MasteringA&P®

iP2

VIDEO TUTORS

learning|**catalytics**™

A&P**Flix**™

DYNAMIC STUDY MODULES

...NEEDS AN INTEGRATED EXPERIENCE

OWNER'S MANUAL

HOW TO USE THIS BOOK

WELCOME TO HUMAN PHYSIOLOGY!

As you begin your study of the human body, one of your main tasks will be to construct for yourself a global view of the body, its systems, and the many processes that keep the systems working. This "big picture" is what physiologists call the integration of systems, and it is a key theme in this book. To integrate information, however, you must do more than simply memorize it. You need to truly understand it and be able to use it to solve problems that you have never encountered before. If you are headed for a career in the health professions, you will do this in the clinics. If you plan a career in biology, you will solve problems in the laboratory, field, or classroom. Analyzing, synthesizing, and evaluating information are skills you need to develop while you are in school, and I hope that the features of this book will help you with this goal.

One of my aims is to provide you not only with information about how the human body functions but also with tips for studying and problem solving. Many of these study aids have been developed with the input of my students, so I think you may find them particularly helpful.

On the following pages, I have put together a brief tour of the special features of the book, especially those that you may not have encountered previously in textbooks. Please take a few minutes to read about them so that you can make optimum use of the book as you study.

In this edition, we have added some new features to help you study. Each chapter now begins with a list of Learning Outcomes to guide you as you read the chapter. We have also added Video Tutors, online video clips that

I created with the assistance of some of my students. Look for the QR codes in the figures with associated Video Tutor clips, and watch Kevin and Michael as they demonstrate physiology in action.

Pattern recognition is important for all healthcare professionals, so you can begin to develop this skill by learning the key concepts of physiology that repeat over and over as you study different organ systems. Chapter 1 includes two special *Focus On* features: one on concept mapping, a study strategy that is also used for decision-making in the clinics, and one on constructing and interpreting graphs. The Chapter 1 Running Problem introduces you to effective ways to find information on the Internet.

Be sure to look for the Essentials and Review figures throughout the book. These figures distill the basics about a topic onto one or two pages, much as the Anatomy Summaries do. My students tell me they find them particularly useful for review when there isn't time to go back and read all the text.

We have also retained the four approaches to learning physiology that proved so popular since this book was first published in 1998.

1. Cellular and Molecular Physiology

Most physiological research today is being done at the cellular and molecular level, and there have been many exciting developments in molecular medicine and physiology in the 10 years since the first edition. For example, now scientists are paying more attention to primary cilia, the single cilium that occurs on most cells of the body. Primary cilia are thought to play a role in some kidney and other diseases. Look for similar links between molecular and cellular biology, physiology, and medicine throughout the book.

2. Physiology as a Dynamic Field

Physiology is a dynamic discipline, with numerous unanswered questions that merit further investigation and research. Many of the "facts" presented in this text are really only our current theories, so you should be prepared to change your mental models as new information emerges from scientific research.

3. An Emphasis on Integration

The organ systems of the body do not work in isolation, although we study them one at a time. To emphasize the integrative nature of physiology, three chapters (Chapters 13, 20, and 25) focus on how the physiological processes of multiple organ systems coordinate with each other, especially when homeostasis is challenged.

4. A Focus on Problem Solving

One of the most valuable life skills students should acquire is the ability to think critically and use information to solve problems. As you study physiology, you should be prepared to practice these skills. You will find a number of features in this book, such as the Concept Check questions and Figure and Graph Questions, that are designed to challenge your critical thinking and analysis skills. In each chapter, read the Running Problem as you work through the text and see if you can apply what you're reading to the clinical scenario described in the problem.

Also, be sure to look at the back of the text, where we have combined the index and glossary to save time when you are looking up unfamiliar words. The appendices have the answers to the Concept Check questions, Figure and Graph Questions, and end-of-chapter questions, as well as reviews of physics, logarithms, and basic genetics. The back end papers include a periodic table of the elements, diagrams of anatomical positions of the body, and tables with conversions and normal values of blood components. Take

a few minutes to look at all these features so that you can make optimum use of them.

It is my hope that by reading this book, you will develop an integrated view of physiology that allows you to enter your chosen profession with respect for the complexity of the human body and a clear vision of the potential of physiological and biomedical research. May you find physiology as fun and exciting I do. Good luck with your studies!

Warmest regards,
Dr. Dee (as my students call me)
silverthorn@utexas.edu

INTEGRATED MEDIA

NEW! **10 PHYS IN ACTION! VIDEO TUTORS AND COACHING ACTIVITIES** help students master challenging physiological concepts through laboratory demonstrations and real-world examples. The Phys in Action! Video Tutors are QR code–accessible from the text, while the Phys in Action! Coaching Activities can track student assessment in MasteringA&P.

FIG. 15.14 **ESSENTIALS**

Cardiovascular Control

The intrinsic rate of the heartbeat is modulated by sympathetic and parasympathetic neurons. Blood vessel diameter is under tonic control by the sympathetic division.

(a) CNS Control of the Heart and Blood Vessels

Medullary cardiovascular control center

Parasympathetic neurons

Sensory neuron

Change in blood pressure

Carotid and aortic baroreceptors

KEY

○ Stimulus
▢ Sensor
▢ Afferent pathway
▢ Integrating center
▢ Output signal
▢ Target
▢ Tissue response
▢ Systemic response

Sympathetic neurons

SA node

Ventricles

Veins

Arterioles

PHYS IN ACTION!

Q FIGURE QUESTION
Name the neurotransmitters and receptors for each of the target tissues.

(b) Baroreceptor Reflex

This map shows the reflex response to an increase in mean arterial pressure.

↑ Blood pressure

↑ Firing of baroreceptors in carotid arteries and aorta ⊖

Sensory neurons

Cardiovascular control center in medulla oblongata

↓ Sympathetic output → *less NE released*

↑ Parasympathetic output → *more ACh on muscarinic receptor*

α-receptor → Arteriolar smooth muscle → Vasodilation

β₁-receptor → Ventricular myocardium → ↓ Force of contraction

β₁-receptor → SA node → ↓ Heart rate

↓ Peripheral resistance

↓ Cardiac output

↓ Blood pressure — *Negative feedback*

iP2

NEW! INTERACTIVE PHYSIOLOGY 2.0 (IP 2.0) COACHING ACTIVITIES

help students dive deeper into complex physiological processes using the engaging Interactive Physiology tutorial program. **The next generation of IP is available for Cardiac Output, Resting Membrane Potential, Electrical Activity of the Heart, Factors Affecting Blood Pressure, Cardiac Cycle, and Generation of an Action Potential.**

Reset the activity to earn additional medals by selecting Heart Rate and Stroke Volume combinations that produce cardiac outputs typical of walking, jogging, or running.

A&PFlix™

A&P FLIX COACHING ACTIVITIES

bring difficult A&P concepts to life with dramatic three-dimensional representations of important processes and carefully developed step-by-step explanations.

GROUP & INDIVIDUAL LEARNING

NEW! DYNAMIC STUDY MODULES

offer a personalized reading experience of the chapter content. As students answer questions to master the chapter content, they receive detailed feedback with text and art from the book itself. Dynamic Study Modules help students acquire, retain, and recall information faster and more efficiently than ever before.

learning | catalytics™

LEARNING CATALYTICS

is a "bring your own device" (laptop, smartphone, or tablet) student engagement, assessment, and classroom intelligent system. Instructors can assess students in real time using open-ended tasks to probe student understanding and facilitate peer-to-peer learning.

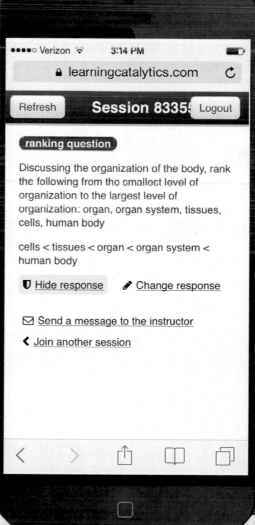

●●●●○ Verizon 🛜   3:14 PM   🔋

🔒 learningcatalytics.com   ↻

Refresh   **Session 83355**   Logout

ranking question

Discussing the organization of the body, rank the following from the smallest level of organization to the largest level of organization: organ, organ system, tissues, cells, human body

cells < tissues < organ < organ system < human body

🛡 Hide response   ✎ Change response

✉ Send a message to the instructor

‹ Join another session

INTEGRATED ART

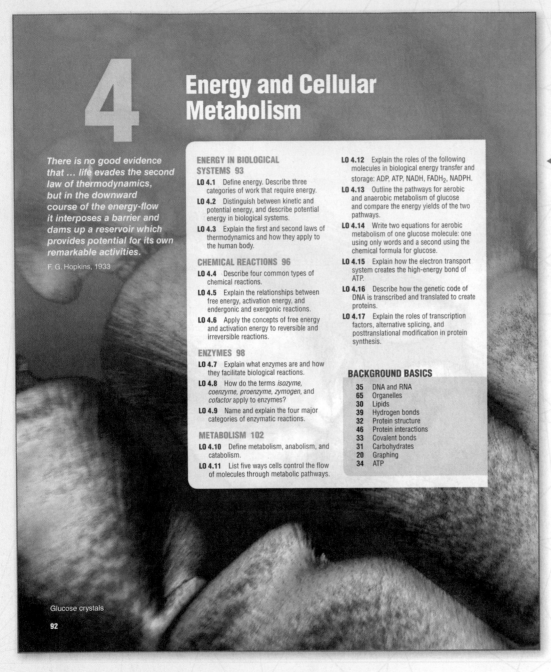

4 Energy and Cellular Metabolism

There is no good evidence that ... life evades the second law of thermodynamics, but in the downward course of the energy-flow it interposes a barrier and dams up a reservoir which provides potential for its own remarkable activities.

F. G. Hopkins, 1933

ENERGY IN BIOLOGICAL SYSTEMS 93

LO 4.1 Define energy. Describe three categories of work that require energy.

LO 4.2 Distinguish between kinetic and potential energy, and describe potential energy in biological systems.

LO 4.3 Explain the first and second laws of thermodynamics and how they apply to the human body.

CHEMICAL REACTIONS 96

LO 4.4 Describe four common types of chemical reactions.

LO 4.5 Explain the relationships between free energy, activation energy, and endergonic and exergonic reactions.

LO 4.6 Apply the concepts of free energy and activation energy to reversible and irreversible reactions.

ENZYMES 98

LO 4.7 Explain what enzymes are and how they facilitate biological reactions.

LO 4.8 How do the terms *isozyme*, *coenzyme*, *proenzyme*, *zymogen*, and *cofactor* apply to enzymes?

LO 4.9 Name and explain the four major categories of enzymatic reactions.

METABOLISM 102

LO 4.10 Define metabolism, anabolism, and catabolism.

LO 4.11 List five ways cells control the flow of molecules through metabolic pathways.

LO 4.12 Explain the roles of the following molecules in biological energy transfer and storage: ADP, ATP, NADH, FADH$_2$, NADPH.

LO 4.13 Outline the pathways for aerobic and anaerobic metabolism of glucose and compare the energy yields of the two pathways.

LO 4.14 Write two equations for aerobic metabolism of one glucose molecule: one using only words and a second using the chemical formula for glucose.

LO 4.15 Explain how the electron transport system creates the high-energy bond of ATP.

LO 4.16 Describe how the genetic code of DNA is transcribed and translated to create proteins.

LO 4.17 Explain the roles of transcription factors, alternative splicing, and posttranslational modification in protein synthesis.

BACKGROUND BASICS

35	DNA and RNA
65	Organelles
30	Lipids
39	Hydrogen bonds
32	Protein structure
46	Protein interactions
33	Covalent bonds
31	Carbohydrates
20	Graphing
34	ATP

Glucose crystals

92

← *NEW!* **LEARNING OUTCOMES** appear at the beginning of every chapter to provide a learning framework that allows students to easily test their knowledge after reading.

NEW! **THE DIGESTIVE SYSTEM AND REPRODUCTION AND DEVELOPMENT CHAPTERS** have been revised to include the latest research and to better foster student understanding and application.

NEW! OVER 40 NEW AND UPDATED PROCESS ART FIGURES

blend narrative explanations and clear visuals that better enable students to visualize difficult concepts and physiological processes.

FIG. 17.5 Airway epithelium

(a) Epithelial cells lining the airways and submucosal glands secrete saline and mucus.

(b) Cilia move the mucus layer toward the pharynx, removing trapped pathogens and particulate matter.

Ciliated epithelium

Movement of mucus

Mucus layer

Lumen of airway

Submucosal gland

Dust particle

Mucus layer traps inhaled particles.

Watery saline layer allows cilia to push mucus toward pharynx.

Cilia

Goblet cell secretes mucus.

Nucleus of columnar epithelial cell

Basement membrane

NEW! OVER 30 NEW AND UPDATED ESSENTIALS ART FIGURES

concisely present important concepts for students to master in each chapter.

FIG. 21.14 ESSENTIALS

The Pancreas

Anatomy of the Exocrine and Endocrine Pancreas

(a) The exocrine pancreas secretes digestive enzymes and sodium bicarbonate.

Pancreatic duct

Pancreas

Small intestine

Pancreatic islet cells secrete hormones that enter the blood

Pancreatic acini form the exocrine portion of the pancreas.

Capillary

Acinar cells secrete digestive enzymes.

Duct cells secrete NaHCO₃ that enters the digestive tract.

Lumen

Activation of Pancreatic Zymogens

(b) Inactive enzymes secreted by the pancreas are activated in a cascade. Trypsinogen is activated to trypsin by brush border enteropeptidase, and trypsin then activates other pancreatic enzymes.

Lumen of small intestine

Pancreatic duct

Pancreatic secretions (include inactive zymogens)

ZYMOGENS
- Chymotrypsinogen
- Procarboxypeptidase
- Procolipase
- Prophospholipase

Trypsinogen

Enteropeptidase in brush border activates trypsin.

Trypsin

activates

ACTIVATED ENZYMES
- Chymotrypsin
- Carboxypeptidase
- Colipase
- Phospholipase

Intestinal mucosa

Bicarbonate Secretion

(c) Bicarbonate secretion in the pancreas and duodenum

Lumen of pancreas or intestine

Pancreatic duct cell or duodenal cell

Interstitial fluid

Capillary

$H_2O + CO_2$

CO_2

❶ CA

HCO_3^-

$HCO_3^- + H^+$

Na^+

Cl^-

Cl^-

Cl^-

❷

Na^+

ATP

K^+

CFTR channel

Na^+

❷

$2 Cl^-$

K^+

Top Ten Ways to Succeed in Classes that Use Active Learning

By Manila Svinicki, Ph.D., Director
University of Texas Center for Teaching Effectiveness

1. Make the switch from an authority-based conception of learning to a self-regulated conception of learning. Recognize and accept your own responsibility for learning.

2. Be willing to take risks and go beyond what is presented in class or the text.

3. Be able to tolerate ambiguity and frustration in the interest of understanding.

4. See errors as opportunities to learn rather than failures. Be willing to make mistakes in class or in study groups so that you can learn from them.

5. Engage in active listening to what's happening in class.

6. Trust the instructor's experience in designing class activities and participate willingly if not enthusiastically.

7. Be willing to express an opinion or hazard a guess.

8. Accept feedback in the spirit of learning rather than as a reflection of you as a person.

9. Prepare for class physically, mentally, and materially (do the reading, work the problems, etc.).

10. Provide support for your classmate's attempts to learn. The best way to learn something well is to teach it to someone who doesn't understand.

Dr. Dee's Eleventh Rule:
DON'T PANIC! Pushing yourself beyond the comfort zone is scary, but you have to do it in order to improve.

Word Roots for Physiology

a- or **an-** without, absence
anti- against
-ase signifies an enzyme
auto self
bi- two
brady- slow
cardio- heart
cephalo- head
cerebro- brain
contra- against
-crine a secretion
crypt- hidden
cutan- skin
-cyte or **cyto-** cell
de- without, lacking
di- two
dys- difficult, faulty
-elle small
-emia blood
endo- inside or within
epi- over
erythro- red
exo- outside
extra- outside
gastro- stomach
-gen, -genie produce
gluco-, glyco- sugar or sweet
hemi- half
hemo- blood
hepato- liver
homo- same
hydro- water
hyper- above or excess
hypo- beneath or deficient

inter- between
intra- within
-itis inflammation of
kali- potassium
leuko- white
lipo- fat
lumen inside of a hollow tube
-lysis split apart or rupture
macro- large
micro- small
mono- one
multi- many
myo- muscle
oligo- little, few
para- near, close
patho-, -pathy related to disease
peri- around
poly- many
post- after
pre- before
pro- before
pseudo- false
re- again
retro- backward or behind
semi- half
sub- below
super- above, beyond
supra- above, on top of
tachy rapid
trans- across, through

HUMAN PHYSIOLOGY

AN INTEGRATED APPROACH

SEVENTH EDITION

Dee Unglaub Silverthorn, Ph.D.
UNIVERSITY OF TEXAS, AUSTIN

WITH CONTRIBUTIONS BY

Bruce R. Johnson, Ph.D.
CORNELL UNIVERSITY

AND

William C. Ober, M.D.
ILLUSTRATION COORDINATOR

Claire E. Ober, R.N.
ILLUSTRATOR

Andrew C. Silverthorn, M.D.
CLINICAL CONSULTANT

PEARSON

Senior Acquisitions Editor: *Kelsey Churchman*
Project Manager: *Lauren Beebe*
Program Manager: *Chriscelle Palaganas*
Development Editor: *Anne A. Reid*
Director of Development: *Barbara Yien*
Assistant Editor: *Ashley Williams*
Rights and Permissions Manager: *Rachel Youdelman*
Photo Rights and Permissions:
  Lauren McFalls, Lumina Datamatics, Inc.
Program Management Team Lead: *Mike Early*
Project Management Team Lead: *Nancy Tabor*

Production Manager: *Andrea Stefanowicz*
Compositor: *Lumina Datamatics, Inc.*
Art House: *Imagineeringart.com Inc.*
Interior Design Manager: *Derek Bacchus*
Exterior Design Manager: *Mark Ong*
Interior Designer: *Elise Lansdon*
Cover Designer: *Tandem Creative, Inc.*
Illustrators: *William C. Ober, Claire E. Ober, Anita Impagliazzo*
Senior Procurement Specialist: *Stacey Weinberger*
Senior Marketing Manager: *Allison Rona*

Cover Photo: Fibroblast cells, fluorescent micrograph
Credit: Dr. Torsten Wittmann/Science Source

Library of Congress Cataloging-in-Publication Data
Silverthorn, Dee Unglaub
    Human physiology: an integrated approach / Dee Unglaub Silverthorn, Ph.D., University of Texas, Austin; with contributions by Bruce R. Johnson, Ph.D., Cornell University and William C. Ober, M.D., illustration coordinator; Claire E. Ober, R.N., illustrator; Andrew C. Silverthorn, M.D., clinical consultant.—Seventh edition.
        pages cm
    Includes index.
    ISBN 978-0-321-98122-6
    ISBN 0-321-98122-7
    1. Human physiology.   I. Johnson, Bruce R.   II. Ober, William C.   III. Garrison, Claire W.
    IV. Silverthorn, Andrew C. V. Title.
    QP34.5.S55 2016
    612—dc23

                                                                                          2014033529

            ISBN 10: 0-321-98122-7; ISBN 13: 978-0-321-98122-6 (Student edition)
          ISBN 10: 0-13-404719-2; ISBN 13: 978-0-13-404719-5 (Instructor's Review Copy)

DEE UNGLAUB SILVERTHORN studied biology as an undergraduate at Newcomb College of Tulane University, where she did research on cockroaches. For graduate school, she switched to studying crabs and received a Ph.D. in marine science from the Belle W. Baruch Institute for Marine and Coastal Sciences at the University of South Carolina. Her research interest is epithelial transport, and most recently work in her laboratory has focused on transport properties of the chick allantoic membrane. Her teaching career started in the Physiology Department at the Medical University of South Carolina but over the years she has taught a wide range of students, from medical and college students to those still preparing for higher education. At the University of Texas–Austin, she teaches physiology in both lecture and laboratory settings, and instructs graduate students on developing teaching skills in the life sciences. She has received numerous teaching awards and honors, including a 2011 UT System Regents' Outstanding Teaching Award, the

Michael Chirillo, Dee Silverthorn, and Kevin Christmas

2009 Outstanding Undergraduate Science Teacher Award from the Society for College Science Teachers, the American Physiological Society's Claude Bernard Distinguished Lecturer and Arthur C. Guyton Physiology Educator of the Year, and multiple awards from UT–Austin, including the Burnt Orange Apple Award. The first edition of her textbook won the 1998 Robert W. Hamilton Author Award for best textbook published in 1997–1998 by a University of Texas faculty member. Dee was the president of the Human Anatomy and Physiology Society in 2012–2013, has served as editor-in-chief of *Advances in Physiology Education*, and is currently chair of the American Physiological Society Book Committee. She works with members of the International Union of Physiological Sciences to improve physiology education in developing countries, and this book has been translated into seven languages. Her free time is spent creating multimedia fiber art and enjoying the Texas hill country with her husband, Andy, and their dogs.

About the Illustrators

William C. Ober, M.D. (*art coordinator and illustrator*) received his undergraduate degree from Washington and Lee University and his M.D. from the University of Virginia. He also studied in the Department of Art as Applied to Medicine at Johns Hopkins University. After graduation, Dr. Ober completed a residency in Family Practice and later was on the faculty at the University of Virginia in the Department of Family Medicine and in the Department of Sports Medicine. He also served as Chief of Medicine of Martha Jefferson Hospital in Charlottesville, VA. He is currently a visiting Professor of Biology at Washington & Lee University, where he has taught several courses and led student trips to the Galapagos Islands. He was part of the Core Faculty at Shoals Marine Laboratory, where he taught Biological Illustration for 22 years. The textbooks illustrated by Medical & Scientific Illustration have won numerous design and illustration awards.

Claire E. Ober, R.N. (*illustrator*) practiced pediatric and obstetric nursing before turning to medical illustration as a full-time career. She returned to school at Mary Baldwin College where she received her degree with distinction in studio art. Following a five-year apprenticeship, she has worked as Dr. Ober's partner in Medical and Scientific Illustration since 1986. She was also on the Core Faculty at Shoals Marine Laboratory and co-taught Biological Illustration at both Shoals Marine Lab and at Washington and Lee University.

About the Clinical Consultant

Andrew C. Silverthorn, M.D. is a graduate of the United States Military Academy (West Point). He served in the infantry in Vietnam, and upon his return entered medical school at the Medical University of South Carolina in Charleston. He was chief resident in family medicine at the University of Texas Medical Branch, Galveston, and is currently a family physician in solo practice in Austin, Texas. When Andrew is not busy seeing patients, he may be found on the golf course or playing with his two rescue dogs, Molly and Callie.

About the Contributor

Bruce Johnson is a Senior Research Associate in the Department of Neurobiology and Behavior at Cornell University. He earned biology degrees at Florida State University (B.A.), Florida Atlantic University (M.S.), and at the Marine Biological Laboratory in Woods Hole (Ph.D.) through the Boston University Marine Program. At Cornell he teaches a laboratory course entitled Principles of Neurophysiology. He is a coauthor of Crawdad: a CD-ROM lab manual for Neurophysiology and the Laboratory Manual for Physiology, and he continues development of model preparations for student neuroscience laboratories. Bruce has directed and taught in neuroscience faculty workshops sponsored by NSF (Crawdad), ADInstruments (Crawdad and CrawFly), and the Faculty for Undergraduate Neuroscience (FUN), and in neuroscience courses at the Universities of Copenhagen and Cologne, the Marine Biological Laboratory, and the Shoals Marine Laboratory. He has received outstanding educator and distinguished teaching awards at Cornell University, and the FUN Educator of the Year Award. He is a past president of FUN and presently the Associate Editor for the *Journal of Undergraduate Neuroscience Education*. Bruce's research addresses the cellular and synaptic mechanisms of motor network plasticity.

DEDICATION | This edition is dedicated to all my graduate teaching assistants, from Carol, Jan, and Peter to Kevin, Michael, and Sarah, and everyone in between. Their enthusiasm for teaching and creative ideas have been a continual inspiration.

CONTENTS IN BRIEF

Owner's Manual

New to This Edition xxi

Acknowledgments xxiii

UNIT 1 — Basic Cell Processes: Integration and Coordination

1 Introduction to Physiology 1
2 Molecular Interactions 28
3 Compartmentation: Cells and Tissues 58
4 Energy and Cellular Metabolism 92
5 Membrane Dynamics 122
6 Communication, Integration, and Homeostasis 165

UNIT 2 — Homeostasis and Control

7 Introduction to the Endocrine System 196
8 Neurons: Cellular and Network Properties 226
9 The Central Nervous System 274
10 Sensory Physiology 309
11 Efferent Division: Autonomic and Somatic Motor Control 358
12 Muscles 377
13 Integrative Physiology I: Control of Body Movement 417

UNIT 3 — Integration of Function

14 Cardiovascular Physiology 435
15 Blood Flow and the Control of Blood Pressure 477
16 Blood 511
17 Mechanics of Breathing 534
18 Gas Exchange and Transport 564
19 The Kidneys 589
20 Integrative Physiology II: Fluid and Electrolyte Balance 618

UNIT 4 — Metabolism, Growth, and Aging

21 The Digestive System 654
22 Metabolism and Energy Balance 693
23 Endocrine Control of Growth and Metabolism 729
24 The Immune System 753
25 Integrative Physiology III: Exercise 786
26 Reproduction and Development 800

APPENDICES

APPENDIX A
Answers A-1

APPENDIX B
Physics and Math A-36

APPENDIX C
Genetics A-39

PHOTO CREDITS C-1

GLOSSARY/INDEX GI-1

CONTENTS

UNIT 1   Basic Cell Processes: Integration and Coordination

CHAPTER 1
Introduction to Physiology 1

Physiology Is an Integrative Science 2
RUNNING PROBLEM What to Believe? 2
   EMERGING CONCEPTS The Changing World of Omics 3
Function and Mechanism 4
Themes in Physiology 5
   FOCUS ON . . . Mapping 6
   Theme 1: Structure and Function Are Closely Related 8
   Theme 2: Living Organisms Need Energy 9
   Theme 3: Information Flow Coordinates Body Functions 9
   Theme 4: Homeostasis Maintains Internal Stability 9
Homeostasis 9
   What Is the Body's Internal Environment? 10
   Homeostasis Depends on Mass Balance 10
   Excretion Clears Substances from the Body 12
   Homeostasis Does Not Mean Equilibrium 13
Control Systems and Homeostasis 13
   Local Control Is Restricted to a Tissue 13
   Reflex Control Uses Long-Distance Signaling 14
   Response Loops Begin with a Stimulus 14
   Feedback Loops Modulate the Response Loop 15
   Negative Feedback Loops Are Homeostatic 15
   Positive Feedback Loops Are Not Homeostatic 16
   Feedforward Control Allows the Body to Anticipate Change 17
   Biological Rhythms Result from Changes in a Setpoint 17
The Science of Physiology 18
   Good Scientific Experiments Must Be Carefully Designed 18
   FOCUS ON . . . Graphs 20
   The Results of Human Experiments Can Be Difficult to Interpret 22
   Human Studies Can Take Many Forms 23

CHAPTER SUMMARY 25 | REVIEW QUESTIONS 26

CHAPTER 2
Molecular Interactions 28

RUNNING PROBLEM Chromium Supplements 29
Molecules and Bonds 29
   Most Biomolecules Contain Carbon, Hydrogen, and Oxygen 29
   Electrons Have Four Important Biological Roles 33
   Covalent Bonds between Atoms Create Molecules 33
   Noncovalent Bonds Facilitate Reversible Interactions 39
Noncovalent Interactions 40
   Hydrophilic Interactions Create Biological Solutions 40
   Molecular Shape Is Related to Molecular Function 40
   Hydrogen Ions in Solution Can Alter Molecular Shape 41
Protein Interactions 46
   Proteins Are Selective about the Molecules They Bind 46
   Protein-Binding Reactions Are Reversible 47
   Binding Reactions Obey the Law of Mass Action 47
   The Dissociation Constant Indicates Affinity 48
   Multiple Factors Alter Protein Binding 48
   The Body Regulates the Amount of Protein in Cells 51
   Reaction Rate Can Reach a Maximum 51
Chemistry Review Quiz 54

CHAPTER SUMMARY 55 | REVIEW QUESTIONS 56

CHAPTER 3
Compartmentation: Cells and Tissues 58

RUNNING PROBLEM Pap Tests Save Lives 59
Functional Compartments of the Body 59
   The Lumens of Some Organs Are Outside the Body 59
   Functionally, the Body Has Three Fluid Compartments 61
Biological Membranes 61
   The Cell Membrane Separates Cell from Environment 61
   Membranes Are Mostly Lipid and Protein 62
   Membrane Lipids Create a Hydrophobic Barrier 62
   Membrane Proteins May Be Loosely or Tightly Bound to the Membrane 62
   BIOTECHNOLOGY Liposomes for Beauty and Health 64
   Membrane Carbohydrates Attach to Both Lipids and Proteins 64
Intracellular Compartments 65
   Cells Are Divided into Compartments 65
   The Cytoplasm Includes Cytosol, Inclusions, Fibers, and Organelles 65
   Inclusions Are in Direct Contact with the Cytosol 65
   Cytoplasmic Protein Fibers Come in Three Sizes 68
   Microtubules Form Centrioles, Cilia, and Flagella 68

The Cytoskeleton Is a Changeable Scaffold 68

EMERGING CONCEPTS *Single Cilia Are Sensors* 69

Motor Proteins Create Movement 69

Organelles Create Compartments for Specialized Functions 70

The Nucleus Is the Cell's Control Center 72

Tissues of the Body 72

Extracellular Matrix Has Many Functions 72

Cell Junctions Hold Cells Together to Form Tissues 72

Epithelia Provide Protection and Regulate Exchange 76

Connective Tissues Provide Support and Barriers 80

BIOTECHNOLOGY *Grow Your Own Cartilage* 82

Muscle and Neural Tissues Are Excitable 84

Tissue Remodeling 84

Apoptosis Is a Tidy Form of Cell Death 84

Stem Cells Can Create New Specialized Cells 85

FOCUS ON . . . The Skin 86

Organs 87

CHAPTER SUMMARY 88 | REVIEW QUESTIONS 90

CHAPTER 4

Energy and Cellular Metabolism 92

RUNNING PROBLEM **Tay-Sachs Disease: A Deadly Inheritance 93**

Energy in Biological Systems 93

Energy Is Used to Perform Work 94

Energy Comes in Two Forms: Kinetic and Potential 94

Energy Can Be Converted from One Form to Another 95

Thermodynamics Is the Study of Energy Use 95

Chemical Reactions 96

Energy Is Transferred between Molecules during Reactions 96

Activation Energy Gets Reactions Started 96

Energy Is Trapped or Released during Reactions 96

Net Free Energy Change Determines Reaction Reversibility 98

Enzymes 98

Enzymes Are Proteins 99

Reaction Rates Are Variable 99

BIOTECHNOLOGY *Seeing Isozymes 100*

Enzymes May Be Activated, Inactivated, or Modulated 100

Enzymes Lower Activation Energy of Reactions 100

Enzymatic Reactions Can Be Categorized 101

Metabolism 102

Cells Regulate Their Metabolic Pathways 103

ATP Transfers Energy between Reactions 104

Catabolic Pathways Produce ATP 105

One Glucose Molecule Can Yield 30–32 ATP 105

Anaerobic Metabolism Makes 2 ATP 110

Proteins Are the Key to Cell Function 111

DNA Guides the Synthesis of RNA 112

Alternative Splicing Creates Multiple Proteins from One DNA Sequence 114

mRNA Translation Links Amino Acids 115

EMERGING CONCEPTS *Purple Petunias and RNAi 116*

Protein Sorting Directs Proteins to Their Destination 116

Proteins Undergo Posttranslational Modification 116

CHAPTER SUMMARY 119 | REVIEW QUESTIONS 120

CHAPTER 5

Membrane Dynamics 122

Homeostasis Does Not Mean Equilibrium 123

RUNNING PROBLEM **Cystic Fibrosis 123**

Osmosis and Tonicity 125

The Body Is Mostly Water 125

The Body Is in Osmotic Equilibrium 125

Osmolarity Describes the Number of Particles in Solution 126

Tonicity Describes the Volume Change of a Cell 127

Transport Processes 132

Cell Membranes Are Selectively Permeable 132

Diffusion 134

Lipophilic Molecules Cross Membranes by Simple Diffusion 136

Protein-Mediated Transport 137

Membrane Proteins Have Four Major Functions 137

Channel Proteins Form Open, Water-Filled Passageways 139

Carrier Proteins Change Conformation to Move Molecules 140

Facilitated Diffusion Uses Carrier Proteins 142

Active Transport Moves Substances against Their Concentration Gradients 143

Carrier-Mediated Transport Exhibits Specificity, Competition, and Saturation 145

Vesicular Transport 147

Phagocytosis Creates Vesicles Using the Cytoskeleton 147

Endocytosis Creates Smaller Vesicles 148

Exocytosis Releases Molecules Too Large for Transport Proteins 148

CLINICAL FOCUS *LDL: The Lethal Lipoprotein 148*

Epithelial Transport 150

Epithelial Transport May Be Paracellular or Transcellular 150

Transcellular Transport of Glucose Uses Membrane Proteins 151

Transcytosis Uses Vesicles to Cross an Epithelium 152

The Resting Membrane Potential 153

Electricity Review 153

The Cell Membrane Enables Separation of Electrical Charge in the Body 153

All Living Cells Have a Membrane Potential 156

The Resting Membrane Potential Is Due Mostly to Potassium 157

Changes in Ion Permeability Change the Membrane Potential 158

Integrated Membrane Processes: Insulin Secretion 159

CHAPTER SUMMARY 161 | REVIEW QUESTIONS 162

CHAPTER 6

Communication, Integration, and Homeostasis 165

Cell-To-Cell Communication 166

RUNNING PROBLEM **Diabetes Mellitus: A Growing Epidemic 166**

Gap Junctions Create Cytoplasmic Bridges 166

Contact-Dependent Signals Require Cell-to-Cell Contact 166

Local Communication Uses Paracrine and Autocrine Signals 168

Long-Distance Communication May Be Electrical or Chemical 168

Cytokines May Act as Both Local and Long-Distance Signals 168

Signal Pathways 169

Receptor Proteins Are Located Inside the Cell or on the Cell Membrane 169

Membrane Proteins Facilitate Signal Transduction 171

The Most Rapid Signal Pathways Change Ion Flow through Channels 172

Most Signal Transduction Uses G Proteins 174

Many Lipophobic Hormones Use GPCR-cAMP Pathways 174

G Protein-Coupled Receptors Also Use Lipid-Derived Second Messengers 174

Receptor-Enzymes Have Protein Kinase or Guanylyl Cyclase Activity 176

Integrin Receptors Transfer Information from the Extracellular Matrix 176

Novel Signal Molecules 176

Calcium Is an Important Intracellular Signal 177

Gases Are Ephemeral Signal Molecules 178

CLINICAL FOCUS *From Dynamite to Medicine 178*

Some Lipids Are Important Paracrine Signals 179

BIOTECHNOLOGY *Calcium Signals Glow in the Dark 179*

Modulation of Signal Pathways 180

Receptors Exhibit Saturation, Specificity, and Competition 180

One Ligand May Have Multiple Receptors 180

Up- and Down-Regulation Enable Cells to Modulate Responses 181

Cells Must Be Able to Terminate Signal Pathways 182

Many Diseases and Drugs Target the Proteins of Signal Transduction 182

Homeostatic Reflex Pathways 182

Cannon's Postulates Describe Regulated Variables and Control Systems 182

Long-Distance Pathways Maintain Homeostasis 183

Control Systems Vary in Their Speed and Specificity 187

Complex Reflex Control Pathways Have Several Integrating Centers 189

CHAPTER SUMMARY 193 | REVIEW QUESTIONS 194

UNIT 2 Homeostasis and Control

CHAPTER 7

Introduction to the Endocrine System 196

Hormones 197

RUNNING PROBLEM **Graves' Disease 197**

Hormones Have Been Known Since Ancient Times 197

What Makes a Chemical a Hormone? 198

CLINICAL FOCUS *Diabetes: The Discovery of Insulin 198*

Hormones Act by Binding to Receptors 199

Hormone Action Must Be Terminated 199

The Classification of Hormones 201

Most Hormones Are Peptides or Proteins 201

Steroid Hormones Are Derived from Cholesterol 204

Some Hormones Are Derived from Single Amino Acids 206

Control of Hormone Release 206

The Endocrine Cell Is the Sensor in Simple Endocrine Reflexes 206

Many Endocrine Reflexes Involve the Nervous System 207

Neurohormones Are Secreted into the Blood by Neurons 207

The Pituitary Gland Is Actually Two Fused Glands 209

The Posterior Pituitary Stores and Releases Two Neurohormones 209

The Anterior Pituitary Secretes Six Hormones 209

A Portal System Connects the Hypothalamus and Anterior Pituitary 211

Anterior Pituitary Hormones Control Growth, Metabolism, and Reproduction 213

Feedback Loops Are Different in the Hypothalamic-Pituitary Pathway 213

Hormone Interactions 215

In Synergism, the Effect of Interacting Hormones Is More than Additive 215

A Permissive Hormone Allows Another Hormone to Exert Its Full Effect 216

Antagonistic Hormones Have Opposing Effects 216

Endocrine Pathologies 216

Hypersecretion Exaggerates a Hormone's Effects 216

Hyposecretion Diminishes or Eliminates a Hormone's Effects 217

Receptor or Second Messenger Problems Cause Abnormal Tissue Responsiveness 217

Diagnosis of Endocrine Pathologies Depends on the Complexity of the Reflex 218

Hormone Evolution 218

FOCUS ON . . . The Pineal Gland 221

CHAPTER SUMMARY 223 | REVIEW QUESTIONS 224

CHAPTER 8
Neurons: Cellular and Network Properties 226

RUNNING PROBLEM **Mysterious Paralysis** 227

Organization of the Nervous System 227

Cells of the Nervous System 229

Neurons Carry Electrical Signals 229

Establishing Synapses Depends on Chemical Signals 232

Glial Cells Provide Support for Neurons 233

Can Stem Cells Repair Damaged Neurons? 235

Electrical Signals in Neurons 236

The Nernst Equation Predicts Membrane Potential for a Single Ion 236

The GHK Equation Predicts Membrane Potential Using Multiple Ions 237

Ion Movement Creates Electrical Signals 238

Gated Channels Control the Ion Permeability of the Neuron 238

Current Flow Obeys Ohm's Law 239

CLINICAL FOCUS *Mutant Channels* 239

Graded Potentials Reflect Stimulus Strength 240

Action Potentials Travel Long Distances 242

Na^+ and K^+ Move across the Membrane during Action Potentials 243

One Action Potential Does Not Alter Ion Concentration Gradients 245

Axonal Na^+ Channels Have Two Gates 245

Action Potentials Will Not Fire during the Absolute Refractory Period 245

Action Potentials Are Conducted 246

Larger Neurons Conduct Action Potentials Faster 247

Conduction Is Faster in Myelinated Axons 249

Chemical Factors Alter Electrical Activity 251

BIOTECHNOLOGY *The Body's Wiring* 251

Cell-to-Cell Communication in The Nervous System 253

Neurons Communicate at Synapses 253

Neurons Secrete Chemical Signals 254

Neurotransmitters Are Highly Varied 254

CLINICAL FOCUS *Myasthenia Gravis* 256

BIOTECHNOLOGY *Of Snakes, Snails, Spiders, and Sushi* 257

Neurotransmitters Are Released from Vesicles 257

Stronger Stimuli Release More Neurotransmitter 260

Integration of Neural Information Transfer 260

Postsynaptic Responses May Be Slow or Fast 261

Pathways Integrate Information from Multiple Neurons 263

Synaptic Activity Can Be Modified 266

Long-Term Potentiation Alters Synapses 267

Disorders of Synaptic Transmission Are Responsible for Many Diseases 268

CHAPTER SUMMARY 269 | REVIEW QUESTIONS 271

CHAPTER 9
The Central Nervous System 274

Emergent Properties of Neural Networks 275

RUNNING PROBLEM **Infantile Spasms** 275

Evolution of Nervous Systems 275

Anatomy of the Central Nervous System 277

The CNS Develops from a Hollow Tube 277

The CNS Is Divided into Gray Matter and White Matter 277

Bone and Connective Tissue Support the CNS 280

The Brain Floats in Cerebrospinal Fluid 280

The Blood-Brain Barrier Protects the Brain 282

CLINICAL FOCUS *Diabetes: Hypoglycemia and the Brain* 283

Neural Tissue Has Special Metabolic Requirements 283

The Spinal Cord 284

The Brain 285

The Brain Stem Is the Oldest Part of the Brain 285

The Cerebellum Coordinates Movement 288

The Diencephalon Contains the Centers for Homeostasis 288

The Cerebrum Is the Site of Higher Brain Functions 289

Brain Function 290

The Cerebral Cortex Is Organized into Functional Areas 291

The Spinal Cord and Brain Integrate Sensory Information 291

Sensory Information Is Processed into Perception 294

The Motor System Governs Output from the CNS 294

The Behavioral State System Modulates Motor Output 294

Why Do We Sleep? 296

Physiological Functions Exhibit Circadian Rhythms 297

Emotion and Motivation Involve Complex Neural Pathways 298

Moods Are Long-Lasting Emotional States 299

Learning and Memory Change Synaptic Connections in the Brain 299

Learning Is the Acquisition of Knowledge 300

Memory Is the Ability to Retain and Recall Information 300

Language Is the Most Elaborate Cognitive Behavior 302

Personality Is a Combination of Experience and Inheritance 304

CHAPTER SUMMARY 305 | REVIEW QUESTIONS 307

CHAPTER 10
Sensory Physiology 309

RUNNING PROBLEM **Ménière's Disease** 310

General Properties of Sensory Systems 310

Receptors Are Sensitive to Particular Forms of Energy 311

Sensory Transduction Converts Stimuli into Graded Potentials 312

A Sensory Neuron Has a Receptive Field 312

The CNS Integrates Sensory Information 313

Coding and Processing Distinguish Stimulus Properties 314

Somatic Senses 317

Pathways for Somatic Perception Project to the Cortex and Cerebellum 317

Touch Receptors Respond to Many Different Stimuli 319

Temperature Receptors Are Free Nerve Endings 320

Nociceptors Initiate Protective Responses 320

CLINICAL FOCUS *Natural Painkillers* 324

Chemoreception: Smell and Taste 324

Olfaction Is One of the Oldest Senses 324

Taste Is a Combination of Five Basic Sensations 325

Taste Transduction Uses Receptors and Channels 327

The Ear: Hearing 329

Hearing Is Our Perception of Sound 329

Sound Transduction Is a Multistep Process 331

The Cochlea Is Filled with Fluid 331

Sounds Are Processed First in the Cochlea 335

Auditory Pathways Project to the Auditory Cortex 335

Hearing Loss May Result from Mechanical or Neural Damage 335

BIOTECHNOLOGY *Artificial Ears* 337

The Ear: Equilibrium 337

The Vestibular Apparatus Provides Information about Movement and Position 337

The Semicircular Canals Sense Rotational Acceleration 337

The Otolith Organs Sense Linear Acceleration and Head Position 339

Equilibrium Pathways Project Primarily to the Cerebellum 339

The Eye and Vision 340

The Skull Protects the Eye 340

CLINICAL FOCUS *Glaucoma 340*

Light Enters the Eye through the Pupil 342

The Lens Focuses Light on the Retina 343

Phototransduction Occurs at the Retina 345

EMERGING CONCEPTS *Melanopsin 346*

Photoreceptors Transduce Light into Electrical Signals 348

Signal Processing Begins in the Retina 350

CHAPTER SUMMARY 354 | REVIEW QUESTIONS 356

CHAPTER 11
Efferent Division: Autonomic and Somatic Motor Control 358

RUNNING PROBLEM **A Powerful Addiction** 359

The Autonomic Division 359

Autonomic Reflexes Are Important for Homeostasis 360

Antagonistic Control Is a Hallmark of the Autonomic Division 361

Autonomic Pathways Have Two Efferent Neurons in Series 361

Sympathetic and Parasympathetic Branches Originate in Different Regions 363

The Autonomic Nervous System Uses a Variety of Chemical Signals 364

Autonomic Pathways Control Smooth and Cardiac Muscle and Glands 364

Autonomic Neurotransmitters Are Synthesized in the Axon 365

Autonomic Receptors Have Multiple Subtypes 366

The Adrenal Medulla Secretes Catecholamines 367

Autonomic Agonists and Antagonists Are Important Tools in Research and Medicine 367

Primary Disorders of the Autonomic Nervous System Are Relatively Uncommon 369

CLINICAL FOCUS *Diabetes: Autonomic Neuropathy 369*

Summary of Sympathetic and Parasympathetic Branches 369

The Somatic Motor Division 371

A Somatic Motor Pathway Consists of One Neuron 371

The Neuromuscular Junction Contains Nicotinic Receptors 373

CHAPTER SUMMARY 374 | REVIEW QUESTIONS 375

CHAPTER 12
Muscles 377

RUNNING PROBLEM **Periodic Paralysis** 378

Skeletal Muscle 379

Skeletal Muscles Are Composed of Muscle Fibers 379

Myofibrils Are Muscle Fiber Contractile Structures 382

Muscle Contraction Creates Force 383

Actin and Myosin Slide Past Each Other during Contraction 385

Myosin Crossbridges Move Actin Filaments 385

Calcium Signals Initiate Contraction 386

Myosin Heads Step along Actin Filaments 386

BIOTECHNOLOGY *Watching Myosin Work 388*

Acetylcholine Initiates Excitation-Contraction Coupling 388

Skeletal Muscle Contraction Requires a Steady Supply of ATP 391

Fatigue Has Multiple Causes 392

Skeletal Muscle Is Classified by Speed and Fatigue Resistance 393

Resting Fiber Length Affects Tension 395

Force of Contraction Increases with Summation 396

A Motor Unit Is One Motor Neuron and Its Muscle Fibers 396

Contraction Force Depends on the Types and Numbers of Motor Units 397

Mechanics of Body Movement 398

Isotonic Contractions Move Loads; Isometric Contractions Create Force without Movement 398

Bones and Muscles around Joints Form Levers and Fulcrums 400

Muscle Disorders Have Multiple Causes 402

Smooth Muscle 403

Smooth Muscle Is More Variable Than Skeletal Muscle 404

Smooth Muscle Lacks Sarcomeres 406

Myosin Phosphorylation Controls Contraction 406

MLCP Controls Ca^{2+} Sensitivity 407

Calcium Initiates Smooth Muscle Contraction 407

Some Smooth Muscles Have Unstable Membrane Potentials 410

Chemical Signals Influence Smooth Muscle Activity 410

Cardiac Muscle 412

CHAPTER SUMMARY 413 | REVIEW QUESTIONS 414

CHAPTER 13

Integrative Physiology I: Control of Body Movement 417

Neural Reflexes 418

Neural Reflex Pathways Can Be Classified in Different Ways 418

RUNNING PROBLEM Tetanus 418

Autonomic Reflexes 420

Skeletal Muscle Reflexes 420

Golgi Tendon Organs Respond to Muscle Tension 421

Muscle Spindles Respond to Muscle Stretch 421

CLINICAL FOCUS *Reflexes and Muscle Tone 421*

Stretch Reflexes and Reciprocal Inhibition Control Movement around a Joint 424

Flexion Reflexes Pull Limbs Away from Painful Stimuli 424

The Integrated Control of Body Movement 426

Movement Can Be Classified as Reflex, Voluntary, or Rhythmic 427

The CNS Integrates Movement 428

EMERGING CONCEPTS *Visualization Techniques in Sports 430*

Symptoms of Parkinson's Disease Reflect Basal Ganglia Function 430

Control of Movement in Visceral Muscles 431

CHAPTER SUMMARY 433 | REVIEW QUESTIONS 434

UNIT 3 Integration of Function

CHAPTER 14

Cardiovascular Physiology 435

RUNNING PROBLEM Myocardial Infarction 436

Overview of the Cardiovascular System 436

The Cardiovascular System Transports Materials throughout the Body 437

The Cardiovascular System Consists of the Heart, Blood Vessels, and Blood 437

Pressure, Volume, Flow, and Resistance 439

The Pressure of Fluid in Motion Decreases over Distance 439

Pressure Changes in Liquids without a Change in Volume 440

Blood Flows from Higher Pressure to Lower Pressure 440

Resistance Opposes Flow 440

Velocity Depends on the Flow Rate and the Cross-Sectional Area 442

Cardiac Muscle and the Heart 443

The Heart Has Four Chambers 443

Heart Valves Ensure One-Way Flow in the Heart 447

Cardiac Muscle Cells Contract without Innervation 447

Calcium Entry Is a Feature of Cardiac EC Coupling 449

Cardiac Muscle Contraction Can Be Graded 450

Myocardial Action Potentials Vary 451

The Heart as a Pump 454

Electrical Signals Coordinate Contraction 454

Pacemakers Set the Heart Rate 455

CLINICAL FOCUS *Fibrillation 457*

The Electrocardiogram Reflects Electrical Activity 457

The Heart Contracts and Relaxes during a Cardiac Cycle 461

CLINICAL FOCUS *Gallops, Clicks, and Murmurs 464*

Pressure-Volume Curves Represent One Cardiac Cycle 464

Stroke Volume Is the Volume of Blood Pumped per Contraction 466

Cardiac Output Is a Measure of Cardiac Performance 466

The Autonomic Division Modulates Heart Rate 466

Multiple Factors Influence Stroke Volume 468

Contractility Is Controlled by the Nervous and Endocrine Systems 469

EDV and Arterial Blood Pressure Determine Afterload 471

EMERGING CONCEPTS *Stem Cells for Heart Disease 471*

CHAPTER SUMMARY 473 | REVIEW QUESTIONS 475

CHAPTER 15

Blood Flow and the Control of Blood Pressure 477

RUNNING PROBLEM **Essential Hypertension 478**

The Blood Vessels 479

Blood Vessels Contain Vascular Smooth Muscle 479

Arteries and Arterioles Carry Blood Away from the Heart 480

Exchange Takes Place in the Capillaries 480

Blood Flow Converges in the Venules and Veins 481

Angiogenesis Creates New Blood Vessels 481

Blood Pressure 482

Blood Pressure Is Highest in Arteries and Lowest in Veins 482

Arterial Blood Pressure Reflects the Driving Pressure for Blood Flow 483

Blood Pressure Is Estimated by Sphygmomanometry 484

Cardiac Output and Peripheral Resistance Determine Mean Arterial Pressure 484

Changes in Blood Volume Affect Blood Pressure 485

CLINICAL FOCUS *Shock 487*

Resistance in the Arterioles 487

Myogenic Autoregulation Adjusts Blood Flow 488

Paracrine Signals Influence Vascular Smooth Muscle 489

The Sympathetic Branch Controls Most Vascular Smooth Muscle 490

Distribution of Blood to the Tissues 492

Regulation of Cardiovascular Function 492

The Baroreceptor Reflex Controls Blood Pressure 493

Orthostatic Hypotension Triggers the Baroreceptor Reflex 495

Other Systems Influence Cardiovascular Function 495

Exchange at the Capillaries 496

Velocity of Blood Flow Is Lowest in the Capillaries 497

Most Capillary Exchange Takes Place by Diffusion and Transcytosis 497

Capillary Filtration and Absorption Take Place by Bulk Flow 497

The Lymphatic System 498

Edema Results from Alterations in Capillary Exchange 500

Cardiovascular Disease 501

Risk Factors Include Smoking and Obesity 501

Atherosclerosis Is an Inflammatory Process 502

CLINICAL FOCUS *Diabetes and Cardiovascular Disease 502*

EMERGING CONCEPTS *Inflammatory Markers for Cardiovascular Disease 504*

Hypertension Represents a Failure of Homeostasis 504

CHAPTER SUMMARY 506 | REVIEW QUESTIONS 507

CHAPTER 16

Blood 511

RUNNING PROBLEM **Blood Doping in Athletes 512**

Plasma and the Cellular Elements of Blood 512

Plasma Is Extracellular Matrix 512

Cellular Elements Include RBCs, WBCs, and Platelets 514

Blood Cell Production 514

Blood Cells Are Produced in the Bone Marrow 515

Hematopoiesis Is Controlled by Cytokines 516

Colony-Stimulating Factors Regulate Leukopoiesis 516

Thrombopoietin Regulates Platelet Production 517

Erythropoietin Regulates RBC Production 518

Red Blood Cells 518

Mature RBCs Lack a Nucleus 518

FOCUS ON . . . Bone Marrow 519

Hemoglobin Synthesis Requires Iron 520

RBCs Live about Four Months 522

RBC Disorders Decrease Oxygen Transport 522

CLINICAL FOCUS *Diabetes: Hemoglobin and Hyperglycemia 522*

Platelets 523

Hemostasis and Coagulation 524

Hemostasis Prevents Blood Loss from Damaged Vessels 524

Platelet Activation Begins the Clotting Process 526

Coagulation Converts a Platelet Plug into a Clot 526

Anticoagulants Prevent Coagulation 528

CHAPTER SUMMARY 531 | REVIEW QUESTIONS 532

CHAPTER 17

Mechanics of Breathing 534

The Respiratory System 535

RUNNING PROBLEM **Emphysema 535**

Bones and Muscles of the Thorax Surround the Lungs 536

Pleural Sacs Enclose the Lungs 536

Airways Connect Lungs to the External Environment 537

The Airways Warm, Humidify, and Filter Inspired Air 537

Alveoli Are the Site of Gas Exchange 540

Pulmonary Circulation Is High-Flow, Low-Pressure 540

CLINICAL FOCUS *Congestive Heart Failure 542*

Gas Laws 542

Air Is a Mixture of Gases 542

Gases Move Down Pressure Gradients 542

Boyle's Law Describes Pressure-Volume Relationships 544

Ventilation 544

Lung Volumes Change during Ventilation 544

During Ventilation, Air Flows because of Pressure Gradients 546

Inspiration Occurs When Alveolar Pressure Decreases 546

Expiration Occurs When Alveolar Pressure Increases 547

Intrapleural Pressure Changes during Ventilation 549

Lung Compliance and Elastance May Change in Disease States 550

Surfactant Decreases the Work of Breathing 551

Airway Diameter Determines Airway Resistance 552

Rate and Depth of Breathing Determine the Efficiency of Breathing 553

Alveolar Gas Composition Varies Little during Normal Breathing 554

Ventilation and Alveolar Blood Flow Are Matched 556

Auscultation and Spirometry Assess Pulmonary Function 557

CHAPTER SUMMARY 560 | **REVIEW QUESTIONS** 561

CHAPTER 18

Gas Exchange and Transport 564

RUNNING PROBLEM **High Altitude** 565

Gas Exchange in the Lungs and Tissues 565

Lower Alveolar P_{O_2} Decreases Oxygen Uptake 566

Diffusion Problems Cause Hypoxia 567

BIOTECHNOLOGY *The Pulse Oximeter 569*

Gas Solubility Affects Diffusion 569

Gas Transport in the Blood 571

Hemoglobin Binds to Oxygen 572

Oxygen Binding Obeys the Law of Mass Action 572

Hemoglobin Transports Most Oxygen to the Tissues 573

P_{O_2} Determines Oxygen-Hb Binding 573

Oxygen Binding Is Expressed as a Percentage 574

EMERGING CONCEPTS *Blood Substitutes 574*

Several Factors Affect O_2-Hb Binding 575

Carbon Dioxide Is Transported in Three Ways 577

Regulation of Ventilation 580

Neurons in the Medulla Control Breathing 581

CO_2, Oxygen, and pH Influence Ventilation 582

Protective Reflexes Guard the Lungs 584

Higher Brain Centers Affect Patterns of Ventilation 584

CHAPTER SUMMARY 586 | **REVIEW QUESTIONS** 587

CHAPTER 19

The Kidneys 589

Functions of the Kidneys 590

RUNNING PROBLEM **Gout** 590

Anatomy of the Urinary System 591

The Urinary System Consists of Kidneys, Ureters, Bladder, and Urethra 591

The Nephron Is the Functional Unit of the Kidney 591

Overview of Kidney Function 594

Kidneys Filter, Reabsorb, and Secrete 594

The Nephron Modifies Fluid Volume and Osmolarity 594

Filtration 596

The Renal Corpuscle Contains Filtration Barriers 597

EMERGING CONCEPTS *Diabetes: Diabetic Nephropathy 597*

Capillary Pressure Causes Filtration 598

GFR Is Relatively Constant 600

GFR Is Subject to Autoregulation 600

Hormones and Autonomic Neurons Also Influence GFR 602

Reabsorption 602

Reabsorption May Be Active or Passive 602

Renal Transport Can Reach Saturation 604

BIOTECHNOLOGY *Artificial Kidneys 605*

Peritubular Capillary Pressures Favor Reabsorption 606

Secretion 607

Competition Decreases Penicillin Secretion 608

Excretion 608

Clearance Is a Noninvasive Way to Measure GFR 609

Clearance Helps Us Determine Renal Handling 611

Micturition 613

CHAPTER SUMMARY 614 | **REVIEW QUESTIONS** 616

CHAPTER 20

Integrative Physiology II: Fluid and Electrolyte Balance 618

Fluid and Electrolyte Homeostasis 619

RUNNING PROBLEM **Hyponatremia** 619

ECF Osmolarity Affects Cell Volume 619

Multiple Systems Integrate Fluid and Electrolyte Balance 619

Water Balance 620

Daily Water Intake and Excretion Are Balanced 620

The Kidneys Conserve Water 621

The Renal Medulla Creates Concentrated Urine 622

CLINICAL FOCUS *Diabetes: Osmotic Diuresis 623*

Vasopressin Controls Water Reabsorption 623

Blood Volume and Osmolarity Activate Osmoreceptors 625

The Loop of Henle Is a Countercurrent Multiplier 627

Sodium Balance and ECF Volume 629

Aldosterone Controls Sodium Balance 630

Low Blood Pressure Stimulates Aldosterone Secretion 630

ANG II Has Many Effects 632

Natriuretic Peptides Promote Na$^+$ and Water Excretion 634

Potassium Balance 634

EMERGING CONCEPTS *WNK Kinases and Hypertension 634*

Behavioral Mechanisms in Salt and Water Balance 636

Drinking Replaces Fluid Loss 636

Low Na$^+$ Stimulates Salt Appetite 636

Avoidance Behaviors Help Prevent Dehydration 636

Integrated Control of Volume and Osmolarity 637

Osmolarity and Volume Can Change Independently 637

Dehydration Triggers Homeostatic Responses 638

Acid-Base Balance 641

pH Changes Can Denature Proteins 641

Acids and Bases in the Body Come from Many Sources 642

pH Homeostasis Depends on Buffers, Lungs, and Kidneys 642

Buffer Systems Include Proteins, Phosphate Ions, and HCO$_3^-$ 643

Ventilation Can Compensate for pH Disturbances 643

Kidneys Use Ammonia and Phosphate Buffers 645

The Proximal Tubule Secretes H$^+$ and Reabsorbs HCO$_3^-$ 645

The Distal Nephron Controls Acid Excretion 646

Acid-Base Disturbances May Be Respiratory or Metabolic 647

CHAPTER SUMMARY 651 | REVIEW QUESTIONS 652

UNIT 4 Metabolism, Growth, and Aging

CHAPTER 21

The Digestive System 654

Anatomy of the Digestive System 655

RUNNING PROBLEM **Cholera in Haiti 655**

The Digestive System Is a Tube 655

The GI Tract Wall Has Four Layers 658

Digestive Function and Processes 659

We Secrete More Fluid Than We Ingest 661

Digestion and Absorption Make Food Usable 661

Motility: GI Smooth Muscle Contracts Spontaneously 661

GI Smooth Muscle Exhibits Different Patterns of Contraction 663

CLINICAL FOCUS *Diabetes: Delayed Gastric Emptying 663*

Regulation of GI Function 664

The Enteric Nervous System Can Act Independently 664

GI Peptides Include Hormones, Neuropeptides, and Cytokines 665

Integrated Function: The Cephalic Phase 668

Chemical and Mechanical Digestion Begins in the Mouth 668

Saliva Is an Exocrine Secretion 668

Swallowing Moves Food from Mouth to Stomach 668

Integrated Function: The Gastric Phase 669

The Stomach Stores Food 669

Gastric Secretions Protect and Digest 670

The Stomach Balances Digestion and Defense 672

Integrated Function: The Intestinal Phase 672

Intestinal Secretions Promote Digestion 673

The Pancreas Secretes Enzymes and Bicarbonate 674

The Liver Secretes Bile 675

FOCUS ON . . . The Liver 677

Most Digestion Occurs in the Small Intestine 678

Bile Salts Facilitate Fat Digestion 678

Carbohydrates Are Absorbed as Monosaccharides 680

CLINICAL FOCUS *Lactose Intolerance 680*

Proteins Are Digested into Small Peptides and Amino Acids 681

Some Larger Peptides Can Be Absorbed Intact 682

Nucleic Acids Are Digested into Bases and Monosaccharides 682

The Intestine Absorbs Vitamins and Minerals 682

The Intestine Absorbs Ions and Water 683

Regulation of the Intestinal Phase 684

EMERGING CONCEPTS *Taste Receptors in the Gut 684*

The Large Intestine Concentrates Waste 685

EMERGING CONCEPTS *The Human Microbiome Project 686*

Diarrhea Can Cause Dehydration 686

Immune Functions of the GI Tract 687

M Cells Sample Gut Contents 687

Vomiting Is a Protective Reflex 687

CHAPTER SUMMARY 689 | REVIEW QUESTIONS 690

CHAPTER 22

Metabolism and Energy Balance 693

Appetite and Satiety 694

RUNNING PROBLEM **Eating Disorders 694**

BIOTECHNOLOGY *Discovering Peptides: Research in Reverse 695*

Energy Balance 696

Energy Input Equals Energy Output 696

CLINICAL FOCUS *Estimating Fat—The Body Mass Index 697*

Oxygen Consumption Reflects Energy Use 697

Many Factors Influence Metabolic Rate 698

Energy Is Stored in Fat and Glycogen 698

Metabolism 699

Ingested Energy May Be Used or Stored 699

Enzymes Control the Direction of Metabolism 700

Fed-State Metabolism 701

Carbohydrates Make ATP 701

Amino Acids Make Proteins 701

Fats Store Energy 703

CLINICAL FOCUS *Antioxidants Protect the Body* 703

Plasma Cholesterol Predicts Heart Disease 705

Fasted-State Metabolism 705

Glycogen Converts to Glucose 706

Proteins Can Be Used to Make ATP 706

Lipids Store More Energy than Glucose or Protein 707

Homeostatic Control of Metabolism 708

The Pancreas Secretes Insulin and Glucagon 708

The Insulin-to-Glucagon Ratio Regulates Metabolism 709

Insulin Is the Dominant Hormone of the Fed State 710

Insulin Promotes Anabolism 711

Glucagon Is Dominant in the Fasted State 714

Diabetes Mellitus Is a Family of Diseases 714

Type 1 Diabetics Are Prone to Ketoacidosis 715

Type 2 Diabetics Often Have Elevated Insulin Levels 718

Metabolic Syndrome Links Diabetes and Cardiovascular Disease 718

Multiple Hormones Influence Metabolism 719

Regulation of Body Temperature 720

Body Temperature Balances Heat Production, Gain, and Loss 720

Body Temperature Is Homeostatically Regulated 721

Movement and Metabolism Produce Heat 722

The Body's Thermostat Can Be Reset 723

CHAPTER SUMMARY 726 | **REVIEW QUESTIONS** 727

CHAPTER 23

Endocrine Control of Growth and Metabolism 729

Review of Endocrine Principles 730

RUNNING PROBLEM **Hyperparathyroidism** 730

BIOTECHNOLOGY *Mutant Mouse Models* 730

Adrenal Glucocorticoids 731

The Adrenal Cortex Secretes Steroid Hormones 732

Cortisol Secretion Is Controlled by ACTH 732

Cortisol Is Essential for Life 732

Cortisol Is a Useful Therapeutic Drug 734

Cortisol Pathologies Result from Too Much or Too Little Hormone 734

CRH and ACTH Have Additional Physiological Functions 735

EMERGING CONCEPTS *Melanocortins and the Agouti Mouse* 735

Thyroid Hormones 736

Thyroid Hormones Contain Iodine 736

TSH Controls the Thyroid Gland 738

Thyroid Pathologies Affect Quality of Life 738

Growth Hormone 740

Growth Hormone Is Anabolic 741

CLINICAL FOCUS *New Growth Charts* 742

Growth Hormone Is Essential for Normal Growth 742

Genetically Engineered hGH Raises Ethical Questions 742

Tissue and Bone Growth 743

Tissue Growth Requires Hormones and Paracrine Signals 743

Bone Growth Requires Adequate Dietary Calcium 743

Calcium Balance 745

Plasma Calcium Is Closely Regulated 745

Three Hormones Control Calcium Balance 746

Calcium and Phosphate Homeostasis Are Linked 748

Osteoporosis Is a Disease of Bone Loss 749

CHAPTER SUMMARY 750 | **REVIEW QUESTIONS** 751

CHAPTER 24

The Immune System 753

RUNNING PROBLEM **HPV: To Vaccinate or Not?** 754

Overview 754

Pathogens of the Human Body 754

Bacteria and Viruses Require Different Defense Mechanisms 755

Viruses Can Replicate Only inside Host Cells 756

The Immune Response 756

Anatomy of the Immune System 757

Lymphoid Tissues Are Everywhere 757

FOCUS ON . . . The Spleen 759

Leukocytes Mediate Immunity 759

Innate Immunity: Nonspecific Responses 762

Barriers Are the Body's First Line of Defense 763

Phagocytes Ingest Foreign Material 763

NK Cells Kill Infected and Tumor Cells 763

Cytokines Create the Inflammatory Response 764

Acquired Immunity: Antigen-Specific Responses 765

Lymphocytes Mediate the Acquired Immune Response 766

B Lymphocytes Become Plasma Cells and Memory Cells 766

Antibodies Are Proteins Secreted by Plasma Cells 768

Antibodies Work outside Cells 768

FOCUS ON . . . The Thymus Gland 770

T Lymphocytes Use Contact-Dependent Signaling 770

Immune Response Pathways 772

Bacterial Invasion Causes Inflammation 772

Viral Infections Require Intracellular Defense 772

Specific Antigens Trigger Allergic Responses 775

MHC Proteins Allow Recognition of Foreign Tissue 776

The Immune System Must Recognize "Self" 778

Immune Surveillance Removes Abnormal Cells 779

BIOTECHNOLOGY *Engineered Antibodies* 779

Neuro-Endocrine-Immune Interactions 779

Stress Alters Immune System Function 781

Modern Medicine Includes Mind-Body Therapeutics 781

CHAPTER SUMMARY 783 | REVIEW QUESTIONS 784

CHAPTER 25

Integrative Physiology III: Exercise 786

RUNNING PROBLEM **Malignant Hyperthermia** 787

Metabolism and Exercise 787

Hormones Regulate Metabolism during Exercise 789

Oxygen Consumption Is Related to Exercise Intensity 789

Several Factors Limit Exercise 789

Ventilatory Responses to Exercise 790

Cardiovascular Responses to Exercise 791

Cardiac Output Increases during Exercise 791

Muscle Blood Flow Increases during Exercise 792

Blood Pressure Rises Slightly during Exercise 792

The Baroreceptor Reflex Adjusts to Exercise 793

Feedforward Responses to Exercise 793

Temperature Regulation during Exercise 794

Exercise and Health 795

Exercise Lowers the Risk of Cardiovascular Disease 795

Type 2 Diabetes Mellitus May Improve with Exercise 795

Stress and the Immune System May Be Influenced by Exercise 795

CHAPTER SUMMARY 798 | REVIEW QUESTIONS 798

CHAPTER 26

Reproduction and Development 800

RUNNING PROBLEM **Infertility** 801

Sex Determination 801

Sex Chromosomes Determine Genetic Sex 801

Sexual Differentiation Occurs Early in Development 802

CLINICAL FOCUS *X-Linked Inherited Disorders* 803

Basic Patterns of Reproduction 806

CLINICAL FOCUS *Determining Sex* 806

Gametogenesis Begins *in Utero* 807

The Brain Directs Reproduction 807

Environmental Factors Influence Reproduction 810

Male Reproduction 810

Testes Produce Sperm and Hormones 811

Spermatogenesis Requires Gonadotropins and Testosterone 815

Male Accessory Glands Contribute Secretions to Semen 815

Androgens Influence Secondary Sex Characteristics 815

Female Reproduction 816

The Ovary Produces Eggs and Hormones 816

A Menstrual Cycle Lasts about One Month 817

Hormonal Control of the Menstrual Cycle Is Complex 817

Hormones Influence Female Secondary Sex Characteristics 823

Procreation 824

The Human Sexual Response Has Four Phases 824

The Male Sex Act Includes Erection and Ejaculation 824

Sexual Dysfunction Affects Males and Females 824

Contraceptives Are Designed to Prevent Pregnancy 825

Infertility Is the Inability to Conceive 827

Pregnancy and Parturition 827

Fertilization Requires Capacitation 827

The Developing Embryo Implants in the Endometrium 827

The Placenta Secretes Hormones during Pregnancy 829

Pregnancy Ends with Labor and Delivery 830

The Mammary Glands Secrete Milk during Lactation 832

Growth and Aging 834

Puberty Marks the Beginning of the Reproductive Years 834

Menopause and Andropause Are a Consequence of Aging 834

CHAPTER SUMMARY 835 | REVIEW QUESTIONS 837

APPENDICES

APPENDIX A Answers A-1

APPENDIX B Physics and Math A-36

APPENDIX C Genetics A-39

PHOTO CREDITS C-1

GLOSSARY/INDEX GI-1

NEW TO THIS EDITION

The seventh edition of *Human Physiology: An Integrated Approach* builds upon the thorough coverage of integrative and molecular physiology topics that have been the foundation of this book since its first publication. The text has been revised with extensive content updates, particularly in the areas of neurobiology and reproductive physiology. Chapter 21 on the digestive system and Chapter 26 on reproductive physiology have been reorganized.

Continuing the revision of the art introduced last edition, we created additional Review and Essentials figures that students can use for quick review as well as new Anatomy Summaries and concept maps. Figures from previous editions that were significantly modified or eliminated are still available to instructors on the Instructor's DVD.

A new feature in the book is the addition of online Video Tutors that can be accessed by using the QR codes found in related figures. These short videos explain complex topics or show physiology in action in research and medicine. Finally, every chapter now begins with a list of Learning Outcomes (LO) for the chapter.

CHAPTER-BY-CHAPTER CONTENT UPDATES

Chapter 1
- Added DOI numbers to literature citation information

Chapter 2
- Updated information on chromium supplements
- Expanded discussion of proteins

Chapter 3
- Expanded discussion of fluid compartments
- Updated Running problem on HPV and Pap smears to reflect latest guidelines for management of abnormal tests
- New discussion of retinoids

Chapter 4
- New Figure Question
- Updated information on Tay Sachs testing

Chapter 5
- New Essentials figure on membrane potential
- Video Tutor walking through the problems in Figure 5.4
- Video Tutor demonstrations of diffusion (Fig. 5.6) and membrane potential (Fig. 5.23)
- New figures for osmotic pressure and transport across membranes
- New concept map of fluid compartments
- Updated transporter gene families and transporter classification

- New quantitative Figure Questions for equilibrium potential
- New Figure Questions for insulin secretion
- Updated information on caveolae

Chapter 6
- Updated information on cytokines
- New discussion of catalytic receptors

Chapter 7
- Updated information on:
  - prolactin
  - short-loop negative feedback
- New discussion of tertiary pathologies

Chapter 8
- Expanded discussion of electrical synapses
- New Essentials figure on integration of synaptic signaling

Chapter 9
- BRAIN and Human Connectome
- Updated information on:
  - sleep
  - circadian rhythms
  - jet lag and shift work disorder
  - serotonin/norepinephrine reuptake inhibitors
  - treatment of infantile spasms (West syndrome)
  - language processing
- New table on brain imaging techniques, including CLARITY
- New discussion of PTSD Cranial nerve mnemonic

Chapter 10
- Updated discussion of
  - pain
  - olfactory processing
  - taste transduction
  - Merkel cell mechanotransduction
- New: olfactory (Bowman's) glands

Chapter 11
- New information on dysautonomia
- New Anatomy Summary for autonomic nervous system
- Updated information on drug treatment for tobacco use disorder

Chapter 12
- Updated information on:
  - skeletal muscle fiber types
  - store-operated calcium channels (STIM1, Orai1)
- New figure on fast- and slow-twitch muscles

Chapter 13

- Updated information on Golgi tendon organ function

Chapter 14

- Updated Running Problem to include cardiac stents and treatment for heart attacks
- Updated info on stem cells for cardiac problems
- Video Tutor showing electrocardiogram recording (Fig. 14.15)

Chapter 15

- Video Tutor demonstrating orthostatic hypotension and the baroreceptor reflex (Fig. 15.14)
- New Figure Questions and Quantitative Question
- Updated information on hypertension

Chapter 16

- New Essentials figure on the complete blood count
- Updated information on:
  - ferritin
  - platelets and thrombocytopenia
  - contact activation pathway
  - cell injury pathway
- New Concept Check questions

Chapter 17

- New information on pulmonary function testing and FEV_1 (forced expiratory volume in one second)
- Video Tutor of pulmonary function testing (Fig. 17.7)
- Video Tutor of lung model and negative intrapleural pressure (Fig. 17.10)
- Video Tutor showing effect of breathing pattern on expired CO_2 (Fig. 17.13)
- New table on gas laws
- Updated information on:
  - COPD
  - velocity of air flow
- New algorithm for calculating lung volumes
- New end-of-chapter questions

Chapter 18

- Video Tutor demonstration of oxygen transport by hemoglobin (Fig. 18.8)
- Updated information on:
  - central regulation of ventilation
  - central chemoreceptors
  - blood substitutes
- Expanded discussion and questions on the Fick equation

Chapter 19

- Expanded histology of the nephron
- Video Tutor demonstrating renal clearance (Fig. 19.13)
- New quantitative and graphing end-of-chapter questions
- New information on primary cilia as flow sensors
- Expanded "useful equations" table

Chapter 20

- New quantitative problem on osmotic diuresis
- Emerging Concepts box on WNK kinases

Chapter 21

- Major reorganization of topics
- New Essentials figures on GI motility, The Pancreas
- New Figure Questions and Concept Checks
- New Emerging Concepts box on the human microbiome project

Chapter 22

- New information on
  - uncoupling protein 1 (UCP1) in brown fat
  - SGLT2 inhibitors
- New discussion of hormones and metabolism
- Updated discussion on control of food intake
- New quantitative problem on BMI
- New questions on diabetes

Chapter 23

- Updated information on
  - POMC and food intake update
- New information on
  - monocarboxylate transporter for thyroid hormones
  - mechanical stress, primary cilia, and bone remodeling
  - calbindin in Ca^{2+} absorption
- New Graph Question

Chapter 24

- New Anatomy Summary on the immune system
- Updated information on HPV
- New concept map for the immune system
- Expanded discussion of antibodies
- New section on Rh blood groups

Chapter 25

- Video Tutor demonstrating the effect of exercise on cardiovascular and pulmonary function (Figs. 25.5, 25.8)
- New Running Problem on malignant hyperthermia

Chapter 26

- New art for hormonal control of spermatogenesis and follicular development
- Updated information:
  - developmental timeline of ovarian follicles
  - reproductive aging
- New Running problem: primary ovarian insufficiency (POI, also called premature ovarian failure) and treatment with *in vitro* activation (IVA)
- Expanded information on environmental estrogens

ACKNOWLEDGMENTS

Writing, editing, and publishing a textbook is a group project that requires the talent and expertise of many people. No one scientist has the detailed background needed in all areas to write a book of this scope, and I am indebted to all my colleagues who have so generously shared their expertise in each edition. I particularly want to acknowledge Bruce Johnson, Cornell University, Department of Neurobiology and Behavior, a superb neurobiologist and educator, who once again ensured that the chapters on neurobiology are accurate and reflect the latest developments in that rapidly changing field.

Many other people devoted their time and energy to making this book a reality, and I would like to thank them all, collectively and individually. I apologize in advance to anyone whose name I have omitted.

Reviewers

I am particularly grateful to the instructors who reviewed one or more chapters of the last edition. There were many suggestions in their thoughtful reviews that I was unable to include in the text, but I appreciate the time and thought that went into their comments. The reviewers for this edition include:

Catherine L. Carpenter, University of California, Los Angeles
Karen M. Chooljian, California State University, Fresno
Patricia J. Clark, Indiana University - Purdue University Indianapolis
Margaret T. Flemming, Austin Community College – Rio Grande Campus
Jill M. Tall Gifford, Youngstown State University
David Harris, University of Central Florida College of Medicine
James D. Herman, Texas A&M University College of Veterinary Medicine & Biomedical Sciences
Anthony Jones, Tallahassee Community College
Daniel Kueh, Emory University
David Kurjiaka, Grand Valley State University
Nathan H. Lents, John Jay College
Cheryl L. Neudauer, Minneapolis Community & Technical College
Byron Noordewier, Northwestern College
Rudy M. Ortiz, University of California, Merced
Mike Reutter, Normandale Community College
Ruy Tchao, University of the Sciences
Chad M. Wayne, University of Houston

Many other instructors and students took time to write or e-mail queries or suggestions for clarification, for which I thank them. I am always delighted to have input, and I apologize that I do not have room to acknowledge them all individually.

Specialty Reviews

No one can be an expert in every area of physiology, and I am deeply thankful for my friends and colleagues who reviewed entire chapters or answered specific questions. Even with their help, there may be errors, for which I take full responsibility. The specialty reviewers for this edition were:

James Bryant, University of Texas, Austin
Douglas Eaton, Emory University
Christina Karatzaferi, University of Thessaly, Greece
Michael Levitzky, Louisiana State University Health Science Center, New Orleans
Donald R. McCrimmon, Northwestern University Feinberg School of Medicine
Stephen Schoech, University of Memphis
Frank Powell, University of California, San Diego
Gary Sieck, Mayo Clinic, Rochester
John West, University of California, San Diego

Photographs

I would like to thank the following colleagues who generously provided micrographs from their research:

Kristen Harris, University of Texas
Flora M. Love, University of Texas
Jane Lubisher, University of Texas
Young-Jin Son, University of Texas

Supplements

Damian Hill once again worked with me to revise and improve the Instructor Resource Manual and Student Workbook that accompany the book. I believe that supplements should reflect the style and approach of the text, so I am grateful that Damian has continued to be my alter-ego for so many editions.

I would also like to thank my colleagues who helped with the test bank and media supplements for this edition: Paul Wagner, Tracy Wagner, Cheryl Neudauer, Chad Wayne, Michelle Harrison, and Myriam Alhadeff Feldman.

The Development and Production Team

Writing a manuscript is only a first step in the long and complicated process that results in a bound book with all its ancillaries. The team that works with me on book development deserves a lot of credit for the finished product. In this edition, Bill and Claire Ober, my art coauthors, created thoughtful updates to the art to make it more user-friendly. Once again, Yvo Riezebos designed a striking cover that reflects how science is really art. Anne A. Reid, my long-time developmental editor, was a joy to work with, and helps ensure that what I write can be understood by students.

The team at Pearson Education worked tirelessly to see this edition move from manuscript to bound book. My acquisitions editor, Kelsey Volker Churchman, was always supportive and ready to help. She also continued what has become a tradition with my revisions by delivering a baby girl as we were completing the production process. Ashley Williams, assistant editor, kept track of everyone and everything for us. Chriscelle Palaganas, Program Manager, provided excellent guidance and support throughout the whole production process.

The task of coordinating production fell to Pearson Project Manager Lauren Beebe. Andrea Stefanowicz handled composition and project management, and Project Manager Cynthia Mutheardy at the art house, Imagineering, managed the team that prepared Bill and Claire's art for production. Kristin Piljay was the photo researcher who found the wonderful new photos that appear in this edition. Annie Wang was the assistant media producer who kept my supplements authors on task and on schedule. Christy Lesko is the director of marketing who works with the excellent sales teams at Pearson Education and Pearson International, and Allison Rona is the Senior Marketing Manager for the anatomy and physiology list.

Special Thanks

This edition I owe a special debt of gratitude to my graduate teaching assistants: Kevin Christmas, a Ph.D. student in exercise physiology, and Michael Chirillo, an M.D.-Ph.D. student who is doing his doctoral research at UT-Austin and his medical studies at University of Texas Health Science Center at Houston. (They are shown with me in the photograph in the About the Author section. Michael is on the left, Kevin is on the right.) In addition to working with my classes of 240 students each semester, they played a major role in helping create and performing in the new Video Tutors that debut with this edition. My thanks also go to Ryan Kelley, the videographer, who in his free time is a radio-TV-film major at UT-Austin.

As always, I would like to thank my students and colleagues who looked for errors and areas that needed improvement. I've learned that awarding one point of extra credit for being the first student to report a typo works really well. My graduate teaching assistants over the years have all played a huge role in my teaching, and their input has helped shape how I teach. Many of them are now faculty members themselves. In addition to Kevin and Michael, I would particularly like to thank:

Ari Berman, Ph.D.
Lawrence Brewer, Ph.D.
Lynn Cialdella Kam, M.S., M.B.A., Ph.D.

Sarah Davies, Ph.D.
Peter English, Ph.D.
Carol C. Linder, Ph.D.
Karina Loyo-Garcia, Ph.D.
Jan M. Machart, Ph.D.
Tonya Thompson, M.D.
Patti Thorn, Ph.D.
Justin Trombold, Ph.D.
Kurt Venator, Ph.D.
Kira Wenstrom, Ph.D.

Finally, special thanks to my colleagues in the American Physiological Society and the Human Anatomy & Physiology Society, whose experiences in the classroom have enriched my own understanding of how to teach physiology. I would also like to recognize a special group of friends for their continuing support: Penelope Hansen (Memorial University, St. John's), Mary Anne Rokitka (SUNY Buffalo), Rob Carroll (East Carolina University School of Medicine), Cindy Gill (Hampshire College), and Joel Michael (Rush Medical College), as well as Ruth Buskirk, Jeanne Lagowski, Jan M. Machart and Marilla Svinicki (University of Texas).

As always, I thank my family and friends for their patience, understanding, and support during the chaos that seems inevitable with book revisions. The biggest thank you goes to my husband Andy, whose love, support, and willingness to forgo home-cooked meals on occasion help me meet my deadlines.

A Work in Progress

One of the most rewarding aspects of writing a textbook is the opportunity it has given me to meet or communicate with other instructors and students. In the years since the first edition was published, I have heard from people around the world and have had the pleasure of hearing how the book has been incorporated into their teaching and learning.

Because science textbooks are revised every 3 or 4 years, they are always works in progress. I invite you to contact me or my publisher with any suggestions, corrections, or comments about this edition. I am most reachable through e-mail at silverthorn@utexas.edu. You can reach my editor at the following address:

Applied Sciences
Pearson Education
1301 Sansome Street
San Francisco, CA 94111

Dee U. Silverthorn
University of Texas
Austin, Texas

Introduction to Physiology

PHYSIOLOGY IS AN INTEGRATIVE SCIENCE 2

LO 1.1 Define physiology.

LO 1.2 List the levels of organization from atoms to the biosphere.

LO 1.3 Name the 10 physiological organ systems of the body and give their functions.

FUNCTION AND MECHANISM 4

LO 1.4 Distinguish between mechanistic explanations and teleological explanations.

THEMES IN PHYSIOLOGY 5

LO 1.5 List and give examples of the four major themes in physiology.

HOMEOSTASIS 9

LO 1.6 Define homeostasis. What happens when homeostasis fails?

LO 1.7 Name and describe the two major compartments of the human body.

LO 1.8 Explain the law of mass balance and how it applies to the body's load of a substance.

LO 1.9 Define mass flow using mathematical units and explain how it relates to mass balance.

LO 1.10 Define clearance and give an example.

LO 1.11 Distinguish between equilibrium and steady state.

CONTROL SYSTEMS AND HOMEOSTASIS 13

LO 1.12 List the three components of a control system and give an example.

LO 1.13 Explain the relationship between a regulated variable and its setpoint.

LO 1.14 Compare local control, long-distance control, and reflex control.

LO 1.15 Explain the relationship between a response loop and a feedback loop.

LO 1.16 Compare negative feedback, positive feedback, and feedforward control. Give an example of each.

LO 1.17 Explain what happens to setpoints in biological rhythms and give some examples.

THE SCIENCE OF PHYSIOLOGY 18

LO 1.18 Explain and give examples of the following components of scientific research: independent and dependent variables, experimental control, data, replication, variability.

LO 1.19 Compare and contrast the following types of experimental study designs: blind study, double-blind study, crossover study, prospective and retrospective studies, cross-sectional study, longitudinal study, meta-analysis.

LO 1.20 Define placebo and nocebo effects and explain how they may influence the outcome of experimental studies.

Thermography of the human body. Warmer areas are red; cooler are blue.

Welcome to the fascinating study of the human body! For most of recorded history, humans have been interested in how their bodies work. Early Egyptian, Indian, and Chinese writings describe attempts by physicians to treat various diseases and to restore health. Although some ancient remedies, such as camel dung and powdered sheep horn, may seem bizarre, we are still using others, such as blood-sucking leeches and chemicals derived from medicinal plants. The way we use these treatments has changed through the centuries as we learned more about the human body.

There has never been a more exciting time in human physiology. **Physiology** is the study of the normal functioning of a living organism and its component parts, including all its chemical and physical processes. The term *physiology* literally means "knowledge of nature." Aristotle (384–322 B.C.E.) used the word in this broad sense to describe the functioning of all living organisms, not just of the human body. However, Hippocrates (ca. 460–377 B.C.E.), considered the father of medicine, used the word *physiology* to mean "the healing power of nature," and thereafter the field became closely associated with medicine. By the sixteenth century in Europe, physiology had been formalized as the study of the vital functions of the human body. Today, the term is again used to refer to the study of animals and plants.

Today, we benefit from centuries of work by physiologists who constructed a foundation of knowledge about how the human body functions. Since the 1970s, rapid advances in the fields of cellular and molecular biology have supplemented this work. A few decades ago, we thought that we would find the key to the secret of life by sequencing the human *genome*, which is the collective term for all the genetic information contained in the DNA of a species. However, this deconstructionist view of biology has proved to have its limitations, because living organisms are much more than the simple sum of their parts.

PHYSIOLOGY IS AN INTEGRATIVE SCIENCE

Many complex systems—including those of the human body—possess **emergent properties,** which are properties that cannot be predicted to exist based only on knowledge of the system's individual components. An emergent property is not a property of any single component of the system, and it is greater than the simple sum of the system's individual parts. Emergent properties result from complex, nonlinear interactions of the different components.

For example, suppose someone broke down a car into its nuts and bolts and pieces and laid them out on a floor. Could you predict that, properly assembled, these bits of metal and plastic would become a vehicle capable of converting the energy in gasoline into movement? Who could predict that the right combination of elements into molecules and assemblages of molecules would result in a living organism? Among the most complex emergent properties in humans are emotion, intelligence, and other aspects of brain function. None of these properties can be predicted from knowing the individual properties of nerve cells.

RUNNING **PROBLEM** | **What to Believe?**

Jimmy had just left his first physiology class when he got the text from his mother: *Please call. Need to ask you something.* His mother seldom texted, so Jimmy figured it must be important. "Hi, Mom! What's going on?"

"Oh, Jimmy, I don't know what to do. I saw the doctor this morning and he's telling me that I need to take insulin. But I don't want to! My type of diabetes doesn't need insulin. I think he's just trying to make me see him more by putting me on insulin. Don't you think I'm right?"

Jimmy paused for a moment. "I'm not sure, Mom. He's probably just trying to do what's best for you. Didn't you talk to him about it?"

"Well, I tried but he didn't have time to talk. You're studying these things. Can't you look it up and see if I really need insulin?"

"I guess so. Let me see what I can find out." Jimmy hung up and thought. "Now what?"

2   5   9   12   16   19   23

When the Human Genome Project (*www.genome.gov*) began in 1990, scientists thought that by identifying and sequencing all the genes in human DNA, they would understand how the body worked. However, as research advanced, scientists had to revise their original idea that a given segment of DNA contained one gene that coded for one protein. It became clear that one gene may code for many proteins. The Human Genome Project ended in 2003, but before then researchers had moved beyond genomics to *proteomics,* the study of proteins in living organisms.

Now scientists have realized that knowing that a protein is made by a particular cell does not always tell us the significance of that protein to the cell, the tissue, or the functioning organism. The exciting new areas in biological research are called functional genomics, systems biology, and integrative biology, but fundamentally these are all fields of physiology. The **integration of function** across many **levels of organization** is a special focus of physiology. (To *integrate* means to bring varied elements together to create a unified whole.)

FIGURE 1.1 illustrates levels of organization ranging from the molecular level all the way up to populations of different species living together in *ecosystems* and in the *biosphere*. The levels of organization are shown along with the various sub-disciplines of chemistry and biology related to the study of each organizational level. There is considerable overlap between the different fields of study, and these artificial divisions vary according to who is defining them. Notice, however, that physiology includes multiple levels, from molecular and cellular biology to the ecological physiology of populations.

At all levels, physiology is closely tied to anatomy. The structure of a cell, tissue, or organ must provide an efficient physical base for its function. For this reason, it is nearly impossible to study the physiology of the body without understanding the

EMERGING CONCEPTS

The Changing World of Omics

If you read the scientific literature, it appears that contemporary research has exploded into an era of "omes" and "omics." What is an "ome"? The term apparently derives from the Latin word for a mass or tumor, and it is now used to refer to a collection of items that make up a whole, such as a genome. One of the earliest uses of the "ome" suffix in biology is the term *biome*, meaning all organisms living in a major ecological region, such as the marine biome or the desert biome. A genome is all the genes in an organism, and a proteome includes all the proteins in that organism.

The related adjective "omics" describes the research related to studying an "ome." Adding "omics" to a root word has become the cutting-edge way to describe a research field. For example, *proteomics*, the study of proteins in living organisms, is now as important as *genomics*, the sequencing of DNA (the genome). If you search the Internet, you will find the *transcriptome* (RNA), *metabolomics* (study of metabolic pathways), and *pharmacogenomics* (the influence of genetics on the body's response to drugs). There is even a journal named *OMICS!*

The Physiome Project (*www.physiome.org* and *physiomeproject.org*) is an organized international effort to coordinate molecular, cellular, and physiological information about living organisms into an Internet database. Scientists around the world apply this information in their own research efforts to create better drugs or genetic therapies for curing and preventing disease. The Physiome Project is an ambitious undertaking that promises to integrate information from diverse areas of research so that we can improve our understanding of the complex processes we call life.

underlying anatomy. Because of the interrelationship of anatomy and physiology, you will find Anatomy Summaries throughout the book. These special review features illustrate the anatomy of the physiological systems at different levels of organization.

At the most basic level of organization shown in Figure 1.1, atoms of elements link together to form molecules. Collections of molecules in living organisms form **cells,** the smallest unit of structure capable of carrying out all life processes. A lipid and protein barrier called the **cell membrane** (also called the *plasma membrane*) separates cells from their external environment. Simple organisms are composed of only one cell, but complex organisms have many cells with different structural and functional specializations.

Collections of cells that carry out related functions are called **tissues** {*texere,* to weave}. Tissues form structural and functional units known as **organs** {*organon,* tool}, and groups of organs integrate their functions to create **organ systems.** Chapter 3 reviews the anatomy of cells, tissues and organs.

The 10 physiological organ systems in the human body are illustrated in **FIGURE 1.2**. Several of the systems have alternate names, given in parentheses, that are based on the organs of the system rather than the function of the system. The **integumentary system** {*integumentum,* covering}, composed of the skin, forms a protective boundary that separates the body's internal environment from the external environment (the outside world). The **musculoskeletal system** provides support and body movement.

Four systems exchange materials between the internal and external environments. The **respiratory (pulmonary) system** exchanges gases; the **digestive (gastrointestinal) system** takes up nutrients and water and eliminates wastes; the **urinary (renal) system** removes excess water and waste material; and the **reproductive system** produces eggs or sperm.

The remaining four systems extend throughout the body. The **circulatory (cardiovascular) system** distributes materials by pumping blood through vessels. The **nervous** and **endocrine systems** coordinate body functions. Note that the figure shows them as a continuum rather than as two distinct systems. Why? Because the lines between these two systems have blurred as we have learned more about the integrative nature of physiological function.

The one system not illustrated in Figure 1.2 is the diffuse **immune system,** which includes but is not limited to the anatomical structures known as the *lymphatic system.* The specialized cells of the immune system are scattered throughout the body.

FIG. 1.1   Levels of organization and the related fields of study

FIG. 1.2 Organ systems of the human body and their integration

System Name	Includes	Representative Functions	The Integration between Systems of the Body
Circulatory	Heart, blood vessels, blood	Transport of materials between all cells of the body	
Digestive	Stomach, intestine, liver, pancreas	Conversion of food into particles that can be transported into the body; elimination of some wastes	
Endocrine	Thyroid gland, adrenal gland	Coordination of body function through synthesis and release of regulatory molecules	
Immune	Thymus, spleen, lymph nodes	Defense against foreign invaders	
Integumentary	Skin	Protection from external environment	
Musculoskeletal	Skeletal muscles, bone	Support and movement	
Nervous	Brain, spinal cord	Coordination of body function through electrical signals and release of regulatory molecules	
Reproductive	Ovaries and uterus, testes	Perpetuation of the species	
Respiratory	Lungs, airways	Exchange of oxygen and carbon dioxide between the internal and external environments	
Urinary	Kidneys, bladder	Maintenance of water and solutes in the internal environment; waste removal	

This schematic figure indicates relationships between systems of the human body. The interiors of some hollow organs (shown in white) are part of the external environment.

They protect the internal environment from foreign substances by intercepting material that enters through the intestines and lungs or through a break in the skin. In addition, immune tissues are closely associated with the circulatory system.

Traditionally, physiology courses and books are organized by organ system. Students study cardiovascular physiology and regulation of blood pressure in one chapter, and then study the kidneys and control of body fluid volume in a different chapter. In the functioning human, however, the cardiovascular and renal systems communicate with each other, so that a change in one is likely to cause a reaction in the other. For example, body fluid volume influences blood pressure, while changes in blood pressure alter kidney function because the kidneys regulate fluid volume. In this book, you will find several chapters devoted to topics of integrated function.

Developing skills to help you understand how the different organ systems work together is just as important as memorizing facts. One way physiologists integrate information is by using visual representations of physiological processes called maps. The Focus on Mapping feature in this chapter helps you learn how to make maps. The first type of map, shown in **FIGURE 1.3a**, is a schematic representation of structure or function. The second

type of map, shown in Figure 1.3b, diagrams a physiological process as it proceeds through time. These maps are called *flow charts* or *process maps*. The end-of-chapter questions throughout the book feature lists of selected terms that you can use to practice mapping.

FUNCTION AND MECHANISM

We define physiology as the normal functioning of the body, but physiologists are careful to distinguish between function and mechanism. The **function** of a physiological system or event is the "why" of the system or event: Why does a certain response help an animal survive in a particular situation? In other words, what is the *adaptive significance* of this event for this animal?

For example, humans are large, mobile, terrestrial animals, and our bodies maintain relatively constant water content despite living in a dry, highly variable external environment. Dehydration is a constant threat to our well-being. What processes have evolved in our anatomy and physiology that allow us to survive in this hostile environment? One is the production of highly concentrated urine by the kidney, which allows the body to conserve water. This statement tells us *why* we produce concentrated urine but does not tell us *how* the kidney accomplishes that task.

Thinking about a physiological event in terms of its adaptive significance is the **teleological approach** to science. For example, the teleological answer to the question of why red blood cells transport oxygen is "because cells need oxygen and red blood cells bring it to them." This answer explains *why* red blood cells transport oxygen—their function—but says nothing about *how* the cells transport oxygen.

In contrast, most physiologists study physiological processes, or **mechanisms**—the "how" of a system. The **mechanistic approach** to physiology examines process. The mechanistic answer to the question "How do red blood cells transport oxygen?" is "Oxygen binds to hemoglobin molecules in the red blood cells." This very concrete answer explains exactly how oxygen transport occurs but says nothing about the significance of oxygen transport to the animal.

Students often confuse these two approaches to thinking about physiology. Studies have shown that even medical students tend to answer questions with teleological explanations when the more appropriate response would be a mechanistic explanation.[1] Often, they do so because instructors ask why a physiological event occurs when they really want to know how it occurs. Staying aware of the two approaches will help prevent confusion.

Although function and mechanism seem to be two sides of the same coin, it is possible to study mechanisms, particularly at the cellular and subcellular level, without understanding their function in the life of the organism. As biological knowledge becomes more complex, scientists sometimes become so involved in studying complex processes that they fail to step back and look at the significance of those processes to cells, organ systems, or the animal. Conversely, it is possible to use teleological thinking incorrectly by saying, "Oh, in this situation the body needs to do this." *This* may be a good solution, but if a mechanism for doing *this* doesn't exist, the situation cannot be corrected.

Applying the concept of integrated functions and mechanisms is the underlying principle in **translational research,** an approach sometimes described as "bench to bedside." Translational research uses the insights and results gained from basic biomedical research on mechanisms to develop treatments and strategies for preventing human diseases. For example, researchers working on rats found that a chemical from the pancreas named *amylin* reduced the rats' food intake. These findings led directly to a translational research study in which human volunteers injected a synthetic form of amylin and recorded their subsequent food intake, but without intentionally modifying their lifestyle.[2] The drug suppressed food intake in humans, and was later approved by the Food and Drug Administration for treatment of diabetes mellitus.

At the systems level, we know about most of the mechanics of body function from centuries of research. The unanswered questions today mostly involve integration and control of these mechanisms, particularly at the cellular and molecular levels. Nevertheless, explaining what happens in test tubes or isolated cells can only partially answer questions about function. For this reason, animal and human trials are essential steps in the process of applying basic research to treating or curing diseases.

THEMES IN PHYSIOLOGY

"Physiology is not a science or a profession but a point of view."[3] Physiologists pride themselves on relating the mechanisms they study to the functioning of the organism as a whole. For students, being able to think about how multiple body systems integrate their function is one of the more difficult aspects of learning physiology. To develop expertise in physiology, you must do more than simply memorize facts and learn new terminology. Researchers have found that the ability to solve problems requires a conceptual framework, or "big picture," of the field.

This book will help you build a conceptual framework for physiology by explicitly emphasizing the basic biological concepts, or themes, that are common to all living organisms. These concepts form patterns that repeat over and over, and you will begin to recognize them when you encounter them in specific contexts. Pattern recognition is an important skill in healthcare professions and it will also simplify learning physiology.

RUNNING **PROBLEM**

When Jimmy got back to his room, he sat down at his computer and went to the Internet. He typed *diabetes* in his search box—and came up with 77.2 million results. "That's not going to work. What about *insulin*?" 14 million results. "How in the world am I going to get any answers?" He clicked on the first sponsored ad, for a site called *type2-diabetes-info.com*. That might be good. His mother had type 2 diabetes. But it was for a pharmaceutical company trying to sell him a drug. "What about this? *WhyInsulin .com*—That might give some answers." But it, too, was trying to sell something. "Maybe my physiology prof can help me. I'll ask tomorrow."

Q1: *What search terms could Jimmy have used to get fewer results?*

[1] D. R. Richardson. A survey of students' notions of body function as teleologic or mechanistic. *Advan Physiol Educ* 258: 8–10, Jun 1990. (*http://advan .physiology.org*)

[2] S. R. Smith *et al*. Pramlintide treatment reduces 24-h caloric intake and meal sizes and improves control of eating in obese subjects: a 6-wk translational research study. *Am J Physiol Endocrinol Metab* 293: E620–E627, 2007.

[3] R. W. Gerard. *Mirror to Physiology: A Self-Survey of Physiological Science.* Washington, DC: American Physiology Society, 1958.

FIG. 1.3 FOCUS ON . . .

Mapping

Mapping is a nonlinear way of organizing material based on the theory that individuals have their own unique ways of integrating new material with what they already know. Mapping is a useful study tool because creating a map requires higher-level thinking about the relationships between pieces of information. Studies have shown that when people interact with information by organizing it in their own way before they load it into memory, their understanding and retention of information improves.

Mapping is not just a study technique. Experts in a field make maps when they are trying to integrate newly acquired information into their knowledge base, and they may create two or three versions of a map before they are satisfied that it represents their understanding. Scientists map out the steps in their experiments. Healthcare professionals create maps to guide them while diagnosing and treating patients.

A map can take a variety of forms but usually consists of terms (words or short phrases) linked by arrows to indicate associations. You can label the connecting arrows to describe the type of linkage between the terms (structure/function, cause/effect) or with explanatory phrases ("is composed of"). You may also choose to use different colors for arrows and terms to represent different categories of ideas. Maps in physiology usually focus either on the relationships between anatomical structures and physiological processes (*structure/function* maps, Fig. 1.3a) or on normal homeostatic control pathways and responses to abnormal (pathophysiological) events (*process maps*, or *flow charts*, Fig. 1.3b). If appropriate, a map may also include graphs, diagrams, or pictures.

Many maps appear in this textbook, and they may serve as the starting point for your own maps. However, the real benefit of mapping comes from preparing maps yourself. By mapping information on your own, you think about the relationships between terms, organize concepts into a hierarchical structure, and look for similarities and differences between items. Interaction with the material in this way helps you process it into long-term memory instead of simply memorizing bits of information and forgetting them.

Some people do not like the messiness of hand-drawn maps. There are several electronic ways of making maps, including PowerPoint and free and commercial software programs.

PowerPoint

1. Select the completely blank slide from FORMAT - SLIDE LAYOUT.
2. Use AUTOSHAPES to create boxes/ovals and arrows. To format the autoshape, right-click on it after you have drawn it. You can change fill color and line color.
3. Use INSERT - TEXT BOX to label your arrows and put terms inside your shapes.

Software

Free concept mapping software is available from IHMC Cmap-Tools at *http://cmap.ihmc.us*. Or search for the term *free concept map* to find other resources on the Web. A popular commercial program for mapping is Inspiration (*www.inspiration.com*).

Getting Started with Mapping

1. First, **select the terms or concepts to map.** (In every chapter of this text, the end-of-chapter questions include at least one list of terms to map.) Sometimes it is helpful to write the terms on individual slips of paper or on 1-by-2-inch sticky notes so that you can rearrange the map more easily.

2. Usually the most difficult part of mapping is **deciding where to begin.** Start by grouping related terms in an organized fashion. You may find that you want to put some terms into more than one group. Make a note of these terms, as they will probably have several arrows pointing to them or leading away from them.

3. Now try to **create some hierarchy with your terms.** You may arrange the terms on a piece of paper, on a table, or on the floor. In a structure/function map, start at the top with the most general, most important, or overriding concept— the one from which all the others stem. In a process map, start with the first event to occur.

   Next, either break down the key idea into progressively more specific parts using the other concepts or follow the event through its time course. Use arrows to point the direction of linkages and include horizontal links to tie related concepts together. The downward development of the map will generally mean either an increase in complexity or the passage of time.

   You may find that some of your arrows cross each other. Sometimes you can avoid this by rearranging the terms on the map. Labeling the linking arrows with explanatory words may be useful. For example,

   $$\text{channel proteins} \xrightarrow{\text{form}} \text{open channels}$$

   Color can be very effective on maps. You can use colors for different types of links or for different categories of terms. You may also add pictures and graphs that are associated with specific terms in your map.

Types of Maps

(a) A map showing structure/function relationships

(b) A process map, or flow chart

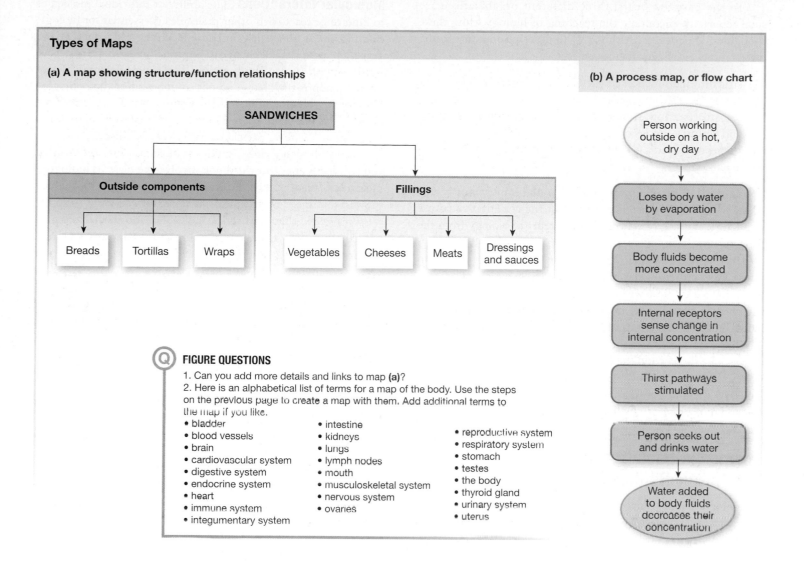

SANDWICHES

Outside components
- Breads
- Tortillas
- Wraps

Fillings
- Vegetables
- Cheeses
- Meats
- Dressings and sauces

Person working outside on a hot, dry day
↓
Loses body water by evaporation
↓
Body fluids become more concentrated
↓
Internal receptors sense change in internal concentration
↓
Thirst pathways stimulated
↓
Person seeks out and drinks water
↓
Water added to body fluids decreases their concentration

Ⓠ FIGURE QUESTIONS

1. Can you add more details and links to map **(a)**?
2. Here is an alphabetical list of terms for a map of the body. Use the steps on the previous page to create a map with them. Add additional terms to the map if you like.

- bladder
- blood vessels
- brain
- cardiovascular system
- digestive system
- endocrine system
- heart
- immune system
- integumentary system
- intestine
- kidneys
- lungs
- lymph nodes
- mouth
- musculoskeletal system
- nervous system
- ovaries
- reproductive system
- respiratory system
- stomach
- testes
- the body
- thyroid gland
- urinary system
- uterus

4. Once you have created your map, **sit back and think about it.** Are all the items in the right place? You may want to move them around once you see the big picture. Revise your map to fill in the picture with new concepts or to correct wrong links. Review by recalling the main concept and then moving to the more specific details. Ask yourself questions like, What is the cause? effect? What parts are involved? What are the main characteristics?

5. Science is a collaborative field. A useful way to study with a map is to **trade maps with a classmate** and try to understand each other's maps. Your maps will almost certainly not look the same! It's OK if they are different. Remember that your map reflects the way you think about the subject, which may be different from the way someone else thinks about it. Did one of you put in something the other forgot? Did one of you have an incorrect link between two items?

6. **Practice making maps.** The study questions in each chapter will give you some ideas of what you should be mapping. Your instructor can help you get started.

In the past few years, three different organizations issued reports to encourage the teaching of biology using these fundamental concepts. Although the descriptions vary in the three reports, five major themes emerge:

1. structure and function across all levels of organization
2. energy transfer, storage, and use
3. information flow, storage, and use within single organisms and within a species of organism
4. homeostasis and the control systems that maintain it
5. evolution

In addition, all three reports emphasize the importance of understanding how science is done and of the quantitative nature of biology. **TABLE 1.1** lists the core concepts in biology from the three reports.

In this book, we focus on the four themes most related to physiology: structure-function relationships, biological energy use, information flow within an organism, and homeostasis and the control systems that maintain it. The first six chapters introduce the fundamentals of these themes. You may already be familiar with some of them from earlier biology and chemistry classes. The themes and their associated concepts, with variations, then appear over and over in subsequent chapters of this book. Look for them in the summary material at the end of the chapters and in the end-of-chapter questions as well.

Theme 1: Structure and Function Are Closely Related

The integration of structure and function extends across all levels of organization, from the molecular level to the intact body. This theme subdivides into two major ideas: molecular interactions and compartmentation.

Molecular Interactions The ability of individual molecules to bind to or react with other molecules is essential for biological function. A molecule's function depends on its structure and shape, and even a small change to the structure or shape may have significant effects on the function. The classic example of this phenomenon is the change in one amino acid of the hemoglobin protein. (Hemoglobin is the oxygen-carrying pigment of the blood.) This one small change in the protein converts normal hemoglobin to the form associated with sickle cell disease.

Many physiologically significant molecular interactions that you will learn about in this book involve the class of biological molecules called *proteins*. Functional groups of proteins include *enzymes* that speed up chemical reactions, *signal molecules* and the *receptor proteins* that bind signal molecules, and specialized proteins that function as biological pumps, filters, motors, or transporters. [Chapter 2 describes molecular interactions involving proteins in more detail.]

Interactions between proteins, water, and other molecules influence cell structure and the mechanical properties of cells and tissues. Mechanical properties you will encounter in your study of physiology include *compliance* (ability to stretch), *elastance* (stiffness or the ability to return to the unstretched state), strength, flexibility, and fluidity (*viscosity*).

Compartmentation **Compartmentation** is the division of space into separate compartments. Compartments allow a cell, a tissue, or an organ to specialize and isolate functions. Each level of organization is associated with different types of compartments. At the macroscopic level, the tissues and organs of the body form discrete functional compartments, such as body cavities or the insides of hollow organs. At the microscopic level, cell membranes separate cells from the fluid surrounding them and also create tiny compartments within the cell called organelles. [Compartmentation is the theme of Chapter 3.]

TABLE 1.1	**Biology Concepts**		
Scientific Foundations for Future Physicians (HHMI and AAMC)[1]	**Vision and Change (NSF and AAAS)**[2]	**The 2010 Advanced Placement Biology Curriculum (College Board)**[3]
Structure/function from molecules to organisms	Structure and function (anatomy and physiology)	Relationship of structure to function
Physical principles applied to living systems. Chemical principles applied to living systems	Pathways and transformations of energy and matter	Energy transfer
Biomolecules and their functions	Information flow, exchange, and storage	Continuity and change
Organisms sense and control their internal environment and respond to external change	Systems	Regulation ("a state of dynamic balance")
Evolution as an organizing principle	Evolution	Evolution

[1]*Scientific Foundations for Future Physicians*. Howard Hughes Medical Institute (HHMI) and the Association of American Medical Colleges (AAMC), 2009. *www.aamc.org/scientificfoundations*
[2]*Vision and Change: A Call to Action*. National Science Foundation (NSF) and American Association for the Advancement of Science (AAAS). 2011. *http://visionandchange.org/finalreport*. The report mentioned the integration of science and society as well.
[3]*College Board AP Biology Course Description*, The College Board, 2010. *http://apcentral.collegeboard.com/apc/public/repository/ap-biology-course-description.pdf*. The AP report also included "Interdependence in Nature" and "Science, Technology and Society" as two of their eight themes.

Theme 2: Living Organisms Need Energy

Growth, reproduction, movement, homeostasis—these and all other processes that take place in an organism require the continuous input of energy. Where does this energy come from, and how is it stored? We will answer those questions and describe some of the ways that energy in the body is used for building and breaking down molecules [in Chapter 4.] In subsequent chapters, you will learn how energy is used to transport molecules across cell membranes and to create movement.

Theme 3: Information Flow Coordinates Body Functions

Information flow in living systems ranges from the transfer of information stored in DNA from generation to generation (genetics) to the flow of information within the body of a single organism. At the organismal level, information flow includes translation of DNA's genetic code into proteins responsible for cell structure and function.

In the human body, information flow between cells *coordinates function. Cell-to-cell communication* uses chemical signals, electrical signals, or a combination of both. Information may go from one cell to its neighbors (local communication) or from one part of the body to another (long-distance communication). [Chapter 6 discusses chemical communication in the body.]

When chemical signals reach their target cells, they must get their information into the cell. Some molecules are able to pass through the barrier of the cell membrane, but signal molecules that cannot enter the cell must pass their message across the cell membrane. [How molecules cross biological membranes is the topic of Chapter 5.]

Theme 4: Homeostasis Maintains Internal Stability

Organisms that survive in challenging habitats cope with external variability by keeping their internal environment relatively stable, an ability known as **homeostasis** {*homeo-*, similar + *-stasis*, condition}. Homeostasis and regulation of the internal environment are key principles of physiology and underlying themes in each chapter of this book. The next section looks in detail at the key elements of this important theme.

HOMEOSTASIS

The concept of a relatively stable internal environment is attributed to the French physician Claude Bernard in the mid-1800s. During his studies of experimental medicine, Bernard noted the stability of various physiological functions, such as body temperature, heart rate, and blood pressure. As the chair of physiology at the University of Paris, he wrote "C'est la fixité du milieu intérieur qui est la condition d'une vie libre et indépendante." (It is the constancy of the internal environment that is the condition for

RUNNING **PROBLEM**

After his second physiology class, Jimmy introduced himself to his professor and explained his problem. The professor's first suggestion was simple: try to narrow the search. "One of the best ways to search is to combine terms using the connector AND. If you remember set theory from your math class, the connector AND will give you the intersection of the sets. In other words, you'll get only the results that occur in both sets."

Seemed simple enough. Jimmy went back to the Internet and tried *diabetes and insulin*. That search still had 43.6 million results but some of the results on the first page after the ads looked pretty good: *mayoclinic.com* and *diabetes.org*. Now he was getting somewhere.

Q2: *What kinds of web sites should Jimmy be looking for in his results list, and how can he recognize them?*

2   5   **9**   12   16   19   23

a free and independent life.)[4] This idea was applied to many of the experimental observations of his day, and it became the subject of discussion among physiologists and physicians.

In 1929, an American physiologist named Walter B. Cannon wrote a review for the American Physiological Society.[5] Using observations made by numerous physiologists and physicians during the nineteenth and early twentieth centuries, Cannon proposed a list of variables that are under homeostatic control. We now know that his list was both accurate and complete. Cannon divided his variables into what he described as environmental factors that affect cells (osmolarity, temperature, and pH) and "materials for cell needs" (nutrients, water, sodium, calcium, other inorganic ions, oxygen, as well as "internal secretions having general and continuous effects"). Cannon's "internal secretions" are the hormones and other chemicals that our cells use to communicate with one another.

In his essay, Cannon created the word *homeostasis* to describe the regulation of the body's internal environment. He explained that he selected the prefix *homeo-* (meaning *like* or *similar*) rather than the prefix *homo-* (meaning *same*) because the internal environment is maintained within a range of values rather than at an exact fixed value. He also pointed out that the suffix *–stasis* in this instance means a *condition*, not a state that is static and unchanging. Cannon's homeostasis, therefore, is a state of maintaining "a similar condition," similar to Claude Bernard's relatively constant internal environment.

Some physiologists contend that a literal interpretation of *stasis* {a state of standing} in the word *homeostasis* implies a static, unchanging state. They argue that we should use the word *homeodynamics* instead, to reflect the small changes constantly taking

---

[4] C. Bernard. *Introduction á l'étude de la medicine*, Paris: J.-B. Baillière, 1865. (*www.gutenberg.org/ebooks/16234*).

[5] W. B. Cannon. Organization for physiological homeostasis. *Physiol Rev* 9: 399–443, 1929.

place in our internal environment {*dynamikos,* force or power}. Whether the process is called homeostasis or homeodynamics, the important concept to remember is that the body monitors its internal state and takes action to correct disruptions that threaten its normal function.

If the body fails to maintain homeostasis of the critical variables listed by Walter Cannon, then normal function is disrupted and a disease state, or **pathological** condition {*pathos,* suffering}, may result. Diseases fall into two general groups according to their origin: those in which the problem arises from internal failure of some normal physiological process, and those that originate from some outside source. Internal causes of disease include the abnormal growth of cells, which may cause cancer or benign tumors; the production of antibodies by the body against its own tissues (autoimmune diseases); and the premature death of cells or the failure of cell processes. Inherited disorders are also considered to have internal causes. External causes of disease include toxic chemicals, physical trauma, and foreign invaders such as viruses and bacteria.

In both internally and externally caused diseases, when homeostasis is disturbed, the body attempts to compensate (**FIG. 1.4**). If the compensation is successful, homeostasis is restored. If compensation fails, illness or disease may result. The study of body functions in a disease state is known as **pathophysiology.** You will encounter many examples of pathophysiology as we study the various systems of the body.

One very common pathological condition in the United States is **diabetes mellitus,** a metabolic disorder characterized by abnormally high blood glucose concentrations. Although we speak of diabetes as if it were a single disease, it is actually a whole family of diseases with various causes and manifestations. You will learn more about diabetes in the focus boxes scattered throughout the chapters of this book. The influence of this one disorder on many systems of the body makes it an excellent example of the integrative nature of physiology.

What Is the Body's Internal Environment?

Claude Bernard wrote of the "constancy of the internal environment," but why is constancy so essential? As it turns out, most cells in our bodies are not very tolerant of changes in their surroundings. In this way they are similar to early organisms that lived in tropical seas, a stable environment where salinity, oxygen content, and pH vary little and where light and temperature cycle in predictable ways. The internal composition of these ancient creatures was almost identical to that of seawater. If environmental conditions changed, conditions inside the primitive organisms changed as well. Even today, marine invertebrates cannot tolerate significant changes in salinity and pH, as you know if you have ever maintained a saltwater aquarium.

In both ancient and modern times, many marine organisms relied on the constancy of their external environment to keep their internal environment in balance. In contrast, as organisms evolved and migrated from the ancient seas into estuaries, then into freshwater environments and onto the land, they encountered highly variable external environments. Rains dilute the salty water of estuaries, and organisms that live there must cope with the influx of water into their body fluids. Terrestrial organisms, including humans, face the challenge of dehydration—constantly losing internal water to the dry air around them. Keeping the internal environment stable means balancing that water loss with appropriate water intake.

But what exactly is the internal environment of the body? For multicellular animals, it is the watery internal environment that surrounds the cells, a "sea within" the body called the **extracellular fluid** {*extra-,* outside of} (**FIG. 1.5**). Extracellular fluid (ECF) serves as the transition between an organism's external environment and the **intracellular fluid (ICF)** inside cells {*intra-,* within}. Because extracellular fluid is a buffer zone between cells and the outside world, elaborate physiological processes have evolved to keep its composition relatively stable.

When the extracellular fluid composition varies outside its normal range of values, compensatory mechanisms activate and try to return the fluid to the normal state. For example, when you drink a large volume of water, the dilution of your extracellular fluid triggers a mechanism that causes your kidneys to remove excess water and protect your cells from swelling. Most cells of multicellular animals do not tolerate much change. They depend on the constancy of extracellular fluid to maintain normal function.

Homeostasis Depends on Mass Balance

In the 1960s, a group of conspiracy theorists obtained a lock of Napoleon Bonaparte's hair and sent it for chemical analysis in

FIG. 1.4   Homeostasis

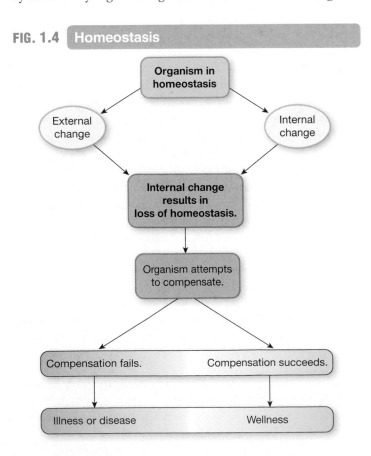

Organism in homeostasis

External change

Internal change

Internal change results in loss of homeostasis.

Organism attempts to compensate.

Compensation fails.

Compensation succeeds.

Illness or disease

Wellness

FIG. 1.5 | The body's internal and external environments

(a) Extracellular fluid is a buffer between cells and the outside world.

Cells contain intracellular fluid (ICF).

The cell membrane separates cells from the ECF.

Cells are surrounded by the extracellular fluid (ECF).

Cells ← ECF ← External environment

(b) A box diagram represents the ECF, ICF, and external environment as three separate compartments.

External environment ↔ Extracellular fluid (ECF) ↔ Intracellular fluid (ICF)

Q **FIGURE QUESTION**
Put a * on the cell membrane of the box diagram.

an attempt to show that he died from arsenic poisoning. Today, a group of students sharing a pizza joke about the garlic odor on their breath. At first glance these two scenarios appear to have little in common, but in fact Napoleon's hair and "garlic breath" both demonstrate how the human body works to maintain the balance that we call homeostasis.

The human body is an open system that exchanges heat and materials with the outside environment. To maintain homeostasis,

the body must maintain mass balance. The **law of mass balance** says that if the amount of a substance in the body is to remain constant, any gain must be offset by an equal loss (**FIG. 1.6a**). The amount of a substance in the body is also called the body's **load**, as in "sodium load."

For example, water loss to the external environment (output) in sweat and urine must be balanced by water intake from the external environment plus metabolic water production (input).

FIG. 1.6 | Mass balance

(a) Mass balance in an open system requires input equal to output.

Input

To maintain constant level, output must equal input.

Output

(b) Mass balance in the body

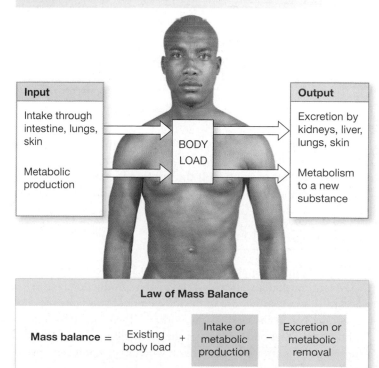

Input		Output
Intake through intestine, lungs, skin	BODY LOAD	Excretion by kidneys, liver, lungs, skin
Metabolic production		Metabolism to a new substance

Law of Mass Balance

Mass balance = Existing body load + Intake or metabolic production − Excretion or metabolic removal

The concentrations of other substances, such as oxygen and carbon dioxide, salts, and hydrogen ions (pH), are also maintained through mass balance. The following equation summarizes the law of mass balance:

> Total amount of = intake + production −
> substance x in the body   excretion − metabolism

Most substances enter the body from the outside environment, but some (such as carbon dioxide) are produced internally through metabolism (Fig. 1.6b). In general, water and nutrients enter the body as food and drink absorbed through the intestine. Oxygen and other gases and volatile molecules enter through the lungs. A few lipid-soluble chemicals make their way to the internal environment by penetrating the barrier of the skin.

To maintain mass balance, the body has two options for output. The simplest option is simply to excrete the material. **Excretion** is defined as the elimination of material from the body, usually through the urine, feces, lungs, or skin. For example, carbon dioxide (CO_2) produced during metabolism is excreted by the lungs. Many foreign substances that enter the body, such as drugs or artificial food additives, are excreted by the liver and kidneys. (Any foreign substance in the body is called a *xenobiotic*, from the Greek word *xenos*, a stranger.)

A second output option for maintaining mass balance is to convert the substance to a different substance through metabolism. Nutrients that enter the body become the starting point for metabolic pathways that convert the original nutrient to a different molecule. However, converting the original nutrient to something different then creates a new mass balance disturbance by adding more of the new substance, or *metabolite*, to the body. (*Metabolite* is the general term for any product created in a metabolic pathway.)

Scientists use **mass flow** to follow material throughout the body. For substance x, the equation for mass flow is

> Mass flow = concentration of x × volume flow
> (amount x/min) = (amount x/vol) × (vol/min)

Mass flow can be used to determine the rate of intake, output, or production of x.

For example, suppose a person is given an intravenous (IV) infusion of glucose solution that has a concentration of 50 grams of glucose per liter of solution. If the infusion is given at a rate of 2 milliliters per minute, the mass flow of glucose into the body is:

$$\frac{50 \text{ g glucose}}{1000 \text{ mL solution}} \times 2 \text{ mL solution/min} = 0.1 \text{ g glucose/min}$$

The rate of glucose input into the body is 0.1 g glucose/min.

Mass flow applies not only to the entry, production, and removal of substances but also to the movement of substances from one compartment in the body to another. When materials enter the body, they first become part of the extracellular fluid. Where a substance goes after that depends on whether or not it can cross the barrier of the cell membrane and enter the cells.

RUNNING **PROBLEM**

Jimmy called his mother with the news that he had found some good information on the Mayo Clinic and American Diabetes Association web sites. According to both those organizations, someone with type 2 diabetes might begin to require insulin as the disease progressed. But his mother was still not convinced that she needed to start insulin injections.

"My friend Ahn read that some doctors say that if you eat a high-fiber diet, you won't need any other treatment for diabetes."

"Mom, that doesn't sound right to me."

"But it must be," Jimmy's mother replied. "It says so in The Doctors' Medical Library."

Q3: *Go to The Doctors' Medical Library at www.medical-library .net and search for the article called "Fiber" by typing the word into the Search box or by using the alphabetical listing of Library Articles. What does Dr. Kennedy, the author of the article, say about high-fiber diet and diabetes?*

Q4: *Should Jimmy's mother believe what it says on this web site? How can Jimmy find out more about who created the site and what their credentials are?*

Excretion Clears Substances from the Body

It is relatively easy to monitor how much of a substance enters the body from the outside world, but it is more difficult to track molecules inside the body to monitor their excretion or metabolism. Instead of directly measuring the substance, we can follow the rate at which the substance disappears from the blood, a concept called **clearance.** Clearance is usually expressed as a volume of blood *cleared* of substance x per unit of time. For this reason, clearance is only an indirect measure of how substance x is eliminated. For example, urea is a normal metabolite produced from protein metabolism. A typical value for urea clearance is 70 mL plasma cleared of urea per minute, written as 70 mL/min.

The kidney and the liver are the two primary organs that clear solutes from the body. *Hepatocytes* {*hepaticus*, pertaining to the liver + *cyte*, cell}, or liver cells, metabolize many different types of molecules, especially xenobiotics such as drugs. The resulting metabolites may be secreted into the intestine for excretion in the feces or released into the blood for removal by the kidneys. Pharmaceutical companies testing chemicals for their potential use as therapeutic drugs must know the clearance of the chemical before they can develop the proper dosing schedule.

Clearance also takes place in tissues other than the liver and kidneys. Saliva, sweat, breast milk, and hair all contain substances that have been cleared from the body. Salivary secretion of the hormone *cortisol* provides a simple noninvasive source of hormone for monitoring chronic stress.

An everyday example of clearance is "garlic breath," which occurs when volatile lipid-soluble garlic compounds in the blood pass into the airways and are exhaled. The lungs also clear ethanol in the blood, and exhaled alcohol is the basis of the "breathalyzer" test used by law enforcement agencies. Drugs and alcohol secreted into breast milk are potentially dangerous because a breast-feeding infant will ingest these substances.

The 1960s analysis of Napoleon Bonaparte's hair tested it for arsenic because hair follicles help clear some compounds from the body. The test results showed significant concentrations of the poison in his hair, but the question remains whether Napoleon was murdered, poisoned accidentally, or died from stomach cancer.

CONCEPT CHECK

1. If a person eats 12 milligrams (mg) of salt in a day and excretes 11 mg of it in the urine, what happened to the remaining 1 mg?

2. Glucose is metabolized to CO_2 and water. Explain the effect of glucose metabolism on mass balance in the body.

Homeostasis Does Not Mean Equilibrium

When physiologists talk about homeostasis, they are speaking of the stability of the body's *internal environment*—in other words, the stability of the extracellular fluid compartment (ECF). One reason for focusing on extracellular fluid homeostasis is that it is relatively easy to monitor by taking a blood sample. When you centrifuge blood, it separates into two parts: **plasma,** the fluid component, plus the heavier blood cells. Plasma is part of the extracellular fluid compartment, and its composition can be easily analyzed. It is much more difficult to follow what is taking place in the intracellular fluid compartment (ICF), although cells do maintain *cellular homeostasis*.

In a state of homeostasis, the composition of both body compartments is relatively stable. This condition is a dynamic **steady state.** The modifier *dynamic* indicates that materials are constantly moving back and forth between the two compartments. In a steady state, there is no *net* movement of materials between the compartments.

Steady state is not the same as **equilibrium** {*aequus,* equal + *libra,* balance}, however. Equilibrium implies that the composition of the body compartments is identical. If we examine the composition of the ECF and ICF, we find that the concentrations of many substances are different in the two compartments (**FIG. 1.7**). For example, sodium (Na^+) and chloride (Cl^-) are far more concentrated in the ECF than in the ICF, while potassium (K^+) is most concentrated in the ICF. Because of these concentration differences, the two fluid compartments are not at equilibrium. Instead the ECF and ICF exist in a state of relatively stable **disequilibrium** {*dis-* is a negative prefix indicating the opposite of the base noun}. For living organisms, the goal of homeostasis is to maintain the dynamic steady states of the body's compartments, not to make the compartments the same.

FIG. 1.7 Steady-state disequilibrium

The body compartments are in a dynamic steady state but are not at equilibrium. Ion concentrations are very different in the extracellular fluid compartment (ECF) and the intracellular fluid compartment (ICF).

CONTROL SYSTEMS AND HOMEOSTASIS

To maintain homeostasis, the human body monitors certain key functions, such as blood pressure and blood glucose concentration, that must stay within a particular operating range if the body is to remain healthy. These important **regulated variables** are kept within their acceptable (normal) range by physiological control mechanisms that kick in if the variable ever strays too far from its **setpoint,** or optimum value. There are two basic patterns of control mechanisms: local control and long-distance reflex control.

In their simplest form, all **control systems** have three components (**FIG. 1.8**): (1) an input signal; (2) a controller, or **integrating center** {*integrare,* to restore}, that integrates incoming information and initiates an appropriate response; and (3) an output signal that creates a response. Long-distance reflex control systems are more complex than this simple model, however, as they may include input from multiple sources and have output that acts on multiple targets.

Local Control Is Restricted to a Tissue

The simplest form of control is **local control,** which is restricted to the tissue or cell involved (**FIG. 1.9**). In local control, a relatively isolated change occurs in a tissue. A nearby cell or group of cells senses the change in their immediate vicinity and responds, usually by releasing a chemical. The response is restricted to the region where the change took place—hence the term *local control.*

FIG. 1.8 A simple control system

FIG. 1.9 A comparison of local control and reflex control

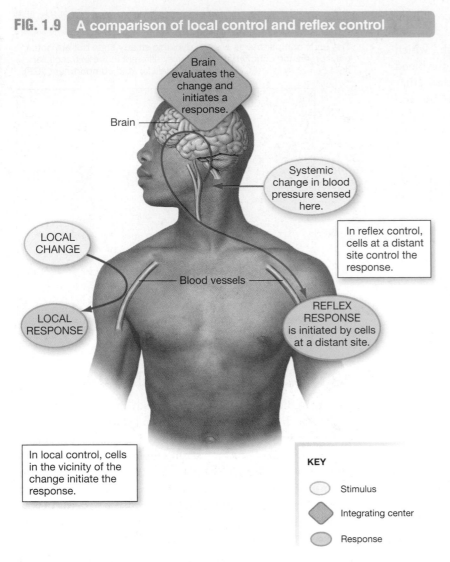

Brain evaluates the change and initiates a response.

Brain

Systemic change in blood pressure sensed here.

In reflex control, cells at a distant site control the response.

LOCAL CHANGE

Blood vessels

LOCAL RESPONSE

REFLEX RESPONSE is initiated by cells at a distant site.

In local control, cells in the vicinity of the change initiate the response.

KEY

Stimulus

Integrating center

Response

One example of local control can be observed when oxygen concentration in a tissue decreases. Cells lining the small blood vessels that bring blood to the area sense the lower oxygen concentration and respond by secreting a chemical signal. The signal molecule diffuses to nearby muscles in the blood vessel wall, bringing them a message to relax. Relaxation of the muscles widens (*dilates*) the blood vessel, which increases blood flow into the tissue and brings more oxygen to the area.

Reflex Control Uses Long-Distance Signaling

Changes that are widespread throughout the body, or *systemic* in nature, require more complex control systems to maintain homeostasis. For example, maintaining blood pressure to drive blood flow throughout the body is a systemic issue rather than a local one. Because blood pressure is body-wide, maintaining it requires long-distance communication and coordination. We will use the term *reflex control* to mean any long-distance pathway that uses the nervous system, endocrine system, or both.

A physiological reflex can be broken down into two parts: a response loop and a feedback loop (**FIG. 1.10**). As with the simple control system just described, a **response loop** has three primary components: an *input signal*, an *integrating center* to integrate the signal, and an *output signal*. These three components can be expanded into the following sequence of seven steps to form a pattern that is found with slight variations in all reflex pathways:

Stimulus → sensor → input signal →

integrating center →

output signal → target → response

The input side of the response loop starts with a *stimulus*—the change that occurs when the regulated variable moves out of its desirable range. A specialized **sensor** monitors the variable. If the sensor is activated by the stimulus, it sends an input signal to the integrating center. The integrating center evaluates the information coming from the sensor and initiates an output signal. The output signal directs a target to carry out a response. If successful, the response brings the regulated variable back into the desired range.

In mammals, integrating centers are usually part of the nervous system or endocrine system. Output signals may be chemical signals, electrical signals, or a combination of both. The targets activated by output signals can be any cell of the body.

Response Loops Begin with a Stimulus

To illustrate response loops, let's apply the concept to a simple nonbiological example. Think about an aquarium whose heater is programmed to maintain the water temperature (the regulated variable) at 30 °C (Fig. 1.10). The room temperature is 25 °C. The desired water temperature (30 °C) is the *setpoint* for the regulated variable.

Assume that initially the aquarium water is at room temperature, 25 °C. When you turn the control box on, you set the response loop in motion. The thermometer (sensor) registers a temperature of 25 °C. It sends this information through a wire (input signal) to the control box (integrating center). The control box is programmed to evaluate the incoming temperature signal, compare it with the setpoint for the system (30 °C), and "decide" whether a response is needed to bring the water temperature up to the setpoint. The control box sends a signal through another wire (output signal) to the heater (the target), which turns on and starts heating the water (response). This sequence—from stimulus to response—is the response loop.

This aquarium example involves a variable (temperature) controlled by a single control system (the heater). We can also describe a system that is under dual control. For example, think of a house that has both heating and air conditioning. The owner would like the house to remain at 70 °F (about 21 °C). On chilly

FIG. 1.10 The steps in a reflex pathway

In the aquarium example shown, the control box is set to maintain a water temperature of 30±1 °C.

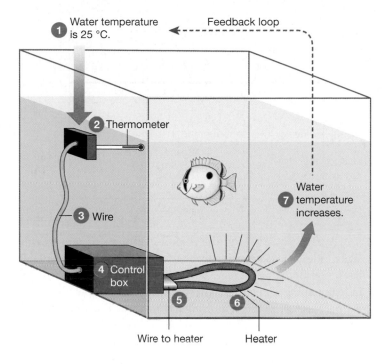

1 Water temperature is 25 °C.

Feedback loop

2 Thermometer

3 Wire

4 Control box

Wire to heater   Heater

Water temperature increases.

7

5   6

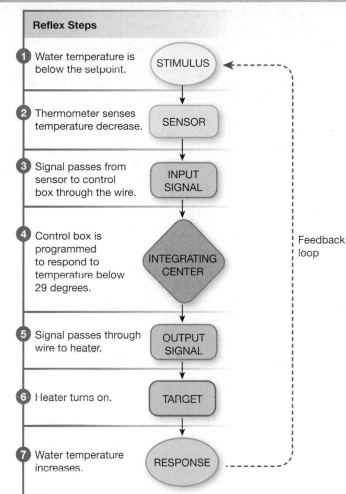

Reflex Steps

1 Water temperature is below the setpoint.   STIMULUS

2 Thermometer senses temperature decrease.   SENSOR

3 Signal passes from sensor to control box through the wire.   INPUT SIGNAL

4 Control box is programmed to respond to temperature below 29 degrees.   INTEGRATING CENTER

5 Signal passes through wire to heater.   OUTPUT SIGNAL

6 Heater turns on.   TARGET

7 Water temperature increases.   RESPONSE

Feedback loop

autumn mornings, when the temperature in the house falls, the heater turns on to warm the house. Then, as the day warms up, the heater is no longer needed and turns off. When the sun heats the house above the setpoint, the air conditioner turns on to cool the house back to 70 °F. The heater and air conditioner have *antagonistic control* over house temperature because they work in opposition to each other. Similar situations occur in the human body when two branches of the nervous system or two different hormones have opposing effects on a single target.

CONCEPT CHECK 3. What is the drawback of having only a single control system (a heater) for maintaining aquarium water temperature in some desired range?

Feedback Loops Modulate the Response Loop

The response loop is only the first part of a reflex. For example, in the aquarium just described, the sensor sends temperature information to the control box, which recognizes that the water is too cold. The control box responds by turning on the heater to warm

the water. Once the response starts, what keeps the heater from sending the temperature up to, say, 50 °C?

The answer is a **feedback loop,** where the response "feeds back" to influence the input portion of the pathway. In the aquarium example, turning on the heater increases the temperature of the water. The sensor continuously monitors the temperature and sends that information to the control box. When the temperature warms up to the maximum acceptable value, the control box shuts off the heater, thus ending the reflex response.

Negative Feedback Loops Are Homeostatic

For most reflexes, feedback loops are homeostatic—that is, designed to keep the system at or near a setpoint so that the regulated variable is relatively stable. How well an integrating center succeeds in maintaining stability depends on the *sensitivity* of the system. In the case of our aquarium, the control box is programmed to have a sensitivity of ± 1 °C. If the water temperature drops from 30 °C to 29.5 °C, it is still within the acceptable range, and no response occurs. If the water temperature drops below 29 °C (30 − 1), the control box turns the heater on (**FIG. 1.11**). As the water heats up, the control box constantly receives information about the water

RUNNING **PROBLEM**

After reading the article on fiber, Jimmy decided to go back to his professor for help. "How can I figure out who to believe on the Internet? Isn't there a better way to get health information?"

"Well, the sites you found from the Mayo Clinic and the American Diabetes Association are fine for general information aimed at the lay public. But if you want to find the same information that scientists and physicians read, you should search using MEDLINE, the database published by the U.S. National Library of Medicine. PubMed is the free public-access version (*www.pubmed.gov*). This database lists articles that are **peer-reviewed,** which means that the research described has gone through a screening process in which the work is critiqued by an anonymous panel of two or three scientists who are qualified to judge the science. Peer review acts as a kind of quality control because a paper that does not meet the standards of the reviewers will be rejected by the editor of the journal."

Q5: *Jimmy went to PubMed and typed in his search terms:* type 2 diabetes and insulin therapy. *Repeat his search. Compare the number of results to the Google searches.*

2 — 5 — 9 — 12 — **16** — 19 — 23

temperature from the sensor. When the water reaches 31 °C (30 ± 1), the upper limit for the acceptable range, the feedback loop causes the control box to turn the heater off. The water then gradually cools off until the cycle starts all over again. The end result is a regulated variable that *oscillates* {*oscillare,* to swing} around the setpoint.

In physiological systems, some sensors are more sensitive than others. For example, the sensors that trigger reflexes to conserve water activate when blood concentration increases only 3% above normal, but the sensors for low oxygen in the blood will not respond until oxygen has decreased by 40%.

FIG. 1.11 **Oscillation around the setpoint**

Most functions that maintain homeostasis have a setpoint, or normal value. The response loop that controls the function activates when the function moves outside a predetermined normal range.

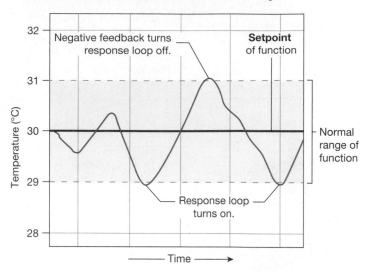

A pathway in which the response opposes or removes the signal is known as **negative feedback** (**FIG. 1.12a**). Negative feedback loops *stabilize* the regulated variable and thus aid the system in maintaining homeostasis. In the aquarium example, the heater warms the water (the response) and removes the stimulus (low water temperature). With loss of the stimulus for the pathway, the response loop shuts off. *Negative feedback loops can restore the normal state but cannot prevent the initial disturbance.*

Positive Feedback Loops Are Not Homeostatic

A few reflex pathways are not homeostatic. In a **positive feedback loop,** the response *reinforces* the stimulus rather than decreasing or removing it. In positive feedback, the response sends

FIG. 1.12 **Negative and positive feedback**

(a) Negative feedback: The response counteracts the stimulus, shutting off the response loop.

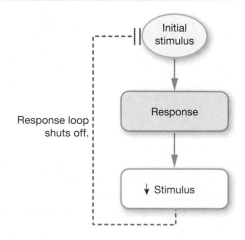

(b) Positive feedback: The response reinforces the stimulus, sending the variable farther from the setpoint.

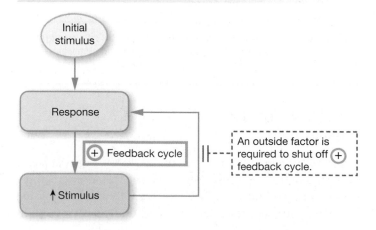

FIG. 1.13 A positive feedback loop

Baby drops lower in uterus to initiate labor.

Cervical stretch

causing   *stimulates*

Push baby against cervix

Oxytocin release

Positive feedback loop

causes

Uterine contractions

Delivery of baby stops the cycle.

the regulated variable even farther from its normal value. This initiates a vicious cycle of ever-increasing response and sends the system temporarily out of control (Fig. 1.12b). Because positive feedback escalates the response, this type of feedback requires some intervention or event outside the loop to stop the response.

One example of a positive feedback loop involves the hormonal control of uterine contractions during childbirth (**FIG. 1.13**). When the baby is ready to be delivered, it drops lower in the uterus and begins to put pressure on the *cervix,* the opening of the uterus. Sensory signals from the cervix to the brain cause release of the hormone *oxytocin,* which causes the uterus to contract and push the baby's head even harder against the cervix, further stretching it. The increased stretch causes more oxytocin release, which causes more contractions that push the baby harder against the cervix. This cycle continues until finally the baby is delivered, releasing the stretch on the cervix and stopping the positive feedback loop.

CONCEPT CHECK

4. Does the aquarium heating system in Figure 1.10 operate using positive feedback or negative feedback?

Feedforward Control Allows the Body to Anticipate Change

Negative feedback loops can stabilize a function and maintain it within a normal range but are unable to prevent the change that triggered the reflex in the first place. A few reflexes have evolved that enable the body to predict that a change is about to occur and start the response loop in anticipation of the change. These anticipatory responses are called **feedforward control.**

An easily understood physiological example of feedforward control is the salivation reflex. The sight, smell, or even the thought of food is enough to start our mouths watering in expectation of eating the food. This reflex extends even further, because the same stimuli can start the secretion of hydrochloric acid as the stomach anticipates food on the way. One of the most complex feedforward reflexes appears to be the body's response to exercise [discussed in Chapter 25].

Biological Rhythms Result from Changes in a Setpoint

As discussed earlier, each regulated variable has a normal range within which it can vary without triggering a correction. In physiological systems, the setpoints for many regulated variables are different from person to person, or may change for the same individual over a period of time. Factors that influence an individual's setpoint for a given variable include normal biological rhythms, inheritance, and the conditions to which the person has become accustomed.

Regulated variables that change predictably and create repeating patterns or cycles of change are called biological rhythms, or *biorhythms*. The timing of many biorhythms coincides with a predictable environmental change, such as daily light–dark cycles or the seasons. Biological rhythms reflect changes in the setpoint of the regulated variable.

For example, all animals exhibit some form of daily biological rhythm, called a **circadian rhythm** {*circa,* about + *dies,* day}. Humans have circadian rhythms for many body functions, including blood pressure, body temperature, and metabolic processes. Body temperature peaks in the late afternoon and declines dramatically in the early hours of the morning (**FIG. 1.14a**). Have you ever been studying late at night and noticed that you feel cold? This is not because of a drop in environmental temperature but because your thermoregulatory reflex has turned down your internal thermostat.

One of the interesting correlations between circadian rhythms and behavior involves body temperature. Researchers found that self-described "morning people" have temperature rhythms that cause body temperature to climb before they wake up in the morning, so that they get out of bed prepared to face the world. On the other hand, "night people" may be forced by school and work schedules to get out of bed while their body temperature is still at its lowest point, before their bodies are prepared for activity. These night people are still going strong and working

FIG. 1.14 Circadian rhythms in humans

(a) Body temperature is lowest in the early morning and peaks in the late afternoon and early evening. Data from W. E. Scales *et al., J Appl Physiol* 65(4): 1840–1846, 1998.

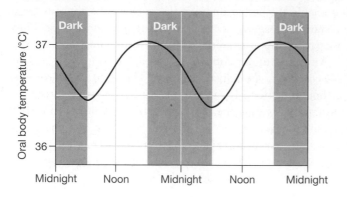

(b) Plasma cortisol is lowest during sleep and peaks shortly after awakening. Data from L. Weibel *et al., Am J Physiol Endocrinol Metab* 270: E608–E613, 1996.

productively in the early hours of the morning, when the morning people's body temperatures are dropping and they are fast asleep.

Many hormones in humans have blood concentrations that fluctuate predictably in a 24-hour cycle. Cortisol, growth hormone, and the sex hormones are among the most noted examples. A cortisol concentration in a 9:00 A.M. sample might be nearly twice as high as one taken in the early afternoon (Fig. 1.14b).

If a patient has a suspected abnormality in hormone secretion, it is therefore important to know when hormone levels are measured. A concentration that is normal at 9:00 A.M. is high at 2:00 P.M. One strategy for avoiding errors due to circadian fluctuations is to collect information for a full day and calculate an average value over 24 hours. For example, cortisol secretion is estimated indirectly by measuring all urinary cortisol metabolites excreted in 24 hours.

What is the adaptive significance of functions that vary with a circadian rhythm? Our best answer is that biological rhythms create an anticipatory response to a predictable environmental variable. There are seasonal rhythms of reproduction in many organisms. These rhythms are timed so that the offspring have food and other favorable conditions to maximize survival.

Circadian rhythms cued by the light–dark cycle may correspond to rest-activity cycles. These rhythms allow our bodies to anticipate behavior and coordinate body processes accordingly. You may hear people who are accustomed to eating dinner at 6:00 P.M. say that they cannot digest their food if they wait until 10:00 P.M. to eat because their digestive system has "shut down" in anticipation of going to bed.

Some variability in setpoints is associated with changing environmental conditions rather than biological rhythms. The adaptation of physiological processes to a given set of environmental conditions is known as **acclimatization** when it occurs naturally. If the process takes place artificially in a laboratory setting, it is called **acclimation.** Each winter, people in the upper latitudes of the northern hemisphere go south in February, hoping to escape the bitter subzero temperatures and snows of the northern

climate. As the northerners walk around in 40 °F (about 4 °C) weather in short-sleeve shirts, the southerners, all bundled up in coats and gloves, think that the northerners are crazy: the weather is cold! The difference in behavior is due to different temperature acclimatization, a difference in the setpoint for body temperature regulation that is a result of prior conditioning.

Biorhythms and acclimatization are complex processes that scientists still do not completely understand. Some rhythms arise from special groups of cells in the brain and are reinforced by information about the light–dark cycle that comes in through the eyes. Some cells outside the nervous system generate their own rhythms. Research in simpler animals such as flies is beginning to explain the molecular basis for biological rhythms. [We discuss the cellular and molecular basis for circadian rhythms in Chapter 9.]

THE SCIENCE OF PHYSIOLOGY

How do we know what we know about the physiology of the human body? The first descriptions of physiology came from simple observations. But physiology is an experimental science, one in which researchers generate **hypotheses** {*hypotithenai,* to assume; singular *hypothesis*}, or logical guesses, about how events take place. They test their hypotheses by designing experiments to collect evidence that supports or disproves their hypothesis, and they publish the results of their experiments in the scientific literature. Healthcare providers look in the scientific literature for evidence from these experiments to help guide their clinical decision-making. Critically evaluating the scientific evidence in this manner is a practice known as *evidence-based medicine*. Observation and experimentation are the key elements of **scientific inquiry.**

Good Scientific Experiments Must Be Carefully Designed

A common type of biological experiment either removes or alters some variable that the investigator thinks is an essential part of an observed phenomenon. That altered variable is the **independent variable.** For example, a biologist notices that birds at a feeder seem to eat more in the winter than in the summer. She generates a hypothesis that cold temperatures cause birds to increase their food intake. To test her hypothesis, she designs an experiment in which she keeps birds at different temperatures and monitors how much food they eat. In her experiment, temperature, the manipulated element, is the independent variable. Food intake, which is hypothesized to be dependent on temperature, becomes the **dependent variable.**

CONCEPT CHECK

5. Students in the laboratory run an experiment in which they drink different volumes of water and measure their urine output in the hour following drinking. What are the independent and dependent variables in this experiment?

An essential feature of any experiment is an experimental **control.** A control group is usually a duplicate of the experimental group in every respect except that the independent variable is not changed from its initial value. For example, in the bird-feeding experiment, the control group would be a set of birds maintained at a warm summer temperature but otherwise treated exactly like the birds held at cold temperatures. The purpose of the control is to ensure that any observed changes are due to the manipulated variable and not to changes in some other variable. For example, suppose that in the bird-feeding experiment food intake increased after the investigator changed to a different food. Unless she had a control group that was also fed the new food, the investigator could not determine whether the increased food intake was due to temperature or to the fact that the new food was more palatable.

During an experiment, the investigator carefully collects information, or **data** {plural; singular *datum,* a thing given}, about the effect that the manipulated (independent) variable has on the observed (dependent) variable. Once the investigator feels that she has sufficient information to draw a conclusion, she begins to analyze the data. Analysis can take many forms and usually includes statistical analysis to determine if apparent differences are statistically significant. A common format for presenting data is a graph (**FIG. 1.15**).

If one experiment supports the hypothesis that cold causes birds to eat more, then the experiment should be repeated to ensure that the results were not an unusual one-time event. This step is called **replication.** When the data support a hypothesis in multiple experiments, the hypothesis may become a working **model.** A model with substantial evidence from multiple investigators supporting it may become a **scientific theory.**

Most information presented in textbooks like this one is based on models that scientists have developed from the best available experimental evidence. On occasion, investigators publish new experimental evidence that does not support a current model. In that case, the model must be revised to fit the available evidence. For this reason, you may learn a physiological "fact" while using this textbook, but in 10 years that "fact" may be inaccurate because of what scientists have discovered in the interval.

For example, in 1970, students learned that the cell membrane was a "butter sandwich," a structure composed of a layer of fats sandwiched between two layers of proteins. In 1972, however, scientists presented a very different model of the membrane, in which globules of proteins float within a double layer of fats. As a result, students who had learned the butter sandwich model had to revise their mental model of the membrane.

Where do our scientific models for human physiology come from? We have learned much of what we know from experiments on animals ranging from fruit flies and squid to rats. In many instances, the physiological processes in such animals are either identical to those taking place in humans or else similar enough that we can extrapolate from the animal model to humans. It is important to use nonhuman models because experiments using human subjects can be difficult to perform.

However, not all studies done on animals can be applied to humans. For example, an antidepressant that Europeans had used

RUNNING **PROBLEM**

"Hi, professor. I'm back again." Most of the articles Jimmy found in PubMed were too focused on single experiments. And he didn't really understand the technical terms the authors used. "Is there any way to find papers that are not so complicated?"

"Yes, there are several ways. Many journals publish **review articles** that contain a synopsis of recent research on a particular topic. When you are just beginning to learn about a topic, it is best to begin with review articles. PubMed will have a link on the Results page that takes you directly to the review articles in your results. Another place to look for basic information is MedlinePlus, another resource from the National Library of Medicine (*www.medlineplus.gov*). Or try Google Scholar (*scholar.google .com*)."

Jimmy decided to try MedlinePlus because the PubMed and Google Scholar results seemed too technical for his simple question. On the MedlinePlus site, he entered *type 2 diabetes and insulin therapy* into the search box. After reading a few of the articles he found linked there, he called his mother. "Hey, Mom! I found the answer to your question!"

Q6: *Repeat Jimmy's search in MedlinePlus and look for links to articles on type 2 diabetes published by the National Institutes of Health (NIH), National Library of Medicine (NLM), or the Centers for Disease Control and Prevention (CDC). Based on what you read in those articles, what did Jimmy tell his mother about her need to take insulin for her type 2 diabetes?*

Q7: *What about the article that said eating a high-fiber diet could help? On the MedlinePlus results pages, look for articles on alternative treatments for diabetes published by the National Center for Complementary and Alternative Medicine. Do these articles mention dietary fiber?*

FIG. 1.15 **FOCUS ON . . .**

Graphs

Graphs are pictorial representations of the relationship between two (or more) variables, plotted in a rectangular region (Fig. 1.15a). We use graphs to present a large amount of numerical data in a small space, to emphasize comparisons between variables, or to show trends over time. A viewer can extract information much more rapidly from a graph than from a table of numbers or from a written description. A well-constructed graph should contain (in very abbreviated form) everything the reader needs to know about the data, including the purpose of the experiment, how the experiment was conducted, and the results.

All scientific graphs have common features. The independent variable (the variable manipulated by the experimenter) is graphed on the horizontal x-axis. The dependent variable (the variable measured by the experimenter) is plotted on the vertical y-axis. If the experimental design is valid and the hypothesis is correct, changes in the independent variable (x-axis) will cause changes in the dependent variable (y-axis). In other words, y is a function of x. This relationship can be expressed mathematically as $y = f(x)$.

Each axis of a graph is divided into units represented by evenly spaced tick marks on the axis. A label tells what variable the axis represents (time, temperature, amount of food consumed) and in what units it is marked (days, degrees Celsius, grams per day). The intersection of the two axes is called the *origin*. The origin usually, but not always, has a value of zero for both axes. A graph should have a title or legend that describes what the graph represents. If multiple groups are shown on one graph, the lines or bars representing the groups may have labels, or a key may show what group each symbol or color represents.

Most graphs you will encounter in physiology display data either as bars (bar graphs or histograms), as lines (line graphs), or as dots (scatter plots). Four typical types of graphs are shown in Figure 1.15b–e. **Bar graphs** (Fig. 1.15b) are used when the independent variables are distinct entities. A histogram (Fig. 1.15c) is a specialized bar graph that shows the distribution of one variable over a range. The x-axis is divided into units (called "bins" in some computer graphing programs), and the y-axis indicates how many pieces of data are associated with each bin.

Line graphs (Fig. 1.15d) are commonly used when the independent variable on the x-axis is a continuous phenomenon, such as time, temperature, or weight. Each point on the graph may represent the average of a set of observations. Because the independent variable is a continuous function, the points on the graph can be connected with a line (point-to-point connections or a mathematically calculated "best fit" line or curve). Connecting the points allows the reader to **interpolate,** or estimate values between the measured values.

Scatter plots (Fig. 1.15e) show the relationship between two variables, such as time spent studying for an exam and performance on that exam. Usually each point on the plot represents one member of a test population. Individual points on a scatter plot are never connected by a line, but a "best fit" line or curve may indicate a trend in the data.

Here are some questions to ask when you are trying to extract information from a graph:

1. What variable does each axis represent?
2. What is the relationship between the variables represented by the axes? This relationship can usually be expressed by substituting the labels on the axes into the following statement: y varies with x. For example, in graph (b), the canaries' daily food intake varied with the type of diet.
3. Are any trends apparent in the graph? For line graphs and scatter plots, is the line horizontal (no change in the dependent variable when the independent variable changes), or does it have a slope? Is the line straight or curved? For bar graphs, are the bars the same height or different heights? If different heights, is there a trend in the direction of height change?

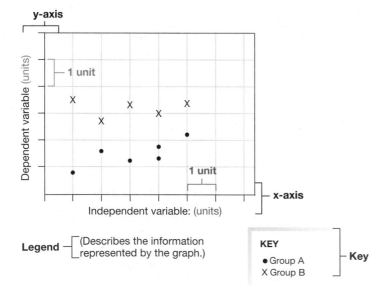

(a) Standard features of a graph. The standard features of a graph include units and labels on the axes, a key, and a figure legend.

CONCEPT CHECK

6. Students in a physiology laboratory collected heart rate data on one another. In each case, heart rate was measured first for the subject at rest and again after the subject had exercised using a step test. Two findings from the experiment were (1) that heart rate was greater with exercise than at rest, and (2) that female subjects had higher resting heart rates than male subjects.

   (a) What was the independent variable in this experiment? What was the dependent variable?

   (b) Draw a graph and label each axis with the correct variable. Draw trend lines or bars that might approximate the data collected.

(b) Bar graph. Each bar shows a distinct variable. The bars are lined up side by side along one axis so that they can be easily compared with one another. Scientific bar graphs traditionally have the bars running vertically.

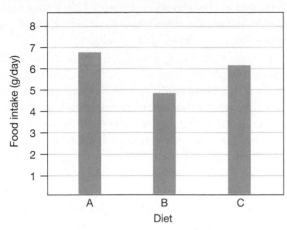

Q GRAPH QUESTION

Which food did the canaries prefer?

Canaries were fed one of three diets and their food intake was monitored for three weeks.

(c) Histogram. A histogram quantifies the distribution of one variable over a range of values.

Q GRAPH QUESTION

How many students took the quiz?

The distribution of student scores on a 10-point quiz is plotted on a histogram.

(d) Line graph. The x-axis frequently represents time; the points represent averaged observations. The points may be connected by lines, in which case the slope of the line between two points shows the rate at which the variable changed.

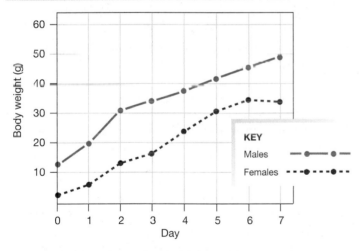

KEY

Males

Females

Male and female mice were fed a standard diet and weighed daily.

Q GRAPH QUESTION

When did male mice increase their body weight the fastest?

(e) Scatter plot. Each point represents one member of a test population. The individual points of a scatter plot are never connected by lines, but a best fit line may be estimated to show a trend in the data, or better yet, the line may be calculated by a mathematical equation.

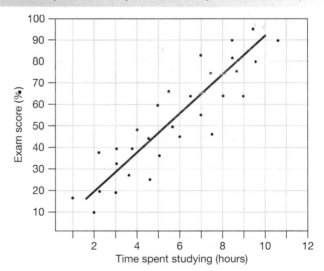

Student scores were directly related to the amount of time they spent studying.

Q GRAPH QUESTIONS

For graphs (d) and (e), answer the following:
• What was the investigator trying to determine?

• What are the independent and dependent variables?
• What are the results or trends indicated by the data?

safely for years was undergoing stringent testing required by the U.S. Food and Drug Administration before it could be sold in this country. When beagles took the drug for a period of months, the dogs started dying from heart problems. Scientists were alarmed until further research showed that beagles have a unique genetic makeup that causes them to break down the drug into a more toxic substance. The drug was perfectly safe in other breeds of dogs and in humans, and it was subsequently approved for human use.

The Results of Human Experiments Can Be Difficult to Interpret

There are many reasons it is difficult to carry out physiological experiments in humans, including variability, psychological factors, and ethical considerations.

Variability Human populations have tremendous genetic and environmental **variability**. Although physiology books usually present *average* values for many physiological variables, such as blood pressure, these average values simply represent a number that falls somewhere near the middle of a wide range of values. Thus, to show significant differences between experimental and control groups in a human experiment, an investigator would have to include a large number of identical subjects.

However, getting two groups of people who are *identical* in every respect is impossible. Instead, the researcher must attempt to recruit subjects who are *similar* in as many aspects as possible. You may have seen newspaper advertisements requesting research volunteers: "Healthy males between 18 and 25, nonsmokers, within 10% of ideal body weight, to participate in a study. . . ." Researchers must take into account the variability inherent in even a select group of humans when doing experiments with human subjects. This variability may affect the researcher's ability to interpret the significance of data collected on that group.

One way to reduce variability within a test population, whether human or animal, is to do a **crossover study.** In a crossover study, each individual acts both as experimental subject and as control. Thus, each individual's response to the treatment can be compared with his or her own control value. This method is particularly effective when there is wide variability within a population.

For example, in a test of blood pressure medication, investigators might divide subjects into two groups. Group A takes an inactive substance called a **placebo** (from the Latin for "I shall be pleasing") for the first half of the experiment, then changes to the experimental drug for the second half. Group B starts with the experimental drug, and then changes to the placebo. This scheme enables the researcher to assess the effect of the drug on each individual. In other words, each subject acts as his or her own control. Statistically, the data analysis can use methods that look at the changes within each individual rather than at changes in the collective group data.

Psychological Factors Another significant variable in human studies is the psychological aspect of administering a treatment. If you give someone a pill and tell the person that it will help alleviate some problem, there is a strong possibility that the pill will have exactly that effect, even if it contains only sugar or an inert substance. This well-documented phenomenon is called the **placebo effect.** Similarly, if you warn people that a drug they are taking may have specific adverse side effects, those people will report a higher incidence of the side effects than a similar group of people who were not warned. This phenomenon is called the **nocebo effect,** from the Latin *nocere,* to do harm. The placebo and nocebo effects show the ability of our minds to alter the physiological functioning of our bodies.

In setting up an experiment with human subjects, we must try to control for the placebo and nocebo effects. The simplest way to do this is with a **blind study,** in which the subjects do not know whether they are receiving the treatment or the placebo. Even this precaution can fail; however, if the researchers assessing the subjects know which type of treatment each subject is receiving. The researchers' expectations of what the treatment will or will not do may color their measurements or interpretations.

To avoid this outcome, researchers often use **double-blind studies.** A third party, not involved in the experiment, is the only one who knows which group is receiving the experimental treatment and which group is receiving the control treatment. The most sophisticated experimental design for minimizing psychological effects is the **double-blind crossover study.** In this type of study, the control group in the first half of the experiment becomes the experimental group in the second half, and vice versa, but no one involved knows who is taking the active treatment.

Ethical Considerations Ethical questions arise when humans are used as experimental subjects, particularly when the subjects are people suffering from a disease or other illness. Is it ethical to withhold a new and promising treatment from the control group? A noteworthy example occurred some years ago when researchers were testing the efficacy of a treatment for dissolving blood clots in heart attack victims. The survival rate among the treated patients was so much higher that testing was halted so that members of the control group could also be given the experimental drug.

In contrast, tests on some anticancer agents have shown that the experimental treatments were less effective in stopping the spread of cancer than were the standard treatments used by the controls. Was it ethical to undertreat patients in the experimental group by depriving them of the more effective current medical practice? Most studies now are evaluated continually over the course of the study to minimize the possibility that subjects will be harmed by their participation.

In 2002, a trial on hormone replacement therapy in postmenopausal women was halted early when investigators realized that women taking a pill containing two hormones were developing cardiovascular disease and breast cancer at a higher rate than women on placebo pills. On the other hand, the women receiving hormones also had *lower* rates of colon cancer and bone fractures. The investigators decided that the risks associated with taking the hormones exceeded the potential benefits, and they stopped the study. To learn more about this clinical trial and the pros and cons of hormone replacement therapy, go to *www.nlm.nih.gov/ medlineplus/hormonereplacementtherapy.html,* the web site of the U.S. National Library of Medicine.

Human Studies Can Take Many Forms

Almost daily, the newspapers carry articles about clinical trials studying the efficacy of drugs or other medical treatments. Many different aspects of experimental design can affect the validity and applicability of the results of these trials. For example, some trials are carried out for only a limited time on a limited number of people, such as studies conducted for the U.S. Food and Drug Administration's drug-approval process. In several instances in the past few years, drugs approved as a result of such studies have later been withdrawn from the market when extended use of the drug uncovered adverse side effects, including deaths.

Longitudinal studies are designed to be carried out for a long period of time. One of the most famous longitudinal studies is the Framingham Heart Study (*www.framingham.com/heart*), started in 1948 and still ongoing. Framingham is a **prospective study** {*prospectus,* outlook, looking forward} that recruited healthy people and has been following them for years to identify factors that contribute to the development of cardiovascular disease. This study has already made important contributions to healthcare, and it continues today with the adult children and grandchildren of the original participants.

Additional study designs you may encounter in the literature include cross-sectional and retrospective studies. **Cross-sectional studies** survey a population for the prevalence of a disease or condition. Data from cross-sectional studies identify trends to be investigated further, such as whether age group or socioeconomic status is associated with a higher risk of developing the condition being surveyed. **Retrospective studies** {*retro,* backward + *spectare,* to look} match groups of people who all have a particular disease to a similar but healthy control group. The goal of these studies is to determine whether development of the disease can be associated with a particular variable.

Often, the results of one or more published studies do not agree with the conclusions of other published studies. In some cases, the reason for the disagreement turns out to be a limitation of the experimental design, such as a small number of subjects who may not be representative of larger populations. In other cases, the disagreement may be due to small but potentially significant differences in the experimental designs of the different studies.

One way scientists attempt to resolve contradictory results is to perform a **meta-analysis** of the data {*meta-,* at a higher level}.

A meta-analysis combines all the data from a group of similar studies and uses sophisticated statistical techniques to extract significant trends or findings from the combined data. For example, multiple studies have been done to assess whether glucosamine and chondroitin, two dietary supplements, can improve degenerative joint disease. However, the individual studies had small numbers of subjects (<50) and used different dosing regimens. A meta-analysis using statistical methods is one way to compare the results from these studies.[6]

The difficulty of using human subjects in experiments is one of the reasons scientists use animals to develop many of our scientific models. Since the 1970s, physiological research has increasingly augmented animal experimentation with techniques developed by cellular biologists and molecular geneticists. As we have come to understand the fundamentals of chemical signaling and communication in the body, we have unlocked the mysteries of many processes. In doing so, we also have come closer to being able to treat many diseases by correcting their cause rather than simply treating their symptoms.

More and more, medicine is turning to therapies based on interventions at the molecular level. A classic example is the treatment of cystic fibrosis, an inherited disease in which the mucus of the lungs and digestive tract is unusually thick. For many years, patients with this condition had few treatment options, and most died at a young age. However, basic research into the mechanisms by which salt and water move across cell membranes provided clues to the underlying cause of cystic fibrosis: a defective protein in the membrane of certain cells. Once molecular geneticists found the gene that coded for that protein, the possibility of repairing the defective protein in cystic fibrosis patients became a reality. Without the basic research into how cells and tissues carry out their normal function, however, this treatment would never have been developed.

As you read this book and learn what we currently know about how the human body works, keep in mind that many of the ideas presented in it reflect models that represent our current understanding and are subject to change. There are still many questions in physiology waiting for investigators to find the answers.

---

[6]See, for example, S. Wandel *et al.* Effects of glucosamine, chondroitin, or placebo in patients with osteoarthritis of hip or knee: network meta-analysis. *Br Med J* 341: c4675–c4676, 2010.

RUNNING **PROBLEM CONCLUSION** | What to Believe?

One skill all physiology students should acquire is the ability to find information in the scientific literature. In today's world, the scientific literature can be found both in print, in the form of books and periodicals, and on the Web. However, unless a book has a recent publication date, it may not be the most up-to-date source of information.

Many students begin their quest for information on a subject by searching the Internet. Be cautious! Anyone can create a web page and publish information on the Web. There is no screening process comparable to peer review in journals, and the reader of a web page must decide how valid the information is. Web sites published by recognized universities and not-for-profit organizations are likely to have good information, but you should view an article about vitamins on the web page of a health food store with a skeptical eye unless the article cites published peer-reviewed research.

—Continued next page

RUNNING **PROBLEM CONCLUSION** *Continued*

Question	Answer and Commentary
Q1: *What search terms could Jimmy have used to get fewer results?*	Including more words in a web search is the best way to narrow the results list. For example, Jimmy could have searched for *insulin therapy diabetes*. That search would have produced about 9.3 million results. Being more specific about his mother's type of diabetes might help. A search for *insulin therapy for type 2 diabetes* has about 7.2 million results. That's still a lot of web pages to look at!
Q2: *What kinds of web sites should Jimmy be looking for in his results list, and how can he recognize them?*	The best web sites for health information come from organizations that are part of the scientific and healthcare communities, such as the National Institutes of Health (NIH), nonprofit groups dedicated to supporting research on a particular disease (The American Diabetes Association, *diabetes.org*), or clinics and universities where scientists and physicians are actively investigating causes and treatments for diseases. Treat commercial web sites that end in **.com* with extra caution.
Q3: *In The Doctors' Medical Library article called "Fiber," what does Dr. Kennedy say about high-fiber diet and diabetes?*	Dr. Kennedy claims that some patients with type 2 diabetes can be "successfully treated" by eating a high-fiber diet. (The classification of type 2 diabetes as "adult onset" is obsolete.)
Q4: *How can Jimmy find out more about who created the site and what their credentials are?*	To learn more about who created a web site and why, look for links at the bottom of the page for HOME or ABOUT US. On the home page for The Doctors' Medical Library, you will learn that the site promotes reader education. The link on Ron Kennedy, M.D., does not give you any information about his medical training or other credentials.
Q5: *Compare the number of results from the PubMed search to those for the Google searches.*	The number of results will depend on when you do the search because new articles are added constantly. But the number will probably be fewer than 25,000, much less than the millions of results that came up following a Google search.
Q6: *What did Jimmy tell his mother about her need to take insulin for her type 2 diabetes?*	The articles published by these national organizations all say that people with type 2 diabetes may need to take insulin. Patients should always listen to their healthcare providers and ask questions if they are uncertain about what they should be doing.
Q7: *Do the articles from NCCAM mention dietary fiber as an alternative treatment for diabetes?*	The NCCAM article lists a number of alternative treatments that people have tried. It also says that, so far, there is no scientific evidence supporting the use of dietary supplements for treating diabetes. Patients should never stop their conventional treatments when using complementary treatments, and they should always inform their healthcare providers about any vitamins or dietary supplements they are taking.

Citation Formats

When you find an article either in print or on the Web, you should write down the full citation. Citation formats for papers vary slightly from source to source but will usually include the following elements (with the punctuation shown):

Author(s). Article title. *Journal Name* volume (issue): inclusive pages, year of publication.

For example:

Echevarria M and Ilundain AA. Aquaporins. *J Physiol Biochem* 54(2): 107–118, 1998.

In many citations, the journal name is abbreviated using standard abbreviations. For example, the *American Journal of Physiology* is abbreviated as *Am J Physiol*. (One-word titles, such as *Science*, are never abbreviated.) For each calendar year, the issues of a given journal are assigned a **volume** number. The first issue of a given volume is designated **issue** 1, the second is issue 2, and so on.

Citing sources from the Web requires a different format. Here is one suggested format:

Author/Editor (if known). Revision or copyright date (if available). Title of web page [Publication medium]. Publisher of web page. *URL* [Date accessed].

For example:

Patton G (editor). 2005. Biological Journals and Abbreviations. [Online]. National Cancer Institute. *http://home.ncifcrf.gov/research/bja* [accessed April 10, 2005].

Unlike print resources, web pages are not permanent and frequently disappear or move. If you access a print journal on the Web, you should give the print citation, not the URL. You might also give the article's unique DOI (digital object identifier) number.

RUNNING **PROBLEM CONCLUSION** | *Continued*

Citing the Work of Others

Copying or paraphrasing material from another source without acknowledging that source is academic dishonesty. Word-for-word quotations placed within quotation marks are rarely used in scientific writing. Instead, we summarize the contents of the source paper and acknowledge the source, as follows:

Some rare forms of epilepsy are known to be caused by mutations in ion channels (Mulley *et al.*, 2003).

When a paper has three or more authors, we use the abbreviation *et al.*—from the Latin *et alii*, meaning "and others"—to save space in the body of the text. All authors' names are given in the full citation, which is usually included within a References section at the end of the paper. Reference lists are often arranged alphabetically by the last name of the paper's first author.

2   5   9   12   16   19   **23**

VISUALIZE, PRACTICE, AND MASTER PHYSIOLOGY WITH MASTERING A&P

- Practice-on-the-go with Dynamic Study Modules
- Visualize important processes with Interactive Physiology and A&P Flix
- Check your understanding with Video Tutors and practice quizzes!

CHAPTER SUMMARY

1. **Physiology** is the study of the normal functioning of a living organism and its component parts. (p. 2)

Physiology Is an Integrative Science

2. Many complex functions are **emergent properties** that cannot be predicted from the properties of the individual component parts. (p. 2)

3. Physiologists study the many **levels of organization** in living organisms, from molecules to populations of one species. (p. 2; Fig. 1.1)

4. The **cell** is the smallest unit of structure capable of carrying out all life processes. (p. 3)

5. Collections of cells that carry out related functions make up **tissues** and **organs.** (p. 3)

6. The human body has 10 physiological organ systems: **integumentary, musculoskeletal, respiratory, digestive, urinary, immune, circulatory, nervous, endocrine,** and **reproductive.** (p. 3; Fig. 1.2)

Function and Mechanism

7. The **function** of a physiological system or event is the "why" of the system. The **mechanism** by which events occur is the "how" of a system. The **teleological approach** to physiology explains why events happen; the **mechanistic approach** explains how they happen. (p. 5)

8. Translational research applies the results of basic physiological research to medical problems. (p. 5)

Themes in Physiology

9. The four key themes in physiology are structure/function relationships, such as **molecular interactions** and **compartmentation;** biological energy use; information flow within the body; and homeostasis. (p. 8)

Homeostasis

10. **Homeostasis** is the maintenance of a relatively constant internal environment. Variables that are regulated to maintain homeostasis include temperature, pH, ion concentrations, oxygen, and water. (p. 9)

11. Failure to maintain homeostasis may result in illness or disease. (p. 10; Fig. 1.4)

12. The body's internal environment is the **extracellular fluid.** (p. 10; Fig. 1.5)

13. The human body as a whole is adapted to cope with a variable external environment, but most cells of the body can tolerate much less change. (p. 10)

14. The **law of mass balance** says that if the amount of a substance in the body is to remain constant, any input must be offset by an equal loss. (p. 11; Fig. 1.6)

15. Input of a substance into the body comes from metabolism or from the outside environment. Output occurs through metabolism or **excretion.** (p. 11; Fig. 1.6)

16. The rate of intake, production, or output of a substance x is expressed as **mass flow,** where mass flow = concentration × volume flow. (p. 11)

17. **Clearance** is the rate at which a material is removed from the blood by excretion, metabolism, or both. The liver, kidneys, lungs, and skin all clear substances from the blood. (p. 12)

18. Cells and the extracellular fluid both maintain homeostasis, but they are not identical in composition. Their stable condition is a dynamic **steady state.** (p. 13)

19. Most solutes are concentrated in either one compartment or the other, creating a state of **disequilibrium.** (p. 13; Fig. 1.7)

Control Systems and Homeostasis

20. **Regulated variables** have a **setpoint** and a normal range. (p. 13; Fig. 1.11)

21. The simplest homeostatic control takes place at the tissue or cell level and is known as **local control.** (p. 13; Fig. 1.9)

22. **Control systems** have three components: an input signal, an **integrating center,** and an output signal. (p. 13; Fig. 1.8)

23. Reflex pathways can be broken down into **response loops** and **feedback loops.** A response loop begins when a **stimulus** is sensed by a **sensor.** The sensor is linked by the input signal to the **integrating center** that decides what action to take. The output signal travels from the integrating center to a **target** that carries out the appropriate **response.** (p. 14; Fig. 1.10)

24. In **negative feedback,** the response opposes or removes the original stimulus, which in turn stops the response loop. (p. 15; Fig. 1.12a)

25. In **positive feedback** loops, the response reinforces the stimulus rather than decreasing or removing it. This destabilizes the system until some intervention or event outside the loop stops the response. (p. 16; Figs. 1.12b, 1.13)

26. **Feedforward control** allows the body to predict that a change is about to occur and start the response loop in anticipation of the change. (p. 17)

27. Regulated variables that change in a predictable manner are called biological rhythms. Those that coincide with light–dark cycles are called **circadian rhythms.** (p. 17; Fig. 1.14)

The Science of Physiology

28. Observation and experimentation are the key elements of **scientific inquiry.** A **hypothesis** is a logical guess about how an event takes place. (p. 18)

29. In scientific experimentation, the factor manipulated by the investigator is the **independent variable,** and the observed factor is the **dependent variable.** All well-designed experiments have **controls** to ensure that observed changes are due to the experimental manipulation and not to some outside factor. (p. 19)

30. **Data,** the information collected during an experiment, are analyzed and presented, often as a graph. (p. 19; Fig. 1.15)

31. A **scientific theory** is a hypothesis supported by data from multiple sources. When new evidence does not support a theory or a model, the theory or model must be revised. (p. 19)

32. Animal experimentation is important because of the tremendous **variability** within human populations and because it is difficult to control human experiments. In addition, ethical questions may arise when using humans as experimental subjects. (p. 22)

33. To control many experiments, some subjects take an inactive substance known as a **placebo. Placebo** and **nocebo effects,** in which changes take place even if the treatment is inactive, may affect experimental outcomes. (p. 22)

34. In a **blind study,** the subjects do not know whether they are receiving the experimental treatment or a placebo. In a **double-blind study,** a third party removed from the experiment is the only one who knows which group is the experimental group and which is the control. In a **crossover study,** the control group in the first half of the experiment becomes the experimental group in the second half, and vice versa. (p. 22)

35. **Meta-analysis** of data combines data from many studies to look for trends. (p. 23)

REVIEW QUESTIONS

In addition to working through these questions and checking your answers on p. A-1, review the Learning Outcomes at the beginning of this chapter.

Level One Reviewing Facts and Terms

1. Define physiology. Describe the relationship between physiology and anatomy.

2. Name the different levels of organization in the biosphere.

3. Name the 10 systems of the body and give their major function(s).

4. What does "Physiology is an integrative science" mean?

5. Define homeostasis. Name some regulated variables that are maintained through homeostasis.

6. Name four major themes in physiology.

7. Put the following parts of a reflex in the correct order for a physiological response loop: input signal, integrating center, output signal, response, sensor, stimulus, target.

8. The name for daily fluctuations of body functions such as blood pressure, temperature, and metabolic processes is a(n) _____.

Level Two Reviewing Concepts

9. **Mapping exercise:** Make a large map showing the organization of the human body. Show all levels of organization in the body (see Fig. 1.1) and all 10 organ systems. Try to include functions of all components on the map and remember that some structures may share functions. (*Hint:* Start with the human body as the most important term. You may also draw the outline of a body and make your map using it as the basis.)

10. Distinguish between the items in each group of terms.

   (a) tissues and organs
   (b) *x*-axis and *y*-axis on a graph
   (c) dependent and independent variables
   (d) teleological and mechanistic approaches
   (e) the internal and external environments for a human
   (f) blind, double-blind, and crossover studies
   (g) the target and the sensor in a control system

11. Name as many organs or body structures that connect directly with the external environment as you can.

12. Which organ systems are responsible for coordinating body function? For protecting the body from outside invaders? Which systems exchange material with the external environment, and what do they exchange?

13. Explain the differences among positive feedback, negative feedback, and feedforward mechanisms. Under what circumstances would each be advantageous?

Level Three Problem Solving

14. A group of biology majors went to a mall and asked passersby, "Why does blood flow?" These are some of the answers they received. Which answers are teleological and which are mechanistic? (Not all answers are correct, but they can still be classified.)

    (a) Because of gravity
    (b) To bring oxygen and food to the cells
    (c) Because if it didn't flow, we would die
    (d) Because of the pumping action of the heart

15. Although dehydration is one of the most serious physiological obstacles that land animals must overcome, there are others. Think of as many as you can, and think of various strategies that different terrestrial animals have to overcome these obstacles. (*Hint:* Think of humans, insects, and amphibians; also think of as many different terrestrial habitats as you can.)

Level Four Quantitative Problems

16. A group of students wanted to see what effect a diet deficient in vitamin D would have on the growth of baby guppies. They fed the guppies a diet low in vitamin D and measured fish body length every third day for three weeks. Their data looked like this:

Day	0	3	6	9	12	15	18	21
Average body length (mm)	6	7	9	12	14	16	18	21

    (a) What was the dependent variable and what was the independent variable in this experiment?
    (b) What was the control in this experiment?
    (c) Make a fully labeled graph with a legend, using the data in the table.
    (d) During what time period was growth slowest? Most rapid? (Use your graph to answer this question.)

17. You performed an experiment in which you measured the volumes of nine slices of potato, then soaked the slices in solutions of different salinities for 30 minutes. At the end of 30 minutes, you again measured the volumes of the nine slices. The changes you found were:

Percent Change in Volume after 30 Minutes			
Solution	Sample 1	Sample 2	Sample 3
Distilled water	10%	8%	11%
1% salt (NaCl)	0%	-0.5%	1%
9% salt (NaCl)	-8%	-12%	-11%

    (a) What was the independent variable in this experiment? What was the dependent variable?
    (b) Can you tell from the information given whether or not there was a control in this experiment? If there was a control, what was it?
    (c) Graph the results of the experiment using the most appropriate type of graph.

18. At the end of the semester, researchers measured an intermediate-level class of 25 male weight lifters for aerobic fitness and midarm muscle circumference. The relationship between those two variables is graphed here.

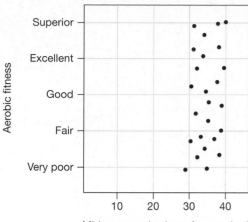

    (a) What kind of graph is this?
    (b) What question were the investigators asking?
    (c) In one sentence, summarize the relationship between the two variables plotted on the graph.

19. Answer the questions after the following article summary.

    A study[7] was carried out on human volunteers to see whether two procedures performed during arthroscopic surgery {*arthro-*, joint + *scopium*, to look at} are effective in relieving knee pain associated with osteoarthritis, or degenerative joint disease {*osteon*, bone + *arthro-*, joint + *-itis*, inflammation}. The volunteers were up to 75 years old and were recruited from a Veterans Affairs Medical Center. They were 93% male and 60% white. One-third of the subjects had placebo operations—that is, they were given anesthesia and their knees were cut open, but the remainder of the treatment procedure was not done. The other two-thirds of the subjects had one of the two treatment procedures performed. Subjects were followed for two years. They answered questions about their knee pain and function and were given an objective walking and stair-climbing test. At the end of the study, the results showed no significant difference in knee function or perception of pain between subjects getting one of the standard treatments and those getting the placebo operation.

    (a) Do you think it is ethical to perform placebo surgeries on humans who are suffering from a painful condition, even if the subjects are informed that they might receive the placebo operation and not the standard treatment?
    (b) Give two possible explanations for the decreased pain reported by the placebo operation subjects.
    (c) Analyze and critique the experimental design of this study. Are the results of this study applicable to everyone with knee pain?
    (d) Was this study a blind, double-blind, or double-blind crossover design?
    (e) Why do you think the investigators felt it was necessary to include a placebo operation in this study?

---

[7]J. B. Moseley *et al.* A controlled trial of arthroscopic surgery for osteoarthritis of the knee. *N Eng J Med* 347(2): 81–88, 2002.

Answers to Concept Checks, Figure and Graph Questions, and end-of-chapter Review Questions can be found in Appendix A [A-1].

2

Molecular Interactions

Science regards man as an aggregation of atoms temporarily united by a mysterious force called the life-principle.

H. P. Blavatsky, 1877. In *Isis Unveiled: A Master-Key to the Mysteries of Ancient and Modern Science and Theology, Vol. I: Science*

MOLECULES AND BONDS 29

LO 2.1 Compare and contrast the composition, structure, and functions of the four major groups of biomolecules.

LO 2.2 Describe four important biological roles of electrons.

LO 2.3 Describe and compare the different types of covalent and noncovalent bonds.

NONCOVALENT INTERACTIONS 40

LO 2.4 Contrast the structure and solubility of polar and nonpolar molecules.

LO 2.5 Describe the covalent and noncovalent interactions that contribute to molecular shape, and explain how

molecular shape is related to molecular function.

LO 2.6 Define pH in words and mathematically, and explain the differences between acids, bases, and buffers.

PROTEIN INTERACTIONS 46

LO 2.7 List seven important functions of soluble proteins in the body.

LO 2.8 Explain the meanings of affinity, specificity, saturation, and competition in protein-ligand binding.

LO 2.9 Explain the different methods by which modulators alter protein binding or protein activity.

Crystals of ATP

Nearly 100 years ago two scientists, Aleksander Oparin in Russia and John Haldane in England, speculated on how life might have arisen on a primitive Earth whose atmosphere consisted mainly of hydrogen, water, ammonia, and methane. Their theories were put to the test in 1953, when a 23-year-old scientist named Stanley Miller combined these molecules in a closed flask and boiled them for a week while periodically discharging flashes of electricity through them, simulating lightning. At the end of his test, Miller found amino acids had formed in the flask. With this simple experiment, he had shown that it was possible to create organic molecules, usually associated with living creatures, from nonliving inorganic precursors.

Miller's experiments were an early attempt to solve one of the biggest mysteries of biology: How did a collection of chemicals first acquire the complex properties that we associate with living creatures? We still do not have an answer to this question. Numerous scientific theories have been proposed, ranging from life arriving by meteor from outer space to molecules forming in deep ocean hydrothermal vents. No matter what their origin, the molecules associated with living organisms have the ability to organize themselves into compartments, replicate themselves, and act as *catalysts* to speed up reactions that would otherwise proceed too slowly to be useful.

The human body is far removed from the earliest life forms, but we are still a collection of chemicals—dilute solutions of dissolved and suspended molecules enclosed in compartments with lipid-protein walls. Strong links between atoms, known as chemical bonds, store and transfer energy to support life functions. Weaker interactions between and within molecules create distinctive molecular shapes and allow biological molecules to interact reversibly with each other.

This chapter introduces some of the fundamental principles of molecular interactions that you will encounter repeatedly in your study of physiology. The human body is more than 50% water, and because most of its molecules are dissolved in this water, we will review the properties of aqueous solutions. If you would like to refresh your understanding of the key features of atoms, chemical bonds, and biomolecules, you will find a series of one- and two-page review features that encapsulate biochemistry

RUNNING **PROBLEM** | **Chromium Supplements**

"Lose weight while gaining muscle," the ads promise. "Prevent heart disease." "Stabilize blood sugar." What is this miracle substance? It's chromium picolinate, a nutritional supplement being marketed to consumers looking for a quick fix. Does it work, though, and is it safe? Some athletes, like Stan—the star running back on the college football team—swear by it. Stan takes 500 micrograms of chromium picolinate daily. Many researchers, however, are skeptical and feel that the necessity for and safety of chromium supplements have not been established.

(29) (39) (40) (41) (46) (48) (53)

as it pertains to physiology. You can test your knowledge of basic chemistry and biochemistry with a special review quiz at the end of the chapter.

MOLECULES AND BONDS

There are more than 100 known elements on Earth, but only three—oxygen, carbon, and hydrogen—make up more than 90% of the body's mass. These three plus eight additional elements are considered *major essential elements.* An additional 19 *minor essential elements* are required in trace amounts. A periodic table showing major and minor essential elements is located inside the back cover of the book.

Most Biomolecules Contain Carbon, Hydrogen, and Oxygen

Molecules that contain carbon are known as **organic molecules,** because it was once thought that they all existed in or were derived from plants and animals. Organic molecules associated with living organisms are also called **biomolecules.** There are four major groups of biomolecules: carbohydrates, lipids, proteins, and nucleotides. The body uses the first three groups for energy and as the building blocks of cellular components. The fourth group, the nucleotides, includes DNA, RNA, ATP, and cyclic AMP. DNA and RNA are the structural components of genetic material. ATP (adenosine triphosphate) and related molecules carry energy, while cyclic AMP (adenosine monophosphate; cAMP) and related compounds regulate metabolism.

Each group of biomolecules has a characteristic composition and molecular structure. Lipids are mostly carbon and hydrogen (**FIG. 2.1**). Carbohydrates are primarily carbon, hydrogen, and oxygen, in the ratio CH_2O (**FIG. 2.2**). Proteins and nucleotides contain nitrogen in addition to carbon, hydrogen, and oxygen (**FIGS. 2.3** and **2.4**). Two amino acids, the building blocks of proteins, also contain sulfur.

Not all biomolecules are pure protein, pure carbohydrate, or pure lipid, however. **Conjugated proteins** are protein molecules combined with another kind of biomolecule. For example, proteins combine with lipids to form **lipoproteins.** Lipoproteins are found in cell membranes and in the blood, where they act as carriers for less soluble molecules, such as cholesterol.

Glycosylated molecules are molecules to which a carbohydrate has been attached. Proteins combined with carbohydrates form **glycoproteins.** Lipids bound to carbohydrates become **glycolipids.** Glycoproteins and glycolipids, like lipoproteins, are important components of cell membranes [see Chapter 3].

Many biomolecules are **polymers,** large molecules made up of repeating units {*poly-*, many + *-mer,* a part}. For example, glycogen and starch are both glucose polymers. They differ in the way the glucose molecules attach to each other, as you can see at the bottom of Figure 2.2.

Some combinations of elements, known as **functional groups,** occur repeatedly in biological molecules. The atoms in a functional group tend to move from molecule to molecule as a single unit.

FIG. 2.1 REVIEW

Biochemistry of Lipids

Lipids are biomolecules made mostly of carbon and hydrogen. Most lipids have a backbone of **glycerol** and 1–3 **fatty acids**. An important characteristic of lipids is that they are nonpolar and therefore not very soluble in water. Lipids can be divided into two broad categories.

• **Fats** are solid at room temperature. Most fats are derived from animal sources.

• **Oils** are liquid at room temperature. Most plant lipids are oils.

Fatty Acids

Fatty acids are long chains of carbon atoms bound to hydrogens, with a carboxyl (–COOH) or "acid" group at one end of the chain.

Palmitic acid, a saturated fatty acid

Saturated fatty acids have no double bonds between carbons, so they are "saturated" with hydrogens. The more saturated a fatty acid is, the more likely it is to be solid at room temperature.

Oleic acid, a monounsaturated fatty acid

Monounsaturated fatty acids have one double bond between two of the carbons in the chain. For each double bond, the molecule has two fewer hydrogen atoms attached to the carbon chain.

Linolenic acid, a polyunsaturated fatty acid

Polyunsaturated fatty acids have two or more double bonds between carbons in the chain.

Formation of Lipids

 Glycerol is a simple 3-carbon molecule that makes up the backbone of most lipids.

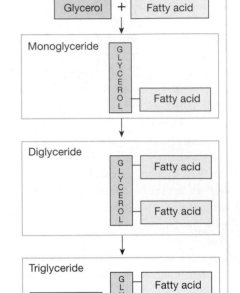

Glycerol plus one fatty acid produces a **monoglyceride**.

Glycerol plus two fatty acids produces a **diglyceride**.

Glycerol plus three fatty acids produces a **triglyceride** (triacylglycerol). More than 90% of lipids are in the form of triglycerides.

Lipid-Related Molecules

In addition to true lipids, this category includes three types of lipid-related molecules.

Eicosanoids

Eicosanoids {*eikosi*, twenty} are modified 20-carbon fatty acids with a complete or partial carbon ring at one end and two long carbon chain "tails."

Prostaglandin E₂ (PGE₂)

Eicosanoids, such as thromboxanes, leukotrienes, and prostaglandins, act as regulators of physiological functions.

Steroids

Steroids are lipid-related molecules whose structure includes four linked carbon rings.

Cholesterol is the primary source of steroids in the human body.

Cortisol

Phospholipids

Phospholipids have 2 fatty acids and a phosphate group (–H₂PO₄). Cholesterol and phospholipids are important components of animal cell membranes.

Phosphate group

FIG. 2.2 REVIEW

Biochemistry of Carbohydrates

Carbohydrates are the most abundant biomolecule. They get their name from their structure, literally carbon {*carbo-*} with water {*hydro-*}. The general formula for a carbohydrate is $(CH_2O)n$ or $C_nH_{2n}O_n$, showing that for each carbon there are two hydrogens and one oxygen. Carbohydrates can be divided into three categories: **monosaccharides**, **disaccharides**, and complex glucose polymers called **polysaccharides.**

Monosaccharides

Monosaccharides are simple sugars. The most common monosaccharides are the building blocks of complex carbohydrates and have either five carbons, like ribose, or six carbons, like glucose.

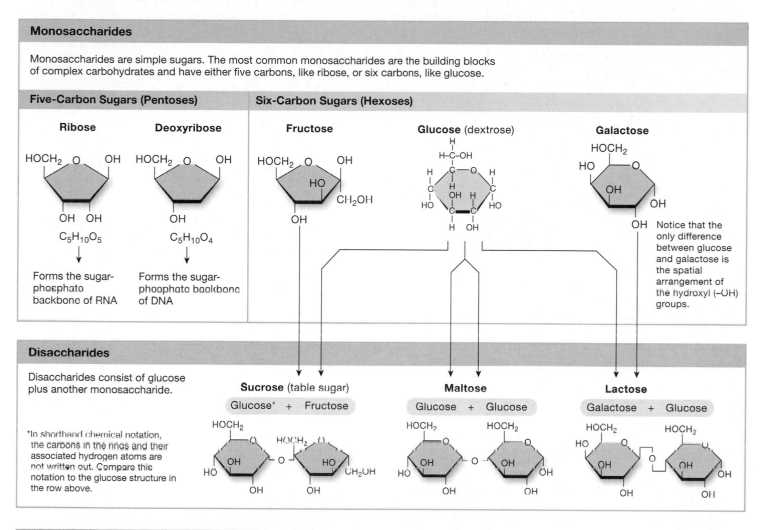

Five-Carbon Sugars (Pentoses)

Ribose

$C_5H_{10}O_5$

Forms the sugar-phosphate backbone of RNA

Deoxyribose

$C_5H_{10}O_4$

Forms the sugar-phosphate backbone of DNA

Six-Carbon Sugars (Hexoses)

Fructose

Glucose (dextrose)

Galactose

Notice that the only difference between glucose and galactose is the spatial arrangement of the hydroxyl (–OH) groups.

Disaccharides

Disaccharides consist of glucose plus another monosaccharide.

*In shorthand chemical notation, the carbons in the rings and their associated hydrogen atoms are not written out. Compare this notation to the glucose structure in the row above.

Sucrose (table sugar)

Glucose* + Fructose

Maltose

Glucose + Glucose

Lactose

Galactose + Glucose

Polysaccharides

Polysaccharides are glucose polymers. All living cells store glucose for energy in the form of a polysaccharide.

Animals

Plants

Yeasts and bacteria

Chitin**

in invertebrate animals

Glycogen

Glucose molecules

Cellulose**

Humans cannot digest cellulose and obtain its energy, even though it is the most abundant polysaccharide on earth.

Starch

Dextran

Digestion of starch or glycogen yields maltose.

**Chitin and cellulose are structural polysaccharides.

31

FIG. 2.3 REVIEW

Biochemistry of Proteins

Proteins are polymers of smaller building-block molecules called **amino acids.**

Amino Acids

All amino acids have a carboxyl group (–COOH), an amino group (–NH$_2$), and a hydrogen attached to the same carbon. The fourth bond of the carbon attaches to a variable "R" group.

The nitrogen (N) in the amino group makes proteins our major dietary source of nitrogen.

The R groups differ in their size, shape, and ability to form hydrogen bonds or ions. Because of the different R groups, each amino acid reacts with other molecules in a unique way.

In a **peptide bond,** the amino group of one amino acid joins the carboxyl group of the other, with the loss of water.

Amino Acids in Natural Proteins

Twenty different amino acids commonly occur in natural proteins. The human body can synthesize most of them, but at different stages of life some amino acids must be obtained from diet and are therefore considered *essential amino acids*. Some physiologically important amino acids are listed below.

Amino Acid	Three-Letter Abbreviation	One-Letter Symbol
Arginine	Arg	R
Aspartic acid (aspartate)*	Asp	D
Cysteine	Cys	C
Glutamic acid (glutamate)*	Glu	E
Glutamine	Gln	Q
Glycine	Gly	G
Tryptophan	Trp	W
Tyrosine	Tyr	Y

Note:

A few amino acids do not occur in proteins but have important physiological functions.

- *Homocysteine*: a sulfur-containing amino acid that in excess is associated with heart disease
- *γ-amino butyric acid* (gamma-amino butyric acid) or *GABA*: a chemical made by nerve cells
- *Creatine*: a molecule that stores energy when it binds to a phosphate group

*The suffix –ate indicates the anion form of the acid.

Structure of Peptides and Proteins

Primary Structure

The 20 protein-forming amino acids assemble into polymers called peptides. The sequence of amino acids in a peptide chain is called the **primary structure.** Just as the 26 letters of our alphabet combine to create different words, the 20 amino acids can create an almost infinite number of combinations.

Peptides range in length from two to two million amino acids:

- **Oligopeptide** {*oligo-*, few}: 2–9 amino acids
- **Polypeptide:** 10–100 amino acids
- **Proteins:** >100 amino acids

Sequence of amino acids

Secondary Structure

Covalent bond angles between amino acids determine secondary structure.

α-helix   β-strands form sheets

Tertiary Structure

Tertiary structure is the protein's three-dimensional shape.

Fibrous proteins
Collagen

Globular proteins

Quaternary Structure

Multiple subunits combine with noncovalent bonds. Hemoglobin molecules are made from four globular protein subunits.

Hemoglobin

For example, *hydroxyl groups,* —OH, common in many biological molecules, are added and removed as a group rather than as single hydrogen or oxygen atoms. Amino groups, —NH$_2$, are the signature of amino acids. The phosphate group, —H$_2$PO$_4$, plays a role in many important cell processes, such as energy transfer and protein regulation. Addition of a phosphate group is called *phosphorylation*; removal of a phosphate group is *dephosphorylation*.

The most common functional groups are listed in **TABLE 2.1**.

CONCEPT CHECK

1. List three major essential elements found in the human body.
2. What is the general formula of a carbohydrate?
3. What is the chemical formula of an amino group? Of a carboxyl group?

Electrons Have Four Important Biological Roles

An atom of any element has a unique combination of protons and electrons that determines the element's properties (**FIG. 2.5**). We are particularly interested in the electrons because they play four important roles in physiology:

1. **Covalent bonds.** The arrangement of electrons in the outer energy level (*shell*) of an atom determines an element's ability to bind with other elements. Electrons shared between atoms form strong covalent bonds that create molecules.
2. **Ions.** If an atom or molecule gains or loses one or more electrons, it acquires an electrical charge and becomes an **ion.**

Ions are the basis for electrical signaling in the body. Ions may be single atoms, like the sodium ion Na$^+$ and chloride ion Cl$^-$. Other ions are combinations of atoms, such as the bicarbonate ion HCO$_3^-$. Important ions of the body are listed in **TABLE 2.2**.

3. **High-energy electrons.** The electrons in certain atoms can capture energy from their environment and transfer it to other atoms. This allows the energy to be used for synthesis, movement, and other life processes. The released energy may also be emitted as radiation. For example, bioluminescence in fireflies is visible light emitted by high-energy electrons returning to their normal low-energy state.
4. **Free radicals.** Free radicals are unstable molecules with an unpaired electron. They are thought to contribute to aging and to the development of certain diseases, such as some cancers. Free radicals and high-energy electrons are discussed later.

The role of electrons in molecular bond formation is discussed in the next section. There are four common bond types, two strong and two weak. Covalent and ionic bonds are strong bonds because they require significant amounts of energy to make or break. Hydrogen bonds and van der Waals forces are weaker bonds that require much less energy to break. Interactions between molecules with different bond types are responsible for energy use and transfer in metabolic reactions as well as a variety of other reversible interactions.

Covalent Bonds between Atoms Create Molecules

Molecules form when atoms share pairs of electrons, one electron from each atom, to create **covalent bonds.** These strong bonds require the input of energy to break them apart. It is possible to predict how many covalent bonds an atom can form by knowing how many unpaired electrons are in its outer shell, because an atom is most stable when all of its electrons are paired (**FIG. 2.6**).

For example, a hydrogen atom has one unpaired electron and one empty electron place in its outer shell. Because hydrogen has only one electron to share, it always forms one covalent bond, represented by a single line (—) between atoms. Oxygen has six electrons in an outer shell that can hold eight.

TABLE 2.1 | Common Functional Groups

Notice that oxygen, with two electrons to share, sometimes forms a double bond with another atom.

	Shorthand	Bond Structure
Amino	—NH$_2$	—N with H, H
Carboxyl (acid)	—COOH	—C with =O, OH
Hydroxyl	—OH	—O—H
Phosphate	—H$_2$PO$_4$	—O—P(=O) with OH, OH

TABLE 2.2 | Important Ions of the Body

Cations		Anions	
Na$^+$	Sodium	Cl$^-$	Chloride
K$^+$	Potassium	HCO$_3^-$	Bicarbonate
Ca^{2+}	Calcium	HPO$_4^{2-}$	Phosphate
H$^+$	Hydrogen	SO$_4^{2-}$	Sulfate
Mg^{2+}	Magnesium		

FIG. 2.4 **REVIEW**

Nucleotides and Nucleic Acids

Nucleotides are biomolecules that play an important role in energy and information transfer. Single nucleotides include the energy-transferring compounds **ATP** (adenosine triphosphate) and **ADP** (adenosine diphosphate), as well as **cyclic AMP,** a molecule important in the transfer of signals between cells. **Nucleic acids** (or nucleotide polymers) such as **RNA** and **DNA** store and transmit genetic information.

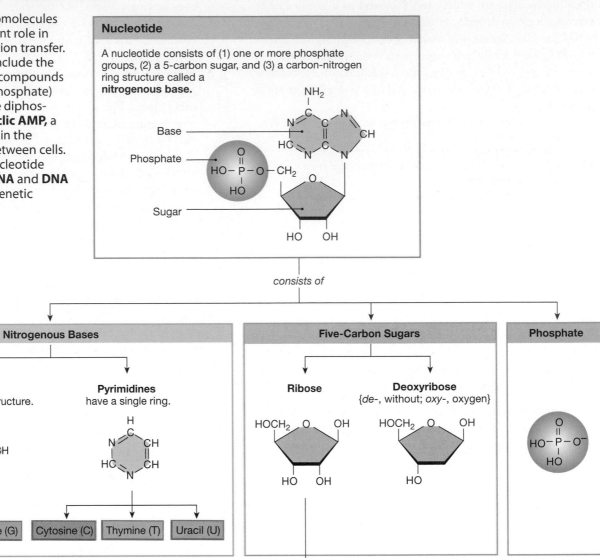

Nucleotide

A nucleotide consists of (1) one or more phosphate groups, (2) a 5-carbon sugar, and (3) a carbon-nitrogen ring structure called a **nitrogenous base.**

Base

Phosphate

Sugar

consists of

Nitrogenous Bases

Purines
have a double ring structure.

Pyrimidines
have a single ring.

| Adenine (A) | Guanine (G) | Cytosine (C) | Thymine (T) | Uracil (U) |

Five-Carbon Sugars

Ribose

Deoxyribose
{*de*-, without; *oxy*-, oxygen}

Phosphate

Adenine + Ribose

Adenosine

Single Nucleotide Molecules

Single nucleotide molecules have two critical functions in the human body: (1) Capture and transfer energy in high-energy electrons or phosphate bonds, and (2) aid in cell-to-cell communication.

Nucleotide	consists of	Base	+	Sugar	+	Phosphate Groups	+	Other Component	Function
ATP	=	Adenine	+	Ribose	+	3 phosphate groups			Energy capture and transfer
ADP	=	Adenine	+	Ribose	+	2 phosphate groups			
NAD	=	Adenine	+	2 Ribose	+	2 phosphate groups	+	Nicotinamide	
FAD	=	Adenine	+	Ribose	+	2 phosphate groups	+	Riboflavin	
cAMP	=	Adenine	+	Ribose	+	1 phosphate group			Cell-to-cell communication

Nucleic acids (nucleotide polymers) function in information storage and transmission. The sugar of one nucleotide links to the phosphate of the next, creating a chain of alternating sugar–phosphate groups. The sugar–phosphate chains, or backbone, are the same for every nucleic acid molecule. Nucleotide chains form strands of DNA and RNA.

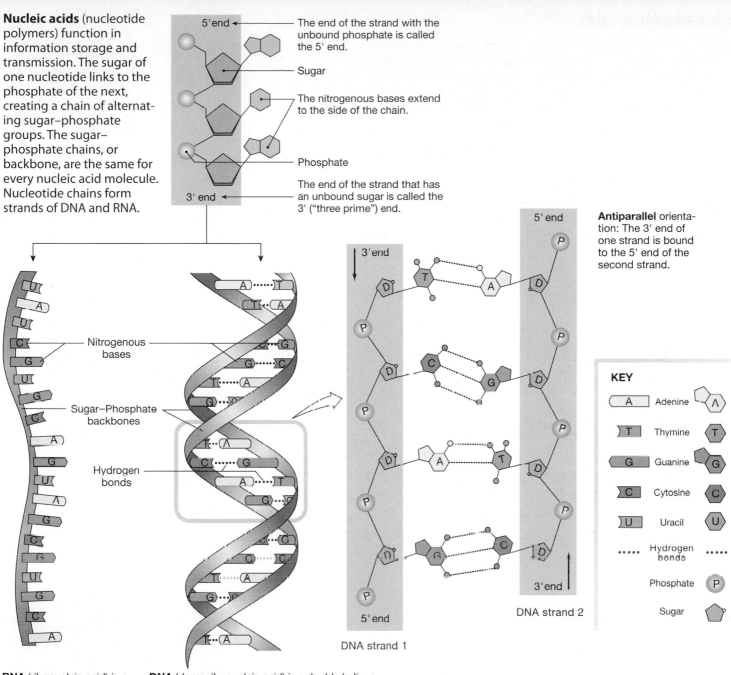

5'end — The end of the strand with the unbound phosphate is called the 5' end.

Sugar

The nitrogenous bases extend to the side of the chain.

Phosphate

3' end — The end of the strand that has an unbound sugar is called the 3' ("three prime") end.

Antiparallel orientation: The 3' end of one strand is bound to the 5' end of the second strand.

3'end

5' end

Nitrogenous bases

Sugar–Phosphate backbones

Hydrogen bonds

DNA strand 1

DNA strand 2

3' end

5' end

KEY

A	Adenine	A
T	Thymine	T
G	Guanine	G
C	Cytosine	C
U	Uracil	U
.....	Hydrogen bonds	.....
	Phosphate	P
	Sugar	

RNA (ribonucleic acid) is a single–strand nucleic acid with ribose as the sugar in the backbone, and four bases—adenine, guanine, cytosine, and uracil.

DNA (deoxyribonucleic acid) is a double helix, a three-dimensional structure that forms when two DNA strands link through hydrogen bonds between complementary base pairs. Deoxyribose is the sugar in the backbone, and the four bases are adenine, guanine, cytosine, and thymine.

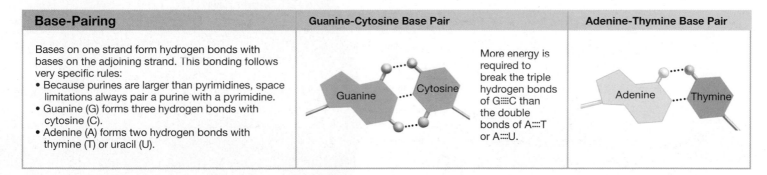

Base-Pairing

Bases on one strand form hydrogen bonds with bases on the adjoining strand. This bonding follows very specific rules:
- Because purines are larger than pyrimidines, space limitations always pair a purine with a pyrimidine.
- Guanine (G) forms three hydrogen bonds with cytosine (C).
- Adenine (A) forms two hydrogen bonds with thymine (T) or uracil (U).

Guanine-Cytosine Base Pair

Guanine — Cytosine

More energy is required to break the triple hydrogen bonds of G≡C than the double bonds of A꞊T or A꞊U.

Adenine-Thymine Base Pair

Adenine — Thymine

FIG. 2.5 **REVIEW**

Atoms and Molecules

Elements are the simplest type of matter. There are over 100 known elements,* but only three—oxygen, carbon, and hydrogen—make up more than 90% of the body's mass. These three plus eight additional elements are *major essential elements*. An additional 19 *minor essential elements* are required in trace amounts. The smallest particle of any element is an **atom** {*atomos*, indivisible}. Atoms link by sharing electrons to form molecules.

* A periodic table of the elements can be found inside the back cover of the book.

Major Essential Elements	Minor Essential Elements
H, C, O, N, Na, Mg, K, Ca, P, S, Cl	Li, F, Cr, Mn, Fe, Co, Ni, Cu, Zn, Se, Y, I, Zr, Nb, Mo, Tc, Ru, Rh, La

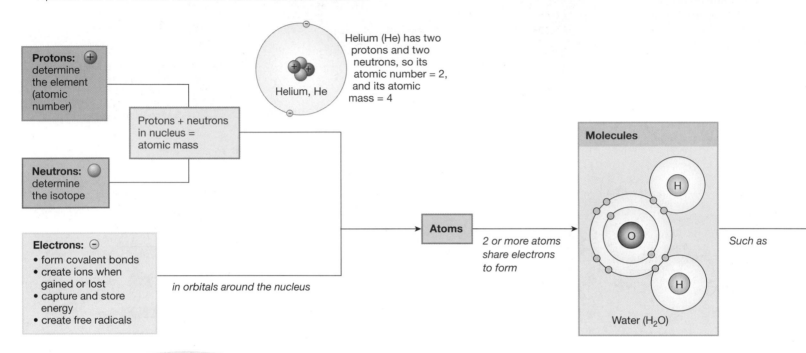

Protons: ⊕ determine the element (atomic number)

Neutrons: ○ determine the isotope

Protons + neutrons in nucleus = atomic mass

Helium (He) has two protons and two neutrons, so its atomic number = 2, and its atomic mass = 4

Helium, He

Electrons: ⊖
• form covalent bonds
• create ions when gained or lost
• capture and store energy
• create free radicals

in orbitals around the nucleus

Atoms

2 or more atoms share electrons to form

Molecules

H
O
H

Water (H_2O)

Such as

Isotopes and Ions

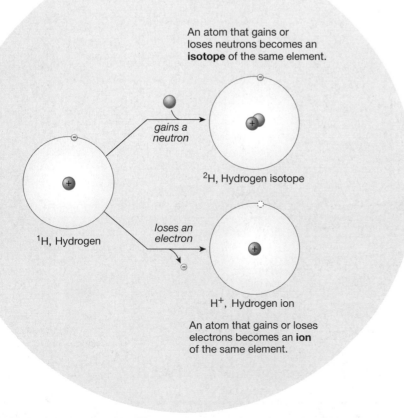

An atom that gains or loses neutrons becomes an **isotope** of the same element.

gains a neutron

^2H, Hydrogen isotope

^1H, Hydrogen

loses an electron

H^+, Hydrogen ion

An atom that gains or loses electrons becomes an **ion** of the same element.

Proteins

Amino acids → Amino acid sequence → α-helix or β-strand → Globular or fibrous shape → **Proteins**

| Ala | Val | Ser | Lys | Arg | Trp |

Amino acid sequence

Carbohydrates

Monosaccharides → Disaccharides → Polysaccharides → **Carbohydrates**

Polysaccharide

Polysaccharides → Glycogen
Polysaccharides → Starch
Polysaccharides → Cellulose

Biomolecules

Glycoproteins

Lipoproteins

Glycolipids

Lipids

Glycerol
Fatty acids

Monoglycerides → Diglycerides → Triglycerides → **Lipids**

Oleic acid, a fatty acid

Lipid-related molecules

Phospholipids

Eicosanoids

Steroids

Nucleotides

cAMP, cGMP

ATP, ADP, FAD, NAD

RNA, DNA

DNA molecule

FIG. 2.6 **REVIEW**

Molecular Bonds

When two or more atoms link by sharing electrons, they make units known as **molecules.** The transfer of electrons from one atom to another or the sharing of electrons by two atoms is a critical part of forming **bonds,** the links between atoms.

Bonds

Covalent Bonds

Covalent bonds result when atoms share electrons. These bonds require the most energy to make or break.

(a) Nonpolar Molecules

Nonpolar molecules have an even distribution of electrons. For example, molecules composed mostly of carbon and hydrogen tend to be nonpolar.

Fatty acid

Hydrogen

Carbon

(b) Polar Molecules

Polar molecules have regions of partial charge (δ^+ or δ^-). The most important example of a polar molecule is water.

Negative pole

δ^- δ^-

δ^+ δ^+

Positive pole

Water molecule

O

H H

$H\!:\!\ddot{O}\!:\!H$ = $H^{\diagdown O \diagup} H$ = H_2O

Noncovalent Bonds

(c) Ionic Bonds

Ionic bonds are electrostatic attractions between ions. A common example is sodium chloride.

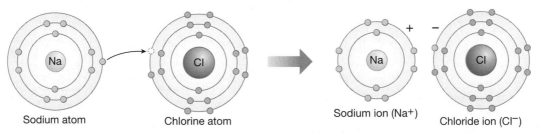

Sodium atom   Chlorine atom

Na   Cl

+   −

Na   Cl

Sodium ion (Na^+)   Chloride ion (Cl^-)

Sodium gives up its one weakly held electron to chlorine, creating sodium and chloride ions, Na^+ and Cl^-.

The sodium and chloride ions both have stable outer shells that are filled with electrons. Because of their opposite charges, they are attracted to each other and, in the solid state, the ionic bonds form a sodium chloride (NaCl) crystal.

(d) Hydrogen Bonds

Hydrogen bonds form between a hydrogen atom and a nearby oxygen, nitrogen, or fluorine atom. So, for example, the polar regions of adjacent water molecules allow them to form hydrogen bonds with one another.

Hydrogen bonding

Hydrogen bonding between water molecules is responsible for the surface tension of water.

(e) Van der Waals Forces

Van der Waals forces are weak, nonspecific attractions between atoms.

RUNNING **PROBLEM**

What is chromium picolinate? Chromium (Cr) is an essential element that has been linked to normal glucose metabolism. In the diet, chromium is found in brewer's yeast, broccoli, mushrooms, and apples. Because chromium in food and in chromium chloride supplements is poorly absorbed from the digestive tract, a scientist developed and patented the compound chromium picolinate. Picolinate, derived from amino acids, enhances chromium uptake at the intestine. The recommended adequate intake (AI) of chromium for men age 19–50 is 35 µg/day. (For women, it is 25 µg/day.) As we've seen, Stan takes more than 10 times this amount.

Q1: *Locate chromium on the periodic table of the elements. What is chromium's atomic number? Atomic mass? How many electrons does one atom of chromium have? Which elements close to chromium are also essential elements?*

That means oxygen can form two covalent bonds and fill its outer shell with electrons. If adjacent atoms share two pairs of electrons rather than just one pair, a **double bond,** represented by a double line ($=$), results. If two atoms share three pairs of electrons, they form a triple bond.

Polar and Nonpolar Molecules

Some molecules develop regions of partial positive and negative charge when the electron pairs in their covalent bonds are not evenly shared between the linked atoms. When electrons are shared unevenly, the atom(s) with the stronger attraction for electrons develops a slight negative charge (indicated by δ^-), and the atom(s) with the weaker attraction for electrons develops a slight positive charge (δ^+). These molecules are called **polar molecules** because they can be said to have positive and negative ends, or poles. Certain elements, particularly nitrogen and oxygen, have a strong attraction for electrons and are often found in polar molecules.

A good example of a polar molecule is water (H_2O). The larger and stronger oxygen atom pulls the hydrogen electrons toward itself. This pull leaves the two hydrogen atoms of the molecule with a partial positive charge, and the single oxygen atom with a partial negative charge from the unevenly shared electrons (Fig. 2.6b). Note that the net charge for the entire water molecule is zero. The polarity of water makes it a good solvent, and all life as we know it is based on watery, or *aqueous,* solutions.

A **nonpolar molecule** is one whose shared electrons are distributed so evenly that there are no regions of partial positive or negative charge. For example, molecules composed mostly of carbon and hydrogen, such as the fatty acid shown in Figure 2.6a, tend to be nonpolar. This is because carbon does not attract electrons as strongly as oxygen does. As a result, the carbons and hydrogens share electrons evenly, and the molecule has no regions of partial charge.

Noncovalent Bonds Facilitate Reversible Interactions

Ionic bonds, hydrogen bonds, and van der Waals forces are noncovalent bonds. They play important roles in many physiological processes, including pH, molecular shape, and the reversible binding of molecules to each other.

Ionic Bonds Ions form when one atom has such a strong attraction for electrons that it pulls one or more electrons completely away from another atom. For example, a chlorine atom needs only one electron to fill the last of eight places in its outer shell, so it pulls an electron from a sodium atom, which has only one weakly held electron in its outer shell (Fig. 2.6c). The atom that gains electrons acquires one negative charge (-1) for each electron added, so the chlorine atom becomes a chloride ion Cl^-. Negatively charged ions are called **anions.**

An atom that gives up electrons has one positive charge (+1) for each electron lost. For example, the sodium atom becomes a sodium ion Na^+. Positively charged ions are called **cations.**

Ionic bonds, also known as *electrostatic attractions,* result from the attraction between ions with opposite charges. (Remember the basic principle of electricity that says that opposite charges attract and like charges repel.) In a crystal of table salt, the solid form of ionized NaCl, ionic bonds between alternating Na^+ and Cl^- ions hold the ions in a neatly ordered structure.

Hydrogen Bonds A **hydrogen bond** is a weak attractive force between a hydrogen atom and a nearby oxygen, nitrogen, or fluorine atom. No electrons are gained, lost, or shared in a hydrogen bond. Instead, the oppositely charged regions in polar molecules are attracted to each other. Hydrogen bonds may occur between atoms in neighboring molecules or between atoms in different parts of the same molecule. For example, one water molecule may hydrogen-bond with as many as four other water molecules. As a result, the molecules line up with their neighbors in a somewhat ordered fashion (Fig. 2.6d).

Hydrogen bonding between molecules is responsible for the **surface tension** of water. Surface tension is the attractive force between water molecules that causes water to form spherical droplets when falling or to bead up when spilled onto a nonabsorbent surface (Fig. 2.6d). The high cohesiveness {*cohaesus,* to cling together} of water is due to hydrogen bonding and makes it difficult to stretch or deform, as you may have noticed in trying to pick up a wet glass that is "stuck" to a slick table top by a thin film of water. The surface tension of water influences lung function [described in Chapter 17].

Van der Waals Forces Van der Waals forces are weak, nonspecific attractions between the nucleus of any atom and the electrons of nearby atoms. Two atoms that are weakly attracted to each other by van der Waals forces move closer together until they are so close that their electrons begin to repel one another. Consequently, van der Waals forces allow atoms to pack closely together and occupy a minimum amount of space. A single van der Waals attraction between atoms is very weak.

CONCEPT CHECK

4. Are electrons in an atom or molecule most stable when they are paired or unpaired?

5. When an atom of an element gains or loses one or more electrons, it is called a(n) _____ of that element.

6. Match each type of bond with its description:

(a) covalent bond	1. weak attractive force between hydrogen and oxygen or nitrogen
(b) ionic bond	2. formed when two atoms share one or more pairs of electrons
(c) hydrogen bond	3. weak attractive force between atoms
(d) van der Waals force	4. formed when one atom loses one or more electrons to a second atom

NONCOVALENT INTERACTIONS

Many different kinds of noncovalent interactions can take place between and within molecules as a result of the four different types of bonds. For example, the charged, uncharged, or partially charged nature of a molecule determines whether that molecule can dissolve in water. Covalent and noncovalent bonds determine molecular shape and function. Finally, noncovalent interactions allow proteins to associate reversibly with other molecules, creating functional pairings such as enzymes and substrates, or signal receptors and molecules.

Hydrophilic Interactions Create Biological Solutions

Life as we know it is established on water-based, or *aqueous*, solutions that resemble dilute seawater in their ionic composition. The adult human body is about 60% water. Na^+, K^+, and Cl^- are the main ions in body fluids, with other ions making up a lesser proportion. All molecules and cell components are either dissolved or suspended in these solutions. For these reasons, it is useful to understand the properties of solutions, which are reviewed in **FIGURE 2.7**.

The degree to which a molecule is able to dissolve in a solvent is the molecule's **solubility**: the more easily a molecule dissolves, the higher its solubility. Water, the biological solvent, is polar, so molecules that dissolve readily in water are polar or ionic molecules whose positive and negative regions readily interact with water. For example, if NaCl crystals are placed in water, polar regions of the water molecules disrupt the ionic bonds between sodium and chloride, which causes the crystals to dissolve (**FIG. 2.8a**). Molecules that are soluble in water are said to be **hydrophilic** {*hydro-*, water + *-philic*, loving}.

In contrast, molecules such as oils that do not dissolve well in water are said to be **hydrophobic** {*-phobic*, hating}. Hydrophobic substances are usually nonpolar molecules that cannot form hydrogen bonds with water molecules. The lipids (fats and oils) are the most hydrophobic group of biological molecules.

RUNNING **PROBLEM**

One advertising claim for chromium is that it improves the transfer of glucose—the simple sugar that cells use to fuel all their activities—from the bloodstream into cells. In people with diabetes mellitus, cells are unable to take up glucose from the blood efficiently. It seemed logical, therefore, to test whether the addition of chromium to the diet would enhance glucose uptake in people with diabetes. In one Chinese study, diabetic patients receiving 500 micrograms of chromium picolinate twice a day showed significant improvement in their diabetes, but patients receiving 100 micrograms or a placebo did not.

Q2: *If people have a chromium deficiency, would you predict that their blood glucose level would be lower or higher than normal? From the results of the Chinese study, can you conclude that all people with diabetes suffer from a chromium deficiency?*

29 39 **40** 41 46 48 53

When placed in an aqueous solution, lipids do not dissolve. Instead they separate into distinct layers. One familiar example is salad oil floating on vinegar in a bottle of salad dressing. Before hydrophobic molecules can dissolve in body fluids, they must combine with a hydrophilic molecule that will carry them into solution.

For example, cholesterol, a common animal fat, is a hydrophobic molecule. Fat from a piece of meat dropped into a glass of warm water will float to the top, undissolved. In the blood, cholesterol will not dissolve unless it binds to special water-soluble carrier molecules. You may know the combination of cholesterol with its hydrophilic carriers as HDL-cholesterol and LDL-cholesterol, the "good" and "bad" forms of cholesterol associated with heart disease.

Some molecules, such as the phospholipids, have both polar and nonpolar regions (Fig. 2.8b). This dual nature allows them to associate both with each other (hydrophobic interactions) and with polar water molecules (hydrophilic interactions). Phospholipids are the primary component of biological membranes.

CONCEPT CHECK

7. Which dissolve more easily in water, polar molecules or nonpolar molecules?

8. A molecule that dissolves easily is said to be hydro_____ic.

9. Why does table salt (NaCl) dissolve in water?

Molecular Shape Is Related to Molecular Function

A molecule's shape is closely related to its function. Molecular bonds—both covalent bonds and weak bonds—play a critical role in determining molecular shape. The three-dimensional shape of a molecule is difficult to show on paper, but many molecules have characteristic shapes due to the angles of covalent bonds between the atoms. For example, the two hydrogen atoms of the water molecule shown in Figure 2.6b are attached to the oxygen with a bond angle of 104.5°. Double bonds in long carbon chain fatty acids cause the chains to kink or bend, as shown by the three-dimensional model of oleic acid in Figure 2.5.

Weak noncovalent bonds also contribute to molecular shape. The complex double helix of a DNA molecule (Fig. 2.4) results both from covalent bonds between adjacent bases in each strand and the hydrogen bonds connecting the two strands of the helix.

Proteins have the most complex and varied shapes of all the biomolecules. The two common **secondary structures** for poly-peptide chains are the α-**helix** (alpha-helix) spiral and β-**strands** (beta-strands) (Fig. 2.3). The covalent bond angles between amino acids create the spiral of the α-helix or the zigzag shape of β-strands. Adjacent β-strands associate into sheet-like structures that are stabilized by hydrogen bonding, shown as dotted lines (...) in Figure 2.3. The sheet configuration is very stable and occurs in many proteins destined for structural uses. Proteins with other functions may have a mix of β-strands and α-helices.

The **tertiary structure** of a protein is its three-dimensional shape, created through spontaneous folding as the result of co-valent bonds and noncovalent interactions. Proteins are catego-rized into two large groups based on their shape: globular and fibrous (see Fig. 2.3). **Globular proteins** have amino acid chains that fold back on themselves to create a complex tertiary struc-ture containing pockets, channels, or protruding knobs. The ter-tiary structure of globular proteins arises partly from the angles of covalent bonds between amino acids and partly from hydrogen bonds, van der Waals forces, and ionic bonds that stabilize the molecule's shape.

In addition to covalent bonds between adjacent amino ac-ids, covalent **disulfide** (S–S) **bonds** play an important role in the shape of many globular proteins (Fig. 2.8c). The amino acid *cys-teine* contains sulfur as part of a *sulfhydryl group* (—SH). Two cysteines in different parts of the polypeptide chain can bond to each other with a disulfide bond that pulls the sections of chain together.

Fibrous proteins may be β-strands or long chains of α-helices. Fibrous proteins are usually insoluble in water and form important structural components of cells and tissues. Examples of fibrous proteins include *collagen*, found in many types of connec-tive tissue, such as skin, and *keratin*, found in hair and nails.

Hydrogen Ions in Solution Can Alter Molecular Shape

Hydrogen bonding is an important part of molecular shape. However, free hydrogen ions, H^+, in solution can also participate in hydrogen bonding and van der Waals forces. If free H^+ dis-rupts a molecule's noncovalent bonds, the molecule's shape, or *conformation*, can change. A change in shape may alter or destroy the molecule's ability to function.

RUNNING **PROBLEM**

Chromium is found in several ionic forms. The chromium usu-ally found in biological systems and in dietary supplements is the cation Cr^{3+}. This ion is called trivalent because it has a net charge of +3. The hexavalent cation, Cr^{6+}, with a charge of +6, is used in industry, such as in the manufacturing of stainless steel and the chrome plating of metal parts.

Q3: *How many electrons have been lost from the hexavalent ion of chromium? From the trivalent ion?*

29   39   40   **41**   46   48   53

The concentration of free H^+ in body fluids, or *acidity*, is measured in terms of **pH**. **FIGURE 2.9** reviews the chemistry of pH and shows a pH scale with the pH values of various substances. The normal pH of blood in the human body is 7.40, slightly alka-line. Regulation of the body's pH within a narrow range is critical because a blood pH more acidic than 7.00 (pH < 7.00) or more alkaline than 7.70 (pH > 7.70) is incompatible with life.

Where do hydrogen ions in body fluids come from? Some of them come from the separation of water molecules (H_2O) into H^+ and OH^- ions. Others come from **acids,** molecules that release H^+ when they dissolve in water (Fig. 2.9). Many of the molecules made during normal metabolism are acids. For example, carbonic acid is made in the body from CO_2 (carbon dioxide) and water. In solution, carbonic acid separates into a bicarbonate ion and a hydrogen ion:

$$CO_2 + H_2O \rightleftharpoons H_2CO_3 \text{ (carbonic acid)} \rightleftharpoons H^+ + HCO_3$$

Note that when the hydrogen is part of the intact carbonic acid molecule, it does not contribute to acidity. *Only free H^+ contributes to the hydrogen ion concentration.*

We are constantly adding acid to the body through metabo-lism, so how does the body maintain a normal pH? One answer is buffers. A **buffer** is any substance that moderates changes in pH. Many buffers contain anions that have a strong attraction for H^+ molecules. When free H^+ is added to a buffer solution, the buffer's anions bond to the H^+, thereby minimizing any change in pH.

The bicarbonate anion, HCO_3^-, is an important buffer in the human body. The following equation shows how a sodium bicarbonate solution acts as a buffer when hydrochloric acid (HCl) is added. When placed in plain water, hydrochloric acid separates, or dissociates, into H^+ and Cl^- and creates a high H^+ concentration (low pH). When HCl dissociates in a sodium bi-carbonate solution, however, some of the bicarbonate ions com-bine with some of the H to form undissociated carbonic acid. "Tying up" the added H^+ in this way keeps the H^+ concentration of the solution from changing significantly and minimizes the pH change.

FIG. 2.7 REVIEW

Solutions

Life as we know it is established on water-based, or aqueous, solutions that resemble dilute seawater in their ionic composition. The human body is 60% water. Sodium, potassium, and chloride are the main ions in body fluids. All molecules and cell components are either dissolved or suspended in these saline solutions. For these reasons, the properties of solutions play a key role in the functioning of the human body.

Terminology

A **solute** is any substance that dissolves in a liquid. The degree to which a molecule is able to dissolve in a solvent is the molecule's solubility. The more easily a solute dissolves, the higher its **solubility**.

A **solvent** is the liquid into which solutes dissolve. In biological solutions, water is the universal solvent.

A **solution** is the combination of solutes dissolved in a solvent. The **concentration** of a solution is the amount of solute per unit volume of solution.

Concentration = solute amount/volume of solution

Expressions of Solute Amount

- **Mass** (weight) of the solute before it dissolves. Usually given in grams (g) or milligrams (mg).

- **Molecular mass** is calculated from the chemical formula of a molecule. This is the mass of one molecule, expressed in atomic mass units (amu) or, more often, in daltons (Da), where 1 amu = 1 Da.

$$\text{Molecular mass} = \text{SUM}\left[\begin{array}{ccc}\text{atomic mass} & & \text{the number of atoms}\\ \text{of each element} & \times & \text{of each element}\end{array}\right]$$

Example

What is the molecular mass of glucose, $C_6H_{12}O_6$?	**Answer**

Element	# of Atoms	Atomic Mass of Element
Carbon	6	12.0 amu × 6 = 72
Hydrogen	12	1.0 amu × 12 = 12
Oxygen	6	16.0 amu × 6 = 96

Molecular mass of glucose = 180 amu (or Da)

- **Moles** (mol) are an expression of the number of solute molecules, without regard for their weight. One mole = 6.02×10^{23} atoms, ions, or molecules of a substance. One mole of a substance has the same number of particles as one mole of any other substance, just as a dozen eggs has the same number of items as a dozen roses.

- **Gram molecular weight.** In the laboratory, we use the molecular mass of a substance to measure out moles. For example, one mole of glucose (with 6.02×10^{23} glucose molecules) has a molecular mass of 180 Da and weighs 180 grams. The molecular mass of a substance expressed in grams is called the gram molecular weight.

- **Equivalents** (Eq) are a unit used for ions, where 1 equivalent = molarity of the ion × the number of charges the ion carries. The sodium ion, with its charge of +1, has one equivalent per mole. The hydrogen phosphate ion (HPO_4^{2-}) has two equivalents per mole. Concentrations of ions in the blood are often reported in milliequivalents per liter (mEq/L).

Q FIGURE QUESTIONS

1. What are the two components of a solution?
2. The concentration of a solution is expressed as:
   (a) amount of solvent/volume of solute
   (b) amount of solute/volume of solvent
   (c) amount of solvent/volume of solution
   (d) amount of solute/volume of solution
3. Calculate the molecular mass of water, H_2O.
4. How much does a mole of KCl weigh?

Expressions of Volume

Volume is usually expressed as liters (L) or milliliters (mL)
{*milli-*, 1/1000}. A volume convention common in medicine is
the deciliter (dL), which is 1/10 of a liter, or 100 mL.

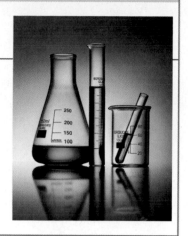

Prefixes

deci- (d)	1/10	1×10^{-1}
milli- (m)	1/1000	1×10^{-3}
micro- (μ)	1/1,000,000	1×10^{-6}
nano- (n)	1/1,000,000,000	1×10^{-9}
pico- (p)	1/1,000,000,000,000	1×10^{-12}

Useful Conversions

• 1 liter of water weighs
  1 kilogram (kg) {*kilo-*, 1000}

• 1 kilogram = 2.2 pounds

Expressions of Concentration

• **Percent solutions.** In a laboratory or pharmacy, scientists cannot measure out solutes by the mole.
Instead, they use the more conventional measurement of weight. The solute concentration may
then be expressed as a percentage of the total solution, or percent solution. A 10% solution
means 10 parts of a solute per 100 parts of total solution. Weight/volume solutions, used for
solutes that are solids, are usually expressed as g/100 mL solution or mg/dL. An out-of-date way
of expressing mg/dL is mg% where % means per 100 parts or 100 mL. A concentration of 20
mg/dL could also be expressed as 20 mg%.

FIGURE QUESTIONS

5. Which solution is more
   concentrated: a 100 mM
   solution of glucose or a 0.1
   M solution of glucose?
6. When making a 5% solution
   of glucose, why don't you
   measure out 5 grams of
   glucose and add it to 100 mL
   of water?

Example

Solutions used for intravenous (IV) infusions are often expressed as percent solutions. How would you make 500 mL of a 5% dextrose (glucose) solution?	**Answer** 5% solution = 5 g glucose dissolved in water to make a final volume of 100 mL solution. 5 g glucose/100 mL = ? g/500 mL 25 g glucose with water added to give a final volume of 500 mL

• **Molarity** is the number of moles of solute in a liter of solution, and is abbreviated as either mol/L
or M. A one molar solution of glucose (1 mol/L, 1 M) contains 6.02×10^{23} molecules of glucose
per liter of solution. It is made by dissolving one mole (180 grams) of glucose in enough water to
make one liter of solution. Typical biological solutions are so dilute that solute concentrations are
usually expressed as **millimoles** per liter (mmol/L or mM).

Example

What is the molarity of a 5% dextrose solution?	**Answer** 5 g glucose/100 mL = 50 g glucose/1000 mL (or 1 L) 1 mole glucose = 180 g glucose 50 g/L × 1 mole/180 g – 0.278 moles/L or 278 mM

FIG. 2.8 **REVIEW**

Molecular Interactions

(a) Hydrophilic Interactions

Molecules that have polar regions or ionic bonds readily interact with the polar regions of water. This enables them to dissolve easily in water. Molecules that dissolve readily in water are said to be **hydrophilic** {*hydro-,* water + *philos,* loving}.

Water molecules interact with ions or other polar molecules to form hydration shells around the ions. This disrupts the hydrogen bonding between water molecules, thereby lowering the freezing temperature of water (freezing point depression).

Cl^-

Hydration shells

Na^+

Glucose molecule

Water molecules

NaCl in solution

Glucose molecule in solution

(b) Hydrophobic Interactions

Because they have an even distribution of electrons and no positive or negative poles, nonpolar molecules have no regions of partial charge, and therefore tend to repel water molecules. Molecules like these do not dissolve readily in water and are said to be **hydrophobic** {*hydro-,* water + *phobos,* fear}. Molecules such as phospholipids have both polar and nonpolar regions that play critical roles in biological systems and in the formation of biological membranes.

Phospholipid molecules have polar heads and nonpolar tails.

Phospholipids arrange themselves so that the polar heads are in contact with water and the nonpolar tails are directed away from water.

Polar head (hydrophilic)

Nonpolar fatty acid tail (hydrophobic)

Molecular models

Stylized model

Water

Hydrophilic head

Hydrophobic tails

Hydrophilic head

Water

This characteristic allows the phospholipid molecules to form bilayers, the basis for biological membranes that separate compartments.

(c) Molecular Shape

Covalent bond angles, ionic bonds, hydrogen bonds, and van der Waals forces all interact to create the distinctive shape of a complex biomolecule. This shape plays a critical role in the molecule's function.

Disulfide bond

KEY

• • • • • Hydrogen bonds or van der Waals forces

+ — Ionic bond

Ionic repulsion

S–S Disulfide bond

FIG. 2.9 **REVIEW**

pH

Acids and Bases

An **acid** is a molecule that contributes H⁺ to a solution.

- The carboxyl group, –COOH, is an acid because in solution it tends to lose its H⁺:

$$R\text{–}COOH \longrightarrow R\text{–}COO^- + H^+$$

A **base** is a molecule that decreases the H⁺ concentration of a solution by combining with free H⁺.

- Molecules that produce hydroxide ions, OH⁻, in solution are bases because the hydroxide combines with H⁺ to form water:

$$R\text{–}OH \longrightarrow R^+ + OH^- \longrightarrow OH^- + H^+ \longrightarrow H_2O$$

- Another molecule that acts as a base is ammonia, NH_3. It reacts with a free H⁺ to form an ammonium ion:

$$NH_3 + H^+ \longrightarrow NH_4^+$$

pH

The concentration of H^+ in body fluids is measured in terms of **pH**.

- The expression pH stands for "power of hydrogen."

1 $pH = -\log [H^+]$

This equation is read as "pH is equal to the negative log of the hydrogen ion concentration." Square brackets are shorthand notation for "concentration" and by convention, concentration is expressed in mEq/L.

- Using the rule of logarithms that says $-\log x = \log(1/x)$, pH equation (1) can be rewritten as:

2 $pH = \log (1/[H^+])$

This equation shows that pH is inversely related to H⁺ concentration. In other words, as the H⁺ concentration goes up, the pH goes down.

Example

What is the pH of a solution whose hydrogen ion concentration [H⁺] is 10^{-7} meq/L?

Answer

$$pH = -\log [H^+]$$
$$pH = -\log [10^{-7}]$$

Using the rule of logs, this can be rewritten as

$$pH = \log (1/10^{-7})$$

Using the rule of exponents that says $1/10^x = 10^{-x}$

$$pH = \log 10^7$$

the log of 10^7 is 7, so the solution has a pH of 7.

Pure water has a pH value of 7.0, meaning its H⁺ concentration is 1×10^{-7} M.

Lemon juice   Tomatoes, grapes   Pancreatic secretions   Household ammonia   1 M NaOH

Stomach acid   Vinegar, cola   Urine (4.5–7)   Saliva   Baking soda   Soap solutions   Chemical hair removers

Extremely acidic — **Extremely basic**

0 1 2 3 4 5 6 6.5 7 7.7 8 8.5 9 10 11 12 13 14

Acidic solutions have gained H⁺ from an acid and have a pH less than 7.

The normal pH of blood in the human body is 7.40. Homeostatic regulation is critical because blood pH less than 7.00 or greater than 7.70 is incompatible with life.

Basic or **alkaline** solutions have an H⁺ concentration lower than that of pure water and have a pH value greater than 7.

The pH of a solution is measured on a numeric scale between 0 and 14. The pH scale is logarithmic, meaning that a change in pH value of 1 unit indicates a 10-fold change in [H⁺]. For example, if a solution changes from pH 8 to pH 6, there has been a 100-fold (10^2 or 10×10) increase in [H⁺].

Q   **FIGURE QUESTIONS**

1. When the body becomes more acidic, does pH increase or decrease?
2. How can urine, stomach acid, and saliva have pH values outside the pH range that is compatible with life and yet be part of the living body?

$$H^+ + Cl^- \quad + \quad HCO_3^- \ + \ Na^+ \rightleftharpoons H_2CO_3 \ + \quad Cl^- + Na^+$$

| Hydrochloric acid | + | Sodium bicarbonate | $\rightleftharpoons$ | Carbonic acid | + | Sodium chloride (table salt) |

CONCEPT CHECK

10. To be classified as an acid, a molecule must do what when dissolved in water?

11. pH is an expression of the concentration of what in a solution?

12. When pH goes up, acidity goes _____.

PROTEIN INTERACTIONS

Noncovalent molecular interactions occur between many different biomolecules and often involve proteins. For example, biological membranes are formed by the noncovalent associations of phospholipids and proteins. Also, glycosylated proteins and glycosylated lipids in cell membranes create a "sugar coat" on cell surfaces, where they assist cell aggregation {*aggregare,* to join together} and adhesion {*adhaerere,* to stick}.

Proteins play important roles in so many cell functions that we can consider them the "workhorses" of the body. Most soluble proteins fall into seven broad categories:

1. **Enzymes.** Some proteins act as **enzymes,** biological catalysts that speed up chemical reactions. Enzymes play an important role in metabolism [discussed in Chapters 4 and 22].

2. **Membrane transporters.** Proteins in cell membranes help move substances back and forth between the intracellular and extracellular compartments. These proteins may form channels in the cell membrane, or they may bind to molecules and carry them through the membrane. [Membrane transporters are discussed in detail in Chapter 5.].

3. **Signal molecules.** Some proteins and smaller peptides act as hormones and other signal molecules. [Different types of signal molecules are described in Chapters 6 and 7.]

4. **Receptors.** Proteins that bind signal molecules and initiate cellular responses are called *receptors*. [Receptors are discussed along with signal molecules in Chapter 6.]

5. **Binding proteins.** These proteins, found mostly in the extracellular fluid, bind and transport molecules throughout the body. Examples you have already encountered include the oxygen-transporting protein *hemoglobin* and the cholesterol-binding proteins, such as LDL (low-density lipoprotein).

6. **Immunoglobulins.** These extracellular immune proteins, also called *antibodies*, help protect the body from foreign invaders and substances. [Immune functions are discussed in Chapter 24.]

7. **Regulatory proteins.** Regulatory proteins turn cell processes on and off or up and down. For example, the regulatory proteins known as *transcription factors* bind to DNA and alter gene expression and protein synthesis. The details of regulatory proteins can be found in cell biology textbooks.

Although soluble proteins are quite diverse, they do share some common features. They all bind to other molecules through

RUNNING **PROBLEM**

The hexavalent form of chromium used in industry is known to be toxic to humans. In 1992, officials at California's Hazard Evaluation System and Information Service warned that inhaling chromium dust, mist, or fumes placed chrome and stainless steel workers at increased risk for lung cancer. Officials found no risk to the public from normal contact with chrome surfaces or stainless steel. In 1995 and 2002, a possible link between the biological trivalent form of chromium (Cr^{3+}) and cancer came from *in vitro* studies {*vitrum,* glass—that is, a test tube} in which mammalian cells were kept alive in tissue culture. In these experiments, cells exposed to moderately high levels of chromium picolinate developed potentially cancerous changes.*

Q4: *From this information, can you conclude that hexavalent and trivalent chromium are equally toxic?*

————————————
*D. M. Stearns *et al*. Chromium(III) picolinate produces chromosome damage in Chinese hamster ovary cells. *FASEB J* 9: 1643–1648, 1995.

D. M. Stearns *et al*. Chromium(III) tris(picolinate) is mutagenic at the hypoxanthine (guanine) phosphoribosyltransferase locus in Chinese hamster ovary cells. *Mutat Res Genet Toxicol Environ Mutagen* 513: 135–142, 2002.

29   39   40   41   **46**   48   53

noncovalent interactions. The binding, which takes place at a location on the protein molecule called a **binding site,** exhibits properties that will be discussed shortly: specificity, affinity, competition, and saturation. If binding of a molecule to the protein initiates a process, as occurs with enzymes, membrane transporters, and receptors, we can describe the activity rate of the process and the factors that modulate, or alter, the rate.

Any molecule or ion that binds to another molecule is called a **ligand** {*ligare,* to bind or tie}. Ligands that bind to enzymes and membrane transporters are also called **substrates** {*sub-*, below + *stratum,* a layer}. Protein signal molecules and protein transcription factors are ligands. Immunoglobulins bind ligands, but the immunoglobulin-ligand complex itself then becomes a ligand [for details, see Chapter 24].

Proteins Are Selective about the Molecules They Bind

The ability of a protein to bind to a certain ligand or a group of related ligands is called **specificity.** Some proteins are very specific about the ligands they bind, while others bind to whole groups of molecules. For example, the enzymes known as *peptidases* bind polypeptide ligands and break apart peptide bonds, no matter which two amino acids are joined by those bonds. For this reason peptidases are not considered to be very specific in their action. In contrast, *aminopeptidases* also break peptide bonds but are more specific. They will bind only to one end of a protein chain (the end with an unbound amino group) and can act only on the terminal peptide bond.

Ligand binding requires *molecular complementarity*. In other words, the ligand and the protein binding site must be complementary, or compatible. In protein binding, when the ligand and protein come close to each other, noncovalent interactions between the ligand and the protein's binding site allow the two molecules to bind. From studies of enzymes and other binding proteins, scientists have discovered that a protein's binding site and the shape of its ligand do not need to fit one another exactly. When the binding site and the ligand come close to each other, they begin to interact through hydrogen and ionic bonds and van der Waals forces. The protein's binding site then changes shape (*conformation*) to fit more closely to the ligand. This **induced-fit model** of protein-ligand interaction is shown in **FIGURE 2.10**.

Protein-Binding Reactions Are Reversible

The degree to which a protein is attracted to a ligand is called the protein's **affinity** for the ligand. If a protein has a high affinity for a given ligand, the protein is more likely to bind to that ligand than to a ligand for which the protein has a lower affinity.

Protein binding to a ligand can be written using the same notation that we use to represent chemical reactions:

$$P + L \rightleftharpoons PL$$

where P is the protein, L is the ligand, and PL is the bound protein-ligand complex. The double arrow indicates that binding is reversible.

Reversible binding reactions go to a state of **equilibrium,** where the rate of binding (P + L → PL) is exactly equal to the rate of unbinding, or *dissociation* (P + L ← PL). When a reaction is at equilibrium, the ratio of the product concentration, or protein-ligand complex [PL], to the reactant concentrations [P][L] is always the same. This ratio is called the **equilibrium constant** K_{eq}, and it applies to all reversible chemical reactions:

$$K_{eq} = \frac{[PL]}{[P][L]}$$

The square brackets [ ] around the letters indicate concentrations of the protein, ligand, and protein-ligand complex.

FIG. 2.10 | The induced-fit model of protein-ligand (L) binding

In this model of protein binding, the binding site shape is not an exact match to the ligands' (L) shape.

Binding Reactions Obey the Law of Mass Action

Equilibrium is a dynamic state. In the living body, concentrations of protein or ligand change constantly through synthesis, breakdown, or movement from one compartment to another. What happens to equilibrium when the concentration of P or L changes? The answer to this question is shown in **FIGURE 2.11**, which begins with a reaction at equilibrium (Fig. 2.11a).

In Figure 2.11b, the equilibrium is disturbed when more protein or ligand is added to the system. Now the ratio of [PL]

FIG. 2.11 | The law of mass action

The **law of mass action** says that when protein binding is at equilibrium, the ratio of the bound and unbound components remains constant.

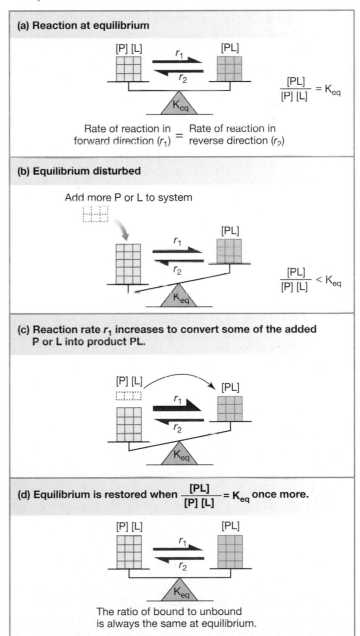

(a) Reaction at equilibrium

$$\frac{[PL]}{[P][L]} = K_{eq}$$

Rate of reaction in forward direction (r_1) = Rate of reaction in reverse direction (r_2)

(b) Equilibrium disturbed

Add more P or L to system

$$\frac{[PL]}{[P][L]} < K_{eq}$$

(c) Reaction rate r_1 increases to convert some of the added P or L into product PL.

(d) Equilibrium is restored when $\frac{[PL]}{[P][L]} = K_{eq}$ **once more.**

The ratio of bound to unbound is always the same at equilibrium.

to [P][L] differs from the K_{eq}. In response, the rate of the binding reaction increases to convert some of the added P or L into the bound protein-ligand complex (Fig. 2.11c). As the ratio approaches its equilibrium value again, the rate of the forward reaction slows down until finally the system reaches the equilibrium ratio once more (Fig. 2.11d). [P], [L], and [PL] have all increased over their initial values, but the equilibrium ratio has been restored.

The situation just described is an example of a reversible reaction obeying the **law of mass action,** a simple relationship that holds for chemical reactions whether in a test tube or in a cell. You may have learned this law in chemistry as *Le Châtelier's principle.* In very general terms, the law of mass action says that when a reaction is at equilibrium, the ratio of the products to the substrates is always the same. If the ratio is disturbed by adding or removing one of the participants, the reaction equation will shift direction to restore the equilibrium condition. (Note that the law of mass action is not the same as mass balance [see Chapter 1, p. 10].)

One example of this principle at work is the transport of steroid hormones in the blood. Steroids are hydrophobic, so more than 99% of hormone in the blood is bound to carrier proteins. The equilibrium ratio [PL]/[P][L] is 99% bound/1% unbound hormone. However, only the unbound or "free" hormone can cross the cell membrane and enter cells. As unbound hormone leaves the blood, the equilibrium ratio is disturbed. The binding proteins then release some of the bound hormone until the 99/1 ratio is again restored. The same principle applies to enzymes and metabolic reactions. Changing the concentration of one participant in a chemical reaction has a chain-reaction effect that alters the concentrations of other participants in the reaction.

CONCEPT CHECK

13. Consider the carbonic acid reaction, which is reversible:

$$CO_2 + H_2O \rightleftharpoons H_2CO_3 \text{ (carbonic acid)} \rightleftharpoons H^+ + HCO_3^-$$

If the carbon dioxide concentration in the body increases, what happens to the concentration of carbonic acid (H_2CO_3)? What happens to the pH?

The Dissociation Constant Indicates Affinity

In protein-binding reactions, the equilibrium constant is a quantitative representation of the protein's binding affinity for the ligand: high affinity for the ligand means a larger K_{eq}. The reciprocal of the equilibrium constant is called the **dissociation constant** (K_d).

$$K_d = \frac{[P][L]}{[PL]}$$

A large K_d indicates low binding affinity of the protein for the ligand and more P and L remaining in the unbound state. Conversely, a small K_d means a higher value for [PL] relative to

[P] and [L], so a small K_d indicates higher affinity of the protein for the ligand.

If one protein binds to several related ligands, a comparison of their K_d values can tell us which ligand is more likely to bind to the protein. The related ligands compete for the binding sites and are said to be **competitors.** Competition between ligands is a universal property of protein binding.

Competing ligands that mimic each other's actions are called **agonists** {*agonist,* contestant}. Agonists may occur in nature, such as *nicotine,* the chemical found in tobacco, which mimics the activity of the neurotransmitter *acetylcholine* by binding to the same receptor protein. Agonists can also be synthesized using what scientists learn from the study of protein–ligand binding sites. The ability of agonist molecules to mimic the activity of naturally occurring ligands has led to the development of many drugs.

CONCEPT CHECK

14. A researcher is trying to design a drug to bind to a particular cell receptor protein. Candidate molecule A has a K_d of 4.9 for the receptor. Molecule B has a K_d of 0.3. Which molecule has the most potential to be successful as the drug?

Multiple Factors Alter Protein Binding

A protein's affinity for a ligand is not always constant. Chemical and physical factors can alter, or *modulate*, binding affinity or can even totally eliminate it. Some proteins must be activated

RUNNING **PROBLEM**

Stan has been taking chromium picolinate because he heard that it would increase his strength and muscle mass. Then a friend told him that the Food and Drug Administration (FDA) said there was no evidence to show that chromium would help build muscle. In one study,* a group of researchers gave high daily doses of chromium picolinate to football players during a two-month training period. By the end of the study, the players who took chromium supplements had not increased muscle mass or strength any more than players who did not take the supplement.

Use Google Scholar (*http://scholar.google.com*) and search for *chromium picolinate*. Look for articles on body composition or muscle strength in humans before you answer the next question. (You should look beyond the first page of results.)

Q5: *Based on the papers you found, the Hallmark et al. study (which did not support enhanced muscle development from chromium supplements), and the studies that suggest that chromium picolinate might cause cancer, do you think that Stan should continue taking chromium picolinate?*

---

*M. A. Hallmark *et al.* Effects of chromium and resistive training on muscle strength and body composition. *Med Sci Sports Exerc* 28(1): 139–144, 1996.

before they have a functional binding site. In this section we discuss some of the processes that have evolved to allow activation, modulation, and inactivation of protein binding.

Isoforms Closely related proteins whose function is similar but whose affinity for ligands differs are called **isoforms** of one another. For example, the oxygen-transporting protein *hemoglobin* has multiple isoforms. One hemoglobin molecule has a quaternary structure consisting of four subunits (see Fig. 2.3). In the developing fetus, the hemoglobin isoform has two α (alpha) chains and two γ (gamma) chains that make up the four subunits. Shortly after birth, fetal hemoglobin molecules are broken down and replaced by adult hemoglobin. The adult hemoglobin isoform retains the two α chain isoforms but has two β (beta) chains in place of the γ chains. Both adult and fetal isoforms of hemoglobin bind oxygen, but the fetal isoform has a higher affinity for oxygen. This makes it more efficient at picking up oxygen across the placenta.

Activation Some proteins are inactive when they are synthesized in the cell. Before such a protein can become active, enzymes must chop off one or more portions of the molecule (**FIG. 2.12a**). Protein hormones (a type of signal molecule) and enzymes are two groups that commonly undergo such *proteolytic activation* {*lysis*, to release}. The inactive forms of these proteins are often identified with the prefix *pro-* {before}: prohormone, proenzyme, proinsulin, for example. Some inactive enzymes have the suffix *-ogen* added to the name of the active enzyme instead, as in *trypsinogen*, the inactive form of trypsin.

The activation of some proteins requires the presence of a **cofactor**, which is an ion or small organic functional group. Cofactors must attach to the protein before the binding site will become active and bind to ligand (Fig. 2.12b). Ionic cofactors include Ca^{2+}, Mg^{2+}, and Fe^{2+}. Many enzymes will not function without their cofactors.

Modulation The ability of a protein to bind a ligand and initiate a response can be altered by various factors, including temperature, pH, and molecules that interact with the protein. A factor that influences either protein binding or protein activity is called a **modulator.** There are two basic mechanisms by which modulation takes place. The modulator either (1) changes the protein's ability to bind the ligand or it (2) changes the protein's activity or its ability to create a response. **TABLE 2.3** summarizes the different types of modulation.

Chemical modulators are molecules that bind covalently or noncovalently to proteins and alter their binding ability or their activity. Chemical modulators may activate or enhance ligand binding, decrease binding ability, or completely inactivate the protein so that it is unable to bind any ligand. Inactivation may be either reversible or irreversible.

Antagonists, also called *inhibitors,* are chemical modulators that bind to a protein and decrease its activity. Many are simply molecules that bind to the protein and block the binding site without causing a response. They are like the guy who slips into the front of the movie ticket line to chat with his girlfriend, the

TABLE 2.3	Factors that Affect Protein Binding
Essential for Binding Activity	
Cofactors	Required for ligand binding at binding site
Proteolytic activation	Converts inactive to active form by removing part of molecule. Examples: digestive enzymes, protein hormones
Modulators and Factors That Alter Binding or Activity	
Competitive inhibitor	Competes directly with ligand by binding reversibly to active site
Irreversible inhibitor	Binds to binding site and cannot be displaced
Allosteric modulator	Binds to protein away from binding site and changes activity; may be inhibitors or activators
Covalent modulator	Binds covalently to protein and changes its activity. Example: phosphate groups
pH and temperature	Alter three-dimensional shape of protein by disrupting hydrogen or S–S bonds; may be irreversible if protein becomes denatured

cashier. He has no interest in buying a ticket, but he prevents the people in line behind him from getting their tickets for the movie.

Competitive inhibitors are reversible antagonists that compete with the customary ligand for the binding site (Fig. 2.12d). The degree of inhibition depends on the relative concentrations of the competitive inhibitor and the customary ligand, as well as on the protein's affinities for the two. The binding of competitive inhibitors is reversible: increasing the concentration of the customary ligand can displace the competitive inhibitor and decrease the inhibition.

Irreversible antagonists, on the other hand, bind tightly to the protein and cannot be displaced by competition. Antagonist drugs have proven useful for treating many conditions. For example, tamoxifen, an antagonist to the estrogen receptor, is used in the treatment of hormone-dependent cancers of the breast.

Allosteric and covalent modulators may be either antagonists or activators. **Allosteric modulators** {*allos*, other + *stereos*, solid (as a shape)} bind reversibly to a protein at a regulatory site away from the binding site, and by doing so change the shape of the binding site. *Allosteric inhibitors* are antagonists that decrease the affinity of the binding site for the ligand and inhibit protein activity (Fig. 2.12e). *Allosteric activators* increase the probability of protein-ligand binding and enhance protein activity (Fig. 2.12c). For example, the oxygen-binding ability of hemoglobin changes with allosteric modulation by carbon dioxide, H^+, and several other factors [see Chapter 18].

Covalent modulators are atoms or functional groups that bind covalently to proteins and alter the proteins' properties. Like allosteric modulators, covalent modulators may either increase or decrease a protein's binding ability or its activity. One of the most common covalent modulators is the phosphate group. Many proteins in the cell can be activated or inactivated when a phosphate group forms a covalent bond with them, the process known as phosphorylation.

FIG. 2.12 ESSENTIALS

Protein Activation and Inhibition

Activation

(a) Proteolytic activation: Protein is inactive until peptide fragments are removed.

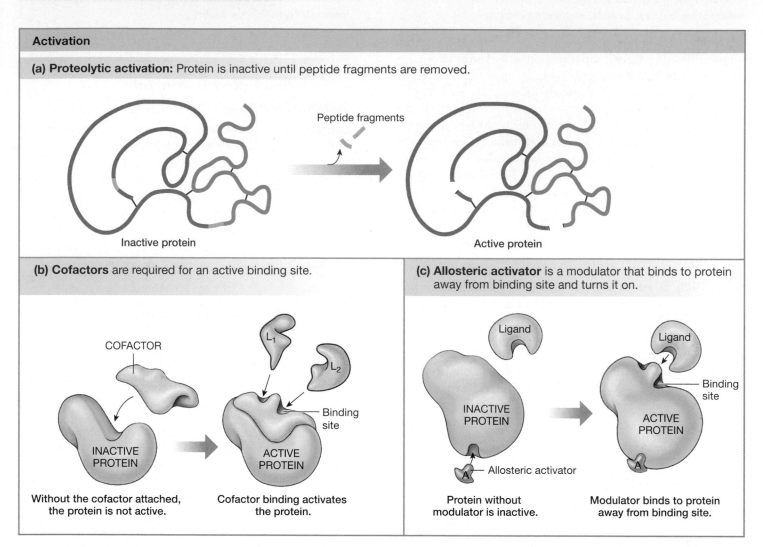

Inactive protein

Peptide fragments

Active protein

(b) Cofactors are required for an active binding site.

COFACTOR

L_1

L_2

Binding site

INACTIVE PROTEIN

ACTIVE PROTEIN

Without the cofactor attached, the protein is not active.

Cofactor binding activates the protein.

(c) Allosteric activator is a modulator that binds to protein away from binding site and turns it on.

Ligand

Ligand

Binding site

INACTIVE PROTEIN

ACTIVE PROTEIN

A — Allosteric activator

A

Protein without modulator is inactive.

Modulator binds to protein away from binding site.

Inhibition

(d) A **competitive inhibitor** blocks ligand binding at the binding site.

Competitive inhibitor

I

L_1

I

L_2

ACTIVE PROTEIN

INACTIVE PROTEIN

(e) Allosteric inhibitor is a modulator that binds to protein away from binding site and inactivates the binding site.

Ligand

Ligand

Binding site

ACTIVE PROTEIN

INACTIVE PROTEIN

Allosteric inhibitor — I

I

Protein without modulator is active.

Modulator binds to protein away from binding site and inactivates the binding site.

One of the best known chemical modulators is the antibiotic penicillin. Alexander Fleming discovered this compound in 1928, when he noticed that *Penicillium* mold inhibited bacterial growth in a petri dish. By 1938, researchers had extracted the active ingredient penicillin from the mold and used it to treat infections in humans. Yet it was not until 1965 that researchers figured out exactly how the antibiotic works. Penicillin is an antagonist that binds to a key bacterial protein by mimicking the normal ligand. Because penicillin forms unbreakable bonds with the protein, the protein is irreversibly inhibited. Without the protein, the bacterium is unable to make a rigid cell wall. With no rigid cell wall, the bacterium swells, ruptures, and dies.

Physical Factors Physical conditions such as temperature and pH (acidity) can have dramatic effects on protein structure and function. Small changes in pH or temperature act as modulators to increase or decrease activity (**FIG. 2.13a**). However, once these factors exceed some critical value, they disrupt the noncovalent bonds holding the protein in its tertiary conformation. The protein loses its shape and, along with that, its activity. When the protein loses its conformation, it is said to be *denatured.*

If you have ever fried an egg, you have watched this transformation happen to the egg white protein *albumin* as it changes from a slithery clear state to a firm white state. Hydrogen ions in high enough concentration to be called acids have a similar effect on protein structure. During preparation of ceviche, the national dish of Ecuador, raw fish is marinated in lime juice. The acidic lime juice contains hydrogen ions that disrupt hydrogen bonds in the muscle proteins of the fish, causing the proteins to become denatured. As a result, the meat becomes firmer and opaque, just as it would if it were cooked with heat.

In a few cases, activity can be restored if the original temperature or pH returns. The protein then resumes its original shape as if nothing had happened. Usually, however, denaturation produces a permanent loss of activity. There is certainly no way to unfry an egg or uncook a piece of fish. The potentially disastrous influence of temperature and pH on proteins is one reason these variables are so closely regulated by the body.

CONCEPT CHECK

15. Match each chemical to its action(s).

(a) Allosteric modulator	1. Bind away from the binding site
(b) Competitive inhibitor	2. Bind to the binding site
(c) Covalent modulator	3. Inhibit activity only
	4. Inhibit or enhance activity

The Body Regulates the Amount of Protein in Cells

The final characteristic of proteins in the human body is that the amount of a given protein varies over time, often in a regulated fashion. The body has mechanisms that enable it to monitor whether it needs more or less of certain proteins. Complex signaling pathways, many of which themselves involve proteins, direct particular cells to make new proteins or to break down (*degrade*) existing proteins. This programmed production of new proteins (receptors, enzymes, and membrane transporters, in particular) is called **up-regulation.** Conversely, the programmed removal of proteins is called **down-regulation.** In both instances, the cell is directed to make or remove proteins to alter its response.

The amount of protein present in a cell has a direct influence on the magnitude of the cell's response. For example, the graph in Figure 2.13b shows the results of an experiment in which the amount of ligand is held constant while the amount of protein is varied. As the graph shows, an increase in the amount of protein present causes an increase in the response.

As an analogy, think of the checkout lines in a supermarket. Imagine that each cashier is an enzyme, the waiting customers are ligand molecules, and people leaving the store with their purchases are products. One hundred customers can be checked out faster when there are 25 lines open than when there are only 10 lines. Likewise, in an enzymatic reaction, the presence of more protein molecules (enzyme) means that more binding sites are available to interact with the ligand molecules. As a result, the ligands are converted to products more rapidly.

Regulating protein concentration is an important strategy that cells use to control their physiological processes. Cells alter the amount of a protein by influencing both its synthesis and its breakdown. If protein synthesis exceeds breakdown, protein accumulates and the reaction rate increases. If protein breakdown exceeds synthesis, the amount of protein decreases, as does the reaction rate. Even when the amount of protein is constant, there is still a steady turnover of protein molecules.

Reaction Rate Can Reach a Maximum

If the concentration of a protein in a cell is constant, then the concentration of the ligand determines the magnitude of the response. Fewer ligands activate fewer proteins, and the response is low. As ligand concentrations increase, so does the magnitude of the response, up to a maximum where all protein binding sites are occupied.

Figure 2.13c shows the results of a typical experiment in which the protein concentration is constant but the concentration of ligand varies. At low ligand concentrations, the response rate is directly proportional to the ligand concentration. Once the concentration of ligand molecules exceeds a certain level, the protein molecules have no more free binding sites. The proteins are fully occupied, and the rate reaches a maximum value. This condition is known as **saturation.** Saturation applies to enzymes, membrane transporters, receptors, binding proteins, and immunoglobulins.

An analogy to saturation appeared in the early days of television on the *I Love Lucy* show. Lucille Ball was working at the conveyor belt of a candy factory, loading chocolates into the little paper cups of a candy box. Initially, the belt moved slowly, and she had no difficulty picking up the candy and putting it into the box.

FIG. 2.13 **ESSENTIALS**

Factors That Influence Protein Activity

(a) Temperature and pH

Temperature and pH changes may disrupt protein structure and cause loss of function.

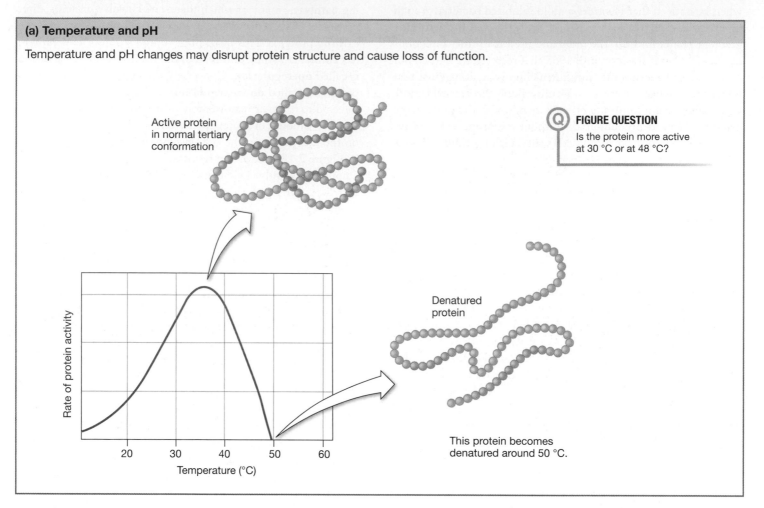

Active protein in normal tertiary conformation

Ⓠ **FIGURE QUESTION**

Is the protein more active at 30 °C or at 48 °C?

Rate of protein activity

Temperature (°C)

20 30 40 50 60

Denatured protein

This protein becomes denatured around 50 °C.

(b) Amount of Protein

Reaction rate depends on the amount of protein. The more protein present, the faster the rate.

Response rate (mg/sec)

0 1 2 3

Protein concentration ⟶

A B C

Ⓠ **GRAPH QUESTIONS**

1. What is the rate when the protein concentration is equal to A?
2. When the rate is 2.5 mg/sec, what is the protein concentration?

In this experiment, the ligand amount remains constant.

(c) Amount of Ligand

If the amount of binding protein is held constant, the reaction rate depends on the amount of ligand, up to the saturation point.

Maximum rate at saturation

Response rate (mg/sec)

0 1 2 3 4

Ligand concentration (mg/mL) ⟶

25 50 75 100 125 150 175

Ⓠ **GRAPH QUESTION**

What is the rate when the ligand concentration is 200 mg/mL?

In this experiment, the amount of binding protein was constant. At the maximum rate, the protein is said to be saturated.

Gradually, the belt brought candy to her more rapidly, and she had to increase her packing speed to keep up. Finally, the belt brought candy to her so fast that she could not pack it all in the boxes because she was working at her maximum rate. That was Lucy's saturation point. (Her solution was to stuff the candy into her mouth as well as into the box!)

In conclusion, you have now learned about the important and nearly universal properties of soluble proteins: shape-function relationships, ligand binding, saturation, specificity, competition, and activation/inhibition. You will revisit these concepts many times as you work through the organ systems of the body. The body's insoluble proteins, which are key structural components of cells and tissues, are covered in the next chapter.

CONCEPT CHECK

16. What happens to the rate of an enzymatic reaction as the amount of enzyme present decreases?

17. What happens to the rate of an enzymatic reaction when the enzyme has reached saturation?

RUNNING **PROBLEM CONCLUSION** | Chromium Supplements

In this running problem, you learned that claims of chromium picolinate's ability to enhance muscle mass have not been supported by evidence from controlled scientific experiments. You also learned that studies suggest that some forms of the biological trivalent form of chromium may be toxic. To learn more about current research, go to PubMed (*www.pubmed.gov*) and search for "*chromium picolinate*" (use the quotation marks). Compare what you find there with the results of a similar Google search. Should you believe everything you read on the Web? Now compare your answers with those in the summary table.

Question	Facts	Integration and Analysis
Q1: *Locate chromium on the periodic table of elements.*	The periodic table organizes the elements according to atomic number.	N/A
What is chromium's atomic number? Atomic mass?	Reading from the table, chromium (Cr) has an atomic number of 24 and an average atomic mass of 52.	N/A
How many electrons does one atom of chromium have?	Atomic number of an element = number of protons in one atom. One atom has equal numbers of protons and electrons.	The atomic number of chromium is 24; therefore, one atom of chromium has 24 protons and 24 electrons.
Which elements close to chromium are also essential elements?	Molybdenum, manganese, and iron.	N/A
Q2: *If people have chromium deficiency, would you predict that their blood glucose level would be lower or higher than normal?*	Chromium helps move glucose from blood into cells.	If chromium is absent or lacking, less glucose would leave the blood and blood glucose would be higher than normal.
From the result of the Chinese study, can you conclude that all people with diabetes suffer from chromium deficiency?	Higher doses of chromium supplements lowered elevated blood glucose levels, but lower doses have no effect. This is only one study, and no information is given about similar studies elsewhere.	We have insufficient evidence from the information presented to draw a conclusion about the role of chromium deficiency in diabetes.
Q3: *How many electrons have been lost from the hexavalent ion of chromium? From the trivalent ion?*	For each electron lost from an ion, a positively charged proton is left behind in the nucleus of the ion.	The hexavalent ion of chromium, Cr^{6+}, has six unmatched protons and therefore has lost six electrons. The trivalent ion, Cr^{3+}, has lost three electrons.
Q4: *From this information, can you conclude that hexavalent and trivalent chromium are equally toxic?*	The hexavalent form is used in industry and, when inhaled, has been linked to an increased risk of lung cancer. Enough studies have shown an association that California's Hazard Evaluation System and Information Service has issued warnings to chromium workers. Evidence to date for toxicity of trivalent chromium in chromium picolinate comes from studies done on isolated cells in tissue culture.	Although the toxicity of Cr^{6+} is well established, the toxicity of Cr^{3+} has not been conclusively determined. Studies performed on cells *in vitro* may not be applicable to humans. Additional studies need to be performed in which animals are given reasonable doses of chromium picolinate for an extended period of time.

—*Continued next page*

RUNNING **PROBLEM CONCLUSION** | *Continued*

Question	Facts	Integration and Analysis
Q5: *Based on the study that did not support enhanced muscle development from chromium supplements and the studies that suggest that chromium picolinate might cause cancer, do you think Stan should continue taking picolinate?*	No research evidence supports a role for chromium picolinate in increasing muscle mass or strength in humans. Other research suggests that chromium picolinate may cause cancerous changes in isolated cells.	The evidence presented suggests that for Stan, there is no benefit from taking chromium picolinate, and there may be risks. Using risk–benefit analysis, the evidence supports stopping the supplements. However, the decision is Stan's personal responsibility. He should keep himself informed of new developments that would change the risk-benefit analysis.

29 39 40 41 46 48 **53**

CHEMISTRY REVIEW QUIZ

Use this quiz to see what areas of chemistry and basic biochemistry you might need to review. Answers are on p. A-2. The title above each set of questions refers to a review figure on this topic.

Atoms and Molecules (Fig. 2.5)

Match each subatomic particle in the left column with all the phrases in the right column that describe it. A phrase may be used more than once.

1. electron	(a) one has atomic mass of 1 amu
2. neutron	(b) found in the nucleus
3. proton	(c) negatively charged
	(d) changing the number of these in an atom creates a new element
	(e) adding or losing these makes an atom into an ion
	(f) gain or loss of these makes an isotope of the same element
	(g) determine(s) an element's atomic number
	(h) contribute(s) to an element's atomic mass

4. Isotopes of an element have the same number of _____ and _____, but differ in their number of _____. Unstable isotopes emit energy called _____.

5. Name the element associated with each of these symbols: C, O, N, and H.

6. Write the one- or two-letter symbol for each of these elements: phosphorus, potassium, sodium, sulfur, calcium, and chlorine.

7. Use the periodic table of the elements on the inside back cover to answer the following questions:

   (a) Which element has 30 protons?
   (b) How many electrons are in one atom of calcium?
   (c) Find the atomic number and average atomic mass of iodine. What is the letter symbol for iodine?

8. A magnesium ion, Mg^{2+}, has (*gained/lost*) two (*protons/neutrons/electrons*).

9. H^+ is also called a proton. Why is it given that name?

10. Use the periodic table of the elements on the inside back cover to answer the following questions about an atom of sodium.

   (a) How many electrons does the atom have?
   (b) What is the electrical charge of the atom?
   (c) How many neutrons does the average atom have?
   (d) If this atom loses one electron, it would be called a(n) *anion/cation*.
   (e) What would be the electrical charge of the substance formed in (d)?
   (f) Write the chemical symbol for the ion referred to in (d).
   (g) What does the sodium atom become if it loses a proton from its nucleus?
   (h) Write the chemical symbol for the atom referred to in (g).

11. Write the chemical formulas for each molecule depicted. Calculate the molecular mass of each molecule.

(a)

(b) $O=C=O$

(c)

(d)

Lipids (Fig. 2.1)

12. Match each lipid with its best description:

(a) triglyceride	1. most common form of lipid in the body
(b) eicosanoid	2. liquid at room temperature, usually from plants
(c) steroid	
(d) oil	3. important component of cell membrane
(e) phospholipids	4. structure composed of carbon rings
	5. modified 20-carbon fatty acid

13. Use the chemical formulas given to decide which of the following fatty acids is most unsaturated: (a) $C_{18}H_{36}O_2$ (b) $C_{18}H_{34}O_2$ (c) $C_{18}H_{30}O_2$

Carbohydrates (Fig. 2.2)

14. Match each carbohydrate with its description:

(a) starch	1. monosaccharide
(b) chitin	2. disaccharide, found in milk
(c) glucose	3. storage form of glucose for animals
(d) lactose	4. storage form of glucose for plants
(e) glycogen	5. structural polysaccharide of invertebrates

Proteins (Fig. 2.3)

15. Match these terms pertaining to proteins and amino acids:

(a) the building blocks of proteins	1. essential amino acids
(b) must be included in our diet	2. primary structure
(c) protein catalysts that speed the rate of chemical reactions	3. amino acids
	4. globular proteins
(d) sequence of amino acids in a protein	5. enzymes
	6. tertiary structure
(e) protein chains folded into a ball-shaped structure	7. fibrous proteins

16. What aspect of protein structure allows proteins to have more versatility than lipids or carbohydrates?

17. Peptide bonds form when the _____ group of one amino acid joins the _____ of another amino acid.

Nucleotides (Fig. 2.4)

18. List the three components of a nucleotide.

19. Compare the structure of DNA with that of RNA.

20. Distinguish between purines and pyrimidines.

VISUALIZE, PRACTICE, AND MASTER PHYSIOLOGY WITH MASTERING A&P
- Practice-on-the-go with Dynamic Study Modules
- Visualize important processes with Interactive Physiology and A&P Flix
- Check your understanding with Video Tutors and practice quizzes!

CHAPTER SUMMARY

This chapter introduces the *molecular interactions* between biomolecules, water, and ions that underlie many of the key themes in physiology. These interactions are an integral part of *information flow, energy storage and transfer,* and the *mechanical properties* of cells and tissues in the body.

Molecules and Bonds

1. The four major groups of **biomolecules** are carbohydrates, lipids, proteins, and nucleotides. They all contain carbon, hydrogen, and oxygen. (p. 29; Figs. 2.1, 2.2, 2.3, 2.4)

2. Proteins, lipids, and carbohydrates combine to form glycoproteins, glycolipids, or lipoproteins. (p. 29; Fig. 2.5)

3. Electrons are important for covalent and ionic bonds, energy capture and transfer, and formation of free radicals. (p. 33)

4. **Covalent bonds** form when adjacent atoms share one or more pairs of electrons. (p. 33; Fig. 2.6)

5. **Polar molecules** have atoms that share electrons unevenly. When atoms share electrons evenly, the molecule is **nonpolar**. (p. 39; Fig. 2.6)

6. An atom that gains or loses electrons acquires an electrical charge and is called an **ion**. (p. 39; Fig. 2.6)

7. **Ionic bonds** are strong bonds formed when oppositely charged ions are attracted to each other. (p. 39)

8. Weak **hydrogen bonds** form when hydrogen atoms in polar molecules are attracted to oxygen, nitrogen, or fluorine atoms. Hydrogen bonding among water molecules is responsible for the surface tension of water. (p. 39; Fig. 2.6)

9. **Van der Waals forces** are weak bonds that form when atoms are attracted to each other. (p. 39)

Noncovalent Interactions

iP Fluids and Electrolytes: Acid-Base Homeostasis

10. The universal solvent for biological solutions is water. (p. 40; Figs. 2.7, 2.8a)

11. The ease with which a molecule dissolves in a solvent is called its **solubility** in that solvent. **Hydrophilic** molecules dissolve easily in water, but **hydrophobic** molecules do not. (p. 40)

12. Molecular shape is created by covalent bond angles and weak noncovalent interactions within a molecule. (p. 40; Fig. 2.8)

13. Free H^+ in solution can disrupt a molecule's noncovalent bonds and alter its ability to function. (p. 41)

14. The **pH** of a solution is a measure of its hydrogen ion concentration. The more acidic the solution, the lower its pH. (p. 41; Fig. 2.9)

15. **Buffers** are solutions that moderate pH changes. (p. 41)

Protein Interactions

16. Most water-soluble proteins serve as enzymes, membrane transporters, signal molecules, receptors, binding proteins, immunoglobulins, or transcription factors. (p. 46)

17. **Ligands** bind to proteins at a binding site. According to the **induced-fit model** of protein binding, the shapes of the ligand and binding site do not have to match exactly. (p. 46; Fig. 2.10)

18. Proteins are specific about the ligands they will bind. The attraction of a protein to its ligand is called the protein's **affinity** for the ligand. The **equilibrium constant** (K_{eq}) and the **dissociation constant** (K_d) are quantitative measures of a protein's affinity for a given ligand. (p. 47)

19. Reversible binding reactions go to equilibrium. If equilibrium is disturbed, the reaction follows the **law of mass action** and shifts in the direction that restores the equilibrium ratio. (p. 47; Fig. 2.11)

20. Ligands may compete for a protein's binding site. If competing ligands mimic each other's activity, they are **agonists.** (p. 48)

21. Closely related proteins having similar function but different affinities for ligands are called **isoforms** of one another. (p. 49)

22. Some proteins must be activated, either by **proteolytic activation** or by addition of **cofactors.** (p. 49; Fig. 2.12)

23. **Competitive inhibitors** can be displaced from the binding site, but irreversible **antagonists** cannot. (p. 49; Fig. 2.12)

24. **Allosteric modulators** bind to proteins at a location other than the binding site. **Covalent modulators** bind with covalent bonds. Both types of modulators may activate or inhibit the protein. (p. 49; Fig. 2.12)

25. Extremes of temperature or pH will **denature** proteins. (p. 51; Fig. 2.13)

26. Cells regulate their proteins by **up-regulation** or **down-regulation** of protein synthesis and destruction. The amount of protein directly influences the magnitude of the cell's response. (p. 51; Fig. 2.13)

27. If the amount of protein (such as an enzyme) is constant, the amount of ligand determines the cell's response. If all binding proteins (such as enzymes) become **saturated** with ligand, the response reaches its maximum. (p. 51; Fig. 2.13)

REVIEW QUESTIONS

In addition to working through these questions and checking your answers on p. A-3, review the Learning Outcomes at the beginning of this chapter.

Level One Reviewing Facts and Terms

1. List the four kinds of biomolecules. Give an example of each kind that is relevant to physiology.

2. True or false? All organic molecules are biomolecules.

3. When atoms bind tightly to one another, such as H_2O or O_2, one unit is called a(n) _____.

4. An atom of carbon has four unpaired electrons in an outer shell with space for eight electrons. How many covalent bonds will one carbon atom form with other atoms?

5. Fill in the blanks with the correct bond type.

   In a(n) _____ bond, electrons are shared between atoms. If the electrons are attracted more strongly to one atom than to the other, the molecule is said to be a(n) _____ molecule. If the electrons are evenly shared, the molecule is said to be a(n) _____ molecule.

6. Name two elements whose presence contributes to a molecule becoming a polar molecule.

7. Based on what you know from experience about the tendency of the following substances to dissolve in water, predict whether they are polar or nonpolar molecules: table sugar, vegetable oil.

8. A negatively charged ion is called a(n) _____, and a positively charged ion is called a(n) _____.

9. Define the pH of a solution. If pH is less than 7, the solution is _____; if pH is greater than 7, the solution is _____.

10. A molecule that moderates changes in pH is called a _____.

11. Proteins combined with fats are called _____, and proteins combined with carbohydrates are called _____.

12. A molecule that binds to another molecule is called a(n) _____.

13. Match these definitions with their terms (not all terms are used):

(a) the ability of a protein to bind one molecule but not another	1. irreversible inhibition
	2. induced fit
(b) the part of a protein molecule that binds the ligand	3. binding site
	4. specificity
(c) the ability of a protein to alter shape as it binds a ligand	5. saturation

14. An ion, such as Ca^{2+} or Mg^{2+}, that must be present in order for an enzyme to work is called a(n) _____.

15. A protein whose structure is altered to the point that its activity is destroyed is said to be _____.

Level Two Reviewing Concepts

16. Mapping exercise: Make the list of terms into a map describing solutions.

• concentration	• nonpolar molecule
• equivalent	• polar molecule
• hydrogen bond	• solubility
• hydrophilic	• solute
• hydrophobic	• solvent
• molarity	• water
• mole	

17. A solution in which $[H^+] = 10^{-3}$ M is (acidic/basic), whereas a solution in which $[H^+] = 10^{-10}$ M is (acidic/basic). Give the pH for each of these solutions.

18. Name three nucleotides or nucleic acids, and tell why each one is important.

19. You know that two soluble proteins are isoforms of each other. What can you predict about their structures, functions, and affinities for ligands?

20. You have been asked to design some drugs for the purposes described next. Choose the desirable characteristic(s) for each drug from the numbered list.

(a) Drug A must bind to an enzyme and enhance its activity.	1. antagonist
	2. competitive inhibitor
(b) Drug B should mimic the activity of a normal nervous system signal molecule.	3. agonist
	4. allosteric activator
(c) Drug C should block the activity of a membrane receptor protein.	5. covalent modulator

Level Three Problem Solving

21. You have been summoned to assist with the autopsy of an alien being whose remains have been brought to your lab. The chemical analysis returns with 33% C, 40% O, 4% H, 14% N, and 9% P. From this information, you conclude that the cells contain nucleotides, possibly even DNA or RNA. Your assistant is demanding that you tell him how you knew this. What do you tell him?

22. The harder a cell works, the more CO_2 it produces. CO_2 is carried in the blood according to the following equation:

$$CO_2 + H_2O \rightleftharpoons H_2CO_3 \rightleftharpoons H^+ + HCO_3^-$$

What effect does hard work by your muscle cells have on the pH of the blood?

Level Four Quantitative Problems

23. Calculate the amount of NaCl you would weigh out to make one liter of 0.9% NaCl. Explain how you would make a liter of this solution.

24. A 1.0 M NaCl solution contains 58.5 g of salt per liter. (a) How many molecules of NaCl are present in 1 L of this solution? (b) How many millimoles of NaCl are present? (c) How many equivalents of Na^+ are present? (d) Express 58.5 g of NaCl per liter as a percent solution.

25. How would you make 200 mL of a 10% glucose solution? Calculate the molarity of this solution. How many millimoles of glucose are present in 500 mL of this solution? (*Hint:* What is the molecular mass of glucose?)

26. The graph shown below represents the binding of oxygen molecules (O_2) to two different proteins, myoglobin and hemoglobin, over a range of oxygen concentrations. Based on the graph, which protein has the higher affinity for oxygen? Explain your reasoning.

3

Compartmentation: Cells and Tissues

Cells are organisms, and entire animals and plants are aggregates of these organisms.

Theodor Schwann, 1839

FUNCTIONAL COMPARTMENTS OF THE BODY 59

LO 3.1 Name and describe the major body cavities and compartments.

BIOLOGICAL MEMBRANES 61

LO 3.2 Explain the four major functions of the cell membrane.

LO 3.3 Draw and label the fluid mosaic model of the cell membrane and describe the functions of each component.

LO 3.4 Compare a phospholipid bilayer to a micelle and a liposome.

INTRACELLULAR COMPARTMENTS 65

LO 3.5 Map the organization of a typical animal cell.

LO 3.6 Draw, name, and list the functions of organelles found in animal cells.

LO 3.7 Compare the structures and functions of the three families of cytoplasmic protein fibers.

LO 3.8 Compare and contrast cilia and flagella.

LO 3.9 Describe five major functions of the cytoskeleton.

LO 3.10 Name the three motor proteins and explain their functions.

LO 3.11 Describe the organization and function of the nucleus.

LO 3.12 Explain how protein synthesis uses compartmentation to separate different steps of the process.

TISSUES OF THE BODY 72

LO 3.13 Describe the structure and functions of extracellular matrix.

LO 3.14 Describe the role of proteins in the three major categories of cell junctions.

LO 3.15 Compare the structures and functions of the four tissue types.

LO 3.16 Describe the anatomy and functions of the five functional categories of epithelia.

LO 3.17 Compare the anatomy and functions of the seven main categories of connective tissue.

LO 3.18 Use structural and functional differences to distinguish between the three types of muscle tissue.

LO 3.19 Describe the structural and functional differences between the two types of neural tissue.

TISSUE REMODELING 84

LO 3.20 Explain the differences between apoptosis and necrosis.

LO 3.21 Distinguish between pluripotent, multipotent, and totipotent stem cells.

ORGANS 87

LO 3.22 List as many organs as you can for each of the 10 physiological organ systems.

BACKGROUND BASICS

Units of measure: inside back cover
8 Compartmentation
10 Extracellular fluid
40 Hydrophobic molecules
32 Proteins
41 pH
33 Covalent and noncovalent interactions

Pancreatic cell

What makes a compartment? We may think of something totally enclosed, like a room or a box with a lid. But not all compartments are totally enclosed ... think of the modular cubicles that make up many modern workplaces. And not all functional compartments have walls ... think of a giant hotel lobby divided into conversational groupings by careful placement of rugs and furniture. Biological compartments come with the same type of anatomic variability, ranging from totally enclosed structures such as cells to functional compartments without visible walls.

The first living compartment was probably a simple cell whose intracellular fluid was separated from the external environment by a wall made of phospholipids and proteins—the cell membrane. **Cells** are the basic functional unit of living organisms, and an individual cell can carry out all the processes of life.

As cells evolved, they acquired intracellular compartments separated from the intracellular fluid by membranes. Over time, groups of single-celled organisms began to cooperate and specialize their functions, eventually giving rise to multicellular organisms. As multicellular organisms evolved to become larger and more complex, their bodies became divided into various functional compartments.

Compartments are both an advantage and a disadvantage for organisms. On the advantage side, compartments separate biochemical processes that might otherwise conflict with one another. For example, protein synthesis takes place in one subcellular compartment while protein degradation is taking place in another. Barriers between compartments, whether inside a cell or inside a body, allow the contents of one compartment to differ from the contents of adjacent compartments. An extreme example is the intracellular compartment called the *lysosome*, with an internal pH of 5 [Fig. 2.9, p. 45]. This pH is so acidic that if the lysosome ruptures, it severely damages or kills the cell that contains it.

The disadvantage to compartments is that barriers between them can make it difficult to move needed materials from one compartment to another. Living organisms overcome this problem with specialized mechanisms that transport selected substances across membranes. [Membrane transport is the subject of Chapter 5.]

In this chapter, we explore the theme of compartmentation by first looking at the various compartments that subdivide the human body, from body cavities to the subcellular compartments called organelles. We then examine how groups of cells with similar functions unite to form the tissues and organs of the body. Continuing the theme of molecular interactions, we also look at how different molecules and fibers in cells and tissues give rise to their *mechanical properties:* their shape, strength, flexibility, and the connections that hold tissues together.

FUNCTIONAL COMPARTMENTS OF THE BODY

The human body is a complex compartment separated from the outside world by layers of cells. Anatomically, the body is divided into three major body **cavities:** the *cranial cavity* (commonly referred to as the *skull*), the *thoracic cavity* (also called the *thorax*), and the *abdominopelvic cavity* (**FIG. 3.1a**). The cavities are separated from one another by bones and tissues, and they are lined with *tissue membranes.*

The cranial cavity {*cranium*, skull} contains the brain, our primary control center. The thoracic cavity is bounded by the spine and ribs on top and sides, with the muscular *diaphragm* forming the floor. The thorax contains the heart, which is enclosed in a membranous *pericardial sac* {*peri-*, around + *cardium*, heart}, and the two lungs, enclosed in separate *pleural sacs.*

The *abdomen* and *pelvis* form one continuous cavity, the *abdominopelvic cavity.* A tissue lining called the *peritoneum* lines the abdomen and surrounds the organs within it (stomach, intestines, liver, pancreas, gallbladder, and spleen). The kidneys lie outside the abdominal cavity, between the peritoneum and the muscles and bones of the back, just above waist level. The pelvis contains reproductive organs, the urinary bladder, and the terminal portion of the large intestine.

In addition to the body cavities, there are several discrete fluid-filled anatomical compartments. The blood-filled vessels and heart of the circulatory system form one compartment. Our eyes are hollow fluid-filled spheres subdivided into two compartments, the aqueous and vitreous humors. The brain and spinal cord are surrounded by a special fluid compartment known as cerebrospinal fluid (CSF). The membranous sacs that surround the lungs (*pleural sacs*) and the heart (*pericardial sac*) also contain small volumes of fluid (Fig. 3.1a).

The Lumens of Some Organs Are Outside the Body

All hollow organs, such as heart, lungs, blood vessels, and intestines, create another set of compartments within the body.

RUNNING **PROBLEM** | Pap Tests Save Lives

Dr. George Papanicolaou has saved the lives of millions of women by popularizing the Pap test, a screening method that detects the early signs of cervical cancer. In the past 50 years, deaths from cervical cancer have dropped dramatically in countries that routinely use the Pap test. In contrast, cervical cancer is a leading cause of death in regions where Pap test screening is not routine, such as Africa and Central America. If detected early, cervical cancer is one of the most treatable forms of cancer. Today, Jan Melton, who had an abnormal Pap test a year ago, returns to Dr. Baird, her family physician, for a repeat test. The results will determine whether she needs to undergo further testing for cervical cancer.

59 61 70 79 85 87

FIG. 3.1 **ESSENTIALS**

Levels of Organization: Body Compartments

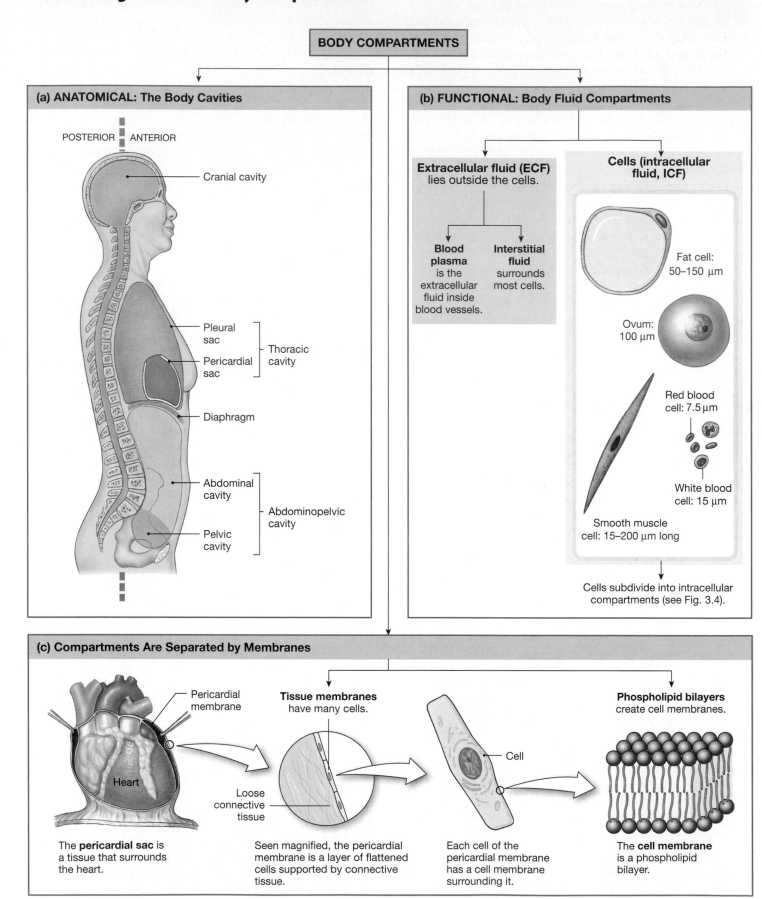

BODY COMPARTMENTS

(a) ANATOMICAL: The Body Cavities

POSTERIOR | ANTERIOR

- Cranial cavity
- Pleural sac ⎫
- Pericardial sac ⎭ Thoracic cavity
- Diaphragm
- Abdominal cavity ⎫
- ⎬ Abdominopelvic cavity
- Pelvic cavity ⎭

(b) FUNCTIONAL: Body Fluid Compartments

Extracellular fluid (ECF) lies outside the cells.

Blood plasma is the extracellular fluid inside blood vessels.

Interstitial fluid surrounds most cells.

Cells (intracellular fluid, ICF)

- Fat cell: 50–150 μm
- Ovum: 100 μm
- Red blood cell: 7.5 μm
- White blood cell: 15 μm
- Smooth muscle cell: 15–200 μm long

Cells subdivide into intracellular compartments (see Fig. 3.4).

(c) Compartments Are Separated by Membranes

- Pericardial membrane
- Heart

The **pericardial sac** is a tissue that surrounds the heart.

Tissue membranes have many cells.

- Loose connective tissue

Seen magnified, the pericardial membrane is a layer of flattened cells supported by connective tissue.

- Cell

Each cell of the pericardial membrane has a cell membrane surrounding it.

Phospholipid bilayers create cell membranes.

The **cell membrane** is a phospholipid bilayer.

The interior of any hollow organ is called its **lumen** {*lumin*, window}. A lumen may be wholly or partially filled with air or fluid. For example, the lumens of blood vessels are filled with the fluid we call blood.

For some organs, the lumen is essentially an extension of the external environment, and material in the lumen is not truly part of the body's internal environment until it crosses the wall of the organ. For example, we think of our digestive tract as being "inside" our body, but in reality its lumen is part of the body's external environment [see Fig. 1.2, p. 4]. An analogy would be the hole through a bead. The hole passes through the bead but is not actually inside the bead.

An interesting illustration of this distinction between the internal environment and the external environment in a lumen involves the bacterium *Escherichia coli.* This organism normally lives and reproduces inside the large intestine, an internalized compartment whose lumen is continuous with the external environment. When *E. coli* is residing in this location, it does not harm the host. However, if the intestinal wall is punctured by disease or accident and *E. coli* enters the body's internal environment, a serious infection can result.

Functionally, the Body Has Three Fluid Compartments

In physiology, we are often more interested in functional compartments than in anatomical compartments. Most cells of the body are not in direct contact with the outside world. Instead, their external environment is the extracellular fluid [Fig. 1.5, p. 11]. If we think of all the cells of the body together as one unit, we can then divide the body into two main fluid compartments: (1) the *extracellular fluid* (ECF) outside the cells and (2) the *intracellular fluid* (ICF) within the cells (Fig. 3.1b). The dividing wall between ECF and ICF is the cell membrane. The extracellular fluid subdivides further into **plasma,** the fluid portion of the blood, and **interstitial fluid** {*inter-,* between + *stare,* to stand}, which surrounds most cells of the body.

RUNNING **PROBLEM**

Cancer is a condition in which a small group of cells starts to divide uncontrollably and fails to differentiate into specialized cell types. Cancerous cells that originate in one tissue can escape from that tissue and spread to other organs through the circulatory system and the lymph vessels, a process known as *metastasis.*

Q1: *Why does the treatment of cancer focus on killing the cancerous cells?*

59　61　70　79　85　87

BIOLOGICAL MEMBRANES

The word *membrane* {*membrana*, a skin} has two meanings in biology. Before the invention of microscopes in the sixteenth century, a membrane always described a tissue that lined a cavity or separated two compartments. Even today, we speak of *mucous membranes* in the mouth and vagina, the *peritoneal membrane* that lines the inside of the abdomen, the *pleural membrane* that covers the surface of the lungs, and the *pericardial membrane* that surrounds the heart. These visible membranes are tissues: thin, translucent layers of cells.

Once scientists observed cells with a microscope, the nature of the barrier between a cell's intracellular fluid and its external environment became a matter of great interest. By the 1890s, scientists had concluded that the outer surface of cells, the **cell membrane,** was a thin layer of lipids that separated the aqueous fluids of the interior and outside environment. We now know that cell membranes consist of microscopic double layers, or *bilayers,* of phospholipids with protein molecules inserted in them.

In short, the word *membrane* may apply either to a tissue or to a phospholipid-protein boundary layer (Fig. 3.1c). One source of confusion is that tissue membranes are often depicted in book illustrations as a single line, leading students to think of them as if they were similar in structure to the cell membrane. In this section, you will learn more about the phospholipid membranes that create compartments for cells.

The Cell Membrane Separates Cell from Environment

There are two synonyms for the term *cell membrane: plasma membrane* and *plasmalemma.* We will use the term *cell membrane* in this book rather than *plasma membrane* or *plasmalemma* to avoid confusion with the term *blood plasma.* The general functions of the cell membrane include:

1. **Physical isolation.** The cell membrane is a physical barrier that separates intracellular fluid inside the cell from the surrounding extracellular fluid.

2. **Regulation of exchange with the environment.** The cell membrane controls the entry of ions and nutrients into the cell, the elimination of cellular wastes, and the release of products from the cell.

3. **Communication between the cell and its environment.** The cell membrane contains proteins that enable the cell to recognize and respond to molecules or to changes in its external environment. Any alteration in the cell membrane may affect the cell's activities.

4. **Structural support.** Proteins in the cell membrane hold the *cytoskeleton,* the cell's interior structural scaffolding, in place to maintain cell shape. Membrane proteins also create specialized junctions between adjacent cells or between cells and the *extracellular matrix* {*extra-,* outside}, which is extracellular material that is synthesized and secreted by the cells. (**Secretion** is the process by which a cell releases a substance into the extracellular space.) Cell-cell and cell-matrix junctions stabilize the structure of tissues.

How can the cell membrane carry out such diverse functions? Our current model of cell membrane structure provides the answer.

Membranes Are Mostly Lipid and Protein

In the early decades of the twentieth century, researchers trying to decipher membrane structure ground up cells and analyzed their composition. They discovered that all biological membranes consist of a combination of lipids and proteins plus a small amount of carbohydrate. However, a simple and uniform structure did not account for the highly variable properties of membranes found in different types of cells. How could water cross the cell membrane to enter a red blood cell but not be able to enter certain cells of the kidney tubule? The explanation had to lie in the molecular arrangement of the proteins and lipids in the various membranes.

The ratio of protein to lipid varies widely, depending on the source of the membrane (TBL. 3.1). Generally, the more metabolically active a membrane is, the more proteins it contains. For example, the inner membrane of a mitochondrion, which contains enzymes for ATP production, is three-quarters protein.

This chemical analysis of membranes was useful, but it did not explain the structural arrangement of lipids and proteins in a membrane. Studies in the 1920s suggested that there was enough lipid in a given area of membrane to create a double layer. The bilayer model was further modified in the 1930s to account for the presence of proteins. With the introduction of electron microscopy, scientists saw the cell membrane for the first time. The 1960s model of the membrane, as seen in in electron micrographs, was a "butter sandwich"—a clear layer of lipids sandwiched between two dark layers of protein.

By the early 1970s, freeze-fracture electron micrographs had revealed the actual three-dimensional arrangement of lipids and proteins within cell membranes. Because of what scientists learned from looking at freeze-fractured membranes, S. J. Singer and G. L. Nicolson in 1972 proposed the **fluid mosaic model** of the membrane. FIGURE 3.2 highlights the major features of this contemporary model of membrane structure.

The lipids of biological membranes are mostly phospholipids arranged in a bilayer so that the phosphate heads are on the membrane surfaces and the lipid tails are hidden in the center of the membrane (Fig. 3.2b). The cell membrane is studded with protein molecules, like raisins in a slice of bread, and the extracellular surface has glycoproteins and glycolipids. All cell membranes are of relatively uniform thickness, about 8 nm.

Membrane Lipids Create a Hydrophobic Barrier

Three main types of lipids make up the cell membrane: phospholipids, sphingolipids, and cholesterol. Phospholipids are made of a glycerol backbone with two fatty acid chains extending to one side and a phosphate group extending to the other [p. 33]. The glycerol-phosphate head of the molecule is polar and thus hydrophilic. The fatty acid "tail" is nonpolar and thus hydrophobic.

When placed in an aqueous solution, phospholipids orient themselves so that the polar heads of the molecules interact with the water molecules while the nonpolar fatty acid tails "hide" by putting the polar heads between themselves and the water. This arrangement can be seen in three structures: the micelle, the liposome, and the phospholipid bilayer of the cell membrane (Fig. 3.2a). **Micelles** are small droplets with a single layer of phospholipids arranged so that the interior of the micelle is filled with hydrophobic fatty acid tails. Micelles are important in the digestion and absorption of fats in the digestive tract.

Liposomes are larger spheres with bilayer phospholipid walls. This arrangement leaves a hollow center with an aqueous core that can be filled with water-soluble molecules. Biologists think that a liposome-like structure was the precursor of the first living cell. Today, liposomes are being used as a medium to deliver drugs and cosmetics through the skin.

Phospholipids are the major lipid of membranes, but some membranes also have significant amounts of **sphingolipids.** Sphingolipids also have fatty acid tails, but their heads may be either phospholipids or glycolipids. Sphingolipids are slightly longer than phospholipids.

Cholesterol is also a significant part of many cell membranes. Cholesterol molecules, which are mostly hydrophobic, insert themselves between the hydrophilic heads of phospholipids (Fig. 3.2b). Cholesterol helps make membranes impermeable to small water-soluble molecules and keeps membranes flexible over a wide range of temperatures.

Membrane Proteins May Be Loosely or Tightly Bound to the Membrane

According to some estimates, membrane proteins may be nearly one-third of all proteins coded in our DNA. Each cell has between 10 and 50 different types of proteins inserted into its membranes. Membrane proteins can be described in several different ways. **Integral proteins** are tightly bound to the membrane, and

TABLE 3.1	Composition of Selected Membranes		
Membrane	**Protein**	**Lipid**	**Carbohydrate**
Red blood cell membrane	49%	43%	8%
Myelin membrane around nerve cells	18%	79%	3%
Inner mitochondrial membrane	76%	24%	0%

FIG. 3.2 ESSENTIALS

The Cell Membrane

(a) Membrane Phospholipids

Membrane phospholipids form bilayers, micelles, or liposomes. They arrange themselves so that their nonpolar tails are not in contact with aqueous solutions such as extracellular fluid.

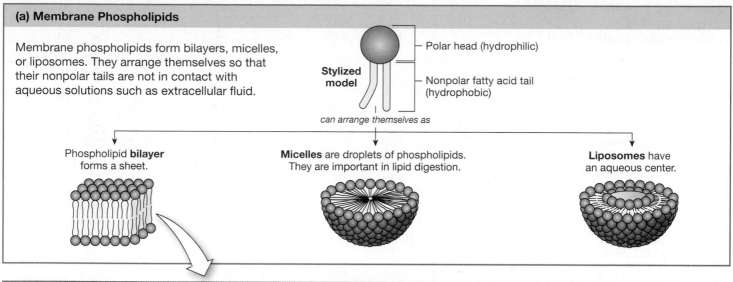

Stylized model
— Polar head (hydrophilic)
— Nonpolar fatty acid tail (hydrophobic)

can arrange themselves as

Phospholipid **bilayer** forms a sheet.

Micelles are droplets of phospholipids. They are important in lipid digestion.

Liposomes have an aqueous center.

(b) The Fluid Mosaic Model of Biological Membranes

Peripheral proteins can be removed without disrupting the integrity of the membrane.

Phospholipid heads face the aqueous intracellular and extracellular compartments.

Lipid tails form the interior layer of the membrane.

Cholesterol molecules insert themselves into the lipid layer.

Glycoprotein

Transmembrane proteins cross the lipid bilayer.

Lipid-anchored proteins

Cytoplasm

Peripheral protein

Cytoskeleton proteins

This membrane-spanning protein crosses the membrane seven times.

Carbohydrate

Extracellular fluid

COOH

Cell membrane

Intracellular fluid NH$_2$

Phosphate

Cytoplasmic loop

(c) Concept Map of Cell Membrane Components

Cell Membrane

consists of

Cholesterol | Phospholipids, Sphingolipids | Carbohydrates | Proteins

together form

Lipid bilayer

together form

Glycolipids | Glycoproteins

together form

functions as

Selective barrier between cytosol and external environment

whose functions include

Structural stability | Cell recognition | Immune response

the only way they can be removed is by disrupting the membrane structure with detergents or other harsh methods that destroy the membrane's integrity. Integral proteins include transmembrane proteins and lipid-anchored proteins.

Peripheral proteins {*peripheria*, circumference} attach to other membrane proteins by noncovalent interactions [p. 40] and can be separated from the membrane by chemical methods that do not disrupt the integrity of the membrane. Peripheral proteins include some enzymes as well as structural binding proteins that anchor the *cytoskeleton* (the cell's internal "skeleton") to the membrane (Fig. 3.2b).

Transmembrane proteins {*trans-*, across} are also called *membrane-spanning* proteins because the protein's chains extend all the way across the cell membrane (Fig. 3.2b). When a protein crosses the membrane more than once, loops of the amino acid chain protrude into the cytoplasm and the extracellular fluid. Carbohydrates may attach to the extracellular loops, and phosphate groups may attach to the intracellular loops. Phosphorylation or dephosphorylation of proteins is one way cells alter protein function [p. 33].

Transmembrane proteins are classified into families according to how many transmembrane segments they have. Many physiologically important membrane proteins have seven transmembrane segments, as shown in Figure 3.2c. Others cross the membrane only once or up to as many as 12 times.

Membrane-spanning proteins are integral proteins, tightly but not covalently bound to the membrane. The 20–25 amino acids in the protein chain segments that pass through the bilayer are nonpolar. This allows those amino acids to create strong noncovalent interactions with the lipid tails of the membrane phospholipids, holding them tightly in place.

Some membrane proteins that were previously thought to be peripheral proteins are now known to be **lipid-anchored proteins** (Fig. 3.2b). Some of these proteins are covalently bound to lipid tails that insert themselves into the bilayer. Others, found only on the external surface of the cell, are held by a **GPI anchor** that consists of a membrane lipid plus a sugar-phosphate chain. (GPI stands for *glycosylphosphatidylinositol*.) Many lipid-anchored proteins are associated with membrane sphingolipids, leading to the formation of specialized patches of membrane called *lipid rafts* (**FIG. 3.3**). The longer tails of the sphingolipids elevate the lipid rafts over their phospholipid neighbors.

According to the original fluid mosaic model of the cell membrane, membrane proteins could move laterally from location to location, directed by protein fibers that run just under the membrane surface. However, researchers have learned that this is not true of all membrane proteins. Some integral proteins are anchored to cytoskeleton proteins (Fig. 3.2b) and are, therefore, immobile. The ability of the cytoskeleton to restrict the movement of integral proteins allows cells to develop *polarity*, in which different faces of the cell have different proteins and therefore different properties. This is particularly important in the cells of the transporting epithelia, as you will see in multiple tissues in the body.

Membrane Carbohydrates Attach to Both Lipids and Proteins

Most membrane carbohydrates are sugars attached either to membrane proteins (glycoproteins) or to membrane lipids (glycolipids). They are found exclusively on the external surface of the cell, where they form a protective layer known as the **glycocalyx**

FIG. 3.3 **Lipid rafts are made of sphingolipids**

Sphingolipids (orange) are longer than phospholipids and stick up above the phospholipids of the membrane (black). A lipid-anchored enzyme, placental alkaline phosphatase (yellow), is almost always associated with a lipid raft. Image courtesy of D. E. Saslowsky, J. Lawrence, X. Ren, D. A. Brown, R. M. Henderson, and J. M. Edwardson. Placental alkaline phosphatase is efficiently targeted to rafts in supported lipid bilayers. *J Biol Chem* 277: 26966–26970, 2002.

{*glyco-,* sweet + *kalyx,* husk or pod}. Glycoproteins on the cell surface play a key role in the body's immune response. For example, the ABO blood groups are determined by the number and composition of sugars attached to membrane sphingolipids.

CONCEPT CHECK

1. Name three types of lipids found in cell membranes.
2. Describe three types of membrane proteins and how they are associated with the cell membrane.
3. Why do phospholipids in cell membranes form a bilayer instead of a single layer?
4. How many phospholipid bilayers will a substance cross passing into a cell?

Figure 3.2c is a summary map organizing the structure of the cell membrane.

INTRACELLULAR COMPARTMENTS

Much of what we know about cells comes from studies of simple organisms that consist of one cell. But humans are much more complex, with trillions of cells in their bodies. It has been estimated that there are more than 200 different types of cells in the human body, each cell type with its own characteristic structure and function.

During development, cells specialize and take on specific shapes and functions. Each cell in the body inherits identical genetic information in its DNA, but no one cell uses all this information. During **differentiation,** only selected genes become active, transforming the cell into a specialized unit. In most cases, the final shape and size of a cell and its contents reflect its function. Figure 3.1b shows five representative cells in the human body. These mature cells look very different from one another, but they all started out alike in the early embryo, and they retain many features in common.

Cells Are Divided into Compartments

We can compare the structural organization of a cell to that of a medieval walled city. The city is separated from the surrounding countryside by a high wall, with entry and exit strictly controlled through gates that can be opened and closed. The city inside the walls is divided into streets and a diverse collection of houses and shops with varied functions. Within the city, a ruler in the castle oversees the everyday comings and goings of the city's inhabitants. Because the city depends on food and raw material from outside the walls, the ruler negotiates with the farmers in the countryside. Foreign invaders are always a threat, so the city ruler communicates and cooperates with the rulers of neighboring cities.

In the cell, the outer boundary is the cell membrane. Like the city wall, it controls the movement of material between the cell interior and the outside by opening and closing "gates" made of protein. The inside of the cell is divided into compartments rather than into shops and houses. Each of these compartments

has a specific purpose that contributes to the function of the cell as a whole. In the cell, DNA in the nucleus is the "ruler in the castle," controlling both the internal workings of the cell and its interaction with other cells. Like the city, the cell depends on supplies from its external environment. It must also communicate and cooperate with other cells to keep the body functioning in a coordinated fashion.

FIGURE 3.4a is an overview map of cell structure. The cells of the body are surrounded by the dilute salt solution of the extracellular fluid. The cell membrane separates the inside environment of the cell (the intracellular fluid) from the extracellular fluid.

Internally the cell is divided into the *cytoplasm* and the *nucleus.* The cytoplasm consists of a fluid portion, called *cytosol;* insoluble particles called *inclusions;* insoluble protein fibers; and membrane-bound structures collectively known as *organelles.* Figure 3.4 shows a typical cell from the lining of the small intestine. It has most of the structures found in animal cells.

The Cytoplasm Includes Cytosol, Inclusions, Fibers, and Organelles

The **cytoplasm** includes all material inside the cell membrane except for the nucleus. The cytoplasm has four components:

1. **Cytosol** {*cyto-,* cell + sol(uble)}, or intracellular fluid: The cytosol is a semi-gelatinous fluid separated from the extracellular fluid by the cell membrane. The cytosol contains dissolved nutrients and proteins, ions, and waste products. The other components of the cytoplasm—inclusions, fibers, and organelles—are suspended in the cytosol.
2. **Inclusions** are particles of insoluble materials. Some are stored nutrients. Others are responsible for specific cell functions. These structures are sometimes called the *nonmembranous organelles.*
3. Insoluble **protein fibers** form the cell's internal support system, or **cytoskeleton.**
4. **Organelles**—"little organs"—are membrane-bound compartments that play specific roles in the overall function of the cell. For example, the organelles called mitochondria (singular, *mitochondrion*) generate most of the cell's ATP, and the organelles called lysosomes act as the digestive system of the cell. The organelles work in an integrated manner, each organelle taking on one or more of the cell's functions.

Inclusions Are in Direct Contact with the Cytosol

The inclusions of cells do not have boundary membranes and so are in direct contact with the cytosol. Movement of material between inclusions and the cytosol does not require transport across a membrane. Nutrients are stored as glycogen granules and lipid droplets. Most inclusions with functions other than nutrient storage are made from protein or combinations of RNA and protein.

Cell Structure

(a) This is an overview map of cell structure. The *cell membrane* separates the inside environment of the cell (the intracellular fluid) from the extracellular fluid. Internally the cell is divided into the *cytoplasm* and the *nucleus.* The cytoplasm consists of a fluid portion, called the *cytosol;* membrane-bound structures called *organelles;* insoluble particles called *inclusions;* and protein fibers that create the *cytoskeleton.*

THE CELL

is composed of

Extracellular fluid

(b) Cytoskeleton

Microvilli increase cell surface area. They are supported by microfilaments.

Microfilaments form a network just inside the cell membrane.

Microtubules are the largest cytoskeleton fiber.

Intermediate filaments include myosin and keratin.

(c) Peroxisomes

Peroxisomes contain enzymes that break down fatty acids and some foreign materials.

(d) Lysosomes

Lysosomes are small, spherical storage vesicles that contain powerful digestive enzymes.

(e) Centrioles

Centrioles are made from microtubules and direct DNA movement during cell division.

Centrioles

FIG. 3.4 REVIEW

(f) Cell Membrane

The **cell membrane** is a phospholipid bilayer studded with proteins that act as structural anchors, transporters, enzymes, or signal receptors. Glycolipids and glycoproteins occur only on the extracellular surface of the membrane. The cell membrane acts as both a gateway and a barrier between the cytoplasm and the extracellular fluid.

(g) Mitochondria

Outer membrane
Intermembrane space
Cristae
Matrix

Mitchondria are spherical to elliptical organelles with a double wall that creates two separate compartments within the organelle. The inner **matrix** is surrounded by a membrane that folds into leaflets called **cristae.** The **intermembrane space,** which lies between the two membranes, plays an important role in ATP production. Mitochondria are the site of most ATP synthesis in the cell.

(h) Golgi Apparatus and Vesicles

Vesicle
Cisternae

The **Golgi apparatus** consists of a series of hollow curved sacs called **cisternae** stacked on top of one another and surrounded by vesicles. The Golgi apparatus participates in protein modification and packaging.

(i) Endoplasmic Reticulum (ER) and Ribosomes

Rough ER
Ribosomes
Smooth ER

The **endoplasmic reticulum (ER)** is a network of interconnected membrane tubes that are a continuation of the outer nuclear membrane. **Rough endoplasmic reticulum** has a granular appearance due to rows of ribosomes dotting its cytoplasmic surface. **Smooth endoplasmic reticulum** lacks ribosomes and appears as smooth membrane tubes. The rough ER is the main site of protein synthesis. The smooth ER synthesizes lipids and, in some cells, concentrates and stores calcium ions.

(j) Nucleus

Nuclear envelope
Nucleolus
Nuclear pores

The **nucleus** is surrounded by a double-membrane **nuclear envelope.** Both membranes of the envelope are pierced here and there by **pores** to allow communication with the cytoplasm. The outer membrane of the nuclear envelope connects to the endoplasmic reticulum membrane. In cells that are not dividing, the nucleus appears filled with randomly scattered granular material composed of DNA and proteins. Usually a nucleus also contains from one to four larger dark-staining bodies of DNA, RNA, and protein called **nucleoli.**

Ribosomes (Fig. 3.4i) are small, dense granules of RNA and protein that manufacture proteins under the direction of the cell's DNA [see Chapter 4 for details]. **Fixed ribosomes** attach to the cytosolic surface of organelles. **Free ribosomes** are suspended free in the cytosol. Some free ribosomes form groups of 10 to 20 known as **polyribosomes.** A ribosome that is fixed one minute may release and become a free ribosome the next. Ribosomes are most numerous in cells that synthesize proteins for export out of the cell.

Cytoplasmic Protein Fibers Come in Three Sizes

The three families of cytoplasmic protein fibers are classified by diameter and protein composition (**TBL. 3.2**). All fibers are polymers of smaller proteins. The thinnest are **actin fibers,** also called *microfilaments.* Somewhat larger **intermediate filaments** may be made of different types of protein, including *keratin* in hair and skin, and *neurofilament* in nerve cells. The largest protein fibers are the hollow **microtubules,** made of a protein called **tubulin.** A large number of *accessory proteins* are associated with the cell's protein fibers.

The insoluble protein fibers of the cell have two general purposes: structural support and movement. Structural support comes primarily from the cytoskeleton. Movement of the cell or of elements within the cell takes place with the aid of protein fibers and a group of specialized enzymes called *motor proteins.* These functions are discussed in more detail in the sections that follow.

Microtubules Form Centrioles, Cilia, and Flagella

The largest cytoplasmic protein fibers, the microtubules, create the complex structures of centrioles, cilia, and flagella, which are all involved in some form of cell movement. The cell's *microtubule-organizing center,* the **centrosome,** assembles tubulin molecules into microtubules. The centrosome appears as a region of darkly staining material close to the cell nucleus. In most animal cells, the centrosome contains two **centrioles,** shown in the typical cell of Figure 3.4e.

Each centriole is a cylindrical bundle of 27 microtubules, arranged in nine triplets. In cell division, the centrioles direct the movement of DNA strands. Cells that have lost their ability to undergo cell division, such as mature nerve cells, lack centrioles.

Cilia are short, hair-like structures projecting from the cell surface like the bristles of a brush {singular, *cilium,* Latin for eyelash}. Most cells have a single short cilium, but cells lining the upper airways and part of the female reproductive tract are covered with cilia. In these tissues, coordinated ciliary movement creates currents that sweep fluids or secretions across the cell surface.

The surface of a cilium is a continuation of the cell membrane. The core of *motile,* or moving, cilia contains nine pairs of microtubules surrounding a central pair (**FIG. 3.5b**). The microtubules terminate just inside the cell at the *basal body*. These cilia beat rhythmically back and forth when the microtubule pairs in their core slide past each other with the help of the motor protein *dynein.*

Flagella have the same microtubule arrangement as cilia but are considerably longer {singular, *flagellum,* Latin for whip}. Flagella are found on free-floating single cells, and in humans the only flagellated cell is the male sperm cell. A sperm cell has only one flagellum, in contrast to ciliated cells, which may have one surface almost totally covered with cilia (Fig. 3.5a). The wavelike movements of the flagellum push the sperm through fluid, just as undulating contractions of a snake's body push it headfirst through its environment. Flagella bend and move by the same basic mechanism as cilia.

The Cytoskeleton Is a Changeable Scaffold

The cytoskeleton is a flexible, changeable three-dimensional scaffolding of actin microfilaments, intermediate filaments, and microtubules that extends throughout the cytoplasm. Some cytoskeleton protein fibers are permanent, but most are synthesized or disassembled according to the cell's needs. Because of the cytoskeleton's changeable nature, its organizational details are complex and we will not discuss the details.

The cytoskeleton has at least five important functions.

1. **Cell shape.** The protein scaffolding of the cytoskeleton provides mechanical strength to the cell and in some cells plays an important role in determining the shape of the cell. Figure 3.4b shows how cytoskeletal fibers help support **microvilli** {*micro-,* small + *villus,* tuft of hair}, fingerlike extensions of the cell membrane that increase the surface area for absorption of materials.

TABLE 3.2	Diameter of Protein Fibers in the Cytoplasm		
	Diameter	**Type of Protein**	**Functions**
Microfilaments	7 nm	Actin (globular)	Cytoskeleton; associates with myosin for muscle contraction
Intermediate Filaments	10 nm	Keratin, neurofilament protein (filaments)	Cytoskeleton, hair and nails, protective barrier of skin
Microtubules	25 nm	Tubulin (globular)	Movement of cilia, flagella, and chromosomes; intracellular transport of organelles; cytoskeleton

FIG. 3.5 Cilia and flagella

(a) Cilia

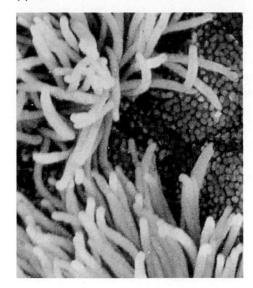

(b) Cilia and flagella have 9 pairs of microtubules surrounding a central pair.

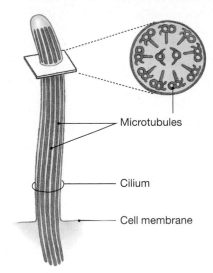

Microtubules

Cilium

Cell membrane

(c) The beating of cilia and flagella creates fluid movement.

Fluid movement

Flagellum

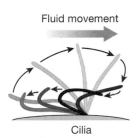

Fluid movement

Cilia

2. **Internal organization.** Cytoskeletal fibers stabilize the positions of organelles. Figure 3.4b illustrates organelles held in place by the cytoskeleton. Note, however, that this figure is only a snapshot of one moment in the cell's life. The interior arrangement and composition of a cell are dynamic, changing from minute to minute in response to the needs of the cell, just as the inside of the walled city is always in motion. One disadvantage of the static illustrations in textbooks is that they are unable to represent movement and the dynamic nature of many physiological processes.

3. **Intracellular transport.** The cytoskeleton helps transport materials into the cell and within the cytoplasm by serving as an intracellular "railroad track" for moving organelles. This function is particularly important in cells of the nervous system, where material must be transported over intracellular distances as long as a meter.

4. **Assembly of cells into tissues.** Protein fibers of the cytoskeleton connect with protein fibers in the extracellular space, linking cells to one another and to supporting material outside the cells. In addition to providing mechanical strength to the tissue, these linkages allow the transfer of information from one cell to another.

5. **Movement.** The cytoskeleton helps cells move. For example, the cytoskeleton helps white blood cells squeeze out of blood vessels and helps growing nerve cells send out long extensions as they elongate. Cilia and flagella on the cell membrane are able to move because of their microtubule cytoskeleton. Special motor proteins facilitate movement and intracellular transport by using energy from ATP to slide or step along cytoskeletal fibers.

EMERGING CONCEPTS

Single Cilia Are Sensors

Cilia in the body are not limited to the airways and the female reproductive tract. Scientists have known for years that most cells of the body contain a single, stationary, or *nonmotile,* cilium, but they thought that these solitary **primary cilia** were mostly evolutionary remnants and of little significance. Primary cilia differ structurally from motile cilia because they lack the central pair of microtubules found in motile cilia (a 9 + 0 arrangement instead of 9 + 2; see Fig. 3.5). Researchers in recent years have learned that primary cilia actually serve a function. They can act as sensors of the external environment, passing information into the cell. For example, primary cilia in photoreceptors of the eye help with light sensing, and primary cilia in the kidney sense fluid flow. Using molecular techniques, scientists have found that these small, insignificant hairs play critical roles during embryonic development as well. Mutations to ciliary proteins cause disorders (*ciliopathies*) ranging from polycystic kidney disease and loss of vision to cancer. The role of primary cilia in other disorders, including obesity, is currently a hot topic in research.

Motor Proteins Create Movement

Motor proteins are proteins that convert stored energy into directed movement. Three groups of motor proteins are associated with the cytoskeleton: myosins, kinesins, and dyneins. All three groups use energy stored in ATP to propel themselves along cytoskeleton fibers.

Myosins bind to actin fibers and are best known for their role in muscle contraction [Chapter 12]. **Kinesins** and **dyneins** assist the movement of vesicles along microtubules. Dyneins also associate with the microtubule bundles of cilia and flagella to help create their whiplike motion.

CONCEPT CHECK

5. Name the three sizes of cytoplasmic protein fibers.
6. How would the absence of a flagellum affect a sperm cell?
7. What is the difference between cytoplasm and cytosol?
8. What is the difference between a cilium and a flagellum?
9. What is the function of motor proteins?

Most motor proteins are made of multiple protein chains arranged into three parts: two heads that bind to the cytoskeleton fiber, a neck, and a tail region that is able to bind "cargo," such as an organelle that needs to be transported through the cytoplasm (**FIG. 3.6**). The heads alternately bind to the cytoskeleton fiber, then release and "step" forward using the energy stored in ATP.

Organelles Create Compartments for Specialized Functions

Organelles are subcellular compartments separated from the cytosol by one or more phospholipid membranes similar in structure to the cell membrane. The compartments created by organelles allow the cell to isolate substances and segregate functions. For example, an organelle might contain substances that could be harmful to the cell, such as digestive enzymes. Figures 3.4g, 3.4h, and 3.4i show the four major groups of organelles: mitochondria, the Golgi apparatus, the endoplasmic reticulum, and membrane-bound spheres called **vesicles** {*vesicula*, bladder}.

Mitochondria Mitochondria {singular, *mitochondrion*; *mitos*, thread+*chondros*, granule} are unique organelles in several ways. First, they have an unusual double wall that creates two separate compartments within the mitochondrion (Fig. 3.4g). In the center, inside the inner membrane, is a compartment called the **mitochondrial matrix** {*matrix*, female animal for breeding}. The matrix contains enzymes, ribosomes, granules, and surprisingly, its own unique DNA. This **mitochondrial DNA** has a different nucleotide sequence from that found in the nucleus. Because mitochondria have their own DNA, they can manufacture some of their own proteins.

Why do mitochondria contain DNA when other organelles do not? This question has been the subject of intense scrutiny. According to the *prokaryotic endosymbiont theory*, mitochondria are the descendants of bacteria that invaded cells millions of years ago. The bacteria developed a mutually beneficial relationship with their hosts and soon became an integral part of the host cells. Supporting evidence for this theory is the fact that our mitochondrial DNA, RNA, and enzymes are similar to those in bacteria but unlike those in our own cell nuclei.

The second compartment inside a mitochondrion is the **intermembrane space,** which lies between the outer and inner mitochondrial membranes. This compartment plays an important role in mitochondrial ATP production, so the number of mitochondria in a cell is directly related to the cell's energy needs. For example, skeletal muscle cells, which use a lot of energy, have many more mitochondria than less active cells, such as adipose (fat) cells.

Another unusual characteristic of mitochondria is their ability to replicate themselves even when the cell to which they belong is not undergoing cell division. This process is aided by the mitochondrial DNA, which allows the organelles to direct their own duplication. Mitochondria replicate by budding, during which small daughter mitochondria pinch off from an enlarged parent. For instance, exercising muscle cells that experience increased energy

---

FIG. 3.6 | Motor proteins

Motor protein chains form a tail that binds organelles or other cargo, a neck, and two heads that "walk" along the cytoskeleton using energy from ATP.

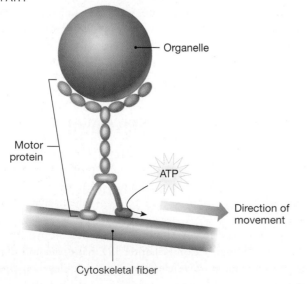

Organelle

Motor protein

ATP

Direction of movement

Cytoskeletal fiber

RUNNING **PROBLEM**

During a Pap test for cervical cancer, tissue is sampled from the cervix (neck) of the uterus with a collection device that resembles a tiny brush. The cells are rinsed off the brush into preservative fluid that is sent to a laboratory. There the sample is processed onto a glass slide that will be examined first by a computer, then by a trained cytologist. The computer and cytologist look for *dysplasia* {*dys-*, abnormal + *-plasia*, growth or cell multiplication}, a change in the size and shape of cells that is suggestive of cancerous changes. Cancer cells can usually be recognized by a large nucleus surrounded by a relatively small amount of cytoplasm. Jan's first Pap test showed all the hallmarks of dysplasia.

Q2: *What is happening in cancer cells that explains the large size of their nucleus and the relatively small amount of cytoplasm?*

59 61 70 79 85 87

demands over a period of time may meet the demand for more ATP by increasing the number of mitochondria in their cytoplasm.

The Endoplasmic Reticulum

The **endoplasmic reticulum, or ER,** is a network of interconnected membrane tubes with three major functions: synthesis, storage, and transport of biomolecules (Fig. 3.4i). The name *reticulum* comes from the Latin word for *net* and refers to the netlike arrangement of the tubules. Electron micrographs reveal that there are two forms of endoplasmic reticulum: **rough endoplasmic reticulum** (RER) and **smooth endoplasmic reticulum** (SER).

The rough endoplasmic reticulum is the main site of protein synthesis. Proteins are assembled on ribosomes attached to the cytoplasmic surface of the rough ER, then inserted into the rough ER lumen, where they undergo chemical modification.

The smooth endoplasmic reticulum lacks attached ribosomes and is the main site for the synthesis of fatty acids, steroids, and lipids [p. 30]. Phospholipids for the cell membrane are produced here, and cholesterol is modified into steroid hormones, such as the sex hormones estrogen and testosterone. The smooth ER of liver and kidney cells detoxifies or inactivates drugs. In skeletal muscle cells, a modified form of smooth ER stores calcium ions (Ca^{2+}) to be used in muscle contraction.

The Golgi Apparatus

The **Golgi apparatus** (also known as the Golgi complex) was first described by Camillo Golgi in 1898 (Fig. 3.4h). For years, some investigators thought that this organelle was just a result of the fixation process needed to prepare tissues for viewing under the light microscope. However, we now know from electron microscope studies that the Golgi apparatus is indeed a discrete organelle. It consists of a series of hollow curved sacs, called *cisternae,* stacked on top of one another like a series of hot water bottles and surrounded by vesicles. The Golgi apparatus receives proteins made on the rough ER, modifies them, and packages them into the vesicles.

Cytoplasmic Vesicles

Membrane-bound cytoplasmic vesicles are of two kinds: secretory and storage. **Secretory vesicles** contain proteins that will be released from the cell. The contents of most **storage vesicles,** however, never leave the cytoplasm.

Lysosomes {*lysis,* dissolution + *soma,* body} are small storage vesicles that appear as membrane-bound granules in the cytoplasm (Fig. 3.4d). Lysosomes act as the digestive system of the cell. They use powerful enzymes to break down bacteria or old organelles, such as mitochondria, into their component molecules. Those molecules that can be reused are reabsorbed into the cytosol, while the rest are dumped out of the cell. As many as 50 types of enzymes have been identified from lysosomes of different cell types.

Because lysosomal enzymes are so powerful, early workers puzzled over the question of why these enzymes do not normally destroy the cell that contains them. What scientists discovered was that lysosomal enzymes are activated only by very acidic conditions, 100 times more acidic than the normal acidity level in the cytoplasm. When lysosomes first pinch off from the Golgi apparatus, their interior pH is about the same as that of the cytosol, 7.0–7.3. The enzymes are inactive at this pH. Their inactivity

serves as a form of insurance. If the lysosome breaks or accidentally releases enzymes, they will not harm the cell.

However, as the lysosome sits in the cytoplasm, it accumulates H^+ in a process that uses energy. Increasing concentrations of H^+ decrease the pH inside the vesicle to 4.8–5.0, and the enzymes are activated. Once activated, lysosomal enzymes can break down biomolecules inside the vesicle. The lysosomal membrane is not affected by the enzymes.

The digestive enzymes of lysosomes are not always kept isolated within the organelle. Occasionally, lysosomes release their enzymes outside the cell to dissolve extracellular support material, such as the hard calcium carbonate portion of bone. In other instances, cells allow their lysosomal enzymes to come in contact with the cytoplasm, leading to self-digestion of all or part of the cell. When muscles *atrophy* (shrink) from lack of use or the uterus diminishes in size after pregnancy, the loss of cell mass is due to the action of lysosomes.

The inappropriate release of lysosomal enzymes has been implicated in certain disease states, such as the inflammation and destruction of joint tissue in *rheumatoid arthritis.* In the inherited conditions known as *lysosomal storage diseases,* lysosomes are ineffective because they lack specific enzymes. One of the best-known lysosomal storage diseases is the fatal inherited condition known as *Tay-Sachs disease.* Infants with Tay-Sachs disease have defective lysosomes that fail to break down glycolipids. Accumulation of glycolipids in nerve cells causes nervous system dysfunction, including blindness and loss of coordination. Most infants afflicted with Tay-Sachs disease die in early childhood.

Peroxisomes are storage vesicles that are even smaller than lysosomes (Fig. 3.4c). For years, they were thought to be a kind of lysosome, but we now know that they contain a different set of enzymes. Their main function appears to be to degrade long-chain fatty acids and potentially toxic foreign molecules.

Peroxisomes get their name from the fact that the reactions that take place inside them generate hydrogen peroxide (H_2O_2), a toxic molecule. The peroxisomes rapidly convert this peroxide to oxygen and water using the enzyme *catalase.* Peroxisomal disorders disrupt the normal processing of lipids and can severely disrupt neural function by altering the structure of nerve cell membranes.

CONCEPT CHECK

10. What distinguishes organelles from inclusions?

11. What is the anatomical difference between rough endoplasmic reticulum and smooth endoplasmic reticulum? What is the functional difference?

12. How do lysosomes differ from peroxisomes?

13. Apply the physiological theme of compartmentation to organelles in general and to mitochondria in particular.

14. Microscopic examination of a cell reveals many mitochondria. What does this observation imply about the cell's energy requirements?

15. Examining tissue from a previously unknown species of fish, you discover a tissue containing large amounts of smooth endoplasmic reticulum in its cells. What is one possible function of these cells?

The Nucleus Is the Cell's Control Center

The nucleus of the cell contains DNA, the genetic material that ultimately controls all cell processes. Figure 3.4j illustrates the structure of a typical nucleus. Its boundary, or **nuclear envelope,** is a two-membrane structure that separates the nucleus from the cytoplasmic compartment. Both membranes of the envelope are pierced here and there by round holes, or **pores.**

Communication between the nucleus and cytosol occurs through the **nuclear pore complexes,** large protein complexes with a central channel. Ions and small molecules move freely through this channel when it is open, but transport of large molecules such as proteins and RNA is a process that requires energy. Specificity of the transport process allows the cell to restrict DNA to the nucleus and various enzymes to either the cytoplasm or the nucleus.

In electron micrographs of cells that are not dividing, the nucleus appears filled with randomly scattered granular material, or **chromatin,** composed of DNA and associated proteins. Usually a nucleus also contains from one to four larger dark-staining bodies of DNA, RNA, and protein called **nucleoli** {singular, *nucleolus,* little nucleus}. Nucleoli contain the genes and proteins that control the synthesis of RNA for ribosomes.

The process of protein synthesis, modification, and packaging in different parts of the cell is an excellent example of how compartmentation allows separation of function, as shown in **FIGURE 3.7**. RNA for protein synthesis is made from DNA templates in the nucleus ❶, then transported to the cytoplasm through the nuclear pores ❷. In the cytoplasm, proteins are synthesized on ribosomes that may be free inclusions ❸ or attached to the rough endoplasmic reticulum ❹. The newly made protein is compartmentalized in the lumen of the rough ER ❺, where it is modified before being packaged into a vesicle ❻. The vesicles fuse with the Golgi apparatus, allowing additional modification of the protein in the Golgi lumen ❼. The modified proteins leave the Golgi packaged in either storage vesicles ❾ or secretory vesicles whose contents will be released into the extracellular fluid ❿. The molecular details of protein synthesis are discussed elsewhere [see Chapter 4].

TISSUES OF THE BODY

Despite the amazing variety of intracellular structures, no single cell can carry out all the processes of the mature human body. Instead, cells assemble into the larger units we call tissues. The cells in tissues are held together by specialized connections called *cell junctions* and by other support structures. Tissues range in complexity from simple tissues containing only one cell type, such as the lining of blood vessels, to complex tissues containing many cell types and extensive extracellular material, such as connective tissue. The cells of most tissues work together to achieve a common purpose.

The study of tissue structure and function is known as **histology** {*histos,* tissue}. Histologists describe tissues by their physical features: (1) the shape and size of the cells, (2) the arrangement of the cells in the tissue (in layers, scattered, and so on), (3) the way cells are connected to one another, and (4) the amount of extracellular material present in the tissue. There are four primary tissue types in the human body: epithelial, connective, muscle, and *neural,* or nerve. Before we consider each tissue type specifically, let's examine how cells link together to form tissues.

Extracellular Matrix Has Many Functions

Extracellular matrix (usually just called *matrix*) is extracellular material that is synthesized and secreted by the cells of a tissue. For years, scientists believed that matrix was an inert substance whose only function was to hold cells together. However, experimental evidence now shows that the extracellular matrix plays a vital role in many physiological processes, ranging from growth and development to cell death. A number of disease states are associated with overproduction or disruption of extracellular matrix, including chronic heart failure and the spread of cancerous cells throughout the body (*metastasis*).

The composition of extracellular matrix varies from tissue to tissue, and the mechanical properties, such as elasticity and flexibility, of a tissue depend on the amount and consistency of the tissue's matrix. Matrix always has two basic components: proteoglycans and insoluble protein fibers. **Proteoglycans** are glycoproteins, which are proteins covalently bound to polysaccharide chains [p. 29]. Insoluble protein fibers such as *collagen, fibronectin,* and *laminin* provide strength and anchor cells to the matrix. Attachments between the extracellular matrix and proteins in the cell membrane or the cytoskeleton are ways cells communicate with their external environment.

The amount of extracellular matrix in a tissue is highly variable. Nerve and muscle tissue have very little matrix, but the connective tissues, such as cartilage, bone, and blood, have extensive matrix that occupies as much volume as their cells. The consistency of extracellular matrix can vary from watery (blood and lymph) to rigid (bone).

Cell Junctions Hold Cells Together to Form Tissues

During growth and development, cells form *cell-cell adhesions* that may be transient or that may develop into more permanent **cell junctions. Cell adhesion molecules (CAMs)** are membrane-spanning proteins responsible both for cell junctions and for transient cell adhesions (**TBL. 3.3**). Cell-cell or cell-matrix adhesions mediated by CAMs are essential for normal growth and development. For example, growing nerve cells creep across the extracellular matrix with the help of *nerve-cell adhesion molecules,* or *NCAMs.* Cell adhesion helps white blood cells escape from the circulation and move into infected tissues, and it allows clumps of platelets to cling to damaged blood vessels. Because cell adhesions are not permanent, the bond between those CAMs and matrix is weak.

Stronger cell junctions can be grouped into three broad categories by function: communicating junctions, occluding junctions

FIG. 3.7   Protein synthesis demonstrates subcellular compartmentation

Nucleus

DNA

mRNA

Nuclear pore

Endoplasmic reticulum

Ribosome

Targeted proteins

Growing amino-acid chain

Peroxisome

Mitochondrion

Cytosolic protein

Transport vesicle

Golgi apparatus

Retrograde Golgi-ER transport

Golgi

Lysosome or storage vesicle

Golgi apparatus

Secretory vesicle

Cytosol

Cell membrane

Extracellular fluid

1 mRNA is transcribed from genes in the DNA.

2 mRNA leaves the nucleus and attaches to cytosolic ribosomes, initiating protein synthesis.

3 Some proteins are released by free ribosomes into the cytosol or are targeted to specific organelles.

4 Ribosomes attached to the rough endoplasmic reticulum direct proteins destined for packaging into the lumen of the rough ER.

5 Proteins are modified as they pass through the lumen of the ER.

6 Transport vesicles move the proteins from the ER to the Golgi apparatus.

7 Golgi cisternae migrate toward the cell membrane.

8 Some vesicles bud off the cisternae and move in a retrograde or backward fashion.

9 Some vesicles bud off to form lysosomes or storage vesicles.

10 Other vesicles become secretory vesicles that release their contents outside the cell.

TABLE 3.3	Major Cell Adhesion Molecules (CAMs)
Name	**Examples**
Cadherins	Cell-cell junctions such as adherens junctions and desmosomes. Calcium-dependent.
Integrins	Primarily found in cell-matrix junctions. These also function in cell signaling.
Immunoglobulin superfamily CAMs	NCAMs (nerve-cell adhesion molecules). Responsible for nerve cell growth during nervous system development.
Selectins	Temporary cell-cell adhesions.

{*occludere,* to close up}, and anchoring junctions (**FIG. 3.8**). In animals, the communicating junctions are gap junctions. The occluding junctions of vertebrates are tight junctions that limit movement of materials between cells. The three major types of junctions are described next.

1. **Gap junctions** are the simplest cell-cell junctions (Fig. 3.8b). They allow direct and rapid cell-to-cell communication through cytoplasmic bridges between adjoining cells. Cylindrical proteins called *connexins* interlock to create passageways that look like hollow rivets with narrow channels through their centers. The channels are able to open and close, regulating the movement of small molecules and ions through them.

   Gap junctions allow both chemical and electrical signals to pass rapidly from one cell to the next. They were once thought to occur only in certain muscle and nerve cells, but we now know they are important in cell-to-cell communication in many tissues, including the liver, pancreas, ovary, and thyroid gland.

2. **Tight junctions** are occluding junctions that restrict the movement of material between the cells they link (Fig. 3.8c). In tight junctions, the cell membranes of adjacent cells partly fuse together with the help of proteins called *claudins* and *occludins,* thereby making a barrier. As in many physiological processes, the barrier properties of tight junctions are dynamic and can be altered depending on the body's needs. Tight junctions may have varying degrees of "leakiness."

   Tight junctions in the intestinal tract and kidney prevent most substances from moving freely between the external and internal environments. In this way, they enable cells to regulate what enters and leaves the body. Tight junctions also create the so-called *blood-brain barrier* that prevents many potentially harmful substances in the blood from reaching the extracellular fluid of the brain.

3. **Anchoring junctions** (Fig. 3.8d) attach cells to each other (cell-cell anchoring junctions) or to the extracellular matrix (cell-matrix anchoring junctions). In vertebrates, cell-cell anchoring junctions are created by CAMs called **cadherins,** which connect with one another across the intercellular space. Cell-matrix junctions use CAMs called **integrins.** Integrins

are membrane proteins that can also bind to signal molecules in the cell's environment, transferring information carried by the signal across the cell membrane into the cytoplasm.

Anchoring junctions contribute to the mechanical strength of the tissue. They have been compared to buttons or zippers that tie cells together and hold them in position within a tissue. Notice how the interlocking cadherin proteins in Figure 3.8d resemble the teeth of a zipper.

The protein linkage of anchoring cell junctions is very strong, allowing sheets of tissue in skin and lining body cavities to resist damage from stretching and twisting. Even the tough protein fibers of anchoring junctions can be broken, however. If you have shoes that rub against your skin, the stress can shear the proteins connecting the different skin layers. When fluid accumulates in the resulting space and the layers separate, a *blister* results.

Tissues held together with anchoring junctions are like a picket fence, where spaces between the pickets (the cells) allow materials to pass from one side of the fence to the other. Movement of materials between cells is known as the **paracellular** pathway. In contrast, tissues held together with tight junctions are more like a solid brick wall: Very little can pass from one side of the wall to the other between the bricks.

Cell-cell anchoring junctions take the form of either adherens junctions or desmosomes. **Adherens junctions** link actin fibers in adjacent cells together, as shown in the micrograph in Figure 3.8e. **Desmosomes** {*desmos,* band + *soma,* body} attach to intermediate filaments of the cytoskeleton. Desmosomes are the strongest cell-cell junctions. In electron micrographs they can be recognized by the dense glycoprotein bodies, or *plaques,* that lie just inside the cell membranes in the region where the two cells connect (Fig. 3.8d, e). Desmosomes may be small points of contact between two cells (spot desmosomes) or bands that encircle the entire cell (belt desmosomes).

There are also two types of cell-matrix anchoring junctions. **Hemidesmosomes** {*hemi-,* half} are strong junctions that anchor intermediate fibers of the cytoskeleton to fibrous matrix proteins such as laminin. **Focal adhesions** tie intracellular actin fibers to different matrix proteins, such as fibronectin.

The loss of normal cell junctions plays a role in a number of diseases and in metastasis. Diseases in which cell junctions are destroyed or fail to form can have disfiguring and painful symptoms, such as blistering skin. One such disease is *pemphigus,* a condition in which the body attacks some of its own cell junction proteins (www.pemphigus.org).

The disappearance of anchoring junctions probably contributes to the metastasis of cancer cells throughout the body. Cancer cells lose their anchoring junctions because they have fewer cadherin molecules and are not bound as tightly to neighboring cells. Once a cancer cell is released from its moorings, it secretes protein-digesting enzymes known as *proteases.* These enzymes, especially those called *matrix metalloproteinases (MMPs),* dissolve the extracellular matrix so that escaping cancer cells can invade adjacent tissues or enter the bloodstream. Researchers are investigating ways of blocking MMP enzymes to see if they can prevent metastasis.

FIG. 3.8 **ESSENTIALS**

Cell Junctions

(a) Cell junctions connect one cell with another cell (or to surrounding matrix) with membrane-spanning proteins called **cell adhesion molecules,** or **CAMs.** This map shows the many ways cell junctions can be categorized.

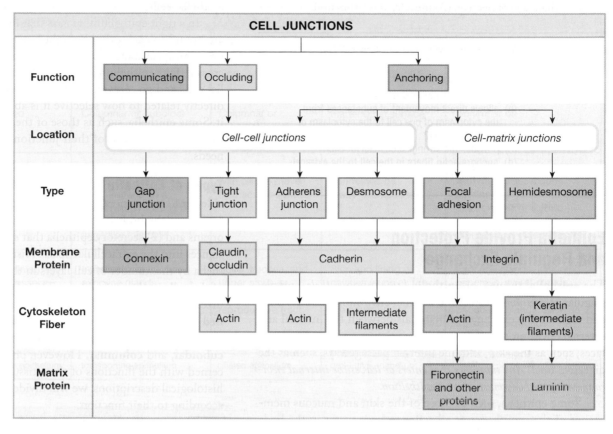

	CELL JUNCTIONS					
Function	Communicating	Occluding		Anchoring		
Location		Cell-cell junctions			Cell-matrix junctions	
Type	Gap junction	Tight junction	Adherens junction	Desmosome	Focal adhesion	Hemidesmosome
Membrane Protein	Connexin	Claudin, occludin	Cadherin		Integrin	
Cytoskeleton Fiber		Actin	Actin	Intermediate filaments	Actin	Keratin (intermediate filaments)
Matrix Protein					Fibronectin and other proteins	Laminin

Cell junctions can be grouped into three categories: **(b) Gap junctions** which allow direct cell to cell communication, **(c) tight junctions** that block movement of material between cells, and **(d) anchoring junctions** that hold cells to one another and to the extracellular matrix.

(b) Gap junctions are communicating junctions.

- Cytosol
- Connexin proteins
- Intercellular space
- Cell membrane
- Cell 1
- Cell 2

(c) Tight junctions are occluding junctions.

- Claudin and occludin proteins
- Cell 1
- Cell 2

(d) A desmosome is a cell-to-cell anchoring junction.

- Cadherin proteins
- Cell membrane
- Plaque glycoproteins
- Intercellular space
- Intermediate filament

Heart muscle has gap junctions that allow chemical and electrical signals to pass rapidly from one cell to the next.

- Clusters of gap junctions

Freeze fracture of cell membrane

- Tight junctions prevent movement between cells.
- Adherens junction
- Desmosomes anchor cells to each other.

(e) Cells may have several types of junctions, as shown in this micrograph of two adjacent intestinal cells.

FIG. 3.10 ESSENTIALS

Types of Epithelia

(a) Exchange Epithelium

The thin, flat cells of exchange epithelium allow movement through and between the cells.

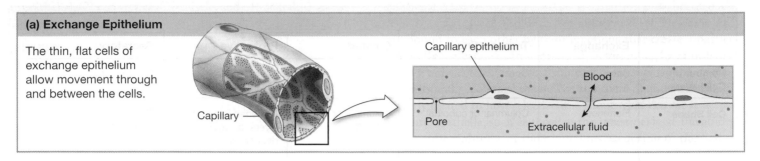

Capillary

Capillary epithelium

Blood

Pore

Extracellular fluid

(b) Transporting Epithelium

Transporting epithelia selectively move substances between a lumen and the ECF.

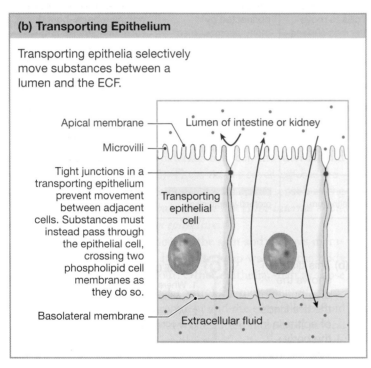

Apical membrane

Lumen of intestine or kidney

Microvilli

Tight junctions in a transporting epithelium prevent movement between adjacent cells. Substances must instead pass through the epithelial cell, crossing two phospholipid cell membranes as they do so.

Transporting epithelial cell

Basolateral membrane

Extracellular fluid

(c) Ciliated Epithelium

Beating cilia create fluid currents that sweep across the epithelial surface.

Cilia

Microvilli

SEM of the epithelial surface of an airway

Golgi apparatus

Nucleus

Mitochondrion

Basal lamina

(d) Protective Epithelium

Protective epithelia have many stacked layers of cells that are constantly being replaced. This figure shows layers in skin (see Fig. 3.15, *Focus On: The Skin*).

Epithelial cells

Section of skin showing cell layers.

(e) Secretory Epithelium

Secretory epithelial cells make and release a product. Exocrine secretions, such as the mucus shown here, are secreted outside the body. The secretions of endocrine cells (hormones) are released into the blood.

Mucus

SEM of goblet cell

Golgi apparatus

Nucleus

Goblet cells secrete mucus into the lumen of hollow organs such as the intestine.

Transporting Epithelia The **transporting epithelia** actively and selectively regulate the exchange of nongaseous materials, such as ions and nutrients, between the internal and external environments. These epithelia line the hollow tubes of the digestive system and the kidney, where lumens open into the external environment [p. 4]. Movement of material from the external environment across the epithelium to the internal environment is called *absorption*. Movement in the opposite direction, from the internal to the external environment, is called *secretion*.

Transporting epithelia can be identified by the following characteristics (Fig. 3.10b):

1. **Cell shape.** Cells of transporting epithelia are much thicker than cells of exchange epithelia, and they act as a barrier as well as an entry point. The cell layer is only one cell thick (a simple epithelium), but cells are cuboidal or columnar.

2. **Membrane modifications.** The **apical membrane,** the surface of the epithelial cell that faces the lumen, has tiny finger-like projections called *microvilli* that increase the surface area available for transport. A cell with microvilli has at least 20 times the surface area of a cell without them. In addition, the **basolateral membrane,** the side of the epithelial cell facing the extracellular fluid, may also have folds that increase the cell's surface area.

3. **Cell junctions.** The cells of transporting epithelia are firmly attached to adjacent cells by moderately tight to very tight junctions. This means that to cross the epithelium, material must move into an epithelial cell on one side of the tissue and out of the cell on the other side.

4. **Cell organelles.** Most cells that transport materials have numerous mitochondria to provide energy for transport processes [discussed further in Chapter 5]. The properties of transporting epithelia differ depending on where in the body the epithelia are located. For example, glucose can cross the epithelium of the small intestine and enter the extracellular fluid but cannot cross the epithelium of the large intestine.

The transport properties of an epithelium can be regulated and modified in response to various stimuli. Hormones, for example, affect the transport of ions by kidney epithelium. You will learn more about transporting epithelia when you study the kidney and digestive systems.

Ciliated Epithelia **Ciliated epithelia** are nontransporting tissues that line the respiratory system and parts of the female reproductive tract. The surface of the tissue facing the lumen is covered with cilia that beat in a coordinated, rhythmic fashion, moving fluid and particles across the surface of the tissue (Fig. 3.10c). Injury to the cilia or to their epithelial cells can stop ciliary movement. For example, smoking paralyzes the ciliated epithelium lining the respiratory tract. Loss of ciliary function contributes to the higher incidence of respiratory infection in smokers, when the mucus that traps bacteria can no longer be swept out of the lungs by the cilia.

Protective Epithelia The **protective epithelia** prevent exchange between the internal and external environments and protect areas subject to mechanical or chemical stresses. These epithelia are stratified tissues, composed of many stacked layers of cells (Fig. 3.10d). Protective epithelia may be toughened by the secretion of *keratin* {*keras,* horn}, the same insoluble protein abundant in hair and nails. The *epidermis* {*epi,* upon + *derma,* skin} and linings of the mouth, pharynx, esophagus, urethra, and vagina are all protective epithelia.

Because protective epithelia are subjected to irritating chemicals, bacteria, and other destructive forces, the cells in them have a short life span. In deeper layers, new cells are produced continuously, displacing older cells at the surface. Each time you wash your face, you scrub off dead cells on the surface layer. As skin ages, the rate of cell turnover declines. *Retinoids,* a group of chemicals derived from vitamin A, speed up cell division and surface shedding so treated skin develops a more youthful appearance.

Secretory Epithelia **Secretory epithelia** are composed of cells that produce a substance and then secrete it into the extracellular space. Secretory cells may be scattered among other epithelial cells, or they may group together to form a multicellular **gland.** There are two types of secretory glands: exocrine and endocrine.

Exocrine glands release their secretions to the body's external environment {*exo-,* outside + *krinein,* to secrete}. This may be onto the surface of the skin or onto an epithelium lining one of the internal passageways, such as the airways of the lung or the lumen of the intestine (Fig. 3.10e). In effect, an exocrine secretion

RUNNING **PROBLEM**

Many kinds of cancer develop in epithelial cells that are subject to damage or trauma. The cervix consists of two types of epithelia. Columnar secretory epithelium with mucus-secreting glands lines the inside of the cervical canal. A protective stratified squamous epithelium covers the outside of the cervix. At the opening of the cervix, these two types of epithelia come together. In many cases, infections caused by the human papillomavirus (HPV) cause the cervical cells to develop dysplasia. Dr. Baird ran an HPV test on Jan's first Pap smear, and it was positive for the virus. Today she is repeating the tests to see if Jan's dysplasia and HPV infection have persisted.

Q3: *What other kinds of damage or trauma are cervical epithelial cells normally subjected to?*

Q4: *Which of the two types of cervical epithelia is more likely to be affected by physical trauma?*

Q5: *The results of Jan's first Pap test showed atypical squamous cells of unknown significance (ASC-US). Were these cells more likely to come from the secretory portion of the cervix or from the protective epithelium?*

59 61 70 79 85 87

leaves the body. This explains how some exocrine secretions, like stomach acid, can have a pH that is incompatible with life [Fig. 2.9, p. 45].

Most exocrine glands release their products through open tubes known as **ducts.** Sweat glands, mammary glands in the breast, salivary glands, the liver, and the pancreas are all exocrine glands.

Exocrine gland cells produce two types of secretions. **Serous secretions** are watery solutions, and many of them contain enzymes. Tears, sweat, and digestive enzyme solutions are all serous exocrine secretions. **Mucous secretions** (also called **mucus**) are sticky solutions containing glycoproteins and proteoglycans. Some exocrine glands contain more than one type of secretory cell, and they produce both serous and mucous secretions. For example, the salivary glands release mixed secretions.

Goblet cells, shown in Figure 3.10e, are single exocrine cells that produce mucus. Mucus acts as a lubricant for food to be swallowed, as a trap for foreign particles and microorganisms inhaled or ingested, and as a protective barrier between the epithelium and the environment.

Unlike exocrine glands, **endocrine glands** are ductless and release their secretions, called **hormones,** into the body's extracellular compartment (Fig. 3.9b). Hormones enter the blood for distribution to other parts of the body, where they regulate or coordinate the activities of various tissues, organs, and organ systems. Some of the best-known endocrine glands are the pancreas, the thyroid gland, the gonads, and the pituitary gland. For years, it was thought that all hormones were produced by cells grouped together into endocrine glands. We now know that isolated endocrine cells occur scattered in the epithelial lining of the digestive tract, in the tubules of the kidney, and in the walls of the heart.

FIGURE 3.11 shows the epithelial origin of endocrine and exocrine glands. During embryonic development, epithelial cells grow downward into the supporting connective tissue. Exocrine glands remain connected to the parent epithelium by a duct that transports the secretion to its destination (the external environment). Endocrine glands lose the connecting cells and secrete their hormones into the bloodstream.

FIG. 3.11 Development of endocrine and exocrine glands

During development, the region of epithelium destined to become glandular tissue divides downward into the underlying connective tissue.

Epithelium

Connective tissue

Exocrine

Endocrine

Duct

Connecting cells disappear

Exocrine secretory cells

Endocrine secretory cells

Blood vessel

A hollow center, or lumen, forms in exocrine glands, creating a duct that provides a passageway for secretions to move to the surface of the epithelium.

Endocrine glands lose the connecting bridge of cells that links them to the parent epithelium. Their secretions go directly into the bloodstream.

CONCEPT CHECK

18. List the five functional types of epithelia.
19. Define secretion.
20. Name two properties that distinguish endocrine glands from exocrine glands.
21. The basal lamina of epithelium contains the protein fiber laminin. Are the overlying cells attached by focal adhesions or hemidesmosomes?
22. You look at a tissue under a microscope and see a simple squamous epithelium. Can it be a sample of the skin surface? Explain.
23. A cell of the intestinal epithelium secretes a substance into the extracellular fluid, where it is picked up by the blood and carried to the pancreas. Is the intestinal epithelium cell an endocrine or an exocrine cell?

Connective Tissues Provide Support and Barriers

Connective tissues, the second major tissue type, provide structural support and sometimes a physical barrier that, along with specialized cells, helps defend the body from foreign invaders such as bacteria. The distinguishing characteristic of connective tissues is the presence of extensive extracellular matrix containing widely scattered cells that secrete and modify the matrix (**FIG. 3.12**). Connective tissues include blood, the support tissues for the skin and internal organs, and cartilage and bone.

Structure of Connective Tissue The extracellular matrix of connective tissue is a **ground substance** of proteoglycans and water in which insoluble protein fibers are arranged, much like pieces of fruit suspended in a gelatin salad. The consistency of ground substance is highly variable, depending on the type of connective tissue (Fig. 3.12a). At one extreme is the watery matrix of blood, and at the other extreme is the hardened matrix

FIG. 3.12 **ESSENTIALS**

Connective Tissue

(a) Map of Connective Tissue Components

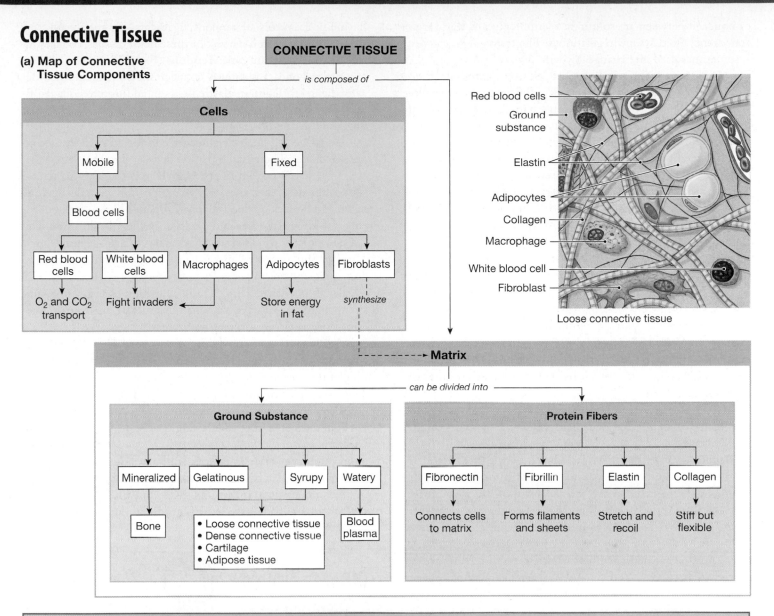

CONNECTIVE TISSUE

is composed of

Cells

Mobile — Fixed

Blood cells

Red blood cells | White blood cells | Macrophages | Adipocytes | Fibroblasts

O₂ and CO₂ transport — Fight invaders — Store energy in fat — *synthesize*

Red blood cells
Ground substance
Elastin
Adipocytes
Collagen
Macrophage
White blood cell
Fibroblast

Loose connective tissue

Matrix

can be divided into

Ground Substance

Mineralized | Gelatinous | Syrupy | Watery

Bone

• Loose connective tissue
• Dense connective tissue
• Cartilage
• Adipose tissue

Blood plasma

Protein Fibers

Fibronectin | Fibrillin | Elastin | Collagen

Connects cells to matrix | Forms filaments and sheets | Stretch and recoil | Stiff but flexible

(b) Types of Connective Tissue

Tissue Name	Ground Substance	Fiber Type and Arrangement	Main Cell Types	Where Found
Loose connective tissue	Gel; more ground substance than fibers or cells	Collagen, elastic, reticular; random	Fibroblasts	Skin, around blood vessels and organs, under epithelia
Dense, irregular connective tissue	More fibers than ground substance	Mostly collagen; random	Fibroblasts	Muscle and nerve sheaths
Dense, regular connective tissue	More fibers than ground substance	Collagen; parallel	Fibroblasts	Tendons and ligaments
Adipose tissue	Very little ground substance	None	Brown fat and white fat	Depends on age and sex
Blood	Aqueous	None	Blood cells	In blood and lymph vessels
Cartilage	Firm but flexible; hyaluronic acid	Collagen	Chondroblasts	Joint surfaces, spine, ear, nose, larynx
Bone	Rigid due to calcium salts	Collagen	Osteoblasts and osteocytes	Bones

of bone. In between are solutions of proteoglycans that vary in consistency from syrupy to gelatinous. The term *ground substance* is sometimes used interchangeably with *matrix*.

Connective tissue cells lie embedded in the extracellular matrix. These cells are described as *fixed* if they remain in one place and as *mobile* if they can move from place to place. **Fixed cells** are responsible for local maintenance, tissue repair, and energy storage. **Mobile cells** are responsible mainly for defense. The distinction between fixed and mobile cells is not absolute, because at least one cell type is found in both fixed and mobile forms.

Extracellular matrix is nonliving, but the connective tissue cells constantly modify it by adding, deleting, or rearranging molecules. The suffix *-blast* {*blastos,* sprout} on a connective tissue cell name often indicates a cell that is either growing or actively secreting extracellular matrix. **Fibroblasts,** for example, are connective tissue cells that secrete collagen-rich matrix. Cells that are actively breaking down matrix are identified by the suffix *-clast* {*klastos,* broken}. Cells that are neither growing, secreting matrix components, nor breaking down matrix may be given the suffix *-cyte,* meaning "cell." Remembering these suffixes should help you remember the functional differences between cells with similar names, such as the osteoblast, osteocyte, and osteoclast, three cell types found in bone.

In addition to secreting proteoglycan ground substance, connective tissue cells produce matrix fibers. Four types of fiber proteins are found in matrix, aggregated into insoluble fibers. **Collagen** {*kolla,* glue + *-genes,* produced} is the most abundant protein in the human body, almost one-third of the body's dry weight. Collagen is also the most diverse of the four protein types, with at least 12 variations. It is found almost everywhere connective tissue is found, from the skin to muscles and bones. Individual collagen molecules pack together to form collagen fibers, flexible but inelastic fibers whose strength per unit weight exceeds that of steel. The amount and arrangement of collagen fibers help determine the mechanical properties of different types of connective tissues.

Three other protein fibers in connective tissue are elastin, fibrillin, and fibronectin. **Elastin** is a coiled, wavy protein that returns to its original length after being stretched. This property is known as *elastance.* Elastin combines with the very thin, straight fibers of **fibrillin** to form filaments and sheets of elastic fibers. These two fibers are important in elastic tissues such as the lungs, blood vessels, and skin. As mentioned earlier, **fibronectin** connects cells to extracellular matrix at focal adhesions. Fibronectins also play an important role in wound healing and in blood clotting.

Types of Connective Tissue Figure 3.12b compares the properties of different types of connective tissue. The most common types are loose and dense connective tissue, adipose tissue, blood, cartilage, and bone. By many estimates, connective tissues are the most abundant of the tissue types as they are a component of most organs.

Loose connective tissues (**FIG. 3.13a**) are the elastic tissues that underlie skin and provide support for small glands. **Dense connective tissues** (irregular and regular) provide strength or

flexibility. Examples are tendons, ligaments, and the sheaths that surround muscles and nerves. In these dense tissues, collagen fibers are the dominant type. **Tendons** (Fig. 3.13c) attach skeletal muscles to bones. **Ligaments** connect one bone to another. Because ligaments contain elastic fibers in addition to collagen fibers, they have a limited ability to stretch. Tendons lack elastic fibers and so cannot stretch.

Cartilage and bone together are considered supporting connective tissues. These tissues have a dense ground substance that contains closely packed fibers. **Cartilage** is found in structures such as the nose, ears, knee, and windpipe. It is solid, flexible, and notable for its lack of blood supply. Without a blood supply, nutrients and oxygen must reach the cells of cartilage by diffusion. This is a slow process, which means that damaged cartilage heals slowly.

The fibrous extracellular matrix of **bone** is said to be *calcified* because it contains mineral deposits, primarily calcium salts, such as calcium phosphate (Fig. 3.13b). These minerals give the bone strength and rigidity. We examine the structure and formation of bone along with calcium metabolism later [Chapter 23].

Adipose tissue is made up of **adipocytes,** or fat cells. An adipocyte of **white fat** typically contains a single enormous lipid droplet that occupies most of the volume of the cell (Fig. 3.13e). This is the most common form of adipose tissue in adults.

Brown fat is composed of adipose cells that contain multiple lipid droplets rather than a single large droplet. This type of fat has been known for many years to play an important role in temperature regulation in infants. Until recently it was thought to be almost completely absent in adults. However, modern imaging techniques such as combined CT and PET scans have revealed that adults do have brown fat [discussed in more detail in Chapter 22].

BIOTECHNOLOGY

Grow Your Own Cartilage

Have you torn the cartilage in your knee playing basketball or some other sport? Maybe you won't need surgery to repair it. Replacing lost or damaged cartilage has moved from the realm of science fiction into reality. Researchers have developed a process in which they take a cartilage sample from a patient and put it into a tissue culture medium to reproduce. Once the culture has grown enough *chondrocytes*—the cells that synthesize the extracellular matrix of cartilage—the mixture is sent back to a physician, who surgically places the cells in the patient's knee at the site of cartilage damage. Once returned to the body, the chondrocytes secrete matrix and help repair the damaged cartilage. Because the person's own cells are grown and reimplanted, there is no tissue rejection. A different method for cartilage repair being used outside the United States is treatment with stem cells derived from bone marrow. Both therapies have proved to be effective treatments for selected cartilage problems.

FIG. 3.13 **ESSENTIALS**

Types of Connective Tissue

(a) Loose Connective Tissue

Loose connective tissue is very flexible, with multiple cell types and fibers.

Collagen fibers

Fibroblasts are cells that secrete matrix proteins.

Free macrophage

Elastic fibers

Ground substance is the matrix of loose connective tissue.

Light micrograph of loose connective tissue

(b) Bone and Cartilage

Hard bone forms when osteo-blasts deposit calcium phosphate crystals in the matrix. Cartilage has firm but flexible matrix secreted by cells called chondrocytes.

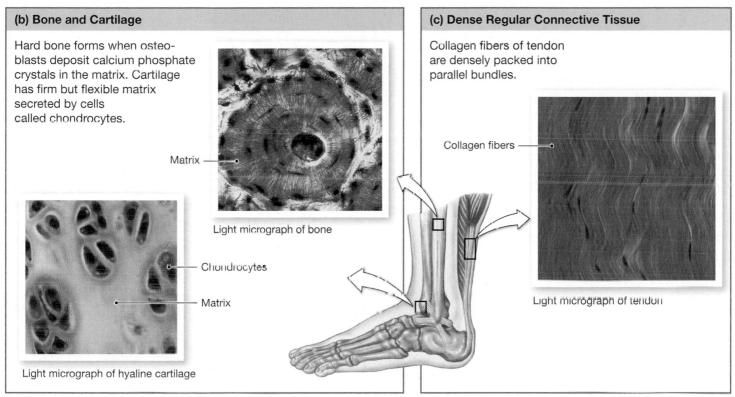

Matrix

Light micrograph of bone

Chondrocytes

Matrix

Light micrograph of hyaline cartilage

(c) Dense Regular Connective Tissue

Collagen fibers of tendon are densely packed into parallel bundles.

Collagen fibers

Light micrograph of tendon

(d) Blood

Blood consists of liquid matrix (*plasma*) plus red and white blood cells and the cell fragments called platelets.

Red blood cell
Platelet

White Blood Cells
— Lymphocyte
— Neutrophil
— Eosinophil

Light micrograph of a blood smear

(e) Adipose Tissue

In white fat, the cell cytoplasm is almost entirely filled with lipid droplets.

Nucleus

Lipid droplets

Light micrograph of adipose tissue

Blood is an unusual connective tissue that is characterized by its watery extracellular matrix called *plasma*. Plasma consists of a dilute solution of ions and dissolved organic molecules, including a large variety of soluble proteins. Blood cells and cell fragments are suspended in the plasma (Fig. 3.13d), but the insoluble protein fibers typical of other connective tissues are absent. [We discuss blood in Chapter 16.]

CONCEPT CHECK

24. What is the distinguishing characteristic of connective tissues?
25. Name four types of protein fibers found in connective tissue matrix and give the characteristics of each.
26. Name six types of connective tissues.
27. Blood is a connective tissue with two components: plasma and cells. Which of these is the matrix in this connective tissue?
28. Why does torn cartilage heal more slowly than a cut in the skin?

Muscle and Neural Tissues Are Excitable

The third and fourth of the body's four tissue types—muscle and neural—are collectively called the *excitable tissues* because of their ability to generate and propagate electrical signals called *action potentials*. Both of these tissue types have minimal extracellular matrix, usually limited to a supportive layer called the *external lamina*. Some types of muscle and nerve cells are also notable for their gap junctions, which allow the direct and rapid conduction of electrical signals from cell to cell.

Muscle tissue has the ability to contract and produce force and movement. The body contains three types of muscle tissue: cardiac muscle in the heart; smooth muscle, which makes up most internal organs; and skeletal muscle. Most skeletal muscles attach to bones and are responsible for gross movement of the body. [We discuss muscle tissue in more detail in Chapter 12.]

Neural tissue has two types of cells. **Neurons,** or nerve cells, carry information in the form of chemical and electrical signals from one part of the body to another. They are concentrated in the brain and spinal cord but also include a network of cells that extends to virtually every part of the body. **Glial cells,** or neuroglia, are the support cells for neurons. [We discuss the anatomy of neural tissue in Chapter 8.] A summary of the characteristics of the four tissue types can be found in **TABLE 3.4**.

TISSUE REMODELING

Most people associate growth with the period from birth to adulthood. However, cell birth, growth, and death continue throughout a person's life. The tissues of the body are constantly remodeled as cells die and are replaced.

Apoptosis Is a Tidy Form of Cell Death

Cell death occurs two ways, one messy and one tidy. In **necrosis,** cells die from physical trauma, toxins, or lack of oxygen when their blood supply is cut off. Necrotic cells swell, their organelles deteriorate, and finally the cells rupture. The cell contents released this way include digestive enzymes that damage adjacent cells and trigger an inflammatory response. You see necrosis when you have a red area of skin surrounding a scab.

TABLE 3.4	Characteristics of the Four Tissue Types			
	Epithelial	**Connective**	**Muscle**	**Nerve**
Matrix Amount	Minimal	Extensive	Minimal	Minimal
Matrix Type	Basal lamina	Varied—protein fibers in ground substance that ranges from liquid to gelatinous to firm to calcified	External lamina	External lamina
Unique Features	No direct blood supply	Cartilage has no blood supply	Able to generate electrical signals, force, and movement	Able to generate electrical signals
Surface Features of Cells	Microvilli, cilia	N/A	N/A	N/A
Locations	Covers body surface; lines cavities and hollow organs, and tubes; secretory glands	Supports skin and other organs; cartilage, bone, and blood	Makes up skeletal muscles, hollow organs, and tubes	Throughout body; concentrated in brain and spinal cord
Cell Arrangement and Shapes	Variable number of layers, from one to many; cells flattened, cuboidal, or columnar	Cells not in layers; usually randomly scattered in matrix; cell shape irregular to round	Cells linked in sheets or elongated bundles; cells shaped in elongated, thin cylinders; heart muscle cells may be branched	Cells isolated or networked; cell appendages highly branched and/or elongated

In contrast, cells that undergo *programmed cell death,* or **apoptosis** {ap-oh-TOE-sis or a-pop-TOE-sis; *apo-,* apart, away + *ptosis,* falling}, do not disrupt their neighbors when they die. Apoptosis, also called cell suicide, is a complex process regulated by multiple chemical signals. Some signals keep apoptosis from occurring, while other signals tell the cell to self-destruct. When the suicide signal wins out, chromatin in the nucleus condenses, and the cell pulls away from its neighbors. It shrinks, then breaks up into tidy membrane-bound *blebs* that are gobbled up by neighboring cells or by wandering cells of the immune system.

Apoptosis is a normal event in the life of an organism. During fetal development, apoptosis removes unneeded cells, such as half the cells in the developing brain and the webs of skin between fingers and toes. In adults, cells that are subject to wear and tear from exposure to the outside environment may live only a day or two before undergoing apoptosis. For example, it has been estimated that the intestinal epithelium is completely replaced with new cells every two to five days.

CONCEPT CHECK

29. What are some features of apoptosis that distinguish it from cell death due to injury?

Stem Cells Can Create New Specialized Cells

If cells in the adult body are constantly dying, where do their replacements come from? This question is still being answered and is one of the hottest topics in biological research today. The following paragraphs describe what we currently know.

All cells in the body are derived from the single cell formed at conception. That cell and those that follow reproduce themselves by undergoing the cell division process known as **mitosis** [see Appendix C]. The very earliest cells in the life of a human

The day after Jan's visit, the computerized cytology analysis system rapidly scans the cells on the slide of Jan's cervical tissue, looking for abnormal cell size or shape. The computer is programmed to find multiple views for the cytologist to evaluate. The results of Jan's two Pap tests are shown in **FIGURE 3.14**.

Q6: *Has Jan's dysplasia improved or worsened? What evidence do you have to support your answer?*

Q7: *Use your answer to question 6 to predict whether Jan's HPV infection has persisted or been cleared by her immune system.*

59　61　70　79　**85**　87

being are said to be **totipotent** {*totus,* entire} because they have the ability to develop into any and all types of specialized cells. Any totipotent cell has the potential to become a functioning organism.

After about day 4 of development, the totipotent cells of the embryo begin to specialize, or *differentiate.* As they do so, they narrow their potential fates and become **pluripotent** {*plures,* many}. Pluripotent cells can develop into many different cell types but not all cell types. An isolated pluripotent cell cannot develop into an organism.

As differentiation continues, pluripotent cells develop into the various tissues of the body. As the cells specialize and mature, many lose the ability to undergo mitosis and reproduce themselves. They can be replaced, however, by new cells created from **stem cells,** less specialized cells that retain the ability to divide.

Undifferentiated stem cells in a tissue that retain the ability to divide and develop into the cell types of that tissue are said to be

---

FIG. 3.14　Pap smears of cervical cells

(a) Jan's abnormal Pap test.

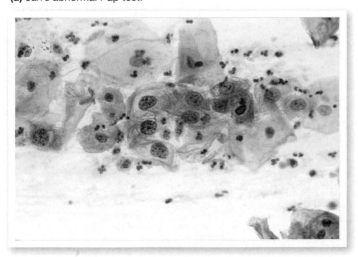

(b) Jan's second Pap test. Are these cells normal or abnormal?

FIG. 3.15 **FOCUS ON . . .**

The Skin

The Layers of the Skin

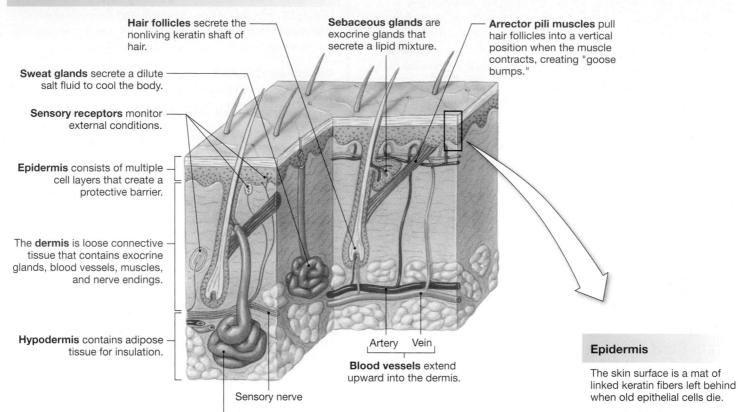

Hair follicles secrete the nonliving keratin shaft of hair.

Sebaceous glands are exocrine glands that secrete a lipid mixture.

Arrector pili muscles pull hair follicles into a vertical position when the muscle contracts, creating "goose bumps."

Sweat glands secrete a dilute salt fluid to cool the body.

Sensory receptors monitor external conditions.

Epidermis consists of multiple cell layers that create a protective barrier.

The **dermis** is loose connective tissue that contains exocrine glands, blood vessels, muscles, and nerve endings.

Hypodermis contains adipose tissue for insulation.

Artery   Vein

Blood vessels extend upward into the dermis.

Sensory nerve

Apocrine glands in the genitalia, anus, axillae (*axilla*, armpit), and eyelids release waxy or viscous milky secretions in response to fear or sexual excitement.

CLINICAL FOCUS

Melanoma Is a Serious Form of Skin Cancer

Melanoma occurs when melanocytes become malignant, often following repeated exposure to UV light. One study found that people who used tanning beds were 24% more likely to develop melanoma.

Epidermis

The skin surface is a mat of linked keratin fibers left behind when old epithelial cells die.

Phospholipid matrix acts as the skin's main waterproofing agent.

Surface keratinocytes produce keratin fibers.

Desmosomes anchor epithelial cells to each other.

Epidermal cell

Melanocytes contain the pigment melanin.

Basal lamina

Connection between Epidermis and Dermis

Hemidesmosomes tie epidermal cells to fibers of the basal lamina.

Basal lamina or basement membrane is an acellular layer between epidermis and dermis.

multipotent {*multi*, many}. Some of the most-studied multipotent adult stem cells are found in bone marrow and give rise to blood cells. However, all adult stem cells occur in very small numbers. They are difficult to isolate and do not thrive in the laboratory.

Biologists once believed that nerve and muscle cells, which are highly specialized in their mature forms, could not be replaced when they died. Now research indicates that stem cells for these tissues do exist in the body. However, naturally occurring neural and muscle stem cells are so scarce that they cannot replace large masses of dead or dying tissue that result from diseases such as strokes or heart attacks. Consequently, one goal of stem cell research is to find a source of pluripotent or multipotent stem cells that could be grown in the laboratory. If stem cells could be grown in larger numbers, they could be implanted to treat damaged tissues and *degenerative* diseases, those in which cells degenerate and die.

One example of a degenerative disease is Parkinson's disease, in which certain types of nerve cells in the brain die. Embryos and fetal tissue are rich sources of stem cells, but the use of embryonic stem cells is controversial and poses many legal and ethical questions. Some researchers hope that adult stem cells will show **plasticity,** the ability to specialize into a cell of a type different from the type for which they were destined.

There are still many challenges facing us before stem cell therapy becomes a standard medical treatment. One is finding a good source of stem cells. A second major challenge is determining the chemical signals that tell stem cells when to differentiate and what type of cell to become. And even once these two challenges are overcome and donor stem cells are implanted, the body may recognize that the new cells are foreign tissue and try to reject them.

Stem cell research is an excellent example of the dynamic and often controversial nature of science. For the latest research findings, as well as pending legislation and laws regulating stem cell research and use, check authoritative web sites, such as that sponsored by the U.S. National Institutes of Health (*http://stemcells.nih.gov/Pages/Default.aspx*).

ORGANS

Groups of tissues that carry out related functions may form structures known as **organs.** The organs of the body contain the four types of tissue in various combinations. The skin is an excellent example of an organ that incorporates all four types of tissue into an integrated whole. We think of skin as a thin layer that covers the external surfaces of the body, but in reality it is the heaviest single organ, at about 16% of an adult's total body weight! If it were flattened out, it would cover a surface area of between 1.2 and 2.3 square meters, about the size of a couple of card-table tops. Its size and weight make skin one of the most important organs of the body.

The functions of the skin do not fit neatly into any one chapter of this book, and this is true of some other organs as well. We will highlight several of these organs in special organ *Focus* features throughout the book. These illustrated boxes discuss the structure and functions of these versatile organs so that you can gain an appreciation for the way different tissues combine for a united purpose. The first of these features, *Focus On: The Skin*, appears as **FIGURE 3.15**.

As we consider the systems of the body in the succeeding chapters, you will see how diverse cells, tissues, and organs carry out the processes of the living body. Although the body's cells have different structures and different functions, they have one need in common: a continuous supply of energy. Without energy, cells cannot survive, let alone carry out all the other processes of daily living. Next, we look at energy in living organisms and how cells capture and use the energy released by chemical reactions.

RUNNING **PROBLEM CONCLUSION** | Pap Tests Save Lives

In this running problem, you learned that the Pap test can detect the early cell changes that precede cervical cancer. The diagnosis is not always simple because the change in cell cytology from normal to cancerous occurs along a continuum and can be subject to individual interpretation. In addition, not all cell changes are cancerous. The human papillomavirus (HPV), a common sexually transmitted infection, can also cause cervical dysplasia. In most cases, the woman's immune system overcomes the virus within two years, and the cervical cells revert to normal. A small number of women with persistent HPV infections have a higher risk of developing cervical cancer, however. Studies indicate that 98% of cervical cancers are associated with HPV infection. To learn more about the association between HPV and cervical cancer, go to the National Cancer Institute homepage (*www.cancer.gov*) and search for HPV. This site also contains information about cervical cancer. To check your understanding of the running problem, compare your answers with the information in the following summary table.

Question	Facts	Integration and Analysis
Q1: *Why does the treatment of cancer focus on killing the cancerous cells?*	Cancerous cells divide uncontrollably and fail to coordinate with normal cells. Cancerous cells fail to differentiate into specialized cells.	Unless removed, cancerous cells will displace normal cells. This may cause destruction of normal tissues. In addition, because cancerous cells do not become specialized, they cannot carry out the same functions as the specialized cells they displace.

—Continued next page

RUNNING **PROBLEM CONCLUSION** | *Continued*

Question	Facts	Integration and Analysis
Q2: *What is happening in cancer cells that explains the large size of their nucleus and the relatively small amount of cytoplasm?*	Cancerous cells divide uncontrollably. Dividing cells must duplicate their DNA prior to cell division, and this DNA duplication takes place in the nucleus, leading to the large size of that organelle [Appendix C].	Actively reproducing cells are likely to have more DNA in their nucleus as they prepare to divide, so their nuclei tend to be larger. Each cell division splits the cytoplasm between two daughter cells. If division is occurring rapidly, the daughter cells may not have time to synthesize new cytoplasm, so the amount of cytoplasm is less than in a normal cell.
Q3: *What other kinds of damage or trauma are cervical epithelial cells normally subjected to?*	The cervix is the passageway between the uterus and vagina.	The cervix is subject to trauma or damage, such as might occur during sexual intercourse and childbirth.
Q4: *Which of its two types of epithelia is more likely to be affected by trauma?*	The cervix consists of secretory epithelium with mucus-secreting glands lining the inside and protective epithelium covering the outside.	Protective epithelium is composed of multiple layers of cells and is designed to protect areas from mechanical and chemical stress [p. 79]. Therefore, the secretory epithelium with its single-cell layer is more easily damaged.
Q5: *Jan's first Pap test atypical squamous cells of unknown significance (ASCUS). Were these cells more likely to come from the secretory portion of the cervix or from the protective epithelium?*	Secretory cells are columnar epithelium. Protective epithelium is composed of multiple cell layers.	Protective epithelium with multiple cell layers has cells that are flat (stratified squamous epithelium). The designation ASC refers to these protective epithelial cells.
Q6: *Has Jan's dysplasia improved or worsened? What evidence do you have to support your answer?*	The slide from Jan's first Pap test shows abnormal cells with large nuclei and little cytoplasm. These abnormal cells do not appear in the second test.	The disappearance of the abnormal cells indicates that Jan's dysplasia has resolved. She will return in another year for a repeat Pap test. If it shows no dysplasia, her cervical cells have reverted to normal.
Q7: *Use your answer to question 6 to predict whether Jan's HPV infection has persisted or been cleared by her immune system.*	The cells in the second Pap test appear normal.	Once Jan's body fights off the HPV infection, her cervical cells should revert to normal. Her second HPV test should show no evidence of HPV infection.

59 61 70 79 85 **87**

VISUALIZE, PRACTICE, AND MASTER PHYSIOLOGY WITH MASTERING A&P
- Practice-on-the-go with Dynamic Study Modules
- Visualize important processes with Interactive Physiology and A&P Flix
- Check your understanding with Video Tutors and practice quizzes!

MasteringA&P®

CHAPTER SUMMARY

Cell biology and histology illustrate one of the major themes in physiology: *compartmentation*. In this chapter, you learned how a cell is subdivided into two main compartments—the nucleus and the cytoplasm. You also learned how cells form tissues that create larger compartments within the body. A second theme in this chapter is the *molecular interactions* that create the *mechanical properties* of cells and tissues. Protein fibers of the cytoskeleton and cell junctions, along with the molecules that make up the extracellular matrix, form the "glue" that holds tissues together.

Functional Compartments of the Body

iP Fluids and Electrolytes: Introduction to Body Fluids

1. The **cell** is the functional unit of living organisms. (p. 59)

2. The major human body cavities are the cranial cavity (skull), thoracic cavity (thorax), and abdominopelvic cavity. (p. 60; Fig. 3.1a)

3. The **lumens** of some hollow organs are part of the body's external environment. (p. 61)

4. The body fluid compartments are the extracellular fluid (ECF) outside the cells and the intracellular fluid (ICF) inside the cells. The ECF can be subdivided into **interstitial fluid** bathing the cells and **plasma,** the fluid portion of the blood. (p. 60; Fig. 3.1b)

Biological Membranes

5. The word *membrane* is used both for cell membranes and for tissue membranes that line a cavity or separate two compartments. (p. 60; Fig. 3.1c)

6. The **cell membrane** acts as a barrier between the intracellular and extracellular fluids, provides structural support, and regulates exchange and communication between the cell and its environment. (p. 61)

7. The **fluid mosaic model** of a biological membrane shows it as a **phospholipid bilayer** with proteins inserted into the bilayer. (p. 62; Fig. 3.2b)

8. Membrane lipids include phospholipids, **sphingolipids,** and cholesterol. **Lipid-anchored proteins** attach to membrane lipids. (p. 62)

9. **Transmembrane proteins** are **integral proteins** tightly bound to the phospholipid bilayer. **Peripheral proteins** attach less tightly to either side of the membrane. (p. 64; Fig. 3.2b, c)

10. Carbohydrates attach to the extracellular surface of cell membranes. (p. 64)

Intracellular Compartments

11. The cytoplasm consists of semi-gelatinous **cytosol** with dissolved nutrients, ions, and waste products. Suspended in the cytosol are the other components of the cytoplasm: insoluble **inclusions** and fibers, which have no enclosing membrane, and **organelles,** which are membrane-enclosed bodies that carry out specific functions. (p. 65; Fig. 3.4a)

12. **Ribosomes** are inclusions that take part in protein synthesis. (p. 67)

13. Insoluble protein fibers come in three sizes: **actin fibers** (also called **microfilaments**), **intermediate filaments,** and **microtubules.** (p. 68; Tbl. 3.2)

14. **Centrioles** that aid the movement of chromosomes during cell division, **cilia** that move fluid or secretions across the cell surface, and **flagella** that propel sperm through body fluids are made of microtubules. (p. 68; Figs. 3.4e, 3.5)

15. The changeable **cytoskeleton** provides strength, support, and internal organization; aids transport of materials within the cell; links cells together; and enables motility in certain cells. (p. 68; Fig. 3.4b)

16. **Motor proteins** such as **myosins, kinesins,** and **dyneins** associate with cytoskeleton fibers to create movement. (p. 69; Fig. 3.6)

17. Membranes around organelles create compartments that separate functions. (p. 70)

18. **Mitochondria** generate most of the cell's ATP. (p. 70; Fig. 3.4g)

19. The **smooth endoplasmic reticulum** is the primary site of lipid synthesis. The **rough endoplasmic reticulum** is the primary site of protein synthesis. (p. 71; Fig. 3.4i)

20. The **Golgi apparatus** packages proteins into vesicles. **Secretory vesicles** release their contents into the extracellular fluid. (p. 71; Fig. 3.4h)

21. **Lysosomes** and **peroxisomes** are small **storage vesicles** that contain digestive enzymes. (p. 71; Figs. 3.4c, d)

22. The **nucleus** contains DNA, the genetic material that ultimately controls all cell processes, in the form of **chromatin.** The double-membrane **nuclear envelope** surrounding the nucleus has **nuclear pore complexes** that allow controlled chemical communication between the nucleus and cytosol. **Nucleoli** are nuclear areas that control the synthesis of RNA for ribosomes. (p. 72; Fig. 3.4j)

23. Protein synthesis is an example of how the cell separates functions by isolating them to separate compartments within the cell (p. 72; Fig. 3.7)

Tissues of the Body

iP Muscular: Anatomy Review—Skeletal Muscle Tissue

24. There are four primary tissue types in the human body: epithelial, connective, muscle, and neural. (p. 72)

25. **Extracellular matrix** secreted by cells provides support and a means of cell-cell communication. It is composed of proteoglycans and insoluble protein fibers. (p. 72)

26. Animal cell junctions fall into three categories. **Gap junctions** allow chemical and electrical signals to pass directly from cell to cell. **Tight junctions** restrict the movement of material between cells. **Anchoring junctions** hold cells to each other or to the extracellular matrix. (p. 74; Fig. 3.8)

27. Membrane proteins called **cell adhesion molecules** (CAMs) are essential in cell adhesion and in anchoring junctions. (p. 72; Tbl. 3.3)

28. **Desmosomes** and **adherens junctions** anchor cells to each other. **Focal adhesions** and **hemidesmosomes** anchor cells to matrix. (p. 74; Fig. 3.8)

29. **Epithelial tissues** protect the internal environment, regulate the exchange of material, or manufacture and secrete chemicals. There are five functional types found in the body: exchange, transporting, ciliated, protective, and secretory. (p. 76; Fig. 3.9)

30. **Exchange epithelia** permit rapid exchange of materials, particularly gases. **Transporting epithelia** actively regulate the selective exchange of nongaseous materials between the internal and external environments. **Ciliated epithelia** move fluid and particles across the surface of the tissue. **Protective epithelia** help prevent exchange between the internal and external environments. The **secretory epithelia** release secretory products into the external environment or the blood. (p. 76; Fig. 3.10)

31. **Exocrine glands** release their secretions into the external environment through **ducts. Endocrine glands** are ductless glands that release their secretions, called **hormones,** directly into the extracellular fluid. (p. 79; Fig. 3.9b)

32. **Connective tissues** have extensive extracellular matrix that provides structural support and forms a physical barrier. (p. 80; Fig. 3.12)

33. **Loose connective tissues** are the elastic tissues that underlie skin. **Dense connective tissues,** including **tendons** and **ligaments,**

have strength or flexibility because they are made of collagen. **Adipose tissue** stores fat. The connective tissue we call **blood** is characterized by a watery matrix. **Cartilage** is solid and flexible and has no blood supply. The fibrous matrix of **bone** is hardened by deposits of calcium salts. (p. 82; Fig. 3.13)

34. Muscle and neural tissues are called excitable tissues because of their ability to generate and propagate electrical signals called action potentials. **Muscle tissue** has the ability to contract and produce force and movement. There are three types of muscle: cardiac, smooth, and skeletal. (p. 84)

35. **Neural tissue** includes **neurons,** which use electrical and chemical signals to transmit information from one part of the body to another, and support cells known as **glial cells** (neuroglia). (p. 84)

Tissue Remodeling

36. Cell death occurs by **necrosis,** which adversely affects neighboring cells, and by **apoptosis,** programmed cell death that does not disturb the tissue. (p. 84)

37. **Stem cells** are cells that are able to reproduce themselves and differentiate into specialized cells. Stem cells are most plentiful in embryos but are also found in the adult body. (p. 85)

Organs

38. **Organs** are formed by groups of tissues that carry out related functions. The organs of the body contain the four types of tissues in various ratios. For example, skin is largely connective tissue. (p. 87)

REVIEW QUESTIONS

In addition to working through these questions and checking your answers on p. A-4, review the Learning Outcomes at the beginning of this chapter.

Level One Reviewing Facts and Terms

1. List the four general functions of the cell membrane.

2. In 1972, Singer and Nicolson proposed the fluid mosaic model of the cell membrane. According to this model, the membrane is composed of a bilayer of _____ and a variety of embedded _____, with _____ on the extracellular surface.

3. What are the two primary types of biomolecules found in the cell membrane?

4. Define and distinguish between inclusions and organelles. Give an example of each.

5. Define cytoskeleton. List five functions of the cytoskeleton.

6. Match each term with the description that fits it best:

(a) cilia	1. in human cells, appears as single, long, whiplike tail
(b) centriole	2. short, hairlike structures that beat to produce currents in fluids
(c) flagellum	3. a bundle of microtubules that aid in mitosis
(d) centrosome	4. the microtubule-organizing center

7. Exocrine glands produce watery secretions (such as tears or sweat) called _____ secretions, or stickier solutions called _____ secretions.

8. Match each organelle with its function:

(a) endoplasmic reticulum	1. powerhouse of the cell where most ATP is produced
(b) Golgi apparatus	2. degrades long-chain fatty acids and toxic foreign molecules
(c) lysosome	3. network of membranous tubules that synthesize biomolecules
(d) mitochondrion	4. digestive system of cell, degrading or recycling components
(e) peroxisome	5. modifies and packages proteins into vesicles

9. What process activates the enzymes inside lysosomes?

10. _____ glands release hormones, which enter the blood and regulate the activities of organs or systems.

11. List the four major tissue types. Give an example and location of each.

12. The largest and heaviest organ in the body is the _____.

13. Match each protein to its function. Functions in the list may be used more than once.

(a) cadherin	1. membrane protein used to form cell junctions
(b) CAM	
(c) collagen	2. matrix glycoprotein used to anchor cells
(d) connexin	
(e) elastin	3. protein found in gap junctions
(f) fibrillin	
(g) fibronectin	4. matrix protein found in connective tissue
(h) integrin	
(i) occludin	

14. What types of glands can be found within the skin? Name the secretion of each type.

15. The term *matrix* can be used in reference to an organelle or to tissues. Compare the meanings of the term in these two contexts.

Level Two Reviewing Concepts

16. List, compare, and contrast the three types of cell junctions and their subtypes. Give an example of where each type can be found in the body and describe its function in that location.

17. Which would have more rough endoplasmic reticulum: pancreatic cells that manufacture the protein hormone insulin, or adrenal cortex cells that synthesize the steroid hormone cortisol?

18. A number of organelles can be considered vesicles. Define *vesicle* and describe at least three examples.

19. Explain why a stratified epithelium offers more protection than a simple epithelium.

20. **Mapping exercise:** Transform this list of terms into a map of cell structure. Add functions where appropriate.

• actin	• microfilament
• cell membrane	• microtubule
• centriole	• mitochondria
• cilia	• nonmembranous organelle
• cytoplasm	• nucleus
• cytoskeleton	• organelle
• cytosol	• peroxisome
• extracellular matrix	• ribosome
• flagella	• rough ER
• Golgi apparatus	• secretory vesicle
• intermediate filament	• smooth ER
• keratin	• storage vesicle
• lysosome	• tubulin

21. Sketch a short series of columnar epithelial cells. Label the apical and basolateral borders of the cells. Briefly explain the different kinds of junctions found on these cells.

22. Arrange the following compartments in the order a glucose molecule entering the body at the intestine would encounter them: interstitial fluid, plasma, intracellular fluid. Which of these fluid compartments is/are considered extracellular fluid(s)?

23. Explain how inserting cholesterol into the phospholipid bilayer of the cell membrane decreases membrane permeability.

24. Compare and contrast the structure, locations, and functions of bone and cartilage.

25. Differentiate between the terms in each set below:
    (a) lumen and wall
    (b) cytoplasm and cytosol
    (c) myosin and keratin

26. When a tadpole turns into a frog, its tail shrinks and is reabsorbed. Is this an example of necrosis or apoptosis? Defend your answer.

27. Match the structures from the chapter to the basic physiological themes in the right column and give an example or explanation for each match. A structure may match with more than one theme.

(a) cell junctions	1. communication
(b) cell membrane	2. molecular interactions
(c) cytoskeleton	3. compartmentation
(d) organelles	4. mechanical properties
(e) cilia	5. biological energy use

28. In some instances, the extracellular matrix can be quite rigid. How might developing and expanding tissues cope with a rigid matrix to make space for themselves?

Level Three Problem Solving

29. One result of cigarette smoking is paralysis of the cilia that line the respiratory passageways. What function do these cilia serve? Based on what you have read in this chapter, why is it harmful when they no longer beat? What health problems would you expect to arise? How does this explain the hacking cough common among smokers?

30. Cancer is abnormal, uncontrolled cell division. What property of epithelial tissues might (and does) make them more prone to developing cancer?

31. What might happen to normal physiological function if matrix metalloproteinases are inhibited by drugs?

Answers to Concept Checks, Figure and Graph Questions, and end-of-chapter Review Questions can be found in Appendix A [A-1].

4

Energy and Cellular Metabolism

There is no good evidence that … life evades the second law of thermodynamics, but in the downward course of the energy-flow it interposes a barrier and dams up a reservoir which provides potential for its own remarkable activities.

F. G. Hopkins, 1933. "Some Chemical Aspects of Life," presidential address to the 1933 meeting of British Association for the Advancement of Science.

ENERGY IN BIOLOGICAL SYSTEMS 93

LO 4.1   Define energy. Describe three categories of work that require energy.

LO 4.2   Distinguish between kinetic and potential energy, and describe potential energy in biological systems.

LO 4.3   Explain the first and second laws of thermodynamics and how they apply to the human body.

CHEMICAL REACTIONS 96

LO 4.4   Describe four common types of chemical reactions.

LO 4.5   Explain the relationships between free energy, activation energy, and endergonic and exergonic reactions.

LO 4.6   Apply the concepts of free energy and activation energy to reversible and irreversible reactions.

ENZYMES 98

LO 4.7   Explain what enzymes are and how they facilitate biological reactions.

LO 4.8   How do the terms *isozyme, coenzyme, proenzyme, zymogen,* and *cofactor* apply to enzymes?

LO 4.9   Name and explain the four major categories of enzymatic reactions.

METABOLISM 102

LO 4.10   Define metabolism, anabolism, and catabolism.

LO 4.11   List five ways cells control the flow of molecules through metabolic pathways.

LO 4.12   Explain the roles of the following molecules in biological energy transfer and storage: ADP, ATP, NADH, $FADH_2$, NADPH.

LO 4.13   Outline the pathways for aerobic and anaerobic metabolism of glucose and compare the energy yields of the two pathways.

LO 4.14   Write two equations for aerobic metabolism of one glucose molecule: one using only words and a second using the chemical formula for glucose.

LO 4.15   Explain how the electron transport system creates the high-energy bond of ATP.

LO 4.16   Describe how the genetic code of DNA is transcribed and translated to create proteins.

LO 4.17   Explain the roles of transcription factors, alternative splicing, and posttranslational modification in protein synthesis.

BACKGROUND BASICS

35   DNA and RNA
65   Organelles
30   Lipids
39   Hydrogen bonds
32   Protein structure
46   Protein interactions
33   Covalent bonds
31   Carbohydrates
20   Graphing
34   ATP

Glucose crystals

Christine Schmidt, Ph.D., and her graduate students seed isolated endothelial cells onto an engineered matrix and watch them grow. They know that if their work is successful, the tissue that results might someday help replace a blood vessel in the body. Just as a child playing with building blocks assembles them into a house, the bioengineer and her students create tissue from cells. In both cases someone familiar with the starting components, building blocks or cells, can predict what the final product will be: blocks make buildings; cells make tissues.

Why then can't biologists, knowing the characteristics of nucleic acids, proteins, lipids, and carbohydrates, explain how combinations of these molecules acquire the remarkable attributes of a living cell? How can living cells carry out processes that far exceed what we would predict from understanding their individual components? The answer is *emergent properties* [p. 2], those distinctive traits that cannot be predicted from the simple sum of the component parts. For example, if you came across a collection of metal pieces and bolts from a disassembled car motor, could you predict (without prior knowledge) that, given an energy source and properly arranged, this collection could create the power to move thousands of pounds?

The emergent properties of biological systems are of tremendous interest to scientists trying to explain how a simple compartment, such as a phospholipid liposome [p. 62], could have evolved into the first living cell. Pause for a moment and see if you can list the properties of life that characterize all living creatures. If you were a scientist looking at pictures and samples sent back from Mars, what would you look for to determine whether life exists there?

Now compare your list with the one in **TABLE 4.1**. Living organisms are highly organized and complex entities. Even a one-celled bacterium, although it appears simple under a microscope, has incredible complexity at the chemical level of organization. It

TABLE 4.1	Properties of Living Organisms
1. Have a complex structure whose basic unit of organization is the cell
2. Acquire, transform, store, and use energy
3. Sense and respond to internal and external environments
4. Maintain homeostasis through internal control systems with feedback
5. Store, use, and transmit information
6. Reproduce, develop, grow, and die
7. Have emergent properties that cannot be predicted from the simple sum of the parts
8. Individuals adapt and species evolve

uses intricately interconnected biochemical reactions to acquire, transform, store, and use energy and information. It senses and responds to changes in its internal and external environments and adapts so that it can maintain homeostasis. It reproduces, develops, grows, and dies; and over time, its species evolves.

Energy is essential for the processes we associate with living things. Without energy for growth, repair, and maintenance of the internal environment, a cell is like a ghost town filled with buildings that are slowly crumbling into ruin. Cells need energy to import raw materials, make new molecules, and repair or recycle aging parts. The ability of cells to extract energy from the external environment and use that energy to maintain themselves as organized, functioning units is one of their most outstanding characteristics. In this chapter, we look at the cell processes through which the human body obtains energy and maintains its ordered systems. You will learn how protein interactions [p. 46] apply to enzyme activity and how the subcellular compartments [p. 8] separate various steps of energy metabolism.

ENERGY IN BIOLOGICAL SYSTEMS

Energy cycling between the environment and living organisms is one of the fundamental concepts of biology. All cells use energy from their environment to grow, make new parts, and reproduce. Plants trap radiant energy from the sun and store it as chemical-bond energy through the process of photosynthesis (**FIG. 4.1**). They extract carbon and oxygen from carbon dioxide, nitrogen from the soil, and hydrogen and oxygen from water to make biomolecules such as glucose and amino acids.

Animals, on the other hand, cannot trap energy from the sun or use carbon and nitrogen from the air and soil to synthesize biomolecules. They must import chemical-bond energy by ingesting the biomolecules of plants or other animals. Ultimately, however, energy trapped by photosynthesis is the energy source for all animals, including humans.

Animals extract energy from biomolecules through the process of *respiration*, which consumes oxygen and produces carbon dioxide and water. If animals ingest more energy than they need

RUNNING **PROBLEM** | **Tay-Sachs Disease: A Deadly Inheritance**

In many American ultra-orthodox Jewish communities—in which arranged marriages are the norm—the rabbi is entrusted with an important, life-saving task. He keeps a confidential record of individuals known to carry the gene for Tay-Sachs disease, a fatal, inherited condition that strikes one in 3600 American Jews of Eastern European descent. Babies born with this disease rarely live beyond age 4, and there is no cure. Based on the family trees he constructs, the rabbi can avoid pairing two individuals who carry the deadly gene.

Sarah and David, who met while working on their college newspaper, are not orthodox Jews. Both are aware, however, that their Jewish ancestry might put any children they have at risk for Tay-Sachs disease. Six months before their wedding, they decide to see a genetic counselor to determine whether they are carriers of the gene for Tay-Sachs disease.

93　99　101　104　111　118

FIG. 4.1 Energy transfer in the environment

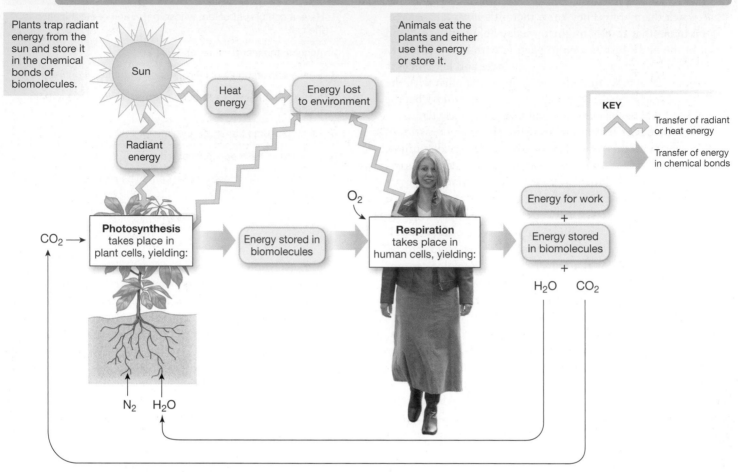

Plants trap radiant energy from the sun and store it in the chemical bonds of biomolecules.

Animals eat the plants and either use the energy or store it.

Sun

Heat energy

Energy lost to environment

Radiant energy

KEY

Transfer of radiant or heat energy

Transfer of energy in chemical bonds

O_2

Energy for work
+
Energy stored in biomolecules

CO_2 →

Photosynthesis takes place in plant cells, yielding:

Energy stored in biomolecules

Respiration takes place in human cells, yielding:

Energy stored in biomolecules
+
H_2O CO_2

N_2 H_2O

---

for immediate use, the excess energy is stored in chemical bonds, just as it is in plants. Glycogen (a glucose polymer) and lipid molecules are the main energy stores in animals [p. 31]. These storage molecules are available for use at times when an animal's energy needs exceed its food intake.

CONCEPT CHECK

1. Which biomolecules always include nitrogen in their chemical makeup?

Energy Is Used to Perform Work

All living organisms obtain, store, and use energy to fuel their activities. **Energy** can be defined as the capacity to do work, but what is *work*? We use this word in everyday life to mean various things, from hammering a nail to sitting at a desk writing a paper. In biological systems, however, the word means one of three specific things: chemical work, transport work, or mechanical work.

Chemical work is the making and breaking of chemical bonds. It enables cells and organisms to grow, maintain a suitable internal environment, and store information needed for reproduction and other activities. Forming the chemical bonds of a protein is an example of chemical work.

Transport work enables cells to move ions, molecules, and larger particles through the cell membrane and through the membranes of organelles in the cell. Transport work is particularly useful for creating **concentration gradients,** distributions of molecules in which the concentration is higher on one side of a membrane than on the other. For example, certain types of endoplasmic reticulum [p. 71] use energy to import calcium ions from the cytosol. This ion transport creates a high calcium concentration inside the organelle and a low concentration in the cytosol. If calcium is then released back into the cytosol, it creates a "calcium signal" that causes the cell to perform some action, such as muscle contraction.

Mechanical work in animals is used for movement. At the cellular level, movement includes organelles moving around in a cell, cells changing shape, and cilia and flagella beating [p. 68]. At the macroscopic level in animals, movement usually involves muscle contraction. Most mechanical work is mediated by motor proteins that make up certain intracellular fibers and filaments of the cytoskeleton [p. 68].

Energy Comes in Two Forms: Kinetic and Potential

Energy can be classified in various ways. We often think of energy in terms we deal with daily: thermal energy, electrical energy, mechanical energy. We speak of energy stored in chemical bonds. Each type of energy has its own characteristics. However, all types of energy share an ability to appear in two forms: as kinetic energy or as potential energy.

Kinetic energy is the energy of motion {*kinetikos,* motion}. A ball rolling down a hill, perfume molecules spreading through the air, electric charge flowing through power lines, heat warming a frying pan, and molecules moving across biological membranes are all examples of bodies that have kinetic energy.

Potential energy is stored energy. A ball poised at the top of a hill has potential energy because it has the potential to start moving down the hill. A molecule positioned on the high-concentration side of a concentration gradient stores potential energy because it has the potential energy to move down the gradient. In chemical bonds, potential energy is stored in the position of the electrons that form the bond [p. 33]. [To learn more about kinetic and potential energy, see Appendix B.]

A key feature of all types of energy is the ability of potential energy to become kinetic energy and vice versa.

Energy Can Be Converted from One Form to Another

Recall that a general definition of energy is the capacity to do work. Work always involves movement and therefore is associated with kinetic energy. Potential energy can also be used to perform work, but the potential energy must first be converted to kinetic energy. The conversion from potential energy to kinetic energy is never 100% efficient, and a certain amount of energy is lost to the environment, usually as heat.

The amount of energy lost in the transformation depends on the *efficiency* of the process. Many physiological processes in the human body are not very efficient. For example, 70% of the energy used in physical exercise is lost as heat rather than transformed into the work of muscle contraction.

FIGURE 4.2 summarizes the relationship of kinetic energy and potential energy:

1. Kinetic energy of the moving ball is transformed into potential energy as work is used to push the ball up the ramp (Fig. 4.2a).
2. Potential energy is stored in the stationary ball at the top of the ramp (Fig. 4.2b). No work is being performed, but the capacity to do work is stored in the position of the ball.
3. The potential energy of the ball becomes kinetic energy when the ball rolls down the ramp (Fig. 4.2c). Some kinetic energy is lost to the environment as heat due to friction between the ball and the air and ramp.

In biological systems, potential energy is stored in concentration gradients and chemical bonds. It is transformed into kinetic energy when needed to do chemical, transport, or mechanical work.

Thermodynamics Is the Study of Energy Use

Two basic rules govern the transfer of energy in biological systems and in the universe as a whole. The **first law of thermodynamics,** also known as the *law of conservation of energy,* states that the total amount of energy in the universe is constant. The universe is considered to be a *closed system*—nothing enters and nothing leaves. Energy can be converted from one type to another, but the total amount of energy in a closed system never changes.

The human body is not a closed system, however. As an *open system,* it exchanges materials and energy with its surroundings. Because our bodies cannot create energy, they import it from outside in the form of food. By the same token, our bodies lose energy, especially in the form of heat, to the environment. Energy that stays within the body can be changed from one type to another or can be used to do work.

FIG. 4.2   Kinetic and potential energy

(a) Work is used to push a ball up a ramp. Kinetic energy of movement up the ramp is being stored in the potential energy of the ball's position.

(b) The ball sitting at the top of the ramp has potential energy, the potential to do work.

(c) The ball rolling down the ramp is converting the potential energy to kinetic energy. However, the conversion is not totally efficient, and some energy is lost as heat due to friction between the ball, ramp, and air.

Kinetic energy

Potential energy

Kinetic energy

The **second law of thermodynamics** states that natural spontaneous processes move from a state of order (nonrandomness) to a condition of randomness or disorder, also known as **entropy.** Creating and maintaining order in an open system such as the body requires the input of energy. Disorder occurs when open systems lose energy to their surroundings without regaining it. When this happens, we say that the entropy of the open system has increased.

The ghost-town analogy mentioned earlier illustrates the second law. When people put all their energy into activities away from town, the town slowly falls into disrepair and becomes less organized (its entropy increases). Similarly, without continual input of energy, a cell is unable to maintain its ordered internal environment. As the cell loses organization, its ability to carry out normal functions disappears, and it dies.

In the remainder of this chapter, you will learn how cells obtain energy from and store energy in the chemical bonds of biomolecules. Using chemical reactions, cells transform the potential energy of chemical bonds into kinetic energy for growth, maintenance, reproduction, and movement.

CONCEPT CHECK	2.	Name two ways animals store energy in their bodies.
	3.	What is the difference between potential energy and kinetic energy?
	4.	What is entropy?

CHEMICAL REACTIONS

Living organisms are characterized by their ability to extract energy from the environment and use it to support life processes. The study of energy flow through biological systems is a field known as **bioenergetics** {*bios*, life + *en-*, in + *ergon*, work}. In a biological system, chemical reactions are a critical means of transferring energy from one part of the system to another.

Energy Is Transferred between Molecules during Reactions

In a **chemical reaction,** a substance becomes a different substance, usually by the breaking and/or making of covalent bonds. A reaction begins with one or more molecules called **reactants** and ends with one or more molecules called **products** (TBL. 4.2). In this discussion, we consider a reaction that begins with two reactants and ends with two products:

$$A + B \rightarrow C + D$$

The speed with which a reaction takes place, the **reaction rate,** is the disappearance rate of the reactants (A and B) or the appearance rate of the products (C and D). Reaction rate is measured as change in concentration during a certain time period and is often expressed as molarity per second (M/sec).

TABLE 4.2	**Chemical Reactions**	
Reaction Type	**Reactants (Substrates)**	**Products**
Combination	A + B $\longrightarrow$	C
Decomposition	C $\longrightarrow$	A + B
Single displacement*	L + MX $\longrightarrow$	LX + M
Double displacement*	LX + MY $\longrightarrow$	LY + MX

*X and Y represent atoms, ions, or chemical groups.

The purpose of chemical reactions in cells is either to transfer energy from one molecule to another or to use energy stored in reactant molecules to do work. The potential energy stored in the chemical bonds of a molecule is known as the **free energy** of the molecule. Generally, complex molecules have more chemical bonds and therefore higher free energies.

For example, a large glycogen molecule has more free energy than a single glucose molecule, which in turn has more free energy than the carbon dioxide and water from which it was synthesized. The high free energy of complex molecules such as glycogen is the reason that these molecules are used to store energy in cells.

To understand how chemical reactions transfer energy between molecules, we should answer two questions. First, how do reactions get started? The energy required to initiate a reaction is known as the *activation energy* for the reaction. Second, what happens to the free energy of the products and reactants during a reaction? The difference in free energy between reactants and products is the *net free energy change of the reaction.*

Activation Energy Gets Reactions Started

Activation energy is the initial input of energy required to bring reactants into a position that allows them to react with one another. This "push" needed to start the reaction is shown in **FIGURE 4.3a** as the little hill up which the ball must be pushed before it can roll by itself down the slope. A reaction with low activation energy proceeds spontaneously when the reactants are brought together. You can demonstrate a *spontaneous reaction* by pouring a little vinegar onto some baking soda and watching the two react to form carbon dioxide. Reactions with high activation energies either do not proceed spontaneously or else proceed too slowly to be useful. For example, if you pour vinegar over a pat of butter, no observable reaction takes place.

Energy Is Trapped or Released during Reactions

One characteristic property of any chemical reaction is the free energy change that occurs as the reaction proceeds. The products of a reaction have either a lower free energy than the reactants

FIG. 4.3 Activation energy in exergonic and endergonic reactions

(a) Activation energy is the "push" needed to start a reaction.

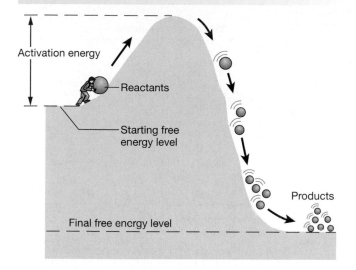

(b) Exergonic reactions release energy because the products have less energy than the reactants.

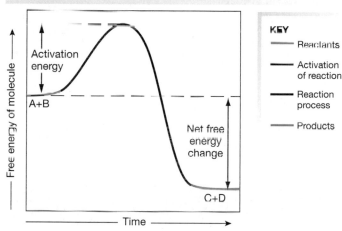

KEY
— Reactants
— Activation of reaction
— Reaction process
— Products

(c) Endergonic reactions trap some activation energy in the products, which then have more free energy than the reactants.

or a higher free energy than the reactants. A change in free energy level means that the reaction has either released or trapped energy.

If the free energy of the products is lower than the free energy of the reactants, as in Figure 4.3b, the reaction releases energy and is called an **exergonic reaction** {*ex-*, out + *ergon*, work}. The energy released by an exergonic, or *energy-producing*, reaction may be used by other molecules to do work or may be given off as heat. In a few cases, the energy released in an exergonic reaction is stored as potential energy in a concentration gradient.

An important biological example of an exergonic reaction is the combination of ATP and water to form ADP, inorganic phosphate (P_i) and H^+. Energy is released during this reaction when the high-energy phosphate bond of the ATP molecule is broken:

$$ATP + H_2O \rightarrow ADP + P_i + H^+ + energy$$

Now contrast the exergonic reaction of Figure 4.3b with the reaction represented in Figure 4.3c. In the latter, products retain part of the activation energy that was added, making their free energy greater than that of the reactants. These reactions that require a net input of energy are said to be **endergonic** {*end(o)*, within + *ergon*, work}, or *energy-utilizing*, reactions.

Some of the energy added to an endergonic reaction remains trapped in the chemical bonds of the products. These energy-consuming reactions are often *synthesis* reactions, in which complex molecules are made from smaller molecules. For example, an endergonic reaction links many glucose molecules together to create the glucose polymer glycogen. The complex glycogen molecule has more free energy than the simple glucose molecules used to make it.

If a reaction traps energy as it proceeds in one direction ($A + B \rightarrow C + D$), it releases energy as it proceeds in the reverse direction ($C + D \rightarrow A + B$). (The naming of forward and reverse directions is arbitrary.) For example, the energy trapped in the bonds of glycogen during its synthesis is released when glycogen is broken back down into glucose.

Coupling Endergonic and Exergonic Reactions

Where does the activation energy for metabolic reactions come from? The simplest way for a cell to acquire activation energy is to couple an exergonic reaction to an endergonic reaction. Some of the most familiar coupled reactions are those that use the energy released by breaking the high-energy bond of ATP to drive an endergonic reaction:

$$E + F \xrightarrow[\text{ATP} \quad \text{ADP} + P_i]{} G + H$$

In this type of coupled reaction, the two reactions take place simultaneously and in the same location, so that the energy from ATP can be used immediately to drive the endergonic reaction between reactants E and F.

However, it is not always practical for reactions to be directly coupled like this. Consequently, living cells have developed ways

FIG. 4.4 **Energy in biological reactions**

Energy released by exergonic reactions can be trapped in the high-energy electrons of NADH, FADH$_2$, or NADPH. Energy that is not trapped is given off as heat.

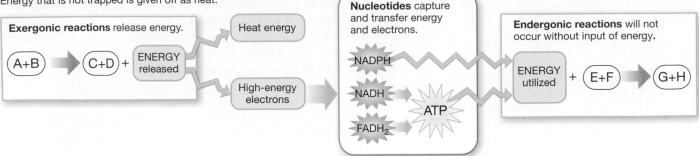

to trap the energy released by exergonic reactions and save it for later use. The most common method is to trap the energy in the form of high-energy electrons carried on nucleotides [p. 33]. The nucleotide molecules NADH, FADH$_2$, and NADPH all capture energy in the electrons of their hydrogen atoms (**FIG. 4.4**). NADH and FADH$_2$ usually transfer most of this energy to ATP, which can then be used to drive endergonic reactions.

Net Free Energy Change Determines Reaction Reversibility

The net free energy change of a reaction plays an important role in determining whether that reaction can be reversed, because the net free energy change of the forward reaction contributes to the activation energy of the reverse reaction. A chemical reaction that can proceed in both directions is called a **reversible reaction.** In a reversible reaction, the forward reaction $A + B \rightarrow C + D$ and its reverse reaction $C + D \rightarrow A + B$ are both likely to take place. If a reaction proceeds in one direction but not the other, it is an **irreversible reaction.**

For example, look at the activation energy of the reaction $C + D \rightarrow A + B$ in **FIGURE 4.5**. This reaction is the reverse of the reaction shown in Figure 4.3b. Because a lot of energy was released in the forward reaction $A + B \rightarrow C + D$, the activation energy of the reverse reaction is substantial (Fig. 4.5). As you will recall, the larger the activation energy, the less likely it is that the reaction will proceed spontaneously. Theoretically, all reactions can be reversed with enough energy input, but some reactions release so much energy that they are essentially irreversible.

In your study of physiology, you will encounter a few irreversible reactions. However, most biological reactions are reversible: if the reaction $A + B \rightarrow C + D$ is possible, then so is the reaction $C + D \rightarrow A + B$. Reversible reactions are shown with arrows that point in both directions: $A + B \rightleftarrows C + D$. One of the main reasons that many biological reactions are reversible is that they are aided by the specialized proteins known as enzymes.

CONCEPT CHECK

5. What is the difference between endergonic and exergonic reactions?

6. If you mix baking soda and vinegar together in a bowl, the mixture reacts and foams up, releasing carbon dioxide gas. Name the reactant(s) and product(s) in this reaction.

7. Do you think the reaction of question 6 is endergonic or exergonic? Do you think it is reversible? Defend your answers.

ENZYMES

Enzymes are proteins that speed up the rate of chemical reactions. During these reactions, the enzyme molecules are not changed in any way, meaning they are biological *catalysts*. Without enzymes, most chemical reactions in a cell would go so slowly

FIG. 4.5 **Some reactions have large activation energies**

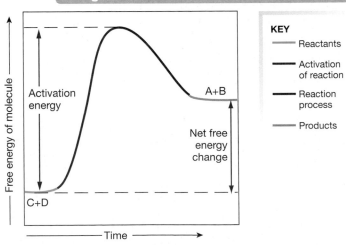

KEY
— Reactants
— Activation of reaction
— Reaction process
— Products

Q GRAPH QUESTION
Is this an endergonic or exergonic reaction?

that the cell would be unable to live. Because an enzyme is not permanently changed or used up in the reaction it catalyzes, we might write it in a reaction equation this way:

$$A + B + enzyme \rightarrow C + D + enzyme$$

This way of writing the reaction shows that the enzyme participates with reactants A and B but is unchanged at the end of the reaction. A more common shorthand for enzymatic reactions shows the name of the enzyme above the reaction arrow, like this:

$$A + B \xrightarrow{enzyme} C + D$$

In enzymatically catalyzed reactions, the reactants A and B are called **substrates.**

Enzymes Are Proteins

Most enzymes are large proteins with complex three-dimensional shapes, although recently researchers discovered that RNA can sometimes act as a catalyst. Like other proteins that bind to substrates, protein enzymes exhibit specificity, competition, and saturation [p. 46].

A few enzymes come in a variety of related forms (isoforms) and are known as **isozymes** {*iso-*, equal} of one another. Isozymes are enzymes that catalyze the same reaction but under different conditions or in different tissues. The structures of related isozymes are slightly different from one another, which causes the variability in their activity. Many isozymes have complex structures with multiple protein chains.

For example, the enzyme *lactate dehydrogenase* (LDH) has two kinds of subunits, named H and M, that are assembled into *tetramers*—groups of four. LDH isozymes include H_4, H_2M_2, and

M_4. The different LDH isozymes are tissue specific, including one found primarily in the heart and a second found in skeletal muscle and the liver.

Isozymes have an important role in the diagnosis of certain medical conditions. For example, in the hours following a heart attack, damaged heart muscle cells release enzymes into the blood. One way to determine whether a person's chest pain was indeed due to a heart attack is to look for elevated levels of heart isozymes in the blood. Some diagnostically important enzymes and the diseases of which they are suggestive are listed in **TABLE 4.3**.

Reaction Rates Are Variable

We measure the rate of an enzymatic reaction by monitoring either how fast the products are synthesized or how fast the substrates are consumed. Reaction rate can be altered by a number of factors, including changes in temperature, the amount of enzyme present, and substrate concentrations [p. 52]. In mammals, we consider temperature to be essentially constant. This leaves enzyme amount and substrate concentration as the two main variables that affect reaction rate.

In protein-binding interactions, if the amount of protein (in this case, enzyme) is constant, the reaction rate is proportional to the substrate concentration. One strategy cells use to control reaction rates is to regulate the amount of enzyme in the cell. In the absence of appropriate enzyme, many biological reactions go very slowly or not at all. If enzyme is present, the rate of the reaction is proportional to the amount of enzyme and the amount of substrate, unless there is so much substrate that all enzyme binding sites are saturated and working at maximum capacity [p. 51].

This seems simple until you consider a reversible reaction that can go in both directions. In that case, what determines in which direction the reaction goes? The answer is that reversible reactions go to a state of *equilibrium*, where the rate of the reaction in the forward direction $(A + B \rightarrow C + D)$ is equal to the rate of the reverse reaction $(C + D \rightarrow A + B)$. At equilibrium,

RUNNING **PROBLEM**

Tay-Sachs disease is a devastating condition. Normally, lysosomes in cells contain enzymes that digest old, worn-out parts of the cell. In Tay-Sachs and related lysosomal storage diseases, genetic mutations result in lysosomal enzymes that are ineffective or absent. Tay-Sachs disease patients lack *hexosaminidase A*, an enzyme that digests glycolipids called *gangliosides*. As a result, gangliosides accumulate in nerve cells in the brain, causing them to swell and function abnormally. Infants with Tay-Sachs disease slowly lose muscle control and brain function. There is currently no treatment or cure for Tay-Sachs disease, and affected children usually die before age 4.

Q1: *Hexosaminidase A is also required to remove gangliosides from the light-sensitive cells of the eye. Based on this information, what is another symptom of Tay-Sachs disease besides loss of muscle control and brain function?*

93　99　101　104　111　118

TABLE 4.3 | Diagnostically Important Enzymes

Elevated blood levels of these enzymes are suggestive of the pathologies listed.

Enzyme	Related Diseases
Acid phosphatase*	Cancer of the prostate
Alkaline phosphatase	Diseases of bone or liver
Amylase	Pancreatic disease
Creatine kinase (CK)	Myocardial infarction (heart attack), muscle disease
Lactate dehydrogenase (LDH)	Tissue damage to heart, liver, skeletal muscle, red blood cells

*A newer test for a molecule called prostate specific antigen (PSA) has replaced the test for acid phosphatase in the diagnosis of prostate cancer.

BIOTECHNOLOGY

Seeing Isozymes

One way to determine which isozymes are present in a tissue sample is to use a technique known as **electrophoresis.** In this technique, a solution derived from the tissue sample is placed at one end of a container filled with a polyacrylamide polymer gel. An electric current passed through the gel causes the negatively charged proteins to move toward the positively charged end of the gel. The rate at which the proteins move depends on their size, their shape, and the electrical charge on their amino acids. As proteins move along the gel at different rates, they separate from one another and appear as individual bands of color when stained with a dye called Coomassie blue or with silver. Electrophoresis can separate mixtures of charged macromolecules, such as proteins and DNA.

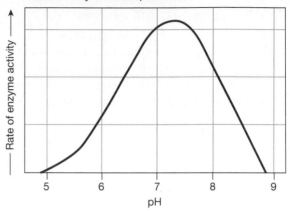

FIG. 4.6 **pH affects enzyme activity**

Most enzymes in humans have optimal activity near the body's internal pH of 7.4.

Rate of enzyme activity (y-axis)

pH (x-axis): 5, 6, 7, 8, 9

Q **FIGURE QUESTION**
If the pH falls from 8 to 7.4, what happens to the activity of the enzyme?

there is no net change in the amount of substrate or product, and the ratio [C][D]/[A][B] is equal to the reaction's equilibrium constant, K_{eq} [p. 47].

If substrates or products are added or removed by other reactions in a pathway, the reaction rate increases in the forward or reverse direction as needed to restore the ratio [C][D]/[A][B]. According to the law of mass action, the ratio of [C] and [D] to [A] and [B] is always the same at equilibrium.

Enzymes May Be Activated, Inactivated, or Modulated

Enzyme activity, like the activity of other soluble proteins, can be altered by various factors. Some enzymes are synthesized as inactive molecules (*proenzymes* or *zymogens*) and activated on demand by proteolytic activation [p. 49]. Others require the binding of inorganic cofactors, such as Ca^{2+} or Mg^{2+} before they become active.

Organic cofactors for enzymes are called **coenzymes.** Coenzymes do not alter the enzyme's binding site as inorganic cofactors do. Instead, coenzymes act as receptors and carriers for atoms or functional groups that are removed from the substrates during the reaction. Although coenzymes are needed for some metabolic reactions to take place, they are not required in large amounts.

Many of the substances that we call **vitamins** are the precursors of coenzymes. The water-soluble vitamins, such as the B vitamins, vitamin C, folic acid, biotin, and pantothenic acid, become coenzymes required for various metabolic reactions. For example, vitamin C is needed for adequate collagen synthesis.

Enzymes may be inactivated by inhibitors or by becoming denatured. Enzyme activity can be modulated by chemical factors or by changes in temperature and pH. **FIGURE 4.6** shows how

enzyme activity can vary over a range of pH values. By turning reactions on and off or by increasing and decreasing the rate at which reactions take place, a cell can regulate the flow of biomolecules through different synthetic and energy-producing pathways.

CONCEPT CHECK

8. What is a biological advantage of having multiple isozymes for a given reaction rather than only one form of the enzyme?

9. The four protein chains of an LDH isozyme are an example of what level of protein structure? (a) primary (b) secondary (c) tertiary (d) quaternary

Enzymes Lower Activation Energy of Reactions

How does an enzyme increase the rate of a reaction? In thermodynamic terms, it lowers the activation energy, making it more likely that the reaction will start (**FIG. 4.7**). Enzymes accomplish this by binding to their substrates and bringing them into the best position for reacting with each other. Without enzymes, the reaction would depend on random collisions between substrate molecules to bring them into alignment.

The rate of a reaction catalyzed by an enzyme is much more rapid than the rate of the same reaction taking place without the enzyme. For example, consider *carbonic anhydrase,* which facilitates conversion of CO_2 and water to carbonic acid. This enzyme plays a critical role in the transport of waste CO_2 from cells to lungs. Each molecule of carbonic anhydrase takes one second to catalyze the conversion of 1 million molecules of CO_2 and water to carbonic acid. In the absence of enzyme, it takes more than a minute for one molecule of CO_2 and water to be converted to

FIG. 4.7 Enzymes lower the activation energy of reactions

In the absence of enzyme, the reaction (curved dashed line) would have much greater activation energy.

KEY
- Reactants
- Activation of reaction
- Reaction process
- Products

carbonic acid. Without carbonic anhydrase and other enzymes in the body, biological reactions would go so slowly that cells would be unable to live.

Enzymatic Reactions Can Be Categorized

Most reactions catalyzed by enzymes can be classified into four categories: oxidation-reduction, hydrolysis-dehydration, exchange-addition-subtraction, and ligation reactions. **TABLE 4.4** summarizes these categories and gives common enzymes for different types of reactions.

An enzyme's name can provide important clues to the type of reaction the enzyme catalyzes. Most enzymes are instantly

RUNNING **PROBLEM**

Tay-Sachs disease is a *recessive* genetic disorder caused by a defect in the gene that directs synthesis of hexosaminidase A. *Recessive* means that for a baby to be born with Tay-Sachs disease, it must inherit two defective genes, one from each parent. People with one Tay-Sachs gene and one normal gene are called *carriers* of the disease. Carriers do not develop the disease but can pass the defective gene on to their children. People who have two normal genes have normal amounts of hexosaminidase A in their blood. Carriers have lower-than-normal levels of the enzyme, but this amount is enough to prevent excessive accumulation of gangliosides in cells.

Q2: *How could you test whether Sarah and David are carriers of the Tay-Sachs gene?*

93   99   **101**   104   111   118

recognizable by the suffix -*ase*. The first part of the enzyme's name (everything that precedes the suffix) usually refers to the type of reaction, to the substrate upon which the enzyme acts, or to both. For example, *glucokinase* has glucose as its substrate, and as a *kinase* it will add a phosphate group [p. 33] to the substrate. Addition of a phosphate group is called **phosphorylation.**

A few enzymes have two names. These enzymes were discovered before 1972, when the current standards for naming enzymes were first adopted. As a result, they have both a new name and a commonly used older name. Pepsin and trypsin, two digestive enzymes, are examples of older enzyme names.

Oxidation-Reduction Reactions Oxidation-reduction **reactions** are the most important reactions in energy extraction and transfer in cells. These reactions transfer electrons from one molecule to another. A molecule that gains electrons is said to be **reduced.** One way to think of this is to remember that adding negatively charged electrons *reduces* the electric charge on the molecule. Conversely, molecules that lose electrons are said to be **oxidized.** Use the mnemonic OIL RIG to remember what happens: <u>O</u>xidation <u>I</u>s <u>L</u>oss (of electrons), <u>R</u>eduction <u>I</u>s <u>G</u>ain.

Hydrolysis-Dehydration Reactions Hydrolysis and dehydration reactions are important in the breakdown and synthesis of large biomolecules. In **dehydration reactions** {*de-*, out + *hydr-*, water}, a water molecule is one of the products. In many dehydration reactions, two molecules combine into one, losing water in the process. For example, the monosaccharides glucose and fructose join to make one sucrose molecule [p. 31]. In the process, one substrate molecule loses a hydroxyl group -OH and the other substrate molecule loses a hydrogen to create water, H_2O. When a dehydration reaction results in the synthesis of a new molecule, the process is known as *dehydration synthesis*.

In a **hydrolysis reaction** {*hydro,* water + *lysis,* to loosen or dissolve}, a substrate changes into one or more products through the addition of water. In these reactions, the covalent bonds of the water molecule are broken ("lysed") so that the water reacts as a hydroxyl group OH^- and a hydrogen ion H^+. For example, an amino acid can be removed from the end of a peptide chain through a hydrolysis reaction.

When an enzyme name consists of the substrate name plus the suffix -*ase,* the enzyme causes a hydrolysis reaction. One example is *lipase,* an enzyme that breaks up large lipids into smaller lipids by hydrolysis. A *peptidase* is an enzyme that removes an amino acid from a peptide.

Addition-Subtraction-Exchange Reactions An **addition reaction** adds a functional group to one or more of the substrates. A **subtraction reaction** removes a functional group from one or more of the substrates. Functional groups are exchanged between or among substrates during **exchange reactions.**

For example, phosphate groups may be transferred from one molecule to another during addition, subtraction, or exchange reactions. The transfer of phosphate groups is an important means

TABLE 4.4	Classification of Enzymatic Reactions	
Reaction Type	**What Happens**	**Representative Enzymes**

1. Oxidation-reduction
  (a) Oxidation
 	
  (b) Reduction
	Add or subtract electrons
	  Transfer electrons from donor to oxygen
	  Remove electrons and H+
	  Gain electrons
		Class:* oxidoreductase
		Oxidase
		Dehydrogenase
		Reductase

2. Hydrolysis-dehydration
  (a) Hydrolysis
  (b) Dehydration
	Add or subtract a water molecule
	  Split large molecules by adding water
	  Remove water to make one large molecule from several smaller ones
		Class:* hydrolase
		Peptidases, saccharidases, lipases
		Dehydratases

3. Transfer chemical groups
 	
  (a) Exchange reaction
 	
  (b) Addition
 	
  (c) Subtraction
	Exchange groups between molecules
	Add or subtract groups
	  Phosphate
	  Amino group (*transamination*)
	  Phosphate (*phosphorylation*)
	  Amino group (*amination*)
	  Phosphate (*dephosphorylation*)
	  Amino group (*deamination*)
		Class:* transferases
		Class:* lyases
		Kinase
		Transaminase
		Phosphorylase
		Aminase
		Phosphatase
		Deaminase

4. Ligation	Join two substrates using energy from ATP	**Class:*** ligases Synthetase

*Enzyme classes as defined by the Nomenclature Committee of the International Union of Biochemistry and Molecular Biology, *www.chem.qmul.ac.uk/iubmb/enzyme.*

of covalent modulation [p. 49], turning reactions on or off or increasing or decreasing their rates. Several types of enzymes catalyze reactions that transfer phosphate groups. **Kinases** transfer a phosphate group from a substrate to an ADP molecule to create ATP, or from an ATP molecule to a substrate. For example, creatine kinase transfers a phosphate group from creatine phosphate to ADP, forming ATP and leaving behind creatine.

The addition, subtraction, and exchange of amino groups [p. 32] are also important in the body's use of amino acids. Removal of an amino group from an amino acid or peptide is a **deamination** reaction. Addition of an amino group is **amination,** and the transfer of an amino group from one molecule to another is **transamination.**

Ligation Reactions Ligation reactions join two molecules together using enzymes known as *synthetases* and energy from ATP. An example of a ligation reaction is the synthesis of *acetyl coenzyme A* (acetyl CoA) from fatty acids and coenzyme A. Acetyl CoA is an important molecule in the body, as you will learn in the next section.

CONCEPT CHECK

10. Name the substrates for the enzymes lactase, peptidase, lipase, and sucrase.

11. Match the reaction type or enzyme in the left column to the group or particle involved.
    - a. kinase
    - b. oxidation
    - c. hydrolysis
    - d. transaminase

    1. amino group
    2. electrons
    3. phosphate group
    4. water

METABOLISM

Metabolism refers to all chemical reactions that take place in an organism. These reactions (1) extract energy from nutrient biomolecules (such as proteins, carbohydrates, and lipids) and (2) either synthesize or break down molecules. Metabolism is often divided into **catabolism,** reactions that release energy through the breakdown of large biomolecules, and **anabolism,** energy-utilizing reactions that result in the synthesis of large biomolecules. Anabolic and catabolic reactions take place simultaneously in cells throughout the body, so that at any given moment, some biomolecules are being synthesized while others are being broken down.

The energy released from or stored in the chemical bonds of biomolecules during metabolism is commonly measured in kilocalories (kcal). A **kilocalorie** is the amount of energy needed to raise the temperature of 1 liter of water by 1 degree Celsius. One kilocalorie is the same as a Calorie, with a capital C, used for quantifying the energy content of food. One kilocalorie is also equal to 1000 calories (small c).

Much of the energy released during catabolism is trapped in the high-energy phosphate bonds of ATP or in the high-energy electrons of NADH, $FADH_2$, or NADPH. Anabolic reactions then transfer energy from these temporary carriers to the covalent bonds of biomolecules.

Metabolism is a network of highly coordinated chemical reactions in which the activities taking place in a cell at any given moment are matched to the needs of the cell. Each step in a metabolic pathway is a different enzymatic reaction, and the reactions of a pathway proceed in sequence. Substrate A is changed into product B, which then becomes the substrate for the next reaction in the pathway. B is changed into C, and so forth:

$$A \rightarrow B \rightarrow C \rightarrow D$$

We call the molecules of the pathway **intermediates** because the products of one reaction become the substrates for the next. You will sometimes hear metabolic pathways called *intermediary metabolism*. Certain intermediates, called *key intermediates,* participate in more than one pathway and act as the branch points for channeling substrate in one direction or another. Glucose, for instance, is a key intermediate in several metabolic pathways.

In many ways, a group of metabolic pathways is similar to a detailed road map (**FIG. 4.8**). Just as a map shows a network of roads that connect various cities and towns, you can think of metabolism as a network of chemical reactions connecting various intermediate products. Each city or town is a different chemical intermediate. One-way roads are irreversible reactions, and big cities with roads to several destinations are key intermediates. Just as there may be more than one way to get from one place to another, there can be several pathways between any given pair of chemical intermediates.

Cells Regulate Their Metabolic Pathways

How do cells regulate the flow of molecules through their metabolic pathways? They do so in five basic ways:

1. By controlling enzyme concentrations
2. By producing modulators that change reaction rates
3. By using two different enzymes to catalyze reversible reactions
4. By compartmentalizing enzymes within intracellular organelles
5. By maintaining an optimum ratio of ATP to ADP

We discussed the effects of changing enzyme concentration in the discussion of protein-binding reactions: as enzyme concentration increases, the reaction rate increases [p. 51]. The sections that follow examine the remaining four items on the list.

Enzyme Modulation Modulators, which alter the activity of a protein, were introduced in the discussion of protein binding [p. 49]. For enzymes, the production of modulators is frequently controlled by hormones and other signals coming from outside the cell. This type of outside regulation is a key element in the integrated control of the body's metabolism following a meal or during periods of fasting between meals.

In addition, some metabolic pathways have their own built-in form of modulation, called **feedback inhibition.** In this form of modulation, the end product of a pathway, shown as Z in **FIGURE 4.9**, acts as an inhibitory modulator of the pathway. As the pathway proceeds and Z accumulates, the enzyme catalyzing the conversion of A to B is inhibited. Inhibition of the enzyme slows down production of Z until the cell can use it up. Once the levels of Z fall, feedback inhibition on enzyme 1 is removed and

---

FIG. 4.8   Metabolic pathways resemble a road map

Cities on the map are equivalent to intermediates in metabolism. In metabolism, there may be more than one way to go from one intermediate to another, just as on the map, there may be many ways to get from one city to another.

(a) Section of Road Map

(b) Metabolic Pathways Drawn Like a Road Map

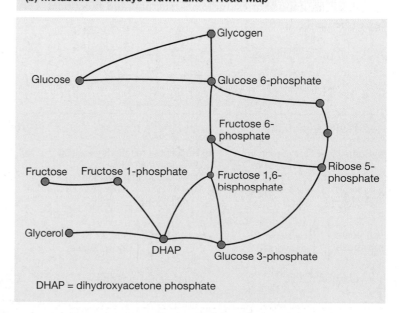

DHAP = dihydroxyacetone phosphate

FIG. 4.9 Feedback inhibition

The accumulation of end product Z inhibits the first step of the pathway. As the cell consumes Z in another metabolic reaction, the inhibition is removed and the pathway resumes.

the pathway starts to run again. Because Z is the end product of the pathway, this type of feedback inhibition is sometimes called *end-product inhibition.*

Reversible Reactions Cells can use reversible reactions to regulate the rate and direction of metabolism. If a single enzyme can catalyze the reaction in either direction, the reaction will go to a state of equilibrium, as determined by the law of mass action (**FIG. 4.10a**). Such a reaction, therefore, cannot be closely regulated except by modulators and by controlling the amount of enzyme.

However, if a reversible reaction requires two different enzymes, one for the forward reaction and one for the reverse reaction, the cell can regulate the reaction more closely (Fig. 4.10b). If no enzyme for the reverse reaction is present in the cell, the reaction is irreversible (Fig. 4.10c).

Compartmentalizing Enzymes in the Cell Many enzymes of metabolism are isolated in specific subcellular compartments. Some, like the enzymes of carbohydrate metabolism, are dissolved in the cytosol, whereas others are isolated within specific organelles. Mitochondria, endoplasmic reticulum, Golgi apparatus, and lysosomes all contain enzymes that are not found in the cytosol. This separation of enzymes means that the pathways controlled by the enzymes are also separated. That allows the cell to control metabolism by regulating the movement of substrate from one cellular compartment to another. The isolation of enzymes within organelles is an important example of structural and functional compartmentation [p. 8].

Ratio of ATP to ADP The energy status of the cell is one final mechanism that can influence metabolic pathways. Through complex regulation, the ratio of ATP to ADP in the cell determines whether pathways that result in ATP synthesis are turned on or off. When ATP levels are high, production of ATP decreases. When ATP levels are low, the cell sends substrates through pathways that result in more ATP synthesis. In the next section, we look further into the role of ATP in cellular metabolism.

ATP Transfers Energy between Reactions

The usefulness of metabolic pathways as suppliers of energy is often measured in terms of the net amount of ATP the pathways can yield. ATP is a nucleotide containing three phosphate groups [p. 34]. One of the three phosphate groups is attached to ADP by a covalent bond in an energy-requiring reaction. Energy is stored in this **high-energy phosphate bond** and then released when the bond is broken during removal of the phosphate group. This relationship is shown by the following reaction:

$$ADP + P_i + energy \rightleftharpoons ADP \sim P (= ATP)$$

RUNNING **PROBLEM**

In 1989, researchers discovered three genetic mutations responsible for Tay-Sachs disease. This discovery paved the way for a new carrier screening test that detects the presence of one of the three genetic mutations in blood cells rather than testing for lower-than-normal hexosaminidase A levels. David and Sarah will undergo this genetic test.

Q3: Why might the genetic test for mutations in the Tay-Sachs gene be more accurate than the test that detects decreased amounts of hexosaminidase A?

Q4: Can you think of a situation in which the enzyme test might be more accurate than the genetic test?

FIG. 4.10 Enzymes control reversibility of metabolic reactions

Reversible Reactions		Irreversible Reactions
(a) Some reversible reactions use one enzyme for both directions.	**(b)** Reversible reactions requiring two enzymes allow more control over the reaction.	**(c)** Irreversible reactions lack the enzyme for the reverse direction.
CO_2 + H_2O	Glucose + PO_4	Glucose + PO_4
carbonic anhydrase ↓ ↑ carbonic anhydrase	hexokinase ↓ ↑ glucose 6-phosphatase	hexokinase ↓
Carbonic acid	Glucose 6-phosphate	Glucose 6-phosphate

Q **FIGURE QUESTION**
What is the difference between a kinase and a phosphatase? (*Hint:* See Tbl. 4.4.)

The squiggle ~ indicates a high-energy bond, and P_i is the abbreviation for an inorganic phosphate group. Estimates of the amount of free energy released when a high-energy phosphate bond is broken range from 7 to 12 kcal per mole of ATP.

ATP is more important as a carrier of energy than as an energy-storage molecule. For one thing, cells can contain only a limited amount of ATP. A resting adult human needs 40 kg (88 pounds!) of ATP to supply the energy required to support one day's worth of metabolic activity, far more than our cells could store. Instead, the body acquires most of its daily energy requirement from the chemical bonds of complex biomolecules. Metabolic reactions transfer that chemical bond energy to the high-energy bonds of ATP, or in a few cases, to the high-energy bonds of the related nucleotide *guanosine triphosphate*, **GTP.**

The metabolic pathways that yield the most ATP molecules are those that require oxygen—the **aerobic,** or *oxidative,* pathways. **Anaerobic** {*an-*, without + *aer,* air} pathways, which are those that can proceed without oxygen, also produce ATP molecules but in much smaller quantities. The lower ATP yield of anaerobic pathways means that most animals (including humans) are unable to survive for extended periods on anaerobic metabolism alone. In the next section, we consider how biomolecules are metabolized to transfer energy to ATP.

CONCEPT CHECK

12. Name five ways in which cells regulate the movement of substrates through metabolic pathways.

13. In which part of an ATP molecule is energy trapped and stored? In which part of a NADH molecule is energy stored?

14. What is the difference between aerobic and anaerobic pathways?

Catabolic Pathways Produce ATP

FIGURE 4.11 summarizes the catabolic pathways that extract energy from biomolecules and transfer it to ATP. Aerobic production of ATP from glucose commonly follows two pathways: **glycolysis** {*glyco-*, sweet + *lysis*, dissolve} and the **citric acid cycle** (also known as the tricarboxylic acid cycle). The citric acid cycle was first described by Hans A. Krebs, so it is sometimes called the *Krebs cycle.* Because Dr. Krebs described other metabolic cycles, we will avoid confusion by using the term *citric acid cycle.*

Carbohydrates enter glycolysis in the form of glucose (top of Fig. 4.11). Lipids are broken down into glycerol and fatty acids [p. 30], which enter the pathway at different points: glycerol feeds into glycolysis, and fatty acids are metabolized to acetyl CoA. Proteins are broken down into amino acids, which also enter at various points. Carbons from glycolysis and other nutrients enter the citric acid cycle, which makes a never-ending circle. At each turn, the cycle adds carbons and produces ATP, high-energy electrons, and carbon dioxide.

Both glycolysis and the citric acid cycle produce small amounts of ATP directly, but their most important contribution to ATP synthesis is trapping energy in electrons carried by

NADH and $FADH_2$. These compounds transfer the electrons to the **electron transport system** (ETS) in the mitochondria. The electron transport system, in turn, uses energy from those electrons to make the high-energy phosphate bond of ATP. At various points, the process produces carbon dioxide and water. Cells can use the water, but carbon dioxide is a waste product and must be removed from the body.

Because glucose is the only molecule that follows both pathways in their entirety, in this chapter, we look at only glucose catabolism.

- **FIGURE 4.12** summarizes the key steps of glycolysis, the conversion of glucose to pyruvate.

- **FIGURE 4.13** shows how pyruvate is converted to acetyl CoA and how carbons from acetyl CoA go through the citric acid cycle.

- **FIGURE 4.14** illustrates the energy-transferring pathway of the electron transport system.

We will examine protein and lipid catabolism and synthetic pathways for lipids and glucose when we look at the fate of the nutrients we eat [Chapter 22].

The aerobic pathways for ATP production are a good example of compartmentation within cells. The enzymes of glycolysis are located in the cytosol, and the enzymes of the citric acid cycle are in the mitochondria. Within mitochondria, concentration of H^+ in the intermembrane compartment stores the energy needed to make the high-energy bond of ATP.

CONCEPT CHECK

15. Match each component on the left to the molecule(s) it is part of:

    (a) amino acids   1. carbohydrates
    (b) fatty acids   2. lipids
    (c) glycerol   3. polysaccharides
    (d) glucose   4. proteins
                         5. triglycerides

16. Do endergonic reactions release energy or trap it in the products?

One Glucose Molecule Can Yield 30–32 ATP

Recall from Figure 4.11 that the aerobic metabolism of one glucose molecule produces carbon dioxide, water, and 30–32 ATP. Let's review the role of glycolysis and the citric acid cycle in that ATP production.

In glycolysis (Fig. 4.12), metabolism of one glucose molecule $C_6H_{12}O_6$ has a net yield of two 3-carbon pyruvate molecules, 2 ATPs, and high-energy electrons carried on 2 NADH:

$$\text{Glucose} + 2\,NAD^+ + 2\,ADP + 2\,P_i \rightarrow$$
$$2\,\text{Pyruvate} + 2\,ATP + 2\,NADH + 2\,H^+ + 2\,H_2O$$

FIG. 4.11 **ESSENTIALS**

ATP Production

The catabolic pathways that extract energy from biomolecules and transfer it to ATP are summarized in this overview figure of aerobic respiration of glucose.

Glycolysis and the **citric acid cycle** produce small amounts of ATP directly, but their most important contributions to ATP synthesis are high-energy electrons carried by NADH and FADH2 to the electron transport system in the mitochondria.

Aerobic Metabolism of Glucose

The energy production from one glucose molecule can be summarized in the following two equations.

$$Glucose + O_2 + ADP + P_i \longrightarrow CO_2 + H_2O + ATP$$

$$C_6H_{12}O_6 + 6\ O_2 \xrightarrow[\ \ 30\text{--}32\ ADP + P_i\ \ \ \ \ \ 30\text{--}32\ ATP\ \ ]{} 6\ CO_2 + 6\ H_2O$$

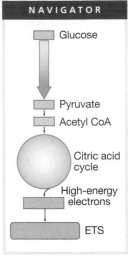

NAVIGATOR

This icon represents the different steps in the metabolic summary figure. Look for it in the figures that follow to help you navigate your way through metabolism.

In the next phase, the conversion of pyruvate to acetyl CoA produces one NADH (Fig. 4.13). Carbons from one acetyl CoA going through the citric acid cycle trap energy in 3 NADH molecules, 1 FADH$_2$ and 1 ATP. These steps happen twice for each glucose, giving a total yield of 8 NADH, 2 FADH$_2$, and 2 ATP for the pyruvate-citric acid cycle phase of glucose metabolism.

In the final step, high-energy electrons of NADH and FADH$_2$ passing along the proteins of the electron transport system use their energy to concentrate H$^+$ in the intermembrane compartment of the mitochondria (Fig. 4.14). When the H$^+$ move down their concentration gradient through a channel in the ATP synthase, the energy released is transferred to the high-energy phosphate bond of ATP. On average, the NADH and FADH$_2$ from one glucose produce 26–28 ATPs.

When we tally the maximum potential energy yield for the catabolism of one glucose molecule through aerobic pathways, the total comes to 30–32 ATP (**FIG. 4.15b**). These numbers are the *potential* maximum because often the mitochondria do not

FIG. 4.12 **ESSENTIALS**

Glycolysis

During glycolysis, one molecule of glucose is converted by a series of enzymatically catalyzed reactions into two pyruvate molecules, producing a net release of energy.

NAVIGATOR

Glucose

Pyruvate

KEY

● = Carbon
○ = Oxygen
Ⓟ = Phosphate group

(side groups not shown)

GLUCOSE

ATP
ADP
Glucose 6-phosphate

1 Glucose is phosphorylated to glucose 6-phosphate. (The "6" in glucose 6-phosphate tells you that the phosphate group has been attached to carbon 6 of the glucose molecule.)

2
Fructose 6-phosphate
ATP
ADP
3
Fructose 1,6 bisphosphate

Key Features of Glycolysis

- In glycolysis, one 6-carbon molecule of glucose becomes two 3-carbon pyruvate molecules.
- Two steps of glycolysis require energy input from ATP. Other steps trap energy in ATP and the high-energy electrons of NADH.
- Glycolysis does not require oxygen. It is the common pathway for aerobic and anaerobic catabolism of glucose.

Dihydroxyacetone phosphate **4**

2 Glyceraldehyde 3-phosphate 2

NAD⁺
NADH
5 Steps 5–9 occur twice for each glucose that begins the pathway.

2 1, 3-Bisphosphoglycerate 2

ADP
ATP
6

2 3-Phosphoglycerate 2

7

2 2-Phosphoglycerate 2

H₂O
8

2 Phosphoenol pyruvate 2

ADP
ATP
9 Pyruvate is the branch point for aerobic and anaerobic metabolism of glucose.

2 Pyruvate 2

Q **FIGURE QUESTIONS**

1. Overall, is glycolysis an endergonic or exergonic pathway?

2. Which steps of glycolysis
   (a) use ATP?
   (b) make ATP or NADH?
   (c) are catalyzed by kinases?
   (d) are catalyzed by dehydrogenases?
   (*Hint:* See Tbl. 4.4.)

3. What is the net energy yield (ATP and NADH) for one glucose?

FIG. 4.13 **ESSENTIALS**

Pyruvate, Acetyl CoA, and the Citric Acid Cycle

If the cell has adequate oxygen, each 3-carbon pyruvate formed during glycolysis reacts with coenzyme A (CoA) to form one **acetyl CoA** and one carbon dioxide (CO_2).

The 2-carbon **acyl unit** of acetyl CoA enters the citric acid cycle pathway, allowing coenzyme A to recycle and react with another pyruvate.

The citric acid cycle makes a never-ending circle, adding carbons from acetyl CoA with each turn of the cycle and producing ATP, high-energy electrons, and carbon dioxide.

NAVIGATOR

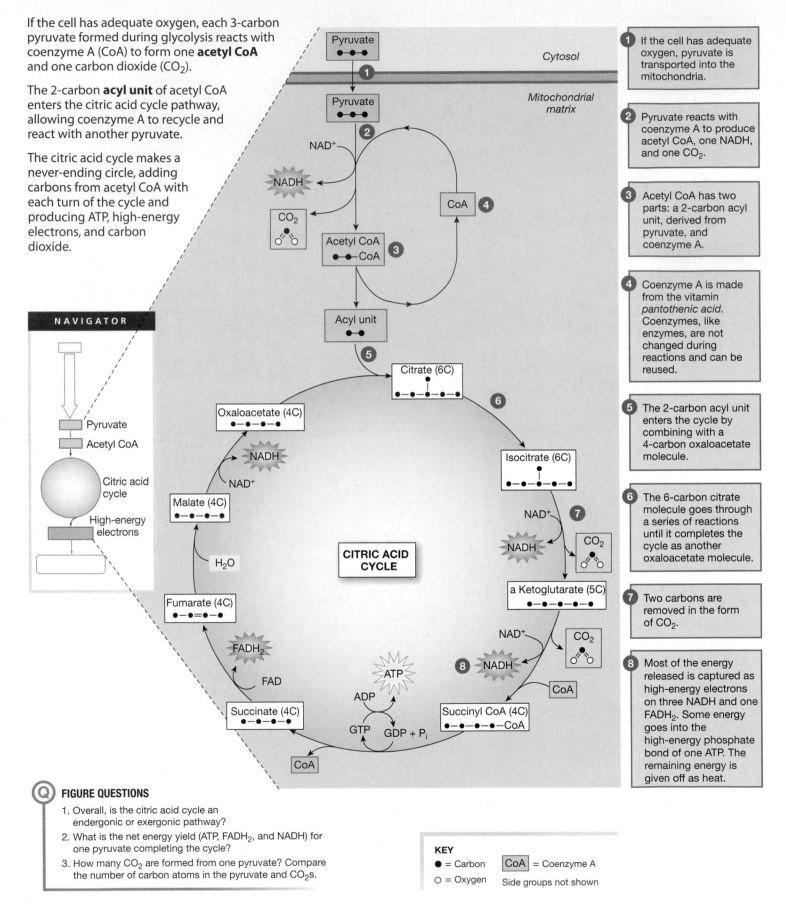

1. If the cell has adequate oxygen, pyruvate is transported into the mitochondria.

2. Pyruvate reacts with coenzyme A to produce acetyl CoA, one NADH, and one CO_2.

3. Acetyl CoA has two parts: a 2-carbon acyl unit, derived from pyruvate, and coenzyme A.

4. Coenzyme A is made from the vitamin *pantothenic acid*. Coenzymes, like enzymes, are not changed during reactions and can be reused.

5. The 2-carbon acyl unit enters the cycle by combining with a 4-carbon oxaloacetate molecule.

6. The 6-carbon citrate molecule goes through a series of reactions until it completes the cycle as another oxaloacetate molecule.

7. Two carbons are removed in the form of CO_2.

8. Most of the energy released is captured as high-energy electrons on three NADH and one $FADH_2$. Some energy goes into the high-energy phosphate bond of one ATP. The remaining energy is given off as heat.

(Q) **FIGURE QUESTIONS**

1. Overall, is the citric acid cycle an endergonic or exergonic pathway?

2. What is the net energy yield (ATP, $FADH_2$, and NADH) for one pyruvate completing the cycle?

3. How many CO_2 are formed from one pyruvate? Compare the number of carbon atoms in the pyruvate and CO_2s.

KEY
● = Carbon   [CoA] = Coenzyme A
○ = Oxygen   Side groups not shown

FIG. 4.14 **ESSENTIALS**

The Electron Transport System

The final step in aerobic ATP production is energy transfer from high-energy electrons of NADH and FADH$_2$ to ATP. This energy transfer requires mitochondrial proteins known as the **electron transport system (ETS),** located in the inner mitochondrial membrane. ETS proteins include enzymes and iron-containing **cytochromes.** The synthesis of ATP using the ETS is called **oxidative phosphorylation** because the system requires oxygen to act as the final acceptor of electrons and H$^+$. The **chemiosmotic theory** says that potential energy stored by concentrating H$^+$ in the intermembrane space is used to make the high-energy bond of ATP.

| 1 NADH and FADH$_2$ release high-energy electrons and H$^+$ to the ETS. NAD$^+$ and FAD are coenzymes that recycle. | 2 Energy released when pairs of high-energy electrons pass along the transport system is used to concentrate H$^+$ from the mitochondrial matrix in the intermembrane space. The H$^+$ concentration gradient is a source of potential energy. | 3 By the end of the ETS, the electrons have given up their stored energy. | 4 Each pair of electrons released by the ETS combines with two H$^+$ and an oxygen atom, creating a molecule of water, H$_2$O. | 5 H$^+$ flow back into the matrix through a protein known as **ATP synthase.** As the H$^+$ move down their concentration gradient, the synthase transfers their kinetic energy to the high-energy phosphate bond of ATP. Because energy conversions are never completely efficient, a portion of the energy is released as heat. | 6 Each three H$^+$ that shuttle through the ATP synthase make a maximum of one ATP. |

Q FIGURE QUESTIONS

1. What is phosphorylation? What is phosphorylated in oxidative phosphorylation?
2. Is the movement of electrons through the electron transport system endergonic or exergonic?
3. What is the role of oxygen in oxidative phosphorylation?

FIG. 4.15 Energy yields from catabolism of one glucose molecule

(a) Anaerobic Metabolism $C_6H_{12}O_6 \longrightarrow 2\,C_3H_5O_3^- + 2\,H^+$

One glucose metabolized anaerobically yields only 2 ATP.

1 Glucose	NADH	FADH$_2$	ATP	CO$_2$
GLYCOLYSIS	2		4	
			-2	
2 Pyruvate → 2 Lactate	-2			

| TOTALS | 0 NADH | | 2 ATP | |

Ⓠ FIGURE QUESTIONS

1. How many NADH enter the electron transport system when glucose is metabolized to lactate?
2. Some amino acids can be converted to pyruvate. If one amino acid becomes one pyruvate, what is the ATP yield from aerobic metabolism of that amino acid?

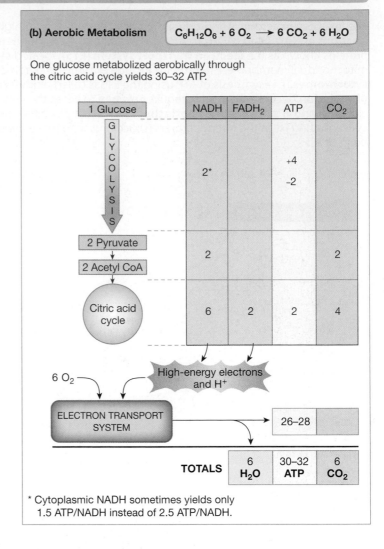

(b) Aerobic Metabolism $C_6H_{12}O_6 + 6\,O_2 \longrightarrow 6\,CO_2 + 6\,H_2O$

One glucose metabolized aerobically through the citric acid cycle yields 30–32 ATP.

1 Glucose	NADH	FADH$_2$	ATP	CO$_2$
GLYCOLYSIS	2*		+4	
			-2	
2 Pyruvate → 2 Acetyl CoA	2			2
Citric acid cycle	6	2	2	4

6 O$_2$ → High-energy electrons and H$^+$ → ELECTRON TRANSPORT SYSTEM → | 26–28 | |

TOTALS	6 H$_2$O	30–32 ATP	6 CO$_2$

* Cytoplasmic NADH sometimes yields only 1.5 ATP/NADH instead of 2.5 ATP/NADH.

work up to capacity. There are various reasons for this, including the fact that a certain number of H$^+$ ions leak from the intermembrane space back into the mitochondrial matrix without producing an ATP.

A second source of variability in the number of ATP produced per glucose comes from the two cytosolic NADH molecules produced during glycolysis. These NADH molecules are unable to enter mitochondria and must transfer their electrons through membrane carriers. Inside a mitochondrion, some of these electrons go to FADH$_2$, which has a potential average yield of only 1.5 ATP rather than the 2.5 ATP made by mitochondrial NADH. If cytosolic electrons go to mitochondrial NADH instead, they produce two additional ATP molecules.

Anaerobic Metabolism Makes 2 ATP

The metabolism of glucose just described assumes that the cells have adequate oxygen to keep the electron transport

system functioning. But what happens to a cell whose oxygen supply cannot keep pace with its ATP demand, such as often happens during strenuous exercise? In that case, the metabolism of glucose shifts from aerobic to anaerobic metabolism, starting at pyruvate (**FIG. 4.16**).

In anaerobic glucose metabolism, pyruvate is converted to lactate instead of being transported into the mitochondria:

Pyruvate is a branch point for metabolic pathways, like a hub city on a road map. Depending on a cell's needs and oxygen content, pyruvate can be shuttled into the citric acid cycle or diverted into lactate production until oxygen supply improves.

The conversion of pyruvate to lactate changes one NADH back to NAD$^+$ when a hydrogen atom and an electron are transferred to the lactate molecule. As a result, the net energy yield for

FIG. 4.16 Aerobic and anaerobic metabolism

Pyruvate is the branch point between aerobic and anaerobic metabolism of glucose.

KEY

● = Carbon | CoA | = Coenzyme A

○ = Oxygen H and –OH not shown

the anaerobic metabolism of one glucose molecule is 2 ATP and 0 NADH (Fig. 4.15a), a very puny yield when compared to the 30–32 ATP/glucose that result from aerobic metabolism (Fig. 4.15b). The low efficiency of anaerobic metabolism severely limits its usefulness in most vertebrate cells, whose metabolic energy demand is greater than anaerobic metabolism can provide. Some cells, such as exercising muscle cells, can tolerate anaerobic metabolism for a limited period of time. Eventually, however, they must shift back to aerobic metabolism. [Aerobic and anaerobic metabolism in muscle are discussed further in Chapters 12 and 25.]

CONCEPT CHECK

17. How is the separation of mitochondria into two compartments essential to ATP synthesis?

18. Lactate dehydrogenase acts on lactate by (adding or removing?) a(n) _____ and a(n) _____. This process is called (oxidation or reduction?).

19. Describe two differences between aerobic and anaerobic metabolism of glucose.

Proteins Are the Key to Cell Function

As you have seen, proteins are the molecules that run a cell from day to day. Protein enzymes control the synthesis and breakdown of carbohydrates, lipids, structural proteins, and signal molecules. Protein transporters and pores in the cell membrane and in organelle membranes regulate the movement of molecules into and out of compartments. Other proteins form the structural skeleton of cells and tissues. In these and other ways, protein synthesis is critical to cell function.

The power of proteins arises from their tremendous variability and specificity. Protein synthesis using 20 amino acids can be compared to creating a language with an alphabet of 20 letters. The "words" vary in length from three letters to hundreds of letters, spelling out the structure of thousands of different proteins with different functions. A change in one amino acid during protein synthesis can alter the protein's function, just as changing one letter turns the word "foot" into "food."

The classic example of an amino acid change causing a problem is sickle cell disease. In this inherited condition, when the amino acid valine replaces one glutamic acid in the protein chain, the change alters the shape of hemoglobin. As a result, red blood cells containing the abnormal hemoglobin take on a crescent (sickle) shape, which causes them to get tangled up and block small blood vessels.

The Protein "Alphabet" One of the mysteries of biology until the 1960s was the question of how only four nitrogenous bases in the DNA molecule—adenine (A), guanine (G), cytosine (C), and thymine (T)—could code for more than 20 different amino acids. If each base controlled the synthesis of one amino acid, a cell could make only four different amino acids. If pairs of bases represented different amino acids, the cell could make 4^2 or 16 different amino acids. Because we have 20 amino acids, this is still not satisfactory. If triplets of bases were the codes for different molecules, however, DNA could create 4^3 or 64 different amino acids. These triplets, called **codons,** are indeed the way information is encoded in DNA and RNA. **FIGURE 4.17** shows the genetic code as it appears in one form of RNA. Remember that RNA substitutes the base uracil (U) for the DNA base thymine [p. 35].

RUNNING PROBLEM

David and Sarah had their blood drawn for the genetic test several weeks ago and have been anxiously awaiting the results. Today, they returned to the hospital to hear the news. The tests show that Sarah carries the gene for Tay-Sachs disease but David does not. This means that although some of their children may be carriers of the Tay-Sachs gene like Sarah, none of the children will develop the disease.

Q5: *The Tay-Sachs gene is a recessive gene (t). If Sarah is a carrier of the gene (Tt) but David is not (TT), what is the chance that any child of theirs will be a carrier? (Consult a general biology or genetics text if you need help solving this problem.)*

FIG. 4.17 **The genetic code as it appears in the codons of mRNA**

The three-letter abbreviations to the right of the brackets indicate the amino acid each codon represents. The start and stop codons are also marked.

First base of codon	Second base of codon				Third base of codon
	U	**C**	**A**	**G**	
U	UUU ⎤ Phe UUC ⎦ UUA ⎤ Leu UUG ⎦	UCU ⎤ UCC ⎥ Ser UCA ⎥ UCG ⎦	UAU ⎤ Tyr UAC ⎦ UAA **Stop** UAG **Stop**	UGU ⎤ Cys UGC ⎦ UGA **Stop** UGG Trp	U C A G
C	CUU ⎤ CUC ⎥ Leu CUA ⎥ CUG ⎦	CCU ⎤ CCC ⎥ Pro CCA ⎥ CCG ⎦	CAU ⎤ His CAC ⎦ CAA ⎤ Gln CAG ⎦	CGU ⎤ CGC ⎥ Arg CGA ⎥ CGG ⎦	U C A G
A	AUU ⎤ AUC ⎥ Ile AUA ⎦ AUG Met **Start**	ACU ⎤ ACC ⎥ Thr ACA ⎥ ACG ⎦	AAU ⎤ Asn AAC ⎦ AAA ⎤ Lys AAG ⎦	AGU ⎤ Ser AGC ⎦ AGA ⎤ Arg AGG ⎦	U C A G
G	GUU ⎤ GUC ⎥ Val GUA ⎥ GUG ⎦	GCU ⎤ GCC ⎥ Ala GCA ⎥ GCG ⎦	GAU ⎤ Asp GAC ⎦ GAA ⎤ Glu GAG ⎦	GGU ⎤ GGC ⎥ Gly GGA ⎥ GGG ⎦	U C A G

Of the 64 possible triplet combinations, one DNA codon (TAC) acts as the initiator or "start codon" that signifies the beginning of a coding sequence. Three codons serve as terminator or "stop codons" that show where the sequence ends. The remaining 60 triplets all code for amino acids. Methionine and tryptophan have only one codon each, but the other amino acids have between two and six different codons each. Thus, like letters spelling words, the DNA base sequence determines the amino acid sequence of proteins.

Unlocking DNA's Code How does a cell know which of the thousands of bases present in its DNA sequence to use in making a protein? It turns out that the information a cell needs to make a particular protein is contained in a segment of DNA known as a gene. What exactly is a gene? The definition keeps changing, but for this text we will say that a **gene** is a region of DNA that contains the information needed to make a functional piece of RNA, which in turn can make a protein.

FIGURE 4.18 shows the five major steps from gene to RNA to functional protein. First, a section of DNA containing a gene must be activated so that its code can be read ❶. Genes that are continuously being read and converted to RNA messages are said to be *constitutively active*. Usually these genes code for proteins that are essential to ongoing cell functions. Other genes are *regulated*—that is, their activity can be turned on (*induced*) or turned off (*repressed*) by regulatory proteins.

Once a gene is activated, the DNA base sequence of the gene is used to create a piece of RNA in the process known as **transcription** {*trans*, over + *scribe*, to write} (Fig. 4.18 ❷). Human cells have three major forms of RNA: **messenger RNA** (mRNA), **transfer RNA** (tRNA), and **ribosomal RNA** (rRNA). Messenger

RNA is processed in the nucleus after it is made ❸. It may either undergo *alternative splicing* (discussed shortly) before leaving the nucleus or be "silenced" and destroyed by enzymes through *RNA interference*. Processed mRNA leaves the nucleus and enters the cytosol. There it works with tRNA and rRNA to direct **translation**, the assembly of amino acids into a protein chain ❹.

Newly synthesized proteins are then subject to **posttranslational modification** (Fig. 4.18 ❺). They fold into complex shapes, may be split by enzymes into smaller peptides, or have various chemical groups added to them. The remainder of this chapter looks at transcription, RNA processing, translation, and posttranslational modification in more detail.

DNA Guides the Synthesis of RNA

The first steps in protein synthesis are compartmentalized within the nucleus because DNA is a very large molecule that cannot pass through the nuclear envelope. Transcription uses DNA as a template to create a small single strand of RNA that can leave the nucleus (**FIG. 4.19**). The synthesis of RNA from the double-stranded DNA template requires an enzyme known as **RNA polymerase,** plus magnesium or manganese ions and energy in the form of high-energy phosphate bonds:

A **promoter** region that precedes the gene must be activated before transcription can begin. Regulatory-protein **transcription factors** bind to DNA and activate the promoter. The active promoter tells the RNA polymerase where to bind to the DNA (Fig. 4.19 ❶). The polymerase moves along the DNA molecule and "unwinds" the double strand by breaking the hydrogen bonds between paired bases ❷. One strand of DNA, called the *template strand*, serves as the guide for RNA synthesis ❸. The promoter region is not transcribed into RNA.

During transcription, each base in the DNA template strand pairs with the complementary RNA base (G-C, C-G, T-A, A-U). This pairing of complementary bases is similar to the process by which a double strand of DNA forms [see Appendix C for a review of DNA synthesis]. For example, a DNA segment containing the base sequence AGTAC is transcribed into the RNA sequence UCAUG.

As the RNA bases bind to the DNA template strand, they also bond with one another to create a single strand of RNA. During transcription, bases are linked at an average rate of 40 per second. In humans, the largest RNAs may contain as many as 5000 bases, and their transcription may take more than a minute—a long time for a cellular process. When RNA polymerase reaches the stop codon, it stops adding bases to the growing RNA strand and releases the strand (Fig. 4.19 ❹).

FIG. 4.18 **ESSENTIALS**

Overview of Protein Synthesis

The major steps required to convert the
genetic code of DNA into a functional protein.

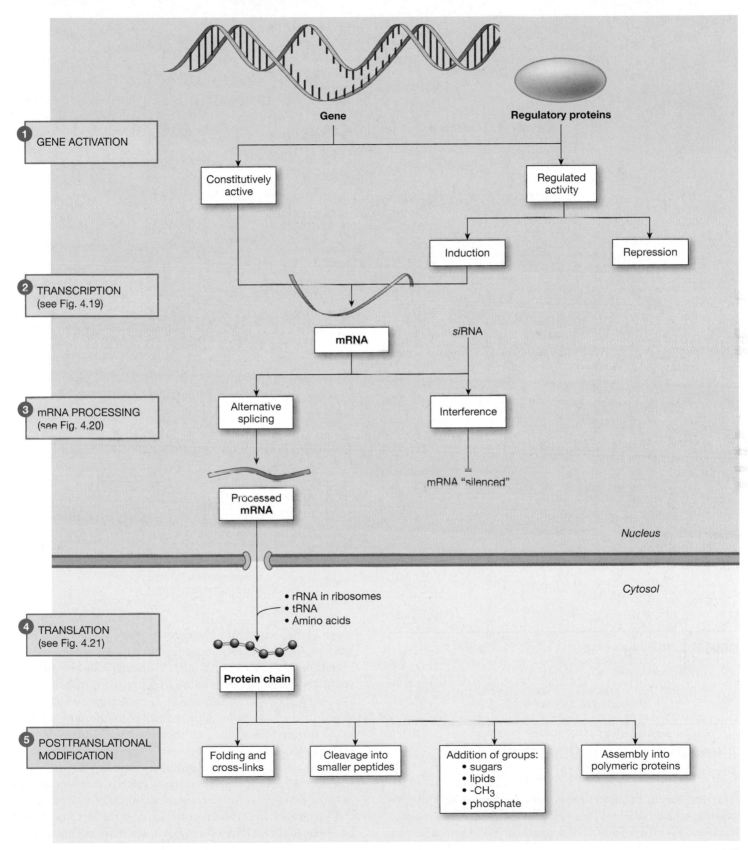

Gene   **Regulatory proteins**

1 GENE ACTIVATION

Constitutively active   Regulated activity

Induction   Repression

2 TRANSCRIPTION (see Fig. 4.19)

mRNA   *si*RNA

3 mRNA PROCESSING (see Fig. 4.20)

Alternative splicing   Interference

mRNA "silenced"

Processed **mRNA**

Nucleus

Cytosol

• rRNA in ribosomes
• tRNA
• Amino acids

4 TRANSLATION (see Fig. 4.21)

Protein chain

5 POSTTRANSLATIONAL MODIFICATION

Folding and cross-links   Cleavage into smaller peptides   Addition of groups:
• sugars
• lipids
• -CH_3
• phosphate   Assembly into polymeric proteins

FIG. 4.19 Transcription

A gene is a segment of DNA that can produce a functional piece of RNA, which in turn can make a protein. Base pairing is the same as in DNA synthesis, except that the base uracil (U) substitutes for thymine (T).

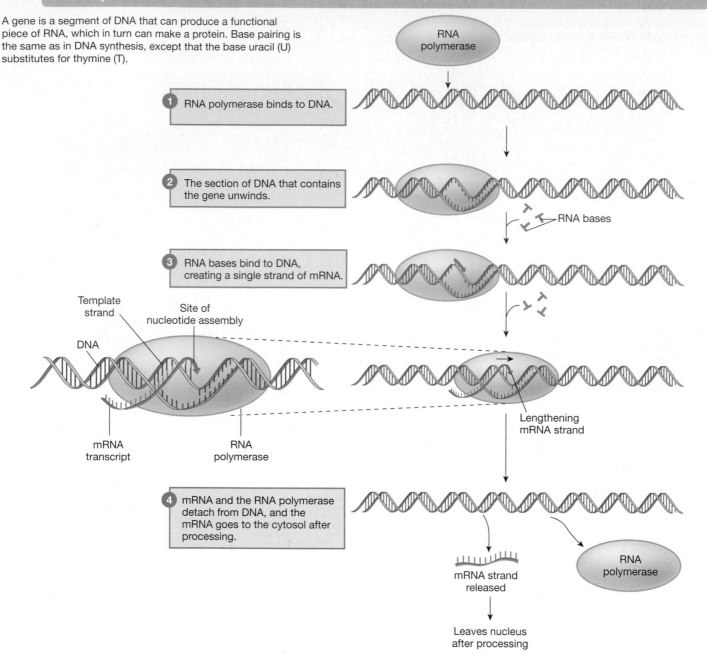

RNA polymerase

1 RNA polymerase binds to DNA.

2 The section of DNA that contains the gene unwinds.

RNA bases

3 RNA bases bind to DNA, creating a single strand of mRNA.

Template strand

Site of nucleotide assembly

DNA

mRNA transcript

RNA polymerase

Lengthening mRNA strand

4 mRNA and the RNA polymerase detach from DNA, and the mRNA goes to the cytosol after processing.

RNA polymerase

mRNA strand released

Leaves nucleus after processing

CONCEPT CHECK

20. Use the genetic code in Figure 4.17 to write the DNA codons that correspond to the three mRNA stop codons.

21. What does the name RNA polymerase tell you about the function of this enzyme?

Alternative Splicing Creates Multiple Proteins from One DNA Sequence

The next step in the process of protein synthesis is **mRNA processing,** which takes two forms (Fig. 4.18 ❸). In *RNA interference*, newly synthesized mRNA is inactivated or destroyed before it can be translated into proteins (see the Emerging Concepts box). In **alternative splicing,** enzymes clip segments out of the middle or off the ends of the mRNA strand. Other enzymes then splice the remaining pieces of the strand back together.

Alternative splicing is necessary because a gene contains both segments that encode proteins (**exons**) and noncoding segments called **introns** (**FIG. 4.20**). That means the mRNA initially made from the gene's DNA contains noncoding segments that must be removed before the mRNA leaves the nucleus. The result of alternative splicing is a smaller piece of mRNA that now contains only the coding sequence for a specific protein.

One advantage of alternative splicing is that it allows a single base sequence on DNA to code for more than one protein. The

FIG. 4.20 **mRNA processing**

In mRNA processing, segments of the newly created mRNA strand called introns are removed. The remaining exons are spliced back together to form the mRNA that codes for a functional protein.

The promoter segment of DNA is not transcribed into RNA.

Removing different introns from mRNA allows a single gene to code for multiple proteins. For protein #1, introns A, C, G, and I were removed. For protein #2, segments B, D, F, and H became the introns.

---

designation of segments as coding or noncoding is not fixed for a given gene. Segments of mRNA that are removed one time can be left in the next time, producing a finished mRNA with a different sequence. The closely related forms of a single enzyme known as *isozymes* are probably made by alternative splicing of a single gene.

After mRNA has been processed, it exits the nucleus through nuclear pores and goes to ribosomes in the cytosol. There mRNA directs the construction of protein.

CONCEPT CHECK   22. Explain in one or two sentences the relationship of mRNA, nitrogenous bases, introns, exons, mRNA processing, and proteins.

mRNA Translation Links Amino Acids

Protein synthesis requires cooperation and coordination among all three types of RNA: mRNA, rRNA, and tRNA. Upon arrival in the cytosol, processed mRNA binds to ribosomes, which are small particles of protein and several types of rRNA [p. 35]. Each ribosome has two subunits, one large and one small, that come together when protein synthesis begins (**FIG. 4.21 ❸**). The small ribosomal subunit binds the mRNA, then adds the large subunit so that the mRNA is sandwiched in the middle. Now the ribosome-mRNA complex is ready to begin translation.

During translation, the mRNA codons are matched to the proper amino acid. This matching is done with the assistance of a tRNA molecule (Fig. 4.21 ❹). One region of each tRNA contains a three-base sequence called an **anticodon** that is complementary to an mRNA codon. A different region of the tRNA molecule binds to a specific amino acid.

As translation begins, the anticodons of tRNAs carrying amino acids attach to the complementary codons of ribosomal mRNA. For example, a tRNA with anticodon sequence UUU carries the amino acid lysine. The UUU anticodon pairs with an AAA codon, one of two codons for lysine, on mRNA. The pairing between mRNA and tRNA puts newly arrived amino acids into the correct orientation to link to the growing peptide chain.

Dehydration synthesis links amino acids by creating a *peptide bond* between the amino group (-NH₂) of the newly arrived amino acid and the carboxyl end (-COOH) of the peptide chain [p. 32]. Once this happens, mRNA releases the "empty" tRNA. The tRNA can then attach to another amino acid molecule with the aid of a cytosolic enzyme and ATP.

EMERGING CONCEPTS

Purple Petunias and RNAi

Who could have guessed that research to develop a deep purple petunia would lead the way to one of the most exciting new areas of molecular biology research? **RNA interference (RNAi)** was first observed in 1990, when botanists who introduced purple pigment genes into petunias ended up with plants that were white or striped with white instead of the deeper purple color they expected. This observation did not attract attention until 1998, when scientists doing research in animal biology and medicine had similar problems in experiments on a nematode worm. Now RNAi is one of the newest tools in biotechnology research.

In very simple terms, RNA "silencing" of mRNA is a naturally occurring event accomplished through the production or introduction of short RNA molecules. These short RNAs bind to mRNA and keep it from being translated. They may even target the mRNA for destruction.

RNAi is a naturally occurring RNA processing mechanism that may have evolved as a means of blocking the replication of RNA viruses. Now researchers are using it to selectively block the production of single proteins within a cell. The scientists' ultimate goal is to create technologies that can be used for the diagnosis and treatment of disease.

When the last amino acid has been joined to the newly synthesized peptide chain, the termination stage has been reached (Fig. 4.21 ❺). The mRNA, the peptide, and the ribosomal subunits separate. The ribosomes are ready for a new round of protein synthesis, but the mRNA is broken down by enzymes known as *ribonucleases*. Some forms of mRNA are broken down quite rapidly, while others may linger in the cytosol and be translated many times.

Protein Sorting Directs Proteins to Their Destination

One of the amazing aspects of protein synthesis is the way specific proteins go from the ribosomes directly to where they are needed in the cell, a process called *protein sorting*. Many newly made proteins carry a *sorting signal,* an address label that tells the cell where the protein should go. Some proteins that are synthesized on cytosolic ribosomes do not have sorting signals. Without a "delivery tag," they remain in the cytosol when they are released from the ribosome [Fig. 3.7, p. 73].

The sorting signal is a special segment of amino acids known as a **signal sequence.** The signal sequence tag directs the protein to the proper organelle, such as the mitochondria or peroxisomes, and allows it to be transported through the organelle membrane.

Peptides synthesized on ribosomes attached to the rough endoplasmic reticulum have a signal sequence directs them through the membrane of the rough ER and into the lumen of this organelle. Once a protein enters the ER lumen, enzymes remove the signal sequence.

Proteins Undergo Posttranslational Modification

The amino acid sequence that comes off a ribosome is the primary structure of a newly synthesized protein [p. 32], but not the final form. The newly made protein can now form different types of covalent and noncovalent bonds, a process known as **posttranslational modification.** Cleavage of the amino acid chain, attachment of molecules or groups, and cross-linkages are three general types of posttranslational modification. More than 100 different types of posttranslational modification have been described so far.

In some common forms of posttranslational modification, the amino acid chain can:

1. fold into various three-dimensional shapes. Protein folding creates the tertiary structure of the protein.
2. create cross-links between different regions of its amino acid chain
3. be cleaved (split) into fragments
4. add other molecules or groups
5. assemble with other amino acid chains into a polymeric (many-part) protein. Assembly of proteins into polymers creates the quaternary structure of the protein.

Protein Folding Peptides released from ribosomes are free to take on their final three-dimensional shape. Each peptide first forms its secondary structure, which may be an α-helix or a β-strand [p. 32]. The molecule then folds into its final shape when hydrogen bonds, covalent bonds, and ionic bonds form between amino acids in the chain. Studies show that some protein folding takes place spontaneously, but it is often facilitated by helper proteins called *molecular chaperones*.

The three-dimensional shape of proteins is often essential for proper function. Misfolded proteins, along with other proteins the cell wishes to destroy, are tagged with a protein called *ubiquitin* and sent to *proteasomes*, cylindrical cytoplasmic enzyme complexes that break down proteins.

Cross-Linkage Some protein folding is held in place by relatively weak hydrogen bonds and ionic bonds. However, other proteins form strong covalent bonds between different parts of the amino acid chain. These bonds are often disulfide bonds (S—S) between two cysteine amino acids, which contain sulfur atoms. For example, the three chains of the digestive enzyme chymotrypsin are held together by disulfide bonds.

Cleavage Some biologically active proteins, such as enzymes and hormones, are synthesized initially as inactive molecules that

FIG. 4.21 Translation

Translation matches the codons of RNA with amino acids to create a protein.

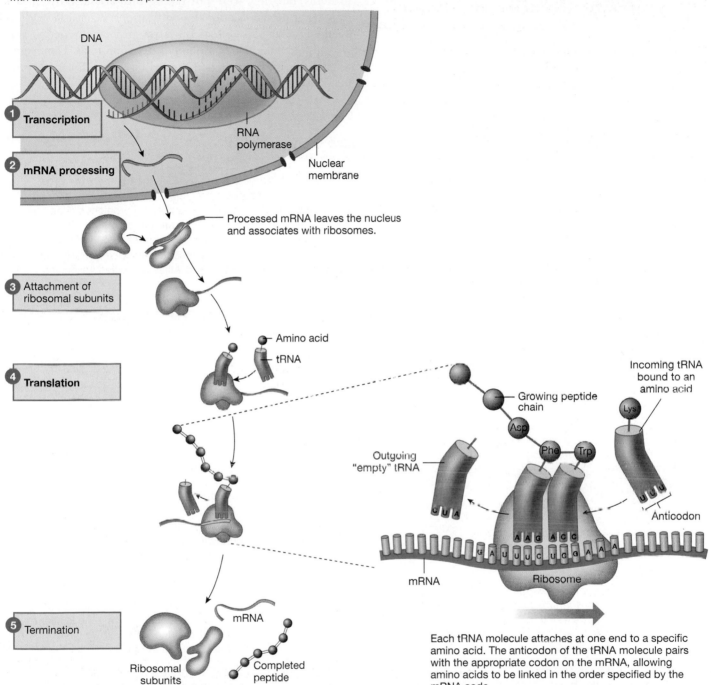

1 Transcription

2 mRNA processing

DNA

RNA polymerase

Nuclear membrane

Processed mRNA leaves the nucleus and associates with ribosomes.

3 Attachment of ribosomal subunits

Amino acid

tRNA

4 Translation

5 Termination

Ribosomal subunits

mRNA

Completed peptide

Growing peptide chain

Asp

Phe   Trp

Lys

Outgoing "empty" tRNA

Incoming tRNA bound to an amino acid

Anticodon

mRNA   Ribosome

Each tRNA molecule attaches at one end to a specific amino acid. The anticodon of the tRNA molecule pairs with the appropriate codon on the mRNA, allowing amino acids to be linked in the order specified by the mRNA code.

must have segments removed before they become active. The enzyme chymotrypsin must have two small peptide fragments removed before it can catalyze a reaction [Fig. 2.12a, p. 50]. Post-translational processing also activates some peptide hormones.

Addition of Other Molecules or Groups Proteins can be modified by the addition of sugars (glycosylation) to create gly-coproteins, or by combination with lipids to make lipoproteins [p. 29]. The two most common chemical groups added to proteins are phosphate groups, PO_4^{2-} and methyl groups, $-CH_3$. (Addition of a methyl group is called *methylation*.)

Assembly into Polymeric Proteins Many complex proteins have a quaternary structure with multiple subunits, in which protein chains assemble into dimers, trimers, or tetramers. One example is the enzyme lactate dehydrogenase (described on p. 99). Another example is the hemoglobin molecule, with four protein chains [Fig. 2.3, p. 32].

CONCEPT CHECK

23. What is the removal of a phosphate group called?
24. List three general types of posttranslational modification of proteins.
25. Is hemoglobin a monomer, dimer, trimer, or tetramer?

The many ways that proteins can be modified after synthesis add to the complexity of the human body. We must know not only the sequence of a protein but also how it is processed, where the protein occurs in or outside the cell, and what it does. Scientists working on the Human Genome Project initially predicted that our DNA would code for about 30,000 proteins, but they were not taking into account alternative splicing or posttranslational modifications. Scientists working on the Human Proteomics Initiative are now predicting that we will find more than a million different proteins. The magnitude of this project means that it will continue for many years into the future.

RUNNING **PROBLEM CONCLUSION** | Tay-Sachs Disease

In this running problem, you learned that Tay-Sachs disease is an incurable, recessive genetic disorder in which the enzyme that breaks down gangliosides in cells is missing. One in 27 Americans of Eastern European Jewish descent in the United States carries the gene for this disorder. Other high-risk populations include French Canadians, Louisiana "Cajuns," and Irish Americans. By one estimate, about one person in every 250 in the general American population is a carrier of the Tay-Sachs gene.

You have also learned that a blood test can detect the presence of genetic mutations that cause this deadly disease. Check your understanding of this running problem by comparing your answers to those in the summary table. To read more on Tay-Sachs disease, see the NIH reference page (*www .ninds.nih.gov/disorders/taysachs/taysachs.htm*) or the web site of the National Tay-Sachs & Allied Diseases Association (*www.ntsad.org*).

Question	Facts	Integration and Analysis
Q1: *What is another symptom of Tay-Sachs disease besides loss of muscle control and brain function?*	Hexosaminidase A breaks down gangliosides. In Tay-Sachs disease, this enzyme is absent, and gangliosides accumulate in cells, including light-sensitive cells of the eye, and cause them to function abnormally.	Damage to light-sensitive cells of the eye could cause vision problems and even blindness.
Q2: *How could you test whether Sarah and David are carriers of the Tay-Sachs gene?*	Carriers of the gene have lower-than-normal levels of hexosaminidase A.	Run tests to determine the average enzyme levels in known carriers of the disease (i.e., people who are parents of children with Tay-Sachs disease) and in people who have little likelihood of being carriers. Compare the enzyme levels of suspected carriers such as Sarah and David with the averages for the known carriers and noncarriers.
Q3: *Why might the genetic test for mutations in the Tay-Sachs gene be more accurate than the test that detects decreased amounts of hexosaminidase A?*	The genetic test detects three mutations in the gene. The enzyme test analyzes levels of the enzyme produced by the gene.	The genetic test is a direct way to test if a person is a carrier. The enzyme test is an indirect indicator. It is possible for factors other than a defective gene to alter a person's enzyme level. Can you think of some? (Answer in Appendix A, p. A-4.)
Q4: *Can you think of a situation in which the enzyme activity test might be more accurate than the genetic test?*	The genetic test looks for three mutations in the Tay-Sachs gene.	There are more than three mutations that can cause Tay-Sachs disease. If the person does not have one of the three mutations being tested, the result will appear to be normal.
Q5: *The Tay-Sachs gene is a recessive gene (t). What is the chance that any child of a carrier (Tt) and a noncarrier (TT) will be a carrier? What are the chances that a child of two carriers will have the disease or be a carrier?*	Mating of Tt × TT results in the following offspring: TT, Tt, TT, Tt. Mating of Tt × Tt results in the following offspring: TT, Tt, Tt, tt.	If only one parent is a carrier, each child has a 50% chance of being a carrier (Tt). If both parents are carriers, there is a 25% chance that a child will have Tay-Sachs disease and a 50% chance a child will be a carrier.

93   99   101   104   111   118

VISUALIZE, PRACTICE, AND MASTER PHYSIOLOGY WITH MASTERING A&P
- Practice-on-the-go with Dynamic Study Modules
- Visualize important processes with Interactive Physiology and A&P Flix
- Check your understanding with Video Tutors and practice quizzes!

CHAPTER SUMMARY

The major theme of this chapter is *energy in biological systems* and how it is acquired, transferred, and used to do biological work. Energy is stored in large biomolecules such as fats and glycogen and is extracted from them through the processes of metabolism. Extracted energy is often stored temporarily in the high-energy phosphate bonds of ATP. Reactions and processes that require energy often use ATP as the energy source. This is a pattern you will see repeated as you learn more about the organ systems of the body.

Other themes in the chapter involve two kinds of *structure-function relationships:* molecular interactions and compartmentation. *Molecular interactions* are important in enzymes, where the ability of an enzyme to bind to its substrate influences the enzyme's activity or in protein synthesis, where nucleic acids direct the assembly of amino acids into larger molecules. *Compartmentation* of enzymes allows cells to direct energy flow by separating functions. Glycolysis takes place in the cytosol of the cell, but the citric acid cycle is isolated within mitochondria, requiring transport of substrates across the mitochondrial membrane. Modulation of enzyme activity and the separation of pathways into subcellular compartments are essential for organizing and separating metabolic processes.

Energy in Biological Systems

1. **Energy** is the capacity to do work. **Chemical work** enables cells and organisms to grow, reproduce, and carry out normal activities. **Transport work** enables cells to move molecules to create concentration gradients. **Mechanical work** is used for movement. (p. 94)

2. **Kinetic energy** is the energy of motion. **Potential energy** is stored energy. (p. 95; Fig. 4.2)

Chemical Reactions

3. A **chemical reaction** begins with one or more **reactants** and ends with one or more **products** (Tbl. 4.2). **Reaction rate** is measured as the change in concentration of products with time. (p. 96)

4. The energy stored in the chemical bonds of a molecule and available to perform work is the **free energy** of the molecule. (p. 96)

5. **Activation energy** is the initial input of energy required to begin a reaction. (p. 96; Fig. 4.3)

6. **Exergonic reactions** are energy-producing. **Endergonic reactions** are energy-utilizing. (p. 96; Fig. 4.3)

7. Metabolic pathways couple exergonic reactions to endergonic reactions. (p. 97; Fig. 4.4)

8. Energy for driving endergonic reactions is stored in ATP. (p. 97)

9. **Reversible reactions** can proceed in both directions. **Irreversible reactions** can go in one direction but not the other. The net free energy change of a reaction determines whether that reaction is reversible. (p. 98)

Enzymes

10. **Enzymes** are biological catalysts that speed up the rate of chemical reactions without themselves being changed. In reactions catalyzed by enzymes, the reactants are called **substrates.** (pp. 98, 99)

11. Like other proteins that bind ligands, enzymes exhibit saturation, specificity, and competition. Related isozymes may have different activities. (p. 99)

12. Some enzymes are produced as inactive precursors and must be activated. This may require the presence of a **cofactor.** Organic cofactors are called **coenzymes.** (p. 100)

13. Enzyme activity is altered by temperature, pH, and modulator molecules. (p. 100)

14. Enzymes work by lowering the activation energy of a reaction. (p. 100; Fig. 4.7)

15. Most reactions can be classified as **oxidation-reduction, hydrolysis-dehydration, addition-subtraction-exchange,** or **ligation** reactions. (pp. 101, 102; Tbl. 4.4)

Metabolism

16. All the chemical reactions in the body are known collectively as **metabolism. Catabolic reactions** release energy and break down large biomolecules. **Anabolic reactions** require a net input of energy and synthesize large biomolecules. (p. 102)

17. Cells regulate the flow of molecules through their metabolic pathways by (1) controlling enzyme concentrations, (2) producing allosteric and covalent modulators, (3) using different enzymes to catalyze reversible reactions, (4) isolating enzymes in intracellular organelles, or (5) maintaining an optimum ratio of ATP to ADP. (p. 103)

18. **Aerobic pathways** require oxygen and yield the most ATP. **Anaerobic pathways** can proceed without oxygen but produce ATP in much smaller quantities. (p. 105)

ATP Production

iP Muscular: Muscle Metabolism

19. Through **glycolysis,** one molecule of glucose is converted into two pyruvate molecules, and yields 2 ATP, 2 NADH, and 2 H^+. Glycolysis does not require the presence of oxygen. (p. 105; Fig. 4.12)

20. **Aerobic metabolism** of pyruvate through the **citric acid cycle** yields ATP, carbon dioxide, and high-energy electrons captured by NADH and $FADH_2$. (p. 108; Fig. 4.13)

21. **High-energy electrons** from NADH and $FADH_2$ give up their energy as they pass through the **electron transport system.** Their energy is trapped in the high-energy bonds of ATP. (p. 105; Fig. 4.14)

22. Maximum energy yield for aerobic metabolism of one glucose molecule is 30–32 ATP. (p. 105; Fig. 4.15)

23. In **anaerobic metabolism,** pyruvate is converted into lactate, with a net yield of 2 ATP for each glucose molecule. (p. 110; Fig. 4.15)

24. Protein synthesis is controlled by nuclear **genes** made of DNA. The code represented by the base sequence in a gene is transcribed into a complementary base code on **RNA. Alternative splicing** of mRNA in the nucleus allows one gene to code for multiple proteins. (pp. 112, 114; Figs. 4.18, 4.19, 4.20)

25. mRNA leaves the nucleus and goes to the cytosol where, with the assistance of **transfer RNA** and **ribosomal RNA,** it assembles amino acids into a designated sequence. This process is called **translation.** (p. 112; Fig. 4.21)

26. **Posttranslational modification** converts the newly synthesized protein to its finished form. (p. 116)

REVIEW QUESTIONS

In addition to working through these questions and checking your answers on p. A-5, review the Learning Outcomes at the beginning of this chapter.

Level One Reviewing Facts and Terms

1. List the three basic forms of work and give a physiological example of each.

2. Explain the difference between potential energy and kinetic energy.

3. State the two laws of thermodynamics in your own words.

4. The sum of all chemical processes through which cells obtain and store energy is called _____.

5. In the reaction $CO_2 + H_2O \rightarrow H_2CO_3$, water and carbon dioxide are the reactants, and H_2CO_3 is the product. Because this reaction is catalyzed by an enzyme, it is also appropriate to call water and carbon dioxide _____. The speed at which this reaction occurs is called the reaction _____, often expressed as molarity/second.

6. _____ are protein molecules that speed up chemical reactions by (increasing or decreasing?) the activation energy of the reaction.

7. Match each definition in the left column with the correct term from the right column (you will not use all the terms):

1. reaction that can run either direction	(a) exergonic
2. reaction that releases energy	(b) endergonic
3. ability of an enzyme to catalyze one reaction but not another	(c) activation energy
	(d) reversible
4. boost of energy needed to get a reaction started	(e) irreversible
	(f) specificity
	(g) free energy
	(h) saturation

8. Since 1972, enzymes have been designated by adding the suffix _____ to their name.

9. Organic molecules that must be present in order for an enzyme to function are called _____. The precursors of these organic molecules come from _____ in our diet.

10. In an oxidation-reduction reaction, in which electrons are moved between molecules, the molecule that gains an electron is said to be _____, and the one that loses an electron is said to be _____.

11. The removal of H_2O from reacting molecules is called _____. Using H_2O to break down polymers, such as starch, is called _____.

12. The removal of an amino group $-NH_2$ from a molecule (such as an amino acid) is called _____. Transfer of an amino group from one molecule to the carbon skeleton of another molecule (to form a different amino acid) is called _____.

13. In metabolism, _____ reactions release energy and result in the breakdown of large biomolecules, and _____ reactions require a net input of energy and result in the synthesis of large biomolecules. In what units do we measure the energy of metabolism?

14. Metabolic regulation in which the last product of a metabolic pathway (the end product) accumulates and slows or stops reactions earlier in the pathway is called _____.

15. Explain how H^+ movement across the inner mitochondrial membrane results in ATP synthesis.

16. List two carrier molecules that deliver high-energy electrons to the electron transport system.

Level Two Reviewing Concepts

17. Create maps using the following terms.

  Map 1: Metabolism

• acetyl CoA	• glycolysis
• ATP	• high-energy electrons
• citric acid cycle	• lactate
• CO_2	• mitochondria
• cytosol	• NADH
• electron transport system	• oxygen
• $FADH_2$	• pyruvate
• glucose	• water

  Map 2: Protein synthesis

• alternative splicing	• ribosome
• base pairing	• RNA polymerase
• bases (A, C, G, T, U)	• RNA processing
• DNA	• start codon
• exon	• stop codon
• gene	• template strand
• intron	• transcription
• promoter	• transcription factors
• mRNA	• translation
• tRNA	

18. When bonds are broken during a chemical reaction, what are the three possible fates for the potential energy found in those bonds?

19. Match the metabolic processes with the letter of the biological theme that best describes the process:

a. Biological energy use	1. Glycolysis takes place in the cytosol; oxidative phosphorylation takes place in mitochondria.
b. Compartmentation	
c. Molecular interactions	
	2. The electron transport system traps energy in a hydrogen ion concentration gradient.
	3. Proteins are modified in the endoplasmic reticulum.
	4. Metabolic reactions are often coupled to the reaction $ATP \rightarrow ADP + P_i$.
	5. Some proteins have S–S bonds between nonadjacent amino acids.
	6. Enzymes catalyze biological reactions.

20. Explain why it is advantageous for a cell to store or secrete an enzyme in an inactive form.

21. Compare the following: (a) the energy yield from the aerobic breakdown of one glucose to CO_2 and H_2O, and (b) the energy yield from one glucose going through anaerobic glycolysis ending with lactate. What are the advantages of each pathway?

22. Briefly describe the processes of transcription and translation. Which organelles are involved in each process?

23. On what molecule does the anticodon appear? Explain the role of this molecule in protein synthesis.

24. Is the energy of ATP's phosphate bond an example of potential or kinetic energy?

25. If ATP releases energy to drive a chemical reaction, would you suspect the activation energy of that reaction to be large or small? Explain.

Level Three Problem Solving

26. Given the following strand of DNA: (1) Find the first start codon in the DNA sequence. *Hint:* The start codon in mRNA is AUG. (2) For the triplets that follow the start codon, list the sequence of corresponding mRNA bases. (3) Give the amino acids that correspond to those mRNA triplets. (See Fig. 4.17.)

DNA: C G C T A C A A G T C A C G T A C C G T A A C G A C T

mRNA:

Amino acids:

Level Four Quantitative Problems

27. The graph shows the free energy change for the reaction $A + B \rightarrow D$. Is this an endergonic or exergonic reaction?

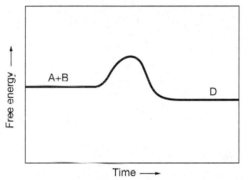

28. If the protein-coding portion of a piece of processed mRNA is 450 bases long, how many amino acids will be in the corresponding polypeptide? (*Hint:* The start codon is translated into an amino acid, but the stop codon is not.)

5 Membrane Dynamics

Organisms could not have evolved without relatively impermeable membranes to surround the cell constituents.

E. N. Harvey, in H. Davson and J. F. Danielli's *The Permeability of Natural Membranes*, 1952

OSMOSIS AND TONICITY 125

LO 5.1 Explain how the body can be in osmotic equilibrium but electrical and chemical disequilibrium.

LO 5.2 Describe the distribution of body water among compartments and the effect of age and sex on total body water.

LO 5.3 Compare and contrast molarity, osmolarity, osmolality, osmotic pressure, and tonicity.

LO 5.4 List the rules for determining osmolarity and tonicity of a solution.

TRANSPORT PROCESSES 132

LO 5.5 Compare bulk flow to solute movement across membranes.

LO 5.6 Create a map to compare simple diffusion, protein-mediated transport, and vesicular transport across membranes.

DIFFUSION 134

LO 5.7 Explain the differences between diffusion in an open system and diffusion across biological membranes.

PROTEIN-MEDIATED TRANSPORT 137

LO 5.8 Compare movement through channels to movement on facilitated diffusion and active transport carriers.

LO 5.9 Apply the principles of specificity, competition, and saturation to carrier-mediated transport.

VESICULAR TRANSPORT 147

LO 5.10 Compare phagocytosis, endocytosis, and exocytosis.

EPITHELIAL TRANSPORT 150

LO 5.11 Explain transcellular transport, paracellular transport, and transcytosis as they apply to epithelial transport.

THE RESTING MEMBRANE POTENTIAL 153

LO 5.12 Explain what it means for a cell to have a resting membrane potential difference.

LO 5.13 Explain how changes in ion permeability change membrane potential, giving examples.

INTEGRATED MEMBRANE PROCESSES: INSULIN SECRETION 159

LO 5.14 Describe the sequence of membrane transport-associated steps that link increased blood glucose to insulin secretion from pancreatic beta cells.

BACKGROUND BASICS

39 Polar and nonpolar molecules
32 Protein and lipid structure
72 Cell junctions
43 Molarity and solutions
62 Membrane structure
68 Cytoskeleton
76 Types of epithelia
99 Enzymes

Small intestine cells

n 1992, the medical personnel at isolated Atoifi Hospital in the Solomon Islands of the South Pacific were faced with a dilemma. A patient was vomiting and needed intravenous (IV) fluids, but the hospital's supply had run out, and it would be several days before a plane could bring more. Their solution was to try something they had only heard about—make an IV of coconut water, the sterile solution that forms in the hollow center of developing coconuts. For two days, the patient received a slow drip of fluid into his veins directly from young coconuts suspended next to his bed. He soon recovered and was well enough to go home.*

No one knows who first tried coconut water as an IV solution, although stories have been passed down that both the Japanese and the British used it in the Pacific Theater of Operations during World War II. Choosing the appropriate IV solution is more than a matter of luck, however. It requires a solid understanding of the body's compartments and of the ways different solutes pass between them.

Homeostasis Does Not Mean Equilibrium

The body has two distinct fluid compartments: the cells and the fluid that surrounds the cells (**FIG. 5.1**). The extracellular fluid (ECF) outside the cells is the buffer between the cells and the environment outside the body. Everything that enters or leaves most cells passes through the ECF.

Water is essentially the only molecule that moves freely between cells and the extracellular fluid. Because of this free movement of water, the extracellular and intracellular compartments reach a state of **osmotic equilibrium** {*osmos*, push or thrust}, in which the fluid concentrations are equal on the two sides of the cell membrane. (Concentration is expressed as amount of solute per volume of solution [Fig. 2.7, p. 42].) Although the overall concentrations of the ECF and intracellular fluid (ICF) are equal, some solutes are more concentrated in one of the two body compartments than in the other (Fig. 5.1d). This means the body is in a state of **chemical disequilibrium.**

Figure 5.1d shows the uneven distribution of major solutes among the body fluid compartments. For example, sodium, chloride, and bicarbonate (HCO_3^-) ions are more concentrated in extracellular fluid than in intracellular fluid. Potassium ions are more concentrated inside the cell. Calcium (not shown in the figure) is more concentrated in the extracellular fluid than in the cytosol, although many cells store Ca^{2+} inside organelles such as the endoplasmic reticulum and mitochondria.

Even the extracellular fluid is not at equilibrium between its two subcompartments, the plasma and the interstitial fluid (IF) [p. 61]. Plasma is the liquid matrix of blood and is found inside the circulatory system. Proteins and other large anions are concentrated in the plasma but cannot cross the leaky exchange epithelium of blood vessels [p. 76], so they are mostly absent from the interstitial fluid (Fig. 5.1d). On the other hand, smaller molecules and ions such as Na^+ and Cl^- are small enough to pass freely between the endothelial cells and therefore have the same concentrations in plasma and interstitial fluid.

The concentration differences of chemical disequilibrium are a hallmark of a living organism, as only the continual input of energy keeps the body in this state. If solutes leak across the cell membranes dividing the intracellular and extracellular compartments, energy is required to return them to the compartment they left. For example, K^+ ions that leak out of the cell and Na^+ ions that leak into the cell are returned to their original compartments by an energy-utilizing enzyme known as the *Na^+-K^+-ATPase*, or the sodium-potassium pump. When cells die and cannot use energy, they obey the second law of thermodynamics [p. 95] and return to a state of randomness that is marked by loss of chemical disequilibrium.

Many body solutes mentioned so far are ions, and for this reason we must also consider the distribution of electrical charge between the intracellular and extracellular compartments. The body as a whole is electrically neutral, but a few extra negative ions are found in the intracellular fluid, while their matching positive ions are located in the extracellular fluid. As a result, the inside of cells is slightly negative relative to the extracellular fluid. This ionic imbalance results in a state of **electrical disequilibrium.** Changes in this disequilibrium create electrical signals. We discuss this topic in more detail later in this chapter.

In summary, note that homeostasis is not the same as equilibrium. The intracellular and extracellular compartments of the body are in osmotic equilibrium, but in chemical and electrical disequilibrium. Furthermore, osmotic equilibrium and the two disequilibria are dynamic *steady states*. The goal of homeostasis is to maintain the dynamic steady states of the body's compartments.

In the remainder of this chapter, we discuss these three steady states, and the role transport mechanisms and the selective permeability of cell membranes play in maintaining these states.

RUNNING **PROBLEM** | Cystic Fibrosis

Over 100 years ago, midwives performed an unusual test on the infants they delivered: The midwife would lick the infant's forehead. A salty taste meant that the child was destined to die of a mysterious disease that withered the flesh and robbed the breath. Today, a similar "sweat test" will be performed in a major hospital—this time, with state-of-the-art techniques—on Daniel Biller, an 18-month-old with a history of weight loss and respiratory problems. The name of the mysterious disease? Cystic fibrosis.

123   133   139   152   153   160

*D. Campbell-Falck *et al.* The intravenous use of coconut water. *Am J Emerg Med* 18: 108–111, 2000.

FIG. 5.1 ESSENTIALS

Body Fluid Compartments

(a) The body fluids are in two compartments: the extracellular fluid (ECF) and intracellular fluid (ICF). The ECF and ICF are in osmotic equilibrium but have very different chemical composition.

BODY FLUID COMPARTMENTS

Cells (Intracellular fluid, ICF)

Intracellular fluid is 2/3 of the total body water volume.

Extracellular Fluid (ECF)

Extracellular fluid is 1/3 of the total body water volume. The ECF consists of:

Interstitial fluid lies between the circulatory system and the cells.

Blood plasma is the liquid matrix of blood.

Material moving into and out of the ICF must cross the cell membrane.

Substances moving between the plasma and interstitial fluid must cross the leaky exchange epithelium of the capillary wall.

KEY
- ☐ Intracellular fluid
- ☐ Interstitial fluid
- ☐ Plasma

(b) This figure shows the compartment volumes for the "standard" 70-kg man.

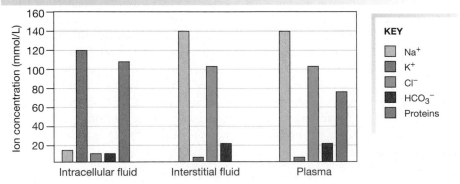

28 L

14 L

Plasma (25% of ECF)

Interstitial Fluid (75% of ECF)

Intracellular fluid (ICF)

Extracellular fluid (ECF)

Percent of total body water

Q GRAPH QUESTIONS

1. Using the ECF volume shown in (b), calculate the volumes of the plasma and interstitial fluid.
2. What is this person's total body water volume?
3. Use your answers from the two questions above to calculate the percentage of total body water in the plasma and interstitial fluid.
4. A woman weighs 121 pounds. Using the standard proportions for the fluid compartments, calculate her ECF, ICF, and plasma volumes. (2.2 lb = 1 kg. 1 kg water = 1 L)

(c) Fluid compartments are often illustrated with box diagrams like this one.

Plasma | Interstitial fluid | Intracellular fluid

ECF 1/3 | ICF 2/3

Cell membrane

(d) The body compartments are in a state of chemical disequilibrium. The cell membrane is a selectively permeable barrier between the ECF and ICF.

Ion concentration (mmol/L)

Intracellular fluid | Interstitial fluid | Plasma

KEY
- ☐ Na^+
- ☐ K^+
- ☐ Cl^-
- ☐ HCO_3^-
- ☐ Proteins

Q GRAPH QUESTIONS

5. How does the ion composition of plasma differ from that of the IF?
6. What ions are concentrated in the ECF? In the ICF?

1. Using what you learned about the naming conventions for enzymes [p. 99], explain what the name Na^+-K^+-$ATPase$ tells you about this enzyme's actions.

2. The intracellular fluid can be distinguished from the extracellular fluid by the ICF's high concentration of _____ ions and low concentration of _____, _____, and _____ ions.

3. In clinical situations, we monitor homeostasis of various substances such as ions, blood gases, and organic solutes by taking a blood sample and analyzing its plasma. For each of the following substances, predict whether knowing its plasma concentration also tells you its concentration in the ECF and the ICF. Defend your answer.

   (a) Na^+

   (b) K^+

   (c) water

   (d) proteins

OSMOSIS AND TONICITY

The distribution of solutes in the body depends on whether a substance can cross cell membranes. Water, on the other hand, is able to move freely in and out of nearly every cell in the body by traversing water-filled ion channels and special water channels created by the protein *aquaporin* (AQP). In this section, we examine the relationship between solute movement and water movement across cell membranes. A sound understanding of this topic provides the foundation for the clinical use of intravenous (IV) fluid therapy.

The Body Is Mostly Water

Water is the most important molecule in the human body because it is the solvent for all living matter. As we look for life in distant parts of the solar system, one of the first questions scientists ask about a planet is, "Does it have water?" Without water, life as we know it cannot exist.

How much water is in the human body? Because one individual differs from the next, there is no single answer. However, in human physiology we often speak of standard values for physiological functions based on "the 70-kg man." These standard values are derived from data published in the mid-twentieth century by The International Commission on Radiological Protection (ICRP). The ICRP was setting guidelines for permissible radiation exposure, and they selected a young (age 20–30) white European male who weighed 70 kilograms (kg) or 154 pounds as their "reference man," or "standard man." In 1984, Reference Man was joined by Reference Woman, a young, 58-kg (127.6 lb) female. The U.S. population is getting larger and heavier, however, and in 1990, the equivalent Reference Man had grown to 77.5 kg and was 8 cm taller.

The 70-kg Reference Man has 60% of his total body weight, or 42 kg (92.4 lb), in the form of water. Each kilogram of water has a volume of 1 liter, so his **total body water** is 42 liters. This is the equivalent of 21 two-liter soft drink bottles!

Adult women have less water per kilogram of body mass than men because women have more adipose tissue. Large fat droplets

in adipose tissue occupy most of the cell's volume, displacing the more aqueous cytoplasm [see Fig. 3.13e, p. 83]. Age also influences body water content. Infants have relatively more water than adults, and water content decreases as people grow older than 60.

TABLE 5.1 shows water content as a percentage of total body weight in people of various ages and both sexes. In clinical practice, it is necessary to allow for the variability of body water content when prescribing drugs. Because women and older people have less body water, they will have a higher concentration of a drug in the plasma than will young men if all are given an equal dose per kilogram of body mass.

The distribution of water among body compartments is less variable. When we look at the relative volumes of the body compartments, the intracellular compartment contains about two-thirds (67%) of the body's water (Fig. 5.1b, c). The remaining third (33%) is split between the interstitial fluid (which contains about 75% of the extracellular water) and the plasma (which contains about 25% of the extracellular water).

4. If the 58-kg Reference Woman has total body water equivalent to 50% of her body weight, what is (a) her total body water volume, (b) her ECF and ICF volumes, and (c) her plasma volume?

The Body Is in Osmotic Equilibrium

Water is able to move freely between cells and the extracellular fluid and distributes itself until water concentrations are equal throughout the body—in other words, until the body is in a state of osmotic equilibrium. The movement of water across a membrane in response to a solute concentration gradient is called **osmosis.** In osmosis, water moves to dilute the more concentrated solution. Once concentrations are equal, net movement of water stops.

Look at the example shown in **FIGURE 5.2** in which a selectively permeable membrane separates two compartments of equal volume. The membrane is permeable to water but does not allow glucose to cross. In ❶, compartments A and B contain equal volumes of glucose solution. Compartment B has more solute (glucose) per volume of solution and therefore is the more concentrated solution. A concentration gradient across the membrane

TABLE 5.1	Water Content as Percentage of Total Body Weight by Age and Sex	
Age	**Male**	**Female**
Infant	65%	65%
1–9	62%	62%
10–16	59%	57%
17–39	61%	51%
40–59	55%	47%
60+	52%	46%

Adapted from I. S. Edelman and J. Leibman, Anatomy of body water and electrolytes, *Am J Med* 27(2): 256–277, 1959.

FIG. 5.2   Osmosis and osmotic pressure

1 Two compartments are separated by a membrane that is permeable to water but not glucose. Solution B is more concentrated than solution A.

Selectively permeable membrane

Glucose molecules

A   B

2 Water moves by osmosis into the more concentrated solution. Osmosis stops when concentrations are equal.

Volume increased

H_2O

Volume decreased

A   B

3 Compartment A is pure water, and compartment B is a glucose solution. **Osmotic pressure** is the pressure that must be applied to oppose osmosis.

Force is applied to exactly oppose osmosis from A to B.

H_2O

Pure water

H_2O

A   B

exists for glucose. However, because the membrane is not permeable to glucose, glucose cannot move to equalize its distribution.

Water, by contrast, can cross the membrane freely. It will move by osmosis from compartment A, which contains the dilute glucose solution, to compartment B, which contains the more concentrated glucose solution. Thus, water moves to dilute the more concentrated solution (Fig. 5.2 **2**).

How can we make quantitative measurements of osmosis? One method is shown in Figure 5.2 **3**. The solution to be measured is placed in compartment B with pure water in compartment A. Because compartment B has a higher solute concentration than compartment A, water will flow from A to B. However, by pushing down on the piston, you can keep water from entering compartment B. The pressure on the piston that exactly opposes the osmotic movement of water into compartment B is known as the **osmotic pressure** of solution B. The units for osmotic pressure, just as with other pressures in physiology, are *atmospheres* (atm) or *millimeters of mercury* (mm Hg). A pressure of 1 mm Hg is equivalent to the pressure exerted on a 1-cm^2 area by a 1-mm-high column of mercury.

Osmolarity Describes the Number of Particles in Solution

Another way to predict the osmotic movement of water quantitatively is to know the concentrations of the solutions with which we are dealing. In chemistry, concentrations are often expressed as *molarity* (*M*), which is defined as number of moles of dissolved solute per liter of solution (mol/L). Recall that one *mole* is 6.02×10^{23} molecules [Fig. 2.7, p. 42].

However, using molarity to describe biological concentrations can be misleading. The important factor for osmosis is the number of osmotically active *particles* in a given volume of solution, not the number of molecules. Because some molecules dissociate into ions when they dissolve in a solution, the number of particles in solution is not always the same as the number of molecules.

For example, one glucose molecule dissolved in water yields one particle, but one NaCl dissolved in water theoretically yields two ions (particles): Na$^+$ and Cl$^-$. Water moves by osmosis in response to the total concentration of all *particles* in the solution. The particles may be ions, uncharged molecules, or a mixture of both.

Consequently, for biological solutions we express the concentration as **osmolarity**, the number of osmotically active particles (ions or intact molecules) per liter of solution. Osmolarity is expressed in *osmoles* per liter (osmol/L or OsM) or, for very dilute physiological solutions, milliosmoles/liter (mOsM). To convert between molarity and osmolarity, use the following equation:

$$\text{molarity (mol/L)} \times \text{particles/molecule (osmol/mol)} = \text{osmolarity (osmol/L)}$$

Let us look at two examples, glucose and sodium chloride, and compare their molarities with their osmolarities.

One mole of glucose molecules dissolved in enough water to create 1 liter of solution yields a 1 molar solution (1 M). Because glucose does not dissociate in solution, the solution has only one mole of osmotically active particles:

$$\text{1 M glucose} \times \text{1 osmole/mole glucose} = \text{1 OsM glucose}$$

Sodium chloride, however, dissociates when placed in solution. At body temperature, a few NaCl ions fail to separate, so instead of 2 ions per NaCl, the *dissociation factor* is about 1.8. Thus, one mole of NaCl dissociates in solution to yield 1.8 moles of particles (Na$^+$, Cl$^-$, and NaCl). The result is a 1.8 OsM solution:

$$\text{1 mole NaCl/L} \times \text{1.8 osmol/mol NaCl} = \text{1.8 osmol/L NaCl}$$

Osmolarity describes only the number of particles in the solution. It says nothing about the composition of the particles. A 1 OsM solution could be composed of pure glucose or pure Na^+ and Cl^- or a mixture of all three solutes.

The normal osmolarity of the human body ranges from 280 to 296 milliosmoles per liter (mOsM). In this book, to simplify calculations, we will round that number up slightly to 300 mOsM.

A term related to osmolarity is osmolality. **Osmolality** is concentration expressed as osmoles of solute per kilogram of water. Because biological solutions are dilute and little of their weight comes from solute, physiologists often use the terms *osmolarity* and *osmolality* interchangeably. Osmolality is usually used in clinical situations because it is easy to estimate people's body water content by weighing them.

Clinicians estimate a person's fluid loss in dehydration by equating weight loss to fluid loss. Because 1 liter of pure water weighs 1 kilogram, a decrease in body weight of 1 kg (or 2.2 lbs.) is considered equivalent to the loss of 1 liter of body fluid. A baby with diarrhea can easily be weighed to estimate its fluid loss. A decrease of 1.1 lbs. (0.5 kg) of body weight is assumed to mean the loss of 500 mL of fluid. This calculation provides a quick estimate of how much fluid needs to be replaced.

CONCEPT CHECK

5. A mother brings her baby to the emergency room because he has lost fluid through diarrhea and vomiting for two days. The staff weighs the baby and finds that he has lost 2 lbs. If you assume that the reduction in weight is due to water loss, what volume of water has the baby lost (2.2 lbs. = 1 kg)?

Comparing Osmolarities of Two Solutions

Osmolarity is a property of every solution. You can compare the osmolarities of different solutions as long as the concentrations are expressed in the same units—for example, as milliosmoles per liter. If two solutions contain the same number of solute particles per unit volume, we say that the solutions are **isosmotic** {*iso-*, equal}. If solution A has a higher osmolarity (contains more particles per unit volume, is more concentrated) than solution B, we say that solution A is **hyperosmotic** to solution B. In the same example, solution B, with fewer osmoles per unit volume, is **hyposmotic** to solution A. **TABLE 5.2** shows some examples of comparative osmolarities.

Osmolarity is a *colligative* property of solutions, meaning it depends strictly on the *number* of particles per liter of solution.

Osmolarity says nothing about what the particles are or how they behave. Before we can predict whether osmosis will take place between any two solutions divided by a membrane, we must know the properties of the membrane and of the solutes on each side of it.

If the membrane is permeable only to water and not to any solutes, water will move by osmosis from a less concentrated (hyposmotic) solution into a more concentrated (hyperosmotic) solution, as illustrated in Figure 5.2. Most biological systems are not this simple, however. Biological membranes are selectively permeable and allow some solutes to cross in addition to water. To predict the movement of water into and out of cells, you must know the *tonicity* of the solution, explained in the next section.

Tonicity Describes the Volume Change of a Cell

Tonicity {*tonikos*, pertaining to stretching} is a physiological term used to describe a solution and how that solution would affect cell volume if the cell were placed in the solution and allowed to come to equilibrium (**TBL. 5.3**).

- If a cell placed in the solution gains water at equilibrium and swells, we say that the solution is **hypotonic** to the cell.
- If the cell loses water and shrinks at equilibrium, the solution is said to be **hypertonic.**
- If the cell in the solution does not change size at equilibrium, the solution is **isotonic.**

By convention, we always describe the tonicity of the solution relative to the cell. How, then, does tonicity differ from osmolarity?

1. Osmolarity describes the number of solute particles dissolved in a volume of solution. It has units, such as osmoles/liter. The osmolarity of a solution can be measured by a machine called an *osmometer*. Tonicity has no units; it is only a comparative term.

2. Osmolarity can be used to compare any two solutions, and the relationship is reciprocal (solution A is hyperosmotic to solution B; therefore, solution B is hyposmotic to solution A). Tonicity always compares a solution and a cell, and by convention, tonicity is used to describe only the solution—for example, "Solution A is hypotonic to red blood cells."

TABLE 5.2	Comparing Osmolarities	
Solution A = 1 OsM Glucose	**Solution B = 2 OsM Glucose**	**Solution C = 1 OsM NaCl**
A is hyposmotic to B	B is hyperosmotic to A	C is isosmotic to A
A is isosmotic to C	B is hyperosmotic to C	C is hyposmotic to B

TABLE 5.3	Tonicity of Solutions	
Solution	**Cell Behavior When Placed in the Solution**	**Description of the Solution Relative to the Cell**
A	Cell swells	Solution A is hypotonic
B	Cell doesn't change size	Solution B is isotonic
C	Cell shrinks	Solution C is hypertonic

3. Osmolarity alone does not tell you what happens to a cell placed in a solution. Tonicity by definition tells you what happens to cell volume at equilibrium when the cell is placed in the solution.

This third point is the one that is most confusing to students. Why can't osmolarity be used to predict tonicity? The reason is that the tonicity of a solution depends not only on its concentration (osmolarity) but also on the *nature* of the solutes in the solution.

By nature of the solutes, we mean whether the solute particles can cross the cell membrane. If the solute particles (ions or molecules) can enter the cell, we call them **penetrating solutes.** We call particles that cannot cross the cell membrane **nonpenetrating solutes.** Tonicity depends on the concentration of nonpenetrating solutes only. Let's see why this is true.

First, some preliminary information. The most important nonpenetrating solute in physiology is NaCl. If a cell is placed in a solution of NaCl, the Na^+ and Cl^- ions do not enter the cell. This makes NaCl a nonpenetrating solute. (In reality, a few Na^+ ions may leak across, but they are immediately transported back to the extracellular fluid by the Na^+-K^+-ATPase. For this reason, NaCl is considered a *functionally* nonpenetrating solute.)

By convention, we assume that cells are filled with other types of nonpenetrating solutes. In other words, the solutes inside the cell are unable to leave so long as the cell membrane remains intact. Now we are ready to see why osmolarity alone cannot be used to predict tonicity.

Suppose you know the composition and osmolarity of a solution. How can you figure out the tonicity of the solution without actually putting a cell in it? The key lies in knowing *the relative concentrations of nonpenetrating solutes in the cell and in the solution.* Water will always move until the concentrations of nonpenetrating solutes in the cell and the solution are equal.

Here are the rules for predicting tonicity:

1. *If the cell has a higher concentration of nonpenetrating solutes than the solution,* there will be net movement of water into the cell. The cell swells, and the solution is *hypotonic.*
2. *If the cell has a lower concentration of nonpenetrating solutes than the solution,* there will be net movement of water out of the cell. The cell shrinks, and the solution is *hypertonic.*
3. *If the concentrations of nonpenetrating solutes are the same in the cell and the solution,* there will be no net movement of water at equilibrium. The solution is *isotonic* to the cell.

How does tonicity relate to osmolarity? **FIGURE 5.3** shows the possible combinations of osmolarity and tonicity, and why osmolarity alone cannot predict tonicity. There is one exception to this statement: A hyposmotic solution is always hypotonic, no matter what its composition. The cell will always have a higher concentration of nonpenetrating solutes than the solution, and water will move into the cell (rule 1 above).

FIG. 5.3 The relationship between osmolarity and tonicity

The osmolarity of a solution is not an accurate predictor of its tonicity.

TONICITY	OSMOLARITY		
	Hyposmotic	**Isosmotic**	**Hyperosmotic**
Hypotonic	√	√	√
Isotonic		√	√
Hypertonic			√

As you can see in Figure 5.3, an isosmotic solution may be isotonic or hypotonic. It can never be hypertonic because it can never have a higher concentration of nonpenetrating solutes than the cell. If all solutes in the isosmotic solution are nonpenetrating, then the solution is also isotonic. If there are any penetrating solutes in the isosmotic solution, the solution will be hypotonic.

Hyperosmotic solutions may be hypertonic, isotonic, or hypotonic. Their tonicity depends on the relative concentration of nonpenetrating solutes in the solution compared to the cell, as described previously.

Often tonicity is explained using a single cell that is placed into a solution, but here we will use a more physiologically appropriate system: a two-compartment box model that represents the total body divided into ECF and ICF (see Fig. 5.1c). To simplify the calculations, we will use a 3-liter body, with 2 liters in the ICF and 1 liter in the ECF. We assume that the starting osmolarity is 300 mOsM (0.3 OsM) and that solutes in each compartment are nonpenetrating (NP) and cannot move into the other compartment. By defining volumes and concentrations, we can use the equation *solute/volume = concentration* ($S/V = C$) to mathematically determine changes to volumes and osmolarity. *Concentration is osmolarity.*

Always begin by defining the starting conditions. This may be the person's normal state or it may be the altered state that you are trying to return to normal. An example of this would be trying to restore normal volume and osmolarity in a person who has become dehydrated through sweat loss.

FIGURE 5.4 shows the starting conditions for the 3-liter body both as a compartment diagram and in a table. The table format allows you to deal with an example mathematically if you know the volumes and concentration of the body and of the solution added or lost.

The body's volumes and concentration will change as the result of adding or losing solutes, water, or both—the law of mass balance [p. 10]. Additions to the body normally come through the ingestion of food and drink. In medical situations, solutions can be added directly to the ECF through intravenous (IV) infusions.

Significant solute and water loss may occur with sweating, vomiting and diarrhea, or blood loss.

Once you have defined the starting conditions, you add or subtract volume and solutes to find the body's new osmolarity. The final step is to determine whether the ECF and ICF volumes change as a result of the water and solute gain or loss. In this last step, you must separate the added solutes into penetrating solutes and nonpenetrating solutes.

In our examples, we use three solutes: NaCl, urea, and glucose. NaCl is considered nonpenetrating. Any NaCl added to the body remains in the ECF. Urea is freely penetrating and behaves as if the cell membranes dividing the ECF and ICF do not exist. An added load of urea distributes itself until the urea concentration is the same throughout the body.

Glucose (also called *dextrose*) is an unusual solute. Like all solutes, it first goes into the ECF. Over time, however, 100% of added glucose will enter the cells. When glucose enters the cells, it is phosphorylated to glucose 6-phosphate (G-6-P) and cannot leave the cell again. So although glucose enters cells, it is not freely penetrating because it stays in the cell and adds to the cell's nonpenetrating solutes.

Giving someone a glucose solution is the same as giving them a slow infusion of pure water because glucose 6-phosphate is the first step in the aerobic metabolism of glucose [p. 105]. The end products of aerobic glucose metabolism are CO_2 and water.

The examples shown in Figure 5.4 walk you through the process of adding and subtracting solutions to the body. Ask the following questions when you are evaluating the effects of a solution on the body:

1. What is the osmolarity of this solution relative to the body? (Tbl. 5.2)
2. What is the tonicity of this solution? (Use Fig. 5.3 to help eliminate possibilities.) To determine tonicity, compare the concentration of the nonpenetrating solutes in the solution to the body concentration. (All body solutes are considered to be nonpenetrating.)

For example, consider a solution that is 300 mOsM—isosmotic to a body that is 300 mOsM. The solution's tonicity depends on the concentration of nonpenetrating solutes in the solution. If the solution is 300 mOsM NaCl, the solution's nonpenetrating solute concentration is equal to that of the body. When the solution mixes with the ECF, the ECF nonpenetrating concentration and osmolarity do not change. No water will enter or leave the cells (the ICF compartment), and the solution is isotonic. You can calculate this for yourself by working through Example 1 in Figure 5.4.

Now suppose the 300 mOsM solution has urea as its only solute. Urea is a penetrating solute, so this solution has zero nonpenetrating solutes. When the 300 mOsM urea solution mixes with the ECF, the added volume of the urea solution dilutes the nonpenetrating solutes of the ECF. (S/V = C: The

same amount of NP solute in a larger volume means a lower NP concentration.)

Now the nonpenetrating concentration of the ECF is less than 300 mOsM. The cells still have a nonpenetrating solute concentration of 300 mOsM, so water moves into the cells to equalize the nonpenetrating concentrations. (Rule: Water moves into the compartment with the higher concentration of NP solutes.) The cells gain water and volume. This means the urea solution is hypotonic, even though it is isosmotic.

Example 2 in Figure 5.4 shows how combining penetrating and nonpenetrating solutes can complicate the situation. This example asks you to describe the solution's osmolarity and tonicity based on its composition before you do the mathematical calculations. This skill is important for clinical situations, when you will not know exact body fluid volumes for the person needing an IV. **TABLE 5.4** lists some rules to help you distinguish between osmolarity and tonicity.

Understanding the difference between osmolarity and tonicity is critical to making good clinical decisions about intravenous (IV) fluid therapy. The choice of IV fluid depends on how the clinician wants the solutes and water to distribute between the extracellular and intracellular fluid compartments. If the problem is dehydrated cells, the appropriate IV solution is hypotonic because the cells need fluid. If the situation requires fluid that remains in the extracellular fluid to replace blood loss, an isotonic IV solution is used. In medicine, the tonicity of a solution is usually the most important consideration.

TABLE 5.5 lists some common IV solutions and their approximate osmolarity and tonicity relative to the normal human cell. What about the coconut water described at the start of the chapter? Chemical analysis shows that it is not an ideal IV solution, although it is useful for emergencies. It is isosmotic to human plasma but is hypotonic, with Na^+ concentrations much lower than normal ECF $[Na^+]$ and high concentrations of glucose and fructose, along with amino acids.

TABLE 5.4	Rules for Osmolarity and Tonicity
1. Assume that all intracellular solutes are nonpenetrating.	
2. Compare osmolarities before the cell is exposed to the solution. (At equilibrium, the cell and solution are always isosmotic.)	
3. Tonicity of a solution describes the volume change of a cell at equilibrium (Tbl. 5.3).	
4. Determine tonicity by comparing nonpenetrating solute concentrations in the cell and the solution. Net water movement is into the compartment with the higher concentration of nonpenetrating solutes.	
5. Hyposmotic solutions are always hypotonic.	

FIG. 5.4 **ESSENTIALS**

Osmolarity and Tonicity

For all problems, define your starting conditions. Assume that all initial body solutes are nonpenetrating (NP) and will remain in either the ECF or ICF.

Use the equation

$$\text{Solute/volume = concentration}$$
$$(S/V = C)$$

to solve the problems. You will know two of the three variables and can calculate the third.

Remember that body compartments are in osmotic equilibrium. Once you know the total body's osmolarity (concentration), you also know the ECF and ICF osmolarity because they are the same.

◄— ECF —►	◄————— ICF —————►
300 mosmol NP	600 mosmol NP
1 L	2 L

Starting Condition:

We have a 3-liter body that is 300 mOsM. The ECF is 1 liter and the ICF is 2 liters.

Use **S/V = C** to find out how much solute is in each of the two compartments. Rearrange the equation to solve for S: **S = CV.**

① $S_{ICF} = 300 \text{ mosmol/L} \times 2\text{ L} = 600 \text{ mosmol NP solute in the ICF}$

We can also do these calculations using the following table format. This table has been filled in with the values for the starting body. Remember that the ECF + ICF must always equal the total body values, and that once you know the total body osmolarity, you know the ECF and ICF osmolarity.

② $S_{ECF} = 300 \text{ mosmol/L} \times 1\text{ L} = 300 \text{ mosmol NP solute in the ECF}$

	Total Body	ECF	ICF
Solute (mosmoles)	900 mosmol	300 mosmol	600 mosmol
Volume (L)	3 L	1 L	2 L
Osmolarity (mOsM)	300 mOsM	300 mOsM	300 mOsM

To see the effect of adding a solution or losing fluid, start with this table and add or subtract volume and solute as appropriate. *You cannot add and subtract concentrations. You must use volumes and solute amounts.*

- Work the total body column first, adding or subtracting solutes and volume. Once you calculate the new total body osmolarity, carry that number across the bottom row to the ECF and ICF columns. (The compartments are in osmotic equilibrium.)

- Distribute nonpenetrating solutes to the appropriate compartment. NaCl stays in the ECF. Glucose goes into the cells. Use **V = S/C** to calculate the new compartment volumes.

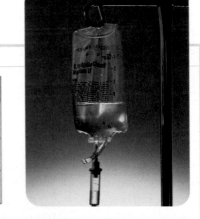

In the tables below and on the following page, the yellow boxes indicate the unknowns that must be calculated.

Example 1

Add an IV solution of 1 liter of 300 mOsM NaCl to this body. This solution adds 1 liter of volume and 300 mosmoles of NaCl.

Answer

Work total body first. Add solute and volume, then calculate new osmolarity (yellow box).

	Total Body
Solute (mosmoles)	900 + 300 = 1200 mosmol
Volume (L)	3 + 1 = 4 L
Osmolarity (mOsM)	1200/4 = 300 mOsM

PHYS IN ACTION!

Carry the new osmolarity across to the ECF and ICF boxes (arrows). All of the added NaCl will stay in the ECF, so add that solute amount to the ECF box. ICF solute amount is unchanged. Use V = S/C to calculate the new ECF and ICF volumes (yellow boxes).

	Total Body	ECF	ICF
Solute (mosmoles)	1200 mosmol	300 + 300 = 600	600 mosmol
Volume (L)	4 L	2 L	2 L
Osmolarity (mOsM)	300 mOsM ⟶	300 mOsM ⟶	300 mOsM

The added solution was isosmotic (300 mOsM) and its nonpenetrating concentration was the same as that of the body's (300 mOsM NP). You would predict that the solution was isotonic, and that is confirmed with these calculations, which show no water entering or leaving the cells (no change in ICF volume).

Example 2

Add 2 liters of a 500 mOsM solution. The solution is equal parts NaCl (nonpenetrating) and urea (penetrating), so it has 250 mosmol/L NaCl and 250 mosmol/L urea.

Answer

This solution has both penetrating and nonpenetrating solutes, but only nonpenetrating solutes contribute to tonicity and cause water to shift between compartments.

Before working this problem, answer the following questions:

(a) This solution is _____ osmotic to the 300 mOsM body.

(b) What is the concentration of nonpenetrating solutes [NP] in the solution? _____

(c) What is the [NP] in the body? _____

(d) Using the rules for tonicity in Table 5.4, will there be water movement into or out of the cells? If so, in what direction?

(e) Based on your answer in (d), this solution is _____ tonic to this body's cells.

Now work the problem using the starting conditions table as your starting point.

> What did you add? 2 L of (250 mosmol/L urea and 250 mosmol/L NaCl) = 2 liters of volume + 500 mosmol urea + 500 mosmol NaCl.
>
> Urea does not contribute to tonicity, so we will set the 500 mosmol of urea aside and only add the volume and NaCl in the first step:

Step 1: Add 2 liters and 500 mosmoles NaCl. Do total body column first.

	Total Body
Solute (mosmoles)	900 + 500 = 1400 mosmol
Volume (L)	3 + 2 = 5 L
Osmolarity (mOsM)	1400/5 = 280 mOsM

Step 2: Carry the new osmolarity across to ECF and ICF. All NaCl remains in the ECF so add that solute to the ECF column. Calculate new ECF and ICF volumes.

- Notice that ICF volume + ECF volume = total body volume.

	Total Body	ECF	ICF
Solute (mosmoles)	1400 mosmol	300 + 500 = 800	600
Volume (L)	5 L	2.857 L	2.143 L
Osmolarity (mOsM)	280 mOsM ➤	280 mOsM ⟶	280 mOsM

Step 3: Now add the reserved urea solute to the whole body solute to get the final osmolarity. That osmolarity carries over to the ECF and ICF compartments. Urea will distribute itself throughout the body until its concentration everywhere is equal, but it will not cause any water shift between ECF and ICF, so the ECF and ICF volumes remain as they were in Step 2.

	Total Body	ECF	ICF
Solute (mosmoles)	1400 + 500 = 1900		
Volume (L)	5 L	2.857 L	2.143 L
Osmolarity (mOsM)	1900/5 = 380 mOsM →	380 mOsM ⟶	380 mOsM

Answer the following questions from the values in the table:

(f) What happened to the body osmolarity after adding the solution? _____ This result means the added solution was _____ osmotic to the body's starting osmolarity.

(g) What happened to the ICF volume? _____ This means the added solution was _____ tonic to the cells.

Compare your answers in (f) and (g) to your answers for (a)–(e). Do they match? They should.

If you know the starting conditions of the body and you know the composition of a solution you are adding, you should be able to describe the solution's osmolarity and tonicity relative to the body by asking the questions in (a)–(e). Now test yourself by working Concept Check questions 8 and 9.

TABLE 5.5	Intravenous Solutions		
Solution	**Also Known as**	**Osmolarity**	**Tonicity**
0.9% saline*	Normal saline	Isosmotic	Isotonic
5% dextrose** in 0.9% saline	D5–normal saline	Hyperosmotic	Isotonic
5% dextrose in water	D5W	Isosmotic	Hypotonic
0.45% saline	Half-normal saline	Hyposmotic	Hypotonic
5% dextrose in 0.45% saline	D5–half-normal saline	Hyperosmotic	Hypotonic

*Saline = NaCl.
**Dextrose = glucose.

CONCEPT CHECK

6. Which of the following solutions has/have the most water per unit volume: 1 M glucose, 1 M NaCl, or 1 OsM NaCl?

7. Two compartments are separated by a membrane that is permeable to water and urea but not to NaCl. Which way will water move when the following solutions are placed in the two compartments? (*Hint*: Watch the units!)

Compartment A	Membrane	Compartment B
(a) 1 M NaCl	|	1 OsM NaCl
(b) 1 M urea	|	2 M urea
(c) 1 OsM NaCl	|	1 OsM urea

8. Use the same 3-liter, 300 mOsM body as in Figure 5.4 for this problem.

   Add 1 liter of 260 mOsM glucose to the body and calculate the new body volumes and osmolarity once all the glucose has entered the cells and been phosphorylated. Before you do the calculations, make the following predictions: This solution is _____ osmotic to the body and is _____ tonic to the body's cells.

9. Use the same 3-liter, 300 mOsM body as in Figure 5.4 for this problem. A 3-liter person working in the hot sun loses 500 mL of sweat that is equivalent to a 130 mOsM NaCl solution. Assume all NaCl loss comes from the ECF.

   (a) The sweat lost is osmotic to the body. This means that the osmolarity of the body after the sweat loss will (*increase/decrease/not change?*).

   (b) As a result of this sweat loss, the body's cell volume will (*increase/decrease/not change?*).

   (c) Using the table, calculate what happens to volume and osmolarity as a result of this sweat loss. Do the results of your calculations match your answers in (a) and (b)?

10. You have a patient who lost 1 liter of blood, and you need to restore volume quickly while waiting for a blood transfusion to arrive from the blood bank.

   (a) Which would be better to administer: 5% dextrose in water or 0.9% NaCl in water? (*Hint*: Think about how these solutes distribute in the body.) Defend your choice.

   (b) How much of your solution of choice would you have to administer to return blood volume to normal?

TRANSPORT PROCESSES

Water moves freely between body compartments, but what about other body components? Humans are large complex organisms, and the movement of material within and between body compartments is necessary for communication. This movement requires a variety of transport mechanisms. Some require an outside source of energy, such as that stored in the high-energy bond of ATP [p. 104], while other transport processes use only the kinetic or potential energy already in the system [p. 95]. Movement between compartments usually means a molecule must cross one or more cell membranes. Movement within a compartment is less restricted. For this reason, biological transport is another theme that you will encounter repeatedly as you study the organ systems.

The most general form of biological transport is the **bulk flow** of fluids within a compartment. Although many people equate **fluids** with liquids, in physics both gases and liquids are considered fluids because they flow. The main difference between the two fluids is that gases are compressible because their molecules are so far apart in space. Liquids, especially water, are not compressible. (Think of squeezing on a water balloon.)

In bulk flow, a *pressure gradient* causes fluid to flow from regions of higher pressure to regions of lower pressure. As the fluid flows, it carries with it all of its component parts, including substances dissolved or suspended in it. Blood moving through the circulatory system is an excellent example of bulk flow. The heart acts as a pump that creates a region of high pressure, pushing plasma with its dissolved solutes and the suspended blood cells through the blood vessels. Air flow in the lungs is another example of bulk flow that you will encounter as you study physiology.

Other forms of transport are more specific than bulk flow. When we discuss them, we must name the molecule or molecules that are moving. Transport mechanisms you will learn about in the following sections include diffusion, protein-mediated transport, and vesicular transport.

Cell Membranes Are Selectively Permeable

Many materials move freely within a body compartment, but exchange between the intracellular and extracellular compartments is restricted by the cell membrane. Whether or not a substance enters a cell depends on the properties of the cell membrane and those of the substance. Cell membranes are **selectively permeable,** which means that some molecules can cross them but others cannot.

The lipid and protein composition of a given cell membrane determines which molecules will enter the cell and which will leave [p. 62]. If a membrane allows a substance to pass through it, the membrane is said to be **permeable** to that substance {*permeare,* to pass through}. If a membrane does not allow a substance to pass, the membrane is said to be **impermeable** {*im-,* not} to that substance.

Membrane permeability is variable and can be changed by altering the proteins or lipids of the membrane. Some molecules, such as oxygen, carbon dioxide, and lipids, move easily across most cell membranes. On the other hand, ions, most polar molecules, and very large molecules (such as proteins), enter cells with more difficulty or may not enter at all.

Two properties of a molecule influence its movement across cell membranes: the size of the molecule and its lipid solubility. Very small molecules and those that are lipid soluble can cross directly through the phospholipid bilayer. Larger and less lipid-soluble molecules usually do not enter or leave a cell unless the cell has specific membrane proteins to transport these molecules across the lipid bilayer. Very large lipophobic molecules cannot be transported on proteins and must enter and leave cells in vesicles [p. 71].

There are multiple ways to categorize how molecules move across membranes. One scheme, just described, separates movement according to physical requirements: whether it moves by diffusion directly through the phospholipid bilayer, crosses with the aid of a membrane protein, or enters the cell in a vesicle (**FIG. 5.5**).

A second scheme classifies movement according to its energy requirements. **Passive transport** does not require the input of energy other than the potential energy stored in a concentration gradient. **Active transport** requires the input of energy from some outside source, such as the high-energy phosphate bond of ATP.

The following sections look at how cells move material across their membranes. The principles discussed here also apply to movement across intracellular membranes, when substances move between organelles.

RUNNING **PROBLEM**

Daniel's medical history tells a frightening story of almost constant medical problems since birth: recurring bouts of respiratory infections, digestive ailments, and, for the past six months, a history of weight loss. Then, last week, when Daniel began having trouble breathing, his mother rushed him to the hospital. A culture taken from Daniel's lungs raised a red flag for cystic fibrosis: The mucus from his airways was unusually thick and dehydrated. In cystic fibrosis, this thick mucus causes life-threatening respiratory congestion and provides a perfect breeding ground for infection-causing bacteria.

Q1: *In people with cystic fibrosis, movement of sodium chloride into the lumen of the airways is impaired. Why would failure to move NaCl into the airways cause the secreted mucus to be thick? (Hint: Remember that water moves into hyperosmotic regions.)*

123   **133**   139   152   153   160

---

123   133   139   152   153   160

FIG. 5.5 | **Transport across membranes**

Movement of substances across membranes can be classified by the energy requirements of transport (in parentheses) or by the physical pathway (through the membrane layer, through a membrane protein, or in a vesicle).

TRANSPORT ACROSS MEMBRANES

Active

Passive

Vesicular transport (ATP)
- Exocytosis
- Endocytosis
- Phagocytosis

Protein-Mediated
- Direct or primary active transport (ATPases)
- Indirect or secondary active transport (concentration gradient created by ATP)
- Facilitated diffusion (concentration gradient)
- Ion channel (electrochemical gradient)
- Aquaporin channel (osmosis)

Simple diffusion (concentration gradient)

DIFFUSION

Passive transport across membranes uses the kinetic energy [p. 95] inherent in molecules and the potential energy stored in concentration gradients. Gas molecules and molecules in solution constantly move from one place to another, bouncing off other molecules or off the sides of any container holding them. When molecules start out concentrated in one area of an enclosed space, their motion causes them to spread out gradually until they distribute evenly throughout the available space. This process is known as diffusion.

Diffusion {*diffundere,* to pour out} may be defined as the movement of molecules from an area of higher concentration of the molecules to an area of lower concentration of the molecules.* If you leave a bottle of cologne open and later notice its fragrance across the room, it is because the aromatic molecules in the cologne have diffused from where they are more concentrated (in the bottle) to where they are less concentrated (across the room).

Diffusion has the following seven properties:

1. *Diffusion is a passive process.* By *passive,* we mean that diffusion does not require the input of energy from some outside source. Diffusion uses only the kinetic energy possessed by all molecules.

2. *Molecules move from an area of higher concentration to an area of lower concentration.* A difference in the concentration of a substance between two places is called a **concentration gradient,** also known as a *chemical gradient.* We say that molecules diffuse *down the gradient,* from higher concentration to lower concentration.

    The rate of diffusion depends on the magnitude of the concentration gradient. The larger the concentration gradient, the faster diffusion takes place. For example, when you open a bottle of cologne, the rate of diffusion is most rapid as the molecules first escape from the bottle into the air. Later, when the cologne has spread evenly throughout the room, the rate of diffusion has dropped to zero because there is no longer a concentration gradient.

3. *Net movement of molecules occurs until the concentration is equal everywhere.* Once molecules of a given substance have distributed themselves evenly, the system reaches equilibrium and diffusion stops. Individual molecules are still moving at equilibrium, but for each molecule that exits an area, another one enters. The *dynamic equilibrium* state in diffusion means that the concentration has equalized throughout the system but molecules continue to move.

4. *Diffusion is rapid over short distances but much slower over long distances.* Albert Einstein studied the diffusion of molecules in solution and found that the time required for a molecule

to diffuse from point A to point B is proportional to the square of the distance from A to B. In other words, if the distance doubles from 1 to 2, the time needed for diffusion increases from 1^2 to 2^2 (from 1 to 4).

What does the slow rate of diffusion over long distances mean for biological systems? In humans, nutrients take five seconds to diffuse from the blood to a cell that is 100 μm from the nearest capillary. At that rate, it would take years for nutrients to diffuse from the small intestine to cells in the big toe, and the cells would starve to death.

To overcome the limitations of diffusion over distance, organisms use various transport mechanisms that speed up the movement of molecules. Most multicellular animals have some form of circulatory system to bring oxygen and nutrients rapidly from the point at which they enter the body to the cells.

5. *Diffusion is directly related to temperature.* At higher temperatures, molecules move faster. Because diffusion results from molecular movement, the rate of diffusion increases as temperature increases. Generally, changes in temperature do not significantly affect diffusion rates in humans because we maintain a relatively constant body temperature.

6. *Diffusion rate is inversely related to molecular weight and size.* Smaller molecules require less energy to move over a distance and therefore diffuse faster. Einstein showed that friction between the surface of a particle and the medium through which it diffuses is a source of resistance to movement. He calculated that diffusion is inversely proportional to the radius of the molecule: the larger the molecule, the slower its diffusion through a given medium. The experiment in **FIGURE 5.6** shows that the smaller and lighter potassium iodide (KI) molecules diffuse more rapidly through the agar gel than the larger and heavier Congo red molecules.

7. *Diffusion can take place in an open system or across a partition that separates two compartments.* Diffusion of cologne within a room is an example of diffusion taking place in an open system. There are no barriers to molecular movement, and the molecules spread out to fill the entire system. Diffusion can also take place between two compartments, such as the intracellular and extracellular compartments, but only if the partition dividing the two compartments allows the diffusing molecules to cross.

    For example, if you close the top of an open bottle of cologne, the molecules cannot diffuse out into the room because neither the bottle nor the cap is permeable to the cologne. However, if you replace the metal cap with a plastic bag that has tiny holes in it, you will begin to smell the cologne in the room because the bag is permeable to the molecules. Similarly, if a cell membrane is permeable to a molecule, that molecule can enter or leave the cell by diffusion. If the membrane is not permeable to that particular molecule, the molecule cannot cross.

TABLE 5.6 summarizes these points.

---

*Some texts use the term *diffusion* to mean any random movement of molecules, and they call molecular movement along a concentration gradient *net diffusion.* To simplify matters, we will use the term *diffusion* to mean movement down a concentration gradient.

FIG. 5.6 | Diffusion experiment

(a) Wells in an agar gel plate are filled with two dyes of equal concentration: potassium iodide (KI, 166 daltons) and Congo red (697 daltons).

(b) Ninety minutes later, the smaller and lighter KI has diffused through the gel to stain a larger area.

KI — Congo red

Time = 0 minutes   Time = 90 minutes

An important point to note: ions do not move by diffusion, even though you will read and hear about ions "diffusing across membranes." Diffusion is random molecular motion down a *concentration* gradient. Ion movement is influenced by *electrical* gradients because of the attraction of opposite charges and repulsion of like charges. For this reason, ions move in response to combined electrical and concentration gradients, or *electrochemical gradients*. This electrochemical movement is a more complex process than diffusion resulting solely from a concentration gradient, and the two processes should not be confused. We discuss ions and electrochemical gradients in more detail at the end of this chapter.

In summary, diffusion is the passive movement of uncharged molecules down their concentration gradient due to random molecular movement. Diffusion is slower over long distances and slower for large molecules. When the concentration of the diffusing molecules is the same throughout a system, the system has come to chemical equilibrium, although the random movement of molecules continues.

CONCEPT CHECK

11. If the distance over which a molecule must diffuse triples from 1 to 3, diffusion takes how many times as long?

TABLE 5.6 | Rules for Diffusion of Uncharged Molecules

General Properties of Diffusion

1. Diffusion uses the kinetic energy of molecular movement and does not require an outside energy source.

2. Molecules diffuse from an area of higher concentration to an area of lower concentration.

3. Diffusion continues until concentrations come to equilibrium. Molecular movement continues, however, after equilibrium has been reached.

4. Diffusion is faster
   — along higher concentration gradients.
   — over shorter distances.
   — at higher temperatures.
   — for smaller molecules.

5. Diffusion can take place in an open system or across a partition that separates two systems.

Simple Diffusion across a Membrane

6. The rate of diffusion through a membrane is faster if
   — the membrane's surface area is larger.
   — the membrane is thinner.
   — the concentration gradient is larger.
   — the membrane is more permeable to the molecule.

7. Membrane permeability to a molecule depends on
   — the molecule's lipid solubility.
   — the molecule's size.
   — the lipid composition of the membrane.

Lipophilic Molecules Cross Membranes by Simple Diffusion

Diffusion across membranes is a little more complicated than diffusion in an open system. Only lipid-soluble (lipophilic) molecules can diffuse through the phospholipid bilayer. Water and the many vital nutrients, ions, and other molecules that dissolve in water are lipo*phobic* as a rule: they do not readily dissolve in lipids. For these substances, the hydrophobic lipid core of the cell membrane acts as a barrier that prevents them from crossing.

Lipophilic substances that can pass through the lipid center of a membrane move by diffusion. Diffusion directly across the phospholipid bilayer of a membrane is called **simple diffusion** and has the following properties in addition to the properties of diffusion listed earlier.

1. *The rate of diffusion depends on the ability of the diffusing molecule to dissolve in the lipid layer of the membrane.* Another way to say this is that the diffusion rate depends on how permeable the membrane is to the diffusing molecules. Most molecules in solution can mingle with the polar phosphate-glycerol heads of the bilayer [p. 62], but only nonpolar molecules that are lipid-soluble (lipophilic) can dissolve in the central lipid core of the membrane. As a rule, only lipids, steroids, and small lipophilic molecules can move across membranes by simple diffusion.

   One important exception to this statement concerns water. Water, although a polar molecule, may diffuse slowly across some phospholipid membranes. For years, it was thought that the polar nature of the water molecule prevented it from moving through the lipid center of the bilayer, but experiments done with artificial membranes have shown that the small size of the water molecule allows it to slip between the lipid tails in some membranes.

   How readily water passes through the membrane depends on the composition of the phospholipid bilayer. Membranes with high cholesterol content are less permeable to water than those with low cholesterol content, presumably because the lipid-soluble cholesterol molecules fill spaces between the fatty acid tails of the lipid bilayer and thus exclude water. For example, the cell membranes of some sections of the kidney are essentially impermeable to water unless the cells insert special water channel proteins into the phospholipid bilayer. Most water movement across membranes takes place through protein channels.

2. *The rate of diffusion across a membrane is directly proportional to the surface area of the membrane.* In other words, the larger the membrane's surface area, the more molecules can diffuse across per unit time. This fact may seem obvious, but it has important implications in physiology. One striking example of how a change in surface area affects diffusion is the lung disease emphysema. As lung tissue breaks down and is destroyed, the surface area available for diffusion of

oxygen decreases. Consequently, less oxygen can move into the body. In severe cases, the oxygen that reaches the cells is not enough to sustain any muscular activity and the patient is confined to bed.

The rules for simple diffusion across membranes are summarized in Table 5.6. They can be combined mathematically into an equation known as **Fick's law of diffusion,** a relationship that involves the factors just mentioned for diffusion across membranes plus the factor of concentration gradient. In an abbreviated form, Fick's law says that the diffusion rate increases when surface area, the concentration gradient, or the membrane permeability increase:

$$\frac{\text{rate of}}{\text{diffusion}} \propto \frac{\text{surface}}{\text{area}} \times \frac{\text{concentration}}{\text{gradient}} \times \frac{\text{membrane}}{\text{permeability}}$$

FIGURE 5.7 illustrates the principles of Fick's law.

Membrane permeability is the most complex of the terms in Fick's law because several factors influence it:

1. the size (and shape, for large molecules) of the diffusing molecule. As molecular size increases, membrane permeability decreases.
2. the lipid-solubility of the molecule. As lipid solubility of the diffusing molecule increases, membrane permeability to the molecule increases.
3. the composition of the lipid bilayer across which it is diffusing. Alterations in lipid composition of the membrane change how easily diffusing molecules can slip between the individual phospholipids. For example, cholesterol molecules in membranes pack themselves into the spaces between the fatty acids tails and retard passage of molecules through those spaces [Fig. 3.2, p. 63], making the membrane less permeable.

We can rearrange the Fick equation to read:

$$\frac{\text{diffusion rate}}{\text{surface area}} = \frac{\text{concentration}}{\text{gradient}} \times \frac{\text{membrane}}{\text{permeability}}$$

This equation now describes the flux of a molecule across the membrane, because **flux** is defined as the diffusion rate per unit surface area of membrane:

$$\text{flux} = \text{concentration gradient} \times \text{membrane permeability}$$

In other words, the flux of a molecule across a membrane depends on the concentration gradient and the membrane's permeability to the molecule.

Remember that the principles of diffusion apply to all biological membranes, not just to the cell membrane. Diffusion of materials in and out of organelles follows the same rules.

FIG. 5.7 | Fick's law of diffusion

Diffusion of an uncharged solute across a membrane is proportional to the concentration gradient of the solute, the membrane surface area, and the membrane permeability to the solute.

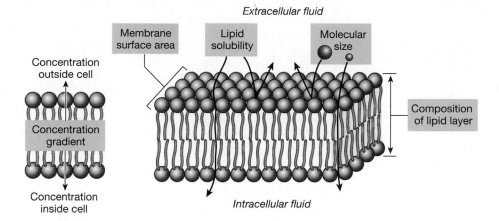

Factors affecting rate of diffusion through a cell membrane:

- Lipid solubility
- Molecular size
- Concentration gradient
- Membrane surface area
- Composition of lipid layer

Fick's Law of Diffusion
Rate of diffusion ∝ surface area × concentration gradient × membrane permeability

Membrane Permeability
Membrane permeability ∝ $\dfrac{\text{lipid solubility}}{\text{molecular size}}$
Changing the composition of the lipid layer can increase or decrease membrane permeability.

CONCEPT CHECK

12. Where does the energy for diffusion come from?

13. Which is more likely to cross a cell membrane by simple diffusion: a fatty acid molecule or a glucose molecule?

14. What happens to the flux of molecules in each of the following cases?

    (a) Molecular size increases.

    (b) Concentration gradient increases.

    (c) Surface area of membrane decreases.

15. Two compartments are separated by a membrane that is permeable only to water and to yellow dye molecules. Compartment A is filled with an aqueous solution of yellow dye, and compartment B is filled with an aqueous solution of an equal concentration of blue dye. If the system is left undisturbed for a long time, what color will compartment A be: yellow, blue, or green? (Remember, yellow plus blue makes green.) What color will compartment B be?

16. What keeps atmospheric oxygen from diffusing into our bodies across the skin? (*Hint:* What kind of epithelium is skin?)

PROTEIN-MEDIATED TRANSPORT

In the body, simple diffusion across membranes is limited to lipophilic molecules. The majority of molecules in the body are either lipophobic or electrically charged and therefore cannot cross membranes by simple diffusion. Instead, the vast majority of solutes cross membranes with the help of membrane proteins, a process we call **mediated transport.**

If mediated transport is passive and moves molecules down their concentration gradient, and if net transport stops when concentrations are equal on both sides of the membrane, the process is **facilitated diffusion** (Fig. 5.5). If protein-mediated transport requires energy from ATP or another outside source and moves a substance against its concentration gradient, the process is known as *active transport.*

Membrane Proteins Have Four Major Functions

Protein-mediated transport across a membrane is carried out by membrane-spanning transport proteins. For physiologists, classifying membrane proteins by their function is more useful than classifying them by their structure. Our functional classification scheme recognizes four broad categories of membrane proteins: (1) structural proteins, (2) enzymes, (3) receptors, and (4) transport proteins. **FIGURE 5.8** is a map comparing the structural and functional classifications of membrane proteins. These groupings are not completely distinct, and as you will learn, some membrane proteins have more than one function, such as receptor-channels and receptor-enzymes.

Structural Proteins The **structural proteins** of membranes have three major roles.

FIG. 5.8 Map of membrane proteins

Functional categories of membrane proteins include transporters, structural proteins, enzymes, and receptors.

---

1. They help create cell junctions that hold tissues together, such as tight junctions and gap junctions [Fig. 3.8, p. 75].

2. They connect the membrane to the cytoskeleton to maintain the shape of the cell [Fig. 3.2, p. 63]. The microvilli of transporting epithelia are one example of membrane shaping by the cytoskeleton [Fig. 3.4b, p. 66].

3. They attach cells to the extracellular matrix by linking cytoskeleton fibers to extracellular collagen and other protein fibers [p. 72].

Enzymes Membrane **enzymes** catalyze chemical reactions that take place either on the cell's external surface or just inside the cell. For example, enzymes on the external surface of cells lining the small intestine are responsible for digesting peptides and carbohydrates. Enzymes attached to the intracellular surface of many cell membranes play an important role in transferring signals from the extracellular environment to the cytoplasm [see Chapter 6].

Receptors Membrane **receptor proteins** are part of the body's chemical signaling system. The binding of a receptor with its ligand usually triggers another event at the membrane (**FIG. 5.9**). Sometimes the ligand remains on the cell surface, and the receptor-ligand complex triggers an intracellular response. In other instances, the receptor-ligand complex is brought into the cell in a vesicle [p. 71]. Membrane receptors also play an important role in some forms of vesicular transport, as you will learn later in this chapter.

Transport Proteins The fourth group of membrane proteins—**transport proteins**—moves molecules across membranes. There are several different ways to classify transport proteins. Scientists have discovered that the genes for most membrane transport proteins belong to one of two gene "superfamilies": the *ATP-binding cassette (ABC) superfamily* or the *solute carrier (SLC) superfamily*. The ABC family proteins use ATP's energy to transport small molecules or ions across membranes. The 52 families of the SLC superfamily include most facilitated diffusion transporters as well as some active transporters.

A second way to classify transport* recognizes two main types of transport proteins: channels and carriers (**FIG. 5.10**). **Channel proteins** create water-filled passageways that directly link the intracellular and extracellular compartments. **Carrier proteins,** also just called *transporters,* bind to the substrates that they carry but never form a direct connection between the intracellular fluid and extracellular fluid. As Figure 5.10 shows, carriers are open to one side of the membrane or the other, but not to both at once the way channel proteins are.

Why do cells need both channels and carriers? The answer lies in the different properties of the two transport proteins. Channel proteins allow more rapid transport across the membrane but generally are limited to moving small ions and water.

---

*The Transporter Classification System, *www.tcdb.org*

FIG. 5.9 Membrane receptors bind extracellular ligands

Extracellular fluid

Ligand binds to a cell membrane receptor protein.

Receptor-ligand complex triggers intracellular response.

Receptor-ligand complex brought into the cell

Cell membrane

Receptor

Events in the cell

Cytoplasmic vesicle

Intracellular fluid

Carriers, while slower, can move larger molecules than channels can. There is some overlap between the two types, both structurally and functionally. For example, the aquaporin protein AQP has been shown to act both as a water channel and as a carrier for certain small organic molecules.

Channel Proteins Form Open, Water-Filled Passageways

Channel proteins are made of membrane-spanning protein subunits that create a cluster of cylinders with a tunnel or *pore* through the center. Nuclear pore complexes [p. 72] and gap junctions [Fig. 3.8b, p. 75] can be considered very large forms of channels. In this book, we restrict use of the term *channel* to smaller channels whose centers are narrow, water-filled pores (**FIG. 5.11**). Movement through these smaller channels is mostly restricted to water and ions. When water-filled ion channels are open, tens of millions of ions per second can whisk through them unimpeded.

Channel proteins are named according to the substances that they allow to pass. Most cells have **water channels** made from a protein called *aquaporin*. In addition, more than 100 types of **ion channels** have been identified. Ion channels may be specific for one ion or may allow ions of similar size and charge to pass. For example, there are Na^+ channels, K^+ channels, and nonspecific *monovalent* ("one-charge") cation channels that transport Na^+, K^+ and lithium ions Li^+. Other ion channels you will encounter frequently in this text are Ca^{2+} channels and Cl^- channels. Ion channels come in many subtypes, or *isoforms*.

The selectivity of a channel is determined by the diameter of its central pore and by the electrical charge of the amino acids that line the channel. If the channel amino acids are positively charged, positive ions are repelled and negative ions can pass

through the channel. On the other hand, a cation channel must have a negative charge that attracts cations but prevents the passage of Cl^- or other anions.

Channel proteins are like narrow doorways into the cell. If the door is closed, nothing can go through. If the door is open, there is a continuous passage between the two rooms connected by the doorway. The open or closed state of a channel is determined by regions of the protein molecule that act like swinging "gates."

According to current models, channel "gates" take several forms. Some channel proteins have gates in the middle of the protein's pore. Other gates are part of the cytoplasmic side of the membrane protein. Such a gate can be envisioned as a ball on a chain that swings up and blocks the mouth of the channel (Fig. 5.10a). One type of channel in neurons has two different gates.

Channels can be classified according to whether their gates are usually open or usually closed. **Open channels** spend most of their time with their gate open, allowing ions to move back and forth across the membrane without regulation. These gates may occasionally flicker closed, but for the most part these channels behave as if they have no gates. Open channels are sometimes called either *leak channels* or pores, as in *water pores*.

Gated channels spend most of their time in a closed state, which allows these channels to regulate the movement of ions through them. When a gated channel opens, ions move through the channel just as they move through open channels. When a gated channel is closed, which it may be much of the time, it allows no ion movement between the intracellular and extracellular fluid.

What controls the opening and closing of gated channels? For **chemically gated channels,** the gating is controlled by intracellular messenger molecules or extracellular ligands that bind to the channel protein. **Voltage-gated channels** open and close when the electrical state of the cell changes. Finally, **mechanically gated channels** respond to physical forces, such as increased temperature or pressure that puts tension on the membrane and pops the channel gate open. You will encounter many variations of these channel types as you study physiology.

RUNNING **PROBLEM**

Cystic fibrosis is a debilitating disease caused by a defect in a membrane channel protein that normally transports chloride ions (Cl^-). The channel—called the **cystic fibrosis transmembrane conductance regulator,** or CFTR—is located in epithelia lining the airways, sweat glands, and pancreas. A gate in the CFTR channel opens when the nucleotide ATP binds to the protein. In the lungs, this open channel transports Cl^- out of the epithelial cells and into the airways. In people with cystic fibrosis, CFTR is nonfunctional or absent. As a result, chloride transport across the epithelium is impaired, and thickened mucus is the result.

Q2: *Is the CFTR a chemically gated, a voltage-gated, or a mechanically gated channel protein?*

123 133 **139** 152 153 160

FIG. 5.10 **ESSENTIALS**

Membrane Transporters

Membrane transporters are membrane-spanning proteins that help move lipophobic molecules across membranes.

Close-up views of transporters are shown in the top two rows and distant views in the bottom row. Primary active transport is indicated by *ATP* on the protein.

CONCEPT CHECK

17. Positively charged ions are called _____, and negatively charged ions are called _____.

Carrier Proteins Change Conformation to Move Molecules

The second type of transport protein is the carrier protein (Fig. 5.10b). Carrier proteins bind with specific substrates and carry them across the membrane by changing conformation. Small organic molecules (such as glucose and amino acids) that are too

large to pass through channels cross membranes using carriers. Ions such as Na^+ and K^+ may move by carriers as well as through channels. Carrier proteins move solutes and ions into and out of cells as well as into and out of intracellular organelles, such as the mitochondria.

Some carrier proteins move only one kind of molecule and are known as **uniport carriers.** However, it is common to find carriers that move two or even three kinds of molecules. A carrier that moves more than one kind of molecule at one time is called a **cotransporter.** If the molecules being transported are moving in the same direction, whether into or out of the cell, the carrier proteins are **symport carriers** {*sym-,* together + *portare,* to carry}. (Sometimes,

FIG. 5.11 | The structure of channel proteins

Many channels are made of multiple protein subunits that assemble in the membrane. Hydrophilic amino acids in the protein line the channel, creating a water-filled passage that allows ions and water to pass through.

One protein subunit of channel

Channel through center of membrane protein

Channel through center of membrane protein (viewed from above)

the term *cotransport* is used in place of *symport.*) If the molecules are being carried in opposite directions, the carrier proteins are **antiport carriers** {*anti*, opposite + *portare*, to carry}, also called *exchangers.* Symport and antiport carriers are shown in Figure 5.10b.

Carriers are large, complex proteins with multiple subunits. The conformation change required of a carrier protein makes this mode of transmembrane transport much slower than movement through channel proteins. A carrier protein can move only 1,000 to 1,000,000 molecules per second, in contrast to tens of millions of ions per second that move through a channel protein.

Carrier proteins differ from channel proteins in another way: carriers never create a continuous passage between the inside and outside of the cell. If channels are like doorways, then carriers are like revolving doors that allow movement between inside and outside without ever creating an open hole. Carrier proteins can transport molecules across a membrane in both directions, like a revolving door at a hotel, or they can restrict their transport to one direction, like the turnstile at an amusement park that allows you out of the park but not back in.

One side of the carrier protein always creates a barrier that prevents free exchange across the membrane. In this respect, carrier proteins function like the Panama Canal (**FIG. 5.12**). Picture the canal with only two gates, one on the Atlantic side and one on the Pacific side. Only one gate at a time is open.

When the Atlantic gate is closed, the canal opens into the Pacific. A ship enters the canal from the Pacific, and the gate closes behind it. Now the canal is isolated from both oceans with the ship trapped in the middle. Then the Atlantic gate opens, making the canal continuous with the Atlantic Ocean. The ship sails out of the gate and off into the Atlantic, having crossed the barrier of the land without the canal ever forming a continuous connection between the two oceans.

Movement across the membrane through a carrier protein is similar (Fig. 5.12b). The molecule being transported binds to the carrier on one side of the membrane (the extracellular side in our example). This binding changes the conformation of the carrier protein so that the opening closes. After a brief transition in which both sides are closed, the opposite side of the carrier opens to the other side of the membrane. The carrier then releases the transported molecule into the opposite compartment, having brought it through the membrane without creating a continuous connection between the extracellular and intracellular compartments.

Carrier proteins can be divided into two categories according to the energy source that powers the transport. As noted earlier,

FIG. 5.12 | Carrier proteins

(a) Carrier proteins, like the canal illustrated, never form a continuous passageway between the extracellular and intracellular fluids.

Closed gate

Pacific Ocean — Atlantic Ocean

Passage open to one side

Pacific Ocean — Atlantic Ocean

Transition state with both gates closed

Pacific Ocean — Atlantic Ocean

Passage open to other side

(b) The ligand binding sites change affinity when the protein conformation changes.

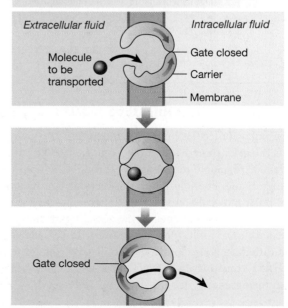

Extracellular fluid — Intracellular fluid

Molecule to be transported

Gate closed

Carrier

Membrane

Gate closed

FIG. 5.13   Facilitated diffusion of glucose into cells

(a) Facilitated diffusion brings glucose into the cell down its concentration gradient using a GLUT transporter.

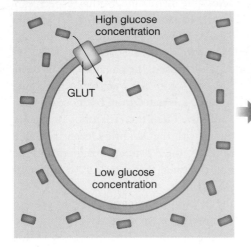

(b) Diffusion reaches equilibrium when the glucose concentrations inside and outside the cell are equal.

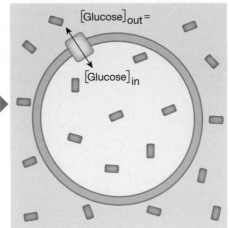

(c) In most cells, conversion of imported glucose into glucose 6-phosphate (G-6-P) keeps intracellular glucose concentrations low so that diffusion never reaches equilibrium.

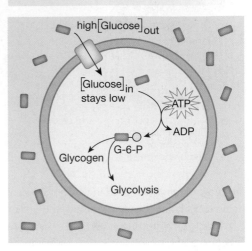

facilitated diffusion is protein-mediated transport in which no outside source of energy except a concentration gradient is needed to move molecules across the cell membrane. Active transport is protein-mediated transport that requires an outside energy source, either ATP or the potential energy stored in a concentration gradient that was created using ATP. We will look first at facilitated diffusion.

CONCEPT CHECK

18. Name four functions of membrane proteins.
19. Which kinds of particles pass through open channels?
20. Name two ways channels differ from carriers.
21. If a channel is lined with amino acids that have a net positive charge, which of the following ions is/are likely to move freely through the channel? Na^+, Cl^-, K^+, Ca^{2+}
22. Why can't glucose cross the cell membrane through open channels?

Facilitated Diffusion Uses Carrier Proteins

Some polar molecules appear to move into and out of cells by diffusion, even though we know from their chemical properties that they are unable to pass easily through the lipid core of the cell membrane. The solution to this seeming contradiction is that these polar molecules cross the cell membrane by facilitated diffusion, with the aid of specific carriers. Sugars and amino acids are examples of molecules that enter or leave cells using facilitated diffusion. For example, the family of carrier proteins known as **GLUT transporters** move glucose and related hexose sugars across membranes.

Facilitated diffusion has the same properties as simple diffusion (see Tbl. 5.6). The transported molecules move down their concentration gradient, the process requires no input of outside energy, and net movement stops at equilibrium, when the concentration inside the cell equals the concentration outside the cell (**FIG. 5.13**):

$$[\text{glucose}]_{ECF} = [\text{glucose}]_{ICF}{}^*$$

Facilitated diffusion carriers always transport molecules down their concentration gradient. If the gradient reverses, so does the direction of transport.

Cells in which facilitated diffusion takes place can avoid reaching equilibrium by keeping the concentration of substrate in the cell low. With glucose, for example, this is accomplished by phosphorylation (Fig. 5.13c). As soon as a glucose molecule enters the cell on the GLUT carrier, it is phosphorylated to glucose 6-phosphate, the first step of glycolysis [p. 107]. Addition of the phosphate group prevents build-up of glucose inside the cell and also prevents glucose from leaving the cell.

CONCEPT CHECK

23. Liver cells (hepatocytes) are able to convert glycogen to glucose, thereby making the intracellular glucose concentration higher than the extracellular glucose concentration. In what direction do the hepatic GLUT2 transporters carry glucose when this occurs?

*In this book, the presence of brackets around a solute's name indicates concentration.

Active Transport Moves Substances against Their Concentration Gradients

Active transport is a process that moves molecules *against* their concentration gradient—that is, from areas of lower concentration to areas of higher concentration. Rather than creating an equilibrium state, where the concentration of the molecule is equal throughout the system, active transport creates a state of *dis*equilibrium by making concentration differences more pronounced. Moving molecules against their concentration gradient requires the input of outside energy, just as pushing a ball up a hill requires energy [see Fig. 4.2, p. 95]. The energy for active transport comes either directly or indirectly from the high-energy phosphate bond of ATP.

Active transport can be divided into two types. In **primary (direct) active transport,** the energy to push molecules against their concentration gradient comes directly from the high-energy phosphate bond of ATP. **Secondary (indirect) active transport** uses potential energy [p. 95] stored in the concentration gradient of one molecule to push other molecules against their concentration gradient. All secondary active transport ultimately depends on primary active transport because the concentration gradients that drive secondary transport are created using energy from ATP.

The mechanism for both types of active transport appears to be similar to that for facilitated diffusion. A substrate to be transported binds to a membrane carrier and the carrier then changes conformation, releasing the substrate into the opposite compartment. Active transport differs from facilitated diffusion because the conformation change in the carrier protein requires energy input.

Primary Active Transport Because primary active transport uses ATP as its energy source, many primary active transporters are known as **ATPases.** You may recall that the suffix *-ase* signifies an enzyme, and the stem (ATP) is the substrate upon which the enzyme is acting [p. 101]. These enzymes hydrolyze ATP to ADP and inorganic phosphate (P_i), releasing usable energy in the process. Most of the ATPases you will encounter in your study of physiology are listed in **TABLE 5.7**. ATPases are sometimes called *pumps,* as in the sodium-potassium pump, or Na^+-K^+-ATPase, mentioned earlier in this chapter.

FIG. 5.14 The sodium-potassium pump, Na^+-K^+-ATPase

The Na^+-K^+-ATPase uses energy from ATP to pump Na^+ out of the cell and K^+ into the cell.

Intracellular fluid:
Low [Na^+]
High [K^+]

Extracellular fluid:
High [Na^+]
Low [K^+]

Na^+ ATP K^+ *

*In this book, carrier proteins that hydrolyze ATP have the letters ATP written on the membrane protein.

The sodium-potassium pump is probably the single most important transport protein in animal cells because it maintains the concentration gradients of Na^+ and K^+ across the cell membrane (**FIG. 5.14**). The transporter is arranged in the cell membrane so that it pumps 3 Na^+ out of the cell and 2 K^+ into the cell for each ATP consumed. In some cells, the energy needed to move these ions uses 30% of all the ATP produced by the cell. **FIGURE 5.15** illustrates the current model of how the Na^+-K^+-ATPase works.

Secondary Active Transport The sodium concentration gradient, with Na^+ concentration high in the extracellular fluid and low inside the cell, is a source of potential energy that the cell can harness for other functions. For example, nerve cells use the sodium gradient to transmit electrical signals, and epithelial cells use it to drive the uptake of nutrients, ions, and water. Membrane transporters that use potential energy stored in concentration gradients to move molecules are called *secondary active transporters.*

Secondary active transport uses the kinetic energy of one molecule moving down its concentration gradient to push other molecules against their concentration gradient. The cotransported molecules may go in the same direction across the membrane (symport) or in opposite directions (antiport). The most

TABLE 5.7	Primary Active Transporters
Names	**Type of Transport**
Na^+-K^+-ATPase or sodium-potassium pump	Antiport
Ca^{2+}-ATPase	Uniport
H^+-ATPase or proton pump	Uniport
H^+-K^+-ATPase	Antiport

FIG. 5.15 Mechanism of the Na⁺-K⁺-ATPase

This figure presents one model of how the Na⁺-K⁺-ATPase uses energy and inorganic phosphate (P$_i$) from ATP to move ions across a membrane. Phosphorylation and dephosphorylation of the ATPase change its conformation and the binding sites' affinity for ions.

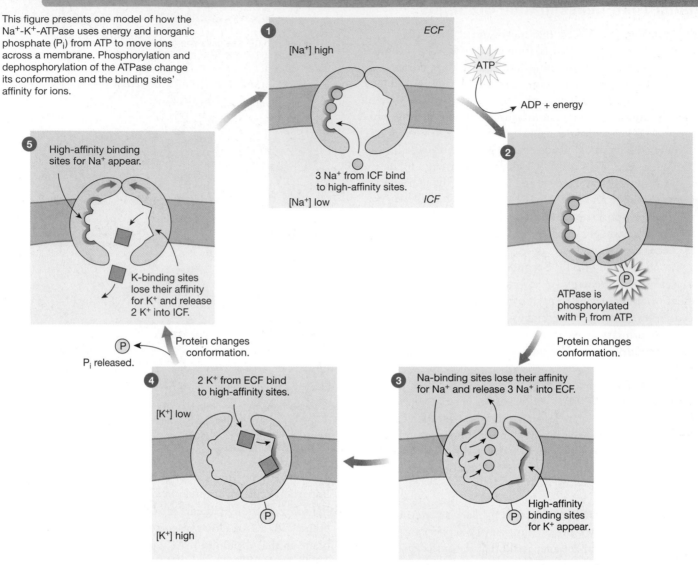

1 ECF

[Na⁺] high

ATP

ADP + energy

3 Na⁺ from ICF bind to high-affinity sites.

[Na⁺] low   ICF

2 ATPase is phosphorylated with P$_i$ from ATP.

Protein changes conformation.

3 Na-binding sites lose their affinity for Na⁺ and release 3 Na⁺ into ECF.

High-affinity binding sites for K⁺ appear.

4 2 K⁺ from ECF bind to high-affinity sites.

[K⁺] low

[K⁺] high

5 High-affinity binding sites for Na⁺ appear.

K-binding sites lose their affinity for K⁺ and release 2 K⁺ into ICF.

Protein changes conformation.

P$_i$ released.

---

common secondary active transport systems are driven by the sodium concentration gradient.

As one Na⁺ moves into the cell, it either brings one or more molecules with it or trades places with molecules exiting the cell. The major Na⁺-dependent transporters are listed in **TABLE 5.8**. Notice that the cotransported substances may be either other ions or uncharged molecules, such as glucose. As you study the different systems of the body, you will find these secondary active transporters taking part in many physiological processes.

The mechanism of the **Na⁺-glucose secondary active transporter** (SGLT) is illustrated in **FIGURE 5.16**. Both Na⁺ and glucose bind to the SGLT protein on the extracellular fluid side. Sodium binds first and causes a conformational change in the protein that creates a high-affinity binding site for glucose **1**. When glucose binds to SGLT **2**, the protein changes conformation again and opens its channel to the intracellular fluid side **3**. Sodium is released to the ICF as it moves down its concentration gradient. The loss of Na⁺ from the protein changes the binding site

TABLE 5.8	Examples of Secondary Active Transporters
Symport Carriers	**Antiport Carriers**
Sodium-Dependent Transporters	
Na⁺-K⁺-2Cl⁻ (NKCC)	Na⁺-H⁺ (NHE)
Na⁺-glucose (SGLT)	Na⁺-Ca²⁺ (NCX)
Na⁺-Cl⁻	
Na⁺-HCO₃⁻	
Na⁺-amino acids (several types)	
Na⁺-bile salts (small intestine)	
Na⁺-choline uptake (nerve cells)	
Na⁺-neurotransmitter uptake (nerve cells)	
Nonsodium-Dependent Transporters	
H⁺-peptide symporter (pepT)	HCO₃⁻-Cl⁻

FIG. 5.16 Sodium-glucose cotransport

The SGLT transporter uses the potential energy stored in the Na^+ concentration gradient to move glucose against its concentration gradient.

1 Na$^+$ binds to carrier.

Lumen of intestine or kidney

Intracellular fluid

Na$^+$

[Na$^+$] high

Glu

[glucose] low

SGLT protein

[Na$^+$] low
[glucose] high

2 Na$^+$ binding creates a high-affinity site for glucose.

Na$^+$

Glu

Lumen

ICF

3 Glucose binding changes carrier conformation so that binding sites now face the ICF.

Na$^+$

Glu

Lumen

ICF

4 Na$^+$ is released into cytosol, where [Na$^+$] is low. Release changes glucose-binding site to low affinity. Glucose is released.

Na$^+$

[Na$^+$] low

Glu

[glucose] high

Lumen

ICF

for glucose back to a low-affinity site, so glucose is released and follows Na$^+$ into the cytoplasm **4**. The net result is the entry of glucose into the cell against its concentration gradient, coupled to the movement of Na$^+$ into the cell down its concentration gradient. The SGLT transporter can only move glucose into cells because glucose must follow the Na$^+$ gradient.

In contrast, GLUT transporters are reversible and can move glucose into or out of cells depending on the concentration gradient. For example, when blood glucose levels are high, GLUT transporters on liver cells bring glucose into those cells. During times of fasting, when blood glucose levels fall, liver cells convert their glycogen stores to glucose. When the glucose concentration inside the liver cells builds up and exceeds the glucose concentration in the plasma, glucose leaves the cells on the reversible GLUT transporters. GLUT transporters are found on all cells of the body.

If GLUT transporters are everywhere, then why does the body need the SGLT Na$^+$-glucose symporter? The simple answer is that both SGLT and GLUT are needed to move glucose from one side of an epithelium to the other. Consequently, SGLT transporters are found on certain epithelial cells, such as intestinal and kidney cells, that bring glucose into the body from the external environment. We discuss the process of transepithelial transport of glucose later in this chapter.

CONCEPT CHECK

24. Name two ways active transport by the Na$^+$-K$^+$-ATPase (Fig. 5.15) differs from secondary transport by the SGLT (Fig. 5.16).

Carrier-Mediated Transport Exhibits Specificity, Competition, and Saturation

Both passive and active forms of carrier-mediated transport demonstrate specificity, competition, and saturation—three properties that result from the binding of a substrate to a protein [p. 46].

Specificity Specificity refers to the ability of a transporter to move only one molecule or only a group of closely related molecules [p. 46]. One example of specificity is found in the GLUT family of transporters, which move 6-carbon sugars (*hexoses*), such as glucose, mannose, galactose, and fructose [p. 31], across cell membranes. GLUT transporters have binding sites that recognize and transport hexoses, but they will not transport the disaccharide maltose or any form of glucose that is not found in nature (**FIG. 5.17b**). For this reason we can say that GLUT transporters are specific for naturally occurring 6-carbon monosaccharides.

For many years, scientists assumed that there must be different isoforms of the glucose-facilitated diffusion carrier because they had observed that glucose transport was regulated by hormones in some cells but not in others. However, it was not until the 1980s that the first glucose transporter was isolated. To date, 14 SCL2A (GLUT) genes have been identified. The important GLUT proteins you will encounter in this book include GLUT1, found in most cells of the body; GLUT2, found in liver and in kidney and intestinal epithelia; GLUT3, found in neurons; GLUT4, the insulin-regulated transporter of skeletal

FIG. 5.17   Transporter saturation and competition

(a) The GLUT transporter brings glucose across cell membranes.

(b) Maltose is a competitive inhibitor that binds to the GLUT transporter but is not itself carried across the membrane.

Extracellular fluid

Glucose

GLUT transporter

Intracellular fluid

Glucose

Maltose

(c) Saturation. This graph shows that transport can reach a maximum rate when all the carrier binding sites are filled with substrate.

Transport maximum

Rate of transport into cell →

Extracellular substrate concentration →

Transport rate is proportional to substrate concentration until the carriers are saturated.

Q GRAPH QUESTION
How could the cell increase its transport rate in this example?

(d) Competition. This graph shows glucose transport rate as a function of glucose concentration. In one experiment, only glucose was present. In the second experiment, a constant concentration of galactose was present.

Glucose transport rate

Glucose only

Glucose and galactose (1 mM)

5   10   15

Glucose concentration (mM)

Q GRAPH QUESTION
Can you tell from this graph if galactose is being transported?

muscle; and GLUT5, the intestinal fructose transporter. The restriction of different GLUT transporters to different tissues is an important feature in the metabolism and homeostasis of glucose.

Competition The property of competition is closely related to specificity. A transporter may move several members of a related group of substrates, but those substrates compete with one another for binding sites on the transporter. For example, GLUT transporters move the family of hexose sugars, but each different GLUT transporter has a "preference" for one or more hexoses, based on its binding affinity.

The results of an experiment demonstrating competition are shown in Figure 5.17d. The graph shows glucose transport rate as a function of glucose concentration. The top line (red) shows transport when only glucose is present. The lower line (black) shows that glucose transport decreases if galactose is also present. Galactose competes for binding sites on the GLUT transporters and displaces some glucose molecules. With fewer glucose molecules able to bind to the GLUT protein, the rate of glucose transport into the cell decreases.

Sometimes, the competing molecule is not transported but merely blocks the transport of another substrate. In this case, the competing molecule is a *competitive inhibitor* [p. 49]. In the glucose transport system, the disaccharide maltose is a competitive inhibitor (Fig. 5.17b). It competes with glucose for the binding site, but once bound, it is too large to be moved across the membrane.

Competition between transported substrates has been put to good use in medicine. An example involves gout, a disease caused by elevated levels of uric acid in the plasma. One method of decreasing uric acid in plasma is to enhance its excretion in the urine. Normally, the kidney's *organic anion transporter* (OAT) reclaims urate (the anion form of uric acid) from the urine and returns the acid to the plasma. However, if an organic acid called probenecid is administered to the patient, OAT binds to probenecid instead of to uric acid, preventing the reabsorption of urate. As a result, more urate leaves the body in the urine, lowering the uric acid concentration in the plasma.

Saturation The rate of substrate transport depends on the substrate concentration and the number of carrier molecules, a property that is shared by enzymes and other binding proteins [p. 51]. For a fixed number of carriers, however, as substrate concentration increases, the transport rate increases up to a maximum, the point at which all carrier binding sites are filled with substrate. At this point, the carriers are said to have reached saturation. At saturation, the carriers are working at their maximum rate, and a further increase in substrate concentration has no effect. Figure 5.17c represents saturation graphically.

As an analogy, think of the carriers as doors into a concert hall. Each door has a maximum number of people that it can allow to enter the hall in a given period of time. Suppose that all the doors together can allow a maximum of 100 people per minute to enter the hall. This is the maximum transport rate, also called the **transport maximum.** When the concert hall is empty, three maintenance people enter the doors every hour. The transport rate is 3 people/60 minutes, or 0.05 people/minute, well under the maximum. For a local dance recital, about 50 people per minute go through the doors, still well under the maximum.

When the most popular rock group of the day appears in concert, however, thousands of people gather outside. When the doors open, thousands of people are clamoring to get in, but the doors allow only 100 people/minute into the hall. The doors are working at the maximum rate, so it does not matter whether there are 1,000 or 3,000 people trying to get in. The transport rate saturates at 100 people/minute.

How can cells increase their transport capacity and avoid saturation? One way is to increase the number of carriers in the membrane. This would be like opening more doors into the concert hall. Under some circumstances, cells are able to insert additional carriers into their membranes. Under other circumstances, a cell may withdraw carriers to decrease movement of a molecule into or out of the cell.

All forms of carrier-mediated transport show specificity, competition, and saturation, but as you learned earlier in the chapter, they also differ in one important way: passive mediated transport—better known as facilitated diffusion—requires no input of energy from an outside source. Active transport requires energy input from ATP, either directly or indirectly.

VESICULAR TRANSPORT

What happens to the many macromolecules that are too large to enter or leave cells through protein channels or carriers? They move in and out of the cell with the aid of bubble-like *vesicles* [p. 71] created from the cell membrane. Cells use two basic processes to import large molecules and particles: phagocytosis and endocytosis. Some scientists consider phagocytosis to be a type of endocytosis, but mechanistically the two processes are different. Material leaves cells by the process known as exocytosis, a process that is similar to endocytosis run in reverse.

CONCEPT CHECK

25. What would you call a carrier that moves two substrates in opposite directions across a membrane?

26. In the concert-hall door analogy, we described how the maximum transport rate might be increased by increasing the number of doors leading into the hall. Using the same analogy, can you think of another way a cell might increase its maximum transport rate?

Phagocytosis Creates Vesicles Using the Cytoskeleton

If you studied *Amoeba* in your biology laboratory, you may have watched these one-cell creatures ingest their food by surrounding it and enclosing it within a vesicle that is brought into the cytoplasm. **Phagocytosis** {*phagein*, to eat + *cyte*, cell + *-sis*, process} is the actin-mediated process by which a cell engulfs a bacterium or other particle into a large membrane-bound vesicle called a **phagosome** {*soma*, body} (**FIG. 5.18**). The phagosome pinches off from the cell membrane and moves to the interior of the cell, where it fuses with a lysosome [p. 71],

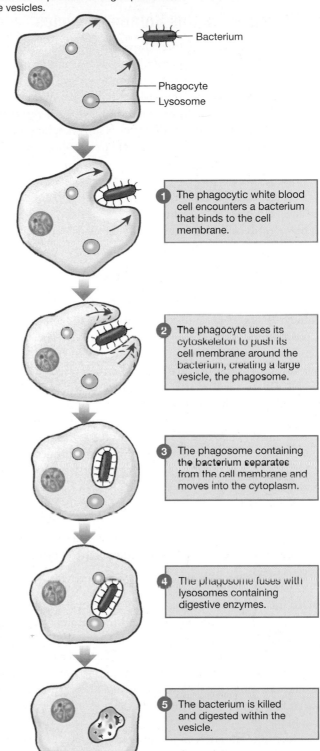

FIG. 5.18 **Phagocytosis**

Phagocytosis uses actin microfilaments and myosin motor proteins to engulf particles in large vesicles.

— Bacterium

— Phagocyte
— Lysosome

1 The phagocytic white blood cell encounters a bacterium that binds to the cell membrane.

2 The phagocyte uses its cytoskeleton to push its cell membrane around the bacterium, creating a large vesicle, the phagosome.

3 The phagosome containing the bacterium separates from the cell membrane and moves into the cytoplasm.

4 The phagosome fuses with lysosomes containing digestive enzymes.

5 The bacterium is killed and digested within the vesicle.

whose digestive enzymes destroy the bacterium. Phagocytosis requires energy from ATP for the movement of the cytoskeleton and for the intracellular transport of the vesicles. In humans, phagocytosis occurs in certain types of white blood cells called *phagocytes,* which specialize in "eating" bacteria and other foreign particles.

Endocytosis Creates Smaller Vesicles

Endocytosis, the second process by which large molecules or particles move into cells, differs from phagocytosis in two important ways. First, in endocytosis the membrane surface indents rather than pushes out. Second, the vesicles formed from endocytosis are much smaller. In addition, some endocytosis is *constitutive;* that is, it is an essential function that is always taking place. In contrast, phagocytosis must be triggered by the presence of a substance to be ingested.

Endocytosis is an active process that requires energy from ATP. It can be nonselective, allowing extracellular fluid to enter the cell—a process called **pinocytosis** {*pino-,* drink}—or it can be highly selective, allowing only specific molecules to enter the cell. In receptor-mediated endocytosis, a ligand binds to a membrane receptor protein to activate the process.

Receptor-Mediated Endocytosis Receptor-mediated **endocytosis** takes place in regions of the cell membrane known as **coated pits,** indentations where the cytoplasmic side of the membrane has high concentrations of protein. The most common protein found in coated pits is *clathrin,* illustrated in **FIGURE 5.19.** In the first step of the process, extracellular ligands that will be brought into the cell bind to their membrane receptors ❶. The receptor-ligand complex migrates along the cell surface until it encounters a coated pit ❷. Once the receptor-ligand complex is in the coated pit, the membrane draws inward, or *invaginates* ❸, then pinches off from the cell membrane and becomes a cytoplasmic vesicle. The clathrin molecules are released and recycle back to the membrane ❹. In the vesicle, the receptor and ligand separate, leaving the ligand inside an *endosome* ❺. The endosome moves to a lysosome if the ligand is to be destroyed, or to the Golgi complex if the ligand is to be processed ❻.

Meanwhile, the ligand's membrane-bound receptors may be reused in a process known as **membrane recycling.** The vesicle with the receptors moves to the cell membrane ❼ and fuses with it ❽. The vesicle membrane then is incorporated back into the cell membrane by exocytosis ❾. Notice in Figure 5.19 that the cytoplasmic face of the membrane remains the same throughout endocytosis and recycling. The extracellular surface of the cell membrane becomes the inside face of the vesicle membrane.

Receptor-mediated endocytosis transports a variety of substances into the cell, including protein hormones, growth factors, antibodies, and plasma proteins that serve as carriers for iron and cholesterol. Elevated plasma cholesterol levels and cardiovascular disease are associated with abnormalities in receptor-mediated removal of cholesterol from the blood (see Clinical Focus box on LDL: The Lethal Lipoprotein).

Caveolae Some endocytosis uses small flask-shaped indentations called **caveolae** ("little caves") rather than clathrin-coated pits to concentrate and bring receptor-bound molecules into the cell. Caveolae are membrane regions with lipid rafts [p. 64], membrane receptor proteins, and specialized membrane proteins named *caveolins* and *cavins.* The receptors in caveolae are lipid-anchored proteins [p. 64]. In many cells, caveolae appear as small indented pockets on the cell membrane, which is how they acquired their name.

Caveolae have several functions: to concentrate and internalize small molecules, to help in the transfer of macromolecules across the capillary endothelium, and to participate in cell signaling. Caveolae appear to be involved in some disease processes, including viral and parasitic infections. Two forms of the disease *muscular dystrophy* are associated with abnormalities in the protein caveolin. Scientists are currently trying to discover more details about the role of caveolae in normal physiology and pathophysiology.

Exocytosis Releases Molecules Too Large for Transport Proteins

Exocytosis is the opposite of endocytosis. In exocytosis, intracellular vesicles move to the cell membrane, fuse with it (Fig. 5.19 ❽), and then release their contents to the extracellular fluid ❾. Cells use exocytosis to export large lipophobic molecules, such

CLINICAL FOCUS

LDL: The Lethal Lipoprotein

"Limit the amount of cholesterol in your diet!" has been the recommendation for many years. So why is too much cholesterol bad for you? After all, cholesterol molecules are essential for membrane structure and for making steroid hormones (such as the sex hormones). But elevated cholesterol levels in the blood also lead to heart disease. One reason some people have too much cholesterol in their blood (*hypercholesterolemia*) is not diet but the failure of cells to take up the cholesterol. In the blood, hydrophobic cholesterol is bound to a lipoprotein carrier molecule to make it water soluble. The most common form of carrier is *low-density lipoprotein* (LDL). When the LDL-cholesterol complex (LDL-C) binds to LDL receptors, it can then enter the cell in a vesicle. When people do not have adequate numbers of LDL receptors on their cell membranes, LDL-C remains in the blood. Hypercholesterolemia due to high levels of LDL-C predisposes these people to develop **atherosclerosis,** also known as hardening of the arteries {*atheroma,* a tumor + *skleros,* hard + *-sis,* condition}. In this condition, the accumulation of cholesterol in blood vessels blocks blood flow and contributes to heart attacks.

FIG. 5.19 ESSENTIALS

Endocytosis, Exocytosis, and Membrane Recycling

Membrane removed from the cell surface by endocytosis is recycled back to the cell surface by exocytosis.

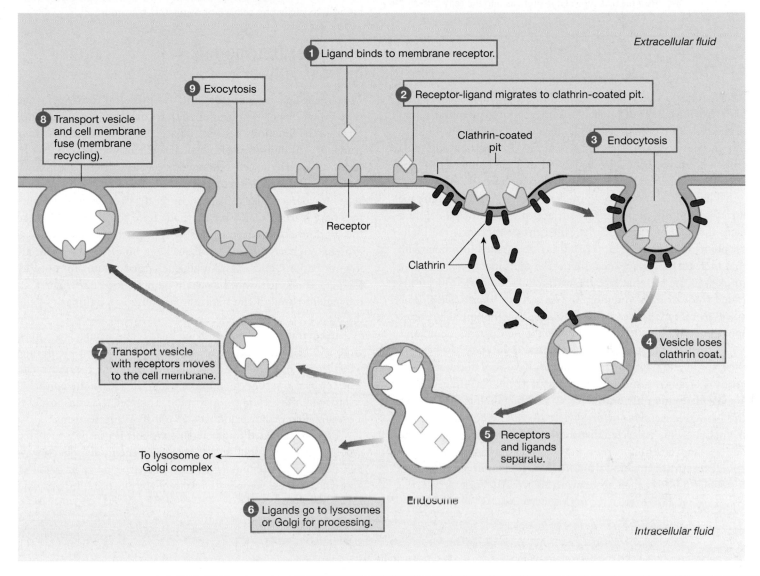

The following labeled steps are shown in the figure:

Extracellular fluid

1. Ligand binds to membrane receptor.
2. Receptor-ligand migrates to clathrin-coated pit.
3. Endocytosis
4. Vesicle loses clathrin coat.
5. Receptors and ligands separate.
6. Ligands go to lysosomes or Golgi for processing.
7. Transport vesicle with receptors moves to the cell membrane.
8. Transport vesicle and cell membrane fuse (membrane recycling).
9. Exocytosis

Clathrin-coated pit

Receptor

Clathrin

Endosome

To lysosome or Golgi complex

Intracellular fluid

as proteins synthesized in the cell, and to get rid of wastes left in lysosomes from intracellular digestion.

The process by which the cell and vesicle membranes fuse is similar in a variety of cell types, from neurons to endocrine cells. Exocytosis involves two families of proteins: *Rabs,* which help vesicles dock onto the membrane, and *SNAREs,* which facilitate membrane fusion. In regulated exocytosis, the process usually begins with an increase in intracellular Ca^{2+} concentration that acts as a signal. The Ca^{2+} interacts with a calcium-sensing protein, which in turn initiates secretory vesicle docking and fusion. When the fused area of membrane opens, the vesicle contents diffuse into the extracellular fluid while the vesicle membrane stays behind and becomes part of the cell membrane. Exocytosis, like endocytosis, requires energy in the form of ATP.

Exocytosis takes place continuously in some cells, making it a *constitutive* process. For example, goblet cells [p. 80] in the intestine continuously release mucus by exocytosis, and fibroblasts in connective tissue release collagen [p. 82]. In other cell types, exocytosis is an intermittent process that is initiated by a signal. In many endocrine cells, hormones are stored in secretory vesicles in the cytoplasm and released in response to a signal from outside the cell. Cells also use exocytosis to insert proteins into the cell membrane, as shown in Figure 5.19. You will encounter many examples of exocytosis in your study of physiology.

27. How does phagocytosis differ from endocytosis?
28. Name the two membrane protein families associated with endocytosis.
29. How do cells move large proteins into the cell? Out of the cell?

EPITHELIAL TRANSPORT

All the transport processes described in the previous sections deal with the movement of molecules across a single membrane, that of the cell. However, molecules entering and leaving the body or moving between certain compartments within the body must cross a layer of epithelial cells [p. 79] that are connected to one another by adhesive junctions and tight junctions [p. 74].

The tight junctions of epithelia separate the cell membrane into two regions, or poles. The surface of the epithelial cell that faces the lumen of an organ is called the *apical* {apex, the highest point} membrane (**FIG. 5.20**). It is often folded into microvilli that increase its surface area. Below the tight junctions, the three surfaces of the cell that face the extracellular fluid are collectively called the *basolateral* membrane {basal, base + latus, side}. The apical membrane is also called the *mucosal* membrane. The corresponding term for the basolateral membrane is *serosal* membrane.

Transporting epithelial cells are said to be *polarized* because their apical and basolateral membranes have very different properties. Certain transport proteins, such as the Na$^+$-K$^+$-ATPase, are usually found only on the basolateral membrane. Others, like the Na$^+$-glucose symporter SGLT, are restricted to the apical membrane. This polarized distribution of transporters allows the one-way movement of certain molecules across the epithelium.

Transport of material from the lumen of an organ to the extracellular fluid is called **absorption** (Fig. 5.20). For example, the intestinal epithelium absorbs digested nutrients. When material moves from the ECF to the lumen, the process is called *secretion*. For example, the salivary glands secrete saliva to help moisten the food you eat. Note that the term *secretion* is also used more broadly to mean the release of a substance from a cell.

Epithelial Transport May Be Paracellular or Transcellular

Movement across an epithelium, or **epithelial transport,** may take place either as **paracellular transport** {para-, beside} through the junctions between adjacent cells or as **transcellular transport** through the epithelial cells themselves (Fig. 5.20). In "tight" epithelia, the cell-cell junctions act as barriers to minimize the unregulated diffusion of material between the cells, so there is very little paracellular transport. In recent years, however, scientists have learned that some epithelia have the ability to change the "tightness" of their junctions. It appears that some junctional proteins such as *claudins* can form large holes or pores that allow water, ions, and a few small uncharged solutes to move by the paracellular pathway. In certain pathological states, increased movement through the paracellular route is a hallmark of the disease.

In contrast, substances moving by the transcellular route must cross two cell membranes. Molecules cross the first membrane when they move into the epithelial cell from one compartment. They cross the second membrane when they leave the epithelial cell to enter the second compartment. Transcellular transport uses a combination of active and passive transport mechanisms.

Protein-mediated transcellular transport is usually a two-step process, with one "uphill" step that requires energy and one "downhill" step in which the molecule moves passively down its gradient. You will see these steps in the example of glucose transport that

---

FIG. 5.20 Transporting epithelia are polarized

The apical membrane and the basolateral membrane are the two poles of the cell. Polarized epithelia have different transport proteins on apical and basolateral membranes. This allows selective directional transport across the epithelium. Transport from lumen to ECF is called **absorption.** Transport from ECF to lumen is called **secretion.**

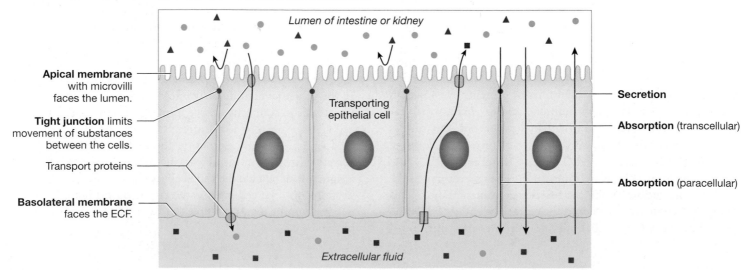

follows. Molecules that are too large to be moved by membrane proteins can be transported across the cell in vesicles.

The cells of transporting epithelia can alter their permeability by selectively inserting or withdrawing membrane proteins. Transporters pulled out of the membrane may be destroyed in lysosomes, or they may be stored in vesicles inside the cell, ready to be reinserted into the membrane in response to a signal (another example of membrane recycling). Most epithelial transport you will study in this book involves the transporting epithelia of intestine and kidney, which are specialized to selectively transport molecules into and out of the body.

Transcellular Transport of Glucose Uses Membrane Proteins

The absorption of glucose from the lumen of the kidney tubule or intestine to the extracellular fluid is an important example of directional movement across a transporting epithelium. Transepithelial movement of glucose involves three transport systems: (1) the SGLT-mediated secondary active transport of glucose with Na^+ from the lumen into the epithelial cell at the apical membrane, followed by the movement of Na^+ and glucose out of the cell and into the extracellular fluid on separate transporters; (2) sodium moves out by primary active transport via a Na^+-K^+-ATPase; and (3) glucose leaves the cell by facilitated diffusion on GLUT carriers.

FIGURE 5.21 shows the process in detail. The glucose concentration in the transporting epithelial cell is higher than the glucose concentration in the lumen of the kidney or intestine. For this reason, moving glucose from the lumen into the cell requires the input of energy—in this case, energy stored in the Na^+ concentration gradient. Sodium ions in the lumen bind to the SGLT carrier, as previously described (see Fig. 5.16), and bring glucose with them into the cell. The energy needed to move glucose against its concentration gradient comes from the kinetic energy of Na^+ moving down its concentration gradient (Fig. 5.21 **1**).

Once glucose is in the epithelial cell, it leaves by moving down its concentration gradient on the facilitated diffusion GLUT transporter in the basolateral membrane (Fig. 5.21 **2**). Na^+ is pumped out of the cell on the basolateral side using Na^+-K^+-ATPase **3**. This step requires energy provided by ATP

FIG. 5.21   Transepithelial absorption of glucose

Absorbing glucose from intestinal or kidney tubule lumen involves indirect (secondary) active transport of glucose across the apical membrane and glucose diffusion across the basolateral membrane.

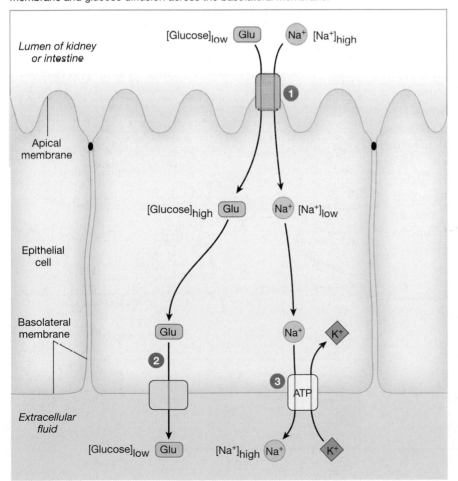

1 **Na^+-glucose symporter** brings glucose into cell against its gradient using energy stored in the Na^+ concentration gradient.

2 **GLUT transporter** transfers glucose to ECF by facilitated diffusion.

3 **Na^+-K^+-ATPase** pumps Na^+ out of the cell, keeping ICF Na^+ concentration low.

Q **FIGURE QUESTIONS**

1. Match each transporter to its location.
   1. GLUT
   2. Na^+-glucose symporter
   3. Na^+-K^+-ATPase

   Choose either
   (a) apical membrane
   (b) basolateral membrane

2. Is glucose movement across the basolateral membrane active or passive? Explain.

3. Why doesn't Na^+ movement at the apical membrane require ATP?

RUNNING **PROBLEM**

The sweat test that Daniel will undergo analyzes levels of NaCl in sweat. Sweat—a mixture of ions and water—is secreted into ducts by the epithelial cells of sweat glands. As sweat moves toward the skin's surface through the ducts, CFTR and Na^+ channels move Cl^- and Na^+ out of the sweat and back into the body. The duct cells are not permeable to water, so that normal reabsorption of NaCl creates sweat with a low salt content. However, without functioning CFTR channels in the epithelium, salt is not reabsorbed. In cystic fibrosis, salt concentrations in the sweat can be four times the normal amount.

Q3: *Based on the information given, is CFTR protein on the apical or basolateral surface of the sweat gland epithelium?*

123   133   139   **152**   153   160

because sodium is more concentrated in the extracellular fluid than in the cell.

The removal of Na^+ from the cell is essential if glucose is to continue to be absorbed from the lumen. The potential energy to run the SGLT symporter comes from the sodium concentration gradient, which depends on low intracellular concentrations of Na^+. If the basolateral Na^+-K^+-ATPase is poisoned with *ouabain* (pronounced wah-bane—a compound related to the heart drug digitalis), Na^+ that enters the cell cannot be pumped out. The Na^+ concentration inside the cell gradually increases until it is equal to that in the lumen. Without a sodium gradient, there is no energy source to run the SGLT symporter, and the absorption of glucose across the epithelium stops.

Transepithelial transport can use ion movement through channels in addition to carrier-mediated transport. For example, the apical membrane of a transporting epithelium may use the Na^+-K^+-$2Cl^-$ (NKCC) symporter to bring K^+ into the cell against its concentration gradient, using energy from the Na^+ gradient. Because the K^+ concentration inside the cell is higher than in the extracellular fluid, K^+ can move out of the cell on the basolateral side through open K^+ leak channels. Na^+ must be pumped out by Na^+-K^+-ATPase. By this simple mechanism, the body can absorb Na^+ and K^+ at the same time from the lumen of the intestine or the kidney.

CONCEPT CHECK

30. Why does Na^+ movement from the cytoplasm to the extracellular fluid require energy?

31. Ouabain, an inhibitor of the Na^+-K^+-ATPase, cannot pass through cell membranes. What would happen to the transepithelial glucose transport shown in Figure 5.21 if ouabain were applied to the apical side of the epithelium? To the basolateral side of the epithelium?

32. Which GLUT transporter is illustrated in Figure 5.21?

Transcytosis Uses Vesicles to Cross an Epithelium

Some molecules, such as proteins, are too large to cross epithelia on membrane transporters. Instead they are moved across epithelia by **transcytosis,** which is a combination of endocytosis, vesicular transport across the cell, and exocytosis (**FIG. 5.22**). In this process, the molecule is brought into the epithelial cell via receptor-mediated endocytosis. The resulting vesicle attaches to microtubules in the cell's cytoskeleton and is moved across the cell by a process known as **vesicular transport.** At the opposite side of the epithelium, the contents of the vesicle are expelled into the interstitial fluid by exocytosis.

Transcytosis makes it possible for large proteins to move across an epithelium and remain intact. It is the means by which infants absorb maternal antibodies in breast milk. The antibodies are absorbed on the apical surface of the infant's intestinal epithelium and then released into the extracellular fluid.

CONCEPT CHECK

33. If you apply a poison that disassembles microtubules to a capillary endothelial cell, what happens to transcytosis?

Now that we have considered how solutes move between the body's compartments, we will examine how the transport of ions creates an electrical disequilibrium between the intracellular and extracellular compartments.

FIG. 5.22   Transcytosis across the capillary endothelium

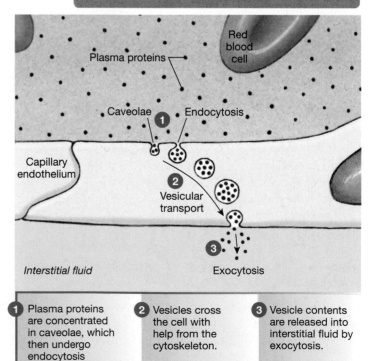

1 Plasma proteins are concentrated in caveolae, which then undergo endocytosis and form vesicles.

2 Vesicles cross the cell with help from the cytoskeleton.

3 Vesicle contents are released into interstitial fluid by exocytosis.

Three days after Daniel's sweat test, the lab returns the grim results: salt levels in his sweat are more than twice the normal concentration. Daniel is diagnosed with cystic fibrosis. Now, along with antibiotics to prevent lung infections and therapy to loosen the mucus in his airways, Daniel must begin a regimen of pancreatic enzymes to be taken whenever he eats, for the rest of his life. In cystic fibrosis, thick mucus in the pancreatic ducts blocks the secretion of digestive enzymes into the intestine. Without artificial enzymes, he would starve.

Q4: *Why will Daniel starve if he does not take artificial pancreatic enzymes?*

123　133　139　152　**153**　160

THE RESTING MEMBRANE POTENTIAL

Many of the body's solutes, including organic compounds such as pyruvate and lactate, are ions and, therefore, carry a net electrical charge. Potassium (K^+) is the major cation within cells, and sodium (Na^+) dominates the extracellular fluid (see Fig. 5.1, p. 124). On the anion side, chloride ions (Cl^-) mostly remain with Na^+ in the extracellular fluid. Phosphate ions and negatively charged proteins are the major anions of the intracellular fluid.

Overall, the body is electrically neutral: for every cation, there is a matching anion. However, ions are not distributed evenly between the ECF and the ICF (**FIG. 5.23a**). The intracellular compartment contains some anions that do not have matching cations, giving the cells a net negative charge. At the same time, the extracellular compartment has the "missing" cations, giving the ECF a net positive charge. One consequence of this uneven distribution of ions is that the intracellular and extracellular compartments are not in electrical equilibrium. Instead, the two compartments exist in a state of *electrical disequilibrium* [p. 123].

The concept of electrical disequilibrium traditionally is taught in chapters on nerve and muscle function because those tissues generate electrical signals known as action potentials. Yet one of the most exciting discoveries in physiology is the realization that other kinds of cells also use electrical signals for communication. In fact, all living organisms, including plants, use electrical signals! This section reviews the basic principles of electricity and discusses what creates electrical disequilibrium in the body. The chapter ends with a look at how the endocrine beta cells of the pancreas use electrical signaling to trigger insulin secretion.

Electricity Review

Atoms are electrically neutral [p. 36]. They are composed of positively charged protons, negatively charged electrons, and uncharged neutrons, but in balanced proportions, so that an atom is neither positive nor negative. The removal or addition of electrons to an atom creates the charged particles we know as ions. We have discussed several ions that are important in the human body, such as Na^+, K^+, and H^+. For each of these positive ions, somewhere in the body there is a matching electron, usually found as part of a negative ion. For example, when Na^+ in the body enters in the form of NaCl, the "missing" electron from Na^+ can be found on the Cl^-.

Remember the following important principles when you deal with electricity in physiological systems:

1. The **law of conservation of electrical charge** states that the net amount of electrical charge produced in any process is zero. This means that for every positive charge on an ion, there is an electron on another ion. Overall, the human body is electrically neutral.

2. Opposite charges (+ and –) are attracted to each other. The protons and electrons in an atom exhibit this attraction. Like charges (two charges of the same type, such as +/+, or –/–) repel each other.

3. Separating positive charges from negative charges requires energy. For example, energy is required to separate the protons and electrons of an atom.

4. When separated positive and negative charges can move freely toward each other, the material through which they move is called a **conductor.** Water is a good conductor of electrical charge. When separated charges cannot move through the material that separates them, the material is known as an **insulator.** The phospholipid bilayer of the cell membrane is a good insulator, as is the plastic coating on electrical wires.

The word *electricity* comes from the Greek word *elektron*, meaning "amber," the fossilized resin of trees. The Greeks discovered that if they rubbed a rod of amber with cloth, the amber acquired the ability to attract hair and dust. This attraction (called static electricity) arises from the separation of electrical charge that occurs when electrons move from the amber atoms to the cloth. To separate these charged particles, energy (work) must be put into the system. In the case of the amber, work was done by rubbing the rod. In the case of biological systems, the work is usually done by energy stored in ATP and other chemical bonds.

The Cell Membrane Enables Separation of Electrical Charge in the Body

In the body, separation of electrical charge takes place across the cell membrane. This process is shown in Figure 5.23b. The diagram shows an artificial cell system. The cell is filled with positive K^+ and large negative ions. The cell is placed in an aqueous solution of sodium chloride that has dissociated into Na^+ and Cl^-. The phospholipid bilayer of the artificial cell, like the membrane

FIG. 5.23 **ESSENTIALS**

Membrane Potential

The electrical disequilibrium that exists between the extracellular fluid (ECF) and intracellular fluid (ICF) of living cells is called the **membrane potential difference** (V_m), or membrane potential for short. The membrane potential results from the uneven distribution of electrical charge (i.e., ions) between the ECF and ICF.

PHYS IN ACTION!

(a) In illustrations, this uneven distribution of charge is often shown by the charge symbols clustered on each side of the cell membrane.

The ECF has a slight excess of cations (+).

The ICF has a slight excess of anions (−).

Creation of a Membrane Potential in an Artificial System

What creates the membrane potential?
1. Ion concentration gradients between the ECF and ICF
2. The selectively permeable cell membrane

To show how a membrane potential difference can arise from ion concentration gradients and a selectively permeable membrane, we will use an artificial cell system where we can control the membrane's permeability to ions and the composition of the ECF and ICF.

(b) When we begin, the cell has no membrane potential: The ECF (composed of Na^+ and Cl^- ions) and the ICF (K^+ and large anions, A^-) are electrically neutral.

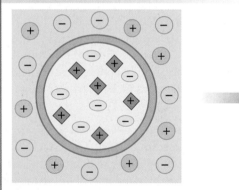

The system is in chemical disequilibrium, with concentration gradients for all four ions. The cell membrane acts as an insulator to prevent free movement of ions between the ICF and ECF.

(c) Now we insert a leak channel for K^+ into the membrane, making the cell freely permeable to K^+.

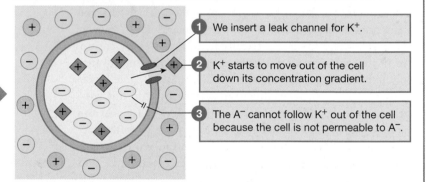

1. We insert a leak channel for K^+.

2. K^+ starts to move out of the cell down its concentration gradient.

3. The A^- cannot follow K^+ out of the cell because the cell is not permeable to A^-.

The transfer of just one K^+ from the cell to the ECF creates an electrical disequilibrium: the ECF has a net positive charge (+1) while the ICF has a net negative charge (−1). The cell now has a membrane potential difference, with the inside of the cell negative relative to the outside.

How much K^+ will leave the cell?

If K^+ was uncharged, like glucose, it would diffuse out of the cell until the concentration outside $[K]_{out}$ equaled the concentration inside $[K]_{in}$. But K^+ is an ion, so we must consider its electrical gradient. Remember the rule for movement along electrical gradients: Opposite charges attract, like charges repel.

(d) As additional K^+ ions leave the cell, going down their concentration gradient, the inside of the cell becomes more negative and the outside becomes more positive.

4. Additional K^+ leaves the cell

5. Now the negative charge inside the cell begins to attract ECF K^+ back into the cell: an electrical gradient in the opposite direction from the concentration gradient.

KEY

+ Sodium ion

◆ Potassium ion

− Chloride ion

− Large anion

Electrochemical Equilibrium

For any given concentration gradient $[Ion]_{out} - [Ion]_{in}$ across a cell membrane, there is a membrane potential difference (i.e., electrical gradient) that exactly opposes ion movement down the concentration gradient. At this membrane potential, the cell is at *electrochemical equilibrium*: There is no net movement of ion across the cell membrane.

 FIGURE QUESTIONS

1. If the cell in (e) was made freely permeable to only Na^+, which way would the Na^+ move? Would the membrane potential become positive or negative?

2. If it became freely permeable to only Cl^-, which way would Cl^- move? Would the membrane potential become positive or negative?

(e) In this example, the concentration gradient sending K^+ out of the cell is exactly opposed by the electrical gradient pulling K^+ into the cell. This is shown by the arrows that are equal in length but opposite in direction.

Efflux due to concentration gradient

Influx due to electrical gradient

Equilibrium Potential

For any ion, the membrane potential that exactly opposes a given concentration gradient is known as the **equilibrium potential** (E_{ion}). To calculate the equilibrium potential for any concentration gradient, we use the Nernst equation:

$$E_{ion} = \frac{61}{z} \log \frac{[Ion]_{out}}{[Ion]_{in}}$$

where z is the charge on the ion. (i.e., $K^+ = +1$)

The Nernst equation is used for a cell that is freely permeable to only one ion at a time. Living cells, however, have limited permeability to several ions. To calculate the actual membrane potential of cells, we use a multi-ion equation called the Goldman-Hodgkin-Katz equation [discussed in Chapter 8].

Approximate Values for Mammalian Cells		
	ICF	ECF
K^+	150	5
Na^+	15	145
Cl^-	10	108

Using these values for K^+ and the Nernst equation, the E_K is −90 mV.

 FIGURE QUESTIONS (You will need the log function on a calculator.)

3. Calculate the equilibrium potential for Na^+ (E_{Na}).

4. Calculate the E_{Cl}.

Measuring Membrane Potential

(f) In the first example, you saw that the membrane potential results from excess cations in the ECF and excess anions in the ICF. To measure this difference, we can place electrodes in the cell and surrounding fluid (equivalent to the ECF).

Intracellular fluid   Extracellular fluid

On a number line, the ECF would be at +1 and the ICF at −1.

Absolute charge scale

Intracellular fluid   Extracellular fluid

In real life, we cannot measure absolute numbers of ions, however. Instead, we measure the difference between the two electrodes. By convention, the ECF is set at 0 mV (the ground). This gives the ICF a relative charge of −2.

Relative charge scale extracellular fluid set to 0.

Intracellular fluid   Extracellular fluid

of a real cell, is not permeable to ions, so it acts as an insulator and prevents the ions from moving. Water can freely cross this cell membrane, making the extracellular and intracellular osmotic concentrations equal.

In Figure 5.23b, both the cell and the solution are electrically neutral, and the system is in electrical equilibrium. However, it is not in chemical equilibrium. There are concentration gradients for all four types of ions in the system, and they would all diffuse down their respective concentration gradients if they could cross the cell membrane.

In Figure 5.23c, a leak channel for K^+ is inserted into the membrane. Now the cell is permeable to K^+, but only to K^+. Because of the K^+ concentration gradient, K^+ moves out of the cell. The negative ions in the cell attempt to follow the K^+ because of the attraction of positive and negative charges. But because the membrane is impermeable to negative ions, the anions remain trapped in the cell.

As soon as the first positive K^+ leaves the cell, the electrical equilibrium between the extracellular fluid and intracellular fluid is disrupted: the cell's interior has developed a net charge of −1 while the cell's exterior has a net charge of +1. The movement of K^+ out of the cell down its concentration gradient has created an **electrical gradient**—that is, a difference in the net charge between two regions. In this example, the inside of the cell has become negative relative to the outside.

If the only force acting on K^+ were the concentration gradient, K^+ would leak out of the cell until the K^+ concentration inside the cell equaled the K^+ concentration outside. The loss of positive ions from the cell creates an electrical gradient, however. The combination of electrical and concentration gradients is called an **electrochemical gradient.**

Because opposite charges attract each other, the negative proteins inside the cell try to pull K^+ back into the cell (Fig. 5.23d). At some point in this process, the electrical force attracting K^+ into the cell becomes equal in magnitude to the chemical concentration gradient driving K^+ out of the cell. At that point, net movement of K^+ across the membrane stops (Fig. 5.23e). The rate at which K^+ ions move out of the cell down the concentration gradient is exactly equal to the rate at which K^+ ions move into the cell down the electrical gradient. The system has reached *electrochemical equilibrium.*

For any given concentration gradient of a single ion, the membrane potential that exactly opposes the concentration gradient is known as the **equilibrium potential,** or E_{ion} (where the subscript *ion* is replaced by the symbol for whichever ion we are looking at). For example, when the concentration gradient is 150 mM K^+ inside and 5 mM K^+ outside the cell, the equilibrium potential for potassium, or E_K is −90 mV.

The equilibrium potential for any ion at 37 °C (human body temperature) can be calculated using the **Nernst equation:**

$$E_{ion} = \frac{61}{z} \log \frac{[ion]_{out}}{[ion]_{in}}$$

where 61 is 2.303 RT/F at 37 °C*

z is the electrical charge on the ion (+1 for K^+),

$[ion]_{out}$ and $[ion]_{in}$ are the ion concentrations outside and inside the cell, and

E_{ion} is measured in mV.

The Nernst equation assumes that the cell in question is freely permeable to only the ion being studied. This is not the usual situation in living cells, however, as you will learn shortly.

All Living Cells Have a Membrane Potential

As the beginning of this chapter pointed out, all living cells are in chemical and electrical disequilibrium with their environment. This electrical disequilibrium, or electrical gradient between the extracellular fluid and the intracellular fluid, is called the **resting membrane potential difference,** or **membrane potential** for short. Although the name sounds intimidating, we can break it apart to see what it means.

1. The *resting* part of the name comes from the fact that an electrical gradient is seen in all living cells, even those that appear to be without electrical activity. In these "resting" cells, the membrane potential has reached a steady state and is not changing.

2. The *potential* part of the name comes from the fact that the electrical gradient created by active transport of ions across the cell membrane is a form of stored, or potential, energy, just as concentration gradients are a form of potential energy. When oppositely charged molecules come back together, they release energy that can be used to do work, in the same way that molecules moving down their concentration gradient can do work [see Appendix B]. The work done using electrical energy includes opening voltage-gated membrane channels and sending electrical signals.

3. The *difference* part of the name is to remind you that the membrane potential represents a difference in the amount of electrical charge inside and outside the cell. The word *difference* is usually dropped from the name, as noted earlier, but it is important for remembering what a membrane potential means.

In living systems, we cannot measure absolute electrical charge, so we describe electrical gradients on a relative scale instead. Figure 5.23f compares the two scales. On an absolute scale, the extracellular fluid in our simple example has a net charge of +1 from the positive ion it gained, and the intracellular fluid has a net charge of −1 from the negative ion that was left behind. On the number line shown, this is a difference of two units.

In real life, because we cannot measure electrical charge as numbers of electrons gained or lost, we use a device that measures the *difference* in electrical charge between two points. This device artificially sets the net electrical charge of one side of the membrane to 0 and measures the net charge of the second side relative

---

*R is the ideal gas constant, T is absolute temperature, and F is the Faraday constant. For additional information, see Appendix B.

to the first. In our example, resetting the extracellular fluid net charge to 0 on the number line gives the intracellular fluid a net charge of –2. We call the ICF value the resting membrane potential (difference) of the cell.

The equipment for measuring a cell's membrane potential is depicted in **FIGURE 5.24**. *Electrodes* are created from hollow glass tubes drawn to very fine points. These *micropipettes* are filled with a liquid that conducts electricity and then connected to a *voltmeter*, which measures the electrical difference between two points in units of either volts (V) or millivolts (mV). A *recording electrode* is inserted through the cell membrane into the cytoplasm of the cell. A *reference electrode* is placed in the external bath, which represents the extracellular fluid.

In living systems, by convention, the extracellular fluid is designated as the *ground* and assigned a charge of 0 mV (Fig. 5.23f). When the recording electrode is placed inside a living cell, the voltmeter measures the membrane potential—in other words, the electrical difference between the intracellular fluid and the extracellular fluid. A recorder connected to the voltmeter can make a recording of the membrane potential versus time.

For resting nerve and muscle cells, the voltmeter usually records a membrane potential between –40 and –90 mV, indicating that the intracellular fluid is negative relative to the extracellular fluid (0 mV). (Throughout this discussion, remember that the extracellular fluid is not really neutral because it has excess positive charges that exactly balance the excess negative charges inside the cell, as shown in Fig. 5.23. The total body remains electrically neutral at all times.)

The Resting Membrane Potential Is Due Mostly to Potassium

Which ions create the resting membrane potential in animal cells? The artificial cell shown in Figure 5.23c used a potassium channel to allow K^+ to leak across a membrane that was otherwise impermeable to ions. But what processes go on in living cells to create an electrical gradient?

In reality, living cells are not permeable to only one ion. They have open channels and protein transporters that allow ions to move between the cytoplasm and the extracellular fluid. Instead

---

FIG. 5.24 Measuring membrane potential

In the laboratory, a cell's membrane potential is measured by placing one electrode inside the cell and a second in the extracellular bath.

A recording electrode is placed inside the cell.

The voltmeter measures the difference in electrical charge between the inside of a cell and the surrounding solution. This value is the **membrane potential difference**, or **V_m**.

Input

Cell –70 mV

0 mV

Saline bath

The ground (⏚) or reference electrode is placed in the bath and given a value of 0 millivolts (mV)

Output

The membrane potential can change over time.

Membrane potential difference (V_m)

V_m

V_m

Hyperpolarization

Depolarization

If the membrane potential becomes less negative than the resting potential, the cell depolarizes.

Repolarization

If the membrane potential becomes more negative, the cell hyperpolarizes.

Membrane potential (mV)

Time (msec) →

of the Nernst equation, we use a related equation called the *Goldman equation* that considers concentration gradients of the permeable ions and the relative permeability of the cell to each ion. [For more detail on the Goldman equation, see Chapter 8.]

The real cell illustrated in **FIGURE 5.25** has a resting membrane potential of –70 mV. Most cells are about 40 times more permeable to K^+ than to Na^+. As a result, a cell's resting membrane potential is closer to the E_K of –90 mV than to the E_{Na} of +60 mV. A small amount of Na^+ leaks into the cell, making the inside of the cell less negative than it would be if Na^+ were totally excluded. Additional Na^+ that leaks in is promptly pumped out by the Na^+-K^+-ATPase. At the same time, K^+ ions that leak out of the cell are pumped back in. The pump contributes to the membrane potential by pumping 3 Na^+ out for every 2 K^+ pumped in. Because the Na^+-K^+-ATPase helps maintain the electrical gradient, it is called an *electrogenic* pump.

Not all ion transport creates an electrical gradient. Many transporters, like the Na^+-K^+-2 Cl^- (NKCC) symporter, are electrically neutral. Some make an even exchange: for each charge that enters the cell, the same charge leaves. An example is the HCO_3^--Cl^- antiporter of red blood cells, which transports these ions in a one-for-one, electrically neutral exchange. Electrically neutral transporters have little effect on the resting membrane potential of the cell.

FIG. 5.25   The resting membrane potential of cells

Most cells in the human body are about 40 times more permeable to K^+ than to Na^+, and the resting membrane potential is about –70 mV. The Na-K-ATPase helps maintain the resting membrane potential by removing Na^+ that leaks into the cell and returning K^+ that has leaked out.

Intracellular fluid
–70 mV

Extracellular fluid
0 mV

Ⓠ **FIGURE QUESTIONS**
1. What force(s) promote(s) Na^+ leak into the cell?
2. What force(s) promote(s) K^+ leak out of the cell?

CONCEPT CHECK   34. What would happen to the resting membrane potential of a cell poisoned with ouabain (an inhibitor of the Na^+-K^+-ATPase)?

Changes in Ion Permeability Change the Membrane Potential

As you have just learned, two factors influence a cell's membrane potential: (1) the concentration gradients of different ions across the membrane and (2) the permeability of the membrane to those ions. If the cell's permeability to an ion changes, the cell's membrane potential changes. We monitor changes in membrane potential using the same recording electrodes that we use to record resting membrane potential.

Figure 5.24 shows a recording of membrane potential plotted against time. The extracellular electrode is set at 0 mV, and the intracellular electrode records the membrane potential difference. The membrane potential (V_m) begins at a steady resting value of –70 mV. When the trace moves upward (becomes less negative), the potential difference between the inside of the cell and the outside (0 mV) is less, and the cell is said to have *depolarized*. A return to the resting membrane potential is termed *repolarization*. If the resting potential becomes more negative, we say the cell has *hyperpolarized*.

A major point of confusion when we talk about changes in membrane potential is the use of the phrases "the membrane potential decreased" or "the membrane potential increased." Normally, we associate "increase" with becoming more positive and "decrease" with becoming more negative—the opposite of what is happening in our cell discussion. The best way to avoid trouble is to speak of the membrane potential becoming more or less negative or the cell depolarizing or hyperpolarizing. Another way to avoid confusion is to add the word *difference* after *membrane potential*. If the membrane potential *difference* is *increasing*, the value of V_m must be moving away from the ground value of 0 and becoming *more negative*. If the membrane potential *difference* is *decreasing*, the value of V_m is moving closer to the ground value of 0 mV and is becoming *less negative*.

What causes changes in membrane potential? In most cases, membrane potential changes in response to movement of one of four ions: Na^+, Ca^{2+}, Cl^-, and K^+. The first three ions are more concentrated in the extracellular fluid than in the cytosol, and the resting cell is minimally permeable to them. If a cell suddenly becomes more permeable to any one of these ions, then those ions will move down their electrochemical gradient into the cell. Entry of Ca^{2+} or Na^+ depolarizes the cell (makes the membrane potential more positive). Entry of Cl^- hyperpolarizes the cell (makes the membrane potential more negative).

Most resting cells are fairly permeable to K^+ but making them more permeable allows even more K^+ to leak out. The cell hyperpolarizes until it reaches the equilibrium potential for K^+. Making the cell *less* permeable to K^+ allows fewer K^+ ions to leak out of the cell. When the cell retains K^+, it becomes more positive and depolarizes. You will encounter instances of all these permeability changes as you study physiology.

It is important to learn that a significant change in membrane potential requires the movement of very few ions. *The concentration gradient does not have to reverse to change the membrane potential.* For example, to change the membrane potential by 100 mV (the size of a typical electrical signal passing down a neuron), only one of every 100,000 K^+ must enter or leave the cell. This is such a tiny fraction of the total number of K^+ ions in the cell that the concentration gradient for K^+ remains essentially unchanged.

INTEGRATED MEMBRANE PROCESSES: INSULIN SECRETION

The movement of Na^+ and K^+ across cell membranes has been known to play a role in generating electrical signals in excitable tissues for many years. You will study these processes in detail when you learn about the nervous and muscular systems. Recently, however, we have come to understand that small changes in membrane potential act as signals in nonexcitable tissues, such as endocrine cells. One of the best-studied examples of this process involves the beta cell of the pancreas. Release of the hormone insulin by beta cells demonstrates how membrane processes—such as facilitated diffusion, exocytosis, and the opening and closing of ion channels by ligands and membrane potential—work together to regulate cell function.

The endocrine beta cells of the pancreas synthesize the protein hormone insulin and store it in cytoplasmic secretory vesicles [p. 71]. When blood glucose levels increase, such as after a meal, the beta cells release insulin by exocytosis. Insulin then directs other cells of the body to take up and use glucose, bringing blood concentrations down to pre-meal levels.

A key question about the process that went unanswered until recently was, "How does a beta cell 'know' that glucose levels have gone up and that it needs to release insulin?" The answer, we have now learned, links the beta cell's metabolism to its electrical activity.

FIGURE 5.26a shows a beta cell at rest. Recall from earlier sections in this chapter that gated membrane channels can be opened or closed by chemical or electrical signals. The beta cell has two such channels that help control insulin release. One is a **voltage-gated Ca^{2+} channel.** This channel is closed at the cell's resting membrane potential (**5** in Fig. 5.26a). The other is a K^+ leak channel (that is, the channel is usually open) that closes when ATP binds to it. It is called an **ATP-gated K^+ channel (K_{ATP} channel).** In the resting cell, when glucose concentrations are low, the cell makes less ATP **1**–**3**. There is little ATP to bind to the K_{ATP} channel, and the channel remains open,

FIG. 5.26 Insulin secretion and membrane transport

(a) Beta cell at rest. The K_{ATP} channel is open, and the cell is at its resting membrane potential.

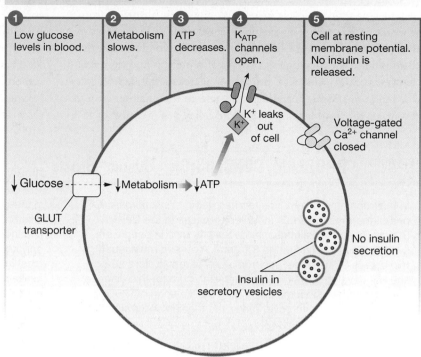

1. Low glucose levels in blood.
2. Metabolism slows.
3. ATP decreases.
4. K_{ATP} channels open.
5. Cell at resting membrane potential. No insulin is released.

K^+ leaks out of cell

Voltage-gated Ca^{2+} channel closed

$\downarrow$ Glucose → $\downarrow$ Metabolism → $\downarrow$ ATP

GLUT transporter

No insulin secretion

Insulin in secretory vesicles

(b) Beta cell secretes insulin. Closure of K_{ATP} channel depolarizes cell, triggering exocytosis of insulin.

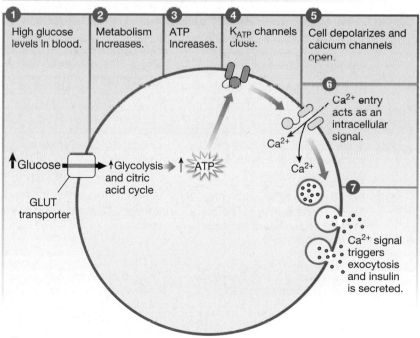

1. High glucose levels in blood.
2. Metabolism increases.
3. ATP increases.
4. K_{ATP} channels close.
5. Cell depolarizes and calcium channels open.
6. Ca^{2+} entry acts as an intracellular signal.
7. Ca^{2+} signal triggers exocytosis and insulin is secreted.

$\uparrow$ Glucose → $\uparrow$ Glycolysis and citric acid cycle → $\uparrow$ ATP

Ca^{2+}

GLUT transporter

Q FIGURE QUESTIONS

1. Which step shows facilitated diffusion?
2. What kind of gating do the beta cell ion channels have?
3. Does insulin secretion in (b) require energy input from ATP?
4. Why is insulin released by exocytosis and not through a carrier or channel?

allowing K^+ to leak out of the cell ④. At the resting membrane potential, the voltage-gated Ca^{2+} channels are closed, and there is no insulin secretion ⑤.

Figure 5.26b shows a beta cell secreting insulin. Following a meal, plasma glucose levels increase as glucose is absorbed from the intestine ①. Glucose reaching the beta cell diffuses into the cell with the aid of a GLUT transporter. Increased glucose in the cell stimulates the metabolic pathways of glycolysis and the citric acid cycle [p. 106], and ATP production increases ②, ③. When ATP binds to the K_{ATP} channel, the gate to the channel closes, preventing K^+ from leaking out of the cell ④. Retention of K^+

depolarizes the cell ⑤, which then causes the voltage-sensitive Ca^{2+} channels to open ⑥. Calcium ions enter the cell from the extracellular fluid, moving down their electrochemical gradient. The Ca^{2+} binds to proteins that initiate exocytosis of the insulin-containing vesicles, and insulin is released into the extracellular space ⑦.

The discovery that cells other than nerve and muscle cells use changes in membrane potential as signals for physiological responses altered our traditional thinking about the role of the resting membrane potential. Next, we will look at other types of signals that the body uses for communication and coordination.

RUNNING **PROBLEM CONCLUSION** | Cystic Fibrosis

In this running problem, you learned about cystic fibrosis (CF), one of the most common inherited diseases in the United States. By some estimates, more than 10 million people are symptomless carriers of the CF gene. A person must inherit two copies of the gene, one from each parent, before he or she will develop CF. Although there is no cure for this disease, treatments have become better, and the life span of CF patients continues to improve. Today, the median survival age is in the early 40s.

Cystic fibrosis is caused by a defect in the CFTR channel protein, which regulates the transport of Cl^- into and out of epithelial cells. Because CFTR channels are found in the epithelial cell membranes of several organs—the sweat glands, lungs, and pancreas—cystic fibrosis may affect many different body processes. Interestingly, the CFTR chloride channel is a member of the ABC transport family and is the only known ion channel in that gene superfamily. Some of the most interesting animal research on cystic fibrosis uses genetically altered mice, called CF mice. These model animals are bred to have CFTR channels with altered functions corresponding to the mutations of the CFTR gene in humans.

To learn more about current research in this disease, go to the Cystic Fibrosis Foundation web site (*www.cff.org*) and click the *Research Overview* tab. To check your understanding of the running problem, compare your answers with the information in the following table.

Question	Facts	Integration and Analysis
Q1: *Why would failure to transport NaCl into the airways cause the secreted mucus to be thick?*	If NaCl is secreted into the lumen of the airways, the solute concentration of the airway fluid increases. Water moves into compartments with higher osmolarity.	Normally, movement of NaCl creates an osmotic gradient so that water also enters the airway lumen, creating a saline solution that thins the thick mucus. If NaCl cannot be secreted into the airways, there will be no fluid movement to thin the mucus.
Q2: *Is the CFTR a chemically gated, a voltage-gated, or a mechanically gated channel protein?*	Chemically gated channels open when a ligand binds to them. CFTRs open when ATP binds to the channel protein.	ATP is a chemical ligand, which means CFTRs are chemically gated channel proteins.
Q3: *Based on the information given, is the CFTR protein on the apical or basolateral surface of the sweat gland epithelium?*	In normal people, the CFTR channels transport Cl^- from sweat into epithelial cells.	The epithelial surface that faces the lumen of the sweat gland, which contains sweat, is the apical membrane. Therefore, the CFTR proteins are on the apical surface.
Q4: *Why will Daniel starve if he does not take artificial pancreatic enzymes?*	The pancreas secretes mucus and digestive enzymes into ducts that empty into the small intestine. In cystic fibrosis, mucus in the ducts is thick because of lack of Cl^- and fluid secretion. This thick mucus blocks the ducts and prevents digestive enzymes from reaching the small intestine.	Without digestive enzymes, Daniel cannot digest the food he eats. His weight loss over the past six months suggests that this has already become a problem. Taking artificial enzymes will enable him to digest his food.

CHAPTER SUMMARY

Several key themes come together in this chapter. You learned how the cell membrane creates distinct intracellular and extracellular compartments, illustrating the theme of *compartmentation*. The contents of the intracellular and extracellular compartments differ, but *homeostasis* keeps them in a dynamic steady state. Movement of materials between and within compartments is necessary for *communication* and is accomplished by *bulk flow* and *biological transport*. Flow of solutes and water across cell membranes occurs in response to osmotic, chemical (concentration), or electrical gradients. The cell membrane creates resistance to flow that can be overcome by inserting membrane proteins that act as channels or carriers. Biological transport in the body requires *energy* from concentration gradients or chemical bonds. Finally, the binding of substrates to transporters demonstrates the theme of *protein interactions*.

Osmosis and Tonicity

iP Fluids and Electrolytes: Introduction to Body Fluids

1. Most solutes are concentrated in either one compartment or the other, creating a state of **chemical disequilibrium.** (p. 123; Fig. 5.1)

2. Cations and anions are not distributed equally between the body compartments, creating a state of **electrical disequilibrium.** (p. 123)

3. Water moves freely between the cells and extracellular fluid, resulting in a state of **osmotic equilibrium.** (p. 123)

4. The movement of water across a membrane in response to a concentration gradient is called **osmosis.** (p. 125)

5. We express the concentration of biological solutions as **osmolarity,** the number of particles (ions or intact molecules) per liter of solution, in units of milliosmoles per liter (mOsM). (p. 126)

6. **Tonicity** of a solution describes the cell volume change that occurs at equilibrium if the cell is placed in that solution. Cells swell in **hypotonic solutions** and shrink in **hypertonic solutions.** If the cell does not change size at equilibrium, the solution is **isotonic.** (p. 127; Tbl. 5.3)

7. The osmolarity of a solution cannot be used to determine the tonicity of the solution. The relative concentrations of **nonpenetrating solutes** in the cell and in the solution determine tonicity. **Penetrating solutes** contribute to the osmolarity of a solution but not to its tonicity. (p. 128; Figs. 5.3, 5.4; Tbl. 5.4)

Diffusion

8. In **bulk flow,** a pressure gradient moves a fluid along with its dissolved and suspended materials. (p. 132)

9. The cell membrane is a selectively permeable barrier that restricts free exchange between the cell and the interstitial fluid. The movement of a substance across a membrane depends on the permeability of the membrane to that substance. (p. 133)

10. Movement of molecules across membranes can be classified either by energy requirements or by the physical means the molecule uses to cross the membrane. (p. 133; Fig. 5.5)

11. Lipid-soluble substances can diffuse through the phospholipid bilayer. Less lipid-soluble molecules require the assistance of a membrane protein or vesicle to cross the membrane. (p. 133)

12. **Passive transport** does not require the input of energy. (p. 133)

13. **Diffusion** is the passive movement of molecules down a chemical (concentration) gradient from an area of higher concentration to an area of lower concentration. Net movement stops when the system reaches **equilibrium,** although molecular movement continues. (p. 134; Tbl. 5.6)

14. Diffusion rate depends on the magnitude of the concentration gradient. Diffusion is slow over long distances, is directly related to temperature, and is inversely related to molecular size. (p. 134)

15. **Simple diffusion** across a membrane is directly proportional to membrane surface area, concentration gradient, and membrane permeability, and inversely proportional to membrane thickness. (p. 136; Fig. 5.7)

Protein-Mediated Transport

16. Most molecules cross membranes with the aid of membrane proteins. (p. 137)

17. Membrane proteins have four functional roles: **structural proteins** maintain cell shape and form cell junctions; **membrane-associated enzymes** catalyze chemical reactions and help transfer signals across the membrane; **receptor proteins** are part of the body's signaling system; and **transport proteins** move many molecules into or out of the cell. (pp. 137, 138; Fig. 5.8)

18. **Channel proteins** form water-filled channels that link the intracellular and extracellular compartments. **Gated channels** regulate movement of substances through them by opening and closing. Gated channels may be regulated by ligands, by the electrical state of the cell, or by physical changes such as pressure. (pp. 138, 139; Fig. 5.10)

19. **Carrier proteins** never form a continuous connection between the intracellular and extracellular fluid. They bind to substrates, then change conformation. (p. 138; Fig. 5.12)

20. Protein-mediated diffusion is called **facilitated diffusion.** It has the same properties as simple diffusion. (p. 137; Tbl. 5.6; Fig. 5.13)

21. **Active transport** moves molecules against their concentration gradient and requires an outside source of energy. In **primary (direct) active transport,** the energy comes directly from ATP. **Secondary (indirect) active transport** uses the potential energy stored in a concentration gradient and is indirectly driven by energy from ATP. (pp. 133, 143)

22. The most important primary active transporter is the **sodium-potassium-ATPase** (Na^+-K^+-ATPase), which pumps Na^+ out of the cell and K^+ into the cell. (p. 143; Figs. 5.14, 5.15)

23. Most secondary active transport systems are driven by the sodium concentration gradient. (p. 143; Tbl. 5.8; Fig. 5.16)

24. All carrier-mediated transport demonstrates **specificity, competition,** and **saturation.** Specificity refers to the ability of a transporter to move only one molecule or a group of closely related molecules. Related molecules may compete for a single transporter. Saturation occurs when a group of membrane transporters are working at their maximum rate. (pp. 145, 146; Fig. 5.17)

Vesicular Transport

25. Large macromolecules and particles are brought into cells by phagocytosis or endocytosis. Material leaves cells by exocytosis. When vesicles that come into the cytoplasm by endocytosis are returned to the cell membrane, the process is called **membrane recycling.** (p. 147; Figs 5.18, 5.19)

26. In **receptor-mediated endocytosis,** ligands bind to membrane receptors that concentrate in **coated pits** or **caveolae.** (pp. 147, 148; Fig. 5.19)

27. In **exocytosis,** the vesicle membrane fuses with the cell membrane before releasing its contents into the extracellular space. Exocytosis requires ATP. (p. 148)

Epithelial Transport

28. Transporting epithelia have different membrane proteins on their **apical** and **basolateral** surfaces. This polarization allows one-way movement of molecules across the epithelium. (p. 150; Figs. 5.20, 5.21)

29. Molecules cross epithelia by moving between the cells by the **paracellular** route or through the cells by the **transcellular** route. (p. 150; Fig. 5.20)

30. Larger molecules cross epithelia by **transcytosis,** which includes **vesicular transport.** (pp. 151, 152; Fig. 5.22)

The Resting Membrane Potential

iP Nervous I: The Membrane Potential

31. Although the total body is electrically neutral, diffusion and active transport of ions across the cell membrane create an **electrical gradient,** with the inside of cells negative relative to the extracellular fluid. (p. 156; Fig. 5.23)

iP2 Resting Membrane Potential

32. The electrical gradient between the extracellular fluid and the intracellular fluid is known as the **resting membrane potential difference.** (p. 156)

33. The movement of an ion across the cell membrane is influenced by the **electrochemical gradient** for that ion. (p. 156)

34. The membrane potential that exactly opposes the concentration gradient of an ion is known as the **equilibrium potential** (E_{ion}). The equilibrium potential for any ion can be calculated using the Nernst equation. (p. 156; Fig. 5.23)

35. In most living cells, K^+ is the primary ion that determines the resting membrane potential. (p. 158)

36. Changes in membrane permeability to ions such as K^+, Na^+, Ca^{2+}, or Cl^- alter membrane potential and create electrical signals. (p. 158)

Integrated Membrane Processes: Insulin Secretion

37. The use of electrical signals to initiate a cellular response is a universal property of living cells. Pancreatic beta cells release insulin in response to a change in membrane potential. (p. 159; Fig. 5.26)

REVIEW QUESTIONS

In addition to working through these questions and checking your answers on p. A-6, review the Learning Outcomes at the beginning of this chapter.

Level One Reviewing Facts and Terms

1. List the four functions of membrane proteins, and give an example of each.

2. Distinguish between active transport and passive transport.

3. Which of the following processes are examples of active transport, and which are examples of passive transport? Simple diffusion, phagocytosis, facilitated diffusion, exocytosis, osmosis, endocytosis.

4. List four factors that increase the rate of diffusion in air.

5. List the three physical methods by which materials enter cells.

6. A cotransporter is a protein that moves more than one molecule at a time. If the molecules are moved in the same direction, the transporters are called _____ carriers; if the molecules are transported in opposite directions, the transporters are called _____ carriers. A transport protein that moves only one substrate is called a(n) _____ carrier.

7. The two types of active transport are _____, which derives energy directly from ATP, and _____, which couples the kinetic energy of one molecule moving down its concentration gradient to the movement of another molecule against its concentration gradient.

8. A molecule that moves freely between the intracellular and extracellular compartments is said to be a(n) _____ solute. A molecule that is not able to enter cells is called a(n) _____ solute.

9. Rank the following individuals in order of how much body water they contain as a percentage of their body weight, from highest to lowest: (a) a 25-year-old, 74-kg male; (b) a 25-year-old, 50-kg female; (c) a 65-year-old, 50-kg female; and (d) a 1-year-old, 11-kg male toddler.

10. What determines the osmolarity of a solution? In what units is body osmolarity usually expressed?

11. What does it mean if we say that a solution is hypotonic to a cell? Hypertonic to the same cell? What determines the tonicity of a solution relative to a cell?

12. Match the membrane channels with the appropriate descriptions. Answers may be used once, more than once, or not at all.

(a) chemically gated channel	1. channel that spends most of its time in the open state
(b) open pore	2. channel that spends most of its time in a closed state
(c) voltage-gated channel	3. channel that opens when resting membrane potential changes
(d) mechanically gated channel	4. channel that opens when a ligand binds to it
	5. channel that opens in response to membrane stretch
	6. channel through which water can pass

13. In your own words, state the four principles of electricity important in physiology.

14. Match each of the following items with its primary role in cellular activity.

(a) Na^+-K^+-ATPase	1. ion channel
(b) protein	2. extracellular cation
(c) unit of measurement for membrane potential	3. source of energy
	4. intracellular anion
(d) K^+	5. intracellular cation
(e) Cl^-	6. millivolts
(f) ATP	7. electrogenic pump
(g) Na^+	8. extracellular anion
	9. milliosmoles

15. The membrane potential at which the electrical gradient exactly opposes the concentration gradient for an ion is known as the _____.

16. A material that allows free movement of electrical charges is called a(n) _____, whereas one that prevents this movement is called a(n) _____.

Level Two Reviewing Concepts

17. Create a map of transport across cell membranes using the following terms. You may add additional terms if you wish.

- active transport
- carrier
- caveolae
- channel
- clathrin-coated pit
- concentration gradient
- electrochemical gradient
- exocytosis
- facilitated diffusion
- glucose
- GLUT transporter
- ion
- large polar molecule
- ligand
- Na^+-K^+-ATPase
- osmosis
- passive transport
- phospholipid bilayer
- receptor-mediated endocytosis
- secondary active transport
- simple diffusion
- small polar molecule
- transcytosis
- vesicle
- vesicular transport
- water

18. Draw a large rectangle to represent the total body volume. Using the information in Figure 5.1b, divide the box proportionately into compartments to represent the different body compartments. Use the information in Figure 5.1d and add solutes to the compartments. Use large letters for solutes with higher concentrations, and small letters for solutes with low concentrations. Label the cell membranes and the endothelial membrane.

19. What factors influence the rate of diffusion across a membrane? Briefly explain each one.

20. Define the following terms and explain how they differ from one another: specificity, competition, saturation. Apply these terms in a short explanation of facilitated diffusion of glucose.

21. Red blood cells are suspended in a solution of NaCl. The cells have an osmolarity of 300 mOsM, and the solution has an osmolarity of 250 mOsM. (a) The solution is (hypertonic, isotonic, or hypotonic) to the cells. (b) Water would move (into the cells, out of the cells, or not at all).

22. Two compartments are separated by a membrane that is permeable to glucose but not water. Each compartment is filled with 1 M glucose. After six hours, compartment A contains 1.5 M glucose and compartment B contains 0.5 M glucose. What kind of transport occurred? Explain.

23. A 2 M NaCl solution is placed in compartment A and a 2 M glucose solution is placed in compartment B. The compartments are separated by a membrane that is permeable to water but not to NaCl or glucose. Complete the following statements. Defend your answers.
    (a) The salt solution is _____ osmotic to the glucose solution.
    (b) True or false? Water will move from one compartment to another. If water moves, it will move from compartment _____ to compartment _____.

24. Explain the differences between a chemical gradient, an electrical gradient, and an electrochemical gradient.

Level Three Problem Solving

25. Sweat glands secrete into their lumen a fluid that is identical to interstitial fluid. As the fluid moves through the lumen on its way to the surface of the skin, the cells of the sweat gland's epithelium make the fluid hypotonic by removing Na^+ and leaving water behind. Design an epithelial cell that will reabsorb Na^+ but not water. You may place water pores, Na^+ leak channels, K^+ leak channels, and the Na^+-K^+-ATPase in the apical membrane, basolateral membrane, or both.

26. Insulin is a hormone that promotes the movement of glucose into many types of cells, thereby lowering blood glucose concentration. Propose a mechanism that explains how this occurs, using your knowledge of cell membrane transport.

27. The following terms have been applied to membrane carriers: specificity, competition, saturation. Why can these terms also be applied to enzymes? What is the major difference in how enzymes and carriers carry out their work?

28. Integral membrane glycoproteins have sugars added as the proteins pass through the lumen of the endoplasmic reticulum and Golgi complex [p. 115]. Based on this information, where would you predict finding the sugar "tails" of the proteins: on the cytoplasmic side of the membrane, the extracellular side, or both? Explain your reasoning.

29. NaCl is a nonpenetrating solute and urea is a penetrating solute for cells. Red blood cells are placed in each of the solutions below. The intracellular concentration of nonpenetrating solute is 300 mOsM. What will happen to the cell volume in each solution? Label each solution with all the terms that apply: hypertonic, isotonic, hypotonic, hyperosmotic, hyposmotic, isosmotic. Watch units! Assume 1 M NaCl = 2 OsM for simplicity.

    (a) 150 mM NaCl plus 150 mM urea
    (b) 100 mM NaCl plus 50 mM urea
    (c) 100 mM NaCl plus 100 mM urea
    (d) 150 mM NaCl plus 100 mM urea
    (e) 100 mM NaCl plus 150 mM urea

Level Four Quantitative Problems

30. The addition of dissolved solutes to water lowers the freezing point of water. A 1 OsM solution depresses the freezing point of water by 1.86 °C. If a patient's plasma shows a freezing-point depression of 0.55 °C, what is her plasma osmolarity? (Assume that 1 kg water = 1 L.)

31. The patient in the previous question is found to have total body water volume of 42 L, ECF volume of 12.5 L, and plasma volume of 2.7 L.

    (a) What is her intracellular fluid (ICF) volume? Her interstitial fluid volume?
    (b) How much solute (osmoles) exists in her whole body? ECF? ICF? plasma? (*Hint:* concentration = solute amount/volume of solution)

32. What is the osmolarity of half-normal saline (= 0.45% NaCl)? [p. 126] Assume that all NaCl molecules dissociate into two ions.

33. If you give 1 L of half-normal saline (see question 32) to the patient in question 31, what happens to each of the following at equilibrium? (*Hint:* NaCl is a nonpenetrating solute.)

    (a) her total body volume
    (b) her total body osmolarity
    (c) her ECF and ICF volumes
    (d) her ECF and ICF osmolarities

34. The following graph shows the results of an experiment in which a cell was placed in a solution of glucose. The cell had no glucose in it at the beginning, and its membrane can transport glucose. Which of the following processes is/are illustrated by this experiment?

    (a) diffusion
    (b) saturation
    (c) competition
    (d) active transport

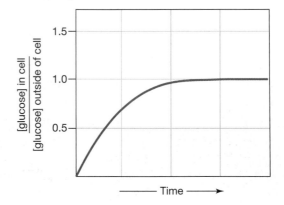

6

Communication, Integration, and Homeostasis

Future progress in medicine will require a quantitative understanding of the many interconnected networks of molecules that comprise our cells and tissues, their interactions, and their regulation.

Overview of the NIH Roadmap, September 30, 2003. *NIH Announces Strategy to Accelerate Medical Research Progress.*

CELL-TO-CELL COMMUNICATION 166

LO 6.1 Describe three forms of local communication and two forms of long-distance communication.

SIGNAL PATHWAYS 169

LO 6.2 Explain the general sequence of events that follow lipophilic ligand binding to intracellular receptors.

LO 6.3 Describe the general sequence of events that follow lipophobic ligand binding to a cell surface receptor.

LO 6.4 Name and describe four major groups of cell surface receptors.

LO 6.5 Explain how cascades and signal amplification play a role in signal transduction.

NOVEL SIGNAL MOLECULES 176

LO 6.6 List five ways calcium acts as an intracellular messenger.

LO 6.7 Describe the advantages and disadvantages of gaseous second messenger molecules.

MODULATION OF SIGNAL PATHWAYS 180

LO 6.8 Apply the concepts of specificity, competition, affinity, and saturation to receptors and their ligands.

LO 6.9 Explain the role of up-regulation, down-regulation, and pathway termination in modulating cell responses to receptors and their ligands.

HOMEOSTATIC REFLEX PATHWAYS 182

LO 6.10 List Cannon's four postulates of homeostatic control and give an example of each.

LO 6.11 List the seven steps of a reflex control pathway in the order in which they occur.

LO 6.12 Compare the speed, specificity, types of signals, and duration of action in neural and endocrine reflexes. How is stimulus intensity coded in each type of reflex?

LO 6.13 Describe some examples of complex reflex pathways with more than one integrating center.

BACKGROUND BASICS

9	Homeostasis
34	Nucleotides
46	Protein interactions
72	Cell junctions
72	Extracellular matrix
80	Endocrine glands
62	Membrane structure
62	Membrane proteins
134	Diffusion
148	Exocytosis

DNA microarrays

In 2003, the United States National Institutes of Health (NIH) embarked on an ambitious project to promote translation of basic research into new medical treatments and strategies for disease prevention. Contributors to the NIH Common Fund Programs (*http://commonfund.nih.gov/bbpn/index*) are compiling information on biological pathways in an effort to understand how cells communicate with one another and maintain the body in a healthy state. In this chapter, we examine the basic patterns of cell-to-cell communication and see how the coordination of function resides in chemical and electrical signals. Each cell in the body can communicate with most other cells. To maintain homeostasis, the body uses a combination of diffusion across small distances; widespread distribution of molecules through the circulatory system; and rapid, specific delivery of messages by the nervous system.

CELL-TO-CELL COMMUNICATION

In recent years, the amount of information available about cell-to-cell communication has mushroomed as a result of advances in research technology. Signal pathways that once seemed fairly simple and direct are now known to be incredibly complex networks and webs of information transfer. In the sections that follow, we distill what is known about cell-to-cell communication into some basic patterns that you can recognize when you encounter them again in your study of physiology. As with many rapidly changing fields, these patterns reflect our current understanding and are subject to modification as scientists learn more about the incredibly complex network of chemical signals that control life processes.

By most estimates, the human body is composed of about 75 *trillion* cells. Those cells face a daunting task—to communicate with one another in a manner that is rapid and yet conveys a tremendous amount of information. Surprisingly, there are only two basic types of physiological signals: electrical and chemical. **Electrical signals** are changes in a cell's membrane potential [p. 153]. **Chemical signals** are molecules secreted by cells into the extracellular fluid. The cells that respond to electrical or chemical signals are called **target cells,** or **targets** for short.

Chemical signals are responsible for most communication within the body. Chemical signals act as *ligands* that bind to proteins to initiate a response. Protein binding of chemical signals obeys the general rules for protein interactions, including *specificity, affinity, competition,* and *saturation* [p. 46].

Our bodies use four basic methods of cell-to-cell communication (**FIG. 6.1**). **Local communication** includes (1) **gap junctions,** which allow direct cytoplasmic transfer of electrical and chemical signals between adjacent cells; (2) **contact-dependent signals,** which occur when surface molecules on one cell membrane bind to surface molecules on another cell's membrane; and (3) chemicals that diffuse through the extracellular fluid to act on cells close by. **Long-distance communication** (4) uses a combination of chemical and electrical signals carried by nerve cells and chemical signals transported in the blood. A given molecule can function as a chemical signal by more than one method. For example, a molecule can act close to the cell that released it (local communication) as well as in distant parts of the body (long-distance communication).

Gap Junctions Create Cytoplasmic Bridges

The simplest form of cell-to-cell communication is the direct transfer of electrical and chemical signals through *gap junctions,* protein channels that create cytoplasmic bridges between adjacent cells (Fig. 6.1a). A gap junction forms from the union of membrane-spanning proteins, called *connexins,* on two adjacent cells [p. 74]. The united connexins create a protein channel (*connexon*) that can open and close. When the channel is open, the connected cells function like a single cell that contains multiple nuclei (a *syncytium*).

When gap junctions are open, ions and small molecules such as amino acids, ATP, and cyclic AMP diffuse directly from the cytoplasm of one cell to the cytoplasm of the next. Larger molecules cannot pass through gap junctions. In addition, gap junctions are the only means by which electrical signals can pass *directly* from cell to cell. Movement of molecules and electrical signals through gap junctions can be modulated or shut off completely.

Gap junctions are not all alike. Scientists have discovered more than 20 different isoforms of connexins that may mix or match to form gap junctions. The variety of connexin isoforms allows gap junction selectivity to vary from tissue to tissue. In mammals, gap junctions are found in almost every cell type, including heart muscle, some types of smooth muscle, lung, liver, and neurons of the brain.

Contact-Dependent Signals Require Cell-to-Cell Contact

Some cell-to-cell communication requires that surface molecules on one cell membrane bind to a membrane protein of another cell (Fig. 6.1b). Such *contact-dependent signaling* occurs in the

RUNNING **PROBLEM** | **Diabetes Mellitus: A Growing Epidemic**

It is 8:00 A.M. and Marvin Garcia, age 20, is hungry. He came to his family physician's office before breakfast to have a fasting blood glucose test as part of a routine physical examination. In this test, blood is drawn after an overnight fast, and the glucose concentration in the blood is measured. Because he knows he is in good condition, Marvin isn't worried about the results. He is surprised, then, when the nurse practitioner in the doctor's office calls two days later. "Your fasting blood sugar is a bit elevated, Marvin. It is 130 milligrams per deciliter, and normal is 100 or less. Does anyone in your family have diabetes?" "Well, yeah—my dad has it. What exactly is diabetes?"

166   169   182   183   187   189   191

FIG. 6.1 **ESSENTIALS**

Communication in the Body

Cell-to-cell communication uses chemical and electrical
signaling to coordinate function and maintain homeostasis.

LOCAL COMMUNICATION

Receptor

(a) Gap junctions form
direct cytoplasmic
connections between
adjacent cells.

(b) Contact-dependent signals
require interaction between
membrane molecules on
two cells.

(c) Autocrine signals act on the same cell
that secreted them. **Paracrine signals**
are secreted by one cell and diffuse to
adjacent cells.

LONG-DISTANCE COMMUNICATION

Long-distance signaling may be electrical signals passing
along neurons or chemical signals that travel through the
circulatory system.

Endocrine System

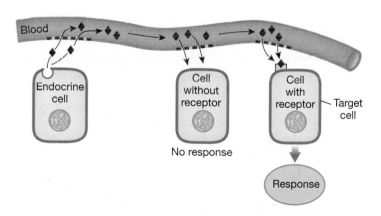

(d) Hormones are secreted by endocrine glands or cells into the blood.
Only target cells with receptors for the hormone respond to the signal.

Nervous System

(e) Neurotransmitters are chemicals secreted by neurons that diffuse
across a small gap to the target cell.

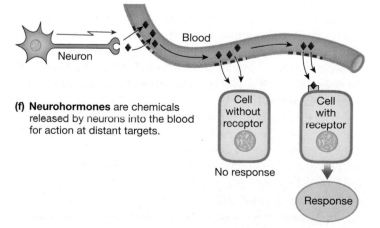

(f) Neurohormones are chemicals
released by neurons into the blood
for action at distant targets.

immune system and during growth and development, such as when nerve cells send out long extensions that must grow from the central axis of the body to the *distal* (distant) ends of the developing limbs. **Cell adhesion molecules (CAMs)** first known for their role in cell-to-cell adhesion [p. 72], have now been shown to act as receptors in cell-to-cell signaling. CAMs are linked to the cytoskeleton or to intracellular enzymes. Through these linkages, CAMs transfer signals in both directions across cell membranes.

Local Communication Uses Paracrine and Autocrine Signals

Local communication takes place through paracrine and autocrine signaling. A **paracrine signal** {*para-*, beside + *krinen*, to secrete} is a chemical that acts on cells in the immediate vicinity of the cell that secreted the signal. A chemical signal that acts on the cell that secreted it is called an **autocrine signal** {*auto-*, self}. In some cases, a molecule may act as both an autocrine signal and a paracrine signal.

Paracrine and autocrine signal molecules reach their target cells by diffusing through the interstitial fluid (Fig. 6.1c). Because distance is a limiting factor for diffusion, the effective range of paracrine signals is restricted to adjacent cells. A good example of a paracrine molecule is *histamine*, a chemical released from damaged cells. When you scratch yourself with a pin, the red, raised *wheal* that results is due in part to the local release of histamine from the injured tissue. The histamine acts as a paracrine signal, diffusing to capillaries in the immediate area of the injury and making them more permeable to white blood cells and antibodies in the plasma. Fluid also leaves the blood vessels and collects in the interstitial space, causing swelling around the area of injury.

Several important classes of molecules act as local signals. *Cytokines* are regulatory peptides, and *eicosanoids* [p. 30] are lipid-derived paracrine and autocrine signal molecules. We discuss cytokines and eicosanoids in more detail later.

Long-Distance Communication May Be Electrical or Chemical

All cells in the body can release paracrine signals, but most long-distance communication between cells takes place through the nervous and endocrine systems. The endocrine system communicates by using **hormones** {*hormon*, to excite}, chemical signals that are secreted into the blood and distributed all over the body by the circulation. Hormones come in contact with most cells of the body, but only those cells with receptors for the hormone are target cells (Fig. 6.1d).

The nervous system uses a combination of chemical signals and electrical signals to communicate over long distances. An electrical signal travels along a nerve cell (*neuron*) until it reaches the very end of the cell, where it is translated into a chemical signal secreted by the neuron. Chemicals secreted by neurons are called **neurocrine molecules.**

If a neurocrine molecule diffuses from the neuron across a narrow extracellular space to a target cell and has a rapid-onset effect, it is called a **neurotransmitter** (Fig. 6.1e). If a neurocrine acts more slowly as an autocrine or paracrine signal, it is called a **neuromodulator.** If a neurocrine molecule diffuses into the blood for body-wide distribution, it is called a **neurohormone** (Fig. 6.1f). The similarities between neurohormones and classic hormones secreted by the endocrine system bridge the gap between the nervous and endocrine systems, making them a functional continuum rather than two distinct systems.

Cytokines May Act as Both Local and Long-Distance Signals

Cytokines are among the most recently identified communication molecules. Initially the term *cytokine* referred only to peptides that modulate immune responses, but in recent years the definition has been broadened to include a variety of regulatory peptides. Most of these peptides share a similar structure of four or more α-helix bundles [p. 32]. Families of cytokines include *interferons, interleukins, colony-stimulating factors,* and *growth factors.*

Cytokines are associated primarily with immune responses, such as inflammation, but they also control cell development and cell differentiation. In development and differentiation, cytokines usually function as autocrine or paracrine signals. In stress and inflammation, some cytokines may act on relatively distant targets and may be transported through the circulation just as hormones are.

How do cytokines differ from classic hormones? Cytokines are not produced by specialized epithelial cells the way hormones are. Instead, any nucleated cell can secrete cytokines at some point in its lifespan. Cytokines are made on demand, in contrast to protein or peptide hormones are that made in advance and stored in the endocrine cell until needed. Also, the signal pathways for cytokines are usually different from those for hormones. However, the distinction between cytokines and hormones is sometimes blurry. For example, erythropoietin, the molecule that controls synthesis of red blood cells, is by tradition considered a hormone but functionally fits the definition of a cytokine.

CONCEPT CHECK

1. Match the communication method on the left with its property on the right.

   (a) autocrine signal   Communication is:
   (b) cytokine   1. electrical
   (c) gap junction   2. chemical
   (d) hormone   3. both electrical and chemical
   (e) neurohormone
   (f) neurotransmitter
   (g) paracrine signal

2. Which signal molecules listed in the previous question are transported through the circulatory system? Which are released by neurons?

3. A cat sees a mouse and pounces on it. Do you think the internal signal to pounce could have been transmitted by a paracrine signal? Give two reasons to explain why or why not.

SIGNAL PATHWAYS

Chemical signal molecules are secreted by cells into the extracellular compartment. This is not a very specific way for these signals to find their targets because substances that diffuse through interstitial fluid or that travel through the blood come in contact with many cells. Yet cells do not respond to every signal that reaches them.

Why do some cells respond to a chemical signal while other cells ignore it? The answer lies in the target cell's **receptor proteins** [p. 138]. *A cell can respond to a particular chemical signal only if the cell has the appropriate receptor protein to bind that signal* (Fig. 6.1d).

If a target cell has the receptor for a signal molecule, binding of the signal molecule to the receptor protein initiates a response. All signal pathways share the following features (**FIG. 6.2**):

1. The signal molecule is a *ligand* that binds to a protein receptor. The ligand is also known as a *first messenger* because it brings information to the target cell.
2. Ligand-receptor binding activates the receptor.
3. The receptor in turn activates one or more intracellular signal molecules.
4. The last signal molecule in the pathway creates a response by modifying existing proteins or initiating the synthesis of new proteins.

In the following sections, we describe some basic signal pathways. They may seem complex at first, but they follow patterns that you will encounter over and over as you study the systems of the body. Most physiological processes, from the beating of your heart to learning and memory, use some variation of these pathways. One of the wonders of physiology is the fundamental importance of these signal pathways and the way they have been conserved in animals ranging from worms to humans.

Receptor Proteins Are Located Inside the Cell or on the Cell Membrane

Protein receptors for signal molecules play an important role in physiology and medicine. About half of all drugs currently in use act on receptor proteins. Target cell receptor proteins may be found in the nucleus, in the cytosol, or on the cell membrane as integral proteins. Where a chemical signal binds to its receptor largely depends on whether that signal molecule is lipophilic or lipophobic (**FIG. 6.3**).

Lipophilic signal molecules enter cells by simple diffusion through the phospholipid bilayer of the cell membrane [p. 62]. Once inside, they bind to *cytosolic receptors* or *nuclear receptors* (Fig. 6.3a). Activation of intracellular receptors often turns on a gene and directs the nucleus to make new mRNA (transcription, [p. 112]). The mRNA then provides a template for synthesis of new proteins (translation, [p. 112]). This process is relatively slow and the cell's response may not be noticeable for an hour or longer. In some instances, the activated receptor can also turn off, or *repress*, gene activity. Many lipophilic signal molecules that follow this pattern are hormones.

Lipophobic signal molecules are unable to enter cells by simple diffusion through the cell membrane. Instead, these signal molecules remain in the extracellular fluid and bind to receptor proteins on the cell membrane (Fig. 6.3b). (Some lipophilic signal molecules also bind to cell membrane receptors in addition to their intracellular receptors.) In general, the response time for pathways linked to membrane receptor proteins is very rapid: responses can be seen within milliseconds to minutes.

FIG. 6.2 | Signal pathways

Most signal pathways consist of the 5 steps shown. Use the shapes and colors of the steps shown here to identify the pattern in later illustrations.

RUNNING **PROBLEM**

Later that day in the physician's office, the nurse practitioner explains diabetes to Marvin. Diabetes mellitus is a family of metabolic disorders caused by defects in the homeostatic pathways that regulate glucose metabolism. Several forms of diabetes exist, and some can be inherited. One form, called *type 1 diabetes mellitus,* occurs when endocrine cells of the pancreas stop making insulin, a protein hormone involved in blood glucose homeostasis. In another form, type 2 diabetes mellitus, insulin may be present in normal or above-normal levels, but the insulin-sensitive cells of the body do not respond normally to the hormone.

Q1: *In which type of diabetes is the target cell's signal pathway for insulin more likely to be defective?*

Q2: *Insulin is a protein hormone. Would you expect to find its receptor on the cell surface or in the cytoplasm of the target cells?*

FIG. 6.3 Target cell receptors may be on the cell surface or inside the cell

(a) Intracellular Signal Receptors

Receptor in cytosol

Receptor in nucleus

Lipophilic signal molecules diffuse through the cell membrane.

Binding to cytosolic or nuclear receptors triggers

Slower responses related to changes in gene activity

(b) Cell Membrane Receptors

Extracellular signal molecule binds to a cell membrane receptor.

Binding triggers

Rapid cellular responses

(c) Four Categories of Membrane Receptors

ECF

Extracellular signal molecules

Channel

Receptor

Receptor

Integrin

Cell membrane

G protein

Enzyme

Anchor protein

ICF

Cytoskeleton

Receptor-channel	**G protein–coupled receptor**	**Receptor-enzyme**	**Integrin receptor**
Ligand binding opens or closes the channel.	Ligand binding to a G protein–coupled receptor opens an ion channel or alters enzyme activity.	Ligand binding to a receptor-enzyme activates an intracellular enzyme.	Ligand binding to integrin receptors alters enzymes or the cytoskeleton.

Catalytic receptors

We can group membrane receptors into four major categories, illustrated in Figure 6.3c. The simplest receptors are chemically gated (*ligand-gated*) ion channels called *receptor-channels* [p. 139]. Ligand binding opens or closes the channel and alters ion flow across the membrane.

Three other receptor types are shown in Figure 6.3c: *G protein-coupled receptors*, *receptor-enzymes*, and *integrin receptors*.

For all three, information from the signal molecule must be passed across the membrane to initiate an intracellular response. This transmission of information from one side of a membrane to the other using membrane proteins is known as *signal transduction*. We will take a closer look at basic signal transduction before returning to the four receptor types that participate in it.

4. List four components of signal pathways.
5. Name three cellular locations of receptors.

Membrane Proteins Facilitate Signal Transduction

Signal transduction is the process by which an extracellular signal molecule activates a membrane receptor that in turn alters intracellular molecules to create a response. The extracellular signal molecule is the first messenger, and the intracellular molecules form a *second messenger system*. The term *signal transduction* comes from the verb *to transduce,* meaning "to lead across" {*trans,* across + *ducere,* to lead}.

A **transducer** is a device that converts a signal from one form into a different form. For example, the transducer in a radio converts radio waves into sound waves (**FIG. 6.4**). In biological systems, membrane proteins act as transducers. They convert the message of extracellular signals into intracellular messenger molecules that trigger a response.

The basic pattern of a biological signal transduction pathway is shown in **FIGURE 6.5a** and can be broken down into the following events.

1. An extracellular signal molecule (the *first messenger*) binds to and activates a membrane receptor.

2. The activated membrane receptor turns on its associated proteins and starts an intracellular cascade of **second messengers.**

3. The last second messenger in the cascade acts on intracellular targets to create a response.

Figure 6.5b details the intracellular events in basic signal transduction pathways:

1. Membrane receptors and their associated proteins usually either
   (a) activate **protein kinases,** which are enzymes that transfer a phosphate group from ATP to a protein [p. 102]. Phosphorylation is an important biochemical method of regulating cellular processes.
   (b) activate amplifier enzymes that create intracellular second messengers.

2. Second messenger molecules in turn
   (a) alter the gating of ion channels. Opening or closing ion channels creates electrical signals by altering the cell's membrane potential [p. 158].
   (b) increase intracellular calcium. Calcium binding to proteins changes their function, creating a cellular response.
   (c) change enzyme activity, especially of protein kinases or **protein phosphatases,** enzymes that remove a phosphate group. The phosphorylation or *dephosphorylation* of a protein can change its configuration and create a response.

3. The proteins modified by calcium binding and phosphorylation are responsible for the cell's response to the signal. Examples of responses include increased or decreased enzyme activity and opening or closing of gated ion channels.

Cascades **FIGURE 6.6a** shows how the steps of a signal transduction pathway form a **cascade.** A signaling cascade starts when a stimulus (the signal molecule) converts inactive molecule A (the receptor) to an active form. Active A then converts inactive molecule B into active B, active molecule B in turn converts inactive molecule C into active C, and so on, until at the final step a substrate is converted into a product. Many intracellular signal pathways are cascades. Blood clotting is an important example of an extracellular cascade.

Amplification In signal transduction pathways, the original signal is not only transformed but also amplified {*amplificare,* to make larger}. In a radio, the radio wave signal is also amplified. In cells, **signal amplification** turns one signal molecule into multiple second messenger molecules (Fig. 6.6b).

The process begins when the first messenger ligand combines with its receptor. The receptor-ligand complex turns on an **amplifier enzyme.** The amplifier enzyme activates several molecules, which in turn each activate several more molecules as the cascade proceeds. By the end of the process, the effects of the

FIG. 6.4 | **Signal transduction**

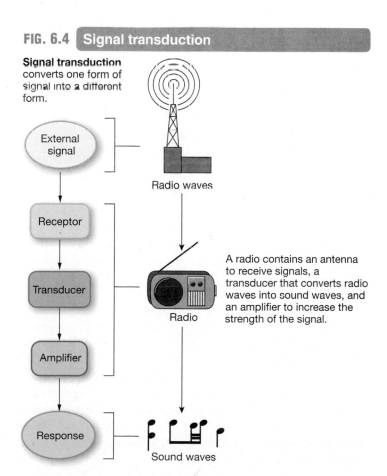

Signal transduction converts one form of signal into a different form.

External signal

Receptor

Transducer

Amplifier

Response

Radio waves

Radio

A radio contains an antenna to receive signals, a transducer that converts radio waves into sound waves, and an amplifier to increase the strength of the signal.

Sound waves

FIG. 6.5 Biological signal transduction

(a) Basic Signal Transduction

(b) Transduction Pathways

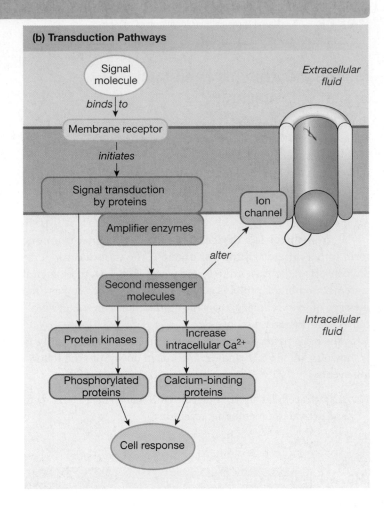

ligand have been amplified much more than if there were a 1:1 ratio between each step.

Amplification gives the body "more bang for the buck" by enabling a small amount of ligand to create a large effect. The most common amplifier enzymes and second messengers are listed in the table in Figure 6.6c.

In the sections that follow, we will examine in more detail the four major types of membrane receptors (see Fig. 6.3c). Keep in mind that these receptors may be responding to any of the different kinds of signal molecules—hormones, neurohormones, neurotransmitters, cytokines, or paracrine and autocrine signals.

CONCEPT CHECK

6. What are the four steps of signal transduction?

7. What happens during amplification? In Figure 6.6b, amplification of one signal molecule binding to the receptor results in how many small dark blue intracellular signal molecules?

8. Why do steroid hormones not require signal transduction and second messengers to exert their action? (*Hint:* Are steroids lipophobic or lipophilic? [p. 30])

The Most Rapid Signal Pathways Change Ion Flow through Channels

The simplest receptors are ligand-gated ion channels. Most of these receptors are neurotransmitter receptors found in nerve and muscle. The activation of **receptor-channels** initiates the most rapid intracellular responses of all receptors. When an extracellular ligand binds to the receptor-channel protein, a channel gate opens or closes, altering the cell's permeability to an ion. Increasing or decreasing ion permeability rapidly changes the cell's membrane potential [p. 158], creating an electrical signal that alters voltage-sensitive proteins (**FIG. 6.7**).

One example of a receptor-channel is the acetylcholine-gated monovalent ("one-charge") cation channel of skeletal muscle. The neurotransmitter *acetylcholine* released from an adjacent neuron binds to the acetylcholine receptor and opens the channel. Both Na^+ and K^+ flow through the open channel, K^+ leaving the cell and Na^+ entering the cell along their electrochemical gradients. The sodium gradient is stronger, however, so net entry of positively charged Na^+ depolarizes the cell. In skeletal muscle, this cascade of intracellular events results in muscle contraction.

FIG. 6.6 **ESSENTIALS**

Signal Transduction: Cascades and Amplification

(a) Signal transduction pathways form a **cascade.**

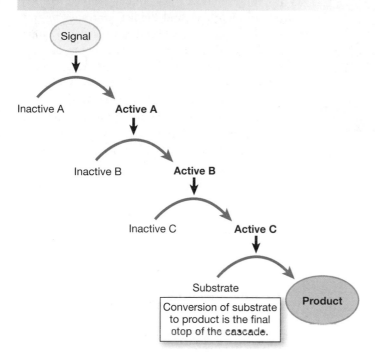

Conversion of substrate to product is the final step of the cascade.

(b) Signal amplification allows a small amount of signal to have a large effect.

Receptor-ligand complex activates an amplifier enzyme (AE).

One ligand is amplified into many intracellular molecules.

(c) Second messenger pathways

Second Messenger	Made from	Amplifier enzyme	Linked to	Action	Effects
Nucleotides					
cAMP	ATP	Adenylyl cyclase (membrane)	GPCR*	Activates protein kinases, especially PKA. Binds to ion channels.	Phosphorylates proteins. Alters channel opening.
cGMP	GTP	Guanylyl cyclase (membrane)	Receptor-enzyme	Activates protein kinases, especially PKG.	Phosphorylates proteins.
		Guanylyl cyclase (cytosol)	Nitric oxide (NO)	Binds to ion channels.	Alters channel opening.
Lipid-Derived*					
IP₃	Membrane phospholipids	Phospholipase C (membrane)	GPCR	Releases Ca²⁺ from intracellular stores.	See Ca²⁺ effects below.
DAG				Activates protein kinase C.	Phosphorylates proteins.
Ions					
Ca²⁺				Binds to calmodulin. Binds to other proteins.	Alters enzyme activity. Exocytosis, muscle contraction, cytoskeleton movement, channel opening.

*GPCR = G protein–coupled receptor. IP₃ = Inositol trisphosphate. DAG = diacylglycerol.

FIG. 6.7 Signal transduction using ion channels

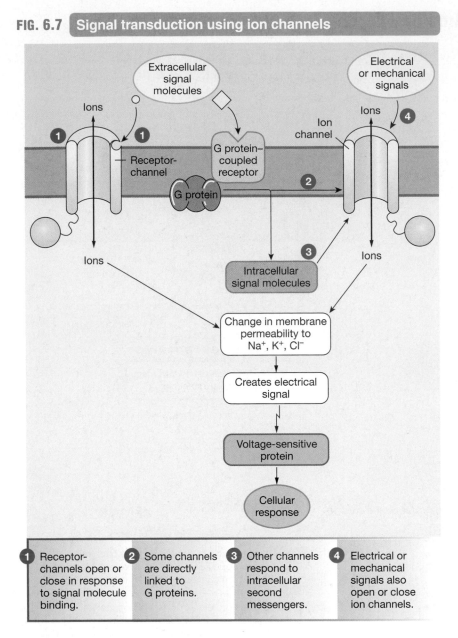

1	2	3	4
Receptor-channels open or close in response to signal molecule binding.	Some channels are directly linked to G proteins.	Other channels respond to intracellular second messengers.	Electrical or mechanical signals also open or close ion channels.

phospholipid bilayer seven times (see Fig. 6.3c). The cytoplasmic tail of the receptor protein is linked to a three-part membrane transducer molecule known as a **G protein.** Hundreds of G protein-coupled receptors have been identified, and the list continues to grow. The types of ligands that bind to G protein-coupled receptors include hormones, growth factors, olfactory molecules, visual pigments, and neurotransmitters. In 1994, Alfred G. Gilman and Martin Rodbell received a Nobel Prize for the discovery of G proteins and their role in cell signaling (see *http://nobelprize.org/nobel_prizes/medicine/laureates/1994*).

G proteins get their name from the fact that they bind guanosine nucleotides [p. 34]. Inactive G proteins are bound to guanosine diphosphate (GDP). Exchanging the GDP for guanosine triphosphate (GTP) activates the G protein. When G proteins are activated, they either (1) open an ion channel in the membrane or (2) alter enzyme activity on the cytoplasmic side of the membrane.

G proteins linked to amplifier enzymes make up the bulk of all known signal transduction mechanisms. The two most common amplifier enzymes for G protein-coupled receptors are adenylyl cyclase and phospholipase C. The pathways for these amplifier enzymes are described next.

Many Lipophobic Hormones Use GPCR-cAMP Pathways

The **G protein-coupled adenylyl cyclase-cAMP system** was the first identified signal transduction pathway (**FIG. 6.8a**). It was discovered in the 1950s by Earl Sutherland when he was studying the effects of hormones on carbohydrate metabolism. This discovery proved so significant to our understanding of signal transduction that in 1971 Sutherland was awarded a Nobel Prize for his work.

The G protein-coupled adenylyl cyclase-cAMP system is the signal transduction system for many protein hormones. In this system, *adenylyl cyclase* is the amplifier enzyme that converts ATP to the second messenger molecule *cyclic AMP* (cAMP). Cyclic AMP then activates *protein kinase A* (PKA), which in turn phosphorylates other intracellular proteins as part of the signal cascade.

G Protein-Coupled Receptors Also Use Lipid-Derived Second Messengers

Some G protein-coupled receptors are linked to a different amplifier enzyme: phospholipase C (Fig. 6.8b). When a signal molecule activates this G protein-coupled pathway, **phospholipase C (PLC)** converts a membrane phospholipid (*phosphatidylinositol bisphosphate*) into two lipid-derived second messenger molecules: diacylglycerol and inositol trisphosphate.

Receptor-channels are only one of several ways to trigger ion-mediated cell signaling. Some ion channels are linked to G protein-coupled receptors. When a ligand binds to the G protein receptor, the G protein pathway opens or closes the channel.

Finally, some membrane ion channels are not associated with membrane receptors at all. Voltage-gated channels can be opened directly with a change in membrane potential. Mechanically gated channels open with pressure or stretch on the cell membrane [p. 139]. Intracellular molecules, such as cAMP or ATP, can open or close non-receptor-linked ligand-gated channels. The ATP-gated K^+ channels of the pancreatic beta cell are an example [Fig. 5.26, p. 159].

Most Signal Transduction Uses G Proteins

The **G protein-coupled receptors (GPCRs)** are a large and complex family of membrane-spanning proteins that cross the

FIG. 6.8   G protein-coupled signal transduction

(a) GPCR-Adenylyl Cyclase Signal Transduction and Amplification

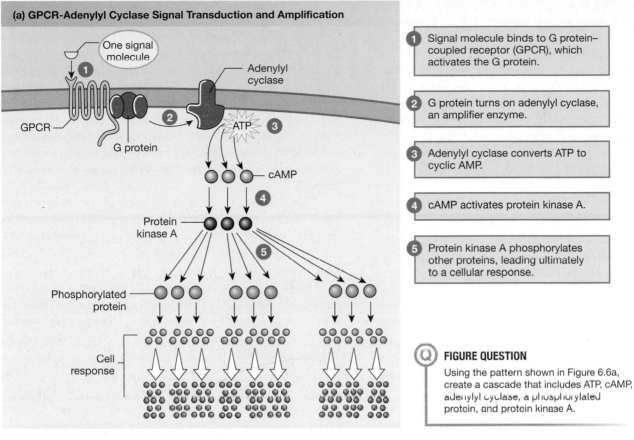

1. Signal molecule binds to G protein–coupled receptor (GPCR), which activates the G protein.

2. G protein turns on adenylyl cyclase, an amplifier enzyme.

3. Adenylyl cyclase converts ATP to cyclic AMP.

4. cAMP activates protein kinase A.

5. Protein kinase A phosphorylates other proteins, leading ultimately to a cellular response.

Q FIGURE QUESTION

Using the pattern shown in Figure 6.6a, create a cascade that includes ATP, cAMP, adenylyl cyclase, a phosphorylated protein, and protein kinase A.

(b) GPCR-Phospholipase C Signal Transduction

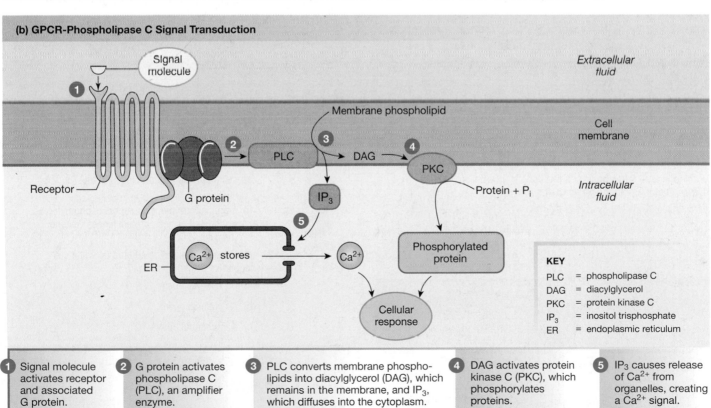

KEY

PLC = phospholipase C
DAG = diacylglycerol
PKC = protein kinase C
IP$_3$ = inositol trisphosphate
ER = endoplasmic reticulum

1. Signal molecule activates receptor and associated G protein.

2. G protein activates phospholipase C (PLC), an amplifier enzyme.

3. PLC converts membrane phospholipids into diacylglycerol (DAG), which remains in the membrane, and IP$_3$, which diffuses into the cytoplasm.

4. DAG activates protein kinase C (PKC), which phosphorylates proteins.

5. IP$_3$ causes release of Ca^{2+} from organelles, creating a Ca^{2+} signal.

Diacylglycerol (DAG) is a nonpolar diglyceride that remains in the lipid portion of the membrane and interacts with **protein kinase C (PKC),** a Ca^{2+}-activated enzyme associated with the cytoplasmic face of the cell membrane. PKC phosphorylates cytosolic proteins that continue the signal cascade.

Inositol trisphosphate (IP$_3$) is a water-soluble messenger molecule that leaves the membrane and enters the cytoplasm. There it binds to a calcium channel on the endoplasmic reticulum (ER). IP$_3$ binding opens the Ca^{2+} channel, allowing Ca^{2+} to diffuse out of the ER and into the cytosol. Calcium is itself an important signal molecule, as discussed later.

Receptor-Enzymes Have Protein Kinase or Guanylyl Cyclase Activity

Receptor-enzymes have two regions: a receptor region on the extracellular side of the cell membrane, and an enzyme region on the cytoplasmic side (see Fig. 6.3c). In some instances, the receptor region and enzyme region are parts of the same protein molecule. In other cases, the enzyme region is a separate protein.

Ligand binding to the receptor activates the enzyme. The enzymes of receptor-enzymes are either protein kinases, such as *tyrosine kinase* (**FIG. 6.9**), or *guanylyl cyclase*, the amplifier enzyme that converts GTP to **cyclic GMP (cGMP)** [p. 34]. Because of the association of these receptors with enzymes, they are now grouped into a receptor family called *catalytic receptors.*

Ligands for receptor-enzymes include the hormone insulin as well as many cytokines and growth factors. The insulin receptor protein has its own intrinsic tyrosine kinase activity. In contrast, most cytokine receptor proteins do not have intrinsic enzyme activity. Instead, cytokine binding activates a cytosolic enzyme called *Janus family tyrosine kinase,* usually abbreviated as *JAK kinase.*

FIG. 6.9 | Receptor-enzymes: The tyrosine kinase receptor

Tyrosine kinase (TK) transfers a phosphate group from ATP to a tyrosine (an amino acid) of a protein.

Integrin Receptors Transfer Information from the Extracellular Matrix

The membrane-spanning proteins called *integrins* [p. 74] mediate blood clotting, wound repair, cell adhesion and recognition in the immune response, and cell movement during development. On the extracellular side of the membrane, integrin receptors bind either to proteins of the extracellular matrix [p. 72] or to ligands such as antibodies and molecules involved in blood clotting. Inside the cell, integrins attach to the cytoskeleton via *anchor proteins* (Fig. 6.3c). Ligand binding to the receptor causes integrins to activate intracellular enzymes or alter the organization of the cytoskeleton. Integrin receptors are also classified as catalytic receptors.

The importance of integrin receptors is illustrated by inherited conditions in which the receptor is absent. In one condition, platelets—cell fragments that play a key role in blood clotting—lack an integrin receptor. As a result, blood clotting is defective in these individuals.

FIGURE 6.10 is a summary map of basic signal transduction, showing the general relationships among first messengers, membrane receptors, second messengers, and cell responses. The modified proteins that control cell responses can be broadly grouped into four categories:

1. metabolic enzymes
2. motor proteins for muscle contraction and cytoskeletal movement
3. proteins that regulate gene activity and protein synthesis
4. membrane transport and receptor proteins

If you think this list includes almost everything a cell does, you're right!

CONCEPT CHECK

9. Name the four categories of membrane receptors.
10. What is the difference between a first messenger and a second messenger?
11. Place the following terms in the correct order for a signal transduction pathway:
    (a) cell response, receptor, second messenger, ligand
    (b) amplifier enzyme, cell response, phosphorylated protein, protein kinase, second messenger
12. In each of the following situations, will a cell depolarize or hyperpolarize?
    (a) Cl^- channel opens
    (b) K^+ channel opens
    (c) Na^+ channel opens

NOVEL SIGNAL MOLECULES

The following sections introduce you to some unusual signal molecules that are important in physiology and medicine. They include an ion (Ca^{2+}), three gases, and a family of lipid-derived

FIG. 6.10 ESSENTIALS

Summary Map of Signal Transduction

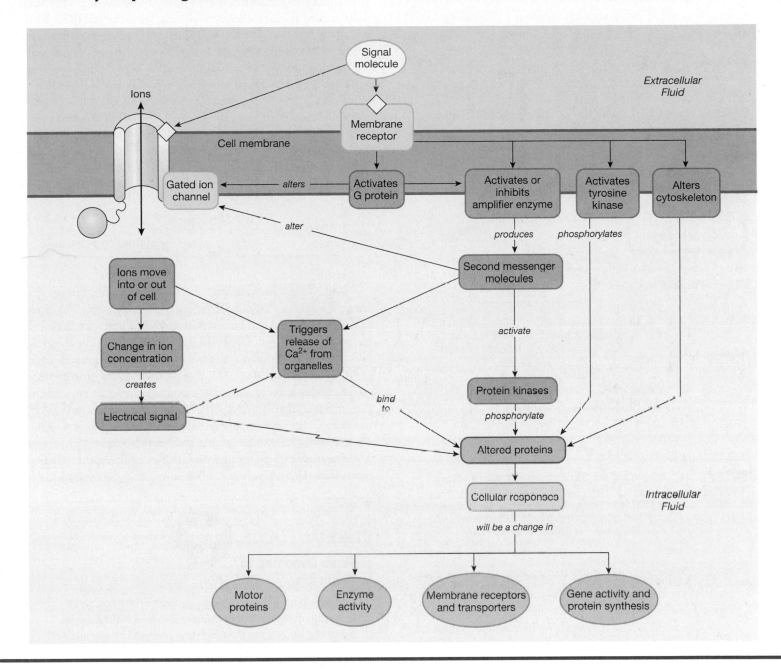

messengers. The processes controlled by these signal molecules have been known for years, but the control signals themselves were discovered only relatively recently.

Calcium Is an Important Intracellular Signal

Calcium ions are the most versatile ionic messengers (FIG. 6.11). Calcium enters the cell through Ca^{2+} channels that may be voltage-gated, ligand-gated, or mechanically gated. Calcium can also be released from intracellular compartments by second messengers, such as IP_3. Most intracellular Ca^{2+} is stored in the

endoplasmic reticulum [p. 71], where it is concentrated by active transport.

Release of Ca^{2+} into the cytosol (from any of the sources just mentioned) creates a Ca^{2+} signal, or Ca^{2+} "spark," that can be recorded using special Ca^{2+}-imaging techniques (see Biotechnology box on calcium signals). The calcium ions combine with cytoplasmic calcium-binding proteins to exert various effects. Several types of calcium-dependent events occur in the cell:

1. Ca^{2+} binds to the protein **calmodulin**, found in all cells. Calcium binding alters enzyme or transporter activity or the gating of ion channels.

FIG. 6.11 Calcium as an intracellular messenger

CONCEPT CHECK

13. The extracellular fluid Ca^{2+} concentration averages 2.5 mmol/L. Free cytosolic Ca^{2+} concentration is about 0.001 mmol/L. If a cell is going to move calcium ions from its cytosol to the extracellular fluid, will it use passive or active transport? Explain.

2. Calcium binds to other regulatory proteins and alters movement of contractile or cytoskeletal proteins such as microtubules. For example, Ca^{2+} binding to the regulatory protein *troponin* initiates muscle contraction in a skeletal muscle cell.

3. Ca^{2+} binds to regulatory proteins to trigger exocytosis of secretory vesicles [p. 148]. For example, the release of insulin from pancreatic beta cells occurs in response to a calcium signal.

4. Ca^{2+} binds directly to ion channels to alter their gating state. An example of this target is a Ca^{2+}-activated K^+ channel found in nerve cells.

5. Ca^{2+} entry into a fertilized egg initiates development of the embryo.

Gases Are Ephemeral Signal Molecules

Soluble gases are short-acting paracrine/autocrine signal molecules that act close to where they are produced. The best-known gaseous signal molecule is **nitric oxide (NO),** but carbon monoxide and hydrogen sulfide, two gases better known for their noxious effects, can also act as local signals.

For years, researchers knew of a short-lived signal molecule produced by the endothelial cells lining blood vessels. They initially named it *endothelial-derived relaxing factor* (EDRF). This molecule diffuses from the endothelium into adjacent smooth muscle cells, causing the muscle to relax and dilate the blood vessel. Scientists took years to identify EDRF as nitric oxide because it is rapidly broken down, with a half-life of only 2 to 30 seconds. (*Half-life* is the time required for the signal to lose half of its activity.) As a result of this difficult work on NO in the cardiovascular system, Robert Furchgott, Louis Ignarro, and Ferid Murad received the 1998 Nobel Prize for physiology and medicine.

In tissues, NO is synthesized by the action of the enzyme *nitric oxide synthase* (NOS) on the amino acid arginine:

$$Arginine + O_2 \xrightarrow{\text{nitric oxide synthase}} NO + citrulline \text{ (an amino acid)}$$

The NO produced in this reaction diffuses into target cells, where it binds to intracellular proteins. In many cases, NO binds to the cytosolic form of guanylyl cyclase and causes formation of the second messenger cGMP. In addition to relaxing blood vessels, NO in the brain acts as a neurotransmitter and a neuromodulator.

Carbon monoxide (CO), a gas known mostly for its toxic effects, is also a signal molecule produced in minute amounts by certain cells. Like NO, CO activates guanylyl cyclase and cGMP, but it may also work independently to exert its effects. Carbon monoxide targets smooth muscle and neural tissue.

The newest gaseous signal molecule to be described is **hydrogen sulfide (H_2S).** Hydrogen sulfide also acts in the cardiovascular

CLINICAL FOCUS

From Dynamite to Medicine

Who would have thought that a component of smog and a derivative of dynamite would turn out to be a biological messenger? Certainly not the peer reviewers who initially rejected Louis Ignarro's attempts to publish his research findings on the elusive gas nitric oxide (NO). The ability of nitrate-containing compounds to relax blood vessels had been known for more than 100 years, ever since workers in Alfred Nobel's dynamite factory complained of headaches caused by nitrate-induced vasodilation. And since the 1860s, physicians have used nitroglycerin to relieve *angina,* heart pain that results from constricted blood vessels. Even today, heart patients carry little nitroglycerin tablets to slide under their tongues when angina strikes. Still, it took years of work to isolate nitric oxide, the short-lived gas that is the biologically active molecule derived from nitroglycerin. Despite our modern technology, direct research on NO is still difficult. Many studies look at its influence indirectly by studying the location and activity of nitric oxide synthase (NOS), the enzyme that produces NO.

system to relax blood vessels. Garlic is a major dietary source of the sulfur-containing precursors, which may explain studies suggesting that eating garlic has protective effects on the heart.

Some Lipids Are Important Paracrine Signals

One of the interesting developments from sequencing the human genome and using genes to find proteins has been the identification of *orphan receptors,* receptors that have no known ligand. Scientists are trying to work backward through signal pathways to find the ligands that bind to these orphan receptors. It was from this type of research that investigators recognized the importance and universality of *eicosanoids,* lipid-derived paracrine signals that play important roles in many physiological processes.

All eicosanoid signal molecules are derived from arachidonic acid, a 20-carbon fatty acid. The synthesis process is a network called the *arachidonic acid cascade* (**FIG. 6.12**). For simplicity, we will break the cascade into steps.

Arachidonic acid is produced from membrane phospholipids by the action of an enzyme, **phospholipase A2 (PLA2).** The activity of PLA2 is controlled by hormones and other signals. Arachidonic acid itself may act directly as a second messenger, altering ion channel activity and intracellular enzymes. It may also be converted into one of several classes of eicosanoid paracrine signals. These lipid-soluble molecules can diffuse out of the cell and combine with receptors on neighboring cells to exert their action.

There are two major groups of arachidonic acid-derived paracrine molecules to be aware of:

1. **Leukotrienes** are molecules produced by the action of the enzyme *lipoxygenase* on arachidonic acid {*leuko-,* white + *triene,* a molecule with three double bonds between carbon atoms}. Leukotrienes are secreted by certain types of white blood cells. They play a significant role in asthma, a

FIG. 6.12 The arachidonic acid cascade

lung condition in which the smooth muscle of the airways constricts, making it difficult to breathe, and in the severe allergic reaction known as *anaphylaxis.* For this reason, pharmaceutical companies have been actively developing drugs to block leukotriene synthesis or action.

2. **Prostanoids** are molecules produced when the enzyme **cyclooxygenase (COX)** acts on arachidonic acid. Prostanoids include **prostaglandins** and **thromboxanes.** These eicosanoids act on many tissues of the body, including smooth

BIOTECHNOLOGY

Calcium Signals Glow in the Dark

If you have ever run your hand through a tropical ocean at night and seen the glow of bioluminescent jellyfish, you've seen a calcium signal. Aequorin, a protein complex isolated from jellyfish such as the *Chrysaora fuscescens* shown here, is one of the molecules that scientists use to monitor the presence of calcium ions. When aequorin combines with calcium, it releases light that can be measured by electronic detection systems. Since the first use of aequorin in 1967, researchers have been designing increasingly sophisticated indicators that allow them to follow calcium signals in cells. With the help of molecules called fura, Oregon green, BAPTA, and chameleons, we can now watch calcium ions diffuse through gap junctions and flow out of intracellular organelles.

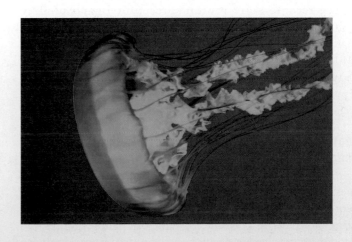

muscle in various organs, platelets, kidney, and bone. In addition, prostaglandins are involved in sleep, inflammation, pain, and fever.

The nonsteroidal anti-inflammatory drugs (NSAIDs), such as aspirin and ibuprofen, help prevent inflammation by inhibiting COX enzymes and decreasing prostaglandin synthesis. However, NSAIDs are not specific and may have serious unwanted side effects, such as bleeding in the stomach. The discovery that COX is made as two isozymes, COX1 and COX2, enabled the design of drugs that target a specific COX isozyme. By inhibiting only COX2, the enzyme that produces inflammatory prostaglandins, physicians hoped to treat inflammation with fewer side effects. However, studies have shown that some patients who take COX2 inhibitors and other NSAIDs have increased risk of heart attacks and strokes, so these drugs are not recommended for long-term use.

Eicosanoids are not the only known lipid signal molecules. Lipids called *sphingolipids* also act as extracellular signals to help regulate inflammation, cell adhesion and migration, and cell growth and death. Like the eicosanoids, sphingolipids combine with G protein-coupled receptors in the membranes of their target cells.

CONCEPT CHECK **14.** One drug blocks leukotriene action in its target cells. A different drug blocks leukotriene synthesis. Use what you have learned about leukotrienes, signal molecules, and signal transduction to predict what these drugs are doing to have those effects.

MODULATION OF SIGNAL PATHWAYS

As you have just learned, signal pathways in the cell can be very complex. Variations among related families of receptors add to the complexity.

Receptors Exhibit Saturation, Specificity, and Competition

Because receptors are proteins, receptor-ligand binding exhibits the general protein-binding characteristics of specificity, competition, and saturation [discussed in Chapter 2, p. 46]. Similar protein-binding reactions occur in enzymes [Chapter 4, p. 99] and transporters [Chapter 5, p. 138]. Receptors, like enzymes and transporters, also come as families of related *isoforms* [p. 49].

Specificity and Competition: Multiple Ligands for One Receptor Receptors have binding sites for their ligands, just as enzymes and transporters do. As a result, different ligand molecules with similar structures may be able to bind to the same receptor. A classic example of this principle involves two neurocrine molecules responsible for fight-or-flight responses: the neurotransmitter *norepinephrine* and its cousin the neurohormone

epinephrine (also called *adrenaline*). Both molecules bind to a class of receptors called *adrenergic receptors.* (*Adrenergic* is the adjective relating to adrenaline.) The ability of adrenergic receptors to bind these two signal molecules, but not others, demonstrates the specificity of the receptors.

Epinephrine and norepinephrine also compete with each other for receptor binding sites. Adrenergic receptors come in two major isoforms designated alpha (α) and beta (β). The α isoform has a higher binding affinity for norepinephrine, and the β_2 isoform has a higher affinity for epinephrine.

Agonists and Antagonists When a ligand combines with a receptor, one of two events follows. Either the ligand activates the receptor and elicits a response, or the ligand occupies the binding site and prevents the receptor from responding (**FIG. 6.13**). A competing ligand that binds and elicits a response is known as an **agonist** of the primary ligand. Competing ligands that bind and block receptor activity are called **antagonists** of the primary ligand.

Pharmacologists use the principle of competing agonists [p. 48] to design drugs that are longer-acting and more resistant to degradation than the *endogenous* ligand produced by the body {*endo-*, within + *-genous*, developing}. One example is the family of modified estrogens (female sex hormones) in birth control pills. These drugs are agonists of naturally occurring estrogens but have chemical groups added to protect them from breakdown and extend their active life.

One Ligand May Have Multiple Receptors

To complicate matters, different cells may respond differently to a single kind of signal molecule. How can one chemical trigger response A in tissue 1 and response B in tissue 2? For most signal molecules, *the target cell response depends on its receptor or its associated intracellular pathways, not on the ligand.*

FIG. 6.13 **Receptor agonists and antagonists**

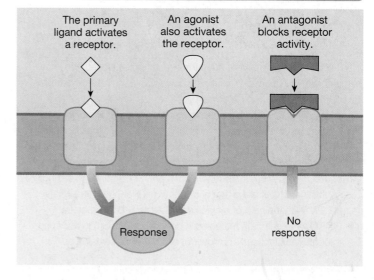

The primary ligand activates a receptor.

An agonist also activates the receptor.

An antagonist blocks receptor activity.

Response

No response

For many years physiologists were unable to explain the observation that a single signal molecule could have different effects in different tissues. For example, the neurohormone epinephrine dilates blood vessels in skeletal muscle but constricts blood vessels in the intestine. How can that one chemical have opposite effects? The answer became clear when scientists discovered that epinephrine was binding to different adrenergic receptor isoforms in the two tissues.

The cellular response that follows activation of a receptor depends on which isoform of the receptor is involved. For example, the α- and β_2-adrenergic receptors for epinephrine are isoforms of each other. When epinephrine binds to α-receptors on intestinal blood vessels, the vessels constrict (**FIG. 6.14**). When epinephrine binds to β_2-receptors on certain skeletal muscle blood vessels, the vessels dilate. The responses of the blood vessels depend on the receptor isoforms and their signal transduction pathways, not on epinephrine. Many drugs now are designed to be specific for only one receptor isoform.

CONCEPT CHECK

15. What do receptors, enzymes, and transporters have in common that explains why they all exhibit saturation, specificity, and competition?

16. Insulin increases the number of glucose transporters on a skeletal muscle cell but not on the membrane of a liver cell. List two possible mechanisms that could explain how this one hormone can have these two different effects.

Up- and Down-Regulation Enable Cells to Modulate Responses

Saturation of proteins refers to the fact that protein activity reaches a maximum rate because cells contain limited numbers of protein molecules [p. 51]. Saturation can be observed with enzymes, transporters, and receptors. A cell's ability to respond to a chemical signal therefore can be limited by the number of receptors for that signal.

A single cell contains between 500 and 100,000 receptors on the surface of its cell membrane, with additional receptors in the cytosol and nucleus. In any given cell, the number of receptors changes over time. Old receptors are withdrawn from the membrane by endocytosis and are broken down in lysosomes. New receptors are inserted into the membrane by exocytosis. Intracellular receptors are also made and broken down. This flexibility permits a cell to vary its responses to chemical signals depending on the extracellular conditions and the internal needs of the cell.

What happens when a signal molecule is present in the body in abnormally high concentrations for a sustained period of time? Initially the increased signal level creates an enhanced response. As this enhanced response continues, the target cells may attempt to bring their response back to normal by either down-regulation or desensitization of the receptors for the signal [p. 51].

Down-regulation is a decrease in receptor number. The cell can physically remove receptors from the membrane through endocytosis [Fig. 5.19, p. 149]. A quicker and more easily reversible way to decrease cell response is *desensitization*, which can be achieved by binding a chemical modulator to the receptor protein. For example, the β-adrenergic receptors described in the previous section can be desensitized by phosphorylation of the receptor.

The result of decreased receptor number or desensitization is a diminished response of the target cell even though the concentration of the signal molecule remains high. Down-regulation and desensitization are one explanation for the development of *drug tolerance*, a condition in which the response to a given dose decreases despite continuous exposure to the drug.

In the opposite situation, when the concentration of a ligand decreases, the target cell may use up-regulation in an attempt to keep its response at a normal level. In **up-regulation,** the target cell inserts more receptors into its membrane. For example, if a neuron is damaged and unable to release normal amounts of neurotransmitter, the target cell may up-regulate its receptors. More receptors make the target cell more responsive to whatever neurotransmitters are present. Up-regulation is also programmed during development as a mechanism that allows cells to vary their responsiveness to growth factors and other signal molecules.

FIG. 6.14 Target response depends on the target receptor

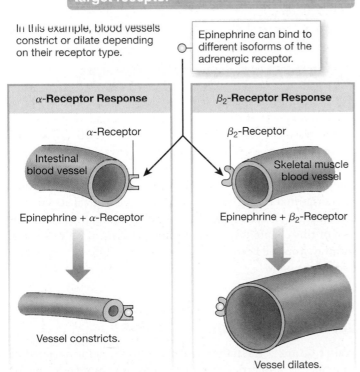

In this example, blood vessels constrict or dilate depending on their receptor type.

Epinephrine can bind to different isoforms of the adrenergic receptor.

α-Receptor Response

α-Receptor

Intestinal blood vessel

Epinephrine + α-Receptor

Vessel constricts.

β_2-Receptor Response

β_2-Receptor

Skeletal muscle blood vessel

Epinephrine + β_2-Receptor

Vessel dilates.

CONCEPT CHECK

17. To decrease a receptor's binding affinity, a cell might (select all that apply):
    (a) synthesize a new isoform of the receptor
    (b) withdraw receptors from the membrane
    (c) insert new receptors into the membrane
    (d) use a covalent modulator [*Hint:* p. 49]

RUNNING **PROBLEM**

"My dad takes insulin shots for his diabetes," Marvin says. "What does insulin do?" The nurse practitioner replies that normally insulin helps most cells take up and use glucose. In both types of diabetes, however, fasting blood glucose concentrations are elevated because the cells are not taking up and using glucose normally. If people with type 1 diabetes are given shots of insulin, their blood glucose levels decline. If people with type 2 diabetes are given insulin, blood glucose levels may change very little.

Q3: *In which form of diabetes are the insulin receptors more likely to be up-regulated?*

166 169 **182** 183 187 189 191

Cells Must Be Able to Terminate Signal Pathways

In the body, signals turn on and off, so cells must be able to tell when a signal is over. This requires that signaling processes have built-in termination mechanisms. For example, to stop the response to a calcium signal, a cell removes Ca^{2+} from the cytosol by pumping it either back into the endoplasmic reticulum or out into the extracellular fluid.

Receptor activity can be stopped in a variety of ways. The extracellular ligand can be degraded by enzymes in the extracellular space. An example is the breakdown of the neurotransmitter acetylcholine. Other chemical messengers, particularly neurotransmitters, can be removed from the extracellular fluid through transport into neighboring cells. A widely used class of antidepressant drugs called *selective serotonin reuptake inhibitors,* or SSRIs, extends the active life of the neurotransmitter serotonin by slowing its removal from the extracellular fluid.

Once a ligand is bound to its receptor, activity can also be terminated by endocytosis of the receptor-ligand complex [Fig. 5.19, p. 149]. After the vesicle is in the cell, the ligand is removed, and the receptors can be returned to the membrane by exocytosis.

Many Diseases and Drugs Target the Proteins of Signal Transduction

As researchers learn more about cell signaling, they are realizing how many diseases are linked to problems with signal pathways. Diseases can be caused by alterations in receptors or by problems with G proteins or second messenger pathways (see **TBL. 6.1** for some examples). A single change in the amino acid sequence of a receptor protein can alter the shape of the receptor's binding site, thereby either destroying or modifying its activity.

Pharmacologists are using information about signaling mechanisms to design drugs to treat disease. Some of the alphabet soup of drugs in widespread use are "beta blockers"

| TABLE 6.1 | Some Diseases or Conditions Linked to Abnormal Signaling Mechanisms |

Genetically Inherited Abnormal Receptors

Receptor	Physiological Alteration	Disease or Condition that Results
Vasopressin receptor (X-linked defect)	Shortens half-life of the receptor	Congenital diabetes insipidus
Calcium sensor in parathyroid gland	Fails to respond to increase in plasma Ca^{2+}	Familial hypercalcemia
Rhodopsin receptor in retina of eye	Improper protein folding	Retinitis pigmentosa

Toxins Affecting Signal Pathways

Toxin	Physiological Effect	Condition that Results
Bordetella pertussis toxin	Blocks inhibition of adenylyl cyclase (i.e., keeps it active)	Whooping cough
Cholera toxin	Blocks enzyme activity of G proteins; cell keeps making cAMP	Ions secreted into lumen of intestine, causing massive diarrhea

(β-adrenergic receptor blockers), and calcium-channel blockers for treating high blood pressure; SERMs (selective estrogen receptor modulators) for treating estrogen-dependent cancers; and H2 (histamine type 2) receptor antagonists for decreasing acid secretion in the stomach. You may encounter many of these drugs again if you study the systems in which they are effective.

HOMEOSTATIC REFLEX PATHWAYS

The cellular signal mechanisms just described are often just one small component of the body's signaling systems that maintain homeostasis. For local control mechanisms, a relatively isolated change occurs in a cell or tissue, and the chemical paracrine or autocrine signals released there are the entire pathway. In more complicated *reflex control pathways* [p. 14], information must be transmitted throughout the body using chemical signals or a combination of electrical and chemical signaling. In the last section of this chapter, we look at some patterns of reflex control pathways you will encounter as you study the different organ systems of the body.

Cannon's Postulates Describe Regulated Variables and Control Systems

Walter Cannon, the father of American physiology, described a number of properties of homeostatic control systems in the

1920s based on his observations of the body in health and disease states.* This was decades before scientists had any idea of how these control systems worked at the cellular and subcellular levels.

Cannon's four postulates are:

1. **The nervous system has a role in preserving the "fitness" of the internal environment.** *Fitness* in this instance means conditions that are compatible with normal function. The nervous system coordinates and integrates blood volume, blood osmolarity, blood pressure, and body temperature, among other regulated variables. (In physiology, a regulated variable is also known as a **parameter** {*para-*, beside + *meter*, measure}).

2. **Some systems of the body are under tonic control {*tonos*, tone}.** To quote Cannon, "An agent may exist which has a moderate activity which can be varied up and down." *Tonic control* is like the volume control on a radio. The radio is always on, but by turning a single knob, you can make the sound level louder or softer. This is one of the more difficult concepts in physiology because we have a tendency to think of responses as being either off or on rather than a response always on that can increase or decrease.

   A physiological example of a tonically controlled system is the minute-to-minute regulation of blood vessel diameter by the nervous system. Increased input from the nervous system decreases vessel diameter, and decreased input from the nervous system increases diameter (**FIG. 6.15a**). In this example, it is the amount of neurotransmitter that determines the vessel's response: more neurotransmitter means a stronger response.

3. **Some systems of the body are under antagonistic control.** Cannon wrote, "When a factor is known which can shift a homeostatic state in one direction, it is reasonable to look for a factor or factors having an opposing effect." Systems that are not under tonic control are usually under *antagonistic control,* either by hormones or the nervous system.

   In pathways controlled by the nervous system, neurons from different divisions may have opposing effects. For example, chemical signals from the sympathetic division increase heart rate, but chemical signals from the parasympathetic division decrease it (Fig. 6.15b).

   When chemical signals have opposing effects, they are said to be antagonistic to each other. For example, insulin and glucagon are antagonistic hormones. Insulin decreases the glucose concentration in the blood and glucagon increases it.

4. **One chemical signal can have different effects in different tissues.** Cannon observed correctly that "homeostatic agents antagonistic in one region of the body may be cooperative in another region." However, it was not until scientists learned about cell receptors that the basis for the seemingly contradictory actions of some hormones or nerves became clear. As you learned earlier in this chapter, a single chemical signal can have different effects depending on the receptor and intracellular pathway of the target cell. For example, epinephrine constricts or dilates blood vessels, depending on whether the vessel has α- or β_2-adrenergic receptors (Fig. 6.14).

The remarkable accuracy of Cannon's postulates, now confirmed with cellular and molecular data, is a tribute to the observational skills of scientists in the nineteenth and early twentieth centuries.

CONCEPT CHECK

18. What is the difference between tonic control and antagonistic control?
19. How can one chemical signal have opposite effects in two different tissues?

Long-Distance Pathways Maintain Homeostasis

Long-distance reflex pathways are traditionally considered to involve two control systems: the nervous system and the endocrine system. However, cytokines [p. 168] can participate in some long-distance pathways. During stress and systemic inflammatory responses, cytokines work with the nervous and endocrine systems to integrate information from all over the body.

Reflex pathway **response loops** have three major components: *input, integration,* and *output* [p. 14]. These three components can be subdivided into seven more detailed steps, as shown next (**FIG. 6.16**):

Stimulus $\longrightarrow$ sensor or receptor $\longrightarrow$ input signal $\longrightarrow$
integrating center $\longrightarrow$
output signal $\longrightarrow$ target $\longrightarrow$ response

---

RUNNING **PROBLEM**

"Why is elevated blood glucose bad?" Marvin asks. "The elevated blood glucose itself is not bad after a meal," says the nurse practitioner, "but when it is high after an overnight fast, it suggests that there is something wrong with the way your body is handling its glucose metabolism." When a normal person absorbs a meal containing carbohydrates, blood glucose levels increase and stimulate insulin release. When cells have taken up the glucose from the meal and blood glucose levels fall, secretion of another pancreatic hormone, glucagon, increases. Glucagon increases blood glucose concentrations to keep the level within the homeostatic range.

Q4: *The homeostatic regulation of blood glucose levels by the hormones insulin and glucagon is an example of which of Cannon's postulates?*

---

*W. B. Cannon. Organization for physiological homeostasis. *Physiological Reviews* 9: 399–443, 1929.

FIG. 6.15 Tonic and antagonistic control patterns

TONIC CONTROL

(a) Tonic control regulates physiological parameters in an up-down fashion. The signal is always present but changes in intensity.

Electrical signals from neuron

Time ⟶

Moderate signal rate results in a blood vessel of intermediate diameter.

Change in signal rate

Decreased signal rate ⟶ vessel dilates

Time ⟶

If the signal rate decreases, the blood vessel dilates.

Increased signal rate ⟶ vessel constricts

Time ⟶

If the signal rate increases, the blood vessel constricts.

ANTAGONISTIC CONTROL

(b) Antagonistic control uses different signals to send a parameter in opposite directions. In this example, antagonistic neurons control heart rate: some speed it up, while others slow it down.

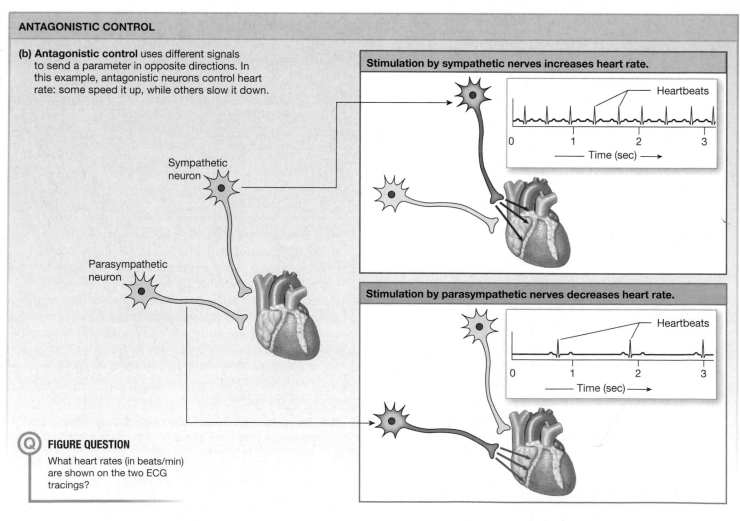

Sympathetic neuron

Parasympathetic neuron

Stimulation by sympathetic nerves increases heart rate.

Heartbeats

0   1   2   3

⟵ Time (sec) ⟶

Stimulation by parasympathetic nerves decreases heart rate.

Heartbeats

0   1   2   3

⟵ Time (sec) ⟶

Q **FIGURE QUESTION**

What heart rates (in beats/min) are shown on the two ECG tracings?

FIG. 6.16 Steps in a reflex pathway

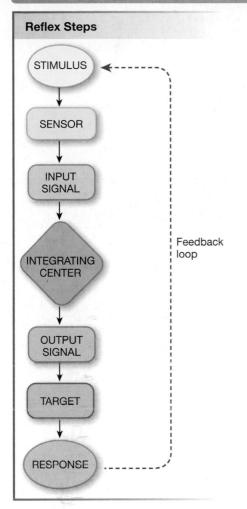

Reflex Steps

STIMULUS

SENSOR

INPUT SIGNAL

INTEGRATING CENTER

Feedback loop

OUTPUT SIGNAL

TARGET

RESPONSE

INPUT:

A **stimulus** is the disturbance or change that sets the pathway in motion. The stimulus may be a change in temperature, oxygen content, blood pressure, or any one of a myriad of other regulated variables.

A **sensor** or sensory receptor continuously monitors its environment for a particular variable.

When activated by a change, the sensor sends an **input (afferent) signal** to the integrating center for the reflex.

INTEGRATION:

The **integrating center** compares the input signal with the **setpoint,** or desired value of the variable. If the variable has moved out of the acceptable range, the integrating center initiates an output signal.

OUTPUT:

The **output (efferent) signal** is an electrical and/or chemical signal that travels to the target.

The **target,** or **effector** {*effectus,* the carrying out of a task} is the cell or tissue that carries out the appropriate **response** to bring the variable back within normal limits.

You will encounter many variations in the number of steps shown. For example, some endocrine reflexes lack a sensor and input signal. Some neural pathways have multiple output signals. Many reflexes have multiple targets and responses.

Now let's look in more detail at each of the reflex steps.

Sensors In the first step in a physiological response loop, a stimulus activates a sensor or receptor. Notice that this is a new and different application of the word *receptor.* Like many other terms in physiology, *receptor* can have different meanings (**FIG. 6.17**). The sensory receptors of a neural reflex are not protein receptors that bind to signal molecules, like those involved in signal transduction. Rather, neural receptors are specialized cells, parts of cells, or complex multicellular receptors (such as the eye) that respond to changes in their environment.

There are many sensory receptors in the body, each located in the best position to monitor the variable it detects. The eyes, ears, and nose are receptors that sense light, sound and motion, and odors, respectively. Your skin is covered with less complex receptors that sense touch, temperature, vibration, and pain. Other sensors are internal: receptors in the joints of the skeleton that send information to the brain about body position, or blood pressure and oxygen receptors in blood vessels that monitor conditions in the circulatory system.

Sensory receptors involved in neural reflexes are divided into central receptors and peripheral receptors. *Central receptors* are located in the brain or are closely linked to the brain. An example is the brain's chemoreceptor for carbon dioxide. *Peripheral receptors* reside elsewhere in the body and include the skin receptors and internal receptors just described.

All sensors have a **threshold,** the minimum stimulus needed to set the reflex response in motion. If a stimulus is below the threshold, no response loop is initiated.

You can demonstrate threshold in a sensory receptor easily by touching the back of your hand with a sharp, pointed object, such as a pin. If you touch the point to your skin lightly enough, you can see the contact between the point and your skin even though you do not feel anything. In this case, the stimulus (pressure from the point of the pin) is below threshold, and the pressure receptors of the skin are not responding. As you press harder, the stimulus reaches threshold, and the receptors respond by sending a signal to the brain, causing you to feel the pin.

Endocrine reflexes that are not associated with the nervous system do not use sensory receptors to initiate their pathways. Instead, endocrine cells act both as sensor and integrating center for the reflex. For example, a pancreatic beta cell sensing and responding directly to changes in blood glucose concentrations is an endocrine cell that is both sensor and integrating center [Fig. 5.26, p. 159].

FIG. 6.17 ESSENTIALS

Multiple Meanings of the Word *Receptor*

The word *receptor* may mean a protein that binds to a ligand. Receptor can also mean a specialized cell or structure for transduction of stimuli into electrical signals (a *sensory receptor* or *sensor*). Sensory receptors are classified as central or peripheral, depending on whether they are found in the brain or outside the brain.

RECEPTORS

can be

Cell Membrane or Intracellular Receptor Proteins

Sensors: Specialized Cells or Structures That Convert Various Stimuli into Electrical Signals

Central receptors are in or close to the brain.

Peripheral receptors lie outside the brain.

Eyes (vision)

Ears (hearing, equilibrium)

Nose (smell)

Tongue (taste)

Central chemoreceptors, osmoreceptors, and thermoreceptors

Chemoreceptor (pH, gases, chemicals)

Osmoreceptor (osmolarity)

Thermoreceptor (temperature)

Baroreceptor (pressure)

Proprioceptor (body position)

Other mechanoreceptors (pain, vibration, touch)

Input Signal The input signal in a reflex varies depending on the type of reflex. In a neural pathway, such as the pin touch above, the input signal is electrical and chemical information transmitted by a sensory neuron. In an endocrine reflex, there is no input pathway because the stimulus acts directly on the endocrine cell, which serves as both sensor and integrating center.

Integrating Center The integrating center in a reflex pathway is the cell that receives information about the regulated variable and can initiate an appropriate response. In endocrine reflexes, the integrating center is the endocrine cell. In neural reflexes, the integrating center usually lies within the *central nervous system* (CNS), which is composed of the brain and the spinal cord.

If information is coming from a single stimulus, it is a relatively simple task for an integrating center to compare that information with the setpoint and initiate a response if appropriate. Integrating centers really "earn their pay," however, when two or more conflicting signals come in from different sources. The center

must evaluate each signal on the basis of its strength and importance and must come up with an appropriate response that integrates information from all contributing receptors. This is similar to the kind of decision-making you must do when on one evening your parents want to take you to dinner, your friends are having a party, there is a television program you want to watch, and you have a major physiology test in three days. It is up to you to rank those items in order of importance and decide what you will do.

Output Signals Output signal pathways are relatively simple. In the nervous system, the output signal is always the electrical and chemical signals transmitted by an efferent neuron. Because all electrical signals traveling through the nervous system are identical, the distinguishing characteristic of the signal is the anatomical pathway of the neuron—the route through which the neuron delivers its signal. For example, the vagus nerve carries neural signals to the heart, and the phrenic nerve carries neural signals to the diaphragm. Output pathways in the nervous system are given the anatomical name of the nerve that carries the signal. For example, we speak of the vagal control of heart rate (*vagal* is the adjective for *vagus*).

In the endocrine system, the anatomical routing of the output signal is always the same—all hormones travel in the blood to their target. Hormonal output pathways are distinguished by the chemical nature of the signal and are therefore named for the hormone that carries the message. For example, the output signal for a reflex integrated through the endocrine pancreas will be either the hormone insulin or the hormone glucagon, depending on the stimulus and the appropriate response.

Targets The targets of reflex control pathways are the cells or tissues that carry out the response. The targets of neural pathways may be any type of muscle, endocrine or exocrine glands, or adipose tissue. Targets of an endocrine pathway are the cells that have the proper receptor for the hormone.

Responses There are multiple levels of response for any reflex control pathway. Let's use the example of a neurotransmitter acting on a blood vessel, as shown in Figure 6.15a. The *cellular response* takes place in the target cell. In this example, the blood vessel smooth muscle contracts in response to neurotransmitter binding. The next level is the *tissue or organ response*. In our example, contraction of smooth muscles in the blood vessel wall decreases the diameter of the blood vessel and decreases flow through this blood vessel. Finally, the more general *systemic response* describes what those specific cellular and tissue events mean to the organism as a whole. In this example, when the blood vessels constrict, the systemic response is an increase in blood pressure.

Now that you have been introduced to the basic parts of a reflex control pathway, we can turn to an analysis of the two primary control systems, the nervous system and the endocrine system.

CONCEPT CHECK
20. What is the difference between local control and reflex control?
21. Name the seven steps in a reflex control pathway in their correct order.

RUNNING **PROBLEM**

Marvin is fascinated by body's ability to keep track of glucose. "How does the pancreas know which hormone to secrete?" he wonders. Special cells in the pancreas called beta cells monitor blood glucose concentrations, and they release insulin when blood glucose increases after a meal. Insulin acts on many tissues of the body so that they take up and use glucose.

Q5: *In the insulin reflex pathway, name the stimulus, the sensor, the integrating center, the output signal, the target(s), and the response(s).*

Control Systems Vary in Their Speed and Specificity

Physiological reflex control pathways are mediated by the nervous system, the endocrine system, or a combination of the two (**FIG. 6.18**). Reflexes mediated solely by the nervous system or solely by the endocrine system are relatively simple, but some pathways combine neural and endocrine reflexes and can be quite complex. In the most complex pathways, signals pass through three different integrating centers before finally

FIG. 6.18 Simple and complex reflexes

This figure compares simple reflexes with one integrating center to a complex pathway with two integrating centers.

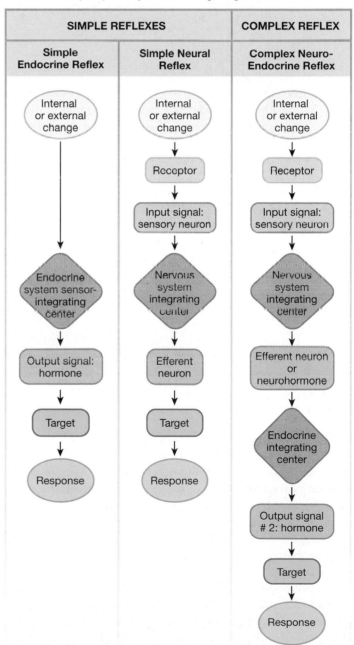

TABLE 6.2	Comparison of Neural and Endocrine Control	
Property	Neural Reflex	Endocrine Reflex
Specificity	Each neuron terminates on a single target cell or on a limited number of adjacent target cells.	Most cells of the body are exposed to a hormone. The response depends on which cells have receptors for the hormone.
Nature of the signal	Electrical signal that passes through neuron, then chemical neurotransmitters that carry the signal from cell to cell. In a few cases, signals pass from cell to cell through gap junctions.	Chemical signals secreted in the blood for distribution throughout the body.
Speed	Very rapid.	Distribution of the signal and onset of action are much slower than in neural responses.
Duration of action	Usually very short. Responses of longer duration are mediated by neuromodulators.	Responses usually last longer than neural responses.
Coding for stimulus intensity	Each signal is identical in strength. Stimulus intensity is correlated with increased frequency of signaling.	Stimulus intensity is correlated with amount of hormone secreted.

reaching the target tissue. With so much overlap between pathways controlled by the nervous and endocrine systems, it makes sense to consider these systems as parts of a continuum rather than as two discrete systems.

Why does the body need different types of control systems? To answer that question, let us compare endocrine control with neural control. Five major differences are summarized in **TABLE 6.2** and discussed next.

Specificity Neural control is very specific because each neuron has a specific target cell or cells to which it sends its message. Anatomically, we can isolate a neuron and trace it from its origin to where it terminates on its target.

Endocrine control is more general because the chemical messenger is released into the blood and can reach virtually every cell in the body. As you learned in the first half of this chapter, the body's response to a specific hormone depends on which cells have receptors for that hormone and which receptor type they have. Multiple tissues in the body can respond to a hormone simultaneously.

Nature of the Signal The nervous system uses both electrical and chemical signals to send information throughout the body. Electrical signals travel long distances through neurons, releasing chemical signals (neurotransmitters) that diffuse across the small gap between the neuron and its target. In a limited number of instances, electrical signals pass directly from cell to cell through gap junctions.

The endocrine system uses only chemical signals: hormones secreted into the blood by endocrine glands or cells. Neuroendocrine pathways represent a hybrid of the neural and endocrine reflexes. In a neuroendocrine pathway, a neuron creates an electrical signal, but the chemical released by the neuron is a neurohormone that goes into the blood for general distribution.

CONCEPT CHECK

22. In the simple neural reflex shown in Figure 6.18, which box or boxes represent the brain and spinal cord? (b) Which box or boxes represent the central and peripheral sense organs? (c) In the simple neural reflex, add a dashed line connecting boxes to show how a negative feedback loop would shut off the reflex [p. 15].

Speed Neural reflexes are much faster than endocrine reflexes. The electrical signals of the nervous system cover great distances very rapidly, with speeds of up to 120 m/sec. Neurotransmitters also create very rapid responses, on the order of milliseconds.

Hormones are much slower than neural reflexes. Their distribution through the circulatory system and diffusion from capillary to receptors take considerably longer than signals through neurons. In addition, hormones have a slower onset of action. In target tissues, the response may take minutes to hours before it can be measured.

Why do we need the speedy reflexes of the nervous system? Consider this example. A mouse ventures out of his hole and sees a cat ready to pounce on him and eat him. A signal must go from the mouse's eyes and brain down to his feet, telling him to run back into the hole. If his brain and feet were only 5 micrometers (5 μm = 1/200 millimeter) apart, it would take a chemical signal 20 milliseconds (msec) to diffuse across the space and the mouse could escape. If the brain and feet were 50 μm (1/20 millimeter) apart, diffusion would take 2 seconds and the mouse might get caught. But because the head and tail of a mouse are *centimeters* apart, it would take a chemical signal *three weeks* to diffuse from the mouse's head to his feet. Poor mouse!

Even if the distribution of the chemical signal were accelerated by help from the circulatory system, the chemical message would still take 10 seconds to get to the feet, and the mouse would become cat food. The moral of this tale is that reflexes requiring a speedy response are mediated by the nervous system because they are so much more rapid.

Duration of Action Neural control is of shorter duration than endocrine control. The neurotransmitter released by a neuron combines with a receptor on the target cell and initiates a response. The response is usually very brief, however, because the neurotransmitter is rapidly removed from the vicinity of the receptor by various mechanisms. To get a sustained response, multiple repeating signals must be sent through the neuron.

Endocrine reflexes are slower to start, but they last longer. Most of the ongoing, long-term functions of the body, such as metabolism and reproduction, fall under the control of the endocrine system.

Coding for Stimulus Intensity As a stimulus increases in intensity, control systems must have a mechanism for conveying this information to the integrating center. The signal strength from any one neuron is constant in magnitude and therefore cannot reflect stimulus intensity. Instead, the frequency of signaling through the afferent neuron increases. In the endocrine system, stimulus intensity is reflected by the amount of hormone released: the stronger the stimulus, the more hormone is released.

Complex Reflex Control Pathways Have Several Integrating Centers

FIGURE 6.19 summarizes variations in the neural, neuroendocrine, and endocrine reflex control pathways.

In a simple neural reflex, all the steps of a reflex pathway are present, from sensor to target (Fig. 6.19 ❶). The neural reflex is represented in its simplest form by the knee jerk (or patellar tendon) reflex. A blow to the knee (the stimulus) activates a stretch receptor. A signal travels through an afferent sensory neuron to the spinal cord (the integrating center). If the blow is strong enough (exceeds threshold), a signal travels from the spinal cord

RUNNING **PROBLEM**

"OK, just one more question," says Marvin. "You said that people with diabetes have high blood glucose levels. If glucose is so high, why can't it just leak into the cells?"

Q6: *Why can't glucose simply leak into cells when the blood glucose concentration is higher than the intracellular glucose concentration?*

Q7: *What do you think happens to insulin secretion when blood glucose levels fall? What kind of feedback loop is operating?*

166 169 182 183 187 **189** 191

through an efferent neuron to the muscles of the thigh (the target or effector). In response, the muscles contract, causing the lower leg to kick outward (the knee jerk).

In a simple endocrine reflex pathway (Fig. 6.19 ❻), some of the steps of the reflex pathway are combined. The endocrine cell acts as both sensor and integrating center; there is no input pathway. The endocrine cell itself monitors the regulated variable and is programmed to initiate a response when the variable goes out of an acceptable range. The output pathway is the hormone, and the target is any cell having the appropriate hormone receptor.

An example of a simple endocrine reflex is secretion of the hormone insulin in response to changes in blood glucose level. The pancreatic beta cells that secrete insulin monitor blood glucose concentrations by using ATP production in the cell as an indicator of glucose availability [Fig. 5.26, p. 159]. When blood glucose increases, intracellular ATP production exceeds the threshold level, and the beta cells respond by secreting insulin into the blood. Any target cell in the body that has insulin receptors responds to the hormone and initiates processes that take glucose out of the blood. The removal of the stimulus acts in a negative feedback manner: the response loop shuts off when blood glucose levels fall below a certain concentration.

CONCEPT CHECK 23. Match the following terms for parts of the knee jerk reflex to the parts of the simple neural reflex shown in Figure 6.19 ❶: blow to knee, leg muscles, neuron to leg muscles, sensory neuron, brain and spinal cord, stretch receptor, muscle contraction.

The neuroendocrine reflex, shown in Figure 6.19 ❷, is identical to the neural reflex except that the neurohormone released by the neuron travels in the blood to its target, just like a hormone. A simple neuroendocrine reflex is the release of breast milk in response to a baby's suckling. The baby's mouth on the nipple stimulates sensory signals that travel through sensory neurons to the brain (integrating center). An electrical signal in the efferent neuron triggers the release of the neurohormone oxytocin from the brain into the circulation. Oxytocin is carried to the breast, where it causes contraction of smooth muscles in the breast (the target), resulting in the ejection of milk.

In complex pathways, there may be more than one integrating center. Figure 6.19 shows three examples of complex neuroendocrine pathways. The simplest of these, Figure 6.19 ❸, combines a neural reflex with a classic endocrine reflex. An example of this pattern can be found in the control of insulin release.

The pancreatic beta cells monitor blood glucose concentrations directly (Fig. 6.19 ❻), but they are also controlled by the nervous system. During a meal, the presence of food in the stomach stretches the wall of the digestive tract and sends input signals to the brain. The brain in turn sends excitatory output signals to the beta cells, telling them to release insulin. These signals take place even before the food has been absorbed and blood glucose levels have gone up (a *feedforward reflex* [p. 17]). This pathway therefore has two integrating centers (the brain and the beta cells).

FIG. 6.19 **ESSENTIALS**

Reflex Pathway Patterns

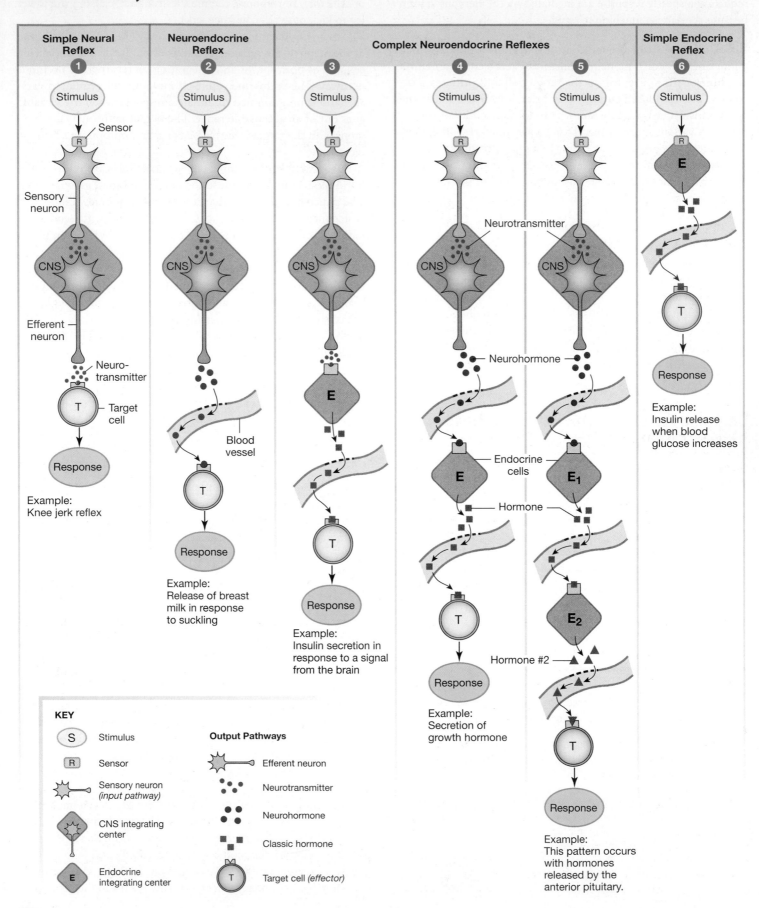

Simple Neural Reflex ❶

Stimulus
Sensor — R
Sensory neuron
CNS
Neurotransmitter
Efferent neuron
Neurotransmitter
T — Target cell
Response

Example: Knee jerk reflex

Neuroendocrine Reflex ❷

Stimulus
R
CNS
Blood vessel
T
Response

Example: Release of breast milk in response to suckling

Complex Neuroendocrine Reflexes

❸
Stimulus
R
CNS
E
T
Response

Example: Insulin secretion in response to a signal from the brain

❹
Stimulus
R
CNS
Neurotransmitter
Neurohormone
E
Endocrine cells
Hormone
T
Response

Example: Secretion of growth hormone

❺
Stimulus
R
CNS
E_1
E_2
Hormone #2
T
Response

Example: This pattern occurs with hormones released by the anterior pituitary.

Simple Endocrine Reflex ❻

Stimulus
R
E
T
Response

Example: Insulin release when blood glucose increases

KEY

S — Stimulus
R — Sensor
Sensory neuron (input pathway)
CNS integrating center
E — Endocrine integrating center

Output Pathways

Efferent neuron
Neurotransmitter
Neurohormone
Classic hormone
T — Target cell (effector)

TABLE 6.3	Comparison of Neural, Neuroendocrine, and Endocrine Reflexes		
	Neural	**Neuroendocrine**	**Endocrine**
Sensor	Special and somatic sensory receptors	Special and somatic sensory receptors	Endocrine cell
Input Signal	Sensory neuron	Sensory neuron	None
Integrating Center	Brain or spinal cord	Brain or spinal cord	Endocrine cell
Output Signal	Efferent neuron (electrical signal and neurotransmitter)	Efferent neuron (electrical signal and neurohormone)	Hormone
Target(s)	Muscles and glands, some adipose tissue	Most cells of the body	Most cells of the body
Response	Contraction and secretion primarily; may have some metabolic effects	Change in enzymatic reactions, membrane transport, or cell proteins	Change in enzymatic reactions, membrane transport, or cell proteins

There are a number of complex reflex pathways, not all of which are shown in Figure 6.19. One (Fig. 6.19 **4**) uses a neurohormone to control the release of a classic hormone. The secretion of growth hormone is an example of this pathway. The most complex neuroendocrine pathways, shown as Figure 6.19 **5**, include a neurohormone and two classic hormones. This pattern is typical of some hormones released by the anterior pituitary, an endocrine gland located just below the brain [see Chapter 7 for details].

CONCEPT CHECK

24. Match the following terms with the appropriate parts of the simple neuroendocrine reflex in Fig. 6.19 (terms may be used more than once): food in stomach following a meal, brain and spinal cord, endocrine cells of pancreas, stretch receptors, efferent neuron to pancreas, insulin, adipose cell, blood, sensory neuron.

In describing complex neuroendocrine reflex pathways, we identify only one receptor and input pathway, as indicated in Figure 6.19 **5**. In the three complex pathways shown, the brain is the first integrating center and the neurohormone is the first output pathway. In Figure 6.19 **5**, the endocrine target (E_1) of the neurohormone is the second integrating center, and its hormone is the second output pathway. The second endocrine gland in the pathway (E_2) is the third integrating center, and its hormone is the third output pathway. The target of the last signal in the sequence is the effector.

TABLE 6.3 compares the various steps in neural, neuroendocrine, and endocrine reflexes. In the remainder of the text, we use the general patterns shown in Figure 6.19 as a tool for classifying complex reflex pathways. Endocrine and neural pathways play key roles in the maintenance of homeostasis.

RUNNING **PROBLEM CONCLUSION** | Diabetes Mellitus

Marvin underwent further tests and was diagnosed with type 2 diabetes. With careful attention to his diet and with a regular exercise program, he managed to keep his blood glucose levels under control. Diabetes is a growing epidemic in the United States, with nearly 26 million diabetics in the United States in 2013 (about 8% of the population). Even scarier is the estimate that another 79 million people are considered "prediabetic"—at significant risk of becoming diabetic. You will learn more about diabetes as you work through the chapters in this book. To learn more about diabetes now, see the American

Diabetes Association web site (*www.diabetes.org*) or the Centers for Disease Control and Prevention (*www.cdc.gov/diabetes*).

In this running problem, you learned about glucose homeostasis and how it is maintained by insulin and glucagon. The disease diabetes mellitus is an indication that glucose homeostasis has been disrupted. Check your understanding of this running problem by comparing your answers to the information in the summary table.

Question	Facts	Integration and Analysis
Q1: *In which type of diabetes is the signal pathway for insulin more likely to be defective?*	Insulin is a peptide hormone that uses membrane receptors linked to second messengers to transmit its signal to cells. People with type 1 diabetes lack insulin; people with type 2 diabetes have normal-to-elevated insulin levels.	Normal or high insulin levels suggest that the problem is not with amount of insulin but with the action of the insulin at the cell. The problem in type 2 diabetes could be a defective signal transduction mechanism.

—Continued next page

RUNNING **PROBLEM CONCLUSION** | *Continued*

Question	Facts	Integration and Analysis
Q2: *Insulin is a protein hormone. Would you expect to find its receptor on the cell surface or in the cytoplasm of the target cells?*	Lipophilic signal molecules have intracellular receptors. Lipophobic molecules have cell membrane receptors.	Proteins are lipophobic so protein hormones like insulin have cell surface receptors.
Q3: *In which form of diabetes are the insulin receptors more likely to be up-regulated?*	Up-regulation of receptors usually occurs if a signal molecule is present in unusually low concentrations [p. 51]. In type 1 diabetes, insulin is not secreted by the pancreas.	In type 1 diabetes, insulin levels are low. Therefore, type 1 is more likely to cause up-regulation of the insulin receptors.
Q4: *The homeostatic regulation of blood glucose levels by the hormones insulin and glucagon is an example of which of Cannon's postulates?*	Cannon's postulates describe the role of the nervous system in maintaining homeostasis, and the concepts of tonic activity, antagonistic control, and different effects of signals in different tissues.	Insulin decreases blood glucose levels, and glucagon increases them. Therefore, the two hormones are an example of an antagonistic control.
Q5: *In the insulin pathway, name the stimulus, the sensor, the integrating center, the output signal, the target(s), and the response(s).*	See the steps of reflex pathways [p. 183].	Stimulus: increase in blood glucose levels; sensor: beta cells of the pancreas that sense the change; integrating center: beta cells; output signal: insulin; targets: any tissues of the body that respond to insulin; responses: cellular uptake and use of glucose.
Q6: *Why can't glucose simply leak into cells when the blood glucose concentration is higher than the intracellular glucose concentration?*	Glucose is lipophobic. Simple diffusion goes across the phospholipid bilayer. Facilitated diffusion uses protein carriers [p. 142].	Because glucose is lipophobic, it cannot cross the membrane by simple diffusion. It must go by facilitated diffusion. If a cell lacks the necessary carriers, facilitated diffusion cannot take place.
Q7: *What do you think happens to the rate of insulin secretion when blood glucose levels fall? What kind of feedback loop is operating?*	The stimulus for insulin release is an increase in blood glucose levels. In negative feedback, the response offsets the stimulus. In positive feedback, the response enhances the stimulus.	An increase in blood glucose concentration stimulates insulin release; therefore, a decrease in blood glucose should decrease insulin release. In this example, the response (lower blood glucose) offsets the stimulus (increased blood glucose), so a negative feedback loop is operating.

(166)(169)(182)(183)(187)(189)(**191**)

VISUALIZE, PRACTICE, AND MASTER PHYSIOLOGY WITH MASTERING A&P

- Practice-on-the-go with Dynamic Study Modules
- Visualize important processes with Interactive Physiology and A&P Flix
- Check your understanding with Video Tutors and practice quizzes!

MasteringA&P®

CHAPTER SUMMARY

Two major themes in physiology stand out in this chapter: *the control of homeostasis* and *communication*. The sensors, integrating centers, and targets of physiological control systems are described in the context of reflex control pathways, which vary from simple to complex. Functional control systems require efficient communication that uses various combinations of chemical and electrical signals. Those signals that cannot enter the cell must use membrane receptor proteins and signal transduction to transfer their information into the cell. The interaction of signal molecules with protein receptors illustrates another fundamental theme of physiology, *molecular interactions*.

Cell-to-Cell Communication

1. There are two basic types of physiological signals: chemical and electrical. Chemical signals are the basis for most communication within the body. (p. 166)

2. There are four methods of cell-to-cell communication: (1) direct cytoplasmic transfer through gap junctions, (2) contact-dependent signaling, (3) local chemical communication, and (4) long-distance communication. (p. 166; Fig. 6.1)

3. **Gap junctions** are protein channels that connect two adjacent cells. When they are open, chemical and electrical signals pass directly from one cell to the next. (p. 166)

4. **Contact-dependent signals** require direct contact between surface molecules of two cells. (p. 166)

5. Local communication uses **paracrine signals,** chemicals that act on cells close to the cell that secreted the paracrine. A chemical that acts on the cell that secreted it is called an **autocrine signal.** The activity of paracrine and autocrine signal molecules is limited by diffusion distance. (p. 168)

6. Long-distance communication uses **neurocrine molecules** and electrical signals in the nervous system, and **hormones** in the endocrine system. Only cells that possess receptors for a hormone will be **target cells.** (pp. 166, 168)

7. **Cytokines** are regulatory peptides that control cell development, differentiation, and the immune response. They serve as both local and long-distance signals. (p. 168)

Signal Pathways

8. Chemical signals bind to **receptors** and change intracellular signal molecules that direct the response. (p. 169)

9. Lipophilic signal molecules enter the cell and combine with cytoplasmic or nuclear receptors. Lipophobic signal molecules and some lipophilic molecules combine with membrane receptors. (p. 169; Fig. 6.3)

10. **Signal transduction** pathways use membrane receptor proteins and intracellular second messenger molecules to translate signal information into an intracellular response. (p. 171; Fig. 6.5a)

11. Some signal transduction pathways activate **protein kinases.** Others activate **amplifier enzymes** that create **second messenger** molecules. (p. 171; Fig. 6.5b)

12. Signal pathways create intracellular **cascades** that amplify the original signal. (p. 171; Fig. 6.6a)

13. **Ligand-gated ion channels** open or close to create electrical signals. (p. 170; Fig. 6.7)

14. **G proteins** linked to amplifier enzymes are the most prevalent signal transduction system. **G protein-coupled receptors** also alter ion channels. (p. 174; Fig. 6.8)

15. The **G protein-coupled adenylyl cyclase-cAMP-protein kinase A** pathway is the most common pathway for protein and peptide hormones. (p. 174; Fig. 6.8a)

16. In the **G protein-coupled phospholipase C** pathway, the amplifier enzyme **phospholipase C (PLC)** creates two second messengers: **inositol trisphosphate (IP₃)** and **diacylglycerol (DAG).** IP_3 causes Ca^{2+} release from intracellular stores. Diacylglycerol activates **protein kinase C (PKC).** (pp. 174, 176; Fig. 6.8b)

17. **Receptor-enzymes** activate protein kinases, such as **tyrosine kinase** (Fig. 6.9), or the amplifier enzyme **guanylyl cyclase,** which produces the second messenger **cGMP.** (p. 176)

18. **Integrin** receptors link the extracellular matrix to the cytoskeleton. (p. 176; Fig. 6.3c)

Novel Signal Molecules

19. Calcium is an important signal molecule that binds to **calmodulin** to alter enzyme activity. It also binds to other cell proteins to alter movement and initiate exocytosis. (p. 177; Fig. 6.11)

20. **Nitric oxide (NO), carbon monoxide (CO),** and **hydrogen sulfide (H₂S)** are short-lived gaseous signal molecules. NO activates guanylyl cyclase directly. (pp. 178, 179)

21. The arachidonic acid cascade creates lipid signal molecules, such as **leukotrienes, prostaglandins,** and **thromboxanes.** (p. 179; Fig. 6.12)

Modulation of Signal Pathways

22. The response of a cell to a signal molecule is determined by the cell's receptor for the signal. (p. 180)

23. Receptors come in related forms called **isoforms.** One ligand may have different effects when binding to different isoforms. (p. 180; Fig. 6.13)

24. A receptor may have multiple ligands. Receptor **agonists** mimic the action of a signal molecule. Receptor **antagonists** block the signal pathway. (p. 180; Fig. 6.14)

25. Receptor proteins exhibit specificity, competition, and saturation. (p. 180)

26. Cells exposed to abnormally high concentrations of a signal for a sustained period of time attempt to bring their response back to normal through down-regulation or by desensitization. In **down-regulation,** the cell decreases the number of receptors. In *desensitization,* the cell decreases the binding affinity of the receptors. **Up-regulation** is the opposite of down-regulation and involves increasing the number of receptors for a signal. (p. 181)

27. Cells have mechanisms for terminating signal pathways, such as removing the signal molecule or breaking down the receptor-ligand complex. (p. 182)

28. Many diseases have been linked to defects in various aspects of signal pathways, such as missing or defective receptors. (p. 182; Tbl. 6.1)

Homeostatic Reflex Pathways

29. Walter Cannon first stated four basic postulates of homeostasis: (1) The nervous system plays an important role in maintaining homeostasis. (2) Some parameters are under **tonic control,** which allows the parameter to be increased or decreased by a single signal (Fig. 6.15a). (3) Other parameters are under **antagonistic control,** in which one hormone or neuron increases the parameter while another decreases it (Fig. 6.15b). (4) Chemical signals can have different effects in different tissues of the body, depending on the type of receptor present at the target cell. (p. 183; Fig. 6.14)

30. In **reflex control pathways,** an integrating center makes the decision to respond to a change. A chemical or electrical signal to the target cell or tissue then initiates the response. Long-distance reflex pathways involve the nervous and endocrine systems and cytokines. (p. 182)

31. Neural control is faster and more specific than endocrine control but is usually of shorter duration. Endocrine control is less specific and slower to start but is longer lasting and is usually amplified. (p. 187; Tbl. 6.2)

32. Many reflex pathways are complex combinations of neural and endocrine control mechanisms. (p. 189; Figs. 6.18, 6.19)

REVIEW QUESTIONS

In addition to working through these questions and checking your answers on p. A-8, review the Learning Outcomes at the beginning of this chapter.

Level One Reviewing Facts and Terms

1. What are the two routes for long-distance signal delivery in the body?

2. Which two body systems maintain homeostasis by monitoring and responding to changes in the environment?

3. What two types of physiological signals does the body use to send messages? Of these two types, which is available to all cells?

4. In a signal transduction pathway, the signal ligand, also called the first messenger, binds to a(n) _____, which activates and changes intracellular _____.

5. The three main amplifier enzymes are (a) _____, which forms cAMP; (b) _____, which forms cGMP; and (c) _____, which converts a phospholipid from the cell's membrane into two different second messenger molecules.

6. An enzyme known as protein kinase adds the functional group _____ to its substrate, by transferring it from a(n) _____ molecule.

7. Distinguish between central and peripheral receptors.

8. Receptors for signal pathways may be found in the _____, _____, or _____ of the cell.

9. Down-regulation results in a(n) _____ (increased or decreased?) number of receptors in response to a prolonged signal.

10. List two ways a cell may decrease its response to a signal.

11. In a negative feedback loop, the response moves the system in the _____ (same/opposite) direction as the stimulus moves it.

Level Two Reviewing Concepts

12. Explain the relationships of the terms in each of the following sets. Give a physiological example or location if applicable.

(a) gap junctions, connexins, connexon
(b) autocrine signal, paracrine signal
(c) cytokine, neurotransmitter, neurohormone, neuromodulator, hormone
(d) Receptor agonist, receptor antagonist, antagonistic control pathways
(e) transduction, amplification, cascade

13. List and compare the four classes of membrane receptors for signal pathways. Give an example of each.

14. Who was Walter Cannon? Restate his four postulates in your own words.

15. Arrange the following terms in the order of a reflex and give an anatomical example of each step when applicable: input signal, integrating center, output signal, response, sensor, stimulus, target.

16. Compare and contrast the advantages and disadvantages of neural versus endocrine control mechanisms.

17. Would the following reflexes have positive or negative feedback?

(a) glucagon secretion in response to declining blood glucose
(b) increasing milk release and secretion in response to baby's suckling
(c) urgency in emptying one's urinary bladder
(d) sweating in response to rising body temperature

18. Identify the target tissue or organ for each example in question 17.

19. Now identify the integrating center for examples (a), (c), and (d) in question 17.

Level Three Problem Solving

20. In each of the following situations, identify the components of the reflex.

(a) You are sitting quietly at your desk, studying, when you become aware of the bitterly cold winds blowing outside at 30 mph, and you begin to feel a little chilly. You start to turn up the thermostat, remember last month's heating bill, and reach for an afghan to pull around you instead. Pretty soon you are toasty warm again.

(b) While you are strolling through the shopping district, the aroma of cinnamon sticky buns reaches you. You inhale appreciatively, but remind yourself that you're not hungry, because you had lunch just an hour ago. You go about your business, but 20 minutes later you're back at the bakery, sticky bun in hand, ravenously devouring its sweetness, saliva moistening your mouth.

21. A researcher is studying the smooth muscle of the respiratory system airways. When she exposes the airways to the neurotransmitter acetylcholine, the smooth muscle contracts. When she exposes the airways to the neurohormone epinephrine, the airways relax.

    (a) The phenomenon just described is an example of _____ control.
    (b) What distinguishes a neurotransmitter from a neurohormone?
    (c) Which chemical messenger is secreted in higher concentrations: acetylcholine or epinephrine? Defend your answer.

Level Four Quantitative Problems

22. In a signal cascade for rhodopsin, a photoreceptor molecule, one rhodopsin activates 1,000 molecules of transducin, the next molecule in the signal cascade. Each transducin activates one phosphodiesterase, and each phosphodiesterase converts 4,000 cGMP to GMP.

    (a) What is the name of the phenomenon described in this paragraph?
    (b) Activation of one rhodopsin will result in the production of how many GMP molecules?

Answers to Concept Checks, Figure and Graph Questions, and end-of-chapter Review Questions can be found in Appendix A [A-1].

7

Introduction to the Endocrine System

The separation of the endocrine system into isolated subsystems must be recognized as an artificial one, convenient from a pedagogical point of view but not accurately reflecting the interrelated nature of all these systems.

Howard Rasmussen, in *Williams' Textbook of Endocrinology*, 1974

HORMONES 197

LO 7.1 Explain the four criteria that make a chemical signal a hormone.

LO 7.2 Explain what the cellular mechanism of action of a hormone is.

THE CLASSIFICATION OF HORMONES 201

LO 7.3 List three chemical classes of hormones and give an example of each.

LO 7.4 Compare endocrine cells' synthesis, storage, and release of peptide and steroid hormones.

LO 7.5 Compare the location of hormone receptors and the cellular mechanisms of action of peptide and steroid hormones.

LO 7.6 Compare the three main groups of amine hormones.

CONTROL OF HORMONE RELEASE 206

LO 7.7 Describe the role of the nervous system in endocrine reflexes.

LO 7.8 Compare the structure and function of the anterior and posterior pituitaries.

LO 7.9 List the six anterior pituitary hormones, the hormones that control their release, and their primary targets.

LO 7.10 Compare long-loop negative feedback for anterior pituitary hormones to the negative feedback loops for insulin and parathyroid hormone.

HORMONE INTERACTIONS 215

LO 7.11 Explain permissiveness, synergism, and functional antagonism as they apply to hormones.

ENDOCRINE PATHOLOGIES 216

LO 7.12 Name the three most common types of endocrine pathologies.

LO 7.13 Explain how negative feedback can be used to determine the location of a problem with one gland in a two- or three-gland pathway.

HORMONE EVOLUTION 218

LO 7.14 Explain how comparative endocrinology is useful for understanding human physiology.

BACKGROUND BASICS

169 Receptors
 32 Peptides and proteins
188 Comparison of endocrine and nervous systems
171 Signal transduction
 30 Steroids
 46 Specificity

Gamma scan of a goiter of the thyroid gland

avid was seven years old when the symptoms first appeared. His appetite at meals increased, and he always seemed to be hungry. Despite eating more, however, he was losing weight. When he started asking for water instead of soft drinks, David's mother became concerned, and when he wet the bed three nights in a row, she knew something was wrong. The doctor listened to David's symptoms and ordered tests to determine the glucose concentrations of David's blood and urine. The test results confirmed the diagnosis: David had diabetes mellitus. In David's case, the disease was due to lack of insulin, a hormone produced by the pancreas. David was placed on insulin injections, a treatment he would continue for the rest of his life.

One hundred years ago, David would have died not long after the onset of symptoms. The field of **endocrinology,** the study of hormones, was then in its infancy. Most hormones had not been discovered, and the functions of known hormones were not well understood. There was no treatment for diabetes, no birth control pill for contraception. Babies born with inadequate secretion of thyroid hormone did not grow or develop normally.

Today, all that has changed. We have identified a long and growing list of hormones. The endocrine diseases that once killed or maimed can now be controlled by synthetic hormones and sophisticated medical procedures. Although physicians do not hesitate to use these treatments, we are still learning exactly how hormones act on their target cells. This chapter provides an introduction to the basic principles of hormone structure and function. You will learn more about individual hormones as you encounter them in your study of the various systems.

HORMONES

As you have learned, hormones are chemical messengers secreted into the blood by specialized epithelial cells. Hormones are responsible for many functions that we think of as long-term, ongoing functions of the body. Processes that fall mostly under hormonal control include metabolism, regulation of the internal environment (temperature, water balance, ions), and reproduction, growth, and development. Hormones act on their target cells in one of three basic ways: (1) by controlling the rates of enzymatic reactions, (2) by controlling the transport of ions or molecules across cell membranes, or (3) by controlling gene expression and the synthesis of proteins.

Hormones Have Been Known Since Ancient Times

Although the scientific field of endocrinology is relatively young, diseases of the endocrine system have been documented for more than a thousand years. Evidence of endocrine abnormalities can even be seen in ancient art. For example, one pre-Colombian statue of a woman shows a mass on the front of her neck (**FIG. 7.1**). The mass is an enlarged thyroid gland, or *goiter,* a common condition high in the Andes, where the dietary iodine needed to make thyroid hormones was lacking.

The first association of endocrine structure and function was probably the link between the testes and male sexuality. Castration of animals and men was a common practice in both Eastern and Western cultures because it decreased the sex drive and rendered males infertile.

In 1849, A. A. Berthold used this knowledge to perform the first classic experiment in endocrinology. He removed the testes from roosters and observed that the castrated birds had smaller combs, less aggressiveness, and less sex drive than uncastrated roosters. If the testes were surgically placed back into the donor

FIG. 7.1   An endocrine disorder in ancient art

This pre-Colombian stone carving of a woman shows a mass at her neck. This mass is an enlarged thyroid gland, a condition known as goiter. It was considered a sign of beauty among the people who lived high in the Andes mountains.

RUNNING **PROBLEM** | Graves' Disease

The ball slid by the hole and trickled off the green: another bogey. Ben Crenshaw's golf game was falling apart. The 33-year-old professional had won the Masters Tournament only a year ago, but now something was not right. He was tired and weak, had been losing weight, and felt hot all the time. He attributed his symptoms to stress, but his family thought otherwise. At their urging, he finally saw a physician. The diagnosis? Graves' disease, which results in an overactive thyroid gland.

(197) (206) (215) (216) (218) (220) (221)

rooster or into another castrated bird, normal male behavior and comb development resumed. Because the reimplanted testes were not connected to nerves, Berthold concluded that the glands must secrete something into the blood that affected the entire body.

Experimental endocrinology did not receive much attention, however, until 1889, when the 72-year-old French physician Charles Brown-Séquard made a dramatic announcement of his sexual rejuvenation after injecting himself with extracts made from bull testes ground up in water. An international uproar followed, and physicians on both sides of the Atlantic began to inject their patients with extracts of many different endocrine organs, a practice known as *organotherapy*.

We now know that the increased virility Brown-Séquard reported was most likely a placebo effect because testosterone is a hydrophobic steroid that cannot be extracted by an aqueous preparation. His research opened the door to hormone therapy, however, and in 1891, organotherapy had its first true success: A woman was treated for low thyroid hormone levels with glycerin extracts of sheep thyroid glands.

As the study of "internal secretions" grew, Berthold's experiments became a template for endocrine research. Once a gland or structure was suspected of secreting hormones, the classic steps for identifying an endocrine gland became:

1. **Remove the suspected gland.** This is equivalent to inducing a state of *hormone deficiency*. If the gland does produce hormones, the animal should start to exhibit anatomical, behavioral, or physiological abnormalities.

2. **Replace the hormone.** This can be done by placing the gland back in the animal or administering an extract of the gland. This *replacement therapy* should eliminate the symptoms of hormone deficiency.

3. **Create a state of hormone excess.** Take a normal animal and implant an extra gland or administer extract from the gland to see if symptoms characteristic of *hormone excess* appear.

Once a gland is identified as a potential source of hormones, scientists purify extracts of the gland to isolate the active substance. They test for hormone activity by injecting animals with the purified extract and monitoring for a response.

Hormones identified by this technique are sometimes called *classic hormones*. They include hormones of the pancreas, thyroid, adrenal glands, pituitary, and gonads, all discrete endocrine glands that could be easily identified and surgically removed. Not all hormones come from identifiable glands, however, and we have been slower to discover them. For example, many hormones involved in digestion are secreted by endocrine cells scattered throughout the wall of the stomach or intestine, which has made them difficult to identify and isolate. The Anatomy Summary in **FIGURE 7.2** lists the major hormones of the body, the glands or cells that secrete them, and the major effects of each hormone.

What Makes a Chemical a Hormone?

In 1905, the term *hormone* was coined from the Greek verb meaning "to excite or arouse." The traditional definition of a **hormone**

CLINICAL FOCUS

Diabetes: The Discovery of Insulin

Diabetes mellitus, the metabolic condition associated with pathologies of insulin function, has been known since ancient times. Detailed clinical descriptions of insulin-deficient diabetes were available to physicians, but they had no means of treating the disease. Patients invariably died. However, in a series of classic experiments, Oskar Minkowski at the University of Strasbourg (Germany) pinpointed the relationship between diabetes and the pancreas. In 1889, Minkowski surgically removed the pancreas from dogs (*pancreatectomy*) and noticed that the dogs developed symptoms of diabetes. He also found that implanting pieces of pancreas under the dogs' skin would prevent development of diabetes. Subsequently, in 1921 Frederick G. Banting and Charles H. Best (Toronto, Canada) identified an antidiabetic substance in pancreas extracts. Banting and Best and others injected pancreatic extracts into diabetic animals and found that the extracts reversed the elevated blood glucose levels of diabetes. From there, it was a relatively short process until, in 1922, purified insulin was used in the first clinical trials. Science had found a treatment for a once-fatal disease.

is a chemical secreted by a cell or group of cells into the blood for transport to a distant target, where it exerts its effect at very low concentrations. However, as scientists learn more about chemical communication in the body, this definition is continually being challenged.

Hormones Are Secreted by a Cell or Group of Cells Traditionally, the field of endocrinology has focused on chemical messengers secreted by endocrine *glands*, the discrete and readily identifiable tissues derived from epithelial tissue [p. 80]. However, we now know that molecules that act as hormones are secreted not only by classic endocrine glands but also by isolated endocrine cells (hormones of the *diffuse endocrine system*), by neurons (*neurohormones*), and occasionally by cells of the immune system (*cytokines*).

Hormones Are Secreted into the Blood Secretion is the movement of a substance from inside a cell to the extracellular fluid or directly into the external environment. According to the traditional definition of a hormone, hormones are secreted into the blood. However, the term *ectohormone* {*ektos*, outside} has been given to signal molecules secreted into the external environment.

Pheromones {*pherein*, to bring} are specialized ectohormones that act on other organisms of the same species to elicit a physiological or behavioral response. For example, sea anemones secrete alarm pheromones when danger threatens, and ants release trail pheromones to attract fellow workers to food sources. Pheromones are also used to attract members of the opposite sex for mating. Sex pheromones are found throughout the animal kingdom, in animals from fruit flies to dogs.

But do humans have pheromones? This question is still a matter of debate. Some studies have shown that human *axillary* (armpit) sweat glands secrete volatile steroids related to sex hormones that may serve as human sex pheromones. In one study, when female students were asked to rate the odors of T-shirts worn by male students, each woman preferred the odor of men who were genetically dissimilar from her. In another study, female axillary secretions rubbed on the upper lip of young women altered the timing of their menstrual cycles. Putative human pheromones are now sold as perfume advertised for attracting the opposite sex, as you will see if you do a web search for *human pheromone*. How humans may sense pheromones is discussed later [see Chapter 10].

Hormones Are Transported to a Distant Target
By the traditional definition, a hormone must be transported by the blood to a distant target cell. Experimentally, this property is sometimes difficult to demonstrate. Molecules that are suspected of being hormones but not fully accepted as such are called *candidate hormones*. They are usually identified by the word *factor*. For example, in the early 1970s, the hypothalamic regulating hormones were known as "releasing factors" and "inhibiting factors" rather than releasing and inhibiting hormones.

Currently, **growth factors,** a large group of substances that influence cell growth and division, are being studied to determine if they meet all the criteria for hormones. Although many growth factors act locally as *autocrine* or *paracrine signals* [p. 168], most do not seem to be distributed widely in the circulation. A similar situation exists with the lipid-derived signal molecules called *eicosanoids* [p. 30].

Complicating the classification of signal molecules is the fact that a molecule may act as a hormone when secreted from one location but as a paracrine or autocrine signal when secreted from a different location. For example, in the 1920s, scientists discovered that *cholecystokinin* (CCK) in extracts of intestine caused contraction of the gallbladder. For many years thereafter, CCK was known only as an intestinal hormone. Then in the mid-1970s, CCK was found in neurons of the brain, where it acts as a neurotransmitter or neuromodulator. In recent years, CCK has gained attention because of its possible role in controlling hunger.

Hormones Exert Their Effect at Very Low Concentrations
One hallmark of a hormone is its ability to act at concentrations in the nanomolar (10^{-9} M) to picomolar (10^{-12} M) range. Some chemical signals transported in the blood to distant targets are not considered hormones because they must be present in relatively high concentrations before an effect is noticed. For example, histamine released during severe allergic reactions may act on cells throughout the body, but its concentration exceeds the accepted range for a hormone.

As researchers discover new signal molecules and new receptors, the boundary between hormones and nonhormonal signal molecules continues to be challenged, just as the distinction

between the nervous and endocrine systems has blurred. Many *cytokines* [p. 168] seem to meet many of the criteria of a hormone. However, experts in cytokine research do not consider cytokines to be hormones because cytokines are synthesized and released on demand, in contrast to classic peptide hormones, which are made in advance and stored in the parent endocrine cell. A few cytokines—for example, *erythropoietin,* the molecule that controls red blood cell production—were classified as hormones before the term *cytokine* was coined, contributing to the overlap between the two groups of signal molecules.

Hormones Act by Binding to Receptors

All hormones bind to target cell receptors and initiate biochemical responses. These responses are the **cellular mechanism of action** of the hormone. As you can see from the table in Figure 7.2, one hormone may act on multiple tissues. To complicate matters, the effects may vary in different tissues or at different stages of development. Or a hormone may have no effect at all in a particular cell. Insulin is an example of a hormone with varied effects. In muscle and adipose tissues, insulin alters glucose transport proteins and enzymes for glucose metabolism. In the liver, it modulates enzyme activity but has no direct effect on glucose transport proteins. In the brain and certain other tissues, glucose metabolism is totally independent of insulin.

CONCEPT CHECK

1. Name the membrane transport process by which glucose moves from the extracellular fluid into cells.

The variable responsiveness of a cell to a hormone depends primarily on the cell's receptor and signal transduction pathways [p. 180]. If there are no hormone receptors in a tissue, its cells cannot respond. If tissues have different receptors and receptor-linked pathways for the same hormone, they will respond differently.

Hormone Action Must Be Terminated

Signal activity by hormones and other chemical signals must be of limited duration if the body is to respond to changes in its internal state. For example, insulin is secreted when blood glucose concentrations increase following a meal. As long as insulin is present, glucose leaves the blood and enters cells. However, if insulin activity continues for too long, blood glucose levels can fall so low that the nervous system becomes unable to function properly—a potentially fatal situation. Normally, the body avoids this situation in several ways: by limiting insulin secretion, by removing or inactivating insulin circulating in the blood, and by terminating insulin activity in target cells.

In general, hormones in the bloodstream are *degraded* (broken down) into inactive metabolites by enzymes found primarily in the liver and kidneys. The metabolites are then excreted in either the bile or the urine. The rate of hormone breakdown is

FIG. 7.2 **ANATOMY SUMMARY**

Hormones

Location	Hormone	Primary Target(s)
Pineal gland	Melatonin [A]	Brain, other tissues
Hypothalamus (N)	Trophic hormones [P, A] (see Fig. 7.9)	Anterior pituitary
Posterior pituitary (N)	Oxytocin [P]	Breast and uterus
	Vasopressin (ADH) [P]	Kidney
Anterior pituitary (G)	Prolactin [P]	Breast
	Growth hormone (somatotropin) [P]	Liver Many tissues
	Corticotropin (ACTH) [P]	Adrenal cortex
	Thyrotropin (TSH) [P]	Thyroid gland
	Follicle-stimulating hormone [P]	Gonads
	Luteinizing hormone [P]	Gonads
Thyroid gland	Triiodothyronine and thyroxine [A]	Many tissues
	Calcitonin [P]	Bone
Parathyroid gland	Parathyroid hormone [P]	Bone, kidney
Thymus gland	Thymosin, thymopoietin [P]	Lymphocytes
Heart (C)	Atrial natriuretic peptide [P]	Kidneys
Liver (C)	Angiotensinogen [P]	Adrenal cortex, blood vessels
	Insulin-like growth factors [P]	Many tissues
Stomach and small intestine (C)	Gastrin, cholecystokinin, secretin, and others [P]	GI tract and pancreas
Pancreas (G)	Insulin, glucagon, somatostatin, pancreatic polypeptide [P]	Many tissues
Adrenal cortex (G)	Aldosterone [S]	Kidney
	Cortisol [S]	Many tissues
	Androgens [S]	Many tissues
Adrenal medulla (N)	Epinephrine, norepinephrine [A]	Many tissues
Kidney (C)	Erythropoietin [P]	Bone marrow
	1,25 Dihydroxy-vitamin D_3 (calciferol) [S]	Intestine
Skin (C)	Vitamin D_3 [S]	Intermediate form of hormone
Testes (male) (G)	Androgens [S]	Many tissues
	Inhibin [P]	Anterior pituitary
Ovaries (female) (G)	Estrogen, progesterone [S]	Many tissues
	Inhibin [P]	Anterior pituitary
	Relaxin (pregnancy) [P]	Uterine muscle
Adipose tissue (C)	Leptin, adiponectin, resistin [P]	Hypothalamus, other tissues
Placenta (pregnant females only) (C)	Estrogen, progesterone [S]	Many tissues
	Chorionic somatomammotropin [P]	Many tissues
	Chorionic gonadotropin [P]	Corpus luteum

KEY

G = gland
C = endocrine cells
N = neurons

P = peptide
S = steroid
A = amino acid–derived

Main Effect(s)
Circadian rhythms; immune function; antioxidant
Release or inhibit pituitary hormones
Milk ejection; labor and delivery; behavior
Water reabsorption
Milk production
Growth factor secretion
Growth and metabolism
Cortisol release
Thyroid hormone synthesis
Egg or sperm production; sex hormone production
Sex hormone production; egg or sperm production
Metabolism, growth, and development
Plasma calcium levels (minimal effect in humans)
Regulates plasma Ca^{2+} and phosphate levels
Lymphocyte development
Increases Na^+ excretion
Aldosterone secretion; increases blood pressure
Growth
Assist digestion and absorption of nutrients
Metabolism of glucose and other nutrients
Na^+ and K^+ homeostasis
Stress response
Sex drive in females
Fight-or-flight response
Red blood cell production
Increases calcium absorption
Precursor of 1,25 dihydroxy-vitamin D_3
Sperm production, secondary sex characteristics
Inhibits FSH secretion
Egg production, secondary sex characteristics
Inhibits FSH secretion
Relaxes muscle
Food intake, metabolism, reproduction
Fetal, maternal development
Metabolism
Hormone secretion

indicated by a hormone's **half-life** in the circulation, the amount of time required to reduce the concentration of hormone by one-half. Half-life is one indicator of how long a hormone is active in the body.

Hormones bound to target membrane receptors have their activity terminated in several ways. Enzymes that are always present in the plasma can degrade peptide hormones bound to cell membrane receptors. In some cases, the receptor-hormone complex is brought into the cell by endocytosis, and the hormone is then digested in lysosomes [Fig. 5.19, p. 149]. Intracellular enzymes metabolize hormones that enter cells.

CONCEPT CHECK

2. What is the suffix in a chemical name that tells you a molecule is an enzyme? [*Hint:* p. 101] Use that suffix to name an enzyme that digests peptides.

THE CLASSIFICATION OF HORMONES

Hormones can be classified according to different schemes. The scheme used in Figure 7.2 groups them according to their source. A different scheme divides hormones into those whose release is controlled by the brain and those whose release is not controlled by the brain. Another scheme groups hormones according to whether they bind to G protein-coupled receptors, tyrosine kinase-linked receptors, or intracellular receptors, and so on.

A final scheme divides hormones into three main chemical classes: peptide/protein hormones, steroid hormones, and amino acid-derived, or amine, hormones (**TBL. 7.1**). The peptide/protein hormones are composed of linked amino acids. The steroid hormones are all derived from cholesterol [p. 30]. The amino acid-derived hormones, also called *amine hormones*, are modifications of single amino acids, either tryptophan or tyrosine.

CONCEPT CHECK

3. What is the classic definition of a hormone?

4. Based on what you know about the organelles involved in protein and steroid synthesis [p. 71], what would be the major differences between the organelle composition of a steroid-producing cell and that of a protein-producing cell?

Most Hormones Are Peptides or Proteins

The peptide/protein hormones range from small peptides of only three amino acids to larger proteins and glycoproteins. Despite the size variability among hormones in this group, they are usually called peptide hormones for the sake of simplicity. You can remember which hormones fall into this category by exclusion: If a hormone is not a steroid hormone and not an amino acid derivative, then it must be a peptide or protein.

Peptide Hormone Synthesis, Storage, and Release The synthesis and packaging of peptide hormones into membrane-bound secretory vesicles is similar to that of other proteins. The

TABLE 7.1	Comparison of Peptide, Steroid, and Amino Acid-Derived Hormones			
	Peptide Hormones	**Steroid Hormones**	**Amine Hormones (Tyrosine Derivatives)**	
			Catecholamines	**Thyroid Hormones**
Synthesis and Storage	Made in advance; stored in secretory vesicles	Synthesized on demand from precursors	Made in advance; stored in secretory vesicles	Made in advance; precursor stored in secretory vesicles
Release from Parent Cell	Exocytosis	Simple diffusion	Exocytosis	Transport protein
Transport in Blood	Dissolved in plasma	Bound to carrier proteins	Dissolved in plasma	Bound to carrier proteins
Half-Life	Short	Long	Short	Long
Location of Receptor	Cell membrane	Cytoplasm or nucleus; some have membrane receptors also	Cell membrane	Nucleus
Response to Receptor-Ligand Binding	Activation of second messenger systems; may activate genes	Activation of genes for transcription and translation; may have nongenomic actions	Activation of second messenger systems	Activation of genes for transcription and translation
General Target Response	Modification of existing proteins and induction of new protein synthesis	Induction of new protein synthesis	Modification of existing proteins	Induction of new protein synthesis
Examples	Insulin, parathyroid hormone	Estrogen, androgens, cortisol	Epinephrine, norepinephrine, dopamine	Thyroxine (T_4)

initial peptide that comes off the ribosome is a large inactive protein known as a preprohormone (**FIG. 7.3 ❶**). **Preprohormones** contain one or more copies of a peptide hormone, a *signal sequence* that directs the protein into the lumen of the rough endoplasmic reticulum, and other peptide sequences that may or may not have biological activity.

As the inactive preprohormone moves through the endoplasmic reticulum, the signal sequence is removed, creating a smaller, still-inactive molecule called a **prohormone** (Fig. 7.3 ❹). In the Golgi complex, the prohormone is packaged into secretory vesicles along with *proteolytic* {*proteo-*, protein + *lysis*, rupture} enzymes that chop the prohormone into active hormone and other fragments. This process is called *post-translational modification* [p. 115].

The secretory vesicles containing peptides are stored in the cytoplasm of the endocrine cell until the cell receives a signal for secretion. At that time, the vesicles move to the cell membrane and release their contents by calcium-dependent exocytosis [p. 148]. All of the peptide fragments created from the prohormone are released together into the extracellular fluid, in a process known as *co-secretion* (Fig. 7.3 ❺).

Post-Translational Modification of Prohormones Studies of prohormone processing have led to some interesting discoveries. Some prohormones, such as that for *thyrotropin-releasing hormone* (TRH), contain multiple copies of the hormone (Fig. 7.3a). Another interesting prohormone is *pro-opiomelanocortin* (Fig. 7.3b). This prohormone splits into three active peptides

plus an inactive fragment. In some instances, even the fragments are clinically useful. For example, proinsulin is cleaved into active insulin and an inactive fragment known as *C-peptide* (Fig. 7.3c). Clinicians measure the levels of C-peptide in the blood of diabetics to monitor how much insulin the patient's pancreas is producing.

Transport in the Blood and Half-Life of Peptide Hormones
Peptide hormones are water soluble and therefore generally dissolve easily in the extracellular fluid for transport throughout the body. The half-life for peptide hormones is usually quite short, in the range of several minutes. If the response to a peptide hormone must be sustained for an extended period of time, the hormone must be secreted continually.

Cellular Mechanism of Action of Peptide Hormones
Because peptide hormones are lipophobic, they are usually unable to enter the target cell. Instead, they bind to surface membrane receptors. The hormone-receptor complex initiates the cellular response by means of a *signal transduction* system (**FIG. 7.4**). Many peptide hormones work through cAMP second messenger systems [p. 173]. A few peptide hormone receptors, such as that of insulin, have tyrosine kinase activity [p. 176] or work through other signal transduction pathways.

The response of cells to peptide hormones is usually rapid because second messenger systems modify existing proteins. The changes triggered by peptide hormones include opening or closing membrane channels and modulating metabolic enzymes or transport proteins. Researchers have recently discovered that

FIG. 7.3 **ESSENTIALS**

Peptide Hormone Synthesis and Processing

Peptide hormones are made as large, inactive preprohormones that include a signal sequence, one or more copies of the hormone, and additional peptide fragments.

(a) Preprohormones

PreproTRH (thyrotropin-releasing hormone) has six copies of the 3-amino acid hormone TRH.

Preprohormone

PreproTRH (242 amino acids)

processes to

6 TRH
(3 amino acids each)

+

Other peptide fragments

+

Signal sequence

(b) Prohormones

Prohormones, such as pro-opiomelanocortin, the prohormone for ACTH, may contain several peptide sequences with biological activity.

Pro-opiomelanocortin

processes to

ACTH   γ lipotropin   β endorphin

+

Peptide fragment

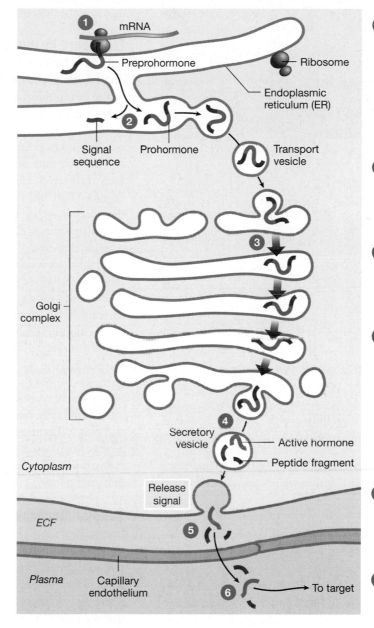

mRNA

Preprohormone

Ribosome

Signal sequence   Prohormone

Endoplasmic reticulum (ER)

Transport vesicle

Golgi complex

Secretory vesicle

Active hormone

Peptide fragment

Cytoplasm

Release signal

ECF

Plasma   Capillary endothelium

To target

1. Messenger RNA on the ribosomes binds amino acids into a peptide chain called a **preprohormone.** The chain is directed into the ER lumen by a **signal sequence** of amino acids.

2. Enzymes in the ER chop off the signal sequence, creating an inactive **prohormone.**

3. The prohormone passes from the ER through the Golgi complex.

4. Secretory vesicles containing enzymes and prohormone bud off the Golgi. The enzymes chop the prohormone into one or more active peptides plus additional peptide fragments.

5. The secretory vesicle releases its contents by exocytosis into the extracellular space.

6. The hormone moves into the circulation for transport to its target.

(c) Prohormones Process to Active Hormone Plus Peptide Fragments

The peptide chain of insulin's prohormone folds back on itself with the help of disulfide (S—S) bonds. The prohormone cleaves to insulin and C-peptide.

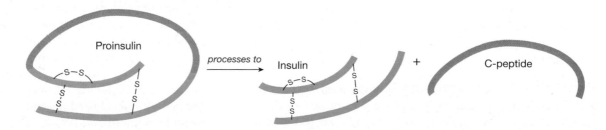

Proinsulin   *processes to*   Insulin   +   C-peptide

203

FIG. 7.4 Peptide hormone receptors and signal transduction

Peptide hormones (H) cannot enter their target cells and must combine with membrane receptors (R) that initiate signal transduction processes.

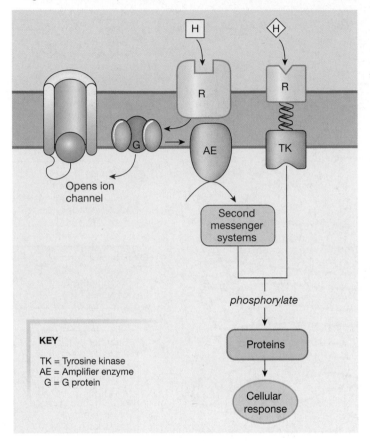

Opens ion channel

phosphorylate

Second messenger systems

Proteins

Cellular response

KEY

TK = Tyrosine kinase
AE = Amplifier enzyme
G = G protein

some peptide hormones also have longer-lasting effects when their second messenger systems activate genes and direct the synthesis of new proteins.

Steroid Hormones Are Derived from Cholesterol

Steroid hormones have a similar chemical structure because they are all derived from cholesterol (**FIG. 7.5a**). Unlike peptide hormones, which are made in tissues all over the body, steroid hormones are made in only a few organs. The **adrenal cortex,** the outer portion of the adrenal glands {*cortex,* bark}, makes several types of steroid hormones. One **adrenal gland** sits atop each kidney {*ad-,* upon + *renal,* kidney}. The gonads produce sex steroids (estrogens, progesterone, and androgens), and the skin can make vitamin D. In pregnant women, the placenta is also a source of steroid hormones.

Steroid Hormone Synthesis and Release Cells that secrete steroid hormones have unusually large amounts of smooth endoplasmic reticulum, the organelle in which steroids are synthesized. Steroids are lipophilic and diffuse easily across membranes, both

out of their parent cell and into their target cell. This property also means that steroid-secreting cells cannot store hormones in secretory vesicles. Instead, they synthesize their hormone as it is needed. When a stimulus activates the endocrine cell, precursors in the cytoplasm are rapidly converted to active hormone. The hormone concentration in the cytoplasm rises, and the hormones move out of the cell by simple diffusion.

Transport in the Blood and Half-Life of Steroid Hormones Like their parent cholesterol, steroid hormones are not very soluble in plasma and other body fluids. For this reason, most of the steroid hormone molecules found in the blood are bound to protein carrier molecules (Fig. 7.5b **1**). Some hormones have specific carriers, such as *corticosteroid-binding globulin.* Others simply bind to general plasma proteins, such as *albumin.*

The binding of a steroid hormone to a carrier protein protects the hormone from enzymatic degradation and results in an extended half-life. For example, **cortisol,** a hormone produced by the adrenal cortex, has a half-life of 60–90 minutes. (Compare this with epinephrine, an amino acid-derived hormone whose half-life is measured in seconds.)

Although binding steroid hormones to protein carriers extends their half-life, it also blocks their entry into target cells. The carrier-steroid complex remains outside the cell because the carrier proteins are lipophobic and cannot diffuse through the membrane. Only an unbound hormone molecule can diffuse into the target cell (Fig. 7.5b **2**). As unbound hormone leaves the plasma, the carriers obey the law of mass action and release hormone so that the ratio of unbound to bound hormone in the plasma remains constant [the K_d; p. 48].

Fortunately, hormones are active in minute concentrations, and only a tiny amount of unbound steroid is enough to produce a response. As unbound hormone leaves the blood and enters cells, additional carriers release their bound steroid so that some unbound hormone is always in the blood and ready to enter a cell.

Cellular Mechanism of Action of Steroid Hormones The best-studied steroid hormone receptors are found within cells, either in the cytoplasm or in the nucleus. The ultimate destination of steroid receptor-hormone complexes is the nucleus, where the complex acts as a *transcription factor,* binding to DNA and either activating or *repressing* (turning off) one or more genes (Fig. 7.5b **3**). Activated genes create new mRNA that directs the synthesis of new proteins. Any hormone that alters gene activity is said to have a *genomic effect* on the target cell.

When steroid hormones activate genes to direct the production of new proteins, there is usually a lag time between hormone-receptor binding and the first measurable biological effects. This lag can be as much as 90 minutes. Consequently, steroid hormones do not mediate reflex pathways that require rapid responses.

In recent years, researchers have discovered that several steroid hormones, including estrogens and aldosterone, have cell membrane receptors linked to signal transduction pathways, just as peptide hormones do. These receptors enable those steroid hormones to initiate rapid **nongenomic responses** in addition to

FIG. 7.5 **ESSENTIALS**

Steroid Hormones

Most steroid hormones are made in the adrenal cortex or gonads (ovaries and testes). Steroid hormones are not stored in the endocrine cell because of their lipophilic nature. They are made on demand and diffuse out of the endocrine cell.

(a) Cholesterol is the parent compound for all steroid hormones.

KEY

DHEA = dehydroepiandrosterone

= intermediate compounds whose names have been omitted for simplicity.

*Each step is catalyzed by an enzyme, but only two enzymes are shown in this figure.

(b) Steroid hormones act primarily on intracellular receptors.

1. Most hydrophobic steroids are bound to plasma protein carriers. Only unbound hormones can diffuse into the target cell.

2. Steroid hormone receptors are in the cytoplasm or nucleus.

2a. Some steroid hormones also bind to membrane receptors that use second messenger systems to create rapid cellular responses.

3. The receptor-hormone complex binds to DNA and activates or represses one or more genes.

4. Activated genes create new mRNA that moves back to the cytoplasm.

5. Translation produces new proteins for cell processes.

their slower genomic effects. With the discovery of nongenomic effects of steroid hormones, the functional differences between steroid and peptide hormones seem almost to have disappeared.

Some Hormones Are Derived from Single Amino Acids

The amino acid-derived, or amine, hormones are small molecules created from either tryptophan or tyrosine, both notable for the carbon ring structures in their R-groups [p. 32]. The pineal gland hormone **melatonin** is derived from tryptophan (see *Focus On: The Pineal Gland*, Fig. 7.16) but the other amino acid-derived hormones—the catecholamines and thyroid hormones—are made from tyrosine (**FIG. 7.6**). Catecholamines are a modification of a single tyrosine molecule. The thyroid hormones combine two tyrosine molecules with iodine atoms.

Despite a common precursor, the two families of tyrosine-based hormones have little in common. The **catecholamines** (epinephrine, norepinephrine, and dopamine) are neurohormones that bind to cell membrane receptors the way peptide hormones do. The **thyroid hormones,** produced by the butterfly-shaped thyroid gland in the neck, behave more like steroid hormones, with intracellular receptors that activate genes.

CONCEPT CHECK

5. What are the three chemical classes of hormones?

6. The steroid hormone aldosterone has a short half-life for a steroid hormone—only about 20 minutes. What would you predict about the degree to which aldosterone is bound to blood proteins?

CONTROL OF HORMONE RELEASE

Some hormones have clear stimuli that initiate their release, such as insulin secreted in response to increasing blood glucose concentrations. Other hormones have less obvious stimuli or are

RUNNING **PROBLEM**

Shaped like a butterfly, the thyroid gland straddles the trachea just below the Adam's apple. The thyroid gland concentrates iodine, an element found in food (most notably as an ingredient added to salt), and combines it with the amino acid tyrosine to make two thyroid hormones—thyroxine and triiodothyronine. These thyroid hormones perform many important functions in the body, including the regulation of growth and development, oxygen consumption, and the maintenance of body temperature.

Q1: *To which of the three classes of hormones do the thyroid hormones belong?*

Q2: *If a person's diet is low in iodine, predict what happens to thyroxine production.*

197　206　215　216　218　220　221

secreted continuously, often with a circadian rhythm [p. 17]. The sections that follow examine some of the most common control pathways for hormones. This discussion is not all-inclusive, and you will encounter a few hormones that do not fit exactly into these patterns.

Reflex pathways are one convenient way to classify hormones and simplify learning the steps that regulate their secretion. All reflex pathways have similar components: a stimulus, a sensor, an input signal, integration of the signal, an output signal, one or more targets, and a response [Fig. 6.16, p. 185]. In endocrine and neuroendocrine reflexes, the output signal is a hormone or a neurohormone.

The Endocrine Cell Is the Sensor in Simple Endocrine Reflexes

The simplest reflex control pathways in the endocrine system are those in which an endocrine cell directly senses a stimulus and responds by secreting its hormone [Fig. 6.19, pathway 6, p. 190]. In this type of pathway, the endocrine cell acts as both sensor and integrating center. The hormone is the output signal, and the response usually serves as a *negative feedback* signal that turns off the reflex [Fig. 1.12a, p. 16].

Parathyroid hormone (PTH), which controls calcium homeostasis, is an example of a hormone that uses a simple endocrine reflex. PTH is secreted by four small parathyroid glands in the neck. The parathyroid endocrine cells monitor plasma Ca^{2+} concentration with the aid of G protein-coupled Ca^{2+} receptors on their cell membranes. When a certain number of receptors are bound to Ca^{2+}, PTH secretion is inhibited. If the plasma Ca^{2+} concentration falls below a certain level and fewer Ca^{2+} receptors are bound, inhibition ceases and the parathyroid cells secrete PTH (**FIG. 7.7a**). Parathyroid hormone travels through the blood to act on bone, kidney, and intestine, initiating responses that increase the concentration of Ca^{2+} in the plasma. The increase in plasma Ca^{2+} is a negative feedback signal that turns off the reflex, ending the release of parathyroid hormone.

Other hormones that follow a simple endocrine reflex pattern include the classic hormones insulin and glucagon. Pancreatic endocrine cells are sensors that monitor blood glucose concentration [p. 159]. If blood glucose increases, the pancreatic beta cells respond by secreting insulin (Fig. 7.7b). Insulin travels through the blood to its target tissues, which increase their glucose uptake and metabolism. Glucose moving into cells decreases the blood concentration, which acts as a negative feedback signal that turns off the reflex, ending release of insulin.

Hormones can be released by more than one pathway, however. For example, insulin secretion can also be triggered by signals from the nervous system or by a hormone secreted from the digestive tract after a meal is eaten (Fig. 7.7b). The pancreatic beta cells—the integrating center for these reflex pathways—therefore must evaluate input signals from multiple sources when "deciding" whether to secrete insulin.

FIG. 7.6   Amine hormones

Most amine hormones are derived from the amino acid tyrosine.

Tyrosine
is the parent amino acid for catecholamines and thyroid hormones.

FIGURE QUESTION
Determine how each catecholamine molecule differs from the tyrosine molecule.

Catecholamines
are made by modifying the side groups of tyrosine.

Dopamine

Norepinephrine

Epinephrine

Thyroid hormones
are synthesized from two tyrosines and iodine (I) atoms.

Thyroxine (Tetraiodothyronine, T_4)

Triiodothyronine (T_3)

CONCEPT CHECK

7. In the blood glucose example, the increase in blood glucose corresponds to which step of a reflex pathway? Insulin secretion and the decrease in blood glucose correspond to which steps?

8. Which insulin release pathway in Figure 7.7b is a simple endocrine reflex? Which is a complex endocrine reflex? Which is a combination neural-endocrine reflex?

9. Glucagon is released from alpha cells in the pancreas when blood glucose levels decrease. Glucagon acts on multiple target tissues to increase blood glucose. Draw a reflex pathway to match this description.

Many Endocrine Reflexes Involve the Nervous System

The nervous system and the endocrine system overlap in both structure and function [see Fig. 6.19, pathways 3–5, p. 190]. Stimuli integrated by the central nervous system influence the release of many hormones through efferent neurons, as previously described for insulin. In addition, specialized groups of neurons secrete neurohormones, and two endocrine structures are incorporated in the anatomy of the brain: the pineal gland (see Fig. 7.16, p. 221) and the pituitary gland.

One of the most fascinating links between the brain and the endocrine system is the influence of emotions over hormone secretion and function. Physicians for centuries have recorded instances in which emotional state has influenced health or normal physiological processes. Women today know that the timing of their menstrual periods may be altered by stressors such as travel or final exams. The condition known as "failure to thrive" in infants can often be linked to environmental or emotional stress that increases secretion of some pituitary hormones and decreases production of others. The interactions among stress, the endocrine system, and the immune system are receiving intense study by scientists [Chapter 24].

Neurohormones Are Secreted into the Blood by Neurons

As noted previously, neurohormones are chemical signals released into the blood by a neuron [p. 168]. The human nervous

FIG. 7.7 Simple endocrine pathways

(a) A Simple Endocrine Reflex: Parathyroid Hormone

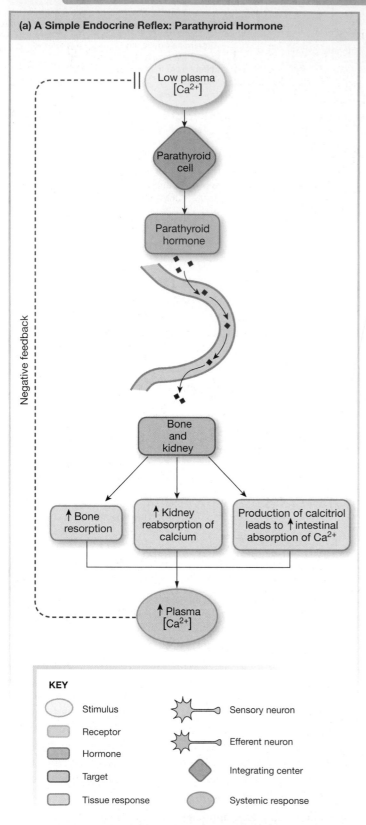

(b) Multiple Pathways for Insulin Secretion

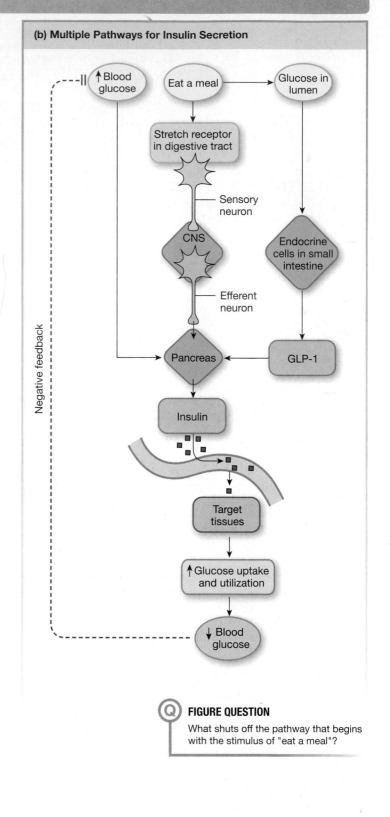

KEY

Stimulus	Sensory neuron
Receptor	Efferent neuron
Hormone	Integrating center
Target	Systemic response
Tissue response	

FIGURE QUESTION

What shuts off the pathway that begins with the stimulus of "eat a meal"?

system produces three major groups of neurohormones: (1) catecholamines (described earlier) made by modified neurons in the adrenal medulla, (2) hypothalamic neurohormones secreted from the posterior pituitary, and (3) hypothalamic neurohormones that control hormone release from the anterior pituitary. Because the latter two groups of neurohormones are associated with the pituitary gland, we describe that important endocrine structure next.

CONCEPT CHECK 10. Catecholamines belong to which chemical class of hormone?

The Pituitary Gland Is Actually Two Fused Glands

The **pituitary gland** is a lima bean–sized structure that extends downward from the brain, connected to it by a thin stalk and cradled in a protective pocket of bone (**FIG. 7.8a**). The first accurate description of the function of the pituitary gland came from Richard Lower (1631–1691), an experimental physiologist at Oxford University. Using observations and some experiments, he theorized that substances produced in the brain passed down the stalk into the gland and from there into the blood.

Lower did not realize that the pituitary gland is actually two different tissue types that merged during embryonic development. The **anterior pituitary** is a true endocrine gland of epithelial origin, derived from embryonic tissue that formed the roof of the mouth [Fig. 3.11, p. 80]. It is also called the *adenohypophysis* {*adeno-*, gland + *hypo-*, beneath + *phyein*, to grow}, and its hormones are *adenohypophyseal* secretions. The **posterior pituitary,** or *neurohypophysis*, is an extension of the neural tissue of the brain. It secretes neurohormones made in the *hypothalamus*, a region of the brain that controls many homeostatic functions.

The Posterior Pituitary Stores and Releases Two Neurohormones

The posterior pituitary is the storage and release site for two neurohormones: oxytocin and vasopressin (Fig. 7.8c). The neurons producing oxytocin and vasopressin are clustered together in areas of the hypothalamus known as the *paraventricular* and *supraoptic nuclei*. (A cluster of nerve cell bodies in the central nervous system is called a *nucleus*.) Each neurohormone is made in a separate cell type, and the synthesis and processing follow the standard pattern for peptide hormones described earlier in this chapter.

Once the neurohormones are packaged into secretory vesicles, the vesicles are transported to the posterior pituitary through long extensions of the neurons called *axons*. After vesicles reach the axon terminals, they are stored there, waiting for the release signal.

When a stimulus reaches the hypothalamus, an electrical signal passes from the neuron cell body in the hypothalamus to the *distal* (distant) end of the cell in the posterior pituitary. Depolarization of the axon terminal opens voltage-gated Ca^{2+} channels, and Ca^{2+} enters the cell. Calcium entry triggers exocytosis and the vesicle contents are released into the circulation. [Compare to insulin release, Fig. 5.26, p. 159.] Once in the blood, the neurohormones travel to their targets.

The two posterior pituitary neurohormones are composed of nine amino acids each. **Vasopressin,** also known as *antidiuretic hormone* or *ADH*, acts on the kidneys to regulate water balance in the body. In women, **oxytocin** released from the posterior pituitary controls the ejection of milk during breast-feeding and contractions of the uterus during labor and delivery.

A few neurons release oxytocin as a neurotransmitter or neuromodulator onto neurons in other parts of the brain. A number of animal experiments plus some human experiments indicate that oxytocin plays an important role in social, sexual, and maternal behaviors. One recent study suggests that *autism*, a developmental disorder in which patients are unable to form normal social relationships, may be related to defects in the normal oxytocin-modulated pathways of the brain.

CONCEPT CHECK 11. What intracellular structure is used for transport of secretory vesicles within the cell?
12. Name the membrane process by which the contents of secretory vesicles are released into the extracellular fluid.

The Anterior Pituitary Secretes Six Hormones

As late as 1889, it was being said in reviews of physiological function that the pituitary was of little or no use to higher vertebrates! By the early 1900s, however, researchers had discovered that animals with their anterior pituitary glands surgically removed were unable to survive more than a day or two. This observation, combined with the clinical signs associated with pituitary tumors, made scientists realize that the anterior pituitary is a major endocrine gland that secretes not one but six physiologically significant hormones: prolactin (PRL), thyrotropin (TSH), adrenocorticotropin (ACTH), growth hormone (GH), follicle-stimulating hormone (FSH), and luteinizing hormone (LH) (Fig. 7.8b).

Secretion of all the anterior pituitary hormones is controlled by hypothalamic neurohormones. The pathways can become quite complex because some hypothalamic neurohormones alter the secretion of several anterior pituitary hormones. In this book, we focus only on the primary targets for the hypothalamic hormones.

The anterior pituitary hormones, their primary hypothalamic neurohormones, and their targets are illustrated in **FIGURE 7.9**. The hypothalamic neurohormones that control release of anterior pituitary hormones are usually identified as *releasing hormones* (e.g., thyrotropin-releasing hormone) or *inhibiting hormones* (e.g., growth hormone-inhibiting hormone). For many years after they were first discovered, the hypothalamic hormones were called *factors*, as in corticotropin-releasing factor.

Notice that of the six anterior pituitary hormones, prolactin acts only on a nonendocrine target (the breast). The remaining five hormones have another endocrine gland or cell as one of their targets. These hormones that control the secretion of other hormones are known as **trophic hormones.**

The adjective *trophic* comes from the Greek word *trophikós*, which means "pertaining to food or nourishment" and refers to the manner in which the trophic hormone "nourishes" the target

FIG. 7.8 **ESSENTIALS**

The Pituitary Gland

The pituitary is actually two glands with different embryological origins that fused during development.

(a) The pituitary gland sits in a protected pocket of bone, connected to the brain by a thin stalk.

ANTERIOR ←——→ POSTERIOR

HYPOTHALAMUS

Infundibulum is the stalk that connects the pituitary to the brain.

Sphenoid bone

Posterior pituitary is an extension of the neural tissue.

Anterior pituitary is a true endocrine gland of epithelial origin.

(b) The **anterior pituitary** is a true endocrine gland that secretes six classic hormones. Neurohormones from the hypothalamus control release of the anterior pituitary hormones. The hypothalamic hormones reach the anterior pituitary through a specialized region of the circulation called a portal system.

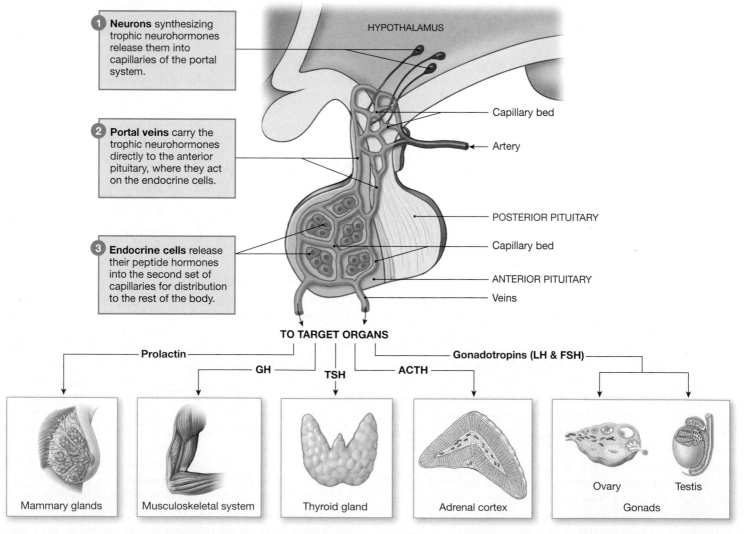

1 **Neurons** synthesizing trophic neurohormones release them into capillaries of the portal system.

2 **Portal veins** carry the trophic neurohormones directly to the anterior pituitary, where they act on the endocrine cells.

3 **Endocrine cells** release their peptide hormones into the second set of capillaries for distribution to the rest of the body.

HYPOTHALAMUS

Capillary bed

Artery

POSTERIOR PITUITARY

Capillary bed

ANTERIOR PITUITARY

Veins

TO TARGET ORGANS

Prolactin

GH

TSH

ACTH

Gonadotropins (LH & FSH)

Mammary glands

Musculoskeletal system

Thyroid gland

Adrenal cortex

Ovary   Testis

Gonads

(c) The **posterior pituitary** is an extension of the brain that secretes neurohormones made in the hypothalamus.

HYPOTHALAMUS

1 Neurohormone is made and packaged in cell body of neuron.

2 Vesicles are transported down the cell.

3 Vesicles containing neurohormone are stored in posterior pituitary.

— POSTERIOR PITUITARY

— Vein

4 Neurohormones are released into blood.

Oxytocin

Ile — Gln
Tyr   Asp
Cys ·· Cys
           Gly
Pro — Leu

Vasopressin

Phe — Gln
Tyr   Asp
Cys ·· Cys
           Gly
Pro — Arg

Mammary glands and uterus

Kidneys

cell. Trophic hormones often have names that end with the suffix *-tropin*, as in *gonadotropin*.* The root word to which the suffix is attached is the target tissue: The gonadotropins are hormones that are trophic to the gonads.

You should be aware that many of the hypothalamic and anterior pituitary hormones have multiple names as well as standardized abbreviations. For example, hypothalamic **somatostatin** (SS) is also called *growth hormone-inhibiting hormone* (GHIH), or in older scientific papers, *somatotropin release-inhibiting hormone* (SRIH). The table in Figure 7.9 lists the hypothalamic and anterior pituitary abbreviations and current alternate names.

A Portal System Connects the Hypothalamus and Anterior Pituitary

Most hormones in the body are secreted into the blood and become rapidly diluted as they distribute throughout the 5 L of blood volume. To avoid dilution, the hypothalamic neurohormones destined for the anterior pituitary enter a special modification of the circulatory system called a portal system. A **portal system** consisting of two sets of capillaries connected in series (one after the other) by a set of small veins (Fig. 7.8b). Hypothalamic neurohormones enter the blood at the first set of capillaries and go directly through the *portal veins* to the second capillary bed in the anterior pituitary, where they diffuse out to reach their target cells. In this way, a small amount of hormone remains concentrated in the tiny volume of portal blood while it goes directly to its target. This arrangement allows a small number of neurosecretory neurons in the hypothalamus to control the anterior pituitary.

The minute amounts of hormone secreted into the hypothalamic portal system posed a great challenge to the researchers who first isolated these hormones. Roger Guillemin and Andrew Schally had to work with huge amounts of tissue to obtain enough hormone to analyze. Guillemin and his colleagues processed more than 50 tons of sheep hypothalami, and a major meat packer donated more than 1 million pig hypothalami to Schally and his associates. For the final analysis, they needed 25,000 hypothalami to isolate and identify the amino acid sequence of just 1 mg of thyrotropin-releasing hormone (TRH), a tiny peptide made of three amino acids (see Fig. 7.3a). For their discovery, Guillemin and Schally shared a Nobel Prize in 1977 (see *http://nobelprize.org/nobel_prizes/medicine/laureates/1977*).

The hypothalamic-anterior pituitary portal system is more formally known as the *hypothalamic-hypophyseal portal system*. There are two additional portal systems in the body that you will encounter as you study physiology: one in the kidneys, and one in the digestive tract.

---

*A few hormones whose names end in *-tropin* do not have endocrine cells as their targets. For example, melanotropin acts on pigment-containing cells in many animals.

FIG. 7.9 **ESSENTIALS**

Hormones of the Hypothalamic-Anterior Pituitary Pathway

The hypothalamus secretes releasing hormones (-RH) and inhibiting hormones (-IH) that act on endocrine cells of the anterior pituitary to influence secretion of their hormones. Alternate names and abbreviations for the hormones are shown in the table below the figure.

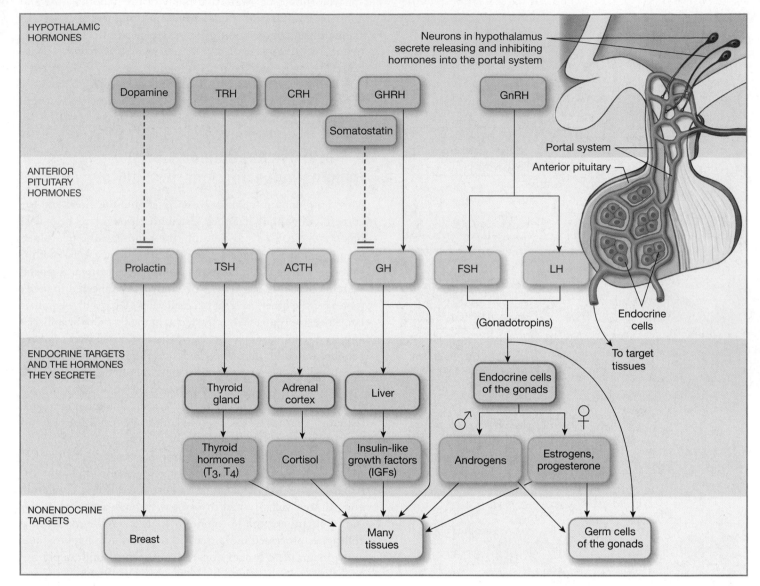

Anterior Pituitary Hormone	Hypothalamic Releasing Hormone	Hypothalamic Inhibiting Hormone
Prolactin (PRL)		Dopamine (PIH)
Thyrotropin, Thyroid-stimulating hormone (TSH)	Thyrotropin-releasing hormone (TRH)	
Adrenocorticotropin, Adrenocorticotrophic hormone (ACTH)	Corticotropin-releasing hormone (CRH)	
Growth hormone (GH), Somatotropin	GHRH (dominant)	Somatostatin (SS), also called growth hormone-inhibiting hormone (GHIH)
Gonadotropins: Follicle-stimulating hormone (FSH) Luteinizing hormone (LH)	Gonadotropin-releasing hormone (GnRH)	

Anterior Pituitary Hormones Control Growth, Metabolism, and Reproduction

The hormones of the anterior pituitary control so many vital functions that the pituitary is often called the master gland of the body. In general, we can say that the anterior pituitary hormones control metabolism, growth, and reproduction, all very complex processes.

One anterior pituitary hormone, **prolactin** (PRL), controls milk production (*lactation*) in the female breast. **Growth hormone** (GH; also called *somatotropin*) affects metabolism of many tissues in addition to stimulating hormone production by the liver (**FIG. 7.10**). Prolactin and growth hormone are the only two anterior pituitary hormones with hypothalamic release-inhibiting hormones, as you can see in Figure 7.9. We discuss prolactin and growth hormone in detail later [Chapters 26 and 23, respectively].

The remaining four anterior pituitary hormones all have another endocrine gland as their primary target. **Follicle-stimulating hormone (FSH)** and **luteinizing hormone (LH),** known collectively as the **gonadotropins,** were originally named for their effects on the ovaries, but both hormones act on the testes as well. **Thyroid-stimulating hormone (TSH)** (or *thyrotropin*) controls hormone synthesis and secretion in the thyroid gland. **Adrenocorticotrophic hormone (ACTH)** (or *adrenocorticotropin*) acts on certain cells of the adrenal cortex to control synthesis and release of the steroid hormone cortisol.

CONCEPT CHECK

13. Match the general reflex pathway patterns shown in Figure 6.19 [p. 190] to:

     (a) the hypothalamic neurohormone—prolactin—breast pattern just described

     (b) the growth hormone pathway in Figure 7.10

14. What is the target tissue of a hypothalamic neurohormone secreted into the hypothalamic-hypophyseal portal system?

Feedback Loops Are Different in the Hypothalamic-Pituitary Pathway

The pathways in which anterior pituitary hormones act as trophic hormones are among the most complex endocrine reflexes because they involve three integrating centers: the hypothalamus, the anterior pituitary, and the endocrine target of the pituitary hormone (**FIG. 7.11a**). Feedback in these complex pathways follows a different pattern. Instead of the response acting as the negative feedback signal, the hormones themselves are the feedback signal.

In hypothalamic-pituitary pathways, the dominant form of feedback is **long-loop negative feedback,** where the hormone secreted by the peripheral endocrine gland "feeds back" to suppress secretion of its anterior pituitary and hypothalamic hormones (Fig. 7.11a). In pathways with two or three hormones in sequence, the "downstream" hormone usually feeds back to suppress the hormone(s) that controlled its secretion. A major

FIG. 7.10 The growth hormone pathway

Hypothalamic growth hormone-releasing hormone (GHRH) stimulates growth hormone (GH) secretion. Growth hormone acts directly on many body tissues but also influences liver production of insulin-like growth factors (IGFs or somatomedins), another group of hormones that regulate growth.

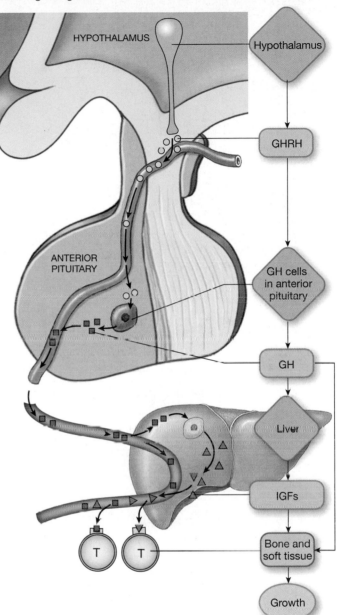

HYPOTHALAMUS

ANTERIOR PITUITARY

Hypothalamus → GHRH → GH cells in anterior pituitary → GH → Liver → IGFs → Bone and soft tissue → Growth

exception to long-loop negative feedback is the ovarian hormones estrogen and progesterone, where feedback alternates between positive and negative [Chapter 26].

Some pituitary hormones also exhibit short-loop negative feedback and ultra-short-loop feedback. In **short-loop negative feedback,** a pituitary hormone feeds back to decrease hormone secretion by the hypothalamus. Prolactin, growth hormone, and ACTH exhibit short-loop negative feedback. There can also be *ultra-short-loop feedback* in the pituitary and hypothalamus, in which a hormone acts as an autocrine or paracrine signal to influence the cell that secreted it. The short-loop feedback

FIG. 7.11   Negative feedback in complex endocrine pathways

(a) In complex endocrine pathways, the hormones of the pathway serve as negative feedback signals. Short-loop negative feedback occurs with prolactin, growth hormone, and ACTH.

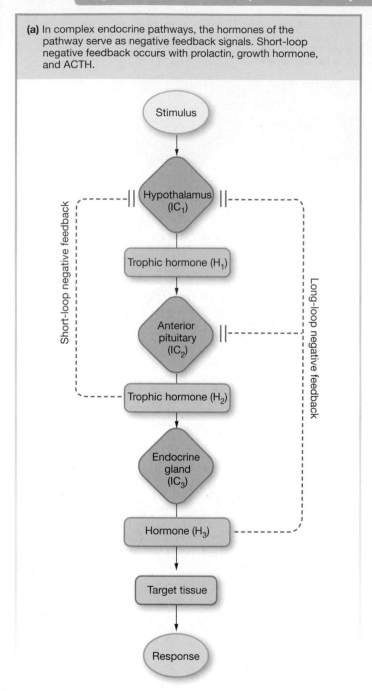

(b) Control Pathway for Cortisol Secretion
Cortisol is a steroid hormone secreted by the adrenal cortex.
ACTH = corticotropin or adrenocorticotropic hormone;
CRH = corticotropin-releasing hormone.

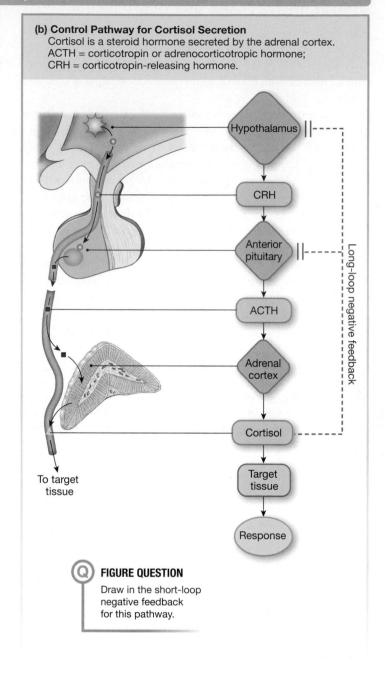

FIGURE QUESTION
Draw in the short-loop negative feedback for this pathway.

pathways are usually secondary to the more significant long-loop pathways.

The hormones of the *hypothalamic-pituitary-adrenal (HPA) pathway* provide a good example of feedback loops (Fig. 7.11b). Cortisol secreted from the adrenal cortex feeds back to suppress secretion of hypothalamic corticotropin-releasing hormone (CRH) and adrenocorticotropin (ACTH) from the anterior pituitary. ACTH also exerts short-loop negative feedback on the secretion of CRH.

One reason hormones must be the feedback signal in these complex endocrine reflexes is that for most anterior pituitary hormone pathways, there is no single response that the body can easily monitor. The hormones act on multiple tissues and have different, often subtle, effects in different tissues. There is no single parameter, such as blood glucose concentration, that can serve as the signal for negative feedback.

With a hormone-based system of negative feedback, the hormones in a pathway normally stay within the range needed

Thyroid hormone production is regulated by thyroid-stimulating hormone (TSH), a hormone secreted by the anterior pituitary. The production of TSH is in turn regulated by the neurohormone thyrotropin-releasing hormone (TRH) from the hypothalamus.

Q3: *In a normal person, when thyroid hormone levels in the blood increase, will negative feedback increase or decrease the secretion of TSH?*

Q4: *In a person with a hyperactive gland that is producing too much thyroid hormone, would you expect the level of TSH to be higher or lower than in a normal person?*

for an appropriate response. Feedback patterns are important in the diagnosis of endocrine pathologies, discussed later in the chapter.

HORMONE INTERACTIONS

One of the most complicated aspects of endocrinology is the way hormones interact at their target cells. It would be simple if each endocrine reflex were a separate entity and if each cell were under the influence of only a single hormone. In many instances, however, cells and tissues are controlled by multiple hormones that may be present at the same time. In addition, multiple hormones acting on a single cell can interact in ways that cannot be predicted by knowing the individual effects of the hormone. In this section, we examine three types of hormone interaction: synergism, permissiveness, and antagonism.

In Synergism, the Effect of Interacting Hormones Is More than Additive

Sometimes different hormones have the same effect on the body, although they may accomplish that effect through different cellular mechanisms. One example is the hormonal control of blood glucose levels. Glucagon from the pancreas is the hormone primarily responsible for elevating blood glucose levels, but it is not the only hormone that has that effect. Cortisol raises blood glucose concentration, as does epinephrine.

What happens if two of these hormones are present in a target cell at the same time, or if all three hormones are secreted at the same time? You may expect their effects to be additive. In other words, if a given amount of epinephrine elevates blood glucose 5 mg/100 mL blood, and glucagon elevates blood glucose 10 mg/100 mL blood, you may expect both hormones

acting at the same time to elevate blood glucose 15 mg/100 mL blood (5 + 10).

Frequently, however, two (or more) hormones interact at their targets so that the combination yields a result that is greater than additive (1 + 2 > 3). This type of interaction is called **synergism.** For our epinephrine/glucagon example, a synergistic reaction would be:

- epinephrine   elevates blood   5 mg/100 mL blood
                           glucose

- glucagon   elevates blood   10 mg/100 mL blood
                           glucose

- epinephrine +   elevate blood   22 mg/100 mL blood
  glucagon   glucose

In other words, the combined effect of the two hormones is greater than the sum of the effects of the two hormones individually.

An example of synergism involving epinephrine, glucagon, and cortisol is shown in **FIGURE 7.12**. The cellular mechanisms that underlie synergistic effects are not always clear, but with peptide hormones, synergism is often linked to overlapping effects on target cell second messenger systems.

Synergism is not limited to hormones. It can occur with any two (or more) chemicals in the body. Pharmacologists have developed drugs with synergistic components. For example, the effectiveness of the antibiotic penicillin is enhanced by the presence of clavulanic acid in the same pill.

FIG. 7.12 **Synergism**

In synergism, the combined effect of hormones is greater than additive.

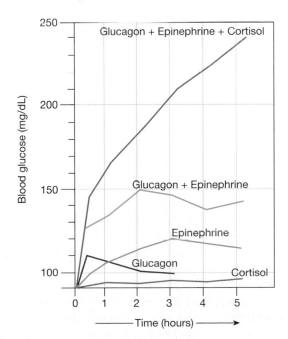

A Permissive Hormone Allows Another Hormone to Exert Its Full Effect

In **permissiveness,** one hormone cannot fully exert its effects unless a second hormone is present, even though the second hormone has no apparent action (2 + 0 > 2). For example, maturation of the reproductive system is controlled by gonadotropin-releasing hormone from the hypothalamus, gonadotropins from the anterior pituitary, and steroid hormones from the gonads. However, if thyroid hormone is not present in sufficient amounts, maturation of the reproductive system is delayed. Because thyroid hormone by itself cannot stimulate maturation of the reproductive system, thyroid hormone is considered to have a permissive effect on sexual maturation.

The results of this interaction can be summarized as follows:

- thyroid hormone alone = no development of reproductive system
- reproductive hormones alone = delayed development of reproductive system
- reproductive hormones with adequate thyroid hormone = normal development of reproductive system

The molecular mechanisms responsible for permissiveness are not well understood in most instances.

Antagonistic Hormones Have Opposing Effects

In some situations, two molecules work against each other, one diminishing the effectiveness of the other. This tendency of one substance to oppose the action of another is called *antagonism.* Antagonism may result when two molecules compete for the same receptor [p. 49]. When one molecule binds to the receptor but does not activate it, that molecule acts as a *competitive inhibitor,* or antagonist, to the other molecule. This type of receptor antagonism has been put to use in the development of pharmaceutical compounds, such as the estrogen receptor antagonist *tamoxifen,* which is used to treat breast cancers that are stimulated by estrogen.

In endocrinology, two hormones are considered *functional antagonists* if they have opposing physiological actions. For example, both glucagon and growth hormone raise the concentration of glucose in the blood, and both are antagonistic to insulin, which lowers the concentration of glucose in the blood. Hormones with antagonistic actions do not necessarily compete for the same receptor. Instead, they may act through different metabolic pathways, or one hormone may decrease the number of receptors for the opposing hormone. For example, evidence suggests that growth hormone decreases the number of insulin receptors, providing part of its functional antagonistic effects on blood glucose concentration.

The synergistic, permissive, and antagonistic interactions of hormones make the study of endocrinology both challenging and

RUNNING **PROBLEM**

Ben Crenshaw was diagnosed with Graves' disease, one form of hyperthyroidism. The goal of treatment is to reduce thyroid hormone activity, and Ben's physician offered him several alternatives. One treatment involved drugs that prevent the thyroid gland from using iodine. Another treatment was a single dose of radioactive iodine that destroys the thyroid tissue. A third treatment was surgical removal of all or part of the thyroid gland. Ben elected initially to use the thyroid-blocking drug. Several months later he was given radioactive iodine.

Q5: *Why is radioactive iodine (rather than some other radioactive element, such as cobalt) used to destroy thyroid tissue?*

197 206 215 **216** 218 220 221

intriguing. With this brief survey of hormone interactions, you have built a solid foundation for learning more about hormone interactions.

ENDOCRINE PATHOLOGIES

As one endocrinologist said, "There are no good or bad hormones. A balance of hormones is important for a healthy life.... Unbalance leads to diseases."[*] We can learn much about the normal functions of a hormone by studying the diseases caused by hormone imbalances. There are three basic patterns of endocrine pathology: hormone excess, hormone deficiency, and abnormal responsiveness of target tissues to a hormone.

To illustrate endocrine pathologies, we will use a single example, that of cortisol production by the adrenal cortex (see Fig. 7.11b). This is a complex reflex pathway that starts with the secretion of corticotropin-releasing hormone (CRH) from the hypothalamus. CRH stimulates release of adrenocorticotropin (ACTH) from the anterior pituitary. ACTH in turn controls the synthesis and release of cortisol from the adrenal cortex. As in other homeostatic reflex pathways, negative feedback shuts off the pathway. As cortisol increases, it acts as a negative feedback signal, causing the pituitary and hypothalamus to decrease their output of ACTH and CRH, respectively.

Hypersecretion Exaggerates a Hormone's Effects

If a hormone is present in excessive amounts, the normal effects of the hormone are exaggerated. Most instances of hormone excess are due to **hypersecretion.** There are numerous causes of hypersecretion, including benign tumors (*adenomas*) and cancerous

---

[*]W. König, preface to *Peptide and Protein Hormones,* New York: VCH Publishers, 1993.

tumors of the endocrine glands. Occasionally, nonendocrine tumors secrete hormones.

Any substance coming from outside the body is referred to as *exogenous* {*exo-,* outside}, and sometimes a patient may exhibit signs of hypersecretion as the result of medical treatment with an exogenous hormone or agonist. In this case, the condition is said to be *iatrogenic,* or physician-caused {*iatros,* healer + *-gen,* to be born}. It seems simple enough to correct the hormone imbalance by stopping treatment with the exogenous hormone, but this is not always the case.

In our example, exogenous cortisol administered as a drug acts as a negative feedback signal, just as cortisol produced within the body would, shutting off the production of CRH and ACTH (**FIG. 7.13**). Without the trophic "nourishing" influence of ACTH, the body's own cortisol production shuts down. If the pituitary remains suppressed and the adrenal cortex is deprived of ACTH long enough, the cells of both glands shrink and lose their ability to manufacture ACTH and cortisol. The loss of cell mass is known as **atrophy** {*a-,* without + *trophikós,* nourishment}. If the cells of an endocrine gland atrophy because of exogenous hormone administration, they may be very slow or totally unable to regain normal function when the treatment with exogenous hormone is stopped.

As you may know, steroid hormones such as cortisol can be used to treat poison ivy and severe allergies. However, when treatment is complete, the dosage must be tapered off gradually to allow the pituitary and adrenal gland to work back up to normal hormone production. As a result, packages of steroid pills direct patients ending treatment to take six pills one day, five the day after that, and so on. Low-dose, over-the-counter steroid creams usually do not pose a risk of feedback suppression when used as directed.

FIG. 7.13   Negative feedback from exogenous hormone

Exogenous hormone has the same negative feedback effect as endogenous hormone.

↓CRH (Hypothalamus)

↓ACTH (Anterior pituitary)

Exogenous cortisol

↓Cortisol (Adrenal cortex)

Target tissue

Response

Hyposecretion Diminishes or Eliminates a Hormone's Effects

Symptoms of hormone deficiency occur when too little hormone is secreted (**hyposecretion**). Hyposecretion may occur anywhere along the endocrine control pathway, in the hypothalamus, pituitary, or other endocrine glands. For example, hyposecretion of thyroid hormone may occur if there is insufficient dietary iodine for the thyroid gland to manufacture the iodinated hormone. The most common cause of hyposecretion pathologies is atrophy of the gland due to some disease process.

Negative feedback pathways are affected in hyposecretion, but in the opposite direction from hypersecretion. The absence of negative feedback causes trophic hormone levels to rise as the trophic hormones attempt to make the defective gland increase its hormone output. For example, if the adrenal cortex atrophies as a result of tuberculosis, cortisol production diminishes. The hypothalamus and anterior pituitary sense that cortisol levels are below normal, so they increase secretion of CRH and ACTH, respectively, in an attempt to stimulate the adrenal gland into making more cortisol.

Receptor or Second Messenger Problems Cause Abnormal Tissue Responsiveness

Endocrine diseases do not always arise from problems with endocrine glands. They may also be triggered by changes in the responsiveness of target tissues to the hormones. In these situations, the target tissues show abnormal responses even though hormone levels may be within the normal range. Changes in the target tissue response are usually caused by abnormal interactions between the hormone and its receptor or by alterations in signal transduction pathways.

Down-Regulation If hormone secretion is abnormally high for an extended period of time, target cells may *down-regulate* (decrease the number of) their receptors in an effort to diminish their responsiveness to excess hormone. **Hyperinsulinemia** {*hyper-,* elevated + insulin + *-emia,* in the blood} is a classic example of down-regulation in the endocrine system. In this disorder, sustained high levels of insulin in the blood cause target cells to remove insulin receptors from the cell membrane. Patients suffering from hyperinsulinemia may show signs of diabetes despite their high blood insulin levels.

Receptor and Signal Transduction Abnormalities Many forms of inherited endocrine pathologies can be traced to problems with hormone action in the target cell. Endocrinologists once believed that these problems were rare, but they are being recognized more frequently as scientists increase their understanding of receptors and signal transduction mechanisms.

Some pathologies are due to problems with the hormone receptor [Tbl. 6.1, p. 182]. If a mutation alters the protein sequence of the receptor, the cellular response to receptor-hormone binding may be altered. In other mutations, the receptors may be absent or completely nonfunctional. For example, in *androgen insensitivity syndrome,* androgen receptors are nonfunctional in the male fetus because of a genetic mutation. As a result, androgens produced by the developing fetus are unable to influence development of the genitalia. The result is a child who appears to be female but lacks a uterus and ovaries.

Genetic alterations in signal transduction pathways can lead to symptoms of hormone excess or deficiency. In the disease called *pseudohypoparathyroidism* {*pseudo-,* false + *hypo-,* decreased + parathyroid + *-ism,* condition or state of being}, patients show signs of low parathyroid hormone even though blood levels of the hormone are normal or elevated. These patients have inherited a defect in the G protein that links the hormone receptor to the cAMP amplifier enzyme, adenylyl cyclase. Because the signal transduction pathway does not function, target cells are unable to respond to parathyroid hormone, and signs of hormone deficiency appear.

Diagnosis of Endocrine Pathologies Depends on the Complexity of the Reflex

Diagnosis of endocrine pathologies may be simple or complicated, depending on the complexity of the reflex. For example, consider a simple endocrine reflex, such as that for parathyroid hormone. If there is too much or too little hormone, the problem can arise in only one location: the parathyroid glands (see Fig.7.7a). However, with complex hypothalamic-pituitary-endocrine gland reflexes, the diagnosis can be much more difficult.

If a pathology (deficiency or excess) arises in the last endocrine gland in a complex reflex pathway, the problem is considered to be a **primary pathology.** For example, if a tumor in the adrenal cortex begins to produce excessive amounts of cortisol, the resulting condition is called *primary hypersecretion.* If dysfunction occurs in the anterior pituitary, the problem is a **secondary pathology.** For example, if the pituitary is damaged because of head trauma and ACTH secretion diminishes, the resulting cortisol deficiency is considered to be *secondary hyposecretion* of cortisol. Pathologies of hypothalamic trophic hormones occur infrequently; they would be considered *tertiary* hyposecretion and hypersecretion.

The diagnosis of pathologies in complex endocrine pathways depends on understanding negative feedback in the control pathway. **FIGURE 7.14** shows three possible causes of excess cortisol secretion. To determine which is the correct *etiology* (cause) of the disease in a particular patient, the clinician must assess the levels of the three hormones in the control pathway.

If cortisol levels are high but levels of both trophic hormones are low, the problem is a primary disorder (Fig. 7.14a). There are two possible explanations for elevated cortisol: endogenous cortisol hypersecretion or the exogenous administration of cortisol for therapeutic reasons (see Fig. 7.13). In both cases, high levels of cortisol act as a negative feedback signal that shuts off production of CRH and ACTH. The pattern of high cortisol with low trophic hormone levels points to a primary disorder.

When the problem is endogenous—an adrenal tumor that is secreting cortisol in an unregulated fashion—the normal control pathways are totally ineffective. Although negative feedback shuts off production of the trophic hormones, the tumor is not dependent on them for cortisol production, so cortisol secretion continues in their absence. The tumor must be removed or suppressed before cortisol secretion can be controlled.

Figure 7.14b shows a secondary hypersecretion of cortisol due to an ACTH-secreting tumor of the pituitary. The high levels of ACTH cause high cortisol production, but in this example the high cortisol level has a negative feedback effect on the hypothalamus, decreasing production of CRH. The combination of low CRH and high ACTH isolates the problem to the pituitary. This pathology is responsible for about two-thirds of cortisol hypersecretion *syndromes* {*syn-,* together + *-drome,* running; a combination of symptoms characteristic of a particular pathology}.

If the problem is overproduction of CRH by the hypothalamus (Fig. 7.14c), CRH levels are higher than normal. High CRH in turn causes high ACTH, which in turn causes high cortisol. This is, therefore, tertiary hypersecretion arising from a problem in the hypothalamus. In clinical practice, hypothalamic hypersecretion pathologies are rare.

FIGURE 7.15 shows two possible etiologies for hyposecretion of cortisol. You can apply your understanding of negative feedback in the hypothalamic-pituitary control pathway to predict whether the levels of CRH, ACTH, and cortisol will be high or low in each case.

HORMONE EVOLUTION

Chemical signaling is an ancient method for communication and the maintenance of homeostasis. As scientists sequence the genomes of diverse species, they are discovering that in many cases hormone structure and function have changed amazingly little from the most primitive vertebrates through the mammals. In fact, hormone signaling pathways that were once considered exclusive to vertebrates, such as those for thyroid hormones and insulin, have now been shown to play physiological or developmental roles in invertebrates such as echinoderms and insects. This *evolutionary conservation* of hormone function is also demonstrated by the fact that some hormones from other organisms

RUNNING **PROBLEM**

Graves' disease is one form of thyroid gland hyperactivity. For this reason, people with Graves' disease have elevated thyroxine levels in the blood. Their TSH levels are very low.

Q6: *If levels of TSH are low and thyroxine levels are high, is Graves' disease a primary disorder or a secondary disorder (one that arises as a result of a problem with the anterior pituitary)? Explain your answer.*

197 — 206 — 215 — 216 — **218** — 220 — 221

FIG. 7.14 Hypercortisolism

Trophic hormone levels help isolate the source of pathology in hypercortisolism.

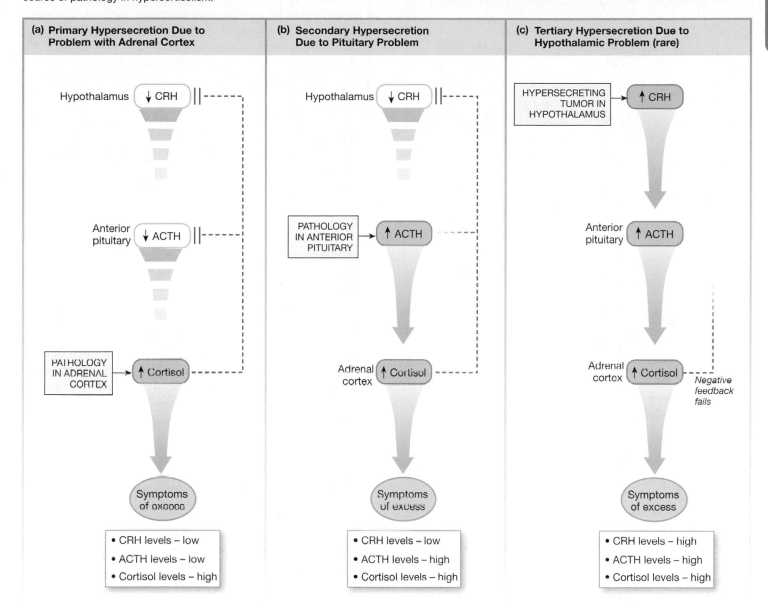

(a) Primary Hypersecretion Due to Problem with Adrenal Cortex

- CRH levels – low
- ACTH levels – low
- Cortisol levels – high

(b) Secondary Hypersecretion Due to Pituitary Problem

- CRH levels – low
- ACTH levels – high
- Cortisol levels – high

(c) Tertiary Hypersecretion Due to Hypothalamic Problem (rare)

- CRH levels – high
- ACTH levels – high
- Cortisol levels – high

have biological activity when administered to humans. By studying which portions of a hormone molecule do not change from species to species, scientists have acquired important clues to aid in the design of agonist and antagonist drugs.

The ability of nonhuman hormones to work in humans was a critical factor in the birth of endocrinology. When Best and Banting discovered insulin in 1921 and the first diabetic patients were treated with the hormone, the insulin was extracted from cow, pig, or sheep pancreases. Before the mid-1980s, slaughterhouses were the major source of insulin for the medical profession. Now, with genetic engineering, the human gene for insulin has been inserted into bacteria, which then synthesize the hormone, providing us with a plentiful source of human insulin.

Although many hormones have the same function in most vertebrates, a few hormones that play a significant role in the physiology of lower vertebrates seem to be evolutionarily "on their way out" in humans. Calcitonin is a good example of such a hormone. It plays a role in calcium metabolism in fish, but apparently has no significant influence on daily calcium balance in adult humans. Neither calcitonin deficiency nor calcitonin excess is associated with any pathological condition or symptom.

Although calcitonin is not a significant hormone in humans, the calcitonin gene does code for a biologically active protein. In the brain, cells process mRNA from the calcitonin gene to make a peptide known as *calcitonin gene–related peptide* (CGRP), which acts as a neurotransmitter. CGRP can act as a powerful dilator of

FIG. 7.15 Hypocortisolism

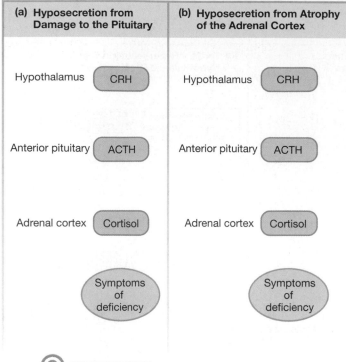

(a) Hyposecretion from Damage to the Pituitary	(b) Hyposecretion from Atrophy of the Adrenal Cortex
Hypothalamus — CRH	Hypothalamus — CRH
Anterior pituitary — ACTH	Anterior pituitary — ACTH
Adrenal cortex — Cortisol	Adrenal cortex — Cortisol
Symptoms of deficiency	Symptoms of deficiency

Q FIGURE QUESTION

For each condition, use arrows to indicate whether levels of the three hormones in the pathway will be increased, decreased, or unchanged. Draw in negative feedback loops where functional.

blood vessels, and CGRP receptor antagonists are being studied for their ability to treat migraine headaches, which occur when cerebral blood vessels dilate (vasodilation). The ability of one gene to produce multiple peptides is one reason research is shifting from genomics to physiology and *proteomics* (the study of the role of proteins in physiological function).

Some endocrine structures that are important in lower vertebrates are *vestigial* {*vestigium*, trace} in humans, meaning that in humans these structures are present as minimally functional glands. For example, *melanocyte-stimulating hormone* (MSH) from the intermediate lobe of the pituitary controls pigmentation in reptiles and amphibians. However, adult humans have only a vestigial intermediate lobe and normally do not have measurable levels of MSH in their blood.

In the research arena, *comparative endocrinology*—the study of endocrinology in nonhuman organisms—has made significant

RUNNING PROBLEM

Researchers have learned that Graves' disease is an autoimmune disorder in which the body fails to recognize its own tissue. In this condition, the body produces antibodies that mimic TSH and bind to the TSH receptor, turning it on. This false signal "fools" the thyroid gland into overproducing thyroid hormone. More women than men are diagnosed with Graves' disease, perhaps because of the influence of female hormones on thyroid function. Stress and other environmental factors have also been implicated in hyperthyroidism.

Q7: *Antibodies are proteins that bind to the TSH receptor. From that information, what can you conclude about the cellular location of the TSH receptor?*

Q8: *In Graves' disease, why doesn't negative feedback shut off thyroid hormone production before it becomes excessive?*

contributions to our quest to understand the human body. Many of our models of human physiology are based on research carried out in fish or frogs or rats, to name a few. For example, the pineal gland hormone *melatonin* (**FIG. 7.16**) was discovered through research using tadpoles. Many small nonhuman vertebrates have short life cycles that facilitate studying aging or reproductive physiology. Genetically altered mice (transgenic or knockout mice) have provided researchers valuable information about proteomics.

Opponents of animal research argue that scientists should not experiment with animals at all and should use only cell cultures and computer models. Cell cultures and models are valuable tools and can be helpful in the initial stages of medical research, but at some point new drugs and procedures must be tested on intact organisms prior to clinical trials in humans. Responsible scientists follow guidelines for appropriate animal use and limit the number of animals killed to the minimum needed to provide valid data.

In this chapter, we have examined how the endocrine system with its hormones helps regulate the slower processes in the body. As you will see, the nervous system takes care of the more rapid responses needed to maintain homeostasis.

FIG. 7.16 **FOCUS ON ...**

The Pineal Gland

Corpus callosum Thalamus

The **pineal gland** is a pea-sized structure buried deep in the brain of humans. Nearly 2000 years ago, this "seat of the soul" was thought to act as a valve that regulated the flow of vital spirits and knowledge into the brain. By 1950, however, scientists had decided that it was a vestigial structure with no known function.

Melatonin is an amino acid–derived hormone made from tryptophan.

Melatonin is the "darkness hormone," secreted at night as we sleep. It is the chemical messenger that transmits information about light-dark cycles to the brain center that governs the body's biological clock.

(Based on J. Arendt, Melatonin, *Clin Endocrinol* 29: 205–229, 1988.)

About 1957, one of the wonderful coincidences of scientific research occurred. An investigator heard about a factor in beef pineal glands that could lighten the skin of amphibians. Using the classical methodology of endocrinology, he obtained pineal glands from a slaughterhouse and started making extracts. His biological assay consisted of dropping pineal extracts into bowls of live tadpoles to see if their skin color blanched. Several years and hundreds of thousands of pineal glands later, he had isolated a small amount of melatonin.

Fifty years later, we are still learning about the functions of melatonin in humans. In addition to its role in sleep-wake cycles and the body's internal clock, scientists have evidence that melatonin is a powerful antioxidant. Some studies using mouse models of Alzheimer's disease suggest that melatonin may help slow the progression of the disease. Melatonin has also been linked to sexual function, the onset of puberty, and depression in the darker winter months (seasonal affective disorder, or SAD). In 2011, there were over 100 active clinical trials in the United States testing the efficacy of melatonin in treating disorders associated with sleep disturbances and depression.

In 2009, European authorities approved the use of a melatonin receptor agonist, *agomelatine*, for treating major depression. The U.S. Food and Drug Administration has been slower to approve the drug, and it is currently being tested in Phase II and Phase III clinical trials in the United States. Phase II trials are usually placebo-controlled, double-blind studies. Phase III trials include more patients and some uncontrolled studies. Some Phase III studies are "open-label," meaning that the patients and healthcare providers know what drug is being administered.

RUNNING **PROBLEM CONCLUSION** | Graves' Disease

In this running problem, you learned that in Graves' disease, thyroid hormone levels are high because an immune-system protein mimics TSH. You also learned that the thyroid gland concentrates iodine for synthesis of thyroid hormones, and that radioactive iodine can concentrate in the gland and destroy the thyroid cells. Ben Crenshaw's treatment for Graves' disease was successful. He went on to win the Masters Tournament for a second time in 1995 and he still plays golf professionally today.

Graves' disease is the most common form of hyperthyroidism. Other famous people who have suffered from it include former U.S. President George H. W. Bush and First Lady Barbara Bush. To learn more about Graves' disease and other thyroid conditions, visit the Endocrine Society's Hormone Foundation web site at *www.hormone.org* or the American Thyroid Association at *www.thyroid.org*. Check your answers to the problem questions by comparing them to the information in the following summary table.

—Continued next page

RUNNING **PROBLEM CONCLUSION** *Continued*

Question	Facts	Integration and Analysis
Q1: *To which of the three classes of hormones do thyroid hormones belong?*	The three classes are peptides, steroids, and amino-acid derivatives.	Thyroid hormones are made from the amino acid tyrosine, making them amino-acid derivatives.
Q2: *If a person's diet is low in iodine, predict what happens to thyroxine production.*	The thyroid gland concentrates iodine and combines it with the amino acid tyrosine to make thyroid hormones.	If iodine is lacking in the diet, a person is unable to make thyroid hormones.
Q3: *In a normal person, when thyroid hormone levels in the blood increase, will negative feedback increase or decrease the secretion of TSH?*	Negative feedback shuts off response loops.	Normally negative feedback decreases TSH secretion.
Q4: *In a person with a hyperactive gland that is producing too much thyroid hormone, would you expect the level of TSH to be higher or lower than in a normal person?*	Thyroid hormone is the negative feed-back signal.	If thyroid hormone is high, you would expect strong negative feedback and even lower levels of TSH.
Q5: *Why is radioactive iodine (rather than some other radioactive element, such as cobalt) used to destroy thyroid tissue?*	The thyroid gland concentrates iodine to make thyroid hormones.	Radioactive iodine is concentrated in the thyroid gland and therefore selectively destroys that tissue. Other radioactive elements distribute more widely throughout the body and may harm normal tissues.
Q6: *If levels of TSH are low and thyroxine levels are high, is Graves' disease a primary disorder or a secondary disorder (one that arises as a result of a problem with the anterior pituitary or the hypothalamus)? Explain your answer.*	In secondary hypersecretion disorders, you would expect the levels of the anterior pituitary trophic hormones to be elevated.	In Graves' disease, TSH from the anterior pituitary is very low. Therefore, the over-secretion of thyroid hormones is not the result of elevated TSH. This means that Graves' disease is a primary disorder that is caused by a problem in the thyroid gland itself.
Q7: *Antibodies are proteins that bind to the TSH receptor. From that information, what can you conclude about the cellular location of the TSH receptor?*	Receptors may be membrane receptors or intracellular receptors. Proteins cannot cross the cell membrane.	The TSH receptor is a membrane receptor. It uses the cAMP second messenger pathway for signal transduction.
Q8: *In Graves' disease, why doesn't negative feedback shut off thyroid hormone production before it becomes excessive?*	In normal negative feedback, increasing levels of thyroid hormone shut off TSH secretion. Without TSH stimulation, the thyroid stops producing thyroid hormone.	In Graves' disease, high levels of thyroid hormone have shut off endogenous TSH production. However, the thyroid gland still produces hormone in response to the binding of antibody to the TSH receptor. In this situation, negative feedback fails to correct the problem.

(197)(206)(215)(216)(218)(220)(**221**)

VISUALIZE, PRACTICE, AND MASTER PHYSIOLOGY WITH MASTERING A&P

• Practice-on-the-go with Dynamic Study Modules
• Visualize important processes with Interactive Physiology and A&P Flix
• Check your understanding with Video Tutors and practice quizzes!

MasteringA&P®

CHAPTER SUMMARY

This chapter introduced you to the endocrine system and the role it plays in *communication* and *control* of physiological processes. As you have seen before, the *compartmentalization of the body* into intracellular and extracellular compartments means that special mechanisms are required for signals to pass from one compartment to the other. The chapter also presented basic patterns that you will encounter again as you study various organ systems: differences among the three chemical classes of hormones, reflex pathways for hormones, types of hormone interactions, and endocrine pathologies.

Hormones

iP Endocrine System: Endocrine System Review

1. The specificity of a hormone depends on its receptors and their associated signal transduction pathways. (p. 199)

2. A **hormone** is a chemical secreted by a cell or group of cells into the blood for transport to a distant target, where it is effective at very low concentrations. (p. 198)

3. **Pheromones** are chemical signals secreted into the external environment. (p. 198)

4. Hormones bind to receptors to initiate responses known as the **cellular mechanism of action.** (p. 199)

5. Hormone activity is limited by terminating secretion, removing hormone from the blood, or terminating activity at the target cell. (p. 199)

6. The rate of hormone breakdown is indicated by a hormone's **half-life.** (p. 201)

The Classification of Hormones

iP Endocrine System: Biochemistry, Secretion and Transport of Hormones, and the Actions of Hormones on Target Cells

7. There are three types of hormones: **peptide/protein hormones,** composed of three or more amino acids; **steroid hormones,** derived from cholesterol; and **amino acid-derived hormones,** derived from either tyrosine (e.g., catecholamines and thyroid hormones) or tryptophan (e.g., melatonin). (p. 201; Tbl. 7.1)

8. Peptide hormones are made as inactive **preprohormones** and processed to **prohormones.** Prohormones are chopped into active hormone and peptide fragments that are co-secreted (p. 202; Fig. 7.3)

9. Peptide hormones dissolve in the plasma and have a short half-life. They bind to surface receptors on their target cells and initiate rapid cellular responses through signal transduction. In some instances, peptide hormones also initiate synthesis of new proteins. (p. 202; Fig. 7.4)

10. Steroid hormones are synthesized as they are needed. They are hydrophobic, and most steroid hormones in the blood are bound to protein carriers. Steroids have an extended half-life. (p. 204; Fig. 7.5)

11. Traditional steroid receptors are inside the target cell, where they turn genes on or off and direct the synthesis of new proteins. Cell response is slower than with peptide hormones. Steroid hormones may bind to membrane receptors and have nongenomic effects. (p. 204; Fig. 7.5)

12. Amine hormones may behave like typical peptide hormones or like a combination of a steroid hormone and a peptide hormone. (p. 207; Fig. 7.6)

Control of Hormone Release

iP Endocrine System: The Hypothalamic-Pituitary Axis

13. Classic endocrine cells act as both sensor and integrating center in the simple reflex pathway. (p. 206; Fig. 7.7)

14. Many endocrine reflexes involve the nervous system, either through **neurohormones** or through neurons that influence hormone release. (p. 207)

15. The pituitary gland is composed of the anterior pituitary (a true endocrine gland) and the posterior pituitary (an extension of the brain). (p. 209; Fig. 7.8a)

16. The posterior pituitary releases two neurohormones, oxytocin and vasopressin, that are made in the hypothalamus. (p. 209; Fig. 7.8c)

17. **Trophic hormones** control the secretion of other hormones. (p. 209)

18. Hypothalamic releasing hormones and inhibiting hormones control the secretion of anterior pituitary hormones. (p. 209; Fig. 7.9)

19. The hypothalamic trophic hormones reach the pituitary through the **hypothalamic-hypophyseal portal system.** (p. 211; Fig. 7.9)

20. There are six anterior pituitary hormones: prolactin, growth hormone, follicle-stimulating hormone, luteinizing hormone, thyroid-stimulating hormone, and adrenocorticotrophic hormone. (p. 213; Fig. 7.9)

21. In complex endocrine reflexes, hormones of the pathway act as negative feedback signals. (p. 213; Fig. 7.11)

Hormone Interactions

22. If the combination of two or more hormones yields a result that is greater than additive, the interaction is **synergism.** (p. 215; Fig. 7.12)

23. If one hormone cannot exert its effects fully unless a second hormone is present, the second hormone is said to be **permissive** to the first. (p. 216)

24. If one hormone opposes the action of another, the two are **antagonistic** to each other. (p. 216)

Endocrine Pathologies

25. Diseases of hormone excess are usually due to **hypersecretion.** Symptoms of hormone deficiency occur when too little hormone is secreted (**hyposecretion**). **Abnormal tissue responsiveness** may result from problems with hormone receptors or signal transduction pathways. (pp. 216, 217)

26. **Primary pathologies** arise in the last endocrine gland in a reflex. A **secondary pathology** is a problem with the anterior pituitary trophic hormones. (p. 218; Fig. 7.14)

Hormone Evolution

27. Many human hormones are similar to hormones found in other vertebrate animals. (p. 220)

REVIEW QUESTIONS

In addition to working through these questions and checking your answers on p. A-9, review the Learning Outcomes at the beginning of this chapter.

Level One Reviewing Facts and Terms

1. The study of hormones is called _____.

2. List the three basic ways hormones act on their target cells.

3. List five endocrine glands, and name one hormone secreted by each. Give one effect of each hormone you listed.

4. Match the following researchers with their experiments:

(a) Lower	1. isolated trophic hormones from the hypothalami of pigs and sheep
(b) Berthold	
(c) Guillemin and Schally	2. claimed sexual rejuvenation after injections of testicular extracts
(d) Brown-Séquard	
(e) Banting and Best	3. isolated insulin
	4. accurately described the function of the pituitary gland
	5. studied comb development in castrated roosters

5. Put the following steps for identifying an endocrine gland in order:

    (a) Purify the extracts and separate the active substances.
    (b) Perform replacement therapy with the gland or its extracts and see if the abnormalities disappear.
    (c) Implant the gland or administer the extract from the gland to a normal animal and see if symptoms characteristic of hormone excess appear.
    (d) Put the subject into a state of hormone deficiency by removing the suspected gland, and monitor the development of abnormalities.

6. For a chemical to be defined as a hormone, it must be secreted into the _____ for transport to a(n) _____ and take effect at _____ concentrations.

7. What is meant by the term *half-life* in connection with the activity of hormone molecules?

8. Metabolites are inactivated hormone molecules, broken down by enzymes found primarily in the _____ and _____, to be excreted in the _____ and _____, respectively.

9. Candidate hormones often have the word _____ as part of their name.

10. List and define the three chemical classes of hormones. Name one hormone in each class.

11. Decide if each of the following characteristics applies best to peptide hormones, steroid hormones, both classes, or neither class.

    (a) are lipophobic and must use a signal transduction system
    (b) have a short half-life, measured in minutes
    (c) often have a lag time of 90 minutes before effects are noticeable
    (d) are water-soluble, and thus easily dissolve in the extracellular fluid for transport
    (e) most hormones belong to this class
    (f) are all derived from cholesterol
    (g) consist of three or more amino acids linked together
    (h) are released into the blood to travel to a distant target organ
    (i) are transported in the blood bound to protein carrier molecules
    (j) are all lipophilic, so diffuse easily across membranes

12. Why do steroid hormones usually take so much longer to act than peptide hormones?

13. When steroid hormones act on a cell nucleus, the hormone-receptor complex acts as a(n) _____ factor, binds to DNA, and activates one or more _____, which create mRNA to direct the synthesis of new _____.

14. Researchers have discovered that some cells have additional steroid hormone receptors on their _____, enabling a faster response.

15. Melatonin is made from the amino acid _____, and the catecholamines and thyroid hormones are made from the amino acid _____.

16. A hormone that controls the secretion of another hormone is known as a(n) _____ hormone.

17. In reflex control pathways involving trophic hormones and multiple integrating centers, the hormones themselves act as _____ _____ signals, suppressing trophic hormone secretion earlier in the reflex.

18. What characteristic defines neurohormones?

19. List the two hormones secreted by the posterior pituitary gland. To what chemical class do they belong?

20. What is the hypothalamic-hypophyseal portal system? Why is it important?

21. List the six hormones of the anterior pituitary gland; give an action of each. Which ones are trophic hormones?

22. Explain long-loop negative feedback.

23. When two hormones work together to create a result that is greater than additive, that interaction is called _____. When hormone A must both be present to achieve full expression of hormone B, that interaction is called _____. When hormone activities oppose each other, that effect is called _____.

Level Two Reviewing Concepts

24. Compare and contrast the terms in each of the following sets:

    (a) paracrine signal, hormone, cytokine
    (b) primary and secondary endocrine pathologies
    (c) hypersecretion and hyposecretion
    (d) anterior and posterior pituitary

25. Compare and contrast the three chemical classes of hormones.

26. Map the following groups of terms. Add terms if you like.

    List 1

• co-secretion	• preprohormone
• endoplasmic reticulum	• prohormone
• exocytosis	• secretory vesicle
• Golgi complex	• signal sequence
• hormone receptor	• synthesis
• peptide hormone	• target cell response

List 2

• ACTH	• oxytocin
• anterior pituitary	• peptide/protein
• blood	• portal system
• endocrine cell	• posterior pituitary
• gonadotropins	• prolactin
• growth hormone	• releasing hormone
• hypothalamus	• trophic hormone
• inhibiting hormone	• TSH
• neurohormone	• vasopressin
• neuron	

Level Three Problem Solving

27. The terms *specificity, receptors,* and *down-regulation* can be applied to many physiological situations. Do their meanings change when applied to the endocrine system? What chemical and physical characteristics do hormones, enzymes, transport proteins, and receptors have in common that makes specificity important?

28. Dexamethasone is a drug used to suppress the secretion of adrenocorticotrophic hormone (ACTH) from the anterior pituitary. Two patients with hypersecretion of cortisol are given dexamethasone. Patient A's cortisol secretion falls to normal as a result, but patient B's cortisol secretion remains elevated. Draw maps of the reflex pathways for these two patients (see Fig. 7.11b for a template) and use the maps to determine which patient has primary hypercortisolism. Explain your reasoning.

29. Some early experiments for male birth control pills used drugs that suppressed gonadotropin (FSH and LH) release. However, men given these drugs stopped taking them because the drugs decreased testosterone secretion, which decreased the men's sex drive and caused impotence.

    (a) Use the information given in Figure 7.9 to draw the GnRH-FSH/LH-testosterone reflex pathway. Use the pathway to show how suppressing gonadotropins decreases sperm production and testosterone secretion.
    (b) Researchers subsequently suggested that a better treatment would be to give men extra testosterone. Draw another copy of the reflex pathway to show how testosterone could suppress sperm production without the side effect of impotence.

Level Four Quantitative Problems

30. The following graph represents the disappearance of a drug from the blood as the drug is metabolized and excreted. Based on the graph, what is the half-life of the drug?

31. The following graph shows plasma TSH concentration in three groups of subjects. Which pattern would be consistent with the following pathologies? Explain your reasoning.

    (a) primary hypothyroidism
    (b) primary hyperthyroidism
    (c) secondary hyperthyroidism

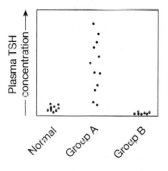

32. Based on what you have learned about the pathway for insulin secretion, draw and label a graph showing the effect of plasma glucose concentration on insulin secretion.

Answers to Concept Checks, Figure and Graph Questions, and end-of-chapter Review Questions can be found in Appendix A [A-1].

8

Neurons: Cellular and Network Properties

The future of clinical neurology and psychiatry is intimately tied to that of molecular neural science.

Eric R. Kandel, James H. Schwartz, and Thomas M. Jessell, in the preface to their book, *Principles of Neural Science*, 2000

ORGANIZATION OF THE NERVOUS SYSTEM 227

LO 8.1 Map the organization of the nervous system in detail.

CELLS OF THE NERVOUS SYSTEM 229

LO 8.2 Draw and describe the parts of a neuron and give their functions.

LO 8.3 Describe the parts of a synapse and their functions.

LO 8.4 Name the types and functions of glial cells.

ELECTRICAL SIGNALS IN NEURONS 236

LO 8.5 Explain in words how the Goldman-Hodgkin-Katz equation relates to the membrane potential of a cell.

LO 8.6 Explain the relationships between the following terms: current flow, conductance, resistance, Ohm's law.

LO 8.7 Compare and contrast graded potentials and action potentials.

LO 8.8 Explain the changes in ion permeability and ion flow that take place during an action potential.

LO 8.9 Describe and compare absolute and relative refractory periods.

LO 8.10 Explain the role of myelin in the conduction of action potentials.

CELL-TO-CELL COMMUNICATION IN THE NERVOUS SYSTEM 253

LO 8.11 Distinguish between electrical and chemical synapses.

LO 8.12 List and give examples of the seven groups of neurocrine secretions.

LO 8.13 Describe different patterns for neurotransmitter synthesis, recycling, release, and termination of action.

INTEGRATION OF NEURAL INFORMATION TRANSFER 260

LO 8.14 Describe the role of the following in synaptic communication: ionotropic and metabotropic receptors, neurotransmitters and neuromodulators, fast and slow synaptic potentials, excitatory and inhibitory postsynaptic potentials.

LO 8.15 Compare temporal and spatial summation.

LO 8.16 Compare presynaptic and postsynaptic inhibition.

LO 8.17 Explain the mechanism of long-term potentiation mediated by AMPA and NMDA receptors.

BACKGROUND BASICS

183 Reflex pathways
16 Positive feedback
65 Organelles
72 Matrix
139 Gated channels
74 Gap junctions
148 Exocytosis
168 Neurohormones
184 Antagonistic control
154 Resting membrane potential
155 Equilibrium potential
153 Bioelectricity

Purkinje cells (red) and glial cells (green) in the cerebellum

In an eerie scene from a science fiction movie, white-coated technicians move quietly through a room filled with bubbling cylindrical fish tanks. As the camera zooms in on one tank, no fish can be seen darting through aquatic plants. The lone occupant of the tank is a gray mass with a convoluted surface like a walnut and a long tail that appears to be edged with beads. Floating off the beads are hundreds of fine fibers, waving softly as the oxygen bubbles weave through them. This is no sea creature. . . . It is a brain and spinal cord, removed from its original owner and awaiting transplantation into another body. Can this be real? Is this scenario possible? Or is it just the creation of an imaginative movie screenwriter?

The brain is regarded as the seat of the soul, the mysterious source of those traits that we think of as setting humans apart from other animals. The brain and spinal cord are also integrating centers for homeostasis, movement, and many other body functions. They are the control center of the **nervous system,** a network of billions of nerve cells linked together in a highly organized manner to form the rapid control system of the body.

Nerve cells, or **neurons,** carry electrical signals rapidly and, in some cases, over long distances. They are uniquely shaped cells, and most have long, thin extensions, or **processes,** that can extend up to a meter in length. In most pathways, neurons release chemical signals, called **neurotransmitters,** into the extracellular fluid to communicate with neighboring cells. In a few pathways, neurons are linked by *gap junctions* [p. 74], allowing electrical signals to pass directly from cell to cell.

Using electrical signals to release chemicals from a cell is not unique to neurons. For example, pancreatic beta cells generate an electrical signal to initiate exocytosis of insulin-containing storage vesicles [p. 159]. Single-celled protozoa and plants also employ electrical signaling mechanisms, in many cases using the same types of ion channels as vertebrates do. Scientists sequencing ion channel proteins have found that many of these channel proteins have been highly conserved during evolution, indicating their fundamental importance.

Although electrical signaling is universal, sophisticated neural networks are unique to animal nervous systems. Reflex pathways in the nervous system do not necessarily follow a straight line from one neuron to the next. One neuron may influence multiple neurons, or many neurons may affect the function of a single neuron. The intricacy of neural networks and their neuronal components underlies the emergent properties of the nervous system. **Emergent properties** are complex processes, such as consciousness, intelligence, and emotion that cannot be predicted from what we know about the properties of individual nerve cells and their specific connections. The search to explain emergent properties makes neuroscience one of the most active research areas in physiology today.

Neuroscience, like many other areas of science, has its own specialized language. In many instances, multiple terms describe a single structure or function, which potentially can lead to confusion. **TABLE 8.1** lists some neuroscience terms used in this book, along with their common synonyms, which you may encounter in other publications.

ORGANIZATION OF THE NERVOUS SYSTEM

The nervous system can be divided into two parts (**FIG. 8.1**). The **central nervous system (CNS)** consists of the **brain** and the **spinal cord.** The **peripheral nervous system (PNS)** consists of **sensory (afferent) neurons** and **efferent neurons.** Information flow through the nervous system follows the basic pattern of a reflex: stimulus → sensor → input signal → integrating center → output signal → target → response [p. 183].

Sensory receptors throughout the body continuously monitor conditions in the internal and external environments. These sensors send information along sensory neurons to the CNS,

RUNNING **PROBLEM** | Mysterious Paralysis

"Like a polio ward from the 1950s" is how Guy McKhann, M.D., a neurology specialist at the Johns Hopkins School of Medicine, describes a ward of Beijing Hospital that he visited on a trip to China in 1986. Dozens of paralyzed children—some attached to respirators to assist their breathing—filled the ward to overflowing. The Chinese doctors thought the children had Guillain-Barré syndrome (GBS), a rare paralytic condition, but Dr. McKhann wasn't convinced. There were simply too many stricken children for the illness to be the rare Guillain-Barré syndrome. Was it polio—as some of the Beijing staff feared? Or was it another illness, perhaps one that had not yet been discovered?

227　229　231　251　253　257　267　268

TABLE 8.1	Synonyms in Neuroscience
Term Used in This Book	**Synonym(s)**
Action potential	AP, spike, nerve impulse, conduction signal
Autonomic nervous system	Visceral nervous system
Axon	Nerve fiber
Axonal transport	Axoplasmic flow
Axon terminal	Synaptic knob, synaptic bouton, presynaptic terminal
Axoplasm	Cytoplasm of an axon
Cell body	Cell soma, nerve cell body
Cell membrane of an axon	Axolemma
Glial cells	Neuroglia, glia
Interneuron	Association neuron
Rough endoplasmic reticulum	Nissl substance, Nissl body
Sensory neuron	Afferent neuron, afferent

FIG. 8.1 **ESSENTIALS**

The Organization of the Nervous System

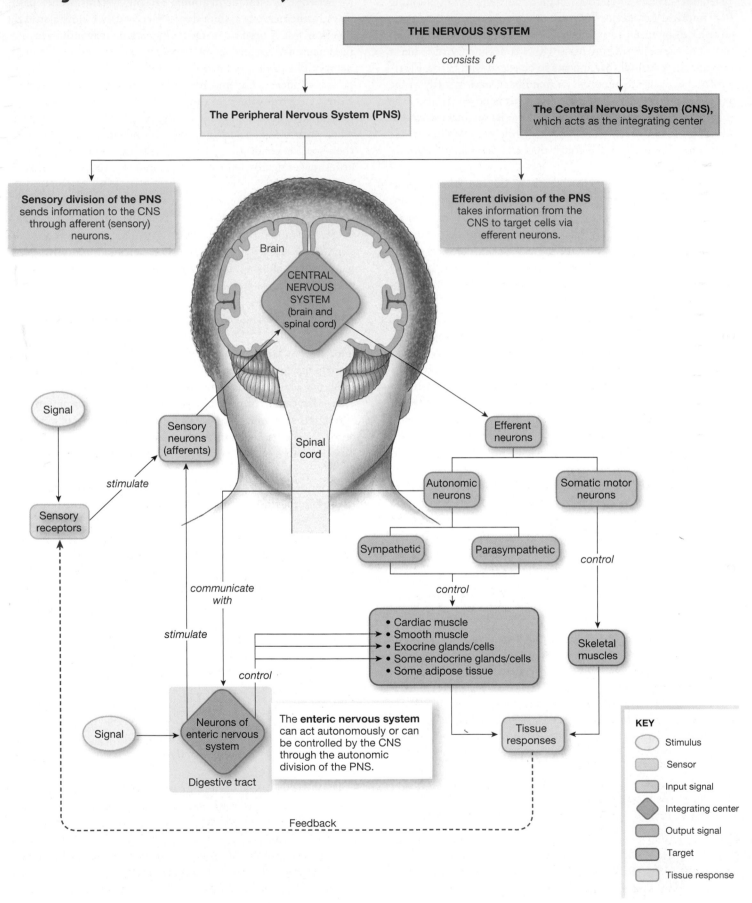

THE NERVOUS SYSTEM

consists of

The Peripheral Nervous System (PNS)

The Central Nervous System (CNS), which acts as the integrating center

Sensory division of the PNS sends information to the CNS through afferent (sensory) neurons.

Efferent division of the PNS takes information from the CNS to target cells via efferent neurons.

Brain

CENTRAL NERVOUS SYSTEM (brain and spinal cord)

Spinal cord

Signal

Sensory neurons (afferents)

stimulate

Sensory receptors

Efferent neurons

Autonomic neurons

Somatic motor neurons

Sympathetic

Parasympathetic

control

communicate with

stimulate

control

• Cardiac muscle
• Smooth muscle
• Exocrine glands/cells
• Some endocrine glands/cells
• Some adipose tissue

Skeletal muscles

control

Signal

Neurons of enteric nervous system

Digestive tract

The **enteric nervous system** can act autonomously or can be controlled by the CNS through the autonomic division of the PNS.

Tissue responses

Feedback

KEY

⬭ Stimulus

▢ Sensor

▢ Input signal

◆ Integrating center

▢ Output signal

▢ Target

▢ Tissue response

which is the integrating center for neural reflexes. CNS neurons integrate information that arrives from the sensory division of the PNS and determine whether a response is needed.

If a response is needed, the CNS sends output signals that travel through efferent neurons to their targets, which are mostly muscles and glands. Efferent neurons are subdivided into the **somatic motor division,** which controls skeletal muscles, and the **autonomic division,** which controls smooth and cardiac muscles, exocrine glands, some endocrine glands, and some types of adipose tissue. Terminology used to describe efferent neurons can be confusing. The expression *motor neuron* is sometimes used to refer to all efferent neurons. However, clinically, the term *motor neuron* (or *motoneuron*) is often used to describe somatic motor neurons that control skeletal muscles.

The autonomic division of the PNS is also called the *visceral nervous system* because it controls contraction and secretion in the various internal organs {*viscera,* internal organs}. Autonomic neurons are further divided into **sympathetic** and **parasympathetic branches,** which can be distinguished by their anatomical organization and by the chemicals they use to communicate with their target cells. Many internal organs receive innervation from both types of autonomic neurons, and it is common for the two divisions to exert *antagonistic control* over a single target [p. 184].

In recent years, a third division of the nervous system has received considerable attention. The **enteric nervous system** is a network of neurons in the walls of the digestive tract. It is frequently controlled by the autonomic division of the nervous system, but it is also able to function autonomously as its own integrating center. You will learn more about the enteric nervous system when you study the digestive system.

It is important to note that the CNS can initiate activity without sensory input, such as when you decide to text a friend. Also, the CNS need not create any measurable output to the efferent divisions. For example, thinking and dreaming are complex higher-brain functions that can take place totally within the CNS.

CONCEPT CHECK

1. Organize the following terms describing functional types of neurons into a map or outline: afferent, autonomic, brain, central, efferent, enteric, parasympathetic, peripheral, sensory, somatic motor, spinal, sympathetic.

CELLS OF THE NERVOUS SYSTEM

The nervous system is composed primarily of two cell types: neurons—the basic signaling units of the nervous system—and support cells known as *glial cells* (or glia or neuroglia).

Neurons Carry Electrical Signals

The neuron, or nerve cell, is the functional unit of the nervous system. (A *functional unit* is the smallest structure that can carry out the functions of a system.) Neurons are uniquely shaped cells with long processes that extend outward from the *nerve cell body.*

RUNNING **PROBLEM**

Guillain-Barré syndrome is a relatively rare paralytic condition that strikes after a viral infection or an immunization. There is no cure, but usually the paralysis slowly disappears, and lost sensation slowly returns as the body repairs itself. In classic Guillain-Barré, patients can neither feel sensations nor move their muscles.

Q1: *Which division(s) of the nervous system may be involved in Guillain-Barré syndrome?*

227   229   231   251   253   257   267   268

These processes are usually classified as either **dendrites,** which receive incoming signals, or **axons,** which carry outgoing information. The shape, number, and length of axons and dendrites vary from one neuron to the next, but these structures are an essential feature that allows neurons to communicate with one another and with other cells. Neurons may be classified either structurally or functionally (**FIG. 8.2**).

Structurally, neurons are classified by the number of processes that originate from the cell body. The model neuron that is commonly used to teach how a neuron functions is *multipolar,* with many dendrites and branched axons (Fig. 8.2c). Multipolar neurons in the CNS look different from multipolar efferent neurons (Fig. 8.2d). In other structural neuron types, the axons and dendrites may be missing or modified. *Pseudounipolar* neurons have the cell body located off one side of a single long process that is called the axon (Fig. 8.2a). (During development, the dendrites fused and became part of the axon.) *Bipolar* neurons have a single axon and single dendrite coming off the cell body (Fig. 8.2b). *Anaxonic* neurons lack an identifiable axon but have numerous branched dendrites (Fig. 8.2c).

Because physiology is concerned chiefly with function, however, we will classify neurons according to their functions: sensory (afferent) neurons, interneurons, and efferent (somatic motor and autonomic) neurons. Sensory neurons carry information about temperature, pressure, light, and other stimuli from sensory receptors to the CNS. Peripheral sensory neurons are pseudounipolar, with cell bodies located close to the CNS and very long processes that extend out to receptors in the limbs and internal organs. In these sensory neurons, the cell body is out of the direct path of signals passing along the axon (Fig. 8.2a). In contrast, sensory neurons in the nose and eye are much smaller bipolar neurons. Signals that begin at the dendrites travel through the cell body to the axon (Fig. 8.2b).

Neurons that lie entirely within the CNS are known as **interneurons** (short for *interconnecting neurons*). They come in a variety of forms but often have quite complex branching processes that allow them to communicate with many other neurons (Fig. 8.2c, d). Some interneurons are quite small compared to the model neuron.

Efferent neurons, both somatic motor and autonomic, are generally very similar to the neuron in Figure 8.2e. The axons

FIG. 8.2 **ESSENTIALS**

Neuron Anatomy

Multipolar
efferent
neuron

Functional Categories

Sensory Neurons

Somatic senses

Neurons for
smell and vision

Dendrites

Schwann
cell

Axon

Interneurons of CNS

Axon

Dendrites

Axon

Efferent Neurons

Dendrites

Axon

Axon
terminal

Collaterals

Structural Categories

Pseudounipolar	Bipolar	Anaxonic	Multipolar	
(a) Pseudounipolar neurons have a single process called the axon. During development, the dendrite fused with the axon.	**(b)** Bipolar neurons have two relatively equal fibers extending off the central cell body.	**(c)** Anaxonic CNS interneurons have no apparent axon.	**(d)** Multipolar CNS interneurons are highly branched but lack long extensions.	**(e)** A typical multipolar efferent neuron has five to seven dendrites, each branching four to six times. A single long axon may branch several times and end at enlarged axon terminals.

(f) Parts of a Neuron

Nucleus

Axon
hillock

Axon (initial
segment)

Myelin sheath

Postsynaptic
neuron

Dendrites

Cell
body

Presynaptic
axon terminal

Synaptic
cleft

Postsynaptic
dendrite

Synapse: The
region where an
axon terminal
communicates with
its postsynaptic
target cell

Input
signal → Integration → Output signal

may divide several time into branches called **collaterals** {*col-*, with + *lateral,* something on the side}. Efferent neurons have enlarged endings called **axon terminals.** Many autonomic neurons also have enlarged regions along the axon called **varicosities** [see Fig. 11.7, p. 365]. Both axon terminals and varicosities store and release neurotransmitter.

The long axons of both afferent and efferent peripheral neurons are bundled together with connective tissue into cordlike fibers called **nerves** that extend from the CNS to the targets of the component neurons. Nerves that carry only afferent signals are called **sensory nerves,** and those that carry only efferent signals are called **motor nerves.** Nerves that carry signals in both directions are **mixed nerves.** Many nerves are large enough to be seen with the naked eye and have been given anatomical names. For example, the *phrenic nerve* runs from the spinal cord to the muscles of the diaphragm.

The Cell Body Is the Control Center

The **cell body** (*cell soma*) of a neuron resembles a typical cell, with a nucleus and all organelles needed to direct cellular activity [p. 65]. An extensive cytoskeleton extends outward into the axon and dendrites. The position of the cell body varies in different types of neurons, but in most neurons the cell body is small, generally making up one tenth or less of the total cell volume. Despite its small size, the cell body with its nucleus is essential to the well-being of the cell because it contains DNA that is the template for protein synthesis [p. 112].

Dendrites Receive Incoming Signals

Dendrites {*dendron,* tree} are thin, branched processes that receive incoming information from neighboring cells (Fig. 8.2f). Dendrites increase the surface area of a neuron, allowing it to communicate with multiple other neurons. The simplest neurons have only a single dendrite. At the other extreme, neurons in the brain may have multiple dendrites with incredibly complex branching (Fig. 8.2d). A dendrite's surface area can be expanded even more by the presence of **dendritic spines** that vary from thin spikes to mushroom-shaped knobs (see Fig. 8.24c, p. 264).

The primary function of dendrites in the peripheral nervous system is to receive incoming information and transfer it to an integrating region within the neuron. Within the CNS, dendrite function is more complex. Dendritic spines can function as independent compartments, sending signals back and forth with other neurons in the brain. Many dendritic spines contain polyribosomes and can make their own proteins.

Dendritic spines can change their size and shape in response to input from neighboring cells. Changes in spine morphology are associated with learning and memory as well as with various pathologies, including genetic disorders that cause mental retardation and degenerative diseases such as Alzheimer's disease. Because of these associations, dendritic spines are a hot topic in neuroscience research.

Axons Carry Outgoing Signals

Most peripheral neurons have a single axon that originates from a specialized region of the cell

RUNNING **PROBLEM**

In classic GBS, the disease affects both sensory and somatic motor neurons. Dr. McKhann observed that although the Beijing children could not move their muscles, they could feel a pin prick.

Q2: *Do you think the paralysis found in the Chinese children affected both sensory (afferent) and somatic motor neurons? Why or why not?*

227   229   **231**   251   253   257   267   268

body called the **axon hillock** (Fig. 8.2f). Axons vary in length from more than a meter to only a few micrometers. They often branch sparsely along their length, forming collaterals. In our model neuron, each collateral ends in a bulbous axon terminal that contains mitochondria and membrane-bound vesicles filled with *neurocrine* molecules [p. 168].

The primary function of an axon is to transmit outgoing electrical signals from the integrating center of the neuron to target cells at the end of the axon. At the distal end of the axon, the electrical signal usually causes secretion of a chemical messenger molecule. In some CNS neurons, electrical signals pass directly to the next neuron through gap junctions that connect the two cells.

CONCEPT CHECK

2. Where do neurohormone-secreting neurons terminate?

3. What is the difference between a nerve and a neuron?

Axons are specialized to convey chemical and electrical signals. The axon cytoplasm is filled with many types of fibers and filaments but lacks ribosomes and endoplasmic reticulum. For this reason, proteins destined for the axon or the axon terminal must be synthesized on the rough endoplasmic reticulum in the cell body. The proteins are then moved down the axon by a process known as **axonal transport.**

Slow axonal transport moves material by **axoplasmic or cytoplasmic flow** from the cell body to the axon terminal. Material moves at a rate of only 0.2–2.5 mm/day, which means that slow transport can be used only for components that are not consumed rapidly by the cell, such as enzymes and cytoskeleton proteins.

Fast axonal transport moves organelles at rates of up to 400 mm (about 15.75 in.) per day (**FIG. 8.3**). The neuron uses stationary microtubules as tracks along which transported vesicles and mitochondria "walk" with the aid of attached foot-like motor proteins [p. 69]. These motor proteins alternately bind and unbind to the microtubules with the help of ATP, stepping their organelles along the axon in a stop-and-go fashion. Fast axonal transport goes in two directions. Forward (or *anterograde*) transport moves vesicles and mitochondria from the cell body to the

FIG. 8.3 | Fast axonal transport

Axonal transport moves proteins and organelles between cell body and axon terminal.

1 Peptides are synthesized on rough ER and packaged by the Golgi apparatus.

2 Fast axonal transport walks vesicles and mitochondria along microtubule network.

3 Vesicle contents are released by exocytosis.

5 Retrograde fast axonal transport

4 Synaptic vesicle recycling

6 Old membrane components digested in lysosomes

Rough endoplasmic reticulum

Soma

Golgi apparatus

Lysosome

Synaptic vesicle

axon terminal. Backward (or *retrograde*) transport returns old cellular components from the axon terminal to the cell body for recycling. There is evidence that nerve growth factors and some viruses also reach the cell body by fast retrograde transport.

Establishing Synapses Depends on Chemical Signals

The region where an axon terminal meets its target cell is called a **synapse** {*syn-*, together + *hapsis*, to join}. The neuron that delivers a signal to the synapse is known as the **presynaptic cell,** and the cell that receives the signal is called the **postsynaptic cell** (Fig. 8.2f). The narrow space between two cells is called the **synaptic cleft.** Although illustrations make the synaptic cleft look like an empty gap, it is filled with extracellular matrix whose fibers hold the presynaptic and postsynaptic cells in position.

The vast majority of synapses in the body are *chemical synapses,* where the presynaptic cell releases a chemical signal that diffuses across the synaptic cleft and binds to a membrane receptor on the postsynaptic cell. The human CNS also contains **electrical synapses,** where the presynaptic and postsynaptic cells are connected by gap junction channels [p. 74]. Gap junctions allow electrical current to flow directly from cell to cell. Communication at electrical synapses is bidirectional as well as faster than at chemical synapses.

During embryonic development, how can billions of neurons in the brain find their correct targets and make synapses? How can a somatic motor neuron in the spinal cord find the correct pathway to form a synapse with its target muscle in the big toe? The answer lies with chemical signals used by the developing embryo, ranging from factors that control differentiation of stem cells into neurons and glia to those that direct an elongating axon to its target.

The axons of embryonic nerve cells send out special tips called **growth cones** that extend through the extracellular compartment until they find their target cell (**FIG. 8.4**). In experiments where target cells are moved to an unusual location in the embryo, the axons in many instances are still able to find their targets by "sniffing out" the target's chemical scent. Growth cones depend on many different types of signals to find their way: growth factors, molecules in the extracellular matrix, and membrane proteins on the growth cones and on cells along the path. For example, *integrins* [p. 74] on the growth cone membrane bind to *laminins,* protein fibers in the extracellular matrix. *Nerve-cell adhesion molecules* (NCAMs) [p. 72] interact with membrane proteins of other cells.

Once an axon reaches its target cell, a synapse forms. However, synapse formation must be followed by electrical and chemical activity, or the synapse will disappear. The survival of neuronal pathways depends on **neurotrophic factors** {*trophikos,* nourishment} secreted by neurons and glial cells. There is still much we have to learn about this complicated process, and it is an active area of physiological research.

FIG. 8.4   The growth cone of a developing axon

The growing tip of a developing axon (blue) is a flattened region filled with microtubules (green) and actin filaments (red and yellow) that continuously assemble at their distal ends, extending the tip of the axon as it seeks its target.

This "use it or lose it" scenario is most dramatically reflected by the fact that the infant brain is only about one-fourth the size of the adult brain. Further brain growth is due not to an increase in the number of cells but to an increase in size and number of axons, dendrites, and synapses. This development depends on electrical signaling between sensory pathways, interneurons, and efferent neurons.

Babies who are neglected or deprived of sensory input may experience delayed development ("failure to thrive") because of the lack of nervous system stimulation. On the other hand, there is no evidence that extra stimulation in infancy enhances intellectual development, despite a popular movement to expose babies to art, music, and foreign languages before they can even walk.

Once synapses form, they are not fixed for life. Variations in electrical activity can cause rearrangement of the synaptic connections, a process that continues throughout life. Maintaining synapses is one reason that older adults are urged to keep learning new skills and information.

CONCEPT CHECK   4. Draw a chain of three neurons that synapse on one another in sequence. Label the presynaptic and postsynaptic ends of each neuron, the cell bodies, dendrites, axons, and axon terminals.

Glial Cells Provide Support for Neurons

Glial cells {*glia*, glue} are the unsung heroes of the nervous system, outnumbering neurons by 10–50 to 1. For many years,

scientists thought that the primary function of glial cells was physical support, and that glial cells had little influence on information processing. That view has changed. Although glial cells do not participate directly in the transmission of electrical signals over long distances, they do communicate with neurons and provide important biochemical support.

The peripheral nervous system has two types of glial cells—Schwann cells and satellite cells—and the CNS has four types: oligodendrocytes, microglia, astrocytes, and ependymal cells (**FIG. 8.5a**).

Myelin-Forming Glia   Neural tissue secretes very little extracellular matrix [p. 72], so glial cells provide structural stability to neurons by wrapping around them. **Schwann cells** in the PNS and **oligodendrocytes** in the CNS support and insulate axons by forming **myelin**, a substance composed of multiple concentric layers of phospholipid membrane (Fig. 8.5c). In addition to providing support, the myelin acts as insulation around axons and speeds up their signal transmission.

Myelin forms when the glial cells wrap around an axon, squeezing out the glial cytoplasm so that each wrap becomes two membrane layers (Fig. 8.5d). As an analogy, think of wrapping a deflated balloon tightly around a pencil. Some neurons have as many as 150 wraps (300 membrane layers) in the myelin sheath that surrounds their axons. Gap junctions connect the membrane layers and allow the flow of nutrients and information from layer to layer.

One difference between oligodendrocytes and Schwann cells is the number of axons each cell wraps around. In the CNS, one oligodendrocyte branches and forms myelin around portions of several axons (Fig. 8.5b). In the peripheral nervous system, one Schwann cell associates with one axon.

Schwann Cells   A single axon may have as many as 500 different Schwann cells along its length. Each Schwann cell wraps around a 1–1.5 mm segment of the axon, leaving tiny gaps, called the **nodes of Ranvier**, between the myelin-insulated areas (Fig. 8.5c). At each node, a tiny section of axon membrane remains in direct contact with the extracellular fluid. The nodes play an important role in the transmission of electrical signals along the axon, as you will learn later.

Satellite Cells   The second type of PNS glial cell, the **satellite cell,** is a nonmyelinating Schwann cell (Fig. 8.5a). Satellite cells form supportive capsules around nerve cell bodies located in ganglia. A **ganglion** {cluster or knot} is a collection of nerve cell bodies found outside the CNS. Ganglia appear as knots or swellings along a nerve. (A cluster of nerve cell bodies inside the CNS, the equivalent of a peripheral ganglion, is called a **nucleus** {plural, *nuclei*}.)

Astrocytes   Astrocytes {*astron*, a star} are highly branched glial cells that by some estimates make up about half of all cells in the brain (Fig. 8.5a, b). They come in several subtypes and form a functional network by communicating with one another through gap junctions. Astrocytes have multiple roles. Some astrocytes

FIG. 8.5 **ESSENTIALS**

Glial Cells

(a) Glial Cells and Their Functions

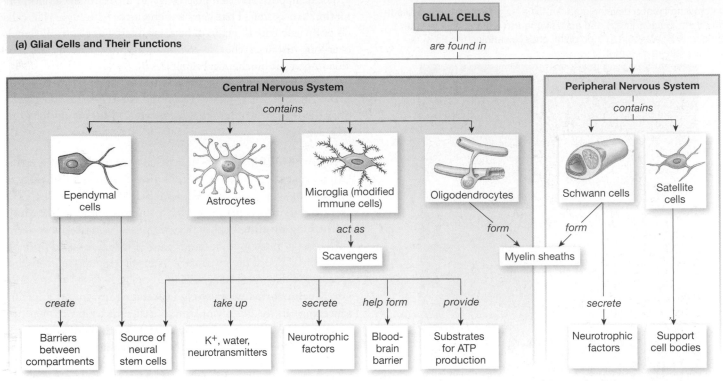

GLIAL CELLS

are found in

Central Nervous System — *contains*

- Ependymal cells — *create* → Barriers between compartments
- Astrocytes → Source of neural stem cells
  - *take up* → K⁺, water, neurotransmitters
  - *secrete* → Neurotrophic factors
  - *help form* → Blood-brain barrier
  - *provide* → Substrates for ATP production
- Microglia (modified immune cells) — *act as* → Scavengers
- Oligodendrocytes — *form* → Myelin sheaths

Peripheral Nervous System — *contains*

- Schwann cells — *form* → Myelin sheaths — *secrete* → Neurotrophic factors
- Satellite cells → Support cell bodies

(b) Glial Cells of the Central Nervous System

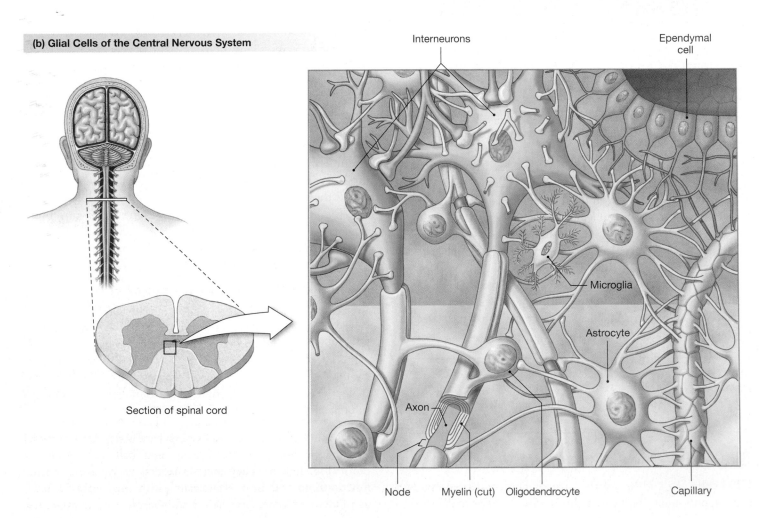

Interneurons

Ependymal cell

Microglia

Astrocyte

Axon

Node   Myelin (cut)   Oligodendrocyte   Capillary

Section of spinal cord

(c) Each Schwann cell forms myelin around a small segment of one axon.

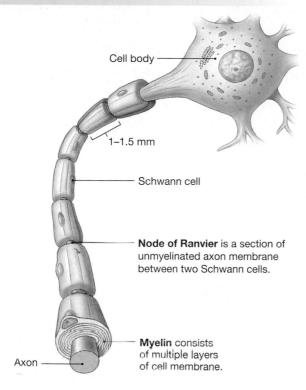

Cell body

1–1.5 mm

Schwann cell

Node of Ranvier is a section of unmyelinated axon membrane between two Schwann cells.

Myelin consists of multiple layers of cell membrane.

Axon

(d) Myelin Formation in the Peripheral Nervous System

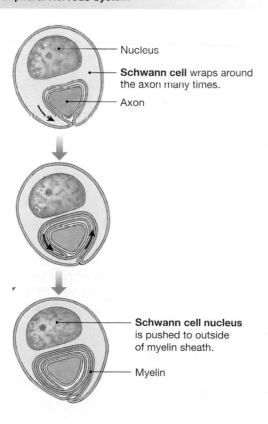

Nucleus

Schwann cell wraps around the axon many times.

Axon

Schwann cell nucleus is pushed to outside of myelin sheath.

Myelin

are closely associated with synapses, where they take up and release chemicals. Astrocytes also provide neurons with substrates for ATP production, and they help maintain homeostasis in the CNS extracellular fluid by taking up K^+ and water. Finally, the ends of some astrocyte processes surround blood vessels and become part of the *blood-brain barrier* that regulates the movement of materials between blood and extracellular fluid.

Microglia The glial cells known as **microglia** are actually not neural tissue. They are specialized immune cells that reside permanently in the CNS (Fig. 8.5a, b). When activated, they remove damaged cells and foreign invaders. However, it now appears that microglia are not always helpful. Activated microglia sometimes release damaging *reactive oxygen species (ROS)* that form free radicals. The *oxidative stress* caused by ROS is believed to contribute to neurodegenerative diseases, such as *amyotrophic lateral sclerosis* (*ALS*, also known as Lou Gehrig's disease).

Ependymal Cells The final class of glial cells is the **ependymal cells,** specialized cells that create a selectively permeable epithelial layer, the *ependyma*, that separates the fluid compartments of the CNS (Fig. 8.5a, b). The ependyma is one source of **neural stem cells** [p. 85], immature cells that can differentiate into neurons and glial cells.

All glial cells communicate with neurons and with one another primarily through chemical signals. Glial-derived growth and *trophic* (nourishing) factors help maintain neurons and guide them during repair and development. Glial cells in turn respond to neurotransmitters and neuromodulators secreted by neurons. Glial cell function is an active area of neuroscience research, and scientists are still exploring the roles these important cells play in the nervous system.

CONCEPT CHECK

5. What is the primary function of each of the following: myelin, microglia, ependymal cells?

6. Name the two glial cell types that form myelin. How do they differ from each other?

Can Stem Cells Repair Damaged Neurons?

Neurons grow when we are young, but what happens when adult neurons are injured? The responses of mature neurons to injury are similar in many ways to the growth of neurons during development. Both processes rely on a combination of chemical and electrical signals.

When a neuron is damaged, if the cell body dies, the entire neuron dies. If the cell body is intact and only the axon is severed, the cell body and attached segment of axon survive (**FIG. 8.6**). The section of axon separated from the cell body usually degenerates slowly and dies because axons lack the cellular organelles to make essential proteins.

What are the cellular events that follow damage to a neuron? First, the axon cytoplasm leaks out at the injury site until membrane is recruited to seal the opening. The segment of axon still attached to the cell body swells as organelles and filaments brought in by axonal transport accumulate. Schwann cells near the injury site send chemical signals to the cell body to tell it that an injury has occurred.

In the distal segment of the axon, synaptic transmission ceases almost immediately. The axon, deprived of its protein source, slowly begins to collapse. The myelin sheath around the distal axon also begins to unravel. Scavenger microglia or phagocytes ingest and clear away the debris. This process may take a month or longer.

If the severed axon belongs to a somatic motor neuron, death of the distal {distant} axon results in permanent paralysis of the skeletal muscles *innervated* by the neuron. (The term *innervated* means "controlled by a neuron.") If the damaged neuron is a sensory neuron, the person may experience loss of sensation (numbness or tingling) in the region previously innervated by the neuron.

Under some conditions, axons in the peripheral nervous system can regenerate and reestablish their synaptic connections. Schwann cells secrete neurotrophic factors that keep the cell body alive and stimulate regrowth of the axon. The growing tip of a regenerating axon behaves much like the growth cone of a developing axon, following chemical signals in the extracellular matrix along its former path until the axon forms a new synapse with its target cell. Sometimes, the loss of the distal axon is permanent, however, and the pathway is destroyed.

Regeneration of axons in the central nervous system is less likely to occur naturally. CNS glial cells tend to seal off and scar the damaged region, and damaged CNS cells secrete factors that inhibit axon regrowth. Many scientists are studying the mechanisms of axon growth and inhibition in the hopes of finding treatments that can restore function to victims of spinal cord injury and degenerative neurological disorders.

Scientists once believed that if a neuron died, it could never be replaced. The discovery of neural stem cells changed that view. During early development, an undifferentiated cell layer called *neuroepithelium* lines the lumen of the neural tube, a structure that will later become the brain and spinal cord. As development proceeds, some cells migrate out of the neuroepithelium and differentiate into neurons. However, some neural stem cells remain unspecialized, waiting until they are called upon to replace damaged cells. Neural stem cells seem to be most concentrated in a few specific areas of the brain (hippocampus and lateral ventricle walls).

When neural stem cells receive the correct signals, they transform into neurons and glial cells. Scientists are working intensely to learn how to control this transformation, in the hope that stem cell transplants can reverse the loss of function that comes with injury and degenerative neurological diseases.

ELECTRICAL SIGNALS IN NEURONS

Nerve and muscle cells are described as *excitable tissues* because of their ability to propagate electrical signals rapidly in response to a stimulus. We now know that many other cell types generate electrical signals to initiate intracellular processes [see insulin secretion, p. 159], but the ability of nerve and muscle cells to send a constant electrical signal over long distance is characteristic of electrical signaling in these tissues.

The Nernst Equation Predicts Membrane Potential for a Single Ion

Recall that all living cells have a resting membrane potential difference (V_m) [p. 154] that represents the separation of electrical charge across the cell membrane. Two factors influence the membrane potential:

1. *The uneven distribution of ions* across the cell membrane. Normally, sodium (Na^+), chloride (Cl^-), and calcium (Ca^{2+}) are more concentrated in the extracellular fluid than in the cytosol. Potassium (K^+) is more concentrated in the cytosol than in the extracellular fluid.
2. *Differing membrane permeability to those ions.* The resting cell membrane is much more permeable to K^+ than to Na^+ or Ca^{2+}. This makes K^+ the major ion contributing to the resting membrane potential.

The *Nernst equation* describes the membrane potential that would result if the membrane were permeable to only one ion [p. 156]. For any given ion concentration gradient, this membrane potential is called the *equilibrium potential* of the ion (E_{ion}):

$$E_{ion \text{ (in mV)}} = \frac{61}{z} \log \frac{[ion]_{out}}{[ion]_{in}}$$

where:

61 is 2.303 RT/F at 37 °C

z is the electrical charge on the ion (+1 for K^+), and

$[ion]_{out}$ and $[ion]_{in}$ are the ion concentrations outside and inside the cell.

(R is the ideal gas constant, T is absolute temperature, and F is the Faraday constant. For additional information on these values, see Appendix B.)

When we use the estimated intracellular and extracellular concentrations for K^+ (**TBL. 8.2**) in the Nernst equation, the equation predicts a potassium equilibrium potential, or E_K of –90 mV. However, an average value for the resting membrane potential of neurons is –70 mV (inside the cell relative to outside), more positive than predicted by the potassium equilibrium potential. This means that other ions must be contributing to the membrane potential. Neurons at rest are slightly permeable to Na^+, and the leak of positive Na^+ into the cell makes the resting membrane potential slightly more positive than it would be if the cell were permeable only to K^+.

FIG. 8.6 Peripheral neuron injury

When an axon is cut, the section attached to the cell body continues to live.

The section of the axon distal to the cut begins to disintegrate.

Site of injury

Connective tissue   Myelin

Under some circumstances, the proximal axon may regrow through the existing sheath of Schwann cells and reform a synapse with the proper target.

Disintegrating distal axon

Proximal axon

CONCEPT CHECK

7. Given the values in Table 8.2, use the Nernst equation to calculate the equilibrium potential for Ca^{2+}. Express the concentrations as powers of 10 and use your knowledge of logarithms [p. A-38] to try the calculations without a calculator.

The GHK Equation Predicts Membrane Potential Using Multiple Ions

In living systems, several different ions contribute to the membrane potential of cells. The **Goldman-Hodgkin-Katz (GHK) equation** calculates the membrane potential that results from the contribution of all ions that can cross the membrane. The GHK equation includes membrane permeability values because the permeability of an ion influences its contribution to the membrane potential. If the membrane is not permeable to a particular ion, that ion does not affect the membrane potential.

For mammalian cells, we assume that Na^+, K^+, and Cl^- are the three ions that influence membrane potential in resting cells. Each ion's contribution to the membrane potential is proportional

to its ability to cross the membrane. The GHK equation for cells that are permeable to Na^+, K^+, and Cl^- is

$$V_m = 61 \log \frac{P_k [K^+]_{out} + P_{Na} [Na^+]_{out} + P_{Cl} [Cl^-]_{in}}{P_K [K^+]_{in} + P_{Na} [Na^+]_{in} + P_{Cl} [Cl^-]_{out}}$$

where:

V_m is the resting membrane potential in mV at 37 °C

61 is 2.303 RT/F at 37 °C

P is the relative permeability of the membrane to the ion shown in the subscript, and

$[ion]_{out}$ and $[ion]_{in}$ are the ion concentrations outside and inside the cell.

Although this equation looks quite intimidating, it can be simplified into words to say: Resting membrane potential (V_m) is determined by the combined contributions of the (concentration gradient × membrane permeability) for each ion.

TABLE 8.2	Ion Concentrations and Equilibrium Potentials		
Ion	**Extracellular Fluid (mM)**	**Intracellular Fluid (mM)**	**E_{ion} at 37 °C**
K^+	5 mM (normal: 3.5–5)	150 mM	–90 mV
Na^+	145 mM (normal: 135–145)	15 mM	+60 mV
Cl^-	108 mM (normal: 100–108)	10 mM (normal: 5–15)	–63 mV
Ca^{2+}	1 mM	0.0001 mM	See Concept Check question 7.

If the membrane is not permeable to an ion, the permeability value for that ion is zero, and the ion drops out of the equation. For example, cells at rest normally are not permeable to Ca^{2+} and, therefore, Ca^{2+} is not part of the GHK equation.

The GHK equation predicts resting membrane potentials based on given ion concentrations and membrane permeabilities. Notice that if permeabilities for Na^+ and Cl^- are zero, the equation reverts back to the Nernst equation for K^+. The GHK equation explains how the cell's slight permeability to Na^+ makes the resting membrane potential more positive than the E_K determined with the Nernst equation. The GHK equation can also be used to predict what happens to membrane potential when ion concentrations or membrane permeabilities change.

Ion Movement Creates Electrical Signals

The resting membrane potential of living cells is determined primarily by the K^+ concentration gradient and the cell's resting permeability to K^+, Na^+, and Cl^-. A change in either the K^+ concentration gradient or ion permeabilities changes the membrane potential. If you know numerical values for ion concentrations and permeabilities, you can use the GHK equation to calculate the new membrane potential.

In medicine, you usually will not have numerical values, however, so it is important to be able to think conceptually about the relationship between ion concentrations, permeabilities, and membrane potential. For example, at rest, the cell membrane of a neuron is only slightly permeable to Na^+. If the membrane suddenly increases its Na^+ permeability, Na^+ enters the cell, moving down its electrochemical gradient [p. 156]. The addition of positive Na^+ to the intracellular fluid *depolarizes* the cell membrane and creates an electrical signal.

Graph of membrane potential changes

The movement of ions across the membrane can also *hyperpolarize* a cell. If the cell membrane suddenly becomes more permeable to K^+, positive charge is lost from inside the cell, and the cell becomes more negative (hyperpolarizes). A cell may also hyperpolarize if negatively charged ions, such as Cl^-, enter the cell from the extracellular fluid.

CONCEPT CHECK

8. Would a cell with a resting membrane potential of –70 mV depolarize or hyperpolarize in the following cases? (You must consider both the concentration gradient and the electrical gradient of the ion to determine net ion movement.)
   (a) Cell becomes more permeable to Ca^{2+}.
   (b) Cell becomes less permeable to K^+.
9. Would the cell membrane depolarize or hyperpolarize if a small amount of Na^+ leaked into the cell?

It is important to understand that a change in membrane potential from –70 mV to a positive value, such as +30 mV *does not mean that the ion concentration gradients have reversed!* A significant change in membrane potential occurs with the movement of very few ions. For example, to change the membrane potential by 100 mV, only 1 of every 100,000 K^+ must enter or leave the cell. This is such a tiny fraction of the total number of K^+ in the cell that the intracellular concentration of K^+ remains essentially unchanged even though the membrane potential has changed by 100 mV.

To appreciate how a tiny change can have a large effect, think of getting one grain of beach sand into your eye. There are so many grains of sand on the beach that the loss of one grain is not significant, just as the movement of one K^+ across the cell membrane does not significantly alter the concentration of K^+. However, the electrical signal created by moving a few K^+ across the membrane has a significant effect on the cell's membrane potential, just as getting that one grain of sand in your eye creates significant discomfort.

Gated Channels Control the Ion Permeability of the Neuron

How does a cell change its ion permeability? The simplest way is to open or close existing channels in the membrane. Neurons contain a variety of gated ion channels that alternate between open and closed states, depending on the intracellular and extracellular conditions [p. 139]. A slower method for changing membrane permeability is for the cell to insert new channels into the membrane or remove some existing channels.

Ion channels are usually named according to the primary ion(s) they allow to pass through them. There are four major types of selective ion channels in the neuron: (1) Na^+ channels, (2) K^+ channels, (3) Ca^{2+} channels, and (4) Cl^- channels. Other channels are less selective, such as the *monovalent cation channels* that allow both Na^+ and K^+ to pass.

The ease with which ions flow through a channel is called the channel's **conductance** (G) {*conductus*, escort}. Channel conductance varies with the gating state of the channel and with the channel protein isoform. Some ion channels, such as the K^+ *leak channels* that are the major determinant of resting membrane potential, spend most of their time in an open state. Other channels have gates that open or close in response to particular stimuli. Most gated channels fall into one of three categories [p. 139]:

1. **Mechanically gated ion channels** are found in sensory neurons and open in response to physical forces such as pressure or stretch.
2. **Chemically gated ion channels** in most neurons respond to a variety of ligands, such as extracellular neurotransmitters and neuromodulators or intracellular signal molecules.
3. **Voltage-gated ion channels** respond to changes in the cell's membrane potential. Voltage-gated Na^+ and K^+ channels play an important role in the initiation and conduction of electrical signals along the axon.

Not all voltage-gated channels behave in exactly the same way. The voltage for channel opening varies from one channel type to another. For example, some channels we think of as leak channels are actually voltage-gated channels that remain open in the voltage range of the resting membrane potential.

The speed with which a gated channel opens and closes also differs among different types of channels. Channel opening to allow ion flow is called channel *activation*. For example, Na^+ channels and K^+ channels of axons are both activated by cell depolarization. The Na^+ channels open very rapidly, but the K^+ channels are slower to open. The result is an initial flow of Na^+ across the membrane, followed later by K^+ flow.

Many channels that open in response to depolarization close only when the cell repolarizes. The gating portion of the channel protein has an electrical charge that moves the gate between open and closed positions as membrane potential changes. This is like a spring-loaded door: It opens when you push on it, then closes when you release it.

Some channels also spontaneously *inactivate*. Even though the activating stimulus that opened them continues, the channel "times out" and closes. This is similar to doors with an automatic timed open-close mechanism. The door opens when you hit the button, then after a certain period of time, it closes itself, whether you are still standing in the doorway or not. An inactivated channel returns to its normal closed state shortly after the membrane repolarizes. The specific mechanisms underlying channel inactivation vary with different channel types.

Each major channel type has several to many subtypes with varying properties, and the list of subtypes gets longer each year. Within each subtype there may be multiple isoforms that express different opening and closing *kinetics* {*kinetikos*, moving} as well as associated proteins that modify channel properties. In addition, channel activity can be modulated by chemical factors that bind to the channel protein, such as phosphate groups.

Current Flow Obeys Ohm's Law

When ion channels open, ions may move into or out of the cell. The flow of electrical charge carried by an ion is called the ion's **current,** abbreviated I_{ion}. The direction of ion movement depends on the *electrochemical* (combined electrical and concentration) gradient of the ion. Potassium ions usually move out of the

CLINICAL FOCUS

Mutant Channels

Ion channels are proteins, and like other proteins they may lose or change function if their amino acid sequence is altered. **Channelopathies** {*pathos,* suffering} are inherited diseases caused by mutations in ion channel proteins. The most common channelopathy is cystic fibrosis, which results from defects in chloride channel function [see Chapter 5 Running Problem]. Because ion channels are so closely linked to the electrical activity of cells, many channelopathies manifest themselves as disorders of the excitable tissues (nerve and muscle). By studying defective ion channels, scientists have now shown that some disease states are actually families of related diseases with different causes but similar symptoms. For example, the condition known as *long Q-T syndrome* (LQTS; named for changes in the electrocardiogram test) is a cardiac problem characterized by an irregular heartbeat {*arrhythmia*; a-, without}, fainting, and sometimes sudden death. Scientists have identified eight different gene mutations in K^+, Na^+, or Ca^{2+} channels that result in various subtypes of LQTS. Other well-known channelopathies include some forms of epilepsy and malignant hyperthermia.

cell. Na^+, Cl^-, and Ca^{2+} usually flow into the cell. The net flow of ions across the membrane depolarizes or hyperpolarizes the cell, creating an electrical signal.

Current flow, whether across a membrane or inside a cell, obeys a rule known as **Ohm's Law**. Ohm's Law says that current flow (I) is directly proportional to the electrical potential difference (in volts, V) between two points and inversely proportional to the resistance (R) of the system to current flow: $I = V \times 1/R$ or $I = V/R$. In other words, as resistance R increases, current flow I decreases. (You will encounter a variant of Ohm's Law when you study fluid flow in the cardiovascular and respiratory systems.)

Resistance in biological flow is the same as resistance in everyday life: It is a force that opposes flow. Electricity is a form of energy and, like other forms of energy it dissipates as it encounters resistance. As an analogy, think of rolling a ball along the floor. A ball rolled across a smooth wood floor encounters less resistance than a ball rolled across a carpeted floor. If you throw both balls with the same amount of energy, the ball that encounters less resistance retains energy longer and travels farther along the floor.

In biological electricity, resistance to current flow comes from two sources: the resistance of the cell membrane (R_m) and the internal resistance of the cytoplasm (R_i). The phospholipid bilayer of the cell membrane is normally an excellent insulator, and a membrane with no open ion channels has very high resistance and low conductance. If ion channels open, ions (current) flow across the membrane if there is an electrochemical gradient for them. Opening ion channels therefore decreases the membrane resistance.

The internal resistance of most neurons is determined by the composition of the cytoplasm and the diameter of the cell. Cytoplasmic composition is relatively constant. Internal resistance decreases as cell diameter increases. The membrane resistance and internal resistance together determine how far current will flow through a cell before the energy is dissipated and the current dies. The combination of the two resistances is called the *length constant* for a given neuron.

Voltage changes across the membrane can be classified into two basic types of electrical signals: graded potentials and action potentials (**TBL. 8.3**). **Graded potentials** are variable-strength signals that travel over short distances and lose strength as they travel through the cell. They are used for short-distance communication. If a depolarizing graded potential is strong enough when it reaches an integrating region within a neuron, the graded potential initiates an action potential. **Action potentials** are very brief, large depolarizations that travel for long distances through a neuron without losing strength. Their function is rapid signaling over long distances, such as from your toe to your brain.

Graded Potentials Reflect Stimulus Strength

Graded potentials in neurons are depolarizations or hyperpolarizations that occur in the dendrites and cell body or, less frequently, near the axon terminals. These changes in membrane potential are called "graded" because their size, or *amplitude* {*amplitudo*, large}, is directly proportional to the strength of the triggering event. A large stimulus causes a strong graded potential, and a small stimulus results in a weak graded potential.

In neurons of the CNS and the efferent division, graded potentials occur when chemical signals from other neurons open chemically gated ion channels, allowing ions to enter or leave the neuron. Mechanical stimuli (such as stretch) or chemical stimuli open ion channels in some sensory neurons. Graded potentials may also occur when an open channel closes, decreasing the movement of ions through the cell membrane. For example, if K^+ leak channels close, fewer K^+ leave the cell. The retention of K^+ depolarizes the cell.

CONCEPT CHECK

10. Match each ion's movement with the type of graded potential it creates.

   (a) Na^+ entry   1. depolarizing

   (b) Cl^- entry   2. hyperpolarizing

   (c) K^+ exit

   (d) Ca^{2+} entry

FIGURE 8.7a shows a graded potential that begins when a stimulus opens monovalent cation channels on the cell body of a neuron. Sodium ions move into the neuron, bringing in electrical energy. The positive charge carried in by the Na^+ spreads as a wave of depolarization through the cytoplasm, just as a stone thrown into water creates ripples or waves that spread outward from the point of entry. The wave of depolarization that moves through the cell is known as **local current flow**. By convention, current in biological systems is the net movement of *positive* electrical charge.

The strength of the initial depolarization in a graded potential is determined by how much charge enters the cell, just as the

---

TABLE 8.3	Comparison of Graded Potential and Action Potential in Neurons	
	Graded Potential	**Action Potential**
Type of Signal	Input signal	Regenerating conduction signal
Occurs Where?	Usually dendrites and cell body	Trigger zone through axon
Types of Gated Ion Channels Involved	Mechanically, chemically, or voltage-gated channels	Voltage-gated channels
Ions Involved	Usually Na^+, K^+, Ca^{2+}	Na^+ and K^+
Type of Signal	Depolarizing (e.g., Na^+) or hyperpolarizing (e.g., Cl^-)	Depolarizing
Strength of Signal	Depends on initial stimulus; can be summed	All-or-none phenomenon; cannot be summed
What Initiates the Signal?	Entry of ions through gated channels	Above-threshold graded potential at the trigger zone opens ion channels
Unique Characteristics	No minimum level required to initiate	Threshold stimulus required to initiate
	Two signals coming close together in time will sum	Refractory period: two signals too close together in time cannot sum
	Initial stimulus strength is indicated by frequency of a series of action potentials	

FIG. 8.7 **ESSENTIALS**

Graded Potentials

(a) Graded potentials decrease in strength as they spread out from the point of origin.

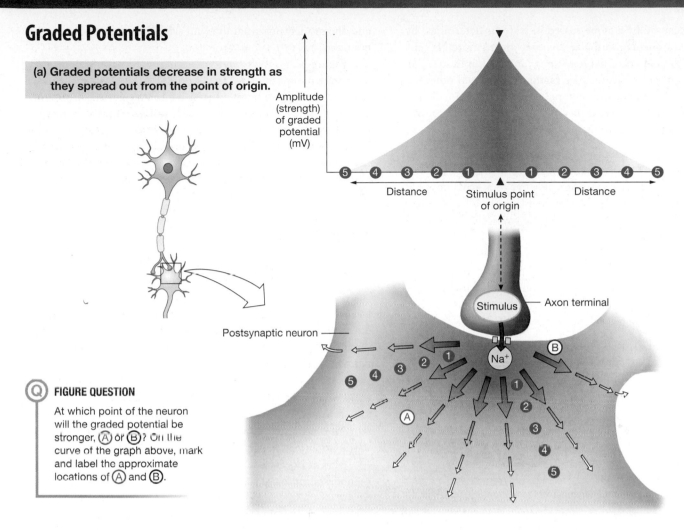

Amplitude (strength) of graded potential (mV)

Distance — Stimulus point of origin — Distance

Postsynaptic neuron

Stimulus — Axon terminal

Na⁺

Q **FIGURE QUESTION**

At which point of the neuron will the graded potential be stronger, Ⓐ or Ⓑ? On the curve of the graph above, mark and label the approximate locations of Ⓐ and Ⓑ.

(b) Subthreshold Graded Potential

A graded potential starts above threshold (T) at its initiation point but decreases in strength as it travels through the cell body. At the trigger zone, it is below threshold and, therefore, does not initiate an action potential.

Stimulus — Synaptic terminal

Cell body

Trigger zone

Axon — No action potential

−40
−55
−70
mV | Stimulus
Time

−40
−55
−70
mV
Time

−40
−55
−70
mV
Graded potential below threshold
Time

(c) Suprathreshold Graded Potential

A stronger stimulus at the same point on the cell body creates a graded potential that is still above threshold by the time it reaches the trigger zone, so an action potential results.

Stimulus

Trigger zone

Action potential

−40
−55
−70
mV | Stimulus
Time — T

−40
−55
−70
mV
Time — T

−40
−55
−70
mV
Graded potential above threshold
Time — T

size of waves caused by a stone tossed in water is determined by the size of the stone. If more Na^+ channels open, more Na^+ enters, and the graded potential has higher initial amplitude. The stronger the initial amplitude, the farther the graded potential can spread through the neuron before it dies out.

Why do graded potentials lose strength as they move through the cytoplasm? Two factors play a role:

1. *Current leak.* The membrane of the neuron cell body has open leak channels that allow positive charge to leak out into the extracellular fluid. Some positive ions leak out of the cell across the membrane as the depolarization wave moves through the cytoplasm, decreasing the strength of the signal moving down the cell.

2. *Cytoplasmic resistance.* The cytoplasm provides resistance to the flow of electricity, just as water creates resistance that diminishes the waves from the stone. The combination of current leak and cytoplasmic resistance means that the strength of the signal inside the cell decreases over distance.

Graded potentials that are strong enough eventually reach the region of the neuron known as the **trigger zone.** In efferent neurons and interneurons, the trigger zone is the *axon hillock* and the very first part of the axon, a region known as the **initial segment.** In sensory neurons, the trigger zone is immediately adjacent to the receptor, where the dendrites join the axon (see Fig. 8.2).

CONCEPT CHECK   **11.** Identify the trigger zones of the neurons illustrated in Figure 8.2, if possible.

The trigger zone is the integrating center of the neuron and contains a high concentration of voltage-gated Na^+ channels in its membrane. If graded potentials reaching the trigger zone depolarize the membrane to the threshold voltage, voltage-gated Na^+ channels open, and an action potential begins. If the depolarization does not reach threshold, the graded potential simply dies out as it moves into the axon.

Because depolarization makes a neuron more likely to fire an action potential, depolarizing graded potentials are considered to be *excitatory.* A hyperpolarizing graded potential moves the membrane potential farther from the threshold value and makes the neuron less likely to fire an action potential. Consequently, hyperpolarizing graded potentials are considered to be *inhibitory.*

Figure 8.7b shows a neuron with three recording electrodes placed at intervals along the cell body and trigger zone. A single stimulus triggers a *subthreshold* graded potential, one that is below threshold by the time it reaches the trigger zone. Although the cell is depolarized to –40 mV at the site where the graded potential begins, the current decreases as it travels through the cell body. As a result, the graded potential is below threshold by the time it reaches the trigger zone. (For the typical mammalian neuron, threshold is about –55 mV.) The stimulus is not strong enough to depolarize the cell to threshold at the trigger zone,

and the graded potential dies out without triggering an action potential.

Figure 8.7c shows *suprathreshold* graded potential, one that is strong enough to cause an action potential. A stronger initial stimulus on the cell body initiates a stronger depolarization and current flow. Although this graded potential also diminishes with distance as it travels through the neuron, its higher initial strength ensures that it is above threshold at the trigger zone. In this example, the graded potential triggers an action potential. The ability of a neuron to respond to a stimulus and fire an action potential is called the cell's **excitability.**

Action Potentials Travel Long Distances

Action potentials, also known as *spikes,* are electrical signals of uniform strength that travel from a neuron's trigger zone to the end of its axon. In action potentials, voltage-gated ion channels in the axon membrane open sequentially as electrical current passes down the axon. As a result, additional Na^+ entering the cell reinforce the depolarization, which is why an action potential does not lose strength over distance the way a graded potential does. Instead, the action potential at the end of an axon is identical to the action potential that started at the trigger zone: a depolarization of about 100 mV amplitude. The high-speed movement of an action potential along the axon is called **conduction** of the action potential.

Action potentials are sometimes called **all-or-none** phenomena because they either occur as a maximal depolarization (if the stimulus reaches threshold) or do not occur at all (if the stimulus is below threshold). The strength of the graded potential that initiates an action potential has no influence on the amplitude of the action potential.

When we talk about action potentials, it is important to realize that there is no single action potential that moves through the cell. The action potential that occurs at the trigger zone is like the movement in the first domino of a series of dominos standing on end (**FIG. 8.8a**). As the first domino falls, it strikes the next, passing on its kinetic energy. As the second domino falls, it passes kinetic energy to the third domino, and so on. If you could take a snapshot of the line of falling dominos, you would see that as the first domino is coming to rest in the fallen position, the next one is almost down, the third one most of the way down, and so forth, until you reach the domino that has just been hit and is starting to fall.

In an action potential, a wave of electrical energy moves down the axon. Instead of getting weaker over distance, action potentials are replenished along the way so that they maintain constant amplitude. As the action potential passes from one part of the axon to the next, the membrane's energy state is reflected in the membrane potential of each region. If we were to insert a series of recording electrodes along the length of an axon and start an action potential at the trigger zone, we would see a series of overlapping action potentials, each in a different part of the waveform, just like the dominos that are frozen in different positions (Fig. 8.8b).

FIG. 8.8 Conduction of an action potential

(a) The conduction of an action potential down an axon is similar to energy passed along a series of falling dominos. In this snapshot, each domino is in a different phase of falling. In the axon, each section of membrane is in a different phase of the action potential.

(b) A wave of electrical current passes down the axon.

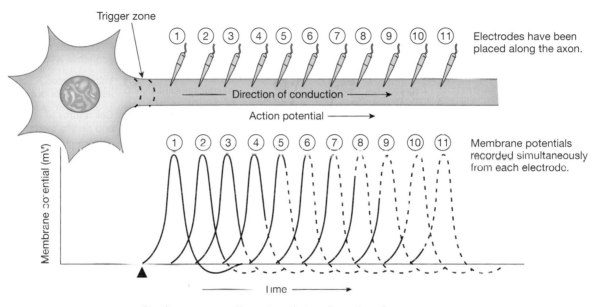

Simultaneous recordings show that each section of axon is experiencing a different phase of the action potential.

CONCEPT CHECK **12.** What is the difference between conductance and conduction in neurons?

Na⁺ and K⁺ Move across the Membrane during Action Potentials

What is happening to the axon membrane when an action potential takes place? As you saw in Figure 8.8b, a suprathreshold (above-threshold) stimulus at the trigger zone initiates the action potential. Conduction of the action potential along the axon requires only a few types of ion channels: voltage-gated Na^+ channels and voltage-gated K^+ channels, plus some leak channels that help set the resting membrane potential. The explanation of action potential generation that follows is typical of an unmyelinated PNS neuron. For their description of this simple but elegant mechanism, A. L. Hodgkin and A. F. Huxley won a Nobel Prize in 1963.

Action potentials begin when voltage-gated ion channels open, altering membrane permeability (P) to Na^+ (P_{Na}) and K^+ (P_K).

FIGURE 8.9 shows the voltage and ion permeability changes that take place in one section of membrane during an action potential. Before and after the action potential, at **1** and **2**, the neuron is at its resting membrane potential of –70 mV. The action potential itself can be divided into three phases: a rising phase, a falling phase, and the after-hyperpolarization phase.

Rising Phase of the Action Potential The rising phase is due to a sudden temporary increase in the cell's permeability to Na^+. An action potential begins when a graded potential reaching the trigger zone depolarizes the membrane to threshold (–55 mV) **3**. As the cell depolarizes, voltage-gated Na^+ channels open, making the membrane much more permeable to Na^+. Na^+ then flows into the cell, down its concentration gradient and attracted by the negative membrane potential inside the cell.

The addition of positive charge to the intracellular fluid further depolarizes the cell (shown by the steep rising phase on the graph **4**). In the top third of the rising phase, the inside of the cell has become more positive than the outside, and the membrane potential has reversed polarity. This reversal is represented

FIG. 8.16   Saltatory conduction

(a) Action potentials appear to jump from one node of Ranvier to the next. Only the nodes have voltage-gated Na⁺ channels.

(b) Demyelinating diseases reduce or block conduction when current leaks out of the previously insulated regions between the nodes.

K⁺ levels and the conduction of action potentials is the most straightforward and easiest to understand, as well as one of the most clinically significant.

The concentration of K⁺ in the blood and interstitial fluid is the major determinant of the resting potential of all cells [p. 158]. If K⁺ concentration in the blood moves out of the normal range of 3.5–5 mmol/L, the result is a change in the resting membrane potential of cells (**FIG. 8.17**). This change is not important to most cells, but it can have serious consequences to the body as a whole because of the relationship between resting potential and the excitability of nervous and muscle tissue.

At normal K⁺ levels, subthreshold graded potentials do not trigger action potentials, and suprathreshold graded potentials do (Fig. 8.17a, b). An increase in blood K⁺ concentration—**hyperkalemia** {*hyper-*, above + *kalium*, potassium + *-emia*, in the blood}—shifts the resting membrane potential of a neuron closer to threshold and causes the cells to fire action potentials in response to smaller graded potentials (Fig. 8.17c).

If blood K⁺ concentration falls too low—a condition known as **hypokalemia**—the resting membrane potential of the cells hyperpolarizes, moving farther from threshold. In this case, a stimulus strong enough to trigger an action potential when the resting potential is the normal –70 mV does not reach the threshold value (Fig. 8.17d). This condition shows up as muscle weakness because the neurons that control skeletal muscles are not firing normally.

Hypokalemia and its resultant muscle weakness are one reason that sport drinks supplemented with Na⁺ and K⁺ were developed. When people sweat excessively, they lose both salts and water. If they replace this fluid loss with pure water, the K⁺ remaining in the blood is diluted, causing hypokalemia.

FIG. 8.17 Potassium and cell excitability

Potassium is the ion primarily responsible for the resting membrane potential.

FIGURE QUESTION

The E_K of −90 mV is based on ECF [K$^+$] = 5 mM and ICF [K$^+$] = 150 mM. Use the Nernst equation to calculate E_K when ECF [K$^+$] is (a) 2.5 mM and (b) 6 mM.

Normal plasma [K$^+$] is 3.5–5 mM.

(a) When blood K$^+$ is in the normal range (normokalemia), a subthreshold graded potential does not fire an action potential.

(b) In normokalemia, a suprathreshold (above threshold) stimulus will fire an action potential.

(c) Hyperkalemia, increased blood K$^+$ concentration, brings the membrane closer to the threshold. Now a stimulus that would normally be subthreshold can trigger an action potential.

Hyperkalemia depolarizes cells.

Hypokalemia hyperpolarizes cells.

(d) Hypokalemia, decreased blood K$^+$ concentration, hyperpolarizes the membrane and makes the neuron less likely to fire an action potential in response to a stimulus that would normally be above the threshold.

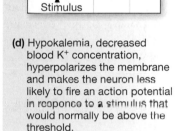

By replacing sweat loss with a dilute salt solution, a person can prevent potentially dangerous drops in blood K$^+$ levels. Because of the importance of K$^+$ to normal function of the nervous system, potassium homeostasis mechanisms keep blood K$^+$ concentrations within a narrow range.

CELL-TO-CELL COMMUNICATION IN THE NERVOUS SYSTEM

Information flow through the nervous system using electrical and chemical signals is one of the most active areas of neuroscience research today because so many devastating diseases affect this process. The specificity of neural communication depends on

RUNNING PROBLEM

Dr. McKhann decided to perform nerve conduction tests on some of the paralyzed children in Beijing Hospital. He found that although the rate of conduction along the children's nerves was normal, the strength of the summed action potentials traveling down the nerve was greatly diminished.

Q4: *Is the paralytic illness that affected the Chinese children a demyelinating condition? Why or why not?*

227 229 231 251 **253** 257 267 268

several factors: the signal molecules secreted by neurons, the target cell receptors for these chemicals, and the anatomical connections between neurons and their targets, which occur in regions known as synapses.

Neurons Communicate at Synapses

Each synapse has two parts: (1) the axon terminal of the *presynaptic cell* and (2) the membrane of the *postsynaptic cell* (Fig. 8.2f). In a neural reflex, information moves from presynaptic cell to postsynaptic cell. The postsynaptic cells may be neurons or nonneuronal cells. In most neuron-to-neuron synapses, the presynaptic axon terminals are next to either the dendrites or the cell body of the postsynaptic neuron.

In general, postsynaptic neurons with many dendrites also have many synapses. A moderate number of synapses is 10,000, but some cells in the brain are estimated to have 150,000 or more synapses on their dendrites! Synapses can also occur on the axon and even at the axon terminal of the postsynaptic cell.

Synapses are classified as electrical or chemical depending on the type of signal that passes from the presynaptic cell to the postsynaptic one.

Electrical Synapses **Electrical synapses** pass an electrical signal, or current, directly from the cytoplasm of one cell to another through the pores of gap junction proteins. Information can flow in both directions through most gap junctions, but in some current can flow in only one direction (a *rectifying synapse*).

Electrical synapses occur mainly in neurons of the CNS. They are also found in glial cells, in cardiac and smooth muscle, and in nonexcitable cells that use electrical signals, such as the pancreatic beta cell. The primary advantage of electrical synapses is rapid and bidirectional conduction of signals from cell to cell to synchronize activity within a network of cells. Gap junctions also allow chemical signal molecules to diffuse between adjacent cells.

Chemical Synapses The vast majority of synapses in the nervous system are **chemical synapses,** which use neurocrine molecules to carry information from one cell to the next. At chemical synapses, the electrical signal of the presynaptic cell is converted into a neurocrine signal that crosses the synaptic cleft and binds to a receptor on its target cell.

Neurons Secrete Chemical Signals

The number of molecules identified as neurocrine signals is large and growing daily. Neurocrine chemical composition is varied, and these molecules may function as neurotransmitters, neuromodulators, or neurohormones [p. 168]. Neurotransmitters and neuromodulators act as *paracrine signals,* with target cells located close to the neuron that secretes them. Neurohormones, in contrast, are secreted into the blood and distributed throughout the body.

The distinction between neurotransmitter and neuromodulator depends on the receptor to which the chemical is binding, as many neurocrine molecules can act in both roles. Generally, if a molecule primarily acts at a synapse and elicits a rapid response, we call it a neurotransmitter, even if it can also act as a neuromodulator. Neuromodulators act at both synaptic and nonsynaptic sites and are slower acting. Some neuromodulators and neurotransmitters also act on the cell that secretes them, making them *autocrine* signals as well as paracrine signals.

Neurocrine Receptors The neurocrine receptors found in chemical synapses can be divided into two categories: receptorchannels, which are ligand-gated ion channels, and G proteincoupled receptors (GPCR) [p. 174]. Receptor-channels mediate rapid responses by altering ion flow across the membrane, so they are also called **ionotropic receptors.** Some ionotropic receptors are specific for a single ion, such as Cl^-, but others are less specific, such as the *nonspecific monovalent cation channel.*

G protein-coupled receptors mediate slower responses because the signal must be transduced through a second messenger system. GPCRs for neuromodulators are described as **metabotropic receptors.** Some metabotropic GPCRs regulate the opening or closing of ion channels.

All neurotransmitters except nitric oxide bind to specific receptor types. Each receptor type may have multiple subtypes, allowing one neurotransmitter to have different effects in different tissues. Receptor subtypes are distinguished by combinations of letter and number subscripts. For example, serotonin (5-HT) has at least 20 receptor subtypes that have been identified, including 5-HT_{1A} and 5-HT_4.

The study of neurotransmitters and their receptors has been greatly simplified by two advances in molecular biology. The genes for many receptor subtypes have been cloned, allowing researchers to create mutant receptors and study their properties. In addition, researchers have discovered or synthesized a variety of agonist and antagonist molecules [p. 49] that mimic or inhibit neurotransmitter activity by binding to the receptors (**TBL. 8.4**).

Neurotransmitters Are Highly Varied

The array of neurocrine molecules in the body and their many receptor types is truly staggering (Tbl. 8.4). Neurocrine molecules can be informally grouped into seven classes according to their structure: (1) acetylcholine, (2) amines, (3) amino acids, (4) peptides, (5) purines, (6) gases, and (7) lipids. CNS neurons release many different chemical signals, including some polypeptides known mostly for their hormonal activity, such as hypothalamic releasing hormones, oxytocin, and vasopressin [p. 209]. In contrast, the PNS secretes only three major neurocrine molecules: the neurotransmitters acetylcholine and norepinephrine, and the neurohormone epinephrine. Some PNS neurons co-secrete additional molecules, such as ATP, which we will mention when they are functionally important.

Acetylcholine Acetylcholine (ACh), in a chemical class by itself, is synthesized from choline and acetyl coenzyme A (acetyl CoA). Choline is a small molecule also found in membrane phospholipids. Acetyl CoA is the metabolic intermediate that links glycolysis to the citric acid cycle [p. 108]. The synthesis of ACh from these two precursors is a simple enzymatic reaction that takes place in the axon terminal. Neurons that secrete ACh and receptors that bind ACh are described as **cholinergic.**

Cholinergic receptors come in two main subtypes: **nicotinic,** named because *nicotine* is an agonist, and **muscarinic,** for which *muscarine*, a compound found in some fungi, is an agonist. Cholinergic nicotinic receptors are receptor-channels found on skeletal muscle, in the autonomic division of the PNS, and in the CNS. Nicotinic receptors are monovalent cation channels through which both Na^+ and K^+ can pass. Sodium entry into cells exceeds K^+ exit because the electrochemical gradient for Na^+ is stronger. As a result, net Na^+ entry depolarizes the postsynaptic cell and makes it more likely to fire an action potential.

Cholinergic muscarinic receptors come in five related subtypes. They are all G protein-coupled receptors linked to second messenger systems. The tissue response to activation of a muscarinic receptor varies with the receptor subtype. These receptors occur in the CNS and on targets of the autonomic parasympathetic division of the PNS.

Amines The amine neurotransmitters are all active in the CNS. Like the amine hormones [p. 206], these neurotransmitters are derived from single amino acids. **Serotonin,** also called *5-hydroxytryptamine* or 5-HT, is made from the amino acid tryptophan. *Histamine*, made from histidine, plays a role in allergic responses in addition to serving as a neurotransmitter.

| TABLE 8.4 | Major Neurocrines* | | | |

Chemical	Receptor	Type	Receptor Location	Key Agonists, Antagonists, and Potentiators**
Acetylcholine (ACh)	Cholinergic			
	Nicotinic (nAChR)	ICR‡ (Na^+, K^+)	Skeletal muscles, autonomic neurons, CNS	**Agonist:** nicotine **Antagonists:** curare, α-bungarotoxin
	Muscarinic (M)	GPCR	Smooth and cardiac muscle, endocrine and exocrine glands, CNS	**Agonist:** muscarine **Antagonist:** atropine
Amines				
Norepinephrine (NE) Epinephrine (E)	Adrenergic (α, β)	GPCR	Smooth and cardiac muscle, glands, CNS	**Antagonists:** α-receptors: ergotamine, phentolamine. β-receptors: propranolol
Dopamine (DA)	Dopamine (D)	GPCR	CNS	**Agonist:** bromocriptine **Antagonists:** antipsychotic drugs
Serotonin (5-hydroxytryptamine, 5-HT)	Serotonergic (5-HT)	ICR (Na^+, K^+), GPCR	CNS	**Agonist:** sumatriptan **Antagonist:** LSD
Histamine	Histamine (H)	GPCR	CNS	**Antagonists:** ranitidine (Zantac®) and cimetidine (Tagamet®)
Amino Acids				
Glutamate	Glutaminergic ionotropic (iGluR)			
	AMPA	ICR (Na^+, K^+)	CNS	**Agonist:** quisqualate
	NMDA	ICR (Na^+, K^+)	CNS	**Potentiator:** serine
	Glutaminergic metabotropic (mGluR)	GPCR	CNS	**Potentiator:** glycine
GABA (γ-aminobutyric acid)	GABA	ICR (Cl^-), GPCR	CNS	**Antagonist:** picrotoxin **Potentiators:** alcohol, barbiturates
Glycine	Glycine (GlyR)	ICR (Cl^-)	CNS	**Antagonist:** strychnine
Purines				
Adenosine	Purine (P)	GPCR	CNS	
Gases				
Nitric oxide (NO)	None	N/A	N/A	

*This table does not include the numerous peptides that can act as neurocrines.
**This list does not include many chemicals that are used as agonists and antagonists in physiological research.
‡ICR = ion channel-receptor; GPCR = G protein-coupled receptor; AMPA = α-amino-3-hydroxy-5-methylisoxazole-4-proprionic acid; NMDA = N-methyl-D-aspartate; LSD = lysergic acid diethylamide; N/A = not applicable.

The amino acid tyrosine is converted to **dopamine, norepinephrine,** and **epinephrine.** Norepinephrine is the major neurotransmitter of the PNS autonomic sympathetic division. All three tyrosine-derived molecules can also function as neurohormones.

Neurons that secrete norepinephrine are called **adrenergic neurons,** or, more properly, **noradrenergic neurons.** The adjective *adrenergic* does not have the same obvious link to its neurotransmitter as *cholinergic* does to *acetylcholine.* Instead, the adjective derives from the British name for epinephrine,

CLINICAL FOCUS

Myasthenia Gravis

What would you think was wrong if suddenly your eyelids started drooping, you had difficulty watching moving objects, and it became difficult to chew, swallow, and talk? What disease attacks these skeletal muscles but leaves the larger muscles of the arms and legs alone? The answer is {*myo-*, muscle + *asthenes*, weak + *gravis*, severe}, an autoimmune disease in which the body fails to recognize the acetylcholine (ACh) receptors on skeletal muscle as part of "self." The immune system then produces antibodies to attack the receptors. The antibodies bind to the ACh receptor protein and change it in some way that causes the muscle cell to pull the receptors out of the membrane and destroy them. This destruction leaves the muscle with fewer ACh receptors in the membrane. Even though neurotransmitter release is normal, the muscle target has a diminished response that exhibits as muscle weakness. Currently medical science does not have a cure for myasthenia gravis, although various drugs can help control its symptoms. To learn more about this disease, visit the web site for the Myasthenia Gravis Foundation of America at *www.myasthenia.org*.

adrenaline. In the early part of the twentieth century, British researchers thought that sympathetic neurons secreted adrenaline (epinephrine), hence the modifier *adrenergic.* Although our understanding has changed, the name persists. Whenever you see reference to "adrenergic control" of a function, you must make the connection to a neuron secreting norepinephrine.

Adrenergic receptors are divided into two classes: α (alpha) and β (beta), with multiple subtypes of each. Like cholinergic muscarinic receptors, adrenergic receptors are linked to G proteins. The two subtypes of adrenergic receptors work through different second messenger pathways. The action of epinephrine on β-receptors in dog liver led E. W. Sutherland to the discovery of cyclic AMP and the concept of second messenger systems as transducers of extracellular messengers [p. 173].

CONCEPT CHECK

18. When pharmaceutical companies design drugs, they try to make a given drug as specific as possible for the particular receptor subtype they are targeting. For example, a drug might target adrenergic β_1-receptors rather than all adrenergic α and β receptors. What is the advantage of this specificity?

Amino Acids Several amino acids function as neurotransmitters in the CNS. **Glutamate** is the primary excitatory neurotransmitter of the CNS, and **aspartate** is an excitatory neurotransmitter in selected regions of the brain. *Excitatory neurotransmitters* depolarize their target cells, usually by opening ion channels that allow flow of positive ions into the cell.

The main inhibitory neurotransmitter in the brain is **gamma-aminobutyric acid (GABA).** The *inhibitory neurotransmitters* hyperpolarize their target cells by opening Cl^- channels and allowing Cl^- to enter the cell.

Glutamate also acts as a neuromodulator. The action of glutamate at a particular synapse depends on which of its receptor types occurs on the target cell. Metabotropic glutaminergic receptors act through GPCRs. Two ionotropic glutamate receptors are receptor-channels.

AMPA receptors are ligand-gated monovalent cation channels similar to nicotinic acetylcholine channels. Glutamate binding opens the channel, and the cell depolarizes because of net Na^+ influx. AMPA receptors are named for their agonist α-*amino-3-hydroxy-5-methylisoxazole-4-proprionic acid.*

NMDA receptors are named for the glutamate agonist N-*methyl-D-aspartate.* They are unusual for several reasons. First, they are nonselective cation channels that allow Na^+, K^+, and Ca^{2+} to pass through the channel. Second, channel opening requires both glutamate binding and a change in membrane potential. The NMDA receptor-channel's action is described in the section on long-term potentiation later in this chapter.

Glycine and the amino acid D-*serine* potentiate, or enhance, the excitatory effects of glutamate at one type of glutamate receptor. D-serine is made and released by glial cells as well as neurons, which illustrates the role that glial cells can play in altering synaptic communication.

Peptides The nervous system secretes a variety of peptides that act as neurotransmitters and neuromodulators in addition to functioning as neurohormones. These peptides include **substance P,** involved in some pain pathways, and the **opioid peptides** (**enkephalins** and **endorphins**) that mediate pain relief, or *analgesia* {*an-,* without + *algos,* pain}. Peptides that function as both neurohormones and neurotransmitters include *cholecystokinin (CCK), vasopressin (AVP),* and *atrial natriuretic peptide (ANP).* Many peptide neurotransmitters are co-secreted with other neurotransmitters.

Purines *Adenosine, adenosine monophosphate* (AMP), and *adenosine triphosphate* (ATP) can all act as neurotransmitters. These molecules, known collectively as *purines* [p. 35], bind to *purinergic* receptors in the CNS and on other excitable tissues such as the heart. The purines all bind to G protein-coupled receptors.

Gases One of the most interesting neurotransmitters is *nitric oxide* (NO), an unstable gas synthesized from oxygen and the amino acid arginine. Nitric oxide acting as a neurotransmitter diffuses freely into a target cell rather than binding to a membrane receptor [p. 178]. Once inside the target cell, nitric oxide binds to target proteins. With a half-life of only 2–30 seconds, nitric oxide is elusive and difficult to study. It is also released from cells other than neurons and often acts as a paracrine signal.

Recent work suggests that *carbon monoxide* (CO) and hydrogen sulfide (H_2S), both known as toxic gases, are produced by the body in tiny amounts to serve as neurotransmitters.

BIOTECHNOLOGY

Of Snakes, Snails, Spiders, and Sushi

What do snakes, marine snails, and spiders have to do with neurophysiology? They all provide neuroscientists with compounds for studying synaptic transmission, extracted from the neurotoxic venoms these creatures use to kill their prey. The Asian snake *Bungarus multicinctus* provides us with α-bungarotoxin, a long-lasting poison that binds tightly to nicotinic acetylcholine receptors. The fish-hunting cone snail, *Conus geographus,* and the funnel web spider, *Agelenopsis aperta,* use toxins that block different types of voltage-gated Ca^{2+} channels. One of the most potent poisons known, however, comes from the Japanese puffer fish, a highly prized delicacy whose flesh is consumed as sushi. The puffer has tetrodotoxin (TTX) in its gonads. This neurotoxin blocks Na^+ channels on axons and prevents the transmission of action potentials, so ingestion of only a tiny amount can be fatal. The Japanese chefs who prepare the puffer fish, or *fugu,* for consumption are carefully trained to avoid contaminating the fish's flesh as they remove the toxic gonads. There's always some risk involved in eating *fugu,* though—one reason that traditionally the youngest person at the table is the first to sample the dish.

Lipids Lipid neurocrine molecules include several eicosanoids [p. 30] that are the endogenous ligands for *cannabinoid receptors.* The CB_1 cannabinoid receptor is found in the brain, and the CB_2 receptor is found on immune cells. The receptors were named for one of their exogenous ligands, Δ^9-*tetrahydrocannabinoid* (THC), which comes from the plant *Cannabis sativa,* more commonly known as marijuana. Lipid neurocrine signals all bind to G protein-coupled receptors.

Neurotransmitters Are Released from Vesicles

When we examine the axon terminal of a presynaptic cell with an electron microscope, we find many small **synaptic vesicles** filled with neurotransmitter that is released on demand (**FIG. 8.18**). Some vesicles are "docked" at active zones along the membrane closest to the synaptic cleft, waiting for a signal to release their contents. Other vesicles act as a reserve pool, clustering close to the docking sites. Axon terminals also contain mitochondria to produce ATP for metabolism and transport. In this section, we discuss general patterns of neurotransmitter synthesis, storage, release, and termination of action.

Neurotransmitter Synthesis Neurotransmitter synthesis takes place both in the nerve cell body and in the axon terminal. Polypeptides must be made in the cell body because axon terminals do not have the organelles needed for protein synthesis.

FIG. 8.18 **A chemical synapse**

The axon terminal contains mitochondria and synaptic vesicles filled with neurotransmitter. The postsynaptic membrane has receptors for neurotransmitter that diffuses across the synaptic cleft.

Schwann cell

Axon terminal

Mitochondrion

Vesicles with neurotransmitter

Synaptic cleft

Muscle fiber

Protein synthesis follows the usual pathway [p. 112]. The large *propeptide* that results is packaged into vesicles along with the enzymes needed to modify it. The vesicles then move from the cell body to the axon terminal by fast axonal transport. Inside the vesicle, the propeptide is broken down into smaller active peptides—a pattern similar to the preprohormone-prohormone-active hormone process in endocrine cells [p. 202]. For example, one propeptide contains the amino acid sequences for three active peptides that are co-secreted: ACTH, gamma (γ-)lipotropin, and beta (β-)endorphin.

RUNNING **PROBLEM**

Dr. McKhann then asked to see autopsy reports on some of the children who had died of their paralysis at Beijing Hospital. In the reports, pathologists noted that the patients had normal myelin but damaged axons. In some cases, the axon had been completely destroyed, leaving only a hollow shell of myelin.

Q5: *Do the results of Dr. McKhann's investigation suggest that the Chinese children had classic GBS? Why or why not?*

227   229   231   251   253   **257**   267   268

Smaller neurotransmitters, such as acetylcholine, amines, and purines, are synthesized and packaged into vesicles in the axon terminal. The enzymes needed for their synthesis are made in the cell body and released into the cytosol. The dissolved enzymes are then brought to axon terminals by slow axonal transport.

CONCEPT CHECK

19. Which organelles are needed to synthesize proteins and package them into vesicles?
20. What is the function of mitochondria in a cell?
21. How do mitochondria get to the axon terminals?

Neurotransmitter Release Neurotransmitters in the axon terminal are stored in vesicles, so their release into the synaptic cleft takes place by exocytosis [p. 148]. From what we can tell, exocytosis in neurons is similar to exocytosis in other types of cells, but much faster. Neurotoxins that block neurotransmitter release, including tetanus and botulinum toxins, exert their action by inhibiting specific proteins of the cell's exocytotic apparatus.

FIGURE 8.19a shows how neurotransmitters are released by exocytosis. When the depolarization of an action potential reaches the axon terminal, the change in membrane potential sets off a sequence of events ❶. The axon terminal membrane has voltage-gated Ca^{2+} channels that open in response to depolarization ❷. Calcium ions are more concentrated in the extracellular fluid than in the cytosol, and so they move into the cell.

Ca^{2+} entering the cell binds to regulatory proteins and initiates exocytosis ❸. The membrane of the synaptic vesicle fuses with the cell membrane, aided by multiple membrane proteins. The fused area opens, and neurotransmitter inside the synaptic vesicle moves into the synaptic cleft ❹. The neurotransmitter molecules diffuse across the gap to bind with membrane receptors on the postsynaptic cell. When neurotransmitters bind to their receptors, a response is initiated in the postsynaptic cell ❺. Each synaptic vesicle contains the same amount of neurotransmitter, so measuring the magnitude of the target cell response is an indication of how many vesicles released their content.

In the classic model of exocytosis, the membrane of the vesicle becomes part of the axon terminal membrane [Fig. 5.19, p. 149]. To prevent a large increase in membrane surface area, the membrane is recycled by endocytosis of vesicles at regions away from the active sites (Fig. 8.3). The recycled vesicles are then re-filled with newly made neurotransmitter.

The transporters that concentrate neurotransmitter into vesicles are H^+-dependent antiporters [p. 141]. The vesicles use H^+-ATPases to concentrate H^+ inside the vesicle, then exchange H^+ for the neurotransmitter.

Recently, a second model of secretion has emerged. In this model, called the **kiss-and-run pathway,** synaptic vesicles fuse to the presynaptic membrane at a complex called the **fusion pore.** This fusion opens a small channel that is just large enough for neurotransmitter to pass through. Then, instead of opening the fused area wider and incorporating the vesicle membrane into the cell membrane, the vesicle pulls back from the fusion pore and returns to the pool of vesicles in the cytoplasm.

CONCEPT CHECK

22. In an experiment on synaptic transmission, a synapse was bathed in a Ca^{2+}-free medium that was otherwise equivalent to extracellular fluid. An action potential was triggered in the presynaptic neuron. Although the action potential reached the axon terminal at the synapse, the usual response of the postsynaptic cell did not occur. What conclusion did the researchers draw from these results?

23. Classify the H^+-neurotransmitter exchange as facilitated diffusion, primary active transport, or secondary active transport. Explain your reasoning.

Termination of Neurotransmitter Activity A key feature of neural signaling is its short duration, due to the rapid removal or inactivation of neurotransmitter in the synaptic cleft. Recall that ligand binding to a protein is reversible and goes to a state of equilibrium, with a constant ratio of unbound to bound ligand [p. 47]. If unbound neurotransmitter is removed from the synapse, the receptors release bound neurotransmitter, terminating its activity, to keep the ratio of unbound/bound transmitter constant.

Removal of unbound neurotransmitter from the synaptic cleft can be accomplished in various ways (Fig. 8.19b). Some neurotransmitter molecules simply diffuse away from the synapse, becoming separated from their receptors. Other neurotransmitters are inactivated by enzymes in the synaptic cleft. For example, acetylcholine (ACh) in the extracellular fluid is rapidly broken down into choline and acetyl CoA by the enzyme **acetylcholinesterase (AChE)** in the extracellular matrix and in the membrane of the postsynaptic cell (**FIG. 8.20**). Choline from degraded ACh is transported back into the presynaptic axon terminal on a Na^+-dependent cotransporter. Once back in the axon terminal, it can be used to make new acetylcholine.

Many neurotransmitters are removed from the extracellular fluid by transport back into the presynaptic cell or into adjacent neurons or glia. For example, norepinephrine action is terminated when the intact neurotransmitter is transported back into the presynaptic axon terminal. Norepinephrine uptake uses a Na^+-dependent cotransporter. Once back in the axon terminal, norepinephrine is either transported back into vesicles or broken down by intracellular enzymes such as *monoamine oxidase (MAO)*, found in mitochondria. Neurotransmitters and their components can be recycled to refill empty synaptic vesicles.

CONCEPT CHECK

24. One class of antidepressant drugs is called selective serotonin reuptake inhibitors (SSRIs). What do these drugs do to serotonin activity at the synapse?

25. How does the axon terminal make acetyl CoA for acetylcholine synthesis? [*Hint:* p. 108]

26. Is Na^+-dependent neurotransmitter reuptake facilitated diffusion, primary active transport, or secondary active transport? Explain your reasoning.

FIG. 8.19 **ESSENTIALS**

Synaptic Communication

Cell-to-cell communication uses chemical and electrical
signaling to coordinate function and maintain homeostasis.

(a) Neurotransmitter Release

Action potential
arrives at
axon terminal.

Synaptic vesicle
with neurotransmitter
molecules

Docking protein

Ca^{2+}

Synaptic
cleft

Receptor

Postsynaptic cell

Voltage-gated
Ca^{2+} channel

Cell
response

1. An action potential depolarizes
the axon terminal.

2. The depolarization opens voltage-
gated Ca^{2+} channels, and Ca^{2+}
enters the cell.

3. Calcium entry triggers exocytosis
of synaptic vesicle contents.

4. Neurotransmitter diffuses across
the synaptic cleft and binds with
receptors on the postsynaptic cell.

5. Neurotransmitter binding initiates
a response in the postsynaptic
cell.

(b) Neurotransmitter Termination

Neurotransmitter action terminates when the chemicals are broken down,
are taken up into cells, or diffuse away from the synapse.

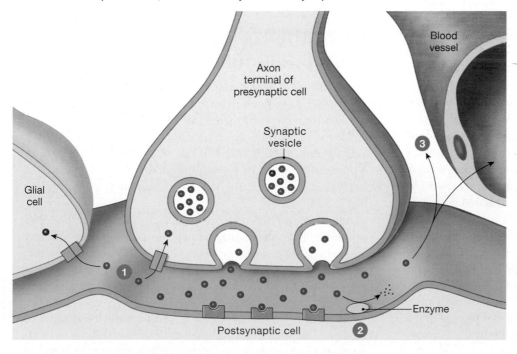

Blood
vessel

Axon
terminal of
presynaptic cell

Synaptic
vesicle

Glial
cell

Enzyme

Postsynaptic cell

1. Neurotransmitters can be returned
to axon terminals for reuse or
transported into glial cells.

2. Enzymes inactivate
neurotransmitters.

3. Neurotransmitters can diffuse
out of the synaptic cleft.

FIG. 8.20   Synthesis and recycling of acetylcholine

1 **Acetylcholine** (ACh) is made from choline and acetyl CoA.

2 In the synaptic cleft, ACh is rapidly broken down by the enzyme **acetylcholinesterase.**

3 Choline is transported back into the axon terminal by cotransport with Na⁺.

4 Recycled choline is used to make more ACh.

Stronger Stimuli Release More Neurotransmitter

A single action potential arriving at the axon terminal releases a constant amount of neurotransmitter. Neurons therefore can use the frequency of action potentials to transmit information about the duration and strength of the stimuli that activated them. Duration of a stimulus is coded by the duration of a series of repeated action potentials. A stronger stimulus causes more action potentials per second to arrive at the axon terminal, which in turn may result in more neurotransmitter release.

For example, let's consider how a sensory neuron tells the CNS the intensity of an incoming stimulus. An above-threshold graded potential reaching the trigger zone of the sensory neuron does not trigger just one action potential. Instead, even a small graded potential that is above threshold triggers a burst of action potentials (**FIG. 8.21a**). As graded potentials increase in strength (amplitude), they trigger more frequent action potentials (Fig. 8.21b).

Electrical signaling patterns in the CNS are more variable. Brain neurons show different electrical personalities by firing action potentials in a variety of patterns, sometimes spontaneously, without an external stimulus to bring them to threshold. For example, some neurons are *tonically active* [p. 183], firing regular trains of action potentials (beating pacemakers). Other neurons exhibit *bursting*, bursts of action potentials rhythmically alternating with intervals of quiet (rhythmic pacemakers).

These different firing patterns in CNS neurons are created by ion channel variants that differ in their activation and inactivation voltages, opening and closing speeds, and sensitivity to neuromodulators. This variability makes brain neurons more dynamic and complicated than the simple somatic motor neuron we use as our model.

INTEGRATION OF NEURAL INFORMATION TRANSFER

Communication between neurons is not always a one-to-one event as we have been describing. Frequently, the axon of a presynaptic neuron branches, and its collaterals (branches) synapse on multiple target neurons. This pattern is known as **divergence** (**FIG. 8.22a**). On the other hand, when a group of presynaptic neurons provide input to a smaller number of postsynaptic neurons, the pattern is known as **convergence** (Fig. 8.22b).

Combination of convergence and divergence in the CNS may result in one postsynaptic neuron with synapses from as many as 10,000 presynaptic neurons (Fig. 8.22c). For example, the Purkinje neurons of the CNS have highly branched dendrites so that they can receive information from many neurons (Fig. 8.22d).

In addition, we now know that the traditional view of chemical synapses as sites of one-way communication, with all messages moving from presynaptic cell to postsynaptic cell, is

FIG. 8.21 **Coding the strength of a stimulus**

The frequency of action potential firing indicates the strength of a stimulus.

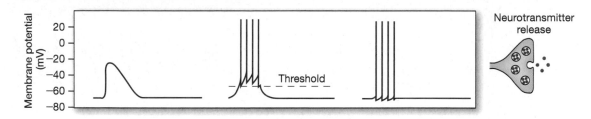

(a) Weak stimulus releases little neurotransmitter.

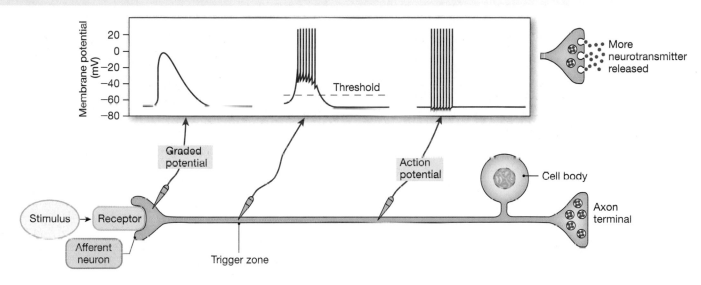

(b) Strong stimulus causes more action potentials and releases more neurotransmitter.

not always correct. In the brain, there are some synapses where cells on both sides of the synaptic cleft release neurotransmitters that act on the opposite cell. Perhaps more importantly, we have learned that many postsynaptic cells "talk back" to their presynaptic neurons by sending neuromodulators that bind to presynaptic receptors. Variations in synaptic activity play a major role in determining how communication takes place in the nervous system.

The ability of the nervous system to change activity at synapses is called **synaptic plasticity** {*plasticus*, that which may be molded}. Short-term plasticity may enhance activity at the synapse (facilitation) or decrease it (depression). For example, in some cases of sustained activity at a synapse, neurotransmitter release decreases over time because the axon cannot replenish its neurotransmitter supply rapidly enough, resulting in synaptic depression.

Sometimes changes at the synapse persist for significant periods of time (long-term depression or long-term potentiation). In the sections that follow, we examine some of the ways that communication at synapses can be modified.

Postsynaptic Responses May Be Slow or Fast

A neurotransmitter combining with its receptor sets in motion a series of responses in the postsynaptic cell (**FIG. 8.23**). Neurotransmitters that bind to G protein-coupled receptors linked to second messenger systems initiate slow postsynaptic responses.

Some second messengers act from the cytoplasmic side of the cell membrane to open or close ion channels. Changes in membrane potential resulting from these alterations in ion flow are called **slow synaptic potentials** because the response of the second messenger pathway takes longer than the direct opening or closing of a channel. In addition, the response itself lasts longer, usually seconds to minutes.

Slow postsynaptic responses are not limited to altering the open state of ion channels. Neurotransmitters acting on GPCRs may also modify existing cell proteins or regulate the production of new cell proteins. These types of slow response have been linked to the growth and development of neurons and to the mechanisms underlying long-term memory.

FIG. 8.22 **ESSENTIALS**

Divergence and Convergence

(a) In a **divergent pathway,** one presynaptic neuron branches to affect a larger number of postsynaptic neurons.

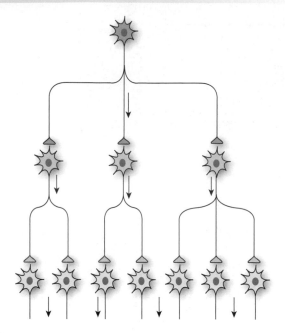

(b) In a **convergent pathway,** many presynaptic neurons provide input to influence a smaller number of postsynaptic neurons.

Q **FIGURE QUESTION**

The pattern of divergence in (a) is similar to _____ in a second messenger system.

(c) The cell body of a somatic motor neuron is nearly covered with synapses providing input from other neurons.

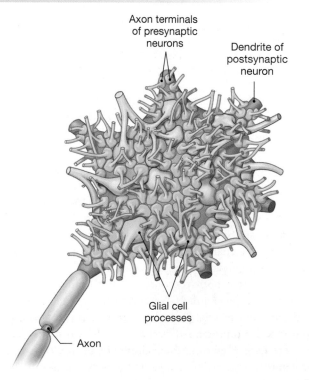

Axon terminals of presynaptic neurons

Dendrite of postsynaptic neuron

Glial cell processes

Axon

(d) The highly branched dendrites of a Purkinje cell (neuron) demonstrate convergence of signals from many synapses onto a cell body.

Highly branched dendrites projecting into the gray matter of the cerebellum

Cell body of Purkinje cell

Light micrograph of Purkinje cells in cerebellum

FIG. 8.23 ESSENTIALS

Fast and Slow Postsynaptic Responses

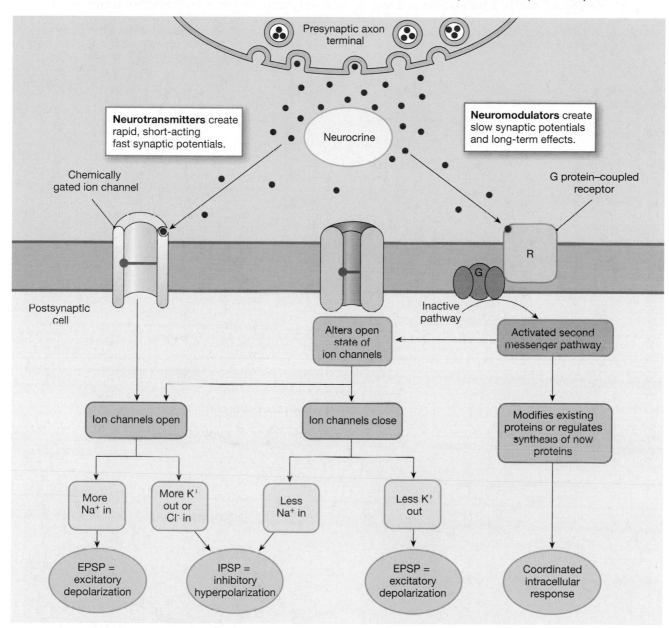

Fast responses are mediated by ion channels.

Slow responses are mediated by G protein–coupled receptors.

Presynaptic axon terminal

Neurotransmitters create rapid, short-acting fast synaptic potentials.

Neurocrine

Neuromodulators create slow synaptic potentials and long-term effects.

Chemically gated ion channel

G protein–coupled receptor

R

Postsynaptic cell

G

Inactive pathway

Alters open state of ion channels

Activated second messenger pathway

Ion channels open

Ion channels close

Modifies existing proteins or regulates synthesis of new proteins

More Na⁺ in

More K⁺ out or Cl⁻ in

Less Na⁺ in

Less K⁺ out

EPSP = excitatory depolarization

IPSP = inhibitory hyperpolarization

EPSP = excitatory depolarization

Coordinated intracellular response

Fast synaptic responses are always associated with the opening of ion channels. In the simplest response, the neurotransmitter binds to and opens a receptor-channel on the postsynaptic cell, allowing ions to move between the postsynaptic cell and the extracellular fluid. The resulting change in membrane potential is called a **fast synaptic potential** because it begins quickly and lasts only a few milliseconds.

If the synaptic potential is depolarizing, it is called an **excitatory postsynaptic potential (EPSP)** because it makes the cell more likely to fire an action potential. If the synaptic potential is hyperpolarizing, it is called an **inhibitory postsynaptic potential (IPSP)** because

hyperpolarization moves the membrane potential away from threshold and makes the cell less likely to fire an action potential.

Pathways Integrate Information from Multiple Neurons

When two or more presynaptic neurons converge on the dendrites or cell body of a single postsynaptic cell, the response of the postsynaptic cell is determined by the summed input from the presynaptic neurons. **FIGURE 8.24c** shows the three-dimensional

FIG. 8.24 **ESSENTIALS**

Integration of Synaptic Signaling

Temporal Summation

Temporal summation occurs when two graded potentials from one presynaptic neuron occur close together in time.

(a) No summation. Two subthreshold graded potentials will not initiate an action potential if they are far apart in time.

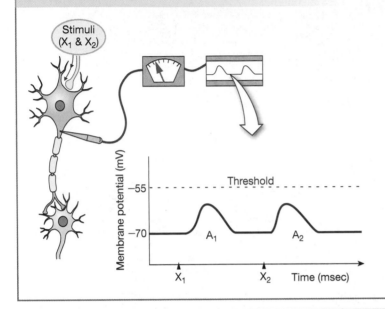

(b) Summation causing action potential. If two subthreshold potentials arrive at the trigger zone within a short period of time, they may sum and initiate an action potential.

Spatial Summation

Spatial summation occurs when the currents from nearly simultaneous graded potentials combine.

(c) Multiple presynaptic neurons provide input on the dendrites and cell body of the postsynaptic neurons.

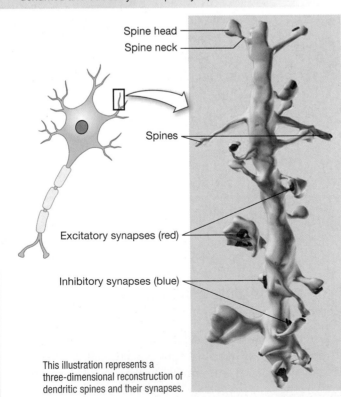

Spine head
Spine neck

Spines

Excitatory synapses (red)

Inhibitory synapses (blue)

This illustration represents a three-dimensional reconstruction of dendritic spines and their synapses.

(d) Summation of several subthreshold signals results in an action potential.

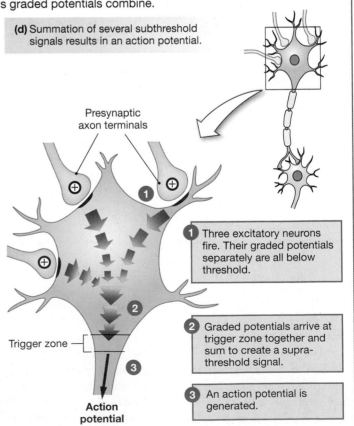

Presynaptic axon terminals

Trigger zone

Action potential

1. Three excitatory neurons fire. Their graded potentials separately are all below threshold.

2. Graded potentials arrive at trigger zone together and sum to create a supra-threshold signal.

3. An action potential is generated.

Synaptic Inhibition

(e) One inhibitory postsynaptic potential (IPSP) sums with two excitatory postsynaptic potentials (EPSPs) to prevent an action potential in the postsynaptic cell.

Excitatory neuron

Inhibitory neuron

Ⓠ FIGURE QUESTIONS
1. Identify examples of divergence and convergence in each part of this figure.
2. Using part (g) as a model, draw an example where the target of one collateral has no response due to postsynaptic inhibition of the target cell.

Trigger zone

1 One inhibitory and two excitatory neurons fire.

2 The summed potentials are below threshold, so no action potential is generated.

No action potential

(f) In **global presynaptic inhibiton,** all targets of the postsynaptic neuron are inhibited equally.

1 Excitatory and inhibitory presynaptic neurons fire.

2 Summed signal in postsynaptic neuron is below threshold.

3 No action potential initiated at trigger zone.

EPSP + IPSP

Excitatory neuron

Inhibitory neuron modifies signal.

Target cells may be other neurons, muscles, or glands.

Collaterals

4 No response occurs in any target cell.

Target cell — No response
Target cell — No response
Target cell — No response

4

(g) In **selective presynaptic inhibiton,** an inhibitory neuron synapses on one collateral of the presynaptic neuron and selectively inhibits one target.

1 An excitatory neuron fires.

2 An action potential is generated.

2 Action potential

Presynaptic axon terminal

Inhibitory neuron

3 An inhibitory neuron fires, blocking neurotransmitter release at one synapse.

3

Neurotransmitter released

Target cell — Response
Target cell — Response
Target cell — No response

reconstruction of dendritic spines of a postsynaptic neuron, with numerous excitatory and inhibitory synapses providing input. The summed input from these synapses determines the activity of the postsynaptic neuron.

The combination of several nearly simultaneous graded potentials is called **spatial summation.** The word *spatial* {*spatium*, space} refers to the fact that the graded potentials originate at different locations (spaces) on the neuron.

Figure 8.24d illustrates spatial summation when three presynaptic neurons releasing excitatory neurotransmitters ("excitatory neurons") converge on one postsynaptic neuron. Each neuron's EPSP is too weak to trigger an action potential by itself, but if the three presynaptic neurons fire simultaneously, the sum of the three EPSPs is above threshold and creates an action potential.

Spatial summation is not always excitatory. If summation prevents an action potential in the postsynaptic cell, the summation is called **postsynaptic inhibition.** This occurs when presynaptic neurons release inhibitory neurotransmitter. For example, Figure 8.24e shows three presynaptic neurons, two excitatory and one inhibitory, converging on a postsynaptic cell. The neurons fire, creating one IPSP and two EPSPs that sum as they reach the trigger zone. The IPSP counteracts the two EPSPs, creating an integrated signal that is below threshold. As a result, no action potential is generated at the trigger zone.

Temporal Summation Summation of graded potentials does not always require input from more than one presynaptic neuron. Two subthreshold graded potentials from the same presynaptic neuron can be summed if they arrive at the trigger zone close enough together in time. Summation that occurs from graded potentials overlapping in time is called **temporal summation** {*tempus*, time}. Let's see how this can happen.

Figure 8.24a shows recordings from an electrode placed in the trigger zone of a neuron. A stimulus (X_1) starts a subthreshold graded potential on the cell body at the time marked on the *x*-axis. The graded potential reaches the trigger zone and depolarizes it, as shown on the graph (A_1), but not enough to trigger an action potential. A second stimulus (X_2) occurs later, and its subthreshold graded potential (A_2) reaches the trigger zone sometime after the first. The interval between the two stimuli is so long that the two graded potentials do not overlap. Neither potential by itself is above threshold, so no action potential is triggered.

In Figure 8.24b, the two stimuli occur closer together in time. As a result, the two subthreshold graded potentials arrive at the trigger zone at almost the same time. The second graded potential adds its depolarization to that of the first, causing the trigger zone to depolarize to threshold.

In many situations, graded potentials in a neuron incorporate both temporal and spatial summation. The summation of graded potentials demonstrates a key property of neurons: *postsynaptic integration*. When multiple signals reach a neuron, postsynaptic integration creates a signal based on the relative strengths and durations of the signals. If the integrated signal is above threshold, the neuron fires an action potential. If the integrated signal is below threshold, the neuron does not fire.

CONCEPT CHECK

27. In Figure 8.24e, assume the postsynaptic neuron has a resting membrane potential of –70 mV and a threshold of –55 mV. If the inhibitory presynaptic neuron creates an IPSP of –5 mV and the two excitatory presynaptic neurons have EPSPs of 10 and 12 mV, will the postsynaptic neuron fire an action potential?

28. In the graphs of Figure 8.24a, b, why doesn't the membrane potential change at the same time as the stimulus?

Synaptic Activity Can Be Modified

The examples of synaptic integration we just discussed all took place on the postsynaptic side of a synapse, but the activity of presynaptic cells can also be altered, or *modulated*. When a modulatory neuron terminates on a presynaptic cell, the IPSP or EPSP created by the modulatory neuron can alter the action potential reaching the axon terminals of the presynaptic cell and modulate neurotransmitter release. In *presynaptic facilitation*, input from an excitatory neuron increases neurotransmitter release by the presynaptic cell.

If modulation of a neuron decreases its neurotransmitter release, the modulation is called *presynaptic inhibition*. Presynaptic inhibition may be global or selective. In global presynaptic inhibition (Fig. 8.24f), input on the dendrites and cell body of a neuron decreases neurotransmitter release by all collaterals and all target cells of the neuron are affected equally.

In selective modulation, one collateral can be inhibited while others remain unaffected. Selective presynaptic alteration of neurotransmitter release provides a more precise means of control than global modulation. For example, Figure 8.24g shows selective presynaptic modulation of a single collateral's axon terminal so that only its target cell fails to respond.

Synaptic activity can also be altered by changing the target (postsynaptic) cell's responsiveness to neurotransmitter. This may be accomplished by changing the structure, affinity, or number of neurotransmitter receptors. Modulators can alter all of these parameters by influencing the synthesis of enzymes, membrane transporters, and receptors. Most neuromodulators act through second messenger systems that alter existing channels, and their effects last much longer than do those of neurotransmitters. One signal molecule can act as either a neurotransmitter or a neuromodulator, depending upon its receptor (Fig. 8.23).

CONCEPT CHECK

29. Why are axon terminals sometimes called "biological transducers"?

Dr. McKhann suspected that the disease afflicting the Chinese children—which he named acute motor axonal polyneuropathy (AMAN)—might be triggered by a bacterial infection. He also thought that the disease initiated its damage of axons at neuromuscular junctions, the synapses between somatic motor neurons and skeletal muscles.

Q6: *Based on information provided in this chapter, name other diseases involving altered synaptic transmission.*

227 229 231 251 253 257 **267** 268

Long-Term Potentiation Alters Synapses

Two of the "hot topics" in neurobiology today are **long-term potentiation** (LTP) {*potentia*, power} and *long-term depression* (LTD), processes in which activity at a synapse brings about sustained changes in the quality or quantity of synaptic connections. Many times changes in synaptic transmission, such as the facilitation and inhibition we just discussed, are of limited duration.

However, if synaptic activity persists for longer periods, the neurons may adapt through LTP and LTD. Our understanding of LTP and LTD is changing rapidly, and the mechanisms may not be the same in different brain areas. The descriptions below reflect some of what we currently know about long-term adaptations of synaptic transmission.

A key element in long-term changes in the CNS is the amino acid glutamate, the main excitatory neurotransmitter in the CNS. As you learned previously, glutamate has two types of receptor-channels: AMPA receptors and NMDA receptors. The NMDA receptor has an unusual property. First, at resting membrane potentials, the NMDA channel is blocked by both a gate and a Mg^{2+} ion. Glutamate binding opens the ligand-activated gate, but ions cannot flow past the Mg^{2+}. However, if the cell depolarizes, the Mg^{2+} blocking the channel is expelled, and then ions can flow through the channel. Thus, the NMDA channel opens only when the receptor is bound to glutamate *and* the cell is depolarized.

In long-term potentiation, when presynaptic neurons release glutamate, the neurotransmitter binds to both AMPA and NMDA receptors on the postsynaptic cell (**FIG. 8.25 ❶**). Binding to the AMPA receptor opens a cation channel, and net Na^+ entry depolarizes the cell ❷. Simultaneously, glutamate binding to

FIG. 8.25 **Long-term potentiation**

❶ Glutamate binds to AMPA and NMDA channels.

❷ Net Na^+ entry through AMPA channels depolarizes the postsynaptic cell.

❸ Depolarization ejects Mg^{2+} from NMDA receptor-channel and opens channel.

❹ Ca^{2+} enters cytoplasm through NMDA channel.

❺ Ca^{2+} activates second messenger pathways.

❻ Paracrine from postsynaptic cell enhances glutamate release.

the NMDA receptor opens the channel gate, and depolarization of the cell creates electrical repulsion that knocks the Mg^{2+} out of the NMDA channel ❸. Once the NMDA channel is open, Ca^{2+} enters the cytosol ❹.

The Ca^{2+} signal initiates second messenger pathways ❺. As a result of these intracellular pathways, the postsynaptic cell becomes more sensitive to glutamate, possibly by inserting more glutamate receptors in the postsynaptic membrane [up-regulation, p. 51]. In addition the postsynaptic cell releases a paracrine that acts on the presynaptic cell to enhance glutamate release ❻.

Long-term depression seems to have two components: a change in the number of postsynaptic receptors and a change in the isoforms of the receptor proteins. In the face of continued neurotransmitter release from presynaptic neurons, the postsynaptic neurons withdraw AMPA receptors from the cell membrane by endocytosis [p. 148], a process similar to down-regulation of receptors in the endocrine system [p. 51]. In addition, different protein subunits are inserted into the AMPA receptor proteins, changing current flow through the ion channels.

Researchers believe that long-term potentiation and depression are related to the neural processes for learning and memory, and to changes in the brain that occur with clinical depression and other mental illnesses. The clinical link makes LTP and LTD hot topics in neuroscience research.

CONCEPT CHECK

30. Why would depolarization of the membrane drive Mg^{2+} from the channel into the extracellular fluid?

Disorders of Synaptic Transmission Are Responsible for Many Diseases

Synaptic transmission is the most vulnerable step in the process of signaling through the nervous system. It is the point at which many things go wrong, leading to disruption of normal function. Yet, at the same time, the receptors at synapses are exposed to the extracellular fluid, making them more accessible to drugs than intracellular receptors are. In recent years, scientists have linked a variety of nervous system disorders to problems with synaptic transmission. These disorders include Parkinson's disease, schizophrenia, and depression. The best understood diseases of the synapse are those that involve the *neuromuscular junction* between somatic motor neurons and skeletal muscles. One example of neuromuscular junction pathology is *myasthenia gravis*. Diseases resulting from synaptic transmission problems within the CNS have proved more difficult to study because they are more difficult to isolate anatomically.

Drugs that act on synaptic activity, particularly synapses in the CNS, are the oldest known and most widely used of all pharmacological agents. Caffeine, nicotine, and alcohol are common drugs in many cultures. Some of the drugs we use to treat conditions such as schizophrenia, depression, anxiety, and epilepsy act by influencing events at the synapse. In many disorders arising in the CNS, we do not yet fully understand either the cause of the disorder or the drug's mechanism of action. This subject is one major area of pharmacological research, and new classes of drugs are being formulated and approved every year.

RUNNING **PROBLEM CONCLUSION** | Mysterious Paralysis

In this running problem you learned about acute motor axonal polyneuropathy (AMAN), a baffling paralytic illness that physicians thought might be a new disease. Although its symptoms resemble those of classic GBS, AMAN is not a demyelinating disease. It affects only somatic motor neurons. However, in both classic GBS and AMAN, the body's immune system makes antibodies against nervous system components. This similarity led experts eventually to conclude that AMAN is a subtype of GBS. The classic form of GBS has been renamed acute inflammatory demyelinating polyneuropathy (AIDP). AIDP is more common in Europe and North America, while

AMAN is the predominant form of GBS in China, Japan, and South America. A significant number of patients with AMAN develop their disease following a gastrointestinal illness caused by the bacterium *Campylobacter jejuni,* and experts suspect that antibodies to the bacterium also attack glycolipids called gangliosides in the axonal membrane. To learn more about the link between *Campylobacter* and GBS, see *"Campylobacter* Species and Guillain-Barré Syndrome," *Clin Microbiol Rev* 11: 555–567, July 1998 (*http://cmr.asm.org/content/11/3/555.full*). Check your understanding of this problem by comparing your answers to the information in the following summary table.

Question	Facts	Integration and Analysis
Q1: *Which division(s) of the nervous system may be involved in GBS?*	The nervous system is divided into the central nervous system (CNS) and the afferent (sensory) and efferent subdivisions of the peripheral nervous system. Efferent neurons are either somatic motor neurons, which control skeletal muscles, or autonomic neurons, which control glands and smooth and cardiac muscle.	Patients with GBS can neither feel sensations nor move their muscles. This suggests a problem in both afferent and somatic motor neurons. However, it is also possible that there is a problem in the CNS integrating center. You do not have enough information to determine which division is affected.

RUNNING **PROBLEM CONCLUSION** | *Continued*

Question	Facts	Integration and Analysis
Q2: *Do you think the paralysis found in the Chinese children affected both sensory (afferent) and somatic motor neurons? Why or why not?*	The Chinese children can feel a pin prick but cannot move their muscles.	Sensory (afferent) function is normal if they can feel the pin prick. Paralysis of the muscles suggests a problem with somatic motor neurons, with the CNS centers controlling movement, or with the muscles themselves.
Q3: *In GBS, what would you expect the results of a nerve conduction test to be?*	Nerve conduction tests measure conduction speed and strength. In GBS, myelin around neurons is destroyed.	Myelin insulates axons and increases speed. Without myelin, ions leak out of the axon. Thus, in GBS you would expect decreased conduction speed or blocked conduction.
Q4: *Is the paralytic illness that affected the Chinese children a demyelinating condition? Why or why not?*	Nerve conduction tests showed normal conduction speed but decreased strength of the summed action potentials.	Myelin loss should decrease conduction speed as well as blocked conduction. Therefore, this illness is probably not a demyelinating disease.
Q5: *Do the results of Dr. McKhann's investigation suggest that the Chinese children had classic GBS? Why or why not?*	Autopsy reports on children who died from the disease showed that the axons were damaged but the myelin was normal.	Classic GBS is a demyelinating disease that affects both sensory and motor neurons. The Chinese children had normal sensory function, and nerve conduction tests and histological studies indicated normal myelin. Therefore, it was reasonable to conclude that the disease was not classic GBS.
Q6: *Based on information provided in this chapter, name other diseases involving altered synaptic transmission.*	Synaptic transmission can be altered by blocking neurotransmitter release from the presynaptic cell, by interfering with the action of neurotransmitter on the target cell, or by removing neurotransmitter from the synapse.	Parkinson's disease, depression, schizophrenia, and myasthenia gravis are related to problems with synaptic transmission.

227 229 231 251 253 257 267 **268**

VISUALIZE, PRACTICE, AND MASTER PHYSIOLOGY WITH MASTERING A&P

- Practice-on-the-go with Dynamic Study Modules
- Visualize important processes with Interactive Physiology and A&P Flix
- Check your understanding with Video Tutors and practice quizzes!

MasteringA&P®

CHAPTER SUMMARY

This chapter introduces the nervous system, one of the major control systems responsible for maintaining *homeostasis*. The divisions of the nervous system correlate with the steps in a reflex pathway. Sensory receptors monitor regulated variables and send input signals to the central nervous system through sensory (afferent) neurons. Output signals, both electrical and chemical, travel through the efferent divisions (somatic motor and autonomic) to their targets throughout the body. Information transfer and *communication* depend on electrical signals that pass along neurons, on *molecular interactions* between signal molecules and their receptors, and on signal transduction in the target cells.

1. The **nervous system** is a complex network of neurons that form the rapid control system of the body. (p. 227)

2. **Emergent properties** of the nervous system include consciousness, intelligence, and emotion. (p. 227)

Organization of the Nervous System

3. The nervous system is divided into the **central nervous system (CNS),** composed of the **brain** and **spinal cord,** and the **peripheral nervous system (PNS).** (p. 227; Fig. 8.1)

4. The peripheral nervous system has **sensory (afferent) neurons** that bring information into the CNS, and **efferent neurons** that carry information away from the CNS back to various parts of the body. (p. 227)

5. The efferent neurons include **somatic motor neurons,** which control skeletal muscles, and **autonomic neurons,** which control smooth and cardiac muscles, glands, and some adipose tissue. (p. 229)

6. Autonomic neurons are subdivided into **sympathetic** and **parasympathetic** branches. (p. 229)

Cells of the Nervous System

iP Nervous System I: Anatomy Review

7. Neurons have a **cell body** with a nucleus and organelles to direct cellular activity, **dendrites** to receive incoming signals, and an **axon** to transmit electrical signals from the cell body to the **axon terminal.** (pp. 229, 231; Fig. 8.2)

8. **Interneurons** are neurons that lie entirely within the CNS. (p. 229; Fig. 8.2c, d)

9. Material is transported between the cell body and axon terminal by **axonal transport.** (p. 232; Fig. 8.3)

10. The region where an axon terminal meets its target cell is called a **synapse.** The target cell is called the **postsynaptic cell,** and the neuron that releases the chemical signal is known as the **presynaptic cell.** The region between these two cells is the **synaptic cleft.** (p. 232; Fig. 8.2f)

11. Developing neurons find their way to their targets by using chemical signals. (p. 232)

12. **Glial cells** provide physical support and communicate with neurons. **Schwann cells** and **satellite cells** are glial cells associated with the peripheral nervous system. **Oligodendrocytes, astrocytes, microglia,** and **ependymal cells** are glial cells found in the CNS. Microglia are modified immune cells that act as scavengers. (pp. 233, 235; Fig. 8.5)

13. Schwann cells and oligodendrocytes form insulating **myelin sheaths** around neurons. The **nodes of Ranvier** are sections of uninsulated membrane occurring at intervals along the length of an axon. (p. 233; Fig. 8.5c)

14. **Neural stem cells** that can develop into new neurons and glia are found in the ependymal layer as well as in other parts of the nervous system. (p. 235)

Electrical Signals in Neurons

iP Nervous System I: The Membrane Potential; Ion Channels; The Action Potential

15. The **Nernst equation** describes the membrane potential of a cell that is permeable to only one ion. (p. 236)

16. Membrane potential is influenced by the concentration gradients of ions across the membrane and by the permeability of the membrane to those ions. (p. 237)

17. The **Goldman-Hodgkin-Katz (GHK) equation** predicts membrane potential based on ion concentration gradients and membrane permeability for multiple ions. (p. 237)

18. The permeability of a cell to ions changes when ion channels in the membrane open and close. Movement of only a few ions significantly changes the membrane potential. (p. 238)

19. Gated ion channels in neurons open or close in response to chemical or mechanical signals or in response to depolarization of the cell membrane. Channels also close through inactivation. (p. 238)

20. Current flow (I) obeys **Ohm's Law:** I = voltage/resistance. **Resistance** to current flow comes from the cell membrane, which is a good insulator, and from the cytoplasm. **Conductance** (G) is the reciprocal of resistance: G = 1/R. (pp. 238, 239)

21. **Graded potentials** are depolarizations or hyperpolarizations whose strength is directly proportional to the strength of the triggering event. Graded potentials lose strength as they move through the cell. (p. 240; Tbl. 8.3; Fig. 8.7)

22. The wave of depolarization that moves through a cell is known as **local current flow.** (p. 240)

iP2 Generation of an Action Potential

23. **Action potentials** are rapid electrical signals that travel undiminished in amplitude (strength) down the axon from the cell body to the axon terminals. (p. 240)

24. Action potentials begin in the **trigger zone** if a single graded potential or the sum of multiple graded potentials exceeds the **threshold** voltage. (p. 242; Fig. 8.7c)

25. Depolarizing graded potentials make a neuron more likely to fire an action potential. Hyperpolarizing graded potentials make a neuron less likely to fire an action potential. (p. 242)

26. Action potentials are uniform, **all-or-none** depolarizations that can travel undiminished over long distances. (p. 242)

27. The rising phase of the action potential is due to increased Na^+ permeability. The falling phase of the action potential is due to increased K^+ permeability. (p. 243; Fig. 8.9)

28. The voltage-gated Na^+ channels of the axon have a fast **activation gate** and a slower **inactivation gate.** (p. 245; Fig. 8.10)

29. Very few ions cross the membrane during an action potential. The Na^+-K^+-ATPase eventually restores Na^+ and K^+ to their original compartments. (p. 245)

30. Once an action potential has begun, there is a brief period of time known as the **absolute refractory period** during which a second action potential cannot be triggered, no matter how large the stimulus. Because of this, action potentials cannot be summed. (p. 246; Fig. 8.12)

31. During the **relative refractory period,** a higher-than-normal graded potential is required to trigger an action potential. (p. 246)

32. The myelin sheath around an axon speeds up conduction by increasing membrane resistance and decreasing current leakage. Larger-diameter axons conduct action potentials faster than smaller-diameter axons do. (p. 249)

33. The apparent jumping of action potentials from node to node is called **saltatory conduction.** (p. 249; Fig. 8.16)

34. Changes in blood K^+ concentration affect resting membrane potential and the conduction of action potentials. (p. 252; Fig. 8.17)

Cell-to-Cell Communication in the Nervous System

iP Nervous System II: Anatomy Review; Synaptic Transmission; Ion Channels

35. In **electrical synapses,** an electrical signal passes directly from the cytoplasm of one cell to another through gap junctions. **Chemical**

synapses use neurotransmitters to carry information from one cell to the next, with the neurotransmitters diffusing across the synaptic cleft to bind with receptors on target cells. (pp. 253, 254)

36. Neurotransmitters come in a variety of forms. **Cholinergic** neurons secrete **acetylcholine. Adrenergic neurons** secrete **norepinephrine. Glutamate, GABA, serotonin, adenosine,** and **nitric oxide** are other major neurotransmitters. (pp. 254, 255, 256; Tbl. 8.4)

37. Neurotransmitter receptors are either ligand-gated ion channels (ionotropic receptors) or G protein-coupled receptors (metabotropic receptors). (p. 254)

38. Neurotransmitters are synthesized in the cell body or in the axon terminal. They are stored in **synaptic vesicles** and are released by exocytosis when an action potential reaches the axon terminal. (p. 257; Fig. 8.19a)

39. Neurotransmitter action is rapidly terminated by reuptake into cells, diffusion away from the synapse, or enzymatic breakdown. (p. 258; Fig. 8.19b)

40. Information about the strength and duration of a stimulus is conveyed by the amount of neurotransmitter released. Increased frequency of action potentials releases more neurotransmitter. (p. 260; Fig. 8.21)

Integration of Neural Information Transfer

iP **Nervous System II: Synaptic Potentials & Cellular Integration**

41. When a presynaptic neuron synapses on a larger number of postsynaptic neurons, the pattern is known as **divergence.** When several presynaptic neurons provide input to a smaller number of postsynaptic neurons, the pattern is known as **convergence.** (p. 260; Fig. 8.22)

42. Synaptic transmission can be modified in response to activity at the synapse, a process known as **synaptic plasticity.** (p. 261)

43. G protein-coupled receptors either create **slow synaptic potentials** or modify cell metabolism. Ion channels create **fast synaptic potentials.** (pp. 261, 263; Fig. 8.23)

44. The summation of simultaneous graded potentials from different neurons is known as **spatial summation.** The summation of graded potentials that closely follow each other sequentially is called **temporal summation.** (p. 266; Fig. 8.24)

45. **Presynaptic modulation** of an axon terminal allows selective modulation of collaterals and their targets. **Postsynaptic modulation** occurs when a modulatory neuron synapses on a postsynaptic cell body or dendrites. (p. 266; Fig. 8.24)

46. **Long-term potentiation** and **long-term depression** are mechanisms by which neurons change the strength of their synaptic connections. (p. 267; Fig. 8.25)

REVIEW QUESTIONS

In addition to working through these questions and checking your answers on p. A-10, review the Learning Outcomes at the beginning of this chapter.

Level One Reviewing Facts and Terms

1. List the three functional classes of neurons, and explain how they differ structurally and functionally.

2. Somatic motor neurons control _____, and _____ neurons control smooth and cardiac muscles, glands, and some adipose tissue.

3. Autonomic neurons are classified as either _____ or _____ neurons.

4. Match each term with its description:

(a) axon	1. process of a neuron that receives incoming signals
(b) dendrite	
(c) afferent	2. sensory neuron, transmits information to CNS
(d) efferent	3. long process that transmits signals to the target cell
(e) trigger zone	4. region of neuron where action potential begins
	5. neuron that transmits information from CNS to the rest of the body

5. Name the two primary cell types found in the nervous system.

6. Draw a typical neuron and label the cell body, axon, dendrites, nucleus, trigger zone, axon hillock, collaterals, and axon terminals. Draw mitochondria, rough endoplasmic reticulum, Golgi complex, and vesicles in the appropriate sections of the neuron.

7. Axonal transport refers to the

   (a) release of neurotransmitters into the synaptic cleft.
   (b) use of microtubules to send secretions from the cell body to the axon terminal.
   (c) movement of organelles and cytoplasm up and down the axon.
   (d) movement of the axon terminal to synapse with a new postsynaptic cell.
   (e) none of these

8. Match the numbers of the appropriate characteristics with the two types of potentials. Characteristics may apply to one or both types.

(a) action potential	1. all-or-none
(b) graded potential	2. can be summed
	3. amplitude decreases with distance
	4. exhibits a refractory period
	5. amplitude depends on strength of stimulus
	6. has no threshold

9. Arrange the following events in the proper sequence:

   (a) Efferent neuron reaches threshold and fires an action potential.
   (b) Afferent neuron reaches threshold and fires an action potential.
   (c) Effector organ responds by performing output.
   (d) Integrating center reaches decision about response.
   (e) Sensory organ detects change in the environment.

10. List the four major types of ion channels found in neurons. Are they chemically gated, mechanically gated, or voltage-gated?

11. Match the glial cell(s) on the right to the functions on the left. There may be more than one correct answer for each function.

(a) modified immune cells	1. astrocytes
(b) help form the blood-brain barrier	2. ependymal cells
	3. microglia
(c) form myelin	4. oligodendrocytes
(d) separate CNS fluid compartments	5. satellite cells
	6. Schwann cells
(e) found in peripheral nervous system	
(f) found in ganglia	

12. An action potential is (circle all correct answers)

    (a) a reversal of the Na^+ and K^+ concentrations inside and outside the neuron.
    (b) the same size and shape at the beginning and end of the axon.
    (c) initiated by inhibitory postsynaptic graded potentials.
    (d) transmitted to the distal end of a neuron and causes release of neurotransmitter.

13. Choose from the following ions to fill in the blanks correctly: Na^+, K^+, Ca^{2+}, Cl^-

    (a) The resting cell membrane is more permeable to _____ than to _____. Although _____ contribute little to the resting membrane potential, they play a key role in generating electrical signals in excitable tissues.
    (b) The concentration of _____ is 12 times greater outside the cell than inside.
    (c) The concentration of _____ is 30 times greater inside the cell than outside.
    (d) An action potential occurs when _____ enter the cell.
    (e) The resting membrane potential is due to the high _____ permeability of the cell.

14. What is the myelin sheath?

15. List two factors that enhance conduction speed.

16. List three ways neurotransmitters are removed from the synapse.

17. Draw and label a graph of an action potential. Below the graph, draw the positioning of the K^+ and Na^+ channel gates during each phase.

Level Two Reviewing Concepts

18. What causes the depolarization phase of an action potential? (Circle all that apply.)

    (a) K^+ leaving the cell through voltage-gated channels
    (b) K^+ being pumped into the cell by the Na^+-K^+-ATPase
    (c) Na^+ being pumped into the cell by the Na^+-K^+-ATPase
    (d) Na^+ entering the cell through voltage-gated channels
    (e) opening of the Na^+ channel inactivation gate

19. Name any four neurotransmitters, their receptor(s), and tell whether the receptor is an ion channel or a GPCR.

20. Create a map showing the organization of the nervous system using the following terms, plus any terms you choose to add:

• afferent signals	• neuron
• astrocyte	• neurotransmitter
• autonomic division	• oligodendrocyte
• brain	• parasympathetic division
• CNS	• peripheral division
• efferent neuron	• satellite cell
• ependymal cell	• Schwann cell
• glands	• sensory division
• glial cells	• somatic motor division
• integration	• spinal cord
• interneuron	• stimulus
• microglia	• sympathetic division
• muscles	• target

21. Arrange the following terms to describe the sequence of events after a neurotransmitter binds to a receptor on a postsynaptic neuron. Terms may be used more than once or not at all.

    (a) action potential fires at axon hillock
    (b) trigger zone reaches threshold
    (c) cell depolarizes
    (d) exocytosis
    (e) graded potential occurs
    (f) ligand-gated ion channel opens
    (g) local current flow occurs
    (h) saltatory conduction occurs
    (i) voltage-gated Ca^{2+} channels open
    (j) voltage-gated K^+ channels open
    (k) voltage-gated Na^+ channels open

22. Match the best term (hyperpolarize, depolarize, repolarize) to the following events. The cell in question has a resting membrane potential of –70 mV.

    (a) membrane potential changes from –70 mV to –50 mV
    (b) membrane potential changes from –70 mV to –90 mV
    (c) membrane potential changes from +20 mV to –60 mV
    (d) membrane potential changes from –80 mV to –70 mV

23. A neuron has a resting membrane potential of –70 mV. Will the neuron hyperpolarize or depolarize when each of the following events occurs? (More than one answer may apply; list all those that are correct.)

    (a) Na^+ enters the cell
    (b) K^+ leaves the cell
    (c) Cl^- enters the cell
    (d) Ca^{2+} enters the cell

24. If all action potentials within a given neuron are identical, how does the neuron transmit information about the strength and duration of the stimulus?

25. The presence of myelin allows an axon to (choose all correct answers)

    (a) produce more frequent action potentials.
    (b) conduct impulses more rapidly.
    (c) produce action potentials of larger amplitude.
    (d) produce action potentials of longer duration.

26. Define, compare, and contrast the following concepts:

    (a) threshold, subthreshold, suprathreshold, all-or-none, overshoot, undershoot
    (b) graded potential, EPSP, IPSP
    (c) absolute refractory period, relative refractory period
    (d) afferent neuron, efferent neuron, interneuron
    (e) sensory neuron, somatic motor neuron, sympathetic neuron, autonomic neuron, parasympathetic neuron
    (f) fast synaptic potential, slow synaptic potential
    (g) temporal summation, spatial summation
    (h) convergence, divergence

Level Three Problem Solving

27. If human babies' muscles and neurons are fully developed and functional at birth, why can't they focus their eyes, sit up, or learn to crawl within hours of being born? (*Hint:* Muscle strength is not the problem.)

28. The voltage-gated Na^+ channels of a neuron open when the neuron depolarizes. If depolarization opens the channels, what makes them close when the neuron is maximally depolarized?

29. One of the pills that Jim takes for high blood pressure caused his blood K^+ level to decrease from 4.5 mM to 2.5 mM. What happens to the resting membrane potential of his liver cells? (Circle all that are correct.)

    (a) decreases
    (b) increases
    (c) does not change
    (d) becomes more negative
    (e) becomes less negative
    (f) fires an action potential
    (g) depolarizes
    (h) hyperpolarizes
    (i) repolarizes

30. Characterize each of the following stimuli as being mechanical, chemical, or thermal:

    (a) bath water at 106 °F
    (b) acetylcholine
    (c) a hint of perfume
    (d) epinephrine
    (e) lemon juice
    (f) a punch on the arm

31. An unmyelinated axon has a much greater requirement for ATP than a myelinated axon of the same diameter and length. Can you explain why?

Level Four Quantitative Problems

32. The GHK equation is sometimes abbreviated to exclude chloride, which plays a minimal role in membrane potential for most cells. In addition, because it is difficult to determine absolute membrane permeability values for Na^+ and K^+, the equation is revised to use the ratio of the two ion permeabilities, expressed as $\alpha = P_{Na}/P_K$:

$$V_m = 61 \log \frac{[K^+]_{out} + \alpha [Na^+]_{out}}{[K^+]_{in} + \alpha [Na^+]_{in}}$$

Thus, if you know the relative membrane permeabilities of the two ions and their intracellular (ICF) and extracellular (ECF) concentrations, you can predict the membrane potential for a cell.

    (a) A resting cell has an alpha value of 0.025 and the following ion concentrations:

$$Na^+\!: ICF = 5\ mM,\ ECF = 135\ mM$$
$$K^+\!: ICF = 150\ mM,\ ECF = 4\ mM$$

What is the cell's membrane potential?

    (b) The Na^+ permeability of the cell in (a) suddenly increases so that $\alpha = 20$. Now what is the cell's membrane potential?
    (c) Mrs. Nguyen has high blood pressure, and her physician puts her on a drug whose side effect decreases her plasma (ECF) K^+ from 4 mM to 2.5 mM. Using the other values in (a), now what is the membrane potential?
    (d) The physician prescribes a potassium supplement for Mrs. Nguyen, who decides that if two pills are good, four must be better. Her plasma (ECF) K^+ now goes to 6 mM. What happens to the membrane potential?

33. In each of the following scenarios, will an action potential be produced? The postsynaptic neuron has a resting membrane potential of –70 mV.

    (a) Fifteen neurons synapse on one postsynaptic neuron. At the trigger zone, 12 of the neurons produce EPSPs of 2 mV each, and the other three produce IPSPs of 3 mV each. The threshold for the postsynaptic cell is –50 mV.
    (b) Fourteen neurons synapse on one postsynaptic neuron. At the trigger zone, 11 of the neurons produce EPSPs of 2 mV each, and the other three produce IPSPs of 3 mV each. The threshold for the postsynaptic cell is –60 mV.
    (c) Fifteen neurons synapse on one postsynaptic neuron. At the trigger zone, 14 of the neurons produce EPSPs of 2 mV each, and the other one produces an IPSP of 9 mV. The threshold for the postsynaptic cell is –50 mV.

Answers to Concept Checks, Figure and Graph Questions, and end-of-chapter Review Questions can be found in Appendix A [A-1].

9

The Central Nervous System

Neuronal assemblies have important properties that cannot be explained by the additive qualities of individual neurons.

O. Hechter, in *Biology and Medicine into the 21st Century*, 1991

EMERGENT PROPERTIES OF NEURAL NETWORKS 275

LO 9.1 Explain and give examples of emergent properties of neural systems in humans and other organisms.

EVOLUTION OF NERVOUS SYSTEMS 275

LO 9.2 Describe how nervous systems increase in complexity from Cnidarians to mammals.

ANATOMY OF THE CENTRAL NERVOUS SYSTEM 277

LO 9.3 Describe how a hollow neural tube develops into the ventricles and seven major divisions of the CNS.

LO 9.4 Define grey matter, white matter, tracts, and nuclei in the CNS.

LO 9.5 Starting at the skull and moving inward, name the membranes and other structures that enclose the brain.

LO 9.6 Explain the formation, distribution, and functions of cerebrospinal fluid.

LO 9.7 Describe the structure and functions of the blood-brain barrier.

THE SPINAL CORD 284

LO 9.8 Explain how the following structures are organized in the spinal cord: ascending and descending tracts, columns, dorsal root ganglia, dorsal and ventral horns, dorsal and ventral roots, propriospinal tracts, spinal nerves

THE BRAIN 285

LO 9.9 Name the major subdivisions of the cerebrum, cerebellum, diencephalon, and brain stem, and give their major functions.

BRAIN FUNCTION 290

LO 9.10 Name the four lobes of the cerebral cortex and explain which sensory, motor, or association areas are associated with each lobe.

LO 9.11 Explain the behavioral state system and how it is related to the diffuse modulatory systems and the reticular activating system.

LO 9.12 Describe the stages of sleep.

LO 9.13 Describe motivation and emotion and how they are related to brain function.

LO 9.14 Explain the role of the following in learning and memory: short-term memory, memory traces, working memory, associative and nonassociative learning, and habituation and sensitization.

LO 9.15 Explain the roles of Wernicke's area and Broca's area in written and spoken language.

BACKGROUND BASICS

Directions of the body: inside back cover
51 Up-regulation
136 Diffusion through membranes
72 Cell-to-cell junctions
232 Synapses
233 Glial cells
79 Transporting epithelium
243 Ionic basis of action potentials
168 Neurotransmitters and neuromodulators
209 Posterior pituitary
261 Fast and slow synaptic potentials
267 Long-term potentiation

Genetically modified neurons in mouse

BRAIN is not just an organ. In 2013, it became the acronym for an ambitious new research initiative, Brain Research through Advancing Innovative Neurotechnologies, funded through the National Institutes of Health (NIH). BRAIN and a related initiative, the NIH-funded Human Connectome project (*www.humanconnectomeproject.org*), are large-scale research programs whose goal is to map the structural and functional organization of the human brain in health and disease. Once we better understand how the human brain works, the possibilities for treating brain disorders become limitless. Researchers already have implantable electrodes that can reduce severe depression and even allow paralyzed subjects to control external objects. Why not wireless devises to restore memory loss or to wipe out distressing memories in posttraumatic stress syndrome? These projects are years in the future, but as scientists work toward them, we are learning more and more about the complex circuits of the brain and how they function.

EMERGENT PROPERTIES OF NEURAL NETWORKS

Neurons in the nervous system link together to form circuits that have specific functions. The most complex circuits are those of the brain, in which billions of neurons are linked into intricate networks that converge and diverge, creating an infinite number of possible pathways. Signaling within these pathways creates thinking, language, feeling, learning, and memory—the complex behaviors that make us human. Some neuroscientists have proposed that the functional unit of the nervous system be changed from the individual neuron to neural networks because even the most basic functions require circuits of neurons.

How is it that combinations of neurons linked together into chains or networks collectively possess emergent properties not found in any single neuron? We do not yet have an answer to this

RUNNING **PROBLEM** | Infantile Spasms

At four months of age, Ben could roll over, hold up his head, and reach for things. At seven months, he was nearly paralyzed and lay listlessly in his crib. He had lost his abilities so gradually that it was hard to remember when each one had slipped away, but his mother could remember exactly when it began. She was preparing to feed him lunch one day when she heard a cry from the highchair where Ben was sitting. As she watched, Ben's head dropped to his chest, came back up, then went hurtling toward his lap, smacking into his highchair tray. Ben's mother snatched him up into her arms, and she could feel him still convulsing against her shoulder. This was the first of many such spells that came with increasing frequency and duration.

275 283 298 300 302 304

question. Some scientists seek to answer it by looking for parallels between the nervous system and the integrated circuits of computers.

Computer programs have been written that attempt to mimic the thought processes of humans. This field of study, called *artificial intelligence,* has created some interesting programs, such as the "psychiatrist" programmed to respond to typed complaints with appropriate comments and suggestions. We are nowhere near creating a brain as complex as that of a human, however, or even one as complex as that of Hal, the computer in the classic movie *2001: A Space Odyssey.*

Probably one reason computers cannot yet accurately model brain function is that computers lack *plasticity,* the ability to change circuit connections and function in response to sensory input and past experience [p. 261]. Although some computer programs can change their output under specialized conditions, they cannot begin to approximate the plasticity of human brain networks, which easily restructure themselves as the result of sensory input, learning, emotion, and creativity. In addition, we now know that the brain can add new connections when neural stem cells differentiate. Computers cannot add new circuits to themselves.

How can simply linking neurons together create **affective behaviors,** which are related to feeling and emotion, and **cognitive behaviors** {*cognoscere,* to get to know} related to thinking? In their search for the organizational principles that lead to these behaviors, scientists seek clues in the simplest animal nervous systems.

EVOLUTION OF NERVOUS SYSTEMS

All animals have the ability to sense and respond to changes in their environment. Even single-cell organisms such as *Paramecium* are able to carry out the basic tasks of life: finding food, avoiding becoming food, finding a mate. Yet, these unicellular organisms have no obvious brain or integrating center. They use the resting membrane potential that exists in living cells and many of the same ion channels as more complex animals to coordinate their daily activities.

Some of the first multicellular animals to develop neurons were members of the phylum *Cnidaria,* the jellyfish and sea anemones. Their nervous system is a *nerve net* composed of sensory neurons, connective interneurons, and motor neurons that innervate muscles and glands (**FIG. 9.1a**). These animals respond to stimuli with complex behaviors, yet without input from an identifiable control center. If you watch a jellyfish swim or a sea anemone maneuver a piece of shrimp into its mouth, it is hard to imagine how a diffuse network of neurons can create such complex coordinated movements. However, the same basic principles of neural communication apply to jellyfish and humans. Electrical signals in the form of action potentials, and chemical signals passing across synapses, are the same in all animals. It is only in the number and organization of the neurons that one species differs from another.

FIG. 9.1 Evolution of the nervous system

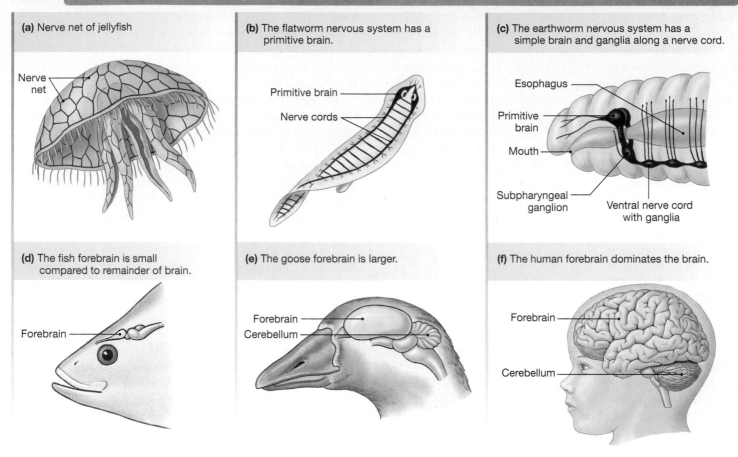

(a) Nerve net of jellyfish

Nerve net

(b) The flatworm nervous system has a primitive brain.

Primitive brain

Nerve cords

(c) The earthworm nervous system has a simple brain and ganglia along a nerve cord.

Esophagus

Primitive brain

Mouth

Subpharyngeal ganglion

Ventral nerve cord with ganglia

(d) The fish forebrain is small compared to remainder of brain.

Forebrain

(e) The goose forebrain is larger.

Forebrain
Cerebellum

(f) The human forebrain dominates the brain.

Forebrain

Cerebellum

CONCEPT CHECK

1. Match each of the following terms with the appropriate neuron type(s).

   (a) afferent neuron
   (b) efferent signal
   (c) integrating center
   (d) input signal
   (e) output signal

   1. interneuron
   2. motor neuron
   3. sensory neuron

In the primitive flatworms, we see the beginnings of a nervous system as we know it in higher animals, although in flatworms the distinction between central nervous system and peripheral nervous system is not clear. Flatworms have a rudimentary brain consisting of a cluster of nerve cell bodies concentrated in the head, or *cephalic* region. Two large nerves called *nerve cords* come off the primitive brain and lead to a nerve network that innervates distal regions of the flatworm body (Fig. 9.1b).

The segmented worms, or annelids, such as the earthworm, have a more advanced central nervous system (Fig. 9.1c). Clusters of cell bodies are no longer restricted to the head region, as they are in flatworms, but also occur in fused pairs, called *ganglia* (singular *ganglion*) [p. 233], along a nerve cord. Because each segment of the worm contains a ganglion, simple reflexes can be integrated within a segment without input from the brain. Reflexes that do not require integration in the brain also occur in higher animals and are called **spinal reflexes** in humans and other vertebrates.

Annelids and higher invertebrates have complex reflexes controlled through neural networks. Researchers use leeches (a type of annelid) and *Aplysia,* a type of shell-less mollusk, to study neural networks and synapse formation because the neurons in these species are 10 times larger than human brain neurons, and because the networks have the same organization of neurons from animal to animal. The neural function of these invertebrates provides a simple model that we can apply to more complex vertebrate networks.

Nerve cell bodies clustered into brains persist throughout the more advanced phyla and become increasingly more complex. One advantage to cephalic brains is that in most animals, the head is the part of the body that first contacts the environment as the animal moves. For this reason, as brains evolved, they became associated with specialized cephalic receptors, such as eyes for vision and chemoreceptors for smell and taste.

In the higher arthropods, such as insects, specific regions of the brain are associated with particular functions. More complex brains are associated with complex behaviors, such as the ability of social insects like ants and bees to organize themselves into colonies, divide labor, and communicate with one another. The octopus (a cephalopod mollusk) has the most sophisticated brain development among the invertebrates, as well as the most sophisticated behavior.

In vertebrate brain evolution, the most dramatic change is seen in the *forebrain* region {*fore,* in front}, which includes the **cerebrum** {*cerebrum,* brain; adjective *cerebral*}. In fish, the forebrain is a small bulge dedicated mainly to processing olfactory information about odors in the environment (Fig. 9.1d). In birds and rodents, part of the forebrain has enlarged into a cerebrum with a smooth surface (Fig. 9.1e).

In humans, the cerebrum is the largest and most distinctive part of the brain, with deep grooves and folds (Fig. 9.1f). More than anything else, the cerebrum is what makes us human. All evidence indicates that it is the part of the brain that allows reasoning and cognition.

The other brain structure whose evolution is obvious in the vertebrates is the *cerebellum,* a region of the *hindbrain* devoted to coordinating movement and balance. Birds (Fig. 9.1e) and humans (Fig. 9.1f) both have well-developed cerebellar structures. The cerebellum, like the cerebrum, is readily identifiable in these animals by its grooves and folds.

In this chapter, we begin with an overview of CNS anatomy and functions. We then look at how neural networks create the higher brain functions of thought and emotion.

ANATOMY OF THE CENTRAL NERVOUS SYSTEM

The vertebrate central nervous system (CNS) consists of the brain and the spinal cord. As you learned in the previous section, brains increase in complexity and degree of specialization as we move up the phylogenetic tree from fish to humans. However, if we look at the vertebrate nervous system during development, a basic anatomical pattern emerges. In all vertebrates, the CNS consists of layers of neural tissue surrounding a fluid-filled central cavity lined with epithelium.

The CNS Develops from a Hollow Tube

In the very early embryo, cells that will become the nervous system lie in a flattened region called the **neural plate.** As development proceeds (at about day 20 of human development), neural plate cells along the edge migrate toward the midline (**FIG. 9.2a**).

By about day 23 of human development, the neural plate cells have fused with each other, creating a **neural tube** (Fig. 9.2b). *Neural crest cells* from the lateral edges of the neural plate now lie dorsal to the neural tube. The lumen of the neural tube will remain hollow and become the central cavity of the CNS.

The cells lining the neural tube will either differentiate into the epithelial *ependyma* [p. 235] or remain as undifferentiated *neural stem cells.* The outer cell layers of the neural tube will become the neurons and glia of the CNS. Neural crest cells will become the sensory and motor neurons of the peripheral nervous system.

By week 4 of human development, the anterior portion of the neural tube has begun to specialize into the regions of the brain (Fig. 9.2c). Three divisions are obvious: a **forebrain,** a **midbrain,** and a **hindbrain.** The tube posterior to the hindbrain will become the spinal cord. At this stage, the portion of the forebrain that will become the cerebrum is not much larger than the other regions of the brain.

As development proceeds, the growth of the cerebrum begins to outpace that of the other regions (Fig. 9.2d). By week 6, the CNS has formed the seven major divisions that are present at birth. Six of these regions are in the brain—(1) the cerebrum, (2) the *diencephalon,* (3) the midbrain, (4) and (5) the cerebellum and *pons,* (6) the *medulla oblongata*—and the seventh is the spinal cord. The cerebrum and diencephalon develop from the forebrain. The cerebellum, pons, and medulla oblongata are divisions of the hindbrain.

By week 6 the central cavity (lumen) of the neural tube has begun to enlarge into the hollow **ventricles** {*ventriculus,* belly} of the brain. There are two *lateral ventricles* (the first and second) and two *descending ventricles* (the third and fourth). The central cavity of the neural tube also becomes the *central canal* of the spinal cord.

By week 11, the cerebrum is noticeably enlarged (Fig. 9.2e), and at birth, the cerebrum is the largest and most obvious structure we see when looking at a human brain (Fig. 9.2f). The fully developed cerebrum surrounds the diencephalon, midbrain, and pons, leaving only the cerebellum and medulla oblongata visible below it. Because of the flexion (bending) of the neural tube early in development (see Fig. 9.2c), some directional terms have different meanings when applied to the brain (Fig. 9.2g).

The CNS Is Divided into Gray Matter and White Matter

The central nervous system, like the peripheral nervous system, is composed of neurons and supportive glial cells. Interneurons are those neurons completely contained within the CNS. Sensory (afferent) and efferent neurons link interneurons to peripheral receptors and effectors.

When viewed on a macroscopic level, the tissues of the CNS are divided into gray matter and white matter (**FIG. 9.3c**). **Gray matter** consists of unmyelinated nerve cell bodies, dendrites, and axons. The cell bodies are assembled in an organized fashion in both the brain and the spinal cord. They form layers in some parts of the brain and in other parts cluster into groups of neurons that have similar functions. Clusters of cell bodies in the brain and spinal cord are known as *nuclei.* Nuclei are usually identified by specific names—for example, the *lateral geniculate nucleus,* where visual information is processed.

White matter is mostly myelinated axons and contains very few cell bodies. Its pale color comes from the myelin sheaths that surround the axons. Bundles of axons that connect different regions of the CNS are known as **tracts.** Tracts in the central nervous system are equivalent to nerves in the peripheral nervous system.

FIG. 9.2 **ESSENTIALS**

Development of the Human Nervous System

(a) Day 20

In the 20-day embryo (dorsal view), neural plate cells (purple) migrate toward the midline. Neural crest cells migrate with the neural plate cells.

Neural crest

Neural plate

(b) Day 23

By day 23 of embryonic development, neural tube formation is almost complete.

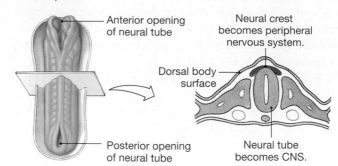

Anterior opening of neural tube

Neural crest becomes peripheral nervous system.

Dorsal body surface

Posterior opening of neural tube

Neural tube becomes CNS.

(c) 4 Weeks

A 4-week human embryo showing the anterior end of the neural tube which has specialized into three brain regions.

Hindbrain

Forebrain Midbrain Spinal cord

Lumen of neural tube

(d) 6 Weeks

At 6 weeks, the neural tube has differentiated into the brain regions present at birth. The central cavity (lumen) shown in the cross section will become the ventricles of the brain (see Fig. 9.4).

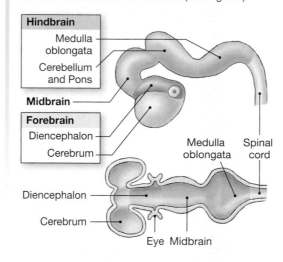

Hindbrain
Medulla oblongata
Cerebellum and Pons

Midbrain

Forebrain
Diencephalon
Cerebrum

Medulla oblongata   Spinal cord

Diencephalon

Cerebrum

Eye Midbrain

(e) 11 Weeks

By 11 weeks of embryonic development, the growth of the cerebrum is noticeably more rapid than that of the other divisions of the brain.

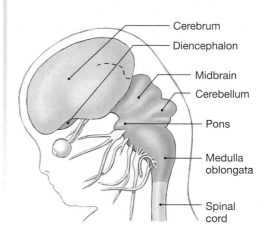

Cerebrum

Diencephalon

Midbrain

Cerebellum

Pons

Medulla oblongata

Spinal cord

(f) 40 Weeks

At birth, the cerebrum has covered most of the other brain regions. Its rapid growth within the rigid confines of the cranium forces it to develop a convoluted, furrowed surface.

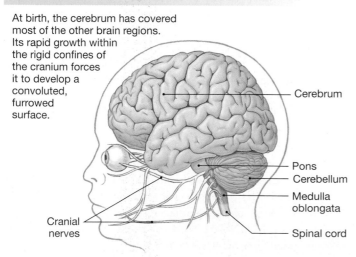

Cerebrum

Pons
Cerebellum
Medulla oblongata

Cranial nerves

Spinal cord

(g) Child

The directions "dorsal" and "ventral" are different in the brain because of flexion in the neural tube during development.

Dorsal (superior)

Rostral

Caudal

Rostral

Ventral (inferior)

Ventral (anterior)

Dorsal (posterior)

Caudal

FIG. 9.3 **ANATOMY SUMMARY**

The Central Nervous System

(a) Posterior View of the CNS

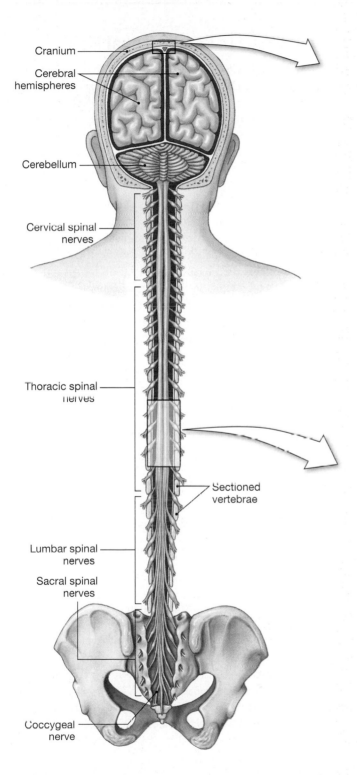

Cranium

Cerebral hemispheres

Cerebellum

Cervical spinal nerves

Thoracic spinal nerves

Sectioned vertebrae

Lumbar spinal nerves

Sacral spinal nerves

Coccygeal nerve

(b) Sectional View of the Meninges

The meninges and extracellular fluid cushion the delicate brain tissue.

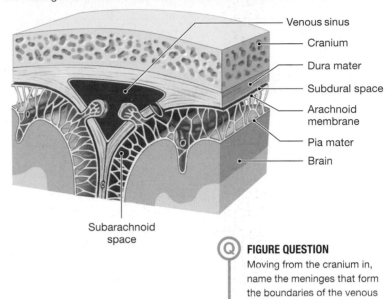

Venous sinus

Cranium

Dura mater

Subdural space

Arachnoid membrane

Pia mater

Brain

Subarachnoid space

Q **FIGURE QUESTION**

Moving from the cranium in, name the meninges that form the boundaries of the venous sinus and the subdural and subarachnoid spaces.

(c) Posterior View of Spinal Cord and Vertebra

Central canal

Gray matter

White matter

Spinal nerve

Spinal cord

Body of vertebra

Pia mater

Arachnoid membrane

Dura mater

Meninges

Autonomic ganglion

Spinal nerve

The consistency of the brain and spinal cord is soft and jellylike. Although individual neurons and glial cells have highly organized internal cytoskeletons that maintain cell shape and orientation, neural tissue has minimal extracellular matrix and must rely on external support for protection from trauma. This support comes in the form of an outer casing of bone, three layers of connective tissue membrane, and fluid between the membranes (Fig. 9.3b, c).

CONCEPT CHECK

2. Name the four kinds of glial cells found in the CNS, and describe the function(s) of each [p. 233].

Bone and Connective Tissue Support the CNS

In vertebrates, the brain is encased in a bony **skull,** or **cranium** (Fig. 9.3a), and the spinal cord runs through a canal in the **vertebral column.** The body segmentation that is characteristic of many invertebrates can still be seen in the bony **vertebrae** (singular *vertebra*), which are stacked on top of one another and separated by disks of connective tissue. Nerves of the peripheral nervous system enter and leave the spinal cord by passing through notches between the stacked vertebrae (Fig. 9.3c).

Three layers of membrane, collectively called the **meninges** {singular *meninx*, membrane}, lie between the bones and tissues of the central nervous system. These membranes help stabilize the neural tissue and protect it from bruising against the bones of the skeleton. Starting from the bones and moving toward the neural tissue, the membranes are (1) the dura mater, (2) the arachnoid membrane, and (3) the pia mater (Fig. 9.3b, c).

The **dura mater** {*durare*, to last + *mater*, mother} is the thickest of the three membranes (think *durable*). It is associated with veins that drain blood from the brain through vessels or cavities called *sinuses*. The middle layer, the **arachnoid** {*arachnoides*, cobweblike} **membrane,** is loosely tied to the inner membrane, leaving a *subarachnoid space* between the two layers. The inner membrane, the **pia mater** {*pius*, pious + *mater*, mother}, is a thin membrane that adheres to the surface of the brain and spinal cord. Arteries that supply blood to the brain are associated with this layer.

The final protective component of the CNS is extracellular fluid, which helps cushion the delicate neural tissue. The cranium has an internal volume of 1.4 L, of which about 1 L is occupied by the cells. The remaining volume is divided into two distinct extracellular compartments: the blood (100–150 mL), and the *cerebrospinal fluid* and interstitial fluid (250–300 mL). The cerebrospinal fluid and interstitial fluid together form the extracellular environment for neurons. Interstitial fluid lies inside the pia mater. Cerebrospinal fluid is found in the ventricles and in the space between the pia mater and the arachnoid membrane. The cerebrospinal and interstitial fluid compartments communicate with each other across the leaky junctions of the pial membrane and the ependymal cell layer lining the ventricles.

CONCEPT CHECK

3. What is a ganglion? What is the equivalent structure in the CNS?

4. Peripheral nerves are equivalent to what organizational structure in the CNS?

The Brain Floats in Cerebrospinal Fluid

Cerebrospinal fluid (CSF) is a salty solution that is continuously secreted by the **choroid plexus,** a specialized region on the walls of the ventricles (**FIG. 9.4b**). The choroid plexus is remarkably similar to kidney tissue and consists of capillaries and a transporting epithelium [p. 79] derived from the ependyma. The choroid plexus cells selectively pump sodium and other solutes from plasma into the ventricles, creating an osmotic gradient that draws water along with the solutes (Fig. 9.4c).

From the ventricles, cerebrospinal fluid flows into the **subarachnoid space** between the pia mater and the arachnoid membrane, surrounding the entire brain and spinal cord in fluid (Fig. 9.4b). The cerebrospinal fluid flows around the neural tissue and is finally absorbed back into the blood by special **villi** {singular *villus*, shaggy hair} on the arachnoid membrane in the cranium (Fig. 9.4d). The rate of fluid flow through the central nervous system is sufficient to replenish the entire volume of cerebrospinal fluid about three times a day.

Cerebrospinal fluid serves two purposes: physical protection and chemical protection. The brain and spinal cord float in the thin layer of fluid between the membranes. The buoyancy of cerebrospinal fluid reduces the weight of the brain nearly 30-fold. Lighter weight translates into less pressure on blood vessels and nerves attached to the CNS.

The cerebrospinal fluid also provides protective padding. When there is a blow to the head, the CSF must be compressed before the brain can hit the inside of the cranium. However, water is minimally compressible, which helps CSF cushion the brain. For a dramatic demonstration of the protective power of cerebrospinal fluid, shake a block of tofu (representing the brain) in an empty jar. Then shake a second block of tofu in a jar completely filled with water to see how cerebrospinal fluid safeguards the brain.

In addition to physically protecting the delicate tissues of the CNS, cerebrospinal fluid creates a closely regulated extracellular environment for the neurons. The choroid plexus is selective about which substances it transports into the ventricles, and, as a result, the composition of cerebrospinal fluid is different from that of the plasma. The concentration of K^+ is lower in the cerebrospinal fluid, and the concentration of H^+ is higher than in plasma. The concentration of Na^+ in CSF is similar to that in the blood. Cerebrospinal fluid normally contains very little protein and no blood cells.

Cerebrospinal fluid exchanges solutes with the interstitial fluid of the CNS and provides a route by which wastes can be removed. Clinically, a sample of cerebrospinal fluid is presumed to be an indicator of the chemical environment in the brain. This

FIG. 9.4 **ANATOMY SUMMARY**

Cerebrospinal Fluid

(a) Ventricles of the Brain

The lateral ventricles consist of the first and second ventricles. The third and fourth ventricles extend through the brain stem and connect to the central canal that runs through the spinal cord. Compare the frontal view to the cross section in Fig. 9.10a.

Lateral ventricles
Third ventricle
Fourth ventricle
Cerebellum
Central canal
Spinal cord

Lateral view

Frontal view

(b) Cerebrospinal Fluid Secretion

Cerebrospinal fluid is secreted into the ventricles and flows throughout the subarachnoid space, where it cushions the central nervous system.

Arachnoid villi

Choroid plexus of third ventricle

Pia mater

Arachnoid membrane

Sinus

Choroid plexus of fourth ventricle

Spinal cord

Central canal

Subarachnoid space

Arachnoid membrane

Dura mater

(c) Choroid Plexus

The choroid plexus transports ions and nutrients from the blood into the cerebrospinal fluid.

Capillary

Ependymal cells

Water   Ions, vitamins, nutrients

Cerebrospinal fluid in third ventricle

(d) Cerebrospinal Fluid Reabsorption

Cerebrospinal fluid is reabsorbed into the blood at fingerlike projections of the arachnoid membrane called villi.

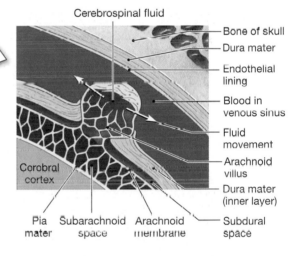

Cerebrospinal fluid

Bone of skull
Dura mater
Endothelial lining
Blood in venous sinus
Fluid movement
Arachnoid villus
Dura mater (inner layer)
Subdural space

Cerebral cortex

Pia mater   Subarachnoid space   Arachnoid membrane

Q FIGURE QUESTIONS

1. Physicians may extract a sample of cerebrospinal fluid when they suspect an infection in the brain. Where is the least risky and least difficult place for them to insert a needle through the meninges? (See Fig. 9.4b.)

2. The aqueduct of Sylvius is the narrow passageway between the third and fourth ventricles. What happens to CSF flow if the aqueduct becomes blocked by infection or tumor, a condition known as aqueductal stenosis {*stenos*, narrow}? On a three-dimensional imaging study of the brain, how would you distinguish aqueductal stenosis from a blockage of CSF flow in the subarachnoid space near the frontal lobe?

sampling procedure, known as a *spinal tap* or *lumbar puncture,* is generally done by withdrawing fluid from the subarachnoid space between vertebrae at the lower end of the spinal cord. The presence of proteins or blood cells in cerebrospinal fluid suggests an infection.

CONCEPT CHECK

5. If the concentration of H^+ in cerebrospinal fluid is higher than that in the blood, what can you say about the pH of the CSF?

6. Why is rupturing a blood vessel running between the meninges potentially a surgical emergency?

7. Is cerebrospinal fluid more like plasma or more like interstitial fluid? Defend your answer.

The Blood-Brain Barrier Protects the Brain

The final layer of protection for the brain is a functional barrier between the interstitial fluid and the blood. This barrier is necessary to isolate the body's main control center from potentially harmful substances in the blood and from blood-borne pathogens such as bacteria. To achieve this protection, most of the 400 miles of brain capillaries create a functional **blood-brain barrier** (FIG. 9.5). Although not a literal barrier, the highly selective permeability of brain capillaries shelters the brain from toxins and from fluctuations in hormones, ions, and neuroactive substances such as neurotransmitters in the blood.

Why are brain capillaries so much less permeable than other capillaries? In most capillaries, leaky cell-cell junctions and pores allow free exchange of solutes between the plasma and interstitial fluid [p. 76]. In brain capillaries, however, the endothelial cells form tight junctions with one another, junctions that prevent solute movement between the cells. Tight junction formation apparently is induced by paracrine signals from adjacent astrocytes whose foot processes surround the capillary. As a result, it is the brain tissue itself that creates the blood-brain barrier.

The selective permeability of the blood-brain barrier can be attributed to its transport properties. The capillary endothelium uses selected membrane carriers and channels to move nutrients and other useful materials from the blood into the brain interstitial fluid. Other transporters move wastes from the interstitial fluid into the plasma. Any water-soluble molecule that is not transported on one of these carriers cannot cross the blood-brain barrier.

FIG. 9.5 The blood-brain barrier

(a) This cerebral angiogram shows the extensive blood supply to the brain, which has about 400 miles of capillaries.

(b) Neurons are protected from harmful substances in the blood because brain capillaries are not leaky.

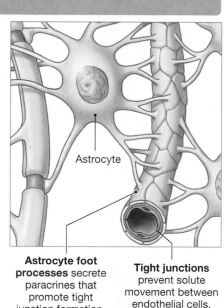

Astrocyte

Astrocyte foot processes secrete paracrines that promote tight junction formation.

Tight junctions prevent solute movement between endothelial cells.

Anterior cerebral artery

Posterior cerebral artery

Middle cerebral artery

Circle of Willis

Internal carotid artery

Capillary lumen

Basal lamina

One interesting illustration of how the blood-brain barrier works is seen in *Parkinson's disease,* a neurological disorder in which brain levels of the neurotransmitter dopamine are too low because dopaminergic neurons are either damaged or dead. Dopamine administered in a pill or injection is ineffective because it is unable to cross the blood-brain barrier. The dopamine precursor *L-dopa,* however, is transported across the cells of the blood-brain barrier on an amino acid transporter [p. 138]. Once neurons have access to L-dopa in the interstitial fluid, they metabolize it to dopamine, thereby allowing the deficiency to be treated.

The blood-brain barrier effectively excludes many water-soluble substances, but smaller lipid-soluble molecules can diffuse through the cell membranes [p. 136]. This is one reason some antihistamines make you sleepy but others do not. Older antihistamines were lipid-soluble amines that readily crossed the blood-brain barrier and acted on brain centers controlling alertness. The newer drugs are much less lipid soluble and as a result do not have the same sedative effect.

A few areas of the brain lack a functional blood-brain barrier, and their capillaries have leaky endothelium like most of the rest of the body. In these areas of the brain, the function of adjacent neurons depends in some way on direct contact with the blood. For instance, the hypothalamus releases neurosecretory hormones that must pass into the capillaries of the *hypothalamic-hypophyseal portal system* for distribution to the anterior pituitary [p. 211].

RUNNING **PROBLEM**

> Ben was diagnosed with infantile spasms, or West syndrome, a form of epilepsy characterized by the onset of head-drop seizures at four to seven months and by arrested or deteriorating mental development. Ben was started on a month-long regimen of adrenocorticotropin (ACTH) [p. 213] shots plus an anti-epileptic drug called vigabatrin to control the seizures. Scientists are unsure why ACTH is so effective in controlling this type of seizure. They have found that, among its effects, it increases myelin formation, increases blood-brain barrier integrity, and enhances binding of the neurotransmitter GABA at synapses. Vigabatrin prolongs synaptic activity of GABA by slowing its breakdown. As expected, Ben's seizures disappeared completely before the month of treatment ended, and his development began to return to a normal level.
>
> **Q1:** How might a leaky blood-brain barrier lead to a cascade of action potentials that trigger a seizure?
>
> **Q2:** GABA opens Cl⁻ channels on the postsynaptic cell. What does this do to the cell's membrane potential? Does GABA make the cell more or less likely to fire action potentials?
>
> **Q3:** Why is it important to limit the duration of ACTH therapy, particularly in very young patients? [p. 217]

(275) **283** (298) (300) (302) (304)

CLINICAL FOCUS

Diabetes:
Hypoglycemia and the Brain

Neurons are picky about their food. Under most circumstances, the only biomolecule that neurons use for energy is glucose. Surprisingly, this can present a problem for diabetic patients, whose problem is too much glucose in the blood. In the face of sustained hyperglycemia (elevated blood glucose), the cells of the blood-brain barrier down-regulate [p. 51] their glucose transporters. Then, if the patient's blood glucose level falls below normal because of excess insulin or failing to eat, the neurons of the brain may not be able to take up glucose fast enough to sustain their electrical activity. The individual may exhibit confusion, irritability, and slurred speech as brain function begins to fail. Prompt administration of sugar, either by mouth or intravenous infusion is necessary to prevent permanent damage. In extreme cases, hypoglycemia can cause coma or even death.

Another region that lacks the blood-brain barrier is the vomiting center in the medulla oblongata. These neurons monitor the blood for possibly toxic foreign substances, such as drugs. If they sense something harmful, they initiate a vomiting reflex. Vomiting removes the contents of the digestive system and helps eliminate ingested toxins.

Neural Tissue Has Special Metabolic Requirements

A unique property of the central nervous system is its specialized metabolism. Neurons require a constant supply of oxygen and glucose to make ATP for active transport of ions and neurotransmitters. Oxygen passes freely across the blood-brain barrier, and membrane transporters move glucose from the plasma to the brain's interstitial fluid. Unusually low levels of either substrate can have devastating results on brain function.

Because of its high demand for oxygen, the brain receives about 15% of the blood pumped by the heart. If blood flow to the brain is interrupted, brain damage occurs after only a few minutes without oxygen. Neurons are equally sensitive to lack of glucose. Under normal circumstances, the only energy source for neurons is glucose.

By some estimates, the brain is responsible for about half of the body's glucose consumption. Consequently, the body uses several homeostatic pathways to ensure that glucose concentrations in the blood always remain adequate to meet the brain's demand. If homeostasis fails, progressive **hypoglycemia** (low blood glucose levels) leads to confusion, unconsciousness, and eventually death.

Now that you have a broad overview of the central nervous system, we will examine the structure and function of the spinal cord and brain in more detail.

CONCEPT CHECK

8. Oxidative phosphorylation takes place in which organelle?

9. Name the two metabolic pathways for aerobic metabolism of glucose. What happens to NADH produced in these pathways?

10. In the late 1800s, the scientist Paul Ehrlich injected blue dye into the bloodstream of animals. He noticed that all tissues except the brain stained blue. He was not aware of the blood-brain barrier, so what conclusion do you think he drew from his results?

11. In a subsequent experiment, a student of Ehrlich's injected the dye into the cerebrospinal fluid of the same animals. What do you think he observed about staining in the brain and in other body tissues?

THE SPINAL CORD

The spinal cord is the major pathway for information flowing back and forth between the brain and the skin, joints, and muscles of the body. In addition, the spinal cord contains neural networks responsible for locomotion. If the spinal cord is severed, there is loss of sensation from the skin and muscles as well as *paralysis,* loss of the ability to voluntarily control muscles.

The spinal cord is divided into four regions: *cervical, thoracic, lumbar,* and *sacral,* named to correspond to the adjacent vertebrae (see Fig. 9.3a). Each spinal region is subdivided into segments, and each segment gives rise to a bilateral pair of **spinal nerves.** Just before a spinal nerve joins the spinal cord, it divides into two branches called **roots** (**FIG. 9.6a**).

The **dorsal root** of each spinal nerve is specialized to carry incoming sensory information. The **dorsal root ganglia,** swellings found on the dorsal roots just before they enter the cord (Fig. 9.6b), contain cell bodies of sensory neurons. The **ventral root** carries information from the CNS to muscles and glands.

In cross section, the spinal cord has a butterfly- or H-shaped core of gray matter and a surrounding rim of white matter. Sensory fibers from the dorsal roots synapse with interneurons in the **dorsal horns** of the gray matter. The dorsal horn cell bodies are organized into two distinct nuclei, one for somatic information and one for visceral information (Fig. 9.6b).

The **ventral horns** of the gray matter contain cell bodies of motor neurons that carry efferent signals to muscles and glands. The ventral horns are organized into somatic motor and autonomic nuclei. Efferent fibers leave the spinal cord via the ventral root.

The white matter of the spinal cord is the biological equivalent of fiber-optic cables that telephone companies use to carry our communications systems. White matter can be divided into a number of **columns** composed of tracts of axons that transfer information up and down the cord. **Ascending tracts** take sensory information to the brain. They occupy the dorsal and external lateral portions of the spinal cord (Fig. 9.6c). **Descending tracts** carry mostly efferent (motor) signals from the brain to the cord. They occupy the ventral and interior lateral portions of the white matter. **Propriospinal tracts** {*proprius,* one's own} are those that remain within the cord.

FIG. 9.6 **Organization of the spinal cord**

The spinal cord contains nuclei with cell bodies of efferent neurons and tracts of axons going to and from the brain.

(a) One segment of spinal cord, ventral view, showing its pair of nerves

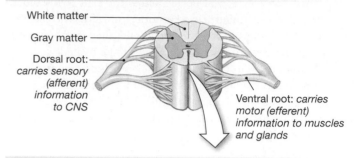

White matter

Gray matter

Dorsal root: *carries sensory (afferent) information to CNS*

Ventral root: *carries motor (efferent) information to muscles and glands*

(b) Gray matter consists of sensory and motor nuclei.

Visceral sensory nuclei

Somatic sensory nuclei

Dorsal root ganglion

Dorsal horn

Lateral horn

Ventral horn

Ventral root

Autonomic efferent nuclei

Somatic motor nuclei

(c) White matter in the spinal cord consists of tracts of axons carrying information to and from the brain.

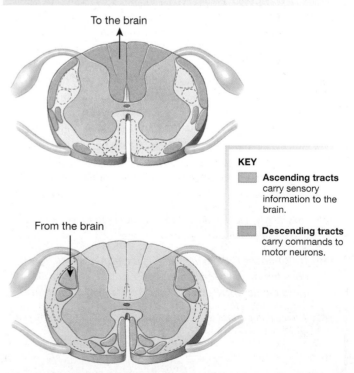

To the brain

From the brain

KEY

Ascending tracts carry sensory information to the brain.

Descending tracts carry commands to motor neurons.

FIG. 9.7 Spinal reflexes

In a spinal reflex, sensory information entering the spinal cord is acted on without input from the brain. However, sensory information about the stimulus may be sent to the brain.

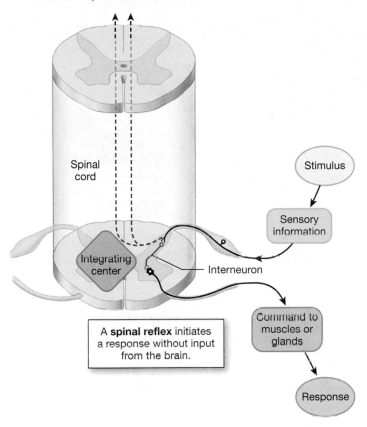

Spinal cord

Stimulus

Sensory information

Integrating center

Interneuron

Command to muscles or glands

A **spinal reflex** initiates a response without input from the brain.

Response

The spinal cord can function as a self-contained integrating center for simple *spinal reflexes*, with signals passing from a sensory neuron through the gray matter to an efferent neuron (**FIG. 9.7**). In addition, spinal interneurons may route sensory information to the brain through ascending tracts or bring commands from the brain to motor neurons. In many cases, the interneurons also modify information as it passes through them. Reflexes play a critical role in the coordination of body movement.

CONCEPT CHECK

12. What are the differences between horns, roots, tracts, and columns of the spinal cord?

13. If a dorsal root of the spinal cord is cut, what function will be disrupted?

THE BRAIN

Thousands of years ago, Aristotle declared that the heart was the seat of the soul. However, most people now agree that the brain is the organ that gives the human species its unique attributes. The challenge facing today's scientists is to understand how circuits formed by millions of neurons result in complex behaviors such as speaking, writing a symphony, or creating imaginary worlds for an interactive computer game. Brain function may be the ultimate

emergent property [p. 2]. The question remains whether we will ever be able to decipher how emotions such as happiness and love arise from the chemical and electrical signals passing along circuits of neurons.

It is possible to study the brain at many levels of organization. The most reductionist view looks at the individual neurons and at what happens to them in response to chemical or electrical signals. A more integrative study might look at groups of neurons and how they interact with one another in *circuits, pathways,* or networks. The most complicated approach starts with a behavior or physiological response and works backward to dissect the neural circuits that create the behavior or response.

For centuries, studies of brain function were restricted to anatomical descriptions. However, when we study the brain we see no tidy 1:1 relationship between structure and function. An adult human brain has a mass of about 1400 g and contains an estimated 85 billion neurons. When you consider that each one of these billions of neurons may receive as many as 200,000 synapses, the number of possible neuronal connections is mind boggling. To complicate matters even more, those synapses are not fixed and are constantly changing.

A basic principle to remember when studying the brain is that one function, even an apparently simple one such as bending your finger, will involve multiple brain regions (as well as the spinal cord). Conversely, one brain region may be involved in several functions at the same time. In other words, understanding the brain is not simple and straightforward.

FIGURE 9.8 is an Anatomy Summary to follow as we discuss major brain regions, moving from the most primitive to the most complex. Of the six major divisions of the brain present at birth (see Fig. 9.2e), only the medulla, cerebellum, and cerebrum are visible when the intact brain is viewed in profile. The remaining three divisions (diencephalon, midbrain, and pons) are covered by the cerebrum.

The Brain Stem Is the Oldest Part of the Brain

The **brain stem** is the oldest and most primitive region of the brain and consists of structures that derive from the embryonic midbrain and hindbrain. The brain stem can be divided into white matter and gray matter, and in some ways, its anatomy is similar to that of the spinal cord. Some ascending tracts from the spinal cord pass through the brain stem, while other ascending tracts synapse there. Descending tracts from higher brain centers also travel through the brain stem on their way to the spinal cord.

Pairs of peripheral nerves branch off the brain stem, similar to spinal nerves along the spinal cord (Fig. 9.8f). Eleven of the 12 **cranial nerves** (numbers II–XII) originate along the brain stem. (The first cranial nerve, the olfactory nerve, enters the forebrain.) Cranial nerves carry sensory and motor information for the head and neck (**TBL. 9.1**).

The cranial nerves are described according to whether they include sensory fibers, efferent fibers, or both (mixed nerves). For

FIG. 9.8 ANATOMY SUMMARY

Central Nervous System

Anatomy of the Brain

(a) Lateral View of the CNS

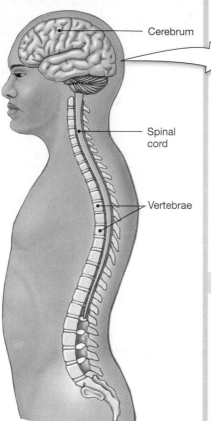

- Cerebrum
- Spinal cord
- Vertebrae

(b) Lateral View of Brain

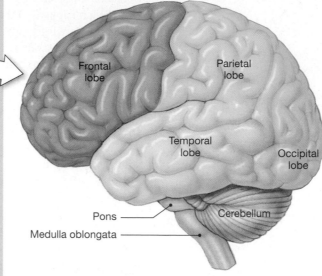

- Frontal lobe
- Parietal lobe
- Temporal lobe
- Occipital lobe
- Pons
- Cerebellum
- Medulla oblongata

(c) Mid-Sagittal View of Brain

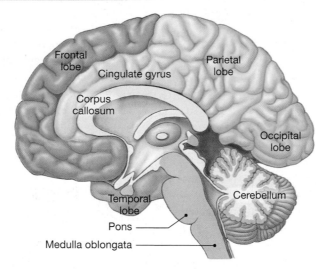

- Frontal lobe
- Cingulate gyrus
- Corpus callosum
- Parietal lobe
- Occipital lobe
- Cerebellum
- Temporal lobe
- Pons
- Medulla oblongata

(d) Diencephalon

- Thalamus
- Pineal gland
- Hypothalamus
- Pituitary gland

Functions of the Cerebrum

Cerebral cortex
See Figure 9.13.

Sensory areas
- Perception [Fig. 10.3]

Motor areas
- Skeletal muscle movement

Association areas
- Integration of information and direction of voluntary movement [Ch. 13]

Basal ganglia (not shown)
See Figure 9.10.

- Movement [Ch. 13]

Limbic System (not shown)
See Figure 9.11.

Amygdala
- Emotion
- Memory

Hippocampus
- Learning
- Memory

Functions of the Cerebellum

- Movement coordination [Ch. 13]

Functions of the Diencephalon

Thalamus
- Integrating center and relay station for sensory and motor information

Pineal gland
- Melatonin secretion [Fig. 7.16]

Hypothalamus
See Table 9.2.

- Homeostasis [Ch. 11]
- Behavioral drives

Pituitary gland
- Hormone secretion [Fig. 7.8]

example, cranial nerve X, the **vagus nerve** {*vagus,* wandering}, is a mixed nerve that carries both sensory and motor fibers for many internal organs. An important component of a clinical neurological examination is testing the functions controlled by these nerves.

The brain stem contains numerous discrete groups of nerve cell bodies, or nuclei. Many of these nuclei are associated with the **reticular formation,** a diffuse collection of neurons that extends throughout the brain stem. The name *reticular* means "network" and comes from the crisscrossed axons that branch profusely up into superior sections of the brain and down into the spinal cord. Nuclei in the brain stem are involved in many basic processes, including arousal and sleep, muscle tone and stretch reflexes, coordination of breathing, blood pressure regulation, and modulation of pain.

CONCEPT CHECK

14. Are the following white matter or gray matter? (a) ascending tracts, (b) reticular formation, (c) descending tracts.

15. Using the information from Table 9.1, describe the types of activities you might ask a patient to perform if you wished to test the function of each cranial nerve.

16. In anatomical directional terminology, the cerebrum, which is located next to the top of the skull, is said to be _____ to the brain stem.

Starting at the spinal cord and moving toward the top of the skull, the brain stem consists of the medulla oblongata, the pons, and the midbrain (Fig. 9.8f). Some authorities include the cerebellum as part of the brain stem. The diamond-shaped fourth ventricle runs through the interior of the brain stem and connects to the central canal of the spinal cord (see Fig. 9.4a).

Medulla The **medulla oblongata,** frequently just called the *medulla* {*medulla,* marrow; adjective *medullary*}, is the transition from the spinal cord into the brain proper (Fig. 9.8f). Its white matter includes ascending **somatosensory tracts** {*soma,* body} that bring sensory information to the brain, and descending **corticospinal tracts** that convey information from the cerebrum to the spinal cord.

About 90% of corticospinal tracts cross the midline to the opposite side of the body in a region of the medulla known as the **pyramids.** As a result of this crossover, each side of the brain controls the opposite side of the body. Gray matter in the medulla includes nuclei that control many involuntary functions, such as blood pressure, breathing, swallowing, and vomiting.

Pons The **pons** {*pons,* bridge; adjective *pontine*} is a bulbous protrusion on the ventral side of the brain stem above the medulla and below the midbrain. Because its primary function is to act as a relay station for information transfer between the cerebellum and cerebrum, the pons is often grouped with the cerebellum. The pons also coordinates the control of breathing along with centers in the medulla.

(e) The Skull

(f) Lateral View of Brain Stem

Functions of the Brain Stem

Midbrain
- Eye movement

Pons
- Relay station between cerebrum and cerebellum
- Coordination of breathing [Fig. 18.14]

Medulla oblongata
- Control of involuntary functions [Fig. 11.3]

Reticular formation (not shown)
See Figure 9.16.
- Arousal
- Sleep
- Muscle tone
- Pain modulation

TABLE 9.1	The Cranial Nerves		
Number	**Name**	**Type**	**Function**
I	Olfactory	Sensory	Olfactory (smell) information from nose
II	Optic	Sensory	Visual information from eyes
III	Oculomotor	Motor	Eye movement, pupil constriction, lens shape
IV	Trochlear	Motor	Eye movement
V	Trigeminal	Mixed	Sensory information from face, mouth; motor signals for chewing
VI	Abducens	Motor	Eye movement
VII	Facial	Mixed	Sensory for taste; efferent signals for tear and salivary glands, facial expression
VIII	Vestibulocochlear	Sensory	Hearing and equilibrium
IX	Glossopharyngeal	Mixed	Sensory from oral cavity, baro- and chemoreceptors in blood vessels; efferent for swallowing, parotid salivary gland secretion
X	Vagus	Mixed	Sensory and efferents to many internal organs, muscles, and glands
XI	Spinal accessory	Motor	Muscles of oral cavity, some muscles in neck and shoulder
XII	Hypoglossal	Motor	Tongue muscles

Mnemonic for remembering the cranial nerves in order: **O**h **O**nce **O**ne **T**akes **T**he **A**natomy **F**inal, **V**ery **G**ood **V**acations **S**ound **H**eavenly.

Midbrain The third region of the brain stem, the **midbrain,** or *mesencephalon* {*mesos,* middle}, is a relatively small area that lies between the lower brain stem and the diencephalon. The primary function of the midbrain is control of eye movement, but it also relays signals for auditory and visual reflexes.

The Cerebellum Coordinates Movement

The **cerebellum** is the second largest structure in the brain (Fig. 9.8a–c). It is located inside the base of the skull, just above the nape of the neck. The name *cerebellum* {adjective *cerebellar*} means "little brain," and, indeed, most of the nerve cells in the brain are in the cerebellum. The specialized function of the cerebellum is to process sensory information and coordinate the execution of movement. Sensory input into the cerebellum comes from somatic receptors in the periphery of the body and from receptors for equilibrium and balance located in the inner ear. The cerebellum also receives motor input from neurons in the cerebrum. [See Chapters 10 and 13 for additional information.]

The Diencephalon Contains the Centers for Homeostasis

The **diencephalon,** or "between-brain," lies between the brain stem and the cerebrum. It is composed of two main sections, the thalamus and the hypothalamus, and two endocrine structures, the pituitary and pineal glands (**FIG. 9.9**).

Most of the diencephalon is occupied by many small nuclei that make up the **thalamus** {*thalamus,* bedroom; adjective *thalamic*}. The thalamus receives sensory fibers from the optic tract,

FIG. 9.9 The diencephalon

The diencephalon lies between the brain stem and the cerebrum. It consists of thalamus, hypothalamus, pineal gland, and pituitary gland.

Corpus callosum

Thalamus

Hypothalamus

Pineal gland

Anterior pituitary

Posterior pituitary

ears, and spinal cord as well as motor information from the cerebellum. It projects fibers to the cerebrum, where the information is processed.

The thalamus is often described as a relay station because almost all sensory information from lower parts of the CNS passes through it. Like the spinal cord, the thalamus can modify information passing through it, making it an integrating center as well as a relay station.

The **hypothalamus** lies beneath the thalamus. Although the hypothalamus occupies less than 1% of total brain volume, it is the center for homeostasis and contains centers for various behavioral drives, such as hunger and thirst. Output from the hypothalamus also influences many functions of the autonomic division of the nervous system, as well as a variety of endocrine functions (**TBL. 9.2**).

The hypothalamus receives input from multiple sources, including the cerebrum, the reticular formation, and various sensory receptors. Output from the hypothalamus goes first to the thalamus and eventually to multiple effector pathways.

Two important endocrine structures are located in the diencephalon: the pituitary gland and the pineal gland [p. 221]. The posterior pituitary (*neurohypophysis*) is a down-growth of the hypothalamus and secretes neurohormones that are synthesized in hypothalamic nuclei. The anterior pituitary (*adenohypophysis*) is a true endocrine gland. Its hormones are regulated by hypothalamic neurohormones secreted into the hypothalamic-hypophyseal portal system. Later in this chapter, we discuss the pineal gland, which secretes the hormone melatonin.

17. Starting at the spinal cord and moving up, name the subdivisions of the brain stem.

18. What are the four primary structures of the diencephalon?

The Cerebrum Is the Site of Higher Brain Functions

As noted earlier in the chapter, the cerebrum is the largest and most distinctive part of the human brain and fills most of the cranial cavity. It is composed of two hemispheres connected primarily at the **corpus callosum** (Figs. 9.8c and 9.9), a distinct structure formed by axons passing from one side of the brain to the other. This connection ensures that the two hemispheres communicate and cooperate with each other. Each cerebral hemisphere is divided into four lobes, named for the bones of the skull under which they are located: *frontal, parietal, temporal,* and *occipital* (Fig. 9.8b, c, e).

The surface of the cerebrum in humans and other primates has a furrowed, walnut-like appearance, with grooves called *sulci* {singular *sulcus*, a furrow} dividing convolutions called *gyri* {singular *gyrus*, a ring or circle}. During development, the cerebrum grows faster than the surrounding cranium, causing the tissue to fold back on itself to fit into a smaller volume. The degree of folding is directly related to the level of processing of which the brain is capable. Less-advanced mammals, such as rodents, have brains with a relatively smooth surface. The human brain, on the other hand, is so convoluted that if it were inflated enough to smooth the surfaces, it would be three times as large and would need a head the size of a beach ball.

Gray Matter and White Matter Cerebral gray matter can be divided into three major regions: the cerebral cortex, the basal ganglia, and the limbic system. The **cerebral cortex** {*cortex,* bark or rind; adjective *cortical,* plural *cortices*} is the outer layer of the cerebrum, only a few millimeters thick (**FIG. 9.10a**). Neurons of the cerebral cortex are arranged in anatomically distinct vertical columns and horizontal layers (Fig. 9.10b). It is within these layers that our higher brain functions arise.

The second region of cerebral gray matter consists of the **basal ganglia** (Fig. 9.10a), which are involved in the control of movement. The basal ganglia are also called the *basal nuclei*. Neuroanatomists prefer to reserve the term *ganglia* for clusters of nerve cell bodies outside the CNS, but the term basal ganglia is commonly used in clinical settings.

The third region of the cerebrum is the **limbic system** {*limbus,* a border}, which surrounds the brain stem (**FIG. 9.11**). The limbic system represents probably the most primitive region of the cerebrum. It acts as the link between higher cognitive functions, such as reasoning, and more primitive emotional responses, such as fear. The major areas of the limbic system are the **amygdala** and **cingulate gyrus,** which are linked to emotion and memory, and the **hippocampus,** which is associated with learning and memory.

TABLE 9.2 | Functions of the Hypothalamus

1. Activates sympathetic nervous system
   - Controls catecholamine release from adrenal medulla (as in fight-or-flight reaction)
   - Helps maintain blood glucose concentrations through effects on endocrine pancreas
   - Stimulates shivering and sweating

2. Maintains body temperature

3. Controls body osmolarity
   - Motivates thirst and drinking behavior
   - Stimulates secretion of vasopressin [p. 209]

4. Controls reproductive functions
   - Directs secretion of oxytocin (for uterine contractions and milk release)
   - Directs trophic hormone control of anterior pituitary hormones FSH and LH [p. 213]

5. Controls food intake
   - Stimulates satiety center
   - Stimulates feeding center

6. Interacts with limbic system to influence behavior and emotions

7. Influences cardiovascular control center in medulla oblongata

8. Secretes trophic hormones that control release of hormones from anterior pituitary gland

FIG. 9.10   Gray matter of the cerebrum

The cerebral cortex and basal ganglia are two of the three regions of gray matter in the cerebrum. The third region, the limbic system, is detailed in Figure 9.11. The frontal view shown here is similar to the sectional view obtained using modern diagnostic imaging techniques.

Q FIGURE QUESTION

The section through this brain is a section through the _____ plane.

(a) coronal
(b) lateral
(c) frontal
(d) transverse
(e) sagittal

(a) Section through the brain showing the basal ganglia

Corpus callosum

Lateral ventricle

Basal ganglia

Tracts of white matter

Tip of lateral ventricle

Gray matter of cerebral cortex

(b) Cell bodies in the cerebral cortex form distinct layers and columns.

Outer surface of the cerebral cortex

Layers 1 2 3 4 5 6

Gray matter

White matter

White matter in the cerebrum is found mostly in the interior (Fig. 9.10a). Bundles of fibers allow different regions of the cortex to communicate with one another and transfer information from one hemisphere to the other, primarily through the

FIG. 9.11   The limbic system

The limbic system includes the amygdala, hippocampus, and cingulate gyrus. Anatomically, the limbic system is part of the gray matter of the cerebrum. The thalamus is shown for orientation purposes and is not part of the limbic system.

Cingulate gyrus plays a role in emotion.

Thalamus

Hippocampus is involved in learning and memory.

Amygdala is involved in emotion and memory.

corpus callosum. According to some estimates, the corpus callosum may have as many as 200 million axons passing through it! Information entering and leaving the cerebrum goes along tracts that pass through the thalamus (with the exception of olfactory information, which goes directly from olfactory receptors to the cerebrum).

CONCEPT CHECK

19. Name the anatomical location in the brain where neurons from one side of the body cross to the opposite side.

20. Name the divisions of the brain in anatomical order, starting from the spinal cord.

BRAIN FUNCTION

From a simplistic view, the CNS is an information processor much like a computer. For many functions, it follows a basic reflex pathway [p. 14]. The brain receives sensory input from the internal and external environments, integrates and processes the information, and, if appropriate, creates a response (**FIG. 9.12a**). What makes the brain more complicated than this simple reflex pathway, however, is its ability to generate information and output signals *in the absence of external input*. Modeling this intrinsic input requires a more complex diagram.

Larry Swanson of the University of Southern California presents one approach to modeling brain function in his book

FIG. 9.12 Simple and complex pathways in the brain

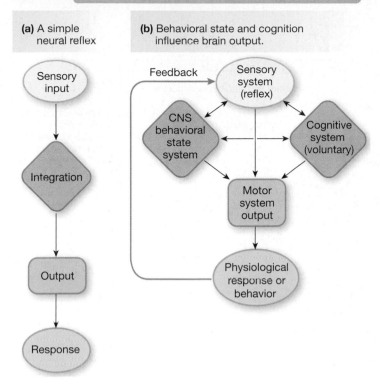

(a) A simple neural reflex

Sensory input → Integration → Output → Response

(b) Behavioral state and cognition influence brain output.

Feedback → Sensory system (reflex)

CNS behavioral state system

Cognitive system (voluntary)

Motor system output

Physiological response or behavior

Brain Architecture: Understanding the Basic Plan (2nd edition, Oxford University Press, 2011). He describes three systems that influence output by the motor systems of the body: (1) the **sensory system,** which monitors the internal and external environments and initiates reflex responses; (2) a **cognitive system** that resides in the cerebral cortex and is able to initiate voluntary responses; and (3) a **behavioral state system,** which also resides in the brain and governs sleep-wake cycles and other intrinsic behaviors. Information about the physiological or behavioral responses created by motor output feeds back to the sensory system, which in turn communicates with the cognitive and behavioral state systems (Fig. 9.12b).

In most of the physiological organ systems of the body that you will study, simple reflex pathways initiated through the sensory system and executed by motor output are adequate to explain homeostatic control mechanisms. However, the cognitive and behavioral state systems remain potential sources of influence. At its simplest, this influence may take the form of voluntary behaviors, such as breath-holding, that override automatic functions. More subtle and complicated interactions include the effect of emotions on normal physiology, such as stress-induced heart palpitations, and the role of circadian rhythms in jet lag and shift work.

In the sections that follow, we take a brief look at sensory and motor systems in the brain. We conclude this chapter with a discussion of some aspects of the behavioral state system and the cognitive system, such as circadian rhythms, sleep-wake cycles, emotion, learning, and memory.

The Cerebral Cortex Is Organized into Functional Areas

The cerebral cortex serves as an integrating center for sensory information and a decision-making region for many types of motor output. If we examine the cortex from a functional viewpoint, we can divide it into three specializations: (1) **sensory areas** (also called sensory fields), which receive sensory input and translate it into perception (awareness); (2) **motor areas,** which direct skeletal muscle movement; and (3) **association areas** (association cortices), which integrate information from sensory and motor areas and can direct voluntary behaviors (**FIG. 9.13**). Information passing along a pathway is usually processed in more than one of these areas.

The functional areas of the cerebral cortex do not necessarily correspond to the anatomical lobes of the brain. For one thing, functional specialization is not symmetrical across the cerebral cortex: each lobe has special functions not shared by the matching lobe on the opposite side. This **cerebral lateralization** of function is sometimes referred to as *cerebral dominance*, more popularly known as left brain–right brain dominance (**FIG. 9.14**). Language and verbal skills tend to be concentrated on the left side of the brain, with spatial skills concentrated on the right side. The left brain is the dominant hemisphere for right-handed people, and it appears that the right brain is the dominant hemisphere for many left-handed people.

Even these generalizations are subject to change, however. Neural connections in the cerebrum, like those in other parts of the nervous system, exhibit a certain degree of plasticity. For example, if a person loses a finger, the regions of motor and sensory cortex previously devoted to control of the finger do not go dormant. Instead, adjacent regions of the cortex extend their functional fields and take over the parts of the cortex that are no longer used by the absent finger. Similarly, skills normally associated with one side of the cerebral cortex can be developed in the other hemisphere, as when a right-handed person with a broken hand learns to write with the left hand.

Much of what we know about functional areas of the cerebral cortex comes from study of patients who have either inherited neurological defects or suffered wounds in accidents or war. In some instances, surgical lesions made to treat some medical condition, such as uncontrollable epilepsy, have revealed functional relationships in particular brain regions. Imaging techniques such as *positron emission tomography* (PET) scans provide noninvasive ways for us to watch the human brain at work (**TBL. 9.3**).

The Spinal Cord and Brain Integrate Sensory Information

The sensory system monitors the internal and external environments and sends information to neural integrating centers, which in turn initiate appropriate responses. In its simplest form, this pathway is the classic reflex, illustrated in Figure 9.12a. The simplest reflexes can be integrated in the spinal cord, without input from higher brain centers (see Fig. 9.7). However, even simple

FIG. 9.13 Functional areas of the cerebral cortex

The cerebral cortex contains sensory areas for perception, motor areas that direct movement, and association areas that integrate information.

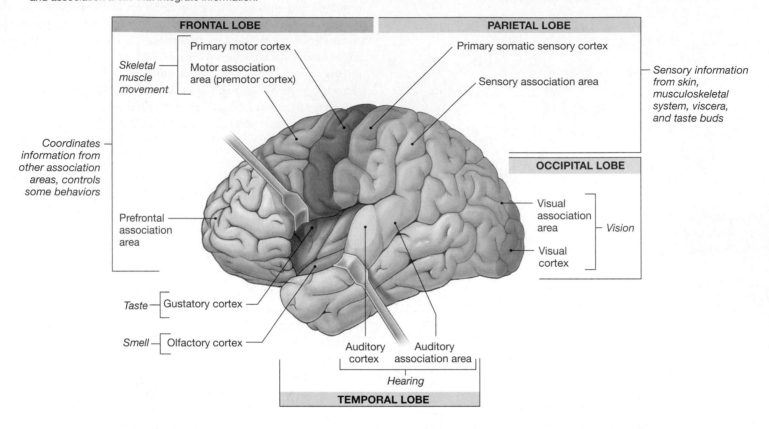

FRONTAL LOBE
- Primary motor cortex
- *Skeletal muscle movement* — Motor association area (premotor cortex)
- *Coordinates information from other association areas, controls some behaviors* — Prefrontal association area
- *Taste* — Gustatory cortex
- *Smell* — Olfactory cortex

PARIETAL LOBE
- Primary somatic sensory cortex
- Sensory association area
- *Sensory information from skin, musculoskeletal system, viscera, and taste buds*

OCCIPITAL LOBE
- Visual association area — *Vision*
- Visual cortex

Auditory cortex   Auditory association area
Hearing

TEMPORAL LOBE

TABLE 9.3	Selected Brain Imaging Techniques
In Vitro Techniques	
Horseradish peroxidase (HRP)	HRP enzyme is brought into axon terminals by endocytosis and transported by retrograde axonal transport to the cell body and dendrites. Completion of the enzyme-substrate reaction makes the entire neuron visible under a microscope.
Brainbow mice	Transgenic mice in which fluorescent proteins have been inserted into the neurons. Neurons light up in a rainbow of colors depending on which proteins they are expressing. (See chapter opener image.) *http://jaxmice.jax.org/jaxnotes/510/510n.html*
CLARITY: Clear, lipid-exchanged, anatomically rigid, imaging/immunostaining-compatible tissue hydrogel	Intact brain samples are made transparent by a technique that removes lipids and embeds the sample in a plastic matrix. Allows easier three-dimensional reconstructions of neural networks. *www.nature.com/news/see-through-brains-clarify-connections-1.12768*
In Vivo Imaging of Living Brain Activity	
Electroencephalography (EEG)	Brain electrical activity from many neurons is measured by electrodes placed on the scalp (see Fig. 9.17a).
Positive emission tomography (PET)	Glucose is tagged with a radioactive substance that emits positively charged particles. Metabolically active cells using glucose light up more (see Fig. 9.20). *www.nature.com/jcbfm/webfocus/mri/index.html*
Functional magnetic resonance imaging (fMRI)	Active brain tissue has increased blood flow and uses more oxygen. Hydrogen nuclei in water create a magnetic signal that indicates more active regions. *www.nature.com/news/brain-imaging-fmri-2-0-1.10365*

FIG. 9.14 | **Cerebral lateralization**

The distribution of functional areas in the two cerebral hemispheres is not symmetrical.

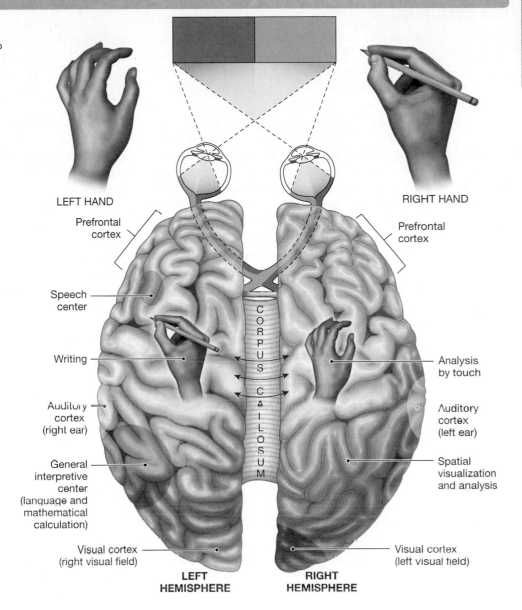

LEFT HAND

RIGHT HAND

Prefrontal cortex

Prefrontal cortex

Speech center

Writing

Analysis by touch

Auditory cortex (right ear)

Auditory cortex (left ear)

General interpretive center (language and mathematical calculation)

Spatial visualization and analysis

Visual cortex (right visual field)

Visual cortex (left visual field)

C O R P U S   C A L L O S U M

LEFT HEMISPHERE

RIGHT HEMISPHERE

Q FIGURE QUESTIONS

1. What would a person see if a stroke destroyed all function in the right visual cortex?
2. What is the function of the corpus callosum?
3. Many famous artists, including Leonardo da Vinci and Michelangelo, were left-handed. How is this related to cerebral lateralization?

spinal reflexes usually send sensory information to the brain, creating perception of the stimulus. Brain functions dealing with perception are the most difficult to study because they require communication between the subject and the investigator—the subject must be able to tell the investigator what he or she is seeing, hearing, or feeling.

Sensory information from the body travels in ascending pathways to the brain. Information about muscle and joint position and movement goes to the cerebellum as well as to the cerebral cortex, allowing the cerebellum to assist with automatic subconscious coordination of movement. Most sensory information continues on to the cerebral cortex, where five sensory areas process information.

The **primary somatic sensory cortex** (also called the *somatosensory cortex*) in the parietal lobe is the termination point of pathways from the skin, musculoskeletal system, and viscera

(Fig. 9.13). The somatosensory pathways carry information about touch, temperature, pain, itch, and body position. Damage to this part of the brain leads to reduced sensitivity of the skin on the opposite side of the body because sensory fibers cross to the opposite side of the midline as they ascend through the spine or medulla.

The special senses of vision, hearing, taste, and olfaction (smell) each have different brain regions devoted to processing their sensory input (Fig. 9.13). The **visual cortex,** located in the occipital lobe, receives information from the eyes. The **auditory cortex,** located in the temporal lobe, receives information from the ears. The **olfactory cortex,** a small region in the temporal lobe, receives input from chemoreceptors in the nose. The **gustatory cortex,** deeper in the brain near the edge of the frontal lobe, receives sensory information from the taste buds. [The sensory systems are described in detail in Chapter 10.]

Sensory Information Is Processed into Perception

Once sensory information reaches the appropriate cortical area, information processing has just begun. Neural pathways extend from sensory areas to appropriate association areas, which integrate somatic, visual, auditory, and other stimuli into *perception,* the brain's interpretation of sensory stimuli.

Often the perceived stimulus is very different from the actual stimulus. For instance, photoreceptors in the eye receive light waves of different frequencies, but we perceive the different wave energies as different colors. Similarly, the brain translates pressure waves hitting the ear into sound and interprets chemicals binding to chemoreceptors as taste or smell.

One interesting aspect of perception is the way our brain fills in missing information to create a complete picture, or translates a two-dimensional drawing into a three-dimensional shape (**FIG. 9.15**). Thus, we sometimes perceive what our brains expect to perceive. Our perceptual translation of sensory stimuli allows the information to be acted upon and used in voluntary motor control or in complex cognitive functions such as language.

The Motor System Governs Output from the CNS

The motor output component of the nervous system is associated with the efferent division of the nervous system [Fig. 8.1, p. 228]. Motor output can be divided into three major types: (1) skeletal muscle movement, controlled by the somatic motor division; (2) neuroendocrine signals, which are neurohormones secreted into the blood by neurons located primarily in the hypothalamus and adrenal medulla; and (3) *visceral* responses, the actions of smooth and cardiac muscle or endocrine and exocrine glands. Visceral responses are governed by the autonomic division of the nervous system.

Information about skeletal muscle movement is processed in several regions of the CNS. Simple stimulus-response pathways, such as the knee jerk reflex, are processed either in the spinal cord or in the brain stem. Although these reflexes do not require

integration in the cerebral cortex, they can be modified or overridden by input from the cognitive system.

Voluntary movements are initiated by the cognitive system and originate in the **primary motor cortex** and **motor association area** in the frontal lobes of the cerebrum (Fig. 9.13). These regions receive input from sensory areas as well as from the cerebellum and basal ganglia. Long output neurons called *pyramidal cells* project axons from the motor areas through the brain stem to the spinal cord. Other pathways go from the cortex to the basal ganglia and lower brain regions. Descending motor pathways cross to the opposite side of the body, which means that damage to a motor area manifests as paralysis or loss of function on the opposite side of the body. [Chapter 13 discusses motor pathways in more detail.]

Neuroendocrine and visceral responses are coordinated primarily in the hypothalamus and medulla. The brain stem contains the control centers for many of the automatic life functions we take for granted, such as breathing and blood pressure. It receives sensory information from the body and relays motor commands to peripheral muscles and glands.

The hypothalamus contains centers for temperature regulation, eating, and control of body osmolarity, among others. The responses to stimulation of these centers may be neural or hormonal reflexes or a behavioral response. Stress, reproduction, and growth are also mediated by the hypothalamus by way of multiple hormones. You will learn more about these reflexes in later chapters as we discuss the various systems of the body.

Sensory input is not the only factor determining motor output by the brain. The behavioral state system can modulate reflex pathways, and the cognitive system exerts both voluntary and involuntary control over motor functions.

The Behavioral State System Modulates Motor Output

The behavioral state system is an important modulator of sensory and cognitive processing. Many neurons in the behavioral state system are found in regions of the brain outside the cerebral cortex, including parts of the reticular formation in the brain stem, the hypothalamus, and the limbic system.

The neurons collectively known as the **diffuse modulatory systems** originate in the reticular formation in the brain stem and project their axons to large areas of the brain (**FIG. 9.16**). There are four modulatory systems that are generally classified according to the neurotransmitter they secrete: *noradrenergic* (norepinephrine), *serotonergic* (serotonin), *dopaminergic* (dopamine), and *cholinergic* (acetylcholine). The diffuse modulatory systems regulate brain function by influencing attention, motivation, wakefulness, memory, motor control, mood, and metabolic homeostasis.

One function of the behavioral state system is control of levels of consciousness and sleep-wake cycles. **Consciousness** is the body's state of arousal or awareness of self and environment. Experimental evidence shows that the **reticular activating system,** a diffuse collection of neurons in the reticular formation, plays an essential role in keeping the "conscious brain" awake.

FIG. 9.15 Perception

The brain has the ability to interpret sensory information to create the perception of (a) shapes or (b) three-dimensional objects.

(a) What shape do you see?   **(b)** What is this object?

FIG. 9.16 | Diffuse modulatory systems

The neurons collectively known as the diffuse modulatory systems originate in the reticular formation of the brain stem and project their axons to large areas of the brain. The four systems are named for their neurotransmitters.

(a) Noradrenergic (Norepinephrine)

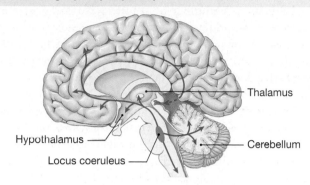

Thalamus

Hypothalamus

Locus coeruleus

Cerebellum

Functions:	Attention, arousal, sleep-wake cycles, learning, memory, anxiety, pain, and mood
Neurons Originate:	Locus coeruleus of the pons
Neurons Terminate:	Cerebral cortex, thalamus, hypothalamus, olfactory bulb, cerebellum, midbrain, spinal cord

(b) Serotonergic (Serotonin)

To basal nuclei

Raphe nuclei

Functions:	1. Lower nuclei: Pain, locomotion 2. Upper nuclei: Sleep-wake cycle; mood and emotional behaviors, such as aggression and depression
Neurons Originate:	Raphe nuclei along brain stem midline
Neurons Terminate:	1. Lower nuclei project to spinal cord 2. Upper nuclei project to most of brain

(c) Dopaminergic (Dopamine)

Prefrontal cortex

To basal nuclei

Substantia nigra

Ventral tegmental area

Functions:	1. Motor control 2. "Reward" centers linked to addictive behaviors
Neurons Originate:	1. Substantia nigra in midbrain 2. Ventral tegmentum in midbrain
Neurons Terminate:	1. Cortex 2. Cortex and parts of limbic system

(d) Cholinergic (Acetylcholine)

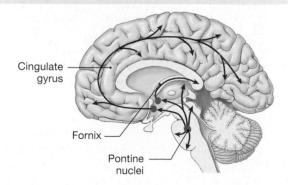

Cingulate gyrus

Fornix

Pontine nuclei

Functions:	Sleep-wake cycles, arousal, learning, memory, sensory information passing through thalamus
Neurons Originate:	Base of cerebrum; pons and midbrain
Neurons Terminate:	Cerebrum, hippocampus, thalamus

If connections between the reticular formation and the cerebral cortex are disrupted surgically, an animal becomes comatose. Other evidence for the importance of the reticular formation in states of arousal comes from studies showing that general anesthetics depress synaptic transmission in that region of the brain. Presumably, blocking ascending pathways between the reticular formation and the cerebral cortex creates a state of unconsciousness.

One way to define arousal states is by the pattern of electrical activity created by the cortical neurons. The measurement of brain activity is recorded by a procedure known as **electroencephalography.** Surface electrodes placed on or in the scalp detect depolarizations of the cortical neurons in the region just under the electrode. The complete cessation of brain waves is one of the clinical criteria for determining death.

Why Do We Sleep?

In humans, our major rest period is marked by a behavior known as **sleep,** defined as an easily reversible state of inactivity characterized by lack of interaction with the external environment. Most mammals and birds show the same stages of sleep as humans, telling us that sleep is a very ancient property of vertebrate brains. Depending on how sleep is defined, it appears that even invertebrates such as flies go through rest periods that could be described as sleep.

Why we need to sleep is one of the unsolved mysteries in neurophysiology, and a question that may have more than one answer. Some explanations that have been proposed include to conserve energy, to avoid predators, to allow the body to repair itself, and to process memories. Some of the newest research indicates that sleep is important for clearing wastes out of the cerebrospinal fluid, particularly some of the proteins that build up in degenerative neurological diseases such as Alzheimer's.

There is good evidence supporting the link between sleep and memory. A number of studies have demonstrated that sleep deprivation impairs our performance on tasks and tests, one reason for not pulling "all-nighters." At the same time, 20–30 minute "power naps" have also been shown to improve memory, and they can help make up a sleep deficit.

Physiologically, what distinguishes being awake from various stages of sleep? From studies, we know that the sleeping brain consumes as much oxygen as the awake brain, so sleep is a metabolically active state. Sleep is divided into four stages, each marked by identifiable, predictable events associated with characteristic somatic changes and EEG patterns.

In awake states, many neurons are firing but not in a coordinated fashion (**FIG. 9.17a**). An *electroencephalogram*, or **EEG,** of the waking-alert (eyes open) state shows a rapid, irregular pattern with no dominant waves. In awake-resting (eyes closed) states, sleep, or coma, electrical activity of the neurons begins to synchronize into waves with characteristic patterns. The more synchronous the firing of cortical neurons, the larger the amplitude of the waves. Accordingly, the awake-resting state is characterized by low-amplitude, high-frequency waves.

As the person falls asleep and the state of arousal lessens, the frequency of the waves decreases. The two major sleep phases are slow-wave sleep and rapid eye movement sleep. **Slow-wave sleep** (also called **deep sleep or non-REM sleep,** stage 4) is indicated on the EEG by the presence of *delta waves*, high-amplitude,

---

FIG. 9.17   Electroencephalograms (EEGs) and the sleep cycle

(a) Recordings of electrical activity in the brain during awake-resting and sleep periods show characteristic patterns.

Alpha waves	Awake, eyes closed	
REM	Stage 1	
	Stage 2	
	Stage 3	
Slow-wave sleep: delta waves	Stage 4	

⟶ Time ⟶

(b) The deepest sleep occurs in the first three hours.

Awake
Stage 1
Stage 2
Stage 3
Stage 4

1 2 3 4 5 6 7
Time of sleep (hr)

KEY

Amplitude

Frequency

Ⓠ **FIGURE QUESTIONS**

1. Which EEG pattern has the fastest frequency? The greatest amplitude?
2. In a 20–30 minute "power nap," what sleep stages will the napper experience?

low-frequency waves of long duration that sweep across the cerebral cortex (Fig. 9.17a). During this phase of the sleep cycle, sleepers adjust body position without conscious commands from the brain to do so.

In contrast, **rapid eye movement (REM) sleep** (stage 1) is marked by an EEG pattern closer to that of an awake person, with low-amplitude, high-frequency waves. During REM sleep, brain activity inhibits motor neurons to skeletal muscles, paralyzing them. Exceptions to this pattern are the muscles that move the eyes and those that control breathing. The control of homeostatic functions is depressed during REM sleep, and body temperature falls toward ambient temperature.

REM sleep is the period during which most dreaming takes place. The eyes move behind closed lids, as if following the action of the dream. Sleepers are most likely to wake up spontaneously from periods of REM sleep.

A typical eight-hour sleep consists of repeating cycles, as shown in Figure 9.17b. In the first hour, the person moves from wakefulness into a deep sleep (stage 4; first blue area in Fig. 9.17b). The sleeper then cycles between deep sleep and REM sleep (stage 1), with stages 2–3 occurring in between. Near the end of an eight-hour sleep period, a sleeper spends the most time in stage 2 and REM sleep, until finally awakening for the day.

If sleep is a neurologically active process, what is it that makes us sleepy? The possibility of a sleep-inducing factor was first proposed in 1913, when scientists found that cerebrospinal fluid from sleep-deprived dogs could induce sleep in normal animals. Since then, a variety of sleep-inducing factors have been identified. Curiously, many of them are also substances that enhance the immune response, such as interleukin-1, interferon, serotonin, and tumor necrosis factor. As a result of this finding, some investigators have suggested that one answer to the puzzle of the biological reason for sleep is that we need to sleep to enhance our immune response. Whether or not that is a reason for why we sleep, the link between the immune system and sleep induction may help explain why we tend to sleep more when we are sick.

Another clue to what makes us sleepy come from studies on *caffeine* and its methylxanthine cousins *theobromine* and *theophylline* (found in chocolate and tea). These chemicals are probably the most widely consumed psychoactive drugs, known since ancient times for their stimulant effect. Molecular research has revealed that the methylxanthines are receptor antagonists for *adenosine*, a molecule composed of the nitrogenous base adenine plus the sugar ribose [p. 35]. The discovery that the stimulant effect of caffeine comes from its blockade of adenosine receptors has led scientists to investigate adenosine's role in sleep-wake cycles. Evidence suggests that adenosine accumulates in the extracellular fluid during waking hours, increasingly suppressing activity of the neurons that promote wakefulness.

Sleep disorders are relatively common, as you can tell by looking at the variety of sleep-promoting agents available over the counter in drugstores. Among the more common sleep disorders are *insomnia* (the inability to go to sleep or remain asleep long enough to awake refreshed), sleep apnea, and sleepwalking. *Sleep apnea* {*apnoos*, breathless} is a condition in which the sleeper awakes when the airway muscles relax to the point of obstructing normal breathing.

Sleepwalking, or *somnambulism* {*somnus*, sleep + *ambulare*, to walk}, is a sleep behavior disorder that for many years was thought to represent the acting out of dreams. However, most dreaming occurs during REM sleep (stage 1), while sleepwalking takes place during deep sleep (stage 4). During sleepwalking episodes, which may last from 30 seconds to 30 minutes, the subject's eyes are open and registering the surroundings. The subject is able to avoid bumping into objects, can negotiate stairs, and in some cases is reported to perform such tasks as preparing food or folding clothes. The subject usually has little if any conscious recall of the sleepwalking episode upon awakening.

Sleepwalking is most common in children, and the frequency of episodes declines with age. There is also a genetic component, as the tendency to sleepwalk runs in families. To learn more about the different sleep disorders, see the NIH web site for the National Center for Sleep Disorder Research (*www.nhlbi.nih.gov/about/ncsdr*).

CONCEPT CHECK

21. During sleep, relay neurons in the thalamus reduce information reaching the cerebrum by altering their membrane potential. Are these neurons more likely to have depolarized or hyperpolarized? Explain your reasoning.

Physiological Functions Exhibit Circadian Rhythms

All organisms (even plants) have alternating daily patterns of rest and activity. These alternating activity patterns, like many other biological cycles, generally follow a 24-hour light-dark cycle and are known as *circadian rhythms* [p. 17]. When an organism is placed in conditions of constant light or darkness, these activity rhythms persist, apparently cued by an internal clock.

In mammals, the primary "clock" resides in networks of neurons located in the **suprachiasmatic nucleus** (SCN) of the hypothalamus, with secondary clocks influencing the behavior of different tissues. A very simple interpretation of how the biological clock works is that clock cycling results from a complex feedback loop in which specific genes turn on and direct protein synthesis. The proteins accumulate, turn off the genes, and then are themselves degraded. As the proteins disappear, the genes turn back on and the cycle begins again. The SCN clock has intrinsic activity that is synchronized with the external environment by sensory information about light cycles received through the eyes.

Circadian rhythms in humans can be found in most physiological functions and usually correspond to the phases of our sleep-wake cycles. For example, body temperature and cortisol secretion both cycle on a daily basis [Fig. 1.14, p. 18]. Melatonin from the pineal gland also is strongly linked to light-dark cycling: melatonin is sometimes called the "darkness hormone" because its secretion increases in the evening. The suprachiasmatic nucleus has melatonin receptors, supporting the hypothesis that melatonin can modulate clock cycling.

Disruption of circadian rhythms, such as occurs with shift work and jet lag, can have detrimental effects on mental and physical health. Sleep disturbances, depression, seasonal affective depressive disorder, diabetes, and obesity have all been linked to abnormal circadian rhythms. Jet lag, which occurs when people shift their light-dark cycles by travel across time zones, is a common manifestation of the effect of circadian rhythms on daily function. Melatonin treatments and exposure to natural daylight in the new location are the only treatments shown to have any significant effect on jet lag.

Emotion and Motivation Involve Complex Neural Pathways

Emotion and motivation are two aspects of brain function that probably represent an overlap of the behavioral state system and cognitive system. The pathways involved are complex and form closed circuits that cycle information among various parts of the brain, including the hypothalamus, limbic system, and cerebral cortex. We still do not understand the underlying neural mechanisms, and this is a large and active area of neuroscience research.

Emotions are difficult to define. We know what they are and can name them, but in many ways they defy description. One characteristic of emotions is that they are difficult to voluntarily turn on or off. The most commonly described emotions, which arise in different parts of the brain, are anger, aggression, sexual feelings, fear, pleasure, contentment, and happiness.

The limbic system, particularly the region known as the *amygdala*, is the center of emotion in the human brain. Scientists have learned about the role of this brain region through experiments in humans and animals. When the amygdala is artificially stimulated in humans, as it might be during surgery for epilepsy, patients report experiencing feelings of fear and anxiety. Experimental lesions that destroy the amygdala in animals cause the animals to become tamer and to display hypersexuality. As a result, neurobiologists believe that the amygdala is the center for basic instincts such as fear and aggression.

The pathways for emotions are complex (**FIG. 9.18**). Sensory stimuli feeding into the cerebral cortex are constructed in the brain to create a representation (perception) of the world. After information is integrated by the association areas, it is passed on to the limbic system. Feedback from the limbic system to the cerebral cortex creates awareness of the emotion, while descending pathways to the hypothalamus and brain stem initiate voluntary behaviors and unconscious responses mediated by autonomic, endocrine, immune, and somatic motor systems.

RUNNING **PROBLEM**

About six months after the start of treatment, Ben's head-drop seizures returned, and his development began to decline once again. An EEG following Ben's relapse did not demonstrate the erratic wave patterns specific to infantile spasms but did show abnormal activity in the right cortex. A neurologist ordered a positron emission tomography (PET) scan to determine the focus of Ben's seizure activity.

Ben received an injection of radioactively labeled glucose. He was then placed in the center of a PET machine lined with radiation detectors that created a map of his brain showing areas of high and low radioactivity. Those parts of his brain that were more active absorbed more glucose and thus emitted more radiation when the radioactive compound began to decay.

Q4: *What is the rationale for using radioactively labeled glucose (and not some other nutrient) for the PET scan?*

275 283 **298** 300 302 304

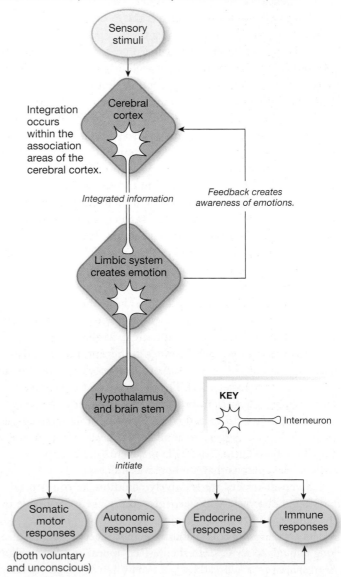

FIG. 9.18 | **Emotions affect physiology**

The association between stress and increased susceptibility to viruses is an example of an emotionally linked immune response.

Sensory stimuli

Cerebral cortex

Integration occurs within the association areas of the cerebral cortex.

Integrated information

Feedback creates awareness of emotions.

Limbic system creates emotion

Hypothalamus and brain stem

KEY

Interneuron

initiate

Somatic motor responses

Autonomic responses

Endocrine responses

Immune responses

(both voluntary and unconscious)

The physical result of emotions can be as dramatic as the pounding heart of a fight-or-flight reaction or as insidious as the development of an irregular heartbeat. The links between mind and body are difficult to study and will take many years of research to understand.

Motivation is defined as internal signals that shape voluntary behaviors. Some of these behaviors, such as eating, drinking, and having sex, are related to survival. Others, such as curiosity and having sex (again), are linked to emotions. Some motivational states are known as **drives** and generally have three properties in common: (1) they create an increased state of CNS arousal or alertness, (2) they create goal-oriented behavior, and (3) they are capable of coordinating disparate behaviors to achieve that goal.

Motivated behaviors often work in parallel with autonomic and endocrine responses in the body, as you might expect with behaviors originating in the hypothalamus. For example, if you eat salty popcorn, your body osmolarity increases. This stimulus acts on the thirst center of the hypothalamus, motivating you to seek something to drink. Increased osmolarity also acts on an endocrine center in the hypothalamus, releasing a hormone that increases water retention by the kidneys. In this way, one stimulus triggers both a motivated behavior and a homeostatic endocrine response.

Some motivated behaviors can be activated by internal stimuli that may not be obvious even to the person in whom they are occurring. Eating, curiosity, and sex drive are three examples of behaviors with complex stimuli underlying their onset. We may eat, for example, because we are hungry or because the food looks good or because we do not want to hurt someone's feelings. Many motivated behaviors stop when the person has reached a certain level of satisfaction, or **satiety,** but they may also continue *despite* feeling satiated.

Pleasure is a motivational state that is being intensely studied because of its relationship to *addictive behaviors,* such as drug use. Animal studies have shown that pleasure is a physiological state that is accompanied by increased activity of the neurotransmitter dopamine in certain parts of the brain. Drugs that are addictive, such as cocaine and nicotine, act by enhancing the effectiveness of dopamine, thereby increasing the pleasurable sensations perceived by the brain. As a result, use of these drugs rapidly becomes a learned behavior.

Interestingly, not all behaviors that are addictive are pleasurable. For example, there are a variety of compulsive behaviors that involve self-mutilation, such as pulling out hair by the roots. Fortunately, many behaviors can be modulated, given motivation.

Moods Are Long-Lasting Emotional States

Moods are similar to emotions but are longer-lasting, relatively stable subjective feelings related to one's sense of well-being. Moods are difficult to define at a neurobiological level, but evidence obtained in studying and treating mood disorders suggests that mood disturbances reflect changes in CNS function, such as abnormal neurotransmitter release or reception in different brain regions.

Mood disorders are estimated to be the fourth leading cause of illness in the world today. **Depression** is a mood disturbance that affects nearly 10% of the United States population each year. It is characterized by sleep and appetite disturbances and alterations of mood and libido that may seriously affect the person's ability to function at school or work or in personal relationships. Many people do not realize that depression is not a sign of mental or moral weakness, or that it can be treated successfully with drugs and psychotherapy. (For detailed information about depression, go to *www.nlm.nih.gov/medlineplus/depression.html.*)

The drug therapy for depression has changed in recent years, but all the major categories of antidepressant drugs alter some aspect of synaptic transmission. The older *tricyclic antidepressants,* such as amitriptyline, block reuptake of norepinephrine into the presynaptic neuron, thus extending the active life of the neurotransmitter. The antidepressants known as *selective serotonin reuptake inhibitors* (SSRIs) and *serotonin/norepinephrine reuptake inhibitors* (SNRIs) slow down the removal of serotonin and norepinephrine from the synapse. As a result of uptake inhibition, the neurotransmitter lingers in the synaptic cleft longer than usual, increasing transmitter-dependent activity in the postsynaptic neuron. Other antidepressant drugs alter brain levels of dopamine. The effectiveness of these different classes of antidepressant drugs suggests that norepinephrine, serotonin, and dopamine are all involved in brain pathways for mood and emotion.

Interestingly, patients need to take antidepressant drugs for several weeks before they experience their full effect. This delay suggests that the changes taking place in the brain are long-term modulation of pathways rather than simply enhanced fast synaptic responses. Several studies in humans and animal models provide evidence that antidepressants promote the growth of new neurons, which would also explain the delayed onset of full action.

The causes of major depression are complex and probably involve a combination of genetic factors, the serotonergic and noradrenergic diffuse modulatory systems, trophic factors such as *brain-derived neurotrophic factor* (BDNF), and stress. The search to uncover the biological basis of disturbed brain function is a major focus of neuroscience research today.

Some research into brain function has become quite controversial, particularly that dealing with sexuality and the degree to which behavior in general is genetically determined in humans. We will not delve deeply into any of these subjects because they are complex and would require lengthy explanations to do them justice. Instead, we will look briefly at some of the recent models proposed to explain the mechanisms that are the basis for higher cognitive functions.

Learning and Memory Change Synaptic Connections in the Brain

For many years, motivation, learning, and memory (all of which are aspects of the cognitive state) were considered to be in the realm of psychology rather than biology. Neurobiologists in decades past were more concerned with the network and cellular aspects of neuronal function. In recent years, however, the two fields have overlapped more and more. Scientists have discovered that the underlying basis for cognitive function seems to be explainable in terms of cellular events that influence plasticity—events such as long-term potentiation [p. 267]. The ability of neurons to change their responsiveness or alter their connections with experience is fundamental to the two cognitive processes of learning and memory.

Learning Is the Acquisition of Knowledge

How do you know when you have learned something? Learning can be demonstrated by behavioral changes, but behavioral changes are not required in order for learning to occur. Learning can be internalized and is not always reflected by overt behavior while the learning is taking place. Would someone watching you read your textbook or listen to a professor's lecture be able to tell whether you had learned anything?

Learning can be classified into two broad types: associative and nonassociative. **Associative learning** occurs when two stimuli are associated with each other, such as Pavlov's classic experiment in which he simultaneously presented dogs with food and rang a bell. After a period of time, the dogs came to associate the sound of the bell with food and began to salivate in anticipation of food whenever the bell was rung. Another form of associative learning occurs when an animal associates a stimulus with a given behavior. An example would be a mouse that gets a shock each time it touches a certain part of its cage. It soon associates that part of the cage with an unpleasant experience and avoids the area.

Nonassociative learning is a change in behavior that takes place after repeated exposure to a single stimulus. This type of learning includes habituation and sensitization, two adaptive behaviors that allow us to filter out and ignore background stimuli while responding more sensitively to potentially disruptive stimuli. In **habituation,** an animal shows a decreased response to an irrelevant stimulus that is repeated over and over. For example, a sudden loud noise may startle you, but if the noise is repeated over and over again, your brain begins to ignore it. Habituated responses allow us to filter out stimuli that we have evaluated and found to be insignificant.

Sensitization is the opposite of habituation, and the two behaviors combined help increase an organism's chances for survival. In sensitization learning, exposure to a noxious or intense stimulus causes an enhanced response upon subsequent exposure. For example, people who become ill while eating a certain food may find that they lose their desire to eat that food again. Sensitization is adaptive because it helps us avoid potentially harmful stimuli. At the same time, sensitization may be maladaptive if it leads to the hypervigilant state known as *post-traumatic stress disorder* (PTSD).

Memory Is the Ability to Retain and Recall Information

Memory is the ability to retain and recall information. Memory is a very complex function, but scientists have tried to classify it in different ways. We think of several types of memory: short-term and long-term, reflexive and declarative. Processing for different types of memory appears to take place through different pathways. With noninvasive imaging techniques such as MRI and PET scans, researchers have been able to track brain activity as individuals learned to perform tasks.

Memories are stored throughout the cerebral cortex in pathways known as **memory traces.** Some components of memories are stored in the sensory cortices where they are processed. For example, pictures are stored in the visual cortex, and sounds in the auditory cortex.

Learning a task or recalling a task already learned may involve multiple brain circuits that work in parallel. This *parallel processing* helps provide backup in case one of the circuits is damaged. It is also believed to be the means by which specific memories are generalized, allowing new information to be matched to stored information. For example, a person who has never seen a volleyball will recognize it as a ball because the volleyball has the same general characteristics as all other balls the person has seen.

In humans, the hippocampus seems to be an important structure in both learning and memory. Patients who have part of the hippocampus destroyed to relieve a certain type of epilepsy also have trouble remembering new information. When given a list of words to repeat, they can remember the words as long as their attention stays focused on the task. If they are distracted, however, the memory of the words disappears, and they must learn the list again. Information stored in long-term memory before the operation is not affected. This inability to remember newly acquired information is a defect known as **anterograde amnesia** {*amnesia, oblivion*}.

Memory has multiple levels of storage, and our memory bank is constantly changing (**FIG. 9.19**). When a stimulus comes into the CNS, it first goes into **short-term memory,** a limited storage area that can hold only about 7 to 12 pieces of information at a time. Items in short-term memory disappear unless an

RUNNING **PROBLEM**

Ben's halted development is a feature unique to infantile spasms. The abnormal portions of the brain send out continuous action potentials during frequent seizures and ultimately change the interconnections of brain neurons. The damaged portions of the brain harm normal portions to such an extent that medication or surgery should be started as soon as possible. If intervention is not begun early, the brain can be permanently damaged and development will never recover.

Q5: *The brain's ability to change its synaptic connections as a result of neuronal activity is called* _____.

275 283 298 **300** 302 304

FIG. 9.19 Memory processing

New information goes into short-term memory but is lost unless processed and stored in long-term memory.

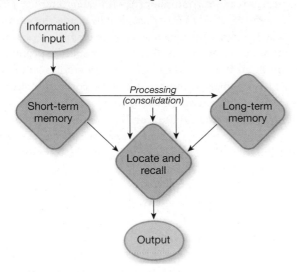

effort, such as repetition, is made to put them into a more permanent form.

Working memory is a special form of short-term memory processed in the prefrontal lobes. This region of the cerebral cortex is devoted to keeping track of bits of information long enough to put them to use in a task that takes place after the information has been acquired. Working memory in these regions is linked to long-term memory stores, so that newly acquired information can be integrated with stored information and acted on.

For example, suppose you are trying to cross a busy road. You look to the left and see that there are no cars coming for several blocks. You then look to the right and see that there are no cars coming from that direction either. Working memory has stored the information that the road to the left is clear, and so using this stored knowledge about safety, you are able to conclude that there is no traffic from either direction and it is safe to cross the road.

In people with damage to the prefrontal lobes of the brain, this task becomes more difficult because they are unable to recall whether the road is clear from the left once they have looked away to assess traffic coming from the right. Working memory allows us to collect a series of facts from short- and long-term memory and connect them in a logical order to solve problems or plan actions.

Long-term memory is a storage area capable of holding vast amounts of information. Think of how much information humans needed to remember in centuries past, when books were scarce and most history was passed down by word of mouth. Wandering bards and troubadours kept long epic poems and ballads, such as *The Odyssey* and *Beowulf,* stored in their memory banks, to be retrieved at will.

The processing of information that converts short-term memory into long-term memory is known as **consolidation** (Fig. 9.19). Consolidation can take varying periods of time, from seconds to minutes. Information passes through many

intermediate levels of memory during consolidation, and in each of these stages, the information can be located and recalled.

As scientists studied the consolidation of short-term memory into long-term memory, they discovered that the process involves changes in neuronal excitability or synaptic connections in the circuits involved in learning. In some cases, new synapses form; in others, the effectiveness of synaptic transmission is altered through long-term potentiation or through long-term depression. These changes are evidence of plasticity and show us that the brain is not "hard-wired."

Long-term memory has been divided into two types that are consolidated and stored using different neuronal pathways (**TBL. 9.4**). **Reflexive (implicit) memory,** which is automatic and does not require conscious processes for either creation or recall, involves the amygdala and the cerebellum. Information stored in reflexive memory is acquired slowly through repetition. Motor skills fall into this category, as do procedures and rules.

For example, you do not need to think about putting a period at the end of each sentence or about how to pick up a fork. Reflexive memory has also been called *procedural memory* because it generally concerns how to do things. Reflexive memories can be acquired through either associative or nonassociative learning processes, and these memories are stored.

Declarative (explicit) memory, on the other hand, requires conscious attention for its recall. Its creation generally depends on the use of higher-level cognitive skills such as inference, comparison, and evaluation. The neuronal pathways involved in this type of memory are in the temporal lobes. Declarative memories deal with knowledge about ourselves and the world around us that can be reported or described verbally.

Sometimes, information can be transferred from declarative memory to reflexive memory. The quarterback on a football team is a good example. When he learned to throw the football as a small boy, he had to pay close attention to gripping the ball and coordinating his muscles to throw the ball accurately. At that point of learning to throw the ball, the process was in declarative memory and required conscious effort as the boy analyzed his movements.

TABLE 9.4	Types of Long-Term Memory
Reflexive (Implicit) Memory	**Declarative (Explicit) Memory**
Recall is automatic and does not require conscious attention	Recall requires conscious attention
Acquired slowly through repetition	Depends on higher-level thinking skills such as inference, comparison, and evaluation
Includes motor skills and rules and procedures	Memories can be reported verbally
Procedural memories can be demonstrated	

With repetition, however, the mechanics of throwing the ball were transferred to reflexive memory: they became a reflex that could be executed without conscious thought. That transfer allowed the quarterback to use his conscious mind to analyze the path and timing of the pass while the mechanics of the pass became automatic. Athletes often refer to this automaticity of learned body movements as *muscle memory.*

Memory is an individual thing. We process information on the basis of our experiences and perception of the world. Because people have widely different experiences throughout their lives, it follows that no two people will process a given piece of information in the same way. If you ask a group of people about what happened during a particular event such as a lecture or an automobile accident, no two descriptions will be identical. Each person processed the event according to her or his own perceptions and experiences. Experiential processing is important to remember when studying in a group situation, because it is unlikely that all group members learn or recall information the same way.

Memory loss and the inability to process and store new memories are devastating medical conditions. In younger people, memory problems are usually associated with trauma to the brain from accidents. In older people, strokes and progressive *dementia* {*demens,* out of one's mind} are the main causes of memory loss.

Alzheimer's disease is a progressive neurodegenerative disease of cognitive impairment that accounts for about half the cases of dementia in the elderly. Alzheimer's is characterized by memory loss that progresses to a point where the patient does not recognize family members. Over time, even the personality changes, and in the final stages, other cognitive functions fail so that patients cannot communicate with caregivers.

Diagnosis of Alzheimer's is usually made through the patient's declining performance on cognitive function examinations. Scientists are studying whether tests for specific proteins in the cerebrospinal fluid or advanced imaging studies can reveal if a person has the disease, but the data are inconclusive at this stage. Currently, the only definitive diagnosis of Alzheimer's comes after death, when brain tissue can be examined for neuronal degeneration, extracellular plaques made of β-*amyloid protein,* and intracellular tangles of *tau,* a protein that is normally associated with microtubules.

The presence of amyloid plaques and tau tangles is diagnostic, but the underlying cause of Alzheimer's is unclear. There is a known genetic component, and other theories include oxidative stress and chronic inflammation. Currently there is no proven prevention or treatment, although drugs that are acetylcholine agonists or acetylcholinesterase inhibitors slow the progression of the disease.

By one estimate, Alzheimer's affects about 5.2 million Americans, with the number expected to rise as Baby Boomers age. The forecast of 16 million Americans with Alzheimer's by the year 2050 has put this disease in the forefront of neurobiological research.

Although pathological memory loss is a concern, the ability to forget is also important for our mental health. Post-traumatic stress disorder is an example of where forgetting would be beneficial.

Language Is the Most Elaborate Cognitive Behavior

One of the hallmarks of an advanced nervous system is the ability of one member of a species to exchange complex information with other members of the same species. Although found predominantly in birds and mammals, this ability also occurs in certain insects that convey amazingly detailed information by means of sound (crickets), touch and sight (bees), and odor (ants). In humans, the exchange of complex information takes place primarily through spoken and written language. Because language is considered the most elaborate cognitive behavior, it has received considerable attention from neurobiologists.

Language skills require the input of sensory information (primarily from hearing and vision), processing in various centers in the cerebral cortex, and the coordination of motor output for vocalization and writing. In most people, the centers for language ability are found in the left hemisphere of the cerebrum. Even 70% of people who are either left-handed (right-brain dominant) or ambidextrous use their left brain for speech. The ability to communicate through speech has been divided into two processes: the combination of different sounds to form words (vocalization) and the combination of words into grammatically correct and meaningful sentences.

RUNNING **PROBLEM**

The PET scan revealed two abnormal spots, or loci (plural of locus), on Ben's right hemisphere, one on the parietal lobe and one overlapping a portion of the primary motor cortex. Because the loci triggering Ben's seizures were located on the same hemisphere and were in the cortex, Ben was a candidate for a hemispherectomy, removal of the cortex of the affected hemisphere. Surgeons removed 80% of his right cerebral cortex, sparing areas crucial to vision, hearing, and sensory processing. Normally the motor cortex would be spared as well, but in Ben's case a seizure locus overlapped much of the region.

Q6: *In which lobes are the centers for vision, hearing, and sensory processing located?*

Q7: *Which of Ben's abilities might have suffered if his left hemisphere had been removed instead?*

Q8: *By taking only the cortex of the right hemisphere, what parts of the cerebrum did surgeons leave behind?*

Q9: *Why were the surgeons careful to spare Ben's right lateral ventricle?*

(275)(283)(298)(300)**(302)**(304)

The model presented here is a simplified version of what scientists now know is a very complex function that involves many regions of the cerebral cortex. Traditionally, integration of spoken language in the human brain has been attributed to two regions in the cerebral cortex: **Wernicke's area** at the junction of the parietal, temporal, and occipital lobes and **Broca's area** in the posterior part of the frontal lobe, close to the motor cortex (**FIG. 9.20**). Most of what we know about these areas comes from studies of people with brain lesions (because nonhuman animals are not capable of speech). Even primates that communicate on the level of a small child through sign language and other visual means do not have the physical ability to vocalize the sounds of human language.

Input into the language areas comes from either the visual cortex (reading) or the auditory cortex (listening). Sensory input from either cortex goes first to Wernicke's area, then to Broca's area. After integration and processing, output from Broca's area to the motor cortex initiates a spoken or written action.

If damage occurs to Wernicke's area, a person may have difficulty understanding spoken or visual information. The person's own speech may be nonsense because the person is unable to retrieve words. This condition is known as **receptive aphasia** {*a-*, not + *phatos*, spoken} because the person is unable to understand sensory input.

Damage to Broca's area causes an **expressive aphasia,** or *Broca aphasia.* People with Broca aphasia understand simple, unambiguous spoken and written language but have difficulty interpreting complicated sentences with several elements linked together. This difficulty appears to be a deficit in short term memory. These people also have difficulty speaking or writing in

FIG. 9.20 Language processing

People with damage to Wernicke's area do not understand spoken or written communication.
Those with damage to Broca's area understand but are unable to respond appropriately.

(a) Speaking a Written Word

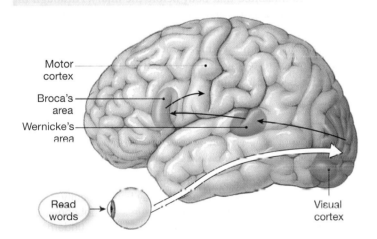

Motor cortex
Broca's area
Wernicke's area
Read words
Visual cortex

(b) Speaking a Heard Word

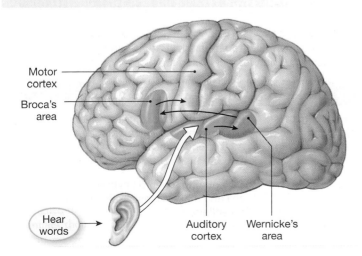

Motor cortex
Broca's area
Hear words
Auditory cortex
Wernicke's area

(c) PET Scan of the Brain at Work

In PET scans, neurons take up radio-labeled glucose. The most active areas show up as red-yellow regions.

Hearing words
Seeing words
Max
Speaking words
Generating words
Min

Q FIGURE QUESTION

In the image above, the brain area active in seeing words is in the _____ lobe, and the brain area active during word generation is in the _____ lobe.

Evolution of Nervous Systems

3. Nervous systems evolved from a simple network of neurons to complex brains. (p. 275; Fig. 9.1)

4. The **cerebrum** is responsible for thought and emotion. (p. 277)

Anatomy of the Central Nervous System

5. The central nervous system consists of layers of cells around a fluid-filled central cavity and develops from the **neural tube** of the embryo. (p. 277; Fig. 9.2)

6. The **gray matter** of the CNS consists of unmyelinated nerve cell bodies, dendrites, and axon terminals. The cell bodies either form layers in parts of the brain or else cluster into groups known as **nuclei.** (p. 277)

7. Myelinated axons form the **white matter** of the CNS and run in bundles called **tracts.** (p. 277)

8. The brain and spinal cord are encased in the **meninges** and the bones of the **cranium** and vertebrae. The meninges are the **pia mater,** the **arachnoid membrane,** and the **dura mater.** (p. 280; Fig. 9.3)

9. The **choroid plexus** secretes **cerebrospinal fluid** (CSF) into the **ventricles** of the brain. Cerebrospinal fluid cushions the tissue and creates a controlled chemical environment. (pp. 277, 281; Fig. 9.4)

10. Tight junctions in brain capillaries create a **blood-brain barrier** that prevents possibly harmful substances in the blood from entering the interstitial fluid. (p. 282; Fig. 9.5)

11. The normal fuel source for neurons is glucose, which is why the body closely regulates blood glucose concentrations. (p. 283)

The Spinal Cord

12. Each segment of the spinal cord is associated with a pair of **spinal nerves.** (p. 284)

13. The **dorsal root** of each spinal nerve carries incoming sensory information. The **dorsal root ganglia** contain the nerve cell bodies of sensory neurons. (p. 284; Fig. 9.6)

14. The **ventral roots** carry information from the central nervous system to muscles and glands. (p. 284)

15. **Ascending tracts** of white matter carry sensory information to the brain, and **descending tracts** carry efferent signals from the brain. **Propriospinal tracts** remain within the spinal cord. (p. 284)

16. **Spinal reflexes** are integrated in the spinal cord. (p. 285; Fig. 9.7)

The Brain

17. The brain has six major divisions: cerebrum, diencephalon, midbrain, cerebellum, pons, and medulla oblongata. (p. 285; Fig. 9.8)

18. The **brain stem** is divided into medulla oblongata, pons, and midbrain (mesencephalon). **Cranial nerves** II to XII originate here. (p. 285; Fig. 9.8f; Tbl. 9.1)

19. The **reticular formation** is a diffuse collection of neurons that play a role in many basic processes. (p. 287)

20. The **medulla oblongata** contains **somatosensory** and **corticospinal tracts** that convey information between the cerebrum and spinal cord. Most tracts cross the midline in the **pyramid** region. The medulla contains control centers for many involuntary functions. (p. 287)

21. The **pons** acts as a relay station for information between the cerebellum and cerebrum. (p. 287)

22. The **midbrain** controls eye movement and relays signals for auditory and visual reflexes. (p. 288)

23. The **cerebellum** processes sensory information and coordinates the execution of movement. (p. 288)

24. The **diencephalon** is composed of the thalamus and hypothalamus. The **thalamus** relays and modifies sensory and motor information going to and from the cerebral cortex. (p. 288; Fig. 9.9)

25. The **hypothalamus** contains centers for behavioral drives and plays a key role in homeostasis by its control over endocrine and autonomic function. (p. 289; Tbl. 9.2)

26. The **pituitary gland** and **pineal gland** are endocrine glands located in the diencephalon. (p. 288)

27. The cerebrum is composed of two hemispheres connected at the **corpus callosum.** Each cerebral hemisphere is divided into **frontal, parietal, temporal,** and **occipital lobes.** (p. 289)

28. Cerebral gray matter includes the **cerebral cortex,** basal ganglia, and limbic system. (p. 289; Fig. 9.10)

29. The **basal ganglia** help control movement. (p. 289)

30. The **limbic system** acts as the link between cognitive functions and emotional responses. It includes the **amygdala** and **cingulate gyrus,** linked to emotion and memory, and the **hippocampus,** associated with learning and memory. (p. 289; Fig. 9.11)

Brain Function

31. Three brain systems influence motor output: a **sensory system,** a **cognitive system,** and a **behavioral state system.** (p. 291; Fig. 9.12)

32. Higher brain functions, such as reasoning, arise in the cerebral cortex. The cerebral cortex contains three functional specializations: **sensory areas, motor areas,** and **association areas.** (p. 291; Fig. 9.13)

33. Each hemisphere of the cerebrum has developed functions not shared by the other hemisphere, a specialization known as **cerebral lateralization.** (p. 291; Fig. 9.14)

34. Sensory areas receive information from sensory receptors. The **primary somatic sensory cortex** processes information about touch, temperature, and other somatic senses. The **visual cortex, auditory cortex, gustatory cortex,** and **olfactory cortex** receive information about vision, sound, taste, and odors, respectively. (p. 293)

35. **Association areas** integrate sensory information into perception. **Perception** is the brain's interpretation of sensory stimuli. (p. 294)

36. Motor output includes skeletal muscle movement, neuroendocrine secretion, and visceral responses. (p. 294)

37. Motor areas direct skeletal muscle movement. Each cerebral hemisphere contains a **primary motor cortex** and **motor association area.** (p. 294)

38. The **behavioral state system** controls states of arousal and modulates the sensory and cognitive systems. (p. 294)

39. The **diffuse modulatory systems** of the reticular formation influence attention, motivation, wakefulness, memory, motor control, mood, and metabolic homeostasis. (p. 294; Fig. 9.16)

40. The **reticular activating system** keeps the brain **conscious,** or aware of self and environment. Electrical activity in the brain varies with levels of arousal and can be recorded by **electroencephalography.** (pp. 294, 296; Fig. 9.17)

41. **Circadian rhythms** are controlled by an internal clock in the **suprachiasmatic nucleus** of the hypothalamus. (p. 297)

42. **Sleep** is an easily reversible state of inactivity with characteristic stages. The two major phases of sleep are **REM (rapid eye**

movement) sleep and **slow-wave sleep** (non-REM sleep). The physiological reason for sleep is uncertain. (pp. 296, 297)

43. The limbic system is the center of **emotion** in the human brain. Emotional events influence physiological functions. (p. 298; Fig. 9.18)

44. **Motivation** arises from internal signals that shape voluntary behaviors related to survival or emotions. Motivational **drives** create goal-oriented behaviors. (p. 299)

45. **Moods** are long-lasting emotional states. Many mood disorders can be treated by altering neurotransmission in the brain. (p. 299)

46. Learning is the acquisition of knowledge about the world around us. **Associative learning** occurs when two stimuli are associated with each other. **Nonassociative learning** is a change in behavior that takes place after repeated exposure to a single stimulus. (p. 300)

47. In **habituation,** an animal shows a decreased response to a stimulus that is repeated over and over. In **sensitization,** exposure to a noxious or intense stimulus creates an enhanced response on subsequent exposure. (p. 300)

48. **Memory** has multiple levels of storage and is constantly changing. Information is first stored in **short-term memory** but disappears unless consolidated into long-term memory. (p. 300; Fig. 9.19)

49. **Long-term memory** includes **reflexive memory,** which does not require conscious processes for its creation or recall, and **declarative memory,** which uses higher-level of cognitive skills for formation and requires conscious attention for its recall. (p. 301; Tbl. 9.4)

50. The **consolidation** of short-term memory into long-term memory appears to involve changes in the synaptic connections of the circuits involved in learning. (p. 301)

51. Language is considered the most elaborate cognitive behavior. The integration of spoken language in the human brain involves information processing in **Wernicke's area** and **Broca's area.** (p. 303; Fig. 9.20)

REVIEW QUESTIONS

In addition to working through these questions and checking your answers on p. A-11, review the Learning Outcomes at the beginning of this chapter.

Level One Reviewing Facts and Terms

1. The ability of human brains to change circuit connections and function in response to sensory input and past experience is known as _____.

2. _____ behaviors are related to feeling and emotion. _____ behaviors are related to thinking.

3. The part of the brain called the _____ is what makes us human, allowing human reasoning and cognition.

4. In vertebrates, the central nervous system is protected by the bones of the _____ and _____.

5. Name the meninges, beginning with the layer next to the bones.

6. List and explain the purposes of cerebrospinal fluid (CSF). Where is CSF made?

7. Compare the CSF concentration of each of the following substances with its concentration in the blood plasma.

   (a) H^+
   (b) Na^+
   (c) K^+

8. The only fuel source for neurons under normal circumstances is _____. Low concentration of this fuel in the blood is termed _____. To synthesize enough ATP to continually transport ions, the neurons consume large quantities of _____. To supply these needs, about _____% of the blood pumped by the heart goes to the brain.

9. What is the blood-brain barrier, and what is its function?

10. How are gray matter and white matter different from each other, both anatomically and functionally?

11. Name the cerebral cortex areas that (a) direct perception, (b) direct movement, and (c) integrate information and direct voluntary behaviors.

12. What does *cerebral lateralization* refer to? What functions tend to be centered in each hemisphere?

13. Match each of the following areas with its function.

(a) medulla oblongata	1. coordinates execution of movement
(b) pons	2. is composed of the thalamus and hypothalamus
(c) midbrain	3. controls arousal and sleep
(d) reticular formation	4. fills most of the cranium
(e) cerebellum	5. contains control centers for blood pressure and breathing
(f) diencephalon	6. relays and modifies information going to and from the cerebrum
(g) thalamus	7. transfers information to the cerebellum
(h) hypothalamus	8. contains integrating centers for homeostasis
(i) cerebrum	9. relays signals and visual reflexes, plus eye movement

14. Name the 12 cranial nerves in numerical order and their major functions.

15. Name and define the two major phases of sleep. How are they different from each other?

16. List several homeostatic reflexes and behaviors influenced by output from the hypothalamus. What is the source of emotional input into this area?

17. The _____ region of the limbic system is believed to be the center for basic instincts (such as fear) and learned emotional states.

18. What are the broad categories of learning? Define habituation and sensitization. What anatomical structure of the cerebrum is important in both learning and memory?

19. What two centers of the cortex are involved in integrating spoken language?

Level Two Reviewing Concepts

20. Map the following terms describing CNS anatomy. You may draw pictures or add terms if you wish.

• arachnoid membrane	• ascending tracts
• blood-brain barrier	• brain
• capillaries	• cell bodies
• cerebrospinal fluid	• cervical nerves
• choroid plexus	• cranial nerves
• descending tracts	• dorsal root
• dorsal root ganglion	• dura mater
• ependyma	• gray matter
• lumbar nerves	• meninges
• nuclei	• pia mater
• propriospinal tracts	• sacral nerves
• spinal cord	• thoracic nerves
• ventral root	• ventricles
• vertebral column	• white matter

21. Trace the pathway that the cerebrospinal fluid follows through the nervous system.

22. What are the three brain systems that regulate motor output by the CNS?

23. Explain the role of Wernicke's and Broca's areas in language.

24. Compare and contrast the following concepts:

    (a) diffuse modulatory systems, reticular formation, limbic system, and reticular activating system
    (b) different forms of memory
    (c) nuclei and ganglia
    (d) tracts, nerves, horns, nerve fibers, and roots

25. Replace each question mark in the following table with the appropriate word(s):

Cerebral Area	Lobe	Functions
Primary somatic sensory cortex	?	Receives sensory information from peripheral receptors
?	Occipital	Processes information from the eyes
Auditory cortex	Temporal	?
?	Temporal	Receives input from chemoreceptors in the nose
Motor cortices	?	?
Association areas	NA	?

26. Given the wave shown below, draw (a) a wave having a lower frequency, (b) a wave having a larger amplitude, (c) a wave having a higher frequency. (*Hint:* See Fig. 9.17, p. 296.)

27. What properties do motivational states have in common?

28. What changes occur at synapses as memories are formed?

Level Three Problem Solving

29. Mr. Andersen, a stroke patient, experiences expressive aphasia. His savvy therapist, Cheryl, teaches him to sing to communicate his needs. What signs did he exhibit before therapy? How do you know he did not have receptive aphasia? Using what you have learned about cerebral lateralization, hypothesize why singing worked for him.

30. A study was done in which 40 adults were taught about the importance of using seat belts in their cars. At the end of the presentation, all participants scored at least 90% on a comprehensive test covering the material taught. The people were also secretly videotaped entering and leaving the parking lot of the class site. Twenty subjects entered wearing their seat belts; 22 left wearing them. Did learning occur? What is the relationship between learning and actually buckling the seat belts?

31. In 1913, Henri Pieron kept a group of dogs awake for several days. Before allowing them to sleep, he withdrew cerebrospinal fluid from the sleep-deprived animals. He then injected this CSF into normal, rested dogs. The recipient dogs promptly went to sleep for periods ranging from two hours to six hours. What conclusion can you draw about the possible source of a sleep-inducing factor? What controls should Pieron have included?

32. A 2002 study* presented the results of a prospective study [p. 23] done in Utah. The study began in 1995 with cognitive assessment of 1889 women whose mean age was 74.5 years. Investigators asked about the women's history of taking calcium, multivitamin supplements, and postmenopausal hormone replacement therapy (estrogen or estrogen/progesterone). Follow-up interviews in 1998 looked for the development of Alzheimer's disease in the study population. Data showed that 58 of 800 women who had not used hormone replacement therapy developed Alzheimer's, compared with 26 of 1066 women who had used hormones.

    (a) Can the researchers conclude from the data given that hormone replacement therapy decreases the risk of developing Alzheimer's? Should other information be factored into the data analysis?
    (b) How applicable are these findings to American women as a whole? What other information might you want to know about the study subjects before you draw any conclusions?

33. A young woman having a seizure was brought into the emergency room. Her roommate said that the woman had taken the street drug Ecstasy the night before and that she had been drinking a lot of water. A blood test showed that her plasma Na^+ was very low: 120 mM (normal 135–145), and her plasma osmolality was 250 mosmoles/kg (normal 280–296). Why would her low osmolality and low Na^+ concentration disrupt her brain function and cause seizures?

---

*P. P. Zandi *et al.* Hormone replacement therapy and incidence of Alzheimer disease in older women: The Cache County study. *JAMA* 288: 2123–2129, 2002 Nov. 6.

Answers to Concept Checks, Figure and Graph Questions, and end-of-chapter Review Questions can be found in Appendix A [A-1].

10

Sensory Physiology

Nature does not communicate with man by sending encoded messages.

Oscar Hechter, in *Biology and Medicine into the 21st Century,* 1991

GENERAL PROPERTIES OF SENSORY SYSTEMS 310

LO 10.1 Describe the different types of receptors for somatic and special senses.

LO 10.2 Explain how receptors convert physical stimuli into electrical signals using the following terms: transduction, threshold, adequate stimulus, receptive field, receptor potential.

LO 10.3 Explain how the central nervous system is able to determine modality, location, intensity, and duration of a stimulus.

LO 10.4 Explain how tonic and phasic receptors adapt to a continuous stimulus.

SOMATIC SENSES 317

LO 10.5 Trace the pathways for somatic sensation from receptor to the somatosensory cortex.

LO 10.6 Describe the different types of somatosensory receptors.

LO 10.7 Explain how pain and itch are mediated by nociceptors, and describe the neural pathways for pain.

CHEMORECEPTION: SMELL AND TASTE 324

LO 10.8 Describe the receptors, sensory transduction, and neural pathways for olfaction.

LO 10.9 Describe the receptors, sensory transduction, and neural pathways for the five primary taste sensations.

THE EAR: HEARING 329

LO 10.10 Trace the anatomical pathway sound energy follows from the air until it becomes an action potential in a primary sensory neuron.

LO 10.11 Describe the anatomical pathway for sound transmission from the cochlea to the auditory cortex.

LO 10.12 Explain how hair cells convert sound energy into an action potential.

THE EAR: EQUILIBRIUM 337

LO 10.13 Explain how otoliths and the cupula convey information about movement and head position to the vestibular nerve.

THE EYE AND VISION 340

LO 10.14 Describe the structures of the eye and the role of each structure in vision.

LO 10.15 Trace the pathway for vision from the retina to the visual cortex.

LO 10.16 Explain how photoreceptors convert light energy into action potentials.

LO 10.17 Explain signal processing in the retina and in the visual cortex.

BACKGROUND BASICS

266 Summation
171 Second messenger systems
242 Threshold
174 G proteins
261 Plasticity
183 Tonic control
154 Membrane potential
240 Graded potentials
257 Neurotransmitter release

Retina blood vessels and nerve cells

magine floating in the dark in an indoor tank of buoyant salt water: there is no sound, no light, and no breeze. The air and water are the same temperature as your body. You are in a sensory deprivation chamber, and the only sensations you are aware of come from your own body. Your limbs are weightless, your breath moves in and out effortlessly, and you feel your heart beating. In the absence of external stimuli, you turn your awareness inward to hear what your body has to say.

In decades past, flotation tanks for sensory deprivation were a popular way to counter the stress of a busy world. These facilities are hard to find now, but they illustrate the role of the afferent division of the nervous system: to provide us with information about the environment outside and inside our bodies. Sometimes we perceive sensory signals when they reach a level of conscious awareness, but other times they are processed completely at the subconscious level (**TBL. 10.1**). Stimuli that usually do not reach conscious awareness include changes in muscle stretch and tension as well as a variety of internal parameters that the body monitors to maintain homeostasis, such as blood pressure and pH. The responses to these stimuli constitute many of the subconscious reflexes of the body, and you will encounter them in later chapters as we explore the processes that maintain physiological homeostasis.

In this chapter, we are concerned primarily with sensory stimuli whose processing reaches the conscious level of perception. These stimuli are associated with the **special senses** of vision, hearing, taste, smell, and equilibrium, and the **somatic senses** of touch, temperature, pain, itch, and proprioception. **Proprioception,** which is defined as the awareness of body movement and position in space, is mediated by muscle and joint sensory receptors called **proprioceptors** and may be either

RUNNING **PROBLEM** | Ménière's Disease

On December 23, 1888, Vincent Van Gogh, the legendary French painter, returned to his room in a boardinghouse in Arles, France, picked up a knife, and cut off his own ear. A local physician, Dr. Felix Ray, examined Van Gogh that night and wrote that the painter had been "assailed by auditory hallucinations" and in an effort to relieve them, "mutilated himself by cutting off his ear." A few months later, Van Gogh committed himself to a lunatic asylum. By 1890, Van Gogh was dead by his own hand. Historians have postulated that Van Gogh suffered from epilepsy, but some American neurologists disagree. They concluded that the painter's strange attacks of dizziness, nausea, and overwhelming tinnitus (ringing or other sounds in the ears), which he described in desperate letters to his relatives, are more consistent with Ménière's disease, a condition that affects the inner ear. Today, Anant, a 20-year-old college student, will be examined by an otolaryngologist (ear-nose-throat specialist) to see if his periodic attacks of severe dizziness and nausea are caused by the same condition that might have driven Van Gogh to suicide.

310   314   331   339   343   348   353

TABLE 10.1 | Information Processing by the Sensory Division

Stimulus Processing Usually Conscious

Special Senses	Somatic Senses
Vision	Touch
Hearing	Temperature
Taste	Pain
Smell	Itch
Equilibrium	Proprioception

Stimulus Processing Usually Subconscious

Somatic Stimuli	Visceral Stimuli
Muscle length and tension	Blood pressure
Proprioception	Distension of gastrointestinal tract
	Blood glucose concentration
	Internal body temperature
	Osmolarity of body fluids
	Lung inflation
	pH of cerebrospinal fluid
	pH and oxygen content of blood

unconscious or conscious. If you close your eyes and raise your arm above your head, you are aware of the arm's position because of proprioceptor activation.

We first consider general properties of sensory pathways. We then look at the unique receptors and pathways that distinguish the different sensory systems from one another.

GENERAL PROPERTIES OF SENSORY SYSTEMS

All sensory pathways have certain elements in common. They begin with a stimulus, in the form of physical energy that acts on a sensory receptor. The receptor is a *transducer* that converts the stimulus into an intracellular signal, which is usually a change in membrane potential. If the stimulus is above threshold, action potentials pass along a sensory neuron to the central nervous system (CNS), where incoming signals are integrated. Some stimuli pass upward to the cerebral cortex, where they reach conscious perception, but others are acted on subconsciously, without our awareness. At each synapse along the pathway, the nervous system can modulate and shape the sensory information.

Sensory systems in the human body vary widely in complexity. The simplest systems are single sensory neurons with branched dendrites that function as receptors, such as pain and itch receptors. The most complex systems include multicellular sense organs, such as the ear and the eye. The cochlea of the ear contains about 16,000 sensory receptors and more than a million associated parts, and the human eye has about 126 million sensory receptors.

Receptors Are Sensitive to Particular Forms of Energy

Receptors in the sensory system vary widely in complexity, ranging from the branched endings of a single sensory neuron to complex, highly organized cells such as photoreceptors. The simplest receptors consist of a neuron with naked ("free") nerve endings (**FIG. 10.1a**). In more complex receptors, the nerve endings are encased in connective tissue capsules (Fig. 10.1b). The axons of both simple and complex receptors may be myelinated or unmyelinated.

The special senses have the most highly specialized receptors. The receptors for smell are neurons, but the other four special senses use non-neural receptor cells that synapse onto sensory neurons. The *hair cell* of the ear, shown in Figure 10.1c, is an example of a non-neural receptor. When activated, the hair cell releases a chemical signal that initiates an action potential in the associated sensory neuron. Both neural and non-neural receptors develop from the same embryonic tissue.

Non-neural *accessory structures* are critical to the operation of many sensory systems. For example, the lens and cornea of the eye help focus incoming light onto photoreceptors. The hairs on our arms help **somatosensory receptors** sense movement in the air millimeters above the skin surface. Accessory structures often enhance the information-gathering capability of the sensory system.

Receptors can be divided into four major groups, based on the type of stimulus to which they are most sensitive (**TBL. 10.2**).

FIG. 10.1 Simple, complex, and nonneural sensory receptors

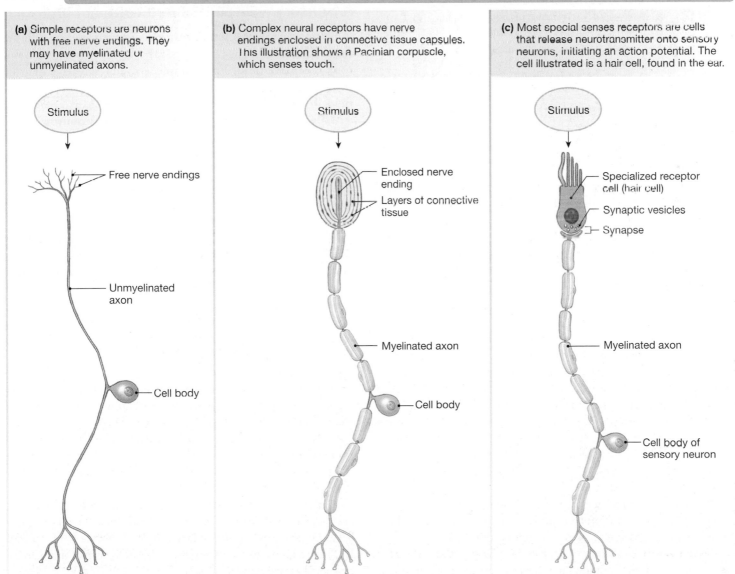

(a) Simple receptors are neurons with free nerve endings. They may have myelinated or unmyelinated axons.

Stimulus

Free nerve endings

Unmyelinated axon

Cell body

(b) Complex neural receptors have nerve endings enclosed in connective tissue capsules. This illustration shows a Pacinian corpuscle, which senses touch.

Stimulus

Enclosed nerve ending

Layers of connective tissue

Myelinated axon

Cell body

(c) Most special senses receptors are cells that release neurotransmitter onto sensory neurons, initiating an action potential. The cell illustrated is a hair cell, found in the ear.

Stimulus

Specialized receptor cell (hair cell)

Synaptic vesicles

Synapse

Myelinated axon

Cell body of sensory neuron

TABLE 10.2	Types of Sensory Receptors
Type of Receptor	**Examples of Stimuli**
Chemoreceptors	Oxygen, pH, various organic molecules such as glucose
Mechanoreceptors	Pressure (baroreceptors), cell stretch (osmoreceptors), vibration, acceleration, sound
Photoreceptors	Photons of light
Thermoreceptors	Varying degrees of heat

Chemoreceptors respond to chemical ligands that bind to the receptor (taste and smell, for example). **Mechanoreceptors** respond to various forms of mechanical energy, including pressure, vibration, gravity, acceleration, and sound (hearing, for example). **Thermoreceptors** respond to temperature, and **photoreceptors** for vision respond to light.

CONCEPT CHECK

1. What advantage do myelinated axons provide?
2. What accessory role does the outer ear (the pinna) play in the auditory system?
3. For each of the somatic and visceral stimuli listed in Table 10.1, which of the following receptor types is the appropriate transducer: mechano-, chemo-, photo-, or thermoreceptors?

Sensory Transduction Converts Stimuli into Graded Potentials

How do receptors convert diverse physical stimuli, such as light or heat, into electrical signals? The first step is **transduction,** the conversion of stimulus energy into information that can be processed by the nervous system [p. 171]. In many receptors, the opening or closing of ion channels converts mechanical, chemical, thermal, or light energy directly into a change in membrane potential. Some sensory transduction mechanisms include signal transduction and second messenger systems that initiate the change in membrane potential.

Each sensory receptor has an **adequate stimulus,** a particular form of energy to which it is most responsive. For example, thermoreceptors are more sensitive to temperature changes than to pressure, and mechanoreceptors respond preferentially to stimuli that deform the cell membrane. Although receptors are specific for one form of energy, they can respond to most other forms if the intensity is high enough. Photoreceptors of the eye respond most readily to light, for instance, but a blow to the eye may cause us to "see stars," an example of mechanical energy of sufficient force to stimulate the photoreceptors.

Sensory receptors can be incredibly sensitive to their preferred form of stimulus. For example, a single photon of light stimulates certain photoreceptors, and a single *odorant* molecule may activate the chemoreceptors involved in the sense of smell. The minimum stimulus required to activate a receptor is known as the **threshold,** just as the minimum depolarization required to trigger an action potential is called the threshold [p. 242].

How is a physical or chemical stimulus converted into a change in membrane potential? The stimulus opens or closes ion channels in the receptor membrane, either directly or indirectly (through a second messenger). In most cases, channel opening results in net influx of Na^+ or other cations into the receptor, depolarizing the membrane. In a few cases, the response to the stimulus is hyperpolarization when K^+ leaves the cell. In the case of vision, the stimulus (light) closes cation channels to hyperpolarize the receptor.

The change in sensory receptor membrane potential is a graded potential [p. 240] called a **receptor potential.** In some cells, the receptor potential initiates an action potential that travels along the sensory fiber to the CNS. In other cells, receptor potentials influence neurotransmitter secretion by the receptor cell, which in turn alters electrical activity in an associated sensory neuron.

A Sensory Neuron Has a Receptive Field

Somatic sensory and visual neurons are activated by stimuli that fall within a specific physical area known as the neuron's **receptive field.** For example, a touch-sensitive neuron in the skin responds to pressure that falls within its receptive field. In the simplest case, one receptive field is associated with one sensory neuron (the **primary sensory neuron** in the pathway), which in turn synapses on one CNS neuron (the **secondary sensory neuron**). (Primary and secondary sensory neurons are also known as *first-order* and *second-order neurons.*) Receptive fields frequently overlap with neighboring receptive fields.

In addition, sensory neurons of neighboring receptive fields may exhibit *convergence* [p. 260], in which multiple presynaptic neurons provide input to a smaller number of postsynaptic neurons (**FIG. 10.2**). Convergence allows multiple simultaneous subthreshold stimuli to sum at the postsynaptic (secondary) neuron. When multiple primary sensory neurons converge on a single secondary sensory neuron, their individual receptive fields merge into a single, large *secondary receptive field*, as shown in Figure 10.2a.

The size of secondary receptive fields determines how sensitive a given area is to a stimulus. For example, sensitivity to touch is demonstrated by a **two-point discrimination test.** In some regions of skin, such as that on the arms and legs, two pins placed within 20 mm of each other are interpreted by the brain as a single pinprick. In these areas, many primary neurons converge on a single secondary neuron, so the secondary receptive field is very large (Fig. 10.2a).

In contrast, more sensitive areas of skin, such as the fingertips, have smaller receptive fields, with as little as a 1:1 relationship between primary and secondary sensory neurons (Fig. 10.2b). In these regions, two pins separated by as little as 2 mm can be perceived as two separate touches.

FIG. 10.2 Receptive fields of sensory neurons

(a) Convergence creates large receptive fields.

(b) Small receptive fields are found in more sensitive areas.

Compass with points separated by 20 mm

The receptive fields of three primary sensory neurons overlap to form one large secondary receptive field.

When fewer neurons converge, secondary receptive fields are much smaller.

Skin surface

Skin surface

Primary sensory neurons

Convergence of primary neurons allows simultaneous subthreshold stimuli to sum at the secondary sensory neuron and initiate an action potential.

Secondary sensory neurons

Two stimuli that fall within the same secondary receptive field are perceived as a single point, because only one signal goes to the brain. Therefore, there is no two-point discrimination.

The two stimuli activate separate pathways to the brain. The two points are perceived as distinct stimuli and hence there is **two-point discrimination.**

The CNS Integrates Sensory Information

Sensory information from much of the body enters the spinal cord and travels through ascending pathways to the brain. Some sensory information goes directly into the brain stem via the cranial nerves [p. 288]. Sensory information that initiates visceral reflexes is integrated in the brain stem or spinal cord and usually does not reach conscious perception. An example of an unconscious visceral reflex is the control of blood pressure by centers in the brain stem.

Each major division of the brain processes one or more types of sensory information (**FIG. 10.3**). For example, the midbrain receives visual information, and the medulla oblongata receives input for sound and taste. Information about balance and equilibrium is processed primarily in the cerebellum. These pathways, along with those carrying somatosensory information, project to the thalamus, which acts as a relay and processing station before passing the information on to the cerebrum.

Only *olfactory* {*olfacere,* to sniff} information is not routed through the thalamus. The sense of smell, a type of chemoreception, is considered one of the oldest senses, and even the most primitive vertebrate brains have well-developed regions for processing olfactory information. Information about odors travels

from the nose through the first cranial nerve [p. 288] and *olfactory bulb* to the olfactory cortex in the cerebrum. Perhaps it is because of this direct input to the cerebrum that odors are so closely linked to memory and emotion. Most people have experienced encountering a smell that suddenly brings back a flood of memories of places or people from the past.

One interesting aspect of CNS processing of sensory information is the **perceptual threshold,** the level of stimulus intensity necessary for you to be aware of a particular sensation. Stimuli bombard your sensory receptors constantly, but your brain can filter out and "turn off" some stimuli. You experience a change in perceptual threshold when you "tune out" the radio while studying or when you "zone out" during a lecture. In both cases, the noise is adequate to stimulate sensory neurons in the ear, but neurons higher in the pathway dampen the perceived signal so that it does not reach the conscious brain.

Decreased perception of a stimulus, or *habituation,* is accomplished by *inhibitory modulation* [p. 263]. Inhibitory modulation diminishes a suprathreshold stimulus until it is below the perceptual threshold. It often occurs in the secondary and higher neurons of a sensory pathway. If the modulated stimulus suddenly becomes important, such as when the professor asks you a

FIG. 10.3   Sensory pathways in the brain

Most pathways pass through the thalamus on their way to the cerebral cortex.

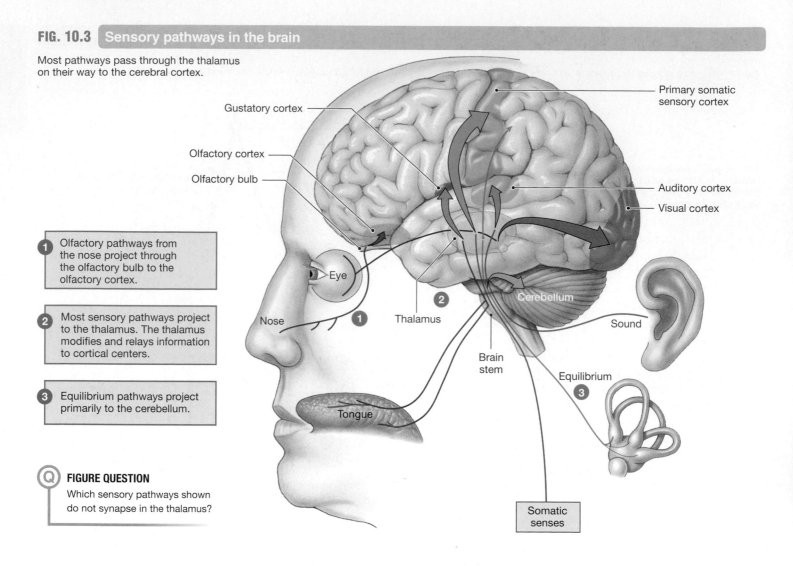

1. Olfactory pathways from the nose project through the olfactory bulb to the olfactory cortex.

2. Most sensory pathways project to the thalamus. The thalamus modifies and relays information to cortical centers.

3. Equilibrium pathways project primarily to the cerebellum.

Gustatory cortex

Olfactory cortex

Olfactory bulb

Eye

Nose

Tongue

Primary somatic sensory cortex

Auditory cortex

Visual cortex

Cerebellum

Thalamus

Sound

Brain stem

Equilibrium

Somatic senses

Q **FIGURE QUESTION**
Which sensory pathways shown do not synapse in the thalamus?

question, you can consciously focus your attention and overcome the inhibitory modulation. At that point, your conscious brain seeks to retrieve and recall recent sound input from your subconscious so that you can answer the question.

RUNNING **PROBLEM**

Ménière's disease—named for its discoverer, the nineteenth-century French physician Prosper Ménière—is associated with a build-up of fluid in the inner ear and is also known as *endolymphatic hydrops* {*hydro-,* water}. Symptoms of Ménière's disease include episodic attacks of vertigo, nausea, and tinnitus, accompanied by hearing loss and a feeling of fullness in the ears. *Vertigo* is a false sensation of spinning movement that patients often describe as dizziness.

Q1: *In which part of the brain is sensory information about equilibrium processed?*

310   314   331   339   343   348   353

Coding and Processing Distinguish Stimulus Properties

If all stimuli are converted to action potentials in sensory neurons and all action potentials are identical, how can the CNS tell the difference between, say, heat and pressure, or between a pinprick to the toe and one to the hand? The attributes of the stimulus must somehow be preserved once the stimulus enters the nervous system for processing. This means that the CNS must distinguish four properties of a stimulus: (1) its nature, or **modality,** (2) its location, (3) its intensity, and (4) its duration.

Sensory Modality   The modality of a stimulus is indicated by which sensory neurons are activated and by where the pathways of the activated neurons terminate in the brain. Each receptor type is most sensitive to a particular modality of stimulus. For example, some neurons respond most strongly to touch; others respond to changes in temperature. Each sensory modality can be subdivided into qualities. For instance, color vision is divided into red, blue, and green according to the wavelengths that most strongly stimulate the different visual receptors.

In addition, the brain associates a signal coming from a specific group of receptors with a specific modality. This 1:1 association of a receptor with a sensation is called **labeled line coding.** Stimulation of a cold receptor is always perceived as cold, whether the actual stimulus was cold or an artificial depolarization of the receptor. The blow to the eye that causes us to "see" a flash of light is another example of labeled line coding.

Location of the Stimulus The location of a stimulus is also coded according to which receptive fields are activated. The sensory regions of the cerebrum are highly organized with respect to incoming signals, and input from adjacent sensory receptors is processed in adjacent regions of the cortex. This arrangement preserves the topographical organization of receptors on the skin, eye, or other regions in the processing centers of the brain.

For example, touch receptors in the hand project to a specific area of the cerebral cortex. Experimental stimulation of that area of the cortex during brain surgery is interpreted as a touch to the hand, even though there is no contact. Similarly, one type of the *phantom limb pain* reported by amputees occurs when secondary sensory neurons in the spinal cord become hyperactive, resulting in the sensation of pain in a limb that is no longer there.

Auditory information is an exception to the localization rule, however. Neurons in the ears are sensitive to different frequencies of sound, but they have no receptive fields and their activation provides no information about the location of the sound. Instead, the brain uses the timing of receptor activation to compute a location, as shown in **FIGURE 10.4**.

A sound originating directly in front of a person reaches both ears simultaneously. A sound originating on one side reaches the closer ear several milliseconds before it reaches the other ear. The brain registers the difference in the time it takes for the sound stimuli to reach the two sides of the auditory cortex and uses that information to compute the sound's source.

Lateral inhibition, which increases the contrast between activated receptive fields and their inactive neighbors, is another way of isolating the location of a stimulus. **FIGURE 10.5** shows this process for a pressure stimulus to the skin. A pin pushing on the skin activates three primary sensory neurons, each of which releases neurotransmitters onto its corresponding secondary neuron.

However, the three secondary neurons do not all respond in the same fashion. The secondary neuron closest to the stimulus (neuron B) suppresses the response of the secondary neurons lateral to it (i.e., on either side), where the stimulus is weaker, and simultaneously allows its own pathway to proceed without interference. The inhibition of neurons farther from the stimulus enhances the contrast between the center and the sides of the receptive field, making the sensation more easily localized. In the visual system, lateral inhibition sharpens our perception of visual edges.

The pathway in Figure 10.5 also is an example of **population coding,** the way multiple receptors function together to send the CNS more information than would be possible from a single receptor. By comparing the input from multiple receptors, the CNS can make complex calculations about the quality and spatial and temporal characteristics of a stimulus.

CONCEPT CHECK

4. In Figure 10.5, what kind(s) of ion channel might open in neurons A and C that would depress their responsiveness: Na^+, K^+, Ca^{2+}, or Cl^-?

Intensity of the Stimulus The intensity of a stimulus cannot be directly calculated from a single sensory neuron action potential because a single action potential is "all-or-none." Instead, stimulus intensity is coded in two types of information: the number of receptors activated (another example of population coding) and the frequency of action potentials coming from those receptors, called *frequency coding*.

Population coding for intensity occurs because the threshold for the preferred stimulus is not the same for all receptors. Only the most sensitive receptors (those with the lowest thresholds) respond to a low-intensity stimulus. As a stimulus increases in intensity, additional receptors are activated. The CNS then translates the number of active receptors into a measure of stimulus intensity.

For individual sensory neurons, intensity discrimination begins at the receptor. If a stimulus is below threshold, the primary sensory neuron does not respond. Once stimulus intensity exceeds threshold, the primary sensory neuron begins to fire action potentials. As stimulus intensity increases, the receptor potential amplitude (strength) increases in proportion, and the frequency of action potentials in the primary sensory neuron increases, up to a maximum rate (**FIG. 10.6**).

FIG. 10.4   Localization of sound

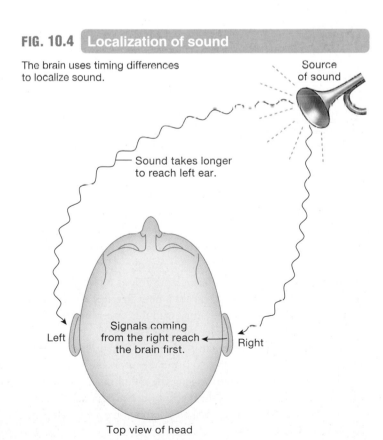

The brain uses timing differences to localize sound.

Source of sound

Sound takes longer to reach left ear.

Left

Signals coming from the right reach the brain first.

Right

Top view of head

FIG. 10.5 | Lateral inhibition

Lateral inhibition enhances contrast and makes a stimulus easier to perceive. The responses of primary sensory neurons A, B, and C are proportional to the intensity of the stimulus in each receptor field. Secondary sensory neuron B inhibits secondary neurons A and C, creating greater contrast between B and its neighbors.

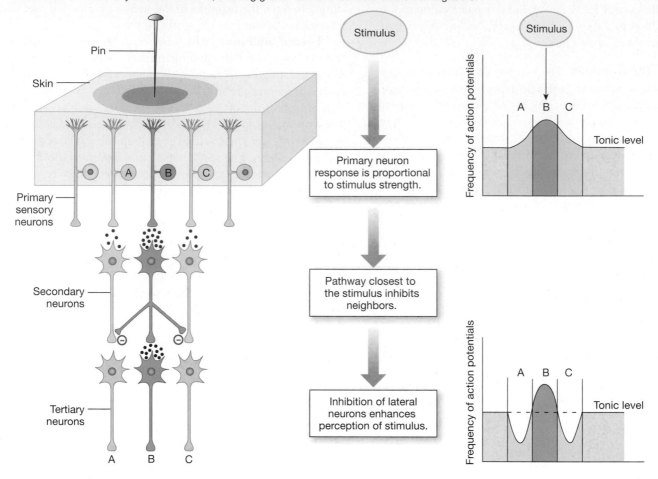

Duration of the Stimulus

The duration of a stimulus is coded by the duration of action potentials in the sensory neuron. In general, a longer stimulus generates a longer series of action potentials in the primary sensory neuron. However, if a stimulus persists, some receptors **adapt,** or cease to respond. Receptors fall into one of two classes, depending on how they adapt to continuous stimulation.

Tonic receptors are slowly adapting receptors that fire rapidly when first activated, then slow and maintain their firing as long as the stimulus is present (**FIG. 10.7a**). Pressure-sensitive baroreceptors, irritant receptors, and some tactile receptors and proprioceptors fall into this category. In general, the stimuli that activate tonic receptors are parameters that must be monitored continuously by the body.

In contrast, **phasic receptors** are rapidly adapting receptors that fire when they first receive a stimulus but cease firing if the strength of the stimulus remains constant (Fig. 10.7b). Phasic receptors are attuned specifically to *changes* in a parameter. Once a stimulus reaches a steady intensity, phasic receptors adapt to the new steady state and turn off. This type of response allows the

body to ignore information that has been evaluated and found not to threaten homeostasis or well-being.

Our sense of smell is an example of a sense that uses phasic receptors. For example, you can smell your cologne when you put it on in the morning, but as the day goes on your olfactory receptors adapt and are no longer stimulated by the cologne molecules. You no longer smell the fragrance, yet others may comment on it.

Adaptation of phasic receptors allows us to filter out extraneous sensory information and concentrate on what is new, different, or essential. In general, once adaptation of a phasic receptor has occurred, the only way to create a new signal is to either increase the intensity of the excitatory stimulus or remove the stimulus entirely and allow the receptor to reset.

The molecular mechanism for sensory receptor adaptation depends on the receptor type. In some receptors, K^+ channels in the receptor membrane open, causing the membrane to repolarize and stopping the signal. In other receptors, Na^+ channels quickly inactivate. In yet other receptors, biochemical pathways alter the receptor's responsiveness.

FIG. 10.6 | Coding for stimulus intensity and duration

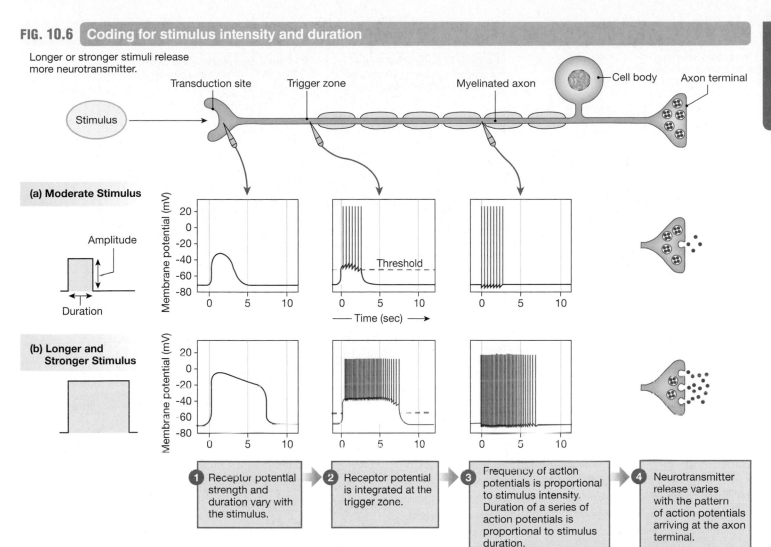

Longer or stronger stimuli release more neurotransmitter.

Transduction site   Trigger zone   Myelinated axon   Cell body   Axon terminal

Stimulus

(a) Moderate Stimulus

Amplitude

Duration

Membrane potential (mV)

Threshold

Time (sec)

(b) Longer and Stronger Stimulus

Membrane potential (mV)

1. Receptor potential strength and duration vary with the stimulus.

2. Receptor potential is integrated at the trigger zone.

3. Frequency of action potentials is proportional to stimulus intensity. Duration of a series of action potentials is proportional to stimulus duration.

4. Neurotransmitter release varies with the pattern of action potentials arriving at the axon terminal.

Accessory structures may also decrease the amount of stimulus reaching the receptor. In the ear, for example, tiny muscles contract and dampen the vibration of small bones in response to loud noises, thus decreasing the sound signal before it reaches auditory receptors.

To summarize, the specificity of sensory pathways is established in several ways:

1. Each receptor is most sensitive to a particular type of stimulus.
2. A stimulus above threshold initiates action potentials in a sensory neuron that projects to the CNS.
3. Stimulus intensity and duration are coded in the pattern of action potentials reaching the CNS.
4. Stimulus location and modality are coded according to which receptors are activated or (in the case of sound) by the timing of receptor activation.
5. Each sensory pathway projects to a specific region of the cerebral cortex dedicated to a particular receptive field. The brain can then tell the origin of each incoming signal.

CONCEPT CHECK

5. How do sensory receptors communicate the intensity of a stimulus to the CNS?
6. What is the adaptive significance of irritant receptors that are tonic instead of phasic?

SOMATIC SENSES

There are four somatosensory modalities: touch, proprioception, temperature, and *nociception*, which includes pain and itch. [We discuss details of proprioception in Chapter 13.]

Pathways for Somatic Perception Project to the Cortex and Cerebellum

Receptors for the somatic senses are found both in the skin and in the viscera. Receptor activation triggers action potentials in the associated primary sensory neuron. In the spinal cord, many primary sensory neurons synapse onto interneurons that serve as the

FIG. 10.7   Receptor adaptation

Receptors adapt to a sustained stimulus.

(a) Tonic receptors are slowly adapting receptors that respond for the duration of a stimulus.

(b) Phasic receptors rapidly adapt to a constant stimulus and turn off.

Stimulus

Receptor

Receptor potential

Axon of sensory neuron

Action potentials in sensory neuron

Stimulus

Time →

Time →

secondary sensory neurons. The location of the synapse between a primary neuron and a secondary neuron varies according to the type of receptor (**FIG. 10.8**).

Neurons associated with receptors for nociception, temperature, and coarse touch synapse onto their secondary neurons shortly after entering the spinal cord. In contrast, most fine touch, vibration, and proprioceptive neurons have very long axons that project up the spinal cord all the way to the medulla.

All secondary sensory neurons cross the midline of the body at some point, so that sensations from the left side of the body are processed in the right hemisphere of the brain and vice versa. The secondary neurons for nociception, temperature, and coarse touch cross the midline in the spinal cord, then ascend to the brain. Fine touch, vibration, and proprioceptive neurons cross the midline in the medulla.

In the thalamus, all secondary sensory neurons synapse onto **tertiary sensory neurons,** which in turn project to the somatosensory region of the cerebral cortex. In addition, many sensory pathways send branches to the cerebellum so that it can use the information to coordinate balance and movement.

The **somatosensory cortex** [p. 293] is the part of the brain that recognizes where ascending sensory tracts originate. Each sensory tract has a corresponding region of the cortex, its *sensory field.* All sensory pathways for the left hand terminate in one area, all pathways for the left foot terminate in another area, and so on (**FIG. 10.9**). Within the cortical region for a particular body part, columns of neurons are devoted to particular types of receptors.

For example, a cortical column activated by cold receptors in the left hand may be found next to a column activated by pressure

receptors in the skin of the left hand. This columnar arrangement creates a highly organized structure that maintains the association between specific receptors and the sensory modality they transmit.

Some of the most interesting research about the somatosensory cortex has been done on patients during brain surgery for epilepsy. Because the brain has no pain fibers, this type of surgery can be performed with the patient awake under local anesthesia. The surgeon stimulates a particular region of the brain and asks the patient about sensations that occur. The ability of the patient to communicate with the surgeon during this process has expanded our knowledge of brain regions tremendously.

Experiments can also be done on nonhuman animals by stimulating peripheral receptors and monitoring electrical activity in the cortex. We have learned from these experiments that the more sensitive a region of the body is to touch and other stimuli, the larger the corresponding region in the cortex. Interestingly, the size of the regions is not fixed. If a particular body part is used more extensively, its topographical region in the cortex will expand. For example, people who are visually handicapped and learn to read Braille with their fingertips develop an enlarged region of the somatosensory cortex devoted to the fingertips.

In contrast, if a person loses a finger or limb, the portion of the somatosensory cortex devoted to the missing structure begins to be taken over by sensory fields of adjacent structures. Reorganization of the somatosensory cortex "map" is an example of the remarkable plasticity [p. 261] of the brain. Unfortunately, sometimes the reorganization is not perfect and

FIG. 10.8 | Somatosensory pathways

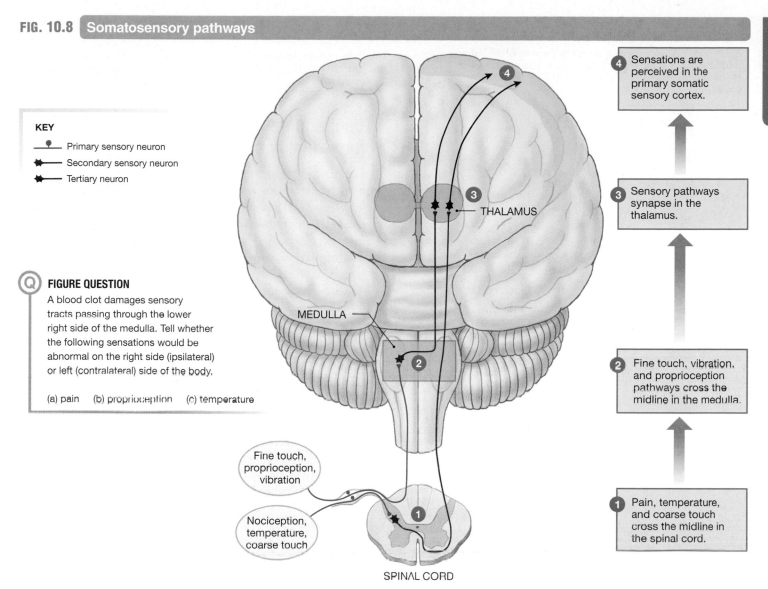

KEY

— Primary sensory neuron

— Secondary sensory neuron

— Tertiary neuron

Q **FIGURE QUESTION**

A blood clot damages sensory tracts passing through the lower right side of the medulla. Tell whether the following sensations would be abnormal on the right side (ipsilateral) or left (contralateral) side of the body.

(a) pain   (b) proprioception   (c) temperature

Fine touch, proprioception, vibration

Nociception, temperature, coarse touch

SPINAL CORD

MEDULLA

THALAMUS

4 Sensations are perceived in the primary somatic sensory cortex.

3 Sensory pathways synapse in the thalamus.

2 Fine touch, vibration, and proprioception pathways cross the midline in the medulla.

1 Pain, temperature, and coarse touch cross the midline in the spinal cord.

	Primary Sensory	**Secondary Sensory**	**Synapse with...**	**Tertiary Sensory**
Fine Touch, Proprioception, Vibration	Primary sensory neuron synapses in the medulla.	Secondary sensory neuron crosses midline of body in medulla.	Synapse with tertiary sensory neuron in the thalamus.	Tertiary sensory neuron terminates in somatosensory cortex.
Irritants, Temperature, Coarse Touch	Primary sensory neuron synapses in dorsal horn of spinal cord.	Secondary sensory neuron crosses midline of body in spinal cord.		

can result in sensory sensations, including pain, that the brain interprets as being located in the missing limb (phantom limb pain).

Contemporary research in this field now uses noninvasive imaging techniques, such as *functional magnetic resonance imaging* (fMRI) and *positron emission tomography* (PET) scans to watch brains at work. Both techniques measure the metabolic activity of neurons, so that more active areas of neuronal activity become highlighted and can be associated with their location. [See Fig. 9.20c for PET scans of the brain.]

Touch Receptors Respond to Many Different Stimuli

Touch receptors are among the most common receptors in the body. These receptors respond to many forms of physical contact, such as stretch, steady pressure, fluttering or stroking movement, vibration, and texture. They are found both in the skin (**FIG. 10.10**) and in deeper regions of the body.

Touch receptors in the skin come in many forms. Some are free nerve endings, such as those that respond to noxious stimuli.

FIG. 10.9 The somatosensory cortex

Each body part is represented next to the area of the sensory cortex that processes stimuli for that body part. This mapping was created by two neurosurgeons, W. Penfield and T. Rasmussen, in 1950 and is called a homunculus (little man).

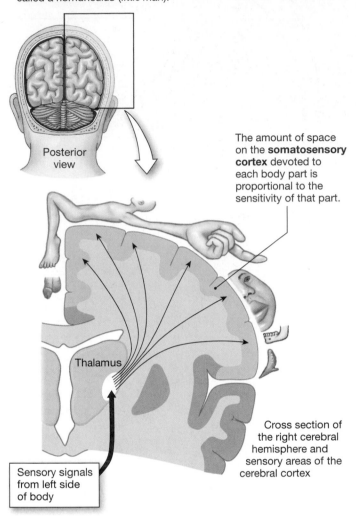

Posterior view

The amount of space on the **somatosensory cortex** devoted to each body part is proportional to the sensitivity of that part.

Thalamus

Cross section of the right cerebral hemisphere and sensory areas of the cerebral cortex

Sensory signals from left side of body

Others are more complex. Most touch receptors are difficult to study because of their small size. However, **Pacinian corpuscles,** which respond to vibration, are some of the largest receptors in the body, and much of what we know about somatosensory receptors comes from studies on these structures.

Pacinian corpuscles are composed of nerve endings encapsulated in layers of connective tissue (see Fig. 10.1b). They are found in the subcutaneous layers of skin and in muscles, joints, and internal organs. The concentric layers of connective tissue in the corpuscles create large receptive fields.

Pacinian corpuscles respond best to high-frequency vibrations, whose energy is transferred through the connective tissue capsule to the nerve ending, where the energy opens mechanically gated ion channels [p. 139]. Recent research using knockout mice indicates that another sensory receptor, the Merkel receptor, also uses mechanically gated ion channels to respond to touch.

Pacinian corpuscles are rapidly adapting phasic receptors, and this property allows them to respond to a change in

touch but then ignore it. For example, you notice your shirt when you first put it on, but the touch receptors soon adapt. Properties of the remaining touch receptors depicted in Figure 10.10—Meissner's corpuscles, Ruffini corpuscles, and Merkel receptors—are summarized in the table of that figure.

Temperature Receptors Are Free Nerve Endings

Temperature receptors are free nerve endings that terminate in the subcutaneous layers of the skin. **Cold receptors** are sensitive primarily to temperatures lower than body temperature. **Warm receptors** are stimulated by temperatures in the range extending from normal body temperature (37 °C) to about 45 °C. Above that temperature, pain receptors are activated, creating a sensation of painful heat. Thermoreceptors in the brain play an important role in thermoregulation.

The receptive field for a thermoreceptor is about 1 mm in diameter, and the receptors are scattered across the body. There are considerably more cold receptors than warm ones. Temperature receptors slowly adapt between 20 and 40 °C. Their initial response tells us that the temperature is changing, and their sustained response tells us about the ambient temperature. Outside the 20–40 °C range, where the likelihood of tissue damage is greater, the receptors do not adapt. Thermoreceptors use a family of cation channels called *transient receptor potential* or TRP channels to initiate an action potential.

Nociceptors Initiate Protective Responses

Nociceptors {*nocere,* to injure} are neurons with free nerve endings (Fig. 10.1a) that respond to a variety of strong noxious stimuli (chemical, mechanical, or thermal) that cause or have the potential to cause tissue damage. Nociceptors are found in the skin, joints, muscles, bones and various internal organs, but not in the central nervous system. Activation of nociceptor pathways initiates adaptive, protective responses. For example, discomfort from overuse of muscles and joints warns us to take it easy and avoid additional damage to these structures. Afferent signals from nociceptors are carried to the CNS in two types of primary sensory fibers: Aδ (A-delta) *fibers,* and *C fibers* (**TBL. 10.3**). The most common sensation carried by these pathways is perceived as pain, but when histamine or some other stimulus activates a subtype of C fiber, we perceive the sensation we call itch.

Pain is a subjective perception, the brain's interpretation of sensory information transmitted along pathways that begin at nociceptors. Pain is highly individual and multidimensional, and may vary with a person's emotional state. The discussion here is limited to the sensory processing of nociception.

Fast pain, described as sharp and localized, is rapidly transmitted to the CNS by the small, myelinated Aδ fibers. **Slow pain,** described as duller and more diffuse, is carried on small, unmyelinated C fibers. The timing distinction between the two is most obvious when the stimulus originates far from the CNS, such as when you stub your toe. You first experience a quick stabbing

FIG. 10.10 | Sensory receptors in the skin

Merkel receptors sense steady pressure and texture.

Meissner's corpuscle responds to flutter and stroking movements.

Hair

Free nerve ending

Free nerve ending of hair root senses hair movement.

Hair root

Pacinian corpuscle senses vibration.

Ruffini corpuscle responds to skin stretch.

Free nerve ending of nociceptor responds to noxious stimuli.

Sensory nerves carry signals to spinal cord.

Receptor	Stimulus	Location	Structure	Adaptation
Free Nerve Endings	Temperature, noxious stimuli, hair movement	Around the hair roots and under surface of skin	Unmyelinated nerve endings	Variable
Meissner's Corpuscles	Flutter, stroking	Superficial layers of skin	Encapsulated in connective tissue	Rapid
Pacinian Corpuscles	Vibration	Deep layers of skin	Encapsulated in connective tissue	Rapid
Ruffini Corpuscles	Stretch of skin	Deep layers of skin	Enlarged nerve endings	Slow
Merkel Receptors	Steady pressure, texture	Superficial layers of skin	Epidermal cell synapsing with enlarged nerve ending	Slow

sensation (fast pain), followed shortly by a dull throbbing (slow pain).

Itch (*pruritus*) comes only from nociceptors in the skin and is characteristic of many rashes and other skin conditions. However, itch can also be a symptom of a number of systemic diseases, including multiple sclerosis, hyperparathyroidism, and diabetes mellitus. The higher pathways for itch are not as well understood as the pathways for pain, but there is an antagonistic interaction between the two sensations. When something itches, we scratch it, creating a mildly painful sensation that seems to interrupt the

TABLE 10.3 | Classes of Somatosensory Nerve Fibers

Fiber Type	Fiber Characteristics	Speed of Conduction	Associated with
Aβ (beta)	Large, myelinated	30–70 m/sec	Mechanical stimuli
Aδ (delta)	Small, myelinated	12–30 m/sec	Cold, fast pain, mechanical stimuli
C	Small, unmyelinated	0.5–2 m/sec	Slow pain, heat, cold, mechanical stimuli

itch sensation. Many of the opioid painkillers, such as morphine, relieve pain but in some people they also induce the side effect of itching.

Nociceptor Pathways Protective nociceptor reflexes begin with activation of the neuron's free nerve endings. Ion channels responding to a variety of chemical, mechanical, and thermal stimuli create graded potentials that trigger action potentials if the stimulus is strong enough. Many of these channels are *transient receptor potential (TRP) channels,* in the same channel family as the thermoreceptor channels.

For example, *vanilloid receptors* ($TRPV_1$ channels) respond to damaging heat from a stove or other source, as well as to *capsaicin,* the chemical that makes hot chili peppers burn your mouth. At the opposite end of the temperature spectrum, researchers have identified a related channel, *TRPM8,* that responds both to cold and to menthol, one reason mint-flavored foods feel cool.

Chemicals that mediate inflammatory responses at the site of tissue injury can activate nociceptors or sensitize them by lowering their activation threshold. Local chemicals released upon tissue injury include K^+, histamine, and prostaglandins released from damaged cells; serotonin released from platelets activated by tissue damage; and the peptide **substance P,** which is secreted by primary sensory neurons. Increased sensitivity to pain at sites of tissue damage is called **inflammatory pain.**

The primary sensory neurons from nociceptors terminate in the dorsal horn of the spinal cord (see Fig. 10.8). Nociceptor activation can follow two pathways: (1) reflexive protective responses that are integrated at the level of the spinal cord (spinal reflexes [p. 285]), and (2) ascending pathways to the cerebral cortex that become conscious sensation (pain or itch). The primary nociceptor neurons synapse onto interneurons for spinal reflex responses or onto secondary sensory neurons that project to the brain.

Nociception responses integrated in the spinal cord initiate rapid unconscious protective reflexes that automatically remove a stimulated area from the source of the stimulus. For example, if you accidentally touch a hot stove, an automatic **withdrawal reflex** causes you to pull back your hand even before you are aware of the heat. The lack of higher control in many protective reflexes has been demonstrated in the classic "spinal frog" preparation, in which the animal's brain has been destroyed. If the frog's foot is placed in a beaker of hot water, the withdrawal reflex causes the leg to contract and move the foot away from the stimulus. The frog is unable to feel pain because the brain, which translates sensory input into perception, is not functional, but its protective spinal reflexes are intact.

The ascending pathways for nociception are similar to other somatosensory pathways (see Fig. 10.8). The secondary sensory neurons cross the body's midline in the spinal cord and ascend to the thalamus and sensory areas of the cortex. The pathways also send branches to the limbic system and hypothalamus. As a result, pain may be accompanied by emotional distress (suffering) and a variety of autonomic reactions, such as nausea, vomiting, or sweating.

Pain can be felt in skeletal muscles (*deep somatic pain*) as well as in the skin. Muscle pain during exercise is associated with the onset of anaerobic metabolism and is often perceived as a burning sensation in the muscle (as in "go for the burn!"). Some investigators have suggested that the exercise-induced metabolite responsible for the burning sensation is K^+, known to enhance the pain response. Muscle pain from **ischemia** (lack of adequate blood flow that reduces oxygen supply) also occurs during *myocardial infarction* (heart attack).

Pain in the heart and other internal organs (*visceral pain*) is often poorly localized and may be felt in areas far removed from the site of the stimulus (**FIG. 10.11a**). For example, the pain of cardiac ischemia may be felt in the neck and down the left shoulder and arm. This **referred pain** apparently occurs because visceral and somatic sensory pain inputs converge on a single ascending tract (Fig. 10.11b). According to this model, when painful stimuli arise in visceral receptors, the brain is unable to distinguish visceral signals from the more common signals arising from somatic receptors. As a result, it interprets the pain as coming from the somatic regions rather than the viscera.

Chronic pain of one sort or another affects millions of people in this country every year. This type of pain is often much greater than nociceptor activation would indicate and reflects damage to or long-term changes in the nervous system. Chronic pain is a **pathological pain** and is also called *neuropathic pain.* One of the most common forms of neuropathic pain is *diabetic neuropathy,* which develops as a consequence of chronically elevated blood glucose concentrations. Scientists do not yet fully understand what causes glucose neurotoxicity or neuropathic pain, which makes its treatment difficult.

Pain Modulation Our perception of pain is subject to modulation at several levels in the nervous system. It can be magnified by past experiences or suppressed in emergencies when survival depends on ignoring injury. In such emergencies, descending pathways that travel through the thalamus inhibit nociceptor neurons in the spinal cord. Artificial stimulation of these inhibitory pathways is one of the newer techniques being used to control chronic pain.

Pain can also be suppressed in the dorsal horn of the spinal cord, before the stimuli are sent to ascending spinal tracts. Normally, tonically active inhibitory interneurons in the spinal cord inhibit ascending pathways for pain (**FIG. 10.12a**). C fibers from nociceptors synapse on these inhibitory interneurons. When activated by a noxious stimulus, the C fibers simultaneously excite the ascending path and block the tonic inhibition (Fig. 10.12b). This action allows the pain signal from the C fiber to travel unimpeded to the brain.

In the **gate control theory** of pain modulation, $A\beta$ fibers carrying sensory information about mechanical stimuli help block pain transmission (Fig. 10.12c). The $A\beta$ fibers synapse on the inhibitory interneurons and *enhance* the interneuron's inhibitory activity. If simultaneous stimuli reach the inhibitory neuron from the $A\beta$ and C fibers, the integrated response is partial inhibition of the ascending pain pathway so that pain perceived by the

FIG. 10.11 | Referred pain

(a) Pain in internal organs is often sensed on the surface of the body, a sensation known as **referred pain.**

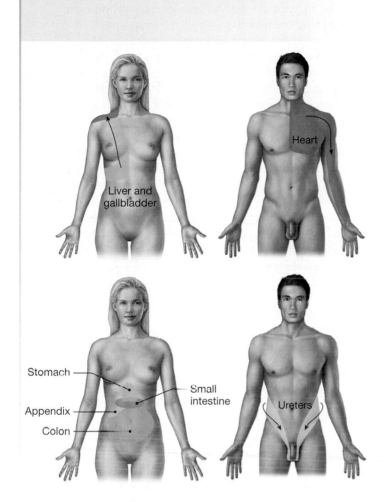

Heart

Liver and gallbladder

Stomach

Appendix

Colon

Small intestine

Ureters

(b) One theory of referred pain says that nociceptors from several locations converge on a single ascending tract in the spinal cord. Pain signals from the skin are more common than pain from internal organs, and the brain associates activation of the pathway with pain in the skin. Adapted from H.L. Fields, *Pain* (McGraw Hill, 1987).

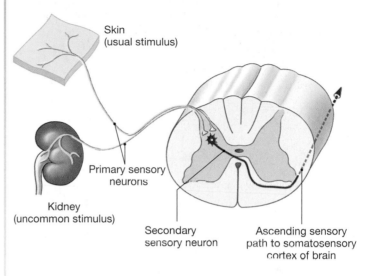

Skin (usual stimulus)

Primary sensory neurons

Kidney (uncommon stimulus)

Secondary sensory neuron

Ascending sensory path to somatosensory cortex of brain

Q FIGURE QUESTION

A man goes to his physician and complains of pain that radiates down his left arm. This suggests to the physician that the man may have a problem with what organ?

---

brain is lessened. The gate control theory explains why rubbing a bumped elbow or shin lessens your pain: the tactile stimulus of rubbing activates Aβ fibers and helps decrease the sensation of pain.

The pharmacological alleviation of pain is of considerable interest to health professionals. **Analgesic drugs** {*analgesia*, pain-lessness} range from aspirin to potent opioids such as morphine. Aspirin inhibits prostaglandins, decreases inflammation, and presumably slows the transmission of pain signals from the site of injury. The opioid drugs act directly on CNS *opioid receptors* that are part of an analgesic system that responds to endogenous opioid molecules [p. 257]. Activation of opioid receptors blocks pain perception by decreasing neurotransmitter release from primary sensory neurons and by postsynaptic inhibition of the secondary sensory neurons.

The endogenous opioids include three families: endorphins, enkephalins, and dynorphins. **Enkephalins** and **dynorphins** are secreted by neurons associated with pain pathways. The endogenous opioid β**-endorphin** is produced from the same prohormone as ACTH (adrenocorticotropin) in neuroendocrine cells of

the hypothalamus [Fig. 7.3b, p. 213]. Although opioid drugs are effective at relieving pain, a person taking them for long periods of time may develop tolerance and need larger and larger doses to obtain the same effect.

As a result, scientists are exploring alternative drugs and strategies for pain relief. Some chronic pain may be caused by sensitization of nociceptive nerve endings near a site of injury when the body releases chemicals in response to the damage. Non-narcotic anti-inflammatory drugs such as aspirin and COX2 inhibitors can often relieve pain, but even over-the-counter doses may have adverse side effects. New research is focused on blocking TRP channels in the sensitized nociceptor nerve endings.

For people with severe chronic pain, possible treatments include electrically stimulating inhibitory pain pathways to the brain, or in extreme cases, surgically severing sensory nerves at the dorsal root. Acupuncture can also be effective, although the physiological reason for its effectiveness is not clear. The leading theory on how acupuncture works proposes that properly placed acupuncture needles trigger the release of endorphins by the brain.

FIG. 10.12 The gate control model

In the gate control model of pain modulation, nonpainful stimuli can diminish the pain signal.

(a) In absence of input from C fibers, a tonically active inhibitory interneuron suppresses pain pathway.

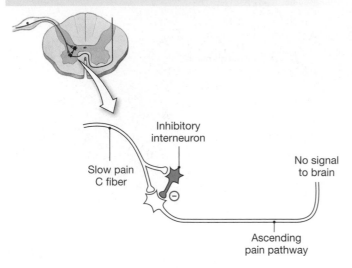

Inhibitory interneuron

Slow pain C fiber

No signal to brain

⊖

Ascending pain pathway

(b) With strong pain, C fiber stops inhibition of the pathway, allowing a strong signal to be sent to the brain.

Noxious stimulus

C fiber

⊖

⊕

Inhibition stops

Strong noxious stimulus to brain

(c) Pain can be modulated by simultaneous somatosensory input.

Aβ fiber

Touch or nonpainful stimulus

Noxious stimulus

C fiber

⊖   ⊕

⊕   ⊖

Noxious stimulus decreased

CONCEPT CHECK

7. What is the adaptive advantage of a spinal reflex?

8. Rank the speed of signal transmission through the following fiber types, from fastest to slowest: (a) small diameter, myelinated fiber; (b) large diameter, myelinated fiber; (c) small diameter, unmyelinated fiber.

9. Your sense of smell uses phasic receptors. What other receptors (senses) adapt to ongoing stimuli?

CLINICAL FOCUS

Natural Painkillers

Many drugs we use today for pain relief are derivatives of plant or animal molecules. One of the newest painkillers in this group is *ziconotide*, a synthetic compound related to the poison that South Pacific cone snails use to kill fish. This drug works by blocking calcium channels on nociceptive neurons. Ziconotide, approved in 2004 for the treatment of severe chronic pain, is highly toxic. To minimize systemic side effects, it must be injected directly into the cerebrospinal fluid surrounding the spinal cord. Ziconotide relieves pain but may also cause hallucinations and other psychiatric symptoms, so it is a last-resort treatment. Other painkilling drugs from biological sources include aspirin, derived from the bark of willow trees (genus *Salix*), and opiate drugs such as morphine and codeine that come from the opium poppy, *Papaver somniferum*. These drugs have been used in Western and Chinese medicine for centuries, and even today you can purchase willow bark as an herbal remedy.

CHEMORECEPTION: SMELL AND TASTE

The five special senses—smell, taste, hearing, equilibrium, and vision—are concentrated in the head region. Like somatic senses, the special senses rely on receptors to transform information about the environment into patterns of action potentials that can be interpreted by the brain. Smell and taste are both forms of *chemoreception,* one of the oldest senses from an evolutionary perspective. Unicellular bacteria use chemoreception to sense their environment, and primitive animals without formalized nervous systems use chemoreception to locate food and mates. It has been hypothesized that chemoreception evolved into chemical synaptic communication in animals.

Olfaction Is One of the Oldest Senses

Imagine waking up one morning and discovering a whole new world around you, a world filled with odors that you never dreamed existed—scents that told you more about your surroundings than you ever imagined from looking at them. This is exactly what happened to a young patient of Dr. Oliver Sacks (an account is in *The Man Who Mistook His Wife for a Hat and Other Clinical Tales*). Or imagine skating along the sidewalk without a helmet, only to fall and hit your head. When you regain consciousness, the world has lost all odor: no smell of grass or perfume or garbage. Even your food has lost much of its taste, and you now eat only to survive because eating has lost its pleasure.

We do not realize the essential role that our sense of smell plays in our lives until a head cold or injury robs us of the ability to smell. **Olfaction** allows us to discriminate among billions of different odors. Even so, our noses are not nearly as sensitive as those of many other animals whose survival depends on olfactory cues. The **olfactory bulb,** the extension of the forebrain that receives input from the primary olfactory neurons, is much better developed in vertebrates whose survival is more closely linked to chemical monitoring of their environment (**FIG. 10.13a**).

Olfactory Pathways The human olfactory system consists of an *olfactory epithelium* lining the nasal cavity, with embedded primary sensory neurons called **olfactory sensory neurons.** Axons of the olfactory sensory neurons form the *olfactory nerve,* or cranial nerve I [p. 288]. The olfactory nerve synapses with secondary sensory neurons in the *olfactory bulb,* located on the underside of the frontal lobe (Fig. 10.13b). Secondary and higher-order neurons project from the olfactory bulb through the *olfactory tract* to the *olfactory cortex* (Fig. 10.13a). The olfactory tract, unlike most other sensory pathways, bypasses the thalamus.

This arrangement seems quite simple, but complex processing takes place before signals pass on to the cortex. Evidence now suggests that modulation of incoming sensory information begins in the olfactory epithelium. Additional processing takes place in the olfactory bulb. Some descending modulatory pathways from the cortex terminate in the olfactory bulb, and there are reciprocal modulatory connections within and between the two branches of the olfactory bulb.

Ascending pathways from the olfactory bulb also lead to the amygdala and hippocampus, parts of the limbic system involved with emotion and memory. The link between smell, memory, and emotion is one amazing aspect of olfaction. A special cologne or the aroma of food can trigger memories and create a wave of nostalgia for the time, place, or people with whom the aroma is associated. In some way that we do not understand, the processing of odors through the limbic system creates deeply buried olfactory memories. Particular combinations of olfactory receptors become linked to other patterns of sensory experience so that stimulating one pathway stimulates them all.

The Olfactory Epithelium Olfactory sensory neurons in humans are concentrated in a 3-cm^2 patch of **olfactory epithelium** high in the nasal cavity (Fig. 10.13a). Olfactory sensory neurons have a single dendrite that extends down from the cell body to the surface of the olfactory epithelium, and a single axon that extends up to the olfactory bulb. Olfactory sensory neurons, unlike other neurons in the body, have very short lives, with a turnover time of about two months (Fig. 10.13c).

Stem cells in the basal layer of the olfactory epithelium are continuously dividing to create new neurons. The axon of each newly formed neuron must then find its way to the olfactory bulb and make the proper synaptic connections. To give us insight into how developing neurons find their targets, scientists are studying how these neurons manage to repeat the same connection each time.

In rodents, an accessory olfactory structure in the nasal cavity, the **vomeronasal organ (VNO),** is known to be involved in behavioral responses to sex pheromones [p. 198]. Anatomical and genetic studies in humans suggest that humans do not have

a functional VNO, but experiments with compounds believed to act as human pheromones support the hypothesis that humans may communicate with chemical signals.

Olfactory Signal Transduction The surface of the olfactory epithelium is composed of the knobby terminals of the olfactory sensory neuron dendrites, each knob branching into multiple nonmotile cilia (Fig. 10.13c). The cilia are embedded in a layer of mucus that is produced by *olfactory (Bowman's) glands* in the epithelium and basal lamina. *Odorant molecules* must first dissolve in and penetrate the mucus before they can bind to an **olfactory receptor** protein on the olfactory cilia. Each olfactory receptor is sensitive to a limited range of odorants.

Olfactory receptors are G protein-linked membrane receptors [p. 174]. Olfactory receptor genes form the largest known gene family in vertebrates (about 1000 genes, or 3–5% of the genome), but only about 400 olfactory receptor proteins are expressed in humans. The combination of most odorant molecules with their olfactory receptors activates a special G protein, G_{olf}, which in turn increases intracellular cAMP. The increase in cAMP concentration opens cAMP-gated cation channels, depolarizing the cell. If the graded receptor potential that results is strong enough, it triggers an action potential that travels along the sensory neuron's axon to the olfactory bulb.

What is occurring at the cellular and molecular levels that allows us to discriminate between thousands of different odors? Current research suggests that each individual olfactory sensory neuron contains a single type of olfactory receptor that responds to a limited range of odorant molecules. The axons of cells with the same receptors converge on a few secondary neurons in the olfactory bulb, which then can modify the information before sending it on to the olfactory cortex. The brain uses information from hundreds of olfactory sensory neurons in different combinations to create the perception of many different smells, just as combinations of letters create different words. This is another example of population coding in the nervous system [p. 315].

CONCEPT CHECK

10. Create a map or diagram of the olfactory pathway from an olfactory sensory neuron to the olfactory cortex.

11. Create a map or diagram that starts with a molecule from the environment binding to its olfactory receptor in the nose and ends with neurotransmitter release from the primary olfactory neuron.

12. The dendrites are which part of an olfactory sensory neuron?

13. Are olfactory neurons pseudounipolar, bipolar, or multipolar? [*Hint:* See Fig. 8.2, p. 230.]

Taste Is a Combination of Five Basic Sensations

Our sense of taste, or **gustation,** is closely linked to olfaction. Indeed, much of what we call the taste of food is actually the aroma, as you know if you have ever had a bad cold. Although

FIG. 10.13 ANATOMY SUMMARY

The Olfactory System

(a) Olfactory Pathways

The olfactory epithelium lies high within the nasal cavity, and its olfactory neurons project to the olfactory bulb. Sensory input at the receptors is carried through the olfactory cortex to the cerebral cortex and the limbic system.

Cerebral cortex

Limbic system

Olfactory bulb → Olfactory tract → Olfactory cortex

Cranial Nerve I

Olfactory neurons in olfactory epithelium

(b) The olfactory neurons synapse with secondary sensory neurons in the olfactory bulb.

Olfactory bulb

Secondary sensory neurons

Bone

Olfactory sensory neurons

Olfactory epithelium

(c) Olfactory neurons in the olfactory epithelium live only about two months. They are replaced by new neurons whose axons must find their way to the olfactory bulb.

Olfactory neuron axons (cranial nerve I) carry information to olfactory bulb.

Lamina propria

Basal cell layer includes stem cells that replace olfactory neurons.

Olfactory sensory neuron

Supporting cell

Olfactory cilia (dendrites) contain odorant receptors.

Mucous layer: Odorant molecules must dissolve in this layer.

Capillary

Olfactory (Bowman's) gland

Developing olfactory neuron

Q FIGURE QUESTION

Multiple primary neurons in the epithelium synapse on one secondary neuron in the olfactory bulb. This pattern is an example of what principle?

smell is sensed by hundreds of receptor types, taste is currently believed to be a combination of five sensations: sweet, sour (acid), salty, bitter, and **umami,** a taste associated with the amino acid glutamate and some nucleotides. Umami, a name derived from the Japanese word for "deliciousness," is a basic taste that enhances the flavor of foods. It is the reason that monosodium glutamate (MSG) is used as a food additive in some countries.

Each of the five currently recognized taste sensations is associated with a physiological process. Sour taste is triggered by the presence of H^+ and salty by the presence of Na^+. The concentrations of these two ions in body fluids are closely regulated because of the roles they play in pH balance and extracellular fluid volume. The other three taste sensations result from organic molecules. Sweet and umami are associated with nutritious food. Bitter taste is recognized by the body as a warning of possibly toxic components. If something tastes bitter, our first reaction is often to spit it out.

Taste Pathways The receptors for taste are located primarily on **taste buds** clustered together on the surface of the tongue (**FIG. 10.14a**). One taste bud is composed of 50–150 **taste receptor cells (TRCs),** along with support cells and regenerative *basal cells.* Taste receptors are also scattered through other regions of the oral cavity, such as the palate.

For a substance (*tastant*) to be tasted, it must first dissolve in the saliva and mucus of the mouth. Dissolved taste ligands then interact with an apical membrane protein (receptor or channel) on the taste receptor cell (Fig. 10.14b). Interaction of the taste ligand with a membrane protein initiates a signal transduction cascade that ends with release of chemical messenger molecules from the TRC. The details of signal transduction for the five taste sensations are still controversial, due partly to the fact that some of the mechanisms appear to be different in humans and mice, the primary model organism for mammalian taste research.

Chemical signals released from taste receptor cells activate primary sensory neurons (*gustatory neurons*) whose axons run through cranial nerves VII, IX, and X to the medulla, where they synapse. Sensory information then passes through the thalamus to the gustatory cortex (see Fig. 10.3). Central processing of sensory information compares the input from multiple taste receptor cells and interprets the taste sensation based on which populations of neurons are responding most strongly (another example of population coding). Signals from the sensory neurons also initiate behavioral responses, such as feeding, and feedforward responses [p. 17] that activate the digestive system.

Taste Transduction Uses Receptors and Channels

The details of taste receptor cell signal transduction, once believed to be relatively straightforward, are more complex than scientists initially thought. Sweet, bitter, and umami tastes are associated with activation of G protein-coupled receptors. In contrast, salty and sour transduction mechanisms both appear to be mediated by ion channels.

Taste buds contain four morphologically distinct cell types designated I, II, and III, plus *basal cells.* Type I cells are glia-like *support cells. Type II* cells, or *receptor cells,* and *type III* cells, or *presynaptic cells,* are taste receptor cells.

Each taste receptor cell is a non-neural polarized epithelial cell [p. 150] tucked down into the epithelium so that only a tiny tip protrudes into the oral cavity through a *taste pore* (Fig. 10.14a). In a given bud, tight junctions link the apical ends of adjacent cells together, limiting movement of molecules between the cells. The apical membrane of a TRC is modified into microvilli to increase the amount of surface area in contact with the environment.

Sweet, Bitter, and Umami Tastes The *type II* **taste receptor cells** respond to sweet, bitter, and umami sensations. These cells express multiple G protein-coupled receptors (GPCR) on their apical surfaces (Fig. 10.14b). Sweet and umami tastes are associated with T1R receptors with different combinations of subunits. Bitter taste uses about 30 variants of T2R receptors.

The type II cell receptors activate a special G protein called **gustducin,** which in turn activates multiple signal transduction pathways. Some of these pathways release Ca^{2+} from intracellular stores, while others open cation channels and allow Ca^{2+} to enter the cell. Calcium signals then initiate ATP release from the type II cells.

ATP in type II cells is not released through secretory vesicles. Instead it leaves the cell through gap junction-like channels. ATP then acts as a paracrine signal on both sensory neurons and neighboring presynaptic cells. This communication between neighboring taste receptor cells creates complex interactions.

Sour Taste The *type III presynaptic cells* respond to sour tastes. Models of transduction mechanisms for sour tastes are complicated by the fact that increasing H^+, the sour taste signal, also changes pH. There is evidence that H^+ acts on ion channels of the presynaptic cell from both extracellular and intracellular sides of the membrane. The intracellular pathways remain uncertain. Ultimately, H^+-mediated depolarization of the presynaptic cell results in serotonin release by exocytosis. Serotonin in turn excites the primary sensory neuron.

Salt Taste The cells responsible for salt taste have not been definitively identified, but some evidence suggests that salt reception may reside in the type 1 support cells. Signal transduction for salt taste in humans is equally unclear, complicated by the fact that mice have two different mechanisms but humans appear to have only one. In the current model for salty taste, Na^+ enters the taste receptor cell through an apical ion channel, such as the epithelial Na^+ channel (ENaC, pronounced ēē-knack). Sodium entry depolarizes the cell, setting off a series of events that culminate with the primary sensory neuron firing an action potential.

The mechanisms of taste transduction are a good example of how our models of physiological function must periodically be revised as new research data are published. For many years, the

FIG. 10.14 ESSENTIALS

Taste

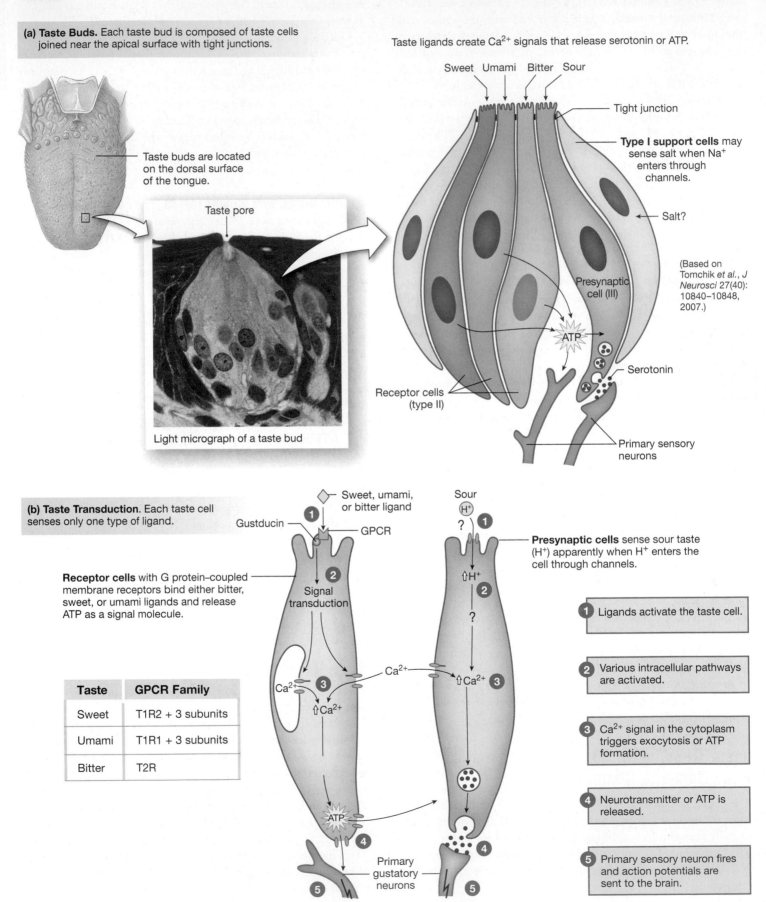

(a) Taste Buds. Each taste bud is composed of taste cells joined near the apical surface with tight junctions.

Taste buds are located on the dorsal surface of the tongue.

Taste pore

Light micrograph of a taste bud

Taste ligands create Ca^{2+} signals that release serotonin or ATP.

Sweet Umami Bitter Sour

Tight junction

Type I support cells may sense salt when Na^+ enters through channels.

Salt?

(Based on Tomchik *et al.*, *J Neurosci* 27(40): 10840–10848, 2007.)

Presynaptic cell (III)

ATP

Serotonin

Receptor cells (type II)

Primary sensory neurons

(b) Taste Transduction. Each taste cell senses only one type of ligand.

Sweet, umami, or bitter ligand

Gustducin

GPCR

Receptor cells with G protein–coupled membrane receptors bind either bitter, sweet, or umami ligands and release ATP as a signal molecule.

Signal transduction

Ca^{2+}

$\Uparrow Ca^{2+}$

Ca^{2+}

ATP

Sour

H^+

?

Presynaptic cells sense sour taste (H^+) apparently when H^+ enters the cell through channels.

$\Uparrow H^+$

?

Ca^{2+}

$\Uparrow Ca^{2+}$

① Ligands activate the taste cell.

② Various intracellular pathways are activated.

③ Ca^{2+} signal in the cytoplasm triggers exocytosis or ATP formation.

④ Neurotransmitter or ATP is released.

⑤ Primary sensory neuron fires and action potentials are sent to the brain.

Primary gustatory neurons

Taste	GPCR Family
Sweet	T1R2 + 3 subunits
Umami	T1R1 + 3 subunits
Bitter	T2R

widely held view of taste transduction was that an individual taste receptor cell could sense more than one taste, with cells differing in their sensitivities. However, gustation research using molecular biology techniques and knockout mice currently indicates that each taste receptor cell is sensitive to only one taste.

Nontraditional Taste Sensations The sensations we call taste are not all mediated by the traditional taste receptors. For years, physiologists thought fat in the diet was appealing because of its texture, and food experts use the phrase "mouth feel" to describe the sensation of eating something fatty, such as ice cream, that seems to coat the inside of the mouth. But now it appears that the tongue may have taste receptors for fats.

Research in rodents has identified a membrane receptor called *CD36* that lines taste pores and binds fats. Activation of this receptor helps trigger the feedforward digestive reflexes that prepare the digestive system for a meal. Currently evidence is lacking for a similar receptor in humans, but "fatty" may turn out to be a sixth taste sensation. Other candidates for new taste sensations include carbonation (dissolved CO_2) and Ca^{2+}, another essential element obtained through the diet.

Some additional taste sensations are related to somatosensory pathways rather than taste receptor cells. Nerve endings in the mouth have TRP receptors and carry spicy sensations through the *trigeminal nerve* (CN V). *Capsaicin* from chili peppers, menthol from mint, and molecules in cinnamon, mustard oil, and many Indian spices activate these receptors to add to our appreciation of the food we eat.

And what would you say to the idea of taste buds in your gut? Scientists have known for years that the stomach and intestines have the ability to sense the composition of a meal and secrete appropriate hormones and enzymes. Now it appears that gut chemoreception is being mediated by the same receptors and signal transduction mechanisms that occur in taste buds on the tongue. Studies have found the T1R receptor proteins for sweet and umami tastes as well as the G protein gustducin in various cells of rodent and human intestines.

An interesting psychological aspect of taste is the phenomenon named **specific hunger.** Humans and other animals that are lacking a particular nutrient may develop a craving for that substance. **Salt appetite,** representing a lack of Na^+ in the body, has been recognized for years. Hunters have used their knowledge of this specific hunger to stake out salt licks because they know that animals will seek them out. Salt appetite is directly related to Na^+ concentration in the body and cannot be assuaged by ingestion of other cations, such as Ca^{2+} or K^+. Other appetites, such as cravings for chocolate, are more difficult to relate to specific nutrient needs and probably reflect complex mixtures of physical, psychological, environmental, and cultural influences.

CONCEPT CHECK

14. With what essential nutrient is the umami taste sensation associated?

15. Map or diagram the neural pathway from a presynaptic taste receptor cell to the gustatory cortex.

THE EAR: HEARING

The ear is a sense organ that is specialized for two distinct functions: hearing and equilibrium. It can be divided into external, middle, and inner sections, with the neurological elements housed in and protected by structures in the inner ear. The vestibular complex of the inner ear is the primary sensor for equilibrium. The remainder of the ear is used for hearing.

The *external ear* consists of the outer ear, or **pinna,** and the **ear canal** (**FIG. 10.15**). The pinna is another example of an important accessory structure to a sensory system, and it varies in shape and location from species to species, depending on the animals' survival needs. The ear canal is sealed at its internal end by a thin membranous sheet of tissue called the **tympanic membrane,** or *eardrum.*

The tympanic membrane separates the external ear from the *middle ear,* an air-filled cavity that connects with the pharynx through the **Eustachian tube.** The Eustachian tube is normally collapsed, sealing off the middle ear, but it opens transiently to allow middle ear pressure to equilibrate with atmospheric pressure during chewing, swallowing, and yawning. Colds or other infections that cause swelling can block the Eustachian tube and result in fluid buildup in the middle ear. If bacteria are trapped in the middle ear fluid, the ear infection known as *otitis media* {*oto-*, ear + *-itis*, inflammation + *media*, middle} results.

Three small bones of the middle ear conduct sound from the external environment to the inner ear: the **malleus** {hammer}, the **incus** {anvil}, and the **stapes** {stirrup}. The three bones are connected to one another with the biological equivalent of hinges. One end of the malleus is attached to the tympanic membrane, and the stirrup end of the stapes is attached to a thin membrane that separates the middle ear from the inner ear.

The inner ear consists of two major sensory structures. The *vestibular apparatus* with its *semicircular canals* is the sensory transducer for our sense of equilibrium, described in the following section. The **cochlea** of the inner ear contains sensory receptors for hearing. On external view the cochlea is a membranous tube that lies coiled like a snail shell within a bony cavity. Two membranous disks, the **oval window** (to which the stapes is attached) and the **round window,** separate the liquid-filled cochlea from the air-filled middle ear. Branches of cranial nerve VIII, the *vestibulocochlear nerve,* lead from the inner ear to the brain.

Hearing Is Our Perception of Sound

Hearing is our perception of the energy carried by *sound waves,* which are pressure waves with alternating peaks of compressed air and valleys in which the air molecules are farther apart (**FIG. 10.16a**). The classic question about hearing is, "If a tree falls in the forest with no one to hear, does it make a noise?" The physiological answer is no, because noise, like pain, is a perception that results from the brain's processing of sensory information. A falling tree emits sound waves, but there is no noise unless someone or something is present to process and perceive the wave energy as sound.

FIG. 10.15 ANATOMY SUMMARY

The Ear

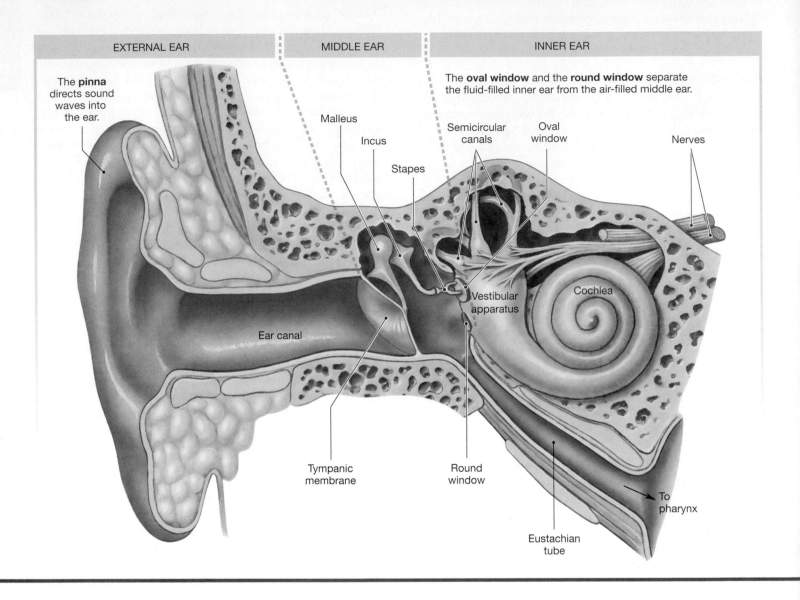

EXTERNAL EAR | MIDDLE EAR | INNER EAR

The **pinna** directs sound waves into the ear.

The **oval window** and the **round window** separate the fluid-filled inner ear from the air-filled middle ear.

Malleus

Incus

Stapes

Semicircular canals

Oval window

Nerves

Cochlea

Vestibular apparatus

Ear canal

Tympanic membrane

Round window

To pharynx

Eustachian tube

Sound is the brain's interpretation of the frequency, amplitude, and duration of sound waves that reach our ears. Our brains translate **frequency** of sound waves (the number of wave peaks that pass a given point each second) into the **pitch** of a sound. Low-frequency waves are perceived as low-pitched sounds, such as the rumble of distant thunder. High-frequency waves create high-pitched sounds, such as the screech of fingernails on a blackboard.

Sound wave frequency (Fig. 10.16b) is measured in waves per second, or **hertz (Hz).** The average human ear can hear sounds over the frequency range of 20–20,000 Hz, with the most acute hearing between 1000–3000 Hz. Our hearing is not as acute as that of many other animals, just as our sense of smell is less acute. Bats listen for ultra-high-frequency sound waves (in the kilohertz range) that bounce off objects in the dark. Elephants and some birds can hear sounds in the infrasound (very low frequency) range.

Loudness is our interpretation of sound intensity and is influenced by the sensitivity of an individual's ear. The intensity of a sound wave is a function of the wave height, or **amplitude** (Fig. 10.16b). Intensity is measured on a logarithmic scale in units called **decibels (dB).** Each 10 dB increase represents a 10-fold increase in intensity.

Normal conversation has a typical noise level of about 60 dB. Sounds of 80 dB or more can damage the sensitive hearing receptors of the ear, resulting in hearing loss. A typical heavy metal rock concert has noise levels around 120 dB, an intensity that puts listeners in immediate danger of damage to their hearing. The amount of damage depends on the duration and frequency of the noise as well as its intensity.

CONCEPT CHECK 16. What is a kilohertz?

FIG. 10.16 | Sound waves

(a) Sound waves alternate peaks of compressed air and valleys where the air is less compressed.

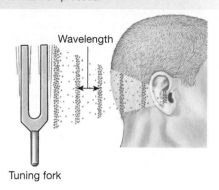

Wavelength

Tuning fork

(b) Sound waves are distinguished by their frequency, measured in hertz (Hz), and amplitude, measured in decibels (dB).

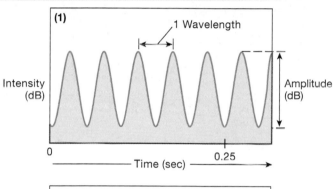

(1)

1 Wavelength

Intensity (dB)

Amplitude (dB)

0 — Time (sec) — 0.25

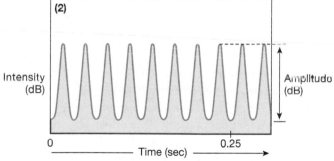

(2)

Intensity (dB)

Amplitude (dB)

0 — Time (sec) — 0.25

Q **FIGURE QUESTIONS**

1. What are the frequencies of the sound waves in graphs (1) and (2) in Hz (waves/second)?
2. Which set of sound waves would be interpreted as having lower pitch?

Sound Transduction Is a Multistep Process

Hearing is a complex sense that involves multiple transductions. Energy from sound waves in the air becomes mechanical vibrations, then fluid waves in the cochlea. The fluid waves open ion channels in *hair cells,* the sensory receptors for hearing. Ion flow into hair cells creates electrical signals that release

RUNNING **PROBLEM**

Anant reports to the otolaryngologist that he never knows when his attacks of dizziness will strike and that they last from 10 minutes to an hour. They often cause him to vomit. He also reports that he has a persistent low buzzing sound in one ear and that he does not seem to hear low tones as well as he could before the attacks started. The buzzing sound (tinnitus) often gets worse during his dizzy attacks.

Q2: *Subjective tinnitus occurs when an abnormality somewhere along the anatomical pathway for hearing causes the brain to perceive a sound that does not exist outside the auditory system. Starting from the ear canal, name the auditory structures in which problems may arise.*

310 314 **331** 339 343 348 353

neurotransmitter (chemical signal), which in turn triggers action potentials in the primary auditory neurons.

These transduction steps are shown in **FIGURE 10.17**. Sound waves striking the outer ear are directed down the ear canal until they hit the tympanic membrane and cause it to vibrate (first transduction). The tympanic membrane vibrations are transferred to the malleus, the incus, and the stapes, in that order. The arrangement of the three connected middle ear bones creates a "lever" that multiplies the force of the vibration (*amplification*) so that very little sound energy is lost due to friction. If noise levels are so high that there is danger of damage to the inner ear, small muscles in the middle ear can pull on the bones to decrease their movement and thereby dampen sound transmission to some degree.

As the stapes vibrates, it pulls and pushes on the thin tissue of the oval window, to which it is attached. Vibrations at the oval window create waves in the fluid-filled channels of the cochlea (second transduction). As waves move through the cochlea, they push on the flexible membranes of the *cochlear duct* and bend sensory **hair cells** inside the duct. The wave energy dissipates back into the air of the middle ear at the round window.

Movement of the cochlear duct opens or closes ion channels on hair cell membranes, creating electrical signals (third transduction). These electrical signals alter neurotransmitter release (fourth transduction). Neurotransmitter binding to the primary auditory neurons initiates action potentials (fifth transduction) that send coded information about sound through the *cochlear branch of the vestibulocochlear nerve* (cranial nerve VIII) and the brain.

The Cochlea Is Filled with Fluid

As we have just seen, the transduction of wave energy into action potentials takes place in the cochlea of the inner ear. Uncoiled, the cochlea can be seen to be composed of three parallel, fluid-filled channels: (1) the **vestibular duct,** or *scala vestibuli* {*scala,* stairway;

FIG. 10.17 Sound transmission through the ear

1 Sound waves strike the tympanic membrane and become vibrations.

2 The sound wave energy is transferred to the three bones of the middle ear, which vibrate.

3 The stapes is attached to the membrane of the oval window. Vibrations of the oval window create fluid waves within the cochlea.

4 The fluid waves push on the flexible membranes of the cochlear duct. Hair cells bend and ion channels open, creating an electrical signal that alters neurotransmitter release.

5 Neurotransmitter release onto sensory neurons creates action potentials that travel through the cochlear nerve to the brain.

6 Energy from the waves transfers across the cochlear duct into the tympanic duct and is dissipated back into the middle ear at the round window.

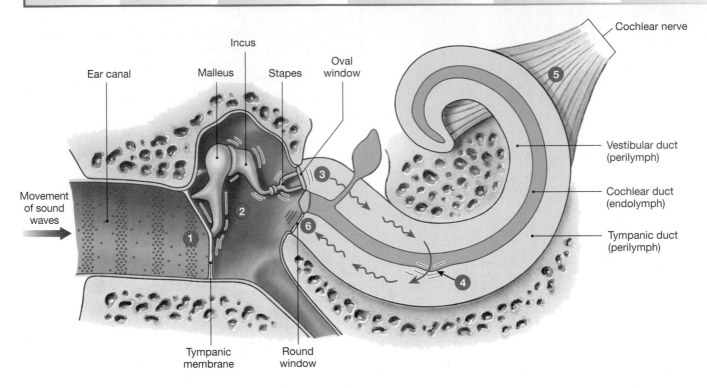

vestibulum, entrance}; (2) the central **cochlear duct,** or *scala media* {*media,* middle}; and (3) the **tympanic duct,** or *scala tympani* {*tympanon,* drum} (**FIG. 10.18**). The vestibular and tympanic ducts are continuous with each other, and they connect at the tip of the cochlea through a small opening known as the **helicotrema** {*helix,* a spiral + *trema,* hole}. The cochlear duct is a dead-end tube, but it connects to the vestibular apparatus through a small opening.

The fluid in the vestibular and tympanic ducts is similar in ion composition to plasma and is known as **perilymph.** The cochlear duct is filled with **endolymph** secreted by epithelial cells in the duct. Endolymph is unusual because it is more like intracellular fluid than extracellular fluid in composition, with high concentrations of K^+ and low concentrations of Na^+.

The cochlear duct contains the **organ of Corti,** composed of hair cell receptors and support cells. The organ of Corti sits on the **basilar membrane** and is partially covered by the **tectorial membrane** {*tectorium,* a cover}, both flexible tissues that move in response to fluid waves passing through the vestibular duct (Fig. 10.18). As the waves travel through the cochlea, they displace basilar and tectorial membranes, creating up-and-down oscillations that bend the hair cells.

Hair cells, like taste receptor cells, are non-neural receptor cells. The apical surface of each hair cell is modified into

50–100 stiffened cilia known as **stereocilia,** arranged in ascending height (**FIG. 10.19a**). The stereocilia of the hair cells are embedded in the overlying tectorial membrane. If the tectorial membrane moves, the underlying cilia do as well.

When hair cells move in response to sound waves, their stereocilia flex, first one way, then the other. The stereocilia are attached to each other by protein bridges called *tip links.* The tip links act like little springs and are connected to gates that open and close ion channels in the cilia membrane. When the hair cells and cilia are in a neutral position, about 10% of the ion channels are open, and there is a low tonic level of neurotransmitter released onto the primary sensory neuron.

When waves deflect the tectorial membrane so that cilia bend toward the tallest members of a bundle, the tip links pop more channels open, so cations (primarily K^+ and Ca^{2+}) enter the cell, which then depolarizes (Fig. 10.19b). Voltage-gated Ca^{2+} channels open, neurotransmitter release increases, and the sensory neuron increases its firing rate. When the tectorial membrane pushes the cilia away from the tallest members, the springy tip links relax and all the ion channels close. Cation influx slows, the membrane hyperpolarizes, less transmitter is released, and sensory neuron firing decreases (Fig. 10.19c).

The vibration pattern of waves reaching the inner ear is thus converted into a pattern of action potentials going to the CNS.

FIG. 10.18 **ANATOMY SUMMARY**

The Cochlea

Oval window

Saccule

Vestibular duct

Cochlear duct

Organ of Corti

Cochlea

Uncoiled

Round window

Tympanic duct

Basilar membrane

Helicotrema

Bony cochlear wall

Vestibular duct

Cochlear duct

Tectorial membrane

Organ of Corti

The cochlear nerve transmits action potentials from the primary auditory neurons to cochlear nuclei in the medulla, on their way to the auditory cortex.

Basilar membrane

Tympanic duct

Fluid wave

Cochlear duct

Tectorial membrane

Hair cell

Tympanic duct

The movement of the tectorial membrane moves the cilia on the hair cells.

Basilar membrane

Nerve fibers of cochlear nerve

FIG. 10.19 | Signal transduction in hair cells

The stereocilia of hair cells have "trap doors" that close off ion channels. These openings are controlled by protein-bridge tip links connecting adjacent cilia.

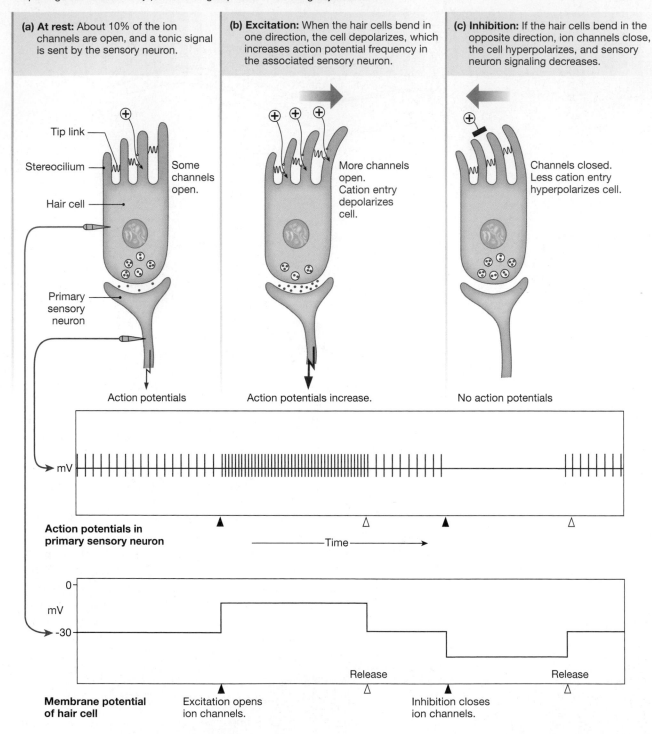

(a) At rest: About 10% of the ion channels are open, and a tonic signal is sent by the sensory neuron.

(b) Excitation: When the hair cells bend in one direction, the cell depolarizes, which increases action potential frequency in the associated sensory neuron.

(c) Inhibition: If the hair cells bend in the opposite direction, ion channels close, the cell hyperpolarizes, and sensory neuron signaling decreases.

Tip link
Stereocilium
Some channels open.
Hair cell
Primary sensory neuron

More channels open. Cation entry depolarizes cell.

Channels closed. Less cation entry hyperpolarizes cell.

Action potentials

Action potentials increase.

No action potentials

mV

Action potentials in primary sensory neuron

Time

0
mV
-30

Release

Release

Membrane potential of hair cell

Excitation opens ion channels.

Inhibition closes ion channels.

Because tectorial membrane vibrations reflect the frequency of the incoming sound wave, the hair cells and sensory neurons must be able to respond to sounds of nearly 20,000 waves per second, the highest frequency audible by a human ear.

CONCEPT CHECK

17. Normally when cation channels on a cell open, either Na^+ or Ca^{2+} enters the cell. Why does K^+ rather than Na^+ enter hair cells when their cation channels open?

Sounds Are Processed First in the Cochlea

The auditory system processes sound waves so that they can be discriminated by location, pitch, and loudness. Localization of sound is a complex process that requires sensory input from both ears coupled with sophisticated computation by the brain (see Fig. 10.4). In contrast, the initial processing for pitch and loudness takes place in the cochlea of each ear.

Coding sound for pitch is primarily a function of the basilar membrane. This membrane is stiff and narrow near its attachment between the round and oval windows but widens and becomes more flexible near its distal end (**FIG. 10.20a**).

FIG. 10.20   Sensory coding for pitch

Coding for pitch is a function of the basilar membrane.

(a) The basilar membrane has variable sensitivity to sound wave frequency along its length.

High frequency (high pitch)

Low frequency (low pitch)

Basilar membrane

Stiff region near round window

Flexible region near helicotrema (distal end)

(b) The frequency of sound waves determines the displacement of the basilar membrane. The location of active hair cells creates a code that the brain translates as information about the pitch of sound.

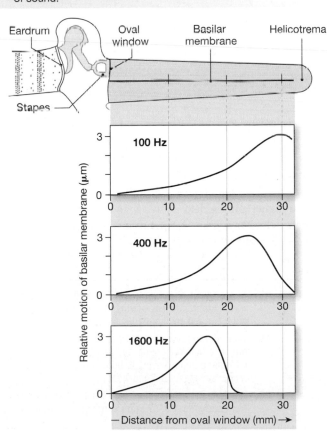

Eardrum   Oval window   Basilar membrane   Helicotrema

Stapes

Relative motion of basilar membrane (μm)

100 Hz

400 Hz

1600 Hz

— Distance from oval window (mm) →

High-frequency waves entering the vestibular duct create maximum displacement of the basilar membrane close to the oval window and consequently are not transmitted very far along the cochlea. Low-frequency waves travel along the length of the basilar membrane and create their maximum displacement near the flexible distal end.

This differential response to frequency transforms the temporal aspect of frequency (number of sound waves per second) into spatial coding for pitch by location along the basilar membrane (Fig. 10.20b). A good analogy is a piano keyboard, where the location of a key tells you its pitch. The spatial coding of the basilar membrane is preserved in the auditory cortex as neurons project from hair cells to corresponding regions in the brain. Loudness is coded by the ear in the same way that signal strength is coded in somatic receptors. The louder the noise, the more rapidly action potentials fire in the sensory neuron.

Auditory Pathways Project to the Auditory Cortex

Once the cochlea transforms sound waves into electrical signals, sensory neurons transfer this information to the brain. The cochlear (auditory) nerve is a branch of cranial nerve VIII, the *vestibulocochlear nerve* [p. 288]. Primary auditory neurons project from the cochlea to *cochlear nuclei* in the medulla oblongata (**FIG. 10.21**). Some of these neurons carry information that is processed into the timing of sound, and others carry information that is processed into the sound quality.

From the medulla, secondary sensory neurons project to two higher nuclei, one *ipsilateral* (on the same side of the body) and one *contralateral* (on the opposite side). Splitting sound signals between two ascending tracts means that each side of the brain gets information from both ears. These ascending tracts then synapse in nuclei in the midbrain and thalamus before projecting to the auditory cortex (see Fig. 10.3). Collateral pathways take information to the reticular formation and the cerebellum.

The localization of a sound source is an integrative task that requires simultaneous input from both ears. Unless sound is coming from directly in front of a person, it will not reach both ears at the same time (see Fig. 10.4). The brain records the time differential for sound arriving at the ears and uses complex computation to create a three-dimensional representation of the sound source.

Hearing Loss May Result from Mechanical or Neural Damage

There are three forms of hearing loss: conductive, central, and sensorineural. In *conductive hearing loss,* sound cannot be transmitted either through the external ear or the middle ear. The causes of conductive hearing loss range from an ear canal plugged with earwax (*cerumen*), to fluid in the middle ear from an infection, to diseases or trauma that impede vibration of the malleus, incus, or stapes. Correction of conductive hearing loss includes microsurgical techniques in which the bones of the middle ear can be reconstructed.

FIG. 10.21 | The auditory pathways

Sound is processed so that information from each ear goes to both sides of the brain.

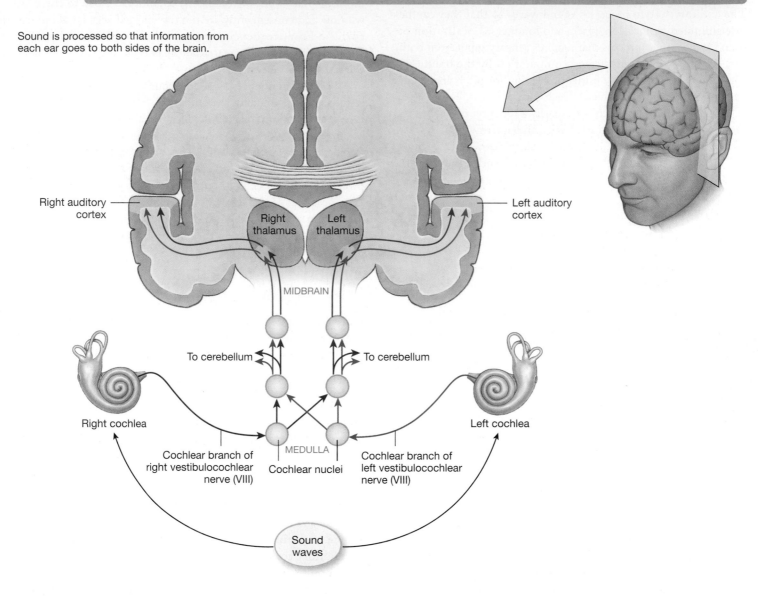

Central hearing loss results either from damage to the neural pathways between the ear and cerebral cortex or from damage to the cortex itself, as might occur from a stroke. This form of hearing loss is relatively uncommon.

Sensorineural hearing loss arises from damage to the structures of the inner ear, including death of hair cells as a result of loud noises. The loss of hair cells in mammals is currently irreversible. Birds and lower vertebrates, however, are able to regenerate hair cells to replace those that die. This discovery has researchers exploring strategies to duplicate the process in mammals, including transplantation of neural stem cells and gene therapy to induce nonsensory cells to differentiate into hair cells.

Therapy that replaces hair cells would be an important advance. The incidence of hearing loss in younger people is increasing because of prolonged exposure to rock music and environmental noises. Ninety percent of hearing loss in the elderly—called *presbycusis* {*presbys*, old man + *akoustikos*, able to be heard}—is sensorineural. Currently, the primary treatment for

sensorineural hearing loss is the use of hearing aids, but amazing results have been obtained with cochlear implants attached to tiny computers (see Biotechnology box).

Hearing is probably our most important social sense. Suicide rates are higher among deaf people than among those who have lost their sight. More than any other sense, hearing connects us to other people and to the world around us.

CONCEPT CHECK

18. Map or diagram the pathways followed by a sound wave entering the ear, starting in the air at the outer ear and ending on the auditory cortex.

19. Why is somatosensory information projected to only one hemisphere of the brain but auditory information is projected to both hemispheres? (*Hint:* See Figs. 10.4 and 10.8.)

20. Would a cochlear implant help a person who suffers from nerve deafness? From conductive hearing loss?

Artificial Ears

One technique used to treat sensorineural hearing loss is the cochlear implant. The newest cochlear implants have multiple components. Externally, a microphone, tiny computerized speech processor, and transmitter fit behind the ear like a conventional hearing aid. The speech processor is a transducer that converts sound into electrical impulses. The transmitter converts the processor's electrical impulses into radio waves and sends these signals to a receiver and 8–24 electrodes, which are surgically placed under the skin. The electrodes take electrical signals directly into the cochlea or to the auditory nerve, bypassing any damaged areas. After surgery, recipients go through therapy so that they can learn to understand the sounds they hear. Cochlear implants have been remarkably successful for many profoundly deaf people, allowing them to hear loud noises and modulate their own voices. In the most successful cases, individuals can even use the telephone. To learn more about cochlear implants, visit the web site of the National Institute for Deafness and Other Communication Disorders (*www.nidcd.nih.gov/health/hearing*).

THE EAR: EQUILIBRIUM

Equilibrium is a state of balance, whether the word is used to describe ion concentrations in body fluids or the position of the body in space. The special sense of equilibrium has two components: a dynamic component that tells us about our movement through space, and a static component that tells us if our head is not in its normal upright position. Sensory information from the inner ear and from joint and muscle proprioceptors tells our brain the location of different body parts in relation to one another and to the environment. Visual information also plays an important role in equilibrium, as you know if you have ever gone to one of the 360° movie theaters where the scene tilts suddenly to one side and the audience tilts with it!

Our sense of equilibrium is mediated by hair cells lining the fluid-filled vestibular apparatus of the inner ear. These non-neural receptors respond to changes in rotational, vertical, and horizontal acceleration and positioning. The hair cells function just like those of the cochlea, but gravity and acceleration rather than sound waves provide the force that moves the stereocilia. Vestibular hair cells have a single long cilium called a **kinocilium** {*kinein,* to move} located at one side of the ciliary bundle. The kinocilium creates a reference point for the direction of bending.

When the cilia bend, tip links between them open and close ion channels. Movement in one direction causes the hair cells to depolarize; with movement in the opposite direction, they hyperpolarize. This is similar to what happens in cochlear hair cells (see Fig. 10.19).

The Vestibular Apparatus Provides Information about Movement and Position

The **vestibular apparatus,** also called the *membranous labyrinth,* is an intricate series of interconnected fluid-filled chambers. (In Greek mythology the labyrinth was a maze that housed a monster called the Minotaur.) In humans, the vestibular apparatus consists of two saclike **otolith organs**—the **saccule** and the **utricle**—along with three **semicircular canals** that connect to the utricle at their bases (**FIG. 10.22a**). The otolith organs tell us about *linear acceleration* and head position. The three semicircular canals sense *rotational acceleration* in various directions.

The vestibular apparatus, like the cochlear duct, is filled with high-K^+, low-Na^+ endolymph secreted by epithelial cells. Like cerebrospinal fluid, endolymph is secreted continuously and drains from the inner ear into the venous sinus in the dura mater of the brain.

If endolymph production exceeds the drainage rate, buildup of fluid in the inner ear may increase fluid pressure within the vestibular apparatus. Excessive accumulation of endolymph is believed to contribute to *Ménière's disease,* a condition marked by episodes of dizziness and nausea. If the organ of Corti in the cochlear duct is damaged by fluid pressure within the vestibular apparatus, hearing loss may result.

The Semicircular Canals Sense Rotational Acceleration

The three semicircular canals of the vestibular apparatus monitor rotational acceleration. They are oriented at right angles to one another, like three planes that come together to form the corner of a box (Fig. 10.22a). The horizontal canal monitors rotations that we associate with turning, such as an ice skater's spin or shaking your head left and right to say "no." The posterior canal monitors left-to-right rotation, such as the rotation when you tilt your head toward your shoulders or perform a cartwheel. The superior canal is sensitive to forward and back rotation, such as nodding your head front to back or doing a somersault.

At one end of each canal is an enlarged chamber, the **ampulla** {bottle}, which contains a sensory structure known as a **crista** {a crest; plural *cristae*}. The crista consists of hair cells and a gelatinous mass, the **cupula** {small tub}, that stretches from floor to ceiling of the ampulla, closing it off (Fig. 10.22b). Hair cell cilia are embedded in the cupula.

How is rotation sensed? As the head turns, the bony skull and the membranous walls of the labyrinth move, but the fluid within the labyrinth cannot keep up because of *inertia* (the tendency of a body at rest to remain at rest). In the ampullae, the drag of endolymph bends the cupula and its hair cells in the direction *opposite* to the direction in which the head is turning.

For an analogy, think of pulling a paintbrush (a cupula attached to the wall of a semicircular canal) through sticky wet paint (the endolymph) on a board. If you pull the brush to the right, the drag of the paint on the bristles bends them to the

FIG. 10.22 **ESSENTIALS**

Equilibrium

The vestibular apparatus of the inner ear responds to changes in the body's position in space. The cristae are sensory receptors for rotational acceleration. The maculae are sensory receptors for linear acceleration and head position.

SEMICIRCULAR CANALS

Superior
Horizontal
Posterior

Cochlea

Cristae within ampulla

Utricle

Saccule

Maculae

(a) Semicircular Canals

The posterior canal of the vestibular apparatus senses the tilt of the head toward the right or left shoulder.

Left ←→ right

The superior canal senses rotation of the head from front to back, such as that which occurs when nodding "yes."

The horizontal canal senses rotation of the head as it turns left or right, such as that which occurs when shaking the head "no."

(b) Crista

Movement of the endolymph pushes on the gelatinous cupula and activates the hair cells.

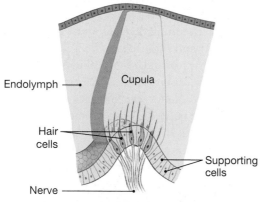

Endolymph
Cupula
Hair cells
Supporting cells
Nerve

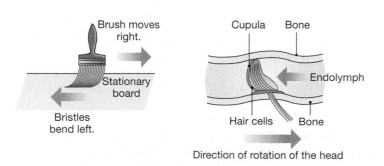

Brush moves right.

Stationary board

Bristles bend left.

Cupula   Bone

Endolymph

Hair cells   Bone

Direction of rotation of the head

When the head turns right, endolymph pushes the cupula to the left.

(c) Macula

Otoliths are crystals that move in response to gravitational forces.

Gelatinous otolith membrane

Hair cells

Nerve fibers

Head in neutral position

Macula   Gravity

Head tilted posteriorly

Gravity

Otolith

RUNNING **PROBLEM**

Although many vestibular disorders can cause the symptoms Anant is experiencing, two of the most common are positional vertigo and Ménière's disease. In *positional vertigo,* calcium crystals normally embedded in the otolith membrane of the maculae become dislodged and float toward the semicircular canals. The primary symptom of positional vertigo is brief episodes of severe dizziness brought on by a change in position, such as moving to the head-down yoga position called "downward-facing dog." People with positional vertigo often say they feel dizzy when they lie down or turn over in bed.

Q3: *When a person with positional vertigo changes position, the displaced crystals float toward the semicircular canals. Why would this cause dizziness?*

Q4: *Compare the symptoms of positional vertigo and Ménière's disease. On the basis of Anant's symptoms, which condition do you think he has?*

310 — 314 — 331 — **339** — 343 — 348 — 353

left (Fig. 10.22b). In the same way, the inertia of the fluid in the semicircular canal pulls the cupula and the cilia of the hair cells to the left when the head turns right.

If rotation continues, the moving endolymph finally catches up. Then if head rotation stops suddenly, the fluid has built up momentum and cannot stop immediately. The fluid continues to rotate in the direction of the head rotation, leaving the person with a turning sensation. If the sensation is strong enough, the person may throw his or her body in the direction opposite the direction of rotation in a reflexive attempt to compensate for the apparent loss of equilibrium.

The Otolith Organs Sense Linear Acceleration and Head Position

The two otolith organs, the utricle {*utriculus,* little bag} and saccule {little sac}, are arranged to sense linear forces. Their sensory structures, called **maculae,** consist of hair cells, a gelatinous mass known as the **otolith membrane,** and calcium carbonate and protein particles called **otoliths** {*oto,* ear + *lithos,* stone}.

The hair cell cilia are embedded in the otolith membrane, and otoliths bind to matrix proteins on the surface of the membrane (Fig. 10.22c). If gravity or acceleration cause the otoliths to slide forward or back, the gelatinous otolith membrane slides with them, bending the hair cell cilia and setting off a signal. For example, the maculae are horizontal when the head is in its normal upright position. If the head tips back, gravity displaces the otoliths, and the hair cells are activated.

The maculae of the utricle sense forward acceleration or deceleration as well as head tilt. In contrast, the maculae of the saccule are oriented vertically when the head is erect, which makes them sensitive to vertical forces, such as dropping downward in an elevator. The brain analyzes the pattern of depolarized and hyperpolarized hair cells to compute head position and direction of movement.

Equilibrium Pathways Project Primarily to the Cerebellum

Vestibular hair cells, like those of the cochlea, are tonically active and release neurotransmitter onto primary sensory neurons of the **vestibular nerve** (a branch of cranial nerve VIII, the vestibulocochlear nerve). Those sensory neurons either synapse in the *vestibular nuclei* of the medulla or run without synapsing to the cerebellum, which is the primary site for equilibrium processing (**FIG. 10.23**). Collateral pathways run from the medulla to

FIG. 10.23 **Equilibrium pathways**

Vestibular apparatus

Vestibular branch of vestibulocochlear nerve (VIII)

Cerebral cortex

Thalamus

Reticular formation

Cerebellum

Vestibular nuclei of medulla

Somatic motor neurons controlling eye movements

the cerebellum or upward through the reticular formation and thalamus.

There are some poorly defined pathways from the medulla to the cerebral cortex, but most integration for equilibrium occurs in the cerebellum. Descending pathways from the vestibular nuclei go to certain motor neurons involved in eye movement. These pathways help keep the eyes locked on an object as the head turns.

CONCEPT CHECK

21. The stereocilia of hair cells are bathed in endolymph, which has a very high concentration of K^+ and a low concentration of Na^+. When ion channels in the stereocilia open, which ions move in which direction to cause depolarization?

22. Why does hearing decrease if an ear infection causes fluid buildup in the middle ear?

23. When dancers perform multiple turns, they try to keep their vision fixed on a single point ("spotting"). How does spotting keep a dancer from getting dizzy?

THE EYE AND VISION

The eye is a sensory organ that functions much like a camera. It focuses light on a light-sensitive surface (the retina) using a lens and an aperture or opening (the pupil) whose size can be adjusted to change the amount of entering light. **Vision** is the process through which light reflected from objects in our environment is translated into a mental image. This process can be divided into three steps:

1. Light enters the eye, and the lens focuses the light on the retina.
2. Photoreceptors of the retina transduce light energy into an electrical signal.
3. Neural pathways from retina to brain process electrical signals into visual images.

The Skull Protects the Eye

The external anatomy of the eye is shown in **FIGURE 10.24**. Like sensory elements of the ears, the eyes are protected by a bony cavity, the *orbit*, which is formed by facial bones of the skull. Accessory structures associated with the eye include six *extrinsic eye muscles*, skeletal muscles that attach to the outer surface of the eyeball and control eye movements. Cranial nerves III, IV, and VI innervate these muscles.

The upper and lower *eyelids* close over the anterior surface of the eye, and the *lacrimal apparatus,* a system of glands and ducts, keeps a continuous flow of tears washing across the exposed surface so that it remains moist and free of debris. Tear secretion is stimulated by parasympathetic neurons from cranial nerve VII.

The **pupil** is an opening through which light can pass into the interior of the eye. Pupil size varies with the contraction and relaxation of a ring of smooth *pupillary muscle.* The pupil appears as the black spot inside the colored ring of pigment known as the **iris.** The pigments and other components of the iris determine eye color.

FIG. 10.24   External anatomy of the eye

Lacrimal gland secretes tears.

Muscles attached to external surface of eye control eye movement.

Upper eyelid

Sclera

Pupil

Iris

Lower eyelid

The **orbit** is a bony cavity that protects the eye.

Nasolacrimal duct drains tears into nasal cavity.

The eye itself is a hollow sphere divided into two compartments (chambers) separated by a lens (**FIG. 10.25**). The **lens,** suspended by ligaments called **zonules,** is a transparent disk that focuses light. The anterior chamber in front of the lens is filled with **aqueous humor** {*humidus,* moist}, a low-protein, plasmalike fluid secreted by the *ciliary epithelium* supporting the lens. Behind the lens is a much larger chamber, the **vitreous chamber,**

CLINICAL FOCUS

Glaucoma

The eye disease *glaucoma*, characterized by degeneration of the optic nerve, is the leading cause of blindness worldwide. Many people associate glaucoma with increased *intraocular* (within the eyeball) pressure, but scientists have discovered that increased pressure is only one risk factor for the disease. A significant number of people with glaucoma have normal intraocular pressure, and not everyone with elevated pressure develops glaucoma. Many cases of elevated eye pressure are associated with excess aqueous humor, a fluid that is secreted by the ciliary epithelium near the lens. Normally, the fluid drains out through the canal of Schlemm in the anterior chamber of the eye, but if outflow is blocked, the aqueous humor accumulates, causing pressure to build up inside the eye. Treatments to decrease intraocular pressure include drugs that inhibit aqueous humor production and surgery to reopen the canal of Schlemm. Research suggests that the optic nerve degeneration in glaucoma may be due to nitric oxide or apoptosis-inducing factors, and studies in these areas are underway.

FIG. 10.25 **ANATOMY SUMMARY**

The Eye

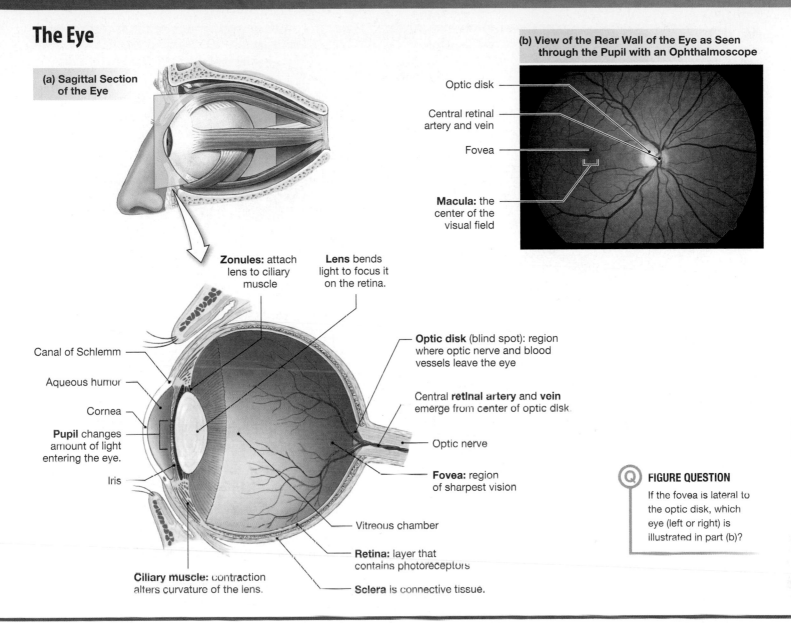

(a) Sagittal Section of the Eye

(b) View of the Rear Wall of the Eye as Seen through the Pupil with an Ophthalmoscope

Optic disk

Central retinal artery and vein

Fovea

Macula: the center of the visual field

Zonules: attach lens to ciliary muscle

Lens bends light to focus it on the retina.

Canal of Schlemm

Aqueous humor

Cornea

Pupil changes amount of light entering the eye.

Iris

Ciliary muscle: contraction alters curvature of the lens.

Optic disk (blind spot): region where optic nerve and blood vessels leave the eye

Central **retinal artery** and **vein** emerge from center of optic disk.

Optic nerve

Fovea: region of sharpest vision

Vitreous chamber

Retina: layer that contains photoreceptors

Sclera is connective tissue.

Q **FIGURE QUESTION**

If the fovea is lateral to the optic disk, which eye (left or right) is illustrated in part (b)?

---

filled mostly with the **vitreous body** {*vitrum,* glass; also called the *vitreous humor*}, a clear, gelatinous matrix that helps maintain the shape of the eyeball. The outer wall of the eyeball, the **sclera,** is composed of connective tissue.

Light enters the anterior surface of the eye through the **cornea,** a transparent disk of tissue that is a continuation of the sclera. After passing through the opening of the pupil, light strikes the lens, which has two convex surfaces. The cornea and lens together bend incoming light rays so that they focus on the **retina,** the light-sensitive lining of the eye that contains the photoreceptors.

When viewed through the pupil with an ophthalmoscope {*ophthalmos,* eye}, the retina is seen to be crisscrossed with small arteries and veins that radiate out from one spot, the **optic disk** (Fig. 10.25b). The optic disk is the location where neurons of the visual pathway form the **optic nerve** (cranial nerve II) and exit the eye. Lateral to the optic disk is a small

dark spot, the *fovea.* The fovea and a narrow ring of tissue surrounding it, the *macula,* are the regions of the retina with the most acute vision.

Neural pathways for the eyes are illustrated in **FIGURE 10.26**. The optic nerves from the eyes go to the **optic chiasm** in the brain, where some of the fibers cross to the opposite side. After synapsing in the **lateral geniculate body** (*lateral geniculate nucleus*) of the thalamus, the vision neurons of the tract terminate in the occipital lobe at the **visual cortex.** Collateral pathways go from the thalamus to the midbrain, where they synapse with efferent neurons of cranial nerve III that control the diameter of the pupils.

CONCEPT CHECK 24. What functions does the aqueous humor serve?

FIG. 10.26   Pathways for vision and the pupillary reflex

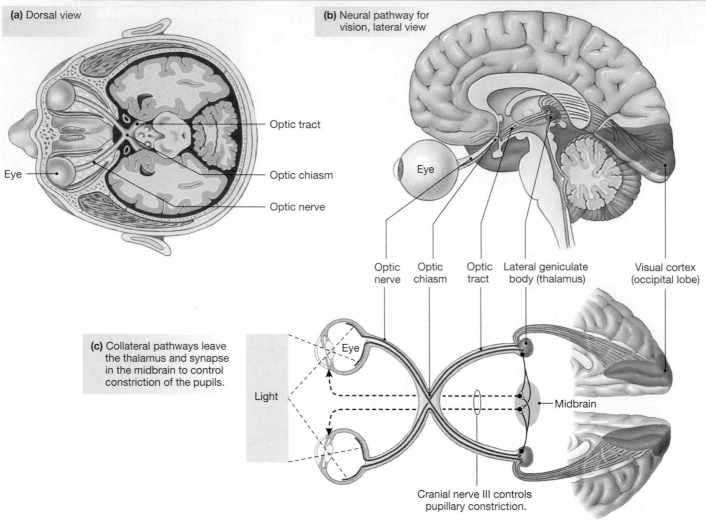

(a) Dorsal view

Optic tract

Eye

Optic chiasm

Optic nerve

(b) Neural pathway for vision, lateral view

Eye

Optic nerve

Optic chiasm

Optic tract

Lateral geniculate body (thalamus)

Visual cortex (occipital lobe)

(c) Collateral pathways leave the thalamus and synapse in the midbrain to control constriction of the pupils.

Eye

Light

Midbrain

Cranial nerve III controls pupillary constriction.

Light Enters the Eye through the Pupil

In the first step of the visual pathway, light from the environment enters the eye. Before it strikes the retina, however, the light is modified two ways. First, the amount of light that reaches photoreceptors is modulated by changes in the size of the pupil. Second, the light is focused by changes in the shape of the lens.

The human eye functions over a 100,000-fold range of light intensity. Most of this ability comes from the sensitivity of the photoreceptors, but the pupils assist by regulating the amount of light that falls on the retina. In bright sunlight, the pupils narrow to about 1.5 mm in diameter when a parasympathetic pathway constricts the circular pupillary muscles. In the dark, the opening of the pupil dilates to 8 mm, a 28-fold increase in pupil area. Dilation occurs when radial muscles lying perpendicular to the circular muscles contract under the influence of sympathetic neurons.

Testing **pupillary reflexes** is a standard part of a neurological examination. Light hitting the retina in one eye activates the reflex. Signals travel through the optic nerve to the thalamus, then

CONCEPT CHECK

25. Use the neural pathways in Figure 10.26 to answer the following questions.
    (a) Why does shining light into one eye cause pupillary constriction in both eyes?
    (b) If you shine a light in the left eye and get pupillary constriction in the right eye but not in the left eye, what can you conclude about the afferent path from the left eye to the brain? About the efferent pathways to the pupils?

26. Parasympathetic fibers constrict the pupils, and sympathetic fibers dilate them. The two autonomic divisions can be said to have _____ effects on pupil diameter.

to the midbrain, where efferent neurons constrict the pupils in *both* eyes (Fig. 10.26c). This response is known as the **consensual reflex** and is mediated by parasympathetic fibers running through cranial nerve III.

In addition to regulating the amount of light that hits the retina, the pupils create what is known as **depth of field.** A simple example comes from photography. Imagine a picture of a puppy sitting in the foreground amid a field of wildflowers. If only the puppy and the flowers immediately around her are in focus, the picture is said to have a shallow depth of field. If the puppy and the wildflowers all the way back to the horizon are in focus, the picture has full depth of field. Full depth of field is created by constricting the pupil (or the diaphragm on a camera) so that only a narrow beam of light enters the eye. In this way, a greater depth of the image is focused on the retina.

The Lens Focuses Light on the Retina

The physics that describes the behavior and properties of light is a field known as **optics.** When light rays pass from air into a medium of different density, such as glass or water, they bend, or **refract.** Light entering the eye is refracted twice: first when it passes through the cornea, and again when it passes through the lens. About two-thirds of the total refraction (bending) occurs at the cornea and the remaining one-third occurs at the lens. Here, we consider only the refraction that occurs as light passes through the lens because the lens is capable of changing its shape to focus light.

When light passes from one medium into another, the angle of refraction (how much the light rays bend) is influenced by two factors: (1) the difference in density of the two media and (2) the angle at which the light rays meet the surface of the medium into which it is passing. For light passing through the lens of the eye, we assume that the density of the lens is the same as the density of the air and thus ignore this factor. The angle at which light meets the face of the lens depends on the curvature of the lens surface and the direction of the light beam.

Imagine parallel light rays striking the surface of a transparent lens. If the lens surface is perpendicular to the rays, the light passes through without bending. If the surface is not perpendicular, however, the light rays bend. Parallel light rays striking a *concave* lens, such as that shown in **FIGURE 10.27a**, are refracted into a wider beam. Parallel rays striking a *convex* lens bend inward and focus to a point—*convex lenses converge* light waves (Fig. 10.27b). You can demonstrate the properties of a convex lens by using a magnifying glass to focus sunlight onto a piece of paper or other surface.

When parallel light rays pass through a convex lens, the single point where the rays converge is called the **focal point** (Fig. 10.27b). The distance from the center of a lens to its focal point is known as the **focal length** (or *focal distance*) of the lens. For any given lens, the focal length is fixed. For the focal length to change, the shape of the lens must change.

When light from an object passes through the lens of the eye, the focal point and object image must fall precisely on the retina if the object is to be seen in focus. In Figure 10.27c, parallel light rays strike a lens whose surface is relatively flat. For this lens, the focal point falls on the retina. The object is therefore in focus. For the normal human eye, any object that is 20 feet or more from the eye creates parallel light rays and will be in focus when the lens is flatter.

RUNNING **PROBLEM**

The otolaryngologist strongly suspects that Anant has Ménière's disease, with excessive endolymph in the vestibular apparatus and cochlea. Many treatments are available, beginning with simple dietary changes. For now, the physician suggests that Anant limit his salt intake and take diuretics, drugs that cause the kidneys to remove excess fluid from the body.

Q5: *Why is limiting salt (NaCl) intake suggested as a treatment for Ménière's disease? (Hint: What is the relationship between salt, osmolarity, and fluid volume?)*

310   314   331   339   **343**   348   353

What happens, though, when an object is closer than 20 feet to the lens? In that case, the light rays from the object are not parallel and strike the lens at an oblique angle that changes the distance from the lens to the object's image (Fig. 10.27d). The focal point now lies behind the retina, and the object image becomes fuzzy and out of focus.

To keep a near object in focus, the lens must become more rounded to increase the angle of refraction (Fig. 10.27e). Making a lens more convex shortens its focal length. In this example, rounding the lens causes light rays to converge on the retina instead of behind it, and the object comes into focus.

The process by which the eye adjusts the shape of the lens to keep objects in focus is known as **accommodation,** and the closest distance at which it can focus an object is known as the **near point of accommodation.** You can demonstrate changing focus with the *accommodation reflex* easily by closing one eye and holding your hand up about 8 inches in front of your open eye, fingers spread apart.

Focus your eye on some object in the distance that is visible between your fingers. Notice that when you do so, your fingers remain visible but out of focus. Your lens is flattened for distance vision, so the focal point for near objects falls behind the retina. Those objects appear out of focus. Now shift your gaze to your fingers and notice that they come into focus. The light rays reflecting off your fingers have not changed their angle, but your lens has become more rounded, and the light rays now converge on the retina.

How can the lens, which is clear and does not have any muscle fibers in it, change shape? The answer lies in the **ciliary muscle,** a ring of smooth muscle that surrounds the lens and is attached to it by the inelastic ligaments called zonules (Fig. 10.27f). If no tension is placed on the lens by the ligaments, the lens assumes its natural rounded shape because of the elasticity of its capsule. If the ligaments pull on the lens, it flattens out and assumes the shape required for distance vision.

Tension on the ligaments is controlled by the ciliary muscle. When the ciliary muscle is relaxed, the ring is more open and the lens is pulled into a flatter shape (Fig. 10.27g). When this circular muscle contracts, the muscle ring gets smaller, releasing tension on the ligaments so that the lens rounds (Fig. 10.27h).

FIG. 10.27 **ESSENTIALS**

Optics of the Eye

Light passing through a curved surface will bend or refract.

(a) A **concave lens** scatters light rays.

Concave lens

Parallel
light rays

(b) A **convex lens** causes light rays to converge.

Convex lens   Focal point

Parallel
light rays

|← Focal length →|

The **focal length** of the lens is the distance
from the center of the lens to the **focal point.**

For clear vision, the focal point must fall on the retina.

(c) Parallel light rays pass through
a flattened lens, and the focal
point falls on the retina.

Focal length

Light
from
distant
source

Light from

distant source

Lens flattened
for distant vision

Focal length

(d) For close objects, the light rays are no longer parallel.
The lens and its focal length have not changed, but
the object is seen out of focus because the light
beam is not focused on the retina.

Image distance

Lens

Object

Object
image

Object
distance (P)

Image distance (Q)

Focal length of lens (F)

(e) To keep an object in focus as it
moves closer, the lens becomes
more rounded.

Focal length

Lens rounded
for close vision

Image distance
now equals focal length

Changes in lens shape are controlled by the ciliary muscle.

(f) The lens is attached to the ciliary
muscle by inelastic ligaments (zonules).

Ciliary
muscle

Lens

Ligaments

Cornea

Iris

(g) When ciliary muscle is relaxed, the
ligaments pull on and flatten the lens.

Ciliary muscle
relaxed

Lens flattened

Ligaments
pulled tight

Cornea

(h) When ciliary muscle contracts, it
releases tension on the ligaments
and the lens becomes more rounded.

Ciliary muscle
contracted

Lens rounded

Ligaments
slacken

Common vision defects can be corrected with external lenses.

(i) Hyperopia, or far-sightedness, occurs when the focal point falls behind the retina.

Hyperopia (corrected with a convex lens)

(j) Myopia, or near-sightedness, occurs when the focal point falls in front of the retina.

Myopia (corrected with a concave lens)

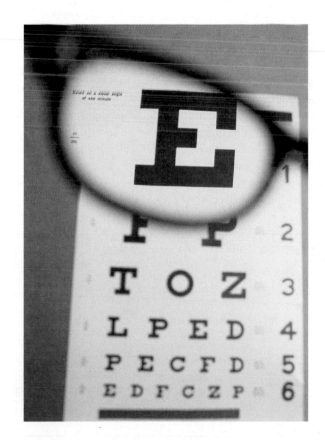

Young people can focus on items as close as 8 cm, but the accommodation reflex diminishes from the age of 10 on. By age 40, accommodation is only about half of what it was at age 10. By age 60, many people lose the reflex completely because the lens has lost flexibility and remains in its flatter shape for distance vision. The loss of accommodation, **presbyopia,** is the reason most people begin to wear reading glasses in their 40s.

Two other common vision problems are near-sightedness and far-sightedness. Near-sightedness, or **myopia,** occurs when the focal point falls in front of the retina (Fig. 10.27j). Far-sightedness, or **hyperopia,** occurs when the focal point falls behind the retina (Fig. 10.27i). These vision problems are caused by abnormally curved or flattened corneas or by eyeballs that are too long or too short. Placing a lens with the appropriate curvature in front of the eye changes the refraction of light entering the eye and corrects the problem. A third common vision problem, **astigmatism,** is usually caused by a cornea that is not a perfectly shaped dome, resulting in distorted images.

CONCEPT CHECK

27. If a person's cornea, which helps focus light, is more rounded than normal (has a greater curvature), is this person more likely to be hyperopic or myopic? (*Hint:* See Fig. 10.27.)

28. The relationship between the focal length of a lens (F), the distance between an object and the lens (P), and the distance from the lens to the object's image (Q) is expressed as $1/F = 1/P + 1/Q$.

  (a) If the focal length of a lens does not change but an object moves closer to the lens, what happens to the image distance Q?

  (b) If an object moves closer to the lens and the image distance Q must stay the same for the image to fall on the retina, what must happen to the focal length F of the lens? For this change in F to occur, should the lens become flatter or more rounded?

29. (a) Explain how convex and concave corrective lenses change the refraction of light.

  (b) Which type of corrective lens should be used for myopia, and why? For hyperopia?

Phototransduction Occurs at the Retina

In the second step of the visual pathway, photoreceptors of the retina convert light energy into electrical signals. Light energy is part of the electromagnetic spectrum, which ranges from high-energy, very-short-wavelength waves such as X-rays and gamma rays to low-energy, lower-frequency microwaves and radio waves (**FIG. 10.28**). However, our brains can perceive only a small portion of this broad energy spectrum. For humans, **visible light** is limited to electromagnetic energy with waves that have a frequency of $4.0–7.5 \times 10^{14}$ cycles per second (hertz, Hz) and a wavelength of 400–750 nanometers (nm). Electromagnetic energy is measured in units called *photons.*

Our unaided eyes see visible light but do not respond to ultraviolet and infrared light, whose wavelengths border the ends of

FIG. 10.28 The electromagnetic spectrum

our visible light spectrum. On the other hand, the eyes of some other animals can see these wavelengths. For example, bees use ultraviolet "runways" on flowers to guide them to pollen and nectar.

Phototransduction is the process by which animals convert light energy into electrical signals. In humans, phototransduction takes place when light hits the retina, the sensory organ of the eye (**FIG. 10.29**). The retina develops from the same embryonic tissue as the brain, and (as in the cortex of the brain) neurons in the retina are organized into layers. There are five types of neurons in the retinal layers: photoreceptors, bipolar cells, ganglion cells, amacrine cells, and horizontal cells (Fig. 10.29f).

Backing the photosensitive portion of the human retina is a dark **pigment epithelium** layer. Its function is to absorb any light rays that escape the photoreceptors, preventing distracting light from reflecting inside the eye and distorting the visual image. The black color of these epithelial cells comes from granules of the pigment *melanin*.

Photoreceptors are the neurons that convert light energy into electrical signals. There are two main types of photoreceptors, rods and cones, as well as a recently discovered photoreceptor that is a modified ganglion cell (see Emerging Concepts Box: Melanopsin). You might expect photoreceptors to be on the surface of the retina facing the vitreous chamber, where light will strike them first, but the retinal layers are actually in reverse order. The photoreceptors are the bottom layer, with their photosensitive tips against the pigment epithelium. Most light entering the eye must pass through several relatively transparent layers of neurons before striking the photoreceptors.

One exception to this organizational pattern occurs in a small region of the retina known as the **fovea** {pit}. This area is free of neurons and blood vessels that would block light reception, so photoreceptors receive light directly, with minimal scattering

EMERGING CONCEPTS

Melanopsin

Circadian rhythms in mammals are cued by light entering the eyes. For many years, scientists believed that rods and cones of the retina were the primary photoreceptors linked to the *suprachiasmatic nucleus* (SCN), the brain center for circadian rhythms. However, in 1999, researchers found that transgenic mice lacking both rods and cones still had the ability to respond to changing light cues, suggesting that an additional photoreceptor must exist in the retina. Now scientists believe they have found it: a subset of retinal ganglion cells that contain an opsin-like pigment called *melanopsin* (mRGCs). Axons from these mRGC ganglion cells project to the SCN as well as to brain areas that control the pupillary reflex. It appears that these newly identified photoreceptors join rods and cones as the light-sensing cells of the mammalian retina, and scientists may have to revise the traditional models of visual processing. To learn more, see C. Sedwick, Melanopsin ganglion cells: A different way of seeing things. *PLoS Biol* 8(12): e1001003, 2010 (*www.plosbiology.org*).

(Fig. 10.29d). As noted earlier, the fovea and the **macula** immediately surrounding it are the areas of most acute vision, and they form the center of the visual field.

When you look at an object, the lens focuses the object image on the fovea. For example, in Figure 10.29b, the eye is focused on the green-yellow border of the color bar. Light from that section of the visual field falls on the fovea and is in sharp focus. Notice also that the image falling on the retina is upside down. Subsequent visual processing by the brain reverses the image again so that we perceive it in the correct orientation.

Sensory information about light passes from the photoreceptors to **bipolar neurons,** then to a layer of **ganglion cells** (Fig. 10.29e). The axons of ganglion cells form the optic nerve, which leaves the eye at the optic disk. Because the optic disk has no photoreceptors, images projected onto this region cannot be seen, creating what is called the eye's **blind spot.**

CONCEPT CHECK

30. Some vertebrate animals that see well in very low light lack a pigment epithelium and instead have a layer called the *tapetum lucidum* behind the retina. What property might this layer have that would enhance vision in low light?

31. How is the difference in visual acuity between the fovea and the edge of the visual field similar to the difference in touch discrimination between the fingertips and the skin of the arm?

32. Macular degeneration is the leading cause of blindness in Americans over the age of 55. Impaired function of the macula causes vision loss in which part of the visual field?

FIG. 10.29 **ANATOMY SUMMARY**

The Retina

(a) Dorsal view of a section of the right eye

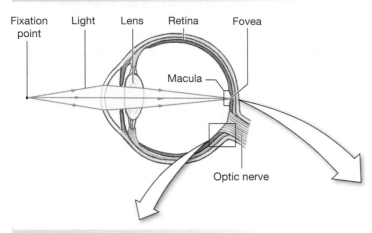

Fixation point
Light
Lens
Retina
Fovea
Macula
Optic nerve

(b) The projected image is upside down on the retina. Visual processing in the brain reverses the image.

Fovea

(c) Axons from the retina exit via the optic nerve.

Optic nerve

Sclera

The choroid layer contains blood vessels.

Pigment epithelium

Neural cells of retina

(d) Light strikes the photoreceptors in the fovea directly because overlying neurons are pushed aside.

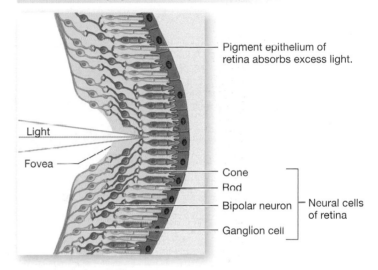

Pigment epithelium of retina absorbs excess light.

Light

Fovea

Cone
Rod
Bipolar neuron — Neural cells of retina
Ganglion cell

(e) Convergence in the retina

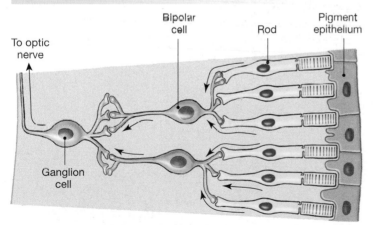

Bipolar cell
Rod
Pigment epithelium

To optic nerve

Ganglion cell

(f) Retinal photoreceptors are organized into layers.

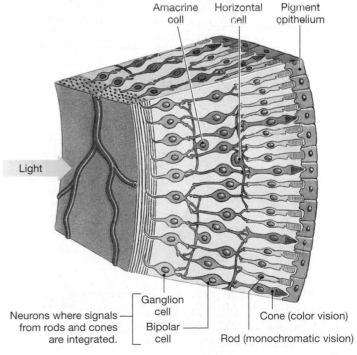

Amacrine cell
Horizontal cell
Pigment epithelium

Light

Neurons where signals from rods and cones are integrated.

Ganglion cell
Bipolar cell

Cone (color vision)
Rod (monochromatic vision)

Q **FIGURE QUESTION**

How many rods converge on the ganglion cell in (e)?

RUNNING **PROBLEM**

Anant's condition does not improve with the low-salt diet and diuretics, and he continues to suffer from disabling attacks of vertigo with vomiting. In severe cases of Ménière's disease, surgery is sometimes performed when less invasive treatments have failed. In one surgical procedure for the disease, a drain is inserted to relieve pressure in the endolymph by removing some of the fluid. If that fails to provide relief, as a last resort the vestibular nerve can be severed. This surgery is difficult to perform, as the vestibular nerve lies near many other important nerves, including facial nerves and the auditory nerve. Patients who undergo this procedure are advised that the surgery can result in deafness if the cochlear nerve is inadvertently severed.

Q6: *Why would severing the vestibular nerve alleviate Ménière's disease?*

310 314 331 339 343 **348** 353

Photoreceptors Transduce Light into Electrical Signals

There are two main types of photoreceptors in the eye: rods and cones. **Rods** function well in low light and are used in night vision, when objects are seen in black and white rather than in color. They outnumber cones by a 20:1 ratio, except in the fovea, which contains only cones.

Cones are responsible for *high-acuity* vision and color vision during the daytime, when light levels are higher. *Acuity* means keenness and is derived from the Latin *acuere*, meaning "to sharpen." The fovea, which is the region of sharpest vision, has a very high density of cones.

The two types of photoreceptors have the same basic structure (**FIG. 10.30**): (1) an outer segment whose tip touches the pigment epithelium of the retina, (2) an inner segment that contains the cell nucleus and organelles for ATP and protein synthesis, and (3) a basal segment with a synaptic terminal that releases glutamate onto bipolar cells.

FIG. 10.30   Photoreceptors: Rods and cones

PIGMENT EPITHELIUM

The dark pigment epithelium absorbs extra light and prevents that light from reflecting back and distorting vision.

Old disks at tip are phagocytized by pigment epithelial cells.

Melanin granules

OUTER SEGMENT
Light transduction takes place in the outer segment of the photoreceptor using visual pigments in membrane disks.

Disks

Connecting stalks

Disks

INNER SEGMENT
Location of major organelles and metabolic operations, such as photopigment synthesis and ATP production

Mitochondria

Cone   Rods

Rhodopsin molecule

Retinal

Opsin

SYNAPTIC TERMINAL
Synapses with bipolar cells.

Bipolar cell

LIGHT

In the outer segment, the cell membrane has deep folds that form disk-like layers. Toward the tip of the outer segments in rods, these layers actually separate from the cell membrane and form free-floating membrane disks. In the cones, the disks stay attached.

Light-sensitive **visual pigments** are bound to the disk membranes in outer segments of photoreceptors. These visual pigments are transducers that convert light energy into a change in membrane potential. Rods have one type of visual pigment, **rhodopsin.** Cones have three different pigments that are closely related to rhodopsin.

The visual pigments of cones are excited by different wavelengths of light, allowing us to see in color. White light is a combination of colors, as demonstrated when you separate white light by passing it through a prism. The eye contains cones for red, green, and blue light. Each cone type is stimulated by a range of light wavelengths but is most sensitive to a particular wavelength (**FIG. 10.31**). Red, green, and blue are the three primary colors that make the colors of visible light, just as red, blue, and yellow are the three primary colors that make different colors of paint.

The color of any object we are looking at depends on the wavelengths of light reflected by the object. Green leaves reflect green light, and bananas reflect yellow light. White objects reflect most wavelengths. Black objects absorb most wavelengths, which is one reason they heat up in sunlight while white objects stay cool.

Our brain recognizes the color of an object by interpreting the combination of signals coming to it from the three different color cones. The details of color vision are still not fully understood, and there is some controversy about how color is processed in the cerebral cortex. **Color-blindness** is a condition in which a person inherits a defect in one or more of the three types of cones and has difficulty distinguishing certain colors. Probably the best-known form of color-blindness is red-green, in which people have trouble telling red and green apart.

CONCEPT CHECK 33. Why is our vision in the dark in black and white rather than in color?

Phototransduction The process of phototransduction is similar for rhodopsin (in rods) and the three color pigments (in cones). Rhodopsin is composed of two molecules: **opsin,** a protein embedded in the membrane of the rod disks, and **retinal,** a vitamin A derivative that is the light-absorbing portion of the pigment (see Fig. 10.30). In the absence of light, retinal binds snugly into a binding site on the opsin (**FIG. 10.32**). When activated by as little as one photon of light, retinal changes shape to a new configuration. The activated retinal no longer binds to opsin and is released from the pigment in the process known as **bleaching.**

How does rhodopsin bleaching lead to action potentials traveling through the optical pathway? To understand the pathway, we must look at other properties of the rods. Electrical signals in cells occur as a result of ion movement between the intracellular and extracellular compartments. Rods contain three main types of cation channels: **cyclic nucleotide-gated (CNG) channels** that allow Na^+ and Ca^{2+} to enter the rod, K^+ channels that allow K^+ to leak out of the rod, and voltage-gated Ca^{2+} channels in the synaptic terminal that help regulate exocytosis of neurotransmitter.

When a rod is in darkness and rhodopsin is not active, cyclic GMP (cGMP) levels in the rod are high, and both CNG and K^+ channels are open (Fig. 10.32 **1**). Sodium and Ca^{2+} ion influx is greater than K^+ efflux, so the rod stays depolarized to an average membrane potential of –40 mV (instead of the more usual –70 mV). At this slightly depolarized membrane potential, the voltage-gated Ca^{2+} channels are open and there is tonic (continuous) release of the neurotransmitter glutamate from the synaptic portion of the rod onto the adjacent bipolar cell.

When light activates rhodopsin, a second-messenger cascade is initiated through the G protein **transducin** (Fig. 10.32 **2**). (Transducin is closely related to gustducin, the G protein found in type II taste receptor cells.) The transducin second-messenger cascade decreases the concentration of cGMP, which closes the CNG channels. As a result, cation influx slows or stops.

With decreased cation influx and continued K^+ efflux, the inside of the rod hyperpolarizes, and glutamate release onto the

FIG. 10.31 Light absorption by visual pigments

There are three types of cone pigment, each with a characteristic light absorption spectrum. Rods are for black and white vision in low light.

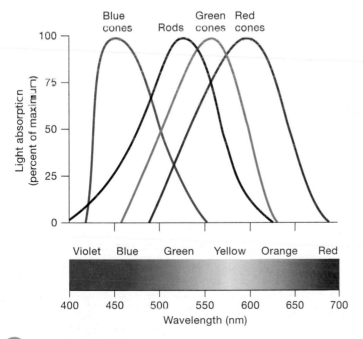

GRAPH QUESTIONS
1. Which pigment absorbs light over the broadest spectrum of wavelengths?
2. Over the narrowest?
3. Which cone pigment absorbs the most light at 500 nm?

FIG. 10.32   Phototransduction in rods

Rods contain the visual pigment rhodopsin. When activated by light, rhodopsin separates into opsin and retinal.

1 In darkness, rhodopsin is inactive, cGMP is high, and CNG and K^+ channels are open.

Pigment epithelium cell

Disk

Transducin (G protein)

Inactive rhodopsin (opsin and retinal)

cGMP levels high

CNG channel open

Ca^{2+} Na^+

K^+

Membrane potential in dark = −40mV

Rod

Tonic release of neurotransmitter onto bipolar neurons

2 Light bleaches rhodopsin. Opsin decreases cGMP, closes CNG channels, and hyperpolarizes the cell.

Activated retinal   Opsin (bleached pigment)   Activates transducin

Cascade

Decreased cGMP

CNG channel closes

Ca^{2+} Na^+

K^+

Membrane hyperpolarizes to −70 mV

Light

Neurotransmitter release decreases in proportion to amount of light.

3 In the recovery phase, retinal recombines with opsin.

Retinal converted to inactive form

Retinal recombines with opsin to form rhodopsin.

Q FIGURE QUESTION

One rod contains about 10,000 CNG channels open in the dark. One photon of light activates one rhodopsin. Each rhodopsin activates 800 transducin. Each transducin cascade removes 6 cGMP. A decrease of 24 cGMP closes one CNG channel. How many photons are needed to close all the CNG channels in one rod?

---

bipolar neurons decreases. Bright light closes all CNG channels and stops all neurotransmitter release. Dimmer light causes a response that is graded in proportion to the light intensity.

After activation, retinal diffuses out of the rod and is transported into the pigment epithelium. There it reverts to its inactive form before moving back into the rod and being reunited with opsin (Fig. 10.32 **3**). The recovery of rhodopsin from bleaching can take some time and is a major factor in the slow adaptation of the eyes when moving from bright light into the dark.

CONCEPT CHECK **34.** Draw a map or diagram to explain phototransduction. Start with bleaching and end with release of neurotransmitter.

Signal Processing Begins in the Retina

We now move from the cellular mechanism of light transduction to the processing of light signals by the retina and brain, the third and final step in our vision pathway. Signal processing in the retina is an excellent example of convergence [p. 260], in which multiple neurons synapse onto a single postsynaptic cell (**FIG. 10.33a**). Depending on location in the retina, as many as 15 to 45 photoreceptors may converge on one bipolar neuron.

Multiple bipolar neurons in turn innervate a single ganglion cell, so that the information from hundreds of millions of retinal photoreceptors is condensed down to a mere 1 million axons leaving the eye in each optic nerve. Convergence is minimal in the fovea, where some photoreceptors have a 1:1 relationship with their bipolar neurons, and greatest at the outer edges of the retina.

FIG. 10.33 Visual fields

(a) Multiple photoreceptors converge on one ganglion cell.

Horizontal and amacrine cells influence communication at the rod-bipolar or bipolar-ganglion synapses.

Rod   Horizontal cell   Amacrine cell

Pigment epithelium

Ganglion cell

Bipolar cell

To optic nerve

(b) A group of adjacent photoreceptors form the visual field for one ganglion cell. This illustration shows an on-center, off-surround field.

Visual fields have centers (yellow) and outer surrounds (gray).

Bipolar cells are either activated or inhibited by light, depending on their type.

Ganglion cells respond most strongly when there is good contrast of light intensity between the center and the surround.

(c) The retina uses contrast rather than absolute light intensity for better detection of weak stimuli.

Visual Field Type	Field Is On-Center/Off-Surround	Field Is Off-Center/On-Surround
On-center, off-surround Bright light onto center	Ganglion cell is excited by light in the center of the visual field.	Ganglion cell is inhibited by light in the center of the visual field.
Off-center, on-surround Bright light onto surround Bright light onto surround	Ganglion cell is inhibited by light on the surround of the visual field.	Ganglion cell is excited by light on the surround of the visual field.
Both field types Diffuse light on both center and surround	Ganglion cell responds weakly.	Ganglion cell responds weakly.

Signal processing in the retina is modulated by input from two additional sets of cells that we will not discuss (Fig. 10.29f). **Horizontal cells** synapse with photoreceptors and bipolar cells. **Amacrine cells** modulate information flowing between bipolar cells and ganglion cells.

Bipolar Cells Glutamate release from photoreceptors onto bipolar neurons begins signal processing. There are two types of bipolar cells, *light-on* (ON bipolar cells) and *light-off* (OFF bipolar cells). ON bipolar cells are activated in the light when glutamate secretion by photoreceptors decreases. In the dark, ON bipolar cells are inhibited by glutamate release. OFF bipolar cells are excited by glutamate release in the dark. In the light, with less glutamate, OFF bipolar cells are inhibited. By using different glutamate receptors, one stimulus (light) creates two different responses with a single neurotransmitter.

Whether glutamate is excitatory or inhibitory depends on the type of glutamate receptor on the bipolar neuron. ON bipolar cells have a metabotropic glutamate receptor called *mGluR6* that hyperpolarizes the cell when the receptor binds glutamate in the dark. When mGluR6 is not activated, the ON bipolar cell depolarizes. OFF bipolar cells have an ionotropic glutamate receptor that opens ion channels and depolarizes the OFF bipolar cell in the dark. Bipolar cell signal processing is also modified by input from the horizontal and amacrine cells.

Ganglion Cells Bipolar cells synapse with ganglion cells, the next neurons in the pathway. We know more about ganglion cells because they lie on the surface of the retina, where their axons are the most accessible to researchers. Extensive studies have been done in which researchers stimulated the retina with carefully placed light and evaluated the response of the ganglion cells.

Each ganglion cell receives information from a particular area of the retina. These areas, known as **visual fields**, are similar to receptive fields in the somatic sensory system [p. 293]. The visual field of a ganglion cell near the fovea is quite small. Only a few photoreceptors are associated with each ganglion cell, and so visual acuity is greatest in these areas. At the edge of the retina, multiple photoreceptors converging onto a single ganglion cell results in vision that is not as sharp (Fig. 10.33a).

An analogy for this arrangement is to think of pixels on your computer screen. Assume that two screens have the same number of "photoreceptors," as indicated by a maximal screen resolution of 1280 × 1024 pixels. If screen A has one photoreceptor becoming one "ganglion cell" pixel, the actual screen resolution is 1280 × 1024, and the image is very clear. If eight photoreceptors on screen B converge into one ganglion cell pixel, then the actual screen resolution falls to 160 × 128, resulting in a very blurry and perhaps indistinguishable image.

Visual fields of ganglion cells are roughly circular (unlike the irregular shape of somatic sensory receptive fields) and are divided into sections: a round center and its doughnut-shaped **surround** (Fig. 10.33b). This organization allows each ganglion cell to use contrast between the center and its surround to interpret visual information. Strong contrast between the center and surround elicits a strong excitatory response (a series of action potentials) or a strong inhibitory response (no action potentials) from the ganglion cell. Weak contrast between center and surround gets an intermediate response.

There are two types of ganglion cell visual fields. In an *on-center/off-surround field*, the associated ganglion cell responds most strongly when light is brightest in the center of the field (Fig. 10.33c). If light is brightest in the off-surround region of the field, the on-center/off-surround field ganglion cell is inhibited and stops firing action potentials. The reverse happens with *off-center/on-surround fields*.

FIG. 10.34 Binocular vision

The left visual field of each eye is projected to the visual cortex on the right side of the brain, and the right visual field is projected to the left visual cortex. Objects seen by both eyes fall within the binocular zone and are perceived in three dimensions. Objects seen with only one eye fall outside the binocular zone and are perceived in only two dimensions.

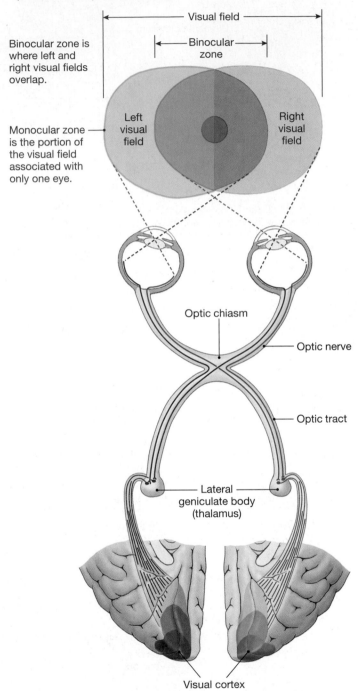

Binocular zone is where left and right visual fields overlap.

Monocular zone is the portion of the visual field associated with only one eye.

Visual field

Binocular zone

Left visual field

Right visual field

Optic chiasm

Optic nerve

Optic tract

Lateral geniculate body (thalamus)

Visual cortex

What happens if light is uniform across a visual field? In that case, the ganglion cell responds weakly. Thus, the retina uses *contrast* rather than absolute light intensity to recognize objects in the environment. One advantage of using contrast is that it allows better detection of weak stimuli.

Scientists have now identified multiple types of ganglion cells in the primate retina. The two predominant types, which account for 80% of retinal ganglion cells, are M cells and P cells. Large *magnocellular* ganglion cells, or **M cells,** are more sensitive to information about movement. Smaller *parvocellular* ganglion cells, or **P cells,** are more sensitive to signals that pertain to form and fine detail, such as the texture of objects in the visual field. A recently discovered subtype of ganglion cell, the *melanopsin retinal ganglion cell,* apparently also acts as a photoreceptor to relay information about light cycles to the suprachiasmatic nucleus, which controls circadian rhythms [p. 17].

Processing Beyond the Retina Once action potentials leave ganglion cell bodies, they travel along the optic nerves to the CNS for further processing. As noted earlier, the optic nerves enter the brain at the optic chiasm. At this point, some nerve fibers from each eye cross to the other side of the brain for processing. **FIGURE 10.34** shows how information from the right side of each eye's visual field is processed on the left side of the brain, and information from the left side of the field is processed on the right side of the brain.

The central portion of the visual field, where left and right sides of each eye's visual field overlap, is the **binocular zone.** The two eyes have slightly different views of objects in this region, and the brain processes and integrates the two views to create three-dimensional representations of the objects. Our sense of depth perception—that is, whether one object is in front of or behind another—depends on binocular vision. Objects that fall within the visual field of only one eye are in the **monocular zone** and are viewed in two dimensions.

Once axons leave the optic chiasm, some fibers project to the midbrain, where they participate in control of eye movement or coordinate with somatosensory and auditory information for balance and movement (see Fig. 10.26). Most axons, however, project to the lateral geniculate body of the thalamus, where the optic fibers synapse onto neurons leading to the visual cortex in the occipital lobe.

The lateral geniculate body is organized in layers that correspond to the different parts of the visual field, which means that information from adjacent objects is processed together. This **topographical organization** is maintained in the visual cortex, with the six layers of neurons grouped into vertical columns. Within each portion of the visual field, information is further sorted by form, color, and movement.

The cortex merges monocular information from the two eyes to give us a binocular view of our surroundings. Information from on/off combinations of ganglion cells is translated into sensitivity to line orientation in the simplest pathways, or into color, movement, and detailed structure in the most complex. Each of these attributes of visual stimuli is processed through a separate pathway, creating a network whose complexity we are just beginning to unravel.

RUNNING **PROBLEM CONCLUSION** | Ménière's Disease

Anant was told about the surgical options but elected to continue medical treatment for a little longer. Over the next two months, his Ménière's disease gradually resolved. The cause of Ménière's disease is still unknown, which makes treatment difficult. To learn more about treatments that are available to alleviate Ménière's disease, do an Internet search. Now check your understanding of this running problem by comparing your answers to those in the summary table.

Question	Facts	Integration and Analysis
Q1: In which part of the brain is sensory information about equilibrium processed?	The major equilibrium pathways project to the cerebellum. Some information is also processed in the cerebrum.	N/A
Q2: Subjective tinnitus occurs when an abnormality occurs somewhere along the anatomical pathway for hearing. Starting from the ear canal, name the auditory structures in which problems may arise.	The middle ear consists of malleus, incus, and stapes, bones that vibrate with sound. The hearing portion of the inner ear consists of hair cells in the fluid-filled cochlea. The cochlear (auditory) nerve leads to the brain.	Subjective tinnitus could arise from a problem with any of the structures named. Abnormal bone growth can affect the middle ear bones. Excessive fluid accumulation in the inner ear will affect the hair cells. Neural defects may cause the cochlear nerve to fire spontaneously, creating the perception of sound.
Q3: When a person with positional vertigo changes position, the displaced crystals float toward the semicircular canals. Why would this cause dizziness?	The ends of the semicircular canals contain sensory cristae, each crista consisting of a cupula with embedded hair cells. Displacement of the cupula creates a sensation of rotational movement.	If the floating crystals displace the cupula, the brain will perceive movement that is not matched to sensory information coming from the eyes. The result is vertigo, an illusion of movement.

—Continued next page

RUNNING **PROBLEM CONCLUSION** | *Continued*

Question	Facts	Integration and Analysis
Q4: *Compare the symptoms of positional vertigo and Ménière's disease. On the basis of Anant's symptoms, which condition do you think he has?*	The primary symptom of positional vertigo is brief dizziness following a change in position. Ménière's disease combines vertigo with tinnitus and hearing loss.	Anant complains of dizzy attacks typically lasting up to an hour that come on without warning, making it more likely that Anant has Ménière's disease.
Q5: *Why is limiting salt (NaCl) intake suggested as a treatment for Ménière's disease?*	Ménière's disease is characterized by too much endolymph in the inner ear. Endolymph is an extracellular fluid.	Reducing salt intake should also reduce the amount of fluid in the extracellular compartment because the body will retain less water. Reduction of ECF volume may decrease fluid accumulation in the inner ear.
Q6: *Why would severing the vestibular nerve alleviate Ménière's disease?*	The vestibular nerve transmits information about balance and rotational movement from the vestibular apparatus to the brain.	Severing the vestibular nerve prevents false information about body rotation from reaching the brain, thus alleviating the vertigo of Ménière's disease.

310 314 331 339 343 348 **353**

VISUALIZE, PRACTICE, AND MASTER PHYSIOLOGY WITH MASTERING A&P

• Practice-on-the-go with Dynamic Study Modules
• Visualize important processes with Interactive Physiology and A&P Flix
• Check your understanding with Video Tutors and practice quizzes!

MasteringA&P®

CHAPTER SUMMARY

We all live in the same world, but different animals perceive the world differently. Dogs hear sounds we can't, for instance, and nocturnal animals have better night vision than we do. An animal can perceive only those stimuli for which it has sensory receptors. In this chapter, you explored sensory receptors in the human body and learned how each type is designed to enable us to perceive different aspects of the world around us.

Despite the unique characteristics of each sense, basic patterns emerge for sensory transduction and perception. *Molecular interactions* between signal molecules and ion channels or G protein-coupled receptors initiate many sensory pathways. Neural and non-neural sensory receptors convert chemical, mechanical, thermal, and light *energy* into electrical signals that pass along sensory neurons to CNS *control centers*. The brain processes and filters incoming signals, sometimes acting on sensory *information* without that information ever reaching conscious awareness. Many of the visceral reflexes you will study are unconscious responses to sensory input.

General Properties of Sensory Systems

1. Sensory stimuli are divided into the **special senses** of vision, hearing, taste, smell, and equilibrium, and the **somatic senses** of touch, temperature, pain, itch, and proprioception. (p. 310)

2. Sensory pathways begin with a stimulus that is converted by a receptor into an electrical potential. (p. 310)

3. If the stimulus is above threshold, action potentials pass along a sensory neuron to the central nervous system. We become aware of some stimuli but are never conscious of others. (p. 310; Tbl. 10.1)

4. Sensory receptors vary from free nerve endings to encapsulated nerve endings to specialized receptor cells. (p. 311; Fig. 10.1)

5. There are four types of sensory receptors, based on the stimulus to which they are most sensitive: **chemoreceptors, mechanoreceptors, thermoreceptors,** and **photoreceptors.** (p. 312; Tbl. 10.2)

6. Each receptor type has an **adequate stimulus,** a particular form of energy to which it is most responsive. (p. 312)

7. A stimulus that is above **threshold** creates a graded potential in the receptor. (p. 312)

8. Multiple sensory neurons may converge on one secondary neuron and create a single large **receptive field.** (p. 312; Fig. 10.2)

9. Sensory information from the spinal cord projects to the thalamus, then on to the sensory areas of the cerebral cortex. Olfactory information does not pass through the thalamus. (p. 313; Fig. 10.3)

10. The central nervous system is able to modify our level of awareness of sensory input. The **perceptual threshold** is the level of stimulus intensity necessary for us to be aware of a particular sensation. (p. 313)

11. The **modality** of a signal and its location are indicated by which sensory neurons are activated. The association of a receptor with a specific sensation is called **labeled line coding.** (p. 314)

12. Localization of auditory information depends on the timing of receptor activation in each ear. (p. 315; Fig. 10.4)

13. **Lateral inhibition** enhances the contrast between the center of the receptive field and the edges of the field. In **population coding,** the brain uses input from multiple receptors to calculate location and timing of a stimulus. (p. 315; Fig. 10.5)

14. Stimulus intensity is coded by the number of receptors activated and by the frequency of their action potentials. (p. 315; Fig. 10.6)

15. For **tonic receptors,** the sensory neuron fires action potentials as long as the **receptor potential** is above threshold. **Phasic receptors** respond to a change in stimulus intensity but adapt if the strength of the stimulus remains constant. (p. 316; Fig. 10.7)

Somatic Senses

16. There are four somatosensory modalities: touch, proprioception, temperature, and nociception. (p. 317)

17. **Secondary sensory neurons** cross the midline so that one side of the brain processes information from the opposite side of the body. Ascending sensory tracts terminate in the **somatosensory cortex.** (p. 318; Fig. 10.8)

18. Touch receptors come in many varieties. Temperature receptors sense heat and cold. (p. 319; Fig. 10.10)

19. **Nociceptors** are free nerve endings that respond to chemical, mechanical, or thermal stimuli. Their activation is perceived as pain and itch. (p. 320)

20. Some responses to irritants, such as the withdrawal reflex, are protective **spinal reflexes.** (p. 322)

21. **Referred pain** from internal organs occurs when multiple primary sensory neurons converge onto a single ascending tract. (p. 322; Fig. 10.11)

22. **Fast pain** is transmitted rapidly by small, myelinated fibers. **Slow pain** is carried by small, unmyelinated fibers. Pain may be modulated either by descending pathways from the brain or by **gating** mechanisms in the spinal cord. (p. 320; Fig. 10.12, Tbl. 10.3)

Chemoreception: Smell and Taste

23. Chemoreception is divided into the special senses of smell (**olfaction**) and taste (**gustation**). (pp. 324, 325)

24. **Olfactory sensory neurons** in the nasal cavity are bipolar neurons whose pathways project directly to the olfactory cortex. (p. 325; Fig. 10.13)

25. **Olfactory receptors** are G protein-coupled membrane proteins. (p. 325)

26. Taste is a combination of five sensations: sweet, sour, salty, bitter, and **umami.** (p. 327)

27. **Taste receptor cells** are nonneural cells with membrane channels or receptors that interact with taste ligands. This interaction creates an intracellular Ca^{2+} signal that ultimately activates the primary sensory neuron. (p. 327; Fig. 10.14)

The Ear: Hearing

28. **Hearing** is our perception of the energy carried by sound waves. Sound transduction turns air waves into mechanical vibrations, then fluid waves, chemical signals, and finally action potentials. (p. 329; Fig. 10.17)

29. The **cochlea** of the inner ear contains three parallel, fluid-filled ducts. The **cochlear duct** contains the **organ of Corti,** which contains **hair cell** receptors. (pp. 329, 331, 332; Fig. 10.18)

30. When sound bends hair cell cilia, the hair cell membrane potential changes and alters release of neurotransmitter onto sensory neurons. (p. 332; Fig. 10.19)

31. The initial processing for pitch, loudness, and duration of sound takes place in the cochlea. Localization of sound is a higher function that requires sensory input from both ears and sophisticated computation by the brain. (p. 335; Figs. 10.20, 10.4)

32. The auditory pathway goes from cochlear nerve to medulla, pons, midbrain, and thalamus before terminating in the auditory cortex. Information from both ears goes to both sides of the brain. (p. 335; Fig. 10.21)

The Ear: Equilibrium

33. **Equilibrium** is mediated through hair cells in the **vestibular apparatus** and **semicircular canals** of the inner ear. Gravity and acceleration provide the force that moves the cilia. (p. 337; Fig. 10.22)

The Eye and Vision

34. **Vision** is the translation of reflected light into a mental image. Photoreceptors of the **retina** transduce light energy into an electrical signal that passes to the visual cortex for processing. (pp. 340, 341)

35. The amount of light entering the eye is altered by changing the size of the pupil. (p. 342)

36. Light waves are focused by the lens, whose shape is adjusted by contracting or relaxing the **ciliary muscle.** (p. 343; Fig. 10.27)

37. Light is converted into electrical energy by the **photoreceptors** of the retina. Signals pass through bipolar neurons to ganglion cells, whose axons form the optic nerve. (p. 346; Fig. 10.29)

38. The **fovea** has the most acute vision because it has the smallest receptive fields. (p. 346)

39. **Rods** are responsible for monochromatic nighttime vision. **Cones** are responsible for high-acuity vision and color vision during the daytime. (p. 348; Fig. 10.30)

40. Light-sensitive **visual pigments** in photoreceptors convert light energy into a change in membrane potential. The visual pigment in rods is **rhodopsin.** Cones have three different visual pigments. (p. 349; Fig. 10.31)

41. Rhodopsin is composed of **opsin** and **retinal.** In the absence of light, retinal binds snugly to opsin. (p. 349; Fig. 10.32)

42. When light bleaches rhodopsin, **retinal** is released and **transducin** begins a second-messenger cascade that hyperpolarizes the rod so that it releases less glutamate onto the bipolar neurons. (p. 349)

43. Signals pass from photoreceptors through bipolar neurons to ganglion cells, with modulation by horizontal and amacrine cells. (p. 350; Fig. 10.33)

44. Ganglion cells called **M cells** convey information about movement. Ganglionic **P cells** transmit signals that pertain to the form and texture of objects in the visual field. (p. 353)

45. Information from one side of the visual field is processed on the opposite side of the brain. Objects must be seen by both eyes to appear three-dimensional. (p. 353; Fig. 10.34)

REVIEW QUESTIONS

In addition to working through these questions and checking your answers on p. A-13, review the Learning Outcomes at the beginning of this chapter.

Level One Reviewing Facts and Terms

1. What is the role of the afferent division of the nervous system?

2. Define proprioception.

3. What are the common elements of all sensory pathways?

4. List and briefly describe the four major types of somatic receptors based on the type of stimulus to which they are most sensitive.

5. The receptors of each primary sensory neuron pick up information from a specific area, known as the _____.

6. Match the brain area with the sensory information processed there:

(a) sounds	1. midbrain
(b) odors	2. cerebrum
(c) visual information	3. medulla
(d) taste	4. cerebellum
(e) equilibrium	5. none of the above

7. The conversion of stimulus energy into a change in membrane potential is called _____. The form of energy to which a receptor responds is called its _____ _____. The minimum stimulus required to activate a receptor is known as the _____.

8. When a sensory receptor membrane depolarizes (or hyperpolarizes in a few cases), the change in membrane potential is called the _____ potential. Is this a graded potential or an all-or-none potential?

9. Explain what is meant by adequate stimulus to a receptor.

10. The organization of sensory regions in the _____ of the brain preserves the topographical organization of receptors on the skin, eye, or other regions. However, there are exceptions to this rule. In which sense(s) does the brain rely on the timing of receptor activation to determine the location of the initial stimulus?

11. What is lateral inhibition?

12. Define tonic receptors and list some examples. Define phasic receptors and give some examples. Which type adapts?

13. Heart pain perceived as coming from the neck and down the left arm is an example of _____ pain.

14. What are the five basic tastes? What is the adaptive significance of each taste sensation?

15. The unit of sound wave measurement is _____, which is a measure of the frequency of sound waves per second. The loudness, or intensity, of a sound is a function of the _____ of the sound waves and is measured in _____. The range of hearing for the average human ear is from _____ to _____ [units], with the most acute hearing in the range of _____ to _____ [units].

16. Which structure of the inner ear codes sound for pitch? Define spatial coding.

17. Loud noises cause action potentials to: (choose all correct answers)

   (a) fire more frequently.
   (b) have higher amplitudes.
   (c) have longer refractory periods.

18. Once sound waves have been transformed into electrical signals in the cochlea, sensory neurons transfer information to the _____, from which collaterals then take the information to the _____ and _____. The main auditory pathway synapses in the _____ and _____ before finally projecting to the _____ in the _____.

19. The parts of the vestibular apparatus that tell our brain about our movements through space are the _____, which sense rotation, and the _____ organs, which respond to linear forces.

20. List the following structures in the sequence in which a beam of light entering the eye will encounter them: (a) aqueous humor, (b) cornea, (c) lens, (d) pupil, (e) retina.

21. The three primary colors of vision are _____, _____, and _____. White light containing these colors stimulates photoreceptors called _____. Lack of the ability to distinguish some colors is called _____.

22. List six types of cells found in the retina, and briefly describe their functions.

Level Two Reviewing Concepts

23. Compare and contrast the following:

   (a) the special senses with the somatic senses
   (b) different types of touch receptors with respect to structure, size, and location
   (c) transmission of sharp localized pain with transmission of dull and diffuse pain (include the particular fiber types involved as well as the presence or absence of myelin in your discussion)
   (d) the forms of hearing loss
   (e) convergence of retinal neurons with convergence of primary somatic sensory neurons

24. Draw three touch receptors having overlapping receptive fields (see Fig. 10.2) and number the fields 1–3. Draw a primary and secondary sensory neuron for each receptor so that they have separate ascending pathways to the cortex. Use the information in your drawing to answer this question: How many different regions of the skin can the brain distinguish using input from these three receptors?

25. Describe the neural pathways that link pain with emotional distress, nausea, and vomiting.

26. Trace the neural pathways involved in olfaction. What is G_{olf}?

27. Compare the current models of signal transduction in taste buds for salty/sour ligands and sweet/bitter/umami ligands.

28. Put the following structures in the order in which a sound wave would encounter them: (a) pinna, (b) cochlear duct, (c) stapes, (d) ion channels, (e) oval window, (f) hair cells/stereocilia, (g) tympanic membrane, (h) incus, (i) vestibular duct, (j) malleus

29. Draw the structures and receptors of the vestibular apparatus for equilibrium. Label the components. Briefly describe how they function to notify the brain of movement.

30. Explain how accommodation by the eye occurs. What is the loss of accommodation called?

31. List four common visual problems, and explain how they occur.

32. Explain how the intensity and duration of a stimulus are coded so that the stimulus can be interpreted by the brain. (Remember, action potentials are all-or-none phenomena.)

33. Make a table of the special senses. In the first row, write these stimuli: sound, standing on the deck of a rocking boat, light, a taste, an aroma. In row 2, describe the location of the receptor for each sense. In row 3, describe the structure or properties of each receptor. In a final row, name the cranial nerve(s) that convey(s) each sensation to the brain. [p. 288]

34. Map the following terms related to vision. Add terms if you wish.

Map 1

• accommodation reflex	• depth of field	• lens
• binocular vision	• field of vision	• macula
• blind spot	• focal point	• optic chiasm
• ciliary muscle	• fovea	• optic disk
• cornea	• iris	• optic nerve
• cranial nerve III	• lateral geniculate	• phototransduction
• pupillary reflex	• visual cortex	• zonules
• retina	• visual field	

Map 2: The Retina

• amacrine cells	• ganglion cells	• pigment epithelium
• bipolar cells	• horizontal cells	• retinal
• bleaching	• melanin	• rhodopsin
• cGMP	• melanopsin	• rods
• cones	• opsin	• transducin

Level Three Problem Solving

35. You are prodding your blindfolded lab partner's arm with two needle probes (with her permission). Sometimes she can tell you are using two probes. But when you probe less sensitive areas, she thinks there is just one probe. Which sense are you testing? Which receptors are being stimulated? Explain why she sometimes feels only one probe.

36. Consuming alcohol depresses the nervous system and vestibular apparatus. In a sobriety check, police officers use this information to determine if an individual is intoxicated. What kinds of tests can you suggest that would show evidence of this inhibition?

37. Often, children are brought to medical attention because of speech difficulties. If you were a clinician, which sense would you test first in such patients, and why?

38. A clinician shines a light into a patient's left eye, and neither pupil constricts. Shining the light into the right eye elicits a normal consensual reflex. What problem in the reflex pathway could explain these observations?

39. An optometrist wishes to examine a patient's retina. Which of the following classes of drugs might dilate the pupil? Explain why you did or did not select each choice.

(a) a sympathomimetic [*mimicus,* imitate]
(b) a muscarinic antagonist
(c) a cholinergic agonist
(d) an anticholinesterase
(e) a nicotinic agonist

40. The iris of the eye has two sets of antagonistic muscles, one for dilation and one for constriction. One set of muscles is radial (radiating from the center of the pupil), and the other set is circular. Draw an iris and pupil, and arrange the muscles so that contraction of one set causes pupillary constriction and contraction of the other set causes dilation.

41. As people age, their ability to see at night decreases. What changes in the retina might explain this?

Level Four Quantitative Problems

42. The relationship between focal length (F) of a lens, object distance (P), and image distance or focal point (Q) is $1/F = 1/P + 1/Q$. Assume the distance from lens to retina is 20 mm.

(a) For a distant object, P = infinity (∞) and $1/\infty = 0$. If Pavi sees a distant object in focus, what is the focal length of her lens in meters?

(b) If the object moves to 1 foot in front of Pavi's lens and the lens does not change shape, what is the image distance (1 in. = 2.54 cm)? What must happen to Pavi's lens for the closer image to come into focus?

11

Efferent Division: Autonomic and Somatic Motor Control

Because a number of cells in the autonomic nervous system act in conjunction, they have relinquished their independence to function as a coherent whole.

Otto Appenzeller and Emilio Oribe, in *The Autonomic Nervous System*, 1997

THE AUTONOMIC DIVISION 359

LO 11.1   Describe the physiological role of the autonomic division and its branches.

LO 11.2   Compare and contrast the anatomy and chemical communication of the sympathetic and parasympathetic branches.

LO 11.3   Describe the synthesis and breakdown of autonomic neurotransmitters.

LO 11.4   Describe the structure and secretions of the adrenal medulla.

THE SOMATIC MOTOR DIVISION 371

LO 11.5   Describe the structure of the neuromuscular junction.

LO 11.6   Compare the anatomy, neurotransmitters and receptors of the somatic motor, sympathetic, and parasympathetic divisions.

BACKGROUND BASICS

170   Membrane receptors
168   Neurotransmitters
171   Second messenger systems
206   Catecholamines
181   Up- and down-regulation
184   Tonic and antagonistic control
228   Organization of the nervous system
229   Neuron structure
232   Synapses
231   Nerves
240   Action potentials
261   Slow synaptic potentials

Muscle fibers with motor neurons terminating at neuromuscular junctions

The picnic lunch was wonderful. You are now dozing on the grass in the warm spring sunlight as you let the meal digest. Suddenly you feel something moving across your lower leg. You open your eyes, and as they adjust to the bright light, you see a four-foot-long snake slithering over your foot. More by instinct than reason, you fling the snake into the grass while scrambling to a safe perch on top of the nearby picnic table. You are breathing heavily, and your heart is pounding.

In less than a second, your body has gone from a state of quiet rest and digestion to a state of panic and frantic activity. How did this happen? The answer is a reflex fight-or-flight reaction, integrated and coordinated through the central nervous system (CNS), then carried out by the efferent division of the peripheral nervous system (PNS). Efferent neurons carry rapid commands from the CNS to the muscles and glands of the body through *nerves,* or bundles of axons. Some nerves, called *mixed nerves,* also carry sensory information through afferent fibers [p. 231].

The efferent division of the peripheral nervous system can be subdivided into **somatic motor neurons,** which control skeletal muscles, and **autonomic neurons,** which control smooth muscle, cardiac muscle, many glands, and some adipose tissue. The somatic and autonomic divisions are sometimes called the voluntary and involuntary divisions of the nervous system, respectively. However, this distinction does not always hold true. Most movement controlled by somatic pathways requires conscious thought, but some skeletal muscle reflexes, such as swallowing and the knee jerk reflex, are involuntary. And autonomic reflexes are mainly involuntary, but a person can use biofeedback training to learn to modulate some autonomic functions, such as heart rate and blood pressure.

We begin our study of the efferent division of the PNS by looking at the autonomic division. Then we consider the somatic motor division, as preparation for learning about muscles [Chapter 12].

RUNNING **PROBLEM** | A Powerful Addiction

Every day, more than 1.3 billion people around the world intentionally absorb a chemical that kills about 5 million people each year. Why would people knowingly poison themselves? If you've guessed that the chemical is nicotine, you already know part of the answer. One of more than 4000 chemicals found in tobacco, nicotine is highly addictive. So powerful is this addiction that fewer than 20% of tobacco users are able to quit smoking the first time they try. Shanika, a smoker for six years, is attempting for the second time to stop smoking. The odds are in her favor this time, however, because she has made an appointment with her physician to discuss all the options available to help her break her addiction to nicotine and smoking.

359   361   363   367   371   373

THE AUTONOMIC DIVISION

The autonomic division of the efferent nervous system (or *autonomic nervous system* for short) is also known in older writings as the *vegetative nervous system,* reflecting the observation that its functions are not under voluntary control. The word *autonomic* comes from the same roots as *autonomous,* meaning *self-governing.* Another name for the autonomic division is *visceral nervous system* because of its control over internal organs.

The autonomic division is subdivided into **sympathetic** and **parasympathetic branches** (often called the *sympathetic* and *parasympathetic nervous systems*). Some parts of the sympathetic branch were first described by the Greek physician Claudius Galen (ca. 130–200 C.E.), who is famous for his compilation of anatomy, physiology, and medicine as they were known during his time. As a result of his dissections, Galen proposed that "animal spirits" flowed from the brain to the tissues through hollow nerves, creating "sympathy" between the different parts of the body. Galen's "sympathy" later gave rise to the name for the sympathetic branch. The prefix *para-,* for the parasympathetic branch, means *beside* or *alongside.*

The sympathetic and parasympathetic branches can be distinguished anatomically, but there is no simple way to separate the actions of the two branches on their targets. They are distinguished best by the type of situation in which they are most active. The picnic scene that began the chapter illustrates the two extremes at which the sympathetic and parasympathetic branches function. If you are resting quietly after a meal, the parasympathetic branch is dominant, taking command of the routine, quiet activities of day-to-day living, such as digestion. Consequently, parasympathetic neurons are sometimes said to control "rest and digest" functions.

In contrast, the sympathetic branch dominates in stressful situations, such as the potential threat from the snake. One of the most dramatic examples of sympathetic action is the **fight-or-flight** response, in which the brain triggers massive simultaneous sympathetic discharge throughout the body. As the body prepares to fight or flee, the heart speeds up; blood vessels to muscles of the arms, legs, and heart dilate; and the liver starts to produce glucose to provide energy for muscle contraction. Digestion becomes a low priority when life and limb are threatened, and so blood is diverted from the gastrointestinal tract to skeletal muscles.

The massive sympathetic discharge that occurs in fight-or-flight situations is mediated through the hypothalamus and is a total-body response to a crisis. If you have ever been scared by the squealing of brakes or a sudden sound in the dark, you know how rapidly the nervous system can influence multiple body systems. Most sympathetic responses are not the all-out response of a fight-or-flight reflex, however, and more importantly, activating one sympathetic pathway does not automatically activate them all.

The role of the sympathetic nervous system in mundane daily activities is as important as a fight-or-flight response. For example, one key function of the sympathetic branch is control

FIG. 11.1 The autonomic division

Homeostasis is a dynamic balance between the autonomic branches.

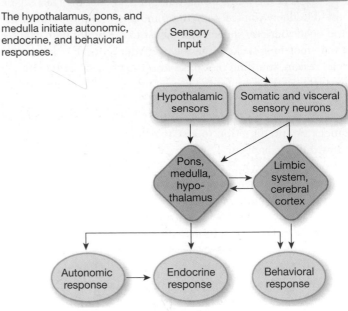

Rest-and-digest: Parasympathetic activity dominates.

Fight-or-flight: Sympathetic activity dominates.

of blood flow to the tissues. Most of the time, autonomic control of body function "seesaws" back and forth between the sympathetic and parasympathetic branches as they cooperate to fine-tune various processes (**FIG. 11.1**). Only occasionally, as in the fight-or-flight example, does the seesaw move to one extreme or the other.

CONCEPT CHECK

1. The afferent division of the nervous system has what two components?

2. The central nervous system consists of the _____ and the _____.

Autonomic Reflexes Are Important for Homeostasis

The autonomic nervous system works closely with the endocrine system and the behavioral state system [p. 291] to maintain homeostasis in the body. Sensory information from somatosensory and visceral receptors goes to homeostatic control centers in the hypothalamus, pons, and medulla (**FIG. 11.2**). These centers monitor and regulate important functions such as blood pressure, temperature control, and water balance (**FIG. 11.3**).

The hypothalamus also contains neurons that act as sensors, such as *osmoreceptors,* which monitor osmolarity, and thermoreceptors, which monitor body temperature. Motor output from the hypothalamus and brain stem creates autonomic responses, endocrine responses, and behavioral responses such as drinking, food-seeking, and temperature regulation (getting out of the heat, putting on a sweater). These behavioral responses are integrated in brain centers responsible for motivated behaviors and control of movement.

In addition, sensory information integrated in the cerebral cortex and limbic system can create emotions that influence autonomic output, as Figure 11.2 illustrates. Blushing, fainting at the sight of a hypodermic needle, and "butterflies in the stomach" are all examples of emotional influences on autonomic functions. Understanding the autonomic and hormonal control of organ systems is the key to understanding the maintenance of homeostasis in virtually every system of the body.

FIG. 11.2 Integration of autonomic function

The hypothalamus, pons, and medulla initiate autonomic, endocrine, and behavioral responses.

Sensory input → Hypothalamic sensors / Somatic and visceral sensory neurons → Pons, medulla, hypothalamus / Limbic system, cerebral cortex → Autonomic response → Endocrine response / Behavioral response

Some autonomic reflexes are capable of taking place without input from the brain. These *spinal reflexes* [Fig. 9.7, p. 285] include urination, defecation, and penile erection—body functions that can be influenced by descending pathways from the brain but do not require this input. For example, people with spinal cord injuries that disrupt communication between the brain and spinal cord may retain some spinal reflexes but lose the ability to sense or control them.

FIG. 11.3 Autonomic control centers

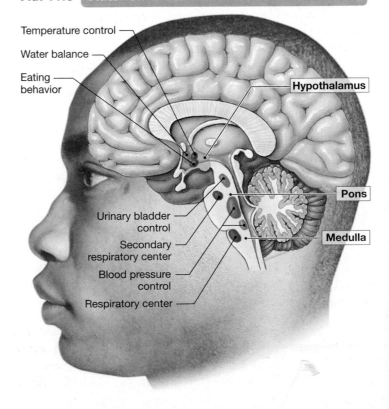

Temperature control
Water balance
Eating behavior
Hypothalamus
Urinary bladder control
Secondary respiratory center
Blood pressure control
Respiratory center
Pons
Medulla

RUNNING **PROBLEM**

Neuroscientists have learned that addictive behaviors develop because certain chemicals act as positive reinforcers in the brain, creating physical and psychological dependence. Nicotine is an addictive drug that enhances dopamine release in the brain's reward centers and creates pleasurable sensations. Over time, the brain also begins to associate the social aspects of cigarette smoking with pleasure, a conditioned response that makes quitting difficult. If smokers do stop smoking, they may suffer from unpleasant physical withdrawal symptoms, including lethargy, hunger, and irritability.

Q1: *To avoid withdrawal symptoms, people continue to smoke, resulting in chronically elevated nicotine levels in their blood. Nicotine binds to nicotinic acetylcholine receptors (nAChR). What is the usual response of cells that are chronically exposed to elevated concentrations of a signal molecule?* [Hint: p. 181]

(359) (**361**) (363) (367) (371) (373)

Antagonistic Control Is a Hallmark of the Autonomic Division

The sympathetic and parasympathetic branches of the autonomic nervous system display all four of Walter Cannon's properties of homeostasis: (1) preservation of the fitness of the internal environment, (2) up-down regulation by tonic control, (3) antagonistic control, and (4) chemical signals with different effects in different tissues [p. 182]

Many internal organs are under *antagonistic control,* in which one autonomic branch is excitatory and the other branch is inhibitory (see the table on the right side of Fig. 11.5). For example, sympathetic innervation increases heart rate, while parasympathetic stimulation decreases it. Consequently, heart rate can be regulated by altering the relative proportions of sympathetic and parasympathetic control.

Exceptions to dual antagonistic innervation include the sweat glands and the smooth muscle in most blood vessels. These tissues are innervated only by the sympathetic branch and rely strictly on tonic (up-down) control.

The two autonomic branches are usually antagonistic in their control of a given target tissue, but they sometimes work cooperatively on different tissues to achieve a common goal. For example, blood flow for penile erection is under control of the parasympathetic branch, while muscle contraction for sperm ejaculation is directed by the sympathetic branch.

In some autonomic pathways, the neurotransmitter receptor determines the response of the target tissue. For instance, most blood vessels contain one type of *adrenergic receptor* [p. 256] that causes smooth muscle contraction (vasoconstriction). However, some blood vessels also contain a second type of adrenergic receptor that causes smooth muscle relaxation (vasodilation). Both receptors are activated by the catecholamines norepinephrine and epinephrine [p. 206]. In these blood vessels, the adrenergic receptor, not the chemical signal, determines the response [p. 182].

CONCEPT CHECK **3.** Define homeostasis.

Autonomic Pathways Have Two Efferent Neurons in Series

All autonomic pathways (sympathetic and parasympathetic) consist of two neurons in a series (**FIG. 11.4**). The first neuron, called the **preganglionic neuron,** originates in the central nervous system (CNS) and projects to an **autonomic ganglion** outside the CNS. There, the preganglionic neuron synapses with the second neuron in the pathway, the **postganglionic neuron.** This neuron has its cell body in the ganglion and projects its axon to the target tissue. (A *ganglion* is a cluster of nerve cell bodies that lie outside the CNS. The equivalent in the CNS is a *nucleus* [p. 233].)

Divergence [p. 260] is an important feature of autonomic pathways. On average, one preganglionic neuron entering a ganglion synapses with 8 or 9 postganglionic neurons. Some synapse on as many as 32 neurons! Each postganglionic neuron may then innervate a different target, meaning that a single signal from the CNS can affect a large number of target cells simultaneously.

In the traditional view of the autonomic division, autonomic ganglia were simply a way station for the transfer of signals from preganglionic neurons to postganglionic neurons. We now know, however, that ganglia are more than a simple collection of axon terminals and nerve cell bodies: they also contain neurons that lie completely within them. These neurons enable the autonomic ganglia to act as mini integrating centers, receiving sensory input from the periphery of the body and modulating outgoing autonomic signals to target tissues. Presumably, this arrangement means that a reflex could be integrated totally within a ganglion, with no involvement of the CNS. That pattern of control is known to exist in the enteric nervous system [p. 228], which is discussed with the digestive system [Chapter 21].

FIG. 11.4 **Autonomic pathways**

Autonomic pathways consist of two neurons that synapse in an autonomic ganglion.

Preganglionic neuron   Postganglionic neuron

CNS   Autonomic ganglion   Target tissue

FIG. 11.5   Sympathetic and parasympathetic divisions

The Autonomic Nervous System

The autonomic nervous system can be divided
into two divisions: the sympathetic division
and the parasympathetic division.

KEY
● —— Sympathetic
● —— Parasympathetic

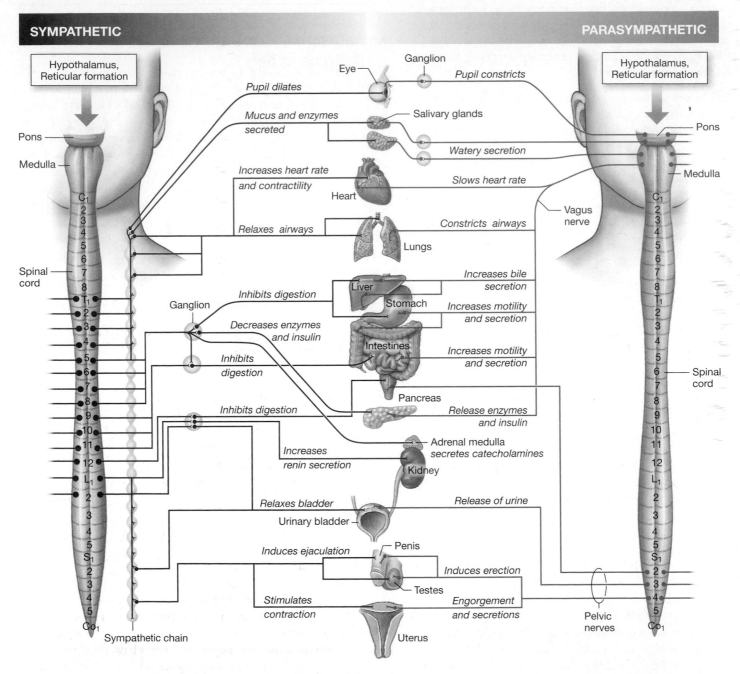

Characteristic	Sympathetic		Parasympathetic
Origin in the CNS	Thoracic and lumbar segments		Brainstem and sacral segments
Ganglion Location	Close to spinal cord		On or close to targets
Pathways	Short preganglionic, long postganglionic neurons		Long preganglionic, short postganglionic neurons

Sympathetic and Parasympathetic Responses

Although the sympathetic and parasympathetic divisions frequently innervate the same organs and tissues, they often have opposing effects.

[handwritten: m1 - m5, muscarinic receptor]

Effector Organ	Sympathetic Response	Adrenergic Receptor	Parasympathetic Response **
Pupil of eye	Dilates	α	Constricts
Salivary glands	Mucus, enzymes	α and β_2	Watery secretion
Heart	Increases rate and force of contraction	β_1	Slows rate *[handwritten: m₂]*
Arterioles and veins	Constricts / Dilates	α / β_2	— —
Lungs	Bronchioles dilate	β_2*	Bronchioles constrict
Digestive tract	Decreases motility and secretion	α, β_2	Increases motility and secretion
Exocrine pancreas	Decreases enzyme secretion	α	Increases enzyme secretion
Endocrine pancreas	Inhibits insulin secretion	α	Stimulates insulin secretion
Adrenal medulla	Secretes catecholamines	—	— —
Kidney	Increases renin secretion	β_1	— —
Urinary bladder	Urinary retention	α, β_2	Release of urine
Adipose tissue	Fat breakdown	β_3	— —
Male and female sex organs	Ejaculation (male)	α	Erection
Uterus	Depends on stage of cycle	α, β_2	Depends on stage of cycle
Lymphoid tissue	Generally inhibitory	α, β_2	— —
	*Hormonal epinephrine only		**All parasympathetic responses are mediated by muscarinic receptors.

[handwritten in margin: ACh]

Ⓠ FIGURE QUESTIONS

1. What is an advantage of having ganglia in the sympathetic chain linked to each other?
2. Which organs have antagonistic control by sympathetic and parasympathetic divisions? Which have cooperative control, with sympathetic and parasympathetic division each contributing to a function?

RUNNING **PROBLEM**

Shanika's doctor congratulates her for trying once more to stop smoking. He explains that quitting is most likely to be successful if the smoker uses a combination of behavioral modification strategies and drug therapy. Currently, there are three types of pharmacological treatments used for nicotine addiction: nicotine replacement, bupropion, and varenicline. Bupropion inhibits reuptake of the monoamines (dopamine, serotonin, and norepinephrine) by neurons, mimicking the effects of nicotine. Varenicline binds to nicotinic cholinergic receptors (nAChR). Nicotinic receptors are found throughout the nervous system, and evidence suggests that activation of nAChR by nicotine in certain regions of the brain plays a key role in nicotine addiction.

Q2: *Cholinergic receptors are classified as either nicotinic or muscarinic, on the basis of the agonist molecules that bind to them. What happens to a postsynaptic cell when nicotine rather than ACh binds to a nicotinic cholinergic receptor?*

359   361   **363**   367   371   373

Sympathetic and Parasympathetic Branches Originate in Different Regions

How, then, do the two autonomic branches differ anatomically? The main anatomical differences are (1) the pathways' point of origin in the CNS and (2) the location of the autonomic ganglia. As **FIGURE 11.5** shows, most sympathetic pathways (red) originate in the thoracic and lumbar regions of the spinal cord. *Sympathetic ganglia* are found primarily in two ganglion chains that run along either side of the bony vertebral column, with additional ganglia along the descending aorta. Long nerves (axons of postganglionic neurons) project from the ganglia to the target tissues. Because most sympathetic ganglia lie close to the spinal cord, sympathetic pathways generally have short preganglionic neurons and long postganglionic neurons.

Many parasympathetic pathways (shown in blue in Fig. 11.5) originate in the brain stem, and their axons leave the brain in several cranial nerves [p. 288]. Other parasympathetic pathways originate in the sacral region (near the lower end of the spinal cord) and control pelvic organs. In general, parasympathetic ganglia are located either on or near their target organs. Consequently, parasympathetic preganglionic neurons have long axons, and parasympathetic postganglionic neurons have short axons.

Parasympathetic innervation goes primarily to the head, neck, and internal organs. The major parasympathetic tract is the **vagus nerve** (cranial nerve X), which contains about 75% of all parasympathetic fibers. This nerve carries both sensory information from internal organs to the brain and parasympathetic output from the brain to organs.

Vagotomy, a procedure in which the vagus nerve is surgically cut, was an experimental technique used in the nineteenth and early twentieth centuries to study the effects of the autonomic nervous system on various organs. For a time, vagotomy was the preferred treatment for stomach ulcers because removal of parasympathetic innervation to the stomach decreased the secretion of stomach acid. However, this procedure had many unwanted side effects and has been abandoned in favor of drug therapies that treat the problem more specifically.

CONCEPT CHECK

4. A nerve that carries both sensory and motor information is called a(n) _____ nerve.
5. Name the four regions of the spinal cord in order, starting from the brain stem.

The Autonomic Nervous System Uses a Variety of Chemical Signals

Chemically, the sympathetic and parasympathetic branches can be distinguished by their neurotransmitters and receptors, using the following rules and **FIGURE 11.6**:

1. Both sympathetic and parasympathetic preganglionic neurons release acetylcholine (ACh) onto *nicotinic cholinergic receptors* (nAChR) on the postganglionic cell [p. 257].
2. Most postganglionic sympathetic neurons secrete norepinephrine (NE) onto *adrenergic receptors* on the target cell.
3. Most postganglionic parasympathetic neurons secrete acetylcholine onto *muscarinic cholinergic receptors* (mAChR) on the target cell.

However, there are some exceptions to these rules. A few sympathetic postganglionic neurons, such as those that terminate on sweat glands, secrete ACh rather than norepinephrine. These neurons are therefore called *sympathetic cholinergic neurons*.

A small number of autonomic neurons secrete neither norepinephrine nor acetylcholine and are known as *nonadrenergic, noncholinergic neurons*. Some of the chemicals they use as neurotransmitters include substance P, somatostatin, vasoactive intestinal peptide (VIP), adenosine, nitric oxide, and ATP. The nonadrenergic, noncholinergic neurons are assigned to either the sympathetic or parasympathetic branch according to where their preganglionic fibers leave the nerve cord.

Autonomic Pathways Control Smooth and Cardiac Muscle and Glands

The targets of autonomic neurons are smooth muscle, cardiac muscle, many exocrine glands, a few endocrine glands, lymphoid tissues, and some adipose tissue. The synapse between a postganglionic autonomic neuron and its target cell is called the **neuroeffector junction** (recall that targets are also called effectors).

The structure of an autonomic synapse differs from the model synapse [Fig. 8.2f, p. 230]. Autonomic postganglionic axons end with a series of swollen areas at their distal ends, like beads spaced out along a string (**FIG. 11.7a**). Each of these swellings, known as

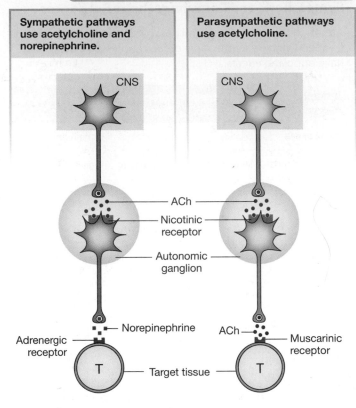

FIG. 11.6 **Sympathetic and parasympathetic neurotransmitters and receptors**

Sympathetic pathways use acetylcholine and norepinephrine.

Parasympathetic pathways use acetylcholine.

FIGURE QUESTIONS
1. Identify all:
   - cholinergic neurons
   - adrenergic neurons
   - preganglionic neurons
   - postganglionic neurons
2. Which pathway will have longer preganglionic neurons? (*Hint:* See Fig. 11.5.)

a **varicosity** {*varicosus*, abnormally enlarged or swollen}, contains vesicles filled with neurotransmitter.

The branched ends of the axon lie across the surface of the target tissue, but the underlying target cell membrane does not possess clusters of neurotransmitter receptors in specific sites. Instead, the neurotransmitter is simply released into the interstitial fluid to diffuse to wherever the receptors are located. The result is a less-directed form of communication than that which occurs between a somatic motor neuron and a skeletal muscle. The diffuse release of autonomic neurotransmitter means that a single postganglionic neuron can affect a large area of target tissue.

The release of autonomic neurotransmitters is subject to modulation from a variety of sources. For example, sympathetic varicosities contain receptors for hormones and for paracrine signals such as histamine. These modulators may either facilitate or inhibit neurotransmitter release. Some preganglionic neurons co-secrete neuropeptides along with acetylcholine. The peptides

FIG. 11.7 Autonomic synapses

(a) Autonomic varicosities release neurotransmitter over the surface of target cells.

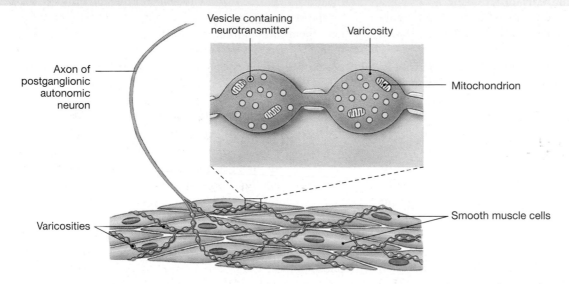

(b) Norepinephrine (NE) release and removal at a sympathetic neuroeffector junction

1. Action potential arrives at the varicosity.

2. Depolarization opens voltage-gated Ca^{2+} channels.

3. Ca^{2+} entry triggers exocytosis of synaptic vesicles.

4. NE binds to adrenergic receptor on target.

5. Receptor activation ceases when NE diffuses away from the synapse.

6. NE is removed from the synapse.

7. NE can be taken back into synaptic vesicles for re-release.

8. NE is metabolized by monoamine oxidase (MAO).

act as neuromodulators, producing slow synaptic potentials that modify the activity of postganglionic neurons [p. 261].

Autonomic Neurotransmitters Are Synthesized in the Axon

The primary autonomic neurotransmitters, acetylcholine and norepinephrine, can be synthesized in the axon varicosities (Fig. 11.7b).

Both are small molecules easily synthesized by cytoplasmic enzymes. Neurotransmitter made in the varicosities is packaged into synaptic vesicles for storage.

Neurotransmitter release follows the pattern found in other cells: depolarization—calcium signal—exocytosis [p. 159]. When an action potential arrives at the varicosity, voltage-gated Ca^{2+} channels open, Ca^{2+} enters the neuron, and the synaptic vesicle contents are released by exocytosis. Once neurotransmitters are released into the synapse, they either diffuse through the interstitial fluid until they encounter a receptor on the target cell or drift away from the synapse.

The concentration of neurotransmitter in the synapse is a major factor in autonomic control of a target: more neurotransmitter means a longer or stronger response. The concentration of neurotransmitter in a synapse is influenced by its rate of breakdown or removal (Fig. 11.7b). Neurotransmitter activation of its receptor terminates when the neurotransmitter (1) diffuses away, (2) is metabolized by enzymes in the extracellular fluid, or (3) is actively transported into cells around the synapse. The uptake of neurotransmitter by varicosities allows neurons to reuse the chemicals.

These steps are shown for norepinephrine in Figure 11.7b. Norepinephrine is synthesized in the varicosity from the amino acid tyrosine. Once released into the synapse, norepinephrine may combine with an adrenergic receptor on the target cell, diffuse away, or be transported back into the varicosity. Inside the neuron, recycled norepinephrine is either repackaged into vesicles or broken down by **monoamine oxidase** (MAO), the main enzyme responsible for degradation of catecholamines. [See Fig. 8.20, p. 260 for a similar figure on acetylcholine.]

TABLE 11.1 compares the characteristics of the two primary autonomic neurotransmitters.

Autonomic Receptors Have Multiple Subtypes

The autonomic nervous system uses only a few neurotransmitters but it diversifies its actions by having multiple receptor subtypes with different second messenger pathways. The sympathetic division has two types of adrenergic receptors with multiple subtypes. The parasympathetic division uses five varieties of muscarinic cholinergic receptors.

Sympathetic Receptors Sympathetic pathways secrete catecholamines that bind to adrenergic receptors on their target cells. Adrenergic receptors come in two varieties: α (alpha) and β (beta), with several subtypes of each. **Alpha receptors**—the most common sympathetic receptor—respond strongly to norepinephrine and only weakly to epinephrine (**TBL. 11.2**).

The three main subtypes of beta receptors differ in their affinity for catecholamines. β_1-**receptors** respond equally strongly to norepinephrine and epinephrine. β_2-**receptors** are more sensitive to epinephrine than to norepinephrine. Interestingly, the β_2-receptors are not innervated (no sympathetic neurons terminate near them), which limits their exposure to the neurotransmitter norepinephrine. β_3-**receptors,** which are found primarily on adipose tissue, are innervated and more sensitive to norepinephrine than to epinephrine.

Adrenergic Receptor Pathways All adrenergic receptors are G protein-coupled receptors rather than ion channels [p. 174]. This means that the target cell response is slower to start and can persist for a longer time than is usually associated with the nervous system. The long-lasting metabolic effects of some autonomic pathways result from modification of existing proteins or from the synthesis of new proteins.

The different adrenergic receptor subtypes use different second messenger pathways (Tbl. 11.2). α_1-**receptors** activate phospholipase C, creating inositol trisphosphate (IP$_3$) and diacylglycerol (DAG) [Fig. 6.8b, p. 175]. DAG initiates a cascade that phosphorylates proteins. IP$_3$ opens Ca^{2+} channels, creating intracellular Ca^{2+} signals. In general, activation of α_1-receptors causes muscle contraction or secretion by exocytosis. α_2-receptors decrease intracellular cyclic AMP and cause smooth muscle relaxation (gastrointestinal tract) or decreased secretion (pancreas).

β-receptors all increase cyclic AMP and trigger the phosphorylation of intracellular proteins. The target cell response then depends on the receptor subtype and the specific downstream pathway in the target cell. For example, activation of β_1-receptors enhances cardiac muscle contraction, but activation of β_2-receptors relaxes smooth muscle in many tissues.

Parasympathetic Pathways As a rule, parasympathetic neurons release ACh at their targets. As noted earlier, the neuroeffector junctions of the parasympathetic branch have muscarinic cholinergic receptors [p. 254]. Muscarinic receptors are all G protein-coupled receptors. The activation of these receptors initiates second messenger pathways, some of which open K$^+$ or Ca^{2+} channels. The tissue response to activation of a muscarinic receptor varies with the receptor subtype, of which there are at least five.

TABLE 11.1	Postganglionic Autonomic Neurotransmitters	
	Sympathetic Division	**Parasympathetic Division**
Neurotransmitter	Norepinephrine (NE)	Acetylcholine (ACh)
Receptor Types	α- and β-adrenergic	Nicotinic and muscarinic cholinergic
Synthesized from	Tyrosine	Acetyl CoA + choline
Inactivation Enzyme	Monoamine oxidase (MAO) in mitochondria of varicosity	Acetylcholinesterase (AChE) in synaptic cleft
Varicosity Membrane Transporters for	Norepinephrine	Choline

TABLE 11.2	Properties of Adrenergic Receptors		
Receptor	**Found in**	**Sensitivity**	**Effect on Second Messenger**
α_1	Most sympathetic target tissues	NE > E*	Activates phospholipase C
α_2	Gastrointestinal tract and pancreas	NE > E	Decreases cAMP
β_1	Heart muscle, kidney	NE = E	Increases cAMP
β_2	Certain blood vessels and smooth muscle of some organs	E > NE	Increases cAMP
β_3	Adipose tissue	NE > E	Increases cAMP

*NE = norepinephrine, E = epinephrine

CONCEPT CHECK

6. In what organelle is most intracellular Ca^{2+} stored?
7. What enzyme (a) converts ATP to cAMP? (b) does cAMP activate? [Fig. 6.8a, p. 175]

The Adrenal Medulla Secretes Catecholamines

The **adrenal medulla** {*ad-*, upon + *renal*, kidney; *medulla*, marrow} is a specialized neuroendocrine tissue associated with the sympathetic nervous system. During development, the neural tissue destined to secrete the catecholamines norepinephrine and epinephrine splits into two functional entities: the sympathetic branch of the nervous system, which secretes norepinephrine, and the adrenal medulla, which secretes epinephrine primarily.

The adrenal medulla forms the core of the *adrenal glands*, which sit atop the kidneys (**FIG. 11.8a**). Like the pituitary gland, each adrenal gland is actually two glands of different embryological origin that fused during development (Fig. 11.8b). The outer portion, the *adrenal cortex*, is a true endocrine gland of epidermal origin that secretes steroid hormones [p. 80]. The adrenal medulla, which forms the small core of the gland, develops from the same embryonic tissue as sympathetic neurons and is a neurosecretory structure.

The adrenal medulla is often described as a *modified sympathetic ganglion*. Preganglionic sympathetic neurons project from the spinal cord to the adrenal medulla, where they synapse (Fig. 11.8c). However, the postganglionic neurons lack the axons that would normally project to target cells. Instead, the axonless cell bodies, called *chromaffin cells*, secrete the neurohormone epinephrine directly into the blood. In response to alarm signals from the CNS, the adrenal medulla releases large amounts of epinephrine for general distribution throughout the body as part of a fight-or-flight response.

CONCEPT CHECK

8. Is the adrenal medulla most like the anterior pituitary or the posterior pituitary? Explain.
9. Predict whether chromaffin cells have nicotinic or muscarinic ACh receptors.

Autonomic Agonists and Antagonists Are Important Tools in Research and Medicine

The study of the two autonomic branches has been greatly simplified by advances in molecular biology. The genes for many autonomic receptors and their subtypes have been cloned, allowing researchers to create mutant receptors and study their properties. In addition, researchers have either discovered or synthesized a variety of agonist and antagonist molecules (**TBL. 11.3**). Direct agonists and antagonists combine with the target receptor to mimic or block neurotransmitter action. Indirect agonists and antagonists act by altering secretion, reuptake, or degradation of neurotransmitters.

For example, cocaine is an indirect agonist that blocks the reuptake of norepinephrine into adrenergic nerve terminals,

RUNNING PROBLEM

The action of nicotine on nAChR is complicated. Normally, chronic exposure of cells to a receptor agonist such as ACh or nicotine causes the cells to down-regulate their receptors. However, one research study that examined brains at autopsy found that smokers have more nAChR receptors on their cell membranes than nonsmokers do. This increase in receptor numbers, or up-regulation [p. 181], usually occurs when cells are chronically exposed to receptor antagonists.

Q3: *Although ACh and nicotine have been shown in short-term studies to be nAChR agonists, continued exposure of the receptors to ACh has been shown to close, or desensitize, the channel. Speculate why this could explain the up-regulation of nAChR observed in smokers.*

Q4: *Name another ion channel you have studied that opens in response to a stimulus but inactivates and closes shortly thereafter [p. 245].*

359   361   363   **367**   371   373

FIG. 11.8 | The adrenal medulla

The adrenal medulla secretes epinephrine into the blood.

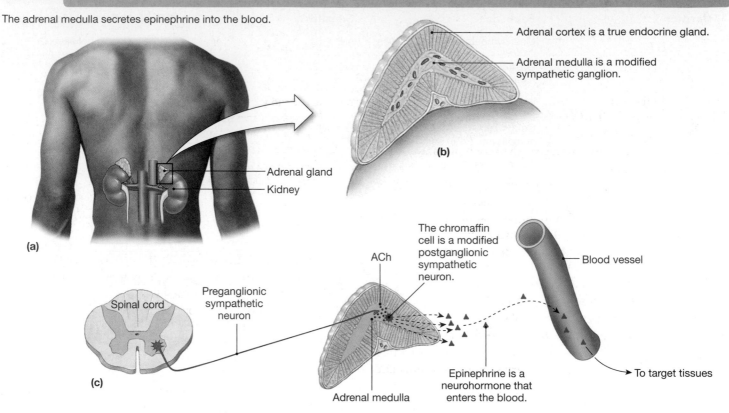

(a)

(b)

Adrenal cortex is a true endocrine gland.

Adrenal medulla is a modified sympathetic ganglion.

Adrenal gland

Kidney

(c)

Spinal cord

Preganglionic sympathetic neuron

ACh

The chromaffin cell is a modified postganglionic sympathetic neuron.

Blood vessel

Epinephrine is a neurohormone that enters the blood.

Adrenal medulla

To target tissues

thereby extending norepinephrine's excitatory effect on the target. This is demonstrated by the toxic effect of cocaine on the heart, where sympathetic-induced vasoconstriction of the heart's blood vessels can result in a heart attack. Cholinesterase inhibitors, also called *anticholinesterases,* are indirect agonists that block ACh degradation and extend the active life of each ACh molecule. The toxic *organophosphate insecticides,* such as parathion and malathion, are anticholinesterases. They kill insects by causing sustained contraction of their breathing muscles so that they are unable to breathe.

TABLE 11.3 | Agonists and Antagonists of Neurotransmitter Receptors

Receptor Type	Neurotransmitter	Agonist	Antagonists	Indirect Agonists/ Antagonists
Cholinergic	Acetylcholine			AChE* *inhibitors:* neostigmine
Muscarinic		Muscarine	Atropine, scopolamine	
Nicotinic		Nicotine	α-bungarotoxin (muscle only), TEA (tetraethylammonium; ganglia only), curare	
Adrenergic	Norepinephrine (NE), epinephrine			*Stimulate NE release:* ephedrine, amphetamines; *Prevents NE uptake:* cocaine
Alpha (α)		Phenylephrine	"Alpha-blockers"	
Beta (β)		Isoproterenol, albuterol	"Beta-blockers": propranolol (β_1 and β_2), metoprolol (β_1 only)	

*AChE = acetylcholinesterase

Many drugs used to treat depression are indirect agonists that act either on membrane transporters for neurotransmitters (tricyclic antidepressants and selective serotonin reuptake inhibitors) or on their metabolism (monoamine oxidase inhibitors). The older antidepressant drugs that act on norepinephrine transport and metabolism (tricyclics and MAO inhibitors) may have side effects related to their actions in the autonomic nervous system, including cardiovascular problems, constipation, urinary difficulty, and sexual dysfunction {*dys-*, abnormal or ill}. The serotonin reuptake inhibitors have fewer autonomic side effects. Some of the newest drugs influence the action of both norepinephrine and serotonin.

Many new drugs have been developed from studies of agonists and antagonists. The discovery of α- and β-adrenergic receptors led to the development of drugs that block only one of the two receptor types. The drugs known as beta-blockers have given physicians a powerful tool for treating high blood pressure, one of the most common disorders in the United States today. Early α-adrenergic receptor antagonists had many unwanted side effects, but now pharmacologists can design drugs to target specific receptor subtypes. For example, tamsulosin (Flomax®) blocks alpha-1A adrenergic receptors (ADRA1A) found largely on smooth muscle of the prostate gland and bladder. Relaxing these muscles helps relieve the urinary symptoms of prostatic enlargement.

Primary Disorders of the Autonomic Nervous System Are Relatively Uncommon

Diseases and malfunction of the autonomic nervous system are relatively rare. Direct damage (trauma) to hypothalamic control centers may disrupt the body's ability to regulate water balance or temperature. Generalized sympathetic dysfunction, or *dysautonomia*, may result from systemic diseases such as cancer and diabetes mellitus. There are also some conditions, such as *multiple system atrophy*, in which the CNS control centers for autonomic functions degenerate.

In many cases of sympathetic dysfunction, the symptoms are manifested most strongly in the cardiovascular system, when diminished sympathetic input to blood vessels results in abnormally low blood pressure. Other prominent symptoms of sympathetic pathology include urinary *incontinence* {*in-*, unable + *continere*, to contain}, which is the loss of bladder control, or *impotence*, which is the inability to achieve or sustain a penile erection.

Occasionally, patients suffer from primary autonomic failure when sympathetic neurons degenerate. In the face of continuing diminished sympathetic input, target tissues up-regulate [p. 181], putting more receptors into the cell membrane to maximize the cell's response to available norepinephrine. This increase in receptor abundance leads to *denervation hypersensitivity*, a state in which the administration of exogenous adrenergic agonists causes a greater-than-expected response.

CLINICAL FOCUS

Diabetes: Autonomic Neuropathy

Primary disorders of the autonomic division are rare, but the secondary condition known as **diabetic autonomic neuropathy** is quite common. This complication of diabetes often begins as a sensory neuropathy, with tingling and loss of sensation in the hands and feet. In some patients, pain is the primary symptom. About 30% of diabetic patients go on to develop autonomic neuropathies, manifested by dysfunction of the cardiovascular, gastrointestinal, urinary, and reproductive systems (abnormal heart rate, constipation, incontinence, impotence). The cause of diabetic neuropathies is unclear. Patients who have chronically elevated blood glucose levels are more likely to develop neuropathies, but the underlying metabolic pathway has not been identified. Other contributing factors for neuropathy include oxidative stress and autoimmune reactions. Currently, there is no prevention for diabetic neuropathies other than controlling blood glucose levels, and no cure. The only recourse for patients is taking drugs that treat the symptoms.

Summary of Sympathetic and Parasympathetic Branches

As you have seen in this discussion, the branches of the autonomic nervous system share some features but are distinguished by others. Many of these features are summarized in **FIGURE 11.9** and compared in **TABLE 11.4**.

1. Both sympathetic and parasympathetic pathways consist of two neurons (preganglionic and postganglionic) in series. One exception to this rule is the adrenal medulla, in which postganglionic sympathetic neurons have been modified into a neuroendocrine organ.

2. All preganglionic autonomic neurons secrete acetylcholine onto nicotinic receptors. Most sympathetic neurons secrete norepinephrine onto adrenergic receptors. Most parasympathetic neurons secrete acetylcholine onto muscarinic receptors.

3. Sympathetic pathways originate in the thoracic and lumbar regions of the spinal cord. Parasympathetic pathways leave the CNS at the brain stem and in the sacral region of the spinal cord.

4. Most sympathetic ganglia are located close to the spinal cord (are *paravertebral*). Parasympathetic ganglia are located close to or in the target tissue.

5. The sympathetic branch controls functions that are useful in stress or emergencies (fight-or-flight). The parasympathetic branch is dominant during rest-and-digest activities.

FIG. 11.9 **ESSENTIALS**

Efferent Divisions of the Nervous System

SOMATIC MOTOR PATHWAY

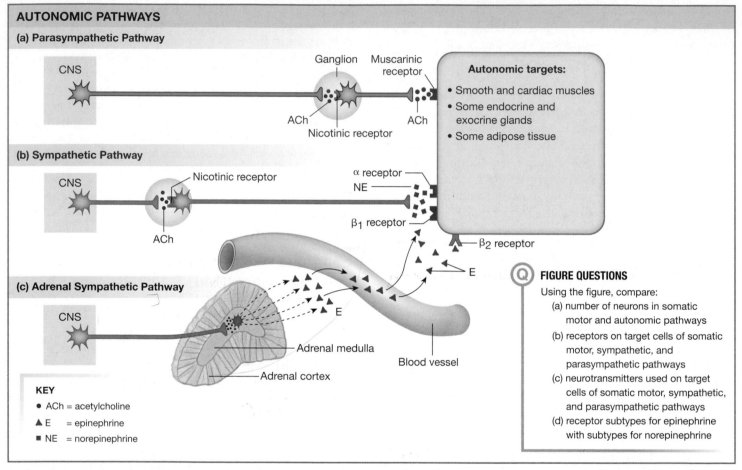

AUTONOMIC PATHWAYS

(a) Parasympathetic Pathway

CNS

Ganglion Muscarinic receptor

ACh Nicotinic receptor ACh

Autonomic targets:
- Smooth and cardiac muscles
- Some endocrine and exocrine glands
- Some adipose tissue

(b) Sympathetic Pathway

CNS Nicotinic receptor ACh

α receptor
NE
β₁ receptor
β₂ receptor
E

(c) Adrenal Sympathetic Pathway

CNS

Adrenal medulla
Adrenal cortex
Blood vessel
E

KEY
- • ACh = acetylcholine
- ▲ E = epinephrine
- ■ NE = norepinephrine

SOMATIC MOTOR PATHWAY
CNS ACh Nicotinic receptor Target: skeletal muscle

Q **FIGURE QUESTIONS**

Using the figure, compare:
(a) number of neurons in somatic motor and autonomic pathways
(b) receptors on target cells of somatic motor, sympathetic, and parasympathetic pathways
(c) neurotransmitters used on target cells of somatic motor, sympathetic, and parasympathetic pathways
(d) receptor subtypes for epinephrine with subtypes for norepinephrine

Comparison of Somatic Motor and Autonomic Divisions

	SOMATIC MOTOR	AUTONOMIC
Number of neurons in efferent path	1	2
Neurotransmitter/receptor at neuron-target synapse	ACh/nicotinic	ACh/muscarinic or NE/α- or β-adrenergic
Target tissue	Skeletal muscle	Smooth and cardiac muscle; some endocrine and exocrine glands; some adipose tissue
Neurotransmitter released from	Axon terminals	Varicosities and axon terminals
Effects on target tissue	Excitatory only: muscle contracts	Excitatory or inhibitory
Peripheral components found outside the CNS	Axons only	Preganglionic axons, ganglia, postganglionic neurons
Summary of function	Posture and movement	Visceral function, including movement in internal organs and secretion; control of metabolism

TABLE 11.4	Comparison of Sympathetic and Parasympathetic Branches	
	Sympathetic	**Parasympathetic**
Point of CNS Origin	First thoracic to second lumbar segments	Midbrain, medulla, and second to fourth sacral segments
Location of Peripheral Ganglia	Primarily in paravertebral sympathetic chain; three outlying ganglia located alongside descending aorta	On or near target organs
Structure of Region from which Neurotransmitter Is Released	Varicosities	Varicosities
Neurotransmitter at Target Synapse	Norepinephrine (adrenergic neurons)	ACh (cholinergic neurons)
Inactivation of Neurotransmitter at Synapse	Uptake into varicosity, diffusion	Enzymatic breakdown, diffusion
Neurotransmitter Receptors on Target Cells	Adrenergic	Muscarinic
Ganglionic Synapse	ACh on nicotinic receptor	ACh on nicotinic receptor
Neuron-Target Synapse	NE on α- or β-adrenergic receptor	ACh on muscarinic receptor

THE SOMATIC MOTOR DIVISION

Somatic motor pathways, which control skeletal muscles, differ from autonomic pathways both anatomically and functionally (see the table in Fig. 11.9). Somatic motor pathways have a single neuron that originates in the CNS and projects its axon to the target tissue, which is always a skeletal muscle. Somatic pathways are always excitatory, unlike autonomic pathways, which may be either excitatory or inhibitory.

A Somatic Motor Pathway Consists of One Neuron

The cell bodies of somatic motor neurons are located either in the ventral horn of the spinal cord [p. 284] or in the brain, with a long single axon projecting to the skeletal muscle target (Fig. 11.9). These myelinated axons may be a meter or more in length, such as the somatic motor neurons that innervate the muscles of the foot and hand.

Somatic motor neurons branch close to their targets. Each branch divides into a cluster of enlarged axon terminals that lie on the surface of the skeletal muscle fiber (**FIG. 11.10a**). This branching structure allows a single motor neuron to control many muscle fibers at one time.

The synapse of a somatic motor neuron on a muscle fiber is called the **neuromuscular junction,** or NMJ (Fig. 11.10b). Like all other synapses, the NMJ has three components: (1) the motor neuron's presynaptic axon terminal filled with synaptic vesicles and mitochondria, (2) the synaptic cleft, and (3) the postsynaptic membrane of the skeletal muscle fiber.

In addition, the neuromuscular junction includes extensions of Schwann cells that form a thin layer covering the top of the axon terminals. For years, it was thought that this cell layer simply provided insulation to speed up the conduction of the action potential, but we now know that Schwann cells secrete a variety of chemical signal molecules. These signal molecules play a critical role in the formation and maintenance of neuromuscular junctions.

On the postsynaptic side of the neuromuscular junction, the muscle cell membrane that lies opposite the axon terminal is modified into a **motor end plate,** a series of folds that look like shallow gutters (Fig. 11.10b, c). Along the upper edge of each gutter, nicotinic ACh receptor (nAChR) channels cluster together in an active zone. Between the axon and the muscle, the synaptic cleft is filled with a fibrous matrix whose collagen fibers hold the axon terminal and the motor end plate in the proper alignment. The matrix also contains **acetylcholinesterase** (AChE), the enzyme that rapidly deactivates ACh by degrading it into acetyl and choline [p. 258].

RUNNING **PROBLEM**

After discussing her options with her doctor, Shanika decides to try the nicotine patch, one form of nicotine replacement therapy. These adhesive patches allow the former smoker to gradually decrease nicotine levels in the body, preventing withdrawal symptoms during the time the cells are down-regulating their receptors back to the normal number. When Shanika reads the package insert prior to applying her first nicotine patch, she notices a warning to keep the patches away from children. An overdose of nicotine (highly unlikely when the patch is used as directed) could result in complete paralysis of the respiratory muscles (the diaphragm and the skeletal muscles of the chest wall).

Q5: *Why might excessive levels of nicotine cause respiratory paralysis?*

359　361　363　367　**371**　373

FIG. 11.10 **ESSENTIALS**

Somatic Motor Neurons and the Neuromuscular Junction

(a) The **neuromuscular junction** consists of axon terminals, motor end plates on the muscle membrane, and Schwann cell sheaths.

Somatic motor neuron branches at its distal end.

Skeletal muscle fiber

Motor end plate

(b) The **motor end plate** is a region of muscle membrane that contains high concentrations of ACh receptors.

Schwann cell sheath

Axon terminal

Mitochondrion

Motor end plate

(c) Neuromuscular junction

Synaptic vesicle (ACh)

Presynaptic membrane

Synaptic cleft

Nicotinic ACh receptors

Postsynaptic membrane is modified into a motor end plate.

(d) An action potential arrives at the axon terminal, causing voltage-gated Ca^{2+} channels to open. Calcium entry causes synaptic vesicles to fuse with the presynaptic membrane and release ACh into the synaptic cleft.

Synaptic vesicle (ACh)

Ca^{2+} Ca^{2+}

ACh

Voltage-gated Ca^{2+} channel

Acetyl + choline

AChE

Nicotinic receptor

Skeletal muscle fiber

Acetylcholine (ACh) is metabolized by acetylcholinesterase (AChE).

(e) The nicotinic cholinergic receptor binds two ACh molecules, opening a nonspecific monovalent cation channel. The open channel allows Na^+ and K^+ to pass. Net Na^+ influx depolarizes the muscle fiber.

Na^+

K^+

ACh

K^+

$+ + + + +$ $-$ $+ + + + +$

$- - - -$ $- - - -$

$- - - -$ $- - - -$

$+ + + +$ $+ + + +$

Na^+

Closed channel

Open channel

10. Is the ventral horn of the spinal cord, which contains the cell bodies of somatic motor neurons, gray matter or white matter?

The Neuromuscular Junction Contains Nicotinic Receptors

As in all neurons, action potentials arriving at the axon terminal open voltage-gated Ca^{2+} channels in the membrane. Calcium diffuses into the cell down its electrochemical gradient, triggering the release of ACh-containing synaptic vesicles. Acetylcholine diffuses across the synaptic cleft and combines with nicotinic receptor channels (nAChR) on the skeletal muscle membrane (Fig. 11.10d).

The nAChR channels of skeletal muscle are similar but not identical to the nicotinic ACh receptors found on neurons. This difference is illustrated by the fact that the snake toxin *α-bungarotoxin* binds to nicotinic skeletal muscle receptors but not to those in autonomic ganglia. Both muscle and neuronal nAChR proteins have five subunits encircling the central pore. However, skeletal muscle has α, β, δ, and ε subunit isoforms, while neuronal nAChR has only the α and β isoforms. The α and β isoforms of nAChR can become desensitized and their channels closed with extended exposure to ACh or other agonists.

Nicotinic cholinergic receptors are chemically gated ion channels with two binding sites for ACh (Fig. 11.10e). When ACh binds to the receptor, the channel gate opens and allows monovalent cations to flow through. In skeletal muscle, net Na^+ entry into the muscle fiber depolarizes it, triggering an action potential that causes contraction of the skeletal muscle cell.

Acetylcholine acting on a skeletal muscle's motor end plate is always excitatory and creates muscle contraction. There is no antagonistic innervation to relax skeletal muscles. Instead, relaxation occurs when the somatic motor neurons are inhibited in the CNS, preventing ACh release. You will learn later about how inhibition of somatic motor pathways controls body movement.

Somatic motor neurons do more than simply create contractions: They are necessary for muscle health. "Use it or lose it" is a cliché that is very appropriate to the dynamics of muscle mass because disrupting synaptic transmission at the neuromuscular junction has devastating effects on the entire body. Without communication between the motor neuron and the muscle, the skeletal muscles for movement and posture weaken, as do the skeletal muscles for breathing. In the most severe cases, loss of respiratory function can be fatal unless the patient is placed on artificial ventilation. *Myasthenia gravis,* a disease characterized by loss of ACh receptors, is the most common disorder of the neuromuscular junction.

11. Compare gating and ion selectivity of acetylcholine receptor-channels in the motor end plate with that of ion channels along the axon of a somatic motor neuron.

12. A nonsmoker who chews nicotine-containing gum might notice an increase in heart rate, a function controlled by sympathetic neurons. Postganglionic sympathetic neurons secrete norepinephrine, not ACh, so how could nicotine affect heart rate?

13. Patients with myasthenia gravis have a deficiency of ACh receptors on their skeletal muscles and have weak muscle function as a result. Why would administration of an anticholinesterase drug (one that inhibits acetylcholinesterase) improve muscle function in these patients?

RUNNING **PROBLEM CONCLUSION** | A Powerful Addiction

Shanika is determined to stop smoking this time because her grandfather, a smoker for many years, was just diagnosed with lung cancer. When the patch alone does not stop her craving for a cigarette, Shanika's physician adds bupropion pills to her treatment. In addition, Shanika attends behavioral modification classes, where she learns to avoid situations that make her likely to smoke and to substitute other activities, such as chewing gum, for smoking. After six months, Shanika proudly informs her family that she thinks she has kicked the habit.

Controlled studies of the drug bupropion (Zyban®) show that it nearly doubles the rate of smoking cessation compared to placebo, and this drug is now considered a first choice for therapy. The nAChR agonist varenicline (Chantix®) may help break the nicotine addiction, but it carries a risk of serious adverse cardiovascular side effects, which has decreased its use. Two drugs that act on cannabinoid receptors [p. 257] were effective in clinical trials but were withdrawn from the market after people taking them exhibited serious psychological side effects. A vaccine against nicotine is currently in clinical trials in the United States. To learn more about nicotine addiction and smoking cessation programs, see Medline Plus (*www .nlm.nih.gov/medlineplus*). Check your understanding of this running problem by comparing your answers to the information in the following summary table.

Question	Facts	Integration and Analysis
Q1: *What is the usual response of cells that are chronically exposed to elevated concentrations of a signal molecule?*	A cell exposed to elevated concentrations of a signal molecule will decrease (down-regulate) its receptors for that molecule.	Down-regulation of receptors allows a cell to respond normally even if the concentration of ligand is elevated.

—Continued next page

RUNNING **PROBLEM CONCLUSION** | *Continued*

Question	Facts	Integration and Analysis
Q2: *What happens to a postsynaptic cell when nicotine rather than ACh binds to a nicotinic cholinergic receptor?*	Nicotine is an agonist of ACh. Agonists mimic the activity of a ligand.	Nicotine binding to a nAChR will open ion channels in the postsynaptic cell, and the cell will depolarize. This is the same effect that ACh binding creates.
Q3: *Although ACh and nicotine have been shown in short-term studies to be nAChR agonists, continued exposure of the receptors to ACh has been shown to close, or desensitize, the channel. Speculate why this could explain the up-regulation of nAChR observed in smokers.*	Chronic exposure to an agonist usually causes down-regulation. Chronic exposure to an antagonist usually causes up-regulation. nAChR channels open with initial exposure to agonists but close with continued exposure.	Although nicotine is a short-term agonist, it appears to be having the same effect as an antagonist during long-term exposure. With both antagonism and the desensitization described here, the cell's activity decreases. The cell subsequently up-regulates the number of receptors in an attempt to restore activity.
Q4: *Name another ion channel you have studied that opens in response to a stimulus but inactivates and closes shortly thereafter.*	The voltage-gated Na^+ channel of the axon first opens, then closes when an inactivation gate shuts.	N/A
Q5: *Why might excessive levels of nicotine cause respiratory paralysis?*	Nicotinic receptors are found at the neuromuscular junction that controls skeletal muscle contraction. The diaphragm and chest wall muscles that regulate breathing are skeletal muscles.	The nicotinic receptors of the neuromuscular junction are not as sensitive to nicotine as are those of the CNS and autonomic ganglia. However, excessively high amounts of nicotine will activate the nAChR of the motor end plate, causing the muscle fiber to depolarize and contract. The continued presence of nicotine keeps these ion channels open, and the muscle remains depolarized. In this state, the muscle is unable to contract again, resulting in paralysis.

(359)(361)(363)(367)(371)(**373**)

VISUALIZE, PRACTICE, AND MASTER PHYSIOLOGY WITH MASTERING A&P
- Practice-on-the-go with Dynamic Study Modules
- Visualize important processes with Interactive Physiology and A&P Flix
- Check your understanding with Video Tutors and practice quizzes!

MasteringA&P®

CHAPTER SUMMARY

The autonomic and somatic motor divisions are the output pathways of the peripheral nervous system. *Communication* among the sensory and efferent divisions and the CNS depends primarily on chemical signaling and *molecular interactions* between neurotransmitters and their receptors. *Homeostasis* requires constant surveillance of body parameters by the nervous system, working in conjunction with the endocrine and immune systems. As you learn about the function of other body systems, you will continue to revisit the principles of communication and coordination.

The Autonomic Division

1. The efferent division of the peripheral nervous system consists of **somatic motor neurons,** which control skeletal muscles, and **autonomic neurons,** which control smooth muscle, cardiac muscle, many glands, lymphoid tissue, and some adipose tissue. (p. 359)

2. The autonomic division is subdivided into a **sympathetic branch** and a **parasympathetic branch.** (p. 359; Tbl. 11.4)

3. The maintenance of homeostasis within the body is a balance of autonomic control, endocrine control, and behavioral responses. (p. 359; Fig. 11.2)

4. The autonomic division is controlled by centers in the hypothalamus, pons, and medulla. Some autonomic reflexes are spinal reflexes. Many of these can be modulated by input from the brain. (p. 360; Fig. 11.3)

5. The two autonomic branches demonstrate Cannon's properties of homeostasis: maintenance of the internal environment, tonic control, antagonistic control, and variable tissue responses. (p. 361)

6. All autonomic pathways are composed of a **preganglionic neuron** from the CNS that synapses with a **postganglionic neuron** in an **autonomic ganglion.** Autonomic ganglia can modulate and integrate information passing through them. (p. 361; Fig. 11.4)

7. Most sympathetic pathways originate in the thoracic and lumbar regions of the spinal cord. Most sympathetic ganglia lie either close to the spinal cord or along the descending aorta. (p. 363; Fig. 11.5)

8. Parasympathetic pathways originate in the brain stem or the sacral region of the spinal cord. Parasympathetic ganglia are located on or near their target organs. (p. 363; Fig. 11.5)

9. The primary autonomic neurotransmitters are **acetylcholine** and **norepinephrine.** All preganglionic neurons secrete ACh onto **nicotinic cholinergic receptors.** As a rule, postganglionic sympathetic neurons secrete norepinephrine onto **adrenergic receptors,** and postganglionic parasympathetic neurons secrete ACh onto **muscarinic cholinergic receptors.** (p. 364; Fig. 11.6; Tbl. 11.1)

10. The synapse between an autonomic neuron and its target cells is called the **neuroeffector junction.** (p. 364)

11. Postganglionic autonomic axons end with **varicosities** from which neurotransmitter is released. (p. 364; Figs. 11.7, 11.8)

12. The **adrenal medulla** secretes epinephrine and is controlled by sympathetic preganglionic neurons. (p. 367; Fig. 11.8)

13. Adrenergic receptors are G protein-coupled receptors. Alpha receptors respond most strongly to norepinephrine. β_1**-receptors** respond equally to norepinephrine and epinephrine. β_2**-receptors** are not associated with sympathetic neurons and respond most strongly to epinephrine. β_3**-receptors** respond most strongly to norepinephrine. (p. 366; Fig. 11.9; Tbl. 11.2)

14. Cholinergic muscarinic receptors are also G protein-coupled receptors. (p. 366)

The Somatic Motor Division

15. Somatic motor pathways, which control skeletal muscles, have a single neuron that originates in the CNS and terminates on a skeletal muscle. Somatic motor neurons are always excitatory and cause muscle contraction. (p. 371; Fig. 11.9)

16. A single **somatic motor neuron** controls many muscle fibers at one time. (p. 371)

17. The synapse of a somatic motor neuron on a muscle fiber is called the **neuromuscular junction.** The muscle cell membrane is modified into a **motor end plate** that contains a high concentration of nicotinic ACh receptors. (p. 371; Fig. 11.10)

18. ACh binding to nicotinic receptor opens cation channels. Net Na^+ entry into the muscle fiber depolarizes the fiber. Acetylcholine in the synapse is broken down by the enzyme **acetylcholinesterase.** (p. 371; Fig. 11.10)

REVIEW QUESTIONS

In addition to working through these questions and checking your answers on p. A-14, review the Learning Outcomes at the beginning of this chapter.

Level One Reviewing Facts and Terms

1. Name the two efferent divisions of the peripheral nervous system. What type of effectors does each control?

2. The autonomic nervous system is sometimes called the _____ nervous system. Why is this an appropriate name? List some functions controlled by the autonomic nervous system.

3. What are the two branches of the autonomic nervous system? How are these branches distinguished from each other anatomically and physiologically?

4. Which neurosecretory endocrine gland is closely allied to the sympathetic branch?

5. Neurons that secrete acetylcholine are described as _____ neurons, whereas those that secrete norepinephrine are called either _____ or _____ neurons.

6. List four things that can happen to autonomic neurotransmitters after they are released into a synapse.

7. The main enzyme responsible for catecholamine degradation is _____, abbreviated as _____.

8. What is acetylcholinesterase? Describe its action.

9. Somatic motor pathways

(a) are excitatory or inhibitory?
(b) are composed of a single neuron or a preganglionic and a postganglionic neuron?
(c) synapse with glands or with smooth, cardiac, or skeletal muscle?

10. What kind of receptor is found on the postsynaptic cell in a neuromuscular junction?

Level Two Reviewing Concepts

11. What is the advantage of divergence of neural pathways in the autonomic nervous system?

12. Compare and contrast:

(a) neuroeffector junctions and neuromuscular junctions
(b) alpha, beta, muscarinic, and nicotinic receptors. Describe where each is found and the ligands that bind to them.

13. Compare and contrast

(a) autonomic ganglia and CNS nuclei
(b) the adrenal medulla and the posterior pituitary gland
(c) axon terminals and varicosities

14. **Concept map:** Use the following terms to make a map comparing the somatic motor division and the sympathetic and parasympathetic branches of the autonomic division. You may add terms.

- acetylcholine
- alpha receptor
- beta receptor
- cholinergic receptor
- endocrine gland
- ganglion
- nicotinic receptor
- one-neuron pathway
- skeletal muscle
- somatic motor division
- two-neuron pathway

- adipose tissue
- autonomic division
- cardiac muscle
- efferent division
- exocrine gland
- muscarinic receptor
- norepinephrine
- parasympathetic branch
- smooth muscle
- sympathetic branch

15. If a target cell's receptor is (use items in left column), the neuron(s) releasing neurotransmitter onto the receptor must be (use all appropriate items from the right column).

(a) nicotinic cholinergic
(b) adrenergic α
(c) muscarinic cholinergic
(d) adrenergic β

1. somatic motor neuron
2. autonomic preganglionic neuron
3. sympathetic postganglionic neuron
4. parasympathetic postganglionic neuron

16. Ganglia contain the cell bodies of (choose all that apply)

(a) somatic motor neurons
(b) preganglionic autonomic neurons
(c) interneurons
(d) postganglionic autonomic neurons
(e) sensory neurons

Level Three Problem Solving

17. If nicotinic receptor channels allow both Na^+ and K^+ to flow through, why does Na^+ influx exceed K^+ efflux? [*Hint:* p. 238]

18. You have discovered a neuron that innervates an endocrine cell in the intestine. To learn more about this neuron, you place a marker substance at the endocrine cell synapse. The marker is taken into the neuron and transported in a vesicle by retrograde axonal transport to the nerve cell body.

(a) By what process is the marker probably taken into the axon terminal?
(b) The nerve cell body is found in a ganglion very close to the endocrine cell. To which branch of the peripheral nervous system does the neuron probably belong? (Be as specific as you can.)
(c) Which neurotransmitter do you predict will be secreted by the neuron onto the endocrine cell?

19. The Huaorani Indians of South America use blowguns to shoot darts poisoned with curare at monkeys. Curare is a plant toxin that binds to and inactivates nicotinic ACh receptors. What happens to a monkey struck by one of these darts?

Level Four Quantitative Problems

20. The U.S. Centers for Disease Control and Prevention (CDC) conduct biennial Youth Risk Behavior Surveys (YRBS) in which they ask high school students to self-report risky behaviors such as alcohol consumption and smoking. The graphs that follow were created from data in the report on cigarette smoking among American high school students. *Current smoking* is defined as smoking cigarettes on at least one day in the 30 days preceding the survey. (*http://www.cdc.gov/mmwr/pdf/ss/ss6104.pdf*)

(a) What can you say about cigarette smoking among high school students in the period from 1991 to 2011?
(b) Which high school students are most likely to be smokers? Least likely to be smokers?

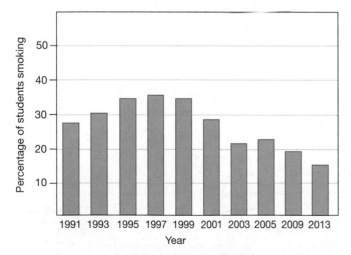

Percentage of students who reported current smoking (1991–2013)

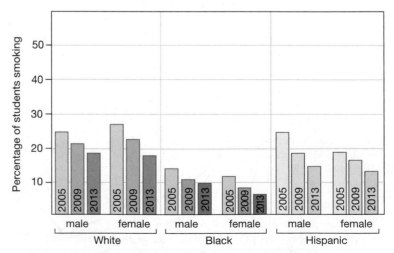

Percentage of students in 2005, 2009, and 2013 who reported current smoking, separated by sex and race/ethnicity*

*Other race/ethnic groups are not shown because the numbers were too small for meaningful statistical analysis.

Answers to Concept Checks, Figure and Graph Questions, and end-of-chapter Review Questions can be found in Appendix A [p. A-1].

12

Muscles

A muscle is ... an engine, capable of converting chemical energy into mechanical energy. It is quite unique in nature, for there has been no artificial engine devised with the great versatility of living muscle.

Ralph W. Stacy and John A. Santolucito, in *Modern College Physiology*, 1966

SKELETAL MUSCLE 379

LO 12.1 Draw and label a series of diagrams to show the different levels of organization of skeletal muscle.

LO 12.2 Diagram the sliding filament theory of contraction.

LO 12.3 Diagram the molecular events of excitation-contraction coupling and the contractile cycle.

LO 12.4 Discuss the different possible causes for muscle fatigue.

LO 12.5 Discuss the differences between slow-twitch fibers, fast-twitch oxidative-glycolytic fibers, and fast-twitch glycolytic fibers.

LO 12.6 Explain how muscle length influences force of contraction.

LO 12.7 Distinguish between summation and the different types of tetanus.

LO 12.8 Define a motor unit and explain how skeletal muscles use them to create graded contractions.

MECHANICS OF BODY MOVEMENT 398

LO 12.9 Compare and contrast isometric and isotonic contractions.

LO 12.10 Describe and give examples of how bones and muscles form fulcrums and levers.

SMOOTH MUSCLE 403

LO 12.11 Diagram smooth muscle anatomy.

LO 12.12 Diagram smooth muscle contraction and relaxation.

LO 12.13 Explain slow wave potentials, pacemaker potentials, and pharmacomechanical coupling.

CARDIAC MUSCLE 412

LO 12.14 Compare and contrast cardiac muscle with skeletal and smooth muscle.

BACKGROUND BASICS

82	Tendons
102	Kinases and phosphatases
99	Isozymes
110	Anaerobic and aerobic metabolism
107	Glycolysis
183	Tonic control
178	Nitric oxide
242	Threshold
266	Summation
359	Autonomic neurons
371	Somatic motor neurons
371	Neuromuscular junction

Striated muscle sarcomeres and sarcoplasmic reticulum between the fibers

I t was his first time to be the starting pitcher. As he ran from the bullpen onto the field, his heart was pounding and his stomach felt as if it were tied in knots. He stepped onto the mound and gathered his thoughts before throwing his first practice pitch. Gradually, as he went through the familiar routine of throwing and catching the baseball, his heart slowed and his stomach relaxed. It was going to be a good game.

The pitcher's pounding heart, queasy stomach, and movements as he runs and throws all result from muscle contraction. Our muscles have two common functions: to generate motion and to generate force. Our skeletal muscles also generate heat and contribute significantly to the homeostasis of body temperature. When cold conditions threaten homeostasis, the brain may direct our muscles to shiver, creating additional heat.

The human body has three types of muscle tissue: skeletal muscle, cardiac muscle, and smooth muscle. Most **skeletal muscles** are attached to the bones of the skeleton, enabling these muscles to control body movement. **Cardiac muscle** {*kardia*, heart} is found only in the heart and moves blood through the circulatory system. Skeletal and cardiac muscles are classified as **striated muscles** {*stria*, groove} because of their alternating light and dark bands seen under the light microscope (**FIG. 12.1a, b**).

RUNNING **PROBLEM** | **Periodic Paralysis**

This morning, Paul, age 6, gave his mother the fright of her life. One minute he was happily playing in the backyard with his new beagle puppy. The next minute, after sitting down to rest, he could not move his legs. In answer to his screams, his mother came running and found her little boy unable to walk. Panic-stricken, she scooped him up, brought him into the house, and dialed 9-1-1. But as she hung up the phone and prepared to wait for the paramedics, Paul got to his feet and walked over to her. "I'm OK now, Mom," he announced. "I'm going outside."

378 — 389 — 394 — 403 — 406 — 412

Smooth muscle is the primary muscle of internal organs and tubes, such as the stomach, urinary bladder, and blood vessels. Its primary function is to influence the movement of material into, out of, and within the body. An example is the passage of food through the gastrointestinal tract. Viewed under the microscope, smooth muscle lacks the obvious cross-bands of striated muscles (Fig. 12.1c). Its lack of banding results from

FIG. 12.1   The three types of muscles

(a) Skeletal muscle fibers are large, multinucleate cells that appear striped or striated under the microscope.

Nucleus

Muscle fiber (cell)

Striations

(b) Cardiac muscle fibers are also striated but they are smaller, branched, and uninucleate. Cells are joined in series by junctions called intercalated disks.

Nucleus

Muscle fiber

Intercalated disk

Striations

(c) Smooth muscle fibers are small and lack striations.

Nucleus

Muscle fiber

the less organized arrangement of contractile fibers within the muscle cells.

Skeletal muscles are often described as voluntary muscles, and smooth and cardiac muscle as involuntary. However, this is not a precise classification. Skeletal muscles can contract without conscious direction, and we can learn a certain degree of conscious control over some smooth and cardiac muscle.

Skeletal muscles are unique in that they contract only in response to a signal from a somatic motor neuron. They cannot initiate their own contraction, and their contraction is not influenced directly by hormones.

In contrast, cardiac and smooth muscle have multiple levels of control. Their primary extrinsic control arises through autonomic innervation, but some types of smooth and cardiac muscle can contract spontaneously, without signals from the central nervous system. In addition, the activity of cardiac and some smooth muscle is subject to modulation by the endocrine system. Despite these differences, smooth and cardiac muscle share many properties with skeletal muscle.

In this chapter, we discuss skeletal and smooth muscle anatomy and contraction, and conclude by comparing the properties of skeletal muscle, smooth muscle, and cardiac muscle. All three muscle types have certain properties in common. The signal to initiate muscle contraction is an intracellular calcium signal, and movement is created when a motor protein called *myosin* uses energy from adenosine triphosphate (ATP) to change its conformation. The details of these processes vary with the different muscle types.

SKELETAL MUSCLE

Skeletal muscles make up the bulk of muscle in the body and constitute about 40% of total body weight. They position and move the skeleton, as their name suggests. Skeletal muscles are usually attached to bones by **tendons** made of collagen [p. 82]. The **origin** of a muscle is the end of the muscle that is attached closest to the trunk or to the more stationary bone. The **insertion** of the muscle is the more *distal* {*distantia*, distant} or more mobile attachment.

When the bones attached to a muscle are connected by a flexible joint, contraction of the muscle moves the skeleton. The muscle is called a **flexor** if the centers of the connected bones are brought closer together when the muscle contracts, and the movement is called *flexion*. The muscle is called an **extensor** if the bones move away from each other when the muscle contracts, and the movement is called *extension*.

Most joints in the body have both flexor and extensor muscles, because a contracting muscle can pull a bone in one direction but cannot push it back. Flexor-extensor pairs are called **antagonistic muscle groups** because they exert opposite effects. **FIGURE 12.2** shows a pair of antagonistic muscles in the arm: the *biceps brachii* {*brachion*, arm}, which acts as the flexor, and the *triceps brachii*, which acts as the extensor. When you do a "dumbbell curl" with a weight in your hand, the biceps muscle contracts and the hand and forearm move toward the shoulder. When you lower the weight,

FIG. 12.2 Antagonistic muscles

Antagonistic muscle groups move bones in opposite directions. Muscle contraction can pull on a bone but cannot push a bone away.

(a) Flexion moves bones closer together.

(b) Extension moves bones away from each other.

Triceps muscle relaxes.

Biceps muscle contracts (flexor).

Triceps muscle contracts (extensor).

Biceps muscle relaxes.

the triceps contracts, and the flexed forearm moves away from the shoulder. In each case, when one muscle contracts and shortens, the antagonistic muscle must relax and lengthen.

CONCEPT CHECK

1. Identify as many pairs of antagonistic muscle groups in the body as you can. If you cannot name them, point out the probable location of the flexor and extensor of each group.

Skeletal Muscles Are Composed of Muscle Fibers

Muscles function together as a unit. A skeletal muscle is a collection of muscle cells, or **muscle fibers,** just as a nerve is a collection of neurons. Each skeletal muscle fiber is a long, cylindrical cell with up to several hundred nuclei near the surface of the fiber (see Anatomy Summary, **FIG. 12.3a**). Skeletal muscle fibers are the largest cells in the body, created by the fusion of many individual embryonic muscle cells. Committed stem cells called **satellite cells** lie just outside the muscle fiber membrane. Satellite cells become active and differentiate into muscle when needed for muscle growth and repair.

The fibers in a given muscle are arranged with their long axes in parallel (Fig. 12.3a). Each skeletal muscle fiber is sheathed in connective tissue, with groups of adjacent muscle fibers bundled together into units called **fascicles.** Collagen, elastic fibers, nerves, and blood vessels are found between the fascicles. The entire muscle is enclosed in a connective tissue sheath that is

FIG. 12.3 **ANATOMY SUMMARY**

Skeletal Muscles

(a) Structure of Skeletal Muscle

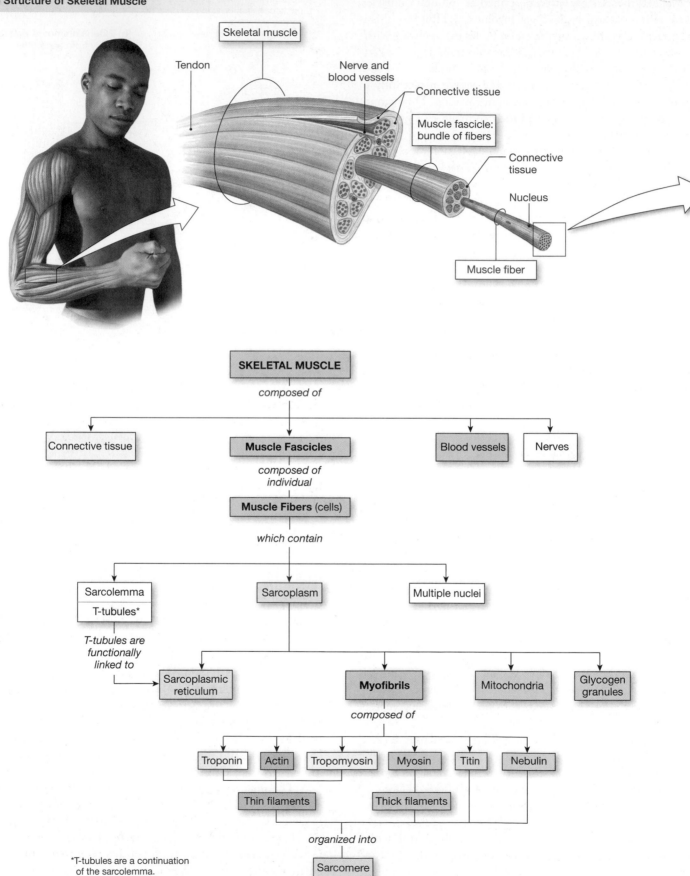

Skeletal muscle

Tendon

Nerve and blood vessels

Connective tissue

Muscle fascicle: bundle of fibers

Connective tissue

Nucleus

Muscle fiber

SKELETAL MUSCLE

composed of

Connective tissue | **Muscle Fascicles** | Blood vessels | Nerves

composed of individual

Muscle Fibers (cells)

which contain

Sarcolemma / T-tubules*

Sarcoplasm

Multiple nuclei

T-tubules are functionally linked to

Sarcoplasmic reticulum | **Myofibrils** | Mitochondria | Glycogen granules

composed of

Troponin | Actin | Tropomyosin | Myosin | Titin | Nebulin

Thin filaments | Thick filaments

organized into

Sarcomere

*T-tubules are a continuation of the sarcolemma.

Ultrastructure of Muscle

(b) Structure of a Skeletal Muscle Fiber

Mitochondria

Sarcoplasmic reticulum

Nucleus

Thick filament

Thin filament

T-tubules

Sarcolemma

Myofibril

(c) Myofibril

A band

Sarcomere

Z disk

Z disk

Myofibril

M line

I band

H zone

(d) Components of a Myofibril

Titin

Z disk

M line

Myosin crossbridges

Z disk

(e) Thick Filaments

M line

Myosin heads

Hinge region

Myosin tail

Myosin Molecule

(f) Thin Filaments

Titin

Troponin

Nebulin

Tropomyosin

G-actin molecule

Actin Chain

continuous with the connective tissue around the muscle fibers and fascicles and with the tendons holding the muscle to underlying bones.

Muscle Fiber Anatomy Muscle physiologists, like neurobiologists, use specialized vocabulary (**TBL. 12.1**). The cell membrane of a muscle fiber is called the **sarcolemma** {*sarkos,* flesh + *lemma,* shell}, and the cytoplasm is called the **sarcoplasm.** The main intracellular structures in striated muscles are **myofibrils** {*myo-,* muscle}, highly organized bundles of contractile and elastic proteins that carry out the work of contraction.

Skeletal muscle fibers also contain extensive **sarcoplasmic reticulum** (SR), a form of modified endoplasmic reticulum that wraps around each myofibril like a piece of lace (Figs. 12.3b, 12.4). The sarcoplasmic reticulum consists of longitudinal tubules with enlarged end regions called the **terminal cisternae** {*cisterna,* a reservoir}. The sarcoplasmic reticulum concentrates and sequesters Ca^{2+} {*sequestrare,* to put in the hands of a trustee} with the help of a *Ca^{2+}-ATPase* in the SR membrane. Calcium release from the SR creates calcium signals that play a key role in contraction in all types of muscle.

The terminal cisternae are adjacent to and closely associated with a branching network of **transverse tubules,** also known as **t-tubules** (**FIG. 12.4**). One t-tubule and its two flanking terminal cisternae are called a *triad.* The membranes of t-tubules are a continuation of the muscle fiber membrane, which makes the lumen of t-tubules continuous with the extracellular fluid.

To understand how this network of t-tubules deep inside the muscle fiber communicates with the outside, take a lump of soft clay and poke your finger into the middle of it. Notice how the outside surface of the clay (analogous to the surface membrane of the muscle fiber) is now continuous with the sides of the hole that you poked in the clay (the membrane of the t-tubule).

T-tubules allow action potentials to move rapidly from the cell surface into the interior of the fiber so that they reach the terminal cisternae nearly simultaneously. Without t-tubules, the action potential would reach the center of the fiber only by conduction of the action potential through the cytosol, a slower and less direct process that would delay the response time of the muscle fiber.

The cytosol between the myofibrils contains many glycogen granules and mitochondria. Glycogen, the storage form of glucose found in animals, is a reserve source of energy. Mitochondria

TABLE 12.1	Muscle Terminology
General Term	**Muscle Equivalent**
Muscle cell	Muscle fiber
Cell membrane	Sarcolemma
Cytoplasm	Sarcoplasm
Modified endoplasmic reticulum	Sarcoplasmic reticulum

FIG. 12.4   T-tubules

T-tubules are extensions of the cell membrane (sarcolemma) that associate with the ends (terminal cisternae) of the sarcoplasmic reticulum.

T-tubule brings action potentials into interior of muscle fiber.

Sarcoplasmic reticulum stores Ca^{2+}.

Sarcolemma

Triad

Thick filament

Thin filament

Terminal cisterna

ontain the enzymes for oxidative phosphorylation of glucose and other biomolecules, so they produce much of the ATP for muscle contraction [p. 109].

Myofibrils Are Muscle Fiber Contractile Structures

One muscle fiber contains a thousand or more myofibrils that occupy most of the intracellular volume, leaving little space for cytosol and organelles (Fig. 12.3b). Each myofibril is composed of several types of proteins organized into repeating contractile structures called *sarcomeres.* Myofibril proteins include the motor protein *myosin,* which forms thick filaments; the microfilament *actin* [p. 68], which creates *thin filaments;* the regulatory proteins *tropomyosin* and *troponin;* and two giant accessory proteins, *titin* and *nebulin.*

Myosin {*myo-,* muscle} is a motor protein with the ability to create movement [p. 69]. Various isoforms of myosin occur in different types of muscle and help determine the muscle's speed of contraction. Each myosin molecule is composed of protein chains that intertwine to form a long tail and a pair of tadpole-like heads (Fig. 12.3e). The rod-like tail is stiff, but the protruding myosin heads have an elastic hinge region where the heads join the rods. This hinge region allows the heads to swivel around their point of attachment.

Each myosin head has two protein chains: a heavy chain and a smaller *light chain.* The heavy chain is the *motor domain* that binds ATP and uses the energy from ATP's high-energy phosphate bond to create movement. Because the motor domain acts as an enzyme, it is considered a **myosin ATPase.** The heavy chain also contains a binding site for actin. In skeletal muscle, about 250 myosin molecules join to create a **thick filament.** Each thick filament is arranged so that the myosin heads are clustered at each end of the filament, and the central region of the filament is a bundle of myosin tails.

Actin {*actum,* to do} is a protein that makes up the **thin filaments** of the muscle fiber. One actin molecule is a globular protein (*G-actin*), represented in Figure 12.3f by a round ball. Usually, multiple G-actin molecules polymerize to form long chains or filaments, called *F-actin.* In skeletal muscle, two F-actin polymers twist together like a double strand of beads, creating the thin filaments of the myofibril.

Most of the time, the parallel thick and thin filaments of the myofibril are connected by myosin **crossbridges** that span the space between the filaments. Each G-actin molecule has a single *myosin-binding site,* and each myosin head has one actin-binding site and one binding site for ATP. Crossbridges form when the myosin heads of thick filaments bind to actin in the thin filaments (Fig. 12.3d). Crossbridges have two states: low-force (relaxed muscles) and high-force (contracting muscles).

Under a light microscope, the arrangement of thick and thin filaments in a myofibril creates a repeating pattern of alternating light and dark bands (Figs. 12.1a, 12.3c). One repeat of the pattern forms a **sarcomere** {*sarkos,* flesh + *-mere,* a unit or segment}, the contractile unit of the myofibril. Each sarcomere has the following elements (**FIG. 12.5**):

1. **Z disks.** One sarcomere is composed of two Z disks and the filaments found between them. Z disks are zigzag protein structures that serve as the attachment site for thin filaments. The abbreviation *Z* comes from *zwischen,* the German word for "between."

2. **I bands.** These are the lightest color bands of the sarcomere and represent a region occupied only by thin filaments. The abbreviation *I* comes from *isotropic,* a description from early microscopists meaning that this region reflects light uniformly under a polarizing microscope. A Z disk runs through the middle of every I band, so each half of an I band belongs to a different sarcomere.

3. **A band.** This is the darkest of the sarcomere's bands and encompasses the entire length of a thick filament. At the outer edges of the A band, the thick and thin filaments overlap. The center of the A band is occupied by thick filaments only. The abbreviation *A* comes from *anisotropic* {*an-,* not}, meaning that the protein fibers in this region scatter light unevenly.

4. **H zone.** This central region of the A band is lighter than the outer edges of the A band because the H zone is occupied by thick filaments only. The *H* comes from *helles,* the German word for "clear."

5. **M line.** This band represents proteins that form the attachment site for thick filaments, equivalent to the Z disk for the thin filaments. Each M line divides an A band in half. *M* is the abbreviation for *mittel,* the German word for "middle."

In three-dimensional array, the actin and myosin molecules form a lattice of parallel, overlapping thin and thick filaments, held in place by their attachments to the Z-disk and M-line proteins, respectively (Fig. 12.5b). When viewed end-on, each thin filament is surrounded by three thick filaments, and six thin filaments encircle each thick filament (Fig. 12.5c, rightmost circle).

The proper alignment of filaments within a sarcomere is ensured by two proteins: titin and nebulin (**FIG. 12.6**). **Titin** is a huge elastic molecule and the largest known protein, composed of more than 25,000 amino acids. A single titin molecule stretches from one Z disk to the neighboring M line. To get an idea of the immense size of titin, imagine that one titin molecule is an 8-foot-long piece of the very thick rope used to tie ships to a wharf. By comparison, a single actin molecule would be about the length and weight of a single eyelash.

Titin has two functions: (1) it stabilizes the position of the contractile filaments and (2) its elasticity returns stretched muscles to their resting length. Titin is helped by **nebulin,** an inelastic giant protein that lies alongside thin filaments and attaches to the Z disk. Nebulin helps align the actin filaments of the sarcomere.

CONCEPT CHECK

2. Why are the ends of the A band the darkest region of the sarcomere when viewed under the light microscope?

3. What is the function of t-tubules?

4. Why are skeletal muscles described as striated?

Muscle Contraction Creates Force

The contraction of muscle fibers is a remarkable process that enables us to create force to move or to resist a load. In muscle physiology, the force created by contracting muscle is called **muscle tension.** The **load** is a weight or force that opposes contraction of a muscle. **Contraction,** the creation of tension in a muscle, is an active process that requires energy input from ATP. **Relaxation** is the release of tension created by a contraction.

FIGURE 12.7 maps the major steps leading up to skeletal muscle contraction.

1. **Events at the neuromuscular junction** convert an acetylcholine signal from a somatic motor neuron into an electrical signal in the muscle fiber [p. 371].

2. **Excitation-contraction (E-C) coupling** is the process in which muscle action potentials initiate calcium signals that in turn activate a contraction-relaxation cycle.

3. At the molecular level, a **contraction-relaxation cycle** can be explained by the *sliding filament theory of contraction.* In intact muscles, one contraction-relaxation cycle is called a muscle *twitch.*

In the sections that follow, we start with the sliding filament theory for muscle contraction. From there, we look at the integrated function of a muscle fiber as it undergoes excitation-contraction coupling. The skeletal muscle section ends with a discussion of the innervation of muscles and how muscles move bones around joints.

CONCEPT CHECK

5. What are the three anatomical elements of a neuromuscular junction?

6. What is the chemical signal at a neuromuscular junction?

FIG. 12.5 **ESSENTIALS**

The Sarcomere

Organization of a Sarcomere

The Z disk (not shown in part (c)) has accessory proteins that link the thin filaments together, similar to the accessory proteins shown for the M line. Myosin heads are omitted for simplicity.

KEY
— Actin
━ Myosin

(a)

(b)

(c)

I band
Actin only

H zone
Myosin only

M line
Myosin linked with accessory proteins

A band
(outer edge)
Actin and myosin overlap

The sarcomere shortens during contraction. As contraction takes place, actin and myosin do not change length but instead slide past one another.

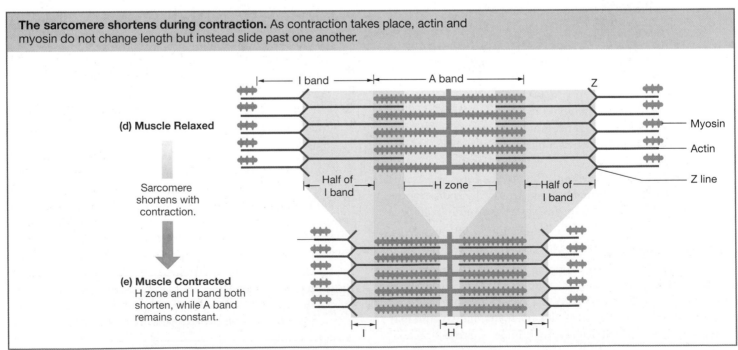

(d) **Muscle Relaxed**

Sarcomere shortens with contraction.

(e) **Muscle Contracted**
H zone and I band both shorten, while A band remains constant.

FIG. 12.6 Titin and nebulin

Titin and nebulin are giant accessory proteins. Titin spans the distance from one Z disk to the neighboring M line. Nebulin, lying along the thin filaments, attaches to a Z disk but does not extend to the M line.

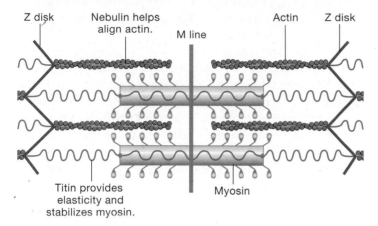

Actin and Myosin Slide Past Each Other during Contraction

In previous centuries, scientists observed that when muscles move a load, they shorten. This observation led to early theories of contraction, which proposed that muscles were made of molecules that curled up and shortened when active, then relaxed and stretched at rest, like elastic in reverse. The theory received support when myosin was found to be a helical molecule that shortened upon heating (the reason meat shrinks when you cook it).

FIG. 12.7 Summary map of muscle contraction

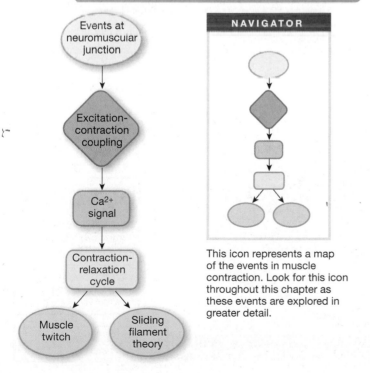

This icon represents a map of the events in muscle contraction. Look for this icon throughout this chapter as these events are explored in greater detail.

In 1954, however, scientists Andrew Huxley and Rolf Niedergerke discovered that the length of the A band of a myofibril remains constant during contraction. Because the A band represents the myosin filament, Huxley and Niedergerke realized that shortening of the myosin molecule could not be responsible for contraction. Subsequently, they proposed an alternative model, the **sliding filament theory of contraction.** In this model, overlapping actin and myosin filaments of fixed length slide past one another in an energy-requiring process, resulting in muscle contraction.

If you examine a myofibril at its resting length, you see that within each sarcomere, the ends of the thick and thin filaments overlap slightly (Fig. 12.5d). In the relaxed state, a sarcomere has a large I band (thin filaments only) and an A band whose length is the length of the thick filament.

When the muscle contracts, the thick and thin filaments slide past each other. The Z disks of the sarcomere move closer together as the sarcomere shortens (Fig. 12.5e). The I band and H zone—regions where actin and myosin do not overlap in resting muscle—almost disappear.

Despite shortening of the sarcomere, the length of the A band remains constant. These changes are consistent with the sliding of thin actin filaments along the thick myosin filaments as the actin filaments move toward the M line in the center of the sarcomere. It is from this process that the sliding filament theory of contraction derives its name.

The sliding filament theory explains how a muscle can contract and create force without creating movement. For example, if you push on a wall, you are creating tension in many muscles of your body without moving the wall. According to the sliding filament theory, tension generated in a muscle fiber is directly proportional to the number of high-force crossbridges between the thick and thin filaments.

Myosin Crossbridges Move Actin Filaments

The movement of myosin crossbridges provides force that pushes the actin filament during contraction. The process can be compared to a competitive sailing team, with many people holding the rope that raises a heavy mainsail. When the order to raise the mainsail comes, each person on the team begins pulling on the rope, hand over hand, grabbing, pulling, and releasing repeatedly as the rope moves past.

In muscle, myosin heads bind to actin molecules, which are the "rope." A calcium signal initiates the **power stroke,** when myosin crossbridges swivel and push the actin filaments toward the center of the sarcomere. At the end of a power stroke, each myosin head releases actin, then swivels back and binds to a new actin molecule, ready to start another contractile cycle. During contraction, the heads do not all release at the same time or the fibers would slide back to their starting position, just as the mainsail would fall if the sailors all released the rope at the same time.

The power stroke repeats many times as a muscle fiber contracts. The myosin heads bind, push, and release actin molecules over and over as the thin filaments move toward the center of the sarcomere.

Myosin ATPase Where does energy for the power stroke come from? The answer is ATP. Myosin converts the chemical bond energy of ATP into the mechanical energy of crossbridge motion.

Myosin is an ATPase (*myosin ATPase*) that hydrolyzes ATP to ADP and inorganic phosphate (P_i). The energy released by ATP hydrolysis is trapped by myosin and stored as potential energy in the angle between the myosin head and the long axis of the myosin filament. Myosin heads in this position are said to be "cocked," or ready to rotate. The potential energy of the cocked heads becomes kinetic energy in the power stroke that moves actin.

Calcium Signals Initiate Contraction

How does a calcium signal turn muscle contraction on and off? The answer is found in **troponin** (TN), a calcium-binding complex of three proteins. Troponin controls the positioning of an elongated protein polymer, **tropomyosin** {*tropos*, to turn}.

In resting skeletal muscle, tropomyosin wraps around actin filaments and partially covers actin's myosin-binding sites (**FIG. 12.8a**). This is tropomyosin's blocking or "off" position. Weak, low-force actin-myosin binding can still take place, but myosin is blocked from completing its power stroke, much as the safety latch on a gun keeps the cocked trigger from being pulled. Before contraction can occur, tropomyosin must be shifted to an "on" position that uncovers the remainder of actin's myosin-binding site.

The off-on positioning of tropomyosin is regulated by troponin. When contraction begins in response to a calcium signal

(❶ in Fig. 12.8b), one protein of the complex—**troponin C**—binds reversibly to Ca^{2+} ❷. The calcium-troponin C complex pulls tropomyosin completely away from actin's myosin-binding sites ❸. This "on" position enables the myosin heads to form strong, high-force crossbridges and carry out their power strokes ❹, moving the actin filament ❺. Contractile cycles repeat as long as the binding sites are uncovered.

For muscle relaxation to occur, Ca^{2+} concentrations in the cytosol must decrease. By the law of mass action [p. 48], when cytosolic calcium decreases, Ca^{2+} unbinds from troponin. In the absence of Ca^{2+}, troponin allows tropomyosin to return to the "off" position, covering most of actin's myosin-binding sites. During the brief portion of the relaxation phase when actin and myosin are not bound to each other, the filaments of the sarcomere slide back to their original positions with the aid of titin and elastic connective tissues within the muscle.

The discovery that Ca^{2+}, not the action potential, is the signal for muscle contraction was the first piece of evidence suggesting that calcium acts as a messenger inside cells. Initially scientists thought that calcium signals occurred only in muscles, but we now know that calcium is an almost universal second messenger [p. 177].

Myosin Heads Step along Actin Filaments

FIGURE 12.9 shows the molecular events of a contractile cycle in skeletal muscle. We will start a cycle with the **rigor state** {*rigere*, to be stiff}, where the myosin heads are tightly bound to G-actin

---

FIG. 12.8   Troponin and tropomyosin

(a) Relaxed state. Myosin head cocked. Tropomyosin partially blocks binding site on actin. Myosin is weakly bound to actin.

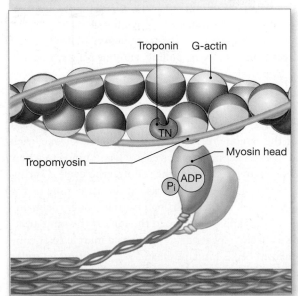

(b) Initation of contraction. A calcium signal initiates contraction.

❶ Ca^{2+} levels increase in cytosol.

❷ Ca^{2+} binds to troponin (TN).

❸ Troponin-Ca^{2+} complex pulls tropomyosin away from actin's myosin-binding site.

❹ Myosin binds strongly to actin and completes power stroke.

❺ Actin filament moves.

FIG. 12.9 The contraction cycle

Tight Binding in the Rigor State

G-actin molecule

Myosin binding sites

Myosin filament

1 ATP binds to myosin. Myosin releases actin.

ATP binds.

NAVIGATOR

Contraction-relaxation

Sliding filament

ADP releases.

4 Myosin releases ADP at the end of the power stroke.

2 Myosin hydrolyzes ATP. Energy from ATP rotates the myosin head to the cocked position. Myosin binds weakly to actin.

ADP

P_i

ADP and P_i remain bound.

The Power Stroke

← Actin filament moves toward M line.

Head swivels.

Myosin releases P_i.

Ca^{2+} signal

3 Power stroke begins when tropomyosin (not shown) moves off the binding site.

molecules. No nucleotide (ATP or ADP) is bound to myosin. In living muscle, the rigor state occurs for only a very brief period. Then:

1 ATP binds and myosin detaches. An ATP molecule binds to the myosin head. ATP-binding decreases the actin-binding affinity of myosin, and myosin releases from actin.

2 ATP hydrolysis provides energy for the myosin head to rotate and reattach to actin. The ATP-binding site on the

myosin head closes around ATP and hydrolyzes it to ADP and inorganic phosphate (P_i). Both ADP and P_i remain bound to myosin as energy released by ATP hydrolysis rotates the myosin head until it forms a 90° angle with the long axis of the filaments. In this cocked position, myosin binds to a new actin that is 1–3 molecules away from where it started.

The newly formed actin-myosin crossbridge is weak and low-force because tropomyosin is partially blocking actin's binding site. However, in this rotated position myosin has stored potential energy, like a stretched spring. The head is cocked, just as someone preparing to fire a gun pulls back or cocks the spring-loaded hammer before firing. Most resting muscle fibers are in this state, cocked and prepared to contract, and just waiting for a calcium signal.

3 The power stroke. The power stroke (*crossbridge tilting*) begins after Ca^{2+} binds to troponin to uncover the rest of the myosin-binding site. The crossbridges transform into strong, high-force bonds as myosin releases P_i. Release of P_i allows the myosin head to swivel. The heads swing toward the M line, sliding the attached actin filament along with them. The power stroke is also called *crossbridge tilting* because the myosin head and hinge region tilt from a 90° angle to a 45° angle.

4 Myosin releases ADP. At the end of the power stroke, myosin releases ADP, the second product of ATP hydrolysis. With ADP gone, the myosin head is again tightly bound to actin in the rigor state. The cycle is ready to begin once more as a new ATP binds to myosin.

The Rigor State The contractile cycle illustrated in Figure 12.9 begins with the rigor state in which no ATP or ADP is bound to myosin. This state in living muscle is normally brief. Living muscle fibers have a sufficient supply of ATP that quickly binds to myosin once ADP is released (step 1). As a result, relaxed muscle fibers remain mostly in step 2.

After death, however, when metabolism stops and ATP supplies are exhausted, muscles are unable to bind more ATP, so they remain in the tightly bound rigor state. In the condition known as *rigor mortis,* the muscles "freeze" owing to immovable crossbridges. The tight binding of actin and myosin persists for a day or so after death, until enzymes released within the decaying fiber begin to break down the muscle proteins.

CONCEPT CHECK

7. Each myosin molecule has binding sites for what molecules?

8. What is the difference between F-actin and G-actin?

9. Myosin hydrolyzes ATP to ADP and P_i. Enzymes that hydrolyze ATP are collectively known as

    _____ .

Although the preceding discussion sounds as if we know everything there is to know about the molecular basis of muscle contraction, in reality this is simply our current model. The process is more complex than presented here. It now appears that myosin can influence Ca^{2+}-troponin binding, depending on whether the myosin is bound to actin in a strong (rigor) state, bound to actin in a weak state, or not bound at all. The details of this influence are still being worked out.

Studying contraction and the movement of molecules in a myofibril has proved very difficult. Many research techniques rely on crystallized molecules, electron microscopy, and other tools that cannot be used with living tissues. Often we can see the thick

---

BIOTECHNOLOGY

Watching Myosin Work

One big step forward in understanding the power stroke of myosin was the development of the *in vitro* motility assay in the 1980s. In this assay, isolated myosin molecules are randomly bonded to a specially coated glass coverslip. A fluorescently labeled actin molecule is placed on top of the myosin molecules. With ATP as a source of energy, the myosin heads bind to the actin and move it across the coverslip, marked by a fluorescent trail as it goes. In even more ingenious experiments, developed in 1995, a single myosin molecule is bound to a tiny bead that elevates it above the surface of the cover slip. An actin molecule is placed on top of the myosin molecule, like the balancing pole of a tightrope walker. As the myosin "motor" moves the actin molecule, lasers measure the nanometer movements and piconewton forces created with each cycle of the myosin head. Because of this technique, researchers can now measure the mechanical work being done by a single myosin molecule! For an animation and movie of the process, visit *http://physiology.med.uvm.edu/warshaw/TechspgInVitro.html.*

---

and thin filaments only at the beginning and end of contraction. Progress is being made, however, and perhaps in the next decade you will see a "movie" of muscle contraction, constructed from photographs of sliding filaments.

CONCEPT CHECK

10. Name an elastic fiber in the sarcomere that aids relaxation.

11. In the sliding filament theory of contraction, what prevents the filaments from sliding back to their original position each time a myosin head releases to bind to the next actin binding site?

Acetylcholine Initiates Excitation-Contraction Coupling

Now let's start at the neuromuscular junction and follow the events leading up to contraction. As you learned earlier in the chapter, this combination of electrical and mechanical events in a muscle fiber is called *excitation-contraction coupling*. E-C coupling has four major events:

1. Acetylcholine (ACh) is released from the somatic motor neuron.
2. ACh initiates an action potential in the muscle fiber.
3. The muscle action potential triggers calcium release from the sarcoplasmic reticulum.
4. Calcium combines with troponin and initiates contraction.

Now let's look at these steps in detail. Acetylcholine released into the synapse at a neuromuscular junction binds to ACh receptor-channels on the motor end plate of the muscle fiber (**FIG. 12.10a ①**) [p. 371]. When the ACh-gated channels open, they allow both Na^+ and K^+ to cross the membrane. However, Na^+ influx exceeds K^+ efflux because the electrochemical driving force is greater for Na^+ [p. 156]. The addition of net positive charge to the muscle fiber depolarizes the membrane, creating an **end-plate potential (EPP)**. Normally, end-plate potentials always reach threshold and initiate a muscle action potential (Fig. 12.10a ②).

The action potential travels across the surface of the muscle fiber and into the t-tubules by the sequential opening of voltage-gated Na^+ channels. The process is similar to the conduction of action potentials in axons, although action potentials in skeletal muscle are conducted more slowly than action potentials in myelinated axons [p. 249].

The action potential that moves down the t-tubules causes Ca^{2+} release from the sarcoplasmic reticulum (Fig. 12.10b ③, ④). Free cytosolic Ca^{2+} levels in a resting muscle are normally quite low, but after an action potential, they increase about 100-fold. As you've learned, when cytosolic Ca^{2+} levels are high, Ca^{2+} binds to troponin, tropomyosin moves to the "on" position ⑤, and contraction occurs ⑥.

At the molecular level, transduction of the electrical signal into a calcium signal requires two key membrane proteins. The t-tubule membrane contains a voltage-sensing **L-type calcium channel** protein ($Ca_v1.1$) called a **dihydropyridine (DHP) receptor** (Fig. 12.10b ③). The DHP receptors, found only in skeletal muscle, are mechanically linked to Ca^{2+} channels in the adjacent sarcoplasmic reticulum. These SR **Ca^{2+} release channels** are also known as **ryanodine receptors (RyR).**

When the depolarization of an action potential reaches a DHP receptor, the receptor changes conformation. The conformation change opens the RyR Ca^{2+} release channels in the sarcoplasmic reticulum (Fig. 12.10b ④). Stored Ca^{2+} then flows down its electrochemical gradient into the cytosol, where it initiates contraction.

Scientists used to believe that the calcium channel we call the DHP receptor did not form an open channel for calcium entry from the ECF. However, in recent years, it has become apparent that there is a small amount of Ca^{2+} movement through the DHP receptor, described as *excitation-coupled Ca^{2+} entry*. However, skeletal muscle contraction will still take place if there is no ECF Ca^{2+} to come through the channel, so the physiological role of excitation-coupled Ca^{2+} entry is unclear.

Relaxation To end a contraction, calcium must be removed from the cytosol. The sarcoplasmic reticulum pumps Ca^{2+} back into its lumen using a **Ca^{2+}-ATPase** [p. 143]. As the free cytosolic Ca^{2+} concentration decreases, the equilibrium between bound and unbound Ca^{2+} is disturbed and calcium releases from troponin. Removal of Ca^{2+} allows tropomyosin to slide back and block actin's myosin-binding site. As the crossbridges release, the muscle fiber relaxes with the help of elastic fibers in the sarcomere and in the connective tissue of the muscle.

Timing of E-C Coupling The graphs in **FIGURE 12.11** show the timing of electrical and mechanical events during E-C coupling. The somatic motor neuron action potential is followed by the skeletal muscle action potential, which in turn is followed by contraction. A single contraction-relaxation cycle in a skeletal muscle fiber is known as a **twitch.** Notice that there is a short delay—the **latent period**—between the muscle action potential and the beginning of muscle tension development. This delay represents the time required for calcium release and binding to troponin.

Once contraction begins, muscle tension increases steadily to a maximum value as crossbridge interaction increases. Tension then decreases in the relaxation phase of the twitch. During relaxation, elastic elements of the muscle return the sarcomeres to their resting length.

A single action potential in a muscle fiber evokes a single twitch (Fig. 12.11, bottom graph). However, muscle twitches vary from fiber to fiber in the speed with which they develop tension (the rising slope of the twitch curve), the maximum tension they achieve (the height of the twitch curve), and the duration of the twitch (the width of the twitch curve). You will learn about factors that affect these parameters in upcoming sections. First, we discuss how muscles produce ATP to provide energy for contraction and relaxation.

CONCEPT CHECK

12. Which part of contraction requires ATP? Does relaxation require ATP?

13. What events are taking place during the latent period before contraction begins?

RUNNING PROBLEM

Paul had experienced mild attacks of muscle weakness in his legs before, usually in the morning. Twice the weakness had come on after exposure to cold. Each attack had disappeared within minutes, and Paul seemed to suffer no lasting effects. On the advice of Paul's family doctor, Mrs. Leong takes her son to see a specialist in muscle disorders, who suspects a condition called periodic paralysis. The periodic paralyses are a family of disorders caused by Na^+ or Ca^{2+} ion channel mutations in the membranes of skeletal muscle fibers. The specialist believes that Paul has a condition in which defective voltage-gated Na^+ channels fail to inactivate after they open.

Q1: *When Na^+ channels on the muscle membrane open, which way does Na^+ move?*

Q2: *What effect would continued movement of Na^+ have on the membrane potential of muscle fibers?*

378 389 394 403 406 412

FIG. 12.10 ESSENTIALS

Excitation-Contraction Coupling and Relaxation

(a) Initiation of Muscle Action Potential

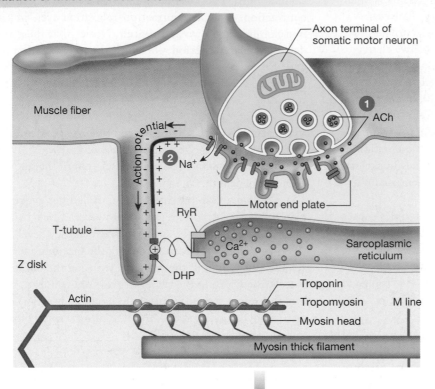

KEY

DHP = dihydropyridine L-type calcium channel

RyR = ryanodine receptor-channel

① Somatic motor neuron releases ACh at neuromuscular junction.

② Net entry of Na⁺ through ACh receptor-channel initiates a muscle action potential.

(b) Excitation-Contraction Coupling

③ Action potential in t-tubule alters conformation of DHP receptor.

④ DHP receptor opens RyR Ca^{2+} release channels in sarcoplasmic reticulum, and Ca^{2+} enters cytoplasm.

⑤ Ca^{2+} binds to troponin, allowing actin-myosin binding.

⑥ Myosin heads execute power stroke.

⑦ Actin filament slides toward center of sarcomere.

(c) Relaxation Phase

⑧ Sarcoplasmic Ca^{2+}-ATPase pumps Ca^{2+} back into SR.

⑨ Decrease in free cytosolic $[Ca^{2+}]$ causes Ca^{2+} to unbind from troponin.

⑩ Tropomyosin re-covers binding site. When myosin heads release, elastic elements pull filaments back to their relaxed position.

FIG. 12.11 Timing of E-C coupling

Action potentials in the axon terminal (top graph) and in the muscle fiber (middle graph) are followed by a muscle twitch (bottom graph).

Muscle fiber

Action potential from CNS

Motor end plate

Axon terminal

Recording electrodes

Muscle action potential

Motor Neuron Action Potential

Neuron membrane potential in mV

+30

-70

Time

Muscle Fiber Action Potential

Muscle fiber membrane potential in mV

+20

-80

2 msec

Time

Development of Tension during One Muscle Twitch

Tension

Latent period | Contraction phase | Relaxation phase

10–100 msec

Time

NAVIGATOR

Neuro-muscular junction (NMJ)

E-C coupling

Muscle twitch

Q **FIGURE QUESTIONS**

Movement of what ion(s) in what direction(s) creates
(a) the neuronal action potential?
(b) the muscle action potential?

Skeletal Muscle Contraction Requires a Steady Supply of ATP

The muscle fiber's use of ATP is a key feature of muscle physiology. Muscles require energy constantly: during contraction for crossbridge movement and release, during relaxation to pump Ca^{2+} back into the sarcoplasmic reticulum, and after E-C coupling to restore Na^+ and K^+ to the extracellular and intracellular compartments, respectively. Where do muscles get the ATP they need for this work?

The amount, or pool, of ATP stored in a muscle fiber at any one time is sufficient for only about eight twitches. As ATP is converted to ADP and P_i during contraction, the ATP pool must be replenished by transfer of energy from other high-energy phosphate bonds or by synthesis of ATP through the slower metabolic pathways of glycolysis and oxidative phosphorylation.

The backup energy source of muscles is **phosphocreatine,** a molecule whose high-energy phosphate bonds are created from creatine and ATP when muscles are at rest (**FIG. 12.12**). When muscles become active, such as during exercise, the high-energy phosphate group of phosphocreatine is quickly transferred to ADP, creating more ATP to power the muscles.

The enzyme that transfers the phosphate group from phosphocreatine to ADP is **creatine kinase (CK),** also known as *creatine phosphokinase* (CPK). Muscle cells contain large amounts of this enzyme. Consequently, elevated blood levels of creatine

FIG. 12.12 Phosphocreatine

Resting muscle stores energy from ATP in the high-energy bonds of phosphocreatine. Working muscle then uses that stored energy.

Muscle at rest

ATP from metabolism + creatine $\xrightarrow{\text{creatine kinase}}$ ADP + phosphocreatine

Working muscle

Phosphocreatine + ADP $\xrightarrow{\text{creatine kinase}}$ Creatine + ATP

needed for

- Myosin ATPase (contraction)
- Ca^{2+}-ATPase (relaxation)
- Na^+-K^+-ATPase (restores ions that cross cell membrane during action potential to their original compartments)

kinase usually indicate damage to skeletal or cardiac muscle. Because the two muscle types contain different isozymes [p. 99], clinicians can distinguish cardiac tissue damage during a heart attack from skeletal muscle damage.

Energy stored in high-energy phosphate bonds is very limited, so muscle fibers must use metabolism of biomolecules to transfer energy from covalent bonds to ATP. Carbohydrates, particularly glucose, are the most rapid and efficient source of energy for ATP production. Glucose is metabolized through glycolysis to pyruvate [p. 107]. In the presence of adequate oxygen, pyruvate goes into the citric acid cycle, producing about 30 ATP for each molecule of glucose.

When oxygen concentrations fall during strenuous exercise, muscle fiber metabolism relies more on *anaerobic glycolysis*. In this pathway, glucose is metabolized to lactate with a yield of only 2 ATP per glucose [p. 110]. Anaerobic metabolism of glucose is a quicker source of ATP but produces many fewer ATP per glucose. When energy demands are greater than the amount of ATP that can be produced through anaerobic glucose metabolism, muscles can function for only a short time without fatiguing.

Muscle fibers also obtain energy from fatty acids, although this process always requires oxygen. During rest and light exercise, skeletal muscles burn fatty acids along with glucose, one reason that modest exercise programs of brisk walking are an effective way to reduce body fat. However, the metabolic process by which fatty acids are converted to acetyl CoA is relatively slow and cannot produce ATP rapidly enough to meet the energy needs of muscle fibers during strenuous exercise. Under these conditions, muscle fibers rely more on glucose.

Proteins normally are not a source of energy for muscle contraction. Most amino acids found in muscle fibers are used to synthesize proteins rather than to produce ATP.

Do muscles ever run out of ATP? You might think so if you have ever exercised to the point of fatigue, the point at which you feel that you cannot continue or your limbs refuse to obey commands from your brain. Most studies show, however, that even intense exercise uses only 30% of the ATP in a muscle fiber. The condition we call fatigue must come from other changes in the exercising muscle.

CONCEPT CHECK

14. According to the convention for naming enzymes, what does the name creatine kinase tell you about this enzyme's function? [*Hint:* p. 101]

15. The reactions in Figure 12.12 show that creatine kinase catalyzes the creatine-phosphocreatine reaction in both directions. What then determines the direction that the reaction goes at any given moment? [*Hint:* p. 48]

Fatigue Has Multiple Causes

The physiological term **fatigue** describes a reversible condition in which an exercising muscle is no longer able to generate or sustain the expected power output. Fatigue is highly variable. It is influenced by the intensity and duration of the contractile activity, by whether the muscle fiber is using aerobic or anaerobic metabolism, by the composition of the muscle, and by the fitness level of the individual. The study of fatigue is complex, and research in this area is complicated by the fact that experiments are done under a wide range of conditions, from "skinned" (sarcolemma removed) single muscle fibers to exercising humans. Although many different factors have been *associated with* fatigue, the factors that *cause* fatigue are still uncertain.

Factors that have been proposed to play a role in fatigue are classified into **central fatigue** mechanisms, which arise in the central nervous system, and **peripheral fatigue** mechanisms, which arise anywhere between the neuromuscular junction and the contractile elements of the muscle (**FIG. 12.13**). Most experimental evidence suggests that muscle fatigue arises from excitation-contraction failure in the muscle fiber rather than from failure of control neurons or neuromuscular transmission.

Central fatigue includes subjective feelings of tiredness and a desire to cease activity. Several studies have shown that this psychological fatigue precedes physiological fatigue in the muscles and therefore may be a protective mechanism. Low pH from acid production during ATP hydrolysis is often mentioned as a possible cause of fatigue, and some evidence suggests that acidosis may influence the sensation of fatigue perceived by the brain. However, homeostatic mechanisms for pH balance maintain blood pH at normal levels until exertion is nearly maximal, so pH as a factor in central fatigue probably applies only in cases of maximal exertion.

Neural causes of fatigue could arise either from communication failure at the neuromuscular junction or from failure of

FIG. 12.13 Muscle fatigue

Muscle fatigue has many possible causes but the strongest evidence supports failure of EC coupling and subsequent events. In recent years, research indicated that lactate accumulation is no longer a likely cause of fatigue.

Types of Fatigue	Process Map	Proposed Mechanisms
Central fatigue	CNS	• Psychological effects • Protective reflexes
	Somatic motor neuron	
Peripheral fatigue	Neuromuscular junction	• ↓ Neurotransmitter release • ↓ Receptor activation
	Excitation-contraction coupling	• Change in muscle membrane potential
	Ca²⁺ signal	• SR Ca²⁺ leak • ↓ Ca²⁺ release • ↓ Ca²⁺-troponin interaction
	Contraction-relaxation	• Depletion theories: PCr, ATP, glycogen • Accumulation theories: H⁺, Pᵢ, lactate

the CNS command neurons. For example, if ACh is not synthesized in the axon terminal fast enough to keep up with neuron firing rate, neurotransmitter release at the synapse decreases. Consequently, the muscle end-plate potential fails to reach the threshold value needed to trigger a muscle fiber action potential, resulting in contraction failure. This type of fatigue is associated with some neuromuscular diseases, but it is probably not a factor in normal exercise.

Fatigue within the muscle fiber (peripheral fatigue) could occur in any of several sites. In extended submaximal exertion, fatigue is associated with the depletion of muscle glycogen stores. Because most studies show that lack of ATP is not a limiting factor, glycogen depletion may be affecting some other aspect of contraction, such as the release of Ca^{2+} from the sarcoplasmic reticulum.

The cause of fatigue in short-duration maximal exertion seems to be different. One theory is based on the increased levels of inorganic phosphate (P_i) produced when ATP and phosphocreatine are used for energy in the muscle fiber. Elevated

cytoplasmic P_i may slow P_i release from myosin and thereby alter the power stroke (see Fig. 12.9 ❹).

Another theory suggests that elevated phosphate levels decrease Ca^{2+} release because the phosphate combines with Ca^{2+} to become calcium phosphate. Some investigators feel that alterations in Ca^{2+} release from the sarcoplasmic reticulum play a major role in fatigue.

Ion imbalances have also been implicated in fatigue. During maximal exercise, K^+ leaves the muscle fiber with each action potential, and as a result K^+ concentrations rise in the extracellular fluid of the t-tubules. The shift in K^+ alters the membrane potential of the muscle fiber. Changes in Na^+-K^+-ATPase activity may also be involved. In short, muscle fatigue is a complex phenomenon with multiple causes that interact with each other.

CONCEPT CHECK 16. If K^+ concentration increases in the extracellular fluid surrounding a cell but does not change significantly in the cell's cytoplasm, the cell membrane (*depolarizes/hyperpolarizes*) and becomes (*more/less*) negative.

Skeletal Muscle Is Classified by Speed and Fatigue Resistance

Skeletal muscle fibers have traditionally been classified on the basis of their speed of contraction and their resistance to fatigue with repeated stimulation. But like so much in physiology, the more scientists learn, the more complicated the picture becomes. The current classification of muscle fiber types depends on the isoform of myosin expressed in the fiber (type 1 or type 2).

Muscle fiber types are not fixed for life. Muscles have plasticity and can shift their type depending on their activity. The currently accepted muscle fiber types in humans include **slow-twitch fibers** (also called *ST* or *type 1*), **fast-twitch oxidative-glycolytic fibers** (*FOG* or *type 2A*), and **fast-twitch glycolytic fibers** (*FG* or *type 2X*). Type 2X was previously identified as type 2B, which is found in other animals but not in humans.

Fast-twitch muscle fibers (type 2) develop tension two to three times faster than slow-twitch fibers (type 1). The speed with which a muscle fiber contracts is determined by the isoform of myosin ATPase present in the fiber's thick filaments. Fast-twitch fibers split ATP more rapidly and can, therefore, complete multiple contractile cycles more rapidly than slow-twitch fibers. This speed translates into faster tension development in the fast-twitch fibers.

The duration of contraction also varies according to fiber type. Twitch duration is determined largely by how fast the sarcoplasmic reticulum removes Ca^{2+} from the cytosol. As cytosolic Ca^{2+} concentrations fall, Ca^{2+} unbinds from troponin, allowing tropomyosin to move into position to partially block the myosin-binding sites. With the power stroke inhibited in this way, the muscle fiber relaxes.

Fast-twitch fibers pump Ca^{2+} into their sarcoplasmic reticulum more rapidly than slow-twitch fibers do, so fast-twitch fibers

RUNNING **PROBLEM**

Two forms of periodic paralysis exist. One form, called *hypoka-lemic periodic paralysis,* is characterized by decreased blood levels of K+ during paralytic episodes. The other form, *hyperka-lemic periodic paralysis (hyperKPP),* is characterized by either normal or increased blood levels of K+ during episodes. Results of a blood test revealed that Paul has the hyperkalemic form.

Q3: *In people with hyperKPP, attacks may occur after a period of exercise (i.e., after a period of repeated muscle contrac-tions). What ion is responsible for the repolarization phase of the muscle action potential, and in which direction does this ion move across the muscle fiber membrane? How might this be linked to hyperKPP?*

(378) (389) **394** (403) (406) (412)

have quicker twitches. The twitches in fast-twitch fibers last only about 7.5 msec, making these muscles useful for fine, quick move-ments, such as playing the piano. Contractions in slow-twitch muscle fibers may last more than 10 times as long. Fast-twitch fibers are used occasionally, but slow-twitch fibers are used almost constantly for maintaining posture, standing, or walking.

The second major difference between muscle fiber types is their ability to resist fatigue. Glycolytic fibers (fast-twitch type 2X)

rely primarily on anaerobic glycolysis to produce ATP. However, the accumulation of H+ from ATP hydrolysis contributes to aci-dosis, a condition implicated in the development of fatigue, as noted previously. As a result, glycolytic fibers fatigue more eas-ily than do oxidative fibers, which do not depend on anaerobic metabolism.

Oxidative fibers rely primarily on oxidative phosphorylation [p. 109] for production of ATP—hence their descriptive name. These fibers, which include type 1 slow-twitch fibers and type 2A fast-twitch oxidative-glycolytic fibers, have more mitochondria (the site of enzymes for the citric acid cycle and oxidative phosphoryla-tion) than glycolytic fibers do. They also have more blood vessels in their connective tissue to bring oxygen to the cells (**FIG. 12.14**).

The efficiency with which muscle fibers obtain oxygen is a factor in their preferred method of glucose metabolism. Oxygen in the blood must diffuse into the interior of muscle fibers in order to reach the mitochondria. This process is facilitated by the presence of **myoglobin,** a red oxygen-binding pigment with a high affinity for oxygen. This affinity allows myoglobin to act as a transfer molecule, bringing oxygen more rapidly to the interior of the fibers. Because oxidative fibers contain more myoglobin, oxy-gen diffusion is faster than in glycolytic fibers. Oxidative fibers are described as *red muscle* because large amounts of myoglobin give them their characteristic color.

In addition to myoglobin, oxidative fibers have smaller diam-eters, so the distance through which oxygen must diffuse before reaching the mitochondria is shorter. Because oxidative fibers have

FIG. 12.14 | **Fast-twitch and slow-twitch muscles**

Slow-twitch oxidative muscle has large amounts of red myoglobin, numerous mitochondria, and extensive capillary blood supply, in contrast to fast-twitch glycolytic muscle.

Slow-Twitch Oxidative Muscle Fibers. Note smaller diameter, darker color due to myoglobin. Fatigue-resistant.

Capillaries

Mitochondria

Cross section of slow-twitch muscle fibers (LM × 170)

Fast-Twitch Glycolytic Muscle Fibers. Larger diameter, pale color. Easily fatigued.

Cross section of fast-twitch muscle fibers (LM × 170)

more myoglobin and more capillaries to bring blood to the cells and are smaller in diameter, they maintain a better supply of oxygen and are able to use oxidative phosphorylation for ATP production.

2X glycolytic fibers, in contrast, are described as *white muscle* because of their lower myoglobin content. These muscle fibers are also larger in diameter than type 1 slow-twitch fibers. The combination of larger size, less myoglobin, and fewer blood vessels means that glycolytic fibers are more likely to run out of oxygen after repeated contractions. Glycolytic fibers therefore rely primarily on anaerobic glycolysis for ATP synthesis and fatigue most rapidly.

Type 2A fast-twitch oxidative-glycolytic fibers exhibit properties of both oxidative and glycolytic fibers. They are smaller than fast-twitch glycolytic fibers and use a combination of oxidative and glycolytic metabolism to produce ATP. Because of their intermediate size and the use of oxidative phosphorylation for ATP synthesis, type 2A fibers are more fatigue resistant than their 2X fast-twitch glycolytic cousins. Type 2A fibers, like type 1 slow-twitch fibers, are classified as red muscle because of their myoglobin content.

Human muscles are a mixture of fiber types, with the ratio of types varying from muscle to muscle and from one individual to another. For example, who would have more fast-twitch fibers in leg muscles, a marathon runner or a high-jumper? Characteristics of the three muscle fiber types are compared in **TABLE 12.2**.

Resting Fiber Length Affects Tension

In a muscle fiber, the tension developed during a twitch is a direct reflection of the length of individual sarcomeres before contraction begins (**FIG. 12.15**). Each sarcomere contracts with optimum force if it is at optimum length (neither too long nor too short) before the contraction begins. Fortunately, the normal resting length of skeletal muscles usually ensures that sarcomeres are at optimum length when they begin a contraction.

At the molecular level, sarcomere length reflects the overlap between the thick and thin filaments (Fig. 12.15). The sliding filament theory predicts that *the tension generated by a muscle fiber is directly proportional to the number of crossbridges formed between the thick and thin filaments*. If the fibers start a contraction at a very long sarcomere length, the thick and thin filaments barely overlap and form few crossbridges (Fig. 12.15e). This means that in the initial part of the contraction, the sliding filaments interact only minimally and therefore cannot generate much force.

At the optimum sarcomere length (Fig. 12.15c), the filaments begin contracting with numerous crossbridges between the thick and thin filaments, allowing the fiber to generate optimum force in that twitch. If the sarcomere is shorter than optimum length at the beginning of the contraction (Fig. 12.15b), the thick and thin fibers have too much overlap before the contraction begins. Consequently, the thick filaments can move the thin filaments only a short distance before the thin actin filaments from opposite ends of the sarcomere start to overlap. This overlap prevents crossbridge formation.

If the sarcomere is so short that the thick filaments run into the Z disks (Fig. 12.15a), myosin is unable to find new binding sites for crossbridge formation, and tension decreases rapidly. Thus, the development of single-twitch tension in a muscle fiber is a passive property that depends on filament overlap and sarcomere length.

TABLE 12.2	Characteristics of Muscle Fiber Types		
	Slow-Twitch Oxidative, Red Muscle (Type 1)	**Fast-Twitch Oxidative-Glycolytic; Red Muscle (Type 2A)**	**Fast-Twitch Glycolytic; White Muscle (Type 2X)**
Speed of Development of Maximum Tension	Slowest	Intermediate	Fastest
Myosin ATPase Activity	Slow	Fast	Fast
Diameter	Small	Medium	Large
Contraction Duration	Longest	Short	Short
Ca^{2+}-ATPase Activity in SR	Moderate	High	High
Endurance	Fatigue resistant	Fatigue resistant	Easily fatigued
Use	Most used: posture	Standing, walking	Least used: jumping; quick, fine movements
Metabolism	Oxidative; aerobic	Glycolytic but becomes more oxidative with endurance training	Glycolytic; more anaerobic than fast-twitch oxidative-glycolytic type
Capillary Density	High	Medium	Low
Mitochondria	Numerous	Moderate	Few
Color	Dark red (myoglobin)	Red	Pale

FIG. 12.15 Length-tension relationships

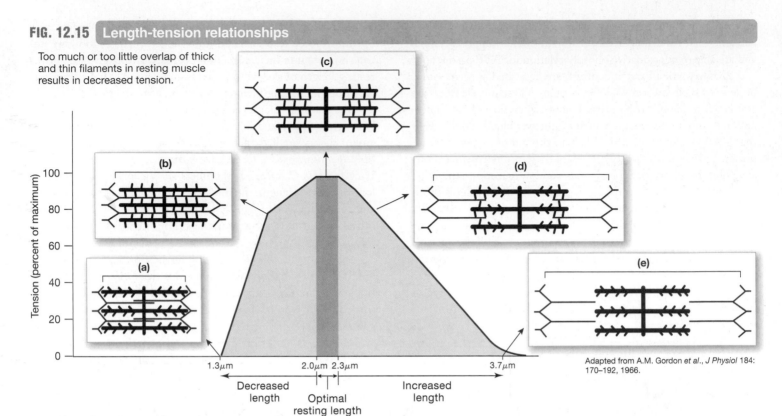

Too much or too little overlap of thick and thin filaments in resting muscle results in decreased tension.

Tension (percent of maximum)

1.3μm   2.0μm 2.3μm   3.7μm

Decreased length   Optimal resting length   Increased length

Adapted from A.M. Gordon *et al.*, *J Physiol* 184: 170–192, 1966.

Force of Contraction Increases with Summation

Although we have just seen that single-twitch tension is determined by the length of the sarcomere, it is important to note that a single twitch does not represent the maximum force that a muscle fiber can develop. The force generated by the contraction of a single muscle fiber can be increased by increasing the rate (frequency) at which muscle action potentials stimulate the muscle fiber.

A typical muscle action potential lasts between 1 and 3 msec, while the muscle contraction may last 100 msec (see Fig. 12.11). If repeated action potentials are separated by long intervals of time, the muscle fiber has time to relax completely between stimuli (**FIG. 12.16a**). If the interval of time between action potentials is shortened, the muscle fiber does not have time to relax completely between two stimuli, resulting in a more forceful contraction (Fig. 12.16b). This process is known as **summation** and is similar to the temporal summation of graded potentials that takes place in neurons [p. 266].

If action potentials continue to stimulate the muscle fiber repeatedly at short intervals (high frequency), relaxation between contractions diminishes until the muscle fiber achieves a state of maximal contraction known as **tetanus.** There are two types of tetanus. In *incomplete,* or *unfused, tetanus,* the stimulation rate of the muscle fiber is not at a maximum value, and consequently the fiber relaxes slightly between stimuli (Fig. 12.16c). In *complete,* or *fused, tetanus,* the stimulation rate is fast enough that the muscle fiber does not have time to relax. Instead, it reaches maximum tension and remains there (Fig. 12.16d).

Thus, it is possible to increase the tension developed in a single muscle fiber by changing the rate at which action potentials occur in the fiber. Muscle action potentials are initiated by the somatic motor neuron that controls the muscle fiber.

CONCEPT CHECK

17. Summation in muscle fibers means that the _____ of the fiber increases with repeated action potentials.

18. Temporal summation in neurons means that the _____ of the neuron increases when two depolarizing stimuli occur close together in time.

A Motor Unit Is One Motor Neuron and Its Muscle Fibers

The basic unit of contraction in an intact skeletal muscle is a **motor unit,** composed of a group of muscle fibers that function together and the somatic motor neuron that controls them (**FIG. 12.17**). When the somatic motor neuron fires an action potential, all muscle fibers in the motor unit contract. Note that although one somatic motor neuron innervates multiple fibers, each muscle fiber is innervated by only a single neuron.

The number of muscle fibers in a motor unit varies. In muscles used for fine motor actions, such as the *extraocular* muscles that move the eyes or the muscles of the hand, one motor unit contains as few as three to five muscle fibers. If one such motor unit is activated, only a few fibers contract, and the muscle response is quite small. If additional motor units are activated, the

FIG. 12.16 Summation of contractions

(a) Single Twitches: Muscle relaxes completely between stimuli (▲).

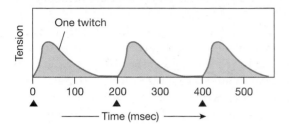

(b) Summation: Stimuli closer together do not allow muscle to relax fully.

NAVIGATOR

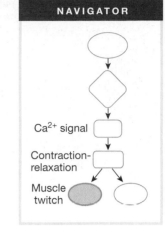

Ca²⁺ signal

Contraction-relaxation

Muscle twitch

(c) Summation Leading to Unfused Tetanus: Stimuli are far enough apart to allow muscle to relax slightly between stimuli.

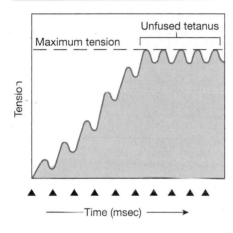

(d) Summation Leading to Complete Tetanus: Muscle reaches steady tension. If muscle fatigues, tension decreases rapidly.

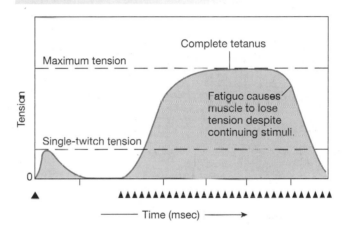

response increases by small increments because only a few more muscle fibers contract with the addition of each motor unit. This arrangement allows fine gradations of movement.

In muscles used for gross motor actions such as standing or walking, each motor unit may contain hundreds or even thousands of muscle fibers. The gastrocnemius muscle in the calf of the leg, for example, has about 2000 muscle fibers in each motor unit. Each time an additional motor unit is activated in these muscles, many more muscle fibers contract, and the muscle response jumps by correspondingly greater increments.

All muscle fibers in a single motor unit are of the same fiber type. For this reason there are fast-twitch motor units and slow-twitch motor units. Which kind of muscle fiber associates with a particular neuron appears to be a function of the neuron. During embryological development, each somatic motor neuron secretes a growth factor that directs the differentiation of all muscle fibers in its motor unit so that they develop into the same fiber type.

Intuitively, it would seem that people who inherit a predominance of one fiber type over another would excel in certain sports. They do, to some extent. Endurance athletes, such as distance runners and cross-country skiers, have a predominance of slow-twitch fibers, whereas sprinters, ice hockey players, and weight lifters tend to have larger percentages of fast-twitch fibers.

Inheritance is not the only determining factor for fiber composition in the body, however, because the metabolic characteristics of muscle fibers have some plasticity. With endurance training, the aerobic capacity of some fast-twitch fibers can be enhanced until they are almost as fatigue-resistant as slow-twitch fibers. Because the conversion occurs only in those muscles that are being trained, a neuromodulator chemical is probably involved. In addition, endurance training increases the number of capillaries and mitochondria in the muscle tissue, allowing more oxygen-carrying blood to reach the contracting muscle and contributing to the increased aerobic capacity of the muscle fibers.

Contraction Force Depends on the Types and Numbers of Motor Units

Within a skeletal muscle, each motor unit contracts in an all-or-none manner. How then can muscles create graded contractions of varying force and duration? The answer lies in the fact that muscles are composed of multiple motor units of different types (Fig. 12.17). This diversity allows the muscle to vary contraction by (1) changing the types of motor units that are active or (2) changing the number of motor units that are responding at any one time.

FIG. 12.17   Motor units

A motor unit consists of one motor neuron and all the muscle fibers it innervates. A muscle may have many motor units of different types.

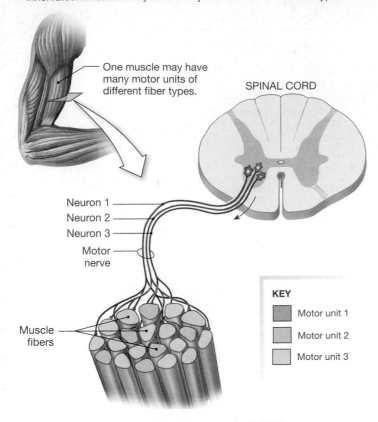

One muscle may have many motor units of different fiber types.

SPINAL CORD

Neuron 1
Neuron 2
Neuron 3

Motor nerve

Muscle fibers

KEY

Motor unit 1

Motor unit 2

Motor unit 3

The force of contraction in a skeletal muscle can be increased by recruiting additional motor units. **Recruitment** is controlled by the nervous system and proceeds in a standardized sequence. A weak stimulus directed onto a pool of somatic motor neurons in the central nervous system activates only the neurons with the lowest thresholds [p. 242]. Studies have shown that these low-threshold neurons control fatigue-resistant slow-twitch fibers, which generate minimal force.

As the stimulus onto the motor neuron pool increases in strength, additional motor neurons with higher thresholds begin to fire. These neurons in turn stimulate motor units composed of fatigue-resistant fast-twitch oxidative-glycolytic fibers. Because more motor units (and, thus, more muscle fibers) are participating in the contraction, greater force is generated in the muscle.

As the stimulus increases to even higher levels, somatic motor neurons with the highest thresholds begin to fire. These neurons stimulate motor units composed of glycolytic fast-twitch fibers. At this point, the muscle contraction is approaching its maximum force. Because of differences in myosin and crossbridge formation, fast-twitch fibers generate more force than slow-twitch fibers do. However, because fast-twitch fibers fatigue more rapidly, it is impossible to hold a muscle contraction at maximum force for an extended period of time. You can demonstrate this by clenching your fist as hard as you can: How long can you hold it before some of the muscle fibers begin to fatigue?

Sustained contractions in a muscle require a continuous train of action potentials from the central nervous system to the muscle. As you learned earlier, however, increasing the stimulation rate of a muscle fiber results in summation of its contractions. If the muscle fiber is easily fatigued, summation leads to fatigue and diminished tension (Fig. 12.16d).

One way the nervous system avoids fatigue in sustained contractions is by **asynchronous recruitment** of motor units. The nervous system modulates the firing rates of the motor neurons so that different motor units take turns maintaining muscle tension. The alternation of active motor units allows some of the motor units to rest between contractions, preventing fatigue.

Asynchronous recruitment prevents fatigue only in submaximal contractions, however. In high-tension, sustained contractions, the individual motor units may reach a state of unfused tetanus, in which the muscle fibers cycle between contraction and partial relaxation. In general, we do not notice this cycling because the different motor units in the muscle are contracting and relaxing at slightly different times. As a result, the contractions and relaxations of the motor units average out and appear to be one smooth contraction. But as different motor units fatigue, we are unable to maintain the same amount of tension in the muscle, and the force of the contraction gradually decreases.

CONCEPT CHECK

19. Which type of runner would you expect to have more slow-twitch fibers, a sprinter or a marathoner?

20. What is the response of a muscle fiber to an increase in the firing rate of the somatic motor neuron?

21. How does the nervous system increase the force of contraction in a muscle composed of many motor units?

MECHANICS OF BODY MOVEMENT

Because one main role of skeletal muscles is to move the body, we now turn to the mechanics of body movement. The term *mechanics* refers to how muscles move loads and how the anatomical relationship between muscles and bones maximizes the work the muscles can do.

Isotonic Contractions Move Loads; Isometric Contractions Create Force without Movement

When we described the function of muscles earlier in this chapter, we noted that they can create force to generate movement but can also create force without generating movement. You can demonstrate both properties with a pair of heavy weights. Pick up one weight in each hand and then bend your elbows so that the weights touch your shoulders. You have just performed an **isotonic contraction** {*iso,* equal + *teinein,* to stretch}. Any contraction that creates force and moves a load is an isotonic contraction.

When you bent your arms at the elbows and brought the weights to your shoulders, the biceps muscles shortened. Now slowly extend your arms, resisting the gravitational forces pulling the weights down. The biceps muscles are again active, but now you are performing a *lengthening (eccentric) contraction*. Lengthening contractions are thought to contribute most to cellular damage after exercise and to lead to delayed muscle soreness.

If you pick up the weights and hold them stationary in front of you, the muscles of your arms are creating tension (force) to overcome the load of the weights but are not creating movement. Contractions that create force without moving a load are called **isometric contractions** {*iso*, equal + *metric*, measurement} or *static contractions*. Isotonic and isometric contractions are illustrated in **FIGURE 12.18**. To demonstrate an isotonic contraction experimentally, we hang a weight (the load) from the muscle and electrically stimulate the muscle. The muscle contracts, lifting the weight. The graph on the right shows the development of force throughout the contraction.

To demonstrate an isometric contraction experimentally, we attach a heavier weight to the muscle, as shown in Figure 12.18b. When the muscle is stimulated, it develops tension, but the force created is not enough to move the load. In isometric contractions, muscles create force without shortening significantly. For example, when your exercise instructor yells at you to "tighten those glutes," your response is isometric contraction of the gluteal muscles in your buttocks.

How can an isometric contraction create force if the length of the muscle does not change significantly? The elastic elements of the muscle provide the answer. All muscles contain elastic fibers in the tendons and other connective tissues that attach muscles to bone, and in the connective tissue between muscle fibers. In muscle fibers, elastic cytoskeletal proteins occur between the myofibrils and as part of the sarcomere. All of these elastic components behave collectively as if they were connected in series (one after the other) to the contractile elements of the muscle.

FIG. 12.18 | **Isotonic and isometric contractions**

(a) Isotonic Contraction. In an isotonic contraction, the muscle contracts, shortens, and creates enough force to move the load.

(b) Isometric Contraction. In an isometric contraction, the muscle contracts but does not shorten. The force created cannot move the load.

Consequently, they are often called the **series elastic elements** of the muscle (**FIG. 12.19**).

When the sarcomeres shorten in the first stages of a contraction, the elastic elements stretch. This stretching of the elastic elements allows the fibers to maintain a relatively constant length even though the sarcomeres are shortening and creating tension (Fig. 12.19 ❷). If the muscle cannot create additional force to move the load, the contraction is isometric. Once the elastic elements have been stretched, if the sarcomeres generate force equal to the load, the muscle shortens in an isotonic contraction and lifts the load.

FIG. 12.19 Series elastic elements in muscle

Muscle has both contractile components (sarcomeres, shown here as a gear and ratchet) and elastic components (shown here as a spring).

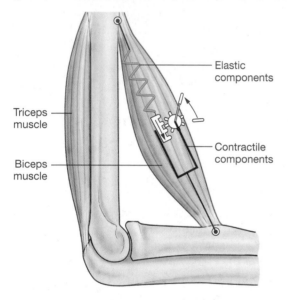

Elastic elements allow isometric contractions.

- In an isometric contraction, sarcomeres shorten, generating force, but elastic elements stretch, allowing muscle length to remain the same.
- In isotonic contractions, sarcomeres shorten more but, because elastic elements are already stretched, the muscles shorten.

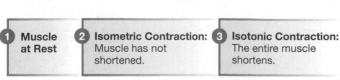

❶ **Muscle at Rest**

❷ **Isometric Contraction:** Muscle has not shortened.

❸ **Isotonic Contraction:** The entire muscle shortens.

Bones and Muscles around Joints Form Levers and Fulcrums

The anatomical arrangement of muscles and bones in the body is directly related to how muscles work. The body uses its bones and joints as levers and fulcrums on which muscles exert force to move or resist a load. A **lever** is a rigid bar that pivots around a point known as the **fulcrum.** In the body, bones form levers, flexible joints form the fulcrums, and muscles attached to bones create force by contracting.

Most lever systems in the body are similar to a fishing pole, like the one shown in **FIGURE 12.20a**. In these lever systems, the fulcrum is located at one end of the lever, the load is near the other end of the lever, and force is applied between the fulcrum and the load. This arrangement maximizes the distance and speed with which the lever can move the load but also requires more force than some other lever systems. Let's see how flexion of the forearm illustrates lever system function.

In the lever system of the forearm, the elbow joint acts as the fulcrum around which rotational movement of the forearm (the lever) takes place (Fig. 12.20b). The biceps muscle is attached at its origin at the shoulder and inserts onto the radius bone of the forearm a few centimeters away from the elbow joint. When the biceps contracts, it creates the upward force F_1 (Fig. 12.20c) as it pulls on the bone. The total rotational force* created by the biceps depends on two things: (1) the force of muscle contraction and (2) the distance between the fulcrum and the point at which the muscle inserts onto the radius.

If the biceps is to hold the forearm stationary and flexed at a 90° angle, the muscle must exert enough upward rotational force to exactly oppose the downward rotational force exerted by gravity on the forearm (Fig. 12.20c). The downward rotational force on the forearm is proportional to the weight of the forearm (F_2) times the distance from the fulcrum to the forearm's center of gravity (the point along the lever at which the forearm load exerts its force). For the arm illustrated in Figure 12.20c, the biceps must exert 6 kg of force to hold the arm at a 90° angle. Because the muscle is not shortening, this is an isometric contraction.

Now what happens if a 7-kg weight is placed in the hand? This weight places an additional load on the lever that is farther from the fulcrum than the forearm's center of gravity. Unless the biceps can create additional upward force to offset the downward force created by the weight, the hand falls. If you know the force exerted by the added weight and its distance from the elbow, you can calculate the additional muscle force needed to keep the arm from dropping the 7-kg weight.

What happens to the force required of the biceps to support a weight if the distance between the fulcrum and the muscle insertion point changes? Genetic variability in the insertion point can have a dramatic effect on the force required to move or resist

---

*In physics, rotational force is expressed as *torque,* and the force of contraction is expressed in newtons (mass × acceleration due to gravity). For simplicity, we ignore the contribution of gravity in this discussion and use the mass unit "kilograms" for force of contraction.

FIG. 12.20 The arm is a lever and fulcrum system

(a) The lever system of the forearm is like that of a fishing pole. The fulcrum is at one end of the lever and the load is at the other end. Force is applied between the fulcrum and the load.

KEY

▲ Fulcrum

⬆ Applied force

⬆ Movement of load

⬭ Load

▬ Lever

Fulcrum

(b) The human forearm acts as a lever. The fulcrum is the elbow joint. The load is gravity acting on the mass of the forearm and hand.

Biceps muscle

Lever

Load

Fulcrum

(c) Force calculations

Biceps contraction creates upward force F_1.

The biceps inserts into the lever 5 cm from the fulcrum.

Rotational force$_{up}$ ∝ biceps force F_1 × 5 cm from the fulcrum.

F_1

$F_2 = 2$ kg

5 cm

15 cm

The weight of the forearm exerts a downward force of 2 kg at its center of gravity, which is 15 cm from the fulcrum.

Rotational force$_{down}$ ∝ load F_2 × 15 cm

∝ 2 kg × 15 cm

To hold the arm stationary at 90 degrees, the rotational force created by the contracting biceps must exactly oppose the downward rotation created by the forearm's weight.

Rotational force$_{up}$ = Rotational force$_{down}$

Biceps force × 5 cm = 2 kg × 15 cm

$$\text{Biceps force} = \frac{30 \text{ kg} \cdot \text{cm}}{5 \text{ cm}}$$

Biceps force = 6 kg

A 7-kg load is added to the hand 25 cm from the elbow.

F_1

D_1 ⊢ 5 cm

D_2 ⊢ 25 cm

ⓠ FIGURE QUESTION

How much additional force must the biceps exert to keep from dropping the weight?

(d) The arm amplifies speed of movement of the load.

Because the insertion of the biceps is close to the fulcrum, a small movement of the biceps becomes a much larger movement of the hand.

When the biceps contracts and shortens 1 cm, the hand moves upward 5 cm.

Lever

5 cm

Fulcrum   1 cm

ⓠ FIGURE QUESTION

If the biceps shortens 1 cm in 1 second, how fast does the hand move upward?

a load. For example, if the biceps in Figure 12.20b inserted 6 cm from the fulcrum instead of 5 cm, it would only need to generate 5 kg of force to offset the weight of the arm. Some studies have shown a correlation between muscle insertion points and success in certain athletic events.

In the example so far, we have assumed that the load is stationary and that the muscle is contracting isometrically. What happens if we want to flex the arm and lift the load? To move the load from its position, the biceps must exert a force that exceeds the force created by the stationary load.

The disadvantage of this type of lever system, where the fulcrum is positioned near one end of the lever, is that the muscle is required to create large amounts of force to move or resist a small load. However, the advantage of this type of lever-fulcrum system is that it maximizes speed and mobility. A small movement of the forearm at the point where the muscle inserts becomes a much larger movement at the hand (Fig. 12.20d). In addition, the two movements occur in the same amount of time, and so the speed of contraction at the insertion point is amplified at the hand. Thus, the lever-fulcrum system of the arm amplifies both the distance the load is moved and the speed at which this movement takes place.

In muscle physiology, the speed with which a muscle contracts depends on the type of muscle fiber (fast-twitch or slow-twitch) and on the load that is being moved. Intuitively, you can see that you can flex your arm much faster with nothing in your hand than you can while holding a 7-kg weight in your hand. The relationship between load and velocity (speed) of contraction in a muscle fiber, determined experimentally, is graphed in **FIGURE 12.21**.

Contraction is fastest when the load on the muscle is zero. When the load on the muscle equals the ability of the muscle to create force, the muscle is unable to move the load and the velocity drops to zero. The muscle can still contract, but the contraction becomes isometric instead of isotonic. Because speed is a function of

load and muscle fiber type, it cannot be regulated by the body except through recruitment of faster muscle fiber types. However, the arrangement of muscles, bones, and joints allows the body to amplify speed so that regulation at the cellular level becomes less important.

CONCEPT CHECK   22. One study found that many world-class athletes have muscle insertions that are farther from the joint than in the average person. Why would this trait translate into an advantage for a weight lifter?

Muscle Disorders Have Multiple Causes

Dysfunction in skeletal muscles can arise from a problem with the signal from the nervous system, from miscommunication at the neuromuscular junction, or from defects in the muscle. Unfortunately, in many muscle conditions, even the simple ones, we do not fully understand the mechanism of the primary defect. As a result, we can treat the symptoms but may not be able to cure the problem.

One common muscle disorder is a "charley horse," or *muscle cramp*—a sustained painful contraction of skeletal muscles. Many muscle cramps are caused by hyperexcitability of the somatic motor neurons controlling the muscle. As the neuron fires repeatedly, the muscle fibers of its motor unit go into a state of painful sustained contraction. Sometimes muscle cramps can be relieved by forcibly stretching the muscle. Apparently, stretching sends sensory information to the central nervous system that inhibits the somatic motor neuron, relieving the cramp.

The simplest muscle disorders arise from overuse. Most of us have exercised too long or too hard and suffered from fatigue or soreness as a result. With more severe trauma, muscle fibers, the connective tissue sheath, or the union of muscle and tendon may tear.

Disuse of muscles can be as traumatic as overuse. With prolonged inactivity, such as may occur when a limb is immobilized in a cast, the skeletal muscles atrophy. Blood supply to the muscle diminishes, and the muscle fibers get smaller. If activity is resumed in less than a year, the fibers usually regenerate. Atrophy of longer than one year is usually permanent. If the atrophy results from somatic motor neuron dysfunction, therapists now try to maintain muscle function by administering electrical impulses that directly stimulate the muscle fibers.

Acquired disorders that affect the skeletal muscle system include infectious diseases, such as influenza, that lead to weakness and achiness, and poisoning by toxins, such as those produced in botulism (*Clostridium botulinus*) and tetanus (*Clostridium tetani*). Botulinum toxin acts by decreasing the release of acetylcholine from the somatic motor neuron. Clinical investigators have successfully used injections of botulinum toxin as a treatment for writer's cramp, a disabling cramp of the hand that apparently arises as a result of hyperexcitability in the distal portion of the somatic motor neuron. Botox® injections are now widely used for cosmetic wrinkle reduction. Botulinum toxin injected under the skin temporarily paralyzes facial muscles that pull the skin into wrinkles.

FIG. 12.21   Load-velocity relationship in skeletal muscle

GRAPH QUESTIONS
1. At what point on the line is contraction isometric?
2. At what point is the muscle contraction at maximum velocity?

RUNNING **PROBLEM**

Paul's doctor explains to Mrs. Leong that the paralytic attacks associated with hyperkalemic periodic paralysis last only a few minutes to a few hours and generally involve only the muscles of the extremities, which become weak and unable to contract (*flaccid paralysis*). "Is there any treatment?" asks Mrs. Leong. The doctor replies that although the inherited condition cannot be cured, attacks may be prevented with drugs. Diuretics, for example, increase the rate at which the body excretes water and ions (including Na^+ and K^+), and these medications have been shown to help prevent attacks of paralysis in people with hyperKPP.

Q4: *Draw a map to explain why a Na^+ channel that does not inactivate results in a muscle that cannot contract (flaccid paralysis).*

378   389   394   **403**   406   412

Inherited muscular disorders are the most difficult to treat. These conditions include various forms of muscular dystrophy as well as biochemical defects in glycogen and lipid storage. In **Duchenne muscular dystrophy,** the structural protein **dystrophin,** which links actin to proteins in the cell membrane, is absent. In muscle fibers that lack dystrophin, extracellular Ca^{2+} enters the fiber through small tears in the membrane or possibly through stretch-activated channels. Calcium entry activates intracellular enzymes, resulting in breakdown of the fiber components. The major symptom of Duchenne dystrophy is progressive muscle weakness, and patients usually die before age 30 from failure of the respiratory muscles.

McArdle's disease, also known as *myophosphorylase deficiency,* is a condition in which the enzyme that converts glycogen to glucose 6-phosphate is absent in muscles. As a result, muscles lack a usable glycogen energy supply, and exercise tolerance is limited.

One way physiologists are trying to learn more about muscle diseases is by using animal models, such as genetically engineered mice that lack the genes for certain muscle proteins. Researchers are trying to correlate the absence of protein with particular disruptions in function.

SMOOTH MUSCLE

Although skeletal muscle has the most muscle mass in the body, cardiac and smooth muscle are more important in the maintenance of homeostasis. Smooth muscle is challenging to describe because smooth muscles in the body have so much functional variability. There are many ways to categorize the different types of smooth muscle, but we will consider three:

1. **By location.** Smooth muscles with widely differing properties are found throughout the animal kingdom. In humans, smooth muscle can be divided into six major groups: *vascular* (blood vessel walls), *gastrointestinal* (walls of digestive tract

and associated organs, such as the gallbladder), *urinary* (walls of bladder and ureters), *respiratory* (airway passages), *reproductive* (uterus in females and other reproductive structures in both females and males), and *ocular* (eye). These muscles have different functions in the body, and their physiology reflects their specialized functions. In contrast, skeletal muscle is relatively uniform throughout the body.

2. **By contraction pattern.** Smooth muscle can be classified by whether it alternates between contraction and relaxation states or whether it is continuously contracted. Muscles that undergo periodic contraction and relaxation cycles are said to be **phasic smooth muscles.** An example would be the wall of the lower esophagus, which contracts only when food passes through it (**FIG. 12.22a**). Some phasic smooth muscles, such as those in the wall of the intestine, cycle rhythmically through contractions alternating with relaxation (Fig. 12.22b).

   Muscles that are continuously contracted are called **tonic smooth muscles** because they are always maintaining some level of muscle tone. The esophageal and urinary bladder **sphincters** {*sphingein,* to close} are examples of tonically contracted muscles that close off the opening to a hollow organ. These sphincters relax when it is necessary to allow material to enter or leave the organ (Fig. 12.22c). The tonic smooth muscle in the walls of some blood vessels maintains an intermediate level of contraction. Under *tonic control* by the nervous system [p. 183], this vascular smooth muscle contracts or relaxes as the situation demands (Fig. 12.22d).

3. **By their communication with neighboring cells.** In some smooth muscles, the cells are electrically connected by gap junctions, and they contract as a coordinated unit. These muscles are called **single-unit smooth muscle,** or *unitary smooth muscle.* In **multiunit smooth muscle,** the cells are not linked electrically and each muscle cell functions independently.

Most smooth muscle is single-unit smooth muscle. Single-unit smooth muscle is also called **visceral smooth muscle** because it forms the walls of internal organs (viscera), such as the intestinal tract. The fibers of single-unit smooth muscle are connected to one another by gap junctions. An electrical signal in one cell spreads rapidly through the entire sheet of tissue to create a coordinated contraction (**FIG. 12.23a**). Because all fibers contract every time, no reserve units are left to be recruited to increase contraction force. Instead, the amount of Ca^{2+} that enters the cell determines the force of contraction, as you will learn in the discussion that follows.

In multiunit smooth muscle, the cells are not linked electrically and they must be stimulated independently to contract. Each individual muscle cell is closely associated with an axon terminal or varicosity (Fig. 12.23b). This arrangement allows fine control of contractions in these muscles through selective activation of individual muscle cells. As in skeletal muscle, increasing the force of contraction requires recruitment of additional fibers.

FIG. 12.22 Smooth muscle contractions

(a) A phasic smooth muscle that is usually relaxed. Example: esophagus.

(b) A phasic smooth muscle that cycles between contraction and relaxation. Example: intestine.

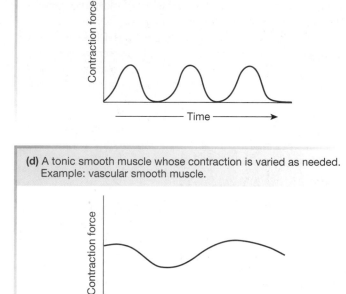

(c) A tonic smooth muscle that is usually contracted. Example: a sphincter that relaxes to allow material to pass.

(d) A tonic smooth muscle whose contraction is varied as needed. Example: vascular smooth muscle.

Multiunit smooth muscle is found in the iris and ciliary muscle of the eye [p. 343], in part of the male reproductive tract, and in the uterus except just prior to labor and delivery. Interestingly, the multiunit smooth muscle of the uterus changes and becomes single-unit during the final stages of pregnancy. Genes for synthesis of gap junction connexin proteins turn on, apparently under the influence of pregnancy hormones. The addition of gap junctions to the uterine muscle cells synchronizes electrical signals, allowing the uterine muscle to contract more effectively while expelling the baby.

Because of the variability in smooth muscle types, we introduce only their general features in this chapter. You will learn properties that are specific to a certain type when you study the different organ systems.

Smooth Muscle Is More Variable Than Skeletal Muscle

Two of the principles that you learned in previous sections for skeletal muscle apply to all smooth muscle. First, force is created by actin-myosin crossbridge interaction between sliding filaments. Second, contraction in smooth muscle, as in skeletal and cardiac muscle, is initiated by an increase in free cytosolic Ca^{2+} concentrations. However, in most other ways smooth muscle function is more complex than skeletal muscle function. Let's examine some differences, starting at the organ level and working to the cellular level.

1. **Smooth muscles must operate over a range of lengths.** Smooth muscle is found predominantly in the walls of hollow organs and tubes, many of which expand and contract as they fill and empty. The bladder, which fills with urine, is an example of a distensible organ. Smooth muscles in organs like this must function efficiently over a range of muscle lengths. In contrast, most skeletal muscles are attached to bone and operate over a narrow range of lengths.

2. **Within an organ, the layers of smooth muscle may run in several directions.** For example, the intestine has one muscle layer that encircles the lumen and a perpendicular layer that runs the length of the intestine. The stomach adds a third layer that is set obliquely to the other two. Contraction in different layers changes the shape of the organ. Sometimes smooth muscles generate force to move material through the lumen of the organ, such as the sequential waves of smooth muscle contraction that move ingested material through the small intestine. In contrast, most skeletal muscles are arranged so that their contraction shortens the muscle.

3. When you compare a single muscle twitch in muscle types, **smooth muscles contract and relax much more slowly** than skeletal or cardiac muscle (**FIG. 12.24**).

4. **Smooth muscle uses less energy to generate and maintain a given amount of force.** Smooth muscles can develop force rapidly but have the ability to slow down their myosin ATPase so that crossbridges cycle slowly as they maintain

FIG. 12.23   Smooth muscle coordination

(a) **Single-unit smooth muscle cells** are connected by gap junctions, and the cells contract as a single unit.

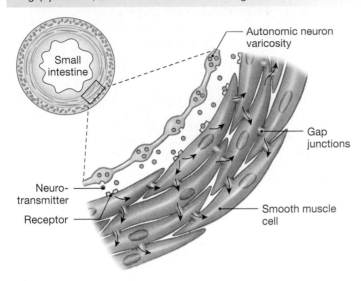

Small intestine

Autonomic neuron varicosity

Gap junctions

Neuro-transmitter

Receptor

Smooth muscle cell

(b) **Multi-unit smooth muscle cells** are not electrically linked, and each cell must be stimulated independently.

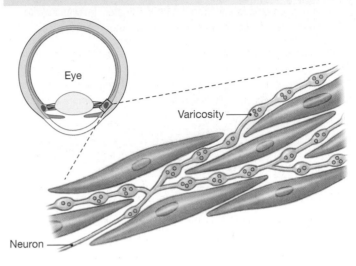

Eye

Varicosity

Neuron

their force. As a result, their use of ATP is lower than that in striated muscles. Smooth muscle has fewer mitochondria than striated muscles and relies more on glycolysis for its ATP production.

5. **Smooth muscle can sustain contractions for extended periods without fatiguing.** This property allows organs such as the bladder to maintain tension despite a continued load. It also allows some smooth muscles to be tonically contracted and maintain tension most of the time.

6. **Smooth muscles have small, spindle-shaped cells with a single nucleus,** in contrast to the large multinucleated fibers of skeletal muscles.

7. In smooth muscle, **the contractile fibers are not arranged in sarcomeres.** Under the microscope, smooth muscle lacks the distinct banding patterns of striated muscle (see Fig. 12.1c).

FIG. 12.24   Duration of muscle twitch in the three types of muscle

Smooth muscles are the slowest to contract and to relax.

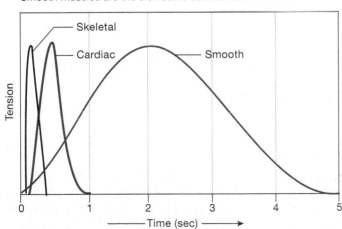

Skeletal

Cardiac

Smooth

Tension

Time (sec)

8. **Contraction in smooth muscle may be initiated by electrical or chemical signals or both.** Skeletal muscle contraction always begins with an action potential in the muscle fiber.

9. **Smooth muscle is controlled by the autonomic nervous system.** Skeletal muscle is controlled by the somatic motor division of the nervous system.

10. **Smooth muscle lacks specialized receptor regions** such as the motor end plates found in skeletal muscle synapses. Instead, receptors are found all over the cell surface. Neurotransmitter is released from autonomic neuron varicosities [p. 364] close to the surface of the muscle fibers and simply diffuses across the cell surface until it finds a receptor.

11. In smooth muscle, **the Ca^{2+} for contraction comes from the extracellular fluid as well as from the sarcoplasmic reticulum.** In skeletal muscle, the Ca^{2+} comes from the sarcoplasmic reticulum.

12. In smooth muscle, **the Ca^{2+} signal initiates a cascade that ends with phosphorylation of myosin light chains and activation of myosin ATPase.** In skeletal muscle, the Ca^{2+} signal binds to troponin to initiate contraction. (Smooth muscle has no troponin.)

With these points in mind, we will now look at some details of smooth muscle function.

CONCEPT CHECK

23. What is the difference in how contraction force is varied in multiunit and single-unit smooth muscle?

24. When the circular muscle layer of the intestine contracts, what happens to the shape of the tube? When the longitudinal layer contracts, what happens to the shape?

RUNNING **PROBLEM**

Three weeks later, Paul had another attack of paralysis, this time at kindergarten after a game of tag. He was rushed to the hospital and given glucose by mouth. Within minutes, he was able to move his legs and arms and asked for his mother.

Q5: *Explain why oral glucose might help bring Paul out of his paralysis.* (Hint: *Glucose stimulates insulin release, and insulin increases Na⁺-K⁺-ATPase activity. What happens to the extracellular K⁺ level when Na⁺-K⁺-ATPase is more active?*)

Smooth Muscle Lacks Sarcomeres

Smooth muscle has the same contractile elements as skeletal muscle—actin and myosin that interact through crossbridges—as well sarcoplasmic reticulum that stores and releases Ca^{2+}. However, details of the structural elements differ in the two muscle types.

Actin and Myosin Actin is more plentiful in smooth muscle than in striated muscle, with an actin-to-myosin ratio of 10–15 to 1, compared with 2–4 to 1 in striated muscle. Smooth muscle actin is associated with tropomyosin, as in skeletal muscle. However, unlike skeletal muscle, smooth muscle lacks troponin.

Smooth muscles have less myosin than skeletal muscle. The less numerous myosin filaments are surrounded by actin filaments and are arranged so that each myosin molecule is in the center of a bundle of 12–15 actin molecules. These contractile units are arranged so that they run parallel to the long axis of the cell.

Myosin filaments in smooth muscle are longer than in skeletal muscle, and the entire surface of the filament is covered by myosin heads (**FIG. 12.25b**). This unique organization enables smooth muscle to stretch more while still maintaining enough overlap to create optimum tension. This is an important property for internal organs, such as the bladder, whose volume varies as it alternately fills and empties.

Smooth muscle cells have an extensive cytoskeleton consisting of intermediate filaments and protein **dense bodies** in the cytoplasm and along the cell membrane. Actin filaments attach to the dense bodies (Fig. 12.25a). Cytoskeleton fibers linking dense bodies to the cell membrane help hold actin in place. Protein fibers in the extracellular matrix tie the smooth muscle cells of a tissue together and transfer force from a contracting cell to its neighbors.

Sarcoplasmic Reticulum The amount of SR in smooth muscle varies from one type of smooth muscle to another. The arrangement of smooth muscle SR is less organized than in skeletal muscle, consisting of a network of tubules that extend from just under the cell membrane into the interior of the cell. There are no t-tubules in smooth muscle, but the SR is closely associated with the membrane invaginations called *caveolae* [p. 148], which apparently participate in cell signaling.

FIG. 12.25 **Smooth muscle organization**

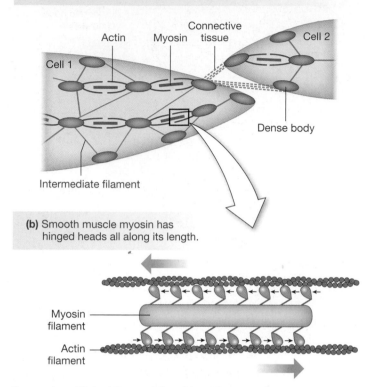

(a) Intermediate filaments and protein dense bodies form a cytoskeleton. Actin attaches to the dense bodies. Each myosin molecule is surrounded by actin filaments.

Actin Myosin Connective tissue Cell 2

Cell 1

Dense body

Intermediate filament

(b) Smooth muscle myosin has hinged heads all along its length.

Myosin filament

Actin filament

Figure courtesy of Marion J. Siegman, Jefferson Medical College.

CONCEPT CHECK

25. The dense bodies that anchor smooth muscle actin are analogous to what structure in a sarcomere? (*Hint:* See Fig. 12.5.)

26. Name two ways smooth muscle myosin differs from skeletal muscle myosin.

27. Name one way actin and its associated proteins differ in skeletal and smooth muscle.

Myosin Phosphorylation Controls Contraction

The molecular events of smooth muscle contraction are similar in many ways to those in skeletal muscle, but some important differences exist. Here is a summary of our current understanding of the key points of smooth muscle contraction. In smooth muscle:

1. An increase in cytosolic Ca^{2+} initiates contraction. This Ca^{2+} is released from the sarcoplasmic reticulum but also enters from the extracellular fluid.
2. Ca^{2+} binds to **calmodulin,** a calcium-binding protein found in the cytosol.
3. Ca^{2+} binding to calmodulin is the first step in a cascade that ends in phosphorylation of myosin light chains.
4. Phosphorylation of myosin light chains enhances myosin ATPase activity and results in contraction. Thus, smooth

muscle contraction is controlled through myosin-linked regulatory processes rather than through tropomyosin.

We begin our discussion with steps 2–4 because those steps are common to all types of smooth muscle. We then go back and look at the different pathways that create Ca^{2+} signals.

FIGURE 12.26 illustrates the steps of smooth muscle contraction. Contraction begins when cytosolic Ca^{2+} concentrations increase following Ca^{2+} entry from the extracellular fluid and Ca^{2+} release from the sarcoplasmic reticulum **1**. The Ca^{2+} ions bind to calmodulin (CaM) **2**, obeying the law of mass action [p. 48]. The Ca^{2+}-calmodulin complex then activates an enzyme called **myosin light chain kinase (MLCK)** **3**.

At the base of the myosin head is a small regulatory protein chain called a **myosin light chain.** Phosphorylation and dephosphorylation of the myosin light chain control contraction and relaxation in smooth muscle. When Ca^{2+}-calmodulin activates MLCK, the enzyme phosphorylates the myosin light protein chains **4**.

Phosphorylation of myosin enhances myosin ATPase activity. When myosin ATPase activity is high, actin binding and crossbridge cycling increase tension in the muscle **5**. The myosin ATPase isoform in smooth muscle is much slower than that in skeletal muscle, which decreases the rate of crossbridge cycling.

Dephosphorylation of the myosin light chain by the enzyme **myosin light chain phosphatase (MLCP)** decreases myosin ATPase activity. Interestingly, dephosphorylation of myosin does not automatically result in relaxation. Under conditions that we do not fully understand, dephosphorylated myosin may remain in an isometric contraction called a **latch state.** This condition maintains tension in the muscle fiber while consuming minimal ATP. It is a significant factor in the ability of smooth muscle to sustain contraction without fatiguing.

Relaxation Because dephosphorylation of myosin does not automatically cause relaxation, it is the ratio of MLCK to MLCP activity that determines the contraction state of smooth muscle. MLCP is always active to some degree in smooth muscle, so the activity of MLCK is often the critical factor. As you learned, MLCK activity depends on Ca^{2+}-calmodulin.

Relaxation in a smooth muscle fiber is a multistep process (Fig. 12.26b). As in skeletal muscle, free Ca^{2+} is removed from the cytosol when Ca^{2+}-ATPase pumps it back into the sarcoplasmic reticulum. In addition, some Ca^{2+} is pumped out of the cell with the help of Ca^{2+}-ATPase and the Na^+-Ca^{2+} exchanger (NCX) [p. 144] **6**.

By the law of mass action, a decrease in free cytosolic Ca^{2+} causes Ca^{2+} to unbind from calmodulin **7**. In the absence of Ca^{2+}-calmodulin, myosin light chain kinase becomes inactive. As MLCK becomes less active, myosin light chain phosphatase dephosphorylates myosin **8**. Myosin ATPase activity decreases **9**, and the muscle relaxes.

MLCP Controls Ca^{2+} Sensitivity

From the earlier discussion, it would appear that calcium and its regulation of MLCK activity is the primary factor responsible for control of smooth muscle contraction. But chemical signals such as neurotransmitters, hormones, and paracrine molecules alter smooth muscle Ca^{2+} **sensitivity** by modulating myosin light chain phosphatase (MLCP) activity. If MLCK and Ca^{2+}-calmodulin are constant but MLCP activity increases, the MLCK/MLCP ratio shifts so that MLCP dominates. Myosin ATPase dephosphorylates and contraction force decreases, even though the cytosolic Ca^{2+} concentration has not changed (**FIG. 12.27**). The contraction process is said to be *desensitized* to calcium—the calcium signal is less effective at causing a contraction. Conversely, signal molecules that *decrease* myosin light chain phosphatase activity make the cell *more sensitive* to Ca^{2+} and contraction force increases even though $[Ca^{2+}]$ has not changed.

Calcium Initiates Smooth Muscle Contraction

We now step back to look in detail at the processes that initiate smooth muscle contraction. Contraction can start with electrical signals—changes in membrane potential—or chemical signals. Contraction caused by electrical signaling is termed *electromechanical coupling*. Contractions initiated by chemical signals without a significant change in membrane potential are called **pharmacomechanical coupling.** Chemical signals may also relax muscle tension without a change in membrane potential. **FIGURE 12.28** is a generalized summary of these pathways.

The Ca^{2+} to initiate contraction comes from two sources: the sarcoplasmic reticulum and the extracellular fluid (Fig. 12.26a). Variable amounts of Ca^{2+} can enter the cytosol from these sources, creating *graded contractions* whose force varies according to the strength of the Ca^{2+} signal.

Sarcoplasmic Ca^{2+} Release The smooth muscle's intracellular Ca^{2+} store is the sarcoplasmic reticulum (SR). SR Ca^{2+} release is mediated both by a ryanodine receptor (RyR) calcium release channel and by an **IP$_3$-receptor channel.** The RyR channel opens in response to Ca^{2+} entering the cell, a process known as **calcium-induced calcium release (CICR).** You will learn more about CICR when you study cardiac muscle.

The IP$_3$ channels open when G protein-coupled receptors activate phospholipase C signal transduction pathways [p. 174]. *Inositol trisphosphate* (IP$_3$) is a second messenger created in that pathway. When IP$_3$ binds to the SR IP$_3$-receptor channel, the channel opens and Ca^{2+} flows out of the SR into the cytosol.

Smooth muscle cells have sufficient SR Ca^{2+} stores for contraction. However, because some Ca^{2+} is lost to the ECF through the membrane pumps, the cells must monitor their SR Ca^{2+} stores. When SR Ca^{2+} stores decrease, a protein sensor (*STIM1*) on the SR membrane interacts with **store-operated Ca^{2+} channels** on the cell membrane. These Ca^{2+} channels, made from the protein *Orai-1*, then open to allow more Ca^{2+} into the cell. The Ca^{2+}-ATPase pumps the cytosolic Ca^{2+} into the SR to replenish its stores.

Cell Membrane Ca^{2+} Entry Store-independent Ca^{2+} entry from the extracellular fluid takes place with the help of membrane

FIG. 12.26 ESSENTIALS

Smooth Muscle Contraction and Relaxation

Smooth muscle contraction and relaxation are similar to those of skeletal muscle, but differ in several important ways: (1) Ca^{2+} comes from the ECF as well as the sarcoplasmic reticulum, (2) an action potential is not required for Ca^{2+} release, (3) there is no troponin, so Ca^{2+} initiates contraction through a cascade that includes phosphorylation of myosin light chains, and (4) an additional step in smooth muscle relaxation is dephosphorylation of myosin light chains by myosin phosphatase.

(a) Smooth Muscle Contraction

Increased cytosolic calcium is the signal for contraction.

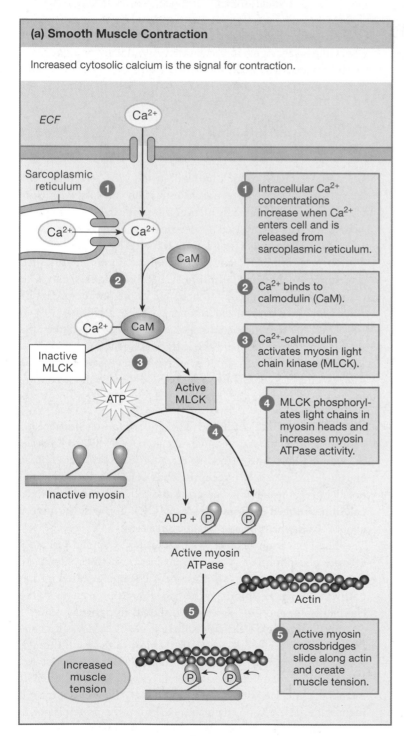

1. Intracellular Ca^{2+} concentrations increase when Ca^{2+} enters cell and is released from sarcoplasmic reticulum.

2. Ca^{2+} binds to calmodulin (CaM).

3. Ca^{2+}-calmodulin activates myosin light chain kinase (MLCK).

4. MLCK phosphorylates light chains in myosin heads and increases myosin ATPase activity.

5. Active myosin crossbridges slide along actin and create muscle tension.

(b) Relaxation in Smooth Muscle

Removal of Ca^{2+} from the cytosol is the first step in relaxation.

6. Free Ca^{2+} in cytosol decreases when Ca^{2+} is pumped out of the cell or back into the sarcoplasmic reticulum.

7. Ca^{2+} unbinds from calmodulin (CaM). MLCK activity decreases.

8. *Myosin phosphatase* (MLCP) removes phosphate from myosin light chains, which decreases myosin ATPase activity.

9. Less myosin ATPase activity results in decreased muscle tension.

FIG. 12.27 Phosphate-mediated Ca²⁺ sensitivity

Changes in phosphatase activity alter myosin's response to Ca²⁺.

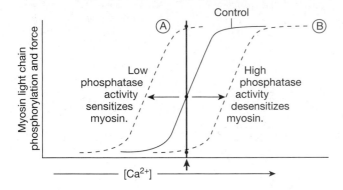

Q GRAPH QUESTION

At the [Ca²⁺] indicated by the red arrow, which graph shows increased myosin light chain phosphorylation?

channels that are voltage-gated, ligand-gated, or mechanically gated [p. 139].

1. Voltage-gated Ca²⁺ channels open in response to a depolarizing stimulus. Action potentials maybe generated in the muscle cell or may enter from neighboring cells via gap junctions. Subthreshold graded potentials may open a few Ca²⁺ channels, allowing small amounts of Ca²⁺ into the cell. This cation entry depolarizes the cell and opens additional voltage-gated Ca²⁺ channels. Sometimes chemical signal molecules open cation channels, and the resulting depolarization opens the Ca²⁺ channels.

2. Ligand-gated Ca²⁺ channels are also known as *receptor-operated calcium channels* or ROCC. These channels open in response to ligand binding and allow enough Ca²⁺ into the cell to induce calcium release from the SR.

3. Stretch-activated channels: Some smooth muscle cells, such as those in blood vessels, contain stretch-activated channels that open when pressure or other force distorts the cell membrane. The exact process is still being worked out, but the cell depolarizes, opening neighboring voltage-gated Ca²⁺ channels. Because contraction in this instance originates from a property of the muscle fiber itself, it is known as a **myogenic contraction.** Myogenic contractions are common in blood vessels that maintain a certain amount of tone at all times.

Although stretch may initiate a contraction, some types of smooth muscle adapt if the muscle cells are stretched for an extended period of time. As the stretch stimulus continues, the Ca²⁺ channels begin to close in a time-dependent fashion. Then, as Ca²⁺ is pumped out of the cell, the muscle relaxes. This adaptation response explains why the bladder develops tension as it fills, then relaxes as it adjusts to the increased volume. (There is a limit to the amount of stretch the muscle can endure, however, and once a critical volume is reached, the urination reflex empties the bladder.)

FIG. 12.28 Membrane potentials vary in smooth muscle

(a) Slow wave potentials fire action potentials when they reach threshold.

(b) Pacemaker potentials always depolarize to threshold.

(c) Pharmacomechanical coupling occurs when chemical signals change muscle tension through signal transduction pathways with little or no change in membrane potential.

CONCEPT CHECK

28. Compare the following aspects of skeletal and smooth muscle contraction:
    (a) signal for crossbridge activation
    (b) source(s) of calcium for the Ca²⁺ signal
    (c) signal that releases Ca²⁺ from the sarcoplasmic reticulum

29. What happens to contraction if a smooth muscle is placed in a saline bath from which all calcium has been removed?

30. Compare Ca²⁺ release channels in skeletal and smooth muscle sarcoplasmic reticulum.

Some Smooth Muscles Have Unstable Membrane Potentials

The role of membrane potentials in smooth muscle contraction is more complex than in skeletal muscle, where contraction always begins in response to an action potential. Smooth muscle exhibits a variety of electrical behaviors: it can hyperpolarize as well as depolarize. Hyperpolarization of the cell decreases the likelihood of contraction. Smooth muscle can also depolarize without firing action potentials. Contraction may take place after an action potential, after a subthreshold graded potential, or without any change in membrane potential.

Many types of smooth muscle display resting membrane potentials that vary between –40 and –80 mV. Cells that exhibit cyclic depolarization and repolarization of their membrane potential are said to have **slow wave potentials** (Fig. 12.28a). Sometimes, the cell simply cycles through a series of subthreshold slow waves. However, if the peak of the depolarization reaches threshold, action potentials fire, followed by contraction of the muscle.

Other types of smooth muscle with oscillating membrane potentials have regular depolarizations that always reach threshold and fire an action potential (Fig. 12.28b). These depolarizations are called **pacemaker potentials** because they create regular rhythms of contraction. Pacemaker potentials are found in some cardiac muscles as well as in smooth muscle. Both slow wave and pacemaker potentials are due to ion channels in the cell membrane that spontaneously open and close.

In pharmacomechanical coupling, the membrane potential of the muscle may not change at all. In the next section, we consider how this occurs.

CONCEPT CHECK

31. How do pacemaker potentials differ from slow wave potentials?

32. When tetrodotoxin (TTX), a poison that blocks Na⁺ channels, is applied to certain types of smooth muscle, it does not alter the spontaneous generation of action potentials. From this observation, what conclusion can you draw about the action potentials of these types of smooth muscle?

Chemical Signals Influence Smooth Muscle Activity

In this section, we look at how smooth muscle function is influenced by neurotransmitters, hormones, or paracrine signals. These chemical signals may be either excitatory or inhibitory, and they modulate contraction by second messenger action at the level of myosin as well as by influencing Ca^{2+} signals (**FIG. 12.29**). One of the interesting properties of smooth muscle is that signal transduction may cause muscle relaxation as well as contraction.

Autonomic Neurotransmitters and Hormones Many smooth muscles are under antagonistic control by both sympathetic and parasympathetic divisions of the autonomic nervous system. Other

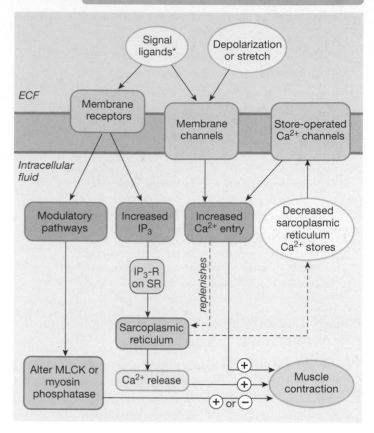

FIG. 12.29   Control of smooth muscle contraction

KEY

IP_3–R = IP_3-activated receptor channel

*Ligands include norepinephrine, ACh, other neurotransmitters, hormones, and paracrines.

smooth muscles, such as those found in blood vessels, are under *tonic control* [p. 183] by only one of the two autonomic branches. In tonic control, the response is graded by increasing or decreasing the amount of neurotransmitter released onto the muscle.

A chemical signal can have different effects in different tissues, depending on the receptor type to which it binds [p. 180]. For this reason, it is important to specify the signal molecule and its receptor and subtype when describing the control of a tissue. For example, the sympathetic neurohormone epinephrine causes smooth muscle contraction when it binds to α-adrenergic receptors but relaxation when it binds to β_2-adrenergic receptors.

Most smooth muscle neurotransmitters and hormones bind to G protein-linked receptors. The second messenger pathways then determine the muscle response: IP_3 triggers contraction and cAMP promotes relaxation.

Pathways that increase IP_3 cause contraction several ways:

- IP_3 opens IP_3 channels on the SR to release Ca^{2+}.

- Diacylglycerol (DAG), another product of the phospholipase C signal pathway, indirectly inhibits myosin

CHAPTER

12

TABLE 12.3	Comparison of the Three Muscle Types		
	Skeletal	**Smooth**	**Cardiac**
Appearance under Light Microscope	Striated	Smooth	Striated
Fiber Arrangement	Sarcomeres	No sarcomeres	Sarcomeres
Location	Attached to bones; a few sphincters close off hollow organs	Forms the walls of hollow organs and tubes; some sphincters	Heart muscle
Tissue Morphology	Multinucleate; large, cylindrical fibers	Uninucleate; small spindle-shaped fibers	Uninucleate; shorter branching fibers
Internal Structure	T-tubule and sarcoplasmic reticulum	No t-tubules; sarcoplasmic reticulum	T-tubule and sarcoplasmic reticulum
Fiber Proteins	Actin, myosin; troponin and tropomyosin	Actin, myosin; tropomyosin	Actin, myosin; troponin and tropomyosin
Control	• Ca^{2+} and troponin • Fibers independent of one another	• Ca^{2+} and calmodulin • Some fibers electrically linked via gap junctions; others independent	• Ca^{2+} and troponin • Fibers electrically linked via gap junctions
Contraction Speed	Fastest	Slowest	Intermediate
Contraction Force of Single Fiber Twitch	Not graded	Graded	Graded
Initiation of Contraction	Requires ACh from motor neuron	Stretch, chemical signals. Can be autorhythmic	Autorhythmic
Neural Control of Contraction	Somatic motor neuron	Autonomic neurons	Autonomic neurons
Hormonal Influence on Contraction	None	Multiple hormones	Epinephrine

phosphatase activity. Increasing the MLCK/MLCP ratio promotes crossbridge activity and muscle tension.

Signals that increase cAMP production cause muscle relaxation through the following mechanisms:

- Free cytosolic Ca^{2+} concentrations decrease when IP_3 channels are inhibited and the SR Ca^{2+}-ATPase is activated.

- K^+ leaking out of the cell hyperpolarizes it and decreases the likelihood of voltage-activated Ca^{2+} entry.

- Myosin phosphatase activity increases, which causes a decrease in muscle tension.

Paracrine Signals Locally released paracrine signal molecules can also alter smooth muscle contraction. For example, asthma is a condition in which smooth muscle of the airways constricts in response to histamine release. This constriction can be reversed by the administration of epinephrine, a neurohormone that relaxes smooth muscle and dilates the airway. Note from this example that not all physiological responses are adaptive or favorable to the body: Constriction of the airways triggered during an asthma attack, if left untreated, can be fatal.

Another important paracrine molecule that affects smooth muscle contraction is *nitric oxide* [p. 178]. This gas is synthesized by the endothelial lining of blood vessels and relaxes adjacent smooth muscle that regulates the diameter of the blood vessels. For many years, the identity of this *endothelium-derived relaxing factor*, or EDRF, eluded scientists even though its presence could be demonstrated experimentally. We know now that EDRF is nitric oxide, an important paracrine signal in many systems of the body.

Because several different signals might reach a muscle fiber simultaneously, smooth muscle fibers must act as integrating centers. For example, sometimes blood vessels receive contradictory messages from two sources: one message signals for contraction, and the other for relaxation. The smooth muscle fibers must integrate the two signals and execute an appropriate response. The complexity of overlapping signal pathways influencing smooth muscle tone can make the tissue difficult to work with in the laboratory.

Although smooth muscles do not have nearly the mass of skeletal muscles, they play a critical role in body function. You will learn more about smooth muscle physiology as you study the different organ systems.

CONCEPT CHECK

33. How can a neuron alter the amount of neurotransmitter it releases? [*Hint:* See Fig. 8.21, p. 261.]
34. Explain how hyperpolarization decreases the likelihood of contraction in smooth muscle.
35. What causes relaxation in skeletal muscle?

CARDIAC MUSCLE

Cardiac muscle, the specialized muscle of the heart, has features of both smooth and skeletal muscle (**TBL. 12.3**). Like skeletal muscle fibers, cardiac muscle fibers are striated and have a sarcomere structure. However, cardiac muscle fibers are shorter than skeletal muscle fibers, may be branched, and have a single nucleus (unlike multinucleate skeletal muscle fibers).

As in single-unit smooth muscle, cardiac muscle fibers are electrically linked to one another. The gap junctions are contained in specialized cell junctions known as *intercalated disks*. Some cardiac muscle, like some smooth muscle, exhibits pacemaker potentials. In addition, cardiac muscle is under sympathetic and parasympathetic control as well as hormonal control. You will learn more about cardiac muscle and how it functions within the heart when you study the cardiovascular system.

RUNNING **PROBLEM CONCLUSION** | Periodic Paralysis

In this running problem, you were introduced to hyperkalemic periodic paralysis (hyperKPP), a condition caused by a genetic defect in voltage-gated Na^+ channels on muscle cell membranes. The periodic paralyses are a family of related disorders caused by muscle ion channel mutations. To learn more about periodic paralyses, see *http://hkpp.org/what-is-periodic-paralysis*. Read the information there to compare the hyperkalemic and hypokalemic forms of the disease. For a more detailed discussion of hyperKPP, read the GeneReviews article at *www.ncbi.nlm.nih.gov/books/NBK1496/*.

Now check your understanding of this running problem by comparing your answers with the information in the following summary table.

Question	Facts	Integration and Analysis
Q1: When Na^+ channels on the muscle membrane open, which way does Na^+ move?	Na^+ is more concentrated in the ECF than in the ICF, and cells have a negative membrane potential.	The electrochemical gradient causes Na^+ to move into cells.
Q2: What effect would continued movement of Na^+ have on the membrane potential of muscle fibers?	The resting membrane potential of cells is negative relative to the extracellular fluid.	The influx of positive charge depolarizes the muscle, and it remains depolarized.
Q3: What ion is responsible for the repolarization phase of the muscle action potential, and in which direction does this ion move across the muscle fiber membrane? How might this be linked to hyperKPP?	In the repolarization phase of the action potential, K^+ leaves the cell.	During repeated contractions, K^+ leaves the muscle fiber, which could contribute to elevated extracellular $[K^+]$ (hyperkalemia).
Q4: Draw a map to explain why a Na^+ channel that does not inactivate results in a muscle that cannot contract (flaccid paralysis).	During an attack, the Na^+ channels remain open and continuously admit Na^+, and the muscle fiber remains depolarized.	If the muscle fiber is unable to repolarize, it cannot fire additional action potentials. The first action potential causes a twitch, but the muscle then goes into a state of flaccid (uncontracted) paralysis.
Q5: Explain why oral glucose might help bring Paul out of his paralysis. (Hint: What happens to the extracellular K^+ level when Na^+-K^+-ATPase is more active?)	The Na^+-K^+-ATPase moves K^+ into cells and Na^+ out of cells.	Providing glucose to cells triggers insulin release. Insulin increases Na^+-K^+-ATPase activity, which removes Na^+ from the cells and helps them repolarize.

378 389 394 403 406 **412**

VISUALIZE, PRACTICE, AND MASTER PHYSIOLOGY WITH MASTERING A&P
- Practice-on-the-go with Dynamic Study Modules
- Visualize important processes with Interactive Physiology and A&P Flix
- Check your understanding with Video Tutors and practice quizzes!

MasteringA&P®

CHAPTER SUMMARY

Muscles provide an excellent system for studying *structure-function* relationships at all levels, from actin, myosin, and sliding filaments in the cell to muscles pulling on bones and joints. *Mechanical properties* of muscles that influence contraction include elastic components, such as the protein titin and the series elastic elements of the intact muscle. *Compartmentation* is essential to muscle function, as demonstrated by the concentration of Ca^{2+} in the sarcoplasmic reticulum and the key role of Ca^{2+} signals in initiating contraction. The *law of mass action* is at work in the dynamics of Ca^{2+}-calmodulin and Ca^{2+}-troponin binding and unbinding. Muscles also show how *biological energy use* transforms stored energy in ATP's chemical bonds to the movement of motor proteins.

Muscles provide many examples of *communication* and *control* in the body. Communication occurs on a scale as small as electrical signals spreading among smooth muscle cells via gap junctions, or as large as a somatic motor neuron innervating multiple muscle fibers. Skeletal muscles are controlled only by somatic motor neurons, but smooth and cardiac muscle have complex regulation that ranges from neurotransmitters to hormones and paracrine molecules.

1. Muscles generate motion, force, and heat. (p. 378)
2. The three types of muscle are **skeletal muscle, cardiac muscle,** and **smooth muscle.** Skeletal and cardiac muscles are **striated muscles.** (p. 378; Fig. 12.1)
3. Skeletal muscles are controlled by somatic motor neurons. Cardiac and smooth muscle are controlled by autonomic innervation, paracrine signals, and hormones. Some smooth and cardiac muscles are autorhythmic and contract spontaneously. (p. 379)

Skeletal Muscle

iP Muscular Physiology

4. Skeletal muscles are usually attached to bones by tendons. The **origin** is the end of the muscle attached closest to the trunk or to the more stationary bone. The **insertion** is the more distal or mobile attachment. (p. 379)
5. At a flexible joint, muscle contraction moves the skeleton. **Flexors** bring bones closer together; **extensors** move bones away from each other. Flexor-extensor pairs are examples of **antagonistic muscle groups.** (p. 379; Fig. 12.2)
6. A skeletal muscle is a collection of **muscle fibers,** large cells with many nuclei. (p. 379; Fig. 12.3)
7. **T-tubules** allow action potentials to move rapidly into the interior of the fiber and release calcium from the **sarcoplasmic reticulum.** (p. 382; Fig. 12.4)
8. **Myofibrils** are intracellular bundles of contractile and elastic proteins. **Thick filaments** are made of **myosin. Thin filaments** are made mostly of **actin. Titin** and **nebulin** hold thick and thin filaments in position. (pp. 382, 383; Figs. 12.3, 12.6)

9. Myosin binds to actin, creating **crossbridges** between the thick and thin filaments. (p. 383; Fig. 12.3d)
10. A **sarcomere** is the contractile unit of a myofibril. It is composed of two **Z disks** and the filaments between them. The sarcomere is divided into **I bands** (thin filaments only), an **A band** that runs the length of a thick filament, and a central **H zone** occupied by thick filaments only. The **M line** and Z disks represent attachment sites for myosin and actin, respectively. (p. 383; Fig. 12.5)
11. The force created by a contracting muscle is called **muscle tension.** The **load** is a weight or force that opposes contraction of a muscle. (p. 383)
12. The **sliding filament theory of contraction** states that during contraction, overlapping thick and thin filaments slide past each other in an energy-dependent manner as a result of actin-myosin crossbridge movement. (p. 385; Fig. 12.5d, e)
13. In relaxed muscle, **tropomyosin** partially blocks the myosin-binding site on actin. To initiate contraction, Ca^{2+} binds to **troponin.** This unblocks the myosin-binding sites and allows myosin to complete its power stroke. (p. 386; Fig. 12.8)
14. During relaxation, the sarcoplasmic reticulum uses a **Ca^{2+}-ATPase** to pump Ca^{2+} back into its lumen. (p. 389)
15. Myosin converts energy from ATP into motion. **Myosin ATPase** hydrolyzes ATP to ADP and P_i. (p. 386; Fig. 12.9)
16. When myosin releases P_i the myosin head moves in the **power stroke.** At the end of the power stroke, myosin releases ADP. The cycle ends in the **rigor state,** with myosin tightly bound to actin. (pp. 385, 386; Fig. 12.9)
17. In **excitation-contraction coupling,** a somatic motor neuron releases ACh, which initiates a skeletal muscle action potential that leads to contraction. (p. 388; Fig. 12.10a)
18. Voltage-sensing Ca^{2+} channels called **DHP receptors** in the t-tubules open RyR **Ca^{2+} release channels** in the sarcoplasmic reticulum. (p. 389; Fig. 12.10b)
19. Relaxation occurs when Ca^{2+} is pumped back into the SR by a Ca^{2+}-ATPase. (p. 389; Fig. 12.10c)
20. A single contraction-relaxation cycle is known as a **twitch.** The **latent period** between the end of the muscle action potential and the beginning of muscle tension development represents the time required for Ca^{2+} release and binding to troponin. (p. 389; Fig. 12.11)
21. Muscle fibers store energy for contraction in **phosphocreatine.** Anaerobic metabolism of glucose is a rapid source of ATP but is not efficient. Aerobic metabolism is very efficient but requires an adequate supply of oxygen to the muscles. (p. 391; Fig. 12.12)
22. **Muscle fatigue** is a reversible condition in which a muscle is no longer able to generate or sustain the expected power output. Fatigue has multiple causes. (p. 392; Fig. 12.13)

23. Skeletal muscle fibers can be classified on the basis of their speed of contraction and resistance to fatigue into **slow-twitch (oxidative) fibers, fast-twitch oxidative-glycolytic fibers,** and **fast-twitch glycolytic fibers.** Oxidative fibers are the most fatigue resistant. (p. 393; Fig. 12.14; Tbl. 12.2)

24. **Myoglobin** is an oxygen-binding pigment that transfers oxygen to the interior of the muscle fiber. (p. 394)

25. The tension of a skeletal muscle contraction is determined by the length of the sarcomeres before contraction begins. (p. 395; Fig. 12.15)

26. Increasing the stimulus frequency causes summation of twitches with an increase of tension. A state of maximal contraction is known as **tetanus.** (p. 396; Fig. 12.16)

27. A **motor unit** is composed of a group of muscle fibers and the somatic motor neuron that controls them. The number of muscle fibers in a motor unit varies, but all fibers in a single motor unit are of the same fiber type. (p. 396; Fig. 12.17)

28. The force of contraction within a skeletal muscle can be increased by **recruitment** of additional motor units. (p. 398)

Mechanics of Body Movement

29. An **isotonic contraction** creates force as the muscle shortens and moves a load. An **isometric contraction** creates force without moving a load. *Lengthening contractions* create force while the muscle lengthens. (pp. 398, 399; Fig. 12.18)

30. Isometric contractions occur because **series elastic elements** allow the fibers to maintain constant length even though the sarcomeres are shortening and creating tension. (p. 400; Fig. 12.19)

31. The body uses its bones and joints as **levers** and **fulcrums.** Most lever-fulcrum systems in the body maximize the distance and speed that a load can be moved but also require that muscles do more work than they would without the lever. (p. 400; Fig. 12.20)

32. Contraction speed is a function of muscle fiber type and load. Contraction is fastest when the load on the muscle is zero. (p. 402; Fig. 12.21)

Smooth Muscle

33. Smooth muscle is slower than skeletal muscle but can sustain contractions for longer without fatiguing. (p. 404; Fig. 12.24)

34. **Phasic muscles** are usually relaxed or cycle through contractions. **Tonic smooth muscle** is usually contracted. (p. 403; Fig. 12.22)

35. **Single-unit smooth muscle** contracts as a single unit when depolarizations pass from cell to cell through gap junctions. In **multiunit smooth muscle,** individual muscle fibers are stimulated independently. (p. 403; Fig. 12.23)

36. Smooth muscle has less myosin than skeletal muscle. Each myosin is associated with about 12–15 actin molecules. Smooth muscle actin lacks troponin. (p. 406; Fig. 12.25)

37. Smooth muscle sarcoplasmic reticulum has both RyR Ca^{2+} release channels and **IP$_3$-receptor channels.** Calcium also enters the cell from the extracellular fluid. (p. 407)

38. In smooth muscle contraction, Ca^{2+} binds to **calmodulin** and activates **myosin light chain kinase (MLCK).** (pp. 406, 407; Fig. 12.26a)

39. MLCK phosphorylates **myosin light protein chains,** which activates myosin ATPase. This allows crossbridge power strokes. (p. 407; Fig. 12.26a)

40. During relaxation, Ca^{2+} is pumped out of the cytosol, and myosin light chains are dephosphorylated by **myosin phosphatase.** (p. 407; Fig. 12.26b)

41. Smooth muscle **calcium sensitivity** can be altered by changing myosin phosphatase activity. (p. 407; Fig. 12.27)

42. In **myogenic contraction,** stretch on the cell depolarizes it and opens membrane Ca^{2+} channels. (p. 409)

43. Unstable membrane potentials in smooth muscle take the form of either **slow wave potentials** or **pacemaker potentials.** (p. 410; Fig. 12.28a, b)

44. In **pharmacomechanical coupling,** smooth muscle contraction initiated by chemical signals can take place without a significant change in membrane potential. (p. 407; Fig. 12.28c)

45. Smooth muscle contraction is influenced by sympathetic and parasympathetic neurons and a variety of hormones and paracrine signals. (p. 410; Fig. 12.29)

Cardiac Muscle

46. Cardiac muscle fibers are striated, have a single nucleus, and are electrically linked through gap junctions. Cardiac muscle shares features with both skeletal and smooth muscle. (p. 411; Tbl. 12.3)

REVIEW QUESTIONS

In addition to working through these questions and checking your answers on p. A-16, review the Learning Outcomes at the beginning of this chapter.

Level One Reviewing Facts and Terms

1. The three types of muscle tissue found in the human body are _____, _____, and _____. Which type is attached to the bones, enabling it to control body movement?

2. Which two muscle types are striated?

3. Which type of muscle tissue is controlled only by somatic motor neurons?

4. Arrange the following skeletal muscle components in order, from outermost to innermost: sarcolemma, connective tissue sheath, thick and thin filaments, myofibrils.

5. The modified endoplasmic reticulum of skeletal muscle is called the _____. Its role is to sequester _____ ions.

6. Which of the following statement(s) is (are) true about skeletal muscles?

   (a) They constitute about 60% of a person's total body weight.
   (b) They position and move the skeleton.
   (c) The insertion of the muscle is more distal or mobile than the origin.
   (d) They are often paired into antagonistic muscle groups called flexors and extensors.

7. T-tubules allow _____ to move to the interior of the muscle fiber.

8. List six proteins that make up the myofibrils. Which protein creates the power stroke for contraction?

9. List the letters used to label the elements of a sarcomere. Which band has a Z disk in the middle? Which is the darkest band? Why? Which element forms the boundaries of a sarcomere? Name the line that divides the A band in half. What is the function of this line?

10. Briefly explain the functions of titin and nebulin.

11. During contraction, the _____ band remains a constant length. This band is composed primarily of _____ molecules. Which components of the sarcomere approach each other during contraction?

12. Explain the sliding filament theory of contraction.

13. Explain the roles of troponin, tropomyosin, and Ca^{2+} in skeletal muscle contraction.

14. Which neurotransmitter is released by somatic motor neurons?

15. What is the motor end plate, and what kinds of receptors are found there? Explain how neurotransmitter binding to these receptors creates an action potential.

16. Match the following characteristics with the appropriate type(s) of muscle.

(a) has the largest diameter	1. fast-twitch glycolytic fibers
(b) uses anaerobic metabolism, thus fatigues quickly	2. fast-twitch oxidative-glycolytic fibers
(c) has the most blood vessels	3. slow-twitch oxidative fibers
(d) has some myoglobin	
(e) is used for quick, fine movements	
(f) is also called red muscle	
(g) uses a combination of oxidative and glycolytic metabolism	
(h) has the most mitochondria	

17. A single contraction-relaxation cycle in a skeletal muscle fiber is known as a(n) _____.

18. List the steps of skeletal muscle contraction that require ATP.

19. The basic unit of contraction in an intact skeletal muscle is the _____. The force of contraction within a skeletal muscle is increased by _____ additional motor units.

20. The two functional types of smooth muscle are _____ and _____.

Level Two Reviewing Concepts

21. Make a map of muscle fiber structure using the following terms. Add terms if you like.

• actin	• myosin
• Ca^{2+}	• nucleus
• cell	• regulatory protein
• cell membrane	• sarcolemma
• contractile protein	• sarcoplasm
• crossbridges	• sarcoplasmic reticulum
• cytoplasm	• titin
• elastic protein	• tropomyosin
• glycogen	• troponin
• mitochondria	• t-tubule
• muscle fiber	

22. How does an action potential in a muscle fiber trigger a Ca^{2+} signal inside the fiber?

23. Muscle fibers depend on a continuous supply of ATP. How do the fibers in the different types of muscle generate ATP?

24. Define muscle fatigue. Summarize factors that could play a role in its development. How can muscle fibers adapt to resist fatigue?

25. Explain how you vary the strength and effort made by your muscles in picking up a pencil versus picking up a full gallon container of milk.

26. Compare and contrast the following in skeletal and smooth muscle:

    (a) cellular anatomy
    (b) neural and chemical control of contraction

27. Arrange the following terms to create a map of skeletal muscle excitation, contraction, and relaxation. Terms may be used more than once. Add terms if you like.

• acetylcholine	• Ca^{2+}-ATPase
• ACh receptor	• calcium-release channels
• actin	• contraction
• action potential	• crossbridge
• ADP	• DHP receptor
• ATP	• end-plate potential
• axon terminal	• exocytosis
• Ca^{2+}	• motor end plate
• myosin	• sarcoplasmic reticulum
• Na^+	• somatic motor neuron
• neuromuscular junction	• tropomyosin
• P_i	• troponin
• power stroke	• t-tubules
• relaxation	• voltage-gated Ca^{2+} channels
• rigor state	

28. What is the role of the sarcoplasmic reticulum in muscular contraction? How can smooth muscle contract when it has so little sarcoplasmic reticulum?

29. Compare and contrast:

    (a) fast-twitch oxidative-glycolytic, fast-twitch glycolytic, and slow-twitch muscle fibers
    (b) a twitch and tetanus
    (c) action potentials in motor neurons and action potentials in skeletal muscles
    (d) temporal summation in motor neurons and summation in skeletal muscles
    (e) isotonic contraction and isometric contraction
    (f) slow-wave and pacemaker potentials
    (g) the source and role of Ca^{2+} in skeletal and smooth muscle contraction

30. Explain the different factors that influence Ca^{2+} entry and release in smooth muscle fibers.

Level Three Problem Solving

31. One way that scientists study muscles is to put them into a state of rigor by removing ATP. In this condition, actin and myosin are strongly linked but unable to move. On the basis of what you know about muscle contraction, predict what would happen to these

muscles in a state of rigor if you (a) added ATP but no free calcium ions; (b) added ATP with a substantial concentration of calcium ions.

32. When curare, a South American Indian arrow poison, is placed on a nerve-muscle preparation, the muscle does not contract when the nerve is stimulated, even though neurotransmitter is still being released from the nerve. Give all possible explanations for the action of curare that you can think of.

33. On the basis of what you have learned about muscle fiber types and metabolism, predict what variations in structure you would find among these athletes:

    (a) a 7-foot, 2-inch-tall, 325-pound basketball player
    (b) a 5-foot, 10-inch-tall, 180-pound steer wrestler
    (c) a 5-foot, 7-inch-tall, 130-pound female figure skater
    (d) a 4-foot, 11-inch-tall, 89-pound female gymnast

Level Four Quantitative Problems

34. Look at the following graph, created from data published in "Effect of ambient temperature on human skeletal muscle metabolism during fatiguing submaximal exercise," *J Appl Physiol* 86(3): 902–908, 1999. What hypotheses might you develop about the cause(s) of muscle fatigue based on these data?

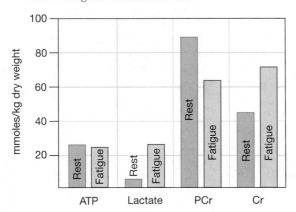

Muscle metabolites in resting muscle and after cycling exercise to fatigue

35. Use the arm in Figure 12.20c to answer the following questions.

    (a) How much force would a biceps muscle inserted 4 cm from the fulcrum need to exert to hold the arm stationary at a 90° angle? How does this force compare with the force needed when the insertion point is 5 cm from the fulcrum?

    (b) Suppose a 7-kg weight band is placed around the wrist 20 cm from the fulcrum. How much force does the biceps inserted 5 cm from the fulcrum need to exert to hold the arm stationary at a 90° angle? How does this force compare with the force needed to keep the arm horizontal in the situation shown in Figure 12.20c, with the same weight in the hand (25 cm from the fulcrum)?

Answers to Concept Checks, Figure and Graph Questions, and end-of-chapter Review Questions can be found in Appendix A [p. A-1].

13

Integrative Physiology I: Control of Body Movement

Extracting signals directly from the brain to directly control robotic devices has been a science fiction theme that seems destined to become fact.

Dr. Eberhard E. Fetz, "Rats Operate Robotic Arm Via Brain Activity," *Science News* 156: 142, 8/28/1999

NEURAL REFLEXES 418

LO 13.1 List four ways to classify neural reflex pathways.

AUTONOMIC REFLEXES 420

LO 13.2 List some examples of autonomic reflexes.

SKELETAL MUSCLE REFLEXES 420

LO 13.3 Diagram the steps of a skeletal muscle reflex, including the following terms: alpha motor neuron, proprioceptor, extrafusal fibers, muscle tone.

LO 13.4 Diagram a stretch reflex with alpha-gamma coactivation in the muscle spindle.

LO 13.5 Use the following terms to explain the patellar tendon reflex: monosynaptic stretch reflex, reciprocal inhibition, myotatic unit.

LO 13.6 Diagram a flexion reflex and its associated crossed extensor reflex.

THE INTEGRATED CONTROL OF BODY MOVEMENT 426

LO 13.7 Compare and contrast reflex, rhythmic, and voluntary movements and their control.

LO 13.8 Describe the role of the following brain structures in the control of movement: basal ganglia, brain stem, cerebellum, motor areas of cerebral cortex, prefrontal cortex, thalamus, spinal cord.

LO 13.9 Describe the anatomy and function of the corticospinal tract.

CONTROL OF MOVEMENT IN VISCERAL MUSCLES 431

BACKGROUND BASICS

183 Reflex pathways
227 Central nervous system
266 Summation of action potentials
398 Isometric contraction
310 Sensory pathways and receptors
240 Graded potentials
183 Tonic control
 82 Tendons

Each dot of a microarray represents one gene. Genes that are active show up in bright colors.

Think of a baseball pitcher standing on the mound. As he looks at the first batter, he receives sensory information from multiple sources: the sound of the crowd, the sight of the batter and the catcher, the smell of grass, the feel of the ball in his hand, and the alignment of his body as he begins his windup. Sensory receptors code this information and send it to the central nervous system (CNS), where it is integrated.

The pitcher acts consciously on some of the information: He decides to throw a fastball. But he processes other information at the subconscious level and acts on it without conscious thought. As he thinks about starting his motion, for instance, he shifts his weight to offset the impending movement of his arm. The integration of sensory information into an involuntary response is the hallmark of a *reflex* [p. 183].

NEURAL REFLEXES

All neural reflexes begin with a stimulus that activates a sensory receptor. The sensor sends information in the form of action potentials through sensory afferent neurons to the CNS [p. 183]. The CNS is the integrating center that evaluates all incoming information and selects an appropriate response. It then initiates action potentials in efferent neurons to direct the response of muscles and glands—the targets.

A key feature of many reflex pathways is *negative feedback* [p. 15]. Feedback signals from muscle and joint receptors keep the CNS continuously informed of changing body position. Some reflexes have a *feedforward* component that allows the body to anticipate a stimulus and begin the response [p. 17]. Bracing yourself in anticipation of a collision is an example of a feedforward response.

Neural Reflex Pathways Can Be Classified in Different Ways

Reflex pathways in the nervous system consist of chains or networks of neurons that link sensory receptors to muscles or glands. Neural reflexes can be classified in several ways (**TBL. 13.1**):

1. *By the efferent division of the nervous system that controls the response.* Reflexes that involve somatic motor neurons and skeletal muscles are known as **somatic reflexes**. Reflexes whose responses are controlled by autonomic neurons are called **autonomic reflexes**.

2. *By the CNS location where the reflex is integrated.* **Spinal reflexes** are integrated in the spinal cord. These reflexes may be modulated by higher input from the brain, but they can occur without that input. Reflexes integrated in the brain are called **cranial reflexes**.

3. *By whether the reflex is innate or learned.* Many reflexes are **innate**; in other words, we are born with them, and they are genetically determined. One example is the knee jerk, or patellar tendon reflex: When the patellar tendon at the lower edge of the kneecap is stretched with a tap from a reflex hammer, the lower leg kicks out. Other reflexes are

418 420 424 427 430 432

acquired through experience [p. 300]. The example of Pavlov's dogs salivating upon hearing a bell is the classic example of a **learned reflex**, also referred to as a **conditioned reflex**.

4. *By the number of neurons in the reflex pathway.* The simplest reflex is a **monosynaptic reflex**, named for the single synapse between the two neurons in the pathway: a sensory afferent neuron (often just called *a sensory afferent*) and an efferent somatic motor neuron (**FIG. 13.1a**). These two neurons synapse in the spinal cord, allowing a signal initiated at the receptor to go directly from the sensory neuron to the motor neuron. (The synapse between the somatic motor neuron and its muscle target is ignored.)

TABLE 13.1	**Classification of Neural Reflexes**

Neural reflexes can be classified by:

1. **Efferent division that controls the effector**
   a. Somatic motor neurons control skeletal muscles.
   b. Autonomic neurons control smooth and cardiac muscle, glands, and adipose tissue.

2. **Integrating region within the central nervous system**
   a. Spinal reflexes do not require input from the brain.
   b. Cranial reflexes are integrated within the brain.

3. **Time at which the reflex develops**
   a. Innate (inborn) reflexes are genetically determined.
   b. Learned (conditioned) reflexes are acquired through experience.

4. **The number of neurons in the reflex pathway**
   a. Monosynaptic reflexes have only two neurons: one afferent (sensory) and one efferent. Only somatic motor reflexes can be monosynaptic.
   b. Polysynaptic reflexes include one or more interneurons between the afferent and efferent neurons. All autonomic reflexes are polysynaptic because they have three neurons: one afferent and two efferent.

FIG. 13.1 **ESSENTIALS**

Neural Reflexes

SKELETAL MUSCLE REFLEXES

(a) A **monosynaptic reflex** has a single synapse between the afferent and efferent neurons.

(b) Polysynaptic reflexes have two or more synapses. This somatic motor reflex has both synapses in the CNS.

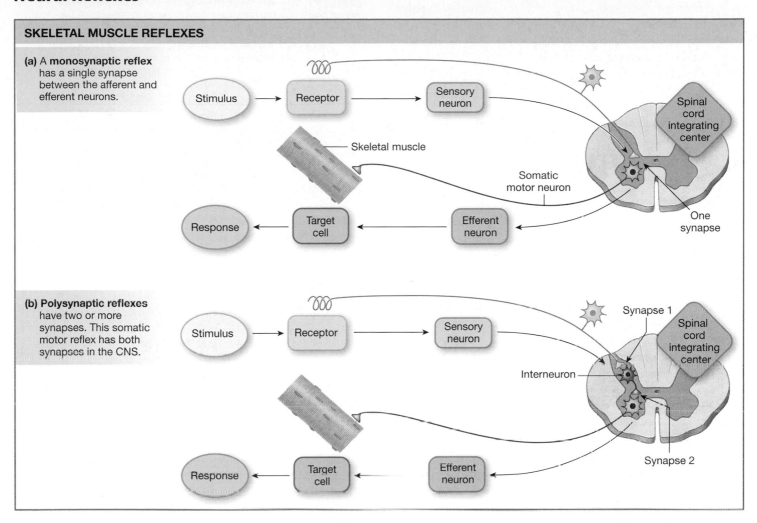

AUTONOMIC REFLEXES

(c) All autonomic reflexes are polysynaptic, with at least one synapse in the CNS and another in the autonomic ganglion.

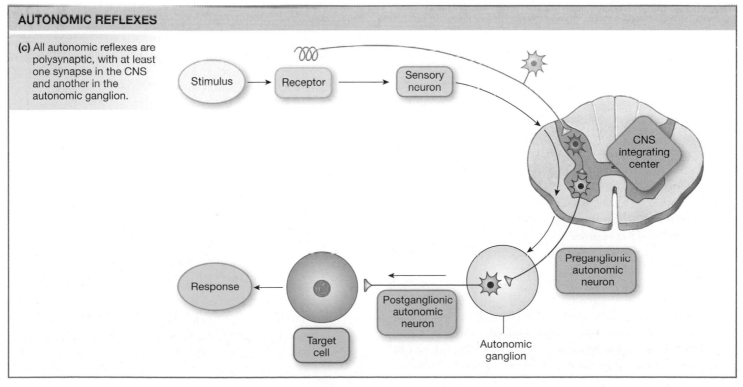

Most reflexes have three or more neurons in the pathway (and at least two synapses), leading to their designation as **polysynaptic reflexes** (Fig. 13.1b, c). Polysynaptic reflexes may be quite complex, with extensive branching in the CNS to form networks involving multiple interneurons. *Divergence* of pathways allows a single stimulus to affect multiple targets [p. 260]. *Convergence* integrates the input from multiple sources to modify the response. The modification in polysynaptic pathways may involve excitation or inhibition [p. 263].

AUTONOMIC REFLEXES

Autonomic reflexes are also known as *visceral reflexes* because they often involve the internal organs of the body. Some visceral reflexes, such as urination and defecation, are spinal reflexes that can take place without input from the brain. However, spinal reflexes are often modulated by excitatory or inhibitory signals from the brain, carried by descending tracts from higher brain centers.

For example, urination may be voluntarily initiated by conscious thought. Or it may be inhibited by emotion or a stressful situation, such as the presence of other people (a syndrome known as "bashful bladder"). Often, the higher control of a spinal reflex is a learned response. The toilet training we master as toddlers is an example of a learned reflex that the CNS uses to modulate the simple spinal reflex of urination.

Other autonomic reflexes are integrated in the brain, primarily in the hypothalamus, thalamus, and brain stem. These regions contain centers that coordinate body functions needed to maintain homeostasis, such as heart rate, blood pressure, breathing, eating, water balance, and maintenance of body temperature [see Fig. 11.3, p. 360]. The brain stem also contains the integrating centers for autonomic reflexes such as salivating, vomiting, sneezing, coughing, swallowing, and gagging.

An interesting type of autonomic reflex is the conversion of emotional stimuli into visceral responses. The limbic system [p. 289]—the site of primitive drives such as sex, fear, rage, aggression, and hunger—has been called the "visceral brain" because of its role in these emotionally driven reflexes. We speak of "gut feelings" and "butterflies in the stomach"—all transformations of emotion into somatic sensation and visceral function. Other emotion-linked autonomic reflexes include urination, defecation, blushing, blanching, and *piloerection,* in which tiny muscles in the hair follicles pull the shaft of the hair erect ("I was so scared my hair stood on end!").

Autonomic reflexes are all polysynaptic, with at least one synapse in the CNS between the sensory neuron and the preganglionic autonomic neuron, and an additional synapse in the ganglion between the preganglionic and postganglionic neurons (Fig. 13.1c).

Many autonomic reflexes are characterized by *tonic activity,* a continuous stream of action potentials that creates ongoing activity in the effector. For example, the tonic control of blood vessels is an example of a continuously active autonomic reflex [p. 183]. You will encounter many autonomic reflexes as you continue your study of the systems of the body.

CONCEPT CHECK

1. List the general steps of a reflex pathway, including the anatomical structures in the nervous system that correspond to each step.
2. If a cell hyperpolarizes, does its membrane potential become more positive or more negative? Does the potential move closer to threshold or farther from threshold?

RUNNING **PROBLEM**

Tetanus {tetanus, a muscle spasm}, also known as lockjaw, is a devastating disease caused by the bacterium *Clostridium tetani.* These bacteria are commonly found in soil and enter the human body through a cut or wound. As the bacteria reproduce in the tissues, they release a protein neurotoxin. This toxin, called *tetanospasmin,* is taken up by somatic motor neurons at the axon terminals. Tetanospasmin then travels along the axons until it reaches the nerve cell body in the spinal cord.

Q1: *a. Tetanospasmin is a protein. By what process is it taken up into neurons? [Hint: p. 148]*
b. By what process does it travel up the axon to the nerve cell body? [Hint: p. 231]

418　420　424　427　430　432

SKELETAL MUSCLE REFLEXES

Although we are not always aware of them, skeletal muscle reflexes are involved in almost everything we do. Receptors that sense changes in joint movements, muscle tension, and muscle length feed this information to the CNS, which responds in one of two ways. If muscle contraction is the appropriate response, the CNS activates somatic motor neurons to the muscle fibers. If a muscle needs to be relaxed to achieve the response, sensory input activates inhibitory interneurons in the CNS, and these interneurons *inhibit* activity in somatic motor neurons controlling the muscle.

Recall that excitation of somatic motor neurons always causes contraction in skeletal muscle [p. 383]. There is no inhibitory neuron that synapses on skeletal muscles to cause them to relax. Instead, *relaxation results from the absence of excitatory input by the somatic motor neuron.* Inhibition and excitation of somatic motor neurons and their associated skeletal muscles must occur at synapses within the CNS.

Skeletal muscle reflexes have the following components:

1. *Sensory receptors,* known as **proprioceptors**, are located in skeletal muscles, joint capsules, and ligaments. Proprioceptors monitor the position of our limbs in space, our movements, and the effort we exert in lifting objects. The input signal from proprioceptors goes to the CNS through sensory neurons.
2. *The central nervous system* integrates the input signal using networks and pathways of *excitatory and inhibitory*

interneurons. In a reflex, sensory information is integrated and acted on subconsciously. However, some sensory information may be integrated in the cerebral cortex and become perception, and some reflexes can be modulated by conscious input.

3. *Somatic motor neurons* carry the output signal. The somatic motor neurons that innervate skeletal muscle contractile fibers are called **alpha motor neurons** (**FIG. 13.2b**).

4. The effectors are contractile skeletal muscle fibers, also known as **extrafusal muscle fibers.** Action potentials in alpha motor neurons cause extrafusal fibers to contract.

Three types of proprioceptors are found in the body: joint receptors, Golgi tendon organs, and muscle spindles. *Joint receptors* are found in the capsules and ligaments around joints in the body. They are stimulated by mechanical distortion that accompanies changes in the relative positioning of bones linked by flexible joints. Sensory information from joint receptors is integrated primarily in the cerebellum.

In the next two sections we examine the function of Golgi tendon organs and muscle spindles, both interesting and unique receptors. These receptors lie inside skeletal muscles and sense changes in muscle length and tension. Their sensory output plays an important role in maintaining body position and movement.

Golgi Tendon Organs Respond to Muscle Tension

The **Golgi tendon organ (GTO)** is a type of receptor found at the junction of tendons and muscle fibers, placed in series with the muscle fibers (Fig. 13.2a). GTOs respond primarily to muscle tension created during the isometric phase of contraction and are relatively insensitive to muscle stretch.

Golgi tendon organs are composed of free nerve endings that wind between collagen fibers inside a connective tissue capsule (Fig. 13.2a). When a muscle contracts, its tendons act as a *series elastic element* during the isometric phase of the contraction [p. 400]. Muscle contraction pulls on collagen fibers within the GTO, pinching sensory endings of the afferent neurons and causing them to fire.

The classic view of Golgi tendon organs was that they were part of a protective reflex initiated by muscle contraction and ending with muscle relaxation. Research has now shown that the Golgi tendon organs primarily provide sensory information to CNS integrating centers. The sensory information from GTOs combines with feedback from muscle spindles and joint receptors to allow optimal motor control of posture and movement.

Muscle Spindles Respond to Muscle Stretch

Muscle spindles are stretch receptors that send information to the spinal cord and brain about muscle length and changes in muscle length. They are small, elongated structures scattered among and arranged parallel to the contractile extrafusal muscle fibers (Fig. 13.2b). With the exception of one muscle in the jaw,

every skeletal muscle in the body has many muscle spindles. For example, a small muscle in the index finger of a newborn human has on average about 50 spindles.

Each muscle spindle consists of a connective tissue *capsule* that encloses a group of small muscle fibers known as **intrafusal fibers** {*intra-,* within + *fusus,* spindle}. Intrafusal muscle fibers are modified so that the ends are contractile but the central region lacks myofibrils (Fig. 13.2b). The noncontractile central region is wrapped by sensory nerve endings that are stimulated by stretch. The contractile ends of the intrafusal fibers have their own innervation from **gamma motor neurons.**

When a muscle is at its resting length, the central region of each muscle spindle is stretched enough to activate the sensory fibers (Fig. 13.2c). As a result, sensory neurons from the spindles are tonically active, sending a steady stream of action potentials to the spinal cord. The sensory neurons synapse directly on alpha motor neurons innervating the muscle in which the spindles lie, creating a monosynaptic reflex as shown in Figure 13.1a. The tonically active sensory neurons mean that tonically active alpha motor neurons are triggering muscle contraction. As a result, even a muscle at rest maintains a certain level of tension known as **muscle tone.**

Muscle spindles are anchored in parallel to the extrafusal muscle fibers. Any movement that increases muscle length also stretches the muscle spindles and causes their sensory fibers to fire more rapidly. Spindle and muscle stretch create a reflex contraction of the muscle to prevent damage from overstretching (**FIG. 13.3**). The reflex pathway in which muscle stretch initiates a contraction response is known as a **stretch reflex.**

An example of how muscle spindles work during a stretch reflex is shown in Figure 13.3. You can demonstrate this yourself with an unsuspecting friend. Have your friend stand with eyes closed, one arm extended with the elbow at 90°, and the hand

---

CLINICAL FOCUS

Reflexes and Muscle Tone

Clinicians use reflexes to investigate the condition of the nervous system and the muscles. For a reflex to be normal, there must be normal conduction through all neurons in the pathway, normal synaptic transmission at the neuromuscular junction, and normal muscle contraction. A reflex that is absent, abnormally slow, or greater than normal (hyperactive) suggests the presence of a pathology. Interestingly, not all abnormal reflexes are caused by neuromuscular disorders. For example, slowed relaxation of the ankle flexion reflex suggests hypothyroidism. (The cellular mechanism linking low thyroid to slow reflexes is not known.) Besides testing reflexes, clinicians assess muscle tone. Even when relaxed and at rest, muscles have a certain resistance to stretch that is the result of continuous (tonic) output by alpha motor neurons. The absence of muscle tone or increased muscle resistance to being stretched by an examiner (increased tone) indicates a problem with the pathways that control muscle contraction.

FIG. 13.2 **ESSENTIALS**

Muscle Spindles and Golgi Tendon Organs

GOLGI TENDON ORGANS

(a) The **Golgi tendon organ** links the muscle and the tendon. It consists of sensory nerve endings interwoven among collagen fibers.

Extrafusal muscle fibers are normal contractile fibers.

Golgi tendon organ

Tendon

Extrafusal muscle fibers

Capsule

Sensory neuron fires when muscle contracts and pulls collagen fibers of the tendon tight.

Tendon

Collagen fiber

MUSCLE SPINDLES

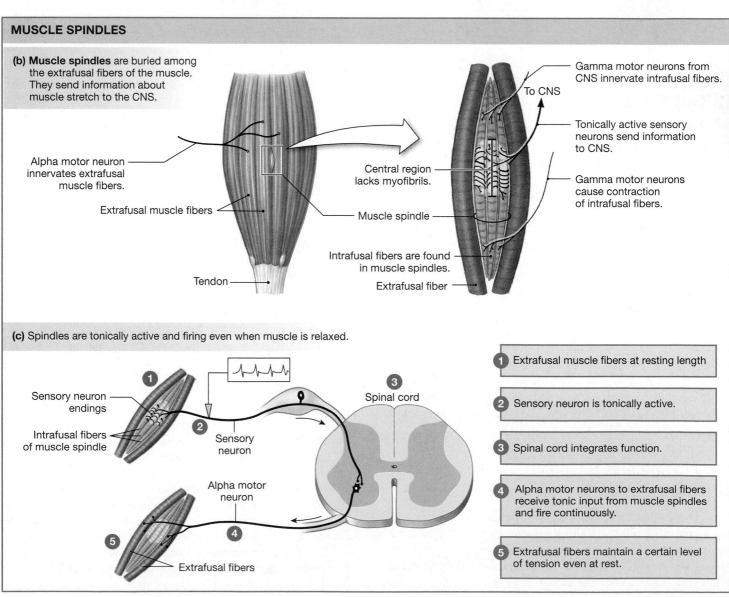

(b) **Muscle spindles** are buried among the extrafusal fibers of the muscle. They send information about muscle stretch to the CNS.

Alpha motor neuron innervates extrafusal muscle fibers.

Extrafusal muscle fibers

Tendon

Gamma motor neurons from CNS innervate intrafusal fibers.

To CNS

Central region lacks myofibrils.

Muscle spindle

Intrafusal fibers are found in muscle spindles.

Extrafusal fiber

Tonically active sensory neurons send information to CNS.

Gamma motor neurons cause contraction of intrafusal fibers.

(c) Spindles are tonically active and firing even when muscle is relaxed.

Sensory neuron endings

Intrafusal fibers of muscle spindle

Sensory neuron

Alpha motor neuron

Extrafusal fibers

Spinal cord

1 Extrafusal muscle fibers at resting length

2 Sensory neuron is tonically active.

3 Spinal cord integrates function.

4 Alpha motor neurons to extrafusal fibers receive tonic input from muscle spindles and fire continuously.

5 Extrafusal fibers maintain a certain level of tension even at rest.

FIG. 13.3   The stretch reflex

Muscle stretch can trigger a stretch reflex. As illustrated below, the addition of a load stretches the muscle and the spindles, creating a reflex contraction.

1 Load added to muscle.

2 Muscle and muscle spindle stretch as arm extends. Muscle spindle afferents fire more frequently.

3 Reflex contraction initiated by muscle stretch restores arm position and prevents damage from over-stretching.

Muscle stretch → Increased afferent signals to spinal cord → Spinal cord → Increased efferent output through alpha motor neurons → Muscle contracts → Firing rate of afferent sensory neuron decreases. Negative feedback

FIGURE QUESTION

Which neuron fires to cause contraction of the extrafusal muscle fibers?

(a) muscle alpha motor neuron

(b) muscle spindle gamma motor neuron

(c) muscle spindle sensory neuron

(d) Golgi tendon organ sensory neuron

(e) none of the above

palm up. Place a small book or other flat weight in the outstretched hand and watch the arm muscles contract to compensate for the added weight.

Now suddenly drop a heavier load, such as another book, onto your friend's hand. The added weight will send the hand downward, stretching the biceps muscle and activating its muscle spindles. Sensory input into the spinal cord then activates the alpha motor neurons of the biceps muscle. The biceps will contract, bringing the arm back to its original position.

CONCEPT CHECK

3. Using the standard steps of a reflex pathway (stimulus, receptor, and so forth), draw a reflex map of the stretch reflex.

Muscle stretch activates muscle spindles, but what happens to spindle activity when a resting muscle contracts and shortens? You might predict that the release of tension on the center of the intrafusal fibers in the absence of gamma motor neuron activity would cause the spindle afferents to slow their firing rate. However, the presence of gamma motor neurons in a normal muscle keeps the muscle spindles active, no matter what the muscle length is.

When alpha motor neurons fire, the muscle shortens, releasing tension on the muscle spindle capsule. To keep the spindle functioning normally, gamma motor neurons innervating the contractile ends of the muscle spindle also fire at the same time (**FIG. 13.4**). The gamma motor neurons cause the spindle intrafusal fibers to contract and shorten. This contraction pulls on the central region of the spindle and maintains stretch on the sensory

FIG. 13.4 Alpha-gamma coactivation

Gamma motor neurons innervate muscle fibers at the ends of muscle spindles. **Alpha-gamma coactivation** keeps the spindles stretched and maintains spindle function when the muscle contracts.

1 Alpha motor neuron fires and gamma motor neuron fires.

2 Muscle and intrafusal fibers both contract.

3 Stretch on centers of intrafusal fibers unchanged. Firing rate of afferent neuron remains constant.

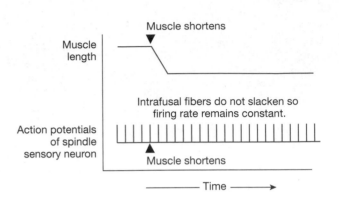

nerve endings. As a result, the spindle remains active even when the muscle contracts. Excitation of gamma motor neurons and alpha motor neurons at the same time is a process known as **alpha-gamma coactivation.**

Stretch Reflexes and Reciprocal Inhibition Control Movement around a Joint

Movement around most flexible joints in the body is controlled by groups of synergistic and antagonistic muscles that act in a coordinated fashion. Sensory neurons from muscle receptors and efferent motor neurons that control the muscle are linked by diverging and converging pathways of interneurons within the spinal cord. The collection of pathways controlling a single joint is known as a **myotatic unit** {*myo-,* muscle + *tasis,* stretching}.

The simplest reflex in a myotatic unit is the monosynaptic stretch reflex, which involves only two neurons: the sensory neuron from the muscle spindle and the somatic motor neuron to the muscle. The patellar tendon reflex is an example of a monosynaptic stretch reflex (**FIG. 13.5**).

To demonstrate this reflex, a person sits on the edge of a table so that the lower leg hangs relaxed. When the patellar tendon below the kneecap is tapped with a small rubber hammer, the tap stretches the quadriceps muscle, which runs up the front of the thigh. This stretching activates muscle spindles and sends action potentials via the sensory fibers to the spinal cord. The sensory neurons synapse directly onto the motor neurons that control contraction of the quadriceps muscle (a monosynaptic reflex). Excitation of the motor neurons causes motor units in the quadriceps to contract, and the lower leg swings forward.

For muscle contraction to extend the leg, the antagonistic flexor muscles must relax, a process known as **reciprocal inhibition.** In the leg, this requires relaxation of the hamstring muscles

running up the back of the thigh. The single stimulus of the tap to the tendon accomplishes both contraction of the quadriceps muscle and reciprocal inhibition of the hamstrings. The sensory fibers branch upon entering the spinal cord. Some of the branches activate motor neurons innervating the quadriceps, while the other branches synapse on inhibitory interneurons. The inhibitory interneurons suppress activity in the motor neurons controlling the hamstrings (a polysynaptic reflex). The result is a relaxation of the hamstrings that allows contraction of the quadriceps to proceed unopposed.

Flexion Reflexes Pull Limbs Away from Painful Stimuli

Flexion reflexes are polysynaptic reflex pathways that cause an arm or leg to be pulled away from a noxious stimulus, such as a pinprick or a hot stove. These reflexes, like the reciprocal inhibition reflex just described, rely on divergent pathways in the spinal

RUNNING **PROBLEM**

Once in the spinal cord, tetanospasmin is released from the motor neuron. It then selectively blocks neurotransmitter release at inhibitory synapses. Patients with tetanus experience muscle spasms that begin in the jaw and may eventually affect the entire body. When the extremities become involved, the arms and legs may go into painful, rigid spasms.

Q2: *Using the reflex pathways diagrammed in Figures 13.5 and 13.6, explain why inhibition of inhibitory interneurons might result in uncontrollable muscle spasms.*

418   420   424   427   430   432

FIG. 13.5 The patellar tendon (knee jerk) reflex

The patellar tendon (knee jerk) reflex illustrates a monosynaptic stretch reflex and reciprocal inhibition of the antagonistic muscle.

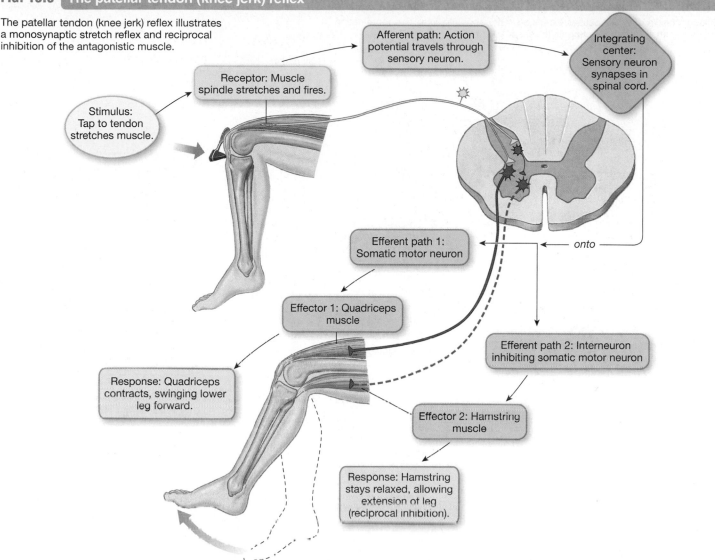

Stimulus: Tap to tendon stretches muscle.

Receptor: Muscle spindle stretches and fires.

Afferent path: Action potential travels through sensory neuron.

Integrating center: Sensory neuron synapses in spinal cord.

Efferent path 1: Somatic motor neuron

onto

Effector 1: Quadriceps muscle

Response: Quadriceps contracts, swinging lower leg forward.

Efferent path 2: Interneuron inhibiting somatic motor neuron

Effector 2: Hamstring muscle

Response: Hamstring stays relaxed, allowing extension of leg (reciprocal inhibition).

cord. **FIGURE 13.6** uses the example of stepping on a tack to illustrate a flexion reflex.

When the foot contacts the point of the tack, nociceptors (pain receptors) in the foot send sensory information to the spinal cord. Here the signal diverges, activating multiple excitatory interneurons. Some of these interneurons excite alpha motor neurons, leading to contraction of the flexor muscles of the stimulated limb. Other interneurons simultaneously activate inhibitory interneurons that cause relaxation of the antagonistic muscle groups. Because of this reciprocal inhibition, the limb is flexed, withdrawing it from the painful stimulus. This type of reflex requires more time than a stretch reflex (such as the knee jerk reflex) because it is a polysynaptic rather than a monosynaptic reflex.

Flexion reflexes, particularly in the legs, are usually accompanied by the **crossed extensor reflex.** The crossed extensor reflex is a postural reflex that helps maintain balance when one

foot is lifted from the ground. The quick withdrawal of the right foot from a painful stimulus (a tack) is matched by extension of the left leg so that it can support the sudden shift in weight (Fig. 13.6). The extensors contract in the supporting left leg and relax in the withdrawing right leg, while the opposite occurs in the flexor muscles.

Note in Figure 13.6 how the one sensory neuron synapses on multiple interneurons. Divergence of the sensory signal permits a single stimulus to control two sets of antagonistic muscle groups as well as to send sensory information to the brain. This type of complex reflex with multiple neuron interactions is more typical of our reflexes than the simple monosynaptic knee jerk stretch reflex.

In the next section, we look at how the CNS controls movements that range from involuntary reflexes to complex, voluntary movement patterns such as dancing, throwing a ball, or playing a musical instrument.

FIG. 13.6 The crossed extensor reflex

A flexion reflex in one limb causes extension in the opposite limb. The coordination of reflexes with postural adjustments is essential for maintaining balance.

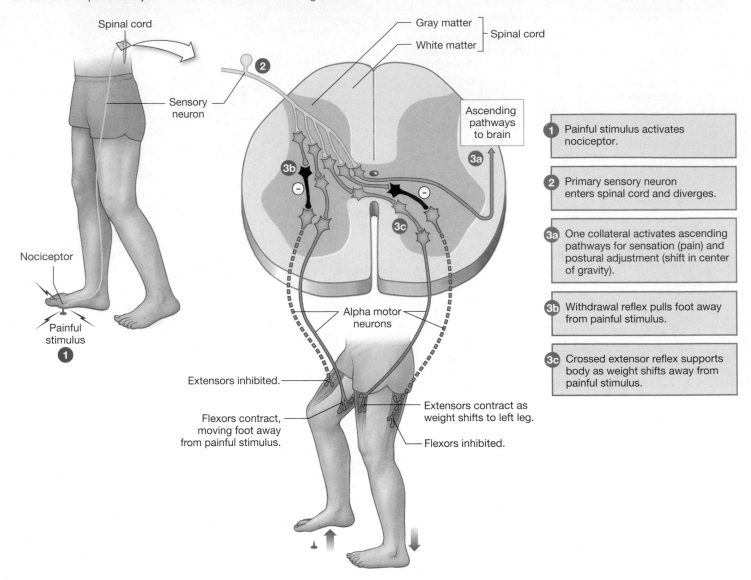

Spinal cord

Sensory neuron

Nociceptor

Painful stimulus

1

Gray matter
White matter
} Spinal cord

Ascending pathways to brain

2

3b

3a

3c

Alpha motor neurons

Extensors inhibited.

Flexors contract, moving foot away from painful stimulus.

Extensors contract as weight shifts to left leg.

Flexors inhibited.

1 Painful stimulus activates nociceptor.

2 Primary sensory neuron enters spinal cord and diverges.

3a One collateral activates ascending pathways for sensation (pain) and postural adjustment (shift in center of gravity).

3b Withdrawal reflex pulls foot away from painful stimulus.

3c Crossed extensor reflex supports body as weight shifts away from painful stimulus.

CONCEPT CHECK

4. Draw a reflex map of the flexion reflex initiated by a painful stimulus to the sole of a foot.

5. Add the crossed extensor reflex in the supporting leg to the map you created in Concept Check 4.

6. As you pick up a heavy weight, which of the following are active in your biceps muscle: alpha motor neurons, gamma motor neurons, muscle spindle afferents, Golgi tendon organ afferent neurons?

7. What distinguishes a stretch reflex from a crossed extensor reflex?

THE INTEGRATED CONTROL OF BODY MOVEMENT

Most of us never think about how our body translates thoughts into action. Even the simplest movement requires proper timing so that antagonistic and synergistic muscle groups contract in the appropriate sequence and to the appropriate degree. In addition, the body must continuously adjust its position to compensate for differences between the intended movement and the actual one. For example, the baseball pitcher steps off the mound to field a ground ball but in doing so slips on a wet patch of grass. His brain quickly compensates for the unexpected change in position

RUNNING **PROBLEM**

Dr. Ling admits Mrs. Evans to the intensive care unit. There, Mrs. Evans is given tetanus antitoxin to deactivate any toxin that has not yet entered motor neurons. She also receives penicillin, an antibiotic that kills the bacteria, and drugs to help relax her muscles. Despite these treatments, by the third day, Mrs. Evans is having difficulty breathing because of spasms in her chest muscles. Dr. Ling calls in the chief of anesthesiology to administer metocurine, a drug similar to curare. Curare and metocurine induce temporary paralysis of muscles by binding to ACh receptors on the motor end plate. Patients receiving metocurine must be placed on respirators that breathe for them. For people with tetanus, however, metocurine can temporarily halt the muscle spasms and allow the body to recover.

Q3: a. Why does the binding of metocurine to ACh receptors on the motor end plate induce muscle paralysis? (Hint: What is the function of ACh in synaptic transmission?)

b. Is metocurine an agonist or an antagonist of ACh?

418 420 424 **427** 430 432

through reflex muscle activity, and he stays on his feet to intercept the ball.

Skeletal muscles cannot communicate with one another directly, and so they send messages to the CNS, allowing the integrating centers to take charge and direct movement. Most body movements are highly integrated, coordinated responses that require input from multiple regions of the brain. Let's examine a few of the CNS integrating centers that are responsible for control of body movement.

Movement Can Be Classified as Reflex, Voluntary, or Rhythmic

Movement can be loosely classified into three categories: reflex movement, voluntary movement, and rhythmic movement

(TBL. 13.2). **Reflex movements** are the least complex and are integrated primarily in the spinal cord (for example, see the knee jerk reflex in Fig. 13.5). However, like other spinal reflexes, reflex movements can be modulated by input from higher brain centers. In addition, the sensory input that initiates reflex movements, such as the input from muscle spindles and Golgi tendon organs, goes to the brain and participates in the coordination of voluntary movements and postural reflexes.

Postural reflexes help us maintain body position as we stand or move through space. These reflexes are integrated in the brain stem. They require continuous sensory input from visual and vestibular (inner ear) sensory systems and from the muscles themselves. Muscle, tendon, and joint receptors provide information about *proprioception*, the positions of various body parts relative to one another. You can tell if your arm is bent even when your eyes are closed because these receptors provide information about body position to the brain.

Information from the vestibular apparatus of the ear and visual cues help us maintain our position in space. For example, we use the horizon to tell us our spatial orientation relative to the ground. In the absence of visual cues, we rely on tactile input. People trying to move in a dark room instinctively reach for a wall or piece of furniture to help orient themselves. Without visual and tactile cues, our orientation skills may fail. The lack of cues is what makes flying airplanes in clouds or fog impossible without instruments. The effect of gravity on the vestibular system is such a weak input when compared with visual or tactile cues that pilots may find themselves flying upside down relative to the ground.

Voluntary movements are the most complex type of movement. They require integration at the cerebral cortex, and they can be initiated at will without external stimuli. Learned voluntary movements improve with practice, and some even become involuntary, like reflexes. Think about learning to ride a bicycle. It may have been difficult at first but once you learned to pedal smoothly and to keep your balance, the movements became automatic. "Muscle memory" is the name dancers and athletes give the ability of the unconscious brain to reproduce voluntary, learned movements and positions.

TABLE 13.2	Types of Movement		
	Reflex	**Voluntary**	**Rhythmic**
Stimulus that Initiates Movement	Primarily external via sensory receptors; minimally voluntary	External stimuli or at will	Initiation and termination voluntary
Example	Knee jerk, cough, postural reflexes	Playing piano	Walking, running
Complexity	Least complex; integrated at level of spinal cord or brain stem with higher center modulation	Most complex; integrated in cerebral cortex	Intermediate complexity; integrated in spinal cord with higher center input required
Comments	Inherent, rapid	Learned movements that improve with practice; once learned, may become subconscious ("muscle memory")	Spinal circuits act as pattern generators; activation of these pathways requires input from brain stem

Rhythmic movements, such as walking or running, are a combination of reflex movements and voluntary movements. Rhythmic movements are initiated and terminated by input from the cerebral cortex, but once activated, networks of CNS interneurons called **central pattern generators** (CPGs) maintain the spontaneous repetitive activity. Changes in rhythmic activity, such as changing from walking to skipping, are also initiated by input from the cerebral cortex.

As an analogy, think of a battery-operated bunny. When the switch is thrown to "on," the bunny begins to hop. It continues its repetitive hopping until someone turns it off (or until the battery runs down). In humans, rhythmic movements controlled by central pattern generators include locomotion and the unconscious rhythm of quiet breathing.

An animal paralyzed by a spinal cord injury is unable to walk because damage to descending pathways blocks the "start walking" signal from the brain to the legs' motor neurons in the spinal cord. However, these paralyzed animals can walk if they are supported on a moving treadmill and given an electrical stimulus to activate the spinal CPG governing that motion. As the treadmill moves the animal's legs, the CPG, reinforced by sensory signals from muscle spindles, drives contraction of the leg muscles.

The ability of central pattern generators to sustain rhythmic movement without continued sensory input has proved important for research on spinal cord injuries. Researchers are trying to take advantage of CPGs and rhythmic reflexes in people with spinal cord injuries by artificially stimulating portions of the spinal cord to restore movement to formerly paralyzed limbs.

The distinctions among reflex, voluntary, and rhythmic movements are not always clear-cut. The precision of voluntary movements improves with practice, but so does that of some reflexes. Voluntary movements, once learned, can become reflexive. In addition, most voluntary movements require continuous input from postural reflexes. **Feedforward reflexes** allow the body to prepare for a voluntary movement, and feedback mechanisms are used to create a smooth, continuous motion. Coordination of movement requires cooperation from many parts of the brain.

The CNS Integrates Movement

Three levels of the nervous system control movement: (1) the spinal cord, which integrates spinal reflexes and contains central pattern generators; (2) the brain stem and cerebellum, which control postural reflexes and hand and eye movements; and (3) the cerebral cortex and basal ganglia [p. 289], which are responsible for voluntary movements. The thalamus relays and modifies signals as they pass from the spinal cord, basal ganglia, and cerebellum to the cerebral cortex (TBL. 13.3).

Reflex movements do not require input from the cerebral cortex. Proprioceptors such as muscle spindles, Golgi tendon organs, and joint capsule receptors provide information to the spinal cord, brain stem, and cerebellum (FIG. 13.7). The brain stem is in charge of postural reflexes and hand and eye movements. It also gets commands from the cerebellum, the part of the brain responsible for "fine-tuning" movement. The result is reflex movement. However, some sensory information is sent through ascending pathways to sensory areas of the cortex, where it can be used to plan voluntary movements.

Voluntary movements require coordination between the cerebral cortex, cerebellum, and basal ganglia. The control of voluntary movement can be divided into three steps: (1) decision-making and planning, (2) initiating the movement, and (3) executing the movement (FIG. 13.8). The cerebral cortex plays a key role in the first two steps. Behaviors such as movement require knowledge of the body's position in space (where am I?), a decision on what movement should be executed (what shall I do?), a plan for executing the movement (how shall I do it?), and the ability to hold the plan in memory long enough to carry it out (now, what was I just doing?). As with reflex movements, sensory feedback is used to continuously refine the process.

TABLE 13.3	Neural Control of Movement		
	Role	**Receives Input from**	**Sends Integrative Output to**
Spinal Cord	Spinal reflexes; locomotor pattern generators	Sensory receptors and brain	Brain stem, cerebellum, thalamus/cerebral cortex
Brain Stem	Posture, hand and eye movements	Cerebellum, visual and vestibular sensory receptors	Spinal cord
Motor Areas of Cerebral Cortex	Planning and coordinating complex movement	Thalamus	Brain stem, spinal cord (corticospinal tract), cerebellum, basal ganglia
Cerebellum	Monitors output signals from motor areas and adjusts movements	Spinal cord (sensory), cerebral cortex (commands)	Brain stem, cerebral cortex (Note: All output is inhibitory.)
Thalamus	Contains relay nuclei that modulate and pass messages to cerebral cortex	Basal ganglia, cerebellum, spinal cord	Cerebral cortex
Basal Ganglia	Motor planning	Cerebral cortex	Cerebral cortex, brain stem

FIG. 13.7 | Integration of muscle reflexes

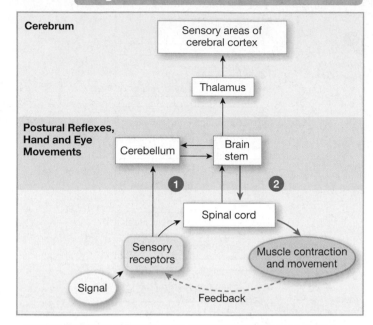

Sensory areas of cerebral cortex

Cerebrum

Thalamus

Postural Reflexes, Hand and Eye Movements

Cerebellum → Brain stem

1

2

Spinal cord

Sensory receptors

Muscle contraction and movement

Signal

Feedback

1. Sensory input (——→) from receptors goes to spinal cord, cerebral cortex, and cerebellum. Signals from the vestibular apparatus go directly to the cerebellum.

2. Postural and spinal reflexes do not require integration in the cortex.
Output signals (——→) initiate movement without higher input.

Let's return to our baseball pitcher and trace the process as he decides whether to throw a fastball or a slow curve. Standing out on the mound, the pitcher is acutely aware of his surroundings: the other players on the field, the batter in the box, and the dirt

FIG. 13.8 | Phases of voluntary movement

Voluntary movements can be divided into three phases: planning, initiation, and execution. Sensory feedback allows the brain to correct for any deviation between the planned movement and the actual movement.

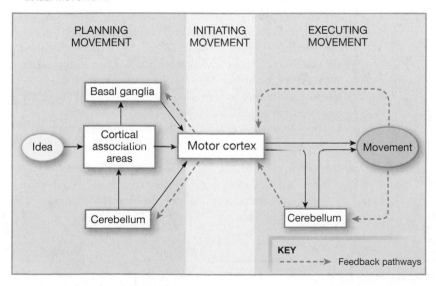

PLANNING MOVEMENT

INITIATING MOVEMENT

EXECUTING MOVEMENT

Basal ganglia

Idea → Cortical association areas → Motor cortex → Movement

Cerebellum

Cerebellum

KEY

- - - - → Feedback pathways

beneath his feet. With the help of visual and somatosensory input to the sensory areas of the cortex, he is aware of his body position as he steadies himself for the pitch (**FIG. 13.9** ❶). Deciding which type of pitch to throw and anticipating the consequences occupy many pathways in his prefrontal cortex and association areas ❷ . These pathways loop down through the basal ganglia and thalamus for modulation before cycling back to the cortex.

Once the pitcher makes the decision to throw a fastball, the motor cortex takes charge of organizing the execution of this complex movement. To initiate the movement, descending information travels from the motor association areas and motor cortex to the brain stem, the spinal cord, and the cerebellum ❸–❹. The cerebellum assists in making postural adjustments by integrating feedback from peripheral sensory receptors. The basal ganglia, which assisted the cortical motor areas in planning the pitch, also provide information about posture, balance, and gait to the brain stem ❺.

The pitcher's decision to throw a fastball now is translated into action potentials that travel down through the **corticospinal tract**, a group of interneurons controlling voluntary movement that run from the motor cortex to the spinal cord, where they synapse directly onto somatic motor neurons (**FIG. 13.10**). Most of these descending pathways cross to the opposite side of the body in a region of the medulla known as the *pyramids*. Consequently, this pathway is sometimes called the *pyramidal tract*.

Neurons from the basal ganglia [p. 289] also influence body movement. These neurons have multiple synapses in the CNS and make up what is sometimes called the *extrapyramidal tract* or the *extrapyramidal system*. It was once believed that the pyramidal and extrapyramidal pathways were separate systems, but we now know that they interact and are not as distinct in their function as was once believed.

As the pitcher begins the pitch, *feedforward postural reflexes* adjust the body position, shifting weight slightly in anticipation of the changes about to occur (**FIG. 13.11**). Through the appropriate divergent pathways, action potentials race to the somatic motor neurons that control the muscles used for pitching: some are excited, others are inhibited. The neural circuitry allows precise control over antagonistic muscle groups as the pitcher flexes and retracts his right arm. His weight shifts onto his right foot as his right arm moves back.

Each of these movements activates sensory receptors that feed information back to the spinal cord, brain stem, and cerebellum, initiating postural reflexes. These reflexes adjust his body position so that the pitcher does not lose his balance and fall over backward. Finally, he releases the ball, catching his balance on the follow-through—another example of postural reflexes mediated through sensory feedback. His head stays erect, and his eyes track the ball as it reaches the batter. Whack! Home run. As the pitcher's eyes follow the ball and he evaluates the result of his pitch, his brain is preparing for the next batter, hoping to use what it has learned from these pitches to improve those to come.

FIG. 13.9 Control of voluntary movements

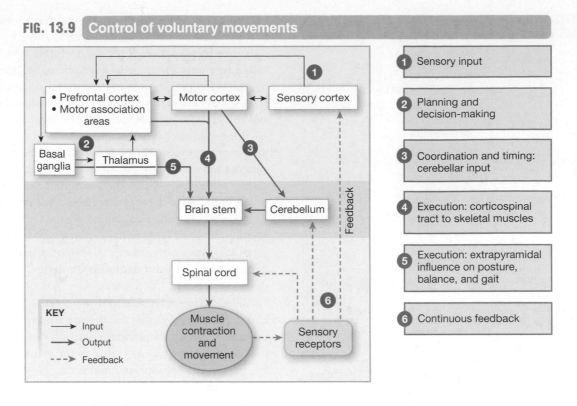

KEY
⟶ Input
⟶ Output
- - -> Feedback

1. Sensory input
2. Planning and decision-making
3. Coordination and timing: cerebellar input
4. Execution: corticospinal tract to skeletal muscles
5. Execution: extrapyramidal influence on posture, balance, and gait
6. Continuous feedback

EMERGING CONCEPTS

Visualization Techniques in Sports

Researchers now believe that *presynaptic facilitation,* in which modulatory input increases neurotransmitter release, is the physiological mechanism that underlies the success of visualization techniques in sports. Visualization, also known as *guided imagery,* enables athletes to maximize their performance by "psyching" themselves, picturing in their minds the perfect vault or the perfect fastball. By pathways that we still do not understand, the mental image conjured up by the cerebral cortex is translated into signals that find their way to the muscles. Guided imagery is also being used in medicine as *adjunct* (supplementary) therapy for cancer treatment and pain management. The ability of the conscious brain to alter physiological function is only one example of the many fascinating connections between the higher brain and the body. To learn more about this, go to *http:// sportsmedicine.about.com* and search for *visualization*.

Symptoms of Parkinson's Disease Reflect Basal Ganglia Function

Our understanding of the role of the basal ganglia in the control of movement has been slow to develop because, for many years, animal experiments yielded little information. Randomly destroying portions of the basal ganglia did not appear to affect research animals. However, research focusing on **Parkinson's disease** (Parkinsonism) in humans has been more fruitful. From studying patients with Parkinson's, scientists have learned that the basal ganglia play a role in cognitive function and memory as well as in the coordination of movement.

Parkinson's disease is a progressive neurological disorder characterized by abnormal movements, speech difficulties, and cognitive changes. These signs and symptoms are associated with loss

RUNNING **PROBLEM**

Four weeks later, Mrs. Evans is ready to go home, completely recovered and showing no signs of lingering effects. Once she could talk, Mrs. Evans, who was born on the farm where she still lived, was able to tell Dr. Ling that she had never had immunization shots for tetanus or any other diseases. "Well, that made you one of only a handful of people in the United States who will develop tetanus this year," Dr. Ling told her. "You've been given your first two tetanus shots here in the hospital. Be sure to come back in six months for the last one so that this won't happen again." Because of national immunization programs begun in the 1950s, tetanus is now a rare disease in the United States. However, in developing countries without immunization programs, tetanus is still a common and serious condition.

Q4: *On the basis of what you know about who receives immunization shots in the United States, predict the age and background of people who are most likely to develop tetanus this year.*

418   420   424   427   **430**   432

FIG. 13.10 The corticospinal tract

Interneurons run directly from the motor cortex to their synapses with somatic motor neurons. Most corticospinal neurons cross the midline at the pyramids.

Primary motor cortex of left cerebral hemisphere

Primary Motor Cortex

Cranial nerves to selected skeletal muscles

MIDBRAIN

Motor nuclei of cranial nerves

MEDULLA OBLONGATA

Most corticospinal pathways cross to the opposite side of the body at the pyramids.

Lateral corticospinal tract

Pyramids

Anterior corticospinal tract

Somatic motor neurons to skeletal muscles

SPINAL CORD

of neurons in the basal ganglia that release the neurotransmitter dopamine. One abnormal sign that most Parkinson patients have is tremors in the hands, arms, and legs, particularly at rest. In addition, they have difficulty initiating movement and walk slowly with stooped posture and shuffling gait. They lose facial expression, fail to blink (the reptilian stare), and may develop depression, sleep disturbances, and personality changes.

FIG. 13.11 Feedforward reflexes and feedback of information during movement

Brain initiates movement. → Body moves. → Posture is disturbed.

Posture adjusted

Feedforward for anticipated postural disturbance

Feedback for unanticipated postural disturbance

The cause of Parkinson's disease is usually not known and appears to be a combination of environmental factors and genetic susceptibility. However, a few years ago, a number of young drug users were diagnosed with Parkinsonism. Their disease was traced to the use of homemade heroin containing a toxic contaminant that destroyed *dopaminergic* (dopamine-secreting) neurons. This contaminant has been isolated and now enables researchers to induce Parkinson's disease in experimental animals so that we have an animal model on which to test new treatments.

The primary current treatment for Parkinson's is administration of drugs designed to enhance dopamine activity in the brain. Dopamine cannot cross the blood-brain barrier, so patients take *L-dopa*, a precursor of dopamine that crosses the blood-brain barrier, then is metabolized to dopamine. Other drug treatments include dopamine agonists and inhibitors of enzymes that break down dopamine, such as MAO [p. 366]. In severe cases, selected parts of the brain may be destroyed to reduce tremors and rigidity.

Experimental treatments include transplants of dopamine-secreting neurons. Proponents of stem cell research feel that Parkinson's may be one of the conditions that would benefit from the transplant of stem cells into affected brains. For more information on Parkinson's treatments, see *www.parkinson.org*, the National Parkinson Foundation.

CONTROL OF MOVEMENT IN VISCERAL MUSCLES

Movement created by contracting smooth and cardiac muscles is very different from that created by skeletal muscles, in large part because smooth and cardiac muscle are not attached to bone. In the internal organs, or viscera, muscle contraction usually changes the shape of an organ, narrowing the lumen of a hollow organ or shortening the length of a tube. In many hollow internal organs, muscle contraction pushes material through the lumen of the organ: the heart pumps blood, the digestive tract moves food, the uterus expels a baby.

Visceral muscle contraction is often reflexively controlled by the autonomic nervous system, but not always. Some types of smooth and cardiac muscle are capable of generating their own action potentials, independent of an external signal. Both the heart and digestive tract have spontaneously depolarizing muscle fibers (often called *pacemakers*) that give rise to regular, rhythmic contractions.

Reflex control of visceral smooth muscle varies from that of skeletal muscle. Skeletal muscles are controlled only by the nervous system, but in many types of visceral muscle, hormones are important in regulating contraction. In addition, some visceral muscle cells are connected to one another by gap junctions that allow electrical signals to pass directly from cell to cell.

Because smooth and cardiac muscle have such a variety of control mechanisms, we will discuss their control as we cover the appropriate organ system for each type of muscle.

RUNNING **PROBLEM CONCLUSION** | Tetanus

In this running problem, you learned about the tetanus toxin tetanospasmin, a potent poison made by the bacterium *Clostridium tetani*. As little as 175 billionths of a gram (175 nanograms) can be fatal to a 70-kg human. Both tetanus toxin and botulinum toxin cause paralysis, but tetanus is a rigid (contracted muscle) paralysis, while botulism is a flaccid (relaxed muscle) paralysis. To learn more about tetanus, visit the web site of the U.S. Centers for Disease Control and Prevention (*www.cdc.gov*). Now check your understanding of this running problem by comparing your answers with the information in the summary table.

Question	Facts	Integration and Analysis
Q1a: *By what process is tetanospasmin taken up into neurons?*	Tetanospasmin is a protein.	Proteins are too large to cross cell membranes by mediated transport. Therefore, tetanospasmin must be taken up by endocytosis [p. 148].
Q1b: *By what process does tetanospasmin travel up the axon to the nerve cell body?*	Substances move from the axon terminal to the cell body by retrograde axonal transport [p. 231].	Tetanospasmin is taken up by endocytosis, so it will be contained in endocytotic vesicles. These vesicles "walk" along microtubules through retrograde axonal transport.
Q2: *Using the reflex pathways diagrammed in Figures 13.5 and 13.6, explain why inhibition of inhibitory interneurons might result in uncontrollable muscle spasms.*	Muscles often occur in antagonistic pairs. When one muscle is contracting, its antagonist must be inhibited.	If the inhibitory interneurons are not functioning, both sets of antagonistic muscles can contract at the same time. This would lead to muscle spasms and rigidity because the bones attached to the muscles would be unable to move in any direction.
Q3a: *Why does the binding of metocurine to ACh receptors on the motor end plate induce muscle paralysis?*	ACh is the somatic motor neuron neurotransmitter that initiates skeletal muscle contraction.	If metocurine binds to ACh receptors, it prevents ACh from binding. Without ACh binding, the muscle fiber will not depolarize and cannot contract, resulting in paralysis.
Q3b: *Is metocurine an agonist or an antagonist of ACh?*	Agonists mimic the effects of a substance; antagonists block the effects of a substance.	Metocurine blocks ACh action; therefore, it is an antagonist.
Q4: *On the basis of what you know about who receives immunization shots in the United States, predict the age and background of people who are most likely to develop tetanus this year.*	Immunizations are required for all children of school age. This practice has been in effect since about the 1950s. In addition, most people who suffer puncture wounds or dirty wounds receive tetanus booster shots when they are treated for those wounds.	Most cases of tetanus in the United States will occur in people over the age of 60 who have never been immunized, in immigrants (particularly migrant workers), and in newborn infants. Another source of the disease is contaminated heroin; injection of the drug under the skin may cause tetanus in drug users who do not receive tetanus booster shots.

418 420 424 427 430 **432**

VISUALIZE, PRACTICE, AND MASTER PHYSIOLOGY WITH MASTERING A&P
- Practice-on-the-go with Dynamic Study Modules
- Visualize important processes with Interactive Physiology and A&P Flix
- Check your understanding with Video Tutors and practice quizzes!

MasteringA&P®

CHAPTER SUMMARY

How many times have you heard people say, "I did it without thinking"? In effect, they were saying that their action was a reflex response. There are many ways to control the functions of muscles and glands of the body, but a neural reflex is the simplest and the fastest.

This chapter discusses how the *nervous system* controls body movement. Postural and spinal reflexes follow the basic pattern of a reflex: sensory input is integrated in the CNS, then acted on when an output signal goes to skeletal muscles. Voluntary movements do not require sensory input to be initiated, but they integrate sensory feedback to ensure smooth execution.

Neural Reflexes

1. A neural reflex consists of the following elements: stimulus, receptor, sensory neurons, integrating center, efferent neurons, effectors (muscles and glands), and response. (p. 418)

2. Neural reflexes can be classified in several ways. **Somatic reflexes** involve somatic motor neurons and skeletal muscles. **Autonomic (or visceral) reflexes** are controlled by autonomic neurons. (p. 418; Tbl. 13.1)

3. **Spinal reflexes** are integrated in the spinal cord. **Cranial reflexes** are integrated in the brain. (p. 418)

4. Many reflexes are innate. Others are acquired through experience. (p. 418)

5. The simplest reflex pathway is a **monosynaptic reflex** with only two neurons. **Polysynaptic reflexes** have three or more neurons in the pathway. (pp. 418, 420; Fig. 13.1)

Autonomic Reflexes

6. Some autonomic reflexes are spinal reflexes that are modulated by input from the brain. Other reflexes needed to maintain homeostasis are integrated in the brain, primarily in the hypothalamus, thalamus, and brain stem. (p. 420)

7. Autonomic reflexes are all polysynaptic, and many are characterized by tonic activity. (p. 420; Fig. 13.1c)

Skeletal Muscle Reflexes

8. Skeletal muscle relaxation must be controlled by the CNS because somatic motor neurons always cause contraction in skeletal muscle. (p. 420)

9. The normal contractile fibers of a muscle are called **extrafusal muscle fibers.** Their contraction is controlled by **alpha motor neurons.** (p. 421; Fig. 13.2)

10. **Golgi tendon organs** are found at the junction of the tendons and muscle fibers. They consist of free nerve endings that wind between collagen fibers. Golgi tendon organs provide information on muscle tension to the CNS. (p. 421; Fig. 13.2a)

11. **Muscle spindles** send information about muscle length to the CNS. These receptors consist of **intrafusal fibers** with sensory neurons wrapped around the noncontractile center. **Gamma motor neurons** innervate the contractile ends of the intrafusal fibers. (p. 421; Fig. 13.2b)

12. Muscle spindles are tonically active stretch receptors. Their output creates tonic contraction of extrafusal muscle fibers. Because of this tonic activity, a muscle at rest maintains a certain level of tension, known as **muscle tone.** (p. 421; Fig. 13.2c)

13. If a muscle stretches, the intrafusal fibers of its spindles stretch and initiate reflex contraction of the muscle. The contraction prevents damage from overstretching. This reflex pathway is known as a **stretch reflex.** (p. 421; Fig. 13.3)

14. When a muscle contracts, **alpha-gamma coactivation** ensures that its muscle spindle remains active. Activation of gamma motor neurons causes contraction of the ends of the intrafusal fibers. This contraction lengthens the central region of the intrafusal fibers and maintains stretch on the sensory nerve endings. (p. 424; Fig. 13.4)

15. The synergistic and antagonistic muscles that control a single joint are known as a **myotatic unit.** When one set of muscles in a myotatic unit contracts, the antagonistic muscles must relax through a reflex known as **reciprocal inhibition.** (p. 424; Fig. 13.5)

16. **Flexion reflexes** are polysynaptic reflexes that cause an arm or leg to be pulled away from a painful stimulus. Flexion reflexes that occur in the legs are usually accompanied by the **crossed extensor reflex,** a postural reflex that helps maintain balance when one foot is lifted from the ground. (pp. 424, 425; Fig. 13.6)

17. **Central pattern generators** are networks of neurons in the CNS that can produce rhythmic motor movements without sensory feedback or higher brain commands. (p. 428)

The Integrated Control of Body Movement

18. Movement can be loosely classified into three categories: reflex movement, voluntary movement, and rhythmic movement. (p. 427; Tbl. 13.2)

19. **Reflex movements** are integrated primarily in the spinal cord. **Postural reflexes** are integrated in the brain stem. (p. 427; Fig. 13.7; Tbl. 13.3)

20. **Voluntary movements** are integrated in the cerebral cortex and can be initiated at will. Learned voluntary movements improve with practice and may even become involuntary, like reflexes. (p. 427; Fig. 13.8)

21. **Rhythmic movements,** such as walking, are a combination of reflexes and voluntary movements. Rhythmic movements can be sustained by central pattern generators. (p. 428)

22. Most signals for voluntary movement travel from cortex to spinal cord through the **corticospinal tract.** Signals from the **basal ganglia** also influence movement through extrapyramidal pathways. (p. 429; Fig. 13.10)

23. **Feedforward reflexes** allow the body to prepare for a voluntary movement; feedback mechanisms are used to create a smooth, continuous motion. (p. 428; Fig. 13.11)

Control of Movement in Visceral Muscles

24. Contraction in smooth and cardiac muscles may occur spontaneously or may be controlled by hormones or by the autonomic division of the nervous system. (p. 431)

REVIEW QUESTIONS

In addition to working through these questions and checking your answers on p. A-17, review the Learning Outcomes at the beginning of this chapter.

Level One Reviewing Facts and Terms

1. All neural reflexes begin with a(n) _____ that activates a receptor.

2. Somatic reflexes involve _____ muscles; _____ (or visceral) reflexes are controlled by autonomic neurons.

3. The pathway pattern that brings information from many neurons into a smaller number of neurons is known as _____.

4. When the axon terminal of a modulatory neuron (cell M) terminates close to the axon terminal of a presynaptic cell (cell P) and decreases the amount of neurotransmitter released by cell P, the resulting type of modulation is called _____. [*Hint:* See p. 265.]

5. Autonomic reflexes are also called _____ reflexes. Why?

6. Some autonomic reflexes are spinal reflexes; others are integrated in the brain. List some examples of each.

7. Which part of the brain transforms emotions into somatic sensation and visceral function? List three autonomic reflexes that are linked to emotions.

8. How many synapses occur in the simplest autonomic reflexes? Where do the synapses occur?

9. List the three types of sensory receptors that convey information for muscle reflexes.

10. Because of tonic activity in neurons, a resting muscle maintains a low level of tension known as _____.

11. Stretching a skeletal muscle causes sensory neurons to (increase/decrease) their rate of firing, causing the muscle to contract, thereby relieving the stretch. Why is this a useful reflex?

12. Match the structure to all correct statements about it.

(a) muscle spindle	1. is strictly a sensory receptor
(b) Golgi tendon organ	2. has sensory neurons that send information to the CNS
(c) joint capsule mechanoreceptor	3. is associated with two types of motor neurons
	4. conveys information about the relative positioning of bones
	5. is innervated by gamma motor neurons
	6. modulates activity in alpha motor neurons

13. The Golgi tendon organ responds primarily to muscle _____.

14. The simplest reflex requires a minimum of how many neurons? How many synapses? Give an example.

15. List and differentiate the three categories of movement. Give an example of each.

Level Two Reviewing Concepts

16. What is the purpose of alpha-gamma coactivation? Explain how it occurs.

17. Modulatory neuron M synapses on the axon terminal of neuron P, just before P synapses with the effector organ. If M is an inhibitory neuron, what happens to neurotransmitter release by P? What effect does M's neurotransmitter have on the postsynaptic membrane potential of P? (*Hint:* Draw this pathway.)

18. At your last physical, your physician checked your patellar tendon reflex by tapping just below your knee while you sat quietly on the edge of the table. (a) What was she checking when she did this test? (b) What would happen if you were worried about falling off the table and were very tense? Where does this additional input to the efferent motor neurons originate? Are these modulatory neurons causing EPSPs or IPSPs [p. 263] at the spinal motor neuron? (c) Your physician notices that you are tense and asks you to count backward from 100 by 3's while she repeats the test. Why would carrying out this counting task enhance your reflex?

Level Three Problem Solving

19. There are several theories about how presynaptic inhibition works at the cellular level. Use what you have learned about membrane potentials and synaptic transmission to explain how each of the following mechanisms would result in presynaptic inhibition:

   (a) Voltage-gated Ca^{2+} channels in axon terminal are inhibited.
   (b) Cl^- channels in axon terminal open.
   (c) K^+ channels in axon terminal open.

20. Andy is working on improving his golf swing. He must watch the ball, swing the club back and then forward, twist his hips, straighten his left arm, then complete the follow-through, where the club arcs in front of him. Which parts of the brain are involved in adjusting how hard he hits the ball, keeping all his body parts moving correctly, watching the ball, and then repeating these actions once he has verified that this swing is successful?

21. It's Halloween, and you are walking through the scariest haunted house around. As you turn a corner and enter the dungeon, a skeleton reaches out and grabs your arm. You let out a scream. Your heart rate quickens, and you feel the hairs on your arm stand on end. (a) What has just happened to you? (b) Where in the brain is fear processed? What are the functions of this part of the brain? Which branch (somatic or autonomic) of the motor output does it control? What are the target organs for this response? (c) How is it possible for your hair to stand on end when hair is made of proteins that do not contract? [*Hint:* See p. 86.] Given that the autonomic nervous system is mediating this reflex response, which type of tissue do you expect to find attached to hair follicles?

22. Using what you have learned about tetanus and botulinum toxins, make a table to compare the two. In what ways are tetanus and botulinum toxin similar? How are they different?

Answers to Concept Checks, Figure and Graph Questions, and end-of-chapter Review Questions can be found in Appendix A [p. A-1].

14

Cardiovascular Physiology

Only in the 17th century did the brain displace the heart as the controller of our actions.

Mary A. B. Brazier, *A History of Neurophysiology in the 19th Century,* 1988

OVERVIEW OF THE CARDIOVASCULAR SYSTEM 436

LO 14.1 Describe the functions of the cardiovascular system and give examples of each function.

LO 14.2 Describe the organization of the cardiovascular system, starting and ending in the aorta.

PRESSURE, VOLUME, FLOW, AND RESISTANCE 439

LO 14.3 Define and explain the relationships among pressure, hydrostatic pressure, pressure gradients, flow, velocity of flow, resistance, and radius as they relate to the cardiovascular system.

CARDIAC MUSCLE AND THE HEART 443

LO 14.4 Describe in detail the internal and external anatomy of the heart.

LO 14.5 Describe the two types of myocardial cells and their arrangement in the heart.

LO 14.6 Describe the membrane proteins and ion movement involved in myocardial excitation-contraction coupling and relaxation.

LO 14.7 Compare and contrast actions potentials of myocardial autorhythmic and contractile cells.

THE HEART AS A PUMP 454

LO 14.8 Describe the conduction of electrical signals through the heart.

LO 14.9 Describe the parts of an electrocardiogram and explain how these electrical events are related to the mechanical events of the cardiac cycle.

LO 14.10 Explain the pressure changes that occur during the cardiac cycle and their relationship to flow through the heart and blood vessels.

LO 14.11 Explain the relationship of heart rate, cardiac output, and stroke volume.

LO 14.12 Explain the role of the autonomic divisions in control of heart rate at the cellular and molecular level.

LO 14.13 Explain how the following factors influence stroke volume: venous return, length-tension relationships, preload, afterload, contractility, skeletal muscle pump, respiratory pump, inotropic agents.

BACKGROUND BASICS

134	Diffusion
378	Striated muscle
74	Desmosomes
383	Excitation-contraction coupling
395	Length-tension relationship in muscle
396	Tetanus in skeletal muscle
382	Muscle contraction
166	Gap junctions
206	Catecholamines
287	Vagus nerve
398	Isometric contraction

Colored electron micrograph of cardiac muscle. Oval mitochondria lie between pink striated muscle fibers.

In the classic movie *Indiana Jones and the Temple of Doom*, the evil priest reaches into the chest of a sacrificial victim and pulls out his heart, still beating. This act was not dreamed up by some Hollywood scriptwriter—it was taken from rituals of the ancient Mayans, who documented this grisly practice in their carvings and paintings. The heart has been an object of fascination for centuries, but how can this workhorse muscle, which pumps 7200 liters of blood a day, keep beating outside the body? To answer that question, let's first consider the role of hearts in circulatory systems.

As life evolved, simple one-celled organisms began to band together, first into cooperative colonies and then into multicelled organisms. In most multicellular animals, only the surface layer of cells is in direct contact with the environment. This body plan presents a problem because diffusion slows exponentially as distance increases [p. 134]. Because of this, oxygen consumption in the interior cells of larger animals exceeds the rate at which oxygen can diffuse from the body surface.

One solution to overcome slow diffusion was the evolutionary development of circulatory systems that move fluid between the body's surface and its deepest parts. In simple animals, muscular activity creates fluid flow when the animal moves. More complex animals have muscular pumps called hearts to circulate internal fluid.

In the most efficient circulatory systems, the heart pumps blood through a closed system of vessels. This one-way circuit steers the blood along a specific route and ensures systematic distribution of gases, nutrients, signal molecules, and wastes. A circulatory system comprising a heart, blood vessels, and blood is known as a **cardiovascular system** {*kardia*, heart + *vasculum*, little vessel}.

Although the idea of a closed cardiovascular system that cycles blood in an endless loop seems intuitive to us today, it has not always been so. **Capillaries** {*capillus*, hair}, the microscopic vessels where blood exchanges material with the interstitial fluid, were not discovered until Marcello Malpighi, an Italian anatomist, observed them through a microscope in the middle of the seventeenth century. At that time, European medicine was still heavily influenced by the ancient belief that the cardiovascular system distributed both blood and air.

Blood was thought to be made in the liver and distributed throughout the body in the veins. Air went from the lungs to the heart, where it was digested and picked up "vital spirits." From the heart, air was distributed to the tissues through vessels called arteries. Anomalies—such as the fact that a cut artery squirted blood rather than air—were ingeniously explained by unseen links between arteries and veins that opened upon injury.

According to this model of the circulatory system, the tissues consumed all blood delivered to them, and the liver had to synthesize new blood continuously. It took the calculations of William Harvey (1578–1657), court physician to King Charles I of England, to show that the weight of blood pumped by the heart in a single hour exceeds the weight of the entire body! Once it became obvious that the liver could not make blood as rapidly as the heart pumped it, Harvey looked for an anatomical route that would allow the blood to recirculate rather than be consumed in the tissues. He showed that valves in the heart and veins created a one-way flow of blood, and that veins carried blood back to the heart, not out to the limbs. He also showed that blood entering the right side of the heart had to go to the lungs before it could go to the left side of the heart.

These studies created a furor among Harvey's contemporaries, leading Harvey to say in a huff that no one under the age of 40 could understand his conclusions. Ultimately, Harvey's work became the foundation of modern cardiovascular physiology. Today, we understand the structure of the cardiovascular system at microscopic and molecular levels that Harvey never dreamed existed. Yet some things have not changed. Even now, with our sophisticated technology, we are searching for "spirits" in the blood, although today we call them by names such as *hormone* and *cytokine*.

OVERVIEW OF THE CARDIOVASCULAR SYSTEM

In the simplest terms, a cardiovascular system is a series of tubes (the blood vessels) filled with fluid (blood) and connected to a pump (the heart). Pressure generated in the heart propels blood through the system continuously. The blood picks up oxygen at the lungs and nutrients in the intestine and then delivers these substances to the body's cells while simultaneously removing cellular wastes and heat for excretion. In addition, the cardiovascular system plays an important role in cell-to-cell communication and in defending the body against foreign invaders. This chapter focuses on an overview of the cardiovascular system and on the heart as a pump. Later, you will learn about the properties of the blood vessels and the homeostatic controls that regulate blood flow and blood pressure.

RUNNING **PROBLEM** | **Myocardial Infarction**

At 9:06 A.M., the blood clot that had silently formed in Walter Parker's left anterior descending coronary artery made its sinister presence known. The 53-year-old advertising executive had arrived at the Dallas Convention Center feeling fine, but suddenly a dull ache started in the center of his chest, and he became nauseated. At first, he brushed it off as the after-effects of the convention banquet the night before. When the chest pain persisted, however, he thought of his family history of heart disease and took an aspirin, remembering a radio ad that said to do this if you were having symptoms of a heart attack. Walter then made his way to the Center's Aid Station. "I'm not feeling very well," he told the emergency medical technician (EMT). The EMT, on hearing Walter's symptoms and seeing his pale, sweaty face, immediately thought of a heart attack. "Let's get you over to the hospital and get this checked out."

436 439 449 455 466 468 472

The Cardiovascular System Transports Materials throughout the Body

The primary function of the cardiovascular system is the transport of materials to and from all parts of the body. Substances transported by the cardiovascular system can be divided into (1) nutrients, water, and gases that enter the body from the external environment, (2) materials that move from cell to cell within the body, and (3) wastes that the cells eliminate (**TBL. 14.1**).

Oxygen enters the body at the exchange surface of the lungs. Nutrients and water are absorbed across the intestinal epithelium. Once all these materials are in the blood, the cardiovascular system distributes them. A steady supply of oxygen for the cells is particularly important because many cells deprived of oxygen become irreparably damaged within a short period of time. For example, about 5–10 seconds after blood flow to the brain is stopped, a person loses consciousness. If oxygen delivery stops for 5–10 minutes, permanent brain damage results. Neurons of the brain have a very high rate of oxygen consumption and cannot meet their metabolic need for ATP by using anaerobic pathways, which have low yields of ATP/glucose [p. 110]. Because of the brain's sensitivity to *hypoxia* {*hypo-*, low + *oxia*, oxygen}, homeostatic controls do everything possible to maintain cerebral blood flow, even if it means depriving other cells of oxygen.

Cell-to-cell communication is a key function of the cardiovascular system. For example, hormones secreted by endocrine glands travel in the blood to their targets. Blood also carries nutrients, such as glucose from the liver and fatty acids from adipose tissue, to metabolically active cells. Finally, the defense team of white blood cells and antibodies patrols the circulation to intercept foreign invaders.

The cardiovascular system also picks up carbon dioxide and metabolic wastes released by cells and transports them to the lungs and kidneys for excretion. Some waste products are transported to the liver for processing before they are excreted in the urine or feces. Heat also circulates through the blood, moving from the body core to body surface, where it dissipates.

The Cardiovascular System Consists of the Heart, Blood Vessels, and Blood

The cardiovascular system is composed of the heart, the blood vessels (also known as the *vasculature*), and the cells and plasma of the blood. Blood vessels that carry blood away from the heart are called **arteries.** Blood vessels that return blood to the heart are called **veins.**

As blood moves through the cardiovascular system, a system of valves in the heart and veins ensures that the blood flows in one direction only. Like the turnstiles at an amusement park, the valves keep blood from reversing its direction of flow. **FIGURE 14.1** is a schematic diagram that shows these components and the route that blood follows through the body. Notice in this illustration, as well as in most other diagrams of the heart, that the right side of the heart is on the left side of the page, which means that the heart is labeled as if you were viewing the heart of a person facing you.

The heart is divided by a central wall, or **septum,** into left and right halves. Each half functions as an independent pump that consists of an **atrium** {*atrium,* central room; plural *atria*} and a **ventricle** {*ventriculus,* belly}. The atrium receives blood returning to the heart from the blood vessels, and the ventricle pumps blood out into the blood vessels. The right side of the heart receives blood from the tissues and sends it to the lungs for oxygenation. The left side of the heart receives newly oxygenated blood from the lungs and pumps it to tissues throughout the body.

Starting in the right atrium in Figure 14.1, trace the path taken by blood as it flows through the cardiovascular system. Note that blood in the right side of the heart is colored blue. This is a convention used to show blood from which the tissues have extracted oxygen. Although this blood is often described as *deoxygenated,* it is not completely devoid of oxygen. It simply has less oxygen than blood going from the lungs to the tissues.

In living people, well-oxygenated blood is bright red, and low-oxygen blood is a darker red. Under some conditions, low-oxygen blood can impart a bluish color to certain areas of the skin, such as around the mouth and under the fingernails. This condition, known as *cyanosis* {*kyanos,* dark blue}, is the reason blue is used in drawings to indicate blood with lower oxygen content.

From the right atrium, blood flows into the right ventricle of the heart. From there it is pumped through the **pulmonary arteries** {*pulmo,* lung} to the lungs, where it is oxygenated. Note the color change from blue to red in Figure 14.1, indicating higher oxygen content after the blood leaves the lungs. From the lungs,

TABLE 14.1	Transport in the Cardiovascular System	
Substance Moved	**From**	**To**
Materials Entering the Body		
Oxygen	Lungs	All cells
Nutrients and water	Intestinal tract	All cells
Materials Moved from Cell to Cell		
Wastes	Some cells	Liver for processing
Immune cells, antibodies, clotting proteins	Present in blood continuously	Available to any cell that needs them
Hormones	Endocrine cells	Target cells
Stored nutrients	Liver and adipose tissue	All cells
Materials Leaving the Body		
Metabolic wastes	All cells	Kidneys
Heat	All cells	Skin
Carbon dioxide	All cells	Lungs

FIG. 14.1 The cardiovascular system

The cardiovascular system is a closed loop. The heart is a pump that circulates blood through the system. Arteries take blood away from the heart, and veins carry blood back to the heart.

Q FIGURE QUESTION

A portal system is two capillary beds joined in series. Identify the two portal systems shown in this figure.

blood travels to the left side of the heart through the **pulmonary veins.** The blood vessels that go from the right ventricle to the lungs and back to the left atrium are known collectively as the **pulmonary circulation.**

Blood from the lungs enters the heart at the left atrium and passes into the left ventricle. Blood pumped out of the left ventricle enters the large artery known as the **aorta.** The aorta branches into a series of smaller and smaller arteries that finally lead into networks of capillaries. Notice at the top of Figure 14.1 the color change from red to blue as the blood passes through the capillaries, indicating that oxygen has left the blood and diffused into the tissues.

After leaving the capillaries, blood flows into the venous side of the circulation, moving from small veins into larger and larger

veins. The veins from the upper part of the body join to form the **superior vena cava.** Those from the lower part of the body form the **inferior vena cava.** The two *venae cavae* empty into the right atrium. The blood vessels that carry blood from the left side of the heart to the tissues and back to the right side of the heart are collectively known as the **systemic circulation.**

Return to Figure 14.1 and follow the divisions of the aorta after it leaves the left ventricle. The first branch represents the *coronary arteries,* which nourish the heart muscle itself. Blood from these arteries flows into capillaries, then into the *coronary veins,* which empty directly into the right atrium at the *coronary sinus.* Ascending branches of the aorta go to the arms, head, and brain. The abdominal aorta supplies blood to the trunk, the legs, and the internal organs such as liver (*hepatic artery*), digestive tract, and the kidneys (*renal arteries*).

Notice two special arrangements of the circulation. One is the blood supply to the digestive tract and liver. Both regions receive well-oxygenated blood through their own arteries, but, in addition, blood leaving the digestive tract goes directly to the liver by means of the *hepatic portal vein*. The liver is an important site for nutrient processing and plays a major role in detoxifying foreign substances. Most nutrients absorbed in the intestine are routed directly to the liver, allowing that organ to process material before it is released into the general circulation. The two capillary beds of the digestive tract and liver, joined by the hepatic portal vein, are an example of a *portal system*.

A second portal system occurs in the kidneys, where two capillary beds are connected in series. A third portal system, discussed earlier but not shown here, is the hypothalamic-hypophyseal portal system, which connects the hypothalamus and the anterior pituitary [p. 211].

CONCEPT CHECK

1. A cardiovascular system has what three major components?

2. What is the difference between (a) the pulmonary and systemic circulations, (b) an artery and a vein, (c) an atrium and a ventricle?

PRESSURE, VOLUME, FLOW, AND RESISTANCE

If you ask people why blood flows through the cardiovascular system, many of them respond, "So that oxygen and nutrients can get to all parts of the body." This is true, but it is a teleological answer, one that describes the purpose of blood flow. In physiology, we are also concerned with how blood flows—in other words, with the mechanisms or forces that create blood flow.

A simple mechanistic answer to "Why does blood flow?" is that liquids and gases flow down **pressure gradients (ΔP)** from regions of higher pressure to regions of lower pressure. For this reason, blood can flow in the cardiovascular system only if one region develops higher pressure than other regions.

In humans, the heart creates high pressure when it contracts. Blood flows out of the heart (the region of highest pressure) into the closed loop of blood vessels (a region of lower pressure). As blood moves through the system, pressure is lost because of friction between the fluid and the blood vessel walls. Consequently, pressure falls continuously as blood moves farther from the heart (**FIG. 14.2**). The highest pressure in the vessels of the cardiovascular system is found in the aorta and systemic arteries as they receive blood from the left ventricle. The lowest pressure is in the venae cavae, just before they empty into the right atrium.

Now let's review the laws of physics that explain the interaction of pressure, volume, flow, and resistance in the cardiovascular system. Many of these principles apply broadly to the flow of all

FIG. 14.2 Blood flows down a pressure gradient

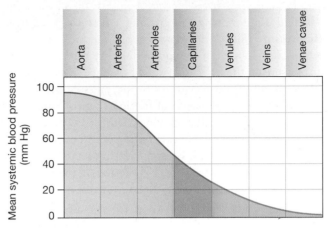

The mean blood pressure of the systemic circulation ranges from a high of 93 mm Hg (millimeters of mercury) in the aorta to a low of a few mm Hg in the venae cavae.

types of liquids and gases, including the flow of air in the respiratory system. However, in this chapter we focus on blood flow and its relevance to the function of the heart.

The Pressure of Fluid in Motion Decreases over Distance

Pressure in a fluid is the force exerted by the fluid on its container. In the heart and blood vessels, pressure is commonly measured

RUNNING **PROBLEM**

When people speak of a "heart attack," they are actually referring to a clot that stops the blood supply to part of the heart, creating a condition known as *ischemia* {*ischien*, to suppress + *-emia*, blood}. In medical terms, a heart attack is called a *myocardial infarction* (MI), referring to an area of heart muscle that is dying because of a lack of blood supply. The clot in Walter's coronary artery had restricted blood flow to part of his left ventricle, and its cells were beginning to die from lack of oxygen. When someone has a heart attack, immediate medical intervention is critical. In the ambulance on the way to the emergency room (ER), the paramedics gave Walter oxygen and a tablet of nitroglycerin, hooked him up to a heart monitor, and started an intravenous (IV) injection of normal (isotonic) saline. With an IV line in place, other drugs could be given rapidly if Walter's condition should suddenly worsen.

Q1: Why did the EMT give Walter oxygen and nitroglycerin? [Hint: p. 179]

Q2: What effect would the injection of isotonic saline have on Walter's extracellular fluid volume? On his intracellular fluid volume? On his total body osmolarity? [Hint: p. 127]

436 439 449 455 466 468 472

in *millimeters of mercury* (mm Hg), where one millimeter of mercury is equivalent to the hydrostatic pressure exerted by a 1-mm-high column of mercury on an area of 1 cm². Some physiological literature reports pressures in *torr* (1 torr = 1 mm Hg) or in *centimeters of water:* 1 cm H_2O = 0.74 mm Hg.

If fluid is not moving, the pressure it exerts is called **hydrostatic pressure (FIG. 14.3a)**, and force is exerted equally in all directions. For example, a column of fluid in a tube exerts hydrostatic pressure on the floor and sides of the tube.

In a system in which fluid is flowing, pressure falls over distance as energy is lost because of friction (Fig. 14.3b). In addition, the pressure exerted by moving fluid has two components: a dynamic, flowing component that represents the kinetic energy of the system, and a lateral component that represents the hydrostatic pressure (potential energy) exerted on the walls of the system. Pressure within our cardiovascular system is usually called hydrostatic pressure even though it is a system in which fluid is in motion. Some textbooks are beginning to replace the term *hydrostatic pressure* with the term *hydraulic pressure*. Hydraulics is the study of fluid in motion.

Pressure Changes in Liquids without a Change in Volume

If the walls of a fluid-filled container contract, the pressure exerted on the fluid in the container increases. You can demonstrate this principle by filling a balloon with water and squeezing the water balloon in your hand. Water is minimally compressible, and so the pressure you apply to the balloon is transmitted throughout the fluid. As you squeeze, higher pressure in the fluid causes parts of the balloon to bulge. If the pressure becomes high enough, the stress on the balloon causes it to pop. The water volume inside the balloon did not change, but the pressure in the fluid increased.

In the human heart, contraction of the blood-filled ventricles is similar to squeezing a water balloon: pressure created by the contracting muscle is transferred to the blood. This high-pressure blood then flows out of the ventricle and into the blood vessels, displacing lower-pressure blood already in the vessels. The pressure created in the ventricles is called the **driving pressure** because it is the force that drives blood through the blood vessels.

When the walls of a fluid-filled container expand, the pressure exerted on the fluid decreases. For this reason, when the heart relaxes and expands, pressure in the fluid-filled chambers falls.

Pressure changes can also take place in the blood vessels. If blood vessels dilate, blood pressure inside the circulatory system falls. If blood vessels constrict, blood pressure in the system increases. Volume changes of the blood vessels and heart are major factors that influence blood pressure in the cardiovascular system.

Blood Flows from Higher Pressure to Lower Pressure

As stated earlier, blood flow through the cardiovascular system requires a pressure gradient. This pressure gradient is analogous to the difference in pressure between two ends of a tube through which fluid flows (Fig. 14.3c). Flow through the tube is directly proportional to (α) the pressure gradient (ΔP):

$$\text{Flow} \, \alpha \, \Delta P \qquad (1)$$

where $\Delta P = P_1 - P_2$. This relationship says that the higher the pressure gradient, the greater the fluid flow.

A pressure gradient is not the same thing as the absolute pressure in the system. For example, the tube in Figure 14.3c has an absolute pressure of 100 mm Hg at each end. However, because there is no pressure gradient between the two ends of the tube, there is no flow through the tube.

On the other hand, two identical tubes can have very different absolute pressures but the same flow. The top tube in Figure 14.3d has a hydrostatic pressure of 100 mm Hg at one end and 75 mm Hg at the other end, which means that the pressure gradient between the ends of the tube is 25 mm Hg. The identical bottom tube has a hydrostatic pressure of 40 mm Hg at one end and 15 mm Hg at the other end. This tube has lower absolute pressure all along its length but the same pressure gradient as the top tube: 25 mm Hg. Because the pressure difference in the two tubes is identical, fluid flow through the tubes is the same.

Resistance Opposes Flow

In an ideal system, a substance in motion would remain in motion. However, no system is ideal because all movement creates friction. Just as a ball rolled across the ground loses energy to friction, blood flowing through blood vessels encounters friction from the walls of the vessels and from cells within the blood rubbing against one another as they flow.

The tendency of the cardiovascular system to oppose blood flow is called the system's **resistance** to flow. *Resistance* (R) is a term that most of us understand from everyday life. We speak of people being resistant to change or taking the path of least resistance. This concept translates well to the cardiovascular system because blood flow also takes the path of least resistance. An increase in the resistance of a blood vessel results in a decrease in the flow through that vessel. We can express that relationship as

$$\text{Flow} \, \alpha \, 1/R \qquad (2)$$

This expression says that flow is inversely proportional to resistance: if resistance increases, flow decreases; and if resistance decreases, flow increases.

What parameters determine resistance? For fluid flowing through a tube, resistance is influenced by three components: the radius of the tube (r), the length of the tube (L), and the **viscosity** (thickness) of the fluid (η, the Greek letter eta). The following equation, derived by the French physician Jean Leonard Marie Poiseuille and known as **Poiseuille's law,** shows the relationship of these factors:

$$R = 8L\eta/\pi \, r^4 \qquad (3)$$

FIG. 14.3 ESSENTIALS

The Physics of Fluid Flow

Pressure in Static and Flowing Fluids

(a) Hydrostatic pressure is the pressure exerted on the walls of the container by the fluid within the container. Hydrostatic pressure is proportional to the height of the water column.

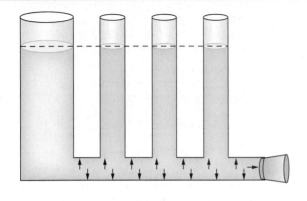

(b) Once fluid begins to flow through the system, pressure falls with distance as energy is lost because of friction. This is the situation in the cardiovascular system.

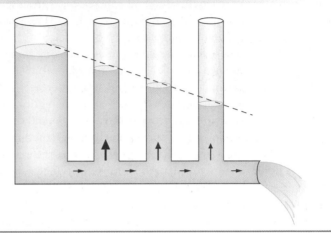

Fluid flow through a tube depends on the pressure gradient.

(c) Fluid flows only if there is a positive pressure gradient (ΔP).

Higher P ——— Flow ——→ Lower P

$$P_1 - P_2 = \Delta P$$

This tube has no pressure gradient, so no flow.

100 mm Hg   100 mm Hg

$\Delta P = 0$, so no flow.

(d) Flow depends on the pressure gradient (ΔP), not on the absolute pressure (P). ΔP is equal in these tubes so flow is the same.

100 mm Hg   75 mm Hg

Flow →

$\Delta P = 100 - 75 = 25$ mm Hg

40 mm Hg   15 mm Hg

Flow →

$\Delta P = 40 - 15 = 25$ mm Hg

Flow is equal.

As the radius of a tube decreases, the resistance to flow increases.

(e)

Radius of A = 1   Radius of B = 2

Volume of A = 1   Volume of B = 16

Resistance $\propto \dfrac{1}{radius^4}$	
Tube A	Tube B
$R \propto \dfrac{1}{1^4}$	$R \propto \dfrac{1}{2^4}$
$R \propto 1$	$R \propto \dfrac{1}{16}$

Flow $\propto \dfrac{1}{resistance}$	
Tube A	Tube B
Flow $\propto \dfrac{1}{1}$	Flow $\propto \dfrac{1}{\frac{1}{16}}$
Flow $\propto 1$	Flow $\propto 16$

(Q) FIGURE QUESTION

If the radius of A changes to 3, the flow through A will be about _____ times the flow through B.

Because the value of $8/\pi$ is a constant, this factor can be removed from the equation, and the relationship can be rewritten as

$$R \propto L\eta/r^4 \tag{4}$$

This expression says that (1) the resistance to fluid flow offered by a tube increases as the length of the tube increases, (2) resistance increases as the viscosity of the fluid increases, but (3) resistance decreases as the tube's radius increases.

To remember these relationships, think of drinking through a straw. You do not need to suck as hard on a short straw as on a long one (the resistance offered by the straw increases with length). Drinking water through a straw is easier than drinking a thick milkshake (resistance increases with viscosity). And drinking the milkshake through a fat straw is much easier than through a skinny cocktail straw (resistance increases as radius decreases).

How significant are tube length, fluid viscosity, and tube radius to blood flow in a normal individual? The length of the systemic circulation is determined by the anatomy of the system and is essentially constant. Blood viscosity is determined by the ratio of red blood cells to plasma and by how much protein is in the plasma. Normally, viscosity is constant, and small changes in either length or viscosity have little effect on resistance. This leaves changes in the radius of the blood vessels as the main variable that affects resistance in the systemic circulation.

Let's return to the example of the straw and the milkshake to illustrate how changes in radius affect resistance. If we assume that the length of the straw and the viscosity of the milkshake do not change, this system is similar to the cardiovascular system—the radius of the tube has the greatest effect on resistance. If we consider only resistance (R) and radius (r) from equation 4, the relationship between resistance and radius can be expressed as

$$R \propto 1/r^4 \tag{5}$$

If the skinny straw has a radius of 1, its resistance is proportional to $1/1^4$ or 1. If the fat straw has a radius of 2, the resistance it offers is $\frac{1}{2}^4$, or 1/16th, that of the skinny straw (Fig. 14.3e). Because flow is inversely proportional to resistance, flow increases 16-fold when the radius doubles.

As you can see from this example, a small change in the radius of a tube has a large effect on the flow of a fluid through that tube. Similarly, a small change in the radius of a blood vessel has a large effect on the resistance to blood flow offered by that vessel. A decrease in blood vessel diameter is known as **vasoconstriction** {*vas,* a vessel or duct}. An increase in blood vessel diameter is called **vasodilation.** Vasoconstriction decreases blood flow through a vessel, and vasodilation increases blood flow through a vessel.

In summary, by combining equations 1 and 2, we get the equation

$$Flow \propto \Delta P/R \tag{6}$$

which, translated into words, says that the flow of blood in the cardiovascular system is directly proportional to the pressure gradient in the system, and inversely proportional to the resistance of the system to flow. If the pressure gradient remains constant, then flow varies inversely with resistance.

CONCEPT CHECK

3. Which is more important for determining flow through a tube: absolute pressure or the pressure gradient?

4. The two identical tubes below have the pressures shown at each end. Which tube has the greater flow? Defend your choice.

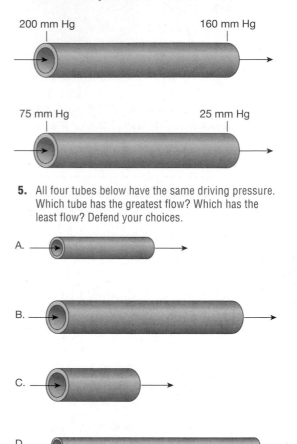

5. All four tubes below have the same driving pressure. Which tube has the greatest flow? Which has the least flow? Defend your choices.

Velocity Depends on the Flow Rate and the Cross-Sectional Area

The word *flow* is sometimes used imprecisely in cardiovascular physiology, leading to confusion. Flow usually means **flow rate,** the volume of blood that passes a given point in the system per unit time. In the circulation, flow is expressed in either liters per minute (L/min) or milliliters per minute (mL/min). For instance, blood flow through the aorta of a 70-kg man at rest is about 5 L/min.

Flow rate should not be confused with **velocity of flow** (or simply *velocity*), the distance a fixed volume of blood travels in a given period of time. Velocity is a measure of *how fast* blood flows past a point. In contrast, flow rate measures *how much* (volume) blood flows past a point in a given period of time. For example, look through the open door at the hallway outside your classroom. The number of people passing the door in 1 minute is the flow rate of people through the hallway. How quickly those people are walking past the door is their velocity.

The relationship between velocity of flow (v), flow rate (Q), and cross-sectional area of the tube (A) is expressed by the equation

$$v = Q/A \qquad (7)$$

which says that the velocity of flow through a tube equals the flow rate divided by the tube's cross-sectional area. In a tube of fixed diameter (and thus fixed cross-sectional area), velocity is directly related to flow rate. In a tube of variable diameter, if the flow rate is constant, velocity varies inversely with the diameter. In other words, velocity is faster in narrow sections, and slower in wider sections.

FIGURE 14.4 shows how the velocity of flow varies as the cross-sectional area of the tube changes. The vessel in the figure has variable width, from narrow, with a cross-sectional area of 1 cm^2, to wide, with a cross-sectional area of 12 cm^2. The flow rate is identical along the length of the vessel: 12 cm^3 per minute (1 cm^3 = 1 cubic centimeter (cc) = 1 mL). This flow rate means that in 1 minute, 12 cm^3 of fluid flow past point X in the narrow section, and 12 cm^3 of fluid flow past point Y in the wide section.

But *how fast* does the fluid need to flow to accomplish that rate? According to equation 7, the velocity of flow at point X is 12 cm/min, but at point Y it is only 1 cm/min. As you can see, fluid flows more rapidly through narrow sections of a tube than through wide sections.

To see this principle in action, watch a leaf as it floats down a stream. Where the stream is narrow, the leaf moves rapidly, carried by the fast velocity of the water. In sections where the stream widens into a pool, the velocity of the water decreases and the leaf meanders more slowly.

In this chapter and the next, we apply the physics of fluid flow to the cardiovascular system. The heart generates pressure when it contracts and pumps blood into the arterial side of the circulation. Arteries act as a pressure reservoir during the heart's relaxation phase, maintaining the *mean arterial pressure* (MAP) that is the primary driving force for blood flow. Mean arterial

pressure is influenced by two parameters: *cardiac output* (the volume of blood the heart pumps per minute) and *peripheral resistance* (the resistance of the blood vessels to blood flow through them):

$$\frac{\text{Mean arterial pressure } \alpha}{\text{cardiac output} \times \text{peripheral resistance}} \qquad (8)$$

We will return to a discussion of peripheral resistance and blood flow later. In the remainder of this chapter, we examine heart function and the parameters that influence cardiac output.

CONCEPT CHECK

6. Two canals in Amsterdam are identical in size, but the water flows faster through one than through the other. Which canal has the higher flow rate?

CARDIAC MUSCLE AND THE HEART

To ancient civilizations, the heart was more than a pump—it was *the seat of the mind*. When ancient Egyptians mummified their dead, they removed most of the viscera but left the heart in place so that the gods could weigh it as an indicator of the owner's worthiness. Aristotle characterized the heart as the most important organ of the body, as well as *the seat of intelligence*. We can still find evidence of these ancient beliefs in modern expressions such as "heartfelt emotions." The link between the heart and mind is still explored today as scientists study the effects of stress and depression on the development of cardiovascular disease.

The heart is the workhorse of the body, a muscle that contracts continually, resting only in the milliseconds-long pause between beats. By one estimate, in 1 minute the heart performs work equivalent to lifting a 5-pound weight up 1 foot. The energy demands of this work require a continuous supply of nutrients and oxygen to the heart muscle.

The Heart Has Four Chambers

The heart is a muscular organ, about the size of a fist. It lies in the center of the *thoracic cavity* (see Anatomy Summary, **FIG. 14.5a, b, c**). The pointed *apex* of the heart angles down to

FIG. 14.4 | Flow rate is not the same as velocity of flow

Flow rate (Q) = 12 cm^3/min

12 cm^3

Flow →

X
A = 1 cm^2

Y
A = 12 cm^2

The narrower the vessel, the faster the velocity of flow.

Velocity (v) = $\dfrac{\text{Flow rate (Q)}}{\text{Cross-sectional area (A)}}$	
At Point X	**At Point Y**
$v = \dfrac{12 \text{ cm}^3/\text{min}}{1 \text{ cm}^2}$	$v = \dfrac{12 \text{ cm}^3/\text{min}}{12 \text{ cm}^2}$
v = 12 cm/min	v = 1 cm/min

FIGURE QUESTION
If the cross-sectional area of this pipe is 3 cm^2, what is the velocity of the flow?

FIG. 14.5 ANATOMY SUMMARY

The Heart

(a) The heart lies in the center of the thorax.

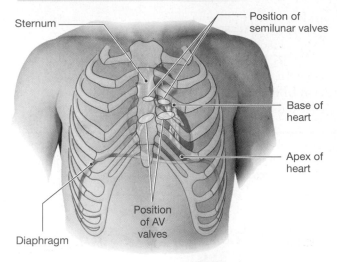

Sternum

Position of semilunar valves

Base of heart

Apex of heart

Position of AV valves

Diaphragm

(b) Vessels that carry well-oxygenated blood are red; those with less well-oxygenated blood are blue.

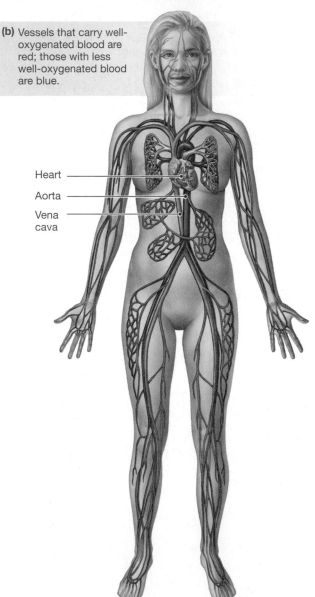

Heart

Aorta

Vena cava

Anatomy of the Thoracic Cavity

(c) The heart is on the ventral side of the thoracic cavity, sandwiched between the lungs.

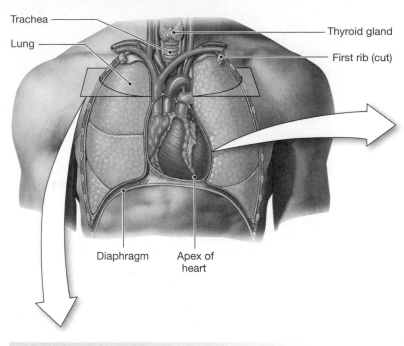

Trachea

Lung

Thyroid gland

First rib (cut)

Diaphragm

Apex of heart

(d) Superior view of transverse plane in (c)

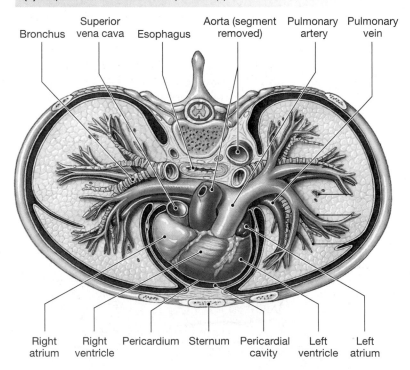

Bronchus

Superior vena cava

Esophagus

Aorta (segment removed)

Pulmonary artery

Pulmonary vein

Right atrium

Right ventricle

Pericardium

Sternum

Pericardial cavity

Left ventricle

Left atrium

Structure of the Heart

(e) The heart is encased within a membranous fluid-filled sac, the pericardium.

Pericardium

Diaphragm

(f) The ventricles occupy the bulk of the heart. The arteries and veins all attach to the base of the heart.

Aorta

Superior vena cava

Pulmonary artery

Auricle of left atrium

Coronary artery and vein

Right atrium

Right ventricle

Left ventricle

(g) One-way flow through the heart is ensured by two sets of valves.

Aorta

Right pulmonary arteries

Superior vena cava

Right atrium

Cusp of right AV (tricuspid) valve

Right ventricle

Inferior vena cava

Septum

Pulmonary semilunar valve

Left pulmonary arteries

Left pulmonary veins

Left atrium

Cusp of left AV (bicuspid) valve

Chordae tendineae

Papillary muscles

Left ventricle

Descending aorta

(h) Myocardial muscle cells are branched, have a single nucleus, and are attached to each other by specialized junctions known as intercalated disks.

Intercalated disks

Myocardial muscle cell

the left side of the body, while the broader *base* lies just behind the breastbone, or *sternum*. Because we usually associate the word *base* with the bottom, remember that the base of a cone is the broad end, and the apex is the pointed end. Think of the heart as an inverted cone with apex down and base up. Within the thoracic cavity, the heart lies on the ventral side, sandwiched between the two lungs, with its apex resting on the diaphragm (Fig. 14.5c).

The heart is encased in a tough membranous sac, the **pericardium** {*peri*, around + *kardia*, heart} (Fig.14.5d, e). A thin layer of clear pericardial fluid inside the pericardium lubricates the external surface of the heart as it beats within the sac. Inflammation of the pericardium (*pericarditis*) may reduce this lubrication to the point that the heart rubs against the pericardium, creating a sound known as a *friction rub*.

The heart itself is composed mostly of cardiac muscle, or **myocardium** {*myo*, muscle + *kardia*, heart}, covered by thin outer and inner layers of epithelium and connective tissue. Seen from the outside, the bulk of the heart is the thick muscular walls of the ventricles, the two lower chambers (Fig. 14.5f). The thinner-walled atria lie above the ventricles.

The major blood vessels all emerge from the base of the heart. The aorta and *pulmonary trunk* (artery) direct blood from the heart to the tissues and lungs, respectively. The venae cavae and pulmonary veins return blood to the heart (**TBL. 14.2**). When the heart is viewed from the front (anterior view), as in Figure 14.5f, the pulmonary veins are hidden behind the other major blood vessels. Running across the surface of the ventricles are shallow grooves containing the **coronary arteries** and **coronary veins,** which supply blood to the heart muscle.

The relationship between the atria and ventricles can be seen in a cross-sectional view of the heart (Fig. 14.5g). As noted earlier, the left and right sides of the heart are separated by a septum, so that blood on one side does not mix with blood on the other side. Although blood flow in the left heart is separated from flow in the right heart, the two sides contract in a coordinated fashion. First, the atria contract together, then the ventricles contract together.

TABLE 14.2	The Heart and Major Blood Vessels	
Blue type indicates structures containing blood with lower oxygen content; red type indicates well-oxygenated blood.		
	Receives Blood from	**Sends Blood to**
Heart		
Right atrium	Venae cavae	Right ventricle
Right ventricle	Right atrium	Lungs
Left atrium	Pulmonary veins	Left ventricle
Left ventricle	Left atrium	Body except for lungs
Vessels		
Venae cavae	Systemic veins	Right atrium
Pulmonary trunk (artery)	Right ventricle	Lungs
Pulmonary vein	Veins of the lungs	Left atrium
Aorta	Left ventricle	Systemic arteries

Blood flows from veins into the atria and from there through one-way valves into the ventricles, the pumping chambers. Blood leaves the heart via the pulmonary trunk from the right ventricle and via the aorta from the left ventricle. A second set of valves guards the exits of the ventricles so that blood cannot flow back into the heart once it has been ejected.

Notice in Figure 14.5g that blood enters each ventricle at the top of the chamber but also leaves at the top. This is because during development, the tubular embryonic heart twists back on itself (**FIG. 14.6b**). This twisting puts the arteries (through which blood leaves) close to the top of the ventricles. Functionally, this means that the ventricles must contract from the bottom up so that blood is squeezed out of the top.

FIG. 14.6 In the embryo, the heart develops from a single tube

(a) Age: embryo, day 25. The heart is a single tube.

(b) By four weeks of development, the atria and ventricles can be distinguished. The heart begins to twist so that the atria move on top of the ventricles.

(c) Age: one year (arteries not shown)

Four fibrous connective tissue rings surround the four heart valves (Fig. 14.5g). These rings form both the origin and insertion for the cardiac muscle, an arrangement that pulls the apex and base of the heart together when the ventricles contract. In addition, the fibrous connective tissue acts as an electrical insulator, blocking most transmission of electrical signals between the atria and the ventricles. This arrangement ensures that the electrical signals can be directed through a specialized conduction system to the apex of the heart for the bottom-to-top contraction.

Heart Valves Ensure One-Way Flow in the Heart

As the arrows in Figure 14.5g indicate, blood flows through the heart in one direction. Two sets of heart valves ensure this one-way flow: one set (the **atrioventricular valves**) between the atria and ventricles, and the second set (the **semilunar valves,** named for their crescent-moon shape) between the ventricles and the arteries. Although the two sets of valves are very different in structure, they serve the same function: preventing the backward flow of blood.

The opening between each atrium and its ventricle is guarded by an atrioventricular (AV) valve (Fig. 14.5g). The AV valve is formed from thin flaps of tissue joined at the base to a connective tissue ring. The flaps are slightly thickened at the edge and connect on the ventricular side to collagenous tendons, the **chordae tendineae** (**FIG. 14.7a, c**).

Most of the chordae fasten to the edges of the valve flaps. The opposite ends of the chordae are tethered to mound-like extensions of ventricular muscle known as the **papillary muscles** {*papilla,* nipple}. These muscles provide stability for the chordae, but they cannot actively open and close the AV valves. The valves move passively when flowing blood pushes on them.

When a ventricle contracts, blood pushes against the bottom side of its AV valve and forces it upward into a closed position (Fig. 14.7a). The chordae tendineae prevent the valve from being pushed back into the atrium, just as the struts on an umbrella keep the umbrella from turning inside out in a high wind. Occasionally, the chordae fail, and the valve is pushed back into the atrium during ventricular contraction, an abnormal condition known as *prolapse.*

The two AV valves are not identical. The valve that separates the right atrium and right ventricle has three flaps and is called the **tricuspid valve** {*cuspis,* point} (Fig. 14.7b). The valve between the left atrium and left ventricle has only two flaps and is called the **bicuspid valve.** The bicuspid is also called the **mitral valve** because of its resemblance to the tall headdress, known as a miter, worn by popes and bishops. You can match AV valves to the proper side of the heart by remembering that the Right Side has the Tricuspid (R-S-T).

The semilunar valves separate the ventricles from the major arteries. The **aortic valve** is between the left ventricle and the aorta, and the **pulmonary valve** lies between the right ventricle and the pulmonary trunk. Each semilunar valve has three cuplike leaflets that snap closed when blood attempting to flow back into

the ventricles fills them (Fig. 14.7c, d). Because of their shape, the semilunar valves do not need connective tendons as the AV valves do.

CONCEPT CHECK

7. What prevents electrical signals from passing through the connective tissue in the heart?
8. Trace a drop of blood from the superior vena cava to the aorta, naming all structures the drop encounters along its route.
9. What is the function of the AV valves? What happens to blood flow if one of these valves fails?

Cardiac Muscle Cells Contract without Innervation

The bulk of the heart is composed of cardiac muscle cells, or myocardium. Most cardiac muscle is contractile, but about 1% of the myocardial cells are specialized to generate action potentials spontaneously. These cells account for a unique property of the heart: its ability to contract without any outside signal. As mentioned in the introduction to this chapter, records tell us of Spanish explorers in the New World witnessing human sacrifices in which hearts torn from the chests of living victims continued to beat for minutes. The heart can contract without a connection to other parts of the body because the signal for contraction is *myogenic,* originating within the heart muscle itself.

The signal for myocardial contraction comes not from the nervous system but from specialized myocardial cells known as **autorhythmic cells.** The autorhythmic cells are also called **pacemakers** because they set the rate of the heartbeat. Myocardial autorhythmic cells are anatomically distinct from contractile cells: autorhythmic cells are smaller and contain few contractile fibers. Because they do not have organized sarcomeres, autorhythmic cells do not contribute to the contractile force of the heart.

Contractile cells are typical striated muscle, however, with contractile fibers organized into sarcomeres [p. 383]. Cardiac muscle differs in significant ways from skeletal muscle and shares some properties with smooth muscle:

1. Cardiac muscle fibers are much smaller than skeletal muscle fibers and usually have a single nucleus per fiber.
2. Individual cardiac muscle cells branch and join neighboring cells end-to-end to create a complex network (Fig. 14.5h and **FIG. 14.8b**). The cell junctions, known as **intercalated disks** {*inter-,* between + *calare,* to proclaim}, consist of interdigitated membranes. Intercalated disks have two components: *desmosomes* and gap junctions [p. 74]. Desmosomes are strong connections that tie adjacent cells together, allowing force created in one cell to be transferred to the adjacent cell.
3. *Gap junctions* in the intercalated disks electrically connect cardiac muscle cells to one another. They allow waves of depolarization to spread rapidly from cell to cell, so that all the heart muscle cells contract almost simultaneously. In this respect, cardiac muscle resembles single-unit smooth muscle.

FIG. 14.7   **Heart valves create one-way flow through the heart**

This longitudinal section shows both the left AV (mitral) valve and the aortic semilunar valve.

This transverse section shows the AV valves as viewed from the atria, and the semilunar valves as viewed from inside the aorta and pulmonary artery.

VENTRICULAR CONTRACTION

(a) Frontal Section

- Aorta
- Left atrium
- Aortic semilunar valve (open)
- Papillary muscles (tense)
- Septum
- Mitral valve
- Chordae tendineae (tense)
- Left ventricle (contracted)

During ventricular contraction, the **AV valves** remain closed to prevent blood flow backward into the atria.

(b) Transverse Section

- Mitral (left AV), or bicuspid, valve
- Fibrous skeleton
- Aortic semilunar valve (open)
- Pulmonary semilunar valve (open)
- Tricuspid (right AV) valve

VENTRICULAR RELAXATION

(c) Frontal Section

- Pulmonary veins
- Semilunar valves
- Mitral valve (open)
- Chordae tendineae (relaxed)
- Papillary muscles (relaxed)
- Left ventricle (dilated)

The **semilunar valves** prevent blood that has entered the arteries from flowing back into the ventricles during ventricular relaxation.

(d) Transverse Section

- Mitral (left AV), or bicuspid, valve (open)
- Semilunar valves (closed)

4. The t-tubules of myocardial cells are larger than those of skeletal muscle, and they branch inside the myocardial cells.
5. Myocardial sarcoplasmic reticulum is smaller than that of skeletal muscle, reflecting the fact that cardiac muscle depends in part on extracellular Ca^{2+} to initiate contraction. In this respect, cardiac muscle resembles smooth muscle.
6. Mitochondria occupy about one-third the cell volume of a cardiac contractile fiber, a reflection of the high energy demand of

FIG. 14.8 | Cardiac muscle

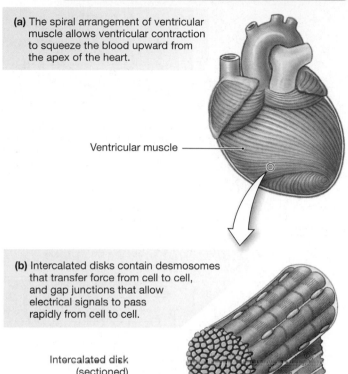

(a) The spiral arrangement of ventricular muscle allows ventricular contraction to squeeze the blood upward from the apex of the heart.

Ventricular muscle

(b) Intercalated disks contain desmosomes that transfer force from cell to cell, and gap junctions that allow electrical signals to pass rapidly from cell to cell.

Intercalated disk (sectioned)

Nucleus

Intercalated disk

Mitochondria

Cardiac muscle cell

Contractile fibers

these cells. By one estimate, cardiac muscle consumes 70–80% of the oxygen delivered to it by the blood, more than twice the amount extracted by other cells in the body.

During periods of increased activity, the heart uses almost all the oxygen brought to it by the coronary arteries. As a result, the only way to get more oxygen to exercising heart muscle is to increase the blood flow. Reduced myocardial blood flow from narrowing of a coronary vessel by a clot or fatty deposit can damage or even kill myocardial cells.

[See Tbl. 12.3, p. 411, for a summary comparison of the three muscle types.]

Calcium Entry Is a Feature of Cardiac EC Coupling

In skeletal muscle, acetylcholine from a somatic motor neuron causes a skeletal muscle action potential to begin

excitation-contraction coupling (EC coupling) [p. 383]. In cardiac muscle, an action potential also initiates EC coupling, but the action potential originates spontaneously in the heart's pacemaker cells and spreads into the contractile cells through gap junctions. Other aspects of cardiac EC coupling are similar to processes you encountered in skeletal and smooth muscle contraction.

FIGURE 14.9 illustrates EC coupling and relaxation in cardiac muscle. An action potential that enters a contractile cell moves across the sarcolemma and into the t-tubules **1**, where it opens voltage-gated L-type Ca^{2+} channels in the cell membrane **2**. Ca^{2+} enters the cell through these channels, moving down its electrochemical gradient. Calcium entry opens *ryanodine receptor Ca^{2+} release channels (RyR)* in the sarcoplasmic reticulum **3**. This process of EC coupling in cardiac muscle is also called **Ca^{2+}-induced Ca^{2+} release** (CICR). When the RyR channels open, stored Ca^{2+} flows out of the sarcoplasmic reticulum and into the cytosol **4**, creating a Ca^{2+} "spark" that can be seen using special biochemical methods [p. 178]. Multiple sparks from different RyR channels sum to create a Ca^{2+} signal **5**.

Calcium released from the sarcoplasmic reticulum provides about 90% of the Ca^{2+} needed for muscle contraction, with the remaining 10% entering the cell from the extracellular fluid. Calcium diffuses through the cytosol to the contractile elements, where the ions bind to troponin and initiate the cycle of crossbridge formation and movement **6**. Contraction takes place by the same type of sliding filament movement that occurs in skeletal muscle [p. 385].

FIG. 14.9 EC coupling in cardiac muscle

This figure shows the cellular events leading to contraction and relaxation in a cardiac contractile cell.

1 Action potential enters from adjacent cell.

2 Voltage-gated Ca²⁺ channels open. Ca²⁺ enters cell.

3 Ca²⁺ induces Ca²⁺ release through ryanodine receptor-channels (RyR).

4 Local release causes Ca²⁺ spark.

5 Summed Ca²⁺ sparks create a Ca²⁺ signal.

6 Ca²⁺ ions bind to troponin to initiate contraction.

7 Relaxation occurs when Ca²⁺ unbinds from troponin.

8 Ca²⁺ is pumped back into the sarcoplasmic reticulum for storage.

9 Ca²⁺ is exchanged with Na⁺ by the NCX antiporter.

10 Na⁺ gradient is maintained by the Na⁺-K⁺-ATPase.

Q FIGURE QUESTION

Using the numbered steps, compare the events shown to EC coupling in skeletal and smooth muscle [see Figs.12.10 and 12.26].

Relaxation in cardiac muscle is generally similar to that in skeletal muscle. As cytoplasmic Ca^{2+} concentrations decrease, Ca^{2+} unbinds from troponin, myosin releases actin, and the contractile filaments slide back to their relaxed position **7**. As in skeletal muscle, Ca^{2+} is transported back into the sarcoplasmic reticulum with the help of a Ca^{2+}-ATPase **8**. However, in cardiac muscle, Ca^{2+} is also removed from the cell via the *Na^+-Ca^{2+} exchanger* (NCX) **9**. One Ca^{2+} moves out of the cell against its electrochemical gradient in exchange for 3 Na^+ entering the cell down their electrochemical gradient. Sodium that enters the cell during this transfer is removed by the Na^+-K^+-ATPase **10**.

Cardiac Muscle Contraction Can Be Graded

A key property of cardiac muscle cells is the ability of a single muscle fiber to execute *graded contractions,* in which the fiber varies the amount of force it generates. (Recall that in skeletal muscle, contraction in a single fiber is all-or-none at any given fiber length.) The force generated by cardiac muscle is proportional to the number of crossbridges that are active. The number of active crossbridges is determined by how much Ca^{2+} is bound to troponin.

If cytosolic Ca^{2+} concentrations are low, some crossbridges are not activated and contraction force is small. If additional Ca^{2+} enters the cell from the extracellular fluid, more Ca^{2+} is released from the sarcoplasmic reticulum. This additional Ca^{2+} binds to troponin, enhancing the ability of myosin to form crossbridges with actin and creating additional force.

Another factor that affects the force of contraction in cardiac muscle is the sarcomere length at the beginning of contraction. In the intact heart, stretch on the individual fibers is a function of how much blood is in the chambers of the heart. The relationship between force and ventricular volume is an important property of cardiac function, and we discuss it in detail later in this chapter.

10. Compare the receptors and channels involved in cardiac EC coupling to those found in skeletal muscle EC coupling. [*Hint:* p. 383]

11. If a myocardial contractile cell is placed in interstitial fluid and depolarized, the cell contracts. If Ca^{2+} is removed from the fluid surrounding the myocardial cell and the cell is depolarized, it does not contract. If the experiment is repeated with a skeletal muscle fiber, the skeletal muscle contracts when depolarized, whether or not Ca^{2+} is present in the surrounding fluid. What conclusion can you draw from the results of this experiment?

12. A drug that blocks all Ca^{2+} channels in the myocardial contractile cell membrane is placed in the solution around the cell. What happens to the force of contraction in that cell?

Myocardial Action Potentials Vary

Cardiac muscle, like skeletal muscle and neurons, is an excitable tissue with the ability to generate action potentials. Each of the two types of cardiac muscle cells has a distinctive action potential that will vary somewhat in shape depending on where in the heart it is recorded. In both autorhythmic and contractile myocardium, Ca^{2+} plays an important role in the action potential, in contrast to the action potentials of skeletal muscle and neurons.

Myocardial Contractile Cells The action potentials of myocardial contractile cells are similar in several ways to those of neurons and skeletal muscle [p. 240]. The rapid depolarization phase of the action potential is the result of Na^+ entry, and the steep repolarization phase is due to K^+ leaving the cell (**FIG. 14.10**). The main difference between the action potential of the myocardial contractile cell and those of skeletal muscle fibers and neurons is that the myocardial cell has a longer action potential due to Ca^{2+} entry. Let's take a look at these longer action potentials. By convention, the action potential phases start with zero.

Phase 4: resting membrane potential. Myocardial contractile cells have a stable resting potential of about –90 mV.

Phase 0: depolarization. When a wave of depolarization moves into a contractile cell through gap junctions, the membrane potential becomes more positive. Voltage-gated Na^+ channels open, allowing Na^+ to enter the cell and rapidly depolarize it. The membrane potential reaches about +20 mV before the Na^+ channels close. These are double-gated Na^+ channels, similar to the voltage-gated Na^+ channels of the axon [p. 245].

Phase 1: initial repolarization. When the Na^+ channels close, the cell begins to repolarize as K^+ leaves through open K^+ channels.

Phase 2: the plateau. The initial repolarization is very brief. The action potential then flattens into a plateau as the result of two events: a decrease in K^+ permeability and an increase in Ca^{2+} permeability. Voltage-gated Ca^{2+} channels activated by depolarization have been slowly opening during phases

FIG. 14.10 Action potential of a cardiac contractile cell

P_X = Permeability to ion X

Phase*	Membrane channels
0	Na^+ channels open
1	Na^+ channels close
2	Ca^{2+} channels open; fast K^+ channels close
3	Ca^{2+} channels close; slow K^+ channels open
4	Resting potential

*The phase numbers are a convention.

Q FIGURE QUESTION

Compare ion movement during this action potential to ion movement of a neuron's action potential [Fig. 8.9].

0 and 1. When they finally open, Ca^{2+} enters the cell. At the same time, some "fast" K^+ channels close. The combination of Ca^{2+} influx and decreased K^+ efflux causes the action potential to flatten out into a plateau.

Phase 3: rapid repolarization. The plateau ends when Ca^{2+} channels close and K^+ permeability increases once more. The "slow" K^+ channels responsible for this phase are similar to those in the neuron: They are activated by depolarization but are slow to open. When the slow K^+ channels open, K^+ exits rapidly, returning the cell to its resting potential (phase 4).

The influx of Ca^{2+} during phase 2 lengthens the total duration of a myocardial action potential. A typical action potential in a neuron or skeletal muscle fiber lasts between 1 and 5 msec. In a contractile myocardial cell, the action potential typically lasts 200 msec or more.

The longer myocardial action potential helps prevent the sustained contraction called *tetanus*. Prevention of tetanus in the heart is important because cardiac muscles must relax between contractions so the ventricles can fill with blood. To understand how a longer action potential prevents tetanus, let's compare

Pressure-Volume Curves Represent One Cardiac Cycle

Another way to describe the cardiac cycle is with a pressure-volume graph, shown in Figure 14.17b. This figure represents the changes in volume (*x*-axis) and pressure (*y*-axis) that occur during one cardiac cycle.

Recall that the flow of blood through the heart is governed by the same principle that governs the flow of all liquids and gases: Flow proceeds from areas of higher pressure to areas of lower pressure. When the heart contracts, the pressure increases and blood flows out of the heart into areas of lower pressure. Figure 14.17b represents pressure and volume changes in the left ventricle, which sends blood into the systemic circulation. The left side of the heart creates higher pressures than the right side, which sends blood through the shorter pulmonary circuit.

The cycle begins at point A. The ventricle has completed a contraction and contains the minimum amount of blood that it will hold during the cycle. It has relaxed, and its pressure is also at its minimum value. Blood is flowing into the atrium from the pulmonary veins.

Once pressure in the atrium exceeds pressure in the ventricle, the mitral valve between the atrium and ventricle opens (Fig. 14.17b, point A). Atrial blood now flows into the ventricle, increasing its volume (point A to point B). As blood flows in, the relaxing ventricle expands to accommodate the entering blood.

Consequently, the volume of the ventricle increases, but the pressure in the ventricle goes up very little.

The last portion of ventricular filling is completed by atrial contraction (point A to B). The ventricle now contains the maximum volume of blood that it will hold during this cardiac cycle (point B). Because maximum filling occurs at the end of ventricular relaxation (diastole), this volume is called the **end-diastolic volume (EDV)**. In a 70-kg man at rest, end-diastolic volume is about 135 mL. However, EDV varies under different conditions. During periods of very high heart rate, for instance, when the ventricle does not have time to fill completely between beats, the end-diastolic value may be less than 135 mL.

When ventricular contraction begins, the mitral (AV) valve closes. With both the AV valve and the semilunar valve closed, blood in the ventricle has nowhere to go. Nevertheless, the ventricle continues to contract, causing the pressure in this chamber to increase rapidly during isovolumic contraction (B → C in Fig. 14.17b). Once ventricular pressure exceeds the pressure in the aorta, the aortic valve opens (point C). Pressure continues to increase as the ventricle contracts further, but ventricular volume decreases as blood is pushed out into the aorta (C → D).

The heart does not empty itself completely of blood each time the ventricle contracts. The volume of blood left in the ventricle at the end of contraction is known as the **end-systolic volume (ESV)**. The ESV (point D) is the minimum volume of blood the ventricle contains during one cycle. An average ESV value in a person at rest is 65 mL, meaning that nearly half of the 135 mL that was in the ventricle at the start of the contraction is still there at the end of the contraction.

At the end of each ventricular contraction, the ventricle begins to relax. As it does so, ventricular pressure decreases. Once pressure in the ventricle falls below aortic pressure, the semilunar valve closes, and the ventricle again becomes a sealed chamber. The remainder of relaxation occurs without a change in blood volume, and so this phase is called *isovolumic relaxation* (Fig. 14.17b, D → A). When ventricular pressure finally falls to the point at which atrial pressure exceeds ventricular pressure, the mitral valve opens and the cycle begins again.

The electrical and mechanical events of the cardiac cycle are summarized together in **FIGURE 14.18**, known as a Wiggers diagram after the physiologist who first created it.

CONCEPT CHECK

27. In Figure 14.17a, at what points in the cycle do EDV and ESV occur?

28. On the Wiggers diagram in Figure 14.18, match the following events to the lettered boxes: (a) end-diastolic volume, (b) aortic valve opens, (c) mitral valve opens, (d) aortic valve closes, (e) mitral valve closes, (f) end-systolic volume

29. Why does atrial pressure increase just to the right of point C in Figure 14.18? Why does it decrease during the initial part of ventricular systole, then increase? Why does it decrease to the right of point D?

30. Why does ventricular pressure shoot up suddenly at point C in Figure 14.18?

FIG. 14.18 | The Wiggers diagram

This diagram follows left heart and aortic pressures, left ventricular volume, and the ECG through one cardiac cycle. The boxed letters refer to Concept Checks 28–30.

RUNNING **PROBLEM**

The electrocardiogram indicated that Walter suffered a myocardial infarction, resulting from blockage of blood vessels nourishing the left ventricle. The exact location of the damage depends on which artery and which branch or branches have become occluded.

Q7: *If the ventricle of the heart is damaged, in which wave or waves of the electrocardiogram would you expect to see abnormal changes?*

436 439 449 455 **466** 468 472

Stroke Volume Is the Volume of Blood Pumped per Contraction

What is the purpose of blood remaining in the ventricles at the end of each contraction? For one thing, the resting end-systolic volume of 65 mL provides a safety margin. With a more forceful contraction, the heart can decrease its ESV, sending additional blood to the tissues. Like many organs of the body, the heart does not usually work "all out."

Stroke volume is the amount of blood pumped by one ventricle during a contraction. It is measured in milliliters per beat and can be calculated as follows:

$$\text{Volume of blood before contraction} - \text{volume of blood after contraction} = \text{stroke volume}$$

$$\text{EDV} - \text{ESV} = \text{stroke volume} \tag{9}$$

For the average contraction in a person at rest:

$$135 \text{ mL} - 65 \text{ mL} = 70 \text{ mL, the normal stroke volume} \tag{10}$$

Stroke volume is not constant and can increase to as much as 100 mL during exercise. Stroke volume, like heart rate, is regulated by mechanisms we discuss later in this chapter.

Cardiac Output Is a Measure of Cardiac Performance

How can we assess the effectiveness of the heart as a pump? One way is to measure **cardiac output (CO),** the volume of blood pumped by one ventricle in a given period of time. Because all blood that leaves the heart flows through the tissues, cardiac output is an indicator of total blood flow through the body. However, cardiac output does not tell us how blood is distributed to various tissues. That aspect of blood flow is regulated at the tissue level.

Cardiac output (CO) can be calculated by multiplying heart rate (beats per minute) by stroke volume (mL per beat, or per contraction):

$$\text{Cardiac output} = \text{heart rate} \times \text{stroke volume} \tag{11}$$

For an average resting heart rate of 72 beats per minute and a stroke volume of 70 mL per beat, we have

$$\text{CO} = 72 \text{ beats/min} \times 70 \text{ mL/beat}$$
$$= 5040 \text{ mL/min (or approx. 5 L/min)} \tag{12}$$

Average total blood volume is about 5 liters. This means that, at rest, one side of the heart pumps all the blood in the body through it in only 1 minute!

Normally, cardiac output is the same for both ventricles. However, if one side of the heart begins to fail for some reason and is unable to pump efficiently, cardiac output becomes mismatched. In that situation, blood pools in the circulation behind the weaker side of the heart.

During exercise, cardiac output may increase to 30–35 L/min. Homeostatic changes in cardiac output are accomplished by varying the heart rate, the stroke volume, or both. Both local and reflex mechanisms can alter cardiac output, as you will see in the sections that follow.

CONCEPT CHECK

31. If the stroke volume of the left ventricle is 250 mL/beat and the stroke volume of the right ventricle is 251 mL/beat, what happens to the relative distribution of blood between the systemic and pulmonary circulation after 10 beats?

The Autonomic Division Modulates Heart Rate

An average resting heart rate in an adult is about 70 beats per minute (bpm). The normal range is highly variable, however. Trained athletes may have resting heart rates of 50 bpm or less, while someone who is excited or anxious may have a rate of 125 bpm or higher. Children have higher average heart rates than adults. Heart rate is initiated by autorhythmic cells in the SA node, but it is modulated by neural and hormonal input.

The sympathetic and parasympathetic branches of the autonomic division influence heart rate through antagonistic control (**FIG. 14.19**). Parasympathetic activity slows heart rate, while sympathetic activity speeds it up.

Parasympathetic Control The parasympathetic neurotransmitter acetylcholine (ACh) slows heart rate. Acetylcholine activates muscarinic cholinergic receptors that influence K^+ and Ca^{2+} channels in the pacemaker cell (Fig. 14.19c). Potassium permeability increases, hyperpolarizing the cell so that the pacemaker potential begins at a more negative value (Fig. 14.19d). At the same time, Ca^{2+} permeability of the pacemaker decreases. Decreased Ca^{2+} permeability slows the rate at which the pacemaker potential depolarizes. The combination of the two effects causes the cell to take longer to reach threshold, delaying the onset of the action potential in the pacemaker and slowing the heart rate.

FIG. 14.19   Autonomic control of heart rate

(a) Stimulation by parasympathetic nerves decreases heart rate.

(b) Stimulation by sympathetic nerves increases heart rate.

Parasympathetic neuron (ACh on M receptor)

Heartbeats

0   1   2   3

Time (sec)

Sympathetic neuron (NE on β_1-receptor)

Heartbeats

0   1   2   3

Time (sec)

SA node

(c) This concept map demonstrates how the parasympathetic and sympathetic neurons alter heart rate through antagonistic control.

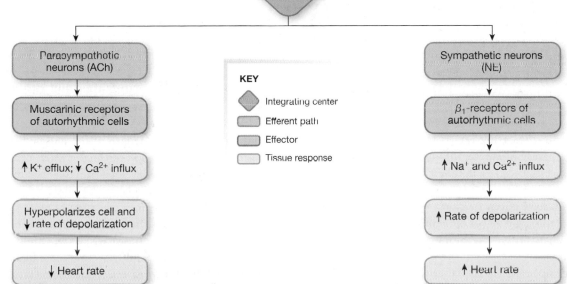

Cardiovascular control center in medulla oblongata

Parasympathetic neurons (ACh)

Muscarinic receptors of autorhythmic cells

↑ K+ efflux; ↓ Ca²⁺ influx

Hyperpolarizes cell and ↓ rate of depolarization

↓ Heart rate

KEY

◇ Integrating center

▭ Efferent path

▭ Effector

▭ Tissue response

Sympathetic neurons (NE)

β_1-receptors of autorhythmic cells

↑ Na+ and Ca²⁺ influx

↑ Rate of depolarization

↑ Heart rate

(d) Parasympathetic stimulation hyperpolarizes the membrane potential of the autorhythmic cell and slows depolarization, slowing down the heart rate.

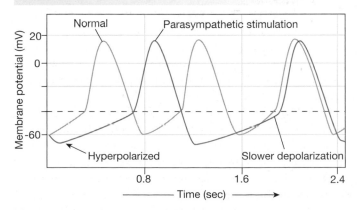

Normal   Parasympathetic stimulation

Membrane potential (mV)

20

0

-60

Hyperpolarized   Slower depolarization

0.8   1.6   2.4

Time (sec)

(e) Sympathetic stimulation and epinephrine depolarize the autorhythmic cell and speed up the pacemaker potential, increasing the heart rate.

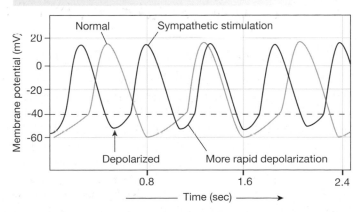

Normal   Sympathetic stimulation

Membrane potential (mV)

20

0

-20

-40

-60

Depolarized   More rapid depolarization

0.8   1.6   2.4

Time (sec)

Sympathetic Control Sympathetic stimulation of pacemaker cells speeds up heart rate (Fig. 14.19b). The catecholamines norepinephrine (from sympathetic neurons) and epinephrine (from the adrenal medulla) increase ion flow through both I_f and Ca^{2+} channels. More rapid cation entry speeds up the rate of the pacemaker depolarization, causing the cell to reach threshold faster and increasing the rate of action potential firing (Fig. 14.19e). When the pacemaker fires action potentials more rapidly, heart rate increases.

Catecholamines exert their effect by binding to and activating β_1-adrenergic receptors on the autorhythmic cells. The β_1-receptors use a cAMP second messenger system to alter the transport properties of the ion channels. In the case of the I_f channels, which are cyclic nucleotide-gated channels, cAMP itself is the messenger. When cAMP binds to open I_f channels, they remain open longer. Increased permeability to Na^+ and Ca^{2+} during the pacemaker potential phase speeds up depolarization and heart rate.

Tonic Control Normally, tonic control of heart rate is dominated by the parasympathetic branch. This control can be shown experimentally by blocking all autonomic input to the heart. When all sympathetic and parasympathetic input is blocked, the spontaneous depolarization rate of the SA node is 90–100 times per minute. To achieve a resting heart rate of 70 beats per minute, tonic parasympathetic activity must slow the intrinsic rate down from 90 bpm.

An increase in heart rate can be achieved in two ways. The simplest method for increasing rate is to decrease parasympathetic activity. As parasympathetic influence is withdrawn from the autorhythmic cells, they resume their intrinsic rate of depolarization, and heart rate increases to 90–100 beats per minute. Sympathetic input is required to increase heart rate above the intrinsic rate. Norepinephrine (or epinephrine) on β_1-receptors speeds up the depolarization rate of the autorhythmic cells and increases heart rate.

Both autonomic branches also alter the rate of conduction through the AV node. Acetylcholine slows the conduction of action potentials through the AV node, thereby increasing AV node delay. In contrast, the catecholamines epinephrine and norepinephrine enhance conduction of action potentials through the AV node and through the conducting system.

Multiple Factors Influence Stroke Volume

Stroke volume, the volume of blood pumped per ventricle per contraction, is directly related to the force generated by cardiac muscle during a contraction. Normally, as contraction force increases, stroke volume increases. In the isolated heart, the force of ventricular contraction is affected by two parameters: the length of muscle fibers at the beginning of contraction and the contractility of the heart. The volume of blood in the ventricle at the beginning of contraction (the end-diastolic volume) determines the length of the muscle. **Contractility** is the intrinsic ability of a cardiac muscle fiber to contract at any given fiber length and is a function of Ca^{2+} interaction with the contractile filaments.

RUNNING **PROBLEM**

Walter's heart attack caused elevation of the S-T segment on his ECG, a type of MI known as STEMI. Early treatment with t-PA dissolved the clots blocking his artery, preventing significant damage to his heart muscle. Walter was transferred to the cardiac care unit, where the cardiologist visited him. "We need to keep an eye on you here for the next few days. There is a possibility that the damage from your heart attack could cause an irregular heartbeat." When Walter was stable, he was taken for a coronary angiogram, a procedure in which an opaque dye visible on X-rays shows where coronary artery lumens have narrowed from atherosclerotic plaques.

Q8: *The beta blocker Walter was given is an antagonist to β_1-adrenergic receptors. What did this drug do to Walter's heart rate? Why is that response helpful following a heart attack?*

Q9: *If Walter's heart attack has damaged the muscle of his left ventricle, what do you predict will happen to his cardiac output?*

436   439   449   455   466   **468**   472

Length-Tension Relationships and the Frank-Starling Law of the Heart In striated muscles, the force created by a muscle fiber is directly related to the length of the sarcomere, as indicated by the initial length of the muscle fiber [p. 395]. The longer the muscle fiber and sarcomere when a contraction begins, the greater the tension developed, up to a maximum (**FIG. 14.20a**).

The length-tension relationship observed in isolated muscles can also be seen in the intact heart: as stretch of the ventricular wall increases, so does the stroke volume (Fig. 14.20b). If additional blood flows into the ventricles, the muscle fibers stretch, then contract more forcefully, ejecting more blood. The degree of myocardial stretch before contraction begins is called the **preload** on the heart because this stretch represents the load placed on cardiac muscles before they contract.

This relationship between stretch and force in the intact heart was first described by a German physiologist, Otto Frank. A British physiologist, Ernest Starling, then expanded on Frank's work. Starling attached an isolated heart-lung preparation from a dog to a reservoir so that he could regulate the amount of blood returning to the heart. He found that in the absence of any nervous or hormonal control, the heart pumped all the blood that returned to it.

The relationship between stretch and force in the intact heart is plotted on a *Starling curve* (Fig. 14.20b). The *x*-axis represents the end-diastolic volume. This volume is a measure of stretch in the ventricles, which in turn determines sarcomere length. The *y*-axis of the Starling curve represents the stroke volume and is an indicator of the force of contraction.

The graph shows that stroke volume is proportional to EDV. As additional blood enters the heart, the heart contracts more forcefully and ejects more blood. This relationship is known as

FIG. 14.20   Length-tension relationships

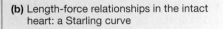

The force (tension) created by a striated muscle is directly related to the starting length of the sarcomere.

(a) Isometric contractions

These data represent tension developed during experiments where muscles were held at a constant length (isometric contraction). The physiological range is the sarcomere length in which the muscle normally functions.

(b) Length-force relationships in the intact heart: a Starling curve

GRAPH QUESTION

What is the maximum stroke volume achieved in this experiment? At what end-diastolic volume is maximum stroke volume first achieved?

(c) Catecholamines increase contractility. Norepinephrine is a positive inotropic agent.

GRAPH QUESTION

At the end-diastolic volume indicated by point A, which heart will create more force: the control heart or the heart under the influence of norepinephrine?

---

the **Frank-Starling law of the heart.** It means that within physiological limits, the heart pumps all the blood that returns to it.

Stroke Volume and Venous Return According to the Frank-Starling law, stroke volume increases as end-diastolic volume increases. End-diastolic volume is normally determined by **venous return,** the amount of blood that enters the heart from the venous circulation. Three factors affect venous return: (1) contraction or compression of veins returning blood to the heart (the skeletal muscle pump), (2) pressure changes in the abdomen and thorax during breathing (the respiratory pump), and (3) sympathetic innervation of veins.

Skeletal muscle pump is the name given to skeletal muscle contractions that squeeze veins (particularly in the legs), compressing them and pushing blood toward the heart. During exercise that involves the lower extremities, the skeletal muscle pump helps return blood to the heart. During periods of sitting or standing motionless, the skeletal muscle pump does not assist venous return.

The **respiratory pump** is created by movement of the thorax during inspiration (breathing in). As the chest expands and the diaphragm moves toward the abdomen, the thoracic cavity enlarges and develops a subatmospheric pressure. This low pressure decreases pressure in the inferior vena cava as it passes through the thorax, which helps draw more blood into the vena cava from veins in the abdomen. The respiratory pump is aided by the higher pressure placed on the outside of abdominal veins when

the abdominal contents are compressed during inspiration. The combination of increased pressure in the abdominal veins and decreased pressure in thoracic veins enhances venous return during inspiration.

Constriction of veins by sympathetic activity is the third factor that affects venous return. When the veins constrict, their volume decreases, squeezing more blood out of them and into the heart. With a larger ventricular volume at the beginning of the next contraction, the ventricle contracts more forcefully, sending the blood out into the arterial side of the circulation. In this manner, sympathetic innervation of veins allows the body to redistribute some venous blood to the arterial side of the circulation.

Contractility Is Controlled by the Nervous and Endocrine Systems

Any chemical that affects contractility is called an **inotropic agent** {*ino,* fiber}, and its influence is called an **inotropic effect.** If a chemical increases the force of contraction, it is said to have a positive inotropic effect. For example, the catecholamines epinephrine and norepinephrine and drugs such as digitalis enhance contractility and are therefore considered to have a positive inotropic effect. Chemicals with negative inotropic effects decrease contractility.

Figure 14.20c illustrates a normal Starling curve (the control curve) along with a curve showing how the stroke volume changes with increased contractility due to norepinephrine. Note that contractility is distinct from the length-tension relationship. A muscle can remain at one length (for example, the end-diastolic volume marked A in Figure 14.20c) but show increased contractility. Contractility increases as the amount of calcium available for contraction increases. Contractility was once considered to be distinct from changes in force resulting from variations in muscle (sarcomere) length. However, it now appears that increasing sarcomere length also makes cardiac muscle more sensitive to Ca^{2+} thus linking contractility to muscle length.

The mechanism by which catecholamines increase Ca^{2+} entry and storage and exert their positive inotropic effect is mapped in **FIGURE 14.21**. The signal molecules bind to and activate β_1-adrenergic receptors [p. 366] on the contractile myocardial cell membrane. Activated β_1-receptors use a cyclic AMP second messenger system to phosphorylate specific intracellular proteins [p. 173]. Phosphorylation of voltage-gated Ca^{2+} channels increases the probability that they will open and stay open longer. More open channels allow more Ca^{2+} to enter the cell.

The catecholamines increase Ca^{2+} storage through the use of a regulatory protein called **phospholamban** (Fig. 14.21).

Phosphorylation of phospholamban enhances Ca^{2+}-ATPase activity in the sarcoplasmic reticulum. The ATPase concentrates Ca^{2+} in the sarcoplasmic reticulum, making more Ca^{2+} available for calcium-induced calcium release. Because more cytosolic Ca^{2+} means more active crossbridges, and because the force of contraction is proportional to the number of active crossbridges, the net result of catecholamine stimulation is a stronger contraction.

In addition to increasing the force of cardiac contraction, catecholamines also shorten the duration of contraction. The enhanced Ca^{2+}-ATPase speeds up removal of Ca^{2+} from the cytosol. This in turn shortens the time that Ca^{2+} is bound to troponin and decreases the active time of the myosin crossbridges. The muscle twitch is therefore briefer.

A different mechanism that enhances contractility can be triggered by the administration of cardiac glycosides, a class of molecules first discovered in the plant *Digitalis purpurea* (purple foxglove). Cardiac glycosides include digitoxin and the related compound *ouabain*, a molecule used to inhibit sodium transport in research studies. Glycosides increase contractility by slowing Ca^{2+} removal from the cytosol (in contrast to the catecholamines just discussed, which

FIG. 14.21 | Catecholamines increase cardiac contraction

Phospholamban is a regulatory protein that alters sarcoplasmic reticulum Ca^{2+}-ATPase activity.

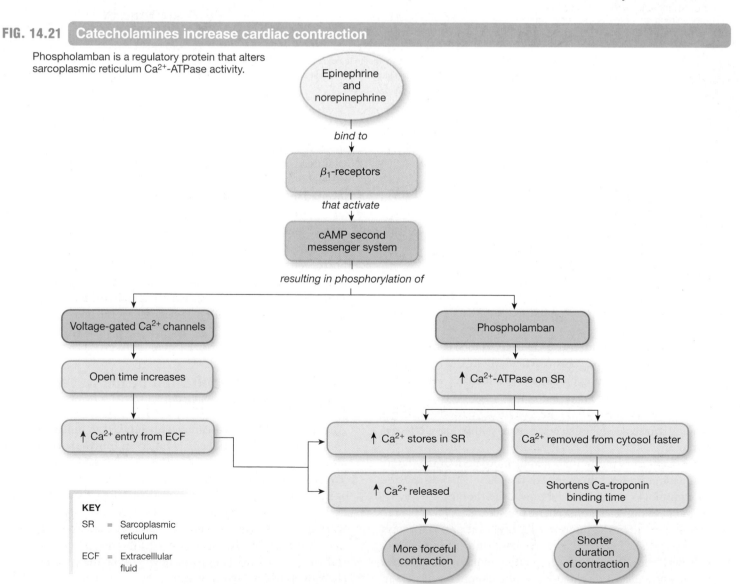

Epinephrine and norepinephrine

bind to

β_1-receptors

that activate

cAMP second messenger system

resulting in phosphorylation of

Voltage-gated Ca^{2+} channels → Open time increases → ↑ Ca^{2+} entry from ECF

Phospholamban → ↑ Ca^{2+}-ATPase on SR → ↑ Ca^{2+} stores in SR → ↑ Ca^{2+} released → More forceful contraction

Ca^{2+} removed from cytosol faster → Shortens Ca-troponin binding time → Shorter duration of contraction

KEY

SR = Sarcoplasmic reticulum

ECF = Extracellular fluid

speed up Ca^{2+} removal). This mechanism is a pharmacological effect and does not occur in the absence of the drug.

Cardiac glycosides have been used since the eighteenth century as a remedy for *heart failure,* a pathological condition in which the heart is unable to contract forcefully. These highly toxic drugs depress Na^{+}-K^{+}-ATPase activity in all cells, not just those of the heart. With depressed Na^{+}-K^{+}-ATPase activity, Na^{+} builds up in the cytosol, and the concentration gradient for Na^{+} across the cell membrane diminishes. This in turn decreases the potential energy available for indirect active transport [p. 143]. In the myocardial cell, cardiac glycosides decrease the cell's ability to remove Ca^{2+} by means of the Na^{+}-Ca^{2+} exchanger. The resultant increase in cytosolic Ca^{2+} causes more forceful myocardial contractions.

CONCEPT CHECK   **32.** Using the myocardial cell in Figure 14.9 as a model, draw a contractile cell and show how catecholamines increase myocardial contractility.

EDV and Arterial Blood Pressure Determine Afterload

Many of the experiments that uncovered the relationship between myocardial stretch and contractile force were conducted using isolated hearts. In the intact animal, ventricular force must

EMERGING CONCEPTS

Stem Cells for Heart Disease

Translating basic scientific research into medicine treatments is the goal of many biomedical scientists. One example is finding stem cells that can repair heart damage. After a heart attack, portions of the myocardium may be so damaged from lack of oxygen that they can no longer contract and contribute to cardiac function. A therapy that could replace dead and damaged cells and restore function would be a dream come true. In 2001, a group of researchers reported that stem cells injected into mice with damaged hearts differentiated into new myocardial cells. This dramatic result prompted rapid translation of the basic research into human clinical trials. By 2008, there were more than 251 clinical trials looking at whether stem cell injections could help impaired cardiac function. But the results have been disappointing and the topic is controversial. Some scientists report that they have been unable to duplicate the 2001 findings that stem cells differentiate into myocardial cells, and several published papers were retracted. As of 2014, the evidence suggests that although the heart may be able to repair itself using stem cells, the process is so slow that it is unlikely to be of therapeutic use.

be used to overcome the resistance created by blood already filling the arterial system. In other words, to eject blood from the ventricle, the heart must create force to displace the blood in the aorta, pushing it farther downstream. The combined load of blood in the ventricle (the EDV) and arterial resistance during ventricular contraction is known as **afterload.**

As an analogy, think of waiters carrying trays of food through a swinging door. A tray is a load equivalent to blood in the ventricles at the beginning of contraction. The door is an additional load that the waiter must push against to leave the kitchen. Normally this additional load is relatively minor. If someone decides to play a prank, however, and piles furniture against the dining room side of the door (increased afterload), the waiter must expend considerably more force to push through the door. Similarly, ventricular contraction must push a load of blood through a semilunar valve and out into the blood-filled arteries.

Increased afterload is found in several pathological situations, including elevated arterial blood pressure and loss of stretchability (*compliance*) in the aorta. To maintain constant stroke volume when afterload increases, the ventricle must increase its force of contraction. This in turn then increases the heart muscle's need for oxygen and ATP production. If increased afterload becomes a chronic situation, the myocardial cells hypertrophy, resulting in increased thickness of the ventricular wall.

Clinically, arterial blood pressure is often used as an indirect indicator of afterload. Other aspects of ventricular function can be assessed noninvasively by echocardiography, an ultrasound procedure in which sound waves are reflected off heart tissue. A common functional index derived from this procedure is the **ejection fraction,** or the percentage of EDV ejected with one contraction (stroke volume/EDV). Using our standard values for the 70-kg man, ejection fraction at rest is 70 mL/135 mL, or 52%. If stroke volume increases to 100 mL with exercise, the ejection fraction increases to 74%.

CONCEPT CHECK   **33.** A person's aortic valve opening has become constricted, creating a condition known as *aortic stenosis.* Which ventricle is affected by this change? What happens to the afterload on this ventricle?

The factors that determine cardiac output are summarized in **FIGURE 14.22.** Cardiac output varies with both heart rate and stroke volume. Heart rate is modulated by the autonomic division of the nervous system and by epinephrine. Stroke volume is a function of the intrinsic length-tension relationship of the Frank-Starling law, as indicated by the end-diastolic volume plus adrenergically mediated changes in contractility. Venous return is a major determinant of EDV and stretch.

The heart is a complex organ with many parts that can malfunction. Next, we examine how cardiac output plays a key role in blood flow through the circulation. You will learn about high blood pressure and atherosclerosis, and how these conditions can cause the heart to fail in its role as a pump.

FIG. 14.22 Stroke volume and heart rate determine cardiac output

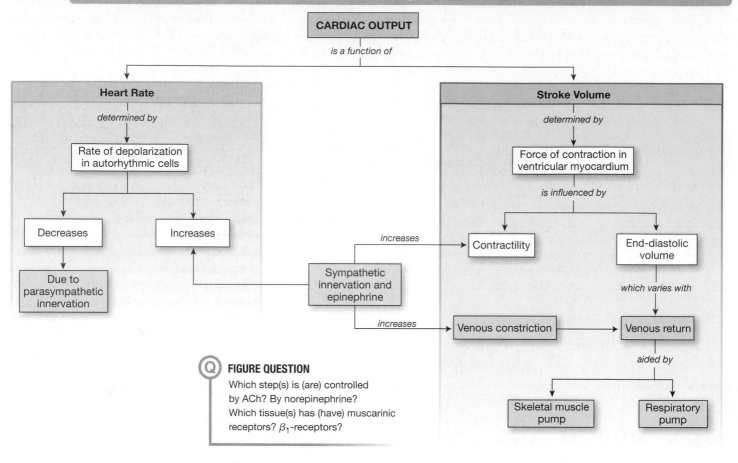

Q **FIGURE QUESTION**

Which step(s) is (are) controlled by ACh? By norepinephrine? Which tissue(s) has (have) muscarinic receptors? β_1-receptors?

RUNNING **PROBLEM** CONCLUSION | Myocardial Infarction

Walter's angiogram showed two small arteries blocked by cholesterol deposits. Blocked arteries like these can be treated either with angioplasty or coronary bypass surgery. In angioplasty, a balloon attached to a tube is passed into the coronary artery and inflated to open the blockage. A small mesh tube called a *stent* is left inside the artery to help keep it from closing up again. In coronary bypass surgery, veins from other parts of the body are grafted onto the heart arteries to provide bypass channels around blocked regions.

Walter's blockages were opened by balloon angioplasty. He returned home with instructions from his doctor for modifying his lifestyle to include a better diet, regular exercise, and no cigarette smoking.

In this running problem, you learned about some current techniques for diagnosing and treating heart attacks. Walter's symptoms are the classic ones, but many women have symptoms that are different. For more information on heart attack symptoms and other cardiovascular diseases, visit *www.americanheart.org,* the American Heart Association web site. Check your understanding of this physiology by comparing your answers with the information in the summary table.

Question	Facts	Integration and Analysis
Q1: *Why did the EMT give Walter oxygen and nitroglycerin?*	In a heart attack, blood flow and oxygen supply to the heart muscle may be blocked. If the heart is not pumping effectively, the brain may not receive adequate oxygen.	Administration of oxygen increases the amount of oxygen that reaches both the heart and the brain. Nitroglycerin is metabolized to nitric oxide, which dilates blood vessels and improves blood flow.
Q2: *What effect would the injection of isotonic saline have on Walter's extracellular fluid volume? On his intracellular fluid volume? On his total body osmolarity?*	An isotonic solution is one that does not change cell volume [p. 127]. Isotonic saline (NaCl) is isosmotic to the body.	The extracellular volume will increase because all of the saline administered will remain in that compartment. Intracellular volume and total body osmolarity will not change.

RUNNING **PROBLEM CONCLUSION** | *Continued*

Question	Facts	Integration and Analysis
Q3: *A related form of creatine kinase is found in skeletal muscle. What are related forms of an enzyme called?*	Related forms of an enzyme are called isozymes.	Although isozymes are variants of the same enzymes, their activity may vary under different conditions, and their structures are slightly different. Cardiac and skeletal muscle isozymes can be distinguished by their different structures.
Q4: *What is troponin, and why would elevated blood levels of troponin indicate heart damage?*	Troponin is the regulatory protein bound to tropomyosin [p. 386]. Ca^{2+} binding to troponin uncovers the myosin-binding site of actin to allow contraction.	Troponin is part of the contractile apparatus of the muscle cell. If troponin escapes from the cell and enters the blood, this is an indication that the cell either has been damaged or is dead.
Q5: *How do electrical signals move from cell to cell in the myocardium?*	Electrical signals pass through gap junctions in intercalated disks [p. 74].	The cells of the heart are electrically linked by gap junctions.
Q6: *What happens to contraction in a myocardial contractile cell if a wave of depolarization passing through the heart bypasses it?*	Depolarization in a muscle cell is the signal for contraction.	If a myocardial cell is not depolarized, it will not contract. Failure to contract creates a nonfunctioning region of heart muscle and impairs the pumping function of the heart.
Q7: *If the ventricle of the heart is damaged, in which wave or waves of the electrocardiogram would you expect to see abnormal changes?*	The P wave represents atrial depolarization. The QRS complex and T wave represent ventricular depolarization and repolarization, respectively.	The QRS complex and the T wave are most likely to show changes after a heart attack. Changes indicative of myocardial damage include enlargement of the Q wave, shifting of the S-T segment off the baseline (elevated or depressed), and inversion of the T wave.
Q8: *What will a beta blocker do to Walter's heart rate? Why is that response helpful following a heart attack?*	A beta blocker is an antagonist to β_1-adrenergic receptors. Activation of β_1-receptors increases heart rate.	A beta blocker therefore decreases heart rate and lowers oxygen demand. Cells that need less oxygen are less likely to die if their blood supply is diminished.
Q9: *If Walter's heart attack has damaged the muscle of his left ventricle, what do you predict will happen to his left cardiac output?*	Cardiac output equals stroke volume times heart rate.	If the ventricular myocardium has been weakened, stroke volume may decrease. Decreased stroke volume in turn decreases cardiac output.

436 439 449 455 466 468 **472**

VISUALIZE, PRACTICE, AND MASTER PHYSIOLOGY WITH MASTERING A&P

- Practice-on-the-go with Dynamic Study Modules
- Visualize important processes with Interactive Physiology and A&P Flix
- Check your understanding with Video Tutors and practice quizzes!

MasteringA&P®

CHAPTER SUMMARY

The cardiovascular system demonstrates many of the basic themes in physiology. Blood flows through vessels as a result of high pressure created during ventricular contraction (*mass flow*). The circulation of blood provides an essential route for *cell-to-cell communication*, particularly for hormones and other chemical signals. Myocardial contraction, like contraction in skeletal and smooth muscle, demonstrates the importance of *molecular interactions, biological energy use,* and the *mechanical properties* of cells and tissues. This chapter also introduced the *control systems* for cardiovascular physiology, a theme that will be expanded in the next chapter.

Overview of the Cardiovascular System

iP Cardiovascular—Anatomy Review: The Heart

1. The human **cardiovascular system** consists of a **heart** that pumps **blood** through a closed system of **blood vessels.** (p. 436; Fig. 14.1)

2. The primary function of the cardiovascular system is the transport of nutrients, water, gases, wastes, and chemical signals to and from all parts of the body. (p. 437; Tbl. 14.1)

3. Blood vessels that carry blood away from the heart are called **arteries.** Blood vessels that return blood to the heart are called **veins. Valves** in the heart and veins ensure unidirectional blood flow. (p. 437; Fig. 14.1)

4. The heart has four chambers: two **atria** and two **ventricles.** (p. 437; Fig. 14.1)

5. The **pulmonary circulation** goes from the right side of the heart to the lungs and back to the heart. The **systemic circulation** goes from the left side of the heart to the tissues and back to the heart. (p. 438; Fig. 14.1)

Pressure, Volume, Flow, and Resistance

6. Blood flows down a **pressure gradient (ΔP),** from the highest pressure in the **aorta** and arteries to the lowest pressure in the **venae cavae** and **pulmonary veins.** (pp. 438, 439; Fig. 14.2)

7. In a system in which fluid is flowing, pressure decreases over distance. (p. 439; Fig. 14.3)

8. The pressure created when the ventricles contract is called the **driving pressure** for blood flow. (p. 440)

9. **Resistance** of a fluid flowing through a tube increases as the length of the tube and the **viscosity** (thickness) of the fluid increase, and as the radius of the tube decreases. Of these three factors, radius has the greatest effect on resistance. (p. 440)

10. If resistance increases, flow rate decreases. If resistance decreases, flow rate increases. (p. 440; Fig. 14.3)

11. Fluid flow through a tube is proportional to the pressure gradient **(ΔP).** A pressure gradient is not the same thing as the absolute pressure in the system. (p. 439; Fig. 14.3)

12. **Flow rate** is the volume of blood that passes one point in the system per unit time. (p. 442)

13. **Velocity** is the distance a volume of blood travels in a given period of time. At a constant flow rate, the velocity of flow through a small tube is faster than the velocity through a larger tube. (p. 442; Fig. 14.4)

Cardiac Muscle and the Heart

iP Cardiovascular: Cardiac Action Potential

14. The heart is composed mostly of cardiac muscle, or **myocardium.** Most cardiac muscle is typical striated muscle. (p. 446; Fig. 14.5h)

15. The signal for contraction originates in **autorhythmic cells** in the heart. Autorhythmic cells are noncontractile myocardium. (p. 447)

16. Myocardial cells are linked to one another by **intercalated disks** that contain gap junctions. The junctions allow depolarization to spread rapidly from cell to cell. (p. 447; Fig. 14.8)

17. In contractile cell excitation-contraction coupling, an action potential opens Ca^{2+} channels. Ca^{2+} entry into the cell triggers the release of additional Ca^{2+} from the sarcoplasmic reticulum through **calcium-induced calcium release.** (p. 449; Fig. 14.9)

18. The force of cardiac muscle contraction can be graded according to how much Ca^{2+} enters the cell. (p. 450)

19. The action potentials of myocardial contractile cells have a rapid depolarization phase created by Na^+ influx, and a steep repolarization phase due to K^+ efflux. The action potential also has a plateau phase created by Ca^{2+} influx. (p. 451; Fig. 14.10)

20. Autorhythmic myocardial cells have an unstable membrane potential called a **pacemaker potential.** The pacemaker potential is due to I_f **channels** that allow net influx of positive charge. (p. 453; Fig. 14.12)

21. The steep depolarization phase of the autorhythmic cell action potential is caused by Ca^{2+} influx. The repolarization phase is due to K^+ efflux. (p. 453; Fig. 14.12)

The Heart as a Pump

iP Cardiovascular: Intrinsic Conduction System

22. Action potentials originate at the **sinoatrial node (SA node)** and spread rapidly from cell to cell in the heart. Action potentials are followed by a wave of contraction. (p. 454; Fig. 14.14)

23. The electrical signal moves from the SA node through the **internodal pathway** to the **atrioventricular node (AV node),** then into the **AV bundle, bundle branches,** terminal **Purkinje fibers,** and myocardial contractile cells. (p. 454; Fig. 14.14)

24. The SA node sets the pace of the heartbeat. If the SA node malfunctions, other autorhythmic cells in the AV node or ventricles take control of heart rate. (p. 457)

iP2 Electrical Activity of the Heart

25. An **electrocardiogram** (ECG) is a surface recording of the electrical activity of the heart. The **P wave** represents atrial depolarization. The **QRS complex** represents ventricular depolarization. The **T wave** represents ventricular repolarization. Atrial repolarization is incorporated in the QRS complex. (p. 457; Fig. 14.15)

26. An ECG provides information on heart rate and rhythm, conduction velocity, and the condition of cardiac tissues. (p. 460)

iP2 Cardiac Cycle

27. One **cardiac cycle** includes one cycle of contraction and relaxation. **Systole** is the contraction phase; **diastole** is the relaxation phase. (pp. 460, 461; Fig. 14.17)

28. Most blood enters the ventricles while the atria are relaxed. Only 20% of ventricular filling at rest is due to atrial contraction. (p. 462)

29. The **AV valves** prevent backflow of blood into the atria. Closure of the AV valves during ventricular contraction set up vibrations that create the **first heart sound.** (p. 462; Figs. 14.7, 14.18)

30. During **isovolumic ventricular contraction,** the ventricular blood volume does not change, but pressure rises. When ventricular pressure exceeds arterial pressure, the **semilunar valves** open, and blood is ejected into the arteries. (p. 462; Fig. 14.18)

31. When the ventricles relax and ventricular pressure falls, the semilunar valves close, creating the **second heart sound.** (p. 462; Fig. 14.18)

32. The amount of blood pumped by one ventricle during one contraction is known as the **stroke volume.** (p. 466)

iP2 Cardiac Output

33. **Cardiac output** is the volume of blood pumped per ventricle per unit time. It is equal to heart rate times stroke volume. The average cardiac output at rest is 5 L/min. (p. 466)

34. Homeostatic changes in cardiac output are accomplished by varying heart rate, stroke volume, or both. (p. 466; Fig. 14.22)

35. Parasympathetic activity slows heart rate; sympathetic activity speeds it up. Norepinephrine and epinephrine act on β_1-receptors to speed up the rate of the pacemaker depolarization. Acetylcholine activates muscarinic receptors to hyperpolarize the pacemakers. (p. 466; Fig. 14.19)

36. The longer a muscle fiber is when it begins to contract, the greater the force of contraction. The **Frank-Starling law of the heart** says that an increase in **end-diastolic volume (EDV)** results in a greater stroke volume. (pp. 464, 469; Fig. 14.20)

37. Epinephrine and norepinephrine increase the force of myocardial contraction when they bind to β_1-adrenergic receptors. They also shorten the duration of cardiac contraction. (p. 468; Fig. 14.21)

38. End-diastolic volume and **preload** are determined by **venous return.** Venous return is affected by skeletal muscle contractions, the respiratory pump, and constriction of veins by sympathetic activity. (pp. 468, 469)

39. **Contractility** of the heart is enhanced by catecholamines and certain drugs. Chemicals that alter contractility are said to have an **inotropic effect.** (pp. 468, 469; Fig. 14.20c)

40. **Afterload** is the load placed on the ventricle as it contracts. Afterload reflects the preload and the effort required to push the blood out into the arterial system. Mean arterial pressure is a clinical indicator of afterload. (p. 471)

41. **Ejection fraction,** the percent of EDV ejected with one contraction (stroke volume/EDV), is one measure for evaluating ventricular function. (p. 471)

REVIEW QUESTIONS

In addition to working through these questions and checking your answers on p. A-19, review the Learning Outcomes at the beginning of this chapter.

Level One Reviewing Facts and Terms

1. What contributions to understanding the cardiovascular system did each of the following people make?
   (a) William Harvey
   (b) Otto Frank and Ernest Starling
   (c) Marcello Malpighi

2. List three functions of the cardiovascular system.

3. Put the following structures in the order in which blood passes through them, starting and ending with the left ventricle:
   (a) left ventricle
   (b) systemic veins
   (c) pulmonary circulation
   (d) systemic arteries
   (e) aorta
   (f) right ventricle

4. The primary factor causing blood to flow through the body is a(n) _____ gradient. In humans, the value of this gradient is highest at the _____ and in the _____. It is lowest in the _____. In a system in which fluid is flowing, pressure decreases over distance because of _____.

5. If vasodilation occurs in a blood vessel, pressure (increases/decreases).

6. The specialized cell junctions between myocardial cells are called _____. These areas contain _____ that allow rapid conduction of electrical signals.

7. Trace an action potential from the SA node through the conducting system of the heart.

8. Distinguish between the two members of each of the following pairs:
   (a) end-systolic volume and end-diastolic volume
   (b) sympathetic and parasympathetic control of heart rate
   (c) diastole and systole
   (d) systemic and pulmonary circulation
   (e) AV node and SA node

9. Match the descriptions with the correct anatomic term(s). Not all terms are used and terms may be used more than once. Give a definition for the unused terms.

(a) tough membranous sac that encases the heart	1. aorta
	2. apex
(b) valve between ventricle and a main artery	3. artery
	4. atria
(c) a vessel that carries blood away from the heart	5. atrium
(d) lower chamber of the heart	6. AV valve
(e) valve between left atrium and left ventricle	7. base
	8. bicuspid valve
(f) primary artery of the systemic circulation	9. endothelium
	10. myocardium
(g) muscular layer of the heart	11. pericardium
(h) narrow end of the heart; points downward	12. semilunar valve
	13. tricuspid valve
(i) valve with papillary muscles	14. ventricle
(j) the upper chambers of the heart	

10. What events cause the two principal heart sounds?

11. What is the proper term for each of the following?
    (a) number of heart contractions per minute
    (b) volume of blood in the ventricle before the heart contracts
    (c) volume of blood that enters the aorta with each contraction
    (d) volume of blood that leaves the heart in 1 minute
    (e) volume of blood in the entire body

Level Two Reviewing Concepts

12. List the events of the cardiac cycle in sequence, beginning with atrial and ventricular diastole. Note when valves open and close. Describe what happens to pressure and blood flow in each chamber at each step of the cycle.

13. Concept maps:

    (a) Create a map showing blood flow through the heart and body. Label as many structures as you can.

    (b) Create a map for control of cardiac output using the following terms. You may add additional terms.

• ACh	• heart rate
• adrenal medulla	• length-tension relationship
• autorhythmic cells	• muscarinic receptor
• β_1-receptor	• norepinephrine
• Ca^{2+}	• parasympathetic neurons
• Ca^{2+}-induced Ca^{2+} release	• respiratory pump
• cardiac output	• skeletal muscle pump
• contractile myocardium	• stroke volume
• contractility	• sympathetic neurons
• force of contraction	• venous return

14. Compare and contrast the structure of a cardiac muscle cell with that of a skeletal muscle cell. What unique properties of cardiac muscle are essential to its function?

15. Explain why contractions in cardiac muscle cannot sum or exhibit tetanus.

16. Correlate the waves of an ECG with mechanical events in the atria and ventricles. Why are there only three electrical events but four mechanical events?

17. Match the following ion movements with the appropriate phrase. More than one ion movement may apply to a single phrase. Some choices may not be used.

(a) slow rising phase of auto-rhythmic cells	1. K^+ from ECF to ICF
(b) plateau phase of contractile cells	2. K^+ from ICF to ECF
	3. Na^+ from ECF to ICF
(c) rapid rising phase of contractile cells	4. Na^+ from ICF to ECF
(d) rapid rising phase of autorhythmic cells	5. Ca^{2+} from ECF to ICF
	6. Ca^{2+} from ICF to ECF
(e) rapid falling phase of contractile cells	
(f) falling phase of autorhythmic cells	
(g) cardiac muscle contraction	
(h) cardiac muscle relaxation	

18. List and briefly explain four types of information that an ECG provides about the heart.

19. Define inotropic effect. Name two drugs that have a positive inotropic effect on the heart.

Level Three Problem Solving

20. Two drugs used to reduce cardiac output are calcium channel blockers and beta (receptor) blockers. What effect do these drugs have on the heart that explains how they decrease cardiac output?

21. Police Captain Jeffers has suffered a myocardial infarction.

    (a) Explain to his (nonmedically oriented) family what has happened to his heart.

    (b) When you analyzed his ECG, you referred to several different leads, such as lead I and lead III. What are leads?

    (c) Why is it possible to record an ECG on the body surface without direct access to the heart?

22. What might cause a longer-than-normal PR interval in an ECG?

23. The following paragraph is a summary of a newspaper article:

    A new treatment for atrial fibrillation due to an excessively rapid rate at the SA node involves a high-voltage electrical pulse administered to the AV node to destroy its autorhythmic cells. A ventricular pacemaker is then implanted in the patient.

    Briefly explain the physiological rationale for this treatment. Why is a rapid atrial depolarization rate dangerous? Why is the AV node destroyed in this procedure? Why must a pacemaker be implanted?

Level Four Quantitative Problems

24. Police Captain Jeffers in question 21 has an ejection fraction (SV divided by EDV) of only 25%. His stroke volume is 40 mL/beat, and his heart rate is 100 beats/min. What are his EDV, ESV, and CO? Show your calculations.

25. If 1 cm water = 0.74 mm Hg:

    (a) Convert a pressure of 120 mm Hg to cm H_2O.

    (b) Convert a pressure of 90 cm H_2O to mm Hg.

26. Calculate cardiac output if stroke volume is 65 mL/beat and heart rate is 80 beats/min.

27. Calculate end-systolic volume if end-diastolic volume is 150 mL and stroke volume is 65 mL/beat.

28. A person has a total blood volume of 5 L. Of this total, assume that 4 L is contained in the systemic circulation and 1 L is in the pulmonary circulation. If the person has a cardiac output of 5 L/min, how long will it take (a) for a drop of blood leaving the left ventricle to return to the left ventricle and (b) for a drop of blood to go from the right ventricle to the left ventricle?

15

Blood Flow and the Control of Blood Pressure

Since 1900, CVD (cardiovascular disease) has been the No. 1 killer in the United States every year but 1918.

American Heart Association, *Heart Disease and Stroke Statistics—2006 Update, A Report From the American Heart Association Statistics Committee and Stroke Statistics Subcommittee*

THE BLOOD VESSELS 479

LO 15.1 Compare and contrast the structure, mechanical properties, and functions of the five major types of blood vessels.

BLOOD PRESSURE 482

LO 15.2 Explain what creates blood pressure and how blood pressure changes as blood flows through the systemic circulation.

LO 15.3 Explain the relationship between blood flow, pressure gradients, and the resistance of the system to flow. Use Poiseuille's Law to explain the factors that influence resistance.

LO 15.4 Describe how blood pressure is estimated using sphygmomanometry.

LO 15.5 Explain the contributions of cardiac output and peripheral resistance to blood pressure. Calculate mean arterial pressure.

LO 15.6 Explain how changes in blood volume affect blood pressure.

RESISTANCE IN THE ARTERIOLES 487

LO 15.7 Define myogenic autoregulation and explain its role in altering local blood flow.

LO 15.8 List and describe the major paracrine molecules involved in local control of blood flow.

LO 15.9 Describe the hormonal and neural control of blood vessel diameter, including significant neurotransmitters and their receptor types.

DISTRIBUTION OF BLOOD TO THE TISSUES 492

LO 15.10 Explain how the body can use local and long-distance signaling to direct blood flow to or away from specific organs or tissues.

REGULATION OF CARDIOVASCULAR FUNCTION 492

LO 15.11 Describe in detail the steps of the baroreceptor reflex, including the stimulus, sensor, input pathway, integrating center(s), output pathways, target(s), cellular response(s), tissue response(s), and systemic response(s). Include all chemical signal molecules and their receptors as well as any feedback loops.

EXCHANGE AT THE CAPILLARIES 496

LO 15.12 Describe the different types of capillaries and where they are found in the body.

LO 15.13 Explain why the velocity of blood flow is lowest in the capillaries.

LO 15.14 Explain the role of diffusion and transcytosis in capillary exchange.

LO 15.15 Explain the forces that influence capillary filtration and absorption.

THE LYMPHATIC SYSTEM 498

LO 15.16 Describe the anatomy and functions of the lymphatic system and how the lymphatics are related to the circulatory and immune systems.

LO 15.17 Explain the pathological factors that might alter capillary exchange and result in edema.

CARDIOVASCULAR DISEASE 501

LO 15.18 List the controllable and uncontrollable risk factors for cardiovascular disease.

LO 15.19 Describe the progression of events that result in atherosclerosis.

LO 15.20 Explain why hypertension represents a failure of homeostasis.

BACKGROUND BASICS

76 Basal lamina
178 Nitric oxide
152 Transcytosis
359 Fight-or-flight response
76 Exchange epithelium
206 Catecholamines
148 Caveolae
35 Diffusion
183 Tonic control
403 Smooth muscle

Blood vessels of the small intestine

nthony was sure he was going to be a physician, until the day in physiology laboratory they studied blood types. When the lancet pierced his fingertip and he saw the drop of bright red blood well up, the room started to spin, and then everything went black. He awoke, much embarrassed, to the sight of his classmates and the teacher bending over him.

Anthony suffered an attack of *vasovagal syncope* (syncope = fainting), a benign and common emotional reaction to blood, hypodermic needles, or other upsetting sights. Normally, homeostatic regulation of the cardiovascular system maintains blood flow, or *perfusion,* to the heart and brain. In vasovagal syncope, signals from the nervous system cause a sudden decrease in blood pressure, and the individual faints from lack of oxygen to the brain. In this chapter you will learn how the heart and blood vessels work together most of the time to prevent such problems.

A simplified model of the cardiovascular system (**FIG. 15.1**) illustrates the key points we discuss in this chapter. This model

RUNNING **PROBLEM** | **Essential Hypertension**

"Doc, I'm as healthy as a horse," says Kurt English, age 56, during his long-overdue annual physical examination. "I don't want to waste your time. Let's get this over with." But to Dr. Cortez, Kurt does not appear to be the picture of health: he is about 30 pounds overweight. When Dr. Cortez asks about his diet, Kurt replies, "Well, I like to eat." Exercise? "Who has the time?" replies Kurt. Dr. Cortez wraps a blood pressure cuff around Kurt's arm and takes a reading. "Your blood pressure is 164 over 100," says Dr. Cortez. "We'll take it again in 15 minutes. If it's still high, we'll need to discuss it further." Kurt stares at his doctor, flabbergasted. "But how can my blood pressure be too high? I feel fine!" he protests.

478 484 489 490 500 505

FIG. 15.1 **A functional model of the cardiovascular system**

This functional model of the cardiovascular system shows the heart and blood vessels as a single closed loop.

The elastic systemic arteries are a pressure reservoir that maintains blood flow during ventricular relaxation.

The arterioles, shown with adjustable screws that alter their diameter, are the site of variable resistance.

Each side of the heart functions as an independent pump.

Exchange between the blood and cells takes place at the capillaries.

Systemic veins serve as an expandable volume reservoir.

Aorta
Aortic valve
Left heart
Left ventricle
Mitral valve
Left atrium
Pulmonary vein
Lungs
Capillaries
Pulmonary artery
Pulmonary valve
Right ventricle
Right heart
Tricuspid valve
Right atrium
Venae cavae
Venules

Q **FIGURE QUESTION**
Are pumps in this model operating in parallel or in series?

shows the heart as two separate pumps, with the right heart pumping blood to the lungs and back to the left heart. The left heart then pumps blood through the rest of the body and back to the right heart.

Blood leaving the left heart enters systemic arteries, shown here as an expandable, elastic region. Pressure produced by contraction of the left ventricle is stored in the elastic walls of arteries and slowly released through *elastic recoil*. This mechanism maintains a continuous driving pressure for blood flow during ventricular relaxation. For this reason, the arteries are known as the *pressure reservoir* {*reservare*, to retain} of the circulatory system.

Downstream from the arteries, small vessels called **arterioles** create a high-resistance outlet for arterial blood flow. Arterioles direct distribution of blood flow to individual tissues by selectively constricting and dilating, so they are known as the site of *variable resistance*. Arteriolar diameter is regulated both by local factors, such as tissue oxygen concentrations, and by the autonomic nervous system and hormones.

When blood flows into the capillaries, their leaky epithelium allows exchange of materials between the plasma, the interstitial fluid, and the cells of the body. At the distal end of the capillaries, blood flows into the venous side of the circulation. The veins act as a *volume reservoir* from which blood can be sent to the arterial side of the circulation if blood pressure falls too low. From the veins, blood flows back to the right heart.

Total blood flow through any level of the circulation is equal to cardiac output. For example, if cardiac output is 5 L/min, blood flow through all the systemic capillaries is 5 L/min. In the same manner, blood flow through the pulmonary side of the circulation is equal to blood flow through the systemic circulation.

THE BLOOD VESSELS

The walls of blood vessels are composed of layers of smooth muscle, elastic connective tissue, and fibrous connective tissue (**FIG. 15.2**). The inner lining of all blood vessels is a thin layer of **endothelium,** a type of epithelium. For years, the endothelium was thought to be simply a passive barrier. However, we now know that endothelial cells secrete many paracrine signals and play important roles in the regulation of blood pressure, blood vessel growth, and absorption of materials. Some scientists have even proposed that endothelium be considered a separate physiological organ system.

In most vessels, layers of connective tissue and smooth muscle surround the endothelium. The endothelium and its adjacent elastic connective tissue together make up the *tunica*

FIG. 15.2   Blood vessel structure

The walls of blood vessels vary in diameter and composition. The bars show the relative proportions of the different tissues. The endothelium and its underlying elastic tissue together form the tunica intima. (Based on A.C. Burton, *Physiol Rev* 34: 619–642, 1954.)

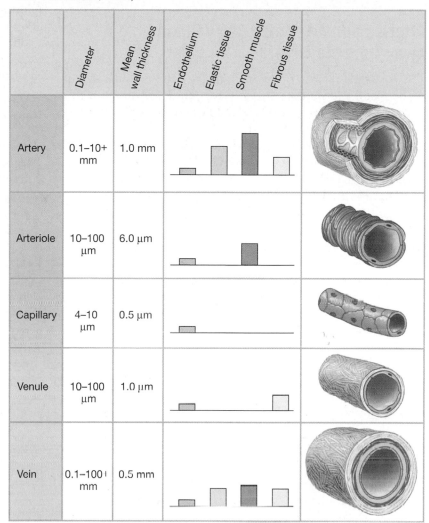

	Diameter	Mean wall thickness	Endothelium / Elastic tissue / Smooth muscle / Fibrous tissue	
Artery	0.1–10+ mm	1.0 mm		
Arteriole	10–100 µm	6.0 µm		
Capillary	4–10 µm	0.5 µm		
Venule	10–100 µm	1.0 µm		
Vein	0.1–100 mm	0.5 mm		

intima, usually called simply the *intima* {*intimus,* innermost}. The thickness of the smooth muscle and connective tissue layers varies in different vessels. The descriptions that follow apply to the vessels of the systemic circulation, although those of the pulmonary circulation are generally similar.

Blood Vessels Contain Vascular Smooth Muscle

The smooth muscle of blood vessels is known as **vascular smooth muscle.** Most blood vessels contain smooth muscle, arranged in either circular or spiral layers. *Vasoconstriction* narrows the diameter of the vessel lumen, and *vasodilation* widens it.

In most blood vessels, smooth muscle cells maintain a state of partial contraction at all times, creating the condition known as *muscle tone* [p. 421]. Contraction of smooth muscle, like that of cardiac muscle, depends on the entry of Ca^{2+} from the extracellular fluid through Ca^{2+} channels [p. 407]. Signal molecules, including neurotransmitters, hormones, and paracrine signals, influence vascular smooth muscle tone. Many vasoactive paracrine molecules are secreted either by endothelial cells lining blood vessels or by tissues surrounding the vessels.

Arteries and Arterioles Carry Blood Away from the Heart

The aorta and major arteries are characterized by walls that are both stiff and springy. Arteries have a thick smooth muscle layer and large amounts of elastic and fibrous connective tissue (Fig. 15.2). Because of the stiffness of the fibrous tissue, substantial amounts of energy are required to stretch the walls of an artery outward, but that energy can be stored by the stretched elastic fibers and released through elastic recoil.

The arteries and arterioles are characterized by a divergent {*divergere*, bend apart} pattern of blood flow. As major arteries divide into smaller and smaller arteries, the character of the wall changes, becoming less elastic and more muscular. The walls of arterioles contain several layers of smooth muscle that contract and relax under the influence of various chemical signals.

Arterioles, along with capillaries and small postcapillary vessels called venules, form the *microcirculation*. Regulation of blood flow through the microcirculation is an active area of physiological research.

Some arterioles branch into vessels known as **metarterioles** {*meta-*, beyond} (**FIG. 15.3**). True arterioles have a continuous smooth muscle layer, but the wall of a metarteriole is only partially surrounded by smooth muscle. Blood flowing through metarterioles can take one of two paths. If muscle rings called **precapillary sphincters** {*sphingein*, to hold tight} are relaxed, blood flowing into a metarteriole is directed into adjoining capillary beds (Fig. 15.3b). If the precapillary sphincters are all constricted, metarteriole blood bypasses the capillaries and goes directly to the venous circulation (Fig. 15.3c).

Exchange Takes Place in the Capillaries

Capillaries are the smallest vessels in the cardiovascular system. They are the primary site of exchange between the blood and the interstitial fluid. A small amount of exchange takes place in the postcapillary venules at the distal ends of the capillaries, but this is insignificant.

To facilitate exchange of materials, capillaries lack smooth muscle and elastic or fibrous tissue reinforcement (Fig. 15.2). Instead, their walls consist of a flat layer of endothelium, one cell thick, supported on an acellular matrix called the *basal lamina* (basement membrane) [p. 76].

Many capillaries are closely associated with cells known as **pericytes** {*peri-*, around}. In most tissues, these highly branched contractile cells surround the capillaries, forming a mesh-like

FIG. 15.3 Capillary beds

(a) The microcirculation

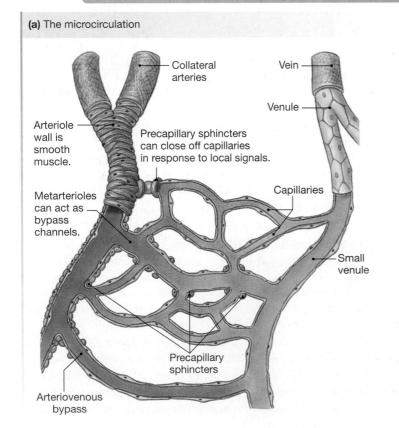

- Collateral arteries
- Vein
- Venule
- Arteriole wall is smooth muscle.
- Precapillary sphincters can close off capillaries in response to local signals.
- Metarterioles can act as bypass channels.
- Capillaries
- Small venule
- Precapillary sphincters
- Arteriovenous bypass

(b) When precapillary sphincters are relaxed, blood flows through all capillaries in the bed.

- Arteriole
- Venule
- Precapillary sphincters relaxed
- Capillaries

(c) If precapillary sphincters constrict, blood flow bypasses capillaries completely and flows through metarterioles.

- Precapillary sphincters constricted

outer layer between the capillary endothelium and the interstitial fluid. Pericytes contribute to the "tightness" of capillary permeability: the more pericytes, the less leaky the capillary endothelium. Cerebral capillaries, for example, are surrounded by pericytes and glial cells, and have tight junctions that create the *blood-brain barrier* [p. 282].

Pericytes secrete factors that influence capillary growth, and they can differentiate to become new endothelial or smooth muscle cells. Loss of pericytes around capillaries of the retina is a hallmark of the disease *diabetic retinopathy,* a leading cause of blindness. Scientists are now trying to determine whether pericyte loss is a cause or consequence of the retinopathy.

Blood Flow Converges in the Venules and Veins

Blood flows from the capillaries into small vessels called **venules.** The smallest venules are similar to capillaries, with a thin exchange epithelium and little connective tissue (Fig. 15.2). They are distinguished from capillaries by their convergent pattern of flow.

Smooth muscle begins to appear in the walls of larger venules. From venules, blood flows into veins that become larger in diameter as they travel toward the heart. Finally, the largest veins, the venae cavae, empty into the right atrium. To assist venous flow, some veins have internal one-way valves (**FIG. 15.4**). These valves, like those in the heart, ensure that blood passing the valve cannot flow backward. Once blood reaches the vena cava, there are no valves.

Veins are more numerous than arteries and have a larger diameter. As a result of their large volume, the veins hold more than half of the blood in the circulatory system, making them the *volume reservoir* of the circulatory system. Veins lie closer to the surface of the body than arteries, forming the bluish blood vessels that you see running just under the skin. Veins have thinner walls than arteries, with less elastic tissue. As a result, they expand easily when they fill with blood.

When you have blood drawn from your arm (*venipuncture*), the technician uses a tourniquet to exert pressure on the blood vessels. Blood flow into the arm through deep high-pressure arteries is not affected, but pressure exerted by the tourniquet stops outflow through the low-pressure veins. As a result, blood collects in the surface veins, making them stand out against the underlying muscle tissue.

Angiogenesis Creates New Blood Vessels

One topic of great interest to researchers is **angiogenesis** {*angeion,* vessel + *gignesthai,* to beget}, the process by which new blood vessels develop, especially after birth. In children, blood vessel growth is necessary for normal development. In adults, angiogenesis takes place as wounds heal and as the uterine lining grows after menstruation. Angiogenesis also occurs with endurance exercise training, enhancing blood flow to the heart muscle and to skeletal muscles.

FIG. 15.4 **Valves ensure one-way flow in veins**

Valves in the veins prevent backflow of blood.

Valve closed

When the skeletal muscles compress the veins, they force blood toward the heart (the skeletal muscle pump).

Valve opened

The growth of malignant tumors is a disease state that requires angiogenesis. As cancer cells invade tissues and multiply, they instruct the host tissue to develop new blood vessels to feed the growing tumor. Without these new vessels, the interior cells of a cancerous mass would be unable to get adequate oxygen and nutrients, and would die.

From studies of normal blood vessels and tumor cells, scientists learned that angiogenesis is controlled by a balance of angiogenic and antiangiogenic cytokines. A number of related growth factors, including *vascular endothelial growth factor* (VEGF) and *fibroblast growth factor* (FGF), promote angiogenesis. These growth factors are *mitogens,* meaning they promote mitosis, or cell division. They are normally produced by smooth muscle cells and pericytes.

Cytokines that inhibit angiogenesis include *angiostatin,* made from the blood protein plasminogen, and *endostatin* {*stasis,* a state of standing still}. Scientists are currently testing antiangiogenic cytokines and inhibitors of angiogenic cytokines as cancer treatments, to see if they can stop angiogenesis and literally starve tumors to death.

In contrast, **coronary heart disease,** also known as *coronary artery disease,* is a condition in which blood flow to the myocardium is decreased by fatty deposits that narrow the lumen of the coronary arteries. In some individuals, new blood vessels develop spontaneously and form *collateral circulation* that supplements flow through the partially blocked artery. Researchers are testing angiogenic cytokines to see if they can duplicate this natural process and induce angiogenesis to replace *occluded* vessels {*occludere*, to close up}.

BLOOD PRESSURE

Ventricular contraction is the force that creates blood flow through the cardiovascular system [p. 436]. As blood under pressure is ejected from the left ventricle, the aorta and arteries expand to accommodate it (**FIG. 15.5a**). When the ventricle relaxes and the aortic valve closes, the elastic arterial walls recoil, propelling the blood forward into smaller arteries and arterioles (Fig. 15.5b). By sustaining the *driving pressure* for blood flow during ventricular relaxation, the arteries keep blood flowing continuously through the blood vessels.

Blood flow obeys the rules of fluid flow [p. 440]. Flow is directly proportional to the pressure gradient between any two points, and inversely proportional to the resistance of the vessels to flow (**TBL. 15.1**). Unless otherwise noted, the discussion that follows is restricted to the events that take place in the systemic circuit. You will learn about pulmonary blood flow when you study the respiratory system.

TABLE 15.1	**Pressure, Flow, and Resistance in the Cardiovascular System**

Flow ∝ ΔP/R

1. Blood flows if a pressure gradient (ΔP) is present.

2. Blood flows from areas of higher pressure to areas of lower pressure.

3. Blood flow is opposed by the resistance (R) of the system.

4. Three factors affect resistance: radius of the blood vessels, length of the blood vessels, and viscosity of the blood. [p. 440]

5. Flow is usually expressed in either liters or milliliters per minute (L/min or mL/min).

6. Velocity of flow is usually expressed in either centimeters per minute (cm/min) or millimeters per second (mm/sec).

7. The primary determinant of velocity (when flow rate is constant) is the total cross-sectional area of the vessel(s). [p. 442]

Blood Pressure Is Highest in Arteries and Lowest in Veins

Blood pressure is highest in the arteries and decreases continuously as blood flows through the circulatory system (**FIG. 15.6**). The decrease in pressure occurs because energy is lost as a result

---

FIG. 15.5 **Arteries are a pressure reservoir**

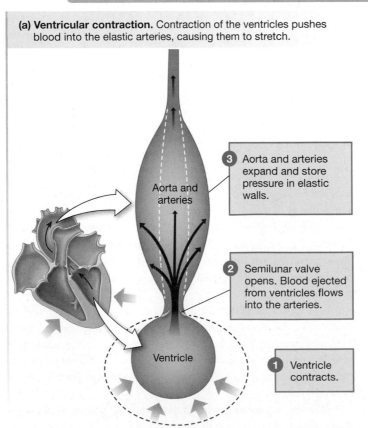

(a) Ventricular contraction. Contraction of the ventricles pushes blood into the elastic arteries, causing them to stretch.

Aorta and arteries

3 Aorta and arteries expand and store pressure in elastic walls.

2 Semilunar valve opens. Blood ejected from ventricles flows into the arteries.

Ventricle

1 Ventricle contracts.

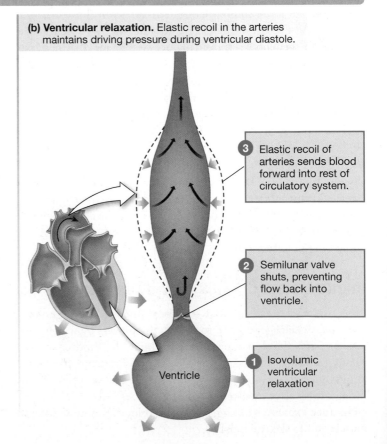

(b) Ventricular relaxation. Elastic recoil in the arteries maintains driving pressure during ventricular diastole.

3 Elastic recoil of arteries sends blood forward into rest of circulatory system.

2 Semilunar valve shuts, preventing flow back into ventricle.

Ventricle

1 Isovolumic ventricular relaxation

FIG. 15.6 Systemic circulation pressures

Pressure waves created by ventricular contraction travel into the blood vessels. Pressure in the arterial side of the circulation cycles but the pressure waves diminish in amplitude with distance and disappear at the capillaries.

Pulse pressure = systolic pressure – diastolic pressure

Mean arterial pressure = diastolic pressure + 1/3 (pulse pressure)

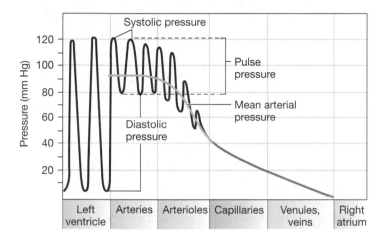

By the time blood reaches the veins, pressure has decreased because of friction, and a pressure wave no longer exists. Venous blood flow is steady rather than *pulsatile* (in pulses), pushed along by the continuous movement of blood out of the capillaries.

Low-pressure blood in veins below the heart must flow "uphill," or against gravity, to return to the heart. Try holding your arm straight down without moving for several minutes and notice how the veins in the back of your hand begin to stand out as they fill with blood. (This effect may be more evident in older people, whose subcutaneous connective tissue has lost elasticity). Then raise your hand so that gravity assists the venous flow and watch the bulging veins disappear.

Blood return to the heart, known as *venous return*, is aided by valves, the *skeletal muscle pump*, and the *respiratory pump* [p. 469]. When muscles such as those in the calf of the leg contract, they compress the veins, which forces blood upward past the valves. While your hand is hanging down, try clenching and unclenching your fist to see the effect muscle contraction has on distention of the veins.

CONCEPT CHECK

1. Would you expect to find valves in the veins leading from the brain to the heart? Defend your answer.

2. If you check the pulse in a person's carotid artery and left wrist at the same time, would the pressure waves occur simultaneously? Explain.

3. Who has the higher pulse pressure, someone with blood pressure of 90/60 or someone with blood pressure of 130/95?

of the resistance to flow offered by the vessels. Resistance to blood flow also results from friction between the blood cells.

In the systemic circulation, the highest pressure occurs in the aorta and results from pressure created by the left ventricle. Aortic pressure reaches an average high of 120 mm Hg during ventricular systole (**systolic pressure**), then falls steadily to a low of 80 mm Hg during ventricular diastole (**diastolic pressure**). Notice that pressure in the ventricle falls to only a few mm Hg as the ventricle relaxes, but diastolic pressure in the large arteries remains relatively high. The high diastolic pressure in arteries reflects the ability of those vessels to capture and store energy in their elastic walls.

The rapid pressure increase that occurs when the left ventricle pushes blood into the aorta can be felt as a **pulse,** or pressure wave, transmitted through the fluid-filled arteries. The pressure wave travels about 10 times faster than the blood itself. Even so, a pulse felt in the arm is occurring slightly after the ventricular contraction that created the wave.

The amplitude of the pressure wave decreases over distance because of friction, and the wave finally disappears at the capillaries (Fig. 15.6). **Pulse pressure,** a measure of the strength of the pressure wave, is defined as systolic pressure minus diastolic pressure:

Systolic pressure – diastolic pressure = pulse pressure (1)

For example, in the aorta:

120 mm Hg – 80 mm Hg = 40 mm Hg pressure (2)

Arterial Blood Pressure Reflects the Driving Pressure for Blood Flow

Arterial blood pressure, or simply "blood pressure," reflects the driving pressure created by the pumping action of the heart. Because ventricular pressure is difficult to measure, it is customary to assume that arterial blood pressure reflects ventricular pressure. As you just learned, arterial pressure is pulsatile, so we use a single value—the **mean arterial pressure (MAP)**—to represent driving pressure. MAP is represented graphically in Figure 15.6.

Mean arterial pressure is estimated as diastolic pressure plus one-third of pulse pressure:

MAP = diastolic P + 1/3 (systolic P – diastolic P) (3)

For a person whose systolic pressure is 120 and diastolic pressure is 80:

MAP = 80 mm Hg + 1/3 (120 – 80 mm Hg)
= 93 mm Hg (4)

Mean arterial pressure is closer to diastolic pressure than to systolic pressure because diastole lasts twice as long as systole.

Abnormally high or low arterial blood pressure can be indicative of a problem in the cardiovascular system. If blood pressure falls too low (*hypotension*), the driving force for blood flow is unable to overcome opposition by gravity. In this instance, blood flow and oxygen supply to the brain are impaired, and the person may become dizzy or faint.

On the other hand, if blood pressure is chronically elevated (a condition known as *hypertension,* or high blood pressure), high pressure on the walls of blood vessels may cause weakened areas to rupture and bleed into the tissues. If a rupture occurs in the brain, it is called a *cerebral hemorrhage* and may cause the loss of neurological function commonly called a *stroke*. If a weakened area ruptures in a major artery, such as the descending aorta, rapid blood loss into the abdominal cavity causes blood pressure to fall below the critical minimum. Without prompt treatment, rupture of a major artery is fatal.

CONCEPT CHECK

4. The formula given for calculating MAP applies to a typical resting heart rate of 60–80 beats/min. If heart rate increases, would the contribution of systolic pressure to mean arterial pressure decrease or increase, and would MAP decrease or increase?

5. Peter's systolic pressure is 112 mm Hg, and his diastolic pressure is 68 mm Hg (written 112/68). What is his pulse pressure? His mean arterial pressure?

Blood Pressure Is Estimated by Sphygmomanometry

We estimate arterial blood pressure in the radial artery of the arm using a *sphygmomanometer,* an instrument consisting of an inflatable cuff and a pressure gauge {*sphygmus*, pulse + *manometer*, an instrument for measuring pressure of a fluid}. The cuff encircles the upper arm and is inflated until it exerts pressure higher than the systolic pressure driving arterial blood. When cuff pressure exceeds arterial pressure, blood flow into the lower arm stops (**FIG. 15.7a**).

Now pressure on the cuff is gradually released. When cuff pressure falls below systolic arterial blood pressure, blood begins to flow again. As blood squeezes through the still-compressed

RUNNING **PROBLEM**

Kurt's second blood pressure reading is 158/98. Dr. Cortez asks him to take his blood pressure at home daily for two weeks and then return to the doctor's office. When Kurt comes back with his diary, the story is the same: his blood pressure continues to average 160/100. After running some tests, Dr. Cortez concludes that Kurt has high blood pressure or *hypertension,* like nearly one out of every three adult Americans. If not controlled, hypertension can lead to heart failure, stroke, and kidney failure.

Q1: *Why are people with high blood pressure at greater risk for having a hemorrhagic (or bleeding) stroke?*

478 — 484 — 489 — 490 — 500 — 505

artery, a thumping noise called a **Korotkoff sound** can be heard with each pressure wave (Fig. 15.7b). Korotkoff sounds are caused by the turbulent flow of blood through the compression. Once the cuff pressure no longer compresses the artery, flow smooths out and the sounds disappear (Fig. 15.7c).

The pressure at which a Korotkoff sound is first heard represents the highest pressure in the artery and is recorded as the systolic pressure. The point at which the Korotkoff sounds disappear is the lowest pressure in the artery and is recorded as the diastolic pressure. By convention, blood pressure is written as systolic pressure over diastolic pressure.

For years, the "average" value for blood pressure has been stated as 120/80. Like many average physiological values, however, these numbers are subject to wide variability, both from one person to another and within a single individual from moment to moment. A systolic pressure that is consistently over 140 mm Hg at rest, or a diastolic pressure that is chronically over 90 mm Hg, is considered a sign of hypertension in an otherwise healthy person.

Furthermore, the guidelines published in the 2003 JNC 7 Report* recommended that individuals maintain their blood pressure *below* 120/80. Persons whose systolic pressure is consistently in the range of 120–139 or whose diastolic pressure is in the range of 80–89 were considered to be prehypertensive and should be counseled on lifestyle modification strategies to reduce their blood pressure.

Cardiac Output and Peripheral Resistance Determine Mean Arterial Pressure

Mean arterial pressure is the driving force for blood flow, but what determines mean arterial pressure? Arterial pressure is a balance between blood flow into the arteries and blood flow out of the arteries. If flow in exceeds flow out, blood volume in the arteries increases, and mean arterial pressure increases. If flow out exceeds flow in, volume decreases and mean arterial pressure falls.

Blood flow into the aorta is equal to the cardiac output of the left ventricle. Blood flow out of the arteries is influenced primarily by **peripheral resistance,** defined as the resistance to flow offered by the arterioles (**FIG. 15.8a**). Mean arterial pressure (MAP) then is proportional to cardiac output (CO) times resistance (R) of the arterioles:

$$MAP \propto CO \times R_{arterioles} \qquad (5)$$

Let's consider how this works. If cardiac output increases, the heart pumps more blood into the arteries per unit time. If resistance to blood flow out of the arteries does not change, flow into the arteries is greater than flow out, blood volume in the arteries increases, and arterial blood pressure increases.

---

*Seventh Report of the Joint National Committee on Prevention, Detection, Evaluation, and Treatment of High Blood Pressure, National Institutes of Health. *www.nhlbi.nih.gov/files/docs/guidelines/express.pdf.* JNC 8 was published in early 2014 and its guidelines can be accessed at *JAMA* 311(5): 507–520, 2014.

FIG. 15.7 Sphygmomanometry

Arterial blood pressure is measured with a sphygmomanometer (an inflatable cuff plus a pressure gauge) and a stethoscope. The inflation pressure shown is for a person whose blood pressure is 120/80.

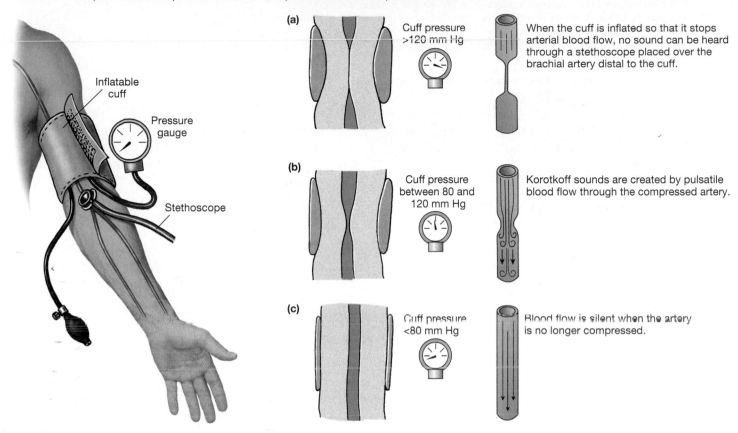

(a) Cuff pressure >120 mm Hg

When the cuff is inflated so that it stops arterial blood flow, no sound can be heard through a stethoscope placed over the brachial artery distal to the cuff.

Inflatable cuff

Pressure gauge

Stethoscope

(b) Cuff pressure between 80 and 120 mm Hg

Korotkoff sounds are created by pulsatile blood flow through the compressed artery.

(c) Cuff pressure <80 mm Hg

Blood flow is silent when the artery is no longer compressed.

In another example, suppose cardiac output remains unchanged but peripheral resistance increases. Flow into arteries is unchanged, but flow out is decreased. Blood again accumulates in the arteries, and the arterial pressure again increases. Most cases of hypertension are believed to be caused by increased peripheral resistance without changes in cardiac output.

Two additional factors can influence arterial blood pressure: the distribution of blood in the systemic circulation and the total blood volume. The relative distribution of blood between the arterial and venous sides of the circulation can be an important factor in maintaining arterial blood pressure. Arteries are low-volume vessels that usually contain only about 11% of total blood volume at any one time. Veins, in contrast, are high-volume vessels that hold about 60% of the circulating blood volume at any one time.

The veins act as a *volume reservoir* for the circulatory system, holding blood that can be redistributed to the arteries if needed. If arterial blood pressure falls, increased sympathetic activity constricts veins, decreasing their holding capacity. Venous return sends blood to the heart, which according to the Frank–Starling law of the heart, pumps all the venous return out to the systemic side of the circulation [p. 469]. Thus, constriction of the veins redistributes blood to the arterial side of the circulation and raises mean arterial pressure.

Changes in Blood Volume Affect Blood Pressure

Although the volume of the blood in the circulation is usually relatively constant, changes in blood volume can affect arterial blood pressure (Fig. 15.8b). If blood volume increases, blood pressure increases. When blood volume decreases, blood pressure decreases.

To understand the relationship between blood volume and pressure, think of the circulatory system as an elastic balloon filled with water. If only a small amount of water is in the balloon, little pressure is exerted on the walls, and the balloon is soft and flabby. As more water is added to the balloon, more pressure is exerted on the elastic walls. If you fill a balloon close to the bursting point, you risk popping the balloon. The best way to reduce this pressure is to remove some of the water.

Small increases in blood volume occur throughout the day due to ingestion of food and liquids, but these increases usually do not create long-lasting changes in blood pressure because of homeostatic compensations. Adjustments for increased blood volume are primarily the responsibility of the kidneys. If blood volume increases, the kidneys restore normal volume by excreting excess water in the urine (**FIG. 15.9**).

FIG. 15.8 ESSENTIALS

Mean Arterial Pressure

(a) Mean arterial pressure (MAP) is a function of cardiac output and resistance in the arterioles (peripheral resistance). MAP illustrates mass balance: the volume of blood in the arteries is determined by input (cardiac output) and flow out (altered by changing peripheral resistance). As arterial volume increases, pressure increases. In this model, the ventricle is represented by a syringe. The variable diameter of the arterioles is represented by adjustable screws.

Cardiac output   **Mean arterial pressure**   Variable resistance

Left ventricle

Elastic arteries

Arterioles

Mean arterial pressure α cardiac output × resistance

Q FIGURE QUESTIONS

1. If arterioles constrict, what happens to blood flow out of the arteries? What happens to MAP?
2. If cardiac output decreases, what happens to arterial blood volume? What happens to MAP?
3. If veins constrict, what happens to blood volume in the veins? What happens to volume in the arteries and to MAP?

(b) Factors that influence mean arterial pressure

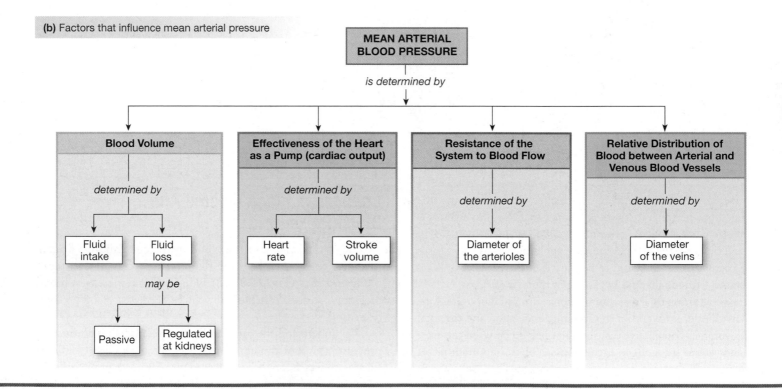

MEAN ARTERIAL BLOOD PRESSURE

is determined by

Blood Volume	Effectiveness of the Heart as a Pump (cardiac output)	Resistance of the System to Blood Flow	Relative Distribution of Blood between Arterial and Venous Blood Vessels
determined by	*determined by*	*determined by*	*determined by*
Fluid intake / Fluid loss	Heart rate / Stroke volume	Diameter of the arterioles	Diameter of the veins

may be → Passive / Regulated at kidneys

Compensation for decreased blood volume is more difficult and requires an integrated response from the kidneys and the cardiovascular system. If blood volume decreases, *the kidneys cannot restore the lost fluid*. The kidneys can only *conserve* blood volume and thereby prevent further decreases in blood pressure.

The only way to restore lost fluid volume is through drinking or intravenous infusions. This is an example of mass balance: Volume lost to the external environment must be replaced from the external environment. Cardiovascular compensation for decreased blood volume includes vasoconstriction and increased sympathetic stimulation of the heart to increase cardiac output [Fig. 14.22, p. 472]. However, there are limits to the effectiveness of cardiovascular compensation—if fluid loss is too great, the body cannot maintain adequate blood pressure. Typical events that might cause significant changes in blood volume include dehydration, hemorrhage, and ingestion of a large quantity of fluid.

FIG. 15.9 Compensation for increased blood volume

Blood pressure control includes rapid responses from the cardiovascular system and slower responses by the kidneys.

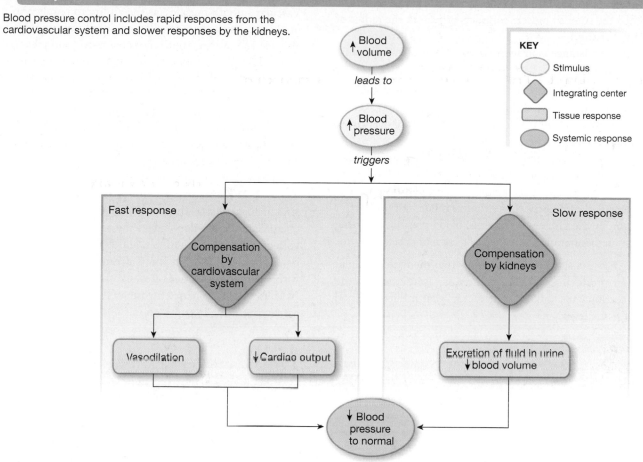

KEY

- Stimulus
- Integrating center
- Tissue response
- Systemic response

CLINICAL FOCUS

Shock

Shock is a broad term that refers to generalized, severe circulatory failure. Shock can arise from multiple causes: failure of the heart to maintain normal cardiac output (*cardiogenic shock*), decreased circulating blood volume (*hypovolemic shock*), bacterial toxins (*septic shock*), and miscellaneous causes, such as the massive immune reactions that cause *anaphylactic shock*. No matter what the cause, the results are similar: low cardiac output and falling peripheral blood pressure. When tissue perfusion can no longer keep up with tissue oxygen demand, the cells begin to sustain damage from inadequate oxygen and from the buildup of metabolic wastes. Once this damage occurs, a positive feedback cycle begins. The shock becomes progressively worse until it becomes irreversible, and the patient dies. The management of shock includes administration of oxygen, fluids, and norepinephrine, which stimulates vasoconstriction and increases cardiac output. If the shock arises from a cause that is treatable, such as a bacterial infection, measures must also be taken to remove the precipitating cause.

Figure 15.8b summarizes the four key factors that influence mean arterial blood pressure.

RESISTANCE IN THE ARTERIOLES

Peripheral resistance is one of the two main factors influencing blood pressure. According to Poiseuille's Law [p. 440], resistance to blood flow (R) is directly proportional to the length of the tubing through which the fluid flows (L) and to the viscosity (η) of the fluid, and inversely proportional to the fourth power of the tubing radius (r):

$$R \propto L\eta/r^4 \qquad (6)$$

Normally, the length of the systemic circulation and the blood's viscosity are relatively constant. That leaves only the radius of the blood vessels as the primary resistance to blood flow:

$$R \propto 1/r^4 \qquad (7)$$

The arterioles are the main site of variable resistance in the systemic circulation and contribute more than 60% of the total resistance to flow in the system. Resistance in arterioles is variable because of the large amounts of smooth muscle in the arteriolar walls. When the smooth muscle contracts or relaxes, the radius of the arterioles changes.

Arteriolar resistance is influenced by both local and systemic control mechanisms:

1. *Local control of arteriolar resistance* matches tissue blood flow to the metabolic needs of the tissue. In the heart and skeletal muscle, these local controls often take precedence over reflex control by the central nervous system.
2. *Sympathetic reflexes* mediated by the CNS maintain mean arterial pressure and govern blood distribution for certain homeostatic needs, such as temperature regulation.
3. *Hormones*—particularly those that regulate salt and water excretion by the kidneys—influence blood pressure by acting directly on the arterioles and by altering autonomic reflex control.

TABLE 15.2 lists significant chemicals that mediate arteriolar resistance by producing vasoconstriction or vasodilation. In the following sections, we look at some factors that influence blood flow at the tissue level.

Myogenic Autoregulation Adjusts Blood Flow

Vascular smooth muscle has the ability to regulate its own state of contraction, a process called **myogenic autoregulation.** In the absence of autoregulation, an increase in blood pressure increases blood flow through an arteriole. However, when smooth muscle fibers in the wall of the arteriole stretch because of increased blood pressure, the arteriole constricts. This vasoconstriction increases the resistance offered by the arteriole, automatically decreasing blood flow through the vessel. With this simple and direct response to pressure, arterioles have limited ability to regulate their own blood flow.

How does myogenic autoregulation work at the cellular level? When vascular smooth muscle cells in arterioles are stretched, mechanically gated channels in the muscle membrane open. Cation entry depolarizes the cell. The depolarization opens voltage-gated Ca^{2+} channels, and Ca^{2+} flows into the cell down its electrochemical gradient. Calcium entering the cell combines with calmodulin and activates myosin light chain kinase [p. 406]. MLCK in turn increases myosin ATPase activity and crossbridge activity, resulting in contraction.

TABLE 15.2	Chemicals Mediating Vasoconstriction and Vasodilation		
Chemical	**Physiological Role**	**Source**	**Type**
Vasoconstriction			
Norepinephrine (α-receptors)	Baroreceptor reflex	Sympathetic neurons	Neurotransmitter
Serotonin	Platelet aggregation, smooth muscle contraction	Neurons, digestive tract, platelets	Paracrine signal, neurotransmitter
Endothelin	Local control of blood flow	Vascular endothelium	Paracrine
Vasopressin	Increases blood pressure in hemorrhage	Posterior pituitary	Neurohormone
Angiotensin II	Increases blood pressure	Plasma hormone	Hormone
Vasodilation			
Epinephrine (β_2-receptors)	Increase blood flow to skeletal muscle, heart, liver	Adrenal medulla	Neurohormone
Acetylcholine	Erection reflex (indirectly through NO production)	Parasympathetic neurons	Neurotransmitter
Nitric oxide (NO)	Local control of blood flow	Endothelium	Paracrine signal
Bradykinin (via NO)	Increases blood flow	Multiple tissues	Paracrine signal
Adenosine	Increases blood flow to match metabolism	Hypoxic cells	Paracrine signal
$\downarrow O_2, \uparrow CO_2, \uparrow H^+, \uparrow K^+$	Increase blood flow to match metabolism	Cell metabolism	Paracrine molecule
Histamine	Increases blood flow	Mast cells	Paracrine signal
Natriuretic peptides (example: ANP)	Reduce blood pressure	Atrial myocardium, brain	Hormone, neurotransmitter
Vasoactive intestinal peptide	Digestive secretion, relax smooth muscle	Neurons	Neurotransmitter, neurohormone

RUNNING **PROBLEM**

Most hypertension is *essential hypertension,* which means high blood pressure that cannot be attributed to any particular cause. "Since your blood pressure is only mildly elevated," Dr. Cortez tells Kurt, "let's see if we can control it with lifestyle changes and a diuretic. You need to reduce salt and fat in your diet, get some exercise, and lose some weight. The diuretic will help your kidneys get rid of excess fluid." "Looks like you're asking me to turn over a whole new leaf," says Kurt. "I'll try it."

Q2: *What is the rationale for reducing salt intake and taking a diuretic to control hypertension? (Hint: Salt causes water retention.)*

478　484　**489**　490　500　505

Paracrine Signals Influence Vascular Smooth Muscle

Local control is an important strategy by which individual tissues regulate their own blood supply. In a tissue, blood flow into individual capillaries can be regulated by the precapillary sphincters described earlier in the chapter. When these small bands of smooth muscle at metarteriole-capillary junctions constrict, they restrict blood flow into the capillaries (see Fig. 15.3). When the sphincters

dilate, blood flow into the capillaries increases. This mechanism provides an additional site for local control of blood flow.

Local regulation also takes place by changing arteriolar resistance in a tissue. This is accomplished by paracrine molecules (including the gases O_2, CO_2, and NO) secreted by the vascular endothelium or by cells to which the arterioles are supplying blood (Tbl. 15.2).

The concentrations of many paracrine molecules change as cells become more or less metabolically active. For example, if aerobic metabolism increases, tissue O_2 levels decrease while CO_2 production goes up. Both low O_2 and high CO_2 dilate arterioles. This vasodilation increases blood flow into the tissue, bringing additional O_2 to meet the increased metabolic demand and removing waste CO_2 (**FIG. 15.10a**). The process in which an increase in blood flow accompanies an increase in metabolic activity is known as **active hyperemia** {*hyper-,* above normal + *(h)aimia,* blood}.

If blood flow to a tissue is occluded {*occludere,* to close up} for a few seconds to a few minutes, O_2 levels fall and metabolic paracrine signals such as CO_2 and H^+ accumulate in the interstitial fluid. Local *hypoxia* {*hypo-,* low + *oxia,* oxygen} causes endothelial cells to synthesize the vasodilator nitric oxide. When blood flow to the tissue resumes, the increased concentrations of NO, CO_2, and other paracrine molecules immediately trigger significant vasodilation. As the vasodilators are metabolized or washed away by the restored tissue blood flow, the radius of the arteriole gradually returns to normal. An increase in tissue blood flow following a period of low perfusion (blood flow) is known as **reactive hyperemia** (Fig. 15.10b).

FIG. 15.10　Hyperemia

Hyperemia is a locally mediated increase in blood flow.

(a) Active hyperemia matches blood flow to increased metabolism.

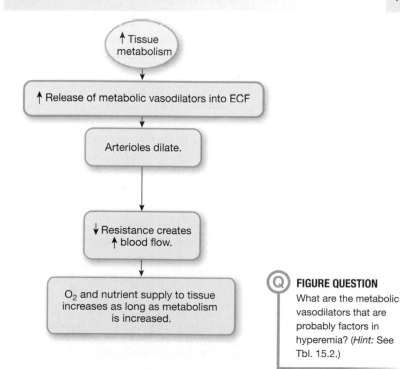

↑ Tissue metabolism

↑ Release of metabolic vasodilators into ECF

Arterioles dilate.

↓ Resistance creates ↑ blood flow.

O_2 and nutrient supply to tissue increases as long as metabolism is increased.

(b) Reactive hyperemia follows a period of decreased blood flow.

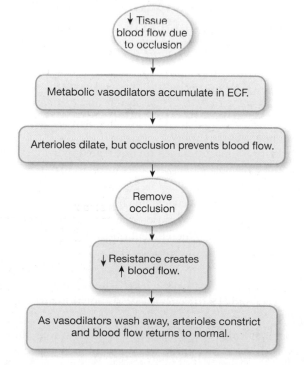

↓ Tissue blood flow due to occlusion

Metabolic vasodilators accumulate in ECF.

Arterioles dilate, but occlusion prevents blood flow.

Remove occlusion

↓ Resistance creates ↑ blood flow.

As vasodilators wash away, arterioles constrict and blood flow returns to normal.

Q **FIGURE QUESTION**

What are the metabolic vasodilators that are probably factors in hyperemia? (*Hint:* See Tbl. 15.2.)

Nitric oxide is probably best known for its role in the male erection reflex: Drugs used to treat erectile dysfunction prolong NO activity. Decreases in endogenous NO activity are suspected to play a role in other medical conditions, including hypertension and *preeclampsia,* the elevated blood pressure that sometimes occurs during pregnancy.

Another vasodilator paracrine signal is the nucleotide **adenosine.** If oxygen consumption in heart muscle exceeds the rate at which oxygen is supplied by the blood, myocardial hypoxia results. In response to low tissue oxygen, the myocardial cells release adenosine. Adenosine dilates coronary arterioles in an attempt to bring additional blood flow into the muscle.

Not all vasoactive paracrine molecules reflect changes in metabolism. For example, *kinins* and *histamine* are potent vasodilators that play a role in inflammation. *Serotonin* (5-HT), previously mentioned as a CNS neurotransmitter [p. 294], is also a vasoconstricting signal molecule released by activated platelets. When damaged blood vessels activate platelets, the subsequent serotonin-mediated vasoconstriction helps slow blood loss. Serotonin agonists called triptans (for example, *sumatriptan*) are drugs that bind to *5-HT$_1$ receptors* and cause vasoconstriction. These drugs are used to treat migraine headaches, which are caused by inappropriate cerebral vasodilation.

CONCEPT CHECK

6. Resistance to blood flow is determined *primarily* by which? (a) blood viscosity, (b) blood volume, (c) cardiac output, (d) blood vessel diameter, or (e) blood pressure gradient (ΔP)

7. The extracellular fluid concentration of K^+ increases in exercising skeletal muscles. What effect does this increase in K^+ have on blood flow in the muscles?

The Sympathetic Branch Controls Most Vascular Smooth Muscle

Smooth muscle contraction in arterioles is regulated by neural and hormonal signals in addition to locally produced paracrine molecules. Among the hormones with significant vasoactive properties are *atrial natriuretic peptide* and *angiotensin II (ANG II).* These hormones also have significant effects on the kidney's excretion of ions and water.

Most systemic arterioles are innervated by sympathetic neurons. A notable exception is arterioles involved in the erection reflex of the penis and clitoris. They are controlled indirectly by parasympathetic innervation. Acetylcholine from parasympathetic neurons causes paracrine release of nitric oxide, resulting in vasodilation.

Tonic discharge of norepinephrine from sympathetic neurons helps maintain tone of arterioles (**FIG. 15.11a**). Norepinephrine binding to α-receptors on vascular smooth muscle causes vasoconstriction. If sympathetic release of norepinephrine decreases, the arterioles dilate. If sympathetic stimulation increases, the arterioles constrict.

Epinephrine from the adrenal medulla travels through the blood and also binds with α-receptors, reinforcing vasoconstriction.

RUNNING PROBLEM

After two months, Kurt returns to the doctor's office for a checkup. He has lost 5 pounds and is walking at least a mile daily, but his blood pressure has not changed. "I swear, I'm trying to do better," says Kurt, "but it's difficult." Because lifestyle changes and the diuretic have not lowered Kurt's blood pressure, Dr. Cortez adds an antihypertensive drug. "This drug, called an ACE inhibitor, blocks production of a chemical called angiotensin II, a powerful vasoconstrictor. This medication should bring your blood pressure back to a normal value."

Q3: *Why would blocking the action of a vasoconstrictor lower blood pressure?*

478 484 489 **490** 500 505

However, α-receptors have a lower affinity for epinephrine and do not respond as strongly to it as they do to norepinephrine [p. 367]. In addition, epinephrine binds to β_2-receptors, found on vascular smooth muscle of heart, liver, and skeletal muscle arterioles. These receptors are not innervated and therefore respond primarily to circulating epinephrine. Activation of vascular β_2-receptors by epinephrine causes vasodilation.

One way to remember which tissues' arterioles have β_2-receptors is to think of a fight-or-flight response to a stressful event [p. 359]. This response includes a generalized increase in sympathetic activity, along with the release of epinephrine. Blood vessels that have β_2-receptors respond to epinephrine by vasodilating. Such β_2-mediated vasodilation enhances blood flow to the heart, skeletal muscles, and liver, tissues that are active during the fight-or-flight response. (The liver produces glucose for muscle contraction.) During fight or flight, increased sympathetic activity at arteriolar α-receptors causes vasoconstriction. The increase in resistance diverts blood from nonessential organs, such as the gastrointestinal tract, to the skeletal muscles, liver, and heart.

The map in Figure 15.11b summarizes the many factors that influence blood flow in the body. The pressure to drive blood flow is created by the pumping heart and captured by the arterial pressure reservoir, as reflected by the mean arterial pressure. Flow through the body as a whole is equal to the cardiac output, but flow to individual tissues can be altered by selectively changing resistance in a tissue's arterioles. In the next section, we consider the relationship between blood flow and arteriolar resistance.

CONCEPT CHECK

8. What happens when epinephrine combines with β_1-receptors in the heart? With β_2-receptors in the heart? (*Hint:* "in the heart" is vague. The heart has multiple tissue types. Which heart tissues possess the different types of β-receptors? [p. 367])

9. Skeletal muscle arterioles have both α- and β-receptors on their smooth muscle. Epinephrine can bind to both. Will these arterioles constrict or dilate in response to epinephrine? Explain.

FIG. 15.11 Resistance and flow

(a) Tonic Control of Arteriolar Diameter

Arteriole Diameter Is Controlled by Tonic Release of Norepinephrine.

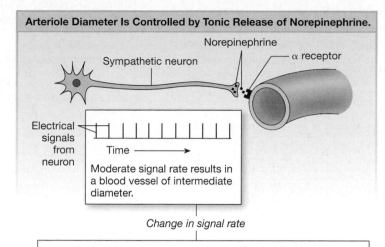

Norepinephrine

α receptor

Sympathetic neuron

Electrical signals from neuron

Time

Moderate signal rate results in a blood vessel of intermediate diameter.

Change in signal rate

↑ Norepinephrine Release onto α Receptors

Time

As the signal rate increases, the blood vessel constricts.

↓ Norepinephrine Release onto α Receptors

Time

As the signal rate decreases, the blood vessel dilates.

(b) Factors Influencing Peripheral Blood Flow

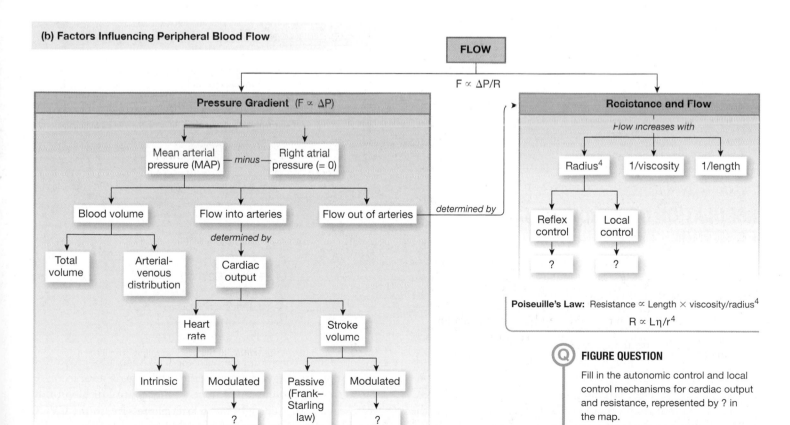

FLOW

$F \propto \Delta P/R$

Pressure Gradient $(F \propto \Delta P)$

Mean arterial pressure (MAP) — *minus* — Right atrial pressure (= 0)

Blood volume

Flow into arteries

Flow out of arteries — *determined by* →

Total volume

Arterial-venous distribution

determined by

Cardiac output

Heart rate

Stroke volume

Intrinsic

Modulated

?

Passive (Frank–Starling law)

Modulated

?

Resistance and Flow

Flow increases with

Radius^4

1/viscosity

1/length

Reflex control

Local control

?

?

Poiseuille's Law: Resistance $\propto$ Length $\times$ viscosity/radius^4

$R \propto L\eta/r^4$

Q FIGURE QUESTION

Fill in the autonomic control and local control mechanisms for cardiac output and resistance, represented by ? in the map.

DISTRIBUTION OF BLOOD TO THE TISSUES

The body's ability to selectively alter blood flow to organs is an important aspect of cardiovascular regulation. The distribution of systemic blood varies according to the metabolic needs of individual organs and is governed by a combination of local control mechanisms and homeostatic reflexes. For example, skeletal muscles at rest receive about 20% of cardiac output. During exercise, when the muscles use more oxygen and nutrients, they receive as much as 85%.

Blood flow to individual organs is set to some degree by the number and size of arteries feeding the organ. **FIGURE 15.12** shows how blood is distributed to various organs when the body is at rest. Usually, more than two-thirds of the cardiac output is routed to the digestive tract, liver, muscles, and kidneys.

Variations in blood flow to individual tissues are possible because the arterioles in the body are arranged in parallel. In other words, all arterioles receive blood at the same time from the aorta (see Fig. 15.1). Total blood flow through *all* the arterioles of the body always equals the cardiac output.

However, the flow through individual arterioles in a branching system of arterioles depends on their resistance (R). The higher the resistance in an arteriole, the lower the blood flow through it. If an arteriole constricts and resistance increases, blood flow through that arteriole decreases (**FIG. 15.13**):

$$\text{Flow}_\text{arteriole} \propto 1/R_\text{arteriole} \qquad (8)$$

In other words, blood is diverted from high-resistance arterioles to lower-resistance arterioles. You might say that blood traveling through the arterioles takes the path of least resistance.

CONCEPT CHECK

10. Use Fig. 15.12 to answer these questions. (a) Which tissue has the highest blood flow per unit weight? (b) Which tissue has the least blood flow, regardless of weight?

REGULATION OF CARDIOVASCULAR FUNCTION

The central nervous system coordinates the reflex control of blood pressure and distribution of blood to the tissues. The main integrating center is in the medulla oblongata. Because of the complexity of the neural networks involved in cardiovascular control, we will simplify this discussion and refer to the CNS network as the **cardiovascular control center** (CVCC).

The primary function of the cardiovascular control center is to ensure adequate blood flow to the brain and heart by maintaining sufficient mean arterial pressure. However, the CVCC also receives input from other parts of the brain, and it has the ability to alter function in a few organs or tissues while leaving others unaffected. For example, thermoregulatory centers in the

FIG. 15.12 Distribution of blood in the body at rest

Blood flow to the major organs is represented in three ways: as a percentage of total flow, as volume per 100 grams of tissue per minute, and as an absolute rate of flow (in L/min).

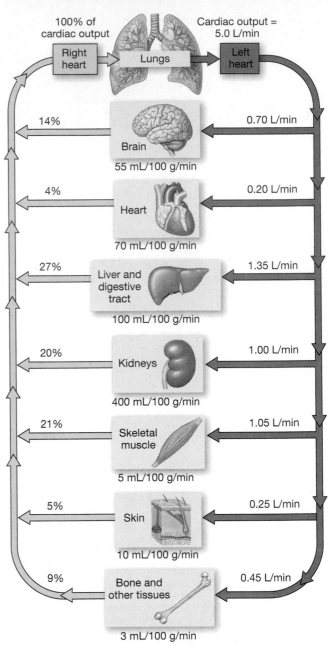

100% of cardiac output — Right heart — Lungs — Left heart — Cardiac output = 5.0 L/min

Organ	%	volume/weight	L/min
Brain	14%	55 mL/100 g/min	0.70 L/min
Heart	4%	70 mL/100 g/min	0.20 L/min
Liver and digestive tract	27%	100 mL/100 g/min	1.35 L/min
Kidneys	20%	400 mL/100 g/min	1.00 L/min
Skeletal muscle	21%	5 mL/100 g/min	1.05 L/min
Skin	5%	10 mL/100 g/min	0.25 L/min
Bone and other tissues	9%	3 mL/100 g/min	0.45 L/min

Q FIGURE QUESTION
What is the rate of blood flow through the lungs?

hypothalamus communicate with the CVCC to alter blood flow to the skin. Brain-gut communication following a meal increases blood flow to the intestinal tract. Reflex control of blood flow to selected tissues changes mean arterial pressure, so the CVCC is constantly monitoring and adjusting its output as required to maintain homeostasis.

FIG. 15.13 | Blood flow through individual blood vessels is determined by the vessel's resistance to flow

(a) Blood flow through four identical vessels (A–D) is equal. Total flow into vessels equals total flow out.

A	1 L/min
B	1 L/min
C	1 L/min
D	1 L/min
Total flow:	4 L/min

(b) When vessel B constricts, resistance of B increases and flow through B decreases. Flow diverted from B is divided among the lower-resistance vessels A, C, and D.

A	$1\frac{1}{4}$ L/min
B	$\frac{1}{4}$ L/min *
C	$1\frac{1}{4}$ L/min
D	$1\frac{1}{4}$ L/min
Total flow unchanged:	4 L/min

FIGURE QUESTIONS

1. You are monitoring blood pressure in the artery at the point indicated by ▲. What happens to arterial blood pressure when vessel B suddenly constricts?

2. Is pressure at the * end of B increased or decreased following constriction?

The Baroreceptor Reflex Controls Blood Pressure

The primary reflex pathway for homeostatic control of mean arterial blood pressure is the **baroreceptor reflex.** The components of the reflex are illustrated in **FIGURE 15.14a**. Stretch-sensitive mechanoreceptors known as **baroreceptors** are located in the walls of the carotid arteries and aorta, where they continuously monitor the pressure of blood flowing to the brain (carotid baroreceptors) and to the body (aortic baroreceptors).

The carotid and aortic baroreceptors are tonically active stretch receptors that fire action potentials continuously at normal blood pressures. When increased blood pressure in the arteries stretches the baroreceptor membrane, the firing rate of the receptor increases. If blood pressure falls, the firing rate of the receptor decreases.

If blood pressure changes, the frequency of action potentials traveling from the baroreceptors to the medullary cardiovascular control center changes. The CVCC integrates the sensory input and initiates an appropriate response. The response of the baroreceptor reflex is quite rapid: changes in cardiac output and peripheral resistance occur within two heartbeats of the stimulus.

Output signals from the cardiovascular control center are carried by both sympathetic and parasympathetic autonomic neurons. As you learned earlier, peripheral resistance is under tonic sympathetic control, with increased sympathetic discharge causing vasoconstriction.

Heart function is regulated by antagonistic control [p. 184]. Increased sympathetic activity increases heart rate, shortens conduction time through the AV node, and enhances the force of myocardial contraction. Increased parasympathetic activity slows heart rate but has only a small effect on ventricular contraction.

The baroreceptor reflex in response to increased blood pressure is mapped in Fig. 15.14b. Baroreceptors increase their firing rate as blood pressure increases, activating the medullary cardiovascular control center. In response, the cardiovascular control center increases parasympathetic activity and decreases sympathetic activity to slow down the heart and dilate arterioles.

When heart rate falls, cardiac output falls. In the vasculature, decreased sympathetic activity causes dilation of arterioles, lowering their resistance and allowing more blood to flow out of the arteries. Because mean arterial pressure is directly proportional to cardiac output and peripheral resistance (MAP $\propto$ CO $\times$ R), the combination of reduced cardiac output and decreased peripheral resistance lowers the mean arterial blood pressure.

It is important to remember that the baroreceptor reflex is functioning all the time, not just with dramatic disturbances in blood pressure, and that it is not an all-or-none response. A change in blood pressure can result in a change in both cardiac output and peripheral resistance or a change in only one of the two variables. Let's look at an example.

For this example, we will use the schematic diagram in **FIGURE 15.15**, which combines the concepts introduced in Figures 15.8 and 15.13. In this model, there are four sets of variable resistance arterioles (A–D) whose diameters can be independently controlled by local or reflex control mechanisms. Baroreceptors in the arteries monitor mean arterial pressure and communicate with the medullary cardiovascular control center.

Suppose arteriole set A constricts because of local control mechanisms. Vasoconstriction increases resistance in A and decreases flow through A. Total peripheral resistance (TPR) across all the arterioles also increases. Using the relationship MAP $\propto$ CO $\times$ TPR, an increase in total resistance results in an increase in mean arterial pressure. The arterial baroreceptors sense the increase in MAP and activate the baroreceptor reflex.

Output from the cardiovascular control center can alter either cardiac output, arteriolar resistance, or both. In this instance, we can assume that blood flow in arteriole sets A–D now matches tissue needs and should remain constant. That means the only option left to decrease MAP is to decrease cardiac output. So efferent signals from the CVCC decrease cardiac output, which in turn brings mean arterial pressure down. Blood pressure

FIG. 15.14 **ESSENTIALS**

Cardiovascular Control

The intrinsic rate of the heartbeat is modulated by sympathetic and parasympathetic neurons. Blood vessel diameter is under tonic control by the sympathetic division.

(a) CNS Control of the Heart and Blood Vessels

KEY

- ⬭ Stimulus
- ▭ Sensor
- ▭ Afferent pathway
- ▭ Integrating center
- ▭ Output signal
- ▭ Target
- ▭ Tissue response
- ⬭ Systemic response

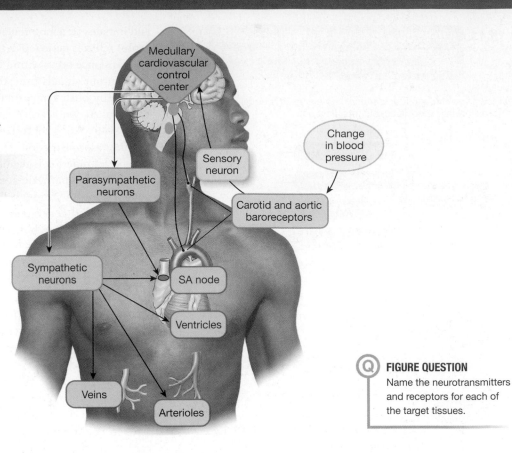

Medullary cardiovascular control center

Parasympathetic neurons

Sensory neuron

Change in blood pressure

Carotid and aortic baroreceptors

Sympathetic neurons

SA node

Ventricles

Veins

Arterioles

Q **FIGURE QUESTION**
Name the neurotransmitters and receptors for each of the target tissues.

(b) Baroreceptor Reflex

This map shows the reflex response to an increase in mean arterial pressure.

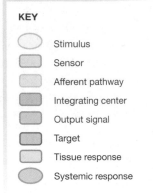

PHYS IN ACTION!

↑ Blood pressure

↑ Firing of baroreceptors in carotid arteries and aorta ⊖

Sensory neurons

Cardiovascular control center in medulla oblongata

↓ Sympathetic output — ↑ Parasympathetic output

less NE released

more ACh on muscarinic receptor

α-receptor — *β₁-receptor* — *β₁-receptor*

Arteriolar smooth muscle

Ventricular myocardium

SA node

Vasodilation

↓ Force of contraction

↓ Heart rate

↓ Peripheral resistance

↓ Cardiac output

↓ Blood pressure

Negative feedback

FIG. 15.15   Integration of resistance changes and cardiac output

1 Arteriole A constricts → Increased resistance ($\uparrow R_A$) → Increased total peripheral resistance ($\uparrow$TPR)

2 $\uparrow$TPR $\times$ Cardiac output (CO) → Increased mean arterial pressure ($\uparrow$ MAP)

3 $\uparrow$ MAP → baroreceptors fire → baroreceptor reflex

Assuming that tissue blood flow is matched to tissue need and does not need to change:

4 Baroreceptor reflex → Decreased cardiac output ($\downarrow$CO)

$\uparrow$TPR $\times$ $\downarrow$CO = MAP restored to normal

homeostasis is restored. In this example, the output signal of the baroreceptor reflex altered cardiac output but did not change peripheral resistance.

Orthostatic Hypotension Triggers the Baroreceptor Reflex

The baroreceptor reflex functions every morning when you get out of bed. When you are lying flat, gravitational forces are distributed evenly up and down the length of your body, and blood is distributed evenly throughout the circulation. When you stand up, gravity causes blood to pool in the lower extremities.

This pooling creates an instantaneous decrease in venous return so that less blood is in the ventricles at the beginning of the next contraction. Cardiac output falls from 5 L/min to 3 L/min, causing arterial blood pressure to decrease. This decrease in blood pressure upon standing is known as *orthostatic hypotension* {*orthos*, upright + *statikos*, to stand}.

Orthostatic hypotension normally triggers the baroreceptor reflex. The result is increased cardiac output and increased peripheral resistance, which together increase mean arterial pressure and bring it back up to normal within two heartbeats.

The skeletal muscle pump also contributes to the recovery by enhancing venous return when abdominal and leg muscles contract to maintain an upright position.

The baroreceptor reflex is not always effective, however. For example, during extended bed rest or in the zero-gravity conditions of space flights, blood from the lower extremities is distributed evenly throughout the body rather than pooled in the lower extremities. This even distribution raises arterial pressure, triggering the kidneys to excrete what the body perceives as excess fluid. Over the course of three days of bed rest or in space, excretion of water leads to a 12% decrease in blood volume.

When the person finally gets out of bed or returns to earth, gravity again causes blood to pool in the legs. Orthostatic hypotension occurs, and the baroreceptors attempt to compensate. In this instance, however, the cardiovascular system is unable to restore normal pressure because of the loss of blood volume. As a result, the individual may become light-headed or even faint from reduced delivery of oxygen to the brain.

CONCEPT CHECK

11. Baroreceptors have stretch-sensitive ion channels in their cell membrane. Increased pressure stretches the receptor cell membrane, opens the channels, and initiates action potentials. What ion probably flows through these channels and in which direction (into or out of the cell)?

12. Use the map in Fig. 15.14b to map the reflex response to orthostatic hypotension.

Other Systems Influence Cardiovascular Function

Cardiovascular function can be modulated by input from peripheral receptors other than the baroreceptors. For example, arterial chemoreceptors activated by low blood oxygen levels increase cardiac output. The cardiovascular control center also has reciprocal communication with centers in the medulla that control breathing.

The integration of function between the respiratory and circulatory systems is adaptive. If tissues require more oxygen, it is supplied by the cardiovascular system working in tandem with the respiratory system. Consequently, increases in breathing rate are usually accompanied by increases in cardiac output.

Blood pressure is also subject to modulation by higher brain centers, such as the hypothalamus and cerebral cortex. The hypothalamus mediates vascular responses involved in body temperature regulation and for the fight-or-flight response. Learned and emotional responses may originate in the cerebral cortex and be expressed by cardiovascular responses such as blushing and fainting.

One such reflex is *vasovagal syncope*, which may be triggered in some people by the sight of blood or a hypodermic needle. (Recall Anthony's experience at the beginning of this chapter.) In this pathway, increased parasympathetic activity and decreased sympathetic activity slow heart rate and cause widespread vasodilation. Cardiac output and peripheral resistance both decrease, triggering a precipitous drop in blood pressure. With insufficient blood to the brain, the individual faints.

Regulation of blood pressure in the cardiovascular system is closely tied to regulation of body fluid balance by the kidneys. Certain hormones secreted from the heart act on the kidneys, while hormones secreted from the kidneys act on the heart and blood vessels. Together, the heart and kidneys play a major role in maintaining homeostasis of body fluids, an excellent example of the integration of organ system function.

CONCEPT CHECK

13. In the classic movie *Jurassic Park,* Dr. Ian Malcolm must flee from the *T. rex.* Draw a reflex map showing the cardiovascular response to his fight-or-flight situation. Remember that fight-or-flight causes epinephrine secretion as well as output from the cardiovascular control center. (*Hints:* What is the stimulus? Fear is integrated in the limbic system.)

EXCHANGE AT THE CAPILLARIES

The transport of materials around the body is only part of the function of the cardiovascular system. Once blood reaches the capillaries, the plasma and the cells exchange materials across the thin capillary walls. Most cells are located within 0.1 mm of the nearest capillary, and diffusion over this short distance proceeds rapidly.

The capillary density in any given tissue is directly related to the metabolic activity of the tissue's cells. Tissues with a higher metabolic rate require more oxygen and nutrients. Those tissues have more capillaries per unit area. Subcutaneous tissue and cartilage have the lowest capillary density. Muscles and glands have the highest. By one estimate, the adult human body has about 50,000 miles of capillaries, with a total exchange surface area of more than 6300 m² , nearly the surface area of two football fields.

Capillaries have the thinnest walls of all the blood vessels, composed of a single layer of flattened endothelial cells supported on a basal lamina (Fig. 15.2). The diameter of a capillary is barely that of a red blood cell, forcing the RBCs to squeeze through in single file. Cell junctions between the endothelial cells vary from tissue to tissue and help determine the "leakiness" of the capillary.

The most common capillaries are **continuous capillaries,** whose endothelial cells are joined to one another with leaky junctions (**FIG. 15.16a**). These capillaries are found in muscle, connective tissue, and neural tissue. The continuous capillaries of the brain have evolved to form the *blood-brain barrier,* with tight junctions that help protect neural tissue from toxins that may be present in the bloodstream [p. 282].

Fenestrated capillaries {*fenestra,* window} have large pores (*fenestrae* or *fenestrations*) that allow high volumes of fluid to pass rapidly between the plasma and interstitial fluid (Fig. 15.16b). These capillaries are found primarily in the kidney and the intestine, where they are associated with absorptive transporting epithelia.

Three tissues—the bone marrow, the liver, and the spleen—do not have typical capillaries. Instead they have modified vessels

---

FIG. 15.16 Capillaries

(a) Continuous capillaries have leaky junctions.

(b) Fenestrated capillaries have large pores.

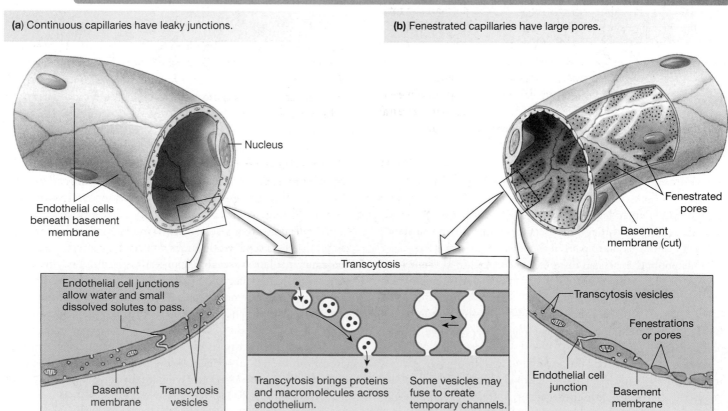

Nucleus

Endothelial cells beneath basement membrane

Endothelial cell junctions allow water and small dissolved solutes to pass.

Basement membrane

Transcytosis vesicles

Transcytosis

Transcytosis brings proteins and macromolecules across endothelium.

Some vesicles may fuse to create temporary channels.

Fenestrated pores

Basement membrane (cut)

Transcytosis vesicles

Fenestrations or pores

Endothelial cell junction

Basement membrane

called **sinusoids** that are as much as five times wider than a capillary. The sinusoid endothelium has fenestrations, and there may be gaps between the cells as well. Sinusoids are found in locations where blood cells and plasma proteins need to cross the endothelium to enter the blood. [Fig. 16.4c, *Focus On: Bone Marrow,* shows blood cells leaving the bone marrow by squeezing between endothelial cells.] In the liver, the sinusoidal endothelium lacks a basal lamina, which allows even more free exchange between plasma and interstitial fluid.

Velocity of Blood Flow Is Lowest in the Capillaries

The rate at which blood flows through the capillaries plays a role in the efficiency of exchange between the blood and the interstitial fluid. At a constant flow rate, velocity of flow is higher in a smaller diameter tube than in a larger one [p. 442]. From this, you might conclude that blood moves very rapidly through the capillaries because they are the smallest blood vessels. However, the primary determinant for velocity is not the diameter of an individual capillary but the *total cross-sectional area* of *all* the capillaries.

What is total cross-sectional area? Imagine circles representing cross sections of all the capillaries placed edge to edge, and you have it. Or think of a package of spaghetti. Each piece of spaghetti has a very small diameter but if you gather many pieces together in your hands, the total area occupied by the ends of the spaghetti pieces is quite large.

This is what happens with capillaries. Even though a single capillary has a tiny diameter, when you put them all together, their summed diameters cover an area much larger than the total cross-sectional areas of all the arteries and veins combined. Because total cross-sectional area of the capillaries is so large, the velocity of flow through them is low.

FIGURE 15.17 compares cross-sectional areas of different parts of the systemic circulation with the velocity of blood flow in each part. The fastest flow is in the relatively small-diameter arterial system. The slowest flow is in the capillaries and venules, which collectively have the largest cross-sectional area. The low velocity of flow through capillaries is a useful characteristic that allows enough time for diffusion to go to equilibrium [p. 134].

Most Capillary Exchange Takes Place by Diffusion and Transcytosis

Exchange between the plasma and interstitial fluid takes place either by movement between endothelial cells (the *paracellular pathway*) or by movement through the cells (*endothelial transport*). Smaller dissolved solutes and gases move by diffusion between or through the cells, depending on their lipid solubility [p. 136]. Larger solutes and proteins move mostly by vesicular transport [p. 147].

The diffusion rate for dissolved solutes is determined primarily by the concentration gradient between the plasma and the interstitial fluid. Oxygen and carbon dioxide diffuse freely across

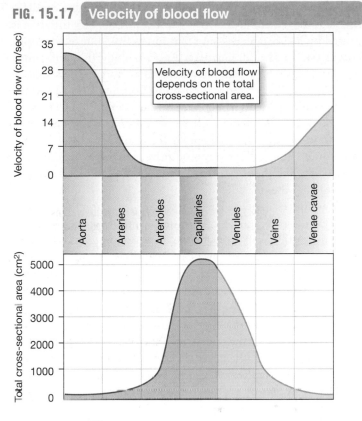

FIG. 15.17 Velocity of blood flow

Velocity of blood flow depends on the total cross-sectional area.

[Graph 1: Velocity of blood flow (cm/sec) on y-axis with values 0, 7, 14, 21, 28, 35. Regions labeled: Aorta, Arteries, Arterioles, Capillaries, Venules, Veins, Venae cavae]

[Graph 2: Total cross-sectional area (cm²) on y-axis with values 0, 1000, 2000, 3000, 4000, 5000]

GRAPH QUESTIONS

1. Is velocity directly proportional to or inversely proportional to cross-sectional area?
2. What effect does changing only the cross-sectional area have on flow rate?

the thin endothelium. Their plasma concentrations reach equilibrium with the interstitial fluid and cells by the time blood reaches the venous end of the capillary. In capillaries with leaky cell junctions, most small dissolved solutes can diffuse freely between the cells or through the fenestrations.

In continuous capillaries, blood cells and most plasma proteins are unable to pass through the junctions between endothelial cells. However, we know that proteins do move from plasma to interstitial fluid and vice versa. In most capillaries, larger molecules (including selected proteins) are transported across the endothelium by *transcytosis* [p. 152]. The endothelial cell surface appears dotted with numerous *caveolae* and noncoated pits that become vesicles for transcytosis. It appears that in some capillaries, chains of vesicles fuse to create open channels that extend across the endothelial cell (Fig. 15.16).

Capillary Filtration and Absorption Take Place by Bulk Flow

A third form of capillary exchange is bulk flow into and out of the capillary. **Bulk flow** refers to the mass movement of fluid as the result of hydrostatic or osmotic pressure gradients. If the direction of bulk flow is into the capillary, the fluid movement is called **absorption.** If the direction of flow is out of the capillary, the fluid movement is known as **filtration.** Capillary filtration is caused by hydrostatic pressure that forces fluid out of the capillary through leaky cell junctions. As an analogy, think of garden "soaker" hoses whose perforated walls allow water to ooze out.

Most capillaries show a transition from net filtration at the arterial end to net absorption at the venous end. There are some exceptions to this rule, though. Capillaries in part of the kidney filter fluid along their entire length, for instance, and some capillaries in the intestine are only absorptive, picking up digested nutrients that have been transported into the interstitial fluid from the lumen of the intestine.

Two forces regulate bulk flow in the capillaries. One is hydrostatic pressure, the lateral pressure component of blood flow that pushes fluid out through the capillary pores [p. 440], and the other is osmotic pressure [p. 126]. These forces are sometimes called *Starling forces,* after the English physiologist E. H. Starling, who first described them (the same Starling as in the Frank–Starling law of the heart). Osmotic pressure is determined by solute concentration of a compartment. The main solute difference between plasma and interstitial fluid is due to proteins, which are present in the plasma but mostly absent from interstitial fluid. The osmotic pressure created by the presence of these proteins is known as **colloid osmotic pressure** (π), also called *oncotic pressure.*

Colloid osmotic pressure is *not* equivalent to the total osmotic pressure in a capillary. It is simply a measure of the osmotic pressure created by proteins. Because the capillary endothelium is freely permeable to ions and other solutes in the plasma and interstitial fluid, these other solutes do not contribute to the osmotic gradient.

Colloid osmotic pressure is higher in the plasma (π_{cap} = 25 mm Hg) than in the interstitial fluid (π_{IF} = 0 mm Hg). Therefore, the osmotic gradient favors water movement by osmosis from the interstitial fluid into the plasma, represented by the red arrows in **FIGURE 15.18b**. For the purposes of our discussion, colloid osmotic pressure is constant along the length of the capillary, at π = 25 mm Hg.

Capillary hydrostatic pressure (P_H), by contrast, decreases along the length of the capillary as energy is lost to friction. Average values for capillary hydrostatic pressure, shown in Fig. 15.18b, are 32 mm Hg at the arterial end of a capillary and 15 mm Hg at the venous end. The hydrostatic pressure of the interstitial fluid P_{IF} is very low, and so we consider it to be essentially zero. This means that water movement due to hydrostatic pressure is directed out of the capillary, as denoted by the blue arrows in Fig. 15.18b, with the pressure gradient decreasing from the arterial end to the venous end.

If we assume that the interstitial hydrostatic and colloid osmotic pressures are zero, as discussed earlier, then the net pressure driving fluid flow across the capillary is determined by the difference between the hydrostatic pressure P_H and the colloid osmotic pressure (π):

$$\text{Net pressure} = P_H - \pi \qquad (9)$$

A positive value for the net pressure indicates net filtration and a negative value indicates net absorption.

Using the hydrostatic and oncotic pressure values given in Fig. 15.18b, we can calculate the following values at the arterial end of a capillary:

$$\begin{aligned} \text{Net pressure} &= P_H \,(32 \text{ mm Hg}) - \pi \,(25 \text{ mm Hg}) \\ &= 7 \text{ mm Hg} \end{aligned} \qquad (10)$$

At the arterial end, P_H is greater than π, so the net pressure is 7 mm Hg of filtration pressure.

At the venous end, where capillary hydrostatic pressure is less:

$$\begin{aligned} \text{Net pressure}_{\text{venous end}} &= (15 \text{ mm Hg} - 25 \text{ mm Hg}) \\ &= -10 \text{ mm Hg} \end{aligned} \qquad (11)$$

At the venous end, π is greater than P_H. The net pressure is –10 mm Hg, favoring absorption. (A negative net pressure indicates absorption.)

Fluid movement down the length of a capillary is shown in Fig. 15.18b. There is net filtration at the arterial end and net absorption at the venous end. If the point at which filtration equals absorption occurred in the middle of the capillary, there would be no net movement of fluid. All volume that was filtered at the arterial end would be absorbed at the venous end. However, filtration is usually greater than absorption, resulting in bulk flow of fluid out of the capillary into the interstitial space.

By most estimates, that bulk flow amounts to about 3 liters per day, which is the equivalent of the entire plasma volume! If this filtered fluid could not be returned to the plasma, the blood would turn into a sludge of blood cells and proteins. Restoring fluid lost from the capillaries to the circulatory system is one of the functions of the lymphatic system, which we discuss next.

CONCEPT CHECK

14. A man with liver disease loses the ability to synthesize plasma proteins. What happens to the colloid osmotic pressure of his blood? What happens to the balance between filtration and absorption in his capillaries?

15. Why did this discussion refer to the colloid osmotic pressure of the plasma rather than the osmolarity of the plasma?

THE LYMPHATIC SYSTEM

The vessels of the lymphatic system interact with three other physiological systems: the cardiovascular system, the digestive system, and the immune system. Functions of the lymphatic system include (1) returning fluid and proteins filtered out of the capillaries to the circulatory system, (2) picking up fat absorbed at the small intestine and transferring it to the circulatory system, and (3) serving as a filter to help capture and destroy foreign pathogens. In this discussion, we focus on the role of the lymphatic system in fluid transport.

FIG. 15.18 | Capillary fluid exchange

(a) A net average of 3 L/day of fluid filters out of the capillaries. The excess water and solutes that filter out of the capillary are picked up by the lymph vessels and returned to the circulation.

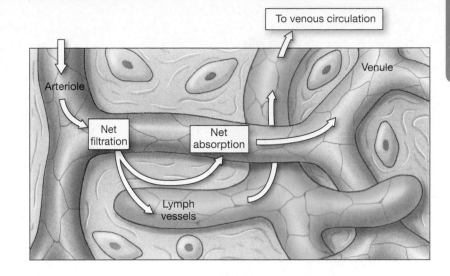

(b) Filtration in systemic capillaries

Net pressure = hydrostatic pressure (P_H) – colloid osmotic pressure (π)

Positive net pressure indicates filtration; negative net pressure indicates absorption.

$P_H = 32$ mm Hg
$\pi = 25$ mm Hg

$\pi = 25$ mm Hg
$P_H = 15$ mm Hg

$\frac{7200 \text{ L}}{\text{day}}$

$P_H > \pi$

$P_H = \pi$

$P_H < \pi$

Net filtration   Net flow out = 3 L/day   Net absorption

KEY

P_H = Hydrostatic pressure forces fluid out of the capillary.

π = Colloid osmotic pressure of proteins within the capillary pulls fluid into the capillary.

Q **FIGURE QUESTION**

Suppose that the hydrostatic pressure (P_H) at the arterial end of a capillary increases from 32 mm Hg to 35 mm Hg. If P_H remains 15 mm Hg at the venous end, does net filtration in this capillary decrease, increase, or stay the same?

The lymphatic system allows the one-way movement of interstitial fluid from the tissues into the circulation. Blind-end lymph vessels (*lymph capillaries*) lie close to all blood capillaries except those in the kidney and central nervous system (Fig. 15.18a). The smallest lymph vessels are composed of a single layer of flattened endothelium that is even thinner than the capillary endothelium.

The walls of these tiny lymph vessels are anchored to the surrounding connective tissue by fibers that hold the thin-walled vessels open. Large gaps between cells allow fluid, interstitial proteins, and particulate matter such as bacteria to be swept into the lymph vessels, also called *lymphatics,* by bulk flow. Once inside the lymphatics, this clear fluid is called simply **lymph.**

Lymph vessels in the tissues join one another to form larger lymphatic vessels that progressively increase in size (**FIG. 15.19**).

These vessels have a system of semilunar valves, similar to valves in the venous circulation. The largest lymph ducts empty into the venous circulation just under the collarbones, where the left and right subclavian veins join the internal jugular veins. At intervals along the way, vessels enter **lymph nodes,** bean-shaped nodules of tissue with a fibrous outer capsule and an internal collection of immunologically active cells, including lymphocytes and macrophages.

The lymphatic system has no single pump like the heart. Lymph flow depends primarily on waves of contraction of smooth muscle in the walls of the larger lymph vessels. Flow is aided by contractile fibers in the endothelial cells, by the one-way valves, and by external compression created by skeletal muscles.

The skeletal muscle pump plays a significant role in lymph flow, as you know if you have ever injured a wrist or ankle.

FIG. 15.19 The lymphatic system

Lymph fluid empties into the venous circulation.

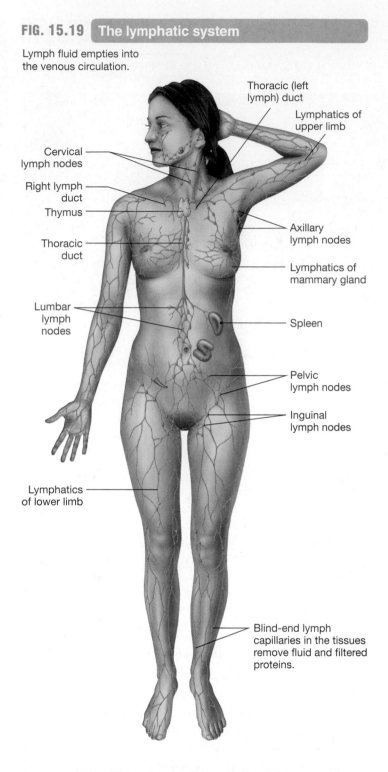

Thoracic (left lymph) duct

Lymphatics of upper limb

Cervical lymph nodes

Right lymph duct

Thymus

Axillary lymph nodes

Thoracic duct

Lymphatics of mammary gland

Lumbar lymph nodes

Spleen

Pelvic lymph nodes

Inguinal lymph nodes

Lymphatics of lower limb

Blind-end lymph capillaries in the tissues remove fluid and filtered proteins.

An immobilized limb frequently swells from the accumulation of fluid in the interstitial space, a condition known as **edema** {*oidema*, swelling}. Patients with edema in an injured limb are told to elevate the limb above the level of the heart so that gravity can assist lymph flow back to the blood.

An important reason for returning filtered fluid to the circulation is the recycling of plasma proteins. The body must maintain a low protein concentration in the interstitial fluid because colloid osmotic pressure is the only significant force that opposes capillary hydrostatic pressure. If proteins move from the plasma to the interstitial fluid, the osmotic pressure gradient that opposes filtration decreases. With less opposition to capillary hydrostatic pressure, additional fluid moves into the interstitial space.

Inflammation is an example of a situation in which the balance of colloid osmotic and hydrostatic pressures is disrupted. Histamine released in the inflammatory response makes capillary walls leakier and allows proteins to escape from the plasma into the interstitial fluid. The local swelling that accompanies a region of inflammation is an example of edema caused by redistribution of proteins from the plasma to the interstitial fluid.

Edema Results from Alterations in Capillary Exchange

Edema is a sign that normal exchange between the circulatory system and the lymphatics has been disrupted. Edema usually arises from one of two causes: (1) inadequate drainage of lymph or (2) blood capillary filtration that greatly exceeds capillary absorption.

Inadequate lymph drainage occurs with obstruction of the lymphatic system, particularly at the lymph nodes. Parasites, cancer, or fibrotic tissue growth caused by therapeutic radiation can block the movement of lymph through the system. For example, *elephantiasis* is a chronic condition marked by gross enlargement of the legs and lower appendages when parasites block the lymph vessels. Lymph drainage may also be impaired if lymph nodes are removed during surgery, a common procedure in the diagnosis and treatment of cancer.

Three factors that disrupt the normal balance between capillary filtration and absorption are:

1. *An increase in capillary hydrostatic pressure.* Increased capillary hydrostatic pressure is usually indicative of elevated venous pressure. An increase in arterial pressure is generally not noticeable at the capillaries because of autoregulation of pressure in the arterioles.

   One common cause of increased venous pressure is *heart failure,* a condition in which one ventricle loses pumping

RUNNING **PROBLEM**

Another few weeks go by, and Kurt again returns to Dr. Cortez for a checkup. Kurt's blood pressure is finally closer to the normal range and has been averaging 135/87. "But, Doc, can you give me something for this dry, hacking cough I've been having? I don't feel bad, but it's driving me nuts." Dr. Cortez explains that a dry cough is an occasional side effect of taking ACE inhibitors. "It is more of a nuisance than anything else, but let's change your medicine. I'd like to try you on a calcium channel blocker instead of the ACE inhibitor."

Q4: *How do calcium channel blockers lower blood pressure?*

478   484   489   490   500   505

power and can no longer pump all the blood sent to it by the other ventricle. For example, if the right ventricle begins to fail but the left ventricle maintains its cardiac output, blood accumulates in the systemic circulation. Blood pressure rises first in the right atrium, then in the veins and capillaries draining into the right side of the heart. When capillary hydrostatic pressure increases, filtration greatly exceeds absorption, leading to edema.

2. *A decrease in plasma protein concentration.* Plasma protein concentrations may decrease as a result of severe malnutrition or liver failure. The liver is the main site for plasma protein synthesis, and these proteins are responsible for the colloid osmotic pressure component (π) of the blood.

3. *An increase in interstitial proteins.* As discussed earlier, excessive leakage of proteins out of the blood decreases the colloid osmotic pressure gradient and increases net capillary filtration.

On occasion, changes in the balance between filtration and absorption help the body maintain homeostasis. For example, if arterial blood pressure falls, capillary hydrostatic pressure also decreases. This change increases fluid absorption. If blood pressure falls low enough, there is net absorption in the capillaries rather than net filtration. This passive mechanism helps maintain blood volume in situations in which blood pressure is very low, such as hemorrhage or severe dehydration.

CONCEPT CHECK

16. If the left ventricle fails to pump normally, blood backs up into what set of blood vessels? Where would you expect edema to occur?

17. Malnourished children who have inadequate protein in their diet often have grotesquely swollen bellies. This condition, which can be described as edema of the abdomen, is called *ascites* (**FIG. 15.20**). Use the information you have just learned about capillary filtration to explain why malnutrition causes ascites.

CARDIOVASCULAR DISEASE

Disorders of the heart and blood vessels, such as heart attacks and strokes, play a role in more than half of all deaths in the United States. The American Heart Association predicted that by 2030, over 40% of the U.S. population will have cardiovascular disease. The direct medical costs for these people are expected to triple, to more than $800 billion. The prevalence of cardiovascular disease is reflected in the tremendous amount of research being done worldwide. The scientific investigations range from large-scale clinical studies that track cardiovascular disease in thousands of people, such as the Framingham (Massachusetts) Heart Study, to experiments at the cellular and molecular levels.

Much of the research at the cellular and molecular levels is designed to expand our understanding of both normal and abnormal function in the heart and blood vessels. Scientists are studying a virtual alphabet soup of transporters and regulators. You have learned about some of these molecules, such as adenosine,

FIG. 15.20   Ascites

This 1960s photo from a Nigerian refugee camp shows ascites (abdominal edema) in a child with protein malnutrition, or kwashiorkor.

endothelin, vascular endothelial growth factor (VEGF), and nitric oxide, in this chapter.

As we increase our knowledge of cardiovascular function, we also begin to understand the actions of drugs that have been used for centuries. A classic example is the cardiac glycoside *digitalis* [p. 470], whose mechanism of action was explained when scientists discovered the role of Na^+-K^+-ATPase. It is a sobering thought to realize that for many therapeutic drugs, we know *what* they do without fully understanding *how* they do it.

Risk Factors Include Smoking and Obesity

Conducting and interpreting research on humans is a complicated endeavor in part because of the difficulty of designing well-controlled experiments [p. 22]. The economic and social importance of cardiovascular disease (CVD) makes it the focus of many studies each year as researchers try to improve treatments and prediction algorithms. (An *algorithm* is a set of rules or a sequence of steps used to solve a problem).

We can predict the likelihood that a person will develop cardiovascular disease during his or her lifetime by examining the various risk factors that the person possesses. The list of risk factors described here is the result of following the medical histories of thousands of people for many years in studies such as the Framingham Heart Study. As more data become available, additional risk factors may be added.

Risk factors are generally divided into those over which the person has no control and those that can be controlled. Medical intervention is aimed at reducing risk from the controllable factors. The risk factors that cannot be controlled include sex, age, and a family history of early cardiovascular disease. As noted earlier in the chapter, *coronary heart disease* (CHD) is a form of cardiovascular disease in which the coronary arteries become blocked by cholesterol deposits and blood clots. Up until middle age, men have a 3–4 times higher risk of developing CHD than do women. After age 55, when most women have entered menopause, the death rate from CHD equalizes in men and women. In general, the risk of coronary heart disease increases as people age. Heredity also plays an important role. If a person has one or more close relatives with this condition, his or her risk is elevated.

Risk factors that can be controlled include cigarette smoking, obesity, sedentary lifestyle, and untreated hypertension. In the United States, smoking-related illnesses such as CHD, lung cancer, and emphysema are the primary preventable cause of death, followed by conditions related to overweight and obesity. Physical inactivity and obesity have been steadily increasing in the United States since 1991, and currently nearly 70% of U.S. adults are either overweight or obese.

Two risk factors for cardiovascular disease—diabetes mellitus and elevated blood lipids—have both an uncontrollable genetic component and a modifiable lifestyle component. Diabetes mellitus is a metabolic disorder that puts a person at risk for developing coronary heart disease by contributing to the development of **atherosclerosis** ("hardening of the arteries"), in which fatty deposits form inside arterial blood vessels. Elevated serum cholesterol and triglycerides also lead to atherosclerosis. The increasing prevalence of these risk factors has created an epidemic in the United States, with one in every 3.4 deaths in 2009 attributed to all forms of cardiovascular disease.

Atherosclerosis Is an Inflammatory Process

Coronary heart disease accounts for the majority of cardiovascular disease deaths and is the single largest killer of Americans, both men and women. Let's look at the underlying cause of this disease: atherosclerosis.

The role of elevated blood cholesterol in the development of atherosclerosis is well established. Cholesterol, like other lipids, is not very soluble in aqueous solutions, such as the plasma. For this reason, when cholesterol in the diet is absorbed from the digestive tract, it combines with lipoproteins to make it more soluble. Clinicians generally are concerned with two of these lipoproteins: **high-density lipoprotein-cholesterol (HDL-C)** complexes and **low-density lipoprotein-cholesterol (LDL-C)** complexes. HDL-C is the more desirable form of blood cholesterol because high levels of HDL-C are associated with lower risk of heart attacks. (Memory aid: "H" in HDL stands for "healthy.")

LDL-C is sometimes called "bad" cholesterol because elevated plasma LDL-C levels are associated with coronary heart disease. (Remember this by associating "L" with "lethal.") Normal levels of LDL-C are not bad, however, because LDL is necessary

for cholesterol transport into cells. LDL-C's binding site—a protein called **apoB**—combines with an LDL receptor found in clathrin-coated pits on the cell membrane. The receptor-LDL-C complex is then brought into the cell by endocytosis. The LDL receptor recycles to the cell membrane, and the endosome fuses with a lysosome. LDL-C's proteins are digested to amino acids, and the freed cholesterol is used to make cell membranes or steroid hormones.

Although LDL is needed for cellular uptake of cholesterol, excess levels of plasma LDL-C lead to atherosclerosis (**FIG. 15.21**). Endothelial cells lining the arteries transport LDL-C into the extracellular space so that it accumulates just under the intima ❶. There, white blood cells called macrophages ingest cholesterol and other lipids to become lipid-filled *foam cells* ❷. Cytokines released by the macrophages promote smooth muscle cell division ❸. This early-stage *lesion* {*laesio*, injury} is called a *fatty streak*.

As the condition progresses, the lipid core grows, and smooth muscle cells reproduce, forming bulging *plaques* that protrude into the lumen of the artery ❹. In the advanced stages of atherosclerosis, the plaques develop hard, calcified regions and fibrous collagen caps ❺–❼. The mechanism by which calcium carbonate is deposited is still being investigated.

Scientists once believed that the occlusion (blockage) of coronary blood vessels by large plaques that triggered blood clots was the primary cause of heart attacks, but that model has been revised. The new model indicates that blood clot formation on plaques is more dependent on the structure of a plaque than on its size. Atherosclerosis is now considered to be an inflammatory

CLINICAL FOCUS

Diabetes and Cardiovascular Disease

Having diabetes is one of the major risk factors for developing cardiovascular disease, and almost two-thirds of people with diabetes will die from cardiovascular problems. In diabetes, cells that cannot use glucose turn to fats and proteins for their energy. The body breaks down fat into fatty acids [p. 30] and dumps them into the blood. Plasma cholesterol levels are also elevated. When LDL-C remains in the blood, the excess is ingested by macrophages, starting a series of events that lead to atherosclerosis. Because of the pivotal role that LDL-C plays in atherosclerosis, many forms of therapy, ranging from dietary modification and exercise to drugs, are aimed at lowering LDL-C levels. Left untreated, blockage of small and medium-sized blood vessels in the lower extremities can lead to loss of sensation and *gangrene* (tissue death) in the feet. Atherosclerosis in larger vessels causes heart attacks and strokes. To learn more about diabetes and the increased risk of cardiovascular disease, visit the web sites of the American Diabetes Association (*www.diabetes.org*) and the American Heart Association (*www.americanheart.org*).

FIG. 15.21 **The development of atherosclerotic plaques**

(a) Normal Arterial Wall

— Endothelial cells
— Elastic connective tissue
— Smooth muscle cells

(b) Fatty Streak

1 LDL-cholesterol accumulates between the endothelium and connective tissue and is oxidized.

2 Macrophages ingest cholesterol and become foam cells.

3 Smooth muscle cells, attracted by macrophage cytokines, begin to divide and take up cholesterol.

(c) Stable Fibrous Plaque

4 A lipid core accumulates beneath the endothelium.

5 Fibrous scar tissue forms to wall off the lipid core.

6 Smooth muscle cells divide and contribute to thickening of the intima.

7 Calcifications are deposited within the plaque.

8 Macrophages may release enzymes that dissolve collagen and convert stable plaques to unstable plaques.

(d) Vulnerable Plaque

9 Platelets that are exposed to collagen become activated and initiate a blood clot.

process in which macrophages release enzymes that convert stable plaques to vulnerable plaques **8** . *Stable plaques* have thick fibrous caps that separate the lipid core from the blood and do not activate platelets. *Vulnerable plaques* have thin fibrous caps that are more likely to rupture, exposing collagen and activating platelets that initiate a blood clot (*thrombus*) **9** .

If a clot blocks blood flow to the heart muscle, a heart attack, or *myocardial infarction*, results. Blocked blood flow in a coronary artery cuts off the oxygen supply to myocardial cells supplied by that artery. The oxygen-starved cells must then rely on anaerobic metabolism [p. 110], which produces lactate. As ATP production declines, the contractile cells are unable to pump Ca^{2+} out of the cell.

The unusually high Ca^{2+} concentration in the cytosol closes gap junctions in the damaged cells. Closure electrically isolates the damaged cells so that they no longer contract, and it forces

EMERGING CONCEPTS

Inflammatory Markers for Cardiovascular Disease

In clinical studies, it is sometimes difficult to determine whether a factor that has a positive correlation with a disease functions in a cause-effect relationship or represents a simple association. For example, two factors associated with higher incidence of heart disease are C-reactive protein and homocysteine. *C-reactive protein* (CRP) is a molecule involved in the body's response to inflammation. In one study, women who had elevated blood CRP levels were more than twice as likely to have a serious cardiovascular problem as women with low CRP. Does this finding mean that CRP is causing cardiovascular disease? Or could it simply be a marker that can be used clinically to predict who is more likely to develop cardiovascular complications, such as a heart attack or stroke?

Similarly, elevated homocysteine levels are associated with an increased incidence of CVD. (*Homocysteine* is an amino acid that takes part in a complicated metabolic pathway that also requires folate and vitamin B_{12} as cofactors). Should physicians routinely measure homocysteine along with cholesterol? Currently, there is little clinical evidence to show that reducing either CRP or homocysteine decreases a person's risk of developing CVD. If these two markers are not indicators for *modifiable* risk factors, should a patient's insurance be asked to pay for the tests used to detect them?

FIG. 15.22   Cardiovascular disease and blood pressure

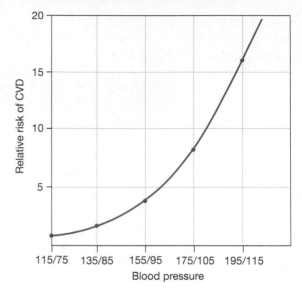

The risk of developing cardiovascular disease doubles with each 20/10 mm Hg increase in blood pressure.

action potentials to find an alternate route from cell to cell. If the damaged area of myocardium is large, the disruption can lead to an irregular heartbeat (*arrhythmia*) and potentially result in cardiac arrest or death.

Hypertension Represents a Failure of Homeostasis

One controllable risk factor for cardiovascular disease is hypertension—chronically elevated blood pressure, with systolic pressures greater than 140 mm Hg or diastolic pressures greater than 90 mm Hg. Hypertension is a common disease in the United States and is one of the most common reasons for visits to physicians and for the use of prescription drugs. High blood pressure is associated with increasing risk of CVD: the risk doubles for each 20/10 mm Hg increase in blood pressure over a baseline value of 115/75 (**FIG. 15.22**).

More than 90% of all patients with hypertension are considered to have *essential* (or *primary*) *hypertension*, with no clear-cut cause other than heredity. Cardiac output is usually normal in these people, and their elevated blood pressure appears to be associated with increased peripheral resistance. Some investigators

have speculated that the increased resistance may be due to a lack of nitric oxide, the locally produced vasodilator formed by endothelial cells in the arterioles. In the remaining 5–10% of hypertensive cases, the cause is more apparent, and the hypertension is considered to be secondary to an underlying pathology. For instance, the cause might be an endocrine disorder that causes fluid retention.

A key feature of hypertension from all causes is adaptation of the carotid and aortic baroreceptors to higher pressure, with subsequent down-regulation of their activity. Without input from the baroreceptors, the cardiovascular control center interprets the high blood pressure as "normal," and no reflex reduction of pressure occurs.

Hypertension is a risk factor for atherosclerosis because high pressure in the arteries damages the endothelial lining of the vessels and promotes the formation of atherosclerotic plaques. In addition, high arterial blood pressure puts additional strain on the heart by increasing afterload [p. 471]. When resistance in the arterioles is high, the myocardium must work harder to push the blood into the arteries.

Amazingly, stroke volume in hypertensive patients remains constant up to a mean blood pressure of about 200 mm Hg, despite the increasing amount of work that the ventricle must perform as blood pressure increases. The cardiac muscle of the left ventricle responds to chronic high systemic resistance in the same way that skeletal muscle responds to a weight-lifting routine. The heart muscle *hypertrophies*, increasing the size and strength of the muscle fibers.

However, if resistance remains high over time, the heart muscle cannot meet the workload and begins to fail: cardiac output by the left ventricle decreases. If cardiac output of the right heart remains normal while the output from the left side decreases, fluid collects in the lungs, creating *pulmonary edema.* At this point, a detrimental positive feedback loop begins. Oxygen exchange in the lungs diminishes because of the pulmonary edema, leading to less oxygen in the blood. Lack of oxygen for aerobic metabolism further weakens the heart, and its pumping effectiveness diminishes even more. Unless treated, this condition, known as *congestive heart failure,* eventually leads to death.

Many of the treatments for hypertension have their basis in the cardiovascular physiology you have learned. For example, calcium entry into vascular smooth muscle and cardiac muscle can be decreased by a class of drugs known as *calcium channel blockers.* These drugs bind to Ca^{2+} channel proteins, making it less likely that the channels will open in response to depolarization. With less Ca^{2+} entry, vascular smooth muscle dilates, while in the heart the depolarization rate of the SA node and the force of contraction decrease.

Vascular smooth muscle is more sensitive than cardiac muscle to certain classes of calcium channel blockers, and it is possible to get vasodilation at drug doses that are low enough to have no effect on heart rate. Other tissues with Ca^{2+} channels, such as neurons, are only minimally affected by calcium channel blockers because their Ca^{2+} channels are of a different subtype.

Other drugs used to treat hypertension include diuretics, which decrease blood volume, and beta-blocking drugs that target β_1-receptors and decrease catecholamine stimulation of cardiac output. Two other groups of antihypertensive drugs, the ACE inhibitors and the angiotensin receptor blockers, act by decreasing the activity of angiotensin, a powerful vasoconstrictor substance. You will learn more about angiotensin when you study the integrated control of blood pressure by the cardiovascular and renal systems. In the future, we may be seeing new treatments for hypertension that are based on other aspects of the molecular physiology of the heart and blood vessels.

RUNNING **PROBLEM** **CONCLUSION** | **Essential Hypertension**

Kurt remained on the calcium channel blocker and diuretic, and after several months his cough went away and his blood pressure stabilized at 130/85—a significant improvement. Kurt's new diet also brought his total blood cholesterol down below 200 mg/dL plasma. By improving two of his controllable risk factors, Kurt decreased his chances of having a

heart attack. To learn more about hypertension and some of the therapies currently used to treat it, visit the web site of the American Heart Association (*www.americanheart .org*). Now check your understanding of this running problem by comparing your answers with the information in the summary table.

Question	Facts	Integration and Analysis
Q1: *Why are people with high blood pressure at greater risk for having a hemorrhagic (or bleeding) stroke?*	High blood pressure exerts force on the walls of the blood vessels.	If an area of blood vessel wall is weakened or damaged, high blood pressure may cause that area to rupture, allowing blood to leak out of the vessel into the surrounding tissues.
Q2: *What is the rationale for reducing salt intake and taking a diuretic to control hypertension?*	Salt causes water retention. Diuretics increase renal fluid excretion.	Blood pressure increases if the circulating blood volume increases. By restricting salt in the diet, a person can decrease retention of fluid in the extracellular compartment, which includes the plasma. Diuretics also help decrease blood volume.
Q3: *Why would blocking the action of a vasoconstrictor lower blood pressure?*	Blood pressure is determined by cardiac output and peripheral resistance.	Resistance is inversely proportional to the radius of the blood vessels. Therefore, if blood vessels dilate as a result of blocking a vasoconstrictor, resistance and blood pressure decrease.
Q4: *How do calcium channel blockers lower blood pressure?*	Calcium entry from the extracellular fluid plays an important role in both smooth muscle and cardiac muscle contraction.	Blocking Ca^{2+} entry through Ca^{2+} channels decreases the force of cardiac contraction and decreases the contractility of vascular smooth muscle. Both of these effects lower blood pressure.

478 484 489 490 500 505

CHAPTER SUMMARY

Blood flow through the cardiovascular system is an excellent example of *mass flow* in the body. Cardiac contraction creates high pressure in the ventricles, and this pressure drives blood through the vessels of the systemic and pulmonary circuits, speeding up cell-to-cell *communication*. Resistance to flow is regulated by *local and reflex control mechanisms* that act on arteriolar smooth muscle and help match tissue perfusion to tissue needs. The *homeostatic* baroreceptor reflex monitors arterial pressure to ensure adequate perfusion of the brain and heart. Capillary *exchange of material* between the plasma and interstitial fluid *compartments* uses several transport mechanisms, including diffusion, transcytosis, and bulk flow.

1. Homeostatic regulation of the cardiovascular system is aimed at maintaining adequate blood flow to the brain and heart. (p. 479)

2. Total blood flow at any level of the circulation is equal to the cardiac output. (p. 479)

The Blood Vessels

iP **Cardiovascular—Anatomy Review: Blood Vessel Structure & Function**

3. Blood vessels are composed of layers of smooth muscle, elastic and fibrous connective tissue, and **endothelium.** (p. 479; Fig. 15.2)

4. **Vascular smooth muscle** maintains a state of muscle tone. (p. 479)

5. The walls of the aorta and major arteries are both stiff and springy. This property allows them to absorb energy and release it through elastic recoil. (p. 480)

6. **Metarterioles** regulate blood flow through capillaries by contraction and dilation of **precapillary sphincters.** (p. 480; Fig. 15.3)

7. Capillaries and postcapillary **venules** are the site of exchange between blood and interstitial fluid. (p. 481)

8. Veins hold more than half of the blood in the circulatory system. Veins have thinner walls with less elastic tissue than arteries, so veins expand easily when they fill with blood. (p. 481)

9. **Angiogenesis** is the process by which new blood vessels grow and develop, especially after birth. (p. 481)

Blood Pressure

iP **Cardiovascular: Measuring Blood Pressure**

10. The ventricles create high pressure that is the driving force for blood flow. The aorta and arteries act as a pressure reservoir during ventricular relaxation. (p. 482; Fig. 15.5)

11. Blood pressure is highest in the arteries and decreases as blood flows through the circulatory system. At rest, desirable **systolic pressure** is 120 mm Hg or less, and desirable **diastolic pressure** is 80 mm Hg or less. (p. 483; Fig. 15.6)

12. Pressure created by the ventricles can be felt as a **pulse** in the arteries. **Pulse pressure** equals systolic pressure minus diastolic pressure. (p. 483)

13. Blood flow against gravity in the veins is assisted by one-way valves and by the respiratory and skeletal muscle pumps. (p. 483; Fig. 15.4)

14. Arterial blood pressure is indicative of the driving pressure for blood flow. **Mean arterial pressure (MAP)** is defined as diastolic pressure + 1/3 (systolic pressure – diastolic pressure). (p. 483)

15. Arterial blood pressure is usually measured with a sphygmomanometer. Blood squeezing through a compressed artery makes **Korotkoff sounds.** (p. 484; Fig. 15.7)

16. Arterial pressure is a balance between cardiac output and **peripheral resistance,** the resistance to blood flow offered by the arterioles. (p. 484; Fig. 15.8)

17. If blood volume increases, blood pressure increases. If blood volume decreases, blood pressure decreases. (p. 485; Fig. 15.9)

18. Venous blood volume can be shifted to the arteries if arterial blood pressure falls. (p. 485; Fig. 15.1)

Resistance in the Arterioles

iP2 **Factors Affecting Blood Pressure**

19. The arterioles are the main site of variable resistance in the systemic circulation. A small change in the radius of an arteriole creates a large change in resistance: $R \propto 1/r^4$. (p. 487)

20. Arterioles regulate their own blood flow through **myogenic autoregulation.** Vasoconstriction increases the resistance offered by an arteriole and decreases the blood flow through the arteriole. (p. 488)

21. Arteriolar resistance is influenced by local control mechanisms that match tissue blood flow to the metabolic needs of the tissue. Vasodilator paracrine molecules include nitric oxide, H^+, K^+, CO_2, prostaglandins, adenosine, and histamine. Low O_2 causes vasodilation. Endothelins are powerful vasoconstrictors. (p. 489; Tbl. 15.2)

22. **Active hyperemia** is a process in which increased blood flow accompanies increased metabolic activity. **Reactive hyperemia** is an increase in tissue blood flow following a period of low perfusion. (p. 489; Fig. 15.10)

23. Most systemic arterioles are under tonic sympathetic control. Norepinephrine causes vasoconstriction. Decreased sympathetic stimulation causes vasodilation. (p. 490)

24. Epinephrine binds to arteriolar α-receptors and causes vasoconstriction. Epinephrine on β_2-receptors, found in the arterioles of the heart, liver, and skeletal muscle, causes vasodilation. (p. 490)

Distribution of Blood to the Tissues

25. Changing the resistance of the arterioles affects mean arterial pressure and alters blood flow through the arteriole. (p. 492; Fig. 15.15)

26. The flow through individual arterioles depends on their resistance. The higher the resistance in an arteriole, the lower the blood flow in that arteriole: $\text{Flow}_{\text{arteriole}} \propto 1/R_{\text{arteriole}}$. (p. 492)

Regulation of Cardiovascular Function

iP Cardiovascular: Blood Pressure Regulation

27. The reflex control of blood pressure resides in the medulla oblongata. **Baroreceptors** in the carotid artery and the aorta monitor arterial blood pressure and trigger the **baroreceptor reflex**. (p. 493; Fig. 15.14)

28. Efferent output from the medullary **cardiovascular control center** goes to the heart and arterioles. Increased sympathetic activity increases heart rate and force of contraction. Increased parasympathetic activity slows heart rate. Increased sympathetic discharge at the arterioles causes vasoconstriction. There is no significant parasympathetic control of arterioles. (p. 492)

29. Cardiovascular function can be modulated by input from higher brain centers and from the respiratory control center of the medulla. (p. 495)

30. The baroreceptor reflex functions each time a person stands up. The decrease in blood pressure upon standing is known as orthostatic hypotension. (p. 495)

Exchange at the Capillaries

iP Cardiovascular: Autoregulation and Capillary Dynamics

31. Exchange of materials between the blood and the interstitial fluid occurs primarily by diffusion. (p. 496)

32. **Continuous capillaries** have leaky junctions between cells but also transport material using transcytosis. Continuous capillaries with tight junctions form the blood-brain barrier. (p. 496; Fig. 15.16)

33. **Fenestrated capillaries** have pores that allow large volumes of fluid to pass rapidly. (p. 496; Fig. 15.16)

34. The velocity of blood flow through the capillaries is slow, allowing diffusion to go to equilibrium. (p. 497; Fig. 15.17)

35. The mass movement of fluid between the blood and the interstitial fluid is **bulk flow.** Fluid movement is called **filtration** if the direction of flow is out of the capillary and **absorption** if the flow is directed into the capillary. (pp. 497, 498; Fig. 15.18)

36. The osmotic pressure difference between plasma and interstitial fluid due to the presence of plasma proteins is the **colloid osmotic pressure.** (p. 498)

The Lymphatic System

iP Fluids & Electrolytes: Electrolyte Homeostasis, Edema

37. About 3 liters of fluid filter out of the capillaries each day. The lymphatic system returns this fluid to the circulatory system. (p. 498; Fig. 15.19)

38. Lymph capillaries accumulate fluid, interstitial proteins, and particulate matter by bulk flow. Lymph flow depends on smooth muscle in vessel walls, one-way valves, and the skeletal muscle pump. (p. 499)

39. The condition in which excess fluid accumulates in the interstitial space is called **edema.** Factors that disrupt the normal balance between capillary filtration and absorption cause edema. (p. 500)

Cardiovascular Disease

40. **Cardiovascular disease** is the leading cause of death in the United States. Risk factors predict the likelihood that a person will develop cardiovascular disease during her or his lifetime. (p. 501)

41. **Atherosclerosis** is an inflammatory condition in which fatty deposits called plaques develop in arteries. If plaques are unstable, they may block the arteries by triggering blood clots. (p. 502; Fig. 15.21)

42. Hypertension is a significant risk factor for the development of cardiovascular disease. (p. 504; Fig. 15.22)

REVIEW QUESTIONS

In addition to working through these questions and checking your answers on p. A-20, review the Learning Outcomes at the beginning of this chapter.

Level One Reviewing Facts and Terms

1. The first priority of blood pressure homeostasis is to maintain adequate perfusion to which two organs?

2. Match the types of systemic blood vessels with the terms that describe them. Each vessel type may have more than one match, and matching items may be used more than once.

a. arterioles	1. store pressure generated by the heart
b. arteries	2. have walls that are both stiff and elastic
c. capillaries	3. carry low-oxygen blood
d. veins	4. have thin walls of exchange epithelium
e. venules	5. act as a volume reservoir
	6. their diameter can be altered by neural input
	7. blood flow slowest through these vessels
	8. have lowest blood pressure
	9. are the main site of variable resistance

3. List the four tissue components of blood vessel walls, in order from inner lining to outer covering. Briefly describe the importance of each tissue.

4. Blood flow to individual tissues is regulated by selective vasoconstriction and vasodilation of which vessels?

5. Aortic pressure reaches a typical high value of _____ (give both numeric value and units) during _____, or contraction of the heart. As the heart relaxes during the event called _____, aortic pressure declines to a typical low value of _____. This blood pressure reading would be written as _____/_____.

6. The rapid pressure increase that occurs when the left ventricle pushes blood into the aorta can be felt as a pressure wave, or _____. What is the equation used to calculate the strength of this pressure wave?

7. List the factors that aid venous return to the heart.

8. What is hypertension, and why is it a threat to a person's health?

9. When measuring a person's blood pressure, at what point in the procedure are you likely to hear Korotkoff sounds?

10. List three paracrine molecules that cause vasodilation. What is the source of each one? In addition to paracrine signals, list two other ways to control smooth muscle contraction in arterioles.

11. What is hyperemia? How does active hyperemia differ from reactive hyperemia?

12. Most systemic arterioles are innervated by the _____ branch of the nervous system. Increased sympathetic input will have what effect on arteriole diameter?

13. Match each event in the left column with all appropriate neurotransmitter(s) *and* receptor(s) from the list on the right.

a. vasoconstriction of intestinal arterioles	1. norepinephrine
	2. epinephrine
b. vasodilation of coronary arterioles	3. acetylcholine
c. increased heart rate	4. β_1-receptor
	5. α-receptor
d. decreased heart rate	6. β_2-receptor
e. vasoconstriction of coronary arterioles	7. nicotinic receptor
	8. muscarinic receptor

14. Which organs receive more than two-thirds of the cardiac output at rest? Which organs have the highest flow of blood on a per unit weight basis?

15. By looking at the density of capillaries in a tissue, you can make assumptions about what property of the tissue? Which tissue has the lowest capillary density? Which tissue has the highest?

16. What type of transport is used to move each of the following substances across the capillary endothelium?

    a. oxygen
    b. proteins
    c. glucose
    d. water

17. With which three physiological systems do the vessels of the lymphatic system interact?

18. Define edema. List some ways in which it can arise.

19. Define the following terms and explain their significance to cardiovascular physiology.

    a. perfusion
    b. colloid osmotic pressure
    c. vasoconstriction
    d. angiogenesis
    e. metarterioles
    f. pericytes

20. The two major lipoprotein carriers of cholesterol are _____ and _____. Which type is bad when present in the body in elevated amounts?

Level Two Reviewing Concepts

21. Calcium channel blockers prevent Ca^{2+} movement through Ca^{2+} channels. Explain two ways this action lowers blood pressure. Why are neurons and other cells unaffected by these drugs?

22. Compare and contrast the following sets of terms:

    a. lymphatic capillaries and systemic capillaries
    b. roles of the sympathetic and parasympathetic branches in blood pressure control
    c. lymph and blood
    d. continuous capillaries and fenestrated capillaries
    e. hydrostatic pressure and colloid osmotic pressure in systemic capillaries

23. **Concept map:** Map all the following factors that influence mean arterial pressure. You may add terms.

• aorta	• parasympathetic neuron
• arteriole	• peripheral resistance
• baroreceptor	• SA node
• blood volume	• sensory neuron
• cardiac output	• stroke volume
• carotid artery	• sympathetic neuron
• contractility	• vein
• heart rate	• venous return
• medulla oblongata	• ventricle

24. Define myogenic autoregulation. What mechanisms have been proposed to explain it?

25. Left ventricular failure may be accompanied by edema, shortness of breath, and increased venous pressure. Explain how these signs and symptoms develop.

Level Three Problem Solving

26. Robert is a 52-year-old nonsmoker. He weighs 180 lbs and stands 5'9" tall, and his blood pressure averaged 145/95 on three successive visits to his doctor's office. His father, grandfather, and uncle all had heart attacks in their early 50s, and his mother died of a stroke at the age of 71.

    a. Identify Robert's risk factors for coronary heart disease.
    b. Does Robert have hypertension? Explain.
    c. Robert's doctor prescribes a drug called a beta blocker. Explain the mechanism by which a beta-receptor-blocking drug may help lower blood pressure.

27. The following figure is a schematic representation of the systemic circulation. Use it to help answer the following questions. (CO = cardiac output, MAP = mean arterial pressure).

a. If resistance in vessels 1 and 2 increases because of the presence of local paracrine signals but cardiac output is unchanged, what happens to MAP? What happens to flow through vessels 1 and 2? Through vessels 3 and 4?

b. Homeostatic compensation occurs within seconds. Draw a reflex map to explain the compensation (stimulus, receptor, and so on).

c. When vessel 1 constricts, what happens to the filtration pressure in the capillaries downstream from that arteriole?

28. The following graphs are recordings of contractions in an isolated frog heart. The intact frog heart is innervated by sympathetic neurons that increase heart rate and by parasympathetic neurons that decrease heart rate. Based on these four graphs, what conclusion can you draw about the mechanism of action of atropine? (Atropine does not cross the cell membrane.)

A
(add epinephrine)

B
(add epinephrine + atropine)

C
(add ACh)

D
(add ACh + atropine)

29. Draw a reflex map that explains Anthony's vasovagal syncope at the sight of blood. Include all the steps of the reflex, and explain whether autonomic pathways are being stimulated or inhibited.

30. A physiologist placed a section of excised arteriole in a perfusion chamber containing saline. When the oxygen content of the saline perfusing (flowing through) the arteriole was reduced, the arteriole dilated. In a follow-up experiment, she used an isolated piece of arteriolar smooth muscle that had been stripped away from the other layers of the arteriole wall. When the oxygen content of the saline was reduced as in the first experiment, the isolated muscle showed no response. What do these two experiments suggest about how low oxygen exerts local control over arterioles?

31. In advanced atherosclerosis, calcified plaques cause the normally elastic aorta and arteries to become stiff and noncompliant. (a) What effect does this change in the aorta have on afterload? (b) If cardiac output remains unchanged, what happens to peripheral resistance and mean arterial pressure?

32. During fetal development, most blood in the pulmonary artery bypasses the lungs and goes into the aorta by way of a channel called the *ductus arteriosus.* Normally, this fetal bypass channel closes during the first day after birth, but each year, about 4000 babies in the United States maintain a *patent* (open) ductus arteriosus and require surgery to close the channel.

a. Use this information to draw an anatomical diagram showing blood flow in an infant with a patent ductus arteriosus.

b. In the fetus, why does most blood bypass the lungs?

c. If the systemic side of the circulatory system is longer than the pulmonary side, which circuit has the greater resistance?

d. If flow is equal in the pulmonary and systemic circulations, which side of the heart must generate more pressure to overcome resistance?

e. Use your answer to (d) to figure out which way blood will flow through a patent ductus arteriosus.

Level Four Quantitative Problems

33. Using the appropriate equation, mathematically explain what happens to blood flow if the *diameter* of a blood vessel increases from 2 mm to 4 mm.

34. Duplicate the calculations that led William Harvey to believe that blood circulated in a closed loop:

a. Take your resting pulse.

b. Assume that your heart at rest pumps 70 mL/beat, and that 1 mL of blood weighs one gram. Calculate how long it would take your heart to pump your weight in blood. (2.2 pounds = 1 kilogram)

35. Calculate the mean arterial pressure (MAP) and pulse pressure for a person with a blood pressure of 115/73.

36. According to the Fick principle, the rate of oxygen consumption by an organ is equal to the blood flow through that organ times the amount of oxygen extracted from the blood as it flows through the organ:

Oxygen consumption = blood flow ×
(arterial O_2 content − venous O_2 content)

(mL O_2 consumed/min) = (mL blood/min × mL O_2/mL blood)

A woman has a total body oxygen consumption rate of 250 mL/min. The oxygen content of blood in her aorta is 200 mL O_2/L blood, the oxygen content of her pulmonary artery blood is 160 mL O_2/L blood. What is her cardiac output?

37. Beau has an average daily heart rate of 75 beats per minute. If his net capillary filtration rate is 3.24 L/day, how much fluid filters from his capillary with each beat of his heart?

38. The solid line on the graph below shows how pressure decreases from the arteries to the right atrium. (a) Which line represents the pressure change that takes place if the arterioles constrict? Explain your reasoning. (b) What will happen to net capillary filtration if pressure changes from line A to line B? Explain.

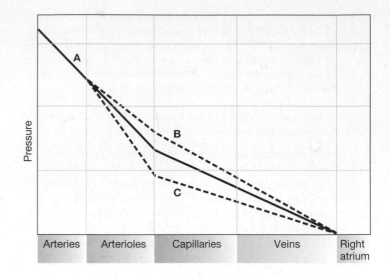

16

Blood

Who would have thought the old man to have had so much blood in him?

William Shakespeare, in *Macbeth,* V, i, 42

PLASMA AND THE CELLULAR ELEMENTS OF BLOOD 512

LO 16.1 Describe the composition of plasma and list the major functions of plasma proteins.

LO 16.2 List the cellular elements of blood, including immature forms and subtypes, and describe the function(s) and distinguishing characteristics of each.

BLOOD CELL PRODUCTION 514

LO 16.3 Describe the differentiation of blood's cellular elements, starting from a pluripotent hematopoietic stem cell and including key cytokines involved in development.

LO 16.4 List the components of a complete blood count.

RED BLOOD CELLS 518

LO 16.5 Compare the structures of immature and mature red blood cells.

LO 16.6 Describe the molecular structure of hemoglobin.

LO 16.7 Create a map of iron metabolism and hemoglobin synthesis.

LO 16.8 Describe the common pathologies of red blood cells.

PLATELETS 523

LO 16.9 Describe the production, structure, and functions of platelets.

HEMOSTASIS AND COAGULATION 524

LO 16.10 Distinguish between hemostasis and coagulation.

LO 16.11 Diagram the key steps of hemostasis, coagulation, and fibrinolysis.

BACKGROUND BASICS

80	Connective tissue
147	Phagocytosis
171	Second messenger cascade
440	Viscosity and resistance
82	Collagen
65	Cell organelles
168	Cytokines

Red blood cells, white blood cells (yellow), and platelets (pink)

Blood, the fluid that circulates in the cardiovascular system, has occupied a prominent place throughout history as an almost mystical fluid. Humans undoubtedly had made the association between blood and life by the time they began to fashion tools and hunt animals. A wounded animal that lost blood would weaken and die if the blood loss was severe enough. The logical conclusion was that blood was necessary for existence. This observation eventually led to the term *lifeblood*, meaning anything essential for existence.

Ancient Chinese physicians linked blood to energy flow in the body. They wrote about the circulation of blood through the heart and blood vessels long before William Harvey described it in seventeenth-century Europe. In China, changes in blood flow were used as diagnostic clues to illness. Chinese physicians were expected to recognize some 50 variations in the pulse. Because blood was considered a vital fluid to be conserved and maintained, bleeding patients to cure disease was not a standard form of treatment.

In contrast, ancient Western civilizations came to believe that disease-causing evil spirits circulated in the blood. The way to remove these spirits was to remove the blood containing them. Because blood was recognized as an essential fluid, however, bloodletting had to be done judiciously. Veins were opened with knives or sharp instruments (*venesection*), or blood-sucking leeches were applied to the skin. In ancient India, people believed that leeches could distinguish between healthy and infected blood.

There is no written evidence that venesection was practiced in ancient Egypt, but the writings of Galen of Pergamum in the second century influenced Western medicine for nearly 2000 years. This early Greek physician advocated bleeding as treatment for many disorders. The location, timing, and frequency of the bleeding depended on the condition, and the physician was instructed to remove enough blood to bring the patient to the point of fainting. Over the years, this practice undoubtedly killed more people than it cured.

What is even more remarkable is the fact that as late as 1923, an American medical textbook advocated bleeding for treating certain infectious diseases, such as pneumonia! Now that we better understand the importance of blood in the immune response, it is doubtful that modern medicine will ever again turn to blood removal as a nonspecific means of treating disease. It is still used, however, for selected *hematological disorders* {*haima*, blood}.

PLASMA AND THE CELLULAR ELEMENTS OF BLOOD

What is this remarkable fluid that flows through the circulatory system? Blood is a connective tissue composed of cellular elements suspended in an extensive fluid matrix called *plasma* [p. 84]. Plasma makes up one-fourth of the extracellular fluid, the internal environment that bathes cells and acts as a buffer between cells and the external environment. Blood is the circulating portion of the extracellular compartment, responsible for carrying material from one part of the body to another.

Total blood volume in a 70-kg man is equal to about 7% of his total body weight, or 0.07×70 kg = 4.9 kg. Thus, if we assume that 1 kg of blood occupies a volume of 1 liter, a 70-kg man has about 5 liters of blood. Of this volume, about 2 liters is composed of blood cells, while the remaining 3 liters is composed of plasma, the fluid portion of the blood. The 58-kg "Reference Woman" [p. 125] has about 4 liters total blood volume.

In this chapter, we present an overview of the components of blood and the functions of plasma, red blood cells, and platelets. You will learn more about hemoglobin when you study oxygen transport in the blood, and more about leukocytes and blood types when you study the immune system.

Plasma Is Extracellular Matrix

Plasma is the fluid matrix of the blood, within which cellular elements are suspended (**FIG. 16.1**). Water is the main component of plasma, accounting for about 92% of its weight. Proteins account for another 7%. The remaining 1% is dissolved organic molecules (amino acids, glucose, lipids, and nitrogenous wastes), ions (Na^+, K^+, Cl^-, H^+, Ca^{2+}, and HCO_3^-), trace elements and vitamins, and dissolved oxygen (O_2) and carbon dioxide (CO_2).

Plasma is identical in composition to interstitial fluid except for the presence of **plasma proteins. Albumins** are the most prevalent type of protein in the plasma, making up about 60% of the total. Albumins and nine other proteins—including *globulins*, the clotting protein *fibrinogen*, and the iron-transporting protein *transferrin*—make up more than 90% of all plasma proteins. The liver makes most plasma proteins and secretes them into the blood. Some globulins, known as *immunoglobulins* or *antibodies*, are synthesized and secreted by specialized blood cells rather than by the liver.

The presence of proteins in the plasma makes the osmotic pressure of the blood higher than that of the interstitial fluid. This osmotic gradient tends to pull water from the interstitial fluid into the capillaries and offset filtration out of the capillaries created by blood pressure [p. 498].

RUNNING PROBLEM | **Blood Doping in Athletes**

Athletes spend hundreds of hours training, trying to build their endurance. For Johann Muehlegg, a cross-country skier at the 2002 Salt Lake City Winter Olympics, it appeared that his training had paid off when he captured three gold medals. On the last day of the Games, however, Olympics officials expelled Muehlegg and stripped him of his gold medal in the 50-kilometer classical race. The reason? Muehlegg had tested positive for a performance-enhancing chemical that increased the oxygen-carrying capacity of his blood. Officials claimed Muehlegg's endurance in the grueling race was the result of blood doping, not training.

512 522 523 527 531

FIG. 16.1 Composition of blood

Blood consists of plasma and cellular elements.

TABLE 16.1	Functions of Plasma Proteins	
Name	**Source**	**Function**
Albumins (multiple types)	Liver	Major contributors to colloid osmotic pressure of plasma; carriers for various substances
Globulins (multiple types)	Liver and lymphoid tissue	Clotting factors, enzymes, antibodies, carriers for various substances
Fibrinogen	Liver	Forms fibrin threads essential to blood clotting
Transferrin	Liver and other tissues	Iron transport

Plasma proteins participate in many functions, including blood clotting and defense against foreign invaders. In addition, they act as carriers for steroid hormones, cholesterol, drugs, and certain ions such as iron (Fe^{2+}). Finally, some plasma proteins act as hormones or as extracellular enzymes. **TABLE 16.1** summarizes the functions of plasma proteins.

Cellular Elements Include RBCs, WBCs, and Platelets

Three main cellular elements are found in blood (Fig. 16.1): **red blood cells (RBCs),** also called **erythrocytes** {*erythros,* red}; **white blood cells (WBCs),** also called **leukocytes** {*leukos,* white}; and **platelets** or *thrombocytes* {*thrombo-,* lump, clot}. White blood cells are the only fully functional cells in the circulation. Red blood cells have lost their nuclei by the time they enter the bloodstream, and platelets, which also lack a nucleus, are cell fragments that have split off a relatively large parent cell known as a **megakaryocyte** {*mega,* extremely large + *karyon,* kernel + *-cyte,* cell}.

Red blood cells play a key role in transporting oxygen from lungs to tissues, and carbon dioxide from tissues to lungs. Platelets are instrumental in *coagulation,* the process by which blood clots prevent blood loss in damaged vessels. White blood cells play a key role in the body's immune responses, defending the body against foreign invaders, such as parasites, bacteria, and viruses. Most white blood cells circulate through the body in the blood, but their work is usually carried out in the tissues rather than in the circulatory system.

Blood contains five types of mature white blood cells: (1) **lymphocytes,** (2) **monocytes,** (3) **neutrophils,** (4) **eosinophils,** and (5) **basophils.** Monocytes that leave the circulation and enter the tissues develop into **macrophages.** Tissue basophils are called **mast cells.**

The types of white blood cells may be grouped according to common morphological or functional characteristics. Neutrophils, monocytes, and macrophages are collectively known as **phagocytes** because they can engulf and ingest foreign particles such as bacteria (phagocytosis) [p. 147]. Lymphocytes are sometimes called **immunocytes** because they are responsible for specific immune responses directed against invaders. Basophils, eosinophils, and neutrophils are called **granulocytes** because they contain cytoplasmic inclusions that give them a granular appearance.

CONCEPT CHECK

1. Name the five types of leukocytes.
2. Why do we say that erythrocytes and platelets are not fully functional cells?
3. On the basis of what you have learned about the origin and role of plasma proteins, explain why patients with advanced liver degeneration frequently suffer from edema [p. 500].

BLOOD CELL PRODUCTION

Where do these different blood cells come from? They are all descendants of a single precursor cell type known as the *pluripotent hematopoietic stem cell* (**FIG. 16.2**). This cell type is found primarily in **bone marrow,** a soft tissue that fills the hollow center of bones. Pluripotent stem cells have the remarkable ability to develop into many different cell types.

As they specialize, they narrow their possible fates. First, they become *uncommitted stem cells,* then *progenitor cells* that are committed to developing into one or perhaps two cell types. Progenitor cells differentiate into red blood cells, lymphocytes, other white blood cells, and megakaryocytes, the parent cells of platelets. It is estimated that only about one out of every 100,000 cells in the bone marrow is an uncommitted stem cell, making it difficult to isolate and study these cells.

In recent years, scientists have been working to isolate and grow uncommitted hematopoietic stem cells to use as replacements in patients whose own stem cells have been killed by cancer chemotherapy. Originally, scientists obtained these stem cells from bone marrow or peripheral blood. Now umbilical cord blood, collected at birth, has been found to be a rich source of hematopoietic stem cells that can be used for transplants in patients with hematological diseases such as leukemia. Public and private cord blood banking programs are active in the United States and Europe, and the American National Marrow Donor Program Registry includes genetic marker information from banked cord blood to help patients find stem cell matches. Currently, researchers are working on techniques for culturing cord blood to increase the number of stem cells in each unit.

FIG. 16.2 Hematopoiesis

Cells below the horizontal line are the predominant forms found circulating in the blood. Cells above the line are found mostly in the bone marrow.

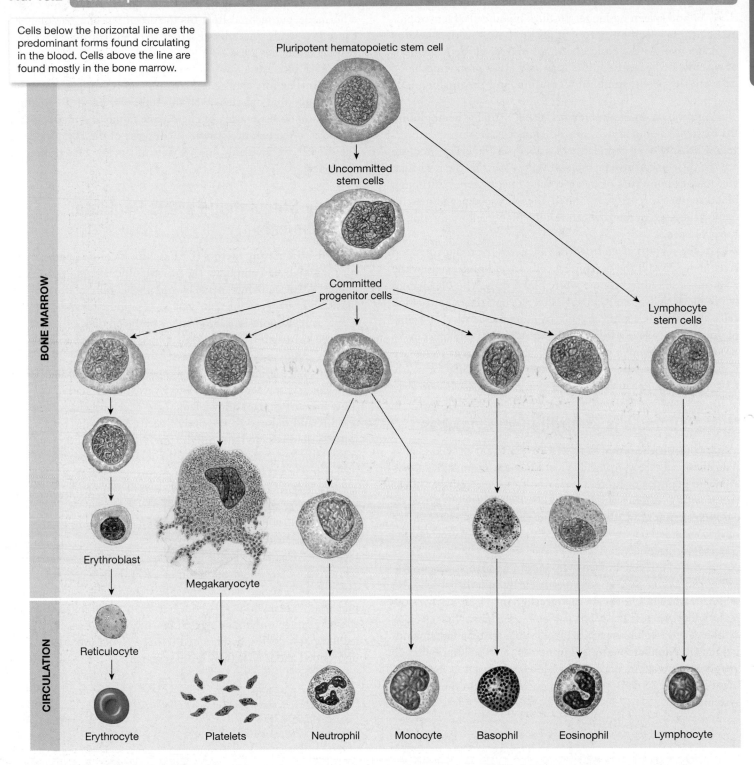

Pluripotent hematopoietic stem cell

Uncommitted stem cells

Committed progenitor cells

Lymphocyte stem cells

BONE MARROW

Erythroblast

Megakaryocyte

CIRCULATION

Reticulocyte

Erythrocyte   Platelets   Neutrophil   Monocyte   Basophil   Eosinophil   Lymphocyte

Blood Cells Are Produced in the Bone Marrow

Hematopoiesis {*haima*, blood + *poiesis*, formation}, the synthesis of blood cells, begins early in embryonic development and continues throughout a person's life. In about the third week of fetal development, specialized cells in the yolk sac of the embryo form clusters. Some of these cell clusters are destined to become the endothelial lining of blood vessels, while others become blood cells. The common embryological origin of the endothelium and blood cells perhaps explains why many cytokines that control hematopoiesis are released by the vascular endothelium.

As the embryo develops, blood cell production spreads from the yolk sac to the liver, spleen, and bone marrow. By birth, the liver and spleen no longer produce blood cells. Hematopoiesis continues in the marrow of all the bones of the skeleton until age five. As the child continues to age, the active regions of marrow decrease. In adults, the only areas producing blood cells are the pelvis, spine, ribs, cranium, and proximal ends of long bones.

Active bone marrow is red because it contains **hemoglobin,** the oxygen-binding protein of red blood cells. Inactive marrow is yellow because of an abundance of adipocytes (fat cells). (You can see the difference between red and yellow marrow the next time you look at bony cuts of meat in the grocery store.) Although blood synthesis in adults is limited, the liver, spleen, and inactive (yellow) regions of marrow can resume blood cell production in times of need.

In the regions of marrow that are actively producing blood cells, about 25% of the developing cells are red blood cells, while 75% are destined to become white blood cells. The life span of white blood cells is considerably shorter than that of red blood cells, and so WBCs must be replaced more frequently. For example, neutrophils have a 6-hour half-life, and the body must make more than *100 million* neutrophils each day in order to replace those that die. Red blood cells, on the other hand, live for nearly four months in the circulation.

Hematopoiesis Is Controlled by Cytokines

What controls the production and development of blood cells? The chemical factors known as cytokines are responsible. *Cytokines* are peptides or proteins released from one cell that affect the growth or activity of another cell [p. 168]. Newly discovered cytokines are often called *factors* and given a modifier that describes their actions: growth factor, differentiating factor, trophic (nourishing) factor.

Some of the best-known cytokines in hematopoiesis are the *colony-stimulating factors,* molecules made by endothelial cells and white blood cells. Others are the **interleukins** {*inter-*, between + *leuko,* white}, such as IL-3. The name *interleukin* was first given to cytokines released by one white blood cell to act on another white blood cell. Numbered interleukin names, such as interleukin-3, are given to cytokines once their amino acid sequences have been identified. Interleukins also play important roles in the immune system.

Another hematopoietic cytokine is *erythropoietin,* which controls red blood cell synthesis. Erythropoietin is usually called a hormone, but technically it fits the definition of a cytokine because it is made on demand rather than stored in vesicles like peptide hormones are.

TABLE 16.2 lists a few of the many cytokines linked to hematopoiesis. The role cytokines play in blood cell production is so complicated that one review on this topic was titled "Regulation of hematopoiesis in a sea of chemokine family members with a plethora of redundant activities"!* Because of the complexity of the subject, we give only an overview of the key hematopoietic cytokines.

Colony-Stimulating Factors Regulate Leukopoiesis

Colony-stimulating factors (CSFs) were identified and named for their ability to stimulate the growth of leukocyte colonies in culture. These cytokines, made by endothelial cells, marrow fibroblasts, and leukocytes, regulate leukocyte production and development, or **leukopoiesis.** CSFs induce both cell division (mitosis) and cell maturation in stem cells. Once a leukocyte matures, it loses its ability to undergo mitosis.

One fascinating aspect of leukopoiesis is that production of new leukocytes is regulated in part by existing white blood cells. This form of control allows leukocyte development to be very specific and tailored to the body's needs. When the body's defense system is called on to fight foreign invaders, both the absolute number of leukocytes and the relative proportions of the different types of leukocytes in the circulation change.

Clinicians often rely on a *differential white cell count* to help them arrive at a diagnosis (**FIG. 16.3**). For example, a person with a bacterial infection usually has a high total number of leukocytes in the blood, with an increased percentage that are neutrophils. Cytokines released by active leukocytes fighting the bacterial infection stimulate the production of additional neutrophils and monocytes. A person with a viral infection may have a high, normal, or low total white cell count but often shows an increase in the percentage of lymphocytes. The complex process by which leukocyte production is matched to need is still not completely understood and is an active area of research.

---

*Broxmeyer H. E. and C. H. Kim, *Exp Hematol* 27(7): 1113–1123, 1999, July.

TABLE 16.2	Cytokines Involved in Hematopoiesis		
Name	**Sites of Production**	**Influences Growth or Differentiation of**
Erythropoietin (EPO)	Kidney cells primarily	Red blood cells
Thrombopoietin (TPO)	Liver primarily	Megakaryocytes
Colony-stimulating factors, interleukins, stem cell factor	Endothelium and fibroblasts of bone marrow, leukocytes	All types of blood cells; mobilizes hematopoietic stem cells

FIG. 16.3 **ESSENTIALS**

The Complete Blood Count (CBC)

A complete blood count, commonly known as a CBC, provides the information in the table below. The numbers shown are the normal ranges of values. In addition, a CBC usually also includes the following information:

- **Mean corpuscular volume (MCV):** the average volume of one red blood cell. A corpuscle is a small unattached cell (diminutive of *corpus*, body)
- **Mean corpuscular hemoglobin (MCH):** amount of hemoglobin per RBC
- **Mean corpuscular hemoglobin concentration (MCHC):** the amount of hemoglobin per volume of one red blood cell

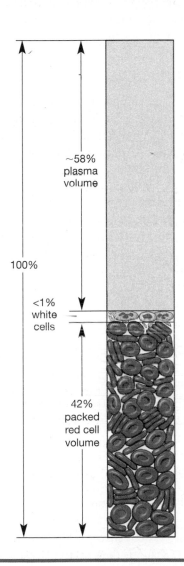

~58% plasma volume

100%

<1% white cells

42% packed red cell volume

Blood Count Normal Ranges of Values

Test	Males	Females
Hematocrit		
Hematocrit is the percentage of total blood volume that is occupied by packed (centrifuged) red blood cells.	40–54%	37–47%
Hemoglobin (g Hb/dL* whole blood)		
The hemoglobin value reflects the oxygen-carrying capacity of red blood cells. (*1 deciliter (dL) = 100 mL)	14–17	12–16
Red Cell Count (cells/μL)		
A machine counts erythrocytes as they stream through a beam of light.	$4.5–6.5 \times 10^6$	$3.9–5.6 \times 10^6$
Total White Count (cells/μL)		
A total white cell count includes all types of leukocytes but does not distinguish between them.	$4–11 \times 10^3$	$4–11 \times 10^3$
Differential White Cell Count		
The differential white cell count presents estimates of the relative proportions of the five types of leukocytes in a thin blood smear stained with biological dyes.		
Neutrophils	50–70%	50–70%
Eosinophils	1–4%	1–4%
Basophils	<1%	<1%
Lymphocytes	20–40%	20–40%
Monocytes	2–8%	2–8%
Platelets (per μL)		
Platelet count is suggestive of the blood's ability to clot.	$150–450 \times 10^3$	$150–450 \times 10^3$

Scientists are working to create a model for the control of leukopoiesis so that they can develop effective treatments for diseases characterized by either a lack or an excess of leukocytes. The *leukemias* are a group of diseases characterized by the abnormal growth and development of leukocytes. In *neutropenias* {*penia*, poverty}, patients have too few leukocytes and are unable to fight off bacterial and viral infections. Researchers hope to find better treatments for both leukemias and neutropenias by unlocking the secrets of how the body regulates cell growth and division.

Thrombopoietin Regulates Platelet Production

Thrombopoietin (TPO) is a glycoprotein that regulates the growth and maturation of megakaryocytes, the parent cells of platelets. (Recall that *thrombocyte* is an alternative name for *platelet*). TPO is produced primarily in the liver. This cytokine was first described in 1958, but its gene was not cloned until 1994. Within a year, genetically engineered TPO was widely available to researchers and physician-scientists hoping to use it to stimulate platelet production in patients with too few platelets, or *thrombocytopenia* {*penia*, poverty}. The first TPO drugs had to be recalled after patients developed adverse side effects but newer TPO agonists are currently in clinical use. Despite having these drugs, scientists still do not understand everything about the basic biology of thrombopoiesis, and research continues.

Erythropoietin Regulates RBC Production

Red blood cell production (**erythropoiesis**) is controlled by the glycoprotein **erythropoietin (EPO),** assisted by several cytokines. Erythropoietin is made primarily in the kidneys of adults. The stimulus for EPO synthesis and release is *hypoxia*, low oxygen levels in the tissues. Hypoxia stimulates production of a transcription factor called *hypoxia-inducible factor 1* (HIF-1), which turns on the EPO gene to increase EPO synthesis. This pathway, like other endocrine pathways, helps the body maintain homeostasis. By stimulating the synthesis of red blood cells, EPO puts more hemoglobin into the circulation to carry oxygen.

The existence of a hormone controlling red blood cell production was first suggested in the 1950s, but two decades passed before scientists succeeded in purifying the substance. One reason for the delay is that EPO is made on demand and not stored, as in an endocrine cell. It took scientists another nine years to identify the amino acid sequence of EPO and to isolate and clone the gene for it. However, an incredible leap was made after the EPO gene was isolated: only two years later, the hormone was produced by recombinant DNA technology and put into clinical use.

In recent years, physicians have been able to prescribe not only genetically engineered EPO, such as epoetin, but also several colony-stimulating factors (sargramostim and filgrastim) that stimulate white blood cell synthesis. Cancer patients in whom hematopoiesis has been suppressed by chemotherapy have benefited from injections of these hematopoietic hormones, but in 2007, the Food and Drug Administration issued new dosing instructions and warnings about an increased risk of blood clots in patients taking erythropoiesis-stimulating agents. Scientists are currently monitoring the safety of CSFs to ensure that they do not increase the likelihood of developing hematological diseases.

CONCEPT CHECK   **4.** Name the cytokine(s) that regulate(s) growth and maturation in (a) erythrocytes, (b) leukocytes, and (c) megakaryocytes.

RED BLOOD CELLS

Erythrocytes, or red blood cells, are the most abundant cell type in the blood. A microliter of blood contains about 5 million red blood cells, compared with only 4000–11,000 leukocytes and 150,000–450,000 platelets. The primary function of red blood cells is to facilitate oxygen transport from the lungs to cells, and carbon dioxide transport from cells to lungs.

The ratio of red blood cells to plasma is indicated clinically by the **hematocrit** and is expressed as a percentage of the total blood volume (Fig. 16.3). Hematocrit is determined by drawing a blood sample into a narrow capillary tube and spinning it in a centrifuge so that the heavier red blood cells go to the bottom of the sealed tube, leaving the thin "buffy layer" of lighter leukocytes and platelets in the middle, and plasma on top.

The column of *packed red cells* is measured, and the hematocrit value is reported as a percentage of the total sample volume. The normal range of hematocrit is 40–54% for a man and 37–47% for a woman. This test provides a rapid and inexpensive way to estimate a person's red cell count because blood for a hematocrit can be collected by simply sticking a finger.

Mature RBCs Lack a Nucleus

In the bone marrow, committed progenitor cells differentiate through several stages into large, nucleated *erythroblasts*. As erythroblasts mature, the nucleus condenses and the cell shrinks in diameter from 20 μm to about 7 μm. In the last stage before maturation, the nucleus is pinched off and phagocytized by bone marrow macrophages. At the same time, other membranous organelles (such as mitochondria) break down and disappear. The final immature cell form, called a *reticulocyte*, leaves the marrow and enters the circulation, where it matures into an erythrocyte in about 24 hours (**FIG. 16.4c**).

Mature mammalian red blood cells in an isotonic solution are biconcave disks, shaped much like jelly doughnuts with the filling squeezed out of the middle (**FIG. 16.5a**). They are simple membranous "bags" filled with enzymes and hemoglobin. Because red blood cells contain no mitochondria, they cannot carry out aerobic metabolism. Glycolysis is their primary source of ATP. Without a nucleus and endoplasmic reticulum to carry out protein synthesis, red blood cells are unable to make new enzymes or to renew membrane components. This inability leads to an increasing loss of membrane flexibility, making older cells more fragile and likely to rupture.

The biconcave shape of the red blood cell is one of its most distinguishing features. The membrane is held in place by a complex cytoskeleton composed of filaments linked to transmembrane attachment proteins (Fig. 16.5b). Despite the cytoskeleton, red blood cells are remarkably flexible, like a partially filled water balloon that can compress into various shapes. This flexibility allows red blood cells to change shape as they squeeze through the narrow capillaries of the circulation.

The disk-like structure of red blood cells also allows them to modify their shape in response to osmotic changes in the blood. In hypertonic media, red blood cells shrink up and develop a spiky surface when the membrane pulls tight against the cytoskeleton (Fig. 16.5c). An erythrocyte placed in a slightly hypotonic medium [p. 127] swells and forms a sphere without disruption of its membrane integrity (Fig. 16.5d).

The **morphology** {*morphe*, form} of red blood cells can provide clues to the presence of disease. Sometimes the cells lose their flattened disk shape and become spherical (*spherocytosis*), a shape

FIG. 16.4 **FOCUS ON...**

Bone Marrow

(a) The bone marrow, hidden within the bones of the skeleton, is easily overlooked as a tissue, although collectively it is nearly the size and weight of the liver!

(b) Marrow is a highly vascular tissue, filled with blood sinuses, widened regions lined with epithelium.

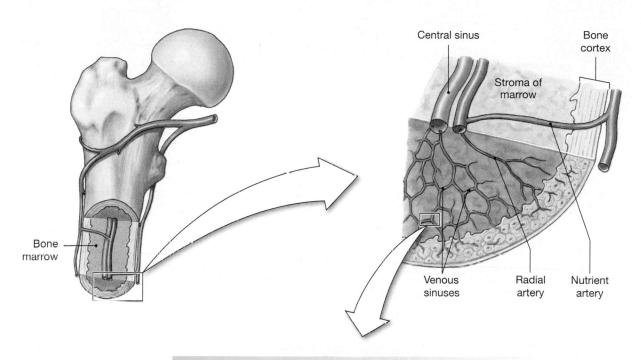

Central sinus

Bone cortex

Stroma of marrow

Bone marrow

Venous sinuses

Radial artery

Nutrient artery

(c) Bone marrow consists of blood cells in different stages of development and supporting tissue known as the **stroma** {mattress}.

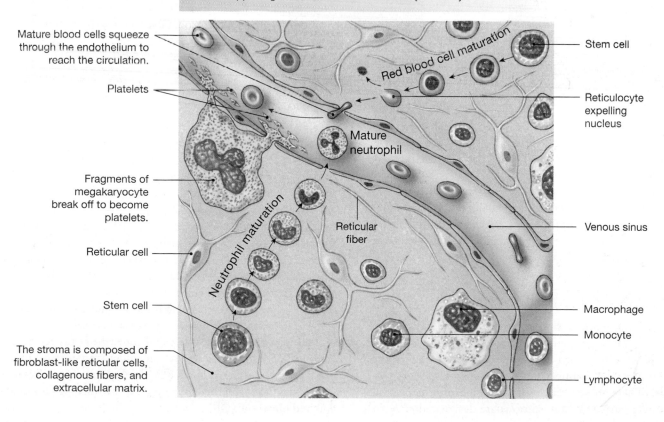

Mature blood cells squeeze through the endothelium to reach the circulation.

Platelets

Fragments of megakaryocyte break off to become platelets.

Reticular cell

Stem cell

The stroma is composed of fibroblast-like reticular cells, collagenous fibers, and extracellular matrix.

Neutrophil maturation

Mature neutrophil

Reticular fiber

Red blood cell maturation

Stem cell

Reticulocyte expelling nucleus

Venous sinus

Macrophage

Monocyte

Lymphocyte

FIG. 16.5   Erythrocytes, or red blood cells (RBCs)

(a) Scanning electron micrographs (SEMs) show biconcave disk shape of RBCs.

(b) The cytoskeleton creates the unique shape of RBCs.

Cross section of RBC

Cytoskeleton filament

Attachment protein

Actin

RBC in isotonic medium

(c) Erythrocytes placed in a hypertonic medium shrink, but the rigid cytoskeleton remains intact, creating a spiky surface. These cells are said to be crenated {*crenatus,* a notch}.

(d) Erythrocytes placed in a hypotonic medium swell and lose their characteristic biconcave disk shape.

(e) Abnormal hemoglobin in sickle cell disease can cause RBCs to change shape.

Sickled RBC

similar to that of the cell in hypotonic medium. In sickle cell anemia, the cells are shaped like a sickle or crescent moon (Fig. 16.5e). In some disease states, the size of red blood cells—the mean red cell volume, also called the **mean corpuscular volume (MCV)**—may be either abnormally large or abnormally small. For example, red blood cells can be abnormally small, or *microcytic,* in iron-deficiency anemia. If they are pale due to lack of red hemoglobin, they are described as *hypochromic* {*chroma,* color}.

Hemoglobin Synthesis Requires Iron

Hemoglobin, the main component of red blood cells, is best known for its role in oxygen transport. Hemoglobin (Hb) is a large, complex protein with four globular protein chains, each of which is wrapped around an iron-containing *heme group* (**FIG. 16.6a**). There are several isoforms of **globin** proteins in hemoglobin. The most common isoforms are designated *alpha* (α),

beta (β), *gamma* (γ), and *delta* (δ), depending on the structure of the chain. Most adult hemoglobin (designated *HbA*) has two alpha chains and two beta chains, as shown. However, a small portion of adult hemoglobin (about 2.5%) has two alpha chains and two delta chains (HbA_2).

The four **heme groups** in a hemoglobin molecule are identical. Each consists of a carbon-hydrogen-nitrogen *porphyrin ring* with an iron atom (Fe) in the center (Fig. 16.6b). About 70% of the iron in the body is found in the heme groups of hemoglobin. Consequently, hemoglobin synthesis requires an adequate supply of iron in the diet (Fig. 16.6c ❶). Most dietary iron comes from red meat, beans, spinach, and iron-fortified bread.

Iron is absorbed in the small intestine by active transport (Fig. 16.6c ❷). A carrier protein called **transferrin** binds iron and transports it in the blood ❸ . The bone marrow takes up iron and uses it to make the heme group of hemoglobin for developing red blood cells ❹ .

FIG. 16.6   Hemoglobin

(a) A hemoglobin molecule is composed of four protein globin chains, each centered around a heme group. In most adult hemoglobin, there are two alpha chains and two beta chains as shown.

α Chain

β Chain

Heme group

(b) Each heme group consists of a porphyrin ring with an iron atom in the center.

Porphyrin ring

R = additional C, H, O groups

(c) Hemoglobin and iron

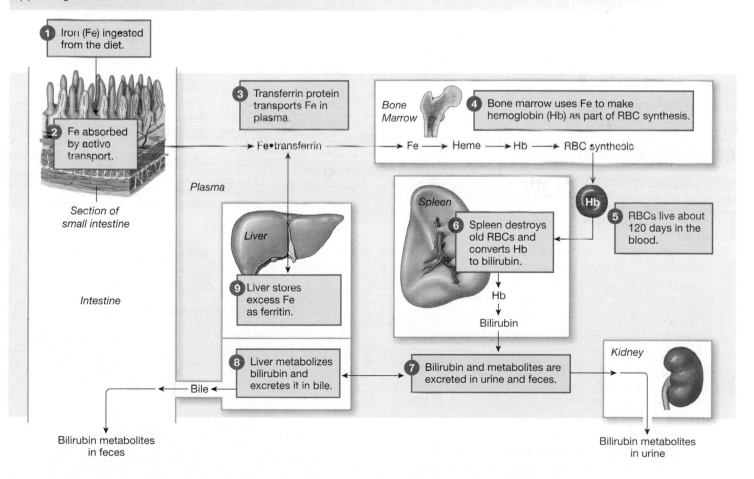

1 Iron (Fe) ingested from the diet.

2 Fe absorbed by active transport.

Section of small intestine

3 Transferrin protein transports Fe in plasma.

Plasma

Intestine

Fe•transferrin

Bone Marrow

4 Bone marrow uses Fe to make hemoglobin (Hb) as part of RBC synthesis.

Fe ⟶ Heme ⟶ Hb ⟶ RBC synthesis

Hb

5 RBCs live about 120 days in the blood.

Liver

Spleen

6 Spleen destroys old RBCs and converts Hb to bilirubin.

Hb

Bilirubin

9 Liver stores excess Fe as ferritin.

8 Liver metabolizes bilirubin and excretes it in bile.

Bile

7 Bilirubin and metabolites are excreted in urine and feces.

Kidney

Bilirubin metabolites in feces

Bilirubin metabolites in urine

Blood doping to increase the oxygen-carrying capacity of the blood has been a problem in endurance sports for more than 30 years. The first sign that Muehlegg might be cheating by this method was the result of a simple blood test for hemoglobin and hematocrit, taken several hours before his 50-km race. Muehlegg's blood hemoglobin level registered above 17.5 g/dL. A repeat test was within acceptable limits, however, and Muehlegg was allowed to race.

Q1: *What is the normal range for Muehlegg's hemoglobin (Fig. 16.3)?*

Q2: *Olympic officials also tested Muehlegg's hematocrit. With blood doping, would you expect a hematocrit value to be lower or higher than normal?*

512   **522**   523   527   531

Excess ingested iron is stored, mostly in the liver. Iron stores are found inside a spherical protein called molecule **ferritin** ❾. The core of the sphere is an iron-containing mineral that can be converted to soluble iron and released when needed for hemoglobin synthesis.

Excess iron in the body is toxic, and poisoning sometimes occurs in children when they ingest too many vitamin pills containing iron. Initial symptoms of iron toxicity are gastrointestinal pain, cramping, and internal bleeding, which occurs as iron corrodes the digestive epithelium. Subsequent problems include liver failure, which can be fatal.

RBCs Live about Four Months

Red blood cells in the circulation live for about 120 ± 20 days. Increasingly fragile older red blood cells may rupture as they try to squeeze through narrow capillaries, or they may be engulfed by scavenging macrophages as they pass through the spleen (Fig. 16.6 ❻). Many components of the destroyed cells are recycled. Amino acids from the globin chains of hemoglobin are incorporated into new proteins, and some iron from the heme groups is reused to make new heme groups.

The spleen and liver convert remnants of the heme groups to a colored pigment called **bilirubin.** Bilirubin is carried by plasma albumin to the liver, where it is metabolized and incorporated into a secretion called **bile** (Fig. 16.6 ❽). Bile is secreted into the digestive tract, and the bilirubin metabolites leave the body in the feces. Small amounts of other bilirubin metabolites are filtered from the blood in the kidneys, where they contribute to the yellow color of urine ❼.

In some circumstances, bilirubin levels in the blood become elevated (*hyperbilirubinemia*). This condition, known as **jaundice,** causes the skin and whites of the eyes to take on a yellow cast. The accumulation of bilirubin can occur from several different causes. Newborns whose fetal hemoglobin is being broken down

and replaced with adult hemoglobin are particularly susceptible to bilirubin toxicity, so doctors monitor babies for jaundice in the first weeks of life. Another common cause of jaundice is liver disease, in which the liver is unable to process or excrete bilirubin.

CONCEPT CHECK
5. Distinguish between (a) heme and hemoglobin, and (b) ferritin and transferrin.

6. Is bile an endocrine secretion or an exocrine secretion?

RBC Disorders Decrease Oxygen Transport

Because hemoglobin plays a critical role in oxygen transport, the red blood cell count and hemoglobin content of the body are important. If hemoglobin content is too low—a condition known as **anemia**—the blood cannot transport enough oxygen to the tissues. People with anemia are usually tired and weak, especially during exercise. The major causes of anemia are summarized in **TABLE 16.3**.

In the *hemolytic anemias* {*lysis*, rupture}, the rate of red blood cell destruction exceeds the rate of red blood cell production. The hemolytic anemias are usually hereditary defects in which the body makes fragile cells. For example, in *hereditary spherocytosis,* the erythrocyte cytoskeleton does not link properly because of defective or deficient cytoskeletal proteins. Consequently, the cells are shaped more like spheres than like biconcave disks. This disruption in the cytoskeleton results in red blood cells that rupture easily and are unable to withstand osmotic changes as well as normal cells can. Several of the hemolytic anemias are acquired diseases, as indicated in Table 16.3.

Some anemias are the result of abnormal hemoglobin molecules. *Sickle cell disease* is a genetic defect in which glutamate, the sixth amino acid in the 146–amino acid beta chain of hemoglobin, is replaced by valine. The result is abnormal hemoglobin (a form referred to as *HbS*) that crystallizes when it gives up its oxygen.

CLINICAL FOCUS

Diabetes: Hemoglobin and Hyperglycemia

One of the goals of diabetes treatment is to keep blood glucose concentrations as close to normal as possible, but how can a clinician tell if a patient has been doing this? One way is to analyze the patient's hemoglobin. Glucose in the plasma binds covalently to hemoglobin, producing a glycohemoglobin known as **hemoglobin A_{1C}** ("A-one-C"). The amount of hemoglobin A_{1C} in the plasma is directly related to hemoglobin's exposure to glucose over the preceding 8–12 weeks. By using this assay, a clinician can monitor long-term fluctuations in blood glucose levels and adjust a diabetic patient's therapy appropriately.

TABLE 16.3 | Causes of Anemia

Accelerated Red Blood Cell Loss

Blood loss: cells are normal in size and hemoglobin content but low in number

Hemolytic anemias: cells rupture at an abnormally high rate

Hereditary

Membrane defects (example: hereditary spherocytosis)

Enzyme defects

Abnormal hemoglobin (example: sickle cell anemia)

Acquired

Parasitic infections (example: malaria)

Drugs

Autoimmune reactions

Decreased Red Blood Cell Production

Defective red blood cell or hemoglobin synthesis in the bone marrow

Aplastic anemia: can be caused by certain drugs or radiation

Inadequate dietary intake of essential nutrients

Iron deficiency (iron is required for heme production)

Folic acid deficiency (folic acid is required for DNA synthesis)

Vitamin B_{12} deficiency (B_{12} is required for DNA synthesis): may be due to lack of intrinsic factor for B_{12} absorption

Inadequate production of erythropoietin

This crystallization pulls the red blood cells into a sickle shape, like a crescent moon (Fig. 16.5e). The sickled cells become tangled with other sickled cells as they pass through the smallest blood vessels, causing the cells to jam up and block blood flow to the tissues. This blockage creates tissue damage and pain from hypoxia.

One treatment for sickle cell disease is the administration of *hydroxyurea*, a compound that inhibits DNA synthesis. Hydroxyurea alters bone marrow function so that immature red blood cells produce the fetal form of hemoglobin (*HbF*) instead of adult hemoglobin. HbF interferes with the crystallization of hemoglobin, so that HbS no longer forms and the red blood cells no longer sickle. In addition, some studies show improvements in sickle cell symptoms before HbF levels increase. One theory of why this happens is based on the finding that hydroxyurea is metabolized to nitric oxide (NO), which causes vasodilation. Inhaled nitric oxide is now being tested as a treatment for sickle cell disease symptoms.

Other anemias result from the failure of the bone marrow to make adequate amounts of hemoglobin. One of the most common examples of an anemia that results from insufficient hemoglobin synthesis is *iron-deficiency anemia*. If iron loss by the body exceeds iron intake, the marrow does not have adequate iron to make heme groups, and hemoglobin synthesis slows.

People with iron-deficiency anemia have either a low red blood cell count (reflected in a low hematocrit) or low hemoglobin content in their blood. Their red blood cells are often smaller than usual (*microcytic* red blood cells), and the lower hemoglobin content may cause the cells to be paler than normal, in which case they are described as being *hypochromic* {*hypo-*, below normal; *chrom-*, color}. Women who menstruate are likely to suffer from iron-deficiency anemia because of iron loss in menstrual blood.

Although the anemias are common, it is also possible to have too many red blood cells. *Polycythemia vera* {*vera*, true} is a stem cell dysfunction that produces too many blood cells, white as well as red. These patients may have hematocrits as high as 60–70% (normal is 37–54%). The increased number of cells causes the blood to become more viscous and thus more resistant to flow through the circulatory system [p. 440].

In *relative polycythemia*, the person's red blood cell number is normal, but the hematocrit is elevated because of low plasma volume. This might occur with dehydration, for example. The opposite problem can also occur. If an athlete overhydrates, the hematocrit may decrease temporarily because of increased plasma volume. In both of these situations, there is no actual pathology involving the red blood cells.

CONCEPT CHECK

7. A person who goes from sea level to a city that is 5000 feet above sea level begins to show an increased hematocrit within days. Draw the reflex pathway that links the hypoxia of high altitude to increased red blood cell production.

PLATELETS

As noted earlier, platelets are cell fragments produced in the bone marrow from huge cells called megakaryocytes. Megakaryocytes develop their formidable size by undergoing DNA replication up to seven times without undergoing nuclear or cytoplasmic division. The result is a *polyploid* cell with multiple copies of its

RUNNING **PROBLEM**

In its earliest form, blood doping was accomplished by blood transfusions, which increased the athlete's oxygen-carrying capacity. One hallmark of a recent blood transfusion is elevated hemoglobin and hematocrit levels. Muehlegg claimed that his elevated hemoglobin was a result of his special diet and of dehydration from diarrhea he had suffered the night before.

Q3: *Explain how diarrhea could cause a temporarily elevated hematocrit.*

Q4: *How might Muehlegg quickly reduce his hematocrit without removing red blood cells?*

512 — 522 — **523** — 527 — 531

FIG. 16.7 Megakaryocytes and platelets

(a) Megakaryocytes are giant cells with multiple copies of DNA in the nucleus.

The edges of the megakaryocyte break off to form cell fragments called platelets.

Platelets

Red blood cell

Endoplasmic reticulum

(b) Inactive platelets are small disk-like cell fragments.

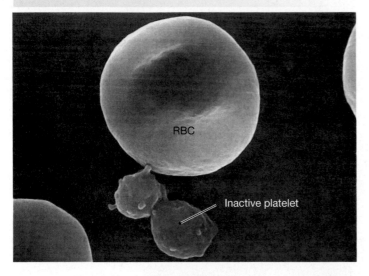

RBC

Inactive platelet

(c) Activated platelets (shown enlarged) develop a spiky outer surface and adhere to each other.

Activated platelet

DNA in its lobed nucleus (**FIG. 16.7**). The outer edges of marrow megakaryocytes extend through the endothelium into the lumen of marrow blood sinuses, where the cytoplasmic extensions fragment into disk-like platelets (Fig. 16.4c).

Platelets are smaller than red blood cells, are colorless, and have no nucleus. Their cytoplasm contains mitochondria, smooth endoplasmic reticulum, and numerous membrane-bound vesicles called *granules* that are filled with a variety of cytokines and growth factors. There are at least three different types of granules. One type of granule has been shown to contain more than 280 different proteins. Some of these proteins you learned about in other contexts, such as VEGF that promotes angiogenesis [p. 481] and matrix metalloproteinases (MMPs) [p. 74].

Platelets are always present in the blood and their typical life span is about 10 days. They are best known for their role in helping stop blood loss, but in recent years, scientists have shown that they also act as immune cells and mediators of the inflammatory response. They apparently help the immune system fight infectious diseases such as malaria, and they may contribute to the inflammatory process of atherosclerosis [p. 502].

The newest role for platelets is their use in *platelet-rich plasma (PRP) therapy*. This treatment became popular after Tiger Woods, the golfer, announced he had used PRP to aid his recovery from knee surgery. Tendons and ligaments have minimal blood supply and are notoriously slow to heal, and the theory behind PRP therapy is that growth factors and cytokines inside platelet granules will promote healing. At this time, however, evidence for PRP efficacy from well-designed, placebo-controlled studies is lacking.

HEMOSTASIS AND COAGULATION

Because of its fluid nature, blood flows freely throughout the circulatory system. However, if there is a break in the "piping" of the system, blood will be lost unless steps are taken. One of the challenges for the body is to plug holes in damaged blood vessels while still maintaining blood flow through the vessel.

It would be simple to block off a damaged blood vessel completely, like putting a barricade across a street full of potholes. However, just as shopkeepers on that street lose business if traffic is blocked, cells downstream from the point of injury die from lack of oxygen and nutrients if the vessel is completely blocked. The body's task is to allow blood flow through the vessel while simultaneously repairing the damaged wall.

This challenge is complicated by the fact that blood in the system is under pressure. If the repair "patch" is too weak, it is blown out by the blood pressure. For this reason, stopping blood loss involves several steps. First, the pressure in the vessel must be decreased long enough to create a secure mechanical seal in the form of a blood clot. Once the clot is in place and blood loss has been stopped, the body's repair mechanisms can take over. Then, as the wound heals, enzymes gradually dissolve the clot while scavenger leukocytes ingest and destroy the debris.

Hemostasis Prevents Blood Loss from Damaged Vessels

Hemostasis {*haima,* blood + *stasis,* stoppage} is the process of keeping blood within a damaged blood vessel (**FIG. 16.8**).

FIG. 16.8 Hemostasis and tissue repair

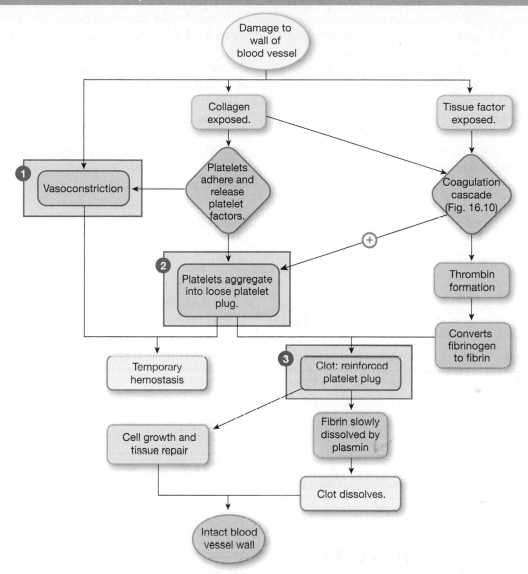

(The opposite of hemostasis is *hemorrhage* {*-rrhagia,* abnormal flow}.) Hemostasis has three major steps: ❶ vasoconstriction, ❷ temporary blockage of a break by a platelet plug, and ❸ *coagulation,* the formation of a clot that seals the hole until tissues are repaired.

The first step in hemostasis is immediate constriction of damaged vessels to decrease blood flow and pressure within the vessel temporarily. When you put pressure on a bleeding wound, you also decrease flow within the damaged vessel. Vasoconstriction normally is caused by paracrine molecules released from the endothelium.

Vasoconstriction is rapidly followed by the second step, mechanical blockage of the hole by a loose **platelet plug.** Plug formation begins with **platelet adhesion,** when platelets *adhere* or stick to exposed collagen in the damaged area. The adhered platelets become activated, releasing cytokines into the area around the injury. These platelet factors reinforce local

vasoconstriction and activate more platelets, which *aggregate* or stick to one another to form a loose platelet plug. Platelets activating more platelets are an example of a positive feedback loop [p. 16].

Simultaneously, exposed collagen and **tissue factor** (a protein-phospholipid mixture) initiate the third step, the formation of a *fibrin* protein mesh that stabilizes the platelet plug to form a **clot.** Fibrin is the end product of a series of enzymatic reactions known as the **coagulation cascade.** Some chemical factors involved in the coagulation cascade also promote platelet adhesion and aggregation in the damaged region. Eventually, as the damaged vessel repairs itself, the clot retracts when fibrin is slowly dissolved by the enzyme *plasmin.*

The body must maintain the proper balance during hemostasis. Too little hemostasis allows excessive bleeding; too much creates a **thrombus,** a blood clot that adheres to the undamaged wall of a blood vessel {*thrombos,* a clot or lump}. A large thrombus can block the lumen of the vessel and stop blood flow.

Hemostasis seems straightforward, but unanswered questions remain at the cellular and molecular levels. Inappropriate blood clotting plays an important role in strokes and heart attacks. Inherited mutations affecting platelet function can lead to inappropriate clotting or excessive bleeding due to failure of hemostasis.

A detailed study of hemostasis involves many chemical factors, some of which play multiple roles and have multiple names. For this reason, learning about hemostasis can be especially challenging. For example, some factors participate in both platelet plug formation and coagulation, and one factor in the cascade activates enzymes for both clot formation and clot dissolution. Because of the complexity of the coagulation cascade, we discuss only a few aspects of hemostasis in additional detail.

Platelet Activation Begins the Clotting Process

When a blood vessel wall is first damaged, exposed collagen and chemicals from endothelial cells activate platelets (**FIG. 16.9 ①**). Normally, the blood vessel's endothelium separates the collagenous matrix fibers from the circulating blood. But when the vessel is damaged, collagen is exposed, and platelets rapidly begin to adhere to it.

Platelets adhere to collagen with the help of *integrins,* membrane receptor proteins that are linked to the cytoskeleton [p. 74].

Binding activates platelets so that they release the contents of their intracellular granules, including *serotonin* (5-hydroxytryptamine), ADP, and **platelet-activating factor (PAF).** PAF sets up a positive feedback loop by activating more platelets.

PAF also initiates pathways that convert platelet membrane phospholipids into **thromboxane A2** [p. 179]. Serotonin and thromboxane A2 are vasoconstrictors. They also contribute to platelet aggregation, along with ADP and PAF (**TBL. 16.4**). The net result is a growing platelet plug that seals the damaged vessel wall.

If platelet aggregation is a positive feedback event, what prevents the platelet plug from continuing to form and spreading beyond the site of injury to other areas of the vessel wall? The answer lies in the fact that platelets do not adhere to normal endothelium. Intact vascular endothelial cells convert their membrane lipids into **prostacyclin,** an eicosanoid [p. 30] that blocks platelet adhesion and aggregation (Fig. 16.9). Nitric oxide released by normal, intact endothelium also inhibits platelets from adhering. The combination of platelet attraction to the injury site and repulsion from the normal vessel wall creates a localized response that limits the platelet plug to the area of damage.

Coagulation Converts a Platelet Plug into a Clot

The third major step in hemostasis, coagulation, is a complex process in which fluid blood forms a gelatinous clot.

FIG. 16.9   Platelet plug formation

Platelets will not adhere to intact endothelium. Damage triggers platelet plug formation where collagen has been exposed.

Lumen of blood vessel

Intact endothelium releases prostacyclin and nitric oxide (NO).

Prevents platelet adhesion

Smooth muscle cells

Collagen subendothelial layer

Exposed collagen in damaged blood vessel wall

ECF

1. Exposed collagen binds and activates platelets.

2. Release of platelet factors

3. Factors attract more platelets.

4. Platelets aggregate into platelet plug.

TABLE 16.4	Factors Involved in Platelet Function			
Chemical Factor	Source	Activated by or Released in Response to	Role in Platelet Plug Formation	Other Roles and Comments
Collagen	Subendothelial extracellular matrix	Injury exposes platelets to collagen	Binds platelets to begin platelet plug	N/A
von Willebrand factor (vWF)	Endothelium, megakaryocytes	Exposure to collagen	Links platelets to collagen	Deficiency or defect causes prolonged bleeding
Serotonin	Secretory vesicles of platelets	Platelet activation	Platelet aggregation	Vasoconstrictor
Adenosine diphosphate (ADP)	Platelet mitochondria	Platelet activation, thrombin	Platelet aggregation	N/A
Platelet-activating factor (PAF)	Platelets, neutrophils, monocytes	Platelet activation	Platelet aggregation	Plays role in inflammation; increases capillary permeability
Thromboxane A2	Phospholipids in platelet membranes	Platelet-activating factor	Platelet aggregation	Vasoconstrictor; eicosanoid
Platelet-derived growth factor (PDGF)	Platelets	Platelet activation	N/A	Promotes wound healing by attracting fibroblasts and smooth muscle cells

RUNNING **PROBLEM**

After the 1984 Olympics Games, where U.S. cyclists reportedly suffered bad side effects following blood transfusions, the International Olympic Committee and other organizations banned blood doping. Then recombinant human EPO (rhEPO) became available in the late 1980s, and athletes started injecting the drug to increase their body's red blood cell production. Subsequently the biotechnology firm Amgen created a longer-acting derivative of EPO named darbepoetin. Athletes using rhEPO and darbepoetin hoped to escape detection by using these natural hormones, but sports organizations worked with scientists to develop methods for detection.

Q5: *Endogenous EPO, rhEPO, and darbepoetin all induce red blood cell synthesis, but they can be distinguished from one another when a urine sample is tested by electrophoresis [p. 100]. Explain how three hormones made from the same gene can all be active yet different enough from one another to be detectable in the laboratory.*

Q6: *One hallmark of illegal EPO use is elevated reticulocytes in the blood. Why would this suggest greater-than-normal EPO activity?*

512 522 523 **527** 531

Coagulation is divided into two pathways that eventually merge into one (**FIG. 16.10**). An **intrinsic pathway** (yellow) begins when damage to the tissue exposes collagen. For this reason, the intrinsic pathway is also known as the *contact activation pathway*. The intrinsic pathway uses proteins already present in the plasma. Collagen activates the first enzyme, factor XII, to begin the cascade.

An **extrinsic pathway** (blue) starts when damaged tissues expose tissue factor, also called *tissue thromboplastin* or factor III. The extrinsic pathway is also called the cell injury pathway or tissue factor pathway. Tissue factor activates factor VII to begin the extrinsic pathway.

The two pathways unite at the **common pathway** (green) to create **thrombin,** the enzyme that converts **fibrinogen** into insoluble **fibrin** polymers. These fibrin fibers become part of the clot.

Coagulation was initially regarded as a cascade similar to second messenger cascades of signal transduction [p. 171]. At each step, an enzyme converts an inactive precursor into an active enzyme, often with the help of Ca^{2+}, membrane phospholipids, or additional factors. We now know, however, that the process is more than a simple cascade. Factors in the intrinsic and extrinsic pathways interact with each other, making coagulation a network rather than a simple cascade. In addition, several positive feedback loops sustain the cascade until one or more of the participating plasma proteins are completely consumed.

The final step of coagulation is the conversion of fibrinogen into fibrin, a reaction catalyzed by the enzyme *thrombin*

FIG. 16.10 The coagulation cascade

Inactive plasma proteins (white boxes) are converted into active enzymes in each step of the pathway.

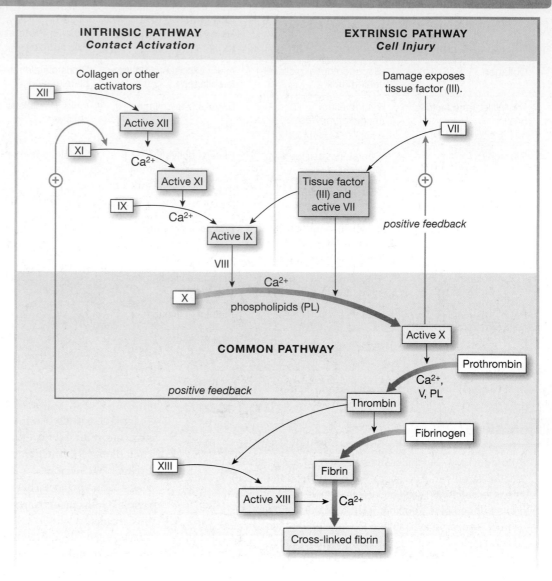

(**FIG. 16.11a**). The fibrin fibers weave through the platelet plug and trap red blood cells within their mesh (Fig. 16.11b). Active factor XIII converts fibrin into a cross-linked polymer that stabilizes the clot.

Clots are only a temporary fix. As the damaged vessel wall slowly repairs itself, the clot disintegrates when fibrin is broken into fragments by the enzyme **plasmin** (Fig. 16.11a). An inactive form of plasmin, **plasminogen,** is part of the clot. After coagulation, thrombin, a factor in the coagulation cascade, works with a second factor called **tissue plasminogen activator (tPA)** to convert inactive plasminogen into plasmin. Plasmin then breaks down fibrin, a process known as **fibrinolysis.**

The large number of factors involved in coagulation and the fact that a single factor may have many different names can be confusing (**TBL. 16.5**). Scientists assigned numbers to the coagulation factors, but the factors are not numbered in the order

in which they participate in the coagulation cascade. Instead, they are numbered according to the order in which they were discovered.

CONCEPT CHECK

8. Using Figure 16.10, draw the positive feedback loop initiated by thrombin. What stops this loop?

Anticoagulants Prevent Coagulation

Once coagulation begins, what keeps it from continuing until the entire circulation has clotted? Two mechanisms limit the extent of blood clotting within a vessel: (1) inhibition of platelet adhesion and (2) inhibition of the coagulation cascade and fibrin production (**TBL. 16.6**). As mentioned earlier, factors such as prostacyclin in the blood vessel endothelium and plasma ensure that

FIG. 16.11 Coagulation and fibrinolysis

(a) Conversion of fibrinogen into fibrin, and subsequent fibrinolysis

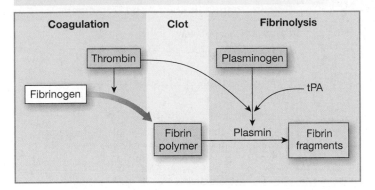

(b) Red blood cells are trapped in the fibrin mesh of a clot.

the platelet plug is restricted to the area of damage (see left side of Fig. 16.9).

In addition, endothelial cells release chemicals known as **anticoagulants,** which prevent coagulation from taking place. Most act by blocking one or more of the reactions in the coagulation cascade. The body produces two anticoagulants, **heparin** and **antithrombin III,** which work together to block active factors IX, X, XI, and XII. **Protein C,** another anticoagulant in the body, inhibits clotting factors V and VIII.

The discovery of the factors controlling coagulation and fibrinolysis was an important step in developing treatments for many diseases related to coagulation problems. For example, heart attacks, more properly called *myocardial infarctions* (MIs), occur when a coronary blood vessel is blocked by a blood clot.

Unless the blockage is removed promptly, the tissue will die or be severely damaged. One option for dissolving blood clots is to use fibrinolytic drugs—such as *streptokinase* (from bacteria) and *tissue plasminogen activator* (tPA)—to dissolve the clots. These drugs are now being combined with *antiplatelet agents* to prevent further platelet plug and clot formation. Some antiplatelet agents act as antagonists to platelet integrin receptors and prevent platelets from adhering to collagen.

Acetylsalicylic acid (aspirin) is an agent that prevents platelet plug formation. It acts by inhibiting the COX enzymes [p. 179] that promote synthesis of the platelet activator thromboxane A2. People who are at risk of developing small blood clots are sometimes told to take one aspirin every other day "to thin the blood." The aspirin does not actually make the blood less viscous, but it does prevent clots from forming by blocking platelet aggregation. Aspirin is now given routinely as emergency treatment for a suspected heart attack.

Anticoagulant drugs may be prescribed for people who are in danger of forming small blood clots that could block off critical vessels in the brain, heart, or lungs. The *coumarin anticoagulants,* such as *warfarin* (Coumadin®), block the action of vitamin K, a cofactor [p. 49] in the synthesis of clotting factors II (thrombin), VII, IX, and X. These anticoagulants were discovered when cattle that developed severe bleeding problems were found to have been eating spoiled sweet clover.

When blood samples are drawn into glass tubes, clotting takes place very rapidly through the contact activation (intrinsic) pathway unless the tube contains an anticoagulant. One of the anticoagulants used for this purpose, EGTA, removes free Ca^{2+} from the plasma. Calcium is an essential clotting factor, so with no Ca^{2+}, no coagulation can occur. In the living body, however, plasma Ca^{2+} levels never decrease to levels that interfere with coagulation.

Several inherited diseases affect the coagulation process. Patients with coagulation disorders bruise easily. In severe forms, spontaneous bleeding may occur throughout the body. Bleeding into the joints and muscles can be painful and disabling. If bleeding occurs in the brain, it can be fatal.

The best-known coagulation disorder is **hemophilia,** a name given to several diseases in which one of the factors in the coagulation cascade is either defective or lacking. Hemophilia A, a factor VIII deficiency, is the most common form, occurring in about 80% of all cases. This disease is a recessive sex-linked trait that usually affects only males.

One exciting development in the treatment of hemophilia is the development of gene therapy for hemophilia B, a deficiency in clotting factor IX. Patients injected with a virus engineered to carry the gene for factor IX started to produce some of the factor on their own, reducing their need for expensive injections of artificial factor IX. To learn more about these clinical trials and the latest treatments for hemophilia, visit the National Hemophilia Foundation web site at *www.hemophilia.org.*

TABLE 16.5	Factors Involved in Coagulation

Chemical Factor	Source	Activated by or Released in Response to	Role in Coagulation	Other Roles and Comments
Collagen	Subendothelial extracellular matrix	Injury that exposes collagen to plasma clotting factors	Starts intrinsic pathway	N/A
von Willebrand factor (vWF)	Endothelium, megakaryocytes	Exposure to collagen	Regulates level of factor VIII	Deficiency or defect causes prolonged bleeding
Kininogen and kallikrein	Liver and plasma	Cofactors normally present in plasma pathway	Cofactors for contact activation of intrinsic pathway	Mediate inflammatory response; enhance fibrinolysis
Tissue factor (tissue thromboplastin or factor III)	Most cells except platelets	Damage to tissue	Starts extrinsic pathway	N/A
Prothrombin and thrombin (factor II)	Liver and plasma	Platelet lipids, Ca^{2+} and factor V	Fibrin production	N/A
Fibrinogen and fibrin (factor I)	Liver and plasma	Thrombin	Form insoluble fibers that stabilize platelet plug	N/A
Fibrin-stabilizing factor (XIII)	Liver, megakaryocytes	Platelets	Cross-links fibrin polymers to make stable mesh	N/A
Ca^{2+} (factor IV)	Plasma ions	N/A	Required for several steps of coagulation cascade	Never a limiting factor
Vitamin K	Diet	N/A	Needed for synthesis of factors II, VII, IX, X	N/A

TABLE 16.6	Endogenous Factors Involved in Fibrinolysis and Anticoagulation

Chemical Factor	Source	Activated by or Released in Response to	Role in Anticoagulation or Fibrinolysis	Other Roles and Comments
Plasminogen and plasmin	Liver and plasma	tPA and thrombin	Dissolves fibrin and fibrinogen	N/A
Tissue plasminogen activator (tPA)	Many tissues	Normally present; levels increase with stress, protein C	Activates plasminogen	Recombinant tPA used clinically to dissolve clots
Antithrombin III	Liver and plasma	N/A	Anticoagulant; blocks factors IX, X, XI, XII, thrombin, kallikrein	Facilitated by heparin; no effect on thrombin despite name
Prostacyclin (prostaglandin I, or PGI2)	Endothelial cells	N/A	Blocks platelet aggregation	Vasodilator

RUNNING **PROBLEM CONCLUSION** | Blood Doping in Athletes

Johann Muehlegg's elevated hemoglobin and hematocrit prior to his 50-km race meant an automatic urine drug test following the race. At the time of the 2002 Olympics, athletes knew that there was a urine test for EPO, but they were not aware that the same test could detect darbepoetin. Both of Muehlegg's urine samples tested positive for darbepoetin, and he was stripped of his 50-km gold metal. The International Olympic Committee tested other athletes for rhEPO at the 2002 Salt Lake City Winter Olympics and had more than 100 positive results. Despite official prohibitions, blood doping in endurance sports remains a major problem.

Now check your understanding of the physiology behind blood doping by comparing your answers with the information in the following table.

Question	Facts	Integration and Analysis
Q1: What is the normal range for Muehlegg's hemoglobin?	Normal hemoglobin range for males is 14–17 g/dL whole blood.	N/A
Q2: With blood doping, would you expect hematocrit value to be lower or higher than normal?	Hematocrit is the percentage of a blood sample volume that is packed red blood cells. A primary function of red blood cells is to carry oxygen.	Blood doping is done to increase oxygen-carrying capacity; therefore, the athlete would want more blood cells. This would mean a higher hematocrit.
Q3: Explain how diarrhea could cause a temporarily elevated hematocrit.	Diarrhea causes dehydration, which is loss of fluid volume. Plasma is the fluid component of blood.	If the total volume of red cells is unchanged but plasma volume decreases with dehydration, the hematocrit will increase.
Q4: How might Muehlegg quickly reduce his hematocrit without removing red blood cells?	Hematocrit = RBC volume/total blood volume (total blood volume = plasma volume + cell volume)	If plasma volume increases, hematocrit will decrease even though red cell volume does not change. By drinking fluids, Muehlegg could increase his plasma volume quickly.
Q5: Explain how endogenous EPO, rhEPO, and darbepoetin made from the same gene can all be active yet different enough from one another to be detectable in the laboratory.	Activity depends on the protein binding to the receptor's binding site. Post-translational modification allows proteins from the same gene to be altered so that they are different from one another.	The three hormones have sites that bind to and activate the EPO receptor, but they have different sizes or charges that cause them to separate during electrophoresis. For example, the glycosylation pattern [p. 29] of rhEPO is different from the pattern in endogenous EPO.
Q6: One hallmark of illegal EPO use is elevated reticulocytes in the blood. Why would this suggest greater-than-normal EPO activity?	Reticulocytes are the final immature stage of red blood cell development. Maturation usually takes place in the marrow.	If red blood cell development becomes more rapid, more reticulocytes may be released into the blood before they have time to mature.

(512)(522)(523)(527)(**531**)

VISUALIZE, PRACTICE, AND MASTER PHYSIOLOGY WITH MASTERING A&P

- Practice-on-the-go with Dynamic Study Modules
- Visualize important processes with Interactive Physiology and A&P Flix
- Check your understanding with Video Tutors and practice quizzes!

MasteringA&P®

CHAPTER SUMMARY

Blood is an interesting tissue, with blood cells and cell fragments suspended in a liquid matrix—the plasma—that forms one of the two extracellular *compartments*. *Exchange* between the plasma and interstitial fluid takes place only in the capillaries. *Bulk flow* of blood through the body depends on the pressure gradient created by the heart. At the same time, high pressure in the blood vessels poses a danger should the wall of a vessel rupture. Collectively, the cellular and protein components of blood carry out hemostasis and coagulation to protect against excessive blood loss. Blood cells are also essential for oxygen transport and defense, as you will learn in later chapters.

Plasma and the Cellular Elements of Blood

1. Blood is the circulating portion of the extracellular compartment. (p. 512)

2. **Plasma,** the liquid matrix of blood, is composed mostly of water, with dissolved proteins, organic molecules, ions, and dissolved gases. (p. 512; Fig. 16.1)

3. The **plasma proteins** include **albumins, globulins,** and the clotting protein **fibrinogen.** They function in blood clotting, defense, and as hormones, enzymes, or carriers for different substances. (p. 512)

4. The cellular elements of blood are **red blood cells (erythrocytes), white blood cells (leukocytes),** and **platelets.** Platelets are fragments of cells called **megakaryocytes.** (p. 514; Fig. 16.1)

5. Blood contains five types of white blood cells: (1) **lymphocytes,** (2) **monocytes,** (3) **neutrophils,** (4) **eosinophils,** and (5) **basophils.** (p. 514; Fig. 16.1)

Blood Cell Production

6. All blood cells develop from a pluripotent hematopoietic stem cell. (p. 514; Fig. 16.2)

7. **Hematopoiesis** begins early in embryonic development and continues throughout a person's life. Most hematopoiesis takes place in the bone marrow. (p. 515; Fig. 16.4)

8. **Colony-stimulating factors** and other cytokines control white blood cell production. **Thrombopoietin** regulates the growth and maturation of megakaryocytes. Red blood cell production is regulated primarily by **erythropoietin.** (pp. 516, 517, 518)

Red Blood Cells

9. Mature mammalian red blood cells are biconcave disks lacking a nucleus. They contain hemoglobin, a red oxygen-carrying pigment. (p. 518; Fig. 16.5)

10. Hemoglobin synthesis requires iron in the diet. Iron is transported in the blood on **transferrin** and stored mostly in the liver on the protein **ferritin.** (pp. 520, 522; Fig. 16.6)

11. When hemoglobin is broken down, some heme groups are converted into **bilirubin,** which is incorporated into **bile** and excreted. Elevated bilirubin concentrations in the blood cause **jaundice.** (p. 522; Fig. 16.6)

Platelets

12. Platelets are cell fragments filled with granules containing clotting proteins and cytokines. Platelets are activated by damage to vascular endothelium. (p. 523; Fig. 16.7)

Hemostasis and Coagulation

13. **Hemostasis** begins with vasoconstriction and the formation of a **platelet plug.** (pp. 524, 525; Fig. 16.8)

14. Exposed collagen triggers **platelet adhesion** and **platelet aggregation.** The platelet plug is converted into a clot when reinforced by **fibrin.** (pp. 525, 527; Fig. 16.8)

15. In the last step of the **coagulation cascade,** fibrin is made from **fibrinogen** through the action of **thrombin.** (pp. 525, 527; Fig. 16.10)

16. As the damaged vessel is repaired, **plasmin** trapped in the platelet plug dissolves fibrin (**fibrinolysis**) and breaks down the clot. (p. 528; Fig. 16.11)

17. Platelet plugs are restricted to the site of injury by **prostacyclin** in the membrane of intact vascular endothelium. **Anticoagulants** limit the extent of blood clotting within a vessel. (pp. 526, 529; Fig. 16.9)

REVIEW QUESTIONS

In addition to working through these questions and checking your answers on p. A-21, review the Learning Outcomes at the beginning of this chapter.

Level One Reviewing Facts and Terms

1. The fluid portion of the blood, called _____, is composed mainly of _____.

2. List the three types of plasma proteins. Name at least one function of each type. Which type is most prevalent in the body?

3. List the cellular elements found in blood, and name at least one function of each.

4. Blood cell production is called _____. When and where does it occur?

5. What role do colony-stimulating factors, cytokines, and interleukins play in blood cell production? How are these chemical signal molecules different? Give two examples of each.

6. List the technical terms for production of red blood cells, production of platelets, and production of white blood cells.

7. The hormone that directs red blood cell synthesis is called _____. Where is it produced, and what is the stimulus for its production?

8. How are the terms *hematocrit* and *packed red cells* related? What are normal hematocrit values for men and women?

9. Distinguish between an erythroblast and an erythrocyte. Give three distinct characteristics of erythrocytes.

10. Which chemical element in the diet is important for hemoglobin synthesis?

11. Define the following terms and explain their significance in hematology.

    (a) jaundice
    (b) anemia
    (c) transferrin
    (d) hemophilia

12. Chemicals that prevent blood clotting from occurring are called _____.

Level Two Reviewing Concepts

13. **Concept maps:** Combine each list of terms into a map. You may add other terms.

List 1

• ADP	• platelet aggregation
• collagen	• platelet plug
• integrins	• positive feedback
• membrane	• serotonin
• phospholipids	• thromboxane A2
• platelet-activating factor	• vasoconstriction
• platelet activation	
• platelet adhesion	

List 2

• clot	• infarct
• coagulation	• plasmin
• fibrin	• plasminogen
• fibrinogen	• polymer
• fibrinolysis	• thrombin

List 3

• bile	• heme
• bilirubin	• hemoglobin
• bone marrow	• intestine
• erythropoietin	• iron
• ferritin	• liver
• globin	• reticulocyte
• hematocrit	• transferrin

14. Distinguish between the intrinsic, extrinsic, and common pathways of the coagulation cascade.

15. Once platelets are activated to aggregate, what factors halt their activity?

Level Three Problem Solving

16. Rachel is undergoing chemotherapy for breast cancer. She has blood cell counts at regular intervals, with these results:

	Normal Count Cells × 10^3/μL	Rachel's Count 10 Days Post-Chemotherapy	Rachel's Count 20 Days Post-Chemotherapy
WBC	4–11	2.6	4.9
RBC	3.9–5.6	3.85	4.2
Platelets	150–450	133	151

At the time of the 10-day test, Jen, Rachel's nurse, notes that Rachel, although pale and complaining of being tired, does not have any bruises on her skin. Jen tells Rachel to eat foods high in protein, take multivitamin tablets containing iron, and stay home and away from crowds as much as possible. How are Jen's observations and recommendations related to the results of the 1-day and 20-day blood tests?

17. Hemochromatosis is an inherited condition in which the body absorbs iron excessively, resulting in an elevated total body load of iron.

    (a) What plasma protein would you expect to be elevated in this disease?
    (b) Which organ(s) would you expect to show damage in this disease?
    (c) Can you think of a simple treatment that could decrease the body's overload of iron in hemochromatosis?

18. Erythropoietin (EPO) was first isolated from the urine of anemic patients who had high circulating levels of the hormone. Although these patients had high concentrations of EPO, they were unable to produce adequate amounts of hemoglobin or red cells. Give some possible reasons why the patients' EPO was unable to correct their anemia.

Level Four Quantitative Problems

19. If we estimate that total blood volume is 7% of body weight, calculate the total blood volume in a 200-lb man and in a 130-lb woman (2.2 lb/kg). What are their plasma volumes if the man's hematocrit is 52% and the woman's hematocrit is 41%?

20. The total blood volume of an average person is 7% of total body weight. Using this figure and the fact that 1 kg of blood occupies a volume of about 1 liter, figure the total red blood cell volume of a 50-kg woman with a hematocrit of 40%.

Answers to Concept Checks, Figure and Graph Questions, and end-of-chapter Review Questions can be found in Appendix A [p. A-1].

17

Mechanics of Breathing

This being of mine, whatever it really is, consists of a little flesh, a little breath, and the part which governs.

Marcus Aurelius Antoninus (121–180 C.E.), *Meditations* (c. 161–180 C.E.) New York: Dutton, 1937

THE RESPIRATORY SYSTEM 535

LO 17.1   List four major functions of the respiratory system.

LO 17.2   Diagram the anatomy of the respiratory system and explain the function of each structure.

GAS LAWS 542

LO 17.3   Explain and express mathematically the relationship between atmospheric pressure, water vapor pressure, and the partial pressures of individual gases.

LO 17.4   Explain the relationship between the pressure of a gas and the volume in which it is contained.

VENTILATION 544

LO 17.5   Define and describe the lung volumes and lung capacities.

LO 17.6   Explain how pressures and lung volumes change during normal breathing, and how that affects air flow in the respiratory system.

LO 17.7   Explain how subatmospheric intrapleural pressure develops and the role it plays in normal breathing.

LO 17.8   Graph the alveolar and intrapleural pressure changes that occur during one respiratory cycle.

LO 17.9   Compare and contrast compliance and elastance in respiratory physiology, giving examples of disease states that demonstrate changes in compliance and/ or elastance.

LO 17.10   Explain the role of surface tension and surfactants in respiratory physiology.

LO 17.11   Map the factors affecting airway resistance, with emphasis on local and reflex control mechanisms involved in bronchodilation and bronchoconstriction.

LO 17.12   Compare and contrast total pulmonary ventilation and alveolar ventilation.

LO 17.13   Explain why gas composition in the alveoli remains relatively constant during normal breathing and how it changes with hyper- and hypoventilation.

LO 17.14   Explain the local control mechanisms by which ventilation and alveolar blood flow are matched.

LO 17.15   Compare obstructive and restrictive lung diseases.

BACKGROUND BASICS

79	Ciliated and exchange epithelia
442	Pressure, volume, flow, and resistance
438	Pulmonary circulation
39	Surface tension
359	Autonomic and somatic motor neurons
442	Velocity of flow

Colored x-ray of the lung showing the branching airways.

Imagine covering the playing surface of a racquetball court (about 75 m²) with thin plastic wrap, then crumpling up the wrap and stuffing it into a 3-liter soft drink bottle. Impossible? Maybe so, if you use plastic wrap and a drink bottle. But the lungs of a 70-kg man have a gas exchange surface the size of that plastic wrap, compressed into a volume that is less than that of the bottle. This tremendous surface area for gas exchange is needed to supply trillions of cells in the body with adequate amounts of oxygen.

Aerobic metabolism in cells depends on a steady supply of oxygen and nutrients from the environment, coupled with the removal of carbon dioxide. In very small aquatic animals, simple diffusion across the body surface meets these needs. Distance limits diffusion rate, however, so most multicelled animals require specialized respiratory organs associated with a circulatory system. Respiratory organs take a variety of forms, but all possess a large surface area compressed into a small space.

Besides needing a large exchange surface, humans and other terrestrial animals face an additional physiological challenge: dehydration. The exchange surface must be thin and moist to allow gases to pass from air into solution, and yet at the same time it must be protected from drying out as a result of exposure to air. Some terrestrial animals, such as the slug (a shell-less snail), meet the challenge of dehydration with behavioral adaptations that restrict them to humid environments and nighttime activities.

A more common solution is anatomical: an internalized respiratory epithelium. Human lungs are enclosed in the chest cavity to control their contact with the outside air. Internalization creates a humid environment for gas exchange with the blood and protects the delicate exchange surface from damage.

Internalized lungs create another challenge, however: how to move air between the atmosphere and an exchange surface deep within the body. Air flow requires a muscular pump to create pressure gradients. More complex respiratory systems therefore consist of two separate components: a muscle-driven pump and a thin, moist exchange surface. In humans, the pump is the musculoskeletal structure of the thorax. The lungs themselves consist of the exchange epithelium and associated blood vessels.

The four primary functions of the respiratory system are:

1. *Exchange of gases between the atmosphere and the blood.* The body brings in O_2 for distribution to the tissues and eliminates CO_2 waste produced by metabolism.
2. *Homeostatic regulation of body pH.* The lungs can alter body pH by selectively retaining or excreting CO_2.
3. *Protection from inhaled pathogens and irritating substances.* Like all other epithelia that contact the external environment, the respiratory epithelium is well supplied with defense mechanisms to trap and destroy potentially harmful substances before they can enter the body.
4. *Vocalization.* Air moving across the vocal cords creates vibrations used for speech, singing, and other forms of communication.

In addition to serving these functions, the respiratory system is also a significant source of water loss and heat loss from the body. These losses must be balanced using homeostatic compensations.

In this chapter, you will learn how the respiratory system carries out these functions by exchanging air between the environment and the interior air spaces of the lungs. This exchange is the *bulk flow* of air, and it follows many of the same principles that govern the bulk flow of blood through the cardiovascular system:

1. Flow takes place from regions of higher pressure to regions of lower pressure.
2. A muscular pump creates pressure gradients.
3. Resistance to air flow is influenced primarily by the diameter of the tubes through which the air is flowing.

Air and blood are both fluids. The primary difference between air flow in the respiratory system and blood flow in the circulatory system is that air is a less viscous, compressible mixture of gases while blood is a noncompressible liquid.

THE RESPIRATORY SYSTEM

The word *respiration* has several meanings in physiology (**FIG. 17.1**). **Cellular respiration** refers to the intracellular reaction of oxygen with organic molecules to produce carbon dioxide, water, and energy in the form of ATP [p. 105]. **External respiration,** the topic of this chapter and the next, is the movement of gases between the environment and the body's cells. External respiration can be subdivided into four integrated processes, illustrated in Figure 17.1:

1. *The exchange of air between the atmosphere and the lungs.* This process is known as **ventilation,** or breathing. **Inspiration** (inhalation) is the movement of air into the lungs. **Expiration**

RUNNING **PROBLEM** | Emphysema

You could hear the sound of her whistling, wheezing breathing preceding her down the hall. "Diagnosis: COPD," reads Edna Wilson's patient chart. COPD—chronic obstructive pulmonary disease—is the name given to diseases in which air exchange is impaired by narrowing of the lower airways. Most people with COPD have emphysema or chronic bronchitis or a combination of the two. Individuals in whom chronic bronchitis predominates are sometimes called "blue bloaters," owing to the bluish tinge of their skin (from low blood oxygen levels) and a tendency to be overweight. In contrast, patients with emphysema have been nicknamed "pink puffers." They tend to be thin, have normal (pink) skin coloration, and often breathe out through pursed lips, which helps open their airways. More than 15 million people in the United States have diagnosed COPD and in 2011 COPD was the third leading cause of death. Most people can avoid the disease simply by not smoking cigarettes. Unfortunately, Edna has been a heavy smoker for 35 of her 47 years.

535   537   550   552   554   559

FIG. 17.1 External respiration

The respiratory and circulatory systems coordinate to move O_2 and CO_2 between the atmosphere and the cells.

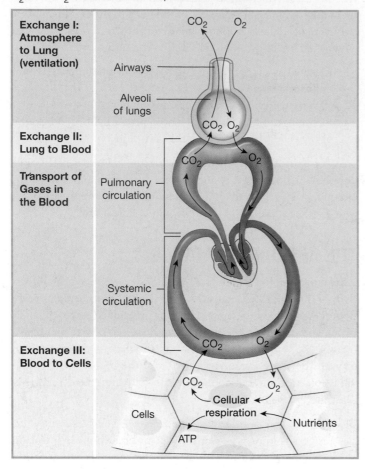

Exchange I:
Atmosphere
to Lung
(ventilation)

Airways

Alveoli
of lungs

Exchange II:
Lung to Blood

Transport of
Gases in
the Blood

Pulmonary
circulation

Systemic
circulation

Exchange III:
Blood to Cells

Cells

Cellular
respiration

Nutrients

ATP

(exhalation) is the movement of air out of the lungs. The mechanisms by which ventilation takes place are collectively called the *mechanics of breathing.*

2. *The exchange of O_2 and CO_2 between the lungs and the blood.*
3. *The transport of O_2 and CO_2 by the blood.*
4. *The exchange of gases between blood and the cells.*

External respiration requires coordination between the respiratory and cardiovascular systems. The **respiratory system** consists of structures involved in ventilation and gas exchange (**FIG. 17.2**):

1. The **conducting system** of passages, or **airways,** that lead from the external environment to the exchange surface of the lungs.
2. The **alveoli** (singular **alveolus**) {*alveus,* a concave vessel}, a series of interconnected sacs and their associated *pulmonary capillaries.* These structures form the exchange surface, where oxygen moves from inhaled air to the blood, and carbon dioxide moves from the blood to air that is about to be exhaled.
3. The bones and muscles of the thorax (chest cavity) and abdomen that assist in ventilation.

The respiratory system can be divided into two parts. The **upper respiratory tract** consists of the mouth, nasal cavity, pharynx, and larynx. The **lower respiratory tract** consists of the trachea, two primary bronchi {*bronchos,* windpipe; singular—bronchus}, their branches, and the lungs. The lower tract is also known as the *thoracic portion* of the respiratory system because it is enclosed in the thorax.

Bones and Muscles of the Thorax Surround the Lungs

The thorax is bounded by the bones of the spine and rib cage and their associated muscles. Together the bones and muscles are called the *thoracic cage.* The ribs and spine (the *chest wall*) form the sides and top of the cage. A dome-shaped sheet of skeletal muscle, the **diaphragm,** forms the floor (Fig. 17.2a).

Two sets of **intercostal muscles,** internal and external, connect the 12 pairs of ribs (Fig. 17.2c). Additional muscles, the **sternocleidomastoids** and the **scalenes,** run from the head and neck to the sternum and first two ribs.

Functionally, the thorax is a sealed container filled with three membranous bags, or sacs. One, the *pericardial sac,* contains the heart. The other two bags, the **pleural sacs,** each surround a lung {*pleura,* rib or side}. The esophagus and thoracic blood vessels and nerves pass between the pleural sacs (Fig. 17.2d).

Pleural Sacs Enclose the Lungs

The **lungs** (Fig. 17.2a, b) consist of light, spongy tissue whose volume is occupied mostly by air-filled spaces. These irregular cone-shaped organs nearly fill the thoracic cavity, with their bases resting on the curved diaphragm. Semi-rigid conducting airways—the bronchi—connect the lungs to the main airway, the trachea.

Each lung is surrounded by a double-walled pleural sac whose membranes line the inside of the thorax and cover the outer surface of the lungs (Fig. 17.2d, **FIG. 17.3**). Each *pleural membrane,* or **pleura,** contains several layers of elastic connective tissue and numerous capillaries. The opposing layers of pleural membrane are held together by a thin film of **pleural fluid** whose total volume is only about 25–30 mL in a 70-kg man. The result is similar to an air-filled balloon (the lung) surrounded by a water-filled balloon (the pleural sac). Most illustrations exaggerate the volume of the pleural fluid, but you can appreciate its thinness if you imagine spreading 25 mL of water evenly over the surface of a 3-liter soft drink bottle.

Pleural fluid serves several purposes. First, it creates a moist, slippery surface so that the opposing membranes can slide across one another as the lungs move within the thorax. Second, it holds the lungs tight against the thoracic wall. To visualize this arrangement, think of two panes of glass stuck together by a thin film of water. You can slide the panes back and forth across each other, but you cannot pull them apart because of the cohesiveness of the water [p. 39]. A similar fluid bond between the two pleural membranes makes the lungs "stick" to the thoracic cage and holds them stretched in a partially inflated state, even at rest.

Airways Connect Lungs to the External Environment

Air enters the upper respiratory tract through the mouth and nose and passes into the **pharynx,** a common passageway for food, liquids, and air {*pharynx,* throat}. From the pharynx, air flows through the **larynx** into the **trachea,** or windpipe (Fig. 17.2a). The larynx contains the **vocal cords,** connective tissue bands that vibrate and tighten to create sound when air moves past them.

The trachea is a semiflexible tube held open by 15 to 20 C-shaped cartilage rings. It extends down into the thorax, where it branches (division 1) into a pair of **primary bronchi,** one *bronchus* to each lung (Fig. 17.2a). Within the lungs, the bronchi branch repeatedly (divisions 2–11) into progressively smaller bronchi (Fig. 17.2e). Like the trachea, the bronchi are semirigid tubes supported by cartilage.

Within the lungs, the smallest bronchi branch to become **bronchioles,** small collapsible passageways with walls of smooth muscle. The bronchioles continue branching (divisions 12–23) until the *respiratory bronchioles* form a transition between the airways and the exchange epithelium of the lung.

The diameter of the airways becomes progressively smaller from the trachea to the bronchioles, but as the individual airways get narrower, their numbers increase geometrically (**FIG. 17.4**). As a result, the total cross-sectional area increases with each division of the airways. Total cross-sectional area is lowest in the upper respiratory tract and greatest in the bronchioles, analogous to the increase in cross-sectional area that occurs from the aorta to the capillaries in the circulatory system [p. 443].

Velocity of air flow is inversely proportional to total cross-sectional area of the airways [p. 442]. This is similar to the velocity of blood flow through different parts of the circulatory system, and means that the velocity of air flow is greatest in the upper airways and slowest in the terminal bronchioles.

CONCEPT CHECK	
	1. What is the difference between cellular respiration and external respiration?
	2. Name the components of the upper respiratory tract and those of the lower respiratory tract.
	3. Name the components (including muscles) of the thoracic cage. List the contents of the thorax.
	4. Which air passages of the respiratory system are collapsible?

The Airways Warm, Humidify, and Filter Inspired Air

During breathing, the upper airways and the bronchi do more than simply serve as passageways for air. They play an important role in conditioning air before it reaches the alveoli. Conditioning has three components:

1. *Warming* air to body temperature (37 °C), so that core body temperature does not change and alveoli are not damaged by cold air;

RUNNING PROBLEM

Edna has not been able to stop smoking, and her COPD is a combination of emphysema and bronchitis. Patients with chronic bronchitis have excessive mucus production and exhibit general inflammation of the entire respiratory tract. The mucus narrows the airways and makes breathing difficult.

Q1: *What does narrowing of the airways do to airway resistance?* [Hint: *Poiseuille's Law, p. 440*]

535   537   550   552   554   559

2. *Adding water vapor* until the air reaches 100% humidity, so that the moist exchange epithelium does not dry out; and

3. *Filtering out foreign material,* so that viruses, bacteria, and inorganic particles do not reach the alveoli.

Inhaled air is warmed by the body's heat and moistened by water evaporating from the mucosal lining of the airways. Under normal circumstances, by the time air reaches the trachea, it has been conditioned to 100% humidity and 37 °C. Breathing through the mouth is not nearly as effective at warming and moistening air as breathing through the nose. If you exercise outdoors in very cold weather, you may be familiar with the ache in your chest that results from breathing cold air through your mouth.

Air is filtered both in the trachea and in the bronchi. These airways are lined with ciliated epithelium whose cilia are bathed in a watery saline layer (**FIG. 17.5**). The saline is produced by epithelial cells when Cl^- secreted into the lumen by apical anion channels draws Na^+ into the lumen through the paracellular pathway (Fig. 17.5c). Movement of solute from the ECF to the lumen creates an osmotic gradient, and water follows the ions into the airways. The *CFTR channel,* whose malfunction causes cystic fibrosis, is one of the anion channels found on the apical surface of this epithelium [p. 150].

A sticky layer of mucus floats over the cilia to trap most inhaled particles larger than 2 μm. The mucus layer is secreted by *goblet cells* in the epithelium (Fig. 17.5b). The cilia beat with an upward motion that moves the mucus continuously toward the pharynx, creating what is called the *mucociliary escalator.* Mucus contains *immunoglobulins* that can disable many pathogens. Once mucus reaches the pharynx, it can be spit out (*expectorated*) or swallowed. For swallowed mucus, stomach acid and enzymes destroy any remaining microorganisms.

Secretion of the watery saline layer beneath the mucus is essential for a functional mucociliary escalator. In the disease *cystic fibrosis,* for example, inadequate ion secretion decreases fluid movement in the airways. Without the saline layer, cilia become trapped in thick, sticky mucus and can no longer move. Mucus cannot be cleared, and bacteria colonize the airways, resulting in recurrent lung infections.

FIG. 17.2 ANATOMY SUMMARY

The Lungs and Thoracic Cavity

(a) The respiratory system is divided into upper and lower regions.

(b) On external view, the right lung is divided into three lobes, and the left lung is divided into two lobes.

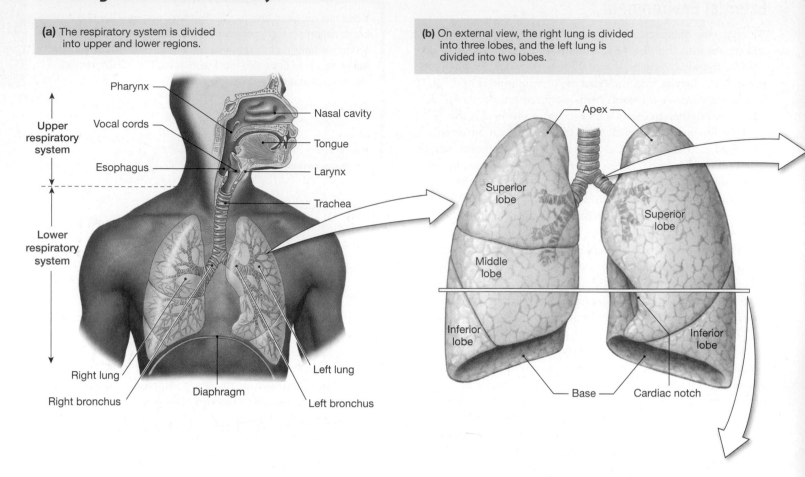

Upper respiratory system

Lower respiratory system

Pharynx
Vocal cords
Esophagus
Nasal cavity
Tongue
Larynx
Trachea
Right lung
Right bronchus
Diaphragm
Left lung
Left bronchus

Apex
Superior lobe
Superior lobe
Middle lobe
Inferior lobe
Inferior lobe
Base
Cardiac notch

(c) Muscles of the thorax, neck, and abdomen create the force to move air during breathing.

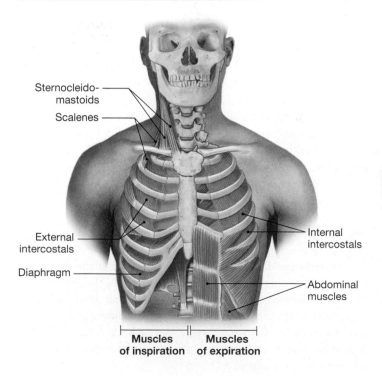

Sternocleido-mastoids
Scalenes
External intercostals
Diaphragm
Internal intercostals
Abdominal muscles

Muscles of inspiration **Muscles of expiration**

(d) **Sectional view of chest.** Each lung is enclosed in two pleural membranes. The esophagus and aorta pass through the thorax between the pleural sacs.

Superior view

Esophagus
Aorta
Pleural membranes
Right lung
Left lung
Heart
Right pleural cavity*
Pericardial cavity*
Left pleural cavity*

*Note: The pericardial cavity and two pleural cavites are filled with small amounts of fluid.

The Bronchi and Alveoli

(e) Branching of airways creates about 80 million bronchioles.

The trachea branches into two primary bronchi.

The primary bronchus divides 22 more times, terminating in a cluster of alveoli.

- Larynx
- Trachea
- Cartilage ring
- Left primary bronchus
- Secondary bronchus
- Bronchiole
- Alveoli

(f) Structure of lung lobule. Each cluster of alveoli is surrounded by elastic fibers and a network of capillaries.

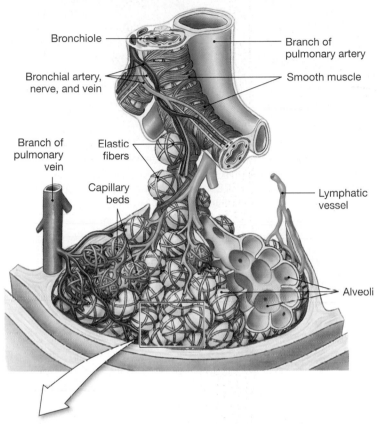

- Bronchiole
- Branch of pulmonary artery
- Bronchial artery, nerve, and vein
- Smooth muscle
- Branch of pulmonary vein
- Elastic fibers
- Capillary beds
- Lymphatic vessel
- Alveoli

(g) Alveolar structure

- Capillary
- Elastic fibers
- Type I alveolar cell for gas exchange
- Endothelial cell of capillary
- Type II alveolar cell (surfactant cell) synthesizes surfactant.
- Limited interstitial fluid
- Alveolar macrophage ingests foreign material.

(h) Exchange surface of alveoli

- Alveolar epithelium
- Nucleus of endothelial cell
- RBC
- Endothelium
- Plasma
- Capillary
- 0.1–1.5 µm
- Alveolar air space
- Surfactant
- Fused basement membranes
- Alveolus

Blue arrow represents gas exchange between alveolar air space and the plasma.

FIG. 17.3 | **The pleural sac**

The pleural sac forms a double membrane surrounding the lung, similar to a fluid-filled balloon surrounding an air-filled balloon.

Air-filled balloon

Fluid-filled balloon

Air space of lung

Pleural membrane

The pleural fluid has a much smaller volume than is suggested by this illustration.

CONCEPT CHECK

5. Cigarette smoking paralyzes cilia in the airways and increases mucus production. Why would these effects cause smokers to develop a cough?

Alveoli Are the Site of Gas Exchange

The air-filled alveoli, clustered at the ends of terminal bronchioles, make up the bulk of lung tissue (Fig. 17.2f, g). Their primary function is the exchange of gases between themselves and the blood.

Each tiny alveolus is composed of a single layer of epithelium (Fig. 17.2g). Two types of epithelial cells are found in the alveoli. About 95% of the alveolar surface area is used for gas exchange and consists of the larger **type I alveolar cells.** These cells are very thin so that gases can diffuse rapidly through them (Fig. 17.2h). In much of the exchange area, a layer of basement membrane fuses the alveolar epithelium to the capillary endothelium. In the remaining area, only a small amount of interstitial fluid is present.

The smaller but thicker **type II alveolar cells** synthesize and secrete a chemical known as **surfactant.** Surfactant mixes with the thin fluid lining of the alveoli to aid lungs as they expand during breathing, as you will see later in this chapter. Type II cells also help minimize the amount of fluid present in the alveoli by transporting solutes, followed by water, out of the alveolar air space.

The thin walls of alveoli do not contain muscle because muscle fibers would block rapid gas exchange. As a result, lung tissue itself cannot contract. However, connective tissue between the alveolar epithelial cells contains many elastin and collagen fibers that create *elastic recoil* when lung tissue is stretched.

The close association of the alveoli with an extensive network of capillaries demonstrates the intimate link between the respiratory and cardiovascular systems. Blood vessels fill 80–90% of the space between alveoli, forming an almost continuous "sheet" of blood in close contact with the air-filled alveoli. The proximity of capillary blood to alveolar air is essential for the rapid exchange of gases.

Pulmonary Circulation Is High-Flow, Low-Pressure

The pulmonary circulation begins with the pulmonary trunk, which receives low-oxygen blood from the right ventricle. The pulmonary trunk divides into two pulmonary arteries, one to each lung [Fig. 14.5, p. 445]. Oxygenated blood from the lungs returns to the left atrium via the pulmonary veins.

At any given moment, the pulmonary circulation contains about 0.5 liter of blood, or 10% of total blood volume. About 75 mL of this amount is found in the capillaries, where gas exchange takes place, with the remainder in pulmonary arteries

FIG. 17.4 | **Airway branching in the lower respiratory tract**

Branching of the Airways

System Name		Name	Division	Diameter (mm)	How Many?	Cross-Sectional Area (cm²)
Conducting system		Trachea	0	15–22	1	2.5
		Primary bronchi	1	10–15	2	
		Smaller bronchi	2		4	
			3			
			4	1–10		
			5			
			6–11		1×10^4	
		Bronchioles	12–23	0.5–1	2×10^4 ↓ 8×10^7	100 ↓ 5×10^3
Exchange surface		Alveoli	24	0.3	$3\text{–}6 \times 10^8$	$>1 \times 10^6$

FIG. 17.5 Airway epithelium

(a) Epithelial cells lining the airways and submucosal glands secrete saline and mucus.

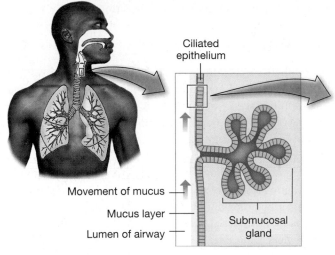

Ciliated epithelium

Movement of mucus

Mucus layer

Lumen of airway

Submucosal gland

(b) Cilia move the mucus layer toward the pharynx, removing trapped pathogens and particulate matter.

Dust particle

Mucus layer traps inhaled particles.

Watery saline layer allows cilia to push mucus toward pharynx.

Cilia

Goblet cell secretes mucus.

Nucleus of columnar epithelial cell

Basement membrane

(c) One model of saline secretion by airway epithelial cells

Saline layer in lumen

Na^+ H_2O Cl^-

Anion channel

Respiratory epithelial cells

K^+ ATP Na^+ Na^+ $2Cl^-$ K^+ K^+

Na^+ H_2O

ECF

1. NKCC brings Cl^- into epithelial cell from ECF.

2. Apical anion channels, including CFTR, allow Cl^- to enter the lumen.

3. Na^+ goes from ECF to lumen by the paracellular pathway, drawn by the electrochemical gradient.

4. NaCl movement from ECF to lumen creates a concentration gradient so water follows into the lumen.

and veins. The rate of blood flow through the lungs is much higher than the rate in other tissues [p. 492] because the lungs receive the entire cardiac output of the right ventricle: 5 L/min. This means that as much blood flows through the lungs in 1 minute as flows through the rest of the body in the same amount of time!

Despite the high flow rate, pulmonary blood pressure is low. Pulmonary arterial pressure averages 25/8 mm Hg, much lower than the average systemic pressure of 120/80 mm Hg. The right ventricle does not have to pump as forcefully to create blood flow through the lungs because resistance of the pulmonary circulation

is low. This low resistance can be attributed to the shorter total length of pulmonary blood vessels and to the distensibility and large total cross-sectional area of pulmonary arterioles.

Normally, the net hydrostatic pressure filtering fluid out of a pulmonary capillary into the interstitial space is low because of low mean blood pressure [p. 497]. The lymphatic system efficiently removes filtered fluid, and lung interstitial fluid volume is usually minimal. As a result, the distance between the alveolar air space and the capillary endothelium is short, and gases diffuse rapidly between them.

CONCEPT CHECK

6. Is blood flow through the pulmonary trunk greater than, less than, or equal to blood flow through the aorta?

7. A person has left ventricular failure but normal right ventricular function. As a result, blood pools in the pulmonary circulation, doubling pulmonary capillary hydrostatic pressure. What happens to net fluid flow across the walls of the pulmonary capillaries?

8. Calculate the mean pressure in a person whose pulmonary arterial pressure is 25/8 mm Hg. [p. 483]

GAS LAWS

Respiratory air flow is very similar in many respects to blood flow in the cardiovascular system because both air and blood are fluids. Their primary difference is that blood is a noncompressible liquid but air is a compressible mixture of gases. **FIGURE 17.6** describes the laws that govern the behavior of gases in air and that provide the basis for the exchange of air between the atmosphere and the alveoli. We will consider the gas laws that govern the solubility of gases in solution when we talk about oxygen transport in blood.

In this book, we report blood pressure and environmental air pressure (**atmospheric pressure**) in millimeters of mercury (mm Hg). Respiratory physiologists sometimes report gas pressures in centimeters of water instead, where 1 mm Hg = 1.36 cm H_2O, or in kiloPascals (kPa), where 760 mm Hg = 101.325 kPa.

At sea level, normal atmospheric pressure is 760 mm Hg. However, in this book we follow the convention of designating atmospheric pressure as 0 mm Hg. Because atmospheric pressure varies with altitude and because very few people live exactly at sea level, this convention allows us to compare pressure differences that occur during ventilation without correcting for altitude.

Air Is a Mixture of Gases

The atmosphere surrounding the earth is a mixture of gases and water vapor. **Dalton's Law** states that the total pressure exerted by a mixture of gases is the sum of the pressures exerted by the individual gases (Fig. 17.6c). For example, in dry air at an atmospheric pressure of 760 mm Hg, 78% of the total pressure is due to N_2, 21% to O_2, and so on.

In respiratory physiology, we are concerned not only with total atmospheric pressure but also with the individual pressures of oxygen and carbon dioxide. The pressure of a single gas in a mixture is known as its **partial pressure (P_{gas})**. The pressure exerted by an individual gas is determined only by its relative abundance in the mixture and is independent of the molecular size or mass of the gas.

The partial pressures of gases in air vary slightly depending on how much water vapor is in the air because the pressure of water vapor "dilutes" the contribution of other gases to the total pressure. The table in Figure 17.6c compares the partial pressures of some gases in dry air and at 100% humidity.

CONCEPT CHECK

9. If nitrogen is 78% of atmospheric air, what is the partial pressure of nitrogen (P_{N_2}) in a sample of dry air that has an atmospheric pressure of 720 mm Hg?

10. The partial pressure of water vapor in inspired air is 47 mm Hg when inhaled air is fully humidified. If atmospheric pressure is 700 mm Hg and oxygen is 21% of the atmosphere at 0% humidity, what is the P_{O_2} of fully humidified air?

Gases Move Down Pressure Gradients

Air flow occurs whenever there is a pressure gradient. Bulk flow of air, like blood flow, is directed from areas of higher pressure to areas of lower pressure. Meteorologists predict the weather by knowing that areas of high atmospheric pressure move in to replace areas of low pressure. In ventilation, bulk flow of air down pressure gradients explains how air is exchanged between the external environment and the lungs. Movement of the thorax during breathing creates alternating conditions of high and low pressure in the lungs.

Diffusion of gases down concentration (partial pressure) gradients applies to single gases. For example, oxygen moves from areas of higher oxygen partial pressure (P_{O_2}) to areas of lower oxygen partial pressure. Diffusion of individual gases is important in the exchange of oxygen and carbon dioxide between alveoli and blood and from blood to cells, as you will learn later.

FIG. 17.6 **ESSENTIALS**

Gas Laws

This figure summarizes the rules that govern the behavior of gases in air. These rules provide the basis for the exchange of air between the external environment and the alveoli.

(a) The Ideal Gas Equation

$$PV = nRT$$

Where P is pressure, V is volume, n is the moles of gas, T is absolute temperature, and R is the universal gas constant, 8.3145 j/mol × K

In the human body, we can assume that the number of moles and temperature are constant. Removing the constants leaves the following equation:

$$V = 1/P$$

This relationship says that if the volume of gas increases, the pressure decreases, and vice versa.

(b) Boyle's Law

Boyle's Law also expresses this inverse relationship between pressure and volume.

$$P_1V_1 = P_2V_2$$

For example, the container on the left is 1 L (V_1) and has a pressure of 100 mm Hg (P_1).

What happens to the pressure when the volume decreases to 0.5 L?

$$100 \text{ mm Hg} \times 1 \text{ L} = P_2 \times 0.5 \text{ L}$$

$$200 \text{ mm Hg} = P_2$$

The pressure has increased ×2.

The Ideal Gas Law and Boyle's Law apply to all gases or mixtures of gases.

$V_1 = 1.0$ L
$P_1 = 100$ mm Hg

$V_2 = 0.5$ L
$P_2 = 200$ mm Hg

(c) Dalton's Law

Dalton's Law says that the total pressure of a mixture of gases is the sum of the pressures of the individual gases. The pressure of an individual gas in a mixture is known as the **partial pressure** of the gas (P_{gas}).

For example, at sea level, atmospheric pressure (P_{atm}) is 760 mm Hg, and oxygen is 21% of the atmosphere. What is the partial pressure of oxygen (P_{O_2})?

To find the partial pressure of any one gas in a sample of dry air, multiply the atmospheric pressure (P_{atm}) by the gas's relative contribution (%) to P_{atm}:

Partial pressure of a gas = P_{atm} × % of gas in atmosphere

$$P_{O_2} = 760 \text{ mm Hg} \times 21\% \text{ oxygen}$$

$$= 760 \text{ mm} \times 0.21 = 160 \text{ mm Hg}$$

The partial pressure of oxygen (P_{O_2}) in dry air at sea level is 160 mm Hg.

The pressure exerted by an individual gas is determined only by its relative abundance in the mixture and is independent of the molecular size or mass of the gas.

In humid air, water vapor "dilutes" the contribution of other gases to the total pressure.

Partial Pressures (P_{gas}) of Atmospheric Gases at 760 mm Hg			
Gas and its percentage in air	P_{gas} in dry 25 °C air	P_{gas} in 25 °C air, 100% humidity	P_{gas} in 37 °C air, 100% humidity
O_2 21%	160 mm Hg	155 mm Hg	150 mm Hg
CO_2 0.03%	0.25 mm Hg	0.24 mm Hg	0.235 mm Hg
Water vapor	0 mm Hg	24 mm Hg	47 mm Hg

To calculate the partial pressure of a gas in humid air, you must first subtract the water vapor pressure from the total pressure. At 100% humidity and 25 °C, water vapor pressure (P_{H_2O}) is 24 mm Hg.

P_{gas} in humid air = (P_{atm} – P_{H_2O}) × % of gas

$$P_{O_2} = (760 - 24) \times 21\% = 155 \text{ mm Hg}$$

Boyle's Law Describes Pressure-Volume Relationships

The pressure exerted by a gas or mixture of gases in a sealed container is created by the collisions of moving gas molecules with the walls of the container and with each other. If the size of the container is reduced, the collisions between the gas molecules and the walls become more frequent, and the pressure rises (Fig. 17.6b). This relationship between pressure and volume was first noted by Robert Boyle in the 1600s and can be expressed by the equation of **Boyle's Law** of gases:

$$P_1V_1 = P_2V_2$$

where P represents pressure and V represents volume.

Boyle's Law states that if the volume of a gas is reduced, the pressure increases. If the volume increases, the pressure decreases.

In the respiratory system, changes in the volume of the chest cavity during ventilation cause pressure gradients that create air flow. When chest volume increases, alveolar pressure falls, and air flows into the respiratory system. When the chest volume decreases, alveolar pressure increases, and air flows out into the atmosphere. This movement of air is bulk flow because the entire gas mixture is moving rather than merely one or two of the gases in the air.

The important gas laws for respiratory physiology are summarized in **TABLE 17.1**.

VENTILATION

The bulk flow exchange of air between the atmosphere and the alveoli is *ventilation,* or *breathing* (Fig. 17.1). A single **respiratory cycle** consists of an inspiration followed by an expiration.

Lung Volumes Change during Ventilation

Physiologists and clinicians assess a person's pulmonary function by measuring how much air the person moves during quiet breathing, then with maximum effort. These **pulmonary function tests** use a **spirometer,** an instrument that measures the volume of air moved with each breath (**FIG. 17.7a**). Most spirometers in clinical use today are small computerized machines rather than the traditional wet spirometer illustrated here.

When a subject is attached to the traditional spirometer through a mouthpiece and the subject's nose is clipped closed, the subject's respiratory tract and the spirometer form a closed system. When the subject breathes in, air moves from the spirometer into the lungs, and the recording pen, which traces a graph on a rotating cylinder, moves up. When the subject exhales, air moves from the lungs back into the spirometer, and the pen moves down.

Lung Volumes The air moved during breathing can be divided into four lung volumes: (1) tidal volume, (2) inspiratory reserve volume, (3) expiratory reserve volume, and (4) residual volume. The numerical values used on the graph in Figure 17.7b represent average volumes for a 70-kg man. The volumes for women are typically less, as shown in Figure 17.7b. Lung volumes vary considerably with age, sex, height, and weight, so clinicians use algorithms based on those parameters to predict lung volumes. (An *algorithm* is an equation or series of steps used to solve a problem.) See question #36 at the end of this chapter for one of these algorithms.

Each of the following paragraphs begins with the instructions you would be given if you were being tested for these volumes.

"Breathe quietly." The volume of air that moves during a single inspiration or expiration is known as the **tidal volume** (V_T). Average tidal volume during quiet breathing is about 500 mL. (It is hard for subjects to breathe normally when they are thinking about their breathing, so the clinician may not give this instruction.)

"Now, at the end of a quiet inspiration, take in as much additional air as you possibly can." The additional volume you inspire above the tidal volume represents your **inspiratory reserve volume (IRV)**. In a 70-kg man, this volume is about 3000 mL, a six-fold increase over the normal tidal volume.

"Now stop at the end of a normal exhalation, then exhale as much air as you possibly can." The amount of air forcefully exhaled after the end of a normal expiration is the **expiratory reserve volume (ERV),** which averages about 1100 mL.

The fourth volume cannot be measured directly. Even if you blow out as much air as you can, air still remains in the lungs and the airways. The volume of air in the respiratory system after maximal exhalation—about 1200 mL—is called the **residual volume (RV)**. Most of this residual volume exists because the lungs are held stretched against the ribs by the pleural fluid.

Lung Capacities The sum of two or more lung volumes is called a *capacity*. The **vital capacity (VC)** is the sum of the inspiratory reserve volume, expiratory reserve volume, and tidal volume. Vital capacity represents the maximum amount of air that can be voluntarily moved into or out of the respiratory system with one breath. Vital capacity decreases with age as muscles weaken and the lungs become less elastic.

TABLE 17.1	**Gas Laws**

1. The total pressure of a mixture of gases is the sum of the pressures of the individual gases (Dalton's Law).

2. To find the partial pressure of a gas (P_{gas}) in dry air, multiply the gas's contribution (%) times the atmospheric pressure (P_{atm}). $P_{gas} = P_{atm} \times \%$ gas

3. To calculate the partial pressure of a gas in humid air, you must first subtract water vapor pressure (P_{H_2O}) from the total pressure (P_{atm}). $P_{gas} = (P_{atm} - P_{H_2O}) \times \%$ gas

4. Gases, singly or in a mixture, move from areas of higher pressure to areas of lower pressure.

5. If the volume of a container of gas changes, the pressure of the gas will change in an inverse manner: $P_1V_1 = P_2V_2$ (Boyle's Law).

FIG. 17.7 Pulmonary function tests

(a) The Spirometer

This figure shows a traditional wet spirometer. The subject inserts a mouthpiece that is attached to an inverted bell filled with air or oxygen. The volume of the bell and the volume of the subject's respiratory tract create a closed system because the bell is suspended in water.

PHYS IN ACTION!

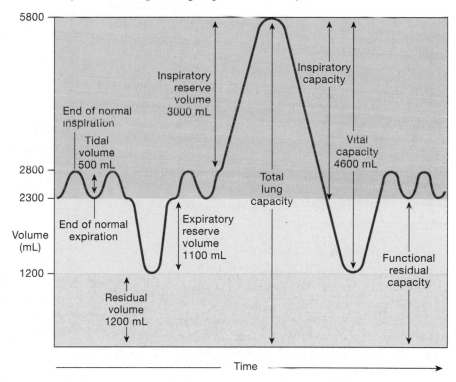

Bell

Air

Water

Inspiration | Expiration | Inspiration | Expiration

0.5

Volume (L)

0

Time

When the subject inhales, air moves into the lungs. The volume of the bell decreases, and the pen rises on the tracing.

(b) Lung Volumes and Capacities

The four lung volumes:

Dead space

RV

ERV

V_T

IRV

KEY

RV = Residual volume
ERV = Expiratory reserve volume
V_T = Tidal volume
IRV = Inspiratory reserve volume

A spirometer tracing showing lung volumes and capacities.

5800

Inspiratory reserve volume 3000 mL

Inspiratory capacity

End of normal inspiration

Tidal volume 500 mL

Vital capacity 4600 mL

2800

Total lung capacity

2300

End of normal expiration

Expiratory reserve volume 1100 mL

Functional residual capacity

1200

Volume (mL)

Residual volume 1200 mL

Time

Capacities are sums of 2 or more volumes.

Inspiratory capacity = V_T + IRV
Vital capacity = V_T + IRV + ERV
Total lung capacity = V_T + IRV + ERV + RV
Functional residual capacity = ERV + RV

Pulmonary Volumes and Capacities*		
	Males	**Females**
Vital capacity — IRV 3000		1900
V_T 500		500
ERV 1100		700
Residual volume 1200		1100
Total lung capacity 5800 mL		4200 mL

*Pulmonary volumes are given for a normal 70-kg man or a 50-kg woman, 28 years old.

To measure vital capacity, the person being tested takes in as much air as possible, then blows it all out as fast as possible. This forced vital capacity test also allows the clinician to measure how fast air leaves the airways in the first second of expiration, a measurement known as the **FEV$_1$**, or **forced expiratory volume in 1 second**. FEV$_1$ decreases in certain lung diseases, such as asthma, and also with age.

Vital capacity plus the residual volume yields the **total lung capacity** (TLC). Other capacities of importance in pulmonary medicine include the **inspiratory capacity** (tidal volume + inspiratory reserve volume) and the **functional residual capacity** (expiratory reserve volume + residual volume).

CONCEPT CHECK

11. How are lung volumes related to lung capacities?
12. Which lung volume cannot be measured directly?
13. If vital capacity decreases with age but total lung capacity does not change, which lung volume must be changing? In which direction?
14. As inhaled air becomes humidified passing down the airways, what happens to the P$_{O2}$ of the air?

During Ventilation, Air Flows because of Pressure Gradients

Breathing is an active process that requires muscle contraction. Air flows into the lungs because of pressure gradients created by a pump, just as blood flows because of the pumping action of the heart. In the respiratory system, muscles of the thoracic cage and diaphragm function as the pump because most lung tissue is thin exchange epithelium. When these muscles contract, the lungs expand, held to the inside of the chest wall by the pleural fluid.

The primary muscles involved in quiet breathing (breathing at rest) are the diaphragm, the external intercostals, and the scalenes. During forced breathing, other muscles of the chest and abdomen may be recruited to assist. Examples of physiological situations in which breathing is forced include exercise, playing a wind instrument, and blowing up a balloon.

As we noted earlier in the chapter, air flow in the respiratory tract obeys the same rule as blood flow:

$$\text{Flow} \propto \Delta P/R$$

This equation means that (1) air flows in response to a pressure gradient (ΔP) and (2) flow decreases as the resistance (R) of the system to flow increases. Before we discuss resistance, let's consider how the respiratory system creates a pressure gradient. The pressure-volume relationships of Boyle's Law provide the basis for pulmonary ventilation.

CONCEPT CHECK

15. Compare the direction of air movement during one respiratory cycle with the direction of blood flow during one cardiac cycle.
16. Explain the relationship between the lungs, the pleural membranes, the pleural fluid, and the thoracic cage.

Inspiration Occurs When Alveolar Pressure Decreases

For air to move into the alveoli, pressure inside the lungs must become lower than atmospheric pressure. According to Boyle's Law, an increase in volume will create a decrease in pressure. During inspiration, thoracic volume increases when certain skeletal muscles of the rib cage and diaphragm contract.

When the diaphragm contracts, it drops down toward the abdomen. In quiet breathing, the diaphragm moves about 1.5 cm, increasing thoracic volume (**FIG. 17.8b**). Contraction of the diaphragm causes between 60% and 75% of the inspiratory volume change during normal quiet breathing.

Movement of the rib cage creates the remaining 25–40% of the volume change. During inhalation, the external intercostal and scalene muscles (see Fig. 17.2c) contract and pull the ribs upward and out (Fig. 17.8b). Rib movement during inspiration has been likened to a pump handle lifting up and away from the pump (the ribs moving up and away from the spine) and to the movement of a bucket handle as it lifts away from the side of a bucket (ribs moving outward in a lateral direction). The combination of these two movements broadens the rib cage in all directions. As thoracic volume increases, pressure decreases, and air flows into the lungs.

For many years, quiet breathing was attributed solely to the action of the diaphragm and the external intercostal muscles. It was thought that the scalenes and sternocleidomastoid muscles were active only during deep breathing. In recent years, however, studies have changed our understanding of how these accessory muscles contribute to quiet breathing.

If an individual's scalenes are paralyzed, inspiration is achieved primarily by contraction of the diaphragm. Observation of patients with neuromuscular disorders has revealed that although the contracting diaphragm increases thoracic volume by moving toward the abdominal cavity, it also tends to pull the lower ribs inward, working against inspiration. In normal individuals, we know that the lower ribs move up and out during inspiration rather than inward. The fact that there is no up-and-out rib motion in patients with paralyzed scalenes tells us that normally the scalenes must be contributing to inspiration by lifting the sternum and upper ribs.

New evidence also downplays the role of the external intercostal muscles during quiet breathing. However, the external intercostals play an increasingly important role as respiratory activity increases. Because the exact contribution of external intercostals and scalenes varies depending on the type of breathing, we group these muscles together and simply call them the *inspiratory muscles*.

Now let's see how alveolar pressure (P$_A$) changes during a single inspiration. Follow the graphs in **FIGURE 17.9** as you read through the process. Remember that atmospheric pressure is assigned a value of 0 mm Hg. Negative numbers designate subatmospheric pressures, and positive numbers denote higher-than-atmospheric pressures.

Time 0. In the brief pause between breaths, alveolar pressure is equal to atmospheric pressure (0 mm Hg at point A$_1$). When pressures are equal, there is no air flow.

FIG. 17.8 | Movement of the thoracic cage and diaphragm during breathing

(a) At rest: Diaphragm is relaxed.

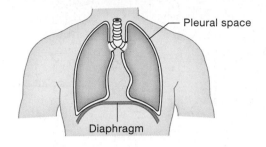

Pleural space

Diaphragm

(b) Inspiration: Thoracic volume increases.

Diaphragm contracts and flattens.

(c) Expiration: Diaphragm relaxes, thoracic volume decreases.

During inspiration, the dimensions of the thoracic cavity increase.

Vertebrae   Rib   Sternum

Side view:

"Pump handle" motion increases anterior-posterior dimension of rib cage. Movement of the handle on a hand pump is analogous to the lifting of the sternum and ribs.

Vertebrae   Rib

Front view:

"Bucket handle" motion increases lateral dimension of rib cage. The bucket handle moving up and out is a good model for lateral rib movement during inspiration.

Sternum

Time 0–2 sec: Inspiration. As inspiration begins, inspiratory muscles contract, and thoracic volume increases. With the increase in volume, alveolar pressure falls about 1 mm Hg below atmospheric pressure (-1 mm Hg, point A_2), and air flows into the alveoli (point C_1 to point C_2). Because the thoracic volume changes faster than air can flow, alveolar pressure reaches its lowest value about halfway through inspiration (point A_2).

As air continues to flow into the alveoli, pressure increases until the thoracic cage stops expanding, just before the end of inspiration. Air movement continues for a fraction of a second longer, until pressure inside the lungs equalizes with atmospheric pressure (point A_3). At the end of inspiration, lung volume is at its maximum for the respiratory cycle (point C_2), and alveolar pressure is equal to atmospheric pressure.

You can demonstrate this phenomenon by taking a deep breath and stopping the movement of your chest at the end of inspiration. (Do not "hold your breath" because doing so closes the opening of the pharynx and prevents air flow.) If you do this correctly, you notice that air flow stops after you freeze the inspiratory movement. This exercise shows that at the end of inspiration, alveolar pressure is equal to atmospheric pressure.

Expiration Occurs When Alveolar Pressure Increases

At the end of inspiration, impulses from somatic motor neurons to the inspiratory muscles cease, and the muscles relax. Elastic recoil of the lungs and thoracic cage returns the diaphragm and rib

FIG. 17.9 Pressure changes during quiet breathing

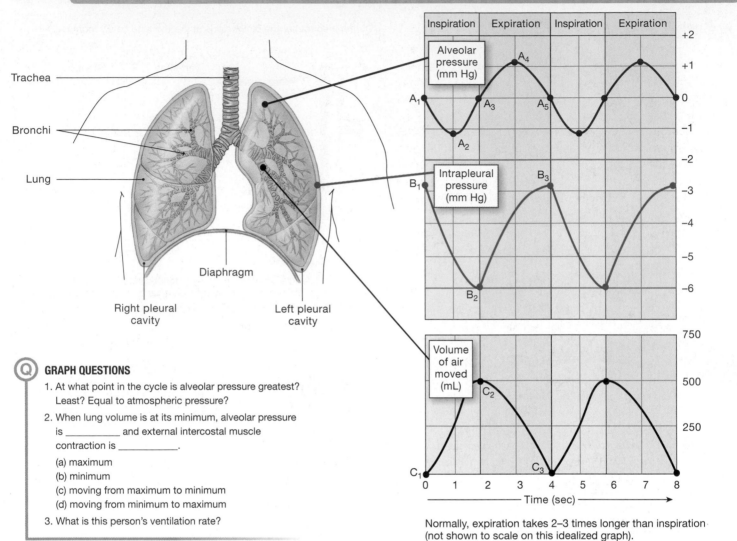

Normally, expiration takes 2–3 times longer than inspiration (not shown to scale on this idealized graph).

GRAPH QUESTIONS

1. At what point in the cycle is alveolar pressure greatest? Least? Equal to atmospheric pressure?

2. When lung volume is at its minimum, alveolar pressure is _____ and external intercostal muscle contraction is _____.
   (a) maximum
   (b) minimum
   (c) moving from maximum to minimum
   (d) moving from minimum to maximum

3. What is this person's ventilation rate?

cage to their original relaxed positions, just as a stretched elastic waistband recoils when released. Because expiration during quiet breathing involves passive elastic recoil rather than active muscle contraction, it is called passive expiration.

Time 2–4 sec: Expiration. As lung and thoracic volumes decrease during expiration, air pressure in the lungs increases, reaching a maximum of about 1 mm Hg above atmospheric pressure (Fig. 17.9, point A_4). Alveolar pressure is now higher than atmospheric pressure, so air flow reverses and air moves out of the lungs.

Time 4 sec. At the end of expiration, air movement ceases when alveolar pressure is again equal to atmospheric pressure (point A_5). Lung volume reaches its minimum for the respiratory cycle (point C_3). At this point, the respiratory cycle has ended and is ready to begin again with the next breath.

The pressure differences shown in Figure 17.9 apply to quiet breathing. During exercise or forced heavy breathing, these values become proportionately larger. **Active expiration** occurs during voluntary exhalations and when ventilation exceeds 30–40 breaths per minute. (Normal resting ventilation rate is 12–20 breaths per minute for an adult.) Active expiration uses the internal intercostal muscles and the abdominal muscles (see Fig. 17.2c), which are not used during inspiration. These muscles are collectively called the *expiratory muscles.*

The internal intercostal muscles line the inside of the rib cage. When they contract, they pull the ribs inward, reducing the volume of the thoracic cavity. To feel this action, place your hands on your rib cage. Forcefully blow as much air out of your lungs as you can, noting the movement of your hands as you do so.

The internal and external intercostals function as antagonistic muscle groups [p. 379] to alter the position and volume of the rib cage during ventilation. The diaphragm, however, has no antagonistic muscles. Instead, abdominal muscles contract during active expiration to supplement the activity of the internal intercostals.

Abdominal contraction pulls the lower rib cage inward and decreases abdominal volume, actions that displace the intestines and liver upward. The displaced viscera push the diaphragm up into the thoracic cavity and passively decrease chest volume even more. The action of abdominal muscles during forced expiration is why aerobics instructors tell you to blow air out as you lift your head and shoulders during abdominal "crunches." The active process of blowing air out helps contract the abdominals, the muscles you are trying to strengthen.

Any neuromuscular disease that weakens skeletal muscles or damages their motor neurons can adversely affect ventilation. With decreased ventilation, less fresh air enters the lungs. In addition, loss of the ability to cough increases the risk of pneumonia and other infections. Examples of diseases that affect the motor control of ventilation include *myasthenia gravis* [p. 256], an illness in which acetylcholine receptors of the motor end plates of skeletal muscles are destroyed, and *polio* (poliomyelitis), a viral illness that paralyzes skeletal muscles.

CONCEPT CHECK

17. Scarlett O'Hara is trying to squeeze herself into a corset with an 18-inch waist. Will she be more successful by taking a deep breath and holding it or by blowing all the air out of her lungs? Why?

18. Why would loss of the ability to cough increase the risk of respiratory infections? (*Hint:* What does coughing do to mucus in the airways?)

Intrapleural Pressure Changes during Ventilation

Ventilation requires that the lungs, which are unable to expand and contract on their own, move in association with the expansion and relaxation of the thorax. As we noted earlier in this chapter, the lungs are enclosed in the fluid-filled pleural sac. The surface of the lungs is covered by the *visceral pleura*, and the portion of the sac that lines the thoracic cavity is called the *parietal pleura* {*paries*, wall}. Cohesive forces of the intrapleural fluid cause the stretchable lung to adhere to the thoracic cage. When the thoracic cage moves during breathing, the lungs move with it.

Subatmospheric Intrapleural Pressure The intrapleural pressure in the fluid between the pleural membranes is normally subatmospheric. This subatmospheric pressure arises during fetal development, when the thoracic cage with its associated pleural membrane grows more rapidly than the lung with its associated pleural membrane. The two pleural membranes are held together by the pleural fluid bond, so the elastic lungs are forced to stretch to conform to the larger volume of the thoracic cavity. At the same time, however, elastic recoil of the lungs creates an inwardly directed force that tries to pull the lungs away from the chest wall (**FIG. 17.10a**). The combination of the outward pull of the thoracic cage and inward recoil of the elastic lungs creates a subatmospheric intrapleural pressure of about −3 mm Hg.

FIG. 17.10   Subatmospheric pressure in the pleural cavity helps keep the lungs inflated

(a) In the normal lung at rest, pleural fluid keeps the lung adhered to the chest wall.

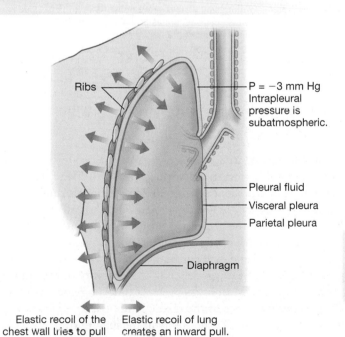

Ribs

P = −3 mm Hg
Intrapleural pressure is subatmospheric.

Pleural fluid
Visceral pleura
Parietal pleura

Diaphragm

Elastic recoil of the chest wall tries to pull the chest wall outward.   Elastic recoil of lung creates an inward pull.

(b) Pneumothorax. If the sealed pleural cavity is opened to the atmosphere, air flows in. The bond holding the lung to the chest wall is broken, and the lung collapses, creating a pneumothorax (air in the thorax).

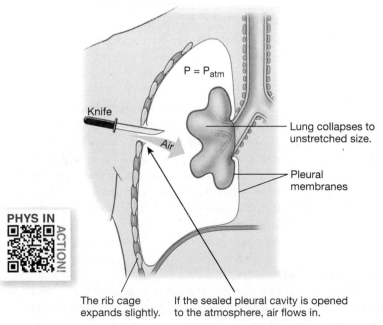

P = P$_{atm}$

Knife

Air

Lung collapses to unstretched size.

Pleural membranes

The rib cage expands slightly.   If the sealed pleural cavity is opened to the atmosphere, air flows in.

PHYS IN ACTION!

RUNNING **PROBLEM**

Edna's COPD began with chronic bronchitis and a morning cough that produced lots of mucus (*phlegm*). Cigarette smoke paralyzes the cilia that sweep debris and mucus out of the airways, and smoke irritation increases mucus production in the airway. Without functional cilia, mucus and debris pool in the airways, leading to a chronic cough. Eventually, smokers may begin to develop emphysema in addition to their bronchitis.

Q2: *Why do people with chronic bronchitis have a higher-than-normal rate of respiratory infections?*

535   537   **550**   552   554   559

You can create a similar situation by half-filling a syringe with water, removing all air, and capping the end with a plugged-up needle or three-way valve. At this point, the pressure inside the barrel is equal to atmospheric pressure. Now hold the syringe barrel (the chest wall) in one hand while you try to withdraw the plunger (the elastic lung pulling away from the chest wall). As you pull on the plunger, the volume inside the barrel increases very slightly, but the cohesive forces between the water molecules cause the water to resist expansion. The pressure inside the barrel, which was initially equal to atmospheric pressure, decreases slightly as you pull on the plunger. If you release the plunger, it snaps back to its resting position, restoring atmospheric pressure inside the syringe. This experiment demonstrates the cohesive forces of water: water will resist being "stretched."

Pneumothorax Now suppose an opening is made between the sealed pleural cavity with its subatmospheric pressure and the atmosphere. In real life, this might happen with a knife thrust between the ribs, a broken rib that punctures the pleural membrane, or any other event that breaks the seal of the pleural cavity. Air moves down pressure gradients, so opening the pleural cavity to the atmosphere allows air to flow into the cavity, just as air enters a vacuum-packed can when you break the seal with a can opener.

Air entering the pleural cavity breaks the fluid bond holding the lung to the chest wall. The chest wall expands outward while the elastic lung collapses to an unstretched state, like a deflated balloon (Fig. 17.10b). This condition, called **pneumothorax** {*pneuma*, air + *thorax*, chest}, results in a collapsed lung that is unable to function normally. Pneumothorax can be due to trauma but can also occur spontaneously when a congenital *bleb*, a weakened section of lung tissue, ruptures, allowing air from inside the lung to enter the pleural cavity.

Correction of a pneumothorax has two components: removing as much air from the pleural cavity as possible with a suction pump, and sealing the hole to prevent more air from entering. Any air remaining in the cavity is gradually absorbed into the blood, restoring the pleural fluid bond and reinflating the lung.

Intrapleural Pressure during the Respiratory Cycle

Pressures in the pleural fluid vary during a respiratory cycle. At the beginning of inspiration, intrapleural pressure is about -3 mm Hg (Fig. 17.9, point B_1). As inspiration proceeds, the pleural membranes and lungs follow the expanding thoracic cage because of the pleural fluid bond, but the elastic lung tissue resists being stretched. The lungs attempt to pull farther away from the chest wall, causing the intrapleural pressure to become even more negative (Fig. 17.9, point B_2).

Because this process is difficult to visualize, let's return to the analogy of the water-filled syringe with the plugged-up needle. You can pull the plunger out a small distance without much effort, but the cohesiveness of the water makes it difficult to pull the plunger out any farther. The increased amount of work you do trying to pull the plunger out is paralleled by the work your inspiratory muscles must do when they contract during inspiration. The bigger the breath, the more work is required to stretch the elastic lung.

By the end of a quiet inspiration, when the lungs are fully expanded, intrapleural pressure falls to around -6 mm Hg (Fig. 17.9, point B_2). During exercise or other powerful inspirations, intrapleural pressure may reach -8 mm Hg or lower.

During expiration, the thoracic cage returns to its resting position. The lungs are released from their stretched position, and the intrapleural pressure returns to its normal value of about -3 mm Hg (point B_3). Notice that intrapleural pressure never equilibrates with atmospheric pressure because the pleural cavity is a sealed compartment.

CONCEPT CHECK

19. A person has periodic spastic contractions of the diaphragm, otherwise known as hiccups. What happens to intrapleural and alveolar pressures when a person hiccups?

20. A stabbing victim is brought to the emergency room with a knife wound between the ribs on the left side of his chest. What has probably happened to his left lung? To his right lung? Why does the left side of his rib cage seem larger than the right side?

Lung Compliance and Elastance May Change in Disease States

Pressure gradients required for air flow are created by the work of skeletal muscle contraction. Normally, about 3–5% of the body's energy expenditure is used for quiet breathing. During exercise, the energy required for breathing increases substantially. The two factors that have the greatest influence on the amount of work needed for breathing are the stretchability of the lungs and the resistance of the airways to air flow.

Adequate ventilation depends on the ability of the lungs to expand normally. Most of the work of breathing goes into overcoming the resistance of the elastic lungs and the thoracic cage to stretching. Clinically, the ability of the lung to stretch is called **compliance.**

Compliance refers to the amount of force that must be exerted in a body to deform it. In the lung, we can express compliance as the change of volume (V) that results from a given force or pressure (P) exerted on the lung: $\Delta V/\Delta P$. A high-compliance lung stretches easily, just as a compliant person is easy to persuade. A low-compliance lung requires more force from the inspiratory muscles to stretch it.

Compliance is the reciprocal of **elastance** (elastic recoil), the ability to resist being deformed. Elastance also refers to the ability of a body to return to its original shape when a deforming force is removed. A lung that stretches easily (high compliance) has probably lost its elastic tissue and will not return to its resting volume when the stretching force is released (low elastance). You may have experienced something like this with old gym shorts. After many washings the elastic waistband is easy to stretch (high compliance) but lacking in elastance, making it impossible for the shorts to stay up around your waist.

Analogous problems occur in the respiratory system. For example, as noted in the Running Problem, emphysema is a disease in which elastin fibers normally found in lung tissue are destroyed. Destruction of elastin results in lungs that exhibit high compliance and stretch easily during inspiration. However, these lungs also have decreased elastance, so they do not recoil to their resting position during expiration.

To understand the importance of elastic recoil to expiration, think of an inflated balloon and an inflated plastic bag. The balloon is similar to the normal lung. Its elastic walls squeeze on the air inside the balloon, thereby increasing the internal air pressure. When the neck of the balloon is opened to the atmosphere, elastic recoil causes air to flow out of the balloon.

The inflated plastic bag, on the other hand, is like the lung of an individual with emphysema. It has high compliance and is easily inflated, but it has little elastic recoil. If the inflated plastic bag is opened to the atmosphere, most of the air remains inside the bag. To get the air out of the bag, you must squeeze it with your hands. Patients with emphysema contract their expiratory muscles (active expiration) to force out air that is not leaving from elastic recoil.

Restrictive Lung Disease A decrease in lung compliance affects ventilation because more work must be expended to stretch a stiff lung. Pathological conditions in which compliance is reduced are called **restrictive lung diseases.** In these conditions, the energy expenditure required to stretch less-compliant lungs can far exceed the normal work of breathing. Two common causes of decreased compliance are inelastic scar tissue formed in *fibrotic lung diseases,* and inadequate alveolar production of surfactant, a chemical that facilitates lung expansion.

Pulmonary **fibrosis** is characterized by the development of stiff, fibrous scar tissue that restricts lung inflation. In *idiopathic* pulmonary fibrosis {*idios,* one's own}, the cause is unknown. Other forms of fibrotic lung disease result from chronic inhalation of fine particulate matter, such as asbestos and silicon, that escapes the mucus lining the airways and reaches the alveoli. Wandering alveolar macrophages (see Fig. 17.2g) then ingest the inhaled particulate matter. If the particles are organic, the macrophages can digest them with lysosomal enzymes. However, if the particles cannot be digested or if they accumulate in large numbers, an inflammatory process ensues. The macrophages then secrete growth factors that stimulate fibroblasts in the lung's connective tissue to produce inelastic collagen. Pulmonary fibrosis cannot be reversed.

Surfactant Decreases the Work of Breathing

For years, physiologists assumed that elastin and other elastic fibers were the primary source of resistance to stretch in the lung. However, studies comparing the work required to expand air-filled and saline-filled lungs showed that air-filled lungs are much harder to inflate. From this result, researchers concluded that lung tissue itself contributes less to resistance than once thought. Some other property of the normal air-filled lung, a property not present in the saline-filled lung, must create most of the resistance to stretch.

This property is the *surface tension* [p. 39] created by the thin fluid layer between the alveolar cells and the air. At any air-fluid interface, the surface of the fluid is under tension, like a thin membrane being stretched. When the fluid is water, surface tension arises because of the hydrogen bonds between water molecules. The water molecules on the fluid's surface are attracted to other water molecules beside and beneath them but are not attracted to gases in the air at the air-fluid interface.

Alveolar surface tension is similar to the surface tension that exists in a spherical bubble, even though alveoli are not perfect spheres. The surface tension created by the thin film of fluid is directed toward the center of the bubble and creates pressure in the interior of the bubble. The **law of LaPlace** is an expression of this pressure. It states that the pressure (P) inside a bubble formed by a fluid film is a function of two factors: the surface tension of the fluid (T) and the radius of the bubble (r). This relationship is expressed by the equation

$$P = 2T/r$$

Notice in **FIGURE 17.11a** that if two bubbles have different diameters but are formed by fluids with the same surface tension, the pressure inside the smaller bubble is greater than that inside the larger bubble.

How does this apply to the lung? In physiology, we can equate the bubble to a fluid-lined alveolus (although alveoli are not perfect spheres). The fluid lining all the alveoli creates surface tension. If the surface tension (T) of the fluid were the same in small and large alveoli, small alveoli would have higher inwardly directed pressure than larger alveoli, and increased resistance to stretch. As a result, more work would be needed to expand smaller alveoli.

Normally, however, our lungs secrete a surfactant that reduces surface tension. Surfactants ("**surf**ace **act**ive **agents**") are molecules that disrupt cohesive forces between water molecules by substituting themselves for water at the surface. For example, that product you add to your dishwasher to aid in the rinse cycle

FIG. 17.11   Law of LaPlace

(a) The two bubbles shown have the same surface tension (T). According to the Law of LaPlace, pressure is greater in the smaller bubble.

(b) Surfactant (●) reduces surface tension (T). In the lungs, smaller alveoli have more surfactant, which equalizes the pressure between large and small alveoli.

Larger bubble
r = 2
T = 3
P = (2 × 3)/2
P = 3

Smaller bubble
r = 1
T = 3
P = (2 × 3)/1
P = 6

Law of LaPlace

$$P = 2T/r$$

P = pressure
T = surface tension
r = radius

According to the law of LaPlace, if two bubbles have the same surface tension, the smaller bubble will have higher pressure.

More surfactant decreases surface tension.

r = 2
T = 2
P = (2 × 2)/2
P = 2

r = 1
T = 1
P = (2 × 1)/1
P = 2

is a surfactant that keeps the rinse water from beading up on the dishes (and forming spots when the water beads dry). In the lungs, surfactant decreases surface tension of the alveolar fluid and thereby decreases resistance of the lung to stretch.

Surfactant is more concentrated in smaller alveoli, making their surface tension less than that in larger alveoli (Fig. 17.11b). Lower surface tension helps equalize the pressure among alveoli of different sizes and makes it easier to inflate the smaller alveoli. With lower surface tension, the work needed to expand the alveoli with each breath is greatly reduced. Human surfactant is a mixture containing proteins and phospholipids, such as *dipalmitoylphosphatidylcholine,* which are secreted into the alveolar air space by type II alveolar cells (see Fig. 17.2g).

Normally, surfactant synthesis begins about the 25th week of fetal development under the influence of various hormones. Production usually reaches adequate levels by the 34th week (about 6 weeks before normal delivery). Babies who are born prematurely without adequate concentrations of surfactant in their alveoli develop *newborn respiratory distress syndrome* (NRDS). In addition to having "stiff" (low-compliance) lungs, NRDS babies also have alveoli that collapse each time they exhale. These infants must use a tremendous amount of energy to expand their collapsed lungs with each breath. Unless treatment is initiated rapidly, about 50% of these infants die. In the past, all physicians could do for NRDS babies was administer oxygen. Today, however, the prognosis for NRDS babies is much better. Amniotic fluid can be sampled to assess whether or not the fetal lungs are producing adequate amounts of surfactant. If they are not, and if delivery cannot be delayed, NRDS babies can be treated with aerosol administration of artificial surfactant until the lungs mature enough to produce their own. The current treatment also includes artificial ventilation that forces air into the lungs (*positive-pressure ventilation*) and keeps the alveoli open.

Airway Diameter Determines Airway Resistance

The other factor besides compliance that influences the work of breathing is the resistance of the respiratory system to air flow. Resistance in the respiratory system is similar in many ways to resistance in the cardiovascular system [p. 442]. Three parameters contribute to resistance (R): the system's length (L), the viscosity of the substance flowing through the system (η), and the radius of the tubes in the system (r). As with flow in the cardiovascular system, Poiseuille's Law relates these factors to one another:

$$R \propto L\eta/r^4$$

Because the length of the respiratory system is constant, we can ignore L in the equation. The viscosity of air is almost constant, although you may have noticed that it feels harder to

RUNNING **PROBLEM**

Emphysema is characterized by a loss of elastin, the elastic fibers that help the alveoli recoil during expiration. Elastin is destroyed by *elastase,* an enzyme released by alveolar macrophages, which must work overtime in smokers to rid the lungs of irritants. People with emphysema have more difficulty exhaling than inhaling. Their alveoli have lost elastic recoil, which makes expiration—normally a passive process—require conscious effort.

Q3: *Name the muscles that patients with emphysema use to exhale actively.*

535   537   550   552   554   559

breathe in a sauna filled with steam than in a room with normal humidity. Water droplets in the steam increase the viscosity of the steamy air, thereby increasing its resistance to flow. Viscosity also changes slightly with atmospheric pressure, decreasing as pressure decreases. A person at high altitude may feel less resistance to air flow than a person at sea level. Despite these exceptions, viscosity plays a very small role in resistance to air flow.

Length and viscosity are essentially constant for the respiratory system. As a result, the radius (or diameter) of the airways becomes the primary determinant of airway resistance. Normally, however, the work needed to overcome resistance of the airways to air flow is much less than the work needed to overcome the resistance of the lungs and thoracic cage to stretch.

Nearly 90% of airway resistance normally can be attributed to the trachea and bronchi, rigid structures with the smallest total cross-sectional area. Because these structures are supported by cartilage and bone, their diameters normally do not change, and their resistance to air flow is constant. However, accumulation of mucus from allergies or infections can dramatically increase resistance. If you have ever tried breathing through your nose when you have a cold, you can appreciate how the narrowing of an upper airway limits air flow!

The bronchioles normally do not contribute significantly to airway resistance because their total cross-sectional area is about 2000 times that of the trachea. Because the bronchioles are collapsible tubes, however, a decrease in their diameter can suddenly turn them into a significant source of airway resistance. **Bronchoconstriction** increases resistance to air flow and decreases the amount of fresh air that reaches the alveoli.

Bronchioles, like arterioles, are subject to reflex control by the nervous system and by hormones. However, most minute-to-minute changes in bronchiolar diameter occur in response to paracrine signals. Carbon dioxide in the airways is the primary paracrine molecule that affects bronchiolar diameter. Increased CO_2 in expired air relaxes bronchiolar smooth muscle and causes **bronchodilation.**

Histamine is a paracrine signal that acts as a powerful bronchoconstrictor. This chemical is released by *mast cells* [p. 514] in response to either tissue damage or allergic reactions. In severe allergic reactions, large amounts of histamine may lead to widespread bronchoconstriction and difficult breathing. Immediate medical treatment in these patients is imperative.

The primary neural control of bronchioles comes from parasympathetic neurons that cause bronchoconstriction, a reflex designed to protect the lower respiratory tract from inhaled irritants. There is no significant sympathetic innervation of the bronchioles in humans. However, smooth muscle in the bronchioles is well supplied with β_2-receptors that respond to epinephrine. Stimulation of β_2-receptors relaxes airway smooth muscle and results in bronchodilation. This reflex is used therapeutically in the treatment of asthma and various allergic reactions characterized by histamine release and bronchoconstriction. **TABLE 17.2** summarizes the factors that alter airway resistance.

CONCEPT CHECK

21. In a normal person, which contributes more to the work of breathing: airway resistance or lung and chest wall elastance?

22. Coal miners who spend years inhaling fine coal dust have much of their alveolar surface area covered with scarlike tissue. What happens to their lung compliance as a result?

23. How does the work required for breathing change when surfactant is not present in the lungs?

24. A cancerous lung tumor has grown into the walls of a group of bronchioles, narrowing their lumens. What has happened to the resistance to air flow in these bronchioles?

25. Name the neurotransmitter and receptor for parasympathetic bronchoconstriction.

Rate and Depth of Breathing Determine the Efficiency of Breathing

You may recall that the efficiency of the heart is measured by the cardiac output, which is calculated by multiplying heart rate by stroke volume. Likewise, we can estimate the effectiveness of ventilation by calculating **total pulmonary ventilation,** the volume

TABLE 17.2	Factors That Affect Airway Resistance	
Factor	**Affected by**	**Mediated by**
Length of the system	Constant; not a factor	
Viscosity of air	Usually constant; humidity and altitude may alter slightly	
Diameter of airways		
Upper airways	Physical obstruction	Mucus and other factors
Bronchioles	Bronchoconstriction	Parasympathetic neurons (muscarinic receptors), histamine, leukotrienes
	Bronchodilation	Carbon dioxide, epinephrine (β_2-receptors)

RUNNING **PROBLEM**

Edna has been experiencing shortness of breath while exercising, so her physician runs some tests, including measuring Edna's lung volumes with spirometry. Part of the test is a forced expiratory volume. With her lungs filled to their maximum with air, Edna is told to blow out as fast and as forcefully as she can. The volume of air that Edna expels in the first second of the test (the forced expiratory volume in one second, or FEV_1) is lower than normal because in COPD, airway resistance is increased. Another test the physician orders is a complete blood count (CBC). The results of this test show that Edna has higher-than-normal red blood cell count and hematocrit [p. 518].

Q4: When Edna fills her lungs maximally, the volume of air in her lungs is known as the _____ capacity. When she exhales all the air she can, the volume of air left in her lungs is the _____.

Q5: Why are Edna's RBC count and hematocrit increased? (Hint: Because of Edna's COPD, her arterial P_{O_2} is low.)

535 537 550 552 **554** 559

of air moved into and out of the lungs each minute (**FIG. 17.12a**). Total pulmonary ventilation, also known as the *minute volume*, is calculated as follows:

$$\text{Total pulmonary ventilation} = \text{ventilation rate} \times \text{tidal volume}$$

The normal ventilation rate for an adult is 12–20 breaths (br) per minute. Using the average tidal volume (500 mL) and the slowest ventilation rate, we get:

$$\text{Total pulmonary ventilation} =$$
$$12 \text{ br/min} \times 500 \text{ mL/br} = 6000 \text{ mL/min} = 6 \text{ L/min}$$

Total pulmonary ventilation represents the physical movement of air into and out of the respiratory tract, but is it a good indicator of how much fresh air reaches the alveolar exchange surface? Not necessarily.

Some air that enters the respiratory system does not reach the alveoli because part of every breath remains in the conducting airways, such as the trachea and bronchi. Because the conducting airways do not exchange gases with the blood, they are known as the **anatomic dead space.** Anatomic dead space averages about 150 mL.

To illustrate the difference between the total volume of air that enters the airways and the volume of fresh air that reaches the alveoli, let's consider a typical breath that moves 500 mL of air during a respiratory cycle (Fig. 17.12b).

1. Start at the end of an inspiration: Lung volume is maximal, and fresh air from the atmosphere fills the upper airways (the dead space).

2. Now exhale: the tidal volume of 500 mL leaves the body. However, the first portion of this 500 mL to exit the airways is the 150 mL of fresh air that had been in the dead space, followed by 350 mL of "stale" air from the alveoli. Even though 500 mL of air exited the alveoli, only 350 mL of that volume left the body. The remaining 150 mL of "stale" alveolar air stays in the dead space.

3. At the end of expiration, lung volume is at its minimum and stale air from alveoli fills the anatomic dead space.

4. With the next inspiration, another 500 mL of fresh air enters the airways. The first air to enter the alveoli is the 150 mL of stale air that was in the anatomic dead space. The remaining 350 mL of air to go into the alveoli is fresh air. The last 150 mL of inspired fresh air again remains in the dead space and never reaches the alveoli.

Thus, although 500 mL of air enters the alveoli with each breath, only 350 mL of that volume is fresh air. The volume of fresh air entering the alveoli equals the tidal volume minus the dead space volume.

Because a significant portion of inspired air never reaches an exchange surface, a more accurate indicator of ventilation efficiency is **alveolar ventilation,** the volume of fresh air that reaches the alveoli each minute. Alveolar ventilation is calculated by multiplying ventilation rate by the volume of fresh air that reaches the alveoli:

$$\text{Alveolar ventilation} =$$
$$\text{ventilation rate} \times (\text{tidal volume} - \text{dead space})$$

Using the same ventilation rate and tidal volume as before, and a dead space of 150 mL, then

$$\text{Alveolar ventilation} =$$
$$12 \text{ br/min} \times (500 - 150 \text{ mL/br}) = 4200 \text{ mL/min}$$

Thus, at 12 breaths per minute, the alveolar ventilation is 4.2 L/min. Although 6 L/min of fresh air enters the respiratory system, only 4.2 L of fresh air reaches the alveoli.

Alveolar ventilation can be drastically affected by changes in the rate or depth of breathing, as you can calculate using the figure question in Figure 17.12. **Maximum voluntary ventilation,** which involves breathing as deeply and quickly as possible, may increase total pulmonary ventilation to as much as 170 L/min. **TABLE 17.3** describes various patterns of ventilation, and **TABLE 17.4** gives normal ventilation values.

Alveolar Gas Composition Varies Little during Normal Breathing

The P_{O_2} and P_{CO_2} in the alveoli change surprisingly little during normal quiet breathing. Alveolar P_{O_2} is fairly constant at 100 mm Hg, and alveolar P_{CO_2} stays close to 40 mm Hg.

Intuitively, you might think that P_{O_2} would increase when fresh air first enters the alveoli, then decrease steadily as oxygen leaves to enter the blood. Instead, we find only very small swings in P_{O_2}. Why? The reasons are that (1) the amount of oxygen that

FIG. 17.12 ESSENTIALS

Ventilation

(a) Total pulmonary ventilation is greater than alveolar ventilation because of dead space.

Total pulmonary ventilation:

> **Total pulmonary ventilation = ventilation rate × tidal volume (V_T)**

For example: 12 breaths/min × 500 mL breath = 6000 mL/min

Alveolar ventilation:

Alveolar ventilation is a better indication of how much fresh air reaches the alveoli. Fresh air remaining in the dead space does not get to the alveoli.

> **Alveolar ventilation = ventilation rate × (V_T – dead space volume V_D)**

If dead space is 150 mL: 12 breaths/min × (500 – 150 mL) = 4200 mL/min

(b) Because the conducting airways do not exchange gases with the blood, they are known as **anatomic dead space**.

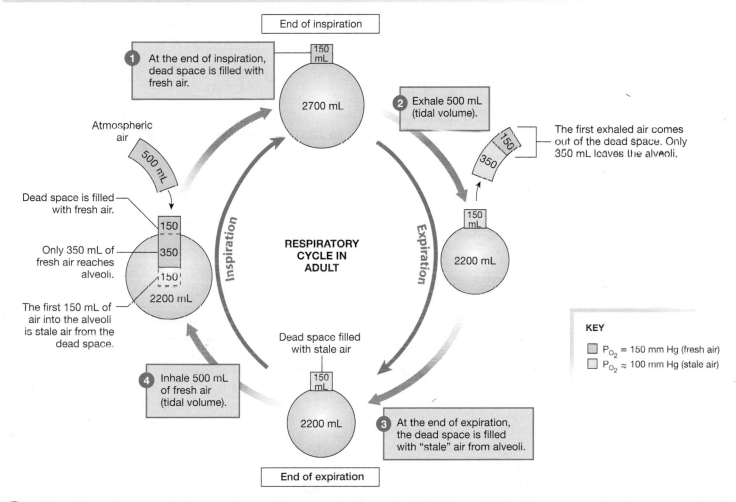

RESPIRATORY CYCLE IN ADULT

End of inspiration

1 At the end of inspiration, dead space is filled with fresh air.

150 mL

2700 mL

2 Exhale 500 mL (tidal volume).

The first exhaled air comes out of the dead space. Only 350 mL leaves the alveoli.

150 350

Atmospheric air

500 mL

Dead space is filled with fresh air.

150

Only 350 mL of fresh air reaches alveoli.

350

150

2200 mL

The first 150 mL of air into the alveoli is stale air from the dead space.

Inspiration

Expiration

150 mL

2200 mL

Dead space filled with stale air

150 mL

4 Inhale 500 mL of fresh air (tidal volume).

2200 mL

3 At the end of expiration, the dead space is filled with "stale" air from alveoli.

End of expiration

KEY

P_{O_2} = 150 mm Hg (fresh air)

P_{O_2} ≈ 100 mm Hg (stale air)

Q FIGURE QUESTION

Complete this table showing the effects of breathing pattern on alveolar ventilation. Assume dead space volume is 150 mL. Which pattern is the most efficient?

Tidal Volume (mL)	Ventilation Rate (breaths/min)	Total Pulmonary Ventilation (mL/min)	Fresh Air to Alveoli (mL)	Alveolar Ventilation (mL/min)
500 (normal)	12 (normal)	6000	350	4200
300 (shallow)	20 (rapid)			
750 (deep)	8 (slow)			

TABLE 17.3	Types and Patterns of Ventilation	
Name	**Description**	**Examples**
Eupnea	Normal quiet breathing	
Hyperpnea	Increased respiratory rate and/or volume in response to increased metabolism	Exercise
Hyperventilation	Increased respiratory rate and/or volume without increased metabolism	Emotional hyperventilation; blowing up a balloon
Hypoventilation	Decreased alveolar ventilation	Shallow breathing; asthma; restrictive lung disease
Tachypnea	Rapid breathing; usually increased respiratory rate with decreased depth	Panting
Dyspnea	Difficulty breathing (a subjective feeling sometimes described as "air hunger")	Various pathologies or hard exercise
Apnea	Cessation of breathing	Voluntary breath-holding; depression of CNS control centers

TABLE 17.4	Normal Ventilation Values in Pulmonary Medicine
Total pulmonary ventilation	6 L/min
Total alveolar ventilation	4.2 L/min
Maximum voluntary ventilation	125–170 L/min
Respiration rate	12–20 breaths/min

enters the alveoli with each breath is roughly equal to the amount of oxygen that enters the blood, and (2) the amount of fresh air that enters the lungs with each breath is only a little more than 10% of the total lung volume at the end of inspiration.

You can see this in Figure 17.12b ❹. In this example, at the end of inspiration only 350 mL out of the total volume of 2700 mL is higher-oxygen fresh air. This comes to about 13% of the total lung volume.

Although alveolar gases do not change much with quiet breathing, changes in alveolar ventilation can significantly affect the amount of fresh air and oxygen that reach the alveoli. **FIGURE 17.13** shows how partial pressures P_{O_2} and P_{CO_2} in the alveoli vary with increased alveolar ventilation (*hyperventilation*) and decreased hypoventilation (*hypoventilation*).

As alveolar ventilation increases during hyperventilation, alveolar P_{O_2} increases and alveolar P_{CO_2} falls. During hypoventilation, when less fresh air enters the alveoli, alveolar P_{O_2} decreases and alveolar P_{CO_2} increases. Carbon dioxide concentrations in the blood are closely linked to the body's pH, and you will learn later how the body uses changes in ventilation to help maintain pH homeostasis.

Ventilation and Alveolar Blood Flow Are Matched

Moving oxygen from the atmosphere to the alveolar exchange surface is only the first step in external respiration. Next, gas exchange must occur across the alveolar-capillary interface. Finally, blood flow (*perfusion*) past the alveoli must be high enough to pick up the available oxygen. Matching the ventilation rate into groups of alveoli with blood flow past those alveoli is a two-part process involving local regulation of both air flow and blood flow.

Alterations in pulmonary blood flow depend almost exclusively on properties of the capillaries and on such local factors as the concentrations of oxygen and carbon dioxide in the lung tissue. Capillaries in the lungs are unusual because they are collapsible. If the pressure of blood flowing through the capillaries falls below a certain point, the capillaries close off, diverting blood to pulmonary capillary beds in which blood pressure is higher.

In a person at rest, some capillary beds in the apex (top) of the lung are closed off because of low hydrostatic pressure. Capillary beds at the base of the lung have higher hydrostatic pressure because of gravity and thus remain open. Consequently, blood flow is diverted toward the base of the lung. During exercise, when blood pressure rises, the closed apical capillary beds open, ensuring that the increased cardiac output can be fully oxygenated as it passes through the lungs. The ability of the lungs to recruit additional capillary beds during exercise is an example of the reserve capacity of the body.

At the local level, the body attempts to match air flow and blood flow in each section of the lung by regulating the diameters of the arterioles and bronchioles. Bronchiolar diameter is mediated primarily by CO_2 levels in exhaled air passing through them (**FIG. 17.14**). An increase in the P_{CO_2} of expired air causes bronchioles to dilate. A decrease in the P_{CO_2} of expired air causes bronchioles to constrict.

Although there is some autonomic innervation of pulmonary arterioles, there is apparently little neural control of pulmonary blood flow. The resistance of pulmonary arterioles to blood flow is regulated primarily by the oxygen content of the interstitial fluid around the arteriole. If ventilation of alveoli in one area of

FIG. 17.13 Alveolar gases

As alveolar ventilation increases, alveolar P_{O_2} increases and P_{CO_2} decreases. The opposite occurs as alveolar ventilation decreases.

Normal ventilation
4.2 L/min

Hypoventilation   Hyperventilation

P_{O_2}

P_{CO_2}

Alveolar partial pressure (P_{gas}) in mm Hg

Alveolar ventilation (L/min)

Q **GRAPH QUESTION**

What are the maximum alveolar P_{O_2}
and minimum P_{CO_2} shown in this graph?

PHYS IN ACTION!

CONCEPT CHECK

26. If a lung tumor decreases blood flow in one small section of the lung to a minimum, what happens to P_{O_2} in the alveoli in that section and in the surrounding interstitial fluid? What happens to P_{CO_2} in that section? What is the compensatory response of the bronchioles in the affected section? Will the compensation bring ventilation in the affected section of the lung back to normal? Explain.

Auscultation and Spirometry Assess Pulmonary Function

Most pulmonary function tests are relatively simple to perform. Auscultation of breath sounds is an important diagnostic technique in pulmonary medicine, just as auscultation of heart sounds is an important technique in cardiovascular diagnosis [p. 464]. Breath sounds are more complicated to interpret than heart sounds, however, because breath sounds have a wider range of normal variation.

Normally, breath sounds are distributed evenly over the lungs and resemble a quiet "whoosh" made by flowing air. When air flow is reduced, such as in pneumothorax, breath sounds may be either diminished or absent. Abnormal sounds include various squeaks, pops, wheezes, and bubbling sounds caused by fluid and secretions in the airways or alveoli. Inflammation of the pleural membrane results in a crackling or grating sound known as a *friction rub*. It is caused by swollen, inflamed pleural membranes rubbing against each other, and it disappears when fluid again separates them.

Diseases in which air flow is diminished because of increased airway resistance are known as **obstructive lung diseases.** When patients with obstructive lower airway diseases are asked to exhale forcefully, air whistling through the narrowed airways creates a wheezing sound that can be heard even without a stethoscope. Depending on the severity of the disease, the bronchioles may even collapse and close off before a forced expiration is completed, reducing both the amount and rate of air flow as measured by a spirometer.

Obstructive lung diseases include asthma, obstructive sleep apnea, emphysema, and chronic bronchitis. The latter two are sometimes called *chronic obstructive pulmonary disease* (COPD) because of their ongoing, or chronic, nature. *Obstructive sleep apnea* {*apnoia,* breathless} results from obstruction of the upper airway, often due to abnormal relaxation of the muscles of the pharynx and tongue that increases airway resistance during inspiration.

Asthma is an inflammatory condition, often associated with allergies, that is characterized by bronchoconstriction and airway edema. Asthma can be triggered by exercise (exercise-induced asthma) or by rapid changes in the temperature or humidity of inspired air. Asthmatic patients complain of "air hunger" and difficulty breathing, or *dyspnea.* The severity of asthma attacks ranges from mild to life threatening. Studies of asthma at the cellular level show that a variety

---

the lung is diminished, as shown in Figure 17.14b, the P_{O_2} in that area decreases, and the arterioles respond by constricting, as shown in Figure 17.14c. This local vasoconstriction is adaptive because it diverts blood away from the under-ventilated region to better-ventilated parts of the lung.

Note that constriction of pulmonary arterioles in response to low P_{O_2} is the opposite of what occurs in the systemic circulation [p. 489]. In the systemic circulation, a decrease in the P_{O_2} of a tissue causes local arterioles to dilate, delivering more oxygen-carrying blood to those tissues that are consuming oxygen. In the lungs, blood is picking up oxygen, so it does not make sense to send more blood to an area with low tissue P_{O_2} due to poor ventilation.

Another important point must be noted here. Local control mechanisms are not effective regulators of air and blood flow under all circumstances. If blood flow is blocked in one pulmonary artery, or if air flow is blocked at the level of the larger airways, local responses that shunt air or blood to other parts of the lung are ineffective because in these cases no part of the lung has normal ventilation or perfusion.

FIG. 17.14 Local control mechanisms attempt to match ventilation and perfusion

(a) Normally, perfusion of blood past alveoli is matched to alveolar ventilation to maximize gas exchange.

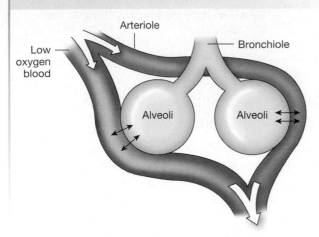

Arteriole

Bronchiole

Low oxygen blood

Alveoli   Alveoli

(b) Ventilation-Perfusion Mismatch Caused by Under-Ventilated Alveoli

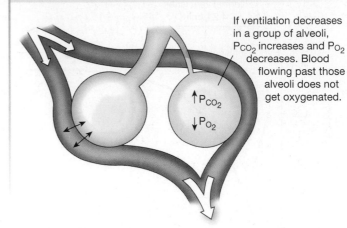

If ventilation decreases in a group of alveoli, P_{CO_2} increases and P_{O_2} decreases. Blood flowing past those alveoli does not get oxygenated.

$\uparrow P_{CO_2}$
$\downarrow P_{O_2}$

(c) Local control mechanisms try to keep ventilation and perfusion matched.

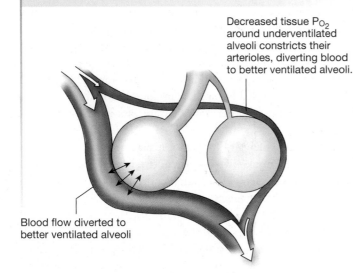

Decreased tissue P_{O_2} around underventilated alveoli constricts their arterioles, diverting blood to better ventilated alveoli.

Blood flow diverted to better ventilated alveoli

(d) Bronchiole diameter is mediated primarily by CO_2 levels in exhaled air passing through them.

Local Control of Arterioles and Bronchioles by Oxygen and Carbon Dioxide

Gas Composition	Bronchioles	Pulmonary Arteries	Systemic Arteries
P_{CO_2} increases	Dilate	(Constrict)*	Dilate
P_{CO_2} decreases	Constrict	(Dilate)	Constrict
P_{O_2} increases	(Constrict)	(Dilate)	Constrict
P_{O_2} decreases	(Dilate)	Constrict	Dilate

*Parentheses indicate weak responses.

Q **FIGURE QUESTIONS**

A blood clot prevents gas exchange in a group of alveoli.

1. What happens to tissue and alveolar gases?
2. What do bronchioles and arterioles do in response?

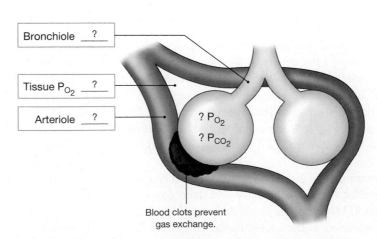

Bronchiole ___?___

Tissue P_{O_2} ___?___

Arteriole ___?___

? P_{O_2}
? P_{CO_2}

Blood clots prevent gas exchange.

of chemical signals may be responsible for inducing asthmatic bronchoconstriction. Among these are acetylcholine, histamine, substance P (a neuropeptide), and leukotrienes secreted by mast cells, macrophages, and eosinophils. *Leukotrienes* are lipid-like bronchoconstrictors that are released during the inflammatory response. Asthma is treated with inhaled and oral medications that include β_2-adrenergic agonists, anti-inflammatory drugs, and leukotriene antagonists.

This completes our discussion of the mechanics of ventilation. Next, we shift focus from the bulk flow of air to the diffusion and transport of oxygen and carbon dioxide as they travel between the air spaces of the alveoli and the cells of the body.

CONCEPT CHECK

27. Restrictive lung diseases decrease lung compliance. How will inspiratory reserve volume change in patients with a restrictive lung disease?

28. Chronic obstructive lung disease causes patients to lose the ability to exhale fully. How does residual volume change in these patients?

RUNNING **PROBLEM CONCLUSION** | **Emphysema**

Edna leaves the office with prescriptions for a mucus-thinning drug, a bronchodilator, and anti-inflammatory drugs to keep her airways as open as possible. She has agreed to try to stop smoking once more and also has a prescription and brochures for that. Unfortunately, the lung changes that take place with COPD are not reversible, and Edna will require treatment for the rest of her life. According to the American Lung Association

(*www.lung.org*), in 2010, COPD cost nearly $50 billion per year in direct medical costs and indirect costs such as lost wages.

In this running problem, you learned about chronic obstructive pulmonary disease. Now check your understanding of the physiology in the problem by comparing your answers with those in the following table.

Question	Facts	Integration and Analysis
Q1: What does narrowing of the airways do to the resistance airways offer to air flow?	The relationship between tube radius and resistance is the same for air flow as for blood flow: as radius decreases, resistance increases [p. 442].	When resistance increases, the body must use more energy to create air flow.
Q2: Why do people with chronic bronchitis have a higher-than-normal rate of respiratory infections?	Cigarette smoke paralyzes the cilia that sweep debris and mucus out of the airways. Without the action of cilia, mucus and trapped particles pool in the airways.	Bacteria trapped in the mucus can multiply and cause respiratory infections.
Q3: Name the muscles that patients with emphysema use to exhale actively.	Normal passive expiration depends on elastic recoil of muscles and elastic tissue in the lungs.	Forceful expiration involves the internal intercostal muscles and the abdominal muscles.
Q4: When Edna fills her lungs maximally, the volume of air in her lungs is known as the _____ capacity. When she exhales all the air she can, the volume of air left in her lungs is the _____.	The maximum volume of air in the lungs is the *total lung capacity.* Air left in the lungs after maximal exhalation is the *residual volume.*	N/A
Q5: Why are Edna's RBC count and hematocrit increased?	Because of Edna's COPD, her arterial P_{O_2} is low. The major stimulus for red blood cell synthesis is hypoxia.	Low arterial oxygen levels trigger EPO release, which increases the synthesis of red blood cells [p. 518]. More RBCs provide more binding sites for oxygen transport.

535 537 550 552 554 **559**

VISUALIZE, PRACTICE, AND MASTER PHYSIOLOGY WITH MASTERING A&P
- Practice-on-the-go with Dynamic Study Modules
- Visualize important processes with Interactive Physiology and A&P Flix
- Check your understanding with Video Tutors and practice quizzes!

MasteringA&P®

CHAPTER SUMMARY

Air flow into and out of the lungs is another example of the principle of *mass flow*. Like blood flow, air flow is bulk flow that requires a pump to create a pressure gradient and that encounters resistance, primarily from changes in the diameter of the tubes through which it flows. The *mechanical properties* of the pleural sacs and elastic recoil in the chest wall and lung tissue are essential for normal ventilation.

1. Aerobic metabolism in living cells consumes oxygen and produces carbon dioxide. (p. 535)

2. Gas exchange requires a large, thin, moist exchange surface; a pump to move air; and a circulatory system to transport gases to the cells. (p. 535)

3. Respiratory system functions include gas exchange, pH regulation, vocalization, and protection from foreign substances. (p. 535)

The Respiratory System

iP **Respiratory: Anatomy Review**

4. **Cellular respiration** refers to cellular metabolism that consumes oxygen. **External respiration** is the exchange of gases between the atmosphere and cells of the body. It includes ventilation, gas exchange at the lung and cells, and transport of gases in the blood. **Ventilation** is the movement of air into and out of the lungs. (p. 535; Fig. 17.1)

5. The **respiratory system** consists of anatomical structures involved in ventilation and gas exchange. (p. 536)

6. The **upper respiratory tract** includes the mouth, nasal cavity, **pharynx,** and **larynx.** The **lower respiratory tract** includes the **trachea, bronchi, bronchioles,** and exchange surfaces of the **alveoli.** (pp. 536, 537; Fig. 17.2a)

7. The thoracic cage is bounded by the ribs, spine, and **diaphragm.** Two sets of **intercostal muscles** connect the ribs. (p. 536; Fig. 17.2c)

8. Each **lung** is contained within a double-membrane **pleural sac** that contains a small quantity of **pleural fluid.** (p. 536; Figs. 17.2d, 17.3)

9. The two **primary bronchi** enter the lungs. Each primary bronchus divides into progressively smaller bronchi and finally into collapsible **bronchioles.** (p. 537; Figs. 17.2e, 17.4)

10. The upper respiratory system filters, warms, and humidifies inhaled air. (p. 537)

11. The alveoli consist mostly of thin-walled **type I alveolar cells** for gas exchange. **Type II alveolar cells** produce surfactant. A network of capillaries surrounds each alveolus. (p. 540; Fig. 17.2f, g)

12. Blood flow through the lungs equals cardiac output. Resistance to blood flow in the pulmonary circulation is low. Pulmonary arterial pressure averages 25/8 mm Hg. (p. 540)

Gas Laws

iP **Respiratory: Pulmonary Ventilation**

13. **Dalton's Law** states that the total pressure of a mixture of gases is the sum of the pressures of the individual gases in the mixture. **Partial pressure** is the pressure contributed by a single gas in a mixture. (p. 542; Fig. 17.6)

14. Bulk flow of air occurs down pressure gradients, as does the movement of any individual gas making up the air. (p. 542)

15. **Boyle's Law** states that as the volume available to a gas increases, the gas pressure decreases. The body creates pressure gradients by changing thoracic volume. (p. 544; Fig. 17.6b)

Ventilation

iP **Respiratory: Pulmonary Ventilation**

16. A single **respiratory cycle** consists of one inspiration followed by one expiration. (p. 544)

17. **Tidal volume** is the amount of air taken in during a single normal inspiration. **Vital capacity** is tidal volume plus **expiratory** and **inspiratory reserve volumes.** Air volume in the lungs at the end of maximal expiration is the **residual volume.** (p. 544; Fig. 17.7b)

18. Air flow in the respiratory system is directly proportional to the pressure gradient, and inversely related to the resistance to air flow offered by the airways. (p. 546)

19. During **inspiration, alveolar pressure** decreases, and air flows into the lungs. Inspiration requires contraction of the inspiratory muscles and the diaphragm. (p. 547; Fig. 17.9)

20. **Expiration** is usually passive, resulting from elastic recoil of the lungs. (p. 548)

21. **Active expiration** requires contraction of the internal intercostal and abdominal muscles. (p. 548)

22. **Intrapleural pressures** are subatmospheric because the pleural cavity is a sealed compartment. (p. 549; Figs. 17.9, 17.10)

23. **Compliance** is a measure of the ease with which the chest wall and lungs expand. Loss of compliance increases the work of breathing. **Elastance** is the ability of a lung to resist stretching or to return to its unstretched state. (p. 551)

24. **Surfactant** decreases surface tension in the fluid lining the alveoli. Reduced surface tension prevents smaller alveoli from collapsing and also makes it easier to inflate the lungs. (p. 552; Fig. 17.11)

25. The diameter of the bronchioles determines how much resistance they offer to air flow. (p. 553)

26. Increased CO_2 in expired air dilates bronchioles. Parasympathetic neurons cause **bronchoconstriction** in response to irritant stimuli. There is no significant sympathetic innervation of bronchioles, but epinephrine causes **bronchodilation.** (p. 553; Tbl. 17.2)

27. **Total pulmonary ventilation** = ventilation rate × tidal volume. **Alveolar ventilation** = ventilation rate × (tidal volume − dead space volume). (p. 554; Fig. 17.12a)

28. Alveolar gas composition changes very little during a normal respiratory cycle. **Hyperventilation** increases alveolar P_{O_2} and decreases alveolar P_{CO_2}. **Hypoventilation** has the opposite effect. (p. 556; Fig. 17.13)

29. Local mechanisms match air flow and blood flow around the alveoli. Increased levels of CO_2 dilate bronchioles, and decreased O_2 constricts pulmonary arterioles. (p. 556; Fig. 17.14)

REVIEW QUESTIONS

In addition to working through these questions and checking your answers on p. A-22, review the Learning Outcomes at the beginning of this chapter.

Level One Reviewing Facts and Terms

1. List four functions of the respiratory system.

2. Give two definitions for the word *respiration*.

3. Which sets of muscles are used for normal quiet inspiration? For normal, quiet expiration? For active expiration? What kind(s) of muscles are the different respiratory muscles (skeletal, cardiac, or smooth)?

4. Give two functions of pleural fluid.

5. Name the anatomical structures that an oxygen molecule passes on its way from the atmosphere to the blood.

6. Diagram the structure of an alveolus, and state the function of each part. How are capillaries associated with an alveolus?

7. Trace the path of the pulmonary circulation. About how much blood is found here at any given moment? What is a typical arterial blood pressure for the pulmonary circuit, and how does this pressure compare with that of the systemic circulation?

8. What happens to inspired air as it is conditioned during its passage through the airways?

9. During inspiration, most of the thoracic volume change is the result of movement of the _____.

10. Describe the changes in alveolar and intrapleural pressure during one respiratory cycle.

11. Refer to the spirogram in the following figure:

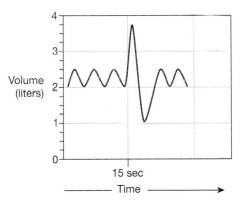

(a) Label tidal volume (V_T), inspiratory and expiratory reserve volumes (IRV and ERV), residual volume (RV), vital capacity (VC), total lung capacity (TLC).
(b) What is the value of each of these volumes and capacities?
(c) What is this person's ventilation rate?

12. Of the three factors that contribute to the resistance of air flow through a tube, which plays the largest role in changing resistance in the human respiratory system?

13. Match the following items with their correct effect on the bronchioles:

(a) histamine	1. bronchoconstriction
(b) epinephrine	2. bronchodilation
(c) acetylcholine	3. no effect
(d) increased P_{CO_2}	

14. What is the function of surfactants in general? In the respiratory system?

15. If a person increases her tidal volume, what would happen to her alveolar P_{O_2}?

Level Two Reviewing Concepts

16. Compare and contrast the terms in each of the following sets:

    (a) compliance and elastance
    (b) inspiration, expiration, and ventilation
    (c) intrapleural pressure and alveolar pressure
    (d) total pulmonary ventilation and alveolar ventilation
    (e) type I and type II alveolar cells
    (f) pulmonary circulation and systemic circulation

17. List the major paracrines and neurotransmitters that cause bronchoconstriction and bronchodilation. What receptors do they act through? (muscarinic, nicotinic, α, β_1, β_2)

18. Compile the following terms into a map of ventilation. Use up arrows, down arrows, greater than symbols (>), and less than symbols (<) as modifiers. You may add other terms.

• abdominal muscles	• inspiratory muscles
• air flow	• internal intercostals
• contract	• P_A
• diaphragm	• P_{atm}
• expiratory muscles	• $P_{intrapleural}$
• external intercostals	• quiet breathing
• forced breathing	• relax
• in, out, from, to	• scalenes

19. Decide whether each of the following parameters will increase, decrease, or not change in the situations given.

    (a) airway resistance with bronchodilation
    (b) intrapleural pressure during inspiration
    (c) air flow with bronchoconstriction
    (d) bronchiolar diameter with increased P_{CO_2}
    (e) tidal volume with decreased compliance
    (f) alveolar pressure during expiration

20. Define the following terms: pneumothorax, spirometer, auscultation, hypoventilation, bronchoconstriction, minute volume, partial pressure of a gas.

21. The cartoon coyote is blowing up a balloon in another attempt to catch the roadrunner. He first breathes in as much air as he can, then blows out all he can into the balloon.

    (a) The volume of air in the balloon is equal to the _____ of the coyote's lungs. This volume can be measured directly by measuring the balloon volume or by adding which respiratory volumes together?
    (b) In 10 years, when the coyote is still chasing the roadrunner, will he still be able to put as much air into the balloon in one breath? Explain.

22. Match the descriptions to the appropriate phase(s) of ventilation:

(a) usually depend(s) on elastic recoil	1. inspiration
(b) is/are easier when lung compliance decreases	2. expiration
(c) is/are driven mainly by positive intrapleural pressure generated by muscular contraction	3. both inspiration and expiration
(d) is usually an active process requiring smooth muscle contraction	4. neither

23. Draw and label a graph showing the P_{O_2} of air in the primary bronchi during one respiratory cycle. (*Hint:* What parameter goes on each axis?)

24. Lung compliance increases but chest wall compliance decreases as we age. In the absence of other changes, would the following parameters increase, decrease, or not change as compliance decreases?

    (a) work required for breathing
    (b) ease with which lungs inflate
    (c) lung elastance
    (d) airway resistance during inspiration

25. Will pulmonary surfactant increase, decrease, or not change the following?

    (a) work required for breathing
    (b) lung compliance
    (c) surface tension in the alveoli

26. A student breathes at a rate of 20 breaths/min, with a tidal volume of 300 mL/breath. If his anatomic dead space is 130 mL, calculate his total pulmonary ventilation rate and his alveolar ventilation rate.

Level Three Problem Solving

27. A 30-year-old computer programmer has had asthma for 15 years. When she lies down at night, she has spells of wheezing and coughing. Over the years, she has found that she can breathe better if she sleeps sitting nearly upright. Upon examination, her doctor finds that she has an enlarged thorax. Her lungs are overinflated on x-ray. Here are the results of her examination and pulmonary function tests. Use the normal values and abbreviations in Fig. 17.7 to help answer the questions.

Ventilation rate: 16 breaths/min	Tidal volume: 600 mL
ERV: 1000 mL	RV: 3500 mL
Inspiratory capacity: 1800 mL	Vital capacity: 2800 mL
Functional residual capacity: 4500 mL	TLC: 6300 mL

After she was given a bronchodilator, her vital capacity increased to 3650 mL.

    (a) What is her minute volume?
    (b) Explain the change in vital capacity with bronchodilators.
    (c) Which other values are abnormal? Can you explain why they might be, given her history and findings?

28. Alveolar air has an average P_{O_2} of 100 mm Hg, but expired air has an average P_{O_2} of 120 mm Hg. If oxygen moves out of the lungs into the body, why is there more oxygen in the expired air?

29. Assume a normal female has a resting tidal volume of 400 mL, a respiratory rate of 13 breaths/min, and an anatomic dead space of 125 mL. When she exercises, which of the following scenarios would be most efficient for increasing her oxygen delivery to the lungs?

    (a) increase respiratory rate to 20 breaths/min but have no change in tidal volume
    (b) increase tidal volume to 550 mL but have no change in respiratory rate
    (c) increase tidal volume to 500 mL and respiratory rate to 15 breaths/min

    Which of these scenarios is most likely to occur during exercise in real life?

Level Four Quantitative Problems

30. A container of gas with a movable piston has a volume of 500 mL and a pressure of 60 mm Hg. The piston is moved, and the new pressure is 150 mm Hg. What is the new volume of the container?

31. You have a mixture of gases in dry air, with an atmospheric pressure of 760 mm Hg. Calculate the partial pressure of each gas if the composition of the air includes:

    (a) 21% oxygen, 78% nitrogen, 0.3% carbon dioxide
    (b) 40% oxygen, 13% nitrogen, 45% carbon dioxide, 2% hydrogen
    (c) 10% oxygen, 15% nitrogen, 1% argon, 25% carbon dioxide

32. Li is a tiny woman, with a tidal volume of 400 mL and a respiratory rate of 12 breaths per minute at rest. What is her total pulmonary ventilation? Just before a physiology exam, her ventilation increases to 18 breaths per minute from nervousness. Now what is her total pulmonary ventilation? Assuming her anatomic dead space is 120 mL, what is her alveolar ventilation in each case?

33. You collected the following data on your classmate Neelesh:

    Minute volume = 5004 mL/min
    Respiratory rate = 3 breaths/15 sec
    Vital capacity = 4800 mL
    Expiratory reserve volume = 1000 mL

    What are Neelesh's tidal volume and inspiratory reserve volume?

34. Use the following figure to help solve this problem. A spirometer with a volume of 1 liter (V_1) is filled with a mixture of oxygen and helium, with the helium concentration being 4 g/L (C_1). Helium does not move from the lungs into the blood or from the blood into the lungs. A subject is told to blow out all the air he possibly can. Once he finishes that exhalation, his lung volume is V_2. He then puts the spirometer tube in his mouth and breathes quietly for several breaths. At the end of that time, the helium is evenly dispersed in the spirometer and the subject's lungs. A measurement shows the new concentration of helium is 1.9 g/L. What was the subject's lung volume at the start of the experiment? (*Hint:* $C_1V_1 = C_2V_2$)

35. The graph shows one lung under two different conditions, A and B. What does this graph show? (a) the effect of lung volume on pressure, or (b) the effect of pressure on lung volume? In which condition does the lung have higher compliance, or is compliance the same in the two situations?

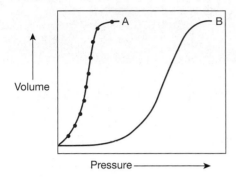

36. Lung volumes and capacities will vary with a person's height and sex. Prediction equations for estimating them have been derived from clinical studies. Use the equations in the table below to estimate your volumes and capacities. What will happen to your predicted vital capacity when you are 70 years old? H = height in cm, where 1 inch = 2.54 cm. A = age in years.

Lung Volume (L)	Subject	Formula
Vital capacity	Men	$(0.06 \times H) - (0.0214 \times A) - 4.65$
	Women	$(0.0491 \times H) - (0.0216 \times A) - 3.59$
Total lung capacity	Men	$(0.0795 \times H) + (0.0032 \times A) - 7.333$
	Women	$(0.059 \times H) - 4.537$
Functional residual capacity	Men	$(0.0472 \times H) + (0.009 \times A) - 5.29$
	Women	$(0.036 \times H) + (0.0031 \times A) - 3.182$
Residual volume	Men	$(0.0216 \times H) + (0.0207 \times A) - 2.84$
	Women	$(0.0197 \times H) + (0.0201 \times A) - 2.421$

Answers to Concept Checks, Figure and Graph Questions, and end-of-chapter Review Questions can be found in Appendix A [p. A-1].

18

Gas Exchange and Transport

The successful ascent of Everest without supplementary oxygen is one of the great sagas of the 20th century.

John B. West, *Climbing with O's*, NOVA Online (*www.pbs.org*)

GAS EXCHANGE IN THE LUNGS AND TISSUES 565

LO 18.1 List three arterial blood parameters that influence ventilation.

LO 18.2 Diagram the normal partial pressures of O_2 and CO_2 in the atmosphere, alveoli, arterial blood, resting cells, and venous blood.

LO 18.3 Describe all the factors that influence gas exchange between the atmosphere and arterial blood.

LO 18.4 Explain the difference between the concentration of a gas in solution and the partial pressure of that gas in solution, using O_2 and CO_2 as examples.

GAS TRANSPORT IN THE BLOOD 571

LO 18.5 Explain how the Fick equation uses mass flow and mass balance to relate cardiac output and cellular oxygen consumption.

LO 18.6 Explain the role of hemoglobin in oxygen transport from the molecular level to the systemic level.

LO 18.7 Describe the relationship between plasma P_{O_2} and oxygen transport.

LO 18.8 Draw the oxyhemoglobin saturation curve, explain the physiological significance of the shape of this curve, and draw the shifts in the curve that result from changes in pH, temperature, and 2,3-BPG.

LO 18.9 Compare and contrast oxygen transport on fetal and adult hemoglobin.

LO 18.10 Write the chemical reaction for the conversion of CO_2 to HCO_3^-, including the enzyme that catalyzes the reaction.

LO 18.11 Map the transport of carbon dioxide in arterial and venous blood, including the exchanges of CO_2 between the blood and the alveoli or cells.

REGULATION OF VENTILATION 580

LO 18.12 Map the reflex control of ventilation including appropriate neurotransmitters and their receptors.

LO 18.13 Diagram the current model for the brainstem neural networks that control breathing.

LO 18.14 Explain the mechanisms by which central and peripheral chemoreceptors monitor CO_2 and O_2 levels.

LO 18.15 Describe the protective reflexes that guard the lungs.

BACKGROUND BASICS

76	Exchange epithelia
41	pH and buffers
48	Law of mass action
134	Simple diffusion
280	Cerebrospinal fluid
285	Structure of the brain stem
282	Blood-brain barrier
259	Autonomic and somatic motor neurons
518	Red blood cells and hemoglobin

Giant liposomes of pulmonary surfactant (40X)

The book *Into Thin Air* by Jon Krakauer chronicles an ill-fated trek to the top of Mt. Everest. To reach the summit of Mt. Everest, climbers must pass through the "death zone" located at about 8000 meters (over 26,000 ft.). Of the thousands of people who have attempted the summit, only about 2000 have been successful, and more than 185 have died. What are the physiological challenges of climbing Mt. Everest (8850 m or 29,035 ft.), and why did it take so many years before humans successfully reached the top? The lack of oxygen at high altitude is part of the answer.

The mechanics of breathing include the events that create bulk flow of air into and out of the lungs. In this chapter, we focus on the two gases most significant to human physiology, oxygen and carbon dioxide, and look at how they move between alveolar air spaces and the cells of the body. The process can be divided into two components: the exchange of gases between compartments, which requires diffusion across cell membranes, and the transport of gases in the blood. **FIGURE 18.1** presents an overview of the topics that we cover in this chapter.

If the diffusion of gases between alveoli and blood is significantly impaired, or if oxygen transport in the blood is inadequate, **hypoxia** (a state of too little oxygen) results. Hypoxia frequently (but not always!) goes hand in hand with **hypercapnia**, elevated concentrations of carbon dioxide. These two conditions are clinical signs, not diseases, and clinicians must gather additional information to pinpoint their cause. **TABLE 18.1** lists several types of hypoxia and some typical causes.

To avoid hypoxia and hypercapnia, the body uses sensors that monitor arterial blood composition. These sensors respond to three regulated variables:

1. *Oxygen.* Arterial oxygen delivery to the cells must be adequate to support aerobic respiration and ATP production.
2. *Carbon dioxide* (CO_2) is produced as a waste product during the citric acid cycle [p. 108]. Excretion of CO_2 by the lungs is important for two reasons: high levels of CO_2 are a central nervous system depressant, and elevated CO_2 causes a state of acidosis (low pH) through the following reaction: $CO_2 + H_2O \rightleftharpoons H_2CO_3 \rightleftharpoons H^+ + HCO_3^-$.
3. *pH.* Maintaining pH homeostasis is critical to prevent denaturation of proteins [p. 51]. The respiratory system monitors plasma pH and uses changes in ventilation to alter pH. We discuss this process later along with renal contributions to pH homeostasis.

The normal values for these three parameters are given in **TABLE 18.2**. In this chapter, we will consider the mechanisms by which oxygen and CO_2 move from the lungs to the cells and back again.

GAS EXCHANGE IN THE LUNGS AND TISSUES

Breathing is the bulk flow of air into and out of the lungs. Once air reaches the alveoli, individual gases such as oxygen and CO_2 diffuse from the alveolar air space into the blood. Recall that diffusion is movement of a molecule from a region of higher concentration to one of lower concentration [p. 134].

When we think of concentrations of solutions, units such as moles/liter and milliosmoles/liter come to mind. However, respiratory physiologists commonly express plasma gas concentrations in partial pressures to establish whether there is a concentration gradient between the alveoli and the blood. Gases move from regions of higher partial pressure to regions of lower partial pressure.

FIG. 18.1 Pulmonary gas exchange and transport

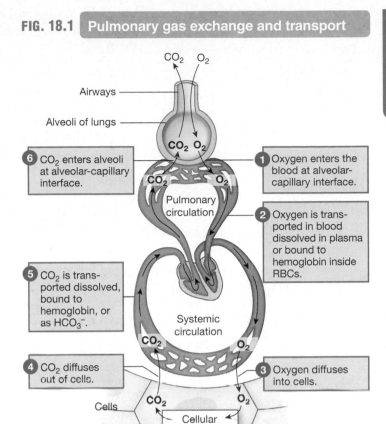

CO₂   O₂

Airways

Alveoli of lungs

CO₂ O₂

6 CO_2 enters alveoli at alveolar-capillary interface.

CO₂   O₂

Pulmonary circulation

1 Oxygen enters the blood at alveolar-capillary interface.

2 Oxygen is transported in blood dissolved in plasma or bound to hemoglobin inside RBCs.

5 CO_2 is transported dissolved, bound to hemoglobin, or as HCO_3^-.

Systemic circulation

CO₂   **O₂**

4 CO_2 diffuses out of cells.

3 Oxygen diffuses into cells.

Cells   **CO₂**   **O₂**

Cellular respiration determines metabolic CO_2 production.

ATP   Nutrients

RUNNING **PROBLEM** | High Altitude

In 1981, a group of 20 physiologists, physicians, and climbers, supported by 42 Sherpa assistants, formed the American Medical Research Expedition to Mt. Everest. The purpose of the expedition was to study human physiology at extreme altitudes, starting with the base camp at 5400 m (18,000 ft) and continuing on to the summit at 8850 m (over 29,000 ft). From the work of these scientists and others, we now have a good picture of the physiology of high-altitude acclimatization [p. 18].

565   567   571   574   579   584   585

TABLE 18.1	Classification of Hypoxias		
Type	**Definition**		**Typical Causes**
Hypoxic hypoxia	Low arterial P_{O_2}		High altitude; alveolar hypoventilation; decreased lung diffusion capacity; abnormal ventilation-perfusion ratio
Anemic hypoxia	Decreased total amount of O_2 bound to hemoglobin		Blood loss; anemia (low [Hb] or altered HbO_2 binding); carbon monoxide poisoning
Ischemic hypoxia	Reduced blood flow		Heart failure (whole-body hypoxia); shock (peripheral hypoxia); thrombosis (hypoxia in a single organ)
Histotoxic hypoxia	Failure of cells to use O_2 because cells have been poisoned		Cyanide and other metabolic poisons

TABLE 18.2	Normal Blood Values in Pulmonary Medicine	
	Arterial	**Venous**
P_{O_2}	95 mm Hg (85–100)	40 mm Hg
P_{CO_2}	40 mm Hg (35–45)	46 mm Hg
pH	7.4 (7.38–7.42)	7.37

FIGURE 18.2 shows the partial pressures of oxygen and CO_2 in air, the alveoli, and inside the body. Normal alveolar P_{O_2} at sea level is about 100 mm Hg. The P_{O_2} of "deoxygenated" venous blood arriving at the lungs is about 40 mm Hg. Oxygen therefore diffuses down its partial pressure (concentration) gradient from the alveoli into the capillaries. Diffusion goes to equilibrium, and the P_{O_2} of arterial blood leaving the lungs is the same as in the alveoli: 100 mm Hg.

When arterial blood reaches tissue capillaries, the gradient is reversed. Cells are continuously using oxygen for oxidative phosphorylation [p. 109]. In the cells of a person at rest, intracellular P_{O_2} averages 40 mm Hg. Arterial blood arriving at the cells has a P_{O_2} of 100 mm Hg. Because P_{O_2} is lower in the cells, oxygen diffuses down its partial pressure gradient from plasma into cells. Once again, diffusion goes to equilibrium. As a result, venous blood has the same P_{O_2} as the cells it just passed.

Conversely, P_{CO_2} is higher in tissues than in systemic capillary blood because of CO_2 production during metabolism (Fig. 18.2). Cellular P_{CO_2} in a person at rest is about 46 mm Hg, compared to an arterial plasma P_{CO_2} of 40 mm Hg. The gradient causes CO_2 to diffuse out of cells into the capillaries. Diffusion goes to equilibrium, and systemic venous blood averages a P_{CO_2} of 46 mm Hg.

At the pulmonary capillaries, the process reverses. Venous blood bringing waste CO_2 from the cells has a P_{CO_2} of 46 mm Hg. Alveolar P_{CO_2} is 40 mm Hg. Because P_{CO_2} is higher in the plasma, CO_2 moves from the capillaries into the alveoli. By the time blood leaves the alveoli, it has a P_{CO_2} of 40 mm Hg, identical to the P_{CO_2} of the alveoli.

In the sections that follow, we will consider some of the other factors that affect the transfer of gases between the alveoli and the body's cells.

CONCEPT CHECK

1. Cellular metabolism review: which of the following three metabolic pathways—glycolysis, the citric acid cycle, and the electron transport system—is *directly* associated with (a) O_2 consumption and with (b) CO_2 production?

2. Why doesn't the movement of oxygen from the alveoli to the plasma decrease the P_{O_2} of the alveoli? [*Hint:* p. 555]

3. If nitrogen is 78% of atmospheric air, what is the partial pressure of this gas when the dry atmospheric pressure is 720 mm Hg?

Lower Alveolar P_{O_2} Decreases Oxygen Uptake

Many variables influence the efficiency of alveolar gas exchange and determine whether arterial blood gases are normal (**FIG. 18.3a**). First, adequate oxygen must reach the alveoli. A decrease in alveolar P_{O_2} means that less oxygen is available to enter the blood. There can also be problems with the transfer of gases between the alveoli and pulmonary capillaries. Finally, blood flow, or *perfusion*, of the alveoli must be adequate [p. 556]. If something impairs blood flow to the lung, then the body is unable to acquire the oxygen it needs. Let's look in more detail at these factors.

There are two possible causes of low alveolar P_{O_2}: either (1) the inspired air has low oxygen content or (2) alveolar ventilation [p. 553] is inadequate.

Composition of the Inspired Air The first requirement for adequate oxygen delivery to the tissues is adequate oxygen intake from the atmosphere. The main factor that affects atmospheric oxygen content is altitude. The partial pressure of oxygen in air decreases along with total atmospheric pressure as you move from sea level (where normal atmospheric pressure is 760 mm Hg) to higher altitudes.

FIG. 18.2 Gases diffuse down concentration gradients

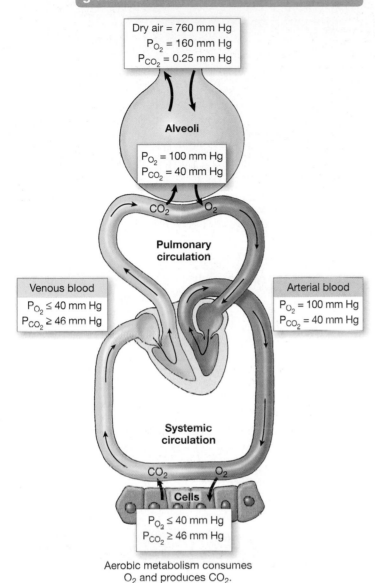

Dry air = 760 mm Hg
P_{O_2} = 160 mm Hg
P_{CO_2} = 0.25 mm Hg

Alveoli

P_{O_2} = 100 mm Hg
P_{CO_2} = 40 mm Hg

CO_2 O_2

Pulmonary circulation

Venous blood
P_{O_2} ≤ 40 mm Hg
P_{CO_2} ≥ 46 mm Hg

Arterial blood
P_{O_2} = 100 mm Hg
P_{CO_2} = 40 mm Hg

Systemic circulation

CO_2 O_2

Cells

P_{O_2} ≤ 40 mm Hg
P_{CO_2} ≥ 46 mm Hg

Aerobic metabolism consumes
O_2 and produces CO_2.

RUNNING PROBLEM

Hypoxia is the primary problem that people experience when ascending to high altitude. High altitude is considered anything above 1500 m (5000 ft), but most pathological responses to altitude occur above 2500 m (about 8000 ft). By one estimate, 25% of people arriving at 2590 m will experience some form of altitude sickness.

Q1: *If water vapor contributes 47 mm Hg to the pressure of fully humidified air, what is the P_{O_2} of inspired air reaching the alveoli at 2500 m, where dry atmospheric pressure is 542 mm Hg? How does this value for P_{O_2} compare with that of fully humidified air at sea level?*

565　**567**　571　574　579　584　585

that can result in alveolar hypoventilation (Fig. 18.3c) include decreased lung compliance [p. 551], increased airway resistance [p. 553], or central nervous system (CNS) depression that slows ventilation rate and decreases depth. Common causes of CNS depression in young people include alcohol poisoning and drug overdoses.

CONCEPT CHECK   4. At the summit of Mt. Everest, an altitude of 8850 m, atmospheric pressure is only 250 mm Hg. What is the P_{O_2} of dry atmospheric air atop Everest? If water vapor added to inhaled air at the summit has a partial pressure of 47 mm Hg, what is the P_{O_2} of the inhaled air when it reaches the alveoli?

Diffusion Problems Cause Hypoxia

If hypoxia is not caused by hypoventilation, then the problem usually lies with some aspect of gas exchange between alveoli and blood. In these situations, alveolar P_{O_2} may be normal, but the P_{O_2} of arterial blood leaving the lungs is low. The transfer of oxygen from alveoli to blood requires diffusion across the barrier created by type I alveolar cells and the capillary endothelium (Fig. 18.3b).

The exchange of oxygen and carbon dioxide across this diffusion barrier obeys the same rules as simple diffusion across a membrane [p. 136]. The diffusion rate is directly proportional to the available surface area, the concentration gradient of the gas, and the permeability of the barrier:

Diffusion rate ∝
surface area × concentration gradient × barrier permeability

From the general rules for diffusion, we can add a fourth factor: *diffusion distance*. Diffusion is inversely proportional to the square of the distance or, in simpler terms—diffusion is most rapid over short distances [p. 134]:

Diffusion rate ∝ $1/distance^2$

For example, Denver, 1609 m above sea level, has an atmospheric pressure of about 628 mm Hg. The P_{O_2} of dry air in Denver is 132 mm Hg, down from 160 mm Hg at sea level. For fully humidified atmospheric air reaching the alveoli, the P_{O_2} is even lower: P_{atm} 628 mm Hg − P_{H_2O} 47 mm Hg = 581 mm Hg × 21% = P_{O_2} of 122 mm Hg, down from 150 mm Hg at sea level. Notice that water vapor pressure at 100% humidity is the same no matter what the altitude, making its contribution to total pressure in the lungs more important as you go higher.

Alveolar Ventilation   Unless a person is traveling, altitude remains constant. If the composition of inspired air is normal but alveolar P_{O_2} is low, then the problem must lie with alveolar ventilation. Low alveolar ventilation is also known as *hypoventilation* and is characterized by lower-than-normal volumes of fresh air entering the alveoli. Pathological changes

FIG. 18.3 Gas exchange in the alveoli

(a) Alveolar gas exchange

Alveolar Gas Exchange

is influenced by

- O_2 reaching the alveoli
  - Composition of inspired air
  - Alveolar ventilation
    - Rate and depth of breathing
    - Airway resistance
    - Lung compliance
- Gas diffusion between alveoli and blood
  - Surface area
  - Diffusion distance
    - Barrier thickness
    - Amount of fluid
- Adequate perfusion of alveoli

(b) Cells form a diffusion barrier between lung and blood.

Alveoli

Capillary

Surfactant
Alveolar epithelium
Fused basement membranes
Nucleus of endothelial cell
Plasma
RBC

O_2 CO_2
O_2 CO_2

Alveolar air space
0.1–1.5 μm
Capillary lumen

(c) Pathologies that cause hypoxia

Diffusion $\propto$ surface area $\times$ barrier permeability/distance^2

Normal Lung	Emphysema	Fibrotic Lung Disease	Pulmonary Edema	Asthma
	Destruction of alveoli means less surface area for gas exchange.	Thickened alveolar membrane slows gas exchange. Loss of lung compliance may decrease alveolar ventilation.	Fluid in interstitial space increases diffusion distance. Arterial P_{CO_2} may be normal due to higher CO_2 solubility in water.	Increased airway resistance decreases alveolar ventilation.
P_{O_2} normal	P_{O_2} normal or low	P_{O_2} normal or low	P_{O_2} normal	P_{O_2} low
P_{O_2} normal	P_{O_2} low	P_{O_2} low	P_{O_2} low	P_{O_2} low

Exchange surface normal
Increased diffusion distance

Bronchioles constricted

Under most circumstances, diffusion distance, surface area, and barrier permeability in the body are constants and are maximized to facilitate diffusion. Gas exchange in the lungs is rapid, blood flow through pulmonary capillaries is slow, and diffusion reaches equilibrium in less than 1 second. This leaves the concentration gradient between alveoli and blood as the primary factor affecting gas exchange in healthy people.

The factors of surface area, diffusion distance, and membrane permeability do come into play with various diseases. Pathological changes that adversely affect gas exchange include (1) a decrease in the amount of alveolar surface area available for gas exchange, (2) an increase in the thickness of the alveolar-capillary exchange barrier, and (3) an increase in the diffusion distance between the alveolar air space and the blood.

Surface Area Physical loss of alveolar surface area can have devastating effects in *emphysema,* a degenerative lung disease most often caused by cigarette smoking (Fig. 18.3c). The irritating effect of smoke chemicals and tar in the alveoli activates alveolar macrophages that release *elastase* and other proteolytic enzymes. These enzymes destroy the elastic fibers of the lung [p. 82] and induce apoptosis of cells, breaking down the walls of the alveoli. The result is a high-compliance/low-elastic recoil lung with fewer and larger alveoli and less surface area for gas exchange.

Diffusion Barrier Permeability Pathological changes in the alveolar-capillary diffusion barrier may alter its properties and slow gas exchange. For example, in fibrotic lung diseases, scar tissue thickens the alveolar wall (Fig. 18.3c). Diffusion of gases through this scar tissue is much slower than normal. However, because the lungs have a built-in reserve capacity, one-third of the exchange epithelium must be incapacitated before arterial P_{O_2} falls significantly.

Diffusion Distance Normally, the pulmonary diffusion distance is small because the alveolar and endothelial cells are thin and there is little or no interstitial fluid between the two cell layers (Fig. 18.3b). However, in certain pathological states, excess fluid increases the diffusion distance between the alveolar air space and the blood. Fluid accumulation may occur inside the alveoli or in the interstitial compartment between the alveolar epithelium and the capillary.

In **pulmonary edema,** accumulation of interstitial fluid increases the diffusion distance and slows gas exchange (Fig. 18.3c). Normally, only small amounts of interstitial fluid are present in the lungs, the result of low pulmonary blood pressure and effective lymph drainage. However, if pulmonary blood pressure rises for some reason, such as left ventricular failure or mitral valve dysfunction, the normal filtration/reabsorption balance at the capillary is disrupted [Fig. 15.18, p. 499].

When capillary hydrostatic pressure increases, more fluid filters out of the capillary. If filtration increases too much, the lymphatics are unable to remove all the fluid, and excess accumulates in the pulmonary interstitial space, creating pulmonary edema. In severe cases, if edema exceeds the tissue's ability to retain it, fluid leaks from the interstitial space into the alveolar air space,

BIOTECHNOLOGY

The Pulse Oximeter

One important clinical indicator of the effectiveness of gas exchange in the lungs is the concentration of oxygen in arterial blood. Obtaining an arterial blood sample is difficult for the clinician and painful for the patient because it means finding an accessible artery. (Most blood is drawn from superficial veins rather than from arteries, which lie deeper within the body.) Over the years, however, scientists have developed instruments that quickly and painlessly measure blood oxygen levels through the surface of the skin on a finger or earlobe. One such instrument, the *pulse oximeter,* clips onto the skin and in seconds gives a digital reading of arterial hemoglobin saturation. The oximeter works by measuring light absorbance of the tissue's hemoglobin at two wavelengths. Another instrument, the *transcutaneous oxygen sensor,* measures dissolved oxygen using a variant of traditional gas-measuring electrodes. Both methods have limitations but are popular because they provide a rapid, noninvasive means of estimating arterial oxygen content.

flooding the alveoli. Normally the inside of the alveoli is a moist surface lined by a very thin (about 2–5 µm) layer of fluid with surfactant (see Fig. 18.3b). With alveolar flooding, this fluid layer can become much thicker and seriously impair gas exchange. Alveolar flooding can also occur with leakage when alveolar epithelium is damaged, such as from inflammation or inhaling toxic gases. If hypoxia due to alveolar fluid accumulation is severe and cannot be corrected by oxygen therapy, the condition may be called *adult respiratory distress syndrome* or ARDS.

CONCEPT CHECK

5. Why would left ventricular failure or mitral valve dysfunction cause elevated pulmonary blood pressure?

6. If alveolar ventilation increases, what happens to arterial P_{O_2}? To arterial P_{CO_2}? To venous P_{O_2} and P_{CO_2}? Explain your answers.

Gas Solubility Affects Diffusion

A final factor that can affect gas exchange in the alveoli is the solubility of the gas. The movement of gas molecules from air into a liquid is directly proportional to three factors: (1) the pressure gradient of the gas, (2) the solubility of the gas in the liquid, and (3) temperature. Because temperature is relatively constant in mammals, we can ignore its contribution in this discussion.

When a gas is placed in contact with water and there is a pressure gradient, gas molecules move from one phase to the other. If gas pressure is higher in the water than in the gaseous phase, then gas molecules leave the water. If gas pressure is higher in the gaseous phase than in water, then the gas dissolves into the water.

For example, consider a container of water exposed to air with a P_{O_2} of 100 mm Hg (**FIG. 18.4a**). Initially, the water has no oxygen dissolved in it (water P_{O_2} = 0 mm Hg). As the air stays in contact with the water, some of the moving oxygen molecules in the air diffuse into the water and dissolve (Fig. 18.4b). This process continues until equilibrium is reached. At equilibrium (Fig. 18.4c), the movement of oxygen from the air into the water is equal to the movement of oxygen from the water back into the air.

We refer to the concentration of oxygen dissolved in the water at any given P_{O_2} as the *partial pressure of the gas in solution*. In our example, therefore, if the air has a P_{O_2} of 100 mm Hg, at equilibrium the water also has a P_{O_2} of 100 mm Hg.

Note that this does *not* mean that the concentration of oxygen is the same in the air and in the water! The concentration of dissolved oxygen also depends on the *solubility* of oxygen in water. The ease with which a gas dissolves in a liquid is the **solubility** of the gas in that liquid. If a gas is very soluble, large numbers of gas molecules go into solution at a low gas partial pressure. With less soluble gases, even a high partial pressure may cause only a few molecules of the gas to dissolve in the liquid.

For example, when P_{O_2} is 100 mm Hg both in the air and the water, air contains 5.2 mmol O_2/L air, but water contains only 0.15 mmol O_2/L water (Fig. 18.4c). As you can see, oxygen is not very soluble in water and, by extension, in any aqueous solution. Its low solubility was a driving force for the evolution of oxygen-carrying molecules in the aqueous solution we call blood.

Now compare oxygen solubility with CO_2 solubility (Fig. 18.4d). Carbon dioxide is 20 times more soluble in water than oxygen is. At a P_{CO_2} of 100 mm Hg, the CO_2 concentration in air is 5.2 mmol CO_2/L air, and its concentration in water is 3 mmol/L water. So although P_{O_2} and P_{CO_2} are both 100 mm Hg in the water, the amount of each gas that dissolves in the water is very different.

Why is solubility important in physiology? The answer is that oxygen's low solubility in aqueous solutions means that very little oxygen can be carried dissolved in plasma. Its low solubility also means oxygen is slower to cross the increased diffusion distance present in pulmonary edema. Diffusion of oxygen into alveolar capillaries does not have time to come to equilibrium before the blood has left the capillaries. The result is decreased arterial P_{O_2} even though alveolar P_{O_2} may be normal.

Carbon dioxide, in contrast, is relatively soluble in body fluids, so increased diffusion distance may not significantly affect

---

FIG. 18.4 Gases in solution

When the temperature remains constant, the amount of a gas that dissolves in a liquid depends on both the solubility of the gas in the liquid and the partial pressure of the gas.

Oxygen Solubility

(a) Initial state: No O_2 in solution

(b) Oxygen dissolves.

(c) At equilibrium, P_{O_2} in air and water are equal. Low O_2 solubility means concentrations are not equal.

P_{O_2} = 100 mm Hg

P_{O_2} = 0 mm Hg

P_{O_2} = 100 mm Hg
$[O_2]$ = 5.20 mmol/L

P_{O_2} = 100 mm Hg
$[O_2]$ = 0.15 mmol/L

CO_2 Solubility

(d) When CO_2 is at equilibrium at the same partial pressure (100 mm Hg), more CO_2 dissolves.

P_{CO_2} = 100 mm Hg
$[CO_2]$ = 5.20 mmol/L

P_{CO_2} = 100 mm Hg
$[CO_2]$ = 3.00 mmol/L

Q **FIGURE QUESTION**

Physiologists also express dissolved gases in blood using the following equation:

$$[Gas]_{diss} = \alpha \, [P_{gas}]$$

α for oxygen is (0.03 mL O_2/L blood)/mm Hg P_{O_2}

α for CO_2 is (0.7 mL CO_2/L blood)/mm Hg P_{CO_2}

If arterial blood has a P_{O_2} of 95 mm Hg and a P_{CO_2} of 40 mm Hg, what are the oxygen and CO_2 concentrations (in mL gas/L blood)?

RUNNING **PROBLEM**

Acute mountain sickness is the mildest illness caused by altitude hypoxia. The primary symptom is a headache that may be accompanied by dizziness, nausea, fatigue, or confusion. More severe illnesses are *high-altitude pulmonary edema* (HAPE) and high-altitude cerebral edema. HAPE is the major cause of death from altitude sickness. It is characterized by high pulmonary arterial pressure, extreme shortness of breath, and sometimes a productive cough yielding a pink, frothy fluid. Treatment is immediate relocation to lower altitude and administration of oxygen.

Q2: *Why would someone with HAPE be short of breath?*

Q3: *Based on what you learned about the mechanisms for matching ventilation and perfusion in the lung [p. 556], can you explain why patients with HAPE have elevated pulmonary arterial blood pressure?*

565 567 **571** 574 579 584 585

CO_2 exchange. In some cases of pulmonary edema, arterial P_{O_2} is low but arterial P_{CO_2} is normal because of the different solubilities of the two gases.

CONCEPT CHECK

7. True or false? Plasma with a P_{O_2} of 40 mm Hg and a P_{CO_2} of 40 mm Hg has the same concentrations of oxygen and carbon dioxide.

8. A saline solution is exposed to a mixture of nitrogen gas and hydrogen gas in which $P_{H_2} = P_{N_2}$. What information do you need to predict whether equal amounts of H_2 and N_2 dissolve in the solution?

GAS TRANSPORT IN THE BLOOD

Now that we have described how gases enter and leave the capillaries, we turn our attention to oxygen and carbon dioxide transport in the blood. Gases that enter the capillaries first dissolve in the plasma. But dissolved gases play only a small part in providing the cells with oxygen. The red blood cells, or *erythrocytes,* have a critical role in ensuring that gas transport between lung and cells is adequate to meet cell needs. Without hemoglobin in the red blood cells, the blood would be unable to transport sufficient oxygen to sustain life (**FIG. 18.5**).

Oxygen transport in the circulation and oxygen consumption by tissues are excellent ways to illustrate the general principles of mass flow and mass balance. *Mass flow* [p. 11] is defined as amount of *x* moving per minute, where mass flow = concentration × volume flow. We can calculate the mass flow of oxygen traveling from lungs to the cells by using the oxygen content of the arterial blood × cardiac output. If arterial blood contains, on average, 200 mL O_2/L blood and the cardiac output is 5 L/min, then oxygen transport to cells is

$$200 \text{ mL } O_2/\text{L blood} \times 5 \text{ L blood/min} = \\ 1000 \text{ mL } O_2/\text{min to cells} \quad (1)$$

FIG. 18.5 Oxygen transport

More than 98% of the oxygen in blood is bound to hemoglobin in red blood cells, and less than 2% is dissolved in plasma.

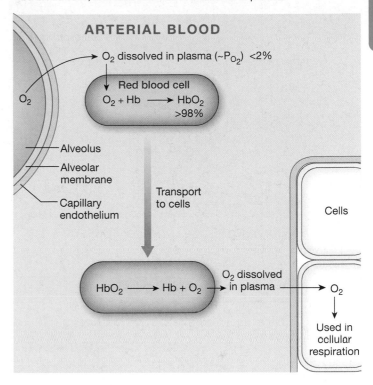

FIGURE QUESTION

How many cell membranes will O_2 cross in its passage between the airspace of the alveolus and binding to hemoglobin?

If we know the mass flow of oxygen in venous blood leaving the cells, we can use the principle of *mass balance* [p. 10] to calculate the uptake and consumption of oxygen by the cells (**FIG. 18.6**):

$$\text{Arterial } O_2 \text{ transport } - \text{ cell use of } O_2 = \text{venous } O_2 \text{ transport} \quad (2)$$

where oxygen transport is expressed as mL O_2 transported by blood per minute.

Using algebra, we can rearrange equation (2) to calculate O_2 use, or *oxygen consumption*, by the cells:

$$\text{Arterial } O_2 \text{ transport} - \text{venous } O_2 \text{ transport} = \\ \text{oxygen consumption} \quad (3)$$

Adolph Fick, the nineteenth-century physiologist who derived Fick's law of diffusion, combined the mass flow equation (1) and mass balance equation (3) to relate oxygen consumption (Q_{O_2}), cardiac output (CO), and blood oxygen content, as shown in Figure 18.6. The result is the **Fick equation:**

$$Q_{O_2} = \\ CO \times (\text{arterial oxygen content} - \text{venous oxygen content}) \quad (4)$$

FIG. 18.6 Mass balance and the Fick equation

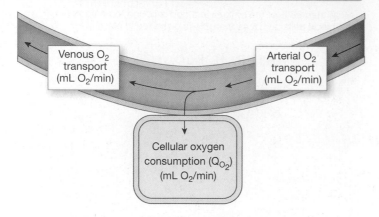

Venous O_2 transport (mL O_2/min)

Arterial O_2 transport (mL O_2/min)

Cellular oxygen consumption (Q_{O_2}) (mL O_2/min)

Mass Balance

Arterial O_2 transport – Q_{O_2} = venous O_2 transport

Rearranges to:

Arterial O_2 transport – venous O_2 transport = Q_{O_2}

Mass Flow

$$O_2 \text{ transport } = \text{ cardiac output (CO)} \times O_2 \text{ concentration}$$
$$(\text{L blood/min}) \qquad (\text{mL } O_2/\text{L blood})$$

Fick Equation

Substitute the mass flow equation for O_2 transport in the mass balance equation:

$$(\text{CO} \times \text{Arterial } [O_2]) - (\text{CO} \times \text{Venous } [O_2]) = Q_{O_2}$$

Using algebra (AB) – (AC) = A(B – C):

$$\text{CO} \times (\text{Arterial } [O_2] - \text{Venous } [O_2]) = Q_{O_2}$$

 FIGURE QUESTION

During exercise, a man consumes 1.8 L of oxygen per minute. His arterial oxygen content is 190 mL O_2/L blood, and the oxygen content of his venous blood is 134 mL O_2/L blood. What is his cardiac output?

The Fick equation (4) can be used to estimate cardiac output or oxygen consumption, assuming that arterial and venous blood gases can be measured.

Hemoglobin Binds to Oxygen

Oxygen transport in the blood has two components: the oxygen that is dissolved in the plasma (the P_{O_2}) and oxygen bound to hemoglobin (Hb). In other words:

Total blood O_2 content = dissolved O_2 + O_2 bound to Hb

As you learned in the previous section, oxygen is only slightly soluble in aqueous solutions, and less than 2% of all oxygen in the blood is dissolved. That means hemoglobin transports more than 98% of our oxygen (Fig. 18.5).

Hemoglobin, the oxygen-binding protein that gives red blood cells their color, binds reversibly to oxygen, as summarized in the equation

$$Hb + O_2 \rightleftharpoons HbO_2$$

Why is hemoglobin an effective oxygen carrier? The answer lies in its molecular structure. Hemoglobin (Hb) is a tetramer with four globular protein chains (*globins*), each centered around an iron-containing *heme* group [p. 521]. The central iron atom of each heme group can bind reversibly with one oxygen molecule. The iron-oxygen interaction is a weak bond that can be easily broken without altering either the hemoglobin or the oxygen.

With four heme groups per hemoglobin molecule, one hemoglobin molecule has the potential to bind four oxygen molecules. Hemoglobin bound to oxygen is known as **oxyhemoglobin**, abbreviated HbO_2. It would be most accurate to show the number of oxygen molecules carried on each hemoglobin molecule—$Hb(O_2)_{1-4}$—but we use the simpler abbreviation of HbO_2 because the number of bound oxygen molecules varies from one hemoglobin molecule to another.

Oxygen Binding Obeys the Law of Mass Action

The hemoglobin binding reaction $Hb + O_2 \rightleftharpoons HbO_2$ obeys the law of mass action [p. 48]. As the concentration of free O_2 increases, more oxygen binds to hemoglobin and the equation shifts to the right, producing more HbO_2. If the concentration of O_2 decreases, the equation shifts to the left. Hemoglobin releases oxygen, and the amount of oxyhemoglobin decreases.

In the blood, the free oxygen available to bind to hemoglobin is dissolved oxygen, indicated by the P_{O_2} of plasma (Fig. 18.5). In pulmonary capillaries, oxygen from the alveoli first dissolves in plasma. Dissolved O_2 then diffuses into the red blood cells, where it can bind to hemoglobin. The hemoglobin acts like a sponge, soaking up oxygen from the plasma until the reaction $Hb + O_2 \rightleftharpoons HbO_2$ reaches equilibrium.

The transfer of oxygen from alveolar air to plasma to red blood cells and onto hemoglobin occurs so rapidly that blood in the pulmonary capillaries normally picks up as much oxygen as the P_{O_2} of the plasma and the number of red blood cells allow.

Once arterial blood reaches the tissues, the exchange process that took place in the lungs reverses. Dissolved oxygen diffuses out of systemic capillaries into cells, which have a lower P_{O_2}. This decreases plasma P_{O_2} and disturbs the equilibrium of the oxygen-hemoglobin binding reaction by removing O_2 from the left side of the equation. The equilibrium shifts to the left according to the law of mass action, causing hemoglobin molecules release their oxygen stores (bottom half of Fig. 18.5).

Like oxygen loading at the lungs, transferring oxygen to the body's cells takes place very rapidly and goes to equilibrium.

The P_{O_2} of the cells determines how much oxygen is unloaded from hemoglobin. As cells increase their metabolic activity, their P_{O_2} decreases, and hemoglobin releases more oxygen to them.

Hemoglobin Transports Most Oxygen to the Tissues

We must have adequate amounts of hemoglobin in our blood or we cannot survive. To understand why, consider the following example.

Assume that a person's total oxygen consumption at rest is about 250 mL O_2/min and cardiac output is 5 L blood/min. How much oxygen must blood contain to meet this demand?

$$250 \text{ mL } O_2/\text{min consumed} = 5 \text{ L blood/min} \times ? \text{ mL } O_2/\text{L blood}$$

To meet the cells' needs for oxygen, the 5 L of blood/min coming to the tissues would need to contain at least 250 mL O_2. This calculates to a blood oxygen concentration of 50 mL O_2/L blood.

The low solubility of oxygen in plasma means that only 3 mL of O_2 can dissolve in the plasma fraction of each liter of arterial blood (**FIG. 18.7a**). The dissolved oxygen delivered to the cells in plasma therefore is

$$3 \text{ mL } O_2/\text{L blood} \times 5 \text{ L blood/min} = 15 \text{ mL } O_2/\text{min}$$

The cells require at least 250 mL O_2/min, so the small amount of oxygen that dissolves in plasma cannot meet the needs of the tissues at rest.

Now let's consider the difference in oxygen delivery if hemoglobin is available. At normal hemoglobin levels, red blood cells carry about 197 mL O_2/L blood (Fig. 18.7b).

$$\begin{aligned} \text{Total blood } O_2 \text{ content} &= \text{dissolved } O_2 + O_2 \text{ bound to Hb} \\ &= 3 \text{ mL } O_2/\text{L blood} + 197 \text{ mL } HbO_2/\text{L blood} \\ &= 200 \text{ mL } O_2/\text{L blood} \end{aligned}$$

If cardiac output remains 5 L/min, total oxygen delivery to cells is 1000 mL/min with hemoglobin present:

$$200 \text{ mL } O_2/\text{L blood} \times 5 \text{ L blood/min} = 1000 \text{ mL } O_2/\text{min}$$

This is four times the oxygen consumption needed by the tissues at rest. The extra O_2 serves as a reserve for times when oxygen demand increases, such as with exercise.

P_{O_2} Determines Oxygen-Hb Binding

The amount of oxygen that binds to hemoglobin depends on two factors: (1) the P_{O_2} in the plasma surrounding the red blood cells and (2) the number of potential Hb binding sites available

---

FIG. 18.7   Hemoglobin increases oxygen transport

(a) Oxygen transport in blood without hemoglobin: Alveolar P_{O_2} – arterial P_{O_2}

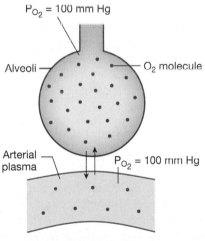

Oxygen dissolves in plasma.

O_2 content of plasma	= 3 mL O_2/L blood
O_2 content of red blood cells	= 0
Total O_2 carrying capacity	3 mL O_2/L blood

(b) Oxygen transport at normal P_{O_2} in blood with hemoglobin

PHYS IN ACTION!

Red blood cells with hemoglobin are carrying 98% of their maximum load of oxygen.

O_2 content of plasma =	3 mL O_2/L blood
O_2 content of red blood cells	= 197 mL O_2/L blood
Total O_2 carrying capacity	200 mL O_2/L blood

(c) Oxygen transport at reduced P_{O_2} in blood with hemoglobin

Red blood cells carrying 50% of their maximum load of oxygen.

O_2 content of plasma =	0.8 mL O_2/L blood
O_2 content of red blood cells	= 99.5 mL O_2/L blood
Total O_2 carrying capacity	100.3 mL O_2/L blood

RUNNING **PROBLEM**

In most people arriving at high altitude, normal physiological responses kick in to help acclimatize the body to the chronic hypoxia. Within two hours of arrival, hypoxia triggers the release of erythropoietin from the kidneys and liver. This hormone stimulates red blood cell production, and as a result, new erythrocytes appear in the blood within days.

Q4: *How does adding erythrocytes to the blood help a person acclimatize to high altitude?*

Q5: *What does adding erythrocytes to the blood do to the viscosity of the blood? What effect will that change in viscosity have on blood flow?*

565   567   571   **574**   579   584   585

in the red blood cells (**FIG. 18.8**). Plasma P_{O_2} is the primary factor determining what percentage of the available hemoglobin binding sites are occupied by oxygen, known as *the percent saturation of hemoglobin*. As you learned in previous sections, arterial P_{O_2} is established by (1) the composition of inspired air, (2) the alveolar ventilation rate, and (3) the efficiency of gas exchange from alveoli to blood. Figure 18.7c shows what happens to O_2 transport when P_{O_2} decreases.

The total number of oxygen-binding sites depends on the number of hemoglobin molecules in red blood cells. Clinically, this number can be estimated either by counting the red blood cells and quantifying the amount of hemoglobin per red blood cell (*mean corpuscular hemoglobin*) or by determining the blood hemoglobin content (g Hb/dL whole blood). Any pathological condition that decreases the amount of hemoglobin in red blood cells or the number of red blood cells will adversely affect the blood's oxygen-transporting capacity.

FIG. 18.8 **Factors controlling oxygen-hemoglobin binding**

People who have lost large amounts of blood need to replace hemoglobin for oxygen transport. A blood transfusion is the ideal replacement for blood loss, but in emergencies this is not always possible. Saline infusions can replace lost blood volume, but saline, like plasma, cannot transport sufficient quantities of oxygen to support cellular respiration. Faced with this problem, researchers are currently testing artificial oxygen carriers to replace hemoglobin. In times of large-scale disasters, these hemoglobin substitutes could eliminate the need to identify a patient's blood type before giving transfusions.

Oxygen Binding Is Expressed as a Percentage

As you just learned, the amount of oxygen bound to hemoglobin at any given P_{O_2} is expressed as the **percent saturation of hemoglobin,** where

(Amount of O_2 bound/maximum that could be bound) $\times$ 100 = percent saturation of hemoglobin

If all binding sites on all hemoglobin molecules are occupied by oxygen, the blood is 100% oxygenated, or *saturated* with oxygen. If half the available binding sites are carrying oxygen, the hemoglobin is 50% saturated, and so on.

EMERGING CONCEPTS

Blood Substitutes

Physiologists have been attempting to find a substitute for blood ever since 1878, when an intrepid physician named T. Gaillard Thomas transfused a patient with whole milk in place of blood. (It helped but the patient died anyway.) Although milk seems an unlikely replacement for blood, it has two important properties: proteins to provide colloid osmotic pressure and molecules (emulsified lipids) capable of binding to oxygen. In the development of hemoglobin substitutes, oxygen transport is the most difficult property to mimic. A hemoglobin solution would seem to be the obvious answer, but hemoglobin that is not compartmentalized in red blood cells behaves differently than native hemoglobin inside RBCs. Investigators have tried polymerizing hemoglobin into more stable molecules, loading hemoglobin polymers into phospholipid liposomes [p. 62], or combining hemoglobin with other compounds. Unfortunately all the products developed to date have adverse side effects, and research on most has been discontinued. To learn more about this research, read "Evaluating the Safety and Efficacy of Hemoglobin-based Blood Substitutes (FDA)" *(www.fda.gov/biologicsbloodvaccines/scienceresearch/biologicsresearchareas/ucm127061.htm).*

The relationship between plasma P_{O_2} and percent saturation of hemoglobin can be explained with the following analogy. The hemoglobin molecules carrying oxygen are like students moving books from an old library to a new one. Each student (a hemoglobin molecule) can carry a maximum of four books (100% saturation). The librarian in charge controls how many books (O_2 molecules) each student will carry, just as plasma P_{O_2} determines the percent saturation of hemoglobin.

The total number of books being carried depends on the number of available students, just as the amount of oxygen delivered to the tissues depends on the number of available hemoglobin molecules. For example, if there are 100 students, and the librarian gives each of them four books (100% saturation), then 400 books are carried to the new library. If the librarian gives three books to each student (decreased plasma P_{O_2}), then only 300 books go to the new library, even though each student could carry four. (Students carrying only three of a possible four books correspond to 75% saturation of hemoglobin.) If the librarian is handing out four books per student but only 50 students show up (fewer hemoglobin molecules at 100% saturation), then only 200 books get to the new library, even though the students are taking the maximum number of books they can carry.

The physical relationship between P_{O_2} and how much oxygen binds to hemoglobin can be studied *in vitro*. Researchers expose samples of hemoglobin to various P_{O_2} levels and quantitatively determine the amount of oxygen that binds. **Oxyhemoglobin saturation curves,** such as the ones shown in **FIGURE 18.9**, are the result of these *in vitro* binding studies. (These curves are also called *dissociation curves*.)

The shape of the HbO_2 saturation curve reflects the properties of the hemoglobin molecule and its affinity for oxygen. If you look at the curve, you find that at normal alveolar and arterial P_{O_2} (100 mm Hg), 98% of the hemoglobin is bound to oxygen (Fig. 18.9a). In other words, as blood passes through the lungs under normal conditions, hemoglobin picks up nearly the maximum amount of oxygen that it can carry.

Notice that the curve is nearly flat at P_{O_2} levels higher than 100 mm Hg (that is, the slope approaches zero). At P_{O_2} above 100 mm Hg, even large changes in P_{O_2} cause only minor changes in percent saturation. In fact, hemoglobin is not 100% saturated until the P_{O_2} reaches nearly 650 mm Hg, a partial pressure far higher than anything we encounter in everyday life.

The flattening of the saturation curve at higher P_{O_2} also means that alveolar P_{O_2} can fall a good bit below 100 mm Hg without significantly lowering hemoglobin saturation. As long as P_{O_2} in the alveoli (and thus in the pulmonary capillaries) stays above 60 mm Hg, hemoglobin is more than 90% saturated and maintains near-normal levels of oxygen transport. However, once P_{O_2} falls below 60 mm Hg, the curve becomes steeper. The steep slope means that a small decrease in P_{O_2} causes a relatively large release of oxygen.

For example, if P_{O_2} falls from 100 mm Hg to 60 mm Hg, the percent saturation of hemoglobin goes from 98 to about 90%, a decrease of 8%. This is equivalent to a saturation change of 2% for each 10 mm Hg change. If P_{O_2} falls further, from 60 to 40 mm Hg,

the percent saturation goes from 90 to 75%, a decrease of 7.5% for each 10 mm Hg. In the 40–20 mm Hg range, the curve is even steeper. Hemoglobin saturation declines from 75 to 35%, a change of 20% for each 10 mm Hg change.

What is the physiological significance of the shape of the saturation curve? In blood leaving systemic capillaries with a P_{O_2} of 40 mm Hg (an average value for venous blood in a person at rest), hemoglobin is still 75% saturated. This means that at the cells, blood releases only one-fourth of the oxygen it is capable of carrying. The oxygen that remains bound serves as a reservoir that cells can draw on if metabolism increases.

When metabolically active tissues use additional oxygen, their cellular P_{O_2} decreases, and hemoglobin releases additional O_2 at the cells. At a P_{O_2} of 20 mm Hg (an average value for exercising muscle), hemoglobin saturation falls to about 35%. With this 20 mm Hg decrease in P_{O_2} (40 mm Hg to 20 mm Hg), hemoglobin releases an additional 40% of the oxygen it is capable of carrying. This is another example of the built-in reserve capacity of the body.

Several Factors Affect O_2-Hb Binding

Any factor that changes the conformation of the hemoglobin protein may affect its ability to bind oxygen. In humans, physiological changes in plasma pH, temperature, and P_{CO_2} all alter the oxygen-binding affinity of hemoglobin. Changes in binding affinity are reflected by changes in the shape of the HbO_2 saturation curve.

Decreased pH, increased temperature, or increased P_{CO_2}, decrease the affinity of hemoglobin for oxygen and shift the oxygen-hemoglobin saturation curve to the right (Fig. 18.9c–e). When these factors change in the opposite direction, binding affinity increases, and the curve shifts to the left. Notice that when the curve shifts in either direction, the changes are much more pronounced in the steep part of the curve. Physiologically, this means that oxygen binding at the lungs (in the 90–100 mm Hg P_{O_2} range) is not greatly affected, but oxygen delivery at the tissues (in the 20–40 mm Hg range) is significantly altered.

Let's examine one situation, the affinity shift that takes place when pH decreases from 7.4 (normal) to 7.2 (more acidic). (The normal range for blood pH is 7.38–7.42, but a pH of 7.2 is compatible with life.) Look at the graph in Figure 18.9c.

At a P_{O_2} of 40 mm Hg (equivalent to a resting cell) and pH of 7.4, hemoglobin is about 75% saturated. At the same P_{O_2}, if the pH falls to 7.2, the percent saturation decreases to about 62%. This means that hemoglobin molecules release 13% more oxygen at pH 7.2 than they do at pH 7.4.

When does the body undergo shifts in blood pH? One situation is with maximal exertion that pushes cells into anaerobic metabolism. Anaerobic metabolism in exercising muscle fibers releases H^+ into the cytoplasm and extracellular fluid. As H^+ concentrations increase, pH falls, the affinity of hemoglobin for oxygen decreases, and the HbO_2 saturation curve shifts to the right. More oxygen is released at the tissues as the blood becomes more acidic (pH decreases). A shift in the hemoglobin saturation curve that results from a change in pH is called the **Bohr effect.**

FIG. 18.9 **ESSENTIALS**

Oxygen-Hemoglobin Binding Curves

Binding Properties of Adult and Fetal Hemoglobin

(a) The oxyhemoglobin saturation curve is determined *in vitro* in the laboratory.

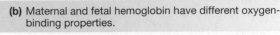

(b) Maternal and fetal hemoglobin have different oxygen-binding properties.

Physical Factors Alter Hemoglobin's Affinity for Oxygen

(c) Effect of pH

(d) Effect of temperature

(e) Effect of P_{CO_2}

(f) Effect of the metabolic compound 2,3-BPG

GRAPH QUESTIONS

1. For the graph in (a):
   (a) When the P_{O_2} is 20 mm Hg, what is the percent O_2 saturation of hemoglobin?
   (b) At what P_{O_2} is hemoglobin 50% saturated with O_2?

2. At a P_{O_2} of 20 mm Hg, how much more oxygen is released at an exercising muscle cell whose pH is 7.2 than at a cell with a pH of 7.4?

3. What happens to oxygen release when the exercising muscle cell warms up?

4. Blood stored in blood banks loses its normal content of 2,3-BPG. Is this good or bad? Explain.

5. Because of incomplete gas exchange across the thick membranes of the placenta, hemoglobin in fetal blood leaving the placenta is 80% saturated with oxygen. What is the P_{O_2} of that placental blood?

6. Blood in the vena cava of the fetus has a P_{O_2} around 10 mm Hg. What is the percent O_2 saturation of maternal hemoglobin at the same P_{O_2}?

An additional factor that affects oxygen-hemoglobin binding is **2,3-bisphosphoglycerate** (2,3-BPG; also called *2,3-diphosphoglycerate* or 2,3-DPG), a compound made from an intermediate of the glycolysis pathway. **Chronic hypoxia** (extended periods of low oxygen) triggers an increase in 2,3-BPG production in red blood cells. Increased levels of 2,3-BPG lower the binding affinity of hemoglobin and shift the HbO_2 saturation curve to the right (Fig. 18.9f). Ascent to high altitude and anemia are two situations that increase 2,3-BPG production.

Changes in hemoglobin's structure also change its oxygen-binding affinity. For example, *fetal hemoglobin (HbF)* has two gamma protein chains in place of the two beta chains found in adult hemoglobin. The presence of gamma chains enhances the ability of fetal hemoglobin to bind oxygen in the low-oxygen environment of the placenta. The altered binding affinity is reflected by the different shape of the fetal HbO_2 saturation curve (Fig. 18.9b). At any given placental P_{O_2}, oxygen released by maternal hemoglobin is picked up by the higher-affinity fetal hemoglobin for delivery to the developing fetus. Shortly after birth, fetal hemoglobin is replaced with the adult form as new red blood cells are made.

FIGURE 18.10 summarizes all the factors that influence the total oxygen content of arterial blood.

CONCEPT CHECK

9. Can a person breathing 100% oxygen at sea level achieve 100% saturation of her hemoglobin?

10. What effect does hyperventilation have on the percent saturation of arterial hemoglobin? [*Hint:* Fig. 17.13, p. 557]

11. A muscle that is actively contracting may have a cellular P_{O_2} of 25 mm Hg. What happens to oxygen binding to hemoglobin at this low P_{O_2}? What is the P_{O_2} of the venous blood leaving the active muscle?

Carbon Dioxide Is Transported in Three Ways

Gas transport in the blood includes carbon dioxide removal from the cells as well as oxygen delivery to cells. Carbon dioxide is a by-product of cellular respiration [p. 105] and is potentially toxic if not excreted (removed from the body). Elevated P_{CO_2} (*hypercapnia*) causes the pH disturbance known as *acidosis*. Extremes of pH interfere with hydrogen bonding of molecules and can denature proteins [p. 51]. Abnormally high P_{CO_2} levels also depress central nervous system function, causing confusion, coma, or even death. For these reasons, CO_2 must be removed, making CO_2 homeostasis an important function of the lungs.

Carbon dioxide is more soluble in body fluids than oxygen is, but the cells produce far more CO_2 than can dissolve in the plasma. Only about 7% of the CO_2 carried by venous blood is dissolved in the plasma. The other 93% diffuses into red blood cells, where 23% binds to hemoglobin ($HbCO_2$) while the remaining 70% is converted to bicarbonate ion (HCO_3^-), as explained next. **FIGURE 18.11** summarizes these three mechanisms of carbon dioxide transport in the blood.

CO_2 and Bicarbonate Ions Most of the CO_2 that enters the blood is transported to the lungs as bicarbonate ions (HCO_3^-) dissolved in the plasma. The conversion of CO_2 to HCO_3^- serves two purposes: (1) it provides an additional way to transport CO_2 from cells to lungs, and (2) HCO_3^- is available to act as a buffer for metabolic acids [p. 41], thereby helping stabilize the body's pH.

How does CO_2 turn into HCO_3^-? The rapid production of HCO_3^- depends on the presence of **carbonic anhydrase (CA)**, an enzyme found concentrated in red blood cells. Let's see how this happens. Dissolved CO_2 in the plasma diffuses into red blood cells,

---

FIG. 18.10 Arterial oxygen

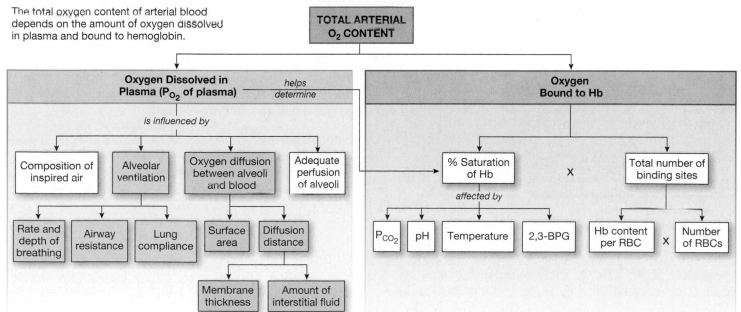

The total oxygen content of arterial blood depends on the amount of oxygen dissolved in plasma and bound to hemoglobin.

FIG. 18.11   Carbon dioxide transport

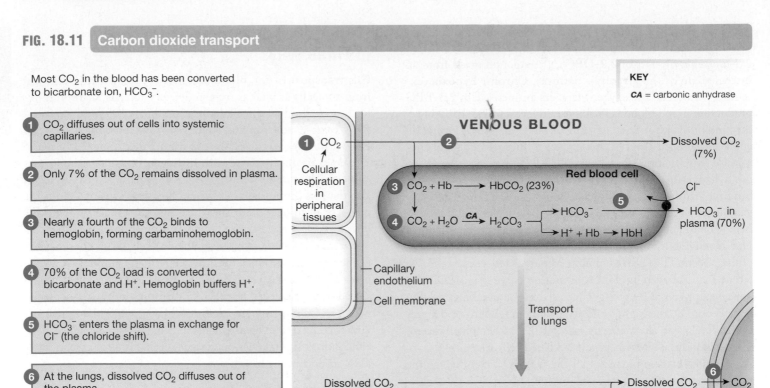

Most CO_2 in the blood has been converted to bicarbonate ion, HCO_3^-.

KEY

CA = carbonic anhydrase

1. CO_2 diffuses out of cells into systemic capillaries.

2. Only 7% of the CO_2 remains dissolved in plasma.

3. Nearly a fourth of the CO_2 binds to hemoglobin, forming carbaminohemoglobin.

4. 70% of the CO_2 load is converted to bicarbonate and H^+. Hemoglobin buffers H^+.

5. HCO_3^- enters the plasma in exchange for Cl^- (the chloride shift).

6. At the lungs, dissolved CO_2 diffuses out of the plasma.

7. By the law of mass action, CO_2 unbinds from hemoglobin and diffuses out of the RBC.

8. The carbonic acid reaction reverses, pulling HCO_3^- back into the RBC and converting it back to CO_2.

where it can react with water in the presence of carbonic anhydrase to form *carbonic acid* (H_2CO_3, top portion of Fig. 18.11). Carbonic acid then dissociates into a hydrogen ion and a bicarbonate ion:

$$CO_2 + H_2O \underset{\text{Carbonic acid}}{\overset{\overset{\text{Carbonic anhydrase}}{}}{\rightleftharpoons}} H_2CO_3 \rightleftharpoons H^+ + HCO_3^-$$

Because carbonic acid dissociates readily, we sometimes ignore the intermediate step and summarize the reaction as:

$$CO_2 + H_2O \rightleftharpoons H^+ + HCO_3^-$$

This reaction is reversible and obeys the law of mass action.

The conversion of carbon dioxide and water to H^+ and HCO_3^- continues until equilibrium is reached. (Water is always in excess in the body, so water concentration plays no role in the dynamic equilibrium of this reaction.) To keep the reaction going, the products (H^+ and HCO_3^-) must be removed from the cytoplasm of the red blood cell. If the product concentrations are kept low, the reaction cannot reach equilibrium. Carbon dioxide will continue to move out of plasma into the red blood cells, which in turn allows more CO_2 to diffuse out of tissues into the blood.

Two separate mechanisms remove free H^+ and HCO_3^-. In the first, bicarbonate leaves the red blood cell on an antiport protein [p. 141]. This transport process, known as the **chloride shift,** exchanges HCO_3^- for Cl^-. The anion exchange maintains the cell's electrical neutrality. The transfer of HCO_3^- into the plasma makes this buffer available to moderate pH changes caused by the production of metabolic acids. Bicarbonate is the most important extracellular buffer in the body.

Hemoglobin and H^+ The second mechanism for keeping product concentrations low removes free H^+ from the red blood cell cytoplasm. Hemoglobin within the red blood cell acts as a buffer and binds hydrogen ions in the reaction

$$H^+ + Hb \rightleftharpoons HbH$$

Hemoglobin's buffering of H^+ is an important step that prevents large changes in the body's pH. If blood P_{CO_2} is elevated much above normal, the hemoglobin buffer cannot soak up all the H^+

produced from the reaction of CO_2 and water. In those cases, excess H^+ accumulates in the plasma, causing the condition known as **respiratory acidosis.** You will learn more about the role of the respiratory system in maintaining pH homeostasis when you study acid-base balance.

Hemoglobin and CO_2 Most carbon dioxide that enters red blood cells is converted to bicarbonate ions, but about 23% of the CO_2 in venous blood binds directly to hemoglobin. At the tissues, when oxygen leaves its binding sites on the hemoglobin molecule, CO_2 binds with free hemoglobin at exposed amino groups $(-NH_2)$, forming **carbaminohemoglobin:**

$$CO_2 + Hb \rightleftharpoons HbCO_2 \text{ (carbaminohemoglobin)}$$

The presence of CO_2 and H^+ facilitates formation of carbaminohemoglobin because both of these factors decrease hemoglobin's binding affinity for oxygen (see Fig. 18.9c and e).

CO_2 Removal at the Lungs When venous blood reaches the lungs, the processes that took place in the systemic capillaries reverse (bottom portion of Fig. 18.11). The P_{CO_2} of the alveoli is lower than that of venous blood in the pulmonary capillaries. In response to this gradient, CO_2 diffuses out of plasma into the alveoli, and the plasma P_{CO_2} begins to fall.

The decrease in plasma P_{CO_2} allows dissolved CO_2 to diffuse out of the red blood cells. As CO_2 levels in the red blood cells decrease, the equilibrium of the CO_2-HCO_3^- reaction is disturbed, shifting toward production of more CO_2:

$$H^+ + HCO_3^- \rightarrow CO_2 + H_2O$$

H^+ unbinds from hemoglobin molecules and HCO_3^- moves back into the red blood cells when the chloride shift reverses. The HCO_3^- and newly released H^+ are converted back into water and CO_2. This newly made CO_2 is then free to diffuse out of the red blood cell into the plasma and from there into the alveoli.

FIGURE 18.12 shows the combined transport of CO_2 and O_2 in the blood. At the alveoli, O_2 diffuses down its pressure gradient, moving from the alveoli into the plasma and then from the plasma into the red blood cells. Hemoglobin binds to O_2, increasing the amount of oxygen that can be transported to the cells.

At the cells, the process reverses. Because P_{O_2} is lower in cells than in the arterial blood, O_2 diffuses from the plasma into the cells. The decrease in plasma P_{O_2} causes hemoglobin to release O_2, making additional oxygen available to enter cells.

Carbon dioxide from aerobic metabolism simultaneously leaves cells and enters the blood, dissolving in the plasma. From there, CO_2 enters red blood cells, where most is converted to HCO_3^- and H^+. The HCO_3^- is returned to the plasma in exchange for a Cl^- while the H^+ binds to hemoglobin. A fraction of the CO_2 that enters red blood cells binds directly to hemoglobin. At the lungs, the process reverses as CO_2 diffuses out of the pulmonary capillaries and into the alveoli.

FIG. 18.12 **Summary of O_2 and CO_2 exchange and transport**

Dry air = 760 mm Hg
P_{O_2} = 160 mm Hg
P_{CO_2} = 0.25 mm Hg

Alveoli

P_{O_2} = 100 mm Hg
P_{CO_2} = 40 mm Hg

CO_2 O_2

CO_2 transport

HCO_3^- = 70%
$HbCO_2$ = 23%
Dissolved CO_2 = 7%

Pulmonary circulation

O_2 transport

HbO_2 > 98%
Dissolved O_2 < 2%
$(\sim P_{O_2})$

Venous blood

$P_{O_2} \leq 40$ mm Hg
$P_{CO_2} \geq 46$ mm Hg

Systemic circulation

Arterial blood

P_{O_2} = 100 mm Hg
P_{CO_2} = 40 mm Hg

CO_2 O_2

Cells

$P_{O_2} \leq 40$ mm Hg
$P_{CO_2} \geq 46$ mm Hg

RUNNING PROBLEM

The usual homeostatic response to high-altitude hypoxia is hyperventilation, which begins on arrival. Hyperventilation enhances alveolar ventilation, but this may not help elevate arterial P_{O_2} levels significantly when atmospheric P_{O_2} is low. However, hyperventilation does lower plasma P_{CO_2}.

Q6: *What happens to plasma pH during hyperventilation? (Hint: Apply the law of mass action to figure out what happens to the balance between CO_2 and $H^+ + HCO_3^-$).*

Q7: *How does this change in pH affect oxygen binding at the lungs when P_{O_2} is decreased? How does it affect unloading of oxygen at the cells?*

565   567   571   574   **579**   584   585

To understand fully how the respiratory system coordinates delivery of oxygen to the lungs with transport of oxygen in the circulation, we now consider the central nervous system control of ventilation and the factors that influence it.

CONCEPT CHECK

12. How would an obstruction of the airways affect alveolar ventilation, arterial P_{CO_2}, and the body's pH?

REGULATION OF VENTILATION

Breathing is a rhythmic process that usually occurs without conscious thought or awareness. In that respect, it resembles the rhythmic beating of the heart. However, skeletal muscles, unlike autorhythmic cardiac muscles, are not able to contract spontaneously. Instead, skeletal muscle contraction must be initiated by somatic motor neurons, which in turn are controlled by the central nervous system.

In the respiratory system, contraction of the diaphragm and other muscles is initiated by a spontaneously firing network of neurons in the brain stem (**FIG. 18.13**). Breathing occurs automatically throughout a person's life but can also be controlled voluntarily, up to a point. Complicated synaptic interactions between neurons in the network create the rhythmic cycles of inspiration and expiration, influenced continuously by sensory input, especially that from chemoreceptors for CO_2, O_2, and H^+. Ventilation pattern depends in large part on the levels of those three substances in the arterial blood and extracellular fluid.

The neural control of breathing is one of the few "black boxes" left in systems-level physiology. As you have learned, the

---

FIG. 18.13 The reflex control of ventilation

Central and peripheral chemoreceptors monitor blood gases and pH. Control networks in the brain stem regulate activity in somatic motor neurons leading to respiratory muscles.

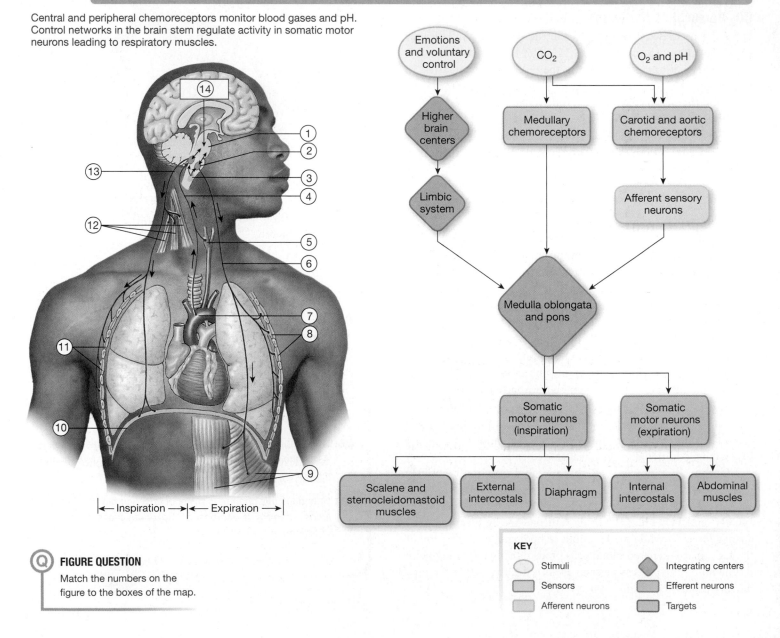

FIGURE QUESTION

Match the numbers on the figure to the boxes of the map.

KEY

- Stimuli
- Sensors
- Afferent neurons
- Integrating centers
- Efferent neurons
- Targets

"facts" presented in a textbook like this are really just our latest models of how the body works [p. 19]. Of all the models presented in this book, the model for neural control of breathing is the one that has changed the most in the past 15 years. We know the major regions of the brain stem that are involved, but the details of the neural networks involved remain elusive. The brain stem network that controls breathing behaves like a *central pattern generator* [p. 428], with intrinsic rhythmic activity that probably arises from *pacemaker neurons* with unstable membrane potentials. Tonic input from CO_2-sensitive and other chemoreceptors adds to the complexity.

Some of our understanding of how ventilation is controlled has come from observing patients with brain damage. Other information has come from animal experiments in which neural connections between major parts of the brain stem are severed, or sections of brain are studied in isolation. Research on CNS respiratory control is difficult because of the complexity of the neural networks and their anatomical locations. In recent years scientists have developed better techniques for studying the system.

The details that follow represent a contemporary model for the control of ventilation. Although some parts of the model are well supported with experimental evidence, other aspects are still under investigation. This model states that:

1. Respiratory neurons in the medulla control inspiratory and expiratory muscles.
2. Neurons in the pons integrate sensory information and interact with medullary neurons to influence ventilation.
3. The rhythmic pattern of breathing arises from a brainstem neural network with spontaneously discharging neurons.
4. Ventilation is subject to continuous modulation by various chemoreceptor- and mechanoreceptor-linked reflexes and by higher brain centers.

Neurons in the Medulla Control Breathing

Classic descriptions of how the brain controls ventilation divided the brain stem into various control centers. More recent descriptions, however, are less specific about assigning function to particular "centers" and instead look at complex interactions between neurons in a network. We know that respiratory neurons are concentrated bilaterally in two areas of the medulla oblongata. **FIGURE 18.14** shows these areas on the left side of the brain stem. One area called the **nucleus tractus solitarius** (NTS) contains the **dorsal respiratory group** (DRG) of neurons that control mostly muscles of inspiration. Output from the DRG goes via the **phrenic nerves** to the diaphragm and via the **intercostal nerves** to the intercostal muscles. In addition, the NTS receives sensory information from peripheral chemo- and mechanoreceptors through the *vagus* and *glossopharyngeal nerves* (cranial nerves X and IX).

Respiratory neurons in the pons receive sensory information from the DRG and in turn influence the initiation and termination of inspiration. The **pontine respiratory groups** (previously called the pneumotaxic center) and other pontine neurons provide tonic input to the medullary networks to help coordinate a smooth respiratory rhythm.

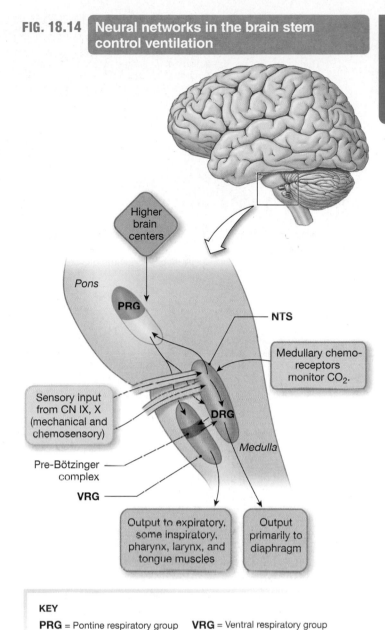

FIG. 18.14 **Neural networks in the brain stem control ventilation**

Higher brain centers

Pons

PRG

NTS

Medullary chemo-receptors monitor CO_2.

Sensory input from CN IX, X (mechanical and chemosensory)

DRG

Medulla

Pre-Bötzinger complex

VRG

Output to expiratory, some inspiratory, pharynx, larynx, and tongue muscles

Output primarily to diaphragm

KEY

PRG = Pontine respiratory group   **VRG** = Ventral respiratory group
DRG = Dorsal respiratory group   **NTS** = Nucleus tractus solitarius

The **ventral respiratory group** (VRG) of the medulla has multiple regions with different functions. One area known as the **pre-Bötzinger complex** contains spontaneously firing neurons that may act as the basic pacemaker for the respiratory rhythm. Other areas control muscles used for active expiration or for greater-than-normal inspiration, such as occurs during vigorous exercise. In addition, nerve fibers from the VRG innervate muscles of the larynx, pharynx, and tongue to keep the upper airways open during breathing. Inappropriate relaxation of these muscles during sleep contributes to *obstructive sleep apnea*, a sleeping disorder associated with snoring and excessive daytime sleepiness.

The integrated action of the respiratory control networks can be seen by monitoring electrical activity in the phrenic nerve and other motor nerves (**FIG. 18.15**). During quiet breathing, a pacemaker initiates each cycle, and inspiratory neurons gradually increase stimulation of the inspiratory muscles. This increase

FIG. 18.15 Neural activity during quiet breathing

During inspiration, the activity of inspiratory neurons increases steadily, apparently through a positive feedback mechanism. At the end of inspiration, the activity shuts off abruptly and expiration takes place through recoil of elastic lung tissue.

Rapid positive feedback loop

Number of active inspiratory neurons

Inspiration shuts off.

Tidal volume (liters)

0.5

0

Inspiration 2 sec | Passive expiration 3 sec | Inspiration 2 sec

← Time →

Q GRAPH QUESTION
What is the ventilation rate of the person in this example?

is sometimes called *ramping* because of the shape of the graph of inspiratory neuron activity. A few inspiratory neurons fire to begin the ramp. The firing of these neurons recruits other inspiratory neurons to fire in an apparent positive feedback loop. As more neurons fire, more skeletal muscle fibers are recruited. The rib cage expands smoothly as the diaphragm contracts.

At the end of inspiration, the inspiratory neurons abruptly stop firing, and the respiratory muscles relax. Over the next few seconds, passive expiration occurs because of elastic recoil of the inspiratory muscles and elastic lung tissue. However, some motor neuron activity can be observed during passive expiration, suggesting that perhaps muscles in the upper airways contract to slow the flow of air out of the respiratory system.

Many neurons of the VRG remain inactive during quiet respiration. They function primarily during forced breathing, when inspiratory movements are exaggerated, and during active expiration. In forced breathing, increased activity of inspiratory neurons stimulates accessory muscles, such as the sternocleidomastoids. Contraction of these accessory muscles enhances expansion of the thorax by raising the sternum and upper ribs.

With active expiration, expiratory neurons from the VRG activate the internal intercostal and abdominal muscles. There seems to be some communication between inspiratory and expiratory neurons, as inspiratory neurons are inhibited during active expiration.

CO₂, Oxygen, and pH Influence Ventilation

Sensory input from central and peripheral chemoreceptors modifies the rhythmicity of the control network to help maintain blood gas homeostasis. Carbon dioxide is the primary stimulus

for changes in ventilation. Oxygen and plasma pH play lesser roles.

The chemoreceptors for oxygen and carbon dioxide are strategically associated with the arterial circulation. If too little oxygen is present in arterial blood destined for the brain and other tissues, the rate and depth of breathing increase. If the rate of CO_2 production by the cells exceeds the rate of CO_2 removal by the lungs, arterial P_{CO_2} increases, and ventilation is intensified to match CO_2 removal to production. These homeostatic reflexes operate constantly, keeping arterial P_{O_2} and P_{CO_2} within a narrow range.

Peripheral chemoreceptors outside the CNS sense changes in the P_{O_2}, pH, and P_{CO_2} of the plasma (Fig. 18.13). The **carotid bodies** in the carotid arteries are the primary peripheral chemoreceptors. They are located close to the baroreceptors involved in reflex control of blood pressure [p. 493]. **Central chemoreceptors** in the brain respond to changes in the concentration of CO_2 in the cerebrospinal fluid. The primary central receptors lie on the ventral surface of the medulla, close to neurons involved in respiratory control.

Peripheral Chemoreceptors When specialized *type 1* or **glomus cells** {*glomus*, a ball-shaped mass} in the carotid bodies are activated by a decrease in P_{O_2} or pH or by an increase in P_{CO_2}, they trigger a reflex increase in ventilation. Under most normal circumstances, oxygen is not an important factor in modulating ventilation because arterial P_{O_2} must fall to less than 60 mm Hg before ventilation is stimulated. This large decrease in P_{O_2} is equivalent to ascending to an altitude of 3000 m. (For reference, Denver is located at an altitude of 1609 m). However, any condition that reduces plasma pH or increases P_{CO_2} will activate the carotid and aortic glomus cells and increase ventilation.

The details of glomus cell function remain to be worked out, but the basic mechanism by which these chemoreceptors respond to low oxygen is similar to the mechanism you learned for insulin release by pancreatic beta cells [p. 159] or taste transduction in taste buds [p. 325].

In all three examples, a stimulus inactivates K^+ channels, causing the receptor cell to depolarize (FIG. 18.16). Depolarization opens voltage-gated Ca^{2+} channels, and Ca^{2+} entry causes exocytosis of neurotransmitter onto the sensory neuron. In the carotid bodies, neurotransmitters initiate action potentials in sensory neurons leading to the brain stem respiratory networks, signaling them to increase ventilation.

Arterial oxygen concentrations do not play a role in the everyday regulation of ventilation because the peripheral chemoreceptors respond only to dramatic changes in arterial P_{O_2}. However, unusual physiological conditions, such as ascending to high

altitude, and pathological conditions, such as chronic obstructive pulmonary disease (COPD), can reduce arterial P_{O_2} to levels that are low enough to activate the peripheral chemoreceptors.

Central Chemoreceptors The most important chemical controller of ventilation is carbon dioxide, mediated both through the peripheral chemoreceptors just discussed and through central chemoreceptors located in the medulla (FIG. 18.17). These receptors set the respiratory pace, providing continuous input into the control network. When arterial P_{CO_2} increases, CO_2 crosses the blood-brain barrier and activates the central chemoreceptors. These receptors signal the control network to increase the rate and depth of ventilation, thereby enhancing alveolar ventilation and removing CO_2 from the blood.

Although we say that the central chemoreceptors monitor CO_2, they actually respond to pH changes in the cerebrospinal fluid (CSF). Carbon dioxide that diffuses across the blood-brain barrier into the CSF is converted to carbonic acid, which dissociates to bicarbonate and H^+. Experiments indicate that the H^+ produced by this reaction is what initiates the chemoreceptor reflex, rather than the increased level of CO_2.

Note, however, that pH changes in the plasma *do not* usually influence the central chemoreceptors directly. Although plasma P_{CO_2} enters the CSF readily, plasma H^+ crosses the blood-brain barrier very slowly and therefore has little direct effect on the central chemoreceptors.

When plasma P_{CO_2} increases, the chemoreceptors initially respond strongly by increasing ventilation. However, if P_{CO_2} remains elevated for several days, ventilation falls back toward normal rates as the chemoreceptor response adapts. The adaptation appears to be due to increased CSF bicarbonate concentrations that buffer the H^+. The mechanism by which bicarbonate increases is not clear.

Even though the central chemoreceptor response adapts to chronically high P_{CO_2}, the response of peripheral chemoreceptor to low arterial P_{O_2} remains intact over time. In some situations, low P_{O_2} becomes the primary chemical stimulus for ventilation. For example, patients with severe chronic lung disease, such as COPD, have chronic hypercapnia and hypoxia. Their arterial P_{CO_2} may rise to 50–55 mm Hg (normal is 35–45) while their P_{O_2} falls to 45–50 mm Hg (normal 75–100). Because these levels are chronic, the chemoreceptor response adapts to the elevated P_{CO_2}. Most of the chemical stimulus for ventilation in this situation then comes from low P_{O_2}, sensed by the carotid body chemoreceptors. If these patients are given too much oxygen, they may stop breathing because their chemical stimulus for ventilation is eliminated.

The central chemoreceptors respond to decreases in arterial P_{CO_2} as well as to increases. If alveolar P_{CO_2} falls, as it does during hyperventilation, plasma P_{CO_2} and cerebrospinal fluid P_{CO_2} follow suit. As a result, central chemoreceptor activity declines, and the control network slows the ventilation rate. When ventilation decreases, carbon dioxide begins to accumulate in alveoli and the plasma. Eventually, the arterial P_{CO_2} rises above the threshold level for the chemoreceptors. At that point, the receptors fire, and the control network again increases ventilation.

FIG. 18.16 Carotid body cells respond to P_{O_2} below 60 mm Hg

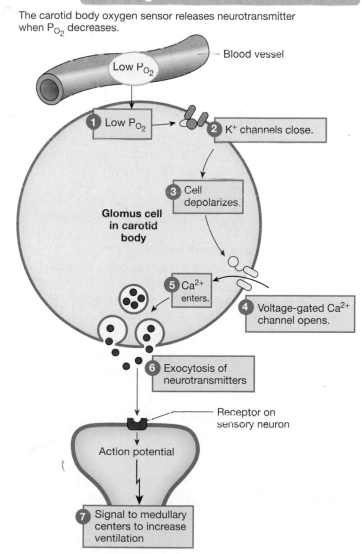

The carotid body oxygen sensor releases neurotransmitter when P_{O_2} decreases.

Blood vessel

Low P_{O_2}

1 Low P_{O_2}

2 K^+ channels close.

3 Cell depolarizes.

Glomus cell in carotid body

4 Voltage-gated Ca^{2+} channel opens.

5 Ca^{2+} enters.

6 Exocytosis of neurotransmitters

Receptor on sensory neuron

Action potential

7 Signal to medullary centers to increase ventilation

FIG. 18.17   **Chemoreceptor response**

Central chemoreceptors monitor CO_2 in cerebrospinal fluid.	Carotid and aortic chemoreceptors monitor CO_2, O_2, and H^+.

KEY

CA = carbonic anhydrase

↑ Plasma P_{CO_2}

Cerebral capillary

Blood-brain barrier   ↑P_{CO_2} H^+

Cerebrospinal fluid $CO_2 + H_2O \rightleftarrows H_2CO_3 \rightleftarrows H^+ + HCO_3^-$
(CA)

Medulla oblongata at brain

Central chemoreceptor

Respiratory control centers

↑Ventilation

↑ Plasma P_{O_2}
↓ Plasma P_{CO_2}

Negative feedback

↑ $CO_2 \longrightarrow$ ↑H^+ + ↑HCO_3^-
(in plasma)

Stimulates peripheral chemoreceptors in carotid and aortic bodies

↓ Plasma P_{O_2}
<60 mm Hg

Sensory neurons

Protective Reflexes Guard the Lungs

In addition to the chemoreceptor reflexes that help regulate ventilation, the body has protective reflexes that respond to physical injury or irritation of the respiratory tract and to overinflation of the lungs. The major protective reflex is *bronchoconstriction*, mediated through parasympathetic neurons that innervate bronchiolar smooth muscle. Inhaled particles or noxious gases stimulate **irritant receptors** in the airway mucosa. The irritant receptors send signals through sensory neurons to integrating centers in the CNS that trigger bronchoconstriction. Protective reflex responses also include coughing and sneezing.

The *Hering–Breuer inflation reflex* was first described in the late 1800s in anesthetized dogs. In these animals, if tidal volume exceeded a certain volume, stretch receptors in the lung signaled the brain stem to terminate inspiration. However, this reflex is difficult to demonstrate in adult humans and does not operate during quiet breathing and mild exertion. Studies on human infants, however, suggest that the Hering-Breuer inflation reflex may play a role in limiting their ventilation volumes.

RUNNING **PROBLEM**

The hyperventilation response to hypoxia creates a peculiar breathing pattern called periodic breathing, in which the person goes through a 10–15-second period of breath-holding followed by a short period of hyperventilation. Periodic breathing occurs most often during sleep.

Q8: *Based on your understanding of how the body controls ventilation, why do you think periodic breathing occurs most often during sleep?*

565   567   571   574   579   **584**   585

Higher Brain Centers Affect Patterns of Ventilation

Conscious and unconscious thought processes also affect respiratory activities. Higher centers in the hypothalamus and cerebrum can alter the activity of the brain stem control network to change ventilation rate and depth. Voluntary control of ventilation falls into this category. Higher brain center control is not a *requirement* for ventilation, however. Even if the brain stem above the pons is severely damaged, essentially normal respiratory cycles continue.

Respiration can also be affected by stimulation of portions of the limbic system. For this reason, emotional and autonomic activities such as fear and excitement may affect the pace and depth of respiration. In some of these situations, the neural pathway goes directly to the somatic motor neurons, bypassing the control network in the brain stem.

Although we can temporarily alter our respiratory performance, we cannot override the chemoreceptor reflexes. Holding your breath is a good example. You can hold your breath voluntarily only until elevated P_{CO_2} in the blood and cerebrospinal fluid activates the chemoreceptor reflex, forcing you to inhale.

Small children having temper tantrums sometimes attempt to manipulate parents by threatening to hold their breath until they die. However, the chemoreceptor reflexes make it impossible for the children to carry out that threat. Extremely strong-willed children can continue holding their breath until they turn blue and pass out from hypoxia, but once they are unconscious, normal breathing automatically resumes.

Breathing is intimately linked to cardiovascular function. The integrating centers for both functions are located in the brain stem, and interneurons project between the two networks, allowing signaling back and forth. The cardiovascular, respiratory, and renal systems all work together to maintain fluid and acid-base homeostasis, as you will see.

RUNNING **PROBLEM CONCLUSION** | High Altitude

On May 29, 1953, Edmund Hillary and Tenzing Norgay of the British Everest Expedition were the first humans to reach the summit of Mt. Everest. They carried supplemental oxygen with them, as it was believed that this feat was impossible without it. In 1978, however, Reinhold Messner and Peter Habeler achieved the "impossible." On May 8, they struggled to the summit using sheer willpower and no extra oxygen. In Messner's words, "I am nothing more than a single narrow gasping lung, floating over the mists and summits." Learn more about these Everest expeditions by doing a Google search for Hillary Everest or Messner Everest.

To learn more about different types of mountain sickness, see "High altitude medicine," *Am Fam Physician* 1998 Apr. 15 (*www.aafp.org/afp/980415ap/harris.html*) and "Altitude Illness" in the Centers for Disease Control and Prevention book *CDC Health Information for International Travel 2014* (*wwwnc.cdc.gov/travel/yellowbook/2014/chapter-2-the-pre-travel-consultation/altitude-illness*).

In this running problem, you learned about normal and abnormal responses to high altitude. Check your understanding of the physiology behind this respiratory challenge by comparing your answers with the information in the following table.

Question	Facts	Integration and Analysis
Q1: What is the P_{O_2} of inspired air reaching the alveoli when dry atmospheric pressure is 542 mm Hg? How does this value for P_{O_2} compare with the P_{O_2} value for fully humidified air at sea level?	Water vapor contributes a partial pressure of 47 mm Hg to fully humidified air. Oxygen is 21% of dry air. Normal atmospheric pressure at sea level is 760 mm Hg.	Correction for water vapor: 542 − 47 = 495 mm Hg × 21% = 104 mm Hg P_{O_2}. In humidified air at sea level, P_{O_2} = 150 mm Hg.
Q2: Why would someone with HAPE be short of breath?	Pulmonary edema increases the diffusion distance for oxygen.	Slower oxygen diffusion means less oxygen reaching the blood, which worsens the normal hypoxia of altitude.
Q3: Based on mechanisms for matching ventilation and perfusion in the lung, why do patients with HAPE have elevated pulmonary arterial blood pressure?	Low oxygen levels constrict pulmonary arterioles.	Constriction of pulmonary arterioles causes blood to collect in the pulmonary arteries behind the constriction. This increases pulmonary arterial blood pressure.
Q4: How does adding erythrocytes to the blood help a person acclimatize to high altitude?	98% of arterial oxygen is carried bound to hemoglobin.	Additional hemoglobin increases the total oxygen-carrying capacity of the blood.
Q5: What does adding erythrocytes to the blood do to the viscosity of the blood? What effect will that change in viscosity have on blood flow?	Adding cells increases blood viscosity.	According to Poiseuille's Law, increased viscosity increases resistance to flow, so blood flow will decrease.
Q6: What happens to plasma pH during hyperventilation?	Apply the law of mass action to the equation $CO_2 + H_2O \rightleftharpoons H^+ + HCO_3^-$.	The amount of CO_2 in the plasma decreases during hyperventilation, which means the equation shifts to the left. This shift decreases H^+, which increases pH (alkalosis).

—Continued next page

RUNNING **PROBLEM CONCLUSION** | Continued

Question	Facts	Integration and Analysis
Q7: *How does this change in pH affect oxygen binding at the lungs when P_{O_2} is decreased? How does it affect unloading of oxygen at the cells?*	See Figure 18.9c.	The left shift of the curve means that, at any given P_{O_2}, more O_2 binds to hemoglobin. Less O_2 will unbind at the tissues for a given P_{O_2}, but P_{O_2} in the cells is probably lower than normal, and consequently there may be no change in unloading.
Q8: *Why do you think periodic breathing occurs most often during sleep?*	Periodic breathing alternates periods of breath-holding (apnea) and hyperventilation.	An awake person is more likely to make a conscious effort to breathe during the breath-holding spells, eliminating the cycle of periodic breathing.

565   567   571   574   579   584   **585**

VISUALIZE, PRACTICE, AND MASTER PHYSIOLOGY WITH MASTERING A&P
- Practice-on-the-go with Dynamic Study Modules
- Visualize important processes with Interactive Physiology and A&P Flix
- Check your understanding with Video Tutors and practice quizzes!

MasteringA&P®

CHAPTER SUMMARY

In this chapter, you learned why climbing Mt. Everest is such a respiratory challenge for the human body, and why people with emphysema experience the same respiratory challenges at sea level. The exchange and transport of oxygen and carbon dioxide in the body illustrate the *mass flow* of gases along concentration gradients. *Homeostasis* of these blood gases demonstrates *mass balance:* The concentration in the blood varies according to what enters or leaves at the lungs and tissues. The *law of mass action* governs the chemical reactions through which hemoglobin binds O_2, and carbonic anhydrase catalyzes the conversion of CO_2 and water to carbonic acid.

Gas Exchange in the Lungs and Tissues

iP Respiratory: Gas Exchange

1. Normal alveolar and arterial P_{O_2} is about 100 mm Hg. Normal alveolar and arterial P_{CO_2} is about 40 mm Hg. Normal venous P_{O_2} is 40 mm Hg, and normal venous P_{CO_2} is 46 mm Hg. (p. 566; Fig. 18.2)

2. Body sensors monitor blood oxygen, CO_2, and pH in an effort to avoid **hypoxia** and **hypercapnia.** (p. 565)

3. Both the composition of inspired air and the effectiveness of alveolar ventilation affect alveolar P_{O_2}. (p. 566)

4. Changes in alveolar surface area, in diffusion barrier thickness, and in fluid distance between the alveoli and pulmonary capillaries can all affect gas exchange efficiency and arterial P_{O_2}. (p. 567; Fig. 18.3)

5. The amount of a gas that dissolves in a liquid is proportional to the partial pressure of the gas and to the **solubility** of the gas in the liquid. Carbon dioxide is 20 times more soluble in aqueous solutions than oxygen is. (p. 570; Fig. 18.4)

Gas Transport in the Blood

iP Respiratory: Gas Transport

6. Gas transport demonstrates mass flow and mass balance. The **Fick equation** relates blood oxygen content, cardiac output, and tissue oxygen consumption. (p. 571; Fig. 18.6)

7. Oxygen is transported dissolved in plasma (<2%) and bound to hemoglobin (>98%). (p. 571; Fig. 18.5)

8. The P_{O_2} of plasma determines how much oxygen binds to hemoglobin. (p. 573; Fig. 18.8)

9. Oxygen-hemoglobin binding is influenced by pH, P_{CO_2}, temperature, and **2,3-bisphosphoglycerate** (2,3-BPG). (p. 577; Fig. 18.9)

10. Venous blood carries 7% of its carbon dioxide dissolved in plasma, 23% as **carbaminohemoglobin,** and 70% as bicarbonate ion in the plasma. (p. 579; Fig. 18.11)

11. **Carbonic anhydrase** in red blood cells converts CO_2 to carbonic acid, which dissociates into H^+ and HCO_3^-. The H^+ then binds to hemoglobin, and HCO_3^- enters the plasma using the **chloride shift.** (pp. 577, 578)

Regulation of Ventilation

iP Respiratory: Control of Respiration

12. Respiratory control resides in networks of neurons in the medulla and pons, influenced by input from central and peripheral sensory receptors and higher brain centers. (p. 580; Fig. 18.13)

13. The medullary **dorsal respiratory group** (DRG) contains mostly inspiratory neurons that control somatic motor neurons to the diaphragm. The **ventral respiratory group** (VRG) includes the **pre-Bötzinger complex** with its apparent pacemakers as well as neurons for inspiration and active expiration. (p. 581; Fig. 18.14)

14. **Peripheral chemoreceptors** in the carotid and aortic bodies monitor P_{O_2}, P_{CO_2}, and pH. P_{O_2} below 60 mm Hg triggers an increase in ventilation. (p. 582; Fig. 18.17)

15. Carbon dioxide is the primary stimulus for changes in ventilation. Chemoreceptors in the medulla and carotid bodies respond to changes in P_{CO_2}. (p. 583; Fig. 18.17)

16. Protective reflexes monitored by peripheral receptors prevent injury to the lungs from inhaled irritants. (p. 584)

17. Conscious and unconscious thought processes can affect respiratory activity. (p. 584)

REVIEW QUESTIONS

In addition to working through these questions and checking your answers on p. A-24, review the Learning Outcomes at the beginning of this chapter.

Level One Reviewing Facts and Terms

1. List five factors that influence the diffusion of gases between alveolus and blood.

2. More than _____% of the oxygen in arterial blood is transported bound to hemoglobin. How is the remaining oxygen transported to the cells?

3. Name four factors that influence the amount of oxygen that binds to hemoglobin. Which of these four factors is the most important?

4. Describe the structure of a hemoglobin molecule. What chemical element is essential for hemoglobin synthesis?

5. The networks for control of ventilation are found in the _____ and _____ of the brain. What do the dorsal and ventral respiratory groups of neurons control? What is a central pattern generator?

6. Describe the chemoreceptors that influence ventilation. What chemical is the most important controller of ventilation?

7. Describe the protective reflexes of the respiratory system.

8. What causes the exchange of oxygen and carbon dioxide between alveoli and blood or between blood and cells?

9. List five possible physical changes that could result in less oxygen reaching the arterial blood.

Level Two Reviewing Concepts

10. Concept map: Construct a map of gas transport using the following terms. You may add other terms.

• alveoli	• hemoglobin saturation
• arterial blood	• oxyhemoglobin
• carbaminohemoglobin	• P_{CO_2}
• carbonic anhydrase	• plasma
• chloride shift	• P_{O_2}
• dissolved CO_2	• pressure gradient
• dissolved O_2	• red blood cell
• hemoglobin	• venous blood

11. In respiratory physiology, it is customary to talk of the P_{O_2} of the plasma. Why is this not the most accurate way to describe the oxygen content of blood?

12. Compare and contrast the following pairs of concepts:

(a) transport of O_2 and CO_2 in arterial blood
(b) partial pressure and concentration of a gas dissolved in a liquid

13. Does HbO_2 binding increase, decrease, or not change with decreased pH?

14. Define hypoxia, COPD, and hypercapnia.

15. Why did oxygen-transporting molecules evolve in animals?

16. Draw and label the following graphs:

(a) the effect of ventilation on arterial P_{O_2}
(b) the effect of arterial P_{CO_2} on ventilation

17. As the P_{O_2} of plasma increases:

(a) what happens to the amount of oxygen that dissolves in plasma?
(b) what happens to the amount of oxygen that binds to hemoglobin?

18. If a person is anemic and has a lower-than-normal level of hemoglobin in her red blood cells, what is her arterial P_{O_2} compared to normal?

19. Create reflex pathways (stimulus, receptor, afferent path, and so on) for the chemical control of ventilation, starting with the following stimuli:

(a) increased arterial P_{CO_2}
(b) arterial $P_{O_2} = 55$ mm Hg

Be as specific as possible regarding anatomical locations. Where known, include neurotransmitters and their receptors.

Level Three Problem Solving

20. Marco tries to hide at the bottom of a swimming hole by breathing in and out through two feet of garden hose, which greatly increases his anatomic dead space. What happens to the following parameters in his arterial blood, and why?

(a) P_{CO_2}
(b) P_{O_2}
(c) bicarbonate ion
(d) pH

21. Which person carries more oxygen in his blood?

(a) one with Hb of 15 g/dL and arterial P_{O_2} of 80 mm Hg
(b) one with Hb of 12 g/dL and arterial P_{O_2} of 100 mm Hg

22. What would happen to each of the following parameters in a person suffering from pulmonary edema?

(a) arterial P_{O_2}
(b) arterial hemoglobin saturation
(c) alveolar ventilation

23. In early research on the control of rhythmic breathing, scientists made the following observations. What hypotheses might the researchers have formulated from each observation?

    (a) *Observation.* If the brain stem is severed below the medulla, all respiratory movement ceases.
    (b) *Observation.* If the brain stem is severed above the level of the pons, ventilation is normal.
    (c) *Observation.* If the medulla is completely separated from the pons and higher brain centers, ventilation becomes irregular but a pattern of inspiration/expiration remains.

24. A hospitalized patient with severe chronic obstructive lung disease has a P_{CO_2} of 55 mm Hg and a P_{O_2} of 50 mm Hg. To elevate his blood oxygen, he is given pure oxygen through a nasal tube. The patient immediately stops breathing. Explain why this might occur.

25. You are a physiologist on a space flight to a distant planet. You find intelligent humanoid creatures inhabiting the planet, and they willingly submit to your tests. Some of the data you have collected are described below.

The graph above shows the oxygen saturation curve for the oxygen-carrying molecule in the blood of the humanoid named Bzork. Bzork's normal alveolar P_{O_2} is 85 mm Hg. His normal cell P_{O_2} is 20 mm Hg, but it drops to 10 mm Hg with exercise.

    (a) What is the percent saturation for Bzork's oxygen-carrying molecule in blood at the alveoli? In blood at an exercising cell?
    (b) Based on the graph, what conclusions can you draw about Bzork's oxygen requirements during normal activity and during exercise?

26. The next experiment on Bzork involves his ventilatory response to different conditions. The data from that experiment are graphed below. Interpret the results of experiments A and C.

27. The alveolar epithelium is an absorptive epithelium and is able to transport ions from the fluid lining of alveoli into the interstitial space, creating an osmotic gradient for water to follow. Draw an alveolar epithelium and label apical and basolateral surfaces, the airspace, and interstitial fluid. Arrange the following proteins on the cell membrane so that the epithelium absorbs sodium and water: aquaporins, Na^+-K^+-ATPase, epithelial Na^+ channel (ENaC). (Remember: Na^+ concentrations are higher in the ECF than in the ICF.)

Level Four Quantitative Problems

28. You are given the following information on a patient.

    Blood volume = 5.2 liters

    Hematocrit = 47%

    Hemoglobin concentration = 12 g/dL whole blood

    Total amount of oxygen carried in blood = 1015 mL

    Arterial plasma = 100 mm Hg

    You know that when plasma P_{O_2} is 100 mm Hg, plasma contains 0.3 mL O_2/dL, and that hemoglobin is 98% saturated. Each hemoglobin molecule can bind to a maximum of four molecules of oxygen. Using this information, calculate the maximum oxygen-carrying capacity of hemoglobin (100% saturated). Units will be mL O_2/g Hb.

29. Adolph Fick, the nineteenth-century physiologist who derived Fick's law of diffusion, also developed the Fick equation that relates oxygen consumption, cardiac output, and blood oxygen content:

    O_2 consumption =
    cardiac output $\times$ (arterial oxygen content $-$ venous oxygen content)

    A person has a cardiac output of 4.5 L/min, an arterial oxygen content of 105 mL O_2/L blood, and a vena cava oxygen content of 50 mL O_2/L blood. What is this person's oxygen consumption?

30. Describe what happens to the oxygen-hemoglobin saturation curve in Figure 18.9a when blood hemoglobin falls from 15 g/dL blood to 10 g/dL blood.

19

The Kidneys

Plasma undergoes modification to urine in the nephron.

Arthur Grollman, in *Clinical Physiology: The Functional Pathology of Disease,* 1957

FUNCTIONS OF THE KIDNEYS 590

LO 19.1 List and describe the six functions of the kidneys.

ANATOMY OF THE URINARY SYSTEM 591

LO 19.2 Trace the anatomical path of a drop of water from Bowman's capsule to urine leaving the body.

LO 19.3 Trace a drop of blood from the renal artery to the renal vein.

LO 19.4 Diagram the anatomical relationship between the vascular and tubular elements of the nephron.

OVERVIEW OF KIDNEY FUNCTION 594

LO 19.5 Describe the three processes of the nephron.

LO 19.6 Diagram the volume and osmolarity changes of filtrate as it passes through each section of the nephron.

FILTRATION 596

LO 19.7 Describe the filtration barriers between the blood and the lumen of the nephron, and explain how they can be modified to control filtration.

LO 19.8 Describe the pressures that promote and oppose glomerular filtration.

LO 19.9 Define glomerular filtration rate and give average values for GFR.

LO 19.10 Explain how GFR can be influenced by local and reflex control mechanisms.

REABSORPTION 602

LO 19.11 Distinguish between transcellular transport and paracellular pathways.

LO 19.12 Describe and give examples of active and passive reabsorption in the proximal tubule.

LO 19.13 Using glucose as an example, create graphs to show filtration, transport maximum, and renal threshold of a substance reabsorbed by protein-mediated transport.

SECRETION 607

LO 19.14 Explain and give examples of the importance of tubular secretion in renal function.

EXCRETION 608

LO 19.15 Explain mathematically and in words the relationship between the excretion of a solute and its renal clearance.

LO 19.16 Explain how clearance can be used as an indirect indicator of renal handling of a solute.

MICTURITION 613

LO 19.17 Diagram the involuntary micturition reflex and include the voluntary control pathway exerted by higher brain centers.

BACKGROUND BASICS

10	Mass balance
41	pH and buffers
51	Saturation
126	Osmolarity and tonicity
137	Membrane transport
146	Competition
79	Transporting epithelium
150	Epithelial transport
152	Transcytosis
497	Capillary filtration
496	Fenestrated capillaries
488	Autoregulation of vascular resistance
391	Phosphocreatine

Resin cast of balllike glomerular capillaries along with their arterioles (gold)

About 100 C.E., Aretaeus the Cappadocian wrote, "Diabetes is a wonderful affection, not very frequent among men, being a melting down of the flesh and limbs into urine.... The patients never stop making water {urinating}, but the flow is incessant, as if from the opening of aqueducts."* Physicians have known since ancient times that **urine,** the fluid waste produced by the kidneys, reflects the functioning of the body. To aid them in their diagnosis of illness, they even carried special flasks for the collection and inspection of patients' urine.

The first step in examining a urine sample is to determine its color. Is it dark yellow (concentrated), pale straw (dilute), red (indicating the presence of blood), or black (indicating the presence of hemoglobin metabolites)? One form of malaria was called *black-water fever* because metabolized hemoglobin from the abnormal breakdown of red blood cells turned victims' urine black or dark red.

Physicians also inspected urine samples for clarity, froth (indicating abnormal presence of proteins), smell, and even taste. Physicians who did not want to taste the urine themselves would allow their students the "privilege" of tasting it for them. A physician without students might expose insects to the urine and study their reaction.

Probably the most famous example of using urine for diagnosis was the taste test for diabetes mellitus, historically known as the *honey-urine disease.* Diabetes is an endocrine disorder characterized by the presence of glucose in the urine. The urine of diabetics tasted sweet and attracted insects, making the diagnosis clear.

Today, we have much more sophisticated tests for glucose in the urine, but the first step of a *urinalysis* is still to examine the color, clarity, and odor of the urine. In this chapter, you will learn why we can tell so much about how the body is functioning by what is present in the urine.

FUNCTIONS OF THE KIDNEYS

If you ask people on the street, "What is the most important function of the kidney?" they are likely to say, "The removal of wastes." Actually, the most important function of the kidney is

---

The Extant Works of Aretaeus the Cappadocian. Edited and translated by Adams, F. London, 1856.

RUNNING **PROBLEM** | Gout

Michael, 43, had spent the past two days on the sofa, suffering from a relentless throbbing pain in his left big toe. When the pain began, Michael thought he had a mild sprain or perhaps the beginnings of arthritis. Then the pain intensified, and the toe joint became hot and red. Michael finally hobbled into his doctor's office, feeling a little silly about his problem. On hearing his symptoms and looking at the toe, the doctor seemed to know instantly what was wrong. "Looks to me like you have gout," said Dr. Garcia.

590 591 608 609 611 613

the homeostatic regulation of the water and ion content of the blood, also called *salt and water balance* or *fluid and electrolyte balance.* Waste removal is important, but disturbances in blood volume or ion levels cause serious medical problems before the accumulation of metabolic wastes reaches toxic levels.

The kidneys maintain normal blood concentrations of ions and water by balancing intake of those substances with their excretion in the urine, obeying the principle of *mass balance* [p. 10]. We can divide kidney function into six general areas:

1. **Regulation of extracellular fluid volume and blood pressure.** When extracellular fluid volume decreases, blood pressure also decreases [p. 485]. If ECF volume and blood pressure fall too low, the body cannot maintain adequate blood flow to the brain and other essential organs. The kidneys work in an integrated fashion with the cardiovascular system to ensure that blood pressure and tissue perfusion remain within an acceptable range.

2. **Regulation of osmolarity.** The body integrates kidney function with behavioral drives, such as thirst, to maintain blood osmolarity at a value close to 290 mOsM. We examine the reflex pathways for regulation of ECF volume and osmolarity later.

3. **Maintenance of ion balance.** The kidneys keep concentrations of key ions within a normal range by balancing dietary intake with urinary loss. Sodium (Na^+) is the major ion involved in the regulation of extracellular fluid volume and osmolarity. Potassium (K^+) and calcium (Ca^{2+}) concentrations are also closely regulated.

4. **Homeostatic regulation of pH.** The pH of plasma is normally kept within a narrow range. If extracellular fluid becomes too acidic, the kidneys remove H^+ and conserve bicarbonate ions (HCO_3^-) which act as a buffer [p. 41]. Conversely, when extracellular fluid becomes too alkaline, the kidneys remove HCO_3^- and conserve H^+. The kidneys play a significant role in pH homeostasis, but they do not correct pH disturbances as rapidly as the lungs do.

5. **Excretion of wastes.** The kidneys remove metabolic waste products and *xenobiotics,* or foreign substances, such as drugs and environmental toxins. Metabolic wastes include *creatinine* from muscle metabolism [p. 391] and the nitrogenous wastes *urea* and *uric acid.* A metabolite of hemoglobin called *urobilinogen* gives urine its characteristic yellow color. Hormones are another endogenous substance the kidneys clear from the blood. Examples of foreign substances that the kidneys actively remove include the artificial sweetener *saccharin* and the anion *benzoate,* part of the preservative *potassium benzoate,* which you ingest each time you drink a diet soft drink.

6. **Production of hormones.** Although the kidneys are not endocrine glands, they play important roles in three endocrine pathways. Kidney cells synthesize *erythropoietin,* the cytokine/hormone that regulates red blood cell synthesis [p. 518]. They also release *renin,* an enzyme that regulates the production of hormones involved in sodium balance and blood

pressure homeostasis. Finally, renal enzymes help convert vitamin D_3 into a hormone that regulates Ca^{2+} balance.

The kidneys, like many other organs in the body, have a tremendous reserve capacity. By most estimates, you must lose nearly three-fourths of your kidney function before homeostasis begins to be affected. Many people function perfectly normally with only one kidney, including the one person in 1000 born with only one kidney (the other fails to develop during gestation) or those people who donate a kidney for transplantation.

CONCEPT CHECK

1. Ion regulation is a key feature of kidney function. What happens to the resting membrane potential of a neuron if extracellular K^+ levels decrease? [p. 252]

2. What happens to the force of cardiac contraction if plasma Ca^{2+} levels decrease substantially? [p. 449]

ANATOMY OF THE URINARY SYSTEM

The **urinary system** is composed of the kidneys and accessory structures (**FIG. 19.1a**). The study of kidney function is called **renal physiology,** from the Latin word *renes,* meaning "kidneys."

The Urinary System Consists of Kidneys, Ureters, Bladder, and Urethra

Let's begin by following the route a drop of water takes on its way from plasma to excretion in the urine. Urine production begins when water and solutes move from plasma into the hollow tubules (*nephrons*) that make up the bulk of the paired **kidneys.** These tubules modify the composition of the fluid as it passes through. The modified fluid, now called *urine,* leaves the kidney and passes into a hollow tube called a **ureter.** There are two ureters, one leading from each kidney to the **urinary bladder.** The bladder expands and fills with urine until, in a reflex called *micturition* or urination, the bladder contracts and expels urine through a single tube, the **urethra.**

The urethra in males exits the body through the shaft of the penis. In females, the urethral opening is found anterior to the openings of the vagina and anus. Because of the shorter length of the female urethra and its proximity to bacteria leaving the large intestine, women are more prone than men to develop bacterial infections of the bladder and kidneys, or **urinary tract infections** (UTIs).

The most common cause of UTIs is the bacterium *Escherichia coli,* a normal inhabitant of the human large intestine. *E. coli* is not harmful while restricted to the lumen of the large intestine, but it is pathogenic {*patho-,* disease + *-genic,* causing} if it gets into the urethra. The most common symptoms of a UTI are pain or burning during urination and increased frequency of urination. A urine sample from a patient with a UTI often contains many red and white blood cells, neither of which is commonly found in normal urine. UTIs are treated with antibiotics.

The Kidneys The kidneys are the site of urine formation. They lie on either side of the spine at the level of the eleventh and twelfth ribs, just above the waist (Fig. 19.1b). Although they are below the diaphragm, they are technically outside the abdominal cavity, sandwiched between the membranous **peritoneum,** which lines the abdomen, and the bones and muscles of the back. Because of their location behind the peritoneal cavity, the kidneys are sometimes described as being *retroperitoneal* {*retro-,* behind}.

The concave surface of each kidney faces the spine. The renal blood vessels, nerves, lymphatics, and ureters all emerge from this surface. **Renal arteries,** which branch off the abdominal aorta, supply blood to the kidneys. **Renal veins** carry blood from the kidneys to the inferior vena cava.

At any given time, the kidneys receive 20–25% of the cardiac output, even though they constitute only 0.4% of total body weight (4.5–6 ounces each). This high rate of blood flow through the kidneys is critical to renal function.

The Nephron Is the Functional Unit of the Kidney

A cross section through a kidney shows that the interior is arranged in two layers: an outer **cortex** and inner **medulla** (Fig. 19.1c). The layers are formed by the organized arrangement of microscopic tubules called **nephrons.** About 80% of the nephrons in a kidney are almost completely contained within the cortex (*cortical* nephrons), but the other 20%—called *juxtamedullary* nephrons {*juxta-,* beside}—dip down into the medulla (Fig. 19.1f, h).

The nephron is the functional unit of the kidney. (A *functional unit* is the smallest structure that can perform all the functions of an organ.) Each of the 1 million nephrons in a kidney is divided into sections (Fig. 19.1i), and each section is closely associated with specialized blood vessels (Fig. 19.1g, h).

RUNNING **PROBLEM**

Gout is a metabolic disease characterized by high blood concentrations of uric acid (*hyperuricemia*). If uric acid concentrations reach a critical level (6.8 mg/dL), monosodium urate precipitates out of solution and forms crystals in peripheral joints, particularly in the feet, ankles, and knees. These crystals trigger an inflammatory reaction and cause periodic attacks of excruciating pain. Uric acid crystals may also form kidney stones in the *renal pelvis* (Fig. 19.1c).

Q1: *Trace the route followed by these kidney stones when they are excreted.*

Q2: *Name the anion formed when uric acid dissociates.*

590 **591** 608 609 611 613

Vascular Elements of the Kidney Blood enters the kidney through the renal artery before flowing into smaller arteries and then into arterioles in the cortex (Fig. 19.1d, e). At this point, the arrangement of blood vessels forms a *portal system,* one of three in the body [p. 439]. Recall that a portal system consists of two capillary beds in series (one after the other).

In the renal portal system, blood flows from renal arteries into an **afferent arteriole.** From the afferent arteriole it goes into the first capillary bed, a ball-like network known as the **glomerulus** {*glomus,* a ball-shaped mass; plural *glomeruli*} (Fig. 19.1g, j). Blood leaving the glomerulus flows into an **efferent arteriole,** then into the second set of capillaries, the **peritubular capillaries** {*peri-,* around} that surround the tubule (Fig. 19.1g). In juxtamedullary nephrons, the long peritubular capillaries that dip into the medulla are called the **vasa recta** (Fig. 19.1h). Finally, peritubular capillaries converge to form venules and small veins, sending blood out of the kidney through the renal vein.

The function of the renal portal system is to filter fluid out of the blood and into the lumen of the nephron at the glomerular capillaries, then to *reabsorb* fluid from the tubule lumen back into the blood at the peritubular capillaries. The forces behind fluid movement in the renal portal system are similar to those that govern filtration of water and molecules out of systemic capillaries in other tissues, as we will describe shortly.

CONCEPT CHECK

3. If net filtration out of glomerular capillaries occurs, then you know that capillary hydrostatic pressure must be (*greater than/less than/equal to*) capillary colloid osmotic pressure. [p. 498]

4. If net reabsorption into peritubular capillaries occurs, then capillary hydrostatic pressure must be (*greater than/less than/equal to*) the capillary colloid osmotic pressure.

Tubular Elements of the Kidney The kidney tubule consists of a single layer of epithelia cells connected together near their apical surface. The apical surfaces are folded into *microvilli* [p. 68] or other area-increasing folds, and the basal side of the polarized epithelium [p. 150] rests on a basement membrane, or *basal lamina*. The cell-cell junctions are mostly tight but some have selective permeability for ions.

The nephron begins with a hollow, ball-like structure called **Bowman's capsule** that surrounds the glomerulus (Fig. 19.1i). The endothelium of the glomerulus is fused to the epithelium of Bowman's capsule so that fluid filtering out of the capillaries passes directly into the lumen of the tubule. The combination of glomerulus and Bowman's capsule is called the **renal corpuscle.**

From Bowman's capsule, filtered fluid flows into the **proximal tubule** {*proximal,* close or near}, then into the **loop of Henle,** a hairpin-shaped segment that dips down toward the medulla and then back up. The loop of Henle is divided into two limbs, a thin **descending limb** and an **ascending limb** with thin and thick segments. The fluid then passes into the **distal tubule** {*distal,* distant or far}. The distal tubules of up to eight nephrons drain into a single larger tube called the **collecting duct.** (The distal

tubule and its collecting duct together form the **distal nephron.**) Collecting ducts pass from the cortex through the medulla and drain into the **renal pelvis** (Fig. 19.1c). From the renal pelvis, the filtered and modified fluid, now called urine, flows into the ureter on its way to excretion.

Notice in Figure 19.1g how the nephron twists and folds back on itself so that the final part of the ascending limb of the loop of Henle passes between the afferent and efferent arterioles. This region is known as the **juxtaglomerular apparatus.** The proximity of the ascending limb and the arterioles allows paracrine communication between the two structures, a key feature of kidney autoregulation. Because the twisted configuration of the nephron makes it difficult to follow fluid flow, we unfold the nephron in many of the remaining figures in this chapter so that fluid flows from left to right across the figure, as in Figure 19.1i.

OVERVIEW OF KIDNEY FUNCTION

Imagine drinking a 12-ounce soft drink every three minutes around the clock: By the end of 24 hours, you would have consumed the equivalent of 90 two-liter bottles. The thought of putting 180 liters of liquid into your intestinal tract is staggering, but that is how much plasma passes into the nephrons every day! Because the average volume of urine leaving the kidneys is only 1.5 L/day, more than 99% of the fluid that enters nephrons must find its way back into the blood, or the body would rapidly dehydrate.

Kidneys Filter, Reabsorb, and Secrete

Three basic processes take place in the nephron: filtration, reabsorption, and secretion (**FIG. 19.2**). **Filtration** is the movement of fluid from blood into the lumen of the nephron. Filtration takes place only in the renal corpuscle, where the walls of glomerular capillaries and Bowman's capsule are modified to allow bulk flow of fluid.

Once the filtered fluid, called *filtrate*, passes into the lumen of the nephron, it becomes part of the body's external environment, just as substances in the lumen of the intestinal tract are part of the external environment [Fig. 1.2, p. 4]. For this reason, anything that filters into the nephron is destined for **excretion,** removal in the urine, unless it is reabsorbed into the body.

After filtrate leaves Bowman's capsule, it is modified by reabsorption and secretion. **Reabsorption** is the process of moving substances in the filtrate from the lumen of the tubule back into the blood flowing through peritubular capillaries. **Secretion** selectively removes molecules from the blood and adds them to the filtrate in the tubule lumen. Although secretion and glomerular filtration both move substances from blood into the tubule, secretion is a more selective process that usually uses membrane proteins to move molecules across the tubule épithelium.

The Nephron Modifies Fluid Volume and Osmolarity

Now let's follow some filtrate through the nephron to learn what happens to it in the various segments (Fig. 19.2). The 180 liters of fluid that filters into Bowman's capsule each day are almost

FIG. 19.2 **ESSENTIALS**

Nephron Function

The four processes of the kidney are:

 = **Filtration:** movement from blood to lumen   **S** = **Secretion:** from blood to lumen

R = **Reabsorption:** from lumen to blood   **E** = **Excretion:** from lumen to outside the body

This model nephron has been untwisted so that fluid flows left to right.

Tubular Elements

Bowman's capsule | Proximal tubule | Loop of Henle | Distal tubule | Collecting duct

Vascular Elements

Efferent arteriole

Glomerulus

Afferent arteriole

Peritubular capillaries

Vasa recta

Filtered
180 L/day
100% volume
300 mOsM

Start of Loop of Henle
54 L/day
30% volume
300 mOsM

End of Loop of Henle
18 L/day
____% volume
100 mOsM

End of Collecting duct
1.5 L/day
____% volume
50–1200 mOsM

To renal vein

To bladder and external environment

Segments of the Nephron and their Functions

Segment of Nephron	Processes
Renal corpuscle (glomerulus + Bowman's capsule)	Filtration of mostly protein-free plasma from the capillaries into the capsule
Proximal tubule	Isosmotic reabsorption of organic nutrients, ions, and water. Secretion of metabolites and xenobiotic molecules such as penicillin.
Loop of Henle	Reabsorption of ions in excess of water to create dilute fluid in the lumen. Countercurrent arrangement contributes to concentrated interstitial fluid in the renal medulla [see Chapter 20].
Distal nephron (distal tubule + collecting duct)	Regulated reabsorption of ions and water for salt and water balance and pH homeostasis.

Q FIGURE QUESTIONS

1. In which segments of the nephron do the following processes take place:
   (a) filtration
   (b) reabsorption
   (c) secretion
   (d) excretion
2. Calculate the percentage of filtered volume that leaves
   (a) the loop of Henle
   (b) the collecting duct

identical in composition to plasma and nearly isosmotic—about 300 mOsM [p. 126]. As this filtrate flows through the proximal tubule, about 70% of its volume is reabsorbed, leaving 54 liters in the lumen.

Reabsorption occurs when proximal tubule cells transport solutes out of the lumen, and water follows by osmosis. Filtrate leaving the proximal tubule has the same osmolarity as filtrate that entered. For this reason, we say that the primary function of the proximal tubule is the isosmotic reabsorption of solutes and water.

Filtrate leaving the proximal tubule passes into the loop of Henle, the primary site for creating dilute urine [by a process discussed in detail in Chapter 20]. As filtrate passes through the loop, proportionately more solute is reabsorbed than water, and the filtrate becomes hyposmotic relative to the plasma. By the time filtrate flows out of the loop, it averages 100 mOsM, and its volume has fallen from 54 L/day to about 18 L/day. Most of the volume originally filtered into Bowman's capsule has been reabsorbed into the capillaries.

From the loop of Henle, filtrate passes into the distal tubule and the collecting duct. In these two segments, the fine regulation of salt and water balance takes place under the control of several hormones. Reabsorption and (to a lesser extent) secretion determine the final composition of the filtrate. By the end of the collecting duct, the filtrate has a volume of about 1.5 L/day and an osmolarity that can range from 50 mOsM to 1200 mOsM. The final volume and osmolarity of urine depend on the body's need to conserve or excrete water and solute. Hormonal control of salt and water balance is discussed in the next chapter.

A word of caution here: It is very easy to confuse *secretion* with *excretion*. Try to remember the origins of the two prefixes. *Se-* means *apart,* as in to separate something from its source. In the nephron, secreted solutes are moved from plasma to tubule lumen. *Ex-* means *out,* or *away,* as in out of or away from the body. Excretion refers to the removal of a substance from the body. Besides the kidneys, other organs that carry out excretory processes include the lungs (CO_2) and intestines (undigested food, bilirubin).

Figure 19.2 summarizes filtration, reabsorption, secretion, and excretion. Filtration takes place in the renal corpuscle as fluid moves from the capillaries of the glomerulus into Bowman's capsule. Reabsorption and secretion occur along the remainder of the tubule, transferring materials between the lumen and the peritubular capillaries. The quantity and composition of the substances being reabsorbed and secreted vary in different segments of the nephron. Filtrate that remains in the lumen at the end of the nephron is excreted as urine.

The amount of any substance excreted in the urine reflects the net handling of that substance during its passage through the nephron (**FIG. 19.3**). The amount excreted is equal to the amount filtered into the tubule, minus any amount reabsorbed into the blood, plus any amount secreted into the tubule lumen:

> Amount excreted = amount filtered − amount reabsorbed
> + amount secreted

FIG. 19.3   Solute movement through the nephron

The urinary excretion of a substance depends on its filtration, reabsorption, and secretion.

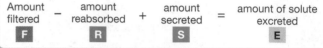

$$\underset{F}{\text{Amount filtered}} - \underset{R}{\text{amount reabsorbed}} + \underset{S}{\text{amount secreted}} = \underset{E}{\text{amount of solute excreted}}$$

Q FIGURE QUESTION

A person filters 720 millimoles of K^+ in a day and secretes 43 millimoles. She excretes 79 millimoles in her urine. What happened to the rest of the K^+ and how much was it?

This equation is a useful way to think about *renal handling* of solutes. Note, however, that not every substance in the plasma is filtered. And substances that are filtered may or may not be reabsorbed or secreted. In the following sections, we look in more detail at the important processes of filtration, reabsorption, secretion, and excretion.

CONCEPT CHECK

5. Name one way in which filtration and secretion are alike. Name one way in which they differ.

6. A water molecule enters the renal corpuscle from the blood and ends up in the urine. Name all the anatomical structures that the molecule passes through on its trip to the outside world.

7. What would happen to the body if filtration continued at a normal rate but reabsorption dropped to half the normal rate?

FILTRATION

The filtration of plasma into the kidney tubule is the first step in urine formation. This relatively nonspecific process creates a filtrate whose composition is like that of plasma minus most of the plasma proteins. Under normal conditions, blood cells remain in the capillary, so that the filtrate is composed of water and dissolved solutes.

When you visualize plasma filtering out of the glomerular capillaries, it is easy to imagine that all the plasma in the capillary moves into Bowman's capsule. However, filtration of all the plasma would leave behind a sludge of blood cells and proteins that could not flow out of the glomerulus. Instead, only about one-fifth of the plasma that flows through the kidneys filters into the nephrons. The remaining four-fifths of the plasma, along with most plasma proteins and blood cells, flows into the peritubular capillaries (**FIG. 19.4**). The percentage of total plasma volume that filters into the tubule is called the **filtration fraction.**

The Renal Corpuscle Contains Filtration Barriers

Filtration takes place in the renal corpuscle (**FIG. 19.5**), which consists of the glomerular capillaries surrounded by Bowman's capsule. Substances leaving the plasma must pass through three *filtration barriers* before entering the tubule lumen: the glomerular capillary endothelium, a basal lamina (basement membrane), and the epithelium of Bowman's capsule (Fig. 19.5d). The details of how these filtration barriers function are still under investigation.

The first barrier is the capillary endothelium. Glomerular capillaries are *fenestrated capillaries* [p. 496] with large pores that allow most components of the plasma to filter through the endothelium. The pores are small enough, however, to prevent blood cells from leaving the capillary. The negatively charged proteins on the pore surfaces also help repel negatively charged plasma proteins.

The second filtration barrier is the **basal lamina,** an acellular layer of extracellular matrix that separates the capillary endothelium from the epithelium of Bowman's capsule (Fig. 19.5d). The basal lamina consists of negatively charged glycoproteins, collagen, and other proteins. The lamina acts like a coarse sieve, excluding most plasma proteins from the fluid that filters through it.

The third filtration barrier is the epithelium of Bowman's capsule. The portion of the capsule epithelium that surrounds each glomerular capillary consists of specialized cells called **podocytes** {*podos*, foot} (Fig. 19.5c). Podocytes have long cytoplasmic extensions called **foot processes** that extend from the main cell body (Fig. 19.5a, b).

Foot processes wrap around the glomerular capillaries and interlace with one another, leaving narrow **filtration slits** closed by a semiporous membrane. The filtration slit membrane contains several unique proteins, including *nephrin* and *podocin*. These proteins were discovered by investigators looking for the gene mutations responsible for two congenital kidney diseases. In these diseases, where nephrin or podocin are absent or abnormal, proteins leak across the glomerular filtration barrier into the urine.

Glomerular **mesangial cells** lie between and around the glomerular capillaries (Fig. 19.5c). Mesangial cells have cytoplasmic bundles of actin-like filaments that enable them to contract and alter blood flow through the capillaries. In addition, mesangial

EMERGING CONCEPTS

Diabetes: Diabetic Nephropathy

End-stage renal failure, in which kidney function has deteriorated beyond recovery, is a life-threatening complication in 30–40% of people with type 1 diabetes and in 10–20% of those with type 2 diabetes. As with many other complications of diabetes, the exact causes of renal failure are not clear. Diabetic nephropathy usually begins with an increase in glomerular filtration. This is followed by the appearance of proteins in the urine (*proteinuria*), an indication that the normal filtration barrier has been altered. In later stages, filtration rates decline. This stage is associated with thickening of the glomerular basal lamina and changes in podocytes and mesangial cells. Abnormal growth of mesangial cells compresses the glomerular capillaries and impedes blood flow, contributing to the decrease in glomerular filtration. At this point, patients must have their kidney function supplemented by dialysis, and eventually they may need a kidney transplant.

FIG. 19.4   The filtration fraction

Only 20% of the plasma that passes through the glomerulus is filtered. Less than 1% of filtered fluid is eventually excreted.

Efferent arteriole   Peritubular capillaries

4 >99% of plasma entering kidney returns to systemic circulation.

80%

2 20% of volume filters.

3 >19% of fluid is reabsorbed.

5 <1% of volume is excreted to external environment.

Afferent arteriole

1 Plasma volume entering afferent arteriole = 100%.

Bowman's capsule

Remainder of nephron

Glomerulus

Q **FIGURE QUESTION**

If 120 mL of plasma filter each minute and the filtration fraction is 20%, what is the daily renal plasma flow?

FIG. 19.5 The renal corpuscle

(a) The epithelium around glomerular capillaries is modified into podocytes.

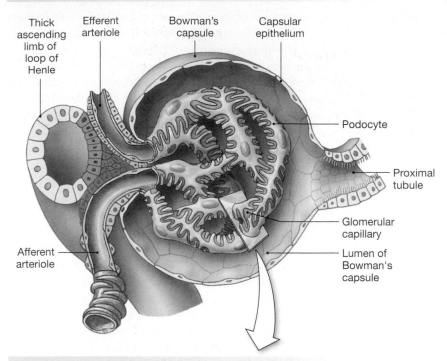

Thick ascending limb of loop of Henle

Efferent arteriole

Bowman's capsule

Capsular epithelium

Podocyte

Proximal tubule

Glomerular capillary

Lumen of Bowman's capsule

Afferent arteriole

(b) Micrograph showing podocyte foot processes around glomerular capillary.

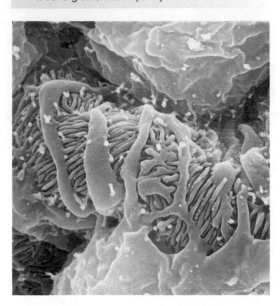

(c) Podocyte foot processes surround each capillary, leaving slits through which filtration takes place. Mesangial cells between the capillaries contract to alter blood flow.

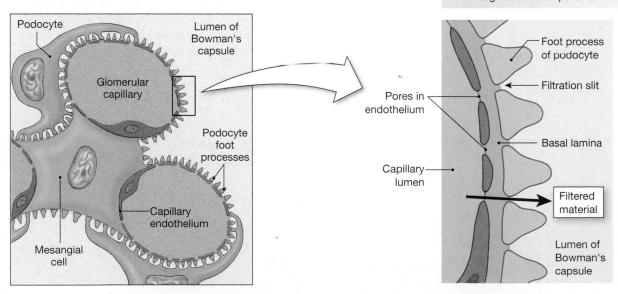

Podocyte

Lumen of Bowman's capsule

Glomerular capillary

Podocyte foot processes

Capillary endothelium

Mesangial cell

(d) The glomerular capillary endothelium, basal lamina, and Bowman's capsule epithelium create a three-layer filtration barrier. Filtered substances pass through endothelial pores and filtration slits.

Foot process of podocyte

Filtration slit

Pores in endothelium

Basal lamina

Capillary lumen

Filtered material

Lumen of Bowman's capsule

cells secrete cytokines associated with immune and inflammatory processes. Disruptions of mesangial cell function have been linked to several disease processes in the kidney.

Capillary Pressure Causes Filtration

What drives filtration across the walls of the glomerular capillaries? The process is similar in many ways to filtration of fluid out of systemic capillaries [p. 497]. The three pressures that influence glomerular filtration—capillary blood pressure, capillary colloid osmotic pressure, and capsule fluid pressure—are summarized in **FIGURE 19.6a**.

1. The *hydrostatic pressure* (P_H) of blood flowing through the glomerular capillaries forces fluid through the leaky endothelium. Capillary blood pressure averages 55 mm Hg and

FIG. 19.6 ESSENTIALS

Glomerular Filtration Rate

Filtration out of glomerular capillaries is similar to filtration in other systemic capillaries. Filtration pressure depends on hydrostatic pressure, and is opposed by colloid osmotic pressure and capsule fluid pressure.

(a) Calculating glomerular filtration pressure

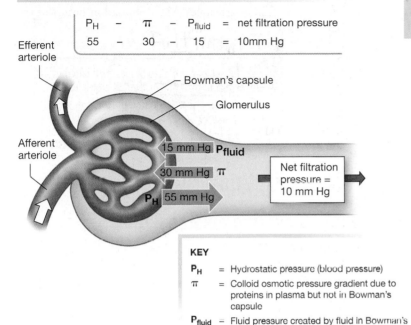

$$P_H - \pi - P_{fluid} = \text{net filtration pressure}$$
$$55 - 30 - 15 = 10 \text{mm Hg}$$

Efferent arteriole

Bowman's capsule

Glomerulus

Afferent arteriole

15 mm Hg P_{fluid}

30 mm Hg π

P_H 55 mm Hg

Net filtration pressure = 10 mm Hg

KEY

P_H = Hydrostatic pressure (blood pressure)

π = Colloid osmotic pressure gradient due to proteins in plasma but not in Bowman's capsule

P_{fluid} = Fluid pressure created by fluid in Bowman's capsule

(b) Autoregulation of glomerular filtration rate takes place over a wide range of blood pressures.

Glomerular filtration rate (L/day)

Autoregulation maintains a nearly constant GFR when mean arterial blood pressure is between 80 and 180 mm Hg.

180

Normal mean blood pressure

0   40   80   120   160   200

Mean arterial blood pressure (mm Hg)

(c) Resistance changes in renal arterioles alter renal blood flow and GFR.

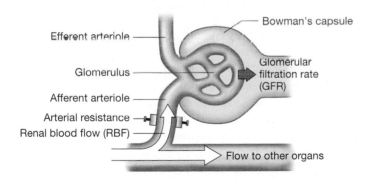

Efferent arteriole

Bowman's capsule

Glomerulus

Afferent arteriole

Arterial resistance

Renal blood flow (RBF)

Glomerular filtration rate (GFR)

Flow to other organs

(d) Vasoconstriction of the afferent arteriole increases resistance and decreases renal blood flow, capillary blood pressure (P_H), and GFR.

Decreased capillary blood pressure ($\downarrow P_H$)

Decreased GFR

Increased resistance in afferent arteriole

Decreased RBF

(e) Increased resistance of efferent arteriole decreases renal blood flow but increases P_H and GFR.

Increased resistance in efferent arteriole

Increased P_H

Increased GFR

Decreased RBF

Q **FIGURE QUESTION**

What happens to capillary blood pressure, GFR, and RBF when the afferent arteriole dilates?

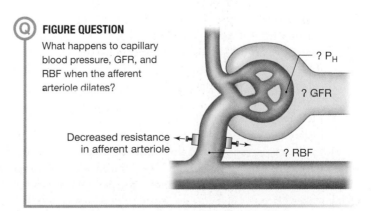

? P_H

? GFR

Decreased resistance in afferent arteriole

? RBF

favors filtration into Bowman's capsule. Although pressure decreases as blood moves through the capillaries, it remains higher than the opposing pressures. Consequently, filtration takes place along nearly the entire length of the glomerular capillaries.

2. The *colloid osmotic pressure* (π) inside glomerular capillaries is higher than that of the fluid in Bowman's capsule. This pressure gradient is due to the presence of proteins in the plasma. The osmotic pressure gradient averages 30 mm Hg and favors fluid movement back into the capillaries.

3. Bowman's capsule is an enclosed space (unlike the interstitial fluid), and so the presence of fluid in the capsule creates a hydrostatic *fluid pressure* (P_{fluid}) that opposes fluid movement into the capsule. Fluid filtering out of the capillaries must displace the fluid already in the capsule lumen. Hydrostatic fluid pressure in the capsule averages 15 mm Hg, opposing filtration.

The net driving force is 10 mm Hg in the direction favoring filtration. This pressure may not seem very high, but when combined with the very leaky nature of the fenestrated glomerular capillaries, it results in rapid fluid filtration into the tubules.

The volume of fluid that filters into Bowman's capsule per unit time is the **glomerular filtration rate (GFR).** Average GFR is 125 mL/min, or 180 L/day, an incredible rate considering that total plasma volume is only about 3 liters. This rate means that the kidneys filter the entire plasma volume 60 times a day, or 2.5 times every hour. If most of the filtrate were not reabsorbed during its passage through the nephron, we would run out of plasma in only 24 minutes of filtration!

GFR is influenced by two factors: the net filtration pressure just described and the filtration coefficient. Filtration pressure is determined primarily by renal blood flow and blood pressure. The **filtration coefficient** has two components: the surface area of the glomerular capillaries available for filtration and the permeability of interface between the capillary and Bowman's capsule. In this respect, glomerular filtration is similar to gas exchange at the alveoli, where the rate of gas exchange depends on partial pressure differences, the surface area of the alveoli, and the permeability of the alveolar-capillary diffusion barrier [p. 565].

GFR Is Relatively Constant

Blood pressure provides the hydrostatic pressure that drives glomerular filtration. Therefore, it might seem reasonable to assume that if blood pressure increased, GFR would increase, and if blood pressure fell, GFR would decrease. That is not usually the case, however. Instead, GFR is remarkably constant over a wide range of blood pressures. As long as mean arterial blood pressure remains between 80 mm Hg and 180 mm Hg, GFR averages 180 L/day (Fig. 19.6b).

GFR is controlled primarily by regulation of blood flow through the renal arterioles. If the overall resistance of the renal arterioles increases, renal blood flow decreases, and blood is diverted to other organs [Fig. 15.13, p. 493]. The effect of increased resistance on GFR, however, depends on *where* the resistance change takes place.

If resistance increases in the *afferent* arteriole (Fig. 19.6d), hydrostatic pressure decreases on the glomerular side of the constriction. This translates into a decrease in GFR. If resistance increases in the *efferent* arteriole, blood "dams up" in front of the constriction, and hydrostatic pressure in the glomerular capillaries increases (Fig. 19.6e). Increased glomerular pressure increases GFR. The opposite changes occur with decreased resistance in the afferent or efferent arterioles. Most regulation occurs at the afferent arteriole.

CONCEPT CHECK

8. Why is the osmotic pressure of plasma in efferent arterioles higher than that in afferent arterioles?

9. If a hypertensive person's blood pressure is 143/107 mm Hg, and mean arterial pressure = diastolic pressure + 1/3 the pulse pressure, what is this person's mean arterial pressure? What is this person's GFR according to Figure 19.6b?

GFR Is Subject to Autoregulation

Autoregulation of glomerular filtration rate is a local control process in which the kidney maintains a relatively constant GFR in the face of normal fluctuations in blood pressure. One important function of GFR autoregulation is to protect the filtration barriers from high blood pressures that might damage them. We do not completely understand the autoregulation process, but several mechanisms are at work. The **myogenic response** is the intrinsic ability of vascular smooth muscle to respond to pressure changes. **Tubuloglomerular feedback** is a paracrine signaling mechanism through which changes in fluid flow through the loop of Henle influence GFR.

Myogenic Response The myogenic response of afferent arterioles is similar to autoregulation in other systemic arterioles. When smooth muscle in the arteriole wall stretches because of increased blood pressure, stretch-sensitive ion channels open, and the muscle cells depolarize. Depolarization opens voltage-gated Ca^{2+} channels, and the vascular smooth muscle contracts [p. 407]. Vasoconstriction increases resistance to flow, and so blood flow through the arteriole diminishes. The decrease in blood flow decreases filtration pressure in the glomerulus.

If blood pressure decreases, the tonic level of arteriolar contraction disappears, and the arteriole becomes maximally dilated. However, vasodilation is not as effective at maintaining GFR as vasoconstriction because normally the afferent arteriole is fairly relaxed. Consequently, when mean blood pressure drops below 80 mm Hg, GFR decreases. This decrease is adaptive in the sense that if less plasma is filtered, the potential for fluid loss in the urine is decreased. In other words, a decrease in GFR helps the body conserve blood volume.

Tubuloglomerular Feedback Tubuloglomerular feedback is a local control pathway in which fluid flow through the tubule influences GFR. The twisted configuration of the nephron, as shown in **FIGURE 19.7a**, causes the final portion of the thick

FIG. 19.7 The juxtaglomerular apparatus

The juxtaglomerular apparatus consists of macula densa and granular cells.
Paracrine signaling between the nephron and afferent arteriole influences GFR.

(a) The nephron loops back on itself so that the ascending limb of the loop of Henle passes between the afferent and efferent arterioles.

(b) The macula densa cells sense distal tubule flow and release paracrines that affect afferent arteriole diameter.

Efferent arteriole

Bowman's capsule

Glomerulus

Ascending limb of loop of Henle

Macula densa cells

Proximal tubule

Endothelium

Granular cells secrete renin, an enzyme involved in salt and water balance.

Afferent arteriole

(c) Tubuloglomerular feedback helps GFR autoregulation.

① GFR increases.

② Flow through tubule Increases.

③ Flow past macula densa increases.

④ Paracrine from macula densa to afferent arteriole

⑤ Afferent arteriole constricts.

Resistance in afferent arteriole increases.

Hydrostatic pressure in glomerulus decreases.

GFR decreases.

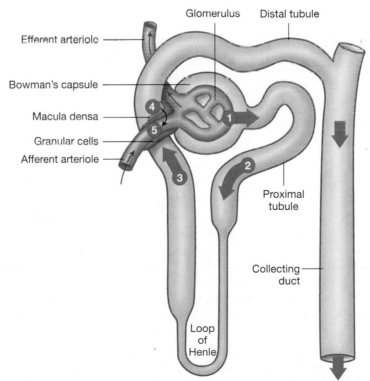

Glomerulus   Distal tubule

Efferent arteriole

Bowman's capsule

Macula densa

Granular cells

Afferent arteriole

Proximal tubule

Collecting duct

Loop of Henle

ascending limb of the loop of Henle to pass between the afferent and efferent arterioles. The tubule and arteriolar walls are modified in the regions where they contact each other and together form the *juxtaglomerular apparatus*.

The modified portion of the tubule epithelium is a plaque of cells called the **macula densa** (Fig. 19.7b). The adjacent wall of the afferent arteriole has specialized smooth muscle cells called **granular cells** (also known as *juxtaglomerular cells* or *JG cells*). The granular cells secrete *renin*, an enzyme involved in salt and water balance [Chapter 20]. When NaCl delivery past the macula densa increases as a result of increased GFR, the macula densa cells send a paracrine message to the neighboring afferent arteriole (Fig. 19.7c). The afferent arteriole constricts, increasing resistance and decreasing GFR.

Experimental evidence indicates that the macula densa cells transport NaCl, and that increases in salt transport initiate tubuloglomerular feedback. Flow can also be sensed in renal tubule cells by the *primary cilia* [p. 69], which are located on the apical surface facing the lumen. The renal primary cilia are known to act as flow sensors as well as signal transducers for normal development.

Paracrine signaling between the macula densa and the afferent arteriole is complex, and the details are still being worked out. Experiments show that multiple paracrine signals, including ATP, adenosine, and nitric oxide, pass from the macula densa to the arteriole as part of tubuloglomerular feedback.

Hormones and Autonomic Neurons Also Influence GFR

Although local mechanisms within the kidney attempt to maintain a constant GFR, the importance of the kidneys in systemic blood pressure homeostasis means that integrating centers outside the kidney can override local controls. Hormones and the autonomic nervous system alter glomerular filtration rate two ways: by changing resistance in the arterioles and by altering the filtration coefficient.

Neural control of GFR is mediated by sympathetic neurons that innervate both the afferent and efferent arterioles. Sympathetic innervation of α-receptors on vascular smooth muscle causes vasoconstriction [p. 490]. If sympathetic activity is moderate, there is little effect on GFR. If systemic blood pressure drops sharply, however, as occurs with hemorrhage or severe dehydration, sympathetically induced vasoconstriction of the arterioles decreases GFR and renal blood flow. This is an adaptive response that helps conserve fluid volume.

A variety of hormones also influence arteriolar resistance. Among the most important are *angiotensin II,* a potent vasoconstrictor, and prostaglandins, which act as vasodilators. These same hormones may affect the filtration coefficient by acting on podocytes or mesangial cells. Podocytes change the size of the glomerular filtration slits. If the slits widen, more surface area is available for filtration, and GFR increases. Contraction of mesangial cells apparently changes the glomerular capillary surface area available for filtration. We still have much to learn about these processes, and physiologists are actively investigating them.

CONCEPT CHECK

10. If systemic blood pressure remains constant but the afferent arteriole of a nephron constricts, what happens to renal blood flow and GFR in that nephron?

11. A person with cirrhosis of the liver has lower-than-normal levels of plasma proteins and consequently a higher-than-normal GFR. Explain why a decrease in plasma protein concentration causes an increase in GFR.

REABSORPTION

Each day, 180 liters of filtered fluid pass from the glomerular capillaries into the tubules, yet only about 1.5 liters are excreted in the urine. Thus, more than 99% of the fluid entering the tubules must be reabsorbed into the blood as filtrate moves through the nephrons. Most of this reabsorption takes place in the proximal tubule, with a smaller amount of reabsorption in the distal segments of the nephrons. Regulated reabsorption in the distal nephron allows the kidneys to return ions and water to the plasma selectively—as needed to maintain homeostasis.

One question you might be asking is, "Why bother to filter 180 L/day and then reabsorb 99% of it? Why not simply filter and excrete the 1% that needs to be eliminated?" There are two reasons. First, many foreign substances are filtered into the tubule but not reabsorbed into the blood. The high daily filtration rate helps clear such substances from the plasma very rapidly.

Once a substance filters into the lumen of Bowman's capsule, it is no longer part of the body's internal environment. The lumen of the nephron is external environment, and anything in the filtrate is destined to leave the body in the urine unless there is a tubule mechanism for reclaiming it. Many small nutrients, such as glucose and citric acid cycle intermediates, are filtered, but the proximal tubule very efficiently reabsorbs them.

Second, filtering ions and water into the tubule simplifies their regulation. If a portion of filtrate that reaches the distal nephron is not needed to maintain homeostasis, it passes into the urine. With a high GFR, this excretion can occur quite rapidly. However, if the ions and water are needed, they are reabsorbed.

Reabsorption May Be Active or Passive

Reabsorption of water and solutes from the tubule lumen to the extracellular fluid depends on active transport. The filtrate flowing out of Bowman's capsule into the proximal tubule has the same solute concentrations as extracellular fluid. To move solute out of the lumen, the tubule cells must therefore use active transport to create concentration or electrochemical gradients. Water osmotically follows solutes as they are reabsorbed.

FIGURE 19.8a is an overview of renal reabsorption. Active transport of Na^+ from the tubule lumen to the extracellular fluid creates a transepithelial electrical gradient in which the lumen is more negative than the ECF. Anions then follow the positively charged Na^+ out of the lumen. The removal of Na^+ and anions from lumen to ECF dilutes the luminal fluid and increases the concentration of the ECF, so water leaves the tubule by osmosis.

FIG. 19.8   **Reabsorption**

(a) Principles Governing the Tubular Reabsorption of Solutes

Some solutes and water move into and then out of epithelial cells (transcellular or epithelial transport); other solutes move through junctions between epithelial cells (the paracellular pathway). Membrane transporters are not shown in this illustration.

Filtrate is similar to interstitial fluid.

1 Na^+

2 Anions

3 H_2O

4 K^+, Ca^{2+}, urea

Tubule lumen   *Tubular epithelium*   *Extracellular fluid*

1 Na^+ is reabsorbed by active transport.

2 Electrochemical gradient drives anion reabsorption.

3 Water moves by osmosis, following solute reabsorption. Concentrations of other solutes increase as fluid volume in lumen decreases.

4 Permeable solutes are reabsorbed by diffusion through membrane transporters or by the paracellular pathway.

(b) Sodium Reabsorption in the Proximal Tubule: Active Transport

This figure shows the epithelial Na^+ channel, ENaC.

Tubule lumen   *Proximal tubule cell*   *Interstitial fluid*

Na^+ reabsorbed

$[Na^+]$ high $[Na^+]$ low $[Na^+]$ high

ENaC

Na^+

Na^+

ATP

K^+

Leak channel

1 Na^+ enters cell through various membrane proteins, moving down its electrochemical gradient.

2 Na^+ is pumped out the basolateral side of cell by the Na^+-K^+-ATPase.

(c) Sodium-Linked Reabsorption: Indirect (secondary) Active Transport

This figure shows glucose, but amino acids, other organic metabolites, and some ions such as phosphate are also absorbed by Na^+-dependent cotransport.

$[Na^+]$ high $[Na^+]$ low   Glucose and Na^+
$[glu]$ low $[glu]$ high   reabsorbed

SGLT

glu   glu $[glu]$ low

GLUT

Na^+ Na^+ $[Na^+]$ high

ATP

K^+

Apical membrane   *Basolateral membrane*

1 Na^+ moving down its electrochemical gradient uses the SGLT protein to pull glucose into the cell against its concentration gradient.

2 Glucose diffuses out the basolateral side of the cell using the GLUT protein.

3 Na^+ is pumped out by Na^+-K^+-ATPase.

The loss of volume from the lumen increases the concentration of solutes (including K^+, Ca^{2+}, and urea) left behind in the filtrate: The same amount of solute in a smaller volume equals higher solute concentration. Once luminal solute concentrations are higher than solute concentrations in the extracellular fluid, the solutes diffuse out of the lumen if the epithelium of the tubule is permeable to them.

Reabsorption involves both transepithelial transport and paracellular transport. In **transepithelial transport** (also called *transcellular transport*), substances cross the apical and basolateral membranes of the tubule epithelial cell [p. 150] to reach the interstitial fluid. In the **paracellular pathway,** substances pass through the cell-cell junction between two adjacent cells. Which route a solute takes depends on the permeability of the epithelial junctions and on the electrochemical gradient for the solute.

For solutes that move by transepithelial transport, their concentration or electrochemical gradients determine their transport mechanisms. Solutes moving down their gradient use open leak channels or facilitated diffusion carriers to cross the cell membrane. Molecules that need to be pushed against their gradient are moved by either primary or indirect (usually secondary) active transport. Sodium is directly or indirectly involved in many instances of both passive and active transport.

Active Transport of Sodium

The active reabsorption of Na^+ is the primary driving force for most renal reabsorption. As noted earlier, filtrate entering the proximal tubule is similar in ion composition to plasma, with a higher Na^+ concentration than is found in cells. Thus, Na^+ in the filtrate can enter tubule cells passively by moving down its electrochemical gradient (Fig. 19.8b). Apical movement of Na^+ uses a variety of symport and antiport transport proteins [p. 141] or open leak channels. In the proximal tubule, the *Na^+-H^+ exchanger* (NHE) plays a major role in Na^+ reabsorption, as does the apical *epithelial Na^+ channel* (ENaC, pronounced ee-knack). Once inside a tubule cell, Na^+ is actively transported out across the basolateral membrane in exchange for K^+ by the Na^+-K^+-ATPase. A basolateral K^+ leak channel prevents K^+ from accumulating in the cell. The end result is Na^+ reabsorption across the tubule epithelium.

Secondary Active Transport: Symport with Sodium

Sodium-linked secondary active transport in the nephron is responsible for the reabsorption of many substances, including glucose, amino acids, ions, and various organic metabolites. Figure 19.8c shows one example: Na^+-dependent glucose reabsorption across the proximal tubule epithelium [Fig. 5.21, p. 151]. The apical membrane contains the Na^+-*glucose cotransporter* (SGLT) that brings glucose into the cytoplasm against its concentration gradient by harnessing the energy of Na^+ moving down its electrochemical gradient. On the basolateral side of the cell, Na^+ is pumped out by the Na^+-K^+-ATPase, while glucose diffuses out with the aid of a facilitated diffusion GLUT transporter.

The same basic pattern holds for many other molecules absorbed by Na^+-dependent transport: an apical symport protein and a basolateral facilitated diffusion carrier or ion exchanger. Other molecules that are reabsorbed by similar mechanisms include amino acids, lactate, citric acid cycle intermediates such as citrate and α-ketoglutarate (αKG), and ions such as phosphate and sulfate. A few of the apical transporters use H^+ in place of Na^+.

Passive Reabsorption: Urea

The nitrogenous waste product urea has no active transporters in the proximal tubule but can move through the epithelial junctions by diffusion if there is a urea concentration gradient. Initially, urea concentrations in the filtrate and extracellular fluid are equal. However, the active transport of Na^+ and other solutes out of the proximal tubule lumen creates a urea concentration gradient by the following process.

When Na^+ and other solutes are reabsorbed from the proximal tubule, the transfer of osmotically active particles makes the extracellular fluid more concentrated than the filtrate remaining in the lumen (see Fig. 19.8a). In response to the osmotic gradient, water moves by osmosis across the epithelium. Up to this point, no urea molecules have moved out of the lumen because there has been no urea concentration gradient.

When water is reabsorbed, the concentration of urea in the lumen increases—the same amount of urea is contained in a smaller volume. Once a concentration gradient for urea exists, urea moves out of the lumen into the extracellular fluid by transport through the cells or by the paracellular pathway.

Endocytosis: Plasma Proteins

Filtration of plasma at the glomerulus normally leaves most plasma proteins in the blood, but some smaller proteins and peptides can pass through the filtration barrier. Most filtered proteins are removed from filtrate in the proximal tubule, with the result that normally only trace amounts of protein appear in urine.

Small as they are, filtered proteins are too large to be reabsorbed by carriers or through channels. Most enter proximal tubule cells by receptor-mediated endocytosis [p. 148] at the apical membrane. Once in the cells, the proteins are digested in lysosomes. The resulting amino acids are transported across the basolateral membrane and absorbed into the blood. The renal digestion of small filtered proteins is actually a significant method by which peptide signal molecules can be removed from the circulation.

Renal Transport Can Reach Saturation

Most transport in the nephron uses membrane proteins and exhibits the three characteristics of mediated transport: saturation, specificity, and competition [p. 146].

Saturation refers to the maximum rate of transport that occurs when all available carriers are occupied by (are saturated with) substrate. At substrate concentrations below the saturation point, transport rate is directly related to substrate concentration (**FIG. 19.9**). At substrate concentrations equal to or above the saturation point, transport occurs at a maximum rate. The transport rate at saturation is the **transport maximum (T_m)** [p. 146].

Glucose reabsorption in the nephron is an excellent example of the consequences of saturation. At normal plasma glucose concentrations, all glucose that enters the nephron is reabsorbed before it reaches the end of the proximal tubule. The tubule epithelium is well supplied with carriers to capture glucose as the filtrate flows past.

But what happens if blood glucose concentrations become excessive, as they do in diabetes mellitus? In that case, glucose is filtered faster than the carriers can reabsorb it. The carriers become saturated and are unable to reabsorb all the glucose that flows through the tubule. As a result, some glucose escapes reabsorption and is excreted in the urine.

Consider the following analogy. Assume that the carriers are like seats on a train at Disney World. Instead of boarding the stationary train from a stationary platform, passengers step onto a moving sidewalk that rolls them past the train. As the passengers see an open seat, they grab it. However, if more people are allowed onto the moving sidewalk than there are seats in the train, some people will not find seats. And because the sidewalk is moving people past the train toward an exit, they cannot wait for the next train. Instead, they end up being transported out the exit.

Glucose molecules entering Bowman's capsule in the filtrate are like passengers stepping onto the moving sidewalk. To be reabsorbed, each glucose molecule must bind to a transporter as the filtrate flows through the proximal tubule. If only a few glucose molecules enter the tubule at a time, each one can find a free transporter and be reabsorbed, just as a small number of people on the moving sidewalk all find seats on the train. However, if glucose molecules filter into the tubule faster than the glucose

BIOTECHNOLOGY

Artificial Kidneys

Many people with severe renal disease depend on *dialysis,* a medical procedure that either supplements or completely replaces their kidney function. Imagine trying to make a machine or develop a procedure that performs the functions of the kidney. What would it have to do? Dialysis is based on diffusion through a semipermeable membrane. Solutes and water pass from a patient's extracellular fluid across the membrane into a dialysis fluid. *Hemodialysis* routes blood from the arm past a membrane in an external dialysis machine. This technique requires attachment to the machine for 3–5 hours three days a week and is used for more severe cases of renal failure. *Peritoneal dialysis* is also called *continuous ambulatory peritoneal dialysis* (CAPD) because it takes place while the patient moves about during daily activities. In CAPD, dialysis fluid is injected into the peritoneal cavity, where it accumulates waste products from the blood for 4–6 hours before being drained out. For more information about dialysis, see the web site for the National Institute of Diabetes and Digestive and Kidney Diseases (*www.niddk.nih.gov*) and search for *dialysis*.

carriers can transport them, some glucose remains in the lumen and is excreted in the urine.

FIGURE 19.10 is a graphic representation of glucose handling by the kidney. Figure 19.10a shows that the filtration rate of glucose from plasma into Bowman's capsule is proportional to the plasma concentration of glucose. Because filtration does not exhibit saturation, the graph continues infinitely in a straight line: The filtrate glucose concentration is always equal to the plasma glucose concentration.

Figure 19.10b plots the reabsorption rate of glucose in the proximal tubule against the plasma concentration of glucose. Reabsorption exhibits a maximum transport rate (T_m) when the carriers reach saturation. Notice that normal plasma glucose concentrations are well below the saturation point.

Figure 19.10c plots the excretion rate of glucose in relation to the plasma concentration of glucose. Remember that excretion equals filtration minus reabsorption $(E = F - R)$. When plasma glucose concentrations are low enough that 100% of the filtered glucose is reabsorbed, no glucose is excreted. Once the carriers reach saturation, glucose excretion begins. The plasma concentration at which glucose first appears in the urine is called the **renal threshold** for glucose.

Figure 19.10d is a composite graph that compares filtration, reabsorption, and excretion of glucose. Recall from our earlier discussion that

$$\text{Amount excreted} = \text{amount filtered} - \text{amount reabsorbed} + \text{amount secreted}$$

FIG. 19.9 | Saturation of mediated transport

The transport rate of a substance is proportional to the plasma concentration of the substance, up to the point at which transporters become saturated. Once saturation occurs, transport rate reaches a maximum. The plasma concentration of substrate at which the transport maximum occurs is called the renal threshold.

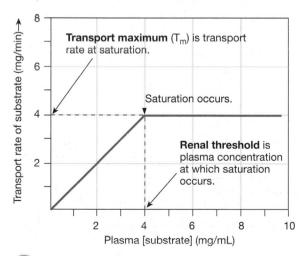

GRAPH QUESTION

What is the transport rate at the following plasma substrate concentrations: 3 mg/mL, 5 mg/mL, 8 mg/mL? At what plasma substrate concentration is the transport rate 2 mg/min?

FIG. 19.10 Glucose handling by the nephron

(a) Filtration

Filtration of glucose is proportional to the plasma concentration. Filtration does not saturate.

(b) Reabsorption

Reabsorption of glucose is proportional to plasma concentration until the transport maximum (T_m) is reached.

(c) Excretion = Filtration – Reabsorption

Glucose excretion is zero until the renal threshold is reached.

(d) Composite graph shows the relationship between filtration, reabsorption, and excretion of glucose.

For glucose, which is not secreted, the equation can be re-written as

| Glucose excreted = glucose filtered – glucose reabsorbed

Under normal conditions, all filtered glucose is reabsorbed. In other words, filtration is equal to reabsorption.

Notice in Figure 19.10d that the lines representing filtration and reabsorption are identical up to the plasma glucose concentration that equals the renal threshold. If filtration equals reabsorption, the algebraic difference between the two is zero, and there is no excretion. Once the renal threshold is reached, filtration begins to exceed reabsorption. Notice on the graph that the filtration and reabsorption lines diverge at this point. The difference between the filtration line and the reabsorption line represents the excretion rate:

$$\text{Excretion} = \underset{\text{(increasing)}}{\text{filtration}} - \underset{\text{(constant)}}{\text{reabsorption}}$$

Excretion of glucose in the urine is called **glucosuria** or **glycosuria** {*-uria,* in the urine} and usually indicates an elevated blood glucose concentration. Rarely, glucose appears in the urine even though the blood glucose concentrations are normal. This situation is due to a genetic disorder in which the nephron does not make enough carriers.

Peritubular Capillary Pressures Favor Reabsorption

The reabsorption we have just discussed refers to the movement of solutes and water from the tubule lumen to the interstitial fluid. How does that reabsorbed fluid then get into the capillary? The answer is that the hydrostatic pressure that exists along the entire length of the peritubular capillaries is less than the colloid osmotic pressure, so the net pressure gradient favors reabsorption (**FIG. 19.11**).

The peritubular capillaries have an average hydrostatic pressure of 10 mm Hg (in contrast to the glomerular capillaries, where hydrostatic pressure averages 55 mm Hg). Colloid osmotic pressure, which favors movement of fluid into the capillaries, is 30 mm Hg. As a result, the pressure gradient in peritubular capillaries is 20 mm Hg, favoring the absorption of fluid into the capillaries. Fluid that is reabsorbed passes from the peritubular capillaries to the venous circulation and returns to the heart.

FIG.19.11 Reabsorption in peritubular capillaries

Lower hydrostatic pressure in peritubular capillaries results in net reabsorption of interstitial fluid.

Efferent arteriole

Bowman's capsule

Peritubular capillaries

π 30

P_H 55

P_F 15

Glomerulus

10

Afferent arteriole

Net driving pressure causes filtration.

Proximal tubule

P_H 10

π 30

20

Pressure gradient favors reabsorption.

KEY

P_H = hydrostatic pressure

π = colloid osmotic pressure

P_F = fluid pressure

All numbers are mm Hg.

Q FIGURE QUESTION

Why is the hydrostatic pressure so much lower in the peritubular capillaries than in the glomerulus?

SECRETION

Secretion is the transfer of molecules from extracellular fluid into the lumen of the nephron (see Fig. 19.2). Secretion, like reabsorption, depends mostly on membrane transport systems. The secretion of K^+ and H^+ by the distal nephron is important in the homeostatic regulation of those ions. In addition, many organic compounds are secreted. These compounds include both metabolites produced in the body and substances brought into the body, or *xenobiotics*.

Secretion enables the nephron to enhance excretion of a substance. If a substance is filtered and not reabsorbed, it is excreted very efficiently. If, however, the substance is filtered into the tubule, not reabsorbed, *and* then more of it is secreted into the tubule from the peritubular capillaries, excretion is even more efficient.

Secretion is an active process because it requires moving substrates against their concentration gradients. Most organic compounds are secreted across the proximal tubule epithelium into the lumen by indirect active transport. Let's look at how the tubule handles the secretion of organic anions (**FIG. 19.12**).

The transporters responsible for organic solute excretion have broad specificity. For example, the **organic anion transporter** (OAT) family, shown in this figure, is able to transport a wide variety of endogenous and exogenous anions, ranging from bile salts to benzoate used as a preservative in soft drinks, salicylate from aspirin, and the artificial sweetener saccharine. Secretion of organic anions on the OAT is an example of *tertiary* active transport, where the use of energy from ATP is two steps removed from the OAT. Let's see how this works.

In the first step of the process, which is direct active transport, the proximal tubule cell uses ATP to maintain the low intracellular concentration of Na^+. In the second step, the Na^+ gradient is then used to concentrate a *dicarboxylate* inside the tubule cell, using a Na^+-*dicarboxylate cotransporter* called the NaDC. The NaDC is found on both apical and basolateral membranes in the proximal tubule.

FIG.19.12 Organic anion secretion

Proximal tubule secretion of organic anions by the organic anion transporter (OAT) is an example of tertiary active transport.

Tubule lumen | Proximal tubule cell | Interstitial fluid

Filtrate

[Na⁺] low

K^+

ATP

Na^+

NaDC ⊕ Na^+

αKG⁻ ⊕ NaDC αKG⁻

Metabolism

αKG⁻ concentrates

A⁻ αKG⁻

? OAT

OA⁻ OA⁻ OA⁻

Apical membrane

Basolateral membrane

④ Organic anions enter the lumen by facilitated diffusion.

① **Direct active transport.** The Na^+-K^+-ATPase keeps intracellular [Na^+] low.

② **Secondary indirect active transport.** The Na^+-dicarboxylate cotransporter (NaDC) concentrates a dicarboxylate inside the cell using energy stored in the [Na^+] gradient.

③ **Tertiary indirect active transport.** The basolateral organic anion transporter (OAT) concentrates organic anions (OA⁻) inside the cell, using the energy stored in the dicarboxylate gradient.

Dicarboxylates are the anion form of dicarboxylic acids, which have two carboxyl (–COOH) groups. Most of the citric acid cycle intermediates, such as citrate, oxaloacetate, and α-ketoglutarate (αKG), are dicarboxylates. Figure 19.12 shows αKG as the dicarboxylate.

The concentration of dicarboxylate inside the tubule cell drives the third step of organic anion secretion. The OAT is an indirect active transporter that uses the dicarboxylate moving out of the cell down its concentration gradient to move an organic anion against its gradient into the cell. In the final step, once the organic anion is concentrated inside the tubule cell, it can use facilitated diffusion to enter the lumen. The apical transporters have not been definitively identified but appear to be anion exchangers.

Competition Decreases Penicillin Secretion

The broad specificity of the organic anion transporters means that different substrates must compete for the transporter binding sites [p. 146]. An interesting and important example of an organic molecule secreted by the OAT is the antibiotic *penicillin*. Many people today take antibiotics for granted, but until the early decades of the twentieth century, infections were a leading cause of death.

In 1928, Alexander Fleming discovered a substance in the bread mold *Penicillium* that retarded the growth of bacteria. But the antibiotic was difficult to isolate, so it did not become available for clinical use until the late 1930s. During World War II, penicillin made a major difference in the number of deaths and amputations caused by infected wounds. The only means of producing penicillin, however, was to isolate it from bread mold, and supplies were limited.

RUNNING **PROBLEM**

Uric acid, the molecule that causes gout, is a normal product of purine metabolism. Increased uric acid production may be associated with cell and tissue breakdown, or it may occur as a result of inherited enzyme defects. Plasma *urate*, the anionic form of uric acid, filters freely into Bowman's capsule but is almost totally reabsorbed in the first part of the proximal tubule. The middle section of the proximal tubule then secretes about half of the reabsorbed urate back into the lumen. Finally, the terminal section of the proximal tubule again reabsorbs some of the urate. The end result is net secretion.

Q3: *Purines are part of which category of biomolecules? Using that information, explain why uric acid levels in the blood go up when cell breakdown increases.*

Q4: *Based on what you have learned about uric acid and urate, predict two ways a person may develop hyperuricemia.*

590 591 **608** 609 611 613

Demand for the drug was heightened by the fact that kidney tubules secrete penicillin. Renal secretion is so efficient at clearing foreign molecules from the blood that within three to four hours after a dose of penicillin has been administered, about 80% has been excreted in the urine. During the war, the drug was in such short supply that it was common procedure to collect the urine from patients being treated with penicillin so that the antibiotic could be isolated and reused.

This solution was not satisfactory, however, and so researchers looked for a way to slow penicillin secretion. They hoped to find a molecule that could compete with penicillin for the organic anion transporter responsible for secretion. That way, when presented with both drugs, the OAT carrier would bind preferentially to the competitor and secrete it, leaving penicillin behind in the blood.

A synthetic compound named *probenecid* was the answer. When probenecid is administered concurrently with penicillin, the transporter removes probenecid preferentially, prolonging the activity of penicillin in the body. Once mass-produced synthetic penicillin became available and supply was no longer a problem, the medical use of probenecid declined.

EXCRETION

Urine output is the result of all the processes that take place in the kidney. By the time fluid reaches the end of the nephron, it bears little resemblance to the filtrate that started in Bowman's capsule. Glucose, amino acids, and useful metabolites are gone, having been reabsorbed into the blood, and organic wastes are more concentrated. The concentrations of ions and water in the urine are highly variable, depending on the state of the body.

Although excretion tells us what the body is eliminating, excretion by itself cannot tell us the details of renal function. Recall that for any substance,

$$\text{Excretion} = \text{filtration} - \text{reabsorption} + \text{secretion}$$

Simply looking at the excretion rate of a substance tells us nothing about how the kidney handled that substance. The excretion rate of a substance depends on (1) the filtration rate of the substance and (2) whether the substance is reabsorbed, secreted, or both, as it passes through the tubule.

Renal handling of a substance and GFR are often of clinical interest. For example, clinicians use information about a person's glomerular filtration rate as an indicator of overall kidney function. And pharmaceutical companies developing drugs must provide the Food and Drug Administration with complete information on how the human kidney handles each new compound.

But how can investigators dealing with living humans assess filtration, reabsorption, and secretion at the level of the individual nephron? They have no way to do this directly: The kidneys are not easily accessible and the nephrons are microscopic. Scientists therefore had to develop a technique that would allow them to assess renal function using only analysis of the urine and blood. To do this, they apply the concept of clearance.

RUNNING **PROBLEM**

Michael found it amazing that a metabolic problem could lead to pain in his big toe. "How do we treat gout?" he asked. Dr. Garcia explained that the treatment includes anti-inflammatory agents, lots of water, and avoidance of alcohol, which can trigger gout attacks. "In addition, I would like to put you on a uricosuric agent, like probenecid, which will enhance renal excretion of urate," replied Dr. Garcia. "By enhancing excretion, we can reduce uric acid levels in your blood and thus provide relief." Michael agreed to try these measures.

Q5: *Urate is reabsorbed by some proximal tubule cells and secreted by others using membrane transporters, one on the apical membrane and one on the basolateral membrane. Could the same transporters be used by cells that reabsorb urate and cells that secrete it? Defend your reasoning.*

Q6: *Uricosuric agents, like urate, are organic acids. Given this fact, explain how uricosuric agents might enhance excretion of urate.*

590   591   608   **609**   611   613

Clearance Is a Noninvasive Way to Measure GFR

Clearance of a solute is the rate at which that solute disappears from the body by excretion or by metabolism [p. 12]. The general equation for clearance is:

$$\text{Clearance of } X = \frac{\text{excretion rate of } X \text{ (mg/min)}}{[X]_{\text{plasma}} \text{ (mg/mL plasma)}}$$

where clearance is mL plasma cleared of X per minute. Notice that the units for clearance are mL plasma and time. Substance X does not appear anywhere in the clearance units.

For any solute that is cleared only by renal excretion, clearance is expressed as the volume of plasma passing through the kidneys that has been totally cleared of that solute in a given period of time. Because this is such an indirect way to think of excretion (how much blood has been cleared of X rather than how much X has been excreted), clearance is often a very difficult concept to grasp.

Before we jump into the mathematical expression of clearance, let's look at an example that shows how clearance relates to kidney function. For our example, we use **inulin,** a polysaccharide isolated from the tuberous roots of a variety of plants. (Inulin is not the same as *insulin*, the protein hormone that regulates glucose metabolism.) Scientists discovered from experiments with isolated nephrons that inulin injected into the plasma filters freely into the nephron. As it passes through the kidney tubule, inulin is neither reabsorbed nor secreted. In other words, 100% of the inulin that filters into the tubule is excreted.

How does this relate to clearance? To answer this question, take a look at **FIGURE 19.13**, which assumes that 100% of a filtered volume of plasma is reabsorbed. (This is not too far off the actual value, which is more than 99%.) In Figure 19.13a, inulin has been injected so that its plasma concentration is 4 inulin molecules per 100 mL plasma. If GFR is 100 mL plasma filtered per minute, we can calculate the filtration rate, or *filtered load*, of inulin using the equation

$$\text{Filtered load of } X = [X]_{\text{plasma}} \times \text{GFR}$$
$$\text{Filtered load of inulin} = (4 \text{ inulin}/100 \text{ mL plasma}) \times 100 \text{ mL plasma filtered/min}$$
$$= 4 \text{ inulin/min filtered}$$

As the filtered inulin and the filtered plasma pass along the nephron, all the plasma is reabsorbed, but all the inulin remains in the tubule. The reabsorbed plasma contains no inulin, so we say it has been totally *cleared* of inulin. The *inulin clearance*, therefore, is 100 mL of plasma cleared/min. At the same time, the excretion rate of inulin is 4 inulin molecules excreted per minute.

What good is this information? For one thing, we can use it to calculate the glomerular filtration rate. Notice from Figure 19.13a that inulin clearance (100 mL plasma cleared/min) is equal to the GFR (100 mL plasma filtered/min). Thus, *for any substance that is freely filtered but neither reabsorbed nor secreted, its clearance is equal to GFR.*

Now let's show mathematically that inulin clearance is equal to GFR. We already know that

$$\text{Filtered load of } X = [X]_{\text{plasma}} \times \text{GFR} \qquad (1)$$

We also know that 100% of the inulin that filters into the tubule is excreted. In other words:

$$\text{Filtered load of inulin} = \text{excretion rate of inulin} \qquad (2)$$

Because of this equality, we can substitute excretion rate for filtered load in equation (1) by using algebra (if A = B and A = C, then B = C):

$$\text{Excretion rate of inulin} = [\text{inulin}]_{\text{plasma}} \times \text{GFR} \qquad (3)$$

This equation can be rearranged to read

$$\text{GFR} = \frac{\text{excretion rate of inulin}}{[\text{inulin}]_{\text{plasma}}} \qquad (4)$$

It turns out that the right side of this equation is identical to the clearance equation for inulin. Thus the general equation for the clearance of any substance X (mL plasma cleared/min) is

$$\text{Clearance of } X = \frac{\text{excretion rate of } X \text{ (mg/min)}}{[X]_{\text{plasma}} \text{ (mg/mL plasma)}} \qquad (5)$$

For inulin:

$$\text{Inulin clearance} = \frac{\text{excretion rate of inulin}}{[\text{inulin}]_{\text{plasma}}} \qquad (6)$$

FIG. 19.13 **ESSENTIALS**

Renal Clearance

These figures show the relationship between clearance and excretion. Each figure represents the events taking place in one minute. For simplicity, 100% of the filtered volume is assumed to be reabsorbed.

PHYS IN
ACTION!

KEY

= 100 mL of plasma

= 100 mL of filtrate

① Plasma concentration is 4/100 mL.

② GFR = 100 mL/min

③ 100 mL plasma is reabsorbed.

④ Clearance depends on renal handling of solute.

(a) Inulin clearance is equal to GFR.

Efferent arteriole

Filtration (100 mL/min) 4 inulin/min

Peritubular capillaries

Glomerulus

②

Afferent arteriole

① Inulin molecules

Nephron

If filtration and excretion are the same, then there is no net reabsorption or secretion, and the clearance of a substance equals the GFR.

③ 100 mL, 0% inulin reabsorbed

④ Inulin clearance = 100 mL/min

100% inulin excreted

4 inulin/min excreted

(b) Glucose clearance: Normally all glucose that filters is reabsorbed.

Filtration (100 mL/min) 4 glucose /min

②

① Glucose molecules

③ 100 mL, 100% glucose reabsorbed

No glucose excreted

④ Glucose clearance = 0 mL/min

0 glucose/min excreted

(c) Urea clearance is an example of net reabsorption. If filtration is greater than excretion, there is net reabsorption.

Filtration (100 mL/min) 4 urea/min

②

① Urea molecules

③ 100 mL, 50% of urea reabsorbed

50% of urea excreted

④ Urea clearance = 50 mL/min

2 urea/min excreted

If clearance of a substance is less than GFR, there is net reabsorption.

(d) Penicillin clearance is an example of net secretion. If excretion is greater than filtration, there is net secretion.

Filtration (100 mL/min) 4 penicillin /min

②

① Penicillin molecules

Some additional penicillin secreted.

③ 100 mL, 0 penicillin reabsorbed

More penicillin is excreted than was filtered.

④ Penicillin clearance = 150 mL/min

6 penicillin/min excreted

If clearance of a substance is greater than GFR, there is net secretion.

TABLE 19.1 | Useful Equations in Renal Physiology

- Excretion = Filtration − Reabsorption + Secretion

- Filtration rate of $X = [X]_{plasma} \times GFR$

- Excretion rate of X = urine flow × $[X]_{urine}$

- Clearance of $X = \dfrac{\text{excretion rate of } X \text{ (mg/min)}}{[X]_{plasma} \text{ (mg/mL plasma)}}$

- When $[X]_{plasma}$ = renal threshold for X, then reabsorption of X = transport maximum for X.

The right sides of equations (4) and (6) are identical, so by using algebra again, we can say that:

$$\text{GFR} = \text{inulin clearance} \qquad (7)$$

So why is this important? For one thing, you have just learned how we can measure GFR in a living human by taking only blood and urine samples. Try the example in Concept Check 12 to see if you understand the preceding discussion. TABLE 19.1 is a summary table of equations you will find useful for renal physiology.

Inulin is not practical for routine clinical applications because it does not occur naturally in the body and must be administered by continuous intravenous infusion. As a result, inulin use is restricted to research. Unfortunately, no substance that occurs naturally in the human body is handled by the kidney exactly the way inulin is handled.

In clinical settings, physicians use creatinine to estimate GFR. **Creatinine** is a breakdown product of phosphocreatine, an energy-storage compound found primarily in muscles [p. 391]. It is constantly produced by the body and need not be administered. Normally, the production and breakdown rates of phosphocreatine are relatively constant, and the plasma concentration of creatinine does not vary much.

Although creatinine is always present in the plasma and is easy to measure, it is not the perfect molecule for estimating GFR because a small amount is secreted into the urine. However, the amount secreted is small enough that, in most people, *creatinine clearance* is routinely used to estimate GFR.

CONCEPT CHECK

12. If plasma creatinine = 1.8 mg/100 mL plasma, urine creatinine = 1.5 mg/mL urine, and urine volume is 1100 mL in 24 hours, what is the creatinine clearance? What is GFR?

Clearance Helps Us Determine Renal Handling

Once we know a person's GFR, we can determine how the kidney handles any solute by measuring the solute's plasma concentration and its excretion rate. If we assume that the solute is freely filtered at the glomerulus, we know from equation (1) that

$$\text{Filtered load of } X = [X]_{plasma} \times \text{GFR}$$

RUNNING **PROBLEM**

Three weeks later, Michael was back in Dr. Garcia's office. The anti-inflammatory drugs and probenecid had eliminated the pain in his toe, but last night he had gone to the hospital with a very painful kidney stone. "We'll have to wait until the analysis comes back," said Dr. Garcia, "but I will guess that it is a uric acid stone. Did you drink as much water as I told you to?" Sheepishly, Michael admitted that he had good intentions but could never find the time at work to drink much water. "You have to drink enough water while on this drug to produce 3 liters or more of urine a day. That's more than three quarts. Otherwise, you may end up with another uric acid kidney stone." Michael remembered how painful the kidney stone was and agreed that this time he would follow instructions to the letter.

Q7: *Explain why not drinking enough water while taking uricosuric agents may cause uric acid crystals to form kidney stones in the urinary tract.*

590 591 608 609 **611** 613

By comparing the filtered load of the solute with its excretion rate, we can tell how the nephron handled that substance (Fig. 19.13). For example, if less of the substance appears in the urine than was filtered, net reabsorption occurred (excreted = filtered − reabsorbed). If more of the substance appears in the urine than was filtered, there must have been net secretion of the substance into the lumen (excreted = filtered + secreted). If the same amount of the substance is filtered and excreted, then the substance is handled like inulin—neither reabsorbed nor secreted. Let's look at some examples.

Suppose that glucose is present in the plasma at 100 mg glucose/dL plasma, and GFR is calculated from creatinine clearance to be 125 mL plasma/min. For these values, equation (1) tells us that

$$\begin{aligned}\text{Filtered load of glucose} &= (100 \text{ mg glucose}/100 \text{ mL plasma}) \\ &\quad \times 125 \text{ mL plasma/min} \\ \text{Filtered load of glucose} &= 125 \text{ mg glucose/min}\end{aligned}$$

There is no glucose in this person's urine, however: glucose excretion is zero. Because glucose was filtered at a rate of 125 mg/min but excreted at a rate of 0 mg/min, it must have been totally reabsorbed.

Clearance values can also be used to determine how the nephron handles a filtered solute. In this method, researchers calculate creatinine or inulin clearance, then compare the clearance of the solute being investigated with the creatinine or inulin clearance. If clearance of the solute is less than the inulin clearance, the solute has been reabsorbed. If the clearance of the solute is higher than the inulin clearance, additional solute has been secreted into the urine. More plasma was cleared of the solute than was filtered, so the additional solute must have been removed from the plasma by secretion.

Figure 19.13 shows filtration, excretion, and clearance of three molecules: glucose, urea, and penicillin. All solutes have the same concentration in the blood entering the glomerulus: 4 molecules/100 mL plasma. GFR is 100 mL/min, and we assume for simplicity that the entire 100 mL of plasma filtered into the tubule is reabsorbed.

For any solute, its clearance reflects how the kidney tubule handles it. For example, 100% of the glucose that filters is reabsorbed, and glucose clearance is zero (Fig. 19.13b). On the other hand, urea is partially reabsorbed: Four molecules filter, but two are reabsorbed and two are excreted (Fig. 19.13c). Consequently, urea clearance is 50 mL plasma per minute. Urea and glucose clearance are both less than the inulin clearance of 100 mL/min, which tells you that urea and glucose have been reabsorbed.

As you learned earlier, penicillin is filtered, not reabsorbed, and additional penicillin molecules are secreted from plasma in the peritubular capillaries. In Figure 19.13d, four penicillin are filtered, but six are excreted. An extra 50 mL of plasma have been cleared of penicillin in addition to the original 100 mL that were filtered. The penicillin clearance therefore is 150 mL plasma cleared per minute. Penicillin clearance is greater than the inulin clearance of 100 mL/min, which tells you that net secretion of penicillin occurs.

Note that a comparison of clearance values tells you only the *net* handling of the solute. It does not tell you if a molecule is both reabsorbed and secreted. For example, nearly all K^+ filtered is reabsorbed in the proximal tubule and loop of Henle, and then a small amount is secreted back into the tubule lumen at the distal nephron. On the basis of K^+ clearance, it appears that only reabsorption occurred.

Clearance calculations are relatively simple because all you need to know are the urine excretion rates and the plasma concentrations for any solute of interest, and both values are easily obtained. If you also know either inulin or creatinine clearance, then you can determine the renal handling of any compound.

FIG.19.14 Micturition

Micturition is a spinal reflex subject to higher brain control.

(a) Bladder at Rest

(b) Micturition

1 Stretch receptors fire.

2 Parasympathetic neurons fire. Motor neurons stop firing.

3 Smooth muscle contracts. Internal sphincter is passively pulled open. External sphincter relaxes.

MICTURITION

Once filtrate leaves the collecting ducts, it can no longer be modified, and its composition does not change. The filtrate, now called urine, flows into the renal pelvis and then down the *ureter* to the bladder with the help of rhythmic smooth muscle contractions. The bladder is a hollow organ whose walls contain well-developed layers of smooth muscle. In the bladder, urine is stored until released in the process known as urination, voiding, or more formally, **micturition** {*micturire*, to desire to urinate}.

The bladder can expand to hold a volume of about 500 mL. The neck of the bladder is continuous with the *urethra*, a single tube through which urine passes to reach the external environment. The opening between the bladder and urethra is closed by two rings of muscle called *sphincters* (**FIG. 19.14a**).

The **internal sphincter** is a continuation of the bladder wall and consists of smooth muscle. Its normal tone keeps it contracted. The **external sphincter** is a ring of skeletal muscle controlled by somatic motor neurons. Tonic stimulation from the central nervous system maintains contraction of the external sphincter except during urination.

Micturition is a simple spinal reflex that is subject to both conscious and unconscious control from higher brain centers. As the bladder fills with urine and its walls expand, stretch receptors send signals via sensory neurons to the spinal cord (Fig. 19.14b). There the information is integrated and transferred to two sets of neurons. The stimulus of a full bladder excites parasympathetic neurons leading to the smooth muscle in the bladder wall. The smooth muscle contracts, increasing the pressure on the bladder contents. Simultaneously, somatic motor neurons leading to the external sphincter are inhibited.

Contraction of the bladder occurs in a wave that pushes urine downward toward the urethra. Pressure exerted by the urine forces the internal sphincter open while the external sphincter relaxes. Urine passes into the urethra and out of the body, aided by gravity.

This simple micturition reflex occurs primarily in infants who have not yet been toilet trained. A person who has been toilet trained acquires a learned reflex that keeps the micturition reflex inhibited until she or he consciously desires to urinate. The learned reflex involves additional sensory fibers in the bladder that signal the degree of fullness. Centers in the brain stem and cerebral cortex receive that information and override the basic micturition reflex by directly inhibiting the parasympathetic fibers and by reinforcing contraction of the external sphincter. When an appropriate time to urinate arrives, those same centers remove the inhibition and facilitate the reflex by inhibiting contraction of the external sphincter.

In addition to conscious control of urination, various subconscious factors can affect the micturition reflex. "Bashful bladder" is a condition in which a person is unable to urinate in the presence of other people despite the conscious intent to do so. The sound of running water facilitates micturition and is often used to help patients urinate if the urethra is irritated from insertion of a *catheter*, a tube inserted into the bladder to drain it passively.

RUNNING **PROBLEM CONCLUSION** | Gout

In this running problem, you learned that gout, which often presents as a debilitating pain in the big toe, is a metabolic problem whose cause and treatment can be linked to kidney function. Urate handling by the kidney is a complex process because urate is both secreted and reabsorbed in different segments of the proximal tubule. Scientists have now identified three different but related transport proteins that are involved in the process: the *organic anion transporter* (OAT), which exchanges anions in an electrically neutral exchange; *urate transporter* 1 (URAT1), which is also an anion exchanger but with high specificity for urate; and *urate transporter* (UAT), an electrogenic uniport urate transporter. The arrangement of these transport proteins on the polarized cell membrane determines whether the cell reabsorbs or secretes urate.

Gout is one of the oldest known diseases and for many years was considered a "rich man's" disease caused by too much rich food and drink. Thomas Jefferson and Benjamin Franklin both suffered from gout. To learn more about its causes, symptoms, and treatments, go to the Mayo Clinic's health information pages (*www.mayoclinic.com*) and search for *gout*. Check your understanding of this running problem by comparing your answers against the information in the summary table.

Question	Facts	Integration and Analysis
Q1: *Trace the route followed by kidney stones when they are excreted.*	Kidney stones often form in the renal pelvis.	From the renal pelvis, a stone passes down the ureter, into the urinary bladder, then into the urethra and out of the body.
Q2: *Name the anion formed when uric acid dissociates.*	The suffix –*ate* is used to identify the anion of organic acids.	The anion of uric acid is urate.

—Continued next page

RUNNING **PROBLEM CONCLUSION** | *Continued*

Question	Facts	Integration and Analysis
Q3: *Purines are part of which category of biomolecules? Using that information, explain why uric acid levels in the blood go up when cell breakdown increases.*	Purines include adenine and guanine, which are components of DNA, RNA, and ATP [p. 34]. When a cell dies, nuclear DNA and other chemical components are broken down.	Degradation of the cell's DNA, RNA, and ATP increases purine production, which in turn increases uric acid production.
Q4: *Based on what you have learned about uric acid and urate, predict two ways a person may develop hyperuricemia.*	Hyperuricemia is a disturbance of mass balance. Uric acid is made from purines. Urate is filtered by the kidneys with net secretion.	Hyperuricemia results either from overproduction of uric acid or from a defect in the renal excretion of urate.
Q5: *Could the same transporters be used by cells that reabsorb urate and cells that secrete it? Defend your reasoning.*	Some transporters move substrates in one direction only but others are reversible. Assume one urate transporter brings urate into the cell and another takes it out.	You could use the same two transporters if you reverse their positions on the apical and basolateral membranes. Cells reabsorbing urate would bring it in on the apical side and move it out on the basolateral. Cells secreting urate would reverse this pattern.
Q6: *Uricosuric agents, like urate, are organic acids. With that information, explain how uricosuric agents might enhance excretion of urate.*	Mediated transport exhibits competition, in which related molecules compete for one transporter. Usually, one molecule binds preferentially and therefore inhibits transport of the second molecule [p. 146].	Uricosuric agents are organic anions, so they may compete with urate for the proximal tubule organic anion transporter. Preferential binding of the uricosuric agents would block urate access to the OAT, leaving urate in the lumen and increasing its excretion.
Q7: *Explain why not drinking enough water while taking uricosuric agents may cause uric acid stones to form in the urinary tract.*	Uric acid stones form when uric acid concentrations exceed a critical level and crystals precipitate.	If a person drinks large volumes of water, the excess water will be excreted by the kidneys. Large amounts of water dilute the urine, thereby preventing the high concentrations of uric acid needed for stone formation.

(590) (591) (608) (609) (611) (**613**)

VISUALIZE, PRACTICE, AND MASTER PHYSIOLOGY WITH MASTERING A&P
- Practice-on-the-go with Dynamic Study Modules
- Visualize important processes with Interactive Physiology and A&P Flix
- Check your understanding with Video Tutors and practice quizzes!

MasteringA&P®

CHAPTER SUMMARY

The urinary system, like the lungs, uses the principle of *mass balance* to maintain homeostasis. The components of urine are constantly changing and reflect the kidney's functions of regulating ions and water and removing wastes. One of the body's three *portal systems*—each of which includes two capillary beds—is found in the kidney. Filtration occurs in the first capillary bed and reabsorption in the second. The *pressure-flow-resistance* relationship you encountered in the cardiovascular and pulmonary systems also plays a role in glomerular filtration and urinary excretion. *Compartmentation* is illustrated by the movement of water and solutes between the internal and external environments as filtrate is modified along the nephron. Reabsorption and secretion of solutes depend on *molecular interactions* and on the *movement of molecules across membranes* of the tubule cells.

Functions of the Kidneys

1. The kidneys regulate extracellular fluid volume, blood pressure, and osmolarity; maintain ion balance; regulate pH; excrete wastes and foreign substances; and participate in endocrine pathways. (p. 590)

Anatomy of the Urinary System

iP Urinary System: Glomerular Filtration

2. The **urinary system** is composed of two kidneys, two ureters, a bladder, and a urethra. (p. 591; Fig. 19.1a)

3. Each **kidney** has about 1 million microscopic **nephrons**. In cross section, a kidney is arranged into an outer **cortex** and inner **medulla**. (p. 591; Fig. 19.1c)

4. Renal blood flow goes from **afferent arteriole** to **glomerulus** to **efferent arteriole** to **peritubular capillaries**. The **vasa recta** capillaries dip into the medulla. (p. 594; Fig. 19.1g, h, j)

5. Fluid filters from the glomerulus into **Bowman's capsule**. From there, it flows through the **proximal tubule, loop of Henle, distal tubule,** and **collecting duct,** then drains into the **renal pelvis.** **Urine** flows through the **ureter** to the **urinary bladder.** (pp. 590, 591, 594; Fig. 19.1b, c, i)

Overview of Kidney Function

6. **Filtration** is the movement of fluid from plasma into Bowman's capsule. **Reabsorption** is the movement of filtered materials from tubule to blood. **Secretion** is the movement of selected molecules from blood to tubule. (p. 594; Fig. 19.2)

7. Average urine volume is 1.5 L/day. Osmolarity varies between 50 and 1200 mOsM. (p. 595; Fig. 19.2)

8. The amount of a solute excreted equals the amount filtered minus the amount reabsorbed plus the amount secreted. (p. 596; Fig. 19.3)

Filtration

iP Urinary System: Glomerular Filtration

9. One-fifth of renal plasma flow filters into the tubule lumen. The percentage of total plasma volume that filters is called the **filtration fraction**. (p. 597; Fig. 19.4)

10. Bowman's capsule epithelium has specialized cells called **podocytes** that wrap around the glomerular capillaries and create **filtration slits**. **Mesangial cells** are associated with the glomerular capillaries. (p. 597; Fig. 19.5a, c)

11. Filtered solutes must pass first through glomerular capillary endothelium, then through a **basal lamina**, and finally through Bowman's capsule epithelium before reaching the lumen of Bowman's capsule. (p. 597; Fig. 19.5d)

12. Filtration allows most components of plasma to enter the tubule but excludes blood cells and most plasma proteins. (p. 596)

13. Hydrostatic pressure in glomerular capillaries averages 55 mm Hg, favoring filtration. Opposing filtration are colloid osmotic pressure of 30 mm Hg and hydrostatic capsule fluid pressure averaging 15 mm Hg. The net driving force is 10 mm Hg, favoring filtration. (p. 598; Fig. 19.6)

14. The **glomerular filtration rate (GFR)** is the amount of fluid that filters into Bowman's capsule per unit time. Average GFR is 125 mL/min, or 180 L/day. (p. 600)

15. Hydrostatic pressure in glomerular capillaries can be altered by changing resistance in the afferent and efferent arterioles. (p. 599; Fig. 19.6c–e)

16. Autoregulation of glomerular filtration is accomplished by a **myogenic response** of vascular smooth muscle in response to pressure changes and by **tubuloglomerular feedback.** When fluid flow through the distal tubule increases, the **macula densa** cells send a paracrine signal to the afferent arteriole, which constricts. (pp. 600, 602; Fig. 19.7c)

17. Reflex control of GFR is mediated through systemic signals, such as hormones, and through the autonomic nervous system. (p. 602)

Reabsorption

iP Urinary System: Early Filtrate Processing

18. Most reabsorption takes place in the proximal tubule. Finely regulated reabsorption takes place in the more distal segments of the nephron. (p. 602)

19. The active transport of Na^+ and other solutes creates concentration gradients for passive reabsorption of urea and other solutes. (p. 602; Fig. 19.8a)

20. Most reabsorption involves transepithelial transport, but some solutes and water are reabsorbed by the paracellular pathway. (p. 604)

21. Glucose, amino acids, ions, and various organic metabolites are reabsorbed by Na^+-linked secondary active transport. (p. 604; Fig. 19.8c)

22. Most renal transport is mediated by membrane proteins and exhibits saturation, specificity, and competition. The **transport maximum T_m** is the transport rate at saturation. (p. 604; Fig. 19.9)

23. The **renal threshold** is the plasma concentration at which a substance first appears in the urine. (p. 605; Fig. 19.9)

24. Peritubular capillaries reabsorb fluid along their entire length. (p. 606; Fig. 19.11)

Secretion

25. Secretion enhances excretion by removing solutes from the peritubular capillaries. K^+, H^+, and a variety of organic compounds are secreted. (p. 607; Fig. 19.12)

26. Molecules that compete for renal carriers slow the secretion of a molecule. (p. 608)

Excretion

27. The excretion rate of a solute depends on (1) its filtered load and (2) whether it is reabsorbed or secreted as it passes through the nephron. (p. 608)

28. **Clearance** describes how many milliliters of plasma passing through the kidneys have been totally cleared of a solute in a given period of time. (p. 609)

29. **Inulin** clearance is equal to GFR. In clinical settings, **creatinine** is used to measure GFR. (p. 611; Fig. 19.13)

30. Clearance can be used to determine how the nephron handles a solute filtered into it. (p. 611; Fig. 19.13)

Micturition

31. The external sphincter of the bladder is skeletal muscle that is tonically contracted except during urination. (p. 613; Fig. 19.14)

32. Micturition is a simple spinal reflex subject to conscious and unconscious control. (p. 613)

33. Parasympathetic neurons cause contraction of the smooth muscle in the bladder wall. Somatic motor neurons leading to the external sphincter are simultaneously inhibited. (p. 613)

REVIEW QUESTIONS

In addition to working through these questions and checking your answers on p. A-25, review the Learning Outcomes at the beginning of this chapter.

Level One Reviewing Facts and Terms

1. List and explain the significance of the five characteristics of urine that can be found by physical examination.

2. List and explain the six major kidney functions.

3. At any given time, what percentage of cardiac output goes to the kidneys?

4. List the major structures of the urinary system in their anatomical sequence, from the kidneys to the urine leaving the body. Describe the function of each structure.

5. Arrange the following structures in the order that a drop of water entering the nephron would encounter them:
   (a) afferent arteriole
   (b) Bowman's capsule
   (c) collecting duct
   (d) distal tubule
   (e) glomerulus
   (f) loop of Henle
   (g) proximal tubule
   (h) renal pelvis

6. Name the three filtration barriers that solutes must cross as they move from plasma to the lumen of Bowman's capsule. What components of blood are usually excluded by these layers?

7. What force(s) promote(s) glomerular filtration? What force(s) oppose(s) it? What is meant by the term *net driving force*?

8. What does the abbreviation GFR stand for? What is a typical numerical value for GFR in milliliters per minute? In liters per day?

9. Identify the following structures, then explain their significance in renal physiology:
   (a) juxtaglomerular apparatus
   (b) macula densa
   (c) mesangial cells
   (d) podocytes
   (e) sphincters in the bladder
   (f) renal cortex

10. In which segment of the nephron does most reabsorption take place? When a molecule or ion is reabsorbed from the lumen of the nephron, where does it go? If a solute is filtered and not reabsorbed from the tubule, where does it go?

11. Match each of the following substances with its mode(s) of transport in proximal tubule reabsorption.

(a) Na$^+$	1. simple diffusion
(b) glucose	2. primary active transport
(c) urea	3. indirect active transport
(d) plasma proteins	4. facilitated diffusion
(e) water	5. movement through open channels
	6. endocytosis
	7. paracellular movement

12. List three solutes secreted into the tubule lumen.

13. What solute that is normally present in the body is used to estimate GFR in humans?

14. What is micturition?

Level Two Reviewing Concepts

15. Map the following terms. You may add terms if you like.

• α-receptor	• afferent arteriole
• autoregulation	• basal lamina
• Bowman's capsule	• capillary blood pressure
• capsule fluid pressure	• colloid osmotic pressure
• efferent arteriole	• endothelium
• epithelium	• GFR
• glomerulus	• JG cells
• macula densa	• mesangial cell
• myogenic autoregulation	• norepinephrine
• paracrine	• plasma proteins
• podocyte	• resistance
• vasoconstriction	

16. Define, compare, and contrast the items in the following sets of terms:
   a. filtration, secretion, and excretion
   b. saturation, transport maximum, and renal threshold
   c. probenecid, creatinine, inulin, and penicillin
   d. clearance, excretion, and glomerular filtration rate

17. What are the advantages of a kidney that filters a large volume of fluid and then reabsorbs 99% of it?

18. If the afferent arteriole of a nephron constricts, what happens to GFR in that nephron? If the efferent arteriole of a nephron constricts, what happens to GFR in that nephron? Assume that no autoregulation takes place.

19. Diagram the micturition reflex. How is this reflex altered by toilet training? How do higher brain centers influence micturition?

20. Antimuscarinic drugs are the accepted treatment for an overactive bladder. Explain why they work for this condition.

Level Three Problem Solving

21. Draw a section of renal tubule epithelium showing three cells joined by cell junctions. Label the apical and basolateral membranes, the tubule lumen, and the extracellular fluid. Use the following written description of proximal tubule processes to draw a model cell.

The proximal tubule cells contain carbonic anhydrase, which promotes the conversion of CO_2 and water to carbonic acid. Carbonic acid then dissociates to H^+ and HCO_3^-. Sodium is reabsorbed by an apical Na^+-H^+ antiporter and a basolateral Na^+-K^+-ATPase. Chloride is passively reabsorbed by movement through the paracellular pathway. Bicarbonate produced in the cytoplasm leaves the cell on a basolateral Na^+-HCO_3^- symporter.

22. You have been asked to study kidney function in a new species of rodent found in the Amazonian jungle. You isolate some nephrons and expose them to inulin. The following graph shows the results of your studies. (a) How is the rodent nephron handling inulin? Is inulin filtered? Is it excreted? Is there net inulin reabsorption? Is there net secretion? (b) On the graph, accurately draw a line indicating the net reabsorption or secretion. (*Hint:* excretion = filtration − reabsorption + secretion)

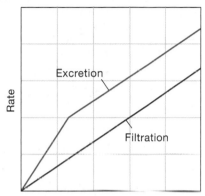

23. Read the box on hemodialysis on p. 605 and see if you can create a model system that would work for dialysis. Draw two compartments (one to represent blood and one to represent dialysis fluid) separated by a semipermeable membrane. In the blood compartment, list normal extracellular fluid solutes and their concentrations (see the table with normal values of blood components inside the back cover of this book). What will happen to the concentrations of these solutes during kidney failure? Which of these solutes should you put in the dialysis fluid, and what should their concentrations be? (*Hint:* Do you want diffusion into the dialysis fluid, out of the dialysis fluid, or no net movement?) How would you change the dialysis fluid if the patient was retaining too much water?

24. Graphing question: You are given a chemical Z and told to determine how it is handled by the kidneys of a mouse. After a series of experiments, you determine that (a) Z is freely filtered; (b) Z is not reabsorbed; (c) Z is actively secreted; and (d) the renal threshold for Z secretion is a plasma concentration of 80 mg/mL plasma, and the transport maximum is 40 mg/min. The mouse GFR is 1 mL/min. On a graph similar to the one in question 22, show how filtration, secretion, and excretion are related. One axis will be plasma concentration of Z (mg/mL) with a range of 0–140, and the other axis will show rates of kidney processes (mg/min) with a range of 0–140.

Level Four   Quantitative Problems

25. If plasma concentration of inulin = 1 mg inulin/mL plasma, plasma concentration of X = 1 mg/mL, and GFR = 125 mL/min: (a) What is the filtration rate of inulin? Of X? (b) What is the excretion rate of inulin? Of X?

26. Darlene weighs 50 kg. Assume that her total blood volume is 8% of her body weight, that her heart pumps her total blood volume once a minute, and that her renal blood flow is 25% of her cardiac output. Calculate the volume of blood that flows through Darlene's kidneys each minute.

27. Dwight was competing for a spot on the Olympic equestrian team. As his horse, Nitro, cleared a jump, the footing gave way, causing the horse to somersault, landing on Dwight and crushing him. The doctors feared kidney damage and ran several tests. Dwight's plasma creatinine level was 2 mg/100 mL. His 24-hour urine specimen had a volume of 1 L and a creatinine concentration of 20 mg/mL. A second specimen taken over the next 24 hours had the same plasma creatinine value and urine volume, but a urine creatinine concentration of 4 mg/mL. How many milligrams of creatinine are in each specimen? What is Dwight's creatinine clearance in each test? What is his GFR? Evaluate these results and comment on Dwight's kidney function.

28. You are a physiologist taking part in an archeological expedition to search for Atlantis. One of the deep-sea submersibles has come back with a mermaid, and you are taking a series of samples from her. You have determined that her GFR is 250 mL/min and that her kidneys reabsorb glucose with a transport maximum of 50 mg/min. What is her renal threshold for glucose? When her plasma concentration of glucose is 15 mg/mL, what is its glucose clearance?

29. If 140 liters of plasma are filtered in a day, and the filtration fraction is 20%:

    (a) What is the renal plasma flow?
    (b) If this person has a hematocrit of 30% [p. 518], what is the renal blood flow?
    (c) If renal blood flow is 20% of this person's cardiac output, what is her cardiac output in L/min?

Answers to Concept Checks, Figure and Graph Questions, and end-of-chapter Review Questions can be found in Appendix A [p. A-1].

20

Integrative Physiology II: Fluid and Electrolyte Balance

At a 10% loss of body fluid, the patient will show signs of confusion, distress, and hallucinations and at 20%, death will occur.

Poul Astrup, in *Salt and Water in Culture and Medicine*, 1993

FLUID AND ELECTROLYTE HOMEOSTASIS   619

LO 20.1   Map an overview of the cardiovascular and renal systems and behaviors that maintain blood volume and blood pressure homeostasis.

WATER BALANCE   620

LO 20.2   Explain how the countercurrent multiplier in the loop of Henle is the key to the regulation of urine concentration.

LO 20.3   Map in detail the reflex pathway through which vasopressin controls water reabsorption in the kidney.

LO 20.4   Diagram the cellular mechanism of action of vasopressin on principal cells.

SODIUM BALANCE AND ECF VOLUME   629

LO 20.5   Map the homeostatic responses to salt ingestion.

LO 20.6   Diagram the cellular mechanism of aldosterone action at principal cells.

LO 20.7   Map the renin-angiotensin-aldosterone system (RAAS), including all the responses initiated by ANG II and aldosterone.

LO 20.8   Describe the release of natriuretic peptides and their effects on sodium and water reabsorption.

POTASSIUM BALANCE   634

LO 20.9   Explain why the regulation of body K^+ levels is essential in maintaining a state of well-being.

BEHAVIORAL MECHANISMS IN SALT AND WATER BALANCE   636

LO 20.10   Describe behavioral mechanisms involved in salt and water balance.

INTEGRATED CONTROL OF VOLUME AND OSMOLARITY   637

LO 20.11   Diagram the appropriate homeostatic compensations for different combinations of volume and osmolarity disturbances.

ACID-BASE BALANCE   641

LO 20.12   Compare and contrast the three mechanisms by which the body copes with minute-to-minute changes in pH.

LO 20.13   Diagram the reflex pathways and cellular mechanisms involved in respiratory compensation of pH changes.

LO 20.14   Diagram the mechanisms by which the kidneys compensate for pH changes.

LO 20.15   Map the causes and compensations involved in each of the four classes of acid-base disturbances (respiratory acidosis, metabolic acidosis, respiratory alkalosis, metabolic alkalosis).

BACKGROUND BASICS

59	Body fluid compartments
32	Protein structure
41	pH and buffers
137	Membrane transport
148	Membrane recycling
150	Polarized epithelial cells
171	Second messenger systems
201	Peptide hormones
209	Posterior pituitary hormones
204	Steroid hormones
493	Control of blood pressure
579	CO_2 excretion by lungs
577	Carbonic anhydrase
126	Osmolarity and tonicity
600	Glomerular filtration rate

Each dot of a microarray represents one gene. Genes that are active show up in bright colors.

The American businesswoman in Tokyo finished her workout and stopped at the snack bar of the fitness club to ask for a sports drink. The attendant handed her a bottle labeled "Pocari Sweat®." Although the thought of drinking sweat is not very appealing, the physiological basis for the name is sound.

During exercise, the body secretes sweat, a dilute solution of water and ions, particularly Na^+, K^+, and Cl^-. To maintain homeostasis, the body must replace any substances it has lost to the external environment. For this reason, the replacement fluid a person consumes after exercise should resemble sweat.

In this chapter, we explore how humans maintain salt and water balance, also known as fluid and electrolyte balance. The homeostatic control mechanisms for fluid and electrolyte balance in the body are aimed at maintaining four parameters: fluid volume, osmolarity, the concentrations of individual ions, and pH.

FLUID AND ELECTROLYTE HOMEOSTASIS

The human body is in a state of constant flux. Over the course of a day, we ingest about 2 liters of fluid that contains 6–15 grams of NaCl. In addition, we take in varying amounts of other electrolytes, including K^+, H^+, Ca^{2+}, HCO_3^-, and phosphate ions (HPO_4^{2-}). The body's task is to maintain *mass balance* [p. 10]: What comes in must be excreted if the body does not need it.

The body has several routes for excreting ions and water. The kidneys are the primary route for water loss and for removal of many ions. Under normal conditions, small amounts of both water and ions are lost in the feces and sweat as well. In addition, the lungs lose water and help remove H^+ and HCO_3^- by excreting CO_2.

Although physiological mechanisms that maintain fluid and electrolyte balance are important, behavioral mechanisms also play an essential role. *Thirst* is critical because drinking is the only normal way to replace lost water. *Salt appetite* is a behavior that leads people and animals to seek and ingest salt (sodium chloride, NaCl).

Why are we concerned with homeostasis of these substances? Water and Na^+ are associated with extracellular fluid volume and

RUNNING **PROBLEM** | Hyponatremia

Lauren was competing in her first ironman distance triathlon, a 140.6-mile race consisting of 2.4 miles of swimming, 112 miles of cycling, and 26.2 miles of running. At mile 22 of the run, approximately 16 hours after starting the race, she collapsed. On being admitted to the medical tent, Lauren complained of nausea, a headache, and general fatigue. The medical staff noted that Lauren's face and clothing were covered in white crystals. When they weighed her and compared that value with her prerace weight recorded at registration, they realized Lauren had gained 2 kg during the race.

619 — 622 — 629 — 637 — 641 — 649

osmolarity. Disturbances in K^+ balance can cause serious problems with cardiac and muscle function by disrupting the membrane potential of excitable cells. Ca^{2+} is involved in a variety of body processes, from exocytosis and muscle contraction to bone formation and blood clotting, and H^+ and HCO_3^- are the ions whose balance determines body pH.

ECF Osmolarity Affects Cell Volume

Why is maintaining osmolarity so important to the body? The answer lies in the fact that water crosses most cell membranes freely. If the osmolarity of the extracellular fluid changes, water moves into or out of cells and changes intracellular volume. If extracellular fluid (ECF) osmolarity decreases as a result of excess water intake, water moves into the cells and they swell. If ECF osmolarity increases as a result of salt intake, water moves out of the cells and they shrink. Cell volume is so important that many cells have independent mechanisms for maintaining it.

For example, renal tubule cells in the medulla of the kidney are constantly exposed to high extracellular fluid osmolarity, yet these cells maintain normal cell volume. They do so by synthesizing organic solutes as needed to make their intracellular osmolarity match that of the medullary interstitial fluid. The organic solutes used to raise intracellular osmolarity include sugar alcohols and certain amino acids. Other cells in the body regulate their volume by changing their ionic composition.

In a few instances, changes in cell volume are believed to act as signals that initiate certain cellular responses. For example, swelling of liver cells activates protein and glycogen synthesis, and shrinkage of these cells causes protein and glycogen breakdown. In many cases, however, inappropriate changes in cell volume—either shrinking or swelling—impair cell function. The brain, encased in the rigid skull, is particularly vulnerable to damage from swelling. In general, maintenance of ECF osmolarity within a normal range is essential to maintain cell volume homeostasis.

Multiple Systems Integrate Fluid and Electrolyte Balance

The process of fluid and electrolyte balance is truly integrative because it involves the respiratory and cardiovascular systems in addition to renal and behavioral responses. Adjustments made by the lungs and cardiovascular system are primarily under neural control and can be made quite rapidly. Homeostatic compensation by the kidneys occurs more slowly because the kidneys are primarily under endocrine and neuroendocrine control. For example, small changes in blood pressure that result from increases or decreases in blood volume are quickly corrected by the cardiovascular control centers in the brain [p. 492]. If volume changes are persistent or of large magnitude, the kidneys step in to help maintain homeostasis.

FIGURE 20.1 summarizes the integrated response of the body to changes in blood volume and blood pressure. Signals from carotid and aortic baroreceptors and atrial volume receptors initiate

FIG. 20.1   Integrated responses to changes in blood volume and blood pressure

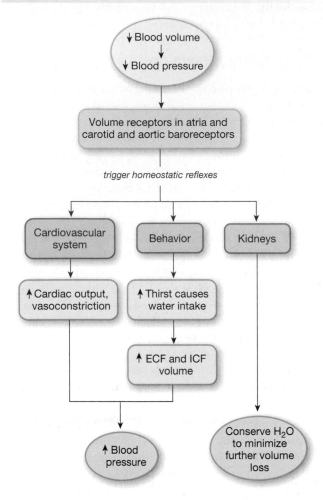

(a) Response to Decreased Blood Pressure and Volume

(b) Response to Elevated Blood Pressure and Volume

a quick neural response mediated through the cardiovascular control center and a slower response elicited from the kidneys. In addition, low blood pressure stimulates thirst. In both situations, renal function integrates with the cardiovascular system to keep blood pressure within a normal range.

Because of the overlap in their functions, a change made by one system—whether renal or cardiovascular—is likely to have consequences that affect the other. Endocrine pathways initiated by the kidneys have direct effects on the cardiovascular system, for instance, and hormones released by myocardial cells act on the kidneys. Sympathetic pathways from the cardiovascular control centers affect not only cardiac output and vasoconstriction but also glomerular filtration and hormone release by the kidneys.

In this way, the maintenance of blood pressure, blood volume, and ECF osmolarity forms a network of interwoven control pathways. This integration of function in multiple systems is one of the more difficult concepts in physiology, but it is also one of the most exciting areas of medicine and physiological research.

WATER BALANCE

Water is the most abundant molecule in the body, constituting about 50% of total body weight in females ages 17 to 39, and 60% of total body weight in males of the same age group. A 60-kg (132-lb) woman contains about 30 liters of body water, and the "standard" 70-kg man contains about 42 liters. Two-thirds of his water (about 28 liters) is inside the cells, about 3 liters are in the plasma, and the remaining 11 liters are in the interstitial fluid [Fig. 5.1, p. 124].

Daily Water Intake and Excretion Are Balanced

To maintain a constant volume of water in the body, we must take in the same amount of water that we excrete: intake must equal output. There are multiple avenues for daily water gain and loss (**FIG. 20.2**). On average, an adult ingests a little more than 2 liters of water in food and drink in a day. Normal metabolism, especially aerobic respiration (glucose + O_2 → CO_2 + H_2O) adds

FIG. 20.2 Water balance in the body

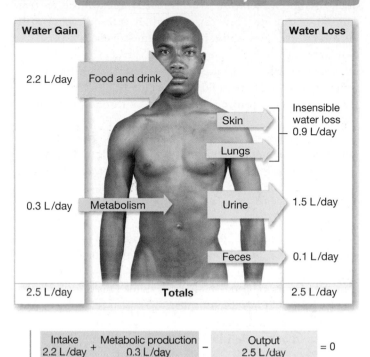

Water Gain		Water Loss
2.2 L/day	Food and drink	
	Skin	Insensible water loss 0.9 L/day
	Lungs	
0.3 L/day	Metabolism	Urine 1.5 L/day
	Feces	0.1 L/day
2.5 L/day	**Totals**	2.5 L/day

$$\underbrace{\text{Intake} \atop 2.2 \text{ L/day}} + \underbrace{\text{Metabolic production} \atop 0.3 \text{ L/day}} - \underbrace{\text{Output} \atop 2.5 \text{ L/day}} = 0$$

about 0.3 liter of water, bringing the total daily intake to approximately 2.5 liters.

Notice that the only means by which water normally enters the body from the external environment is by absorption through the digestive tract. Unlike some animals, we cannot absorb significant amounts of water directly through our skin. If fluids must be rapidly replaced or an individual is unable to eat and drink, fluid can be added directly to the plasma by means of **intravenous (IV) injection,** a medical procedure.

Most water is lost from the body in the urine, which has a daily volume of about 1.5 liters (Fig. 20.2). A small volume of water (about 100 mL) is lost in the feces. Additionally, water leaves the body through **insensible water loss.** This water loss, called *insensible* because we are not normally aware of it, occurs across the skin surface and through exhalation of humidified air. Even though the human epidermis is modified with an outer layer of keratin to reduce evaporative water loss in a terrestrial environment [p. 86], we still lose about 900 mL of water insensibly each day. Thus the 2.5 liters of water we take in are balanced by the 2.5 liters that leave the body. Only water loss in the urine can be regulated.

Although urine is normally the major route of water loss, in certain situations other routes of water loss can become significant. Excessive sweating is one example. Another way in which water is lost is through *diarrhea*, a condition that can pose a major threat to the maintenance of water balance, particularly in infants.

Pathological water loss disrupts homeostasis in two ways. Volume depletion of the extracellular compartment decreases blood pressure. If blood pressure cannot be maintained through homeostatic compensations, the tissues do not get adequate

oxygen. Also, if the fluid lost is hyposmotic to the body (as is the case in excessive sweating), the solutes left behind in the body raise osmolarity, potentially disrupting cell function.

Normally, water balance takes place automatically. Salty food makes us thirsty. Drinking 42 ounces of a soft drink means an extra trip to the bathroom. Salt and water balance is a subtle process that we are only peripherally aware of, like breathing and the beating of the heart.

Now that we have discussed *why* regulation of osmolarity is important, let's see *how* the body accomplishes that goal.

The Kidneys Conserve Water

FIGURE 20.3 summarizes the role of the kidneys in water balance. The main point to remember is that the kidneys can remove excess fluid by excreting it in the urine, but the kidneys cannot replace lost volume. Volume lost to the environment must be replaced from the environment.

The mug represents the body, and its hollow handle represents the kidneys, where body fluid filters into the nephrons. Once fluid filters, it is in the outside environment. Unless it is reabsorbed, it will go into the urine. The volume that leaves can be regulated, as indicated by the little gates at the bottom of the handle.

The normal range for fluid volume in the mug lies between the dashed line and the open top. Fluid in the mug enters the handle (equivalent to being filtered into the kidney) and cycles back into the body of the mug to maintain the mug's volume.

FIG. 20.3 The kidneys conserve volume

Kidneys cannot restore lost volume. They only conserve fluid.

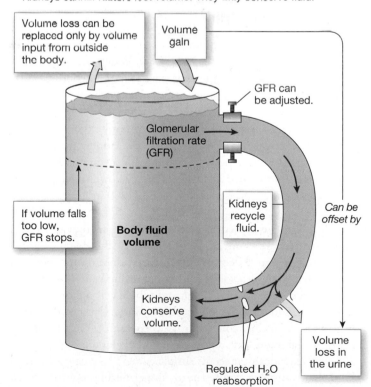

Volume loss can be replaced only by volume input from outside the body.

Volume gain

GFR can be adjusted.

Glomerular filtration rate (GFR)

If volume falls too low, GFR stops.

Body fluid volume

Kidneys recycle fluid.

Can be offset by

Kidneys conserve volume.

Regulated H_2O reabsorption

Volume loss in the urine

RUNNING **PROBLEM**

The medical staff was concerned with Lauren's large weight increase during the race. They asked her to recall what she ate and drank during the race. Lauren reported that to avoid getting dehydrated in the warm weather, she had drunk large quantities of water in addition to sports gel and sports drinks containing carbohydrates and electrolytes.

Q1: *Name the two major body fluid compartments and give the major ions in each compartment.*

Q2: *Based on Lauren's history, give a reason for why her weight increased during the race.*

619 · **622** · 629 · 637 · 641 · 649

If fluid is added to the mug and threatens to overflow, the extra fluid is allowed to drain out of the handle (comparable to excess water excreted in urine). If a small volume is lost from the mug, fluid still flows through the handle, but fluid loss from the handle is turned off to prevent additional fluid loss.

The only way to replace lost fluid is to add water from a source outside the mug. Translating this model to the body underscores the fact that *the kidneys cannot replenish lost water: All they can do is conserve it.* And as shown in the mug model, if fluid loss is severe and volume falls below the dashed line, fluid no longer flows into the handle, just as a major fall in blood volume and blood pressure shuts down renal filtration.

The Renal Medulla Creates Concentrated Urine

The concentration, or osmolarity, of urine is a measure of how much water is excreted by the kidneys. When maintenance of homeostasis requires eliminating excess water, the kidneys produce copious amounts of dilute urine with an osmolarity as low as 50 mOsM. Removal of excess water in urine is known as **diuresis** {*diourein*, to pass in urine}. Drugs that promote the excretion of urine are called *diuretics*. In contrast, if the kidneys need to conserve water, the urine becomes quite concentrated. Specialized mechanisms in the medulla of the kidney allow urine to be up to four times as concentrated as the blood (1200 mOsM versus the blood's 300 mOsM).

The kidneys control urine concentration by varying the amounts of water and Na^+ reabsorbed in the distal nephron (distal tubule and collecting duct). To produce dilute urine, the kidney must reabsorb solute without allowing water to follow by osmosis. This means that the apical tubule cell membranes and cell junctions must not be permeable to water. On the other hand, if urine is to become concentrated, the nephron must be able to reabsorb water but leave solute in the tubule lumen.

Mechanistically, it seems simple enough to create an epithelium that transports solutes but is impermeable to water (dilute

urine)—simply remove all water pores on the apical cell membrane. But mechanistically it seems much more difficult to create concentrated urine. How can the kidney reabsorb water without first reabsorbing solute? At one time, scientists speculated that water was actively transported on carriers, just as Na^+ and other ions are. However, once scientists developed micropuncture techniques for sampling fluid inside kidney tubules, they discovered that water is reabsorbed by osmosis through water pores (*aquaporins*).

The mechanism for absorbing water without solute turned out to be simple: make the collecting duct cells and interstitial fluid surrounding them more concentrated than the fluid flowing into the tubule. Then, if the tubule cells have water pores, water can be absorbed from the lumen without first reabsorbing solute.

This is indeed the situation in the kidney. Through an unusual arrangement of blood vessels and renal tubules, which we discuss later, the renal medulla maintains a high osmotic concentration in its cells and interstitial fluid. This high *medullary interstitial osmolarity* allows urine to be concentrated as it flows through the collecting duct.

Let's follow some filtered fluid through a nephron to see where these changes in osmolarity take place (**FIG. 20.4**). The renal cortex has an interstitial osmolarity of about 300 mOsM. Reabsorption in the proximal tubule is isosmotic [p. 126], and filtrate entering the loop of Henle has an osmolarity of about 300 mOsM (Fig. 20.4 **1**).

As the nephrons dip into the medulla, the interstitial osmolarity steadily increases until it reaches about 1200 mOsM where the collecting ducts empty into the renal pelvis [Fig. 19.1c, p. 592]. Fluid passing through the descending limb of the loop loses water to the interstitium. Tubule fluid at the bottom of the loop will be of the same osmolarity as in the medulla.

In the ascending limb, the permeability of the tubule wall changes. The cells in the thick portion of the ascending limb of the loop have apical surfaces (facing the tubule lumen) that are impermeable to water. These cells do transport ions out of the tubule lumen (Fig. 20.4 **2**), but in this part of the nephron, solute movement is not followed by water movement. The reabsorption of solute without water decreases the concentration of the tubule fluid. Fluid leaving the loop of Henle therefore is hyposmotic, with an osmolarity of around 100 mOsM. The loop of Henle is the primary site where the kidney creates hyposmotic fluid.

Once hyposmotic fluid leaves the loop of Henle, it passes into the distal nephron. Here the water permeability of the tubule cells is variable and under hormonal control (Fig. 20.4 **3**). When the apical membrane of distal nephron cells is not permeable to water, water cannot leave the tubule, and the filtrate remains dilute. A small amount of additional solute can be reabsorbed as fluid passes along the collecting duct, making the filtrate even more dilute. When this happens, the concentration of urine can be as low as 50 mOsM (Fig. 20.4 **4**).

On the other hand, if the body needs to conserve water by reabsorbing it, the tubule epithelium in the distal

FIG. 20.4   Osmolarity changes as fluid flow through the nephron

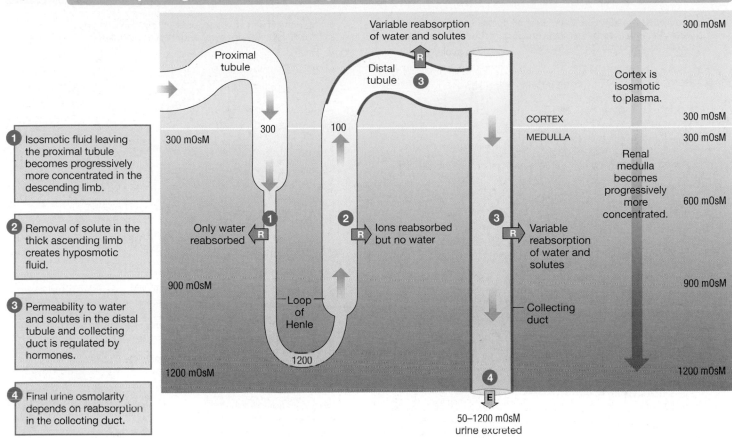

1 Isosmotic fluid leaving the proximal tubule becomes progressively more concentrated in the descending limb.

2 Removal of solute in the thick ascending limb creates hyposmotic fluid.

3 Permeability to water and solutes in the distal tubule and collecting duct is regulated by hormones.

4 Final urine osmolarity depends on reabsorption in the collecting duct.

Proximal tubule

Variable reabsorption of water and solutes

Distal tubule

3

300 mOsM

Cortex is isosmotic to plasma.

CORTEX   300 mOsM

MEDULLA   300 mOsM

Renal medulla becomes progressively more concentrated.

600 mOsM

900 mOsM

1200 mOsM

300 mOsM

300

100

900 mOsM

Only water reabsorbed

1

2

Ions reabsorbed but no water

3

Variable reabsorption of water and solutes

Loop of Henle

1200

1200 mOsM

4

E

Collecting duct

50–1200 mOsM urine excreted

---

nephron must become permeable to water. Under hormonal control, the cells insert water pores into their apical membranes. Once water can enter the epithelial cells, osmosis draws water out of the less concentrated lumen and into the

CLINICAL FOCUS

Diabetes: Osmotic Diuresis

The primary sign of diabetes mellitus is an elevated blood glucose concentration. In untreated diabetics, if blood glucose levels exceed the renal threshold for glucose reabsorption [p. 605], glucose is excreted in the urine. This may not seem like a big deal, but any additional solute that remains in the lumen forces additional water to be excreted, causing *osmotic diuresis*. Suppose, for example, that the nephrons must excrete 300 milliosmoles of NaCl. If the urine is maximally concentrated at 1200 mOsM, the NaCl is excreted in a volume of 0.25 L. However, if the NaCl is joined by 300 milliosmoles of glucose that must be excreted, the volume of urine doubles, to 0.5 L. Osmotic diuresis in untreated diabetics (primarily type 1) causes *polyuria* (excessive urination) and *polydipsia* (excessive thirst) {*dipsios*, thirsty} as a result of dehydration and high plasma osmolarity.

more concentrated interstitial fluid. At maximal water permeability, removal of water from the tubule leaves behind concentrated urine with an osmolarity that can be as high as 1200 mOsM (Fig. 20.4 **4**).

Water reabsorption in the kidneys conserves water and can decrease body osmolarity to some degree when coupled with excretion of solute in the urine. But remember that the kidney's homeostatic mechanisms can do nothing to restore lost fluid volume. Only the ingestion or infusion of water can replace water that has been lost.

Vasopressin Controls Water Reabsorption

How do the distal tubule and collecting duct cells alter their permeability to water? The process involves adding or removing water pores in the apical membrane under the direction of the posterior pituitary hormone **vasopressin** [p. 209]. In most mammals, the nine-amino-acid peptide contains the amino acid arginine, so vasopressin is called *arginine vasopressin* or *AVP*. Because vasopressin causes the body to retain water, it is also known as *antidiuretic hormone (ADH)*.

When vasopressin acts on target cells, the collecting duct epithelium becomes permeable to water, allowing water to move out of the lumen (**FIG. 20.5a**). The water moves by osmosis because osmolarity of tubule cells and the medullary interstitial

FIG. 20.5 Vasopressin makes the collecting duct epithelium permeable to water

(a) With maximal vasopressin, the collecting duct is freely permeable to water. Water leaves by osmosis and is carried away by the vasa recta capillaries. Urine is concentrated.

(b) In the absence of vasopressin, the collecting duct is impermeable to water and the urine is dilute.

CORTEX
MEDULLA
300 mOsM

300 mOsM · 300 mOsM
100 mOsM
H_2O
400 mOsM
H_2O
500 mOsM
600 mOsM · 600 mOsM · 600 mOsM
H_2O
700 mOsM
Vasa recta · Vasa recta
800 mOsM
900 mOsM · 900 mOsM · 900 mOsM
H_2O
1000 mOsM
1200 mOsM · 1200 mOsM
H_2O
1100 mOsM
Collecting duct
1200 mOsM

Urine = 1200 mOsM

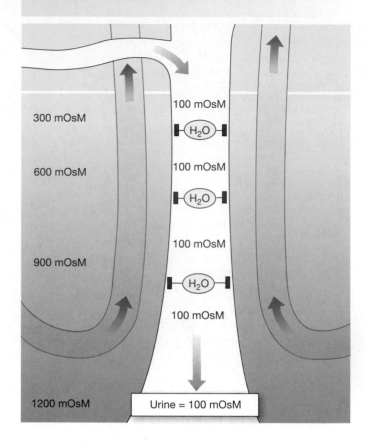

300 mOsM
100 mOsM
H_2O
100 mOsM
600 mOsM
H_2O
100 mOsM
900 mOsM
H_2O
100 mOsM
1200 mOsM

Urine = 100 mOsM

(c) Vasopressin causes insertion of water pores into the apical membrane.

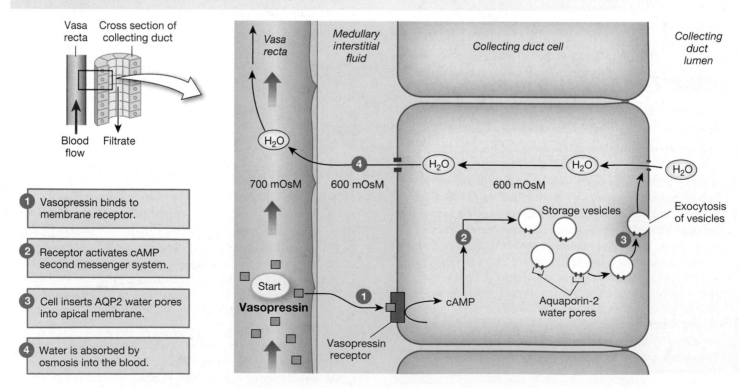

Vasa recta   Cross section of collecting duct

Blood flow   Filtrate

Vasa recta | Medullary interstitial fluid | Collecting duct cell | Collecting duct lumen

H_2O
700 mOsM   600 mOsM   600 mOsM

Storage vesicles
Exocytosis of vesicles
cAMP
Aquaporin-2 water pores
Start
Vasopressin
Vasopressin receptor

1 Vasopressin binds to membrane receptor.

2 Receptor activates cAMP second messenger system.

3 Cell inserts AQP2 water pores into apical membrane.

4 Water is absorbed by osmosis into the blood.

fluid is higher than osmolarity of fluid in the tubule. In the absence of vasopressin, the collecting duct is impermeable to water (Fig. 20.5b). Although a concentration gradient is present across the epithelium, water remains in the tubule, producing dilute urine.

The water permeability of the collecting duct is not an all-or-none phenomenon, as the previous paragraph might suggest. Permeability is variable, depending on how much vasopressin is present. The graded effect of vasopressin allows the body to match urine concentration closely to the body's needs: the more vasopressin is present, the more water is reabsorbed.

One point that is sometimes difficult to remember is that this is not a static system, where the filtrate sits passively in the lumen waiting for solutes and water to be reabsorbed. The collecting duct, like other segments of the nephron, is a flow-through system. If the apical membrane has low water permeability, most of the water in filtrate will pass through the tubule unabsorbed and end up in the urine.

Vasopressin and Aquaporins Most membranes in the body are freely permeable to water. What makes the cells of the distal nephron different? The answer lies with the *water pores* found in these cells. Water pores are **aquaporins,** a family of membrane channels with at least 10 different isoforms that occur in mammalian tissues. The kidney has multiple isoforms of aquaporins, including *aquaporin-2* (AQP2), the water channel regulated by vasopressin.

AQP2 in a collecting duct cell may be found in two locations: on the apical membrane facing the tubule lumen and in the membrane of cytoplasmic storage vesicles (Fig. 20.5c). (Two other isoforms of aquaporins are present in the basolateral membrane, but they are not regulated by vasopressin.) When vasopressin levels and, consequently, collecting duct water permeability are low, the collecting duct cell has few water pores in its apical membrane and stores its AQP2 water pores in cytoplasmic storage vesicles.

When vasopressin arrives at the collecting duct, it binds to its *V2 receptors* on the basolateral side of the cell (step 1 in Fig. 20.5c). Binding activates a G-protein/cAMP second messenger system [p. 173]. Subsequent phosphorylation of intracellular proteins causes the AQP2 vesicles to move to the apical membrane and fuse with it. Exocytosis inserts the AQP2 water pores into the apical membrane. Now the cell is permeable to water. This process, in which parts of the cell membrane are alternately added by exocytosis and withdrawn by endocytosis, is known as **membrane recycling** [Fig. 5.19, p. 149].

Blood Volume and Osmolarity Activate Osmoreceptors

What stimuli control vasopressin secretion? There are three: plasma osmolarity, blood volume, and blood pressure (**FIG. 20.6**). The most potent stimulus for vasopressin release is an increase in plasma osmolarity. Osmolarity is monitored by **osmoreceptors,** stretch-sensitive neurons that increase their firing rate as osmolarity increases. Our current model indicates that when the osmoreceptors shrink, cation channels linked to actin filaments open, depolarizing the cell.

The primary osmoreceptors for vasopressin release are in the hypothalamus. When plasma osmolarity is below the threshold value of 280 mOsM, the osmoreceptors do not fire, and vasopressin release from the pituitary ceases (Fig. 20.6b). If plasma osmolarity rises above 280 mOsM, the osmoreceptors shrink and fire to stimulate release of vasopressin.

Decreases in blood pressure and blood volume are less powerful stimuli for vasopressin release. The primary receptors for decreased volume are stretch-sensitive receptors in the atria. Blood pressure is monitored by the same carotid and aortic baroreceptors that initiate cardiovascular responses [p. 493]. When blood pressure or blood volume is low, these receptors signal the hypothalamus to secrete vasopressin and conserve fluid.

In adults, vasopressin secretion also shows a circadian rhythm, with increased secretion during the overnight hours. As a result of this increase, less urine is produced overnight than during the day, and the first urine excreted in the morning is more concentrated. One theory for the cause of bed-wetting, or **nocturnal enuresis,** in children is that these children have a developmental delay in the normal pattern of increased vasopressin secretion at night. With less vasopressin, the children's urine output stays elevated, causing the bladder to fill to its maximum capacity and empty spontaneously during sleep. Many of these children can be successfully treated with a nasal spray of *desmopressin*, a vasopressin derivative, administered at bedtime.

CONCEPT CHECK

1. Does the apical membrane of a collecting duct cell have more water pores when vasopressin is present or when it is absent?

2. People who inherit vasopressin V2 receptor deficiency will have urine that is dilute or concentrated?

CONCEPT CHECK

3. A scientist monitoring the activity of osmoreceptors notices that infusion of hyperosmotic saline (NaCl) causes increased firing of the osmoreceptors. Infusion of hyperosmotic urea (a penetrating solute) [p. 128] had no effect on the firing rate. If osmoreceptors fire only when cell volume decreases, explain why hyperosmotic urea did not affect them.

4. If vasopressin increases water reabsorption by the nephron, would vasopressin secretion be increased or decreased with dehydration?

5. Experiments suggest that there are peripheral osmoreceptors in the lumen of the upper digestive tract and in the hepatic portal vein [Fig. 14.1, p. 438]. What is the adaptive significance of osmoreceptors in these locations?

FIG. 20.6 **ESSENTIALS**

Vasopressin

High osmolarity or low blood pressure cause vasopressin release.

(a) Control of Vasopressin Secretion

Decreased blood pressure → Carotid and aortic baroreceptors → Sensory neuron to hypothalamus

Decreased atrial stretch due to low blood volume → Atrial stretch receptor → Sensory neuron to hypothalamus

Osmolarity greater than 280 mOsM → Hypothalamic osmoreceptors → Interneurons to hypothalamus

↓

Hypothalamic neurons that synthesize vasopressin

Vasopressin (AVP)
Phe — Gln
Tyr — Asp
Cys ·· Cys
Pro — Arg — Gly

↓

Vasopressin (released from posterior pituitary)

↓

Collecting duct epithelium

↓

Insertion of water pores in apical membrane

↓

Increased water reabsorption to conserve water

Kidneys

KEY

- ⬭ Stimulus
- ▭ Sensor
- ▭ Input signal
- ◆ Integrating center
- ▭ Output signal
- ▭ Target
- ▭ Tissue response
- ⬭ Systemic response

HYPOTHALAMUS

1. AVP is made and packaged in cell body of neuron.
2. Vesicles are transported down the cell.
3. Vesicles containing AVP are stored in posterior pituitary.

Posterior pituitary
Vein

4. AVP is released into blood.

ARGININE VASOPRESSIN (AVP), ANTIDIURETIC HORMONE (ADH)

Origin	Hypothalamic neurons. Released from posterior pituitary
Chemical Nature	9-amino acid peptide
Transport in the Circulation	Dissolved in plasma
Half-Life	15 min
Factors Affecting Release	↑ Osmolarity (hypothalamic osmoreceptors) ↓ Blood pressure or volume (carotid, aortic, atrial receptors)
Target Cells or Tissues	Renal collecting duct
Receptor/Second Messenger	V2 receptor/cAMP
Tissue Action	Increases renal water reabsorption
Action at Cellular-Molecular Level	Inserts AQP water pores in apical membrane

(b) The Effect of Plasma Osmolarity on Vasopressin Secretion

Plasma vasopressin (picomol/L) vs. Plasma osmolarity (mOsM)

FIGURE QUESTIONS

1. What is the threshold osmolarity for vasopressin release?
2. What signal in the AVP neuron triggers exocytosis of AVP-containing vesicles?

The Loop of Henle Is a Countercurrent Multiplier

Vasopressin is the signal for water reabsorption out of the nephron tubule, but the key to the kidney's ability to produce concentrated urine is the high osmolarity of the medullary *interstitium* (interstitial fluid compartment of the kidney). Without it, there would be no concentration gradient for osmotic movement of water out of the collecting duct. What creates this high ECF osmolarity? And why isn't the interstitial fluid osmolarity reduced as water is reabsorbed from the collecting duct and descending limb of the loop of Henle (see Fig. 20.4)? The answers to these questions lie in the anatomical arrangement of the loop of Henle and its associated blood vessels, the vasa recta. Together, these structures form a *countercurrent exchange system.*

Countercurrent Exchange Systems **Countercurrent exchange systems** require arterial and venous blood vessels that pass very close to each other, with their fluid flow moving in opposite directions (the name *countercurrent* reflects the fact that the two flows run *counter to* each other). This anatomical arrangement allows the passive transfer of heat or molecules from one vessel to the other. Because the countercurrent heat exchanger is easier to understand, we first examine how it works and then apply the same principle to the kidney.

The countercurrent heat exchanger in mammals and birds evolved to reduce heat loss from flippers, tails, and other limbs that are poorly insulated and have a high surface-area-to-volume ratio. Without a heat exchanger, warm blood flowing from the body core into the limb would easily lose heat to the surrounding environment (**FIG. 20.7a**). With a countercurrent heat exchanger, warm arterial blood entering the limb transfers its heat to cooler venous blood flowing from the tip of the limb back into the body (Fig. 20.7b). This arrangement reduces the amount of heat lost to the external environment.

The countercurrent exchange system of the kidney works on the same principle, except that it transfers water and solutes instead of heat. However, because the kidney forms a closed system, the solutes are not lost to the environment. Instead, the solutes concentrate in the interstitium. This process is aided by active transport of solutes out of the ascending limb of the loop of Henle, which makes the ECF osmolarity even greater. A countercurrent exchange system in which exchange is enhanced by active transport of solutes is called a **countercurrent multiplier.**

The Renal Countercurrent Multiplier An overview of the countercurrent multiplier system in the renal medulla is shown in Figure 20.7c. The system has two components: loops of Henle that leave the cortex, dip down into the more concentrated environment of the medulla, then ascend into the cortex again, and the peritubular capillaries known as the **vasa recta.** These capillaries, like the loop of Henle, dip down into the medulla and then go back up to the cortex, also forming hairpin loops that act as a countercurrent exchanger.

Although textbooks traditionally show a single nephron with a single loop of capillary (as we do in Fig. 20.7c), each kidney has thousands of collecting ducts and loops of Henle packed between thousands of vasa recta capillaries, blurring the direct association between a nephron and its vascular supply. Functionally, blood flow in the vasa recta moves in the opposite direction from filtrate flow in the loops of Henle, as shown in Figure 20.7c.

Let's follow some fluid as it moves through the loop. Isosmotic filtrate from the proximal tubule first flows into the descending limb of the loop of Henle. The descending limb is permeable to water but does not transport ions. As the loop dips into the medulla, water moves by osmosis from the descending limb into the progressively more concentrated interstitial fluid, leaving solutes behind in the tubule lumen.

The filtrate becomes progressively more concentrated as it moves deeper into the medulla. At the tips of the longest loops of Henle, the filtrate reaches a concentration of 1200 mOsM. Filtrate in shorter loops (which do not extend into the most concentrated regions of the medulla) does not reach such a high concentration.

When the filtrate rounds the hairpin turn at the tip of the loop and enters the ascending limb, the properties of the tubule epithelium change. The tubule epithelium in this segment of the nephron is impermeable to water while actively transporting Na^+, K^+, and Cl^- out of the tubule into the interstitial fluid. The loss of solute from the tubule lumen causes the filtrate osmolarity to decrease steadily, from 1200 mOsM at the bottom of the loop to 100 mOsM at the point where the ascending limb leaves the medulla and enters the cortex. The net result of the countercurrent multiplier in the kidney is to produce hyperosmotic interstitial fluid in the medulla and hyposmotic filtrate leaving the loop of Henle.

Normally, about 25% of all Na^+ and K^+ reabsorption takes place in the ascending limb of the loop. Some transporters responsible for active ion reabsorption in the thick portion of the ascending limb are shown in Figure 20.7d. The *NKCC symporter* uses energy stored in the Na^+ concentration gradient to transport Na^+, K^+, and 2 Cl^- from the lumen into the epithelial cells of the ascending limb. The Na^+-K^+-ATPase removes Na^+ from the cells on the basolateral side of the epithelium, while K^+ and Cl^- leave the cells together on a cotransport protein or through open channels. NKCC-mediated transport can be inhibited by drugs known as "loop diuretics," such as *furosemide* (Lasix).

CONCEPT CHECK

6. Explain why patients taking a loop diuretic that inhibits solute reabsorption excrete greater-than-normal volumes of urine.

7. Loop diuretics that inhibit the NKCC symporter are sometimes called "potassium-wasting" diuretics. Explain why people who are on loop diuretics must increase their dietary K^+ intake.

The Vasa Recta Removes Water It is easy to see how transport of solute out of the ascending limb of the loop of Henle dilutes the filtrate and helps concentrate the interstitial fluid in the medulla. Still, why doesn't the water leaving the descending

FIG. 20.7 Countercurrent mechanisms

A countercurrent heat exchanger

(a) If blood vessels are not close to each other, heat is dissipated to the external environment.

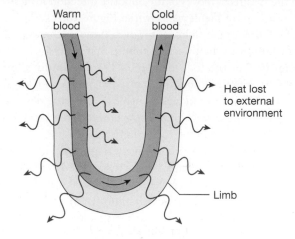

Warm blood   Cold blood

Heat lost to external environment

Limb

(b) Countercurrent heat exchanger allows warm blood entering the limb to transfer heat directly to blood flowing back into the body.

Warm blood   Warm blood

(c) Countercurrent exchange in the vasa recta

Filtrate entering the descending limb becomes progressively more concentrated as it loses water.

Blood in the vasa recta removes water leaving the loop of Henle.

The ascending limb pumps out Na⁺, K⁺, and Cl⁻, and filtrate becomes hyposmotic.

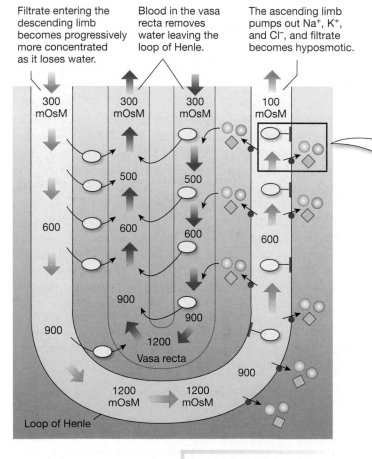

300 mOsM   300 mOsM   300 mOsM   100 mOsM

500   500

600   600   600

900   900

900   1200

Vasa recta

900

1200 mOsM → 1200 mOsM

Loop of Henle

(d) The apical surface of the ascending limb is not permeable to water. Active reabsorption of ions in this region creates a dilute filtrate in the lumen.

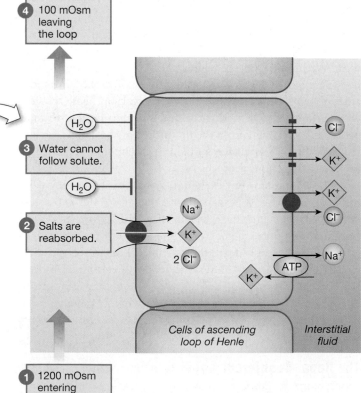

4 100 mOsm leaving the loop

3 Water cannot follow solute.

H_2O

H_2O

2 Salts are reabsorbed.

Na^+
K^+
$2\ Cl^-$

Cl^-
K^+
K^+
Cl^-
Na^+
ATP
K^+

Cells of ascending loop of Henle   *Interstitial fluid*

1 1200 mOsm entering ascending loop of Henle

KEY

H_2O = ⬭ K^+ = ◆

Cl^- = ◯ Na^+ = ◯

limb of the loop (see Fig. 20.7c) *dilute* the interstitial fluid of the medulla? The answer lies in the close anatomical association of the loop of Henle and the peritubular capillaries of the vasa recta, which functions as a countercurrent exchanger.

Water or solutes that leave the tubule move into the vasa recta if an osmotic or concentration gradient exists between the medullary interstitium and the blood in the vasa recta. For example, assume that at the point at which the vasa recta enters the medulla, the blood in the vasa recta is 300 mOsM, isosmotic with the cortex. As the blood flows deeper into the medulla, it loses water and picks up solutes transported out of the ascending limb of the loop of Henle, carrying these solutes farther into the medulla. By the time the blood reaches the bottom of the vasa recta loop, it has a high osmolarity, similar to that of the surrounding interstitial fluid (1200 mOsM).

Then, as blood in the vasa recta flows back toward the cortex, the high plasma osmolarity attracts the water that is being lost from the descending limb, as Figure 20.7c shows. The movement of this water into the vasa recta decreases the osmolarity of the blood while simultaneously preventing the water from diluting the concentrated medullary interstitial fluid.

The end result of this arrangement is that blood flowing through the vasa recta removes the water reabsorbed from the loop of Henle. Without the vasa recta, water moving out of the descending limb of the loop of Henle would eventually dilute the medullary interstitium. The vasa recta thus plays an important part in keeping the medullary solute concentration high.

Urea Increases the Osmolarity of the Medullary Interstitium

The high solute concentration in the medullary interstitium is only partly due to NaCl. Nearly half the solute in this compartment is urea. Where does this urea come from? For many years scientists thought urea crossed cell membranes only by passive transport. However, in recent years, researchers have learned that membrane transporters for urea are present in the collecting duct and loops of Henle. One family of transporters consists of facilitated diffusion carriers, and the other family has Na^+-dependent secondary active transporters. These urea transporters apparently help concentrate urea in the medullary interstitium, where it contributes to the high interstitial osmolarity.

SODIUM BALANCE AND ECF VOLUME

With an average American diet, we ingest a lot of NaCl—about 9 grams per day. This is about 2 teaspoons of salt, or 155 milliosmoles of Na^+ and 155 milliosmoles of Cl^-. Let's see what would happen to our bodies if the kidneys could not get rid of this Na^+.

Our normal plasma Na^+ concentration, measured from a venous blood sample, is 135–145 milliosmoles Na^+ per liter of plasma. Because Na^+ distributes freely between plasma and interstitial fluid, this value also represents our ECF Na^+ concentration. Clinically, it is simple to find ECF values for ions by drawing a blood sample and analyzing the plasma portion.

If we add NaCl to the body to increase the ECF concentration to 155 milliosmoles Na^+/L, how much water would we have

to add to keep the ECF Na^+ concentration at 140 mOsM? One form of an equation asking this question is

$$155\ \text{mosmol}/x\ \text{liters} = 140\ \text{mosmol}/\text{liter}$$
$$x = 1.1\ \text{liters}$$

We would have to add 0.1 liter of water for each liter of ECF volume to compensate for the addition of the Na^+. If we assume normal ECF volume is 14 liters, we would have to add 1.4 L—a 10% gain! Imagine what that volume increase would do to blood pressure.

Suppose, however, that instead of adding water to keep plasma concentrations constant, we add the NaCl but don't drink any water. What happens to osmolarity now? If we assume that normal total body osmolarity is 300 mOsM and that the volume of fluid in the body is 42 L, the addition of 155 milliosmoles of Na^+ and 155 milliosmoles of Cl^- would increase total body osmolarity to 307 mOsM*—a substantial increase. In addition, because NaCl is a nonpenetrating solute, it would stay in the ECF. Higher osmolarity in the ECF would draw water from the cells, shrinking them and disrupting normal cell function.

Fortunately, our homeostatic mechanisms usually maintain mass balance: Anything extra that comes into the body is excreted. **FIGURE 20.8** shows a generalized homeostatic pathway for sodium balance in response to salt ingestion. Here's how it works.

The addition of NaCl to the body raises osmolarity. This stimulus triggers two responses: vasopressin secretion and thirst. Vasopressin release causes the kidneys to conserve water (by reabsorbing water from the filtrate) and concentrate the urine. Thirst

RUNNING **PROBLEM**

The medical staff analyzed Lauren's blood for electrolyte concentrations. Her serum Na^+ concentration was 124 mEq/L. The normal range is 135–145 mEq/L. Lauren's diagnosis was **hyponatremia** {*hypo-*, below + *natri-*, sodium + *-emia*, blood}, defined as a serum Na^+ concentration below 135 mEq/L. Hyponatremia induced by the consumption of large quantities of low-sodium or sodium-free fluid, which is what happened in Lauren's case, is sometimes called *dilutional hyponatremia*.

Q3: *Which body fluid compartment is being diluted in dilutional hyponatremia?*

Q4: *One way to estimate body osmolarity is to double the plasma Na^+ concentration. Estimate Lauren's osmolarity and explain what effect the dilutional hyponatremia has on her cells.*

Q5: *In dilutional hyponatremia, the medical personnel are most concerned about which organ or tissue?*

619   622   **629**   637   641   649

---

*(155 mosmol Na^+ + 155 mosmol Cl^-)/42 L = 7.4 mosmol/L added; 300 mosmol/L initial concentration increased by 7.4 mosmol/L = 307 mOsM final

FIG. 20.8 Homeostatic responses to salt ingestion

Q **FIGURE QUESTION**
Map the cardiovascular reflex pathway represented by the ✱.

the kidney is regulated by the steroid hormone **aldosterone**: the more aldosterone, the more Na$^+$ reabsorption. Because one target of aldosterone is increased activity of the Na$^+$-K$^+$-ATPase, aldosterone also causes K$^+$ secretion (**FIG. 20.9**).

Aldosterone is a steroid hormone synthesized in the adrenal cortex, the outer portion of the adrenal gland that sits atop each kidney [p. 204]. Like other steroid hormones, aldosterone is secreted into the blood and transported on a protein carrier to its target.

The primary site of aldosterone action is the last third of the distal tubule and the portion of the collecting duct that runs through the kidney cortex (the *cortical collecting duct*). The primary target of aldosterone is **principal cells (P cells)** (Fig. 20.9b), the main cell type found in the distal nephron epithelium. Principal cells are arranged much like other polarized transporting epithelial cells, with Na$^+$-K$^+$-ATPase pumps on the basolateral membrane, and various channels and transporters on the apical membrane [p. 79]. In principal cells, the apical membranes contain leak channels for Na$^+$ (called ENaC, for **e**pithelial **Na**$^+$ **c**hannel) and for K$^+$ (called ROMK, for **r**enal **o**uter **m**edulla **K**$^+$ channel).

Aldosterone enters P cells by simple diffusion. Once inside, it combines with a cytoplasmic receptor (Fig. 20.9b ❶). In the early response phase, apical Na$^+$ and K$^+$ channels increase their open time under the influence of an as-yet-unidentified signal molecule. As intracellular Na$^+$ levels rise, the Na$^+$-K$^+$-ATPase pump speeds up, transporting cytoplasmic Na$^+$ into the ECF and bringing K$^+$ from the ECF into the P cell. The net result is a rapid increase in Na$^+$ reabsorption and K$^+$ secretion that does not require the synthesis of new channel or ATPase proteins. In the slower phase of aldosterone action, newly synthesized channels and pumps are inserted into epithelial cell membranes (Fig. 20.9b).

Note that Na$^+$ and water reabsorption are separately regulated in the distal nephron. Water does not automatically follow Na$^+$ reabsorption: Vasopressin must be present to make the distal nephron epithelium permeable to water. In contrast, Na$^+$ reabsorption in the proximal tubule is automatically followed by water reabsorption because the proximal tubule epithelium is always freely permeable to water.

prompts us to drink water or other fluids. The increased fluid intake decreases osmolarity, but the combination of salt and water intake increases both ECF volume and blood pressure. These increases then trigger another series of control pathways, which bring ECF volume, blood pressure, and total-body osmolarity back into the normal range by excreting extra salt and water.

The kidneys are responsible for most Na$^+$ excretion, and normally only a small amount of Na$^+$ leaves the body in feces and perspiration. However, in situations such as vomiting, diarrhea, and heavy sweating, we may lose significant amounts of Na$^+$ and Cl$^-$ through nonrenal routes.

Although we speak of ingesting and losing salt (NaCl), only renal Na$^+$ absorption is regulated. And actually, the stimuli that set the Na$^+$ balance pathway in motion are more closely tied to blood volume and blood pressure than to Na$^+$ levels. Chloride movement usually follows Na$^+$ movement, either indirectly via the electrochemical gradient created by Na$^+$ transport or directly via membrane transporters such as the NKCC transporter of the loop of Henle or the Na$^+$-Cl$^-$ symporter of the distal tubule.

Aldosterone Controls Sodium Balance

The regulation of blood Na$^+$ levels takes place through one of the most complicated endocrine pathways of the body. The reabsorption of Na$^+$ in the distal tubules and collecting ducts of

CONCEPT CHECK

8. In Figure 20.9b, what forces cause Na$^+$ and K$^+$ to cross the apical membrane?

9. If a person experiences hyperkalemia, what happens to resting membrane potential and the excitability of neurons and the myocardium?

10. Laboratory values for ions may be reported as mg/L, mmol/L, or mEq/L. If normal plasma Na$^+$ is 140 mmol/L, what is that concentration expressed as mEq/L? [Fig. 2.7, p. 42].

Low Blood Pressure Stimulates Aldosterone Secretion

What controls physiological aldosterone secretion from the adrenal cortex? There are two primary stimuli: increased extracellular K$^+$ concentration and decreased blood pressure (Fig. 20.9a).

FIG. 20.9 **ESSENTIALS**

Aldosterone

(a) The primary action of aldosterone is renal sodium reabsorption.

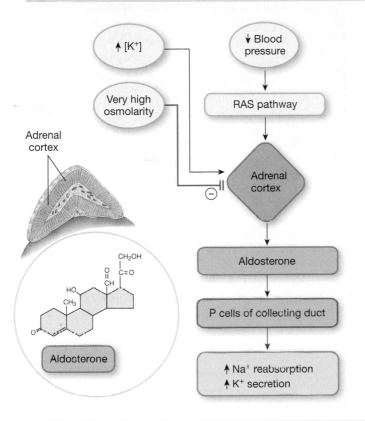

ALDOSTERONE	
Origin	Adrenal cortex
Chemical Nature	Steroid
Biosynthesis	Made on demand
Transport in the Circulation	50–70% bound to plasma protein
Half-Life	15 min
Factors Affecting Release	↓ Blood pressure (via renin) ↑ K^+ (hyperkalemia) Natriuretic poptides inhibit release
Target Cells or Tissues	Renal collecting duct—principal cells
Receptor	Cytosolic mineralocorticoid (MR) receptor
Tissue Action	Increases Na^+ reabsorption and K^+ secretion
Action at Cellular-Molecular Level	Synthesis of new ion channels (ENaC and ROMK) and pumps (Na^+-K^+-ATPase); increased activity of oxisting channels and pumps.

(b) Aldosterone acts on principal cells.

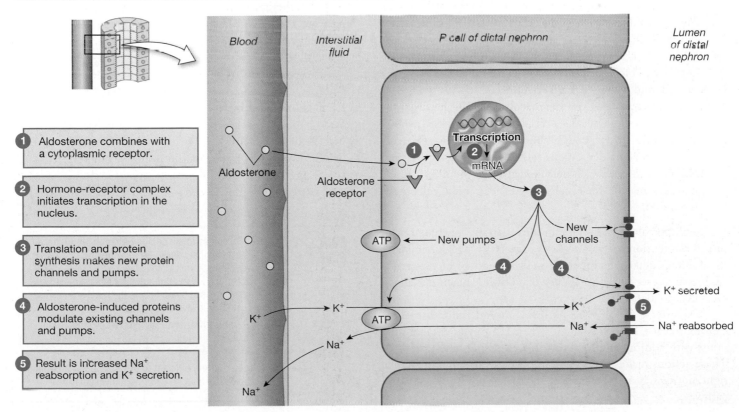

1. Aldosterone combines with a cytoplasmic receptor.

2. Hormone-receptor complex initiates transcription in the nucleus.

3. Translation and protein synthesis makes new protein channels and pumps.

4. Aldosterone-induced proteins modulate existing channels and pumps.

5. Result is increased Na^+ reabsorption and K^+ secretion.

Elevated K^+ concentrations act directly on the adrenal cortex in a reflex that protects the body from hyperkalemia. Decreased blood pressure initiates a complex pathway that results in release of a hormone, **angiotensin II,** that stimulates aldosterone secretion in most situations.

Two additional factors modulate aldosterone release in pathological states: An increase in ECF osmolarity acts directly on adrenal cortex cells to inhibit aldosterone secretion during severe dehydration, and an abnormally large (10–20 mEq/L) decrease in plasma Na^+ can directly stimulate aldosterone secretion.

The Renin-Angiotensin Pathway Angiotensin II (ANG II) is the usual signal controlling aldosterone release from the adrenal cortex. ANG II is one component of the **renin-angiotensin system (RAS),** a complex, multistep pathway for maintaining blood pressure. The RAS pathway begins when juxtaglomerular granular cells in the afferent arterioles of a nephron [p. 602] secrete an enzyme called **renin (FIG. 20.10).** Renin converts an inactive plasma protein, **angiotensinogen,** into **angiotensin I (ANG I).** (The suffix *-ogen* indicates an inactive precursor.) When ANG I in the blood encounters an enzyme called **angiotensin-converting enzyme (ACE),** ANG I is converted into ANG II.

This conversion was originally thought to take place only in the lungs, but ACE is now known to occur on the endothelium of blood vessels throughout the body. When ANG II in the blood reaches the adrenal gland, it causes synthesis and release of aldosterone. Finally, at the distal nephron, aldosterone initiates the intracellular reactions that cause the tubule to reabsorb Na^+.

The stimuli that begin the RAS pathway are all related either directly or indirectly to low blood pressure (Fig. 20.10):

1. The *granular cells* are directly sensitive to blood pressure. They respond to low blood pressure in renal arterioles by secreting renin.
2. *Sympathetic neurons,* activated by the cardiovascular control center when blood pressure decreases, terminate on the granular cells and stimulate renin secretion.
3. *Paracrine feedback*—from the macula densa in the distal tubule to the granular cells—stimulates renin release [p. 600]. When fluid flow through the distal tubule is relatively high, the macula densa cells release paracrine signals that inhibit renin release. When fluid flow in the distal tubule decreases, macula densa cells signal the granular cells to secrete renin.

Sodium reabsorption does not directly raise low blood pressure, but retention of Na^+ increases osmolarity, which stimulates thirst. Fluid intake when the person drinks more water increases ECF volume (see Fig. 20.8). When blood volume increases, blood pressure also increases.

The effects of the RAS pathway are not limited to aldosterone release, however. Angiotensin II is a remarkable hormone with additional effects directed at raising blood pressure. These actions make ANG II an important hormone in its own right, not merely an intermediate step in the aldosterone control pathway.

ANG II Has Many Effects

Angiotensin II has significant effects on fluid balance and blood pressure beyond stimulating aldosterone secretion, underscoring the integrated functions of the renal and cardiovascular systems. ANG II increases blood pressure both directly and indirectly through five additional pathways (Fig. 20.10):

1. *ANG II increases vasopressin secretion.* ANG II receptors in the hypothalamus initiate this reflex. Fluid retention in the kidney under the influence of vasopressin helps conserve blood volume, thereby maintaining blood pressure.
2. *ANG II stimulates thirst.* Fluid ingestion is a behavioral response that expands blood volume and raises blood pressure.
3. *ANG II is one of the most potent vasoconstrictors* known in humans. Vasoconstriction causes blood pressure to increase without a change in blood volume.
4. *Activation of ANG II receptors in the cardiovascular control center increases sympathetic output to the heart and blood vessels.* Sympathetic stimulation increases cardiac output and vasoconstriction, both of which increase blood pressure.
5. *ANG II increases proximal tubule Na^+ reabsorption.* ANG II stimulates an apical transporter, the **Na^+-H^+ exchanger (NHE).** Sodium reabsorption in the proximal tubule is followed by water reabsorption, so the net effect is reabsorption of isosmotic fluid, conserving volume.

Once these blood pressure-raising effects of ANG II became known, it was not surprising that pharmaceutical companies started looking for drugs to block ANG II. Their research produced a new class of antihypertensive drugs called *ACE inhibitors.* These drugs block the ACE-mediated conversion of ANG I to ANG II, thereby helping to relax blood vessels and lower blood pressure. Less ANG II also means less aldosterone release, a decrease in Na^+ reabsorption and, ultimately, a decrease in ECF volume. All these responses contribute to lowering blood pressure.

However, the ACE inhibitors have side effects in some patients. ACE inactivates a cytokine called *bradykinin.* When ACE is inhibited by drugs, bradykinin levels increase, and in some patients this creates a dry, hacking cough. One solution was the development of drugs called *angiotensin receptor blockers (ARBs),* which block the blood pressure–raising effects of ANG II at target cells by binding to AT_1 receptors. Recently another new class of drugs, *direct renin inhibitors,* was approved. Direct

CONCEPT CHECK

11. A man comes to the doctor with high blood pressure. Tests show that he also has elevated plasma renin levels and atherosclerotic plaques that have nearly blocked blood flow through his renal arteries. How does decreased blood flow in his renal arteries cause elevated renin levels?

12. Map the pathways through which elevated renin causes high blood pressure in the man mentioned in Concept Check 11.

13. Why is it more efficient to put ACE in the pulmonary vasculature than in the systemic vasculature?

FIG. 20.10 **ESSENTIALS**

The Renin-Angiotensin System (RAS)

This map outlines the control of aldosterone secretion as well as the blood pressure–raising effects of ANG II. The pathway begins when decreased blood pressure stimulates renin secretion.

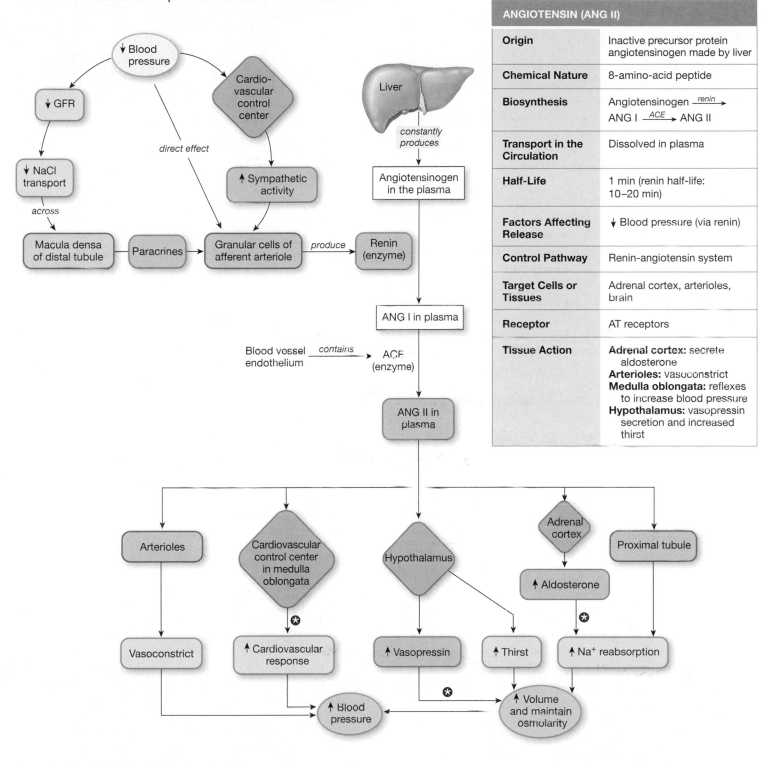

ANGIOTENSIN (ANG II)	
Origin	Inactive precursor protein angiotensinogen made by liver
Chemical Nature	8-amino-acid peptide
Biosynthesis	Angiotensinogen $\xrightarrow{renin}$ ANG I $\xrightarrow{ACE}$ ANG II
Transport in the Circulation	Dissolved in plasma
Half-Life	1 min (renin half-life: 10–20 min)
Factors Affecting Release	↓ Blood pressure (via renin)
Control Pathway	Renin-angiotensin system
Target Cells or Tissues	Adrenal cortex, arterioles, brain
Receptor	AT receptors
Tissue Action	**Adrenal cortex:** secrete aldosterone **Arterioles:** vasoconstrict **Medulla oblongata:** reflexes to increase blood pressure **Hypothalamus:** vasopressin secretion and increased thirst

Q FIGURE QUESTION

Add efferent pathways and/or targets to the pathways marked with a ✲.

renin inhibitors decrease the plasma activity of renin, which in turn blocks production of ANG I and inhibits the entire RAS pathway.

Natriuretic Peptides Promote Na⁺ and Water Excretion

Once it was known that aldosterone and vasopressin increase Na$^+$ and water reabsorption, scientists speculated that other hormones might cause urinary Na$^+$ loss, or **natriuresis** {*natrium,* sodium + *ourein,* to urinate} and water loss (diuresis). If found, these hormones might be used clinically to lower blood volume and blood pressure in patients with essential hypertension [p. 504]. During years of searching, however, evidence for the other hormones was not forthcoming.

Then, in 1981, a group of Canadian researchers found that injections of homogenized rat atria caused rapid but short-lived excretion of Na$^+$ and water in the rats' urine. They hoped they had found the missing hormone, one whose activity would complement that of aldosterone and vasopressin. As it turned out, they had discovered the first *natriuretic peptide* (NP), one member of a family of hormones that appear to be endogenous RAS antagonists (**FIG. 20.11**).

Atrial natriuretic peptide (**ANP;** also known as *atriopeptin*) is a peptide hormone produced in specialized myocardial cells primarily in the atria of the heart. ANP is synthesized as part of a large prohormone that is cleaved into several active hormone fragments [p. 202]. A related hormone, **brain natriuretic peptide** (BNP), is synthesized by ventricular myocardial cells and certain brain neurons. Natriuretic peptides are released by the heart when myocardial cells stretch more than normal. The natriuretic peptides bind to membrane receptor-enzymes that work through a cGMP second messenger system.

ANP is the more important signal molecule in normal physiology. ANP and its co-secreted natriuretic peptides are released when increased blood volume causes increased atrial stretch. At the systemic level, ANP enhances Na$^+$ and water excretion to decrease blood volume. ANP acts at multiple sites. In the kidney it increases GFR by dilating the afferent arterioles, and it directly decreases Na$^+$ reabsorption in the collecting duct.

Natriuretic peptides also act indirectly to increase Na$^+$ and water excretion by suppressing the release of renin, aldosterone, and vasopressin (Fig. 20.11), actions that reinforce the natriuretic-diuretic effect. In addition, natriuretic peptides act directly on the cardiovascular control center of the medulla to lower blood pressure.

BNP is now recognized as an important biological marker for heart failure because production of this substance increases with ventricular dilation and increased ventricular pressure. Hospital emergency departments now use BNP levels to distinguish *dyspnea* (difficulty breathing) in heart failure from other causes. BNP levels are also used as an independent predictor of heart failure and sudden death from cardiac arrhythmias.

POTASSIUM BALANCE

Aldosterone (but not other factors in the RAS pathway) plays a critical role in potassium homeostasis. Only about 2% of the body's K$^+$ load is in the ECF, but regulatory mechanisms keep plasma K$^+$ concentrations within a narrow range (3.5–5 mEq/L). Under normal conditions, mass balance matches K$^+$ excretion to K$^+$ ingestion. If intake exceeds excretion and plasma K$^+$ goes up, aldosterone is released into the blood through the direct effect of hyperkalemia on the adrenal cortex. Aldosterone acting on distal-nephron P cells keeps the cells' apical ion channels open longer and speeds up the Na$^+$-K$^+$-ATPase pump, enhancing renal excretion of K$^+$.

The regulation of body potassium levels is essential in maintaining a state of well-being. Changes in extracellular K$^+$ concentration affect the resting membrane potential of all cells [Fig. 8.17, p. 253]. If plasma (and ECF) K$^+$ concentrations decrease (*hypokalemia*), the concentration gradient between the cell and the ECF becomes larger, more K$^+$ leaves the cell, and the resting membrane potential becomes more negative. If ECF K$^+$ concentrations increase (*hyperkalemia*), the concentration gradient decreases and more K$^+$ remains in the cell, depolarizing it. (Remember that when plasma K$^+$ concentrations change, anions such as Cl$^-$ are also added to or subtracted from the ECF in a 1:1 ratio, maintaining overall electrical neutrality.)

Because of the effect of plasma K$^+$ on excitable tissues, such as the heart, clinicians are always concerned about keeping plasma K$^+$ within its normal range. If K$^+$ falls below 3 mEq/L or rises above 6 mEq/L, the excitable tissues of muscle and nerve begin to show altered function. For example, hypokalemia causes muscle weakness because it is more difficult for hyperpolarized neurons and muscles to fire action potentials. The danger in

EMERGING CONCEPTS

WNK Kinases and Hypertension

We usually think of basic physiology as giving us insight into disease processes, but sometimes things work the other way around. One example from the renal system was the 1964 discovery of a rare inherited form of hypertension (high blood pressure) that was associated with *hyperkalemia* (elevated blood K$^+$). The first thought was that these patients must have hypoaldosteronism, but subsequent tests showed normal hormone levels. It was not until this century, when scientists developed the ability to screen for gene mutations, that they discovered a clue to this unusual form of hypertension. These patients have mutations in the genes that code for a family of proteins called *WNK kinases.* The WNK kinases have important roles in helping aldosterone regulate Na$^+$ and K$^+$ homeostasis in the distal nephron. With this new information, scientists are now studying the WNK kinases and their effects in normal kidneys, hoping to find some novel ways to treat essential hypertension and other diseases.

FIG. 20.11 **ESSENTIALS**

Natriuretic Peptides

Atrial natriuretic peptide (ANP) promotes salt and water excretion. Brain natriuretic peptide (BNP) is a clinical marker for heart failure.

NATRIURETIC PEPTIDES (ANP, BNP)	
Origin	Myocardial cells
Chemical Nature	Peptides. ANP: 28 amino acids, BNP: 32 amino acids
Biosynthesis	Typical peptide. Stored in secretory cells
Transport in the Circulation	Dissolved in plasma
Half-Life	ANP: 2–3 min, BNP: 12 min
Factors Affecting Release	↑ Myocardial stretch. ANP: atrial stretch due to increased blood volume. BNP: ventricular stretch in heart failure
Target Cells or Tissues	ANP: kidney, brain, adrenal cortex primarily
Receptor	NPR receptors. Guanylyl cyclase-linked receptor-enzymes
Systemic Action of ANP	Increase salt and water excretion
Tissue Action	**Afferent arterioles:** vasodilate to increase GFR; inhibit renin secretion **Nephron:** decrease Na^+ and water reabsorption **Adrenal cortex:** inhibit aldosterone secretion **Medulla oblongata:** reflexes to decrease blood pressure **Hypothalamus:** inhibit vasopressin secretion

this condition lies in the failure of respiratory muscles and the heart. Fortunately, skeletal muscle weakness is usually significant enough to lead patients to seek treatment before cardiac problems occur. Mild hypokalemia may be corrected by oral intake of K^+ supplements and K^+-rich foods, such as orange juice and bananas.

Hyperkalemia is a more dangerous potassium disturbance because in this case depolarization of excitable tissues makes them more excitable initially. Subsequently, the cells are unable to repolarize fully and actually become *less* excitable. In this state, they have action potentials that are either smaller than normal or nonexistent. Cardiac muscle excitability affected by changes in plasma K^+ can lead to life-threatening cardiac arrhythmias.

Disturbances in K^+ balance may result from kidney disease, eating disorders, loss of K^+ in diarrhea, or the use of certain types of diuretics that prevent the kidneys from fully reabsorbing K^+. Inappropriate correction of dehydration can also create K^+ imbalance. Consider a golfer playing a round of golf when the temperature was above 100°F. He was aware of the risk of dehydration, so he drank lots of water to replace fluid lost through sweating. The replacement of lost sweat with pure water kept his ECF volume normal but dropped his total blood osmolarity and his K^+ and Na^+ concentrations. He was unable to finish the round of golf because of muscle weakness, and he required medical attention that included ion replacement therapy. A more suitable replacement fluid would have been one of the sports drinks that include salt and K^+.

Potassium balance is also closely tied to acid-base balance, as you will learn in the final section of this chapter. Correction of a pH disturbance requires close attention to plasma K^+ levels. Similarly, correction of K^+ imbalance may alter body pH.

BEHAVIORAL MECHANISMS IN SALT AND WATER BALANCE

Although neural, neuroendocrine, and endocrine reflexes play key roles in salt and water homeostasis, behavioral responses are critical in restoring the normal state, especially when ECF volume decreases or osmolarity increases. Drinking water is normally the only way to restore lost water, and eating salt is the only way to raise the body's Na^+ content. Both behaviors are essential for normal salt and water balance. Clinicians must recognize the absence of these behaviors in patients who are unconscious or otherwise unable to obey behavioral urges, and must adjust treatment accordingly. The study of the biological basis for behaviors, including drinking and eating, is a field known as *physiological psychology*.

Drinking Replaces Fluid Loss

Thirst is one of the most powerful urges known in humans. In 1952, the Swedish physiologist Bengt Andersson showed that stimulating certain regions of the hypothalamus triggered drinking behavior. This discovery led to the identification of hypothalamic osmoreceptors that initiate drinking when body osmolarity rises above 280 mOsM. This is an example of a behavior initiated by an internal stimulus.

It is interesting to note that although increased osmolarity triggers thirst, the act of drinking is sufficient to relieve thirst. The ingested water need not be absorbed in order for thirst to be quenched. As-yet-unidentified receptors in the mouth and pharynx (*oropharynx receptors*) respond to cold water by decreasing thirst and decreasing vasopressin release even though plasma osmolarity remains high. This oropharynx reflex is one reason surgery patients are allowed to suck on ice chips: the ice alleviates their thirst without putting significant amounts of fluid into the digestive system.

A similar reflex exists in camels. When led to water, they drink just enough to replenish their water deficit. Oropharynx receptors apparently act as a feedforward "metering" system that helps prevent wide swings in osmolarity by matching water intake to water need.

In humans, cultural rituals complicate the thirst reflex. For example, we may drink during social events, whether or not we are thirsty. As a result, our bodies must be capable of eliminating fluid ingested in excess of our physiological needs.

Low Na⁺ Stimulates Salt Appetite

Thirst is not the only urge associated with fluid balance. **Salt appetite** is a craving for salty foods that occurs when plasma Na^+ concentrations drop. It can be observed in deer and cattle attracted to salt blocks or naturally occurring salt licks. In humans, salt appetite is linked to aldosterone and angiotensin, hormones that regulate Na^+ balance. The centers for salt appetite are in the hypothalamus close to the center for thirst.

Avoidance Behaviors Help Prevent Dehydration

Other behaviors play a role in fluid balance by preventing or promoting dehydration. Desert animals avoid the heat of the day and become active only at night, when environmental temperatures fall and humidity rises. Humans, especially now that we have air conditioning, are not always so wise.

The midday nap, or *siesta*, is a cultural adaptation in tropical countries that keeps people indoors during the hottest part of the day, thereby helping prevent dehydration and overheating. In the United States, we have abandoned this civilized custom and are active continuously during daylight hours, even when the temperature soars during summer in the South and Southwest. Fortunately, our homeostatic mechanisms usually keep us out of trouble.

CONCEPT CHECK   14. Incorporate the thirst reflex into Figure 20.8.

RUNNING **PROBLEM**

During exercise in the heat, sweating rate and sweat composition are quite variable among athletes and depend partly on how acclimatized the individual is to the heat. Sweat fluid losses can range from less than 0.6 L/h to more than 2.5 L/h, and sweat Na^+ concentrations can range from less than 20 mEq/L to more than 90 mEq/L. The white salt crystals noted on Lauren's face and clothing suggest that she is a "salty sweater" who probably lost a large amount of salt during the race. Follow-up testing revealed that Lauren's sweat Na^+ concentration was 70 mEq/L.

Q6: *Assuming a sweating rate of 1 L/hr, how much Na^+ did Lauren lose during the 16-hour race?*

Q7: *Total body water for a 60-kg female is approximately 30 L, and her ECF volume is 10 L. Based on the information given in the problem so far, calculate how much fluid Lauren probably ingested during the race.*

619   622   629   **637**   641   649

FIG. 20.12   Disturbances in volume and osmolarity

		Osmolarity		
		Decrease	No change	Increase
Volume	Increase	Drinking large amount of water	Ingestion of isotonic saline	Ingestion of hypertonic saline
	No change	Replacement of sweat loss with plain water	Normal volume and osmolarity	Eating salt without drinking water
	Decrease	Incomplete compensation for dehydration	Hemorrhage	Dehydration (e.g., sweat loss or diarrhea)

INTEGRATED CONTROL OF VOLUME AND OSMOLARITY

The body uses an integrated response to correct disruptions of salt and water balance. The cardiovascular system responds to changes in blood volume, and the kidneys respond to changes in blood volume or osmolarity. Maintaining homeostasis throughout the day is a continuous process in which the amounts of salt and water in the body shift, according to whether you just drank a soft drink or sweated through an aerobics class.

In that respect, maintaining fluid balance is like driving a car down the highway and making small adjustments to keep the car in the center of the lane. However, just as exciting movies feature wild car chases, not sedate driving, the exciting part of fluid homeostasis is the body's response to crisis situations, such as severe dehydration or hemorrhage. In this section, we examine more extreme challenges to salt and water balance.

Osmolarity and Volume Can Change Independently

Normally, volume and osmolarity are maintained within an acceptable range through homeostatic control pathways. Under some circumstances, however, fluid loss exceeds fluid gain or vice versa, and the body goes out of balance. Common pathways for fluid loss include excessive sweating, vomiting, diarrhea, and hemorrhage. All these situations may require medical intervention. In contrast, fluid gain is seldom a medical emergency, unless it is addition of water that decreases osmolarity below an acceptable range, as described in this chapter's Running Problem.

Volume and osmolarity of the ECF can each have three possible states: normal, increased, or decreased. The relation of

volume and osmolarity changes can be represented by the matrix in **FIGURE 20.12**. The center box represents the normal state, and the surrounding boxes represent the most common examples of the variations from normal.

In all cases, the appropriate homeostatic compensation for the change acts according to the principle of mass balance: Whatever fluid and solute were added to the body must be removed, or whatever was lost must be replaced. However, perfect compensation is not always possible. Let's begin at the upper right corner of Figure 20.12 and move right to left across each row.

1. *Increased volume, increased osmolarity.* A state of increased volume and increased osmolarity might occur if you ate salty food and drank liquids at the same time, such as popcorn and a soft drink at the movies. The net result could be ingestion of hypertonic saline that increases ECF volume and osmolarity. The appropriate homeostatic response is excretion of hypertonic urine. For homeostasis to be maintained, the osmolarity and volume of the urinary output must match the salt and water input from the popcorn and soft drink.

2. *Increased volume, no change in osmolarity.* Moving one cell to the left in the top row, we see that if the proportion of salt and water in ingested food is equivalent to an isotonic NaCl solution, volume increases but osmolarity does not change. The appropriate response is excretion of isotonic urine whose volume equals that of the ingested fluid.

3. *Increased volume, decreased osmolarity.* This situation would occur if you drank pure water without ingesting any solute. The goal here would be to excrete very dilute urine to maximize water loss while conserving salts. However, because our kidneys cannot excrete pure water, there is always some loss of solute in the urine. In this situation, urinary output cannot exactly match input, and so compensation is imperfect.

4. *No change in volume, increased osmolarity.* This disturbance (middle row, right cell) might occur if you ate salted popcorn without drinking anything. The ingestion of salt without water increases ECF osmolarity and causes some water to shift from cells to the ECF. The homeostatic response is intense thirst, which prompts ingestion of water to dilute the added solute. The kidneys help by creating highly concentrated urine of minimal volume, conserving water while removing excess NaCl. Once water is ingested, the disturbance becomes that described in situation 1 or situation 2.

5. *No change in volume, decreased osmolarity.* This scenario (middle row, left cell) might occur when a person who is dehydrated replaces lost fluid with pure water, like the golfer described earlier. The decreased volume resulting from the dehydration is corrected, but the replacement fluid has no solutes to replace those lost. Consequently, a new imbalance is created.

   This situation led to the development of electrolyte-containing sports beverages. If people working out in hot weather replace lost sweat with pure water, they may restore volume but run the risk of diluting plasma K^+ and Na^+ concentrations to dangerously low levels (*dilutional hypokalemia* and *hyponatremia*, respectively).

6. *Decreased volume, increased osmolarity.* Dehydration is a common cause of this disturbance (bottom row, right cell). Dehydration has multiple causes. During prolonged heavy exercise, water loss from the lungs can double while sweat loss may increase from 0.1 liter to as much as 5 liters! Because the fluid secreted by sweat glands is hyposmotic, the fluid left behind in the body becomes hyperosmotic.

   Diarrhea {*diarrhein*, to flow through}, excessively watery feces, is a pathological condition involving major water and solute loss, this time from the digestive tract. In both sweating and diarrhea, if too much fluid is lost from the circulatory system, blood volume decreases to the point that the heart can no longer pump blood effectively to the brain. In addition, cell shrinkage caused by increased osmolarity disrupts cell function.

7. *Decreased volume, no change in osmolarity.* This situation (bottom row, middle cell) occurs with hemorrhage. Blood loss represents the loss of isosmotic fluid from the extracellular compartment, similar to scooping a cup of seawater out of a large bucketful. If a blood transfusion is not immediately available, the best replacement solution is one that is isosmotic and remains in the ECF, such as isotonic NaCl.

8. *Decreased volume, decreased osmolarity.* This situation (bottom row, left cell) might also result from incomplete compensation of dehydration, but it is uncommon.

Dehydration Triggers Homeostatic Responses

To understand the body's integrated response to changes in volume and osmolarity, you must first have a clear idea of which pathways become active in response to various stimuli. **TABLE 20.1** is a summary of the many pathways involved in the homeostasis

of salt and water balance. For details of individual pathways, refer to the figures cited in Table 20.1.

The homeostatic response to severe dehydration is an excellent example of how the body works to maintain blood volume and cell volume in the face of decreased volume and increased osmolarity. It also illustrates the role of neural and endocrine integrating centers. In severe dehydration, the adrenal cortex receives two opposing signals. One says, "Secrete aldosterone"; the other says, "Do not secrete aldosterone." The body has multiple mechanisms for dealing with diminished blood volume, but high ECF osmolarity causes cells to shrink and presents a more immediate threat to well-being. Thus, faced with decreased volume and increased osmolarity, the adrenal cortex does not secrete aldosterone. (If secreted, aldosterone would cause Na^+ reabsorption, which could worsen the already-high osmolarity associated with dehydration.)

In severe dehydration, compensatory mechanisms are aimed at restoring normal blood pressure, ECF volume, and osmolarity by (1) conserving fluid to prevent additional loss, (2) triggering cardiovascular reflexes to increase blood pressure, and (3) stimulating thirst so that normal fluid volume and osmolarity can be restored. **FIGURE 20.13** maps the interwoven nature of these responses. This figure is complex and intimidating at first glance, so let's discuss it step by step.

At the top of the map (in yellow) are the two stimuli caused by dehydration: decreased blood volume/pressure, and increased osmolarity. Decreased ECF volume causes decreased blood pressure. Blood pressure acts both directly and as a stimulus for several reflex pathways that are mediated through the carotid and aortic baroreceptors and the pressure-sensitive granular cells. Decreased volume is sensed by the atrial volume receptors.

1. *The carotid and aortic baroreceptors signal the cardiovascular control center (CVCC) to raise blood pressure.* Sympathetic output from the CVCC increases while parasympathetic output decreases.
   a. Heart rate goes up as control of the SA node shifts from predominantly parasympathetic to sympathetic.
   b. The force of ventricular contraction also increases under sympathetic stimulation. The increased force of contraction combines with increased heart rate to increase cardiac output.
   c. Simultaneously, sympathetic input causes arteriolar vasoconstriction, increasing peripheral resistance.
   d. Sympathetic vasoconstriction of afferent arterioles in the kidneys decreases GFR, helping conserve fluid.
   e. Increased sympathetic activity at the granular cells of the kidneys increases renin secretion.

2. *Decreased peripheral blood pressure directly decreases GFR.* A lower GFR conserves ECF volume by filtering less fluid into the nephron.

3. *Paracrine feedback causes the granular cells to release renin.* Lower GFR decreases fluid flow past the macula densa. This triggers renin release.

4. *Granular cells respond to decreased blood pressure by releasing renin.* The combination of decreased blood pressure, increased sympathetic input onto granular cells, and signals from the

TABLE 20.1	Responses Triggered by Changes in Volume, Blood Pressure, and Osmolarity		
Stimulus	**Organ or Tissue Involved**	**Response(s)**	**Figure(s)**
Decreased Blood Pressure/Volume			
Direct effects			
	Granular cells	Renin secretion	20.10
	Glomerulus	Decreased GFR	19.6, 20.10
Reflexes			
Carotid and aortic baroreceptors	Cardiovascular control center	Increased sympathetic output, decreased parasympathetic output	15.14b, 20.10
Carotid and aortic baroreceptors	Hypothalamus	Thirst stimulation	20.1a
Carotid and aortic baroreceptors	Hypothalamus	Vasopressin secretion	20.6
Atrial volume receptors	Hypothalamus	Thirst stimulation	20.1a
Atrial volume receptors	Hypothalamus	Vasopressin secretion	20.6
Increased Blood Pressure			
Direct effects			
	Glomerulus	Increased GFR (transient)	19.6, 19.7
	Myocardial cells	Natriuretic peptide secretion	20.11
Reflexes			
Carotid and aortic baroreceptors	Cardiovascular control center	Decreased sympathetic output, increased parasympathetic output	15.14b
Carotid and aortic baroreceptors	Hypothalamus	Thirst inhibition	
Carotid and aortic baroreceptors	Hypothalamus	Vasopressin inhibition	
Atrial volume receptors	Hypothalamus	Thirst inhibition	
Atrial volume receptors	Hypothalamus	Vasopressin inhibition	
Increased Osmolarity			
Direct effects			
Pathological dehydration	Adrenal cortex	Decreased aldosterone secretion	20.13
Reflexes			
Osmoreceptors	Hypothalamus	Thirst stimulation	20.8
Osmoreceptors	Hypothalamus	Vasopressin secretion	20.6
Decreased Osmolarity			
Direct effects			
Pathological hyponatremia	Adrenal cortex	Increased aldosterone secretion	
Reflexes			
Osmoreceptors	Hypothalamus	Decreased vasopressin secretion	

FIG. 20.13   Homeostatic compensation for severe dehydration

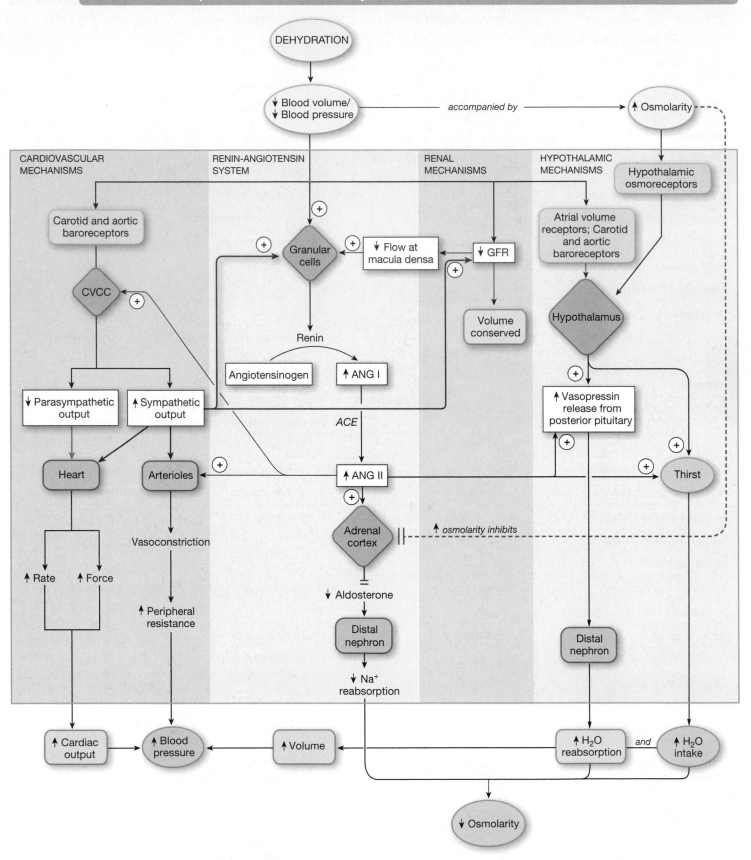

macula densa stimulates renin release and ensures increased production of ANG II.

5. *Decreased blood pressure, decreased blood volume, increased osmolarity, and increased ANG II production all stimulate vasopressin and the thirst centers of the hypothalamus.*

The redundancy in the control pathways ensures that all four main compensatory mechanisms are activated: cardiovascular responses, ANG II, vasopressin, and thirst.

1. *Cardiovascular responses* combine increased cardiac output and increased peripheral resistance to raise blood pressure. Note, however, that this increase in blood pressure does *not necessarily* mean that blood pressure returns to normal. If dehydration is severe, compensation may be incomplete, and blood pressure may remain below normal.

2. *Angiotensin II* has a variety of effects aimed at raising blood pressure, including stimulation of thirst, vasopressin release, direct vasoconstriction, and reinforcement of cardiovascular control center output. ANG II also reaches the adrenal cortex and attempts to stimulate aldosterone release. In severe dehydration, however, Na$^+$ reabsorption worsens the already high osmolarity. Consequently, high osmolarity at the adrenal cortex directly inhibits aldosterone release, blocking the action of ANG II. The RAS pathway in dehydration produces the beneficial blood pressure–enhancing effects of ANG II while avoiding the detrimental effects of Na$^+$ reabsorption. This is a beautiful example of integrated function.

3. *Vasopressin* increases the water permeability of the renal collecting ducts, allowing water reabsorption to conserve fluid. Without fluid replacement, however, vasopressin cannot bring volume and osmolarity back to normal.

4. *Oral (or intravenous) intake of water* in response to thirst is the only mechanism for replacing lost fluid volume and for restoring ECF osmolarity to normal.

The net result of all four mechanisms is (1) restoration of volume by water conservation and fluid intake, (2) maintenance of blood pressure through increased blood volume, increased cardiac output, and vasoconstriction, and (3) restoration of normal osmolarity by decreased Na$^+$ reabsorption and increased water reabsorption and intake.

RUNNING **PROBLEM**

The human body attempts to maintain fluid and sodium balance via several hormonal mechanisms. During exercise sessions, increased sympathetic output causes increased production of aldosterone and vasopressin, which promote the retention of Na$^+$ and water by the kidneys.

Q8: *What would you expect to happen to vasopressin and aldosterone production in response to dilutional hyponatremia?*

619 622 629 637 **641** 649

Using the pathways listed in Table 20.1 and Figure 20.13 as a model, try to create reflex maps for the seven other disturbances of volume and osmolarity shown in Figure 20.12.

ACID-BASE BALANCE

Acid-base balance (also called pH homeostasis) is one of the essential functions of the body. The pH of a solution is a measure of its H$^+$ concentration [p. 41]. The H$^+$ concentration of normal arterial plasma sample is 0.00004 mEq/L, minute compared with the concentrations of other ions. (For example, the plasma concentration of Na$^+$ is about 135 mEq/L.)

Because the body's H$^+$ concentration is so low, it is commonly expressed on a logarithmic pH scale of 0–14, in which a pH of 7.0 is neutral (neither acidic nor basic). If the pH of a solution is below 7.0, the H$^+$ concentration is greater than 1×10^{-7} M and the solution is considered acidic. If the pH is above 7.0, the H$^+$ concentration is lower than 1×10^{-7} M and the solution is considered alkaline (basic).

The normal pH of the body is 7.40, slightly alkaline. A change of 1 pH unit represents a 10-fold change in H$^+$ concentration. [To review the concept of pH, see Fig. 2.9, p. 45. To review logarithms, see Appendix B.]

pH Changes Can Denature Proteins

The normal pH range of plasma is 7.38–7.42. Extracellular pH usually reflects intracellular pH, and vice versa. Because monitoring intracellular conditions is difficult, plasma values are used clinically as an indicator of ECF and whole body pH. Body fluids that are "outside" the body's internal environment, such as those in the lumen of the gastrointestinal tract or kidney tubule, can have a pH that far exceeds the normal range. Acidic secretions in the stomach, for instance, may create a gastric pH as low as 1. The pH of urine varies between 4.5 and 8.5, depending on the body's need to excrete H$^+$ or HCO$_3^-$.

The concentration of H$^+$ in the body is closely regulated. Intracellular proteins, such as enzymes and membrane channels, are particularly sensitive to pH because the function of these proteins depends on their three-dimensional shape. Changes in H$^+$ concentration alter the tertiary structure of proteins by interacting with hydrogen bonds in the molecules, disrupting the proteins' three-dimensional structures and activities [p. 51].

Abnormal pH may significantly affect the activity of the nervous system. If pH is too low—the condition known as **acidosis**—neurons become less excitable, and CNS depression results. Patients become confused and disoriented, then slip into a coma. If CNS depression progresses, the respiratory centers cease to function, causing death.

If pH is too high—the condition known as **alkalosis**—neurons become hyperexcitable, firing action potentials at the slightest signal. This condition shows up first as sensory changes, such as numbness or tingling, then as muscle twitches. If alkalosis is severe, muscle twitches turn into sustained contractions (*tetanus*) that paralyze respiratory muscles.

Disturbances of acid-base balance are associated with disturbances in K^+ balance. This is partly due to a renal transporter that moves K^+ and H^+ ions in an antiport fashion. In acidosis, the kidneys excrete H^+ and reabsorb K^+ using an *H^+-K^+-ATPase*. In alkalosis, the kidneys reabsorb H^+ and excrete K^+. Potassium imbalance usually shows up as disturbances in excitable tissues, especially the heart.

Acids and Bases in the Body Come from Many Sources

In day-to-day functioning, the body is challenged by intake and production of acids more than bases. Hydrogen ions come from both food and internal metabolism. Maintaining mass balance requires that acid intake and production be balanced by acid excretion. Hydrogen balance in the body is summarized in **FIGURE 20.14**.

Acid Input Many metabolic intermediates and foods are organic acids that ionize and contribute H^+ to body fluids.* Examples of organic acids include amino acids, fatty acids, intermediates in the citric acid cycle, and lactate produced by anaerobic metabolism. Metabolic production of organic acids each day generates a significant amount of H^+ that must be excreted to maintain mass balance.

Under extraordinary circumstances, metabolic organic acid production can increase significantly and create a crisis. For example, severe anaerobic conditions, such as circulatory collapse, produce so much lactate that normal homeostatic mechanisms

cannot keep pace, resulting in a state of *lactic acidosis*. In diabetes mellitus, abnormal metabolism of fats and amino acids creates strong acids known as **ketoacids.** These acids cause a state of metabolic acidosis known as *ketoacidosis*.

The biggest source of acid on a daily basis is the production of CO_2 during aerobic respiration. Carbon dioxide is not an acid because it does not contain any hydrogen atoms. However, CO_2 from respiration combines with water to form carbonic acid (H_2CO_3), which dissociates into H^+ and bicarbonate ion, HCO_3^-.

$$CO_2 + H_2O \rightleftharpoons H_2CO_3 \rightleftharpoons H^+ + HCO_3^-$$

This reaction takes place in all cells and in the plasma, but at a slow rate. However, in certain cells of the body, the reaction proceeds very rapidly because of the presence of large amounts of *carbonic anhydrase* [p. 577]. This enzyme catalyzes the conversion of CO_2 and H_2O to H^+ and HCO_3^-.

The production of H^+ from CO_2 and H_2O is the single biggest source of acid input under normal conditions. By some estimates, CO_2 from resting metabolism produces 12,500 mEq of H^+ each day. If this amount of acid were placed in a volume of water equal to the plasma volume, it would create an H^+ concentration of 4167 mEq/L, over one hundred million (10^8) times as concentrated as the normal plasma H^+ concentration of 0.00004 mEq/L!

These numbers show that CO_2 from aerobic respiration has the potential to affect pH in the body dramatically. Fortunately, homeostatic mechanisms normally prevent CO_2 from accumulating in the body.

Base Input Acid-base physiology focuses on acids for good reasons. First, our diet and metabolism have few significant sources of bases. Some fruits and vegetables contain anions that metabolize to HCO_3^-, but the influence of these foods is far outweighed by the contribution of acidic fruits, amino acids, and fatty acids. Second, acid-base disturbances due to excess acid are more common than those due to excess base. For these reasons, the body uses far more resources removing excess acid.

pH Homeostasis Depends on Buffers, Lungs, and Kidneys

How does the body cope with minute-to-minute changes in pH? There are three mechanisms: (1) buffers, (2) ventilation, and (3) renal regulation of H^+ and HCO_3^-. Buffers are the first line of defense, always present and waiting to prevent wide swings in pH. Ventilation, the second line of defense, is a rapid, reflexively controlled response that can take care of 75% of most pH disturbances. The final line of defense lies with the kidneys. They are slower than buffers or the lungs but are very effective at coping with any remaining pH disturbance under normal conditions. Usually these three mechanisms help the body balance acid so effectively that normal body pH varies only slightly. Let's take a closer look at each of them.

FIG. 20.14 pH balance in the body

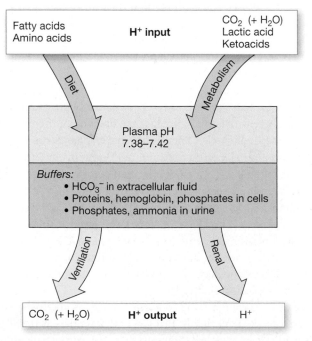

---

*The anion forms of many organic acids end with the suffix –ate, such as pyruvate and lactate.

Buffer Systems Include Proteins, Phosphate Ions, and HCO_3^-

A buffer is a molecule that moderates but does not prevent changes in pH by combining with or releasing H^+ [p. 41]. In the absence of buffers, the addition of acid to a solution causes a sharp change in pH. In the presence of a buffer, the pH change is moderated or may even be unnoticeable. Because acid production is the major challenge to pH homeostasis, most physiological buffers combine with H^+.

Buffers are found both within cells and in the plasma. Intracellular buffers include cellular proteins, phosphate ions (HPO_4^{2-}), and hemoglobin. Hemoglobin in red blood cells buffers the H^+ produced by the reaction of CO_2 with H_2O [Fig. 18.11, p. 578].

Each H^+ ion buffered by hemoglobin leaves a matching bicarbonate ion inside the red blood cell. This HCO_3^- can then leave the red blood cell in exchange for plasma Cl^-, the *chloride shift* [p. 578].

The large amounts of plasma bicarbonate produced from metabolic CO_2 create the most important extracellular buffer system of the body. Plasma HCO_3^- concentration averages 24 mEq/L, which is approximately 600,000 times as concentrated as plasma H^+. Although H^+ and HCO_3^- are created in a 1:1 ratio from CO_2 and H_2O, intracellular buffering of H^+ by hemoglobin is a major reason the two ions do not appear in the plasma in the same concentration. The HCO_3^- in plasma is then available to buffer H^+ from nonrespiratory sources, such as metabolism.

The relationship between CO_2, HCO_3^-, and H^+ in the plasma is expressed by the equation we just looked at:

$$CO_2 + H_2O \rightleftharpoons \underset{\text{carbonic acid}}{H_2CO_3} \rightleftharpoons H^+ + HCO_3^- \qquad (1)$$

According to the law of mass action, any change in the amount of CO_2, H^+, or HCO_3^- in the reaction solution causes the reaction to shift until a new equilibrium is reached. (Water is always in excess in the body and does not contribute to the reaction equilibrium.) For example, if CO_2 increases (red), the equation shifts to the right, creating one additional H^+ and one additional HCO_3^- from each CO_2 and water:

$$\uparrow CO_2 + H_2O \rightarrow H_2CO_3 \rightarrow \uparrow H^+ + \uparrow HCO_3^- \qquad (2)$$

Once a new equilibrium is reached, both H^+ and HCO_3 levels have increased. The addition of H^+ makes the solution more acidic and therefore lowers its pH.

Note that in reaction (2), it does not matter that a bicarbonate molecule has also been produced. HCO_3^- acts as a buffer only when it binds to H^+ and becomes carbonic acid. When the reaction is at equilibrium, as shown here, HCO_3^- will not combine with H^+.

Now suppose H^+ (red) is added to the plasma from some metabolic source, such as lactic acid:

$$CO_2 + H_2O \rightleftharpoons H_2CO_3 \rightleftharpoons \uparrow\uparrow H^+ + HCO_3^- \qquad (3)$$

Adding H^+ disturbs the equilibrium state of the reaction. By the law of mass action [p. 48], adding a molecule to the right side of the equilibrium will send the equation to the left. Now plasma

HCO_3^- can act as a buffer and combine with some of the added H^+. The reaction shifts to the left, converting some of the added H^+ and bicarbonate buffer to CO_2 and H_2O:

$$CO_2 + H_2O \leftarrow H_2CO_3 \leftarrow \uparrow\uparrow H^+ + HCO_3^- \qquad (4)$$

When the equation comes back to equilibrium, H^+ is still elevated, but not as much as it was initially. The concentration of HCO_3^- is decreased because some bicarbonate has been used as a buffer. CO_2 and H_2O have both increased. At equilibrium, the reaction looks like this:

$$\uparrow CO_2 + \uparrow H_2O \rightleftharpoons H_2CO_3 \rightleftharpoons \uparrow H^+ + \downarrow HCO_3^- \qquad (5)$$

The law of mass action is a useful way to think about the relationship between changes in the concentrations of H^+, HCO_3^-, and CO_2 as long as you remember certain qualifications. First, a change in HCO_3^- concentration (as indicated in reaction 2) may not show up clinically as a HCO_3^- concentration outside the normal range. This is because HCO_3^- is 600,000 times more concentrated in the plasma than H^+ is. If both H^+ and HCO_3^- are added to the plasma, you may observe changes in pH but not in HCO_3^- concentration because so much bicarbonate was present initially. Both H^+ and HCO_3^- experience an *absolute* increase in concentration, but because so many HCO_3^- were in the plasma to begin with, the *relative increase* in HCO_3^- goes unnoticed.

As an analogy, think of two football teams playing in a stadium packed with 80,000 fans. If 10 more players (H^+) run out onto the field, everyone notices. But if 10 people (HCO_3^-) come into the stands at the same time, no one pays any attention because there were already so many people watching the game that 10 more make no significant difference.

The relationship between pH, HCO_3^- concentration in mM, and dissolved CO_2 concentration is expressed mathematically by the **Henderson-Hasselbalch equation.** One variant of the equation that is more useful in clinical medicine uses P_{CO_2} instead of dissolved CO_2 concentration:

$$pH = 6.1 + \log[HCO_3^-]/0.03 \times P_{CO_2}$$

If you know a patient's P_{CO_2} and plasma bicarbonate concentration, you can predict the plasma pH.

The second qualification for the law of mass action is that when the reaction shifts to the left and increases plasma CO_2, a nearly instantaneous increase in ventilation takes place in a normal person. If extra CO_2 is ventilated off, arterial P_{CO_2} may remain normal or even fall below normal as a result of hyperventilation.

Ventilation Can Compensate for pH Disturbances

The increase in ventilation just described is a *respiratory compensation* for acidosis. Ventilation and acid-base status are intimately linked, as shown by the equation

$$CO_2 + H_2O \rightleftharpoons H_2CO_3 \rightleftharpoons H^+ + HCO_3^-$$

Fig. 20.15   Respiratory compensation for metabolic acidosis

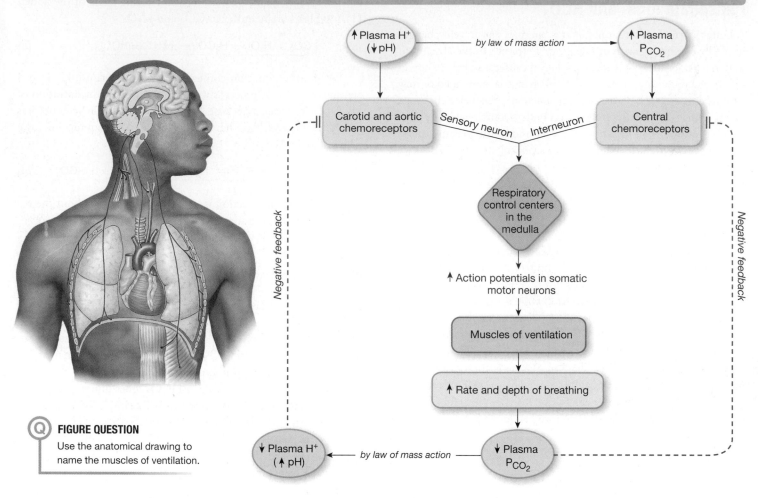

FIGURE QUESTION
Use the anatomical drawing to name the muscles of ventilation.

Changes in ventilation can correct disturbances in acid-base balance, but they can also cause them. Because of the dynamic equilibrium between CO_2 and H^+, any change in plasma P_{CO_2} affects both H^+ and HCO_3^- content of the blood.

Hypoventilation   For example, if a person hypoventilates and P_{CO_2} increases (red), the equation shifts to the right. More carbonic acid is formed, and H^+ goes up, creating a more *acidotic* state:

$$\uparrow CO_2 + H_2O \;\rightarrow\; H_2CO_3 \;\rightarrow\; \uparrow H^+ + \uparrow HCO_3^- \qquad (6)$$

Hyperventilation   On the other hand, if a person hyperventilates, blowing off CO_2 and thereby decreasing the plasma P_{CO_2} (red), the equation shifts to the left, which means that H^+ combines with HCO_3^- and becomes $CO_2 + H_2O$, thereby decreasing the H^+ concentration. Lower H^+ means an increase in pH:

$$\downarrow CO_2 + H_2O \;\leftarrow\; H_2CO_3 \;\leftarrow\; \downarrow H^+ + \downarrow HCO_3^- \qquad (7)$$

In these two examples, you can see that a change in P_{CO_2} affects the H^+ concentration and therefore the pH of the plasma.

Ventilation Reflexes   The body uses ventilation as a homeostatic method for adjusting pH only if a stimulus associated with pH triggers the reflex response. Two stimuli can do so: H^+ and CO_2.

Ventilation is affected directly by plasma H^+ levels primarily through carotid body chemoreceptors (**FIG. 20.15**). These chemoreceptors are located in the carotid arteries along with oxygen sensors and blood pressure sensors [p. 493]. An increase in plasma H^+ stimulates the chemoreceptors, which in turn signal the medullary respiratory control centers to increase ventilation. Increased alveolar ventilation allows the lungs to excrete more CO_2 and convert H^+ to $CO_2 + H_2O$.

The central chemoreceptors of the medulla oblongata cannot respond directly to changes in plasma pH because H^+ does not cross the blood-brain barrier. However, changes in pH change P_{CO_2}, and CO_2 stimulates the central chemoreceptors [Fig. 18.17, p. 584]. Dual control of ventilation through the central and peripheral chemoreceptors helps the body respond rapidly to changes in either pH or plasma CO_2.

15. In equation 6, the amount of HCO_3^- present is increased at equilibrium. Why doesn't this HCO_3^- act as a buffer and prevent acidosis from occurring?

Kidneys Use Ammonia and Phosphate Buffers

The kidneys take care of the 25% of compensation that the lungs cannot handle. They alter pH two ways: (1) directly, by excreting or reabsorbing H^+ and (2) indirectly, by changing the rate at which HCO_3^- buffer is reabsorbed or excreted.

In acidosis, the kidney secretes H^+ into the tubule lumen using direct and indirect active transport (**FIG. 20.16**). Ammonia from amino acids and phosphate ions (HPO_4^{2-}) in the kidney act as buffers, trapping large amounts of H^+ as NH_4^+ and $H_2PO_4^-$. These buffers allow more H^+ to be excreted. Phosphate ions are present in filtrate and combine with H^+ secreted into the nephron lumen:

$$HPO_4^{2-} + H^+ \rightleftharpoons H_2PO_4^-$$

Even with these buffers, urine can become quite acidic, down to a pH of about 4.5. While H^+ is being excreted, the kidneys make new HCO_3^- from CO_2 and H_2O. The HCO_3^- is reabsorbed into the blood to act as a buffer and increase pH.

In alkalosis, the kidney reverses the general process just described for acidosis, excreting HCO_3^- and reabsorbing H^+ in an effort to bring pH back into the normal range. Renal compensations are slower than respiratory compensations, and their effect on pH may not be noticed for 24–48 hours. However, once activated, renal compensations effectively handle all but severe acid-base disturbances.

The cellular mechanisms for renal handling of H^+ and HCO_3^- resemble transport processes in other epithelia. However, these mechanisms involve some membrane transporters that you have not encountered before:

1. The apical *Na⁺-H⁺ exchanger* (NHE) is an indirect active transporter that brings Na^+ into the epithelial cell in exchange for moving H^+ against its concentration gradient into the lumen. This transporter also plays a role in proximal tubule Na^+ reabsorption.
2. The basolateral **Na⁺-HCO₃⁻ symporter** moves Na^+ and HCO_3^- out of the epithelial cell and into the interstitial fluid. This indirect active transporter couples the energy of HCO_3^- diffusing down its concentration gradient to the uphill movement of Na^+ from the cell to the ECF.
3. The **H⁺-ATPase** uses energy from ATP to acidify the urine, pushing H^+ against its concentration gradient into the lumen of the distal nephron. The H^+-ATPase is also called the *proton pump*.
4. The **H⁺-K⁺-ATPase** puts H^+ into the urine in exchange for reabsorbed K^+. This exchange contributes to the potassium imbalance that sometimes accompanies acid-base disturbances.
5. A **Na⁺-NH₄⁺ antiporter** moves NH_4^+ from the cell to the lumen in exchange for Na^+.

FIG. 20.16 **Overview of renal compensation for acidosis**

The kidney secretes H^+, which is buffered in the urine by ammonia and phosphate ions. It reabsorbs bicarbonate to act as an extracellular buffer.

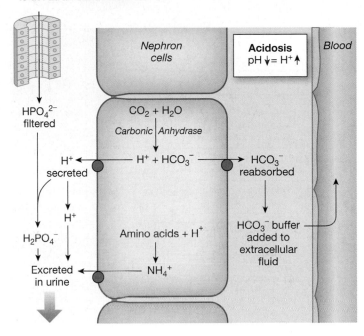

The transporters shown here are generic membrane proteins. For specific transporters involved, see Figure 20.17.

In addition to these transporters, the renal tubule also uses the ubiquitous Na^+-K^+-ATPase and the same HCO_3^--Cl^- antiport protein that is responsible for the chloride shift in red blood cells.

The Proximal Tubule Secretes H^+ and Reabsorbs HCO_3^-

The amount of bicarbonate ion the kidneys filter each day is equivalent to the bicarbonate in a pound of baking soda ($NaHCO_3$)! Most of this HCO_3^- must be reabsorbed to maintain the body's buffer capacity. The proximal tubule reabsorbs most filtered HCO_3^- by indirect methods because there is no apical membrane transporter to bring HCO_3^- into the tubule cell.

FIGURE 20.17 shows the two pathways by which bicarbonate is reabsorbed in the proximal tubule. (The numbers in the following lists correspond to the steps shown in the figure.) By following this illustration, you will see how the transporters listed in the previous section function together.

The first pathway converts filtered HCO_3^- into CO_2, then back into HCO_3^-, which is reabsorbed:

1. H^+ is secreted from the proximal tubule cell into the lumen in exchange for filtered Na^+, which moves from the lumen into the tubule cell. This exchange takes place using the NHE.
2. The secreted H^+ combines with filtered HCO_3^- to form CO_2 in the lumen. This reaction is facilitated by carbonic anhydrase that is bound to the luminal membrane of the tubule cells.

FIG. 20.17   The proximal tubule reabsorbs filtered bicarbonate

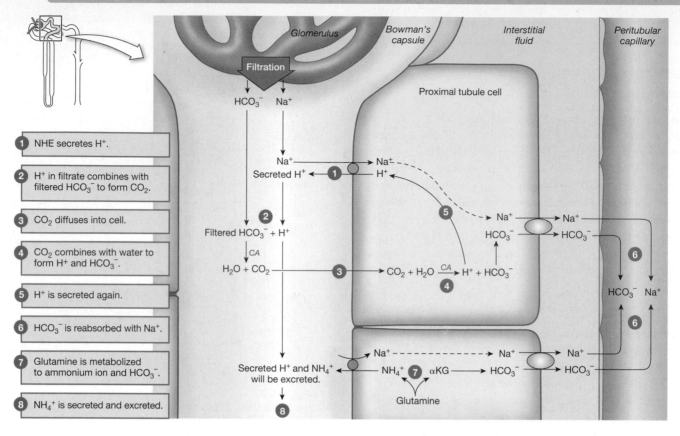

1. NHE secretes H^+.

2. H^+ in filtrate combines with filtered HCO_3^- to form CO_2.

3. CO_2 diffuses into cell.

4. CO_2 combines with water to form H^+ and HCO_3^-.

5. H^+ is secreted again.

6. HCO_3^- is reabsorbed with Na^+.

7. Glutamine is metabolized to ammonium ion and HCO_3^-.

8. NH_4^+ is secreted and excreted.

3. The newly formed CO_2 diffuses from the lumen into the tubule cell.

4. In the cytoplasm, CO_2 combines with water to form H_2CO_3, which dissociates to H^+ and HCO_3^-.

5. The H^+ created in step 4 can be secreted into the lumen again, replacing the H^+ that combined with filtered HCO_3^- in step 2. It can combine with another filtered bicarbonate or be buffered by filtered phosphate ion and excreted.

6. The HCO_3^- created in step 3 is transported out of the cell on the basolateral side of the proximal tubule cell by the HCO_3^--Na^+ symporter.

The net result of this process is reabsorption of filtered Na^+ and HCO_3^-, and secretion of H^+.

A second way to reabsorb bicarbonate and excrete H^+ comes from metabolism of the amino acid glutamine:

7. Glutamine in the proximal tubule cell is metabolized to α-ketoglutarate (αKG) and two amino groups ($-NH_2$). The amino groups become ammonia (NH_3), and the ammonia buffers H^+ to become ammonium ion NH_4^+. The ammonium ion NH_4^+ is transported into the lumen in exchange for Na^+. The α-ketoglutarate molecule is metabolized further to HCO_3^-, which is transported into the blood along with Na^+.

The net result of both pathways shown in Figure 20.17 is secretion of acid (H^+) and reabsorption of buffer in the form of sodium bicarbonate—baking soda, $NaHCO_3$.

The Distal Nephron Controls Acid Excretion

The distal nephron plays a significant role in the fine regulation of acid-base balance. Specialized cells called **intercalated cells (I cells)** interspersed among the principal cells are primarily responsible for acid-base regulation.

Intercalated cells are characterized by high concentrations of carbonic anhydrase in their cytoplasm. This enzyme allows them to rapidly convert CO_2 and water into H^+ and HCO_3^-. The H^+ ions are pumped out of the intercalated cell either by the H^+-ATPase or by the H^+-K^+-ATPase. Bicarbonate leaves the cell by means of the HCO_3^--Cl^- antiport exchanger.

There are two types of intercalated cells, and their transporters are found on different faces of the epithelial cell. During periods of acidosis, type A intercalated cells secrete H^+ and reabsorb bicarbonate. During periods of alkalosis, type B intercalated cells secrete HCO_3^- and reabsorb H^+.

FIGURE 20.18a shows how type A intercalated cells work during acidosis, secreting H^+ and reabsorbing HCO_3^-. The process is similar to H^+ secretion in the proximal tubule except for the specific H^+ transporters. The distal nephron uses apical H^+-ATPase and H^+-K^+-ATPase rather than the Na^+-H^+ antiport protein found in the proximal tubule.

During alkalosis, when the H^+ concentration of the body is too low, H^+ is reabsorbed and HCO_3^- buffer is excreted in the urine (Fig. 20.18b). Once again, the ions are formed from H_2O and CO_2. Hydrogen ions are reabsorbed by transport into the

FIG. 20.18 | Intercalated cells function in acid-base disturbances

(a) Acidosis. Type A intercalated cells in collecting duct function in acidosis. H^+ is excreted; HCO_3^- and K^+ are reabsorbed.

Lumen of collecting duct — Type A intercalated cell — Interstitial space — Blood

[H^+] high

Filtered K^+

$H_2O + CO_2 \leftarrow\!\!-\!\!-\!\!- CO_2 \leftarrow HCO_3^- + H^+$

CA

$H^+ + HCO_3^-$ → HCO_3^- acts as a buffer to ↓[H^+]

ATP → K^+ — Cl^-

H^+

ATP

High [K^+] → K^+ reabsorbed

H^+ excreted in urine

(b) Alkalosis. Type B intercalated cells in collecting duct function in alkalosis. HCO_3^- and K^+ are excreted; H^+ is reabsorbed.

Lumen of collecting duct — Type B intercalated cell — Interstitial space — Blood

[H^+] low

$H_2O + CO_2$

CA

$HCO_3^- \leftarrow HCO_3^- + H^+$ — ATP → H^+

Cl^- ↑H^+

H^+

ATP

K^+ — K^+

Excreted in urine

ECF on the basolateral side of the cell, and HCO_3^- is secreted into the lumen. The polarity of the two types of I cells is reversed, with the same transport processes taking place, but on the opposite sides of the cell.

The H^+-K^+-ATPase of the distal nephron helps create parallel disturbances of acid-base balance and K^+ balance. In acidosis,

when plasma H^+ is high, the kidney secretes H^+ and reabsorbs K^+. For this reason, acidosis is often accompanied by hyperkalemia. (Other non-renal events also contribute to elevated ECF K^+ concentrations in acidosis.) The reverse is true for alkalosis, when blood H^+ levels are low. The mechanism that allows the distal nephron to reabsorb H^+ simultaneously causes it to secrete K^+, with the result that alkalosis goes hand in hand with hypokalemia.

CONCEPT CHECK

16. Why is ATP required for H^+ secretion by the H^+-K^+ transporter but not for the Na^+-H^+ exchanger?

17. In hypokalemia, the intercalated cells of the distal nephron reabsorb K^+ from the tubule lumen. What happens to blood pH as a result?

Acid-Base Disturbances May Be Respiratory or Metabolic

The three compensatory mechanisms (buffers, ventilation, and renal excretion) take care of most variations in plasma pH. But under some circumstances, the production or loss of H^+ or HCO_3^- is so extreme that compensatory mechanisms fail to maintain pH homeostasis. In these states, the pH of the blood moves out of the normal range of 7.38–7.42. If the body fails to keep pH between 7.00 and 7.70, acidosis or alkalosis can be fatal (**FIG. 20.19**).

Acid-base problems are classified both by the direction of the pH change (acidosis or alkalosis) and by the underlying cause (metabolic or respiratory). Changes in P_{CO_2} resulting from hyperventilation or hypoventilation cause pH to shift. These disturbances are said to be of respiratory origin. If the pH problem arises from acids or bases of non-CO_2 origin, the problem is said to be a metabolic problem.

Note that by the time an acid-base disturbance becomes evident as a change in plasma pH, the body's buffers are ineffectual. The loss of buffering ability leaves the body with only two options: respiratory compensation or renal compensation. And if the problem is of respiratory origin, only one homeostatic compensation is available—the kidneys. If the problem is of metabolic origin, both respiratory and renal mechanisms can compensate. Compensation can bring the pH back closer to normal but may not correct the disturbance completely (Fig. 20.19).

The combination of an initial pH disturbance and the resultant compensatory changes is one factor that makes analysis of acid-base disorders in the clinical setting so difficult. In this book, we concentrate on simple scenarios with a single underlying cause. Changes that occur in the four simple acid-base disturbances are listed in **TABLE 20.2**.

Respiratory Acidosis A state of respiratory acidosis occurs when alveolar hypoventilation results in CO_2 retention and elevated plasma P_{CO_2}. Some situations in which this occurs are respiratory depression due to drugs (including alcohol), increased airway resistance in asthma, impaired gas exchange in fibrosis or severe pneumonia, and muscle weakness in muscular dystrophy

FIG. 20.19 Acid-base disturbances may be incompletely compensated

TABLE 20.2	Plasma P_{CO_2}, Ions, and pH in Acid-Base Disturbances			
Disturbance	P_{CO_2}	H^+	pH	HCO_3^-
Acidosis				
Respiratory	↑	↑	↓	↑
Metabolic	Normal* or ↓	↑	↓	↓
Alkalosis				
Respiratory	↓	↓	↑	↓
Metabolic	Normal* or ↑	↓	↑	↑

*These values are different from what you would expect from the law of mass action because almost instantaneous respiratory compensation keeps P_{CO_2} from changing significantly.

and other muscle diseases. The most common cause of respiratory acidosis is *chronic obstructive pulmonary disease* (COPD), which includes emphysema. In emphysema, inadequate alveolar ventilation is compounded by loss of alveolar exchange area.

No matter what the cause of respiratory acidosis, plasma CO_2 levels increase (red), leading to elevated H^+ and HCO_3^-:

$$\uparrow CO_2 + H_2O \rightarrow H_2CO_3 \rightarrow \uparrow H^+ + \uparrow HCO_3^- \qquad (8)$$

The hallmark of respiratory acidosis is decreased pH with elevated bicarbonate levels (Tbl. 20.2). Because the problem is of respiratory origin, the body cannot carry out respiratory compensation. (However, depending on the problem, mechanical ventilation can sometimes be used to assist breathing.)

Any compensation for respiratory acidosis must occur through renal mechanisms that excrete H^+ and reabsorb HCO_3^-. The excretion of H^+ raises plasma pH. Reabsorption of HCO_3^- provides additional buffer that combines with H^+, lowering the H^+ concentration and therefore raising the pH.

In chronic obstructive pulmonary disease, renal compensation mechanisms for acidosis can moderate the pH change, but they may not be able to return the pH to its normal range. If you look at pH and HCO_3^- levels in patients with compensated respiratory acidosis, you find that both those values are closer to normal than they were when the acidosis was at its worst.

Metabolic Acidosis Metabolic acidosis is a disturbance of mass balance that occurs when the dietary and metabolic input of H^+ exceeds H^+ excretion. Metabolic causes of acidosis include *lactic acidosis,* which is a result of anaerobic metabolism, and *keto-acidosis*, which results from excessive breakdown of fats or certain amino acids. The metabolic pathway that produces ketoacids is associated with type 1 diabetes mellitus and with low-carbohydrate diets, like the Atkins diet [Chapter 22]. Ingested substances that cause metabolic acidosis include methanol, aspirin, and ethylene glycol (antifreeze).

Metabolic acidosis is expressed by the equation

$$\uparrow CO_2 + H_2O \leftarrow H_2CO_3 \leftarrow \uparrow H^+ + \downarrow HCO_3^- \qquad (9)$$

Hydrogen ion concentration increases (red) because of the H^+ contributed by the metabolic acids. This increase shifts the equilibrium represented in the equation to the left, increasing CO_2 levels and using up HCO_3^- buffer.

Metabolic acidosis can also occur if the body loses HCO_3^-. The most common cause of bicarbonate loss is diarrhea, during which HCO_3^- is lost from the intestines. The pancreas produces HCO_3^- from CO_2 and H_2O by a mechanism similar to the renal mechanism illustrated in Figure 20.16. The H^+ made at the same time is released into the blood. Normally, the HCO_3^- is released into the small intestine, then reabsorbed into the blood, buffering the H^+. However, if a person is experiencing diarrhea, HCO_3^- is not reabsorbed, and a state of acidosis may result.

Whether HCO_3^- concentration is elevated or decreased is an important criterion for distinguishing metabolic acidosis from respiratory acidosis (Tbl. 20.2).

You would think from looking at equation 9 that metabolic acidosis would be accompanied by elevated P_{CO_2}. However, unless the individual also has a lung disease, respiratory compensation takes place almost instantaneously. Both elevated CO_2 and elevated H^+ stimulate ventilation through the pathways described earlier. As a result, P_{CO_2} decreases to normal or even below-normal levels due to hyperventilation.

Uncompensated metabolic acidosis is rarely seen clinically. Actually, a common sign of metabolic acidosis is hyperventilation, evidence of respiratory compensation occurring in response to the acidosis.

The renal compensations discussed for respiratory acidosis also take place in metabolic acidosis: secretion of H^+ and reabsorption of HCO_3^-. Renal compensations take several days to reach full effectiveness, and so they are not usually seen in recent-onset (acute) disturbances.

Respiratory Alkalosis

States of alkalosis are much less common than acidotic conditions. Respiratory alkalosis occurs as a result of hyperventilation, when alveolar ventilation increases without a matching increase in metabolic CO_2 production. Consequently, plasma P_{CO_2} falls (red), and alkalosis results when the equation shifts to the left:

$$\downarrow CO_2 + H_2O \leftarrow H_2CO_3 \leftarrow \downarrow H^+ + \downarrow HCO_3^- \qquad (10)$$

The decrease in CO_2 shifts the equilibrium to the left, and both plasma H^+ and plasma HCO_3^- decrease. Low plasma HCO_3^- levels in alkalosis indicate a respiratory disorder.

The primary clinical cause of respiratory alkalosis is excessive artificial ventilation. Fortunately, this condition is easily corrected by adjusting the ventilator. The most common physiological cause of respiratory alkalosis is hysterical hyperventilation caused by anxiety. When this is the cause, the neurological symptoms caused by alkalosis can be partially reversed by having the patient breathe into a paper bag. In doing so, the patient rebreathes exhaled CO_2, a process that raises arterial P_{CO_2} and corrects the problem.

Because this alkalosis has respiratory cause, the only compensation available to the body is renal. Filtered bicarbonate is not reabsorbed in the proximal tubule and is secreted in the distal nephron. The combination of HCO_3^- excretion and H^+ reabsorption in the distal nephron decreases the body's HCO_3^- buffer load and increases its H^+, both of which help correct the alkalosis.

Metabolic Alkalosis

Metabolic alkalosis has two common causes: excessive vomiting of acidic stomach contents and excessive ingestion of bicarbonate-containing antacids. In both cases, the resulting alkalosis reduces H^+ concentration (red):

$$\downarrow CO_2 + \downarrow H_2O \rightarrow H_2CO_3 \rightarrow \downarrow H^+ + \uparrow HCO_3^- \qquad (11)$$

The decrease in H^+ shifts the equilibrium to the right, meaning that carbon dioxide (P_{CO_2}) decreases and HCO_3^- goes up.

Just as in metabolic acidosis, respiratory compensation for metabolic alkalosis is rapid. The increase in pH and decrease in P_{CO_2} depress ventilation. Hypoventilation means the body retains CO_2, raising the P_{CO_2} and creating more H^+ and HCO_3^-. This respiratory compensation helps correct the pH problem but elevates HCO_3^- levels even more. However, this respiratory compensation is limited because hypoventilation also causes hypoxia. Once the arterial P_{O_2} drops below 60 mm Hg, hypoventilation ceases.

The renal response to metabolic alkalosis is the same as that for respiratory alkalosis: HCO_3^- is excreted and H^+ is reabsorbed.

This chapter has used fluid balance and acid-base balance to illustrate functional integration in the cardiovascular, respiratory, and renal systems. Changes in body fluid volume, reflected by changes in blood pressure, trigger both cardiovascular and renal homeostatic responses. Disturbances of acid-base balance are met with compensatory responses from both the respiratory and renal systems. Because of the interwoven responsibilities of these three systems, a disturbance in one system is likely to cause disturbances in the other two. Recognition of this fact is an important aspect of treatment for many clinical conditions.

RUNNING **PROBLEM CONCLUSION** | Hyponatremia

In acute cases of dilutional hyponatremia such as Lauren's, the treatment goal is to correct the body's depleted Na^+ load and raise the plasma osmolarity to reduce cerebral swelling. The physicians in the emergency medical tent started a slow intravenous drip of 3% saline and restricted Lauren's oral fluid intake. Over the course of several hours, the combination of Na^+ intake and excretion of dilute urine returned Lauren's plasma Na^+ to normal levels.

Hyponatremia has numerous causes, including inappropriate secretion of antidiuretic hormone, a condition known as SIADH, which stands for **s**yndrome of **I**nappropriate **a**ntidiuretic **h**ormone secretion. (ADH is also called vasopressin.) To learn more about medical causes of hyponatremia, see "Clinical Practice Guideline on Diagnosis and Treatment of Hyponatraemia," *Nephrol. Dial. Transplant. (2014) 29 (suppl 2): i1–i39.* Available at www.eje-online.org/content/170/3/G1.full.pdf. *doi: 10.1093/ndt/gfu040* (free pdf; note the British spelling of hyponatremia).

Question	Facts	Integration and Analysis
Q1: *Name the two major body fluid compartments and give the major ions in each compartment.*	The major compartments are the intracellular fluid (ICF) and extracellular fluid (ECF) compartments. The primary ICF ion is K^+, and the major ECF ions are Na^+ and Cl^-.	N/A
Q2: *Based on Lauren's history, give a reason for why her weight increased during the race.*	Lauren reported drinking lots of water and sports drinks. One liter of pure water has a mass of 1 kg.	Lauren's fluid intake was greater than her fluid loss from sweating. A 2-kg increase in body weight means she drank an excess of about 2 L.

—Continued next page

RUNNING **PROBLEM CONCLUSION** | *Continued*

Question	Facts	Integration and Analysis
Q3: Which body fluid compartment is being diluted in dilutional hyponatremia?	Ingested water distributes itself throughout the ECF and ICF. Sodium is one of the major extracellular cations.	Lauren consumed a large amount of Na-free fluid and therefore diluted her Na^+ stores. However, the body compartments are in osmotic equilibrium so both ECF and ICF have lower osmolarities.
Q4: One way to estimate osmolarity is to double the plasma Na^+ concentration. Estimate Lauren's osmolarity and explain what effect the dilutional hyponatremia has on her cells.	Lauren's plasma Na^+ is 124 mEq/L. For Na^+, 1 mEq = 1 milliosmole. Doubling this value tells you that Lauren's estimated plasma osmolarity is 248 mOsM. Water distributes to maintain osmotic equilibrium.	At the start of the race, Lauren's cells were at 280 mOsM. The water she ingested distributed to maintain osmotic equilibrium, so water entered the ICF from the ECF, resulting in cell swelling.
Q5: In dilutional hyponatremia, the medical personnel are most concerned about which organ or tissue?	All cells in Lauren's body swell as a result of excess water ingestion. The brain is encased in the rigid skull.	The bony skull restricts the swelling of brain tissue, causing neurological symptoms, including confusion, headache, and loss of coordination. With lower Na^+ concentrations, death can result.
Q6: Assuming a sweating rate of 1 L/hr, how much Na^+ did Lauren lose during the 16-hour race?	1 L sweat lost/hr $\times$ 16 hr $\times$ 70 mEq Na^+/L sweat = 1120 mEq Na^+ lost during the 16-hour race.	N/A
Q7: Total body water for a 60-kg female is approximately 30 L, and her ECF volume is 10 L. Based on the information given so far, how much fluid did Lauren ingest during the race?	From the sweating rate given in question 6, you know that Lauren lost 16 L of sweat during the race. You also know that she gained 2 kg in weight. One liter of water weighs 1 kg.	Lauren must have ingested at least 18 L of fluid. You have no information on other routes of fluid loss, such as urine and insensible water lost during breathing.
Q8: What would you expect to happen to vasopressin and aldosterone production in response to dilutional hyponatremia?	Vasopressin secretion is inhibited by a decrease in osmolarity. The usual stimuli for renin or aldosterone release are low blood pressure and hyperkalemia.	Vasopressin secretion decreases with hyponatremia. The usual stimuli for aldosterone secretion are absent, but a pathological decrease in plasma Na^+ of 10 mEq/L can stimulate the adrenal cortex to secrete aldosterone. Thus, Lauren's plasma Na^+ may be low enough to increase her aldosterone secretion.

This problem was developed by Matt Pahnke while he was a kinesiology graduate student at the University of Texas.

(619)(622)(629)(637)(641)(**649**)

VISUALIZE, PRACTICE, AND MASTER PHYSIOLOGY WITH MASTERING A&P

- Practice-on-the-go with Dynamic Study Modules
- Visualize important processes with Interactive Physiology and A&P Flix
- Check your understanding with Video Tutors and practice quizzes!

MasteringA&P®

CHAPTER SUMMARY

Homeostasis of body fluid volume, electrolytes, and pH follows the principle of *mass balance:* To maintain constant amount of a substance in the body, any intake or production must be offset by metabolism or excretion. The *control systems* that regulate these parameters are among the most complicated reflexes of the body because of the overlapping functions of the kidneys, lungs, and cardiovascular system. At the cellular level, however, the *movement of molecules across membranes* follows familiar patterns, as transfer of water and solutes from one *compartment* to another depends on osmosis, diffusion, and protein-mediated transport.

Fluid and Electrolyte Homeostasis

1. The renal, respiratory, and cardiovascular systems control fluid and electrolyte balance. Behaviors such as drinking also play an important role. (p. 619; Fig. 20.1)

2. Pulmonary and cardiovascular compensations are more rapid than renal compensation. (p. 619)

Water Balance

iP Urinary: Early Filtrate Processing

3. Most water intake comes from food and drink. The largest water loss is 1.5 liters/day in urine. Smaller amounts are lost in feces, by evaporation from skin, and in exhaled humidified air. (p. 620; Fig. 20.2)

4. Renal water reabsorption conserves water but cannot restore water lost from the body. (p. 621; Fig. 20.3)

5. To produce dilute urine, the nephron must reabsorb solute without reabsorbing water. To concentrate urine, the nephron must reabsorb water without reabsorbing solute. (p. 622)

6. Filtrate leaving the ascending limb of the loop of Henle is dilute. The final concentration of urine depends on the water permeability of the collecting duct. (p. 622; Fig. 20.4)

7. The hypothalamic hormone **vasopressin** controls collecting duct permeability to water in a graded fashion. When vasopressin is absent, water permeability is nearly zero. (p. 623; Fig. 20.5a, b)

8. Vasopressin causes distal nephron cells to insert **aquaporin** water pores in their apical membrane. (p. 625; Fig. 20.5c)

9. An increase in ECF osmolarity or a decrease in blood pressure stimulates vasopressin release from the posterior pituitary. Osmolarity is monitored by hypothalamic **osmoreceptors**. Blood pressure and blood volume are sensed by receptors in the carotid and aortic bodies, and in the atria, respectively. (p. 625; Fig. 20.6)

10. The loop of Henle is a **countercurrent multiplier** that creates high osmolarity in the medullary interstitial fluid by actively transporting Na^+, Cl^-, and K^+ out of the nephron. This high medullary osmolarity is necessary for formation of concentrated urine as filtrate flows through the collecting duct. (p. 627; Fig. 20.7)

11. The **vasa recta** capillaries form a countercurrent exchanger that carries away water leaving the tubule so that the water does not dilute the medullary interstitium. (p. 627; Fig. 20.7)

12. Urea contributes to the high osmolarity in the renal medulla. (p. 629)

Sodium Balance and ECF Volume

iP Urinary: Late Filtrate Processing

13. The total amount of Na^+ in the body is a primary determinant of ECF volume. (p. 629; Fig. 20.8)

14. The steroid hormone **aldosterone** increases Na^+ reabsorption and K^+ secretion. (p. 631; Fig. 20.9a)

15. Aldosterone acts on **principal cells (P cells)** of the distal nephron. This hormone enhances Na^+-K^+-ATPase activity and increases open time of Na^+ and K^+ leak channels. It also stimulates the synthesis of new pumps and channels. (p. 631; Fig. 20.9b)

16. Aldosterone secretion can be controlled directly at the adrenal cortex. Increased ECF K^+ stimulates aldosterone secretion, but very high ECF osmolarity inhibits it. (p. 632; Fig. 20.9)

17. Aldosterone secretion is also stimulated by **angiotensin II.** In response to signals associated with low blood pressure, granular cells in the kidney secrete **renin,** which converts **angiotensinogen** in the blood to **angiotensin I. Angiotensin-converting enzyme (ACE)** converts ANG I to ANG II. (p. 632; Fig. 20.10)

18. The signals for the release of renin are related either directly or indirectly to low blood pressure. (p. 632; Fig. 20.10)

19. ANG II has additional effects that raise blood pressure, including increased vasopressin secretion, stimulation of thirst, vasoconstriction, and activation of the cardiovascular control center. (p. 632; Fig. 20.10)

20. **Atrial natriuretic peptide (ANP)** and **brain natriuretic peptide** (BNP) enhance Na^+ excretion and urinary water loss by increasing GFR, inhibiting tubular reabsorption of NaCl, and inhibiting the release of renin, aldosterone, and vasopressin. (p. 634; Fig. 20.11)

Potassium Balance

21. Potassium homeostasis keeps plasma K^+ concentrations in a narrow range. **Hyperkalemia** and **hypokalemia** cause problems with excitable tissues, especially the heart. (p. 634)

Behavioral Mechanisms in Salt and Water Balance

22. Thirst is triggered by hypothalamic osmoreceptors and relieved by drinking. (p. 636)

23. **Salt appetite** is triggered by aldosterone and angiotensin. (p. 636)

Integrated Control of Volume and Osmolarity

iP Fluids & Electrolytes: Water Homeostasis

24. Homeostatic compensations for changes in salt and water balance follow the law of mass balance. Fluid and solute added to the body must be removed, and fluid and solute lost from the body must be replaced. However, perfect compensation is not always possible. (p. 637; Tbl. 20.1)

Acid-Base Balance

iP Fluids & Electrolytes: Acid/Base Homeostasis

25. The body's pH is closely regulated because pH affects intracellular proteins, such as enzymes and membrane channels. (p. 641)

26. Acid intake from foods and acid production by the body's metabolic processes are the biggest challenge to body pH. The most significant source of acid is CO_2 from respiration, which combines with water to form carbonic acid (H_2CO_3). (p. 642; Fig. 20.14)

27. The body copes with changes in pH by using buffers, ventilation, and renal secretion or reabsorption of H^+ and HCO_3^-. (p. 642; Fig. 20.14)

28. Bicarbonate produced from CO_2 is the most important extracellular buffer of the body. Bicarbonate buffers organic acids produced by metabolism. (p. 643)

29. Ventilation can correct disturbances in acid-base balance because changes in plasma P_{CO_2} affect both the H^+ content and the HCO_3^- content of the blood. An increase in P_{CO_2} stimulates central chemoreceptors. An increase in plasma H^+ stimulates carotid and aortic chemoreceptors. Increased ventilation excretes CO_2 and decreases plasma H^+. (p. 643; Fig. 20.15)

30. In **acidosis,** the kidneys secrete H^+ and reabsorb HCO_3^-. (p. 641; Figs. 20.16, 20.18a)

31. In **alkalosis,** the kidneys secrete HCO_3^- and reabsorb H^+. (p. 641; Fig. 20.18b)

32. **Intercalated cells** in the collecting duct are responsible for the fine regulation of acid-base balance. Type A intercalated cells are active in acidosis and type B cells active in alkalosis. (p. 646; Fig. 20.18)

REVIEW QUESTIONS

In addition to working through these questions and checking your answers on p. A-26, review the Learning Outcomes at the beginning of this chapter.

Level One Reviewing Facts and Terms

1. What is an electrolyte? Name five electrolytes whose concentrations must be regulated by the body.

2. List five organs and four hormones important in maintaining fluid and electrolyte balance.

3. Compare the routes by which water enters the body with the routes by which the body loses water.

4. List the receptors that regulate osmolarity, blood volume, blood pressure, ventilation, and pH. Where are they located, what stimulates them, and what compensatory mechanisms are triggered by them?

5. How do the two limbs of the loop of Henle differ in their permeability to water? What makes this difference in permeability possible?

6. Which ion is a primary determinant of ECF volume? Which ion is the determinant of extracellular pH?

7. What happens to the resting membrane potential of excitable cells when plasma K^+ concentrations decrease? Which organ is most likely to be affected by changes in K^+ concentration?

8. Appetite for which two substances is important in regulating fluid volume and osmolarity?

9. Write out the words for the following abbreviations: ADH, ANP, ACE, ANG II, JG apparatus, P cell, I cell.

10. Make a list of all the different membrane transporters in the kidney. For each transporter, tell (a) which section(s) of the nephron contain(s) the transporter; (b) whether the transporter is on the apical membrane only, on the basolateral membrane only, or on both; (c) whether it participates in reabsorption only, in secretion only, or in both.

11. List and briefly explain three reasons why monitoring and regulating ECF pH are important. What three mechanisms does the body use to cope with changing pH?

12. Which is more likely to accumulate in the body, acids or bases? List some sources of each.

13. What is a buffer? List three intracellular buffers. Name the primary extracellular buffer.

14. Name two ways the kidneys alter plasma pH. Which compounds serve as urinary buffers?

15. Write the equation that shows how CO_2 is related to pH. What enzyme increases the rate of this reaction? Name two cell types that possess high concentrations of this enzyme.

16. When ventilation increases, what happens to arterial P_{CO_2}? To plasma pH? To plasma H^+ concentration?

Level Two Reviewing Concepts

17. Concept map: Map the homeostatic reflexes that occur in response to each of the following situations:

    (a) decreased blood volume, normal blood osmolarity
    (b) increased blood volume, increased blood osmolarity
    (c) normal blood volume, increased blood osmolarity

18. Figures 20.15 and 20.18a show the respiratory and renal compensations for acidosis. Draw similar maps for alkalosis.

19. Explain how the loop of Henle and vasa recta work together to create dilute renal filtrate.

20. Diagram the mechanism by which vasopressin alters the composition of urine.

21. Make a table that specifies the following for each substance listed: hormone or enzyme? steroid or peptide? produced by which cell or tissue? target cell or tissue? target has what response?

    (a) ANP
    (b) aldosterone
    (c) renin
    (d) ANG II
    (e) vasopressin
    (f) angiotensin-converting enzyme

22. Name the four main compensatory mechanisms for restoring low blood pressure to normal. Why do you think there are so many homeostatic pathways for raising low blood pressure?

23. The interstitial fluid in contact with the basolateral side of collecting duct cells has an extremely high osmolarity, and yet the cells do not shrivel up. How can they maintain normal cell volume in the face of such high ECF osmolarity?

24. Compare and contrast the terms in each set:

 (a) principal cells and intercalated cells
 (b) renin, ANG II, aldosterone, ACE
 (c) respiratory acidosis and metabolic acidosis, including causes and compensations
 (d) water reabsorption in proximal tubule, distal tubule, and ascending limb of the loop of Henle
 (e) respiratory alkalosis and metabolic alkalosis, including causes and compensations

Level Three Problem Solving

25. A 45-year-old man visiting from out of town arrives at the emergency room having an asthma attack caused by pollen.

 (a) Blood drawn before treatment showed the following: $HCO_3^- = 30$ mEq/L (normal: 24), $P_{CO_2} = 70$ mm Hg, pH = 7.24. What is the man's acid-base state? Is this an acute or a chronic situation?
 (b) The man was treated and made a complete recovery. Over the next 10 years, he continued to smoke a pack of cigarettes a day, and a year ago, his family doctor diagnosed chronic obstructive pulmonary disease (emphysema). The man's most recent blood test showed the following: $HCO_3^- = 45$ mEq/L, $P_{CO_2} = 85$ mm Hg, pH = 7.34. What is the man's acid-base state now? Is this an acute or a chronic situation?
 (c) Explain why in his second illness his plasma bicarbonate level and P_{CO_2} are higher than in the first illness but his pH is closer to normal.

26. The U.S. Food and Drug Administration has now approved a new class of drugs called *vasopressin receptor antagonists*. Predict the effect these drugs would have on renal function and describe some clinical situations or diseases in which these drugs might be useful.

27. Karen has bulimia, in which she induces vomiting to avoid weight gain. When the doctor sees her, her weight is 89 lb and her respiration rate is 6 breaths/min (normal 12). Her blood HCO_3^- is 62 mEqE/L (normal: 24–29), arterial blood pH is 7.61, and P_{CO_2} is 61 mm Hg.

 (a) What is her acid-base condition called?
 (b) Explain why her plasma bicarbonate level is so high.
 (c) Why is she hypoventilating? What effect does this have on the pH and total oxygen content of her blood? Explain your answers.

28. Hannah, a 31-year-old woman, decided to have colonic irrigation, a procedure during which large volumes of distilled water were infused into her rectum. During the treatment, she absorbed 3000 mL of water. About 12 hours later, her roommate found her in convulsions and took her to the emergency room. Her blood pressure was 140/90, her plasma Na^+ concentration was 106 mEq/L (normal: 135 mEq/L), and her plasma osmolarity was 270 mOsM. In a concept map or flow chart, diagram all the homeostatic responses her body was using to attempt compensation for the changes in blood pressure and osmolarity.

Level Four Quantitative Problems

29. In extreme dehydration, urine can reach a concentration of 1400 mOsM. If the minimum amount of waste solute that a person must excrete daily is about 600 milliosmoles, what is the minimum urine volume that is excreted in one day?

30. The **Henderson-Hasselbalch equation** is a mathematical expression of the relationship between pH, HCO_3^- concentration, and dissolved CO_2 concentration. One variant of the equation uses P_{CO_2} instead of dissolved CO_2 concentration:

$$pH = 6.1 + \log[HCO_3^-]/0.03 \times P_{CO_2}$$

 (a) If arterial blood has a P_{CO_2} of 40 mm Hg and its HCO_3^- concentration is 24 mM, what is its pH? (Use a log table or calculator with a logarithmic function capability.)
 (b) What is the pH of venous blood with the same HCO_3^- concentration but a P_{CO_2} of 46 mm Hg?

31. Hyperglycemia in a diabetic patient leads to osmotic diuresis and dehydration. Given the following information, answer the questions.

 Plasma glucose = 400 mg/dL

 Normal urine flow = 1 L per day

 GFR = 130 mL/min

 Normal urine osmolarity = 300 mOsM

 Glucose T_m = 400 mg/min

 Molecular mass of glucose = 180 daltons

 Renal plasma flow = 500 mL/min

 (a) How much glucose filters into the nephron each minute?
 (b) How much glucose is reabsorbed each minute?
 (c) How much glucose is excreted in the urine each day?
 (d) Assuming that dehydration causes maximal vasopressin secretion and allows the urine to concentrate to 1200 mOsM, how much additional urine does this diabetic patient excrete in a day?

32. Osmotic diuresis refers to the loss of additional water in urine as a result of unreabsorbed solutes. To see what difference unreabsorbed solutes make, calculate the volumes of filtrate that would be needed for excretion of 150 milliosmoles of NaCl. Then repeat the calculation for a diabetic who is excreting the same 150 mosmol NaCl plus 200 mosmol unreabsorbed glucose.

	Filtrate Concentration	Volume Needed for Excretion of 150 mosmol NaCl	Volume Needed for Excretion of 150 mosmol NaCl + 200 mosmol Glucose
End of loop of Henle	100 mOsM		
End of cortical collecting duct	300 mOsM		
Urine leaving medullary collecting duct	1200 mOsM		

Answers to Concept Checks, Figure and Graph Questions, and end-of-chapter Review Questions can be found in Appendix A [p. A-1].

21

The Digestive System

Give me a good digestion, Lord, and also something to digest.

Anonymous, *A Pilgrim's Grace*

ANATOMY OF THE DIGESTIVE SYSTEM 655

LO 21.1 Trace a piece of undigested food from mouth to anus.

LO 21.2 Describe the four layers of the GI tract wall.

DIGESTIVE FUNCTION AND PROCESSES 659

LO 21.3 Describe the primary function of the digestive system.

LO 21.4 Explain the challenges of autodigestion, mass balance, and defense.

LO 21.5 Describe and compare secretion, digestion, absorption, and motility.

LO 21.6 Describe single-unit smooth muscle, slow wave potentials, tonic and phasic contractions.

LO 21.7 Describe and compare peristalsis, segmentation, and the migrating motor complex.

REGULATION OF GI FUNCTION 664

LO 21.8 Compare the enteric nervous system to the central nervous system.

LO 21.9 Contrast long reflexes, short reflexes, and control involving GI peptides.

LO 21.10 Name the three families of GI hormones and give examples of each.

INTEGRATED FUNCTION: THE CEPHALIC PHASE 668

LO 21.11 Explain feedforward control in digestion.

LO 21.12 Map the processes and control pathways of the cephalic phase.

LO 21.13 Explain the functions of saliva and the process by which it is secreted.

LO 21.14 List the steps of the deglutition (swallowing) reflex.

INTEGRATED FUNCTION: THE GASTRIC PHASE 669

LO 21.15 Explain the three functions of the stomach.

LO 21.16 Map the processes and control pathways of the gastric phase.

LO 21.17 Describe the gastric secretions and their major actions.

INTEGRATED FUNCTION: THE INTESTINAL PHASE 672

LO 21.18 Compare and contrast digestion and motility in the large and small intestine.

LO 21.19 Describe the anatomy and function of the hepatic portal system.

LO 21.20 Describe the major secretions of the pancreas and liver.

LO 21.21 Diagram the cellular mechanisms for secretion or absorption of water and ions.

LO 21.22 Diagram the digestion and absorption of carbohydrates, proteins, and fats.

LO 21.23 Explain the neural and hormonal control of the intestinal phase of digestion.

LO 21.24 Explain the role of bacteria in the gut.

IMMUNE FUNCTIONS OF THE GI TRACT 687

LO 21.25 Describe the GALT.

LO 21.26 Contrast the protective reflexes of vomiting and diarrhea.

BACKGROUND BASICS

16 Positive feedback and feedforward control
29 Biomolecules
62 Micelles
68 Microvilli
72 Cell junctions
150 Transporting epithelia
79 Apical and basolateral membranes
79 Endocrine and exocrine glands
99 Enzymes
113 Protein synthesis and storage
143 Secondary active transport
148 Exocytosis and transcytosis
403 Smooth muscle
439 Portal systems
499 Lymphatics
604 Renal transport
645 Acidification of urine

Cross-section of intestinal villi (outlined in red)

A shotgun wound to the stomach seems an unlikely beginning to the scientific study of digestive processes. But in 1822, at Fort Mackinac, a young Canadian trapper named Alexis St. Martin narrowly escaped death when a gun discharged three feet from him, tearing open his chest and abdomen and leaving a hole in his stomach wall. U.S. Army surgeon William Beaumont attended to St. Martin and nursed him back to health over the next two years.

The gaping wound over the stomach failed to heal properly, leaving a *fistula,* or opening, into the lumen. St. Martin was destitute and unable to care for himself, so Beaumont "retained St. Martin in his family for the special purpose of making physiological experiments." In a legal document, St. Martin even agreed to "obey, suffer, and comply with all reasonable and proper experiments of the said William [Beaumont] in relation to . . . the exhibiting . . . of his said stomach and the power and properties . . . and states of the contents thereof."

Beaumont's observations on digestion and on the state of St. Martin's stomach under various conditions created a sensation. In 1832, just before Beaumont's observations were published, the nature of gastric juice {*gaster,* stomach} and digestion in the stomach was a subject of much debate. Beaumont's careful observations went far toward solving the mystery. Like physicians of old who tasted urine when making a diagnosis, Beaumont tasted the mucous lining of the stomach and the gastric juices. He described them both as "saltish," but mucus was not at all acidic, and gastric fluid was very acidic. Beaumont collected copious amounts of gastric fluid through the fistula, and in controlled experiments he confirmed that gastric fluid digested meat, using a combination of hydrochloric acid and another active factor now known to be the enzyme pepsin.

These observations and others about motility and digestion in the stomach became the foundation of what we know about digestive physiology. Although research today is conducted more at the cellular and molecular level, researchers still create surgical fistulas in experimental animals to observe and sample the contents of the digestive tract.

Why is the digestive system—also referred to as the **gastrointestinal system** {*intestinus,* internal}—of such great interest? The reason is that gastrointestinal diseases today account for nearly 10% of the money spent on healthcare. Many of these conditions, such as heartburn, indigestion, gas, and constipation, are troublesome rather than major health risks, but their significance should not be underestimated. Go into any drugstore and look at the number of over-the-counter medications for digestive disorders to get a feel for the impact digestive diseases have on our society. In this chapter, we examine the gastrointestinal system and the remarkable way in which it transforms the food we eat into nutrients for the body's use.

ANATOMY OF THE DIGESTIVE SYSTEM

The digestive system begins with the oral cavity (mouth and pharynx), which serves as a receptacle for food (**FIG. 21.1a**). Swallowed food enters the **gastrointestinal tract (GI tract)** consisting of esophagus, stomach, small intestine, and large

RUNNING **PROBLEM** | Cholera in Haiti

Brooke was looking for a way to spend her 2013 winter break, so she volunteered to join a disaster relief team going to Haiti. Upon her arrival in the earthquake-devastated country, Brooke was appalled to see the living conditions. Many people were still living in tents with little or no running water and sanitation. To make matters worse, in October 2010, the World Health Organization (WHO) had issued a global outbreak alert for a cholera epidemic. *Vibrio cholerae,* the cholera bacterium, causes vomiting and massive volumes of watery diarrhea in people who consume contaminated food or water. There had been no cholera in Haiti for nearly a hundred years, but in the years since the earthquake, nearly 700,000 cases and over 8000 deaths have been reported.

intestine. The portion of the GI tract running from the stomach to the anus is also called the **gut.**

Digestion, the chemical and mechanical breakdown of food, takes place primarily in the lumen of the gut. Along the way, secretions are added to ingested food by secretory epithelial cells and by *accessory glandular organs* that include salivary glands, the liver, the gallbladder, and the pancreas. The soupy mixture of food and secretions is known as **chyme.**

The GI tract is a long tube with muscular walls lined by secretory and transporting epithelium [p. 150]. At intervals along the tract, rings of muscle function as *sphincters* to separate the tube into segments with distinct functions. Food moves through the tract propelled by waves of muscle contraction.

The products of digestion are absorbed across the intestinal epithelium and pass into the interstitial fluid. From there, they move into the blood or lymph for distribution throughout the body. Any waste remaining in the lumen at the end of the GI tract leaves the body through an opening called the *anus.*

Because the digestive system opens to the outside world, the tract lumen and its contents are actually part of the external environment. (Think of a hole passing through the center of a bead.) [Fig. 1.2, p. 4] This allows an amazing variety of bacteria to live in the lumen, particularly in the large intestine. The arrangement is usually described as a *commensal* relationship, in which the bacteria benefit from having a home and food supply, while the human body is not affected. However, we are discovering ways in which the body does benefit from its bacterial companions. The relationship between humans and their bacterial *microbiome* is a hot topic in physiology today, and you will learn more about it at the end of the chapter.

The Digestive System Is a Tube

In the oral cavity, the first stages of digestion begin with chewing and the secretion of saliva by three pairs of **salivary glands:** *sublingual glands* under the tongue, *submandibular glands* under the

FIG. 21.1 **ANATOMY SUMMARY**

The Digestive System

(a) Overview of the Digestive System

Oral cavity
Salivary glands
Esophagus
Gallbladder
Pancreas
Small intestine
Rectum
Liver
Stomach
Large intestine

(b) Salivary Glands

Parotid
Sublingual
Submandibular

(c) Stomach

Diaphragm
Esophagus
Fundus
Body
Antrum
Pylorus
Rugae: Surface folding increases area

(d) Structure of the Small Intestine

Mesentery
Mucosa
Submucosa
Circular muscle
Longitudinal muscle
Serosa
Plica
Submucosal glands
Villi

Q **FIGURE QUESTION**

Name the accessory glands and organs of the digestive system.

(e) Sectional View of the Stomach

In the stomach, surface area is increased by invaginations called gastric glands.

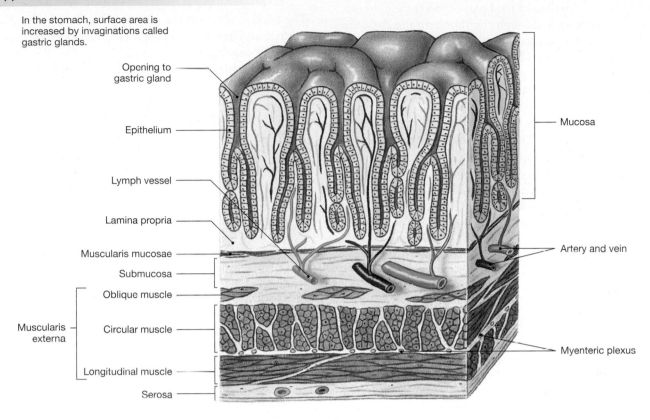

Opening to gastric gland

Epithelium

Lymph vessel

Lamina propria

Muscularis mucosae

Submucosa

Oblique muscle

Muscularis externa

Circular muscle

Longitudinal muscle

Serosa

Mucosa

Artery and vein

Myenteric plexus

(f) Sectional View of the Small Intestine

Intestinal surface area is enhanced by fingerlike villi and invaginations called crypts.

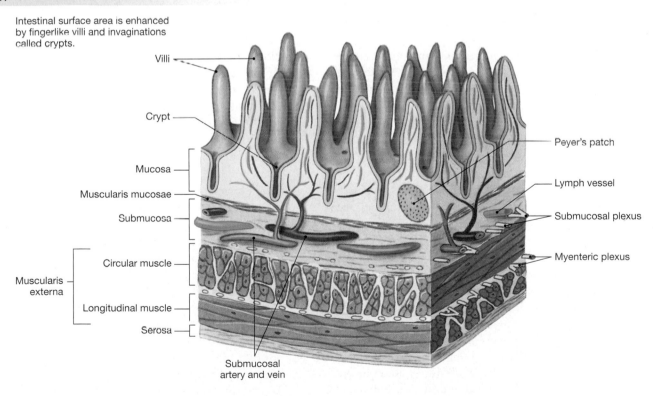

Villi

Crypt

Mucosa

Muscularis mucosae

Submucosa

Circular muscle

Muscularis externa

Longitudinal muscle

Serosa

Submucosal artery and vein

Peyer's patch

Lymph vessel

Submucosal plexus

Myenteric plexus

mandible (jawbone), and *parotid glands* lying near the hinge of the jaw (Fig. 21.1b). Swallowed food passes into the **esophagus,** a narrow tube that travels through the thorax to the abdomen (Fig. 21.1a). The esophageal walls are skeletal muscle initially but transition to smooth muscle about two-thirds of the way down the length. Just below the diaphragm, the esophagus ends at the **stomach,** a baglike organ that can hold as much as 2 liters of food and fluid when fully (if uncomfortably) expanded.

The stomach has three sections: the upper **fundus,** the central **body,** and the lower **antrum** (Fig. 21.1c). The stomach continues digestion that began in the mouth by mixing food with acid and enzymes to create chyme. The **pylorus** {gatekeeper} or opening between the stomach and the **small intestine** is guarded by the **pyloric valve.** This thickened band of smooth muscle relaxes to allow only small amounts of chyme into the small intestine at any one time.

The stomach acts as an intermediary between the behavioral act of eating and the physiological events of digestion and absorption in the intestine. Integrated signals and feedback loops between the intestine and stomach regulate the rate at which chyme enters the duodenum. This ensures that the intestine is not overwhelmed with more than it can digest and absorb.

Most digestion takes place in the small intestine, which has three sections: the **duodenum** (the first 25 cm), **jejunum,** and **ileum** (the latter two together are about 260 cm long). Digestion is carried out by intestinal enzymes, aided by exocrine secretions from two accessory glandular organs: the pancreas and the liver (Fig. 21.1a). Secretions from these two organs enter the initial section of the duodenum through ducts. A tonically contracted sphincter (the *sphincter of Oddi*) keeps pancreatic fluid and bile from entering the small intestine except during a meal.

Digestion finishes in the small intestine, and nearly all digested nutrients and secreted fluids are absorbed there, leaving about 1.5 liters of chyme per day to pass into the **large intestine** (Fig. 21.1a). In the **colon**—the proximal section of the large intestine—watery chyme becomes semisolid **feces** {*faeces*, dregs} as water and electrolytes are absorbed out of the chyme and into the extracellular fluid (ECF).

When feces are propelled into the terminal section of the large intestine, known as the **rectum,** distension of the rectal wall triggers a *defecation reflex*. Feces leave the GI tract through the **anus,** with its **external anal sphincter** of skeletal muscle, which is under voluntary control.

In a living person, the digestive system from mouth to anus is about 450 cm (nearly 15 ft.) long! Of this length, 395 cm (about 13 ft.) consists of the large and small intestines. Try to imagine 13 ft. of rope ranging from 1 to 3 inches in diameter all coiled up inside your abdomen from the belly button down. The tight arrangement of the abdominal organs helps explain why you feel the need to loosen your belt after consuming a large meal.

Measurements of intestinal length made during autopsies are nearly double those given here because after death, the longitudinal muscles of the intestinal tract relax. This relaxation accounts for the wide variation in intestinal length you may encounter in different references.

The GI Tract Wall Has Four Layers

The basic structure of the gastrointestinal wall is similar in the stomach and intestines, although variations exist from one section of the GI tract to another (Fig. 21.1e, f). The gut wall is crumpled into folds to increase its surface area. These folds are called *rugae* in the stomach and *plicae* in the small intestine. The intestinal mucosa also projects into the lumen in small fingerlike extensions known as **villi** (Fig. 21.1f). Additional surface area is added by tubular invaginations of the surface that extend down into the supporting connective tissue. These invaginations are called **gastric glands** in the stomach and **crypts** in the intestine. Some of the deepest invaginations form secretory **submucosal glands** that open into the lumen through ducts.

The gut wall consists of four layers: (1) an inner *mucosa* facing the lumen, (2) a layer known as the *submucosa*, (3) layers of smooth muscle known collectively as the *muscularis externa*, and (4) a covering of connective tissue called the *serosa*.

Mucosa The **mucosa,** the inner lining of the gastrointestinal tract, has three layers: a single layer of **mucosal epithelium** facing the lumen; the **lamina propria,** subepithelial connective tissue that holds the epithelium in place; and the **muscularis mucosae,** a thin layer of smooth muscle. Several structural modifications increase the amount of mucosal surface area to enhance absorption.

1. *The mucosal epithelium* is the most variable feature of the GI tract, changing from section to section. The cells of the mucosa include transporting epithelial cells (called *enterocytes* in the small intestine), endocrine and exocrine secretory cells, and stem cells. At the *mucosal* (apical) surface of the epithelium [p. 79], cells secrete ions, enzymes, mucus, and paracrine molecules into the lumen. On the *serosal* (basolateral) surface of the epithelium, substances being absorbed from the lumen and molecules secreted by epithelial cells enter the ECF.

   The cell-to-cell junctions that tie GI epithelial cells together vary [p. 72]. In the stomach and colon, the junctions form a tight barrier so that little can pass between the cells. In the small intestine, junctions are not as tight. This intestinal epithelium is considered "leaky" because some water and solutes can be absorbed between the cells (the *paracellular pathway*) instead of through them. We now know that these junctions have plasticity and that their "tightness" and selectivity can be regulated to some extent.

   GI *stem cells* are rapidly dividing, undifferentiated cells that continuously produce new epithelium in the crypts and gastric glands. As stem cells divide, the newly formed cells are pushed toward the luminal surface of the epithelium. The average life span of a GI epithelial cell is only a few days, a good indicator of the rough life such cells lead. As with other types of epithelium, the rapid turnover and cell division rate in the GI tract make these organs susceptible to developing cancers. In 2013, cancers of the colon and rectum (colorectal cancer) were the second leading cause of cancer deaths in the United States. The death rate has been steadily falling, however, due to more screening examinations and better treatments.

2. The *lamina propria* is subepithelial connective tissue that contains nerve fibers and small blood and lymph vessels. Absorbed nutrients pass into the blood and lymph here (Fig. 21.1e). This layer also contains wandering immune cells, such as macrophages and lymphocytes, patrolling for invaders that enter through breaks in the epithelium.

   In the intestine, collections of lymphoid tissue adjoining the epithelium form small *nodules* and larger **Peyer's patches** that create visible bumps in the mucosa (Fig. 21.1f). These lymphoid aggregations are a major part of the **gut-associated lymphoid tissue (GALT)**.

3. The *muscularis mucosae*, a thin layer of smooth muscle, separates the lamina propria from the submucosa. Contraction of muscles in this layer alters the effective surface area for absorption by moving the villi back and forth, like the waving tentacles of a sea anemone.

Submucosa The **submucosa** is the middle layer of the gut wall. It is composed of connective tissue with larger blood and lymph vessels running through it (Fig. 21.1e, f). The submucosa also contains the **submucosal plexus** {*plexus*, interwoven}, one of the two major nerve networks of the **enteric nervous system** [p. 229]. The submucosal plexus (also called *Meissner's plexus*) innervates cells in the epithelial layer as well as smooth muscle of the muscularis mucosae.

Muscularis Externa The outer wall of the gastrointestinal tract, the **muscularis externa,** consists primarily of two layers of smooth muscle: an inner circular layer and an outer longitudinal layer (Fig. 21.1d, f). Contraction of the circular layer decreases the diameter of the lumen. Contraction of the longitudinal layer shortens the tube. The stomach has an incomplete third layer of oblique muscle between the circular muscles and the submucosa (Fig. 21.1e).

   The second nerve network of the enteric nervous system, the **myenteric plexus** {*myo-*, muscle + *enteron*, intestine}, lies between the longitudinal and circular muscle layers. The myenteric plexus (also called *Auerbach's plexus*) controls and coordinates the motor activity of the muscularis externa.

Serosa The outer covering of the entire digestive tract, the **serosa,** is a connective tissue membrane that is a continuation of the **peritoneal membrane** (*peritoneum*) lining the abdominal cavity [p. 59]. The peritoneum also forms sheets of **mesentery** that hold the intestines in place so that they do not become tangled as they move.

   The next section is a brief look at the four processes of secretion, digestion, absorption, and motility. Gastrointestinal physiology is a rapidly expanding field, and this textbook does not attempt to be all inclusive. Instead, it focuses on selected broad aspects of digestive physiology.

CONCEPT CHECK

1. Is the lumen of the digestive tract on the apical or basolateral side of the intestinal epithelium? On the serosal or mucosal side?
2. Name the four layers of the GI tract wall, starting at the lumen and moving out.
3. Name the structures a piece of food passes through as it travels from mouth to anus.

DIGESTIVE FUNCTION AND PROCESSES

The primary function of the digestive system is to move nutrients, water, and electrolytes from the external environment into the body's internal environment. To accomplish this, the system uses four basic processes: digestion, absorption, secretion, and motility (**FIG.21.2**). **Digestion** is the chemical and mechanical breakdown of foods into smaller units that can be taken across the intestinal epithelium into the body. **Absorption** is the movement of substances from the lumen of the GI tract to the extracellular fluid. **Secretion** in the GI tract has two meanings. It can mean the movement of water and ions from the ECF to the digestive tract lumen (the opposite of absorption), but it can also mean the release of substances synthesized by GI epithelial cells into either the lumen or the ECF. **Motility** {*movere*, move + *tillis*, characterized by} is movement of material in the GI tract as a result of muscle contraction.

   Although it might seem simple to digest and absorb food, the digestive system faces three significant challenges:

1. *Avoiding autodigestion* The food we eat is mostly in the form of macromolecules, such as proteins and complex carbohydrates, so our digestive systems must secrete powerful enzymes to digest food into molecules that are small enough to be absorbed into the body. At the same time, however, these enzymes must not digest the cells of the GI tract itself

FIG. 21.2 Four processes of the digestive system

Secretion	Movement of material from cells into lumen or ECF
Digestion	Chemical and mechanical breakdown of food into absorbable units
Absorption	Movement of material from GI lumen to ECF
Motility	Movement of material through the GI tract as a result of muscle contraction

← Lumen of digestive tract → | ← Wall → | Interstitial fluid | ← Blood →

Food
SECRETION
DIGESTION
ABSORPTION
MOTILITY

FIG. 21.3 Mass balance in the digestive system

To maintain homeostasis, the volume of fluid entering the GI tract by intake or secretion must equal the volume leaving the lumen.

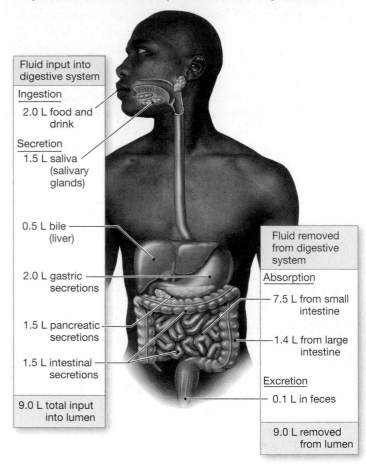

Fluid input into digestive system

Ingestion
2.0 L food and drink

Secretion
1.5 L saliva (salivary glands)

0.5 L bile (liver)

2.0 L gastric secretions

1.5 L pancreatic secretions

1.5 L intestinal secretions

9.0 L total input into lumen

Fluid removed from digestive system

Absorption
7.5 L from small intestine

1.4 L from large intestine

Excretion
0.1 L in feces

9.0 L removed from lumen

(*autodigestion*). If protective mechanisms against autodigestion fail, raw patches known as *peptic ulcers* {*peptos,* digested} develop on the walls of the GI tract.

2. *Mass balance* Another challenge the digestive system faces daily is maintaining mass balance by matching fluid input with output (**FIG. 21.3**). People ingest about 2 liters of fluid a day. In addition, exocrine glands and cells secrete 7 liters or so of enzymes, mucus, electrolytes, and water into the lumen of the GI tract. That volume of secreted fluid is the equivalent of one-sixth of the body's total body water (42 liters), or more than twice the plasma volume of 3 liters. If the secreted fluid could not be reabsorbed, the body would rapidly dehydrate.

3. Normally intestinal reabsorption is very efficient, and only about 100 mL of fluid is lost in the feces. However, vomiting and diarrhea (excessively watery feces) can become an emergency when GI secretions are lost to the environment

instead of being reabsorbed. In severe cases, this fluid loss can deplete extracellular fluid volume to the point that the circulatory system is unable to maintain adequate blood pressure.

4. *Defense* A final challenge the digestive system faces is protecting the body from foreign invaders. It is counterintuitive, but the largest area of contact between the body's internal environment and the outside world is in the lumen of the digestive system. As a result, the GI tract, with a total surface area about the size of a tennis court, faces daily conflict between the need to absorb water and nutrients, and the need to keep bacteria, viruses, and other pathogens from entering the body. To this end, the transporting epithelium of the GI tract is assisted by an array of physiological defense mechanisms, including mucus, digestive enzymes, acid, and the largest collection of lymphoid tissue in the body, the *gut-associated lymphoid tissue (GALT).* By one estimate, 80% of all lymphocytes [p. 514] in the body are found in the small intestine.

The human body meets these sometimes conflicting physiological challenges by coordinating motility and secretion to maximize digestion and absorption.

We Secrete More Fluid than We Ingest

In a typical day, 9 liters of fluid pass through the lumen of an adult's gastrointestinal tract—equal to the contents of three 3-liter soft drink bottles! Only about 2 liters of that volume enter the GI system through the mouth. The remaining 7 liters of fluid come from body water secreted along with ions, enzymes, and mucus (see Fig. 21.3). The ions are transported from the ECF into the lumen. Water then follows the osmotic gradient created by this transfer of solutes from one side of the epithelium to the other. Water moves through the epithelial cells via channels or through leaky junctions between cells (the paracellular pathway).

Gastrointestinal epithelial cells, like those in the kidney, are *polarized* [p. 150], with distinct apical and basolateral membranes. Each cell surface contains membrane proteins for solute and water movement, many of them similar to those of the renal tubule. The arrangement of transport proteins on the apical and basolateral membranes determines the direction of movement of solutes and water across the epithelium.

Digestive Enzymes Digestive enzymes are secreted either by exocrine glands (salivary glands and the pancreas) or by epithelial cells in the stomach and small intestine. Enzymes are proteins, which means that they are synthesized on the rough endoplasmic reticulum, packaged by the Golgi complex into secretory vesicles, and then stored in the cell until needed. On demand, they are released by exocytosis [p. 148]. Many intestinal enzymes remain bound to the apical membranes of intestinal cells, anchored by transmembrane protein "stalks" or lipid anchors [p. 64].

Some digestive enzymes are secreted in an inactive *proenzyme* form known collectively as *zymogens* [p. 100]. Zymogens must be activated in the GI lumen before they can carry out digestion. Synthesizing the enzymes in a nonfunctional form allows them to be stockpiled in the cells that make them without damaging those cells. Zymogen names often have the suffix *–ogen* added to the enzyme name, such as *pepsinogen*.

Mucus *Mucus* is a viscous secretion composed primarily of glycoproteins collectively called **mucins.** The primary functions of mucus are to form a protective coating over the GI mucosa and to lubricate the contents of the gut. Mucus is made in specialized exocrine cell called *mucous cells* in the stomach and salivary glands, and *goblet cells* in the intestine [Fig. 3.10, p. 78]. Goblet cells make up between 10% and 24% of the intestinal cell population.

The signals for mucus release include parasympathetic innervation, a variety of neuropeptides found in the enteric nervous system, and cytokines from immunocytes. Parasitic infections and inflammatory processes in the gut also cause substantial increases in mucus secretion as the body attempts to fortify its protective barrier.

CONCEPT CHECK

4. Define digestion. What is the difference between digestion and metabolism [p. 102]?

5. Why is the digestive system associated with the largest collection of lymphoid tissue in the body?

6. Draw a cell showing (1) an enzyme in a cytoplasmic secretory vesicle, (2) exocytosis of the vesicle, and (3) the enzyme remaining bound to the surface membrane of the cell rather than floating away.

Digestion and Absorption Make Food Usable

Most GI secretions facilitate digestion. The GI system digests macromolecules into absorbable units using a combination of mechanical and chemical breakdown. Chewing and churning create smaller pieces of food with more surface area exposed to digestive enzymes. The pH at which different digestive enzymes function best [p. 100] reflects the location where they are most active. For example, enzymes that act in the stomach work well at acidic pH, and those that are secreted into the small intestine work best at alkaline pH.

Most absorption takes place in the small intestine, with additional absorption of water and ions in the large intestine. Absorption, like secretion, uses many of the same transport proteins as the kidney tubule. Once absorbed, nutrients enter the blood or the lymphatic circulation.

Motility: GI Smooth Muscle Contracts Spontaneously

Motility in the gastrointestinal tract serves two purposes: moving food from the mouth to the anus and mechanically mixing food to break it into uniformly small particles. This mixing maximizes exposure of the particles to digestive enzymes by increasing particle surface area. Gastrointestinal motility is determined by the properties of the GI smooth muscle and is modified by chemical input from nerves, hormones, and paracrine signals.

Most of the gastrointestinal tract is composed of single-unit smooth muscle, with groups of cells electrically connected by gap junctions [p. 405] to create contracting segments. Different regions exhibit different types of contraction. **Tonic contractions** are sustained for minutes or hours. They occur in some smooth muscle sphincters and in the anterior portion of the stomach. **Phasic contractions,** with contraction-relaxation cycles lasting only a few seconds, occur in the posterior region of the stomach and in the small intestine.

Cycles of smooth muscle contraction and relaxation are associated with cycles of depolarization and repolarization known as **slow wave potentials** or simply *slow waves* (**FIG. 21.4a**). Current research indicates that slow waves originate in a network of cells called the **interstitial cells of Cajal** (named after the Spanish neuroanatomist Santiago Ramón y Cajal), or ICCs. These modified smooth muscle cells lie between smooth muscle layers and the intrinsic nerve plexuses, and they may act as an intermediary between the neurons and smooth muscle.

FIG. 21.4 **ESSENTIALS**

Gastrointestinal Motility

(a) Slow waves are spontaneous depolarizations in GI smooth muscle.

Membrane potential (mV)

Slow wave

Action potential

Threshold

Action potentials fire when slow wave potentials exceed threshold.

Force of muscle contraction

The force and duration of muscle contraction are directly related to the amplitude and frequency of action potentials.

⟶ Time ⟶

Q FIGURE QUESTION

Why do the peaks of the contraction waves occur after the peaks of the action potentials?

(b) The **migrating motor complex** (MMC) is a series of contractions that begin in the empty stomach and end in the large intestine.

(c) Peristaltic contractions are responsible for forward movement.

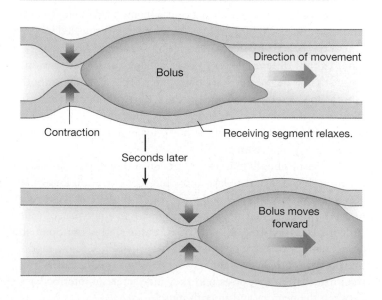

Bolus

Direction of movement

Contraction

Receiving segment relaxes.

Seconds later

Bolus moves forward

(d) Segmental contractions are responsible for mixing.

Alternate segments contract, and there is little or no net forward movement.

It appears that ICCs function as the pacemakers for slow wave activity in different regions of the GI tract, just as cells of the cardiac conduction system act as pacemakers for the heart [p. 455]. Slow wave potentials differ from myocardial pacemaker potentials in that the GI waves occur at a much slower frequency (3–12 waves/min GI versus 60–90 waves/min myocardial). In addition, slow wave frequency varies by region of the digestive tract, ranging from 3 waves/min in the stomach to 12 waves/min in the duodenum.

Slow waves that begin spontaneously in ICCs spread to adjacent smooth muscle layers through gap junctions. Just as in the cardiac conducting system, the fastest pacemaker in a group of ICCs sets the pace for the entire group [p. 456]. The observation that ICCs seem to coordinate GI motility now has researchers working to establish a link between ICCs and functional bowel disorders, such as irritable bowel syndrome and chronic constipation.

One difference between slow waves and cardiac pacemaker potentials is that slow waves do not reach threshold with each cycle, and a slow wave that does not reach threshold will not cause muscle contraction. When a slow wave potential does reach threshold, voltage-gated Ca^{2+} channels in the muscle fiber open, Ca^{2+} enters, and the cell fires one or more action potentials. The depolarization phase of the slow wave action potential, like that in myocardial autorhythmic cells, is the result of Ca^{2+} entry into the cell. In addition, Ca^{2+} entry initiates muscle contraction [p. 407].

Contraction of smooth muscle, like that of cardiac muscle, is graded according to the amount of Ca^{2+} that enters the fiber. The longer the duration of the slow wave, the more action potentials fire, and the greater the contraction force in the muscle. The likelihood of a slow wave firing an action potential depends primarily on input from the enteric nervous system.

GI Smooth Muscle Exhibits Different Patterns of Contraction

Muscle contractions in the gastrointestinal tract occur in three patterns that bring about different types of movement within the tract. Between meals, when the tract is largely empty, a series of contractions begins in the stomach and passes slowly from section to section, each series taking about 90 minutes to reach the large intestine. This pattern, known as the **migrating motor complex,** is a "housekeeping" function that sweeps food remnants and bacteria out of the upper GI tract and into the large intestine (Fig. 21.4b).

Muscle contractions during and following a meal fall into one of two other patterns. (Fig. 21.4) **Peristalsis** {*peri-*, surrounding + *stalsis*, contraction} is progressive waves of contraction that move from one section of the GI tract to the next, just like the human "waves" that ripple around a football stadium or basketball arena. In peristalsis, circular muscles contract just behind a mass, or **bolus,** of food (Fig. 21.4c). This contraction pushes the bolus forward into a *receiving segment*, where the circular muscles are relaxed. The receiving segment then contracts, continuing the forward movement.

Peristaltic contractions push a bolus forward at speeds between 2 and 25 cm/sec. Peristalsis in the esophagus propels material from pharynx to stomach. Peristalsis contributes to food mixing in the stomach but in normal digestion, intestinal peristaltic waves are limited to short distances.

In **segmental contractions,** short (1–5 cm) segments of intestine alternately contract and relax (Fig. 21.4d). In the contracting segments, circular muscles contract while longitudinal muscles relax. These contractions may occur randomly along the intestine or at regular intervals. Alternating segmental contractions churn the intestinal contents, mixing them and keeping them in contact with the absorptive epithelium. When segments contract sequentially, in an oral-to-aboral direction {*ab-*, away}, intestinal contents are propelled short distances.

Motility disorders are among the more common gastrointestinal problems. They range from esophageal spasms and delayed gastric (stomach) emptying to constipation and diarrhea. *Irritable bowel syndrome* is a chronic functional disorder characterized by altered bowel habits and abdominal pain.

CONCEPT CHECK

7. What is the difference between absorption and secretion?

8. How do fats absorbed into the lymphatic system get into the general circulation for distribution to cells? [*Hint:* p. 499]

9. Why are some sphincters of the digestive system tonically contracted?

CLINICAL FOCUS

Diabetes: Delayed Gastric Emptying

Diabetes mellitus has an impact on almost every organ system. The digestive tract is not exempt. One problem that plagues more than a third of all people with diabetes is *gastroparesis,* also called delayed gastric emptying. In these patients, the migrating motor complex is absent between meals, and the stomach empties very slowly after meals. Many patients suffer nausea and vomiting as a result. The causes of diabetic gastroparesis are unclear, but recent studies of animal models and human patients show loss or dysfunction of the interstitial cells of Cajal, which serve as pacemakers and as a link between GI smooth muscle and the enteric and autonomic nervous systems. Adopting the cardiac model of an external pacemaker, scientists are now testing an implantable gastric pacemaker to promote gastric motility in diabetic patients with severe gastroparesis.

REGULATION OF GI FUNCTION

Of the four GI processes, motility and secretion are the primary regulated functions. If food moves through the system too rapidly, there will not be enough time for everything in the lumen to be digested and absorbed. Secretion is regulated so that the appropriate digestive enzymes can break down food into an absorbable form. Digestion in turn depends on motility and secretion.

Scientists used to believe that nutrient absorption was not regulated and that "what you eat is what you get." Now, however, evidence indicates that some nutrient absorption can be altered in response to long-term environmental changes.

The Enteric Nervous System Can Act Independently

The enteric nervous system (ENS) was first recognized more than a century ago, when scientists noted that isolated sections of intestine removed from the body created a reflex wave of peristaltic contraction when pressure in the lumen increased. What they observed was the ability of the ENS to carry out a reflex independent of control by the central nervous system (CNS).

In this respect, the ENS is much like the nerve networks of jellyfish and sea anemones (phylum Cnidaria) [p. 275]. You might have seen sea anemones being fed at an aquarium. As a piece of shrimp or fish drifts close to the tentacles, they begin to wave, picking up chemical "odors" through the water. Once the food contacts the tentacles, it is directed toward the mouth, passed from one tentacle to another until it disappears into the digestive cavity.

This purposeful reflex is accomplished without a brain, eyes, or a nose. The anemone's nervous system consists of a nerve network with sensory neurons, interneurons, and efferent neurons that control the muscles and secretory cells of the anemone's body. The neurons of the Cnidarian network are linked in a way that allows them to integrate information and act on it. In the same way that an anemone captures its food, the human ENS receives stimuli and acts on them. The enteric nervous system controls motility, secretion, and growth of the digestive tract.

Anatomically and functionally, the ENS shares many features with the CNS:

1. *Intrinsic neurons.* The **intrinsic neurons** of the two nerve plexuses of the digestive tract are those neurons that lie completely within the wall of the gut, just as interneurons are completely contained within the CNS. Autonomic neurons that bring signals from the CNS to the digestive system are called **extrinsic neurons.**

2. *Neurotransmitters and neuromodulators.* ENS neurons release more than 30 neurotransmitters and neuromodulators, most of which are identical to molecules found in the brain. These neurotransmitters are sometimes called *nonadrenergic, noncholinergic* to distinguish them from the traditional autonomic neurotransmitters norepinephrine and acetylcholine. Among the best known GI neurotransmitters and neuromodulators are serotonin, vasoactive intestinal peptide, and nitric oxide.

3. *Glial support cells.* The glial cells of neurons within the ENS are more similar to astroglia of the brain than to Schwann cells of the peripheral nervous system.

4. *Diffusion barrier.* The capillaries that surround ganglia in the ENS are not very permeable and create a diffusion barrier that is similar to the blood-brain barrier of cerebral blood vessels.

5. *Integrating center.* As noted earlier, reflexes that originate in the GI tract can be integrated and acted on without neural signals leaving the ENS. For this reason, the neuron network of the ENS is its own integrating center, much like the brain and spinal cord.

It was once thought that if we could explain how the ENS integrates simple behaviors, we could use the system as a model for CNS function. But studying ENS function is difficult because enteric reflexes have no discrete command center. Instead, in an interesting twist, GI physiologists are applying information gleaned from studies of the brain and spinal cord to investigate ENS function. The complex interactions between the enteric and central nervous systems, the endocrine system, and the immune system promise to provide scientists with questions to investigate for many years to come.

Short Reflexes Integrate in the Enteric Nervous System The enteric nerve plexuses in the gut wall act as a "little brain," allowing local reflexes to begin, be integrated, and end completely in the GI tract (**FIG. 21.5**, red arrows). Reflexes that originate within the enteric nervous system and are integrated there without outside input are called **short reflexes.** The submucosal plexus contains sensory neurons that receive signals from the lumen of the gut. The ENS network integrates this sensory information, then initiates responses. The submucosal plexus controls secretion by GI epithelial cells. Myenteric plexus neurons in the muscularis externa influence motility.

Long Reflexes Integrate in the CNS Although the ENS can work in isolation, it also sends sensory information to the CNS and receives input from the CNS through autonomic neurons. A classic neural reflex begins with a stimulus transmitted along a sensory neuron to the CNS, where the stimulus is integrated and acted on. In the digestive system, some classic reflexes originate with sensory receptors in the GI tract, but others originate outside the digestive system (Fig. 21.5, gray arrows). No matter where they originate, digestive reflexes integrated in the CNS are called **long reflexes.**

Long reflexes that originate outside the digestive system include feedforward reflexes [p. 17] and emotional reflexes. These reflexes are called **cephalic reflexes** because they originate in the brain {*cephalicus,* head}. *Feedforward reflexes* begin with stimuli such as the sight, smell, sound, or thought of food, and they prepare the digestive system for food that the brain is anticipating. For example, if you are hungry and smell dinner cooking, your mouth waters and your stomach growls.

FIG. 21.5 Integration of digestive reflexes

→ **Long reflexes** are integrated in the CNS. Some long reflexes originate outside the GI tract, but others originate in the enteric nervous system.

→ **Short reflexes** originate in the enteric nervous system and are carried out entirely within the wall of the gut.

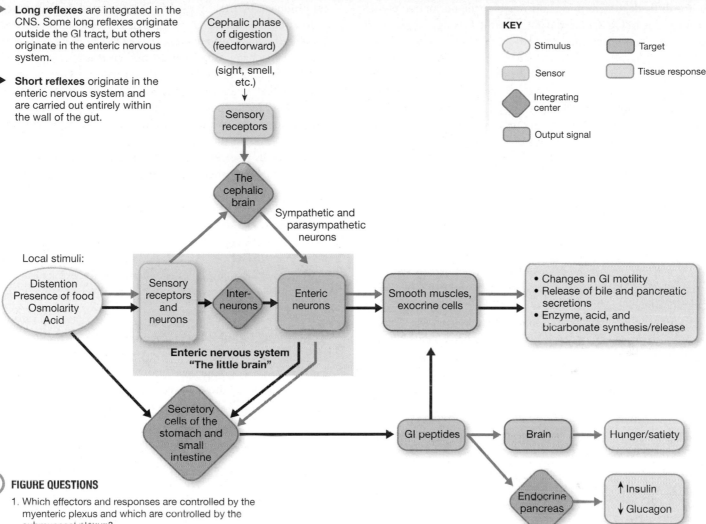

KEY

- Stimulus
- Sensor
- Integrating center
- Output signal
- Target
- Tissue response

Cephalic phase of digestion (feedforward)

(sight, smell, etc.)

Sensory receptors

The cephalic brain

Sympathetic and parasympathetic neurons

Local stimuli:

Distention Presence of food Osmolarity Acid

Sensory receptors and neurons

Inter-neurons

Enteric neurons

Enteric nervous system "The little brain"

Smooth muscles, exocrine cells

- Changes in GI motility
- Release of bile and pancreatic secretions
- Enzyme, acid, and bicarbonate synthesis/release

Secretory cells of the stomach and small intestine

GI peptides

Brain

Hunger/satiety

Endocrine pancreas

↑ Insulin
↓ Glucagon

Q FIGURE QUESTIONS

1. Which effectors and responses are controlled by the myenteric plexus and which are controlled by the submucosal plexus?

2. What type of sensory receptor responds to stretch? to osmolarity? to products of digestion?

Emotional reflexes and their influence on the GI tract illustrate another link between the brain and the digestive system. GI responses to emotions range from traveler's constipation to "butterflies in the stomach" to psychologically induced vomiting and diarrhea.

In long reflexes, the smooth muscle and glands of the GI tract are under autonomic control. In general, we say that the parasympathetic division is excitatory and enhances GI functions, leading to its nickname of "rest and digest." Most parasympathetic neurons to the GI tract are found in the vagus nerve. Sympathetic neurons usually inhibit GI function.

CONCEPT CHECK

10. Excitation of GI function by the parasympathetic division and inhibition by the sympathetic division is an example of what kind of control?

GI Peptides Include Hormones, Neuropeptides, and Cytokines

Peptides secreted by cells of the digestive tract may act as hormones or paracrine signals. Some of these GI peptides were first identified and named in other body systems. Because their names have nothing to do with their function in the gastrointestinal system, learning the terminology can be challenging.

In the digestive system, GI peptides excite or inhibit motility and secretion. Some paracrine peptides are secreted into the lumen, where they combine with receptors on the apical membrane of the luminal epithelium to elicit a response. Others are secreted into the extracellular fluid where they diffuse short distances to act on neighboring cells.

GI peptides also act outside the GI tract, and some of their most interesting actions involve the brain. For example, in experimental studies the GI hormone **cholecystokinin (CCK)** enhances *satiety*, the feeling that hunger has been satisfied. However, CCK is also manufactured by neurons and functions as a neurotransmitter in the brain, so it is difficult to determine how much of the normal satiety response is due to CCK from the gut. Another GI peptide, *ghrelin,* is secreted by the stomach and acts on the brain to increase food intake.

Researchers have now sequenced more than 30 peptides from the GI mucosa, but only some of them are widely accepted as hormones. A few peptides have well-defined paracrine effects, but most fall into a long list of candidate hormones. In addition, we know of nonpeptide regulatory molecules, such as histamine, that function as paracrine signals. Because of the uncertainty associated with the field, we restrict our focus in this chapter to the major regulatory molecules.

GI Hormones GI hormones, like all hormones, are secreted into the blood and transported throughout the body. They act on the GI tract, on accessory organs such as the pancreas, and on distant targets, such as the brain.

The hormones of the gastrointestinal tract occupy an interesting place in the history of endocrinology. In 1902, two English physiologists, W. M. Bayliss and E. H. Starling, discovered that acidic chyme entering the small intestine from the stomach caused the release of pancreatic juices even when all nerves to the pancreas were cut. Because the only communication remaining between intestine and pancreas was the blood supply that ran between them, Bayliss and Starling postulated the existence of some blood-borne (*humoral*) factor released by the intestine.

When duodenal extracts applied directly to the pancreas stimulated secretion, they knew they were dealing with a chemical produced by the duodenum. They named the substance *secretin.* Starling further proposed that the general name *hormone,* from the Greek word meaning "I excite," be given to all humoral agents that act at a site distant from their release.

In 1905, J. S. Edkins postulated the existence of a gastric hormone that stimulated gastric acid secretion. It took more than 30 years for researchers to isolate a relatively pure extract of the gastric hormone, and it was 1964 before the hormone, named *gastrin,* was finally purified.

Why was research on the digestive hormones so slow to develop? A major reason is that GI hormones are secreted by isolated endocrine cells scattered among other cells of the mucosal epithelium. At one time, the only way to obtain these hormones was to make a crude extract of the entire epithelium, a procedure that also liberated digestive enzymes and paracrine molecules made in adjacent cells. For this reason, it was very difficult to tell whether the physiological effect elicited by the extract came from one hormone, from more than one hormone, or from a paracrine signal such as histamine.

GI Hormone Families The gastrointestinal hormones are usually divided into three families. All the members of a family have similar amino acid sequences, and in some cases there is overlap in their ability to bind to receptors. The sources, targets, and effects of the major GI hormones are summarized in **TABLE 21.1**.

TABLE 21.1	The GI Hormones			
	Stimulus for Release	**Primary Target(s)**	**Primary Effect(s)**	**Other Information**
Stomach				
Gastrin (G Cells)	Peptides and amino acids; neural reflexes	ECL cells and parietal cells	Stimulates gastric acid secretion and mucosal growth	Somatostatin inhibits release.
Intestine				
Cholecystokinin (CCK)	Fatty acids and some amino acids	Gallbladder, pancreas, stomach	• Stimulates gallbladder contraction and pancreatic enzyme secretion • Inhibits gastric emptying and acid secretion	• Promotes satiety • Some effects may be due to CCK as a neurotransmitter.
Secretin	Acid in small intestine	Pancreas, stomach	• Stimulates HCO_3^- secretion • Inhibits gastric emptying and acid secretion	
Motilin	Fasting: periodic release every 1.5–2 hours	Gastric and intestinal smooth muscle	Stimulates migrating motor complex	Inhibited by eating a meal
Gastric Inhibitory Peptide (GIP)	Glucose, fatty acids, and amino acids in small intestine	Beta cells of pancreas	• Stimulates insulin release (feedforward mechanism) • Inhibits gastric emptying and acid secretion	
Glucagon-Like Peptide-1 (GLP-1)	Mixed meal that includes carbohydrates or fats in the lumen	Endocrine pancreas	• Stimulates insulin release • Inhibits glucagon release and gastric function	Promotes satiety

The *gastrin family* includes the hormones **gastrin** and *cholecystokinin* (CCK), plus several variants of each. Their structural similarity means that gastrin and CCK can bind to and activate the same CCKB receptor.

The *secretin family* includes **secretin; vasoactive intestinal peptide (VIP),** a nonadrenergic-noncholinergic neurotransmitter; and **GIP,** a hormone known originally as *gastric inhibitory peptide* because it inhibited gastric acid secretion in early experiments. Subsequent studies, however, indicated that GIP administered in lower physiological doses does not block acid secretion. Researchers proposed a new name with the same initials—**glucose-dependent insulinotropic peptide**—that more accurately describes the hormone's action: it stimulates insulin release in response to glucose in the intestinal lumen. However, for the most part *gastric inhibitory peptide* has remained the preferred name.

Another member of the secretin family is the hormone **glucagon-like peptide-1** (GLP-1). GIP and GLP-1 act together as feedforward signals for insulin release, as you will learn when you study the endocrine pancreas [Chapter 22].

The third family of peptides contains those that do not fit into the other two families. The primary member of this group is the hormone **motilin.** Increases in motilin secretion are associated with the migrating motor complex.

In the remainder of this chapter, we integrate motility, secretion, digestion, and absorption as we follow food passing through the GI tract. **FIGURE 21.6** is a summary of the main events that occur in each section of the GI tract. Food processing traditionally is divided into three phases: a cephalic phase, a gastric phase, and an intestinal phase.

FIG. 21.6 Overview of digestive function

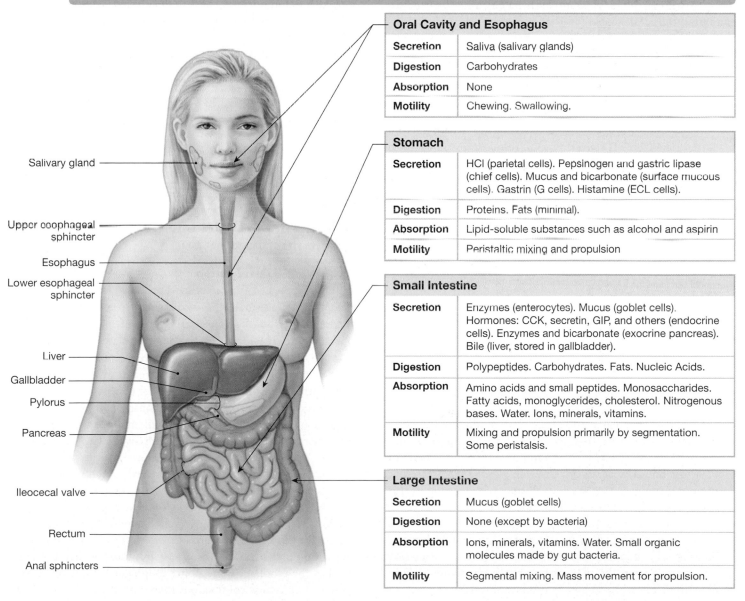

Oral Cavity and Esophagus	
Secretion	Saliva (salivary glands)
Digestion	Carbohydrates
Absorption	None
Motility	Chewing. Swallowing.

Stomach	
Secretion	HCl (parietal cells). Pepsinogen and gastric lipase (chief cells). Mucus and bicarbonate (surface mucous cells). Gastrin (G cells). Histamine (ECL cells).
Digestion	Proteins. Fats (minimal).
Absorption	Lipid-soluble substances such as alcohol and aspirin
Motility	Peristaltic mixing and propulsion

Small Intestine	
Secretion	Enzymes (enterocytes). Mucus (goblet cells). Hormones: CCK, secretin, GIP, and others (endocrine cells). Enzymes and bicarbonate (exocrine pancreas). Bile (liver, stored in gallbladder).
Digestion	Polypeptides. Carbohydrates. Fats. Nucleic Acids.
Absorption	Amino acids and small peptides. Monosaccharides. Fatty acids, monoglycerides, cholesterol. Nitrogenous bases. Water. Ions, minerals, vitamins.
Motility	Mixing and propulsion primarily by segmentation. Some peristalsis.

Large Intestine	
Secretion	Mucus (goblet cells)
Digestion	None (except by bacteria)
Absorption	Ions, minerals, vitamins. Water. Small organic molecules made by gut bacteria.
Motility	Segmental mixing. Mass movement for propulsion.

Salivary gland

Upper esophageal sphincter

Esophagus

Lower esophageal sphincter

Liver

Gallbladder

Pylorus

Pancreas

Ileocecal valve

Rectum

Anal sphincters

INTEGRATED FUNCTION: THE CEPHALIC PHASE

Digestive processes in the body begin before food ever enters the mouth. Simply smelling, seeing, or even *thinking* about food can make our mouths water and our stomachs rumble. These long reflexes that begin in the brain create a feedforward response known as the **cephalic phase** of digestion.

Anticipatory stimuli and the stimulus of food in the oral cavity activate neurons in the medulla oblongata. The medulla in turn sends an efferent signal through autonomic neurons to the salivary glands, and through the vagus nerve to the enteric nervous system. In response to these signals, the stomach, intestine, and accessory glandular organs begin secretion and increase motility in anticipation of the food to come.

Chemical and Mechanical Digestion Begins in the Mouth

When food first enters the mouth, it is met by a flood of the secretion we call *saliva*. Saliva has four important functions:

1. *Soften and moisten food.* The water and mucus in saliva soften and lubricate food to make it easier to swallow. You can appreciate this function if you have ever tried to swallow a dry soda cracker without chewing it thoroughly.
2. *Digestion of starch.* Chemical digestion begins with the secretion of *salivary amylase*. Amylase breaks starch into maltose after the enzyme is activated by Cl^- in saliva. If you chew on an unsalted soda cracker for a long time, you may be able to detect the conversion of the cracker's flour starch to maltose, which is sweeter.
3. *Taste.* Saliva dissolves food so that we can taste it [p. 325].
4. *Defense.* The final function of saliva is defense. *Lysozyme* is an antibacterial salivary enzyme, and salivary *immunoglobulins* disable bacteria and viruses. In addition, saliva helps wash the teeth and keep the tongue free of food particles.

Mechanical digestion of food begins in the oral cavity with chewing. The lips, tongue, and teeth all contribute to the **mastication** {*masticare,* to chew} of food, creating a softened, moistened mass (*bolus*) that can be easily swallowed.

Saliva Is an Exocrine Secretion

Saliva is a complex hyposmotic fluid that contains water, ions, mucus, and proteins such as enzymes and immunoglobulins. Three pairs of salivary glands produce as much as 1.5 liters of saliva a day. Salivary glands are exocrine glands, with secretory epithelium arranged in grape-like clusters of cells called **acini** {*acinus,* grape or berry}. Each acinus surrounds a duct, and the individual ducts join to form larger and larger ducts (like the stems on a bunch of grapes). The main excretory duct of each gland empties into the mouth.

Secretions from the three pairs of salivary glands vary in composition. The parotid glands produce a watery solution of enzymes while sublingual glands produce a mucus-rich saliva. Secretions from the submandibular glands are mixed, with both mucus and enzymes.

The production of saliva is a two-step process. The initial fluid secreted by the acinar cells resembles extracellular fluid in its ionic composition: an isotonic NaCl solution. As this fluid passes through the duct on its way to the oral cavity, epithelial cells along the duct reabsorb NaCl and secrete K^+ and bicarbonate ion until the ion ratio in the duct fluid is more like that of intracellular fluid (high in K^+ and low in Na^+). The apical membranes of the duct cells have very low water permeability, and the net removal of solute from the secreted fluid results in saliva that is hyposmotic to plasma.

Salivation is under autonomic control and can be triggered by multiple stimuli, including the sight, smell, touch, and even thought of food. Parasympathetic innervation is the primary stimulus for secretion of saliva, but there is also some sympathetic innervation to the glands. In ancient China, a person suspected of a crime was sometimes given a mouthful of dry rice to chew during questioning. If he could produce enough saliva to moisten the rice and swallow it, he went free. If his nervous state dried up his salivary reflex, however, he was pronounced guilty. Recent research has confirmed that stress, such as that associated with lying or anxiety from being questioned, decreases the volume of salivary secretion.

CONCEPT CHECK
11. How do mucin, amylase, and immunoglobulins move from salivary gland epithelial cells into the lumen of the gland? (*Hint:* They are all proteins.)

Swallowing Moves Food from Mouth to Stomach

Swallowing, or **deglutition** {*glutire,* to swallow}, is a reflex action that pushes a bolus of food or liquid into the esophagus (**FIG. 21.7**). The stimulus for swallowing is pressure created when the tongue pushes the bolus against the soft palate and the back of the mouth. Pressure from the bolus activates sensory neurons that run through the *glossopharyngeal nerve* (cranial nerve IX) to a swallowing center in the medulla oblongata.

Output from the swallowing center consists of somatic motor neurons that control the skeletal muscles of the pharynx and upper esophagus as well as autonomic neurons that act on the lower portions of the esophagus. As the swallowing reflex begins, the soft palate elevates to close off the nasopharynx. Muscle contractions move the larynx up and forward, which helps close off the trachea and open the upper esophageal sphincter.

As the bolus moves down toward the esophagus, the **epiglottis** folds down, completing closure of the upper airway and preventing food and liquid from entering the airways. At the same time, respiration is briefly inhibited. When the bolus reaches the esophagus, the upper esophageal sphincter relaxes. Waves of peristaltic contractions then push the bolus toward the

FIG. 21.7 Deglutition: The swallowing reflex

Swallowing is integrated in the medulla oblongata. Sensory afferents in cranial nerve IX and somatic motor and autonomic neurons mediate the reflex.

1 Tongue pushes bolus against soft palate and back of mouth, triggering swallowing reflex.

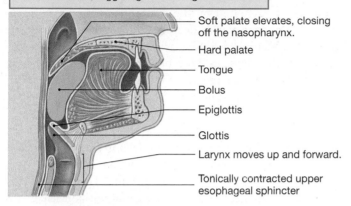

Soft palate elevates, closing off the nasopharynx.

Hard palate

Tongue

Bolus

Epiglottis

Glottis

Larynx moves up and forward.

Tonically contracted upper esophageal sphincter

2 Breathing is inhibited as the bolus passes the closed airway.

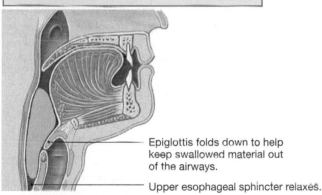

Epiglottis folds down to help keep swallowed material out of the airways.

Upper esophageal sphincter relaxes.

3 Food moves downward into the esophagus, propelled by peristaltic waves and aided by gravity.

stomach, aided by gravity. Gravity is not required, however, as you know if you have ever participated in the party trick of swallowing while standing on your head.

The lower end of the esophagus lies just below the diaphragm and is separated from the stomach by the lower esophageal sphincter. This area is not a true sphincter but a region of relatively high muscle tension that acts as a barrier between the esophagus and the stomach. When food is swallowed, the tension relaxes, allowing the bolus to pass into the stomach.

If the lower esophageal sphincter does not stay contracted, gastric acid and pepsin can irritate the lining of the esophagus, leading to the pain and irritation of *gastroesophageal reflux* {*re-*, backward + *fluxus*, flow}, more commonly called heartburn. During the inspiratory phase of breathing, when the intrapleural pressure falls, the walls of the esophagus expand [p. 549]. This expansion creates subatmospheric pressure in the esophageal lumen and can suck acidic contents out of the stomach if the sphincter is relaxed. The churning action of the stomach when filled with food can also squirt acid back into the esophagus if the sphincter is not fully contracted. *Gastroesophageal reflux disorder* or GERD is one the most common digestive disorders in American society.

INTEGRATED FUNCTION: THE GASTRIC PHASE

About 3.5 liters of food, drink, and saliva enter the fundus of the stomach each day. The stomach has three general functions:

1. **Storage.** The stomach stores food and regulates its passage into the small intestine, where most digestion and absorption take place.
2. **Digestion.** The stomach chemically and mechanically digests food into the soupy mixture of uniformly small particles called chyme.
3. **Defense.** The stomach protects the body by destroying many of the bacteria and other pathogens that are swallowed with food or trapped in airway mucus. At the same time, the stomach must protect itself from being damaged by its own secretions.

Before food even arrives, digestive activity in the stomach begins with the long **vagal reflex** of the cephalic phase (**FIG. 21.8**). Then, once food enters the stomach, stimuli in the gastric lumen initiate a series of short reflexes that constitute the **gastric phase** of digestion.

In gastric phase reflexes, distension of the stomach and the presence of peptides or amino acids in the lumen activate endocrine cells and enteric neurons. Hormones, neurotransmitters, and paracrine molecules then influence motility and secretion.

The Stomach Stores Food

When food arrives from the esophagus, the stomach relaxes and expands to hold the increased volume. This neurally mediated reflex is called *receptive relaxation*. The upper half of the stomach remains relatively quiet, holding food until it is ready to be digested. The storage function of the stomach is perhaps the least

FIG. 21.8 Cephalic and gastric phase reflexes

The sight, smell, and taste of food initiate long reflexes that prepare the stomach for the arrival of food.

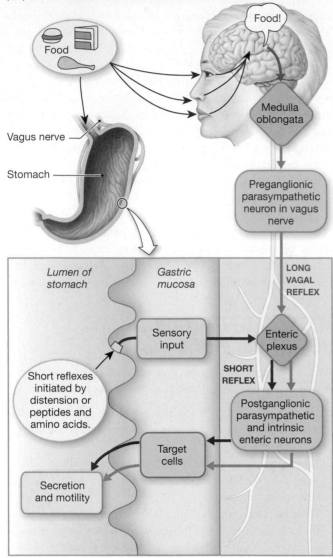

the pylorus into the duodenum. Enhanced gastric motility during a meal is primarily under neural control and is stimulated by distension of the stomach.

Gastric Secretions Protect and Digest

The lumen of the stomach is lined with mucus-producing epithelium punctuated by the openings of *gastric pits*. The pits lead to **gastric glands** deep within the mucosal layer (see Fig. 21.1e). Multiple cell types within the glands produce gastric acid (HCl), enzymes, hormones, and paracrine molecules. The various secretions of gastric mucosa cells, their stimuli for release, and their functions are summarized in **FIGURE 21.9** and described next.

Gastrin Secretion G cells, found deep in the gastric glands, secrete the hormone **gastrin** into the blood. In short reflexes, gastrin release is stimulated by the presence of amino acids and peptides in the stomach and by distension of the stomach. Coffee (even if decaffeinated) also stimulates gastrin release—one reason people with excess acid secretion syndromes are advised to avoid coffee.

Gastrin release is also triggered by neural reflexes. Short reflexes are mediated by an ENS neurotransmitter called **gastrin-releasing peptide** (GRP). In cephalic reflexes, parasympathetic neurons from the vagus nerve stimulate G cells to release gastrin into the blood.

Gastrin's primary action is to promote acid release. It does this directly by acting on parietal cells and indirectly by stimulating histamine release.

Acid Secretion Parietal cells deep in the gastric glands secrete **gastric acid** (HCl) into the lumen of the stomach. Acid secretion in the stomach averages 1–3 liters per day and can create a luminal pH as low as 1. The cytoplasmic pH of the parietal cells is about 7.2, which means the cells are pumping H^+ against a gradient that can be 1.5 million times more concentrated in the lumen.

Gastric acid has multiple functions:

- Acid in the stomach lumen causes release and activation of pepsin, an enzyme that digests proteins.
- Acid triggers somatostatin release from D cells. Somatostatin is discussed later in the section on paracrine signals.
- HCl *denatures* proteins by breaking disulfide and hydrogen bonds that hold the protein in its tertiary structure [p. 32]. Unfolding protein chains make the peptide bonds between amino acids more accessible to digestion by pepsin.
- Gastric acid helps kill bacteria and other ingested microorganisms.
- Acid inactivates salivary amylase, stopping carbohydrate digestion that began in the mouth.

The parietal cell pathway for acid secretion is depicted in Figure 21.9c. The process begins when H^+ from water inside the parietal cell is pumped into the stomach lumen by an

obvious aspect of digestion. However, whenever we ingest more than we need from a nutritional standpoint, the stomach must regulate the rate at which food enters the small intestine.

Without such regulation, the small intestine would not be able to digest and absorb the load presented to it, and significant amounts of unabsorbed chyme would pass into the large intestine. The epithelium of the large intestine is not designed for large-scale nutrient absorption, so most of the chyme would become feces, resulting in diarrhea. This "dumping syndrome" is one of the less pleasant side effects of surgery that removes portions of either the stomach or small intestine.

While the upper stomach is quietly holding food, the lower stomach is busy with digestion. In the distal half of the stomach, a series of peristaltic waves pushes the food down toward the pylorus, mixing food with acid and digestive enzymes. As large food particles are digested to the more uniform texture of chyme, each contractile wave squirts a small amount of chyme through

FIG. 21.9 **ESSENTIALS**

Gastric Secretions

(a) Secretory Cells of the Gastric Mucosa

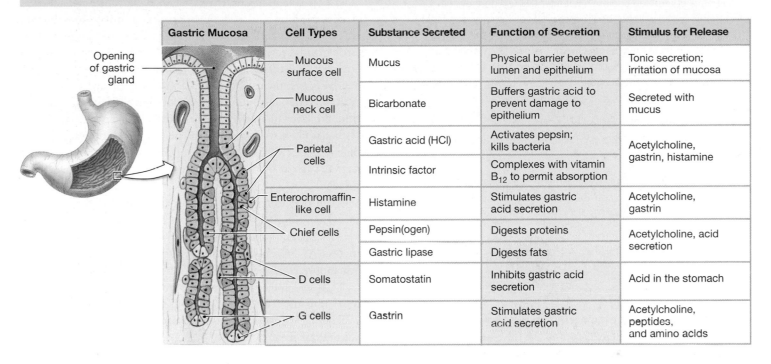

Gastric Mucosa	Cell Types	Substance Secreted	Function of Secretion	Stimulus for Release
Opening of gastric gland	Mucous surface cell	Mucus	Physical barrier between lumen and epithelium	Tonic secretion; irritation of mucosa
	Mucous neck cell	Bicarbonate	Buffers gastric acid to prevent damage to epithelium	Secreted with mucus
	Parietal cells	Gastric acid (HCl)	Activates pepsin; kills bacteria	Acetylcholine, gastrin, histamine
		Intrinsic factor	Complexes with vitamin B$_{12}$ to permit absorption	
	Enterochromaffin-like cell	Histamine	Stimulates gastric acid secretion	Acetylcholine, gastrin
	Chief cells	Pepsin(ogen)	Digests proteins	Acetylcholine, acid secretion
		Gastric lipase	Digests fats	
	D cells	Somatostatin	Inhibits gastric acid secretion	Acid in the stomach
	G cells	Gastrin	Stimulates gastric acid secretion	Acetylcholine, peptides, and amino acids

(b) Mucus-Bicarbonate Barrier

Gastric juice pH ~ 2 — Stomach lumen

The mucus layer is a physical barrier.

HCO_3^- Bicarbonate is a chemical barrier that neutralizes acid. HCO_3^- — Mucus layer

pH ~ 7 at cell surface

Gastric mucous cells — Mucus droplets

Capillary

(c) Acid Secretion in the Stomach

Lumen of stomach

Interstitial fluid

Capillary

H_2O

H^+ ← ATP → $H^+ + OH^-$

CA

K^+ → K^+ → HCO_3^- HCO_3^-

CO_2

Cl^- ← Cl^- ---- Cl^- ← Cl^-

Parietal cell

H^+-K^+-ATPase in exchange for K^+ entering the cell. Cl^- then follows the electrical gradient created by H^+ by moving through open chloride channels. The net result is secretion of HCl by the cell.

By learning the cellular mechanism of parietal cell acid secretion, scientists were able to develop a new class of drugs to treat oversecretion of gastric acid. These drugs, known as *proton pump inhibitors (PPIs),* block activity of the H^+-K^+-ATPase. Generic versions of some PPIs (omeprazole, for example) are available over the counter in the United States.

While acid is being secreted into the lumen, bicarbonate made from CO_2 and the OH^- from water is absorbed into the blood. The buffering action of HCO_3^- makes blood leaving the stomach less acidic, creating an *alkaline tide* that can be measured as a meal is being digested.

RUNNING **PROBLEM**

Brooke, who had always been healthy, was baffled. How could she have contracted cholera? But after discussing the methods of transmission with her healthcare providers, she realized that she hadn't been as careful about consuming only bottled water as she should have been. One of the doctors noticed that Brooke's medical history form listed Nexium® (esomeprazole) among her current medications. "You know, taking Nexium might have also contributed to your contracting cholera."

Q3: *Esomeprazole is a proton pump inhibitor (PPI). For what symptom or condition might Brooke have been taking this drug?*

Q4: *Why might taking a protein pump inhibitor like esomeprazole have increased Brooke's chances of contracting cholera?*

655 659 **672** 675 682 688

Enzyme Secretion The stomach produces two enzymes: pepsin and a gastric lipase. **Pepsin** carries out the initial digestion of proteins. It is particularly effective on collagen and therefore plays an important role in digesting meat.

Pepsin is secreted as the inactive enzyme *pepsinogen* by **chief cells** in the gastric glands. Acid stimulates pepsinogen release from chief cells through a short reflex mediated in the ENS (**FIG. 21.10**). Once in the stomach lumen, pepsinogen is cleaved to active pepsin by the action of H^+, and protein digestion begins.

Gastric lipase is co-secreted with pepsin. Lipases are enzymes that break down triglycerides. However, less than one-third of fat digestion takes place in the stomach.

Paracrine Secretions Paracrine secretions from the gastric mucosa include histamine, somatostatin, and intrinsic factor. **Histamine** is a paracrine signal secreted by **enterochromaffin-like cells (ECL cells)** in response to gastrin or acetylcholine stimulation. Histamine diffuses to its target, the parietal cells, and stimulates acid secretion by combining with H_2 *receptors* on parietal cells (Fig. 21.10). H_2 *receptor antagonists* (cimetidine and ranitidine, for example) that block histamine action are a second class of drugs used to treat acid hypersecretion.

Intrinsic factor is a protein secreted by the same gastric parietal cells that secrete acid. In the lumen of the stomach intrinsic factor complexes with vitamin B_{12}, a step that is needed for the vitamin's absorption in the intestine.

Somatostatin (SS), also known as hypothalamic growth hormone-inhibiting hormone, is secreted by **D cells** in the stomach. Somatostatin is the primary negative feedback signal for gastric phase secretion. It shuts down acid secretion directly and indirectly by decreasing gastrin and histamine secretion. Somatostatin also inhibits pepsinogen secretion (Fig. 21.10).

The Stomach Balances Digestion and Defense

Under normal conditions, the gastric mucosa protects itself from autodigestion by acid and enzymes with a mucus-bicarbonate barrier. **Mucous cells** on the luminal surface and in the neck of gastric glands secrete both substances. The mucus forms a physical barrier, and the bicarbonate creates a chemical buffer barrier underlying the mucus (Fig. 21.9b).

Researchers using microelectrodes have shown that the bicarbonate layer just above the cell surface in the stomach has a pH that is close to 7, even when the pH in the lumen is highly acidic at pH 2. Mucus secretion is increased when the stomach is irritated, such as by the ingestion of aspirin (acetylsalicylic acid) or alcohol.

Even the protective mucus-bicarbonate barrier can fail at times. In *Zollinger-Ellison syndrome,* patients secrete excessive levels of gastrin, usually from gastrin-secreting tumors in the pancreas. As a result, hyperacidity in the stomach overwhelms the normal protective mechanisms and causes a peptic ulcer. In peptic ulcers, acid and pepsin destroy the mucosa, creating holes that extend into the submucosa and muscularis of the stomach and duodenum. *Acid reflux* into the esophagus can erode the mucosal layer there as well.

Excess acid secretion is an uncommon cause of peptic ulcers. By far the most common causes are nonsteroidal anti-inflammatory drugs (NSAIDs), such as aspirin, and *Helicobacter pylori,* a bacterium that creates inflammation of the gastric mucosa.

For many years the primary therapy for excess acid secretion, or *dyspepsia,* was the ingestion of *antacids,* agents that neutralize acid in the gastric lumen. But as molecular biologists discovered the mechanism for acid secretion by parietal cells, the potential for new therapies became obvious. Today, we have two classes of drugs to fight hyperacidity: the H_2 receptor antagonists and proton pump inhibitors that block the H^+-K^+-ATPase.

INTEGRATED FUNCTION: THE INTESTINAL PHASE

Once chyme passes into the small intestine, the **intestinal phase** of digestion begins. Chyme entering the small intestine has undergone relatively little chemical digestion, so its entry must be controlled to avoid overwhelming the small intestine. Motility in the small intestine is also controlled. Intestinal contents are slowly propelled forward by a combination of segmental and peristaltic contractions. These actions mix chyme with enzymes and they expose digested nutrients to the mucosal epithelium for absorption. Forward movement of chyme through the intestine must be slow enough to allow digestion and absorption to go to completion. Parasympathetic innervation and the GI hormones gastrin and CCK promote intestinal motility; sympathetic innervation inhibits it.

About 5.5 liters of food, fluid, and secretions enter the small intestine each day, and about 3.5 liters of hepatic, pancreatic,

FIG. 21.10 Integration of cephalic and gastric phase secretion

The cephalic phase is initiated by the sight, smell, sound, or thought of food or by the presence of food in the mouth. The gastric phase is initiated by the arrival of food in the stomach.

1. Food or cephalic reflexes initiate gastric secretion of gastrin, histamine, and acid.

2. Gastrin stimulates acid secretion by direct action on parietal cells or indirectly through histamine.

3. Acid stimulates short reflex secretion of pepsinogen.

4. Somatostatin release by H⁺ is the negative feedback signal that modulates acid and pepsin release.

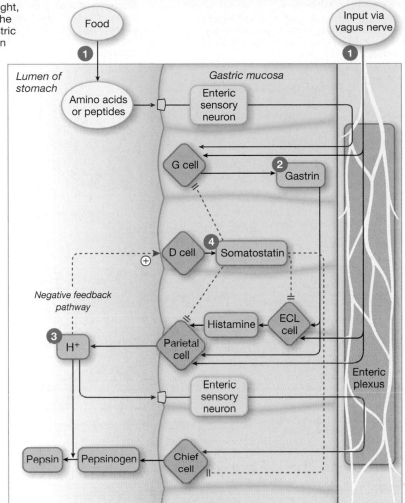

FIGURE QUESTIONS

1. Is the autonomic vagal input sympathetic or parasympathetic?

2. What are the neurotransmitter and receptor for this input?

and intestinal secretions are added there, making a total input of 9 liters into the lumen (see Fig. 21.3). All but about 1.5 liters of this volume is absorbed in the small intestine, mostly in the duodenum and jejunum.

The anatomy of the small intestine facilitates secretion, digestion, and absorption by maximizing surface area (**FIGS. 21.11** and 21.1f). At the macroscopic level, the surface of the lumen is sculpted into fingerlike villi and deep crypts. Most absorption takes place along the villi while fluid and hormone secretion and cell renewal from stem cells occurs in the crypts. On a microscopic level the apical surface of the enterocytes is modified into microvilli whose surfaces are covered with membrane-bound enzymes and a *glycocalyx* coat [p. 64]. The surface of the intestinal epithelium is called the **brush border** from the bristle-like appearance of the microvilli.

Most nutrients absorbed across the intestinal epithelium move into capillaries in the villi for distribution through the circulatory system. The exception is digested fats, most of which pass into lacteals of the lymphatic system. Venous blood from the digestive tract does not go directly back to the heart.

Instead, it passes into the *hepatic portal system* [p. 439]. This specialized region of the circulation has two sets of capillary beds: one that picks up absorbed nutrients at the intestine, and another that delivers the nutrients directly to the liver (**FIG. 21.12**).

The delivery of absorbed materials directly to the liver underscores the importance of that organ as a biological filter. Hepatocytes contain a variety of enzymes, such as the *cytochrome P450* isozymes, that metabolize drugs and xenobiotics and clear them from the bloodstream before they reach the systemic circulation. Hepatic clearance is one reason a drug administered orally must often be given in higher doses than the same drug administered by IV infusion.

Intestinal Secretions Promote Digestion

Each day, the liver, pancreas, and intestine produce more than 3 liters of secretions whose contents are necessary for completing the digestion of ingested nutrients. The added secretions include

FIG. 21.11 The villus and a crypt in the small intestine

Villi and crypts increase the effective surface area of the small intestine. Stem cells in the crypts produce new epithelial cells to replace those that die or are damaged. Most absorption occurs along the villi. Most fluid secretion occurs in the crypts.

- Brush border
- Microvilli

Enterocyte

- Enterocytes transport nutrients and ions.
- Capillaries transport most absorbed nutrients.
- Goblet cells secrete mucus.
- Lacteals transport most fats to the lymph.
- Stem cells divide to replace damaged cells.
- Crypt cells secrete ions and water.
- Endocrine cells secrete hormones.
- Muscularis mucosae

Crypt lumen

Lamina propria

FIG. 21.12 The hepatic portal system

Most nutrients absorbed by the intestine pass through the liver, which serves as a filter that can remove potentially harmful xenobiotics before they get into the systemic circulation.

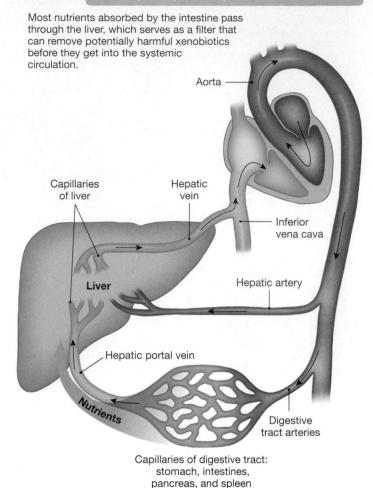

- Aorta
- Capillaries of liver
- Hepatic vein
- Inferior vena cava
- **Liver**
- Hepatic artery
- Hepatic portal vein
- *Nutrients*
- Digestive tract arteries

Capillaries of digestive tract: stomach, intestines, pancreas, and spleen

---

digestive enzymes, bile, bicarbonate, mucus, and an isotonic NaCl solution.

1. *Digestive enzymes* are produced by the intestinal epithelium and the exocrine pancreas. Intestinal brush border enzymes are anchored to the luminal cell membranes and are not swept out of the small intestine as chyme is propelled forward. The control pathways for enzyme release vary but include a variety of neural, hormonal, and paracrine signals. Usually, stimulation of parasympathetic neurons in the vagus nerve enhances enzyme secretion.

2. *Bile* made in the liver and released from the gall bladder is a nonenzymatic solution that facilitates the digestion of fats.

3. *Bicarbonate secretion* into the small intestine neutralizes the highly acidic chyme that enters from the stomach. Most bicarbonate comes from the pancreas and is released in response to neural stimuli and secretin.

4. *Mucus* from intestinal goblet cells protects the epithelium and lubricates the gut's contents.

5. An *isotonic NaCl solution* mixes with mucus to help lubricate the contents of the gut.

Isotonic NaCl Secretion Crypt cells in the small intestine and colon secrete an isotonic NaCl solution in a process similar to the initial step of salivation (**FIG. 21.13**). Chloride from the ECF enters cells via NKCC transporters, then exits into the lumen via an apical gated Cl⁻ channel known as the **cystic fibrosis transmembrane conductance regulator,** or **CFTR channel.** Movement of negatively charged Cl⁻ into the lumen draws Na⁺ down the electrical gradient through leaky cell junctions. Water follows Na⁺ along the osmotic gradient created by redistribution of NaCl. The result is secretion of isotonic saline solution.

The Pancreas Secretes Enzymes and Bicarbonate

The pancreas is an organ that contains both types of secretory epithelium: endocrine and exocrine [p. 79]. Endocrine secretions come from clusters of cells called *islets* and include the hormones insulin and glucagon (**FIG. 21.14**). Exocrine secretions include digestive enzymes and a watery solution of sodium bicarbonate, $NaHCO_3$.

The exocrine portion of the pancreas consists of lobules called *acini,* similar to those of the salivary glands. Ducts from the acini

FIG. 21.13 Isotonic NaCl secretion

Intestinal and colonic crypt cells and salivary gland acini secrete isotonic NaCl solutions.

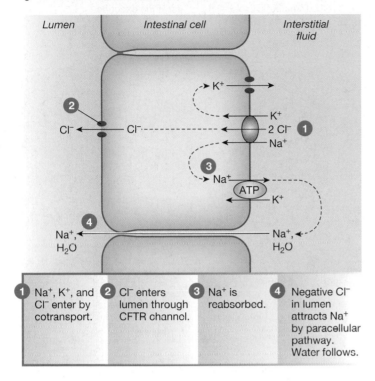

Lumen	Intestinal cell	Interstitial fluid

1. Na^+, K^+, and Cl^- enter by cotransport.
2. Cl^- enters lumen through CFTR channel.
3. Na^+ is reabsorbed.
4. Negative Cl^- in lumen attracts Na^+ by paracellular pathway. Water follows.

empty into the duodenum (Fig. 21.14a). The acinar cells secrete digestive enzymes, and the duct cells secrete the $NaHCO_3$ solution.

Enzyme Secretion   Most pancreatic enzymes are secreted as zymogens that must be activated upon arrival in the intestine. This activation process is a cascade that begins when brush border **enteropeptidase** (previously called *enterokinase*) converts inactive trypsinogen to active trypsin (Fig. 21.14b). Trypsin then converts the other pancreatic zymogens to their active forms.

The signals for pancreatic enzyme release include distension of the small intestine, the presence of food in the intestine, neural signals, and the GI hormone CCK. Pancreatic enzymes enter the intestine in a watery fluid that also contains bicarbonate.

Bicarbonate Secretion   Bicarbonate secretion into the duodenum neutralizes acid entering from the stomach. A small amount of bicarbonate is secreted by duodenal cells, but most comes from the pancreas.

Bicarbonate production requires high levels of the enzyme *carbonic anhydrase*, levels similar to those found in renal tubule cells and red blood cells [pp. 577, 646]. Bicarbonate produced from CO_2 and water is secreted by an apical Cl^--HCO_3^- exchanger (Fig. 21.14c). Hydrogen ions produced along with bicarbonate leave the cell on basolateral Na^+-H^+ exchangers. The H^+ thus reabsorbed into the intestinal circulation helps balance HCO_3^- put into the blood when parietal cells secrete H^+ into the stomach (see Fig. 21.9c).

The chloride for bicarbonate exchange enters the cell on a basolateral NKCC cotransporter and leaves via an apical CFTR channel. Luminal Cl^- then re-enters the cell in exchange for HCO_3^- entering the lumen. Defects in CFTR channel structure or function cause the disease *cystic fibrosis*, and disruption of pancreatic secretion is one hallmark of cystic fibrosis.

In cystic fibrosis, an inherited mutation causes the CFTR channel protein to be defective or absent. As a result, secretion of Cl^- and fluid ceases but goblet cells continue to secrete mucus, resulting in thickened mucus. In the digestive system, the thick mucus clogs small pancreatic ducts and prevents digestive enzyme secretion into the intestine. In airways of the respiratory system, where the CFTR channel is also found, failure to secrete fluid clogs the mucociliary escalator [Fig. 17.5c, p. 541] with thick mucus, leading to recurrent lung infections.

In both the pancreas and intestinal crypts, sodium and water secretion is a passive process, driven by electrochemical and osmotic gradients. The movement of negative ions from the ECF to the lumen creates a lumen-negative electrical gradient that attracts Na^+. Sodium moves down its electrochemical gradient through leaky junctions between the cells. The transfer of Na^+ and HCO_3^- from ECF into the lumen creates an osmotic gradient, and water follows by osmosis. The net result is secretion of a watery sodium bicarbonate solution.

The Liver Secretes Bile

Bile is a nonenzymatic solution secreted from **hepatocytes,** or liver cells (see *Focus On: The Liver,* **FIG. 21.15**). The key components of bile are (1) **bile salts,** which facilitate enzymatic fat digestion, (2) *bile pigments,* such as bilirubin, which are the waste products of hemoglobin degradation, and (3) *cholesterol,* which is excreted in the feces. Drugs and other xenobiotics are cleared from the blood by hepatic processing and are also excreted in bile. Bile salts, which act as detergents to make fats soluble during digestion, are made from steroid **bile acids** combined with amino acids.

RUNNING **PROBLEM**

A hallmark of *Vibrio cholerae* infection is profuse, dilute diarrhea sometimes said to resemble "rice water." The toxin secreted by *Vibrio cholerae* is a protein complex with six subunits. The toxin binds to intestinal cells, and the A subunit is taken into the enterocytes by endocytosis. Once inside the enterocyte, the toxin turns on adenylyl cyclase, which then produces cAMP continuously. Because the CFTR channel of the enterocyte is a cAMP-gated channel, the effect of cholera toxin is to open the CFTR channels and keep them open.

Q5: *Why would continuously open enterocyte CFTR channels cause secretory diarrhea and dehydration in humans?*

655   659   672   **675**   682   688

FIG. 21.14 ESSENTIALS

The Pancreas

Anatomy of the Exocrine and Endocrine Pancreas

(a) The exocrine pancreas secretes digestive enzymes and sodium bicarbonate.

Pancreatic duct

Pancreas

Small intestine

Pancreatic islet cells secrete hormones that enter the blood.

Pancreatic acini form the exocrine portion of the pancreas.

Capillary

Acinar cells secrete digestive enzymes.

Duct cells secrete $NaHCO_3$ that enters the digestive tract.

Lumen

Activation of Pancreatic Zymogens

(b) Inactive enzymes secreted by the pancreas are activated in a cascade. Trypsinogen is activated to trypsin by brush border enteropeptidase, and trypsin then activates other pancreatic enzymes.

Lumen of small intestine

Pancreatic duct

Pancreatic secretions (include inactive zymogens)

ZYMOGENS
- Chymotrypsinogen
- Procarboxypeptidase
- Procolipase
- Prophospholipase

Trypsinogen

Enteropeptidase in brush border activates trypsin.

Trypsin

activates

ACTIVATED ENZYMES
- Chymotrypsin
- Carboxypeptidase
- Colipase
- Phospholipase

Intestinal mucosa

Bicarbonate Secretion

(c) Bicarbonate secretion in the pancreas and duodenum

Lumen of pancreas or intestine | Pancreatic duct cell or duodenal cell | Interstitial fluid | Capillary

$H_2O + CO_2 \leftarrow CO_2$

1 *CA*

$HCO_3^- \quad HCO_3^- + H^+ \quad Na^+$

Cl^-

$Cl^- \quad Na^+$

ATP

2 K^+

CFTR channel

Na^+

2 $2\,Cl^-$

K^+

K^+

3 H_2O, Na^+

1 Cells that produce bicarbonate have high concentrations of carbonic anhydrase (*CA*).

2 Chloride enters cells by indirect active transport and leaves the apical side through a CFTR channel. Cl^- then reenters the cell in exchange for HCO_3^-.

3 Leaky junctions allow paracellular movement of ions and water. Negative ions in the lumen attract Na^+ by the paracellular pathway. Water follows.

FIG. 21.15 FOCUS ON . . .

The Liver

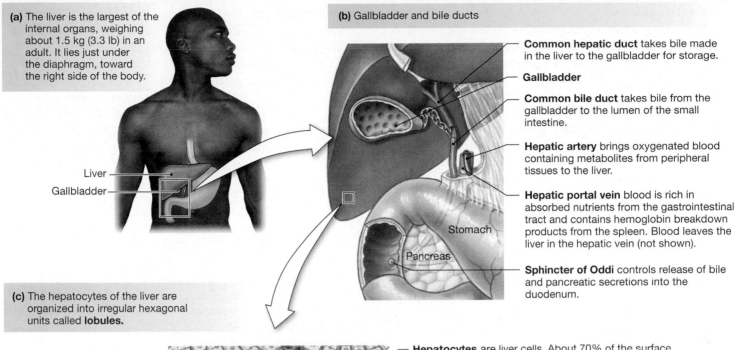

(a) The liver is the largest of the internal organs, weighing about 1.5 kg (3.3 lb) in an adult. It lies just under the diaphragm, toward the right side of the body.

Liver

Gallbladder

(b) Gallbladder and bile ducts

Common hepatic duct takes bile made in the liver to the gallbladder for storage.

Gallbladder

Common bile duct takes bile from the gallbladder to the lumen of the small intestine.

Hepatic artery brings oxygenated blood containing metabolites from peripheral tissues to the liver.

Hepatic portal vein blood is rich in absorbed nutrients from the gastrointestinal tract and contains hemoglobin breakdown products from the spleen. Blood leaves the liver in the hepatic vein (not shown).

Stomach

Pancreas

Sphincter of Oddi controls release of bile and pancreatic secretions into the duodenum.

(c) The hepatocytes of the liver are organized into irregular hexagonal units called **lobules.**

Bile canaliculi

Sinusoid

Hepatic artery

Hepatic portal vein

Bile ductule

Hepatocytes are liver cells. About 70% of the surface area of each hepatocyte faces the sinusoids, maximizing the exchange between the blood and the cells.

Each lobule is centered around a central vein that drains blood into the hepatic vein.

Along its periphery, a lobule is associated with branches of the hepatic portal vein and hepatic artery.

These vessels branch among the hepatocytes, forming **sinusoids** into which the blood flows.

The **bile canaliculi** are small channels into which bile is secreted. The canaliculi coalesce into bile ductules that run through the liver alongside the portal veins.

(d) Blood entering the liver brings nutrients and foreign substances from the digestive tract, bilirubin from hemoglobin breakdown, and metabolites from peripheral tissues of the body. In turn, the liver excretes some of these in the bile and stores or metabolizes others. Some of the liver's products are wastes to be excreted by the kidney; others are essential nutrients, such as glucose. In addition, the liver synthesizes an assortment of plasma proteins.

Absorbed from gastrointestinal tract

- Bilirubin
- Nutrients
- Drugs
- Foreign substances

Hepatic portal vein

Liver

- Glucose and fat metabolism
- Protein synthesis
- Hormone synthesis
- Urea production
- Detoxification
- Storage

Hepatic artery

Metabolites and drugs from peripheral tissues

- Bilirubin
- Metabolites of hormones and drugs
- Nutrients

Secreted into duodenum

- Bile salts
- Bilirubin
- Water, ions
- Phospholipids

Bile duct

Hepatic vein

Metabolites to peripheral tissues

- Glucose
- Plasma proteins: Albumin, clotting factors, angiotensinogen
- Urea
- Vitamin D, somatomedins
- Metabolites for excretion

Bile secreted by hepatocytes travels in hepatic ducts to the **gallbladder,** which stores and concentrates the bile solution. During a meal that includes fats, contraction of the gallbladder sends bile into the duodenum through the **common bile duct.** The gallbladder is an organ that is not essential for normal digestion, and if the duct becomes blocked by hard deposits known as gallstones, the gallbladder can be removed without creating long-term problems.

Bile salts are not altered during fat digestion. When they reach the terminal section of the small intestine (the ileum), they encounter cells that reabsorb them and send them back into the circulation. From there, bile salts return to the liver, where the hepatocytes take them back up and re-secrete them. This recirculation of bile salts is essential to fat digestion because the body's pool of bile salts must cycle from two to five times for each meal. Bilirubin and other wastes secreted in bile cannot be reabsorbed and pass into the large intestine for excretion.

Most Digestion Occurs in the Small Intestine

The intestinal, pancreatic, and hepatic secretion of enzymes and bile is essential for normal digestive function. Although a significant amount of mechanical digestion takes place in the mouth and stomach, chemical digestion of food there is limited to a small amount of starch breakdown and incomplete protein digestion in the stomach. When chyme enters the small intestine, protein digestion stops when pepsin is inactivated at the higher intestinal pH. Pancreatic and brush border enzymes then finish digestion of peptides, carbohydrates, and fats into smaller molecules that can be absorbed.

Bile Salts Facilitate Fat Digestion

Fats and related molecules in the Western diet include triglycerides, cholesterol, phospholipids, long-chain fatty acids, and the fat-soluble vitamins [Fig. 2.1, p. 30]. Nearly 90% of our fat calories come from triglycerides because they are the primary form of lipid in both plants and animals.

Fat digestion is complicated by the fact that most lipids are not particularly water soluble. As a result, the aqueous chyme leaving the stomach contains a coarse emulsion of large fat droplets, which have less surface area than smaller particles. To increase the surface area available for enzymatic fat digestion, the liver secretes bile salts into the small intestine (**FIG. 21.16a**). Bile salts help break down the coarse emulsion into smaller, more stable particles.

Bile salts, like phospholipids of cell membranes, are *amphipathic* {*amphi-,* on both sides + *pathos,* experience}, meaning that they have both a hydrophobic region and a hydrophilic region. The hydrophobic regions of bile salts associate with the surface of lipid droplets while the polar side chains interact with water, creating a stable emulsion of small, water-soluble fat droplets (Fig. 21.16a). You can see a similar emulsion when you shake a bottle of salad dressing to combine the oil and aqueous layers.

Enzymatic fat digestion is carried out by **lipases,** enzymes that remove two fatty acids from each triglyceride molecule. The result is one monoglyceride and two free fatty acids (Fig. 21.16c). The bile salt coating of the intestinal emulsion complicates digestion, however, because lipase is unable to penetrate the bile salts. For this reason, fat digestion also requires **colipase,** a protein cofactor secreted by the pancreas. Colipase displaces some bile salts, allowing lipase access to fats inside the bile salt coating.

Phospholipids are digested by pancreatic *phospholipase.* Free cholesterol is not digested and is absorbed intact.

As enzymatic and mechanical digestion proceed, fatty acids, bile salts, mono- and diglycerides, phospholipids, and cholesterol coalesce to form small disk-shaped **micelles** (Fig. 21.16b) [p. 63]. Micelles then enter the unstirred aqueous layer at the edge of the brush border.

Fat Absorption Lipophilic fats such as fatty acids and monoglycerides are absorbed primarily by simple diffusion. They move out of their micelles and diffuse across the enterocyte membrane into the cells (Fig. 21.16d). Initially scientists believed that cholesterol also diffused across the enterocyte membrane, but the discovery of a drug called *ezetimibe* that inhibits cholesterol absorption suggested that transport proteins were involved. Experiments now indicate that some cholesterol is transported across the brush border membrane on specific, energy-dependent membrane transporters, including one named *NPC1L1,* the protein that is inhibited by ezetimibe.

Once monoglycerides and fatty acids are inside the enterocytes, they move to the smooth endoplasmic reticulum, where they recombine into triglycerides (Fig. 21.16d). The triglycerides then join cholesterol and proteins to form large droplets called **chylomicrons.** Because of their size, chylomicrons must be packaged into secretory vesicles by the Golgi. The chylomicrons then leave the cell by exocytosis.

The large size of chylomicrons also prevents them from crossing the basement membrane of capillaries (Fig. 21.16d). Instead, chylomicrons are absorbed into *lacteals,* the lymph vessels of the villi. Chylomicrons pass through the lymphatic system and finally enter the venous blood just before it flows into the right side of the heart [p. 499].

Some shorter fatty acids (10 or fewer carbons) are not assembled into chylomicrons. These fatty acids can therefore cross the capillary basement membrane and go directly into the blood.

CONCEPT CHECK

12. Do bile salts digest triglycerides into monoglycerides and free fatty acids?

13. Bile acids are reabsorbed in the distal intestine by an apical sodium-dependent bile acid transporter (ASBT) and a basolateral organic anion transporter (OAT). Draw one enterocyte. Label the lumen, ECF, and basolateral and apical sides. Diagram bile acid reabsorption as described.

14. Explain how pH can be used to predict the location where a particular digestive enzyme might be most active.

FIG. 21.16 **ESSENTIALS**

Digestion and Absorption: Fats

Most lipids are hydrophobic and must be emulsified to facilitate digestion in the aqueous environment of the intestine.

(a) Bile salts coat lipids to make emulsions.

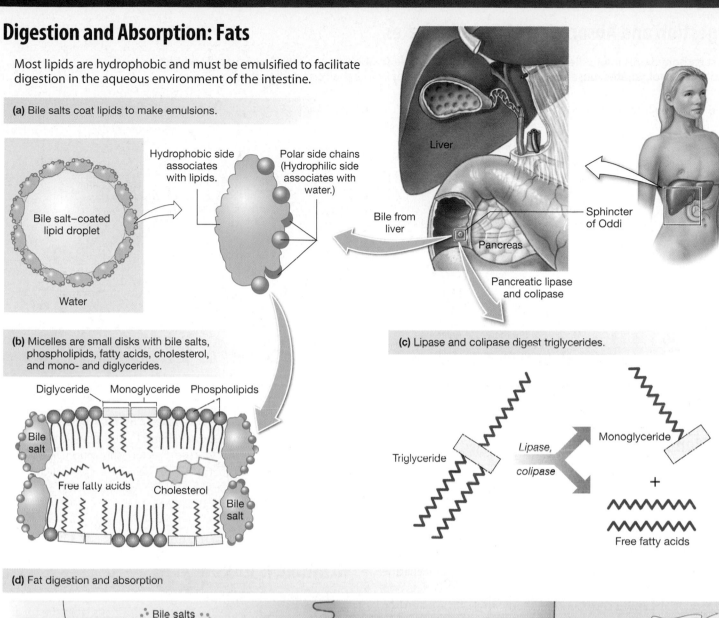

Hydrophobic side associates with lipids.

Polar side chains (Hydrophilic side associates with water.)

Bile salt–coated lipid droplet

Water

Liver

Bile from liver

Sphincter of Oddi

Pancreas

Pancreatic lipase and colipase

(b) Micelles are small disks with bile salts, phospholipids, fatty acids, cholesterol, and mono- and diglycerides.

Diglyceride   Monoglyceride   Phospholipids

Bile salt

Free fatty acids   Cholesterol   Bile salt

(c) Lipase and colipase digest triglycerides.

Triglyceride   *Lipase, colipase*   Monoglyceride

+

Free fatty acids

(d) Fat digestion and absorption

Bile salts recycle

Bile salts

Bile salts

Cholesterol + triglycerides + protein

Lacteal   Lymph to vena cava

3b

4

Micelles

Chylomicron

Golgi apparatus

5

Capillary

1

2

Emulsion

Smooth ER

Large fat droplets from stomach

3a

Lumen of small intestine

Cells of small intestine

Interstitial fluid

1 Bile salts from liver coat fat droplets.

2 Pancreatic lipase and colipase break down fats into monoglycerides and fatty acids stored in micelles.

3a Monoglycerides and fatty acids move out of micelles and enter cells by diffusion.

3b Cholesterol is transported into cells.

4 Absorbed fats combine with cholesterol and proteins in the intestinal cells to form chylomicrons.

5 Chylomicrons are removed by the lymphatic system.

FIG. 21.17 **ESSENTIALS**

Digestion and Absorption of Carbohydrates

Most carbohydrates in our diets are disaccharides and complex carbohydrates. Cellulose is not digestible. All other carbohydrates must be digested to monosaccharides before they can be absorbed.

(a) Carbohydrates break down into monosaccharides.

(b) Carbohydrate absorption in the small intestine

Carbohydrates Are Absorbed as Monosaccharides

About half the calories the average American ingests are in the form of carbohydrates, mainly *starch* and *sucrose* (table sugar). Other dietary carbohydrates include the glucose polymers *glycogen* and *cellulose,* disaccharides such as *lactose* (milk sugar) and *maltose,* and the monosaccharides *glucose* and *fructose* [Fig. 2.2, p. 31]. The enzyme *amylase* breaks long glucose polymers into smaller glucose chains and into the disaccharide maltose (**FIG. 21.17a**).

Starch digestion starts in the mouth with salivary amylase but that enzyme is denatured in the acidic stomach. Pancreatic amylase then resumes digestion of starch into maltose. Maltose and other disaccharides are broken down by intestinal brush-border enzymes known as **disaccharidases** (maltase, sucrase, and lactase). The absorbable end products of carbohydrate digestion are glucose, galactose, and fructose.

Because intestinal carbohydrate absorption is restricted to monosaccharides, all larger carbohydrates must be digested if they are to be used by the body. The complex carbohydrates we can digest are starch and glycogen. We are unable to digest cellulose because we lack the necessary enzymes. As a result, the cellulose in plant matter becomes what is known as dietary *fiber* or *roughage* and is excreted undigested. Similarly, *sucralose* (Splenda®), the artificial sweetener made from sucrose, cannot be digested because chlorine atoms substituted for three hydroxyl groups block enzymatic digestion of the sugar derivative.

Carbohydrate Absorption Intestinal glucose and galactose absorption uses transporters identical to those found in the renal proximal tubule: the apical Na^+-glucose SGLT symporter and the basolateral GLUT2 transporter (Fig. 21.17b). These transporters move galactose as well as glucose. Fructose absorption, however, is not Na^+-dependent. Fructose moves across the apical membrane by facilitated diffusion on the GLUT5 transporter and across the basolateral membrane by GLUT2 [p. 144].

CLINICAL FOCUS

Lactose Intolerance

Lactose, or milk sugar, is a disaccharide composed of glucose and galactose. Ingested lactose must be digested before it can be absorbed, a task accomplished by the intestinal brush border enzyme *lactase*. Generally, lactase is found only in juvenile mammals, except in some humans of European descent. Those people inherit a dominant gene that allows them to produce lactase after childhood. Scientists believe the lactase gene provided a selective advantage to their ancestors, who developed a culture in which milk and milk products played an important role. In cultures in which dairy products are not part of the diet after weaning, most adults lack the gene and synthesize less intestinal lactase. Decreased lactase activity is associated with a condition known as *lactose intolerance*. If a person with lactose intolerance drinks milk or eats dairy products, diarrhea may result. In addition, bacteria in the large intestine ferment lactose to gas and organic acids, leading to bloating and flatulence. The simplest remedy is to remove milk products from the diet, although milk predigested with lactase is available.

FIG. 21.18 **ESSENTIALS**

Digestion and Absorption of Proteins

(a) Proteins are chains of amino acids.

Amino-terminal end | Amino acids | Peptide bonds | Carboxy-terminal end

H_2N — ◼ ◼ ◼ ◼ ◼ ◼ ◼ — COOH

(b) Enzymes for protein digestion

Endopeptidase digests internal peptide bonds.

Endopeptidases include pepsin in the stomach, and trypsin and chymotrypsin in the small intestine.

$+H_2O$

H_2N — ◼ ◼ ◼ ◼ ◼ ◼ ◼ — COOH

2 smaller peptides

H_2N — ◼ ◼ ◼ — COOH H_2N — ◼ ◼ ◼ — COOH

Exopeptidases digest terminal peptide bonds to release amino acids.

Aminopeptidase | Carboxypeptidase

$+H_2O$ | $+H_2O$

H_2N — ◼ ◼ ◼ ◼ ◼ — COOH

Amino acid | Peptide | Amino acid

H_2N—◼—COOH H_2N—◼ ◼ ◼—COOH H_2N—◼—COOH

(c) Peptide absorption

After digestion, proteins are absorbed mostly as free amino acids. A few di- and tripeptides are absorbed. Some peptides larger than tripeptides can be absorbed by transcytosis.

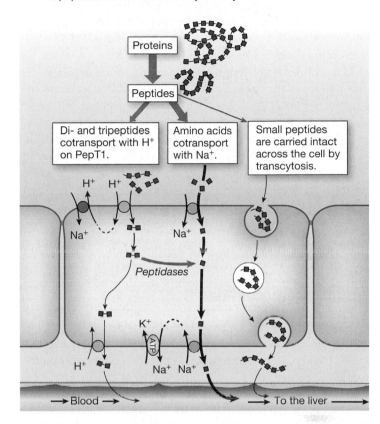

Proteins

Peptides

Di- and tripeptides cotransport with H^+ on PepT1. | Amino acids cotransport with Na^+. | Small peptides are carried intact across the cell by transcytosis.

H^+ H^+

Na^+ Na^+

Peptidases

K^+

ATP

H^+ Na^+ Na^+

→ Blood →   → To the liver →

How are enterocytes able to keep intracellular glucose concentrations high so that facilitated diffusion moves glucose into the extracellular space? In most cells, glucose is the major metabolic substrate for aerobic respiration and is immediately phosphorylated when it enters the cell [p. 142]. However, the metabolism of enterocytes (and proximal tubule cells) apparently differs from that of most other cells. These transporting epithelial cells do not use glucose as their preferred energy source. Current studies indicate that these cells use the amino acid glutamine as their main source of energy, thus allowing absorbed glucose to pass unchanged into the bloodstream.

Proteins Are Digested into Small Peptides and Amino Acids

Unlike carbohydrates, which are ingested in forms ranging from simple to complex, most ingested proteins are polypeptides or larger [Fig. 2.3, p. 32]. Not all proteins are equally digestible by humans, however. Plant proteins are the least digestible. Among the most digestible is egg protein, 85–90% of which is in a form that can be digested and absorbed. Surprisingly, between 30% and 60% of the protein found in the intestinal lumen comes not from ingested food but from the sloughing of dead cells and from protein secretions such as enzymes and mucus.

The enzymes for protein digestion are classified into two broad groups: endopeptidases and exopeptidases (**FIG. 21.18b**). **Endopeptidases,** more commonly called **proteases,** attack peptide bonds in the interior of the amino acid chain and break a long peptide chain into smaller fragments. Proteases are secreted as inactive *proenzymes* (zymogens) from epithelial cells in the stomach, intestine, and pancreas. They are activated once they reach the GI tract lumen. Examples of proteases include **pepsin** secreted in the stomach, and **trypsin** and **chymotrypsin** secreted by the pancreas.

Exopeptidases release single amino acids from peptides by chopping them off the ends, one at a time. *Aminopeptidases* act on the amino-terminal end of the protein; *carboxypeptidases* act at the carboxy-terminal end. The most important digestive exopeptidases are two isozymes of carboxypeptidase secreted by the pancreas. Aminopeptidases play a lesser role in digestion.

15. What activates pepsinogen, trypsinogen, and chymotrypsinogen?

Protein Absorption The primary products of protein digestion are free amino acids, dipeptides, and tripeptides, all of which can be absorbed. Amino acid structure is so variable that multiple amino acid transport systems occur in the intestine. Most free amino acids are carried by Na^+-dependent cotransport proteins similar to those in the proximal tubule of the kidney (Fig. 21.18b). A few amino acid transporters are H^+-dependent.

Dipeptides and tripeptides are carried into enterocytes on the oligopeptide transporter *PepT1* that uses H^+-dependent cotransport (Fig. 21.18c). Once inside the epithelial cell, these *oligopeptides* {*oligos*, little} have two possible fates. Most are digested by cytoplasmic peptidases into individual amino acids, which are then transported across the basolateral membrane and into the circulation. Those oligopeptides that are not digested are transported intact across the basolateral membrane on an H^+-dependent exchanger. The transport system that moves oligopeptides also is responsible for intestinal uptake of certain drugs, including some antibiotics, angiotensin-converting enzyme inhibitors, and thrombin inhibitors.

Some Larger Peptides Can Be Absorbed Intact

Some peptides larger than three amino acids are absorbed by transcytosis [p. 152] after binding to membrane receptors on the luminal surface of the intestine. The discovery that ingested proteins can be absorbed as small peptides has implications in medicine because these peptides may act as *antigens*, substances that stimulate antibody formation and result in allergic reactions. Consequently, the intestinal absorption of peptides may be a significant factor in the development of food allergies and food intolerances.

In newborns, peptide absorption takes place primarily in intestinal crypt cells (Fig. 21.11). At birth, intestinal villi are very small, so the crypts are well exposed to the luminal contents. As the villi grow and the crypts have less access to chyme, the high peptide absorption rates present at birth decline steadily. If parents delay feeding the infant allergy-inducing peptides, the gut has a chance to mature, lessening the likelihood of antibody formation.

One of the most common antigens responsible for food allergies is gluten, a component of wheat. The incidence of childhood gluten allergies has decreased since the 1970s, when parents were cautioned not to feed infants gluten-based cereals until they were several months old.

In another medical application, pharmaceutical companies have developed indigestible peptide drugs that can be given orally instead of by injection. Probably the best-known example is *DDAVP* (1-deamino-8-D-arginine vasopressin), the synthetic analog of vasopressin. If the natural hormone vasopressin is ingested, it is digested rather than absorbed intact. By changing the structure of the hormone slightly, scientists created a synthetic peptide that has the same activity but is absorbed without being digested.

Nucleic Acids Are Digested into Bases and Monosaccharides

The nucleic acid polymers DNA and RNA are only a very small part of most diets. They are digested by pancreatic and intestinal enzymes, first into their component nucleotides and then into nitrogenous bases and monosaccharides [Fig. 2.4, p. 34]. The bases are absorbed by active transport, and the monosaccharides are absorbed by facilitated diffusion and secondary active transport, as other simple sugars are.

The Intestine Absorbs Vitamins and Minerals

In general, the fat-soluble vitamins (A, D, E, and K) are absorbed in the small intestine along with fats—one reason that health professionals are concerned about excessive consumption of "fake fats," such as Olestra, that are not absorbed. The same concern exists with orlistat (Alli®), a lipase inhibitor used for weight loss. Users of these weight-loss aids are advised to take a daily multivitamin to avoid vitamin deficiencies.

The water-soluble vitamins (C and most B vitamins) are absorbed by mediated transport. The major exception is **vitamin B_{12},** also known as *cobalamin* because it contains the element cobalt. We obtain most of our dietary supply of B_{12} from seafood, meat, and milk products. The intestinal transporter for B_{12} is found only in the ileum and recognizes B_{12} only when the vitamin is complexed with a protein called **intrinsic factor,** secreted by the same gastric parietal cells that secrete acid.

RUNNING **PROBLEM**

Rehydrating people with cholera is the key to their survival. Most patients who develop cholera can be treated successfully with oral rehydration salts. However, in about 5% of patients, the dehydration caused by cholera-induced diarrhea can be severe. If left untreated, these patients can die from circulatory collapse as soon as 18 hours after infection. Because Brooke's blood pressure was so low, the medical personnel decided that she needed intravenous (IV) fluids to restore her volume.

Q6: *Recipes for oral rehydration therapy usually include sugar (sucrose) and table salt. Explain how the sugar enhances intestinal absorption of Na^+.*

Q7: *Which type of IV solution would you select for Brooke, and why? Your choices are normal (isotonic) saline, half-normal saline, and 5% dextrose in water (D-5-W).*

655 659 672 675 **682** 688

FIG. 21.19   Ion and water absorption

(a) Iron absorption

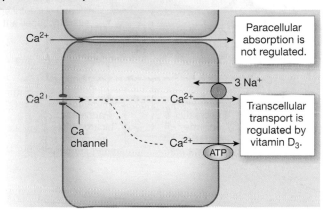

(b) Calcium absorption

Paracellular absorption is not regulated.

Transcellular transport is regulated by vitamin D_3.

(c) Na^+, K^+, Cl^-, and water absorption

1 Na^+ enters cells by multiple pathways.

2 The Na^+-K^+-ATPase pumps Na^+ into the ECF.

Water and K^+ move through the paracellular pathway.

One concern about extended use of drugs that inhibit gastric acid secretion, such as the proton-pump inhibitors discussed earlier, is that they may cause decreased absorption of vitamin B_{12}. In the complete absence of intrinsic factor, severe vitamin B_{12} deficiency causes the condition known as *pernicious anemia*. In this state, red blood cell synthesis (*erythropoiesis*), which depends on vitamin B_{12}, is severely diminished. Lack of intrinsic factor cannot be remedied directly, but patients with pernicious anemia can be given vitamin B_{12} shots.

Iron and Calcium   Mineral absorption usually occurs by active transport. Iron and calcium are two of the few substances whose intestinal absorption is regulated. For both minerals, a decrease in body concentrations of the mineral leads to enhanced uptake at the intestine.

Dietary iron is ingested as heme iron [p. 521] in meat and as ionized iron in some plant products. Heme iron is absorbed by an apical transporter on the enterocyte (**FIG. 21.19a**). Ionized iron Fe^{2+} is actively absorbed by apical cotransport with H^+ on a protein called the *divalent metal transporter 1 (DMT1)*. Inside the cell, enzymes convert heme iron to Fe^{2+} and both pools of ionized iron leave the cell on a transporter called *ferroportin*.

Iron uptake by the body is regulated by a peptide hormone called *hepcidin*. When body stores of iron are high, the liver secretes hepcidin, which binds to ferroportin. Hepcidin binding causes the enterocyte to destroy the ferroportin transporter, which results in decreased iron uptake across the intestine.

Most Ca^{2+} absorption in the gut occurs by passive, unregulated movement through paracellular pathways (Fig. 21.19b). Hormonally regulated transepithelial Ca^{2+} transport takes place in the duodenum. Calcium enters the enterocyte through apical Ca^{2+} channels and is actively transported across the basolateral membrane by either a Ca^{2+}-ATPase or by the Na^+-Ca^{2+} antiporter. Calcium absorption is regulated by vitamin D_3, discussed in Chapter 23.

The Intestine Absorbs Ions and Water

Most water absorption takes place in the small intestine, with an additional 0.5 liter per day absorbed in the colon. The absorption of nutrients moves solute from the lumen of the intestine to the ECF, creating an osmotic gradient that allows water to follow.

Ion absorption into the body also creates the osmotic gradients needed for water movement. Enterocytes in the small intestine and **colonocytes,** the epithelial cells on the luminal surface of the colon, absorb Na^+ using three membrane proteins (Fig. 21.19c): apical Na^+ channels such as ENaC, a Na^+-Cl^- symporter, and the Na^+-H^+ exchanger (NHE). In the small intestine, a significant fraction of Na^+ absorption also takes place through Na^+-dependent organic solute uptake, such as the SGLT and Na^+-amino acid transporters.

On the basolateral side of both enterocytes and colonocytes, the primary transporter for Na^+ is Na^+-K^+-ATPase. Chloride layer but a discontinuous longitudinal muscle layer concentrated into three bands called the **tenia coli.** Contractions of the tenia pull the wall into bulging pockets called **haustra** {*haustrum,* bucket or scoop}.

The mucosa of the colon has two regions, like that of the small intestine. The luminal surface lacks villi and appears smooth. It is composed of colonocytes and mucus-secreting goblet cells. The crypts contain stem cells that divide to produce new epithelium, as well as goblet cells, endocrine cells, and maturing colonocytes.

Motility in the Large Intestine

Chyme that enters the colon continues to be mixed by segmental contractions. Forward movement is minimal during mixing contractions and depends primarily on a unique colonic contraction known as **mass movement.** A wave of contraction decreases the diameter of a segment of colon and sends a substantial bolus of material forward. These contractions occur 3–4 times a day and are associated with eating and distension of the stomach through the *gastrocolic reflex.* Mass movement is responsible for the sudden distension of the rectum that triggers defecation.

The **defecation reflex** removes undigested feces from the body. Defecation resembles urination in that it is a spinal reflex triggered by distension of the organ wall. The movement of fecal material into the normally empty rectum triggers the reflex. Smooth muscle of the **internal anal sphincter** relaxes, and peristaltic contractions in the rectum push material toward the anus. At the same time, the external anal sphincter, which is under voluntary control, is consciously relaxed if the situation is appropriate. Defecation is often aided by conscious abdominal contractions and forced expiratory movements against a closed glottis (the *Valsalva maneuver*).

Defecation, like urination, is subject to emotional influence. Stress may increase intestinal motility and cause psychosomatic diarrhea in some individuals but may decrease motility and cause *constipation* in others. When feces are retained in the colon, either through consciously ignoring a defecation reflex or through decreased motility, continued water absorption creates hard, dry feces that are difficult to expel. One treatment for constipation is glycerin suppositories, small bullet-shaped wads that are inserted through the anus into the rectum. Glycerin attracts water and helps soften the feces to promote defecation.

Digestion and Absorption in the Large Intestine

According to the traditional view of the large intestine, no significant digestion of organic molecules takes place there. However, in recent years, this view has been revised. We now know that the numerous bacteria inhabiting the colon break down significant amounts of undigested complex carbohydrates and proteins through fermentation. The end products include lactate and short-chain fatty acids, such as butyric acid. Several of these products are lipophilic and can be absorbed by simple diffusion. The fatty acids, for example, are used by colonocytes as their preferred energy substrate.

EMERGING CONCEPTS

The Human Microbiome Project

Did you realize that the average human body has many more bacteria living on and in it than it has cells? And that most of these bacteria reside in the gut? Scientists have known for decades about intestinal bacteria and the problems they cause when they leave the external environment of the gut lumen and enter the body proper. Bacterial infections are common if your appendix ruptures or if trauma, such as a stab wound, punctures the wall of the intestine. At the same time, our continued health depends on absorption of vitamins and other nutrients from bacterial metabolism. The relationship between our *microbiota* (the bacteria that inhabit our bodies) and our health has become a topic of research studies in recent years, and data are being collected by an international collaboration known as the Human Microbiome Project (*http:// commonfund.nih.gov/hmp*). Do foods advertised as "probiotics" really do anything? Can bacteria influence whether we gain weight or not? Do they affect fetal development and our susceptibility to disease? We will be learning more about the answers to these questions in the years to come.

Colonic bacteria also produce significant amounts of absorbable vitamins, especially vitamin K. Intestinal gases, such as hydrogen sulfide, that escape from the gastrointestinal tract are a less useful product. Some starchy foods, such as beans, are notorious for their tendency to produce intestinal gas (**flatus**).

Diarrhea Can Cause Dehydration

Diarrhea is a pathological state in which intestinal secretion of fluid is not balanced by absorption, resulting in watery stools. Diarrhea occurs if normal intestinal water absorption mechanisms are disrupted or if there are unabsorbed osmotically active solutes that "hold" water in the lumen. Substances that cause *osmotic diarrhea* include undigested lactose and sorbitol, a sugar alcohol from plants. Sorbitol is used as an "artificial" sweetener in some chewing gums and in foods made for people with diabetes. Another unabsorbed solute that can cause osmotic diarrhea, intestinal cramping, and gas is Olestra, the "fake fat" made from vegetable oil and sugar.

In clinical settings, patients who need to have their bowels cleaned out before surgery or other procedures are often given 4 liters of an isotonic solution of polyethylene glycol and electrolytes to drink. Because polyethylene glycol cannot be absorbed, a large volume of unabsorbed solution passes into the colon, where it triggers copious diarrhea that removes all solid waste from the GI tract.

Secretory diarrheas occur when bacterial toxins, such as cholera toxin from *Vibrio cholerae* and *Escherichia coli* enterotoxin, enhance colonic Cl^- and fluid secretion (see Fig. 21.13). When excessive fluid secretion is coupled with increased motility, diarrhea results. Secretory diarrhea in response to intestinal infection can be viewed as adaptive because it helps flush pathogens out of the lumen. However, it also has the potential to cause dehydration if fluid loss is excessive.

The World Health Organization estimates that in developing countries, 4 million people die from diarrhea each year. In the United States, diarrhea in children causes about 200,000 hospitalizations a year. Oral replacement fluids for treatment of diarrheal salt and water loss can prevent the *morbidity* (illness) and *mortality* (death) associated with diarrhea. Oral rehydration solutions usually contain glucose or sucrose as well as Na^+, K^+, and Cl^- because the inclusion of a sugar enhances Na^+ absorption. If dehydration is severe, intravenous fluid therapy may be necessary.

CONCEPT CHECK

16. In secretory diarrhea, epithelial cells in the intestinal villi may be damaged or may slough off. In these cases, would it be better to use an oral rehydration solution containing glucose or one containing sucrose? Explain your reasoning.

IMMUNE FUNCTIONS OF THE GI TRACT

As you learned at the beginning of the chapter, the GI tract is the largest immune organ in the body. Its luminal surface is continuously exposed to disease-causing organisms, and the immune cells of the GALT must prevent these pathogens from entering the body through delicate absorptive tissues. The first lines of defense are the enzymes and immunoglobulins in saliva and the highly acidic environment of the stomach. If pathogens or toxic materials make it into the small intestine, sensory receptors and the immune cells of the GALT respond. Two common responses are diarrhea, just described, and vomiting.

M Cells Sample Gut Contents

The immune system of the intestinal mucosa consists of immune cells scattered throughout the mucosa, clusters of immune cells in Peyer's patches (see Fig. 21.1f), and specialized epithelial cells called **M cells** that overlie the Peyer's patches. The M cells provide information about the contents of the lumen to the immune cells of the GALT.

The microvilli of M cells are fewer in number and more widely spaced than in the typical intestinal cell. The apical surface of M cells contains clathrin-coated pits [p. 148] with embedded membrane receptors. When antigens bind to these receptors, the M cell uses transcytosis to transport them to its basolateral membrane, where they are released into the interstitial

fluid. Macrophages and lymphocytes [p. 514] are waiting in the extracellular compartment for the M cell to present them with antigens.

If the antigens are substances that threaten the body, the immune cells swing into action. They secrete cytokines to attract additional immune cells that can attack the invaders and cytokines that trigger an inflammatory response. A third response to cytokines is increased secretion of Cl^-, fluid, and mucus to flush the invaders from the GI tract.

In *inflammatory bowel diseases* (such as ulcerative colitis and Crohn's disease), the immune response is triggered inappropriately by the normal contents of the gut. One apparently successful experimental therapy for these diseases involves blocking the action of cytokines released by the gut-associated lymphoid tissues.

How certain pathogenic bacteria cross the barrier created by the intestinal epithelium has puzzled scientists for years. The discovery of M cells may provide the answer. It appears that some bacteria, such as *Salmonella* and *Shigella*, have evolved surface molecules that bind to M cell receptors. The M cells then obligingly transport the bacteria across the epithelial barrier and deposit them inside the body, where the immune system immediately reacts. Both bacteria cause diarrhea, and *Salmonella* also causes fever and vomiting.

Vomiting Is a Protective Reflex

Vomiting, or *emesis*, the forceful expulsion of gastric and duodenal contents from the mouth, is a protective reflex that removes toxic materials from the GI tract before they can be absorbed. However, excessive or prolonged vomiting, with its loss of gastric acid, can cause metabolic alkalosis [p. 647].

The vomiting reflex is coordinated through a vomiting center in the medulla. The reflex begins with stimulation of sensory receptors and is often (but not always) accompanied by nausea. A variety of stimuli from all over the body can trigger vomiting. They include chemicals in the blood, such as cytokines and certain drugs; pain; disturbed equilibrium, such as occurs in a moving car or rocking boat, and emotional stress. Tickling the back of the pharynx can also induce vomiting.

Efferent signals from the vomiting center initiate a wave of reverse peristalsis that begins in the small intestine and moves upward. The motility wave is aided by abdominal contraction that increases intra-abdominal pressure. The stomach relaxes so that the increased pressure forces gastric and intestinal contents back into the esophagus and out of the mouth.

During vomiting, respiration is inhibited. The epiglottis and soft palate close off the trachea and nasopharynx to prevent the *vomitus* from being inhaled (*aspirated*). Should acid or small food particles get into the airways, they could damage the respiratory system and cause *aspiration pneumonia*.

RUNNING **PROBLEM CONCLUSION** | Cholera in Haiti

Vibrio cholerae, the bacterium that causes cholera, was first identified in India in the 1800s. It has caused seven worldwide epidemics in the years since. About 75% of people who become infected with *V. cholerae* have no symptoms, but the remaining 25% develop potentially fatal *secretory diarrhea*. The gut immune systems in most people overcome the infection within about a week. But until that happens, even asymptomatic people shed the bacteria in their feces, which contributes to the spread of the disease. In Haiti, plagued by inadequate sanitation and an earthquake-damaged water supply, cholera spread rapidly. By November 2013, nearly 700,000 cases and more than 8,000 deaths had been reported. Genetic analysis

of the *Vibrio cholera* strain in Haiti suggests that the bacterium was accidentally brought to the island by asymptomatic United Nations peacekeepers from Asia.

To learn more about cholera in Haiti, see the CDC (*www.cdc.gov*) and WHO (*www.who.int*) web sites. If you plan to travel to Haiti or any place with a declared cholera epidemic, visit *www.cdc.gov/travel* and review proper guidelines and procedures on avoiding contact with this potentially lethal bacterium. Now check your understanding of this running problem by comparing your answers to the information in the following summary table.

Question	Facts	Integration and Analysis
Q1: *What would you expect Brooke's ECF volume to be?*	Most fluid in diarrhea has been secreted from the body into the lumen of the GI tract.	Loss of fluid from the body would decrease ECF volume.
Q2: *Why was Brooke experiencing a rapid heartbeat?*	Loss of ECF volume with the diarrhea decreased Brooke's blood pressure.	Decreased blood pressure triggered a baroreceptor reflex [p. 493]. Increased sympathetic and decreased parasympathetic output to the SA node resulted in a faster heart rate.
Q3: *Esomeprazole is a proton pump inhibitor (PPI). For what symptom or condition might Brooke have been taking this drug?*	"Proton pump" is another name for an ATP-dependent H^+ transporter. Stomach acid is secreted by H^+-K^+-ATPase.	A proton pump inhibitor would decrease stomach acid, so Brooke may have been taking the PPI for heartburn or gastroesophageal reflex disorder (GERD).
Q4: *Why might taking a protein pump inhibitor like esomeprazole have increased Brooke's chances of contracting cholera?*	Proton pump inhibitors decrease the acidity in the stomach. Low gastric pH is one of the body's defense mechanisms.	In a less acidic stomach environment, more cholera bacteria might survive passage through the stomach to the small intestine, where they could infect the enterocytes.
Q5: *Why would continuously open enterocyte CFTR channels cause secretory diarrhea and dehydration in humans?*	Chloride leaves enterocytes by the CFTR channel. Na^+ and water follow by the paracellular pathway. See Figure 21.13.	A continuously open CFTR channel means increased secretion of NaCl and water into the lumen, which leads to watery diarrhea. The salt and water come from the ECF, and their loss causes dehydration.
Q6: *Recipes for oral rehydration therapy usually include sugar (sucrose) and table salt. Explain how the sugar enhances intestinal absorption of Na^+.*	Sucrose is digested to glucose and fructose. Glucose is absorbed by Na^+-dependent indirect active transport on the SGLT.	Na^+ uptake by the SGLT provides an additional pathway for Na^+ absorption and speeds the replenishing of fluid loss.
Q7: *Which type of IV solution would you select for Brooke and why? Your choices are normal (isotonic) saline, half-normal saline, and 5% dextrose in water (D-5-W).*	Chloride secretion by enterocytes causes Na^+ and water to follow, with the net result being secretion of isotonic fluid. The replacement fluid should match the fluid loss as closely as possible.	Normal saline (isosmotic) approximates the fluid lost in cholera diarrhea. Half-normal saline would dilute the body's osmolarity. D-5-W is not acceptable because it is equivalent to giving pure water and would not replace the lost NaCl.

This problem was written by Claire Conroy when she was an undergraduate Nutritional Sciences/Pre-Physical Therapy student at the University of Texas at Austin.

655 659 672 675 682 **688**

CHAPTER SUMMARY

The digestive system, like the renal system, plays a key role in *mass balance* in the body. Most material that enters the system, whether by mouth or by secretion, is absorbed before it reaches the end of the GI tract. In pathologies such as diarrhea, in which absorption and secretion are unbalanced, the loss of material through the GI tract can seriously disrupt *homeostasis*. Absorption and secretion in the GI tract provide numerous examples of *movement across membranes,* and most transport processes follow patterns you have encountered in the kidney and other systems. Finally, regulation of GI tract function illustrates the complex interactions that take place between endocrine and neural *control systems* and the immune system.

Anatomy of the Digestive System

iP GI: Anatomy Review

1. Food entering the digestive system passes through the mouth, pharynx, **esophagus, stomach (fundus, body, antrum), small intestine (duodenum, jejunum, ileum), large intestine (colon, rectum),** and **anus.** (p. 658; Fig. 21.1a)

2. The **salivary glands, pancreas,** and **liver** add exocrine secretions containing enzymes and mucus to the lumen. (p. 655; Fig. 21.1a)

3. **Chyme** is a soupy substance created as ingested food is broken down by mechanical and chemical digestion. (p. 655)

4. The wall of the GI tract consists of four layers: mucosa, submucosa, muscle layers, and serosa. (p. 658; Fig. 21.1d)

5. The **mucosa** faces the lumen and consists of epithelium, the **lamina propria,** and the **muscularis mucosae.** The lamina propria contains immune cells. Small **villi** and invaginations increase the surface area. (p. 658; Fig. 21.1e, f)

6. The **submucosa** contains blood vessels and lymph vessels and the **submucosal plexus** of the **enteric nervous system.** (p. 659; Fig. 21.1f)

7. The **muscularis externa** consists of a layer of circular muscle and a layer of longitudinal muscle. The **myenteric plexus** lies between these two muscle layers. (p. 659; Fig. 21.1e, f)

8. The **serosa** is the outer connective tissue layer that is a continuation of the peritoneal membrane. (p. 659; Fig. 21.1d)

Digestive Function and Processes

iP GI: Secretion, Digestion and Absorption, Motility

9. The GI tract moves nutrients, water, and electrolytes from the external environment to the internal environment. (p. 659)

10. **Digestion** is chemical and mechanical breakdown of foods into absorbable units. **Absorption** is transfer of substances from the lumen of the GI tract to the ECF. **Motility** is movement of material through the GI tract. **Secretion** is the transfer of fluid and electrolytes from ECF to lumen or the release of substances from cells. (p. 659; Fig. 21.2)

11. About 2 L of fluid per day enter the GI tract through the mouth. Another 7 L of water, ions, and proteins are secreted by the body. To maintain mass balance, nearly all of this volume is reabsorbed. (p. 661; Fig. 21.3)

12. Many digestive enzymes are secreted as inactive zymogens to prevent autodigestion. (p. 661)

13. For defense from invaders, the GI tract contains the largest collection of lymphoid tissue in the body, the **gut-associated lymphoid tissue (GALT).** (p. 660)

14. GI smooth muscle cells depolarize spontaneously and are electrically connected by gap junctions. Some segments of the gut are **tonically contracted,** but others exhibit **phasic contractions.** (p. 661)

15. Intestinal smooth muscle exhibits spontaneous **slow wave potentials** that originate in the **interstitial cells of Cajal.** (p. 661)

16. When a slow wave reaches threshold, it fires action potentials and the muscle contracts. (p. 663; Fig. 21.4a)

17. Between meals, the **migrating motor complex** moves food remnants from the upper GI tract to the lower regions. (p. 663; Fig. 21.4b)

18. **Peristaltic contractions** are progressive waves of contraction that occur mainly in the esophagus. **Segmental contractions** are mixing contractions. (p. 663; Fig. 21.4c, d)

Regulation of GI Function

iP GI: Control of the Digestive System

19. The **enteric nervous system** can integrate information without input from the CNS. **Intrinsic neurons** lie completely within the ENS. (p. 664)

20. **Short reflexes** originate in the ENS and are integrated there. **Long reflexes** may originate in the ENS or outside it but are integrated in the CNS. (p. 664; Fig. 21.5)

21. Generally, parasympathetic innervation is excitatory for GI function, and sympathetic innervation is inhibitory. (p. 665)

22. GI peptides excite or inhibit motility and secretion. Most stimuli for GI peptide secretion arise from the ingestion of food. (p. 665)

23. GI hormones are divided into the gastrin family (**gastrin, cholecystokinin**), secretin family (**secretin, gastric inhibitory peptide, glucagon-like peptide-1**), and hormones that do not fit into either of those two families (**motilin**). (p. 665, 667; Tbl. 21.1)

The Cephalic Phase

24. The sight, smell, or taste of food initiates GI reflexes in the **cephalic phase** of digestion. (p. 668; Fig. 21.8)

25. Mechanical digestion begins with chewing, or **mastication.** Saliva moistens and lubricates food. Salivary amylase digests carbohydrates. (p. 668)

26. **Saliva** is an exocrine secretion that contains water, ions, mucus, and proteins. Salivation is under autonomic control. (p. 668)

27. Swallowing, or **deglutition,** is a reflex integrated by a medullary center. (p. 668; Fig. 21.7)

The Gastric Phase

28. The stomach stores food, begins protein and fat digestion, and protects the body from swallowed pathogens. (p. 669)

29. The stomach secretes mucus and bicarbonate from **mucous cells,** pepsinogen from **chief cells,** somatostatin from **D cells,** histamine from **ECL cells,** and gastrin from **G cells.** (p. 670, 672; Fig. 21.9a, b; Fig. 21.10)

30. **Parietal cells** in gastric glands secrete hydrochloric acid. (p. 670; Fig. 21.9c)

31. Gastric function is integrated with the cephalic and intestinal phases of digestion. (p. 670; Fig. 21.8, Fig. 21.20)

The Intestinal Phase

32. Most nutrient absorption takes place in the small intestine. The large intestine absorbs water and ions. (p. 673)

33. Most absorbed nutrients go directly to the liver via the hepatic portal system before entering the systemic circulation. (p. 673; Fig. 21.12)

34. Intestinal enzymes are part of the **brush border. Goblet cells** secrete mucus. (p. 673)

35. Intestinal cells secrete an isotonic NaCl solution using the **CFTR chloride channel.** Water and Na^+ follow Cl^- down osmotic and electrochemical gradients. (p. 674; Fig. 21.13)

36. The pancreas secretes a watery $NaHCO_3$ solution from duct cells and inactive digestive enzymes from the acini. (p. 674; Fig. 21.14)

37. Bile made by **hepatocytes** contains **bile salts,** bilirubin, and cholesterol. Bile is stored and concentrated in the gallbladder (p. 675; Fig. 21.15)

38. Fat digestion is facilitated by bile salts. As digestion proceeds, fat droplets form **micelles.** (p. 675; Fig. 21.16)

39. Fat digestion requires the enzyme **lipase** and the cofactor **colipase.** (p. 678; Fig. 21.16)

40. Fat absorption occurs primarily by simple diffusion. Cholesterol is actively transported. (p. 676; Fig. 21.16)

41. **Chylomicrons,** made of monoglycerides, fatty acids, cholesterol, and proteins, are absorbed into the lymph. (p. 678; Fig. 21.16)

42. **Amylase** digests starch to maltose. **Disaccharidases** digest disaccharides to monosaccharides. (p. 680; Fig. 21.17)

43. Glucose absorption uses the SGLT Na^+-glucose symporter and GLUT2 transporter. Fructose uses the GLUT5 and GLUT2 transporters. (p. 680; Fig. 21.17)

44. **Endopeptidases** (also called proteases) break proteins into smaller peptides. **Exopeptidases** remove amino acids from peptides. (p. 681; Fig. 21.18)

45. Amino acids are absorbed via Na^+- or H^+-dependent cotransport. Dipeptides and tripeptides are absorbed via H^+-dependent cotransport. Some larger peptides are absorbed intact via transcytosis. (p. 682; Fig. 21.18)

46. Nucleic acids are digested and absorbed as nitrogenous bases and monosaccharides. (p. 682)

47. Fat-soluble vitamins are absorbed along with fats. Water-soluble vitamins are absorbed by mediated transport. Vitamin B_{12} absorption requires **intrinsic factor** secreted by the stomach. (p. 682)

48. Mineral absorption usually occurs via active transport. Some calcium moves by the paracellular pathway. Ions and water move by the paracellular pathway as well as by membrane proteins. (p. 682; Fig. 21.19)

49. Acid in the intestine, CCK, and secretin delay gastric emptying. (p. 684; Fig. 21.20)

50. Undigested material in the colon moves forward by **mass movement.** The **defecation reflex** is triggered by sudden distension of the rectum. (p. 686; Fig. 21.21)

51. Colonic bacteria use fermentation to digest organic material. (p. 686)

52. Cells of the colon can both absorb and secrete fluid. Excessive fluid secretion or decreased absorption causes diarrhea. (p. 686)

Immune Functions of the GI Tract

53. Protective mechanisms of the GI tract include acid and mucus production, vomiting, and diarrhea. (p. 687)

54. **M cells** sample gut contents and present antigens to cells of the GALT. (p. 687)

55. **Vomiting** is a protective reflex integrated in the medulla. (p. 687)

REVIEW QUESTIONS

In addition to working through these questions and checking your answers on p. A-28, review the Learning Outcomes at the beginning of this chapter.

Level One Reviewing Facts and Terms

1. Match each of the following descriptions with the appropriate term(s):

a.	chyme is produced here	1. colon
b.	organ where most digestion occurs	2. stomach
c.	initial section of small intestine	3. small intestine
d.	this adds exocrine secretions to duodenum via a duct	4. duodenum
		5. ileum
e.	sphincter between stomach and intestine	6. jejunum
		7. pancreas
f.	enzymes produced here	8. pylorus
g.	distension of its walls triggers the defecation reflex	9. rectum
		10. liver

2. For most nutrients, which two processes are not regulated? Which two are continuously regulated? Why do you think these differences exist? Defend your answer.

3. Define the four basic processes of the digestive system and give an example of each.

4. List the four layers of the GI tract walls. What type of tissue predominates in each layer?

5. Describe the functional types of epithelium lining the stomach and intestines.

6. What are Peyer's patches? M cells of the intestine?

7. What purposes does motility serve in the gastrointestinal tract? Which types of tissue contribute to gut motility? Which types of contraction do the tissues undergo?

8. What is a zymogen? What is a proenzyme? List two examples of each.

9. Match each of the following cells with the product(s) it secretes. Items may be used more than once.

a.	parietal cells	1.	enzymes
b.	goblet cells	2.	histamine
c.	brush border cells	3.	mucus
d.	pancreatic cells	4.	pepsinogen
e.	D cells	5.	gastrin
f.	ECL cells	6.	somatostatin
g.	chief cells	7.	HCO_3^-
h.	G cells	8.	HCl
		9.	intrinsic factor

10. How does each of the following factors affect digestion? Briefly explain how and where each factor exerts its effects.
   (a) emulsification
   (b) neural activity
   (c) low pH
   (d) size of food particles

11. Most digested nutrients are absorbed into the _____ of the _____ system, delivering nutrients to the _____ (organ). However, digested fats go into the _____ system because intestinal capillaries have a(n) _____ around them that most lipids are unable to cross.

12. What is the enteric nervous system, and what is its function?

13. What are short reflexes? What types of responses do they regulate? What is meant by the term *long reflex?*

14. What role do paracrines play in digestion? Give specific examples.

Level Two Reviewing Concepts

15. Mapping. **Map 1:** List the three major groups of biomolecules across the top of a large piece of paper. Down the left side of the paper write *mouth, stomach, small intestine.* For each biomolecule in each location, fill in the enzymes that digest the biomolecule, the products of digestion for each enzyme, and the location and mechanisms by which these products are absorbed.

   Map 2: Create a diagram or map using the following terms related to iron absorption:

• DMT1	• heme iron
• endocytosis	• hepcidin
• enterocyte	• Fe^{2+}
• ferroportin	• liver

16. Define, compare, and contrast the following pairs or sets of terms:
   (a) mastication, deglutition
   (b) microvilli, villi
   (c) peristalsis, segmental contractions, migrating motor complex, mass movements
   (d) chyme, feces
   (e) short reflexes, long reflexes
   (f) submucosal plexus, myenteric plexus, vagus nerve
   (g) cephalic, gastric, and intestinal phases of digestion

17. (a) Diagram the cellular mechanisms by which Na^+, K^+, and Cl^- are reabsorbed from the intestine.
   (b) Diagram the cellular mechanisms by which H^+ and HCO_3^- are secreted into the lumen.

18. Compare the enteric nervous system with the cephalic brain. Give some specific examples of neurotransmitters, neuromodulators, and supporting cells in the two.

19. List and briefly describe the actions of the members of each of the three groups of GI hormones.

20. Explain how H_2 receptor antagonists and proton pump inhibitors decrease gastric acid secretion.

Level Three Problem Solving

21. In the disease state called *hemochromatosis,* the hormone hepcidin is either absent or not functional. Use your understanding of iron homeostasis to predict what would happen to intestinal iron uptake and plasma levels of iron in this disease.

22. Erica's baby, Justin, has had a severe bout of diarrhea and is now dehydrated. Is his blood more likely to be acidotic or alkalotic? Why?

23. Mary Littlefeather arrives in her physician's office complaining of severe, steady pain in the upper right quadrant of her abdomen. The pain began shortly after she ate a meal of fried chicken, French fries, and peas. Lab tests and an ultrasound reveal the presence of gallstones in the common bile duct running from the liver, gallbladder, and pancreas into the small intestine.

   (a) Why was Mary's pain precipitated by the meal she ate?
   (b) Which of the following processes will be affected by the gallstones: micelle formation in the intestine, carbohydrate digestion in the intestine, and protein absorption in the intestine. Explain your reasoning.

24. Using what you have learned about epithelial transport, draw a picture of the salivary duct cells and lumen. Arrange the following membrane channels and transporters on the apical and basolateral membranes so that the duct cell absorbs Na^+ and secretes K^+: ENaC, Na^+-K^+-ATPase, and K^+ leak channel. With neural stimulation, the flow rate of saliva can increase from 0.4 mL/min to 2 mL/min. What do you think happens to the Na^+ and K^+ content of saliva at the higher flow rate?

Level Four Quantitative Problems

25. Intestinal transport of the amino acid analog MIT (monoiodotyrosine) can be studied using the "everted sac" preparation. A length of intestine is turned inside out, filled with a solution containing MIT, tied at both ends, and then placed in a bath containing nutrients, salts, and an equal concentration of MIT. Changes in the concentration of MIT are monitored in the bath (mucosal or apical side of the inverted intestine), in the intestinal cells (tissue), and within the sac (serosal or basolateral side of the intestine) over a 240-minute period. The results are displayed in the graph shown here. (Data from Nathans *et al.*, *Biochimica et Biophysica Acta 41:* 271–282, 1960.)

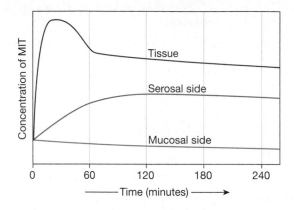

(a) Based on the data shown, is the transepithelial transport of MIT a passive process or an active process?

(b) Which way does MIT move: (1) apical to tissue to basolateral, or (2) basolateral to tissue to apical? Is this movement absorption or secretion?

(c) Is transport across the apical membrane active or passive? Explain your reasoning.

(d) Is transport across the basolateral membrane active or passive? Explain your reasoning.

22

Metabolism and Energy Balance

Probably no single abnormality has contributed more to our knowledge of the intermediary metabolism of animals than has the disease known as diabetes.

Helen R. Downes, *The Chemistry of Living Cells,* 1955

APPETITE AND SATIETY 694

LO 22.1   Diagram the control pathways that influence hunger and satiety.

ENERGY BALANCE 696

LO 22.2   Explain how we measure energy use and metabolic rate in humans.

LO 22.3   Identify the factors that affect metabolic rate.

METABOLISM 699

LO 22.4   Distinguish between anabolic and catabolic pathways, and name as many specific pathways as possible.

LO 22.5   Distinguish between the fed (absorptive) state and the fasted (postabsorptive) state.

LO 22.6   Describe the possible fates of ingested nutrients and indicate which is the most common for each class of biomolecules.

LO 22.7   Create a map that summarizes the balance of nutrient pools and nutrient storage for carbohydrates, proteins, and lipids.

LO 22.8   Explain the regulatory significance of push-pull control.

FED-STATE METABOLISM 701

LO 22.9   Create a summary diagram for anabolic metabolism of carbohydrates, proteins, and lipids in the fed state.

LO 22.10   Explain the relationship between different forms of cholesterol and cardiovascular disease.

FASTED-STATE METABOLISM 705

LO 22.11   Create a summary diagram for catabolic metabolism of carbohydrates, proteins, and lipids in the fasted state.

HOMEOSTATIC CONTROL OF METABOLISM 708

LO 22.12   Explain the roles of insulin and glucagon in the control of metabolism,

LO 22.13   Create a reflex map for insulin, including mechanisms of action where possible.

LO 22.14   Draw a reflex map for glucagon, including mechanisms of action where possible.

LO 22.15   Compare type 1 and type 2 diabetes mellitus. Explain how treatments for diabetes are related to the pathophysiology of the disease.

LO 22.16   Create a map for type 1 diabetes to show the body's responses to elevated plasma glucose in absence of insulin.

REGULATION OF BODY TEMPERATURE 720

LO 22.17   Explain the normal routes of heat gain and loss for the human body.

LO 22.18   Map the homeostatic control of body temperature.

BACKGROUND BASICS

31	Glycogen
82	Brown fat
94	Biological work
10	Mass balance
102	Metabolism
142	GLUT transporters
148	Receptor-mediated endocytosis
176	Tyrosine kinase receptors
201	Peptide hormones
493	Blood pressure control
605	Renal threshold for glucose
676	Exocrine pancreas
665	CCK
667	GIP
678	Chylomicrons

Vitamin C crystal

Magazine covers at grocery store checkout stands reveal a lot about Americans. Headlines screaming "Lose 10 pounds in a week without dieting" or "CCK: the hormone that makes you thin" vie for attention with glossy photographs of fat-laden, high-calorie desserts dripping with chocolate and whipped cream. As one magazine article put it, we are a nation obsessed with staying trim and with eating—two mutually exclusive occupations. But what determines when, what, and how much we eat? The factors influencing food intake are an area of intense research because the act of eating is the main point at which our bodies exert control over energy input.

APPETITE AND SATIETY

The control of food intake is a complex process. The digestive system does not regulate energy intake, so we must depend on behavioral mechanisms, such as hunger and **satiety** {*satis,* enough}, to tell us when and how much to eat. Psychological and social aspects of eating, such as parents who say "Clean your plate," complicate the physiological control of food intake. As a result, we still do not fully understand what governs when, what, and how much we eat. What follows is an overview of this increasingly complex and constantly changing field.

Our current model for behavioral regulation of food intake is based on two hypothalamic centers [Fig. 11.3, p. 360]: a **feeding center** that is tonically active and a **satiety center** that stops food intake by inhibiting the feeding center. Output signals from these centers cause changes in eating behavior and create sensations of hunger and fullness. Animals whose feeding centers are destroyed cease to eat. If the satiety centers are destroyed, animals overeat and become obese {*obesus,* plump or fat}.

Studies using transgenic and knockout mice [p. 730] show that these hypothalamic centers form a complicated neural network whose neurons secrete a variety of chemical messengers. The control of food intake is complex. Higher brain centers, including the cerebral cortex and limbic system, provide input to the hypothalamus. Many different chemical signals influence food intake and satiety, including neuropeptides, "brain-gut" hormones secreted by the GI tract, and chemical signals called **adipocytokines,** which are secreted by adipose tissue.

There are two classic theories for regulation of food intake: the glucostatic theory and the lipostatic theory. Current evidence indicates that these two classic theories are too simple to be the only models, however. The **glucostatic theory** states that glucose metabolism by hypothalamic centers regulates food intake. When blood glucose concentrations decrease, the satiety center is suppressed, and the feeding center is dominant. When glucose metabolism increases, the satiety center inhibits the feeding center.

The **lipostatic theory** of energy balance proposes that a signal from the body's fat stores to the brain modulates eating behavior so that the body maintains a particular weight. If fat stores are increased, eating decreases. In times of starvation, eating increases. Obesity results from disruption of this pathway.

The 1994 discovery of **leptin** {*leptos,* thin}, a protein hormone synthesized in adipocytes, provided evidence for the lipostatic theory. Leptin acts as a negative-feedback signal between adipose tissue and the brain. As fat stores increase, adipose cells secrete more leptin, and food intake decreases.

Leptin is synthesized in adipocytes under control of the *obese (ob) gene.* Mice that lack the *ob* gene (and therefore lack leptin) become obese, as do mice with defective leptin receptors. However, these findings did not translate well to humans, as only a small percentage of obese humans are leptin deficient. The majority of them have *elevated* leptin levels. As scientists have learned, leptin is only part of the story.

A key signal molecule is **neuropeptide Y (NPY),** a brain neurotransmitter that seems to be the stimulus for food intake. In normal-weight animals, leptin inhibits NPY in a negative feedback pathway (**FIG. 22.1**). Other neuropeptides, hormones, and adipocytokines also influence NPY secretion, leptin release by adipocytes, and the hypothalamic centers controlling food intake.

For example, the peptide **ghrelin** is secreted by the stomach during fasting and increases hunger when infused into human subjects. Other peptides, such as the hormones CCK and GLP-1, are released by the gut during a meal and help decrease hunger. Many of these appetite-regulating peptides have functions in addition to control of food intake. Ghrelin promotes release of growth hormone, for instance. The brain peptides called *orexins* appear to play a role in wakefulness and sleep. Our understanding of how all these factors interact is incomplete, and many scientists are studying this topic. Some of the major signal molecules being studied are listed in the table in Figure 22.1.

Appetite and eating are also influenced by sensory input through the nervous system. The simple acts of swallowing and chewing food help create a sensation of fullness. The sight, smell, and taste of food can either stimulate or suppress appetite.

In one interesting study researchers tried to determine whether chocolate craving is attributable to psychological factors or to

RUNNING **PROBLEM** | Eating Disorders

Sara and Nicole had been best friends growing up but had not seen each other for several semesters. When Sara heard that Nicole was in the hospital, she went to visit but wasn't prepared for what she saw. Nicole, who had always been thin, now was so emaciated that Sara hardly recognized her. Nicole had been taken to the emergency room when she fainted in a ballet class. When Dr. Ayani saw Nicole, her concern about her weight was as great as her concern about her broken wrist. At 5'6", Nicole weighed only 95 pounds (normal healthy range for that height is 118–155 pounds). Dr. Ayani suspected an eating disorder, probably anorexia nervosa, and she ordered blood tests to confirm her suspicion.

694   696   698   700   720   724

FIG. 22.1 | Complex chemical signaling controls food intake

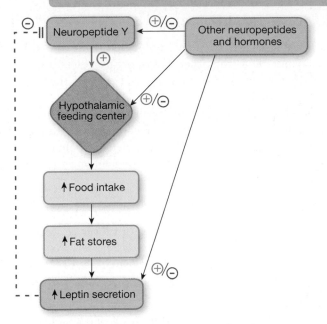

Key Peptides that Modulate Food Intake

Peptide	Source
Increase Food Intake	
Ghrelin	Stomach
NPY and Agouti-related protein (AgRP)	Co-expressed in hypothalamus
Orexins (hypocretins)	Hypothalamus
Decrease Food Intake	
CCK	Small intestine, neurons
Glucagon-like peptide-1 (GLP-1)	Intestines
PYY	Intestines
Leptin	Adipose cells
Corticotropin-releasing hormone (CRH)	Hypothalamus
α-Melanocyte-stimulating hormone (α-MSH)	Hypothalamus
CART (cocaine-and-amphetamine-regulated transcript) and POMC (pro-opiomelanocortin)	Co-expressed in hypothalamus

BIOTECHNOLOGY

Discovering Peptides: Research in Reverse

In the early days of molecular biology, scientists collected tissues that contained active peptides, isolated and purified the peptides, and then analyzed their amino acid sequence. Now, in the era of proteomics, investigators are using reverse techniques to discover new proteins in the body. One group of researchers, for example, found the hunger-inducing *orexin* (or *hypocretin*) peptides by isolating mRNA expressed in a particular region of the hypothalamus. The investigators then used this mRNA to create the amino acid sequence of the prepropeptide. At the same time, a different group of scientists also discovered orexin peptides by working backwards from an "orphan" G protein-coupled receptor [p. 179] to find its peptide ligand. The endogenous hunger hormone *ghrelin* was discovered by a similar method. Pharmacologists testing synthetic peptides to stimulate the release of growth hormone found that their peptides were binding to a previously unknown receptor, and from that receptor they discovered ghrelin.

physiological stimuli, such as chemicals in the chocolate.* Subjects were given either dark chocolate, white chocolate (which contains none of the pharmacological agents of cocoa), cocoa capsules, or placebo capsules. The researchers found that white chocolate was the best substitute for the real thing, which suggests that taste and aroma play a significant role in satisfying chocolate craving.

Psychological factors, such as stress, can also play a significant role in regulating food intake. In another study,** researchers found that subjects who imagined eating 30 M&M's one at a time ate fewer real M&M's than subjects who thought about eating only 3 M&M's. A repeat of the experiment using cheese cubes instead of M&M's had the same result.

The eating disorder *anorexia nervosa* has both psychological and physiological components, which complicates its treatment. And the concept of *appetite* is closely linked to the psychology of eating, which may explain why dieters who crave smooth, cold ice cream cannot be satisfied by a crunchy carrot stick. A considerable amount of money is directed to research on eating behaviors.

CONCEPT CHECK

1. Explain the roles of the satiety and feeding centers. Where are they located?

2. Studies show that most obese humans have elevated leptin levels in their blood. Based on your understanding of endocrine disorders [p. 216], propose some reasons why leptin is not decreasing food intake in these people.

*W. Michener and P. Rozin. Pharmacological versus sensory factors in the satiation of chocolate craving. *Physiol Behav* 56(3): 419–422, 1994.
**C. K. Morewedge *et al.* Thought for food: Imagined consumption reduces actual consumption. *Science* 330: 1530–1533, 2010.

RUNNING **PROBLEM**

Sara and Nicole first met in ballet class at age 11. They were both serious about dance and worked hard to maintain the perfect thin ballet-dancer's body. Both girls occasionally took diet pills or laxatives, and it was almost a contest to see who could eat the least food. Sara was always impressed with Nicole's willpower, but then again, Nicole was a perfectionist. Two years ago, Sara found herself less interested in dance and went away to college. Nicole was accepted into a prestigious ballet company and remained focused on her dancing—and on her body image. Nicole's regimented diet became stricter, and if she felt she'd eaten too much, she would make herself throw up. Her weight kept dropping, but each time she looked in the mirror, she saw a fat girl looking back.

Q1: If you measured Nicole's leptin level, what would you expect to find?

Q2: Would you expect Nicole to have elevated or depressed levels of neuropeptide Y?

694 696 698 700 720 724

ENERGY BALANCE

Once food has been digested and absorbed, the body's chemical reactions—known collectively as metabolism—determine what happens to the nutrients in the food. Are they destined to burn off as heat? Become muscle? Or turn into extra pounds that make it difficult to zip up blue jeans? In this section, we examine energy balance in the body.

Energy Input Equals Energy Output

The *first law of thermodynamics* [p. 95] states that the total amount of energy in the universe is constant. By extension, this statement means that all energy that goes into a biological system, such as the human body, can be accounted for (**FIG. 22.2**). In the body, most stored energy is contained in the chemical bonds of molecules.

We can apply the concept of mass balance to energy balance: changes to the body's energy stores result from the difference between the energy put into the body and the energy used.

$$\text{Total body energy} = \text{energy stored} + \text{energy intake} - \text{energy output} \qquad (1)$$

Energy intake for humans consists of energy in the nutrients we eat, digest, and absorb. *Energy output* is a combination of work performed and energy returned to the environment as heat:

$$\text{Energy output} = \text{work} + \text{heat} \qquad (2)$$

In the human body, at least half the energy released in chemical reactions is lost to the environment as unregulated "waste" heat.

The work in equation (2) takes one of three forms [p. 94]:

1. **Transport work** moves molecules from one side of a membrane to the other. Transport processes bring materials into and out of the body and transfer them between compartments.

2. **Mechanical work** uses intracellular fibers and filaments to create movement. This form of work includes external work, such as movement created by skeletal muscle contraction, and internal work, such as the movement of cytoplasmic vesicles and the pumping of the heart.

3. **Chemical work** is used for growth, maintenance, and storage of information and energy. Chemical work in the body can be subdivided into synthesis and storage. Storage includes both *short-term energy storage* in high-energy phosphate compounds such as ATP and *long-term energy storage* in the chemical bonds of glycogen and fat.

Most of this energy-consuming work in the body is not under conscious control. The only way people can voluntarily increase energy *output* is through body movement, such as walking and exercise. People can control their energy *intake*, however, by watching what they eat.

Although energy balance is a very simple concept, it is a difficult one for people to accept. Behavior modifications, such as eating less and exercising more, are among the most frequent instructions healthcare professionals give to patients. These

FIG. 22.2   **Energy balance in the body**

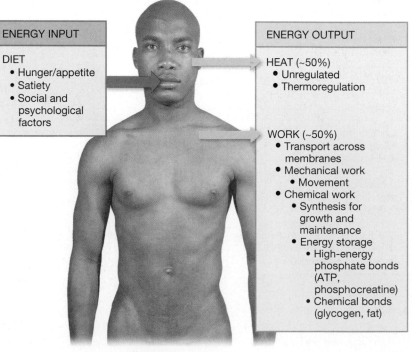

ENERGY INPUT

DIET
- Hunger/appetite
- Satiety
- Social and psychological factors

ENERGY OUTPUT

HEAT (~50%)
- Unregulated
- Thermoregulation

WORK (~50%)
- Transport across membranes
- Mechanical work
  - Movement
- Chemical work
  - Synthesis for growth and maintenance
  - Energy storage
    - High-energy phosphate bonds (ATP, phosphocreatine)
  - Chemical bonds (glycogen, fat)

Estimating Fat— The Body Mass Index

One way to estimate the total amount of energy stored in the body is a person's body weight. Although it is a simplistic approach, we say that when energy intake exceeds energy output, a person gains weight. If energy use exceeds dietary energy input, the body taps into its energy stores and the person loses weight. One current assessment for healthy weight is a measurement known as the **body mass index (BMI)**. To calculate your BMI:

$$\text{weight (lb)} \times 703/\text{height}^2 \text{ (in)} = \text{BMI}$$
$$\text{weight (kg)}/\text{height}^2 \text{ (m)} = \text{BMI}$$

A BMI between 18.5 and 24.9 is considered normal weight. Less than 18.5 is underweight, and more than 24.5 is overweight. A BMI over 30 indicates obesity. Higher risk for a number of diseases, including diabetes, heart disease, and high blood pressure, correlate with higher BMI values. The BMI calculation does not distinguish between fat mass and muscle mass, however, and heavily muscled athletes, such as football players, may have a BMI that seems unhealthy. Muscle tissue weighs more than fat, which explains the discrepancy. BMI calculations also do not allow for differences due to age, gender, and ethnicity. For example, data indicate that Asians with BMIs in the normal weight range may still be at higher risk of certain diseases. Researchers have suggested that a *fat mass index* (fat mass/height^2) is a better health indicator than BMI.

instructions are also among the most difficult for people to follow, and patient compliance is low.

Oxygen Consumption Reflects Energy Use

To compile an energy balance sheet for the human body, we must estimate both the energy content of food (energy intake) and the energy expenditure from heat loss and various types of work (energy output). The most direct way to measure the energy content of food is by **direct calorimetry**. In this procedure, food is burned in an instrument called a *bomb calorimeter*, and the heat released is trapped and measured.

The heat released is a direct measure of the energy content of the burned food and is usually measured in kilocalories (kcal). One **kilocalorie** (kcal) is the amount of heat needed to raise the temperature of 1 liter of water by 1 °C. A kilocalorie is the same as a *Calorie* (with a capital C). Direct calorimetry is a quick way of measuring the total energy content of food, but the *metabolic* energy content of food is slightly less because most foods cannot be fully digested and absorbed.

The caloric content of any food can be calculated by multiplying the number of grams of each component by its metabolic energy content. The metabolic energy content of proteins and carbohydrates is 4 kcal/g. Fats contain more than twice as much energy—9 kcal/g. For example, a plain bagel contains:

2 g fat	2 g fat × 9 kcal/g fat	= 18 kcal
7 g protein	7 g protein × 4 kcal/g protein	= 28 kcal
38 g carbohydrates	38 g carbohydrates × 4 kcal/g CHO	= 152 kcal
	Total calories	= 198 kcal

$$(3)$$

In the United States, you can find the energy content for various foods on the *Nutrition Facts* label of food packages.

Estimating an individual's energy expenditure, or **metabolic rate,** is more complex than figuring the caloric content of ingested food. Applying the *law of mass balance* to energy balance, a person's caloric intake minus heat production is the energy used for chemical, mechanical, and transport work. Heat released by the body can be measured by enclosing a person in a sealed compartment. Practically speaking, however, measuring total body heat release is not a very easy way to measure energy use.

Probably the most common method for estimating metabolic rate is to measure a person's **oxygen consumption,** the rate at which the body consumes oxygen as it metabolizes nutrients. Recall [p. 105] that metabolism of glucose to trap energy in the bonds of ATP is most efficient in the presence of adequate oxygen:

$$C_6H_{12}O_6 + O_2 + ADP + P_i \rightarrow$$
$$CO_2 + H_2O + ATP + \text{heat} \quad (4)$$

Studies have shown that oxygen consumption for different foods is relatively constant at a rate of 1 liter of oxygen consumed for each 4.5–5 kcal of energy released from the food being metabolized. The measurement of oxygen consumption is one form of **indirect calorimetry.**

Another method of estimating metabolic rate is to measure carbon dioxide production, either alone or in combination with oxygen consumption. Equation (4) shows that aerobic metabolism consumes O_2 and produces CO_2. However, the ratio of CO_2 produced to O_2 consumed varies with the composition of the diet. This ratio of CO_2 produced to O_2 consumed is known as the **respiratory quotient (RQ)** or the **respiratory exchange ratio (RER).** RQ varies from a high of 1 for a pure carbohydrate diet to 0.8 for pure protein and 0.7 for pure fat. The average American diet has an RQ of about 0.82.

Metabolic rate is calculated by multiplying oxygen consumption by the number of kilocalories metabolized per liter of oxygen consumed:

$$\text{Metabolic rate (kcal/day)} = \text{L } O_2 \text{ consumed/day} \times \text{kcal/L } O_2 \quad (5)$$

RUNNING **PROBLEM**

When Nicole's blood test results came back, Dr. Ayani immediately wrote orders to start an intravenous infusion and heart monitoring. The laboratory report showed plasma potassium of 2.5 mEq/L (normal: 3.5–5.0 mEq/L), plasma HCO_3^- of 40 mEq/L (normal: 24–29), and plasma pH of 7.52 (normal: 7.38–7.42). Dr. Ayani admitted Nicole to the hospital for further treatment and evaluation, hoping to convince her that she needed help for her anorexia. Anorexia, meaning "no appetite," can have both physiological and psychological origins.

Q3: What is Nicole's K^+ disturbance called? What effect does it have on the resting membrane potential of her cells?

Q4: Why does Dr. Ayani want to monitor Nicole's cardiac function?

Q5: Based on her laboratory blood values, what is Nicole's acid-base status?

694　696　**698**　700　720　724

A mixed diet with an RQ of 0.8 requires 1 liter of O_2 for each 4.80 kcal metabolized. For a 70-kg male whose resting oxygen consumption is 430 L/day, this means:

$$\text{Resting metabolic rate} = 430 \text{ L } O_2/\text{day} \times 4.80 \text{ kcal/L } O_2$$
$$= 2064 \text{ kcal/day} \qquad (6)$$

Many Factors Influence Metabolic Rate

Whether measured by O_2 consumption or by CO_2 production, metabolic rate can be highly variable from one person to another or from day to day in a single individual. An individual's lowest metabolic rate is considered the **basal metabolic rate (BMR).** In reality, metabolic rate would be lowest when an individual is asleep. However, because measuring the BMR of a sleeping person is difficult, metabolic rate is often measured after a 12-hour fast in a person who is awake but resting: a **resting metabolic rate (RMR).**

Other factors that affect metabolic rate in humans include age, sex, amount of lean muscle mass, activity level, diet, hormones, and genetics.

1. **Age and sex.** Adult males have an average BMR of 1 kcal per hour per kilogram of body weight. Adult females have a lower rate than males: 0.9 kcal/hr/kg. The difference arises because women have a higher percentage of adipose tissue and less lean muscle mass. Metabolic rates in both sexes decline with age. Some of this decline is due to decreases in lean muscle mass.

2. **Amount of lean muscle mass.** Muscle has higher oxygen consumption than adipose tissue, even at rest. (Most of the volume of an adipose tissue cell is occupied by metabolically inactive lipid droplets.) This is one reason weight loss advice often includes weight training in addition to aerobic exercise. Weight training adds muscle mass to the body, which increases basal metabolic rate and results in more calories being burned at rest.

3. **Activity level.** Physical activity and muscle contraction increase metabolic rate over the basal rate. Sitting and lying down consume relatively little energy. Competitive rowing and cycling are among the activities that expend the most energy.

4. **Diet.** Resting metabolic rate increases after a meal, a phenomenon termed **diet-induced thermogenesis.** In other words, there is an energetic cost to the digestion and assimilation of food. Diet-induced thermogenesis is related to the type and amount of food ingested. Fats cause relatively little diet-induced thermogenesis, and proteins increase heat production the most. This phenomenon may support the claim of some nutritionists that eating a calorie of fat is different from eating a calorie of protein, although they contain the same amount of energy when measured by direct calorimetry.

5. **Hormones.** Basal metabolic rate is increased by thyroid hormones and by catecholamines (epinephrine and norepinephrine). Some of the peptides that regulate food intake also appear to influence metabolism.

6. **Genetics.** The effect of inherited traits on energy balance can be observed in the variety of normal body types. Some people have very efficient metabolism that converts food energy into energy stored in adipose tissue with little heat loss, while others can eat large amounts of food and never gain weight because their metabolism is less efficient.

Of the factors affecting metabolic rate, a person can voluntarily control only two: energy intake (how much food is eaten) and level of physical activity. If a person's activity includes strength training, which increases lean muscle mass, resting metabolic rate goes up. The addition of lean muscle mass to the body creates additional energy use, which in turn decreases the number of calories that go into storage.

Energy Is Stored in Fat and Glycogen

A person's daily energy requirement, expressed as caloric intake, varies with the needs and activity of the body. For example, during the 2008 Olympics, swimming champion Michael Phelps consumed more than 12,000 kcal per day. On the other hand, a woman engaged in normal activities may require only 2000 kcal/day.

Suppose that our woman's energy requirement could be met by ingesting only glucose. Glucose has an energy content of 4 kcal/g, which means that to get 2000 kcal, she would have to consume 500 g, or 1.1 pounds, of glucose each day. Our bodies cannot absorb crystalline glucose, however, so those 500 g of glucose would have to be dissolved in water. If the glucose were made as an isosmotic 5% solution, the 500 g would have to be dissolved in 10 liters of water—a substantial volume to drink in a day!

Fortunately, we do not usually ingest glucose as our primary fuel. Proteins, complex carbohydrates, and fats also provide energy. The glucose polymer glycogen is a more compact form of energy than an equal number of individual glucose molecules. Glycogen also requires less water for hydration.

For this reason, our cells convert glucose to glycogen for storage. Normally, we keep about 100 g of glycogen in the liver and 200 g in skeletal muscles. But even this 300 g of glycogen can provide only enough energy for 10 to 15 hours. The brain alone requires 150 g of glucose per day.

Consequently, the body keeps most of its energy reserves in compact, high-energy fat molecules. One gram of fat has 9 kcal, more than twice the energy content of an equal amount of carbohydrate or protein. This means that each pound of body fat stores 3500 kcal of energy.

The high caloric content of fat and the histology of fat cells, with minimal cytosol and a large central fat droplet [p. 82], make adipose tissue very efficient at storing large amounts of energy in minimal space. Metabolically, however, the energy in fat is harder to access, and the metabolism of fats is slower than that of carbohydrates.

CONCEPT CHECK

3. Name seven factors that can influence a person's metabolic rate.

4. Why does the body store most of its extra energy in fat and not in glycogen?

5. Complete and balance the following equation for aerobic metabolism of one glucose molecule:
$C_6H_{12}O_6 + O_2 \rightarrow ? + ?$

6. What is the RQ for the balanced equation in Concept Check 5?

METABOLISM

Metabolism is the sum of all chemical reactions in the body. The reactions making up these pathways (1) extract energy from nutrients, (2) use energy for work, and (3) store excess energy so that it can be used later. Metabolic pathways that synthesize large molecules from smaller ones are called **anabolic pathways** {*ana-*, completion + *metabole*, change}. Those that break large molecules into smaller ones are called **catabolic pathways** {*cata-*, down or back}.

The classification of a pathway is its net result, not what happens in any individual step of the pathway. For example, in the first step of *glycolysis* [p. 107], glucose gains a phosphate to become a larger molecule, glucose 6-phosphate. This single reaction is anabolic, but by the end of glycolysis, the initial 6-carbon glucose molecule has been converted to two 3-carbon pyruvate molecules. The breakdown of one glucose to two pyruvate makes glycolysis a catabolic pathway.

In the human body, we divide metabolism into two states. The period of time following a meal, when the products of digestion are being absorbed, used, and stored, is called the **fed state** or the **absorptive state.** This is an anabolic state in which the energy of nutrient biomolecules is transferred to high-energy compounds or stored in the chemical bonds of other molecules.

Once nutrients from a recent meal are no longer in the bloodstream and available for use by the tissues, the body enters what is called the **fasted state** or the **postabsorptive state.** As the pool of available nutrients in the blood decreases, the body taps into its stored reserves. The fasted state is catabolic because cells break down large molecules into smaller molecules. The energy released by breaking chemical bonds of large molecules is used to do work.

Ingested Energy May Be Used or Stored

The biomolecules we ingest are destined to meet one of three fates:

1. **Energy.** Biomolecules can be metabolized immediately, with the energy released from broken chemical bonds trapped in ATP, phosphocreatine, and other high-energy compounds. This energy can then be used to do mechanical work.
2. **Synthesis.** Biomolecules entering cells can be used to synthesize basic components needed for growth and maintenance of cells and tissues.
3. **Storage.** If the amount of food ingested exceeds the body's requirements for energy and synthesis, the excess energy goes into storage in the bonds of glycogen and fat. Storage makes energy available for times of fasting.

The fate of an absorbed biomolecule depends on whether it is a carbohydrate, protein, or fat. **FIGURE 22.3** is a schematic diagram that follows these biomolecules from the diet into the three *nutrient pools* of the body: the free fatty acid pool, the glucose pool, and the amino acid pool. **Nutrient pools** are nutrients that are available for immediate use. They are located primarily in the plasma.

Free fatty acids form the primary pool of fats in the blood. They can be used as an energy source by many tissues but are also easily stored as fat (*triglycerides*) in adipose tissue.

Carbohydrates are absorbed mostly as glucose. Plasma glucose concentration is the most closely regulated of the three nutrient pools because glucose is the only fuel the brain can metabolize, except in times of starvation. Notice the locations of the exit "pipes" on the glucose pool in Figure 22.3. If the glucose pool falls below a certain level, only the brain has access to glucose. This conservation measure ensures that the brain has an adequate energy supply. Just as the circulatory system gives priority to supplying oxygen to the brain, metabolism also gives priority to the brain.

If the body's glucose pool is within the normal range, most tissues use glucose for their energy source. Excess glucose goes into storage as glycogen. The synthesis of glycogen from glucose is known as **glycogenesis.** Glycogen stores are limited, however, and additional excess glucose is converted to fat by **lipogenesis.**

If plasma glucose concentrations decrease, the body converts glycogen to glucose through **glycogenolysis.** The body maintains plasma glucose concentrations within a narrow range by balancing oxidative metabolism, glycogenesis, glycogenolysis, and lipogenesis.

FIG. 22.3 Nutrient pools and metabolism

Based on L. L. Langley, *Homeostasis*.
New York: Reinhold, 1965.

If homeostasis fails and plasma glucose exceeds a critical level, as occurs in diabetes mellitus, excess glucose is excreted in the urine. Glucose excretion takes place only when the *renal threshold* for glucose reabsorption is exceeded [p. 605].

The amino acid pool of the body is used primarily for protein synthesis. However, if glucose intake is low, amino acids can be converted into glucose through the pathways known as **gluconeogenesis.** This word literally means "the birth (*genesis*) of new (*neo*) glucose" and refers to the synthesis of glucose from a non-carbohydrate precursor.

Amino acids are the main source for glucose through the gluconeogenesis pathways, but glycerol from triglycerides can also be used. Both gluconeogenesis and glycogenolysis are important backup sources for glucose during periods of fasting.

Enzymes Control the Direction of Metabolism

How does the body control the shift of metabolism between the fed state and the fasted state? One key feature of metabolic regulation is the use of different enzymes to catalyze forward and reverse reactions. This dual control, sometimes called *push-pull control*, allows close regulation of a reaction's direction.

RUNNING **PROBLEM**

When Nicole was admitted to the hospital, her blood pressure was 80/50, and her pulse was a weak and irregular 90 beats per minute. She weighed less than 85% of the minimal healthy weight for a woman of her height and age. She had an intense fear of gaining weight, even though she was underweight. Her menstrual periods were irregular, she had just suffered a fractured wrist from a fall that normally shouldn't have caused a fracture, and her hair was thinning. When Dr. Ayani questioned Nicole, she admitted that she had been feeling weak during dance rehearsals and had been having difficulty concentrating at times.

Q6: *Based on what you know about heart rate and blood pressure, speculate on why Nicole has low blood pressure with a rapid pulse.*

Q7: *Would you expect Nicole's renin and aldosterone levels to be normal, elevated, or depressed? How might these levels relate to her K^+ disturbance?*

Q8: *Give some possible reasons Nicole had been feeling weak during dance rehearsals.*

FIG. 22.4 Push-pull control

In push-pull control, different enzymes catalyze forward and reverse reactions.

(a) Without regulation of enzymatic activity, the pathway will simply cycle back and forth. There is no net synthesis of substrate A or B.

No net synthesis
of substrate A or B

(b) In fed-state metabolism under the influence of insulin, enzyme activity for the forward reaction increases. Enzymes for glycogen breakdown are inhibited. Net glycogen synthesis results.

Net glycogen
synthesis

(c) In fasted-state metabolism under the influence of glucagon, enzymes that break down glycogen are more active, and enzymes for glycogen synthesis are inhibited. Net glucose synthesis results.

Net glucose
synthesis

FIGURE 22.4 shows how push-pull control can regulate the flow of nutrients through metabolic pathways. In Figure 22.4a, enzyme 1 catalyzes the reaction A → B, and enzyme 2 catalyzes the reverse reaction, B → A. When the activity of the two enzymes is roughly equal, as soon as A is converted into B, B is converted back into A. Turnover of the two substrates is rapid, but there is no net production of either A or B.

To alter the net direction of the reaction, the enzyme activity must change. Enzymes are proteins that bind ligands, so their activity can be modulated [p. 49]. Most modulation of metabolic enzymes is controlled by hormones.

Figure 22.4b represents the series of reactions through which glucose becomes glycogen. During the fed state, the pancreatic hormone insulin stimulates the enzymes promoting glycogenesis and inhibits the enzymes for glycogenolysis. The net result is glycogen synthesis from glucose.

The reverse pattern is shown in Figure 22.4c. In the fasted state, *glucagon,* another pancreatic hormone, is dominant. Glucagon stimulates the enzymes of glycogenolysis while inhibiting the enzymes for glycogenesis. The net result is glucose synthesis from glycogen.

FED-STATE METABOLISM

The fed state following ingestion of nutrients is anabolic: absorbed nutrients are being used for energy, synthesis, and storage. The table at the bottom of **FIGURE 22.5** summarizes the fates of carbohydrates, proteins, and fats in the fed state. In the sections that follow, we examine some of these pathways.

Carbohydrates Make ATP

The most important biochemical pathways for energy production are summarized in Figure 22.5. This figure does not include all of the metabolic intermediates in each pathway [see Chapter 4, p. 107 for detailed pathways]. Instead, it emphasizes the points at which different pathways intersect, because these intersections are often key points at which metabolism is controlled.

Glucose is the primary substrate for ATP production. Glucose absorbed from the intestine enters the hepatic portal vein and is taken directly to the liver, where about 30% of all ingested glucose is metabolized. The remaining 70% continues in the bloodstream for distribution to the brain, muscles, and other organs and tissues.

Glucose moves from interstitial fluid into cells by membrane GLUT transporters [p. 142]. Most glucose absorbed from a meal goes immediately into glycolysis and the citric acid cycle to make ATP. Some glucose is used by the liver for lipoprotein synthesis. Glucose that is not required for energy and synthesis is stored either as glycogen or fat. The body's ability to store glycogen is limited, so most excess glucose is converted to triglycerides and stored in adipose tissue.

Glucose Storage Glycogen, a large polysaccharide, is the main storage form of glucose in the body. Glycogen is a glucose polymer, created by linking many individual glucose molecules together into a branching chain [see Fig. 2.2, p. 31]. A single glycogen particle in the cytoplasm may contain as many as 55,000 linked glucose molecules! Glycogen granules occur as insoluble inclusions in the cytosol of cells [p. 65].

Glycogen is found in all cells of the body, but the liver and skeletal muscles contain especially high concentrations. Glycogen in skeletal muscles provides a ready energy source for muscle contraction. Glycogen in the liver acts as the main source of glucose for the body in periods between meals (the fasted state). It is estimated that the liver keeps about a 4-hour supply of glucose stored as glycogen.

CONCEPT CHECK

7. Are GLUT transporters active or passive transporters?

Amino Acids Make Proteins

Most amino acids absorbed from a meal go to the tissues for protein synthesis [p. 112]. Like glucose, amino acids are taken first to the liver by the hepatic portal system. The liver uses them to synthesize lipoproteins and plasma proteins, such as albumin, clotting factors, and angiotensinogen.

Amino acids not taken up by the liver are used by cells to create structural or functional proteins, such as cytoskeletal elements, enzymes, and hormones. Amino acids are also

FIG. 22.5 **ESSENTIALS**

Biochemical Pathways for Energy Production

(a) Summary of Biochemical Pathways for Energy Production

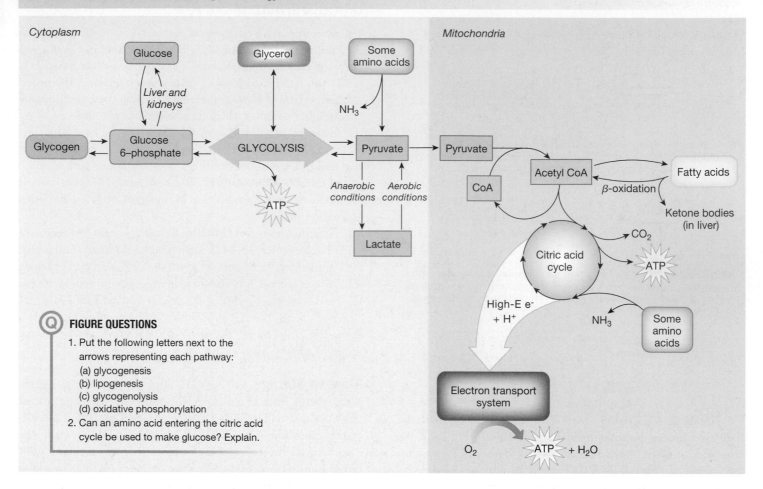

Q FIGURE QUESTIONS

1. Put the following letters next to the arrows representing each pathway:
   (a) glycogenesis
   (b) lipogenesis
   (c) glycogenolysis
   (d) oxidative phosphorylation
2. Can an amino acid entering the citric acid cycle be used to make glucose? Explain.

(b) Fates of Nutrients in Fed-State and Fasted-State Metabolism

Nutrient	Absorbed as	Fed-State Metabolism	Fasted-State Metabolism
Carbohydrates	Glucose primarily; also fructose and galactose	• Used immediately for energy through aerobic pathways* (*glycolysis* and *citric acid cycle*) • Used for lipoprotein synthesis in liver • Stored as glycogen in liver and muscle (*glycogenesis*) • Excess converted to fat and stored in adipose tissue (*lipogenesis*)	• Glycogen polymers broken down (*glycogenolysis*) to glucose in liver and kidney or to glucose 6-phosphate for use in glycolysis
Proteins	Amino acids primarily plus some small peptides	• Most amino acids go to tissues for protein synthesis* • If needed for energy, amino acids converted in liver to intermediates for aerobic metabolism (*deamination*) • Excess converted to fat and stored in adipose tissue (*lipogenesis*)	• Proteins broken down into amino acids • Amino acids deaminated in liver for ATP production or used to make glucose (*gluconeogenesis*)
Fats	Fatty acids, triglycerides and cholesterol	• Stored as triglycerides primarily in the liver and adipose tissue* (*lipogenesis*) • Cholesterol used for steroid synthesis or as a membrane component • Fatty acids used for lipoprotein and eicosanoid synthesis	• Triglycerides broken down into fatty acids and glycerol (*lipolysis*) • Fatty acids used for ATP production through aerobic pathways (*β-oxidation*)

*Primary fate

incorporated into nonprotein molecules, such as amine hormones and neurotransmitters.

If glucose intake is low, amino acids can be used for energy, as described in the next section on fasted-state metabolism. However, if more protein is ingested than is needed for synthesis and energy expenditures, excess amino acids are converted to fat. Some bodybuilders spend large amounts of money on amino-acid supplements advertised to build bigger muscles. But these amino acids do not automatically go into protein synthesis. When amino acid intake exceeds the body's need for protein synthesis, excess amino acids are burned for energy or stored as fat.

Fats Store Energy

Most ingested fats are assembled inside intestinal epithelial cells into lipoprotein and lipid complexes called *chylomicrons* [p. 678]. Chylomicrons leave the intestine and enter the venous circulation via the lymphatic vessels (**FIG. 22.6a**). Chylomicrons consist of cholesterol, triglycerides, phospholipids, and lipid-binding proteins called **apoproteins,** or *apolipoproteins* {*apo-*, derived from}. Once these lipid complexes begin to circulate through the blood, the enzyme **lipoprotein lipase** (lpl) bound to the capillary endothelium of muscles and adipose tissue converts the triglycerides to free fatty acids and glycerol. These molecules may be used for energy by most cells or reassembled into triglycerides for storage in adipose tissue.

Chylomicron remnants that remain in the circulation are taken up and metabolized by the liver (Fig. 22.6a). Cholesterol from the remnants joins the liver's pool of lipids. If cholesterol is in excess, some may be converted to bile salts and excreted in the bile. The remaining cholesterol is added to newly synthesized cholesterol and fatty acids, and packaged into lipoprotein complexes for secretion into the blood.

The lipoprotein complexes that re-enter the blood contain varying amounts of triglycerides, phospholipids, cholesterol, and apoproteins. The more protein a complex contains, the heavier it is, with plasma lipoprotein complexes ranging from *very-low-density lipoprotein* (VLDL) to *high-density lipoprotein* (HDL). The combination of lipids with proteins makes cholesterol more soluble in plasma, but the complexes are unable to diffuse through cell membranes. Instead, they must be brought into cells by *receptor-mediated endocytosis* [p. 148]. The apoproteins in the complexes have specific membrane receptors in different tissues.

Most lipoprotein in the blood is *low-density lipoprotein-cholesterol* (LDL-C) [p. 502]. LDL-C is sometimes known as the "lethal cholesterol" because elevated concentrations of plasma LDL-C are associated with the development of atherosclerosis [p. 535]. LDL-C complexes contain *apoprotein B* (apoB), which combines with receptors that bring LDL-C into most cells of the body. Several inherited forms of *hypercholesterolemia* (elevated plasma cholesterol levels) have been linked to defective forms of apoB. These abnormal apoproteins may help explain the accelerated development of atherosclerosis in people with hypercholesterolemia.

The second most common lipoprotein in the blood is *high-density lipoprotein-cholesterol* (HDL-C). HDL-C is sometimes called the "healthy cholesterol" because HDL is the lipoprotein involved in cholesterol transport out of the plasma. HDL-C contains *apoprotein A* (apoA), which facilitates cholesterol uptake by the liver and other tissues.

Lipid Synthesis Most people get sufficient cholesterol from animal products in the diet, but cholesterol is such an important molecule that the body will synthesize it if the diet is deficient. Even vegetarians who eat no animal products (vegans) have substantial amounts of cholesterol in their cells. The body can make cholesterol from acetyl CoA through a series of reactions. Once the ring structure of cholesterol is synthesized, it is a fairly simple matter for the cell to change cholesterol into hormones and other steroids.

Other fats needed for cell structure and function, such as phospholipids, can also be made from non-lipid precursors during the fed state. Lipids are so diverse that generalizing about their synthesis is difficult. Enzymes in the smooth endoplasmic reticulum and cytosol of cells are responsible for most lipid synthesis. For example, the phosphorylation steps that convert triglycerides into phospholipids take place in the smooth endoplasmic reticulum (ER).

FIG. 22.6 **ESSENTIALS**

Fat Synthesis

(a) Transport and Fate of Dietary Fats

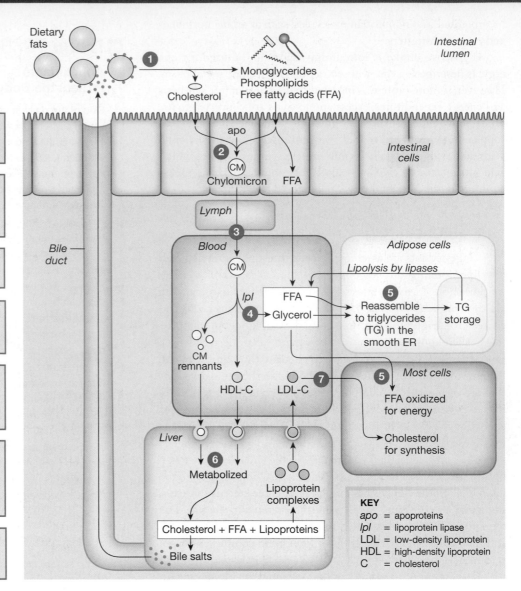

1 Bile salts help break down dietary fats into components that can be absorbed.

2 Intestinal epithelial cells assemble absorbed cholesterol, lipoproteins, and lipid complexes into chylomicrons.

3 Chylomicrons are transported to the blood via the lymphatic vessels.

4 Lipoprotein lipase (lpl) converts triglycerides into free fatty acids and glycerol.

5 Adipose cells reassemble free fatty acids and glycerol into triglycerides for storage. Other cells use free fatty acids for energy production.

6 Chylomicron remnants and HDL-C enter the liver for further processing, creating lipoprotein complexes such as LDL and VLDL. Some of the cholesterol is recycled in new bile salts.

7 LDL-C is transported via the blood to most of the cells, where the cholesterol is used for synthesis.

KEY

apo	=	apoproteins
lpl	=	lipoprotein lipase
LDL	=	low-density lipoprotein
HDL	=	high-density lipoprotein
C	=	cholesterol

(b) Triglyceride and Cholesterol Synthesis from Glucose

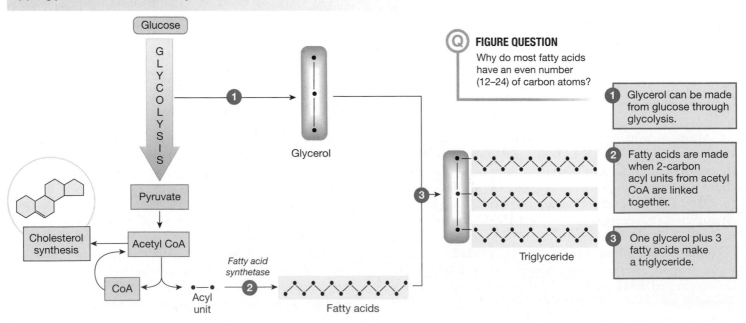

Q FIGURE QUESTION
Why do most fatty acids have an even number (12–24) of carbon atoms?

1 Glycerol can be made from glucose through glycolysis.

2 Fatty acids are made when 2-carbon acyl units from acetyl CoA are linked together.

3 One glycerol plus 3 fatty acids make a triglyceride.

Triglyceride synthesis from excess ingested glucose and protein is an important part of fed-state metabolism. Figure 22.6b shows some pathways for triglyceride synthesis. Glycerol can be made from glucose or from glycolysis intermediates (Fig. 22.6b). Fatty acids are made from acetyl CoA when a cytosolic enzyme called *fatty acid synthetase* links the 2-carbon acyl groups into carbon chains. This process also requires hydrogens and high-energy electrons from NADPH. The combination of glycerol and fatty acids into triglycerides takes place in the smooth endoplasmic reticulum.

Plasma Cholesterol Predicts Heart Disease

Of the nutrients in the plasma, lipids and glucose receive the most attention from health professionals. Abnormal glucose metabolism is the hallmark of diabetes mellitus, described later in this chapter. Abnormal plasma lipids are used as predictors of atherosclerosis and coronary heart disease (CHD) [p. 501].

Tests to measure blood lipids and assess cardiovascular risk range from simple but less accurate finger-stick blood samples to expensive tests on venous blood that look at all sizes of lipoproteins, from VLDL through HDL. As more epidemiological and treatment data are gathered, experts continue to redefine desirable lipid values. The U.S. National Cholesterol Education Panel issued guidelines in 2004 (*www.nhlbi.nih.gov/health-pro/ guidelines/current/cholesterol-guidelines/update-2004.htm*). Emphasis over the years has shifted from concern about total cholesterol levels (<200 mg/dL of plasma is recommended) to the absolute amounts and relative proportions of the various subtypes.

Some studies indicate that elevated LDL-C is the single largest cholesterol risk factor for CHD (**FIG. 22.7**). Oxidized LDL-C in the circulation is taken up by macrophages and leads to the development of atherosclerotic plaques inside blood vessels [Fig. 15.21, p. 503]. The optimum LDL-C value is <100 mg/dL

FIG. 22.7 **The relationship between LDL-C and risk of developing heart disease**

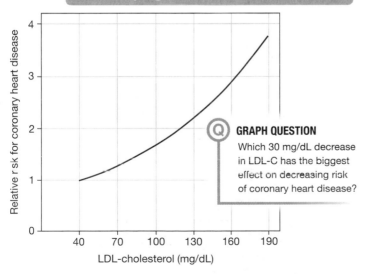

Q GRAPH QUESTION
Which 30 mg/dL decrease in LDL-C has the biggest effect on decreasing risk of coronary heart disease?

Data taken from Grundy *et al.*, *Circulation* 110: 227–239, 2004 July 13.

plasma, but up to 130 mg/dL is acceptable. LDL-C higher than 130 mg/dL increases a person's risk for CHD.

The HDL-C level in plasma has also been used to predict risk of developing atherosclerosis. Like high LDL-C, low HDL-C (<40 mg/dL) is associated with a higher risk of developing CHD. More recently, health professionals have started looking at the *non-HDL cholesterol value* (total cholesterol minus HDL-C) as perhaps a better indicator of CHD risk.

Lifestyle modifications (improved diet, smoking cessation, and exercise) can be very effective in improving lipid profiles but can be difficult for patients to implement and sustain. All the pharmacological therapies developed to treat elevated cholesterol target some aspect of cholesterol absorption and metabolism, underscoring the principle of mass balance in cholesterol homeostasis. Decreasing cholesterol uptake or synthesis and increasing cholesterol clearance by metabolism or excretion are effective mechanisms for reducing the amount of cholesterol in the body.

The drugs known as *bile acid sequestrants {sequestrare,* to put in the hands of a trustee} bind to bile acids in the intestinal lumen, preventing their reabsorption and increasing cholesterol excretion [p. 678]. In the absence of recycled bile salts, the liver increases bile acid synthesis from cholesterol, thereby decreasing plasma cholesterol when hepatic LDL receptors import cholesterol from the blood. A drug-free way to accomplish the same result is to eat more soluble fiber—indigestible fiber found in oatmeal, for example. Soluble fiber traps bile salts and increases their excretion in the feces.

Another lipid-lowering drug, *ezetimibe,* inhibits intestinal cholesterol transport [p. 678]. A different strategy for reducing intestinal cholesterol uptake is adding plant *sterols* (steroid-alcohols) and *stanols* (saturated sterols) to the diet by eating nuts, seeds, cereals, and vegetable oils. Sterols and stanols displace cholesterol in chylomicrons, decreasing cholesterol absorption.

The remaining lipid-lowering drugs all affect cholesterol metabolism in the liver. The drugs called *statins* inhibit the enzyme *HMG CoA reductase,* which mediates cholesterol synthesis in hepatocytes. The *fibrates,* which stimulate a transcription factor called *PPARα* (pronounced *p-par-alpha*), and *niacin* (vitamin B_3, nicotinic acid) decrease LDL-C and increase HDL-C by mechanisms that are not well understood.

CONCEPT CHECK
8. What is a dL (as in mg/dL)?
9. Use your understanding of digestive physiology to predict the common side effects of taking bile acid sequestrants and ezetimibe.

FASTED-STATE METABOLISM

Once all nutrients from a meal have been digested, absorbed, and distributed to various cells, plasma concentrations of glucose begin to fall. This is the signal for the body to shift from fed (absorptive) state to fasted (postabsorptive) state metabolism. Metabolism is under the control of hormones whose goal is to

FIG. 22.14 Insulin and glucagon

Metabolism is controlled by the insulin:glucagon ratio.

(a) Fed State: Insulin Dominates

- ↑ Glucose oxidation
- ↑ Glycogen synthesis
- ↑ Fat synthesis
- ↑ Protein synthesis

(b) Fasted State: Glucagon Dominates

- ↑ Glycogenolysis
- ↑ Gluconeogenesis
- ↑ Ketogenesis

(c) Glucose, Glucagon, and Insulin Levels before and after a Meal

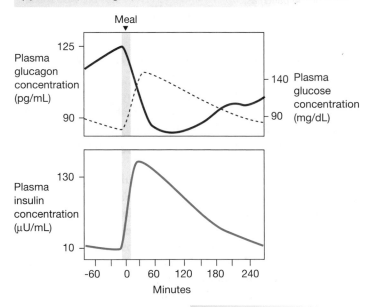

KEY

---- Plasma glucose concentration

—— Plasma glucagon concentration

—— Plasma insulin concentration

Based on data from Unger, *New Engl J Med* 285: 443–449, 1971.

Insulin Is the Dominant Hormone of the Fed State

Insulin is a typical peptide hormone (**TBL. 22.1**). It is synthesized as an inactive prohormone and activated prior to secretion [Fig. 7.3c, p. 203]. Glucose is an important stimulus for insulin secretion, but additional factors may enhance, amplify, or inhibit secretion (**FIG. 22.15**).

1. **Increased plasma glucose.** A major stimulus for insulin release is plasma glucose concentrations greater than 100 mg/dL. Glucose absorbed from the small intestine reaches pancreatic beta cells, where it is taken up by GLUT2 transporters [Fig. 5.26b, p. 159]. With more glucose available as substrate, ATP production increases and ATP-gated K^+ channels close. The cell depolarizes, voltage-gated Ca^{2+} channels open, and Ca^{2+} entry initiates exocytosis of insulin.

2. **Increased plasma amino acids.** Increased plasma amino acid concentrations following a meal also trigger insulin secretion.

3. **Feedforward effects of GI hormones.** Recently it has been shown that as much as 50% of insulin secretion is stimulated

TABLE 22.1	Insulin
Cell of origin	Beta cells of pancreas
Chemical nature	51-amino acid peptide
Biosynthesis	Typical peptide
Transport in the circulation	Dissolved in plasma
Half-life	5 minutes
Factors affecting release	Plasma [glucose] >100 mg/dL; ↑ blood amino acids; GLP-1 (feedforward reflex). Parasympathetic activity amplifies. Sympathetic activity inhibits.
Target cells or tissues	Liver, muscle, and adipose tissue primarily; brain, kidney, and intestine not insulin dependent
Target receptor	Membrane receptor with tyrosine kinase activity; pathways with insulin-receptor substrates
Whole body or tissue action	↓ Plasma [glucose] by ↑ transport into cells or ↑ metabolic use of glucose
Action at cellular level	↑ Glycogen synthesis; ↑ aerobic metabolism of glucose; ↑ protein and triglyceride synthesis
Action at molecular level	Inserts GLUT4 transporters in muscle and adipose cells; alters enzyme activity. Complex signal transduction pathways
Feedback regulation	↓ Plasma [glucose] shuts off insulin release
Other information	Growth hormone and cortisol are antagonistic

FIG. 22.15 Insulin in the fed state

KEY
- Stimulus
- Sensor
- Input signal
- Integrating center
- Output signal
- Target
- Tissue response
- Systemic response

Q **FIGURE QUESTION**
What is the neurotransmitter and receptor for parasympathetic output at ✱?

by the hormone *glucagon-like peptide-1* (GLP-1). GLP-1 and GIP (gastric inhibitory peptide) are *incretin* hormones produced by cells of the ileum and jejunum in response to nutrient ingestion. The incretins travel through the circulation to pancreatic beta cells and may reach them even before the first glucose is absorbed. The anticipatory release of insulin in response to these hormones prevents a sudden surge in plasma glucose concentrations when the meal is absorbed. Other GI hormones, such as CCK and gastrin, amplify insulin secretion.

4. **Parasympathetic activity.** Parasympathetic activity to the GI tract and pancreas increases during and following a meal. Parasympathetic input to beta cells stimulates insulin secretion.

5. **Sympathetic activity.** Insulin secretion is inhibited by sympathetic neurons. In times of stress, sympathetic input to the endocrine pancreas increases, reinforced by catecholamine release from the adrenal medulla (**TBL. 22.2**). Epinephrine and norepinephrine inhibit insulin secretion and switch metabolism to gluconeogenesis to provide extra fuel for the nervous system and skeletal muscles.

Insulin Promotes Anabolism

Like other peptide hormones, insulin combines with a membrane receptor on its target cells (**FIG. 22.16**). The insulin receptor has *tyrosine kinase* activity, which initiates complex intracellular

TABLE 22.2	Adrenal Catecholamines (Epinephrine and Norepinephrine)
Origin	Adrenal medulla (90% epinephrine and 10% norepinephrine)
Chemical nature	Amines made from tyrosine
Biosynthesis	Typical peptide
Transport in the circulation	Some bound to sulfate
Half-life	2 minutes
Factors affecting release	Primarily fight-or-flight reaction through CNS and autonomic nervous system; hypoglycemia
Target cells or tissues	Mainly neurons, pancreatic endocrine cells, heart, blood vessels, adipose tissue
Target receptor	G protein-coupled receptors; α and β subtypes
Second messenger	cAMP for α_2 and all β receptors; IP_3 for α_1 receptors
Whole body or tissue action	↑ Plasma [glucose]; activate fight-or-flight and stress reactions; ↑ glucagon and ↓ insulin secretion
Onset and duration of action	Rapid and brief

cascades whose details are still not completely understood. The activated insulin receptor phosphorylates proteins that include a group known as the **insulin-receptor substrates** (IRS). These proteins act through complicated pathways to influence transport and cellular metabolism. The enzymes that regulate metabolic pathways may be inhibited or activated directly, or their synthesis may be influenced indirectly through transcription factors.

The primary target tissues for insulin are the liver, adipose tissue, and skeletal muscles (Fig. 22.15). The usual target cell response is increased glucose metabolism. In some target tissues, insulin also regulates the GLUT transporters. Other tissues, including the brain and transporting epithelia of the kidney and intestine, are insulin-independent—they do not require insulin for glucose uptake and metabolism.

Insulin lowers plasma glucose in the following four ways:

1. **Insulin increases glucose transport into most, but not all, insulin-sensitive cells.** Adipose tissue and resting skeletal muscle require insulin for glucose uptake (**FIG. 22.17a**). Without insulin, their GLUT4 transporters are withdrawn from the membrane and stored in cytoplasmic vesicles—another example of membrane recycling. When insulin binds to the receptor and activates it, the resulting signal transduction cascade causes the vesicles to move to the cell membrane and insert the GLUT4 transporters by exocytosis (Fig. 22.17b). The cells then take up glucose from the interstitial fluid by facilitated diffusion.

Interestingly, *exercising* skeletal muscle is not dependent on insulin activity for its glucose uptake. When muscles contract, GLUT4 transporters are inserted into the membrane even in the absence of insulin, and glucose uptake increases. The intracellular signal for this is complex but appears to involve Ca^{2+} as well as a variety of intracellular proteins.

FIG. 22.16   Insulin's cellular mechanism of action

1. Insulin binds to tyrosine kinase receptor.
2. Receptor phosphorylates insulin-receptor substrates (IRS).
3. Second messenger pathways alter protein synthesis and existing proteins.
4. Membrane transport is modified.
5. Cell metabolism is changed.

FIG. 22.17 Glucose transport in fed and fasted states

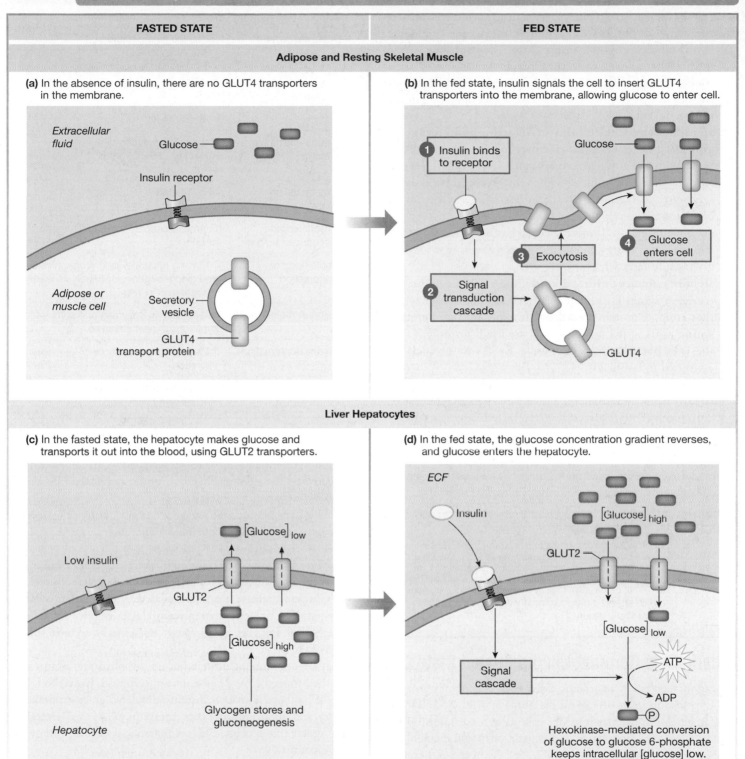

FASTED STATE	FED STATE

Adipose and Resting Skeletal Muscle

(a) In the absence of insulin, there are no GLUT4 transporters in the membrane.

Extracellular fluid

Glucose

Insulin receptor

Adipose or muscle cell

Secretory vesicle

GLUT4 transport protein

(b) In the fed state, insulin signals the cell to insert GLUT4 transporters into the membrane, allowing glucose to enter cell.

1 Insulin binds to receptor

Glucose

2 Signal transduction cascade

3 Exocytosis

4 Glucose enters cell

GLUT4

Liver Hepatocytes

(c) In the fasted state, the hepatocyte makes glucose and transports it out into the blood, using GLUT2 transporters.

[Glucose]$_{low}$

Low insulin

GLUT2

[Glucose]$_{high}$

Glycogen stores and gluconeogenesis

Hepatocyte

(d) In the fed state, the glucose concentration gradient reverses, and glucose enters the hepatocyte.

ECF

Insulin

[Glucose]$_{high}$

GLUT2

[Glucose]$_{low}$

ATP

Signal cascade

ADP

P

Hexokinase-mediated conversion of glucose to glucose 6-phosphate keeps intracellular [glucose] low.

Glucose transport in liver cells (*hepatocytes*) is not *directly* insulin dependent but is influenced by the presence or absence of insulin. Hepatocytes have GLUT2 transporters always present in the cell membrane. In the fasted state, when insulin levels are low, glucose moves *out* of the liver and into the blood to help maintain glucose homeostasis.

In this process (Fig. 22.17c), hepatocytes are converting glycogen stores and amino acids to glucose. Newly formed glucose moves down its concentration gradient out of the cell using GLUT2 facilitated diffusion transporters. If the GLUT transporters were pulled from the hepatocyte membrane during the fasted state, as they are in muscle and adipose tissue, glucose would have no way to leave the hepatocytes.

In the fed state (Fig. 22.17d), insulin activates *hexokinase,* the enzyme that phosphorylates glucose to glucose 6-phosphate. This phosphorylation reaction keeps free intracellular glucose concentrations low relative to plasma concentrations [Fig. 5.13, p. 142]. Now glucose diffuses into the hepatocyte on the GLUT2 transporter operating in the reverse direction.

2. **Insulin enhances cellular utilization and storage of glucose.** Insulin activates enzymes for glucose utilization (*glycolysis*), and for glycogen synthesis (*glycogenesis*). Insulin simultaneously inhibits enzymes for glycogen breakdown (*glycogenolysis*), glucose synthesis (*gluconeogenesis*), and fat breakdown (*lipolysis*) to ensure that metabolism moves in the anabolic direction. If more glucose has been ingested than is needed for energy and synthesis, the excess is made into glycogen or fatty acids.

3. **Insulin enhances utilization of amino acids.** Insulin activates enzymes for protein synthesis and inhibits enzymes that promote protein breakdown. If a meal includes protein, amino acids in the ingested food are used for protein synthesis by both the liver and muscle. Excess amino acids are converted to fatty acids.

4. **Insulin promotes fat synthesis.** Insulin inhibits β-oxidation of fatty acids and promotes conversion of excess glucose or amino acids into triglycerides (*lipogenesis*). Excess triglyceride is stored as lipid droplets in adipose tissue.

In summary, insulin is an anabolic hormone because it promotes glycogen, protein, and fat synthesis. When insulin is absent or deficient, cells go into catabolic metabolism.

CONCEPT CHECK

13. What are the primary target tissues for insulin?
14. Why are glucose metabolism and glucose transport independent of insulin in renal and intestinal epithelium and in neurons?
15. What is the advantage to the body of inhibiting insulin release during a sympathetically mediated fight-or-flight response?

Glucagon Is Dominant in the Fasted State

Glucagon, secreted by pancreatic alpha cells, is generally antagonistic to insulin in its effects on metabolism (**TBL. 22.3**). When plasma glucose concentrations decrease after a meal, insulin secretion slows, and the effects of glucagon on tissue metabolism take on greater significance (Fig. 22.14c). As noted earlier, it is the ratio of insulin to glucagon that determines the direction of metabolism rather than an absolute amount of either hormone.

The function of glucagon is to prevent hypoglycemia, so the primary stimulus for glucagon release is low plasma glucose concentration (**FIG. 22.18**). When plasma glucose falls below 100 mg/dL, glucagon secretion rises dramatically. At glucose concentrations above 100 mg/dL, when insulin is being secreted, glucagon secretion is inhibited and remains at a low but relatively

TABLE 22.3	Glucagon
Cell of origin	Alpha cells of pancreas
Chemical nature	29-amino acid peptide
Biosynthesis	Typical peptide
Transport in the circulation	Dissolved in plasma
Half-life	4–6 minutes
Factors affecting release	Enhanced secretion when plasma [glucose] <65–70 mg/dL; ↑ blood amino acids
Target cells or tissues	Liver primarily
Target receptor/ second messenger	G protein-coupled receptor linked to cAMP
Whole body or tissue action	↑ Plasma [glucose] by glycogenolysis and gluconeogenesis; ↑ lipolysis leads to ketogenesis in liver
Action at molecular level	Alters existing enzymes and stimulates synthesis of new enzymes
Feedback regulation	↑ Plasma [glucose] shuts off glucagon secretion
Other information	Member of secretin family (along with VIP, GIP, and GLP-1)

constant level. The strong relationship between insulin secretion and glucagon inhibition has led to speculation that alpha cells are regulated by some factor linked to insulin rather than by plasma glucose concentrations directly.

The liver is glucagon's primary target tissue (Fig. 22.18). Glucagon stimulates glycogenolysis and gluconeogenesis to increase glucose output. It is estimated that during an overnight fast, 75% of the glucose produced by the liver comes from glycogen stores, and the remaining 25% from gluconeogenesis.

Glucagon release is also stimulated by plasma amino acids. This pathway prevents hypoglycemia after ingestion of a pure protein meal. Let's see how hypoglycemia might occur in the absence of glucagon.

If a meal contains protein but no carbohydrate, amino acids absorbed from the food cause insulin secretion. Even though no glucose has been absorbed, insulin-stimulated glucose uptake increases, and plasma glucose concentrations fall. Unless something counteracts this process, the brain's fuel supply is threatened by hypoglycemia.

Co-secretion of glucagon in this situation prevents hypoglycemia by stimulating hepatic glucose output. As a result, although only amino acids were ingested, both glucose and amino acids are made available to peripheral tissues.

Diabetes Mellitus Is a Family of Diseases

The most common pathology of the pancreatic endocrine system is the family of metabolic disorders known as **diabetes mellitus.**

FIG. 22.18 Endocrine response to hypoglycemia

Glucagon helps maintain adequate plasma glucose levels by promoting glycogenolysis and gluconeogenesis.

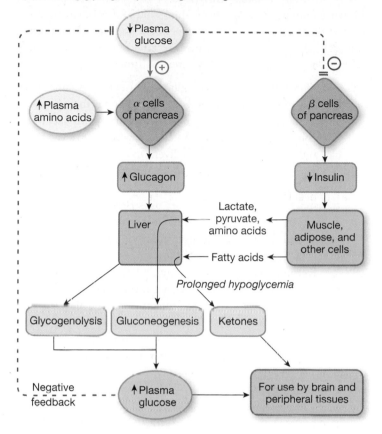

this disease that consists of the "melting down of the flesh . . . into urine," accompanied by terrible thirst that cannot be quenched. The copious production of glucose-laden urine gave the disease its name. *Diabetes* refers to the flow of fluid through a siphon, and *mellitus* comes from the word for honey. In the Middle Ages, diabetes was known as "the pissing evil."

The severe type of diabetes described by Aretaeus is **type 1 diabetes mellitus.** It is a condition of insulin deficiency resulting from beta cell destruction. Type 1 diabetes is most commonly an *autoimmune disease* in which the body fails to recognize the beta cells as "self" and destroys them with antibodies and white blood cells.

The other major variant of diabetes mellitus is **type 2 diabetes.** This type of diabetes is also known as *insulin-resistant diabetes* because in most patients, insulin levels in the blood are normal or even elevated initially. Later in the disease process, many type 2 diabetics become insulin deficient and require insulin injections. Type 2 diabetes is actually a whole family of diseases with a variety of causes.

Diabetes is diagnosed by testing blood glucose. The first indication that someone may have diabetes or prediabetes is often seen in the results of routine health checkup blood tests. These tests are done after the person has fasted for at least 8 hours. A fasting plasma glucose concentration between 100 and 125 mg/dL indicates prediabetes, and a fasting value greater than 125 mg/dL is diagnostic for diabetes.

Another test for diabetes is the 2-hour oral *glucose tolerance test* (**FIG. 22.19**). First, the person's fasting plasma glucose concentration is determined (time 0). Then the person consumes a special drink containing 75 g of glucose dissolved in water. Plasma glucose is measured every 30 minutes for 2 hours.

Normal individuals show a slight increase in plasma glucose concentration soon after drinking glucose, but the level rapidly returns to normal with insulin secretion. In diabetic patients, however, fasting plasma glucose concentrations are above normal and go even higher as glucose is absorbed into the body. In diabetes, plasma glucose remains elevated above 200 mg/dL after 2 hours. This slow response indicates that cells are not taking up and metabolizing glucose normally. Prediabetic patients show an intermediate response, with 2-hour plasma glucose concentrations in the 140–199 mg/dL range.

Elevated fasting glucose levels and abnormal oral glucose tolerance tests simply show that the body's response to an ingested glucose load is not normal. The tests cannot distinguish between problems with insulin synthesis, insulin release, or the responsiveness of target tissues to insulin.

Diabetes is characterized by abnormally elevated plasma glucose concentrations (*hyperglycemia*) resulting from inadequate insulin secretion, abnormal target cell responsiveness [p. 216], or both. Chronic hyperglycemia and its associated metabolic abnormalities cause the many complications of diabetes, including damage to blood vessels, eyes, kidneys, and the nervous system.

Diabetes is reaching epidemic proportions in the United States. In 2014, the U.S. Centers for Disease Control and Prevention estimated that over 29 million people in the United States (9.3% of the population) have diabetes and that more than a quarter of these people were unaware that they have the disease. Another 86 million, or 37% of the population, have prediabetes. Prediabetes is a condition that will likely become diabetes if those people do not alter their eating and exercise habits. Experts attribute the cause of the epidemic to our sedentary lifestyle, ample food, and overweight and obesity, which affect more than 50% of the population.

Diabetes has been known to affect humans since ancient times, and written accounts of the disorder highlight the calamitous consequences of insulin deficiency. Aretaeus the Cappadocian (81–138 C.E.) wrote of the "wonderful"* nature of

CONCEPT CHECK

16. For the oral glucose tolerance test, could you use equal amounts (in grams) of table sugar and expect the same results? Explain.

Type 1 Diabetics Are Prone to Ketoacidosis

Type 1 diabetes is a complex disorder whose onset in genetically susceptible individuals is sometimes preceded by a viral infection.

---

*In the sense of "causing wonder."

FIG. 22.19 Diagnosing diabetes

(a) Normal and Abnormal Glucose Tolerance Tests

At time 0, the subject consumes a drink containing 75 g anhydrous glucose dissolved in water.

Fasting plasma glucose concentration

Time after oral glucose (hr)

KEY

Range for diagnosis of pre-diabetes

(b) Diagnostic Criteria for Diabetes

Condition	Fasting Blood Glucose	After 2-Hour Oral Glucose Tolerance Test
Normal	<100 mg/dL (<5.6 mM)	<140 mg/dL (<7.8 mM)
Pre-diabetes	100–125 mg/dL (5.6–6.9 mM)	140–199 mg/dL (7.8–11 mM)
Diabetes	>125 mg/dL (>6.9 mM)	>199 mg/dL (>11 mM)

Many type 1 diabetics develop their disease in childhood, giving rise to the old name *juvenile-onset diabetes*. About 10% of all diabetics have type 1 diabetes.

Because individuals with type 1 diabetes are insulin deficient, the only treatment is insulin injections. Until the arrival of genetic engineering, most pharmaceutical insulin came from swine, cow, and sheep pancreases. However, once the gene for human insulin was cloned, biotechnology companies began to manufacture artificial human insulin for therapeutic use. In addition, scientists are developing techniques for implanting encapsulated beta cells in the body, in the hope that individuals with type 1 diabetes will no longer need to rely on regular insulin injections.

The events that follow ingestion of food in an insulin-deficient diabetic create a picture of what happens to metabolism in the absence of insulin (**FIG. 22.20**). They also show the integrative nature

of physiology because the problems that arise from abnormal metabolism affect nearly every organ system of the body.

Following a meal, nutrient absorption by the intestine proceeds normally because this process is insulin independent. However, nutrient uptake from the blood and cellular metabolism in many tissues are insulin dependent and therefore severely diminished in the absence of insulin. Lacking nutrients to metabolize, cells go into fasted-state metabolism:

1. *Protein metabolism.* Without glucose for energy or amino acids for protein synthesis, muscles break down their proteins to provide a substrate for ATP production. Amino acids are also converted to pyruvate and lactate, which leave the muscles and are transported to the liver.

2. *Fat metabolism.* Adipose tissue in fasted-state metabolism breaks down its fat stores. Fatty acids enter the blood for transport to the liver. The liver uses β-oxidation to break down fatty acids. However, this organ is limited in its ability to send fatty acids through the citric acid cycle, and the excess fatty acids are converted to ketones.

 Ketone bodies re-enter the circulation and can be used by other tissues (such as muscle and brain) for ATP synthesis. (The breakdown of muscle and adipose tissue in the absence of insulin leads to tissue loss and the "melting down of the flesh" described by Aretaeus.) However, ketones are also metabolic acids, creating a state of ketoacidosis (see point 7).

3. *Glucose metabolism.* In the absence of insulin, glucose remains in the blood, causing hyperglycemia. The liver, unable to metabolize this glucose, initiates fasted-state pathways of glycogenolysis and gluconeogenesis. These pathways produce *additional* glucose from glycogen, amino acids, and glycerol. When the liver dumps this glucose into the blood, hyperglycemia worsens.

 Diabetic hyperglycemia will increase the osmolarity of the blood and create a *hyperglycemic hyperosmolar state*. Plasma glucose may be as high as 600–1200 mg/dL and total osmolarity ranges from 330–380 mOsM. The high osmolarity will trigger vasopressin (ADH) secretion and thirst in an effort to conserve water and return osmolarity back to the normal range [p. 626].

4. *Brain metabolism.* Tissues that are not insulin dependent, such as most neurons in the brain, carry on metabolism as usual. However, neurons in the brain's satiety center *are* insulin sensitive. Therefore, in the absence of insulin, the satiety center is unable to take up plasma glucose. The center perceives the absence of intracellular glucose as starvation and allows the feeding center to increase food intake. The result is *polyphagia* (excessive eating), a classic symptom associated with untreated type 1 diabetes mellitus.

5. *Osmotic diuresis and polyuria.* If the hyperglycemia of diabetes causes plasma glucose concentrations to exceed the *renal threshold* for glucose, glucose reabsorption in the proximal tubule of the kidney becomes saturated [p. 605]. As a result, some filtered glucose is not reabsorbed, and it is excreted in the urine (*glucosuria*).

FIG. 22.20 Acute pathophysiology of type 1 diabetes mellitus

Untreated type 1 diabetes is marked by tissue breakdown, glucosuria, polyuria, polydipsia, polyphagia, and metabolic ketoacidosis.

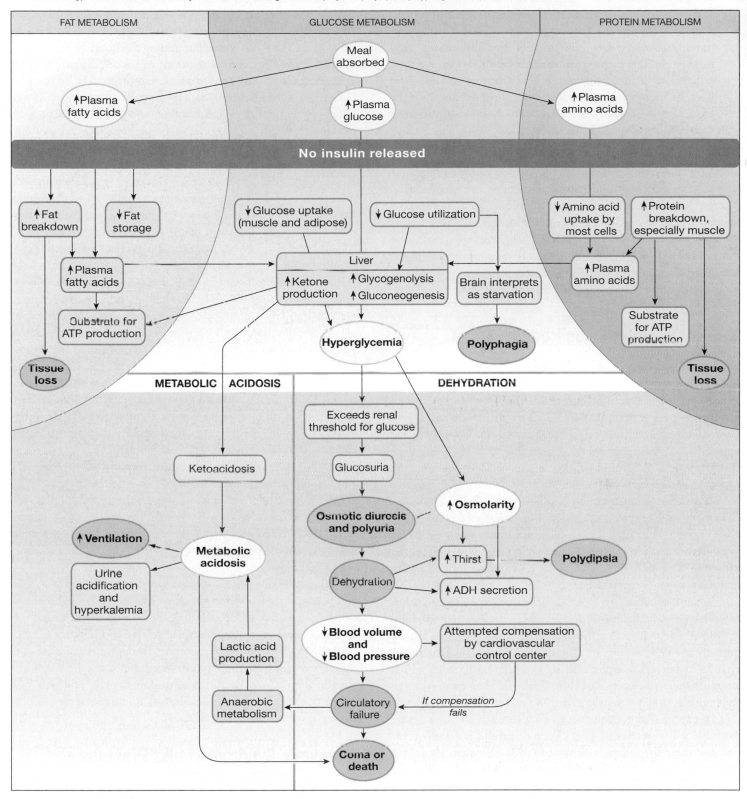

The presence of additional solute in the collecting duct lumen causes less water to be reabsorbed and more to be excreted [see Chapter 20, question 32, p. 653]. This creates large volumes of urine (*polyuria*) and, if unchecked, results in dehydration. The loss of water in the urine due to unreabsorbed solutes is known as **osmotic diuresis.**

6. *Dehydration.* Dehydration caused by osmotic diuresis leads to decreased circulating blood volume and decreased blood

pressure. Low blood pressure triggers homeostatic mechanisms for raising blood pressure, including secretion of vasopressin, thirst that causes constant drinking (*polydipsia*), and cardiovascular compensations [Fig. 20.13, p. 640].

7. *Metabolic acidosis.* Metabolic acidosis in diabetes has two potential sources: anaerobic metabolism and ketone body production. The primary cause of metabolic acidosis in type 1 diabetics is the production of acidic ketone bodies by the liver. Patients in *diabetic ketoacidosis* (DKA) exhibit the signs of metabolic acidosis: increased ventilation, acidification of the urine, and hyperkalemia [p. 648].

Tissues may also go into anaerobic glycolysis (which creates lactate) if low blood pressure decreases blood flow to the point that oxygen delivery to peripheral tissues becomes inadequate. Lactate leaves the cells and enters the blood, contributing to a state of metabolic acidosis. If untreated, the combination of ketoacidosis and hypoxia from circulatory collapse can cause coma and even death. The treatment for a patient in diabetic ketoacidosis is insulin replacement, accompanied by fluid and electrolyte therapy to replenish lost volume and ions.

Type 2 Diabetics Often Have Elevated Insulin Levels

Type 2 diabetics account for 90% of all diabetics. A significant genetic predisposition to develop the disease exists among certain ethnic groups. For example, about 25% of Hispanics over age 45 have diabetes. The disease is more common in people over the age of 40, but there is growing concern about the increased diagnosis of type 2 diabetes in children and adolescents. About 80% of type 2 diabetics are obese.

A common hallmark of type 2 diabetes is **insulin resistance,** demonstrated by the delayed response to an ingested glucose load seen in the 2-hour oral glucose tolerance test. Some type 2 diabetics have both resistance to insulin action and decreased insulin secretion. Others have normal-to-high insulin secretion but decreased target cell responsiveness.

In addition, although type 2 diabetics are hyperglycemic, they often have elevated glucagon levels as well. This seems contradictory until you realize that the pancreatic alpha cells, like muscle and adipose cells, require insulin for glucose uptake. This means that in diabetes, the alpha cells do not take up glucose, which prompts them to secrete glucagon. Glucagon then contributes to hyperglycemia by promoting glycogenolysis and gluconeogenesis.

In type 2 diabetes, the acute symptoms of the disease are not nearly as severe as in type 1 because insulin is usually present, and the cells, although resistant to insulin's action, are able to carry out a certain amount of glucose metabolism. The liver, for example, does not have to turn to ketone production. As a result, ketosis is uncommon in type 2 diabetes.

Nevertheless, overall metabolism is not normal in type 2 diabetes, and patients with this condition develop a variety of physiological problems because of abnormal glucose and fat metabolism. Complications of type 2 diabetes include

atherosclerosis, neurological changes, renal failure, and blindness from diabetic retinopathy. As many as 70% of type 2 diabetics die from cardiovascular disease.

Because many people with type 2 diabetes are asymptomatic when diagnosed, they can be very difficult to treat. People who come in for their yearly checkup feeling fine, only to be told that they have diabetes, can be reluctant to make dramatic lifestyle changes when they do not feel sick. Unfortunately, by the time diabetic symptoms appear, damage to tissues and organs is well under way. Patient compliance at that point can slow the progress of the disease but cannot reverse the pathological changes. The goal of treatment is to correct hyperglycemia to prevent the complications described earlier.

The first therapy recommended for most type 2 diabetics and prediabetics, and for those individuals at high risk of developing the disease, is to exercise and lose weight. For some patients, simply losing weight eliminates their insulin resistance. Exercise decreases hyperglycemia because exercising skeletal muscle does not require insulin for glucose uptake.

Drugs used to treat type 2 diabetes may (1) stimulate beta-cell secretion of insulin, (2) slow the digestion or absorption of carbohydrates in the intestine, (3) inhibit hepatic glucose output, (4) make target tissues more responsive to insulin, or (5) promote glucose excretion in the urine (**TBL. 22.4**). Many of the newest antidiabetic drugs mimic endogenous hormones. For example, *pramlintide* is an analog of **amylin,** a peptide hormone that is co-secreted with insulin. Amylin helps regulate glucose homeostasis following a meal by slowing digestion and absorption of carbohydrates. Amylin also decreases food intake by a central effect on appetite, and it decreases secretion of glucagon.

Other hormone-based therapies recently approved by the FDA are incretin *mimetics* (agonists). *Exendin-4* (Byetta®) is a GLP-1 mimetic derived from a compound found in the venomous saliva of gila monsters. Exendin-4 has four primary effects: It increases insulin production, decreases production of glucagon, slows digestion, and increases satiety. It has also been associated with weight loss.

In normal physiology, the combined actions of amylin, GIP, and GLP-1 create a self-regulating cycle for glucose absorption and fed-state glucose metabolism. Glucose in the intestine following a meal causes feedforward release of GIP and GLP-1 (Fig. 22.15). The two incretins travel through the circulation to the pancreas, where they initiate insulin and amylin secretion. Amylin then acts on the GI tract to slow the rate at which food enters the intestine, while insulin acts on target tissues to promote glucose uptake and utilization.

Metabolic Syndrome Links Diabetes and Cardiovascular Disease

Clinicians have known for years that people who are overweight are prone to develop type 2 diabetes, atherosclerosis, and high blood pressure. The combination of these three conditions has been formalized into a diagnosis called **metabolic syndrome,** which highlights the integrative nature of metabolic pathways.

TABLE 22.4	Drugs for Treating Diabetes	
Drug Class	**Effect**	**Mechanism of Action**
Sulfonylureas and meglitinides	Stimulate insulin secretion	Close beta cell K_{ATP} channels and depolarize the cell
α-Glucosidase inhibitors	Decrease intestinal glucose uptake	Block intestinal enzymes that digest complex carbohydrates
Sodium-glucose cotransporter 2 (SGLT2) inhibitors (e.g., dapagliflozin)	Increase glucose excretion in urine	Inhibit glucose reabsorption in the renal proximal tubule
Biguanides (e.g., metformin)	Reduce plasma glucose by decreasing hepatic gluconeogenesis	Unclear
PPAR activators ("glitazones")	Increase gene transcription for proteins that promote glucose utilization and fatty acid metabolism	Activate PPARγ, a nuclear receptor activator
Amylin analogs (pramlintide)	Reduce plasma glucose	Delay gastric emptying, suppress glucagon secretion, and promote satiety
Incretin (GLP-1) analogs (exendin-4)	Reduce plasma glucose and induce weight loss	Stimulate insulin secretion, reduce glucagon secretion, delay gastric emptying, and promote satiety
DPP4 inhibitors (sitagliptin)	Increase insulin secretion and decrease gastric emptying	Inhibit dipeptidyl peptidase-4, which breaks down GLP-1 and GIP

People with metabolic syndrome meet at least three of the following five criteria: central (visceral) obesity, blood pressure ≥130/85 mm Hg, fasting plasma glucose ≥110 mg/dL, elevated fasting plasma triglyceride levels, and low plasma HDL-C levels. *Central obesity* is defined as a waist circumference >40″ in men and 35″ in women. Women who have an apple-shaped body (widest at the waist) are more prone to developing metabolic syndrome than women who have a pear-shaped body (widest at the hips).

The association between obesity, diabetes, and cardiovascular disease illustrates the fundamental disturbances in cellular metabolism that occur with obesity. One common mechanism known to play a role in both glucose metabolism and lipid metabolism involves the family of nuclear receptors called *peroxisome proliferator-activated receptors* (PPARs, pronounced *p-pars*). Lipids and lipid-derived molecules bind to PPARs, which then turn on a variety of genes. The PPAR subtype called PPARγ (*p-par-gamma*) has been linked to adipocyte differentiation, type 2 diabetes, and foam cells, the endothelial macrophages that have ingested oxidized cholesterol. PPARα, mentioned earlier in the discussion of cholesterol metabolism, is important in hepatic cholesterol metabolism. The PPARs may be important clues to the link between obesity, type 2 diabetes, and atherosclerosis that has evaded scientists for so long.

Multiple Hormones Influence Metabolism

The long-term regulation of metabolism is much more complicated than we are able to present here, and it is still not completely understood. Many of the neuropeptides and hypothalamic neurons mentioned in the discussion of hunger and satiety also have metabolic effects. Hormones such as thyroid hormone, hormones of the cortisol pathway [p. 214], growth

CONCEPT CHECK

17. Why must insulin be administered as a shot and not as an oral pill?

18. Patients admitted to the hospital with acute diabetic ketoacidosis and dehydration are given insulin and fluids that contain K^+ and other ions. The acidosis is usually accompanied by hyperkalemia, so why is K^+ included in the rehydration fluids? (*Hint:* Dehydrated patients may have a high *concentration* of K^+, but their total body fluid volume is low.)

19. In 2006, the FDA approved sitagliptin (Januvia®), a DPP4 inhibitor. This drug blocks action of the enzyme *dipeptidyl peptidase-4*, which breaks down GLP-1 and GIP. Explain how sitagliptin is helpful in treating diabetes.

20. One of the newest drugs for treating diabetes increases urinary excretion of glucose. It does so by inhibiting the Na^+-glucose cotransporter (SGLT) that allows glucose reabsorption in the proximal tubule. What are some potential side effects of increasing urinary glucose excretion? [*Hint:* p. 653]

hormone, and epinephrine modulate metabolic pathways both directly and indirectly through their influence on insulin secretion.

For example, in times of stress, cortisol and circulating epinephrine levels increase. Sympathetic influence on the endocrine pancreas decreases insulin secretion and enhances glucagon secretion. The combined metabolic effects of insulin, cortisol, and glucagon are *synergistic*, or more than additive [Fig. 7.12, p. 215], and blood glucose concentrations rise sharply. When this happens in diabetics who are stressed, they may need to increase their medications to keep their blood glucose levels under control.

The medical staff and Nicole's family decided to talk to her about her weight and eating habits. The disorder Dr. Ayani suspects, **anorexia nervosa** (AN), can have serious physiological consequences. As a result, AN has the highest death rate of any psychiatric disorder, and the mortality rate of young women ages 15–24 with AN is 12 times greater than that of the general population. The most common causes of death are cardiac arrest, electrolyte imbalance, and suicide. AN reportedly affects as many as 3% of females in industrialized nations at some point in their lifetime. (While 90% of AN cases are female, the number of male cases is increasing.) Successful treatment of AN includes providing nutrition, psychotherapy, and family therapy. Current research is investigating the usefulness of neuropeptide Y and other brain peptides in treating anorexia.

Q9: *Why might an NPY agonist help in cases of anorexia?*

694   696   698   700   **720**   724

REGULATION OF BODY TEMPERATURE

Type 2 diabetes is an excellent example of the link between body weight and metabolism. The development of obesity may be linked to the efficiency with which the body converts food energy into cell and tissue components. According to one theory, people who are more efficient in transferring energy from food to fat are the ones who put on weight. In contrast, people who are less metabolically efficient can eat the same number of calories and not gain weight because more food energy is released as heat. Much of what we know about the regulation of energy balance comes from studies on body temperature regulation.

Body Temperature Balances Heat Production, Gain, and Loss

Temperature regulation in humans is linked to metabolic heat production (*thermogenesis*). Humans are **homeothermic** animals, which means our bodies regulate internal temperature within a relatively narrow range. Average body temperature is 37 °C (98.6 °F), with a normal range of 35.5–37.7 °C (96–99.9 °F).

These values are subject to considerable variation, both among individuals and throughout the day in a single individual. The site at which temperature is measured also makes a difference because core body temperature may be higher than temperature at the skin surface. Oral temperatures are about 0.5 °C lower than rectal temperatures.

Several factors affect body temperature in a given individual. Body temperature increases with exercise or after a meal (because of *diet-induced thermogenesis*). Temperature also cycles throughout the day: The lowest (basal) body temperature occurs in the early morning, and the highest occurs in the early evening. Women of reproductive age also exhibit a monthly temperature cycle: Basal body temperatures are about 0.5 °C higher in the second half of the menstrual cycle (after ovulation) than before ovulation.

Heat Gain and Loss Are Balanced Temperature balance in the body, like energy balance, depends on a dynamic equilibrium between input and output (**FIG. 22.21**). Heat input has two components: *internal heat production*, which includes heat from normal metabolism and heat released during muscle contraction, and *external heat input* from the environment through either *radiation* or *conduction*.

All objects with a temperature above absolute zero give off radiant energy (radiation) with infrared or visible wavelengths. This energy can be absorbed by other objects and constitutes **radiant heat gain** for those objects. You absorb radiant energy each time you sit in the sun or in front of a fire. **Conductive heat gain** is the transfer of heat between objects that are in contact with each other, such as the skin and a heating pad or the skin and hot water.

We lose heat from the body in four ways: conduction, radiation, convection, and evaporation. **Conductive heat loss** is the loss of body heat to a cooler object that is touching the body, such as an icepack or a cold stone bench. **Radiant heat loss** from the human body is estimated to account for nearly half the heat lost from a person at rest in a normal room. *Thermography* is a diagnostic imaging technique that measures radiant heat loss. Some cancerous tumors can be visually identified because they have higher metabolic activity and give off more heat than surrounding tissues.

FIG. 22.21   Heat balance in the body

EXTERNAL HEAT INPUT + INTERNAL HEAT PRODUCTION = HEAT LOSS

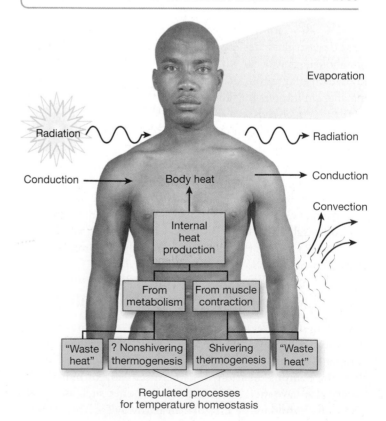

Regulated processes
for temperature homeostasis

Radiant and conductive heat loss is enhanced by **convective heat loss,** the process in which heat is carried away by warm air rising from the body's surface. Convective air currents are created wherever a temperature difference in the air exists: hot air rises and is replaced by cooler air. Convection helps move warmed air away from the skin's surface. Clothing, which traps air and prevents convective air currents, helps retain heat close to the body.

The fourth type of heat loss from the body is **evaporative heat loss,** which takes place as water evaporates at the skin's surface and in the respiratory tract. The conversion of water from the liquid state to the gaseous state requires the input of substantial amounts of heat energy. When water on the body evaporates, it removes heat from the body.

You can demonstrate the effect of evaporative cooling by wetting one arm and letting the water evaporate. As it dries, the wet arm feels much cooler than the rest of your body because heat is being drawn from the arm to vaporize the water. Similarly, the half-liter of water vapor that leaves the body through the lungs and skin each day takes with it a significant amount of body heat. Evaporative heat loss is affected by the humidity of surrounding air: less evaporation occurs at higher humidities.

Conductive, convective, and evaporative heat loss from the body is enhanced by the *bulk flow* of air across the body, such as air moved across the skin by a fan or breeze. The effect of wind on body temperature regulation in the winter is described by the *wind chill factor,* a combination of absolute environmental temperature and the effect of convective heat loss.

Body Temperature Is Homeostatically Regulated

The human body is usually warmer than its environment and therefore loses heat. However, normal metabolism generates enough heat to maintain body temperature when the environmental temperature stays between 27.8–30 °C (82–86 °F). This range is known as the **thermoneutral zone.**

In temperatures above the thermoneutral zone, the body has a net gain of heat because heat production exceeds heat loss. Below the thermoneutral zone, heat loss exceeds heat production. In both cases, the body must use homeostatic compensation to maintain a constant internal temperature.

A human without clothing can thermoregulate at environmental air temperatures between 10–55 °C (50–131°F). Because we are seldom exposed to the higher end of that temperature range, the main physiological challenge in thermal regulation is posed by cold environments.

Humans have been described by some physiologists as tropical animals because we are genetically adapted for life in warm climates. But we have retained a certain amount of genetic flexibility, and the physiological mechanisms by which we thermoregulate have some ability to adapt to changing conditions.

The autonomic control of body temperature regulation is generally considered a function of *thermoregulatory centers* in the hypothalamus. Sensory neurons known as **thermoreceptors** [p. 312] are located peripherally in the skin and centrally in the anterior hypothalamus. These sensors monitor skin temperature and core body temperature, respectively, and send that information to the thermoregulatory center. The hypothalamic "thermostat" then compares the input signals with the desired temperature set-point and coordinates an appropriate physiological response to raise or lower the core temperature (**FIG. 22.22**). Heat loss from the body is promoted by dilation of blood vessels in the skin and by sweating. Heat gain is generated by shivering and possibly by nonshivering thermogenesis.

Alterations in Cutaneous Blood Flow Conserve or Release Heat

Heat loss across the skin surface is regulated by controlling blood flow in *cutaneous* {*cutis,* skin} blood vessels (vessels near the skin's surface). These blood vessels can pick up heat from the environment by convection and transfer it to the body core, or they can lose heat to the surrounding air. Blood flow through cutaneous blood vessels varies from close to zero when heat must be conserved, to nearly one-third of cardiac output when heat must be released to the environment. Local control influences cutaneous blood flow to a certain degree, possibly through vasodilators produced by the vascular endothelium. However, neural regulation is the primary determining factor.

Most arterioles in the body are under tonic sympathetic adrenergic control [p. 490]. If core body temperature falls, the hypothalamus selectively activates sympathetic neurons innervating cutaneous arterioles. The arterioles constrict, increasing their resistance to blood flow and diverting blood to lower-resistance blood vessels in the interior of the body. This response keeps warmer core blood away from the cooler skin surface, thereby reducing heat loss.

In warm temperatures, the opposite happens: Cutaneous arterioles dilate to increase blood flow near the skin surface and enhance heat loss. Only a small fraction of vasodilation results from the withdrawal of tonic sympathetic input. Active cutaneous vasodilation is mediated through **sympathetic cholinergic neurons,** specialized neurons that co-secrete acetylcholine and other molecules. Some proposed mediators of active vasodilation include nitric oxide, substance P, histamine, and prostaglandins. It remains unclear which vasodilator substances are most important in the thermoregulatory response.

CONCEPT CHECK

21. What neurotransmitter and neurotransmitter receptor mediate cutaneous vasoconstriction?

22. What observations might have prompted researchers who discovered sympathetic neurons that secrete ACh to classify them as sympathetic rather than parasympathetic? [*Hint:* p. 259]

Sweat Contributes to Heat Loss

Surface heat loss is enhanced by the evaporation of sweat. By some estimates, the human integument has 2–3 million sweat glands. The highest concentrations are found on the forehead, scalp, axillae (armpits), palms of the hands, and soles of the feet.

FIG. 22.22 Thermoregulatory reflexes

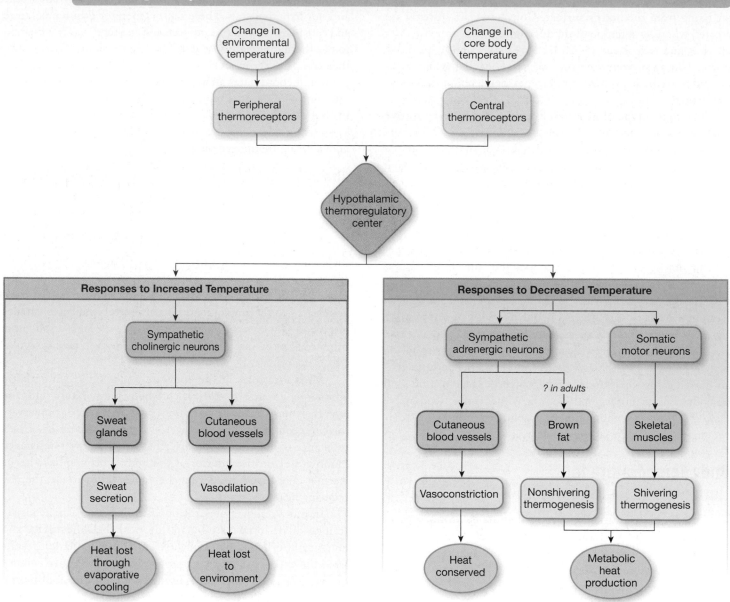

Sweat glands are made of transporting epithelium. Cells deep in the gland secrete an isotonic solution similar to interstitial fluid. As the fluid travels through the duct to the skin, NaCl is reabsorbed, resulting in hypotonic sweat. A typical value for sweat production is 1.5 L/hr. With acclimatization to hot weather, some people sweat at rates of 4–6 L/hr. However, they can maintain this high rate only for short periods unless they are drinking to replace lost fluid volume. Sweat production is regulated by cholinergic sympathetic neurons.

Cooling by evaporative heat loss depends on the evaporation of water from sweat on the skin's surface. Because water evaporates rapidly in dry environments but slowly or not at all in humid ones, the body's ability to withstand high temperatures is directly related to the relative humidity of the air. Meteorologists report the combination of heat and humidity as the *heat index*

or *humidex*. Air moving across a sweaty skin surface enhances evaporation even with high humidity, which is one reason fans are useful in hot weather.

Movement and Metabolism Produce Heat

Heat production by the body falls into two broad categories: (1) unregulated heat production from voluntary muscle contraction and normal metabolic pathways, and (2) regulated heat production for maintaining temperature homeostasis in low environmental temperatures. Regulated heat production is further divided into shivering thermogenesis and nonshivering thermogenesis (Fig. 22.21).

In **shivering thermogenesis,** the body uses shivering (rhythmic tremors caused by skeletal muscle contraction) to generate heat. Signals from the hypothalamic thermoregulatory center initiate these skeletal muscle tremors. Shivering muscle generates five to six times as much heat as resting muscle. Shivering can be partially suppressed by voluntary control.

Nonshivering thermogenesis is metabolic heat production by means other than shivering. In laboratory animals such as the rat, cold exposure significantly increases heat production in *brown adipose tissue* (BAT), also known as brown fat [p. 82]. The mechanism for brown fat heat production is *mitochondrial uncoupling*, induced by BAT-produced *uncoupling protein 1* (UCP1).

In mitochondrial uncoupling, energy flowing through the electron transport system [p. 109] is released as heat rather than being trapped in ATP. Mitochondrial uncoupling in response to cold is promoted by thyroid hormones and by increased sympathetic activity on β_3 adrenergic receptors in brown fat.

The importance of nonshivering thermogenesis in adult humans is becoming a topic of increasing interest. Humans are born with significant amounts of brown fat, found primarily in the *interscapular* area between the shoulder blades. In newborns, nonshivering thermogenesis in this brown fat contributes significantly to raising and maintaining body temperature. We used to believe that as children age, white fat replaced most brown fat. Recently, however, imaging studies used for cancer diagnosis showed that adult humans still have active brown fat. Scientists are now investigating whether increasing brown fat activity might be one way to help people burn calories as heat instead of storing them as fat.

The body's responses to high and low temperatures are summarized in **FIGURE 22.23**. In cold environments, the body tries to reduce heat loss while increasing internal heat production. In hot temperatures, the opposite is true. Notice from Figure 22.23 that voluntary behavioral responses play a significant role in temperature regulation. We reduce activity during hot weather, thereby decreasing muscle heat production. In cold weather, we put on extra clothing, tuck our hands in our armpits, or curl up in a ball to slow heat loss.

The Body's Thermostat Can Be Reset

Variations in body temperature regulation can be either physiological or pathological. Examples of physiological variation include the circadian rhythm of body temperature mentioned earlier, menstrual cycle variations, postmenopausal hot flashes, and fever. These processes share a common mechanism: resetting of the hypothalamic thermostat.

Hot flashes appear to be transient decreases in the thermostat's setpoint caused by the absence of estrogen. When the setpoint is lower, a room temperature that had previously been comfortable suddenly feels too hot. This discomfort triggers the usual thermoregulatory responses to heat, including sweating and cutaneous vasodilation, which leads to flushing of the skin.

For many years, *fever* was thought to be a pathological response to infection, but it is now considered part of the body's normal immune response. Toxins from bacteria and other pathogens trigger the release of chemicals known as **pyrogens** {*pyr*, fire} from various immunocytes. Pyrogens are fever-producing cytokines that also have many other effects.

Experimentally, some interleukins (IL 1, IL-6), some interferons, and tumor necrosis factor have all been shown to induce fever. They do so by resetting the hypothalamic thermostat to a

---

FIG. 22.23   Homeostatic responses to environmental extremes

High environmental temperature ← → **Low environmental temperature**

Maximize Heat Loss

- Vasodilation of cutaneous blood vessels
- Increased sweating
- Behavioral responses
  - Use of fans to increase convective heat loss
  - Immersion in water to increase conductive heat loss
  - Staying out of sun to prevent radiant heat gain
  - Removing clothes

Minimize Heat Production

- Diminished food intake to lessen obligatory heat production
- Behavioral responses
  - Decreased physical activity

Minimize Heat Loss

- Vasoconstriction of cutaneous blood vessels
- Lack of sweating
- Behavioral responses
  - Adding layers of protective clothing
  - Curling up to minimize exposed surface
  - Standing near heat source to enhance radiant heat gain

Maximize Heat Production

- Shivering thermogenesis
- Nonshivering thermogenesis
- Behavioral responses
  - Increased voluntary activity (stamping feet, etc.)

higher setpoint. Normal room temperature feels too cold, and the patient begins to shiver, creating additional heat. Pyrogens may also increase nonshivering thermogenesis, causing body temperature to rise.

The adaptive significance of fever is still unclear, but it seems to enhance the activity of white blood cells involved in the immune response. For this reason, some people question whether patients with a fever should be given aspirin and other fever-reducing drugs simply for the sake of comfort. High fever can be dangerous, however, as a fever of 41 °C (106 °F) for more than a brief period causes brain damage.

Pathological conditions in which body temperature strays outside the normal range include different states of *hyperthermia* and *hypothermia*. Heat exhaustion and heat stroke are the most common forms of **hyperthermia,** a condition in which body temperature rises to abnormally high values. **Heat exhaustion** is marked by severe dehydration and core body temperatures of 37.5–39 °C (99.5–102.2 °F). Patients may experience muscle cramps, nausea, and headache. They are usually pale and sweating profusely. Heat exhaustion often occurs in people who are physically active in hot, humid climates to which they are not acclimatized. It also occurs in the elderly, whose ability to thermoregulate is diminished.

Heat stroke is a more severe form of hyperthermia, with higher core body temperatures. The skin is usually flushed and dry. Immediate and rapid cooling of these patients is important, as enzymes and other proteins begin to denature at temperatures above 41 °C (106 °F). Mortality in heat stroke is nearly 50%.

Malignant hyperthermia, in which body temperature becomes abnormally elevated, is a genetically linked condition. A defective Ca^{2+} channel in skeletal muscle releases too much Ca^{2+} into the cytoplasm. As cell transporters work to move the Ca^{2+} back into mitochondria and the sarcoplasmic reticulum, the heat released from ATP hydrolysis substantially raises body temperature. Some investigators have suggested that a mild version of this process plays a role in nonshivering thermogenesis in mammals.

Hypothermia, a condition in which body temperature falls abnormally low, is also a dangerous condition. As core body temperature falls, enzymatic reactions slow, and the person loses consciousness. When metabolism slows, oxygen consumption also decreases.

Victims of drowning in cold water can sometimes be revived without brain damage if they have gone into a state of hypothermia. This observation led to the development of induced hypothermia for certain surgical procedures, such as heart surgery. The patient is cooled to 21–24 °C (70–75 °F) so that tissue oxygen demand can be met by artificial oxygenation of the blood as it passes through a bypass pump. After surgery is complete, the patient is gradually rewarmed.

CONCEPT CHECK

23. Why must a water bed be heated to allow a person to sleep on it comfortably?

24. Will a person who is exercising outside overheat faster when the air humidity is low or when it is high?

RUNNING **PROBLEM CONCLUSION** | Eating Disorders

Nicole finally agreed to undergo counseling and enter a treatment program for anorexia nervosa. She was lucky—her wrist would heal, and her medical complications could have been much worse. After seeing Nicole and discussing her anorexia, Sara realized that she also needed to see a counselor. Even though she was no longer dancing, Sara still used diet pills, diuretics, and laxatives when she became uncomfortable with her weight, and she had started bingeing and purging—eating much more than normal when she was stressed and then forcing herself to vomit

to avoid gaining any weight. These are the behavioral patterns of bulimia nervosa (BN), a condition that is as serious as AN and that affects an estimated 4% of females. Its physiological effects and treatments are similar to those for AN.

To learn more about anorexia and bulimia, and for help finding a support group, see the National Association of Anorexia Nervosa and Associated Disorders web site at *www.anad.org* or *www.nationaleatingdisorders.org*.

Question	Facts	Integration and Analysis
Q1: *If you measured Nicole's leptin level, what would you expect to find?*	Leptin is a hormone secreted by adipose tissue.	Nicole has little adipose tissue, so she would have a low leptin level.
Q2: *Would you expect Nicole to have elevated or depressed levels of neuropeptide Y?*	NPY is inhibited by leptin. NPY stimulates feeding centers.	Because her leptin level is low, you might predict that NPY would be elevated and feeding stimulated. However, the feeding center is affected by other factors besides NPY (Fig. 22.1). Brain studies of anorexic patients show high levels of CRH, which opposes NPY and depresses feeding.

RUNNING **PROBLEM CONCLUSION** | *Continued*

Question	Facts	Integration and Analysis
Q3: *What is Nicole's K⁺ disturbance called? What effect does it have on the resting membrane potential of her cells?*	Nicole's K⁺ is 2.5 mEq/L, and normal is 3.5–5 mEq/L.	Low plasma K⁺ is called hypokalemia. Hypokalemia causes the membrane potential to hyperpolarize [p. 252].
Q4: *Why does Dr. Ayani want to monitor Nicole's cardiac function?*	Cardiac muscle is an excitable tissue whose activity depends on changes in membrane potential.	Hypokalemia can alter the membrane potential of cardiac autorhythmic and contractile cells and cause a potentially fatal cardiac arrhythmia.
Q5: *Based on her clinical values, what is Nicole's acid-base status?*	Nicole's pH is 7.52, and her plasma HCO₃⁻ is elevated at 40 mEq/L.	Normal pH is 7.38–7.42, so she is in alkalosis. Her elevated HCO₃⁻ indicates a metabolic alkalosis. The cause is probably induced vomiting and loss of HCl from her stomach.
Q6: *Based on what you know about heart rate and blood pressure, speculate on why Nicole has low blood pressure with a rapid pulse.*	Her blood pressure is 80/50 (low), and her pulse is 90 (high).	Normally, increasing the heart rate would increase blood pressure. In this case, the increased pulse is a compensatory attempt to raise her low blood pressure. The low blood pressure probably results from dehydration.
Q7: *Would you expect Nicole's renin and aldosterone levels to be normal, elevated, or depressed? How might these levels relate to her K⁺ disturbance?*	All the primary stimuli for renin secretion are associated with low blood pressure. Renin begins the RAAS pathway that stimulates aldosterone secretion.	Because Nicole's blood pressure is low, you would expect elevated renin and aldosterone levels. Aldosterone promotes renal K⁺ secretion, which would lower her body load of K⁺. She probably also has low dietary K⁺ intake, which contributes to her hypokalemia.
Q8: *Give some possible reasons Nicole had been feeling weak during dance rehearsals.*	In fasted-state metabolism, the body breaks down skeletal muscle.	Loss of skeletal muscle proteins, hypokalemia, and possibly hypoglycemia could all be causes of Nicole's weakness.
Q9: *Why might an NPY agonist help in cases of anorexia?*	NPY stimulates the feeding center.	An NPY agonist might stimulate the feeding center and help Nicole want to eat.

694 696 698 700 720 **724**

VISUALIZE, PRACTICE, AND MASTER PHYSIOLOGY WITH MASTERING A&P

- Practice-on-the-go with Dynamic Study Modules
- Visualize important processes with Interactive Physiology and A&P Flix
- Check your understanding with Video Tutors and practice quizzes!

MasteringA&P®

CHAPTER SUMMARY

Energy balance in the body means that the body's energy intake equals its energy output. The same balance principle applies to metabolism and body temperature. The amount of nutrient in each of the body's nutrient pools depends on intake and output. Glucose *homeostasis* is one of the most important goals of regulated metabolism, for without adequate glucose, the brain is unable to function. Flow of material through the biochemical pathways of metabolism depends on the *molecular interactions* of substrates and enzymes.

Appetite and Satiety

1. The hypothalamus contains a tonically active **feeding center** and a **satiety center** that inhibits the feeding center. (p. 694)

2. Blood glucose concentrations (the **glucostatic theory**) and body fat content (the **lipostatic theory**) influence food intake. (p. 694)

3. Food intake is influenced by a variety of peptides, including **leptin, neuropeptide Y,** and **ghrelin.** (p. 694; Fig. 22.1)

Energy Balance

4. To maintain a constant amount of energy in the body, energy intake must equal energy output. (p. 696; Fig. 22.2)

5. The body uses energy for transport, movement, and chemical work. About half of energy used is released as heat. (p. 696)

6. **Direct calorimetry** measures the energy content of food. (p. 697)

7. The body's **oxygen consumption rate** is the most common method of estimating energy expenditure. (p. 697)

8. The **respiratory quotient (RQ)** or **respiratory exchange ratio (RER)** is the ratio of CO_2 produced to O_2 consumed. RQ varies with diet. (p. 697)

9. **Basal metabolic rate (BMR)** is an individual's lowest metabolic rate. Metabolic rate (kcal/day) = L O_2 consumed/day $\times$ kcal/L O_2. (p. 698)

10. **Diet-induced thermogenesis** is an increase in heat production after eating. (p. 698)

11. Glycogen and fat are the two primary forms of energy storage in the human body. (p. 698)

Metabolism

12. **Metabolism** is all the chemical reactions that extract, use, or store energy. (p. 699; Figs. 22.3, 22.5)

13. **Anabolic pathways** synthesize small molecules into larger ones. **Catabolic pathways** break large molecules into smaller ones. (p. 699)

14. Metabolism is divided into the **fed** (absorptive) **state** and **fasted** (postabsorptive) **state.** The fed state is anabolic; the fasted state is catabolic. (p. 699)

15. **Glycogenesis** is glycogen synthesis. (p. 699; Fig. 22.5)

16. Ingested fats enter the circulation as chylomicrons. **Lipoprotein lipase** removes triglycerides, leaving chylomicron remnants to be taken up and metabolized by the liver. (p. 703; Fig. 22.6)

17. The liver secretes lipoprotein complexes, such as LDL-C. **Apoproteins A** and **B** are the ligands for receptor-mediated endocytosis of lipoprotein complexes. (p. 704; Fig. 22.6)

18. Elevated blood LDL-C and low blood HDL-C are risk factors for coronary heart disease. Therapies for lowering cholesterol decrease cholesterol uptake or synthesis or increase cholesterol clearance. (p. 705)

19. The function of fasted-state metabolism is to maintain adequate plasma glucose concentrations because glucose is normally the only fuel that the brain can metabolize. (p. 705; Fig. 22.8)

20. **Glycogenolysis** is glycogen breakdown. **Gluconeogenesis** is glucose synthesis from noncarbohydrate precursors, especially amino acids. (pp. 699, 700; Figs. 22.9, 22.11)

21. In the fasted state, the liver produces glucose from glycogen and amino acids. **Beta oxidation** of fatty acids forms acidic **ketone bodies.** (pp. 707, 708; Fig. 22.8)

Homeostatic Control of Metabolism

22. Hour-to-hour metabolic regulation depends on the ratio of insulin to glucagon. Insulin dominates the fed state and decreases plasma glucose. Glucagon dominates the fasted state and increases plasma glucose. (p. 709; Fig. 22.14)

23. The **islets of Langerhans** secrete insulin and amylin from beta cells, glucagon from alpha cells, and somatostatin from D cells. (p. 709; Fig. 22.13)

24. Increased plasma glucose and amino acid levels stimulate insulin secretion. GI hormones and parasympathetic input amplify it. Sympathetic signals inhibit insulin secretion. (p. 710; Fig. 22.15)

25. Insulin binds to a tyrosine kinase receptor and activates multiple **insulin-receptor substrates.** (p. 712; Fig. 22.16)

26. Major insulin target tissues are the liver, adipose tissue, and skeletal muscles. Some tissues are insulin independent. (p. 712)

27. Insulin increases glucose transport into muscle and adipose tissue, as well as glucose utilization and storage of glucose and fat. (p. 712; Fig. 22.17)

28. **Glucagon** stimulates glycogenolysis and gluconeogenesis. (p. 709; Fig. 22.18)

29. **Diabetes mellitus** is a family of disorders marked by abnormal secretion or activity of insulin that causes hyperglycemia. In **type 1 diabetes,** pancreatic beta cells are destroyed by antibodies. In **type 2 diabetes,** target tissues fail to respond normally to insulin. (pp. 714, 715)

30. Type 1 diabetes is marked by catabolism of muscle and adipose tissue, glucosuria, polyuria, and metabolic ketoacidosis. Type 2 diabetics have less acute symptoms. In both types, complications include atherosclerosis, neurological changes, and problems with the eyes and kidneys. (p. 715; Fig. 22.20)

31. **Metabolic syndrome** is a condition in which people have central obesity, elevated fasting glucose levels, and elevated lipids. These people are at high risk for developing cardiovascular disease. (p. 718)

Regulation of Body Temperature

32. Body temperature homeostasis is controlled by the hypothalamus. (p. 721)

33. Heat loss from the body takes place by radiation, conduction, convection, and evaporation. Heat loss is promoted by cutaneous vasodilation and sweating. (p. 720; Figs. 22.21, 22.22)

34. Heat is generated by **shivering thermogenesis** and by **nonshivering thermogenesis.** (pp. 722, 723; Fig. 22.21)

REVIEW QUESTIONS

In addition to working through these questions and checking your answers on p. A-29, review the Learning Outcomes at the beginning of this chapter.

Level One Reviewing Facts and Terms

1. Define metabolic, anabolic, and catabolic pathways.

2. List and briefly explain the three forms of biological work.

3. Define a kilocalorie. What is direct calorimetry?

4. What is the respiratory quotient (RQ)? What is a typical RQ value for an American diet?

5. Define basal metabolic rate (BMR). Under what conditions is it measured? Why does the average BMR differ in adult males and females? List at least four factors other than sex that may affect BMR in humans.

6. What are the three general fates of biomolecules in the body?

7. What are the main differences between metabolism in the absorptive and postabsorptive states?

8. What is a nutrient pool? What are the three primary nutrient pools of the body?

9. What is the primary goal of fasted-state metabolism?

10. In what forms is excess energy stored in the body?

11. What are the three possible fates for ingested proteins? For ingested fats?

12. Name the two hormones that regulate glucose metabolism, and explain what effect each hormone has on blood glucose concentrations.

13. Which noncarbohydrate molecules can be made into glucose? What are the pathways called through which these molecules are converted to glucose?

14. Under what circumstances are ketone bodies formed? From what biomolecule are ketone bodies formed? How are they used by the body, and why is their formation potentially dangerous?

15. Name two stimuli that increase insulin secretion, and one stimulus that inhibits insulin secretion.

16. What are the two types of diabetes mellitus? How do their causes and basic symptoms differ?

17. What factors release glucagon? What organ is the primary target of glucagon? What effect(s) do(es) glucagon produce?

18. Define the following terms and explain their physiological significance:

 (a) lipoprotein lipase
 (b) amylin
 (c) ghrelin
 (d) neuropeptide Y
 (e) apoprotein
 (f) leptin
 (g) osmotic diuresis
 (h) insulin resistance

19. What effect does insulin have on:

 (a) glycolysis
 (b) gluconeogenesis
 (c) glycogenesis
 (d) lipogenesis
 (e) protein synthesis

Level Two Reviewing Concepts

20. **Map:** Draw a map that compares the fed state and the fasted state. For each state, compare metabolism in skeletal muscles, the brain, adipose tissue, and the liver. Indicate which hormones are active in each stage and at what points they exert their influence.

21. Examine the graphs of insulin and glucagon secretion in Figure 22.14c. Why have some researchers concluded that the ratio of these two hormones determines whether glucose is stored or removed from storage?

22. Define, compare, and contrast or relate the terms in each of the following sets:

 (a) glucose, glycogenolysis, glycogenesis, gluconeogenesis, glucagon, glycolysis
 (b) shivering thermogenesis, nonshivering thermogenesis, diet-induced thermogenesis
 (c) lipoproteins, chylomicrons, cholesterol, HDL-C, LDL-C, apoproteins
 (d) direct and indirect calorimetry
 (e) conductive heat loss, radiant heat loss, convective heat loss, evaporative heat loss
 (f) absorptive and postabsorptive states

23. Describe (or map) the physiological events that lead to the following signs or symptoms in a type 1 diabetic:

 (a) hyperglycemia
 (b) glucosuria
 (c) polyuria
 (d) ketosis
 (e) dehydration
 (f) severe thirst

24. Both insulin and glucagon are released following ingestion of a protein meal that raises plasma amino acid levels. Why is the secretion of both hormones necessary?

25. Explain the current theory of the control of food intake. Use the following terms in your explanation: hypothalamus, feeding center, satiety center, appetite, leptin, NPY, neuropeptides.

26. Compare human thermoregulation in hot environments and cold environments.

Level Three Problem Solving

27. Scott is a bodybuilder who consumes large amounts of amino acid supplements in the belief that they will increase his muscle mass. He believes that the amino acids he consumes are stored in his body until he needs them. Is Scott correct? Explain.

28. Draw and label a graph showing the effect of insulin secretion on plasma glucose concentration.

29. The signal molecules involved in active cutaneous vasodilation are unclear but it is known that sympathetic cholinergic neurons are involved. In one experiment* scientists used botulinum toxin [p. 402] to block release of chemicals from the sympathetic axon terminal. When they

---

*D. L. Kellogg Jr. *et al.* Cutaneous active vasodilation in humans is mediated by cholinergic nerve cotransmission. *Circ Res* 77: 1222–1228, 1995.

I n 1998, Mark McGwire made news when he hit 70 home runs, surpassing the single-season home run record Roger Maris established in 1961. McGwire also created a firestorm of controversy when he admitted to taking *androstenedione,* a performance-enhancing steroid prohormone banned by the International Olympic Committee and other groups but not by professional baseball. As a result of the controversy, Congress passed the Anabolic Steroids Act of 2004, which made androstenedione and some other steroid prohormones controlled substances available only by prescription.

What is this prohormone, and why is it so controversial? You will learn more about androstenedione in this chapter as we discuss the hormones that play a role in long-term regulation of metabolism and growth. In individuals with normal metabolism, these hormones can be difficult to study because their effects are subtle and their interactions with one another complex. As a result, much of what we know about endocrinology comes from studying pathological conditions in which a hormone is either oversecreted or undersecreted. In recent years, however, advances in molecular biology and the use of transgenic animal models have enabled scientists to learn more about hormone action at the cellular level.

REVIEW OF ENDOCRINE PRINCIPLES

Before we delve into the different hormones, let's do a quick review of some basic principles and patterns of endocrinology.

1. **The hypothalamic-pituitary control system** [p. 210]. Several of the hormones described in this chapter are controlled by hypothalamic and anterior pituitary (*adenohypophyseal*) trophic hormones.
2. **Feedback patterns** [p. 15]. The negative feedback signal for simple endocrine pathways is the systemic response to the hormone. For example, insulin secretion shuts off when blood glucose concentrations decrease. In complex pathways using

the hypothalamic-pituitary control system, the feedback signal may be the hormone itself. In pathological states, endocrine cells may not respond appropriately to feedback signals.

3. **Hormone receptors** [p. 169]. Hormone receptors may be on the cell surface or inside the cell.
4. **Cellular responses** [p. 169]. In general, hormone target cells respond by altering existing proteins or by making new proteins. The historical distinctions between the actions of peptide and steroid hormones no longer apply. Some steroid hormones exert rapid, nongenomic effects, and some peptide hormones alter transcription and translation.
5. **Modulation of target cell response** [p. 217]. The amount of active hormone available to the cell and the number and activity of target cell receptors determine the magnitude of target cell response. Cells may up-regulate or down-regulate their receptors to alter their response. Cells that do not have hormone receptors are nonresponsive.
6. **Endocrine pathologies** [p. 216]. Endocrine pathologies result from (a) excess hormone secretion, (b) inadequate hormone secretion, and (c) abnormal target cell response to the hormone. It now appears that failure of the target cell to respond appropriately to its hormone is a major cause of endocrine disorders.

In the following sections, we first examine adrenal corticosteroids and thyroid hormones, two groups of hormones that

RUNNING **PROBLEM** | Hyperparathyroidism

"Broken bones, kidney stones, abdominal groans, and psychic moans." Medical students memorize this saying when they learn about *hyperparathyroidism,* a disease in which parathyroid glands (see Fig. 23.12, p. 747) work overtime and produce excess parathyroid hormone (PTH). Dr. Spinks suddenly recalls the saying as she examines Prof. Magruder, who has arrived at her office in pain from a kidney stone lodged in his ureter. When questioned about his symptoms, Prof. Magruder also mentions pain in his shin bones, muscle weakness, stomach upset, and a vague feeling of depression. "I thought it was all just the stress of getting my book published," he says. To Dr. Spinks, however, Prof. Magruder's combination of symptoms sounds suspiciously like he might be suffering from hyperparathyroidism.

730 (732) (736) (746) (749) (750)

BIOTECHNOLOGY

Mutant Mouse Models

The use of animal models to study human diseases has become a valuable part of biomedical research. The mouse genome sequencing effort tells us that 99% of the approximately 30,000 mouse genes have direct human *homologs* (equivalents). This means that in many cases, we can use the mouse to understand what genes, and the proteins they encode, do in healthy and sick humans. Sometimes, natural mutations produce animal diseases that resemble human diseases. One example of such mutants is the twitcher mouse, in which normal myelin degenerates owing to an inherited metabolic problem. In other cases, scientists have used biotechnology techniques to create **knockout mice,** which lack specific genes, or to breed **transgenic mice,** which contain extra genes that were inserted artificially. The mouse is the ideal organism for conducting these experiments. It is small, relatively inexpensive, has a short lifespan, and its genome can be easily manipulated. Additionally, its biological processes are similar to those of humans. To learn more about mice in research, read the special edition of the journal *Nature* (*www.nature .com/nature/mousegenome*) or look for information about knockout and transgenic mice in the Mouse Genome Informatics databases from The Jackson Laboratory (*www.informatics .jax.org/mgihome/homepages/humanDisease.shtml*).

influence long-term metabolism. We then consider the endocrine control of growth.

ADRENAL GLUCOCORTICOIDS

The paired adrenal glands sit on top of the kidneys like little caps (**FIG. 23.1**). Each adrenal gland, like the pituitary gland, is two embryologically distinct tissues that merged during development. This complex organ secretes multiple hormones, both neurohormones and classic hormones. The *adrenal medulla* occupies a little over a quarter of the inner mass and is composed of modified sympathetic ganglia that secrete catecholamines (mostly epinephrine) to mediate rapid responses in fight-or-flight situations [p. 359]. The *adrenal cortex* forms the outer three-quarters of the gland and secretes a variety of steroid hormones.

FIG. 23.1   The adrenal gland

(a) The paired adrenal glands sit on top of the kidneys. Each region secretes different hormones.

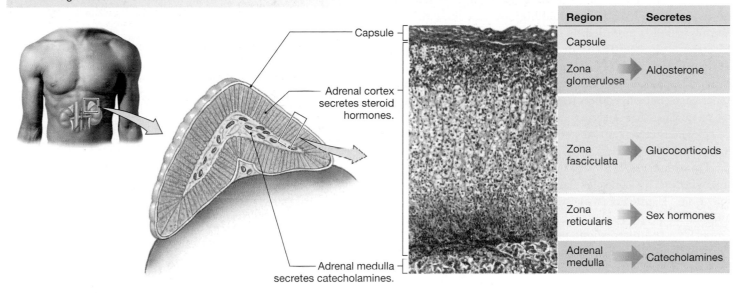

Region	Secretes
Capsule	
Zona glomerulosa	Aldosterone
Zona fasciculata	Glucocorticoids
Zona reticularis	Sex hormones
Adrenal medulla	Catecholamines

(b) Synthesis pathways for steroid hormones

All steroid hormones are synthesized from cholesterol. The blank boxes represent intermediate compounds whose names have been omitted for simplicity. Each step is catalyzed by an enzyme, but only two enzymes are shown in the figure.

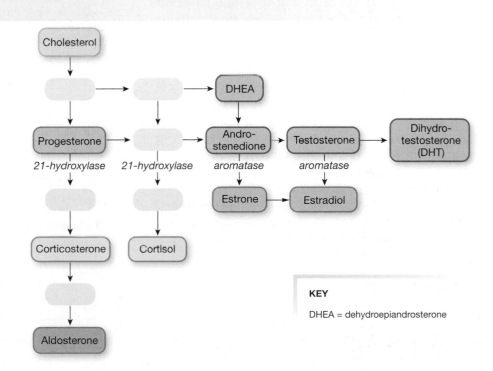

FIGURE QUESTIONS

1. A baby is born with a genetic mutation that results in a deficiency of the enzyme 21-hydroxylase. Based on the role of this enzyme in the pathway illustrated, what symptoms might you predict in the baby?

2. Would men or women have more aromatase activity?

KEY

DHEA = dehydroepiandrosterone

The Adrenal Cortex Secretes Steroid Hormones

The adrenal cortex secretes three major types of steroid hormones: aldosterone (sometimes called a *mineralocorticoid* because of its effect on the minerals sodium and potassium) [p. 631], glucocorticoids, and sex hormones. Histologically, the adrenal cortex is divided into three layers, or zones (Fig. 23.1a). The outer *zona glomerulosa* secretes only aldosterone. The inner *zona reticularis* secretes mostly *androgens*, the sex hormones dominant in men. The middle *zona fasciculata* secretes mostly **glucocorticoids,** named for their ability to increase plasma glucose concentrations. **Cortisol** is the main glucocorticoid secreted by the adrenal cortex.

The generalized synthesis pathway for steroid hormones is shown in Figure 23.1b. All steroid hormones begin with cholesterol, which is modified by multiple enzymes to end up as aldosterone, glucocorticoids, or sex steroids (androgens as well as *estrogens* and *progesterone,* the dominant sex hormones in females). The pathways are the same in the adrenal cortex, gonads, and placenta, but what differs from tissue to tissue is the distribution of enzymes that catalyze the different reactions. For example, the enzyme that makes aldosterone is found in only one of the three adrenal cortex zones.

This chapter opened with the story of baseball player Mark McGwire and his controversial use of the supplement androstenedione. Figure 23.1b shows that this prohormone is one intermediate in the synthesis of testosterone and dihydrotestosterone. One androstenedione precursor, *dehydroepiandrosterone* (DHEA), is used as a dietary supplement. In the United States, purchase of DHEA is not regulated, despite the fact that this substance is metabolically converted to androstenedione and testosterone, both controlled substances whose use is widely banned by sports associations.

The close structural similarity among steroid hormones means that the binding sites on their receptors are also similar, leading to *crossover effects* when one steroid binds to the receptor for a related molecule. For example, *mineralocorticoid receptors* (MRs) for aldosterone are found in the distal nephron. MRs also bind and respond to cortisol, which may be 100 times more concentrated in the blood than aldosterone. What is to keep cortisol from binding to an MR and influencing Na^+ and K^+ excretion? It turns out that renal tubule cells with MRs have an enzyme (11β-*hydroxysteroid dehydrogenase*) that converts cortisol to a less active form with low specificity for the MR. By inactivating cortisol, renal cells normally prevent crossover effects from cortisol. However, crossover activity and the structural similarities of steroid hormones mean that in many endocrine disorders, patients may experience symptoms related to more than one hormone.

CONCEPT CHECK

1. Name the two parts of the adrenal gland and the major hormones secreted by each part.
2. For what hormones is androstenedione a prohormone? (See Fig. 23.1b.) Why might this prohormone give an athlete an advantage?

Cortisol Secretion Is Controlled by ACTH

The control pathway for cortisol secretion is known as the *hypothalamic-pituitary-adrenal (HPA) pathway* (**FIG. 23.2a**). The HPA pathway begins with hypothalamic **corticotropin-releasing hormone (CRH),** which is secreted into the hypothalamic-hypophyseal portal system and transported to the anterior pituitary. CRH stimulates release of **adrenocorticotropic hormone (ACTH)** (or *corticotropin*) from the anterior pituitary. ACTH in turn acts on the adrenal cortex to promote synthesis and release of cortisol. Cortisol then acts as a negative feedback signal, inhibiting ACTH and CRH secretion.

Cortisol secretion is continuous and has a strong diurnal rhythm (Fig. 23.2c). Secretion normally peaks in the morning and diminishes during the night. Cortisol secretion also increases with stress.

Cortisol is a typical steroid hormone and is synthesized on demand [Fig. 7.5, p. 205]. Once synthesized, it diffuses out of adrenal cells into the plasma, where most of it is transported by a carrier protein, *corticosteroid-binding globulin* (CBG or *transcortin*). Unbound hormone is free to diffuse into target cells.

All nucleated cells of the body have cytoplasmic glucocorticoid receptors. The hormone-receptor complex enters the nucleus, binds to DNA, and alters gene expression, transcription, and translation. In general, a tissue's response to glucocorticoid hormones is not evident for 60–90 minutes. However, cortisol's negative feedback effect on ACTH secretion occurs within minutes.

Cortisol Is Essential for Life

Adrenal glucocorticoids are sometimes called the body's stress hormones because of their role in the mediation of long-term stress. Adrenal catecholamines, particularly epinephrine, are responsible for rapid metabolic responses needed in fight-or-flight situations.

RUNNING **PROBLEM**

Hyperparathyroidism causes breakdown of bone and the release of calcium phosphate into the blood. Elevated plasma Ca^{2+} can affect the function of excitable tissues, such as muscles and neurons. Surprisingly, however, most people with hyperparathyroidism have no symptoms. The condition is usually discovered during blood work performed for a routine health evaluation.

Q1: *What role does Ca^{2+} play in the normal functioning of muscles and neurons?*

Q2: *What is the technical term for "elevated levels of calcium in the blood"? (Use your knowledge of word roots to construct this term.)*

730   732   736   746   749   750

FIG. 23.2 The hypothalamic-pituitary-adrenal (HPA) pathway

(a) The control of cortisol secretion

(b) Cortisol

Origin	Adrenal cortex
Chemical Nature	Steroid
Biosynthesis	From cholesterol; made on demand; not stored
Transport in the Circulation	On corticosteroid-binding globulin (made in liver)
Half-Life	60–90 min
Factors Affecting Release	Circadian rhythm of tonic secretion; stress enhances release
Control Pathway	CRH (hypothalamus) ⟶ ACTH (anterior pituitary) ⟶ cortisol (adrenal cortex)
Target Cells or Tissues	Most tissues
Target Receptor	Intracellular
Whole Body or Tissue Reaction	↑ Plasma [glucose]; ↓ immune activity; permissive for glucagon and catecholamines
Action at Cellular Level	↑ Gluconeogenesis and glycogenolysis; ↑ protein catabolism. Blocks cytokine production by immune cells
Action at Molecular Level	Initiates transcription, translation, and new protein synthesis
Feedback Regulation	Negative feedback to anterior pituitary and hypothalamus

(c) The circadian rhythm of cortisol secretion

[Graph: Plasma cortisol concentration (y-axis) vs. time (x-axis: Noon, 6 PM, Midnight, 6 AM, Noon)]

Ⓠ **FIGURE QUESTION**
What do the following abbreviations stand for? ACTH, CRH, MSH

(d) Post-translational processing of POMC creates a variety of active peptides.

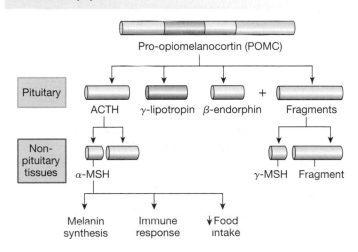

Cortisol is essential for life. Animals whose adrenal glands have been removed die if exposed to any significant environmental stress. The most important metabolic effect of cortisol is its protective effect against *hypoglycemia*. When blood glucose decreases, the normal response is secretion of pancreatic glucagon, which promotes gluconeogenesis and glycogen breakdown [p. 714]. In the absence of cortisol, however, glucagon is unable to respond adequately to a hypoglycemic challenge. Because cortisol is required for full glucagon and catecholamine activity, it is said to have a *permissive effect* on those hormones [p. 216].

Cortisol receptors are found in every tissue of the body, but for many targets we do not fully understand the physiological actions of cortisol. However, we can speculate on these actions based on tissue responses to high levels (*pharmacological doses*) of cortisol administered for therapeutic reasons or associated with hypersecretion.

All the metabolic effects of cortisol are directed at preventing hypoglycemia. Overall, cortisol is catabolic (Fig. 23.2a, b).

1. **Cortisol promotes gluconeogenesis** in the liver. Some glucose produced in the liver is released into the blood, and the rest is stored as glycogen. As a result, cortisol increases blood glucose concentrations.
2. **Cortisol causes the breakdown of skeletal muscle proteins** to provide a substrate for gluconeogenesis.
3. **Cortisol enhances lipolysis** so that fatty acids are available to peripheral tissues for energy use. The glycerol from fatty acids can be used for gluconeogenesis.
4. **Cortisol suppresses the immune system** through multiple pathways. This effect is discussed in more detail next.
5. **Cortisol causes negative calcium balance.** Cortisol decreases intestinal Ca^{2+} absorption and increases renal Ca^{2+} excretion, resulting in net Ca^{2+} loss from the body. In addition, cortisol is catabolic in bone tissue, causing net breakdown of calcified bone matrix. As a result, people who take therapeutic cortisol for extended periods have a higher-than-normal incidence of broken bones.
6. **Cortisol influences brain function.** States of cortisol excess or deficiency cause mood changes as well as memory and learning alterations. Some of these effects may be mediated by hormones in the cortisol release pathway, such as CRH. We discuss this effect of cortisol in more detail next.

CONCEPT CHECK

3. What do the abbreviations HPA and CBG stand for? If there is an alternate name for each term, what is it?
4. You are mountain-biking in Canada and encounter a bear, which chases you up a tree. Is your stress response mediated by cortisol? Explain.
5. The illegal use of anabolic steroids by bodybuilders and athletes periodically receives much attention. Do these illegal steroids include cortisol? Explain.

Cortisol Is a Useful Therapeutic Drug

Cortisol suppresses the immune system by preventing cytokine release and antibody production by white blood cells. It also inhibits the inflammatory response by decreasing leukocyte mobility and migration. These *immunosuppressant effects* of cortisol make it a useful drug for treating a variety of conditions, including bee stings, poison ivy, and pollen allergies. Cortisol also helps prevent rejection of transplanted organs. However, glucocorticoids also have potentially serious side effects because of their metabolic actions. Once *nonsteroidal anti-inflammatory drugs* (NSAIDs), such as ibuprofen, were developed, the use of glucocorticoids for treating minor inflammatory problems was discontinued.

Exogenous administration of glucocorticoids has a negative feedback effect on the anterior pituitary and may shut down ACTH production [Fig. 7.13, p. 217]. Without ACTH stimulation, the adrenal cells that produce cortisol atrophy. For this reason, it is essential that patients taking steroids taper their dose gradually, giving the pituitary and adrenal glands a chance to recover, rather than stopping the drug abruptly.

Cortisol Pathologies Result from Too Much or Too Little Hormone

The most common HPA pathologies result from hormone deficiency and hormone excess. Abnormal tissue responsiveness is an uncommon cause of adrenal steroid disorders.

Hypercortisolism Excess cortisol in the body is called **hypercortisolism.** It can arise from hormone-secreting tumors or from exogenous administration of the hormone. Cortisol therapy with high doses for more than a week has the potential to cause hypercortisolism—also known as *Cushing's syndrome,* after Dr. Harvey Cushing, who first described the condition in 1932.

Most signs of hypercortisolism can be predicted from the normal actions of the hormone. Excess gluconeogenesis causes hyperglycemia, which mimics diabetes. Muscle protein breakdown and lipolysis cause tissue wasting. Paradoxically, excess cortisol deposits extra fat in the trunk and face, perhaps in part because of increased appetite and food intake. The classic appearance of patients with hypercortisolism is thin arms and legs, obesity in the trunk, and a "moon face" with plump cheeks (**FIG. 23.3**). CNS effects of too much cortisol include initial mood elevation followed by depression, as well as difficulty with learning and memory.

Hypercortisolism has three common causes:

1. **An adrenal tumor that autonomously secretes cortisol.** These tumors are not under the control of pituitary ACTH. This condition is an instance of *primary hypercortisolism* [p. 218].
2. **A pituitary tumor that autonomously secretes ACTH.** Excess ACTH prompts the adrenal gland to oversecrete cortisol (*secondary hypercortisolism*). The tumor does not respond to negative feedback. This condition is also called *Cushing's disease* because it was the actual disease described by Dr. Cushing. (Hypercortisolism from any cause is called Cushing's *syndrome.*)
3. **Iatrogenic (physician-caused) hypercortisolism** occurs secondary to cortisol therapy for some other condition.

Hypocortisolism Hyposecretion pathologies are far less common than Cushing's syndrome. *Adrenal insufficiency,* commonly known as **Addison's disease,** is hyposecretion of all adrenal steroid hormones, usually following autoimmune destruction of the adrenal cortex. Hereditary defects in the enzymes needed for

FIG. 23.3 Hypercortisolism (Cushing's syndrome)

(a) Moon face.
A "moon face" with red cheeks is typical in this condition.

(b) Abdominal fat with striations.
Fat also deposits in the trunk. The dark striations result from protein breakdown in the skin.

adrenal steroid production cause several related syndromes collectively known as *congenital adrenal hyperplasia* (see the question in Fig. 23.1). In some of these inherited disorders, excess androgens are secreted because substrate that cannot be made into cortisol or aldosterone is converted to androgens. In newborn girls, excess androgens cause masculinization of the external genitalia, a condition called *adrenogenital syndrome*.

CONCEPT
CHECK

6. For primary, secondary, and iatrogenic hypercortisolism, indicate whether the ACTH level is normal, higher than normal, or lower than normal.

7. Would someone with Addison's disease have normal, low, or high levels of ACTH in the blood?

CRH and ACTH Have Additional Physiological Functions

In recent years, research interest has shifted away from glucocorticoids to CRH and ACTH, the trophic hormones of the HPA pathway. Both peptides are now known to belong to larger families of related molecules, with multiple receptor types found in numerous tissues. Experiments with knockout mice lacking a particular receptor have revealed some of the physiological functions of peptides related to CRH and ACTH.

Two interesting findings of this research are that cytokines secreted by the immune system can stimulate the HPA pathway and that immune cells have receptors for ACTH and CRH. The association between stress and immune function appears to be mediated through CRH and ACTH, and this association provides one explanation for mind-body interactions in which mental state influences physiological function.

CRH Family The CRH family includes CRH and a related brain neuropeptide called *urocortin*. In addition to its involvement in inflammation and the immune response, CRH is known to decrease food intake [Fig. 22.1, p. 695] and has been associated with signals that mark the onset of labor in pregnant women. Additional evidence links CRH to anxiety, depression, and other mood disorders.

POMC and Melanocortins CRH acting on the anterior pituitary stimulates secretion of ACTH. ACTH is synthesized from a large glycoprotein called **pro-opiomelanocortin** (POMC, pronounced *pom-see*). POMC undergoes posttranslational processing to produce a variety of biologically active peptides in addition to ACTH (Fig. 23.2d). In the pituitary, POMC products include **β-endorphin,** an endogenous opioid that binds to receptors that block pain perception [p. 323].

Processing of POMC in nonpituitary tissues creates additional peptides, such as *melanocyte-stimulating hormone* (MSH). **α-MSH** is produced in the brain, where it inhibits food intake, and in the skin, where it acts on **melanocytes.** Melanocytes contain pigments called **melanins** that influence skin color in humans and coat color in rodents (see the Emerging Concepts box).

The MSH hormones plus ACTH have been given the family name **melanocortins.** Five *melanocortin receptors* (MCRs) have been identified. MC2R responds only to ACTH and is the adrenal cortex receptor. MC1R is found in skin melanocytes and responds equally to α-MSH and ACTH. When ACTH is elevated in Addison's disease, the action of ACTH on MC1R leads to increased melanin production and the apparent "tan," or skin darkening, characteristic of this disorder.

EMERGING CONCEPTS

Melanocortins and the *Agouti* Mouse

Much of what we are learning about the functions of the hypothalamic-pituitary-adrenal (HPA) pathway comes from research on mutant mice. One such animal is the *agouti* mouse, a strain that resulted from a spontaneous mutation first described in 1905. *Agouti* mice with one mutated gene overproduce a protein (called *agouti protein*) that gives them a characteristic yellow coat. Agouti protein is an antagonist to the melanocortin receptor MC1R, which controls melanin synthesis in hair. Of more interest to physiologists, however, is the fact that *agouti* mice overeat and develop adult-onset obesity, hyperglycemia, and insulin resistance—in other words, these mice are a model for obesity-related type 2 diabetes. Their agouti protein blocks the MC4R receptor in the brain, a receptor that normally depresses feeding behavior. If the receptor is not functional, the animal overeats and becomes obese. This link between melanocortin receptors, eating behavior, and diabetes has opened up a new area of research on treatments to prevent type 2 diabetes.

Some POMC is made in tissues outside the anterior pituitary. POMC-producing neurons in the hypothalamus are being studied for their role in food intake, energy balance, and thermoregulation. Hypothalamic neurons co-secrete POMC with other peptides known to decrease food intake. Recent investigations suggest that the action of nicotine on these POMC neurons explains why smoking decreases food intake. Other research suggests that these POMC neurons may respond to changes in blood glucose and possibly participate in the glucostatic control mechanism influencing food intake [p. 694].

CONCEPT CHECK
8. What might be the advantage for the body to co-secrete ACTH and β-endorphin?

THYROID HORMONES

The thyroid gland is a butterfly-shaped gland that lies across the trachea at the base of the throat, just below the larynx (**FIG. 23.4a**). It is one of the larger endocrine glands, weighing 15–20 g. The thyroid gland has two distinct endocrine cell types: *C* ("clear") *cells*, which secrete a calcium-regulating hormone called *calcitonin*, and *follicular cells*, which secrete thyroid hormone. We discuss calcitonin later, with calcium homeostasis.

Thyroid Hormones Contain Iodine

Thyroid hormones, like glucocorticoids, have long-term effects on metabolism. Unlike glucocorticoids, however, thyroid hormones are not essential for life: adults can live, although not comfortably, without thyroid hormone or a thyroid gland. Thyroid hormones are essential for normal growth and development in children, and infants born with thyroid deficiency will be developmentally delayed unless treated promptly. Because of the importance of thyroid hormones in children, the United States and Canada test all newborns for thyroid deficiency.

Thyroid hormones are amines derived from the amino acid tyrosine, and they are unusual because they contain the element iodine (Fig. 23.4c). Currently, thyroid hormones are the only known use for iodine in the body, although a few other tissues also concentrate this mineral.

Synthesis of thyroid hormones takes place in the thyroid **follicles** (also called *acini*), spherical structures whose walls are a single layer of epithelial cells (Fig. 23.4b). The hollow center of each follicle is filled with a sticky glycoprotein mixture called **colloid.** The colloid holds a 2–3 month supply of thyroid hormones at any one time.

The follicular cells surrounding the colloid manufacture a glycoprotein called **thyroglobulin** and enzymes for thyroid hormone synthesis (Fig. 23.4c **1**). These proteins are packaged into vesicles, then secreted into the center of the follicle. Follicular cells also actively concentrate dietary iodide, I^-, using the *sodium-iodide symporter* (NIS) **2**. I^- transport into the colloid is mediated by an anion transporter known as *pendrin* (SLC26A4).

As I^- enters the colloid, the enzyme *thyroid peroxidase* removes an electron from the iodide ion and adds iodine to tyrosine

on the thyroglobulin molecule **3**. The addition of one iodine to tyrosine creates **monoiodotyrosine** (MIT). The addition of a second iodine creates **diiodotyrosine** (DIT). MIT and DIT then undergo *coupling reactions*. One MIT and one DIT combine to create the thyroid hormone **triiodothyronine,** or T_3 (Note the change from *tyrosine* to *thyronine* in the name.) Two DIT couple to form **tetraiodothyronine** (T_4, also known as **thyroxine**). At this point, the hormones are still attached to thyroglobulin.

When hormone synthesis is complete, the thyroglobulin–T_3/T_4 complex is taken back into the follicular cells in vesicles **4**. There intracellular enzymes free the hormones T_3 and T_4 from the thyroglobulin protein **5**. For many years, scientists believed that the lipophilic nature of T_3 and T_4 allowed the hormones to diffuse out of the follicular cells and into the plasma, but current evidence indicates that the thyroid hormones move across cell membranes by protein carriers **6**. The transporter for thyroid gland export of T_3 and T_4 has not been fully identified but appears to be one isoform of the *monocarboxylate transporter* (MCT8).

T_3 and T_4 have limited solubility in plasma because they are lipophilic molecules. As a result, thyroid hormones bind to plasma proteins, such as **thyroid-binding globulin** (TBG). Most thyroid hormone in the plasma is in the form of T_4.

Target tissue uptake transporters for thyroid hormones vary from tissue to tissue. They include the monocarboxylate transporters MCT8 and MCT10 as well as one member of the organic anion transporter (OAT) family.

For years it was thought that T_4 was the active hormone but we now know that T_3 is three to five times more active biologically, and that T_3 is the active hormone in target cells. Target cells make about 85% of active T_3 by using enzymes called **deiodinases** to remove an iodine from T_4. Target tissue activation of the hormone adds another layer of control because individual target tissues can alter their exposure to active thyroid hormone by regulating their tissue deiodinase synthesis.

Thyroid receptors, with multiple isoforms, are in the nucleus of target cells. Hormone binding initiates transcription, translation, and synthesis of new proteins.

RUNNING **PROBLEM**

Elevated blood Ca^{2+} leads to high Ca^{2+} concentrations in the kidney filtrate. Calcium-based kidney stones occur when calcium phosphate or calcium oxalate crystals form and aggregate with organic material in the lumen of the kidney tubule. Once Prof. Magruder's kidney stone passes into the urine, Dr. Spinks sends it for a chemical analysis.

Q3: *Only free Ca^{2+} in the blood filters into Bowman's capsule at the nephron. A significant portion of plasma Ca^{2+} cannot be filtered. Use what you have learned about filtration at the glomerulus to speculate on why some plasma Ca^{2+} cannot filter [p. 600].*

730   732   **736**   746   749   750

FIG. 23.4 | Thyroid hormone synthesis

(a) The thyroid gland is a butterfly-shaped gland, located just below the larynx. It secretes thyroid hormones and calcitonin.

(b) Section of thyroid gland. Thyroid hormone synthesis takes place in the colloid of the thyroid follicle.

Larynx

Thyroid gland

Trachea

Follicular cells secrete thyroid hormone.

Thyroid follicle

C cells secrete calcitonin.

Colloid is a glycoprotein.

Capillary

Capsule of connective tissue

(c) Thyroid hormones are made from iodine and tyrosine.

Blood

1 A Na⁺-I⁻ symporter brings I⁻ into the cell. The pendrin transporter moves I⁻ into the colloid.

2 Follicular cell synthesizes enzymes and thyroglobulin for colloid.

6 Free T_3 and T_4 enter the circulation.

5 Intracellular enzymes separate T_3 and T_4 from the protein.

4 Thyroglobulin is taken back into the cell in vesicles.

3 Thyroid peroxidase adds iodine to tyrosine to make T_3 and T_4.

Protein synthesis

Enzymes, Thyroglobulin

Follicular cells

Pendrin

Thyroglobulin

T_3 T_4

Colloid

MIT ← I + tyrosine
DIT ← I + MIT
T_3 ← MIT + DIT
T_4 ← DIT + DIT

KEY
MIT = monoiodotyrosine
DIT = diiodotyrosine
T_3 = triiodothyronine
T_4 = thyroxine

Tyrosine

Triiodothyronine (T_3)
2 tyrosine + 3 I

Thyroxine (T_4)
2 tyrosine + 4 I

Q FIGURE QUESTIONS

1. Identify the apical and basolateral membranes of the follicular cell.
2. What kind of transport brings I⁻ into follicular cells?
3. How does thyroglobulin get into the colloid?
4. How does the cell take thyroglobulin back in?
5. How do T_3 and T_4 leave the cell?

TSH Controls the Thyroid Gland

The control of thyroid hormone secretion follows the typical hypothalamic-pituitary-peripheral endocrine gland pattern (**FIG. 23.5**). **Thyrotropin-releasing hormone (TRH)** from the hypothalamus controls secretion of the anterior pituitary hormone **thyrotropin**, also known as **thyroid-stimulating hormone (TSH)**. TSH in turn acts on the thyroid gland to promote hormone synthesis. The thyroid hormones normally act as a negative feedback signal to prevent oversecretion.

The main function of thyroid hormones in adults is to provide substrates for oxidative metabolism. Thyroid hormones are thermogenic [p. 723] and increase oxygen consumption in most tissues. The exact mechanism is unclear but is at least partly related to changes in ion transport across the cell and mitochondrial membranes. Thyroid hormones also interact with other hormones to modulate protein, carbohydrate, and fat metabolism.

In children, thyroid hormones are necessary for full expression of growth hormone, which means they are essential for normal growth and development, especially of the nervous system. In the first few years after birth, myelin and synapse formation requires T_3 and T_4. Cytological studies suggest that thyroid hormones regulate microtubule assembly, which is an essential part of neuronal growth. Thyroid hormone is also necessary for proper bone growth.

The actions of thyroid hormones are most observable in people who secrete too much or too little hormone. Physiological effects that are subtle in people with normal hormone secretion often become exaggerated in patients with endocrine disorders. Patients with thyroid excess or deficiency may experience decreased tolerance to heat or cold and mood disturbances, in addition to other symptoms.

Thyroid Pathologies Affect Quality of Life

Problems with thyroid hormone secretion can arise either in the thyroid gland or along the control pathway depicted in Figure 23.5. The trophic action of TSH on the thyroid gland

FIG. 23.5 Thyroid hormones

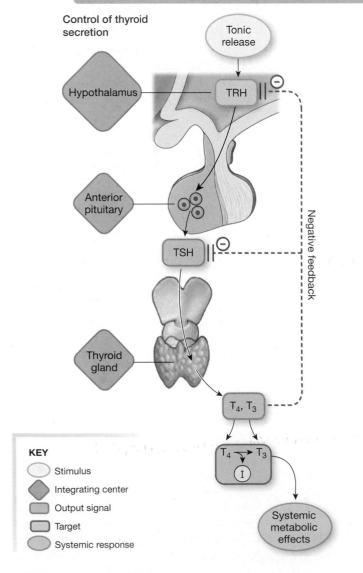

Control of thyroid secretion

KEY
- Stimulus
- Integrating center
- Output signal
- Target
- Systemic response

Thyroid Hormones	
Cell of Origin	Thyroid follicle cells
Chemical Nature	Iodinated amine
Biosynthesis	From iodine and tyrosine. Formed and stored on thyroglobulin in follicle colloid.
Transport in the Circulation	Bound to thyroxine-binding globulin and albumins
Half-Life	6-7 days for thyroxine (T_4); about 1 day for triiodothyronine (T_3)
Factors Affecting Release	Tonic release
Control Pathway	TRH (hypothalamus) $\rightarrow$ TSH (anterior pituitary) $\rightarrow T_3 + T_4$ (thyroid) $\rightarrow T_4$ deiodinates in tissues to form more T_3
Target Cells or Tissues	Most cells of the body
Target Receptor	Nuclear receptor
Whole Body or Tissue Reaction	↑ Oxygen consumption (thermogenesis). Protein catabolism in adults but anabolism in children. Normal development of nervous system
Action at Cellular Level	Increases activity of metabolic enzymes and Na^+-K^+-ATPase
Action at Molecular Level	Production of new enzymes
Feedback Regulation	Free T_3 and T_4 have negative feedback on anterior pituitary and hypothalamus.

FIG. 23.6   Signs of thyroid pathologies

(a) Goiter. Excessive stimulation of the thyroid gland by TSH causes the gland to enlarge (goiter).

(b) Myxedema. In hypothyroid individuals, mucopolysaccharide deposits beneath the skin may cause bags under the eyes.

(c) Exophthalmos. In hyperthyroid states, excessive deposition of mucopolysaccharides in the bony orbit may cause the bulging eyeball called exophthalmos.

causes enlargement, or *hypertrophy*, of follicular cells. In pathological conditions with elevated TSH levels, the thyroid gland will enlarge, a condition known as a **goiter**. A large goiter can weigh hundreds of grams and almost encircle the neck (**FIG. 23.6a**).

Goiters are the result of excess TSH stimulation of the thyroid gland. Simply knowing that someone has a goiter does not tell you what the pathology is, however. Let's see how both hypothyroidism and hyperthyroidism can be associated with goiter.

Hyperthyroidism A person whose thyroid gland secretes too much hormone suffers from **hyperthyroidism**. Excess thyroid hormone causes changes in metabolism, the nervous system, and the heart.

1. Hyperthyroidism increases oxygen consumption and metabolic heat production. Because of the internal heat generated, these patients have warm, sweaty skin and may complain of being intolerant of heat.
2. Excess thyroid hormone increases protein catabolism and may cause muscle weakness. Patients often report weight loss.
3. The effects of excess thyroid hormone on the nervous system include hyperexcitable reflexes and psychological disturbances ranging from irritability and insomnia to psychosis. The mechanism for psychological disturbances is unclear, but morphological changes in the hippocampus and effects on β-adrenergic receptors have been suggested.
4. Thyroid hormones are known to influence β-adrenergic receptors in the heart, and these effects are exaggerated with hypersecretion. A common sign of hyperthyroidism is rapid heartbeat and increased force of contraction due to up-regulation of β_1-receptors on the myocardium [p. 472].

The most common cause of hyperthyroidism is *Graves' disease* (**FIG. 23.7a**). In this condition, the body produces antibodies called **thyroid-stimulating immunoglobulins (TSI).** These antibodies mimic the action of TSH by combining with and activating TSH receptors on the thyroid gland. The result is goiter, hypersecretion of T_3 and T_4 and symptoms of hormone excess.

Negative feedback by the high levels of T_3 and T_4 shuts down the body's TRH and TSH secretion but does nothing to block the TSH-like activity of TSI on the thyroid gland. Graves' disease is often accompanied by **exophthalmos** (Fig. 23.6c), a bug-eyed appearance caused by immune-mediated enlargement of muscles and tissue in the eye socket. The English comic Marty Feldman was known for his wild-eyed appearance caused by exophthalmos.

Thyroid gland tumors are another cause of primary hyperthyroidism. Secondary hyperthyroidism will occur with pituitary tumors secreting TSH.

Hypothyroidism Hyposecretion of thyroid hormones affects the same systems altered by hyperthyroidism.

1. Decreased thyroid hormone secretion slows metabolic rate and oxygen consumption. Patients become intolerant of cold because they are generating less internal heat.
2. Hypothyroidism decreases protein synthesis. In adults, this causes brittle nails, thinning hair, and dry, thin skin. Hypothyroidism also causes accumulation of *mucopolysaccharides* under the skin. These molecules attract water and cause the puffy appearance of *myxedema* (Fig. 23.6b). Hypothyroid children have slow bone and tissue growth and are shorter than normal for their age.
3. Nervous system changes in adults include slowed reflexes, slow speech and thought processes, and feelings of fatigue. Deficient thyroid hormone secretion in infancy causes **cretinism**, a condition marked by decreased mental capacity.
4. The primary cardiovascular change in hypothyroidism is *bradycardia* (slow heart rate).

FIG. 23.7 Thyroid pathologies

(a) Hyperthyroidism due to Graves' disease. In Graves' disease, thyroid-stimulating immune proteins (TSI) bind to thyroid gland TSH receptors and cause the gland to hypertrophy.

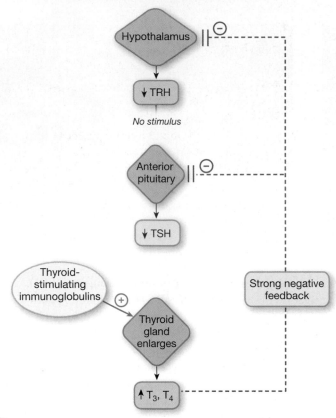

(b) Hypothyroidism due to low iodine. In hypothyroidism caused by iodine deficiency, absence of negative feedback increases TSH secretion and results in goiter.

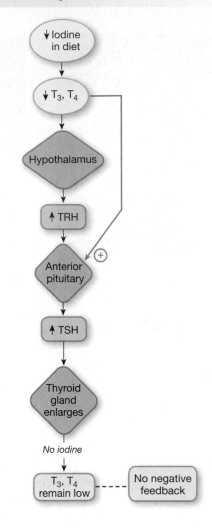

Q FIGURE QUESTION

Draw the pathway for a person with a pituitary tumor that is oversecreting TSH. Would this person be hypothyroid or hyperthyroid? Would this person have a goiter?

Primary hypothyroidism is most commonly caused by a lack of iodine in the diet. Without iodine, the thyroid gland cannot make thyroid hormones (Fig. 23.7b). Low levels of T_3 and T_4 in the blood mean no negative feedback to the hypothalamus and anterior pituitary. In the absence of negative feedback, TSH secretion rises dramatically, and TSH stimulation enlarges the thyroid gland (goiter). Despite hypertrophy, the gland cannot obtain iodine to make hormone, so the patient remains hypothyroid. These patients exhibit the previously described signs of hypothyroidism. The goiter shown in the photograph of Figure 23.6a is probably due to iodine deficiency.

Therapy for thyroid disorders depends on the cause of the problem. Hypothyroidism is treated with oral thyroid hormone. Hyperthyroidism can be treated by surgical removal of all or part of the gland, by destruction of thyroid cells with radioactive iodine, or by drugs that block either hormone synthesis (thiourea drugs) or peripheral conversion of T_4 to T_3 (propylthiouracil).

CONCEPT CHECK

9. A woman who had her thyroid gland removed because of cancer was given pills containing only T_4. Why was this less active form of the hormone an effective treatment for her hypothyroidism?

10. Why would excessive production of thyroid hormone, which uncouples mitochondrial ATP production and proton transport [p. 109], cause a person to become intolerant of heat?

GROWTH HORMONE

Growth in human beings is a continuous process that begins before birth. However, growth rates in children are not steady, with the first two years of life and the adolescent years marked by spurts of rapid growth and development. Normal growth is a complex process that depends on a number of factors:

1. **Growth hormone and other hormones.** Without adequate amounts of growth hormone, children simply fail to grow.

Thyroid hormones, insulin, and the sex hormones at puberty also play both direct and permissive roles. A deficiency in any one of these hormones leads to abnormal growth and development.

2. **An adequate diet** that includes protein, sufficient energy (caloric intake), vitamins, and minerals. Many amino acids can be manufactured in the body from other precursors, but essential amino acids must come from dietary sources. Among the minerals, calcium in particular is needed for proper bone formation.

3. **Absence of chronic stress.** Cortisol from the adrenal cortex is released in times of stress and has significant catabolic effects that inhibit growth. Children who are subjected to stressful environments may exhibit a condition known as *failure to thrive* that is marked by abnormally slow growth.

4. **Genetics.** Each human's potential adult size is genetically determined at conception.

Growth Hormone Is Anabolic

Growth hormone (GH or somatotropin [p. 213]) is released throughout life, although its biggest role is in children. Peak GH secretion occurs during the teenage years. The stimuli for growth hormone release are complex and not well understood, but they include circulating nutrients, stress, and other hormones interacting with a daily rhythm of secretion (**FIG. 23.8**).

The stimuli for GH secretion are integrated in the hypothalamus, which secretes two neuropeptides into the hypothalamic-hypophyseal portal system: **growth hormone-releasing hormone** (GHRH) and *growth hormone-inhibiting hormone,* better known as *somatostatin* (SS). On a daily basis, pulses of GHRH from the hypothalamus stimulate GH release. In adults, the largest pulse of GH release occurs in the first two hours of sleep. It is speculated that GHRH has sleep-inducing properties, but the role of GH in sleep cycles is unclear.

GH is secreted by cells in the anterior pituitary. It is a typical peptide hormone in most respects, except that nearly half the GH in blood is bound to a plasma **growth hormone-binding protein.** The binding protein protects plasma GH from being filtered into the urine and extends its half-life by 12 minutes. Researchers have hypothesized that genetic determination of binding protein concentration plays a role in determining adult height.

FIG. 23.8   Growth hormone

Growth hormone control pathway

Growth Hormone (hGH)	
Origin	Anterior pituitary
Chemical Nature	191-amino acid peptide; several closely related forms
Biosynthesis	Typical peptide
Transport in the Circulation	Half is dissolved in plasma, half is bound to a binding protein whose structure is identical to that of the GH receptor
Half-Life	18 minutes
Factors Affecting Release	Circadian rhythm of tonic secretion; influenced by circulating nutrients, stress, and other hormones in a complex fashion
Control Pathway	GHRH, somatostatin (hypothalamus) → growth hormone (anterior pituitary)
Target Cells or Tissues	Trophic on liver for insulin-like growth factor production; also acts directly on many cells
Target Receptor	Membrane receptor with tyrosine kinase activity
Whole Body or Tissue Reaction (with IGFs)	Bone and cartilage growth; soft tissue growth; ↑ plasma glucose
Action at Cellular Level	Receptor linked to kinases that phosphorylate proteins to initiate transcription

CLINICAL FOCUS

New Growth Charts

When you were growing up, did your family mark your growth each year on a special wall chart? Monitoring growth is an important part of healthcare for children and adolescents, particularly as we see a growing problem with childhood obesity in the United States. In 2000, the U.S. Centers for Disease Control and Prevention (CDC) issued new growth charts for the first time since 1977. In 2006, they recommended that clinicians use an international chart from the World Health Organization for children under two years of age. The old charts were based on 1929–1979 data from mostly bottle-fed, middle-class white children. We now know that breast-fed babies grow more rapidly than bottle-fed infants in the first two months, then more slowly for the remainder of the first year. We also have data showing that babies in lower socioeconomic groups grow more slowly. The new charts take these differences into account and also include body mass index (BMI) information up to age 20. To see the new charts and learn more about monitoring growth in infants and children, visit the CDC web site at *www.cdc.gov/growthcharts*.

The target tissues for GH include both endocrine and nonendocrine cells. GH acts as a trophic hormone to stimulate secretion of **insulin-like growth factors (IGFs)** (formerly called *somatomedins*) from the liver and other tissues. IGFs have a negative feedback effect on growth hormone secretion by acting on the anterior pituitary and on the hypothalamus. IGFs act in concert with growth hormone to stimulate bone and soft tissue growth (Fig. 23.8).

Metabolically, growth hormone and IGFs are anabolic for proteins and promote protein synthesis, an essential part of tissue growth. Growth hormone also acts with IGFs to stimulate bone growth. IGFs are responsible for cartilage growth. GH increases plasma fatty acid and glucose concentrations by promoting fat breakdown and hepatic glucose output.

CONCEPT CHECK   **11.** Which pituitary hormone in addition to GH has two hypothalamic factors that regulate its release?

Growth Hormone Is Essential for Normal Growth

The disorders that reflect the actions of growth hormone are most obvious in children. Severe growth hormone deficiency in childhood leads to **dwarfism**, which can result from a problem either with growth hormone synthesis or with defective GH receptors. Unfortunately, neither bovine nor porcine growth hormone is effective as replacement therapy, as only primate growth hormone is active in humans. Prior to 1985, when genetically engineered human growth hormone became available, donated

human pituitaries harvested at autopsy were the only source of growth hormone. Fortunately, severe growth hormone deficiency is relatively rare. At the opposite extreme, oversecretion of growth hormone in children leads to **gigantism.**

Once bone growth stops in late adolescence, growth hormone cannot further increase height. GH and IGFs can continue to act on cartilage and soft tissues, however. Adults with excessive secretion of growth hormone develop a condition known as **acromegaly,** characterized by lengthening of the jaw, coarsening of facial features, and growth of hands and feet (**FIG. 23.9**). André the Giant, a French wrestler who also had a role in the classic movie *The Princess Bride*, exhibited signs of both giantism (he grew to 7′4″ tall) and acromegaly before his death at age 47.

Genetically Engineered hGH Raises Ethical Questions

When genetically engineered human growth hormone (hGH) became available in the mid-1980s, the medical profession was faced with a dilemma. Obviously, the hormone should be used to treat children who would otherwise be dwarfs, but what about children with only partial GH deficiency or genetically short children with normal GH secretion levels? This question is

FIG. 23.9   Acromegaly

Excess growth hormone secretion in adults causes acromegaly, with lengthening of the jaw, coarsening of the features, and growth in hands and feet. Compare these signs in André the Giant (top) to the features of his co-stars in *The Princess Bride*, Mandy Patinkin (middle) and Wallace Shawn (bottom).

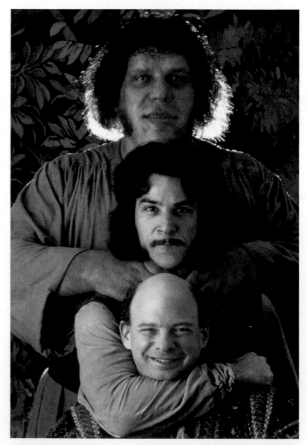

complicated by the difficulty of accurately identifying children with partial growth hormone deficiency. And what about children whose parents want them to be taller? Should these healthy children be given the hormone?

In 2003, the U.S. Food and Drug Administration approved use of a recombinant human growth hormone for treating children with non-GH-deficient short stature, defined as being more than 2.25 standard deviations below the mean height for their age and sex. (This means children in the bottom 1% of their age-sex group.) In clinical trials, daily injections of the drug for two years resulted in an average height increase of 1.3″ (3.3 cm). According to a 2006 analysis in a pediatric medicine journal, the cost for this treatment was more than $52,000 per inch of height gained. Side effects reported during hGH studies include glucose intolerance [p. 715] and *pancreatitis* (inflammation of the pancreas). Long-term risks associated with hGH treatment are unknown, and parents must be made aware that hGH therapy has the potential to create psychological problems in children if the results are less than optimum.

TISSUE AND BONE GROWTH

Growth can be divided into two general areas: soft tissue growth and bone growth. In children, bone growth is usually assessed by measuring height, and tissue growth by measuring weight. Multiple hormones have direct or permissive effects on growth. In addition, we are just beginning to understand how paracrine growth factors interact with classic hormones to influence tissue development and differentiation.

Tissue Growth Requires Hormones and Paracrine Signals

Soft tissue growth requires adequate amounts of growth hormone, thyroid hormone, and insulin. Growth hormone and IGFs are required for tissue protein synthesis and cell division. Under the influence of these hormones, cells undergo both *hypertrophy* (increased cell size) and **hyperplasia** (increased cell number).

Thyroid hormones play a permissive role in growth and contribute directly to nervous system development. At the target tissue level, thyroid hormone interacts synergistically with growth hormone in protein synthesis and nervous system development. Children with untreated hypothyroidism (cretinism) do not grow to normal height even if they secrete normal amounts of growth hormone.

Insulin supports tissue growth by stimulating protein synthesis and providing energy in the form of glucose. Because insulin is permissive for growth hormone, insulin-deficient children fail to grow normally even though they may have normal concentrations of growth and thyroid hormones.

Bone Growth Requires Adequate Dietary Calcium

Bone growth, like soft tissue development, requires the proper hormones and adequate amounts of protein and calcium. Bone

contains calcified extracellular matrix formed when calcium phosphate crystals precipitate and attach to a collagenous lattice support. The most common form of calcium phosphate is **hydroxyapatite,** $Ca_{10}(PO_4)_6(OH)_2$.

Although the large amount of inorganic matrix in bone makes some people think of it as nonliving, bone is a dynamic tissue, constantly being formed and broken down, or *resorbed.* Spaces in the collagen-calcium matrix are occupied by living cells that are well supplied with oxygen and nutrients by blood vessels that run through adjacent channels (**FIG. 23.10**).

Bones generally have two layers: an outer layer of dense **compact bone** and an inner layer of spongy **trabecular bone.** In some bones, a central cavity is filled with bone marrow. Compact bone provides strength and is thickest where support is needed (such as in the long bones of the legs) or where muscles attach. Trabecular bone is less sturdy and has open, cell-filled spaces between struts of calcified lattice.

Bones grow when matrix is deposited faster than it is resorbed. Specialized bone-forming cells called **osteoblasts** produce enzymes and **osteoid,** a mixture of collagen and other proteins to which hydroxyapatite binds. Recent research has found two other proteins, *osteocalcin* and *osteonectin,* that appear to aid in deposition of the calcified matrix.

Bone diameter increases when matrix deposits on the outer surface of the bone. Linear growth of long bones occurs at specialized regions called **epiphyseal plates,** located at each end of the bone shaft (*diaphysis*) (Fig. 23.10b). The side of the plate closer to the end (*epiphysis*) of the bone contains continuously dividing columns of **chondrocytes,** collagen-producing cells of cartilage. As the collagen layer thickens, the older cartilage calcifies and older chondrocytes degenerate, leaving spaces that osteoblasts invade. The osteoblasts then lay down bone matrix on top of the cartilage base.

As new bone is added at the ends, the shaft lengthens. Long bone growth continues as long as the epiphyseal plate is active. When osteoblasts complete their work, they revert to a less active form known as **osteocytes.**

Growth of long bone is under the influence of growth hormone and the insulin-like growth factors. In the absence of these hormones, normal bone growth does not occur. Long bone growth is also influenced by steroid sex hormones. The growth spurt of adolescent boys used to be attributed solely to increased androgen production but it now appears that estrogens play a significant role in pubertal bone growth in both sexes.

In all adolescents, the sex hormones eventually inactivate the epiphyseal plate so that long bones no longer grow. Because the epiphyseal plates of various bones close in a regular, ordered sequence, X-rays that show which plates are open and which have closed can be used to calculate a child's "bone age."

Linear bone growth ceases in adults, but bones are dynamic tissues that undergo continual remodeling throughout life. The *resorption* or breakdown of bone is controlled by **osteoclasts,** large, mobile, multinucleate cells derived from hematopoietic stem cells [p. 514]. Osteoclasts are responsible for dissolving bone.

Osteoclasts attach around their periphery to a section of matrix, much like a suction cup (Fig. 23.10c). The central region of

RUNNING **PROBLEM**

Prof. Magruder's blood work reveals that his Ca^{2+} level is 12.3 mg/dL plasma (normal is 8.5–10.5 mg/dL). These results support the suspected diagnosis of hyperparathyroidism. "Do you take vitamin D or use a lot of antacids?" Dr. Spinks asks. "Those could raise your blood calcium." Prof. Magruder denies using either substance. "Well, we need one more test before we can say conclusively that you have hyperparathyroidism," Dr. Spinks says.

Q4: *What one test could definitively prove that Prof. Magruder has hyperparathyroidism?*

730   732   736   **746**   749   750

Plasma Calcium Is Closely Regulated

Because calcium is critical to so many physiological functions, the body's plasma Ca^{2+} concentration is very closely regulated. Calcium homeostasis follows the principle of mass balance:

Total body calcium = intake − output

1. **Total body Ca^{2+}** is all the calcium in the body, distributed among three compartments (Fig. 23.11):
   (a) *Extracellular fluid.* Ionized Ca^{2+} is concentrated in the ECF. In the plasma, nearly half the Ca^{2+} is bound to plasma proteins and other molecules. The unbound Ca^{2+} is free to diffuse across membranes through open Ca^{2+} channels. Total plasma Ca^{2+} concentration is about 2.5 mM.
   (b) *Intracellular Ca^{2+}.* The concentration of free Ca^{2+} in the cytosol is about 0.001 mM. In addition, Ca^{2+} is concentrated inside mitochondria and the sarcoplasmic reticulum. Electrochemical gradients favor movement of Ca^{2+} into the cytosol when Ca^{2+} channels open.
   (c) *Extracellular matrix (bone).* Bone is the largest Ca^{2+} reservoir in the body, with most bone Ca^{2+} in the form of hydroxyapatite crystals. Bone Ca^{2+} forms a reservoir that can be tapped to maintain plasma Ca^{2+} homeostasis. Usually only a small fraction of bone Ca^{2+} is ionized and readily exchangeable, and this pool remains in equilibrium with Ca^{2+} in the interstitial fluid.

2. **Intake** is the Ca^{2+} ingested in the diet and absorbed in the small intestine. Only about one-third of ingested Ca^{2+} is absorbed, and unlike organic nutrients, Ca^{2+} absorption is hormonally regulated. Many people do not eat enough Ca^{2+}-containing foods, however, and intake may not match output.

   Intestinal calcium absorption is apparently both transcellular and paracellular (between the cells). Transcellular transport is accomplished by entry into the enterocyte through apical Ca^{2+} channels (*TRPV6, also called ECaC*). Once inside the cell, Ca^{2+} binds to a protein called *calbindin* that helps keep free intracellular $[Ca^{2+}]$ low. This is necessary because of the role of free Ca^{2+} as an intracellular signal molecule. On the basolateral side of the cell Ca^{2+} exits through basolateral Na^+-Ca^{2+} exchangers (NCX) and Ca^{2+}-ATPase

transporters. Transcellular absorption is hormonally regulated; paracellular absorption is unregulated.

3. **Output,** or Ca^{2+} loss from the body, occurs primarily through the kidneys, with a small amount excreted in feces. Ionized Ca^{2+} is freely filtered at the glomerulus and then reabsorbed along the length of the nephron. Hormonally regulated reabsorption takes place only in the distal nephron and uses transporters similar to those found in the intestine. There is no paracellular transport in the kidney.

CONCEPT CHECK

14. What does hypercalcemia do to neuronal membrane potential, and why does that effect depress neuromuscular excitability?

15. Draw a picture of a distal tubule cell and label apical and basolateral membranes, lumen, and ECF. Use the description of intestinal Ca^{2+} absorption in list item 2 (Intake) to draw the appropriate transporters.

16. Describe the renal transport of Ca^{2+} from the tubule lumen to the ECF as active, passive, facilitated diffusion, and so on.

Three Hormones Control Calcium Balance

Three hormones regulate the movement of Ca^{2+} between bone, kidney, and intestine: parathyroid hormone, calcitriol (vitamin D_3), and calcitonin (Fig. 23.11). Of these, parathyroid hormone and calcitriol are the most important in adult humans.

The **parathyroid glands,** which secrete parathyroid hormone {*para-*, alongside of}, were discovered in the 1890s by physiologists studying the role of the thyroid gland. These scientists noticed that if they removed all of the thyroid gland from dogs and cats, the animals died in a few days. In contrast, rabbits died only if the little parathyroid "glandules" alongside the thyroid were removed. The scientists then looked for parathyroid glands in dogs and cats and found them tucked away behind the larger thyroid gland. If the parathyroid glands were left behind when the thyroid was surgically removed, the animals lived.

The scientists concluded that the parathyroid glands contained a substance that was essential for life, although the thyroid gland did not. That essential substance was parathyroid hormone. The absence of parathyroid hormones causes hypocalcemic tetany and respiratory paralysis, as mentioned in the section on functions of calcium.

Parathyroid Hormone Four small parathyroid glands lie on the dorsal surface of the thyroid gland (**FIG. 23.12**). They secrete **parathyroid hormone (PTH)** (also called *parathormone*), a peptide whose main function is to increase plasma Ca^{2+} concentrations. The stimulus for PTH release is a decrease in plasma Ca^{2+} monitored by a cell membrane **Ca^{2+}-sensing receptor** (CaSR). The CaSR, a G protein-coupled receptor, was the first membrane receptor identified whose ligand was an ion rather than an organic molecule.

PTH acts on bone, kidney, and intestine to increase plasma Ca^{2+} concentrations (Fig. 23.12). Increased plasma Ca^{2+} acts as negative feedback and shuts off PTH secretion. Parathyroid hormone raises plasma Ca^{2+} in three ways:

FIG. 23.12   Parathyroid glands and parathyroid hormone (PTH)

This illustration shows the dorsal side of the thyroid gland. (Compare with the ventral view in Fig. 23.4a.)

The four parathyroid glands lie hidden behind the thyroid gland.

Left superior thyroid artery

Thyroid

Left parathyroid glands

Left inferior thyroid artery

Left subclavian artery

Pharynx

Right internal jugular vein

Right common carotid artery

Right parathyroid glands

Esophagus

Trachea

Parathyroid Hormone (PTH)	
Origin	Parathyroid glands
Chemical Nature	84-amino acid peptide
Biosynthesis	Continuous production, little stored
Transport in the Circulation	Dissolved in plasma
Half-Life	Less than 20 minutes
Factors Affecting Release	↓ Plasma Ca^{2+}
Target Cells or Tissues	Kidney, bone, intestine
Target Receptor	Membrane receptor acts via cAMP
Whole Body or Tissue Reaction	↑ Plasma Ca^{2+}
Action at Cellular Level	↑ Vitamin D synthesis; ↑ renal reabsorption of Ca^{2+}; ↑ bone resorption
Action at Molecular Level	Rapidly alters Ca^{2+} transport but also initiates protein synthesis in osteolcasts
Onset of Action	2–3 hours for bone, with increased osteoclast activity requiring 1–2 hours. 1–2 days for intestinal absorption. Within minutes for renal transport
Feedback Regulation	Negative feedback by ↑ plasma Ca^{2+}
Other Information	Osteoclasts have no PTH receptors and are regulated by PTH-induced paracrines. PTH is essential for life. Absence causes hypocalcemic tetany

1. **PTH mobilizes calcium from bone.** Increased bone resorption by osteoclasts takes about 12 hours to become measurable. Curiously, although osteoclasts are responsible for dissolving the calcified matrix and would be logical targets for PTH, they do not have PTH receptors. Instead, PTH effects are mediated by a collection of paracrine molecules, including *osteoprotegerin* (OPG) and an osteoclast differentiation factor called *RANKL*. These paracrine factors are receiving intense scrutiny as potential pharmacological agents. In late 2010, a RANKL inhibitor named denosumab was approved for use as treatment for conditions with excessive bone loss.

2. **PTH enhances renal reabsorption of calcium.** As we mentioned previously, regulated Ca^{2+} reabsorption takes place in the distal nephron. PTH simultaneously enhances renal excretion of phosphate by reducing its reabsorption. The opposing effects of PTH on calcium and phosphate are needed to keep their combined concentrations below a critical level. If the concentrations exceed that level, calcium phosphate crystals form and precipitate out of solution. High concentrations of calcium phosphate in the urine are one cause of kidney stones. We discuss additional aspects of phosphate homeostasis later.

3. **PTH indirectly increases intestinal absorption of calcium** through its influence on vitamin D_3, a process described next.

Calcitriol   Intestinal absorption of calcium is enhanced by the action of a hormone known as **1,25-dihydroxycholecalciferol** or $1,25(OH)_2D_3$, also known as **calcitriol** or vitamin D_3 (**FIG. 23.13**). The body makes calcitriol from vitamin D that has been obtained through diet or made in the skin by the action of sunlight on precursors made from acetyl CoA. People who live above 37 degrees of latitude north or below 37 degrees south do not get enough sunlight to make adequate vitamin D except in the summer, and they should consider taking vitamin supplements.

Vitamin D is modified in two steps—first in the liver, then in the kidneys—to make vitamin D_3 or calcitriol. Calcitriol is the primary hormone responsible for enhancing Ca^{2+} uptake from the small intestine. In addition, calcitriol facilitates renal reabsorption of Ca^{2+} and helps mobilize Ca^{2+} out of bone.

The production of calcitriol is regulated at the kidney by the action of PTH. Decreased plasma Ca^{2+} increases PTH secretion, which stimulates calcitriol synthesis. Intestinal and renal absorption of Ca^{2+} raises blood Ca^{2+}, turning off PTH in a negative feedback loop that decreases calcitriol synthesis.

FIG. 23.13 Endocrine control of calcium balance

PTH works with calcitriol to promote bone resorption, intestinal Ca^{2+} absorption, and distal nephron Ca^{2+} reabsorption, all of which tend to elevate plasma Ca^{2+} concentrations.

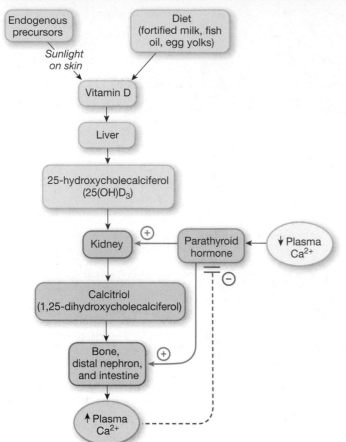

Vitamin D_3 (calcitriol, 1,25-dihydroxycholecalciferol)	
Origin	Complex biosynthesis; see below
Chemical Nature	Steroid
Biosynthesis	Vitamin D formed by sunlight on precursor molecules or ingested in food; converted in two steps (liver and kidney) to $1,25(OH)_2D_3$
Transport in The Circulation	Bound to plasma protein
Stimulus for Synthesis	↓Ca^{2+}. Indirectly via PTH. Prolactin also stimulates synthesis.
Target Cells or Tissues	Intestine, bone, and kidney
Target Receptor	Nuclear
Whole Body or Tissue Reaction	↑Plasma Ca^{2+}
Action at Molecular Level	Stimulates production of calbindin, a Ca^{2+}-binding protein, and of CaSR in parathyroid gland. Associated with intestinal transport by unknown mechanism
Feedback Regulation	↑Plasma Ca^{2+} shuts off PTH secretion

Prolactin, the hormone responsible for milk production in breast-feeding (lactating) women, also stimulates calcitriol synthesis. This action ensures maximal absorption of Ca^{2+} from the diet at a time when metabolic demands for calcium are high.

Calcitonin The third hormone involved with calcium metabolism is **calcitonin,** a peptide produced by the C cells of the thyroid gland (**TBL. 23.1**). Its actions are opposite to those of parathyroid hormone. Calcitonin is released when plasma Ca^{2+} increases. Experiments in animals have shown that calcitonin decreases bone resorption and increases renal calcium excretion.

Calcitonin apparently plays only a minor role in daily calcium balance in adult humans. Patients whose thyroid glands have been removed show no disturbance in calcium balance, and people with thyroid tumors that secrete large amounts of calcitonin also show no ill effects.

Calcitonin has been used medically to treat patients with *Paget's disease,* a genetically linked condition in which osteoclasts are overactive and bone is weakened by resorption. Calcitonin in these patients stabilizes the abnormal bone loss, leading scientists to speculate that this hormone is most important during childhood growth, when net bone deposition is needed, and during pregnancy and lactation, when the mother's body must supply calcium for both herself and her child.

Calcium and Phosphate Homeostasis Are Linked

Phosphate homeostasis is closely linked to calcium homeostasis. Phosphate is the second key ingredient in the hydroxyapatite of bone, $Ca_{10}(PO_4)_6(OH)_2$, and most phosphate in the body is found in bone. However, phosphates have other significant physiological roles, including energy transfer and storage in high-energy phosphate bonds, and activation or deactivation of enzymes, transporters, and ion channels through phosphorylation and dephosphorylation. Phosphates also form part of the DNA and RNA backbone.

Phosphate homeostasis parallels that of Ca^{2+}. Phosphate is absorbed in the intestines, filtered and reabsorbed in the kidneys, and divided between bone, ECF, and intracellular compartments. Vitamin D_3 enhances intestinal absorption of phosphate. Renal excretion is affected by both PTH (which promotes phosphate excretion) and vitamin D_3 (which promotes phosphate reabsorption).

TABLE 23.1	Calcitonin
Cell of Origin	**C Cells of Thyroid Gland (parafollicular cells)**
Chemical Nature	32-amino acid peptide
Biosynthesis	Typical peptide
Transport in the Circulation	Dissolved in plasma
Half-Life	<10 minutes
Factors Affecting Release	↑ plasma $[Ca^{2+}]$
Target Cells or Tissues	Bone and kidney
Target Receptor	G protein-coupled membrane receptor
Whole Body or Tissue Action	Prevents bone resorption. Enhances kidney excretion
Action at Molecular Level	Signal transduction pathways appear to vary during cell cycle
Other Information	Experimentally decreases plasma Ca^{2+} but has little apparent physiological effect in adult humans. Possible effect on skeletal development; possible protection of bone Ca^{2+} stores during pregnancy and lactation

CONCEPT CHECK

17. Name two compounds that store energy in high-energy phosphate bonds.
18. What are the differences between a kinase, a phosphatase, and a phosphorylase?

Osteoporosis Is a Disease of Bone Loss

One of the best-known pathologies of bone function is **osteoporosis,** a metabolic disorder in which bone resorption exceeds bone deposition. The result is fragile, weakened bones that are more easily fractured (Fig. 23.10c). Most bone resorption takes place in spongy trabecular bone, particularly in the vertebrae, hips, and wrists.

Osteoporosis is most common in women after menopause, when estrogen concentrations fall. However, older men also develop osteoporosis. Bone loss and small fractures and compression in the spinal column lead to *kyphosis* {hump-back}, the stooped, hunchback appearance that is characteristic of advanced osteoporosis in the elderly. Osteoporosis is a complex disease with genetic and environmental components. Risk factors include small, thin body type; postmenopausal age; smoking; and low dietary Ca^{2+} intake.

For many years estrogen or estrogen/progesterone *hormone replacement therapy* (HRT) was used to prevent osteoporosis. However, estrogen therapy alone increases the risk of endometrial and possibly other cancers, and some studies suggest that combined estrogen/progesterone HRT might increase risk of heart attacks and strokes. A *selective estrogen receptor modulator (SERM)* called raloxifene has been used to treat osteoporosis.

The most effective drugs for preventing or treating osteoporosis act more directly on bone metabolism. They include *bisphosphonates*, which induce osteoclast apoptosis and suppress bone resorption, and *teriparatide*, a PTH derivative, which stimulates formation of new bone. Teriparatide consists of the first 34 amino acids of the 84-amino acid PTH molecule and must be injected rather than taken orally. Currently clinical studies are investigating whether some combination of bisphosphonates and teriparatide is more effective in combating osteoporosis than either drug alone.

To avoid osteoporosis in later years, young women need to maintain adequate dietary calcium intake and perform weight-bearing exercises, such as running or aerobics, which increase bone density. Loss of bone mass begins by age 30, long before people think they are at risk, and many women suffer from low bone mass (*osteopenia*) before they are aware of a problem. Bone mass testing can help with early diagnosis of osteopenia.

RUNNING **PROBLEM**

The results of Prof. Magruder's last test confirm that he has hyperparathyroidism. He goes on a low-calcium diet, avoiding milk, cheese, and other dairy products, but several months later he returns to the emergency room with another painful kidney stone. Dr. Spinks sends him to an endocrinologist, who recommends surgical removal of the overactive parathyroid glands. "We can't tell which of the parathyroid glands is most active," the specialist says, "and we'd like to leave you with some parathyroid hormone of your own. So I will take out all four glands, but we'll reimplant two of them in the muscle of your forearm. In many patients, the implanted glands secrete just enough PTH to maintain calcium homeostasis. And if they secrete too much PTH, it is much easier to take them out of your arm than do major surgery on your neck again."

Q5: *Why can't Prof. Magruder simply take replacement PTH by mouth? (Hint: PTH is a peptide hormone.)*

RUNNING **PROBLEM CONCLUSION** | Hyperparathyroidism

Prof. Magruder had the surgery, and the implanted glands produced an adequate amount of PTH. He must have his plasma Ca^{2+} levels checked regularly for the rest of his life to ensure that the glands continue to function adequately. To learn more about hyperparathyroidism, see this article in *The American Family Physician: www.aafp.org/afp/20040115/333.html*.

Question	Facts	Integration and Analysis
Q1: What role does Ca^{2+} play in the normal functioning of muscles and neurons?	Calcium triggers neurotransmitter release [p. 258] and uncovers myosin-binding sites on muscle actin filaments [p. 386].	Muscle weakness in hyperparathyroidism is the opposite of what you would predict from knowing the role of Ca^{2+} in muscles and neurons. However, Ca^{2+} also affects the Na^+ permeability of neurons, and this effect leads to muscle weakness and CNS effects.
Q2: What is the technical term for "elevated levels of calcium in the blood"?	Prefix for elevated levels: *hyper-*. Suffix for "in the blood": *-emia*.	*Hypercalcemia*
Q3: Speculate on why some plasma Ca^{2+} cannot filter into Bowman's capsule.	Filtration at the glomerulus is a selective process that excludes blood cells and most plasma proteins [p. 600].	A significant amount of plasma Ca^{2+} is bound to plasma proteins and therefore cannot filter.
Q4: What one test could definitively prove that Prof. Magruder has hyperparathyroidism?	Hyperparathyroidism is excessive secretion of PTH.	A test for PTH concentration in the blood would confirm the diagnosis of hyperparathyroidism.
Q5: Why can't Prof. Magruder simply take replacement PTH by mouth?	PTH is a peptide hormone. Ingested peptides are digested by proteolytic enzymes.	PTH taken orally will be digested and not absorbed intact. Consequently it will not be effective.

730　732　736　746　749　**750**

VISUALIZE, PRACTICE, AND MASTER PHYSIOLOGY WITH MASTERING A&P
- Practice-on-the-go with Dynamic Study Modules
- Visualize important processes with Interactive Physiology and A&P Flix
- Check your understanding with Video Tutors and practice quizzes!

MasteringA&P®

CHAPTER SUMMARY

Endocrinology is based on the physiological principles of *homeostasis* and *control systems*. Each hormone has stimuli that initiate its secretion, and feedback signals that modulate its release. *Molecular interactions* and *communication across membranes* are also essential to hormone activity. In many instances, such as calcium and phosphate homeostasis, the principle of *mass balance* is the focus of homeostatic regulation.

Review of Endocrine Principles

1. Basic components of endocrine pathways include hormone receptors, feedback loops, and cellular responses. (p. 730)

Adrenal Glucocorticoids

2. The **adrenal cortex** secretes **glucocorticoids,** sex steroids, and aldosterone. (pp. 731, 732; Fig. 23.1)

3. **Cortisol** secretion is controlled by hypothalamic **CRH** and **ACTH** from the pituitary. Cortisol is the feedback signal. Cortisol is a typical steroid hormone in its synthesis, secretion, transport, and action. (p. 732; Fig. 23.2)

4. Cortisol is catabolic and essential for life. It promotes gluconeogenesis, breakdown of skeletal muscle proteins and adipose tissue, Ca^{2+} excretion, and suppression of the immune system. (p. 733)

5. **Hypercortisolism** usually results from a tumor or therapeutic administration of the hormone. **Addison's disease** is hyposecretion of all adrenal steroids. (p. 734)

6. CRH and the **melanocortins** have physiological actions in addition to cortisol release. (p. 735; Fig. 23.2d)

Thyroid Hormones

7. The thyroid **follicle** has a hollow center filled with **colloid** containing **thyroglobulin** and enzymes. (p. 736; Fig. 23.4b)

8. Thyroid hormones are made from tyrosine and iodine. **Tetraiodothyronine** (thyroxine, T_4) is converted in target tissues to the more active hormone **triiodothyronine** (T_3). (p. 736; Fig. 23.4)

9. Thyroid hormones are not essential for life, but they influence metabolic rate as well as protein, carbohydrate, and fat metabolism. (p. 738)

10. Thyroid hormone secretion is controlled by **thyrotropin** (**thyroid-stimulating hormone [TSH]**) and **thyrotropin-releasing hormone (TRH).** (p. 738; Fig. 23.5)

Growth Hormone

11. Normal growth requires growth hormone, thyroid hormones, insulin, and sex hormones at puberty. Growth also requires adequate diet and absence of chronic stress. (p. 740)

12. Growth hormone is secreted by the anterior pituitary and stimulates secretion of **insulin-like growth factors (IGFs)** from the liver and other tissues. These hormones promote bone and soft tissue growth. (p. 742; Fig. 23.8)

13. Secretion of growth hormone is controlled by **growth hormone-releasing hormone** (GHRH) and growth hormone-inhibiting hormone (**somatostatin**). (p. 741; Fig. 23.8)

Tissue and Bone Growth

14. **Bone** is composed of **hydroxyapatite** crystals attached to a collagenous support. Bone is a dynamic tissue with living cells. (p. 743)

15. **Osteoblasts** synthesize bone. Long bone growth occurs at **epiphyseal plates,** where **chondrocytes** produce cartilage. (p. 743; Fig. 23.10)

Calcium Balance

iP Fluids & Electrolytes: Electrolyte Homeostasis

16. Calcium acts as an intracellular signal for second messenger pathways, exocytosis, and muscle contraction. It also plays a role in cell junctions, coagulation, and neural function. (p. 745)

17. Ca^{2+} homeostasis balances dietary intake, urinary output, and distribution of Ca^{2+} among bone, cells, and the ECF. (p. 745; Fig. 23.11)

18. Decreased plasma Ca^{2+} stimulates **parathyroid hormone (PTH)** secretion by the **parathyroid glands.** (p. 746; Fig. 23.12)

19. PTH promotes Ca^{2+} resorption from bone, enhances renal Ca^{2+} reabsorption, and increases intestinal Ca^{2+} absorption through its effect on **calcitriol.** (p. 747; Fig. 23.12)

20. Calcitonin from the thyroid gland plays only a minor role in daily calcium balance in adult humans. (p. 748; Tbl. 23.1)

REVIEW QUESTIONS

In addition to working through these questions and checking your answers on p. A-31, review the Learning Outcomes at the beginning of this chapter.

Level One Reviewing Facts and Terms

1. Name the zones of the adrenal cortex and the primary hormones secreted in each zone.

2. For (a) cortisol, (b) growth hormone, (c) parathyroid hormone, and (d) T_3 and T_4: Draw the full control pathway and show feedback where appropriate. Do not use abbreviations.

3. List four conditions that are necessary for people to achieve their full growth. Include five specific hormones known to exert an effect on growth.

4. Name the thyroid hormones. Which one has the highest activity? How and where is most of it produced?

5. Define each of the following terms and explain its physiological significance:
   (a) melanocortins
   (b) osteoporosis
   (c) hydroxyapatite
   (d) mineralocorticoid
   (e) trabecular bone
   (f) POMC
   (g) epiphyseal plates

6. List seven functions of calcium in the body.

7. Make a table showing the effects of cortisol, thyroid hormones, growth hormone, insulin, and glucagon on protein, carbohydrate, and lipid metabolism.

Level Two Reviewing Concepts

8. Mapping exercise: Create a reflex map with feedback loops for each of the following situations:
   (a) hypercortisolism from an adrenal tumor
   (b) hypercortisolism from a pituitary tumor
   (c) hyperthyroidism from a hormone-secreting thyroid tumor
   (d) hypothyroidism from a pituitary problem that decreases TSH synthesis

9. Define, compare, and contrast or relate the terms in each set:
   (a) cortisol, glucocorticoids, ACTH, CRH
   (b) thyroid, C cell, follicle, colloid
   (c) thyroglobulin, tyrosine, iodide, TBG, deiodinase, TSH, TRH
   (d) somatotropin, IGF, GHRH, somatostatin, growth hormone–binding protein
   (e) giantism, acromegaly, dwarfism
   (f) hyperplasia, hypertrophy
   (g) osteoblast, osteoclast, chondrocyte, osteocyte
   (h) vitamin D, calcitriol, 1,25-dihydroxycholecalciferol, calcitonin, estrogen, PTH

10. Based on what you know about the cellular mechanism of action for T_3, would you expect to see tissue response to this hormone within a few minutes or in more than an hour?

11. If average plasma $[Ca^{2+}]$ is 2.5 mmol/L, what is the concentration in mEq/L?

12. Osteoclasts make acid (H^+) from CO_2 and H_2O. They secrete the acid at their apical membrane and put bicarbonate into the ECF. Draw an osteoclast and diagram this process, including enzymes and the appropriate transporters on each membrane. How many different transporters can you think of that could be used to reabsorb bicarbonate?

Level Three Problem Solving

13. Diabetic patients who have surgery, become sick, or are under other physiological stress are told to monitor their blood sugar carefully because they may need to increase their insulin dose temporarily. What is the physiological explanation behind this advice?

14. One diagnostic test to determine the cause of hypercortisolism is a *dexamethasone suppression test*. Dexamethasone blocks secretion of ACTH by the pituitary. The following table shows the results from two patients given a dexamethasone suppression test.

Plasma Cortisol Concentration	Before Test	After Test
Patient A	High	High
Patient B	High	Low

Can you tell from these results where the patients' pathologies originate? Explain for each patient.

15. When blood test results came back last week, someone in the office spilled a cup of coffee on them, smearing the patient names and some of the numbers. One report shows elevated TSH levels, but the thyroid levels are so low they are unreadable. You have three charts waiting for test results on thyroid hormone levels. Your tentative diagnoses, based on physical findings and symptoms, for those three patients are:

Mr. A: primary hypothyroidism
Ms. B: primary hyperthyroidism
Ms. C: secondary hyperthyroidism

(a) Can you tell whose results are on the smeared report, based on the TSH results and the tentative diagnosis?
(b) Can you rule out any of the three people based on those same criteria? Explain.

16. The following graph shows the results of a study done in Boston that compared blood vitamin D levels during summer and winter.

Boston is located at 42 degrees north latitude, and weak sunlight in winter there does not allow skin synthesis of vitamin D. (Data from *Am J Med* 112: 659–662, 2002 Jun 1)

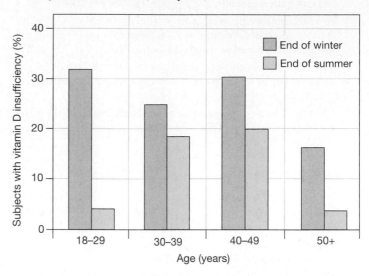

(a) Summarize the results shown in the graph. How many variables are shown in the graph that you must address in your summary?
(b) Based on what you know, how could you explain the results of the study?
(c) Would taking a multivitamin supplement affect the results?

Level Four Quantitative Problems

17. Filterable plasma Ca^{2+} is about 5 mg/L. Assume a man has a GFR of 125 mL of plasma filtered per minute.

(a) How much Ca^{2+} does he filter in a day?
(b) Net dietary Ca^{2+} intake is 170 mg /day. To remain in Ca^{2+} balance, how much Ca^{2+} must he excrete?
(c) What percentage of filtered Ca^{2+} is reabsorbed by the kidney tubule?

18. Draw two graph axes. On graph A, plot the effect of plasma parathyroid hormone concentration on plasma Ca^{2+} concentration. On graph B, plot the effect of plasma Ca^{2+} concentration on plasma parathyroid hormone concentration. Be sure to label the axes of each graph.

Answers to Concept Checks, Figure and Graph Questions, and end-of-chapter Review Questions can be found in Appendix A [p. A-1].

24 The Immune System

Although, at first sight, the immune system may appear to be autonomous, it is connected by innumerable structural and functional bridges with the nervous system and the endocrine system, so as to constitute a multisystem.

Branislav D. Jankovic,
Neuroimmunomodulation: The State of the Art, 1994

OVERVIEW 754

LO 24.1 List the three major functions of the immune system.

LO 24.2 List the three categories of immune pathologies.

PATHOGENS OF THE HUMAN BODY 754

LO 24.3 Describe the differences between bacteria and viruses that require a variety of defense mechanisms.

THE IMMUNE RESPONSE 756

LO 24.4 Describe and differentiate between innate immunity and acquired immunity.

LO 24.5 Describe and differentiate between cell-mediated immunity and humoral immunity.

ANATOMY OF THE IMMUNE SYSTEM 757

LO 24.6 Describe and differentiate between different types of primary lymphoid tissues and secondary lymphoid tissues.

LO 24.7 List and describe leukocytes according to their morphological and functional characteristics

INNATE IMMUNITY: NONSPECIFIC RESPONSES 762

LO 24.8 Identify the physical and chemical barriers that protect the body's internal environment.

LO 24.9 Diagram the nonspecific responses of innate immunity mediated by phagocytes, NK cells, and cytokines.

ACQUIRED IMMUNITY: ANTIGEN-SPECIFIC RESPONSES 765

LO 24.10 Diagram the humoral immune responses of B lymphocytes.

LO 24.11 Describe how antibodies make antigens more visible to the immune system and how they activate other immune cells.

LO 24.12 Diagram the cell-mediated immune responses of T lymphocytes.

IMMUNE RESPONSE PATHWAYS 772

LO 24.13 Map and compare the immune responses in bacterial and viral infections, allergic reactions, and following the transfusion of incompatible blood.

LO 24.14 Describe self-tolerance and how the breakdown of this safeguard can lead to autoimmune diseases.

NEURO-ENDOCRINE-IMMUNE INTERACTIONS 779

LO 24.15 Describe the relationships between the immune, nervous, and endocrine systems.

LO 24.16 Explain how stress affects immunity.

BACKGROUND BASICS

35	DNA, RNA
46	Ligands
76	Types of epithelium
86	Skin
112	Transcription and translation
147	Phagocytosis
168	Cytokines
499	Lymphatic system
516	Colony-stimulating factors
514	Leukocytes
521	Coagulation
537	Mucociliary escalator
723	Pyrogens

Cell infected with poxvirus (650x)

"Laughter is the best medicine." This old saying seemed silly for many years, when our attention was focused on learning as much as possible about the cells and chemicals of the immune system. But even as our list of immune cytokines grows, scientists are coming to recognize that the immune system is only one part of a complex communication network that includes the nervous and endocrine systems. The brain-immune connection has even been given its own name: *psychoneuroimmunology*.

The connections between the immune system and other body systems are becoming more important in our understanding of physiology. For example, we have learned that bacteria are not always harmful, and that the symbiotic bacteria permitted by our immune system to live inside our gut play an important role in metabolism. Inflammation created by the immune system is a factor in many disease states, including atherosclerosis [p. 502]. And on the flip side, the brain can assist or derail the immune system, depending on our psychological state.

The primary job of the immune system is to recognize "self" from "nonself." "Nonself" includes viruses, bacteria, parasites, and other disease-causing entities as well as our own cells that have become defective and threaten to do harm, such as become cancer. Together *specificity* and *memory* in immune cells enable the body to distinguish normal from abnormal and to mount a targeted response.

The body's ability to protect itself is known as **immunity**, from the Latin word *immunis*, meaning *exempt*. The human immune system consists of the lymphoid tissues of the body, the immune cells, and chemicals (both intracellular and secreted) that coordinate and execute immune functions. Most immune functions depend heavily on cell-to-cell communication, particularly local communication by cytokines and contact-dependent signaling that relies on cells binding to surface receptors on other cells [p. 166].

RUNNING **PROBLEM** | **HPV: To Vaccinate or Not?**

Rebecca and her daughter Lizzie arrived at the doctor's office for Lizzie's annual back-to-school checkup. Lizzie is 12 years old and will be in sixth grade in the fall. Halfway through the appointment, Dr. Paul asked Rebecca if Lizzie had started the series of vaccinations to protect her against HPV, the human papillomavirus. "Isn't HPV the virus that causes cervical cancer?" Rebecca asked. When the doctor confirmed that it was, Rebecca said, "No, Lizzie doesn't need that yet. She's only 12 and she's a good girl." "Let's talk about it," Dr. Paul suggested. "The American Academy of Pediatrics, the American Cancer Society, The American Congress of Obstetricians and Gynecologists, and the Centers for Disease Control and Prevention all recommend this vaccine for girls and boys, and now is the time to start it."

754 〔756〕〔766〕〔775〕〔780〕〔782〕

OVERVIEW

The immune system serves three major functions:

1. **It tries to recognize and remove abnormal "self" cells** created when normal cell growth and development go wrong. For example, the diseases we call *cancer* result from abnormal cells that multiply uncontrollably, crowding out normal cells and disrupting function. Scientists believe that cancer cells form on a regular basis but are usually detected and destroyed by the immune system before they get out of control.
2. **It removes dead or damaged cells,** as well as old red blood cells. Scavenger cells of the immune system, such as macrophages, patrol the extracellular compartment, gobbling up and digesting dead or dying cells.
3. **It protects the body from disease-causing invaders** known as **pathogens.** Microorganisms (*microbes*) that act as pathogens include bacteria, viruses, fungi, and one-celled protozoans. Larger pathogens include multicellular **parasites,** such as hookworms and tapeworms. Almost any exogenous molecule or cell has the potential to elicit an immune response. Pollens, chemicals, and foreign bodies are examples of substances to which the body may react. Substances that trigger the body's immune response are called *immunogens*. Immunogens that react with products of the response are known as **antigens.**

Sometimes, the body's immune system fails to perform its normal functions. Pathologies of the immune system generally fall into one of three categories: incorrect responses, overactive responses, or lack of response.

1. *Incorrect responses.* If mechanisms for distinguishing self from nonself fail and the immune system attacks the body's normal cells, an *autoimmune disease* results. Type 1 diabetes mellitus, in which proteins made by immune cells destroy pancreatic beta cells, is an example of an autoimmune disease in humans.
2. *Overactive responses.* Allergies are conditions in which the immune system creates a response that is out of proportion to the threat posed by the antigen. In extreme cases, the systemic effects of allergic responses can be life threatening.
3. *Lack of response.* **Immunodeficiency diseases** arise when some component of the immune system fails to work properly. *Primary immunodeficiency* is a family of genetically inherited disorders that range from mild to severe. *Acquired immunodeficiencies* may occur as a result of infection, such as *acquired immunodeficiency syndrome* (AIDS) caused by the *human immunodeficiency virus* (HIV). Acquired immunodeficiencies may also arise as a side effect of drug or radiation therapy, such as those used to treat cancer.

PATHOGENS OF THE HUMAN BODY

In the United States, our most prevalent infectious diseases are of viral and bacterial origin. Worldwide, parasites are an additional significant public health concern. For example, malaria, a pathogenic protozoan whose life cycle alternates between human and mosquito hosts, was estimated to infect more than 200 million people in 2012, with over 600,000 deaths that year.

Many parasitic organisms, such as the malaria protozoan, are introduced into the body by biting insects. Others enter the body through the digestive tract, brought in by contaminated food and water. Some, such as the fungi that cause valley fever and histoplasmosis, are inhaled. A few, such as the blood fluke *Schistosoma*, burrow through the host's skin. Once in the body, microbes and parasites may enter host cells in an effort to evade the immune response, or they may remain in the extracellular compartment.

Bacteria and Viruses Require Different Defense Mechanisms

Bacteria and viruses differ from each other in several ways. These differences require the body to have a variety of immune responses.

1. **Structure.** Bacteria are cells, with a cell membrane that is usually surrounded by a cell wall (**FIG. 24.1a**). Some *encapsulated* bacteria also produce an additional protective outer layer known as a *capsule*. Viruses are not cells. They consist

FIG. 24.1 Bacteria and viruses

(a) Bacterial Structure

Bacteria come in many shapes, but most are rod-shaped (bacillus), spherical (coccus), or spiral. They may have one or many flagella.

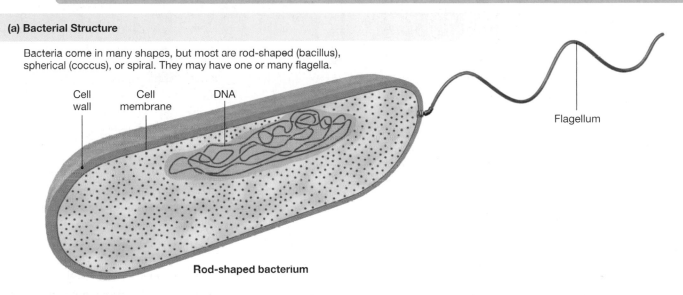

Rod-shaped bacterium

(b) Viral Structure

Viruses may be made of DNA or RNA.
This shows influenza, an RNA virus.

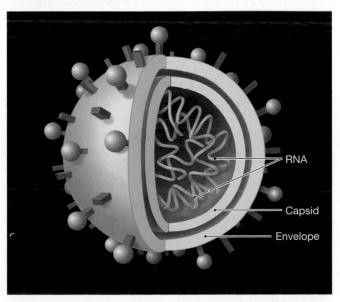

(c) Differences between Bacteria and Viruses

	Bacteria	Viruses
Structure	Cells. Usually surrounded by cell wall.	Not cells. Nucleic acid core with protein coat.
Living Conditions	Most can survive and reproduce outside a host.	Parasitic. Must have a host cell to reproduce.
Susceptibility to Drugs	Most can be killed or inhibited by antibiotics.	Cannot be killed with antibiotics. Some can be inhibited with antiviral drugs.

of nucleic acid (DNA or RNA) enclosed in a coat of viral proteins called a *capsid*. Some viruses add an *envelope* of phospholipid and protein made from the host's cell membrane and incorporate viral proteins into the envelope (Fig. 24.1b).

2. **Living conditions and reproduction.** Most bacteria can survive and reproduce outside a host if they have the required nutrients, temperature, pH, and so on. Viruses *must* use the intracellular machinery of a host cell to replicate. The location of pathogens in both major compartments of the body requires different defense mechanisms for each compartment.

3. **Susceptibility to drugs.** Most bacteria can be killed by the drugs we call **antibiotics.** These drugs act directly on bacteria and destroy them or inhibit their growth. Viruses cannot be killed by antibiotics. A few viral infections can be treated with *antiviral drugs,* which target specific stages of viral replication.

Viruses Can Replicate Only inside Host Cells

The replication cycle of a virus begins when the virus invades the host cell. To cross the human host cell membrane, the virus binds to membrane receptors, triggering endocytosis of the entire virus particle. In an alternative scenario, the virus envelope fuses with the host cell membrane, allowing the core of the virus to enter the cytoplasm.

Once inside the host cell and free of the capsid, the virus's nucleic acid takes over the host cell's resources to make new viral nucleic acid and viral proteins. These components assemble into additional virus particles that are released from the host cell to infect other cells.

Viruses can be released from host cells in one of two ways. (1) The virus causes the host cell to rupture, releasing virus particles into the extracellular fluid (ECF), or (2) virus particles surround themselves with a layer of host cell membrane and then bud off from the surface of the host cell.

Viruses do other kinds of damage to host cells. In taking over a host cell, they may totally disrupt the cell's metabolism so that the cell dies. Some viruses (*Herpes simplex type 1* and varicella-zoster virus, which cause cold sores and chicken pox, respectively) "hide out" in the host cell and replicate only sporadically. Other viruses incorporate their DNA into the host cell DNA. Viruses with this characteristic include HIV and **oncogenic viruses,** which cause cancer.

CONCEPT CHECK

1. Explain the differences between a virus and a bacterium.
2. List the sequence of steps by which a virus replicates itself.

RUNNING **PROBLEM**

Dr. Paul explained to Rebecca that there are over 100 known types of human papillomavirus. HPV is a DNA-based virus that can live only in the thin, flat epithelial cells known as *squamous epithelium {squama,* a scale}. Squamous epithelium is found on the surface of the skin and in mucous membrane, such as those lining the mouth and genital tract. HPV can cause both ordinary warts on the skin and the growths known as genital warts. More importantly, HPV infections cause nearly all cases of cervical cancer. (The *cervix* is neck of the uterus that opens into the vagina.) Two high-risk strains of HPV, designated types 16 and 18, account for about 70% of cervical cancer cases.

Q1: *Why must viruses live and reproduce inside host cells?*

Q2: *How might you test a person to see if she is infected with HPV?*

754   756   766   775   780   782

THE IMMUNE RESPONSE

The body has two lines of defense. Physical and chemical barriers, such as skin, mucus, and stomach acid, first try to keep pathogens out of the body's internal environment (**FIG. 24.2**). If this first line of defense fails, then the internal **immune response** takes over. The internal immune response has four basic steps: (1) *detection* and *identification* of the foreign substance, (2) *communication* with other immune cells to rally an organized response, (3) *recruitment* of assistance and *coordination* of the response among all participants, and (4) *destruction* or *suppression* of the invader.

The immune response is distinguished by its extensive use of chemical signaling. Detection, identification, communication, recruitment, coordination, and the attack on the invader all depend on signal molecules such as cytokines and antibodies. **Cytokines** are protein messengers released by one cell that affect the growth or activity of another cell [p. 168]. **Antibodies,** proteins secreted by certain immune cells, bind antigens and make them more visible to the immune system.

The human immune response is generally divided into two categories: nonspecific innate immunity and specific acquired immunity. **Innate immunity** is present from birth {*innatus,* inborn} and is the body's **nonspecific immune response** to invasion. The membrane receptors that mediate innate immunity have broad specificity and allow some immune cells to respond to molecular signals that are both unique and common to pathogenic microorganisms. An example of a common pathogen-specific signal would be certain components of the bacterial cell wall. Because the nonspecific innate response does not target a particular pathogen, it begins within minutes to hours. **Inflammation,** apparent on the skin as a red, warm, swollen area, is a hallmark reaction of cytokine-mediated innate immunity.

FIG. 24.2 | Physical and chemical barriers

Epithelium

The protective barrier of skin and mucous membranes is the body's first line of defense.

Glandular Secretions

Salivary glands and the glands in airways secrete mucus and immunoglobulins to trap and disable inhaled or ingested pathogens.

Stomach Acidity

The low pH of the stomach helps destroy swallowed pathogens.

Acquired immunity (also called *adaptive immunity*) is directed at specific invaders and for this reason is the body's **specific immune response.** The membrane receptors that mediate acquired immunity are highly specific and can distinguish between different pathogens. One characteristic of acquired immunity is that a specific immune response following first exposure to a pathogen may take days. With repeated exposures, however, the immune system "remembers" prior exposure to the pathogen and reacts more rapidly.

Acquired immunity can be divided into cell-mediated immunity and humoral immunity. *Cell-mediated immunity* uses contact-dependent signaling in which an immune cell binds through receptors to its target cell. *Humoral immunity*, also known as *antibody-mediated immunity,* uses the secreted proteins known as antibodies to carry out the immune response. Antibodies combine with foreign substances to make them more visible to the cells of the immune system. (The term *humoral,* referring to the blood, comes from the ancient Hippocratic school of medicine, which classified the body's fluids into four *humors:* blood, phlegm, black bile, and yellow bile.)

As you will see later in this chapter, the nonspecific and specific immune responses overlap. Although we have described them as if they are separate, in reality they are two interconnected parts of a single process. The innate response is the more rapid response, but it does not target a specific invader. It is reinforced by the antigen-specific acquired response, which amplifies the efficacy of the innate response. Communication and coordination among all the different pathways of immunity are vital for maximum protective effect.

Keep in mind that not all invaders can be destroyed by the body's immune system. In some cases, the best the body can do is control the damage and keep the invader from spreading. Pathogens that are suppressed by the immune system rather than destroyed include the bacterium that causes tuberculosis, which hides inside macrophages in the lung; the malaria parasite, which hides inside liver cells; and the herpes viruses responsible for outbreaks of cold sores or genital lesions, which hide inside cells in the skin.

Because of the huge numbers of molecules involved in human immunity, and because of the complex interactions between different components of the immune system, the field of immunology is continuously expanding. In this chapter, we cover the basics.

ANATOMY OF THE IMMUNE SYSTEM

The immune system is probably the least anatomically identifiable system of the body because most of it is integrated into the tissues of other organs, such as the skin and gastrointestinal tract. Yet, the mass of all immune cells in the body equals the mass of the brain! The immune system has two anatomical components: lymphoid tissues and the cells responsible for the immune response.

Lymphoid Tissues Are Everywhere

Lymphoid tissues are found all over the body (**FIG. 24.3**). The two *primary lymphoid tissues* are the **thymus gland** (see *Focus On: The Thymus Gland,* Fig. 24.10, p. 770) and the **bone marrow** [Fig. 16.4, p. 519], both sites where cells involved in the immune response form and mature. Some types of mature immune cells do not specialize until their first exposure to the pathogen they will fight. These mature but unspecialized immune cells are said to be *naïve* cells {*naif,* natural}.

In the *secondary lymphoid tissues,* mature immune cells interact with pathogens and initiate a response. Secondary tissues are divided into encapsulated tissues and unencapsulated diffuse lymphoid tissues. The **encapsulated lymphoid tissues** are the **spleen** (**FIG. 24.4**) and the **lymph nodes** (Fig. 24.3). Both spleen and lymph nodes have an outer wall formed from fibrous collagenous capsules. The spleen contains immune cells positioned so that they monitor the blood for foreign invaders. Phagocytic cells in the spleen also trap and remove aging red blood cells.

The lymph nodes are part of the lymphatic circulation, which is closely associated with capillaries of the cardiovascular system. Recall that blood pressure creates net flow of fluid out of capillaries and into the interstitial space [p. 497]. The filtered fluid, amounting to about 3 L/day, is picked up by lymph capillaries

FIG. 24.3 ANATOMY SUMMARY

The Immune System

IMMUNE SYSTEM

Lymphoid Tissues

Secondary lymphoid tissues
Active mature immune cells

Encapsulated
• Spleen (see Fig. 24.4)
• Lymph nodes

Unencapsulated (diffuse)
• GALT
• Tonsils
• Other tissues in contact with external environment

Primary lymphoid tissues
Immune cells form and mature here

Thymus gland (see Fig. 24.10)

Bone marrow (see Fig. 16.4)

Immune Cells

Basophil /mast cell

Eosinophil

Neutrophil

Macrophage /monocyte

Lymphocyte

Dendritic cells

The Lymphatic System

Tonsil is diffuse lymphoid tissue.

Thymus produces T lymphocytes.

Lymph nodes ⎫
                ⎬ Encapsulated lymphoid tissues
Spleen   ⎭

Gut-associated lymphoid tissue (GALT) is a diffuse lymphoid tissue.

Bone marrow produces most blood cells.

Lymph vessels

Structure of a Lymph Node

Efferent lymph vessel

Lymph node artery and vein

Capsule

Afferent lymph vessel

Clusters of immune cells intercept pathogens that invade interstitial fluid.

FIG. 24.4 FOCUS ON...

The Spleen

The spleen is about the size of a fist, and is the largest lymphoid organ in the body. It is located in the upper left quadrant of the abdomen close to the stomach.

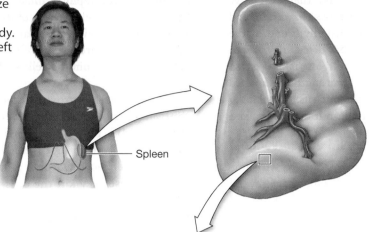

Spleen

The outer surface of the spleen is a connective tissue capsule that extends into the interior to create an open framework that supports the blood vessels and lymphoid tissue.

Vein

Artery

Venous sinuses

Capillary

Darker regions of **red pulp** are closely associated with extensive blood vessels and open venous sinuses. The red pulp contains many macrophages that act as a filter by trapping and destroying foreign material circulating in the blood. In addition, the macrophages ingest old, damaged, and abnormal red blood cells, breaking down their hemoglobin molecules into amino acids, iron, and bilirubin that is transported to the liver for excretion.

Regions of **white pulp** resemble the interior of lymph nodes and are composed mainly of lymphocytes.

Capsule

and passes through the encapsulated lymph nodes on its journey back to the heart.

Inside lymph nodes, clusters of immune cells intercept pathogens that have entered the interstitial fluid through breaks in the skin or through mucous membranes (Fig. 24.3). Once these microbes have been swept into the lymph, immune cells in the lymph nodes help prevent their spread throughout the body. You have probably noticed that if you have a sinus infection or a sore throat, the lymph nodes in your neck become swollen. These sore, swollen nodes result from the presence of active immune cells that have collected in the nodes to fight the infection.

The unencapsulated **diffuse lymphoid tissues** are aggregations of immune cells that appear in other organs of the body (Fig. 24.3). They include the **tonsils** at the posterior nasopharynx; the **gut-associated lymphoid tissue (GALT),** which lies just under the epithelium of the esophagus and intestines [p. 659]; and clusters of lymphoid tissue associated with the skin and the respiratory, urinary, and reproductive tracts. In each case, these tissues contain immune cells positioned to intercept invading pathogens before they get into the general circulation. Because of the large surface area of the digestive tract epithelium, some authorities consider the GALT to be the largest immune organ. Anatomically, the immune system is positioned wherever pathogens are most likely to enter the body.

Leukocytes Mediate Immunity

The white blood cells (WBCs), or **leukocytes,** are the primary cells responsible for the immune responses of the body. Most leukocytes are much larger than red blood cells, and they are not nearly as numerous. A microliter (μL) of whole blood contains about 5 million red blood cells but only about 7000 leukocytes.

Although most leukocytes circulate in the blood, they usually leave the capillaries and function *extravascularly* (outside the vessels). Some types of leukocytes can live out in the tissues for several months, but others may live for only hours or days.

Cytokines Create the Inflammatory Response

Inflammation is a hallmark reaction of innate immunity. Inflammation has three important roles in fighting infection in damaged tissue: (1) attracting immune cells and chemical mediators to the site, (2) producing a physical barrier to retard the spread of infection, and (3) promoting tissue repair once the infection is under control (a non-immunological function).

The inflammatory response is created when activated tissue macrophages release cytokines. These chemicals attract other immune cells, increase capillary permeability, and cause fever. Immune cells attracted to the site in turn release their own cytokines. Some representative examples of chemicals involved in the innate immune response are described next and listed in **TABLE 24.1**, a convenient "scorecard" for following the "players" in these complex processes.

Acute-Phase Proteins In the time immediately following an injury or pathogen invasion (the acute phase), the body responds by increasing the concentration of various plasma proteins. Some of these proteins, produced mostly by the liver, are given the general name of **acute-phase proteins.** They include molecules that act as opsonins by coating pathogens; antiprotease molecules that help prevent tissue damage; and *C-reactive protein* (CRP).

Normally, the levels of acute-phase proteins decline to normal as the immune response proceeds, but in *chronic inflammatory diseases,* such as rheumatoid arthritis, elevated levels of acute-phase proteins may persist. One interesting finding is that increased levels of CRP are associated with increased risk of coronary heart disease. This association exists because atherosclerosis is an inflammatory process that begins when macrophages in blood vessels ingest excess cholesterol and become foam cells [p. 502] that secrete CRP, other cytokines, and growth factors. Elevated CRP is not a specific indicator of atherosclerosis, however, because CRP levels can also be elevated in other acute and chronic inflammatory conditions.

Histamine Histamine is found primarily in the granules of mast cells and basophils, and it is the active molecule that helps initiate the inflammatory response when mast cells degranulate. Histamine's actions bring more leukocytes to the injury site to kill bacteria and remove cellular debris.

TABLE 24.1	Chemicals of the Immune Response

Functional Classes

Acute phase proteins: Liver proteins that act as opsonins and that enhance the inflammatory response

Chemotaxins: Molecules that attract phagocytes to a site of infection

Cytokines: Proteins released by one cell that affect growth or activity of another cell

Opsonins: Proteins that coat pathogens so that phagocytes recognize and ingest them

Pyrogens: Fever-producing substances

Specific Chemicals and Their Functions

Antibodies (immunoglobulins, gamma globulins): Proteins secreted by B lymphocytes that fight specific invaders

Bradykinin: Stimulates pain receptors; vasodilator

Complement: Plasma and cell membrane proteins that act as opsonins, cytolytic agents, and mediators of inflammation

C-reactive protein: Opsonin that activates complement cascade

Granzymes: Cytotoxic enzymes that initiate apoptosis

Heparin: An anticoagulant

Histamine: Vasodilator and bronchoconstrictor released by mast cells and basophils

Interferons (IFN): Cytokines that inhibit viral replication and modulate the immune response

Interleukins (IL): Cytokines secreted by leukocytes to act primarily on other leukocytes; IL-1 mediates inflammatory response and induces fever

Kinins: Plasma proteins that activate to form bradykinin

Lysozyme: An extracellular enzyme that attacks bacteria

Major histocompatibility complex (MHC): Membrane protein complexes involved in cell recognition

Membrane attack complex: A membrane pore protein made in the complement cascade

Perforin: A membrane pore protein that allows granzymes to enter the cell; made by NK and cytotoxic T cells

Superoxide anion (O_2^-): Powerful oxidant in phagocyte lysosomes

T-cell receptors: T lymphocyte receptors that recognize and bind antigen presented by MHC receptors

Tumor necrosis factor (TNF): Cytokines that promote inflammation and can cause cells to self-destruct through apoptosis

Histamine opens pores in capillaries, allowing plasma proteins to escape into the interstitial space. This process causes local edema, or swelling. In addition, histamine dilates blood vessels (vasodilation), increasing blood flow to the area. The result of histamine release is a hot, red, swollen area around a wound or infection site.

Mast cell degranulation is triggered by different cytokines in the immune response. Because mast cells are concentrated under mucous membranes that line the airways and digestive tract, the inhalation or ingestion of certain antigens can trigger histamine release. The resultant edema in the nasal passages leads to one annoying symptom of seasonal pollen allergies: the stuffy nose. Fortunately, pharmacologists have developed a variety of drugs called *antihistamines,* which block the action of histamine at its receptor.

Scientists once believed that histamine was the primary chemical responsible for the bronchoconstriction of asthma and for the severe systemic allergic reaction called *anaphylactic shock* [p. 487]. However, they now know that mast cells release powerful lipid mediators in addition to histamine, including leukotrienes, platelet-activating factor, and prostaglandins. These mediators work with histamine to create the bronchoconstriction and hypotension characteristic of anaphylactic shock.

Interleukins

Interleukins are cytokines initially thought to mediate communication among the body's different types of leukocytes. Scientists have since discovered that many different tissues in the body respond to interleukins. **Interleukin-1 (IL-1)** is secreted by activated macrophages and other immune cells. Its main role is to mediate the inflammatory response, but it also has widespread systemic effects on immune function and metabolism. IL-1 modulates the immune response by:

1. *Altering blood vessel endothelium* to ease passage of white blood cells and proteins during the inflammatory response.
2. *Stimulating production of acute-phase proteins* by the liver.
3. *Inducing fever* by acting on the hypothalamic thermostat [p. 721]. IL-1 is a known pyrogen.
4. *Stimulating cytokine and endocrine secretion* by a variety of other cells.

Bradykinin

Kinins are a group of inactive plasma proteins that participate in a cascade similar to the coagulation cascade [p. 521]. The end product of the kinin cascade is **bradykinin,** a molecule that has the same vasodilator effects as histamine. Bradykinin also stimulates pain receptors, creating the tenderness associated with inflammation. Pain draws the brain's attention to the injury.

Complement Proteins

Complement is a collective term for a group of more than 25 plasma proteins and cell membrane proteins. The *complement cascade* is similar to the blood coagulation cascade. The complement proteins are secreted in inactive forms that are activated as the cascade proceeds. Intermediates of the complement cascade act as opsonins, chemical attractants for leukocytes, and agents that cause mast cell degranulation.

FIG. 24.7   Membrane attack complex creates pores in pathogens

Complement proteins insert themselves into the membrane of a pathogen, creating pores.

Pore of membrane attack complex

H_2O and ions

Water and ions enter the pathogen cell through the pores of the membrane attack complexes.

Cell swells and lyses.

The complement cascade terminates with the formation of **membrane attack complex,** a group of lipid-soluble proteins that insert themselves into the cell membranes of pathogens and virus-infected cells and form giant pores (**FIG. 24.7**). These pores allow water and ions to enter the pathogen cells. As a result, the cells swell and lyse.

CONCEPT CHECK   3. How does the action of histamine on capillary permeability result in swelling?

ACQUIRED IMMUNITY: ANTIGEN-SPECIFIC RESPONSES

Acquired immune responses are *antigen-specific responses* in which the body recognizes a particular foreign substance and selectively reacts to it. Acquired immunity is mediated primarily by lymphocytes.

There are three main types of lymphocytes: B lymphocytes, T lymphocytes, and natural killer (NK) cells. Activated B lymphocytes develop into **plasma cells,** which secrete antibodies. Activated **T lymphocytes** develop either into cells that attack and destroy virus-infected cells (**cytotoxic T** or **TC cells**) or into cells that regulate other immune cells (**helper T** or **TH cells**). NK cells attack and destroy virus-infected cells and tumor cells as part of the innate response discussed previously. NK cells are also cytotoxic cells that release chemicals to destroy antibody-tagged pathogens. All lymphocytes secrete cytokines that act on immune cells, on non-immune cells, and, sometimes, on pathogens.

The process of acquired immunity overlaps with the process of innate immunity. Cytokines released by the inflammatory response attract lymphocytes to the site of an immune reaction.

The lymphocytes release additional cytokines that enhance the inflammatory response.

Acquired immunity can be subdivided into active immunity and passive immunity. **Active immunity** occurs when the body is exposed to a pathogen and produces its own antibodies. Active immunity can occur naturally, when a pathogen invades the body, or artificially, as when we are given vaccinations containing dead or disabled pathogens.

Passive immunity occurs when we acquire antibodies made by another organism. The transfer of antibodies from mother to fetus across the placenta is one example. Injections containing antibodies are another. Travelers going abroad may be injected with *gamma globulin* (antibodies extracted from donated human plasma), but this passive immunity lasts only about three months as the injected proteins degrade and are cleared from the circulation.

Lymphocytes Mediate the Acquired Immune Response

On a microscopic level, all lymphocytes look alike. At the molecular level, however, the cell types can be distinguished from one another by their membrane receptors. Each lymphocyte binds only one particular ligand. All lymphocytes that bind that ligand form a group known as a **clone** {*klon*, a twig} (**FIG. 24.8a**).

If each pathogen that entered the body needs a dedicated type of lymphocyte, there must be millions of different types of lymphocytes ready to combat millions of different pathogens. But how can the body store both the number and variety of lymphocytes needed for adequate defense? As it turns out, the immune system keeps only a few of each type of lymphocyte on hand. If the pathogen these cells fight appears, the cells reproduce to provide the numbers needed.

At an individual's birth, each clone of lymphocytes is represented by only a few cells, called **naïve lymphocytes.** Because the small number of cells in each naïve clone is not enough to fight off foreign invaders, the first exposure to an antigen activates the appropriate clone and stimulates it to divide (Fig. 24.8b). This process, called **clonal expansion,** creates additional cells in the clone. Naïve cells continue to be generated throughout an individual's lifetime.

The newly formed lymphocytes in an expanded clone differentiate into effector cells and memory cells. **Effector cells** carry out the immediate response and then die within a few days. **Memory cells,** in contrast, are long lived and continue reproducing themselves. Second and subsequent exposures to the antigen activate the memory cells and cause rapid clonal expansion, creating a quicker and stronger secondary response to the antigen.

B Lymphocytes Become Plasma Cells and Memory Cells

B lymphocytes (B cells) develop in the bone marrow. (Memory aid: bone starts with B.) Clones of activated B lymphocytes differentiate primarily into specialized cells that secrete antibodies.

RUNNING **PROBLEM**

In 2010, there were nearly 12,000 cases of cervical cancer in the United States, and almost 4000 women died from the disease. The HPV vaccines were developed to prevent HPV infection and decrease the incidence of cervical cancer. The first vaccine, Gardasil®, protects against four different strains of HPV: 16 and 18, associated with cervical cancer, and types 6 and 11, which cause genital warts. A newer vaccine, Cervarix®, protects against only types 16 and 18. The vaccines are composed of viruslike particles (VLPs), which are proteins from the HPV virus capsid. When injected into a person, VLPs initiate an antibody response that has a protective effect against the types of HPV from which they were derived.

Q3: *Antibody production begins with activation of which type of lymphocyte? What type of cell produces antibodies?*

Q4: *In a normal infection, HPV infects cells in the superficial layers of the skin but does not enter the circulation. What kind of immune cell is mostly likely to encounter HPV in the skin?*

754 756 **766** 775 780 782

The word *antibody* describes what the molecules do: work against foreign bodies. Antibodies are also called **immunoglobulins,** and this alternative name describes what the molecules are: globular proteins that participate in the humoral immune response.

Mature B lymphocytes insert antibody molecules into their cell membranes so that the antibodies become surface receptors marking the members of each clone (Fig. 24.8). When a clone of B lymphocytes responds to antigen exposure, some of the effector cells differentiate into plasma cells. Plasma cells do not have antibody proteins bound in their membranes. Instead, they synthesize and secrete additional antibody molecules at incredible rates, estimated to be as high as 2000 molecules per second! Plasma cell antibodies form *humoral immunity,* the soluble antibodies of the plasma.

After each invader has been successfully repulsed, the short-lived plasma cells die off—it could be dangerous to have them continuing to secrete antibody after the antigen is gone. A few memory cells of the clone remain behind, waiting to respond to the next exposure to the same antigen. Figure 24.8 shows B cell responses following the first and second exposures to an antigen.

The **primary immune response** occurs after the initial exposure. Antibody production by plasma cells is slower and lower in magnitude because the body has not encountered the antigen previously. The **secondary immune response,** which takes place after second and subsequent exposures, is quicker and larger because of memory cells that remain behind after the first exposure. Clonal expansion is enhanced by lymphocytes that carry a molecular memory of the first exposure to the antigen, so antibody production begins sooner and reaches higher concentrations.

The existence of a secondary immune response is what allows vaccinations to be an effective protection from disease.

FIG. 24.8 Lymphocyte clones

(a) A clone is a group of lymphocytes that are specific to one antigen.

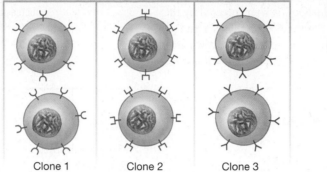

Clone 1   Clone 2   Clone 3

FIGURE QUESTION

Which clone will this antigen activate?

PRIMARY IMMUNE RESPONSE

(b) Exposure to an antigen triggers clonal expansion and the immune response.

(c) B lymphocytes secrete antibodies.

Upon first exposure to an antigen, naive lymphocytes reproduce.

Antigen

Clonal expansion

Effector cells (B lymphocytes) become plasma cells.

Plasma cells   *secrete*   Antibodies

Short-lived effector cells carry out the immediate response.

Clonal expansion leads to immunologic memory.

Memory cells are long lived and continue to reproduce.

PRIMARY IMMUNE RESPONSE

Antibody concentration in plasma

Antibody concentration

First antigen exposure

Time (weeks) ⟶

1   2   3   4

SECONDARY IMMUNE RESPONSE

(d) When memory cells are reexposed to the appropriate antigen, the clone expands more rapidly to create additional effector and memory cells.

Plasma cells   *secrete*   Antibodies

Antibody production in response to the first exposure to an antigen is both slower and weaker than antibody production following subsequent exposures to the same antigen.

SECONDARY IMMUNE RESPONSE

Antibody concentration in plasma

Antibody concentration

Second antigen exposure

Time (weeks) ⟶

1   2   3   4

A vaccine contains an altered pathogen that no longer harms the host but that can be recognized as foreign by immune cells. The altered pathogen triggers creation of memory cells specific to that particular pathogen. If a vaccinated person is later infected by the pathogen, memory cells created by the vaccine produce a stronger and more rapid secondary immune response. We are now learning, however, that immunity after immunization may not last for a person's lifetime.

Antibodies Are Proteins Secreted by Plasma Cells

Antibodies were among the first aspects of the immune system to be discovered, and traditionally their names—agglutinins, precipitins, hemolysins, and so on—indicated what they do. Today, however, antibodies or immunoglobulins (Ig) are divided into five general classes: IgG, IgA, IgE, IgM, and IgD (pronounced *eye-gee-[letter]*). Antibodies are collectively referred to as **gamma globulins.**

IgGs make up 75% of plasma antibody in adults because they are produced in secondary immune responses. Some maternal IgGs cross the placental membrane and give infants immunity in the first few months of life. Some IgGs activate complement.

IgA antibodies are found in external secretions, such as saliva, tears, intestinal and bronchial mucus, and breast milk, where they bind to pathogens and flag them for phagocytosis if they reach the internal environment.

IgEs target gut parasites and are associated with allergic responses. When mast cell receptors bind with IgEs and antigen, the mast cells degranulate and release chemical mediators, such as histamine.

IgM antibodies are associated with primary immune responses and with the antibodies that react to blood group antigens. IgMs strongly activate complement.

IgD antibody proteins appear on the surface of B lymphocytes along with IgM, but the physiological role of IgDs is unclear.

Antibody Proteins The basic antibody molecule has four polypeptide chains linked into a Y shape (**FIG. 24.9a**). The two sides of the Y are identical, with one **light chain** attached to one **heavy chain**. The two arms, or **Fab regions**, form antigen-binding sites that confer the antibody's specificity (Fig. 24.9b).

The stem of the Y-shaped antibody molecule is known as the **Fc region.** The Fc region determines the Ig class to which the antibody belongs. A *hinge region* between the arms and the stem allows flexible positioning of the arms as the antibody binds to the antigen.

In any one antibody molecule, the two light chains are identical and the two heavy chains are identical. However, the chains vary widely among different antibodies, giving the antibody its specificity. Each clone produces a unique antibody. Two classes of immunoglobulins (IgM and IgA) are secreted as polymers: IgM is made up of five Y-shaped antibody molecules, and IgA has from one to four antibody molecules.

CONCEPT CHECK

4. Because antibodies are proteins, they are too large to cross cell membranes on transport proteins or through channels. How then do IgAs and other antibodies become part of external secretions such as saliva, tears, and mucus?

5. A child is stung by a bee for the first time. Why should the parent be particularly alert when the child is stung a second time?

Antibodies Work Outside Cells

Most antibodies are found in the blood, where they make up about 20% of the plasma proteins in a healthy individual. These antibodies are most effective against extracellular pathogens (such as bacteria), some parasites, antigenic macromolecules, and viruses that have not yet invaded their host cells.

Antibody function is summarized in Figure 24.9c. Because antibodies are not toxic, they cannot destroy antigens. Their primary role is to help the immune system react to specific antigens.

In most instances, the antibody binds first to the antigen, forming an antibody-antigen complex, also known as an *immune complex.* Each antibody molecule can bind to two different antigen particles, one on each arm of the antibody (Fig. 24.9c, 1). This creates clumping of antigens, which facilitates their recognition and destruction by the immune system.

In some instances, antibody binding inactivates toxins produced by bacteria. One example where antibodies neutralize a bacterial toxin is infection by *Corynebacterium diphtheria. C. diphtheria* is the bacterium that causes diphtheria, an upper respiratory infection. In this disease, the bacterial toxin kills host cells, leaving ulcers that have a characteristic grayish membrane. Natural immunity to diptheria occurs when the host produces antibodies that disable the toxin.

To develop a vaccine for diphtheria, researchers created an inactivated toxin that did not harm living cells. When administered to a person, the vaccine triggers antibody production without causing any symptoms of the disease. If the vaccinated person is then exposed to *C. diphtheria*, the antibodies are present and ready to neutralize the toxin. As a result, diphtheria has been almost eliminated in countries with good immunization programs.

Fc Receptors The subsequent steps in immune complex-mediated pathogen destruction require that the Fc "stem" regions of the antibody-antigen complex to bind to **Fc receptors** on an immune cell. Activating Fc receptors on neutrophils and macrophages initiates phagocytosis (Fig. 24.9c, 2). In this way, antibodies act as *opsonins*, tagging the immune complex for destruction.

If the Fc receptors are on basophils, mast cells, NK cells, or eosinophils, antibody binding to the receptor triggers degranulation, the release of chemicals stored in the immune cell (Fig. 24.9c, 3). The cellular signaling pathway for degranulation is similar to the release of vesicle contents in endocrine cells, neurons, and other cells: Receptor binding opens Ca^{2+} channels, and Ca^{2+} entry is the signal for exocytosis.

FIG. 24.9 | Antibodies

(a) Antibody Structure

An antibody molecule is composed of two identical light chains and two identical heavy chains, linked by disulfide bonds.

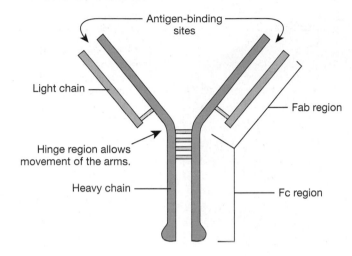

Antigen-binding sites

Light chain

Fab region

Hinge region allows movement of the arms.

Heavy chain

Fc region

(b) Antigen Binding

Antibodies have antigen-binding sites on the Fab regions.

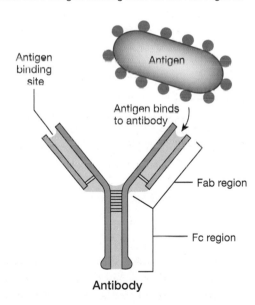

Antigen binding site

Antigen

Antigen binds to antibody

Fab region

Fc region

Antibody

(c) Antibody Functions

1. **Cause antigen clumping and inactivation of bacterial toxins**

Bacterial toxins

2. **Act as opsonins to tag antigens for phagocytosis by neutrophils and macrophages**

Fc receptor

Enhanced phagocytosis

3. **Trigger degranulation**

Antibody-dependent cell-mediated cytotoxicity

NK cell or eosinophil

Mast cell or basophil

4. **Activate complement**

Complement

5. **Activate B lymphocytes**

Plasma cells → Secrete antibodies

Memory cells

If the immune cell that binds the antibody is a cytotoxic eosinophil or NK cell, the chemicals released by the cell destroy the antibody-bound antigen. This nonspecific response of cytotoxic cells to antibody-antigen binding is called **antibody-dependent cell-mediated cytotoxicity.**

The Fc receptors found on mast cells are specific for the Fc region of the IgE antibody family. When IgE antibody-antigen complexes bind to mast cell Fc receptors, the cells degranulate, releasing chemicals that mediate the inflammatory response.

The final immune process activated by the Fc ends of immune complexes is the activation of complement proteins (Fig. 24.9c, 4). Complement proteins in turn trigger mast cell degranulation by binding to a different set of Fc receptors. Complement proteins also serve as opsonins. The ultimate product of the complement cascade is the production

FIG. 24.10 FOCUS ON...

The Thymus Gland

The thymus gland is a two-lobed organ located in the thorax just above the heart.

The thymus gland reaches its greatest size during adolescence. Then it shrinks and is largely replaced by adipose tissue as a person ages.

During development in the thymus, those cells that would be self-reactive are eliminated. Those that do not react with "self" tissues multiply to form clones.

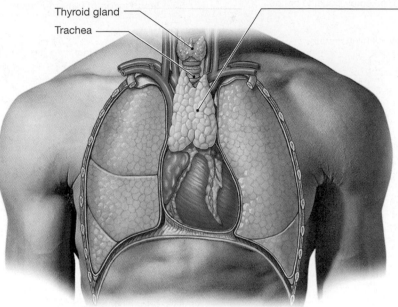

Thyroid gland
Trachea

Thymus

The thymus gland produces:

- T lymphocytes
- Peptides
  thymosin
  thymopoietin
  thymulin

Q FIGURE QUESTION

New T lymphocyte production in the thymus is low in adults, but the number of T lymphocytes in the blood does not decrease. What conclusion(s) about T lymphocytes can you draw from this information?

---

of *membrane attack complex,* which kills pathogens by creating membrane pores (see p. 765).

Non-Fc-Mediated Actions The final role of antibody-antigen binding is to activate B lymphocytes. The surface of every B lymphocyte is covered with as many as 100,000 antibody molecules whose Fc ends are inserted into the lymphocyte membrane (Fig. 24.9c, 5 and Fig. 24.8). This arrangement leaves the Fab regions of the membrane-bound antibodies available to bind free-floating extracellular antigens or bits of antigen on the surface of antigen-presenting cells (Fig. 24.6c). Once antigen binds to the cell's antibodies, B cells are activated and differentiate into plasma cells that secrete more antibodies. Some B cells differentiate into memory cells to await a subsequent invasion.

T Lymphocytes Use Contact-Dependent Signaling

Antibodies are typically effective only against extracellular pathogens because antibodies can bind only to soluble or exposed antigens. Once a pathogen gets inside a host cell, it can no longer be "seen" by the humoral immune system. Defending the body against intracellular pathogens is the role of cytotoxic T lymphocytes, which carry out **cell-mediated immunity.** In this process, cytotoxic T cells bind to cells that display foreign antigen fragments as part of a *major histocompatibility complex* (MHC) on their surface.

T lymphocytes develop in the thymus gland (**FIG. 24.10**). Immature precursor cells migrate from the bone marrow to the thymus. During the development phase, cells that do not react with "self" form clones and insert **T-cell receptors** into their cell membranes (**FIG. 24.11a**). T-cell receptors are not antibodies like the receptors on B lymphocytes, although the proteins are closely related. That means that T cells cannot bind to free-floating antigens as B cells do. Instead, T cell receptors bind only to MHC-antigen complexes on the surface of an antigen-presenting cell.

MHC and Antigen What is MHC? The **major histocompatibility complexes** are a family of membrane protein complexes encoded by a specific set of genes. (These proteins were named when they were discovered to play a role in rejecting foreign tissue following organ or tissue transplants.) We know now that every nucleated cell of the body has MHC proteins on its membrane.

MHC proteins combine with fragments of antigens that have been digested within the cell. The MHC-antigen complex is then inserted into the cell membrane so that the antigen is visible on the extracellular surface (see Fig. 24.6c for an example). Free antigen in the extracellular fluid cannot bind to unoccupied MHC receptors on the cell surface.

FIG. 24.11 T lymphocytes

(a) T Lymphocyte Development

During embryonic development, T lymphocytes insert their T-cell receptors into the membrane.

Multipotent stem cell in bone marrow → T-cell precursor *migrates to* Thymus gland → Cytotoxic T cells → *kill* → MHC Class I target cells

Helper T cells — Bind to MHC-II antigen-presenting cells — *secrete* → Cytokines that activate other immune cells

(b) T Lymphocyte Activation

When T-cell receptors bind to antigen presented on MHC receptors.

MHC-antigen complex

MHC receptor

T-cell receptor

T lymphocyte

1 Cell binds to T lymphocyte.

2 Signal transduction activates T lymphocyte.

Q FIGURE QUESTION
What kind of T cell is shown if the MHC receptor is MHC-I? MHC-II?

There are two types of MHC molecules. **MHC class I molecules** are found on all nucleated human cells. When viruses or bacteria invade the cell, they are digested into peptide fragments and loaded onto MHC-I "platforms." If a cytotoxic T cell (T_C cell) encounters a human host cell with foreign antigen fragment on its MHC-I, the T_C cell recognizes the host cell as a pathogen-infected cell and kills it to prevent it from reproducing (Fig. 24.11a).

MHC class II molecules are found primarily on the antigen-presenting cells (APCs): macrophages, B lymphocytes, and dendritic cells. When an immune cell engulfs and digests an antigen, the fragments are returned to the immune cell membrane combined with MHC-II proteins. If a helper T cell (T_H cell) encounters an APC with a foreign antigen fragment on its MHC-II, the T_H cell responds by secreting cytokines that enhance the immune response.

All MHC proteins are related, but they vary from person to person because of the huge number of alleles (variants of a gene)

people inherit from their parents. There are so many alleles that it is unlikely that any two people other than identical twins inherit exactly the same set. Major histocompatibility complexes are one reason tissues cannot be transplanted from one person to another without first establishing compatibility.

Cytotoxic T Cells As previously described, cytotoxic T (T_C) cells attack and destroy cells that display MHC-I-antigen complexes. Although this destruction may seem to be an extreme response, it prevents the reproduction of intracellular invaders such as viruses, some parasites, and some bacteria.

How do cytotoxic T cells kill their targets? They do so in two ways. First they can release a cytotoxic pore-forming molecule called **perforin** along with **granzymes,** enzymes related to the digestive enzymes trypsin and chymotrypsin. When granzymes enter the target cell through perforin channels, they activate an enzyme cascade that induces the cell to commit suicide (*apoptosis*). Second, cytotoxic T cells instruct target cells to undergo apoptosis by activating **Fas,** a "death receptor" protein on the target cell membrane that is linked to the enzyme cascade.

Helper T (T$_H$) Cells Helper T (T$_H$) cells do not directly attack pathogens and infected cells, but they play an essential role in the immune response by secreting cytokines that activate other immune cells. The cytokines secreted by T$_H$ cells include (1) *interferon-gamma* (IFN-γ), which activates macrophages; (2) *interleukins* that activate antibody production and cytotoxic T lymphocytes; (3) *colony-stimulating factors,* which enhance leukocyte production; and (4) *interleukins* that support the actions of mast cells and eosinophils. Helper T cells also bind to B cells and promote their differentiation into plasma cells and memory B cells. HIV, the virus that causes AIDS, preferentially infects and destroys T$_H$ cells, leaving the host unable to respond to pathogens that otherwise could be easily suppressed.

IMMUNE RESPONSE PATHWAYS

How does the body respond to different kinds of immune challenges? The details depend on the particular challenge, but the basic pattern is the same. The innate response starts first, and it is reinforced by the more specific acquired response. The two pathways are interconnected, so cooperation and communication are essential. In the following sections we examine the body's responses in four situations: an extracellular bacterial infection, a viral infection, an allergic response to pollen, and the transfusion of incompatible blood.

Bacterial Invasion Causes Inflammation

What happens when bacteria invade? As we've seen, if the passive barricades of the skin and mucous membranes fail, bacteria reach the extracellular fluid. There they usually cause an inflammatory response that represents the combined effects of many cells working to get rid of the invader. If bacteria enter the lymph, infection-fighting takes place in lymph nodes as well.

Inflammation is characterized by a red, swollen warm area that is tender or painful. In addition to the nonspecific inflammatory response, lymphocytes attracted to the area produce antibodies keyed to the specific type of bacterium. The entry of bacteria sets off several sets of interrelated reactions (**FIG. 24.12**):

1. *Activity of the complement system.* Components of the bacterial cell wall are antigens that activate the complement system. Some products of the complement cascade are chemical signals (*chemotaxins*) that attract leukocytes from the circulation to help fight the infection. Others act as opsonins to enhance phagocytosis.

   Complement also causes degranulation of mast cells and basophils. Cytokines secreted by mast cells act as additional chemotaxins, attracting more immune cells. Vasoactive chemicals such as histamine dilate blood vessels and increase capillary permeability. The enhanced blood supply to the site creates the redness and warmth of inflammation. Plasma proteins that escape into the interstitial space pull water with them, leading to tissue edema (swelling).

   The complement cascade ends with formation of membrane attack complex molecules that insert themselves into the bacterial wall of unencapsulated bacteria. The subsequent influx of ions and water lyses the bacteria, aided by the enzyme lysozyme. This is a purely chemical process that does not involve immune cells.

2. *Activity of phagocytes.* If the bacteria are not encapsulated, macrophages can begin to ingest the bacteria immediately. But if the bacteria are encapsulated, the capsule hides the bacteria from recognition by the macrophage receptors. Opsonins such as antibodies must coat the capsule before the bacteria can be identified and ingested by phagocytes. Opsonins enhance phagocytosis of bacteria that are not encapsulated. Molecules that act as opsonins include antibodies, complement, and acute-phase proteins.

3. *Role of the acquired immune response.* Some elements of the acquired immune response are called into play in bacterial infections. If antibodies against the bacteria are already present, they enhance the innate response by acting as opsonins and neutralizing bacterial toxins. Antigen-presenting cells that digest the bacteria can then move to secondary lymphoid tissues, where they present antigen to memory cells to initiate more antibody production.

   If the infection is a novel one, some of the bacterial antigens activate naïve B cells with the assistance of APC and helper T cells. Antigen-presenting cells ingest the bacteria and present bacterial fragments to helper T cells to activate them. This triggers cytokine secretion from the T$_H$ cells, B cell clonal expansion, antibody production by plasma cells, and formation of memory B and T$_H$ cells.

4. *Initiation of repair.* If the initial wound damaged blood vessels underlying the skin, platelets and the proteins of the coagulation cascade are also recruited to minimize the damage [p. 521]. Once the bacteria are removed by the immune response, the injured site is repaired under the control of growth factors and other cytokines.

Viral Infections Require Intracellular Defense

What happens when viruses invade the body? First, they encounter an extracellular phase of immune response similar to that described for bacteria. In the early stages of a viral infection, innate immune responses and antibodies can help control the invasion of the virus.

Once the viruses enter body cells, humoral immunity in the form of antibodies is no longer effective. Cytotoxic T lymphocytes (and to a lesser extent, NK cells) are the main defense against intracellular viruses. When these cells recognize infected host cells, they destroy them.

FIG. 24.12 **ESSENTIALS**

Immune Responses to Extracellular Bacteria

Bacterial infections cause inflammation
and trigger specific immune responses.

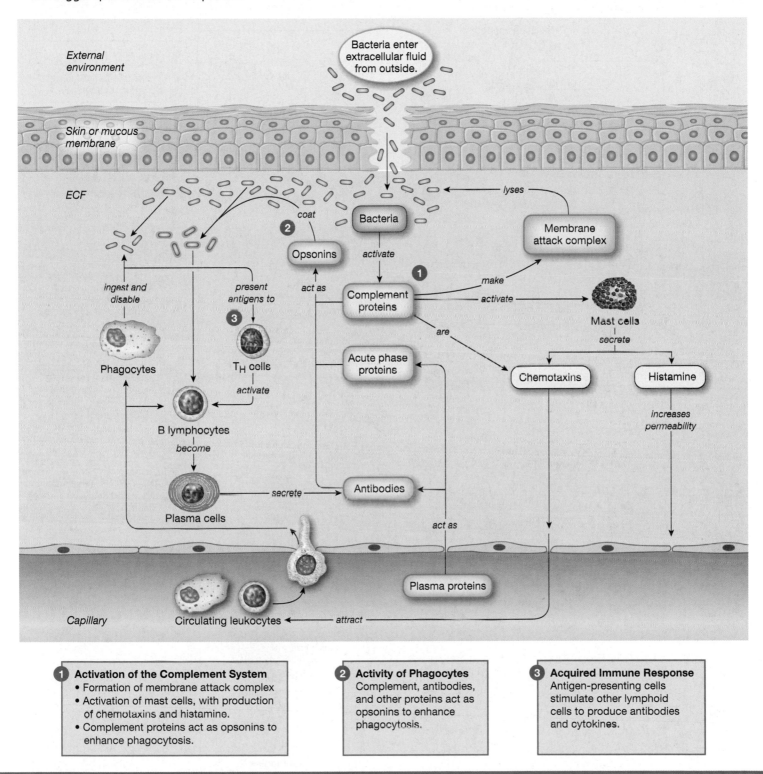

1 **Activation of the Complement System**
- Formation of membrane attack complex
- Activation of mast cells, with production
  of chemotaxins and histamine.
- Complement proteins act as opsonins to
  enhance phagocytosis.

2 **Activity of Phagocytes**
Complement, antibodies,
and other proteins act as
opsonins to enhance
phagocytosis.

3 **Acquired Immune Response**
Antigen-presenting cells
stimulate other lymphoid
cells to produce antibodies
and cytokines.

For years, T cell–mediated immunity and humoral immunity
controlled by B lymphocytes were considered independent processes. We now know that the two types of immunity are linked.
FIGURE 24.13 depicts how these two types of lymphocytes coordinate
to destroy viruses and virus-infected cells. In this figure, we have assumed prior exposure to the virus and the presence of preexisting
antibodies in the circulation. Antibodies can play an important de-
fensive role in the early extracellular stages of a viral infection.

FIG. 24.13 **ESSENTIALS**

Immune Responses to Viruses

This figure assumes prior exposure to the virus and preexisting antibodies.

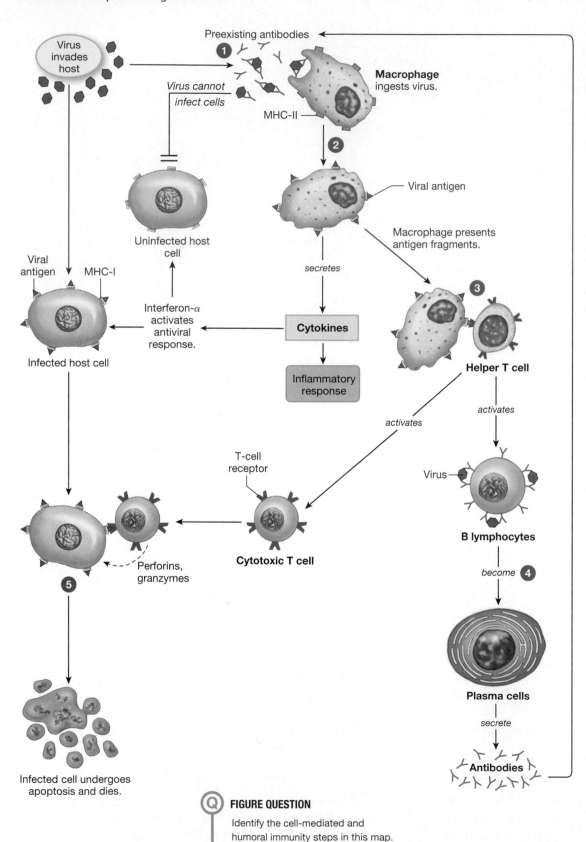

Virus invades host

Preexisting antibodies ①

Macrophage ingests virus.

MHC-II

Virus cannot infect cells

Uninfected host cell

②

Viral antigen

Macrophage presents antigen fragments.

secretes

Viral antigen   MHC-I

Interferon-α activates antiviral response.

Cytokines

Infected host cell

Inflammatory response

③

Helper T cell

activates

activates

activates

T-cell receptor

Virus

B lymphocytes

Cytotoxic T cell

Perforins, granzymes

⑤

become ④

Infected cell undergoes apoptosis and dies.

Plasma cells

secrete

Antibodies

① Antibodies act as opsonins, coating viral particles to make them better targets for macrophages.

② Macrophages ingest virus and insert viral antigens into MHC-II molecules on their membranes. Activated macrophages also secrete cytokines.

③ Helper T cells bind to viral antigen on macrophages and become activated. These activated T_H cells stimulate B lymphocytes and cytotoxic cells.

④ The activated memory B lymphocytes become plasma cells, resulting in additional antibody production.

⑤ Activated cytotoxic T cells attack and destroy infected host cells.

Ⓠ FIGURE QUESTION

Identify the cell-mediated and humoral immunity steps in this map.

1. Antibodies act as opsonins, coating viral particles to make them better targets for antigen-presenting cells such as macrophages. Antibodies can also bind to virus particles, preventing them from entering their target cells. However, once the virus is inside host cells, antibodies are no longer as effective.

2. Macrophages that ingest viruses insert fragments of viral antigen into MHC-II molecules on their membranes. Macrophages also secrete a variety of cytokines. Some of these cytokines initiate the inflammatory response. They produce interferon-α, which causes host cells to make antiviral proteins that keep viruses from replicating. Other macrophage cytokines stimulate NK cells and helper T cells.

3. Helper T cells bind to viral antigen on macrophage MHC-II molecules. Activated T_H cells then secrete cytokines to stimulate B lymphocytes and cytotoxic cells.

4. Previous exposure to the virus created memory B lymphocytes with the viral antibody on their surface. This second exposure to the virus activates the memory cells and promotes development of plasma cells, resulting in additional antibody production.

5. Cytotoxic T cells use viral antigen–MHC-I complexes to recognize infected host cells and kill them. When T_C cells recognize the infected host cell from MHC-I receptors with antigen, they secrete the contents of their granules onto the cell surface. Perforin molecules insert pores into the host cell membrane so that granzymes can enter the cell, inducing it to commit suicide and undergo apoptosis. Destruction of infected host cells is a key step in halting the replication of invading viruses.

NK cells recognize virally infected cells by a different process. Some viruses cause their host cells to withdraw MHC-I receptors from the cell surface in an effort to hide from the immune system. NK cells recognize infected host cells lacking MHC-I complexes and kill them by a process similar to the one described for T_C cells.

Antibodies and Viruses Figure 24.13 shows antibodies from a previous infection protecting the body, but there is no guarantee that antibodies produced during one infection will be effective against the next invasion by the same virus. Why not? The reason is that many viruses mutate constantly, and the protein coat forming the primary antigen may change significantly over time. If the protein coat changes, the antibody may no longer recognize it.

The influenza virus is one virus that changes yearly. Consequently, annual vaccines against influenza must be developed based on virologists' predictions of what mutations will occur. If the predictions do not match the mutations or the prevalent viral strains, getting a flu shot that year cannot keep someone from catching the flu.

The rapid mutation of viruses is also one reason that researchers have not yet developed an effective vaccine against HIV, the virus that causes **acquired immunodeficiency syndrome (AIDS).** HIV infects cells of the immune system, particularly T lymphocytes, monocytes, and macrophages. When HIV wipes out the helper T cells, cell-mediated immunity against the virus is lost. The general loss of an immune response in AIDS leaves these patients susceptible to a variety of viral, bacterial, fungal, and parasitic infections.

Specific Antigens Trigger Allergic Responses

An **allergy** is an inflammatory immune response to a nonpathogenic antigen. The **allergen** is an antigen that is typically not harmful to the body. But if an individual is sensitive to the antigen, the body produces an inappropriate immune response, as if the antigen were a more threatening pathogen such as a parasitic worm. Allergic inflammatory responses can range from mild tissue damage to fatal reactions.

The immune response in allergies is called **sensitivity** or **hypersensitivity** to the antigen. **Immediate hypersensitivity reactions** are mediated by antibodies and occur within minutes of exposure to antigens, which are called **allergens. Delayed hypersensitivity reactions** are mediated by helper T cells and macrophages and may take several days to develop.

Allergens can be practically any exogenous molecule: naturally occurring or synthetic, organic or inorganic. Certain foods, insect venoms, and pollen all trigger immediate hypersensitivity reactions. Allergens can be ingested, inhaled, injected, or simply come in contact with the skin.

Delayed hypersensitivity reactions include contact allergies to copper and other base metals. These are common among people who wear costume jewelry. Poison ivy and poison oak also cause common contact allergies.

Allergies have a strong genetic component, so if parents have a ragweed allergy, chances are good that their children will too. The development of allergies requires exposure to the allergen, a factor that is affected by geographical, cultural, and social conditions.

RUNNING **PROBLEM**

Rebecca cares greatly about the health and well-being of her daughter but is not convinced that her daughter needs the Gardasil immunizations—at least not yet. "Isn't HPV a sexually transmitted disease?" Rebecca asked Dr. Paul. "Why can't Lizzie wait until she is older to get the shots?" Dr. Paul explained that although HPV transmission is usually associated with sexual intercourse, it can be spread by skin-to-skin or hand-to-genital contact. Because the vaccine is a preventive measure rather than a treatment, it is effective only when given before a person has been exposed to HPV.

Q5: *Why are the vaccinations ineffective if the person receiving them already has HPV of the type in the vaccine?*

754 — 756 — 766 — **775** — 780 — 782

FIG. 24.14   Allergic responses

FIRST EXPOSURE

Allergen

MCH-II

1 Allergen ingested and processed by antigen-presenting cell.

2 Antigen-presenting cell activates helper T cell.

3 Activated helper T cell in turn activates B lymphocyte

Helper T cell

activates

B lymphocyte

4 Activated B lymphocytes become plasma cells and memory cells.

becomes

Plasma cell

secretes

Antibodies

5 Memory B and T cells retain memory of exposure to allergen.

6 Upon reexposure to allergen, B and T cells activate more quickly (Fig. 24.8). The body reacts strongly with the release of histamine, cytokines, and other mediators causing allergic symptoms.

REEXPOSURE

Allergen

IgE

IgG

Mast cell

Degranulation

Activated T cells

Activation of complement proteins

Cytokines

Cytokines, histamine, etc

Vasodilation
Bronchoconstriction
↑Vascular permeability

Inflammation

FIGURE 24.14 shows what happens during an immediate hypersensitivity reaction to pollen. The initial step—first exposure or the sensitization phase—is equivalent to the primary immune response discussed previously: the allergen is ingested and processed by an antigen-presenting cell such as a macrophage, which in turn activates a helper T cell (see the right side of Fig. 24.13).

The T_H cell activates B lymphocytes that have bound the allergen. This results in plasma cell production of antibodies (IgE and IgG) to the allergen. The IgE antibodies are immediately bound by their Fc ends to the surface of mast cells and basophils. Memory T and memory B cells store the record of the initial allergen exposure.

Upon re-exposure, equivalent to the secondary immune response shown in Figure 24.8d, the body reacts very strongly and rapidly. The allergen binds to IgE already present on mast cells, triggering the immediate release of histamine, cytokines, and other mediators that cause allergic symptoms (Fig. 24.14).

The type of allergic reaction depends on the antigen antibody, immune cell involved, and location of antigen introduction. Allergen binding to IgE is the most common response to inhaled, ingested, or injected allergens. When allergens bind to IgE antibodies on mast cells, the cells degranulate, releasing histamine and other cytokines. The result is an inflammatory reaction.

The severity of the reaction varies, ranging from localized reactions near the site of allergen entry to systemic reactions such as total body rashes. The most severe IgE-mediated allergic reaction is called **anaphylaxis** {*ana-,* without + *phylax,* guard}. In an anaphylactic reaction, massive release of histamine and other cytokines causes widespread vasodilation, circulatory collapse, and severe bronchoconstriction. Unless treated promptly with epinephrine, anaphylaxis can result in death within as little as 20 minutes.

MHC Proteins Allow Recognition of Foreign Tissue

As surgeons have developed techniques to transplant organs between human beings or from animals to humans, physicians have had to wrestle with the problems of rejection of the host by the donor tissue (known as *graft versus host*) or donor tissue rejection by the recipient's immune system (*host versus graft*). The ubiquitous MHC proteins are the primary tissue antigens that determine whether donated tissue is recognized as foreign by the recipient's immune system.

The MHC proteins are also known as **human leukocyte antigens (HLA)** and are classified according to an international HLA system. A tissue graft or transplanted organ is more likely to be successful if donor and recipient share HLA antigens. Incompatible matches trigger antibody production and activate cytotoxic T cells and T_H cells. Generally it is T cells that are responsible for acute rejection of solid tissue grafts.

One of the most common examples of tissue donation is blood transfusion. Human red blood cell (RBC) membranes contain antigenic proteins and glycoproteins, but RBCs lack the MHC protein markers for recognizing foreign tissue that are found on nucleated cells. In the absence of MHC proteins, two surface proteins—the ABO blood group antigens and the Rh antigens—become the most important causes of a rejection reaction after a blood transfusion. (Rh stands for Rhesus, the type of monkey in which the antigen was first discovered.)

ABO Blood Groups The ABO blood groups consist of four blood types created by combinations of two different glycoprotein antigens (designated A and B) found on the membrane of red blood cells (**FIG. 24.15a**). Each person inherits two alleles (one from each parent) of three possible ABO alleles: A, B, and O (neither A nor B antigen is produced). Because alleles A and B are dominant to allele O, the various combinations of two alleles produce four blood types: A, B, AB, and O (**TBL. 24.2**).

Problems with blood transfusions arise because plasma normally contains antibodies to the ABO group antigens. These antibodies are thought to be produced early in life in response to bacterial antigens or food antigens in the gut. The antibodies can be measured in the blood of infants as early as age 3–6 months.

TABLE 24.2	ABO Blood Group Frequencies in the United States			
	U.S. Population (%)			
Blood Group	**White**	**Black**	**Asian**	**Native American**
O	44	49	43	79
A	43	27	27	16
B	9	20	25	4
AB	4	4	5	<1

People express antibodies to the red blood cell antigen(s) that they do not possess. For this reason, people with blood type A have anti-B antibodies in their plasma, people with blood type B have anti-A antibodies in their plasma, people with no antigens on their red blood cells (blood type O) have both anti-A and anti-B antibodies, and people with both antigens on their red blood cells (blood type AB) have no antibodies to A or B antigens.

FIG. 24.15 ABO blood groups

(a) Characteristics of the ABO Blood Groups

Blood Type	Antigen on Red Blood Cell	Antibodies in Plasma
O	No A or B antigens	"Anti-A" and "anti-B"
A	A antigens	"Anti-B"
B	B antigens	"Anti-A"
AB	A and B antigens	None to A or B

(b) A Mixture of Type O and Type A Blood

When red blood cells with group A antigens on their membranes are mixed with plasma containing antibodies to group A, the antibodies cause the blood cells to clump, or agglutinate.

Q **FIGURE QUESTION**

Each person inherits one allele for ABO blood groups from each parent. A and B are dominant to O but equal if they occur together (blood type AB). Fill in the table showing combinations of inherited alleles. In the shaded blocks, show the blood type that would be expressed.

			Mother		
			A	**B**	**O**
Father	**A**		A A		
			A		
	B				
	O				

How does the body respond to a transfusion of incompatible blood? If a person with blood type O is mistakenly given a transfusion of type A blood, for example, an immune reaction takes place (Fig. 24.15b). The anti-A antibodies of the type O recipient bind to the transfused type A red blood cells, causing them to clump (*agglutinate*). This reaction is easily seen in a blood sample and forms the basis for the blood-typing tests often performed in student laboratories.

Antibody binding also activates the complement system, resulting in production of membrane attack complexes that cause the transfused cells to swell and leak hemoglobin. Free hemoglobin released into the plasma can result in acute renal failure as the kidneys try to filter the large molecules from the blood. Matching donor and recipient blood types is critical prior to giving a blood transfusion.

Rh Blood Groups The Rh blood groups are best known for their role in maternal-fetal interactions. There are at least 49 different Rh group antigens that can be expressed on the surface of red blood cells, but the D antigen is the one usually being referred to when people describe themselves as "Rh-positive" or "Rh-negative." Someone who lacks the Rh D antigen (i.e., is Rh-negative) and is exposed to the D antigen on foreign red blood cells will make antibodies to Rh D antigen.

Production of anti-D antibodies may occur during pregnancy if an Rh-negative mother is pregnant with an Rh-positive fetus. Leakage of the placental barrier can allow fetal blood into the mother's circulation, and the mother's immune system will react to the foreign D antigen. Exposure can also occur during delivery of an Rh-positive child.

Antibodies are transported across the placenta from mother to fetus so that a newborn will have some immunity at birth. If a mother has anti-D antibodies, they will be transferred to the fetus. If the fetus is Rh D-positive, the anti-D antibodies will attack the fetus's red blood cells and destroy them. This destruction causes *hemolytic disease of the newborn* (HDN), also known as *erythroblastosis fetalis*. HDN can also occur if there are other blood type incompatibilies, such as an ABO mismatch.

Rh D incompatibility can be fatal, but treatments have been developed to help prevent HDN. These treatments involve the use of manufactured antibodies that bind to fetal D antigen or to the mother's anti-D antibodies.

CONCEPT CHECK

6. A person with AB blood type is transfused with type O red blood cells. What happens and why?

7. A person with O blood type is transfused with type A blood. What happens? Why?

The Immune System Must Recognize "Self"

One amazing property of immunity is the ability of the body to distinguish its own cells from foreign cells. The lack of immune response by lymphocytes to cells of the body is known as **self-tolerance** and appears to be due to the elimination of self-reactive lymphocytes in a process known as *clonal deletion*. During development, some lymphocyte clones develop with antibodies that can combine with MHC-self-antigen complexes. The primary lymphoid tissues contain self-antigens that can combine with these self-reactive lymphocytes and eliminate them by inducing apoptosis.

When self-tolerance fails, the body makes antibodies against its own components through T cell–activated B lymphocytes. The body's attack on its own cells leads to **autoimmune diseases** {*auto-*, self}. The antibodies produced in autoimmune diseases are specific against a particular antigen and are usually restricted to a particular organ or tissue type. **TABLE 24.3** lists some common autoimmune diseases of humans.

Why does self-tolerance suddenly fail? We still do not know. We have learned, however, that autoimmune diseases often begin in association with an infection. One potential trigger for autoimmune diseases is foreign antigens that are similar to human antigens. When the body makes antibodies to the foreign antigen, those antibodies have enough cross-reactivity with human tissues to do some damage.

One example of an autoimmune disease is type 1 diabetes mellitus, in which the body makes *islet cell antibodies* that destroy pancreatic beta cells but leave other endocrine cells untouched [p. 709]. Another autoimmune condition is the hyperthyroidism of Graves' disease [p. 739]. The body makes *thyroid-stimulating immunoglobulins* that mimic thyroid-stimulating hormone and cause the thyroid gland to oversecrete hormone.

Severe autoimmune diseases are sometimes treated by administering glucocorticosteroids, such as cortisol and its derivatives. Glucocorticoids depress immune system function by suppressing bone marrow production of leukocytes and

TABLE 24.3	**Some Common Autoimmune Diseases in Humans**
Disease	**Antibodies Produced Against**
Graves' disease (hyperthyroidism)	TSH receptor on thyroid cells
Insulin-dependent diabetes mellitus	Pancreatic beta cell antigens
Multiple sclerosis	Myelin of CNS neurons
Myasthenia gravis	Acetylcholine receptor of motor endplate
Rheumatoid arthritis	Collagen
Systemic lupus erythematosus	Intracellular nucleic acid protein complexes (antinuclear antibodies)
Guillain-Barré syndrome (acute inflammatory demyelinating polyneuropathy)	Myelin of peripheral nerves

decreasing the activity of circulating immune cells. Although helpful in suppressing symptoms of the autoimmune disease, large doses of steroids may trigger the same signs and symptoms as hormone oversecretion, causing patients to develop the central obesity and moon face characteristic of Cushing's syndrome [p. 734].

Immune Surveillance Removes Abnormal Cells

What is the role of the human immune system in protecting the body against cancer? This question holds considerable interest for scientists. One theory, known as **immune surveillance,** proposes that cancerous cells develop on a regular basis, but they are detected and destroyed before they can spread. Some evidence supports that theory, but it does not explain why so many cancerous tumors develop every year, or why people with depressed immune function do not develop cancerous tumors in all types of tissues. However, immune surveillance does appear to recognize and control some virus-associated tumors. In addition, some types of endogenous tumors lack surface antigens, which allow NK cells to recognize these cells as abnormal and destroy them. One active area of cancer research is investigating ways to activate the immune system to fight cancerous tumor cells.

BIOTECHNOLOGY

Engineered Antibodies

One goal of cancer treatments is to destroy cancerous cells while minimizing damage to normal body cells. One of the newest cancer treatments is based on the genetic engineering of antibodies against tumor surface proteins, such as the growth factor receptor HER2 that occurs in high density on some breast cancer cells. The engineered antibody then binds to its antigen on the tumor cell. In the case of the HER2 receptor, the anti-HER2 antibody blocks the action of a growth factor that the tumor cell needs to reproduce. Other antibodies have been engineered to carry chemotherapy drugs or radioactive molecules that specifically target cancer cells.

One strategy for developing these antibody treatments involves injecting the antigen into mice, which then produce antibodies against the antigen. Murine (mouse) antibodies injected into humans normally would trigger an immune response against the mouse protein, so scientists have used genetic engineering to create "humanized antibodies." In this procedure, the antigen-binding portion of the mouse Fab region is integrated into the Fab region of a human antibody, resulting in a protein that is less likely to be rejected in the human body.

NEURO-ENDOCRINE-IMMUNE INTERACTIONS

One of the most fascinating and rapidly developing areas in physiology involves the link between the mind and the body. Although serious scientists scoffed at the topic for many years, the interaction of emotions and somatic illnesses has been described for centuries. The Old Testament says, "A merry heart doeth good like a medicine, but a broken spirit drieth the bones" (Proverbs 17:22, King James Version). Most societies have tales of people who lost the will to live and subsequently died without obvious illness, or of people given up for dead who made remarkable recoveries.

Today, the medical field is beginning to recognize the reality of the mind-body connection. Psychosomatic illnesses and the placebo effect have been accepted for many years. Why should we not investigate the possibility that cancer patients can enhance their immune function by visualizing their immune cells gobbling up the abnormal cancer cells?

The study of brain-immune interactions is now a recognized field known as **neuroimmunomodulation,** or *psychoimmunology.* Currently the field is still developing, and the results of many studies raise more questions than they answer. In one experiment, for example, mice were repeatedly injected with a chemical that suppressed the activity of their lymphocytes. During each injection, they were also exposed to the odor of camphor, a chemical that does not affect the immune system. After a period of conditioning, the mice were exposed only to the camphor odor. When the researchers checked the mice's lymphocyte function, they found that the odor of camphor had suppressed the activity of the immune cells, just as the chemical suppressant had done previously. The pathways through which this conditioning occurred are still largely unexplained.

What do we know at this point about the relationship between the immune, nervous, and endocrine systems?

1. **The three systems share common signal molecules and common receptors for those molecules.** Chemicals once thought to be exclusive to a single cell or tissue are being discovered all over the body. Immune cells secrete hormones, and leukocyte cytokines are produced by nonimmune cells. For example, lymphocytes secrete thyrotropin (TSH), ACTH, growth hormone, and prolactin, as well as hypothalamic corticotropin-releasing hormone (CRH).

   Receptors for hormones, neurotransmitters, and cytokines are being discovered everywhere. Neurons in the brain have receptors for cytokines produced by immune cells. Natural killer cells have opiate receptors and beta adrenergic receptors. Everywhere we turn, the nervous, endocrine, and immune systems seem to share chemical signal molecules and their receptors.

2. **Hormones and neuropeptides can alter the function of immune cells.** It has been known for years that increased cortisol levels due to stress were associated with decreased antibody production, depressed lymphocyte proliferation, and diminished activity of natural killer cells. Substance P,

25

Integrative Physiology III: Exercise

One fascinating aspect of the physiology of the human being at work is that it provides basic information about the nature and the range of the functional capacity of different organ systems.

Per-Olof Åstrand and Kaare Rodahl, *Textbook of Work Physiology*, 1977

METABOLISM AND EXERCISE 787

LO 25.1   Identify sources of ATP used by exercising muscle during aerobic and anaerobic exercise and during low-intensity and high-intensity exercise.

LO 25.2   Compare and contrast metabolic pathways for ATP production during aerobic and anaerobic exercise.

LO 25.3   Describe how plasma hormone levels change during exercise, and explain how the hormones' effects on glucose metabolism support exercising muscle.

LO 25.4   Explain how physiologists quantify the intensity of periods of exercise.

LO 25.5   Identify the factors that can limit exercise.

VENTILATORY RESPONSES TO EXERCISE 790

LO 25.6   Describe how oxygen consumption and ventilation change during and after exercise, and explain how these changes take place.

LO 25.7   Explain what happens to arterial and venous blood gases and plasma pH over a range of exercise intensities.

CARDIOVASCULAR RESPONSES TO EXERCISE 791

LO 25.8   Explain what happens to cardiac output, tissue blood flow, and blood pressure over a range of exercise intensities.

LO 25.9   Explain the control mechanisms that influence blood pressure and blood flow, and describe how they change during exercise.

FEEDFORWARD RESPONSES TO EXERCISE 793

LO 25.10   Describe the feedforward reflexes that anticipate homeostatic challenges of exercise.

TEMPERATURE REGULATION DURING EXERCISE 794

LO 25.11   Diagram the homeostatic thermo-regulatory responses and mechanisms that take place during periods of exercise.

EXERCISE AND HEALTH 795

LO 25.12   Describe what has been proved and what has been suggested about the influence of exercise on different aspects of health.

BACKGROUND BASICS

105   Aerobic and anaerobic metabolism
 17   Feedforward reflexes
393   Muscle metabolism
398   Isometric and isotonic contraction
466   Control of heart rate and contractility
493   Control of blood pressure
489   Local control of blood flow
553   Alveolar ventilation
582   Respiratory chemoreceptors
638   Dehydration
709   Insulin
720   Thermoregulation

Each dot of a microarray represents one gene. Genes that are active show up in bright colors.

n July 2008, the world watched with great anticipation as Michael Phelps swam his way to a record eight gold medals at the 2008 Olympics. The work done by Phelps and the hundreds of other Olympic athletes from around the world represents one of the most common challenges to body homeostasis: exercise. Exercise comes in many forms. Distance running, swimming, and cycling are examples of a dynamic endurance exercise. Weight lifting and strength training are examples of resistance training. In this chapter, we examine dynamic exercise as a challenge to homeostasis that is met with an integrated response from multiple body systems.

In many ways, exercise is the ideal example for teaching physiological integration. Everyone is familiar with it, and unlike high-altitude mountaineering or deep-sea diving, it involves no special environmental conditions. Moreover, exercise is a normal physiological state, not a pathological one—although it can be affected by disease and (if excessive) can result in injury.

In addition to being an excellent teaching example, exercise physiology is a very active area of integrative physiology research. The coordinated functioning of multiple body systems is still not well understood in many instances because of complex interactions between neural and local control mechanisms. Researchers use a combination of animal models and studies with human subjects, including elite athletes, in their quest to explain how the body adapts to the metabolic demands of exercise. Note that this

RUNNING **PROBLEM** | **Malignant Hyperthermia**

Zach, age 11, jumped off the school bus and ran into his mother's open arms after a long day at school. "You're a little warm, Zach. Are you feeling ok?" "It was really hot during gym today but I'm ok. Can I go play with Jacob?" "Sure, but be home in an hour." "Ok, Mom. See you later." And with that, Zach was off. Thirty minutes later, the doorbell rang. It was Jacob. "Come quick, something's wrong with Zack! I think he's having a seizure!" Soon, EMS would be rushing Zach to the hospital.

787 790 794 795 797

chapter is not intended to be a complete discussion of exercise physiology. For further information, you might want to consult an exercise physiology textbook.

METABOLISM AND EXERCISE

Exercise begins with skeletal muscle contraction, an active process that requires ATP for energy. But where does the ATP for muscle contraction come from? A small amount is in the muscle fiber when contraction begins (**FIG. 25.1** ❹). As this ATP is used

FIG. 25.1 Overview of muscle metabolism

ATP for muscle contraction is continuously produced by aerobic metabolism of glucose and fatty acids. During short bursts of activity, when ATP demand exceeds the rate of aerobic ATP production, aerobic glycolysis produces ATP, lactate, and H^+.

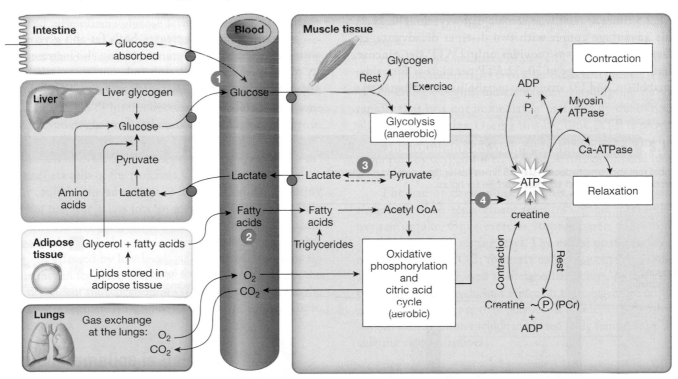

1. Glucose comes from liver glycogen or dietary intake.
2. Fatty acids can be used only in aerobic metabolism.
3. Lactate from anaerobic metabolism can be converted to glucose by the liver.
4. Both aerobic and anaerobic metabolism provide ATP for muscle contraction.

though experiments have shown that arterial P_{O_2} and P_{CO_2} do not change (Fig. 25.6).

How does the feedforward response work? One model says that as exercise begins, proprioceptors in the muscles and joints send information to the motor cortex of the brain. Descending signals from the motor cortex go not only to the exercising muscles but also along parallel pathways to the cardiovascular and respiratory control centers and to the limbic system of the brain.

Output from the limbic system and cardiovascular control center triggers generalized sympathetic discharge. As a result, an immediate slight increase in blood pressure marks the beginning of exercise. Sympathetic discharge causes widespread vasoconstriction, increasing blood pressure. Once exercise has begun, this increase in blood pressure compensates for decreases in blood pressure resulting from muscle vasodilation.

As exercise proceeds, reactive compensations become superimposed on the feedforward changes. For example, when exercise reaches 50% of aerobic capacity, muscle chemoreceptors detect the buildup of H^+, lactate, and other metabolites, and send this information to central command centers in the brain. The command centers then maintain changes in ventilation and circulation that were initiated in a feedforward manner. Thus, the integration of systems in exercise probably involves both common reflex pathways and some unique centrally mediated reflex pathways.

TEMPERATURE REGULATION DURING EXERCISE

As exercise continues, heat released through metabolism creates an additional challenge to homeostasis. Most of the energy released during metabolism is not converted into ATP but instead is released as heat. (Efficiency of energy conversion from organic substrates to ATP is only 20–25%.) With continued exercise, heat production exceeds heat loss, and core body temperature rises. In endurance events, body temperature can reach 40–42 °C (104–108 °F), which we would normally consider a fever.

This rise in body temperature during exercise triggers two thermoregulatory mechanisms: sweating and increased cutaneous blood flow [p. 720]. Both mechanisms help regulate body temperature, but both can also disrupt homeostasis in other ways. While sweating lowers body temperature through evaporative cooling, the loss of fluid from the extracellular compartment can cause dehydration and significantly reduce circulating blood volume. Because sweat is a hypotonic fluid, the extra water loss increases body osmolarity. The combination of decreased ECF volume and increased osmolarity during extended exercise sets in motion the complex homeostatic pathways for overcoming dehydration, including thirst and renal conservation of water [Fig. 20.13, p. 640].

The other thermoregulatory mechanism—increased blood flow to the skin—causes body heat loss to the environment through convection [p. 721]. However, increased sympathetic output during exercise tends to vasoconstrict cutaneous blood vessels, which opposes the thermoregulatory response. The primary

RUNNING PROBLEM

In MH, abnormal ryanodine receptors remain open longer than normal, allowing excess Ca^{2+} to be released from the sarcoplasmic reticulum. The increase in cytosolic Ca^{2+} causes continuous muscle contraction, which increases cellular demand for ATP. Oxygen consumption and CO_2 production increase. If the muscle's demand for ATP exceeds the availability of O_2, metabolism becomes anaerobic. Subsequent production of lactate and other acids create a state of metabolic acidosis. Finally, the elevated cytoplasmic Ca^{2+} activates enzymes that cause muscle breakdown (*rhabdomyolysis*). Dr. Jones ordered blood and urine tests to help pin down a diagnosis.

Q2: *What aspect of MH pathophysiology explains Zach's high fever?*

Q3: *The laboratory tests include plasma K^+ concentration and urine levels of myoglobin, a muscle protein similar to hemoglobin. What do you predict the results of these tests would be in a patient suffering from MH (compared to normal)?*

787 790 **794** 795 797

control of vasodilation in hairy regions of skin, such as trunk and limbs, during exercise appears to come from a sympathetic vasodilator system. Activation of these acetylcholine-secreting sympathetic neurons as body core temperature rises dilates some cutaneous blood vessels without altering sympathetic vasoconstriction in other body tissues.

Although cutaneous vasodilation is essential for thermoregulation, it can disrupt homeostasis by decreasing peripheral resistance and diverting blood flow from the muscles. In the face of these contradictory demands, the body initially gives preference to thermoregulation. However, if central venous pressure falls below a critical minimum, the body abandons thermoregulation in the interest of maintaining blood flow to the brain.

The degree to which the body can adjust to both demands depends on the type of exercise being performed and its intensity and duration. Strenuous exercise in hot, humid environments can severely impair normal thermoregulatory mechanisms and cause *heat stroke,* a potentially fatal condition. Unless prompt measures are taken to cool the body, core temperatures can go as high as 43 °C (109 °F).

It is possible for the body to adapt to repeated exercise in hot environments, however, through **acclimatization.** In this process, physiological mechanisms shift to fit a change in environmental conditions. As the body adjusts to exercise in the heat, sweating begins sooner and doubles or triples in volume, enhancing evaporative cooling. With acclimatization, sweat also becomes more dilute, as salt is reabsorbed from the sweat glands under the influence of increased aldosterone. Salt loss in an unacclimatized person exercising in the heat may reach 30 g NaCl per day, but that value decreases to as little as 3 g after a month of acclimatization.

3. The active vasodilator nerves to the skin secrete ACh but are classified as sympathetic neurons. On what basis were they identified as sympathetic?

EXERCISE AND HEALTH

Physical activity has many positive effects on the human body. The lifestyles of humans have changed dramatically since we were hunter-gatherers, but our bodies still seem to work best with a certain level of physical activity. Several common pathological conditions—including high blood pressure, strokes, and diabetes mellitus—can be improved by physical activity. Even so, developing regular exercise habits is one lifestyle change that many people find difficult to make. In this section, we look at the effects exercise has on several common health conditions.

Exercise Lowers the Risk of Cardiovascular Disease

As early as the 1950s, scientists showed that physically active men have a lower rate of heart attacks than do men who lead sedentary lives. These studies started many investigations into the exact relationship between cardiovascular disease and exercise. Scientists have subsequently demonstrated that exercise has positive benefits for both men and women. These benefits include lowering blood pressure, decreasing plasma triglyceride levels, and raising plasma HDL-cholesterol levels. High blood pressure is a major risk factor for strokes, and elevated triglycerides and low HDL-cholesterol levels are associated with development of atherosclerosis and increased risk of heart attack.

Overall, exercise reduces the risk of death or illness from a variety of cardiovascular diseases, although the exact mechanisms by which this occurs are still unclear. Even such mild exercise as walking has significant health benefits that could reduce the risk of developing cardiovascular diseases or diabetes and the complications of obesity in the estimated 23% of adult Americans who have sedentary lifestyles.

Type 2 Diabetes Mellitus May Improve with Exercise

Regular exercise is now widely accepted as effective in preventing and alleviating type 2 diabetes mellitus and its complications, including microvascular retinopathy [p. 481], diabetic neuropathy [p. 369], and cardiovascular disease [p. 501]. With regular exercise, skeletal muscle fibers up-regulate both the number of GLUT4 glucose transporters and the number of insulin receptors on their membrane. The addition of insulin-independent GLUT4 transporters decreases the muscle's dependence on insulin for glucose uptake. Glucose uptake into the exercising muscle also helps correct the hyperglycemia of diabetes.

Up-regulation of insulin receptors with exercise makes the muscle fibers more sensitive to insulin. A smaller amount of insulin then can achieve a response that previously required more

insulin. Because the cells are responding to lower insulin levels, the endocrine pancreas secretes less insulin. This lessens the stress on the pancreas, resulting in a lower incidence of type 2 diabetes mellitus.

FIGURE 25.9 shows the effects of seven days of exercise on glucose utilization and insulin secretion in men with mild type 2 diabetes. Individuals in the experiment underwent *glucose tolerance tests*, in which they ingested 100 g of glucose after an overnight fast. Their plasma glucose levels were assessed before and for 120 minutes after ingesting the glucose. Simultaneous measurements were made of plasma insulin.

The graph in Figure 25.9a shows glucose tolerance tests in control subjects (blue line) and in the diabetic men before and after exercise (red and green lines, respectively). Figure 25.9b shows concurrent insulin secretion in the three groups. After only seven days of exercise, both the glucose tolerance test and insulin secretion in exercising diabetic subjects had shifted to a pattern that was more like that of the normal control subjects. These results demonstrate the beneficial effect of exercise on glucose transport and metabolism, and support the recommendation that patients with type 2 diabetes maintain a regular exercise program.

Stress and the Immune System May Be Influenced by Exercise

Another health-related topic receiving much attention is the interaction of exercise with the immune system, and "exercise immunology" has become a recognized scientific discipline. Epidemiological studies looking at large populations of people suggest that exercise is associated with a reduced incidence of disease and with increased longevity. Moreover, many people believe

RUNNING **PROBLEM**

Zach's temperature continued to climb and his condition rapidly worsened despite efforts to cool his body with ice packs. Blood test results indicated hyperkalemia, elevated levels of potassium. Although the results of Zach's urinalysis were not available, his urine sample did appear cola-colored, which was indicative of the presence of myoglobin (*myoglobinuria*). This was enough evidence for Dr. Jones to suspect Zach had MH. He quickly ordered dantrolene, a drug that inhibits Ca^{2+} release from the sarcoplasmic reticulum of skeletal muscle. Within minutes of receiving dantrolene, Zach's muscles began to relax and his temperature began to fall. He was admitted to the intensive care unit for continued treatment to correct his hyperkalemia and myoglobinuria, which can lead to kidney failure if left untreated.

Q4: *Why is hyperkalemia dangerous?*

Q5: *Once the Ca^{2+} leak from the sarcoplasmic reticulum stops, how do skeletal muscle cells remove Ca^{2+} from the cytoplasm?*

787 790 794 **795** 797

FIG. 25.9   Exercise improves glucose tolerance and insulin secretion

The experiments tested normal men (blue line), men with type 2 diabetes who had not been exercising (red line), and those same diabetic men after seven days of exercise (green line).

KEY

— Normal controls

— Type 2 diabetes, no exercise

— Type 2 diabetes, after 7 days of exercise

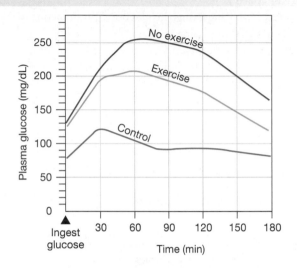

(a) Plasma Glucose During Glucose Tolerance Test

(b) Plasma Insulin During Glucose Tolerance Test

Data from B. R. Seals, *et al. J App Physiol* 56(6): 1521–1525, 1984; and M. A. Rogers, *et al. Diabetes Care* 11: 613–618, 1988.

that exercise boosts immunity, prevents cancer, and helps HIV-infected patients combat AIDS.

However, there are few rigidly controlled research studies providing evidence to support those viewpoints. Indeed, other evidence suggests that strenuous exercise is a form of stress that suppresses the immune response. Immune suppression may be due to corticosteroid release, or it may be due to release of interferon-γ during strenuous exercise.

Researchers have proposed that the relationship between exercise and immunity can be represented by a J-shaped curve (**FIG. 25.10**). People who exercise moderately have slightly more effective immune systems than those who are sedentary, but people who exercise strenuously may experience a decrease in immune function because of the stress of the exercise.

Another area of exercise physiology filled with interesting though contradictory results is the effect of exercise on stress, depression, and other psychological parameters. Research has shown an inverse relationship between exercise and depression: People who exercise regularly are significantly less likely to be clinically depressed than are people who do not exercise regularly. The association exists, but assigning cause and effect to the two parameters is difficult. Are the exercisers less depressed because they exercise? Or do depressed individuals exercise less because they are depressed? What physiological factors are involved?

Many published studies appear to show that regular exercise is effective in reducing depression. But a careful analysis of experimental design suggests that the conclusions of some of those

studies may be overstated. The subjects in many of the experiments were being treated concurrently with drugs or psychotherapy, so it is difficult to attribute improvement in their condition solely to exercise. In addition, participation in exercise studies

FIG. 25.10   Immune function and exercise

Individuals who exercise in moderation have fewer upper respiratory infections (URIs) than sedentary individuals or those who exercise strenuously.

Moderate exercise enhances immunity, but strenuous exercise is a form of stress that depresses immunity.

gives subjects a period of social interaction, another factor that might play a role in the reduction of stress and depression.

The assertion that exercise reduces depression has support from studies showing that exercise increases serotonin in the brain. Drugs that enhance serotonin activity, such as the selective serotonin reuptake inhibitors [p. 299], are currently being used to treat depression, and a way to achieve the same result without drugs would be desirable. A number of clinical trials that look at exercise effects on depression and health are currently underway.

RUNNING **PROBLEM CONCLUSION** | Malignant Hyperthermia

Zach was fortunate to have a doctor familiar with the signs of malignant hyperthermia: A diagnosis of MH is usually not even considered unless the patient is in surgery and under anesthesia. If left untreated, MH proves fatal within hours. Zach spent two days in the hospital but recovered completely due to early administration of dantrolene and treatment for the symptoms of rhabdomyolysis.

Zach and his family were referred to a specialist, who recommended they all undergo a caffeine halothane contracture test (CHCT) to confirm the diagnosis of MH. In a CHCT, a sample of skeletal muscle removed surgically is bathed in a solution containing caffeine and halothane (a common anesthetic). If the muscle contracts while in the solution, the individual is considered to be susceptible to MH. Zach's CHCT and that of his father were both positive. Subsequent genetic testing revealed that Zach and his father both had the defective RyR gene. MH is frequently inherited as an autosomal dominant disorder, which means that a person inheriting one copy of the mutated gene will manifest the condition.

In this running problem, you were introduced to malignant hyperthermia (MH), a genetic disorder that can be difficult to diagnose. To learn more about malignant hyperthermia, see literature from the Malignant Hyperthermia Association of the United States at *www.mhaus.org.*

Now test your understanding of the running problem by checking your answers against the information in this summary table.

Question	Facts	Integration and Analysis
Q1: *Where is the RyR found in skeletal muscle and what is its role in muscle contraction?*	The RyR is a Ca^{2+} channel located on the sarcoplasmic reticulum [p. 389].	Opening the RyR channels allows Ca^{2+} to flow into the cytoplasm and trigger muscle contraction.
Q2: *What aspect of MH pathophysiology explains Zach's high fever?*	The conversion of chemical bond energy to high-energy phosphate bonds in ATP is only about 25% efficient, and the remaining energy is lost as heat.	Muscle contraction requires ATP. ATP is produced through aerobic metabolism, and heat is a byproduct of metabolism.
Q3: *What do you predict the results of plasma $[K^+]$ and urine myoglobin tests would be in a patient suffering from MH?*	Muscle cells contain high concentrations of K^+ and myoglobin.	Muscle breakdown releases K^+ and myoglobin into the ECF, so plasma $[K^+]$ would be elevated and filtered myoglobin would be excreted by the kidneys.
Q4: *Why is hyperkalemia dangerous?*	The ratio of ECF:ICF $[K^+]$ is the primary determinant of the membrane potential.	Hyperkalemia can lead to dangerous cardiac arrhythmias because elevated $[K^+]$ depolarizes cells.
Q5: *How do skeletal muscle cells remove Ca^{2+} from the cytoplasm?*	Calcium is more concentrated in the ECF and sarcoplasmic reticulum than in the cytoplasm.	A Ca^{2+}-ATPase pumps Ca^{2+} back into the sarcoplasmic reticulum.

This problem was written by Douglas Shannon when he was a student at the University of Texas preparing to enter a Physician Assistant program at the University of Texas Medical Branch.

787 790 794 795 **797**

CHAPTER SUMMARY

In this chapter, you learned about exercise and the physiological challenges it presents. *Integration* and *coordination* between the body's *physiological control systems* keep the internal environment relatively constant, despite the challenges to *homeostasis* that exercise presents.

Metabolism and Exercise

1. Exercising muscle requires a steady supply of ATP from metabolism or from conversion of phosphocreatine. (p. 787; Fig. 25.1)

2. Carbohydrates and fats are the primary energy substrates. Glucose can be metabolized through both oxidative and anaerobic pathways, but fatty acid metabolism requires oxygen. (p. 788; Fig. 25.1)

3. Anaerobic **glycolytic metabolism** converts glucose to lactate and H^+. Glycolytic metabolism is 2.5 times more rapid than aerobic pathways but is not as efficient at ATP production. (p. 788; Fig. 25.2)

4. Glucagon, cortisol, catecholamines, and growth hormone influence glucose and fatty acid metabolism during exercise. These hormones favor the conversion of glycogen to glucose. (p. 789)

5. Plasma glucose concentrations rise with exercise, but insulin secretion decreases. This response reduces glucose uptake by most cells, making more glucose available for exercising muscle. (p. 789)

6. The intensity of exercise is indicated by **oxygen consumption** (V_{O_2}). A person's maximal rate of oxygen consumption (V_{O_2max}) is an indicator of that person's ability to perform endurance exercise. (p. 789)

7. Oxygen consumption increases rapidly at the onset of exercise. **Excess postexercise oxygen consumption** (EPOC) is due to ongoing metabolism, increased body temperature, and circulating catecholamines. (p. 789; Fig. 25.4)

8. Muscle mitochondria increase in size and number with endurance training. (p. 790)

9. At maximal exertion, the ability of the cardiovascular system to deliver oxygen and nutrients appears to be the primary limiting factor. (p. 790)

Ventilatory Responses to Exercise

10. Exercise hyperventilation results from feedforward signals from the motor cortex and sensory feedback from peripheral sensory receptors. (p. 790; Fig. 25.5)

11. Arterial P_{O_2}, P_{CO_2}, and pH do not change significantly during mild to moderate exercise. (p. 790; Fig. 25.6)

Cardiovascular Responses to Exercise

12. Cardiac output increases with exercise because of increased venous return and sympathetic stimulation of heart rate and contractility. (p. 791; Fig. 25.7)

13. Blood flow through exercising muscle increases dramatically when skeletal muscle arterioles dilate. Arterioles in other tissues constrict. (p. 792; Fig. 25.7)

14. Decreased tissue O_2 and glucose or increased muscle temperature, CO_2, and acid act as paracrine signals and cause local vasodilation. (p. 793)

15. Mean arterial blood pressure increases slightly as exercise intensity increases. The baroreceptors that control blood pressure change their setpoints during exercise. (p. 793; Fig. 25.8)

Feedforward Responses to Exercise

16. When exercise begins, feedforward responses prevent significant disruption of homeostasis. (p. 793)

Temperature Regulation during Exercise

17. Heat released during exercise is dissipated by sweating and increased cutaneous blood flow. (p. 794)

Exercise and Health

18. Physical activity can help prevent or decrease the risk of developing high blood pressure, strokes, and type 2 diabetes mellitus. (p. 795)

19. Studies suggest that serotonin release during exercise may help alleviate depression. (p. 796)

REVIEW QUESTIONS

In addition to working through these questions and checking your answers on p. A-33, review the Learning Outcomes at the beginning of this chapter.

Level One Reviewing Facts and Terms

1. Name the two muscle compounds that store energy in the form of high-energy phosphate bonds.

2. The most efficient ATP production is through *aerobic/anaerobic* pathways. When these pathways are being used, then *glucose/fatty acids/both/neither* can be metabolized to provide ATP.

3. What are the differences between aerobic and anaerobic metabolism?

4. List three sources of glucose that can be metabolized to ATP, either directly or indirectly.

5. List four hormones that promote the conversion of triglycerides into fatty acids. What effects do these hormones have on plasma glucose levels?

6. What is meant by the term oxygen deficit, and how is it related to excessive postexercise oxygen consumption?

7. What organ system is the limiting factor for maximal exertion?

8. In endurance events, body temperature can reach 40–42 °C. What is normal body temperature? Which two thermoregulatory mechanisms are triggered by this change in temperature during exercise?

Level Two Reviewing Concepts

9. Concept map: Map the metabolic, cardiovascular, and respiratory changes that occur during exercise. Include the signals to and from the nervous system, and show what specific areas signal and coordinate the exercise response.

10. What causes insulin secretion to decrease during exercise, and why is this decrease adaptive?

11. State two advantages and two disadvantages of anaerobic glycolysis.

12. Compare and contrast each of the terms in the following sets of terms, especially as they relate to exercise:

    (a) ATP, ADP, PCr
    (b) myoglobin, hemoglobin

13. Match the following brain areas with the response(s) that each controls. Brain areas may control one response, more than one response, or none at all. Some responses may be associated with more than one brain area.

(a) pons	1. changes in cardiac output
(b) medulla oblongata	2. vasoconstriction
(c) midbrain	3. exercise hyperventilation
(d) motor cortex	4. increased stroke volume
(e) hypothalamus	5. increased heart rate
(f) cerebellum	6. coordination of skeletal muscle movement
(g) brain not involved (i.e., local control)	

14. Specify whether each of the following parameters stays the same, increases, or decreases when a person becomes better conditioned for athletic activities:

    (a) heart rate during exercise
    (b) resting heart rate
    (c) cardiac output during exercise
    (d) resting cardiac output
    (e) breathing rate during exercise
    (f) blood flow to muscles during exercise
    (g) blood pressure during exercise
    (h) total peripheral resistance during exercise

15. Why doesn't increased venous return during exercise overstretch the heart muscle?

16. Diagram the three theories that explain why the normal baroreceptor reflex is absent during exercise.

17. List and briefly discuss the benefits of a lifestyle that includes regular exercise.

18. Explain how exercise decreases blood glucose in type 2 diabetes mellitus.

Level Three Problem Solving

19. You have decided to manufacture a new sports drink that will help athletes, from football players to gymnasts. List at least four different ingredients you would include in your drink, and indicate why each is important for the athlete.

Level Four Quantitative Problems

20. You are a well-conditioned athlete. At rest, your heart rate is 60 beats/minute and your stroke volume is 70 mL/beat. What is your cardiac output? At one point during exercise, your heart rate increases to 120 beats/min. Does your cardiac output increase proportionately? Explain.

21. The following graph shows left ventricular pressure-volume curves in one individual. Curve A is the person sitting at rest. Curve B shows the person's cardiac response to mild exercise on a stationary bicycle. Curve C shows the cardiac response during maximum intensity cycling.

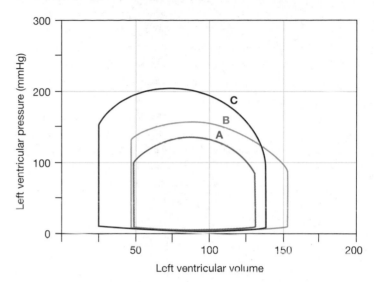

Data from G. D. Plotnick, *et al. Am J Physiol* 251: H1101–H1105, 1986.

    (a) Calculate the stroke volume for each of the curves.
    (b) Given the following cardiac outputs (CO), calculate the heart rates for each condition. $CO_A = 6$ L/min, $CO_B = 10.5$ L/min, $CO_C = 19$ L/min.
    (c) Which exercise curve shows an increase in stroke volume due primarily to increased contractility? Which exercise curve shows an increase in stroke volume due primarily to increased venous return?
    (d) Mechanistically, why did the end-diastolic volume in curve C fall back toward the resting value?

26

Reproduction and Development

Teenagers who received some type of comprehensive sex education were 60 percent less likely to get pregnant or get someone else pregnant.

Amanda Peterson Beadle, *Teen Pregnancies Highest in States with Abstinence-Only Policies*, April 10, 2012. This material was published by ThinkProgress. *http://thinkprogress .org/health/2012/04/10/461402/ teen-pregnancy-sex-education/*.

Seminiferous tubules with sperm (blue)

SEX DETERMINATION 801

LO 26.1 Describe the role of sex chromosomes in sex determination.

LO 26.2 Describe the bipotential reproductive structures of the early embryo.

LO 26.3 Diagram the processes of sexual differentiation in male and female embryonic development.

BASIC PATTERNS OF REPRODUCTION 806

LO 26.4 Describe and compare male and female patterns of gametogenesis.

LO 26.5 Diagram the common hormonal control and feedback pathways for reproductive function.

LO 26.6 Explain the significance of pulsatile GnRH secretion.

LO 26.7 Describe some environmental factors that influence reproductive physiology.

MALE REPRODUCTION 810

LO 26.8 Diagram the internal and external anatomy of the adult male reproductive and accessory structures and give the function(s) of each.

LO 26.9 Diagram the process and timeline of spermatogenesis.

LO 26.10 Explain the hormonal control of spermatogenesis.

LO 26.11 Describe the primary and secondary sex characteristics of the male and the hormones that influence their development.

FEMALE REPRODUCTION 816

LO 26.12 Diagram the internal and external anatomy of the adult female reproductive and accessory structures and give the function(s) of each.

LO 26.13 Diagram and give the timeline for follicular development from primordial follicle to corpus albicans.

LO 26.14 Explain the role of atresia in ovarian function.

LO 26.15 Diagram the ovarian and uterine stages of the menstrual cycle.

LO 26.16 Relate the hormonal control and feedback patterns of the menstrual cycle to different stages of the ovarian and uterine cycles.

LO 26.17 Describe the secondary sex characteristics of the female and the hormones that influence their development.

PROCREATION 824

LO 26.18 Diagram the erection reflex and describe the four phases of the human sexual response.

LO 26.19 Explain the anatomy or physiology of currently available contraceptive methods.

LO 26.20 Describe the common causes of male and female infertility.

PREGNANCY AND PARTURITION 827

LO 26.21 Diagram the process of sperm capacitation and fertilization of an ovum.

LO 26.22 Diagram the process of embryo development from fertilization through implantation in the endometrium.

LO 26.23 Describe the role of placental hormones during pregnancy.

LO 26.24 Describe what we currently understand about the processes of labor and parturition.

LO 26.25 Diagram a mammary gland and the control of milk and colostrum production.

LO 26.26 Diagram the let-down (milk ejection) reflex.

GROWTH AND AGING 834

LO 26.27 Describe how the reproductive systems of males and females change at puberty and with menopause and andropause.

BACKGROUND BASICS

16   Positive and negative feedback
68   Flagella
30   Steroids
48   Agonist/antagonist
181   Up- and down-regulation
179   Prostaglandins
210   Hypothalamic-pituitary axis
213   Prolactin
209   Oxytocin
285   Spinal reflex
723   Hot flashes

Imagine growing up as a girl, then at the age of 12 or so, finding that your voice is deepening and your genitals are developing into those of a man. This scenario actually happens to a small number of men who have a condition known as *pseudohermaphroditism* {*pseudes,* false + *hermaphrodites,* the dual-sex offspring of Hermes and Aphrodite}. These men have the internal sex organs of a male but inherit a gene that causes a deficiency in one of the male hormones. Consequently, they are born with external genitalia that appear feminine, and they are raised as girls. At **puberty** {*pubertas,* adulthood}, the period when a person makes the transition from being nonreproductive to being reproductive, individuals with pseudohermaphroditism begin to secrete more male hormones. As a result, they develop some, but not all, of the characteristics of men. Not surprisingly, a conflict arises: Should these individuals change gender or remain female? Most choose to change and continue life as men.

Reproduction is one area of physiology in which we humans like to think of ourselves as significantly advanced over other animals. We mate for pleasure as well as procreation, and women are always sexually receptive (i.e., not only during fertile periods). But just how different are we?

Like many other terrestrial animals, humans have internal fertilization that allows motile flagellated sperm to remain in an aqueous environment. To facilitate the process, we have mating and courtship rituals, as do other animals. Development is also internal, within the uterus, which protects the growing embryo from dehydration and cushions it in a layer of fluid.

Humans are *sexually dimorphic* {*di-,* two + *morphos,* form}, meaning that males and females are physically distinct. This distinction is sometimes blurred by dress and hairstyle, but these are cultural acquisitions. Everyone agrees that male and female humans are physically dimorphic, but we are still debating just how behaviorally and psychologically dimorphic we are.

Sex hormones play a significant role in the behavior of other mammals, acting on adults as well as influencing the brain of the developing embryo. Their role in humans is more controversial. Human fetuses are exposed to sex hormones while in the uterus, but it is unclear how much influence these hormones have on behavior later in life. Does the preference of little girls for dolls and of little boys for toy guns have a biological basis or a cultural basis? We have no answer yet, but growing evidence suggests that at least part of our brain structure is influenced by sex hormones before we ever leave the womb.

In this chapter, we address the biology of human reproduction and development. The topic is an incredibly complex one, with numerous hormones, cytokines, and paracrine signal molecules interacting in constantly shifting interplay. In recent decades, reproductive research has raised many new questions. Because of the complexity of the topic, this chapter presents only a generalized overview.

We begin our discussion with gametes that fuse to form the fertilized egg, or **zygote.** As the zygote begins to divide (2-cell stage, 4-cell stage, etc.), it becomes first an **embryo** (weeks 0–8 of development), then a **fetus** (eight weeks until birth).

RUNNING PROBLEM | Infertility

Kate and Jon have just about everything to make them happy: successful careers, a loving marriage, a comfortable home. But one thing is missing: After seven years of marriage, they have been unable to have a child. Today, Kate and Jon have their first appointment with Dr. Baker, an infertility specialist. "Finding the cause of your infertility is going to require some painstaking detective work," Dr. Baker explains. She begins her workup of Kate and Jon by asking detailed questions about their reproductive histories. Based on the answers to these questions, she will then order tests to pinpoint the problem.

801   814   817   823   830   834

SEX DETERMINATION

The male and female sex organs consist of three sets of structures: the gonads, the internal genitalia, and the external genitalia. **Gonads** {*gonos,* seed} are the organs that produce **gametes** {*gamein,* to marry}, the eggs and sperm that unite to form new individuals. The male gonads are the **testes** (singular *testis*), which produce **sperm** (*spermatozoa*). The female gonads are the **ovaries,** which produce eggs, or **ova** (singular *ovum*). The undifferentiated gonadal cells destined to produce eggs and sperm are called **germ cells.** The **internal genitalia** consist of accessory glands and ducts that connect the gonads with the outside environment. The **external genitalia** include all external reproductive structures.

Sexual development is programmed in the human genome. Each nucleated cell of the body except eggs and sperm contains 46 chromosomes. This set of chromosomes is called the *diploid number* because the chromosomes occur in pairs: 22 matched, or *homologous,* pairs of **autosomes** plus one pair of **sex chromosomes** (**FIG. 26.1a**). The amount of DNA in a diploid cell is written as 2N, indicating duplicated chromosomes.

The 22 pairs of autosomal chromosomes in our cells direct development of the human body form and of variable characteristics such as hair color and blood type. The two sex chromosomes, designated as either X or Y, contain genes that direct development of internal and external sex organs. The X chromosome is larger than the Y chromosome and includes many genes that are missing from the Y chromosome.

Eggs and sperm are *haploid* (1N) cells with 23 chromosomes, one from each of the 22 matched pairs plus one sex chromosome. When egg and sperm unite, the resulting zygote then contains a unique set of 46 chromosomes, with one chromosome of each matched pair coming from the mother and the other from the father.

Sex Chromosomes Determine Genetic Sex

The sex chromosomes a person inherits determine the genetic sex of that individual. Genetic females are XX, and genetic males are XY (Fig. 26.1b). Females inherit one X chromosome from each

FIG. 26.1 Human chromosomes

(a) Humans have 23 pairs of chromosomes: 22 pairs of autosomes and one pair of sex chromosomes. X and Y chromosomes (lower right) mean that these chromosomes came from a male. The autosomes are arranged in homologous pairs in this figure.

(b) X and Y chromosomes determine sex. Each egg produced by a female (XX) has an X chromosome. Sperm produced by a male (XY) have either an X chromosome or a Y chromosome.

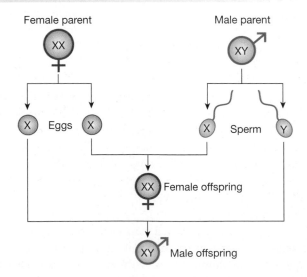

parent. Males inherit a Y chromosome from the father and an X chromosome from the mother. The Y chromosome is essential for development of the male reproductive organs.

If sex chromosomes are abnormally distributed at fertilization, the presence or absence of a Y chromosome determines whether development proceeds along male or female lines. The presence of a Y chromosome means the embryo will become male, even if the zygote also has multiple X chromosomes. For instance, an XXY zygote will become male. A zygote that inherits only a Y chromosome (YO) will die because the larger X chromosome contains essential genes that are missing from the Y chromosome.

In the absence of a Y chromosome, an embryo will develop into a female. For this reason, a zygote that gets only one X chromosome (XO; Turner syndrome) will develop into a female. Two X chromosomes are needed for normal female reproductive function, however.

Once the ovaries develop in a female fetus, one X chromosome in each cell of her body is inactivated and condenses into a clump of nuclear chromatin known as a *Barr body*. (Barr bodies in females can be seen in stained cheek epithelium.) The selection of the X chromosome that becomes inactive during development is random: Some cells will have an active maternal X chromosome and others have an active paternal X chromosome. Because inactivation occurs early in development—before cell division is complete—all cells of a given tissue will usually have the same active X chromosome, either maternal or paternal.

CONCEPT CHECK

1. Name the male and female gonads and gametes.

Sexual Differentiation Occurs Early in Development

The sex of an early embryo is difficult to determine because reproductive structures do not begin to differentiate until the seventh week of development. Before differentiation, the embryonic tissues are considered *bipotential* because they cannot be morphologically identified as male or female.

The bipotential gonad has an outer cortex and an inner medulla (**FIG. 26.2a**). Under the influence of the appropriate developmental signal (described later), the medulla will develop into a testis. In the absence of that signal, the cortex will differentiate into ovarian tissue.

The bipotential internal genitalia consist of two pairs of accessory ducts: **Wolffian ducts** (mesonephric ducts) derived from the embryonic kidney, and **Müllerian ducts** (paramesonephric ducts). As development proceeds along either male or female lines, one pair of ducts develops while the other degenerates (Fig. 26.2a).

The bipotential external genitalia consist of a *genital tubercle, urethral folds, urethral groove,* and *labioscrotal swellings* (Fig. 26.2b). These structures differentiate into the male and female reproductive structures as development progresses.

What directs some single-cell zygotes to become males, and others to become females? Sex determination depends on the presence or absence of the *sex-determining region of the Y chromosome,* or **SRY gene.** In the presence of a functional SRY gene, the

X-Linked Inherited Disorders

Normally, a person inherits two copies of the gene for a given trait: one copy from each parent. However, many genes on the X chromosome, called *X-linked genes,* have no matching gene on the much smaller Y chromosome. Females always get two copies of X-linked genes, so the expression of X-linked traits follows the usual pattern of gene dominance and recessiveness. Males, however, receive only one copy of an X-linked gene—on the X chromosome from their mother—so males *always* exhibit the traits associated with an X-linked gene. If that X-linked gene is defective, male offspring will exhibit the mutation. Among the identified X-linked diseases are Duchenne muscular dystrophy [p. 403], hemophilia [p. 529], and color-blindness.

bipotential gonads develop into testes. In the absence of the SRY gene and under the direction of multiple female-specific genes, the gonads develop into ovaries.

Male Embryonic Development   The SRY gene produces a protein (*testis-determining factor* or TDF) that binds to DNA and activates additional genes, including SOX9, WT1 (Wilms tumor protein), and SF1 (steroidogenic factor). The protein products of these and other genes direct development of the gonadal medulla into a testis (**FIG. 26.3**). Note that testicular development does *not* require male sex hormones such as **testosterone.** The developing embryo cannot secrete testosterone until after the gonads differentiate into testes.

Once the testes differentiate, they begin to secrete three hormones that influence development of the male internal and external genitalia. Testicular **Sertoli cells** secrete glycoprotein **anti-Müllerian hormone** (AMH; also called *Müllerian-inhibiting substance*). Testicular **interstitial (Leydig) cells** secrete **androgens** {*andro-,* male}: testosterone and its derivative **dihydrotestosterone (DHT).** Testosterone and DHT are the dominant steroid hormones in males. Both bind to the same *androgen receptor,* but the two ligands elicit different responses.

In the developing fetus, anti-Müllerian hormone causes the embryonic Müllerian ducts to regress (Fig. 26.2a, male ❷). Testosterone converts the Wolffian ducts into male accessory structures: epididymis, vas deferens, and seminal vesicle (male ❸). Later in fetal development, testosterone controls migration of the testes from the abdomen into the *scrotum,* or scrotal sac. The remaining male sex characteristics, such as differentiation of the external genitalia, are controlled primarily by DHT.

The importance of DHT in male development came to light in studies of the individuals with pseudohermaphroditism

described in the opening of this chapter. These men inherit a defective gene for **5α-reductase,** the enzyme that catalyzes the conversion of testosterone to DHT (**FIG. 26.4**). Despite normal testosterone secretion, these men have inadequate levels of DHT, and as a result the male external genitalia and prostate gland fail to develop fully during fetal development.

At birth, the infants with pseudohermaphroditism appear to be female and are raised as such. However, at puberty, the testes again begin to secrete testosterone, causing masculinization of the external genitalia, pubic hair growth (although scanty facial and body hair), and deepening voice. By studying the 5α-reductase defect in these individuals, scientists have been able to separate the effects of testosterone from those of DHT.

Exposure of nongenital tissues to testosterone during embryonic development is known to have masculinizing effects, such as altering the brain's responsiveness to certain hormones. One controversial aspect of the masculinizing effects of testosterone is its influence on human sexual behavior and gender identity. It is well documented that in many nonhuman mammals, adult sexual behavior depends on the absence or presence of testosterone during critical periods of brain development. However, a similar cause-effect relationship has never been proved in humans. In human behavior, it is very difficult to separate biological influences from environmental factors, and it will probably be years before this question is resolved.

Female Embryonic Development   In female embryos, which have no SRY gene, the cortex of the bipotential gonad develops into ovarian tissue (Fig. 26.2a female ❶). Research indicates that female development is more complex than originally thought, with multiple genes required for the development of functional ovaries.

In the absence of testicular AMH, the Müllerian ducts develop into the upper portion of the **vagina,** the **uterus,** and the **Fallopian tubes,** named after the anatomist Fallopius, who first described them (Fig. 26.2a female ❸). Fallopian tubes are also called *uterine tubes* or *oviducts.* Without testosterone, the Wolffian ducts degenerate (Fig. 26.2a female ❷). Without DHT, the external genitalia take on female characteristics (Fig. 26.2b).

CONCEPT CHECK

2. Where in a target cell would you expect to find receptors for androgens? Where would you expect to find receptors for AMH?

3. Why was King Henry VIII of England wrong to blame his wives when they were unable to produce a male heir to the throne?

4. Which sex will a zygote become if it inherits only one X chromosome (XO)?

5. If the testes are removed from an early male embryo, why does it develop a uterus and Fallopian tubes rather than the normal male accessory structures? Will the embryo have male or female external genitalia? Explain.

FIG. 26.2 **ESSENTIALS**

Sexual Development in the Human Embryo

(a) Development of Internal Organs

Bipotential stage: 6-week embryo
The internal reproductive organs have the potential to develop into male or female structures.

Bipotential stage (6-week embryo)

Müllerian duct — — Wolffian duct

Gonad (bipotential) — — Kidney

IF FEMALE:

Gonadal cortex forms ovary.

Gonadal medulla regresses.

Wolffian duct regresses (testosterone absent).

Müllerian duct becomes Fallopian tube, uterus, cervix, and upper 1/2 of vagina (AMH absent).

IF MALE:

Gonadal cortex regresses.

Gonadal medulla forms a testis.

Wolffian duct forms epididymis, vas deferens, and seminal vesicle (testosterone present).

Müllerian duct regresses (AMH present).

FEMALE

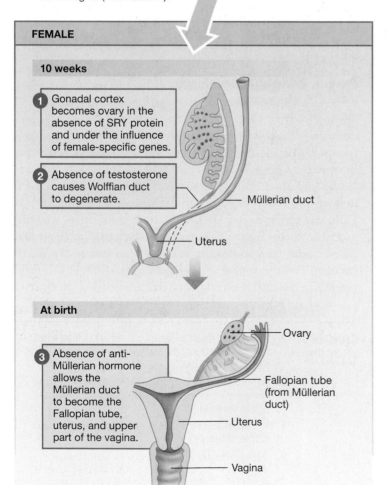

10 weeks

1 Gonadal cortex becomes ovary in the absence of SRY protein and under the influence of female-specific genes.

2 Absence of testosterone causes Wolffian duct to degenerate.

Müllerian duct

Uterus

At birth

3 Absence of anti-Müllerian hormone allows the Müllerian duct to become the Fallopian tube, uterus, and upper part of the vagina.

Ovary

Fallopian tube (from Müllerian duct)

Uterus

Vagina

MALE

10 weeks

1 SRY protein in a male embryo directs the medulla of the bipotential gonad to develop into testis.

Testis

Wolffian duct

2 Anti-Müllerian hormone from testis causes the Müllerian ducts to disappear.

At birth

Prostate

Seminal vesicle

Vas deferens

Testis

Epididymis

3 Testosterone from testis converts Wolffian duct into seminal vesicle, vas deferens, and epididymis. DHT controls prostate development.

(b) Development of External Genitalia

Bipotential stage: The external genitalia of a 6-week embryo cannot be visually identified as male or female.

Bipotential stage
(6-week embryo)

Genital tubercle

Urethral groove

Urethral fold

Labioscrotal swelling

Anus

IF FEMALE:

Genital tubercle forms clitoris.

Urethral folds and grooves form labia minora, opening of vagina and urethra.

Labioscrotal swellings form labia majora.

IF MALE:

Genital tubercle forms glans penis.

Urethral folds and grooves form shaft of penis.

Labioscrotal swellings form shaft of penis and scrotum.

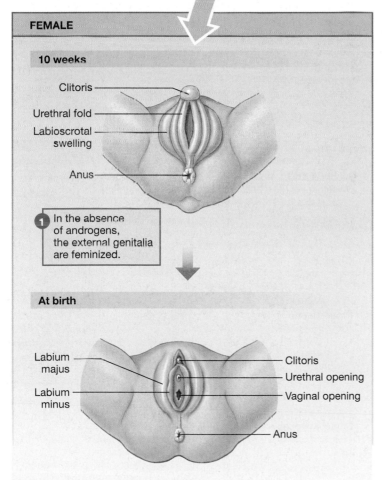

FEMALE

10 weeks

Clitoris

Urethral fold

Labioscrotal swelling

Anus

1 In the absence of androgens, the external genitalia are feminized.

At birth

Labium majus

Labium minus

Clitoris

Urethral opening

Vaginal opening

Anus

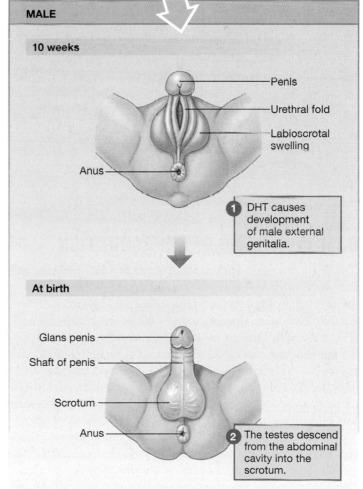

MALE

10 weeks

Penis

Urethral fold

Labioscrotal swelling

Anus

1 DHT causes development of male external genitalia.

At birth

Glans penis

Shaft of penis

Scrotum

Anus

2 The testes descend from the abdominal cavity into the scrotum.

FIG. 26.3 Male embryonic development

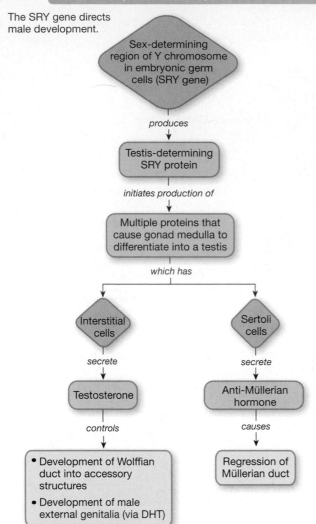

The SRY gene directs male development.

Sex-determining region of Y chromosome in embryonic germ cells (SRY gene)

produces

Testis-determining SRY protein

initiates production of

Multiple proteins that cause gonad medulla to differentiate into a testis

which has

Interstitial cells — *secrete* → Testosterone — *controls* →
- Development of Wolffian duct into accessory structures
- Development of male external genitalia (via DHT)

Sertoli cells — *secrete* → Anti-Müllerian hormone — *causes* → Regression of Müllerian duct

CLINICAL FOCUS

Determining Sex

The first question new parents typically ask about their child is, "Is it a boy or a girl?" Sometimes the answer is not obvious because in approximately 1 in 3000 births, the sex of the child cannot easily be determined. Multiple criteria might be used to establish an individual's sex: genetic, chromosomal, gonadal, morphological, or even psychological characteristics. For example, presence of a Y chromosome with a functional SRY gene could be one criterion for "maleness." However, it is possible for an infant to have a Y chromosome and not appear to be male because of a defect in some aspect of development. Traditionally, sex determination has been based on appearance of the external genitalia at birth, but the idea that individuals should be allowed to choose their sex when they become old enough is gaining ground. The sex a person considers himself or herself to be is called the person's *gender identity.* Learn and read more about the current criteria used to decide a child's sex in cases of ambiguous genitalia in the American Academy of Pediatrics policy statement "Evaluation of the Newborn with Developmental Anomalies of the External Genitalia," *Pediatrics* 106(1): 138–142, 2000 (July).

BASIC PATTERNS OF REPRODUCTION

The testis and ovary both produce hormones and gametes, and they share other similarities, as might be expected of organs having the same origin. However, male and female gametes are very different from each other. Eggs are some of the largest cells in the body [Fig. 3.1, p. 60]. They are nonmotile and must be moved through the reproductive tract on currents created by smooth muscle contraction or the beating of cilia. Sperm, in contrast, are quite small. They are the only flagellated cells in the body and are highly motile so that they can swim up the female reproductive tract in their search for an egg to fertilize.

The timing of gamete production, or **gametogenesis,** is also very different in males and females. Most evidence indicates that women are born with all the eggs, or **oocytes,** they will ever have. During the reproductive years, eggs mature in a cyclic pattern and are released from the ovaries roughly once a month. After about 40 years, female reproductive cycles cease (*menopause*).

FIG. 26.4 Synthesis pathways for steroid hormones

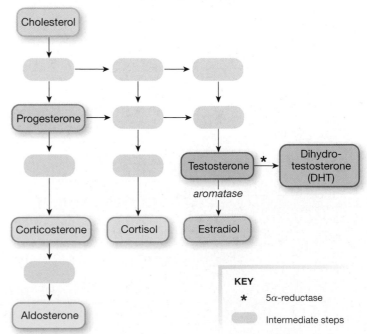

The blank boxes represent intermediate compounds whose names have been omitted for simplicity.

Cholesterol

Progesterone

Corticosterone

Aldosterone

Cortisol

Testosterone — * → Dihydrotestosterone (DHT)

aromatase

Estradiol

KEY
* 5α-reductase
Intermediate steps

Men, in contrast, manufacture sperm continuously from the time they reach reproductive maturity. Sperm production and testosterone secretion diminish with age but do not cease as women's reproductive cycles do.

Gametogenesis Begins in Utero

FIGURE 26.5 compares the male and female patterns of gametogenesis. In both sexes, germ cells in the embryonic gonads first undergo a series of mitotic divisions to increase their numbers ❶. After that, the germ cells are ready to undergo **meiosis,** the cell division process that forms gametes.

In the first step of meiosis ❷, the germ cell's DNA (2N) replicates until each chromosome is duplicated (46 chromosomes duplicated = 92 chromosomes). The cell, now called a **primary spermatocyte** or **primary oocyte,** contains twice the normal amount of DNA (4N). However, cell and chromosomal division do not take place as they do in mitosis. Instead, each duplicated chromosome forms two identical **sister chromatids,** linked together at a region known as the **centromere.** The primary gametes are then ready to undergo meiotic divisions to create haploid cells.

In the **first meiotic division** ❸, one primary gamete divides into two *secondary gametes* (**secondary spermatocyte** or **secondary oocyte**). Each secondary gamete gets one copy of each duplicated autosome plus one sex chromosome (2N). In the **second meiotic division** ❹, the sister chromatids separate. In males, the cells split during the second meiotic division, resulting in two haploid (1N) sperm from each secondary spermatocyte.

In females, the second meiotic division creates one egg and one small cell called a *polar body.* What happens after that depends on whether or not the egg is fertilized.

The timing of mitotic and meiotic divisions is very different in males and females. Let's take a closer look at gametogenesis in each sex.

Male Gametogenesis

At birth, the testes of a newborn boy have not progressed beyond mitosis and contain only immature germ cells (Fig. 26.5 ❶). After birth, the gonads become *quiescent* (relatively inactive) until puberty, the period in the early teen years when the gonads mature.

At puberty, germ cell mitosis resumes. From that point onward, the germ cells, known as **spermatogonia** (singular *spermatogonium*), have two possible fates. Some continue to undergo mitosis throughout the male's reproductive life. Others are destined to start meiosis and become primary spermatocytes ❷.

Each primary spermatocyte creates four sperm. In the first meiotic division ❸, a primary spermatocyte (4N) divides into two secondary spermatocytes (2N). In the second meiotic division ❹, each secondary spermatocyte divides into two spermatids. Each spermatid has 23 single chromosomes, the haploid number (1N) characteristic of a gamete. The spermatids then mature into sperm.

Female Gametogenesis

In the embryonic ovary, germ cells are called **oögonia** (singular *oögonium*) (Fig. 26.5 ❶). Oögonia complete mitosis and the DNA duplication stage of meiosis by the fifth month of fetal development, resulting in **primary oocytes** (4N) ❷. At birth each ovary contains about half a million primary oocytes. The best evidence indicates that at this time, germ cell mitosis ceases and no additional oocytes can be formed.

In the ovary, meiosis does not resume until puberty ❸. If a primary oocyte develops, it divides into two cells, a large **egg (secondary oocyte)** and a tiny **first polar body.** Despite the size difference, the egg and polar body each contain 23 duplicated chromosomes (2N). The first polar body disintegrates.

If the secondary oocyte is selected for ovulation, the second meiotic division takes place just before the egg is released from the ovary ❹. The sister chromatids separate but now meiosis pauses again. The final step of meiosis, in which sister chromatids go to separate cells, does not take place unless the egg is fertilized.

The ovary releases the mature egg during a process known as **ovulation.** If the egg is not fertilized, meiosis never goes to completion, and the egg disintegrates ❺. If fertilization by a sperm occurs, the final step of meiosis takes place ❻. Half the sister chromatids remain in the fertilized egg (zygote), while the other half are released in a **second polar body** (1N). The second polar body, like the first, degenerates. As a result of meiosis, each primary oocyte gives rise to only one egg.

Gametogenesis in both males and females is under the control of hormones from the brain and from endocrine cells in the gonads. Some of these hormones are identical in males and females, but others are different.

CONCEPT CHECK

6. The gametes in a newborn male are at what stage of development? Is it the same in a newborn female?

7. Compare the amount of DNA in the first polar body with the amount of DNA in the second polar body.

8. How many gametes are formed from one primary oocyte? From one primary spermatocyte?

The Brain Directs Reproduction

The reproductive system has some of the most complex control pathways of the body, in which multiple hormones interact in an ever-changing fashion. The endocrine pathways that regulate reproduction begin with secretion of peptide hormones by the hypothalamus and anterior pituitary. These trophic hormones control gonadal secretion of the steroid sex hormones, including androgens, and the so-called female sex hormones **estrogen** and **progesterone.**

These sex steroids are closely related to one another and arise from the same steroid precursors (Fig. 26.4). Both sexes produce all three groups of hormones, but androgens predominate in males, and estrogens and progesterone are dominant in females.

In men, most testosterone is secreted by the testes, but about 5% comes from the adrenal cortex. Testosterone is converted in peripheral tissues to its more potent derivative DHT. Some of

FIG. 26.5   Gametogenesis

Germ cells first duplicate themselves through mitosis. Then, through meiosis, they form gametes with one chromosome from each pair. For simplicity, this figure shows only one of the body's 22 pairs of autosomes in each cell.

the physiological effects attributed to testosterone are actually the result of DHT activity.

Males synthesize some estrogens, but the feminizing effects of these compounds are usually not obvious in males. Both testes and ovaries contain the enzyme **aromatase,** which converts androgens to estrogens, the female sex hormones. A small amount of estrogen is also made in peripheral tissues.

In women, the ovary produces estrogens (particularly **estradiol** and *estrone*) and *progestins,* particularly progesterone. The ovary and the adrenal cortex produce small amounts of androgens.

FIG. 26.6 Hormonal control of reproduction

(a) In both sexes, the brain controls reproduction through GnRH and pituitary gonadotropins (FSH and LH).

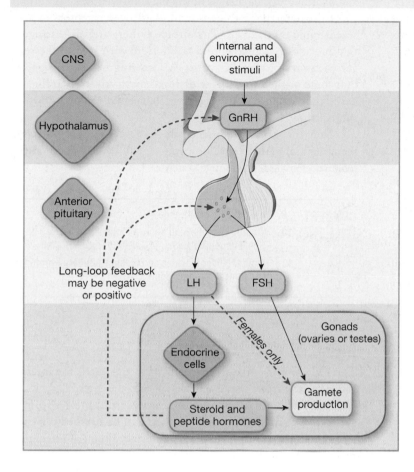

KEY

- Stimulus
- Integrating center
- Output signal
- Target
- Tissue response

GnRH = gonadotropin-releasing hormone

LH = luteinizing hormone

FSH = follicle-stimulating hormone

CNS

Internal and environmental stimuli

Hypothalamus

GnRH

Anterior pituitary

Long-loop feedback may be negative or positive

LH   FSH

Females only

Endocrine cells

Gonads (ovaries or testes)

Gamete production

Steroid and peptide hormones

(b) Feedback effects of sex steroids on gonadotropin release

Steroid Hormone	Effect	Gonadotropin Level
Low estrogen or androgen	Absence of negative feedback	Increases
Moderate estrogen or androgen	Negative feedback	Decreases
High androgen	Negative feedback	Decreases
Sustained high estrogen	Positive feedback	Increases

Control Pathways The hormonal control of reproduction in both sexes follows the basic hypothalamus–anterior pituitary–peripheral gland pattern (**FIG. 26.6**). **Gonadotropin-releasing hormone** (GnRH*) from the hypothalamus controls secretion of two anterior pituitary **gonadotropins: follicle-stimulating hormone (FSH)** and **luteinizing hormone (LH)**. FSH and LH in turn act on the gonads. FSH, along with steroid sex hormones, is required to initiate and maintain gametogenesis. LH acts primarily on endocrine cells, stimulating production of the steroid sex hormones.

Although primary control of gonadal function arises in the brain, the gonads also influence their own function. Both ovary and testis secrete peptide hormones that feed back to act directly on the pituitary. **Inhibins** inhibit FSH secretion, and related peptides called **activins** stimulate FSH secretion. Activins also promote spermatogenesis, oocyte maturation, and development of the embryonic nervous system. These gonadal peptides are produced in nongonadal tissues as well, and their other functions are still being investigated.

AMH, introduced earlier in the discussion of sexual differentiation during development, is also made by cells of both ovary and testis after birth. The inhibins, activins, and AMH are part of a large family of related growth and differentiation factors known as the *transforming growth factor*-β family.

Feedback Pathways The feedback loops of the reproductive system also become quite complex. The feedback pathways for trophic hormones follow the general patterns for negative feedback [p. 15]. Gonadal hormones feed back to alter secretion of GnRH, FSH, and LH in a long-loop response (Fig. 26.6a).

When circulating levels of gonadal steroids are low, the pituitary secretes FSH and LH (Fig. 26.6b). As steroid secretion increases, negative feedback usually inhibits gonadotropin release. Androgens always maintain negative feedback on gonadotropin release: as androgen levels go up, FSH and LH secretion decreases.

However, in an unusual twist, higher concentrations of estrogen can exert either positive or negative feedback. Low levels of estrogen have no feedback effect. Moderate concentrations of estrogen have a negative feedback effect. But if estrogen rises rapidly to a threshold level and remains high for at least 36 hours,

*GnRH is sometimes called *luteinizing hormone releasing hormone (LHRH)* because it was first thought to have its primary effect on LH.

feedback switches from negative to positive, and gonadotropin release (particularly LH) is *stimulated*. The paradoxical effects of estrogen on gonadotropin release play a significant role in the female reproductive cycle, as you will learn later in this chapter. Scientists still do not fully understand the mechanism underlying the change from negative to positive feedback but evidence points to estrogen influencing GnRH release by hypothalamic neurons.

Pulsatile GnRH Release Tonic GnRH release from the hypothalamus occurs in small pulses every 1–3 hours in both males and females. One region of the hypothalamus that contains the GnRH neuron cell bodies has been called a **GnRH pulse generator** because it apparently coordinates the periodic pulsatile secretion of GnRH.

Scientists wondered why tonic GnRH release occurred in pulses rather than in a steady fashion, but several studies have shown the significance of the pulses. Children who suffer from a GnRH deficiency will not mature sexually in the absence of gonadotropin stimulation of the gonads. If treated with steady infusions of GnRH through drug-delivery pumps, these children still fail to mature sexually. But if the pumps are adjusted to deliver GnRH in pulses similar to those that occur naturally, the children will go through puberty. Apparently, steady high levels of GnRH cause down-regulation of the GnRH receptors on gonadotropin cells, making the pituitary unable to respond to GnRH.

This receptor down-regulation is the basis for the therapeutic use of GnRH in treating certain disorders. For example, patients with prostate and breast cancers stimulated by androgens or estrogens may be given GnRH agonists to slow the growth of the cancer cells. It seems paradoxical to give these patients a drug that stimulates secretion of androgens and estrogens, but after a brief increase in FSH and LH, the pituitary becomes insensitive to GnRH. Then FSH and LH secretion decreases, and gonadal output of steroid hormones also falls. In essence, the GnRH agonist creates chemical castration that reverses when the drug is no longer administered.

Environmental Factors Influence Reproduction

Among the least-understood influences on reproductive hormones and gametogenesis are environmental effects. In men, factors that influence gametogenesis are difficult to monitor short of requesting periodic sperm counts. Disruption of the normal reproductive cycle in women is easier to study because physiological uterine bleeding in the menstrual cycle is easily monitored.

Factors that affect reproductive function in women include stress, nutritional status, and changes in the day–night cycle, such as those that occur with travel across time zones or with shift work. The hormone **melatonin** from the pineal gland [p. 221] mediates reproduction in seasonally breeding animals, such as birds and deer, and researchers are investigating whether melatonin also plays a role in seasonal and daily rhythms in humans.

Environmental estrogens are also receiving a lot of attention. These may be naturally occurring compounds, such as the

phytoestrogens of plants, or synthetic compounds that have been released into the environment. Man-made chemicals with estrogenic properties include plastics, pesticides, industrial chemicals, and pharmaceuticals such as hormonal contraceptives.

Environmental estrogens bind to estrogen receptors in both sexes. Some mimic estrogen's effects. Others are anti-estrogens that block estrogen receptors or interfere with second messenger pathways or protein synthesis. Growing evidence suggests that some of these endocrine disruptors can adversely influence developing embryos and even have their effects passed down to subsequent generations.

Now that you have learned the basic patterns of hormone secretion and gamete development, we will look in detail at the male and female reproductive systems.

CONCEPT CHECK

9. What does aromatase do?
10. What do the following abbreviations stand for? (Spelling counts!) FSH, DHT, SRY, LH, GnRH, AMH
11. Name the hypothalamic and anterior pituitary hormones that control reproduction.

MALE REPRODUCTION

The male reproductive system consists of the testes, the internal genitalia (accessory glands and ducts), and the external genitalia. The external genitalia consist of the **penis** and the **scrotum,** a saclike structure that contains the testes. The **urethra** serves as a common passageway for sperm and urine, although not simultaneously. It runs through the ventral aspect of the shaft of the penis (**FIG. 26.7a**) and is surrounded by a spongy column of tissue known as the **corpus spongiosum** {*corpus,* body; plural *corpora*}. The corpus spongiosum and two columns of tissue called the **corpora cavernosa** constitute the erectile tissue of the penis.

The tip of the penis is enlarged into a region called the **glans** that at birth is covered by a layer of skin called the **foreskin,** or **prepuce.** In some cultures, the foreskin is removed surgically in a procedure called **circumcision.** In the United States, circumcision of newborns goes through cycles of popularity. Opponents of circumcision claim that subjecting baby boys to this surgical procedure is unnecessary. Proponents of the procedure claim that it is necessary for good hygiene, and they cite evidence suggesting that the incidence of penile cancer, sexually transmitted diseases, and urinary tract infections is lower in circumcised men. Studies from Africa showed that circumcising heterosexual adult men helped prevent infection with the HIV virus that causes AIDS (acquired immunodeficiency syndrome).

The scrotum is an external sac into which the testes migrate during fetal development. This location outside the abdominal cavity is necessary because normal sperm development requires a temperature that is 2–3 °F lower than core body temperature. Men who have borderline or low sperm counts are advised to switch from jockey-style underwear, which keeps the scrotum close to the body, to boxer shorts, which allow the testes to stay cooler.

The failure of one or both testes to descend is known as **cryptorchidism** {*crypto,* hidden + *orchis,* testicle} and occurs in 1–3% of newborn males. If left alone, about 80% of cryptorchid testes spontaneously descend later. Those that remain in the abdomen through puberty become sterile and are unable to produce sperm.

Although cryptorchid testes lose their sperm-producing potential, they can produce androgens, indicating that hormone production is not as temperature sensitive as sperm production. Because undescended testes are prone to become cancerous, authorities recommend that they be moved to the scrotum with testosterone treatment or, if necessary, surgically.

The male accessory glands and ducts include the **prostate gland,** the **seminal vesicles,** and the **bulbourethral (Cowper's) glands** (Fig. 26.7b). The bulbourethral glands and seminal vesicles empty their secretions into the urethra through ducts. The individual glands of the prostate open directly into the urethral lumen.

The prostate gland is the best known of the three accessory glands because of its medical significance. Cancer of the prostate is the most common form of cancer in men, and *benign prostatic hyperplasia* (BHP) creates problems for many men after age 50. Because the prostate gland completely encircles the urethra, its enlargement causes difficulty in urinating by narrowing the passageway.

Fetal development of the prostate gland, like that of the external genitalia, is under the control of dihydrotestosterone. Discovery of the role of DHT in prostate growth led to the development of *finasteride,* a 5α-reductase inhibitor that blocks DHT production. This drug was the first nonsurgical treatment for benign prostatic hyperplasia.

The Prostate Cancer Prevention Trial (PCPT) was a placebo-controlled study to see if finasteride could also prevent prostate cancer. Nearly 19,000 men participated, with half of them receiving the drug and half receiving a placebo. The trial was stopped a year early after analysis of the results showed that the risk of developing prostate cancer fell by 25% in the men taking the drug.

Testes Produce Sperm and Hormones

The human testes are paired ovoid structures about 5 cm by 2.5 cm (Fig. 26.7a). The word *testis* means "witness" in Latin, but the reason for its application to male reproductive organs is not clear. Testes are also called *testicles.*

The testes have a tough outer fibrous capsule that encloses masses of coiled **seminiferous tubules** clustered into 250–300 compartments (Fig. 26.7c). Interstitial tissue with blood vessels and testosterone-producing **interstitial cells of Leydig** lies between the tubules (Fig. 26.7e). The seminiferous tubules constitute nearly 80% of the testicular mass in an adult. Each individual tubule is 0.3–1 meter long, and, if stretched out and laid end to end, the entire mass would extend for about the length of two and a half football fields.

The seminiferous tubules leave the testis and join the **epididymis** {*epi-,* upon + *didymos,* twin}, a single duct that forms a tightly coiled cord on the surface of the testicular capsule (Fig. 26.7c). The epididymis becomes the **vas deferens** {*vas,*

vessel + *deferre,* to carry away from}, also known as the **ductus deferens.** This duct passes into the abdomen, where it eventually empties into the urethra, the passageway from the urinary bladder to the external environment (see Fig. 26.7a).

Seminiferous Tubules The seminiferous tubules are the site of sperm production and contain two types of cells: spermatogonia in various stages of becoming sperm plus supportive **Sertoli cells** (Fig. 26.7d, e). The developing spermatocytes stack in columns from the outer edge of the tubule to the lumen. Between each column is a single Sertoli cell that extends from the outer edge of the tubule to the lumen. Surrounding the outside of the tubule is a basal lamina (Fig. 26.7e) that acts as a barrier, preventing certain large molecules in the interstitial fluid from entering the tubule but allowing testosterone to enter easily. The basolateral ends of the Sertoli cells rest on the basal lamina, creating a *basal compartment* between the cells and the lamina. The apical ends of the Sertoli cells face the tubule lumen.

Adjacent Sertoli cells in a tubule are linked to each other by tight junctions that form an additional barrier between the lumen of the tubule and the interstitial fluid outside the basal lamina. These tight junctions are sometimes called the **blood-testis barrier** because functionally they behave much like the impermeable capillaries of the blood-brain barrier, restricting movement of molecules between compartments. The basal lamina and tight junctions create three functional compartments: the tubule lumen, the basal compartment on the basolateral side of the Sertoli cells, and the interstitial fluid outside the basal lamina. Because of the barriers between these compartments, the luminal fluid has a composition different from that of interstitial fluid, with low concentrations of glucose and high concentrations of K^+ and steroid hormones.

Sertoli Cells The function of Sertoli cells is to regulate sperm development. Another name for Sertoli cells is *sustentacular cells* because they provide sustenance, or nourishment, for the developing spermatogonia. Sertoli cells manufacture and secrete proteins that range from the hormones inhibin and activin to growth factors, enzymes, and **androgen-binding protein** (ABP). ABP is secreted into the seminiferous tubule lumen, where it binds to testosterone (**FIG. 26.8**). Testosterone bound to protein is less lipophilic and cannot diffuse out of the tubule lumen.

Interstitial Cells Interstitial (Leydig) cells, located in the interstitial tissue between seminiferous tubules (Fig. 26.7d, e), secrete testosterone. They first become active in the fetus, when testosterone is needed to direct development of male characteristics. After birth, the cells become inactive. At puberty, they resume testosterone production. The interstitial cells also convert some testosterone to estradiol.

Sperm Production Spermatogonia, the germ cells that undergo meiotic division to become sperm, are found clustered near the basal ends of the Sertoli cells, just inside the basal lamina of the seminiferous tubules (Fig. 26.7d, e). In this basal compartment, they undergo mitosis to create additional germ cells. Some of the spermatogonia remain here to produce future

FIG. 26.7 ANATOMY SUMMARY

The Male Reproductive System

(a) Reproductive anatomy of the male

Ureter

Urinary bladder

Prostate gland surrounds the urethra.

Seminal vesicle

Urethra

Vas (ductus) deferens transports sperm from testes to urethra.

Bulbourethral gland

Opening to ejaculatory duct

Corpus spongiosum

Dorsal blood vessels

Corpora cavernosa

Corpora cavernosa

Penis

Central artery

Glans

Corpus spongiosum

Prepuce (foreskin)

Testis

Urethra

The scrotum holds the testes outside the abdominal cavity to keep them below body core temperature.

(b) Lateral view

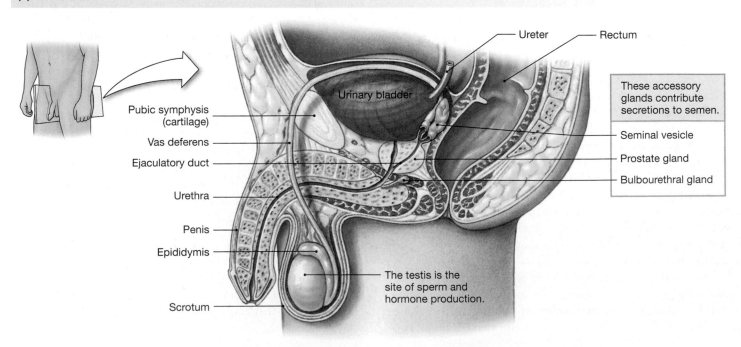

Ureter

Rectum

Urinary bladder

Pubic symphysis (cartilage)

Vas deferens

Ejaculatory duct

Urethra

Penis

Epididymis

Scrotum

These accessory glands contribute secretions to semen.

Seminal vesicle

Prostate gland

Bulbourethral gland

The testis is the site of sperm and hormone production.

(c) Cutaway view of a testis showing coiled tubules

- Head of epididymis
- Seminiferous tubule
- Epididymis
- Vas deferens
- Scrotal cavity

(d) Cross section of a seminiferous tubule

- Interstitial cell
- Capillary
- Sertoli cell
- Lumen
- Spermatogonium

(e) Sperm development

- Mature sperm
- Germ cells
- Spermatozoa
- Spermatid
- Secondary spermatocyte
- Primary spermatocyte
- Spermatogonium
- Capillary
- Lumen of seminiferous tubule
- Luminal fluid composition is high in K⁺ and steroid hormones.
- Sertoli cells secrete proteins to support sperm production.
- Tight junction between Sertoli cells
- Basal lamina
- Fibroblast
- Interstitial tissue
- Interstitial cells secrete testosterone.

(f) Semen is composed of sperm and secretions from the accessory glands.

COMPONENT	FUNCTION	SOURCE
Sperm	Gametes	Seminiferous tubules
Mucus	Lubricant	Bulbourethral glands
Water	Provides liquid medium	All accessory glands
Buffers	Neutralize acidic environment of vagina	Prostate, bulbo-urethral glands
Nutrients	Nourish sperm	
Fructose		Seminal vesicles
Citric acid		Prostate
Vitamin C		Seminal vesicles
Carnitine		Epididymis
Enzymes	Clot semen in vagina, then liquefy the clot	Seminal vesicles and prostate
Zinc	Unknown, possible association with fertility	Unknown
Prostaglandins	Smooth muscle contraction; may aid sperm transport	Seminal vesicles

(g) A sperm consists of a head with enzymes and DNA, a long tail, and mitochondria.

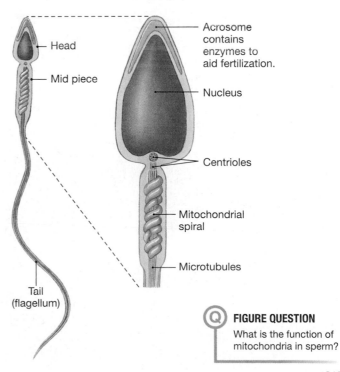

- Head
- Mid piece
- Tail (flagellum)
- Acrosome contains enzymes to aid fertilization.
- Nucleus
- Centrioles
- Mitochondrial spiral
- Microtubules

FIGURE QUESTION

What is the function of mitochondria in sperm?

spermatogonia. Other spermatogonia start meiosis and become primary spermatocytes.

As spermatocytes differentiate into sperm, they move inward toward the tubule lumen, continuously surrounded by Sertoli cells. The tight junctions of the blood-testis barrier break and reform around the migrating cells, ensuring that the barrier remains intact. By the time spermatocytes reach the luminal ends of Sertoli cells, they have divided twice and become spermatids (Fig. 26.5).

Spermatids remain embedded in the apical membrane of Sertoli cells while they complete the transformation into sperm, losing most of their cytoplasm and developing a flagellated tail

(Fig. 26.7g). The chromatin of the nucleus condenses into a dense structure that fills most of the head, while a lysosome-like vesicle called an **acrosome** flattens out to form a cap over the tip of the nucleus. The acrosome contains enzymes essential for fertilization. Mitochondria to produce energy for sperm movement concentrate in the midpiece of the sperm body, along with microtubules that extend into the tail [p. 68]. The result is a small, motile gamete that bears little resemblance to the parent spermatid.

Sperm are released into the lumen of the seminiferous tubule, along with secreted fluid. From there, they are free to move out of the testis. The entire development process—from spermatogonium division until sperm release—takes about 64 days. At any given time, different regions of the tubule contain spermatocytes in different stages of development. The staggering of developmental stages allows sperm production to remain nearly constant at a rate of *200 million* sperm per day. That may sound like an extraordinarily high number, but it is about the number of sperm released in a single ejaculation.

Sperm just released from Sertoli cells are not yet mature and are incapable of swimming. They are pushed out of the tubule lumen by other sperm and by bulk flow of the fluid secreted by Sertoli cells. Sperm complete their maturation in the epididymis during the 12 or so days of their transit time, aided by protein secretions from epididymal cells.

FIG. 26.8 Hormonal control of spermatogenesis

Compare this to the general pattern shown in Figure 26.6.

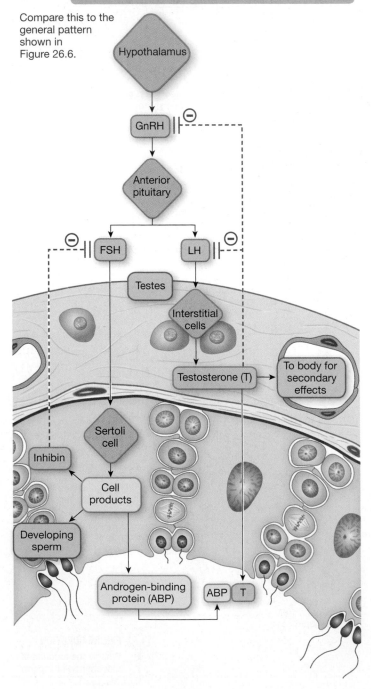

RUNNING **PROBLEM**

Infertility can be caused by problems in either the man or the woman. Sometimes, however, both partners have problems that contribute to their infertility. In general, male infertility is caused by low sperm counts, abnormalities in sperm morphology, or abnormalities in the reproductive structures that carry sperm. Female infertility may be caused by problems in hormonal pathways that govern maturation and release of eggs or by abnormalities of the reproductive structures (cervix, uterus, Fallopian tubes, ovaries). Because tests of male fertility are simpler to perform, Dr. Baker first orders an analysis of Jon's sperm. In this test, trained technicians examine a fresh sperm sample under a microscope. They note the shape and motility of the sperm and estimate the concentration of sperm in the sample.

Q1: *Name (in order) the male reproductive structures that carry sperm from the seminiferous tubules to the external environment.*

Q2: *In what male reproductive structure do sperm reach maturity?*

Q3: *One technique for the treatment of male infertility involves retrieval of sperm from the epididymis. The retrieved sperm can be used to fertilize an egg, which is then implanted in the uterus. Which causes of male infertility might make this treatment necessary?*

801　**814**　817　823　830　834

Spermatogenesis Requires Gonadotropins and Testosterone

The hormonal control of spermatogenesis follows the general pattern described previously: hypothalamic GnRH promotes release of LH and FSH from the anterior pituitary (Fig. 26.8). FSH and LH in turn stimulate the testes. The gonadotropins were named originally for their effect on the female ovary, but the same names have been retained in the male.

GnRH release is pulsatile, peaking every 1.5 hours, and LH release follows the same pattern. FSH levels are not as obviously related to GnRH secretion because FSH secretion is also influenced by inhibin and activin.

FSH targets Sertoli cells. Unlike oocytes, male germ cells do not have FSH receptors. Instead, FSH stimulates Sertoli synthesis of paracrine molecules needed for spermatogonia mitosis and spermatogenesis. In addition, FSH stimulates production of androgen-binding protein and inhibin.

The primary target of LH is the interstitial cells, which produce testosterone. In turn, testosterone feeds back to inhibit LH and GnRH release. Testosterone is essential for spermatogenesis, but its actions appear to be mediated by Sertoli cells, which have androgen receptors. Spermatocytes lack androgen receptors and cannot respond directly to testosterone.

Spermatogenesis is a very difficult process to study *in vivo* or *in vitro*, and the available animal models may not accurately reflect the situation in the human testis. For these reasons, it may be some time before we can say with certainty how testosterone and FSH regulate spermatogenesis.

CONCEPT CHECK

12. What do Sertoli cells secrete? What do interstitial cells secrete?

13. Because GnRH agonists cause down-regulation of GnRH receptors, what would be the advantages and disadvantages of using these drugs as a male contraceptive?

14. Which cells of the testes have receptors for FSH? For LH? For androgens?

Male Accessory Glands Contribute Secretions to Semen

The male reproductive tract has three accessory glands—bulbourethral glands, seminal vesicles, and prostate—whose primary function is to secrete various fluid mixtures (Fig. 26.7b). When sperm leave the vas deferens during ejaculation, they are joined by these secretions, resulting in a sperm-fluid mixture known as **semen.** About 99% of the volume of semen is fluid added from the accessory glands.

Accessory gland contributions to the composition of semen are listed in Figure 26.7f. Semen provides a liquid medium for delivering sperm. The bulbourethral glands contribute mucus for lubrication plus buffers to neutralize the usually acidic environment of the vagina. Seminal vesicles contribute prostaglandins

[p. 179] that appear to influence sperm motility and transport in both male and female reproductive tracts. Prostaglandins were originally believed to come from the prostate gland, and their name was well established by the time their true source was discovered. Both the prostate and seminal vesicles contribute nutrients for sperm metabolism.

In addition to providing a medium for sperm, accessory gland secretions help protect the male reproductive tract from pathogens that might ascend the urethra from the external environment. The secretions physically flush out the urethra and supply immunoglobulins, lysozyme, and other compounds with antibacterial action. One interesting component of semen is zinc. Its role in reproduction is unclear, but concentrations of zinc below a certain level are associated with male infertility.

Androgens Influence Secondary Sex Characteristics

Androgens have a number of effects on the body in addition to gametogenesis. These effects are divided into primary and secondary sex characteristics. **Primary sex characteristics** are the internal sexual organs and external genitalia that distinguish males from females. As you have already learned, androgens are responsible for the differentiation of male genitalia during embryonic development and for their growth during puberty.

The **secondary sex characteristics** are other traits that distinguish males from females. The male body shape is sometimes described as an inverted triangle, with broad shoulders and narrow waist and hips. The female body is usually more pear shaped, with broad hips and narrow shoulders. Androgens are responsible for such typically male traits as beard and body hair growth, muscular development, thickening of the vocal chords with subsequent lowering of the voice, and behavioral effects, such as the sex drive, also called **libido** {*libido,* desire or lust}.

Androgens are anabolic hormones that promote protein synthesis, which gives them their street name of *anabolic steroids*. The illicit use of these drugs by athletes has been widespread despite possible adverse side effects such as liver tumors, infertility, and excessive aggression (*'roid rage*). One of the more interesting side effects is the apparent addictiveness of anabolic steroids. Withdrawal from the drugs may be associated with behavioral changes that include depression, psychosis, or aggression. These psychiatric disturbances suggest that human brain function can be modulated by sex steroids, just as the brain function of other animals can. Fortunately, many side effects of anabolic steroids are reversible once their use is discontinued.

CONCEPT CHECK

15. Diagram the pathways that explain why use of exogenous anabolic steroids might shrink a man's testes and make him temporarily infertile.

FEMALE REPRODUCTION

Female reproduction is an example of a physiological process that is cyclic rather than steady state. The cycles of gamete production in the ovary and the interactions of reproductive hormones and feedback pathways are part of one of the most complex control systems of the human body.

The female external genitalia are collectively known as either the **vulva** or the **pudendum** {*vulva,* womb; *pudere,* to be ashamed}. They are shown in **FIGURE 26.9c**, the view seen by a healthcare provider who is about to do a pelvic exam or take a Pap smear [p. 87].

Starting at the periphery are the **labia majora** {*labium,* lip}, folds of skin that arise from the same embryonic tissue as the scrotum. Within the labia majora are the **labia minora,** derived from embryonic tissues that in the male give rise to the shaft of the penis (see Fig. 26.2b). The **clitoris** is a small bud of erectile, sensory tissue at the anterior end of the vulva, enclosed by the labia minora and an additional fold of tissue equivalent to the foreskin of the penis.

In females, the urethra opens to the external environment between the clitoris and the vagina {*vagina,* sheath}, the cavity that acts as receptacle for the penis during intercourse. At birth, the external opening of the vagina is partially closed by a thin ring of tissue called the **hymen,** or *maidenhead.* The hymen is external to the vagina, not within it, so the normal use of tampons during menstruation will not rupture the hymen. However, it can be stretched by normal activities such as cycling and horseback riding and therefore is not an accurate indicator of a woman's virginity.

Now let's follow the path of sperm deposited in the vagina during intercourse. To continue into the female reproductive tract, sperm must pass through the narrow opening of the **cervix,** the neck of the uterus that protrudes slightly into the upper end of the vagina (Fig. 26.9a). The cervical canal is lined with mucous glands whose secretions create a protective barrier between the vagina and uterus.

Sperm that make it through the cervical canal arrive in the lumen of the uterus, or *womb,* a hollow, muscular organ slightly smaller than a woman's clenched fist. The uterus is the structure in which fertilized eggs implant and develop during pregnancy. It is composed of three tissue layers (Fig. 26.9d): a thin outer connective tissue covering, a thick middle layer of smooth muscle known as the **myometrium,** and an inner layer known as the **endometrium** {*metra,* womb}.

The endometrium consists of an epithelium with glands that dip into a connective tissue layer below. The thickness and character of the endometrium vary during the menstrual cycle. Cells of the epithelial lining alternately proliferate and slough off, accompanied by a small amount of bleeding in the process known as **menstruation** {*menstruus,* monthly}.

Sperm swimming upward through the uterus leave its cavity through openings into the two Fallopian tubes (Fig. 26.9a). The Fallopian tubes are 20–25 cm long and about the diameter of a drinking straw. Their walls have two layers of smooth muscle, longitudinal and circular, similar to the walls of the intestine. A ciliated epithelium lines the inside of the tubes.

Fluid movement created by the cilia and aided by muscular contractions transports an egg along the Fallopian tube toward the uterus. If sperm moving up the tube encounter an egg moving down the tube, fertilization may occur. Pathological conditions in which ciliary function is absent are associated with female infertility and with pregnancies in which the embryo implants in the Fallopian tube rather than the uterus.

The flared open end of the Fallopian tube divides into finger-like projections called **fimbriae** {*fimbriae,* fringe}. The fimbriae (Fig. 26.9a) are held close to the adjacent ovary by connective tissue, which helps ensure that eggs released from the surface of the ovary will be swept into the tube rather than floating off into the abdominal cavity.

The Ovary Produces Eggs and Hormones

The ovary is an elliptical structure, about 2–4 cm long (Fig. 26.9e). It has an outer connective tissue layer and an inner connective tissue framework known as the **stroma** {*stroma,* mattress}. Most of the ovary consists of a thick outer *cortex* filled with ovarian follicles in various stages of development or decline. The small central *medulla* contains nerves and blood vessels.

The ovary, like the testis, produces both gametes and hormones. As mentioned earlier, about 7 million oögonia in the embryonic ovary develop into half a million primary oocytes. Each primary oocyte is surrounded by a single layer of **granulosa cell** precursors and enclosed in a basal lamina, forming a **primordial follicle** (**FIG. 26.10**). Most of these primordial follicles will never develop and will die over a period of years in an apoptosis-like process called *atresia* (hormonally regulated cell death).

Some primordial follicles develop slowly into **primary follicles.** The oocyte enlarges and the granulosa cells begin to divide but remain in a single layer. At puberty, chemical signals cause groups of primary follicles to leave their resting state and enter a period of active growth that may take months. As the growing follicles enlarge, a layer of cells known as the **theca** {*theke,* case or cover} develops outside the basal lamina. At this point, the follicles are known as preantral or **secondary follicles.** Some primary follicles never complete the transition to secondary follicles and are lost through atresia.

As the secondary follicles enlarge, granulosa cells begin to secrete fluid that collects in a central cavity in the follicle known as the **antrum** {*antron,* cave}. Antral fluid contains hormones and enzymes needed for ovulation. At this point, the follicle becomes a **tertiary follicle.** From the pool of early tertiary follicles, only a few follicles survive to reach the final growth stages, and usually only one, called the *dominant follicle,* develops to the point where it releases its egg. The time required for growth from secondary follicle to the selection of a dominant tertiary follicle is estimated to be three months or more.

A Menstrual Cycle Lasts about One Month

Female humans produce gametes in monthly cycles (average 28 days; normal range 24–35 days). These cycles are commonly called **menstrual cycles** because they are marked by a 3–7 day period of bloody uterine discharge known as the **menses** {*menses,* months}, or **menstruation.** The menstrual cycle can be described by following changes that occur in follicles of the ovary, the **ovarian cycle,** or by following changes in the endometrial lining of the uterus, the **uterine cycle. FIGURE 26.11** is a summary figure showing a typical menstrual cycle and its phases.

Notice that the ovarian cycle is divided into three phases:

1. **Follicular phase.** The first part of the ovarian cycle, known as the **follicular phase,** is a period of follicular growth in the ovary. This phase is the most variable in length and lasts from 10 to 21 days.

2. **Ovulation.** Once one or more follicles have ripened, the ovary releases the oocyte(s) during **ovulation.**

3. **Luteal phase.** The phase of the ovarian cycle following ovulation is known as the *postovulatory* or **luteal phase.** The second name comes from the transformation of a ruptured follicle into a **corpus luteum** {*corpus,* body + *luteus,* yellow}, named for its yellow pigment and lipid deposits. The corpus luteum secretes hormones that continue the preparations for pregnancy. If a pregnancy does not occur, the corpus luteum ceases to function after about two weeks, and the ovarian cycle begins again.

The endometrial lining of the uterus also goes through a cycle—the uterine cycle—regulated by ovarian hormones:

1. **Menses.** The beginning of the follicular phase in the ovary corresponds to menstrual bleeding from the uterus.

2. **Proliferative phase.** The latter part of the ovary's follicular phase corresponds to the **proliferative phase** in the uterus, during which the endometrium adds a new layer of cells in anticipation of pregnancy.

3. **Secretory phase.** After ovulation, hormones from the corpus luteum convert the thickened endometrium into a secretory structure. This means that the luteal phase of the ovarian cycle corresponds to the **secretory phase** of the uterine cycle. If no pregnancy occurs, the superficial layers of the secretory endometrium are lost during menstruation as the uterine cycle begins again.

Hormonal Control of the Menstrual Cycle Is Complex

The ovarian and uterine cycles are under the primary control of various hormones:

- GnRH from the hypothalamus
- FSH and LH from the anterior pituitary
- Estrogen, progesterone, inhibin, and AMH from the ovary

RUNNING **PROBLEM**

The results of Jon's sperm analysis are normal. Dr. Baker is therefore able to rule out sperm abnormalities as a cause of Kate and Jon's infertility. Kate, who just turned 30, has a history of irregular menstrual periods. Dr. Baker asks Kate to take her body temperature daily for several months and record the results on a chart. This temperature tracking is intended to determine whether or not she is ovulating. Following ovulation, body temperature rises slightly and remains elevated through the remainder of the menstrual cycle.

Q4: *For which causes of female infertility is temperature tracking useful? For which causes is it not useful?*

801 814 **817** 823 830 834

During the follicular phase of the cycle, estrogen is the dominant steroid hormone (Fig. 26.11). Ovulation is triggered by surges in LH and FSH. In the luteal phase, progesterone is dominant, although estrogen is still present.

Anti-Mullerian hormone (AMH) was first known for its role in male development, but scientists have discovered that AMH is also produced by ovarian follicles in the first part of the ovarian cycle. AMH apparently acts as a brake to keep too many follicles from developing at one time.

Now, we will examine an ovarian cycle in detail.

Early Follicular Phase The first day of menstruation is day 1 of a cycle. This point was chosen to start the cycle because the bleeding of menstruation is an easily observed physical sign. Just before the beginning of each cycle, gonadotropin secretion from the anterior pituitary increases. Under the influence of FSH, a group of tertiary follicles in the ovaries begins to mature (Fig. 26.10 and second row of Fig. 26.11).

As the follicles grow, their granulosa cells (under the influence of FSH) and their thecal cells (under the influence of LH) start to produce steroid hormones (**FIG. 26.12a**). Granulosa cells also begin to secrete AMH. This AMH decreases follicle sensitivity to FSH, which apparently prevents recruitment of additional follicles once one group has started developing. Physicians now use blood AMH levels as an indicator of how many follicles are developing early in a cycle and as a marker for the condition known as *polycystic ovary syndrome (PCOS),* in which ovarian follicles form fluid-filled cysts.

Thecal cells synthesize androgens that diffuse into the neighboring granulosa cells, where aromatase converts them to estrogens (Fig. 26.12a). Gradually increasing estrogen levels in the circulation have several effects. Estrogen exerts negative feedback on pituitary FSH and LH secretion, which prevents the development of additional follicles in the same cycle. At the same time, estrogen stimulates additional estrogen production by the granulosa cells. This positive feedback loop allows the follicles to

FIG. 26.9 **ANATOMY SUMMARY**

The Female Reproductive System

(a) Internal reproductive structures

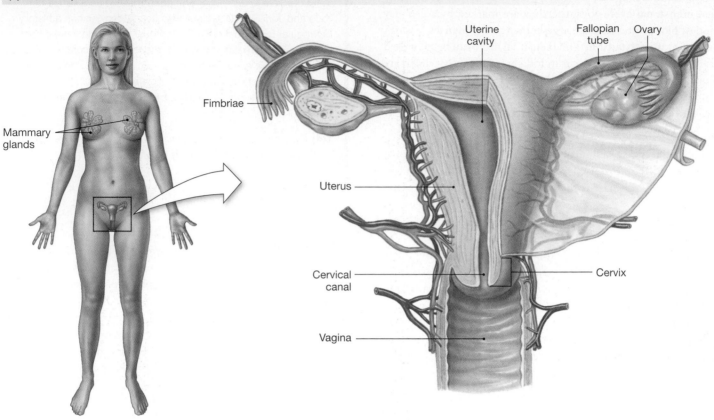

Uterine cavity

Fallopian tube

Ovary

Fimbriae

Mammary glands

Uterus

Cervical canal

Cervix

Vagina

(b) Cross-sectional view of pelvis

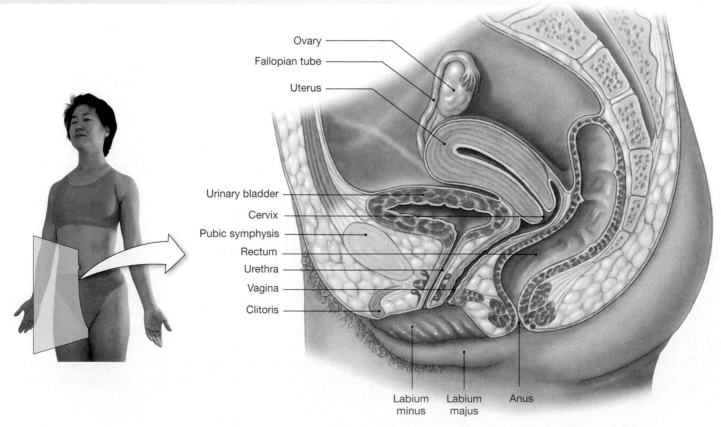

Ovary

Fallopian tube

Uterus

Urinary bladder

Cervix

Pubic symphysis

Rectum

Urethra

Vagina

Clitoris

Labium minus

Labium majus

Anus

(c) Female external genitalia

This is the view seen by a healthcare provider doing a pelvic exam.

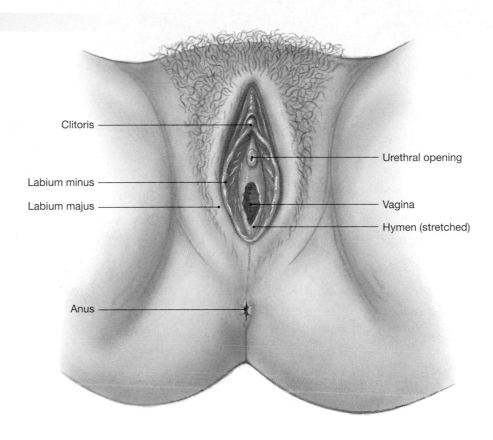

Clitoris

Urethral opening

Labium minus

Labium majus

Vagina

Hymen (stretched)

Anus

(d) Structure of the uterus

Endometrium is glandular epithelium whose structure varies with phases of the menstrual cycle.

Myometrium is smooth muscle

Outer connective tissue

Uterine cavity

Uterine artery

(e) Schematic cross section of an ovary, showing all different stages of follicular development.

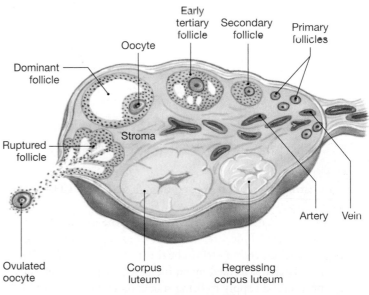

Early tertiary follicle

Secondary follicle

Primary follicles

Oocyte

Dominant follicle

Ruptured follicle

Stroma

Ovulated oocyte

Corpus luteum

Regressing corpus luteum

Artery

Vein

FIG. 26.10 | Follicular development

Growth from a secondary follicle to ovulation requires multiple ovarian cycles.

Theca
Granulosa cells
Surface epithelium
Basal lamina
Antral fluid

Ovulation
Antrum

Primordial follicle
Primary follicle
Secondary follicle
Tertiary follicle
Dominant follicle
Corpus luteum
Corpus albicans

STAGE	PRIMORDIAL FOLLICLE	PRIMARY FOLLICLE	SECONDARY FOLLICLE	TERTIARY FOLLICLE	SELECTED PREOVULATORY FOLLICLE (Dominant follicle)	CORPUS LUTEUM	POST-LUTEAL CORPUS ALBICANS
Activity	Resting	Growing	Growing, preantral	Growing, antral	Rapid growth	Secretes hormones	None
Size	≈ 0.03 mm	≈ 0.04–0.1 mm	≈ 0.1–0.2 mm	≈ 0.2–2 mm	≈ 5 mm up to 20–30 mm		
Ovum	Small primary oocyte (4N DNA)	Enlarges. Primary oocyte	Enlarges. Primary oocyte	Enlarges. Primary oocyte	Meiosis resumes to form secondary oocyte (2N) plus 2N polar body.	None	None
Zona pellucida*		Appears	Present	Present	Present	None	None
Granulosa cells	Precursor cells	Single layer. Cell number increases.	3–6 layers	Multiple layers	Increase in cell numbers	Converted to luteal cells	Luteal cells degenerate
Antrum				Develops within granulosa layer and fills with fluid.	Increases in size	Fills with migrating cells	None
Basal lamina	Present	Present	Present	Present	Present	Disappears	
Theca			Appears and begins to form 2 layers	*Inner layer*: secretory cells and small blood vessels *Outer layer*: connective tissue, smooth muscle, large blood vessels		Converted to luteal cells	Luteal cells degenerate
Vascularization			Appears	Blood vessels in the theca	Increases	Increases	Disappears

*The zona pellucida is a glycoprotein coat that protects the ovum.

continue estrogen production even though FSH and LH levels decrease.

In the uterus, menstruation ends during the early follicular phase (Fig. 26.11). Under the influence of estrogen from developing follicles, the endometrium begins to grow, or proliferate. This period is characterized by an increase in cell number and by enhanced blood supply to bring nutrients and oxygen to the thickening endometrium. Estrogen also causes mucous glands of the cervix to produce clear, watery mucus.

Mid- to Late Follicular Phase As the follicular phase nears its end, ovarian estrogen secretion peaks (Fig. 26.12b). By this point of the cycle, only one follicle is still developing. As the follicular phase ends, granulosa cells of the dominant follicle begin to secrete inhibin and progesterone in addition to estrogen. Estrogen, which had exerted a negative feedback effect on GnRH earlier in the follicular phase, changes to positive feedback, leading to a preovulatory GnRH surge.

Immediately before ovulation, the persistently high levels of estrogen, aided by rising levels of progesterone, enhance pituitary

FIG. 26.11 The menstrual cycle

This 28-day menstrual cycle is divided into phases based on
events in the ovary (ovarian cycle) and in the uterus (uterine cycle).

PHASES OF THE OVARIAN CYCLE	FOLLICULAR PHASE	OVULATION	LUTEAL PHASE

Gonadotrophin levels

LH

FSH

Ovarian cycle

Tertiary follicle — Antrum — Dominant follicle — **Ovulation** — Corpus luteum formation — Mature corpus luteum

Ovarian hormone levels

Progesterone

Estrogen

Inhibin

Uterine cycle

PHASES OF THE UTERINE CYCLE	MENSES	PROLIFERATIVE PHASE	SECRETORY PHASE

Basal body temperature (°C)

36.7

36.4

DAYS   28/0   7   14   21   28/0 responsiveness to GnRH. As a result, LH secretion increases dramatically, a phenomenon known as the *LH surge*. FSH surges also, but to a lesser degree, presumably because it is being suppressed by inhibin and estrogen.

The LH surge is an essential part of ovulation because it triggers the secretion of numerous chemical signals necessary for the final steps of oocyte maturation. Meiosis resumes in the developing follicle with the first meiotic division. This step divides the primary oocyte into a secondary oocyte (2N DNA) and a first polar body (2N), which disintegrates (Fig. 26.5). While this

division is taking place, antral fluid collects and the follicle grows to its greatest size, preparing to release the egg.

High levels of estrogen in the late follicular phase prepare the uterus for a possible pregnancy. The endometrium grows to a thickness of 3–4 mm (Fig. 26.11). Just before ovulation, the cervical glands produce copious amounts of thin, stringy mucus to facilitate sperm entry. The stage is set for ovulation.

Ovulation   About 16–24 hours after LH peaks, ovulation occurs (Fig. 26.11). The mature follicle secretes prostaglandins and

FIG. 26.12   Hormonal control of the menstrual cycle

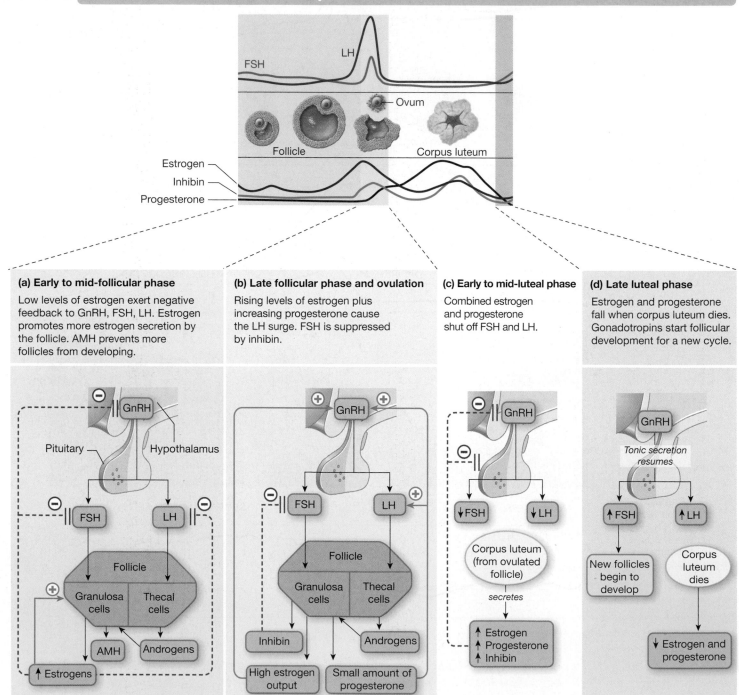

(a) Early to mid-follicular phase
Low levels of estrogen exert negative feedback to GnRH, FSH, LH. Estrogen promotes more estrogen secretion by the follicle. AMH prevents more follicles from developing.

(b) Late follicular phase and ovulation
Rising levels of estrogen plus increasing progesterone cause the LH surge. FSH is suppressed by inhibin.

(c) Early to mid-luteal phase
Combined estrogen and progesterone shut off FSH and LH.

(d) Late luteal phase
Estrogen and progesterone fall when corpus luteum dies. Gonadotropins start follicular development for a new cycle.

proteolytic enzymes such as *matrix metalloproteinases* (MMPs) [p. 74] that dissolve collagen and other components of the connective tissue holding follicular cells together. The prostaglandins may contribute to the follicle wall rupturing at its weakest point. Antral fluid spurts out along with the egg, which is surrounded by two to three layers of granulosa cells. The egg is swept into the Fallopian tube and carried away to be fertilized or die.

Early to Mid-Luteal Phase

After ovulation, follicular thecal cells migrate into the antral space, mingling with the former granulosa cells and filling the cavity. Both cell types then transform into *luteal cells* of the corpus luteum. This process, known as *luteinization,* involves biochemical and morphological changes. The newly formed luteal cells accumulate lipid droplets and glycogen granules in their cytoplasm and begin to secrete hormones. As the luteal phase progresses, the corpus luteum produces steadily increasing amounts of progesterone, estrogen, and inhibin.

Progesterone is the dominant hormone of the luteal phase. Estrogen synthesis diminishes initially, then increases. However, estrogen levels never reach the peak seen before ovulation.

The combination of estrogen and progesterone exerts negative feedback on the hypothalamus and anterior pituitary (Fig. 26.12c). Gonadotropin secretion, further suppressed by luteal inhibin production, remains shut down throughout most of the luteal phase.

Under the influence of progesterone, the endometrium continues its preparation for pregnancy and becomes a secretory structure. Endometrial glands coil, and additional blood vessels grow into the connective tissue layer. Endometrial cells deposit lipids and glycogen in their cytoplasm. These deposits will provide nourishment for a developing embryo while the **placenta,** the fetal-maternal connection, is developing.

Progesterone also causes cervical mucus to thicken. Thicker mucus creates a plug that blocks the cervical opening, preventing bacteria as well as sperm from entering the uterus.

One interesting effect of progesterone is its thermogenic ability. During the luteal phase of an ovulatory cycle, a woman's basal body temperature, taken immediately upon awakening and before getting out of bed, jumps 0.3–0.5 °F and remains elevated until menstruation. Because this change in the temperature setpoint occurs after ovulation, it cannot be used effectively to predict ovulation. However, it is a simple way to assess whether a woman is having ovulatory or *anovulatory* (non-ovulating) cycles.

Late Luteal Phase and Menstruation

The corpus luteum has an intrinsic life span of approximately 12 days. If pregnancy does not occur, the corpus luteum spontaneously undergoes apoptosis. As the luteal cells degenerate, progesterone and estrogen production decrease (Fig. 26.12d). This fall removes the negative feedback signal to the pituitary and hypothalamus, so secretion of FSH and LH increases. The remnants of the corpus luteum become an inactive structure called a **corpus albicans** {*albus,* white}

Maintenance of a secretory endometrium depends on the presence of progesterone. When the corpus luteum degenerates and hormone production decreases, blood vessels in the surface

layer of the endometrium contract. Without oxygen and nutrients, the surface cells die. About two days after the corpus luteum ceases to function, or 14 days after ovulation, the endometrium begins to slough its surface layer, and menstruation begins.

Menstrual discharge from the uterus totals about 40 mL of blood and 35 mL of serous fluid and cellular debris. There are usually few clots of blood in the menstrual flow because of the presence of *plasmin* [p. 528], which breaks up clots. Menstruation continues for 3–7 days, well into the follicular phase of the next ovulatory cycle.

Hormones Influence Female Secondary Sex Characteristics

Estrogens control the development of primary sex characteristics in females, just as androgens control them in males. Estrogens also control the most prominent female secondary sex traits: breast development and the female pattern of fat distribution (hips and upper thighs). Other female secondary sex characteristics, however, are governed by androgens produced in the adrenal cortex. Pubic and axillary (armpit) hair growth and libido (sex drive) are under the control of adrenal androgens.

CONCEPT CHECK

16. Name the phases of the ovarian cycle and the corresponding phases of the uterine cycle.

17. What side effects would you predict in female athletes who take anabolic steroids to build muscles?

18. Aromatase converts testosterone to estrogen. What would happen to the ovarian cycle of a woman given an aromatase inhibitor?

19. On what day of the menstrual cycle will a woman with the following cycle lengths ovulate?

    (a) 28 days   (b) 23 days   (c) 31 days

TABLE 26.1	Efficacy of Various Contraceptive Methods
Method	**Pregnancy Rate with Typical Use***
No contraception	85%
Spermicides	28%
Abstinence during times of predicted fertility	24%
Diaphragm, cervical cap, or sponge	12–24%**
Oral contraceptive pills	9%
Intrauterine devices (IUDs)	<1%
Contraceptive hormone implant	<1%
Male condom	18%
Female condom	21%
Sterilization	<1%

*Rates reflect unintentional pregnancies in the first year of using the method. Data are from www.cdc.gov/reproductivehealth/unintendedpregnancy/contraception.htm (Accessed 1/3/14).
**Lower rates are in women who have never delivered a child.

of interventional contraceptives depends in part on how consistently and correctly they are used (**TBL. 26.1**).

Barrier Methods Contraceptive methods based on chemical or physical barriers are among the earliest recorded means of birth control. Once people made the association between pregnancy and semen, they concocted a variety of physical barriers and *spermicides* {*cida*, killer} to kill sperm. An ancient Egyptian papyrus with the earliest known references to birth control describes the use of vaginal plugs made of leaves, feathers, figs, and alum held together with crocodile and elephant dung. Sea sponges soaked in vinegar and disks of oiled silk have also been used at one time or another. In subsequent centuries, women used douches of garlic, turpentine, and rose petals to rinse the vagina after intercourse. As you can imagine, many of these methods also caused vaginal or uterine infections.

Modern versions of the female barrier include the **diaphragm,** introduced into the United States in 1916. These rubber domes and a smaller version called a *cervical cap* are usually filled with a spermicidal cream, then inserted into the top of the vagina so they cover the cervix. One advantage to the diaphragm is that it is nonhormonal. When used properly and regularly, diaphragms are highly effective (97–99%). However, they are not always used because they must be inserted close to the time of intercourse, and consequently about 20% of women who depend on diaphragms for contraception are pregnant within the first year. Another female barrier contraceptive is the **contraceptive sponge,** which contains a spermicidal chemical.

The male barrier contraceptive is the **condom,** a closed sheath that fits closely over the penis to catch ejaculated semen. Males have used condoms made from animal bladders and intestines for centuries. Condoms lost popularity when oral contraceptives came into widespread use in the 1960s and 1970s, but in recent years, they have regained favor because they combine pregnancy protection with protection from many sexually transmitted diseases. However, latex condoms may cause allergic reactions, and there is evidence that HIV can pass through pores in some condoms currently produced. A female version of the condom is also commercially available. It covers the cervix and completely lines the vagina, providing more protection from sexually transmitted diseases.

Implantation Prevention Some contraceptive methods do not prevent fertilization but do keep a fertilized egg from establishing itself in the endometrium. They include **intrauterine devices (IUDs)** as well as chemicals that change the properties of the endometrium. IUDs are copper-wrapped plastic devices that are inserted into the uterine cavity, where they kill sperm and create a mild inflammatory reaction that prevents implantation. They have low failure rates (0.5% per year) but side effects that range from pain and bleeding to infertility caused by pelvic inflammatory disease and blockage of the Fallopian tubes. Some IUDs contain progesterone-like hormones.

Hormonal Treatments Techniques for decreasing gamete production depend on altering the hormonal milieu of the body. In centuries past, women would eat or drink various plant concoctions for contraception. Some of these substances actually worked because the plants contained estrogen-like compounds. Modern pharmacology has improved on this method, and now women can choose between oral contraceptive pills, injections lasting months, or a vaginal contraceptive ring.

The **oral contraceptives,** also known as *birth control pills,* first became available in 1960. They rely on various combinations of estrogen and progesterone that inhibit gonadotropin secretion from the pituitary. Without adequate FSH and LH, ovulation is suppressed. In addition, progesterones in the contraceptive pills thicken the cervical mucus and help prevent sperm penetration. These hormonal methods of contraception are highly effective when used correctly but also carry some risks, including an increased incidence of blood clots and strokes, especially in women who smoke.

Development of a male hormonal contraceptive has been slow because of undesirable side effects. Contraceptives that block testosterone secretion or action are also likely to decrease the male libido or even cause impotence. Both side effects are unacceptable to men who would be most interested in using the contraceptive. Some early male oral contraceptives irreversibly suppressed sperm production, which was also unacceptable. A number of clinical trials are currently looking at both hormonal and nonhormonal methods for decreasing male fertility.

Contraceptive vaccines are based on antibodies against various components of the male and female reproductive systems, such as antisperm or antiovum antibodies. These contraceptives can be administered as shots. However, clinical trials of human vaccines have been disappointing and vaccines may not be a practical contraceptive for humans.

Infertility Is the Inability to Conceive

While some couples are trying to prevent pregnancy, others are spending thousands of dollars trying to get pregnant. *Infertility* is the inability of a couple to conceive a child after a year of unprotected intercourse. For years, infertile couples had no choice but adoption if they wanted to have a child, but incredible strides have been made in this field since the 1970s. As a result, many infertile couples today are able to have children.

Infertility can arise from a problem in the male, the female, or both. Male infertility usually results from a low sperm count or an abnormally high number of defective sperm. Female infertility can be mechanical (blocked Fallopian tubes or other structural problems) or hormonal, leading to decreased or absent ovulation. One problem involving both partners is that the woman may produce antibodies to her partner's sperm. In addition, not all pregnancies go to a successful conclusion. By some estimates, as many as a third of all pregnancies spontaneously terminate—many within the first weeks, before the woman is even aware that she was pregnant.

Some of the most dramatic advances have been made in the field of **assisted reproductive technology** (ART), strategies in which both sperm and eggs are manipulated. For *in vitro* fertilization, a woman's ovaries are manipulated with hormones to ovulate multiple eggs at one time. The eggs are collected surgically and fertilized outside the body. The developing embryos are then placed in the woman's uterus, which has been primed for pregnancy by hormonal therapy. Because of the expense and complicated nature of the procedure, multiple embryos are usually placed in the uterus at one time, which may result in multiple births. *In vitro* fertilization has allowed some infertile couples to have children, with a 2009 live-birth delivery rate in the United States averaging 37%. ART success varies significantly with age, ranging from 41% for women younger than 35 to 12% for women older than 40.

PREGNANCY AND PARTURITION

Now, we will return to a recently ovulated egg and some sperm deposited in the vagina and follow them through fertilization, pregnancy, and **parturition**, the birth process.

Fertilization Requires Capacitation

Once an egg is released from the ruptured follicle, it is swept into the Fallopian tube by beating cilia. Meanwhile, sperm deposited in the female reproductive tract must go through their final maturation step, **capacitation**, which enables the sperm to swim rapidly and fertilize an egg. The process apparently involves changes in the outer membrane of the sperm head.

Normally, capacitation takes place in the female reproductive tract, which presents a problem for *in vitro* fertilization. Those sperm must be artificially capacitated by placing them in physiological saline supplemented with human serum. Much of what we know about human fertilization has come from infertility research aimed at improving the success rate of *in vitro* fertilization.

Fertilization of an egg by a sperm is the result of a chance encounter, possibly aided by chemical attractants produced by the egg. An egg can be fertilized for only about 12–24 hours after ovulation. Sperm in the female reproductive tract remain viable for 5–6 days. Apparently they bind to the epithelium of the Fallopian tube while awaiting a chemical signal from a newly ovulated egg.

Fertilization normally takes place in the distal part of the Fallopian tube. Of the millions of sperm in a single ejaculation, only about 100 reach this point. To fertilize the egg, a sperm must penetrate both an outer layer of loosely connected granulosa cells (the *corona radiata*) and a protective glycoprotein coat called the **zona pellucida** (**FIG. 26.14b**). To get past these barriers, capacitated sperm release powerful enzymes from the acrosome in the sperm head, a process known as the **acrosomal reaction.** The enzymes dissolve cell junctions and the zona pellucida, allowing the sperm to wiggle their way toward the egg.

The first sperm to reach the egg quickly finds sperm-binding receptors on the oocyte membrane and binds to the egg (Fig. 26.14c). The fusion of sperm and oocyte membranes triggers a chemical reaction called the **cortical reaction** that excludes other sperm. In the cortical reaction, membrane-bound **cortical granules** in the peripheral cytoplasm of the egg release their contents into the space just outside the egg membrane. These chemicals rapidly alter the membrane and surrounding zona pellucida to prevent **polyspermy,** in which more than one sperm fertilizes an egg.

To complete fertilization, the fused section of sperm and egg membrane opens, and the sperm nucleus sinks into the egg's cytoplasm. This signals the egg to resume meiosis and complete its second division. The final meiotic division creates a second polar body, which is ejected. At this point, the 23 chromosomes of the sperm join the 23 chromosomes of the egg, creating a zygote nucleus with a full set of genetic material.

Once an egg is fertilized and becomes a zygote, it begins mitosis as it slowly makes its way along the Fallopian tube to the uterus, where it will settle for the remainder of the **gestation** period {*gestare,* to carry in the womb}.

The Developing Embryo Implants in the Endometrium

The dividing embryo takes four or five days to move through the Fallopian tube into the uterine cavity (Fig. 26.14d). Under the influence of progesterone, smooth muscle of the tube relaxes, and transport proceeds slowly. By the time the developing embryo reaches the uterus, it consists of a hollow ball of about 100 cells called a **blastocyst.**

Fig. 26.14 **ESSENTIALS**

Fertilization

Fertilization must occur within 24 hours of ovulation.

(a) This photograph shows the tremendous difference in the sizes of human sperm and egg.

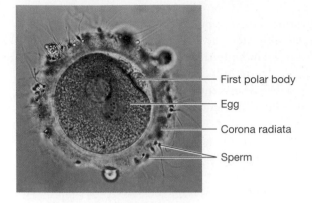

First polar body

Egg

Corona radiata

Sperm

(b) Capacitated sperm release enzymes from their acrosomes in order to penetrate the cells and zona pellucida surrounding the egg.

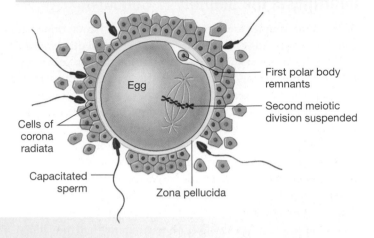

Egg

First polar body remnants

Second meiotic division suspended

Cells of corona radiata

Capacitated sperm

Zona pellucida

(c) The first sperm to fuse with the egg fertilizes it.

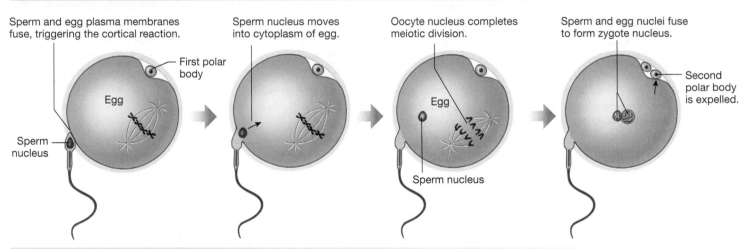

Sperm and egg plasma membranes fuse, triggering the cortical reaction.

Sperm nucleus moves into cytoplasm of egg.

Oocyte nucleus completes meiotic division.

Sperm and egg nuclei fuse to form zygote nucleus.

First polar body

Egg

Sperm nucleus

Egg

Sperm nucleus

Second polar body is expelled.

(d) Timing of ovulation, fertilization, and implantation

2 Day 1: Fertilization

3 Days 2–4: Cell division takes place.

4 Day 4–5: Blastocyst reaches uterus.

Inner cell mass

Zygote

Fallopian tube

Egg

1 Ovulation

Ovary

Uterus

Blastocyst

5 Days 5–9: Blastocyst implants.

Some of the outer layer of blastocyst cells will become the **chorion,** an *extraembryonic membrane* that will enclose the embryo and form the placenta (**FIG. 26.15a**). The inner cell mass of the blastocyst will develop into the embryo and into three other extraembryonic membranes. These membranes include the **amnion,** which secretes *amniotic fluid* in which the developing embryo floats; the **allantois,** which becomes part of the umbilical cord that links the embryo to the mother; and the **yolk sac,** which degenerates early in human development.

Implantation of the blastocyst into the uterine wall normally takes place about seven days after fertilization. The blastocyst secretes enzymes that allow it to invade the endometrium, like a parasite burrowing into its host. As it does so, endometrial cells grow out around the blastocyst until it is completely engulfed.

As the blastocyst continues dividing and becomes an embryo, cells that will become the placenta form fingerlike **chorionic villi** that penetrate into the vascularized endometrium. Enzymes from the villi break down the walls of maternal blood vessels until the villi are surrounded by pools of maternal blood (Fig. 26.15b). The blood of the embryo and that of the mother do not mix, but nutrients, gases, and wastes exchange across the membranes of the villi. Many of these substances move by simple diffusion, but some, such as maternal antibodies, must be transported across the membrane.

The placenta continues to grow during pregnancy until, by delivery, it is about 20 cm in diameter (the size of a small dinner plate). The placenta receives as much as 10% of the total maternal cardiac output. The tremendous blood flow to the placenta is one reason sudden, abnormal separation of the placenta from the uterine wall is a medical emergency.

The Placenta Secretes Hormones during Pregnancy

As the blastocyst implants in the uterine wall and the placenta begins to form, the corpus luteum nears the end of its preprogrammed 12-day life span. Unless the developing embryo sends a hormonal signal, the corpus luteum disintegrates, progesterone and estrogen levels drop, and the embryo is flushed from the body along with the surface layers of endometrium during menstruation. The placenta secretes several hormones that prevent menstruation during pregnancy, including human chorionic gonadotropin, human placental lactogen, estrogen, and progesterone.

Human Chorionic Gonadotropin The corpus luteum remains active during early pregnancy because of **human chorionic gonadotropin** (hCG), a peptide hormone secreted by the chorionic villi and developing placenta. Human chorionic

FIG. 26.15 The placenta

(a) The developing embryo floats in amniotic fluid. It obtains oxygen and nutrients from the mother through the placenta and umbilical cord.

Umbilical cord
Placenta
Extraembryonic membranes enclose the embryo and form the placenta.
Yolk sac
Chorion
Amnion
Amniotic fluid
WEEK 10

(b) Some material is exchanged across placental membranes by diffusion, but other material must be transported.

Umbilical vein carries well-oxygenated blood to the embryo.
Umbilical arteries return embryonic blood to placenta.
Chorionic villi contain embryonic blood vessels.
Umbilical cord
Maternal blood bathes the chorionic villi.
Maternal blood vessels
Amnion

gonadotropin is structurally related to LH, and it binds to LH receptors. Under the influence of hCG, the corpus luteum keeps producing progesterone to keep the endometrium intact.

By the seventh week of development, however, the placenta has taken over progesterone production, and the corpus luteum is no longer needed. At that point, it finally degenerates. Human chorionic gonadotropin production by the placenta peaks at three months of development, then diminishes.

A second function of hCG is stimulation of testosterone production by the developing testes in male fetuses. As you learned in the opening sections of this chapter, fetal testosterone and its metabolite DHT are essential for expression of male characteristics and for descent of the testes into the scrotum before birth.

Human chorionic gonadotropin is the chemical detected by pregnancy tests. Because hCG can induce ovulation in rabbits, years ago rabbits were used for pregnancy testing. If a woman suspected she was pregnant, her urine was injected into a rabbit. The rabbit's ovaries were then inspected for signs of ovulation. It took several days for women to learn the results of this test. Today, with modern biochemical techniques, women can perform their own pregnancy tests in a few minutes in the privacy of their home.

Human Placental Lactogen (hPL) Another peptide hormone produced by the placenta is **human placental lactogen** (hPL), also known as *human chorionic somatomammotropin (hCS)*. This hormone, structurally related to growth hormone and prolactin, was initially believed to be necessary for breast development during pregnancy and for milk production (**lactation**). hPL probably does contribute to lactation, but women who do not make hPL during pregnancy because of a genetic defect still have adequate breast development and milk production. A second role for hPL is alteration of the mother's glucose and fatty acid metabolism to support fetal growth.

RUNNING **PROBLEM**

A few months after receiving the devastating news, Kate and Jon get a call from their doctor. "I have some potentially good news for you! A treatment has finally become available for women with POI and I am hopeful that it might work for you," explains Dr. Baker. In a recently developed procedure known as *in vitro* activation (IVA), the mother's ovaries are removed, cut into strips, and then inserted back into the abdomen after being treated with egg-stimulating drugs. After approximately six weeks, mature eggs are harvested surgically and fertilized *in vitro*. The zygotes are allowed to develop into early embryos before being frozen for later placement into the mother's uterus.

Q6: *Because of the time delay with IVA, the mother must be treated with hormones to prepare the uterine endometrium for implantation of the embryo. What hormones must Kate be given so that her uterus is able to receive an embryo?*

801　814　817　823　**830**　834

Maternal glucose moves across the membranes of the placenta by facilitated diffusion and enters the fetal circulation. During pregnancy, about 4% of women develop *gestational diabetes mellitus (GDM)*, with elevated blood glucose levels caused by insulin resistance, similar to type 2 diabetes. The cause is not clear. After delivery, glucose metabolism in most women with GDM returns to normal, but these mothers and their babies are at higher risk of developing type 2 diabetes later in life.

Estrogen and Progesterone Estrogen and progesterone are produced continuously during pregnancy, first by the corpus luteum under the influence of hCG and then by the placenta. With high circulating levels of these steroid hormones, feedback suppression of the pituitary continues throughout pregnancy, preventing another set of follicles from beginning development.

During pregnancy, estrogen contributes to the development of the milk-secreting ducts of the breasts. Progesterone is essential for maintaining the endometrium and also helps suppress uterine contractions. The placenta makes a variety of other hormones, including inhibin and prorenin, but the function of most of them remains unclear.

Pregnancy Ends with Labor and Delivery

Parturition normally occurs between the 38th and 40th week of gestation. What triggers this process? For many years, researchers developed animal models of the signals that initiate parturition, only to discover recently that there are no good non-primate models that apply to humans. Parturition begins with **labor,** the rhythmic contractions of the uterus that push the fetus out into the world (**FIG. 26.16a**). Signals that initiate these contractions could begin with either the mother or the fetus, or they could be a combination of signals from both.

In many nonhuman mammals, a decrease in estrogen and progesterone levels marks the beginning of parturition. A decrease in progesterone levels is logical, as progesterone inhibits uterine contractions. In humans, however, levels of these hormones do not decrease until labor is well under way.

Another possible labor trigger is oxytocin, the peptide hormone that causes uterine muscle contraction. As a pregnancy nears full term, the number of uterine oxytocin receptors increases. However, studies have shown that oxytocin secretion does not increase until after labor begins. Synthetic oxytocin is often used to induce labor in pregnant women, but it is not always effective. Apparently, the start of labor requires something more than adequate amounts of oxytocin.

Another possibility is that the fetus somehow signals that it has completed development. One theory supported by clinical evidence is that corticotropin-releasing hormone (CRH) secreted by the placenta is the signal to begin labor. (CRH is also a hypothalamic releasing factor that controls release of ACTH from the anterior pituitary.) In the weeks prior to delivery, maternal blood CRH levels increase rapidly. In addition, women with elevated

FIG. 26.16   Parturition: The birth process

(a) Fully developed fetus. As labor begins, the fetus is normally head down in the uterus.

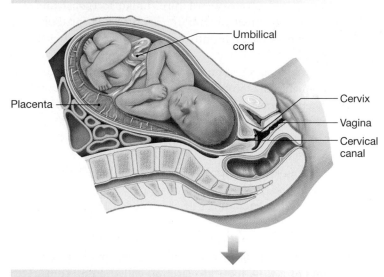

- Umbilical cord
- Placenta
- Cervix
- Vagina
- Cervical canal

(b) Cervical dilation. Uterine contractions push the head against the softened cervix, stretching and dilating it.

(c) Delivery. Once the cervix is fully dilated and stretched, the uterine contractions push the fetus out through the vagina.

(d) The process of labor is controlled by a positive feedback loop that ends with delivery.

Fetus drops lower in uterus.

Cervical stretch ⊕

Oxytocin from posterior pituitary ⊕

Uterine contractions ⊕

Prostaglandins from uterine wall ⊕ ⊕

Delivery of baby stops the cycle.

CRH levels as early as 15 weeks of gestation are more likely to go into premature labor.

Although we do not know for certain what initiates parturition, we do understand the sequence of events. In the days prior to the onset of active labor, the cervix softens ("ripens") and ligaments holding the pelvic bones together loosen as enzymes destabilize collagen in the connective tissue. The control of these processes is not clear and may be due to estrogen or the peptide hormone **relaxin,** which is secreted by ovaries and the placenta.

Once the contractions of labor begin, a positive feedback loop consisting of mechanical and hormonal factors is set into motion. The fetus is normally oriented head down (Fig. 26.16a). At the beginning of labor, it repositions itself lower in the abdomen ("the baby has dropped") and the head pushes on the softened cervix (Fig. 26.16b).

Cervical stretch triggers uterine contractions that move in a wave from the top of the uterus down, pushing the fetus farther into the pelvis. The lower portion of the uterus stays relaxed, and the cervix stretches and dilates. Cervical stretch starts a positive feedback cycle of escalating contractions (Fig. 26.16d). The contractions are reinforced by secretion of oxytocin from the posterior pituitary [p. 209], with continued cervical stretch reinforcing oxytocin secretion.

Prostaglandins are produced in the uterus in response to oxytocin and CRH secretion. Prostaglandins are very effective at causing uterine muscle contractions at any time. They are the primary cause of menstrual cramps and have been used to induce abortion in early pregnancy. During labor and delivery, prostaglandins reinforce the uterine contractions induced by oxytocin (Fig. 26.16d).

As the contractions of labor intensify, the fetus moves down though the vagina and out into the world (Fig. 26.16c), still attached to the placenta. The placenta then detaches from the uterine wall and is expelled a short time later. Uterine contractions clamp the maternal blood vessels and help prevent excessive bleeding, although typically the mother loses about 240 mL of blood.

The Mammary Glands Secrete Milk during Lactation

A newborn has lost its source of maternal nourishment through the placenta and must rely on an external source of food instead. Primates, who normally have only one or two offspring at a time, have two functional mammary glands. A mammary gland is composed of 15–20 milk-secreting lobes (**FIG. 26.17a**). Each lobe branches into lobules, and the lobules terminate in clusters of cells called *alveoli* or *acini*. Each alveolus is composed of secretory epithelium that secretes into a duct, similar to the exocrine secretions of the pancreas [Fig. 21.14, p. 676]. Contractile *myoepithelium* surrounds the alveoli.

During puberty, the breasts begin to develop under the influence of estrogen. The milk ducts grow and branch, and fat is deposited behind the glandular tissue. During pregnancy, the glands develop further under the direction of estrogen, growth hormone, and cortisol. The final development step also requires progesterone, which converts the duct epithelium into a secretory structure. This process is similar to progesterone's effect on the uterus, in which progesterone makes the endometrium into a secretory tissue during the luteal phase.

Although estrogen and progesterone stimulate mammary development, they inhibit secretion of milk. Milk production is stimulated by prolactin from the anterior pituitary [p. 213]. Prolactin is an unusual pituitary hormone in that its secretion is primarily controlled by **prolactin-inhibiting hormone** (PIH) from the hypothalamus. Most evidence suggests that PIH is actually *dopamine,* an amine neurohormone related to epinephrine and norepinephrine [p. 206].

During the later stages of pregnancy, PIH secretion falls, and prolactin reaches levels 10 or more times those found in nonpregnant women. Prior to delivery, when estrogen and progesterone are also high, the mammary glands produce only small amounts of a thin, low-fat secretion called **colostrum.** After delivery, when estrogen and progesterone decrease, the glands produce greater amounts of milk that contains 4% fat and substantial amounts of calcium. Proteins in colostrum and milk include maternal immunoglobulins, secreted into the duct and absorbed by the infant's intestinal epithelium [p. 682]. This process transfers some of the mother's immunity to the infant during its first weeks of life.

Suckling, the mechanical stimulus of the infant nursing at the breast, also inhibits PIH production (Fig. 26.17b). In the absence of PIH, prolactin secretion increases, resulting in milk production. Pregnancy is not a requirement for lactation, however, and some women who have adopted babies have been successful in breast-feeding.

The ejection of milk from the glands, known as the **let-down reflex,** requires the presence of oxytocin from the posterior pituitary. Oxytocin initiates smooth muscle contraction in the uterus and breasts. In the *postpartum* (after delivery) uterus, oxytocin-induced contractions help return the uterus to its prepregnancy size.

In the lactating breast, oxytocin causes contraction of myoepithelial cells surrounding the alveoli and in the walls of the ducts. Contraction creates high pressure in the ducts that sends the milk squirting into the infant's mouth. Although prolactin release requires the mechanical stimulus of suckling, oxytocin release can be stimulated by various cerebral stimuli, including the thought of the child. Many nursing mothers experience inappropriate milk release triggered by hearing someone else's child cry.

Prolactin is related to growth hormone and plays a role in other reproductive and nonreproductive processes. All nonnursing women and men have tonic prolactin secretion that exhibits a diurnal cycle, peaking during sleep. Prolactin is also involved in fertility in both males and females but this function is still being investigated.

FIG. 26.17   Lactation

(a) Mammary glands

Epithelial cells of the mammary glands secrete milk into the ducts of the gland. Contraction of the myoepithelium forces fluid out of the ducts through openings in the nipple.

Pectoralis major muscle
Pectoral fat pad
Suspensory ligaments
Lobes of glandular tissue
Milk duct
Nipple
Areola

Mammary gland lobule

Milk duct
Muscle cells in wall of duct
Epithelial milk-secreting cells
Myoepithelial cells
Mammary gland alveolus

(b) The hormonal control of milk secretion and release

Prolactin controls milk secretion, and oxytocin causes smooth muscle contraction to eject milk.

Sound of child's cry
Higher brain centers
Hypothalamus
Oxytocin neuron
PIH cell
Portal system
↓PIH
Anterior pituitary
Posterior pituitary
Inhibition of prolactin cells is removed.
↑ Prolactin
↑ Oxytocin
Ascending sensory information
Milk secretion
Milk ejected
Smooth muscle contraction
Baby suckling
Mechanoreceptors in nipple

GROWTH AND AGING

The reproductive years begin with the events surrounding puberty and end with decreasing gonadal hormone production.

Puberty Marks the Beginning of the Reproductive Years

In girls, the onset of puberty is marked by budding breasts and the first menstrual period, called **menarche,** a time of ritual significance in many cultures. In the United States, the average age at menarche is 12 years (normal range is considered 8–13 years).

In boys, the onset of puberty is more subtle. The signs include growth and maturation of the external genitalia; development of secondary sex characteristics, such as pubic and facial hair and lowering of voice pitch; change in body shape; and growth in height. The age range for male puberty is 9–14 years.

Puberty requires maturation of the hypothalamic-pituitary control pathway. Before puberty, the child has low levels of both steroid sex hormones and gonadotropins. Because low sex hormone levels normally enhance gonadotropin release, the combination of low steroids and low gonadotropins indicates that the hypothalamus and pituitary are not yet sensitive to steroid levels in the blood.

At puberty, the hypothalamic GnRH-secreting neurons increase their pulsatile secretion of GnRH, which in turn increases gonadotropin release. The signals responsible for the onset of puberty are complex. One theory holds that the genetically programmed maturation of hypothalamic neurons initiates puberty. We know that puberty has a genetic basis because inherited patterns of maturation are common. If a woman did not start her menstrual periods until she was 16, for example, it is likely that her daughters will also have late menarche.

The adipose tissue hormone *leptin* [p. 694] also contributes to the onset of puberty. Undernourished women with little adipose tissue and low leptin levels often stop having menstrual periods (*amenorrhea*), and knockout mice without leptin are infertile. Presumably, improved nutrition over the last century increased individuals' prepubertal fat stores and leptin secretion, which could interact with other factors to initiate puberty.

Menopause and Andropause Are a Consequence of Aging

In nineteenth century America, many people died of acute illnesses while still reproductively active. Now modern medicine has overcome most acute illnesses, and most of us will live well past the time we are likely to have children.

Women's reproductive cycles stop completely at the time known as **menopause.** The physiology of menopause has been well studied. After about 40 years of menstrual cycles, a woman's periods become irregular (*perimenopause*) and finally cease. The cessation of reproductive cycles is due not to the pituitary but to the ovaries, which can no longer respond to gonadotropins. In the absence of negative feedback, gonadotropin levels increase dramatically in an effort to stimulate the ovaries into maturing more follicles.

The absence of estrogen in postmenopausal women leads to symptoms of varying severity. These may include hot flashes [p. 723], atrophy of genitalia and breasts, and osteoporosis as calcium is lost from bones [p. 749]. Hormone replacement therapy (HRT) for women in menopause traditionally consists of estrogen or a combination of estrogen and progesterone. This treatment became controversial around 2002, however, because of some studies that suggested that HRT risks outweigh its benefits. An international 2013 consensus statement by multiple professional societies provides guidelines for the use of HRT.

A newer drug therapy for menopause uses *selective estrogen receptor modulators (SERMs).* These drugs bind with different affinities to the two estrogen receptor subtypes, which allows the drugs to mimic the beneficial effects of estrogen on bone while avoiding the potentially detrimental effects on breasts and uterus.

In men, testosterone production decreases with age, and about half of men over the age of 50 have symptoms of **andropause,** a term coined as the counterpart to menopause. The existence of physiological andropause in men is still controversial because the physical and psychological symptoms of aging in men are not clearly linked to a decline in testosterone, despite a recent trend in advertising that promotes testosterone replacement therapy.

RUNNING **PROBLEM CONCLUSION** | Infertility

Kate and Jon were excited by Dr. Baker's news and decided to move forward with the new treatment. Six weeks after the procedure, three eggs were harvested from Kate's Fallopian tubes and fertilized *in vitro* with Jon's sperm. One at a time, the frozen embryos were thawed, transferred into Kate's uterus, and given time to implant and develop. The first two cycles were unsuccessful, but the last remaining transfer produced a healthy little boy.

In this running problem, you learned how the cause of infertility is diagnosed in a typical couple. To learn more about infertility, see literature from the American Society for Reproductive Medicine at *www.asrm.org* or go to Medline Plus (*www.nlm.nih.gov/medlineplus*) and look under Health Topics. For more on *in vitro* activation, go to *ivafertility.com*.

Now test your understanding of the running problem by checking your answers against the information in this summary table.

RUNNING **PROBLEM CONCLUSION** | *Continued*

Question	Facts	Integration and Analysis
Q1: *Name (in order) the male reproductive structures that carry sperm from the seminiferous tubules to the external environment.*	The male reproductive structures include the testes, accessory glandular organs, a series of ducts, and the external genitalia.	Sperm leaving the seminiferous tubules of the testes pass into the epididymis, then into the vas deferens, and finally exit the body via the urethra.
Q2: *In what male reproductive structure do sperm reach maturity?*	Sperm leaving the seminiferous tubules are not yet mature. They take nearly two weeks to move through the epididymis.	Maturation takes place in the epididymis, aided by secretions from that structure.
Q3: *Which causes of male infertility might make retrieval of sperm from the epididymis necessary?*	The epididymis is the first duct the sperm enter upon leaving the testes.	If the infertility problem is due to blockage or congenital defects in the vas deferens or urethra, removal of sperm from the epididymis might be useful. If the problem is caused by low sperm count or abnormal sperm morphology, this technique would probably not be useful.
Q4: *For which causes of female infertility is temperature tracking useful? For which causes is it not useful?*	Basal body temperature rises slightly following ovulation.	Temperature tracking is a useful way to tell if a woman is ovulating, but it cannot reveal structural or physiological problems in the Fallopian tubes or uterus.
Q5: *What are the levels of FSH and estrogen in a woman with POI, compared to normal?*	The ovaries of women with POI do not produce mature follicles. Estrogen has a negative feedback effect on the anterior pituitary.	Estrogen is secreted by mature follicles and corpora lutea, so estrogen levels will be low. Low estrogen means little negative feedback and FSH levels increase.
Q6: *What hormones must Kate be given so that her uterus is able to receive an embryo?*	Normally the endometrium develops into a secretory structure able to receive an embryo during the luteal phase of the ovarian cycle.	The corpus luteum secretes estrogen and progesterone to prepare the uterus for implantation of an embryo, so these are the hormones Kate is given.

This problem was developed by Douglas Shannon when he was a student at the University of Texas preparing to enter a Physician Assistant program. Reprinted with permission.

801 814 817 823 830 **834**

VISUALIZE, PRACTICE, AND MASTER PHYSIOLOGY WITH MASTERING A&P

- Practice-on-the-go with Dynamic Study Modules
- Visualize important processes with Interactive Physiology and A&P Flix
- Check your understanding with Video Tutors and practice quizzes!

MasteringA&P®

CHAPTER SUMMARY

In this chapter, you learned how the human species perpetuates itself through reproduction. The reproductive system has some of the most complex *control systems* of the body, in which multiple hormones and other signal molecules interact in an ever-changing fashion. *Homeostasis* in the adult reproductive system is anything but steady state, particularly during the female menstrual cycle, when the *feedback effects* of estrogen change from negative to positive and back again. An example of *positive feedback* occurs with oxytocin secretion during labor and delivery. The testis provides a nice example of *compartmentation*, with the lumen of the seminiferous tubules, where sperm develop, isolated from the rest of the extracellular compartment.

Sex Determination

1. The sex organs consist of **gonads, internal genitalia,** and **external genitalia.** (p. 801)

2. **Testes** produce **sperm. Ovaries** produce eggs, or **ova.** Embryonic cells that will produce **gametes** (eggs and sperm) are called **germ cells.** (p. 801)

3. Humans have 23 pairs of chromosomes, or 46 chromosomes. (p. 801; Fig. 26.1a)

4. The genetic sex of an individual depends on the **sex chromosomes:** females are XX, and males are XY. In the absence of a Y chromosome, an embryo will develop into a female. (p. 801; Fig. 26.1b)

5. The **SRY gene** on the Y chromosomes produces SRY protein, a *testis-determining factor* that converts the bipotential gonad into a testis. In the absence of SRY protein, the gonad becomes an ovary. (p. 802)

6. Testicular **Sertoli cells** secrete **anti-Müllerian hormone** (AMH), which causes the **Müllerian ducts** to regress. **Interstitial (Leydig) cells** secrete **testosterone,** which converts **Wolffian ducts** to male accessory structures. **Dihydrotestosterone** (DHT) promotes development of the prostate gland and external genitalia. (pp. 802, 803; Fig. 26.2)

7. Absence of testosterone and AMH causes Müllerian ducts to develop into **Fallopian tubes (oviducts), uterus,** and **vagina.** In females, the Wolffian ducts regress. (p. 803; Fig. 26.2)

Basic Patterns of Reproduction

8. **Gametogenesis** begins with mitotic divisions of **spermatogonia** and **oögonia.** The first step of meiosis creates **primary spermatocytes** and **primary oocytes.** The first meiotic division creates two identical **secondary spermatocytes** in males or a large secondary oocyte (egg) and a tiny **first polar body** in females. (pp. 806, 807; Fig. 26.5)

9. The second meiotic division in males creates haploid **spermatids** that mature into sperm. In females, the second meiotic division does not go to completion unless the egg is fertilized. (p. 807; Fig. 26.5)

10. In both sexes, **gonadotropin-releasing hormone** (GnRH) controls the secretion of **follicle-stimulating hormone** (FSH) and **luteinizing hormone** (LH) from the anterior pituitary. FSH and steroid sex hormones regulate gametogenesis in gonadal gamete-producing cells. LH stimulates production of steroid sex hormones. (p. 809; Fig. 26.6)

11. The steroid sex hormones include **androgens, estrogens,** and **progesterone. Aromatase** converts androgens to estrogens. **Inhibin** inhibits secretion of FSH, and **activin** stimulates FSH secretion. (pp. 807, 808, 809)

12. Gonadal steroids generally suppress secretion of GnRH, FSH, and LH. However, if estrogen rises rapidly above a threshold level for at least 36 hours, its feedback changes to positive and stimulates gonadotropin release. (p. 809)

13. After puberty, tonic GnRH release occurs in small pulses every 1–3 hours from a region of the hypothalamus called a **pulse generator.** (p. 810)

Male Reproduction

14. The **corpus spongiosum** and **corpora cavernosa** make up the erectile tissue of the penis. The **glans** is covered by the **foreskin.** The urethra runs though the **penis.** (p. 810; Fig. 26.7)

15. The testes migrate into the scrotum during fetal development. Failure of one or both testes to descend is known as **cryptorchidism.** (p. 811)

16. The testes consist of **seminiferous tubules** and interstitial tissue containing blood vessels and Leydig cells. The seminiferous tubules join the **epididymis,** which becomes the **vas deferens.** The vas deferens empties into the urethra. (p. 811; Fig. 26.7b)

17. A seminiferous tubule contains spermatogonia, spermatocytes, and Sertoli cells. Tight junctions between Sertoli cells form a **blood-testis barrier.** (p. 811; Fig. 26.7d, e)

18. Spermatogonia in the tubule undergo meiosis, becoming primary spermatocytes, spermatids, and finally sperm in about 64 days. Sperm reach maturity in the epididymis. (p. 811; Fig. 26.7e)

19. Sertoli cells regulate sperm development. They also produce inhibin, activin, growth factors, enzymes, and **androgen-binding protein.** (p. 811; Fig. 26.8)

20. Interstitial cells produce 95% of a male's testosterone. The other 5% comes from the adrenal cortex. (p. 811)

21. FSH stimulates Sertoli cell production of androgen-binding protein, inhibin, and paracrine molecules. Interstitial cells produce testosterone under the direction of LH. (p. 815; Fig. 26.8)

22. The **prostate gland, seminal vesicles,** and **bulbourethral glands** secrete the fluid component of **semen.** (pp. 811, 815)

23. **Primary sex characteristics** are the internal sexual organs and external genitalia. **Secondary sex characteristics** are other features of the body, such as body shape. (p. 815)

Female Reproduction

24. Female external genitalia, called the **vulva** or **pudendum,** are the **labia majora, labia minora,** and **clitoris.** The urethra opening lies between the clitoris and the **vagina.** (p. 816; Fig. 26.9)

25. The uterine tissue layers are outer connective tissue, **myometrium,** and **endometrium.** (p. 816; Fig. 26.9d)

26. Fallopian tubes are lined with ciliated epithelium. The bulk of an ovary consists of ovarian follicles. (p. 816; Fig. 26.9e)

27. Eggs are produced in monthly **menstrual cycles.** (p. 817; Fig. 26.11)

28. In the **ovarian cycle,** the **follicular phase** is a period of follicular growth. **Ovulation** is the release of an egg from its follicle. In the **luteal phase,** the ruptured follicle becomes a **corpus luteum.** (p. 817; Fig. 26.11)

29. The **menses** begin the **uterine cycle.** This is followed by a **proliferative phase,** with endometrial thickening. Following ovulation, the endometrium goes into a **secretory phase.** (p. 817; Fig. 26.11)

30. Follicular **granulosa cells** secrete estrogen. As the follicular phase ends, a surge in LH is necessary for oocyte maturation. (p. 820; Fig. 26.11)

31. The corpus luteum secretes progesterone and some estrogen, which exert negative feedback on the hypothalamus-anterior pituitary. (p. 823; Fig. 26.12)

32. Estrogens and androgen control primary and secondary sex characteristics in females. (p. 823)

Procreation

33. The human sex act is divided into four phases: (1) excitement, (2) plateau, (3) orgasm, and (4) resolution. (p. 824)

34. The male **erection reflex** is a spinal reflex that can be influenced by higher brain centers. Parasympathetic input mediated by

nitric oxide actively vasodilates the penile arterioles. (p. 824; Fig. 26.13)

35. **Emission** is the movement of sperm out of the vas deferens and into the urethra. **Ejaculation** is the expulsion of semen to the external environment. (p. 824)

36. Contraceptive methods include **abstinence, barrier methods, implantation prevention,** and **hormonal treatments.** (pp. 825, 826)

37. Infertility can arise from a problem in the male, the female, or both. *In vitro* fertilization has allowed some infertile couples to have children. (p. 827)

Pregnancy and Parturition

38. Sperm must go through **capacitation** before they can fertilize an egg. (p. 827)

39. Fertilization normally takes place in the Fallopian tube. Capacitated sperm release acrosomal enzymes (the **acrosomal reaction**) to dissolve cell junctions and the **zona pellucida** of the egg. The first sperm to reach the egg fertilizes it. (p. 827; Fig. 26.14)

40. Fusion of egg and sperm membranes initiates a **cortical reaction** that prevents **polyspermy.** (p. 827)

41. The developing embryo is a hollow **blastocyst** when it reaches the uterus. Once the blastocyst implants, it develops extraembryonic membranes. (p. 827; Figs. 26.14d, 26.15)

42. The **chorionic villi** of the placenta are surrounded by pools of maternal blood where nutrients, gases, and wastes are exchanged between mother and embryo. (p. 829; Fig. 26.15)

43. The corpus luteum remains active during early pregnancy because of **human chorionic gonadotropin** (hCG) produced by the developing embryo. (p. 829)

44. The placenta secretes hCG, estrogen, progesterone, and **human placental lactogen.** This last hormone plays a role in maternal metabolism. (p. 830)

45. Estrogen during pregnancy contributes to development of milk-secreting ducts in the breasts. Progesterone is essential for maintaining the endometrium and, along with **relaxin,** helps suppress uterine contractions. (p. 832)

46. **Parturition** normally occurs between the 38th and 40th week of gestation. It begins with **labor** and ends with delivery of the fetus and placenta. A positive feedback loop of oxytocin secretion causes uterine muscle contraction. (pp. 827, 830; Fig. 26.16)

47. Following delivery, the mammary glands produce milk under the influence of prolactin. Milk is released during nursing by oxytocin, causing mammary gland myoepithelial cells to contract. (p. 832; Fig. 26.17)

Growth and Aging

48. In girls, puberty begins with **menarche,** the first menstrual period, at age 8–13 years. The age range for the onset of puberty in boys is 9–14 years. (p. 834)

49. The cessation of reproductive cycles in women is known as the **menopause.** With increasing age, some men exhibit symptoms of testosterone deficiency. (p. 834)

REVIEW QUESTIONS

In addition to working through these questions and checking your answers on p. A-34, review the Learning Outcomes at the beginning of this chapter.

Level One Reviewing Facts and Terms

1. Match each of the following items with all the terms it applies to:

(a) X or Y	1. chromosomes other than sex chromosomes
(b) inactivated X chromosome	2. fertilized egg
(c) XX	3. sperm or ova
(d) XY	4. sex chromosomes
(e) XX or XY	5. germ cells
(f) autosomes	6. male chromosomes
	7. female chromosomes
	8. Barr body

2. The Y chromosome contains a region for male sex determination that is known as the _____ gene.

3. List the functions of the gonads. How do the products of gonadal function differ in males and females?

4. Define each of the following terms and describe its significance to reproductive physiology:
   (a) aromatase
   (b) blood-testis barrier
   (c) androgen-binding protein
   (d) first polar body
   (e) acrosome

5. Trace the anatomical routes to the external environment followed by a newly formed sperm and by an ovulated egg. Name all structures the gametes pass through on their journey.

6. Decide whether each of the following statements is true or false, and defend your answer.
   (a) All testosterone is produced in the testes.
   (b) Only males make androgens and only females make estrogens.
   (c) Anabolic steroid use appears to be addictive, and withdrawal symptoms include psychological disturbances.
   (d) High levels of estrogen in the late follicular phase help prepare the uterus for menstruation.
   (e) Progesterone is the dominant hormone of the luteal phase of the ovarian cycle.

7. What is semen? What are its main components, and where are they produced?

8. List and give a specific example of the various methods of contraception. Which is/are most effective? Least effective?

Level Two Reviewing Concepts

9. Why are X-linked traits exhibited more frequently by males than females?

10. Diagram the hormonal control of gametogenesis in males.

11. Diagram the menstrual cycle, distinguishing between the ovarian cycle and the uterine cycle. Include all relevant hormones.

12. **Concept maps:** Map the following groups of terms. You may add terms.

List 1

• AMH	• spermatids
• DHT	• spermatocytes
• Interstitial cells	• spermatogonia
• Müllerian ducts	• SRY
• Sertoli cells	• testosterone
• sperm	• Wolffian ducts

List 2

• antrum	• granulosa cells
• corpus luteum	• myometrium
• endometrium	• ovum
• follicle	• thecal cells

13. Define and relate each of the following terms in each group:

(a) gamete, zygote, germ cell, embryo, fetus
(b) coitus, erection, ejaculation, orgasm, emission, erogenous zones
(c) capacitation, zona pellucida, acrosomal reaction, cortical reaction, cortical granules
(d) puberty, menarche, menopause, andropause

14. Compare the actions of each of the following hormones in males and females:

(a) FSH
(b) inhibin
(c) activin
(d) GnRH
(e) LH
(f) DHT
(g) estrogen
(h) testosterone
(i) progesterone

15. Compare and contrast the events of the four phases of sexual intercourse in males and in females.

16. Discuss the roles of each of the following hormones in pregnancy, labor and delivery, and mammary gland development and lactation:

(a) human chorionic gonadotropin
(b) luteinizing hormone
(c) human placental lactogen
(d) estrogen
(e) progesterone
(f) relaxin
(g) prolactin

Level Three Problem Solving

17. Down syndrome is a chromosomal defect known as "trisomy" (three copies instead of two) of chromosome 21. The extra chromosome usually comes from the mother. Speculate what causes trisomy, using what you have learned about the events surrounding fertilization.

18. Sometimes the follicle fails to rupture at ovulation, even though it appears to have gone through all stages of development. This condition results in benign ovarian cysts, and the unruptured follicles can be palpated as bumps on the surface of the ovary. If the cysts persist, symptoms of this condition often mimic pregnancy, with missed menstrual periods and tender breasts. Explain what causes these symptoms, using diagrams as needed.

19. An XY individual inherits a mutation that results in completely nonfunctional androgen receptors.

(a) Is this person genetically male or female?
(b) Will this person have functional ovaries, functional testes, or incompletely developed or nonfunctional gonads?
(c) Will this person have Wolffian ducts or their derivatives? Müllerian ducts or their derivatives?
(d) Will this person have the external appearance of a male or a female?

20. The babies of mothers with gestational diabetes mellitus tend to weigh more at birth. They are also at risk of developing hypoglycemia immediately following birth. Use what you have learned about diabetes and insulin to explain these two observations. *Hint:* these babies have normal insulin responses.

Level Four Quantitative Problems

21. The following graph shows the results of an experiment in which normal men were given testosterone over a period of months (indicated by the bar from A to E). Control values of hormones were measured prior to the start of the experiment. From time B to time C, the men were also given FSH. From time D to time E, they were also given LH. Based on the information given, answer the following questions.

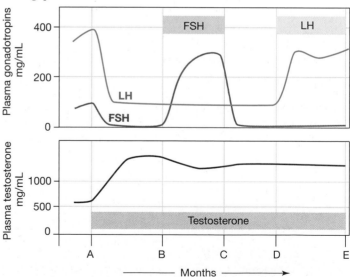

(a) Why did testosterone level increase beginning at point A?
(b) Why did LH and FSH levels decrease beginning at point A?
(c) Predict what happened to the men's sperm production in the A–B interval, the B–C interval, and the D–E interval.

CHAPTER 1

Concept Check Questions

1. It remains in the body.
2. Glucose metabolism adds CO_2 and water to the body, disturbing the mass balance of these two substances. To maintain mass balance, both metabolites must be either excreted or further metabolized.
3. There is no control mechanism for cooling water that is too warm.
4. Negative feedback shuts off the heater.
5. Independent variable is the amount of water the students drink. Dependent variable is urine output.
6. (a) Activity level was independent variable (x-axis), and heart rate was dependent variable (y-axis). (b) A line graph would be appropriate, but a bar graph could also be used if exercise intensity was the same in all subjects. To show the difference between males and females, the graphs should have either separate lines or separate bars for males and females.

Figure Questions

Fig. 1.3: 1. You might include different kinds of breads, meats, and so on, or add a category of sandwich characteristics, such as temperature (hot, cold) or layers (single, club). The sample map shown here is one possible way to map the terms. Notice how some items are linked to more than one other item. Other terms could be included. 2. Sample map below. Maps can vary.

Fig. 1.5: The * goes on the line dividing ECF and ICF.

Fig. 1.15: (b) They preferred diet A. (c) Twenty-four students, (d) Between day 1 and day 2, (e) Graph 1.15d: changes of body weight over time in male and female mice. Graph 1.15e: Is there a relationship between time spent studying for an examination and the student's score? Graph 1.15d: The independent variable was time and the dependent variable was body weight. Graph 1.15e: The independent variable was number of hours spent studying and the dependent variable was the student score.

Graph 1.15d shows that male mice weigh more than female mice from the start of the experiment and that body weight increases with time. The rate of increase is about the same in males and females, indicated by the nearly parallel lines. Graph 1.15e shows that more hours spent studying resulted in higher exam scores.

Review Questions

LEVEL ONE Reviewing Facts and Terms

1. The normal functioning of a living organism. Anatomy is the study of structure.
2. See Fig. 1.1.
3. See Fig. 1.2.
4. Physiology integrates body function across all levels of organization and emphasizes the coordinated function of body systems.
5. The maintenance of internal stability. Examples: body temperature and water balance.
6. Homeostasis and control systems; structure-function relationships; biological energy; communication.
7. Stimulus, sensor, input signal, integrating center, output signal, target, response.
8. Circadian rhythm.

LEVEL TWO Reviewing Concepts

9. Maps are highly individual. Evaluate your map by comparing it to some done by classmates or to ask your instructor for comments.
10. (a) Tissues—collections of cells that carry out related functions. Organs—collections of tissues that form structural and functional units. (b) x-axis—independent variable; y-axis—dependent variable. (c) Independent variable is manipulated to change the dependent variable. (d) Teleological—functional approach, the "why" of a system. Mechanistic approach—physiological mechanisms, the "how" of a system. (e) Internal environment—extracellular fluid; external environment—the world outside the body. (f) Blind study—subjects

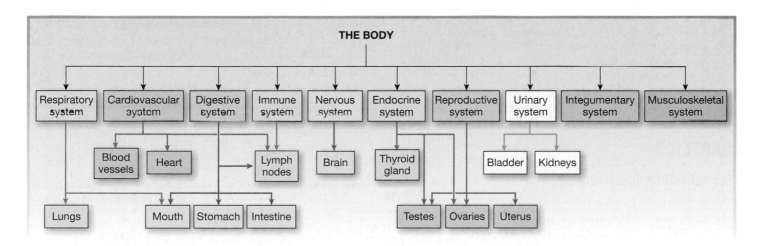

do not know the treatment they are receiving. Double-blind study—neither subjects nor experimenters know which treatment is the active one. Crossover study—each subject serves as both control and experimental. (g) Sensors receive signals. Targets respond to signals.

11. Nasal and oral cavities; external ear; lacrimal (tear) ducts; sweat, sebaceous, and mammary gland ducts; lumens of esophagus, stomach, small and large intestines; ducts of the salivary glands, pancreas, liver and gall bladder; urinary tract, reproductive, and respiratory organs.

12. Coordinate—endocrine and nervous systems. Protection: integumentary, digestive, cardiovascular, and immune systems. Exchange with external environment—respiratory exchanges gases; digestive system takes in nutrients; digestive and urinary eliminate waste products. Integumentary loses water and solutes.

13. Negative feedback—feedback signal turns response loop off; helps maintain homeostasis. Positive feedback—feedback keeps the response loop going; makes a change bigger. Feedforward—starts response loop before the stimulus does; minimizes change.

LEVEL THREE Problem Solving

14. (a) incorrect mechanistic answer, (b) correct teleological answer, (c) correct teleological answer, (d) correct mechanistic answer

15. Other problems: requirement of an aqueous environment for fertilization (internal fertilization in mammals; many other terrestrial animals return to water to breed); aqueous environment for embryonic development (eggs in birds, some reptiles and insects; internal development in mammals, some reptiles and insects); physical support (exoskeletons in insects, internal skeletons in vertebrates)

LEVEL FOUR Quantitative Problems

16. (a) independent—time; dependent—body length, (b) There was no control. (c) Should be a line graph with time in days on x-axis and body length on y-axis. (d) Growth slowest from days 0–3 and most rapid for days 6–9 and days 18–21.

17. (a) independent = solution concentration; dependent = volume change, (b) Volume before soaking provides a baseline but there is no experimental control. (c) A scatter plot with best-fit line would allow you to estimate volume change at intermediate salt concentrations, such as 5%.

18. (a) scatter plot, (b) Is there a relationship between midarm muscle circumference and aerobic fitness? (c) There appears to be no relationship between midarm muscle circumference and aerobic fitness.

19. (a) There is no "correct" answer. For peer critiques of the study, see *New England Journal of Medicine* 347(2): 132–133 and 137–139, 2002, July 11. For a more recent similar study, see article by R. Sihvonen *et al., N Engl J Med* 369: 2515–2524, 2013, Dec. 26. (b) The subjects believed that the surgery had helped (a placebo effect) or other interventions, such as physical therapy, helped. (c) The study is directly applicable to a limited population: male veterans, under age 76, predominantly white, with osteoarthritis or degenerative joint disease. (d) blind study, (e) The investigators were trying to determine whether a placebo effect could account for postsurgical improvement.

CHAPTER 2

Concept Check Questions

1. O, C, H, N, P, Na, K, Ca, Mg, S, Cl
2. $C_nH_{2n}O_n$ or $(CH_2O)_n$
3. Amino group is $—NH_2$. Carboxyl group is $—COOH$.

4. paired
5. ion
6. (a) 2, (b) 4, (c) 1, (d) 3
7. polar
8. hydrophilic
9. Na^+ and Cl^- ions form hydrogen bonds with the polar water molecules. This disrupts the ionic bonds that hold the NaCl crystal together.
10. Dissociate into one or more H^+ plus anions.
11. pH is the concentration of H^+.
12. down
13. Carbonic acid increases and pH decreases.
14. Molecule B is a better candidate because its smaller K_d means higher binding affinity.
15. (a) 1, 4; (b) 2, 3; (c) 4 (can bind anywhere)
16. Reaction rate decreases.
17. Rate is at its maximum.

Figure Questions

Fig. 2.7: 1. solute and solvent, 2. (d), 3. 18 amu or 18 Da, 4. 74.6 g, 5. 0.1 M solution = 100 mM solution, so concentrations are equal. 6. The 5 g of glucose add volume, so if you begin with 100 mL of solvent, you end up with more than 100 mL of solution.

Fig. 2.9: 1. pH decreases. 2. Urine, stomach acid, and saliva are all inside the lumens of hollow organs, where they are not part of the body's internal environment [see Fig. 1.2 on p. 4].

Fig. 2.13: (a) more active at 30 °C, (b) At A, rate is 1 mg/sec. When rate = 2.5 mg/sec, protein concentration is C. (c) At 200 mg/mL, rate is 4 mg/sec.

Chemistry Review Quiz

1. c, e
2. a, b, f, h
3. a, b, d, g, h
4. *protons* and *electrons; neutrons*. Radiation.
5. carbon, oxygen, nitrogen, and hydrogen
6. P, K, Na, S, Ca, Cl
7. (a) zinc, (b) 20, (c) atomic number 53; average atomic mass = 126.9. Iodine = I.
8. Mg^{2+} has *lost* two *electrons*.
9. Loss of hydrogen's one electron leaves behind one proton.
10. (a) 11, (b) zero, (c) 12, (d) cation, (e) +1, (f) Na^+, (g) neon, (h) Ne
11. (a) $C_6H_{12}O_6$ (glucose), m.w. 180; (b) CO_2, m.w. 44; (c) leucine, $C_6H_{13}NO_2$, m.w. 131; (d) $C_3H_7NO_2$ (alanine), m.w. 89
12. (a) 1, (b) 5, (c) 4, (d) 2, (e) 3
13. Unsaturated fatty acids have double bonds between carbons. Each double bond removes two hydrogens from the molecule, therefore, (c) $C_{18}H_{30}O_2$ is the most unsaturated because it has the fewest hydrogens.
14. (a) 4, (b) 5, (c) 1, (d) 2, (e) 3
15. (a) 3, (b) 1, (c) 5, (d) 2, (e) 4, 6
16. Proteins are composed of 20 different amino acids that can be linked in different numbers and an almost infinite number of sequences.
17. *amino; carboxyl* (or vice versa)
18. one or more phosphate groups, a 5-carbon sugar, and a base.

19. DNA: a double-stranded molecule with adenine, guanine, cytosine, and thymine linked in an α-helix; sugar is deoxyribose. RNA: a single-stranded molecule with uracil instead of thymine plus the sugar ribose.

20. Purines have 2 carbon rings. Pyrimidines have 1 carbon ring.

Review Questions

LEVEL ONE Reviewing Facts and Terms

1. Proteins (collagen, hemoglobin, enzymes); carbohydrates (glucose, sucrose); lipids (cholesterol, phospholipids); and nucleic acids (ATP, DNA, RNA).

2. False. All biomolecules are organic molecules, but the reverse is not true.

3. *molecule*

4. One carbon atom needs to share four electrons to fill its outer shell; therefore, it will form four covalent bonds.

5. covalent; polar; nonpolar

6. Oxygen and nitrogen strongly attract electrons and tend to form polar bonds.

7. Table sugar dissolves easily, so it is polar. Vegetable oil does not dissolve in water, so it is nonpolar.

8. *anion, cation*

9. pH = H^+ concentration. pH <7 is *acidic*. pH >7 is *basic* or *alkaline*.

10. *buffer*

11. *lipoproteins; glycoproteins*

12. *ligand*

13. (a) 4, (b) 3, (c) 2

14. *cofactor*

15. *denatured*

LEVEL TWO Reviewing Concepts

16. Check your map with your instructor or your fellow students. Maps will vary.

17. 10^{-3} M = pH 3; acidic. 10^{-10} M = pH 10; basic.

18. ATP: usable energy in a high-energy bond. DNA stores genetic information. RNA translates genetic information into proteins. cAMP: transfer of signals into cells. NAD and FAD transfer energy.

19. Isoforms are structurally similar, with similar functions but differing affinities for ligands. They may function best under different conditions.

20. (a) 4, 5; (b) 3; (c) 2, 1

LEVEL THREE Problem Solving

21. Nucleotides contain all of the elements listed in the right ratio. Carbohydrates have a C:H:O ratio of 1:2:1, so alien does not have enough H. Fats are mostly C and H with little O (not enough H and too much O). Proteins do not have P and have less N relative to C.

22. More CO_2 means more H^+ by the law of mass action. More H^+ means a decrease in pH.

LEVEL FOUR Quantitative Problems

23. 0.9% = 0.9 g/100 mL. Dissolve 9 g NaCl in water to yield 1 L of solution.

24. (a) 6.02×10^{23} molecules of NaCl, (b) 1000 millimoles, (c) 1 equivalent, (d) 5.85% solution

25. 10% glucose = 10g/100 mL or 20 g/200 mL solution. Molarity: 10 g/100 mL = 100 g/L × 1 mole/180g = 0.556 moles/L or

556 millimoles/L (556 mM). 500 mL of 10% glucose would have 50 g glucose × 1 mole/180 g = 278 millimoles glucose.

26. Myoglobin has a higher affinity for O_2 because at lower oxygen concentrations, myoglobin binds more O_2 than hemoglobin does.

CHAPTER 3

Concept Check Questions

1. phospholipids, sphingolipids, and cholesterol

2. Integral proteins are tightly bound to the membrane. Peripheral proteins are loosely bound to membrane components. Proteins may be transmembrane, lipid-anchored, or loosely bound to other proteins.

3. To hide the hydrophobic tails of phospholipids from direct contact with aqueous body fluids

4. one

5. microfilaments (actin fibers), intermediate filaments, and microtubules

6. It would be unable to swim to find an egg to fertilize.

7. Cytoplasm is everything inside the cell membrane except the nucleus. Cytosol is the semi-gelatinous substance in which organelles and inclusions are suspended.

8. Cilia are short, usually are very numerous on a cell, and move fluid or substances across the cell surface. Flagella are longer, usually occur singly on human sperm, and are used to propel a cell through a fluid.

9. Motor proteins use energy to create movement.

10. A membrane separates organelles from the cytosol; inclusions have no membrane.

11. Rough ER has ribosomes attached to the cytoplasmic side of its membrane; smooth ER lacks ribosomes. Rough ER synthesizes proteins; smooth ER synthesizes lipids.

12. Lysosomal enzymes break down bacteria and old organelles. Peroxisomal enzymes break down fatty acids and foreign molecules.

13. The membranes of organelles create compartments that physically isolate their lumens from the cytosol. The double membrane of mitochondria creates two different compartments inside the organelle.

14. It suggests that the cell has a high energy requirement because mitochondria are the site of greatest energy production in the cell.

15. This suggests that the tissue synthesizes large amounts of lipids, fatty acids, or steroids, or that it detoxifies foreign molecules.

16. gap (communicating), tight (occluding), and anchoring

17. (a) tight, (b) gap, (c) anchoring (specifically, desmosome), (d) anchoring (specifically, focal adhesion)

18. protective, secretory, transporting, ciliated, and exchange

19. the process by which a cell releases a substance into its environment

20. Endocrine glands do not have ducts, and they secrete into the blood. Exocrine glands have ducts and secrete into the external environment.

21. Hemidesmosomes (see Fig. 3.8a).

22. No, skin has many layers of cells to protect the internal environment. A one-cell-thick simple squamous epithelium would not be protective.

23. Endocrine cell because it secretes its product into the ECF for distribution in the blood.

24. extensive matrix

25. Collagen provides strength and flexibility; elastin and fibrillin provide elastance; fibronectin helps anchor cells to matrix.

26. bone, cartilage, blood, dense connective tissues (ligaments and tendons), loose connective tissue, and adipose tissue

27. Plasma is the extracellular matrix.

28. Cartilage lacks a blood supply, so oxygen and nutrients needed for repair must reach the cells by diffusion, a slow process.

29. Apoptosis is a tidy form of cell death that removes cells without disrupting their neighbors. By contrast, necrosis releases digestive enzymes that damage neighboring cells.

Figure Questions

Fig. 3.9: Endocrine glands secrete hormones into the blood. Exocrine glands, with ducts, secrete their products outside the body—onto the surface of the skin or into the lumen of an organ that opens into the environment outside the body.

Review Questions

LEVEL ONE Reviewing Facts and Terms

1. Barrier between cell and ECF; regulate exchange of material between cell and ECF; transfer information between the cell and other cells; provide structural support.

2. *phospholipids; proteins; carbohydrates*

3. phospholipids and proteins

4. Inclusions: particles of insoluble material, such as glycogen and ribosomes. Organelles, such as mitochondria and Golgi apparatus, are separated from cytosol by membranes.

5. A flexible, changeable, three-dimensional scaffold of actin, microfilaments, intermediate filaments, and microtubules. Functions: mechanical strength; stabilize position of organelles; transport material; link cells together; movement.

6. (a) 2, (b) 3, (c) 1, (d) 4

7. *serous; mucous*

8. (a) 3, (b) 5, (c) 4, (d) 1, (e) 2

9. very acidic conditions

10. *endocrine*

11. connective tissue (tendons that hold muscles to bones); epithelium (skin); neural tissue (the brain); and muscular tissue (heart and skeletal muscles)

12. *skin*

13. (a) 1, (b) 1, (c) 4, (d) 3, (e) 4, (f) 4, (g) 4, (h) 1, (i) 1

14. sweat glands—sweat; apocrine glands—waxy or milky secretions; sebaceous glands—a mixture of lipids

15. mitochondrial matrix—the internal compartment; tissue matrix—noncellular material found outside cells

LEVEL TWO Reviewing Concepts

16. Anchoring junctions (skin)—allow twisting and stretching of tissue. Tight junctions (epithelia)—prevent movement of materials between cells. Gap junctions (some muscles)—allow material to pass directly from cytoplasm of one cell to another.

17. Rough ER is where proteins are made, so pancreatic cells would have more.

18. Vesicles—membranous spheres. Examples: lysosomes, peroxisomes, secretory vesicles.

19. Stratified has many cell layers for protection; simple epithelium has only one layer.

20. Map: See Fig. 3.2.

21. See Fig. 3.10e. Tight junctions prevent movement of material between cells; leaky junctions allow some material to pass between cells.

22. *intracellular fluid; interstitial fluid; plasma.* Interstitial fluid and plasma are ECF.

23. Cholesterol molecules fill space between phospholipid tails.

24. Bone is rigid due to calcification; cartilage is firm but elastic. Bones are the primary support structure for the body; cartilage forms the ear, nose, larynx, and spine and helps hold bones together at the joints.

25. (a) lumen—hollow inside of an organ or tube; wall—cell layer. (b) cytoplasm—everything inside the cell except the nucleus; cytosol—semi-gelatinous, intracellular fluid. (c) myosin—motor protein filament; keratin—structural protein fiber.

26. Apoptosis—it is a normal part of development.

27. (a) 1 (gap junctions), 2 (tight junction proteins), 4 (strength of desmosomes); (b) 1 (receptors), 2 (enzymes), 3 (barrier), 4 (fluidity), 5 (ATP-dependent transporters); (c) 2 (microtubules direct movement), 4 (strength), 5 (ATP required for actin-myosin interaction); (d) 2 (mRNA binds to ribosomes), 3 (membrane-bounded organelles), 5 (ATP-dependent processes); (e) 2 (microtubules and dynein), 4 (flexibility), 5 (ATP-dependent movement)

28. The matrix can be broken down, then reassembled.

LEVEL THREE Problem Solving

29. Cilia sweep mucus and particles up and out of the airways. When they fail, inhaled pathogens are more likely to reach the lungs, resulting in infections, inflammation, or cancer. The smoker's cough removes the mucus that would normally be swept away by the cilia.

30. Many epithelia are vulnerable to damage and need to be replaced frequently. Cells undergoing frequent mitosis are more likely to develop abnormal cell division.

31. MMPs are enzymes that dissolve the extracellular matrix, so blocking them might inhibit tissue growth and repair.

CHAPTER 4

Concept Check Questions

1. Amino acids and nucleotides

2. in chemical bonds and in concentration gradients

3. Kinetic energy is the energy of motion: Something is happening. Potential energy is stored energy: Something is waiting to happen.

4. a state of randomness or disorder.

5. Endergonic reactions consume energy; exergonic reactions release energy.

6. The reactants are baking soda and vinegar; the product is carbon dioxide.

7. The foaming indicates energy release, so this is an exergonic reaction. The large amount of energy released indicates that the reaction is not readily reversible.

8. Isozymes allow one reaction to be catalyzed under a variety of conditions.

9. (d) quaternary level

10. lactose (lactase), peptides (peptidase), lipids (lipase), and sucrose (sucrase)

11. (a) 3, (b) 2, (c) 4, (d) 1

12. (1) controlling the amount of enzyme, (2) producing allosteric and covalent modulators, (3) using two different enzymes to catalyze reversible reactions, (4) isolating enzymes within intracellular organelles, and (5) altering the ratio of ADP to ATP in the cell

13. Energy is trapped and stored in one of the three phosphate bonds. In NADH, energy is stored in high-energy electrons.

14. Aerobic pathways require sufficient quantities of O_2 in the cell. Anaerobic pathways can run without oxygen.

15. (a) 4; (b) 2, 5; (c) 2, 5; (d) 1, 3

16. Endergonic reactions trap energy in the products.

17. Pumping H^+ into the intermembrane space, creates potential energy in the H^+ concentration. Release of this energy as H^+ pass through the ATP synthase is coupled to ATP synthesis.

18. Lactate dehydrogenase acts by *removing* an *electron* and a *hydrogen atom*. This process is called *oxidation*.

19. Anaerobic metabolism of glucose can proceed in the absence of oxygen; aerobic metabolism requires oxygen. Anaerobic metabolism produces much less ATP per glucose than aerobic metabolism.

20. The DNA triplets are ATT, ATC, and ACT.

21. RNA polymerase makes polymers of RNA.

22. During mRNA processing, nitrogenous base sequences called introns are cut out of the mRNA. The remaining segments, the exons, are spliced back together and provide the code for a protein.

23. dephosphorylation

24. cleavage, addition of groups, and cross-linking

25. It contains four protein chains so it is a tetramer.

Figure Questions

Fig. 4.5: endergonic

Fig. 4.6: Enzyme activity increases.

Fig. 4.10: A kinase moves a phosphate group from one molecule to another. A phosphatase removes a phosphate group.

Fig. 4.12: 1. exergonic; 2. (a) ①, ③; (b) ⑤, ⑥, ⑨; (c) Kinases add a phosphate group, so ①, ③, ⑤. (d) Dehydrogenases remove an electron and a hydrogen atom. In step ⑤, NAD^+ acquires an electron and an H, suggesting that this step is catalyzed by a dehydrogenase. 3. 2 ATP and 2 NADH.

Fig. 4.13: 1. exergonic; 2. 4 NADH, 1 $FADH_2$, and 1 ATP; 3. 3 CO_2, one for each carbon in pyruvate.

Fig. 4.14: 1. Phosphorylation is the addition of a phosphate group. ADP is phosphorylated. 2. exergonic, 3. Oxygen accepts electrons and H^+.

Fig. 4.15: 1. none, 2. 12–13 ATP per pyruvate

Answer to Running Problem Conclusion

Question 3. Other factors that could alter enzyme levels include decreased protein synthesis or increased protein breakdown in the cell. These changes could occur even though the gene was normal.

Review Questions

LEVEL ONE Reviewing Facts and Terms

1. Transport work (moving substances across membranes); chemical work (making proteins); and mechanical work (muscle contraction).

2. Potential energy = stored energy; kinetic energy = energy of motion.

3. First Law: There is a fixed amount of energy in the universe. Second Law: Without input of energy, an open system will become progressively less organized.

4. *metabolism*

5. *substrates; rate*

6. *enzymes; decreasing*

7. 1 (d), 2 (a), 3 (f), 4 (c)

8. *-ase*

9. *coenzymes; vitamins*

10. *reduced; oxidized*

11. *dehydration; hydrolysis*

12. *deamination; transamination*

13. *catabolic; anabolic.* Kilocalories.

14. *feedback inhibition*

15. H^+ transported into the inner compartment stores energy in a concentration gradient. When the ions move back across the membrane, the released energy is trapped in the high-energy bond of ATP.

16. NADH and $FADH_2$

LEVEL TWO Reviewing Concepts

17. Map 1: Start with Fig. 4.11. Map 2: Use Figs. 4.19, 4.20, and 4.21.

18. Perform work, transfer to another molecule, or be released as heat.

19. 1 (b), 2 (a), 3 (b), 4 (a), 5 (c), 6 (c) or (a)

20. When inactive, they cannot harm the cell if accidentally released.

21. Aerobic breakdown = 30–32 ATP; anaerobic breakdown = 2 ATP. Anaerobic is faster and does not require oxygen, but energy yield is lower.

22. Transcription: synthesis of RNA from the sense strand of DNA. Takes place in the nucleus. Translation: conversion of information coded in mRNA into a string of amino acids. Takes place on ribosomes.

23. Anticodons are part of tRNA. Amino acids attach to tRNA.

24. Chemical bond energy is potential energy.

25. If the reaction requires ATP, the activation energy must be large compared to a reaction that does not require ATP.

LEVEL THREE Problem Solving

26. 1. TAC is start codon.
    2. DNA: AAG TCA CGT ACC GTA ACG ACT
       mRNA: UUC AGU GCA UGG CAU UGC UGA
    3. Amino acids: Phe-Ser-Ala-Trp-His-Cys-Stop codon

LEVEL FOUR Quantitative Problems

27. exergonic

28. 149 amino acids

CHAPTER 5

Concept Check Questions

1. The stem preceding *–ase* is the substrate on which the enzyme acts; therefore, ATP is a substrate for this enzyme.

2. ICF has a high K^+ concentration and low Na^+, Cl^-, and Ca^{2+} concentrations.

3. Plasma is part of the ECF, so knowing the plasma concentration of (a) Na^+ and (b) K^+ tells you the ECF concentration of these ions. The ICF ion concentrations are different, so plasma concentration will not tell you the ICF values. (c) The body is in osmotic equilibrium, so water concentration is the same in all compartments. (d) Proteins are in the plasma but not in the interstitial fluid, so plasma protein concentration will not tell you ECF or ICF protein concentrations.

4. (a) total body water = 29 L; (b) ECF = 9.7, ICF = 19.3 L; (c) plasma = 2.4 L

5. The baby has lost 0.91 kg of water, which is 0.91 L.

6. 1 M NaCl = 2 OsM NaCl. The 1 M (= 1 OsM) glucose and 1 OsM NaCl have the most water.

7. (a) Water moves into A because A is 2 OsM. (b) No net movement occurs because urea will diffuse across the membrane until it reaches equilibrium. (c) Water moves into A because A has a higher concentration of nonpenetrating solutes.

8. Prediction: 260 mOsM glucose is hyposmotic and hypotonic. Adding 260 mosmoles glucose and 1 L volume:

	Total Body	ECF	ICF
S (mosmol)	900 + 260 = 1160	300	600 + 260 = 860
V (L)	3 + 1 = 4	1.034	2.966
C (mOsM)	1160/4 = 290 mOsM	290	290

9. (a) Sweat loss is hyposmotic. Body osmolarity will increase. (b) Cell volume will decrease. (c) Losing 0.5 L and 65 mosmoles NaCl. Yes, cell volume decreased and osmolarity increased.

	Total Body	ECF	ICF
S (mosmol)	900 − 65 = 835	300 − 65 = 235	600
V (L)	3 − 0.5 = 2.5	0.704	1.796
C (mOsM)	334 mOsM	334	334

10. (a) The NaCl solution is better, even though both solutions are isosmotic to the body (Tbl. 5.8). Because blood is lost from the extracellular compartment, the best replacement solution would remain in the ECF. For this reason, glucose is not as good a choice because it slowly enters cells, taking water with it. (b) If 1 L has been lost, you should replace at least 1 L.

11. If distance triples, diffusion takes nine times as long.

12. Energy for diffusion comes from molecular motion.

13. Because it is lipophilic, the fatty acid is more likely to cross by simple diffusion.

14. The flux (a) decreases, (b) increases, (c) decreases.

15. Compartment A remains yellow, and compartment B turns green.

16. The skin's thick extracellular matrix is generally impermeable to oxygen. Also, oxygen needs a moist surface to effectively diffuse across a tissue membrane.

17. Positive ions are cations, and negative ions are anions.

18. Membrane proteins serve as structural proteins, receptors, enzymes, and transporters.

19. Ions and water molecules move through open channels.

20. Channel proteins form continuous connections between the two sides of a membrane and transport molecules more quickly.

21. A channel lined with positive charges attracts anions, which in this instance means Cl⁻.

22. Glucose is too large to pass through a channel.

23. The direction of facilitated diffusion of glucose reverses, and glucose leaves the cell.

24. The ATPase is an antiporter, but the SGLT is a symporter. The ATPase requires energy from ATP to change conformation, whereas the SGLT uses energy stored in the Na^+ concentration gradient.

25. An antiporter moves substrates in opposite directions.

26. Larger doors could move more people. This would be analogous to a cell's synthesizing a new isoform of the transporter that would let the transporter move more substrate per second.

27. In phagocytosis, the cytoskeleton pushes the membrane out to engulf a particle in a large vesicle. In endocytosis, the membrane surface indents and the vesicle is much smaller.

28. The proteins associated with endocytosis are clathrin and caveolin.

29. Proteins move into cells by endocytosis and out of cells by exocytosis.

30. Na^+ movement out of the cell requires energy because the direction of ion flow is against the concentration gradient.

31. Ouabain applied to the apical side would have no effect because there are no Na^+-K^+-ATPase molecules on that side. Ouabain applied to the basolateral side would stop the pump. Glucose transport would continue for a time until the Na^+ gradient between the cell and the lumen disappears because Na^+ has entered the cell.

32. The GLUT2 transporter is illustrated.

33. Transcytosis will stop because vesicular transport by the cytoskeleton depends on functioning microtubules.

34. Over time, Na^+ would leak into the cell, and the resting membrane potential would become more positive.

Figure Questions

Fig. 5.1: 1. Plasma is 25% of 14 L = 3.5 L. Interstitial fluid is 75% = 10.5 L. 2. Total body water = 42 L. 3. Plasma is 3.5 L/42 L = 8.3% of total body water; interstitial volume is 10.5 L/42 L= 25%. 4. Total body weight is 121 lb × 1 kg/2.2 lb = 55 kg. If body water = 50% of body weight, total body water = 27.5 L. ICF is 67% of 27.5 L = 18.425 L. ECF is 33% of 27.5 L = 9.075 L. Plasma is 25% of ECF = 2.269 L. 5. Plasma contains proteins and large anions not present in interstitial fluid. 6. The extracellular compartment contains more Na^+, Cl^-, and bicarbonate than the intracellular compartment, and fewer K^+.

Fig. 5.4: Example 2: (a) hyperosmotic; (b) 250 mOsM; (c) 300 mOsM; (d) yes, into the cells; (e) hypotonic; (f) increased, hyperosmotic; (g) increased, hypotonic

Fig. 5.17: (c) The cell could increase transport by adding more transporters to the membrane. (d) You cannot tell if galactose is being transported because the graph only shows glucose transport.

Fig. 5.21: 1. 1 = b, 2 = a, 3 = b; 2. Basolateral glucose transport is passive because the glucose moves down its concentration gradient. 3. Na^+ movement across the apical membrane does not require ATP because Na^+ is moving down its concentration gradient.

Fig. 5.23: 1. Because Na^+ is more concentrated outside the cell, some Na^+ will move into the cell, giving the cell a positive membrane potential. 2. Cl^- also will move into the cell, making the membrane potential negative. 3. E_{Na} = +60 mV. 4. E_{Cl} = –60 mV.

Fig. 5.25: Na^+ leak into the cell is promoted by concentration and electrical gradients. K^+ leak out of the cell is promoted by the concentration gradient.

Fig. 5.26: 1. ①, 2. ligand-gated (ATP-gated), 3. Yes, exocytosis requires energy from ATP. 4. It is too large to go through a carrier or channel.

Review Questions

LEVEL ONE Reviewing Facts and Terms

1. structural proteins (link cell to matrix), transporter proteins (water channels), receptors (hormone receptors), and enzymes (intestinal digestive enzymes)

2. Active: requires direct or indirect use of energy. Passive: uses energy stored in a concentration gradient.

3. Passive: simple and facilitated diffusion, osmosis. Active: phagocytosis, exocytosis, and endocytosis.

4. greater concentration gradient, smaller distance, higher temperature, and smaller molecular size

5. simple diffusion, protein-mediated transport, or vesicular transport

6. *symport; antiport; uniport*

7. *primary (direct)* and *secondary (indirect)*

8. *penetrating; nonpenetrating*

9. (d), (a), (b), (c)

10. Osmolarity: concentration of osmotically active particles, expressed as osmol/L or milliosmoles per liter

11. Hypotonic: net influx of water into the cell at equilibrium. Hypertonic: net water loss at equilibrium. Tonicity is determined by relative concentrations of nonpenetrating solutes in cell versus solution.

12. (a) 2, 4, 6; (b) 1, 6; (c) 2, 3, 6; (d) 2, 5, 6

13. (1) Like charges repel; opposite charges attract. (2) Every positive ion has a matching negative ion. (3) Energy must be used to separate ions or electrons and protons. (4) Conductors allow ions to move through them; insulators keep ions separated.

14. (a) 7; (b) 1, 7; (c) 6; (d) 5; (e) 8; (f) 3; (g) 2

15. *equilibrium potential*

16. *conductor; insulator*

LEVEL TWO Reviewing Concepts

17. Use Figs. 5.5, 5.8, 5.10, and 5.19 to create your map.

18. See Fig. 5.1c and d.

19. Lipid solubility, so that a molecule can pass through the lipid core of the membrane. Diffusion is slower for larger or heavier molecules and faster when there is more membrane surface area.

20. Specificity: Enzyme or transporter works on one molecule or class of molecules. Competition: Similar substrates can compete for the protein binding site. Saturation: Rate reaches a maximum when all binding sites are filled. GLUT is specific for hexose sugars. If glucose and fructose are both present, they compete for GLUT binding sites. If enough sugar is present, transport saturates.

21. (a) *hypotonic*, (b) *into the cells*

22. Active transport. Must use energy to go from a state of equilibrium to one of disequilibrium.

23. (a) *hyperosmotic* (convert molarity to osmolarity), (b) True. Water moves from B to A.

24. Chemical gradient = concentration gradient. Electrical gradient = separation of electrical charge. Electrochemical gradient includes both concentration and electrical gradients.

LEVEL THREE Problem Solving

25. Apical side: Na^+ leak channels but no water pores. Basolateral side: Na^+-K^+-ATPase and K^+ leak channels. May also have water channels.

26. Insulin could increase the number or affinity of GLUT proteins or could act on cell metabolism to keep the intracellular glucose concentration low.

27. Both enzymes and transporters are proteins that bind ligands at a specific binding site. Enzymes alter their substrates. Transporters move substrates unchanged across a membrane.

28. Sugars are added to proteins inside the organelle/vesicle, therefore will face the ECF after being inserted into the membrane.

29. Must convert units from mM to mOsM. (a) hyperosmotic, isotonic; (b) hyposmotic, hypotonic; (c) isosmotic, hypotonic; (d) hyperosmotic, isotonic; (e) hyperosmotic, hypotonic

LEVEL FOUR Quantitative Problems

30. 296 mOsM

31. (a) ICF = 29.5 L; interstitium = 9.8 L. (b) Total solute = 12.432 osmoles; ECF = 3.7 osmoles; ICF = 8.732 osmoles; plasma = 0.799 osmoles.

32. 154 mOsM

33. (a) increases, (b) decreases, (c) increase, (d) decrease

34. Diffusion (a). Cannot be active transport because concentration$_{in}$ never exceeds concentration$_{out}$.

CHAPTER 6

Concept Check Questions

1. 1. no matches; 2. a, b, d, e, f, g; 3. c

2. Cytokines, hormones, and neurohormones travel through the blood. Cytokines, neurohormones, and neurotransmitters are released by neurons.

3. It could not have been a paracrine signal because the eyes are too far away from the legs. Also, the response was too rapid for it to have taken place by diffusion.

4. Signal pathways have a signal molecule, receptor, intracellular signal molecule(s), and target proteins.

5. Receptors are on cell membrane or in cytosol or nucleus.

6. (1) signal molecule binds to receptor that (2) activates a protein that (3) creates second messengers that (4) creates a response.

7. Amplification turns one signal molecule into multiple second messenger molecules. In Fig. 6.6b, 1 ligand amplifies into 18 intracellular molecules.

8. Steroids are lipophilic, so they can enter cells and bind to intracellular receptors.

9. Receptors are either ligand-gated ion channels, receptor-enzymes, G protein–coupled receptors, or integrins.

10. First messengers are extracellular; second messengers are intracellular.

11. (a) ligand, receptor, second messenger, cell response; (b) amplifier enzyme, second messenger, protein kinase, phosphorylated protein, cell response

12. (a) Cl^- channel opens: cell hyperpolarizes, (b) K^+ channel opens: cell hyperpolarizes, (c) Na^+ channel opens: cell depolarizes.

13. The cell must use active transport to move Ca^{2+} against its concentration gradient.

14. A drug that blocks leukotriene action could act at the target cell receptor or at any step downstream. A drug that blocks leukotriene synthesis might inhibit lipoxygenase.

15. They are all proteins.

16. It could be one receptor with different second messenger systems or two different receptor isoforms.

17. Choices (a) and (d) could decrease binding affinity. Changing receptor number would not affect binding affinity.

18. Tonic control usually involves one control system, but antagonistic control uses two.

19. A signal can have opposite effects by using different receptors or different signal pathways.

20. In local control, the stimulus, integration of the signal, and response all take place in or very close to the target cell. With reflex control, integration of the input signal and initiation of a response may take place far from the location where the change occurred.

21. Stimulus, sensor or sensory receptor, input signal (afferent pathway), integrating center, output signal (efferent pathway), target or effector, response (tissue and systemic)

22. (a) The "neural system integrating center" is the brain and spinal cord. (b) "Receptor" represents the sense organs. (c) The dashed line indicating negative feedback runs from "Response" back to "Internal or external change."

23. blow to knee = internal or external change, leg muscles = targets, neuron to leg muscles = efferent neuron, sensory neuron = input signal, brain and spinal cord = CNS integrating center, stretch receptor = sensor or receptor, muscle contraction = response

24. food in stomach = stimulus, brain and spinal cord = CNS integrating center, endocrine cells of pancreas = E (integrating center), stretch receptors = receptor, efferent neuron to pancreas = efferent neuron, insulin = classic hormone, adipose cell = target cell, sensory neuron = afferent neuron. Blood is the anatomical route that hormones use to reach their target but is not part of the reflex pathway.

Figure Questions

Fig. 6.8: A (inactive and active) = adenylyl cyclase; inactive B = ATP; active B = cAMP; C (inactive and active) = protein kinase A; product = phosphorylated protein.

Fig. 6.15: 180 beats/min for the top ECG and 60 beats/min for the bottom ECG.

Review Questions

LEVEL ONE Reviewing Facts and Terms

1. neurons and blood

2. nervous and endocrine systems

3. chemical (available to all cells) and electrical

4. *receptor, targets (effectors)*, or *proteins*

5. (a) *adenylyl cyclase*, (b) *guanylyl cyclase*, (c) *phospholipase C*

6. *phosphate, ATP*

7. Central: located within the central nervous system. Peripheral: found outside the CNS

8. *nucleus, cytosol, cell membrane*

9. *decreased*

10. It may down-regulate receptor number or decrease receptor affinity for the substrate.

11. *opposite*

LEVEL TWO Reviewing Concepts

12. (a) *Gap junctions* connect two cells using protein channels called *connexons*, made from *connexin* subunits. (b) *Paracrine signals* act on nearby cells; *autocrine signals* act on the cell that secretes them. (c) *Cytokines* are peptide autocrine and paracrine signals. *Neurotransmitters, neuromodulators, and neurohormones* are all chemicals secreted by neurons. Neurotransmitters act rapidly on nearby cells; neuromodulators act more slowly. Neurohormones and *hormones* are secreted into the blood for action on distant targets. (d) *Receptor agonists* activate receptors just like the normal ligand; *receptor antagonists* also bind to the receptor but block its activation. *Antagonistic pathways* create responses that oppose each other. (e) *Transduction:* A signal molecule transfers information from ECF to the cytoplasm. *Cascade:* a series of steps. *Amplification:* One signal molecule creates a larger signal.

13. Ligand-gated channels (ATP-gated K^+ channel); integrin receptors (platelet receptors); receptor enzymes (tyrosine kinase receptor); G protein-coupled receptors (adenylyl cyclase/cAMP-linked receptors).

14. The father of American physiology. (1) The nervous system keeps body functions within normal limits. (2) Some functions have tonic control rather than on-off control. (3) Some signals act in opposition to each other. (4) Cell response depends on the cell's receptor for a signal.

15. Stimulus to sensor (sensory receptor) to input signal (sensory nerve) to integrating center. Integrating center (the brain or an endocrine cell) sends an output signal (through nerve or hormone) to target cell (muscles and glands), which reacts to the stimulus with a response.

16. Neural control is faster than endocrine and better for short-acting responses. Endocrine control can affect widely separated tissues with a single signal and better for long-acting responses.

17. (a) negative, (b) positive, (c) negative, (d) negative

18. (a) tissues that respond to glucagon, such as liver; (b) breast; (c) bladder; (d) sweat glands

19. (a) pancreatic endocrine cells that secrete glucagon, (c) and (d) nervous system

LEVEL THREE Problem Solving

20. (a) stimulus = decrease in body temperature to decrease, sensor = temperature receptors, input = sensory neurons, integrating center = brain, output = efferent neurons, targets = muscles used to pull up afghan, response = afghan conserves heat; (b) stimulus = smell of sticky buns, sensor = odor receptors in the nose, input = sensory neurons, integrating center = brain, output = efferent neurons, target = skeletal muscles, response = walk to bakery, buy buns, and eat

21. (a) *antagonistic*, (b) Neurotransmitters act on nearby cells (paracrine action). Neurohormones act on distant targets. (c) Epinephrine is secreted in larger amounts because it will be diluted by the blood volume before reaching its target.

LEVEL FOUR Quantitative Problems

22. (a) amplification and a cascade, (b) (1000 × 4000) or 4,000,000 GMP

CHAPTER 7

Concept Check Questions

1. facilitated diffusion (GLUT transporters).

2. *-ase* indicates an enzyme. A *peptidase* digests peptides.

3. A chemical secreted into the blood that acts on a distant target in very low concentrations

4. steroid-producing cell: extensive smooth endoplasmic reticulum, protein-producing cell: lots of rough endoplasmic reticulum and secretory vesicles

5. peptide, steroid, and amino acid-derived

6. The short half-life suggests that aldosterone is not bound to plasma proteins as much as other steroid hormones are.

7. Increased blood glucose is the stimulus. Insulin secretion is the efferent pathway; decrease in blood glucose is the response.

8. Insulin release by blood glucose is a simple endocrine reflex. Insulin release in response to a digestive hormone is the complex endocrine reflex. Insulin release triggered by a neural signal following a meal is the neural-endocrine reflex.

9. stimulus: decreased blood glucose, sensor/integrating center: pancreatic endocrine cells, efferent path: glucagon, target: multiple target tissues, response: increased blood glucose

10. amino acid-derived hormones

11. Microtubules of the cytoskeleton move secretory vesicles.

12. Contents released by exocytosis.

13. (a) pathway 4; (b) pathway 4 for GH acting directly on targets, and pathway 5 for GH acting on the liver

14. The target is endocrine cells of the anterior pituitary.

Figure Questions

Fig. 7.6: The conversion of tyrosine to dopamine adds a hydroxyl (–OH) group to the 6-carbon ring and changes the carboxyl (–COOH) group to a hydrogen. Norepinephrine is made from dopamine by changing one hydrogen to a hydroxyl group. Epinephrine is made from norepinephrine by changing one hydrogen attached to the nitrogen to a methyl (–CH₃) group.

Fig. 7.7: The pathway begun by eating a meal shuts off when the stretch stimulus disappears as the meal is digested and absorbed from the digestive tract.

Fig. 7.11: In short-loop negative feedback, ACTH feeds back to inhibit hypothalamic release of CRH.

Fig. 7.15: (a) CRH high, ACTH low, cortisol low. No negative feedback loops are functioning. (b) CRH normal/high, ACTH high, cortisol low. Absence of negative feedback by cortisol increases trophic hormones. Short-loop negative feedback from ACTH may keep CRH within the normal range.

Review Questions

LEVEL ONE Reviewing Facts and Terms

1. *endocrinology*

2. Alter the rate of enzymatic reactions, control transport of molecules into and out of cells, or change gene expression and protein synthesis in target cells.

3. See Fig. 7.2.

4. (a) 4, (b) 5, (c) 1, (d) 2, (e) 3

5. (d), (b), (c), (a)

6. *blood, distant target, very low*

7. the time required for half a dose of hormone to disappear from the blood

8. *kidneys* and *liver, urine* and *bile*

9. *factor*

10. Peptides—three or more amino acids; example: insulin. Steroids—derived from cholesterol; example: estrogen. Amino acid–derived—made from single amino acids; example: thyroid hormone

11. (a) peptide, (b) peptide, (c) steroid, (d) peptide, (e) peptide, (f) steroid, (g) peptide, (h) all classes, (i) steroid, (j) steroid

12. Steroid hormones usually initiate new protein synthesis, which takes time. Peptides modify existing proteins.

13. *transcription* factor, *genes, proteins*

14. *cell membrane*

15. *tryptophan, tyrosine*

16. *trophic*

17. *negative feedback*

18. synthesized by and secreted from neurons

19. oxytocin and vasopressin, both peptide neurohormones

20. The portal system is composed of hypothalamic capillaries that take up hormones and deliver them directly to capillaries in the anterior pituitary. The direct connection allows very small amounts of hypothalamic hormone to control the anterior pituitary endocrine cells.

21. See Fig. 7.9.

22. A hormone from a peripheral endocrine gland decreases pituitary and hypothalamic hormone secretion.

23. *synergism, permissiveness, antagonistic*

LEVEL TWO Reviewing Concepts

24. (a) Paracrines—local; cytokines—local or long distance; hormones—long distance. Cytokines—peptides; hormones—peptides, steroids, or amines. Cytokines—made on demand; peptides—made in advance and stored. (b) Primary pathology arises in the last endocrine gland of the pathway. Secondary pathology arises in a gland secreting a trophic hormone. (c) Hypersecretion—too much hormone; hyposecretion—too little hormone. (d) Both secrete peptide hormones. Anterior pituitary gland—true endocrine gland; posterior pituitary—neural tissue.

25. See Tbl. 7.1.

26. Use Fig. 7.3 for List 1 and Figs. 7.8 and 7.9 for List 2.

LEVEL THREE Problem Solving

27. The meanings do not change significantly. Enzymes, hormone receptors, transport proteins, and receptors are all proteins that bind ligands.

28. Patient A—cortisol hypersecretion results from ACTH hypersecretion. When dexamethasone suppresses ACTH secretion, the adrenal gland is no longer stimulated. Cortisol secretion decreases as a result. Patient B—problem in the adrenal gland. His normal negative feedback pathways do not operate, and the adrenal gland continues oversecreting cortisol even though ACTH secretion has been suppressed by dexamethasone.

29. (a) See Fig. 26.8. (b) Both LH and testosterone are needed for gamete formation. Testosterone does not directly suppress gamete formation, but it does have a negative feedback effect and shuts off LH secretion. LH is needed for gamete production, so its absence would suppress gamete synthesis.

LEVEL FOUR Quantitative Problems

30. Half-life is 3 hours.

31. (a) Group A, (b) Group B, (c) Group A

32. *x*-axis—plasma glucose; *y*-axis—insulin secretion. As *X* increases, *Y* increases.

CHAPTER 8

Concept Check Questions

1. Compare your answer to the map in Fig. 8.1.

2. Neurons that secrete neurohormones terminate close to blood vessels so that the neurohormones can enter the circulation.

3. A neuron is a single nerve cell. A nerve is a bundle of axons from many neurons.

4. See Fig. 8.2.

5. Myelin insulates axon membranes. Microglia are scavenger cells in the CNS. Ependymal cells form epithelial barriers between fluid compartments of the CNS.

6. Schwann cells are in the PNS, and each Schwann cell forms myelin around a small portion of one axon. Oligodendrocytes are in the CNS, and one oligodendrocyte forms myelin around axons of several neurons.

7. For Ca^{2+} the electrical charge z is +2; the ratio of ion concentrations is $1/0.0001 = 10,000$ or 10^4. Log of 10^4 is 4 (see Appendix B). Thus, E_{Ca} (in mV) = $(61 \times 4)/(+2) = 122$ mV.

8. (a) depolarize, (b) depolarize

9. depolarize

10. (a) 1, (b) 2, (c) 2, (d) 1

11. The trigger zone for the sensory neurons is close to where the dendrites converge. You cannot tell where the trigger zone is for the anaxonic neuron. For multipolar neurons, the trigger zone is at the junction of the cell body and the axon.

12. Conductance refers to the movement of ions across a cell membrane. Conduction is the rapid, undiminished movement of an electrical signal down the axon of a neuron.

13. (b)

14. The membrane potential depolarizes and remains depolarized.

15. During resetting, the activation gate is closing, and the inactivation gate is opening.

16. The action potential will go in both directions because the Na^+ channels around the stimulation site have not been inactivated by a previous depolarization. See discussion of refractory periods.

17. (a), (c), (b)

18. Because different receptor subtypes work through different signal transduction pathways, targeting drugs to specific receptor subtypes decreases the likelihood of unwanted side effects.

19. Proteins are synthesized on the ribosomes of the rough endoplasmic reticulum; then the proteins are directed into the Golgi apparatus to be packaged into vesicles.

20. Mitochondria are the primary sites of ATP synthesis.

21. Mitochondria reach the axon terminal by fast axonal transport along microtubules.

22. The researchers concluded that some event between arrival of the action potential at the axon terminal and depolarization of the postsynaptic cell is dependent on extracellular Ca^{2+}. We now know that this event is neurotransmitter release.

23. The exchange is secondary active transport because it uses energy stored in the H^+ concentration gradient to concentrate neurotransmitter inside the vesicles.

24. SSRIs decrease reuptake of serotonin into the axon terminal, thereby increasing the time serotonin is active in the synapse.

25. Acetyl CoA is made from pyruvate, the end product of glycolysis, and CoA.

26. Neurotransmitter uptake is secondary active transport because it uses energy stored in the Na^+ concentration gradient to concentrate neurotransmitter inside the axon terminal.

27. The postsynaptic neuron will fire an action potential, because the net effect would be a 17 mV depolarization: –70 mV + 17 mV = –53 mV, which is just above the threshold of –55 mV.

28. The membrane potential does not change at the same time as the stimulus because the depolarization must travel from the point of the stimulus to the recording point.

29. Axon terminals convert (transduce) the electrical action potential signal into a chemical neurotransmitter signal.

30. Membrane depolarization makes the inside of the membrane more positive than the outside. Like charges repel one another, so the more positive membrane potential tends to repel Mg^{2+}.

Figure Questions

Fig. 8.7: The graded potential is stronger at B. On the graph, A is between 3 and 4, and B is about at 1.

Fig. 8.14: (a) 4; (b) 2, 3; (c) 1; (d) 3; (e) 4

Fig. 8.15: Area of 100 giant axons is 50.3344 mm^2, so $r^2 = 16$ mm and r = 4 mm. Diameter = 8 mm.

Fig. 8.17: (a) –108 mV, (b) –85 mV

Fig. 8.22: Amplification

Fig. 8.24: 1. Convergence occurs in (c), (d), (e), (f), and (g). Divergence occurs in (f) and (g). 2. Move the inhibitory neuron from the right collateral terminal to the target cell. Right collateral now releases neurotransmitter but target cell does not respond.

Review Questions

LEVEL ONE Reviewing Facts and Terms

1. Sensory afferents carry messages from sensory receptors to CNS. Their cell bodies are located close to the CNS. Interneurons are completely contained within the CNS and are often extensively branched. Efferents carry signals from the CNS to effectors. They have short, branched dendrites and long axons.

2. *skeletal muscles, autonomic*

3. *sympathetic* or *parasympathetic*

4. (a) 3, (b) 1, (c) 2, (d) 5, (e) 4

5. neurons and glial cells

6. See Figs. 8.2 and 8.3.

7. (c). Answer (b) is only partly correct because not all axonal transport uses microtubules and not all substances moved will be secreted.

8. (a) 1, 4; (b) 2, 3, 5, 6

9. (e) – (b) – (d) – (a) – (c)

10. Na^+ channels (voltage-gated along axon; any type of gating on dendrites); voltage-gated K^+ channels along axon; voltage-gated Ca^{2+} channels in axon terminal; chemically gated Cl^- channels

11. (a) 3; (b) 1; (c) 4, 6; (d) 2; (e) 5, 6; (f) 5

12. (b) and (d)

13. (a) K^+, Na^+, Na^+; (b) Na^+; (c) K^+; (d) Na^+; (e) K^+

14. insulating membranes around neurons that prevent current leak

15. larger axon diameter and the presence of myelin

16. enzymatic degradation, reabsorption, and diffusion

17. See Figs. 8.9, 8.10, and 8.12.

LEVEL TWO Reviewing Concepts

18. (d)

19. See Tbl. 8.4.

20. See Figs. 8.1 and 8.5.

21. (f) – (c) – (g) – (e) – (b) – (k) – (c) – (a) – (h) – (j) – (i) – (d)

22. (a) depolarize, (b) hyperpolarize, (c) repolarize, (d) depolarize

23. (a) depolarize, (b) hyperpolarize, (c) hyperpolarize, (d) depolarize

24. Strength is coded by the frequency of action potentials; duration is coded by the duration of a train of repeated action potentials.

25. (b)

26. (a) Threshold signals trigger action potentials. Suprathreshold also trigger action potentials, but subthreshold do not unless summed. Action potentials are all-or-none events. Overshoot—portion of the action potential above 0 mV. Undershoot—after-hyperpolarization portion of the action potential. (b) Graded potentials may be depolarizing or hyperpolarizing. Graded potential in a postsynaptic cell is an EPSP if depolarizing and an IPSP if hyperpolarizing. (c) No stimulus can trigger another action potential during the absolute refractory period, but a suprathreshold stimulus can trigger an action potential during the relative refractory period. (d) See answer to question 1. (e) Sensory

are afferents; all others are efferents. (f) Fast synaptic potentials result from neurotransmitters altering ion channel gating, occur rapidly, and are short-lived. Slow synaptic potentials are mediated through second messengers, may involve protein modification, and last longer. (g) Temporal summation—multiple stimuli arrive at the trigger zone close together in time. Spatial summation—multiple stimuli from different locations arrive simultaneously at the trigger zone. (h) Divergence—a single neuron branches and its collaterals synapse on multiple targets. Convergence—multiple presynaptic neurons provide input to a smaller number of postsynaptic neurons.

LEVEL THREE Problem Solving

27. All the necessary synapses have not yet been made between neurons or between neurons and effectors.

28. Inactivation gates also respond to depolarization, but they close more slowly than the activation gates open, allowing ions to flow for a short period of time.

29. (b), (d), and (h)

30. (a) thermal, (b) chemical, (c) chemical, (d) chemical, (e) chemical, (f) mechanical

31. Unmyelinated axons have many ion channels, so more ions cross during an action potential and must be returned to their original compartments by the Na^+-K^+-ATPase, using energy from ATP.

LEVEL FOUR Quantitative Problems

32. (a) –80 mV, (b) +63 mV, (c) –86 mV, (d) –73 mV

33. (a) $(12 \times 2\ mV = 24) + (3 \times -3\ mV = -9)$ = signal strength of 15 mV. $V_m = -70 + 15 = -55$ mV. Threshold is –50 so no action potential. (V_m must be equal to or more positive than threshold.) (b) Net signal = $+13\ mV$. $-70 + 13 = -57$. Threshold is –60 so action potential will fire. (c) Net signal = $+19$. $-70 + 19 = -51$. Threshold is –50 so no action potential.

CHAPTER 9

Concept Check Questions

1. (a) 3; (b) 2; (c) 1; (d) 3; (e) 2

2. CNS glial cells are astrocytes, oligodendrocytes, microglia, and ependyma. See Fig. 8.5, p. 234, for functions.

3. Ganglion: A cluster of nerve cell bodies outside the CNS. CNS equivalent is a nucleus.

4. Tracts are the CNS equivalent of peripheral nerves.

5. When H^+ concentration increases, pH decreases, which means CSF pH must be lower than blood pH.

6. Blood will collect in the space between the membranes, pushing on the soft brain tissue under the skull. (This is called a *subdural hematoma.*)

7. CSF is more like interstitial fluid because they both contain little protein and no blood cells.

8. Oxidative phosphorylation takes place in mitochondria.

9. The two pathways are glycolysis and the citric acid cycle. Glucose is metabolized to pyruvate through glycolysis and then enters the citric acid cycle. $NADH_2$ passes high-energy electrons to the electron transport system for ATP synthesis.

10. Ehrlich concluded that some property of brain tissue made it resistant to staining by the dye.

11. The brain stained blue this time, but none of the other body tissues were stained because the dye was unable to cross the blood-brain barrier and enter the bloodstream.

12. Horns are areas of gray matter in the spinal cord. Roots are sections of spinal nerves just before they enter the spinal cord. Tracts are long projections of white matter (axons) that extend up and down the spinal cord. Columns are groups of tracts carrying similar information.

13. Cutting a dorsal root disrupts sensory function.

14. (a) and (c) are white matter, (b) is gray matter.

15. Activities would include moving the eyes, jaw, or tongue and testing taste, smell, and hearing.

16. The cerebrum is dorsal or superior to the brain stem.

17. The three subdivisions of the brain stem are medulla oblongata, pons, and midbrain.

18. The diencephalon is composed of thalamus, hypothalamus, pituitary gland, and pineal gland.

19. Neurons cross from one side of the body to the other at the pyramids in the medulla.

20. The divisions of the brain, starting at the spinal cord, are medulla, pons, cerebellum, midbrain, diencephalon, and cerebrum.

21. Neurons that are sending fewer signals have probably hyperpolarized because they would then require a larger stimulus to initiate an action potential.

Figure Questions

Fig. 9.3: Dura mater completely surrounds the venous sinus and forms one boundary of the subdural space. The arachnoid membrane separates subdural and subarachnoid spaces. Pia mater forms the other boundary of the subarachnoid space.

Fig. 9.4: 1. The easiest access is into the subarachnoid space below the bottom of the spinal cord, where there is less risk of damaging the cord. This is called a spinal tap or lumbar puncture. 2. Blockage of the aqueduct will cause CSF to accumulate in the first, second, and third ventricles. Blockage near the frontal lobe will cause fluid build-up in all the ventricles. You would look for enlargement of the fourth ventricle to help localize the site of the blockage.

Fig. 9.10: (a) and (c) are both correct.

Fig. 9.14: 1. Losing function in the right visual cortex would mean that the person could see nothing in the left visual field, indicated by the red box at the top of the figure. 2. The tracts of the corpus callosum exchange information between the two sides of the cerebrum. 3. Spatial visualization, which is important in art, is primarily a function of the right brain. Many left-handed people are right-brain dominant, and therefore may have more well-developed spatial visualization skills.

Fig. 9.17: 1. Alpha waves have the highest frequency, and delta waves have the greatest amplitude. 2. Stages 1 and 2.

Fig. 9.20: occipital, frontal

Review Questions

LEVEL ONE Reviewing Facts and Terms

1. *plasticity*

2. *affective, cognitive*

3. *cerebrum*

4. *skull, vertebral column*

5. From the bones inward: dura mater, arachnoid membrane, pia mater

6. Buoyancy reduces brain's weight; cushion between the brain and bone; chemical protection by creating a closely regulated ECF for brain cells. CSF is secreted by the choroid plexus.

7. (a) H^+ is higher in CSF. (b) Na^+ is the same in CSF and plasma. (c) K^+ is lower in CSF

8. *glucose, hypoglycemia, oxygen, 15%*

9. Capillaries that are less leaky due to tight junctions between endothelial cells. Function—regulate substances allowed into brain tissue.

10. Gray matter: nerve cell bodies, dendrites, and axon terminals. Forms nuclei or layers in the brain and spinal cord. Information passes from neuron to neuron. White matter: mostly myelinated axons; tracts carry information up and down the spinal cord.

11. (a) Sensory areas—perception. (b) Motor cortex—movement. (c) Association areas integrate information and direct voluntary behavior.

12. Asymmetrical distribution of function between the two lobes of the cerebrum. Left brain—language and verbal functions; right brain—spatial skills.

13. (a) 5, (b) 7, (c) 9, (d) 3, (e) 1, (f) 2, (g) 6, (h) 8, (i) 4

14. See Tbl. 9.1.

15. REM (rapid eye movement) sleep—when most dreaming takes place. Rapid, low-amplitude EEG waves, suppressed muscle movement, and depression of homeostatic function. Slow-wave (deep) sleep—high amplitude, low frequency EEG waves and unconscious body movements.

16. Homeostasis of body temperature and osmolarity, reproduction functions, hunger and satiety, and cardiovascular function. Emotional input from the limbic system.

17. *amygdala*

18. Associative and nonassociative. Habituation—a person responds less and less to a repeated stimulus, sensitization—an enhanced response to a dangerous or unpleasant stimulus. Hippocampus is involved in learning and memory.

19. Broca's area and Wernicke's area

LEVEL TWO Reviewing Concepts

20. Include information from Tbl. 9.1 and Figs. 9.3, 9.4, and 9.6.

21. Secreted into the ventricles and flows into the subarachnoid space around the brain and spinal cord before being reabsorbed by the cerebral arachnoid membrane.

22. sensory system, behavioral state system, and cognitive system

23. Wernicke's area—understand language, Broca's area—produce language

24. (a) Diffuse modulatory—attention, motivation, wakefulness, memory, motor control, mood, and metabolism. Reticular formation—arousal and sleep, muscle tone, breathing, blood pressure, and pain. Reticular activating system—helps maintain consciousness. Limbic system—links higher cognitive functions with more primitive emotions such as fear. (b) Short-term memory—disappears unless consolidated; long-term memory—stored for recall. Long-term includes reflexive, or unconscious, memory, and declarative, or conscious, memory. (c) Nuclei—clusters of nerve cell bodies in the CNS; ganglia—clusters of nerve cell bodies outside the CNS. (d) Tracts—bundles of axons within the CNS. Nerves—bundles of axons outside the CNS. Horns—extensions of spinal cord gray matter that connect to peripheral nerves. Nerve fibers—bundles of axons. Roots—branches of peripheral nerves that enter or exit the spinal cord.

25. Primary somatic sensory cortex—parietal lobe. Visual cortex—processes information from eyes. Auditory cortex—processes information from ears. Olfactory cortex—processes information from nose. Frontal lobe motor cortices—control skeletal muscle movements. Association areas—integrate sensory information into perception.

26. (a) lower frequency: wave peaks farther apart, (b) larger amplitude: taller peaks, (c) higher frequency: peaks closer together.

27. Drives increase arousal, initiate goal-oriented behavior, and coordinate disparate behaviors to achieve goals.

28. new synapses and changes in effectiveness of synaptic transmission

LEVEL THREE Problem Solving

29. Expressive aphasia—could understand people but unable to communicate in any way that made sense. Speech centers are in the left brain. If music centers are in the right brain, then perhaps information from Wernicke's area can be integrated by the right brain so that Mr. Anderson can musically string together words so that they make sense.

30. Learning probably occurred, but need not be translated into behavioral responses. The participants who didn't buckle their seat belts learned that wearing seat belts was important but did not consider this knowledge important enough to act on.

31. Sleep-deprived dogs are producing a substance that induces sleep. Controls: putting CSF from normal dogs into sleep-deprived dogs, CSF from normal dogs into normal dogs, and CSF from sleep-deprived dogs into other sleep-deprived dogs.

32. (a) No, other information that should be taken into consideration include genetics, age, and general health. (b) The application of this study would be limited to women of similar age, background, and health. Other factors you might be interested in would include the ethnicity of the participants, factors as listed in (a), and geographical location.

33. Decreasing ECF osmolality by drinking a lot of water causes water to move into cells. The brain is enclosed in the bony cranium and has limited room in which to expand. If pressure inside the skull increases because of brain swelling, seizures will result.

CHAPTER 10

Concept Check Questions

1. Myelinated axons have a faster conduction velocity than unmyelinated axons.

2. The pinna funnels sound into the ear canal.

3. Muscle length/tension, proprioception = mechanoreception. Pressure, inflation, distension = mechanoreception. Osmolarity = mechanoreception. Temperature = thermoreception. Oxygen, glucose, pH = chemoreception.

4. K^+ and Cl^- channels in neurons A and C are probably opening and causing hyperpolarization.

5. Sensory neurons signal intensity of a stimulus by the rate at which they fire action potentials.

6. Irritant receptors warn the body of danger. If possible, the body should respond in some way that stops the harmful stimulus. Therefore, it is important that signals continue as long as the stimulus is present, meaning the receptors should be tonic rather than phasic.

7. The adaptive advantage of a spinal reflex is a rapid reaction.

8. b, a, c (see Tbl. 10.3).

9. There are many examples, including receptors for taste and touch.

10. olfactory sensory neuron (primary neuron) $\rightarrow$ cranial nerve I $\rightarrow$ secondary neuron in olfactory bulb $\rightarrow$ olfactory tract $\rightarrow$ olfactory cortex in temporal lobe

11. If you need help, use Fig. 10.13 as the basic pattern for creating this map.

12. The knobby terminals of olfactory sensory neurons function as dendrites.

13. Olfactory neurons are bipolar neurons.

14. Umami is associated with ingestion of the amino acid glutamate.

15. presynaptic taste cell → primary sensory neuron through cranial nerves VII, IX, or X → medulla (synapse with secondary neuron) → thalamus → gustatory cortex in parietal lobe

16. A kilohertz is 1000 Hz, which means 1000 waves per second.

17. Endolymph has high $[K^+]$ and low $[Na^+]$ so the electrochemical gradient favors K^+ movement into the cell.

18. Use Figs. 10.15, 10.17, and 10.21 to create your map.

19. Somatosensory information projects to the hemisphere of the brain opposite to the side of the body on which the signal originates. The location of sound is coded by the time a stimulus arrives in each hemisphere, so a signal to both hemispheres is necessary.

20. A cochlear implant would not help people with nerve deafness or conductive hearing loss. It can help only those people with sensorineural hearing loss.

21. K^+ entry into hair cells causes depolarization.

22. When fluid builds up in the middle ear, the eardrum is unable to move freely and cannot transmit sound through the bones of the middle ear as efficiently.

23. When a dancer spots, the endolymph in the ampulla moves with each head rotation but then stops as the dancer holds the head still. This results in less inertia than if the head were continuously turning.

24. The aqueous humor supports the cornea and lens. It also brings nutrients to and removes wastes from the epithelial layer of the cornea, which has no blood supply.

25. (a) The sensory pathway from one eye diverges to activate motor pathways for both pupils. (b) The afferent path and its integration must be functioning because there is an appropriate response on the right side. The motor (efferent) path to the left eye must not be functioning.

26. *antagonistic*

27. A more curved cornea causes light rays to converge more sharply. This causes the focal point to fall in front of the retina, so the person will be myopic.

28. (a) Image distance gets longer. (b) Focal length must decrease, which is accomplished by the lens becoming more rounded.

29. (a) Convex lenses focus a beam of light, and concave lenses scatter a beam of light passing through them. (b) In myopia, the focal point lies in front of the retina so a concave corrective lens increases the focal length and moves the focal point onto the retina. In hyperopia, the focal point lies behind the retina so a convex corrective lens shortens the focal length. This moves the focal point onto the retina.

30. The tapetum lucidum reflects light, which enhances the amount of light hitting the photoreceptors.

31. In both the retina and skin, the finest discrimination occurs in the region with the smallest visual or receptive fields.

32. Damage to the macula, which surrounds the fovea, results in vision loss in the central portion of the visual field. Peripheral vision remains unaffected.

33. Our dark vision is in black and white because only rods (black and white vision), not cones (color vision), are sensitive enough to be stimulated by such low levels of light.

34. Use the information in Figs. 10.30 and 10.32 to create your map.

Figure Questions

Fig. 10.3: The olfactory and some equilibrium pathways do not synapse in the thalamus.

Fig. 10.8: Sensations affected would be contralateral pain and temperature, and ipsilateral proprioception.

Fig. 10.11: his heart

Fig. 10.13: Multiple neurons synapsing on a single neuron is an example of convergence.

Fig. 10.16: Graph (1) shows 20 Hz waves (5 waves in the 0.25-sec interval shown means 20 waves in 1 minute). Graph (2) shows 32 Hz waves. The waves in (1) have the lower pitch because they have the lower frequency.

Fig. 10.25: It shows the right eye.

Fig. 10.29: Six rods converge on the ganglion cell.

Fig. 10.31: The pigment in red cones absorbs light over the broadest spectrum, and blue cones absorb over the narrowest range. At 500 nm, the pigments in blue and green cones absorb light equally.

Fig. 10.32: (10,000 CNG channels × 24 cGMP/channel) × 1 transducin/ 6 cGMP × 1 rhodopsin/800 transducin × 1 photon/rhodopsin = 50 photons needed

Review Questions

LEVEL ONE Reviewing Facts and Terms

1. Carry information from sensory receptors to the CNS.

2. The ability to tell where our body is in space and to sense the relative locations of different body parts.

3. A sensor and a sensory neuron. Could be one cell or two.

4. Mechanoreceptors—pressure, sound, stretch, etc. Chemoreceptors—specific chemicals. Photoreceptors—photons of light. Thermoreceptors—heat and cold.

5. *receptive field*

6. (a) 3; (b) 2; (c) 1, 2; (d) 2, 3; (e) 4

7. *transduction, adequate stimulus, threshold*

8. *Receptor potentials* are graded potentials.

9. Adequate stimulus—form of energy to which a receptor is most sensitive.

10. *cortex.* Exception—hearing.

11. Sensory neurons surrounding a receptive field are inhibited, which enhances contrast between the stimulus and surrounding areas.

12. Tonic receptors, such as for heat, adapt slowly and respond to stimuli that need to be constantly monitored. Phasic receptors adapt rapidly and stop responding unless the stimulus changes. An example is smell.

13. *referred*

14. Sweet and umami indicate nutritious foods, and bitter may contain toxins. Salty (Na^+) and sour (H^+) ions are related to body osmolarity and pH, respectively.

15. Sound waves per second—*hertz (Hz)*. Loudness—a function of the *wave amplitude* and measured in *decibels (dB)*. Range of hearing: *20–20,000 Hz*. Most acute hearing: *1000–3000 Hz*.

16. Basilar membrane. Spatial coding—association of wave frequencies with different areas of the membrane.

17. (a)

18. Signals from cochlea to *medulla*, with collaterals to *reticular formation* and *cerebellum*. Synapses in *midbrain* and *thalamus* before projecting to *auditory cortex* in the *cerebrum*.

19. *semicircular canals*—rotation; *otolith organs*—linear forces

20. (b), (a), (d), (c), (e)

21. *red, blue,* and *green; cones; color-blindness*

22. Rods and cones (photoreceptors), bipolar cells, ganglion cells, horizontal cells, and amacrine cells. Photoreceptors transduce light energy. Remaining cells carry out signal processing.

pathways through the limbic system. Divergent pathways go to the cardiovascular control center, which increases sympathetic output to heart and arterioles. A second descending spinal pathway goes to the adrenal medulla, which releases epinephrine. Epinephrine on β_2-receptors of liver, heart, and skeletal muscle arterioles causes vasodilation of those arterioles. Norepinephrine onto α-receptors in other arterioles causes vasoconstriction. Both catecholamines increase heart rate and force of contraction.

14. Loss of plasma proteins will decrease colloid osmotic pressure. As a result, hydrostatic pressure will have a greater effect in the filtration-absorption balance, and filtration will increase.

15. Using osmotic pressure rather than osmolarity allows a direct comparison between absorption pressure and filtration pressure, both of which are expressed in mm Hg.

16. If the left ventricle fails, blood backs up into the left atrium and pulmonary veins, and then into lung capillaries. Edema in the lungs is known as *pulmonary edema*.

17. Low-protein diets result in a low concentration of plasma proteins. Capillary absorption is reduced while filtration remains constant, resulting in edema and ascites.

Figure Questions

Fig. 15.1: The pumps are arranged in series (one after the other).

Fig. 15.8: 1. Flow decreases and MAP increases. 2. Volume and MAP decrease. 3. Venous volume decreases, arterial volume increases, and MAP increases.

Fig. 15.10: Increased CO_2, H^+, and NO and decreased O_2 are probable factors.

Fig. 15.11: Sympathetic innervation and epinephrine increase heart rate and stroke volume; parasympathetic innervation decreases heart rate. Sympathetic input causes vasoconstriction but epinephrine causes vasodilation in selected arterioles. For paracrine factors that influence arteriolar diameter, see Tbl. 15.2.

Fig. 15.12: Blood flow through the lungs is 5 L/min.

Fig. 15.13: 1. Arterial blood pressure upstream increases. 2. Pressure downstream of the constriction decreases.

Fig. 15.14a: SA node has muscarinic cholinergic receptors for ACh and β_1-receptors for catecholamines. Ventricles have β_1-receptors for catecholamines. Arterioles and veins have α-receptors for norepinephrine.

Fig. 15.17: (a) Velocity of flow is inversely proportional to area: As area increases, velocity decreases. (b) Changing only cross-sectional area has no effect on flow rate because flow rate is determined by cardiac output.

Fig. 15.18: Net filtration will increase as a result of the increased hydrostatic pressure.

Review Questions

LEVEL ONE Reviewing Facts and Terms

1. brain and heart

2. (a) 6, 9; (b) 1, 2; (c) 4, 7; (d) 3, 5, 6, 8; (e) 3, 4

3. endothelium (capillary exchange and paracrine secretion); elastic tissue (recoil); smooth muscle (contraction); fibrous connective tissue (resistance to stretch).

4. arterioles

5. *120 mm Hg; systole; diastole; 80 mm Hg; 120/80*

6. *pulse*. Pulse pressure = systolic P − diastolic P.

7. One-way valves in the veins, skeletal muscle pump, and low pressure in the thorax during breathing

8. Elevated blood pressure can cause a weakened blood vessel to rupture and bleed.

9. Korotkoff sounds occur when cuff pressure is lower than systolic pressure but higher than diastolic pressure.

10. See Tbl. 15.2. Sympathetic neurons (α-receptors) vasoconstrict, and epinephrine on β_2-receptors in certain organs vasodilates.

11. A region of increased blood flow. Active—increased blood flow is in response to an increase in metabolism. Reactive—increase in flow follows a period of decreased blood flow.

12. *Sympathetic* innervation causes vasoconstriction.

13. (a) 1, 5; (b) 2, 6; (c) 1, 2, 4; (d) 3, 8; (e) none of the above

14. Digestive tract, liver, kidneys, and skeletal muscles. Kidneys have the highest flow on a per unit weight basis.

15. Capillary density is proportional to the tissue's metabolic rate. Cartilage—lowest; muscles and glands—highest.

16. (a) diffusion, (b) diffusion or transcytosis, (c) facilitated diffusion, (d) osmosis

17. immune, circulatory, and digestive systems

18. Edema is excess fluid in the interstitial space. Causes include lower capillary colloid osmotic pressure due to decreased plasma proteins or blockage of the lymphatic vessels by a tumor or other pathology.

19. (a) blood flow though a tissue; (b) the contribution of plasma proteins to the osmotic pressure of the plasma; (c) a decrease in blood vessel diameter; (d) growth of new blood vessels, especially capillaries, into a tissue; (e) small vessels between arterioles and venules that can act as bypass channels; (f) cells surrounding the capillary endothelium that regulate capillary leakiness

20. *HDL* and *LDL*. LDL-C is harmful in elevated amounts.

LEVEL TWO Reviewing Concepts

21. Preventing Ca^{2+} entry decreases ability of cardiac and smooth muscles to contract. Decreasing Ca^{2+} entry into autorhythmic cells decreases heart rate. Neurons and other cells are unaffected because they have types of calcium channels not affected by the drugs.

22. (a) Pores of lymphatic capillaries are larger. Lymphatic capillaries have contractile fibers to help fluid flow; systemic capillaries depend on systemic blood pressure for flow. (b) Sympathetic division raises blood pressure by increasing cardiac output and causing vasoconstriction. Parasympathetic division can decrease heart rate. (c) Lymph fluid is similar to blood plasma minus most plasma proteins. Blood also has nearly half its volume occupied by blood cells. (d) Continuous capillaries have smaller pores and regulate the movement of substances better than fenestrated capillaries. Fenestrated can open large gaps to allow proteins and blood cells to pass. (e) Hydrostatic pressure forces fluid out of capillaries; colloid osmotic pressure of plasma proteins draws fluid into capillaries.

23. Use Fig. 15.8 as the starting point.

24. The ability of vascular smooth muscle to regulate its own contraction. Probably results from Ca^{2+} influx when the muscle is stretched.

25. Left ventricular failure causes blood to pool in the lungs, increasing pulmonary capillary hydrostatic pressure. This may cause pulmonary edema and shortness of breath when oxygen has trouble diffusing into the body. Blood backing up into the systemic circulation increases venous pressure.

LEVEL THREE Problem Solving

26. (a) Uncontrollable: male, middle-aged, family history of cardiovascular disease on both sides of his family. Controllable: elevated blood pressure. (b) Yes, because blood pressure was >140 or diastolic pressure >90 on several occasions. It would be useful to confirm that this was not

"white coat hypertension" by having him take his blood pressure for a week or so at locations away from the doctor's office, such as at a drug store. (c) Beta blockers block β_1-receptors in the heart, thus lowering cardiac output and MAP.

27. (a) MAP increases, flow through vessels 1 and 2 decreases, flow through 3 and 4 increases. (b) Pressure increase $\rightarrow$ arterial baroreceptor $\rightarrow$ cardiovascular control center $\rightarrow$ arteriolar vasodilation and decreased CO $\rightarrow$ decreased pressure (c) decreases

28. Atropine is an ACh antagonist, possibly by binding to an ACh receptor.

29. sight of blood $\rightarrow$ cerebral cortex $\rightarrow$ CVCC in the medulla oblongata $\rightarrow$ increased parasympathetic and decreased sympathetic output $\rightarrow$ decreased heart rate and vasodilation $\rightarrow$ decreased blood pressure

30. Cells (endothelium) in the intact wall detect changes in oxygen and communicate these changes to the smooth muscle.

31. (a) increases (b) resistance increases and pressure increases

32. (a) In Fig. 14.1, draw a connection from pulmonary artery to aorta. In Fig. 14.5f, you can see a remnant of the closed ductus as a small ligament connecting the aorta and pulmonary artery. (b) The lungs are not functioning. (c) systemic, (d) left side, (e) from the aorta into the pulmonary artery

LEVEL FOUR Quantitative Problems

33. Increases 16-fold.

34. Answers will vary. For a 50-kg individual with a resting pulse of 70 bpm, she will pump her weight in blood in about 10 minutes.

35. MAP = 87 mm Hg; pulse pressure = 42 mm Hg.

36. 250 mL O_2/min = CO × (200 – 160 mL O_2/L blood). CO = 6.25 L/min.

37. 75 beats/min × 1440 min/day = 108,000 beats/day. 3240 mL filtered/day × day/108,000 beats = 0.03 mL/beat.

38. (a) Line C because increased resistance in the arteriole requires more energy to overcome friction, meaning lower pressure at the distal end. (b) Net filtration will increase because the hydrostatic pressure forcing fluid out of the capillaries is higher but colloid osmotic pressure is unchanged.

CHAPTER 16

Concept Check Questions

1. The five types of leukocytes are lymphocytes, monocytes/macrophages, basophils/mast cells, neutrophils, and eosinophils.

2. Erythrocytes and platelets lack nuclei, which would make them unable to carry out protein synthesis.

3. Liver degeneration reduces the total plasma protein concentration, which reduces the osmotic pressure in the capillaries. This decrease in osmotic pressure increases net capillary filtration and edema results.

4. (a) erythropoietin (EPO), (b) colony-stimulating factors (CSFs), (c) thrombopoietin (TPO)

5. (a) Heme is an iron-containing subunit of a hemoglobin molecule. (b) Ferritin is the liver protein that stores iron. Transferrin is the plasma protein that transports iron in the blood.

6. Bile is an exocrine secretion because it is secreted into the intestine.

7. Low atmospheric oxygen at high altitude $\rightarrow$ low arterial oxygen $\rightarrow$ sensed by kidney cells $\rightarrow$ secrete erythropoietin $\rightarrow$ acts on bone marrow $\rightarrow$ increased production of red blood cells

8. Thrombin promotes production of active factor XI, which creates active IX, which creates active X, which converts prothrombin to more thrombin. The loop stops when prothrombin is used up.

Review Questions

LEVEL ONE Reviewing Facts and Terms

1. *plasma; water*

2. albumins (most prevalent), globulins, and fibrinogen. Functions: Tbl. 16.1.

3. erythrocytes (transport O_2 and CO_2); leukocytes or white blood cells (defense); platelets (clotting)

4. *hematopoiesis*. Embryo—yolk sac, liver, spleen, and bone marrow. At birth—restricted to the bone marrow. By adulthood—only in axial skeleton and proximal ends of long bones.

5. Colony-stimulating factors stimulate hematopoiesis. Cytokines are released by one cell to act on another cell. Interleukins are cytokines released by leukocytes to act on other leukocytes. All influence growth and differentiation of blood cells. Examples: See Tbl. 16.2.

6. RBC—erythropoiesis, WBC—leukopoiesis, platelets—thrombopoiesis

7. *erythropoietin*. Produced primarily in the kidney in response to low oxygen.

8. Hematocrit—percent total blood volume occupied by packed (centrifuged) red cells. Men: 40–54%; women: 37–47%.

9. Erythroblast is an immature, large, nucleated precursor of the erythrocyte. Characteristics: biconcave disk shape, no nucleus, and red color due to hemoglobin.

10. iron

11. (a) yellow color to the skin due to elevated bilirubin, (b) low level of hemoglobin, (c) plasma protein that acts as a carrier for iron, (d) inherited defects of the coagulation cascade, resulting in decreased clotting ability

12. *anticoagulants*

LEVEL TWO Reviewing Concepts

13. List 1: See Figs. 16.8 and 16.9 and Tbl. 16.5. List 2: See Fig. 16.11. List 3: See Fig. 16.6.

14. Intrinsic pathway—exposed collagen and other triggers activate factor XII. Extrinsic pathway—damaged tissue exposes tissue factor (III), which activates factor VII. The two pathways unite at the common pathway to initiate the formation of thrombin. See Fig. 16.10.

15. Activated platelets cannot stick to undamaged regions of endothelium that release prostacyclin and nitric oxide (NO).

LEVEL THREE Problem Solving

16. Rachel is pale and tired because she is anemic. Bruising is a sign that platelet count is low. Proteins and vitamins promote hemoglobin synthesis and the production of new blood cell components. Iron is also necessary for hemoglobin synthesis. Avoid crowds to prevent being exposed to infections because her WBC count and ability to fight infection are decreased. By day 20, all blood counts are back into the low-normal range.

17. (a) transferrin; (b) the liver, which stores iron; (c) withdraw blood. This illustrates mass balance: If input exceeds output, restore body load by increasing output.

18. Some other factor essential for RBC synthesis, such as iron, folic acid, or vitamin B_{12}, must be lacking.

LEVEL FOUR Quantitative Problems

19. 200-lb man: blood 6.4 L and plasma about 3.1 L. 130-lb woman: blood 4.1 L and plasma about 2.4 L.

20. Blood volume is 3.5 L, and total erythrocyte volume is 1.4 L.

CHAPTER 17

Concept Check Questions

1. Cellular respiration is intracellular and uses O_2 and organic substrates to produce ATP. External respiration is exchange and transport of gases between the atmosphere and cells.

2. The upper respiratory tract includes the mouth, nasal cavity, pharynx, and larynx. The lower respiratory tract includes the trachea, bronchi, bronchioles, and exchange surface of lungs.

3. The thoracic cage consists of the rib cage with intercostal muscles, spinal (vertebral) column, and diaphragm. The thorax contains two lungs in pleural sacs, the heart and pericardial sac, esophagus, and major blood vessels.

4. The bronchioles are collapsible.

5. If cilia cannot move mucus, the mucus collecting in the airways triggers a cough reflex to clear out the mucus.

6. Blood flow is approximately equal in the pulmonary trunk and aorta. (Normally, some venous blood leaving the bronchi, pleura, and part of the heart bypasses the pulmonary circulation and drains directly into the left side of the heart. This is called an anatomic shunt.)

7. Increased hydrostatic pressure causes greater net filtration out of capillaries and may result in pulmonary edema.

8. Mean pressure = 8 mm Hg + 1/3(25 − 8) mm Hg = 13.7 mm Hg.

9. 720 mm Hg × 0.78 = 562 mm Hg

10. 700 mm Hg − 47 mm Hg = 653 mm Hg × 21% = 137.1 mm Hg P_{O_2}

11. Lung capacities are the sum of two or more lung volumes.

12. Residual volume cannot be measured directly.

13. If aging individuals have reduced vital capacity while total lung capacity does not change, then residual volume must increase.

14. As air becomes humidified, the P_{O_2} decreases.

15. Air flow reverses direction during a respiratory cycle, but blood flows in a loop and never reverses direction.

16. See Figs. 17.2c and 17.3. The lungs are enclosed in a pleural sac. One pleural membrane attaches to the lung, and the other lines the thoracic cage. Pleural fluid fills the pleural sac.

17. Scarlett will be more successful if she exhales deeply, as this will decrease her thoracic volume and will pull her lower rib cage inward.

18. Inability to cough decreases the ability to expel the potentially harmful material trapped in airway mucus.

19. A hiccup causes a rapid decrease in both intrapleural pressure and alveolar pressure.

20. The knife wound would collapse the left lung if the knife punctured the pleural membrane. Loss of adhesion between the lung and chest wall would release the inward pressure exerted on the chest wall, and the rib cage would expand outward. The right side would be unaffected as the right lung is contained in its own pleural sac.

21. Normally, lung and chest wall elastance contribute more.

22. Compliance decreases.

23. The work of breathing increases.

24. Resistance increases.

25. acetylcholine on muscarinic receptor.

26. P_{O_2} in alveoli in the affected section will increase because O_2 is not leaving the alveoli. P_{CO_2} will decrease because new CO_2 is not entering the alveoli from the blood. Bronchioles constrict when P_{CO_2} decreases (see Fig. 17.14), shunting air to areas of the lung with better blood flow. This compensation cannot restore normal ventilation in this section of lung, and local control is insufficient to maintain homeostasis.

27. IRV decreases.

28. Residual volume increases.

Figure Questions

Fig. 17.9: 1. Alveolar pressure is greatest in the middle of expiration and least in the middle of inspiration. It is equal to atmospheric pressure at the beginning and end of inspiration and expiration. 2. When lung volume is at its minimum, alveolar pressure is (c) moving from maximum to minimum and external intercostal muscle contraction is (b) minimal. 3. 2 br/8 sec = ? br/60 sec = 15 br/min.

Fig. 17.12: Shallow and rapid: total pulmonary ventilation = 6000 mL/min, 150 mL fresh air, alveolar ventilation = 3000 mL/min. Slow and deep: total pulmonary ventilation = 6000 mL/min, 600 mL fresh air, alveolar ventilation = 4800 mL/min. Slow and deep is the most efficient.

Fig. 17.13: Alveolar P_{O_2} goes to 120 mm Hg and P_{CO_2} falls to about 19 mm Hg.

Fig. 17.14: 1. Alveolar P_{O_2} increases and P_{CO_2} decreases in the affected alveoli. Local tissue P_{O_2} increases. 2. This constricts local arterioles, which then shunts blood to better-perfused sections of lung. Bronchioles constrict to divert air to better-perfused alveoli.

Review Questions

LEVEL ONE Reviewing Facts and Terms

1. gas exchange, vocalization, pH regulation, and protection

2. Cellular respiration—oxygen and nutrients are used for energy production. External respiration—gas exchange between atmosphere and cells.

3. Quiet inspiration—external intercostals, scalenes, and diaphragm. Quiet expiration—no significant muscle contraction. Active expiration—internal intercostals and abdominal muscles. These are all skeletal muscles.

4. Pleural fluid reduces friction and holds the lungs tight against the chest wall.

5. Nose and mouth, pharynx, larynx, trachea, main bronchus, secondary bronchi, bronchioles, epithelium of the alveoli, interstitial fluid, and capillary endothelium

6. See Fig. 17.2g and h. Type I—gas exchange; type II—surfactant. Macrophages ingest foreign material. Capillary endothelium is almost fused to the alveolar epithelium, and the space between alveoli is almost filled with capillaries.

7. Right ventricle to pulmonary trunk, to left and right pulmonary arteries, smaller arteries, arterioles, capillaries, venules, small veins, pulmonary veins, left atrium. Contains about 0.5 L of blood. Arterial pressure is 25/8, compared with 120/80 for systemic.

8. Warmed, humidified, and cleaned (filtered)

9. *diaphragm*

10. See Fig. 17.9.

11. (a) See Fig. 17.7. (b) V_T = 0.5 L, IRV = 1.25 L, ERV = 1.0 L. (c) 3 breaths/15 sec × 60 sec/min = 12 br/min.

12. radius of the airways

13. (a) 1, (b) 2, (c) 1, (d) 2

14. Surfactant decreases surface tension of water and makes it easier for lungs to inflate and stay inflated.

15. Increased tidal volume increases alveolar P_{O_2}.

LEVEL TWO Reviewing Concepts

16. (a) Compliance—ability to deform in response to force; elastance—ability to resume original shape after deforming force has been

removed. (b) Ventilation—air exchange between atmosphere and lungs. Inspiration—air movement into lungs. Expiration—air movement out of lungs. (c) Intrapleural pressure—always subatmospheric (except during forced expiration, when it may become positive); alveolar pressures vary from subatmospheric to above atmospheric. (d) Total pulmonary ventilation—volume of air entering or leaving airways in a given period of time. Alveolar ventilation—volume of air entering or leaving alveoli in a given period of time. (e) Type I—thin cells for gas exchange; Type II—synthesize and secrete surfactant. (f) Pulmonary—from right heart to lung and back to left atrium. Systemic—left heart to most tissues and back to right atrium.

17. Bronchoconstrictors: histamine, leukotrienes, acetylcholine (muscarinic); bronchodilators: carbon dioxide, epinephrine (β_2)

18. See Figs. 17.8 and 17.9.

19. (a) decrease, (b) decrease, (c) decrease, (d) increase, (e) decrease, (f) increase

20. Pneumothorax—air in the pleural cavity. Spirometer—device used to measure ventilation. Auscultation—listening for body sounds. Hypoventilation—decreased pulmonary ventilation. Bronchoconstriction—decrease in bronchiole radius. Minute volume—total pulmonary ventilation. Partial pressure of gas—portion of total pressure in a mixture of gases that is contributed by a specific gas.

21. (a) *vital capacity*. Sum of tidal volume and expiratory and inspiratory reserve volumes. (b) No, because lung function decreases with age as elastance and compliance diminish.

22. (a) 2, (b) 2, (c) 4, (d) 4

23. *x*-axis—time; *y*-axis—P_{O_2} During inspiration, the P_{O_2} of the primary bronchi will increase, as fresh air (P_{O_2} = 160 mm Hg) replaces the stale air (P_{O_2} = 100 mm Hg). During expiration, the P_{O_2} will decrease, as the oxygen-depleted air exits the alveoli. The curve will vary from 100 mm Hg to 160 mm Hg.

24. (a) Work increases. (b) Lungs inflate more easily. (c) Elastance decreases. (d) Airway resistance is not affected.

25. (a) decrease, (b) increase, (c) decrease

26. Total pulmonary ventilation = 20 br/min × 300 mL/br = 6000 mL/min. Alveolar ventilation = 20 br/min × (300 mL/br – 130 mL/br) = 3400 mL/br.

LEVEL THREE Problem Solving

27. (a) 9600 mL/min. (b) Dilating bronchioles reduces airway resistance. The patient is able to force more air out of the lungs on expiration, which increases her ERV and decreases her RV. (c) Respiratory rate is normal, but lung volumes are abnormal. Her high RV is confirmed by the X-ray. In obstructive lung diseases such as asthma, the bronchioles collapse on expiration, trapping air in the lungs and resulting in hyperinflation. Her low IRV accounts for most of the low vital capacity and is to be expected in someone with asthma, where the lungs are already overinflated at the beginning of inspiration. Her higher tidal volume may be the result of the effort she must exert to breathe.

28. Expired alveolar air mixes with higher O_2 atmospheric air in the anatomic dead space, increasing the P_{O_2} of air leaving the airways.

29. Resting alveolar ventilation = 3575 mL/min. Exercising: (a) 5500 mL/min (b) 5525 mL/min (c) 5625 mL/min. Increasing both rate and depth has the largest effect and is what would happen in real life.

LEVEL FOUR Quantitative Problems

30. $P_1V_1 = P_2V_2$. New volume = 200 mL.

31. (a) O_2 = 160 mm Hg, nitrogen = 593 mm Hg, CO_2 = 2.3 mm Hg. (b) O_2 = 304 mm Hg, nitrogen = 99 mm Hg, CO_2 = 342 mm Hg, H_2 =

15 mm Hg. (c) O_2 = 76 mm Hg, nitrogen = 114 mm Hg, argon = 8 mm Hg, CO_2 = 190 mm Hg.

32. Total pulmonary ventilation = 4800 mL/min. Before an exam, ventilation is 7200 mL/min. Alveolar ventilation is 3360 mL/min (at rest) and 5040 mL/min (before exam).

33. Tidal volume = 417 mL/breath. IRV = 3383 mL.

34. Lung volume is 1.1 L. (Did you forget to subtract the volume of the spirometer?)

35. (b) The lung in A has the highest compliance.

36. Answers will vary. Vital capacity will decrease significantly at age 70.

CHAPTER 18

Concept Check Questions

1. (a) electron transport system, (b) citric acid cycle

2. The P_{O_2} of the alveoli is constantly being replenished by fresh air. [p. 555]

3. 720 mm Hg × 0.78 N_2 = 561.6 mm Hg

4. Air is 21% oxygen. Therefore, for dry air on Everest, P_{O_2} = 0.21 × 250 mm Hg = 53 mm Hg. Correction for P_{H_2O}: P_{O_2} = (250 mm Hg – 47 mm Hg) × 21% = 203 mm Hg × 0.21 = 43 mm Hg.

5. Blood pools in the lungs because the left heart is unable to pump all the blood coming into it from the lungs. Increased blood volume in the lungs increases pulmonary blood pressure.

6. When alveolar ventilation increases, arterial P_{O_2} increases because more fresh air enters the alveoli. Arterial P_{CO_2} decreases because the low P_{CO_2} of fresh air dilutes alveolar P_{CO_2}. The CO_2 pressure gradient between venous blood and the alveoli increases, causing more CO_2 to leave the blood. Venous P_{O_2} and P_{CO_2} do not change because they are determined by metabolism in the cells.

7. False. Plasma is essentially water, and Fig. 18.4 shows that CO_2 is more soluble in water than O_2.

8. You need to know the solubility of each gas in that solution.

9. Yes. Hemoglobin reaches 100% saturation at 650 mm Hg. If P_{atm} = 760 mm Hg, and the atmosphere is 100% oxygen, then P_{O_2} is 760 mm Hg.

10. The flatness at the top of the P_{O_2} curve tells you that hyperventilation causes only a minimal increase in percent saturation of arterial Hb.

11. As the P_{O_2} falls, more oxygen is released. The P_{O_2} of venous blood leaving the muscle is 25 mm Hg, same as the P_{O_2} of the muscle.

12. An airway obstruction would decrease alveolar ventilation and increase arterial P_{CO_2}. Elevated arterial P_{CO_2} would increase arterial H^+ and decrease pH.

Figure Questions

Fig. 18.4: Oxygen is 2.85 mL/L blood and CO_2 is 28 mL/L blood.

Fig. 18.5: O_2 crosses five cell membranes: two of the alveolar cell, two of the capillary endothelium, and one of the red blood cell.

Fig. 18.6: 32.1 L/min

Fig. 18.9: 1. (a) At P_{O_2} = 20 mm Hg, Hb saturation = 34%. (b) Hemoglobin is 50% saturated with oxygen at a P_{O_2} of 28 mm Hg. 2. When pH falls from 7.4 to 7.2, Hb saturation decreases by 13%, from about 37% saturation to 24%. 3. When an exercising muscle cell warms up, Hb releases more oxygen. 4. Loss of 2, 3-BPG is not good because then hemoglobin binds more tightly to oxygen at the P_{O_2} values found in cells. 5. The P_{O_2} of placental blood is about 28 mm Hg. 6. At a P_{O_2} of 10 mm Hg, maternal blood is only about 8% saturated with oxygen.

Fig. 18.13: 1. pons, 2. medulla oblongata, 3. medullary chemoreceptor, 4. sensory neuron, 5. carotid chemoreceptor, 6. somatic motor neuron (expiration), 7. aortic chemoreceptor, 8. internal intercostals, 9. abdominal muscles, 10. diaphragm, 11. external intercostals, 12. scalenes and sternocleidomastoids, 13. somatic motor neuron (inspiration), 14. limbic system and higher brain centers (emotions and voluntary control)

Fig. 18.15: One breath takes 5 seconds, so there are 12 breaths/min.

Review Questions

LEVEL ONE Reviewing Facts and Terms

1. Pressure gradients, solubility in water, alveolar capillary perfusion, blood pH, temperature.

2. *98%.* Remainder is dissolved in plasma.

3. P_{O_2}, temperature, pH, and the amount of hemoglobin available for binding (most important).

4. Four globular protein chains, each wrapped around a central heme group. Requires iron.

5. *medulla* and *pons.* Dorsal—neurons for inspiration; ventral—neurons for inspiration and active expiration. Central pattern generator—group of neurons that interact spontaneously to control rhythmic contraction of certain muscle groups.

6. Medullary chemoreceptors increase ventilation when P_{CO_2} increases. Carotid body chemoreceptors respond to P_{CO_2}, pH, and P_{O_2} <60 mm Hg. P_{CO_2} is most important.

7. They include irritant-mediated bronchoconstriction and the cough reflex.

8. Partial pressure gradients

9. Decreased atmospheric P_{O_2}, decreased alveolar perfusion, loss of hemoglobin, increased thickness of respiratory membrane, decreased respiratory surface area, increased diffusion distance.

LEVEL TWO Reviewing Concepts

10. Start with Fig. 18.10.

11. Most oxygen is bound to hemoglobin, not dissolved in the plasma.

12. (a) Most O_2 is transported bound to hemoglobin, but most CO_2 is converted to HCO_3^-. (b) Concentration is amount of gas per volume of solution, measured in units such as moles/L. Partial pressure and concentration are proportional, but concentration is affected by the gas solubility and therefore is not the same as partial pressure.

13. decrease

14. Hypoxia—low oxygen inside cells. COPD—chronic obstructive pulmonary disease (includes chronic bronchitis and emphysema). Hypercapnia—elevated CO_2.

15. Oxygen is not very soluble in water, and the metabolic requirement for oxygen in most multicellular animals would not be met without an oxygen-transport molecule.

16. (a) *x*-axis—ventilation in L/min; *y*-axis—arterial P_{O_2}, in mm Hg. See Fig. 18.9. (b) *x*-axis—arterial P_{CO_2} in mm Hg; *y*-axis—ventilation in L/min. As arterial P_{CO_2} increases, ventilation increases. There is a maximum ventilation rate, and the slope of the curve decreases as it approaches this maximum.

17. (a) increases (b) increases

18. Normal, because P_{O_2} depends on the P_{O_2} of the alveoli, not on how much Hb is available for oxygen transport.

19. (a) See Fig. 18.17. (b) See Fig. 18.13.

LEVEL THREE Problem Solving

20. Increased dead space decreases alveolar ventilation. (a) increases, (b) decreases, (c) increases, (d) decreases

21. Person (a) has slightly reduced dissolved O_2 but at P_{O_2} = 80 Hb saturation is still about 95%. Most oxygen is transported on Hb but the increased P_{O_2} of 100 mm Hg cannot compensate for the decreased hemoglobin content.

22. (a) decrease, (b) decrease, (c) decrease

23. (a) Respiratory movements originate above the level of the cut, which could include any area of the brain. (b) Ventilation depends upon signals from the medulla and/or pons. (c) Respiratory rhythm is controlled by the medulla alone, but other important aspects of respiration depend upon signals originating in the pons or higher.

24. With chronic elevated P_{CO_2}, the chemoreceptor response adapts, and CO_2 is no longer a chemical drive for ventilation. The primary chemical signal for ventilation becomes low oxygen (below 60 mm Hg). Thus, when the patient is given O_2, there is no chemical drive for ventilation, and the patient stops breathing.

25. (a) Alveoli—96%; exercising cell—23% (b) At rest, Bzork only uses about 20% of the oxygen that his hemoglobin can carry. With exercise, his hemoglobin releases more than 3/4 of the oxygen it can carry.

26. All three lines show that as P_{CO_2} increases, ventilation increases. Line A shows that a decrease in P_{CO_2} enhances this increase in ventilation (when compared to line B). Line C shows that ingestion of alcohol lessens the effect of increasing P_{CO_2} on ventilation. Because alcohol is a CNS-depressant, we can hypothesize that the pathway that links increased P_{CO_2} and increased ventilation is integrated in the CNS.

27. Apical—faces airspace; basolateral—faces interstitial fluid. Apical side has ENaC and aquaporins; basolateral side has aquaporins and Na^+-K^+-ATPase. Na^+ enters the cell through ENaC, then is pumped out the basolateral side. (Cl^- follows to maintain electrical neutrality.) Translocation of NaCl allows water to follow by osmosis.

LEVEL FOUR Quantitative Problems

28. 1.65 mL O_2/gm Hb

29. 247.5 mL O_2/min

30. Nothing. The percent saturation of Hb is unchanged at any given P_{O_2}. However, with less Hb available, less oxygen will be transported.

CHAPTER 19

Concept Check Questions

1. hyperpolarizes (becomes more negative)

2. Force of contraction decreases.

3. *greater than*

4. *less than*

5. Alike: both represent movement from ECF into the lumen. Filtration is only into Bowman's capsule; secretion takes place along the rest of the tubule.

6. Glomerulus → Bowman's capsule → proximal tubule → loop of Henle → distal tubule → collecting duct → renal pelvis → ureter → urinary bladder → urethra

7. The body would run out of plasma in under an hour.

8. Osmotic pressure is higher in efferent arterioles due to same amount of protein in a smaller volume.

9. Mean arterial pressure is 119 mm Hg and GFR is 180 L/day.

10. Renal blood flow and GFR decrease.

11. With fewer plasma proteins, the plasma has lower-than-normal colloid osmotic pressure opposing GFR, so GFR increases.

12. Creatinine clearance = (1.5 mg creatinine/mL urine × 1.1 L urine/day)/1.8 mg creatinine/100 mL plasma= 92 L/day. GFR = 92 L/day.

Figure Questions

Fig. 19.2: 1. (a) Bowman's capsule, (b) proximal tubule, loop of Henle, distal tubule, collecting duct, (c) proximal tubule, distal tubule, collecting duct, (d) collecting duct. 2. (a) 18/180 = 10%, 1.5/180 = 0.8%

Fig. 19.3: E = F − R + S. 79 mmol/day = 720 − R + 43. R = 684 mmol reabsorbed per day.

Fig. 19.4: 120 mL/min × 1440 min/day = 172,800 mL/day filtered = 172.8 L. 172.8 L = 20% of plasma flow. Plasma flow = 864 L/day.

Fig. 19.6: Capillary blood pressure, GFR, and renal blood flow all increase.

Fig. 19.9: The transport rate at 3 mg/mL is 3 mg/min; at 5 and 8 mg/mL, it is 4 mg/min. The transport rate is 2 mg/min at a plasma concentration of 2 mg/mL.

Fig. 19.11: Pressure is lower because blood flowing out of the glomerulus loses pressure as it moves along the peritubular capillaries.

Review Questions

LEVEL ONE Reviewing Facts and Terms

1. color (concentration), odor (infection or excreted substances), clarity (presence of cells), taste (presence of glucose), and froth (presence of proteins)

2. regulation of extracellular fluid volume (to maintain adequate blood pressure), regulation of osmolarity, maintenance of ion balance (neuron function), regulation of pH (proteins denature if pH not maintained), excretion of wastes and foreign substances (to prevent toxic effects), and production of hormones (that regulate RBC synthesis, Ca^{2+} and Na^+ balance)

3. 20–25%

4. nephrons through ureters to urinary bladder (storage), leaving through the urethra

5. (a), (e), (b), (g), (f), (d), (c), (h)

6. Glomerular capillary endothelium, basal lamina, and epithelium of Bowman's capsule. Blood cells and most plasma proteins are excluded.

7. Capillary hydrostatic pressure promotes filtration. Fluid pressure in Bowman's capsule and colloid osmotic (oncotic) pressure of plasma oppose it. Net driving force is the sum of these pressures.

8. GFR—glomerular filtration rate. 125 mL/min or 180 L/day.

9. (a) Found where distal tubule passes between afferent and efferent arterioles. Composed of macula densa cells in the distal tubule and granular cells in arteriole wall. (b) Macula densa paracrine signals control autoregulation of GFR and renin secretion. (c) Alter the size of filtration slits. (d) Specialized epithelial cells that surround glomerular capillaries. Changes in slit size alter GFR. (e) An internal smooth muscle sphincter that is passively contracted and an external skeletal muscle sphincter that is tonically (actively) contracted. (f) Outer layer of the kidney that contains renal corpuscles, proximal and distal tubules, and parts of the loop of Henle and collecting ducts.

10. 70% occurs in the proximal tubule. Reabsorbed molecules go into the peritubular capillaries and the systemic venous circulation. If filtered and not reabsorbed, a molecule is excreted in the urine.

11. (a) 2, 3, 5; (b) 3, 4; (c) 4, 7; (d) 6; (e) 5, 7

12. penicillin, K^+, and H^+

13. creatinine

14. urination

LEVEL TWO Reviewing Concepts

15. Use Figs. 19.5 to 19.7.

16. (a) Filtration and secretion both move material from blood to tubule lumen, but filtration is a bulk flow process while secretion is a selective process. Excretion is also bulk flow but involves movement from the kidney lumen to the outside world. (b) Saturation—all transporter binding sites are occupied by ligand. Transport maximum—the maximum rate at which carriers are saturated by substrate. Renal threshold—plasma concentration at which saturation occurs. (c) Creatinine and inulin—compounds used to determine GFR. Penicillin and probenecid—xenobiotics that are secreted. (d) Clearance—rate at which plasma is cleared of a substance (mL plasma cleared of substance X/min). GFR—filtration rate of plasma (mL plasma filtered/min). Excretion—removal of urine, mL urine/min.

17. Allows rapid removal of foreign substances that are filtered but not reabsorbed.

18. Afferent arteriole constricts, GFR decreases. Efferent arteriole constricts, GFR increases.

19. See Fig. 19.14. Toilet training allows higher brain centers to inhibit the reflex until an appropriate time. Higher brain centers can also initiate the reflex.

20. Bladder smooth muscle contracts under parasympathetic control, so blocking muscarinic receptors decreases bladder contraction.

LEVEL THREE Problem Solving

21. See Fig. 19.8. Place transporters as described. Cl^- moves between the cells.

22. (a) Inulin is filtered, secreted, and excreted. No evidence for reabsorption is presented. (b) The line indicating net secretion will be close to the filtration line until the slope changes, after which the secretion line is horizontal (no further increase in rate due to saturation).

23. Dialysis fluid should resemble plasma without waste substances, such as urea. This will allow diffusion of solutes and water from the blood into the dialysis fluid, but diffusion will stop at the desired concentration. To remove excess water from the blood, you can make the dialysis fluid more concentrated than plasma.

24. Filtration line: Use several plasma concentrations of Z (0–140 mg Z/mL plasma) × GFR. The line will be a straight line beginning at the origin and extending upward to the right. Secretion reaches its maximum rate of 40 mg/min at 80 mg Z/mL plasma. Plot that point. Draw the secretion line from the origin to that point. Above the renal threshold, secretion rate does not change, so the line becomes horizontal. Excretion line: Add the filtration rate and secretion rate at a number of plasma concentrations of Z.

LEVEL FOUR Quantitative Problems

25. 1 mg X/mL plasma × 125 mL plasma/min = 125 mg X filtered/min. Same values for inulin. Inulin excretion = filtration = 125 mg inulin excreted/min. Cannot say what the excretion rate of X is because there is insufficient information.

26. 1 L/min

27. First specimen clearance = 1000 L plasma/day. Normally, creatinine clearance = GFR. However, this value is not at all realistic for GFR (normal average is 180 L/day). The repeat test has 4000 mg of creatinine and gives a clearance of 200 L/day, which is within normal limits. The abnormal values on the first test were probably a laboratory error. Dwight's kidney function is normal.

28. For any solute that filters: plasma concentration × GFR = filtration rate. At the transport maximum: filtration rate = reabsorption rate of T_m. By substitution: plasma concentration × GFR = T_m. The renal threshold represents the plasma concentration at which the transporters are working at their maximum (T_m). By substitution: renal threshold × GFR = T_m. Mermaid's GFR is 250 mL/min and T_m is 50 mg/min, so renal threshold is 0.2 mg glucose/mL plasma.

Clearance = excretion rate/plasma concentration. At 15 mg glucose/mL plasma, 3750 mg/min filter and 50 mg/min reabsorb, so 3700 mg/min are excreted.

29. (a) 140 L/day is 20% of renal plasma flow (RBF), so plasma flow is 700 L/day. (b) Hematocrit is percent of blood occupied by packed red blood cells; the remainder (70%) is plasma. 700 L/day is 70% of RBF, so RBF is 1000 L/day. (c) If RBF is 20% of cardiac output (CO), then CO = 5000 L/day or 3.47 L/min.

CHAPTER 20

Concept Check Questions

1. More water pores when vasopressin is present.

2. If vasopressin action is suppressed, the urine is dilute.

3. Hyperosmotic NaCl is hypertonic and shrinks the osmoreceptors, but hyperosmotic urea is hypotonic and causes them to swell. Because only cell shrinkage causes firing, osmoreceptors exposed to urea do not fire.

4. Vasopressin levels would increase with dehydration.

5. Osmoreceptors in the lumen of the digestive tract and hepatic portal vein would sense high-osmolarity food or drink that has been ingested and absorbed, before it is in the general circulation. This would allow an anticipatory, or feed-forward, secretion of vasopressin to conserve body water.

6. Solutes that remain in the lumen when the NKCC symporter is inhibited force water to remain in the lumen with them, increasing the urine volume.

7. Diuretics that inhibit the NKCC symporter leave K^+ in the tubule lumen, where it is likely to be excreted, thus increasing urinary K^+ loss.

8. Na^+ and K^+ are moving down their electrochemical gradients.

9. In hyperkalemia, resting membrane potential depolarizes. Excitable tissues fire one action potential but are unable to repolarize to fire a second one.

10. 140 mmol/L = 140 mEq/L

11. Atherosclerotic plaques block blood flow, which decreases pressure in the afferent arteriole and decreases GFR. Both events stimulate renin release.

12. Renin secretion begins a cascade that produces ANG II. ANG II causes vasoconstriction, acts on medullary centers to increase blood pressure, increases production of ADH and aldosterone, and increases thirst, resulting in drinking and an increased fluid volume in the body. All these responses contribute to increased blood pressure.

13. All blood passes through the pulmonary blood vessels with each circuit. Unless ACE was in every systemic blood vessel, some blood might not be exposed to ACE.

14. On the left side of Fig. 20.8, interneurons also lead from hypothalamic osmoreceptors to the hypothalamic thirst centers.

15. The bicarbonate level increases as the reaction shifts to the right as a result of added CO_2. Once a new equilibrium state is achieved, bicarbonate cannot act as a buffer because the system is at equilibrium.

16. In the distal nephron, both H^+ and K^+ are being moved against their concentration gradients, which requires ATP. In the proximal tubule, Na^+ is moving down its concentration gradient, providing the energy to push H^+ against its gradient.

17. When intercalated cells reabsorb K^+, they secrete H^+, and therefore blood pH increases.

Figure Questions

Fig. 20.6: 1. Threshold is 280 mOsM. 2. An action potential arriving at the axon terminal initiates exocytosis.

Fig. 20.8: See Fig. 15.14b, p. 494.

Fig. 20.10: See Fig. 15.14b, p. 494, for the cardiovascular pathway; Fig. 20.9b for the target cell involved in aldosterone action; and Fig. 20.5c for the target cell involved in vasopressin action.

Fig. 20.15: The muscles of inspiration are the diaphragm, the external intercostals, the scalenes, and sternocleidomastoid. Muscles of expiration are the abdominals and internal intercostals.

Review Questions

LEVEL ONE Reviewing Facts and Terms

1. Electrolytes are ions, which can conduct electric current through a solution. Examples: Na^+, K^+, Ca^{2+}, H^+, HPO_4^{2-}, and HCO_3^-.

2. Organs: kidneys, lungs, heart, blood vessels, digestive tract. Hormones: vasopressin or antidiuretic hormone (AVP or ADH), aldosterone, atrial natriuretic peptides (ANP), RAS pathway.

3. Entry: ingested and a small amount from metabolism. Loss: exhaled air, evaporation and perspiration from skin, excreted by kidneys, and in feces.

4. See Tbl. 20.1 and Fig. 20.15.

5. Descending limb: permeable to water but lacks transporters for salts. Ascending limb: impermeable to water but reabsorbs NaCl.

6. ECF volume—Na^+; pH—H^+

7. More K^+ leaves the cell, and membrane potential becomes more negative (hyperpolarizes). The heart is most likely to be affected.

8. salt and water

9. ADH = antidiuretic hormone, ANP = atrial natriuretic peptide, ACE = angiotensin-converting enzyme, ANG II = angiotensin II, JG (apparatus) = juxtaglomerular, P cell = principal cell, I cell = intercalated cell

10. Use Figs. 19.8, 19.12, 20.5c, 20.7d, 20.9b, 20.17, and 20.18.

11. pH alters protein structure (enzyme activity, membrane transporters, neural function). Buffers, renal and respiratory compensation.

12. Acids from CO_2, metabolism, and food are more likely. Sources of bases include some foods.

13. A molecule that moderates changes in pH. Intracellular: proteins, HPO_4^{2-}, and hemoglobin. Extracellular: HCO_3^-.

14. Kidneys excrete or reabsorb H^+ or HCO_3^-. Ammonia and phosphates.

15. $CO_2 + H_2O \rightleftharpoons H_2CO_3 \rightleftharpoons H^+ + HCO_3^-$. Carbonic anhydrase. High in renal tubule cells and RBCs.

16. Arterial P_{CO_2} decreases, pH increases, and plasma H^+ concentration decreases.

LEVEL TWO Reviewing Concepts

17. Use the information in Tbl. 20.1 and compile multiple pathways into a single map similar to Fig. 20.13. Include all steps of the reflex.

18. Combine information from Figs. 20.15 and 20.18b.

19. See Fig. 20.7.

20. See Fig. 20.6.

21. (a) ANP—peptide from atrial myocardial cells. Causes Na^+ and water excretion; inhibits ADH secretion. (b) Aldosterone—steroid from

adrenal cortex. Increases distal nephron Na^+ reabsorption and K^+ excretion. (c) Renin—enzyme from JG cells. Converts plasma angiotensinogen to ANG I. (d) ANG II—peptide hormone made from ANG I. Increases blood pressure by actions on arterioles, brain, and adrenal cortex. (e) Vasopressin—hypothalamic peptide. Increases distal nephron water reabsorption. (f) ACE—enzyme on vascular endothelium. Converts ANG I to ANG II.

22. Vasoconstriction, increased cardiac output, water conservation by kidneys, and increased thirst. If blood pressure falls too low, oxygen supply to the brain will decrease, resulting in damage or death.

23. The cells concentrate organic solutes to increase their internal osmolarity.

24. (a) Both are in the distal nephron. P cells are associated with aldosterone-mediated Na^+ reabsorption; I cells are involved with acid-base regulation. (b) All are parts of the RAS system. Renin and ACE—enzymes; ANG II and aldosterone—hormones. See Fig. 20.10. (c) In both, body pH falls below 7.38. Respiratory—results from CO_2 retention (from any number of causes); metabolic—results from excessive production of metabolic acids. Respiratory compensation—renal H^+ excretion and HCO_3^- retention. Metabolic compensation—increased ventilation, renal H^+ excretion, and HCO_3^- retention. Respiratory—arterial P_{CO_2} is elevated; metabolic—P_{CO_2} usually decreased. (d) Proximal tubule—not regulated; distal nephron—regulated by vasopressin. Ascending limb—impermeable to water. (e) Both—pH goes above 7.42. Metabolic—may be caused by excessive ingestion of bicarbonate-containing antacids or vomiting; respiratory—hyperventilation. Metabolic compensation—decrease ventilation, decreased renal H^+ excretion, increased HCO_3^- excretion. Respiratory compensation—decreased renal H^+ excretion, increased HCO_3^- excretion.

LEVEL THREE Problem Solving

25. (a) acute respiratory acidosis, (b) chronic respiratory acidosis, (c) Renal compensation has increased his pH by H^+ excretion and HCO_3^- reabsorption. His P_{CO_2} is elevated because of his emphysema.

26. These drugs decrease ADH-mediated water reabsorption. Useful in people who secrete too much vasopressin (SIADH, or syndrome of inappropriate ADH secretion) or in hyponatremia, such as the woman in this chapter's Running Problem.

27. (a) Metabolic alkalosis, partially compensated. (b) After vomiting acid (H^+), her body was left with HCO_3^-. (c) Hypoventilation increases P_{CO_2}, HCO_3^-, and H^+. Increased H^+ decreases pH (compensation). Hypoventilation also decreases arterial P_{O_2} and decreases the total oxygen content of blood (see Fig. 17.13).

28. Blood pressure is high, plasma Na^+ and osmolarity are low. Use Tbl. 20.1 to select reflex pathways for the map.

LEVEL FOUR Quantitative Problems

29. 429 mL (600 mosmol/? L = 1400 mosmol/L)

30. (a) $pH = 6.1 + \log [24/(0.03 \times 40)] = 7.40$ (b) 7.34

31. (a) 400 mg glucose/100 mL × 130 mL/min = 520 mg glucose/min filters. (b) Can reabsorb up to T_m so 400 mg/min reabsorbed. (c) Excreted = filtered − reabsorbed = 120 mg/min × 1440 min/day = 172.8 g/day excreted. (d) Convert grams to milliosmoles: 172.8 g × mole/180 g × 1000 mosmol/mole = 960 mosmol glucose excreted/day. Concentration = amount/volume. 1200 mosmol/L = 960 mosmol/? liters. Will require 0.8 L additional volume.

32.

	150 mosmol NaCl	150 mosmol NaCl + 200 mosmol glucose
Loop of Henle	1.5 L	3.5 L
Cortical collecting duct	0.5 L	1.167 L
Urine	0.125 L	0.292 L

CHAPTER 21

Concept Check Questions

1. The lumen is on the apical or mucosal side of the intestinal epithelium.

2. The four layers are mucosa, submucosa, muscularis externa, and serosa.

3. Mouth → pharynx → esophagus → stomach (fundus, body, antrum) → small intestine (duodenum, jejunum, ileum) → large intestine (colon, rectum) → anus

4. Digestion is chemical and mechanical breakdown of food into absorbable units. Digestion takes place in the GI tract lumen, which is external to the body; metabolism takes place in the body's internal environment.

5. Because the GI tract has a large, vulnerable surface area facing the external environment, it needs the immune cells to combat potential invaders.

6. See Fig. 5.19, steps 1, 7, 8, and 9. The main difference is that the pink membrane-bound protein will be an enzyme rather than a receptor.

7. Absorption moves material from the GI lumen into the ECF; secretion moves substances from the cells or the ECF into the lumen.

8. Vessels of the lymphatic system empty into the venous blood just before it returns to the vena cava.

9. Some sphincters are tonically contracted to close off the GI tract from the outside world and to keep material from passing freely from one section of the tract to another.

10. Parasympathetic exciting and sympathetic inhibiting is an example of antagonistic control.

11. They are released by exocytosis.

12. Bile salts do not digest triglycerides. They emulsify them into small particles so that lipase can digest them.

13. The ASBT faces the lumen and brings Na^+ and bile acid into the enterocyte together. The OAT transports bile acid by itself out of the enterocyte and into the ECF.

14. Enzymes function best in a restricted range of pH. Stomach enzymes must be active at acidic pH; salivary and intestinal enzymes work best in alkaline pH.

15. They are activated by trypsin.

16. Damage to epithelial cells means that brush border sucrase may be less effective or absent. In these cases, sucrose digestion would be impaired, so it would be better to use a solution containing glucose because this sugar need not be digested before being absorbed.

Figure Questions

Fig. 21.1: Glands are the salivary glands. Organs are liver, gallbladder, and pancreas.

Fig. 21.5: 1. The myenteric plexus controls smooth muscle contraction; the submucosal plexus controls smooth muscle contraction and endocrine and exocrine secretions by secretory cells. 2. Stretch receptors respond to stretch, osmoreceptors to osmolarity, and chemoreceptors to products of digestion.

Fig. 21.10: 1. The vagal input is parasympathetic, which stimulates digestion. 2. Neurotransmitter is ACh, and receptor is muscarinic.

Review Questions

LEVEL ONE Reviewing Facts and Terms

1. (a) 2; (b) 3; (c) 4; (d) 7, 10; (e) 8; (f) 2, 3, 7; (g) 9

2. *absorption* and *digestion; secretion* and *motility*. By not regulating absorption and digestion, the body ensures that it will always absorb the maximum available nutrients.

3. Digestion—chemical or mechanical breakdown of nutrients (proteins). Absorption—transport from lumen to ECF (water). Secretion—transport from ECF to lumen (enzymes). Motility—movement of material through the digestive tract.

4. Layers (lumen outward): mucosa (epithelium, connective tissue, and smooth muscle), submucosa (connective tissue), musculature (smooth muscle), serosa (connective tissue).

5. Secretory epithelium (endocrine and exocrine) lines the stomach; absorptive epithelium with a few secretory cells lines the intestines.

6. Peyer's patches—nodes of lymphoid tissue. M cells—epithelial cells that transfer information from gut lumen to Peyer's patches.

7. Moves food through the GI tract and helps mix food with secretions. Results from contraction of longitudinal and circular muscle layers to create propulsive peristaltic movements or mixing segmental movements.

8. An inactive digestive proenzyme. Must have a segment of protein chain removed to activate. Examples: pepsinogen-pepsin, trypsinogen-trypsin.

9. (a) 8, 9; (b) 3; (c) 1, 3, 7; (d) 1, 7; (e) 6; (f) 2; (g) 4; (h) 5

10. (a) Increases surface area for enzymes to work; stomach and small intestine. (b) Motility and secretion along the length of the digestive tract. (c) Acidic pH in stomach helps break down food and digest microorganisms. Must be neutralized in the small intestine. (d) Size determines the surface area upon which enzymes can act.

11. *capillaries, hepatic portal system, liver, lymphatic, basement membrane (basal lamina)*

12. ENS: network of neurons within the GI tract that can sense a stimulus, integrate information, and create an appropriate response without integration or input from the CNS. Also interacts with the CNS through sensory and autonomic neurons.

13. Short reflexes—mediated entirely within the ENS; regulate secretion and motility. Long reflexes—GI reflexes integrated in the CNS.

14. Paracrines help mediate secretion and motility. Examples: serotonin (5-HT) and histamine.

LEVEL TWO Reviewing Concepts

15. Map 1: Use Figs. 21.6 to 21.7 and 21.16 to 21.18, then add details. Map 2: Use Fig. 21.19a.

16. (a) Mastication—chewing; deglutition—swallowing. (b) Villi—folds of intestine; microvilli—folds of cell membrane. Both increase surface area. (c) All patterns of GI muscle contraction. Migrating motor complex—move material from stomach to large intestine between meals. Peristalsis—progressive waves of contraction. Segmental contraction—contraction and relaxation of short intestinal segments. Mass movements—push material into rectum, triggering defecation. (d) Chyme—semidigested food and secretions, produced in the stomach. Feces—solid waste material that remains after digestion and absorption are complete; produced in the large intestine. (e) Short reflexes—integrated within the ENS. Long reflexes—integrated within the CNS. (f) Submucosal plexus—ENS in the submucosal layer. Myenteric plexus—ENS that lies between muscle layers of the GI tract wall. Vagus nerve—carries sensory and efferent signals between the brain and ENS. (g) Cephalic phase—digestive reflexes triggered by stimuli received in the brain. Gastric phase—short reflexes that begin with food entering the stomach. Intestinal phase—begins when chyme enters the small intestine.

17. (a) See Fig. 21.19c. (b) See Figs. 21.9c and 21.14c.

18. Both use similar neurotransmitters and neuromodulators (serotonin, VIP, NO). Enteric support cells are similar to CNS astroglia. GI capillaries are not very permeable, like blood-brain barrier. Both act as integrating centers.

19. See Tbl. 21.1 for specific hormones.

20. See Figs. 21.9c and 21.10.

LEVEL THREE Problem Solving

21. Hepcidin causes enterocytes to destroy ferroportin transporters. If hepcidin is absent or not functional, intestinal iron uptake cannot be down-regulated when iron levels become too high, so these patients have elevated plasma levels of iron.

22. severe diarrhea $\rightarrow$ loss of small intestine HCO_3^- $\rightarrow$ metabolic acidosis

23. (a) Ingestion of a fatty meal triggers contraction of the gallbladder to release bile salts, but the blocked bile duct prevented bile secretion, causing pain. (b) Micelle formation—decreased due to lack of bile salts. Carbohydrate digestion—decreased because pancreatic secretions with amylase not able to pass blockage. Protein absorption—decreased slightly because of low pancreatic secretion; however, brush border enzymes also digest protein, so digestion does not stop completely when the bile duct is blocked. Therefore, some digested proteins will be absorbed.

24. Apical membrane has ENaC (Na^+ leak channel) and K^+ leak channels. Basolateral membrane has the Na^+-K^+-ATPase. At high flow, saliva has more Na^+ and less K^+.

LEVEL FOUR Quantitative Problems

25. (a) MIT started out with equal concentrations in both solutions, but by the end of the experiment MIT was more concentrated on the serosal side. Therefore, MIT must be moving by active transport. (b) MIT moves apical to basolateral, which is absorption. (c) Transport across the apical membrane goes from bath into tissue. Tissue MIT is more concentrated than bath. Therefore, this must be active transport. (d) Transport across the basolateral membrane goes from tissue into the sac. Tissue MIT is more concentrated than sac fluid, so this must be passive transport.

CHAPTER 22

Concept Check Questions

1. Feeding center causes an animal to eat, and satiety center causes an animal to stop eating. Both centers are in the hypothalamus.

2. Might be abnormal tissue responsiveness—a target cell with no leptin receptors or defective receptors. There might also be a problem with leptin's signal transduction/second messenger pathway.

3. age, sex, lean muscle mass, activity, diet, hormones, and genetics

4. One g of fat contains more than twice the energy of 1 g of glycogen.

5. $C_6H_{12}O_6 + 6\,O_2 \rightarrow 6\,CO_2 + 6\,H_2O$

6. $RQ = CO_2/O_2 = 6/6 = 1$

7. GLUT transporters are passive facilitated diffusion transporters.

8. dL is the abbreviation for deciliter, or 1/10th of a liter (100 mL).

9. Bile acid sequestrants and ezetimibe leave bile salts and cholesterol in the intestinal lumen to be excreted, so possible side effects are loose, fatty feces and inadequate absorption of fat-soluble vitamins.

10. Glycogenesis is glycogen synthesis; gluconeogenesis is synthesis of glucose from amino acids or glycerol.

11. Amino acids used for energy become pyruvate or enter the citric acid cycle.

12. Plasma cholesterol is bound to a carrier protein and can't diffuse across the cell membrane.

13. The primary target tissues for insulin are liver, skeletal muscle, and adipose tissue.

14. If glucose uptake depended on insulin, the intestine, kidney tubule, and neurons could not absorb glucose in the fasted state. Neurons use glucose exclusively for metabolism and must always be able to take it up.

15. During fight-or-flight, skeletal muscles need glucose for energy. Inhibiting insulin secretion causes the liver to release glucose into the blood and prevents adipose cells from taking it up, making more glucose available for exercising muscle, which does not require insulin for glucose uptake.

16. No, you would not get the same result because you would not be ingesting the same amount of glucose. Table sugar is sucrose: half glucose and half fructose. Most soft drinks are sweetened with high-fructose corn syrup.

17. Insulin is a protein and is digested if administered orally.

18. Although dehydrated patients may have elevated plasma K^+ concentrations, their total *amount* of K^+ is below normal. If fluid volume is restored to normal with no K^+ added, a below-normal K^+ concentration results.

19. Sitagliptin inhibits DPP4, the enzyme that breaks down GLP-1 and GIP. Prolonging the action of these two gut hormones enhances insulin release and slows digestion, which gives cells time to take up and use absorbed glucose.

20. Possible side effects include hypoglycemia from excessive loss of glucose and dehydration from osmotic diuresis.

21. Norepinephrine binds to α-receptors to elicit vasoconstriction.

22. Researchers probably classified the neurons as sympathetic because of where they leave the spinal cord.

23. Water cooler than body temperature draws heat away from the body through conductive heat transfer. If this loss exceeds the body's heat production, the person feels cold.

24. A person exercising in a humid environment loses the benefit of evaporative cooling and is likely to overheat faster.

Figure Questions

Fig. 22.5: 1. (a) next to left arrow from G-6-P to glycogen, (b) next to arrow going from acetyl CoA to fatty acids, (c) next to right arrow from glycogen to G-6-P, (d) with electron transport system. 2. No, amino acids entering the citric acid cycle cannot be used to make glucose because the step from pyruvate to acetyl CoA is not reversible.

Fig. 22.6: Because they are made from 2-carbon acyl units.

Fig. 22.7: The decrease from 190 to 160 mg/dL has the greatest effect.

Fig. 22.10: hydrolysis

Fig. 22.15: acetylcholine on muscarinic receptor

Review Questions

LEVEL ONE Reviewing Facts and Terms

1. Metabolic—all pathways for synthesis or energy production, use, or storage. Anabolic—primarily synthetic; catabolic—break down large molecules into smaller ones.

2. Transport (moving molecules across membranes), mechanical work (movement of muscles), chemical work (protein synthesis).

3. The amount of heat required to raise the temperature of 1 L water by 1°C. In direct calorimetry, food is burned to see how much energy it contains.

4. The ratio of CO_2 produced to O_2 used in cellular metabolism. Typical RQ is 0.82.

5. BMR—an individual's lowest metabolic rate, measured at rest after sleep and a 12-hour fast. Higher in adult males because females have more adipose tissue with a lower respiration rate. Factors that affect BMR: age, physical activity, lean muscle mass, diet, hormones, and genetics.

6. Broken down for energy, used for synthesis, or stored.

7. Absorptive state—anabolic reactions and nutrient storage. Postabsorptive—mobilizes stored nutrients for energy and synthesis.

8. A group of nutrients (glucose, free fatty acids, and amino acids), mostly in the blood, available for cell use.

9. To maintain adequate glucose supply for the brain.

10. glycogen and adipose tissue fat

11. Proteins: protein synthesis, energy, and conversion to fat for storage. Fats: lipid synthesis, energy, and storage as fats.

12. Insulin decreases blood glucose and glucagon increases it.

13. amino acids and glycerol, gluconeogenesis.

14. Excessive breakdown of fatty acids, as occurs in starvation. Can be burned as fuel by neurons and other tissues. Many ketone bodies are strong acids and can cause metabolic acidosis.

15. Increased plasma glucose or amino acids and parasympathetic input stimulate, sympathetic input inhibits.

16. Type 1: absolute lack of insulin. Type 2: cells do not respond normally to insulin. Both: elevated fasting blood glucose levels. Type 1: body uses fats and proteins for fuel. Type 2: not as severe because the cells can use some glucose.

17. Low plasma glucose or increased plasma amino acids stimulate. Primary target—liver, which increases glycogenolysis and gluconeogenesis.

18. (a) Capillary endothelium enzyme that converts triglycerides into free fatty acids and monoglycerides. (b) Co-secreted with insulin; slows gastric emptying and gastric acid secretion. (c) "Hunger hormone" secreted by the stomach. (d) Hypothalamic peptide that increases food intake. (e) Protein components of lipoproteins. Apoprotein B on LDL-C facilitates transport into most cells. (f) "Satiety hormone" produced by adipocytes. (g) Loss of water in the urine due to high amounts of urine solutes. Hyperglycemia causes dehydration through osmotic diuresis. (h) Target cells fail to respond normally to insulin.

19. (a) stimulates, (b) inhibits, (c) stimulates, (d) stimulates, (e) stimulates

LEVEL TWO Reviewing Concepts

20. Use Figs. 22.5 and 22.8. Try different colors for each organ or hormone.

21. Glucagon and insulin cycle according to food intake, but both hormones are always present in some amount. So it appears that the ratio rather than an absolute amount of hormone determines the direction of metabolism.

22. (a) Glucose—monosaccharide. Glycogenolysis—glycogen break-down. Glycogenesis—glycogen production from glucose. Gluconeogenesis—glucose synthesis from amino acids and fats. Glucagon—hormone that increases plasma glucose. Glycolysis—first pathway in glucose metabolism for ATP production. (b) Thermo-genesis—heat production by cells. Shivering thermogenesis—muscle twitches produce heat as a by-product. Nonshivering thermogen-esis occurs in all cells. Diet-induced thermogenesis—heat gener-ated by digestive and anabolic reactions during the absorptive state. (c) Lipoproteins—transport molecules. Chylomicrons—lipoprotein complexes assembled in intestinal epithelium and absorbed into lym-phatic system. Cholesterol—steroid component of cell membranes and precursor to steroid hormones. HDL-C—takes cholesterol into liver cells, where it is metabolized or excreted. LDL-C—elevated concentrations are associated with atherosclerosis. Apoprotein—protein component of lipoproteins. (d) Calorimetry—measurement of energy content and a means of determining metabolic rate. Direct calorimetry—measuring heat production when food is burned. Indirect calorimetry—measures oxygen consumption or CO_2 pro-duction. (e) Conductive heat loss—loss of body heat to a cooler object. Radiant heat loss—loss from production of infrared electro-magnetic waves. Convective heat loss—upward movement of warm air and its replacement by cooler air. Evaporative heat loss—heat lost when water evaporates. (f) Absorptive state—following a meal, when anabolism exceeds catabolism. Postabsorptive state—catabolism exceeds anabolism.

23. (a) Hyperglycemia results from the lack of insulin production and failure of cells to take up and use glucose. (b) Glucosuria results when filtered glucose exceeds the kidney's capacity to reabsorb it. (c) Poly-uria results from osmotic diuresis caused by glucosuria. (d) Ketosis results from increased fatty acid metabolism. (e) Dehydration is a consequence of polyuria due to osmotic diuresis. (f) Severe thirst is a consequence of dehydration.

24. If a person ingests a pure protein meal and only insulin is re-leased, blood glucose concentrations might fall too low. Glucagon co-secretion ensures that blood glucose remains within normal levels.

25. See Fig. 22.1. The satiety center inhibits the feeding center.

26. See Figs. 22.22 and 22.23.

LEVEL THREE Problem Solving

27. Amino acids in excess of what's needed for protein synthesis are stored as glycogen or fat.

28. As insulin secretion (*x*-axis) increases, plasma glucose (*y*-axis) decreases.

29. Some other neurotransmitter besides acetylcholine (which binds to muscarinic receptors) is involved in the vasodilation reflex.

30. (a) [See Fig. 18.9 on p. 576.] Acidosis shifts the curve to the right. Low BPG would shift curve to the left [Fig. 18.9f]. The net effect would be close to normal oxygen binding. (b) As pH normalizes, curve shifts back to the left. With BPG still low, curve would be between the left shift for low BPG and normal. Oxygen release after treatment would therefore be less than normal.

LEVEL FOUR Quantitative Problems

31. (a) Height = 1.555 m and weight = 45.909 kg. BMI = 19, which is normal. (b) Answers vary.

32. Fat: 6 g × 9 kcal/g = 54 kcal. Carbohydrate: 30 g × 4 kcal/g = 120 kcal. Protein: 8 g × 4 kcal/g = 32 kcal. Total = 206 kcal. 54/206 = 26% of calories from fat.

CHAPTER 23

Concept Check Questions

1. The medulla secretes catecholamines (epinephrine, norepinephrine), and cortex secretes aldosterone, glucocorticoids, and sex hormones.

2. Androstenedione is a prohormone for testosterone. Testosterone is anabolic for skeletal muscle, which might give an athlete a strength advantage.

3. HPA = hypothalamic-pituitary-adrenal. CBG = corticosteroid-binding globulin or transcortin.

4. This immediate stress response is too rapid to be mediated by cortisol and must be a fight-or-flight response mediated by the sympathetic nervous system and catecholamines.

5. No, because cortisol is catabolic on muscle proteins.

6. Primary and iatrogenic hypercortisolism: ACTH is lower than normal because of negative feedback. Secondary hypercortisolism: ACTH is higher because of the ACTH-secreting tumor.

7. Addison's disease: high ACTH due to reduced corticosteroid produc-tion and lack of negative feedback.

8. ACTH is secreted during stress. If the stress is a physical one caused by an injury, the endogenous opioid β-endorphin can decrease pain and help the person continue functioning.

9. In peripheral tissues, T_4 is converted to T_3, which is the more active form of the hormone.

10. When mitochondria are uncoupled, energy normally captured in ATP is released as heat. This raises the person's body temperature and causes heat intolerance.

11. prolactin

12. Normal growth and development require growth hormone, thyroid hormone, insulin, and insulin-like growth factors.

13. Their epiphyseal plates have closed.

14. Hypercalcemia hyperpolarizes the membrane potential, which makes it harder for the neuron to fire an action potential.

15. The figure should resemble Fig. 21.19b.

16. The Na^+-Ca^{2+} exchanger is a secondary active transporter, and the Ca^{2+}-ATPase is an active transporter.

17. ATP and phosphocreatine store energy in high-energy phosphate bonds.

18. A kinase transfers a phosphate group from one substrate to another. A phosphatase removes a phosphate group, and a phosphorylase adds one.

Figure Questions

Fig. 23.1: 1. A baby born with deficient 21-hydroxylase would have low aldosterone and cortisol levels and an excess of sex steroids, particularly androgens. Low cortisol would decrease the child's ability to respond to stress. Excess androgen would cause masculinization in female infants. 2. Women, who synthesize more estrogens, would have more aromatase.

Fig. 23.2: ACTH = adrenocorticotropic hormone or corticotropin. CRH = corticotropin-releasing hormone. MSH = melanocyte-stimulating hormone.

Fig. 23.4: 1. The apical membrane faces the colloid, and the basolateral membrane faces the ECF. 2. I^- comes into the cell by secondary active transport (co-transport with Na^+). 3. and 4. Thyroglobulin moves between colloid and cytoplasm by exocytosis and endocytosis. 5. Thyroid hormones leave the cell by an unknown membrane transporter.

Fig. 23.7: A pituitary tumor hypersecreting TSH would cause hyper-thyroidism and an enlarged thyroid gland. The pathway would show decreased TRH resulting from short-loop negative feedback from TSH to the hypothalamus, increased TSH caused by the tumor, elevated thyroid hormones, but no negative feedback from the thyroid hormones to the anterior pituitary because the tumor does not respond to feedback signals.

Review Questions

LEVEL ONE Reviewing Facts and Terms

1. Zona glomerulosa (aldosterone), zona fasciculata (glucocorticoids), zona reticularis (sex steroids, primarily androgens).

2. (a) corticotropin releasing hormone (hypothalamus) → adrenocorticotropic hormone (anterior pituitary) → cortisol (adrenal cortex) feeds back to inhibit secretion of both CRH and ACTH. (b) growth hormone releasing hormone and somatostatin (hypothalamus) → growth hormone (anterior pituitary). Negative feedback by IGFs. (c) decreased blood Ca^{2+} → parathyroid hormone (parathyroid glands) → increases blood Ca^{2+} by increasing reabsorption of bone, among other effects → negative feedback inhibits secretion of PTH. (d) Thyrotropin releasing hormone (hypothalamus) → thyroid stimulating hormone (thyrotropin) (anterior pituitary) → triiodothyronine (T_3) and thyroxine (T_4) (thyroid gland) → negative feedback to hypothalamus and anterior pituitary

3. Conditions: adequate diet, absence of chronic stress, and adequate amounts of thyroid and growth hormones. Other important hormones: insulin, IGFs (somatomedins), and sex hormones at puberty.

4. Triiodothyronine (T_3) and tetraiodothyronine (T_4 or thyroxine). T_3 is the more active; most of it is made from T_4 in peripheral tissues.

5. (a) Include ACTH (cortisol secretion) and MSH (not significant in humans). (b) Bone loss that occurs when bone reabsorption exceeds bone deposition. (c) The inorganic portion of bone matrix, mostly calcium salts. (d) Steroid hormones that regulate minerals, i.e., aldosterone. (e) Spongy bone, with an open latticework. (f) Pro opiomelanocortin, inactive precursor to ACTH and other molecules. (g) Growth zones in long bones, comprised of cartilage.

6. Functions: blood clotting, cardiac muscle excitability and contraction, skeletal and smooth muscle contraction, second messenger systems, exocytosis, tight junctions, strength of bones and teeth

7. In this table, A indicates anabolism, C indicates catabolism, and CHO = carbohydrate.

HORMONE	PROTEIN	CHO	FAT
Cortisol	C (skeletal muscle)	C	C
Thyroid	A (children), C (adults)	C	C
GH	A	C	C
Insulin	A	A	A
Glucagon	C	C	C

LEVEL TWO Reviewing Concepts

8. (a) CRH and ACTH low, cortisol high; (b) CRH low, ACTH and cortisol high; (c) TRH and TSH low, thyroid high; (d) TRH high, TSH and thyroid hormones low

9. (a) Hypothalamic CRH stimulates anterior pituitary secretion of ACTH, which stimulates adrenal cortex (zona fasciculata) secretion of glucocorticoids such as cortisol. (b) Thyroid gland follicle cells secrete colloid from which thyroid hormones are produced; C cells secrete calcitonin. (c) Thyroid hormone synthesis is controlled by TSH, whose release is controlled by TRH. In the thyroid gland,

tyrosine and iodine combine on thyroglobulin to make thyroid hormones. Thyroid-binding globulin (TBG) carries thyroid hormones in the blood. Target cell deiodinase removes iodine from T_4 to make T_3. (d) Growth hormone releasing hormone (GHRH) stimulates anterior pituitary secretion of growth hormone (GH or somatotropin). Somatostatin (GHIH) inhibits production of GH. Growth hormone binding protein binds about half the GH in the blood. Insulin-like growth factors (IGFs) from the liver act with GH to promote growth. (e) Dwarfism results from severe GH deficiency in childhood. Giantism results from hypersecretion of GH during childhood. Acromegaly is lengthening of jaw and growth in hands and feet, caused by hypersecretion of GH in adults. (f) Hyperplasia—increased cell number. Hypertrophy—increased cell size. (g) Osteoblasts—bone cells that secrete organic bone matrix. Osteocytes—inactive form of osteoblasts. Chondrocytes—cartilage cells. Osteoclasts—bone-destroying cells. (h) PTH increases blood Ca^{2+} by stimulating bone and renal reabsorption, and intestinal absorption of Ca^{2+}. Calcitriol (1,25-dihydroxycholecalciferol) is a vitamin D derivative that mediates PTH effect on intestinal absorption of Ca^{2+}. Calcitonin decreases bone reabsorption of Ca^{2+}. Estrogen promotes bone deposition.

10. Thyroid hormones have intracellular receptors, so you expect 60–90 minute onset of action. However, effects on metabolic rate are apparent within a few minutes and are thought to be related to changes in ion transport across cell and mitochondrial membranes.

11. Equivalent = ion's molarity × the number of charges/ion. 2.5 mmoles Ca^{2+} × 2 = 5 mEq Ca^{2+}.

12. See Fig. 23.10c. The cell uses carbonic anhydrase to make H^+ from $CO_2 + H_2O$. Apical membrane: H^+-ATPase secretes H^+ basolateral membrane secretes HCO_3^- with Cl^--HCO_3^- antiporter. Could also be Na^+-HCO_3^- symporter.

LEVEL THREE Problem Solving

13. Physiological stress stimulates secretion of cortisol, which increases blood glucose. Increased insulin opposes this effect.

14. Normal response: dexamethasone → ACTH suppression → decrease in cortisol. Patient A: no response to dexamethasone suggests adrenal hypersecretion that is insensitive to ACTH. Patient B: Dexamethasone decreases cortisol, suggesting that the problem is in the pituitary.

15. Mr. A—elevated TSH. Ms. B—low TSH. Ms. C—elevated TSH. (a) Not possible to determine if the lab slip has the results of Mr. A or Ms. C without knowing the thyroid hormone levels. (b) Ms. B can be ruled out, because her TSH would be low if the tentative diagnosis is correct.

16. (a) People in all age groups showed vitamin D insufficiency at the end of winter. Deficiency was most pronounced in the 18–29 age group and least pronounced in the 50+ age group. At the end of summer, fewer subjects were deficient in vitamin D. Variables: season when blood collected, age group, and percent of people with vitamin D insufficiency. (b) Energy from the sun is required for precursors in the skin to be converted to vitamin D. Days are shorter in the winter and the sun is at a more oblique angle and its rays are weaker. Also, at northern latitudes like Boston, people spend less time outside during the winter. This explains the difference between the two seasons. Fewer than half the people tested were deficient, however, suggesting that most people consumed enough vitamin D. The biggest seasonal difference was in the 18–29 age group, who probably spent more time outside in the summer than members of other groups. (c) Taking multivitamin supplements containing vitamin D should reduce vitamin D insufficiency.

LEVEL FOUR Quantitative Problems

17. (a) 5 mg Ca^{2+}/L plasma × 125 mL plasma filtered/min × 1440 min/day = 900 mg Ca^{2+} filtered/day, (b) To remain in Ca^{2+} balance, he must

excrete 170 mg/day. (c) 900 mg filtered – 170 mg excreted = 730 mg reabsorbed. 730/900 = 81%.

18. Graph A: x-axis = plasma parathyroid hormone concentration, y-axis = plasma Ca^{2+} concentration. Graph line goes up to the right. Graph B: x-axis = plasma Ca^{2+} concentration, y-axis = plasma parathyroid hormone concentration. Graph starts high at low x-axis values and goes down to the right.

CHAPTER 24

Concept Check Questions

1. Bacteria are cells; viruses are not. Most bacteria can reproduce outside a host; viruses cannot. Bacteria can be killed or inhibited by antibiotics; viruses cannot.

2. Virus enters host cell and uses its viral nucleic acids to tell host cell to make viral nucleic acids and proteins. These assemble into new virus particles, which are then released from the host cell.

3. When capillary permeability increases, proteins move from plasma to interstitial fluid. This decreases the colloid osmotic force opposing capillary filtration, and additional fluid accumulates in the interstitial space (swelling or edema).

4. Antibodies can be moved across cells by transcytosis or released from cells by exocytosis.

5. The child developed antibodies to bee venom on first exposure, so if the child will have a severe allergic reaction to bee venom, it will occur on the second exposure. Allergic reactions are mediated by IgE. After the first bee sting, IgE antibodies secreted in response to the venom are bound to the surface of mast cells. At the second exposure, bee venom binding to the IgE causes the mast cells to degranulate, resulting in an allergic reaction that may be severe enough to cause anaphylaxis.

6. The AB recipient has no A or B antibodies and will not react to RBCs of any blood type.

7. The type O recipient's anti-A antibodies will cause agglutination of the type A blood cells.

Figure Questions

Fig. 24.8: The antigen will activate clone 1.

Fig. 24.10: Either the lymphocytes live for a long time or they can reproduce themselves outside the thymus gland.

Fig. 24.11: T_C recognize MHC-I antigen and T_H recognize MHC-II antigen.

Fig. 24.13: Steps 3 and 5 are cell-mediated immunity. Steps 1 and 4 are humoral immunity.

Fig. 24.15:

		Mother		
		A	**B**	**O**
Father	**A**	**AA**	AB	AO
		A	AB	A
	B	AB	BB	BO
		AB	B	B
	O	AO	BO	OO
		A	B	O

Review Questions

LEVEL ONE Reviewing Facts and Terms

1. The body's ability to defend itself against disease-causing pathogens. Memory—immune cells remember prior exposure to an antigen and create a stronger immune response. Specificity—antibodies that target specific antigens.

2. Thymus gland, bone marrow, spleen, lymph nodes, and diffuse lymphoid tissues

3. Protect the body against foreign pathogens; remove dead or damaged tissues and cells; recognize and remove abnormal "self" cells.

4. Viruses bud off the host cell or kill and rupture it. Viruses damage host cells by killing them, taking over their metabolism, or causing them to become cancerous and reproduce uncontrollably.

5. Detect the pathogen, recognize it as foreign, organize a response, and recruit assistance from other cells. If a pathogen cannot be destroyed, it may be suppressed.

6. (a) Severe IgE-mediated allergic reaction with widespread vasodilation, circulatory collapse, and bronchoconstriction. (b) To clump together. When blood cells are exposed to an antibody, an antibody-antigen reaction may cause the blood cells to agglutinate. (c) Outside the blood vessels. Many immune reactions are extravascular. (d) Release of cytoplasmic granule chemicals into the ECF. (e) Opsonins that coat pathogens; released in the early stages of injury or infection. (f) One cell of a clone divides to make many identical cells. (g) Ability of the immune system to find and destroy abnormal cells (especially cancerous).

7. They are all names given to specialized tissue macrophages before scientists recognized that they were the same cell type.

8. It consists of monocytes and macrophages, which ingest and destroy invaders and abnormal cells

9. (a) 5, (b) 1, (c) 3, (d) 6, (e) 2, (f) 4

10. Physical: skin, mucous membranes, respiratory mucociliary escalator. Chemical: lysozymes, opsonins, enzymes, and antibodies.

11. B lymphocytes secrete antibodies; T lymphocytes and NK cells kill infected cells. T lymphocytes bind to antigen presented by MHC complexes; NK cells can also bind to antibodies coating foreign cells.

12. The ability of the body's immune system to ignore the body's own cells. Occurs because T lymphocytes that react with "self" cells are eliminated by clonal deletion. If self-tolerance fails, the body makes antibodies against itself (autoimmune disease).

13. the study of brain-immune interactions

14. Stress—nonspecific stimulus that disturbs homeostasis. Stressor—stimulus that causes stress. General adaptation syndrome—stress response that includes activation of the adrenal glands (fight-or-flight response by adrenal medulla and cortisol secretion by the cortex).

LEVEL TWO Reviewing Concepts

15. Use figures and tables of the chapter to create the map.

16. When lymph nodes trap bacteria, activated immune cells create a localized inflammatory response with swelling and cytokine activation of nociceptors that create the pain sensation.

17. (a) Pathogen—any organism that causes disease. Microbes—microscopic organisms, pathogens or not. Pyrogens—fever-causing chemicals. Antigens—substances that trigger an immune response and react with products of the response. Antibodies—disease-fighting chemicals produced by the body. Antibiotics—drugs that destroy bacteria and fungi. (b) Infection—illness caused by pathogens, especially viruses or bacteria. Inflammation—nonspecific response to cell damage or invaders, including nonpathogens such as a splinter. Allergy—inflammatory response to a nonpathogenic

invader, such as plant pollen. Autoimmune disease—body creates antibodies to its own cells. (c) Allergens—nonpathogenic substances that create allergic reactions. Bacteria—cellular microorganisms. Viruses—acellular parasites that must invade the host's cells to reproduce. (d) Chemotaxins—chemicals that attract immune cells. Cytokines—peptides made on demand and secreted for action on other cells. Opsonins—proteins that coat and tag foreign material so that it can be recognized by the immune system. Interleukins—cytokines initially thought to act only on leukocytes. Interferons—lymphocyte cytokines that aid in the immune response. Bradykinin—paracrine vasodilator. (e) Innate immunity—nonspecific, present from birth; acquired—directed at specific invaders. Acquired can be divided into cell-mediated and humoral (antibodies). (f) Immediate hypersensitivity response—mediated by antibodies; occurs within minutes of exposure to allergen. Delayed—may take several days to develop; mediated by helper T cells and macrophages. (g) All chemicals of the immune response. Membrane attack complex and perforin are membrane pore proteins. Perforins allow granzymes (cytotoxic enzymes) to enter the cell.

18. Histamine—opens pores in capillaries so immune cells and proteins can leave the blood. IL-1—increases capillary permeability, stimulates acute phase proteins, causes fever. Acute phase proteins—opsonins. Bradykinin—vasodilator; stimulates pain receptors. Complement—opsonins, chemotaxins, histamine release, membrane attack complex. Interferon-γ—activates macrophages. These molecules work together so they are not antagonistic. If their combined effect is greater than the sum of their individual effects, they are synergistic.

19. See Fig. 24.9. Fc region—determines antibody class; Fab region—antigen-binding sites that confer the antibody's specificity.

20. See Fig. 24.12.

21. See Fig. 24.13.

22. See Fig. 24.14.

LEVEL THREE Problem Solving

23. Type O—universal donor because these RBCs lack A or B surface antigens and do not trigger an immune response. Type AB—universal recipient because these RBCs have both A and B antigens and no A or B antibodies.

24. Maxie and baby are both OO. Snidley could be either BB or BO. Baby received an O gene from Maxie, and could have received the other O gene from Snidely. Thus, it is possible that Snidley is the father of Maxie's baby.

25. Emotional stress $\rightarrow$ increases cortisol secretion $\rightarrow$ immune system suppression. Also likely that students are spending more time inside and having closer contact with fellow students.

26. Barbara's immune cells recognize her connective tissue as an antigen, and attack it. Autoimmune diseases often begin in association with an infection and are thought to represent cross-reactivity of antibodies that developed because of the infection. Stress reduces the immune system's ability to suppress the autoantibodies and the inflammation they cause.

27. Increase in neutrophils—bacterial infection because neutrophils eat bacteria. Increase in eosinophils—parasitic infection because eosinophils kill parasites.

CHAPTER 25

Concept Check Questions

1. If venous P_{O_2} decreases, P_{O_2} in the cells is also decreasing.

2. The mean blood pressure line lies closer to the diastolic pressure line because the heart spends more time in diastole than systole.

3. The neurons are classified as sympathetic because of where they originate along the spinal cord.

Figure Questions

Fig. 25.6: 1. Arterial P_{O_2} remains constant because pulmonary ventilation is matched to blood flow through the lungs. 2. Although arterial P_{O_2} is constant, oxygen delivery to cells increases due to increased cardiac output (not shown). 3. Venous P_{O_2} drops as exercise increases because cells remove more oxygen from hemoglobin as oxygen consumption increases. 4. Arterial P_{CO_2} does not increase because increased production is matched by increased ventilation. 5. As the person begins to hyperventilate, arterial (and alveolar) P_{CO_2} decreases.

Fig. 25.7: Blood flow to an organ is calculated by multiplying cardiac output (L/min) times the percentage of flow to that organ. When rest and exercise values are compared, actual blood flow decreases only in the kidneys, GI tract, and "other tissues."

Fig. 25.8: Mean arterial pressure is cardiac output $\times$ resistance. If resistance is falling but MAP is increasing, then cardiac output must be increasing.

Review Questions

LEVEL ONE Reviewing Facts and Terms

1. ATP and phosphocreatine

2. *aerobic, both glucose and fatty acids*

3. Aerobic metabolism: requires O_2; glucose goes through glycolysis and citric acid cycle; produces 30–32 ATP/glucose through oxidative phosphorylation. Anaerobic: no O_2 used; glucose undergoes glycolysis to lactate; produces only 2 ATP/glucose.

4. glycogen, plasma glucose, glucose produced through gluconeogenesis

5. Cortisol, growth hormone, epinephrine, and norepinephrine all increase plasma glucose.

6. At the beginning of exercise, muscle ATP use exceeds aerobic ATP production, so cellular stores of ATP are used. This creates an oxygen deficit reflected by increased oxygen consumption after exercise ceases.

7. Cardiovascular system

8. normal—37 °C; sweating and cutaneous vasodilation

LEVEL TWO Reviewing Concepts

9. Look for figures in Chapters [4, 15, 17, 18, 23, and 25].

10. Sympathetic input on pancreatic beta cells decreases insulin secretion. Less insulin $\rightarrow$ liver produces glucose; insulin-sensitive tissues do not take up blood glucose $\rightarrow$ more blood glucose available for brain and exercising muscle (glucose uptake does not require insulin).

11. Advantages: fast; uses readily available glucose. Disadvantages: low ATP yield per glucose; contributes to metabolic acidosis.

12. (a) ATP—energy for muscle contraction. ADP—accepts high-energy phosphate from PCr and becomes ATP. (b) Myoglobin—muscle O_2-binding protein that aids O_2 diffusion from blood to mitochondria. Hemoglobin—RBC O_2-binding pigment that transports O_2 from lungs to cells.

13. (a) 3; (b) 1, 2, 3, 4, 5; (c) 1, 2, 4, 5, 6; (d) 6; (e) no match; (f) 6; (g) 1 (venous return), 4

14. (a) increases, (b) decreases, (c) increases, (d) increases, (e) increases, (f) increases, (g) stays the same, (h) decreases

15. Increased heart rate shortens filling time and helps offset increased end diastolic volume that might be expected from increased venous return.

16. (1) the baroreceptor reflex resets to a higher setpoint, (2) afferent signals from the baroreceptors are being blocked in transit up the spinal cord, or

(3) chemo- and mechanoreceptor input from exercising tissues overrides the baroreceptor input.

17. Regular exercise lowers risk of heart attacks, lowers blood pressure, creates better lipid profiles, and lowers risk of developing type 2 diabetes.

18. Exercising muscle does not require insulin for glucose uptake, so regular exercise can help keep blood glucose levels normal.

LEVEL THREE Problem Solving

19. Water, NaCl, and K$^+$ to replace fluid and ions lost in sweat plus a carbohydrate that is easily absorbed and metabolized to form ATP

LEVEL FOUR Quantitative Problems

20. CO = 60 beats/minute $\times$ 70 mL/beat = 4200 mL/minute. If heart rate doubles, CO goes to 8400 mL/min, or doubles also.

21. Values are approximate. (a) Stroke volumes A: 88 mL, B: 108 mL, C: 114 mL. (b) CO = HR $\times$ SV. HR—A: 68 bpm. B: 97 bpm. C: 167 bpm. (c) Curve C shows increased contractility and curve B shows increased venous return. (d) At HR = 167 bpm, there is less time for ventricular filling, and the end-diastolic volume decreases.

CHAPTER 26

Concept Check Questions

1. Female gonad: ovary; female gamete: egg, or ovum. Male gonad: testis; male gametes: sperm.

2. Primary androgen receptors are in the cytoplasm or nucleus of the target cell. AMH has membrane receptors.

3. The male parent donates the chromosome that determines sex of the zygote; therefore, the wives were not at fault.

4. An XO fetus will be a female because she lacks a Y chromosome.

5. Lack of AMH from the testes allows Müllerian ducts to develop into uterus and Fallopian tubes. External genitalia will be female because there is no DHT for development of male genitalia.

6. A newborn male's gametes are spermatogonia; a newborn female's gametes are primary oocytes.

7. The first polar body has twice as much DNA as the second polar body.

8. Each primary oocyte forms one egg; each primary spermatocyte forms four sperm.

9. Aromatase converts testosterone to estradiol.

10. FSH = follicle-stimulating hormone, DHT = dihydrotestosterone, SRY = sex-determining region of Y chromosome, LH = luteinizing hormone, GnRH = gonadotropin-releasing hormone, AMH = anti-Müllerian hormone.

11. Hypothalamic GnRH, and FSH and LH from the anterior pituitary, control reproduction.

12. Sertoli cells secrete inhibin, activin, androgen-binding protein, enzymes, and growth factors. Interstitial cells secrete testosterone.

13. The advantage is that GnRH agonists decrease FSH and LH, so the testes stop producing sperm. The disadvantage is that the testes also stop producing testosterone, which causes decreased sex drive.

14. FSH receptors—Sertoli cells. LH receptors—interstitial cells. Androgen receptors—Sertoli cells.

15. See Fig. 26.8. Exogenous anabolic steroids (androgens) shut down FSH and LH secretion. In response, the testes shrink and stop producing sperm.

16. Ovarian cycle: follicular phase, ovulation, and luteal phase. The menses and proliferative phases of the uterine cycle correspond to the follicular phase and ovulation; the secretory uterine phase corresponds to the luteal phase.

17. Women who take anabolic steroids may experience growth of facial and body hair, deepening of the voice, increased libido, and irregular menstrual cycles.

18. A woman given an aromatase inhibitor would have decreased estrogen production.

19. Ovulation occurs about 14 days before the end of the cycle, which would be (a) day 14, (b) day 9, or (c) day 17.

Figure Questions

Fig. 26.7: Mitochondria produce ATP to power the flagellum.

Review Questions

LEVEL ONE Reviewing Facts and Terms

1. (a) 3, 4, 5; (b) 8; (c) 2, 7; (d) 2, 6; (e) 2; (f) 1

2. *SRY*

3. Gonads produce gametes and secrete sex hormones. Female gamete—egg (ovum); male—sperm. Female gonadal hormones—estrogen, progesterone, androgens, and inhibin; male—androgens and inhibin.

4. (a) converts androgens to estrogens, (b) tight junctions that prevent free movement of substances between blood and seminiferous tubule lumen, (c) Sertoli cell protein secreted into seminiferous tubule lumen, where it binds and concentrates androgens, (d) formed by the first meiotic division of a primary oocyte; disintegrates and has no function, (e) lysosome-like structure in the head of sperm; contains enzymes essential for fertilization

5. Newly formed sperm: seminiferous tubule → epididymis → ductus (vas) deferens → ejaculatory duct (passing the seminal vesicles, prostate gland, and bulbourethral glands) → urethra. Ovulated egg: fallopian tube → uterine cavity → cervix → vagina

6. (a) False. Some are produced in the adrenal glands of both sexes. (b) False. Both sexes produce both hormones. (c) True. (d) False. High levels of late follicular estrogen help prepare the uterus for implantation of a fertilized ovum. (e) True.

7. A sperm-fluid mixture made mostly by the accessory glands. See Fig. 26.7f for components.

8. The most effective contraception is abstinence. Least effective forms rely on avoiding intercourse during times when the female thinks she might be fertile.

LEVEL TWO Reviewing Concepts

9. Males have one Y chromosome, which often does not have a gene to match one found on the X chromosome. Thus, a male may inherit a recessive X trait and will exhibit it, while a female who inherits the same recessive trait will not exhibit it if her second X chromosome has the dominant gene for the trait.

10. See Fig. 26.8.

11. See Figs. 26.11 and 26.12.

12. List 1: Use Figs. 26.2 to 26.4. List 2: Use Figs. 26.9, 26.11, and 26.12.

13. (a) Gamete—eggs and sperm. Germ cell—cell that will become a gamete. Zygote—formed from the fusion of egg and sperm; undergoes mitosis to become an embryo. In 8th week of pregnancy, embryo becomes a fetus. (b) Coitus—intercourse. Erection—stiffening and enlargement of the penis. Male orgasm—sperm move into the urethra during emission, then out of the body in semen during ejaculation. Erogenous zones—portions of the body with receptors for sexually

arousing stimuli. (c) Capacitation—sperm maturation necessary before it can fertilize an egg. Zona pellucida—protective glycoprotein coat around the ovum. Acrosomal reaction—enzymes help sperm penetrate the zona pellucida. Cortical reaction—granules in egg cytoplasm release their contents at fertilization to change the egg membrane properties. (d) Puberty—time of sexual maturation. Menarche—the first menstrual period. Menopause—female reproductive cycles cease. Andropause—male counterpart to menopause.

14. (a) FSH—stimulates gamete production in both sexes. (b) Inhibin—inhibits FSH secretion. (c) Activin—stimulates FSH secretion. (d) GnRH—stimulates release of FSH and LH. (e) LH—stimulates gonadal sex hormone production; in females, also necessary for gamete maturation. (f) DHT—testosterone metabolite responsible for fetal development of male genitalia. (g) Estrogen—present in both sexes but dominant in females; female gamete formation and some secondary characteristics. (h) Testosterone in males—gamete formation. Both sexes—some secondary sex traits such as hair growth. (i) Progesterone—females only; helps prepare the uterus for pregnancy.

15. The four phases are similar in both sexes. Excitement—penis and clitoris become erect due to increased blood flow. The vagina secretes fluids for lubrication. In male orgasm, ejaculation takes place, while in female orgasm the uterus and vaginal walls contract.

16. (a) hCG—keeps the corpus luteum from dying. (b) LH—no direct role in pregnancy. (c) HPL—regulation of maternal metabolism during pregnancy. (d) Estrogen—breast development; negative feedback signal to prevent new follicles from developing. (e) Progesterone—maintenance of the uterine lining; prevents uterine contractions; mammary gland development. (f) Relaxin—prevents uterine contractions. (g) Prolactin—PIH levels decrease so that prolactin levels will increase, allowing milk production.

LEVEL THREE Problem Solving

17. Normally after fertilization, the second polar body, containing a haploid set of chromosomes, is released from the zygote. If all or some of the second polar body chromosomes are retained, the embryo will have three copies of a chromosome instead of just two.

18. If the unovulated cysts continue to secrete estrogen and do not develop into corpora lutea, the uterine lining will continue to grow and the breasts will develop, just as during pregnancy.

19. (a) male, (b) nonfunctional testes, (c) no ducts of either type, (d) female

20. During pregnancy, the mother's blood glucose is available to the fetus, which metabolizes the extra energy and gains weight. The fetus also up-regulates insulin secretion to handle the glucose coming across the placenta. After birth, when insulin is still high but glucose drops to normal, the baby may become hypoglycemic.

LEVEL FOUR Quantitative Problems

21. (a) Because it was being administered to the subjects. (b) Negative feedback by testosterone. (c) Sperm production decreased in the A–B interval because FSH and LH decreased. It increased toward the end of the B–C interval because FSH allowed sperm production to resume. Sperm production did not increase significantly during the D–E interval.

APPENDIX B | PHYSICS AND MATH

RICHARD D. HILL AND DANIEL BILLER *University of Texas*

Introduction

This appendix discusses selected aspects of **biophysics,** the study of physics as it applies to biological systems. Because living systems are in a continual exchange of force and energy, it is necessary to define these important concepts. According to the seventeenth-century scientist Sir Isaac Newton, a body at rest tends to stay at rest and a body in motion tends to continue moving in a straight line unless the body is acted upon by some force (Newton's First Law).

Newton further defined **force** as an influence, measurable in both intensity and direction, that operates on a body in such a manner as to produce an alteration of its state of rest or motion. Put another way, force gives **energy** to a quantity, or mass, thereby enabling it to do work. In general, a driving force multiplied by a quantity yields energy or work. For example:

$$\text{force} \times \text{distance} = \text{work}$$

Energy exists in two general forms: kinetic energy and potential energy. **Kinetic energy** {*kinein,* to move} is the energy possessed by a mass in motion. **Potential energy** is energy possessed by a mass because of its position. Kinetic energy (*KE*) is equal to one-half the mass (*m*) of a body in motion multiplied by the square of the velocity (*v*) of the body:

$$KE = 1/2 \ mv^2$$

Potential energy (*PE*) is equal to the mass (*m*) of a body multiplied by acceleration due to gravity (*g*) times the height (*h*) of the body above the earth's surface:

$$PE = mgh \text{ where } g = 10 \text{ m/s}^2$$

Both kinetic and potential energy are measured in joules.

Basic Units of Measurement

For physical concepts to be useful in scientific endeavors, they must be measurable and should be expressed in standard units of measurement. Some fundamental units of measure include the following:

Length (*l*): Length is measured in meters (m).
Time (*t*): Time is measured in seconds (s).
Mass (*m*): Mass is measured in kilograms (kg) and is defined as the weight of a body in a gravitational field.
Temperature (*T*): Absolute temperature is measured on the Kelvin (K) scale,

where K = degrees Celsius (°C) + 273.15
and °C = (degrees Fahrenheit − 32)/1.8

Electric current (*I*): Electric current is measured in amperes (A).
Amount of substance (*n*): The amount of a substance is measured in moles (mol).

Using these fundamental units of measure, we can now establish standard units for physical concepts (**TBL. B.1**). Although these are the standard units for these concepts at this time, they are not the only units ever used to describe them. For instance, force can also be measured in dynes, energy can be measured in calories, pressure can be measured in torr or mm Hg, and power can be measured in horsepower. However, all of these units can be converted into a standard unit counterpart and vice versa.

The remainder of this appendix discusses some biologically relevant applications of physical concepts. This discussion includes topics such as bioelectrical principles, osmotic principles, and behaviors of gases and liquids relevant to living organisms.

Bioelectrical Principles

Living systems are composed of different molecules, many of which exist in a charged state. Cells are filled with charged particles such as proteins and organic acids, and ions are in continual flux across the cell membrane. Therefore, electrical forces are important to life.

When molecules gain or lose electrons, they develop positive or negative charges. A basic principle of electricity is that opposite charges attract and like charges repel. A force must act on a charged particle (a mass) to bring about changes in its position. Therefore, there must be a force acting on charged particles to cause attraction or repulsion, and this electrical force can be measured.

TABLE B.1 Standard Units for Physical Concepts

Measured Concept	Standard (SI*) Unit	Mathematical Derivation/ Definition
Force	Newton (N)	$1 \text{ N} = 1 \text{ kg} \cdot \text{m/s}^2$
Energy/Work/ Heat	Joule (J)	$1 \text{ J} = 1 \text{ N} \cdot \text{m}$
Power	Watt (W)	$1 \text{ W} = 1 \text{ J/s}$
Electrical charge	Coulomb (C)	$1 \text{ C} = 1 \text{ A} \cdot \text{s}$
Potential	Volt (V)	$1 \text{ V} = 1 \text{ J/C}$
Resistance	Ohm (Ω)	$1 \ \Omega = 1 \text{ V/A}$
Capacitance	Farad (F)	$1 \text{ F} = 1 \text{ C/V}$
Pressure	Pascal (Pa)	$1 \text{ Pa} = 1 \text{ N/m}^2$

*SI = Système International d'Unités

Electrical force increases as the strength (number) of charges increases, and it decreases as the distance between the charges increases (**FIG. B.1**). This observation has been called **Coulomb's law** and can be written as

$$F = q_1 q_2 / \varepsilon d^2$$

where q_1 and q_2 are the electrical charges (coulombs), d is the distance between the charges (meters), ε is the dielectric constant, and F is the force of attraction or repulsion, depending on the type of charge on the particles.

When opposite charges are separated, a force acts over a distance to draw them together. As the charges move together, work is being done by the charged particles and energy is being released. Conversely, to separate the united charges, energy must be added and work done. If charges are separated and kept apart, they have the potential to do work. This electrical potential is called **voltage.** Voltage is measured in **volts (V).**

If electrical charges are separated and there is a potential difference between them, then the force between the charges allows electrons to flow. Electron flow is called an electric **current.** The **Faraday constant (F)** is an expression of the electrical charge carried by one mole of electrons and is equal to 96,485 coulombs/mole.

The amount of current that flows depends on the nature of the material between the charges. If a material hinders electron flow, then it is said to offer **resistance (R),** measured in ohms. Current is inversely proportional to resistance, such that current decreases as resistance increases. If a material offers high resistance to electron flow, then that material is called an **insulator.** If resistance is low and current flows relatively freely, then the material is called a **conductor.** Current, voltage, and resistance are related by **Ohm's law,** which states:

$$V = IR$$

where V = potential difference in volts
I = current in amperes
R = resistance in ohms

In biological systems, pure water is not a good conductor, but water containing dissolved NaCl is a good conductor because ions provide charges to carry the current. In biological membranes, the lipids have few or no charged groups, so they offer high resistance to current flow across them. Thus, different cells can have different electrical properties depending on their membrane lipid composition and the permeability of their membranes to ions.

Osmotic Principles

Freezing point, vapor pressure, boiling point, and osmotic pressure are properties of solutions collectively called **colligative properties.** These properties depend on the number of solute particles present in a solution. **Osmotic pressure** is the force that drives the diffusion of water across a membrane. Because there are no solutes in pure water, it has no osmotic pressure. However, if one adds a solute like NaCl, the greater the concentration (c) of a solute dissolved in water, the greater the osmotic pressure. The osmotic pressure (π) varies directly with the concentration of solute (number of particles (n) per volume (V) or c):

$$\pi = (n/V)\ RT$$

$$\pi = cRT$$

where R is the ideal gas constant (8.3145 joules/K mol) and T is the absolute temperature in Kelvin. Osmotic pressure can be measured by determining the mechanical pressure that must be applied to a solution so that osmosis ceases.

Water balance in the body is under the control of osmotic pressure gradients (concentration gradients). Most cell membranes allow water to pass freely, primarily through open channels. To control the movement of water, the body either removes these channels from the membrane or alters solute movement that creates concentration gradients.

Relevant Behaviors of Gases and Liquids

The respiratory and circulatory systems of the human body obey the physical laws that govern the behavior of gases and liquids. This section discusses some of the important laws that govern these behaviors and how our body systems utilize these laws.

FIG. B.1 **Electrical force**

ELECTRICAL FORCE

If you separate two opposite charges, there will be an electric force between them.

If you increase the number of charges that are separated, the force increases.

If you increase the distance between the charges, the force decreases.

If charges are separated by some distance d, they have the potential to do work. This electrical potential is called voltage.

If separated charges are allowed to move together, they do work and energy is released. The amount of work done depends on the number of particles and the distance between them.

Work = force × distance

To separate the charged particles, energy must be put into the system and work is done.

FIG. C.1 Phases of cell division

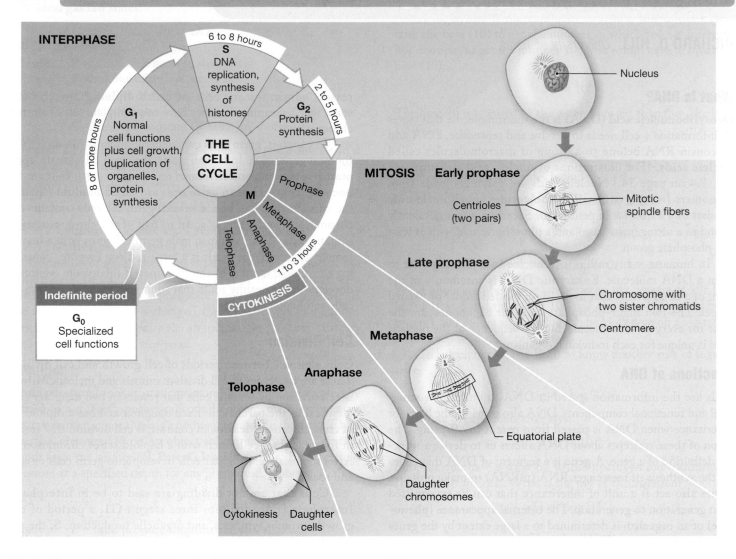

Mitosis Creates Two Identical Daughter Cells

As stated earlier, mitosis is the division of a somatic cell that results in two diploid daughter cells. The DNA of the parent cell first duplicates itself into two complete sets of chromosomes. One set of chromosomes then goes to each daughter cell, and the daughter cells separate.

The four main steps of mitosis are **prophase, metaphase, anaphase,** and **telophase** (Fig. C.1). The entire somatic cell cycle can be remembered by the acronym, IPMAT, in which the "I" stands for interphase and the other letters stand for the steps of mitosis that follow.

Prophase

During prophase, chromatin becomes condensed and microscopically visible as duplicate chromosomes. The duplicated chromosomes form **sister chromatids,** which are joined to each other at the **centromere.** The cell's centriole pair duplicates and the two centriole pairs move to opposing ends of the cell. The **mitotic spindle,** composed of microtubules, assembles between the centriole pairs. The nuclear membrane begins to break down and disappears by the end of prophase.

Metaphase

In metaphase, mitotic spindle fibers extending from the centrioles attach to the centromere of each chromosome. The 46 chromosomes, each consisting of a pair of sister chromatids, line up at the "equator" of the cell.

Anaphase

During anaphase, the spindle fibers pull the sister chromatids apart, so that an identical copy of each chromosome moves toward each pole of the cell. By the end of anaphase, an identical set of 46 chromosomes is present at each pole. At this point, the cell has a total of 92 chromosomes, double the diploid number.

Telophase

The actual division of the parent cell into two daughter cells takes place during telophase. In **cytokinesis,** the cytoplasm divides when an actin contractile ring tightens at the midline of the cell. The result is two separate daughter cells, each with a full diploid set of chromosomes. The spindle fibers disintegrate, nuclear envelopes form around the chromosomes in each cell, and the chromatin returns to its loosely coiled state.

FIG. C.2 | DNA: Levels of organization

(a) Levels of Organization of DNA

(b) DNA Replication

Nondividing cell

Chromatin in nucleus

Histones

Nucleosome

DNA double helix

KEY

Adenine   Cytosine

Guanine   Thymine

Replication fork

DNA polymerase

Leading strand, 5′ end

3′ end of parent DNA

DNA strand

DNA nucleotide

Okazaki fragment

5′ end of parent DNA

DNA Replication

The information stored in DNA is encoded in the nucleotide sequence of the molecule. When nucleotides link together, the phosphate group of one nucleotide bonds covalently to the sugar group of the adjacent nucleotide. The end of the polymer that has an unbound sugar is called the 3′ ("three prime") end. The end of the polymer with the unbound phosphate is called the 5′ end. A DNA molecule has four types of nucleotides, distinguished by their nitrogenous bases.

The nitrogenous bases in nucleic acids are classified as either **purines** or **pyrimidines.** The purine bases are **guanine (G)** and **adenine (A).** The pyrimidine bases are **cytosine (C),** and either **thymine (T),** found in DNA only, or **uracil (U),** found in RNA only. To remember which DNA bases are pyrimidines, look at the first syllable. The word "pyrimidine" and names of the DNA pyrimidine bases all have a "y" in the first syllable.

The "rungs" of the DNA double helix are created when the nitrogenous bases on one DNA strand form hydrogen bonds with nitrogenous bases on the adjoining DNA strand. This phenomenon is called **base-pairing.** The base-pairing rules are as follows:

1. Purines pair only with pyrimidines.
2. Guanine (G) bonds with cytosine (C) in both DNA and RNA.
3. Adenine (A) bonds with thymine (T) in DNA or with uracil (U) in RNA.

The two strands of DNA are bound in **antiparallel** orientation, so that the 3′ end of one strand is bound to the 5′ end of the second strand. This organization has important implications for DNA replication.

DNA Replication Is Semi-Conservative

To be transmitted from one generation to the next, DNA must be replicated. Furthermore, the process of replication must be accurate and fast enough for a living system. The base-pairing rules for nitrogenous bases provide a means for making an appropriate replication system.

In DNA replication, special proteins unzip the DNA double helix and build new DNA by pairing new nucleotide molecules to the two existing DNA strands. The result of this replication is two double-stranded DNA molecules, such that each DNA molecule contains one DNA strand from the template and one newly synthesized DNA strand. This form of replication is called **semi-conservative replication.**

Replication of DNA is bidirectional. A portion of DNA that is "unzipped" and has enzymes performing replication is called a **replication fork** (Fig. C.2b). Replication begins at many points (**origins of replication**), and it continues along both parent strands simultaneously until all the replication forks join.

Nucleotides bond together to form new strands of DNA with the help of an enzyme called **DNA polymerase.** DNA polymerase can add nucleotides only to the 3′ end of a growing strand of DNA. For this reason, DNA is said to replicate in a 5′ to 3′ direction.

The antiparallel orientation of the DNA strands and the directionality of DNA polymerase force replication into two different modes: **leading strand replication** and **lagging strand replication.** The DNA polymerase can replicate continuously along only one parent strand of DNA: the parent strand in the 3′ to 5′ orientation. The DNA replicated continuously is called the **leading strand.**

The DNA replication along the other parent strand is discontinuous because of the strand's 5′ to 3′ orientation. DNA replication on this strand occurs in short fragments called **Okazaki fragments** that are synthesized in the direction away from the replication fork. Another enzyme known as **DNA ligase** later connects these fragments into a continuous strand. The DNA replicated in this way is called the **lagging strand.** Because the 5′ ends of the lagging strand of DNA cannot be replicated by DNA polymerase, a specialized enzyme called **telomerase** has arisen to replicate the 5′ ends.

Much of the accuracy of DNA replication comes from base pairing, but on occasion, mistakes in replication happen. However, several quality control mechanisms are in place to keep the error rate at 1 error/10^9 to 10^{12} base pairs. **Genome** (the entire amount of DNA in an organism) sizes in eukaryotes range from 10^9 to 10^{11} base pairs per genome, so this error rate is low enough to prevent many lethal mutations, yet still allows genetic variation to arise.

Mutations Change the Sequence of DNA

Over the course of a lifetime, there are countless opportunities for mistakes to arise in the replication of DNA. A change in a DNA sequence, such as the addition, substitution, or deletion of a base, is a **point mutation.** If a mutation is not corrected, it may cause a change in the gene product. These changes may be relatively minor, or they may result in dysfunctional gene products that could kill the cell or the organism. Only rarely does a mutation result in a beneficial change in a gene product. Fortunately, our cells contain enzymes that detect and repair damage to DNA.

Some mutations are caused by **mutagens,** factors that increase the rate of mutation. Various chemicals, ionizing radiation such as X-rays and atomic radiation, ultraviolet light, and other factors can behave as mutagens. Mutagens either alter the base code of DNA or interfere with repair enzymes, thereby promoting mutation.

Mutations that occur in body cells are called **somatic mutations.** Somatic mutations are perpetuated in the somatic cells of an individual, but they are not passed on to subsequent generations. However, **germ-line mutations** can also occur. Because these mutations arise in the germ cells of an individual, they are passed on through gametes to future generations.

Oncogenes and Cancer

Proto-oncogenes are normal genes in the genome of an organism that code primarily for protein products that regulate cell growth, cell division, and cell adhesion. Mutations in these proto-oncogenes give rise to **oncogenes** {*onkos*, a mass}, genes that induce uncontrolled cell proliferation and the condition known as **cancer.** The mutations in proto-oncogenes that give rise to cancer-causing oncogenes are often the result of viral activity.

PHOTO CREDITS

NOTE: Page numbers followed by a *t* refer to tables. Page numbers followed by an *f* refer to figures.

A

A band Band of striated muscle sarcomere whose length equals that of the thick filament (Ch 12), 381*f*, 383, 384*f*

abdomen, 59, 61, 538*f*

abdominal cavity, 59, 60*f*

abdominopelvic cavity, 59, 60*f*

abducens nerve, 288*t*

Aβ (A-beta) fibers, 321*t*

abnormal tissue responsiveness, 217

ABO blood group, 777–778, 777*f*, 777*t*

absolute refractory period, 246, 248*f*

absorption Transfer of substances from the lumen of the kidney or gastrointestinal tract to the extracellular space (Ch 3, 5, 15, 21). *See also* digestion

absorptive cell Small intestinal cell. Synonym: enterocyte (Ch 21), 658, 683*f*

abstinence, 825, 826*t*

α-bungarotoxin. *See* alpha-bungarotoxin

accessory nerve (XI), 288*t*

accessory proteins, 68, 384*f*–385*f*

acclimation Physiological adjustment to environmental change in a laboratory setting (Ch 1), 18

acclimatization The adaptation of physiological processes to a given set of environmental conditions (Ch 1), 18
to exercise in heat, 794

accommodation The process by which the eye adjusts the shape of the lens to keep objects in focus (Ch 10), 343

accommodation reflex, 343

acetylcholine (ACh) Neurotransmitter used by neurons of the central and peripheral nervous system (Ch 2, 6, 8, 9, 11, 12, 14, 17), 48, 254, 255*t*, 411*t*
action potentials in cardiac and skeletal muscle, in comparison of, 454*t*
autonomic pathway, 370*f*
diffuse modulatory system, 294, 295*f*
excitation-contraction coupling, 389–391, 390*f*
postganglionic autonomic neurotransmitters, 366*t*
receptor for, 364*f*, 372*f*, 778*t*
sympathetic and parasympathetic branches, comparison of, 371*t*
synthesis and recycling of, 260*f*
vasodilation, 488*t*

acetylcholinesterase (AChE) Enzyme that breaks down acetylcholine in the synapse

(Ch 8, 11), 258, 260*f*, 366*t*, 368*t*, 371, 372*f*

acetyl CoA. *See* acetyl coenzyme A

acetyl coenzyme A (acetyl CoA) metabolic intermediate that links glycolysis and β-oxidation to the citric acid cycle (Ch 4), 102, 105, 106, 108*f*, 260*f*, 787*f*

acid Molecule that ionizes and contributes an H^+ to a solution (Ch 2, 20, 21, 23), 41, 45*f*, 642, 665*f*, 670–671, 684*f*, 744*f*
excretion, 646–467
secretion, 666*t*, 670–671, 671*f*, 673*f*, 684*f*

acid-base balance The homeostatic regulation of body pH (Ch 20), 641–649

acid-base disturbances, 647, 648*f*. *See also* acidosis

acidity, 41

acidosis Extracellular pH less than 7.38 (Ch 18, 20), 577, 647

acid reflux (Ch 21), 672

acini (acinar cells), 668, 831

acquired immunity Immune responses directed at specific invaders and mediated by antibodies (Ch 24), 757, 765–772

acquired immunodeficiency syndrome (AIDS), 754

acromegaly Abnormal growth of cartilage and soft tissues due to excess growth hormone secretion in an adult (Ch 23), 742, 742*f*

acrosomal reaction Release of enzymes from the sperm head when it contacts an egg (Ch 26), 827

acrosome Lysosome-like vesicle of sperm that contains powerful enzymes essential for fertilization (Ch 26), 813*f*, 814, 828*f*

ACTH. *See* adrenocorticotrophic hormone

actin A globular protein (G-actin) that polymerizes to form thin filaments (F-actin) (Ch 3, 12, 14), 68, 75*f*, 382, 383, 384*f*, 385, 450
filament, 233*f*, 385–388, 386*f*, 387*f*, 390*f*, 406*f*
skeletal muscle, 380*f*
smooth muscle, 406, 406*f*, 408*f*, 411*t*

actin-binding site, 382, 383, 386*f*, 387–388

action potential Rapid and uniform electrical signal conducted down a cell membrane (Ch 3, 8, 12, 13, 14, 21) 84, 240, 259*f*
absolute refractory period, 246
autorhythmic cell, 453*f*, 455*f*
cardiac contractile cell, 451*f*
cardiac muscle, 450*f*, 451*f*, 454*t*
conduction of, 246, 247–249, 250*f*
firing frequency, 261*f*
graded potential, differentiated from, 241*f*
ion movement, 243–245
motor neuron, 391*f*

muscle, 390*f*
in neurons, 240*t*, 318*f*, 391*f*, 424*f*
presynaptic and postsynaptic inhibition, 268*f*
refractory period, 245–246, 248*f*
saltatory conduction, 249, 252*f*
in sensory neuron, 318*f*
signal transduction, 334*f*
in skeletal muscle, 454*t*
smooth muscle, 408*f*, 409*f*
in somatic motor neuron, 645*f*
sound transduction, 331
spatial summation, 264*f*
special senses receptor, 311*f*
spindle sensory neuron, 424*f*
stimulus intensity coding and, 317*f*
varicosity, 365*f*

action potential frequency, 334*f*

activation (channel), 239

activation (protein), 49, 50*f*

activation energy (1) The energy required to initiate a reaction. (2) the initial input of energy required to bring reactants into a position that allows them to react with one another (Ch 4), 96, 97*f*
in endergonic reaction, 97–98, 97*f*
in exergonic reaction, 97–98, 97*f*

activation gate Sodium channel gate that opens to initiate an action potential (Ch 8), 245, 246*f*

active expiration, 548

active hyperemia An increase in blood flow that accompanies an increase in metabolism (Ch 15), 489, 489*f*

active immunity, 766

active transport Movement across a membrane that requires the input of energy from ATP (Ch 5, 11, 16, 19, 21, 23), 133, 137, 143, 365*f*, 521*f*, 603*f*, 604, 667*f*, 745*f*
secondary, 143

activin Peptide hormone from the gonads that stimulates FSH secretion (Ch 26), 809

acuity Keenness of vision (Ch 10), 348

acute motor axonal polyneuropathy (AMAN), 267

acute-phase protein Liver proteins that act as opsonins and enhance the inflammatory response (Ch 24), 764*t*, 765, 773*f*

acyl unit, 108*f*

adaptation of receptors Process in which sensory receptors decrease their response to a stimulus over time (Ch 10), 318*f*

adaptive immunity (acquired immunity), 757

addiction, 359, 373

addictive behavior, 299

Addison's disease, 734

addition reaction Reaction in which a functional group is added to one or more of the substrates (Ch 4), 101, 102, 102*t*

Aδ (A-delta) fibers, 320, 321*t*

adenine Nucleotide base found in ADP, ATP, DNA, RNA, and cAMP (Ch 2), 34*f*–35*f*

adenohypophyseal secretions/hormones, 209, 730

adenoma, 216

adenosine Nucleoside composed of adenine and ribose (Ch 2, 8, 9, 15), 34, 255*t*, 256, 297, 488*t*, 490

adenosine diphosphate (ADP) Composed of adenine, ribose, and two phosphates, (2; 3; 4) (Ch 2, 4, 6, 10,16), 34*f*, 37*f*, 97, 104

adenosine monophosphate, cyclic (cyclic AMP/cAMP) Nucleotide that participates in the transfer of signals between the external environment and the cell (Ch 2, 6, 8), 34*f*, 37*f*, 174, 256

adenosine triphosphate (ATP) An energy-storing compound composed of adenine, ribose, and three phosphate groups (Ch 2, 3, 4, 5, 6, 8, 12), 34*f*, 37*f*, 143*f*, 144*f*, 234*f*, 235

aerobic metabolism, 105, 110*f*, 111*f*, 787*f*

anaerobic metabolism, 105, 110–111, 111*f*

electron transport system (ETS), 105, 109*f*

energy transfer between reactions, 104–105

glycolysis, 105, 107*f*

membrane transport uses, 143

motor proteins and, 69, 70*f*

muscle contraction, 386, 387*f*, 391–392

as a neurotransmitter, 256

in taste transduction, 328*f*

adenylyl cyclase-cAMP system The first signal transduction system discovered (Ch 6), 174

adenylyl cyclase Membrane-bound enzyme that converts ATP to cyclic AMP (Ch 6), 174, 175*f*, 182*t*

adequate stimulus The form of energy to which a particular receptor is most responsive (Ch 10) 312

adherens junctions Bands that link action microfilaments in adjacent cells together with the help of cadherins (Ch 3), 74, 74*t*, 75*f*

adipocyte Fat cells (Ch 3, 16, 22), 81*f*, 82, 516, 694

adipocytokines, 694

adipose tissue, 81*f*, 82, 83*f*,

autonomic control of, 370*f*, 418*t*

adrenal cortex Outer portion of adrenal gland that produces steroid hormones (Ch 7, 11, 20, 23), 204, 367, 630, 634, 640*f*, 732

steroid hormone production, 200*f*, 204, 205*f*

adrenal gland Endocrine and neuroendocrine gland that sits on top of the kidney (Ch 7, 11, 18, 23), 204, 367, 368*f*, 592*f*, 731*f*

adrenal glucocorticoids, 731–736

adrenal medulla Modified sympathetic ganglion, the inner portion of the adrenal gland that produces catecholamines (Ch 7,

9, 11, 15, 23), 200*f*, 367, 368*f*, 370*f*, 488*t*, 731, 731*f*

adrenal sympathetic pathway, 370*f*

adrenergic Adjective pertaining to epinephrine (adrenaline) or norepinephrine (Ch 6, 8, 11), 180, 255, 255*t*, 256, 366*t*, 371*t*

adrenergic neuron Neuron that secretes norepinephrine (Ch 8, 11), 255, 371*t*

adrenergic receptor Receptor that binds to norepinephrine or epinephrine, 180, 181*f*, 256, 361, 364, 367*t*, 371*t*

adrenocorticotrophic hormone (ACTH) Anterior pituitary hormone that regulates secretion of cortisol from the adrenal cortex, 210*f*, 212*f*, 213, 214*f*, 217*f*, 219*f*, 220*f*, 732, 733*f*, 734

adrenogenital syndrome, 735

adult respiratory distress syndrome (ARDS), 569

aequorin, 178

aerobic Adjective pertaining to a process that requires oxygen, 105 (Ch 4)

aerobic metabolism, 105, 788*f*

affective behavior Behaviors related to feeling and emotion (Ch 9), 275

afferent arteriole Renal arterioles that bring blood to the glomerulus (Ch 19, 20), 592*f*–593*f*, 594, 599*f*, 601*f*, 607*f*, 633*f*, 634, 635*f*

afferent pathway The pathway that connects a receptor to an integrating center (Ch 6, 13), 185, 425*f*

affinity The degree to which a protein is attracted to its ligand (Ch 2), 47–48

after-hyperpolarization, 244*f*, 250*f*

afterload, 471–473

Agelenopsis aperta, 257

agglutination Clumping of cells together (Ch 24), 777

aging, 834

agomelatine, 221*f*

agonist Molecules that combine with a receptor and mimic a response (Ch 2, 6, 7, 8, 11), 48, 180, 180*f*, 217, 219, 221*f*, 255*t*, 368*t*

agouti mouse, 735

agouti protein, 735

AIDS (acquired immunodeficiency syndrome), 754, 775

air hunger, 556

airway Anatomical structures from mouth to bronchioles that carry air to the alveoli (Ch 11, 17, 18, 24), 565*f*

autonomic and somatic motor control, 362*f*

branching, 540*f*

diameter, 552–553

epithelium, 541*f*

glands, 757*f*

lumen, 541*f*

lungs, external environment, 537

upper, 553*t*

airway resistance, 552–553

albumin Plasma protein made in the liver (Ch 2, 7, 16, 21, 23), 51, 204, 512, 513*f*, 514*t*, 677*f*, 738

albuterol, 368*t*

alcohol, 255*t*, 667*f*

aldosterone A steroid hormone that stimulates Na^+ reabsorption and K^+ secretion in the kidney (Ch 7, 20, 23, 26), 200*f*–201*f*, 630–632, 635*f*, 640*f*

secretion, 630–632, 633*f*, 639*t*

aldosterone-induced proteins, 631*f*

algorithm, 501, 544

alkalosis Extracellular pH greater than 7.42 (Ch 20), 641, 647, 647*f*, 648*f*

allantois Extraembryonic membrane that becomes part of the umbilical cord (Ch 26), 829

allergen Any substance capable of triggering an allergic reaction (Ch 24), 775, 776*f*

allergy (allergic response), 760*f*, 775, 776*f*

all-or-none phenomenon, 242

allosteric activator, 49, 50*f*

allosteric inhibitor, 49, 50*f*

allosteric modulation, 49

allosteric modulator Binds to an enzyme away from the binding site and change the shape of the active site (Ch 2), 49, 49*t*

alpha₁ (α_1) receptors, 367*t*

alpha₂ (α_2) receptor, 367

alpha-adrenergic receptor (α receptor) Membrane receptor that binds to norepinephrine and epinephrine (Ch 6, 8, 11), 180–181, 181*f*, 183, 255*t*, 256, 366, 367*t*, 371*t*

alpha-blockers, 368*t*

alpha-bungarotoxin Snake toxin that is a nicotinic receptor antagonist, 255*t*, 373

alpha (α) chain, 521*f*

alpha-gamma coactivation, 424*f*

alpha (α) helix Spiral configuration formed by some amino acid chains (Ch 2), 32*f*, 37*f*, 41

alpha (α) ketoglutarate (αKG), 646

alpha (α) melanocyte-stimulating hormone (MSH), 695*f*

alpha (α) melanocyte-stimulating hormone (MSH), 735

alpha motor neuron Neurons that innervate extrafusal muscle fibers and cause muscle contraction (Ch 13), 421, 422*f*, 423*f*–424*f*, 426*f*

5-alpha-reductase Enzyme that converts testosterone to DHT (Ch 26), 803

alpha wave Low amplitude, high-frequency brain waves characteristic of the awake-resting state (Ch 9), 296*f*

alternative splicing The processing of mRNA to make different proteins from a single strand of DNA (Ch 4), 114–115, 115*f*

alveolar gas exchange, 565

alveolar macrophage Immune cells that patrol the alveoli (Ch 17), 539*f*

alveolar pressure
    expiration, 547–549
    inspiration, 546–547
alveolar ventilation The volume of fresh air that reaches the alveoli each minute (Ch 17, 18), 553, 556, 557f–558f, 567
alveoli The exchange surface of the lungs, where oxygen and carbon dioxide transfer between air and the blood (Ch 17, 18), 536, 536f, 539f, 540, 552f, 558f
    gas exchange, 540, 568f
    structure, 539f
    surface tension of, 551
Alzheimer's disease (Ch 9), 302
amacrine cell, 347f, 351, 351f
amenorrhea, 834
American Association for the Advancement of Science (AAAS), 8t
American Physiological Society, 9
amination Addition of an amino group to a molecule (Ch 4), 102
amine hormones, 201, 207f
amine neurotransmitter Neurotransmitters made from amino acids, including the catecholamines, histamine, and serotonin (Ch 8), 254–255
amines, 254–256, 255t
amino acid Molecule with a central carbon atom linked to a hydrogen atom, an amino group, a carboxyl group, and a variable group of atoms designated "R." The building blocks of proteins (Ch 2, 4, 16), 32f, 255t, 256, 513f, 642f, 646, 787f
    mRNA translation, 115–116, 117f
amino acid-derived hormones, 200f, 202t, 206
amino group Functional group whose composition is -NH_2 (Ch 2), 32f, 33, 33t
aminopeptidases Digestive enzyme that removes amino acids from the NH_2 terminal end of a peptide (Ch 2, 21), 46, 681, 681f
ammonia, 642f, 645, 645f
ammonium ion (Ch 20, 22), 647f
amnesia, 300
amnion Extraembryonic membrane that secretes amniotic fluid (Ch 26), 829, 829f
amniotic fluid, 829, 829f
amoeba, 147
AMPA receptor Glutamate receptor-channel that allows net Na^+ influx (Ch 8), 255t, 256, 267f
amphetamines, 368t
amphipathic, 678
amplification, 331
amplifier enzyme A membrane enzyme that creates two or more second messengers during signal transduction (Ch 6, 7), 172f–173f, 175f, 177f, 204f
ampulla, 337, 338f
amygdala Portion of the brain linked to emotion and memory (Ch 9), 286f, 289, 290f, 298

amylase Enzyme that digests starch to maltose (Ch 4, 21), 680, 680f
amylin, 5, 709
amyotrophic lateral sclerosis (ALS/Lou Gehrig's disease), 235
anabolism Metabolic pathways that require a net input of energy and that synthesize small molecules into larger ones (Ch 4, 23), 102, 738f, 740f
anaerobic Adjective pertaining to a process that does not require oxygen, 105 (Ch 4)
anaerobic metabolism/pathways, 105, 110–111, 111f, 788f
analgesia, 256
analgesic drugs, 256, 323
anal sphincters, 667f
anaphylactic shock (anaphylaxis), 179, 487, 775, 776
anatomic dead space The portions of the airways that do not exchange gases with the blood (Ch 17), 554
anatomy The study of structure (Ch 1, 7), 2–3, 200f–201f
anaxonic neuron, 229, 230f
anchoring junction Form of cell-cell or cell-matrix junctions (Ch 3), 74, 75f
Andersson, Bengt, 636
Andre the Giant, 742
androgen Steroid hormone produced in the gonads and adrenal cortex; dominant hormone in males (Ch 7, 23, 26), 200f, 212f, 732, 803, 805f, 809f, 815, 822f
androgen-binding protein (ABP) Sertoli cell protein that binds testosterone to keep it in the lumen of the seminiferous tubule (Ch 26), 811, 814f
andropause, 834
androstenedione, 205f, 730, 731f
anemia Pathological state with low hemoglobin (Ch 16), 522, 523t
angina, 179
angiogenesis The process by which new blood vessels develop, especially after birth (Ch 15), 481
angioplasty, 472
angiostatin, 481
angiotensin-converting enzyme (ACE) inhibitor Drug used to treat high blood pressure by blocking ACE (Ch 20), 632, 633f
angiotensin converting enzyme (ACE) Membrane-bound endothelial enzyme that converts ANGI into ANGII (Ch 20), 632
angiotensin I (ANG I), 632, 640f
angiotensin II (ANG II) Trophic hormone that regulates aldosterone secretion; also raises blood pressure and causes thirst and ADH secretion (Ch 15, 19, 20), 488t, 490, 602, 632–634, 633f, 640f, 642
angiotensinogen, 632, 633f, 640f, 677f
angiotensin receptor blockers (ARBs), 632

anion Negatively charged ions (Ch 2, 19), 33t, 39
anisotropic, 383
ankyrin-B, 460
anorexia nervosa (AN), 694–695
antacid, 672
antagonistic control Hormones or neurons with opposing effects on some homeostatic function (Ch 6, 8, 11), 183, 184f, 216, 229, 361
antagonistic muscle groups Flexor-extensor pairs of muscles attached to the same set of bones (Ch 12, 13), 379, 379f, 425f
antagonist One substance opposes the action of another (Ch 2, 6, 8, 11), 49, 180, 180f, 255t, 367, 368t
anterior pituitary gland An endocrine gland in the brain that secretes multiple hormones (Ch 7, 9, 23, 26), 200f, 209–213, 217f, 219f–220f
anterior pituitary hormones, 212f, 213
anterograde amnesia Inability to remember newly acquired information (Ch 9), 300
anterograde axonal transport Fast transport of vesicles and mitochondria from cell body to axon terminal (Ch 8), 231–232
antibiotics, 756
antibody A molecule keyed to a particular pathogen that helps target it for destruction. Synonym: immunoglobulin (Ch 2, 14, 16, 24), 46, 437t, 514, 514t, 756, 764, 768, 773–778
antibody-dependent cell-mediated cytotoxicity Process in which natural killer cells kill a target cell by binding to the Fc portion of antibodies that are coating the cell (Ch 24), 769f
anticholinesterases, 368
anticoagulant Any chemical that inhibits blood coagulation (Ch 16, 24), 528–530, 530t, 764t
anticodon The tRNA base triplet that pairs with the mRNA codon for an amino acid (Ch 4), 115
antidepressants, 299, 369
antidiuretic hormone (ADH, vasopressin) Posterior pituitary hormone that regulates water reabsorption in the kidney (Ch 7, 20), 209, 623, 626f
antigen Substances that trigger an immune response from the body and that can react with products of that response (Ch 24), 754, 768, 775
antigen-presenting cell (APC) Immune cells that ingest and digest pathogens, then insert a fragment of the pathogen into a surface protein (Ch 24), 760, 760f
antigen-specific responses (acquired immunity), 765
antihistamines, 765
antihypertensive drug, 490, 505, 632
anti-inflammatory drugs, 611, 734

anti-Müllerian hormone (Mullerian inhibiting substance) Glycoprotein that causes the Mullerian ducts to degenerate during embryonic development (Ch 26), 803, 804*f*, 806*f*, 817, 822*f*

antiparallel orientation, 35*f*

antiplatelet agents, 529

antiport carrier A membrane transport protein that moves two or more molecules in opposite directions across a membrane (Ch 5), 140*f*, 141, 144*t*

antipsychotic drugs, 255*t*

antiviral drugs, 756

antrum (1) Distal portion of the stomach; (2) Fluid-filled cavity of mature ovarian follicle (Ch 21, 26), 658, 820, 820*f*, 821*f*

anus, 658

anxiety, 295*f*

aorta The main artery taking blood from the left ventricle to the body (Ch 11, 14, 15, 17, 19, 21), 438

aortic body Region of the aortic wall that contains sensory receptors (Ch 18), 584*f*

aortic body baroreceptor Pressure-sensing receptors (Ch 15, 20), 493, 494*f*

aortic body chemoreceptor Receptors that respond to P$_{O2}$ less than 60 mm Hg, decreased pH, or increased P$_{CO2}$ (Ch 18, 20), 582, 584*f*, 645*f*

aortic stenosis, 471

aortic valve The valve between the left ventricle and the aorta (Ch 14, 15), 447, 478*f*

apex
   of heart, 444*f*
   of lung, 556

apical membrane/surface The surface of transporting epithelial cells that faces the lumen of an organ (Ch 3, 5, 10, 19, 20, 25), 79, 150, 151*f*

aplastic anemia, 523*t*

Aplysia, 276

apnea Cessation of breathing (Ch 17), 556*t*

apoB (apoprotein B), 502

apocrine glands, 86*f*

apoptosis (cell suicide) Programmed cell death (Ch 3, 24, 26), 85, 754, 763, 764*t*, 771*f*, 774*f*

appendix, 685, 685*f*

appetite, 695, 696*f*

aquaporin (AQP) Family of membrane water channels (Ch 5, 20), 125, 139, 622, 625

aqueous humor Plasma-like fluid filling the compartment of the eye between the cornea and the lens (Ch 10), 340, 341*f*

aqueous solution Solution in which water is the solvent (Ch 2), 39, 40, 42*f*

arachidonic acid 20-carbon fatty acid precursor of eicosanoid signal molecules (Ch 6), 179, 179*f*

arachnoid membrane The middle membrane layer of the meninges (Ch 9), 279*f*, 280, 281*f*

ARDS. *See* Adult respiratory distress syndrome

Aretaeus the Cappadocian, 590

arginine Amino acid precursor of nitric oxide (Ch 2), 32*f*

arginine vasopressin (AVP), Synonym: vasopressin, antidiuretic hormone, ADH, 623, 626*f*

Aristotle, 2

aromatase An enzyme that converts androgens to estrogens (Ch 26), 808

arrector pili muscles, 86*f*

arrhythmia, cardiac 239, 460, 504

arterial blood pressure, 471–473, 485*f*

arteries, 437, 438*f*–439*f*, 445*f*, 480, 482*f*–483*f*, 486*f*, 491*f*, 495*f*, 497*f*

arteriole The smallest arteries and site of variable resistance in the circulatory system (Ch 11, 14, 15, 19, 20), 479, 593*f*, 633*f*, 640*f*
   resistance, 487–491

arteriovenous bypass, 480*f*

artery Blood vessels that carry blood away from the heart (Ch 14, 15, 21, 24, 26), 437, 438*f*, 479

ascending limb of loop of Henle Portion of the nephron where dilute fluid is produced (Ch 19, 20), 594, 595*f*, 596, 601*f*, 602, 627–629

ascending tract Spinal neurons that carry signals to the brain (Ch 9), 284, 284*f*

ascites (abdominal edema), 501*f*

-*ase* (suffix), 143

aspartate Amino acid that also acts as an excitatory neurotransmitter (Ch 8), 256

assisted reproductive technology (ART), 827

association area Parts of the cerebrum that translate sensory information into perception (Ch 9), 291, 292*f*, 298*f*

association cortices, 291

Association of American Medical Colleges (AAMC), 8*t*

associative learning Learning that occurs by association of two stimuli (Ch 9), 300

asthma Lung disease characterized by bronchoconstriction (Ch 17, 18), 557, 568*f*

astigmatism Blurred vision caused by an irregularly shaped cornea (Ch 10), 345

astrocyte Glial cells in the CNS that contact both neurons and blood vessels (Ch 8, 9), 233–235, 234*f*, 282*f*

asynchronous recruitment Alternation of active motor units to prevent fatigue (Ch 12), 398

AT$_1$ receptors (angiotensin receptors), 632

-*ate* (suffix), 32, 642

atherosclerosis Pathological condition in which lipids and calcium deposit beneath the vascular endothelium (Ch 5, 15), 148, 502–504

atherosclerotic plaque, 503*f*

atmospheric pressure, 542

atomic mass The mass of protons and neutrons in one atom of an element (Ch 2), 36*f*

atomic mass units (amu), 42*f*

atomic number, 36*f*

atoms The smallest particle of an element (Ch 2), 33, 36*f*

ATPases, 143

ATP-binding cassette (ABC) superfamily, 138

ATP-gated K$^+$ channel Channel that closes when the ATP/ADP ratio increases (Ch 5), 159*f*

atresia Apoptosis of ovarian follicles (Ch 26), 816

atrial fibrillation, 459*f*

atrial natriuretic peptide (ANP) Peptide hormone from atria of the heart that increases renal Na$^+$ and water excretion (Ch 7, 8, 15, 20), 200*f*, 256, 490, 634, 635*f*

atrial volume receptor, 620f, 625, 639*t*, 640*f*

atrioventricular bundle (bundle of His), 454, 456*f*, 461*f*

atrioventricular node (AV node) The electrical gateway to the ventricles, located near the floor of the right atrium (Ch 14), 454, 456*f*

atrioventricular node (AV node) delay Slowing of electrical conduction through the AV node that allows atria to complete contraction before the ventricles begin (Ch 14), 455

atrioventricular (AV) valves Heart valves that separate the atria from the ventricles (Ch 14), 444*f*, 445*f*, 447

atrium (plural: atria) Upper chamber of the heart that receives blood from the blood vessels (Ch 14), 437

atrophy, 71, 217

atropine Muscarinic receptor antagonist (Ch 8, 11), 255*t*, 368*t*

auditory association area, 292*f*

auditory cortex, 292*f*–293*f*, 293, 335, 336*f*

auditory information, 315, 353. *See also* sound

auditory neurons, 331, 333*f*

auditory pathway, 335, 336*f*

Auerbach's plexus, 659

auscultation, 464, 557, 559

autism, 209

autocrine signal A local chemical signal that acts on the cell that secreted it (Ch 6, 7, 8), 167*f*, 168, 199, 254

autoimmune disease Diseases in which the immune system creates antibodies against the body's own tissues (Ch 16, 22, 24), 523*t*, 715, 778*t*

autonomic ganglion, 279*f*, 361, 361*f*, 364*f*

autonomic nervous system Efferent division of the nervous system that controls smooth muscle, cardiac muscle, glands, and some adipose tissue (Ch 8, 9, 11, 14), 227*t*, 229, 359, 362*f*, 364, 369, 405
   agonists and antagonists in, 367
   division, 359–371, 466–468
   in heart rate modulation, 467*f*

autonomic neuron Efferent neurons that control smooth muscle, cardiac muscle, many glands, and some adipose tissue (Ch 8, 11, 12, 13, 19, 21), 228*f*, 229, 255*t*, 259, 359

autonomic neuropathy, 369

autonomic neurotransmitters, 366*t*, 410

autonomic pathways, 361, 361*f*, 364, 370*f*

autonomic (visceral) reflex, 360, 418, 418*t*, 419*f*, 420

autorhythmic cell Cardiac cells that spontaneously and rhythmically depolarize and fire action potentials (Ch 12, 14), 411*t*, 447, 455*f*, 467*f*, 472*f*

autosome The 22 pairs of chromosomes that contain information for non-sex-related development (Ch 26), 801, 802*f*, 808

AV valve. *See atrioventricular valve*

axon An extension of a neuron that carries signals to the target cell (Ch 7, 8, 9, 10, 11), 209, 284*f*, 295*f*, 347*f*

postganglionic autonomic neuron, 365*f*, 366*t*

axonal Na⁺ channels, 245

axonal transport Movement of material between the axon terminal and the cell body (Ch 8), 227*t*, 231

axon hillock Region of the axon where it joins the cell body. Often contains the trigger zone (Ch 8), 230 *f*, 231, 242

axon terminal The distal end of a neuron where neurotransmitter is released into a synapse (Ch 8, 10, 11, 12), 227*t*, 231, 241*f*, 246, 257, 259*f*, 372*f*

axoplasm Cytoplasm of the axon (Ch 8), 227*t*

axoplasmic flow Movement of cytoplasm in the axon. Used for slow axonal transport (Ch 8), 227*t*

B

bacteria, 667*f*, 755, 755*f*, 756, 762*f*, 764*t*, 769*f*, 772, 773*f*, 774*f*

Banting, Fredrick G., 198

barbiturate, 255*t*

bar graph, 20*f*, 21*f*

baroreceptor reflex The primary reflex pathway for homeostatic control of blood pressure (Ch 15, 25), 488*t*, 493, 494*f*–495*f*, 793

baroreceptor Stretch-sensitive mechanoreceptors that respond to changes in pressure (Ch 6, 9, 15), 186*f*, 288*t*, 493

Barr body The inactivated X chromosome in each female cell (Ch 26), 802

barrier method Contraception based on putting a physical or chemical barrier between egg and sperm (Ch 26), 826

basal body Cytoplasmic structure where microtubules of cilia and flagella terminate (Ch 3), 68

basal body temperature, 821*f*

basal ganglia Nuclei surrounding the thalamus that help with planning movement (Ch 9, 13), 286*f*, 289, 290*f*, 294, 428*t*, 429*f*, 430

basal lamina An acellular layer of extracellular matrix that lies beneath an epithelium, holding the epithelial cells to underlying cell layers. Synonym: basement membrane (Ch 3, 9, 13, 15, 19, 26), 76, 77*f*, 86*f*, 282*f*, 480, 496, 594, 597, 813*f*, 820, 820*f*

basal metabolic rate (BMR) An individual's lowest metabolic rate (Ch 22), 698

base A molecule that decreases the H⁺ concentration of a solution by combining with free H⁺ (Ch 2, 20, 21), 45*f*, 642, 682

base (nitrogenous) A carbon-nitrogen molecule with a ring structure that is an essential component of a nucleotide (Ch 2), 34*f*

base pairing, 35*f*

basic solution, 45*f*

basilar membrane Membrane that separates the cochlear duct from the tympanic duct. It supports the organ of Corti (Ch 10), 332, 333*f*

basolateral membrane/surface The sides of transporting epithelial cells that face the extracellular fluid. Synonym: serosal membrane (Ch 3, 5, 19, 23), 79, 150, 151*f*, 603*f*, 744*f*

basophil Leukocyte that releases histamine, heparin (Ch 16, 24), 513*f*, 514, 515*f*, 517*f*, 760–761, 760*f*, 764*t*

Bayliss, W. M, 666

Beaumont, William, 655

bed-wetting, 625

behavior, mechanisms, 636

behavioral state system, 291, 291*f*, 294

benign positional vertigo, 339

Bernard, Claude, 9, 10

Berthold, A. A., 197, 198

Best, Charles H., 198

beta (*β*) 1-receptors, 366, 370*f*, 494*f*

beta (*β*) 2-receptors, 181, 181*f*, 366, 370*f*

beta (*β*) 3-receptors, 366

beta-adrenergic receptor (*β* receptor) Sympathetic target cell receptors (Ch 8, 11), 255*t*, 256, 366, 366*t*, 368*t*, 370*f*

beta-amyloid protein, 302

beta-blocker Drugs that are beta-adrenergic receptor antagonists. Used to treat hypertension (Ch 6, 11), 182, 368*t*, 369

beta (*β*)-strands Sheet-like structure formed from some chains of amino acids (Ch 2), 32*f*, 37*f*, 41

beta cells of the pancreas Endocrine cells that secrete insulin (Ch 5, 21, 24), 153, 159, 666*t*, 778*t*

beta (*β*)-endorphin, 323, 735

bicarbonate ion (HCO₃⁻), 33, 33*t*, 41, 123, 577, 642–646

in acid-base balance, 643

renal reabsorption of, 645

in GI tract, 671, 674, 676*f*

biceps brachii muscle, 379

bicuspid valve The left AV valve of the heart. Synonym: mitral valve (Ch 14), 447, 448*f*

bile A solution secreted by the liver and composed primarily of bile acids, bile pigments and cholesterol (Ch 16, 21), 522, 674, 675

bile acid Steroid detergents made from cholesterol by the liver (Ch 21), 675

bile canaliculi Liver channels through which bile is secreted (Ch 21), 677*f*

bile pigments, 675

bile salt Bile acids conjugated with amino acids (Ch 21, 22), 675, 678

bile secretion, 362*f*, 675, 678

bilirubin Breakdown product of heme groups from hemoglobin (Ch 21, 16), 521*f*, 522

binding site Region of an enzyme or transport protein to which the substrate binds (Ch 2) 46, 386*f*, 387*f*, 390*f*

binocular vision Three-dimensional vision from overlapping visual fields of two eyes (Ch 10), 352*f*, 353

bioenergetics The study of energy flow through biological systems (Ch 4), 95

biological membranes, 61–65, 63*f*

biological rhythm The cyclic variation of a biological process, (Ch 1), 17–18, 18*f*

biome, 3

biomolecule Organic molecules associated with living organisms (Ch 2), 29, 37*f*

biorhythms. *See* biological rhythm

biosynthesis, 733

bipolar cell of retina, 347*f*, 351–352, 351*f*

bipolar neuron Neuron with a single axon and single dendrite (Ch 8, 10), 229, 230*f*, 347*f*, 350*f*

bipotential gonad Embryonic tissue that has the potential to develop into either testis or ovary (Ch 26), 802, 804*f*

birth control, 825–827

birth process, 832*f*

2,3-bisphosphoglycerate (2,3-BPG) A metabolite of red blood cells that lowers the binding affinity of hemoglobin for oxygen (Ch 18), 577

bisphosphonates, 749

bitter taste, 327, 328*f*

bladder sphincter, 403

-blast (suffix), 82

blastocyst Early embryo, consisting of a hollow ball of cells (Ch 26), 828*f*, 829

bleaching of visual pigment, 349, 350*f*

bleb Weakened section of the lung tissue (Ch 17), 550

blind spot Region of the retina with no photoreceptors because the optic nerve and blood vessels exit the eye. Synonym: optic disk (Ch 10), 346

blind study An experiment in which the subject does not know if he or she is receiving the experimental treatment (Ch 1), 22

blood The circulating portion of the extracellular fluid (Ch 3, 11, 16, 17, 18), 81*f*, 82, 83*f*, 84, 281*f*, 368*f*, 437–440, 444*f*, 512–514
  blood cell production, 514–518
  cellular elements of, 512–514, 513*f*
  hemostasis and coagulation, 524–531
  loss from damaged vessels, 524–526
  oxygen transport, 572
  plasma, 61
  red blood cell, 518–523
  volume, 625
blood-brain barrier Tight junctions in the brain capillaries that prevent free exchange of many substances between the blood and the cerebrospinal fluid (Ch 3, 8, 9, 18), 74, 234*f*, 235, 282, 481, 496, 583, 584*f*
blood cells, 514–518, 516*t*, 758*f*. *See also specific type*
blood clot, 503*f*, 526
blood count, 517*f*
blood doping, 512, 522, 523, 527, 531
blood flow, 477–510
  arterioles, resistance in the, 487–491
  distribution of blood to the tissues, 492
  during exercise, 792
  velocity of, 497*f*
blood groups (blood types), 777–778, 777*f*
blood pressure The pressure exerted by blood on the walls of the blood vessels. Usually measured in the systemic arteries (Ch 10, 14, 15, 17, 20, 22, 25), 310*t*, 440, 477–510, 488*t*, 590, 792–793
  arterial, 482
  baroreceptor reflex, 493–495
  cardiac output, 484
  exercise, 792–793
  mean arterial pressure (MAP), 484
  peripheral resistance, 484
  responses triggered by changes in, 639*t*
  sphygmomanometry, 484
blood substitutes, 574
blood-testis barrier Tight junctions between Sertoli cells that prevent free exchange between the extracellular fluid and the lumen of the seminiferous tubules (Ch 26), 811
blood type (blood groups), 777–778, 777*f*
blood vessels, 479–482
  angiogenesis, 481
  cardiovascular system, 437–439, 478*f*
  heart, 444*f*
  structure, 479*f*
  vascular smooth muscle, 479
blood volume, 466, 625, 641
B lymphocyte (B cell) White blood cell that secretes antibodies (Ch 24), 760, 760*f*, 763, 764*t*, 765, 766–768
BMR. *See* basal metabolic rate (BMR)
body cavities, 59, 60*f*
body compartments, 59, 60*f*, 8, 13, 13*f*, 60*f*, 61

body fluid compartment, 60*f*, 124*f*. *See also* extracellular fluid
body load, in mass balance, 11, 11*f*
body mass index (BMI), 697
body movement
  autonomic reflexes, 420
  control of, 417–434
  integrated control of, 426–431
  neural reflexes, 418
  skeletal muscle reflexes, 420–426
body temperature Normal human body temperature is 37 °C or 98.6 °F (Ch 1, 9, 10, 17, 22), 17–18, 289*t*, 297, 320, 537, 720–724
body water, 124*f*, 125, 620, 621*f*
Bohr effect The effect of a change in pH on hemoglobin binding of oxygen (Ch 18), 575
bolus A mass (Ch 21), 663, 668, 669*f*
bomb calorimeter Instrument that determines the caloric content of food by combustion (Ch 22), 697
bone Calcified connective tissue (Ch 3, 9, 10, 12, 15, 16, 23), 81*f*, 82, 83*f*, 280, 400–402, 411*t*, 743–745, 747–749
  calcium in, 743–745
  growth, 743–745, 744*f*
  loss, 749
  resorption, 743
bone marrow A soft tissue that fills the hollow centers of bones; site of hematopoiesis (Ch 7, 16, 23, 24), 514–516, 515*f*, 516*t*, 757, 758*f*
Bordetella pertussis toxin, 182*t*
Botox®, 402
botulinum toxin/botulism, 402, 432
Bowman's capsule The initial segment of the renal tubule. Receives filtered fluid from the glomerular capillaries (Ch 19, 20), 593*f*, 594, 597
Boyle, Robert, 544
Boyle's Law If the volume of a gas increases, the pressure decreases, and vice versa. $P_1V_1 = P_2V_2$ (Ch 17), 543*f*, 544
bradycardia Slow heart rate (Ch 14, 23), 460, 739
bradykinin A paracrine vasodilator (Ch 15, 20, 24), 488*t*, 632, 764*t*, 765
brain, 200*f*
  anatomy of the, 285–290, 286*f*
  central nervous system, 227, 228*f*, 279*f*, 281*f*, 284*f*, 285–290
  functions of, 289–304
  hypoglycemia and the, 283
BRAIN (Brain Research through Advancing Innovative Neurotechnologies), 275
Brain Architecture: Understanding the Basic Plan (Swanson), 291
brain-derived neurotrophic factor (BDNF), 299
brain natriuretic peptide (BNP), 635*f*
brain stem Portion of the brain closest to the spinal cord; contains centers for many unconscious body functions (Ch 9, 10, 13), 281*f*, 285, 287*f*, 295*f*

breast feeding (lactation), 749*t*, 830–833, 833*f*
breathalyzer test, 13
breathing, 535. *See also* ventilation
  alveolar gas composition, 554–556
  cessation, 556*t*
  efficiency of, 553–554
  lung volumes, 544
  mechanics of, 536
  neurons, medulla control, 581–582
  rate and depth of, 553–554, 645*f*
  thoracic cage and diaphragm movement, 547*f*
Broca's area Speech center in the frontal lobe (Ch 9), 303, 303*f*
bronchiole Small collapsible airways with smooth muscle walls (Ch 11, 17), 363*f*, 537
bronchoconstriction, 553, 764*t*, 776*f*
  protective reflex, 584
bronchodilation, 553
brown adipose tissue (BAT), 723
brown fat Adipose cells that contain multiple lipid droplets (Ch 3, 23), 81*f*, 82, 723
Brown-Séquard, Charles, 198
brush border Name given to microvilli covering the luminal surface of intestinal and renal tubule epithelia (Ch 21), 680
buffer A molecule that moderates changes in pH (Ch 2, 20, 26), 41, 642, 642*f*, 643, 645*f*, 813*f*
bulbourethral (Cowper's) gland Male accessory gland that produces components of semen (Ch 26), 811, 812*f*
bulk flow Mass movement of water or air as the result of pressure gradients (Ch 5, 15, 17), 132, 161, 497, 535
bundle branch Two branches of the bundle of His that carry electrical signals to each ventricle (Ch 14), 454, 456*f*
bundle of His (atrioventricular bundle) Specialized electrical conducting cells of the heart that carry signals into the ventricles (Ch 14), 454, 456*f*, 461*f*
bungarotoxin, 368*t*, 373
Bungarus multicinctus, 257
bupropion (Zyban®), 363, 373
Bush, Barbara, 221
Bush, George H. W., 221

C

CA. *See* carbonic anhydrase
Ca^{2+}-sensing receptor (CaSR), 746–747, 748*f*
cable properties of axons, 249
cadherin Membrane-spanning protein of adhesive junctions that links two cells together (Ch 3), 74, 74*t*
caffeine, 297
Cajal, interstitial cells of, 661
Cajal, Santiago Ramón, 661
calbindin, 746, 748*f*
calcification, 82, 503*f*
calcitonin gene-related peptide (CGRP) Neuronal peptide that is coded by the same gene as calcitonin (Ch 7), 219

calcitonin Thyroid gland hormone that decreases plasma Ca^{2+} concentrations in lower animals (Ch 7, 23), 200*f*, 736, 737*f*, 745*f*, 747–748, 749*t*

calcitriol (vitamin D_3), 200*f*, 745*f*, 747, 748*f*

calcium (Ca^{2+}), 33*t*, 386, 405, 411*t*, 683, 743
  and phosphate homeostasis, 748–749
  as a signal molecule, 177

calcium ATPase Membrane transporter that moves Ca^{2+} against its concentration gradient (Ch 12, 14), 382, 389

calcium channel blocker Drugs that block Ca^{2+} channels; used to treat high blood pressure (Ch 15), 505

calcium channel Ion channel that allows movement of Ca^{2+} across a membrane (Ch 14), 451*f*

calcium-induced Ca^{2+} release Process in which Ca^{2+} entry into a muscle fiber triggers release of additional Ca^{2+} from the sarcoplasmic reticulum (Ch 12, 14), 407, 449

calcium release channel, 389, 390*f*

calcium signal, 178, 405

calmodulin Intracellular second messenger that binds Ca^{2+} (Ch 6, 12), 177, 178*f*, 406, 408*f*, 411*t*

Campylobacter jejuni, 268

canal of Schlemm, 341*f*

cancer, 61, 69, 79, 754

candidate hormone Molecules that have not been shown to fulfill all the qualifications of a hormone (Ch 7), 199

cannabinoid receptors, 257

Cannabis sativa, 257

Cannon, Walter B. The father of American physiology, 9, 10, 182, 183, 361

Cannon's postulates, 182–183

capacitance of axons, 251

capacitation Changes in sperm that confer the ability to swim rapidly and fertilize an egg (Ch 26), 827, 828*f*

capillary Smallest blood vessels where blood exchanges material with the interstitial fluid (Ch 5, 8, 9, 12, 14, 15, 17, 18, 19, 21, 23, 24, 26), 134, 148, 152*f*, 234*f*, 281*f*, 395*t*, 593*f*, 737*f*, 813*f*
  absorption, 497–498
  cardiovascular system, 436, 438*f*–439*f*
  exchange at the, 496–498, 499*f*
  filtration, 497–498
  hydrostatic pressure, 500, 598, 600
  pressure, 598, 600

capsaicin, 322, 329

capsid, viral, 755*f*, 756

carbaminohemoglobin Hemoglobin with bound carbon dioxide (Ch 18), 579

carbohydrate Molecules of carbon, hydrogen, and oxygen in the ratio CH_2O (Ch 2, 3, 4, 5, 21), 31*f*, 37*f*, 138
  digestion and absorption, 680, 680*f*
  glycolysis and, 105

carbon dioxide (CO_2) Gaseous product of aerobic respiration (Ch 4, 14, 16, 18, 20, 21), 100–101
  and bicarbonate ions, 577–578
  hemoglobin and, 579
  removal at the lungs, 579–580
  solubility, 570, 570*f*
  ventilation, influence on, 582–583

carbonic acid, 578

carbonic anhydrase (CA) Enzyme that catalyzes the conversion of carbon dioxide and water into carbonic acid (Ch 4, 18, 20, 21, 23), 100, 577, 642, 645*f*, 675, 676*f*, 744*f*

carbon monoxide (CO), 179, 256

carboxyl group, 35*t*

carboxypeptidase Enzyme that breaks peptide bonds at the carboxy terminal end of a peptide (Ch 21), 676*f*, 681, 681*f*

cardiac arrhythmia, 460, 634, 636

cardiac cycle The period of time from the end of one heartbeat through the end of the next beat (Ch 14), 460, 461, 463*f*

cardiac glycosides Drugs such as ouabain and digitalis that block the Na^+-K^+-ATPase (Ch 14), 470, 471, 501

cardiac muscle Striated muscle of the heart (Ch 8, 11, 12, 13, 14), 228*f*, 255*t*, 364, 370*f*, 418*t*, 443, 452*f*, 469*f*
  fibers, 378, 378*f*

cardiac output (CO) The amount of blood pumped per ventricle per unit time (Ch 14, 15, 20, 25), 443, 466, 472*f*, 486*f*–487*f*, 491*f*–492*f*, 494*f*–495*f*, 620*f*, 640*f*, 791, 792*f*

cardiogenic shock, 487

cardiovascular control center (CVCC) Neurons in the medulla oblongata that integrate sensory information and direct autonomic responses aimed at maintaining adequate blood pressure (Ch 9, 11, 15, 20), 289*t*, 467*f*, 492, 494*f*–495*f*, 632, 633*f*, 638, 639*t*

cardiovascular disease, 501–505, 795

cardiovascular system The heart and blood vessels (Ch 14, 15, 20), 3, 436–439, 620*f*
  blood, 437–439
  blood vessels, 437–439
  cardiac muscle, 443–454
  function, 495–496
  heart, 437–439, 443–454
  heart as a pump, 454–475
  overview of the, 436–439
  pressure, volume, flow and resistance, 439–443

carotid baroreceptors, 494*f*

carotid bodies, 582

carotid body chemoreceptor Receptor in the carotid artery that responds to low arterial P_{O_2}, decreased pH, or increased P_{CO_2} (Ch 18), 583

carrier-mediated transport, 145

carrier protein Membrane protein that binds to the molecule it transports. Synonym: transporter (Ch 2, 5), 48, 138, 138*f*, 140–142, 140*f*, 143, 143*f*

CART (cocaine-and-amphetamine-regulated transcript), 695*f*

cartilage Firm, elastic connective tissue with collagen fibers (Ch 3, 23), 81*f*, 82, 83*f*, 741*f*, 744*f*

cascade Response in which a series of inactive molecules convert to active forms until a product is formed (Ch 6, 10, 16, 23, 24), 171, 173*f*
  arachidonic acid, 179, 179*f*
  coagulation, 525, 525*f*, 526–529, 530*t*, 745*f*
  complement, 765
  signal transduction pathways, 171–172, 173*f*, 174, 175*f*

castration Removal of the gonads (Ch 7, 26), 197, 810

catabolism Reactions that release energy and result in the breakdown of large biomolecules (Ch 4), 102, 105, 106, 110*f*

catalase Enzyme that converts peroxide to oxygen and water (Ch 3), 71

catalyst A molecule that speeds up the rate of a chemical reaction without itself being changed (Ch 2), 29

catalytic receptors, 170*f*, 176

catecholamine Signal molecule formed from tyrosine; includes epinephrine, norepinephrine, and dopamine (Ch 7, 9, 11, 14, 23), 202*t*, 206, 207*f*, 289*t*, 362*f*–363*f*, 367, 469*f*–470*f*, 731*f*, 733*f*

cation Positively charged ions (Ch 2), 33*t*, 39

caveolae, 148, 152*f*, 406, 497

caveolin, 148

cavins, 148

cavities (body), 59

C cell Thyroid cell that secretes calcitonin (Ch 23), 736, 737*f*, 749, 749*t*

cecum The initial section of the large intestine (Ch 21), 685, 685*f*

cell The basic functional unit of most living organisms (Ch 3, 4, 6, 12, 16, 21), 3, 66*f*, 81*f*, 154*f*. *See also specific cell*
  compartmentation, 65
  intracellular fluid (ICF), 60*f*
  nucleus of the, 72
  organelles, 79

cell adhesion molecule (CAM) Membrane proteins that link cells to each other and to the extracellular matrix (Ch 3, 6), 72, 74*t*, 75*f*, 168

cell body Part of the cell that contains the nucleus and many organelles. Synonym: cell soma (Ch 8, 9, 10), 227*t*, 261*f*–262*f*, 284*f*, 290*f*

cell-cell adhesions, 72, 74*t*

cell-cell contact, 166–168

cell injury pathway, 527

cell junction Membrane proteins and extracellular matrix that hold cells together to form tissues (Ch 3, 5), 72, 75*f*, 79, 138*f*

cell-mediated immunity Immune reaction that requires T lymphocytes to come in contact with the antigen (Ch 24), 757, 770

cell membrane The cell membrane that serves as both a gateway and a barrier for substances moving into and out of the cell (Ch 1, 3, 5, 6, 8, 12, 24), 3, 66f–67f, 73f, 132–133, 139f, 186f, 235f, 382t, 755f
  fluid mosaic model of, 62

cell-to-cell communication Chemical and electrical processes by which cells coordinate their functions (Ch 1, 2, 6, 8), 9, 34f, 166–168, 253–260

cellular mechanism of action of hormone The intracellular events through which the hormone's message is carried out (Ch 7), 199, 202, 204
  peptide hormones, 202–204
  steroid hormones, 204–206, 205f

cellular respiration Intracellular reaction of oxygen with organic molecules to produce CO_2, water, and energy in the form of ATP (Ch 17), 535, 536f

cellulose, 31f, 37f, 680, 680f

cell volume, 127, 128, 619

central canal, 277, 279f, 281f

central chemoreceptor Chemoreceptor in the medulla oblongata that monitors plasma P_{CO_2} (Ch 6, 18, 20), 186f, 583, 645f

central fatigue Subjective feeling of fatigue during exercise (Ch 12), 392, 393f

central hearing loss Hearing loss due to damage to the auditory cortex or the neural pathways between it and the ear (Ch 10), 336

central nervous system (CNS) Brain and spinal cord (Ch 6, 8, 9, 10, 11, 12, 13, 17, 19, 22, 24, 26), 186, 227, 228f, 234f, 255t, 274–305, 274–308, 286f, 364f, 370f, 391f, 612f, 809f
  anatomy of the, 277–283
  cerebrospinal fluid, 280–282, 281f
  gray matter and white matter, divided into, 277
  development, 277

central pattern generator Networks of CNS neurons that function spontaneously to control certain rhythmic muscle movements (Ch 13, 18), 428, 581

central receptor Sensory receptors located in or closely linked to the brain (Ch 6), 186f

centriole, 66f, 68, 69f, 813f

centrosome The cell's microtubule-organizing center (Ch 3), 68

cephalic phase Digestive reflexes triggered by stimuli received in the brain, such as the smell or sight of food (Ch 21), 665f, 668, 670f, 673f

cephalic reflex, 664

cerebellum Portion of the brain that coordinates the execution of movement (Ch 8, 10, 13), 262f, 314f, 317, 336f, 339, 339f, 428t, 429f, 430f

cerebral cortex Outer portion of the cerebrum that carries out higher cognitive functions (Ch 9, 10, 13, 15), 314f, 320f, 326f, 339f, 427t, 428t, 429f, 825f

cerebral hemorrhage, 484

cerebral lateralization Asymmetrical distribution of function between the left and right sides of the brain (Ch 9), 291, 293f

cerebrospinal fluid (CSF) A salty solution that is continuously secreted into the ventricles of the brain (Ch 9, 10, 18), 280–282, 281f, 310t, 583, 584f

cerebrum Largest region of the brain (Ch 9, 13), 277, 278f, 286f–287f, 289–290, 290f, 429f

cerumen, 335

cervical canal, 816, 818f, 832f

cervical cancer, 87–88, 754, 756, 766, 781

cervical spinal cord/nerves, 279f, 284

cervix Neck of the uterus that opens into the vagina (Ch 1, 24, 26), 756, 804f, 816, 818f, 832f

C fibers, 320, 321t

channel kinetics The speed with which channels open, close, or deactivate (Ch 8), 239

channelopathy, 239

channel protein A membrane protein that forms water-filled channels to link intracellular and extracellular compartments (Ch 5), 138, 138f, 139, 140f

channels
  axonal Na^+, 245
  chemoreception, 327–329
  ion, 172f–173f, 174f

chemical bond The physical forces that attract and hold atoms together (Ch 2, 22), 38f, 696–697, 696f

chemical communication, 168

chemical digestion, 668

chemical disequilibrium, 123

chemical equilibrium Reaction in which the forward and reverse rates of the reaction are equal so that there is no net change in the concentrations of products or reactants (Ch 5), 135, 156

chemical gradient, 134

chemically gated channel Channels whose open state is controlled by binding to a chemical ligand (Ch 5, 8), 138f, 139, 239, 240t

chemical modulator, 49

chemical reaction A substance undergoes a chemical change to become a different substance by breaking existent covalent bonds or making new bonds (Ch 2, 4, 22), 47–48, 96, 96t, 696–697

chemical signal, 166, 178f, 185, 188, 232–233, 254, 259f, 364, 405, 410, 411t

chemical synapse Synapse that uses neurotransmitters to pass information to the target cell (Ch 8), 254, 257f

chemical work Making and breaking of chemical bonds. It enables cells

and organisms to grow, maintain a suitable internal environment, and store information needed for reproduction and other activities (Ch 4, 22), 94, 696f, 697

chemiosmotic theory of oxidation phosphorylation Model of mitochondrial ATP production that links H^+ transport across the inner mitochondrial membrane to ATP synthesis (Ch 4), 109f

chemoreception, 324–329

chemoreceptor A sensory receptor that is activated by binding of a chemical substance (Ch 9, 10, 18), 288t, 312, 312t
  central, 583
  peripheral, 582–583

chemotaxin A molecule that attracts cells such as white blood cells (Ch 24), 763, 764t, 772, 773f

chief cell A cell of the stomach that secretes pepsinogen (Ch 21), 671f, 672, 673f

chloride ion (Cl^-), 33t, 42f, 174f, 240t, 628f, 675f, 683f

chloride shift Process in which red blood cells exchange HCO_3^- for Cl^- (Ch 18, 20), 578, 643

cholecystokinin (CCK) Intestinal hormone that regulates digestive function and may play a role in appetite (Ch 7, 8, 21, 22), 199, 200f, 256, 665–666, 666

cholera, 659, 682

cholera toxin, 182t, 655, 675, 688

cholesterol A steroid that serves as the basis for steroid hormones; also a key component of membranes (Ch 2, 3, 4, 5, 7, 15, 21, 22, 26), 30f, 63f, 204, 205f, 675, 679f

cholinergic neuron A neuron secreting acetylcholine (Ch 11), 371t

cholinergic receptor, 254, 260f, 368t

cholinesterase inhibitors, 368

chondrocyte Cells that produce cartilage (Ch 3, 23), 82, 743, 744f

chordae tendineae Collagenous cords that prevent the atrioventricular valves from being pushed back into the atria during ventricular contraction (Ch 14), 445f, 447, 448f

chorion, 829, 829f

chorionic gonadotropin. *See* human chorionic gonadotropin

chorionic villi, 829, 829f

choroid layer, 347f

choroid plexus A transporting epithelium that secretes cerebrospinal fluid (Ch 9), 280, 281f

chromaffin cells Modified postganglionic sympathetic neurons in the adrenal medulla that secrete epinephrine (Ch 11), 367, 368f

chromatid, sister, 807, 808f

chromium picolinate, 29, 39, 40, 46, 48, 53

chromosome, 801–802, 802f, 803, 808f

chronic obstructive pulmonary disease (COPD) Pulmonary diseases characterized by nonreversible decreased air flow through bronchioles; emphysema and chronic bronchitis (Ch 17, 20), 535, 557, 648

Chrysaora fuscescens, 178

chylomicron Large droplets of triglycerides, cholesterol, proteins, and lipoproteins that are synthesized in cells of the small intestine (Ch 21), 678, 679*f*

chyme A soupy substance produced by digestion in the digestive tract (Ch 21), 655, 684*f*

chymotrypsin, 676*f*, 681, 681*f*

chymotrypsinogen, 676*f*

cilia Short, hair-like structures whose movement creates currents that move fluids or secretions across the cell surface (Ch 3, 10, 17), 68, 69*f*, 333*f*, 541*f*

ciliary muscle Muscle in the eye whose contraction slackens zonules and rounds the lens (Ch 10), 341*f*, 343, 344*f*

ciliated epithelia Epithelia covered with cilia that move fluid over the surface (Ch 3), 76, 77*f*, 78*f*, 79

ciliopathies, 69

cimetidine, 255*t*

cingulate gyrus, 286*f*, 289, 290*f*, 295*f*

circadian rhythm Biological rhythm based on a 24-hour cycle (Ch 1, 9, 23), 17–18, 18*f*, 297, 733, 733*f*, 741*f*

circulatory system The heart and blood vessels (Ch 3, 15, 16), 3, 4*f*, 77*f*, 482*f*, 519*f*

circumcision Removal of the foreskin of the penis (Ch 26), 810

citric acid cycle Key metabolic pathway of aerobic respiration. Synonyms: Krebs cycle, tricarboxylic cycle, TCA cycle (Ch 4), 105, 106, 106*f*, 108*f*

CLARITY, 292*t*

classic hormones, 190*f*, 198

-*clast* (suffix), 82

clathrin, 148, 149*f*

clathrin-coated pit, 148, 149*f*

claudin, 74, 75*f*, 150

clearance A measurement of the disappearance of a substance from the blood, expressed as milliliters of plasma cleared of solute per unit time (Ch 1, 19), 12–13, 609, 610*f*, 611–612

clitoris, 805*f*, 816, 818*f*–819*f*

clonal deletion, 778

clonal expansion Reproduction of one type of lymphocyte following exposure to an antigen (Ch 24), 766, 767*f*

clone A group of cells that are genetically identical (Ch 24), 766, 767*f*, 770*f*

closed system nothing enters and nothing leaves (Ch 4) 95

Clostridium botulinus, 402

Clostridium tetani, 402, 420

clot, 517*f*, 524–525, 525*f*, 529*f*

clotting factor, 514*t*, 529, 530*t*, 677*f*

clotting process. *See* coagulation

clotting protein, 437*t*

Cnidaria, 275

coagulation cascade, 525, 525*f*, 526–529, 530*t*, 745*f*

coagulation Process in which fluid blood forms a gelatinous clot (Ch 16), 514, 525, 526–530, 529*f*, 530*t*

coarse touch, 319*f*

coated pit, 148, 149*f*

cobalamin, 682

cocaine, 368*t*

coccygeal nerve, 279*f*

cochlea Coiled structure of ear that contains receptors for hearing (Ch 10), 329, 330*f*, 331–335, 333*f*, 336*f*

cochlear duct (endolymph), 331–334, 333*f*

cochlear nerve, 332*f*–333*f*

cochlear nuclei, 333*f*, 335, 336*f*

coding, stimulus intensity, 189

coenzymes Organic cofactors for enzymes; they do not alter the enzyme's binding site as inorganic cofactors do (Ch 4) 100

cofactor An inorganic or nonprotein organic molecule required for activation of protein (Ch 2), 49, 49*t*, 50*f*

cognitive behavior Behaviors that deal with thought processes rather than emotion (Ch 9), 275, 302

cognitive system, 291, 291*f*

cold receptors, 320

colipase A protein cofactor that allows lipase to break through the bile salt coating of an emulsion (Ch 21), 676*f*, 678

collagen Flexible but inelastic protein fibers of connective tissue (Ch 2, 3, 15, 16, 24, 26), 41, 72, 81*f*, 82, 503*f*, 525*f*, 526*f*, 527*t*, 528*f*, 778*t*

collagen fibers, 422*f*, 519*f*

collateral Branch of an axon (Ch 8), 229–231

collateral circulation, 482

collecting duct Terminal region of the kidney tubule (Ch 19, 20), 593*f*, 594, 595*f*, 601*f*, 624*f*, 626*f*, 631*f*, 647*f*

colligative property, 127

colloid, 736, 737*f*

colloid osmotic pressure (π) Osmotic pressure that due to the presence of plasma proteins that cannot cross the capillary endothelium. Synonym: oncotic pressure (Ch 15, 16, 19), 498, 499*f*, 514*t*, 599*f*, 600, 607*f*

colon Proximal portion of the large intestine. (Ch 10, 21), 323*f*, 658, 685

colonocyte Transporting epithelial cell of the large intestine (Ch 21), 683

colony-stimulating factor (CSF) Cytokines made by endothelial cells and white blood cells that direct the production and development of white blood cells (Ch 16, 24), 168, 516–517, 516*t*, 772

color-blindness, 349

columnar epithelial cell, 76, 84*t*, 541*f*

coma, 296

committed progenitor cells, 515*f*

common bile duct, 677*f*, 678

common pathway of cogulation, 527*t*, 528*f*

communicating junction, 75*f*

communication, 161, 165, 193, 223, 756
  local, 168
  long-distance, 168
  neighboring cells, 403

compact bone, 743, 744*f*

compartment, 70–71, 134, 223, 234*f*

compartmentation The internal division of the body or cell into compartments so that functions can be isolated from one another (Ch 1, 5), 8, 161

competition, 146, 146*f*, 180, 608

competitive inhibitor Molecules that bind to the active site of the enzyme, preventing substrate binding (Ch 2, 5, 7), 49, 49*t*, 50*f*, 146, 216

complement A group of plasma enzymes that are involved in immune function (Ch 24), 764*t*, 765, 765*f*, 768, 769*f*, 772, 773*f*, 776*f*

complete heart block, 457

complete (fused) tetanus Sustained maximal contraction of a muscle in response to repeated stimuli (Ch 12), 396, 397*f*

compliance The ability of the lung or other tissue to stretch (Ch 1, 14, 17), 471, 550

concave lens, 343, 344*f*–345*f*

concentration gradient A difference in the concentration of a substance between two places (Ch 4, 5, 18), 94, 134, 135, 137*f*, 143, 159, 567

concentration The amount of solute per unit volume of solution (Ch 2, 5), 42*f*–43*f*, 128, 134

concept map, 63*f*, 467*f*

conditioned reflex, 418

conductance (G), 238

conducting system of airways, 536

conducting system, electrical in the heart (Ch 14), 454, 456*f*

conduction of action potentials, 242, 243*f*, 246, 250*f*, 321*t*

conductive hearing loss Hearing loss due to failure of sound transmission though outer or middle ear (Ch 10), 335

conductor, 153

cone A photoreceptor for high acuity vision and color vision during the daytime (Ch 10), 347*f*, 348, 348*f*

conformation of proteins, 41, 47

congenital adrenal hyperplasia Hereditary defects in the enzymes needed for adrenal steroid production (Ch 23), 735

congestive heart failure (CHF) Pathological condition in which the left ventricle fails to pump blood adequately, causing backup of fluid into the lungs (Ch 15, 17), 505, 542

conjugated protein Molecules of protein combined with either lipid or carbohydrate (Ch 2), 29 connective tissue, 80, 80*f*–81*f*, 82, 83*f*, 84*t*, 237*f*, 280, 311*f*, 321*f*, 341*f*, 380*f*, 406*f*, 819*f*

connexin Membrane-spanning cylindrical proteins that form gap junctions; capable of opening and closing (Ch 3, 6), 74, 75*f*, 166

connexon The protein channel of a gap junction, made of connexins (Ch 6), 166

consciousness, 294

consensual reflex Light shined in one pupil constricts both pupils (Ch 10), 342

consolidation Process that converts short-term memory to long-term memory (Ch 9), 301

constipation, 686

constitutive/constitutive process Any essential bodily function that is always taking place (Ch 4, 5), 112, 113*f*, 148, 149

contact activation pathway, 527

contact-dependent signals Cell-cell signals that require surface molecule binding between two cells (Ch 6, 24), 166–168, 167*f*, 770–772

continuous capillary Capillary whose endothelial cells are tightly joined (Ch 15), 496, 496*f*

contraction, 362*f*, 379*f*, 383–389, 396–400, 397*f*, 404*f*, 405–409, 469*f*, 470*f*, 472*f*, 494*f*, 787*f*
    cardiac muscle cells, 447–449
    GI smooth muscle, 663
    heart, 461

contralateral On the opposite side from (Ch 9, 10), 304, 335

control Part of an experiment designed to ensure that any observed changes are due to the experimental manipulation and not to an outside factor (Ch 1, 12), 18, 411*t*

control systems, 13, 182–183, 187, 650
    components, 13, 13*f*

Conus geographus, 257

convergence A number of presynaptic neurons provide input to a smaller number of postsynaptic neurons (Ch 8, 10, 13), 260, 262*f*, 312, 313*f*, 347*f*, 420

convex lens, 343, 344*f*–345*f*

cornea The clear covering of the anterior surface of the eye (Ch 10), 341, 344*f*, 346*f*

corona radiata, 827, 828*f*

coronary artery Artery supplying blood to the heart muscle (Ch 14), 436, 439, 445*f*, 455

coronary artery disease, 481

coronary heart disease (ChD), 481, 502

coronary sinus, 438

coronary veins, 438, 446

corpora cavernosa Two columns of erectile tissue in the penis (Ch 26), 810, 812*f*

corpus albicans The remnants of a degenerated corpus luteum (Ch 26), 820*f*–821*f*, 823

corpus callosum The central region where neurons pass from one hemisphere of the cerebrum to the other (Ch 7, 9), 221*f*, 286*f*, 289, 290*f*

corpus luteum Ovarian structure that produces estrogen and progesterone after ovulation (Ch 7, 26), 200*f*, 817, 819*f*–822*f*

corpus spongiosum A column of spongy erectile tissue in the penis (Ch 10, 20, 26), 810, 812*f*

cortex Literally, bark; the outer or surface portion of an organ (Ch 10, 19, 20), 318, 592*f*–593*f*, 624*f*

cortical association areas, 429*f*

cortical collecting duct, 630

cortical granules Cytoplasmic granules in the egg that contain chemicals to prevent polyspermy (Ch 26), 827

cortical nephron, 591–594, 591–596, 592*f*

cortical reaction Chemical reaction that changes the zona pellucida after fertilization so that additional sperm cannot reach the egg (Ch 26), 827

corticospinal tract Neurons from motor cortex to spinal cord (Ch 9, 13), 287, 428*t*, 429, 430*f*, 431*f*

corticosteroid-binding globulin (CBG), 204, 732

corticotropin. *See* adrenocorticotrophic hormone

corticotropin-releasing hormone (CRH) Hypothalamic hormone that regulates secretion of ACTH from the anterior pituitary (Ch 7, 22, 23), 212*f*, 214*f*, 217*f*, 219*f*–220*f*, 695*f*, 732, 733*f*, 735–736

cortisol Steroid hormone from the adrenal cortex that regulates metabolism, particularly during stress (Ch 1, 2, 7, 9, 22, 23, 26), 13, 18, 18*f*, 30*f*, 731–736, 741*f*, 806*f*
    pathologies, 218, 734–735

Corynebacterium diphtheria, 768

co-secretion Secretion of more than one compound from a secretory vesicle (Ch 7), 202

cotransporter A protein that moves more than one kind of molecule at one time (Ch 5, 21), 140, 675, 681*f*

countercurrent exchange system Anatomical arrangement of vessels so that flow in one vessel is in the opposite direction from flow in the adjacent vessel (Ch 20), 627

countercurrent heat exchanger, 628*f*

countercurrent multiplier Anatomical arrangement of the loop of Henle that concentrates solute in the renal medulla (Ch 20), 627–629

covalent bond Bonds created by two atoms that share one or more pairs of electrons (Ch 2), 33, 38*f*

covalent modulator Atoms or functional groups bind to proteins and affect their activity (Ch 2), 49, 49*t*

Cowper's glands, 811

C-peptide, 202, 203*f*

cranial cavity, 59, 60*f*

cranial nerve 12 pairs of peripheral nerves that originate primarily from the brain stem (Ch 9), 278*f*, 285, 287*f*, 288*t*, 431*f*

cranial nerve I, 326*f*

cranial nerve III, 342*f*

cranial nerve IX, 669*f*

cranial reflex, 418, 418*t*

cranium, 279*f*, 280

C-reactive protein (CRP), 504, 764, 764*t*

creatine, 32*f*, 787*f*

creatinine The breakdown product of phosphocreatine (Ch 19), 611

creatine kinase (CK) Enzyme that transfers a high-energy phosphate group from phosphocreatine to ADP (Ch 4, 12), 99*t*, 102, 391, 392*f*

creatine phosphokinase (CPK), 391

crenated, 520*f*

cretinism Congenital hypothyroidism that causes mental retardation and short stature (Ch 23), 739

crista of vestibular apparatus, 337, 338*f*

cristae (mitochondria) Folds of the inner membrane of a mitochondrion (Ch 3, 10), 67*f*, 337, 338*f*

crossbridge Connection formed when mobile myosin heads bind to actin molecules in muscle (Ch 12), 383

crossbridge tilting, 388

crossed extensor reflex A postural reflex that helps maintain balance during flexion reflexes (Ch 13), 425, 426*f*

crossover study Experimental design in which the subjects spend half the time on the experimental treatment and half the time on placebo (Ch 1), 22

cross-sectional area, 442

cross-sectional studies These examine a population for the prevalence of a disease or condition (Ch 1), 23

crypt Deep pockets created by the highly folded surface of the intestine (Ch 21), 657*f*, 658, 674

cryptorchidism Failure of one or both testes to descend into the scrotum (Ch 26), 811

cupula Gelatinous mass in the vestibular apparatus that contains cilia of hair cells (Ch 10), 337, 338*f*

curare, 255*t*, 368*t*

current, 239

current flow in the heart, 458*f*

current leak, 242, 252*f*

Cushing, Harvey, 734

Cushing's syndrome, 734

cyanosis Blue appearance to the skin and mucous membranes due to excessive amounts of reduced hemoglobin (Ch 14), 437

cyclic AMP (cAMP/cyclic adenosine-3', 5'-monophosphate) Nucleotide that participates in the transfer of signals between the external environment and the cell (Ch 2, 6, 11, 20, 23), 37*f*, 173*f*, 175*f*, 176, 182*t*, 350*f*, 367*t*, 470*f*, 624*f*, 747*f*

cyclic nucleotide-gated channels (CNG channels), 349, 350*f*

cyclooxygenase (COX) Enzyme that converts arachidonic acid to prostanoids (Ch 6), 179, 179*f*

cysteine, 41

cystic fibrosis (CF), 123, 160, 537, 674

cystic fibrosis transmembrane regulator (CFTR channel) Nucleotide-gated chloride channel in epithelia that is defective in cystic fibrosis (Ch 5, 21), 139, 674, 675*f*

-*cyte* (suffix), 82

cytochrome P450 isozymes, 673

cytochromes, 109*f*

cytokine Regulatory peptides that control cell development, differentiation, and the immune response (Ch 5, 6, 7, 15, 16, 17, 21, 24), 168, 198–199, 436, 516, 665–666, 733*f*, 756, 761, 776*f*, 779, 780*f*
  inflammatory response, 765
  virus, response to, 771*f*, 774*f*

cytoplasm All material inside the cell membrane except for the nucleus (Ch 3, 7, 8, 10, 12, 22, 26), 65, 66*f*, 68, 68t
  of an axon, 227*t*
  of egg, 828*f*

cytosine, 34*f*–35*f*

cytoskeleton The internal scaffolding of the cell, composed of microfilaments, intermediate filaments, and microtubules (Ch 3, 5, 6, 12, 16), 61, 64, 65, 66*f*, 68–69, 138*f*, 147, 177*f*, 406*f*, 520*f*

cytosol Semi-gelatinous intracellular fluid containing dissolved nutrients, ions, and waste products (Ch 3, 6, 13, 14), 65–68, 66*f*, 73*f*, 178*f*, 386*f*, 408*f*, 470*f*

cytotoxic T cell (TC cell) A lymphocyte that kills its target cells (Ch 24), 760*f*, 764*t*, 765, 770, 771*f*, 772, 774*f*

D

D⁹ -*tetrahydrocannabinoid* (THC), 257

dalton (Da) 1 dalton = 1 atomic mass unit (Ch 2), 42*f*

Dalton's Law The total pressure of a mixture of gases is determined by the sum of the pressures of the individual gases (Ch 17), 542, 543*f*

dantrolene, 795

data Information or facts gathered during an experiment (Ch 1), 19

D cell Pancreatic endocrine cell that secretes somatostatin, 671, 673*f*

dead space Those portions of the respiratory system that do not exchange gases with the blood (Ch 17), 545*f*

deamination Removal of an amino group from a molecule (Ch 4), 102

deci- (d) (prefix), 43*f*

decibel (dB) Measure of sound wave intensity (Ch 10), 330, 331*f*

deciliter (dL) 1/10 of a liter or 100 mL (Ch 2), 43*f*

decision-making, 430*f*

declarative memory Memory that depends on the use of higher level cognitive skills such as inference, comparison, and evaluation. Synonym: explicit memory (Ch 9), 301, 301*t*

defecation reflex, 658, 685*f*, 686

degenerative disease, 87

deglutition Swallowing (Ch 21), 668, 669*f*

degranulation Process in which immune cells release the contents of their granules (Ch 24), 760, 776*f*

dehydration, 636, 638, 641
  diarrhea as a cause for, 686–687
  homeostatic responses, 638–641

dehydration reaction Reaction in which two molecules combine into one, losing water in the process (Ch 4), 101, 102*t*, 115

dehydroepiandrosterone (DHEA), 205*f*, 731*f*, 732

deiodinase Tissue enzyme that converts T4 to T3 by removal of an iodine (Ch 23), 736

delayed gastric emptying, 663

delayed hypersensitivity reaction Allergic reaction mediated by T cells that may take several days to develop (Ch 24), 775

delta wave High-amplitude, low-frequency brain waves of deep sleep (Ch 9), 296, 296*f*

dementia, 302

denaturation, 51, 670

dendrite Thin, branched processes that receive and transfer incoming information to an integrating region within the neuron (Ch 8), 229, 230*f*, 231, 240*t*, 262*f*, 326*f*

dendritic cell Antigen-presenting immune cells with long, thin processes (Ch 24), 760, 760*f*, 763

dendritic spine Projections of the dendrite membrane that increase surface area (Ch 8), 231, 264*f*

denervation hypersensitivity Up-regulation of neurotransmitter receptors following denervation creates greater than expected response to exogenous neurotransmitter (Ch 11), 369

dense bodies Attachment proteins for smooth muscle actin fibers (Ch 12), 406, 406*f*

dense connective tissues, 81*f*, 82, 83*f*

deoxyribose, 31*f*, 34*f*

dependent variable, 19, 20*f*

dephosphorylation Removal of a phosphate group (Ch 2, 6), 33, 171, 408*f*

depolarization A decrease in the membrane potential difference of a cell (Ch 5, 8, 12, 14, 18), 157*f*, 158, 239, 240*t*, 246*f*, 247*f*, 249*f*, 252*f*, 267*f*, 365*f*, 410*f*, 451, 459*f*, 461*f*, 467*f*, 472*f*, 583

depression, 299

depth of field, 342

dermis, 86*f*

descending limb of loop of Henle, 594

descending tract Neurons that carry information from the brain to the spinal cord (Ch 9), 284, 284*f*

desensitization Reversible form of receptor down-regulation achieved using modulators (Ch 6), 181

desmopressin A form of vasopressin (Ch 20), 625

desmosome A type of cell-to-cell junction (Ch 3, 14), 74, 74*t*, 75*f*, 86*f*, 447, 449*f*

dextrose A six-carbon sugar; also known as glucose (Ch 5), 129

diabetes insipidus Disease characterized by lack of vasopressin (Ch 6), 182*t*

diabetes mellitus Disease characterized by lack of or abnormal action of insulin (Ch 1, 6, 7, 10, 15, 19, 20, 21, 24, 25, 26), 10, 166, 191, 369, 502, 590, 795, 830

diabetic autonomic neuropathy Disturbances of neuronal function as a complication of diabetes mellitus (Ch 11), 369

diabetic neuropathy, 322

diabetic retinopathy, 481

diacylglycerol (DAG) A second messenger (Ch 6), 173*f*, 175*f*, 176

diaphragm (muscle) The skeletal muscle that forms the floor of the thoracic cage (Ch 3, 14, 17, 18, 19, 21, 26), 59, 60*f*, 444*f*–445*f*, 536, 592*f*, 656*f*, 826

diaphysis The shaft of a long bone (Ch 23), 743, 744*f*

diarrhea Excessive amounts of watery stool (Ch 6, 20, 21), 182*t*, 621, 638, 686

diastole The time during which cardiac muscle relaxes (Ch 14), 461

diastolic pressure Lowest pressure in the circulatory system, associated with relaxation of the ventricles (Ch 15), 483, 483*f*

dicarboxylate, 607, 607*f*

dicrotic notch, 465*f*

diencephalon Brain portion between brain stem and cerebrum, consisting of thalamus and hypothalamus (Ch 9), 277, 278*f*, 286*f*, 288, 288*f*

differential white cell count, 516, 517*f*

differentiation Developmental process during which cells take on different forms and functions (Ch 3), 65, 85

diffuse endocrine system Hormones secreted by isolated endocrine cells (Ch 7), 198

diffuse lymphoid tissue, 758*f*, 759

diffuse modulatory system Clusters of brain stem neurons that influence large areas of the brain (Ch 9), 294, 295*f*

diffusion Movement of molecules from an area of higher concentration to an area of lower concentration (Ch 5, 11, 26, 15, 17, 18), 133*f*, 134–137, 135*t*, 371*t*, 497
  distance in alveoli, 567, 569
  gas solubility, 569–571

digestion Chemical and mechanical breakdown of foods into smaller units that can be absorbed (Ch 11, 21), 362*f*, 660*f*, 667*f*, 669, 669*f*, 672, 678
  and absorption, 686
  fats, 679*f*
  large intestine, 685–686
  small intestine, 678
digestive enzyme, 661, 674, 676*f*
digestive hormone, 666, 666*t*
digestive reflex, 665*f*
digestive system Those structures involved in ingestion, processing, absorption, and elimination of food (Ch 1, 3, 21, 24), 3, 4*f*, 77*f*
  anatomy of the, 655–659
  cephalic phase, 668–669, 670*f*
  function and processes, 659–663
  gastric phase, 669–672, 670*f*
  gastric secretions protect and digest, 670–672
  GI function, regulation of, 664
  GI tract, immune functions of the, 687
  immune functions of the GI tract, 687
  intestinal phase, 672–687
  overview of the, 656*f*
digitalis, 501
Digitalis purpurea, 470
dihydropyridine (DHP) receptor Voltage-sensing receptors in the t-tubules, linked to Ca^{2+} channels (Ch 12), 389, 390*f*
dihydrotestosterone (DHT), 205*f*, 731*f*, 803, 805*f*–806*f*
diiodotyrosine (DIT), 736, 737*f*
dilutional hyponatremia, 629, 641
dipalmitoylphosphatidylcholine Surfactant in the alveoli that decreases surface tension (Ch 17), 552
dipeptide, 681*f*
2,3-diphosphoglycerate (2,3-DPG). *See* 2,3-bisphosphoglycerate (2,3-BPG)
direct calorimetry A procedure in which food is burned and the heat released is trapped and measured (Ch 22), 697
disaccharidase Enzyme that digests disaccharides (Ch 21), 680
disaccharide Sugar composed of two sugar monomers (Ch 2, 21), 31*f*, 37*f*, 680*f*
disequilibrium, 13, 143
dissociation constant (K_d), 48
dissociation factor, 126
distal, 168, 209, 379
distal nephron The distal tubule and collecting duct (Ch 19, 20, 23), 594, 640*f*, 646–647, 748*f*
distal tubule, 593*f*, 594, 595*f*, 601*f*, 633*f*
disulfide bond A weak bond between two sulfur atoms (Ch 2, 7), 203*f*
diuresis Loss of water in the urine (Ch 20), 622
diuretic A drug that causes water loss in the urine (Ch 20), 622
divalent metal transporter 1 (DMT1), 683
divergence A few presynaptic neurons branch to affect a larger number of

postsynaptic neurons (Ch 8, 11, 13), 260, 262*f*, 361, 420
DNA (deoxyribonucleic acid) Nucleotide that stores genetic information in the nucleus (Ch 2, 3, 4, 7, 16, 24, 26), 34*f*–35*f*, 37*f*, 73*f*, 205*f*, 524*f*, 755*f*, 808*f*, 813*f*
  metabolism, 102–118
docking protein Membrane proteins that connect vesicles to the cell membrane for exocytosis (Ch 8), 259*f*
dopamine (DA) Amine CNS neurotransmitter (Ch 7, 8, 9, 26), 207*f*, 212*f*, 255, 255*t*, 294, 295*f*, 831
dopaminergic (dopamine-secreting) neurons, 431
dorsal horn Region of spinal cord that contains sensory nuclei (Ch 10), 284*f*, 319*f*
dorsal respiratory group (DRG) Medullary neurons that control normal inspiration (Ch 18), 581
dorsal root Branch of a spinal nerve that carries sensory information (Ch 9), 284*f*
dorsal root ganglion A collection of sensory cell bodies found on the dorsal root just before it enters the spinal cord (Ch 9), 284*f*
double-blind crossover study Double-blind experiment in which the subjects switch between experimental treatment and placebo halfway through the study (Ch 1), 22
double-blind study Experimental design in which neither the subject nor the researcher knows who is getting the experimental treatment and who is getting the placebo (Ch 1), 22
double bond Bonds formed when two atoms share two pairs of electrons (Ch 2), 39
down-regulation Decrease in protein number or binding affinity that lessens response (Ch 2, 6, 7), 51, 181, 217
DRG. *See* dorsal respiratory group
drinking, 636
drug susceptibility, 755
drug tolerance, 181
Duchenne muscular dystrophy, 403
ducts Open tubes through which most exocrine glands release their products (Ch 3), 80
ductus deferens, 811
duodenum Initial segment of the small intestine (Ch 21), 658, 676*f*, 677*f*
dura mater Outer membrane of the meninges (Ch 9), 279*f*, 280, 281*f*
dwarfism A condition of short stature caused by inadequate growth hormone during childhood (Ch 23), 742
dynamic equilibrium Equilibrium related to movement through space (Ch 5), 134
dynamite, 179
dynein A motor protein (Ch 3), 68
dynorphins, 323
dysautonomia, 369
dyspepsia, 672
dysplasia, 70

dyspnea A subjective feeling of not being able to breathe or get air (Ch 17), 556*t*, 557
dystrophin Muscle protein that links actin to the cell membrane (Ch 12), 403

E

ear, 186*f*, 330*f*
  artificial, 337
  cochlea, 331–334
  equilibrium, 337–340
  head position, 339
  hearing, 329–336
  linear acceleration, 339
  mechanical damage, 335
  movement and position, 337
  neural damage, 335
  otolith organs, 339
  rotational acceleration, 337
  semicircular canals, 337
  sound perception, 329–330
  vestibular apparatus, 337
ear canal, 329, 330*f*, 332*f*
eardrum, 329, 335*f*
eating disorders, 694
ecosystems, 2, 3*f*
ectohormone Signal molecules secreted to the external environment (Ch 7), 198
edema The accumulation of fluid in the interstitial space (Ch 15), 500–501
effector The cell or tissue that carries out the homeostatic response (Ch 6, 14), 185, 190*f*, 467*f*
efferent arteriole of the kidney, 593*f*, 594, 595*f*–596*f*, 599*f*, 601*f*, 607*f*
efferent division of the nervous system, 228*f*, 358, 418, 418*t*
efferent neuron A peripheral neuron that carries signals from the central nervous system to the target cells, 187*f*, 190*f*, 208*f*, 227, 228*f*, 230*f*, 284*f*, 361, 418*t*, 419*f*
efferent pathway Outgoing signal that goes from the integrating center to an effector (Ch 13, 14), 425*f*, 467*f*
egg, 802*f*, 808*f*, 817, 828*f*. *See also* ovum
eicosanoid Modified 20-carbon fatty acids that act as regulators of physiological functions (Ch 2, 16, 7), 30*f*, 37*f*, 168, 179, 180, 199, 527*t*
Einstein, Albert, 134
Einthoven's triangle The triangle formed by the three lead electrodes of the simple ECG (Ch 14), 458*f*
E_{ion}, 156, 237*t*
ejaculation Semen in the urethra is expelled to the exterior (Ch 11, 26), 362*f* 363*f*, 824
ejaculatory duct, 812*f*
ejection fraction, 471
elastance Ability of a stretched substance to return to its unstretched state (Ch 1, 3), 8, 82, 551
elastic recoil, 479, 482*f*
elastin A coiled, wavy protein that returns to its original length after being stretched (Ch 3), 81*f*, 82

electrical communication, 168

electrical disequilibrium, 123

electrical gradient Uneven distribution of electrical change, especially across a membrane (Ch 5, 14), 135, 156

heart, 449f, 454–455, 456f

electrical signal, 166, 177f–178f, 186f, 236–253, 240t, 407, 491f

ear, 331, 332f

eye, 348–350

electrical synapse Synapse where electrical signals pass directly from cell to cell through gap junctions (Ch 8), 253–254

electricity, 153

electrocardiogram (ECG) A recording of the summed electrical events of the cardiac cycle (Ch 14), 449, 457, 458f

electrochemical gradient The combined concentration and electrical gradients for an ion (Ch 5, 8, 17, 19, 23), 135, 156, 239, 603f, 745f

electrodes, 157, 391f, 458f

electroencephalogram (EEG), 296, 296f

electrogenic pump, 158

electrolyte An ion (Ch 19, 20), 590, 619–620. *See also* ion electrolyte balance; *specific electrolyte*

electron, 33, 36f

electron transport system (ETS), 105, 109f

electrophoresis A technique used to determine which isozymes are present in a tissue sample (Ch 4), 100

electrostatic attraction, 39

element The simplest kind of matter, such as oxygen and carbon (Ch 2), 36f

embryo, 446f, 801, 827–829, 829f

emergent properties Properties that cannot be predicted to exist based only on knowledge of the system's individual components (Ch 1, 4), 2, 93, 227, 275

emesis Vomiting (Ch 21), 687

emission Movement of sperm from vas deferens to the urethra (Ch 26), 824

emotion, 286f, 290t, 295f, 298–299, 298f

emphysema Lung disease characterized by loss of elastance and alveolar surface area (Ch 18), 535, 552, 568f

emulsion Small droplets suspended in a liquid (Ch 21), 679f

ENaC (epithelial sodium channel), 603f, 604, 631f

encapsulated lymphoid tissue Lymph nodes and the spleen (Ch 24), 757, 758f

end-diastolic volume (EDV) The maximum volume of blood that the ventricles hold during a cardiac cycle (Ch 14), 464, 471–473, 472f

endergonic reaction A reaction that requires net input of energy from an outside source (Ch 4), 97–98, 97f, 98f

endocrine control, 188f, **729–752**, 731–743, 748f

endocrine gland A ductless gland or single cell that secretes a hormone (Ch 3, 7, 8, 11),

80, 80f, 214f, 228f, 255t, 368f. *See also* endocrine system

endocrine reflex, 188t, 191t, 206, 207, 208f

endocrine system The cells and tissues of the body that secrete hormones (Ch 1, 6, 7, 9, 11, 14, 21, 22, 23, 24), 3, 4f, 167f, 196–225, 469–471, 780f. *See also* endocrine gland; *specific gland*

disorders of, 217–218, 732

glands, 198

integrating center, 187f, 190f

introduction to, 196

and metabolism, 730–749

principles, review of, 730–731

endocrinology, 197. *See also* endocrine system

endocytosis, 133f, 148, 149f, 152f, 604

endolymphatic hydrops, 314

endolymph High K^+, low Na^+ fluid that fills the cochlear duct of the ear (Ch 10), 332, 338f

endometrium The secretory inner lining of the uterus (Ch 26), 816, 819f, 827

endopeptidase An enzyme that attacks peptide bonds in the interior of an amino acid chain (Ch 21, 22), 681, 681f

endoplasmic reticulum (ER) A network of interconnected membrane tubes in the cytoplasm; site of protein and lipid synthesis (Ch 3, 7, 16), 67f, 71, 203f, 382t, 524f

endorphin, 256, 733f

endosome Vesicle formed by endocytosis (Ch 5), 148

endostatin, 481

endothelial cell, 282f, 496f, 503f, 530t

endothelial-derived relaxing factor (EDRF) Nitric oxide released by endothelial cells; relaxes vascular smooth muscle, 178, 411

endothelin, 488t

endothelium Layer of thin epithelial cells that line the lumen of the heart and blood vessels (Ch 3, 16), 76, 479, 479f, 527t, 530t, 601f

atherosclerotic plaques, 503f

end-plate potential (EPP) Depolarization at the motor end plate due to acetylcholine (Ch 12), 389

end-product inhibition, 104. *See also* feedback inhibition

end-systolic volume (ESV) The amount of blood left in the ventricle at the end of contraction (Ch 14), 464

energy The capacity to do work (Ch 1, 4), 9, 346f

biological transport in the body requires, 161

body, balance in the, 696f

capture and transfer, 34f

conversion, 95

forms of, 94–95

input, 696–697, 696f

intake, 697

kinetic, 95, 95f

law of conservation, 95

output, 696–697, 696f, 697

potential, 95, 95f

requirements, 133f

storage, 696f

substrate, 788, 789f

thermodynamics, 95–96

transfer in environment, 94f

use, 404, 697

work and, 94

enkephalins, 256, 323

enteric nervous system Neurons in the wall of the gastrointestinal tract that are capable of sensing and integrating information and carrying out a response without input from the CNS (Ch 8, 21), 228f, 229, 659, 664, 665f

enteric plexus, 670f, 673f

enterochromaffin-like cell Stomach cells that secrete hormone, 666t, 672, 673f

enterocyte, 658, 683, 683f

enterokinase Old name for enteropeptidase (Ch 21), 675

enteropeptidase Intestinal enzyme that activates trypsin (Ch 21), 675

entropy A condition of randomness or disorder (Ch 4), 96. *See also* second law of thermodynamics

enzyme Protein catalysts that speed up reactions by lowering their activation energy (Ch 1, 2, 4, 5, 6, 7, 8, 11, 15, 16, 21, 22, 23, 24), 8, 46, 98–99, 203f, 503f, 737f, 744f

eosinophil Leukocytes associated with parasitic infections and allergic reactions (Ch 16, 17, 24), 513f, 514, 515f, 517f, 760, 760f, 761, 769f

ependyma Epithelium that lines the brain ventricles and spinal canal (Ch 8, 9), 234f, 35, 277, 281f

epidermis, 79, 86f

epididymis Duct from seminiferous tubules to vas deferens where sperm complete their maturation (Ch 26), 804f, 811, 812f–813f

epiglottis, 668, 669f

epilepsy, 239, 283, 298, 300, 304, 310, 318

epinephrine Catecholamine neurohormone secreted by the adrenal medulla, 180, 181f, 255, 255t, 411t, 488t

epiphyseal plate Region of long bones where active bone growth takes place (Ch 23), 743, 744f

epiphysis The end of a long bone (Ch 23), 743, 744f

epithelia, 76, 210f

structure of, 76

types of, 76, 78f

epithelial Na^+ channel (ENaC), 327, 603f

epithelial tissue, 76, 77f, 84t

epithelial transport Movement of material from one side of an epithelium to the other (Ch 5, 19), 150–152

epithelium Tissue that protects surface of the body, lines hollow organs, and manufactures and secretes substances (Ch 3, 5, 16, 17, 19, 21, 24), 80*f*, 519*f*, 598*f*, 657*f*, 757*f*

equilibrium, 13, 47, 47*f*, 123, 288*t*, 310*t*, 314*f*, 337–340, 338*f*

equilibrium constant K_{eq}, 47

equilibrium (balance) pathways, 339, 339*f*

equilibrium potential (E_{ion}) The membrane potential that exactly opposes the concentration gradient of an ion (Ch 5, 8, 14), 156, 236, 237*t*, 244, 454*t*

equivalent (Eq) Molarity of an ion times the number of charges the ion carries (Ch 2), 42*f*

erectile dysfunction (ED), 824

erection Blood trapped within spongy tissues of the penis causes it to lengthen and harden (Ch 26), 363*f*, 824, 825*f*

erection reflex, 488*t*, 824, 825*f*

erogenous zones, 824

erythroblast A large, nucleated immature red blood cell (Ch 16), 515*f*, 518

erythrocyte Red blood cells that transport oxygen and carbon dioxide between the lungs and the tissues, 514, 515*f*, 517*f*, 520*f*, 571

erythropoiesis Red blood cell production (Ch 16, 21), 518, 683

erythropoietin (EPO) Hormone made in the kidneys that regulates red blood cell production (Ch 7, 16), 199, 200*f*, 516, 516*t*, 518, 523*t*, 590

Escherichia coli, 61, 591, 686

esophagus The passageway connecting the mouth and stomach (Ch 9, 12, 14, 17, 21, 23), 276*f*, 404*f*, 444*f*, 538*f*, 656*f*, 658, 667*f*, 747*f*

essential amino acid Amino acids the human body cannot synthesize and must obtain from the diet (Ch 2), 32*f*

essential element Those elements necessary for life (Ch 2), 36*f*

essential hypertension High blood pressure whose cause is unclear (Ch 15), 489, 505

estradiol Form of estrogen produced when aromatase acts on testosterone (Ch 7, 23, 26), 205*f*, 731*f*, 806*f*, 809

estrogen Steroid hormone produced in ovary and adrenal cortex; dominant steroid in females (Ch 7, 23, 26), 200*f*, 212*f*, 732, 807, 809*f*, 821*f*, 822*f*, 830

estrone, 205*f*, 731*f*

ethics, 22, 742–743

etiology The cause or origin of a disease, 218

eupnea Normal breathing, 556*t*

eustachian tube, 329, 330*f*

evidence-based medicine, 18

excess postexercise oxygen consumption (EPOC) Increased oxygen consumption following exercise that represents metabolism to replace ATP and other stores consumed during exercise (Ch 25), 789

exchange epithelia Thin epithelia designed for easy transfer of material from one compartment to another (Ch 3), 76, 77*f*, 78*f*

exchange reaction Reaction in which a functional groups are exchanged between or among substrates (Ch 4), 101, 102, 102*t*

excitable tissue Neural and muscle tissue that is capable of generating and responding to electrical signals (Ch 3), 84, 236

excitation-contraction coupling The sequence of action potentials and Ca^{2+} release that initiate contraction (Ch 12, 14), 383, 388–391, 390*f*, 391*f*, 450*f*

excitatory neurotransmitter, 256

excitatory postsynaptic potential (EPSPs) Depolarizing graded potentials that make a neuron more likely to fire an action potential (Ch 8), 263

excretion The elimination of material from the body, usually through the urine, feces, lungs, or skin (Ch 1), 12, 596, 608–612, 620–621, 660*f*

exercise, 787–797, 787*f*–788*f*. *See also* muscle contraction
   blood pressure and, 792–793, 793*f*
   cardiac output, 791, 792*f*
   and cardiovascular disease, 795
   cardiovascular response, 791–793
   and diabetes, 795
   feedforward response to, 793–794
   and health, 795–797
   heat stroke, 792–794
   and immune function, 796, 796*f*
   intensity, 788
   limiting factors, 790
   metabolism and, 787–790
   muscle blood flow increases, 792
   oxygen consumption, 789–790, 790*f*
   and stress, 795–797
   temperature regulation during, 794
   type 2 diabetes mellitus, 795
   ventilatory responses to, 790–791, 790*f*

exercise hyperventilation An increase in ventilation that accompanies an increase in metabolic rate. Synonym: hyperpnea (Ch 25), 790–791

exergonic reaction Chemical reaction that releases energy (Ch 4), 97–98, 97*f*, 98*f*

exocrine cell, 228*f*

exocrine gland A gland that releases secretions into the external environment through ducts (Ch 3), 79, 80*f*, 228*f*, 255*t*, 370*f*

exocrine pancreas, 363*f*, 676*f*

exocrine secretion, 668

exocytosis Process in which intracellular vesicles fuse with the cell membrane and release their contents into the extracellular fluid (Ch 18), 133*f*, 159*f*, 178*f*, 203*f*, 232*f*, 259*f*, 328*f*, 365*f*, 583

exopeptidase Enzymes that release single amino acids from peptides by chopping them off the ends (Ch 21, 22), 681, 681*f*

exophthalmos Bulging eyes in hyperthyroidism due to enlargement of tissue in the eye socket (Ch 23), 739, 739*f*

expiration The movement of air out of the lungs (Ch 17, 18), 535, 538*f*, 544, 545*f*, 547–549, 580*f*

expiratory muscles The abdominal muscles and internal intercostals (Ch 17), 548

expiratory reserve volume (ERV) The amount of air that can be exhaled after the end of a normal expiration, 544

expressive aphasia Inability to speak coherently as a result of damage to Broca's area (Ch 9), 303

extensor A muscle that moves bones away from each other when the muscle contracts (Ch 12), 379, 379*f*, 426*f*

external anal sphincter, 658, 685*f*

external ear, 329, 330*f*

external environment The environment surrounding the body (Ch 17), 537, 543*f*, 595*f*

external genitalia, 801, 805*f*–806*f*, 819*f*

external lamina Thin matrix layer supporting nerve and muscle cells (Ch 3), 84

external respiration The interchange of gases between the environment and the body's cells (Ch 17), 535, 536*f*. *See also* breathing; gas exchange

extracellular fluid (ECF) The internal fluid that surrounds the cells (Ch 1, 3), 10, 11*f*, 13, 60*f*, 61, 124*f*, 405, 642*f*, 745, 745*f*
   ion concentrations and equilibrium potentials, 237*t*
   osmolarity, 130*f*–131*f*, 619
   volume, 620*f*, 629–634

extracellular fluid volume and blood pressure, regulation of, 590

extracellular matrix, 61, 72, 176, 512–514, 519*f*, 745*f*, 746

extraembryonic membrane, 829

extrafusal muscle fiber The normal contractile fibers of a muscle (Ch 13), 421, 422*f*

extraocular muscle, 396

extrapyramidal tract or system Neural network associated with basal ganglia that influences body position and movement (Ch 13), 430

extravasation, 763

extrinsic eye muscle, 340

extrinsic pathway Coagulation pathway that starts when damaged tissues expose tissue factor (Ch 16), 527, 528*f*, 530*t*

eye, 186*f*, 278*t*, 288*t*, 340–353, 362*f*–363*f*, 405*f*
   external anatomy of the, 340*f*
   lens, 343–345
   optics of the, 344*f*
   photoreceptors, 348–350
   phototransduction, 345–348
   pupil, 342

retina, 343–348, 349
signal processing, 350
eye movement, 287f, 340f

F

fab region The antigen-binding arms of an antibody molecule (Ch 24), 768, 769f

facial nerve, 288t

facilitated diffusion Movement of molecules across cell membranes in response to a concentration gradient with the aid of a membrane protein (Ch 5, 19), 133f, 137, 142, 142f, 607f

F-actin Long chains or filaments of actin molecules (Ch 12), 383

factor General name given to signal molecules when first discovered (Ch 7, 16), 199, 516

factor V, 530t

factor VII, 530t

factor VIII, 530t

factor X, 530t

factor XI, 530t

factor XII, 530t

FAD (flavin adenine dinucleotide), 34f, 37f

Fallopian tube Tube that transport eggs from the ovary to the uterus. Synonym: oviduct (Ch 26), 803, 804f, 818f

Fallopius, 803

far-sightedness, 345f

Fas A "death receptor" on cell membranes whose activation causes a cell to commit suicide by apoptosis (Ch 24), 772

fascicles, 379

fast axonal transport Rapid movement of particles along an axon using microtubules and kinesin foot proteins (Ch 8), 231–232, 232f

fasting, 666t, 741f. *See also* postabsorptive state

fast pain Sharp, rapidly transmitted pain (Ch 10), 320, 321t

fast synaptic potential Graded potential in postsynaptic cells that begins quickly and lasts only a few milliseconds (Ch 8), 263

fast-twitch fiber Muscle fibers that develop tension rapidly (Ch 12, 14), 393–395, 397, 398, 452f

fast-twitch glycolytic fiber Fast muscle fibers that rely on anaerobic metabolism and therefore fatigue rapidly (Ch 12), 393, 394f, 395t

fast-twitch oxidative-glycolytic fiber Fast muscle fibers that use a combination of aerobic and anaerobic metabolism and therefore do not fatigue as fast as glycolytic fibers (Ch 12), 393, 395t

fat, 666t, 667f, 678, 684f, 696–697
digestion, 678

fatigue, muscle Inability of a muscle to continue to generate or sustain tension (Ch 12), 392, 393f

fatty acid Long chain of carbon atoms bound to hydrogens and terminating with a carboxyl (Ch 2, 20, 21, 25), 30f, 37f, 642f, 666t, 679f, 787f

fatty streak, 502, 503f

Fc receptors, 768–769

Fc region Stem of antibody molecule that binds to receptors on immune cells (Ch 24), 769f

feces, 521f, 621f, 658, 660f, 745f

feedback inhibition The end product of a metabolic pathway acts as an inhibitory modulator of the pathway. Synonym: end-product inhibition, 103, 104f

feedback loop Information about a homeostatic response that is sent back to the integrating center (Ch 1, 7), 15–17, 16f–17f, 213–215
negative, 15–16
positive, 16, 17, 17f

feedforward, 418, 793–794

feedforward control Anticipatory responses that start a response loop in anticipation of a change that is about to occur (Ch 1), 17, 189, 428, 431f, 664, 793–794

feedforward postural reflexes, 429

feeding center Tonically active hypothalamic center that promotes food intake (Ch 22), 694

female embryonic development, 803

female gametogenesis, 807

female reproduction, 816–823, 818f

female secondary sex characteristics, hormones and, 823

fenestrated capillary Capillary with large pores in the endothelium (Ch 15, 19), 496, 496f, 597

ferritin Protein that binds and stores iron in the body (Ch 16), 521f, 522

ferroportin, 683, 683f

fertility-awareness methods, 825

fertilization, 813f, 827, 828f

fetal hemoglobin (HbF), 577

fetus, 801, 804f–805f, 832f

fever, 764t

fibrillation, 457

fibrillin, 81f, 82

fibrin Plasma protein that forms polymer fibers that stabilize platelet plugs (Ch 16), 525, 525f, 527, 528, 529f, 530t

fibrinogen Plasma protein that becomes fibrin in blood clots (Ch 16), 512, 513f, 514t, 525f, 527, 528f, 530t

fibrinolysis Dissolution of fibrin by plasmin (Ch 16), 528, 529f, 530t

fibrin-stabilizing factor (XIII), 530t

fibroblast, 81f, 82, 83f, 516t, 527t, 813f

fibroblast growth factor (FGF), 481

fibronectin A protein fiber that helps connect cells to their extracellular matrix (Ch 3), 72, 75f, 81f, 82

fibrotic lung disease, 568f

fibrous protein, 32f, 41

Fick, Adolph, 571

Fick equation, 571, 572f

Fick's law of diffusion Diffusion through a membrane is directly proportional to the surface area and concentration gradient and inversely proportional to the thickness of the membrane and its resistance (Ch 5), 136, 137f

fight-or-flight response, 359, 360f, 781

filtration (renal) Bulk flow of plasma-like fluid from the glomerular capillaries into Bowman's capsule (Ch 15, 17, 19), 498, 499f, 594, 596–602, 596f
coefficient, 600
glomerular filtration rate, 599f

filtration fraction The percentage of total plasma volume that filters at the glomerulus (Ch 19), 597

filtration slit Opening between podocyte foot processes through which renal filtration takes place (Ch 19), 597, 598f

fimbriae The fringed opening of the Fallopian tube (Ch 26), 817, 818f

first heart sound Sounds created by vibrations from closure of AV valves (Ch 14), 462

first law of thermodynamics The total amount of energy in the universe is constant (Ch 4), 95, 696

first messenger Chemical signal molecules released by cells (Ch 6), 169, 171, 172f

first-order neuron, 312

first polar body, 807, 808f, 828f

fistula, 655

fixed ribosomes Ribosomes attaching to the cytosolic surface of organelles (Ch 3), 68

flaccid paralysis, 403

flagella Long hair-like extensions of the cell whose microtubules create movement (Ch 2, 3, 4), 68, 69f, 755f

flatus Intestinal gas (Ch 21), 686

flavin adenine dinucleotide (FAD) Molecule that captures and transfers energy with high-energy electrons (Ch 2), 34f, 37f

Fleming, Alexander, 51, 608

flexion reflex A polysynaptic reflex that causes an arm or leg to be pulled away from a painful stimulus (Ch 13), 424, 426f

flexor A muscle that brings connected bones closer together when it contracts (Ch 12, 13), 379, 379f, 426f

flow (rate) The volume of blood that passes one point in the system per unit time (Ch 3, 5, 9, 10, 14, 18, 19, 20, 21), 442, 443f

fluid mosaic model Membrane composed of phospholipid bilayer with proteins inserted wholly or partially into the bilayer (Ch 3), 62, 63f

fluid pressure Pressure created by the presence of fluid within an enclosed space (Ch 10, 19), 337, 599f, 600. *See also* hydrostatic pressure

flux, 136

foam cell, 502, 503f

focal adhesion Junction between intracellular actin and matrix proteins (Ch 3), 74, 75f

focal length (focal distance) The distance from the center of a lens to the focal point (Ch 10), 343, 344f

focal point The point where parallel light waves passing through a lens converge (Ch 10), 343, 344*f*–345*f*

follicle, ovarian, 822*f*

follicle colloid, 738, 738*f*

follicle stage, 820, 820*f*

follicle-stimulating hormone (FSH) Anterior pituitary hormone that stimulates gamete production in the gonads, 200*f*, 212*f*, 213

follicular cells, 736, 737*f*

follicular phase Phase of the menstrual cycle during which ovarian follicles mature and prepare to release an egg (Ch 26), 817, 820*f*, 821–823, 821*f*–822*f*

food intake, 289*t*, 621*f*, 660*f*, 665*f*, 666*t*, 669, 669*f*, 673*f*, 695*f*, 733*f*

foot process Long cytoplasmic extension of a podocyte that wraps around glomerular capillaries (Ch 19), 597, 598*f*

force (tension), 404, 469*f*

forced expiratory volume in 1 second (FEV1), 546

forebrain, 276*f*, 277, 278*f*

foreskin, 810

fovea The region of most acute vision and the point on which light is focused when you look at an object (Ch 10), 341, 346, 347*f*

Framingham Heart Study, 23

Frank, Otto, 468

Frank-Starling law of the heart The principle that within physiological limits, the heart will pump all the blood that returns to it (Ch 14), 468, 491*f*

free energy The potential energy stored in the chemical bonds of a molecule (Ch 4), 96

free nerve endings, 311*f*, 320, 321*f*

free radical Unstable molecule with one more unpaired electrons (Ch 2), 33

free ribosome Ribosome suspended free in the cytosol (Ch 3), 68, 73*f*

frequency coding The frequency of action potentials encodes the intensity of a stimulus (Ch 10), 315

friction rub, 446, 557

frontal lobe, 286*f*, 289, 292*f*

fructose, 31*f*, 667*f*, 680, 680*f*, 813*f*

FSH (follicle-stimulating hormone), 809*f*, 814*f*, 820, 821*f*–822*f*

fulcrum, 400–402, 401*f*

functional group Groups of atoms that tend to move from molecule to molecule as a single unit (Ch 2), 29, 33*t*

functional magnetic resonance imaging (fMRI), 292*t*, 319

functional residual capacity, 546

functional unit The smallest structure that can carry out all the functions of a system (Ch 8), 229, 591–594

fundus The upper portion of the stomach (Ch 21), 656*f*, 658

funny current, 453

furosemide, 627

fusion pore Membrane complex through which secretory vesicle contents can be released (Ch 8), 258

G

GABA (gamma-amino butyric acid), 255*t*, 256

G-actin Single globular molecule of actin (Ch 12), 381*f*, 383

galactose A hexose monosaccharide (Ch 2, 5, 21, 22), 31*f*, 146*f*, 680*f*

Galen, Claudius, of Pergamum, 359, 512

gallbladder Organ that stores and concentrates bile (Ch 10, 21), 323*f*, 656*f*, 666*t*, 667*f*, 677*f*, 678

gamete The reproductive cells that unite to form a new individual (Ch 26), 801, 808*f*

gametogenesis Gamete production (Ch 26), 806–807, 808*f*–809*f*

gamma-aminobutyric acid (GABA) Inhibitory neurotransmitter of the CNS (Ch 2), 32*f*

gamma globulin Name given to the immune globulins of plasma (Ch 24), 764*t*, 765, 766. *See also* antibody

gamma motor neuron Small neuron that innervates intrafusal fibers within the muscle spindle (Ch 13), 421, 422*f*, 424*f*

gamma (γ) radiation High energy waves that penetrate matter deeply, like x-rays (Ch 10), 346*f*

ganglion A cluster of nerve cell bodies in the peripheral nervous system. Plural: ganglia (Ch 8, 11), 233, 276, 276*f*, 361, 362*f*, 370*f*, 371*t*

ganglion cells Neurons of the eye whose axons form the optic nerve (Ch 10), 346, 347*f*, 351*f*, 352

gap junction Cytoplasmic bridges between adjacent cells, created by linked membrane proteins (Ch 3), 74, 75*f*, 166, 167*f*, 227, 405*f*, 411*t*, 447, 449*f*, 454*t*, 455*f*

gas exchange, 787*f*

alveoli, 540, 568*f*

lungs, 565–571

tissues, 565–571

gas laws, 544*t*

Boyle's law, 543*f*, 544

Dalton's law, 542, 543*f*

ideal gas equation, 543*f*

partial pressure, 542, 543*f*

gas solubility, 569–571

gas transport, in blood, 571–580

gases, 178–179, 255*t*, 256, 513*f*

in solution, 570*f*

gastric acid (HCl), 670, 674

secretion, 670

gastric emptying, 666*t*

gastric gland, 657*f*, 658

gastric inhibitory peptide (GIP), 666*t*, 684

gastric lipase Stomach enzyme that digests lipids (Ch 21), 667*f*, 672

gastric mucosa, 670*f*, 671*f*, 673*f*

gastric phase, 669, 670*f*

gastrin Hormone secreted by G cells of the stomach that stimulates gastric acid secretion (Ch 7, 21), 200*f*, 666, 666*t*, 671*f*, 673*f*

gastrin family, 667

gastrin-releasing peptide, 670

gastrocolic reflex, 686

gastroesophageal reflux disorder (GERD), 669

gastroileal reflex, 685

gastrointestinal tract (GI tract) Synonym: digestive tract (Ch 7, 10, 11, 21, 25), 200*f*, 310*t*, 367*t*, 655, 658, 677*f*, 687, 792*f*

motility, 665*f*

peptide, 665–667, 665*f*

smooth muscle, 661–663, 662*f*

spontaneous contractions, 661–663

stem cells, 658

wall, 658–659

gate control model of pain, 322, 324*f*

gated channel A channel that opens and closes in response to stimuli (Ch 5, 8), 138*f*, 139, 140*f*, 238–239, 240*t*

G cell Cell of the stomach that secretes gastrin (Ch 21), 670, 673*f*

gene A region of DNA that contains all the information needed to make a functional piece of mRNA, 112, 205*f*

general adaptation syndrome The body's response to stress (Ch 24), 781

genetic code, 112*f*, 741

genetic sex determination, 801–802

genitalia The external reproductive structures (Ch 26), 804*f*–805*f*, 819*f*. *See also specific structure*

genome, 2

genomic effect Any effect that occurs due to altered gene activity, 204

genomics, 3

germ cell Embryonic gonadal cells that produce gametes (Ch 7, 26), 212*f*, 801, 808*f*

gestation, 827

gestational diabetes mellitus, 830

ghrelin, 666, 694, 695*f*

giantism, 742

Gilman, Alfred G., 174

GIP (gastric inhibitory peptide) GI hormone that causes feedforward release of insulin (Ch 21, 22), 667*f*, 684*f*

gland Group of epithelial cells specialized for synthesis and secretion of substances (Ch 3, 7, 9, 13), 79, 200*f*, 284*f*–285*f*, 288*t*, 418*t*

endocrine, 198

glans penis, 805*f*, 810, 812*f*

glaucoma, 340

glial cells Nonexcitable support cells of the central nervous system (Ch 3), 84, 227*t*, 229, 233, 234*f*, 259*f*

globin, 520

globular protein, 32*f*, 41

globular shape, 37*f*

globulin, 512, 513*f*, 514*t*

glomerular capillary, 598*f*, 599*f*

glomerular filtration rate (GFR) The amount of fluid that filters into Bowman's capsule per unit time (Ch 19, 20), 599*f*, 600–602, 601*f*, 608–610, 610*f*, 621*f*, 633*f*, 635*f*, 639*t*, 640*f*
autoregulation, 600–602, 601*f*, 602
hormones and autonomic neurons influence on, 602
and peripheral blood pressure, 639

glomerulus Ball-like network of capillaries in the kidney; site of filtration (Ch 19, 20), 592*f*–593*f*, 594, 595*f*–596*f*, 599*f*, 601*f*, 607*f*, 610*f*, 639*t*, 646*f*

glomus cells Cells of the carotid and aortic body that respond to low oxygen (Ch 18), 582

glossopharyngeal nerves, 288*t*, 581

glottis, 669*f*

glucagon Pancreatic hormone that elevates plasma glucose (Ch 7), 200*f*, 215*f*, 665*f*, 666*t*, 733*f*

glucocorticoids Adrenal steroid hormones such as cortisol that elevate plasma glucose (Ch 23), 731, 731*f*

gluconeogenesis Pathways through which noncarbohydrate precursors, especially amino acids, are converted into glucose (Ch 22, 23, 25), 733*f*, 788

glucose A six-carbon sugar that is a major energy source for the body. Synonym: dextrose (Ch 2, 16), 31*f*, 142*f*, 513*f*, 787*f*
epithelial transport of, 151
renal reabsorption of, 603*f*

glucose clearance, 610*f*

glucose-dependent insulinotropic peptide, 667

glucose tolerance test, 795

glucostatic theory Theory that glucose utilization by the hypothalamic centers regulates food intake (Ch 22), 694

glucosuria (glycosuria) Excretion of glucose in the urine (Ch 19, 22), 606

glutamate Amino acid that also acts as an excitatory neurotransmitter (Ch 8), 255*t*, 256, 267*f*

glutaminergic ionotropic receptor (iGluR), 255*t*

glutaminergic metabotropic receptor (mGluR), 255*t*

GLUT transporter Family of facilitated diffusion carriers for glucose and other hexose sugars (Ch 5, 19, 21, 22), 142, 146*f*, 151*f*, 603*f*, 680*f*

glycerol A simple 3-carbon molecule that is the backbone of fatty acids, 30*f*, 37*f*, 787*f*, (Ch 2)

glycine Amino acid that also acts as an inhibitory neurotransmitter (Ch 8), 255*t*, 256

glycocalyx Glycoproteins on the surface of cells (Ch 3), 64

glycogenolysis The breakdown of glycogen (Ch 22, 23), 733*f*

glycogen Storage polysaccharide found in animal cells (Ch 2, 4, 12, 21, 22), 31*f*, 37*f*, 94, 393*f*, 680, 680*f*, 696*f*

granules, 380*f*

glycolipid Molecule that is a combination of carbohydrate and lipid (Ch 2), 29, 37*f*, 63*f*

glycolysis Metabolic pathway that converts glucose to pyruvate (aerobic) or lactic acid (anaerobic) (Ch 25), 787*f*

glycoprotein Molecule that is a combination of carbohydrate and protein (Ch 2, 3), 29, 63*f*, 737*f*

glycosuria, 606

glycosylated molecule A molecule that has sugar molecules attached to it (Ch 2), 29

glycosylphosphatidylinositol (GPI), 64

GnRH (gonadotropin releasing hormone), 809*f*, 814*f*, 822*f*

GnRH pulse generator, 810

goblet cell Single exocrine cell that produces mucus (Ch 3), 78*f*, 80, 541*f*, 661, 667*f*

goiter Enlarged thyroid gland (Ch 7, 23), 197, 197*f*, 739, 739*f*–740*f*

Goldman-Hodgkin-Katz (GHK) equation Calculates resting membrane potential using membrane permeability and ion concentrations gradients (Ch 5, 8), 155*f*, 158, 237

G$_{olf}$ G protein for olfactory transduction (Ch 10), 325

Golgi, Camillo, 71

Golgi apparatus Organelle that modifies and packages proteins into vesicles (Ch 7), 67*f*, 71, 73*f*, 149*f*, 203*f*, 232*f*, 679*f*

Golgi tendon organ Receptors are found at the junction of the tendons and muscle fibers that responds to contraction of the muscle (Ch 13), 421, 422*f*

gonad The organs (ovaries and testes) that produce gametes (Ch 7, 26), 200*f*, 210*f*, 801, 809*f*

gonadotropin (FSH and LH) Peptide hormones from the anterior pituitary that act on the gonads (Ch 7, 26), 210*f*, 212*f*, 213, 809, 810, 815

gonadotropin-releasing hormone (GnRH) Hypothalamic hormone that stimulates release of gonadotropins from the anterior pituitary (Ch 7, 26), 212*f*, 809

gout, 590, 591, 608, 613–614, 745–746

GPCR-cAMP pathways, 174

GPCR-phospholipase C signal transduction, 175*f*

GPI (glycosylphosphatidylinositol) anchor, 64

G protein Membrane proteins that couple membrane receptors to ion channels or membrane enzymes, 174, 175*f*, 177*f*, 182*t*, 204*f*

G protein-coupled adenylyl cyclase-cAMP system The first signal transduction system discovered (Ch 6), 174

G protein-coupled receptors (GPCR), 173*f*, 174, 175*f*, 255*t*, 263*f*, 328*f*

graded contraction Muscle contraction whose force varies with the amount of Ca^{2+} that enters the cell (Ch 12, 14), 407, 450

graded potential A change in membrane potential whose magnitude is proportional to the stimulus and that decreases with distance as it spreads through the cytoplasm (Ch 8), 240–242, 240*t*, 241*f*, 250*f*, 261*f*

gram molecular weight The molecular mass of a substance expressed in grams (Ch 2), 42*f*

granular cells Specialized cells in the walls of renal arterioles that synthesize and release renin (Ch 19, 20), 601*f*, 602, 632, 633*f*, 639, 639*t*

granulocyte White blood cell whose cytoplasmic inclusions give it a granular appearance: basophils, eosinophils, and neutrophils (Ch 16, 24), 514, 760, 760*f*

granulosa cell Cell of the ovarian follicle that secretes estrogen (Ch 26), 817, 820, 821*f*–822*f*

granzyme Enzyme of cytotoxic T cells that triggers apoptosis in target cells (Ch 24), 764*t*, 765, 771, 774*f*

graphs, 20*f*–21*f*

Graves' disease Hyperthyroid disorder caused by TSH-like antibodies (Ch 7, 23, 24), 197, 221–222, 740*f*, 778*t*

gray matter Nerve cell bodies, dendrites, and axon terminals (Ch 9, 13), 277, 279*f*, 284*f*, 289, 290*f*, 426*f*

ground substance A cellular portion of matrix consisting of glycoproteins and water (Ch 3), 80, 81*f*, 83*f*

growth, 213, 213*f*, 696*f*.
and aging, 834
charts, 742

growth cone, 232, 233*f*

growth factor, 168, 199

growth hormone Protein hormone from the anterior pituitary that controls tissue growth (Ch 7, 23), 211*f*, 220*f*, 221, 222*f*, 223, 224*f*, 741–742, 741*f*
biosynthesis, 747
cellular actions, 747
ethical questions, 742–743
genetically engineered, 742–743
half life, 747
pathway, 213*f*
transport, 747

growth hormone-binding protein, 741

growth hormone control pathway, 741, 741*f*

growth hormone-inhibiting hormone (GHIH), 211, 741. *See also* somatostatin

growth hormone-releasing hormone (GHRH) Hypothalamic hormone that influences growth hormone secretion (Ch 7), 212*f*, 741, 741*f*

guanosine diphosphate (GDP), 174

guanosine triphosphate (GTP), 105

guanylyl cyclase Enzyme that controls formation of cyclic GMP (Ch 6), 176

guanylyl cyclase activity, 176

Guillain-Barré syndrome Rare autoimmune paralytic neural condition with loss of both sensory and motor function (Ch 8, 24), 227, 229, 249, 268, 778*t*

Guillemin, Roger, 211
gustation, 325. *See also* taste
gustatory cortex, 292*f*, 293, 314*f*
gustatory neurons, 327
gustducin, 327, 328*f*
gut-associated lymphoid tissue (GALT)
Immune cells and tissues of the GI tract
(Ch 21, 24), 659, 758*f*, 759
gyrus Convolution of cerebral surface (Ch 9), 289

H

H_2 receptor for histamine, 672
habituation A decreased response to a stimulus
that is repeated over and over (Ch 9, 10),
300, 313
hair cell Sensory cells for transduction of sound
and equilibrium (Ch 10), 311, 331,
332*f*–335*f*, 338*f*
Haldane, John, 29
half-life The amount of time required to reduce
the concentration of hormone by one-half
(Ch 7), 178, 182*t*, 201, 202
HAPE. *See* high-altitude pulmonary edema
Harvey, William, 436, 512
haustra Bulging pockets of the large intestine
wall (Ch 21), 685*f*, 686
HbA (hemoglobin A), 520
HbF (hemoglobin F/fetal hemoglobin), 523
HbS (hemoglobin S) Abnormal hemoglobin of
sickle cell disease (Ch 16), 522
head position, 338*f*, 339
health and exercise, 795–797
hearing, 288*t*, 292*f*, 310*t*, 329–337
hearing loss, 335
heart Muscular organ that serves as the pump
for the circulatory system, 323*f*, 362*f*–363*f*,
478*f*, 492*f*, 640*f*
cardiovascular system, 436–476
conducting system of the, 456*f*
heart contraction, 461–464
heart failure, 471, 500, 635*f*
heart rate, 184*f*, 362*f*–363*f*, 457, 460, 466–468,
467*f*, 472*f*, 486*f*, 491*f*, 494*f*
heart valve Connective tissue valves that prevent
back flow of blood in the heart (Ch 14),
447, 448*f*
heat stroke, 789
Helicobacter pylori, 672
helicotrema, 332, 333*f*, 335*f*
helper T cell Immune cells that secrete cytokines
to help other immune cells (Ch 24), 760,
760*f*, 765, 771, 771*f*, 772, 774*f*, 776*f*
hematocrit Percentage of the total blood volume
that is packed red blood cells (Ch 16),
517*f*, 518
hematopoiesis Blood cell production in the
bone marrow (Ch 16), 515*f*, 516, 516*t*
heme group A carbon-hydrogen-nitrogen
porphyrin ring with an iron atom in the
center (Ch 16), 520, 521*f*, 522, 523
hemidesmosome Strong junction that ties a cell
to matrix (Ch 3), 74, 75*f*, 86*f*
hemoglobin A_{1C}, 522

hemoglobin Oxygen-carrying pigment of red
blood cells (Ch 2, 16, 18, 20), 46, 49, 516,
517*f*, 520, 520*f*, 521*f*, 523*t*, 642*f*
binds oxygen, 572–573
and CO_2, 579
H^+, 578–579
mean corpuscular, 574
molecules, 759*f*
percent saturation, 574
spleen, breakdown products from the, 677*f*
synthesis, 523*t*
hemolytic anemias, 522, 523*t*
hemophilia, 529
hemorrhage Excessive blood loss (Ch 16), 488*t*,
523*t*, 525
hemostasis Process of keeping blood within the
blood vessels by repairing breaks without
compromising the fluidity of the blood
(Ch 16), 524–526, 524–531, 525*f*
Henderson-Hasselbalch equation, 643
heparin An anticoagulant molecule (Ch 16, 24),
529, 530*t*, 761, 764*t*
hepatic artery, 438*f*, 439, 677*f*
hepatic portal system Specialized region of
the circulation that transports material
absorbed at the intestine directly to cells of
the liver (Ch 21), 673, 674*f*
hepatic portal vein, 438*f*, 439, 677*f*, 685*f*
hepatocyte Liver cell (Ch 1, 21, 22), 12, 675,
677*f*
hepcidin, 683
hereditary spherocytosis, 522, 523*t*
Hering-Breuer inflation reflex Reflex to
prevent overinflation of the lungs (Ch 18),
584
hermaphrodites, 801
herpes simplex type 1, 756
hertz (Hz) Measure of sound wave frequency
(Ch 10), 330, 331*f*
hexose A six-carbon sugar (Ch 2), 31*f*, 145
high-altitude pulmonary edema (HAPE), 571
high-density lipoprotein-cholesterol (HDL-C),
502
high-density lipoprotein (HDL) The "good"
plasma carrier for cholesterol (Ch 2), 40,
795
high-energy electrons, 33
high-energy phosphate bond, 104, 696*f*
hindbrain, 277, 278*f*
hinge region, 381*f*, 768, 769*f*
hippocampus Portion of the brain associated
with learning and memory (Ch 8, 9), 236,
286*f*, 289, 290*f*
histamine Paracrine secreted by mast cells
and basophils; acts as a vasodilator and
bronchoconstrictor (Ch 6, 7, 8, 15, 17, 21,
24), 168, 254, 255*t*, 553, 764–765, 764*t*,
773*f*, 776*f*
digestive system, 667*f*, 671*f*, 673*f*
vasoconstrictor, 488*t*, 490
histiocyte Old name for skin macrophages
(Ch 24), 761
histogram, 20*f*, 21*f*

histology The study of tissue structure and
function (Ch 3), 72
histotoxic hypoxia, 566*t*
H^+-K^+-ATPase, 642, 645
Hodgkin, A. L., 243
homeostasis The ability of the body to maintain
a relatively constant internal environment
(Ch 1, 5), 9–10, 10*f*, 123, 161, 165, 183,
193, 259*f*, 286*f*, 288, 360, 360*f*, 504, 650,
748–749
circadian rhythm, 17–18, 18*f*
control systems and, 13–18
feedback loop, 15–17, 16*f*–17*f*
local control, 13–14, 14*f*
mass balance and, 10–12
reflex control, 14, 14*f*
response loop, 14–15, 15*f*
homeostatic reflex pathways, 182–192
homocysteine, 32*f*, 504
homunculus, 320*f*
horizontal canal, 338*f*
horizontal cell, 347*f*, 351, 351*f*
hormone Chemical secreted by a cell or group
of cells into the blood for transport to
a distant target where it acts in very
low concentrations to affect growth,
development, homeostasis, or metabolism
(Ch 3, 6, 7, 12, 15, 19, 22, 23, 24, 25), 80,
167*f*, 168, 187*f*, 410, 411*t*, 436, 437*t*, 779,
780*f*. *See also specific type*
action termination, 199–201
amine, 201, 207*f*
amino acid-derived, 202*t*, 206
antagonistic, 216
blood, secretion into the, 198–199
blood, transport by, 198–199
calcium balance control, 746–748
chemical, 198–199
classification, 201–206
concentration effect, 199
deficiency, 198
degraded, 199
digestive system, 666, 667*f*
endocrine system, 197–201, 208*f*
evolution, 218–222
excess, 198
half-life, 201, 202
hypersecretion effect on, 216–217
hypothalamic-anterior pituitary pathway, 212*f*
inhibiting, 213
interactions, 215–216
male reproduction, 811–814
menstrual cycle, control of the, 817
metabolism, 695*f*, 789
ovarian production of eggs and, 816
peptide, 201, 202*t*, 204*f*
permissive, 216
placenta secretion of, 829
pregnancy, during, 829–830
release, control of, 206–215
releasing, 209
secondary sex characteristics, influence
female, 823

secretion, 219f, 220f, 286f, 670, 829–830
steroid, 202t, 204, 205f, 206
thyroid, 206, 207f
trophic, 211
hormone receptor, 730
hormone replacement therapy (HRT), 749, 834
hormone-response element, 732
host cell, 756, 774f, 776
Howard Hughes Medical Institute (HHMI), 8t
HPV vaccine, 754, 756, 766, 776
human chorionic gonadotropin (hCG)
Hormone secreted by the developing placenta (Ch 26), 830
human chorionic somatomammotropin (hCS), 830
human chromosome, 802, 802f
Human Connectome, 275
human experiments, 22
ethical considerations, 22
variability, 22
Human Genome Project, 2
human immunodeficiency virus (HIV), 754
human leukocyte antigen (HLA) Name for classification of human MHC proteins (Ch 24), 776
human papillomavirus (HPV), 87–88, 754
human placental lactogen (hPL) Peptide placental hormone that influences maternal metabolism. Synonym human chorionic somatomammotropin (hCS) (Ch 7, 26), 830
Human Proteomics Initiative, 118
human sexual response, four phases of, 824
human studies, 23
humoral immunity Immunity conferred by antibodies (Ch 24), 757, 766
humors, 757
humor therapy, 781
Huxley, A. F., 243, 385
hydraulic pressure Pressure exerted by fluid in motion. Used synonymously with hydrostatic pressure in the circulatory system (Ch 14), 440
hydrogen ATPase (H^+-ATPase), Synonym: proton pump, 645
hydrogen bond Weak attractive forces between hydrogens and other atoms, especially oxygen and nitrogen (Ch 2), 38f, 39
hydrogen ion, 41, 645, 646f–647f, 673f, 676f, 681f, 787f
hydrogen sulfide (H_2S), 179
hydrolysis reaction Reaction in which large molecules are broken into smaller ones by addition of water (Ch 4), 101, 102t
hydrophilic molecule Molecules that dissolve readily in water (Ch 2), 40, 44f
hydrophobic molecule Molecules that do not dissolve readily in water (Ch 2), 44f
hydrophobic steroids, 205f
hydrostatic pressure The pressure exerted by a stationary column of fluid in a tube (Ch 14, 15, 19), 440, 441f, 499f, 598, 599f, 600, 601f, 607f
hydroxyapatite Calcium phosphate crystals of bone (Ch 23), 743

hymen (maidenhead), 816, 819f
hyperbilirubinemia, 522
hypercalcemia, 745
hypercapnia Elevated P_{CO_2} in the blood (Ch 18), 565, 577
hypercholesterolemia, 148
hypercortisolism (Cushing's Syndrome), 734, 735f
hyperemia, 489f
hyperglycemia, 522. *See also* diabetes mellitus
hyperinsulinemia, 217
hyperkalemia, 252, 253f, 631f, 636
hyperkalemic periodic paralysis (hyperKPP), 394
hyperopia Far-sightedness (Ch 10), 345, 345f
hyperosmotic, 127
hyperparathyroidism, 730, 732, 746, 750
hyperplasia Increased cell number due to cell division (Ch 23), 743
hyperpnea Increase in ventilation rate to match an increase in metabolic rate. Synonym: exercise hyperventilation (Ch 17), 556t, 790–791
hyperpolarization A membrane potential that is more negative than the resting potential (Ch 5, 8, 14), 158, 157f, 238, 240t, 253f, 454t, 467f
hyperpolarization-activated cyclic nucleotide-gated channel (HCN channel), 453
hypersecretion of hormones, 216
hypersensitivity, 775
hypertension Chronically elevated blood pressure (Ch 15), 478, 484, 504
hyperthyroidism, 739, 778t
hypertonic solution A solution that causes net movement of water out of a cell (Ch 5), 127–128, 128t
hypertrophy An increase in cell size without an increase in cell number (Ch 15, 23), 504, 739, 740f, 743
hyperuricemia, 591
hyperventilation An increase in alveolar ventilation that is not associated with an increase in metabolic rate (Ch 17), 556t
hypocalcemia, 745
hypochromic erythrocytes, 520, 523
hypocretins, 695, 695f
hypoglossal nerve, 288t
hypoglycemia, 283, 733
hypokalemia, 252, 253f, 636
hypokalemic periodic paralysis, 394
hyponatremia, 619, 637–641, 638, 643, 649–650
hypoprolactinemia, 834
hyposmotic, 127,
hypotension, 484
hypothalamic-anterior pituitary pathway, 212f
hypothalamic feeding center, 695f
hypothalamic-hypophyseal portal system
Modified section of the circulation that takes neurohormones directly from the hypothalamus to the anterior pituitary (Ch 7), 211, 283
hypothalamic-pituitary-adrenal (HPA) pathway, 732, 733f

hypothalamus Region of the brain that contains centers for behavioral drives and plays a key role in homeostasis (Ch 11, 20, 22, 23, 26), 360f, 362f, 626f, 695f, 733f, 809f, 814f, 822f, 833f
brain, anatomy of the, 286f
hypothyroidism, 739–740, 740f
hypotonic solution A solution that causes a net influx of water into a cell (Ch 5), 127–128, 128t
hypoventilation A decrease in alveolar ventilation without a change in metabolic rate (Ch 17), 556t, 567
hypovolemic shock, 487
hypoxia Lack of oxygen in the cells (Ch 14, 15, 16, 18), 437, 489, 518, 565
anemic, 566t
cell, 488t
chronic, 577
classification, 566t
diffusion causes, 567–569
histotoxic, 566t
hypoxic, 566t
ischemic, 566t
pathologies causes, 568f
hypoxia-inducible factor 1 (HIF-1), 518
H zone Region of sarcomere with only thick filaments (Ch 12), 381f, 383, 384f

I

iatrogenic condition Physician-caused condition, 217, 461
I band Region of the sarcomere occupied only by thin filaments (Ch 12), 383, 384f
ideal gas equation, 543f
idiopathic pulmonary fibrosis, 551
I_f channel Monovalent cation channels in cardiac autorhythmic cells that contribute to the pacemaker potential (Ch 14), 453
IgA, 768
IgD, 768
IgE, 768, 776f
IgG, 768, 776f
IgM, 768
Ignarro, Louis, 178
ileocecal valve Muscular region whose contraction separates the large and small intestines (Ch 21), 667f, 685, 685f
ileum Distal portion of the small intestine (Ch 21), 658, 685f
immediate hypersensitivity reaction Allergic reaction that occurs within minutes (Ch 24), 775
immune cell, 437t, 733f, 760, 762f, 771f, 779, 780f
immune complex, 768
immune function, 687, 796, 796f
immune response, 298f, 733f, 756–757, 764t, 767f, 773f, 774f
immune surveillance Theory that cancer cells develop regularly but are usually detected and destroyed by the immune system (Ch 24), 779

immune system The cells and tissues and their products that defend the body against invaders (Ch 1, 23, 24), 3, 4*f*, 733*f*, 753–785, 780*f*

    acquired immunity, antigen-specific responses, 765–772

    anatomy of the, 757–762

    cells of, 760*f*

    and exercise, 795–797

    function, and stress, 781

    immune response, 756–757

    immune response pathways, 772–779

    innate immunity, nonspecific response, 762–765

    lymphatic system, 758*f*

    neuro-endocrine-immune interactions, 779–781

    overview of, 754

    pathogens of the human body, 754–756

    recognize "self," 778–779

immunity The ability of the body to protect itself from pathogens (Ch 24), 754

immunocyte General name given to any of the immune cells (Ch 16, 24), 514, 760*f*

immunodeficiency diseases, 754

immunoglobulin Synonym for antibody (Ch 2, 16, 17, 24), 46, 512, 537, 757*f*, 764*t*, 768

immunosuppressant effects, 734

impermeable membrane A membrane that does not allow substances to cross (Ch 5), 133

impotence, 369, 826

inactivation gate The slow gate of the Na⁺ channel that closes to stop ion flow (Ch 8), 245, 246*f*

inclusion Particle of insoluble material in the cytoplasm such as glycogen granules and lipid droplets (Ch 3), 65, 66*f*, 68

incontinence Inability to voluntarily control urination or defecation (Ch 11), 369

incus Middle of the three small bones of the middle ear (Ch 10), 329, 330*f*, 332*f*

independent variable, 19, 20*f*

indirect (secondary) active transport, 143

induced-fit model of protein-ligand activity The active site changes shape to fit either substrate or product molecules (Ch 2), 47, 47*f*

inertia, 337

infantile spasm, 275, 283, 298, 300, 304

inferior vena cava Great vein that returns blood from the lower body to the right atrium (Ch 14, 19, 21), 438, 438*f*

infertility Inability to conceive (Ch 26), 801, 815, 823, 827, 830, 834–835

inflammation A nonspecific reaction of the immune system to a foreign invader (Ch 24), 760*f*, 764*t*, 772, 773*f*, 776*f*

inflammatory bowel diseases, 687

inflammatory pain, 322

inflammatory response, 764–765, 764*t*, 774*f*

information flow, 9

information processing, 310*t*

infundibulum, 210*f*

inheritance, 304

inhibin Peptide hormone from the gonads that inhibits FSH secretion (Ch 7, 26), 200*f*, 809, 814*f*, 821*f*, 822*f*

inhibiting hormone, 209

inhibitory neurotransmitters, 256

inhibitory postsynaptic potential (IPSP) Hyperpolarizing graded potentials that make a neuron less likely to fire an action potential (Ch 8), 263

inhibitory presynaptic neuron, 265*f*

inhibitory synapses, 264*f*

initial segment The axon hillock and first part of an axon; often the location of the neuron's trigger zone (Ch 8), 242

innate immunity The nonspecific responses of the body to invasion by foreign substances (Ch 24), 756, 763

innate reflexes, 418, 418*t*

inner ear Portion of the ear containing the cochlea and hair cells (Ch 10), 330*f*, 338*f*

innervated Controlled by a neuron (Ch 8), 236

inositol trisphosphate (IP₃) A second messenger made from membrane (Ch 6, 12), 173*f*, 175*f*, 407, 410*f*

inotropic agent Any chemical that affects cardiac contractility (Ch 14), 469

input (afferent) signal, 185

insensible water loss Water loss across the skin and in exhaled air of which we are not normally aware (Ch 20), 621

insertion of a muscle, 379

insomnia Inability to sleep well (Ch 9), 297

inspiration The movement of air into the lungs (Ch 17, 18), 535, 538*f*, 544, 545*f*, 547, 580*f*

inspiratory capacity, 546

inspiratory muscles The external intercostals, diaphragm, scalenes, and sternocleidomastoids (Ch 17), 546

inspiratory neuron Somatic motor neurons controlling the inspiratory muscles (Ch 18), 581–582, 582*f*

inspiratory reserve volume (IRV) The volume of air that can be inhaled in addition to a normal inspiration (Ch 17), 544, 545*f*

insulin Pancreatic hormone that decreases plasma glucose concentration (Ch 7), 153, 159*f*, 198, 199, 200*f*, 203*f*, 610*f*, 665*f*, 666*t*

insulin-like growth factors (IGFs), 200*f*, 212*f*, 213*f*, 741*f*, 742

insulin secretion, 159–160, 159*f*, 208*f*, 363*f*, 684*f*, 795

integral protein Proteins tightly bound to the membrane, and the only way they can be removed is by disrupting the membrane structure with detergents or other harsh methods that destroy the membrane's integrity (Ch 3), 62, 138*f*

integrating center The control center that evaluates incoming signal and decides on an appropriate response (Ch 1), 13, 185–186, 208*f*, 285*f*–286*f*, 425*f*, 467*f*, 664

integrin Membrane-spanning proteins that link the cytoskeleton to extracellular matrix proteins (Ch 3, 6, 8, 16), 74, 74*t*, 75*f*, 176, 232, 526

integrin receptor, 170, 170*f*, 176

integumentary system, 3, 4*f*, 77*f*

intercalated cell (I cell) Cell of the collecting duct that transports H⁺ and bicarbonate (Ch 20, 24), 646, 647*f*, 764

intercalated disk Specialized cell junctions in cardiac muscle that contain gap junctions (Ch 12, 14), 378*f*, 412, 445*f*, 447

intercostal muscles Muscles associated with the rib cage; used for breathing, 536

intercostal nerves, 581

interferon Cytokines secreted by lymphocytes (Ch 24), 168, 764*t*, 765

interferon-alpha (IFN-α), 763, 774*f*

interferon-beta (IFN-β), 763

interferon-gamma (IFN-γ), 763, 772

interleukin (IL) Cytokines released by one type of white blood cell to act on another (Ch 16, 24), 168, 516, 516*t*, 764*t*, 765, 772

intermediary metabolism, 103

intermediate filament Cytoplasmic protein fiber made of myosin, keratin, neurofilament, and other proteins (Ch 3), 66*f*, 68, 68*t*, 75*f*, 406*f*

intermembrane space Region between the two outer membranes of a mitochondrion (Ch 3), 67*f*, 70

internal environment The extracellular fluid that surrounds the cells of the body (Ch 1), 10, 11*f*, 13, 183

International Commission on Radiological Protection, 125

Internet resources, 688, 730, 742, 749, 781, 806

interneuron A neuron that is completely contained within the central nervous system (Ch 8, 9, 13, 20, 21), 227*t*, 229, 230*f*, 234*f*, 285*f*, 298*f*, 419*f*, 431*f*, 645*f*, 665*f*

internodal pathway Conduction pathway from the SA node to the AV node (Ch 14), 454, 456*f*

interstitial cells of Cajal (ICC) Modified smooth muscle cells of the digestive tract that appear to initiate slow waves (Ch 21), 661

interstitial fluid Extracellular fluid that surrounds the cells and lies between the cells and the plasma (Ch 3, 5, 17, 18, 19, 20, 24), 60*f*, 61, 124*f*, 152*f*, 539*f*, 603*f*, 607*f*, 646*f*, 758*f*

interventional methods of contraception, 825

intestinal phase, 672–687, 684*f*

intestine, 182*t*, 362*f*, 404*f*, 521*f*, 675*f*–676*f*, 695*f*, 787*f* calcium balance, 747*f*–748*f*

digestive hormones, 665*f*, 666*t*, 679*f*, 683–684

epithelial transport, 150*f*–151*f*

vitamins and minerals, absorption of, 682–683

intima, 479, 503*f*

intracellular compartments, 65–72

intracellular defense, 772–775

intracellular fluid (ICF) Fluid within the cells (Ch 1, 3), 10, 11*f*, 13, 61, 124*f*, 143*f*, 145*f*, 155*f*, 410*f*

intracellular nucleic acid protein complexes (antinuclear antibodies), 778*t*

intracellular signal molecules, 169*f*, 172*f*, 174*f*

intracellular transport, 69

intrafusal fiber Modified muscle fibers of the muscle spindle that lack myofibrils in their central portions (Ch 13), 421, 422*f*, 424*f*

intrapleural pressure Pressure within the pleural fluid (Ch 17), 549–550

pneumothorax, 550

during respiratory cycle, 550

subatmospheric, 549–550

during ventilation, 549–550

intravenous (IV) injection, 439, 621

intravenous solutions, 132*t*

intrinsic enteric neurons, 670*f*

intrinsic factor Protein secreted by gastric parietal cells that is required for vitamin B_{12} absorption in the intestine (Ch 21), 682

intrinsic pathway Coagulation reaction that begins with collagen exposure and uses proteins already present in plasma (Ch 16), 527, 528*f*, 530*t*

inulin clearance, 609, 610*f*, 611

in utero, 806–807

in vitro activation (IVA), 830

in vitro Experiments performed "in glass" (Ch 2), 46

in vitro fertilization Fertilization of an egg outside the body (Ch 18, 24, 26), 575, 761, 815, 827

in vitro motility assay, 388

in vivo, 815

iodine, *207f*, *736*, 737*f*, 738, 738*f*, 740*f*

ion An atom with a net positive or negative charge due to gain or loss of one or more electrons (Ch 2, 16, 20, 21), 33, 36*f*, 154*f*, 182*t*, 281*f*, 392*f*, 513*f*, 603*f*

ion channel, 139, 174*f*, 331, 332*f*, 334*f*, 453*f*, 631*f*

ion concentration gradient, 238, 245

ionic bond A bond between ions attracted to each other by opposite charge (Ch 2), 38*f*, 39

ionotropic receptor Neurotransmitter receptor that alters ion channel function (Ch 8), 254

ion permeability (P_{ion}), 158, 238, 244*f*, 248*f*

IP_3-receptor-channel Ca^{2+} channels in smooth muscle sarcoplasmic reticulum that open in response to IP_3 (Ch 12), 407, 410*f*

ipsilateral On the same side as (Ch 10), 335

iris, 340, 341*f*, 344*f*

iron, 520, 521*f*, 522, 683, 683*f*

iron transport, 514*t*

iron-deficiency anemia, 520, 523

irreversible antagonists, 49

irreversible inhibitor, 49*t*

irreversible reaction If a chemical reaction proceeds in one direction but not the other (Ch 4), 98

irritable bowel syndrome, 663

irritant receptors Stimulated by inhaled particles or noxious gases in the airway mucosa (Ch 18), 584

ischemia Lack of adequate blood flow and oxygen to a tissue (Ch 10, 14), 322, 439

islet cells, 674, 676*f*

isoform Related forms of a molecule (Ch 2, 6), 49, 180, 181*f*

isometric contraction A contraction that creates force without movement (Ch 12, 14), 398–400, 399*f*–400*f*, 469*f*

isoproterenol, 368*t*

isosmotic, 127

isotonic contraction A contraction that creates force and moves a load (Ch 12), 398–400, 399*f*–400*f*

isotonic solution A solution that results in no net water movement when a cell is placed in it (Ch 5), 127–128, 128*t*

isotope Atoms of the same element that have different numbers of neutrons (Ch 2), 36*f*

isovolumic ventricular contraction Phase of the cardiac cycle when the ventricles are contracting but all valves are closed and the volume of blood in them is not changing (Ch 14), 462, 463*f*

isovolumic ventricular relaxation Phase of the cardiac cycle when the ventricles are relaxing but the volume of blood in them is not changing (Ch 14, 15), 462, 463*f*, 464, 482*f*

isozymes Related forms of a single enzyme (Ch 4), 99, 115

itch (pruritus), 310*t*, 321–322

ivabradine, 454

J

JAK kinase. *See* Janus family tyrosine kinase, Janus family tyrosine kinase, 176

jaundice A yellow tint to the skin and sclera due to excessive levels of bilirubin (Ch 16), 522

jejunum The middle section of the small intestine (Ch 21), 658

JG cells (granular), 602

joint, 424

joint receptor Sensory receptors that send information about the relative positioning of bones linked by flexible joints (Ch 13), 421

juxtaglomerular (JG) apparatus Region where the distal tubule of the nephron passes

between afferent and efferent arterioles (Ch 19), 593*f*, 594, 601*f*

juxtamedullary nephrons, 591

K

kallikrein, 530*t*

keratin Insoluble protein prevalent in hair and nails (Ch 2, 3), 41, 68, 75*f*, 79

ketoacidosis A state of acidosis that results from excessive ketone production (Ch 20, 22), 642, 648

ketoacids, 642, 642*f*

kidney, 323*f*, 437*t*, 438*f*, 492*f*, 521*f*, **591–612**, 592*f*

ammonia and phosphate buffers, use of, 645

calcium balance, 745*f*, 747*f*

epithelial transport, 150*f*–151*f*

excretion, 609, 749*t*

filtration, 596–602

function, 590–594, 596

reabsorption, 594, 604–607

secretion, 594, 608

structure of the, 592*f*

tubular elements, 594

urinary system, 591, 592*f*

water balance, 620*f*, 621–622

kilocalorie (kcal or Calorie) Amount of energy needed to raise the temperature of 1 liter of water by 1 °C (Ch 4, 22), 102, 697

kinase An enzyme that adds a phosphate group to the substrate (Ch 4, 23), 102, 741*f*

kinesin A motor protein (Ch 3), 69

kinetic energy The energy of motion (Ch 4), 95, 95*f*

kinin, 490, 764*t*, 765

kinocilium, 337

kiss-and-run pathway Secretion in which the secretory vesicle fuses transiently with the membrane, then pulls away (Ch 8), 258

knockout mouse A mouse whose DNA has been altered to remove ("knock out") a specific gene (Ch 23), 730

Korotkoff sound, 484, 485*f*

Krakauer, Jon, 565

Krebs, Hans A., 105

Krebs cycle, 105

kwashiorkor, 501*f*

kyphosis, 749

L

labeled line coding The 1:1 association of a sensory receptor with a sensation (Ch 10), 315

labia majora Outer lips of the vulva (Ch 26), 805*f*, 816

labia minora Small inner lips of the vulva (Ch 26), 805*f*, 816

labioscrotal swelling, 802, 805*f*

labor and delivery, 830–831, 832*f*

lacrimal apparatus Tear ducts and glands (Ch 10), 340, 340*f*

lactase Enzyme that breaks down the milk sugar lactose (Ch 21), 680, 680*f* lactate, 393*f*, 787*f*

lactate dehydrogenase (LDH), 99

lactation Milk production by the mammary gland (Ch 23, 26), 749*t*, 830–833, 833*f*

lacteal A fingerlike projection of the lymph system that extends into the villi of the intestine (Ch 21), 678, 679*f*

lactic acidosis, 642, 648

lactic acid The end product of anaerobic glycolysis (Ch 20), 642*f*

lactose intolerance, 680

lactose Milk sugar (Ch 2, 21), 31*f*, 680, 680*f*

lamina propria, 326*f*, 657*f*, 658

laminin Insoluble protein fiber in extracellular matrix (Ch 3), 72, 75*f*, 232

Langerhans cell Alternate name for dendritic cell (Ch 24), 760*f*, 761, 763

language, 303

LaPlace, law of Pressure of a fluid sphere equals 2 times the surface tension of the fluid divided by the radius of the sphere (Ch 17), 551, 552*f*

large intestine The terminal portion of the intestine (Ch 21), 656*f*, 658, 660*f*, 662*f*, 667*f*, 685–686, 685*f*

larynx The "voice box" that contains vocal cords (Ch 8, 17, 21, 23), 537, 669*f*, 737*f*

latch state, 407

latent period Delay between the muscle action potential and beginning of muscle tension that represents the time required for Ca^{2+} release and binding to troponin (Ch 12), 389, 391*f*

lateral geniculate body/nucleus Nucleus in the thalamus where optic fibers synapse with neurons going to the visual cortex (Ch 9, 10), 277, 341, 342*f*

lateral inhibition Process in which sensory neurons close to a stimulus are inhibited to intensify the perception of the stimulus (Ch 10), 315, 316*f*

lateral ventricles of the brain, 277, 281*f*, 290*f*

law of conservation of electrical charge The body is electrically neutral (Ch 5), 153

law of conservation of energy The total amount of energy in the universe is constant. Also called the first law of thermodynamics (Ch 4), 95

law of mass action For a reaction at equilibrium, the ratio of substrates to products is always the same (Ch 2, 18, 20), 47–48, 47*f*, 578*f*, 645*f*

law of mass balance If the amount of a substance in the body remains constant, any gain must be offset by an equal loss (Ch 1, 22), 11–12, 11*f*, 697

L-dopa Dopamine precursor that can cross the blood-brain barrier (Ch 9, 14), 283, 431

leads, ECG 457

leak channel Ion channels that spend most of their time in (Ch 5), 139

leaky epithelium, 76

learned reflex, 418, 418*t*

learning, 286*f*, 290*f*, 295*f*, 299–300

Le Châtelier's principle, 48. *See also* law of mass action

left atrium Chamber of the heart that receives blood from the lungs (Ch 14, 15), 438*f*, 444*f*, 445*f*, 478*f*

left ventricle Chamber of the heart that pumps blood to the systemic circulation (Ch 14, 15), 438*f*, 444*f*, 445*f*, 478*f*, 483*f*, 486*f*, 495*f*

length constant, 240

lengthening (eccentric) contraction, 399

length-force relationship, 469*f*

length-tension relationship, 396*f*, 468, 469*f*

lens Portion of the eye that focuses light upon the retina (Ch 10), 340–341, 343–345

leptin Protein hormone from adipocytes that acts as a satiety factor (Ch 22, 26), 694, 695*f*, 834

let-down reflex Neuroendocrine reflex that triggers oxytocin release and ejection of milk from the mammary gland (Ch 26), 831

leukemia, 517

leukocyte White blood cells that defend the body against foreign invaders (Ch 16, 24), 514, 517*f*, 759–761, 760*f*, 764*t*

leukopoiesis, 516

leukotriene Eicosanoid signal molecule; plays a role in the etiology of asthma (Ch 17), 179, 179*f*, 559

levels of organization, 2–3, 3*f*, 60*f*

levers, 400–402

Leydig cell Testicular cells that secrete testosterone (Ch 26), 803, 806*f*, 813*f*, 814, 814*f*

libido Sex drive (Ch 26), 815

ligament Connective tissue that connects one bone to another (3, 10), 82, 344*f*

ligand-gated ion channel Synonym: chemically gated ion channel (Ch 6), 170

ligand The molecule that binds to a protein (Ch 2), 46, 47*f*, 52*f*, 139*f*, 149*f*, 169, 180–181, 180*f*, 186*f*, 328*f*

ligation reaction, 102

light, 342–343, 348

transduction, 348*f*. *See also* phototransduction

light chain (myosin) Small protein chains that make up part of the smooth muscle myosin head (Ch 12, 24), 382, 769*f*

limbic system Region of the cerebrum that acts as the link between higher cognitive functions and more primitive emotional responses (Ch 9, 10, 18), 286*f*, 289, 289*t*, 290*f*, 298*f*, 326*f*

line graph, 20*f*, 21*f*

lipase Enzyme that digests lipids (Ch 4, 21, 22), 678, 679*f*

secretion of, 684*f*

lipid Synonym: fats (Ch 2, 3, 6, 8, 16, 21, 22), 29, 30*f*, 37*f*, 62, 62*t*, 64–65, 179–180, 257, 679*f*, 787*f*

lipid-anchored protein, 64

lipid bilayer, 63*f*

lipid-derived second messengers, 174–176

lipid raft, 64, 64*f*

lipolysis Lipid breakdown (Ch 22, 23), 733*f*

lipophilic molecules Molecules that can diffuse through cell membranes (Ch 5), 136

signal molecules, 169, 170*f*

lipophobic molecules Molecules that cannot diffuse through the phospholipid bilayer (Ch 5, 6), 136, 169, 174

lipoprotein Protein combined with a lipid (Ch 2), 29, 37*f*

liposome Spherical structures with an exterior composed of a phospholipid bilayer, leaving a hollow center with an aqueous core (Ch 3), 62, 63*f*, 64

lipostatic theory Control of food intake is based on a set point for body weight that is set by adipocytes (Ch 22), 694

lipoxygenase Enzyme that converts arachidonic acid to leukotrienes (Ch 6), 179

liver, 200*f*, 212*f*, 213*f*, 323*f*, 362*f*, 492*f*, 514*t*, 516*t*, 521*f*, 530*t*, 633*f*, 656*f*, 660*f*, 679*f*

bile secretion, 675, 677*f*, 678

digestion, 667*f*

gluconeogenesis, 733*f*

load A weight or force that opposes contraction of a muscle (Ch 12, 13), 383, 401*f*, 423*f*

load-velocity relationship, 402*f*

local communication, 166, 167*f*, 168

local control Homeostatic control that takes place strictly at the tissue or cell by using paracrine or autocrine signals (Ch 1), 13, 14, 14*f*

local current flow A wave of electrical current that spreads throughout the cytoplasm (Ch 8), 240, 249*f*

locomotor pattern generator, 428*t*

long-distance communication, 166, 167*f*, 168

longitudinal studies These are designed to be carried out for a long period of time (Ch 1), 23

long-loop negative feedback Negative feedback from a peripheral endocrine gland hormone to the hypothalamus and anterior pituitary (Ch 3, 7), 213, 733*f*

long QT syndrome (LQTS) (Ch 8, 14), 239, 460

long reflex A GI reflex that is integrated in the CNS rather than in the enteric nervous system (Ch 21), 664–665, 665*f*, 670*f*

long-term depression (LTD), 267

long-term memory, 301, 301*f*, 301*t*

long-term potentiation (LTP) Physical changes in a synapse that allow the response of the postsynaptic cell to a constant stimulus to be enhanced (Ch 8), 267–268

loop of Henle Portion of the renal tubule that creates dilute urine and sets up the conditions needed to make concentrated urine (Ch 19, 20), 593*f*, 594, 595*f*, 596, 598*f*, 601*f*, 627–629, 628*f*

loose connective tissue Elastic connective tissues that underlie skin and provide support for small glands (Ch 3), 81*f*, 82, 83*f*

low blood pressure, 626*f*, 630–632

low-density lipoprotein (LDL) The "bad" protein carrier for plasma cholesterol (Ch 5, 15), 148, 503*f*

low-density lipoprotein-cholesterol (LDL-C), 502

Lower, Richard, 209

lower esophageal sphincter, 667*f*

lower respiratory tract, 536
   airway branching, 540*f*

L-type calcium channel, 389, 450*f*

lumbar puncture, 282

lumen The cavity of a hollow tube or organ (Ch 2, 3, 16), 59–61, 73*f*, 150*f*–151*f*, 182*t*, 278*f*, 628*f*, 660*f*, 683*f*

lung(s) Organs where gases are exchanged with the blood (Ch 11, 14, 15, 17, 18, 20), 362*f*–363*f*, 437*t*, 438*f*, 642
   CO$_2$ removal, 579–580
   compliance, 550–551
   elastance, 550–551
   gas exchange, 565–571
   pleural sacs, 536, 540*f*
   restrictive disease, 551
   volumes, 544

lung capacity Sums of two or more lung volumes (Ch 17), 544

luteal cell, 820*f*, 823

luteal phase The portion of the menstrual cycle following ovulation, when the corpus luteum produces estrogen and progesterone (Ch 26), 817, 821*f*–822*f*, 823

luteinization Conversion of the follicle to a corpus luteum (Ch 26), 823

luteinizing hormone (LH) Anterior pituitary hormone that acts on the gonads to influence hormone production (Ch 7, 26), 200*f*, 209, 212*f*, 213, 809, 809*f*, 814*f*, 822*f*

luteinizing hormone releasing hormone, 809n

luteinizing hormone surge, 823

lymph The fluid within the lymphatic system that moves from the tissues to the venous side of the systemic circulation (Ch 15), 499

lymphatic system, 3, 500*f*, 679*f*, 758*f*

lymph capillary Small vessels of the lymph system (Ch 15), 499

lymph node Collections of immune cells that monitor the lymph for pathogens (Ch 15, 24), 757–759, 758*f*

lymphocyte A white blood cell responsible primarily for the acquired immune response (Ch 7, 16, 24), 200*f*, 513*f*, 514, 515*f*, 517*f*, 519*f*, 759, 759*f*, 760*f*, 763, 766, 767*f*. *See also* B lymphocyte; T lymphocyte

lymphocyte clone, 766, 767*f*

lymphoid tissues The tissues of the immune system, including the thymus gland, bone marrow, lymph nodes, and spleen (Ch 11, 16, 24), 363*f*, 514*t*, 757–759, 759*f*. *See also* lymphatic system

lysosomal storage disease, 71

lysosome, 59, 66*f*, 71, 73*f*, 149*f*, 232*f*, 762*f*

lysozyme Antibacterial enzyme found in respiratory tract secretions and tears (Ch 21, 24), 668, 763, 764*t*

M

Macbeth (Shakespeare, William), 824

macrophage Tissue phagocytes that develop from monocytes (Ch 16, 24), 514, 519*f*, 759*f*, 760, 760*f*, 761, 762*f*, 774*f*

macula densa Specialized cells in the distal tubule wall that monitor fluid flow through the tubule (Ch 19, 20), 601*f*, 602, 633*f*, 640*f*

maculae Sensory receptors of the utricle and saccule of the vestibular apparatus (Ch 10), 338*f*, 339, 341, 341*f*, 346

magnocellular ganglion cell (M cell), 353, 687

major essential element, 29, 36*f*

major histocompatibility complex (MHC) Family of membrane protein complexes that participate in the immune response; play a role in foreign tissue rejection (Ch 24), 764*t*, 771, 771*f*, 776–778

male accessory glands The prostate gland, bulbourethral gland, and seminal vesicles (Ch 26), 815

male embryonic development, 803

male gametogenesis, 807

male reproduction, 810–815

male sex act, 824

malignant hyperthermia A genetically linked condition in which body temperature becomes abnormally elevated (Ch 22, 25), 724, 787, 789, 797

malleus The first bone of the middle ear that sits against the tympanic membrane (Ch 10), 329, 330*f*, 332*f*

maltose A disaccharide composed of two glucose molecules (Ch 2, 5, 21), 31*f*, 146*f*, 680, 680*f*

mammary gland The exocrine glands of the breast (Ch 7, 26), 210*f*–211*f*, 818*f*, 831–833, 833*f*

mapping, 4, 6–7*f*

Maris, Roger, 730

mass action, law of For a reaction at equilibrium, the ratio of substrates to products is always the same (Ch 2), 47*f*

mass balance, law of If the amount of a substance in the body remains constant, any gain must be offset by an equal loss (Ch 1), 11–12, 11*f*, 486*f*, 571, 572*f*, 590, 650, 660, 660*f*

mass flow Mass flow equals concentration times volume flow (Ch 1), 12, 571, 572*f*

mass movement Wave of contraction in the large intestine that triggers defecation (Ch 21), 667*f*, 686

mast cell A tissue cell that secretes histamine (Ch 15, 16, 17, 24), 488*t*, 514, 553, 760, 761*f*, 764*t*, 769*f*, 773*f*, 776*f*

mastication Chewing (Ch 21), 668

matrix (extracellular matrix) Extracellular material synthesized and secreted by cells (Ch 3), 67*f*, 72, 81*f*, 82

matrix metalloproteinases (MMPs) Enzymes that dissolve extracellular matrix (Ch 3, 26), 74, 823

maximal rate of oxygen consumption (V$_{O_2max}$), 789

maximum voluntary ventilation The maximum speed and depth at which a person can voluntarily breathe (Ch 17), 554

McArdle's disease, 403

M cell 1. Magnocellular ganglion cells in the retina that transmit information about movement, location, and depth perception; 2. Modified intestinal epithelial cell overlying a Peyer's patch; absorbs intestinal contents by transcytosis (Ch 10, 21), 353, 687

mean arterial pressure (MAP) Average blood pressure in the arteries, estimated as diastolic pressure plus one-third of the pulse pressure (Ch 14, 15, 19, 25), 443, 483*f*, 484, 486*f*, 491*f*, 495*f*, 599*f*, 793*f*

mean corpuscular hemoglobin Average amount of hemoglobin in one red blood cell (Ch 18), 574

mean red cell volume (MCV), 520

mechanical digestion, 668

mechanically gated channel A channel that opens in response to mechanical stimuli such as pressure and heat (Ch 5, 8), 138*f*, 139, 239, 240*t*

mechanical work Used for movement. At the cellular level, movement includes organelles moving around in a cell, cells changing shape, and cilia and flagella beating. Most mechanical work is mediated by motor proteins that make up certain intracellular fibers and filaments of the cytoskeleton (Ch 4), 94, 696

mechanisms This referes to physiological processes or "how" of a system (Ch 1), 5, 144*f*

mechanistic approach The ability to explain the mechanisms that underlie physiological events (Ch 1), 5

mechanoreceptor A sensory receptor that responds to mechanical energy such as touch or pressure (Ch 6, 10, 26), 186*f*, 312, 312*t*, 825*f*, 833*f*

mediated transport Movement across a membrane with the aid of a protein transporter (Ch 5, 19), 137, 605

medulla oblongata Portion of the brain stem that contains centers for breathing, cardiovascular control, swallowing, and other unconscious or involuntary functions (Ch 9, 13, 18, 20, 21), 277, 278*f*, 319*f*, 360*f*, 362*f*, 371*t*, 431*f*, 467*f*, 494*f*, 624*f*, 645*f*
   central nervous system, 277, 278*f*, 286*f*, 287, 287*f*, 289*t*
megakaryocyte Parent cell of platelets, found in bone marrow (Ch 16), 514, 515*f*, 516*t*, 519*f*, 524*f*, 527*t*, 530*t*
meiosis Cell division that produces haploid gametes (Ch 26), 807, 808*f*
Meissner's corpuscle, 321*f*
Meissner's plexus, 659. *See also* submucosal plexus
melanins, 346, 735
melanocortin receptors (MCRs), 735
melanocytes Pigment-containing cells that skin color in humans and coat color in rodents (Ch 3), 86*f*, 735
melanocyte-stimulating hormone (MSH), 220, 733*f*
melanoma, 86*f*
melanopsin retinal ganglion cell (mRGCs), 353
melatonin Hormone secreted by the pineal gland (Ch 7, 26), 200*f*, 206, 220, 221*f*, 810
membrane (1) the phospholipid bilayer that surrounds cells and divides the cytoplasm into compartments, or (2) a thin sheet of connective tissue (Ch 3, 5, 8, 20, 22, 24), 60*f*, 61–62, 62*t*, 132–133, 157*f*, 650. *See also* cell membrane; tissue membrane
   simple diffusion across a, 135*t*
membrane attack complex Proteins produced by immune cells that create membrane pores in the target cells (Ch 24), 764*t*, 765, 765*f*, 773*f*
membrane lipid, 62
membrane potential difference The electrical potential created by living cells due to uneven distribution of ions between the intracellular and extracellular fluids (Ch 5, 8, 10, 12, 14, 21), 154*f*, 155*f*, 156, 157*f*, 158, 236–238, 243*f*–244*f*, 246*f*, 253*f*, 261*f*, 317*f*, 334*f*, 350*f*, 391*f*, 454*t*, 662*f*
   in cardiac and skeletal muscle, 454*t*
   cardiac autorhythmic cells, 453*f*
   cardiac muscle fiber, 452*f*
   terminology, 157*f*
membrane processes, 159–160
membrane protein, 62–64, 133*f*, 137, 138*f*, 151, 171, 762*f*, 764*t*
membrane receptor, 138*f*, 149*f*, 170*f*, 177*f*, 204*f*–205*f*, 410*f*, 747*f*, 762*f*
membrane recycling Process in which cell membrane is withdrawn by endocytosis and stored as vesicles in the cytoplasm until needed. At that time, the vesicle is reinserted into the membrane by exocytosis (Ch 5), 148, 149*f*

membrane-spanning proteins Membrane proteins that are tightly bound into the phospholipid bilayer (Ch 3), 64
membrane surface area, 137*f*
membrane transporter, 46, 140*f*, 177*f*, 603*f*
membranous organelle, 66*f*
memory, 286*f*, 290*f*, 295*f*, 299–302
memory cell Lymphocytes responsible for creating stronger and more rapid immune response following second exposure to an antigen (Ch 24), 760*f*, 766, 767*f*, 769*f*
menarche A woman's first menstrual period (Ch 26), 834
Ménière's disease An inner ear condition characterized by dizziness, ringing, and nausea (Ch 10), 310, 315, 337, 339, 343, 348, 353
meninges Three layers of membrane that lie between the spinal cord and vertebrae, or brain and skull (Ch 9), 279*f*, 280
menopause The time when a woman's menstrual cycles cease (Ch 26), 807, 834
menses, 817, 821*f*
menstrual cycle The cyclic production of eggs and cyclic preparation of the uterus for pregnancy in females (Ch 26), 817, 819*f*–820*f*, 822*f*, 823
menstruation Cyclic sloughing of the endometrial lining (Ch 26), 817
Merkel receptor Skin receptor for steady pressure (Ch 10), 321*f*
mesangial cell Contractile cells in the renal corpuscle that alter glomerular blood flow (Ch 19), 597, 598*f*
mesencephalon, 288
mesonephric ducts. *See* Wolffian duct
mesentery Peritoneal membrane that hold the intestines in place (Ch 21), 656*f*, 659
messenger RNA (mRNA) RNA produced in the nucleus from a DNA template; travels to the cytoplasm to direct the synthesis of new proteins (Ch 4, 7), 112, 112*f*, 114, 115–116, 115*f*, 203*f*
meta-analysis Statistical technique that combines data from multiple studies to look for trends (Ch 1), 23
metabolic acidosis State of acidosis resulting from overproduction of metabolic acids (Ch 20, 22), 645*f*, 648–649
metabolic alkalosis State of alkalosis usually resulting from loss of gastric acid through vomiting or excessive ingestion of alkaline antacids (Ch 20), 649
metabolic energy, 697
metabolic pathway regulation, 103–104, 103*f*
metabolism All the chemical reactions in the body (Ch 4, 25), 102–118, 138*f*, 370*f*, 392*f*, 395*t*, 621*f*, 642*f*
   and exercise, 787–790
   hormones, 213, 787, 789
   muscle, 787*f*–788*f*
   vasodilation, 488*t*

metabolite, 12, 521*f*, 677*f*
metabotropic receptor Neurotransmitter receptor that acts through a second messenger system (Ch 8), 254
metarteriole, 480, 480*f*
metastasis Spread of cancer or another disease throughout the body (Ch 3), 61, 72
methylation, 117
mGluR6, 352
MHC class I molecules, 771, 774*f*
MHC class II molecules, 771, 771*f*, 774*f*, 776*f*
micelle Small droplet of phospholipid, arranged micelle so that the interior is filled with hydrophobic fatty acid tails (Ch 3), 62, 63*f*, 678, 679*f*
micro- (μ) (prefix), 43*f*
microbiome, 655, 686
microcirculation The arterioles, capillaries and venules (Ch 15), 480, 480*f*
microcytic red blood cells, 523
microfilament Thinnest protein fibers in the cytoplasm, made of the protein actin (Ch 3), 66*f*, 68, 68*t*
microglia Macrophages in the CNS (Ch 8), 234*f*, 235
microtubule Tubular fibers made of the protein tubulin (Ch 3, 8, 26), 66*f*, 68, 68*t*, 69*f*, 233*f*, 813*f*
microtubule-organizing center, 68
microvilli Finger-like extensions of the cell membrane that increase the surface area for absorption of material (Ch 3), 66*f*, 68, 79
micturition Urination (Ch 19), 591, 612*f*, 613–614
midbrain, 277, 278*f*, 287*f*, 288, 295*f*, 336*f*, 342*f*, 371*t*, 431*f*
middle ear, 330*f*, 332*f*
migrating motor complex Contractions that move food remnants and bacteria from the stomach to the large intestine between meals (Ch 21), 662*f*, 663, 666*t*
milli- (m) (prefix), 43*f*
milliequivalent (mEq), 42*f*
millimeters of mercury (mm Hg), 126, 439
millimole (mmol), 43*f*
mind-body therapeutics, 781
mineralocorticoid, 732. *See also* aldosterone
mineralocorticoid receptors (MRs), 732
minerals, 667*f*, 682–683
Minkowski, Oscar, 198
minor essential elements, 29, 36*f*
mitochondrial DNA, 70
mitochondrial matrix Central region of a mitochondrion (Ch 3), 70
mitochondria Organelles that generate ATP through oxidative phosphorylation. Singular: mitochondrion (Ch 3, 8, 10, 12, 14, 22, 26), 65, 67*f*, 70, 232*f*, 348*f*, 380*f*–381*f*, 394*f*, 395*t*, 449*f*, 813*f*
   of varicosity, 366*t*
mitosis Cell division that results in two identical diploid daughter cells (Ch 3, 26), 85, 808*f*

mitral valve, 447, 448f, 478f

mixed nerve A nerve that carries both sensory and motor information (Ch 8, 11), 231, 359

M line, 381f, 383, 384f–385f, 387f, 390f

modality The nature of a stimulus (Ch 10), 314

modifiable risk factor, 504

modulation of signal pathway, 180–182

modulator, 49, 49t

molarity (M) Solution concentration expressed as moles of solute per liter of solution (Ch 2), 43f, 126

mole (mol) 6.02 * 10²³ atoms, ions, or molecules of a substance. Avogadro's number of particles (Ch 2), 42f, 126

molecular chaperones Protein that helps a newly-made protein fold into shape (Ch 4), 116

molecular complementarity The physical compatibility of a ligand and its binding site (Ch 2), 47

molecular interaction, 8, 28, 44f, 193

molecular mass The mass of one molecule, expressed in atomic mass units or daltons (Ch 2), 42f

molecular shape, 41, 44f

molecular size, 134, 137f

molecular weight, 134

molecule Two or more atoms linked together by sharing electrons (Ch 1, 2, 4, 5, 20), 36f, 38f, 140, 150, 151, 650

covalent bonds, 33, 39

polar and nonpolar, 38f

monoamine oxidase (MAO) The enzyme that breaks down norepinephrine (Ch 8, 11), 258, 365f, 366, 366t

monocarboxylate transporter, 736

monocular zone The portion of the visual field where vision is two-dimensional (Ch 10), 352f, 353

monocyte Blood cell that is the parent cell of tissue macrophages (Ch 16, 24), 513f, 514, 515f, 517f, 519f, 527t, 760, 760f, 761

monoglyceride, 30f, 37f, 679f

monoiodotyrosine (MIT), 736, 737f

mononuclear phagocyte system Monocytes in the blood and tissue macrophages (Ch 24), 760f, 761

monosaccharide Simple sugars such as glucose (Ch 2, 21), 31f, 37f, 680–681, 680f, 682

monosynaptic reflex Reflex in which there is one synapse between neurons (Ch 13), 418, 418t, 419f

monosynaptic stretch reflex, 424, 425f

monounsaturated fatty acid Fatty acid with one double bond (Ch 2), 30f

monovalent cation channel, 238

mood Relatively stable feelings related to sense of well-being, 295f, 299

motilin GI hormone that stimulates the migrating motor complex (Ch 21), 666t, 667

motility, 363f, 660f, 661–663, 667f, 669f, 686

motivation, 298–299

motor area, 291, 292f, 428t

motor association area, 292f, 294, 430f

motor cortex, 303f, 430f–431f

motor end plate, 371, 372f, 390f–391f, 778t

motor neuron, 229, 396–397, 398f, 411t, 423f, 612f. See also efferent neuron

motor protein Proteins that create movement (Ch 3), 69–70, 70f, 177f

motor unit Group of skeletal muscle fibers and the somatic motor neuron that controls them (Ch 12), 396–398 398f

mouth, 276f, 288t, 668–669

movement, 69, 178f, 286f, 292f, 370f, 426–432, 696f

execution, 429f, 430f

feedback of information during, 431f

initiation, 429f

rhythmic, 427–428, 427t

types of, 427t

mRNA processing, 114–115, 115f, 205f

mRNA translation, 73f, 115–116, 117f

mucin Glycoproteins of mucus (Ch 21), 661

mucociliary escalator The layer of mucus lining the respiratory tract that is moved upward by cilia so that it can be swallowed (Ch 17), 537

mucopolysaccharide, 739, 739f

mucosa The inner lining of the intestinal tract (Ch 21), 656f–657f, 658–659

mucosal membrane, 150

mucous cell Cell that secretes mucus. Synonym: goblet cell (Ch 21), 661, 672

mucous membrane, 61, 757f

mucous secretion, 80. See also mucus

mucus A thick, sticky exocrine secretion containing glycoproteins and proteoglycans (Ch 3, 17, 21, 24, 26), 80, 661, 667f, 674, 757f, 813f

Müllerian duct Embryonic structures that develop into female reproductive structures (Ch 26), 802–803, 804f, 806f

Müllerian inhibiting substance, 803

multiple sclerosis, 249, 778t

multiple system atrophy, 369

multipolar neuron, 229, 230f

multipotent Undifferentiated cells in a tissue that can divide and develop into the specialized cells of that tissue (Ch 3), 87

multi-unit smooth muscle Smooth muscle in which cells are not linked electrically and each muscle fiber is controlled individually (Ch 12), 403

murmurs, heart, 464

muscarine An agonist for cholinergic muscarinic receptors (Ch 8, 11), 254, 255t, 368t

muscarinic receptor One subtype of cholinergic receptor (Ch 8, 11), 254, 255t, 364f, 366t, 368f, 370f, 370t, 467f, 494f

muscle A collection of muscle cells (Ch 3, 9, 10, 12, 13, 14, 15, 17, 22, 23, 25), 84, 284f–285f, 288t, 340f, 377–416, 733f

cardiac, 378, 378f, 405f, 411t, 412, 418t, 443–454, 449f, 452f

contraction, 379f, 383–389, 396–400, 397f, 404f, 405–409, 422f–423f, 424f

disorders, 402

elastic elements in, 400

exercising, 793f

extrafusal fibers, 421

extraocular, 396

fast-twitch glycolytic muscle, 393

fast-twitch oxidative glycolytic, 393

fatigue, 392

fiber types, 395t

intrafusal fibers, 421

smooth, 378, 378f, 403–411, 404f–406f, 411t, 418t, 479f–480f, 488t

spindles, 421, 422f, 423f

stretch, 421–424, 422f, 423f

tension, 409f, 421

tone, 421

types, comparison of the three, 411t

muscle blood flow and exercise, 792

muscle cell. See muscle fiber

muscle contraction Process by which a muscle creates force (Ch 11, 14, 21, 22, 23), 370f, 450–451, 662f, 745f

muscle cramp, 402

muscle fiber A muscle cell (Ch 12), 379

anatomy, 382

muscle memory, 302

muscle metabolism, 787

muscles of ventilation, 645f

muscle spindle Muscle receptors that send information about muscle length (Ch 13), 421, 422f, 423f

muscle tension, 383

muscle tone The basal state of muscle contraction or tension that results from tonic activity of the muscle spindles (Ch 9, 15), 287f, 479

muscular dystrophy, 148

muscularis externa, 657f, 658

muscularis mucosae, 657f, 658, 685f

musculoskeletal system, 3, 4f, 77f, 210f, 292f

mutant mouse model, 730

myasthenia gravis, 256, 549, 778t

myelin Concentric layers of cell membrane that wrap around and insulate axons (Ch 8, 24), 233, 234f–235f, 778t

myelinated axon, 249, 251f

myelin sheath, 230f, 235f, 252f

myenteric plexus Nerve network of the enteric nervous system that lies between the muscle layers (Ch 21), 657f, 659

myocardial action potential, 451–454

myocardial autorhythmic cell, 453

myocardial cell, 455f, 635f, 639, 639t

myocardial contractile cell, 451

myocardial infarction A region of damaged myocardium caused by lack of blood flow (Ch 16), 322, 472, 503, 529

myocardium Cardiac muscle (Ch 14), 446

myoepithelium, 833f

myofibril Bundles of contractile and elastic proteins responsible for muscle contraction (Ch 12), 380f–381f, 382–383, 422f

myogenic autoregulation, 488–489

myogenic contraction, 409

myoglobin Oxygen-binding pigment in muscle that transfers oxygen between cell membrane and mitochondria (Ch 12), 394, 394f, 795

myoglobinuria, 795

myometrium Smooth muscle layer of the uterus (Ch 26), 816, 819f

myophosphorylase deficiency, 403

myopia Near-sightedness (Ch 10), 345, 345f

myosin Forms thick filaments of the myofibril that convert chemical bond energy of ATP into motion (Ch 3), 69, 379, 380f, 381f, 382, 384f, 385–388, 385f, 406, 408f, 409f, 411t, 450f

myosin ATPase, 382, 386, 392f, 405, 787f

myosin-binding site, 383, 386f

myosin light chain, 405, 407, 408f, 409f

myosin light chain kinase (MLCK) Enzyme that phosphorylates light protein chains of myosin in smooth muscle (Ch 12), 407, 410f

myosin light chain phosphatase Enzyme that dephosphorylates light protein chains of myosin in smooth muscle (Ch 12), 407, 408f

myotatic unit Collection of synergistic and antagonistic muscles that act in a coordinated fashion to control a single joint (Ch 13), 424

myxedema, 739, 739f

N

Na⁺-Ca²⁺ exchanger (NCX), 450

NAD (nicotinamide adenine dinucleotide), 34f, 37f

NADH, 98, 98f, 105, 106, 106f, 10, 110f

Na⁺-dicarboxylate cotransporter (NaDC), 607f

Na⁺-glucose secondary active transporter (SGLT), 144, 151f

Na⁺-H⁺ exchanger (NHE), 604, 632, 645, 683

Na⁺-I⁻ symporter (NIS), *See* sodium-iodide symporter 736

naïve immune cell, 757

naïve lymphocyte A lymphocyte that has not yet been exposed to its specific antigen (Ch 24), 766

Na⁺-NH₄⁺ antiport, 645

nano- (n) (prefix), 43f

National Science Foundation (NSF), 8t

natriuresis Sodium (Na⁺) loss in the urine (Ch 20), 634

natriuretic peptide, 488t, 631f, 634–635, 635f, 639, 639t

natural killer cell (NK cell) A type of lymphocyte that apparently attacks certain tumor and virus-infected cells (Ch 24), 760f, 763, 764t, 769f

NCX antiporter, Synonym: Na-Ca exchanger, 450f

near point of accommodation, 343

near-sightedness, 345f

nebulin Inelastic giant protein that aligns filaments of the sarcomere (Ch 12), 380f–381f, 382, 383, 385f

necrosis Cell death due to toxins, physical damage, or lack of oxygen. The dying cell releases enzymes that may damage neighboring cells (Ch 3), 84

negative feedback A homeostatic feedback loop designed to keep the system at or near a setpoint (Ch 1, 7, 13, 23), 16, 16f, 206, 214f, 418, 423f, 733f, 747f

nephrin, 597

nephron Microscopic tubule that is the functional unit of the kidney (Ch 19, 20), 591, 592f, 597f, 601f, 610f, 635

   fluid volume modification, 594, 596

   functional unit of the kidney, 591–594

   structure of the, 594

Nernst equation The equation that determines the equilibrium potential for a single ion based on the ion concentrations inside and outside the cell (Ch 5, 8), 156, 236

nerve A collection of axons running between the central nervous system and the peripheral target cells (Ch 12, 17), 380f

nerve-cell adhesion molecule (NCAM) Membrane proteins in nerve cells that aid cell growth, (Ch 3), 72, 74t, 232

nerve cell body, 227t, 229

nerve cord, 276, 276f

nerve fiber Synonym: axon (Ch 8, 10), 227t, 333f, 338f

nerve net, 275, 276f

nerves, 231, 330f

nervous system Network of billions or trillions of nerve cells linked together in a highly organized manner to form the rapid control system of the body (Ch 1, 6, 7, 8, 14, 23, 26), 3) 3, 4f, 167f, 738f, 780f

   cell-cell communication, 253–260

   cells of the, 229–236

   development, 71

   efferent divisions of the, *359*

   electrical signals in neurons, 236–253

   evolution of, 275, 276f

   neural information transfer, integration of, 260–268

   neural reflex pathways classification, 418

   organization of the, 227–229, 228f

network properties, 226

neural crest cells Embryonic cells that form the peripheral nervous system (Ch 9), 277, 278f

neural network, 275

neural pathway, 263–266, 298–299

neural plate, 277

neural reflex pathway, 418–420

neural stem cell, 234f, 235, 277

neural tissue, 84, 210f, 283

neural tube Embryonic cells that develop into the CNS (Ch 9), 277, 278f

neurocrine Any molecule secreted by a nerve cell (Ch 6, 8), 255t, 263f, 168, 231

neuroeffector junction Synapse between an autonomic neuron and its target muscle or gland (Ch 11), 364

neuroendocrine control Signal molecule secreted by a neuron into the blood, where it functions as a hormone (Ch 6), 190f

neuro-endocrine-immune interaction, 779–781

neuroepithelium, 236

neurofilament Intermediate filament of neurons (Ch 3), 68

neuroglia, 227t. *See also* glial cell

neurohormone A hormone that is produced and secreted by a neuron (Ch 6, 7, 11, 15), 167f, 168, 187f, 190f, 198, 207–209, 210f–211f, 368f, 488t

neurohypophysis, 209

neuroimmunomodulation The ability of the nervous system to influence immune function (Ch 24), 779–780, 780f

neuromodulator Chemicals that alter the response of a neuron more slowly than neurotransmitters (Ch 6, 8), 168, 263f

neuromuscular junction The synapse of a somatic motor neuron and a skeletal muscle fiber (Ch 11, 12), 371, 372f, 383, 385f, 390f–391f, 393f

neuron A nerve cell, capable of generating and transmitting electrical signals (Ch 3, 6, 7, 8, 9, 10), 84, 168, 227, 260, 262f, 282f, 347f, 398f, 405f, 491f, 695f, 745. *See also* neuron; *specific type*

   anatomy, 229

   cellular and network properties, 253–254, 262f, 263–266

   classification of, 229

   electrical signals in, 236–253

   ion permeability of the, 238–239

   stem cell repair of damaged, 235–236

neuropathic pain, 322

neuropeptides, 665, 779–780, 780f

neuropeptide Y Brain neurotransmitter that stimulates food intake (Ch 22), 694, 695f

neurotoxin Chemical that adversely alters neuronal function (Ch 8), 251

neurotransmitter A chemical signal released by a neuron that influences the neuron's target cell (Ch 6, 8, 9, 11, 14), 167f, 168, 190f, 227, 234f, 317f, 328f, 331, 365f, 405f

   molecules, 254–257

   receptors, 254, 368t, 371t

   release, 258, 259f, 332f, 393f

   synthesis, 257–258

   termination of action, 258, 259f

neurotrophic factor Chemicals secreted by Schwann cells that keep damaged neurons alive (Ch 8), 232, 234*f*

neutron Subatomic particle with no charge and mass of 1 amu (Ch 2), 36*f*

neutropenia Low number of neutrophils (Ch 16), 517

neutrophils White blood cells that ingest pathogens and release cytokines (Ch 16, 24), 513*f*, 514, 515f, 517f, 527t, 760*f*, 761, 762*f*

newborn respiratory distress syndrome (NRDS), 552

Nicolson, G. L., 62

nicotinamide adenine dinucleotide (NAD) Molecule that captures and transfers energy with high-energy electrons (Ch 2), 34*f*, 37*f*

nicotine An agonist of cholinergic nicotinic receptors and a chemical found in tobacco (Ch 2, 8, 11), 48, 254, 255*t*, 359, 361, 363, 367, 368*t*, 371, 373

nicotinic cholinergic receptors (nAChR), 364*f*, 366*t*, 368*t*, 370*f*, 371*t*, 372*f*, 254, 255*t*, 366*t*, 364, 372*f*

nicotinic receptor A type of acetylcholine receptor that also responds to nicotine (Ch 11).

Niedergerke, Rolf, 385

NIS. *See* sodium-iodide transporter

nitric oxide (NO) A short-acting paracrine that relaxes smooth muscle; also acts as a neurotransmitter and neuromodulator (Ch 6, 8, 12, 16), 178, 255*t*, 256, 411, 488*t*, 526*f*

nitric oxide synthase (NOS) Enzyme that synthesizes NO from arginine and oxygen (Ch 6), 178

nitrogenous base, 34*f*

NKCC symporter (Na-K-2 Cl symporter), 627

NMDA receptor Glutamate receptor that opens only when the cell is depolarized (Ch 8), 255*t*, 256, 267*f*

Nobel, Alfred, 179

nocebo effect Adverse effect that occurs because the patient expects it to (Ch 1), 22

nociception, 317, 318, 319*f*, 320

nociceptor A sensory receptor associated with pain (Ch 10, 13), 320–324, 321*f*, 323*f*, 426*f*

nocturnal enuresis Involuntary urination, especially bedwetting at night (Ch 20), 625

nodes of Ranvier Unmyelinated regions on myelinated axons (Ch 8), 233, 249

nonadrenergic, noncholinergic neuron A neuron that secretes a neurotransmitter other than ACh or norepinephrine (Ch 11, 20, 21), 364, 635, 664

nonassociative learning, 300

noncovalent bonds, 38*f*, 39

noncovalent interactions, 40–46

nongenomic effect of steroid hormones Actions of steroid hormones that do not require altered gene activity (Ch 7), 204

nonmembranous organelle Cell organelle that is not surrounded by a phospholipid membrane (Ch 3), 65

nonpenetrating solute A solute that cannot cross the cell membrane (Ch 5), 128

nonpolar molecule A molecule whose electrons are distributed so evenly that there are no regions of partial positive or negative charge (Ch 2), 38*f*, 39

nonspecific immune response, 756

nonsteroidal anti-inflammatory drugs (NSAIDs), 734

noradrenergic Adjective related to norepinephrine (noradrenaline) 255, 294, 295*f*

norepinephrine (NE) Primary neurotransmitter of the sympathetic division of the nervous system (Ch 6, 8, 9, 11, 14, 15, 22), 180, 294, 295*f*, 364*f*–365*f*, 491*f*
    efferent division, 366*t*, 370*f*, 371*t*
    as a vasoconstrictor, 488*t*

noxious stimuli, 321*f*

NPC1L1 cholesterol transporter, 678

NPR receptors, 635*f*

NRDS. *See* newborn respiratory distress syndrome

NTS. *See* nucleus tractus solitarius

nuclear envelope, 67*f*, 72

nuclear pores/nuclear pore complexes Protein complexes in the nuclear envelope with a central pore (Ch 3), 72

nuclear receptor, 169, 738*f*

nucleic acid, 34*f*–35*f*, 667*f*, 682

nucleolus, 67*f*, 72

nucleotide, 34*f*–35*f*, 37*f*, 173*f*

nucleus (cell) Central body of a cell that contains DNA (Ch 3, 8, 16), 66*f*, 230*f*, 233, 235*f*, 361, 405, 449*f*, 496*f*, 524*f*, 762*f*
    compartmentation, 65, 66*f*–67*f*, 71, 73*f*

nucleus tractus solitarius (NTS), 581

nutrient, 281*f*, 437*t*, 489*f*, 523*t*, 674*f*, 677*f*, 813*f*

O

O₂. *See* oxygen

OAT. *See* organic anion transporter

obese *(ob)* gene, 694

obesity Excess body fat (Ch 15), 501

obstructive lung diseases, 557

obstructive sleep apnea, 557, 581

occipital lobe, 286*f*, 289, 292*f*, 342*f*

occluding junction A cell-cell junction that prevents movement of material between cells (Ch 3), 75*f*

occludin proteins Proteins in tight junctions, 74

oculomotor nerve, 288*t*

odorant, 312

Ohm's Law, 239–240

oleic acid, 30*f*, 37*f*

olfaction Pertaining to the sense of smell (Ch 10), 324–325

olfactory bulb Part of the brain that receives input from primary olfactory neurons (Ch 9, 10), 295*f*, 313, 314*f*, 325, 326*f*

olfactory cortex, 292*f*, 293, 314*f*, 325, 326*f*

olfactory epithelium, 325, 326*f*

olfactory (Bowman's) glands, 325

olfactory information, 313

olfactory nerve (cranial nerve I), 288*t*, 325

olfactory receptor protein, 325, 326*f*

olfactory sensory neuron, 325, 326*f*

olfactory system, 326*f*

olfactory tract, 325, 326*f*

oligodendrocyte CNS glial cell that forms myelin around several axons (Ch 8), 233, 234*f*

-ome (suffix), 3

-omics (suffix), 3

oncotic pressure, 498

oocyte Developing female germ cells that have begun meiosis (Ch 26), 806–807, 819*f*, 820*f*

oögonia Germ cells of the ovary (Ch 26), 807, 808*f*

Oparin, Aleksander, 29

open system, 134, 136

opioid peptide, 256

opioid receptor, 323

opsin Visual pigment forms from rhodopsin when light strikes it; opsin initiates a signal transduction cascade (Ch 10), 348*f*, 349, 350*f*

opsonins Proteins that coat pathogens to make them targets for immune cells (Ch 24), 763–764, 764*t*, 769*f*, 773*f*

optic chiasm Portion of the brain where some fibers from each eye cross to opposite sides of the brain (Ch 10), 341, 342*f*, 354*f*

optic disk Region of the retina where the optic nerve and blood vessels exit the eye (Ch 10), 341, 343*f*, 346

optic nerve, 340, 341, 341*f*

optic tract Neurons leading from the eyes to the visual cortex (Ch 9, 10), 287*f*, 342*f*, 352*f*

optics The physical relationship between light and lenses (Ch 9, 10), 288*t*, 343

Orai-1, 407

oral cavity, 288*t*, 656*f*, 667*f*

oral contraceptive, 826, 826*t*

orbit Bony cavity that protects the eye (Ch 10), 340, 340*f*

orexin, 694, 695*f*

organ Group of tissues that carries out related functions (Ch 3, 12), 87, 404, 411*t*

organelle Assorted intracellular structures that each take on one or more of the cell's functions (Ch 3), 65, 66*f*, 70–71, 73*f*

organic anion transporter (OAT), 146, 607–608, 607*f*, 613

organic molecules Molecules that contain carbon (Ch 2, 16), 29, 513*f*

organ of Corti Portion of the cochlea that contains the hair cells (Ch 10), 332, 333*f*

organotherapy, 198

organ system Group of organs that work together to perform a certain task (Ch 1), 3, 4*f*

orgasm A series of involuntary muscular contractions during the sex act, accompanied by sensations of intense pleasure (Ch 26), 824

origin of a muscle The end of the muscle attached closest to the trunk or to the more stationary bone (Ch 1, 12), 379

oropharynx receptor An unidentified receptor that monitors oral water intake (Ch 20), 636

orphan receptor One that has no known ligand (Ch 6), 179

orthostatic hypotension Low blood pressure that occurs when going from the supine position to standing up (Ch 15), 495

osmolarity Concentration expressed in osmoles per liter, 127, 310*t*, 594, 596, 625, 631*f*, 633*f*, 638, 640*f*, 665*f*
blood volume and, 625
integrated control of, 637–641
responses triggered by changes in, 639*t*
and tonicity, 125–132, 127*t*–128*t*, 130*f*–131*f*
vasopressin release, 626*f*

osmole, 126

osmometer An instrument for measuring osmolarity of a fluid (Ch 5), 127

osmoreceptor Sensory receptor that monitors extracellular fluid osmolarity (Ch 6, 11, 20, 21), 186*f*, 360, 625, 639

osmosis The movement of water across a membrane in response to a solute concentration gradient (Ch 5, 19), 125–132, 126*f*, 603*f*

osmotic diarrhea, 686

osmotic diuresis Water loss in the urine due to unreabsorbed solute in the tubule lumen (Ch 20, 22), 623

osmotic equilibrium, 123, 125–126

osmotic pressure The pressure that exactly opposes a given concentration gradient (Ch 5), 126, 126*f*

osteoblast Cells that produce bone (Ch 3, 23), 81*f*, 743, 744*f*

osteocalcin, 743

osteoclast Large, mobile, multinucleate cell that is responsible for bone resorption (Ch 23), 743, 747*f*

osteocyte A less active form of osteoblast (Ch 3, 23), 81*f*, 743

osteoid, 743

osteopenia, 749

osteoprotegerin (OPG), 746

otolith Small calcium carbonate crystals whose movement activates hair cells for equilibrium (Ch 10), 338*f*, 339

otolith membrane Gelatinous mass within which otoliths are embedded (Ch 10), 339

otolith organ The utricle and saccule of the vestibular apparatus that sense linear acceleration and head position (Ch 10), 337, 339

ouabain Cardiac glycoside that specifically inhibits the Na$^+$- K$^+$-ATPase (Ch 5, 14), 152, 470

output (efferent) signal, 185, 185*f*, 230*f*, 429*f*

oval window Membrane between the middle ear and cochlea (Ch 10), 329, 330*f*, 332

ovarian cycle The monthly cycle of egg development in the ovary (Ch 26), 817, 821*f*

ovary The female gonad (Ch 7, 26), 200*f*, 205*f*, 210*f*, 801, 804*f*, 808*f*, 809*f*, 817, 818*f*–820*f*, 828*f*

overshoot, 244

oviduct, 803. *See also* Fallopian tube

ovulation Release of a mature egg from its follicle in the ovary (Ch 26), 807, 808*f*, 817, 820*f*–822*f*, 822–823, 828*f*

ovum The female gamete. Synonym: egg. Plural: ova (Ch 26), 801, 820, 822*f*

oxidation-reduction reaction Involves the transfer of electrons or protons (H$^+$) between chemicals (Ch 4), 101, 102*t*

oxidative stress, 235

oxygen, 29, 36*f*, 38*f*, 310*t*, 437*t*
consumption and exercise intensity, 789
hemoglobin binds to, 572–573, 574*f*
solubility, 570

oxygen consumption The disappearance of oxygen during oxidative phosphorylation, when oxygen combines with hydrogen (Ch 22, 23, 25), 571, 697–698, 738*f*, 789–790, 790*f*

oxygen deficit Oxygen needed for metabolism to replace muscle ATP and phosphocreatine reserves (Ch 25), 789

oxygen transport, 512, 518, 520, 522, 571

oxyhemoglobin Hemoglobin bound to oxygen (Ch 18), 572

oxyhemoglobin saturation curves, 575

oxytocin Posterior pituitary hormone that causes uterine and breast smooth muscle contraction (Ch 1, 7, 9, 26), 17, 200*f*, 209, 211*f*, 289*t*, 832*f*

P

pacemaker neurons, 581

pacemaker of the heart The fastest depolarizing cell, usually in the SA node (Ch 13, 14), 431, 447, 455

pacemaker potential Cyclic depolarizations of smooth and cardiac muscle that always reach threshold (Ch 12, 14), 409*f*, 410, 453

Pacinian corpuscle Sensory receptors of skin that sense vibration (Ch 10), 311*f*, 320, 321*f*

Paget's disease, 748

pain The brain's perception of irritating or damaging stimuli (Ch 9, 10), 278*f*, 295*f*, 310*t*, 320, 321*t*, 322–324, 323*f*

pain receptor, 764*t*. *See also* nociceptor

pancreas Digestive organ that secretes enzymes, bicarbonate, and hormones, 362*f*, 367*t*, 674
beta cells of, 159, 666*t* 709
hormones, 200*f*, 709
acini, 676*f*

bicarbonate secretion, 675, 684*f*
enzyme secretion, 675, 666*t*, 684*f*
islet cells, 676*f*, 709

pancreatic polypeptide, 200*f*

Papanicolaou, George, 59

papillary muscle Small muscle in the interior of the ventricles to which the chordae tendineae attach (Ch 14), 445*f*

pap test, 59, 70, 85*f*, 87–88

para- (prefix), 359

paracellular pathway, 74, 150, 497, 603*f*, 604, 658, 676*f*, 683*f*

paracrine signal A chemical secreted by a cell that acts on cells in the immediate vicinity (Ch 6, 8), 167*f*, 168, 179–180, 254

parallel processing One function is carried out by more than one region of the CNS (Ch 9), 300

paralysis, 284

Paramecium, 275

paramesonephric ducts. *See* Müllerian ducts

parameter One of the variables in a system (Ch 6, 10), 183, 316

parasitic infection, 523*t*

parasympathetic branch Division of the autonomic nervous system that is responsible for day-to-day activities (Ch 8, 11), 228*f*, 229, 359, 360*f*, 362*f*, 363–364, 369, 371*t*

parathormone, 746. *See also* parathyroid hormone

parathyroid gland, 182*t*, 200*f*, 746, 747*f*–748*f*

parathyroid hormone (parathormone, or PTH) Hormone from the parathyroid glands that increases plasma Ca^{2+} concentration (Ch 7, 23), 200*f*, 206, 208*f*, 745*f*, 746–747, 747*f*–748*f*

parietal cells Cells of the stomach that secrete hydrochloric acid (Ch 21), 666*t*, 670, 671*f*, 673*f*

parietal lobe, 286*f*, 289, 292*f*

parietal pleura, 549

Parkinson's disease, 283, 430

parotid salivary gland secretion, 288*t*, 656*f*

partial pressure The pressure of a single gas (Ch 18), 565, 566*f*, 542, 543*f*

parturition The birth process (Ch 26), 827–833, 832*f*

parvocellular ganglion cell (P cell), 353

passive immunity, 766

passive transport Movement across a membrane that does not depend on an outside source of energy (Ch 5), 133

patellar tendon (knee jerk) reflex, 425*f*

pathogen Any substance capable of causing disease (Ch 24), 754–756, 757*f*–758*f*, 762*f*, 765*f*, 780*f*

pathogen-associated molecular patterns (PAMPs), 763

pathophysiology The study of body functions in a disease state (Ch 1), 10

pathways Network of interconnected chemical reactions formed by the enzymatic reactions of metabolism (Ch 9, 10, 22), 285

pupillary reflexes, 342*f*
vision, 342*f*
pattern recognition, 5
pattern recognition receptor (PRR), 763
Pavlov, 300
P cell Parvocellular ganglion cells of the retina that transmit information about color, form, and texture (Ch 10, 20), 353, 630, 631*f. See also* principal cell
P_{co_2}, 644*f*
pelvic cavity, 60*f*
pelvis, 59, 438*f*, 818*f*
pendrin transporter (SLC26A4), 736, 737*f*
penetrating solute A solute that freely crosses the cell membrane (Ch 5), 128
Penfield, W., 320*f*
penicillin, 608
Penicillium, 51, 608
penis, 362*f*, 805*f*, 810, 812*f*, 825*f*
pentose, 31*f*
pepsinogen The inactive form of pepsin (Ch 21), 661, 667*f*, 671*f*, 673*f*
pepsin release, 673*f*, 684*f*
PepT1, 682
peptic ulcer, 660
peptidase Enzyme that breaks up peptides into smaller peptides or amino acids (Ch 2), 46
peptide A chain of 2-9 nine amino acid (Ch 2, 8, 20, 21, 22, 23, 24), 32*f*, 232*f*, 256, 635*f*, 666*t*, 670*f*, 671*f*, 673*f*, 681–682, 695, 695*f*, 733*f*, 741*f*, 749*t*, 770*f*
absorption, 681*f*, 682
peptide bond Bond formed between carboxyl group of one amino acid and amino group of another amino acid (Ch 2), 32*f*, 115, 681*f*
peptide hormone Any hormone made of amino acids, including peptides, proteins, and glycoproteins (Ch 7, 24, 26), 200*f*, 202*t*, 203*f*, 204*f*, 780*f*, 809*f*
cellular mechanism of action of, 202–204
membrane receptors and signal transduction for, 204*f*
synthesis, storage, and release, 201–202
synthesis and processing, 203*f*
transport in the blood and half-life of, 202
percent saturation of hemoglobin, 574
percent solution Solution concentration expressed as parts of solute per 100 parts of total solution (Ch 2), 43*f*
perception, 286*f*, 292*f*, 294, 294*f*
perceptual threshold The level of stimulus intensity necessary for awareness (Ch 10), 313
perforin Pore-forming protein secreted by immune cells (Ch 24), 764*t*, 765, 772, 774*f*
perfusion Blood flow to lung tissues (Ch 15), 478, 556, 558
pericardium The connective tissue sac that encloses the heart, 61, 446
pericytes Cells that form a mesh-like outer layer between the capillary endothelium and the interstitial fluid (Ch 15), 480–481

perilymph Fluid within the vestibular and tympanic ducts of the cochlea (Ch 10), 332, 332*f*–333*f*
periodic paralysis, 379, 389, 394, 403, 412
periodic table of the elements, 36*f*
peripheral chemoreceptors Chemoreceptors not found in the CNS (Ch 18), 582–583
peripheral fatigue, 392, 393*f*
peripheral nervous system (PNS) All neurons that lie completely or partially outside the central nervous system (Ch 8, 9), 227, 228*f*, 234*f*–235*f*, 278*f*
efferent division of the, 228*f*, 358
sensory division of the, 228*f*, 310
peripheral protein Proteins attached to membrane-spanning proteins or to the polar regions of membrane phospholipids (Ch 3), 64, 138*f*
peripheral receptor Sensory receptors that are not located in or close to the brain (Ch 6), 186*f*
peripheral resistance Resistance to blood flow created primarily by the arterioles (Ch 14, 15, 20, 25), 484, 486*f*, 494*f*, 640*f*, 793*f*
peristalsis Waves of contraction that move along the gastrointestinal tract (Ch 21), 663
peritoneal membrane Lines the inside of abdominal cavity (Ch 3, 21), 61, 659
peritoneum A membrane that lines the abdomen (Ch 3, 19, 21), 59, 591, 592*f*, 659
peritubular capillaries, 593*f*, 594, 595*f*–596*f*, 597*f*, 607*f*, 610*f*, 646*f*
peritubular capillary pressure, 606
permissiveness One hormone cannot exert its effects fully unless a second hormone is present (Ch 7), 216
pernicious anemia, 683
peroxisome Storage vesicles that contain enzymes to degrade long-chain fatty acids and potentially toxic foreign molecules (Ch 3), 66*f*, 71
personality, 304
PET scan, 303*f*
Peyer's patch Bump of lymphoid tissue visible in the mucosa of the GI tract (Ch 21), 657*f*, 659
pH A measure of the concentration of H⁺; pH = –log [H⁺] (Ch 2, 4, 10, 17, 18, 20, 24), 41, 45*f*, 49*t*, 52*f*, 310*t*, 641
denature proteins, changes can, 41, 641–642
disturbances, 643–644
homeostasis, 642
phagocyte Immune cell that ingests material by phagocytosis (Ch 5, 16, 24), 147–148, 514, 760*f*, 761, 762*f*, 763, 764*t*, 772, 773*f*
phagocytosis The process by which a cell engulfs a particle into a vesicle by using the cytoskeleton to push the membrane around the particle (Ch 5, 24), 133*f*, 147, 147*f*, 761, 762*f*, 769*f*

phagosome The vesicle formed around ingested material during phagocytosis; site of digestion (Ch 5, 24), 147, 762*f*, 763
phantom limb pain, 315
pharmacological doses, 734
pharmacomechanical coupling Contraction that occurs in smooth muscle as a result of a ligand binding; not accompanied by a change in membrane potential (Ch 12), 407, 409*f*
pharynx, 330*f*, 446*f*, 537, 747*f*
phasic contraction, 661
phasic receptor Rapidly adapting receptors that are attuned to changing conditions (Ch 10), 316, 318*f*
phasic smooth muscle, 403, 404*f*
pheromone External hormones secreted to influence others of the same species, 198–199
phosphatase activity, 409*f*
phosphate, 748–749
buffer, 645
group, 33*t*, 34*f*
homeostasis, 748–749
ion, 643, 645*f*
phosphocreatine Muscle molecule that stores energy in high-energy phosphate bonds (Ch 12, 25), 391, 392*f*, 788, 788*f*
phosphodiesterase-5 (PDE-5), 825
phospholamban Regulatory protein in contractile myocardium that alters Ca^{2+}-ATPase activity in the sarcoplasmic reticulum (Ch 14), 470, 470*f*
phospholipase A2 (PLA2) Enzyme that converts membrane phospholipids to arachidonic acid (Ch 6), 179, 179*f*
phospholipase C Enzyme that converts a membrane phospholipid into two different second messenger molecules, DAG and IP_3 (Ch 6, 11), 174, 175*f*, 367*t*
phospholipid bilayers, 60*f*
phospholipid Diglycerides with phosphate attached to the single carbon that lacks a fatty acid (Ch 2, 3, 16), 30*f*, 37*f*, 44*f*, 63*f*, 527*t*, 528*f*, 677*f*, 679*f*
phosphorylation Addition of a phosphate group to a molecule (Ch 2, 12), 33, 49, 405–407, 408*f*–409*f*, 470*f*
photoreceptors Sensory receptors in the eye that respond primarily to light energy (Ch 10), 312, 312*t*, 341*f*, 346, 347*f*, 348–350
phototransduction Conversion of light energy to action potentials (Ch 10), 349–350
phrenic nerve, 231, 581
physiological psychology, 636
physiology The study of the normal functioning of a living organism and its component parts, including all its chemical and physical processes (Ch 1), 2, 3*f*, 298*f*
control systems and homeostasis, 13–18
function versus mechanism, 4–5
homeostasis, 9–18
levels of organization, 2–3, 3*f*

physiological organ systems, 3–4, 3f, 4f
science of, 18–23
themes in, 5, 8–9
Physiome Project, 3
phytoestrogens, 810
P$_i$ (inorganic phosphate), 387f, 393f, 787f
pia mater Inner membrane of the meninges
(Ch 9), 279f, 280, 281f
pico- (p) (prefix), 43f
pigment epithelium, 346, 347f
piloerection Hair standing on end (Ch 13), 420
pineal gland, 200f, 221f, 286f, 288f
pinna The outer ear (Ch 10), 329, 330f
pinocytosis, 148
pitch Physiological interpretation of sound wave
frequency (Ch 10), 330, 331f
pituitary gland Endocrine and neuroendocrine
gland that lies beneath the hypothalamus
(Ch 7), 209, 210f, 219f–220f, 286f, 288f,
733f, 822f
pituitary tumor, 734
placebo An inactive substance used in medical
treatment (Ch 1), 22
placebo effect Response to a placebo treatment
(Ch 1), 22
placenta, 200f, 823, 829f, 832f
plaque Deposition of lipid in arterial walls,
accompanied by smooth muscle
proliferation, scar tissue formation, and
calcification (Ch 3, 15), 74, 503
plasma The fluid portion of the blood (Ch 1, 2,
5, 7, 8, 16, 17, 18, 20, 22, 23), 3), 60f, 61,
81f, 82, 83f, 124f, 203f, 253f, 512–514,
513f, 642f, 745–746, 749t
proteins, 514t
plasma cell Type of lymphocyte that secretes
antibodies (Ch 24), 760, 760f, 765, 767f,
768, 769f, 773f, 776f
plasma membrane The cell membrane that
serves as both a gateway and a barrier for
substances moving into and out of the cell
(Ch 3), 61. *See also* cell membrane
plasma volume, 517f
plasmin Enzyme that breaks down fibrin.
Synonym: fibrinolysin (Ch 16, 26), 525,
525f, 528, 529f, 530t, 823
plasminogen, 528, 529f, 530t
plasticity Ability of adult stem cells to develop
into multiple cell types (Ch 3, 9),
87, 275
plateau phase intermediate phase of the human
sexual response (Ch 26), 824
platelet Cell fragments that participate in
coagulation. Synonym: thrombocyte (Ch 15,
16), 488t, 503f, 513f, 514, 515f, 517–518,
517f, 519f, 523–524, 524f, 525f, 530t
production of, 517–518
platelet plug, 525, 527t, 528, 530t
platelet-activating factor (PAF), 526, 527t
platelet activation, 526, 527t
platelet adhesion Platelets stick to exposed
collagen in wall of damaged blood vessel
(Ch 16), 525, 526f

platelet aggregation Activated platelets stick to
each other (Ch 16), 488t, 527t, 530t
platelet-derived growth factor (PDGF), 527t
platelet-rich plasma (PRP) therapy, 524
pleural fluid, 536
pleura The membranes that line the chest cavity
and cover the outer surface of the lungs
and form the pleural sacs (Ch 17), 59, 60f,
61, 536
plicae Large folds of the intestinal wall (Ch 21),
656f, 658
pluripotent cell A stem cell that can develop
into many but not all cell types, 85, (Ch 3)
pluripotent hematopoietic stem cell, 514, 515f
pneumonia Bacterial or viral lung infection
(Ch 20, 21), 647, 687
pneumothorax Air in the intrapleural space
(Ch 17), 550
P$_{O_2}$, 790f
podocin, 597
podocyte Specialized epithelial cells in Bowman's
capsule that surround each capillary and
form filtration slits (Ch 19), 597, 598f
Poiseuille's law, 440, 491f, 552
polar body, first and second Unused
chromosomes that are discarded from the
egg as it undergoes meiosis (Ch 26), 807
polarity, of cell Cells restrict certain
membrane proteins to particular
regions, thereby creating cells with
different functions in different areas
(Ch 3, 5), 64, 150, 150f
polar molecule Molecules that develop regions
of partial positive and negative charge
when one or more atoms in the molecule
have a strong attraction for electrons
(Ch 2), 38f, 39
polio, 549
polycystic ovary syndrome (PCOS), 817
polycythemia Elevated hematocrit (Ch 16), 523
polymeric proteins, 117–118
polymers Large molecules made up of repeating
units (Ch 2), 29
polymorphonuclear leukocyte, 761. *See also*
neutrophil
polypeptide A chain of 10-100 amino acids
(Ch 2, 21), 32f, 667f
polyploid cell, 523
polyribosomes Free ribosomes forming groups
of 10 to 20 (Ch 3), 68
polysaccharides Complex carbohydrates
composed of glucose polymers; used for
energy storage and structure (Ch 2), 31f, 37f
polysaccharide capsule, 762f
polyspermy Fertilization of an egg by more than
one sperm (Ch 26), 827
polyunsaturated fatty acid A fatty acid with
more than one double bond (Ch 2), 30f
pons Region of the brain stem that contains
centers for respiration and serves as a
relay station (Ch 9, 11, 18), 277, 278f,
286f–287f, 287, 295f, 360f, 362f
pontine respiratory groups, 581

population coding The number of sensory
receptors activated encodes the intensity of
a stimulus (Ch 10), 315
porphyrin, 683f
porphyrin ring, 520, 521f
portal system A specialized region of the
circulation consisting of two capillary
beds directly connected by a set of blood
vessels (Ch 7, 14, 19), 210f, 211, 212f,
439, 593f, 833f
positive feedback loop A feedback loop in
which the response reinforces the stimulus,
triggering a vicious cycle of ever-increasing
response (Ch 1, 8, 16, 26), 16, 17, 17f, 245,
247f, 528f, 832f
positron emission tomography (PET), 291, 319
posterior pituitary gland An extension of
the brain that secretes neurosecretory
hormones made in the hypothalamus
(Ch 7, 9, 15, 20, 26), 200f, 209,
210f–211f, 288f, 488t, 626f, 640f,
832f–833f
postexercise oxygen consumption, 790f
postganglionic neuron Autonomic neuron that
has its cell body in the ganglion and sends
its axon to the target tissue (Ch 11), 361,
361f, 365f, 370f
postsynaptic cell The target cell at a synapse
(Ch 8), 232, 253, 259f–260f, 263f, 265f
postsynaptic inhibition, 265f, 266
postsynaptic integration Multiple signals in a
postsynaptic cell combine to create a single
integrated signal (Ch 8), 266
postsynaptic modulation A modulatory
neuron, usually inhibitory, synapses on the
dendrites or cell body of a post synaptic
cell (Ch 8), 266
post-translational modification Alterations to
a protein molecule made after translation
(Ch 4, 7), 116–118, 202
of prohormones, 202
of POMC, 733f
potassium (K$^+$), 33t, 36f, 157
and cell excitability, 253f
balance, 634, 636
channels, 154, 238, 244f, 350f, 451f
potential difference, 158. *See also* membrane
potential difference
potential energy Stored energy that has the
ability to do work, 95, 95f, (Ch 4)
power stroke Movement of the myosin head
that is the basis for muscle contraction
(Ch 12), 385, 386f, 387f, 388, 390f
P-Q segment, 461f
pre-Bötzinger complex, 581
precapillary sphincter Bands of smooth muscle
that can alter blood flow through capillary
beds (Ch 15), 480, 480f
preeclampsia, 490
prefrontal association area, 292f
prefrontal cortex, 293f, 295f
preganglionic neuron Autonomic neuron that
originates in the central nervous system

and terminates in an autonomic ganglion (Ch 11), 361, 361*f*

pregnancy, 749*t*, 825–827, 829–830

pregnancy prevention, 825

preload The degree of myocardial stretch created by venous return (Ch 14), 468

premature ovarian failure. *See* primary ovarian insufficiency (POI)

premature ventricular contractions (PVCs), 460

premotor cortex, 292*f*

preprohormone Inactive molecule composed of one or more copies of a peptide hormone, a signal sequence, and other peptide sequences that may or may not have biological activity (Ch 7), 202, 203*f*

prepuce (foreskin), 810, 812*f*

presbycusis Age-related hearing loss (Ch 10), 336

presbyopia Loss of the accommodation reflex with aging (Ch 10), 345

pressure, 439–443, 441*f*, 482*t*

changes during quiet breathing, 548*f*

subatmospheric, 549*f*

mm Hg, 465*f*

pressure gradient (ΔP), 132, 439, 439*f*, 441*f*, 482*t*, 491*f*, 546, 607*f*

pressure reservoir, veins as 478*f*, 479, 482*f*

pressure-volume curve, 463*f*, 464–466, 544

presynaptic axon terminal, 230*f*, 264*f*–265*f*

presynaptic cell The cell releasing neurotransmitter into a chemical synapse (Ch 8, 10), 232, 253, 259*f*, 327, 328*f*

presynaptic facilitation Modulation of the presynaptic neuron that enhances neurotransmitter release (Ch 8, 13), 266

presynaptic inhibition, 265*f*, 266

primary active transport The energy for transport comes from the high-energy phosphate bond of ATP (Ch 5), 133*f*, 143, 143*t*

primary bronchi The first two airways created by branching of the trachea (Ch 17), 537

primary cilia, 69

primary endocrine pathology, 218

primary follicle An undeveloped oocyte and its outer layer of granulosa cells (Ch 26), 816, 819*f*–821*f*

primary hypercortisolism, 218, 219*f*, 734

primary hypothyroidism, 740

primary immune response The immune response that occurs with first exposure to a pathogen (Ch 24), 767*f*, 768

primary lymphoid tissue, 757

primary motor cortex Regions of the frontal lobe that coordinate skeletal muscle movements (Ch 9, 13), 292*f*, 294, 431*f*

primary oocyte Oocyte that has duplicated its DNA but not undergone a meiotic division (Ch 26), 807, 808*f*, 820*f*

primary ovarian insufficiency (POI), 823

primary sensory neuron The sensory neuron that takes information from the sensory receptor into the spinal cord (Ch 10), 312, 313*f*, 316*f*, 319*f*, 323*f*, 328*f*

primary sex characteristics The internal sexual organs and external genitalia that distinguish each sex (Ch 26), 816

primary spermatocyte Spermatocyte that has duplicated its DNA but not undergone a meiotic division (Ch 26), 807, 808*f*, 813*f*

primary structure, of protein The sequence of amino acids in the peptide chain (Ch 2), 32*f*

primordial follicle Formed by primary oocyte (Ch 26), 816

principal cell (P cell), 630, 631*f*

procarboxypeptidase, 676*f*

procedural memory, 301

processes of a neuron, 227

process map, 4, 7*f*

procolipase, 676*f*

procreation The act of creating a new being (Ch 26), 824–827

products A reaction ends with one or more molecules (Ch 4), 96, 96*t*

proenzyme An inactive enzyme (Ch 4, 21), 661

progenitor cell, 514, 518

progesterone Female sex hormone produced by the corpus luteum (Ch 7, 26), 200*f*, 205*f*, 212*f*, 731*f*, 806*f*, 807, 821*f*, 822*f*

programmed cell death, 85. *See also* apoptosis

prohormone Inactive protein containing one or more copies of a hormone (Ch 7), 202, 203*f*

proinsulin, 203*f*

prokaryotic endosymbiont theory, 70

prolactin A peptide hormone from the anterior pituitary that controls milk production in the breast (Ch 7, 23, 26), 200*f*, 210*f*, 212*f*, 213, 748*f*, 833*f*, 834

prolactin-inhibiting hormone (PIH) Hypothalamic hormone that inhibits prolactin secretion by the anterior pituitary (Ch 26), 831, 833*f*, 834. *See also* dopamine

proliferative phase Phase of the menstrual cycle when the endometrium grows and thickens (Ch 26), 817, 821*f*

promoter region Section of DNA near the starting end of a gene that must be activated to begin transcription (Ch 4), 112

pro-opiomelanocortin (POMC) Anterior pituitary pro-hormone that is processed into ACTH and other active fragments (Ch 7, 23), 202, 733*f*, 735

propeptide, 257

prophospholipase, 676*f*

proprioception Awareness of body position in space and of the relative location of body parts to each other (Ch 10, 13), 310, 310*t*, 427

proprioceptor, 186*f*, 310, 420

propriospinal tract Tracts of white matter that remain within the cord (Ch 9), 284

prospective study, 23

prostacyclin Eicosanoid in membrane of intact endothelial cells that prevents platelets from adhering (Ch 16), 526, 526*f*, 530*t*

prostaglandin Lipid-derived molecules that act as physiological regulators (Ch 2, 6, 26), 30*f*, 179, 179*f*, 530*t*, 813*f*, 832*f*

prostanoid Eicosanoid signal molecules that include prostaglandins and thromboxanes (Ch 6), 179

Prostate Cancer Prevention Trial (PCPT), 811

prostate gland Male accessory organ that contributes enzymes, nutrients, and other secretions to semen (Ch 26), 804*f*, 811, 812*f*

protease Enzymes that break proteins up into smaller peptides (Ch 3, 21, 22), 74, 681

proteasome Cylindrical cytoplasmic enzyme complex that destroys proteins (Ch 4), 116

protective epithelium, 77*f*, 78*f*, 79

protective reflex, 687

lungs, 584

protein A chain of more than one hundred amino acids (Ch 2, 3, 4, 6, 7, 8, 15, 16, 20, 21, 22, 24) 37*f*, 47*f*, 62, 65, 172*f*, 204*f*, 263*f*, 496*f*, 513*f*, 642*f*, 643, 667*f*, 681–682, 684*f*, 768

absorption, 681*f*, 682

activation, 50*f*

activity, 52*f*

binding, 46–47, 48–49

biochemistry of, 32*f*

cell function, 111–112

cell membrane components, 62*f*

digestion, 681*f*

enzymes, 8

fibrous, 32*f*, 41

functional groups, 8

globular, 32*f*, 41

inhibition, 50*f*

integral, 62

modification, 73*f*

molecular interaction, 8

motor, 69–70, 70*f*

posttranslational modification, 116–118

receptor, 8

signal transduction, 182

structure of, 32*f*

transmembrane, 64

protein-bridge tip links, 334*f*

protein C, 529

protein catabolism, 733*f*, 738*f*

protein folding, 116, 182*t*

protein kinase Enzymes that transfer a phosphate group from ATP to a protein (Ch 6, 14), 171, 172*f*, 177*f*

protein kinase A (PKA), 174, 175*f*

protein kinase C (PKC) Associated membrane enzyme that is activated by DAG and Ca²⁺ (Ch 6), 176

protein-mediated transport, 137–147

protein sorting, 116

protein synthesis, 73*f*, 112, 113*f*

proteinuria, 597

proteoglycans Glycoproteins in extracellular matrix (Ch 3), 72

proteolytic activation, 49, 49*t*, 50*f*

proteomics, 2, 3, 220

prothrombin, 528*f*, 530*t*

proton Subatomic particle with one positive charge and mass of 1 amu (Ch 2), 36*f*

proton pump inhibitors (PPIs), 671

proximal tubule The initial segment of the kidney tubule where most reabsorption takes place (Ch 19, 20), 593*f*, 594, 595*f*, 598*f*, 601*f*, 603*f*, 607*f*, 633*f*, 645–646, 646*f*

PR segment From the end of the P wave to the beginning of the QRS complex (Ch 14), 461*f*

pseudohermaphroditism, 801

pseudohypoparathyroidism, 218

pseudounipolar neuron, 229, 230*f*

PTSD (post-traumatic stress disorder), 300

puberty The period in the early teen years when the gonads mature and produce gametes (Ch 26), 801, 834

pudendum, 816. *See also* vulva

pulmonary artery Blood vessel that carries low oxygen blood from the right heart to the lung (Ch 14, 15, 17), 444*f*, 445*f*, 448*f*, 478*f*

pulmonary circulation That portion of the circulation that carries blood to and from the lungs (Ch 14, 17, 18), 438, 536 low-pressure high-flow, 540–541

pulmonary edema Excessive interstitial fluid volume in the lungs (Ch 15, 18), 505, 568*f*, 569

pulmonary fibrosis, 551

pulmonary function tests, 544

pulmonary system. *See* respiratory (pulmonary) system

pulmonary trunk The single artery that receives blood from the right ventricle; splits into left and right pulmonary arteries (Ch 14), 446

pulmonary valve The semilunar valve between the right ventricle and the pulmonary trunk (Ch 14, 15), 447, 478*f*, 445*f*, 448*f*

pulmonary vein Vessel that carries well-oxygenated blood from the lung to the left heart (Ch 14, 15), 438, 438*f*, 448*f*, 478*f*

pulsatile blood flow, 483, 485*f*

pulsatile GnRH release, 810

pulse Pressure wave that is transmitted through the fluid of the cardiovascular system (Ch 15), 483

pulse generator Region of the hypothalamus that coordinates the pulsatile secretion of GnRH (Ch 26), 810

pulse oximeter, 569

pulse pressure The strength of the pulse wave, defined as the systolic pressure minus the diastolic pressure (Ch 15), 483, 483*f*

pupil of the eye, 340, 340*f*, 341*f*

pupillary reflex Constriction of the pupil in response to light (Ch 10), 342, 342*f*

purine, 34*f*, 255*t*, 256

purinergic receptor A receptor that binds to purines, such as AMP or ATP (Ch 8), 256

Purkinje fiber Specialized myocardial cell that rapidly conducts electrical signals to the apex of the heart (Ch 8, 14), 262*f*, 454, 456*f*

pus, 763

P wave Wave of the ECG that represents atrial depolarization (Ch 14), 457, 459*f*

pylorus The region of increased muscle tone separating the stomach and small intestine (Ch 21), 656*f*, 658, 667*f*

pyramid Region of the medulla where neurons from one side of the body cross to the other (Ch 9, 13), 287, 429, 431*f*

pyramidal cell, 294

pyramidal tract Descending pathways for movement that pass through the pyramids (Ch 13), 429

pyrimidine, 34*f*

pyrogen Fever-causing substances (Ch 22, 24), 761, 764*t*, 765

pyruvate, 105, 106, 106*f*, 107*f*, 108*f*, 110, 110*f*, 111*f*, 787*f*

Q

Q wave First wave of ventricular depolarization (Ch 14), 457, 461*f*

QRS complex Wave complex that represents ventricular depolarization and atrial repolarization (Ch 14), 457, 459*f*, 465*f*

QT interval From the beginning of the Q wave to the end of the T wave. Corresponds to ventricular contraction (Ch 14), 459*f*, 460

quadriceps, 425*f*

quaternary structure, of protein Arrangement of a protein with multiple peptide chains (Ch 2), 32*f*

R

Rasmussen, T., 320*f*

RAS pathway (renin-angiotensin system), 631*f*

reabsorption Movement of filtered material from the lumen of the nephron to the blood (Ch 19, 20), 594, 596*f*, 603*f*, 604–607

of H_2O, 621*f*

peritubular capillary pressures, 606–607

reactants A reaction begins with one or more molecules (Ch 4), 96, 96*t*

reaction rate The speed with which a reaction takes place (Ch 4), 51, 96

reactive hyperemia An increase in tissue blood flow following a period of low perfusion (Ch 15), 489, 489*f*

reactive oxygen species (ROS), 235

receiving segment, 663

receptive aphasia Inability to understand spoken or visual information due to damage to Wernicke's area (Ch 9), 303

receptive field The region within which a sensory neuron can sense a stimulus (Ch 10), 312, 313*f*

receptor A cellular protein that binds to a ligand; (2) A cell or group of cells that continually monitor changes in the internal or external environment (Ch 2, 5, 6, 7, 8, 10,11), 46, 139*f*, 217, 259*f*, 263*f*, 370*f*, 405*f*, 419*f*, 425*f*, 741*f*

chemoreception, 327–329

sensory system, 311

signal transduction, 180, 182*t*

types of, 312*t*

receptor adaptation A repeated stimulus loses its ability to stimulate a receptor (Ch 10), 318*f*

receptor cells (type II) for taste, 327, 328*f*

receptor-channel, 170, 172, 174*f*

receptor-enzyme Membrane proteins that bind ligands on the extracellular side and activate enzymes on the intracellular side (Ch 6), 170*f*, 173*f*, 176

receptor-mediated endocytosis A ligand binds to a membrane protein, which triggers endocytosis of the membrane receptor complex (Ch 5, 22), 138*f*, 148

receptor-operated calcium channels (ROCC), 409

receptor potential Graded potential in a special senses receptor (Ch 10), 312, 317*f*–318*f*

recruitment Addition of motor units to increase the force of contraction in a muscle (Ch 12, 24), 398, 756

rectifying synapse, 253

rectum The distal segment of the large intestine (Ch 19, 21, 26), 592*f*, 656*f*, 658, 667*f*, 685*f*, 686, 812*f*, 818*f*

red blood cell (RBC), Synonym: erythrocyte, 152*f*, 513*f*, 514, 516*t*, 517*f*, 520*f*, 524*f*, 529*f*, 571, 777*f*

accelerated loss, 523*t*

disorders, 522–523

life span, 522

maturation, 519*f*

membrane, 62*t*

nucleus, lack a, 518–520

oxygen-carrying capacity, 517*f*

production, 518, 521*f*, 523*t*

red cell count, 517*f*

red muscle Muscle that has lots of mitochondria and good blood supply so that it can carry out oxidative metabolism (Ch 12), 394, 395*t*

red pulp, 759*f*

referred pain Pain that is felt in a location away from the actual site of the stimulus (Ch 10), 322, 323*f*

reflex Any long-distance pathway that receives input about a change, integrates the information, and uses the nervous system, endocrine system, or both to react appropriately (Ch 1, 7, 11, 13, 15, 20, 21), 218, 418, 639, 664–665, 686

control pathway Any long-distance pathway that uses the nervous system, endocrine system, or both (Ch 1), 14, 14*f*, 182, 206–207

movement, 427–428, 427*t*

reflexive memory Automatic memory that is acquired slowly through repetition and does not require conscious processes for its creation or recall. Synonym: implicit memory (Ch 9), 301, 301*t*

refractory period, 241*t*, 245, 248*f*, 250*f*, 452, 452*f*

regulated variables, 13, 182–183

regulatory protein, 46

relative polycythemia, 523

relative refractory period A period of time immediately following an action potential during which a higher-than-normal graded potential is required to start another action potential (Ch 8), 246, 248*f*

relaxin Peptide hormone secreted by the ovary and placenta to prepare the uterus and pelvis for delivery (Ch 7, 26), 200*f*, 831

REM (rapid eye movement) sleep, 296*f*, 297, 825*f*

renal arteries, 438*f*, 439, 591, 592*f*

renal corpuscle The combination of glomerulus and Bowman's capsule (Ch 19), 594, 597–598, 598*f*

renal (kidney) failure, 597, 604, 778

renal medulla Inner portion of the kidney whose interstitial osmolarity ranges from 300-1200 mOsM (Ch 20), 622–623

renal pelvis, 592*f*, 594

renal system. *See* urinary (renal) system

renin Peptide secreted by juxtaglomerular cells that converts angiotensinogen into angiotensin I (Ch 20), 632, 633*f*, 635*f*, 639, 362*f*–363*f*, 601*f*, 633*f*, 639, 639*t*

renin-angiotensin system (RAS), 632, 633*f*, 640*f*

replication An experiment repeated to ensure that the results were not an unusual one-time event (Ch 1), 19

repolarization Phase during which depolarized membrane returns to its resting potential (Ch 5, 8, 14), 157*f*, 158, 246*f*–247*f*, 451, 454*t*, 461*f*

reproduction, 213, 403
    and aging, 834
    basic patterns of, 806–810
    brain directs, 808–809
    and development, 800–838
    environmental influences on, 810
    female, 816–823
    hormonal control of, 809*f*
    male, 810–815

reproductive system, 3, 4*f*, 77*f*
    female, 818*f*
    male, 812*f*

residual volume (RV) The volume of air left in the lungs following a maximal exhalation (Ch 17), 544

resorption of bone Process in which osteoclasts dissolve the calcium phosphate matrix (Ch 23), 743

respiration Cellular use of oxygen and substrates to produce energy (Ch 4), 93

respiratory acidosis Acidosis due to retention of CO_2 (Ch 18, 20), 579, 647–648

respiratory alkalosis Alkalosis due to hyperventilation that decreases arterial P_{CO_2} (Ch 20), 649

respiratory cycle An inspiration followed by an expiration (Ch 17), 544

respiratory system Those structures involved in ventilation and gas exchange (Ch 1, 3, 17) 3, 77*f*, 536. *See also* lung

response loop Control pathway that begins with the stimulus and ends with the response (Ch 1), 14–15, 15*f*

rest-and-digest, 359, 360*f*

resting membrane potential difference The uneven distribution of ions across a living cell membrane (Ch 5), 156 153–159, 244*f*, 246*f*, 451

restrictive lung diseases, 551

reticular activating system Neurons that contribute to arousal (Ch 9), 294

reticular formation Diffuse groups of neurons that branch from the brain stem into the brain and spinal cord; involved in muscle tone, stretch reflexes, coordination of breathing, blood pressure regulation, and modulation of pain (Ch 9, 10, 11), 287, 287*f*, 295*f*, 339*f*, 362*f*

reticulocyte Immature red blood cell with no nucleus (Ch 16), 515*f*, 518

reticuloendothelial system Old term for tissue macrophages (Ch 24), 761

retina Sensory receptors lining the posterior cavity of the eye (Ch 6, 10), 182*t*, 343–347, 347*f*

retinal The light-absorbing pigment of rhodopsin (Ch 10), 348*f*, 349, 350*f*

retinal processing, 353

retinitis pigmentosa, 182*t*

retinoids A group of chemicals derived from vitamin A, speed up cell division and surface shedding so treated skin develops a more youthful appearance (Ch 3), 79

retrograde transport, 232

retroperitoneal, 591

retrospective studies These studies match groups of people who all have a particular disease to a similar but healthy control group (Ch 1), 23

reversible reaction A chemical reaction that can proceed in both directions (Ch 4), 98, 104, 104*f*

rhabdomyolysis, 794

rheumatoid arthritis, 71, 778*t*

rhodopsin receptor, 182*t*

rhodopsin Visual pigment of rods (Ch 10), 349, 350*f*, 358*f*

rhythmic movement, 427–428, 427*t*

ribonucleases, 116

ribose, 31*f*, 34*f*

ribosome Small dense granules of RNA and protein that assemble amino acids into proteins (Ch 3, 7), 68, 73*f*, 203*f*

right atrium Chamber of the heart that receives systemic venous blood (Ch 14, 15), 438*f*, 444*f*, 445*f*, 446*t*, 478*f*, 483*f*

right ventricle Chamber of the heart that pumps blood to the lungs (Ch 14, 15), 438*f*, 445*f*, 446*f*, 446*t*, 478*f*

rigor mortis, 388

rigor state Tight binding between actin and myosin in the absence of ATP (Ch 12), 386, 387*f*

RNA (ribonucleic acid) Nucleotide that interprets the genetic information stored in DNA and uses it to direct protein synthesis (Ch 2, 4, 24), 34*f*–35*f*, 37*f*, 111–116, 755*f*

RNA interference (RNAi), 116

RNA polymerase Enzyme needed for synthesis of mRNA from DNA (Ch 4), 112

rod Receptors for monochromatic nighttime vision (Ch 10), 347*f*, 348

Rodbell, Martin, 174

ROMK (potassium channel), 630, 631*f*

roots, spinal 284

rough endoplasmic reticulum (RER) Organelle that is the primary site of protein synthesis (Ch 3, 8), 67*f*, 71, 73*f*, 227*t*, 232*f*

round window Membrane between cochlea and middle ear (Ch 10), 329, 330*f*, 332*f*

Ruffini corpuscle, 321*f*

rugae Surface folds in the interior of the stomach (Ch 21), 656*f*, 658

R wave The largest wave of the QRS complex (Ch 14), 461*f*

ryanodine receptor-channel (RyR) Calcium-release channel of sarcoplasmic reticulum in striated muscles (Ch 14), 450*f*, 389, 390*f*, 449, 450*f*

S

S2 (second heart sound), 462

saccule One of the otolith organs of the vestibular apparatus (Ch 10), 333*f*, 337, 338*f*

sacral spinal cord/nerves, 279*f*, 284

saliva Watery enzyme and mucous secretions of the mouth (Ch 21), 668

salivary amylase, 668

salivary gland, 288*t*, 362*f*–363*f*, 655, 656*f*, 660*f*, 667*f*, 757

Salmonella, 687

salt and water balance, 590. *See also* salt balance; water balance

salt appetite, 329, 619, 636

saltatory conduction The apparent leap-frogging of the action potential down myelinated axons (Ch 8), 249, 252*f*

salt taste, 327

SA node, 455*f*–456*f*, 458*f*, 467*f*, 494*f*

sarcolemma The cell membrane of a muscle fiber (Ch 12), 381*f*–382*f*, 382, 382*t*

sarcomere The contractile unit of a myofibril (Ch 12, 14), 380*f*–381*f*, 382, 383, 384*f*, 390*f*, 400*f*, 405–406, 411*t*, 447, 469*f*

sarcoplasmic reticulum, 380*f*, 381*f*–382*f*, 382, 382*t*, 390*f*, 393*f*, 406, 410*f*, 411*t*, 450*f*, 470*f*

satellite cell, 233, 234*f*, 379

satiety A sensation of fullness (Ch 9, 21, 22), 299, 665, 665*f*, 666*t*

satiety center Hypothalamic center that decreases food intake (Ch 22), 694

saturated fatty acid Fatty acid with no double bonds between carbons (Ch 2), 30*f*

saturation All active sites on a given amount of protein are filled with substrate and reaction rate is maximal (Ch 2, 5, 6, 19), 51, 146–147, 146*f*, 180, 604

scala media, 332

scala tympani, 332

scalene muscle Respiratory muscle that lifts the upper rib cage (Ch 17), 536

scanning electron micrographs (SEMs), 520*f*

scatter plot, 20*f*, 21*f*

Schistosoma, 755, 761

schizophrenia, 304

Schmidt, Christine, 93

Schwann cell Cell that forms myelin around a peripheral neuron axon (Ch 8, 11), 230*f*, 233, 234*f*–235*f*, 236, 237*f*, 257*f*, 372*f*

scientific theory A model with substantial evidence from multiple investigators supporting it (Ch 1), 19

sclera, 340*f*, 341, 347*f*

scrotal cavity, 813*f*

scrotum The external sac into which the testes descend so that they can stay cooler than body temperature (Ch 26), 805*f*, 810, 812*f*

sebaceous gland, 86*f*

secondary active transport, 133*f*, 143–145, 144*t*, 604, 667*f*

secondary endocrine pathology An endocrine pathology that arises in a trophic gland of a pathway (Ch 7), 218

secondary gametes, 807

secondary hypercortisolism, 218, 219*f*, 735

secondary immune response The stronger and more rapid immune response that occurs with the second or subsequent exposure to a pathogen (Ch 24), 767*f*, 768

secondary lymphoid tissues, 757

secondary oocyte The ovulated egg which has gone through the first meiotic division (Ch 26), 807, 808*f*, 813*f*, 820*f*

secondary sensory neuron, 312, 313*f*, 316*f*, 319*f*, 323*f*, 326*f*

secondary sex characteristic Features of the body, such as body shape, that distinguish males from females (Ch 26), 816, 824

secondary spermatocyte Spermatocyte that has gone through the first meiotic division (Ch 26), 807, 808*f*

secondary structure, of protein Spatial arrangement of amino acids in the chain (Ch 2), 32*f*, 41

second heart sound Vibrations created when the semilunar valves close (Ch 14), 462

second law of thermodynamics Natural spontaneous processes move from a state of order (nonrandomness) to a condition of randomness or disorder (Ch 4), 96

second meiotic division, 807

second messenger Intracellular molecules that translate the signal from a first messenger into an intracellular response (Ch 6, 7, 10,

11), 171, 172*f*, 173*f*, 204*f*–205*f*, 217, 367*t*, 814*f*, 263*f*, 267*f*, 745*f*

second-order neuron, 312

second polar body, 807, 808*f*, 828*f*

secretin Intestinal hormone that stimulates bicarbonate secretion and pepsin release; inhibits gastric acid (Ch 7, 21), 200*f*, 666, 666*t*, 667*f*, 684*f*

secretion (1) The movement of selected molecules from the blood into the nephron; (2) The process by which a cell releases a substance into the extracellular space (Ch 3, 5, 11, 19, 21), 61, 79, 150, 150*f*, 198, 362*f*–363*f*, 370*f*, 594, 596*f*, 607–608, 660*f*, 667*f*, 669*f*, 670–672, 671*f*

exocrine, 668

secretory diarrhea, 686–687

secretory epithelia Epithelia that secrete hormones or exocrine secretions (Ch 3), 76, 77*f*, 78*f*, 79–80

secretory phase Postovulatory phase of the uterus when it develops into a secretory structure (Ch 26), 817, 821*f*

secretory vesicles, 71, 73*f*, 203*f*

segmental contractions, 662*f*, 663

selective estrogen receptor modulator (SERM), 749, 834

selective serotonin reuptake inhibitor (SSRI), 182, 299

selectively permeable membrane (Ch 5), 132–133

self-reactive cells, 770*f*

self-tolerance The lack of immune response to cells of the body (Ch 24), 778

Selye, Hans, 781

semen Sperm plus secretions from accessory glands (Ch 26), 812*f*–813*f*, 815

semicircular canal (Ch 10), 329, 330*f*, 337–339

semilunar valve Heart valves between the ventricles and major arteries (Ch 14, 15), 463*f*, 464, 482*f*

seminal vesicle Male accessory glands that contribute enzymes and other secretions to semen (Ch 26), 804*f*, 811, 812*f*

seminiferous tubule Region of the testes where sperm and hormones are produced (Ch 26), 811, 813*f*

sense organ, 311

sensitization Exposure to a noxious or intense stimulus creates an enhanced response upon subsequent exposure (Ch 9), 300

sensorineural hearing loss, 336

sensory afferent, 418, 669*f*

sensory coding, 335*f*

sensory cortex, 430*f*

sensory field, 291

sensory modality, 314

sensory neuron A neuron that transmits sensory information to the central nervous system (Ch 6, 7, 8, 9, 10, 13, 15, 18, 20, 26), 187*f*, 190*f*, 208*f*, 227, 227*t*, 228*f*, 230*f*, 284*f*, 311*f*, 312, 313*f*, 334*f*, 418*t*, 419*f*, 422*f*–426*f*, 494*f*, 612*f*, 645*f*, 825*f*

sensory pathway, 314*f*, 319*f*

sensory physiology, 309–357

chemoreception, 324–329

ear, equilibrium, 337–340

ear, hearing, 329–336

eye and vision, 340–353

sensory systems, 310–317

somatic senses, 317–324

sensory receptor, 86*f*, 186*f*, 228*f*, 311*f*, 321*f*, 338*f*, 420, 428*t*, 429*f*, 430*f*, 665*f*. *See also* sensory neuron; *specific type*

types of, 312*t*

sensory system, 291, 291*f*. *See also specific sense*

central nervous system (CNS), 313–314

coding and processing, 314–317

general properties of, 310–317

receptors, 311

sensory neuron, 312

sensory transduction Conversion of a sensory stimulus to an action potential (Ch 10), 312

septic shock, 487

septum A dividing wall, such as between the chambers of the heart (Ch 14), 437

series elastic element Elastic fibers in the muscle that stretch during isometric contraction (Ch 12), 400

serosa Outer surface of the digestive tract created by a continuation of the peritoneum (Ch 21), 656*f*–657*f*, 658

serosal membrane, 150

serotonin A CNS neurotransmitter. Synonym: 5-hydroxytryptamine (5-HT) (Ch 8, 9, 10, 15, 16), 254, 255*t*, 294, 295*f*, 328*f*, 488*t*, 490, 526, 527*t*

serotonin/norepinephrine reuptake inhibitors, 289

serous secretion Watery exocrine solution that often contains enzymes (Ch 3), 80

Sertoli cell Testicular cells that secrete anti-Mullerian hormone and support sperm production (Ch 26), 803, 806*f*, 811, 813*f*, 814, 814*f*

setpoint, 13, 14, 15, 16*f*, 17–18, 185

sex act, 824

sex chromosome, 801, 802*f*

sex determination, 801–806

sexual differentiation, 802–806

sexual dimorphism Males and females have different physical characteristics (Ch 26), 801

sexual dysfunction, 824–825

sexual response, 824

SGLT, 145*f*, 603*f*, 680*f*

SGLT2 inhibitors, 719*t*

shock Generalized, severe circulatory failure (Ch 15), 487

short-loop negative feedback, 213, 809*f*

short reflexes, GI tract 664, 665*f*, 670*f*

short-term energy storage, 696

short-term memory, 301, 301*f*

sickle cell disease, 520*f*, 522, 523

sigmoid colon, 685, 685*f*

signal amplification Process by which a single signal molecule can generate multiple intracellular effector molecules (Ch 6), 173*f*

signal pathway, 169–176, 169*f*, 182*t*

signal processing, 350

signal sequence Initial segment of a newly-made protein that directs the protein to the proper organelle for processing, packaging, and delivery (Ch 4, 7), 116, 202, 203*f*

signal transduction The transmission of information from one side of a membrane to the other using membrane proteins (Ch 6, 7, 10, 24), 171, 171*f*, 173*f*, 174, 177*f*, 182, 202, 204*f*, 217–218, 328*f*, 334*f*, 771*f*

simple epithelium (one cell thick), 76

simple diffusion Diffusion across the phospholipid bilayer of a cell (Ch 5), 133*f*, 135*t*, 136

simple squamous epithelium, 76

Singer, S. J., 62

single-unit smooth muscle Smooth muscle fibers that are electrically coupled by numerous gap junctions (Ch 12), 403, 405*f*.

sinoatrial node (SA node) group of autorhythmic cells in the right atrium of the heart; the main pacemaker of the heart (Ch 14), 454, 455*f*

sinus, 280, 281*f*

sinusoid, 497, 677*f*

sister chromatid, 807, 808*f*

skeletal muscle Striated muscle usually attached to bones; responsible for positioning and movement of the skeleton (Ch 8, 11, 12, 13, 14, 15, 17, 19, 21, 22, 25), 228*f*, 255*t*, 370*f*, 372*f*, 378, 378*f*, 379–398, 405*f*, 411*t*, 418*t*, 430*f*–431*f*, 452*f*, 454*t*, 469*f*, 481*f*, 488*t*, 492*f*, 612*f*, 792*f*

   comparison, 411*t*

   contraction, 391

   fibers, 378*f*, 379–382

   load-velocity relationship in, 402*f*

   speed and fatigue resistance, 393–395

skin, 86*f*, 200*f*, 292*f*, 321*f*, 323*f*, 437*t*, 492*f*, 621*f*, 757*f*, 792*f*

skin cancer, 86*f*

skull, 59, 280, 287*f*, 340

sleep, 287*f*, 296–297

sleep apnea, 297

sleep cycle, 296*f*

sliding filament theory of contraction The current model for muscle contraction in which muscle proteins slide past each other to generate force (Ch 12), 385, 385*f*, 387*f*

slow axonal transport, 231

slow pain, 320, 321*t*

slow synaptic potential Slower onset and longer lasting response of postsynaptic cells to certain neurotransmitters and neuromodulators (Ch 8), 261

slow-twitch oxidative muscle, 393, 394*f*, 395*t*

slow wave potential Cyclic depolarization and repolarization of membrane potential in smooth muscle (Ch 12, 21), 409*f*, 410, 661, 662*f*

slow-wave sleep, 296, 296*f*

small intestine The segment of the gastrointestinal tract where most absorption and digestion take place (Ch 10, 2, 22, 231), 323*f*, 656*f*–657*f*, 658, 660*f*, 665*f*, 666*t*, 667*f*, 674*f*, 676*f*, 677*f*, 678, 679*f*–680*f*, 695*f*, 745*f*

smell, 292*f*, 310*t*, 324–329

smoking, 501

smooth endoplasmic reticulum (SER), 67*f*, 71

smooth muscle, 228*f*, 255*t*, 364, 367*t*, 370*f*, 378, 378*f*, 403–411, 404*f*–406*f*, 411*t*, 418*t*, 479*f*–480*f*, 488*t*, 612*f*, 665*f*, 819*f*

   activity, 410

   contraction, 404*f*, 405, 407–409, 408*f*

   membrane potential (mV), 409*f*, 410

   operate over a range of lengths, 404

sodium, 33*t*, 36*f*

   balance, 629–634

sodium bicarbonate, 676*f*

sodium-calcium exchanger (NCX) 144*t*, 450

sodium ion, 42*f*, 174*f*, 240*t*, 241*f*, 243–245, 247, 252*f*, 328*f*, 350*f*, 372*f*, 390*f*, 451*f*, 603*f*, 628*f*, 634–635, 636, 647*f*, 675*f*, 680*f*–681*f*, 683*f*

   and Ca^{2+} influx, 467*f*

   channel, 244*f*, 246*f*, 250*f*, 328*f*, 451*f*

   channel activation, 247*f*

   channel inactivation gate, 247*f*

   dependent cotransport, 603*f*

sodium-dependent transporter, 144*t*

sodium-hydrogen (Na^+-H^+) exchanger (NHE), 604, 632, 645, 683

sodium-iodide symporter (NIS) Transport protein for uptake of iodide into thyroid gland (Ch 23), 736

sodium-linked reabsorption, 603*f*

sodium-potassium ATPase (Na^+-K^+-ATPase) Active transporter that moves Na^+ out of the cell and K^+ into the cell against their respective concentration gradients (Ch 5, 12, 14, 19, 20, 21, 23), 123, 143*f*–144*f*, 151*f*, 392*f*, 450*f*, 603*f*, 607*f*, 631*f*, 683*f*, 738*f*

soft tissue growth, 741*f*

solubility The ease with which a molecule or gas dissolves in a solution: The more easily a substance dissolves, the higher its solubility (Ch 2), 40, 42*f*, 570

solute Molecules that dissolve in liquid (Ch 2,5, 19), 42*f*, 128, 130*f*–131*f*, 603*f*

solute carrier (SLC) superfamily, 138

solution A solute or combination of solutes dissolved in solvent (Ch 2, 5), 42*f*, 127, 127*t*, 132*t*

solvent The liquid into which solutes dissolve. In biological solutions, water is the solvent (Ch 2), 42*f*

soma, 232*f*. *See also* cell body

somatic motor division Efferent branch of nervous system that controls skeletal muscles (Ch 8, 11), 229, 371

somatic motor neuron Efferent neurons that control skeletal muscles (Ch 8, 10, 11, 12, 13, 18, 21, 22), 228*f*, 262*f*, 339*f*, 359, 372*f*, 390*f*, 393*f*, 411*f*, 418*t*, 419*f*, 421, 431*f*, 669*f*

somatic perception, 317

somatic reflex, 418

somatic senses Touch-pressure, temperature, pain, and proprioception (Ch 8, 9, 10), 230*f*, 314*f*, 317–324

   cerebellum, 317

   conscious stimulus processing, 310*t*

   cortex, 317

   free nerve endings, 320

   itch, 321–322

   nociceptors, 320–324

   pain, 322–324

   protective responses, 320

   somatic perception, 317

   stimuli, 310

   temperature receptors, 320

   touch receptors, 319–320

somatomedin Old name for insulin-like growth factors (Ch 21), 677*f*, 742

somatosensory cortex, 318, 319*f*–320*f*

somatosensory receptor, 311

somatosensory tract Axons carrying sensory information from the body to the brain (Ch 9), 287

somatostatin Hypothalamic hormone that inhibits growth hormone release and gastric paracrine that inhibits gastrin secretion (Ch 7, 21), 200*f*, 211, 212*f*, 741*f*

somatotropin, 200*f*, 213. *See* growth hormone

somatotropin release inhibiting hormone. *See* somatostatin

sorting signal, 116

sound The brain's interpretation of the amplitude, frequency, and duration of sound waves (Ch 10), 314*f*–315*f*, 329–331

sound perception, 329–330

sound transduction, 331

sound wave, 329, 330*f*–331*f*

sour taste, 327, 328*f*

spatial summation Summation of graded potentials from several sources (Ch 8), 264*f*, 266

spatial visualization, 293*f*

special senses Vision, hearing, taste, smell, and equilibrium (Ch 10), 310, 310*t*

specific hunger A craving for a particular substance such as salt (Ch 10), 329

specific immune response, 757

specificity The ability of an enzyme or receptor to bind to a particular molecule or a group of closely related molecules (Ch 2, 5, 6), 46, 145–146, 180, 187

sperm, 801, 802*f*, 808*f*, 811–814, 828*f*

spermatid, 808*f*, 813*f*

spermatogenesis, 814*f*, 815

spermatogonium, 807, 808*f*, 813*f*–814*f*

spermatozoa, 813*f*. *See also* sperm

spherocytosis, 518, 522

sphincter, 403, 404f, 411t, 655. *See also specific type*

sphincter of Oddi, 658, 677f, 679f

sphingolipid, 62, 63f–64f, 180

sphygmomanometer, 484

spinal accessory nerve, 288t

spinal cord, 227, 228f, 234f, 278f–279f, 281f, 284–285, 284f–287f, 319f, 321f, 323f, 362f, 368t, 398f, 422f, 423f, 425f, 426f, 428t, 429f–430f, 825f

spinal nerve, 279f, 284

spinal reflex A simple reflex that can be integrated within the spinal cord without input from the brain (Ch 9, 11, 13, 19), 276, 285, 285f, 360, 612f

spinal tap, 282

spindle, 422f, 423f

spine, dendritic, 264f

spirometer, 544

spleen, 500f, 521f, 677f, 757–759, 758f–759f

spongy bone, 744f. *See also* trabecular bone

spontaneous reaction, 96

squamous epithelium, 756

SRY gene The sex-determining region on the Y chromosome (Ch 26), 803, 806f

SRY protein, 804f

St. Martin, Alexis, 655

stapes The third bone of the inner ear that connects the incus to the oval window (Ch 10), 329, 332f

starch Digestible storage polysaccharide made by plants (Ch 2, 21), 31f, 37f, 680, 680f

Starling, Ernest, 468, 498

Starling curve, 468, 469f

Starling forces, 498

steady state, 13, 13f, 123

stem cell Immature cells that have the ability to differentiate (Ch 3, 8, 10, 14, 15, 21, 24), 85, 87, 235, 326f, 471, 519f, 771f

stenotic valve, 464

stereocilia Stiffened cilia of hair cells in the ear (Ch 10), 332, 334f

sterilization, 825, 826t

sternocleidomastoid muscle Inspiratory muscles that help elevate the upper ribs (Ch 18), 536, 580f

steroid Lipid-related molecules derived from cholesterol (Ch 2, 20, 23), 30f, 37f, 631f, 733, 733f, 748f

steroid hormone Hormones made from cholesterol (Ch 7, 23, 26), 201f, 202t, 204, 205f, 731f, 732, 806f, 809f, 813f

  cellular mechanism of action of, 204

  synthesis and release, 204

steroidogenic factor (SF1), 803

stethoscope, 485f

STIM1, 407

stimuli, 260, 264f, 397f, 424

  mechanical, 321t

  nonpainful, 324f

  noxious, 321f

  sensory transduction, 312

  touch receptors, 319–320

two-point discrimination, 313f

  properties, 314–317

stimulus The disturbance or change that sets a reflex in motion (Ch 1, 6, 7, 8, 9, 10, 13, 14, 20), 14, 185, 190f, 208f, 214f, 240t, 241f, 248f, 285f, 419f, 425f–426f, 452f, 639

  duration, 317f

    intensity and duration, 315–316, 317f, , 189

  location of the, 315

  receptor adaption, 318f

    strength, 240–242

stomach, 200f, 323f, 656f–657f, 658, 662f, 665f, 666t, 667f, 668–669, 669f, 672, 673f, 677f, 680f–681f, 695f

  acidity, 757f

  acid secretion in the, 671f

storage vesicle, 71, 73f, 624f

store-operated Ca²⁺ channel, 407, 410f

stratified epithelium (multiple cell layer), 76

streptokinase An enzyme that dissolves blood clots (Ch 16), 529

stress, 298f, 733f, 741f, 780f, 781

  and exercise, 795–797

stressor An event that causes a stress reaction (Ch 24), 781

stretch receptor, 612f

stretch reflex A reflex pathway in which muscle stretch initiates a contraction response (Ch 13), 421, 423f, 424

striated muscles Muscles that appear to have alternating light and dark bands; includes skeletal and cardiac muscle (Ch 12, 14), 378

striation, 378f, 735

stroke Blockage or rupture of a blood vessel in the brain (Ch 15), 484

stroke volume The amount of blood pumped by one ventricle during one contraction (Ch 14, 15), 486f, 491f

stroma Supporting connective tissue (Ch 16, 26), 519f, 816, 819f

strong pain, 324f

structural protein, 138f

subarachnoid space Fluid-filled space beneath the arachnoid membrane of the skull (Ch 9), 279f, 280, 281f, 282

subatmospheric intrapleural pressure, 549–550

subcellular compartmentation, 73f

subconscious stimulus processing, 310t

subdural space, 279f, 281f

submandibular gland, 655

submucosa, 656f–657f, 659, 685f

substance P, 256, 322

substantia nigra, 295f

substrate The ligand that binds to an enzyme or a membrane transporter (Ch 2, 4, 8), 46, 99, 234f

subthreshold graded potential A graded potential that is not strong enough to trigger an action potential (Ch 8), 241f, 242, 253f, 264f

subtraction reaction Reaction in which a functional group is removed from one or more of the substrates (Ch 4), 101, 102, 102t

suckling, 831

sucrose Disaccharide made from one glucose and one fructose. Synonym: table sugar (Ch 2, 21), 31f, 680, 680f

sulci, 289

sulfate (SO_4^{2-}), 33t

sulfhydryl group (-SH), 41

summation, 264f, 396, 452f

superior vena cava, 438, 438f, 444f–446f

superoxide anion, 764t, 765

suprachiasmatic nucleus Region of the hypothalamus believed to be the center for the biological clock (Ch 9, 10), 297, 346

supraoptic nuclei, 209

suprathreshold graded potential A graded potential that is strong enough to trigger an action potential (Ch 8), 241f, 242

surface area, 136, 569

surface tension, 551

surfactant Chemical that decreases the surface tension of water (Ch 17), 540

sustentacular cell, 814. *See also* Sertoli cell

Sutherland, Earl, 174, 256

swallowing, 288t, 667f, 668–669

sweat gland, 86f

sweating, 637

sweet taste, 328f

sympathetic branch Division of the autonomic nervous system that is responsible for fight-or-flight response (Ch 8, 11, 15), 228f, 229, 359, 360f, 362f, 363–364, 369, 371t, 490

  control, 468

  division, 366t, 494f

sympathetic cholinergic neuron Sympathetic neuron that uses ACh as a neurotransmitter (Ch 11), 364

sympathetic ganglia, 363, 368t,

symport carrier A membrane transport protein that moves two or more molecules in the same direction across a membrane (Ch 5), 140–141, 140f, 144t

synapse Region where a neuron meets its target cell (Ch 8, 10, 13), 230f, 232, 237f, 262f, 264f, 267, 311f, 431f

synaptic activity, 266–267

synaptic cleft The space between the pre- and postsynaptic cells (Ch 8, 11), 230f, 232, 257f, 259f, 366t, 372f

synaptic knob, 227t. *See also* axon terminal

synaptic plasticity, 261

synaptic potential, 265f

synaptic transmission, 268

synaptic vesicle Small secretory vesicles that release neurotransmitter into the synapse (Ch 10, 11, 26), 232f, 257, 257f, 259f–260f, 311f, 365f, 372f

syncytium, 166

syndrome, 218

synergism Interaction of two or more hormones or drugs that yields a result that is more than additive (Ch 7), 215, 215*f*

systemic circulation Portion of the circulation that carries blood to and from most tissues of the body (Ch 14, 17, 18, 21), 438, 439*f*

systole The time when the heart is contracting (Ch 14), 461, 463*f*

systolic pressure The highest pressures in the circulatory system that reflect the pressures created by contraction of the ventricles (Ch 15), 483, 483*f*

T

T₃, 737*f*, 738*f*, 740*f*. *See also* triiodothyronine

T₄, 737*f*, 738*f*, 740*f*. *See also* thyroxine

tachycardia Rapid heart rate (Ch 14), 460, 789

tachypnea Rapid breathing (Ch 17), 556*t*

tamoxifen Drug that is a selective estrogen receptor (Ch 7), 216

tastant, 327

taste, 288*t*, 292*f*, 310*t*, 325–329, 328*f*

taste bud, 292*f*, 327, 328*f*

taste cell, 327, 328*f*

taste pore, 327, 328*f*

taste transduction, 327–329, 328*f*

tau protein, 302

Tay-Sachs disease, 71, 93

T cell, 774*f*. *See also* T lymphocyte

T-cell receptor Membrane receptors of T cells that bind to MHC-antigen complexes (Ch 24), 764*t*, 765, 771, 771*f*, 774*f*

TEA (tetraethylammonium), 368*t*

tear gland, 288*t*

tectorial membrane Membrane in the cochlea whose movement moves cilia of hair cells (Ch 10), 332, 333*f*

teleological approach Describing physiological processes by their purpose rather than their mechanism (Ch 1), 5

temperature, 134, 310*t*, 319*f*, 320

temperature regulation, 794

template strand, 112

temporal lobe, 286*f*, 289, 292*f*

temporal summation Summation of two stimuli that follow one another in time (Ch 8), 264*f*, 266

tendon Connective tissue that attaches skeletal muscle to bone (Ch 3, 12, 13), 82, 379, 380*f*, 422*f*

tenia coli Muscle bands of the large intestine that pull the wall into haustra (Ch 21), 685*f*, 686

tension The force created by a contracting muscle (Ch 12, 14), 391*f*, 405*f*, 452*f*, 468, 469*f*

terminal cisternae The ends of sarcoplasmic reticulum that abut the t-tubules (Ch 12), 382, 382*f*

tertiary active transport, 607, 607*f*

tertiary follicle, 820*f*

tertiary hyposecretion, 218, 219*f*

testes, 200*f*, 210*f*, 362*f*, 801, 809*f*, 811–814, 812*f*, 814*f*

testis The male gonads (Ch 26), 804*f*, 806*f*, 812*f*–813*f*

testis-determining factor (TDF) A protein that activates genes for male development (Ch 26), 803

testosterone Steroid sex hormone, dominant in males (Ch 7, 23, 26), 205*f*, 731*f*, 803, 804*f*, 806*f*, 813*f*–814*f*, 815

tetanus Sustained muscle contraction (Ch 12, 13, 14, 20), 396, 418, 420, 432, 451, 452*f*, 641

tetany, 745

tetraiodothyronine, 736

tetramer Molecule with four subunits (Ch 4), 99

thalamus Portion of the brain that serves as a relay station for information going to and from higher brain centers (Ch 7, 9, 10, 13), 221*f*, 286*f*–287*f*, 288, 288*f*, 290*f*, 295*f*, 314*f*, 319*f*–320*f*, 336*f*, 342*f*, 352*f*, 428*t*, 429*f*, 430*f*

T_H cell, 773*f*. *See also* helper T cell

theca Layer of cells in the follicle that secrete steroid hormones (Ch 26), 816, 820*f*, 821*f*

The Princess Bride, 742

therapeutic drug, 734

thermodynamics. *See also* energy
    first law of, 95
    second law of, 96

thermoreceptor, 186*f*, 312, 312*t*

thick filament An aggregation of myosin in muscle (Ch 12), 380*f*–382*f*, 382, 390*f*, 396*f*

thin filament An actin-containing filament of the myofibril (Ch 12), 380*f*–382*f*, 382, 383, 396*f*

thirst, 619, 620*f*, 633*f*, 639*t*, 640*f*

thoracic cage The ribs, sternum, spine, and attached muscles (Ch 17), 536, 547*f*

thoracic cavity, 59, 60*f*, 443, 444*f*

thorax The body cavity above the diaphragm (Ch 3, 17, 24), 59, 444*f*, 770*f*
    bones and muscles, lungs, 536
    muscles, 538*f*

threshold (1) The minimum depolarization that will initiate an action potential in the trigger zone; (2) The minimum stimulus required to set a reflex response in motion (Ch 6, 8, 10, 12, 21), 185, 241*f*, 250*f*, 261*f*, 264*f*, 312, 409*f*, 662*f*

thrombin Plasma protein that converts fibrinogen into fibrin (Ch 16), 527, 527t, 528*f*, 529*f*, 530*t*

thrombocyte Alternate name for platelets (Ch 16), 514, 517. *See also* platelet

thrombocytopenia, 518

thromboplastin, 530*t*

thrombopoietin (TPO) Cytokine that promotes platelet formation (Ch 16), 516*t*, 517–518

thromboxane, 179, 179*f*

thromboxane A₂, 526, 527*t*

thrombus A blood clot that adheres to the wall of a blood vessel (Ch 16), 503, 525

thymine, 34*f*–35*f*

thymopoietin Hormone made in thymus gland that promotes lymphocyte formation (Ch 7, 24), 200*f*, 770*f*

thymosin Hormone made in thymus gland that promotes lymphocyte formation (Ch 7, 24), 200*f*, 770*f*

thymulin, 770*f*

thymus, 758*f*, 770*f*

thymus gland Immune tissue that produces lymphocytes (Ch 7, 15, 24), 200*f*, 500*f*, 757, 770*f*–771*f*

thyroglobulin Large protein on which thyroid hormones are formed (Ch 23), 736, 737*f*, 738*f*

thyroid follicle, 736, 737*f*, 738, 738*f*

thyroid gland Endocrine gland in the neck that produces thyroid hormones (Ch 7, 14, 23, 24), 200*f*, 210*f*, 212*f*, 736, 737*f*, 738*f*, 740*f*, 747*f*,

thyroid hormone, 202*t*, 206, 207*f*, 212*f*, 737*f*, 740
    control pathway, 738*f*
    and iodine, 736

thyroid pathologies, 738–740, 739*f*–740*f*

thyroid peroxidase, 736, 737*f*

thyroid-binding globulin (TBG), 736

thyroid-stimulating hormone (TSH), 212*f*, 213, 738

thyroid-stimulating immunoglobulin (TSI), 739, 740*f*, 778*t*

thyronine, 736

thyrotropin (TSH), 200*f*, 210*f*, 213, 738, 738*f*–740*f*, 778*t*

thyrotropin-releasing hormone (TRH), 202, 212*f*, 738, 738*f*

thyroxine (T₄), 201*f*, 207*f*, 736, 737*f*

thyroxine-binding globulin (TBG) Plasma protein that serves as carrier for thyroid hormones (Ch 23), 736, 738*f*

tidal volume The volume of air that moves in a single normal inspiration or expiration (Ch 17), 544

tight junction Cell-to-cell junction in epithelia that does not allow much movement of material between the cells (Ch 3, 5, 9, 10, 26), 74, 75*f*, 150*f*, 282*f*, 328*f*, 813*f*

time constant, 251

tip link, 332, 334*f*

tissue A collection of cells, usually held together by cell junctions, that works together to achieve a common purpose (Ch 1, 3, 4, 6, 7, 15, 16 18, 22, 23, 25), 3, 69, 80, 82, 183, 187, 200*f*, 212*f*, 214*f*, 492*f*, 519*f*, 733*f*, 743, 792*f*. *See also specific type*

tissue and bone growth, 743–749

tissue factor A protein-phospholipid mixture released by damaged blood vessel walls (Ch 16), 525, 525*f*, 530*t*

tissue plasminogen activator (tPA) An molecule that promotes dissolution of blood clots (Ch 15, 16), 449, 528, 529*f*, 530*t*

tissue remodeling, 84–87

tissue repair, 525*f*

tissue thromboplastin, 527, 530*t*. *See also* tissue factor

titin Elastic giant protein that maintains spatial structure of myofibrils (Ch 12), 380*f*–381*f*, 382, 383, 385*f*

T lymphocyte (T cell) Immune cells that bind to and kill their target cells (Ch 24), 758*f*, 760*f*, 763, 766, 770*f*–771*f*

T lymphocyte activation, 771*f*

T lymphocyte development, 771*f*

T lymphocyte receptor, 764*t*

toll-like receptors (TLRs), 763

tongue, 186*f*, 288*t*, 314*f*, 328*f*, 669*f*

tonic activity, 420

tonic control Ongoing control that is adjusted up and down (Ch 6, 12, 14, 15), 183, 184*f*, 403, 410, 468, 491*f*, 494*f*

tonic receptor Slowly adapting receptors (Ch 10), 316, 318*f*

tonic smooth muscle, 403, 404*f*

tonicity, 125–132, 127*t*–128*t*, 130*f*–131*f*

tonsil, 758*f*

torque, 400

torr, 440

total cross sectional area, 197, 197*f*

total lung capacity (TLC) Vital capacity plus residual volume (Ch 17), 546

total peripheral resistance (TPR), 495*f*

total pulmonary ventilation The volume of air moved in and out of the lungs each minute (Ch 17), 553

totipotent A stem cell that can develop into a functioning organism, 85

touch, 293*f*, 310*t*, 324*f*

touch receptor, 319–320

trabecular bone Spongy bone with many open spaces (Ch 23), 743, 744*f*

trace element Essential element required in small amounts (Ch 16), 513*f*

trachea Main airway of the respiratory system (Ch 14, 17, 23, 24), 444*f*, 537, 737*f*, 747*f*, 770*f*

tract Bundles of axons in the CNS, generally with a common origin and destination (Ch 9), 277

transamination Transfer of an amino group from one molecule to another (Ch 4), 102

transcellular transport, 150–152, 603*f*, 683*f*

transcription Transfer of information coded in DNA to mRNA (Ch 4, 20, 23), 112, 114*f*, 733*f*, 741*f*

  in the nucleus, 631*f*

transcription factors Regulatory proteins that bind to DNA and alter gene expression (Ch 2, 4), 46, 112, 204

transcytosis A combination of endocytosis, vesicular transport across the cell, and exocytosis; used to move macromolecules across an epithelium (Ch 5, 15, 21), 152*f*, 496*f*, 497, 681, 681*f*

transducer, 171, 171*f*–172*f*, 310

transducin G protein that mediates bitter taste and photoreceptor transduction (Ch 10), 349, 350*f*

transduction Conversion of a signal from one modality to another (Ch 10), 312

transepithelial transport, 151*f*, 603*f*, 604. *See also* epithelial transport

transferrin Plasma protein that binds and transports iron (Ch 16), 512, 514*t*, 520

transforming growth factor-β, 809

transgenic mouse A mouse whose DNA has been altered to contain additional genes that are not normally present (Ch 23), 730

transient receptor potential channel (TRP), 320

transient receptor potential V_1 channel (TRPV$_1$), 322

translation Conversion of the message carried by mRNA into a peptide chain (Ch 4, 7, 20, 23), 115–116, 117*f*, 205*f*, 631*f*, 733*f*

translational research applies basic biomedical research findings to treatment and prevention of human diseases (Ch 1), 5

transmembrane protein, 64

transporter, 138

transporter gene families, 138

transporting epithelia Epithelium whose primary function is the movement of solutes and water between two compartments (Ch 3, 5), 76, 77*f*, 78*f*, 150*f*

transport maximum (T_m) The maximum transport rate that occurs when all carriers are saturated (Ch 5, 19), 146, 604

transport protein, 148

transport vesicle Vesicles that shuttle their contents from endoplasmic reticulum to the Golgi apparatus (Ch 3, 5, 7), 73*f*, 149*f*, 203*f*

transport work This enables cells to move ions, molecules, and larger particles through the cell membrane and through the membranes of organelles in the cell (Ch 4, 22), 94, 696

transverse colon, 685, 685*f*

transverse tubule (t-tubules) Invaginations of the muscle fiber membrane, associated with the sarcoplasmic reticulum (Ch 12), 382

triad One t-tubule with its flanking terminal cisternae (Ch 12), 382, 382*f*

triceps brachii, 379

tricuspid valve The right AV valve of the heart (Ch 14), 447, 448*f*, 478*f*

tricyclic antidepressant, 299

trigeminal nerve, 288*t*, 329

trigger zone The region of the axon where graded potentials are integrated and an action potential begins if the signal is above threshold (Ch 8, 10), 240*t*, 241*f*, 242, 250*f*, 261*f*, 317*f*

triglyceride Lipid composed of one glycerol and three fatty acids. Synonym: triacylglycerol (Ch 2, 21, 22, 25), 30*f*, 37*f*, 679*f*, 787*f*

triiodothyronine (T_3) Most active form of thyroid hormone; produced mostly in

peripheral tissues from T4 (Ch 7, 23), 200*f*, 207*f*, 736, 737*f*

trochlear nerve, 288*t*

trophic hormone Any hormone that controls the secretion of another hormone (Ch 7, 9), 200*f*, 209, 214*f*, 219*f*, 235, 289*t*

-tropin (suffix), 211

tropomyosin A regulatory protein that blocks the myosin binding site on actin (Ch 12), 381*f*, 382, 386, 386*f*, 387*f*, 390*f*, 411*t*

troponin Complex of three proteins associated with tropomyosin (Ch 6, 12, 14), 178, 380*f*–381*f*, 382, 386, 386*f*, 390*f*, 408*f*, 411*t*, 450*f*

troponin C, 386

trypsin Enzyme that digests proteins (Ch 21), 676*f*, 681, 681*f*

trypsinogen Inactive form of trypsin (Ch 3, 21), 49, 676*f*

tryptophan Amino acid from which the hormone melatonin is made (Ch 7), 221*f*

TSH. *See* thyroid-stimulating hormone

T-tubule, 380*f*–382*f*, 382, 390*f*, 411*t*, 450*f*

tubal ligation, 825

tubule, renal, 591. *See also* nephron

tubulin, 68

tubuloglomerular feedback The process by which changes in fluid flow through the distal tubule influence glomerular filtration rate (Ch 19), 600, 601*f*

tumor necrosis factor (TNF), 764*t*, 765

tunica intima, 479

Turner syndrome, 802

T wave ECG wave that represents ventricular repolarization (Ch 14), 457, 459*f*

twitch A single contraction/relaxation cycle in a muscle fiber (Ch 12), 383, 389

two-point discrimination test, 312, 313*f*

tympanic duct (perilymph), 332, 332*f*–333*f*

tympanic membrane, 329, 330*f*, 332*f*

type 1 diabetes mellitus, 169, 715

type 2 diabetes mellitus, 715

type I alveolar cells Thin alveolar cells for gas exchange (Ch 17), 540

type I support cell, 328*f*

type II alveolar cells Alveolar cells that synthesize and secrete surfactant (Ch 17), 540

type II taste cell, 327

type III taste cells, 327

tyrosine Amino acid that is the basis for thyroid hormones and the catecholamines (Ch 7, 11, 22, 23), 207*f*, 365*f*, 366*t*, 737*f*, 738–739, 738*f*

tyrosine kinase (TK) Membrane enzyme that adds a phosphate group to the tyrosine residue of a cytoplasmic protein, enhancing or inhibiting its activity (Ch 6, 7), 176, 177*f*, 204*f*

U

ubiquitin Protein that tags molecules for destruction by proteasomes (Ch 4), 116

umami The taste sensation triggered by glutamate and associated with nutritious food (Ch 10), 327, 328*f*

uncoupling protein 1 (UCP1), in brown fat, 723

unfused tetanus, 396, 397*f*

uniport carrier A membrane transport protein that moves only one kind of molecule (Ch 5), 140, 140*f*

unitary smooth muscle, 403

unsaturated fatty acid Fatty acid with one or more double bonds between carbons (Ch 2), 30*f*

unstable plaque, 503*f*

up- and down-regulation, 181

upper esophageal sphincter, 667*f*, 669*f*–670*f*

upper respiratory tract, 536

up-regulation Increase in protein number or binding affinity that increases the response of the target cell (Ch 2, 6), 51, 181

uracil, 34*f*–35*f*

urate, 609

urate transporter (UAT), 613

urate transporter 1 (URAT1), 613

urea Nitrogenous waste product produced from amino groups (Ch 4, 19, 20, 21), 590, 604, 610*f*, 629, 677*f*

urea and medullary interstitium, 629

ureter Tube that links a kidney to the urinary bladder (Ch 10, 19, 26), 323*f*, 591, 592*f*, 812*f*

urethra Single tube that drains urine from the bladder to the external environment (Ch 19, 26), 591, 592*f*, 805*f*, 811, 812*f*, 818*f*

urethral fold, 802, 802*f*, 805*f*

urethral groove, 802, 805*f*

uric acid Nitrogenous waste product (Ch 19), 608

urinalysis, 590

urinary bladder, 360*f*, 362*f*–363*f*, 591, 592*f*, 595*f*–596*f*, 612*f*, 812*f*, 818*f*

urinary system The kidneys, bladder, and accessory structures (Ch 1, 3, 19), 3, 4*f*, 77*f*, 591–594, 592*f*

urinary tract infections (UTI), 591, 745–746

urine Fluid waste product produced by the kidneys (Ch 11, 15, 16, 19, 20, 22, 23), 590, 620*f*, 622–623, 624*f*

urobilinogen, 590

urocortin, 735

U.S. Food and Drug Administration (FDA), 23, 825

uterine cycle, 817, 821*f*

uterine tube. *See* Fallopian tube

uterus, 200*f*, 211*f*, 362*f*–363*f*, 803, 804*f*, 816, 818*f*–820*f*, 828*f*, 832*f*

utricle One of the otolith organs of the vestibular apparatus (Ch 10), 337, 338*f*

V

V2 receptor for vasopressin, 625, 626*f*

vagina, 803, 804*f*, 818*f*–819*f*, 832*f*

vagotomy Operation that severs the vagus nerve (Ch 11), 363

vagus nerve Cranial nerve that carries sensory information and efferent signals to many internal organs including the heart and GI tract (Ch 9, 11, 14, 21), 287, 362*f*, 363, 454, 670*f*, 673*f*, 187, 288*t*, 581

Valsalva maneuver Abdominal contraction and forced expiratory movement against a closed glottis (Ch 21), 686

valve, heart, 481*f*. *See also specific valve*

valvular stenosis, 462

van der Waals force Weak attractive force that occurs between two polar molecules or a polar molecule and an ion (Ch 2), 38*f*, 39

vanilloid receptor, 322

variable resistance, 479, 486*f*

varicosity Swollen regions along autonomic axons that store and release neurotransmitter (Ch 8, 11), 231, 365*f*, 370*f*, 371*t*, 405*f*

vasa recta Peritubular capillaries in the kidney that dip into the medulla and then go back up to the cortex, forming hairpin loops (Ch 19, 20), 593*f*, 594, 595*f*, 624*f*, 628*f*

vascular elements of the kidney, 594

vascular endothelial growth factor (VEGF) Growth factors that regulate angiogenesis (Ch 15), 481

vascular smooth muscle The smooth muscle of blood vessels (Ch 12, 15), 404*f*, 479, 489, 490

vasculature The blood vessels (Ch 14), 437

vas deferens Tube that carries sperm from the epididymis to the urethra. Synonym: ductus deferens (Ch 26), 804*f*, 811, 812*f*–813*f*

vasectomy, 825

vasoactive intestinal peptide, 488*t*

vasoconstriction Contraction of circular vascular smooth muscle that narrows the lumen of a blood vessel (Ch 14, 15, 16, 19, 20, 22), 442, 479, 488*t*

vasodilation Relaxation of circular vascular smooth muscle that widens the lumen of a blood vessel (Ch 14, 15, 16, 18, 20, 22, 24), 442, 479, 487*f*, 488*t*

vasopressin, 200*f*, 209, 211*f*, 256, 289*t*, 488*t*, 633*f*, 635*f*, 641

   inhibition, 639

   receptor, 182*t*, 624*f*

   release, 640*f*

   secretion, 626*f*, 639, 639*t*

   and water reabsorption, 625

vasopressin, arginine (AVP), 623. *See also* vasopressin

vasovagal syncope Fainting due to a sudden decrease in blood pressure as a result of an emotional stimulus (Ch 15), 478, 495

vegetative nervous system, 359. *See also* autonomic nervous system

vein Blood vessels that return blood to the heart (Ch 11, 14, 15, 17, 21), 363*f*, 444*f*, 445*f*, 478*f*–481*f*, 486*f*, 494*f*, 497*f*, 657*f*

velocity of flow The distance a fixed volume will travel in a given period of time (Ch 14, 15), 442, 443*f*, 482*t*

venae cavae (singular: vena cava), 478*f*, 497*f*

venipuncture, 481

venous blood, 567*f*, 578*f*

venous circulation, 499*f*–500*f*

venous constriction, 472*f*

venous P$_{O_2}$, 791*f*

venous return The amount of blood that enters the heart from the venous circulation (Ch 14, 15), 469, 472*f*, 483

ventilation The movement of air between the atmosphere and the lungs (Ch 17, 18, 20, 25), 535, 643–644, 791*f*

   air flows, 546

   alveolar, 554, 567

   and alveolar blood flow, 556–557

   brain centers, 584–585

   CO_2, oxygen, and pH, 582–583

   and exercise, 790–791, 790*f*

   intrapleural pressure during, 549–550

   local control mechanisms, 558

   lung volumes, 544–546

   maximum voluntary, 554

   neural networks, brain stem control, 581*f*

   pH disturbances, compensates for, 643–644

   pressure gradients, 546

   reflex control of, 580*f*

   regulation of, 580–585

   total pulmonary, 553–554

   types and patterns, 556*t*

ventral horn Region of the spinal cord that contains efferent nuclei (Ch 9), 284*f*

ventral respiratory group (VRG) Medullary neurons for active expiration and greater-than-normal inspiration (Ch 18), 581

ventral root Section of a spinal nerve that carries information from the central nervous system to the muscles and glands (Ch 9), 284*f*

ventricle

   of the brain, 277, 281*f*

   of the heart, 437

ventricular

   contraction, 462

   diastole, 462, 463*f*, 465*f*, 482*f*

   ejection , 462, 463*f*

   end-diastolic volume (mL), *464*

   filling, 462

   systole, 462, 463*f*, 465*f*

vertebrae, 279*f*, 280, 286*f*

vertebral column, 280

vertigo, 314

vesicle A sac-like, membrane-bound organelle used for storage and transport (Ch 3, 5, 7, 8, 14, 15), 70, 73*f*, 133*f*, 147, 211*f*, 232*f*, 257–260, 257*f*, 365*f*, 496*f*, 626*f*, 737*f*

vesicular transport, 147–149, 152*f*

vestibular apparatus Portion of the inner ear that contains sensory receptors for balance and equilibrium (Ch 10, 13), 329, 330*f*, 337, 429*f*